MECHANICAL AND ELECTRICAL EQUIPMENT FOR BUILDINGS

MECHANICAL AND ELECTRICAL EQUIPMENT FOR BUILDINGS

Seventh Edition

Benjamin Stein
Adjunct Associate Professor
B'zalel Institute
School of Environmental Design
Jerusalem, Israel

John S. Reynolds
Professor of Architecture
University of Oregon

William J. McGuinness
Professor of Architecture Emeritus
Pratt Institute

John Wiley & Sons
New York Chichester Brisbane Toronto Singapore

Book and cover designed by Laura C. Ierardi
Production supervised by Cindy Funkhouser
Manuscript edited by Brenda Griffing, Dan Flanagan, and Vivi Danser under the supervision of Martha Cooley

Library of Congress Cataloging in Publication Data:

McGuinness, William J.
Mechanical and electrical equipment for buildings.

Includes indexes.
1. Buildings—Mechanical equipment. 2. Buildings—Electric equipment. I. Stein, Benjamin. II. Reynolds, John, 1938–. III. Title.
TH6010.M38 1986 696 85-17831
ISBN 0-471-86937-6

Printed in the United States of America

10 9 8 7 6

PREFACE

Change and Progress

Since the publication of the first edition of *Mechanical and Electrical Equipment for Buildings* in 1937, each succeeding interval of six to eight years between editions has resulted in what has essentially been, except for unchanging basics, a new book. This edition is no exception. It has been written during a period that has seen pronounced changes in the fields of architectural engineering—changes both in design philosophy and in design execution.

Traditionally, Mechanical and Electrical Equipment for Buildings has been concerned primarily with the relationship between mechanical and electrical systems and the buildings they serve. This edition also emphasizes the role that environmental control systems play in the design of buildings as well as in building performance.

Objectives of the Book

The book is intended as a text for courses in environmental control systems or building control systems in schools of architecture. Because of the large amount of design information and reference data it provides, the book is also appropriate for the practicing professional. Since a background of college-level mathematics and physics is assumed, each subject is logically developed with a minimum of mathematical treatment; emphasis is instead placed on design approach and equipment applications.

The Seventh Edition

This edition extends the philosophy of the sixth edition, emphasizing the themes of energy conservation and the use of renewable energy sources while keeping the reader informed of the major changes in equipment technology wrought by the microprocessor and the computer.

Because of our commitment to the conservation of the earth's resources, we present methods for assessing on-site energy sources and their usage: daylighting, solar heating, passive cooling, solid waste, even rainwater. These renewable energy sources can have important impacts on the building design process; such impacts are discussed in stages, beginning with the most approximate and early "rules of thumb" or building organizational strategies. Because renewable resource usage usually increases first cost, we have included a new appendix to encourage life cycle cost analyses.

Along with our emphasis on the architectural design implications of systems, we have reorganized the approach to system design itself. Throughout the book, sizing information is presented: first the approximate information, then the more detailed. There is a new and very fast rule-of-thumb calculation for heat gain; later, we present the much more detailed heat gain calculations that recognize thermal mass effects through elapsed time. Similarly, both rules of thumb and detailed procedures are presented for solar heating, daylighting, passive cooling, and cisterns.

Many environmental control systems need space; some need outside air, water, fuel, cooling or heating, and so forth. These needs are quantified where possible. We expect environmental control systems to aid the building design process—and the building, in turn, to respond to the needs of such systems.

We recognize that many environmental control systems interact and can be combined to form especially successful spaces. Many buildings are shown that achieve this synergy. A new chapter on the bathroom combines psychology with physiology in exploring light, heat, sound, water, and waste simultaneously.

In the portion of this text devoted to fire safety, the material has been largely rewritten to reflect modern design techniques in compartmentation, sprinklering, and smoke venting. Also included here are a discussion of lightning protection and an extensive discussion of fire alarm systems. This latter involves both principles of fire and smoke detection, and equipment operation. This discussion is followed by design recommendations for various building types which cover both small structures and high-rise buildings.

A streamlined section on modern electrical design for buildings includes treatments of energy and demand control and energy management. The section on illumination and lighting design has been entirely rewritten to emphasize improved daylight utilization, use of efficient and effective light sources and equipment, and energy-conserving design. The latest recommendations of the Illuminating Engineering Society (IES) are given, along with parallel continental and U.S. governmental agency standards, so as to provide a broad view of worldwide design trends. Detailed design examples are developed for both natural and electric lighting design. Cost analyses are included here, as in other sections of the book, to demonstrate the practical aspects of design alternatives.

The signal section deals with the systems and hardware of signalling, communications, and surveillance in buildings. This is followed by a section on transportation that contains updated material on low-energy-use solid-state controls for elevators and escalators, along with our traditional extensive coverage of elevator selection and the design of moving stairs and ramps. The discussion of materials handling in commercial buildings has also been updated.

Noise is our society's most pervasive pollution problem. The section on architectural acoustics discusses noise control at length in addition to standard topics of acoustics design and analysis.

The expanded appendices present detailed tabular data on climate, passive solar design, acoustic design, and metrication. New appendices discuss sunshading, economic analysis, and computer-aided design. Finally, an expanded index—along with a detailed contents and table list—simplify use of this large and data-filled text.

Benjamin Stein
John S. Reynolds
William J. McGuinness

ACKNOWLEDGMENTS

A good many people have contributed to this book. We begin with those who provided early encouragement and continuing inspiration: G.Z. Brown, William McGuinness, and Barbara-Jo Novitski. There are those from whose work we have borrowed at length: J. Douglas Balcomb, Alexander Kira, and Murray Milne, as well as organizations such as ASHRAE (the American Society of Heating, Refrigerating and Air Conditioning Engineers), the American Solar Energy Society, the International Association of Plumbing and Mechanical Officials, the National Fire Protection Association, the American National Standards Institute, and the Carrier Corporation.

We are grateful to the many reviewers who carefully read earlier drafts and made helpful and clarifying suggestions. These include Edward Allen; Dennis A. Andrejko, State University of New York at Buffalo; George T. Balich, Wentworth Institute of Technology; Charles C. Benton, University of California, Berkeley; Dwight Bonner, Columbus Technical Institute; William L. Borner, Harvard Graduate School of Design; Floyd O. Calvert, University of Oklahoma; Robert E. Diedrich, University of Minnesota; Stephen Donahue, John Tyler Community College; Steven C. Easley, Purdue University; James M. Fearing; Walter T. Grondzik, University of Petroleum & Minerals, Saudi Arabia; Bruce Haglund, University of Idaho; Dean R. Heerwagen, University of Washington; Ernest E. Jacks, University of Arkansas; William Jahnke, Kansas State University; Richard M. Kelso, University of Tennessee; Jack Alan Kremers, Kent State University; William P. Lloyd, Mount Hood Community College; Magnus McLetchie, New Hampshire Technical Institute; Murray Milne, University of California at Los Angeles; Fuller Moore, Miami University; Joseph B. Olivieri, Lawrence Institute of Technology; Thurman Potts, Northeast Louisiana University; Alex Rogic, Santa Monica College; Marc Schiler, University of Southern California; Susan Ubbelohde, Florida A & M University; Stephen Vamosi, University of Cincinnati; and John A. Van Deusen.

Many professionals have gone to great lengths to help us clarify details, assemble materials, or express ideas. For this edition, this supporting cast includes Lynda Anderson, John Andrews, Edward Arens, Steve Baker, Paul Beamer, Chris Benton, Donald Bergeson, Eric Bressman, James Brown, Michael Bush, Huber Buehrer, Alan Butler, Gene Clark, Robert Colyer, James Converse, Polly Cooper, Michael Corbett, Karen Crowther, P. O. Fanger, Ken Haggard, Glenn Hezmanlhalch, John Hogen, Seymour Jarmul, Timothy Johnson, Ben Jones, Helen Kessler, Ralph Knowles, Ken Labs, Jane Lidz, Margaret Lyons, David Lung, Kenneth MacLean, Michael Maybaum, Mark Mendell, Marietta Millet, Ric Panciera, Harold Roth, Robert Roth, Edward Rothe, Paul Scanlon, Marc Schiff, Robert Schubert, William Shurcliff, David Soderstrom, Rob Thayer, Joseph Tichenor, Carol Venolia, Donald Wharton, C. Stuart White, Jr., Barry Wollison, and Mary Woolever.

A special role is played by the students in our classes, with whom ideas are tested, rules of thumb developed, early drafts "troubleshot," and examples gathered. The students in the Passive Cooling Seminar and the Solar Heating Seminar at the University of Oregon were especially helpful.

The heaviest and most tedious burdens fall upon supporting staff. At the University of Oregon, these burdens were shouldered by Rosemarie Millet, assisted by several members of the Department of Architecture staff. Much of the manuscript typing was done by Joann Brady. For handling correspondence, typing the manuscript, assisting in proofreading and other production-related items, and continuous encouragement and support, we thank Lila Stein.

We are indebted to the staff of John Wiley for their diligent and highly professional work—to Judy Joseph for managing the project from its inception, assisted by Cindy Zigmund; to Elizabeth Doble, Martha Cooley, and Barbara Mele for editorial work; to Lilly Kaufman and Cindy Funkhouser for production; to Dean Gonzalez for his work on the hundreds of illustrations; and to Laura Ierardi for the design of the book.

CONTENTS

PART I ENERGY OVERVIEW

PART II THERMAL CONTROL

PART IV FIRE PROTECTION

PART V ELECTRICITY

PART VI ILLUMINATION

PART VIII TRANSPORTATION

PART X APPENDICES

TABLES

MECHANICAL AND ELECTRICAL EQUIPMENT FOR BUILDINGS

PART I

ENERGY OVERVIEW

Often mechanical and electrical equipment for buildings is not considered until many important design decisions have been made. Often, too, such equipment is considered to have a corrective function, which permits a building to ''work'' in a climate it essentially ignores.

Part I discusses some topics that encourage designers to *include* both climate and comfort in their early decisions. Chapter 1 discusses the fuel and resource relationship to buildings, from design through demolition. Chapter 2 discusses human comfort, the variety of conditions that seem ''comfortable,'' and some implications for building design of a more broadly defined ''comfort zone.'' Chapter 3 encourages a view of a building site as a collection of renewable resources, to be used and shared in the lighting, heating, and cooling of buildings.

1

ENERGY SOURCES FOR BUILDINGS

Today, some buildings in the United States are labeled as energy gluttons, while others are praised for their low annual energy consumption. Our society uses more energy per capita than most other countries, even some that have "high standards of living" and colder winters than ours. Although buildings are not our only users of energy—automobiles, industry, and agriculture are some others—buildings and their energy-related equipment are the subjects of this book. One way to begin is to examine the path that architecture has followed, from simple shelters that have adapted well to their climate, to sophisticated and tightly controlled internal climates that strive to ignore the conditions outside.

1.1 Energy and Architectural History

Few books on architectural history deal with the influence of changing fuel sources on the development of building. Style—including the proportion of spaces and the elements of the facade—has often been more influential, particularly on exterior appearance. Structural innovation is more visible and more permanent than fuel sources in architecture, and therefore easier to trace back to its earliest forms. Most buildings constructed before the latter part of this century relied heavily on the sun for light and sometimes for heat, and on breezes or massive construction to temper the hot portions of the day. These energy-related distinctions were relatively subtle, particularly among the buildings in given climatic regions. Size and placement of windows were energy related; yet, the size, placement, and, particularly, the shape of the windows were more clearly attributable to prevailing customs of proportion and materials than to a building's energy supply. The amount of sun or breeze admitted through a window, or the amount of building heat lost through it, has less often been cited as an influence on window design.

The impact of energy on buildings in the past has been both considerable and visible, and it is becoming so again. Space heating in buildings beyond that provided by the sun began with the burning of available fuel (Figs. 1.1 and 1.2). From the central open fire—around which all inhabitants slept each night—building heating dispersed to individual fireplaces. This began the continuing choice between concepts of central versus dispersed climate control and is credited for hastening the separation between social classes as well (since the masters got the fireplaces, while the servants still slept around the central fire).

Building heating then proceeded by the gradual use of more imported fuels and less visible equipment, until the obvious architectural impacts (Fig. 1.3) nearly disappeared (Figs. 1.4 and 1.5). Today, solar collection devices, such as large windows, Trombe walls, and/or flat-plate collectors

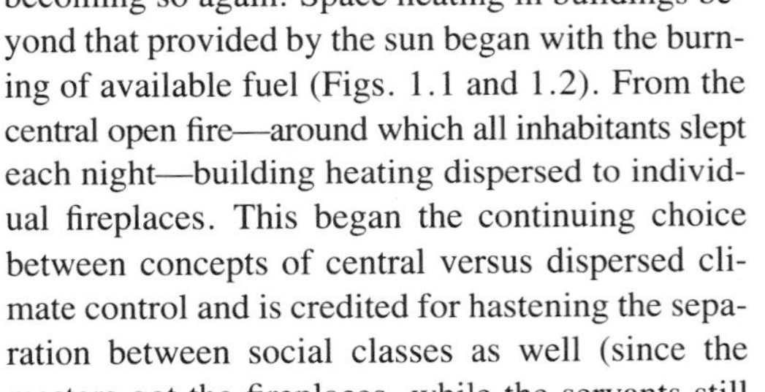

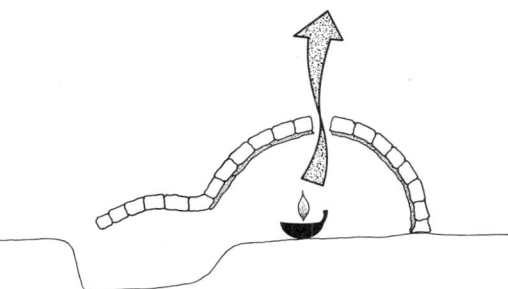

Fig. 1.1 The portable lamp (as used in the igloo) has a slight architectural impact: a storage place is needed for fuel; the design must admit combustion air (to feed oxygen to the flame) and must allow ventilation for the gases of combustion; there is a residue of soot on interior surfaces.

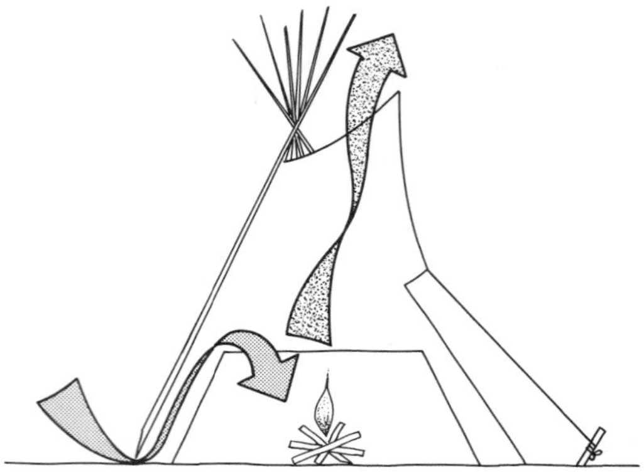

Fig. 1.2 The indoor fire (as used in the tipi) introduces two evident architectural impacts: the adjustable smoke flap, *to encourage venting of smoke by prevailing winds and to minimize the entrance to rain, and the* interior liner, *which forces the cold combustion air to rise along the sides of the tipi, gaining some warmth, before it moves across the occupants on its way to the fire. Again, some fuel storage is needed, and there is a residue of soot indoors.*

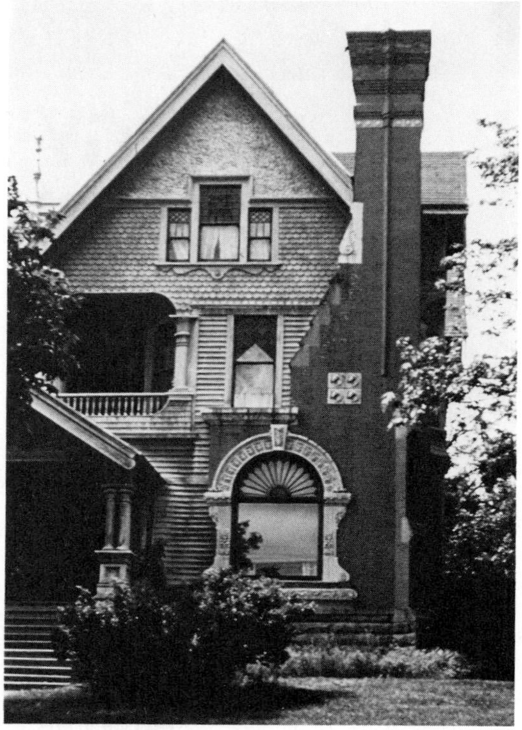

Fig. 1.3 The fireplace and the more efficient wood stove increase the impact of heating on architecture. Their permanent location within a space both allows a massive and visible chimney and determines the placement of other furnishings within the space. As the enclosed fire burns hotter and heats more space, the needs for fuel storage and for providing combustion air intensify. (Photo by William Johnston.)

Fig. 1.4 The furnace (or boiler) begins the trend toward less visible, more automatic heating. These plans show a basement remodeling that resulted from changing fuel sources. (a) *A central large "machine" can be located in a basement along with its fuel storage area. A single chimney exhausts gases, and combustion air can infiltrate the basement without causing undue discomfort upstairs. An extensive ductwork system is threaded through floors and walls to simple grille openings in each room.* (b) *The use of piped-in fuel, such as natural gas, eliminates on-site fuel storage, and furnace sizes diminish with technical developments. With the electric furnace, even the chimney disappears along with the gases of combustion.*

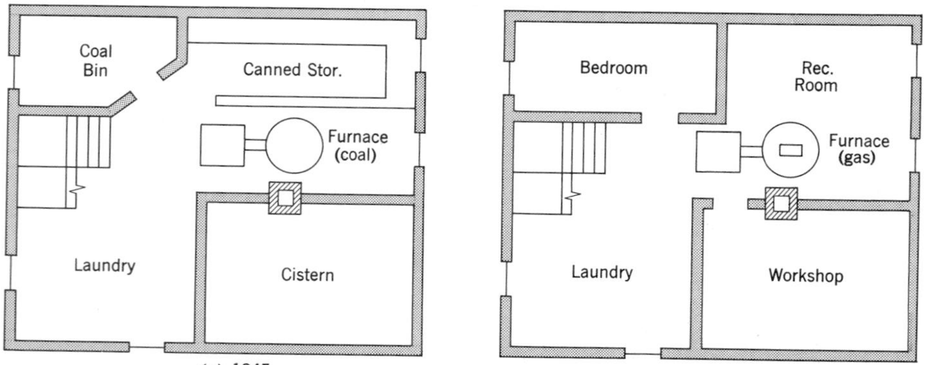

(a) 1945 *(b)* 1955

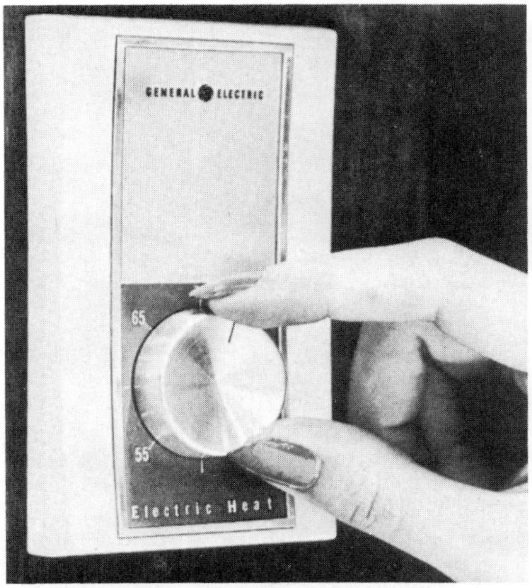

Fig. 1.5 Radiant heat in ceiling or floor, provided through pipes or cables, reduces the visible architectural impact of heating to a thermostat on the wall—and occasional cracks in the heating surfaces!

(Fig. 1.6), have reintroduced the significant architectural impact of heating; their large glass areas, significant thermal mass, and the slopes of collectors combine to produce one of the most visible influences of any heating system on building form.

This is not the only sudden and recent shift in long-term trends concerning energy in buildings. A new awareness of the sources, characteristics, and limitations of energy supplies is resulting in new directions in building design, away from many practices of the recent past.

1.2 Trends, Recent Shifts, and Some Challenges

In the following discussion of energy in today's architecture, two assumptions are made: (1) designers can have a positive impact on society through energy conservation, and (2) buildings that encourage their users to directly experience the natural environment will both facilitate energy conservation and enrich the user's architectural experience.

Fig. 1.6 Solar energy used for space heating brings the visible impact of large glass areas, in this case windows facing south, sloping back to express their relationship to the low winter sun. Sundown House at Sea Ranch, California, David Wright, AIA, architect.

(a) A Trend from Local, Renewable Energy Sources to Imported, Nonrenewable Ones. Renewable fuel sources are those that are available indefinitely but arrive at a relatively fixed rate; the influx of solar energy varies from day to day, but on the average it continues at a steady rate. A woodlot will produce a limited amount of wood per year, but it will do so for centuries if properly managed. A popular analogy for renewable fuel sources is a fixed-but-steady monetary income, such as a salary with no raises.

Using the same analogy, nonrenewable fuel sources are like savings accounts that draw no interest—once spent, they are gone. The nonrenewable resources we utilize can be bought and used in large quantities all at once, which make possible many processes that are more difficult to attain with low-concentration, steady, renewable resources. Thus, the use of nonrenewable fossil and nuclear fuels in high-temperature and high-concentration processes—for example, in power-generating plants—is widespread and is likely to continue until resource depletion is more closely approached.

Plentiful, locally available fuels such as wood and solar energy are cheap and convenient up to a point. As population density increases, however, such supplies dwindle per capita, and locally available quantities can become inadequate. As in the case of firewood in urban areas, prices can also rise. At that point, the conveniences of an established distribution network and a precisely

Fig. 1.7 Solar water heaters are widely used and are available in many versions. This house in Miami, Florida, incorporates a storage tank above the collector, which is enclosed in a chimneylike form. When the lower collector is warmer than the upper storage tank, water circulates without the need for a pump. (Photo by M. Steven Baker.)

measured heat content, as found with electricity, natural gas, and fuel oil, become attractive. The source of pollution that was formerly evident with some local fuels, such as woodsmoke, can be relocated to a distant generating plant. Eventually, of course, the problems of pollution must be faced, wherever they occur.

Even where local fuels are plentiful and "free," imported fuels have often supplanted them. Figure 1.7 shows one of thousands of residential solar water heaters formerly in use in Florida. When automatic washing machines were introduced, their greater appetite for hot water taxed the capacity of these heaters, which had been sized for laundry practices that reused hot soapy water rather than discharging it after one washing cycle. Natural gas utilities promoted the fast temperature recovery of their water heaters, and solar water heating was replaced by natural gas. In another climate, Eskimo families have turned, with government subsidies, from burning seal oil—a locally available, renewable resource—to imported and nonrenewable kerosene for home heating.

This changing pattern in the use of energy sources has been accompanied by a trend in design: away from designing a building so that its external skin and internal organization work with the surrounding climate for "natural" (passive) heating, cooling, and lighting, toward buildings much more dependent on mechanical and electrical equipment for their interior comfort.

Shifts back to renewable and local fuel sources, such as the sun in areas like Florida and California, are occurring rapidly. For example, California has banned new hookups for natural gas swimming pool heaters, creating an early and heavy demand for solar pool heaters. Although the shift in the United States to solar heating has only begun, it is nonetheless highly visible. Figure 1.8 charts the increase in the number of known solar-heated buildings in the United States between 1940 and 1976 (the year Dr. William A. Shurcliff realized that his directory of solar buildings could no longer keep pace with solar development). A similar growth is currently felt in the wood stove industry. Government subsidies have added solar

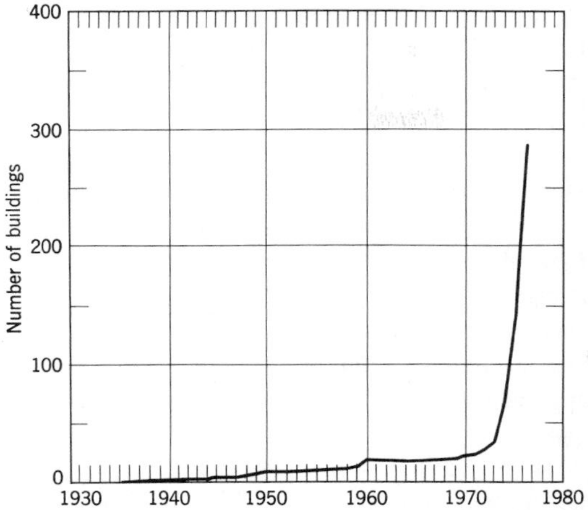

Fig. 1.8 The number of solar-heated buildings in the United States increased rapidly during the 1970s. The 146 new solar buildings (in 1976) exceeded the total number of such buildings in all preceding years. (From Shurcliff, 1977.)

heating to thousands of buildings, just as such subsidies have encouraged mining and drilling for fossil fuels and have supported uranium enrichment.

Buildings designed today may outlast the supplies of the fuels that currently support them. This possibility provides two major challenges for designers:

1. To design our buildings not only to save energy, but so that they can also eventually be weaned away from dependence on nonrenewable fuels.
2. To use energy wisely; to expect only a ''fair share'' of locally available renewable fuels, recognizing that such sources are limited even though they are continually available.

For example, with supplies stretched tight by increasing density, it is tempting to erect a larger solar collector to intercept sunlight that would otherwise be utilized by the neighboring building. The temptation grows stronger as the building is designed to rely more heavily on the sun. The concept of a ''solar envelope'' to protect each building's fair share is discussed in Section 3.4.

(b) The Trend from Labor-Intensive to Energy-Intensive Practices Inside Buildings. All of us know the convenience of energy-driven motors that do what we otherwise would do manually. As appliances that replace human labor become widespread, we design them into our buildings, often without provision for the now-obsolete human-powered practices they replace.

Fig. 1.9 Compare the simple solar clothes dryer with the collectors above. This time- and labor-intensive solution is very different from the common energy-intensive mechanical clothes dryers. (Photo by Douglas Boleyn.)

Figure 1.9 illustrates the effective but time-consuming and labor-intensive practice of solar clothes drying. The use of the energy-intensive alternative, the mechanical clothes dryer, was increasing U.S. energy consumption for this purpose by about 10% per year in the 1970s. A designer who considers weather unpredictability, cultural expectations, and energy scarcity can provide an outdoor clothesline that is both visible and easily accessible from the mechanical dryer.

Another trend to energy-intensive practices is evident in the thermal control of buildings. Figure 1.10 shows the plan of a pioneer house that is dependent on fireplaces and the kitchen's wood stove for heat. Spaces containing fireplaces could be kept warm by practices requiring periodic labor

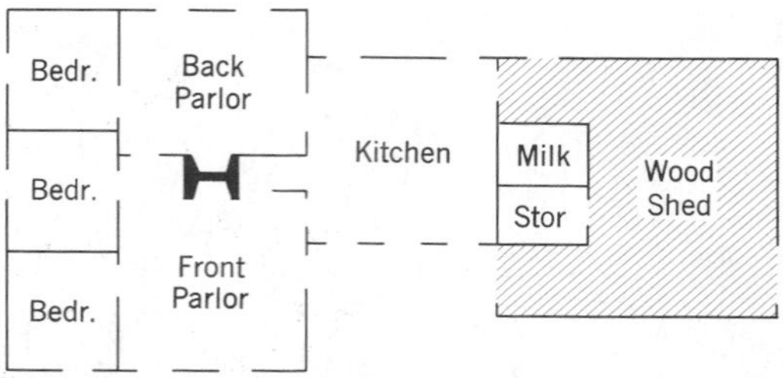

Fig. 1.10 Labor-intensive heat regulation. The house dependent on fireplaces or wood stoves also depends on someone to tend the fire. The warm area near the fire in this early Oregon farmhouse was used for social purposes; the colder spaces at the extremities served for sleeping areas and for storage of food and fuel. (Based on a plan drawn by Philip Dole.)

on the part of the inhabitants (adding wood to the fire or stirring it). Spaces without fireplaces are less controlled, with a temperature somewhere between that of the fireplace rooms and that of the outdoors. Today, the common approach is to provide thermal control via a thermostat, only occasionally adjusted by users. One such device often governs several rooms, keeping them all near one temperature, whether or not they are occupied. In another labor-intensive versus energy-intensive design situation (Fig. 1.11), the labor-intensive approach to snow removal has significant consequences for architectural form.

Recent shifts to labor-intensive practices are scarce. The manually operated thermal shades or shutters of passively heated buildings are the most widespread examples; the increasingly popular wood stove is another.

Labor intensiveness is not so much an energy saver—most controls need very little power—as it is an education. A user who understands how a building gets and conserves its heat in cold weather is more likely to respond by lowering the indoor temperature and tightening up any heat leaks.

Fig. 1.11 Two approaches to dealing with snow at entrances. (a) *The labor-intensive solution utilizes the stair (or ramp); as snow accumulates, "ground level" simply moves up this inclined surface. The person climbing toward the entry, or using a shovel to clear the upper treads, has a labor-intensive task.* (b) *Heated water is piped below the sidewalk to turn snow into water. As temperatures drop below freezing, this solution becomes a particularly energy-intensive procedure.*

(a)

(b)

(c)

Fig. 1.12 Varying provisions for user awareness in schools. The famous early solar-heated school at Wallasey, England (a *and* b) *depends on custodians and users to make the adjustments that alternately admit or exclude direct sun, control the extent of diffused light and outdoor air admitted, and insulate against loss of stored heat at night. (Photos by Reg McDonald.) In a similar climate, a windowless school in the Pacific Northwest* (c) *excludes the unpredictable outdoors. Here the users' consciousness is directed away from energy.*

As we face a continued increase in world population and continued depletion of renewable resources, it seems obvious that labor-intensive alternatives must be made available. The designer's challenge is to make these alternatives attractive and rewarding to use.

(c) The Trend from Energy Awareness to Unconscious Energy Use. With human labor being greatly reduced and the fuel that replaces it being brought into our buildings from far away, we have slipped into an unconscious use of energy. Until 1973, with energy prices declining, the cheap, seemingly endless, and unobtrusive supply of energy encouraged increased use. Even our attitudes toward on-site energy sources, such as the sun, have evolved to exclude user awareness (Fig. 1.12).

With the apparent depletion of nonrenewable energy resources, a "scarcity ethic" has developed. This encourages increased self-reliance in the face of eventually decreasing imports, emphasizes conservation rather than increased production, and brings "user control" of on-site energy sources back among architectural design considerations.

With the widespread availability of computer-controlled heating, cooling, and lighting systems, sophisticated thermal performance is possible without getting people involved (see many examples in Chapter 6). Thus, the overall trend toward user unconsciousness continues, except where visible devices such as sunshading automatically change position during the day.

This is the challenge to the designer: to engage the users in indoor environmental control without making them slaves to a building's thermal regulation.

1.3 Energy Sources and Uses

Before taking a detailed look at the way we use energy in buildings, it is helpful to get perspective on how energy sources are converted to end uses for all purposes. Figure 1.13 shows this energy pattern in the United States from 1960 (during the war in Vietnam) to 1979 (6 years after the start of the OPEC oil embargo). Note that solar energy and wood fuel used in space heating are *not* included; only fuel sources ''measurable'' (by sales) are included here. Several trends are evident; in those 19 years, we gradually required more energy per person (from 243 million to 357 million Btu/person, annually); we gradually wasted a higher percentage of our energy sources in converting them to useful work (from 52 to 57%); and we gradually imported a much higher percentage of the oil we use. The first two trends are directly traceable in large part to our increasing reliance upon electricity; this will be discussed further in Section 1.4.

A note about units of energy: One Btu (British thermal unit) is about the amount of energy expended in burning a standard wooden match; a quad is one quadrillion (10^{15}) Btu's, about equal to 44 million short tons of coal, or 1 trillion cubic feet of natural gas, or 170 million barrels of crude oil, or 8 billion gallons of gasoline, or 63 short tons of dry firewood—or one quadrillion wooden matches. A more scientific definition of energy terms is found in Chapter 4, Table 4.1.

The ''effective use'' figures at the right side of Fig. 1.13 include all our end uses, such as for transportation and industrial processes, as well as that portion of the total (approximately one-third) which is used for all purposes in buildings. Figure 1.14 graphs the approximate percentage of gross-

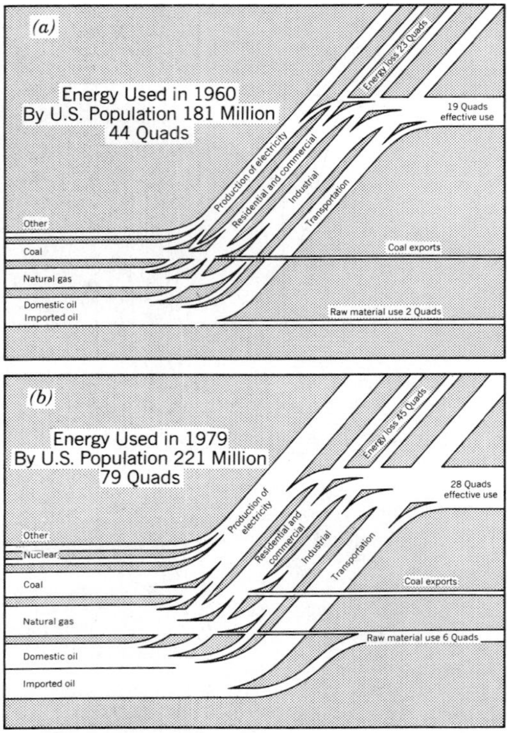

Fig. 1.13 *The flow of energy, from source to end use (''effective use''); graphs show the growing percentage that is wasted in the conversion of energy from one form (low-grade resource) to another (higher grade resource, such as electricity). ''Unmetered'' fuels, such as solar and wood energy used in building heating are not included. One Quad = 10^{15} Btu. (Based on* National Geographic Society, *1981.)*

energy use for various end uses in our society. Transportation, 25% of gross usage, is linked with energy used in buildings; this is particularly evident in the layout of residential areas, which essentially require frequent automobile use. The automobile accounts for about one-half of transportation's share of energy usage. The goal of renewable energy sources for buildings may conflict with the goal of reduced automobile usage, since solar energy is a somewhat thinly spread resource. This tends to limit the density of buildings; decreasing density usually results in the increasing use of transportation, as people travel farther between places of home and work. This conflict has several possible resolutions that may include a merging of work and home sites, a substitution of communication for transportation, or a provision

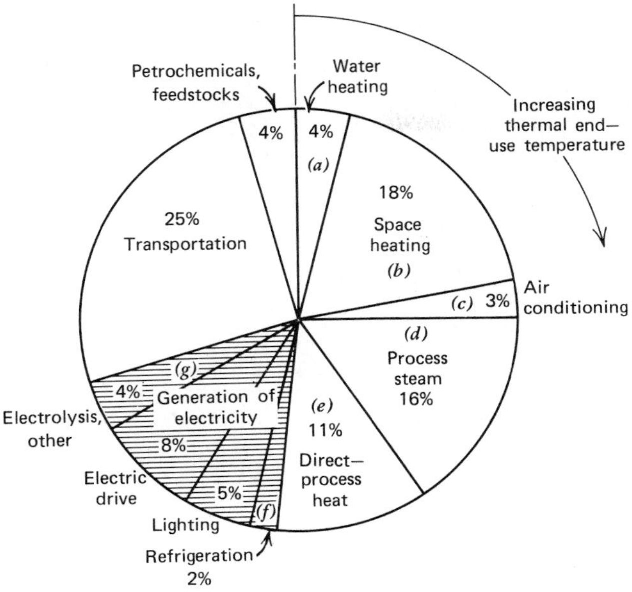

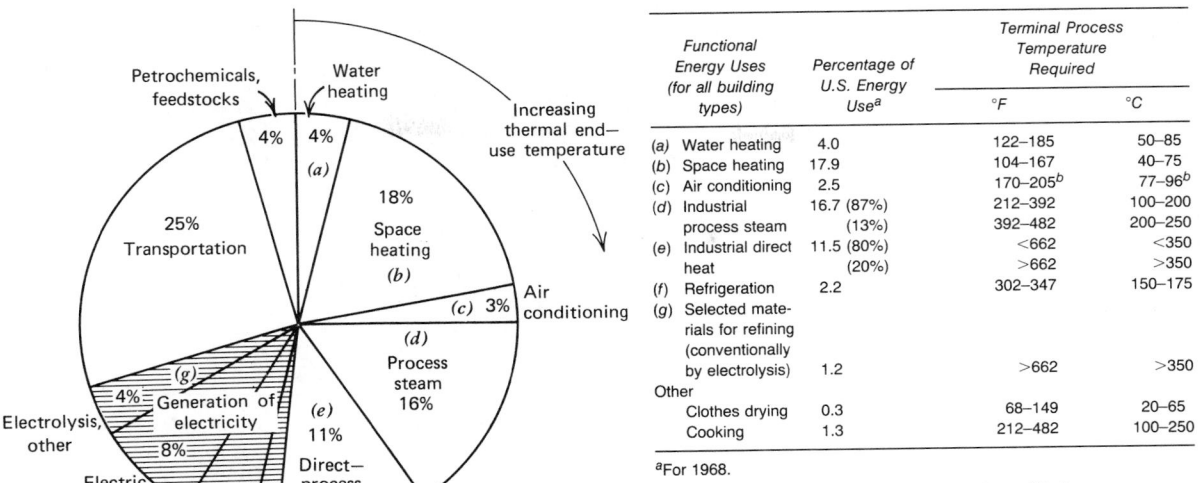

Functional Energy Uses (for all building types)	Percentage of U.S. Energy Use[a]	Terminal Process Temperature Required °F	°C
(a) Water heating	4.0	122–185	50–85
(b) Space heating	17.9	104–167	40–75
(c) Air conditioning	2.5	170–205[b]	77–96[b]
(d) Industrial	16.7 (87%)	212–392	100–200
process steam	(13%)	392–482	200–250
(e) Industrial direct	11.5 (80%)	<662	<350
heat	(20%)	>662	>350
(f) Refrigeration	2.2	302–347	150–175
(g) Selected materials for refining (conventionally by electrolysis)	1.2	>662	>350
Other			
Clothes drying	0.3	68–149	20–65
Cooking	1.3	212–482	100–250

[a]For 1968.

[b]These temperatures are the range for solar air conditioning.

Fig. 1-14 *Percentage and quality of energy use in the United States as of 1968. The categories of water and space heating, air conditioning, refrigeration, and lighting are those most influenced by architectural design. "Quality" is related to terminal process temperatures; "low-grade" uses, at lower temperatures, can be fueled by a variety of energy sources. As temperatures rise, the variety of possible fuels diminishes. At the top of the "high-grade" uses (shaded on graph), electricity is produced by the expenditure of much lower grade fuel. (See also Fig. 1.20) (From E. Lazslo and J. Bierman, Eds.,* Goals in a Global Community, *published by Pergamon Press, Inc., Fairview Park, Elmsford, NY. Copyright © 1977 by State University of New York.)*

for transportation to be powered by renewable resources.

A summary of our U.S. energy sources, their characteristics, and their impacts on architecture and the environment is presented in Table 1.1. Several widely discussed, but still uncertain, energy sources omitted from this table include nuclear fusion, nuclear fission with breeder technology, oil shale, and solar energy from sea-thermal gradients. The uncertainties inherent in these sources include a combination of technical, political, and environmental problems (especially with the nuclear breeder and with oil shale) and involve doubt about whether more energy would eventually be produced by some of these sources than is required to develop them.

Similarly, the known total U.S. quantities of natural gas, crude oil, coal, and uranium in the ground are greater than the portion called *recoverable* reserves; the numbers of Quads of this most easily tapped portion for each of the nonrenewable energy sources are shown in Table 1.1. The remainder of our U.S. supply of these resources is called by names such as "submarginal" (difficult to utilize, expensive to extract, etc.) or "undiscovered."

Despite the significant total Quads of solar radiation that the United States receives each year, we utilize relatively little for energy in buildings. The relatively high initial cost of using such a plentiful but thinly spread resource has discouraged its use, but as the submarginal, nonrenewable resources must be worked to replace the dwindling recoverable portions, the difference in cost between solar and nonrenewable energy sources will shrink.

Probably even before nationwide utilization of solar energy is achieved, the United States is expected to shift away from reliance on gas and oil to coal as its major energy source. This is evident from a comparison of Figs. 1.15 and 1.16 and Table 1.1.

TABLE 1.1 **U.S. Energy Sources**

*Part A. **Nonrenewable Sources***

Nonrenewable Source	*Estimated U.S. Reserves and Resources*	*Notable Characteristics*	*Form in Which Supplied to Consumer*	*Typical Architectural Utilization as Fuel (not including electricity)*	*Architectural Impacts*	*Environmental Impacts*
Coal	U.S. Proven coal resources,[a] total 4577 Quads 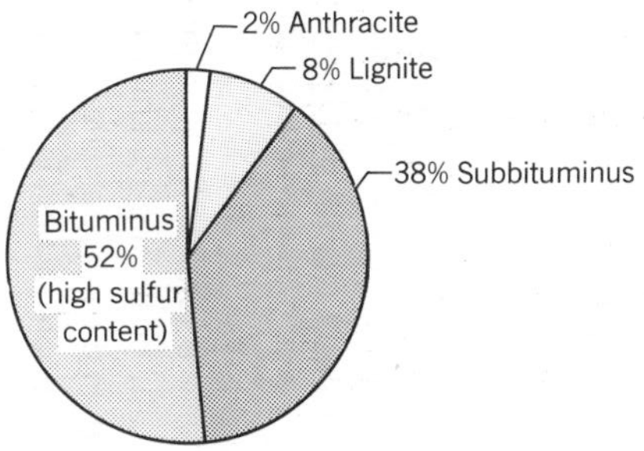 U.S. 1981 use: 16 Quads[b]	• Least flammable fossil fuel to transport, but bulky (less fuel value for its weight) • Coal dust presents mining, transport, and storage problems • Coal gasification can supply cleaner fuel to electric generating plants	• Mostly as electricity • Some as coal	• Some central steam plants (for campuses and other building complexes)	• Combustion air, stack, air pollution control equipment, and cooling tower or pond • District heating possibilities for waste heat from steam plants • Large and "dusty" fuel storage and delivery areas • Ash collection and disposal	Recovery • Strip mining's long-term ecological disruption, including acid drainage to streams • Underground mine safety problems, and subsidence of surface over mines Transport • For slurry pipelines, water pollution Use • Air pollution (particularly with high-sulfur coal); acid rain • Thermal pollution Waste • Landfills for ash disposal

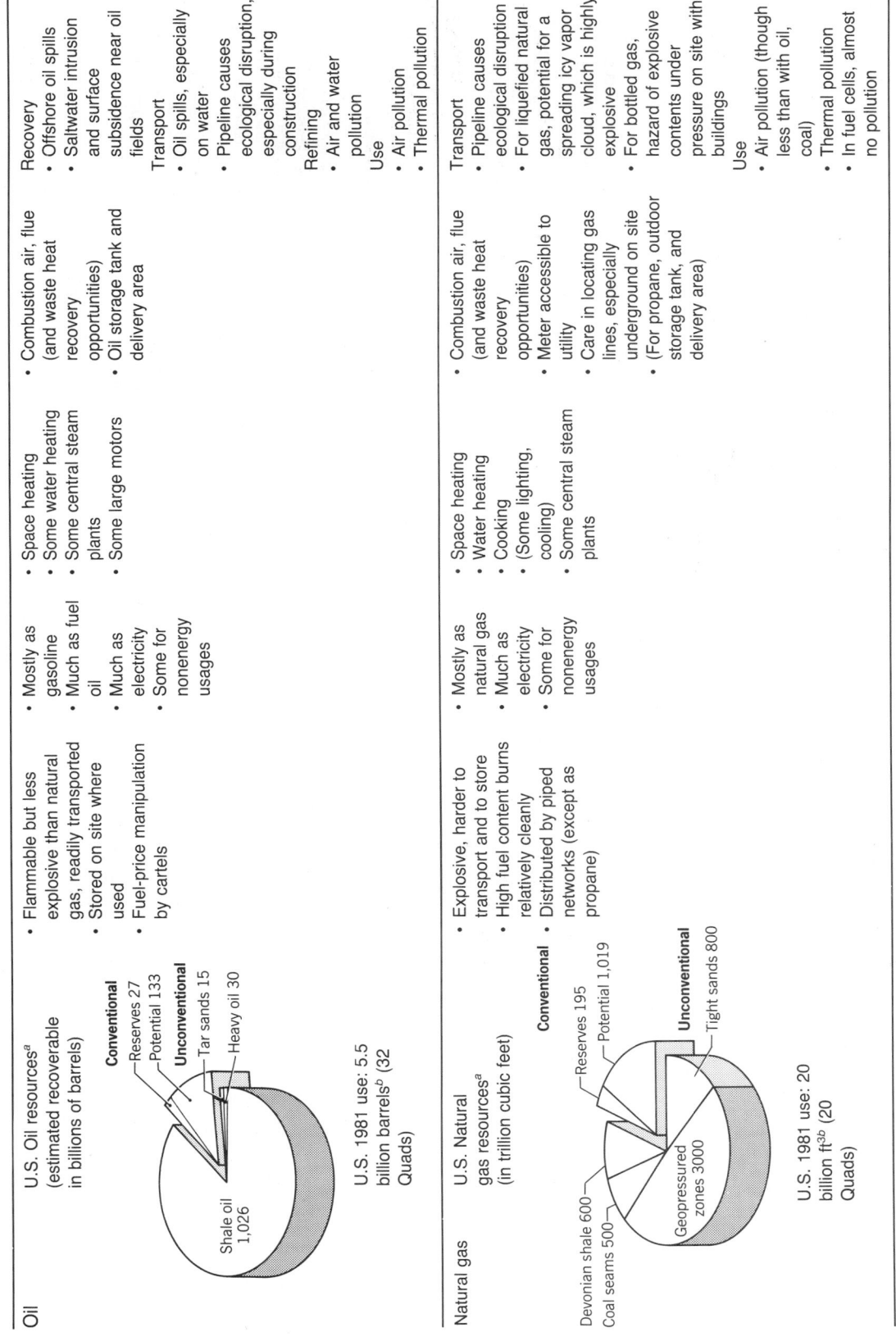

Oil	U.S. Oil resources[a] (estimated recoverable in billions of barrels) 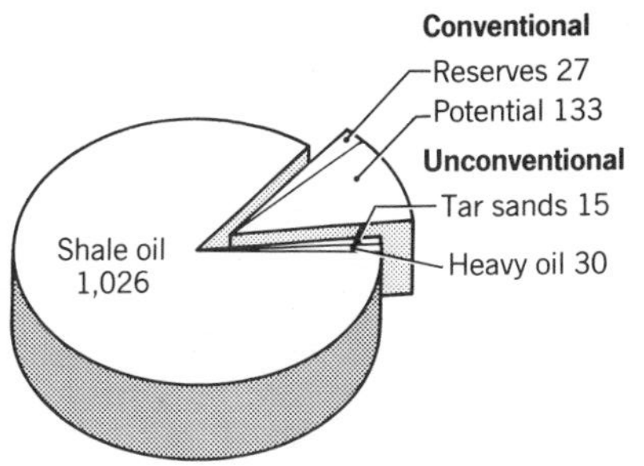 U.S. 1981 use: 5.5 billion barrels[b] (32 Quads)	• Flammable but less explosive than natural gas, readily transported • Stored on site where used • Fuel-price manipulation by cartels	• Mostly as gasoline • Much as fuel oil • Much as electricity • Some for nonenergy usages	• Space heating • Some water heating • Some central steam plants • Some large motors	• Combustion air, flue (and waste heat recovery opportunities) • Oil storage tank and delivery area	Recovery • Offshore oil spills • Saltwater intrusion and surface subsidence near oil fields Transport • Oil spills, especially on water • Pipeline causes ecological disruption, especially during construction Refining • Air and water pollution Use • Air pollution • Thermal pollution
Natural gas	U.S. Natural gas resources[a] (in trillion cubic feet) 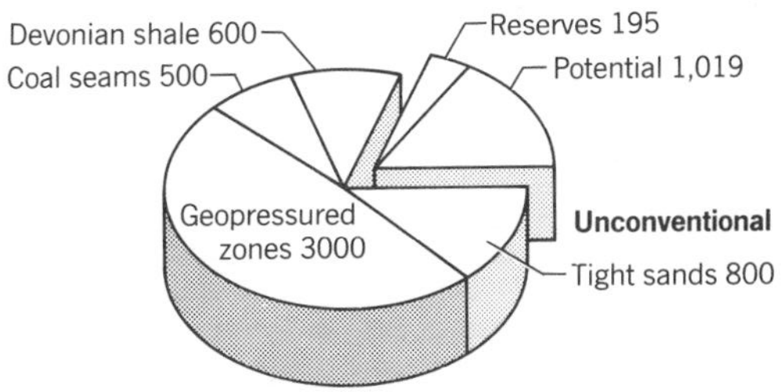 U.S. 1981 use: 20 billion ft^{3b} (20 Quads)	• Explosive, harder to transport and to store • High fuel content burns relatively cleanly • Distributed by piped networks (except as propane)	• Mostly as natural gas • Much as electricity • Some for nonenergy usages	• Space heating • Water heating • Cooking • (Some lighting, cooling) • Some central steam plants	• Combustion air, flue (and waste heat recovery opportunities) • Meter accessible to utility • Care in locating gas lines, especially underground on site • (For propane, outdoor storage tank, and delivery area)	Transport • Pipeline causes ecological disruption • For liquefied natural gas, potential for a spreading icy vapor cloud, which is highly explosive • For bottled gas, hazard of explosive contents under pressure on site with buildings Use • Air pollution (though less than with oil, coal) • Thermal pollution • In fuel cells, almost no pollution

TABLE 1.1 **U.S. Energy Sources (*Continued*)**

Part A. ***Nonrenewable Sources*** **(*Continued*)**

Nonrenewable Source	*Estimated U.S. Reserves and Resources*	*Notable Characteristics*	*Form in Which Supplied to Consumer*	*Typical Architectural Utilization as Fuel (not including electricity)*	*Architectural Impacts*	*Environmental Impacts*
Uranium	U.S. uranium recoverable reserves: 220 Quads[c] U.S. 1981 Use: 2.9 Quads[b]	• Light-water reactors now operating use very little of the energy contained in the enriched fuel • Fuel enrichment itself consumes large amounts of energy • An especially controversial source, involving both environmental and arms control issues • Fuel/price manipulation by cartels	• Almost all as electricity	• (Radioactive materials utilized in some medical and scientific equipment)	• (Radioactive materials require heavy shielding for delivery, storage, use, and waste pickup areas) • "Fallout shelter" designation for qualifying spaces	Recovery • Mining problems similar to coal; less bulky, but with additional problem of radioactive tailings Transport • Radiation shielding and security needs for nuclear wastes Refining • Large amount of energy consumed in fuel enrichment • Radiation shielding and security needs for waste reprocessing Use • Potential for catastrophic radiation release • Radiation shielding and security needs • Thermal pollution Waste • Long-term radiation shielding and security needs
Geothermal energy	U.S. potential[a], estimated 18.5 Quads/year	• Renewable, from earth's core, but millions of years after heat is removed from outermost layers of rock	• Electricity • Hot water	• Direct heating of buildings	• "Hot springs" water often highly corrosive • Source usually must be very near to point of use	• Unknown impacts of large-scale removal of heat from outer crust • Subsidence of surface is possible where water is removed without recharging to the earth

Part B. **Renewable Sources**

Renewable Source	Estimated Maximum U.S. Arrival Rate [Quads (10^{15} Btu) per Year]	Notable Characteristics	Form in Which Supplied to Consumer	Typical Architectural Utilization as Fuel (not including electricity)	Architectural Impacts	Environmental Impacts
Solar radiation	47,000[d] Estimated U.S. yearly capture[d] by year 2000	• Thinly spread, with large daily and seasonal variations in most of the United States		• Daylighting • Space and water heating • Some cooling	• Window and skylight areas for daylighting • Internal arrangement of thermal storage materials in passive solar buildings • Large collectors, storage volumes • Less densely built complexes for solar buildings	• Ecological disruption where large land areas are converted to solar collection • Soil depletion where organic waste is not returned as fertilizer • River ecosystem disruption by hydroelectric dams
Fuel wood	3[d]		• Wood fuel	• Some cooking (wood and gas)		
Farm waste	6[d]	• Methane gas highly explosive	• Methane gas • Gas or oil			
Photosynthesis fuel	15[d]					
Solar heating and cooling of buildings	9[e]		• Heated air or water			
Photovoltaics	7[e]		• Electricity			
Hydropower	9[d] U.S. 1981 use: 3 Quads	• Varies with seasonal river flows	• Electricity			
Wind power	6[e]	• Varies considerably between sites	• Electricity			
Waste paper and plastic incineration	2[f]		• Electricity			• Air pollution • Thermal pollution
Tidal power	3[d]		• Electricity			• Ecological disruption of estuaries

[a]National Geographic Society (1981).

[b]U.S. Department of Energy (1982).

[c]From Fisher (1974). This material was based on U.S. Geologic Survey Circular 650, *Energy Resources of the United States* (1972).

[d]From Starr (1971).

[e]Jewell (1978).

[f]R. Berry and H. Makino, "Energy Thrift in Packaging and Marketing," *Technology Review*, 1974. © 1974, by the Alumni Association of the Massachusetts Institute of Technology, Cambridge.

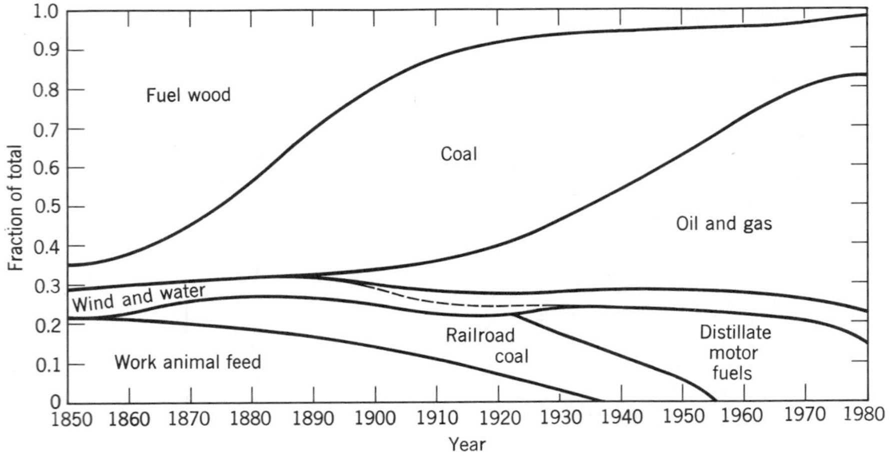

Fig. 1.15 *U.S. fuel input, 1850 to 1980, illustrates the progression from dependence on renewable fuels (wood and work animal) to fossil fuels (coal, then oil and gas). Wind and water power changed their form from "mills" to hydroelectricity between 1890 and 1940. Similarly, though not shown here, much fossil fuel is now converted to electricity before use. (From Fisher, 1974, and Meyers, 1983.)*

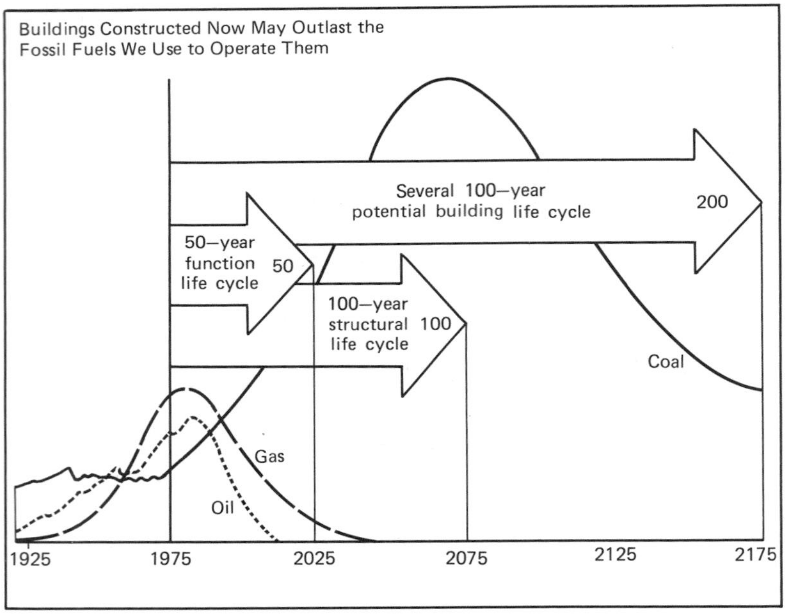

Fig. 1.16 *Life cycles of buildings and of fossil fuel reserves, showing our relative dependence on three major energy sources. In 1925 we used mostly coal, some oil, and very little natural gas. In 1975 natural gas ranked first, followed by oil, then coal. By 2025, we may be almost entirely dependent on coal. (From California, Office of the State Architect, 1976.)*

In our past, we shifted from almost complete reliance on renewable resources (wood and work animals) to nonrenewable (fossil fuel) resources. More recently, we shifted from heavy reliance on coal to reliance on oil and gas. At present, electricity is the dominant form by which the energy of coal is delivered to buildings. Thus part of the prediction for electricity's rapid growth (Section 1.4) can be traced to the relatively plentiful coal reserves in the United States. The waste heat and other environmental impacts associated with the production of electricity are significant and reinforce the necessity of careful and appropriate use of this versatile energy form by the designers of buildings.

The environmental impacts listed in Table 1.1 show only the most obvious or large-scale impacts for each energy source. Although each source has environmentally damaging consequences when used, the relatively mild impacts of some sources make them more environmentally attractive than others. The details of such consequences of energy utilization are not treated further in this text, except for a summary of air pollution in the United States and its impacts on buildings and their occupants, in Chapter 3 (Fig. 3.16).

1.4 Architecture and Energy Usage

(a) The Recent Past. The "energy-rich" decades of the 1950s and 1960s saw a dramatic increase in recommended levels of heating and lighting for buildings. The evolution of higher recommended indoor temperatures is shown in Table 1.2. Lighting levels rose even more rapidly, as a look at past editions of this textbook (Table 1.3) reveals. (Footcandles are defined in Chapter 18.) These lighting levels have been reassessed, as shown by later scaled-down recommendations of the Federal Energy Administration. These recommended increases and their impact on energy consumption reached a peak in the years around 1970, which we could call the "pre-energy-aware-

TABLE 1.2 **Criteria for Thermal Comfort Since 1900**[a]

Date	*Environmental Specifications*	*Dry-Bulb Temperature (Fahrenheit) for 40% rh*
Before 1900	65–70 F DBT	65–70
Early 1900	56 F WBT	70
1914	68 F DBT 40% rh	68
1923	66–72 F DBT, 19–61% rh	66–72
1923	62–69 ET 64 ET (optimum)	68
1925	63–71 ET 66 ET (optimum)	71
1929	66–75 ET 71 ET (optimum)	77
1939	64.8–76.0 F ET 71.8 F ET (optimum)	78
1941	68 ET (optimum)	74
1938–1956	73–77 DBT, 25–60% rh	73–77
1960	77.6 F DBT, 30% rh 76.5 F DBT, 85% rh	77
1965[b]	73–77 DBT, less than 60% rh	73–77
1965[c]	77 F DBT, 70% rh 79.5 F DBT, 20% rh	78
1975[d]	72 F DBT, (30%) winter 78 F DBT, summer	
1980s	See Figs. 2.2–2.4	

[a]DBT = dry bulb temperature, WBT = wet bulb temperature, ET = effective temperature, similar to "operative temperature" defined in Section 2.2.

[b]ASHRAE Standards, Series 55.

[c]For lightly clothed subjects engaged in sedentary activity. Reprinted by permission from *Criteria for Thermal Comfort* by R. G. Nevins, Institute for Environmental Research, Kansas State University.

[d]ASHRAE Standard 90–75.

TABLE 1.3 **Trends in Recommended Minimum Lighting Levels (footcandles)**

Category	*From* Mechanical and Electrical Equipment for Buildings		*Federal Energy Administration (1976)*
	2nd Edition (1945)	*5th Edition (1971)*	
Offices			
Accounting, bookkeeping	30	150	
Regular office work	20	100	50
Conference rooms	10	(Not listed)	30
Corridors, stairs	5	30 (But not less than one fifth the level of the adjacent area)	10
Schools			
Auditoriums	10	30	
Classrooms, regular deskwork	20	70	
Drafting, drawing	30–50	100	
Sewing	50–100	150	
Libraries			
Reading room	20	70	

ness'' period in recent U.S. history. One example of the extent to which increased lighting has been carried is seen in an electric utility headquarters building, renovated and enlarged in the mid-1960s (see Fig. 1.17). Among its features are:

- 300 to 350 footcandles (fc) of light in office areas.
- Up to 550 fc in display areas.
- Up to 600 fc in conference and demonstration rooms.
- 500-watt (W) luminaires in each of 288 window sills, for nighttime facade lighting.

During the design phase of this building, the architect calculated that construction costs would be reduced by about $1 million if lighting levels were cut to 150 fc. This savings would be evident in the lower number of luminaires and in the reduced size of cooling equipment necessary to remove the surplus heat. The utility, however, expressed its interest in leading the trend to higher lighting levels, and retained the 300-plus footcandle level. Some of the heat emitted by this lighting is captured and used when needed to heat the office building (''heat-of-light'' systems are discussed in Chapter 6, Section 6.9). Yet this high a lighting level produces a need for cooling, not for additional heating, in a typical office building on most winter days. The lighting thus provides surplus heat for almost the entire year and requires that still more energy be spent to remove surplus heat. This utility has since discontinued the lighting of its facade at night but continues to provide very high lighting levels for its interiors.

Fig. 1.17 *An office building for an electric utility, designed in the mid-1960s when energy seemed to be in plentiful supply. Extraordinarily high interior lighting levels as well as (now unused) 500-W exterior night lights in* each *window sill are remnants of a less energy-conscious era. (Photo by Stan A. Adams.)*

(b) The Near Future. Most projections of future U.S. energy consumption assume continued, though less rapid, growth in demand. These views of energy supply and demand are constantly shifting, and this year's estimate does not assure next year's performance, nor certainly the next decade's. Consider the impact of the 1973 oil embargo both on the reality of our subsequent consumption and on the revised forecasts for the future. Figure 1.18 plots the predicted impact on an "active conservation program," including improved energy efficiency for automobiles, homes, and office buildings. Compare this 1976 forecast of about 80 Quads for 1981 with the actual total of 74 Quads used (Fig. 1.19). Clearly, energy efficiency in buildings can have an impact on future consumption patterns.

A more detailed look at building energy efficiency is found in many parts of this book, espe-

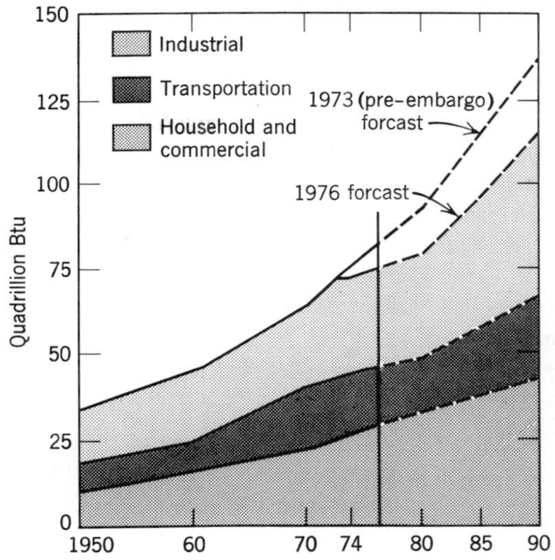

Fig. 1.18 U.S. energy consumption, past and projected future. Energy for buildings is expected to continue to form a large portion of our national energy usage. This forecast was made in 1976. (Based on Federal Energy Administration, 1976.)

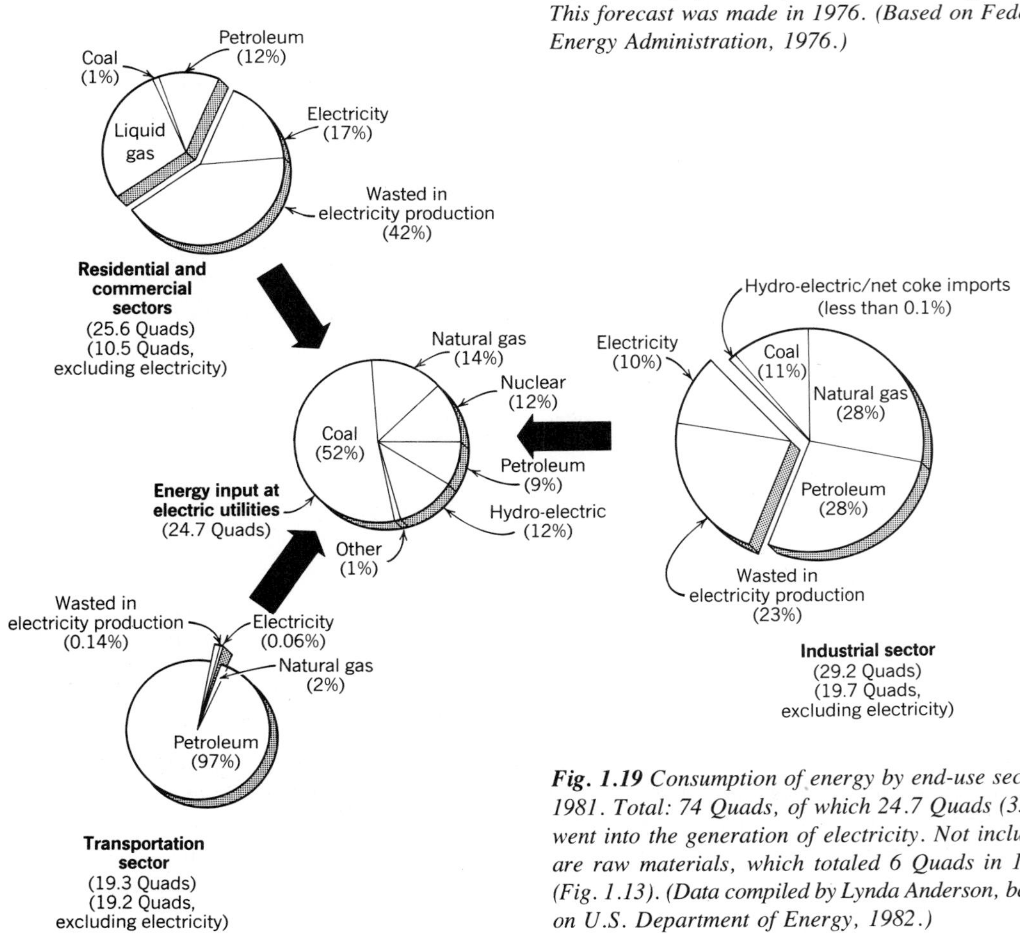

Fig. 1.19 Consumption of energy by end-use sector, 1981. Total: 74 Quads, of which 24.7 Quads (33%) went into the generation of electricity. Not included are raw materials, which totaled 6 Quads in 1979 (Fig. 1.13). (Data compiled by Lynda Anderson, based on U.S. Department of Energy, 1982.)

cially Part II (Thermal Control) and Part VI (Illumination). The relationship between indoor climate and illumination is likely to be particularly influential on building energy consumption, as the office building cited above demonstrates. Electricity is one of the keys to this relationship, for the following reasons.

1. Consumption of electricity is expected to rise about twice as fast as the overall energy demand, and more and more we are expected to use electricity in place of other energy forms. Part of the reason for this is seen in Table 1.1; for some energy sources, conversion to electricity before distribution to buildings is the most convenient option. (A hopeful note was sounded in 1982, the first year in which electricity sales actually declined, by 1.3%. However, the decline was by industrial users; residential and commercial electricity usage increased slightly.)

2. Electricity is almost the only source of illumination in buildings, other than daylight; the pre-1973 annual growth rate (about 7%) of electricity consumption was boosted by the steeply increased illumination recommendations and the cooling that accompanied them.

3. Electricity is a convenient and versatile energy form; it not only serves such high-temperature and highly concentrated (or ''high-grade'') energy tasks as lighting and motive power, but it can also serve the low-temperature simpler (or ''low-grade'') tasks, such as cooking, heating water, and space heating for buildings (see also Fig. 1.14). ''All-electric'' buildings have become commonplace, even though they are subject to paralysis in blackouts, as is any building dependent on a single energy source.

4. Electricity delivers only about one third the total energy that goes into its production; the other two thirds is usually lost as waste heat at the generating plant. See Fig. 1.19 for proportions of energy thus wasted.

Thus, the relationships among greater energy demands, electricity, lighting, and interior climate are complex. In the future, solar (photovoltaic) generation of electricity will remove most of the environmental costs associated with electricity production. Until then, as we increase our use of electricity, we doubly increase our use of the nonrenewable fuels that serve almost all today's electrical generating plants.

The building designer interested in slowing our energy-demand growth rate can begin with a careful utilization of electricity; low-grade energy tasks might be better served by nonelectric fuel sources. As an example, consider the furnace in a residence (Fig. 1.20). The low-grade alternatives to electric-resistance furnaces include not only fossil-fuel-burning furnaces but also renewable heat sources, such as wood and solar energy.

Although the low efficiency of electrical generation is obvious, the central generating plant has some advantages over many fuel-burning furnaces. Air pollution control is more efficient at a central plant, and the massive concentration of its waste heat could be coupled with certain industrial processes or piped to nearby buildings for space heating. (However, since buildings need this waste heat only in winter, a separate, additional cooling solution for the summer must also be provided: two cooling systems are obviously more costly to build.)

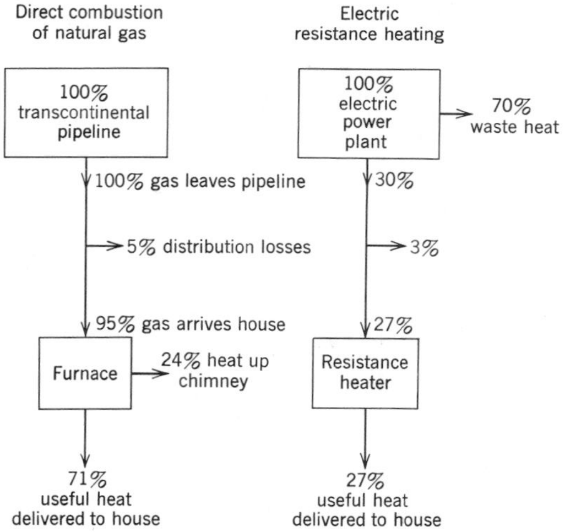

Fig. 1.20 *Efficiency comparisons. For lower grade heating tasks, more of the energy available in fossil fuels is utilized by the furnace, which burns the fuel directly. Much more efficient furnaces are now available (see Chapter 6; see also Fig. 6.57). The electric furnace using resistance heaters is fully efficient in the home, but the higher grade energy it receives has produced waste heat at the generating plant. (From Fisher, 1974.)*

Fig. 1.21 Combining general daylight with specific-task electric light. (a) *Library designed by Alvar Aalto at Mount Angel Abbey (Oregon) utilizes a central north-facing skylight to provide daylight on two levels of the interior. The individual electric lights* (b), *each with a pull chain, can be used when reading fine print makes more light desirable.*

The higher-grade tasks for which electricity is suitable also need careful consideration. The higher the level of electric lighting, the more cooling is needed; depending on the climate, every unit of energy expended in lighting will require another half-unit of energy for removing the surplus heat thus generated. The annual energy growth rate for the cooling of buildings is about 10%—more than three times as fast as our overall energy growth shown previously in Figure 1.18.

One way to reduce the electricity consumed by lighting is to substitute daylight for electric lighting. This is most applicable at the perimeters of buildings; yet, even the interiors of low-rise buildings can be served with daylight for the general or overall illumination, using small individually controlled electric lights only where and when needed (Fig. 1.21).

As with many such substitutions, the designer must consider the tradeoff: will more glass area to admit daylight produce greater heat loss on winter nights and undesired heat gain on summer days? (Calculations of heat gain and heat loss are presented in Part II.) In Chapter 6 techniques are discussed for protecting glass against heat loss with insulating shutters and for designing windows to minimize summer solar gain.

In summary, today's designer is aware of the overall price increases and fuel reserve decreases associated with nonrenewable fuels. The design response, however, is pulled in seemingly opposite directions: a tendency to *close in* the building to conserve energy, versus the notion of *opening* it to daylighting and passive heating and cooling opportunities. This balancing act between conservation and passive design will be a continuing theme in the next five chapters.

1.5 Energy in Building Construction

In the United States, the total energy expended in *constructing* new buildings, as a percentage of the total used for all purposes, appears relatively small (Fig. 1.22). The decisions that designers make regarding the building envelope and the structural and mechanical systems generally have greater energy consequences in building operation over many years than in the shorter construction period. Most of the energy embodied in construction is invested in the manufacture of materials and components (Fig. 1.23).

Some building types are much more "energy intensive" in the construction phase than others, as shown in Fig. 1.24. Single-family residences are relatively low in "energy intensiveness," compared to the total energy invested in them (from Fig. 1.22); such residences constitute almost one third of the total square footage of all new buildings constructed in the years for which these statistics are compiled. The single-family residence and the closely related garden apartments (two- to four-family residential and low-rise residential) are low in energy intensiveness primarily because they utilize so much wood in their construction, and

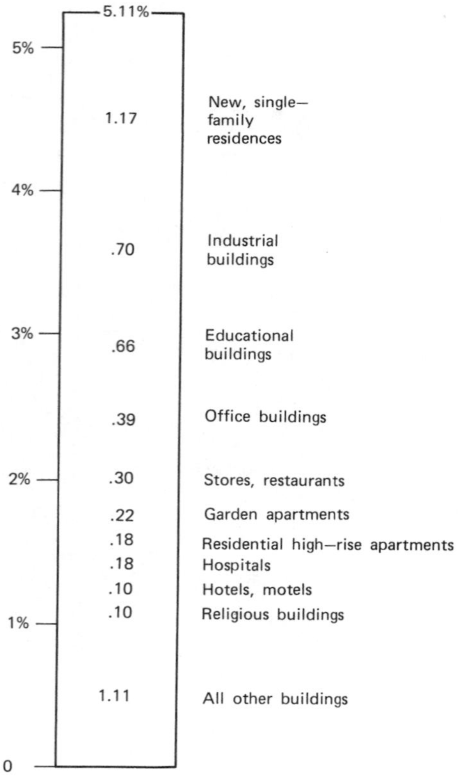

Fig. 1.22 Energy in new building construction, as percentage of U.S. total energy consumption, including energy to manufacture, transport, and erect new buildings. (From B. M. Hannon, et al., 1977.)

Fig. 1.23 Energy embodied in building construction goes largely into the manufacture of building materials and other components. (Based on Hannon, et al., 1977.)

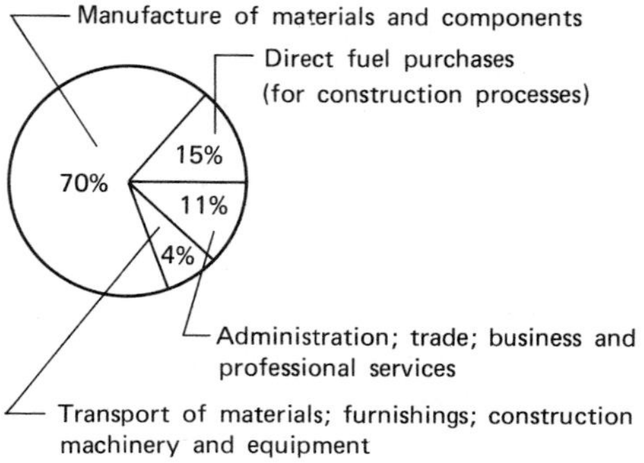

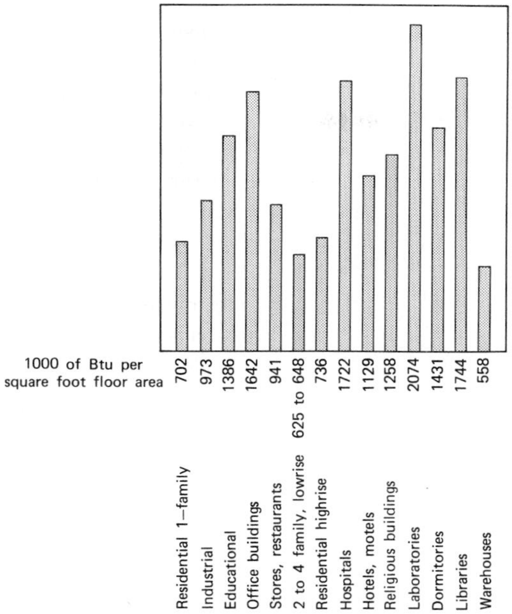

***Fig. 1.24** Energy embodied per square foot of floor area, for various building types. These comparisons of 1000 Btu per unit area are for the building types shown in Fig. 1.22. (From Hannon et al., 1977.)*

wood is a low-energy material. Table 1.4 compares the approximate total energy embodied in selected materials, delivered to the job site. Figure 1.25 illustrates the effects of added insulation on the total energy embodied per square foot of typical residential wall construction. (For the annual energy savings, see Fig. 1.27, following.)

Until recently, building material choices were rarely influenced by the amount of energy invested in those materials. Now that such information is becoming more widely available, energy content can be included with other factors—esthetics, durability, fire resistance, labor intensiveness, and installed cost—that a designer considers in choosing materials for a building.

A related consideration is the monetary cost of the energy-consuming equipment in a building. Table 1.5 indicates the portion of recent construction costs that are assigned to the building elements discussed in this text: plumbing, HVAC (heating, ventilating, air conditioning), and electrical equipment. The thermal and other environmental control systems for buildings are already a major influence on construction costs, as well as on energy consumption in operating the building. As the techniques and equipment for reducing energy consumption are developed through such steps as recovery of waste heat and the substitution of solar energy for fossil fuels, the percentage of mechanical and electrical system cost will probably become an even greater influence on the cost of constructing new buildings.

1.6 Energy in Building Operation

The typical building will consume more energy in its lifetime of operation than is used in its construction. The estimated energy needs for the construction, operation, and demolition of an office building in Albany, New York, are shown in Fig. 1.26.

There are several proposals to establish an upper limit for building energy consumption in operation. In their simplest form, such regulations would specify the maximum allowable energy input per building-area unit per year: British thermal units per square foot per year (Btu/ft^2 year), or megajoules per square meter per year (MJ/m^2 yr). A more complicated set of regulations to define—but relatively easy to apply to the design of walls, roofs, and so on—would establish maximum overall thermal transfer values. This means that the designer would choose a combination of wall materials—for instance, those that did not allow more than a specified rate of heat transfer. These more detailed regulations would also include HVAC system performance (indoor temperatures, controls, ventilation rates, humidity control, zoning, pipe and duct insulation, equipment performance ratings, etc.), hot water service, electrical distribution systems, and lighting. Details of such proposals are found in Section 5.2.

The impact of more energy-efficient buildings can be estimated in several ways. The American Institute of Architects (AIA) estimated in 1976 that a high-priority national energy conservation program could save at least 30% of the energy then used in existing buildings, and 60% of the energy that would have been used by new buildings, had they not been designed with energy conservation in mind. The AIA estimated that by 1990

TABLE 1.4 **Total Energy Embodiment in Selected Building Materials per Unit of Material at Job Site**

Material	*Unit*	*Total Embodied Energy (Btu/unit) at Job Site*
Wood products		
Lumber	Board foot	7,600–9,800
Shingles		7,300
Flooring		10,300–14,300
Mouldings		17,900
Glu-lam		16,700
⅜-in. plywood (softwood)	Square foot	5,000–5,800
Paints	Gallon	437,000–508,500
Asphalt roofing		
Rolls	Square foot	7,800–11,000
Shingles		25,600–29,700
Mineral-surface insulating board siding		67,500
Glass		
Flat glass: double strength	Square foot	15,430
Flat glass: tempered		72,600
Plate and float glass, ⅛ to ¼-in. thick		48,000
Laminated plate glass, ¼-in. and over		212,500
Stone and clay products		
Common brick	Per brick	14,300
Ceramic glazed brick		33,413
Quarry tile	Square foot	51,000
Ceramic mosaic tile and accessories, glazed		63,600–68,700
Concrete block	Per block	31,800
Ready-mix concrete	Cubic yard	2,594,300
Gypsum board-⅜ in.	Square foot	5,300
Mineral wool insulation, 4½-in. thick	Square foot	8,300
Iron and steel		
Steel sheets: 22 gauge	Square foot	29,400
16 gauge		58,800
Galvanized sheets: 22 gauge		49,800
16 gauge		98,500
Steel shapes: W 12 × 65, carbon	Lineal foot	1,217,800
W 12 × 65, alloy		1,749,200
WT 6 × 27, carbon		543,344
WT 6 × 27, alloy		780,400
Reinforcing bars: #2		2,600
#8		41,800
Welded wire mesh: 2 × 4, 14/14	Pound	3,900
2 × 12, 8/8		25,400
Pipe, carbon steel: ½-in. diameter	Lineal foot	21,900
6-in. diameter		489,700
Stainless steel sheets: cold rolled	Pound	138,300
hot rolled		80,800
Aluminum		
¼-in. plate	Square foot	420,700
1-in. plate		1,680,300
⅛-in. sheet		174,800
Standard shapes: 8 I 8.81	Lineal foot	811,800
6 I 5.10		469,900

Source: Hannon et al. (1977).

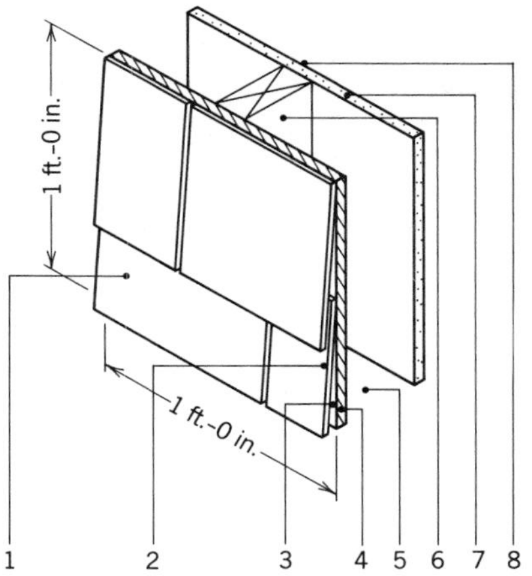

(a)

Construction	*R Value*		*Embodied Energy (Btu/ft²) in Building Section*
1. Outside surface (15-mph wind)	0.17		—
2. Wood shingles (½ in. × 8 in. lapped)	0.87		7,315
3. Building paper (asphalt)	0.15		—
4. Plywood (½ in.)	0.62		7,705
5. 4 in. Airspace	0.97		—
6. 2 in. × 4 in. at 16 in. o.c.	—	4.35	3,486
7. Gypsum wallboard (½ in.)	0.45		6,920
8. Inside surface (still air)	0.68		—
	3.91	4.35	25,426
$U = 1/R =$	0.26	$U =$ 0.23 at framing	
Adjusted U (to account for framing) $=$	0.25		

Addition of Insulation	*R Value*		*Embodied Energy (Btu/ft²) in Building Section*
Add 3½-in. batt insulation	11.00		Add 6,860
Deduct R value of air space	0.97		
	10.03		
Add to above R value	3.91		
	13.79		32,286
$U = 1/R =$	0.07	$U =$ 0.23 at framing	
Adjusted U (to account for framing) $=$	0.085		

Fig. 1.25 *Energy embodied per square foot in typical residential wall constructions.* (a) *With the addition of insulation, frame walls experience a 29% increase in embodied energy—and a 73% decrease in heat loss rate.* (b) *Similarly, insulation added to a brick veneer or frame wall increases embodied energy by only 3%, while decreasing heat loss rate by 72%. See also Fig. 1.27. (From Hannon et al., 1977.)*

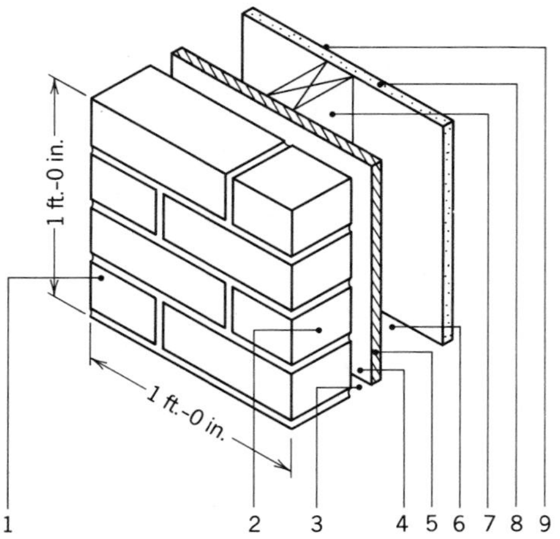

(b)

Construction	*R Value*		*Embodied Energy (Btu/ft²) in Building Section*
1. Outside surface (15-mph wind)	0.17		—
2. Brick and masonry (4 in.)	0.44		105,004
3. 1-in. Air space	0.97		—
4. Building paper (asphalt)	0.15		—
5. Plywood (3/8 in.)	0.47		5,779
6. 4-in. Air space	0.97		—
7. 2 in. × 4 in. at 16 in. o.c.	—	4.35	3,486
8. Gypsum wallboard (3/8 in.)	0.32		5,297
9. Inside surface	0.68		—
	3.98	4.35	119,566
$U = 1/R =$	0.25	$U = 0.23$ at framing	
Adjusted U (to account for framing) $=$	0.24		

Addition of Insulation	*R Value*		*Embodied Energy (Btu/ft²) in Building Section*
Add 3½-in. batt insulation	11.00		Add 6,860
Deduct R value of air space	0.97		
	10.03		
Add to above R value	3.98		
	14.01		126,426
$U = 1/R =$	0.07	$U = 0.23$ at framing	
Adjusted U (to account for framing) $=$	0.085		

Fig. 1.25 *(Continued)*

TABLE 1.5 **Construction Costs for Mechanical and Electrical Equipment**

	Portion of Total Construction Dollar Cost (%)		
Building Type	*Plumbing*	*Heating, Ventilating, Air Conditioning*	*Electrical*
Apartments, mid-rise (4–7 story)	9	4	7
Auditoriums	7	10	9
Banks	4	7	10
Churches	5	10	9
College			
Classrooms and administrative offices	7	15	10
Laboratories	7	17	10
Community centers	7	10	9
Department stores	4	11	12
Hospitals	9	17	14
Libraries	5	10	11
Motels	9	4	10
Offices, mid-rise (5–10 story)	3	11	8
Restaurants	8	12	10
Retail stores	5	9	10
Schools			
High schools	7	10	10
Elementary schools	7	12	10
Supermarkets	6	8	12

Source: © Robert Snow Means Co., Inc. Reproduced with permission from *Building Construction Cost Data 1983.*

the implementation of such a program would save as much energy as our nation could expect from the daily output of any one of our 1990 domestic energy sources. A more specific example is shown in Fig. 1.27.

The building envelope is not the only opportunity for energy conservation. The significant energy savings now available in more efficient appliances and mechanical equipment are detailed in Chapter 6; Fig. 1.28 summarizes such opportunities within residences.

As energy used in operating buildings becomes more expensive and more difficult to obtain, its influence on the design process will become evident over a wider range of the designers' choices. We can expect buildings to be placed not only to take advantage of natural energy sources, but to be so conscious of the energy they consume that significant efforts will be made to utilize formerly wasted heat. It is possible that even a single-family residence will be so thoroughly insulated and so well equipped to recover otherwise wasted heat that a furnace (or other space heating equipment) will become unnecessary. For many brightly lighted office buildings, this has already occurred.

1.7 Energy in Building Reuse or Demolition

Compared to construction and operation, relatively little energy is invested in a building's demolition (see again Fig. 1.26). Yet, another look at Fig. 1.23 is a reminder that about 70% of the total energy invested in construction is embodied in a building's materials and components. If more of a building can be recycled, more energy can be recovered. At present, the recovery of usable materials from demolition is limited. The cost of labor is high, and the cost of energy is relatively low; it is currently easier, quicker, and cheaper to reduce a building to rubble and haul it to a landfill.

Designing for the recycling of buildings is a two-part balancing act. First, the designer should

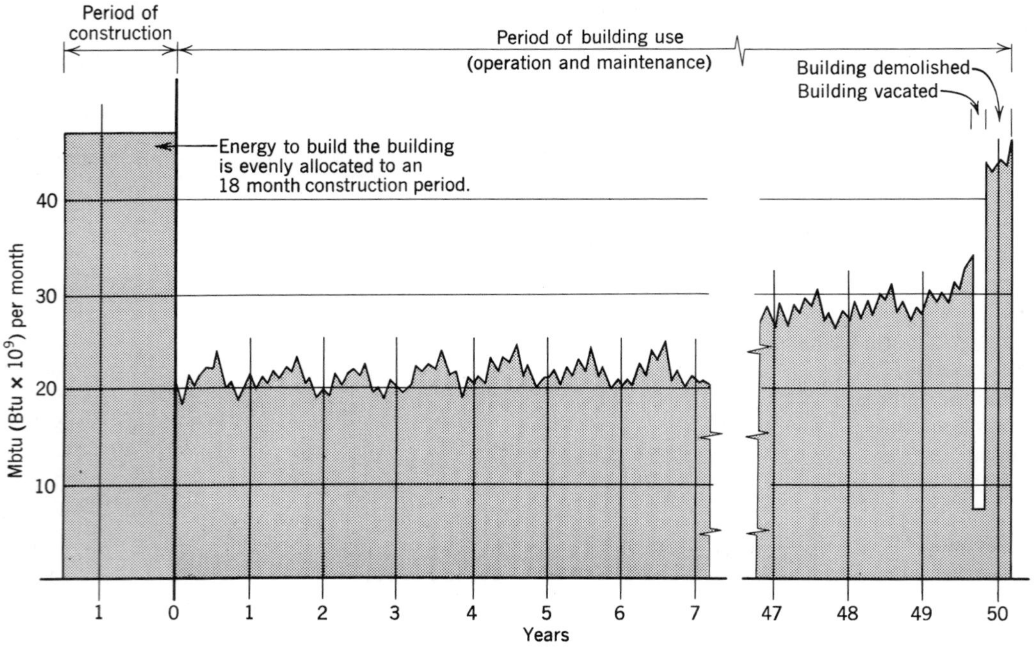

Fig. 1.26 *Energy use in constructing, operating, and demolishing an office building. This shows the approximate energy used over an estimated 50-year life span for a 650,000-ft² office building in Albany, New York. Energy in construction includes total energy embodied in the materials and is shown as constant through the construction period. The peaks and valleys of energy use in each year of operation correspond to the expected weather, peaks during the extremes of winter cold and summer heat, valleys in the milder spring and fall. As the building ages and energy leaks become more numerous, the consumption is assumed to rise gradually. (From* Architecture and Energy, *by Richard G. Stein. Copyright © 1977 by Richard G. Stein. Used by permission of Doubleday & Company, Inc.)*

provide enough flexibility to prolong the useful life of the building by enabling it to easily adapt to changed usage. Flexibility can be expensive to implement physically and can result in a bland "sameness" throughout the structure. The latter characteristic is easier for the designer to change than the former. Second, the design can allow for demounting parts so that the structure can safely remain intact, while reusable materials and components are removed. Yet this can result in heavier buildings, where floor systems are not structurally integral with beams. It also discourages the integration of mechanical and structural systems, as is discussed in Chapter 6. Furthermore, a demountable building is perhaps especially subject to energy leaks, such as cracks widening around self-contained components of the facade.

Some initial design guidelines for recyclable buildings are as follows:

1. Design the structure to be separable from everything else and, of itself, to be easily disassembled. Extensive remodeling is then possible without major structural modifications, and at the end of a building's life, elements of its structure can be reused elsewhere.

2. Design for "breathing room" where possible, between a building and its neighbors, or between major spaces within a building. Some expansion is thus possible without rebuilding. This could include designing the columns and footings to support an extra floor or two, for vertical expansion.

3. Maximize the utilization of on-site (or natural) forces such as sun and wind. The less sophisticated the mechanical and electrical equipment, the less obvious will be the obsolescence of such equipment with the passing of time.

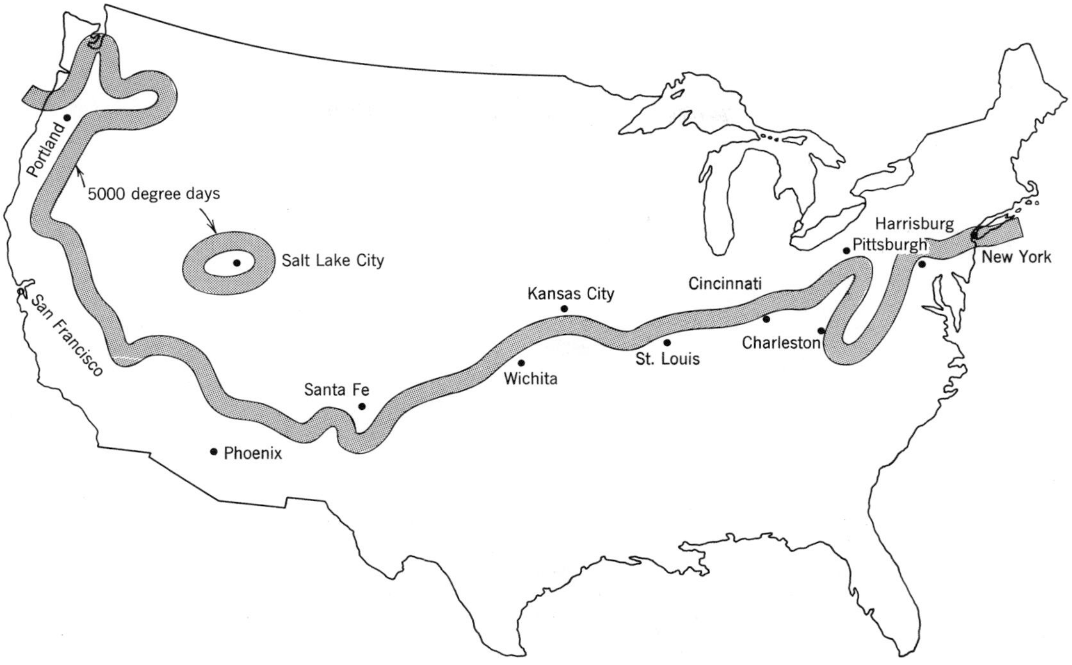

(a)

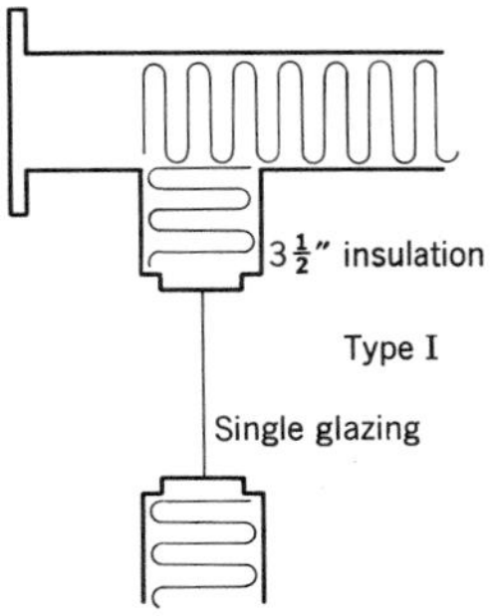

(b)

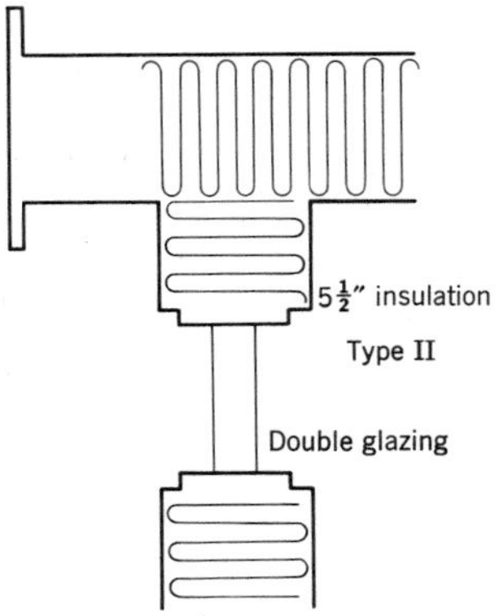

(c)

Fig. 1.27 *Estimated energy for constructing and operating a residence based on a one-story, flat-roofed house with a 1500-ft² floor area, located in an area with about a 5000-degree day heating season, near the band shown in the map* (a). *(Degree days are discussed in Chapter 5.) In areas north of the line, yearly energy consumed—and saved, by the more efficient house—would be even greater.*

	Energy Embodied	*Annual Energy Demand*	*Demand over 20 Years*
Figure 1.27*b*, type I	169 million Btu	109 million Btu	2180 million Btu
Figure 1.27*c*, type II	179 million Btu	77 million Btu	1540 million Btu
Difference between types I and II	10 million Btu more to build type II	32 million Btu less to operate type II yearly	640 million Btu less to operate type II for 20 years

The annual energy demand includes heat lost to cold air infiltrating the house. The extra energy used in building the 5½-in. insulation and the double glazing is quickly repaid—in about one third of one heating season! (Based on Hannon, et al., 1977.)

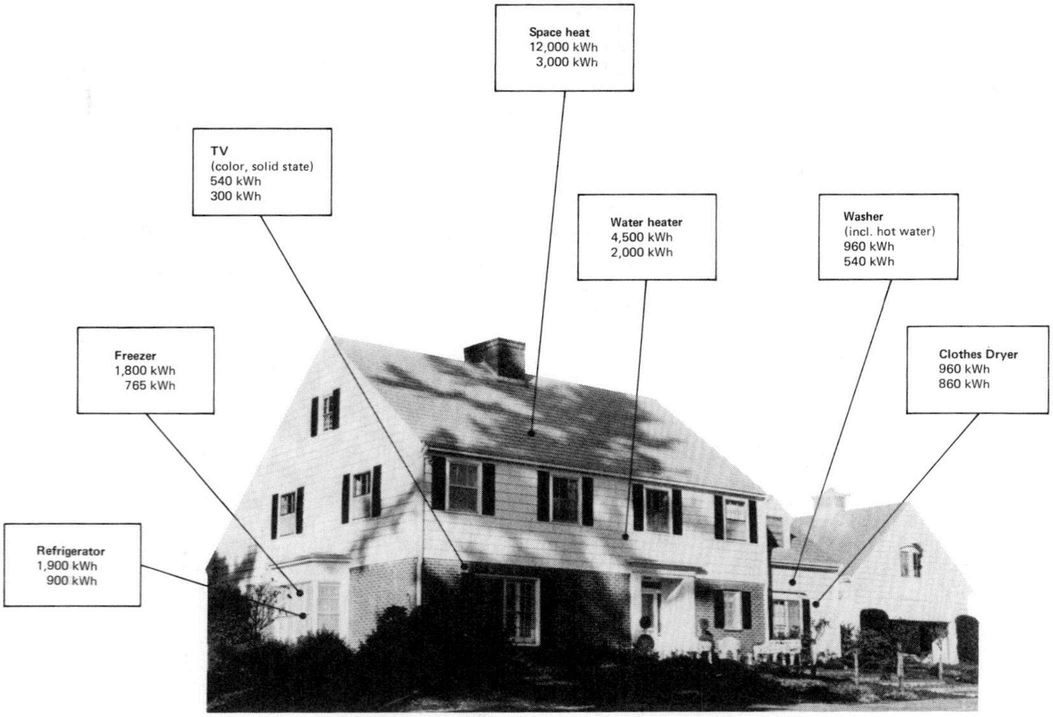

Fig. 1.28 *Appliance and equipment efficiency improvements in residences. Upper kWh figures are "before" and lower ones "after" the initiation of conservation measures. This is an example of energy conservation by "leak plugging," rather than "belt tightening"—the standard of living is not diminished. Space heat (typical for a rainy, cool Pacific Northwest winter) reduction is due to difference between today's Northwest energy codes, compared to those of the mid 1970s. Water heater reduction is due to the use of a heat-pump water heater (see Chapter 9). All other reductions are due to increased efficiency in today's appliances. When all other, more minor electricity uses are added in, the annual regional averages are: currently, 26,860 kWh; in the most efficient available home, 11,715 kWh—a reduction of 56%. (Courtesy of* Northwest Energy News, *Vol. 1, No. 4, July 1982, published by the Northwest Power Planning Council.)*

4. Use the materials and components distinctly; avoid combinations that make recycling of these elements difficult. A steel or plastic pipe embedded in a concrete slab is neither repaired nor recycled easily; some "sandwiches" or panels of building materials do not allow metals, plastics, and other products they contain to be separated for reuse at the end of the panel's life.

Although maximum savings of energy can be realized when building components can be used "as is," even the crushing and reprocessing of some (separated) building materials will save energy when compared to their original manufacture when virgin material; see Chapter 12.

References

Publications of the American Society of Heating, Refrigerating and Air Conditioning Engineers, Atlanta:

ASHRAE Standard 55 is the series "Thermal Environmental Conditions for Human Occupancy."

ASHRAE Standard 90 is the series "Energy Conservation in New Building Design." (ASHRAE Standard 90–75, for example, was published in 1975.)

Banham, R. (1969). *The Architecture of the Well-Tempered Environment,* University of Chicago Press, Chicago, and Architectural Press, London.

Berry, R., and Makino, H. (1974). "Energy Thrift

in Packaging and Marketing,'' *Technology Review*.

California, Office of the State Architect. (1976). *Building Value, Energy Guidelines for State Buildings,* Sacramento.

Commoner, B. (1976). *The Poverty of Power: Energy and the Economic Crisis,* Knopf, New York (distributed by Random House, New York).

Federal Energy Administration. (1976). *National Energy Outlook,* FEA, Washington, D.C.

Fisher, J. (1974). *The Energy Crisis in Perspective,* Wiley, New York.

Hannon, B. M., Stein, R. G., Segal, B. Z., and Serber, D. (1977). *Energy Use for Building Construction,* Center for Advanced Computation, University of Illinois, Champaign-Urbana.

Jewell, W. J. (1978). ''Biomass Fuels—Past, Present, and Future,'' presented at the American Section Conference of the International Solar Energy Society, Denver.

Lazslo, E., and Bierman, J., Editors. (1977). *Goals in a Global Community,* Pergamon Press, Elmsford, NY.

Meyers, R. (1983). *Handbook of Energy Technology and Economics,* Wiley, New York, 1983.

National Geographic Society. (1981). *Special Report; Energy,* National Geographic, Washington, D.C.

Nevins, R. G. *Criteria for Thermal Comfort,* Institute for Environmental Comfort, Kansas State University, Manhattan.

Shurcliff, W. A. (1977). *Solar Heated Buildings: A Brief Survey,* 13th ed.

Starr, C. (1971). ''Energy and Power,'' *Scientific American,* September.

Stein, R. G. (1977). *Architecture and Energy,* Anchor Press-Doubleday, New York.

U.S. Department of Energy. (1982). *Monthly Energy Review,* November.

2

COMFORT, CLIMATE, AND DESIGN STRATEGIES

One of the earliest reasons for building was to create shelter from the climate; to enhance thermal comfort. This chapter introduces this interrelationship between bodies, buildings, and climate by discussing bodily heat flow, then thermal comfort, then the design strategies that are appropriate to various climates. The final topic is the role of building skin elements as they change the outside climate to the one inside.

2.1 The Body

Because we are alive, we are always generating bodily heat. And because the cores of our bodies need to stay within a narrow temperature range, we nearly always need to lose this internally produced heat to our environment. The rate at which we produce heat changes frequently, as does the environment's ability to accept or reject heat. To regulate our bodily heat loss, we have available three common layers between our body cores and our environment: the first skin, our own; the second skin, clothing; and the third skin, a building.

(a) Metabolism. The rate at which we generate heat (our metabolic rate) depends mostly on our level of muscular activity, partly on what we ate and drank (and when), and partly on where we are in our normal daily cycle. Our heat production is measured in metabolic or *met* units (see Table 2.1). One met is defined as 58.2 W/m^2, or 18.4 Btu/h ft^2; it is the energy produced per unit of surface area, by a seated person at rest. (The total heat thus produced for the normal adult is about 117 W, or 400 Btu/h.) The more active we are, the more heat we produce, and our own first skin is the most important regulator of heat flow.

Of the many interactions between our skin and the rest of our body, consider these three: touch, blood, and water. The sensation of *touch* includes pressure and pain, as well as heat and cold. The sensations of heat and cold are produced by contact with surfaces or moving air, as well as by radiant heat. They are frequently the signals for shifts in bodily heat regulation, which is controlled by the thermostat in our brains, called the hypothalamus.

In response to signals from our surface, or to changes in our core temperature, the hypothalamus calls for changes in our *blood* distribution system. If we are too cold, we need to decrease our rate of heat loss, so the flow of blood toward the surface of our skin decreases. Blood carries

TABLE 2.1 **Metabolic Rates for Typical Tasks**

Activity	*Metabolic rate*[a] *(met units)*[b]
Reclining	0.8
Seated, quietly	1.0
Sedentary activity (office, dwelling, lab, school)	1.2
Standing, relaxed	1.2
Light activity, standing (shopping, lab, light industry)	1.6
Medium activity, standing (shop assistant, domestic work, machine work)	2.0
High activity (heavy machine work, garage work)	3.0

Source: Copyright © by the American Society of Heating, Refrigerating and Air Conditioning Engineers, Inc., Atlanta, GA; reprinted by permission from ASHRAE Standard 55-1981.

[a]For whole-body average heat production in watts and Btu per hour, see Table 5.21.

[b]One met = 58.2 W/m^2 = 18.4 Btu/h ft^2.

heat; the less exposure to cool air at the surface, the less heat is lost. This decreased blood flow toward the surface is called vasoconstriction. In this condition, less *water* is forced to the skin surface by our sweat glands, which reduces evaporation (and accompanying heat loss).

If these cold conditions worsen, we get "goose bumps," symptoms of our skin's unsuccessful attempt to create insulation by fluffing up our body hair. Other furred creatures probably find this amusing. Since the added insulation doesn't work, we soon increase our metabolic rate, or burn more fuel, by shivering. At this point, we seek help from our second, then third skins of clothing and architecture.

The opposite occurs when we are too hot; blood flow toward the skin surface increases (vasodilation), and the sweat glands greatly increase their secretion of water and salt to the skin surface. This increases heat loss by evaporation. As we get hotter and wetter, the roles of our second and third skins change. In a hot, humid environment, our first skin needs exposure to air to encourage heat loss, but protection from the sun's radiant heat; we need a simple second or third skin, which acts as a sunshade. In addition, thanks to mechanical equipment, our third skin can create and enclose a less humid body of air. In a hot, arid environment, our second skin may keep us from losing too much valuable water, while our third skin might help us with stored "coolth" from the often chilly night air. Both can also play the vital role of shading.

(b) Migration. An important design principle has been demonstrated here: that of *zoning*. Our bodies strive to maintain, at all costs, a nearly constant core temperature for our vital organs. This most-protected zone takes thermal precedence over the less-vital zone of our extremities, such as arms and legs; next down in priority are our fingers and toes. The most variable thermal zone of all is our skin surface. Similarly, buildings are frequently thermally zoned, and users (paralleling human blood flow) can retreat from—or advance to—the less-protected zones as conditions demand. On a larger scale, migration occurs from one climate to another as seasons change.

(c) Heat Flow. Once the blood and water get our surplus heat to the skin surface, we have four ways to pass it to the environment: *convection* (air molecules pass by our surface, absorbing heat), *conduction* (we touch cooler surfaces, and heat is transferred), *radiation* (when our skin surface is hotter than other surfaces "seen" but not touched, heat is radiated to these cooler surfaces); and *evaporation* (a liquid can evaporate only by removing large amounts of heat from the surface it is leaving). The amount of heat we lose by each of these four methods depends on the interaction of our metabolism, our clothing, and our environment; Fig. 2.1 illustrates the typical situation of a person at rest in a changing environment. As air and surface temperatures approach our own body temperature, we lose the options of convection, conduction, and radiation. Evaporation is essential, so access to dry, moving air is greatly appreciated. As air and surface temperatures fall, evaporation drops while convection, conduction, and particularly radiation increase. Under the "normal" conditions in Fig. 2.1 of about 70 F, the proportions of bodily heat loss per hour are:

Radiation, convection, and conduction	72%
Evaporation	
From skin surface	15%
From lungs (exhaled air)	7%
Warming of air inhaled to lungs	3%
Heat contained in feces and urine expelled	3%

(d) Clothing. Usually clothing acts as an insulating layer and is particularly effective at retarding radiation, convection, and conduction. As air and surface temperatures fall well below our own, we adjust the second skin. The insulating value of clothing is measured in Clo units, 1.0 Clo being equivalent to the typical American man's business suit in 1941, when Clo was born (see Table 2.2). The total Clo of what you're now wearing can be found by summing up the items and multiplying this total by 0.82; it can also be estimated by assuming 0.35 Clo per kilogram (0.15 Clo per pound) of your clothing. Our second skin is just as likely as our third skin to be dominated by considerations of style more than of thermal regulation; we can't always count on clothing—or buildings—to increase our comfort. A basic discussion of the third skin begins in Section 2.5.

Heat generated, Btuh	400	400	400	400	400 (curve 1)
Heat lost by:					
Radiation and convection	350	300	200	100	0 (curve 2)
Evaporation	50	100	200	300	400 (curve 3)
Total, Btuh	400	400	400	400	400 Total (curve 1)

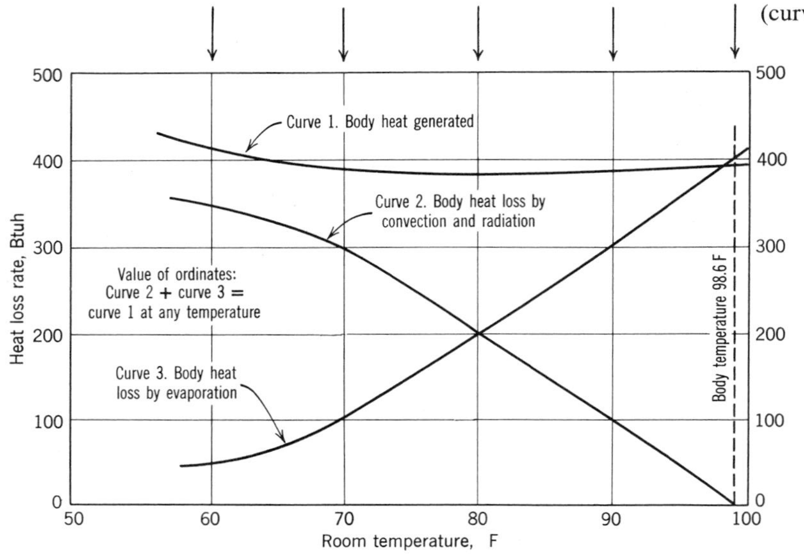

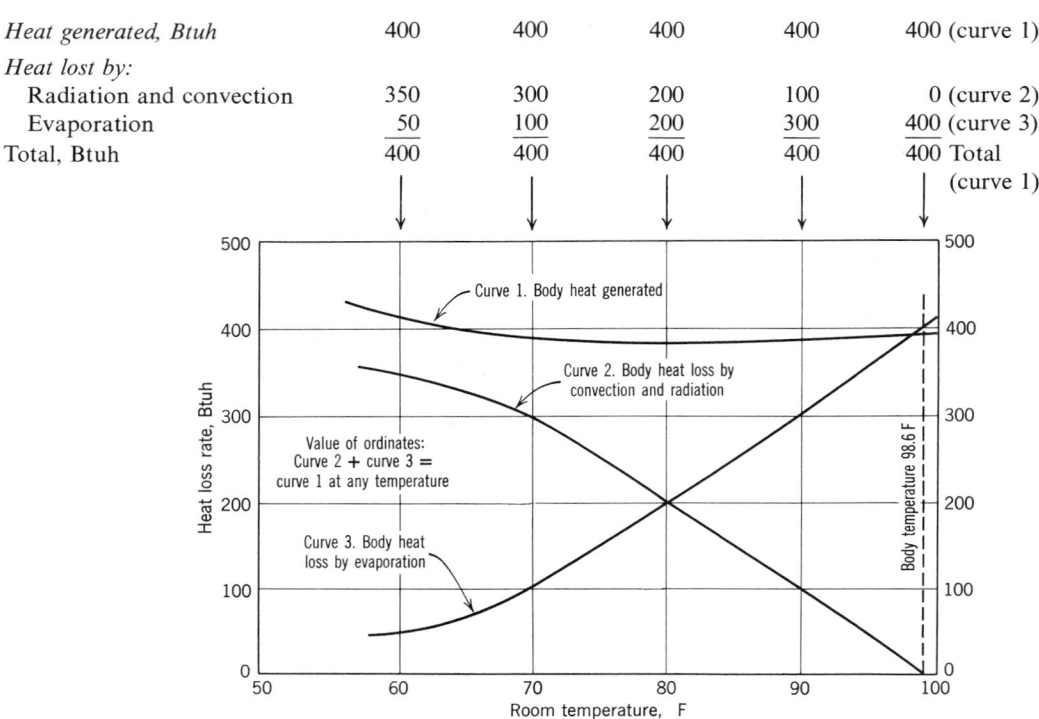

Fig. 2.1 *Heat generated and lost (approximate) by a person at rest (rh fixed at 45%).*

2.2 Comfort

A positive definition of comfort is ''a feeling of well-being.'' However, the more common experience of thermal comfort is a lack of discomfort—or being unconscious of how you are losing heat to your environment. There are three categories of factors that affect comfort: personal, measurable environmental, and psychological. Most *personal* (or physical) factors are under your control; they are the metabolism, migration, and clothing factors just discussed. *Measurable environmental* factors are the familiar tools of the designer and engineer: air temperature, surface temperature, air motion, and humidity. *Psychological* factors are also familiar designers' tools, but they are harder to measure for comfort: color, texture, sound, light, movement, and aroma. These psychological factors are often overlooked as we strive to meet the numerical physical criteria for comfort. However, our primary objective is to heat or cool *people;* buildings are our means to that end. Consider a courtyard in a hot climate. Its fountain *suggests* coolness in the color and texture of its water; running water provides splashing sounds and sparkles of light. Plants provide shade while their leaves sway in the slightest breeze; their blossoms yield a cool fragrance. Then consider a fireplace in a cold climate. The fire's color is intensely warm, it dances and casts a flickering light; it crackles, it yields a smoky aroma. Few textures *seem* hotter than that of glowing coals. Psychological factors can create an impression of comfort even when measurable factors indicate that mild thermal discomfort is expected. In our society, designers are often tempted to take numerical data quite literally. The measurable environmental factors have been extensively tested in laboratories—but such testing is done by *excluding* other factors. A holistic view of designing for comfort will consider numbers as a nonabsolute guide; common sense and the designers' own thermal experience play important roles as well. Lisa Heschong's *Thermal Delight in Architecture* (1979) is an excellent extension of these ideas.

Ultimately, our buildings will be expected to demonstrate success with regard to the measurable environmental factors of comfort, so it is neces-

TABLE 2.2 **Clo Units for Individual Items of Clothing**

Men		*Women*	
Clothing	*Clo*	*Clothing*	*Clo*
Underwear			
Sleeveless	0.06	Bra and Panties	0.05
T Shirt	0.09	Half slip	0.13
Briefs	0.05	Full slip	0.19
Long underwear, upper	0.10	Long underwear, upper	0.10
Long underwear, lower	0.10	Long underwear, lower	0.10
Torso			
Shirt		Blouse	
Light, short sleeved	0.14	Light	0.20
long sleeved	0.22	Heavy	0.29
Heavy, short sleeved	0.25	Dress	
long sleeved (plus 5% for tie or turtleneck)	0.29	Light	0.22
		Heavy	0.70
Vest		Skirt	
Light	0.15	Light	0.10
Heavy	0.29	Heavy	0.22
Trousers		Slacks	
Light	0.26	Light	0.26
Heavy	0.32	Heavy	0.44
Sweater		Sweater	
Light	0.20	Light	0.17
Heavy	0.37	Heavy	0.37
Jacket		Jacket	
Light	0.22	Light	0.17
Heavy	0.49	Heavy	0.37
Footwear			
Socks		Stockings	
Ankle length	0.04	Any length	0.01
Knee high	0.10	Panty hose	0.01
Shoes		Shoes	
Sandals	0.02	Sandals	0.02
Oxfords	0.04	Pumps	0.04
Boots	0.08	Boots	0.08

Source: Copyright © by the American Society of Heating, Refrigerating and Air Conditioning Engineers, Inc., Atlanta, GA; reprinted by permission from ASHRAE Standard 55-1981.

sary to understand how air and surface temperatures, air motion, and humidity are related to heat transfer.

Heat Transferred by:	*Is Primarily Dependent on:*
Conduction	Surface temperature
Convection	Air temperature, then air motion, then humidity
Radiation	Surface temperature
Evaporation	Humidity, air motion, air temperature

From this comparison, it is evident that humidity is of relatively low importance in cold conditions, where heat loss by conduction, convection, and radiation is dominant. But humidity is of primary importance in hot conditions, dominated by evaporative heat loss. This is further evident in comfort studies, which show that skin temperature is an important factor in cold conditions, while skin wettedness (percent covered by water) is of most importance in hot conditions.

A detailed analysis of a space's surface temperature can be performed using the *mean radiant temperature* (MRT). Such a calculation involves determining the temperatures of all surfaces and the position within the space at which to measure MRT, then determining the solid angle that is formed between the measuring position and the outer edges of each surface. These values are averaged to find the mean radiant temperature. However, a more useful term—both for impact on comfort and because it can be physically measured—is *operative temperature* (t_{op}), defined below. For details on MRT calculations, see the *Handbook for Fundamentals* published periodically by the American Society of Heating, Refrigerating and Air Conditioning Engineers (ASHRAE).

The interaction between comfort and those environmental factors can be generally summarized in Fig. 2.2. The "comfort zone" represents combinations of air temperature and relative humidity that most often produce comfort for a seated North American adult in shirtsleeves, in the shade. Surface temperatures are not markedly different from

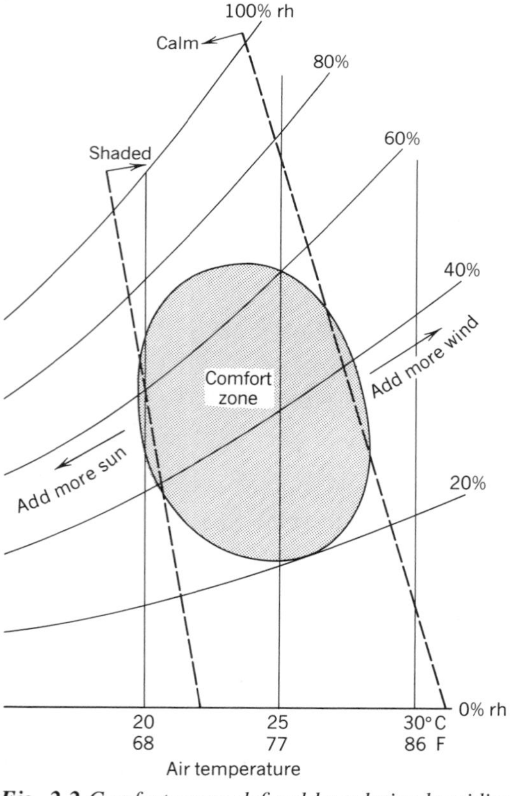

Fig. 2.2 Comfort zones defined by relative humidity and air temperature.

air temperatures in this zone, and minimal air motion is assumed. However, at air temperatures *below* (to the left of) this comfort zone, comfort is still attainable *if* added radiant heat (increased surface temperatures) such as increased exposure to sun is provided. (Also, more activity and/or more clothing is a possible option.) Similarly, at air temperatures *above* (to the right of) this comfort zone, comfort is still attainable *if* added air motion is provided. (Less activity and clothing may also be an option.) In both cases, limits are eventually reached; but the important point is that the basic comfort zone can be extended by utilizing more sun or more wind and, in very dry climates, by adding more moisture to the air.

Designers of buildings can aim for specific comfort goals, especially where the measurable environmental factors are concerned. The following combinations of air and surface temperatures, air motion, and humidity are those recommended by ASHRAE as maintaining comfort for 80% of the users of a space.

Figure 2.3 shows acceptable ranges of air and surface temperatures and humidity for persons wearing typical clothing for both summer and winter. Air motion is considered later in this chapter. The terms used here are defined as follows:

Operative temperature (t_{op}). Essentially an average of the air temperature of a space and the average of the various surface temperatures surrounding the space (or MRT). (To get a working grasp of operative temperature, get a 6-in. diameter metal toilet float, paint it flat black, and drill a small hole in it. Then through a rubber stopper, insert an ordinary air-temperature thermometer about 5 in. long, so that its bulb is near the center of the toilet float. The thermometer will read approximate operative temperature; it will rise dramatically when passed from shade to sun; it will fall slightly in a breeze. Such a device is called a globe thermometer. Ordinary air temperature by itself is often not a good comfort indicator, *especially* in passively heated or cooled spaces where radiant temperature or air motion may be more influential than air temperature.)

Relative humidity (rh). This familiar weather report term is approximately the ratio of the density of water vapor in air to the maximum density

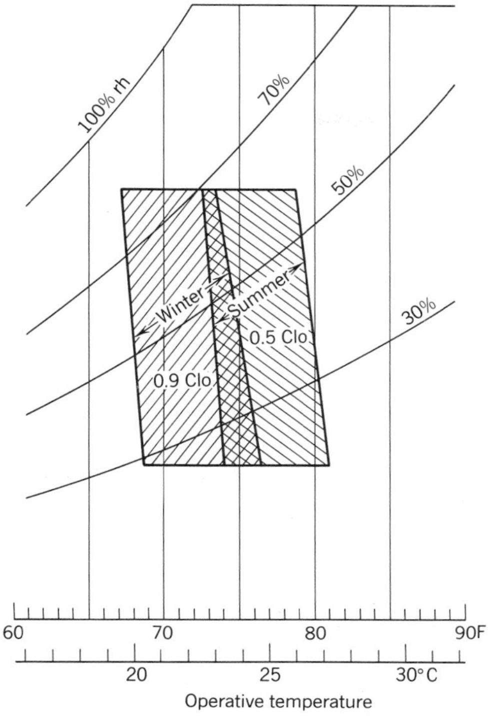

(a)

of water vapor that such air *could* contain, at the same temperature, if it were saturated.

The criteria for operative temperature–relative humidity combinations must be specific for combinations of activity and clothing, since such personal factors are just as important in determining comfort. Also, air motion must be included in any comfort standard. When using *any* standard, remember that conditions can change with the *age* of the users of buildings, as well as the degree of *acclimatization* that they have achieved. Very hot weather is usually more bearable at the end of summer than at its beginning.

Figure 2.3*a* shows the ASHRAE comfort standard range for typical home or office activity and clothing levels, along with fixed air motion conditions. Numerically, these are as follows:

ASHRAE 55—1981:

Activity: 1.2 met, mainly sedentary

Winter: 0.9 Clo, heavy slacks, long-sleeved shirt and sweater. Air motion at maximum = 0.15 m/s (30 fpm). The lower the

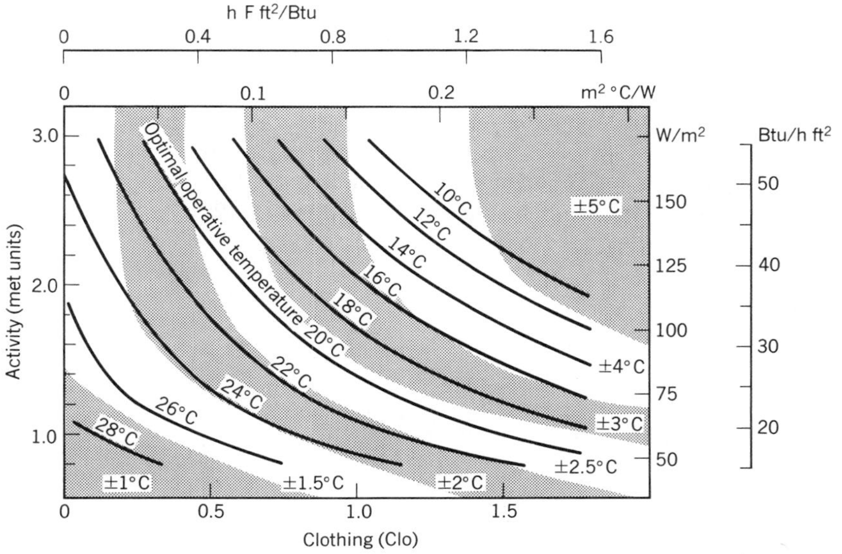

(b)

Fig. 2.3 (a). *The winter and summer comfort zones for light activity (1.2 met) that conform to ASHRAE Standard 55-1981. (Copyright © by the American Society of Heating, Refrigerating and Air Conditioning Engineers, Inc., Atlanta, GA; reprinted by permission from ASHRAE Standard 55-1981.)*
(b) *Relationships between preferred (optimal) operative temperature, clothing, and activity. The shaded bands represent the degree to which departures from the optimal are tolerable; that is, nearly all people will find the departures only "slightly cool" or "slightly warm." (P. O. Fanger, Laboratory of Heating and Air Conditioning, Technical University of Denmark, Copenhagen.)*

temperature, the less air motion is desirable. *Note:* Optimum operative temperature is 22.7°C (71 F).

Summer: 0.5 Clo, light slacks and short-sleeved shirt or blouse. Air motion is assumed at 0.25 m/s (50 fpm); however, this summer zone can be extended by increasing this air motion 0.275 m/s for each degree Kelvin, up to a maximum temperature of 28°C at 0.8 m/s of air motion (or increasing this air motion 30 fpm for each degree Fahrenheit, up to a maximum temperature of 82.5 F at 160 fpm of air motion). At this point, loose paper, hair, and other light objects might start to be blown about (however, see Table 2.3). *Note:* Optimum summer operative temperature is 24.4°C (76 F). If minimal clothing is worn (0.05 Clo), this optimum operative temperature is 27.2°C (81 F).

The converging edges of the "comfort zones" in Fig. 2.2 show that less tolerance of hotter air temperatures is expected at higher relative humidities; it's harder to sweat successfully in a humid environment, because the moisture can't easily evaporate. The horizontal edges of these comfort zones are based on practical humidity limits as follows: *above* the upper edge, excessive indoor moisture with mold problems can be expected. *Below* the lower edge, respiratory discomfort is expected from excessively dry air; coughs and nosebleeds can result. (For those acclimatized to "extreme" conditions, such discomfort is minimized.)

Figure 2.3*b* presents the relationship between clothing, activity, and comfort. It shows the expected tolerance of lower temperatures with increased clothing and activity; importantly, it also shows a greater tolerance of temperatures somewhat different from the optimum with increased clothing and activity. Figure 2.3*b* does not illustrate such subtleties as a preference for a lower operative temperature early in the morning, which increases during the morning. The preferred operative temperature then drops just after lunch.

A different approach to comfort standards has been proposed for buildings that take a more passive approach to heating and cooling. In these buildings, direct sun might add significantly to body warming in winter; strong air currents might be expected as a part of summer body cooling. Arens et al. (1980) demonstrated a wider range of comfort conditions, and considerably more tolerance for summer air motion, than the 1981 ASHRAE standard range. A summary of these proposals is shown in Fig. 2.4. Such graphs are called "bioclimatic" charts because of their interrelation of climate and human comfort factors. Where the users of buildings are expected to adjust to the wider temperature swings associated with passive buildings, these standards should be used. Some differences in the basic assumptions of Arens et al. from ASHRAE 1981 should be noted:

TABLE 2.3 **Indoor Air Velocity and Comfort**

Air Velocity	*Possible "Lower Temperature" Comfort Sensation (between 80 and 90 F; larger numbers correspond to high-humidity areas)*	*Probable Impact*
Up to 50 fpm	No change in comfort sensation	Unnoticed
50–100 fpm	2–3 F lower	Pleasant
100–200 fpm	4–5 F lower	Generally pleasant, but causing a constant awareness of air movement
200–300 fpm	5–7 F lower	From slightly drafty to annoyingly drafty
Above 300 fpm	More than 5–7 F lower	Requires corrective measures if work is to be efficient and health secured

Source: Adapted from Victor Olgyay, *Design with Climate: Bioclimatic Approach to Architectural Regionalism,* Copyright © 1963, Princeton University Press. Reprinted by permission.

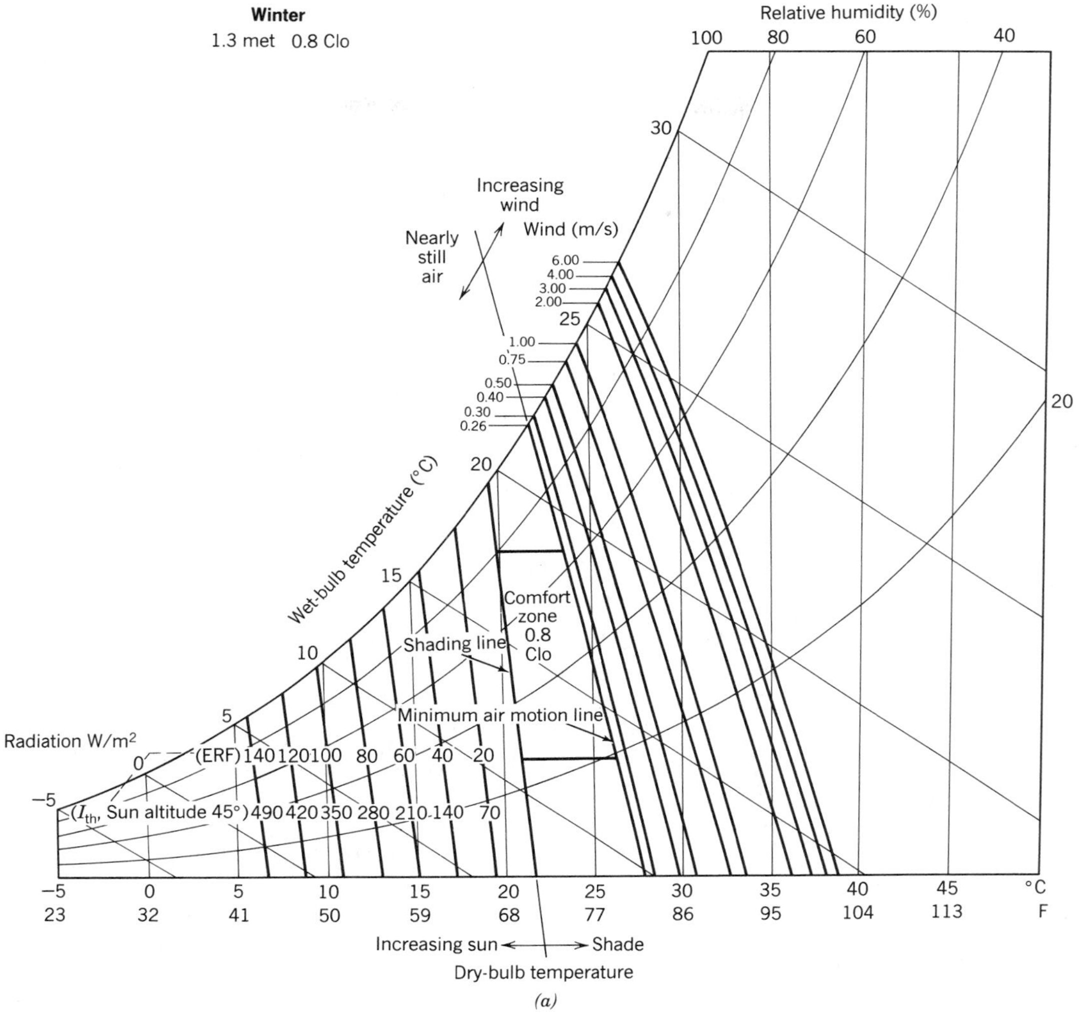

Fig. 2.4 *Comfort zones that encourage passively heated and cooled buildings. The winter comfort zone* (a) *and the summer* (b) *are combined for the year-round zone shown in* (c). *Based on Arens, et al., (1980).*

Activity	1.3 met (ASHRAE, 1.2 met)
Winter	0.8 Clo (ASHRAE, 0.9 Clo)
Summer	0.4 Clo (ASHRAE, 0.5 Clo)

Note particularly that rather than "operative temperature," ordinary air temperature is charted here. Added radiant heat below the shading line can be utilized to extend the comfort zone, as can added air motion above the minimum air motion line. *Radiant heat* to be added to an extended comfort zone in Fig. 2.4 is shown in two quantities: effective radiant field (ERF) is a measure of the net radiant heat flux to the body from all surfaces at temperatures other than air temperature. However, a more convenient figure for designers may be the total solar radiation (or *insolation*) on a horizontal surface, termed I_{TH}. Such values are more readily available, and the added-radiation lines of Fig. 2.4 represents I_{TH} converted to its approximate radiant impact on a human body's surface area, when the sun is at an altitude of 45°. Arens et al. (1980) give further details on the effects of solar radiation on comfort (e.g., at other sun altitudes or for other activity and clothing combinations).

Added air motion is shown for a wide range of velocities; while Table 2.3 indicates that people easily tolerate air motion outdoors up to about

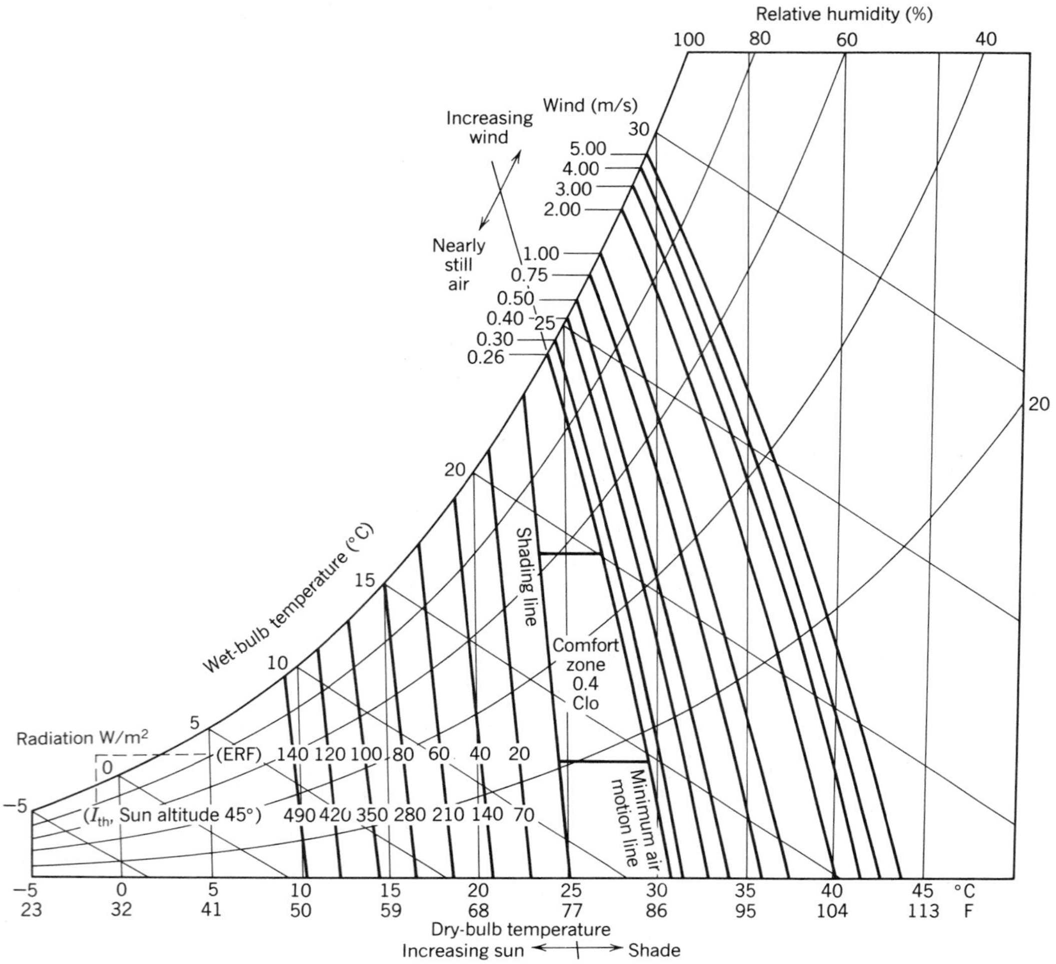

(b)

Fig. 2.4 (continued)

3 m/s (about 600 fpm), indoor studies have indicated 2 m/s (about 400 fpm) maximum for overhead fans. Different comfort zones are also to be expected in other cultures and climates; tolerance for heat is higher in the always-hot cultures of the tropics; cold is much less bothersome to those acclimated to Arctic conditions.

Although the designer can select a range of air and surface temperatures and relative humidity to achieve average comfort conditions in a space, there remains the more personal matter of comfort at each work station or location where people spend longer time periods. The location of heating, cooling, and ventilating components is therefore an important detail. A closer look at the body, air and surface temperatures, and air motion and humidity follows.

The human body is most affected by its thermal environment in its thermally sensitive places. Although we sweat and are sensitive to heat or cold over nearly all our skin surface, we are most thermally sensitive due to concentrations in these places:

Heat receptors: Fingertips, nose, elbows.

Cold receptors: Upper lip, nose, chin, chest, fingers.

So, on a hot and humid day, cool air moving across the face is a particularly strong promise of comfort, while on a cold day, a burst of heat (such as

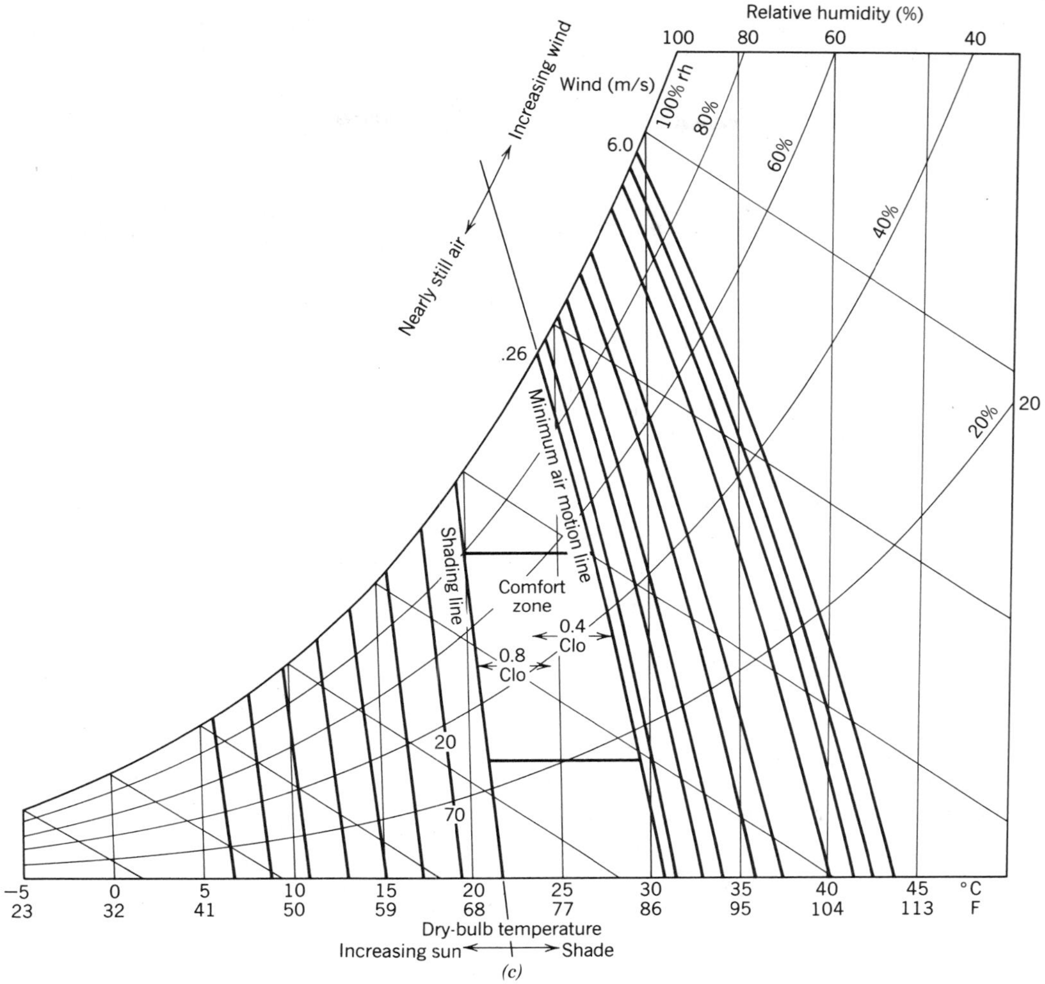

Fig. 2.4 (continued)

radiant heat from a light, a heater, or a cup of coffee) to the face and fingers is quickly effective.

A note here about our sense of touch: our fingertips are really most sensitive to the *rate* at which heat is being *conducted* to colder objects, or from hotter objects. Since the temperature of our skin at fingertips in ordinary conditions is in the high 20s on the Celsius scale (high 80s Fahrenheit), our sense of touch works *against* many passively heated surfaces, which feel ''cool'' even though they are successfully warmer than room air temperature. These surfaces are made of materials that conduct heat rapidly, so that they can soak up solar radiation without overheating the room. This high conductivity makes them eager to accept human warmth as well, and persuades us that they are cooler than in fact they are. Everyone has walked barefoot from a rug to a tile floor; even if a surface temperature thermometer shows the same temperature for both rug and tile, our feet will signal ''colder'' for the tile. This same characteristic works *for* passively *cooled* surfaces; their conductivity persuades us that they are even cooler than our passive design has in fact achieved.

Since warm air is less dense than cold air, it rises; this principle is often applied in locating heating and cooling sources within spaces. Hot air outlets are placed near the floor, so that the hottest incoming air can induce cooler air at the floor to mix with it and rise slowly to warm the center of

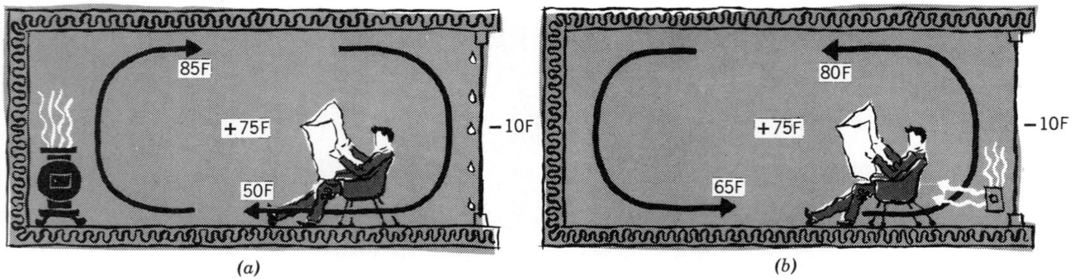

Fig. 2.5 Dealing with outdoor conditions. Convection is inevitable. It can work for or against you. The temperatures are approximate. (a) *The stove is not merely in the wrong place. It* accelerates *the "downslip" of cold air from the glass.* (b) *The convector strip moves air up to warm the glass and provides local radiant warmth. The location of wood stoves should be carefully chosen.*

spaces. Cold air outlets are placed in the ceiling or high in walls, so that the reverse procedure can occur. Again, more detailed thermal factors enter in; Fig. 2.5 shows how heat sources near the floor need also to be near colder exterior surfaces to avoid unpleasant drafts. The overall goal is usually to provide an evenness in the thermal environment.

The body is more prone to such discomfort from large temperature differences between horizontal-facing surfaces (such as a hot radiant ceiling and a cold floor) than between vertical-facing surfaces such as opposite walls. Large hot or cold surfaces, such as are used in passive solar heating and many passive cooling systems, are more likely to produce comfort when used opposite one another across a space; where that is impossible, vertical (wall) surfaces are somewhat preferable to horizontal ones.

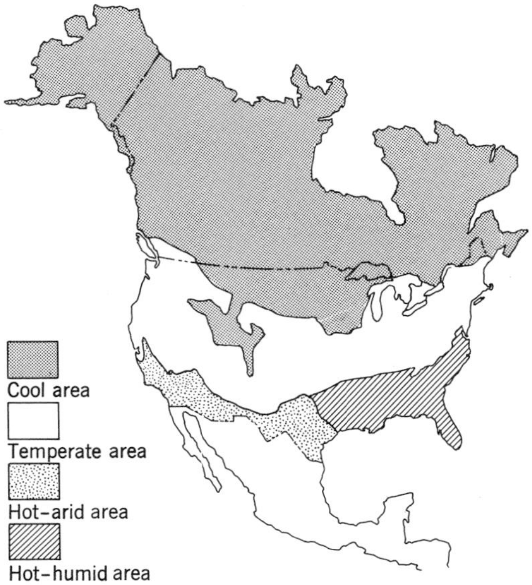

***Fig. 2.6** Regional climate zones of the North American continent. From Victor Olgyay,* Design with Climate: Bioclimatic Approach to Architectural Regionalism, *Copyright © 1963 by Princeton University Press. Reprinted by permission.*

2.3 Climates

Our most familiar names for climates describe their most severe season, as shown in Fig. 2.6. This is a convenient description, but can be misleading for designers. "Cold" climates can have very hot, sometimes humid summer days; hot-arid climates can have bitterly cold winter conditions. Before designing the buildings that modify exterior conditions to provide indoor comfort, we should know when and how much of this modifying is appropriate.

We can utilize the bioclimatic chart of Fig. 2.4 to compare specific climate information with human comfort. Table 2.4 shows the kind of information available at every local climatological station maintained by the National Oceanic and Atmospheric Administration (NOAA) of the U.S. Department of Commerce. These climate summaries, often called LCDs, are available from the National Climatic Data Center (Federal Building, Asheville, NC 28801). In Fig. 2.7, this average information for Dodge City, Kansas, is graphed on the bioclimatic chart.

EXAMPLE. Compare the coldest and hottest average months for Dodge City, Kansas, to the basic

TABLE 2.4 **Normal Data from Annual Summary[a] of Local Climatological Data for Dodge City, Kansas**

	Daily Temperatures (F)				*Relative Humidity (%) at Hour*				*Wind*	
	Normal		*Extreme*							
Month	*Max*	*Min*	*Max*	*Min*	*00*	*06*	*12*	*18*	*Mean Speed (mph)*	*Prevailing Direction*
December	44.6	22.2	86	− 7	72	76	56	60	14.0	South
January	42.6	19.0	78	−12	72	76	58	59	13.6	South
February	47.1	23.2	85	−15	70	75	54	52	14.1	North
June	86.0	61.4	108	41	63	75	44	38	14.4	South
July	91.4	66.9	109	47	66	78	46	42	12.9	South
August	90.4	65.7	107	47	71	80	50	46	12.7	South

[a]Records as of 1979.

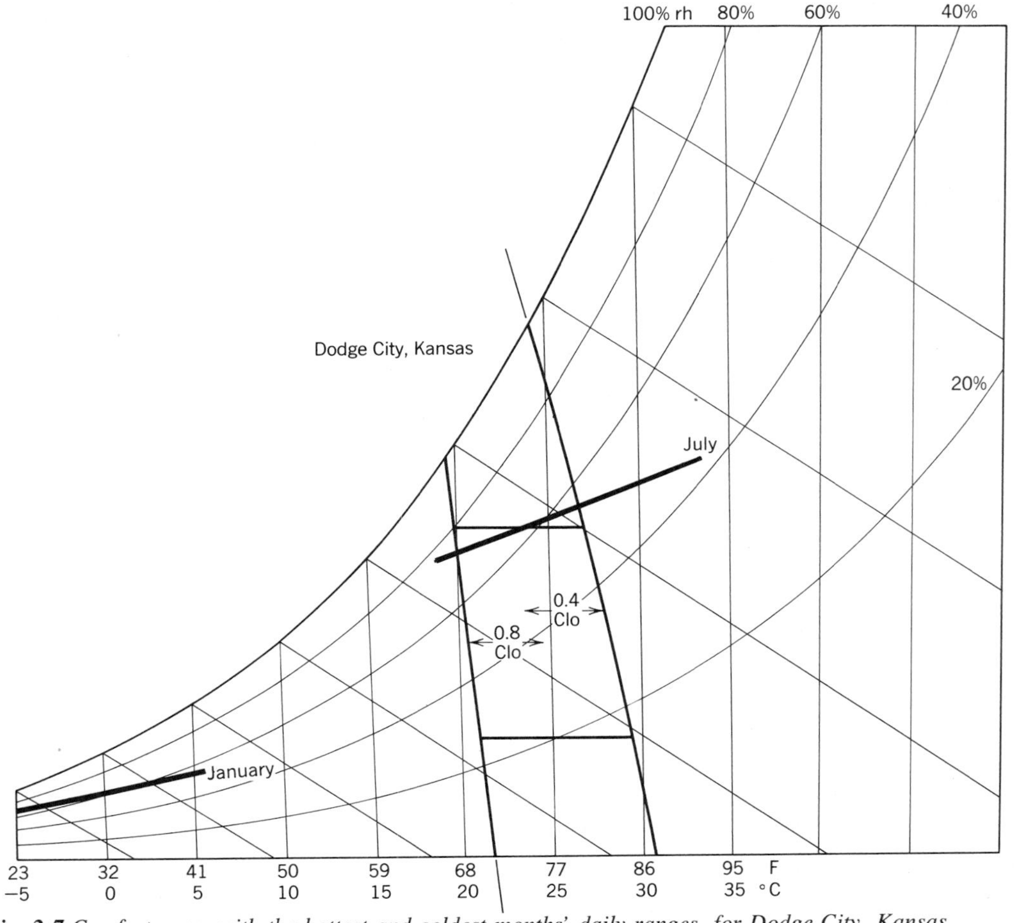

Fig. 2.7 Comfort zone, with the hottest and coldest months' daily ranges, for Dodge City, Kansas.

comfort zone. By inspection of Table 2.4, the coldest month is January. To chart the typical January day, find the approximate relative humidity (rh) for the coldest and the warmest hours. Since rh is listed only at four times, select the probable coldest hour (usually the highest rh), which is

6 A.M.; and the warmest hour (usually lowest rh), which is noon. Then plot these combinations for January:

42.6 F, 58% rh (at 12 noon)
19.0 F, 76% rh (at 6 A.M.)

Similarly, the hottest month is July:

91.4 F, 42% rh (at 6 P.M.)
66.9 F, 70% rh (at 6 A.M.)

These plots, and the average-day line that connects them, are shown in Fig. 2.7.

From the analysis in Fig. 2.7, it is evident that outdoor conditions in the coldest months are not normally within even the sun-extended comfort zone. (Remember, however, that the clothing and activity levels assumed are indoor, sedentary ones.) In the hottest month, most of the cooler half of the normal day falls within the (shaded) comfort zone, although some hours are very humid; in the hotter half of this day, comfort outdoors can be extended by adding air motion (as needed to about 2 m/s). From these averages of the "worst" months, it appears that Dodge City will need well-enclosed buildings in winter, while air motion at 2 m/s (about 4½ mph) seems achievable within buildings where the average summer wind speed is almost 13 mph. Remember that Dodge City is described as a "temperate" climate (Fig. 2.6), despite its *average* range of 19 to 91.4 F!

But what about extreme, rather than average, conditions and their impact on design? How much of the rest of the year is normally within the comfort zone? How many summer design strategies besides open windows might be appropriate, and how reasonable is solar heating in winter?

2.4 Design Strategies

The "building bioclimatic chart" shown in Fig. 2.8 was developed by Milne and Givoni (1979) to relate climate and design strategies. The average monthly data (as in Table 2.4) can be plotted here, thus revealing some appropriate strategies. Although the edges of each strategy zone are shown as lines, these boundaries are much more broad and vague than such lines suggest. (See Milne and Givoni for a discussion of how these boundaries were drawn.)

Before discussing these design strategies, a further elaboration on the bioclimatic chart has occurred. What we have called ordinary air temperature is now called *dry-bulb* temperature; and a new set of lines called *wet-bulb* temperatures have appeared. These terms are derived from the instrument shown in Fig. 2.9, the sling psychrometer.

Dry-bulb temperature (DB). The temperature of the ambient mixture of air and water vapor measured in the normal way with a simple thermometer.

Wet-bulb temperature (WB). The temperature shown by a thermometer with a wetted bulb rotated rapidly in the air to cause evaporation of its moisture. In dry air the moisture evaporates and draws heat out of the thermometer to produce a large *wet-bulb depression* (difference between dry- and wet-bulb temperatures). This is an index of low relative humidity. Slow evaporation when the air is already moisture laden results in a small wet-bulb depression and indicates a condition of high relative humidity.

The wet-bulb lines have many uses, which become more evident when discussing evaporative cooling; for now, they are added because climate data often are given in terms of WB rather than rh. Note that at 100% rh, DB and WB are equal. The design strategies shown in Fig. 2.8 include the following.

Active Solar Heating and Conventional Heating. These are the familiar heating methods that can concentrate heat (from burning fuel, a heat storage tank, electric resistance coils, etc.) on demand. They are frequently used as backup heat to passive solar systems, as well as stand-alone systems.

Passive Solar Heating. The collection and storage of solar heat is integrated within the surfaces of the building. Large (optimum) areas of south-facing glass admit winter sun, which strikes surfaces of high thermal mass (brick, quarry tile, concrete, water containers, etc.). In colder climates, thermal shades or shutters protect the windows against nighttime heat loss. The lines of required insolation shown in Fig. 2.8 indicate the point above which 100% of space heating by passive solar energy is likely; below these lines, some

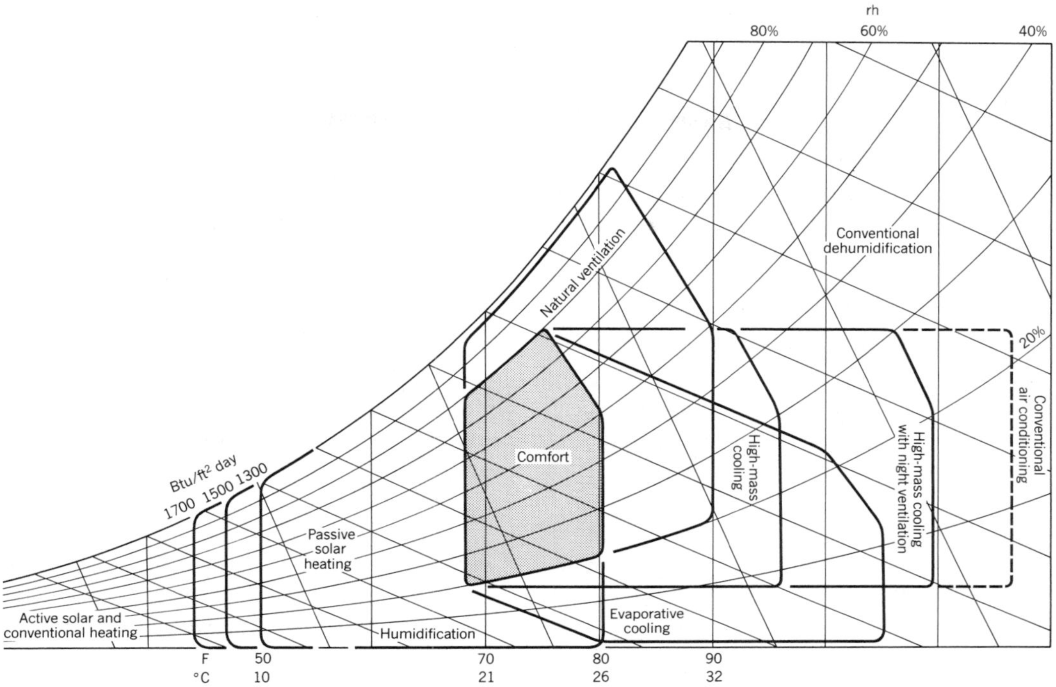

Fig. 2.8 *Design strategies by climate. Buildings usually contain sources of heat. The more heat occurs within the building, the more an* artificially warmer *climate is created. Thus, after plotting outdoor climate data on this chart, consider how* shifting *these plots would affect your choice of design strategy.*

The more:
- *Solar gains allowed inside*
- *Electric lights*
- *Business machines, etc.*

the more your plots move horizontally to the right.

The more:
- *People*
- *Cooking, bathing*
- *Other heat-plus-moisture sources*

the more your plots move up and to the right, following the rh curves. (Reprinted from Milne and Givoni, in Energy Conservation Through Building Design, *edited by Donald Watson, with the permission of McGraw-Hill, Inc.)*

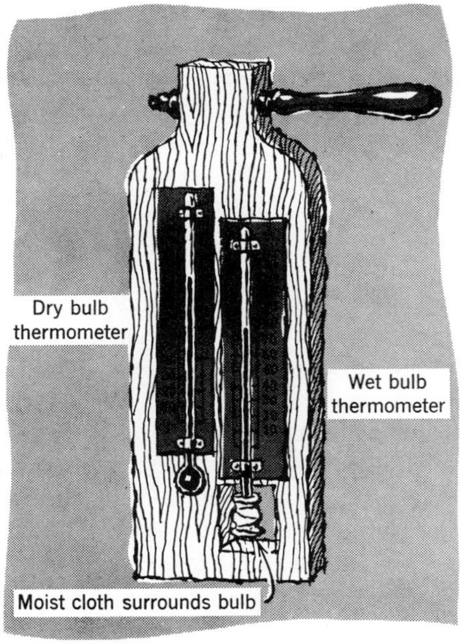

Fig. 2.9 *The Sling psychrometer.*

backup heat will probably be needed. These insolation levels may be checked for your approximate latitude in Tables 5.34 and 5.35 or in Appendix B.

Humidification. In this infrequently occurring zone, moisture should be added to indoor air to avoid respiratory problems and to reduce static electricity.

For the following cooling strategies, the exclusion of direct sun, by shading devices, is assumed.

High-Mass Cooling. This is an especially useful strategy in warm, dry climates, where the extremes of hot days are tempered by the still-cool thermal mass of the building. Cool nights slowly drain away the heat that such mass accumulates during the day. The thermal mass can be on floors, walls, or roofs; the roof has the advantage of radiating to the cold night sky, but should, like all other thermal mass in these conditions, be protected from the sun by day. As in all these cooling strategies, daylighting is usually designed to occur without direct sun, since solar space heating is unneeded in these conditions.

High-Mass Cooling with Night Ventilation. This hot-dry climate design strategy must use the air at comfortable nighttime temperatures to flush away the heat stored in daytime. The fewer the comfortable night hours, the more thermally massive surfaces must be provided to store the day's heat. Also, ventilation must occur more quickly and thoroughly, perhaps with fans. The building switches from a closed condition by day (to exclude sun and hot air) to an open one at night (to allow ventilation to cool the mass). *Note*: Nighttime outdoor temperatures should be *cooler* than the comfort zone, if this strategy is to be effective.

Natural Ventilation Cooling. This is the most obvious strategy suggested by the comfort charts presented earlier, where higher air temperatures were offset by increased air motion. It is most appropriate in more humid, hot climates where temperatures are only slightly lower by night than by day. Buildings should be very open to breezes while simultaneously closed to direct sun. They should be thermally lightweight as well, since night air is not cool enough to remove stored daytime heat.

Evaporative Cooling. This design strategy relies on the principle that when moisture is added to air, the air *increases* in relative humidity while *decreasing* in dry-bulb air temperature. (On the bioclimatic chart, this pattern exactly follows the WB line, upward to the left.) In conditions that are more uncomfortably dry then uncomfortably hot, higher humidity is gladly exchanged for lower air temperature. However, large quantities of both water and outdoor air are needed; fan-driven evaporative coolers (see Section 6.6) are the most common way to provide this kind of cooling.

Conventional Dehumidification and Air Conditioning. These are the familiar air conditioning methods, which rely on machinery and can cool on demand. Buildings utilizing such equipment should be very closed to both wind and direct sun. Thus, conventional air conditioners can be used as backup to passive systems, which imply closed buildings, such as high mass with night ventilation. Where evaporative coolers are used, conventional air conditioners can also be used to back up their cooling.

Returning to the analysis of the climate of Dodge City, the winter maximum of about 43 F, 58% rh indicates that passive solar heating will probably need a backup source almost all day in the coldest month. The summer night minimum of about 67 F, 78% rh is slightly below the comfort zone, while the daytime maximum of about 91 F, 42% rh allows several possible strategies: high thermal mass, high mass with night ventilation, and (only slightly beyond) natural ventilation. To help in making this choice, consider two more factors: temperature extremes and averages for the other hot months.

Temperature extremes are available from sources such as shown back in Table 2.4. A more useful guide, however, is the *design conditions* information, as found in Appendix A. This condition shows the climate near its ''worst,'' while avoiding the freakish temperatures found in the all-time record highs and lows. In Fig. 2.10, this design condition is plotted on the building bioclimatic chart, along with normal conditions for the other two hot months. The design condition indicates high thermal mass to be appropriate; simple natural ventilation is not sufficient (although nighttime ventilation could be useful). Normal conditions for other hot months, however, reinforce the option of simple natural ventilation, along with the options of high mass and night ventilation.

Building function is a potentially large influence on the choice of design strategy. By how much will internally generated heat produce an

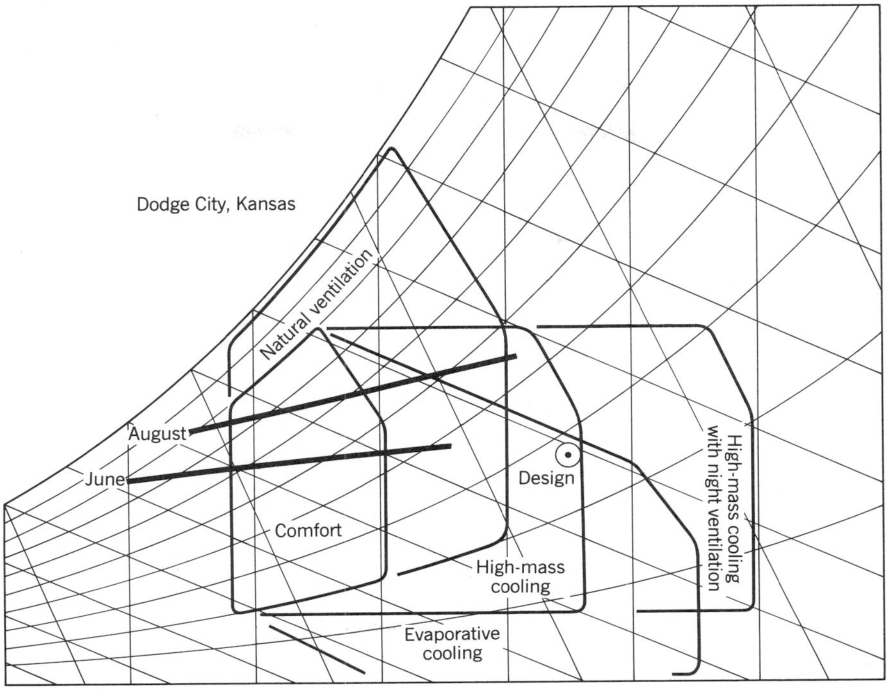

Fig. 2.10 *Hot month daily ranges and design condition, for Dodge City, Kansas, superimposed on design strategy diagram (from Fig. 2.8).*

artificially warmer climate? A Dodge City office building that excludes daylight and utilizes 100% electric lighting may shift its indoor climate plot so dramatically that 100% passive solar heating is possible in winter, and high mass with night ventilation is strained to (or beyond) its limit to provide summer cooling. Such an approach would eliminate both simple natural ventilation and simple high thermal mass as strategies, introducing the probability of conventional air conditioning. An office building that uses only daylight (excluding direct sun except in winter) and allows for the use of outdoor conditions when appropriate would not appreciably shift its indoor plot. A more detailed treatment of the daylight–heat–cool tradeoff for various buildings and climates is given in Section 5.1.

The actual choice of design strategy will be influenced by many factors in addition to climate and function. These include the peculiarities of the site, first cost versus lifetime costs, the desired architectural characteristics of the building's envelope and its internal spaces, and simple personal preference. A closer look at climate could also help in choosing a design strategy.

Climate varies not only from day to day but *during* the day as well. Many buildings are not expected to provide comfort conditions all the time; office buildings, for example, are used almost entirely within a 9- or 10-h period centered around noon. A view of climate as a daily pattern, linked to hours of operation, can reveal a picture quite different from the "monthly average day" that has so far been considered here.

One such approach to climate analysis is illustrated, again for Dodge City, in Fig. 2.11. Brown and Novitski (1981) look for the *least* architectural response necessary to maintain comfort in a given climate. Their studies emphasize the daily *change* in this architectural response. When a typical year in Dodge City was analyzed (using approximately the comfort conditions charted in Fig. 2.4), five standard daily patterns emerged.

Too cold (58 days): Even with sun admitted and wind blocked, comfort outdoors (seated, ordinary winter indoor clothing) is not attainable. (This does not imply that passive solar heating is not viable; only that indoor clothing plus available sun don't add up to outdoor com-

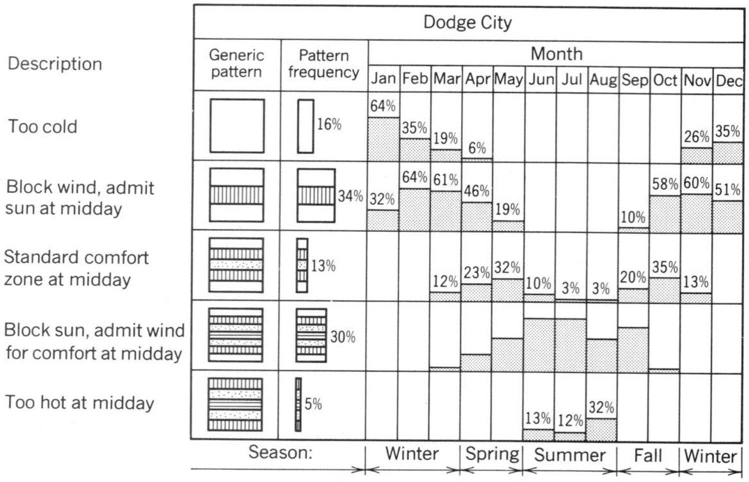

Fig. 2.11 Percentages of time in various modified comfort zones, for Dodge City, Kansas. (From Brown and Novitski, 1981. Reprinted from the Proceedings of the Sixth National Passive Solar Conference *by permission of the publisher, American Solar Energy Society, Boulder.)*

fort.) The building remains closed to outdoor conditions all day, except for solar gains through glass.

Block wind, admit sun at midday (124 days): For several hours, seated comfort outdoors is possible, given sun and blocked wind. The remaining hours are too cold, as above. A wind-protected, sunny outdoor space, probably south-facing, would be comfortable for ordinary indoor activities at midday.

Standard comfort zone at midday (47 days): Most of the daytime hours are within reach of the comfort zone; a few hours of shaded comfort occur at midday. Outdoor spaces protected from the wind and shaded at midday are suitable for most of the day. (A sunny but windy space at midday is also a possibility.)

Block sun, admit wind at midday (110 days): Given rather frequent changes in architectural response, all daylight hours are within reach of the comfort zone. The heat at midday requires added wind as well as shade. Outdoor spaces can be used extensively, with shading devices in use most of the day. Exposure to prevailing wind will be important at midday, while blocking the wind will be needed very early and very late.

Too hot at midday (18 days): For several hours, the air is too hot to provide comfort within acceptable (or available) wind speeds. The rest of the day, the comfort zone is within reach. The building is closed at midday; the hours before and after can open to shaded, wind-exposed outdoor spaces. Still earlier and later, wind must be blocked.

The lesson taught by such an analysis is that a building sensitive to the Dodge City climate will be capable of switching from a closed position to a degree of openness on a daily basis, for 10 months of the year. This is particularly applicable to office buildings and other day-use functions such as schools and stores. This potential interaction between indoor activity and outdoor space is not unique to Dodge City, or to the "temperate" zone of the United States. For example, the same study found that the daily pattern of block wind–admit sun at midday, occurring most frequently (34%) in Dodge City, also occurred 24% of days in the hot-humid climate of Charleston, South Carolina, and 24% of days in the cold climate of Madison, Wisconsin.

Although the detailed analysis presented here for Dodge City is not yet available for a wide range of locations, the same general patterns can be seen by following the daily cycle of the typical monthly plots (refer again to Fig. 2.10).

Another collection of design strategies shown for various climates is found in *Climatic Design* (Watson and Labs, 1983).

You now have some appropriate design strategies identified, and an idea of the extent to which your building can be closed or opened to exterior conditions. Next, the site and the skin or envelope of the building can be more closely examined. Following are some basic factors concerning envelopes; the details of envelope thermal design procedures are then found in Chapter 4. The site is the subject of the next chapter.

2.5 The Building Envelope

The envelope of a building is not merely a set of two-dimensional exterior surfaces; it is a transition space—a theater where the interaction between outdoor forces and indoor conditions can be watched. Some of these interactions include the ways in which sun and daylight are admitted or redirected to the interior, the channeling of breezes and sounds, and the deflection of rain. This transition space, which forms the envelope, is a place where people indoors experience something of what the outdoors is like at the moment, and where people outside get a glimpse of the functions within. The more suited the outdoors to comfort, the more easily indoor activity can move into this outdoor transition space. At entries, where there is a space created in the transition from one environment to another, a person will be especially aware of the difference between outdoors and indoors; an example of entry as space, not just surface, is shown in Fig. 2.12.

The envelope also has a fourth dimension; it changes with time. Daily changes were discussed in the preceding section; seasonal changes have a marked effect on the entry in Fig. 2.12, and a more subtle effect on the east-facing balconies of the apartments in Fig. 2.13. The year-round usable volume of these apartments is increased by making the balcony into a sun porch, as has been done for many apartments in Fig. 2.13. This is a direct response to the prevalence of "block wind–admit sun" conditions in Oregon's long but mild winter. Not all buildings encourage such flexibility; an unchanging envelope can be symbolic of stability and is considered appropriate for some governmental and religious monuments. Generally, the more that users are involved in decisions about how much of the outside to bring inside, the more changeable the building envelope will be. Further examples appear in Figs. 2.20 and 2.21, below.

2.6 Connectors, Filters, and Barriers

Called by their familiar names, the basic components of the envelope are windows, doors, walls, and roofs. On closer inspection we find that windows can include skylights, clerestories, screens, shutters, drapes, blinds, diffusing glass, and reflecting glass—an array of components that determines more exactly how the envelope does its job of making the transition between inside and outside. Norberg-Schulz (1965) suggests that a component can also be thought of by its function in the exchange of energies: as a *filter, connector, barrier,* or *switch*.

> In general, we define a *connector* as a means to establish a direct connection, a *filter* as a means to make the connection indirect (controlled), a *switch* as a regulating connector, and a *barrier* as a separating element. . . . An opaque wall thus serves as a filter to heat and cold, and as a barrier to light. Doors and windows have the character of switches, because they can stop or connect at will.

This approach to the choice of components can be illustrated by two opposite concepts of envelope design: the *open frame* and the *closed shell*. In harsh climates (or where unwanted external influences such as noise or intruding activities abound), the designer frequently conceives of a building's envelope as a closed shell and proceeds to selectively punch holes in it to make limited and special contacts with the outdoors. In the hot-humid regions (or where external conditions are very close to the desired internal ones), the envelope begins as an open structural frame, with pieces of building skin selectively added to modify only a few outdoor forces. The open-frame or the closed-shell approach to envelope design, when combined with material availability and the influence of local culture, can produce a distinct regional architecture (Fig. 2.14).

With a wide range of energy sources, building materials, and mechanical equipment available, it is possible to design connector-dominated buildings anywhere, despite the climate. The conse-

Fig. 2.12 *The envelope is more than a surface. This south-facing entry to an architect's office in Oregon becomes a microclimate that buffers the transition between the indoors and outdoors.* (a) *Three kinds of entry conditions are visible: the awning (over the windows of a restaurant), the gable roof with bare rafters (over the planting in the architect's entry), and the arcade, a second story carried out over a covered walkway, that links shops.* (b) *Detail of architect's entry shown on March 21.* (c) *The change of seasons brings deep shade to this entry in summer and early fall. (Wilmsen & Endicott, Architects.)*

Fig. 2.13 *Envelopes change with time. This east-facing building envelope shows changes between seasons and between years.* (a) *December 21, 8:30* A.M., *the sun is low and shining from the southeast; shading occurs not from balcony overhang but from vertical balcony-divider walls.* (b) *June 21, 8:30* A.M., *the sun is high and nearly due east, so shading occurs only from balcony overhangs and from blinds at the railings.* (c) *An August morning, several years later, showing the conversion of balconies by many of the occupants. (Cascade Manor, Eugene, Oregon; John Graham & Associates, Architects.)*

quences of the resulting energy consumption can be severe. By contrast, if defending against outdoor conditions becomes an overriding consideration, barrier-dominated envelopes can occur in any climate. The resulting fitness for human usage—and the potential of using natural energy sources—can be reduced. The designer's combinations of connectors, filters, and barriers (and the switches that allow these elements to respond to changing conditions) are basic to the design of building exteriors and can give them the liveliness that makes a building an attractive addition to its neighborhood. *Note:* A connector, filter, or barrier for one natural energy may change its role for another: glazing may be a connector to daylight but a barrier to wind.

Connectors are strong indicators that something outside is welcome inside. They are char-

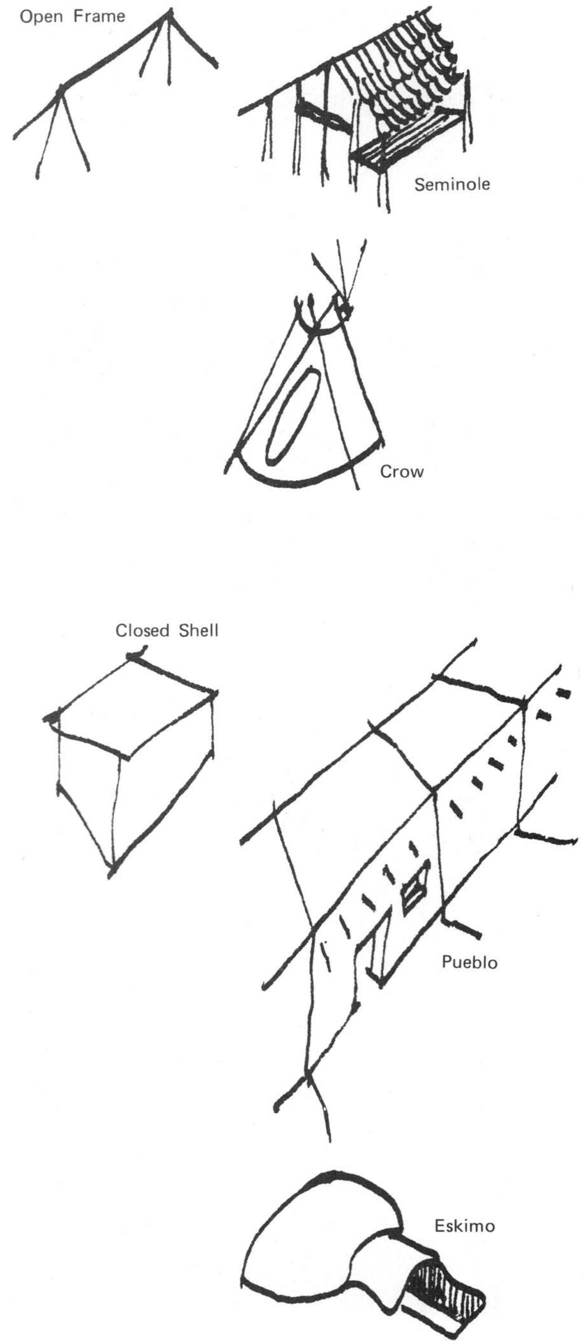

Hot Humid Climate: To the open frame, a barrier roof of local plant materials is added to reject rain and sun. A raised floor avoids damp earth and its creatures and allows breezes to pass over and under its users.

Temperate Climate: This open frame is wrapped in light–filtering animal skin, doubled near the ground. Wind and rain are rejected; protection against cold is provided by users' clothing (blankets) more than by the envelope. The switch at the crown controls smoke (see also Fig. 2.19). Portability of shelter is a cultural factor here.

Arid Climate: The closed shell of mud block is a barrier to wind and sunlight; it filters heat by both delaying and reducing its impact on the interior. Some light and heat are admitted directly by small connectors: the door and window, typically south–facing.
By early morning, the cold interiors are abandoned in favor of rapidly warming south terraces.

Cold Climate: The igloo's closed shell of ice is a filter to light and heat, a barrier to wind. Holes for entry and for smoke are allowed, but sparingly. Fur–bearing hides hung inside can increase thermal comfort for users.

Fig. 2.14 Open frame and closed shell: climate, materials, and culture. The influence of climate is clear, but material availability and cultural patterns are also strong determinants in the choice of envelope design in these examples.

acteristic of regional architecture in milder climates, but sun connectors are dominant in solar-heated buildings anywhere. Connectors, being open to outside influences, are often one position of a switch that in other positions becomes a filter or a barrier. Or, as in Fig. 2.15, a connector is sometimes followed by a combination of filters and barriers somewhere further inside the space that forms a building's envelope.

Filters represent decisions about how much or what kind of outdoor condition is to be admitted. They are found in some form in all building envelopes and in all climates, and they include a wide variety of types. Because they admit desired amounts or qualities of light, air, and sound, they offer an opportunity for an enhanced awareness of selected outside conditions from inside the building. For example, the stained glass of a church enhances the blue of a north sky, or the warm reds and oranges of a sunset; the *texture* of the sky's cloud patterns is not admitted to the interior—only its color comes through. Often the filter is one of the positions of a switch, as in the case of the windows in a famous building by Le Corbusier (Fig. 2.16).

Fig. 2.15 Connectors for a corridor. An old cannery building in the mild climate of San Francisco is converted into a series of shops; a corridor is given a near-outdoors environment by the connectors at the right. Across the corridor, filters give greater internal climate control to the shops beyond. (The Cannery; Joseph Esherick & Associates, Architects. Photo by Richard A. Cooke III.)

Barriers are more drastic in their complete severance of the outdoor–indoor relationship. They are characteristic of regional architecture in harsh climates, but are also common to spaces needing a tightly controlled environment (such as an auditorium). Barriers to rain are an almost universal building feature; barriers to wind are at least seasonally common in all climates, except hot-humid ones. Barriers to sun are more likely to be one position of a switch, unless a building is suffused with electric light or other plentiful sources of internal heat that make solar heat permanently unwelcome. In practice, cultural influences often override those of climate; barriers to sun are erected even in cold, damp environments, such as those of Canadian Pacific (Fig. 2.17).

2.7 Switches and Users' Choice

Switches are a valued component for designers who appreciate that climate and functions change, and that people are unpredictable. As much as a

Fig. 2.16 Filters in various positions of a switch. Corbusier's Pavillon Suisse, 1930–1932, at the University of Paris. (Photo by Nicolai Shur.)

Fig. 2.17 Barriers, but not to heat. These barrier-dominated exteriors, typical of Haida villages in coastal British Columbia and southern Alaska, shed rain, wind, and light. The walls are filters, however, to heat, which passes with relative ease through the thick planks. (In winter, the narrow spaces between the houses can be enclosed, producing a more heat-conserving ''row-house'' configuration.) The central fires of each house are fed by plentiful wood fuel, and firelight is preferred over daylight for illuminating the elaborate masks used in Haida ceremonies. (Photo courtesy of the Milwaukee Public Museum.)

designer might prefer that a building have a certain identity, unchanging in its expression of a set of ideas, time will pass and changes will occur. Switches allow the designer to give control of the building to its users. If the designer has carefully considered the range of choices that the switch should provide, successful user control will be possible. Buildings whose skins are plentifully supplied with switches become continuing demonstrations that architecture is a performing art, not just sculpture.

In Section 2.4, switches were called on to alternately admit or block the sun or wind, so that a building's envelope might fully utilize the climate rather than simply exclude it. Passive solar heating systems rely on switches to throttle down the incoming sun on warmer days; especially in cold climates, such systems sometimes rely on switches to insulate windows by night (Fig. 2.18). Ventilating switches can be integral to a building's envelope in ways other than as sets of opening windows, as in Fig. 2.19.

Daylighting switches are perhaps our most common and visible ones. Awnings might block direct sun at some times, admit it at others. Opaque drapes might expose all the window to incoming daylight on dark winter days, but only part of the window on bright summer days, blocking the window entirely at night. Translucent curtains might change bright sun into a diffuse light for the interior, or be drawn back to allow strong shadows to be cast inside the room. Electric lighting can be turned off while daylight is plentiful.

Visually, switches are a particularly promising source of three- and four-dimensional interest on the building's exterior. For the buildings in Fig. 2.20, on which facade would daily and seasonal changes be most visually evident? If it is easier to imagine humans working behind the windows of one building than of the other, might that suggest a more satisfactory work environment?

Switches allow change, and thus encourage interaction between users and their environments. This is usually satisfying to the users, who are

Fig. 2.18 *Thermal switches in New Mexico.* (a) *The north wall is mostly a barrier.* (b) *The entire south wall can be a connector to sunlight, or* (c) *it can be isolated to varying degrees from the outside by operating switches. (David Wright, AIA, architect. Photos by Edward Mazria.)*

able to select the desired exposure to climate at that moment. Yet without automation, supervision, or training in their uses, switches can be detrimental to system performance. Examples are a thermal shade left to cover a passive solar collecting window on a cold, sunny morning, a vent left open during the hottest hours in a high-mass, night ventilation building, or an awning rolled up to leave a window exposed to summer sun. People in buildings often utilize switches in unexpected ways (see Fig. 2.21). Conventionally air conditioned buildings typically do away with ventilating switches for users (operating windows), so that the system will function with a closely controlled amount of outdoor air. Thermally efficient as this practice might be, it can also be the source of widespread user dissatisfaction with air conditioned spaces. Sealed windows greatly curtail peo-

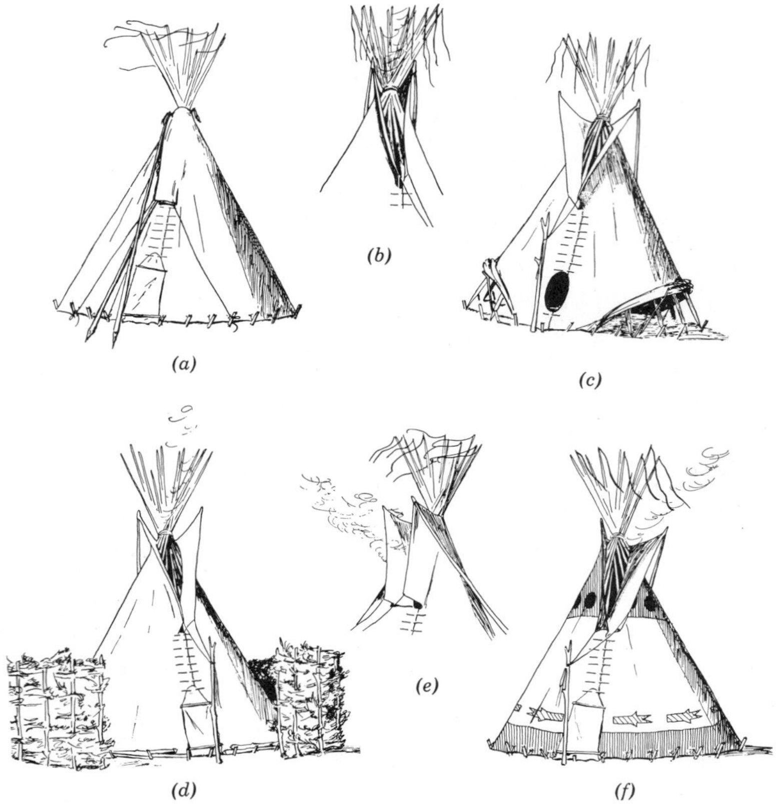

Fig. 2.19 *Thermal switches on the Great Plains. Six adjustments to the lightweight, translucent, and portable tipi are shown; the tipis in these diagrams are east facing, with their backs to the prevailing westerly winds.* (a) *In severe rainstorms, the smoke flaps can be closed.* (b) *For ordinary conditions of west wind, the smoke flaps block the wind, thus creating a suction at the opening to draw out the smoke.* (c) *For hot weather, breezes are admitted under the cover at the ground.* (d) *For extremely cold weather, a temporary windbreak can be added.* (e) *For the unusual wind or* (f) *southwest wind, the smokeflaps are manipulated to block this wind, thus encouraging smoke draw-out, as in* (b). *(From* The Indian Tipi: Its History, Construction, and Use, *by Reginald and Gladys Laubin. Copyright 1957 by the University of Oklahoma Press.)*

ple's contact with the sounds, smells, and breezes of the outdoors. This is frequently beneficial in urban areas, yet on beautiful days it can be very frustrating. A lack of switches contributes to a feeling of helplessness about the internal environment.

Passive heating and cooling techniques are especially reliant on switches, hence on the understanding and cooperation of the people within these spaces. These users often must base their actions of the moment on what will be needed later, by manipulating thermal switches. This practice, called "thermal sailing," is similar to actions of outdoor workers in the far north, who learn to unbutton their coats in the cold early hours of the workday *before* they begin to sweat and rebutton in the relative warmth of the late afternoon before the rapidly falling temperatures near dusk can chill the skin. Misjudgments in passive solar-heated homes can result in extraordinarily high temperatures on a sunny day, or nights without stored solar heat. For the building closely related to its climate, the design of switches is also the design of an educational process for users. The challenge is to involve, but not enslave, the users in the management of their environment. Auto-

Fig. 2.20 *Sun control for Boston offices.* (a) *Switches both inside and outside the glass are evident, and the windows themselves are operable. Awnings are more likely to be fully extended as the windows sit higher up in the canyon of the street. (New England Merchants' Bank; Shepley Rutan & Coolidge, Architects; demolished 1966.)* (b) *Three-dimensional filters in the form of overhangs dominate the south facade (right) of this building at Boston University; the west windows (where overhangs are less effective) have internal switches, shown here in a variety of positions. (Law and Education Building; Sert, Jackson and Gourley, Architects.)* (c) *The two-dimensional filter of reflective glass sheathes all faces of this office tower, sending a beam of reflected sunlight to the neighborhood below. The switches in this case are thermostats; users are not involved in the outside–inside decisions. (John Hancock Headquarters; I. M. Pei & Partners, Architects. Photo by Stephen Tang.)*

mated controls are a partial answer to this challenge; switches that are easy and fun to use are another.

References

Publications of the American Society of Heating, Refrigerating, and Air Conditioning Engineers, Atlanta:

ASHRAE *Handbook of Fundamentals,* Publications of the American Society of Heating, Refrigerating and Air Conditioning Engineers: 1981.

ASHRAE Standard 55-1981, *Thermal Environmental Conditions for Human Occupancy,* 1981.

Arens, E., Gonzalez, R., Berglund, L., McNall, P., and Zeren, L. (1980). ''A New Bioclimatic Chart for Passive Solar Design.'' In Vol. 5.2, *Proceed-*

Fig. 2.21 *Which is spring, which is summer?* (a) *The awnings for this office building in Eugene, Oregon, are joined by overhangs and side walls (or fins) to provide sun control for these south-facing windows.* (b) *The summer sun, at a high angle is readily blocked by the overhang; the awnings can be rolled up with less risk of direct glare into the users' eyes. (Moreland-Unruh-Smith, Architects.)*

ings of the Fifth National Passive Solar Conference, American Section of the International Solar Energy Society, Boulder, CO., pp. 1202–1206.

Brown, G. Z., and Novitski, B. J. (1981). ''A Design Methodology Based on Climate Characteristics.'' In *Proceedings of the Sixth National Passive Solar Conference,* American Section of the International Solar Energy Society, Boulder, CO., pp. 372–376.

Heschong, L. (1979). *Thermal Delight in Architecture,* MIT Press, Cambridge, MA.

Laubin, R., and Laubin, G. (1975). *The Indian Tipi: Its History, Construction and Use,* University of Oklahoma Press, Norman, OK.

Milne, M., and Givoni, B. (1979). ''Architectural Design Based on Climate.'' In *Energy Conservation Through Building Design,* D. Watson, Editor, McGraw-Hill, New York.

Norberg-Schulz, C. (1965). *Intentions in Architecture.* MIT Press, Cambridge, MA.

Olgyay, Victor (1963). *Design with Climate, Bioclimatic Approach to Architectural Regionalism,* Princeton University Press, Princeton, NJ.

3

SITES AND RESOURCES

In the first chapter, the supply of ultimately limited nonrenewable energy was contrasted with the supply of timeless but limited-delivery renewable resources. In the second chapter, climate was examined for its suitability for indoor comfort, so that buildings might open themselves often to this free energy available on site, rather than retreating permanently within an indoor climate dependent on purchased energy. This chapter looks at microclimates; the localized climates near buildings, where inside meets outside.

A designer's early decisions in site planning will influence the later choices of both the building's mechanical and electrical equipment and its overall consumption of energy. If the site is seen as a collection of resources (sun, wind, water, plants) and also as a part of the environment we all share, the buildings we design can approach self-sufficiency of energy supply, without limiting the availability of local energy resources for neighboring buildings. The use of on-site resources not only can reduce the amount of energy needed to maintain the interior climate, it also can produce outdoor spaces that become especially pleasant to use. Such spaces can direct winter sun to a glass wall while blocking the wind, or funnel the summer breeze through shade to an open window. Site planning is greatly influenced by economic considerations, zoning regulations, and adjacent developments, all of which can interfere with the design of a site to utilize the sun and the wind. Integration of all these concerns at the site-planning stage is the first step in adapting a building to its climate.

The microclimate is influenced by the interaction of both site characteristics and climate characteristics:

Site	*Climate*
Soil type	Sun
Ground surface	Temperature
Topography (profile)	Humidity
Vegetation	Precipitation
Water presence	Air motion
View	Air quality
Human activity (heat, noise, etc.)	

This chapter looks briefly at some results of site–climate interactions.

3.1 Climates Within Climates

The climate at a particular site can be quite different from the averages that are published for the whole region. This is particularly evident where a neighboring hill blocks wind or winter sun, or an adjacent lake cools summer breezes or adds a damp chill to winter air. It may be less obvious that urban sites can be under the influence of an urban subclimate, a ''heat island'' that differs from the conditions at the official local climatological station. (These stations are often at a more rural site, such as an outlying airport.) The effects of this urban heat island are summarized in Table 3.1.

The most obvious reason for the city's year-round warmth is its concentration of *heat sources:* the air conditioners, furnaces, and internal combustion engines in cars and buildings. The *rain* that falls on the city and countryside can be an effective cooling mechanism, especially as water evaporates from wet surfaces; but the materials of streets and buildings are usually designed to shed water quickly and thoroughly, so evaporative cooling for these surfaces is minimized.

The city also changes the overall cooling action of the *wind* by channeling it into narrow streets. The geometry of high vertical walls and narrow streets also increases the summer heat in cities as the high sun is reflected downward and is absorbed by the often rocklike street and building surfaces.

Table 3.1 **Average Changes in Climatic Elements Caused by Urbanization**

Element	*Comparison with Rural Environment*
Contaminants	
Condensation nuclei and particulates	10 times more
Gaseous admixtures	5 to 25 times more
Cloudiness	
Cover	5 to 10% more
Fog, winter	100% more
Fog, summer	30% more
Precipitation	
Totals	5 to 10% more
Days with less than 5 mm	10% more
Snowfall	5% less
Relative humidity	
Winter	2% less
Summer	8% less
Radiation	
Global	15 to 20% less
Ultraviolet, winter	30% less
Ultraviolet, summer	5% less
Sunshine duration	5 to 15% less
Temperature	
Annual mean	0.5 to 1.0°C more
Winter minima (average)	1 to 2°C more
Heating degree days	10% less
Wind speed	
Annual mean	20 to 30% less
Extreme gusts	10 to 20% less
Calms	5 to 20% more

Source: H. E. Landsberg, "Climates and Urban Planning," *Urban Climates,* World Meteorological Organization, 1970. WMO Technical Note No. 108, p. 372.

In winter, however, this geometry puts most urban surfaces at a solar disadvantage, since the low sun strikes only the upper portion of south-facing walls.

A more subtle climate influence is *contaminated air;* small particles in the city's air can keep some sunlight from reaching the city; yet it can help to keep the city's heat from radiating outward. These particles also form the nucleus of fog droplets; Table 3.1 shows that up to twice as much fog occurs in the city in winter as in the surrounding countryside. Trees and greenery are not as available in the city to act as crude filters to airborne dust.

The city thus changes its climate from that of its surroundings—to the city's decided disadvantage in summer. In winter, the city's additional warmth reduces the need for heating buildings. The typical means of providing additional winter heating (by fossil-fuel-burning furnaces and power plants) contribute to airborne particles and urban fog. A switch to solar-assisted heating would diminish this pollution—and with less air pollution, more sun would reach a solar collector.

Site and urban planning responses to these urban climate characteristics can sometimes lead in contradictory directions. For example, the provision of greenways within cities would bring softer ground surfaces cooled in summer by shading, breezes, and evaporation. However its winter impact might be an increase in fog via evaporation of retained water; or perhaps fog would be discouraged by the increased wind speeds facilitated by such greenways. The summer impact of the greenway is also not entirely positive; increased evaporation could mean higher relative humidity—but with lower temperature. On balance, the greenway within a city seems to be a positive and esthetic step in ameliorating the urban climate, but the complexity of such a question deserves further research.

While considering how the sun, wind, and other natural forces can be utilized on a site for the benefit of a building, it is important to remember the need to protect the access of others to these same resources. This is effectively illustrated in Garrett Hardin's "Tragedy of the Commons" (1968), often reprinted since its original publication in *Science*. In his example, the commons are meadows publicly owned and shared by many herders. Each herder realizes that his personal wealth will increase as animals are added to his herd, so all herders increase their livestock. But the meadow capacity is not increased; overgrazing occurs, and as a result the commons become unable to support any animals. The following discussions of the resources of sun, air, and water found on a site are influenced both by the "private" needs of a building and the "public" patterns of these resources, which should remain accessible to all.

3.2 Analyzing the Site

One of the designer's first concerns is to recognize the resources that exist on and around a site (or in a place) and to decide how best to integrate these resources into the final design, while making the design a successful addition to the larger patterns of its surroundings. Schematic site plans are typically used as a kind of inventory; overlaid ''bubble diagrams'' can test possible design arrangements that can relate rooms and functions to their surroundings in the plan. Sun and wind conditions (in both summer and winter), noise sources, and water runoff patterns can be included in this schematic plan. It is particularly important to identify microclimates on the site, the places that have special characteristics differing from the regional climate. Microclimates can present opportunities to utilize the expanded comfort zone discussed in Section 2.4; they can sometimes represent building sites where less energy is consumed because the winter is warmer, or the summer cooler. Or microclimates can be problem areas to be avoided for building or outdoor activity, if possible, or where special design measures need to be taken to correct their difficulties.

The site's collection of microclimates is not limited to those visible in plan, on the surface. Conditions of privacy and accessibility, of view, heat, light, air motion, sound, and water, all change with vertical distance from the surface (Fig. 3.1*a*).

To minimize energy consumption in constructing and using buildings, and to integrate them with their surroundings, the conditions best suited to various functions should approach the characteristics of the ''layer'' of the site in which they are located. Consequently, both horizontal and vertical site analyses are needed. A lecture hall, needing both an isolated and a closely controlled environment, is a candidate for the subsurface layer. Electrical equipment, which benefits from cool environments, also is suitable to the subsurface layer.

One building that derives its form from these layers is Boston's City Hall (Fig. 3.1*b*). Activities with the most frequent public interaction are located near the skylit, high-ceilinged lobby on surface levels, whereas special ceremonial functions are elevated to distinctive forms in the near-surface layer. The city offices with less frequent public contact occupy several floors in the sky layer, and storage and mechanical functions are in the subsurface layer. (An esthetic equivalent of this horizontal layering is architect Louis Sullivan's concept of a facade as a ''base, body, and capital.'')

3.3 Site Design Strategies

In the preceding chapter, appropriate building design strategies were identified by analyzing climate data. Sites can also be organized to aid in the heating, lighting, cooling, and noise characteristics of buildings; a few of the most common site elements are analyzed by Watson and Labs (1981) for their thermal contribution in Fig. 3.2. Although housing is suggested in these graphics, they apply to any function that can utilize the extended comfort zone. Later in this chapter, several of these approaches are discussed in more detail.

One example of how the information in Fig. 3.1 can be applied (and manipulated) in the planning of a building and its site is a house designed by Frank Lloyd Wright nearly 40 years ago (Fig. 3.3). The direct gain of solar heat through its windows in winter makes this an early example of ''passive solar heating'' (by contrast, ''active solar heating'' includes collectors and a separate storage volume for solar heat). This house, known as the Solar Hemicycle, was built in 1948 near Madison, Wisconsin, which lies between the cool and temperate climate zones. Winter heating is the dominant thermal influence in this area. The house stands on a hilltop site, particularly vulnerable to winter winds. In construction, earth was scooped from in front of the south face of the house and bermed against the entire curved north wall, almost to roof level. This gives winter wind protection from northeast to northwest and provides further insulation behind the almost windowless north wall. The north wall is made of stone, which absorbs and stores the winter solar heat that comes in through the floor-to-ceiling, southeast-to-southwest-facing glass. The concrete floor also stores solar heat in winter. The impression that this house and site are sun collectors is heightened by an entrance tunnel at the northeast end of the house, which leads from the parking area through the berm wall and onto the sunny, protected south terrace and front door.

This house is longer in the east–west direction (about 1 : 3). This elongation is typical of passive solar designs, which are able to store and use the winter solar heat gain, and thus profit from having long south-facing glass walls to act as collectors. Had the large windows been well insulated at night, and the roof and walls insulated up to the present

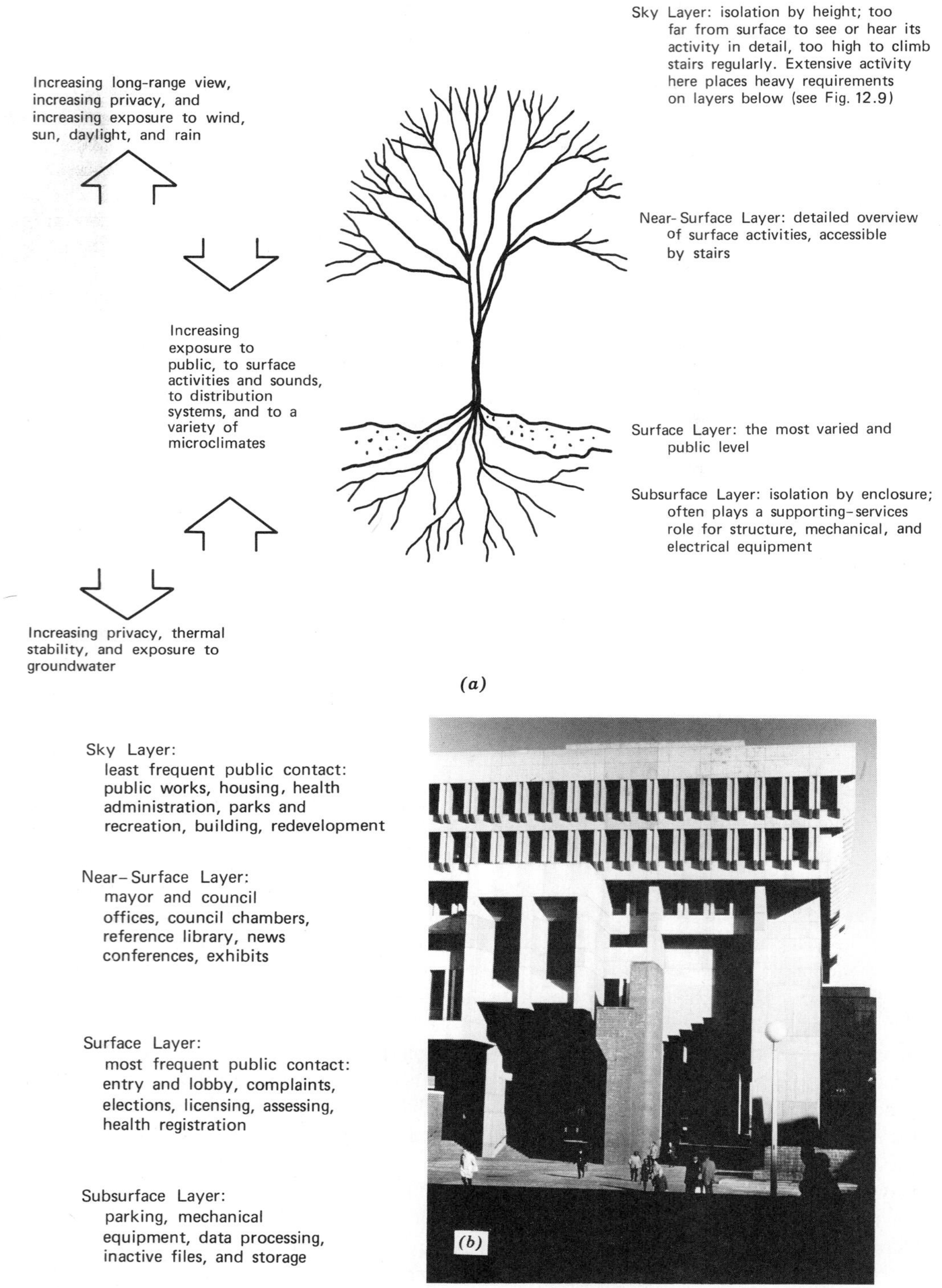
Increasing long-range view, increasing privacy, and increasing exposure to wind, sun, daylight, and rain
Increasing exposure to public, to surface activities and sounds, to distribution systems, and to a variety of microclimates
Increasing privacy, thermal stability, and exposure to groundwater
Sky Layer: isolation by height; too far from surface to see or hear its activity in detail, too high to climb stairs regularly. Extensive activity here places heavy requirements on layers below (see Fig. 12.9)
Near-Surface Layer: detailed overview of surface activities, accessible by stairs
Surface Layer: the most varied and public level
Subsurface Layer: isolation by enclosure; often plays a supporting-services role for structure, mechanical, and electrical equipment
(a)
Sky Layer:
least frequent public contact: public works, housing, health administration, parks and recreation, building, redevelopment
Near-Surface Layer:
mayor and council offices, council chambers, reference library, news conferences, exhibits
Surface Layer:
most frequent public contact: entry and lobby, complaints, elections, licensing, assessing, health registration
Subsurface Layer:
parking, mechanical equipment, data processing, inactive files, and storage
(b)

Fig. 3.1 (a) *Characteristics of horizontal layers for site analysis.* (b) *Layers and form: Boston City Hall, 1969. (Kallmann, McKinnell and Knowles, Architects. Photo by M. D. Ross.)*

	Winter: Minimize Conductive Heat Flow	Winter: Minimize External Heat Flow	Winter: Promote Solar Gain	Winter: Minimize Infiltration	Summer: Delay Periodic Heat Flow	Summer: Promote Ventilation	Summer: Minimize Solar Gain	Summer: Promote Radiant Cooling	Summer: Promote Evaporative Cooling
Wind breaks		●		●					
Plants and water							●		●
Indoor/outdoor rooms			●			●			
Earth sheltering	●	●		●	●		●		●
Solar walls and windows			●						
Thermal envelope	●			●	●			●	
Sun shading							●	●	
Natural ventilation						●			

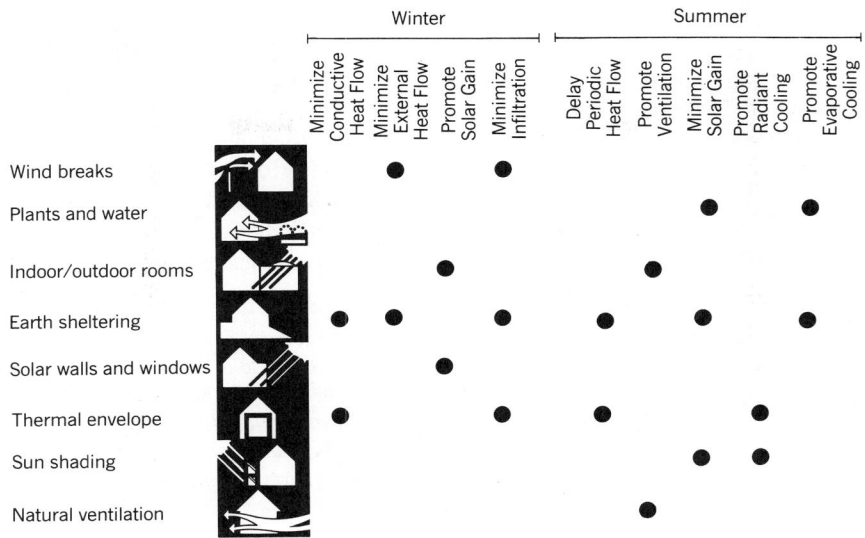

***Fig. 3.2** Generic bioclimatic design concepts. (Reprinted from* Passive Cooling *by permission of the publisher, American Solar Energy Society, Inc.)*

standards, this house would be a very up-to-date example of passive solar heating.

In the Solar Hemicycle house, protection from summer overheating is provided by an overhang along the south glass walls, the cool thermal mass of the north stone/berm wall combination (which receives no direct sun in summer), and the high windows in both north and south walls, which allow warmed air to rise and escape. Over the years since these early photographs were taken, the growth of some plants around the south rim of the scooped pocket has diminished the summer "heat trap" effect of this pleasant outdoor space.

Almost all of today's energy sources have the sun as an ancestor. The earth receives a very small percentage of the sun's daily energy output; the amount of solar energy available to each site varies both seasonally and daily. Typically, the closer the sun to a position directly overhead, the more solar energy reaches the site. Direct sunlight's radiant energy is not its only resource; direct sun is our most intense light source, whose usage is complicated by its slow but constant change of position. Indirect sun, as on an overcast day, is a diffuse and readily utilized light source. Daylighting was the dominant method of lighting buildings until the middle of this century.

3.4 Direct Sun and Daylight

(a) Access to Light and Sun. The value of daylight (and air) to buildings has long been recognized in zoning laws, which require that minimum distances (setbacks) be maintained between a building and the property line in lower density areas. Height restrictions often accompany these setbacks, defining a maximum "buildable volume" in which a building can grow (Fig. 3.4). As buildings become taller and density increases, daylight reflecting down between them is diminished; in response, the maximum buildable volume approximates a pyramid (Figs. 3.4 and 3.5).

When *direct sun* in winter is desirable at the ground floor of each site, this pyramid changes shape and its volume decreases (Fig. 3.4*c*); this is due to the low angle of the winter sun, which is readily blocked by taller objects south of a given window. This most-restricted pyramid (called the "solar envelope") is at present rarely achieved, but various proposals to guarantee access to direct winter sun for solar collection are under development on federal, state, and local levels. The most restrictive feature of the solar envelope is the low slope of its northern face, usually corresponding to the altitude of the sun above the horizontal

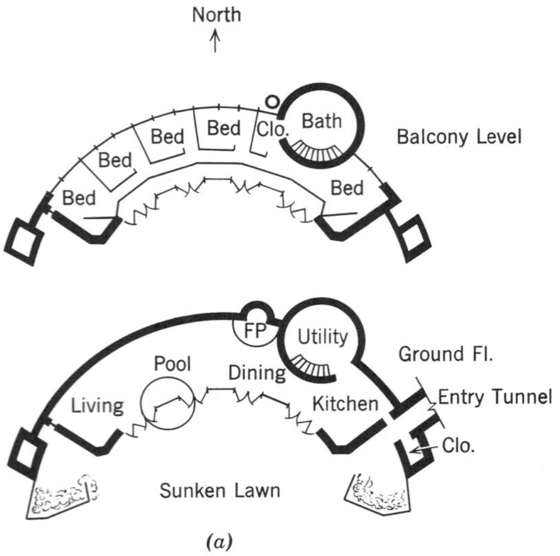

at about 2 hours from noon on December 21. This allows 4 hours of access to direct sun on even the shortest day.

The impact of solar zoning on site utilization can be considerable. Figure 3.6 presents further development of the solar envelope based on the work of Knowles (1981).

For daylight access, building surfaces can be almost as important as building geometry. The im-

Fig. 3.3 *An early passive solar-heated home, Frank Lloyd Wright's Solar Hemicycle near Madison, Wisconsin (between the cool and temperate zones), designed in the early 1940s, built in 1948.* (a) *A representative floor plan.* (b) *An early view of the northeast berm wall.* (c) *The passive solar-heated interior. (Photos by Ezra Stoller. Copyright © ESTO.)*

Fig. 3.4 Access to light. Some methods of compromising private optima (e.g., maximum rentable floor space) with public optima (e.g., daylight at street level).

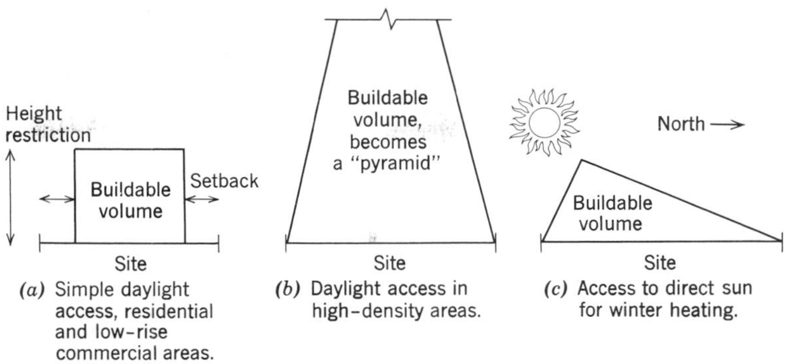

Fig. 3.5 Some results of zoning for "pyramids" in densely built areas.

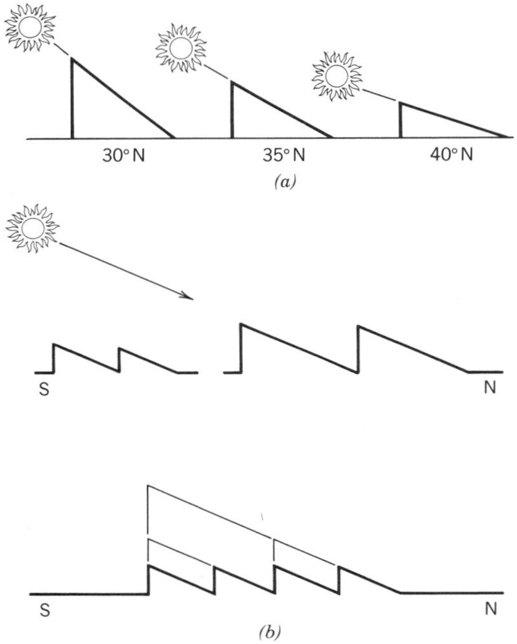

Fig. 3.6 These solar envelopes are refinements of the solar access "pyramid" of Fig. 3.4 and 3.5. (a) *The slope of the solar envelope changes with latitude.* (b) *The larger the site, the greater the volume of the solar envelope. (Reprinted by permission from Knowles, R.,* Sun, Rhythm, Form, *© 1981, MIT Press.)*

portance of reflected light from vertical and horizontal surfaces will be apparent in the daylighting calculations in Chapter 19; lighter colored surfaces produce more internal light, especially in crowded urban conditions where a view of the sky from windows is not common. An increase in daylight will result for all surfaces on an urban street, if the building and site surfaces are light colors. This can conflict with the use of planting for shading and evaporative cooling, although lighter colored ground cover plants are available. While daylight reflected in a diffused way from building and ground surfaces is a potential benefit, the harsh specular reflections from mirror surfaces are often unwelcome (see Section 3.4*d*).

(b) Charting the Sun. The details of sun position and solar utilization are found in later chapters on solar heating (Chapters 5, 6) and daylighting (Chapter 19). Sun position and intensity data

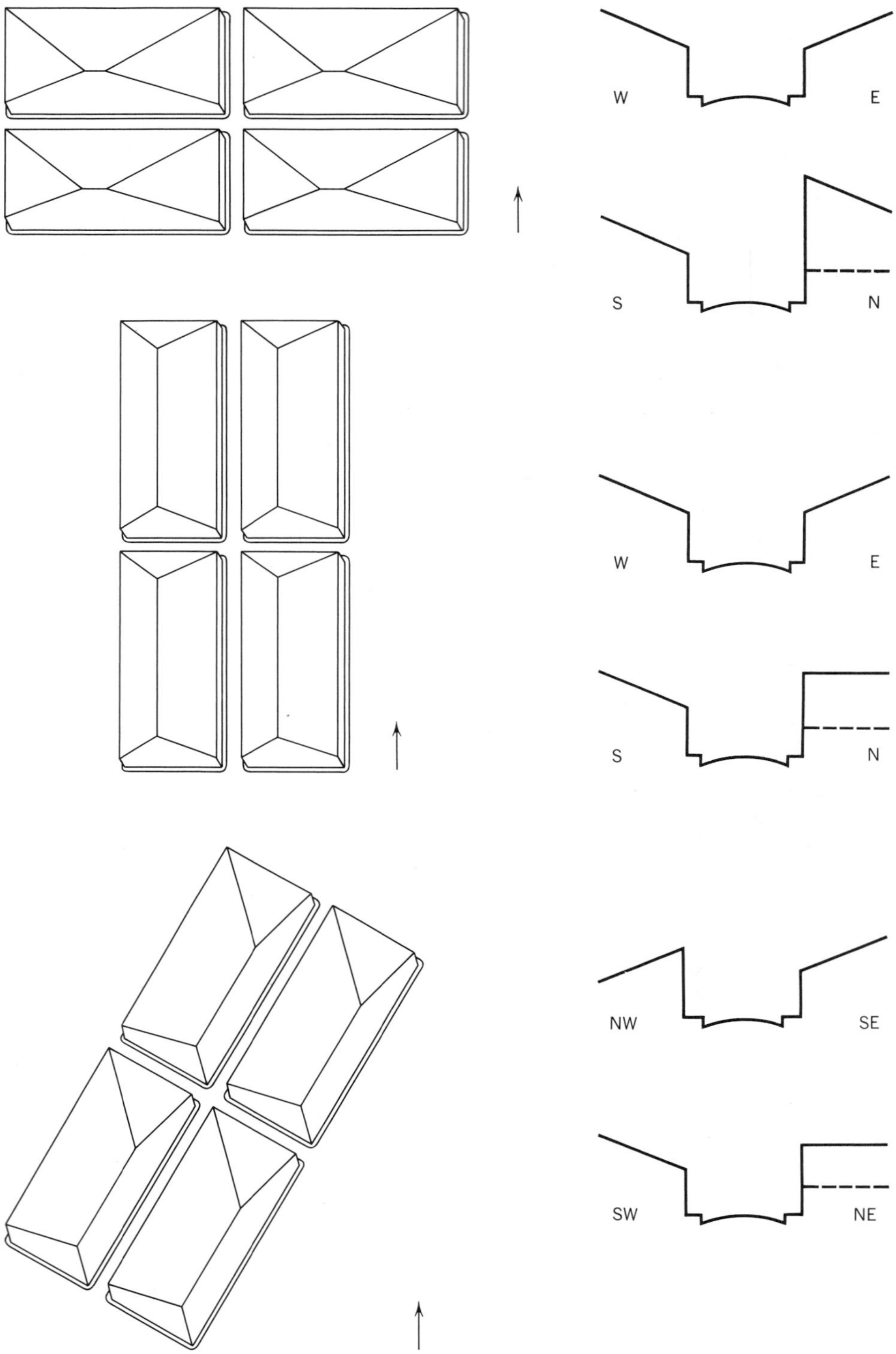

(c)

Fig. 3.6 (c) *Solar envelopes for various orientations of individual sites. (Reprinted by permission of Knowles, R.,* Sun, Rhythm, Form, *© 1981, MIT Press.*

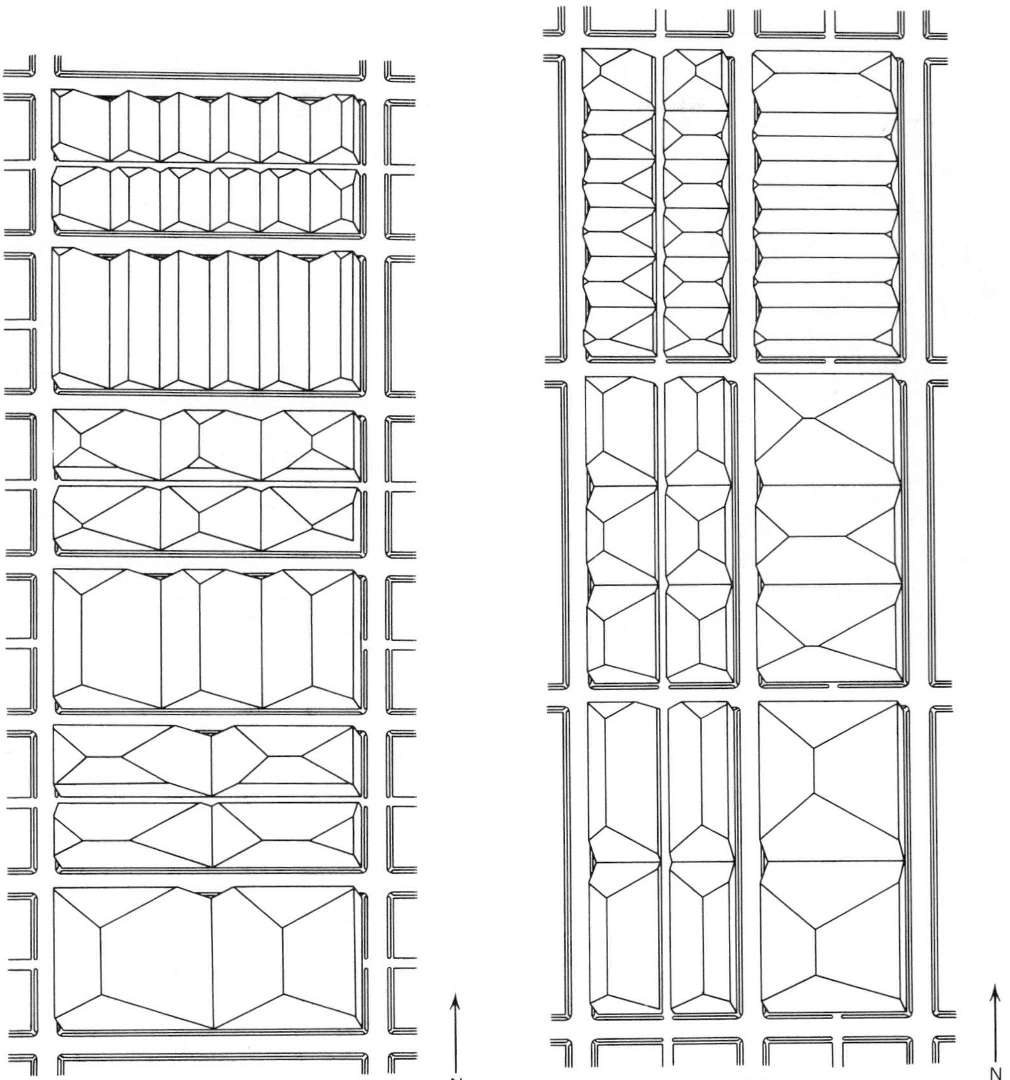

(d) (e)

Fig. 3.6 (d) *Solar envelopes for east–west elongated blocks.* (e) *Solar envelopes for north–south elongated blocks. (Reprinted by permission of Knowles, R.,* Sun, Rhythm, Form, *© 1981, MIT Press.)*

are found in Appendices B and C. For site planning, one of the earliest frequent tasks is to determine whether direct sun reaches a building (or playground, etc.) in the winter. Various devices are widely available for making such determinations; examples of a very simple tool, and a more complex tool are shown in Fig. 3.7. From the resulting graphs of skylines, a site can be easily evaluated for its accessibility to winter sun (or summer shade).

(c) Sun and Shadows: Model Technique. The graphic techniques just discussed have a limitation: each graph applies to only *one* particular location. To study multiple locations, multiple graphs must be constructed. By contrast, a three-dimensional model used in conjunction with a sun shadow plot (Fig. 3.8) can yield the three-dimensional sun penetration patterns as they change over time, for *any* location, indoors or out, on the model. Models are initially time-consuming to build, but

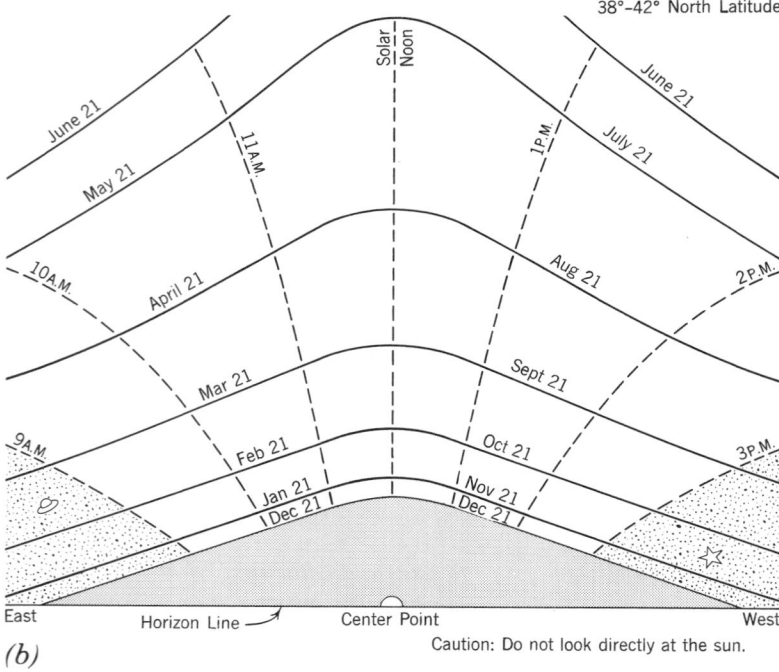

Fig. 3.7 *Tools for charting the sun.* (a) *The expensive (about $180), accurate, and durable Solar Pathfinder includes a compass, bubble level, and tripod. It yields a 360° view of all potential obstructions to sun; it is as useful for siting roof ponds as it is for solar windows. (Courtesy of Bernard M. Haines, Solar Pathways, Inc., Glenwood Springs, CO.)* (b) *The cheap (about $13) and quick Solar Card gives a picture of winter sun obstructions from about 9* A.M. *to 3* P.M. *The user must furnish a compass, and must have the ability to keep the card perfectly vertical while sketching the skyline on it. Its best performance is for siting south-facing windows. (Courtesy of Dan Reif and SolarVision, Inc., Harrisville, NH 03450.)*

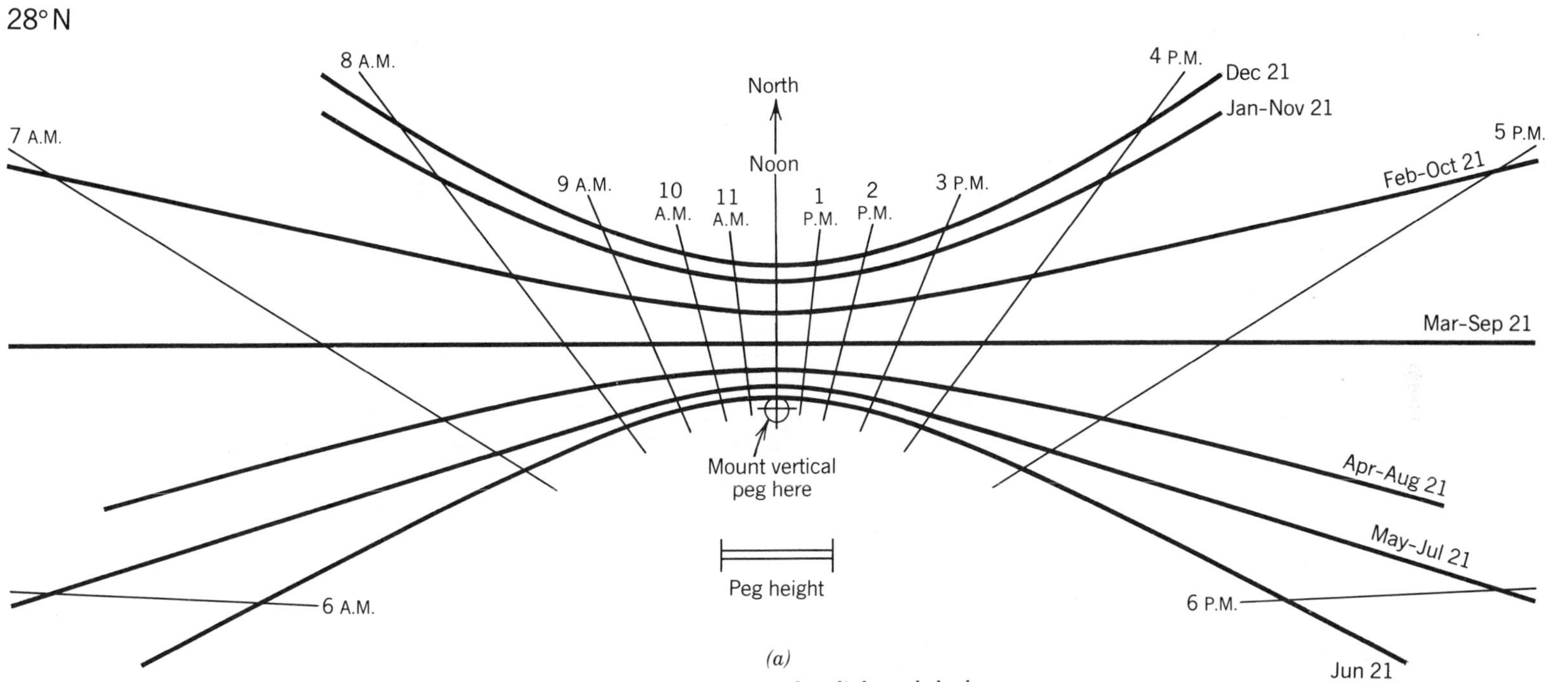

(a)

Fig. 3.8 (a–f) *Sun peg shadow charts. These charts will show the exact position of sunlight and shadow on a model of any scale, on any date, at any time of day between shortly after sunrise and shortly before sunset.*

1. *Find the chart nearest your latitude using Fig. 3.8g.*
2. *Make a copy of this chart. (Don't worry if the copier changes the chart size, because the peg height will be changed in the same proportion.)*
3. *Construct a "peg" that will stand—and remain—perfectly vertical, and whose finished height above the chart surface corresponds exactly to the "peg height" shown on your copy of the chart for your latitude.*
4. *Mount your copy of the chart on the model to be tested. It is important that the chart be perfectly horizontal over its entire surface, and that the north arrow on the chart point to true north for the model.*
5. *Mount your vertical peg at the location shown on the chart. Be sure it's vertical, and will stay vertical.*
6. *Choose a test time and date. Take your model out into direct sunlight. Then* tilt *the model until the shadow of the peg points toward the intersection of the chosen time's line and the chosen date's curve. When the end of the peg touches this intersection, your model will show the* same sun–shadow patterns *as would occur on the time and date at the intersection you chose.*

(Reprinted by permission from Brown, Reynolds, and Ubbelohde, InsideOut: Design Procedures for Passive Environmental Technologies, *John Wiley and Sons, © 1982.)*

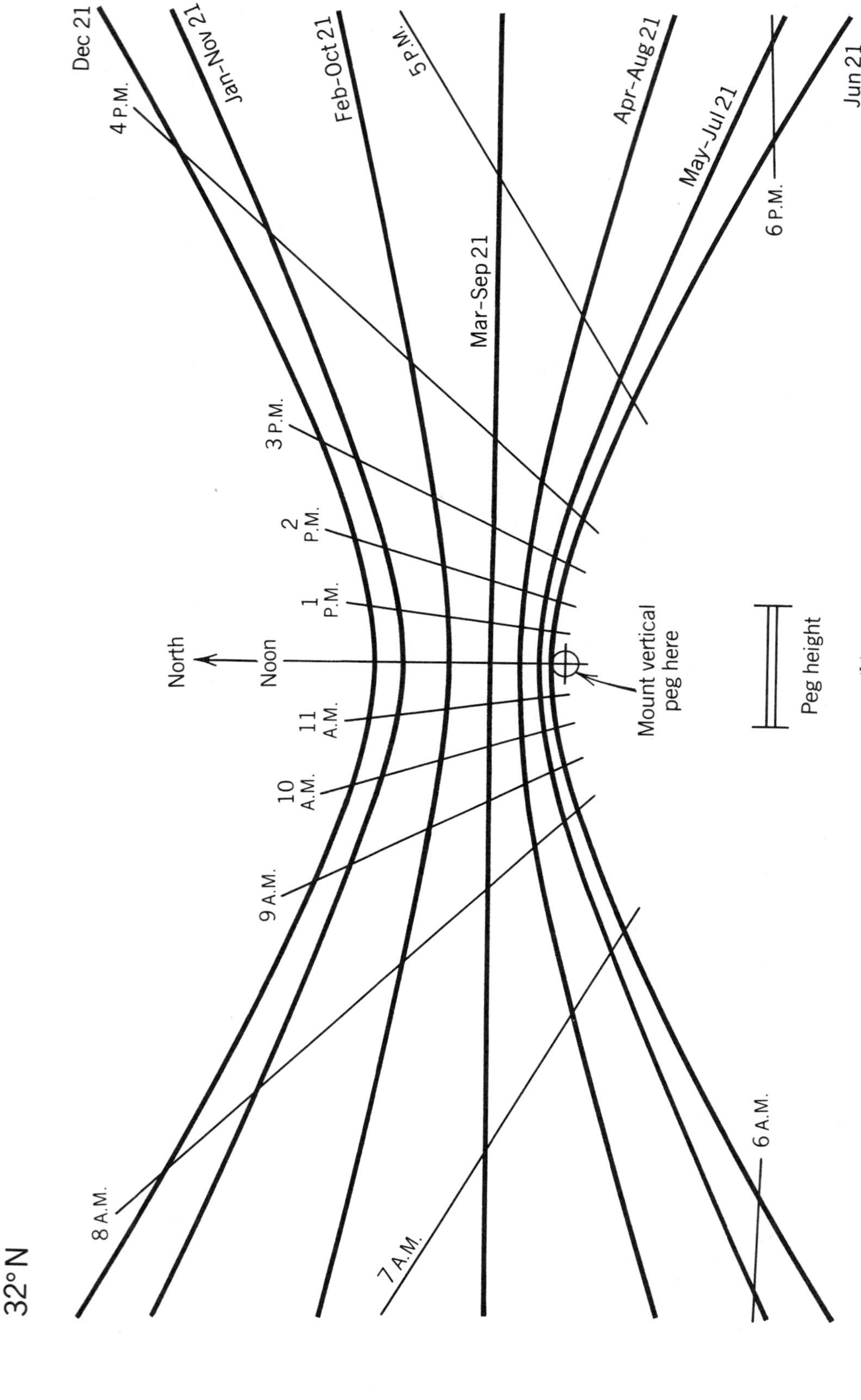
32° N
Dec 21
Jan–Nov 21
Feb–Oct 21
Mar–Sep 21
Apr–Aug 21
May–Jul 21
Jun 21
4 P.M.
5 P.M.
6 P.M.
3 P.M.
2 P.M.
1 P.M.
Noon
11 A.M.
10 A.M.
9 A.M.
8 A.M.
7 A.M.
6 A.M.
North
Mount vertical peg here
Peg height

(b)

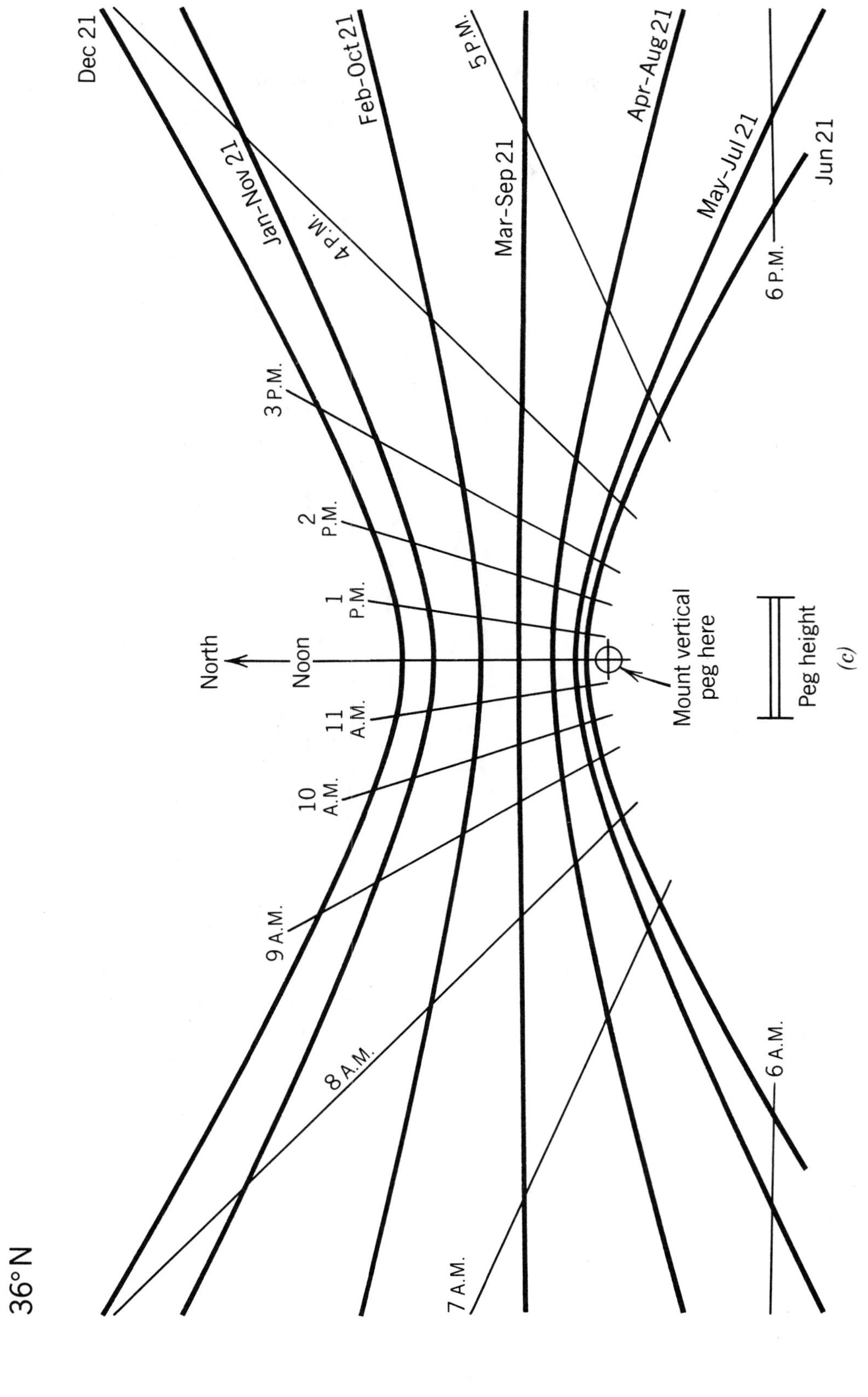
36° N
Dec 21
Jan-Nov 21
Feb-Oct 21
Mar-Sep 21
Apr-Aug 21
May-Jul 21
Jun 21
North
Noon
7 A.M.
8 A.M.
9 A.M.
10 A.M.
11 A.M.
1 P.M.
2 P.M.
3 P.M.
4 P.M.
5 P.M.
6 A.M.
6 P.M.
Mount vertical peg here
Peg height

(c)

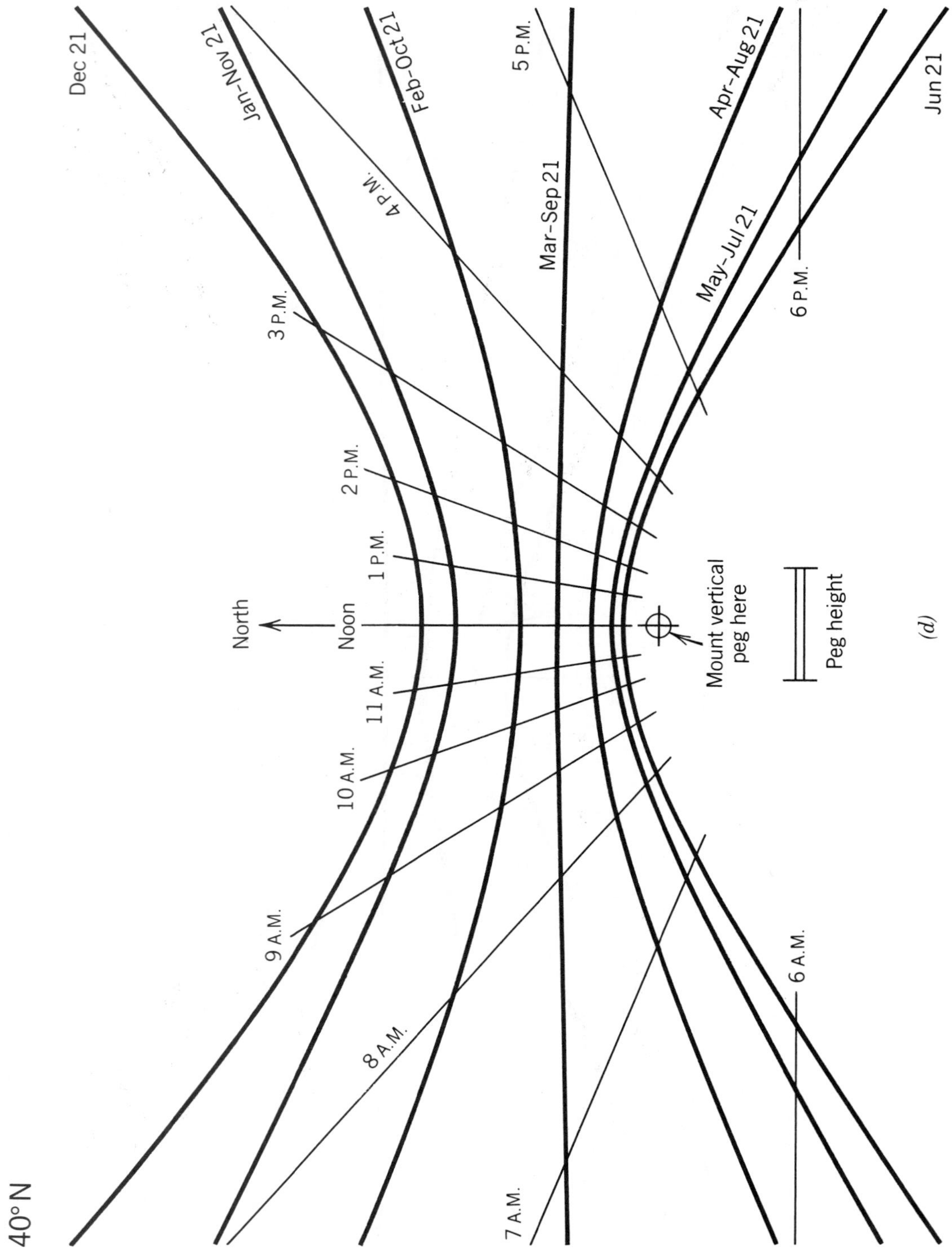
40° N
Dec 21
Jan-Nov 21
Feb-Oct 21
Mar-Sep 21
Apr-Aug 21
May-Jul 21
Jun 21
North
Noon
7 A.M.
8 A.M.
9 A.M.
10 A.M.
11 A.M.
1 P.M.
2 P.M.
3 P.M.
4 P.M.
5 P.M.
6 A.M.
6 P.M.
Mount vertical peg here
Peg height

(d)

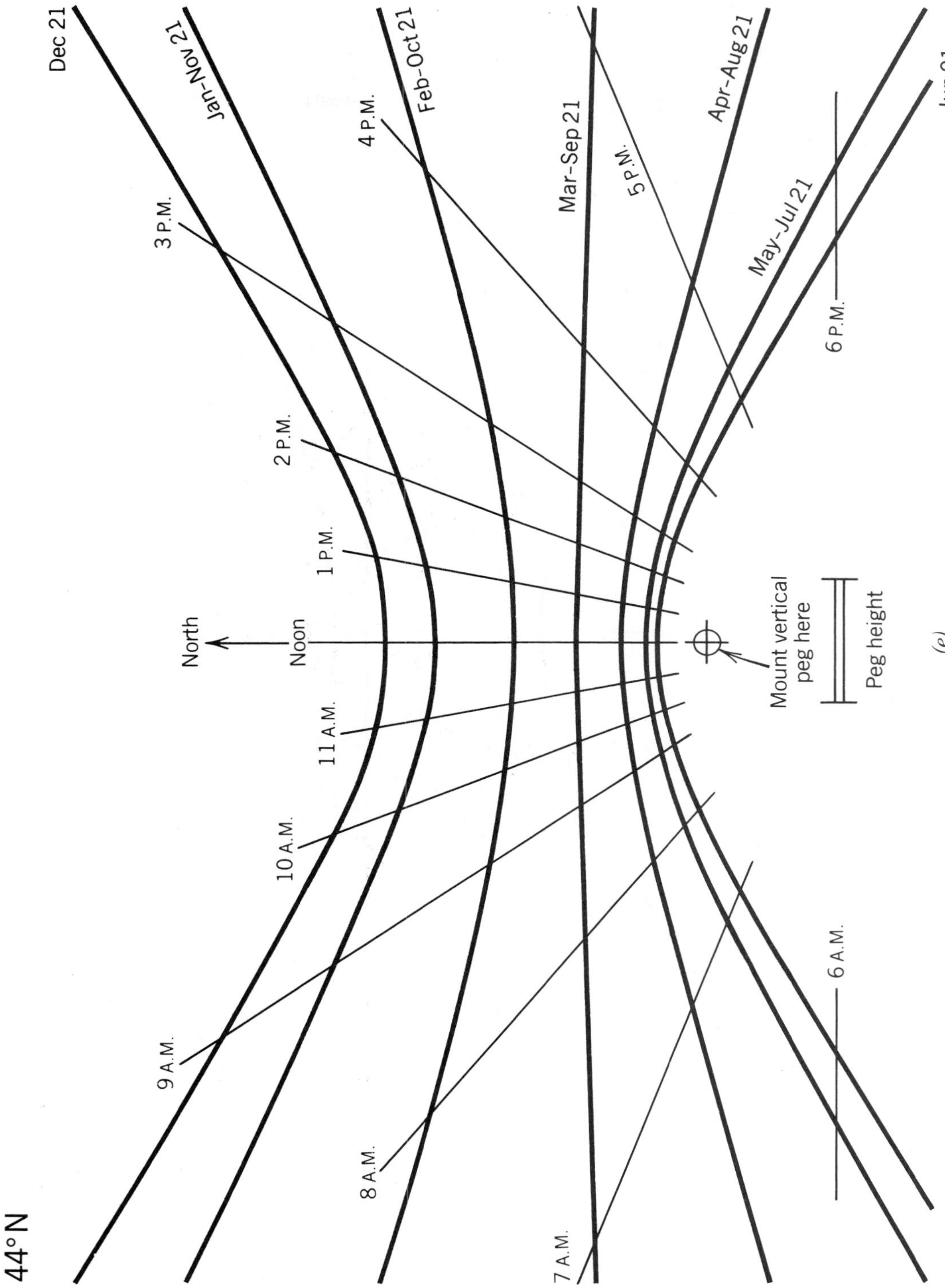
Dec 21
Jan–Nov 21
Feb–Oct 21
Mar–Sep 21
Apr–Aug 21
May–Jul 21
Jun 21
North
Noon
1 P.M.
2 P.M.
3 P.M.
4 P.M.
5 P.M.
6 P.M.
11 A.M.
10 A.M.
9 A.M.
8 A.M.
7 A.M.
6 A.M.
Mount vertical peg here
Peg height
44° N

(e)

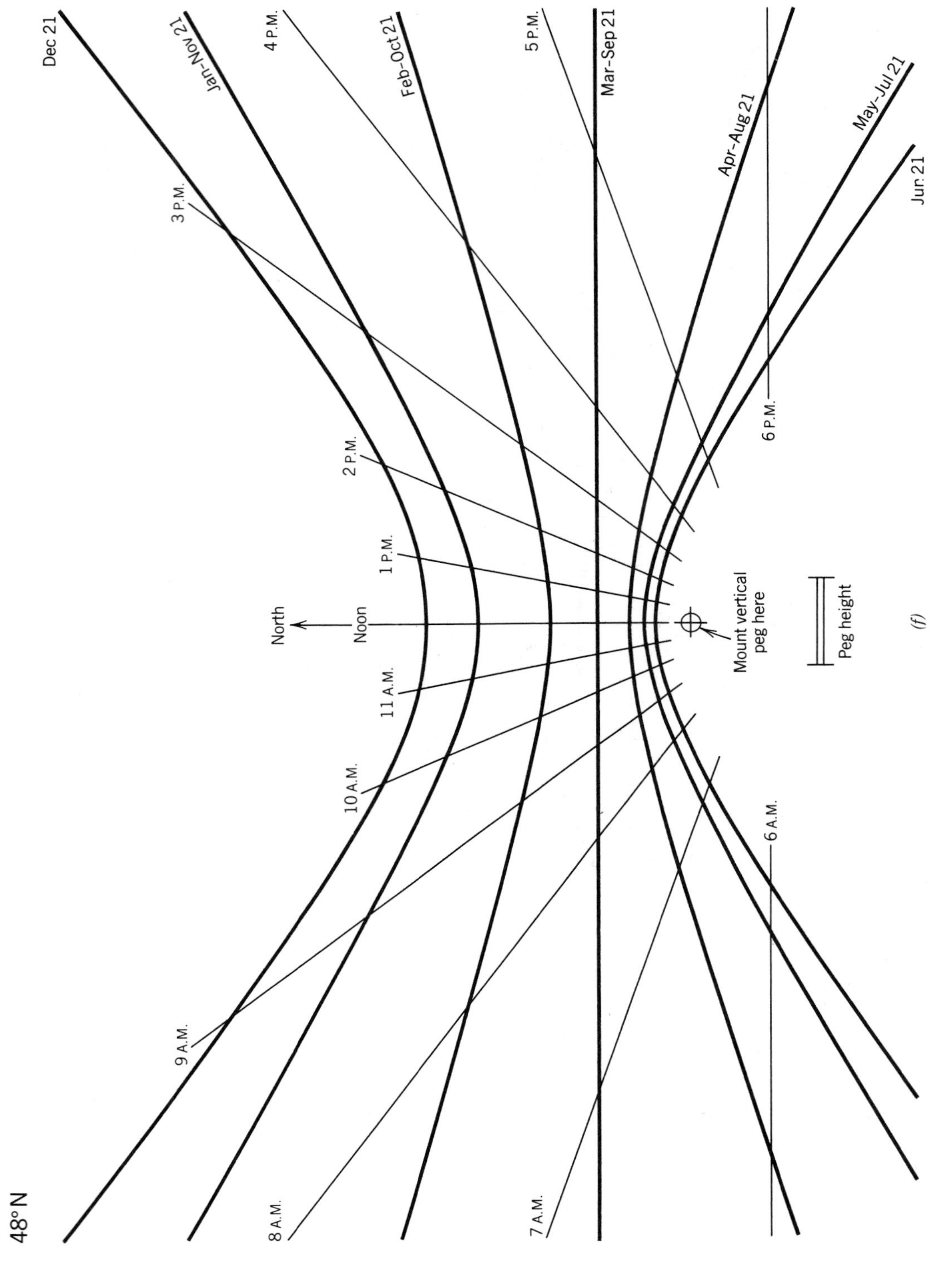

48° N
Dec 21
Jan–Nov 21
Feb–Oct 21
Mar–Sep 21
Apr–Aug 21
May–Jul 21
Jun 21
North
Noon
1 P.M.
2 P.M.
3 P.M.
4 P.M.
5 P.M.
6 P.M.
11 A.M.
10 A.M.
9 A.M.
8 A.M.
7 A.M.
6 A.M.
Mount vertical peg here
Peg height

(f)

(g)

Fig. 3.8 (g) *Map of latitudes across North America.*

can save time when considering alternate locations on a site, alternate window and space combinations, alternate shading devices, and so on. Perhaps most important to the designer, models suggest three-dimensional alternative solutions, because you are testing *space,* not merely plan or section.

However, if obstructions to needed sun exist far from the site (such as nearby mountains), you might want to rely on graphs rather than to try to include such large obstructions on your model.

In Fig. 3.8, ''sun peg shadow plots'' are found for six latitudes. A copy of such plots, correctly attached to a model *of any scale,* will allow the designer to quickly determine exact sun penetration patterns at many times, for any date. These plots are one of the most important tools for the solar designer, both early and late in the design process.

(d) Controlling Solar Reflections. The use of highly reflective (or ''mirror'') glass to reduce heat gain in office buildings, and the rapid spread of solar collectors on walls and sloped roofs, has increased the frequency of annoying solar reflections from buildings (Fig. 3.9). The farther the sun's rays are from being perpendicular to any surface, the more of the sun's light is reflected, rather than absorbed, by that surface. Thus, the intensity of the reflection is greatest when the sun's rays are nearly parallel to a surface. Fortunately, this is the poorest time for solar collection through that surface; curtailed reflections need not mean curtailed collections.

The path or direction of the reflection is a ''mirror image'' of the sun ray's path. Figure 3.10 shows the intensity (percentage of sunlight striking a surface that is reflected) and the path (on horizontal ground) of reflections from a vertical south wall at 44°N latitude. Such an analysis can be used to minimize the most intense solar reflections.

Since these most intense reflections occur where the sun's rays are nearly parallel to a wall, they are easily blocked. For example, foliage (Fig. 3.11) can intercept reflected sunlight. In another approach, the designer can call for external projections around windows (Fig. 3.12).

3.5 Sound and Air

Sound and air are considered together because they are so difficult to separate. Many buildings that could be opened to ventilation or cooling by breezes

Fig. 3.9 *Reflections. Mirror-glass windows in a newer office building (left) in Milwaukee, Wisconsin, cast strong reflections on the north-facing wall of the older hotel next door. Although this reflected heat might be welcome in winter, the glare is intense. In summer, the older building is particularly disadvantaged.*

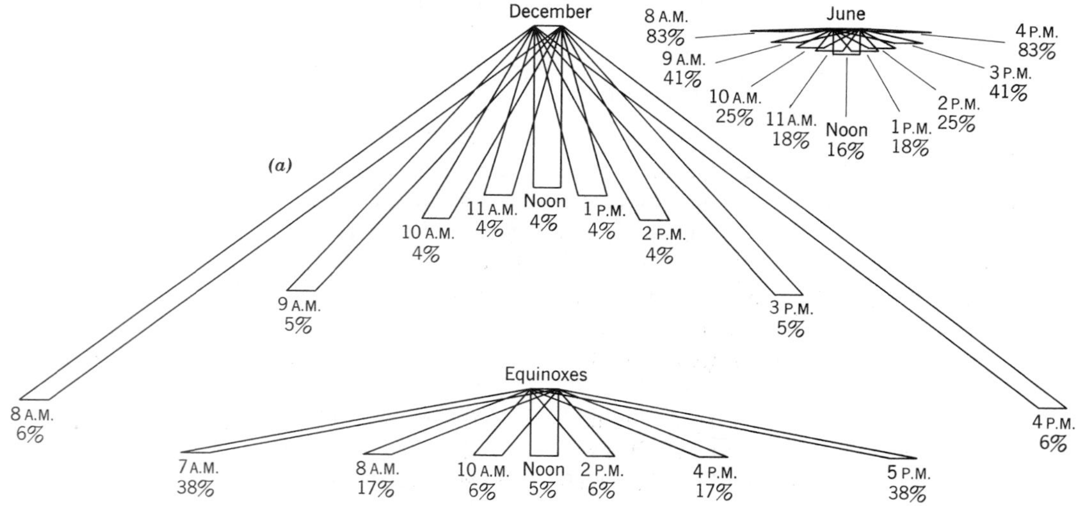

Fig. 3.10 (a) *Reflections in plan view showing the sizes of the reflections from a vertical south-facing wall (at 44°N latitude) whose east–west length is half its height. Although the December reflections reach farthest from this vertical glass wall, they are the weakest in intensity; only 8% of the sunlight striking the glass is reflected at 8* A.M. *and 4* P.M. *Conversely, the June reflections plunge quickly into the ground, but at great intensity (82% at 8* A.M. *and 4* P.M.*).* (b) *Reflections at noon, in section, for solstices and equinox. (Copyright © Michael D. Lee, Anchorage, Alaska. Reprinted by permission.)*

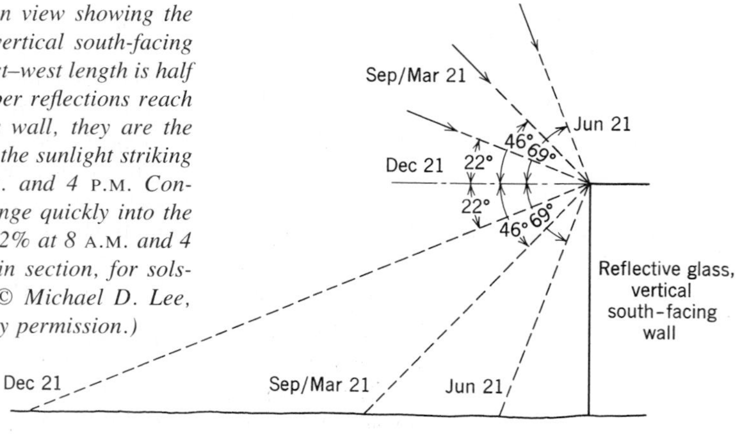

(b)

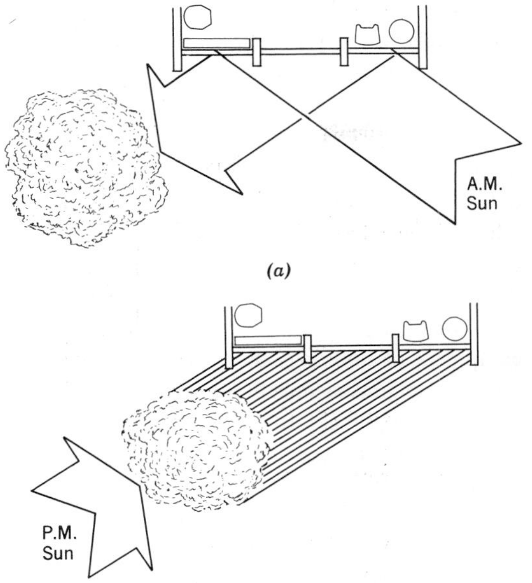

Fig. 3.11 *Selective protection from reflections.* (a) *The tree standing west of this south window wall does not interfere with solar access during the best hours for solar gain (around noon), nor does it prevent early morning sun from entering the windows. The reflections of early morning are intercepted by the tree, before they can escape to annoy elsewhere.* (b) *The late afternoon sun is blocked by the tree, before either heat gain or reflections can occur.*

Fig. 3.12 *The "eggcrate" shading devices on the southwest corner of this University of Oregon building prevent reflections by blocking the sun from either side of the glass. However, different shading devices are appropriate for different orientations of facades.*

rely instead on forced ventilation because of the noise that would accompany the breeze through an open window. Polluted air is another potential deterrent to "natural" ventilation. Almost any device that reduces sound will also reduce the velocity of the breeze, as is true of most filtering devices to remove dirt particles.

(a) Noise. The building in Fig. 3.13 is unusual both in providing an opportunity for wind ventilation—air moves freely below it as well as around and over it—and in the extraordinary intensity of noise that is carried through such air.

Two characteristics of cities (see Section 3.1) contribute to increased noise at street level: *hard*

Fig. 3.13 *Apartment buildings in a series straddle the approach ramps to New York's George Washington Bridge. These buildings were the scene of a study linking noise level with reading disabilities (Cohen et al., 1973).*

surfaces reflect rather than absorb sound, and *parallel walls* intensify noise between them rather than dissipating it. Increasing horizontal and vertical distance from urban noise sources affects the outdoor noise level on the street in different ways. See the graphs of sound level versus distance (and height) of Fig. 3.14. Although building surfaces are generally hard for durability in weathering, softer and multiplaned materials (such as plants) are desirable from a public noise reduction viewpoint. Their impact on measured noise may be slight, but visually softer surfaces reinforce our perception of acoustically softer environments (much as the sound of running water reinforces our perception of cool environments). Furthermore, the sounds made by plants as wind moves through them can help mask the unwanted noises of the street beyond. Fountains are especially useful sources of such *masking sound;* they can be kept running as long as the noise persists, and they enhance the cooling function of natural ventilation, especially in drier climates.

Where site conditions allow, barriers to street noise can be installed that cast ''sound shadows'' on the surface of the site (Fig. 3.15). Such barriers do little to reduce noise levels at upper windows, but surface activities can be given much lower noise levels, especially very near the barrier. Many cities now require such barriers between new housing developments and highways or railroads.

Another urban noise source is the mechanical equipment of buildings themselves. Many of the noise complaints against buildings involve air conditioning equipment (the compressive refrigeration cycle and its year-round utilization as a heat pump are described in Chapter 6). Noise is generated both by the compressor and by the great quantity of exterior air that must be rapidly pushed through outdoor coils (further heating outdoor air so that the indoors may be cooled). When densely packed buildings are forced to rely on mechanical cooling, such closeness makes the noise of the systems even more annoying. The machine's ap-

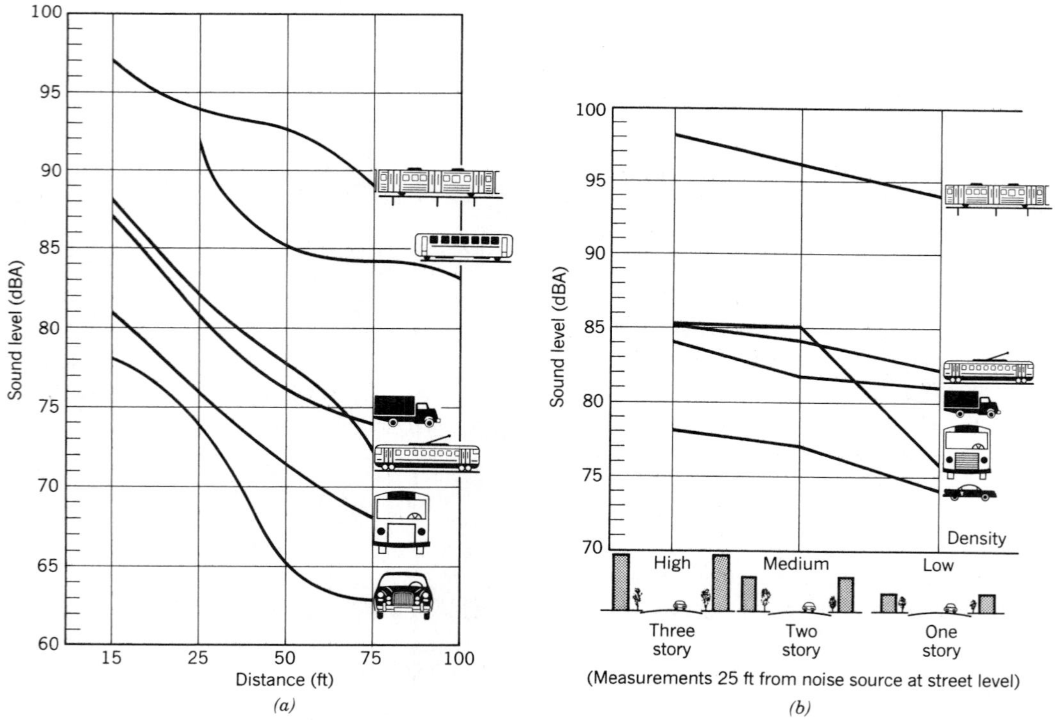

***Fig. 3.14** Predicting noise levels outdoors.* (a) *Distance as a factor in noise intensity.* (b) *Building height as a factor in noise propagation. (From Bragdon, Clifford R.,* Noise Pollution: The Unquiet Crisis, *University of Pennsylvania Press, 1971. Reprinted by permission.)*

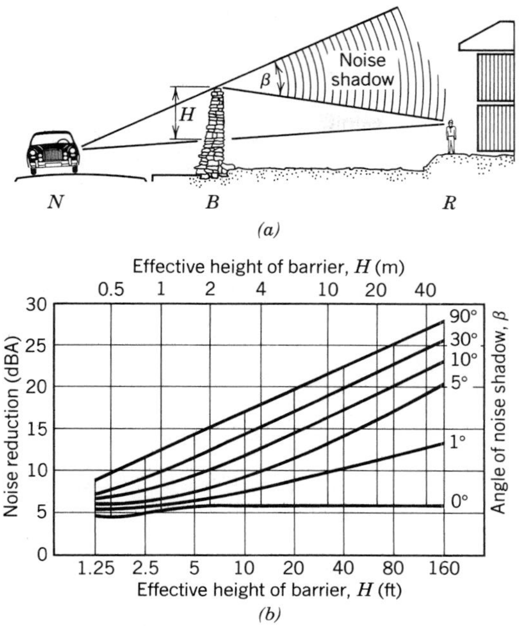

Fig. 3.15 Barriers to outdoor noise sources. (a) *To determine the approximate reduction in noise in decibels (dBA) due to an outdoor barrier, construct a section locating the noise source* (N), *the solid barrier* (B), *and the receiver's location* (R). *On this section, determine effective height* (H) *of the barrier, and the diffraction angle* (β) *with the resulting "noise shadow." Enter the graph* (b) *with* H *and* β; *where their lines intersect, determines the noise reduction in dBA (left margin). Note significant noise reduction from simply breaking the line of sight* (β = *1°). (From Doelle, Leslie,* Environmental Acoustics, *McGraw-Hill Book Company, 1972. Reprinted by permission.)*

petite for outdoor air is so enormous that attempts to surround it with noise shields can greatly hinder its efficient operation and shorten its life.

In residential neighborhoods, the greater distance between buildings might be expected to lessen these difficulties. Yet the much lower ambient (or background) noise level of residential areas is one of their more desirable characteristics, and an intruding compressor can trigger lawsuits from neighbors, who formerly enjoyed cool—and quiet—night breezes.

(b) Air Pollution. While one of the deterrents to operable windows is the dirt that they could admit to the interior, such air pollution is a threat to people and their buildings in other ways, as well. Figure 3.16 summarizes the impacts of air pollution in the United States and indicates hope for improvement, provided we continue to enforce the 1970 Clean Air Act. Buildings (and the electric generating plants that supply them) are major contributors to the pollution of the "fresh air" needed for ventilation. It is encouraging that pollution levels of particulates have declined since 1970.

Sulfur oxide levels have declined slightly nationwide, although "acid rain" remains a serious problem; there has been a decrease in sulfur dioxide in the urban centers, counterbalanced by an increase in rural areas from sources such as electric generating plants that burn fossil fuel. Carbon monoxide levels have changed little nationwide, but the concentration of this pollutant in urban centers has declined. The pollutants that cause smog continue to be troublesome, since levels of nitrogen dioxide have increased because of the increased total automobile miles traveled, and because of increasing electrical generation in fossil-fueled plants.

Carbon dioxide is not a pollutant, but a necessary and widely debated constituent of air. Will the amount of carbon dioxide continue to increase in our atmosphere primarily as a result of the burning of fossil fuels? (Coal is projected as a prime future U.S. energy source, see Table 1.1.) If carbon dioxide does increase, will it warm the earth by discouraging our outward radiation of heat, as does the glass cover on a solar collector? Or will particulate matter, which accompanies carbon dioxide from the furnaces of buildings and generating plants, reflect incoming sunlight and nullify the warming effect of carbon dioxide?

(c) Wind Control. For most buildings, wind (like sun) changes from friend to enemy with the change of seasons. Control of wind often means utilizing wind-sheltered areas in winter, while encouraging increased wind speeds for the building in summer. Outdoor spaces, as seen in Section 2.4, can benefit from this block-to-admit changeover on a daily basis. The generalized patterns of wind flow around thin windbreaks and thicker buildings (Fig. 3.17) help us understand where shelter and increased flows occur. However, they

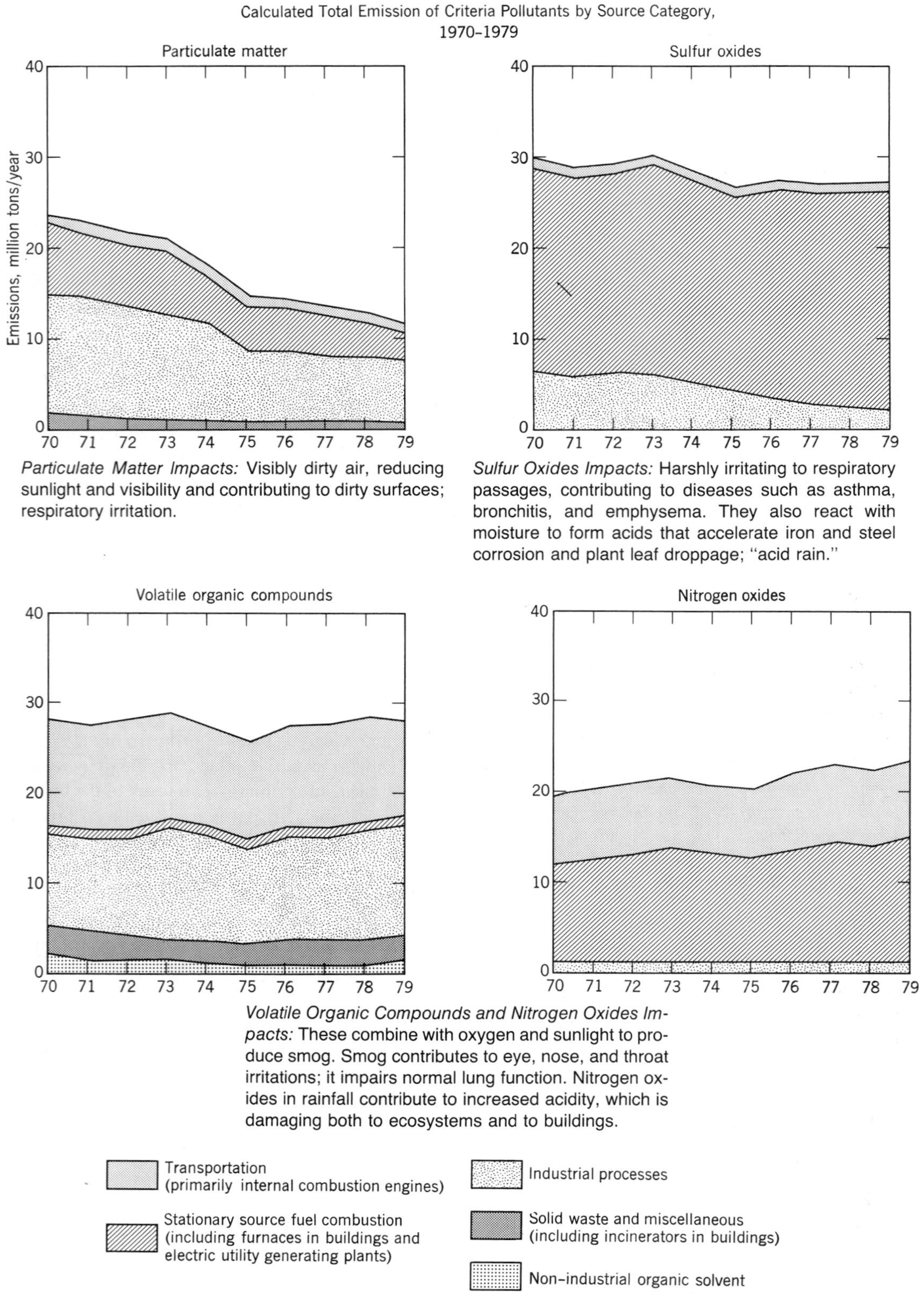

Calculated Total Emission of Criteria Pollutants by Source Category, 1970–1979
Particulate matter
Sulfur oxides
Emissions, million tons/year
40
30
20
10
0
70 71 72 73 74 75 76 77 78 79
Particulate Matter Impacts: Visibly dirty air, reducing sunlight and visibility and contributing to dirty surfaces; respiratory irritation.
Sulfur Oxides Impacts: Harshly irritating to respiratory passages, contributing to diseases such as asthma, bronchitis, and emphysema. They also react with moisture to form acids that accelerate iron and steel corrosion and plant leaf droppage; "acid rain."
Volatile organic compounds
Nitrogen oxides
Volatile Organic Compounds and Nitrogen Oxides Impacts: These combine with oxygen and sunlight to produce smog. Smog contributes to eye, nose, and throat irritations; it impairs normal lung function. Nitrogen oxides in rainfall contribute to increased acidity, which is damaging both to ecosystems and to buildings.
Transportation (primarily internal combustion engines)
Industrial processes
Stationary source fuel combustion (including furnaces in buildings and electric utility generating plants)
Solid waste and miscellaneous (including incinerators in buildings)
Non-industrial organic solvent

Fig. 3.16 *Air pollution in the United States, 1970–1979. Calculated total emissions of pollutants by source category. (From U.S. Environmental Protection Agency, 1980.)*

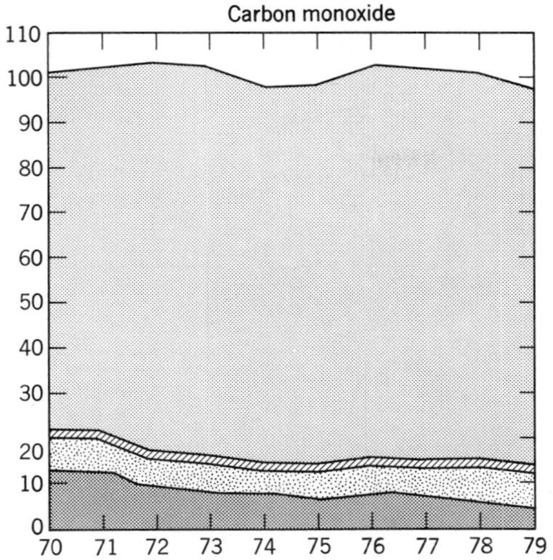

Carbon Monoxide Impacts: Tends to cause suffocation by interfering with the blood's ability to supply oxygen to the body; as a result, heart and respiratory systems are forced to work harder. Symptoms include headache, loss of vision, decreased muscular coordination, nausea, and abdominal pain.

***Fig. 3.16** (continued)*

Fig. 3.17 (below) *Approximate patterns of wind around objects.* (a) *Difference in barrier height.* (b) *Difference in barrier distance.* (c) *Difference in windbreak distance.* (d) *Reduction in wind speed due to density. (Reproduced with the permission of the American Institute of Architects, © 1981, AIA.)*

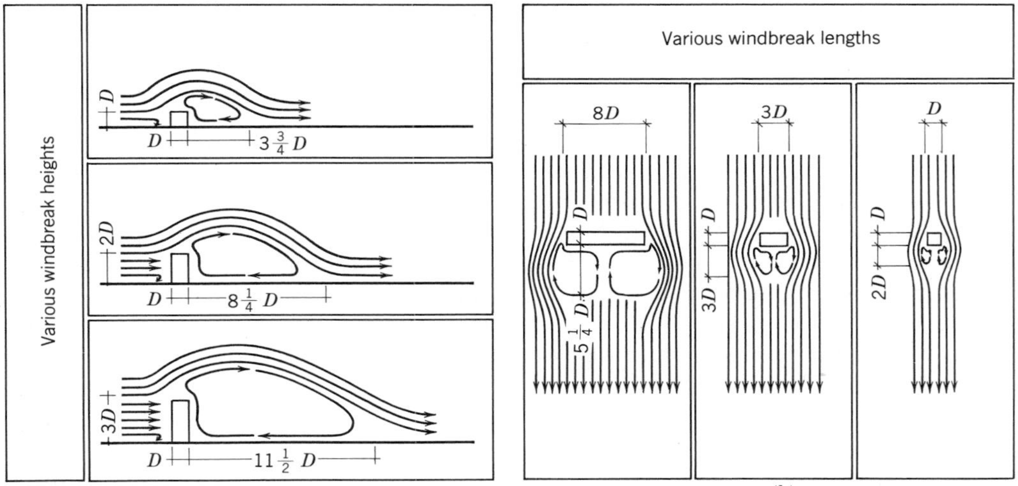

(a) *(b)*

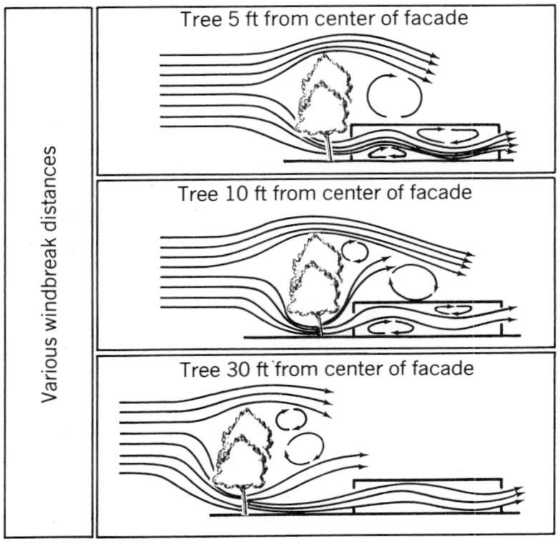

(c)

Wind Speed Reduction (%)				
Density of Belt	*Average Over First 50 Yd*	*Average Over First 100 Yd*	*Average Over First 150 Yd*	*Average Over First 300 Yd*
Very open	18	24	25	18
Open	54	46	37	20
Medium	60	56	48	28
Dense	66	55	44	25
Very dense	66	48	37	20

(d)

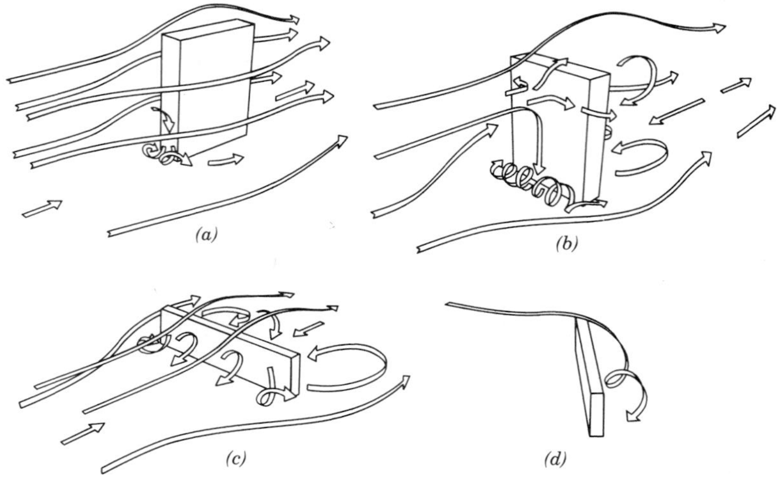

Fig. 3.18 *Wind patterns around single buildings.* (a) *Tall slender buildings; height greater than 2.5 times width.* (b) *Tall, rather wide buildings; height between 2.5 and 0.6 times width.* (c) *Long buildings; height less than 0.6 times width.* (d) *Long buildings; wind at 45° to face. (From W. J. Beranek, "General Rules of the Determination of Wind Environment,"* Wind Engineering, *J. E. Cermak, ed., Vol. 1, © 1980, Pergamon Press Ltd., reprinted by permission.)*

are much more complicated than they first appear and are highly influenced by objects upstream, to the sides, and downstream of the wind-directing object being analyzed. Wind tunnel tests are far more reliable than these patterns; unfortunately, such tests are expensive and still fraught with opportunity for misprediction. With these warnings in mind, the site can be analyzed graphically for seasonal wind utilization.

Wind will ultimately return to its original flow pattern after encountering an obstacle such as a windbreak or a building. Before it reaches the obstacle, it slows, builds pressure, and turns upward or sideways; it increases its speed as it passes, and reduced or negative pressure results at the sides and behind the obstacle. These pressure differences, flow patterns, and the size and shape of wind-protected areas behind the obstacle are all utilizable for control of air motion, inside and out.

Wind around buildings is a complex matter; some rules of thumb for shelter areas were shown in Fig. 3.17, while Figs. 3.18 and 3.19 show some wind behavior to be expected in some typical building combinations. Beranek (1980) has discussed methods for charting the shelter areas in these building groups.

Fig. 3.19 *Wind patterns among building clusters.* (a) *Increased wind speeds can occur, especially where wind is at 45° to the face. With the bar effect, the downward spinning wind behind these buildings can reach 1.4 times the speed of the average wind.* (b) *"Venturi" effect: Where few obstructions occur upwind or downwind from the narrow neck of this plan illustrating the Venturi effect, wind speeds through the neck can reach 1.3 times the average, up to heights of 30 m, and 1.6 times the average at about 50 m height.* (c) *The "gap effect" begins to occur with perpendicular winds at buildings with more than 5 stories (15 m) height; by 7 stories, wind speeds 1.2 times the average can occur through the gaps; by 60 stories, gap wind speeds can be 1.5 times the average.* (d) *For higher buildings, increased wind speeds occur at the corners (localized within a radius, equal to the width of the building* d, *around the corner). Where height is 15 m, wind speed can reach 1.2 times the average; for heights above 35 m, wind speed can be 1.5 times the average. Where two towers approach each other, increased wind around corners and between the towers can go as high as 2.2 times the average for towers 100 m high.* (e) *Increased wind speed and turbulence within the (shaded) wake of buildings can be especially serious for towers where at heights from 16 to 30 stories, wind speeds can reach 1.4 to 2.2 times the average. (From J. Gandemer, "Wind Environment Around Buildings,"* Wind Effects on Buildings and Structures, *K. J. Eaton, ed., © 1977, Cambridge University Press. Reprinted by permission.)*

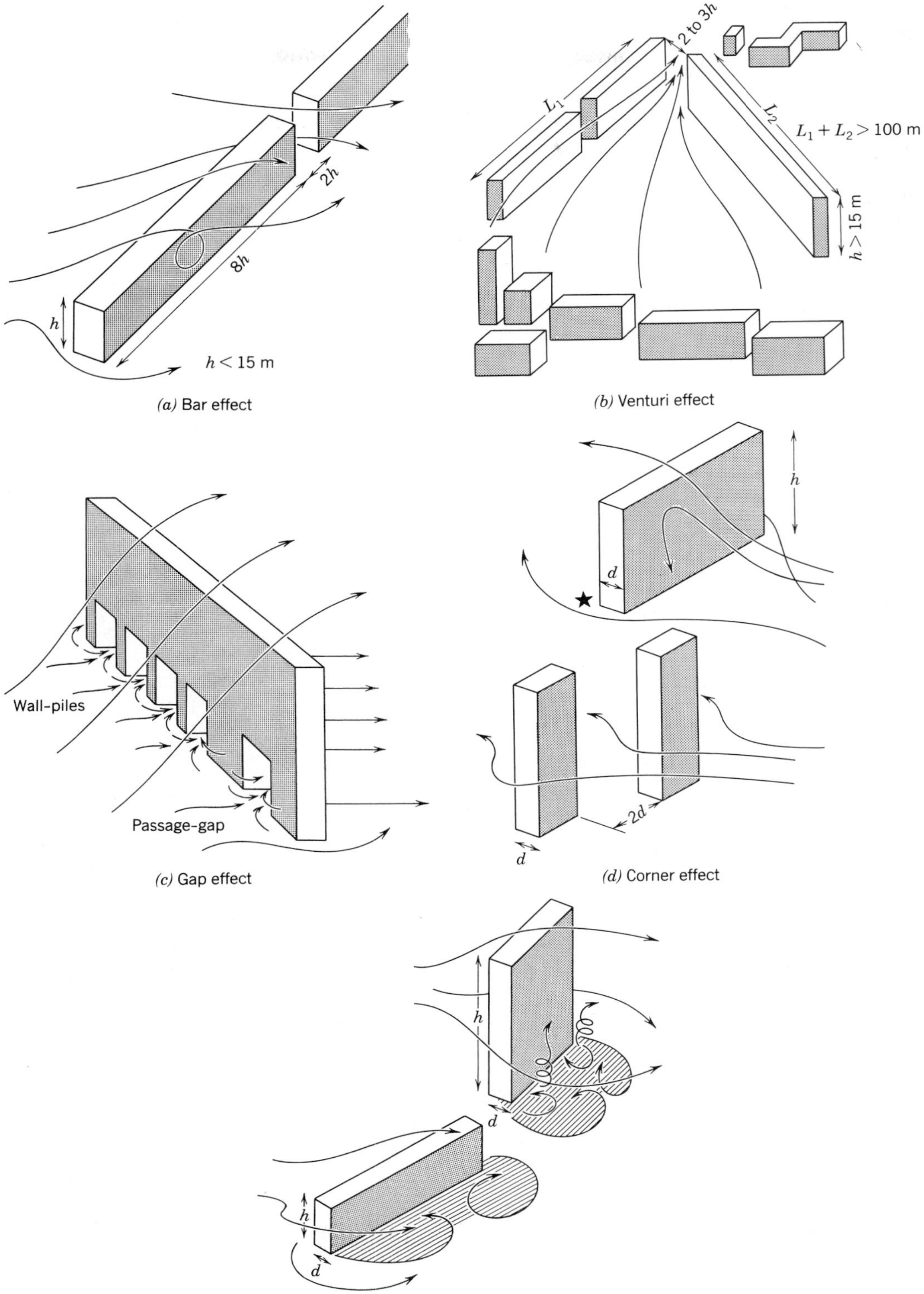

(a) Bar effect

(b) Venturi effect

(c) Gap effect

(d) Corner effect

(e) Wake effect

(d) Ventilation and Cooling. A distinction must be made between two common architectural introductions of outdoor air: ventilation and cooling. *Ventilation* involves the provision of "fresh air" to interiors to replenish the oxygen used by people and to help carry away their by-products of carbon dioxide and bodily odors. Ventilation is desirable all year round; recommended minimum rates for providing fresh air are found in Chapter 4 (Table 4.19). *Cooling* (with outdoor air) replaces heated indoor air with cooler outdoor air. Thus cooling by breezes is a seasonal opportunity, limited to times when the outdoor air temperature is lower than the indoor air temperature. Cooling can require far greater quantities of air than ventilation, and its influence on building siting and window size and placement is considerable. Figure 3.20 illustrates the differing approaches to these two usages.

Two examples of contemporary buildings that use the wind appear in Figs. 3.21 and 3.22*a*. The first served as a summer exhibition space in Montreal, Canada, and relied on prevailing winds to remove the air heated by display lighting and crowds of people. The second is a faculty office building in England, whose windows can be manipulated to provide either a small amount of ventilation air or a more thorough scouring for cooling purposes. The building that relies on prevailing wind for cooling must be sited with attention to wind direction. Again, one building should not be erected to obstruct the next building's access to breeze. As seen previously, obstacles upstream from intake openings, or downstream near the outlets, can

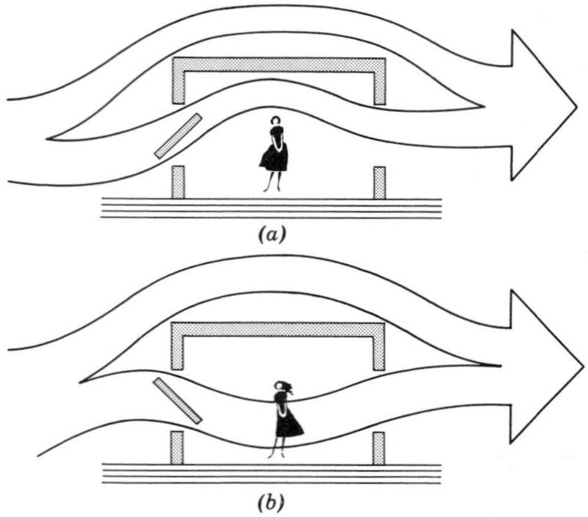

Fig. 3.20 *Ventilation, with and without cooling. The window's placement and size can influence the stream of air within a space, providing either ventilation alone or adding a cooling function to moving air.* (a) *Ventilation: the pivoting window sends the breeze toward the ceiling, where it replaces inside air. It has minimum contact with people below.* (b) *Ventilating and cooling: this position of the window directs the breeze toward the floor, where it encounters people and provides a direct cooling effect from air motion.*

Fig. 3.21 *Cooling with the wind. The air inside this exhibition hall is heated by lights and crowds of people. It then rises and is sucked out of the large clerestory opening by the prevailing winds. Note that these winds do not blow* into *this high opening, but encourage outward flow of the overheated interior's air. (African Place, at Expo '67, Montreal; John Andrews, architect.)*

Fig. 3.22 (a) *These faculty offices and seminar rooms in England have ventilation and cooling options.* (b) *They can admit a small amount of winter ventilating air or a large amount of air to directly cool the interior. The prefabricated window unit* (c) *is hoisted into position during construction* (d). *(Leicester University's Attenborough Building; Arup Associates, Architects, Engineers and Quantity Surveyors, London.)*

substantially reduce the velocity—and thereby the cooling effect—of the wind. But although wind cools people in hot weather primarily by speeding the evaporation of sweat from skin, it can itself become an irritant at high velocity (Tables 2.3 and 3.2). Manual controls of openings are a necessary part of natural ventilation equipment. And finally, the proper size and placement of openings, with a ventilation path through the building, must be provided.

Fig. 3.22 (continued)

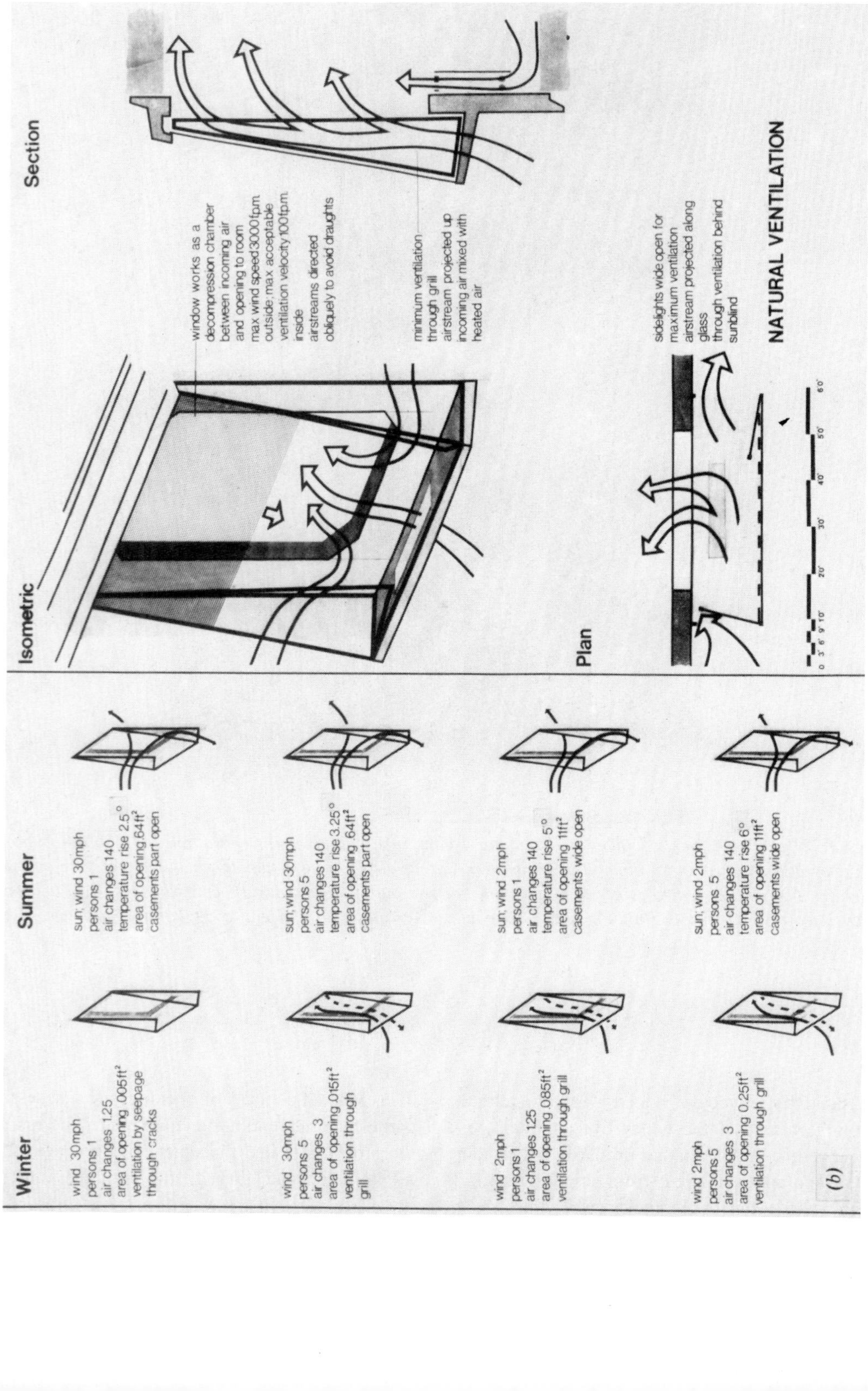
Winter
wind 30mph
persons 1
air changes 1.25
area of opening .005ft²
ventilation by seepage
through cracks
wind 30mph
persons 5
air changes 3
area of opening .015ft²
ventilation through
grill
wind 2mph
persons 1
air changes 1.25
area of opening .085ft²
ventilation through grill
wind 2mph
persons 5
air changes 3
area of opening 0.25ft²
ventilation through grill
Summer
sun; wind 30mph
persons 1
air changes 140
temperature rise 2.5°
area of opening .64ft²
casements part open
sun; wind 30mph
persons 5
air changes 140
temperature rise 3.25°
area of opening .64ft²
casements part open
sun; wind 2mph
persons 1
air changes 140
temperature rise 5°
area of opening 11ft²
casements wide open
sun; wind 2mph
persons 5
air changes 140
temperature rise 6°
area of opening 11ft²
casements wide open
Isometric
Section
window works as a decompression chamber between incoming air and opening to room
max wind speed 3000fpm. outside; max acceptable ventilation velocity 100fpm. inside
airstreams directed obliquely to avoid draughts
minimum ventilation through grill
airstream projected up
incoming air mixed with heated air
Plan
sidelights wide open for maximum ventilation
airstream projected along glass
through ventilation behind sunblind
NATURAL VENTILATION
0 3' 6' 9' 10' 20' 30' 40' 50' 60'
(b)

Fig. 3.22 (continued)

(e) Wind, Daylight, and Sun. When daylight and wind ventilation are desired, they combine to limit the width of buildings. This is particularly evident in multistory office buildings, where increasing urban density and reliance upon electric lighting and cooling have changed the form of buildings considerably in the past few decades (Fig. 3.23). Another impact of combined wind and sun is shown in Fig. 3.24, which represents the factors a designer would weigh in determining the orientation of a clerestory window.

TABLE 3.2 **Beaufort Scale (lower speeds only)**

Beaufort Number	*Speed 6 m (19.7 ft) Above Ground*			*Description of Effects Outdoors*	
	m/s	*fpm*	*mph*	*On Land*	*Over Water*
0	0.3	Less than 88	Less than 1	Smoke rises; no perceptible movement	Smooth sea
1	0.6–1.7	88–264	1–3	Smoke drift shows wind direction; tree leaves barely move	Scalelike ripples
2	1.8–3.3	352–616	4–7	Wind felt on face; leaves rustle	Small wavelets
3	3.4–5.2	704–968	8–11	Leaves, twigs in constant motion; hair is disturbed; wind extends light flag	Large wavelets; occasional white foam crests
4	5.3–7.4	1056–1408	12–16	Small branches move; dust rises; hair disarranged	Small waves become longer

Source: Reprinted from *Passive Cooling* by permission of the publisher, American Solar Energy Society, Inc.

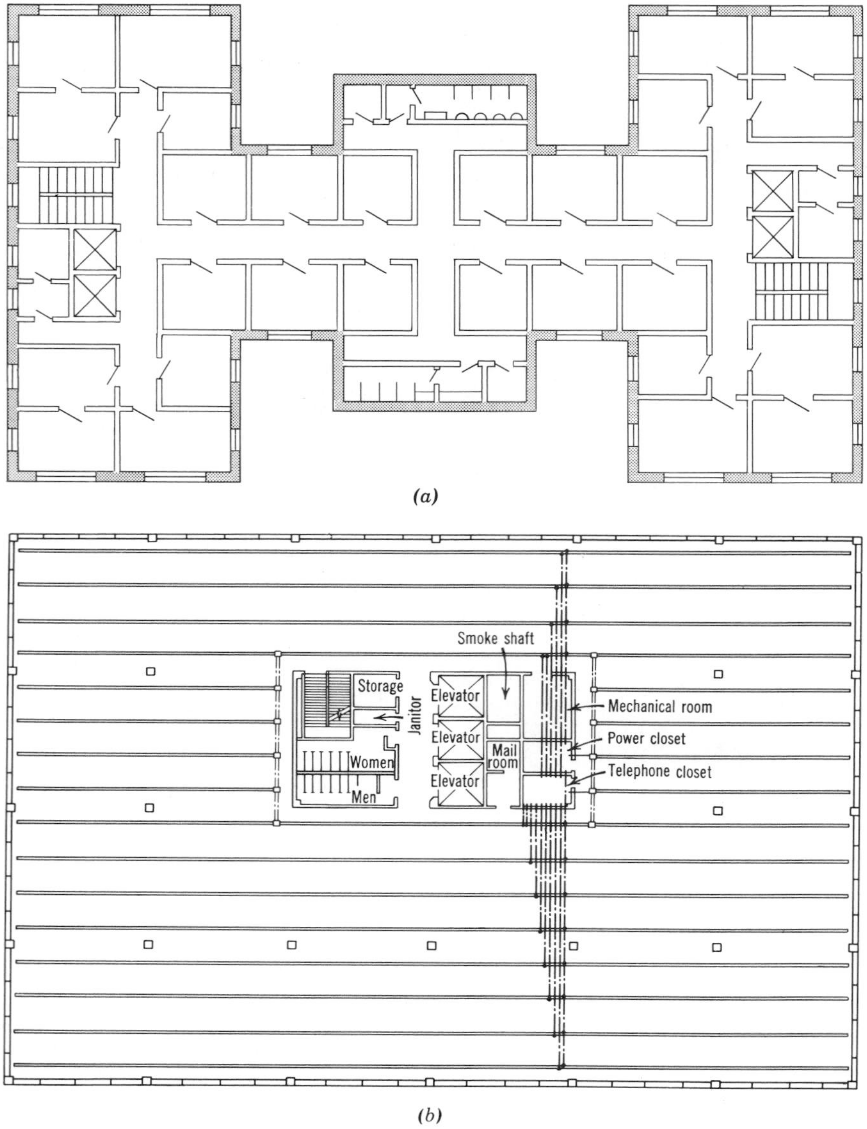

Fig. 3.23 In contrast to building plan (a), *which uses daylight and wind ventilation in each office, office building* (b) *receives cooler, filtered air, is less subject to the noise of the city, typically provides both constant light and temperature throughout, and provides for more rentable floor space on the site. It also allows less daylight to reach street level, requires much more electricity (though probably less heating fuel), and thus contributes more heat (and possibly noise from mechanical equipment) to the city all year round.*

3.6 Rain and Groundwater

Most buildings interact with water in three forms: rainwater, groundwater, and potable water (brought to and taken from urban sites by utilities). A detailed treatment of water within buildings is found in Part III, Water and Waste, and a close look at the utilization of rainwater is found in Chapter 8.

Like solar energy, *rainwater* is a diffuse, intermittent, and often seasonal resource. As a source of water, it is most often collected and used where other water sources are scarce, or of poor quality.

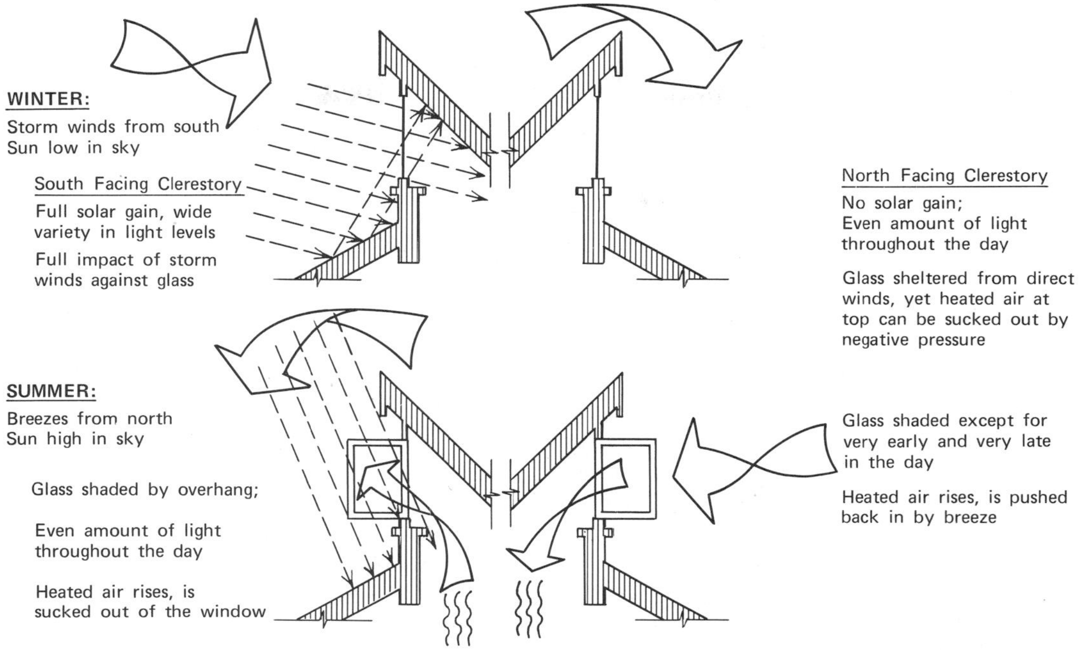

Fig. 3.24 *Which way for a clerestory? Some relative advantages of north versus south orientation for a clerestory window/shed roof combination. (Seasonal wind directions shown prevail in the Pacific Northwest.)*

Rain also has an influence on building design: heavy rains and pitched roofs have long been found in the same locales. Overhangs may extend further beyond walls exposed to storm winds; even gutter and downspout details can become a design feature, as shown in the examples in Chapter 8. A building that reflects the combined influences of daylight, wind, and rain is shown in Fig. 3.25.

Rainwater's impact on site design can be thought of as similar to that of wind. Once on the surface, it will flow downhill in a wide path, and buildings that obstruct this path must make provisions for diverting it. Slight, shallow ditches called swales are frequently used in such diversion; orientation of the building to the slope can affect surface water diversion, as shown in Fig. 3.26.

Surface water can be used to advantage in ther-

Fig. 3.25 *Rain, wind, sun, daylight, and design. This covered outdoor tennis facility at the University of Oregon, Eugene, was designed to ward off the rain-bearing south winds in winter. Direct sun is also unwelcome; instead, north skylight is admitted along with reflected light from roof surfaces. Cool outdoor temperatures are maintained, appropriate to strenuous activity. Gutters at the lower edge of each roof plane carry away rain; the courts stay dry for all but a few days of the year. (Unthank, Seder, Poticha Architects.)*

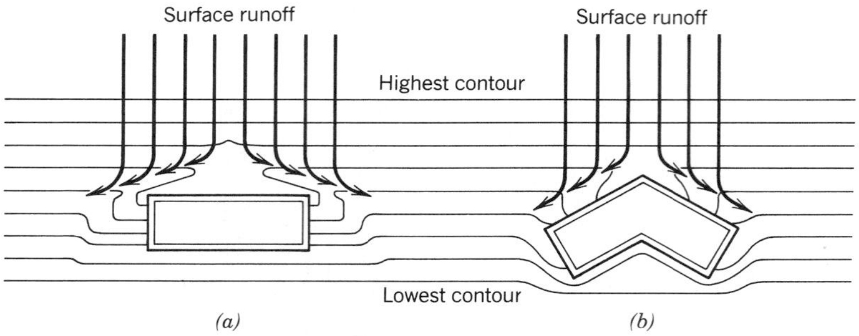

Fig. 3.26 Rain as surface flow. (a) *Where buildings intercept surface water, provisions for diversion are necessary. The building oriented as in* (b) *needs less elaborate provisions.*

mal, acoustic, and daylighting roles. Hot, dry breezes that pass over water surfaces (and especially through misty sprays above ponds) gain substantial moisture while undergoing a drop in dry-bulb temperature, as seen in the preceding discussion on evaporative cooling (Section 2.4). Such air can provide improved comfort in hot, dry conditions. If the water body also provides the sound of running water, it can serve to mask unwanted sounds such as freeway traffic, or conversations in adjacent offices. Surface water has a more complex role in daylighting, due to the reflection characteristics of water. The surface of water is highly reflective to light that strikes it at *low* angles of incidence; the setting sun off the west coast can throw a blinding sheet of light across the ocean surface. Conversely, sunlight near noon on summer strikes a water body at *high* angles of incidence, nearly perpendicular at southern U.S. latitudes. Water is highly absorptive of light at these angles, reflecting relatively little light. Therefore, water bodies east, south, and west of our buildings can provide increased reflected light on sunny winter days, and somewhat decreased light in summer, relative to grass surfaces. However, on heavily *overcast* days, when the sky is uniformly gray (and least daylight is available), water surfaces will not be particularly helpful. Furthermore, the reflection of sun off water tends to be in sparkling patches of always-changing light. This can be fascinating, or it can be annoying, either as glare to the eye or as distraction from a visual task. Reflected off a matte-finish ceiling, such dancing light might be welcome; directly on eyes or work surface, it easily becomes a problem.

Generally avoided by designers where possible, *groundwater* is a threat to foundations and underground spaces. This avoidance carries over into site planning, and marshy places are usually unwelcome near buildings. Urban dryness is intensified as ponds are drained and streams are piped away. However, architects have used groundwater as a heat sink, discharging building heat to groundwater in summer and removing the groundwater heat for the building's winter benefit. The quantity of groundwater available to act thus as a heat sink varies with subsurface conditions, and little information is yet available to indicate its limits. More buildings discharging more heat will eventually raise groundwater temperatures; but how many degrees, and for how long, remains unknown in many areas. Another risk is that of pollution, through increased handling of groundwater.

The ability of water to conduct and store heat, which encourages its use as a heat sink, also makes it a thermal enemy of heat storage tanks or bins in solar buildings. There is little point in collecting solar energy to store in an underground tank for later use if groundwater is allowed to rob the tank of its heat. On the other hand, groundwater helps storage tanks that are to be kept as cool as possible. Such tanks provide cool water to buildings at peak-heat hours of hot days and can greatly lessen the demand for electricity to run conventional air conditioners.

Snow has special implications; it delays runoff, provides a blanket of thermal insulation, absorbs sound, and reflects more light than almost any other naturally occurring surface. Wind patterns can deposit snow much more thickly in some places, for better or worse. Snow hampers the moving of external devices such as thermal shutters (or roof pond covers) and can create a disabling glare when it reflects the low winter sun into windows at eye level.

3.7 Plants

Plants have several roles: they are part of the water cycle; they turn carbon dioxide into oxygen by day; they provide organic matter suitable for composting (eating, in some cases); and they help us tell time both by growth and by change with the seasons. Our associations with plants are mostly pleasant ones, and they contribute to our enjoyment of the places where they grow.

Plants are also of immediate practical value to buildings because they enhance privacy, slow the winter wind, reduce the glare of strong daylight, or prevent summer sun from entering and overheating buildings. In this latter role, plants are particularly noteworthy, because they enhance a feeling of coolness when breezes rustle or sway their leaves, and especially because they respond more to *cycles of outdoor temperature* than to those of *sun position*. Unlike fixed sunscreens on buildings, plants thus provide deepest shade in the hottest weather.

To illustrate this contrast, a fixed sunscreen (here an overhang for south-facing windows) is shown in Fig. 3.27. Such sunscreens block the sun for some portion of the year, centered on June 21; that is, maximum shade is provided at the summer solstice. A typical approach to such sunscreens in the U.S. temperate zone is to shade at least half a south window in a residence (or all the window in an office with internal heat sources), from March 21 to September 21, or from equinox to equinox. Yet, March is on the average a colder month than September; March 21 is the last day of winter, while September 21 is the last day of summer. Full solar radiation is more welcome in early spring than in early fall, yet sun position is identical at these times. In contrast to fixed sunscreens, deciduous plants do most of their shading from the middle of June to early October, giving windows access to solar radiation through much of the spring (Figs. 3.28 and 3.29).

Fig. 3.27 *Fixed overhang sun control. This south-facing corridor in Oregon is in the open air all year round. The low winter sun fills the space, and some of its heat is stored by the tile floor. However, in cold weather the offices that it connects are disadvantaged by exposure to cold air. In summer, little of the corridor is exposed to sun. Sun control is identical in the spring and fall.*

Deciduous trees have a potential solar heating disadvantage in that certain species (such as some oaks) hold onto their leaves well into the heating season, a tendency increased by fertilizing or irrigating near the tree. Others have a dense branch structure, blocking a surprisingly high percentage of solar radiation even when bare. A listing of shading density for various trees is found in Moffat and Schiler (1981). Agricultural extension services in most areas can provide information on late-defoliating trees. For solar collection, avoiding trees or large shrubs within the area shown in Fig. 3.30 is recommended.

Many large buildings have a relatively short heating season, since they have constant internal heat sources. So, the choice of trees, shrubs, and vines for the improvement of a building's micro-

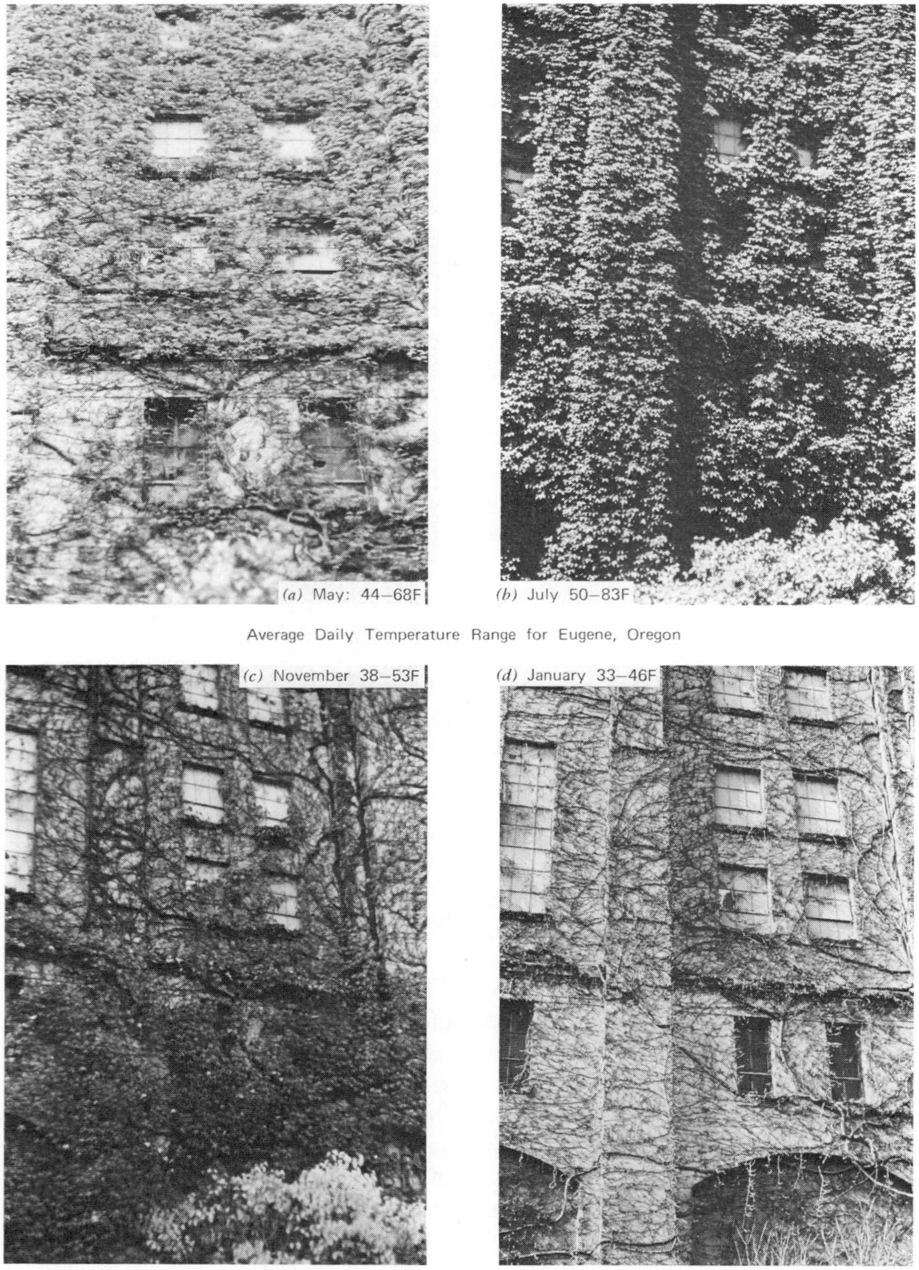

Fig. 3.28 *Deciduous vines, temperature, and sun position. The sun's path through the sky is identical in late May and late July (see Fig. 3.8)* (a *and* b). *Identical lower sun paths occur in November and January* (c *and* d). *This deciduous vine responds more to the temperature of its Oregon climate than to the sun's position at 44°N latitude, which makes it particularly useful as a potential sun control device. (From Reynolds, 1976.)*

(a)
(b)

Fig. 3.29 A deciduous tree and the equinoxes. The sun position (4:45 P.M. *sun time) is identical on September 21* (a) *and March 21* (b); *yet, this hour's average temperature (Eugene, Oregon) is 75F in September and 53F in March. The second-floor overhang provides identical shading, while the tree shades only in September. (From Reynolds, 1976.)*

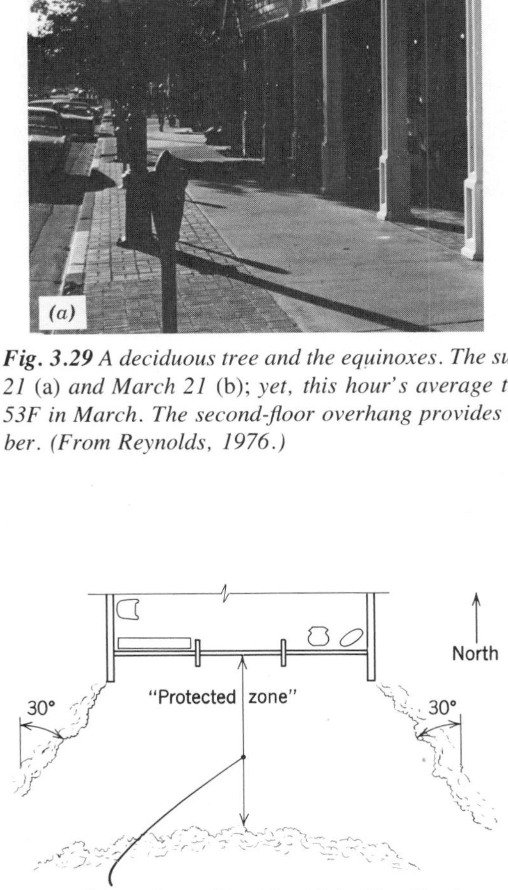
North
"Protected zone"
30°
30°
This distance is equal to at least twice the ultimate height of the trees of shrubs beyond.

Fig. 3.30 Protecting access to winter sun, given a lawn or terrace of minimum size and shape to the south of solar collecting surfaces. Deciduous plants within this "protected zone" should be avoided, unless they are very low growing or are a reliably early-defoliating species. Summer sun protection for these south windows is best provided by flexible architectural controls, such as awnings or hanging screens.

climate can best be made after an analysis of building heating and cooling needs by month.

Clearly, a tree or vine that unfolds its leaves early in spring and holds them until late fall is advantageous to these internal-heat-dominated buildings.

References

American Solar Energy Society (1981). *Passive Cooling*, ASES, Boulder, CO.

Beranek, W. J. (1980). "General Rules for the Determination of Wind Environment." In *Wind Engineering*, J. E. Cermak, Editor, Pergamon Press, New York.

Bragdon, C. (1970). *Noise Pollution, The Unquiet Crisis*, University of Pennsylvania Press, Philadelphia.

Brown, G. Z., Reynolds, J. S., and Ubbelohde, M. S. (1982). *Inside Out: Design Procedures for*

Passive Environmental Technologies, John Wiley, New York.

Cohen, S., Glass, D. C., and Singer, J. E. (1973). ''Apartment noise, auditory discrimination, and reading ability in children,'' *Journal of Experimental Social Psychology,* Vol. 9.

Doelle, Leslie (1972). *Environmental Acoustics,* McGraw-Hill, New York.

Gandemer, J. (1977). ''Wind Environment Around Buildings: Aerodynamic Concepts.'' In *Wind Effects on Buildings and Structures,* K. J. Eaton, Editor, Cambridge University Press, New York.

Gandemer, J. (1981). ''The Aerodynamic Characteristics of Windbreaks, Resulting in Empirical Design Rules,'' *Journal of Wind Engineering and Industrial Aerodynamics,* Vol. 1, July. Elsevier Scientific Publishing Co., Amsterdam.

Knowles, R. (1981). *Sun, Rhythm, Form,* MIT Press, Cambridge, MA.

Landsberg, H. E. (1970). ''Climates and Urban Planning.'' In *Urban Climates,* World Meteorological Organization Technical Note No. 108, Geneva.

Lowry, W. P. (1967). ''The Climate of Cities,'' *Scientific American,* August.

Moffat, A. S., and Schiler, M. (1981). *Landscape Design that Saves Energy,* William Morrow, New York.

Olgyay, A. and Olgyay, V. (1957). *Solar Control and Shading Devices,* Princeton University Press, Princeton, NJ.

Reynolds, J. S. (1976). *Solar Energy for Pacific Northwest Buildings,* Center for Environmental Research, University of Oregon, Eugene.

U.S. Environmental Protection Agency (1980). *Trends in the Quality of the Nation's Air.* U.S. EPA, Washington, D.C.

Watson, D., and Labs, K. (1981). *Climatic Design for Home Building.* Distributed by the authors, c/o Box 401, Guilford, CT 06437.

PART II

THERMAL CONTROL

The next three chapters deal with the topics of heating and cooling. The objective continues to be the *comfort of people,* for whom we heat and cool buildings; thermal comfort criteria were discussed in Chapter 2. In Chapter 4, the basic theory of heat flow is discussed; whether we are calculating heat gain in summer or heat loss in winter, the process is based on the same principles of heat flow. Chapter 5 presents criteria for building performance, followed by detailed heat loss and heat gain procedures. Chapter 6 introduces some of the extensive array of available systems and equipment for heating and cooling.

It is important to keep our perspective during the process of calculating thermal loads for buildings. Throughout these lengthy chapters, keep the following points in mind:

1. Human comfort is the ultimate objective of our calculations and resulting architectural manipulations; this comfort is not attainable by numbers alone. Most of us have experienced situations in which psychological factors seemed to have at least as much influence on comfort as temperature, humidity, or air motion. Achieving numerical adequacy is important, but it is not by itself a guarantee of comfort.

2. These calculations are a rare opportunity to numerically evaluate a building design. When the numbers are all cranked out, sit back and take a larger view. What design strategies do they suggest? We've worked hard to get them; now let them work for us.

3. There are two specific objectives to our calculations, which yield different kinds of answers and require different amounts of work. (a) How big should a device be to furnish enough heating (or cooling) in the coldest (or hottest) conditions we are likely to encounter for our building? (b) How much energy will be used by our building in a typical season?

For the first question, we deal with extreme "design" conditions, as discussed in Section 2.4. We calculate our hourly winter heat loss (or summer gain) under these special conditions, then adjust our buildings, probably recalculate, and finally size our equipment according to these calculations. We have, in effect, a snapshot of the building's performance at a special high-stress time—one that occurs rarely in the building's lifetime.

By contrast, the second question looks at normal conditions, as they change over a long period of time. Our calculations for design-hour performance can be used, but must be adjusted for normal and continuous-change conditions. These calculations take much longer, and are more suited to automatic computation (Appendix H) than simple design-hour answers. Once completed, we again have the opportunity to adjust our buildings. This time adjustments are aimed more at reducing long-term energy use than at equipment sizing. We have, in effect, a series of photos that become a movie of our building's performance over time, under normal circumstances. We get another perspective on architecture as a performing art.

4. Buildings, like our bodies, exchange heat with the outside environment in two distinct ways: *through the envelope,* akin to heat exchange through human skin, and with *incoming fresh air,* akin to our heat exchange with the air we inhale. When calculating envelope (skin) losses or gains, materials, areas, and rates of heat flow through the envelope are used. When calculating fresh air (lung) losses or gains, volumes of space and rates of fresh air exchange are used. For heavily insulated buildings in cold climates, it is not unusual for "lung" losses to exceed "skin" heat losses. Skin and lung losses (or gains) must be added together to determine the total rate of a building's heat exchanges with its environment; both can be manipulated through design to improve a building's thermal performance.

5. Buildings nearly always experience internal heat production; this is helpful in meeting cold weather demands, but harmful in hot weather. For most buildings, it will be helpful to manipulate the envelope to *minimize* internal gains (e.g., to reduce the need for electric lights by specifying larger, better-placed windows), since the reduction in internal heat is so helpful in summer, and can so often be replaced by an increased reliance on passive solar heating in winter.

6. An occupied building can be very different thermally from an unoccupied one. Many buildings (such as offices and schools) experience high internal gains and rates of fresh air for only about a third of a typical day, and only 5 days per week; this amounts to less than a quarter of the year. It is important to remember that although we are designing to provide thermal comfort while a building is being used, we cannot depart too far from comfortable interior temperatures during the night (or weekend) without straining the heating and cooling systems early in the morning. Said another way, one hour's surplus heat can be another hour's needed warmth.

4

HEAT FLOW

Most of this chapter discusses the theory of heat flow; applications are generally found in Chapter 5. Heat flow through the skin—by convection, conduction, and radiation—is contrasted to heat flow due to ''breathing''—by infiltration and provisions for fresh air. Finally, there is psychrometry, which combines the properties of air and moisture.

4.1 Convection, Conduction, and Radiation

Whenever an object is at a temperature different from its surroundings, heat flows from the hotter to the colder. Chapter 2 discussed the human body's means of disposing of its surplus heat to the cooler environment. Buildings, like bodies, experience heat loss to, and heat gain from, the environment by *convection* (molecules to cool air absorb heat from a warm surface, rise, and carry it away), *conduction* (heat is transferred directly from molecules of hot building surfaces to the molecules of cooler solids in contact with the building, such as earth or water), and *radiation* (heat flows in electromagnetic waves from hotter surfaces to detached, distant colder ones, through any transparent medium, even empty space). Evaporation is also involved (from wet surfaces), but much less for most buildings than for our bodies.

This combination of heat flow by convection, conduction, and radiation is illustrated in Fig. 4.1 through some typical combinations of materials. Notice that multiple air spaces and reflective surfaces are an inexpensive but particularly useful way to slow the flow of heat. Some of the most effective insulating materials therefore combine multiple dead-air spaces and layers of reflective films. Heat flow through the various components of a building's skin involves both heat flow through solids and heat flow through films and layers of air.

4.2 Heat Flow Through Solids

Certain terms are fundamental to any discussion of heat flow; Table 4.1 presents terms of energy, terms of power (energy used over a given time span), and terms of heat flow rate.

Each individual material has a characteristic rate at which heat will flow through it. For such individual, or homogeneous, solids this rate is called its *conductivity,* designated as *k*; the number of British thermal units per hour (Btu/h) that flow through 1 square foot (ft^2) of material, 1 in. thick, when the temperature drop through this material is 1F° (under conditions of steady heat flow). The units are Btu/h ft F. The SI equivalent is the rate at which watts flow through 1 square meter (m^2) of material, when the temperature drop is 1K° (equal to 1C°), so units are W/m K (or W/m°C; these terms are used interchangeably in the tables of this chapter). Conductivity is an important factor in passive heating or cooling design, which depends heavily upon the rate at which heat is conducted through a material from its surface; when we touch the solid surface, we sense the material's conductivity. Conductivity for each solid is established by tests and is published as a basic rating.

Many solids (common brick, wood siding, thermal batt insulation, gypsum board, etc.) are widely available in standard thicknesses. For such materials, it is useful to know the rate of heat flow for the standard thickness instead of the rate per inch. *Conductance,* designated as *C*, is the number of Btu per hour that flow through 1 square foot of a given thickness of material when the temperature drop is 1F°. The units are Btu/h ft^2 F. The SI units of conductance are W/m^2 K.

Conductivity and conductance are compared in

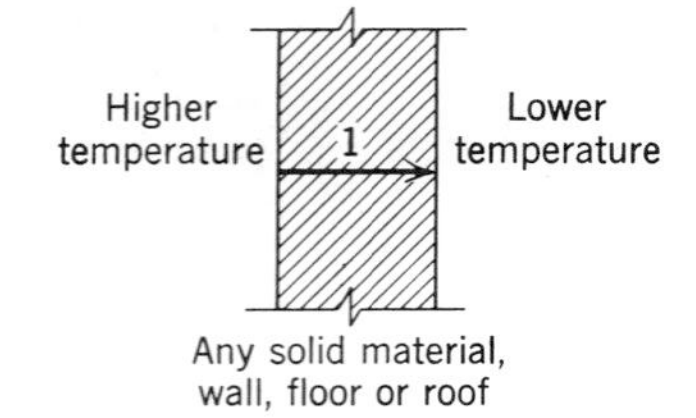

A single solid material illustrates the transfer of heat from the warmer to the cooler particles by conduction (1).

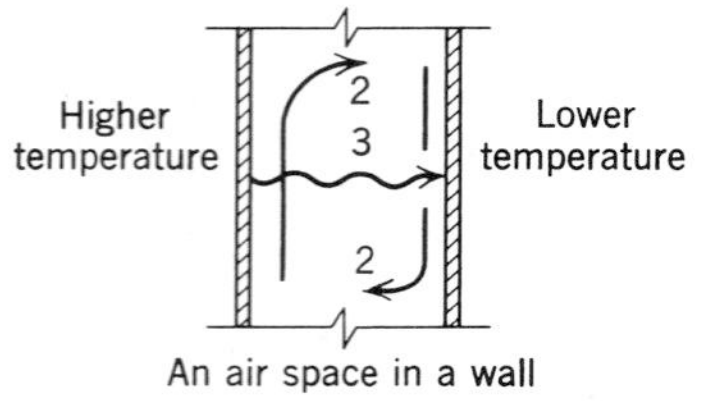

As air is warmed by the warmer side of the air space it rises. As it falls down along the cooler side it transfers heat to this surface (2). Radiant energy (3) is transferred from the warmer to the cooler surface. The rate depends upon the relative temperature of the surfaces and upon their emissive and absorptive qualities. Direction is always from the warmer to the cooler surface.

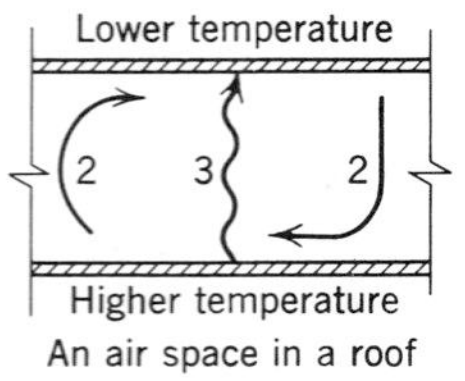

The convective action (2) in the air space of a roof is similar to that in a wall although the height through which the air rises and falls is usually less. The radiant transfer is up in this case because its direction is always to the cooler surface.

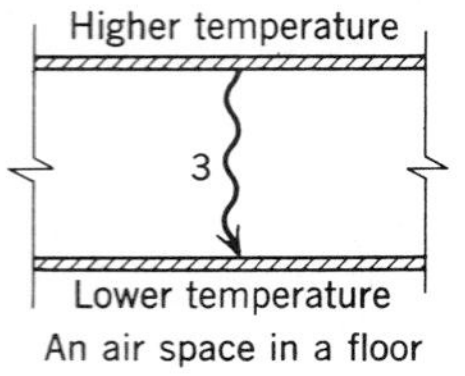

When the higher temperature is at the top of a horizontal air space the warm air is trapped at the top and, being less dense than the cooler air at the bottom, will not flow down to transfer its heat to the cooler surface. This results in little flow by convection. The radiant transfer in this case is down because that is the direction from the warmer surface to the cooler.

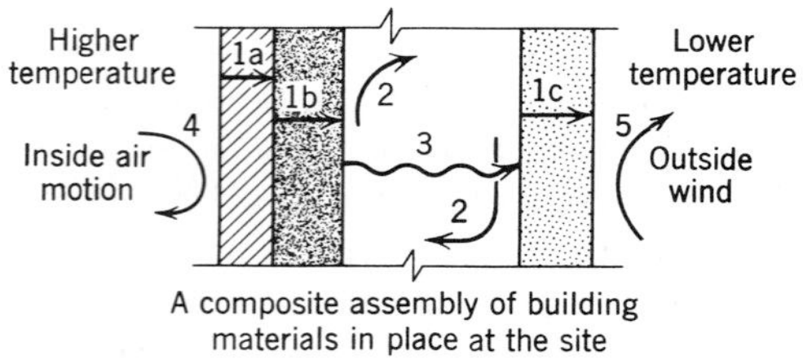

This example of a wall in place illustrates the several methods by which heat is lost through a composite assembly of materials. Conduction at varying rates in different materials is accounted for in 1a, 1b, 1c. Convection currents (2) and radiation (3) carry the heat across the air space.

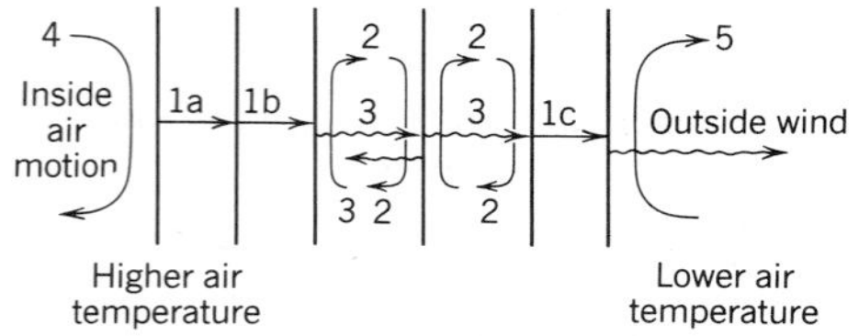

Heat is conducted from the room air by warm air currents that strike the inside wall. Heat is conducted away from the exterior surface of the wall by the action of the wind.

Fig. 4.1 *Nature of heat flow through materials, air spaces, and assembled structures. Thermal action is identified by the following numbers: 1, conduction; 2, convection; 3, radiation; 4, inside surface conductance; and 5, outside surface conductance.*

Table 4.1 and Fig. 4.2, which include another especially useful term, *resistance*. Resistance, designated R, tells us how effective a solid is as an insulator. The reciprocal of conductivity, R is measured in *hours* needed for 1 Btu to flow through a given thickness of a solid when the temperature

TABLE 4.1 **Heat Symbols, Terms, and Definitions**[a]

Conventional Units			*SI (Système International) Units*		
Symbol	*Term*	*Definition or Usage; Conversions*	*Symbol*	*Term*	*Definition or Usage; Conversions*
Part A. ***Energy***					
Btu	British thermal unit	The amount of heat required to raise 1 pound of water by 1F°. (It roughly corresponds to the energy released in burning an ordinary wooden match.) 1 Btu = 1.055 kilojoules	J	joule	Newton-meter, or a force of one newton (1 N) acting through a distance of one meter. In terms of heat, a joule is 1/4.184 of the amount of heat required to raise a gram of water by 1C°.
			kJ	kilojoule	1000 joules; 1 kJ = 0.9478 Btu
Part B. ***Power***					
Btu/h	British thermal units per hour	Commonly used to express the total heat loss or gain of a building, and to express the size of heating or cooling equipment. 1 Btu/h = 0.2929 W	W	watt	The power required to produce energy at the rate of one joule per second. Commonly used to specify size for light bulbs, furnaces, air conditioners, and other heat producers. 1 W = 3.412 Btu/h
hp	horsepower	1 hp = 746 W			
Part C. ***Rate of Heat Flow***					
Btu/h ft² F		1 Btu/h ft² F = 5.67 W/m² K 1 W/m² K = 0.176 Btu/h ft² F These are the units of conductance and of overall heat transmission for the following terms: *C* (conductance). The rate of heat flow through a homogeneous material (or combination of materials) of a stated thickness. *a* (air space conductance). The rate of heat flow through an air space bounded by two surfaces. *h* (film or surface conductance coefficient). The rate of heat flow from a surface due to air (or other fluid) motion against the surface. (Identical to *f*, the film conductance.) *U* (overall coefficient of heat transmission). The overall rate of heat flow through any combination of materials, air layers, and air spaces. It is equal to the reciprocal of the sum of all resistances *R* (see below) that are involved in this combination. This is the term used directly in building envelope heat loss or gain calculations.	W/m² K (or W/m² °C)		
Btu/h ft F	*k*	(conductivity). The rate of heat flow through a homogeneous material, per unit of thickness.	W/m K (or W/m °C)		

TABLE 4.1 **Heat Symbols, Terms, and Definitions**[a] (***Continued***)

Conventional Units		*Definitions or Usage; Conversions*	*SI Units*
*Part C. **Rate of Heat Flow*** (***Continued***)			
h ft^2 F/Btu	*R*	(resistance). A measure of resistance to the passage of heat; the reciprocal of conductance.	m^2 K/W
	ε	(emittance). The ratio of the radiant flux from a given surface to that of a blackbody (a "perfect" emitter) at the same temperature.	
	E	(effective emittance). The combined effect of the emittance of parallel surfaces bounding an air space.	

[a]A more complete listing of conventional–SI conversion units is found in Appendix I.

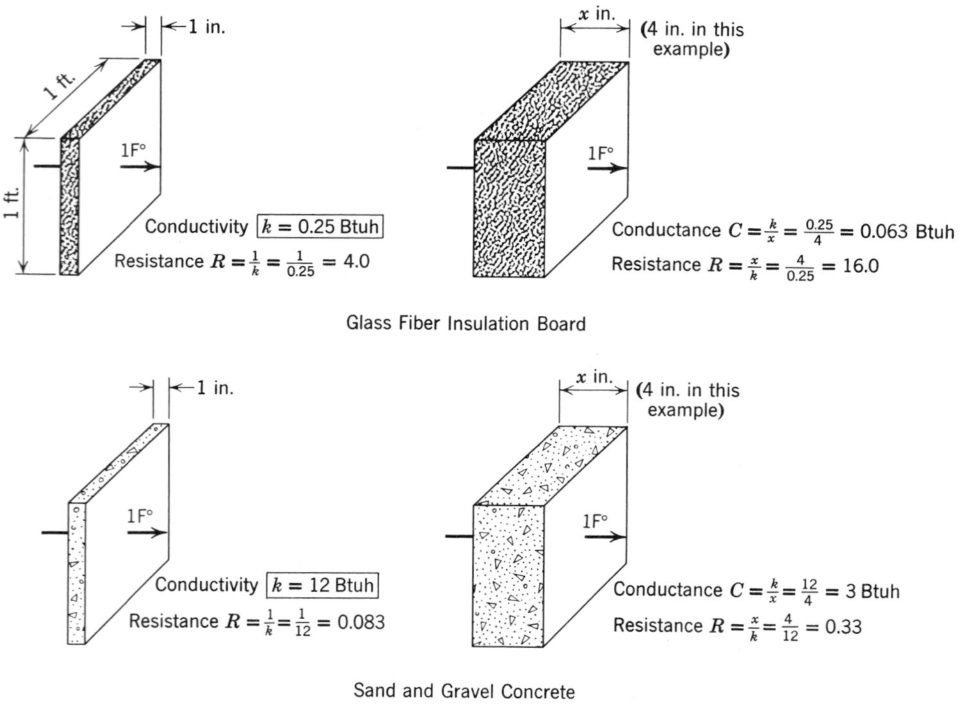

Fig. 4.2 *Sample conductivities* (k) *for 1-in. thickness, conductances* (C) *for any thickness (4 in. in this example), and resistance* (R) *for two materials. Glass fiber is a material of low conductivity; concrete is a material of high conductivity.* Note: *Standard unit of area is 1 square foot; standard unit temperature differential is 1 degree Fahrenheit (shown as 1F°).*

difference is 1F°; its units are h ft^2 F/Btu. In SI, units are m^2 K/W. Resistance is especially useful when comparing insulating materials, since the greater the *R* value, the more effective the insulator. Resistances and other important thermal properties are listed for many materials in Table 4.2. For a nonhomogeneous material, such as concrete block, conductance is a better indicator of performance than resistance.

This discussion has so far concentrated on heat flow by conduction, but convection and radiation are also involved. Both these latter methods become more evident when looking at heat flow through air.

TABLE 4.2 **Thermal Properties of Typical Building and Insulating Materials (design values)**[a]

NOTE: The customary units for conductivity (*k*), [conductance (*C*), and resistance (*R*), either per inch (1/*k*)] or for thickness stated (1/*C*), are given in Table 4.1. Values are for a mean temperature of 75 F unless noted by an asterisk (*), which indicates that such a value has been reported at 45 F. The SI units for resistance (last two columns) were calculated by taking the values from the two resistance columns under Customary Unit, multiplying by the factor 1/*k*(*r*/in.) and 1/*C*(*R*) for the appropriate conversion factor. Author's note: Actual (on-site) resistance values frequently are lower than the test-cell-determined "design" values listed in this table.

Description	Customary Unit						SI Unit	
	Density (lb/ft³)	Conductivity (*k*)	Conductance (*C*)	Resistance[b] (*R*) Per inch thickness (1/*k*)	Resistance[b] (*R*) For thickness listed (1/*C*)	Specific Heat, Btu/(lb)(deg F)	Resistance[b] (*R*) (m·K)/W	Resistance[b] (*R*) (m²·K)/W
BUILDING BOARD								
Boards, Panels, Subflooring, Sheathing								
Woodboard Panel Products								
Asbestos-cement board	120.	4.0	—	*0.25*	—	0.24	1.73	
Asbestos-cement board . . . 0.125 in.	120	—	33.00	—	*0.03*			*0.005*
Asbestos-cement board . . . 0.25 in.	120	—	16.50	—	*0.06*			*0.01*
Gypsum or plaster board . . . 0.375 in.	50	—	3.10	—	*0.32*	0.26		*0.06*
Gypsum or plaster board . . . 0.5 in.	50	—	2.22	—	*0.45*			*0.08*
Gypsum or plaster board . . . 0.625 in.	50	—	1.78	—	*0.56*			*0.10*
Plywood (Douglas Fir)[c]	34	0.80	—	*1.25*	—	0.29	8.66	
Plywood (Douglas Fir) . . . 0.25 in.	34	—	3.20	—	*0.31*			*0.05*
Plywood (Douglas Fir) . . . 0.375 in.	34	—	2.13	—	*0.47*			*0.08*
Plywood (Douglas Fir) . . . 0.5 in.	34	—	1.60	—	*0.62*			*0.11*
Plywood (Douglas Fir) . . . 0.625 in.	34	—	1.29	—	*0.77*			*0.19*
Plywood or wood panels . . . 0.75 in.	34	—	1.07	—	*0.93*	0.29		*0.16*
Vegetable Fiber Board								
Sheathing, regular density . . . 0.5 in.	18	—	0.76	—	*1.32*	0.31		*0.23*
. . . 0.78125 in.	18	—	0.49	—	*2.06*			*0.36*
Sheathing intermediate density . . . 0.5 in.	22	—	0.82	—	*1.22*	0.31		*0.21*
Nail-base sheathing . . . 0.5 in.	25	—	0.88	—	*1.14*	0.31		*0.20*
Shingle backer . . . 0.375 in.	18	—	1.06	—	*0.94*	0.31		*0.17*
Shingle backer . . . 0.3125 in.	18	—	1.28	—	*0.78*			*0.14*
Sound deadening board . . . 0.5 in.	15	—	0.74	—	*1.35*	0.30		*0.24*
Tile and lay-in panels, plain or acoustic	18	0.40	—	*2.50*	—	0.14	17.33	
. . . 0.5 in.	18	—	0.80	—	*1.25*			*0.22*
. . . 0.75 in.	18	—	0.53	—	*1.89*			*0.33*
Laminated paperboard	30	0.50	—	*2.00*	—	0.33	13.86	
Homogeneous board from repulped paper	30	0.50	—	*2.00*	—	0.28	13.86	
Hardboard								
Medium density	50	0.73	—	*1.37*	—	0.31	9.49	

(continued)

TABLE 4.2 **Thermal Properties of Typical Building and Insulating Materials (design values)[a]** *(Continued)*

Description	Customary Unit						SI Unit	
	Density (lb/ft³)	Conductivity (k)	Conductance (C)	Resistance[b] (R): Per inch thickness (1/k)	Resistance[b] (R): For thickness listed (1/C)	Specific Heat, Btu/(lb)(deg F)	Resistance[b] (R): (m·K)/W	Resistance[b] (R): (m²·K)/W
High density, service temp. service underlay	55	0.82	—	*1.22*	—	0.32	8.46	
High density, std. tempered	63	1.00	—	*1.00*	—	0.32	6.93	
Particleboard								
Low density	37	0.54	—	*1.85*	—	0.31	12.82	
Medium density	50	0.94	—	*1.06*	—	0.31	7.35	
High density	62.5	1.18	—	*0.85*	—	0.31	5.89	
Underlayment ... 0.625 in.	40	—	1.22	—	*0.82*	0.29		*0.14*
Wood subfloor ... 0.75 in.		—	1.06	—	*0.94*	0.33		*0.17*
BUILDING MEMBRANE								
Vapor—permeable felt	—	—	16.70	—	*0.06*			*0.01*
Vapor—seal, 2 layers of mopped 15-lb felt	—	—	8.35	—	*0.12*			*0.02*
Vapor—seal, plastic film	—	—	—	—	Negl.			
FINISH FLOORING MATERIALS								
Carpet and fibrous pad	—	—	0.48	—	*2.08*	0.34		*0.37*
Carpet and rubber pad	—	—	0.81	—	*1.23*	0.33		*0.22*
Cork tile ... 0.125 in.	—	—	3.60	—	*0.28*	0.48		*0.05*
Terrazzo ... 1 in.	—	—	12.50	—	*0.08*	0.19		*0.01*
Tile—asphalt, linoleum, vinyl, rubber	—	—	20.00	—	*0.05*	0.30		*0.01*
vinyl asbestos						0.24		
ceramic						0.19		
Wood, hardwood finish ... 0.75 in.			1.47		*0.68*			0.12
INSULATING MATERIALS								
Blanket and Batt[d]								
Mineral Fiber, fibrous form processed from rock, slag, or glass								
approx.[e] 3–3.5 in.	0.3–2.0	—	0.091	—	*11*[d]			*1.94*
approx.[e] 5.50–6.5	0.3–2.0	—	0.053	—	*19*[d]			*3.35*
approx.[e] 6–7 in.	0.3–2.0		0.045		*22*[d]			*3.87*
approx.[e] 8.5–9 in.	0.3–2.0		0.033		*30*[d]			*5.28*
approx. 12 in.	0.3–2.0	—	0.026		*38*[d]			*6.69*
Board and Slabs								
Cellular glass	8.5	0.35	—	*2.86*	—	0.18	*19.81*	
Glass fiber, organic bonded	4–9	0.25	—	*4.00*	—	0.23	*27.72*	
Expanded perlite, organic bonded	1.0	0.36	—	*2.78*	—	0.30	*19.26*	
Expanded rubber (rigid)	4.5	0.22	—	*4.55*	—	0.40	*31.53*	

Expanded polystyrene extruded								
Cut cell surface	1.8	0.25	—	*4.00*	—	0.29	*27.72*	
Smooth skin surface	1.8–3.5	0.20	—	*5.00*	—	0.29	*34.65*	
Expanded polystyrene, molded beads	1.0	0.26	—	—	—	—	*26.3*	*3.8*
	1.25	0.25	—	—	—	—	*27.8*	*4.0*
	1.5	0.24	—	—	—	—	*29.1*	*4.2*
	1.75	0.24	—	—	—	—	*29.1*	*4.2*
	2.0	0.23	—	—	—	—	*29.8*	*4.3*
Cellular polyurethane[f] (R-11 exp.)(unfaced)	1.5	0.16	—	*6.25*	—	0.38	*43.82*	
(Thickness 1 in. or greater)	2.5							
(Thickness 1 in. or greater—high resistance to gas permeation facing)	1.5	0.14						
Foil-faced, glass fiber-reinforced cellular								
Polyisocyanurate (R-11 exp.)[n]	2	0.14	—	*7.04*	—	0.22	*48.79*	
Nominal 0.5 in.		—	0.278	—	*3.6*			*0.63*
Nominal 1.0 in.		—	0.139	—	*7.2*			*1.27*
Nominal 2.0 in.		—	0.069	—	*14.4*			*2.53*
Mineral fiber with resin binder	15	0.29	—	*3.45*	—	0.17	23.91	
Mineral fiberboard, wet felted								
Core or roof insulation	16–17	0.34	—	*2.94*	—		20.38	
Acoustical tile	18	0.35	—	*2.86*	—	0.19	19.82	
Acoustical tile	21	0.37	—	*2.70*	—		18.71	
Mineral fiberboard, wet molded								
Acoustical tile[g]	23	0.42	—	*2.38*	—	0.14	16.49	
Wood or cane fiberboard								
Acoustical tile[g] 0.5 in.	—	—	0.80	—	*1.25*	0.31		0.22
Acoustical tile[g] 0.75 in.	—	—	0.53	—	*1.89*			0.33
Interior finish (plank, tile)	15	0.35	—	*2.86*	—	0.32	19.82	
Cement fiber slabs (shredded wood with Portland cement binder	25-27	0.50–0.53	—	*2.0–1.89*	—	—	*13.87*	
Cement fiber slabs (shredded wood with magnesia oxysulfide binder)	22	0.57	—	*1.75*	—	0.31	*12.16*	
LOOSE FILL								
Cellulosic insulation (milled paper or wood pulp)	2.3–3.2	0.27–0.32	—	*3.13–3.70*	--	0.33	*21.69–25.64*	
Sawdust or shavings	8.0–15.0	0.45	—	*2.22*	—	0.33	*15.39*	
Wood fiber, softwoods	2.0–3.5	0.30	—	*3.33*	—	0.33	*23.08*	
Perlite, expanded			—	*2.70*	—	0.26	*18.71*	
	2.0–4.1	0.27–0.31	3.7–3.3					
	4.1–7.4	0.31–0.36	3.3–2.8					
	7.4–11.0	0.36–0.42	2.8–2.4					

(continued)

THERMAL CONTROL

TABLE 4.2 **Thermal Properties of Typical Building and Insulating Materials (design values)**[a] (*Continued*)

Description	Density (lb/ft³)	Conductivity (k)	Conductance (C)	Customary Unit Resistance[b] (R): Per inch thickness (1/k)	Customary Unit Resistance[b] (R): For thickness listed (1/C)	Specific Heat, Btu/(lb)(deg F)	SI Unit Resistance[b] (R): (m·K)/W	SI Unit Resistance[b] (R): (m²·K)/W
Mineral fiber (rock, slag or glass)								
approx.[e] 3.75–5 in.	0.6–2.0	—	—		11	0.17		*1.94*
approx.[e] 6.5–8.75 in.	0.6–2.0	—	—		19			*3.35*
approx.[e] 7.5–10 in.	0.6–2.0	—	—		22			*3.87*
approx.[e] 10.25–13.75 in.	0.6–2.0	—	—		30			*5.28*
Vermiculite, exfoliated	7.0–8.2	0.47	—	*2.13*	—	3.20	*14.76*	
	4.0–6.0	0.44	—	*2.27*	—		*15.73*	
Roof Insulation[h]								
Preformed, for use above deck								
Different roof insulations are available in different thicknesses to provide the design *C* values listed.[h] Consult individual manufacturers for actual thickness of their material.			0.36 to 0.05		2.7 *to* 20		—	*0.49 to 3.52*
MASONRY MATERIALS								
Concretes								
Cement mortar	116	5.0	—	*0.20*	—		*1.39*	
Gypsum-fiber concrete 87.5% gypsum, 12.5% wood chips	51	1.66	—	*0.60*	—	0.21	*4.16*	
Lightweight aggregates including expanded shale, clay or slate; expanded slags; cinders; pumice; vermiculite; also cellular concretes	120	5.2	—	*0.19*	—		*1.32*	
	100	3.6	—	*0.28*	—		*1.94*	
	80	2.5	—	*0.40*	—		*2.77*	
	60	1.7	—	*0.59*	—		4.09	
	40	1.15	—	*0.86*	—		*5.96*	
	30	0.90	—	*1.11*	—		*7.69*	
	20	0.70		*1.43*			*9.91*	
Perlite, expanded	40	0.93		*1.08*			*7.48*	
	30	0.71		*1.41*			*9.77*	
	20	0.50		2.00		0.32	*13.86*	
Sand and gravel or stone aggregate (oven dried)	140	9.0	—	*0.11*		0.22	*0.76*	
Sand and gravel or stone aggregate (not dried)	140	12.0	—	*0.08*			*0.55*	
Stucco	116	5.0	—	*0.20*			*1.39*	
MASONRY UNITS								
Brick, common[i]	120	5.0	—	*0.20*	—	0.19	*1.39*	
Brick, face[i]	130	9.0	—	*0.11*	—		*0.76*	

Clay tile, hollow:								
1 cell deep . . . 3 in.	—	—	1.25	—	*0.80*	0.21		*0.14*
1 cell deep . . . 4 in.	—	—	0.90	—	*1.11*			*0.20*
2 cells deep . . . 6 in.	—	—	0.66	—	*1.52*			*0.27*
2 cells deep . . . 8 in.	—	—	0.54	—	*1.85*			*0.33*
2 cells deep . . . 10 in.	—	—	0.45	—	*2.22*			*0.39*
3 cells deep . . . 12 in.	—	—	0.40	—	*2.50*			*0.44*
Concrete blocks, three oval core:								
Sand and gravel aggregate . . . 4 in.	—	—	1.40	—	*0.71*	0.22		*0.13*
. . . 8 in.	—	—	0.90	—	*1.11*			*0.20*
. . . 12 in.	—	—	0.78	—	*1.28*			*0.23*
Cinder aggregate . . . 3 in.	—	—	1.16	—	*0.86*	0.21		*0.15*
. . . 4 in.	—	—	0.90	—	*1.11*			*0.20*
. . . 8 in.	—	—	0.58	—	*1.72*			*0.30*
. . . 12 in.	—	—	0.53	—	*1.89*			*0.33*
Lightweight aggregate . . . 3 in.	—	—	0.79	—	*1.27*	0.21		*0.22*
(expanded shale, clay, slate . . . 4 in.	—	—	0.67	—	*1.50*			*0.26*
or slag; pumice) . . . 8 in.	—	—	0.50	—	*2.00*			*0.35*
. . . 12 in.	—	—	0.44	—	*2.27*			*0.40*
Concrete blocks, rectangular core.*[j]								
Sand and gravel aggregate								
2 core, 8 in. 36 lb.[k]* . . .	—	—	0.96	—	*1.04*	0.22		*0.18*
Same with filled cores[l]* . . .	—	—	0.52	—	*1.93*	0.22		*0.34*
Lightweight aggregate (expanded shale, clay, slate or slag, pumice):								
3 core, 6 in. 19 lb.[k]* . . .	—	—	0.61	—	*1.65*	0.21		*0.29*
Same with filled cores[l]* . . .	—	—	0.33	—	*2.99*			*0.53*
2 core, 8 in. 24 lb.[k]* . . .	—	—	0.46	—	*2.18*			*0.38*
Same with filled cores[l]* . . .	—	—	0.20	—	*5.03*			*0.89*
3 core, 12 in. 38 lb.[k]* . . .	—	—	0.40	—	*2.48*			*0.44*
Same with filled cores[l]* . . .	—	—	0.17	—	*5.82*			*1.02*
Stone, lime or sand . . .	—	12.50	—	*0.08*	—	0.19	*0.55*	
Gypsum partition tile:								
3 × 12 × 30 in. solid . . .	—	—	0.79	—	*1.26*	*0.19*		*0.22*
3 × 12 × 30 in. 4-cell . . .	—	—	0.74	—	*1.35*			*0.24*
4 × 12 × 30 in. 3-cell . . .	—	—	0.60	—	*1.67*			*0.29*
METALS								
(See ASHRAE Handbook of Fundamentals)								
PLASTERING MATERIALS								
Cement plaster, sand aggregate . . .	116	5.0	—	*0.20*	—	0.20	*1.39*	
Sand aggregate . . . 0.375 in.	—	—	13.3	—	*0.08*	0.20		*0.01*
Sand aggregate . . . 0.75 in.	—	—	6.66	—	*0.15*	0.20		*0.03*
Gypsum plaster:								
Lightweight aggregate . . . 0.5 in.	45	—	3.12	—	*0.32*			*0.06*
Lightweight aggregate . . . 0.625 in.	45	—	2.67	—	*0.39*			*0.07*
Lightweight agg. on metal lath . . . 0.75 in.	—	—	2.13	—	*0.47*			*0.08*
Perlite aggregate . . .	45	1.5	—	*0.67*	—	0.32	*4.64*	
Sand aggregate . . .	105	5.6	—	*0.18*	—	0.20	*1.25*	
Sand aggregate . . . 0.5 in.	105	—	11.10	—	*0.09*			*0.02*
Sand aggregate . . . 0.625 in.	105	—	9.10	—	*0.11*			*0.02*
Sand aggregate on metal lath . . . 0.75 in.	—	—	7.70	—	*0.13*			*0.02*
Vermiculite aggregate . . .	45	1.7	—	*0.59*	—		*4.09*	

(continued)

TABLE 4.2 **Thermal Properties of Typical Building and Insulating Materials (design values)**[a] *(Continued)*

Description	Density (lb/ft³)	Conductivity (k)	Conductance (C)	Customary Unit Resistance[b] (R): Per inch thickness (1/k)	Customary Unit Resistance[b] (R): For thickness listed (1/C)	Specific Heat, Btu/(lb) (deg F)	SI Unit Resistance[b] (R): (m·K)/W	SI Unit Resistance[b] (R): (m²·K)/W
ROOFING								
Asbestos-cement shingles	120	—	4.76	—	*0.21*	0.24		*0.04*
Asphalt roll roofing	70	—	6.50	—	*0.15*	0.36		*0.03*
Asphalt shingles	70	—	2.27	—	*0.44*	0.30		0.08
Built-up roofing 0.375 in.	70	—	3.00	—	*0.33*	0.35		*0.06*
Slate 0.5 in.	—	—	20.00	—	*0.05*	0.30		*0.01*
Wood shingles, plain and plastic film faced	—	—	1.06	—	*0.94*	0.31		*0.17*
SIDING MATERIALS (ON FLAT SURFACE)								
Shingles								
Asbestos-cement	120	—	4.75	—	*0.21*			*0.04*
Wood, 16 in., 7.5 exposure	—	—	1.15	—	*0.87*	0.31		*0.15*
Wood, double, 16-in., 12-in. exposure	—	—	0.84	—	*1.19*	0.28		*0.21*
Wood, plus insul. backer board, 0.3125 in.	—	—	0.71	—	*1.40*	0.31		*0.25*
Siding								
Asbestos-cement, 0.25 in., lapped	—	—	4.76	—	*0.21*	0.24		*0.04*
Asphalt roll siding	—	—	6.50	—	*0.15*	0.35		*0.03*
Asphalt insulating siding (0.5 in. bed.)	—	—	0.69	—	*1.46*	0.35		*0.26*
Hardboard siding, 0.4375 in.	40	1.49	—	*0.67*		0.28	*4.65*	
Wood, drop, 1 × 8 in.	—	—	1.27	—	*0.79*	0.28		*0.14*
Wood, bevel, 0.5 × 8 in., lapped	—	—	1.23	—	*0.81*	0.28		*0.14*
Wood, bevel, 0.75 × 10 in., lapped	—	—	0.95	—	*1.05*	0.28		*0.18*
Wood, plywood, 0.375 in., lapped	—	—	1.59	—	*0.59*	0.29		*0.10*
Aluminum or Steel[m], over sheathing								
Hollow-backed	—	—	1.61	—	*0.61*	0.29		*0.11*
Insulating-board backed nominal 0.375 in.	—	—	0.55	—	*1.82*	0.32		*0.32*
Insulating-board backed nominal 0.375 in., foil backed			0.34		*2.96*			*0.52*
Architectural glass	—	—	10.00	—	*0.10*	0.20		*0.02*
WOODS[o,p]								
Maple, oak, and similar hardwoods	45	1.10	—	*0.91*	—	0.30	*6.31*	—
Fir, pine, etc.	32	0.80	—	*1.25*	—	0.33	*8.66*	—
0.75 in.	32	—	1.06	—	*0.94*		—	*0.17*
1.5 in.		—	0.53	—	*1.88*		—	*0.33*
2.5 in.		—	0.32	—	*3.12*		—	*0.55*
3.5 in.		—	0.23	—	*4.38*		—	*0.77*
5.5 in.		—	0.14	—	*7.14*		—	*1.26*
7.25 in.		—	-.11	—	*9.09*		—	*1.60*
9.25 in.		—	0.09	—	*11.11*		—	*1.96*
11.25 in.		—	0.07	—	*14.28*		—	*2.15*

Source: Copyright © by the American Society of Heating, Refrigerating and Air Conditioning Engineers, Inc., Atlanta, GA. Reprinted by permission from *1981 Handbook of Fundamentals.*

[a]Representative values for dry materials were selected by ASHRAE TC 4.4, Thermal Insulation and Moisture Retarders (Total Thermal Performance Design Criteria). They are intended as design (not specification) values for materials in normal use. Insulation materials in actual service may have thermal values which vary from design values depending on their in-situ properties such as density and moisture content. For properties of a particular product, use the value supplied by the manufacturer or by unbiased tests.

[b]Resistance values are the reciprocals of C before rounding off C to two decimal places.

[c]Also see Insulating Materials, Board.

[d]Does not include paper backing and facing, if any. Where insulation forms a boundary (reflective or otherwise) of an air space, see Tables 4.3 and 4.4 for the insulating value of air space for the appropriate effective emittance and temperature conditions of the space.

[e]Conductivity varies with fiber diameter. Insulation is produced in different densities; therefore, there is a wide variation in thickness for the same R value among manufacturers. No effort should be made to relate any specific R value to any specific thickness. Commercial thicknesses generally available range from 2 to 8.5.

[f]Values are for aged, unfaced, board stock. For change in conductivity with age of expanded urethane, see manufacturers' data.

[g]Insulating values of acoustical tile vary, depending on density of the board and on type, size, and depth of perforations.

[h]ASTM C-855-77 recognizes the specification of roof insulation on the basis of the C values shown. Roof insulation is made in thicknesses to meet these values.

[i]Face brick and common brick do not always have these specific densities. When density is different from that shown, there will be a change in thermal conductivity.

[j]Data on rectangular core concrete blocks differ from the above data on oval core blocks, due to core configuration, different mean temperatures, and possibly differences in unit weights. Weight data on the oval core blocks tested are not available.

[k]Weights of units approximately 7.625 in. high and 15.75 in. long. These weights are given as a means of describing the blocks tested, but conductance values are all for 1 ft^2 of area.

[l]Vermiculite, perlite, or mineral wool insulation. Where insulation is used, vapor barriers or other precautions must be considered to keep insulation dry.

[m]Values for metal siding applied over flat surfaces vary widely, depending on amount of ventilation of air space beneath the siding; whether air space is reflective or nonreflective; and on thickness, type, and application of insulating backing-board used. Values given are averages for use as design guides, and were obtained from several guarded hotbox tests (ASTM C236) or calibrated hotbox (BSS 77) on hollow-backed types and types made using backing-boards of wood fiber, foamed plastic, and glass fiber. Departures of $\pm 50\%$ or more from the values given may occur.

[n]Time-aged values for board stock with gas-barrier quality (0.001 in. thickness or greater) aluminum foil facers on two major surfaces.

[o]Forest Products Laboratory Wood Handbook, U.S. Dept. of Agriculture #72, 1974, Tables 3 and 4.

[p]L. Adams: Supporting cryogenic equipment with wood (*Chemical Engineering,* May 17, 1971).

TABLE 4.3 **Surface Conductances (Btu/h ft[2] F) and Resistances for Air[a]**

		Surface Emittance					
		Non-reflective		*Reflective*			
		ε = 0.90		*ε = 0.20*		*ε = 0.05*	
Position of Surface	*Direction of Heat Flow*	h_i	R	h_i	R	h_i	R
Still air							
Horizontal.........	Upward	1.63	0.61	0.91	1.10	0.76	1.32
Sloping (45°)	Upward	1.60	0.62	0.88	1.14	0.73	1.37
Vertical	Horizontal	1.46	0.68	0.74	1.35	0.59	1.70
Sloping (45°)	Downward	1.32	0.76	0.60	1.67	0.45	2.22
Horizontal.........	Downward	1.08	0.92	0.37	2.70	0.22	4.55
		h_o	R	h_o	R	h_o	R
Moving air (any position)							
15-mph wind (for winter)	Any	6.00	0.17				
7.5-mph wind (for summer)	Any	4.00	0.25				

NOTE: A surface cannot take credit for both an air space resistance value and a surface resistance value. No credit for an air space value can be taken for any surface facing an air space of less than 0.5 in.

Source: Copyright © by the American Society of Heating, Refrigerating and Air Conditioning Engineers, Inc., Atlanta, GA. Reprinted by permission, from *1981 Handbook of Fundamentals.*

[a]Conductances are for surfaces of the stated emittance facing virtual blackbody surroundings at the same temperature as ambient air. Values are based on a surface-air temperature difference of 10 F° and for surface temperature of 70 F.

TABLE 4.4 **Thermal Resistances of Plane[a] Air Spaces**

SECTION A

All resistance values expressed in $ft^2 \cdot F \cdot h/Btu$

Values apply only to air spaces of uniform thickness bounded by plane, smooth, parallel surfaces with no leakage of air to or from the space. These conditions are not normally present in standard building construction. When accurate values are required, use overall *U*-factors determined for your particular construction through calibrated hot box (BSS-77) or guarded hot box (ASTM-C-236) testing. Thermal resistance values for multiple air spaces must be based on careful estimates of mean temperature differences for each air space.

Position of Air Space	Direction of Heat Flow	Air Space Mean Temp,[b] (F)	Air Space Temp Diff,[b] (deg F)	0.5-in. Air Space[d] Value of $E^{b,c}$ 0.03	0.05	0.2	0.5	0.82	0.75-in. Air Space[d] Value of $E^{b,c}$ 0.03	0.05	0.2	0.5	0.82
		90	10	2.13	2.03	1.51	0.99	0.73	2.34	2.22	1.61	1.04	0.75
		50	30	1.62	1.57	1.29	0.96	0.75	1.71	1.66	1.35	0.99	0.77
		50	10	2.13	2.05	1.60	1.11	0.84	2.30	2.21	1.70	1.16	0.87
Horiz.	Up ↑	0	20	1.73	1.70	1.45	1.12	0.91	1.83	1.79	1.52	1.16	0.93
		0	10	2.10	2.04	1.70	1.27	1.00	2.23	2.16	1.78	1.31	1.02
		−50	20	1.69	1.66	1.49	1.23	1.04	1.77	1.74	1.55	1.27	1.07
		−50	10	2.04	2.00	1.75	1.40	1.16	2.16	2.11	1.84	1.46	1.20
		90	10	2.44	2.31	1.65	1.06	0.76	2.96	2.78	1.88	1.15	0.81
		50	30	2.06	1.98	1.56	1.10	0.83	1.99	1.92	1.52	1.08	0.82
45°		50	10	2.55	2.44	1.83	1.22	0.90	2.90	2.75	2.00	1.29	0.94
Slope	Up ↗	0	20	2.20	2.14	1.76	1.30	1.02	2.13	2.07	1.72	1.28	1.00
		0	10	2.63	2.54	2.03	1.44	1.10	2.72	2.62	2.08	1.47	1.12
		−50	20	2.08	2.04	1.78	1.42	1.17	2.05	2.01	1.76	1.41	1.16
		−50	10	2.62	2.56	2.17	1.66	1.33	2.53	2.47	2.10	1.62	1.30
		90	10	2.47	2.34	1.67	1.06	0.77	3.50	3.24	2.08	1.22	0.84
		50	30	2.57	2.46	1.84	1.23	0.90	2.91	2.77	2.01	1.30	0.94
		50	10	2.66	2.54	1.88	1.24	0.91	3.70	3.46	2.35	1.43	1.01
Vertical	Horiz. →	0	20	2.82	2.72	2.14	1.50	1.13	3.14	3.02	2.32	1.58	1.18
		0	10	2.93	2.82	2.20	1.53	1.15	3.77	3.59	2.64	1.73	1.26
		−50	20	2.90	2.82	2.35	1.76	1.39	2.90	2.83	2.36	1.77	1.39
		−50	10	3.20	3.10	2.54	1.87	1.46	3.72	3.60	2.87	2.04	1.56
		90	10	2.48	2.34	1.67	1.06	0.77	3.53	3.27	2.10	1.22	0.84
		50	30	2.64	2.52	1.87	1.24	0.91	3.43	3.23	2.24	1.39	0.99
45°		50	10	2.67	2.55	1.89	1.25	0.92	3.81	3.57	2.40	1.45	1.02
Slope	Down ↘	0	20	2.91	2.80	2.19	1.52	1.15	3.75	3.57	2.63	1.72	1.26
		0	10	2.94	2.83	2.21	1.53	1.15	4.12	3.91	2.81	1.80	1.30
		−50	20	3.16	3.07	2.52	1.86	1.45	3.78	3.65	2.90	2.05	1.57
		−50	10	3.26	3.16	2.58	1.89	1.47	4.35	4.18	3.22	2.21	1.66
		90	10	2.48	2.34	1.67	1.06	0.77	3.55	3.29	2.10	1.22	0.85
		50	30	2.66	2.54	1.88	1.24	0.91	3.77	3.52	2.38	1.44	1.02
		50	10	2.67	2.55	1.89	1.25	0.92	3.84	3.59	2.41	1.45	1.02
Horiz.	Down ↓	0	20	2.94	2.83	2.20	1.53	1.15	4.18	3.96	2.83	1.81	1.30
		0	10	2.96	2.85	2.22	1.53	1.16	4.25	4.02	2.87	1.82	1.31
		−50	20	3.25	3.15	2.58	1.89	1.47	4.60	4.41	3.36	2.28	1.69
		−50	10	3.28	3.18	2.60	1.90	1.47	4.71	4.51	3.42	2.30	1.71

(continued)

TABLE 4.4 **Thermal Resistances of Plane[a] Air Spaces** *(Continued)*

Position of Air Space	Direction of Heat Flow	Air Space Mean Temp,[b] (F)	Air Space Temp Diff,[b] (deg F)	1.5-in. Air Space[d] Value of $E^{b,c}$ 0.03	1.5-in. 0.05	1.5-in. 0.2	1.5-in. 0.5	1.5-in. 0.82	3.5-in. Air Space[d] Value of $E^{b,c}$ 0.03	3.5-in. 0.05	3.5-in. 0.2	3.5-in. 0.5	3.5-in. 0.82
Horiz	Up ↑	90	10	2.55	2.41	1.71	1.08	0.77	2.84	2.66	1.83	1.13	0.80
		50	30	1.87	1.81	1.45	1.04	0.80	2.09	2.01	1.58	1.10	0.84
		50	10	2.50	2.40	1.81	1.21	0.89	2.80	2.66	1.95	1.28	0.93
		0	20	2.01	1.95	1.63	1.23	0.97	2.25	2.18	1.79	1.32	1.03
		0	10	2.43	2.35	1.90	1.38	1.06	2.71	2.62	2.07	1.47	1.12
		−50	20	1.94	1.91	1.68	1.36	1.13	2.19	2.14	1.86	1.47	1.20
		−50	10	2.37	2.31	1.99	1.55	1.26	2.65	2.58	2.18	1.67	1.33
45° Slope	Up ↗	90	10	2.92	2.73	1.86	1.14	0.80	3.18	2.96	1.97	1.18	0.82
		50	30	2.14	2.06	1.61	1.12	0.84	2.26	2.17	1.67	1.15	0.86
		50	10	2.88	2.74	1.99	1.29	0.94	3.12	2.95	2.10	1.34	0.96
		0	20	2.30	2.23	1.82	1.34	1.04	2.42	2.35	1.90	1.38	1.06
		0	10	2.79	2.69	2.12	1.49	1.13	2.98	2.87	2.23	1.54	1.16
		−50	20	2.22	2.17	1.88	1.49	1.21	2.34	2.29	1.97	1.54	1.25
		−50	10	2.71	2.64	2.23	1.69	1.35	2.87	2.79	2.33	1.75	1.39
Vertical	Horiz. →	90	10	3.99	3.66	2.25	1.27	0.87	3.69	3.40	2.15	1.24	0.85
		50	30	2.58	2.46	1.84	1.23	0.90	2.67	2.55	1.89	1.25	0.91
		50	10	3.79	3.55	2.39	1.45	1.02	3.63	3.40	2.32	1.42	1.01
		0	20	2.76	2.66	2.10	1.48	1.12	2.88	2.78	2.17	1.51	1.14
		0	10	3.51	3.35	2.51	1.67	1.23	3.49	3.33	2.50	1.67	1.23
		−50	20	2.64	2.58	2.18	1.66	1.33	2.82	2.75	2.30	1.73	1.37
		−50	10	3.31	3.21	2.62	1.91	1.48	3.40	3.30	2.67	1.94	1.50
45° Slope	Down ↘	90	10	5.07	4.55	2.56	1.36	0.91	4.81	4.33	2.49	1.34	0.90
		50	30	3.58	3.36	2.31	1.42	1.00	3.51	3.30	2.28	1.40	1.00
		50	10	5.10	4.66	2.85	1.60	1.09	4.74	4.36	2.73	1.57	1.08
		0	20	3.85	3.66	2.68	1.74	1.27	3.81	3.63	2.66	1.74	1.27
		0	10	4.92	4.62	3.16	1.94	1.37	4.59	4.32	3.02	1.88	1.34
		−50	20	3.62	3.50	2.80	2.01	1.54	3.77	3.64	2.90	2.05	1.57
		−50	10	4.67	4.47	3.40	2.29	1.70	4.50	4.32	3.31	2.25	1.68
Horiz.	Down ↓	90	10	6.09	5.35	2.79	1.43	0.94	10.07	8.19	3.41	1.57	1.00
		50	30	6.27	5.63	3.18	1.70	1.14	9.60	8.17	3.86	1.88	1.22
		50	10	6.61	5.90	3.27	1.73	1.15	11.15	9.27	4.09	1.93	1.24
		0	20	7.03	6.43	3.91	2.19	1.49	10.90	9.52	4.87	2.47	1.62
		0	10	7.31	6.66	4.00	2.22	1.51	11.97	10.32	5.08	2.52	1.64
		−50	20	7.73	7.20	4.77	2.85	1.99	11.64	10.49	6.02	3.25	2.18
		−50	10	8.09	7.52	4.91	2.89	2.01	12.98	11.56	6.36	3.34	2.22

SECTION B. Reflectivity and Emittance Values of Various Surfaces and Effective Emittances of Air Spaces

Surface	Reflectivity in Percent	Average *Emittance* ε	Effective *Emittance E* of Air Space	
			One surface emittance ε; the other 0.90	Both surfaces emittances ε
Aluminum foil, bright	92 to 97	0.05	0.05	0.03
Aluminum sheet	80 to 95	0.12	0.12	0.06
Aluminum coated paper, polished	75 to 84	0.20	0.20	0.11
Steel, galvanized, bright...	70 to 80	0.25	0.24	0.15
Aluminum paint	30 to 70	0.50	0.47	0.35
Building materials: wood, paper, masonry, nonmetallic paints	5 to 15	0.90	0.82	0.82
Regular glass	5 to 15	0.84	0.77	0.72

Source: Copyright © by the American Society of Heating, Refrigerating and Air Conditioning Engineers, Inc., Atlanta, GA. Reprinted by permission from *1981 Handbook of Fundamentals.*

[a]Thermal resistance values were determined from the relation $R = 1/C$, where $C = h_c + Eh_r$, h_c is the conduction-convection coefficient, Eh_r is the radiation coefficient $\cong 0.00686E\,[(t_m + 460)/100]^3$, and t_m is the mean temperature of the air space. Values for h_c were determined from research data (National Bureau of Standards), such as those presented in 1954 in Housing Research Paper No. 32 (HRP No. 32) by the Housing and Home Finance Agency (Government Printing Office, Washington, D.C.). For interpolation from Table 4.4 to air space thicknesses less than 0.5 in. (as in insulating window glass), assume:

$$h_c = 0.159(1 + 0.0016t_m)/l,$$

where l is the thickness in inches, and h_c is assumed to represent heat transfer by conduction alone through air.

[b]Interpolation is permissible for other values of mean temperature, temperature differences, and effective emittance E. Interpolation and moderate extrapolation for air spaces greater than 3.5 in. are also permissible.

[c]Effective emittance of the space E is given by $1/E = 1/e_1 + 1/e_2 - 1$, *where* e_1 and e_2 are the emittances of the surfaces of the air space.

[d]Credit for an air space resistance value cannot be taken more than once and only for the boundary conditions established.

[e]Resistances of horizontal spaces with heat flow downward are substantially independent of temperature difference.

TABLE 4.5 **Coefficients of Transmission, U (Btu/h ft^2 F), of Frame Walls**

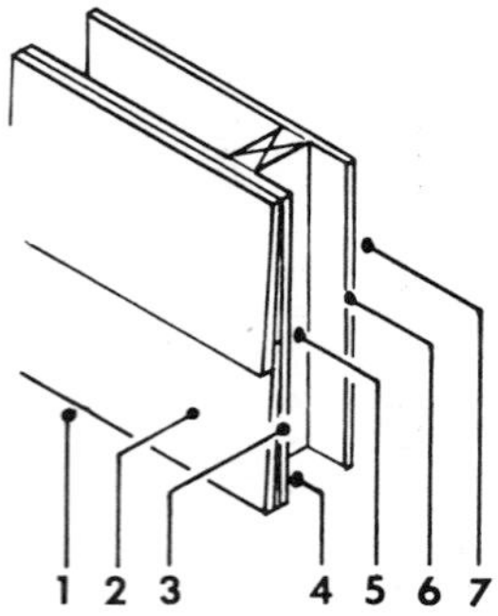

Base Case Resistance: Between Framing, R_i	Base Case Resistance: At Framing, R_s	Construction	Resistance[a]: New Item 4 ("Construction"): Between Framing, R_i	Resistance[a]: New Item 4 ("Construction"): At Framing, R_s
0.17	0.17	1. Outside surface (15-mph wind)	0.17	0.17
0.81	0.81	2. Siding, wood, 0.5 in. × 8 in. lapped (average)	0.81	0.81
1.32	1.32	3. Sheathing, 0.5-in. vegetable fiber board	1.32	1.32
1.01	—	4. Nonreflective air space, 3.5 in. (50 F mean; 10 F temperature difference)	11.00	—
—	4.35	5. Nominal 2-in. × 4-in. wood stud	—	4.35
0.45	0.45	6. Gypsum wallboard, 0.5 in.	0.45	0.45
0.68	0.68	7. Inside surface (still air)	0.68	0.68
4.44	7.78	*Total thermal resistance, R*	14.43	7.78

Base Case

$$U_i = \frac{1}{4.44} = 0.225$$

New Item 4

$$U_i = \frac{1}{14.43} = 0.069$$

To adjust U values for the effect of framing, see the example below. With 15% framing (typical of 2-in. × 4-in. studs at 16-in. o.c.) these adjusted U values are, respectively:

$U_{av} = 0.199$ (12% less heat loss) $\qquad$ $U_{av} = 0.081$ (17% more heat loss)

EXAMPLE. To adjust U values for the effect of framing, average the U_i and U_s values in proportion to their areas. For the constructions above, with 15% of the wall area in framing and 85% in insulation:

$$U_i = \frac{1}{4.44} = 0.225 \qquad U_i = \frac{1}{14.43} = 0.069$$

$$U_s = \frac{1}{7.78} = 0.128 \qquad U_s = \frac{1}{7.78} = 0.128$$

$$U_{av} = 0.85(0.225) + 0.15(0.128) = 0.199 \qquad U_{av} = 0.85(0.069) + 0.15(0.128) = 0.081$$

Source: Copyright © by the American Society of Heating, Refrigerating and Air Conditioning Engineers, Inc., Atlanta, GA. Reprinted by permission from *1981 Handbook of Fundamentals.*

[a]When air space replaced with 3.5-in. R-11 blanket insulation.

TABLE 4.6 **Coefficients of Transmission, *U* (Btu/h ft² F), of Solid Masonry Walls**

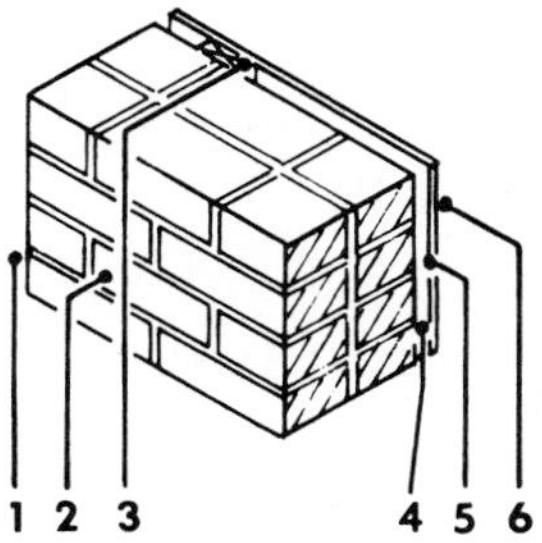

Base Case Resistance: Between Furring, R_i	Base Case Resistance: At Furring, R_s	*Construction*	*Resistance[a]: New Item 4 ("Construction")*
0.17	0.17	1. Outside surface (15-mph wind)	0.17
1.60	1.60	2. Common brick, 8 in.	1.60
—	0.94	3. Nominal 1-in. × 3-in. vertical furring	—
1.01	—	4. Nonreflective air space, 0.75 in. (50 F mean; 10 F° temperature difference)	5.00
0.45	0.45	5. Gypsum wallboard, 0.5 in.	0.45
0.68	0.68	6. Inside surface (still air)	0.68
3.91	3.84	*Total thermal resistance, R*	$R_i = 7.90 = R_s$

Base Case

$$U_i = \frac{1}{3.91} = 0.256$$

New Item 4

$$U = \frac{1}{7.90} = 0.127$$

To adjust *U* values for the effect of furring strips, see the example given in Table 4.5. With 20% furring (typical of 1-in. × 3-in. vertical furring on masonry at 16-in. o.c.):

$$U_{av} = 0.257 \quad \text{(about 1\% more heat loss)}$$

Source: Copyright © by the American Society of Heating, Refrigerating and Air Conditioning Engineers, Inc., Atlanta, GA. Reprinted by permission from *1981 Handbook of Fundamentals.*

[a]When furring strips and air space replaced with 1-in. expanded polystyrene extruded, smooth skin surface, 2.2 lb/ft³.

TABLE 4.7 **Coefficients of Transmission, *U* (Btu/h ft² F), of Masonry Cavity Walls**

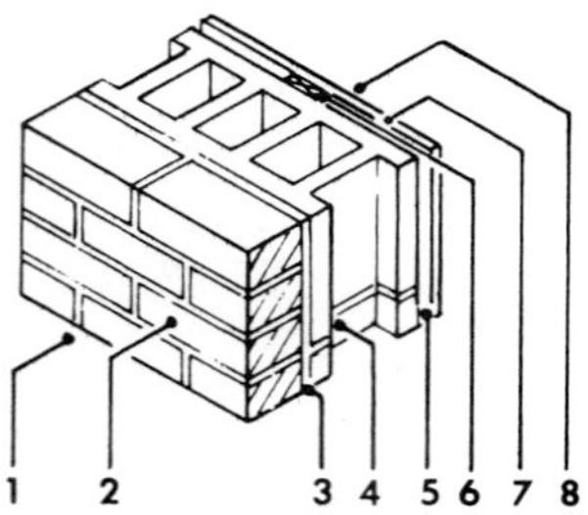

Base Case Resistance: Between Furring, R_i	Base Case Resistance: At Furring, R_s	Construction	Resistance[c]: New Items 3 and 7 ("Construction")
0.17	0.17	1. Outside surface (15-mph wind)	0.17
0.80	0.80	2. Common brick, 4 in.	0.80
1.10[a]	1.10[a]	3. Nonreflective air space, 2.5 in. (30 F mean; 10 F° temperature difference)	5.32[b]
0.71	0.71	4. Concrete block, three-oval core, stone and gravel aggregate, 4 in.	0.71
1.01	—	5. Nonreflective air space 0.75 in. (50 F mean; 10 F° temperature difference)	—
—	0.94	6. Nominal 1-in. × 3-in vertical furring	—
0.45	0.45	7. Gypsum wallboard, 0.5 in.	0.11
0.68	0.68	8. Inside surface (still air)	0.68
4.92	4.85	*Total thermal resistance, R*	$R_i = R_s = 7.79$

Base Case

$$U_i = \frac{1}{4.92} = 0.203$$

New Items 3 and 7

$$U = \frac{1}{7.79} = 0.128$$

To adjust U values for the effect of furring strips, see the example given in Table 4.5; with 20% furring (typical of 1-in. × 3-in. vertical furring on masonry at 16-in. o.c.):

$$U_{av} = 0.204 \qquad \text{(about 1\% more heat loss)}$$

[a]Interpolated from vertical air space values in Table 4.4.

[b]Calculated from vermiculite R values in Table 4.2.

[c]Resistance when furring strips and gypsum wallboard replaced with 0.625-in. plaster (sand aggregate) applied directly to concrete block and 2.5-in. air space filled with vermiculite insulation, 7–8.2 lb/ft³.

Source: Copyright © by the American Society of Heating, Refrigerating and Air Conditioning Engineers, Inc., Atlanta, GA. Reprinted by permission from *1981 Handbook of Fundamentals.*

TABLE 4.8 **Coefficients of Transmission, U (Btu/h ft^2 F), of Wood Construction Flat Roofs and Ceilings: Winter Conditions, Upward Flow**

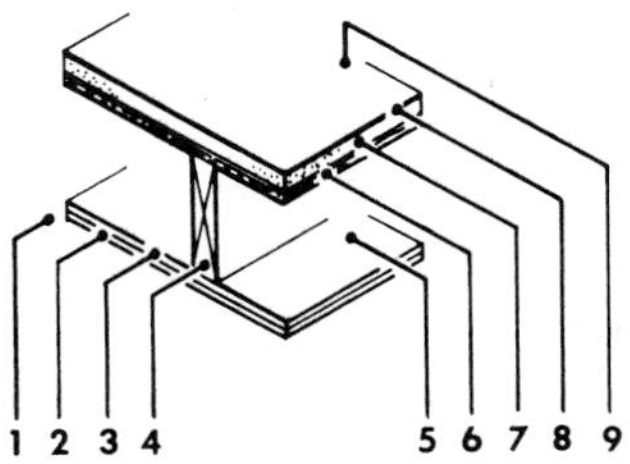

Base Case Resistance			*Resistance[c]: New Items 5 and 7 ("Construction")*	
Between Joists, R_i	*At Joists,* R_s	*Construction (heat flow up)*	*Between Joists,* R_i	*At Joists,* R_s
0.61	0.61	1. Inside surface (still air)	0.61	0.61
1.25	1.25	2. Acoustical tile, fiberboard, 0.5 in.	1.25	1.25
0.45	0.45	3. Gypsum wallboard, 0.5 in.	0.45	0.45
—	9.06	4. Nominal 2-in. × 8-in. ceiling joists	—	9.06
0.93[a]	—	5. Nonreflective air space, 7.25 in. (50 F mean; 10 F° temperature difference)	1.05[b]	—
0.78	0.78	6. Plywood deck, 0.625 in.	0.78	0.78
1.39	1.39	7. Rigid roof deck insulation, $C = 0.72$ ($R = 1/C$)	19.00	—
0.33	0.33	8. Built-up roof	0.33	0.33
0.17	0.17	9. Outside surface (15-mph wind)	0.17	0.17
5.91	14.04	*Total thermal resistance, R*	23.64	12.65

Base Case

$$U_i = \frac{1}{5.91} = 0.169$$

New Items 5 and 7

$$U_i = \frac{1}{23.64} = 0.042$$

To adjust U-values for the effect of framing, see the example given in Table 4.5. With 10% framing (typical of 2-in. joists at 16-in. o.c.), these adjusted U_{av} values are, respectively:

$U_{av} = 0.159$ (6% less heat loss) $U_{av} = 0.046$ (10% more heat loss)

[a]Use largest air space (3.5 in.) value shown in Table 4.4.

[b]Interpolated value (0 F mean; 10 F° temperature difference).

[c]When roof deck insulation replaced and 7.25-in. air space partially filled with 6-in. R-19 blanket insulation and 1.25-in. air space.

Source: Copyright © by the American Society of Heating, Refrigerating and Air Conditioning Engineers, Inc., Atlanta, GA. Reprinted by permission from *1981 Handbook of Fundamentals*.

TABLE 4.9 **Coefficients of Transmission, *U* (Btu/h ft² F), of Flat Masonry Roofs with Built-Up Roofing, With and Without Suspended Ceilings: Winter Conditions, Upward Flow**

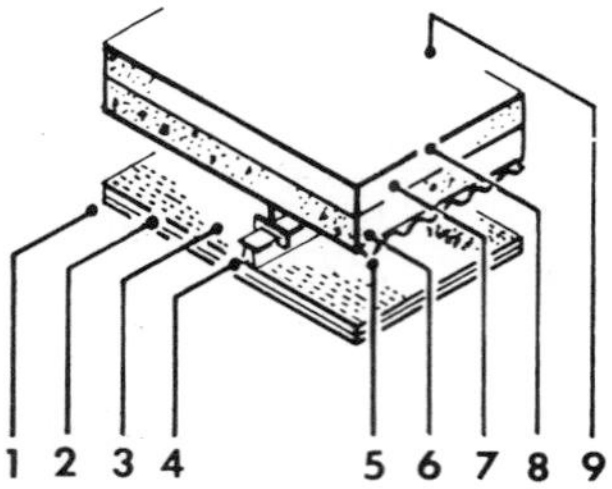

Base Case Resistance R	Construction (heat flow up)	Resistance R[c]: New Item 7 ("Construction")
0.61	1. Inside surface (still air)	0.61
0.47	2. Metal lath and lightweight aggregate plaster, 0.75 in.	0.47
0.93[a]	3. Nonreflective air space, greater than 3.5 in. (50 F mean; 10 F° temperature difference)	0.93[a]
0[b]	4. Metal ceiling suspension system with metal hanger rods	0[b]
0	5. Corrugated metal deck	0
2.22	6. Concrete slab, lightweight aggregate, 2 in. (30 lb/ft³)	2.22
—	7. Rigid roof deck insulation (none)	4.17
0.33	8. Built-up roofing, 0.375 in.	0.33
0.17	9. Outside surface (15-mph wind)	0.17
4.73	*Total Thermal Resistance, R*	8.90

Base Case

$$U = \frac{1}{4.73} = 0.211$$

New Item 7

$$U = \frac{1}{8.90} = 0.112$$

[a]Use largest air space (3.5 in.) value shown in Table 4.4.

[b]Area of hanger rods is negligible in relation to ceiling area.

[c]When rigid roof deck insulation added, $C = 0.24$ ($R = 1/C = 4.17$).

Source: Copyright © by the American Society of Heating, Refrigerating and Air Conditioning Engineers, Inc., Atlanta, GA. Reprinted by permission from *1981 Handbook of Fundamentals.*

TABLE 4.10 **Coefficients of Transmission, *U* (Btu/h ft² F), of Metal Construction Flat Roofs and Ceilings: Winter Conditions, Upward Flow**

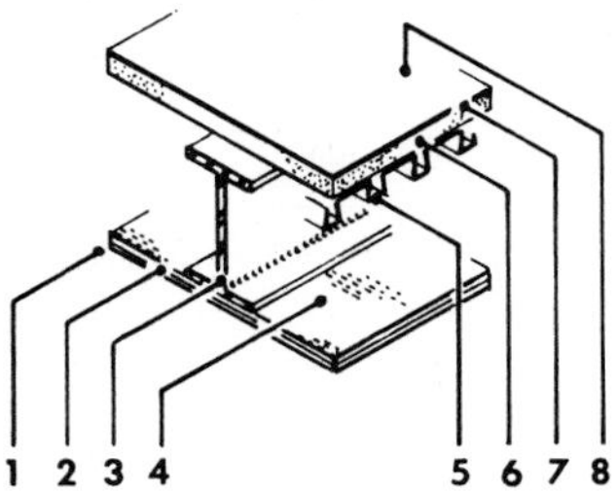

Base Case Resistance, R	Construction (heat flow up)	Resistance R[c]: New Items 2 and 6 ("Construction")
0.61	1. Inside surface (still air)	0.61
0.13	2. Metal lath and sand aggregate plaster, 0.75 in.	0.47
0.00[a]	3. Structural beam	0.00[a]
0.93[b]	4. Nonreflective air space (50 F mean; 10 F° temperature difference)	0.93[b]
0.00[a]	5. Metal deck	0.00[a]
4.17	6. Rigid roof deck insulation, $C = 0.24 (R = 1/C)$	2.78
0.33	7. Built-up roofing, 0.375 in.	0.33
0.17	8. Outside surface (15-mph wind)	0.17
6.34	*Total Thermal Resistance, R*	5.29

Base Case

$$U = \frac{1}{6.34} = 0.158$$

New Items 2 and 6

$$U = \frac{1}{5.29} = 0.189$$

[a]If thermal effects of structural beams and metal deck are to be considered, see *ASHRAE 1981 Handbook of Fundamentals,* Chapter 23.

[b]Use larger air space (3.5 in.) value shown in Table 4.4.

[c]When rigid roof deck insulation ($C = 0.24$) and sand aggregate plaster replaced with rigid roof deck insulation ($C = 0.36$) and lightweight aggregate plaster, 0.75 in., on metal lath.

Source: Copyright © by the American Society of Heating, Refrigerating and Air Conditioning Engineers, Inc., Atlanta, GA. Reprinted by permission from *1981 Handbook of Fundamentals.*

TABLE 4.11 **Coefficients of Transmission, *U* (Btu/h ft² F), of 45° Pitched Roofs**

*Part A. **Reflective Air Space***

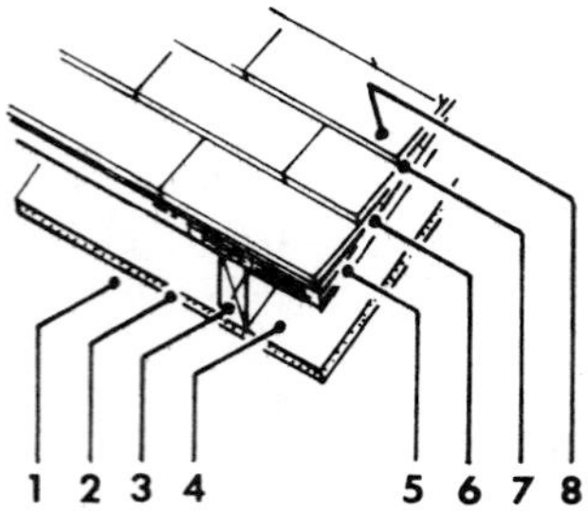

Resistance for Heat Flow Up: Winter Conditions — Between Rafters, R_i	Resistance for Heat Flow Up: Winter Conditions — At Rafters, R_s	Construction	Resistance for Heat Flow Down: Summer Conditions — Between Rafters, R_i	Resistance for Heat Flow Down: Summer Conditions — At Rafters, R_s
0.62	0.62	1. Inside surface (still air)	0.76	0.76
0.45	0.45	2. Gypsum wallboard 0.5 in., foil backed	0.45	0.45
—	4.35	3. Nominal 2-in. × 4-in. ceiling rafter	—	4.35
2.17	—	4. 45° slope reflective air space, 3.5 in. (50 F mean, 30 F° temperature difference) E = 0.05	4.33	—
0.77	0.77	5. Plywood sheathing, 0.625 in.	0.77	0.77
0.06	0.06	6. Permeable felt building membrane	0.06	0.06
0.44	0.44	7. Asphalt shingle roofing	0.44	0.44
0.17	0.17	8. Outside surface (15-mph wind)	0.25[a]	0.25[a]
4.68	6.86	*Total Thermal Resistance, R*	7.06	7.08

Heat Flow Up

$$U_i = \frac{1}{4.68} = 0.213$$

Heat Flow Down

$$U_i = \frac{1}{7.06} = 0.141$$

To adjust U values for the effect of framing, see the example given in Table 4.5. With 10% framing (typical of 2-in. rafters at 16-in. o.c.), these adjusted U_{av} values are, respectively:

$U_{av} = 0.206$ (3% less heat loss) $U_{av} = 0.141$ (unchanged)

*Part B. **Nonreflective Air Space***

Resistance for Heat Flow Up: Winter Conditions — Between Rafters, R_i	Resistance for Heat Flow Up: Winter Conditions — At Rafters, R_s	Construction	Resistance for Heat Flow Down: Summer Conditions — Between Rafters, R_i	Resistance for Heat Flow Down: Summer Conditions — At Rafters, R_s
0.62	0.62	1. Inside surface (still air)	0.76	0.76
0.45	0.45	2. Gypsum wallboard, 0.5 in.	0.45	0.45
—	4.35	3. Nominal 2-in. × 4-in. ceiling rafter	—	4.35
0.96	—	4. 45° slope, nonreflective air space, 3.5 in. (50 F mean, 10 F° temperature difference)	0.90[b]	—

*Part B. **Nonreflective Air Space (Continued)***

Resistance for Heat Flow Up: Winter Conditions — Between Rafters, R_i	Resistance for Heat Flow Up: Winter Conditions — At Rafters, R_s		Resistance for Heat Flow Down: Summer Conditions — Between Rafters, R_i	Resistance for Heat Flow Down: Summer Conditions — At Rafters, R_s
0.77	0.77	5. Plywood sheathing, 0.625 in.	0.77	0.77
0.06	0.06	6. Permeable felt building membrane	0.06	0.06
0.44	0.44	7. Asphalt shingle roofing	0.44	0.44
0.17	0.17	8. Outside surface	0.25[a]	0.25[a]
3.47	6.86	*Total Thermal Resistance, R*	3.63	7.08

Heat Flow Up

$$U_i = \frac{1}{3.47} = 0.287$$

Heat Flow Down

$$U_i = \frac{1}{3.63} = 0.275$$

Adjusted for 10% framing, as above:

$U_{av} = 0.273$ (5% less heat loss) $U_{av} = 0.262$ (5% less heat loss)

[a]Outside wind velocity 7.5 mph.

[b]Air space value of 90 F mean, 10F° temperature difference.

Source: Copyright © by the American Society of Heating, Refrigerating and Air Conditioning Engineers, Inc., Atlanta, GA. Reprinted by permission from *1981 Handbook of Fundamentals.*

TABLE 4.12 **Determination of *U* Value Resulting from Addition of Insulation to the Total Area of Any Given Building Section**

Given Building Section Property[a,b]		*Added R*[c,d,e]						
		R = *4*	R = *6*	R = *8*	R= *12*	R = *16*	R = *20*	R = *24*
U	R	U	U	U	U	U	U	U
1.00	1.00	0.20	0.14	0.11	0.08	0.06	0.05	0.04
0.90	1.11	0.20	0.14	0.11	0.08	0.06	0.05	0.04
0.80	1.25	0.19	0.14	0.11	0.08	0.06	0.05	0.04
0.70	1.43	0.18	0.13	0.11	0.07	0.06	0.05	0.04
0.60	1.67	0.18	0.13	0.10	0.07	0.06	0.05	0.04
0.50	2.00	0.17	0.13	0.10	0.07	0.06	0.05	0.04
0.40	2.50	0.15	0.12	0.10	0.07	0.05	0.04	0.04
0.30	3.33	0.14	0.11	0.09	0.07	0.05	0.04	0.04
0.20	5.00	0.11	0.09	0.08	0.06	0.05	0.04	0.03
0.10	10.00	0.07	0.06	0.06	0.05	0.04	0.03	0.03
0.08	12.50	0.06	0.05	0.05	0.04	0.04	0.03	0.03

[a]For *U*- or *R*-values not shown in the table, interpolate as necessary.

[b]Enter column 1 with *U* or *R* of the design building section.

[c]Under appropriate column heading for added *R*, find *U* value of resulting design section.

[d]If the insulation occupies previously considered air space, an adjustment must be made in the given building section *R* value.

Source: Copyright © by the American Society of Heating, Refrigerating and Air Conditioning Engineers, Inc., Atlanta, GA. Reprinted by permission from *1981 Handbook of Fundamentals.*

TABLE 4.13 **Determination of *U* Value Resulting from Addition of Insulation to Uninsulated Roof Deck**

U Value of Roof without Roof-Deck Insulation[a]	*Conductance C of Roof Deck Insulation*					
	0.12	*0.15*	*0.19*	*0.24*	*0.36*	*0.72*
	U	U	U	U	U	U
0.10	0.05	0.06	0.07	0.07	0.08	0.09
0.15	0.07	0.08	0.08	0.09	0.11	0.12
0.20	0.08	0.09	0.10	0.11	0.13	0.16
0.25	0.08	0.09	0.11	0.12	0.15	0.19
0.30	0.09	0.10	0.12	0.13	0.16	0.21
0.35	0.09	0.11	0.12	0.14	0.18	0.24
0.40	0.09	0.11	0.13	0.15	0.19	0.26
0.50	0.10	0.12	0.14	0.16	0.21	0.30
0.60	0.10	0.12	0.14	0.17	0.23	0.33
0.70	0.10	0.12	0.15	0.18	0.24	0.35

[a]Interpolation or mild extrapolation may be used.

Source: Copyright © by the American Society of Heating, Refrigerating and Air Conditioning Engineers, Inc., Atlanta, GA. Reprinted by permission from *1981 Handbook of Fundamentals.*

TABLE 4.14 **Approximate Thickness (in.) of Insulation for Thermal Resistances, *R* (ft² F h/Btu)**

Thermal Resistance of Insulation	*Batts or Blankets*		*Loose Fill*			*Boards and Slabs*	
	Glass Fiber	*Rock Wool*	*Glass Fiber*	*Rock Wool*	*Cellulosic*	*Polyurethane*	*Cellular Glass*
R-7	2¼–2¾	2	3–4	2–3	2	1	2⅝
R-11	3½–4	3	5	4	3	1¾	4¼
R-13	3⅝	3½	6	4–5	4	2	5
R-19	6–6½	5¼	8–9	6–7	5	3	7¼
R-22	6½	6	10	7–8	6	3½	8⅜
R-30	9½–10½	9	13–14	10–11	8	4¾	11⅜
R-38	12–13	10½	17–18	13–14	10–11	6	14½

Source: Copyright © by the American Society of Heating, Refrigerating and Air Conditioning Engineers, Inc., Atlanta, GA. Reprinted by permission from *Cooling and Heating Load Calculation Manual,* 1979.

TABLE 4.15 **Heat Flow Below Grade**[a]

Part A. ***Basement Walls***

Heat Loss (W/m^2 K)				Depth		Path Length Through Soil		Heat Loss (Btu/h ft^2 F)			
Uninsulated	R = *0.73*	R = *1.47*	R = *2.20*	*m*	*(ft)*	*m*	*(ft)*	*Uninsulated*	R = *4.17*	R = *8.34*	R = *12.51*
2.33	0.86	0.53	0.38	0–0.30	(0–1)	0.20	(0.68)	0.410	0.152	0.093	0.067
1.26	0.66	0.45	0.36	0.30–0.61	(1–2)	0.69	(2.27)	0.222	0.116	0.079	0.059
0.88	0.53	0.38	0.30	0.61–0.91	(2–3)	1.18	(3.88)	0.155	0.094	0.068	0.053
0.67	0.45	0.34	0.27	0.91–1.22	(3–4)	1.68	(5.52)	0.119	0.079	0.060	0.048
0.54	0.39	0.30	0.25	1.22–1.52	(4–5)	2.15	(7.05)	0.096	0.069	0.053	0.044
0.45	0.34	0.27	0.23	1.52–1.83	(5–6)	2.64	(8.65)	0.079	0.060	0.048	0.040
0.39	0.30	0.25	0.21	1.83–2.13	(6–7)	3.13	(10.28)	0.069	0.054	0.044	0.037

Part B. ***Basement Floors***

Heat Loss (W/m^2 K), Width of House in Meters				Depth of Foundation Wall Below Grade	Heat Loss (Btu/h ft^2 F), Width of House in Feet			
6	*7.3*	*8.5*	*9.7*	*m (ft)*	*20*	*24*	*28*	*32*
0.18	0.16	0.15	0.13	1.50 (5)	0.032	0.029	0.026	0.023
0.17	0.15	0.14	0.12	1.83 (6)	0.030	0.027	0.025	0.022
0.16	0.15	0.13	0.12	2.10 (7)	0.029	0.026	0.023	0.021

[a]Only heat losses are assumed, since the interior temperature is normally above the ground temperature.

Source: Copyright © by the American Society of Heating, Refrigerating and Air Conditioning Engineers, Inc., Atlanta, GA. Reprinted by permission from *1981 Handbook of Fundamentals.*

TABLE 4.16 **Heat Flow Coefficients of Slab Floor Construction, F_2: Units are W/K per meter of perimeter (Btu/h F per foot of perimeter)**

Construction	Insulation	Degree Days (65 F base) 7433	5350	2950
8-in. block wall, brick facing	Uninsulated	1.07 (0.62)	1.17 (0.68)	1.24 (0.72)
	R = 0.95 (5.4) from edge to footer	0.83 (0.48)	0.86 (0.50)	0.97 (0.56)
4-in. block wall, brick facing	Uninsulated	1.38 (0.80)	1.45 (0.84)	1.61 (0.93)
	R = 0.95 (5.4) from edge to footer	0.81 (0.47)	0.85 (0.49)	0.93 (0.54)
Metal stud wall, stucco	Uninsulated	1.99 (1.15)	2.07 (1.20)	2.32 (1.34)
	R = 0.95 (5.4) from edge to footer	0.88 (0.51)	0.92 (0.53)	1.00 (0.58)
Poured concrete wall, with duct near perimeter[a]	Uninsulated	3.18 (1.84)	3.67 (2.12)	4.72 (2.73)
	R = 0.95 (5.4) from edge to footer, 0.91 m (3 ft) under floor	1.11 (0.64)	1.24 (0.72)	1.56 (0.90)

[a]Weighted average temperature of the heating duct was assumed at 43°C (110 F) during the heating season [outdoor air temperature less than 18°C (65 F)].

NOTE: To use this table: $q = F_2 P \Delta t$

where q = heat loss through perimeter [W (Btu/h)]
F_2 = heat loss coefficients, from above
P = perimeter of exposed edge of floor slab [m (ft)]
Δt = temperature difference between indoors and outdoors air

Do not assume additional heat losses from the slab to the earth below (as would be calculated from Table 4.15). Heat gains are never assumed to occur, either through perimeter or from the earth below.

Source: Copyright © by the American Society of Heating, Refrigerating and Air Conditioning Engineers, Inc., Atlanta, GA. Reprinted by permission from *1981 Handbook of Fundamentals.*

TABLE 4.17 **Coefficients of Transmission, *U*, of Windows, Sliding Patio Doors, and Skylights: Units are W/m² °C (Btu/h ft² F)**

Part A. ***Exterior[a] Vertical Panels***

	No Storm Sash				Glass Outdoor Storm Sash 25-mm (1-in.) Air Space[b]			
	No Shade		Indoor Shade		No Shade		Indoor Shade	
	Winter*	Summer**	Winter*	Summer**	Winter*	Summer**	Winter*	Summer**
Flat glass[c]								
Single glass, clear	6.2 (1.10)	5.9 (1.04)	4.7 (0.83)	4.6 (0.81)	2.3 (0.40)	2.8 (0.50)	2.5 (0.44)	2.8 (0.49)
Single glass, low-emittance coating[d] e = 0.40	5.2 (0.91)	5.1 (0.90)	3.9 (0.68)	4.0 (0.70)	2.5 (0.44)	3.4 (0.60)	2.1 (0.37)	3.1 (0.55)
Insulating glass; Double[c]								
5-mm (3/16-in.) air space[f]	3.5 (0.62)	3.7 (0.65)	3.0 (0.52)	3.3 (0.58)	2.1 (0.37)	2.3 (0.40)	1.7 (0.29)	2.1 (0.37)
6-mm (1/4-in.) air space[f]	3.3 (0.59)	3.5 (0.61)	2.7 (0.48)	3.1 (0.55)	2.0 (0.35)	2.2 (0.39)	1.6 (0.28)	2.0 (0.36)
13-mm (1/2-in.) air space[g]	2.8 (0.49)	3.2 (0.56)	2.4 (0.42)	3.0 (0.52)	1.8 (0.32)	2.2 (0.39)	1.4 (0.25)	2.1 (0.30)
13-mm (1/2-in.) air space Low-emittance coating[h]								
e = 0.60	2.4 (0.43)	2.9 (0.51)	2.2 (0.38)	2.7 (0.48)	1.7 (0.30)	2.0 (0.36)	1.4 (0.24)	2.0 (0.35)
e = 0.40	2.2 (0.38)	2.6 (0.48)	2.0 (0.36)	2.5 (0.43)	1.5 (0.27)	1.9 (0.39)	1.3 (0.22)	1.8 (0.35)
e = 0.20	1.8 (0.32)	2.2 (0.38)	1.7 (0.30)	2.1 (0.37)	1.4 (0.24)	1.7 (0.30)	1.1 (0.20)	1.6 (0.28)
Insulating glass; triple[e]								
6-mm (1/4-in.) air space[f]	2.2 (0.39)	2.5 (0.44)	1.8 (0.31)	2.3 (0.40)	1.5 (0.27)	1.8 (0.32)	1.3 (0.22)	1.7 (0.30)
13-mm (1/2-in.) air space[i]	1.8 (0.31)	2.2 (0.39)	1.5 (0.26)	2.0 (0.36)	1.3 (0.23)	1.8 (0.31)	1.1 (0.19)	1.7 (0.29)

	Glass Indoor Storm Sash 25-mm (1-in.) Air Space[b]				Acrylic Indoor Storm Sash 25-mm (1-in.) Air Space[b]			
	No Shade		Indoor Shade		No Shade		Indoor Shade	
	Winter*	Summer**	Winter*	Summer**	Winter*	Summer**	Winter*	Summer**
Flat glass[c]								
Single glass, clear	2.8 (0.50)	2.8 (0.50)	2.5 (0.44)	2.8 (0.49)	2.7 (0.48)	2.7 (0.48)	2.4 (0.42)	2.7 (0.47)

(continued)

TABLE 4.17 **Coefficients of Transmission, *U*, of Windows, Sliding Patio Doors, and Skylights: Units are W/m² °C (Btu/h ft² F) (*Continued*)**

	Glass Indoor Storm Sash 25-mm (1-in.) Air Space[b]				Acrylic Indoor Storm Sash 25-mm (1-in.) Air Space[b]			
	No Shade		Indoor Shade		No Shade		Indoor Shade	
	*Winter**	*Summer***	*Winter**	*Summer***	*Winter**	*Summer***	*Winter**	*Summer***
Single glass, low-emittance coating[d] e = 0.40	2.4 (0.42)	2.6 (0.45)	2.0 (0.36)	2.3 (0.40)	2.3 (0.41)	2.6 (0.45)	2.0 (0.35)	2.3 (0.40)
Insulating glass; double[e]								
5-mm (3/16-in.) air space[f]	2.1 (0.37)	2.3 (0.40)	1.7 (0.29)	2.0 (0.36)	2.0 (0.35)	2.2 (0.39)	1.6 (0.28)	2.0 (0.35)
6-mm (1/4-in.) air space[f]	2.0 (0.35)	2.2 (0.39)	1.6 (0.28)	2.0 (0.36)	1.9 (0.34)	2.2 (0.38)	1.5 (0.27)	1.9 (0.34)
13-mm (1/2-in.) air space[g]	1.8 (0.31)	2.2 (0.38)	1.4 (0.25)	2.0 (0.35)	1.7 (0.30)	2.1 (0.37)	1.4 (0.24)	1.9 (0.33)
13-mm (1/2-in.) air space								
Low-emittance coating[h]								
e = 0.60	1.7 (0.29)	2.0 (0.36)	1.4 (0.24)	1.9 (0.33)	1.6 (0.28)	2.0 (0.35)	1.3 (0.23)	1.8 (0.32)
e = 0.40	1.5 (0.27)	1.9 (0.33)	1.3 (0.22)	1.8 (0.31)	1.5 (0.26)	1.8 (0.32)	1.3 (0.22)	1.7 (0.30)
e = 0.20	1.4 (0.25)	1.7 (0.29)	1.1 (0.20)	1.5 (0.26)	1.4 (0.24)	1.6 (0.28)	1.1 (0.20)	1.5 (0.27)
Insulating glass; triple[e]								
6-mm (1/4-in.) air space[f]	1.5 (0.27)	1.8 (0.32)	1.3 (0.22)	1.7 (0.30)	1.5 (0.26)	1.8 (0.31)	1.3 (0.22)	1.7 (0.29)
13-mm (1/2-in.) air space[i]	1.3 (0.23)	1.7 (0.30)	1.1 (0.19)	1.6 (0.28)	1.3 (0.22)	1.7 (0.29)	1.0 (0.18)	1.6 (0.28)

Part B. ***Exterior[a] Horizontal Panels (Skylights)***

Description	*Winter[j]**	*Summer[k]****
Flat glass[g]		
Single glass	7.0 (1.23)	4.7 (0.83)
Insulating glass; double[e]		
5-mm (3/16-in.) air space[f]	4.0 (0.70)	3.2 (0.57)
6-mm (1/4-in.) air space[f]	3.7 (0.65)	3.1 (0.54)
13-mm (1/2-in.) air space[g]	3.4 (0.59)	2.8 (0.49)
13-mm (1/2-in.) air space		
Low-emittance coating[h]		
e = 0.20	2.7 (0.48)	2.0 (0.36)
e = 0.40	3.0 (0.52)	2.4 (0.42)
e = 0.60	3.2 (0.56)	2.6 (0.46)
Plastic domes[l]		
Single walled	6.5 (1.15)	4.5 (0.80)
Double walled	4.0 (0.70)	2.6 (0.46)

Part C. ***Adjustment Factors for Various Window, Sliding Patio Door, and Skylight Types (multiply U in Parts A and B by these factors)***

Product Description	*Single Glass*	*Double Insulating Glass*	*Triple Insulating Glass*	*Storm Sash Applied over Single Glass*	*Storm Sash Applied over Double or Triple Insulating Glass*
All glass	1.00	1.00	1.00	1.00	1.00
Wood frame	0.85–0.95	0.90–1.00	0.95–1.00	0.90–1.00	0.95–1.00
Metal frame	1.10–1.00	1.30–1.20	1.50–1.30	1.40–1.20	1.50–1.30
Thermally improved metal frame	0.90–1.00	0.95–1.15	1.00–1.25	0.90–1.20	0.95–1.25

Part D. ***Exterior Vertical Transparent Acrylic and Polycarbonate Sheeting***

	U at Various Thicknesses				
	3 mm (1/8 in.)	*5 mm (3/16 in.)*	*6 mm (1/4 in.)*	*10 mm (3/8 in.)*	*12 mm (1/2 in.)*
	For Winter Heat Loss: 24 Km/h (15 mph) Wind Velocity				
Single-glazed[m]	6.1 (1.07)	5.7 (1.01)	5.4 (0.96)	5.0 (0.88)	4.6 (0.81)
Reflective[n]	— —	— —	5.0 (0.88)	— —	— —
Double-glazed[m] 6-mm (1/4-in.) air space	3.1 (0.54)	2.9 (0.51)	2.8 (0.49)	— —	— —
Double-glazed[m] 12-mm (1/2-in.) air space	2.7 (0.47)	2.6 (0.45)	2.4 (0.43)	— —	— —
	For Summer Heat Gain: 12 Km/h (7.5 mph) Wind Velocity				
Single glazed[m]	5.4 (0.95)	5.2 (0.91)	5.0 (0.88)	4.7 (0.82)	4.3 (0.76)
Reflective[n]	— —	— —	4.7 (0.83)	— —	— —
Double glazed[m] 6-mm (1/4-in.) air space	3.2 (0.56)	3.0 (0.53)	2.9 (0.51)	— —	— —
Double glazed[m] 12-mm (1/2-in.) air space	2.8 (0.50)	2.7 (0.48)	2.6 (0.46)	— —	— —

Source: Copyright © by the American Society of Heating, Refrigerating and Air Conditioning Engineers, Inc., Atlanta GA. Reprinted by permission from *1981 Handbook of Fundamentals.*

TABLE 4.17 **Coefficients of Transmission, U, of Windows, Sliding Patio Doors, and Skylights: Units are W/m^2 °C (Btu/h ft^2 F) (*Continued*)**

*Part E. **Passive Solar Collecting Components***

Component	*U, Winter*[o] *No Night Insulation*	*With Night Insulation*[p] *R-4*	*R-9*
Direct gain[q]	3.1 (0.55)	1.9 (0.33)	1.6 (0.28)
Trombe wall, 18 in. thick[q]	1.2 (0.22)	0.9 (0.16)	0.7 (0.13)
Water wall[q]	1.9 (0.33)	1.2 (0.22)	1.1 (0.19)
Selective transmitter film, with glass spaced ½ in.[r]	1.6 (0.29) (not used with night insulation)		

[a]See Part C for appropriate adjustments for various windows and sliding patio doors. Window manufacturers should be consulted for specific data.

[b]3-mm (⅛-in.) glass or acrylic as noted, 25 to 100-mm (1 to 4-in.) air space.

[c]Hemispherical emittance of uncoated glass surface = 0.84, coated glass surface as specified.

[d]Coating on second surface (i.e., room side of glass).

[e]Double and triple refer to number of lights of glass.

[f]3-mm (⅛-in.) glass.

[g]6-mm (¼-in.) glass.

[h]Coating on either glass surface 2 or 3 for winter, and for surface 2 for summer U factors.

[i]Window design 6-mm (¼-in.) glass, 3-mm (⅛-in.) glass, and 6-mm (¼-in.) glass.

[j]For heat flow up.

[k]For heat flow down.

[l]Based on area of opening, not total surface area.

[m]Hemispherical emittance = 0.86.

[n]Aluminum metallized polyester film on plastic.

[o]Summer day U can be developed, from the designer's assumptions about the hours for which the insulation is in place.

[p]Insulation assumed in place from 5:30 P.M. to 7:30 A.M., solar time. These U values are thus *average,* rather than instantaneous, values, unlike the others throughout this table.

[q]From Balcomb, et al. (1980).

[r]From Johnson (1981).

*24 km/h (15 mph) outdoor air velocity; −18°C (0 F) outdoor air; 21°C (70 F) inside air temperature, natural convection.

**12 km/h (7.5 mph) outdoor air velocity; 32°C (89 F) outdoor air; 24°C (75 F) inside air, natural convection; solar radiation 782 W/m^2 (248.3 Btu/h ft^2).

The reciprocal of the above U factors is the thermal resistance, R for each type of glazing. If tightly drawn drapes (heavy close weave), closed venetian blinds, or closely fitted roller shades are used internally, the additional R is approximately 0.05 m^2 °C/W (0.29 ft^2 h F/Btu). If miniature louvered solar screens are used in close proximity to the outer fenestration surface, the additional R is approximately 0.04 m^2 °C/W (0.24 ft^2 h F/Btu).

EXAMPLE: Find the winter U factor for uncoated double insulating glass [13-mm (0.5-in.) air space] when (1) external miniature louvered sun screens are used; and (2) tightly woven drapes are added.

SOLUTION: Winter R for 13-mm (0.5-in.) air space double insulating glass = 1/2.8 = 0.36 (2.04); added resistance for the miniature louvered sun screen = 0.04 (0.24); so total R = 0.40 (2.28) and U factor = 2.5 W/m^2 °C (0.44 Btu/h ft^2 F). Adding the tightly woven drape R = 0.05 (0.29), total R = 0.45 (2.57), so U = 2.2 W/m^2 °C (0.39 Btu/h ft^2 F).

TABLE 4.18 **Coefficients of Transmission, *U*, of Doors (Btu/h ft^2 F)**

Part A. ***For Wood Doors***[a]

Door Thickness (in.)[d]	Description	Winter[b] No Storm Door	Winter[b] Wood Storm Door[e]	Winter[b] Metal Storm Door[f]	Summer[c] No Storm Door
1–3/8	Hollow-core flush door	0.47	0.30	0.32	0.45
1–3/8	Solid-core flush door	0.39	0.26	0.28	0.38
1–3/8	Panel door with 7/16-in. panels	0.57	0.33	0.37	0.54
1–3/4	Hollow core flush door with single glazing[g]	0.46	0.29	0.32	0.44
		0.56	0.33	0.36	0.54
1–3/4	Solid-core flush door	0.33	0.28	0.25	0.32
	With single glazing[g]	0.46	0.29	0.32	0.44
	With insulating glass[g]	0.37	0.25	0.27	0.36
1–3/4	Panel door with 7/16-in. panels[h]	0.54	0.32	0.36	0.52
	With single glazing[i]	0.67	0.36	0.41	0.63
	With insulating glass[i]	0.50	0.31	0.34	0.48
1–3/4	Panel door with 1 1/8-in. panels[h]	0.39	0.26	0.28	0.38
	With single glazing[i]	0.61	0.34	0.38	0.58
	With insulating glass[i]	0.44	0.28	0.31	0.42
2–1/4	Solid core flush door	0.27	0.20	0.21	0.26
	With single glazing[g]	0.41	0.27	0.29	0.40
	With insulating glass[g]	0.33	0.23	0.25	0.32

Part B. ***For Steel Doors, 1 3/4 in. Thick***

Core	Steel Door	No Storm Door
Mineral fiber (2 lb/ft^3)	0.59	0.58
Solid urethane core with thermal break	0.40	0.39
Solid polystyrene with thermal break	0.47	0.46

[a]Values for doors are based on nominal 3′8-in. × 6′8-in. door size. Interpolation and moderate extrapolation are permitted for glazing areas and door thicknesses other than those specified.

[b]15 mph outdoor air velocity; 0 F outdoor air; 70 F inside air temperature, natural convection.

[c]7.5 mph outdoor air velocity; 89 F outdoor air; 75 F inside air temperature, natural convection.

[d]Nominal thickness.

[e]Values for wood storm door are approximately 50% glass area.

[f]Values for metal storm door are for any percent of glass area.

[g]17% exposed glass area; insulating glass contains 0.25-in. air space.

[h]55% panel area.

[i]33% glass area; insulating glass contains 0.25-in. air space.

Source: Copyright © by the American Society of Heating, Refrigerating and Air Conditioning Engineers, Inc., Atlanta, GA. Reprinted by permission from *1981 Handbook of Fundamentals.*

4.3 Heat Flow Through Air

Where surfaces of solids are exposed to air, heat transfer takes place both by convection and by radiation. (Evaporation also can occur, occasionally with thermally significant results, as in the case of roof ponds for passive cooling.) Convection is highly dependent on air motion, so wind outdoors must be considered. Also, since warm air rises and cold air falls, vertical surfaces that encourage this kind of airflow will exchange heat faster than the same surfaces placed horizontally, unless the direction of heat flow is *upward* through this horizontal air layer, as was evident in Fig. 4.1.

When air motion along surfaces is minimal, a surprisingly effective insulating layer of air is created. The resistance of a layer of still air along a vertical surface is numerically equal to that of $\frac{1}{2}$-in. plywood, for example. When this air layer is disturbed, its resistance drops quickly; with a 15-mph (6.7 m/s) wind, resistance drops to about one quarter of the still-air value (see Table 4.3). Similar drops in resistance will occur when forced-air grilles are located immediately above or below windows.

These surface layers of air are often evaluated by their conductance, the opposite of resistance, because we often wish to *encourage* heat transfer between solids and air. For example, passive heating and cooling systems are dependent on large building surfaces as heat exchangers. Active (conventional) systems typically concentrate the heat exchange within mechanical equipment.

Surface conductances are termed h_i for interior air layers and h_o for exterior or outside layers; sometimes f_i and f_o are used instead. Like other conductances, they are expressed in Btu per hour per square foot for 1F° temperature difference (in SI units, watts per square meter for 1K°). The variations of resistances and surfaces conductances can be seen in Table 4.3.

Radiation is also highly influenced by surface characteristics; shiny materials are much less able to radiate than are common rough building materials. This characteristic is called *emittance,* the *ratio* of the radiation emitted by a given material to that emitted by a blackbody at the same temperature. The impact of shiny versus ordinary surfaces can be seen in Tables 4.3 and 4.4, which deal with air layers and air spaces; the lower the emittance, the lower the radiative heat exchange. For most materials, emittance is related to absorptance; a highly absorptive (low reflectance) material will usually have high emittance as well.

The combination of dead air and reflective surfaces produces some of our most effective insulating products, especially when they are made of lightweight materials of low conductivity. Glass fiber, cellular glass, expanded styrenes (foamed plastics), and mineral fibers all exhibit the characteristics of enclosing vast numbers of dead-air spaces per unit volume. When bonded to reflective films and properly installed (facing an air space), high resistance to heat flow is achieved. Resistances to heat flow can be compared between various building materials, including insulation products (Table 4.2) and air films and spaces (Tables 4.3 and 4.4).

4.4 Heat Flow Through the Building Envelope

When the process of heat flow is understood, the calculations can begin. Initially, the *hourly* heat loss or heat gain is calculated, since these rates can later be used either to establish equipment sizing (design conditions) or energy consumption (normal conditions). To calculate hourly heat flow through a building's envelope, these factors are necessary:

1. The *rate* at which heat flows through the various assemblies of materials that make up the envelope.
2. The *area* of each of these assemblies.
3. The *temperature difference* between inside and outside for the hour being calculated.

The *areas* of walls, roofs, windows, and floors of various kinds are determined from preliminary architectural design drawings; final versions of these drawings should be subject to changes suggested by these calculations. The *temperature differences* for design conditions are listed in Appendix A; for energy consumption, see Section 5.5. The *rates* at which heat flows are discussed in this section.

The variety of terms used so far to express heat flow is potentially bewildering. These terms are but parts of a larger picture; what is needed is *one* final, overall expression of the steady-state rate at which heat flows through architectural skin elements (walls, roofs, floors, etc.) This is provided by the *U value,* where *U* is thermal transmittance, again expressed in the familiar terms of Btu/h ft^2 F (W/m^2 K). Since *U* values are both common and rather complex, considerable space is devoted here to listing them for familiar constructions of walls, floors, roofs, windows, and doors. See Tables 4.5 to 4.18. *U* values are also used in some design criteria, presented in Section 5.2.

These *U* values are also readily calculated for a particular (floor, roof, wall, etc.) by finding the resistances of each of its materials, its air layers, and its internal air spaces, then adding all these resistances (obtaining ΣR) and finding the reciprocal:

$$U = \frac{1}{\Sigma R}$$

This procedure is illustrated in Fig. 4.3, and examples of typical constructions are shown in Tables 4.5 through 4.11. The coefficients in these tables are expressed in Btu/h ft^2 F (i.e., per degree Fahrenheit difference in temperature between the air on the two sides) and are based on an outside wind velocity of 15 mph, except as noted in Table 4.11.

Tables 4.12 and 4.13 show the effects of adding insulation to constructions; Table 4.14 shows *R* value equivalents of common insulation types. Tables 4.15 and 4.16 deal with the special case of floor slabs on grade; Tables 4.17 and 4.18 show *U* values for windows and doors.

The *U* value must be determined for each element (or construction type) of the building envelope; then the hourly conductive heat loss (or gain) through a building's skin can be calculated as follows:

$$q = \Sigma(U \cdot A)\, \Delta t$$

where *q* is the total heat exchange conducted through building's skin (Btu/h or W)

U and *A* (*U* values and areas) are specific to each skin element

Σ indicates that all *UA* for these elements are to be added together

Δt is the temperature difference between indoors and outdoors. For *design* conditions, use outdoor temperature from Appendix A. For *average* conditions, use average temperatures (such as from Appendix B, Part 3, or local climatological data).

One important use of this quantity, *q*, will be shown in Section 4.6.

Another use of this overall thermal transmittance *q* is in determining temperatures at various points within walls, roofs, or floors. This is particularly useful information for surfaces such as windows on cold days, which have a strong influence on human comfort. It is also useful for avoiding moisture problems as discussed in the next section.

The gradual change of temperature through a wall, roof, or floor from inside to outside is known as its *thermal gradient;* it can be charted by proportioning the amount of temperature change to the amount of thermal resistance at any point. This procedure is illustrated in Fig. 4.4.

In the special case of basement walls, the Δt is affected by the temperature of the earth, which changes only slowly. For these special cases, a combination of U and Δt is appropriate; these values are presented in Table 4.15. For floor slabs

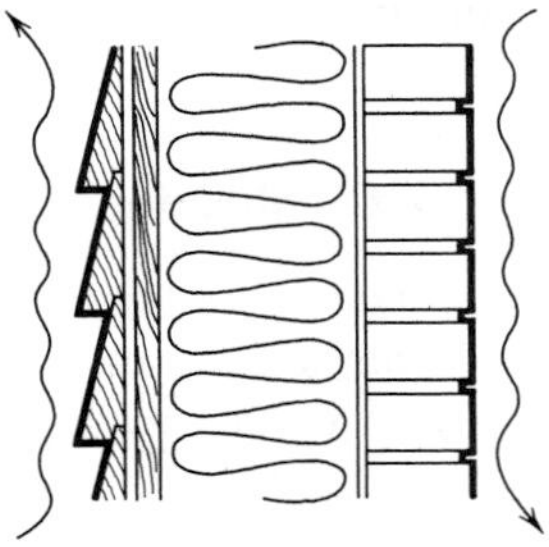

Component	R	*Reference Table*
Inside air layer	0.68	Table 4.3
Common brick	0.20	Table 4.2
Nominal 6-in. batt fiberglass	19.00	Table 4.2
½-in. Plywood	0.62	Table 4.2
1-in. Wood siding	0.79	Table 4.2
Outside air layer	0.17	Table 4.3
ΣR	21.46	

$$U = \frac{1}{\Sigma R} = \frac{1}{21.46} = 0.046$$

Fig. 4.3 *Procedure for determining* U *values (overall thermal transmittance).*

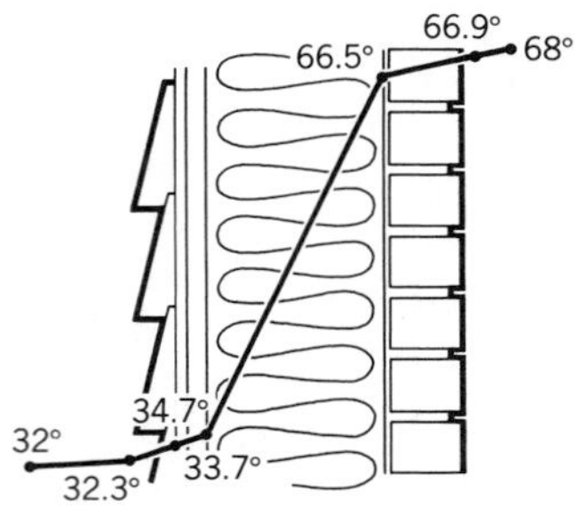

R	Component	ΣR *from Interior*	*Temperature Drop from Interior (F°)*	*Temperature at Outer Edge of Component (F)*
0.68	Inside air layer	0.68	0.68/21.46 × 36 = 1.1	66.9
0.20	Common brick	0.88	0.88/21.46 × 36 = 1.5	66.5
19.00	Nominal 6-in. insulation	19.88	19.88/21.46 × 36 = 33.3	34.7
0.62	½-in. Plywood	20.5	20.5/21.46 × 36 = 34.3	33.7
0.79	1-in. Wood siding	21.29	21.29/21.46 × 36 = 35.7	32.3
0.17	Outside air	21.46		32

Fig. 4.4 *Procedure for calculating the thermal gradient through construction. Assume that outside and inside temperatures are 32 and 68F (0 and 20°C):* $\Delta t = 68 - 32 = 36F°$.

on grade, the only significant heat flow is through the perimeter in winter; see Table 4.16.

4.5 Moisture and Infiltration

Buildings, like our bodies, exchange moisture and air with the environment, as well as exchanging heat. Although most of this moisture exchange occurs during the exchange of fresh air, some moisture is exchanged through a building's skin. This can cause problems, either in hot, humid climates or very cold ones.

In hot-humid conditions, cool inside surfaces are often encountered—for example, the ceiling directly below a roof pond used for passive cooling. As hot and humid air contacts such a surface, *condensation* can occur; the moisture vapor in the air condenses to form visible droplets of water on the ceiling. The result can be mildly annoying as water drips on the head, or serious as water stains occur, and, eventually, mold grows on surfaces. For these reasons, such large cooled surfaces are often avoided in hot-humid climates. Even in less humid but hot conditions, condensation can occur within the rock beds that store heat for passively solar heated or cooled buildings, creating problems of odor and bacterial growth.

In cold climates, cold interior surfaces also occur, especially at windows. Although the air indoors may not be particularly humid (40–50% rh is common), it contains enough moisture to permit condensation on cold surfaces. (For details, see the discussion on psychrometry in Section 4.7.) Again, mild annoyance or more serious damage can result. A much less visible moisture threat occurs *within* walls, ceilings, or floors. Almost all common building materials, including gypsum board, concrete, clay masonry, and wood, are easily permeated by moisture. Most surface finishes are also permeable. In cold climates, the air outside contains relatively little moisture, even though the rh may be high. By contrast, inside air contains much more moisture per unit of volume, despite its probably lower rh. (This is evident from the psychrometric chart, Fig. 4.6, below.) The result is a flow of vapor from high vapor pressure to low vapor pressure (typically warm to cold).

The problem with such a flow occurs when the temperature within the wall (floor, etc.) drops low enough for this vapor to condense. Insulation can then become wet and thereby less effective, since water conducts heat far better than the air pockets it has filled. If wet insulation compacts, these air pockets are permanently lost. Worse yet, moisture damage can occur, such as dry rot in wood structural members.

The usual remedy for such a potential problem is to install a *vapor barrier* within the building envelope; since very low permeability is desired, these barriers are commonly of plastic film installed with as few holes as possible. Since the moister air on the warm side is the source of the problem, this barrier needs to be installed as *close to the warm side* as possible—typically, just behind interior surfaces (gypsum board, wood floors, etc.). A less desirable approach is to use vinyl wallpaper or vapor barrier paints on interior surfaces. These cannot give the around-the-corners, wrapped protection that is obtained with properly installed plastic films. This disadvantage is shared by aluminum-foil-faced insulation; it is very effective thermally, but less effective as a vapor barrier. (Also, the aluminum foil must face an air space if it is to be effective.)

A substantial benefit of plastic films is that they reduce airflow through construction. Outdoor air is always infiltrating a building, gradually replacing the indoor air. This unintentional source of fresh air becomes a problem when temperatures outside are very different from those inside, especially when strong winds force outdoor air indoors fast enough to produce noticeably cold (or hot) drafts. Some fresh air is always desirable in buildings, but so is user control of how and where it is admitted. Therefore, the moisture-tight and infiltration-tight characteristics of plastic film vapor barriers are usually beneficial. When good vapor barriers are installed, the smaller air-change values that accompany "tight" construction may be assumed in the following calculations of heat flow due to infiltration and ventilation.

4.6 Heat Flow by Ventilation

Two common terms for outdoor air that enters a building are:

Infiltration. The "accidental" influx of outdoor air due to air leakage through the building skin.

Ventilation. The deliberate, designed introduction of outdoor air.

Winter heat loss (and summer heat gain in closed, cooled buildings) occurs when "fresh" outdoor air enters the building to replace "stale" indoor air. This heat exchange, analogous to human lung heat losses or gains, must also be calculated when sizing heating or cooling equipment, or when estimating energy use per season.

It is important to provide at least a minimum amount of fresh air indoors, both for comfort and for health. Odors and a sense of staleness can be uncomfortable, and dangerous buildups of radon gas and other pollutants such as formaldehyde can be produced within buildings. These pollutants are easily removed with air changes through rooms. Table 4.19 lists recommended design outdoor airflow rates.

The calculation for the heat lost (or gained) by the introduction of outdoor air into spaces is:

$$q_v = (V)(1.08)(\Delta t)$$

where q_v = sensible heat exchange due to ventilation (Btu/h)

V = volume flow rate, in cubic feet per minute (cfm) of outdoor air introduced

Δt = temperature difference between outdoor and indoor air (F°)

1.08 = a constant derived from the density of air at 0.075 lb/ft^3 under "average" conditions, multiplied by the specific heat of air (heat required to raise 1 lb of air 1 F°), which is 0.24 Btu/lb F, and by 60 min/h. The units of this constant are Btu min/ft^3 F h.

In SI units, this equation becomes

$$q_v = (V)(1200)(\Delta t)$$

where q_v is in watts, V is in liters per second, 1200 is the approximate volumetric specific heat capacity of air (J sec/m^3 °C h), and Δt is in degrees Celsius (or K).

The next task is to determine how much air, V (in cfm or L/s), should be assumed for this calculation. Four basic approaches that represent various degrees of design control over ventilation are considered here.

TABLE 4.19 **Recommended Outdoor Air Requirements for Ventilation**

NOTE: A blank (—) indicates absence of data only; it should not be assumed that smoking (or nonsmoking) is sanctioned in an area so designated.

Part A. ***Commercial Facilities (offices, stores, shops, hotels, sports facilities, etc.)***

	Estimated Occupancy (persons per 1000 ft² or 100 m² floor area) Use only when design occupancy is not known	Outdoor Air Requirements per Person or Area				
		Smoking	Non-smoking	Smoking	Non-smoking	Comments
Dry cleaners and laundries		*cfm/person*		*L/s person*		
Commercial	10	—	15	—	7.75	
Storage/pickup areas	30	35	10	17.5	5	
Coin-operated laundries	20	35	15	17.5	7.5	Dry cleaning processes may require more air.
Coin-operated dry cleaning	20	—	15	—	7.5	
Food and Beverage Services						
Dining rooms	70	35	7	17.5	3.5	
Kitchens	20	—	10	—	5	
Cafeterias, fast-food facilities	100	35	7	17.5	3.5	
Bars and cocktail lounges	100	50	10	25	5	
Garages, auto repair shops, service stations		*cfm/ft² floor*		*L/s m² floor*		
Parking garages (enclosed)	—	1.5	1.5	7.5	7.5	Distribution must consider worker location and concentration of running engines; stands where engines are run must incorporate systems for positive engine exhaust withdrawal.
Auto repair workrooms (general)	—	1.5	1.5	7.5	7.5	
Hotels, Motels, Resorts, Dormitories, and Correctional Facilities		*cfm/room*		*L/s room*		See also Food and Beverage Services, Merchandising, and Barber and beauty shops, Garages.
Bedrooms (single, double)	5	30	15	15	7.5	Independent of room size.
Living rooms (suites)	20	50	25	25	12.5	(Independent of room size.)
Baths, toilets (attached to bedrooms)		50	50	25	25	Independent of room size; installed capacity for intermittent use.

		cfm/person		*L/s person*		
Lobbies	30	15	5	7.5	2.5	
Conference rooms (small)	50	35	7	17.5	3.5	
Assembly rooms (large)	120	35	7	17.5	3.5	
Gambling casinos	120	35	7	17.5	3.5	
Offices						
Office space	7	20	5	10	2.5	
Meeting and waiting spaces	60	35	7	17.5	3.5	
Public spaces		*cfm/ft² floor*		*L/s m² floor*		
Corridors and utility rooms		0.02	0.02	0.10	0.10	
		cfm/stall or urinal		*L/s stall or urinal*		
Public restrooms	100	75	—	37.5	—	
		cfm/locker		*L/s locker*		
Locker and dressing rooms	50	35	15	17.5	7.5	
Retail stores		*cfm/person*		*L/s person*		
Sales floor and showrooms						
Basement and street floors	30	25	5	12.5	2.5	
Upper floors	20	25	5	12.5	2.5	
Storage areas (serving sales and storerooms)	15	25	5	12.5	2.5	
Dressing rooms	—	25	5	12.5	3.5	
Malls and arcades	20	10	5	5	2.5	
Shipping and receiving areas	10	10	5	5	2.5	
Warehouses	5	10	5	5	2.5	
Elevators	—	—	15	—	7.5	
Smoking rooms	70	50	—	25	—	
Specialty shops						
Barber and beauty shops	25	35	20	17.5	10	
Reducing salons, health spas (exercise rooms)	20	—	15	—	7.5	
Florists	10	25	5	12.5	2.5	Ventilation to optimize plant growth may dictate requirements.
Greenhouses	1	—	5	—	2.5	

(continued)

THERMAL CONTROL

TABLE 4.19 **Recommended Outdoor Air Requirements for Ventilation** (*Continued*)

	Estimated Occupancy (persons per 1000 ft² or 100 m² floor area) Use only when design occupancy is not known	*Outdoor Air Requirements per Person or Area*				
		Smoking	*Non-smoking*	*Smoking*	*Non-smoking*	*Comments*
Speciality shops (continued)		*cfm/person*		*L/s person*		
Show repair shops (combined workrooms/trade areas)	10	15	10	7.5	5	
		cfm/ft² floor		*L/s m² floor*		
Pet shops	—	1	1	5	5	
Sports and amusement facilities		*cfm/person*		*L/s person*		
Ballrooms and discos	100	35	7	17.5	3.5	
Bowling alleys (seating area)	70	35	7	17.5	3.5	
Playing floors (e.g., gymnasiums, ice arenas)	30	—	20	—	10	When internal combustion engines are operated for maintenance of playing surfaces, increased ventilation rates will be required.
Spectator areas	150	35	7	17.5	3.5	
Game rooms (e.g., cards and billiards rooms)	70	35	7	17.5	3.5	
Swimming Pools		*cfm/ft² area*		*L/s m² area*		
Pool and deck areas	—	—	0.5	—	2.5	Higher values may be required for humidity control.
		cfm/person		*L/s person*		
Spectator areas	70	35	7	17.5	3.5	
Theaters						
Ticket booths	—	20	5	10	2.5	
Lobbies, foyers, and lounges; auditoriums in motion picture theaters, lecture, concert, and opera halls	150	35	7	17.5	3.5	

Stages, TV, and movie studios	70	—	10	—	5	Special ventilation will be needed to eliminate special stage effects (e.g., dry ice vapors, mists.)
Transportation						
Waiting rooms, ticket and baggage areas, corridors and gate areas, platforms, concourses	150	35	7	17.5	3.5	Ventilation within vehicles will require special consideration.
Workrooms						
Meat processing rooms	10	—	5	—	2.5	Spaces maintained at low temperatures (−10 to +50 F, or −23 to +10°C) are not covered by these requirements unless the occupancy is continuous. Ventiliation from adjoining spaces is permissible. When the occupancy is intermittent, infiltration will normally exceed the ventilation requirement.
Pharmacists' workroom	20	—	7	—	3.5	
Bank vaults	10	—	5	—	2.5	
Photo studios						
Camera room, stages	10	—	5	—	2.5	
Darkrooms	10	—	20	—	10	
		cfm/ft² floor		*L/s m² floor*		
Duplicating and printing rooms		—	0.5	—	2.5	Installed equipment must incorporate positive exhaust and control (as required) of undesirable contaminants (toxic or otherwise).

(continued)

TABLE 4.19 **Recommended Outdoor Air Requirements for Ventilation** (*Continued*)

Part B. ***Institutional Facilities***

	Estimated Occupancy (persons per 1000 ft^2 or 100 m^2 floor area): Use only when design occupancy is not known	Outdoor Air Requirements				Comments
		Smoking	Non-smoking	Smoking	Non-smoking	
Educational facilities		*cfm/person*		*L/s person*		
Classrooms	50	25	5	12.5	2.5	Special contaminant control systems may be required for processes or functions including laboratory animal occupancy.
Laboratories	30	—	10	—	5	
Training shops	30	35	7	17.5	3.5	
Music rooms	50	35	7	17.5	3.5	
Libraries	20	—	5	—	2.5	
Hospital, nursing and convalescent homes						Special requirements or codes and pressure relationships may determine minimum ventilation rates and filter efficiency.
		cfm/bed		*L/s bed*		
Patient rooms	10	35	7	17.5	3.5	
		cfm/person		*L/s person*		
Medical procedure areas	10	35	7	17.5	3.5	
Operating rooms, delivery rooms	20	—	40	—	20	Procedures generating contaminants may require higher rates.
Recovery and intensive care rooms	20	—	15	—	7.5	
Autopsy rooms	20	—	100	—	50	Air shall not be recirculated into other spaces.
Physical therapy areas	20	—	15	—	7.5	

Part C. **Residential Facilities (private dwelling places, single or multiple, low or high rise)**

	Outdoor Air Requirements		Comments
			Operable windows or mechanical ventilation systems shall be provided for use when occupancy is greater than usual conditions or when unusual contaminant levels are generated within the space.
	cfm/room	*L/s room*	
General living areas	10	5	Ventilation rate is independent of room size (General living areas through Baths, toilets)
Bedrooms	10	5	
All other rooms	10	5	
Kitchens	100	50	Installed capacity for intermittent use. (Kitchens; Baths, toilets)
Baths, toilets	50	25	
	cfm/car space	*L/s car space*	
Garages (separate for each dwelling unit)	*100*	*50*	
	cfm/ft² floor	*L/s m² floor*	
Garages (common for several units)	1.5	7.5	

Part D. **Industrial Facilities**

Activity Level[a]	Outdoor Air Requirements: Smoking	Outdoor Air Requirements: Non-smoking	Outdoor Air Requirements: Smoking	Outdoor Air Requirements: Non-smoking	Comments
					Occupational safety laws in various states usually regulate process ventilation requirements. The following list gives the requirements for the occupants only, assuming that the ventilated air is of acceptable quality. Air of this quality may be included as part of the process ventilation.
	cfm/person		*L/s person*		
High activity (2.5 met)	35	20	17.5	10	Mining, foundry, etc.
Medium activity (2.0 met)	35	10	17.5	5	Automotive repair, assembly line, etc.
Low activity (1.5 met)	35	7	17.5	3.5	Laboratory work, light assembly, etc.

[a]1.0 met = sedentary activity level = 18.4 Btu/hr ft^2 body surface (58.2 W/m^2).

Source: Copyright © by the American Society of Heating, Refrigerating and Air Conditioning Engineers, Inc., Atlanta, GA. Reprinted by permission from *1981 Handbook of Fundamentals.*

TABLE 4.20 **Estimated Overall Infiltration Rates for Small Buildings**

*Part A. **Construction Types***

Construction Type	*Description*
Tight	New buildings where there is close supervision of workmanship and special precautions are taken to prevent infiltration. Descriptions for tight windows and doors are given in Table 4.21.
Medium	Building is constructed using conventional construction procedures. Medium-fitting windows and doors are described in Table 4.21.
Loose	Buildings constructed with poor workmanship or older buildings where joints have separated. Loose windows and doors are described in Table 4.21.

*Part B. **Design Infiltration Rate (ACH) for Winter; Heating; Wind Speed = 15 mph***

Type of Construction	*Winter Outdoor Design Temperature (F)*									
	50	*40*	*30*	*20*	*10*	*0*	*−10*	*−20*	*−30*	*−40*
Tight	0.4	0.5	0.6	0.6	0.7	0.8	0.8	0.9	0.9	1.0
Medium	0.6	0.7	0.8	0.9	1.0	1.1	1.2	1.2	1.3	1.4
Loose	0.8	0.9	1.0	1.2	1.3	1.4	1.5	1.6	1.8	1.9

Part C. **Design Infiltration Rate (ACH) for Summer; Cooling; Wind Speed = 7.5 mph**

Type of Construction	Summer Outdoor Design Temperature (F)					
	85	90	95	100	105	110
Tight	0.3	0.3	0.3	0.4	0.4	0.4
Medium	0.4	0.4	0.5	0.5	0.5	0.6
Loose	0.4	0.5	0.6	0.6	0.7	0.8

Part D. **Infiltration per Square Foot of Floor Area**

Ceiling Height (ft)	Air Changes per Hour																	
	0.3	0.4	0.5	0.6	0.7	0.8	0.9	1.0	1.1	1.2	1.3	1.4	1.5	1.6	1.7	1.8	1.9	2.0
	cfm/ft²																	
7.5	0.04	0.05	0.06	0.08	0.09	0.10	0.11	0.13	0.14	0.15	0.16	0.18	0.19	0.20	0.21	0.23	0.24	0.25
8	0.04	0.05	0.07	0.08	0.09	0.11	0.12	0.13	0.15	0.16	0.17	0.19	0.20	0.21	0.23	0.24	0.26	0.27
8.5	0.04	0.06	0.07	0.09	0.10	0.11	0.13	0.14	0.16	0.17	0.18	0.20	0.21	0.23	0.24	0.26	0.27	0.28
9	0.05	0.06	0.08	0.09	0.11	0.12	0.14	0.15	0.17	0.18	0.20	0.21	0.23	0.24	0.26	0.27	0.29	0.30
	Btu/h ft² F																	
7.5	0.04	0.05	0.07	0.08	0.09	0.11	0.12	0.14	0.15	0.16	0.18	0.20	0.20	0.22	0.23	0.24	0.26	0.27
8	0.04	0.06	0.07	0.09	0.10	0.12	0.13	0.14	0.16	0.17	0.19	0.22	0.22	0.23	0.24	0.26	0.27	0.29
8.5	0.05	0.06	0.08	0.09	0.11	0.12	0.14	0.15	0.17	0.18	0.20	0.23	0.23	0.24	0.26	0.28	0.29	0.30
9	0.05	0.06	0.08	0.10	0.11	0.13	0.15	0.16	0.18	0.19	0.21	0.24	0.24	0.26	0.28	0.29	0.31	0.32

Source: Copyright © American Society of Heating, Refrigerating and Air Conditioning Engineers, Inc., Atlanta, GA. Reprinted by permission from *Cooling and Heating Load Calculation Manual,* 1979.

TABLE 4.21 **Approximate Infiltration Through Doors and Windows of Small Buildings**

Part A. ***Converting Wind Speed to Velocity Head Factor***

NOTE: Typical design assumptions:
Winter wind V_w = 15 mph = velocity head factor of 0.105
Summer wind V_w = 7.5 mph = velocity head factor of 0.028

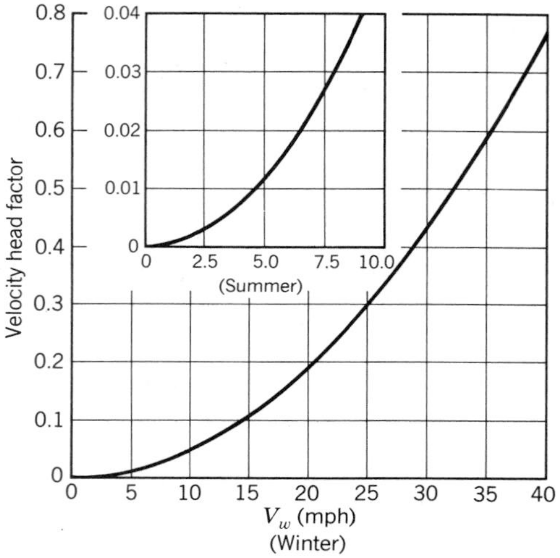

Part B. ***Infiltration Rates for Velocity Head Factors***

NOTE: Enter this graph with velocity head factor (from Part A) to find infiltration rate in cfm/ft of crack (using values of *k* found in Part C or D).

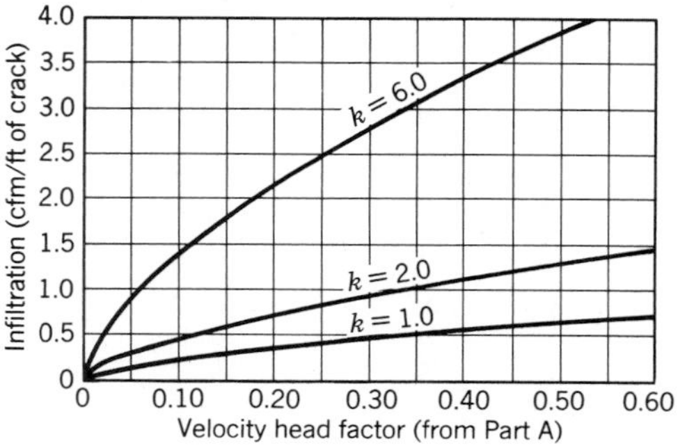

Part C. ***Classifications of Windows for Infiltration***

Window Fit	*Wood Double-Hung (Locked)*	*Other Types*
Tight, $k = 1.0$	Weather-stripped; average gap (1/64-in. crack)	Wood casement and awning windows; weather-stripped.
		Metal casement windows; weather-stripped.
Average, $k = 2.0$	Nonweather-stripped; average gap (1/64-in. crack) or weather-stripped; large gap (3/32-in. crack)	All types of vertical and horizontal sliding windows; weather-stripped. If average gap (1/64-in. crack), this could be tight-fitting window.
		Metal casement windows; non-weather-stripped. If large gap (3/32-in. crack), this could be a loose-fitting window.
Loose, $k = 6.0$	Non-weather-stripped; large gap (3/32-in. crack)	Vertical and horizontal sliding windows; non-weather-stripped.

Part D. ***Classification of Residential-type Doors for Infiltration***

Door Fit	*Comments*
Tight, $k = 1.0$	Very small perimeter gap and perfect fit weather-stripping—often characteristic of new doors
Average, $k = 2.0$	Small perimeter gap having stop trim fitting properly around door; weather-stripped
Loose, $k = 6.0$	Large perimeter gap having poor fitting stop trim; weather-stripped or Small perimeter gap; no weather-stripping

Source: Copyright © American Society of Heating, Refrigerating and Air Conditioning Engineers, Inc., Atlanta, GA. Reprinted by permission from *Cooling and Heating Load Calculation Manual,* 1979.

(a) Natural Infiltration by Approximation. For residences and small commercial buildings without significant internal heat sources, many building codes do not require mechanical or forced ventilation. In some codes, if the openable (or operable) window area is 5% of the floor area of each room in these buildings, adequate natural ventilation is assumed to be achievable. The two methods that follow are typically used with design condition calculations; that is, to find worst-hour loss (or gain) when windows and doors are closed.

There are two approximation methods available: the *air-change* method and the *crack method;* the air-change method is very quick (but tends to overestimate); a glance at Table 4.20 is sufficient to find an assumed number of air changes per hour (ACH) for a space; then,

$$V = \frac{(\text{ACH})(\text{room volume})}{60\ \text{min/h}}$$

where V is in cfm.

The crack method takes longer; it assumes that data on window and door construction and wind velocities are known. This method assumes that doors and openable windows represent all the cracks by which outdoor air infiltrates a closed room under worst-hour conditions. Table 4.21 shows how many cfm per foot of crack should be assumed, *on the windward exposure(s) only,* to arrive at a total cfm. To convert to SI units, use:

$$\text{cfm} \times 0.4719 = \text{L/s}$$

Often, designers aim at removing excess indoor heat by deliberately admitting cooler outdoor air. This procedure is usually applied in passive cooling situations utilizing an "open" building, where outdoor air "as is" is desired indoors.

(b) Cooling by Natural Ventilation. It is important to remember that natural ventilation cooling—whether by windows or by stack effect—will only work when the *outside is cooler than the inside*. More details of natural cooling calculations are found in Chapter 5. For window ventilation, the quantity of outdoor air admitted is called Q (analogous to V in the preceding paragraphs):

$$Q = C_v A v$$

where Q = volume flow rate of air (cfm)

A = area of openable windows on inlet side or sides (ft^2). *Note:* Outlet side(s) should have at least this much openable area as well.

C_v = effectiveness factor that adjusts for different wind orientations (dimensionless): 0.5–0.6 for winds perpendicular to window openings; 0.25–0.35 for wind diagonal to window openings

v = velocity of wind in *feet per minute*.
feet per minute = (miles per hour)(88)
feet per minute = (meters per second)(196.86)

In SI units,

$$Q = 1000\ C_v A v$$

where Q is in liters per second, A is in square meters, and v is in meters per second.

(c) Natural Ventilation Through Stack Effect. Another way of deliberately ventilating and cooling buildings is to take advantage of the lighter weight of hot air, which allows it to rise. When there is a difference in height between inlet openings low in walls (or in floors) and outlets through roofs, *and* when outdoor air is cooler than indoor air, natural ventilation will occur through the *stack effect* of warm air rising and leaving through the higher openings. For stack effect,

$$Q = CA\sqrt{\frac{h(t_i - t_o)}{t_i}}$$

where Q = airflow (cfm)

C = constant of proportionality = 313. *Note:* This assumes a value of 65% of the maximum theoretical flow, due to limited effectiveness of actual openings. With less favorable conditions (due to indirect paths from openings to the stack, etc.), the effectiveness drops to 50%, and C = 240.

A = area of cross section through stack, or outlets (ft^2). *Note:* Inlet area must be at least equal to this amount.

h = height difference, between inlets and outlets (ft)

t_i = (higher) temperature inside, within the height h, (F)

t_o = (lower) temperature outside (F)

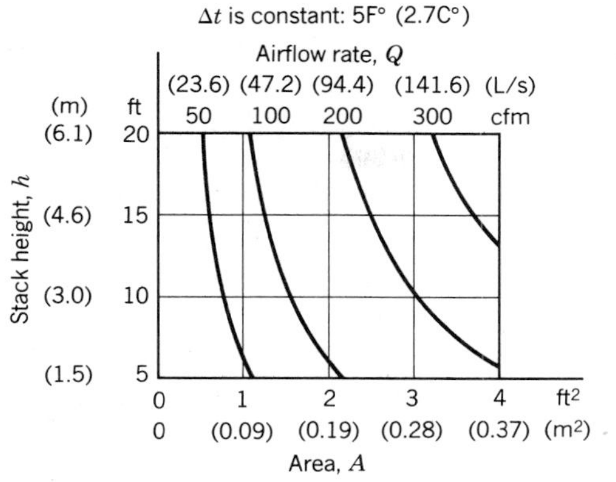

Fig. 4.5 Approximate rate of natural ventilation due to stack effect. Indoor temperatures are near the comfort zone, and Δt *is assumed to be 5F° (2.7C°).*

In SI units, Q is in cubic meters per second, $C = 91.1$ at 65% effectiveness (70 at 50% effectiveness), A is in square meters, h is in meters, and t_i and t_o are in degrees kelvin.

Clearly, the taller the stack and the greater the temperature difference, the more air will flow. Increased area of openings is even more influential. Figure 4.5 presents solutions for some small stack effect combinations in a convenient graph.

When *both* wind and stack effect are used together, their combined flow rate effect is *less* than the simple sum of their individual flow rates:

$$Q_{ws} = \sqrt{(Q_w^2 + Q_s^2)}$$

where Q_{ws} = flow rate of the combined effects of wind and stack ventilation
Q_w = flow rate from wind effect
Q_s = flow rate from stack effect

(d) Forced Ventilation. Fans can be used to forcibly introduce the desired amount of outdoor air directly into spaces. Fan manufacturers list their capacity in cubic feet per hour (cfh) (or in cfm). This outdoor air can be blown into (or sucked out of) spaces, or it can be mixed with air being recirculated so that the different temperature of outdoor air is less noticeable. An important energy conservation opportunity arises with forced ventilation: a heat exchanger may be used. Outgoing and incoming airstreams can be kept as separate airstreams, but heat can be transferred from one stream to the other. Thus, incoming very cold outdoor air can be given the heat, but not the pollutants, of outgoing warm indoor air; this flow of heat can be reversed in the hot, humid summer. (Our bodies do the same thing with arteries and veins, where arms and legs intersect the trunk. That is, in cold weather, cooler blood from the extremities is warmed before it enters the thermally protected core of the body.) The equipment involved in heat exchange is discussed in Sections 6.6 and 6.8.

To approximate the size of the fan, where Q is the desired flow rate (from Table 4.19):

$$Q = \text{(cfm outdoor air person) (number of people)};$$

or

$$Q = (\text{cfm/ft}^2 \text{ floor area}) (\text{ft}^2 \text{ floor area}).$$

In SI units, substitute liters per second for cfm, and square meters for square feet.

4.7 Psychrometry

Before going into the next chapters' details of hourly heat gain and seasonal energy use, it is necessary to better understand the interaction of heat, moisture, and air. Moisture is a relatively minor component in winter comfort and usually can be safely ignored in calculating heat losses. By contrast, moisture is highly influential on heat gain, both for comfort and in calculations.

Air, moisture, and heat interactions are rather complex; as air temperature rises, its capacity to hold moisture rises also; and warmer air becomes less dense. These combined interactions are described by *psychrometry,* the study of moist air. Fortunately, these interactions can be combined in a single chart (Figs. 4.6*a* and *b*).

The basis of the psychrometric chart has been developed in Sections 2.2, 2.3, and 2.4, so that relative humidity (rh) and dry-bulb (DB) and wet-bulb (WB) temperatures are familiar terms. They are combined in the basic chart of Fig. 4.7, where the term ''saturation line,'' at 100% rh, is also introduced. This is also called the ''dew point'' (DP) because dew will form (water vapor will condense) when saturated air touches any surface at or below the air's dew point temperature. This condensation is sometimes undesirable (as within walls or on ceiling, air duct, or glass surfaces), but often desirable, as on air conditioner coils,

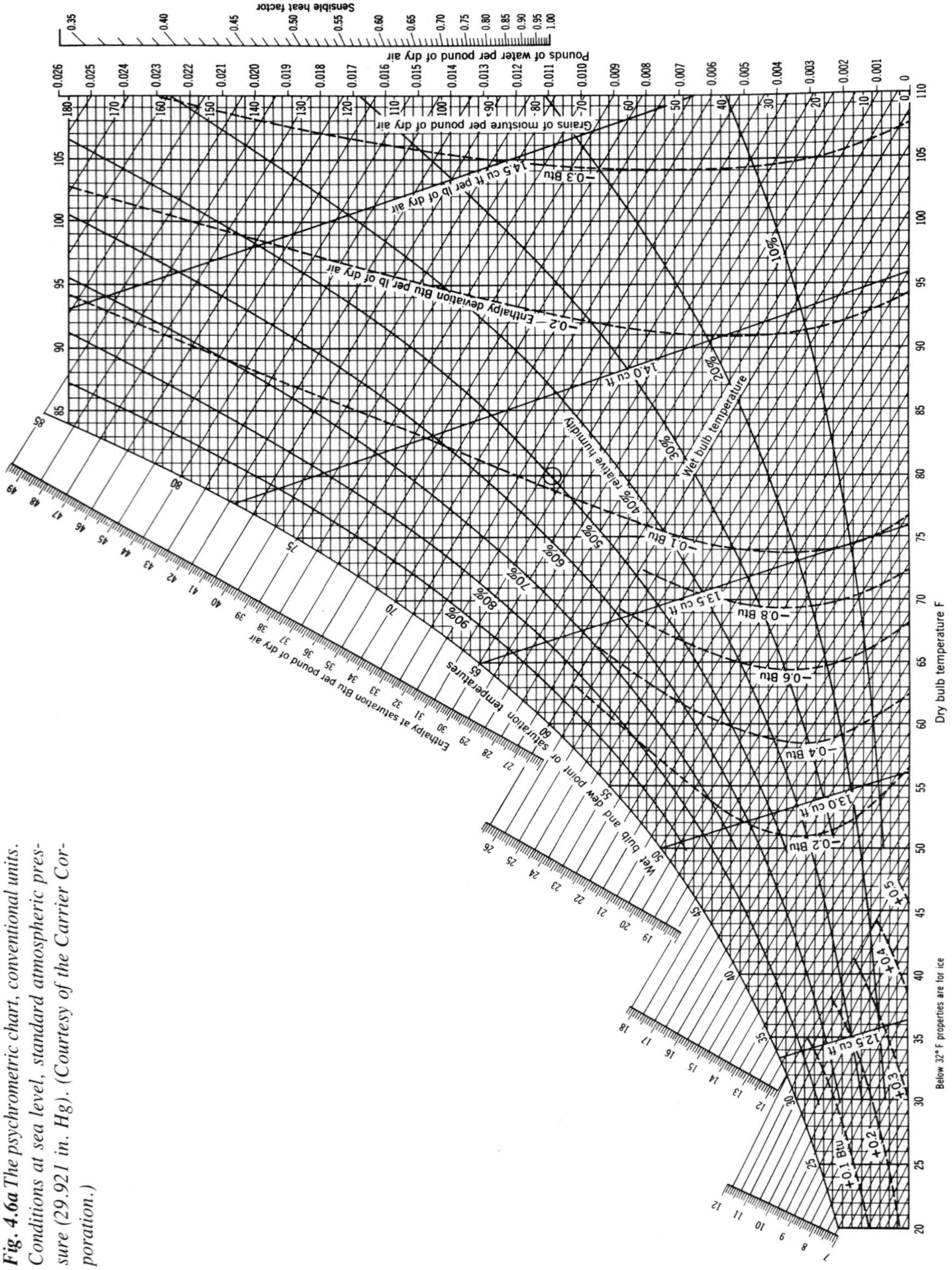

Fig. 4.6a The psychrometric chart, conventional units. Conditions at sea level, standard atmospheric pressure (29.921 in. Hg). (Courtesy of the Carrier Corporation.)

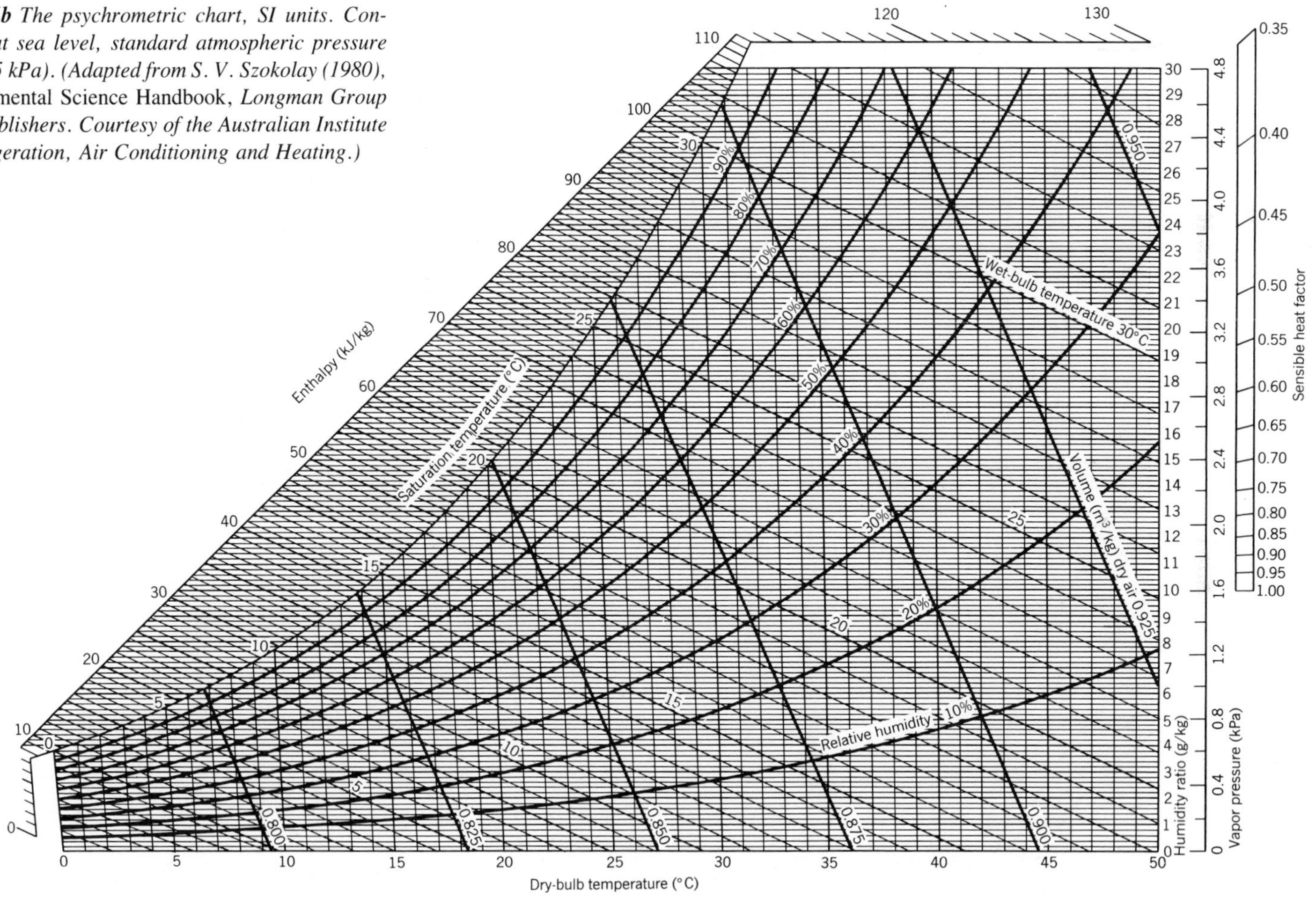

Fig. 4.6b The psychrometric chart, SI units. Conditions at sea level, standard atmospheric pressure (101.325 kPa). (Adapted from S. V. Szokolay (1980), Environmental Science Handbook, *Longman Group Ltd., Publishers. Courtesy of the Australian Institute of Refrigeration, Air Conditioning and Heating.)*

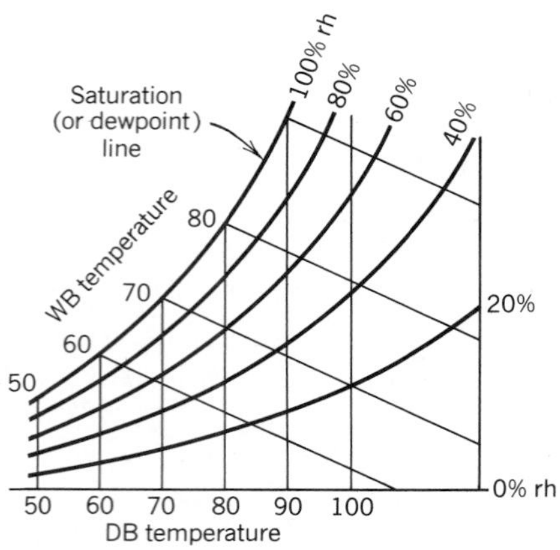

Fig. 4.7 *Some basic components of the psychrometric chart.*

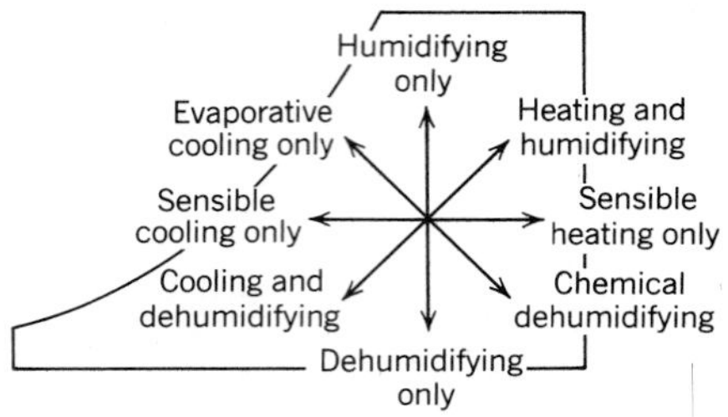

Fig. 4.8 *Climatic-conditioning processes, expressed on the psychrometric chart. [Adapted from "Architectural Design Based on Climate," by M. Milne and B. Givoni, in Watson (ed.),* Energy Conservation Through Building Design. *Reprinted with the permission of the publisher, McGraw-Hill, Inc.]*

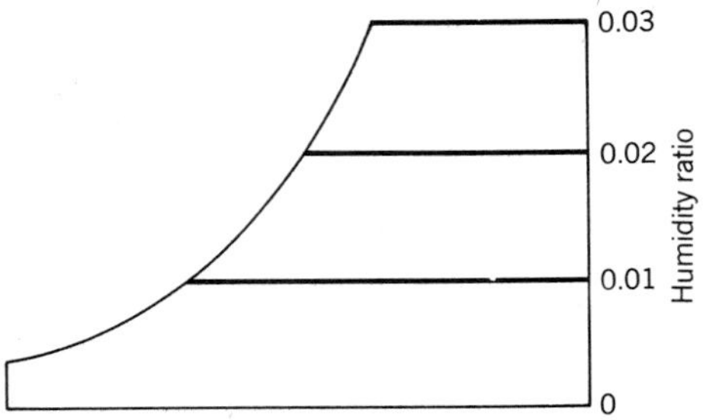

Fig. 4.9 *Humidity ratio on the psychrometric chart: conventional units, lb moisture/lb dry air; SI units, kg moisture/kg dry air.*

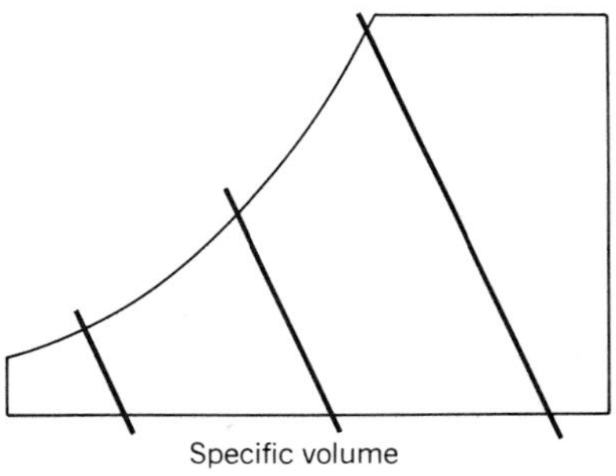

Fig. 4.10 *Specific volume on the psychrometric chart: conventional units, ft³/lb dry air; SI units, m³/kg dry air.*

where a reduction of the moisture content in the air is deliberate.

The psychrometric chart may be used to graph a wide variety of processes, which are summarized in Fig. 4.8. To understand these processes, we must add to the basic chart of Fig. 4.7. The first addition is *humidity ratio,* which indicates the amount of moisture by weight within a given weight of air. Air treatment processes that travel along these horizontal lines of constant humidity ratio (Fig. 4.9) are the familiar processes of simple heating (air passing through the heating coil of a furnace, or through a solar collector) and simple (sensible) cooling (air passes through the cooling coil of an air conditioner, before saturation). The humidity ratio will be used later in calculating latent heat gains from outdoor air (see Section 5.9).

The next addition shows how the *density* of air varies as its temperature and moisture content vary. These lines are those of *specific volume,* the reciprocal of density, a useful quantity in air conditioning calculations and helpful in understanding how the stack effect (Section 4.6) works. The specific volume is given in ft^3/lb (m^3/kg) of dry air. It is evident from these lines in Fig. 4.10 that a pound of hot air is larger (has more volume) than a pound of cold air. This larger volume per pound increases buoyancy; thus hot air rises while cold air sinks.

The next addition (Fig. 4.11) contains some of the most important information of all: *enthalpy,* the total amount of both *sensible* and *latent* heat in the air–moisture mixture.

Sensible heat is the kind of heat that *increases the temperature* of air; the glowing coil of an electric range is adding sensible heat to the air of the kitchen.

Latent heat is the kind of heat that is present in *increased moisture* in air; the boiling water in a pan on top of the electric range is evaporating, and as it does it increases the latent heat in the

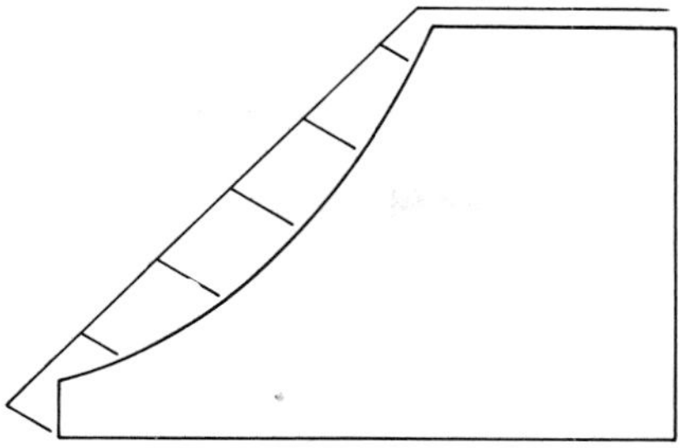

Fig. 4.11 *Enthalpy on the psychrometric chart: conventional units, Btu/lb; SI units, kJ/kg.*

kitchen air. This moisture has the potential to condense, thereby releasing its heat of vaporization.

(Note that 180 Btu is required to raise the temperature of a pound of water from 32 to 212 F; 1061 Btu is required to then evaporate the pound of water, without raising its temperature.)

Enthalpy is the sum of the sensible and latent heat content of an air–moisture mixture, relative to the sensible plus latent heat in air at 0 F (0°C in SI units) at standard atmospheric pressure. Its units are Btu/lb of dry air (in SI units, kJ/kg of dry air).

Enthalpy lines almost parallel those of WB temperature. To avoid further complicating the visual appearance of the psychrometric chart, enthalpy lines (Fig. 4.11) are shown as solid lines beyond the saturation point only.

Perhaps the most familiar air treatment process that travels along lines of constant enthalpy is evaporative cooling, where increased moisture and lower air temperature are obtained *without changing the enthalpy* (total heat content) of the air. There is indeed a drop in sensible heat as the temperature drops, but in terms of heat content, it is matched by an increase in latent heat as the moisture content increases. From the standpoint of human comfort, very hot, very dry air is improved by becoming less hot and more humid, up to a point. Thus evaporative cooling "works" in hot dry climates even though it doesn't change the total heat (enthalpy) of the air at all. The opposite process is chemical dehumidifying, where decreased moisture content is obtained at the price of increased temperature; again, no change in total heat (enthalpy) occurs.

The work to be done by mechanical air conditioning equipment is measured by the total change in enthalpy that must occur within the air that is treated by such equipment. The psychrometric chart is used to accurately size an air conditioner. Although both detailed calculations of heat gain and discussions of mechanical equipment occur later (in the next two chapters), an example of how this chart is used to determine air conditioner size is presented here.

First, consider the problem of determining the total change in enthalpy. Assume outdoor conditions of 90 F dry bulb, 76 F wet bulb, with desired indoor conditions at 75 F DB, 50% rh (which translates to 62.7 F WB). In the simple case of cooling 100% outside air, the total heat to be removed is determined as shown in Fig. 4.12. (Once-through systems cooling 100% outdoor air are rare, largely because they require substantial energy. They do occur in hospital surgical rooms and in laboratories with many fume hoods.) How much total heat will be removed, in getting from 90 F DB, 76 F WB to 75 F DB, 62.7 F WB?

For every pound of "dry" air (it is not really dry, but the values are based on the weight of the air alone) that is cooled and dehumidified, 39.6 − 28.3 = 11.3 Btu must be extracted. Similarly, for every pound of air so treated, 0.0162 − 0.0093 = 0.0069 lb of condensed moisture must be disposed of.

Each pound of moist air contains heat that is, by custom, measured above the value of 0 F. It consists of the sensible heat of the air and the water vapor, and the latent heat of the water vapor. Using 1061 Btu as an average value of latent heat per pound of moisture, one may check the enthalpy values of Fig. 4.12 as shown in Table 4.22.

The actual cooling process is more complex, as shown in Fig. 4.13. The conditioned air must be introduced to the space at conditions of temperature and relative humidity lower in value than those of the indoor conditions so that the entering air may "soak up" heat and moisture and leave through the return grills no worse than the design characteristics of 75 F and 50% relative humidity. So another set of conditions lower in dry bulb and humidity ratio will be established, and the values will depend on the rates at which the air is introduced and the amounts of heat and moisture to be

THERMAL CONTROL

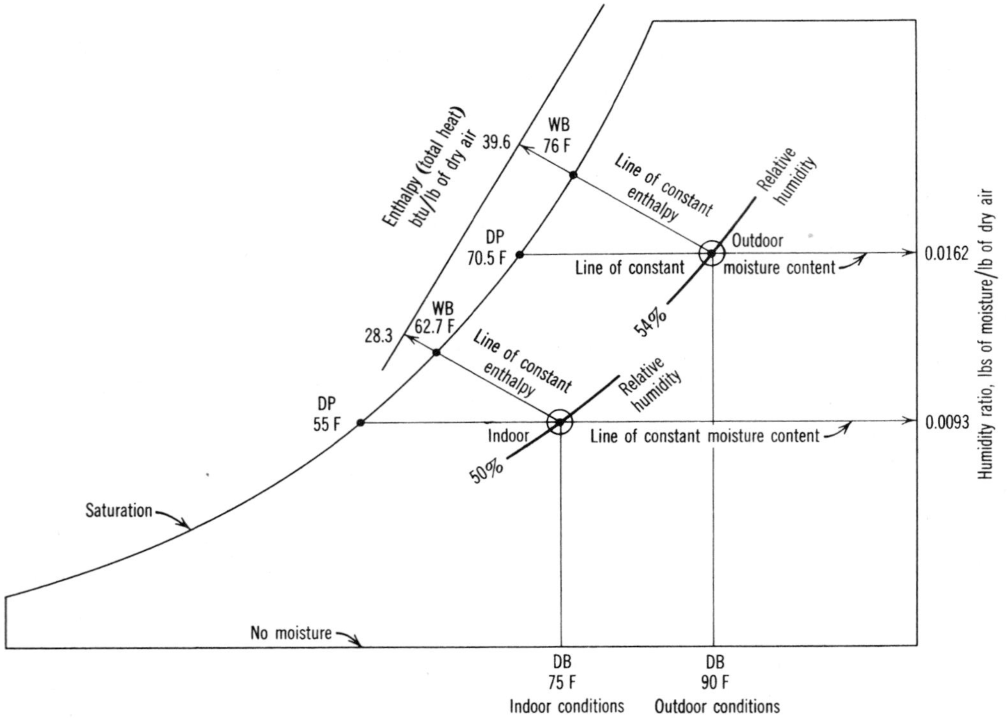

Fig. 4.12 Use of the psychrometric chart to determine the change in enthalpy between given outdoor and indoor air conditions.

TABLE 4.22 **Verification of the Enthalpy Values of Fig. 4.12**

Item	*Indoor Conditions*		*Outdoor Conditions*	
Latent heat, vapor	0.0093 × 1061[a] =	9.90	0.0162 × 1061[a] =	17.25
Sensible heat, vapor	0.0093 × 0.444 × 75 =	0.31	0.0162 × 0.444 × 90 =	0.65
Sensible heat, air (1 lb)	1 × 0.241 × 75 =	18.10	1 × 0.241 × 90 =	21.70
Enthalpy (Total heat)		28.30		39.60

[a]Latent heat of vaporization, Btu per pound (approximate).

absorbed by the air passing through. The heat and moisture are, of course, the sensible and latent heat gain of the space to be conditioned.

Within the cooling equipment, the outdoor air follows a complex path; the lines labeled 1, 2, and 3 in Fig. 4.13 trace the cooling and dehumidifying steps. Air is cooled without loss of moisture until it reaches a temperature at which it is saturated (its dew point). The cooling then continues. The air that had been losing sensible heat in step 1 continues to lose it in step 2, but additionally, moisture is condensed, which also requires the extraction of heat. When step 3 begins (it is known as a reheating process), the remaining smaller amount of moisture and the air are both heated to 65 F, 40% rh—slightly below the conditions to be maintained in the room. The changes in the heat content of the air and moisture at the various stages are measured along the enthalpy line.

The psychrometric chart can also be used to track the process of heating outdoor air in winter (Fig. 4.14). The chart shows that air at low temperatures in winter often has a humidity ratio so low that this moisture content would be unsatisfactory when the mixture was warmed to acceptable room temperature. Moisture must be added. Often this is in the form of a warm water spray. For our purpose, however, consider an adiabatic

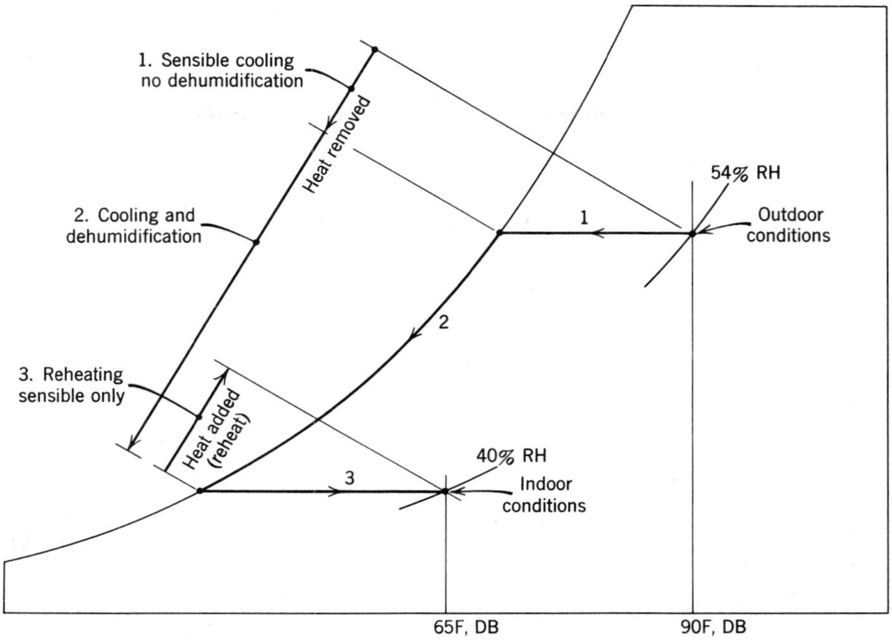

Fig. 4.13 The process of cooling and dehumidifying outdoor air, summer conditions.

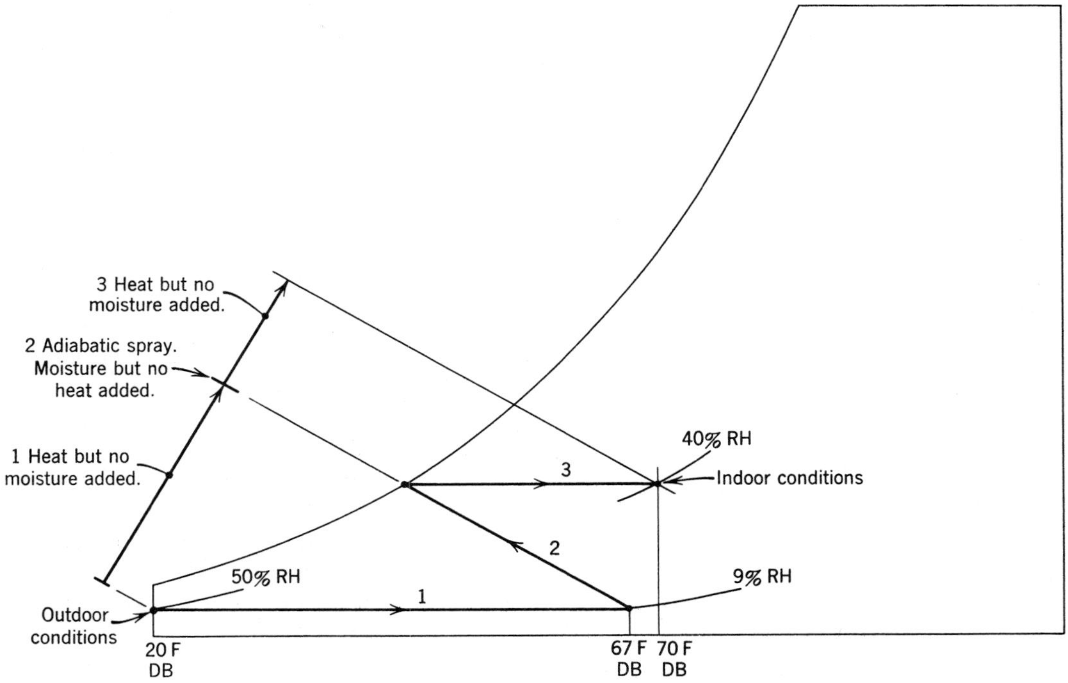

Fig. 4.14 The process of heating and humidifying outdoor air, winter conditions.

spray (no gain or loss of total heat) accomplished by first warming the outdoor air and vapor to a predetermined temperature, in this case 67 F (step 1). Then it is sprayed with water at room temperature to saturation. During this process (step 2), the added water that is evaporated mechanically draws sensible heat from the air, cooling it. The water acquires an equal amount of heat for its change of state to vapor—latent heat. Thus the value at the enthalpy scale does not change. The saturated mixture is then heated (step 3) with no addition of moisture until, at 70 F, it has a reasonably suitable relative humidity. Again the heat changes are read on the enthalpy scale. The action in Fig. 4.13 is known as cooling and dehumidifying with a reheat process. Figure 4.14 illustrates a preheat, spray, and reheat arrangement.

EXAMPLE 4.1. Find the total heat to be removed, and thus the refrigeration capacity required, for a dance hall. The design conditions are:

Room conditions (summer)	75 F DB, 50% RH
Number of occupants	80 people
Activity	Dancing
Ventilation required	35 cfm per person (Table 4.19, Part A)
Conditions, outdoor air	95 F DB, 75 F WB

Heat Gains in the Room	*Sensible heat, SH (Btu/h)*	*Latent Heat, LH (Btu/h)*
80 people dancing (see Table 5.21)		
80 × 405 Btu/h	32,400	
80 × 875 Btu/h		70,000
Total transmission and solar gain, and lights, equipment, etc.	67,600	(none)
	Room Sensible Heat, (RSH) = 100,000	Room Latent Heat, (RLH) = 70,000

Total heat gains in room: 170,000 Btu/h (RSH + RLH)

SOLUTION. First, determine the portion of the heat gain that is due to sensible heat gain, called the *sensible heat factor,* SHF:

$$\text{SHF} = \frac{\text{RSH}}{\text{RSH} + \text{RLH}} = \frac{100{,}000}{170{,}000} = 0.59$$

On the psychrometric chart (Fig. 4.15*a*), draw a line between the fixed "bulls-eye" (80 DB, 50% rh) and the value of 0.59 on the SHF scale, at the upper right edge of the chart. This is called the "SHF line."

Point A is the condition of the "used" air within the dance hall, as it is returned for reprocessing: 75 F DB, 50% rh (62.5 F WB). Next, decide how much cooler the supply air should be than the return air. To avoid uncomfortable drafts, this supply temperature is usually 20F° or less below the space's air temperature. In this case, choose 15F°. Then the quantity of air required to cool the room will be:

$$\text{cfm} = \frac{\text{RSH}}{1.08\Delta t} = \frac{100{,}000\ \text{Btu/h}}{1.08(15)} = 6200\ \text{cfm}$$

(The factor 1.08 is the constant explained in the natural ventilation formulas in Section 4.6.)

The portion of this supply air that must be *outdoor* air is found by reference to Table 4.19, Part A, as follows:

$$80\ \text{people} \times 35\ \text{cfm/person} = 2800\ \text{cfm}$$

So, the percentage of outdoor air is:

$$\frac{2800}{6200} = 45\%$$

Now, several important points can be located on the chart. *Point B* is the condition of the air entering the rooms; it has been decided that it will be 15F° cooler than 75 F, which places point *B* somewhere on the 60 F DB line. To determine exactly where, draw a dashed line through point *A*, parallel to the "SHF line," and extend it until it crosses the vertical 60 F DB line. This occurs at 60 DB, 53 WB; enthalpy (h_B) = 22.0 Btu/lb. *Point D* is the condition of the outdoor air, given at 95 F DB, 75 F WB.

Point C (Fig. 4.15*b*) represents the mixture of 45% outdoor air and 55% return air that is brought to the cooling equipment for treatment and distribution back to the dance hall. Connect points *A* (the return air) and *D* (outdoor air); then plot *C* at 45% of the distance from *A* to *D*. This occurs at

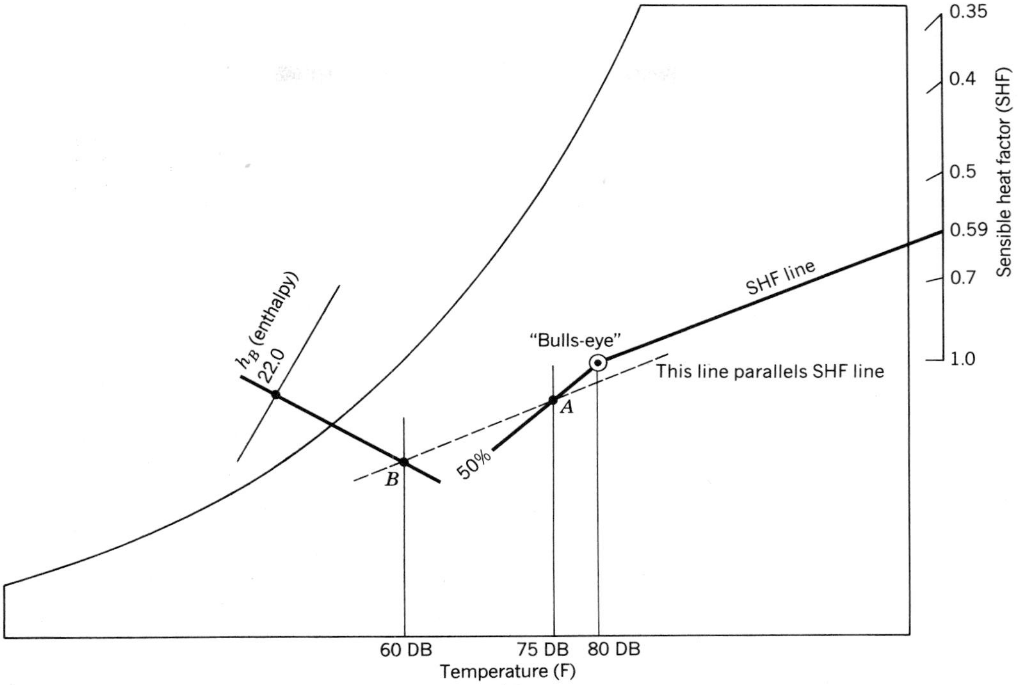

(a)

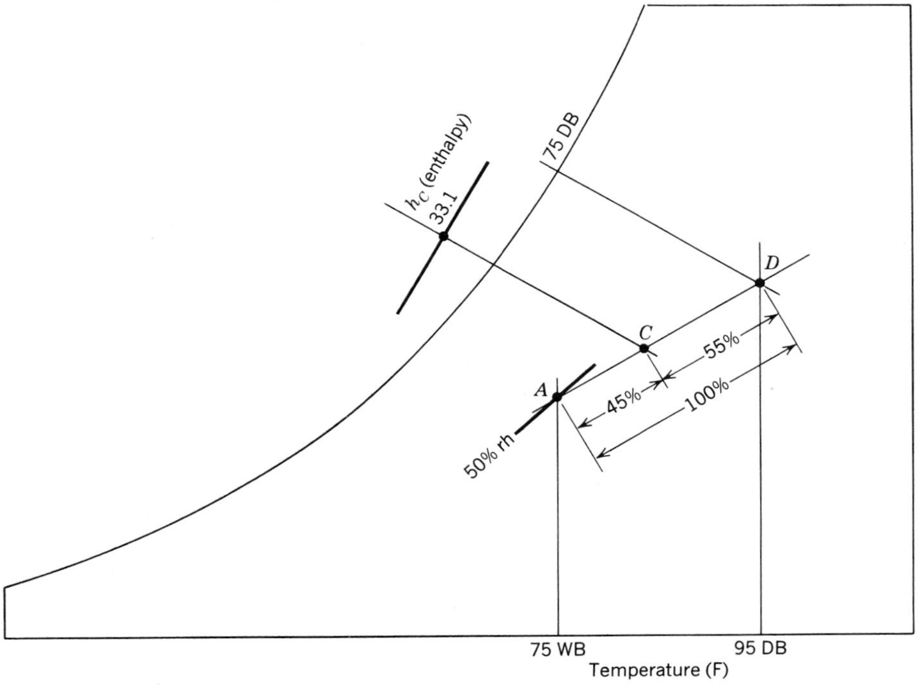

(b)

Fig. 4.15 *Sizing cooling equipment using the psychrometric chart. (a) Finding the conditions for the supply air. (b) Finding the conditions for the return air–outdoor air mixture. (c) Points* A, B, C, *and* D, *shown within the cooling equipment and the building.*

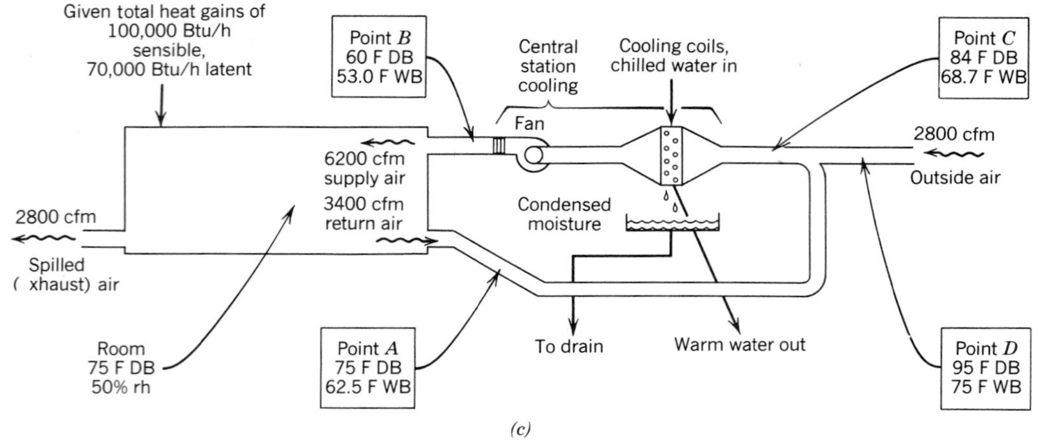

Fig. 4.15 (continued)

84 F DB, 68.7 F WB; enthalpy (h_C) = 33.1 Btu/lb.

The cooling equipment must remove the grand total heat (GTH) according to the formula:

$$\text{GTH} = 4.5 \times \text{cfm} \times (h_C - h_B)$$

(where 4.5 is a constant = 60 min/h × 0.075 lb/ft^3 average air density). So, in this example:

$$\text{GTH} = 4.5 \times 6200 \text{ cfm} \times (33.1 - 22.0) = 309{,}690 \text{ Btu/h}$$

The size of the required refrigeration unit is specified in "tons," where 1 ton = 12,000 Btu/h.

refrigeration required

$$= \frac{309{,}690 \text{ Btu/h}}{12{,}000 \text{ Btu/h ton}} = 25.8 \text{ tons}$$

Note: If smoking were not permitted, the outdoor air requirements could drop to 7 cfm per person.

$$80 \text{ people} \times 7 \text{ cfm/person} = 560 \text{ cfm}$$

The percentage of outdoor air becomes 560/6200 = 9%.

Point *C* then moves to about 77 F DB, 64 F WB, at which point h_C = about 29.4 Btu/lb.

The refrigeration required then becomes:

$$4.5 \times 6200 \times (29.4 - 22.0)$$

$$= \frac{206{,}460 \text{ Btu/h}}{12{,}000 \text{ Btu/h ton}} = 17.2 \text{ tons}$$

This represents a sizable first-cost saving in equipment size, and of course a dramatic energy savings over the life of the dance hall. (Dancers would smell less smoke, and more sweat.)

References

Publications of the American Society of Heating, Refrigerating and Air-Conditioning Engineers, Atlanta:

ASHRAE *Cooling and Heating Load Calculation Manual.*

ASHRAE *Handbook of Fundamentals.*

ASHRAE *Standard 62-1981,* Ventilation for Acceptable Indoor Air Quality, 1981.

Balcomb, J. D., Barley, D., McFarland, R., Perry, J., Jr., Wray, W., and Noll, S. (1980). *Passive Solar Design Handbook, Volume Two: Passive Solar Design Analysis,* U.S. Department of Energy, Washington, D.C.

Balcomb, J. D., Jones, R. W. (ed.), Kosiewicz, C. E., Lazarus, G. S., McFarland, R. D., and Wray, W. O. (1983). *Passive Solar Design Handbook, Volume 3,* American Solar Energy Society, Boulder, CO.

Johnson, T. (1981). *Solar Architecture, the Direct Gain Approach,* McGraw-Hill, New York.

Milne, M. and Givoni, B. (1979). "Architectural Design Based on Climate", in D. Watson, Editor, *Energy Conservation Through Building Design,* McGraw-Hill, New York.

Szokolay, S. V. (1980). *Environmental Science Handbook,* Longman Group, London.

5

DESIGNING FOR HEATING AND COOLING

Chapters 1–4 have provided the necessary preparation for design: a perspective on national energy resources, an understanding of human comfort, an analysis of the climate resources available on site, a list of the general design strategies appropriate to various climates, and a discussion of the basics of heat transfer calculations. Hence, we now know much about the variables outside a building, about the desired conditions inside, and about individual components of a building's skin. In this chapter, all these variables are integrated into the process of designing for heating and cooling with the important related factor of daylighting. The design of mechanical support systems is introduced in Chapter 6.

Chapter 5 has two major parts: Sections 5.1–5.3 accompany preliminary design, where quick decisions are made based on broad guidelines. Sections 5.4–5.9 accompany final design, where more detail is known and more accuracy is appropriate.

5.1 Organizing the Problem

How should your building's skin respond to the sometimes conflicting needs for heating, cooling, and daylighting? Heating and cooling design strategies were related to climate in Sections 2.3 and 2.4, in which it became evident that internal heat gains can shift the appropriate design strategies for a building toward cooling, perhaps eliminating heating needs entirely. Typically, much of this internal heat is provided by electric lighting. Daylighting can replace electric lighting for most of the typical working day in most building types—*if* the building is designed to allow daylight to reach most of the interior. Daylighting is accompanied by large glass areas, which increase a building's heating needs in winter; in such buildings, however, less heat from electric lighting is available to fill those needs. Another complication is that adequate daylight requires much larger glass areas under dim winter skies than under bright summer skies. If the glass area is sized for winter daylighting, then excessive daylight—and along with it, excessive heat—will be admitted in summer. With proper controls, daylighting can reduce summer cooling loads, relative to electric lights, but it will usually increase winter heating loads. Where passive solar heating or surplus heat from another source is readily available, this tradeoff is attractive. One of the earliest and most difficult questions for the designer is what relative weights to give to considerations for heating, cooling, and daylighting.

The most desirable balance among the energy needs for heating, cooling, and daylighting will vary both by building type and by climate. Table 5.1 shows the usual ranking of these three energy usages, as well as of power and process energy use and domestic hot water. The rankings are shown for six typical building types and three categories of winter heating needs. The table indicates only rank, not size; ranks 1 and 2 may be very close in size, or quite different.

The ranking of probable energy use for your building in its climate is the first step toward the design of the building's envelope and mechanical support systems. In the next step, we look at the building as a collection of thermal conditions, or zones.

The thermal zoning of a building recognizes that different envelope and support systems may be required within the building. The more carefully zoning is considered in these early design stages, the better will be the thermal performance and the lower will be the annual energy consumption. (Also, the less likely it will be that all sides of a building will have an identical appearance.)

TABLE 5.1 **Ranking the Energy Use in Buildings**

Rank 1 = most energy used in typical building of this type, relative to other uses

	Categories of Energy Usage				
Building Types	*Heating and Ventilating*	*Cooling and Ventilating*	*Lighting*	*Power and Process*[a]	*Domestic Hot Water*
Schools					
A	4	3	1	5	2
B	1	4	2	5	3
C	1	4	2	5	3
Colleges					
A	5	2	1	4	3
B	1	3	2	5	4
C	1	5	2	4	3
Office buildings					
A	3	1	2	4	5
B	1	3	2	4	5
C	1	3	2	4	5
Stores					
A	3	1	2	4	5
B	2	3	1	4	5
C	1	3	2	3	5
Religious buildings					
A	3	2	1	4	5
B	1	3	2	4	5
C	1	3	2	4	5
Hospitals					
A	4	1	2	5	3
B	1	3	4	5	2
C	1	5	3	4	2

Source: F. Dubin and G. C. Long, *Energy Conservation Standards for Building Design, Construction and Operation,* © 1978, McGraw-Hill, New York.

[a]Extensive use of computer terminals in workplaces may increase the ranking of this category.

A = climate with fewer than 2500 heating degree days.
B = climate with 2500 to 5500 degree days.
C = climate with 5500 to 9500 degree days.
(See Appendix B for degree day information.)

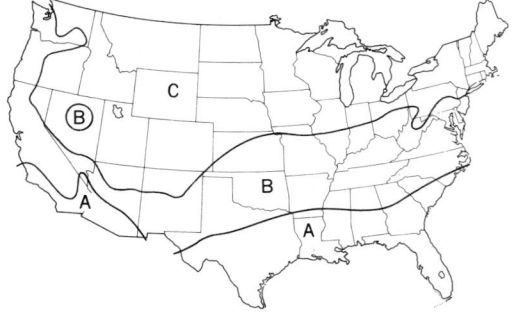

Zoning is most often influenced by the following factors.

1. *Function*. Particularly important because of the variations in internal heat gains between functions, this factor may also influence the vertical organization of a building, as was shown in Fig. 3.1. Comfort conditions may vary considerably between functions; air temperatures can be lower for a strenuous activity than for a sedentary activity, or heat tolerance may be greater for some activities (kitchens) than for others.

2. *Schedule*. This factor, closely related to function, can influence both the envelope and the support system. An activity scheduled only between 9 A.M. and 4 P.M. can often be entirely

daylit, at a time when the outside temperatures are the warmest of the daily cycle. By contrast, activity that takes place only from 9 P.M. to 4 A.M. will be entirely dependent on electric lighting, whose heat can be used to overcome the chill of the outside temperatures during these hours in winter. (In the summer, such heat can be flushed away with the cool outside night air in many U.S. locations). Support systems are often divided by scheduling considerations: if one activity has operating hours different from those of the remainder of the building, a separate mechanical system is often provided. This saves energy, because large equipment will not be underused to provide heating or cooling for only one zone.

3. *Orientation*. The degree of exposure to daylight, direct sun, and wind is obviously important to thermal zoning. Consider a block-square office building on a cold, sunny, and windy day. Perimeter spaces with direct sun through the windows may gain more heat than is lost, and thus need cooling. This might be done by the opening of windows, but too much cold air (especially on the windy side of a building) may make the workers near the windows uncomfortable. Perimeter spaces without direct sun may have a net heat loss, due to heat loss through glass, infiltration, and a lack of electric lights (since daylight is adequate). These spaces will need heat from a mechanical support system. Interior (no daylight) spaces are overheated by electric lights, since they cannot lose heat. These spaces will need cooling from the support system.

Zoning considerations are among the most important influences on building form and external design, along with the familiar esthetic, social, legal, economic, and technical influences that combine in a tug-of-war familiar to the building designer. Some of the most basic design concepts related to thermal zoning are building form, building envelope, and support system design.

Building form, at its simplest, can be reduced to questions of tall or short, thick or thin. Figures 5.1 and 5.2 compare these form variations to the relative importance of heating, cooling, and daylighting. It is apparent that thicker, taller buildings have more floor space away from climate influences; being electrically lit rather than daylit, they generate heat and need cooling all year. These buildings are called *internal load dominated* (ILD) (Fig. 5.1). In contrast, thinner buildings in which nearly all spaces have an exterior wall will need heating in cold weather and cooling in hot weather; electric lights by day are largely unnecessary. These buildings are called *skin load dominated* (SLD) (Fig. 5.2).

By comparing the rankings of energy use by function and climate given in Table 5.1 with the forms shown in Figs. 5.1 and 5.2, we can identify a range of design opportunities and their attendant thermal problems. The ultimate choice of building form is determined by a combination of design issues; the energy use issue can help in the selection process. Once the building form has been chosen, the functions can be distributed according to typical architectural criteria, including the thermal zoning considerations mentioned above.

In the selection of a building form, some particularly important questions accompany each energy use. Because the question of daylighting versus electric lighting frequently is so influential in determining whether heating or cooling will be the dominant concern, we begin with daylighting.

For *daylighting:*

1. What will be the relative emphasis on sidelighting (characterized by uneven distribution and glare in the visual field but little glare on horizontal surfaces) and toplighting (the reverse characteristics)?
2. What role will direct sun play in daylighting? In winter, can heat without glare be admitted?
3. How can seasonal adjustments be made in the size of daylight openings?
4. To what extent will daily changes in daylighting control be necessary?
5. How can adequate daylight be admitted in an even way, such that unwelcome dark-appearing places are avoided?

For *heating:*

1. Can the sun be used to heat spaces? If so, how will south wall design be affected?
2. How can openings in walls facing other directions be kept to a minimum without daylight being shut out?
3. What role will direct sun through south glass or skylights play in daylighting?

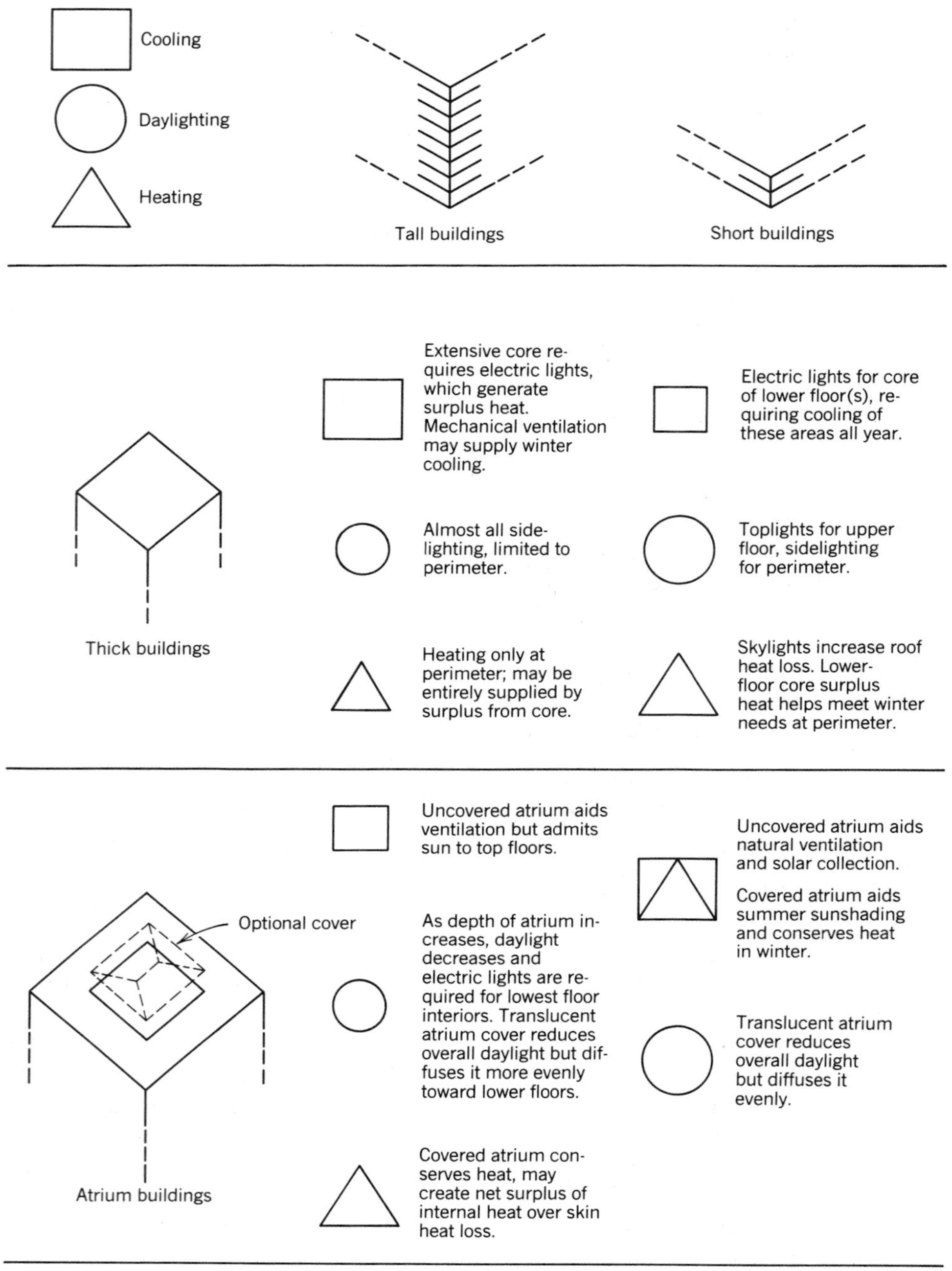

Fig. 5.1. *Internal-load-dominated (ILD) buildings: the relative importance of cooling, daylighting, and heating varies with building form, as do opportunities for the use of on-site or natural energy sources. See also Table 5.1.*

4. How can daylight be admitted but the chilling effects of large, cold glass surfaces be minimized?

5. How can incoming fresh air be warmed before it chills the people sitting near the fresh air opening?

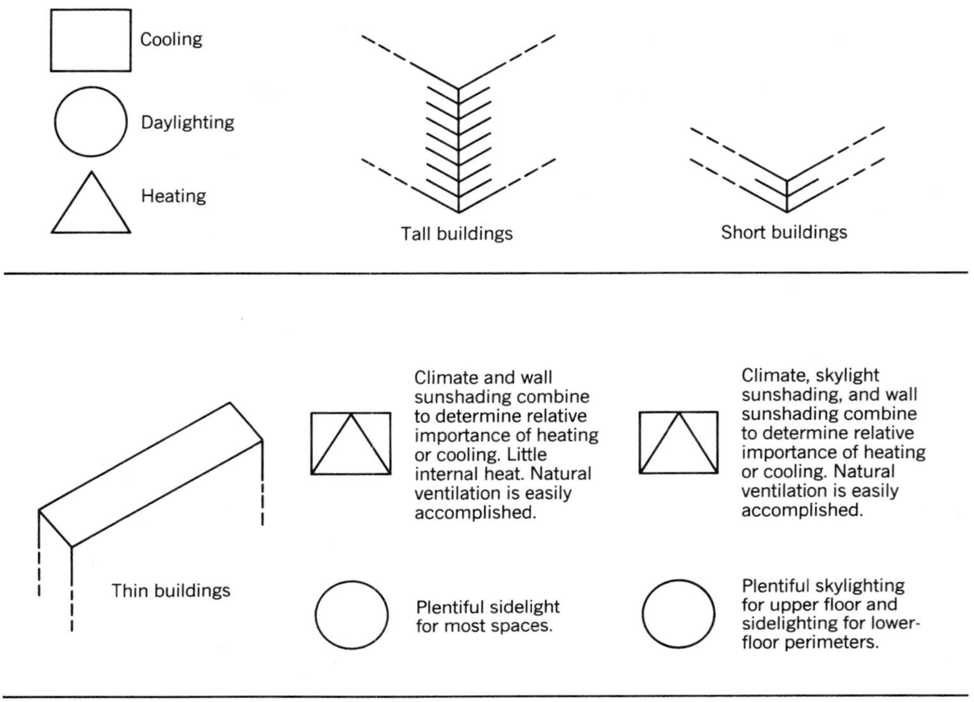

Fig. 5.2. *Skin-load-dominated (SLD) buildings: the relative importance of cooling, daylighting, and heating varies with building form, as do opportunities for the use of on-site or natural energy sources. See also Table 5.1.*

6. Is there surplus heat elsewhere in the building that can be used to help warm perimeter spaces?

For *cooling:*

1. Will the strategy be to open the building to breeze or to close the building for "coolth" retention, or to use a combination of these alternatives (open by night, closed by day)?
2. How can direct sun be kept out of the building? Can east and west windows be minimized and adequate daylight still provided?
3. How can adequate daylight be admitted for winter conditions without overlighting (and thus overheating) for summer conditions?
4. When can cooling be provided by outdoor air, rather than by a refrigeration cycle?
5. Can the operation of refrigeration machinery be concentrated during the coldest (nighttime) hours?
6. How can incoming fresh air be cooled before it warms the people sitting near the fresh air opening?
7. Can the structure of the building be used to absorb heat by day, then be flushed with night air in climates with cool nights?

Building envelope, the next design step, involves relating the climate to the perimeter of the building zones in the design of the building's skin. Each skin element provides an opportunity for thermal and luminous exchange between inside and out; heating, cooling, ventilating, and daylighting devices can be mixed as needed. Figure 5.3 shows some of the most common of these devices for varying orientations. Section 5.2 gives numerical criteria for sizing these skin elements.

Support systems are considered in detail in the next chapter. At this design stage, the most important concept is that of the "distribution tree." Mechanical support systems often produce heating and cooling in one place, then distribute them to other building spaces according to respective needs. The distribution tree is the means for delivering heating and cooling: the "roots" are the machines that provide heat and cold, the "trunk" is the

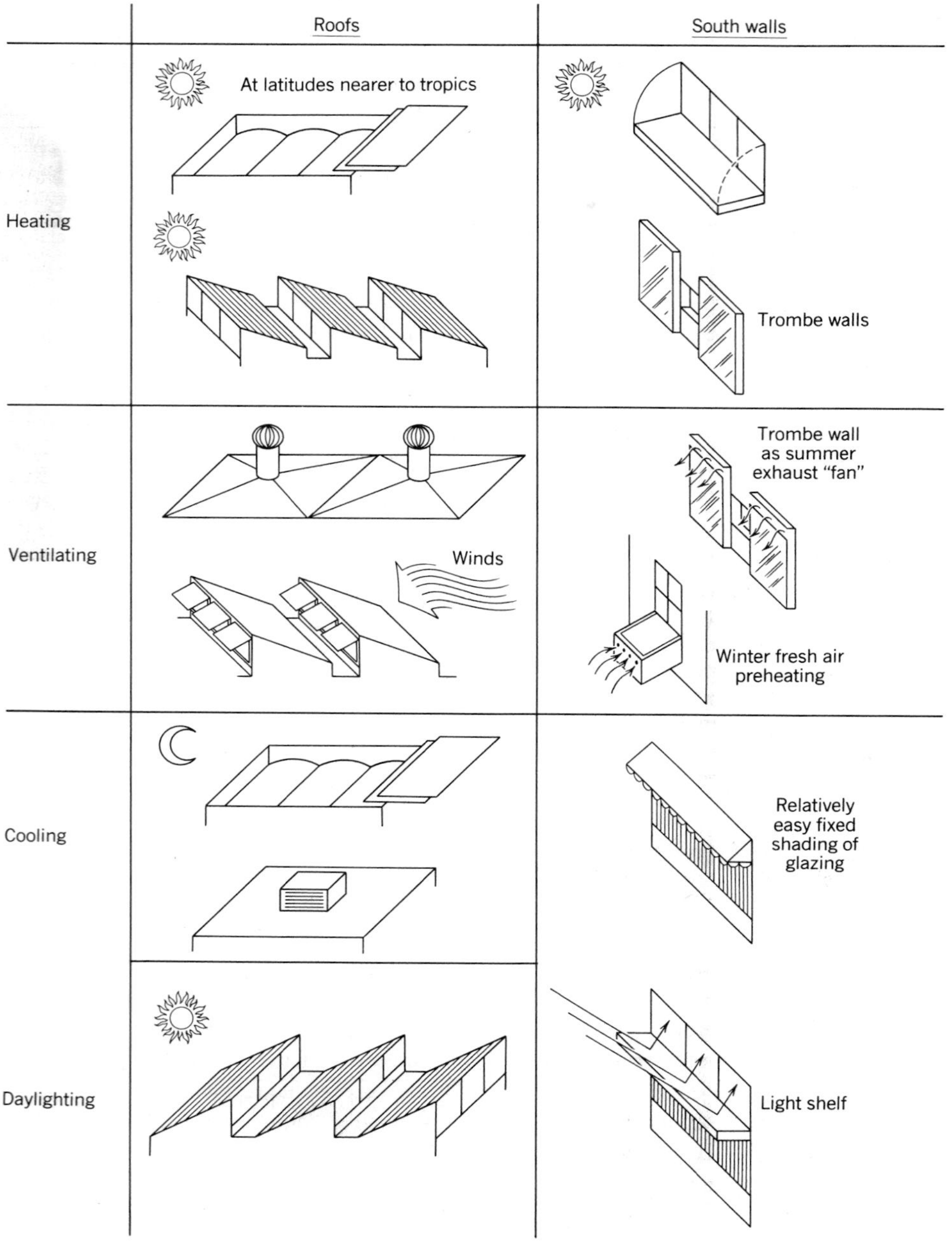

Fig. 5.3. The components of a building's envelope can be used both to conserve energy and to admit on-site or natural energy sources. Some of the many opportunities for heating, ventilating, cooling, and daylighting are shown here.

main duct or pipe from the mechanical equipment to the zone to be served, and the "branches" are the many smaller ducts or pipes that lead to individual spaces.

For now, the questions to be answered about distribution trees for buildings are How many? What kind? and Where? A building can have one giant distribution tree, several medium-sized trees, or an orchard of much smaller trees. At one extreme, a large mechanical room is the scene of all heating

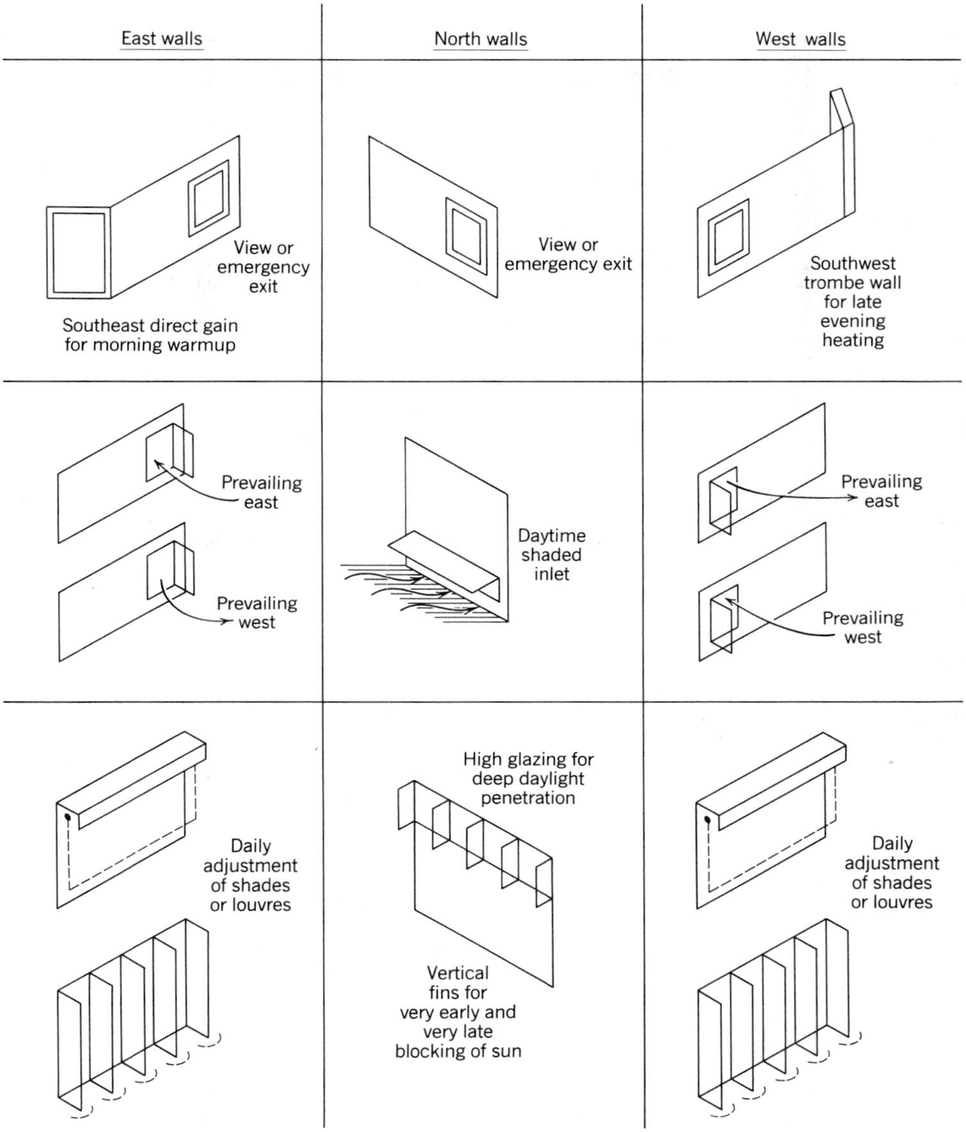

Fig. 5.3. (continued)

and cooling production; leading from this room is a very large trunk duct with perhaps hundreds of branches. At the other extreme, each zone has its own mechanical equipment (such as a rooftop heat pump), with short trunks and relatively few branches on each tree.

What kind of distribution tree? Most simply, air (ducts) or water (pipes). Air distribution trees are bulky and therefore likely to have major visual impacts unless they are concealed above ceilings or within vertical chases. Water distribution trees consume much less space (a given volume of water carries vastly more heat than does the same volume of air at the same temperature) and can be easily integrated within structural members, such as columns. Both air and water trees can be sources of noise.

Where does the distribution tree fit in? On the exterior, it can lend a three-dimensional organizational structure to a facade, as can be seen in the Blue Cross–Blue Shield office building in Boston (see Fig. 6.110). Exterior trees can take up a smaller amount of rentable floor space but require expensive cladding and are subject to heat losses and gains, which could increase energy usage. Interior trees are often combined with other continuous vertical spaces, such as elevator shafts and stairways. If the choice is an exterior distribution tree, its potential contribution to facade performance should be considered. For example, the distribution tree might act as a sunshade or as a light reflector.

To carry the tree analogy to its logical conclusion, consider the ''leaves,'' the points of interchange between the piped or ducted heating or cooling and the spaces served. One example is a large, bulky device such as a fan-coil unit on the exterior wall, below windows. In contrast, a perforated ceiling system with thousands of small holes acting as a widely spread grille is essentially invisible.

Distribution trees are rich in form possibilities; their variations are enormous. For now, assume that the choices of how many, what kind, and where have tentatively been made and that the details and variations are to be considered later. The designer who has chosen the basic form, types of envelope components, and a support system distribution tree can now consider questions of size. The remainder of this chapter deals with such questions.

5.2 Guidelines and Criteria

After approximating the degree to which lighting, heating, and cooling are necessary for your building and climate, you can begin the detailed design of the envelope. Chapter 4 presented the procedures for determining the rate at which heat flows through the many parts of the envelope; now you must make the *choice* of which components to combine for your building.

The information in this section can be used either before preliminary design, as guidelines to window sizing and roof/wall/floor insulation, or after preliminary design, as criteria for measurement of the predicted performance of a building's envelope. Several kinds of design guidelines are available for energy-conserving construction. Presented here are some of the most widely used guidelines, ranging from very rough—but quickly formulated—criteria to very detailed criteria that require computer processing.

(a) Daylighting. When a building is designed to rely heavily on daylighting, a prime design concern is the *daylight factor* (DF), which is expressed as a percentage of the outdoor light that is available indoors.

$$\text{DF} = \frac{\text{indoor illumination from daylight}}{\text{outdoor illumination}} \times 100\%$$

Chapter 19 presents detailed procedures for determining the DF for any point within a building. For now, some simple target DFs are presented in Table 5.2. Table 5.3 gives the simplest design rules of thumb that yield these DF results. Typically, these rules of thumb compare the window area to the floor area for each daylit space.

The daylight factors listed in Table 5.2 will provide sufficient light during most of the daylight hours on overcast winter days. Obviously, much more light will be available on summer days—more light than is needed, bringing heat along with it. When sizing windows and skylights, remember that controlling direct sun is necessary and that less opening area is needed in summer than in winter. Further details on sunshading and daylighting control devices can be found in Section 6.2.

(b) Electric Lighting. To control the electricity usage (and the building overheating that frequently results) that accompanies electric lights, a

TABLE 5.2 **Recommended Daylight Factors**[a]

Task	*DF*
Ordinary seeing tasks, such as reading, filing, and easy office work	1.5–2.5%
Moderately difficult tasks, such as prolonged reading, stenographic work, normal machine tool work	2.5–4.0%
Difficult, prolonged tasks, such as drafting, proofreading poor copy, fine machine work, and fine inspection	4.0–8.0%

Source: Millet and Bedrick (1980).

[a]Use the smaller DF values for southern latitudes with plentiful winter daylight.

TABLE 5.3 **Daylight Factor (DF) Design Rules of Thumb for Overcast Sky Conditions**

For spaces with sidelighting[a,b,c]

$$DF_{av} = 0.2\left(\frac{\text{window area}}{\text{floor area}}\right)$$

$$DF_{min} = 0.1\left(\frac{\text{window area}}{\text{floor area}}\right)$$

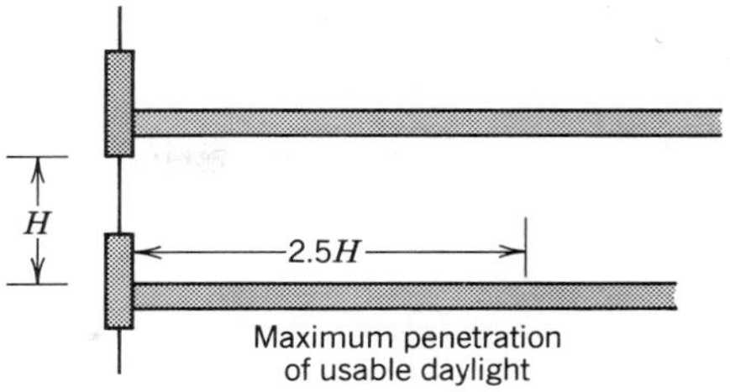

For spaces with toplighting[c,d]

Vertical monitors:

$$DF_{av} = 0.2\left(\frac{\text{skylight glazing area}}{\text{floor area}}\right)$$

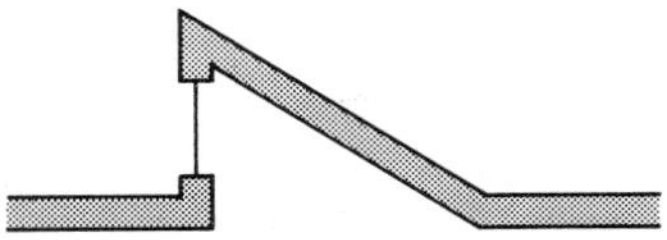

North-facing sawtooth:

$$DF_{av} = 0.33\left(\frac{\text{skylight glazing area}}{\text{floor area}}\right)$$

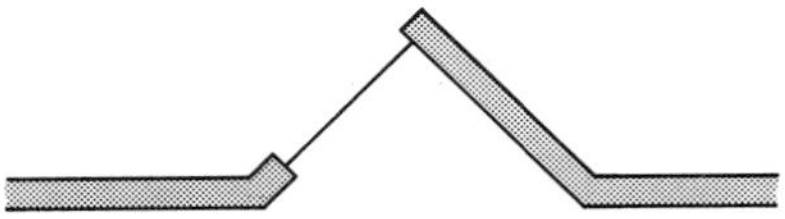

Horizontal skylights:

$$DF_{av} = 0.5\left(\frac{\text{skylight glazing area}}{\text{floor area}}\right)$$

Source: Millet and Bedrick (1980).

[a]Assumes windows in one wall of a room with relatively light-colored surfaces.

[b]Window height/room depth relationships based on the works of R. G. Hopkinson and others at the British Research Station.

[c]For sunny, clear winter sky conditions, use same formula for north, east or west glazing. For south or horizontal glazing, divide each formula's constant by 3.

[d]Assumes an even distribution of such skylights in the roof, so that an even distribution of light results in room below; thus, only average DF are listed.

"power budget" is established. This budget sets a *maximum* of installed lighting, expressed in watts of electric lighting per square foot of floor area. Several state building codes specify lighting power budgets; check the applicable code before using the recommendations given in Table 5.4. The initial approximations of electric lighting include the recommended lighting intensity levels in Tables 18.3 and 18.7 and the power relationships in Fig. 20.2, page 1008 (Part VI). To use Table 5.4, first find the power density allowable for the building occupancy type, then multiply the power density by an area factor (see Table 5.5) that accounts for area and volume of the space to be lighted. The result is the maximum allowable lighting-heat gain, in watts per square foot. As with daylighting, the design of electric lighting is far more involved and subtle than these simple rules of thumb indicate. See Chapters 18–20 for the details on these lighting methods.

(c) Heating: Whole-Building Criteria. The recommended maximum rates of heat loss [in Btu per degree days (DD) per square foot] shown in Table 5.6 were developed for energy conservation in residential or small commercial buildings. Many states have adopted similar criteria as part of their building codes; be sure to check the applicable code before using these numbers. The heat loss rates are shown for two conditions, as follows.

1. Conventional (nonpassively solar heated) small buildings. The overall rate of Btu/DD ft^2 is based on *total* heat loss, including all portions of the envelope *and* infiltration. To determine the whole-building heat loss rate, list for each en-

THERMAL CONTROL

TABLE 5.4 **Electric Lighting Power Densities**

Electric lighting energy consumption for building interiors should stay below these maximum levels, based on the gross floor area of the building. Multiply these power densities by the area factors given in Table 5.5 to obtain the power budget.

Occupancy Areas	Power Densities		
	W/ft²	Btu/h ft²	W/m²
Common areas			
Boiler room	0.6	2.0	6.46
Conference room	1.0	3.4	10.76
Corridor	0.5	1.7	5.38
Dining (fast service)	2.1	7.2	22.60
Dining (leisure)	1.6	5.5	17.22
Electrical equipment room	0.5	1.7	5.38
General assembly (auditorium)	0.8	2.7	8.61
Kitchen	1.3	4.4	14.00
Laboratories	2.4	8.2	25.83
Library room	1.7	5.8	18.30
Lobby, reception, waiting	0.8	2.7	8.61
Locker room and shower	0.5	1.7	5.38
Mail room	2.1	7.2	22.60
Material handling (bulk)	0.6	2.0	6.46
Mechanical equipment room	0.5	1.7	5.38
Stairs	0.5	1.7	5.38
Storerooms or warehouse			
Inactive	0.2	0.7	2.15
Active bulky	0.4	1.4	4.31
Active medium	0.5	1.7	5.38
Switchboard and control room	1.3	4.4	14.00
Toilet and washroom	0.6	2.0	6.46
Utility room, general	0.4	1.4	4.31
Office			
Accounting	2.4	8.2	25.83
Drafting	3.5	12.0	37.67
Filing (active)	1.5	5.1	16.15
Filing (inactive)	0.6	2.0	6.46
Graphic arts	2.3	7.9	24.76
Office machine operation			
Computer machinery	1.3	4.4	14.00
Duplicating machines	0.6	2.0	6.46
EDP I/O terminal (internally illuminated)	0.6	2.0	6.46
EDP I/O terminal (room illuminated)	1.3	4.4	14.00
Typing and reading	1.7	5.8	18.30
Commercial and institutional			
Armories			
Drill	0.5	1.7	5.38
Exhibitions	0.6	2.0	6.46
Seating area	0.4	1.4	4.31
Art galleries	1.2	4.1	12.92
Bakeries			
Hand decorating	2.6	8.9	27.99
Mixing and filling	1.0	3.4	10.76
Banks			
Lobby, general	1.7	5.8	18.30

TABLE 5.4 **Electric Lighting Power Densities (*continued*)**

Occupancy Areas	Power Densities W/ft²	Btu/h ft²	W/m²
Posting and keypunch	3.5	12.0	37.67
Tellers' stations	3.5	12.0	37.67
Bar (lounge)	0.8	2.7	8.61
Barber shops and beauty parlors	2.9	9.9	31.22
Church and synagogues, main worship area	1.7	5.8	18.30
Cleaning and pressing			
General processing	1.1	3.8	11.84
Pressing	3.3	11.3	35.52
Repair and alteration, inspection and spotting	5.8	19.8	62.43
Club and lodge rooms	0.8	2.7	8.61
Courtrooms	0.7	2.4	7.53
Depots, air terminals, and stations			
Baggage checkroom	1.0	3.4	10.76
Concourse (main thruway)	0.6	2.0	6.46
Platforms	0.5	1.7	5.38
Ticket counter	1.7	5.8	18.30
Waiting and lounge area	0.6	2.0	6.46
Hotels			
Bathrooms (public)	1.0	3.4	10.76
Entrance foyer	0.8	2.7	8.61
Lobby, general	0.8	2.7	8.61
Laundries			
Fine hand ironing	2.1	7.2	22.60
Ironing, weighing, listing, marking	1.0	3.4	10.76
Machine and press finishing, sorting	1.3	4.4	14.00
Washing	0.6	2.0	6.46
Library			
Audio listening areas, general	0.6	2.0	6.46
Audiovisual areas	1.3	4.4	14.00
Book stacks (active)	0.7	2.4	7.53
Book stacks (inactive)	0.4	1.4	4.31
Book repair and binding	1.4	4.8	15.07
Card files	2.4	8.2	25.83
Cataloging	1.7	5.8	18.30
Microfilm areas	1.7	5.8	18.30
Reading areas	1.7	5.8	18.30
Municipal building, fire and police			
Fire engine room	0.6	2.0	6.46
Firemen's dormitory	1.4	4.8	15.07
Identification records	3.5	12.0	37.67
Jail cells	0.6	2.0	6.46
Recreation room	0.7	2.4	7.53
Nursing homes			
Administrative and lobby areas	1.1	3.8	11.84
Chapel or quiet areas, general	0.7	2.4	7.53

THERMAL CONTROL

TABLE 5.4 **Electric Lighting Power Densities (*continued*)**

Occupancy Areas	Power Densities W/ft²	Btu/h ft²	W/m²
Nurses' stations	1.1	3.8	11.84
Occupational therapy	1.0	3.4	10.76
Pharmacy area, general	1.2	4.1	12.92
Physical therapy	1.4	4.8	15.07
Recreation area	1.1	3.8	11.84
Post offices			
Lobby	0.6	2.0	6.46
Sorting, mailing, etc.	2.1	7.2	22.60
Printing			
Composing	3.0	10.2	32.29
Electrotyping	1.8	6.1	19.38
Photoengraving	1.8	6.1	19.38
Schools			
Art	2.3	7.9	24.76
Classrooms	1.7	5.8	18.30
Dormitories	1.1	4.8	11.84
Drafting	2.4	8.2	25.83
Home economics	1.1	4.8	11.84
Laboratories	2.1	7.2	22.60
Lecture	1.7	5.8	18.30
Music	1.3	4.4	14.00
Sewing	3.0	10.2	32.29
Shops	2.1	7.2	22.60
Study halls or typing	1.7	5.8	18.30
Service stations, auto	0.6	2.0	6.46
Stores			
Alteration and fitting	4.3	14.7	46.28
Circulation	0.7	2.4	7.53
Merchandise	2.9	9.9	31.22
Sales transaction	1.4	4.8	15.07
Show windows	6.5	22.2	69.97
Stockrooms	0.6	2.0	6.46
Wrapping and packing	1.0	3.4	10.76
Theaters and motion picture houses	0.8	2.7	8.61
Upholstering—automobile, coach, furniture	2.6	8.9	27.99
Sports			
Seating area—all sports	0.4	1.4	4.31
Badminton			
Club	0.5	1.7	5.38
Recreational	0.4	1.4	4.31
Tournament	0.6	2.0	6.46
Basketball			
College and professional	1.1	3.8	11.84
College and intramural, high school	0.6	2.0	6.46
Bowling			
Approach area	0.4	1.4	4.31
Lanes	0.5	1.7	5.38

TABLE 5.4 **Electric Lighting Power Densities (*continued*)**

Occupancy Areas	*Power Densities*		
	W/ft²	*Btu/h ft²*	*W/m²*
Boxing or wrestling (ring)			
Amateur	1.9	6.5	20.45
Championship or professional	3.8	13.0	40.90
Gymnasiums (refer to individual sports listed)			
Exhibitions, matches	1.1	3.8	11.84
General exercising and recreation	0.6	2.0	6.46
Handball			
Club	0.6	2.0	6.46
Recreational	0.5	1.7	5.38
Tournament	1.1	3.8	11.84
Ice hockey			
Amateur	1.1	3.8	11.84
College or professional	2.1	7.2	22.60
Recreational	0.5	1.7	5.38
Skating rinks	0.4	1.4	4.31
Swimming			
Exhibitions	1.0	3.4	10.76
Recreational	0.6	2.0	6.46
Tennis			
Professional (class I)	2.1	7.2	22.60
Club (class II)	1.4	4.8	15.07
Recreational (class III)	1.1	3.8	11.84
Tennis, table			
Club	0.6	2.0	6.46
Recreational	0.5	1.7	5.38
Tournament	1.1	3.8	11.84
Volleyball	0.5	1.7	5.38

Exterior Areas		
Parking lots (open): watts per space: 20 private, 30 public		
Entrances without canopy	30.0 W/ft	98.43 W/m
Exits, w/without canopy	20.0 W/ft	65.62 W/m
Loading doors	20.0 W/ft	65.62 W/m
Driveways		
Private (based on 2-lane width)	2.0 W/ft	6.56 W/m
Public (based on 2-lane width)	3.0 W/ft	9.84 W/m
Loading areas	0.3 W/ft²	3.23 W/m²
Entrances with canopy		
Decorative (retail, hotel, theater, etc.)	10.0 W/ft²	107.64 W/m²
Utilitarian (hospital, office, industrial, etc.)	4.0 W/ft²	43.06 W/m²

Source: Pacific Northwest Laboratory, *Draft Recommendations for Energy Conservation Standards and Guidelines for New Commercial Buildings,* prepared August 1983 for U.S. Department of Energy with the assistance of ASHRAE.

TABLE 5.5 **Area Factors (for use with the power densities from Table 5.4)**

Area		Ceiling Height, ft (m)									
(ft.²)	*(m²)*	*8 (2.4)*	*8.5 (2.6)*	*9 (2.7)*	*10 (3.0)*	*11 (3.4)*	*12 (3.7)*	*14 (4.3)*	*16 (4.9)*	*18 (5.5)*	*20 (6.1)*
50	4.6	2.00									
60	5.6	1.90	2.00								
70	6.5	1.80	1.92	2.00							
80	7.4	1.72	1.82	1.94							
90	8.4	1.66	1.75	1.85	2.00						
100	9.3	1.61	1.69	1.79	1.98						
110	10.2	1.56	1.64	1.73	1.91						
120	11.1	1.53	1.60	1.68	1.85	2.00					
130	12.1	1.50	1.57	1.64	1.80	1.97					
140	13.0	1.47	1.54	1.61	1.76	1.92					
150	13.9	1.44	1.51	1.57	1.72	1.87	2.00				
160	14.9	1.42	1.48	1.55	1.68	1.83	1.99				
170	15.8	1.40	1.46	1.52	1.65	1.79	1.94				
180	16.7	1.39	1.44	1.50	1.62	1.75	1.90				
190	17.7	1.37	1.42	1.48	1.60	1.72	1.86				
200	18.6	1.36	1.41	1.46	1.57	1.70	1.83				
220	20.4	1.33	1.38	1.43	1.53	1.65	1.77	2.00			
240	22.3	1.31	1.35	1.40	1.50	1.60	1.72	1.97			
260	24.2	1.29	1.33	1.38	1.47	1.57	1.67	1.91			
280	26.0	1.27	1.31	1.36	1.44	1.54	1.64	1.86			
300	27.9	1.26	1.30	1.34	1.42	1.51	1.60	1.81	2.00		
350	32.5	1.23	1.26	1.30	1.37	1.45	1.54	1.72	1.93		
400	37.2	1.20	1.23	1.27	1.34	1.41	1.48	1.65	1.83	2.00	
450	41.8	1.18	1.21	1.24	1.31	1.37	1.44	1.59	1.76	1.94	
500	46.5	1.17	1.19	1.22	1.28	1.34	1.41	1.55	1.70	1.87	2.00
550	51.1	1.15	1.18	1.21	1.26	1.32	1.38	1.51	1.65	1.81	1.98
600	55.7	1.14	1.16	1.19	1.24	1.30	1.35	1.48	1.61	1.75	1.91
700	65.0	1.12	1.14	1.17	1.21	1.26	1.31	1.42	1.54	1.67	1.81
800	74.3	1.10	1.13	1.15	1.19	1.24	1.28	1.38	1.49	1.60	1.73

900	84	1.09	1.11	1.13	1.17	1.21	1.26	1.35	1.45	1.55	1.66
1,000	93	1.08	1.10	1.12	1.16	1.19	1.24	1.32	1.41	1.51	1.61
1,500	139	1.05	1.06	1.07	1.10	1.13	1.17	1.23	1.30	1.37	1.45
2,000	186	1.02	1.04	1.05	1.07	1.10	1.13	1.18	1.24	1.30	1.36
2,500	232	1.01	1.02	1.03	1.05	1.08	1.10	1.15	1.20	1.25	1.30
3,000	279	1.00	1.01	1.02	1.04	1.06	1.08	1.12	1.17	1.21	1.26
4,000	372		1.00	1.00	1.02	1.04	1.05	1.09	1.13	1.17	1.20
5,000	465				1.01	1.02	1.04	1.07	1.10	1.13	1.17
6,000	557				1.00	1.01	1.02	1.05	1.08	1.11	1.14
7,000	650					1.00	1.01	1.04	1.07	1.09	1.12
8,000	743						1.01	1.03	1.06	1.08	1.11
9,000	836						1.00	1.02	1.05	1.07	1.09
10,000	929							1.02	1.04	1.06	1.08
20,000	1,858							1.00	1.00	1.01	1.03
30,000	2,787									1.00	1.00

Source: Pacific Northwest Laboratory, *Draft Recommendations for Energy Conservation Standards and Guidelines for New Commercial Buildings,* prepared August 1983 for U.S. Department of Energy with the assistance of ASHRAE.

TABLE 5.6 **Overall Heat Loss Criteria**

Annual Heating Degree Days (base 65 F)	Maximum Heat Loss (BTU/DD ft²) Conventional Buildings	Passively Solar Heated Buildings, Exclusive of Solar Wall
Less than 1000	9	7.6
1000–3000	8	6.6
3000–5000	7	5.6
5000–7000	6	4.6
Greater than 7000	5	3.6

Source: Balcomb et al. (1980).

velope component (roof, walls, floor, windows, etc.) the U value (Chapter 4) and the total exposed area A, then multiply $U \times A$. (For slab floors on grade, see Chapter 4 for determining perimeter heat losses.) For infiltration, determine the number of air changes per hour (ACH) under winter design conditions (Section 4.6) and multiply this infiltration (or fresh air) rate by a constant that accounts for density and specific heat:

$$\left.\begin{array}{l}\text{ACH (volume, ft}^3) \times 0.018 = \\ \textit{or} \\ \text{ACH (volume, m}^3) \times 0.33 = \end{array}\right\} \textit{UA} \text{ for infiltration}$$

Add the envelope UA values to those for infiltration, multiply by 24 h/day to account for degree days (DD), and divide by the building's total heated floor area:

$$\frac{(UA_{\text{envelope}} + UA_{\text{infiltration}}) \times 24\text{ h}}{\text{total heated floor area (ft}^2)} = \text{Btu/DD ft}^2$$

2. Passively solar heated buildings. Here, the overall rate of Btu/DD ft² *excludes* the solar collecting portion(s) of the envelope; otherwise, it is also based on total heat loss from all other portions of the envelope, and it includes infiltration. The equation used to determine the overall rate is

$$\frac{(UA_{\text{envelope, except south glass}} + UA_{\text{infiltration}}) \times 24\text{ h}}{\text{total heated floor area (ft}^2)} = \text{Btu/DD ft}^2$$

One of the biggest unknowns in this procedure is the assumed rate of infiltration. A carefully designed and constructed small building can easily achieve a rate of 0.75 ACH; with increased attention to infiltration (vapor) barrier installation, caulking of all cracks, and so on, rates of 0.33 ACH have been demonstrated.

(d) Heating and Cooling: Criteria for Components of the Envelope. The preceding whole-building, single-number criteria have several limitations; they do not satisfactorily account for differences in building size (and the accompanying differences in the likely rates of internal heat gains), nor do they distinguish between the thermal roles of roofs and walls, which can be quite different seasonally. The following criteria are taken from ASHRAE Standard 90A–1980, Energy Conservation in New Building Design. This standard is frequently updated; see the discussion in Section 5.2.

An important basis of these criteria is the concept of an average U value for all portions of vertical walls, called U_{overall} walls or U_o walls; similarly, an average roof U value is called U_o roofs. This value is calculated in the same way as any other weighted average:

$$U_o \text{ walls} = \frac{(U_{\text{opaque wall}} \times A_{\text{opaque wall}}) + (U_{\text{window}} \times A_{\text{window}}) + \cdots \text{ etc.}}{A_o \text{ walls}}$$

where A_o includes *all* the areas of all vertical exposed surfaces (walls, windows, doors, etc.)—in other words, the sum of all the areas listed in the numerator of the equation. Note that *each* wall component that is thermally different from other wall components is treated separately in the equation; thus $(U_{\text{wall}} \times A_{\text{wall}})$ might become $(U_{\text{masonry wall}} \times A_{\text{masonry wall}})$ plus $(U_{\text{frame wall}} \times A_{\text{frame wall}})$, and so on.

Similarly,

$$U_o \text{ roofs} = \frac{(U_{\text{opaque roof}} \times A_{\text{opaque roof}}) + (U_{\text{skylight}} \times A_{\text{skylight}}) + \ldots \text{ etc.}}{A_o \text{ roofs}}$$

where A_o roofs is, again, the sum of all horizontal or sloped surface areas, from the numerator.

These criteria are listed separately for two major types of buildings, residential (type A) and all other (type B). Within each type, differences are shown between larger and smaller buildings. Type A buildings are shown first (item 1), followed by two items common to both types A and B buildings—namely, exposed floors (item 2) and infiltration (item 3). The more complex criteria for type B are shown last (item 4).

1. *Residential (type A) buildings*. This includes detached one- and two-family dwellings (A_1) and all other residential buildings that are *three stories or less* (A_2) including multifamily dwellings, hotels, and motels. (For four or more floors, see type B criteria.)

Roof/ceiling: Figure 5.4 shows the *maximum* value of U_o roofs for buildings of types A_1 and A_2 that are heated and/or mechanically cooled. For heating degree days, see Appendix B. There is an exception for "cathedral" ceilings, where the finished interior surface is essentially the exposed underside of the structural roof deck. In this case, for any geographic location, U_o roofs maximum = 0.08 Btu/ft^2 h F (0.45 W/m^2 °C; units are interchangeable with W/m^2 K).

Walls: Figure 5.5 shows the *maximum* value of U_o walls for buildings of types A_1 and A_2 that are heated and/or mechanically cooled. In the special case of warmer climates below 500 heating degree days Fahrenheit (278 DD Celsius), buildings that are mechanically cooled have a maximum U_o wall as follows: type A_1, 0.30 Btu/ft^2 h F (1.70 W/m^2 °C); type A_2, 0.38 Btu/ft^2 h F (2.15 W/m^2 °C).

Slab floors on grade: Figure 5.6 shows the minimum required thermal resistance (*R* value) for *perimeter insulation,* which extends downward from the top of the slab for a minimum distance of 24 in. (0.6 m). Alternatively, such insulation should extend from the top of the slab to the bottom, then horizontally below the slab for a minimum distance of 24 in. (0.6 m).

2. *Exposed floors*. The criteria for floors apply both to type A and type B buildings, and two conditions are covered here. Heated and/or mechanically cooled spaces with floors that are exposed to *exterior* conditions should have, at maximum, a floor U_o as shown in Fig. 5.4 (same as for roofs). However, for floors over protected, but unheated spaces, the maximum U_o is as shown in Fig. 5.7.

3. *Infiltration*. For both A and B building types, air leakage rates are based on a pressure differential equivalent to the effect of a 25-mph (11.1-m/s) wind. Technically, this is a pressure differential of 1.57 lb/ft^2 (75 Pa).

Windows should have a maximum air infiltration rate of 0.5 cfm per foot (7.74×10^{-4} m^3/s per meter) of sash crack.

Doors for residential living units should have a maximum air infiltration rate of 0.5 cfm per square foot (2.54×10^{-3} m^3/s per square meter) of door area.

Doors for other usages should have a maximum air infiltration rate of 11 cfm per linear foot (1.7×10^{-2} m^3/s per linear meter) of door crack.

Other joints in the exterior envelope that can be sources of air leakage (around window and door frames, between walls and foundation, walls and roof, at utility penetrations, etc.) should be caulked, gasketed, weather stripped, or otherwise sealed.

4. *All other (type B) buildings*. Since many nonresidential buildings have plentiful sources of internal heat gain, the insulative value of walls may not be particularly helpful; such buildings may need to lose heat for much of the cold season, rather than hold it inside. For buildings with especially high internal gains or particularly long hours of operation, a more complicated whole-year energy analysis is desirable. For the more ordinary nonresidential building, however, these heating- and cooling-based criteria will assist in achieving energy-conserving envelope design.

Floors were presented in item 2 above.

Infiltration was presented in item 3 above.

Roofs: Figure 5.8 shows the maximum value of U_o roofs for any type B building that is heated or mechanically cooled. For type B buildings that are mechanically cooled, a more complex analysis is also necessary, to account for the impact of direct sun on roof surfaces. This analysis is based on an *overall thermal trans-*

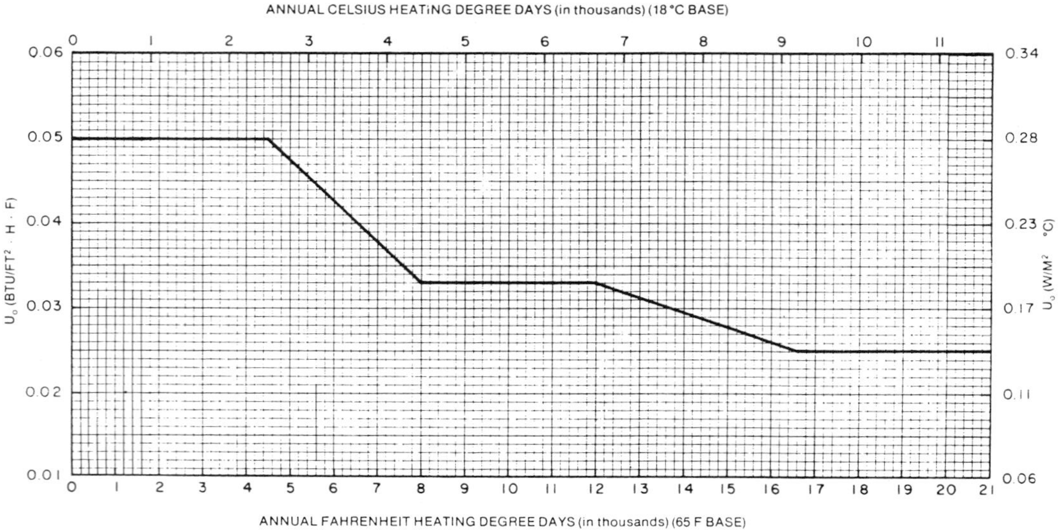

Fig. 5.4. U_o *values for roofs/ceilings, type* A_1 *and* A_2 *(small residential) buildings. Copyright © by the American Society of Heating, Refrigerating and Air Conditioning Engineers, Inc., Atlanta, GA. Reprinted by permission from* ASHRAE Standard 90A-1980.

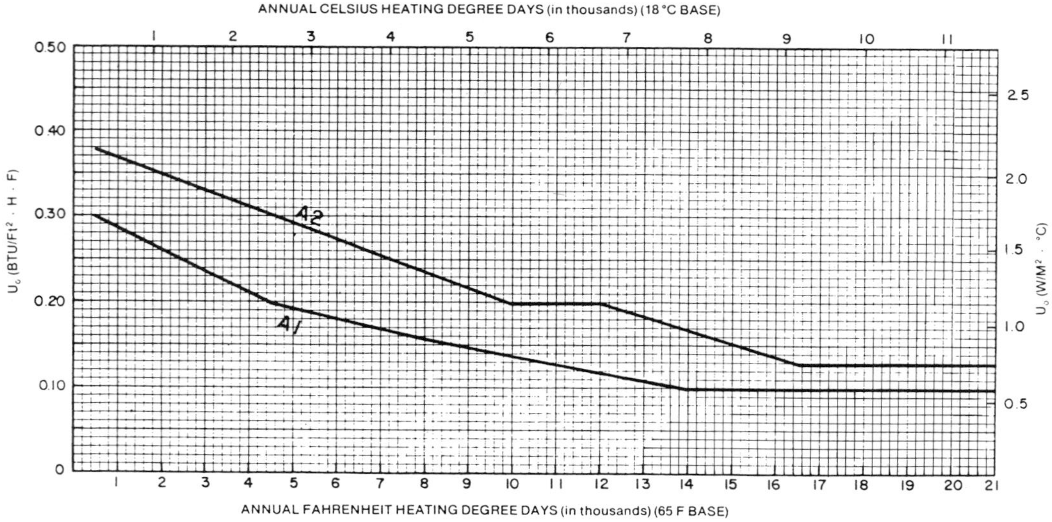

Fig. 5.5. U_o *values for walls, type* A_1 *and* A_2 *(small residential) buildings. Copyright © by the American Society of Heating, Refrigerating and Air Conditioning Engineers, Inc., Atlanta, GA. Reprinted by permission from* ASHRAE Standard 90A-1980.

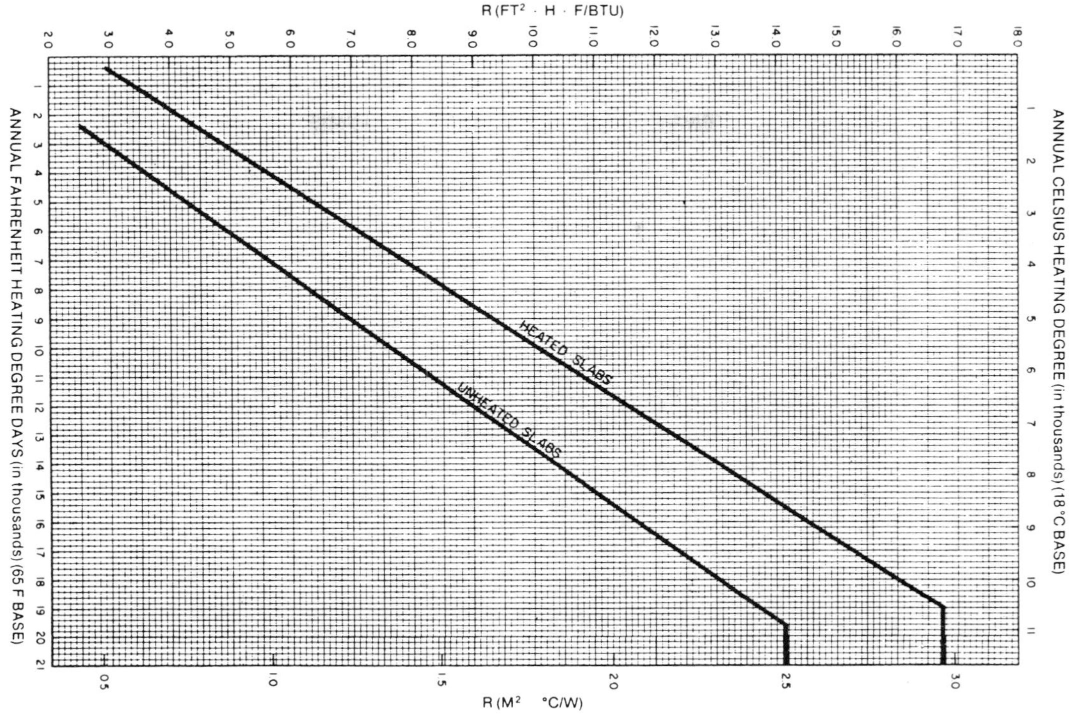

Fig. 5.6. R values for slab floors on grade. Copyright © by the American Society of Heating, Refrigerating and Air Conditioning Engineers, Inc., Atlanta, GA. Reprinted by permission from ASHRAE Standard 90A-1980.

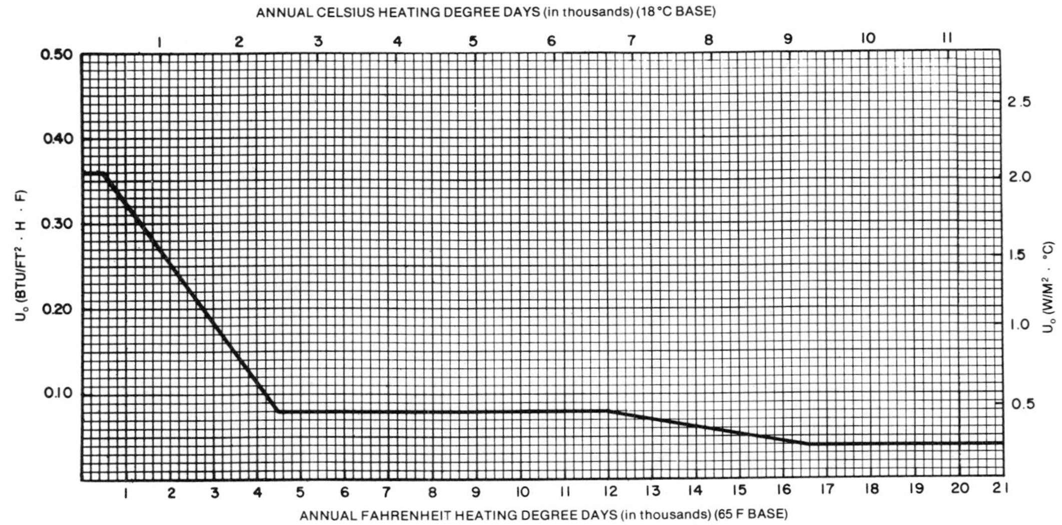

Figure 5.7. U_o *values for floors over unheated spaces. Copyright © by the American Society of Heating, Refrigerating and Air Conditioning Engineers, Inc., Atlanta, GA. Reprinted by permission from* ASHRAE Standard 90A-1980.

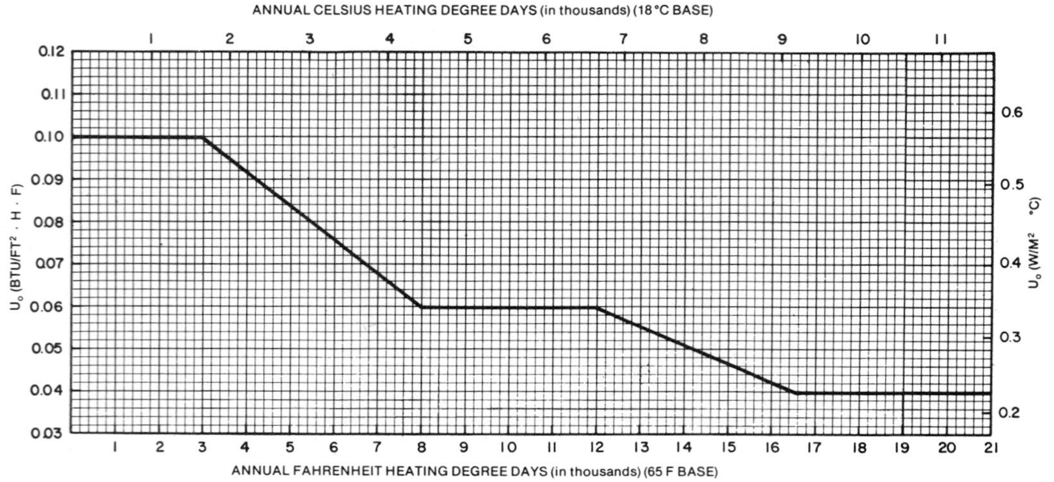

Fig. 5.8. U_o *values for roofs/ceilings, type B buildings. Copyright © by the American Society of Heating, Refrigerating and Air Conditioning Engineers, Inc., Atlanta, GA. Reprinted by permission from* ASHRAE Standard 90A-1980.

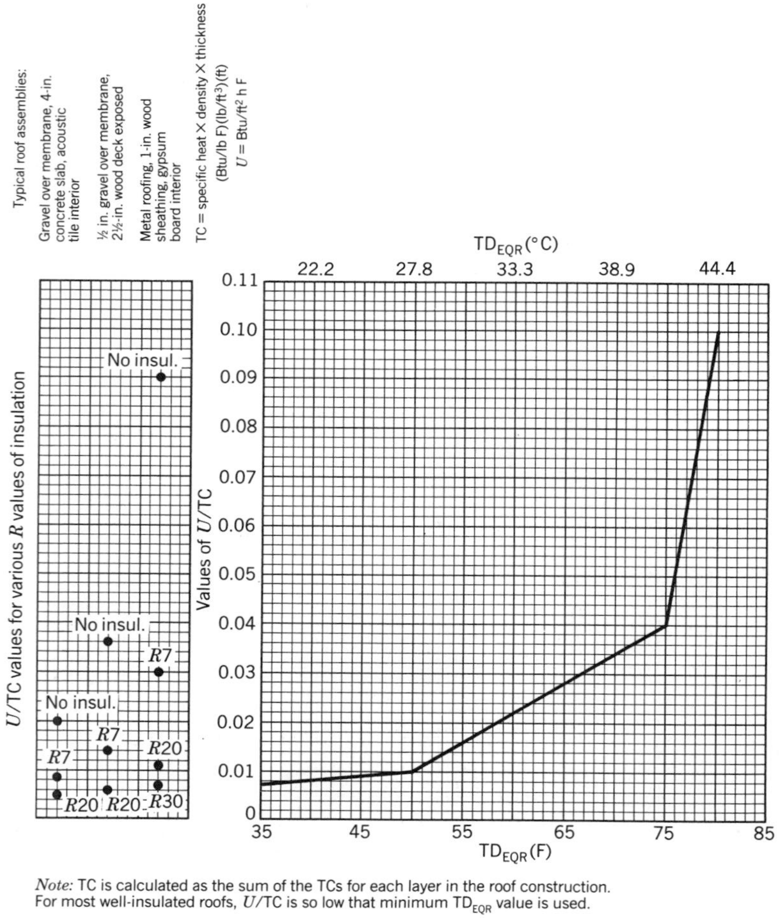

Note: TC is calculated as the sum of the TCs for each layer in the roof construction. For most well-insulated roofs, *U*/TC is so low that minimum TD_{EQR} value is used.

Fig. 5.9. Values of TD_{EQR}*, along with representative roof/ceiling assemblies. Copyright © by the American Society of Heating, Refrigerating and Air Conditioning Engineers, Inc., Atlanta, GA. Reprinted by permission from* ASHRAE Standard 90A-1980.

fer value for roofs ($OTTV_r$). In conventional units, *maximum* $OTTV_r$ is 8.5, where

$$OTTV_r = \frac{U_r A_r \, TD_{EQR} + (138 A_s SC_s) + U_s A_s \, \Delta T_s}{A_o}$$

In SI units, *maximum* $OTTV_r$ is 26.8, where

$$OTTV_r = \frac{U_r A_r TD_{EQR} + (434.7 A_s SC_s) + U_s A_s \, \Delta T_s}{A_o}$$

where

U_r and A_r are, as before, the U value and area, respectively, of the opaque roof/ceiling construction.

TD_{EQR} is a temperature difference factor accounting for a roof's thermal mass, given in Fig. 5.9; for most well-insulated roofs, TD_{EQR} = 35F° (19.4C°).

Note: If more than one type of roof is used, each type is calculated: $U_{r1}A_{r1}TD_{EQR1} + U_{r2}A_{r2}TD_{EQR2} + \cdots$.

A_s is the area of skylights.

SC_s is the shading coefficient of the skylight (Tables 5.37, 5.38, 5.41, 5.42).

U_s is the U value of the skylight.

ΔT_s is the design temperature difference between inside and outside.

A_o is the overall area of all roof and skylight surfaces.

Walls: Figure 5.10 shows the maximum value

THERMAL CONTROL

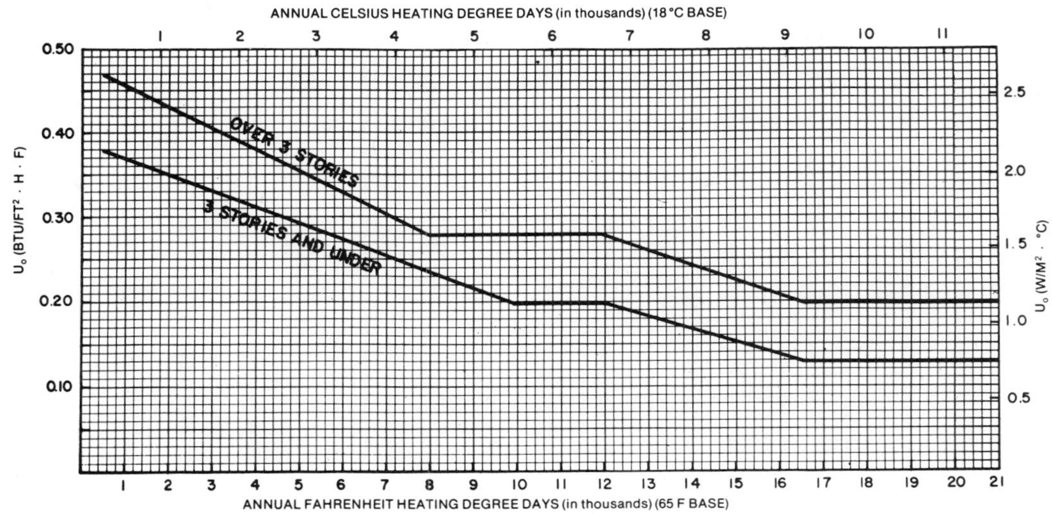

Fig. 5.10. *U_o values for walls, type B buildings. Copyright © by the American Society of Heating, Refrigerating and Air Conditioning Engineers, Inc., Atlanta, GA. Reprinted by permission from* ASHRAE Standard 90A-1980.

of U_o walls for any type B building that is mechanically heated. For type B buildings that are mechanically cooled, the more complex overall thermal transfer procedure is necessary. Figure 5.11 shows the *maximum* values of $OTTV_w$, calculated as

$$OTTV_w = \frac{U_w A_w TD_{EQ} + (A_f SF\ SC) + U_f A_f\ \Delta T}{A_o}$$

where

U_w and A_w are, as before, the U value and area, respectively, of the opaque wall construction.

TD_{EQ} is a temperature difference factor accounting for wall thermal mass, given in Fig. 5.12.

Note: If more than one type of wall is used, each type is calculated: $U_{w1}A_{w1}TD_{EQ1} + U_{w2}A_{w2}TD_{EQ2} + \cdots$.

U_f and A_f are the U value and area, respectively, of the fenestration (windows).

SF is a "solar factor," given in Fig. 5.13.

SC is the shading coefficient of the windows; see the discussion in Section 5.8 and Tables 5.37–5.42.

ΔT is the temperature difference between inside and outside.

A_o is the overall area of all exterior wall and window surfaces.

EXAMPLE 5.1. Use the guidelines given in this section to make preliminary design decisions. A five-story office building in Denver is to be designed for the potential of passive cooling:

Denver: heating DD approx. 6000 (Appendix B)
latitude 39°N

Table 5.1 shows that Denver is in climate C; heating and ventilation are likely to be the most

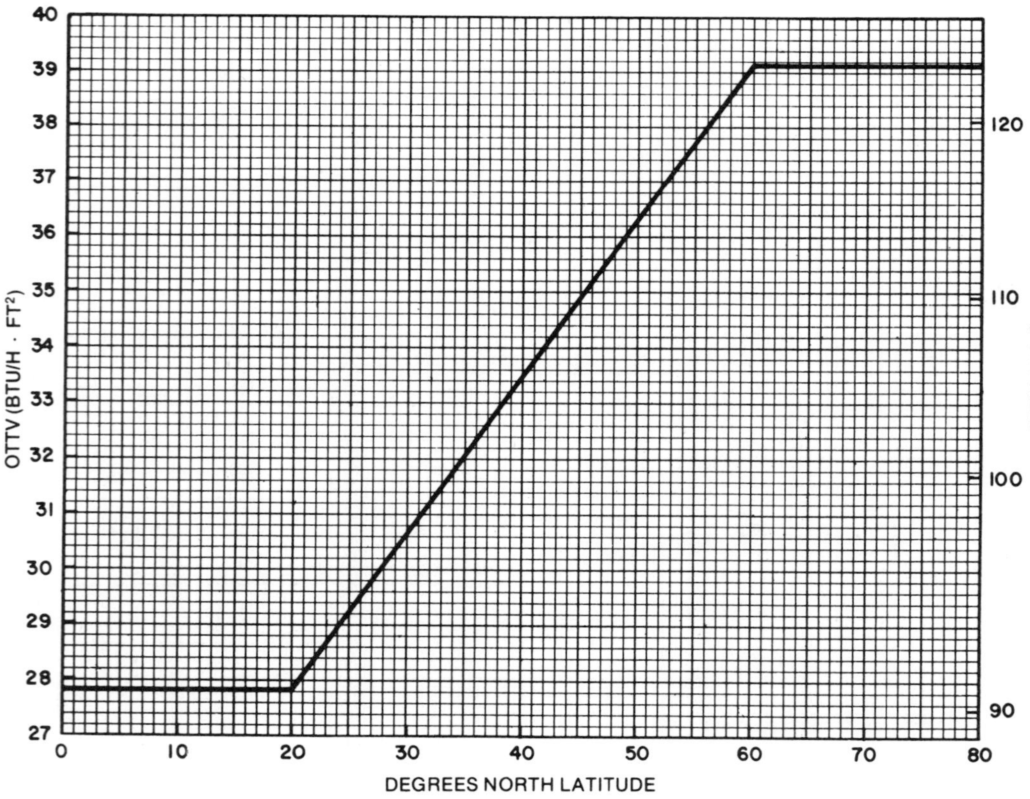

Fig. 5.11. Overall thermal transfer values ($OTTV_w$) for walls of type B buildings that are mechanically cooled. Copyright © by the American Society of Heating, Refrigerating and Air Conditioning Engineers, Inc., Atlanta, GA. Reprinted by permission from ASHRAE Standard 90A-1980.

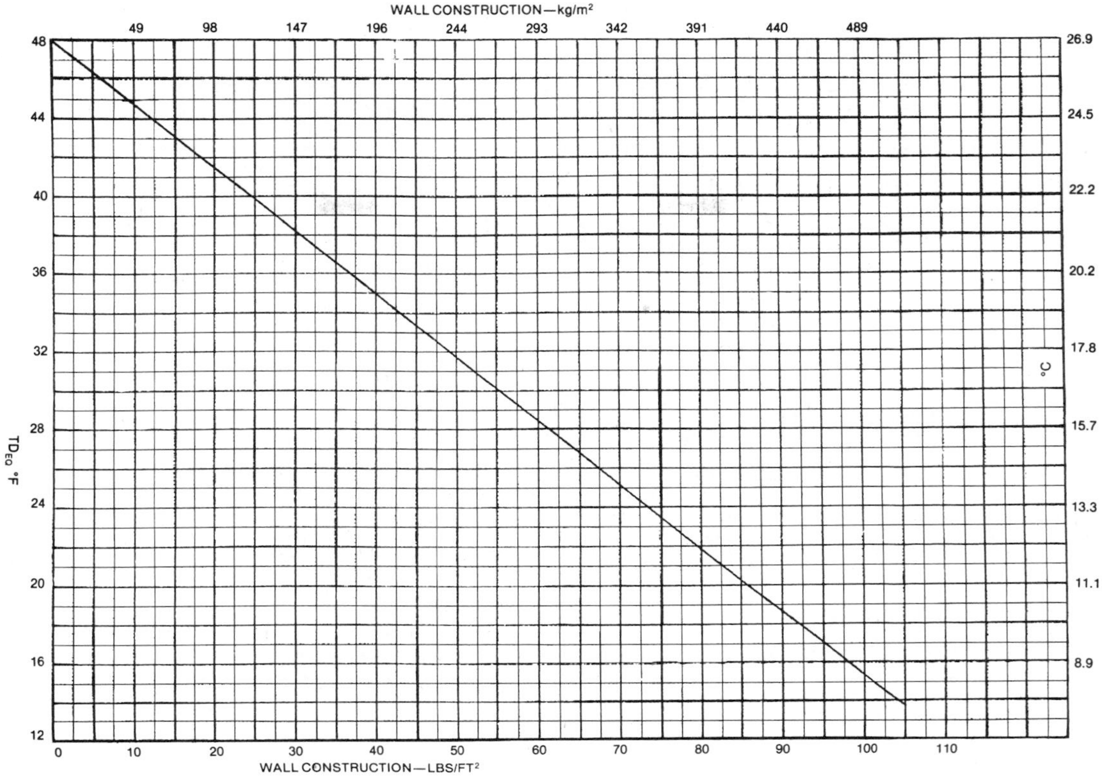

(a)

EXTERIOR WALL ASSEMBLIES FOR ADDITIONAL INFORMATION CONSULT MANUFACTURERS' LITERATURE AND TRADE ASSOCIATIONS		WALL THICKNESS (NOMINAL) (IN.)	WEIGHT (PSF)
C. M. U.* (INSULATED)	C. M. U. INSULATION INT. WALL FIN.	8 + 12 +	60 90
C. M. U.* AND BRICK VENEER (INSULATED)	BRICK VENEER C. M. U. INSULATION INT. WALL FIN.	4 + 4 + 4 + 8 +	75 100
CAVITY	BRICK VENEER CAVITY (MIN. 2") INSULATION (WATER REPELLENT) C. M. U. INT. WALL FIN.	4 + 2 + 4 4 + 2 + 8	75 100
C. M. U.* AND STUCCO (INSULATED)	STUCCO C. M. U. INSULATION INT. WALL FIN.	8 +	67
WOOD STUD	EXT. WALL FIN. SHEATHING WITH MOISTURE BARRIER WOOD STUD INSULATION WITH VAPOR BARRIER INT. WALL FIN	4 6	12 16
BRICK VENEER	BRICK VENEER SHEATHING WITH MOISTURE BARRIER METAL STUD AT 16" O.C. INSULATION WITH VAPOR BARRIER INT. WALL FIN.	4 + 4	54
INSULATED SANDWICH PANEL	METAL SKIN AIRSPACE INSULATING CORE METAL SKIN	5	6
CONCRETE (INSULATED)	CONCRETE INSULATION INT. WALL FIN.	8 +	97
PRECAST CONCRETE	CONCRETE (REINFORCED) INSULATION INT. WALL FINISH	2 + 4 +	23 46

*C.M.U. = CONCRETE MASONRY UNIT

(b)

Fig. 5.12. *Figuring the temperature difference (TD_{EQ}) for calculating $OTTV_w$ for walls of mechanically cooled buildings, type B.* (a) *Values of TD_{EQ} as they vary with wall weight. Copyright © by the American Society of Heating, Refrigerating and Air Conditioning Engineers, Inc., Atlanta, GA. Reprinted by permission from* ASHRAE Standard 90A-1980. (b) *Typical wall assemblies and their weights. From AIA, Ramsey and Sleeper* Architectural Graphic Standards, *7th ed., John Wiley & Sons, Inc., 1981. Reprinted by permission.*

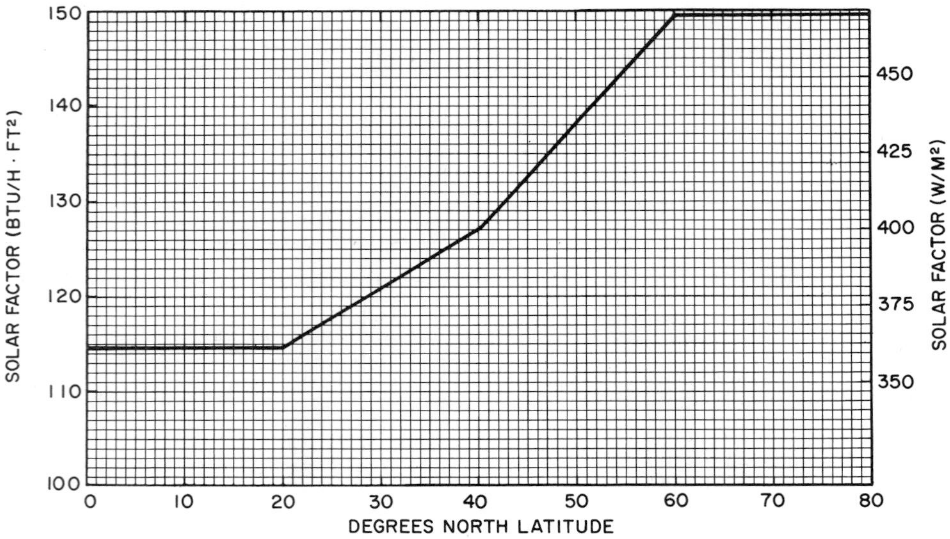

Fig. 5.13. Solar factor (SF) values, for use in calculating $OTTV_w$ *for walls of mechanically cooled buildings, type B. Copyright © by the American Society of Heating, Refrigerating and Air Conditioning Engineers, Inc., Atlanta, GA. Reprinted by permission from* ASHRAE Standard 90A-1980.

energy intensive, followed by lighting, then by cooling (including the cooling of ventilating air). Since solar heating is a strong possibility in Denver's winter, and since lighting energy usage is high, daylighting would seem a high priority. From Table 5.2, select a daylight factor (DF = 3.5% to cover nearly all office lighting situations). Then, assuming sidelighting only (for a five-story building), from Table 5.3:

$$\text{if } DF_{av}3.5 = 0.2\,\frac{\text{window area}}{\text{floor area}}$$

$$\text{then window area} = \frac{3.5\%}{0.2} = 17.5\% \text{ of floor area}$$

Assuming a cross section that keeps the office floor area within 2.5H of the side walls for good daylight penetration (from Table 5.3), and assuming office areas 25 ft deep:

$$\frac{25}{2.5} = 10 \text{ ft window height above floor}$$

The building's total width is 25 ft office + 10 ft circulation/storage + 25 ft office = 60 ft, and floor-to-floor height is 10 ft + 2 ft (for structure above window) = 12 ft. (See Fig. 5.14 for cross section and area tabulations.)

The relative wall area is thus (2 walls × 12 ft tall)/60-ft-wide floor = 40% of the floor area. And since windows are 17.5% of the floor area, the windows are 17.5%/40% = 44% of the overall wall area. With these basic relationships set, the analysis for thermal criteria can proceed.

A long, thin building will be helpful for both sidelighting and cross ventilation; if the long sides face north–south, sunshading problems will be minimized and winter sun can help at least half the building. For this analysis, we will concentrate on these long window walls; the shorter and nearly windowless end walls can be assumed to be safely within any thermal criteria that are met by the window walls.

Walls: To determine the U value of opaque walls with such window areas, refer to Fig. 5.10.

Wall U_o maximum is 0.33 for type B, over three stories, at 6000 DD (Fig. 5.10). If ordinary double-glazed windows are used, U = 0.59 (winter, ¼-in. air space, no inside shade) (from Table 4.17*a*). Required winter wall insulation U_w can now be found:

$$0.33 = \frac{(56\% \times U_w) + (44\% \times 0.59)}{100\%}$$

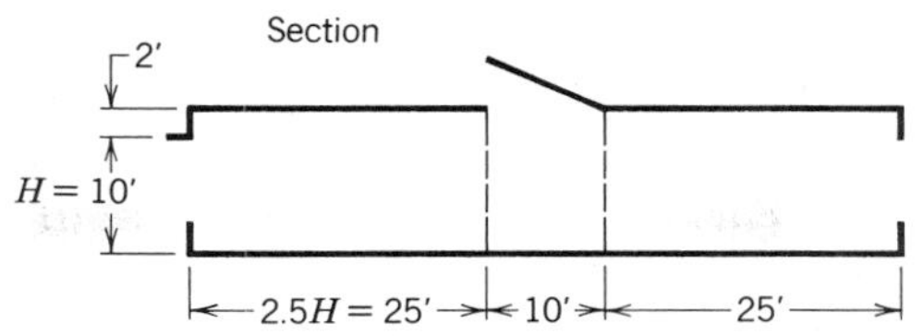

Component	Percentage of Floor Area	Percentage of Overall North–South Wall Area
North plus south wall, overall area	40	(100)
North plus south glass area	17.5	44
North plus south opaque wall area	22.5	56
(For top floor only:)		
Skylight	4	
Opaque roof area	96	

Note: End walls (east and west) are considerably shorter than north and south walls, and nearly windowless. Their influence on heat gain or heat loss is therefore minor and can be ignored in a preliminary rule-of-thumb analysis.

Fig. 5.14. *Cross section for office building in Example 5.1.*

$$U_w = \frac{0.33 - 0.26}{0.56} = 0.125$$

Note that the approximate R value of the wall insulation is simply $1/U = 1/0.125 = 8$. The $R = 8$ can be obtained with approximately 2 in. of either fiberglass or extruded polystyrene (see Table 4.2).

However, the wall should also meet summer cooling requirements, for which OTTV_w must be calculated. Hence, decisions must be made about the wall construction thermal mass and about window shading devices.

Select a masonry cavity wall that allows some thermal mass *indoors,* to aid in passive heating and cooling—for example, the cavity wall shown in Fig. 5.12*b*, for which

$$\text{weight} = 75 \text{ lb/ft}^2 \text{ (Fig. 5.12}b\text{)}$$

$$\text{TD}_{\text{EQ}} = 23.5 \text{ F}° \text{ (Fig. 5.12}a\text{)}$$

Select canvas exterior awnings for summer shading:

$$\text{Shading coefficient} = 0.25 \text{ (Table 5.42)}$$

The "solar factor" for Denver's 39°N latitude is 126 (Fig. 5.13) and the maximum allowable OTTV_w is 33.2 (Fig. 5.11). Assume a rather low ΔT for passive cooling, such as 15F°. The summer U for the double-glazed windows is 0.61 (Table 4.17*a*).

$$\text{actual OTTV}_w = \frac{U_w A_w \text{TD}_{\text{EQ}} + (A_f \text{SF SC}) + U_f A_f \,\Delta T}{A_o}$$

$$= \frac{0.125(56\%)23.5 + (44\%)126 \times 0.25 + 0.61(44\%)15}{100\%}$$

$$= 1.65 + 13.86 + 4.03$$

$$= 19.54$$

which is well below the 33.2 maximum.

Roofs: Although sidelighting has been assumed to provide a DF_{av} of 3.5, the topmost floor could be further daylighted at its interior by small skylights. Since the *minimum* DF (Table 5.3) from sidelighting is 0.1 (window area)/(floor area), we have

$$0.1 \times 17.5\% = 1.75\%$$

A skylight producing an additional 0.75 DF is desirable, to obtain a minimum DF of 2.5 (for average office tasks) at the interior of the topmost floor. From Table 5.3, select a vertical monitor (which is easily shaded and less likely to leak):

$$DF_{av} = 0.2 \frac{\text{skylight area}}{\text{floor area}}$$

$$0.75 = 0.2 \frac{\text{skylight area}}{\text{floor area}}$$

So

$$\text{skylight area} = \frac{0.75}{0.2} = 3.75\% \text{ of floor area}$$

Say the skylight area is 4% of the topmost floor area (which is also 4% of the flat roof's area).

Roof U_o maximum is 0.074 for 6000 DD (Fig. 5.8). Assuming that the vertical glazing of the skylight is the same as the windows, $U = 0.59$ winter, then the required opaque roof insulation U_{r1} can be found:

$$0.074 = \frac{(96\% \times U_r) + (4\% \times 0.59)}{100\%}$$

$$U_r = \frac{0.074 - 0.024}{0.96} = 0.05$$

Note that the approximate R value of the roof insulation is simply

$$\frac{1}{U} = \frac{1}{0.05} = 20$$

The roof should also meet summer cooling requirements, especially in this cooling-dominated example; calculation of $OTTV_r$ is needed, as are decisions as to the roof's mass and insulation. Select a 4-in. concrete slab, which will be useful in thermal storage for passive cooling; cover it with at least $R20$ insulation. The TD_{EQR} for this construction is 35F° (minimum listed in Fig. 5.9). For skylight shading, stay with the canvas awnings whose shading coefficient is 0.25; summer U is 0.61, as before. The maximum allowable $OTTV_r$ is 8.5, as defined.

$$OTTV_r = \frac{U_r A_r TD_{EQR} + 138 A_s SC_s + U_s A_s \Delta T_s}{A_o}$$

$$\text{actual } OTTV_r = \frac{0.05(96\%)35 + 138(4\%)0.25 + 0.61(4\%)15}{100\%}$$

$$= 1.68 + 1.38 + 0.366$$

$$= 3.43$$

which is well below the 8.5 maximum.

(e) Yearly Total Energy: Whole-Building Criteria. When computer assistance is available, the total yearly average consumption of energy can be estimated for the combination of heating, cooling, ventilation, lights, vertical transportation, and domestic hot water in each new building. By varying the proposed design, we can estimate the resulting changes in this total energy consumption. This approach, like the earlier single-number criteria for building heating (Table 5.6), has the advantage of allowing *many combinations* of energy conservation measures, so long as they produce acceptable final results. They also encourage the use of renewable energy, by exempting such sources from budget limitations. Obviously, the calculations that produce this result are lengthy.

Late in 1979, the U.S. Department of Energy published proposed *Building Energy Performance Standards* (BEPS), summaries of which appear in Table 5.7, for commercial and multifamily residential buildings. The objectives of these standards are to achieve the maximum practical improvements in energy efficiency and to increase the use of renewable sources of energy.

Energy budget levels were proposed for 16 frequently built types of commercial and multifamily residential buildings, in 78 metropolitan areas of the 48 contiguous states. (Restaurants and industrial buildings proved too diverse to include, as did university buildings. Single-family houses are covered in detail but are excluded from this presentation.) These energy budget levels are expressed in MBtu/ft^2 year (MBtu = Btu × 10^3; ft^2 = gross floor area). They can be used as targets during the design process, much as cost per square foot is used.

Comparison of a proposed building design's performance to these standards requires the use of one of several computer programs (Appendix H). This establishes the total average yearly MBtu required by the design. Separate totals can be kept for electricity, oil, natural gas, and solar energy. These separate totals can then be *weighted,* to reflect the relative value to the nation of the conservation of scarce and nonrenewable fuel resources. The proposed weighting factors are

MBtu natural gas, × 1.0.

MBtu oil, × 1.20.

MBtu electricity, × 3.08.

TABLE 5.7 **Proposed Energy Budget Levels for Commercial and Multifamily Buildings**

NOTE: These total MBtu/ft^2 year include all energy for heating, cooling, ventilation, domestic hot water, lighting, and vertical transportation. Numbers in **bold face** indicate the total energy purchased from utilities ("building line design energy"); numbers in parentheses indicate the total purchased energy, multiplied by weighting factors for oil and electricity, in the appropriate mix for building type and location.

State	*Standard Metropolitan Statistical Area (SMSA)*	*Clinic*	*Community Center*	*Gymnasium*	*Hospital*	*Hotel/Motel*	*Multifamily High Rise*	*Multifamily Low Rise*	*Nursing Home*	*Office, Large*	*Office, Small*	*School, Elementary*	*School, Secondary*	*Shopping Center*	*Store*	*Theater/Auditorium*	*Warehouse*
Minnesota	Minneapolis	**58**	**45**	**59**	**140**	**74**	**58**	**45**	**72**	**51**	**48**	**50**	**57**	**82**	**64**	**65**	**38**
		(142)	(109)	(144)	(335)	(180)	(140)	(110)	(175)	(123)	(117)	(122)	(138)	(198)	(155)	(157)	(93)
Missouri	St. Louis	**52**	**43**	**53**	**140**	**68**	**50**	**44**	**66**	**47**	**43**	**41**	**50**	**75**	**59**	**58**	**28**
		(133)	(110)	(136)	(353)	(175)	(128)	(112)	(163)	(119)	(109)	(105)	(128)	(192)	(150)	(149)	(72)
District of	Washington	**50**	**42**	**51**	**140**	**66**	**47**	**43**	**64**	**45**	**41**	**37**	**47**	**72**	**56**	**56**	**24**
Columbia		(127)	(107)	(129)	(353)	(169)	(120)	(109)	(164)	(115)	(104)	(96)	(121)	(185)	(144)	(142)	(63)
Florida	Miami	**52**	**48**	**55**	**140**	**69**	**45**	**50**	**68**	**48**	**43**	**35**	**48**	**74**	**61**	**60**	**14**
		(152)	(142)	(161)	(406)	(203)	(133)	(147)	(201)	(140)	(125)	(103)	(141)	(219)	(179)	(178)	(41)
Texas	Dallas	**51**	**45**	**53**	**140**	**67**	**46**	**46**	**66**	**46**	**41**	**36**	**48**	**73**	**58**	**40**	**19**
		(131)	(116)	(136)	(358)	(175)	(119)	(119)	(171)	(120)	(107)	(94)	(124)	(190)	(152)	(150)	(50)
California	San Diego	**43**	**39**	**44**	**140**	**60**	**39**	**40**	**58**	**41**	**35**	**28**	**41**	**65**	**51**	**49**	**15**
		(114)	(103)	(117)	(364)	(158)	(104)	(106)	(153)	(107)	(92)	(75)	(107)	(172)	(134)	(128)	(40)
Oregon	Portland	**47**	**38**	**47**	**140**	**63**	**45**	**39**	**60**	**42**	**38**	**35**	**45**	**69**	**53**	**51**	**26**
		(119)	(98)	(120)	(353)	(161)	(116)	(99)	(154)	(108)	(97)	(91)	(115)	(176)	(135)	(131)	(66)
Massachusetts	Boston	**51**	**41**	**52**	**140**	**67**	**50**	**42**	**65**	**45**	**42**	**41**	**49**	**74**	**57**	**57**	**30**
		(125)	(101)	(126)	(338)	(165)	(121)	(102)	(159)	(111)	(102)	(99)	(121)	(181)	(140)	(139)	(72)

Source: BEPS (1979).

MBtu solar is *excluded,* because it is a renewable resource.

(f) Later Criteria Developments. Recently there has been an effort to merge the best features of ASHRAE Standard 90A-1980 and of BEPS. This procedure would require a refinement of the limited ASHRAE 90A-1980 envelope criteria but would not require the computer processing that is essential for BEPS. The 90A-1980 envelope criteria do not take into account some important energy factors, such as daylighting, orientation, internal loads, building configuration, external sunshading, and internal thermal mass.

Although the final results of this effort were not published in time to be included in this book, a sample will show how the ASHRAE 90A-1980 wall criteria procedure might be improved. Under this proposal, walls would have to meet criteria based on separate requirements for annual heating, annual cooling, and peak cooling. Each criterion is a function of conduction gain and losses, solar gain through glass, and gains from electric lighting (modified by daylighting). To test a wall's compliance with the peak cooling criterion, for example, the following procedure might be used.

1. A ''budget'' is established, based on internal loads, configuration and size of building, cooling design temperature, and glass area (as a percentage of wall area). This budget is a target figure reached by manipulating the parts of a lengthy equation, described as follows.

2. For each wall orientation, the peak cooling criterion equation is solved; the sum of the walls' criteria must fall within the target figure established by the budget. The peak cooling criterion equation contains factors that account for

Mass.
Overall U value (U_o).
Fraction of wall in opaque construction.
Fraction of wall in glass.
Peak solar load for that orientation.
Internal shading coefficient.
External shading.
Power installed for electric lighting.
Percentage of electric lights controlled (adjusted as daylight changes).

It is likely that this more detailed but more accurate procedure will eventually replace ASHRAE 90A-1980 as the basis for the energy standards in the building codes for most of the United States.

5.3 Rules of Thumb for Preliminary Design

With the aid of the guidelines provided in this section, the designer can make early decisions that will lead to energy-efficient buildings that, where possible, utilize on-site energy sources. From the criteria given in the preceding section and these guidelines, a rough estimate of energy performance can often be made. Remember that the simpler the rule of thumb, the cruder the result: the largest divergence from actual performance can be expected from the easiest shortcut! Rules of thumb for daylighting and electric lighting were presented along with the criteria in Section 5.2; the similar guidelines that follow cover various approaches to heating and cooling.

(a) Passive Solar Heating. The energy conservation criteria for passively solar heated buildings were presented in Table 5.6. Given that the U_o wall criterion is difficult to meet with a high percentage of window area, the first design question becomes, what mix of passive collecting area and opaque wall insulation will meet these criteria?

Passive solar heating and energy conservation have a complex relationship. Relative to ''conventional'' buildings, passively solar heated buildings usually conserve purchased energy; yet, buildings that aim at very high percentages of solar heating can use more *total* energy than is used by those with smaller window areas and ''superinsulated'' walls, floors, and roofs. Designers interested primarily in saving purchased energy may aim at lower solar percentages and more insulation; those interested in buildings that closely relate to climate and climatic changes may aim at higher solar percentages, along with higher thermal masses and, probably, greater ranges of indoor temperature.

The solar savings fraction, SSF, is used to evaluate a building's solar performance. The solar savings is the extent to which a solar design reduces a building's auxiliary heat requirement relative to a ''reference'' building—one that has, instead of a solar wall, an energy-neutral wall that experiences neither solar gain nor heat loss; other-

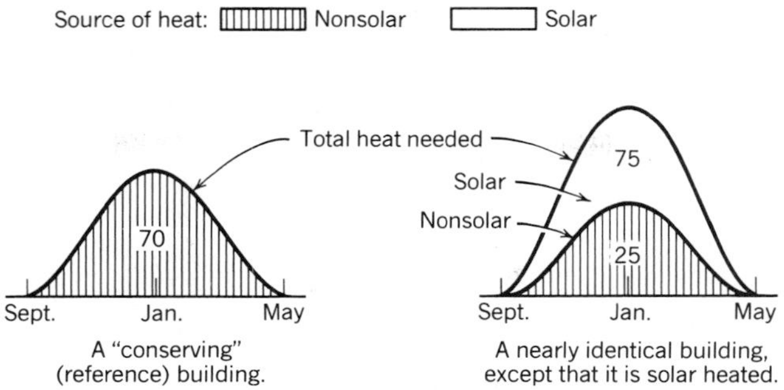

Fig. 5.15. *The solar savings fraction (SSF) compares the auxiliary heat needed by solar-heated buildings to that needed by a nonsolar but energy-conserving building that is otherwise similar, called the "reference" building. In this example, the solar building needs 25 units of auxiliary heat, whereas the reference building needs 70 units. The difference is 70 − 25 = 45, or 64% of the reference 70 units. Therefore, SSF = 64%. (Note, however, that the solar building is 75% solar heated.) From Brown, Reynolds, and Ubbelohde* InsideOut: Design Procedures for Passive Environmental Technologies, *© 1982, John Wiley & Sons, New York. Reprinted by permission.*

wise, the solar building and the reference building are identical. The solar savings fraction compares the auxiliary energy needed by the solar building to the auxiliary energy needed by the reference building, as illustrated in Fig. 5.15. Remember that the SSF is *not* the percentage of the solar building's heat supplied by the sun; typically, the sun provides a much *higher* fraction of a building's heat than does the SSF. Rather, the SSF is more a measure of the solar building's *conservation* advantage.

A starting point for the solar/conservation mix is the combination of solar savings fraction (SSF) and optimum wall insulation shown in the maps of Fig. 5.16. (For a more precise method of optimizing the blend of costs of passive solar heating and wall insulation, see Chapter 2 of Balcomb et al., 1983.) From this range of SSF values, a range of areas of south glass can be determined for passive solar design in the many locations given in Table 5.8. The table shows these areas as percentages of the total floor areas of solar-heated buildings and shows the SSF ranges for both uninsulated and night-insulated solar openings.

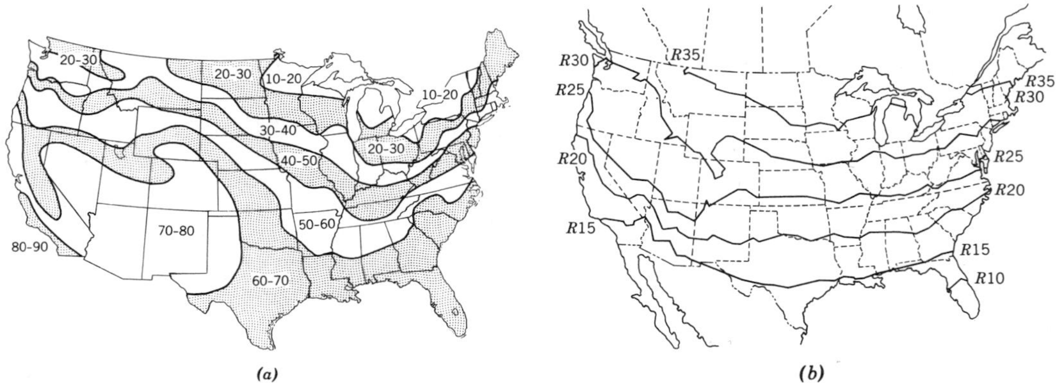

Fig. 5.16. (a) *Recommended ranges of SSF.* (b) *Optimum wall insulation* R *factors for the SSF values in Fig. 5.16a. Combination of these* R *factors with the SSF values will produce a near-optimum mix of solar and conservation under specific conditions, and can be used as a general guide. (Specifically optimum for a semienclosed sunspace with 50° sloped double glazing and a masonry common wall with the house; no night insulation is used. Incremental costs assumed are $0.05/R/ft^2$ for wall insulation and $\$13.80/ft^2$ for the entire passive system.) Note that these wall* R *values usually exceed the criteria from ASHRAE Standard 90A-1980. Reprinted from* Passive Solar Design Handbook, *vol. 3, © 1983, by permission of the publisher, the American Solar Energy Society, Inc.*

THERMAL CONTROL

TABLE 5.8 **Rules of Thumb for Passive Solar Glazing Area**

Location	Area of Solar Glazing[a] as Percentage of Floor Area		Approximate SSF's			
			No Night Insulation		With R9 Night Insulation[b]	
	Low	High	Low	High	Low	High
Birmingham, Alabama	0.09	0.18	22	37	34	58
Mobile, Alabama	0.06	0.12	26	44	34	60
Montgomery, Alabama	0.07	0.15	24	41	34	59
Phoenix, Arizona	0.06	0.12	37	60	48	75
Prescott, Arizona	0.10	0.20	29	48	44	72
Tucson, Arizona	0.06	0.12	35	57	45	73
Winslow, Arizona	0.12	0.24	30	47	48	74
Yuma, Arizona	0.04	0.09	43	66	51	78
Fort Smith, Arkansas	0.10	0.20	24	39	38	64
Little Rock, Arkansas	0.10	0.19	23	38	37	62
Bakersfield, California	0.08	0.15	31	50	42	67
Daggett, California	0.07	0.15	35	56	46	73
Fresno, California	0.09	0.17	29	46	41	65
Long Beach, California	0.05	0.10	35	58	44	72
Los Angeles, California	0.05	0.09	36	58	44	72
Mount Shasta, California	0.11	0.21	24	38	42	67
Needles, California	0.06	0.12	39	61	49	76
Oakland, California	0.07	0.15	35	55	46	72
Red Bluff, California	0.09	0.18	29	46	41	65
Sacramento, California	0.09	0.18	29	47	41	66
San Diego, California	0.04	0.09	37	61	46	74
San Francisco, California	0.06	0.13	34	54	45	71
Santa Maria, California	0.05	0.11	31	53	42	69
Colorado Springs, Colorado	0.12	0.24	27	42	47	74
Denver, Colorado	0.12	0.23	27	43	47	74
Eagle, Colorado	0.14	0.29	25	35	53	77
Grand Junction, Colorado	0.13	0.27	29	43	50	76
Pueblo, Colorado	0.11	0.23	29	45	48	75
Hartford, Connecticut	0.17	0.35	14	19	40	64
Wilmington, Delaware	0.15	0.29	19	30	39	63
Washington, D.C.	0.12	0.23	18	28	37	61
Apalachicola, Florida	0.05	0.10	28	47	36	61
Daytona Beach, Florida	0.04	0.08	30	51	36	63
Jacksonville, Florida	0.05	0.09	27	47	35	62
Miami, Florida	0.01	0.02	27	48	31	54
Orlando, Florida	0.03	0.06	30	52	37	63
Tallahassee, Florida	0.05	0.11	26	45	35	60
Tampa, Florida	0.03	0.06	30	52	36	63
West Palm Beach, Florida	0.01	0.03	30	51	34	59
Atlanta, Georgia	0.08	0.17	22	36	34	58
Augusta, Georgia	0.08	0.16	24	40	35	60
Macon, Georgia	0.07	0.15	25	41	35	59
Savannah, Georgia	0.06	0.13	25	43	35	60
Boise, Idaho	0.14	0.28	27	38	48	71
Lewiston, Idaho	0.15	0.29	22	29	44	65
Pocatello, Idaho	0.13	0.26	25	35	51	74
Chicago, Illinois	0.17	0.35	17	23	43	67
Moline, Illinois	0.20	0.39	17	22	46	70
Springfield, Illinois	0.15	0.30	19	28	42	67

TABLE 5.8 **Rules of Thumb for Passive Solar Glazing Area (*continued*)**

	Area of Solar Glazing[a] as Percentage of Floor Area		*Approximate SSF's*			
			No Night Insulation		*With R9 Night Insulation[b]*	
Location	*Low*	*High*	*Low*	*High*	*Low*	*High*
Evansville, Indiana	0.14	0.27	19	29	37	61
Fort Wayne, Indiana	0.16	0.33	13	17	37	60
Indianapolis, Indiana	0.14	0.28	15	21	37	60
South Bend, Indiana	0.18	0.35	12	15	39	61
Burlington, Iowa	0.18	0.36	20	27	47	71
Des Moines, Iowa	0.21	0.43	19	25	50	75
Mason City, Iowa	0.22	0.44	18	19	56	79
Sioux City, Iowa	0.23	0.46	20	24	53	76
Dodge City, Kansas	0.12	0.23	27	42	46	73
Goodland, Kansas	0.13	0.27	26	39	47	74
Topeka, Kansas	0.14	0.28	24	35	45	71
Wichita, Kansas	0.14	0.28	26	41	45	72
Lexington, Kentucky	0.13	0.27	17	26	35	58
Louisville, Kentucky	0.13	0.27	18	27	35	59
Baton Rouge, Louisiana	0.06	0.12	26	43	34	59
Lake Charles, Louisiana	0.06	0.11	24	41	32	57
New Orleans, Louisiana	0.05	0.11	27	46	35	61
Shreveport, Louisiana	0.08	0.15	26	43	36	61
Caribou, Maine	0.25	0.50	— NR	—[c]	53	74
Portland, Maine	0.17	0.34	14	17	45	69
Baltimore, Maryland	0.14	0.27	19	30	38	62
Boston, Massachusetts	0.15	0.29	17	25	40	64
Alpena, Michigan	0.21	0.42	— NR	—	47	69
Detroit, Michigan	0.17	0.34	13	17	39	61
Flint, Michigan	0.15	0.31	11	12	40	62
Grand Rapids, Michigan	0.19	0.38	12	13	39	61
Sault Ste. Marie, Michigan	0.25	0.50	— NR	—	50	70
Traverse City, Michigan	0.18	0.36	— NR	—	42	62
Duluth, Minnesota	0.25	0.50	— NR	—	50	70
International Falls, Minnesota	0.25	0.50	— NR	—	47	66
Minneapolis–St. Paul, Minnesota	0.25	0.50	— NR	—	55	76
Rochester, Minnesota	0.24	0.49	— NR	—	54	76
Jackson, Mississippi	0.08	0.15	24	40	34	59
Meridian, Mississippi	0.08	0.15	23	39	34	58
Columbia, Missouri	0.13	0.26	20	30	41	66
Kansas City, Missouri	0.14	0.29	22	32	44	70
Saint Louis, Missouri	0.15	0.29	21	33	41	65
Springfield, Missouri	0.13	0.26	22	34	40	65
Billings, Montana	0.16	0.32	24	31	53	76
Cut Bank, Montana	0.24	0.49	22	23	62	81
Dillon, Montana	0.16	0.32	24	32	54	77
Glasgow, Montana	0.25	0.50	— NR	—	55	75
Great Falls, Montana	0.18	0.37	23	28	56	77
Helena, Montana	0.20	0.39	21	25	55	77
Lewistown, Montana	0.19	0.38	21	25	54	76
Miles City, Montana	0.23	0.47	21	23	60	80
Missoula, Montana	0.18	0.36	15	16	47	68
Grand Island, Nebraska	0.18	0.36	24	33	51	76
North Omaha, Nebraska	0.20	0.40	21	29	51	76

TABLE 5.8 **Rules of Thumb for Passive Solar Glazing Area** (*continued*)

	Area of Solar Glazing[a] as Percentage of Floor Area		*Approximate SSF's*			
			No Night Insulation		*With R9 Night Insulation[b]*	
Location	*Low*	*High*	*Low*	*High*	*Low*	*High*
North Platte, Nebraska	0.17	0.34	25	36	50	76
Scotts Bluff, Nebraska	0.16	0.31	24	36	49	74
Elko, Nevada	0.12	0.25	27	39	52	76
Ely, Nevada	0.12	0.23	27	41	50	77
Las Vegas, Nevada	0.09	0.18	35	56	48	75
Lovelock, Nevada	0.13	0.25	32	48	53	78
Reno, Nevada	0.11	0.22	31	48	49	76
Tonopah, Nevada	0.11	0.23	31	48	51	77
Winnemucca, Nevada	0.13	0.26	28	42	49	75
Concord, New Hampshire	0.17	0.34	13	15	45	68
Newark, New Jersey	0.13	0.25	19	29	39	64
Albuquerque, New Mexico	0.11	0.22	29	47	46	73
Clayton, New Mexico	0.10	0.20	28	45	45	73
Farmington, New Mexico	0.12	0.24	29	45	49	76
Los Alamos, New Mexico	0.11	0.22	25	40	44	72
Roswell, New Mexico	0.10	0.19	30	49	45	73
Truth or Consequences, New Mexico	0.09	0.17	32	51	46	73
Tucumcari, New Mexico	0.10	0.20	30	48	45	73
Zuñi, New Mexico	0.11	0.21	27	43	45	73
Albany, New York	0.21	0.41	13	15	43	66
Binghamton, New York	0.15	0.30	— NR	—	35	56
Buffalo, New York	0.19	0.37	— NR	—	36	57
Massena, New York	0.25	0.50	— NR	—	50	71
New York (Central Park), NY	0.15	0.30	16	25	36	59
Rochester, New York	0.18	0.37	— NR	—	37	58
Syracuse, New York	0.19	0.38	— NR	—	37	59
Asheville, North Carolina	0.10	0.20	21	35	36	61
Cape Hatteras, North Carolina	0.09	0.17	24	40	36	60
Charlotte, North Carolina	0.08	0.17	23	38	36	60
Greensboro, North Carolina	0.10	0.20	23	37	37	63
Raleigh–Durham, North Carolina	0.09	0.19	22	37	36	61
Bismarck, North Dakota	0.25	0.50	— NR	—	56	77
Fargo, North Dakota	0.25	0.50	— NR	—	51	72
Minot, North Dakota	0.25	0.50	— NR	—	52	72
Akron–Canton, Ohio	0.15	0.31	12	16	35	57
Cincinnati, Ohio	0.12	0.24	15	23	35	57
Cleveland, Ohio	0.15	0.31	11	14	34	55
Columbus, Ohio	0.14	0.28	13	18	35	57
Dayton, Ohio	0.14	0.28	14	20	36	59
Toledo, Ohio	0.17	0.34	13	17	38	61
Youngstown, Ohio	0.16	0.32	— NR	—	34	54
Oklahoma City, Oklahoma	0.11	0.22	25	41	41	67
Tulsa, Oklahoma	0.11	0.22	24	38	40	65
Astoria, Oregon	0.09	0.19	21	34	37	60
Burns, Oregon	0.13	0.25	23	32	47	71
Medford, Oregon	0.12	0.24	21	32	38	60
North Bend, Oregon	0.09	0.17	25	42	38	64
Pendleton, Oregon	0.14	0.27	22	30	43	64
Portland, Oregon	0.13	0.26	21	31	38	60

TABLE 5.8 **Rules of Thumb for Passive Solar Glazing Area** (*continued*)

	Area of Solar Glazing[a] *as Percentage of Floor Area*		*Approximate SSF's*			
			No Night Insulation		*With R9 Night Insulation*[b]	
Location	*Low*	*High*	*Low*	*High*	*Low*	*High*
Redmond, Oregon	0.13	0.27	26	38	47	71
Salem, Oregon	0.12	0.24	21	32	37	59
Allentown, Pennsylvania	0.15	0.29	16	24	39	63
Erie, Pennsylvania	0.17	0.34	— NR	—	35	55
Harrisburg, Pennsylvania	0.13	0.26	17	26	38	62
Philadelphia, Pennsylvania	0.15	0.29	19	29	38	62
Pittsburgh, Pennsylvania	0.14	0.28	12	16	33	55
Wilkes Barre–Scranton, PA	0.16	0.32	13	18	37	60
Providence, Rhode Island	0.15	0.30	17	24	40	64
Charleston, South Carolina	0.07	0.14	25	41	34	59
Columbia, South Carolina	0.08	0.17	25	41	36	61
Greenville–Spartanburg, SC	0.08	0.17	23	38	36	60
Huron, South Dakota	0.25	0.50	— NR	—	58	79
Pierre, South Dakota	0.22	0.43	21	23	58	80
Rapid City, South Dakota	0.15	0.30	23	32	51	76
Sioux Falls, South Dakota	0.22	0.45	18	19	57	79
Chattanooga, Tennessee	0.09	0.19	19	32	33	56
Knoxville, Tennessee	0.09	0.18	20	33	33	56
Memphis, Tennessee	0.09	0.19	22	36	36	60
Nashville, Tennessee	0.10	0.21	19	30	33	55
Abilene, Texas	0.09	0.18	29	47	41	68
Amarillo, Texas	0.11	0.22	29	46	45	72
Austin, Texas	0.06	0.13	27	46	37	63
Brownsville, Texas	0.03	0.06	27	46	32	57
Corpus Christi, Texas	0.05	0.09	29	49	36	63
Dallas, Texas	0.08	0.17	27	44	38	64
Del Rio, Texas	0.06	0.12	30	50	39	66
El Paso, Texas	0.09	0.17	32	53	45	72
Fort Worth, Texas	0.09	0.17	26	44	38	64
Houston, Texas	0.06	0.11	25	43	34	59
Laredo, Texas	0.05	0.09	31	52	39	64
Lubbock, Texas	0.09	0.19	30	49	44	72
Lufkin, Texas	0.07	0.14	26	43	35	61
Midland–Odessa, Texas	0.09	0.18	32	52	44	72
Port Arthur, Texas	0.06	0.11	26	44	34	60
San Angelo, Texas	0.08	0.15	29	48	40	67
San Antonio, Texas	0.06	0.12	28	48	38	64
Sherman, Texas	0.10	0.20	25	41	38	64
Waco, Texas	0.08	0.15	27	45	38	64
Wichita Falls, Texas	0.10	0.20	27	45	41	67
Bryce Canyon, Utah	0.13	0.25	26	39	52	78
Cedar City, Utah	0.12	0.24	28	43	48	75
Salt Lake City, Utah	0.13	0.26	27	39	48	72
Burlington, Vermont	0.22	0.43	— NR	—	46	68
Norfolk, Virginia	0.09	0.19	23	38	37	62
Richmond, Virginia	0.11	0.22	21	34	37	61
Roanoke, Virginia	0.11	0.23	21	34	37	61
Olympia, Washington	0.12	0.23	20	29	38	59
Seattle–Tacoma, Washington	0.11	0.22	21	30	39	59

TABLE 5.8 **Rules of Thumb for Passive Solar Glazing Area (*continued*)**

	Area of Solar Glazing[a] *as Percentage of Floor Area*		*Approximate SSF's*			
			No Night Insulation		*With R9 Night Insulation*[b]	
Location	*Low*	*High*	*Low*	*High*	*Low*	*High*
Spokane, Washington	0.20	0.39	20	24	48	68
Yakima, Washington	0.18	0.36	24	31	49	70
Charleston, West Virginia	0.13	0.25	16	24	32	54
Huntington, West Virginia	0.13	0.25	17	27	34	57
Eau Claire, Wisconsin	0.25	0.50	— NR	—	53	75
Green Bay, Wisconsin	0.23	0.46	— NR	—	53	75
La Crosse, Wisconsin	0.21	0.43	— NR	—	52	75
Madison, Wisconsin	0.20	0.40	15	17	51	74
Milwaukee, Wisconsin	0.18	0.35	15	18	48	71
Casper, Wyoming	0.13	0.26	27	39	53	78
Cheyenne, Wyoming	0.11	0.21	25	39	47	74
Rock Springs, Wyoming	0.14	0.28	26	38	54	79
Sheridan, Wyoming	0.16	0.31	22	30	52	75
Canada						
Edmonton, Alberta	0.25	0.50	— NR	—	54	72
Suffield, Alberta	0.25	0.50	28	30	67	85
Nanaimo, British Columbia	0.13	0.26	26	35	45	66
Vancouver, British Columbia	0.13	0.26	20	28	40	60
Winnipeg, Manitoba	0.25	0.50	— NR	—	54	74
Dartmouth, Nova Scotia	0.14	0.28	17	24	45	70
Moosonee, Ontario	0.25	0.50	— NR	—	48	67
Ottawa, Ontario	0.25	0.50	— NR	—	59	80
Toronto, Ontario	0.18	0.36	17	23	44	68
Normandin, Quebec	0.25	0.50	— NR	—	54	74

Source: Balcomb et al. (1980).

[a]Double-glazed, due-south-facing solar openings are assumed. Higher-percentage glazings, with correspondingly higher SSF's, can, of course, be designed.

[b]Night insulation in place from 5:30 P.M. to 7:30 A.M., solar time.

[c]NR = Not recommended.

Another early design question involves the amount of thermal mass necessary to store the solar heat admitted by day. Table 5.9 details the simple relationship between SSF and weight of water or masonry. The *distribution* of the thermal mass is also important, however. In Trombe wall and water wall systems (Table 5.16, below), the thermal mass usually sits in full sun for the entire day, often just inside the glazing. In direct-gain systems, this thermal mass should be within (or should enclose) the direct-gain-heated space, and the exposed surface area of the mass should be at least three times the glazing area. Masonry surfaces are not thermally effective beyond a depth of 4 to 6 in. Note that thermal storage is relatively unimportant at low SSF's, but as the SSF increases, so does the relative proportion of thermal mass to solar glazing.

Other types of thermal mass include phase change materials and rock beds. Thin, horizontal tiles packed with phase change materials can store great quantities of heat with the phase change from solid to liquid. This change can be formulated to occur in the low 70s F, to prevent overheating of the space. As a preliminary guide, the phase-change tile surface area equals one to three times the area of solar opening. Additional information on these materials can be found in Johnson (1981).

TABLE 5.9 **Rules of Thumb for Thermal Mass**

Expected Solar Savings Fraction	*Recommended Effective Thermal Storage per Square Foot of Solar Collection Area*	
	Water (lb)	*Masonry (lb)*
10%	6	30
20%	12	60
30%	18	90
40%	24	120
50%	30	150
60%	36	180
70%	42	210
80%	48	240
90%	54	270

Source: Balcomb et al. (1980).

Rock beds are frequently used to store the excess heat that sun spaces often generate. The general guidelines cited by Mazria (1979) are

Rock bed volume, in cubic feet of rock per square foot of solar opening:

cold climates: ¾ to 1½

temperate climates: 1½ to 3

Rock bed surface area, in contact with floor above:

cold climates: 75 to 100% of floor area above

temperate climates: 50 to 75% of floor area above

Yet another early design question involves orientation: How important is it that the solar opening face due south? The general recommendation is that this orientation be *within 30° of south.* In Volume 2 of *The Passive Solar Design Handbook,* the average penalties for off-south orientation were listed as follows.

5% decrease in SSF at 18° east or 30° west of true south.

10% decrease in SSF at 28° east or 40° west of true south.

20% decrease in SSF at 42° east or 54° west of true south.

See Section 5.6 for more detailed coverage of passive heating performance, including the question of the expected annual auxiliary energy needed by a passively solar-heated building.

EXAMPLE **5.2.** A small office building in Omaha, Nebraska, is to be passively solar heated. The maps in Fig. 5.16 suggest an optimum SSF range of 40 to 50 and wall insulation of *R*30 to *R*35. Table 5.8 shows that an Omaha building with a south glass area equal to 20% of its floor area can expect an SSF of 51 if its solar openings are insulated at night.

Select 20% of floor area in south glass, with night insulation; SSF = 51.

Thermal mass can be estimated from Table 5.9; for SSF = 50, either 30 lb of water or 150 lb of masonry (per square foot of glass) should be provided. Note that 150 lb of masonry is provided by 3 ft^2 of 4-in. brick.

Roof ponds provide another approach to passive solar heating, one not covered by the preceding guidelines. In general, roof ponds are used in warmer, less humid areas of the southern United States, where snow will not impede the movement of roof insulating panels and the winter sun is higher in the sky than at northern latitudes. They are frequently sized for their *summer cooling* performance; for this region of the country, a pond sized for cooling will usually be adequate to absorb the needed winter sun. As a check for the pond's heating capacity, Mazria (1979) recommends the following guidelines.

Roof pond area = 85 to 100% of floor area for winter average outdoor temperatures of 25 to 35 F.

Roof pond area = 60 to 90% of floor area for winter average outdoor temperatures of 35 to 45 F.

These figures are for roof ponds that have two layers of enclosing material between the water and the sky (i.e., are "double glazed") and movable night insulation.

The roof-pond-cooling rule of thumb, which usually governs, is discussed later in this section.

(b) Active Solar Heating. In contrast to passive systems, which incorporate sun collection and storage as part of a building's walls, floors, or ceilings, active solar heating uses mechanical equipment to collect and store solar energy. The most common early design questions for such systems involve the area of the solar collectors, their tilt and azimuth, and the size of the thermal storage. (Domestic hot water, or DHW, solar heating information is given in Section 9.4.)

For active solar space-heating (SH) systems, the rules of thumb are more complex. Building heating needs vary by function and climate. The percentage of space heating that can economically be provided with active solar systems is another big variable. Nevertheless, a rough guide is desirable as a design starting point. For collector area,

collector area = the smaller of the two floor area percentages listed in Table 5.8

This should provide a portion of the annual heating load somewhere in the range of the listed SSF for night insulation passive systems (also from Table 5.8). Larger arrays of collectors can be designed, of course, but they rarely will be economically attractive.

For collector tilt and azimuth optima,

optimum tilt = latitude plus 10° to 15°; optimum azimuth is from due south to 15° W of south

where the tilt angle is measured up from horizontal. The orientation somewhat west of south is attractive in climates with frequent morning fog. Also, because air temperatures are higher in the afternoon, collectors lose less heat then and therefore operate more efficiently.

The rules of thumb for storage size are

2 gal of water storage per square foot of collector, or 0.5 to 0.75 ft^3 of rock bed per square foot of collector

The large arrays of collectors necessary for space heating must be served by correspondingly large pipes or ducts. Whereas pipe size rarely influences design, the air ducts for air-type collectors can consume large amounts of space. Therefore, the following flow rates are typical.

water flow rate of 0.25 to 0.5 gpm per square foot of collector, or air flow rate of 2 cfm per square foot of collector

Approximate pipe sizes can be determined from Chapter 9, and approximate duct sizing is shown in Section 5.3*d*, page 199.

(c) Passive Cooling. The rules of thumb for passive cooling are much newer and less tested than are those for passive heating. Design guidelines for cooling are further complicated by the fact that cooling loads frequently are more related to individual building characteristics than to climate: sunshading and internal heat gains are particularly influential on cooling loads. Thus, the following rules of thumb are especially crude. It is important that the designer *first* check the match between the climate and the cooling strategy, as was done in Chapter 2 using Fig. 2.8. Second, since these rules of thumb are expressed in heat to be removed per unit of floor area, it is necessary to estimate the extent of the heat gain problem. Detailed calculations for heat gain are presented in Sections 5.7 and 5.8. Table 5.10 gives a quick *approximation*. Many buildings (restaurants, factories, stores selling heating appliances, etc.) have special heat sources within. For these unusually heat-loaded situations, Table 5.10 will be inadequate. As a starting point for preliminary passive-cooling sizing for typical buildings, however, it should be helpful.

Note that "open" buildings, such as those that are naturally ventilated, do not have heat gains from infiltration, because they assumedly maintain internal temperatures that are slightly *above* exterior temperatures. However, these buildings do experience heat gains through windows, walls, and

TABLE 5.10 **Approximate Heat Gains**

Part A. ***Internal Heat Sources—People and Equipment***

Function	Area per Person (ft^2)	Sensible Heat Gain (Btu/h ft^2 of Floor Area)		
		People[a]	Equipment[b]	Total
Office	100	2.5	3.4	5.9
School: elementary	100	2.5	3.4	5.9
School: secondary, college	150	1.7	3.4	5.1
Hospital	100	2.5	Varies	2.5 plus
Clinic	50	5.0	Varies	5.0 plus
Assembly: theater[c]	15	15.3	—	15.3
Assembly: arena[c]	15	16.7	—	16.7
Restaurant	25	11.0	Varies	11.0 plus
Mercantile	50	5.0	Varies	5.0 plus
Warehouse	1000	0.4	—	0.4
Hotels, nursing homes	300	0.8	3.4	4.2
Apartments[d]	300	0.8	(see note *d*)	(see note *d*)

Part B. ***Internal Heat Sources—Lighting, Daylight and Electric***

Function	Sensible Heat Gain (Btu/h ft^2 of Floor Area)[e]		
	$DF < 1$	$1 < DF < 4$	$DF > 4$
Office	5.1	2.0	0.5
School: elementary	6.3–6.8	2.5–2.7	0.6–0.7
School: secondary, college	6.3–6.8	2.5–2.7	0.6–0.7
Hospital	6.8	2.7	0.7
Clinic	6.8	2.7	0.7
Assembly: theater[c]	3.8	1.5	0.4
Assembly: arena[c]	3.8	1.5	0.4
Restaurant	6.3	2.5	0.6
Mercantile	5.1–6.8	2.0–2.7	0.5–0.7
Warehouse	2.4	1.0	0.2
Hotels, nursing homes	6.8	2.7	0.7
Apartments[d]	Up to 6.8	Up to 2.7	Up to 0.7

Part C. ***Heat Gains through Envelope[f] (Btu/h ft^2 of Floor Area)***

		Outdoor Design Temperature	
		90 F	100 F
I. Gains through externally shaded windows: Find $\dfrac{\text{total window area}}{\text{total floor area}}$,	then multiply by	16	21
II. Gains through opaque walls: Find $\dfrac{\text{total opaque wall area}}{\text{total floor area}} \times (U_{wall})$,	then multiply by	15	25
III. Gains through roofs: Find $\dfrac{\text{total opaque roof area}}{\text{total floor area}} \times (U_{roof})$,	then multiply by	35	45

TABLE 5.10 **Approximate Heat Gains** (*continued*)

*Part D. **Summary Gains (Btu/h ft² of Floor Area)***

I. Passive cooling systems for "open" buildings:
- Cross ventilation
- Stack ventilation
- Nighttime or "open" hours of thermal mass/night ventilation
- Total: add Parts A, B, and C gains to obtain total cooling load

II. Passive cooling system for "closed" buildings:
- Roof ponds
- Evaporative cooling
- Daytime or "closed" hours of thermal mass/night ventilation
- Total: add Parts A, B, C and Part E (below) gains, to obtain total cooling load

*Part E. **Gains from Infiltration/Ventilation of "Closed" Buildings (Btu/h ft² of Floor Area)***

	Outdoor Design Temperature	
	90 F	*100 F*
Find $\frac{\text{total window + opaque wall area}}{\text{total floor area}}$, then multiply by	1.0	1.9
OR		
Find $\frac{\text{known total cfm of outdoor air}}{\text{total floor area}}$, then multiply by	16.0	27.0

[a]Adapted from Buehrer (1978).

[b]The usual load of 1 W/ft² is assumed here. However, heavy use of computers can produce loads of up to 6 W/ft².

[c]Gains listed for these functions are only for the seating areas, not for lobbies, stage areas, kitchens, and so on.

[d]Residential internal gains often assumed at 225 Btu/h per occupant plus 1200 Btu/h total from appliances. See Section 5.7.

[e]Adapted from Northwest Power Planning Council, *Maximum Lighting Standards,* 1983.

[f]Averaged from the more specific data found in Table 5.19 (Section 5.7).

roofs, due to solar impacts on these surfaces. For "closed" buildings, heat gain from infiltration or ventilation must be added, since these structures maintain internal temperatures lower than outside temperatures.

Cross ventilation inlet areas, expressed as percentage of total floor area, are related to wind speed and resulting heat removal in Fig. 5.17. Remember that an equal (or greater) area of outlet openings must also be provided. The assumptions about wind direction and indoor–outdoor temperature differences that were used to produce these guidelines are explained in the figure.

The ΔT of 3F° used in Fig. 5.17 is deliberately kept small, to encourage "open" strategies in milder summer climates. Thus, an interior temperature of 83 F, which is comfortable if sufficient air motion, a lower-percentage relative humidity, and comfortable surface temperatures are present, would be obtainable with an outside temperature of 80 F. However, a greater ΔT is often appropriate—for example, for spring or fall cooling of office buildings, or for summer cooling of factories or kitchens where internal temperatures may remain in the low 90s. In such cases, find the percentage of inlet area required, then multiply by the ratio

$$\frac{3\text{F}^\circ}{\text{actual } \Delta T}$$

to obtain required cross-ventilation areas for any specific temperature difference (see Example 5.3).

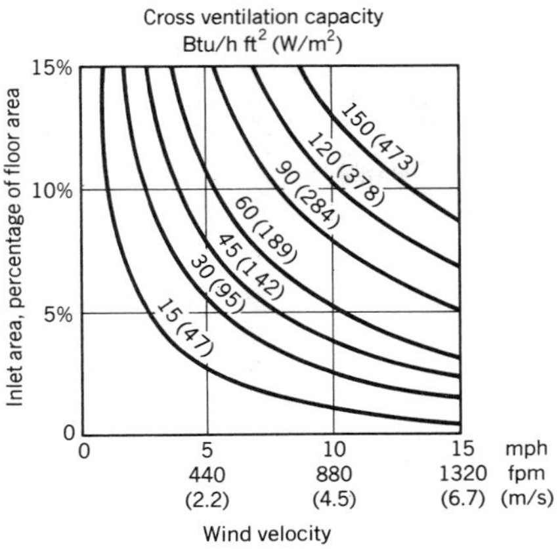

Fig. 5.17. Cross ventilation rule of thumb, for heat removed per unit floor area, and relationship of area of inlet openings and wind speed. Total inlet opening area is expressed as a percent of total floor area. Note: *Outlet areas must* also *be at least equal to this area. The figure assumes that the internal temperature is 3F° (1.7C°) above the exterior temperature and that wind is not quite perpendicular to the inlet opening, for a wind effectiveness factor of 0.4 (see Section 4.6).*

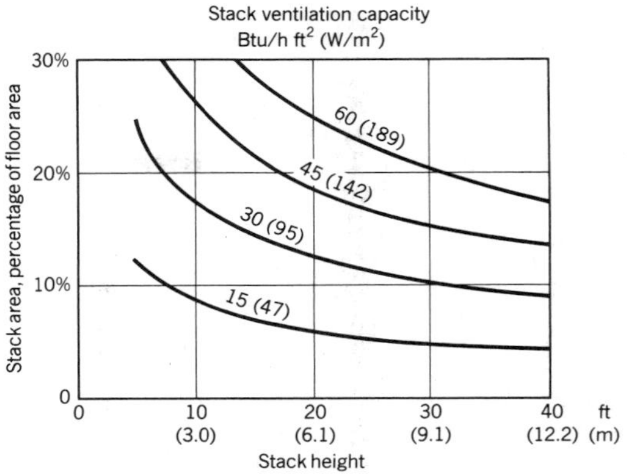

Fig. 5.18. Stack ventilation rule of thumb, for heat removed per unit floor area, and relationship area of stack and height of stack. Total stack area is expressed as a percent of total floor area. Note: *Stack area refers to minimum area of inlets* and *of cross section through the vertical stack* and *of outlets. The figure assumes an internal temperature 3F° (1.7C°) above the exterior temperature.*

Stack ventilation inlet areas, expressed as percentage of total floor area, are related to stack height and the resulting heat removal in Fig. 5.18. Remember that an equal (or greater) area of outlet openings, as well as at least an equal cross-sectional area through the vertical stack, are also required. The assumptions about indoor–outdoor temperature differences that were used to produce this guideline are explained in the figure.

Adjustment of the ΔT for this rule of thumb is similar to the procedure for cross ventilation. It requires multiplication of the required percentage of stack areas by the ratio

$$\sqrt{\frac{3F^\circ}{\text{actual } \Delta T}}$$

to obtain the required stack ventilation areas for any specific temperature difference (see Example 5.3).

Night ventilation of thermal mass is shown in Fig. 5.19, in which climate data are related to two representative types of thermally massive building. For each type, the graphs show the daily Btu per square foot of floor area that can be stored.

The climate data (maximum summer design DB temperature and mean daily range) are given in Appendix A. These data also allow calculation of the *minimum* summer design DB temperature; that is, maximum design DB temperature minus mean daily range. This minimum temperature is of interest here because the thermal mass of the building will be lowered toward (but not quite to) it during night ventilation. For high daily range climates, the lowest temperature obtained by the thermal mass will be one quarter of the mean daily range *above* the minimum air temperature (for lower daily range climates—30 F° or less—one fifth of the mean daily range).

EXAMPLE. Sacramento, California:

highest DB = 98 F (Appendix A)
mean daily range = 36 (Appendix A)
lowest DB = 98 − 36 = 62 F
approximate lowest mass temperature: ¼ × 36 = 9; 62 + 9 = 71 F

Oklahoma City, Oklahoma:

highest DB = 97 F
mean daily range = 23
lowest DB = 97 − 23 = 74 F
approximate lowest mass temperature: ⅕ × 23 = 4.6; 74 + 4.6 = 78.6 F

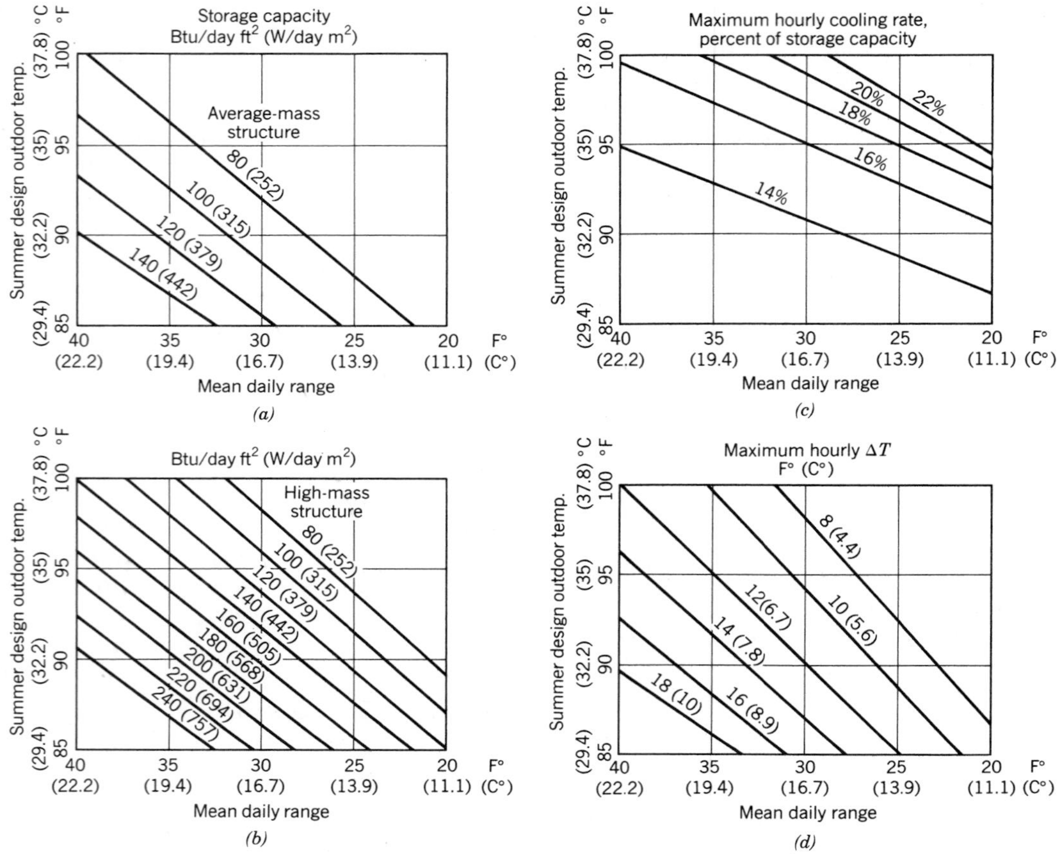

Fig. 5.19. Night ventilation of thermal mass rule of thumb, for Btu/day removed per ft^2 of floor area. Parts a *and* b *also relate daily maximum air temperature, mean daily outdoor temperature range, and degree of thermal massiveness of the building. Assumptions: "average" mass structure* (a) *has 1 ft^2 of surface exposed per ft^2 of floor area, of a 4-in. ordinary-density concrete slab; no other thermal mass is included. "High"-mass structure* (b) *has 2 ft^2 of surface exposed per ft^2 of floor area, of 3-in. ordinary-density concrete (or, alternatively, both sides of a 6-in. concrete wall or slab exposed); no other thermal mass is included. Both buildings go into "open" mode when outdoor temperature drops below 80 F (therefore, the highest mass temperature is assumed to be 80 F). Part* c *shows the approximate percent of the total daily stored heat that is removed during the hour of maximum cooling. Part* d *shows the* ΔT *between outside air and inside mass that exists during the hour of maximum cooling. Required ventilation rates can be determined from* c *and* d.

From these data, it appears that Sacramento, with a comfortably low 71 F mass temperature, is a likely site for night ventilation of heat stored in the building's mass. However, Oklahoma City's 78.6 F lowest mass temperature is at the middle of the comfort zone, a climate in which this passive strategy will be harder to achieve. In either location, performance studies can best be carried out with the details of this mass cooling procedure, found in Section 5.9.

Before these rules of thumb can be utilized, the number of hours for which the building must be "closed" must be determined. During these hours, heat will be stored in the structure, up to the maximum indicated in Fig. 5.19. Typically, these buildings will go into the "closed" mode (allowing minimum ventilation only) at 6 A.M. for 100 F maximum, or 8 A.M. for 85 F maximum, remaining in the closed mode until the outdoor temperature drops below 80 F. (To approximate this hour, assume that the midpoint outdoor temperature, between daily high and daily low, occurs around 10 P.M.) Thus, the typical office building remains "closed" during the eight to nine hours of summer occupancy. All the heat generated during the "closed" mode—eight to nine daytime hours—must be stored in the structure:

$$\frac{\text{heat to be stored/}}{\text{ft}^2\text{ floor area}} = \begin{bmatrix}\text{hours occupied}\\ \text{in ``closed''}\\ \text{mode}\end{bmatrix} \times \begin{bmatrix}\text{heat gain, Btu/h ft}^2\text{ floor}\\ \text{(from Table 5.10)}\end{bmatrix}$$

Figure 5.19 shows two types of thermally massive buildings and explains the details of their construction. The "average" mass building is represented by a building with an exposed concrete floor 4-in. thick. The "high" mass building is similar to the typical passively solar heated, direct-gain building, or to a multistory building with an exposed concrete structure, in which both sides of the floor slab are available for thermal storage.

Having found the amount of heat that can be stored each day, the designer next must solve the problem of removing the heat by night. Either natural or forced ventilation can be used. The ventilation rate is determined by the "best hour" of cooling during the night: that is, the hour during which the temperature difference between inside mass and outside air is greatest, hence the most heat is removed. This "best-hour" information can also be found in Fig. 5.19*c*,*d*. The ventilation equations presented in Section 4.6 can then be used to find the volume of air needed, or the cross or stack ventilation rules of thumb can be used to size the openings.

EXAMPLE 5.3, PART A. Find the approximate cooling performance in night ventilation of thermal mass for the Denver office building described in Example 5.1 (see also Fig. 5.14).

Denver: max temperature = 91 F
mean daily range = 28 (Appendix A)

From these data, the nightly minimum temperature is 91 − 28 = 63 F, and if the internal mass temperature gets to within one fifth of the mean daily range, or 5.6, then 63 + 6 = 69 F is the expected minimum mass temperature. This sounds quite comfortable!

You can now calculate the estimated heat gain, using Parts A–E of Table 5.10 as indicated.

(5.10*a*) People, equipment heat gain (office buildings)	5.9 Btu/h ft²
(5.10*b*) Lighting heat gain (DF_{av} = 3.5)	2.0
(5.10*c*) Envelope heat gains:	
I. For shaded windows,	
$\frac{17.5\%\ \text{window}}{\text{floor}} \times 16 = 2.8$	2.8
II. For opaque walls (U = 0.22),	
$\frac{22.5\%\ \text{opaque wall}}{\text{floor}} \times 15 = 3.4$;	
3.4 × 0.22 = 0.7	0.7
III. For roof and skylights, gain is confined to the top floor. The roof area is equal to the top floor area; the U_o for roof/skylight combination is 0.07:	
$\frac{100\%\ \text{roof-skylight}}{\text{floor}} \times 35 \times 0.07 = 1.1$	(1.1)
(5.10*e*) Since the building will be "closed" during hot days, ventilation gains must be added. From Table 4.19,	
$\frac{7\ \text{persons}}{1000\ \text{ft}^2\ \text{floor area}} \times 5\ \text{cfm/person} = 0.035\ \text{cfm/ft}^2$;	
0.035 × 16 = 0.56	0.6
(5.10*d*) Summary gains	12.0 Btu/h ft² (13.1 Btu/h ft² for top floor)

Note that this calculation does *not* include the thermal impact of the nearly windowless end walls. Assuming the building were to be five times as long as it is wide, then

$$\text{end walls} = \frac{\text{side walls at 40\% floor area}}{5} = 8\%\ \text{floor area}$$

added opaque wall gains $= 8\% \times 15 \times 0.22$

$= 0.3$ Btu/h ft^2 added load

The total approximate heat gain is therefore about 12 Btu/h ft^2. We can assume that the building's mass reaches the same temperature as the outdoor air at about 8 A.M. (as explained above), at which point the building goes into "closed" mode. It will "open" again at about 80 F, which, if Denver's 91 F high occurs between 2 and 4 P.M., will occur well after the 5 P.M. close of work. Therefore, the estimated nine hours of occupancy will occur during "closed" mode, and storage must total 9 h $\times$ 12 Btu/h ft^2 = 108 Btu/ft^2 each day.

Figure 5.19*a* shows that the "average" mass building at 91 F maximum and 28 mean daily range will store only about 80 Btu/ft^2; hence, it has inadequate thermal mass. Figure 5.19*b* shows that the high mass structure will store about 140 Btu/ft^2, which is more than this building's approximate gain of 108 Btu/ft^2.

The next preliminary design consideration involves the ventilation rate for the nighttime flushing of heat from the thermal mass. Figure 5.19*c* shows that during the hour of maximum cooling, 14% of the total day's stored heat will be removed; 14% $\times$ 108 Btu/ft^2 = 15.1 Btu/h ft^2. Figure 5.19*d* shows that the expected ΔT is about 11F°. Consider three options: cross ventilation, stack ventilation, or fan-forced ventilation.

Cross ventilation can be estimated from Fig. 5.17; however, the rule of thumb illustrated there is based on ΔT = 3F°. The graph indicates a required inlet area of about 3% for the 15.1 Btu/h ft^2 cooling load, at a low wind velocity of 5 mph.

$$3\% \times \frac{3\text{F}^\circ}{11\text{F}^\circ \text{ actual}} = \text{about } 1\%$$

If Denver always had at least a 5-mph wind all night, an inlet (or outlet) area of only 1% of the floor area would be adequate. The primary disadvantage of exclusive reliance on cross ventilation is the variability of wind speed; on a perfectly calm night, there could be little or no cooling. Obstructions to cross ventilation (such as partitions) also pose a problem.

Stack ventilation, which works as long as there is a temperature difference between inside and outside, might be a better choice for this application. Stack height, of course, will vary for each floor; with 12-ft floor heights, the upper four floors will provide a minimum 4 $\times$ 12 = 48 ft stack for the first floor. For the upper floors, stacks added above the roof would be needed. (Alternatively, because the top two floors get the strongest wind, stacks for lower floors and cross ventilation for upper floors could be used.)

For analysis purposes, assume that a 50-ft stack is available for the lowest floor and a 10-ft stack for the highest floor. Because the graph is based on 3F° ΔT, the percentage stack areas will be adjusted:

$$\sqrt{\frac{3\text{F}^\circ}{11\text{F}^\circ \text{ actual}}} = 0.52$$

For the lowest floor, with a 50-ft stack, Fig. 5.18 indicates that to remove 15.1 Btu/h ft^2, inlet/stack/outlet areas of 5% $\times$ 0.52, or about 3%, of the floor area would be adequate. (However, *each* of the upper floors must also give up 3% of its area, to allow the stack to pass up through the building.) Then,

2nd floor, with 40-ft stack: 3% inlet/floor

3rd floor, with 30-ft stack: 4% inlet/floor

4th floor, with 20-ft stack: 4% inlet/floor

For the highest floor, with 10-ft stack, the required cooling rate is actually higher than the average rate, due to the added roof gains. This factor must be considered when stack size is chosen. Figure 5.18 indicates that inlet/stack/outlet areas of about 5% of the floor area would be adequate, so an increase to 5% or so seems appropriate at this preliminary stage.

It appears that stack ventilation would impose the following requirements on floor area allocation:

1st floor (ground): none

2nd floor: 3% for 1st floor stack = 3% of floor area

3rd floor: 3% for 1st-floor + 3% for 2nd-floor stacks = 6% of floor area

4th floor: 3% + 3% + 4% for 3rd-floor stack = 10% of floor area

5th floor (top): 3% + 3% + 4% + 4% for 4th-floor stack = 14% of floor area

Clearly, stack ventilation can impose a floor area penalty on multistory buildings, unless a central atrium is utilized or stacks are placed on the exterior.

Forced ventilation is often chosen for night flushing because the delivery rate can be closely controlled, the air can be filtered, and the airstream can be directed over the mass surface in the most thermally advantageous way. It also allows much greater flexibility for partitions, and so on. Its disadvantages, of course, are the energy required by the fans and the expense of the ductwork.

From Section 4.6,

$$q_v = V\left(\frac{1.08}{60 \text{ min}}\right)(\Delta T)$$

$$15.1 \text{ Btu/h ft}^2 = V(0.018)(11\text{F}^\circ)$$

$$V = \frac{15.1}{0.018 \times 11} = 76.4 \text{ cfh/ft}^2 \text{ floor area}$$

Since the building's height is 12 ft between floors, 76.4 cfh/12 cf = about six air changes per hour, at maximum hour of cooling. The duct sizes for such a system are considered in Example 5.3, Part B.

Roof ponds sized for cooling frequently are nearly equal in area to the floors of the buildings they cool. Average pond depth is between 3 and 6 in. A more detailed and precise procedure for determining pond area is presented in Section 5.9, but here is the quick approximation:

pond maximum temperature: 80 F

pond minimum temperature = minimum nighttime DB (= max. daytime temp − mean daily range, from Appendix A)

pond ΔT = pond maximum − pond minimum

pond's allowable daily heat stored (from building), Btu/day ft² = 0.7 × pond ΔT × pond depth in feet, not to exceed 0.75 × 62.5 lb/ft³ water (assuming 70% of the pond's heat gain from the building below and 30% through the insulated panels above the pond)

required size of pond (ft² pond per ft² floor area) =

$$\frac{\text{building total heat gain per day (Btu/day ft}^2\text{ floor area)}}{\text{pond allowable heat stored per day (Btu/day ft}^2\text{ floor area)}}$$

EXAMPLE 5.4, PART A. An office building in Albuquerque, New Mexico, is to be cooled by a roof pond 4-in. deep. An hourly heat gain of 15 Btu/h ft² is assumed (somewhat more than the Denver office building of Example 5.3). For Albuquerque,

maximum temperature = 94 F (Appendix A)

mean daily range = 27 (Appendix A)

minimum (night) temperature = 94 − 27 = 67 F

Therefore,

pond ΔT = 80 maximum − 67 minimum = 13 F°

pond storage capacity for building heat = 0.7(13)(0.33 ft) (62.5 lb/ft³) = 188 Btu/day ft²

The required pond size, then, is

$$\frac{15 \text{ Btu/h ft}^2 \times 9 \text{ h/day}}{188 \text{ Btu/day ft}^2} = 0.72 \text{ ft}^2/\text{ft}^2 \text{ floor area}$$

Therefore, a 4-in. pond covering 75% of the one-story building's floor area will be approximately large enough to cool the building.

Earth contact tempering, another cooling approach, is still in a very early development stage. The most direct application of this approach would be an underground building with uninsulated concrete walls set against the soil. The problem, of course, is that winter heat loss will exceed summer loss; if solar heat can be admitted (e.g., through skylights) to counterbalance the increased winter loss, uninsulated walls sized for the desired summer loss become more attractive. (However, except in dry climates, condensation on walls may pose a seasonal or even year-round problem.)

The long-term potential of the earth as a heat sink is lessened by the fact that soils are relatively slow heat conductors. For example, we ignore heat loss through concrete slabs to the earth below, calculating instead the heat losses through the slab's exposed perimeter. A factor accounting for lower long-term heat flow rate is included in the following rule of thumb.

To estimate the cooling potential of uninsulated walls against the earth, proceed as follows.

1. Determine average underground earth tem-

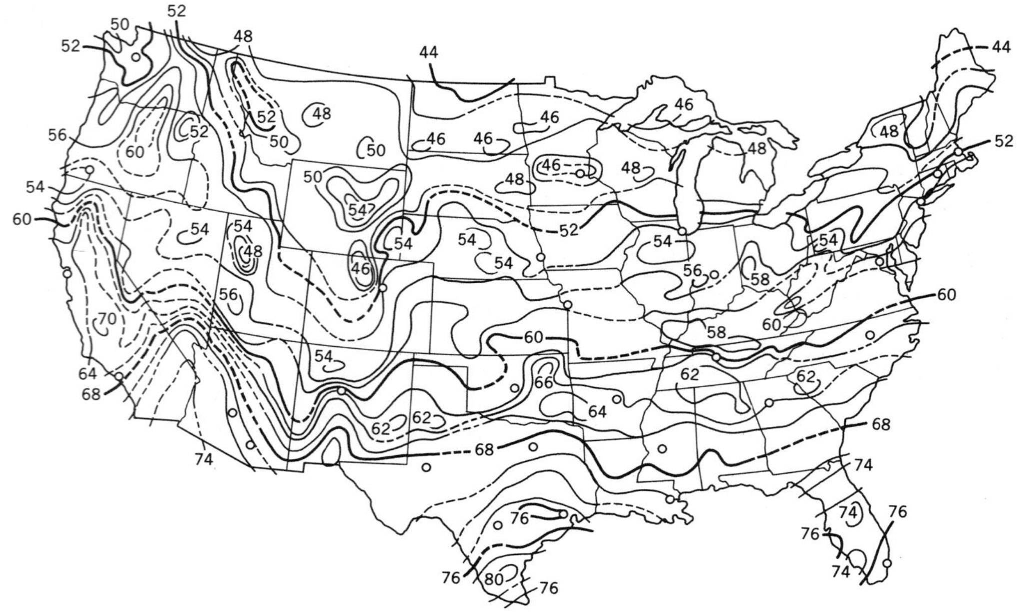

Fig. 5.20. Distribution of well water temperature in the United States, adapted from the National Water Well Association. These temperatures may be assumed to be the yearly average ground temperature. Reprinted from Passive Cooling *by permission of the publisher, the American Solar Energy Society, Inc.*

perature (about equal to well water temperature) from the map in Fig. 5.20.

2. Decide whether the soil outside your underground wall is dry, average, or wet during the cooling period.
3. Adjust the average earth temperature by comparing it to the average amplitude of surface temperature, using the map in Fig. 5.21. These amplitudes, or seasonal variations in earth temperature, diminish to near zero at about these depths:

Dry soil	14 ft
Average soil	18 ft
Wet soil	22 ft

Thus, the *summer* ground temperature at the surface will be equal to the average ground temperature (Fig. 5.20) *plus* the amplitude (Fig. 5.21); but at the depths listed above, the ground temperature remains essentially the same. Considering this variation, estimate the average ground temperature outside the cooling wall, at its average depth below the surface.

4. Determine the temperature difference that makes cooling possible:

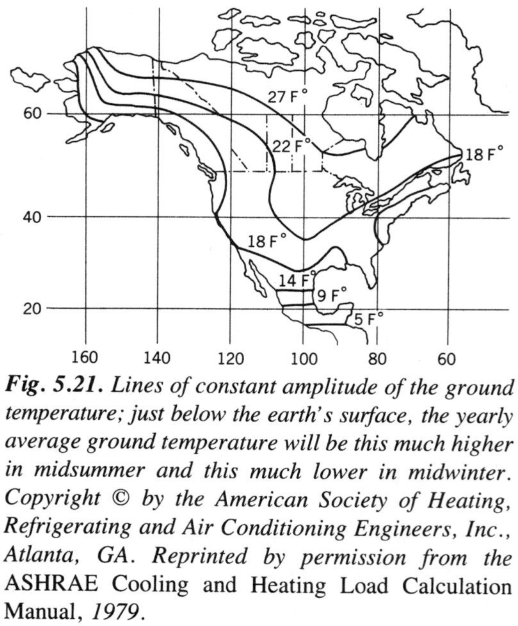

Fig. 5.21. Lines of constant amplitude of the ground temperature; just below the earth's surface, the yearly average ground temperature will be this much higher in midsummer and this much lower in midwinter. Copyright © by the American Society of Heating, Refrigerating and Air Conditioning Engineers, Inc., Atlanta, GA. Reprinted by permission from the ASHRAE Cooling and Heating Load Calculation Manual, *1979.*

$$\Delta T = \text{inside temperature} - \text{average depth summer ground temperature}$$

5. Determine the long-term cooling rate:

$$\text{Btu/h ft}^2 \text{ wall surface} = (\Delta T)(U \text{ wall})(C)$$

where C is a factor that accounts for the long-term effects of heat flow from the building to the soil through an uninsulated wall:

$C = 0.11$ for dry soil (soil conductivity

$k = 0.25$ Btu/ft h F)

$C = 0.28$ for average soil ($k = 0.75$)

$C = 0.44$ for wet soil ($k = 1.5$)

A more thorough discussion of this procedure can be found in Watson and Labs (1983).

Earth tubes are devices for cooling the incoming ventilating air through earth contact, before it enters the building. In general, a small amount of air (perhaps equal to the minimum fresh air requirements—see Table 4.19) is brought in through several tubes. These tubes are 8–12 in. in diameter, are buried at a depth of 5–10 ft, and are 100–200 ft long.

The lowest temperature to which air can be cooled in such tubes will *approach* earth temperature; the slower the air moves through the tube, the more time there will be for cooling. The air temperature might come within 4F° of the earth temperature, under good conditions.

As a rule of thumb for earth tubes:

Air velocity within each tube will be 500 fpm.

Then, air volume, cfm = (500 fpm) × (tube cross section, ft^2).

Sensible cooling will be about 1.3 Btu/h F per foot of length (maximum length about 300 ft).

For more detailed information, see Frances, C. E., "Earth Cooling Tubes, Case Studies . . ." in Bowen et al. (1981).

EXAMPLE 5.5. A one-floor underground office building is being considered for Boise, Idaho. To maximize the use of the wall as a heat sink, the wall:floor ratio is set at 1:2 (as would be obtained in a cross section 10 ft high × 40 ft wide, with both walls underground). Uninsulated concrete walls 8 in. thick will be set at an average depth of 8 ft below ground level (including 3 ft soil depth over the roof). The U for these walls is the reciprocal of the resistances:

8 in. concrete R =

0.08×8 in. $= 0.64$ (Table 4.2)

Inside air film $R = \underline{0.68}$ (Table 4.3)

Then, $0.64 + 0.68 = 1.32$

$$U = \frac{1}{1.32} = 0.76$$

The skylights to be used are south-facing, triple-glazed, vertical monitors providing an average 3.0 DF. From Table 5.3,

$$\text{DF}_{\text{av}}\ 3.0 = 0.2\left(\frac{\text{skylight area}}{\text{floor area}}\right)$$

skylight area $= 3 \div 0.2 =$ 15% of the floor area

For Boise,

maximum summer temperature = 94 F (Appendix A)

(coincident Wet Bulb = 64 F)

winter heating DD = about 5800

winter design temperature = 10 F

Therefore, the roof U_o should be, at most, 0.077 (Fig. 5.8) to meet heating and cooling criteria. (We assume that the thermal mass of a sod roof will more than meet the OTTV_r criterion!) Skylight winter $U = 0.39$ (Table 4.17*a*).

$$U_o \text{ roof} = \frac{(U_{\text{opaque roof}})(85\%) + (0.39)(15\%)}{100\%} = 0.077 \text{ max}$$

$$U_{\text{opaque roof}} = \frac{0.077 - 0.059}{0.85} = \frac{0.018}{0.85} = 0.02 \text{ maximum}$$

At 5 cfm per person, ventilation will meet the requirements of Table 4.19, assuming no smoking. Assuming (also from Table 4.19) that each seven persons occupy 1000 ft^2 of floor area, ventilation becomes $(5 \times 7)/1000 = 0.035$ cfm/ft^2 floor.

The approximate heat gain is calculated as follows.

Internal (people, equip., lights) (Table 5.10)	= 7.9 Btu/h ft^2
From shaded skylights: 15% × (interpolated) 18	= 2.7 Btu/h ft^2
Through roof (at most, since sodded): 100% × 0.02 × (interpolated) 39	= 0.8 Btu/h ft^2

(No wall gains; these are *losses* to heat sink of earth.)

Ventilation: $0.035\ \text{cfh/ft}^2 \times$ (interpolated) 20 $= \underline{0.7\ \text{Btu/h ft}^2}$

Total gain = $12.1\ \text{Btu/h ft}^2$

To find the cooling rate, begin with the average ground temperature, which, for Boise, is 55 F (from Fig. 5.20). Assume that the soil is dry outside the walls. The amplitude at the surface for Boise is about 20F° (from Fig. 5.21), so in summer the surface temperature is 55 + 20 = 75 F. In dry soil, the temperature diminishes to the Boise average of 55 F at a depth of 14 ft. Since the average depth of the wall is 8 ft, assume that this outside-the-wall earth temperature in summer is about halfway between 75 and 55 F, or 65 F.

Consider an inside air temperature of 80 F, which would be quite comfortable when accompanied by cool wall surfaces; a lower MRT will allow a higher air temperature. The long-term cooling available through these walls is

$$\text{Btu/h ft}^2\ \text{wall} = (80 - 65)(0.76)(0.11\ \text{long-term for dry soil}) = 1.25\ \text{Btu/h ft}^2\ \text{wall}$$

Or, translated to floor area terms,

$$1.25 \times \frac{1\ \text{wall area}}{2\ \text{floor area}} = 0.68\ \text{Btu/h ft}^2\ \text{floor}$$

or about 5% of the cooling requirements. Note that the wall's inner surface temperature will be approximately

$$80\ \text{F} - \left[\frac{(0.68R\ \text{of air film})(15\text{F}^\circ\ \Delta T)}{1.32\ \text{total}\ R}\right] = 80 - 7.7 = 72.3\ \text{F}$$

(procedure from Fig. 4.4). Because 72.3 F is well above the wet-bulb 64 F design outdoor temperature, summer condensation therefore would not seem to be a problem.

In *winter*, however, the heat gains from the envelope disappear; the internal gains at 7.9 Btu/h ft² remain during occupied hours. To calculate envelope losses, assume an internal temperature of 75 F (remember those cold walls). At worst, winter conditions will be based on 75 inside − 10 outside = 65F° ΔT.

Skylights: $15\% \times 0.39 \times 65 = 3.8\ \text{Btu/h ft}^2$

Roof: $100\% \times 0.02 \times 65 = 1.3\ \text{Btu/h ft}^2$

Ventilation: $0.035 \times 65 = \underline{2.3}\ \text{Btu/h ft}^2$

Total: $7.4\ \text{Btu/h ft}^2$ loss, at worst

The earth temperature outside the walls now swings to the lower side of the average earth temperature. With the midpoint between the 55 F average ground temperature and the 35 F winter surface temperature being 45 F, wall heat losses become

$$(75 - 45)(0.76)(0.11) = 2.5\ \text{Btu/h ft}^2\ \text{wall}$$

or

$$2.5\left(\frac{1\ \text{wall area}}{2\ \text{floor area}}\right) = 1.3\ \text{Btu/h ft}^2\ \text{floor, average winter.}$$

Total envelope losses therefore are 7.4 + 1.3 = 8.7 Btu/h ft², under worst conditions. Comparison of the winter gain to the winter loss gives

$$9\ \text{h gain} \times 7.9\ \text{Btu/h ft}^2 = 71\ \text{Btu/day ft}^2$$

$$24\ \text{h loss} \times 8.7\ \text{Btu/h ft}^2 = \underline{209\ \text{Btu/day ft}^2}$$
$$138\ \text{Btu/day ft}^2\ \text{heat loss, at worst}$$

Solar radiation through the south skylights will help:

Boise January insolation = 927 Btu/ft² day (Appendix B)

927 × 15% = 139 Btu/day ft² floor, about equal to the net daily heat loss under worst conditions

However, there may be condensation problems with the cold wall surfaces, as well as radiant heat loss discomfort.

(d) Equipment Space Allocations. The design impact of the passive systems described above would be felt throughout a building—in a certain percentage of the floor area that must be provided in south glass, or inlets and outlets, and in the emphasis on spreading thermal mass throughout the structure. In contrast, a conventional heating and cooling system has a *concentrated* impact on a building: a central mechanical room and the distribution tree connecting it with a building are the initial design impacts.

One or a few central mechanical room(s) can serve many floors, or a smaller mechanical room on each floor can be provided. Each mechanical room should have both a central location relative to the area it serves and direct access to the outside—contradictory requirements in many cases.

Central locations minimize distribution tree size; access to the outdoors facilitates the use of outdoor air as a heat source (winter) or sink (summer) and allows equipment to be installed or removed in later remodelings. Mechanical rooms also need relatively high ceilings (12 ft clear is a typical minimum, 20 ft clear a typical maximum). Table 5.11 presents the approximate space requirements for conventional mechanical systems. (For a more detailed look at space requirements, see Table 6.14.)

Air Ducts. Air duct sizes are included, in general, in the space allocations of Table 5.11, but the approximate cross section is frequently of interest early in the design process. Duct depths can help determine floor heights; duct cross sections influence the sizes and shapes of the vertical cores that serve multistory buildings. An early approximation of duct size can be obtained as follows.

1. Determine the quantity of air to be distributed through the largest duct, using Table 5.12, or the air changes per hour, from a calculation of night ventilation of thermal mass (e.g., Example 5.3). This will usually be expressed in cubic feet per hour (cfh).
2. Convert cfh to cfm:
$$\frac{\text{cfh} \times 1\ \text{h}}{60\ \text{min}} = \text{cfm}$$
3. Find the recommended velocity of this air within the duct from Table 5.13, expressed in feet per minute (fpm)

TABLE 5.11 **Mechanical Equipment Space Requirements**

Part A.[a] ***In general, the mechanical and electrical equipment requires 6% to 9% of the total floor area of most buildings.***

Part B.[b] ***Expressed per ton (= 12,000 Btu/h, or 3516 W) of equipment capacity, the mechanical equipment room requirements are***

Heating/cooling equipment, 2.8 to 4.5 ft²/ton.

Air handling equipment, 5.0 to 6.7 ft²/ton.

Part C[c]

Type of System (See Section 6.7 for Descriptions)	Percentage of Gross Floor Area Required for Mechanical Spaces[d]: Larger Buildings		Smaller (<10,000 ft²) Buildings, or Heavier Cooling Loads
All-air systems[e]	3%	to	8%
Air & water systems[e]	3%	to	6%
All-water systems	1.5%	to	5%
Smaller package units throughout building, or on roof	1%	to	3%

[a]*ASHRAE Systems Handbook,* 1980.

[b]Andrews (1966).

[c]M. David Egan, *Concepts in Thermal Comfort,* p. 114. © 1975, Prentice-Hall, Inc. Reprinted by permission.

[d]A much more detailed sizing guide is shown in Tables 6.14 and 6.15.

[e]Separate rooms for air-handling equipment may require an additional 2% to 4% of the building's total occupied area.

TABLE 5.12 **Suggested Air Changes per Hour (ACH)**

Includes both fresh air and recirculated air; acceptable air motion is the objective. For minimum *fresh* air, see Table 4.19.

Function of Space	*(ACH)*
Machinery spaces, maintenance shops (such as for electrical equipment), school and industrial shops, etc.	8 to 12
Cafeterias, restaurants, offices, reception areas, hospitals, residences, garages, etc.	6 to 20
Churches, libraries, reading rooms, bowling alleys, retail shops and stores, etc.	15 to 25
Auditoriums, theaters, classrooms, kitchens, conference rooms, etc.	10 to 30

Source: M. David Egan, *Concepts in Thermal Comfort,* p. 15. © 1975, Prentice-Hall, Inc. Reprinted by permission.

4. The approximate required cross-sectional area of the duct A is then

$$A_{\text{in.}^2} = \frac{\text{volume of air (cfm)}}{\text{velocity (fpm)}} \times 144 \text{ in.}^2/\text{ft}^2 \times \text{friction allowance}$$

where the friction allowances are

round ducts = 1.0 (may be neglected)

nearly square ducts (ratio of width to depth, 1:1):

small (< 1000 cfm) = 1.10

large (> 1000 cfm) = 1.05

thin rectangular ducts (ratio of width to depth, 1:5) = 1.25

TABLE 5.13 **Recommended and Maximum Air Velocities for Ducts**

	Recommended Velocities (fpm)		
Designation	*Residences*	*Schools, Theaters, Public Buildings*	*Industrial Buildings*
Outdoor air intakes[a]	500	500	500
Filters[a]	250	300	350
Heating coils[a,b]	450	500	600
Cooling coils[a]	450	500	600
Air washers[a]	500	500	500
Fan outlets	1000–1600	1300–2000	1600–2400
Main ducts[b]	700–900	1000–1300	1200–1800
Branch ducts[b]	600	600–900	800–1000
Branch risers[b]	500	600–700	800
	Maximum Velocities (fpm)		
Outdoor air intakes[a]	800	900	1200
Filters[a]	300	350	350
Heating coils[a,b]	500	600	700
Cooling coils[a]	450	500	600
Air washers[a]	500	500	500
Fan outlets	1700	1500–2200	1700–2800
Main ducts[b]	800–1200	1100–1600	1300–2200
Branch ducts[b]	700–1000	800–1300	1000–1800
Branch risers[b]	650–800	800–1200	1000–1600

Source: Copyright © by the American Society of Heating, Refrigerating and Air-conditioning Engineers, Inc., Atlanta, GA. Reprinted by permission from *ASHRAE Handbook Systems and Equipment,* 1967.

[a]These velocities are for total face area, not the net free area; other velocities in table are for net free area.

[b]For low-velocity systems only.

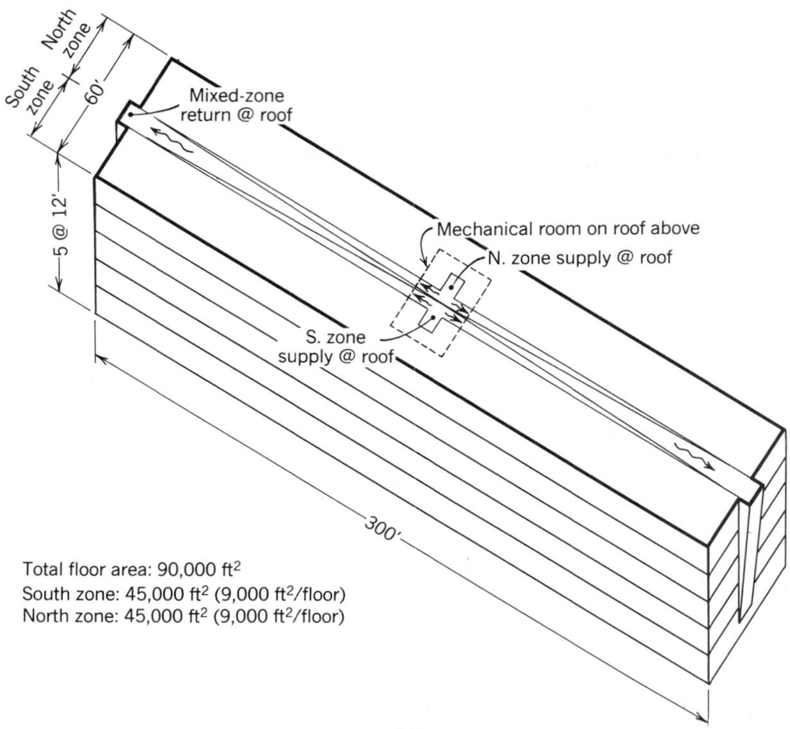

(a)

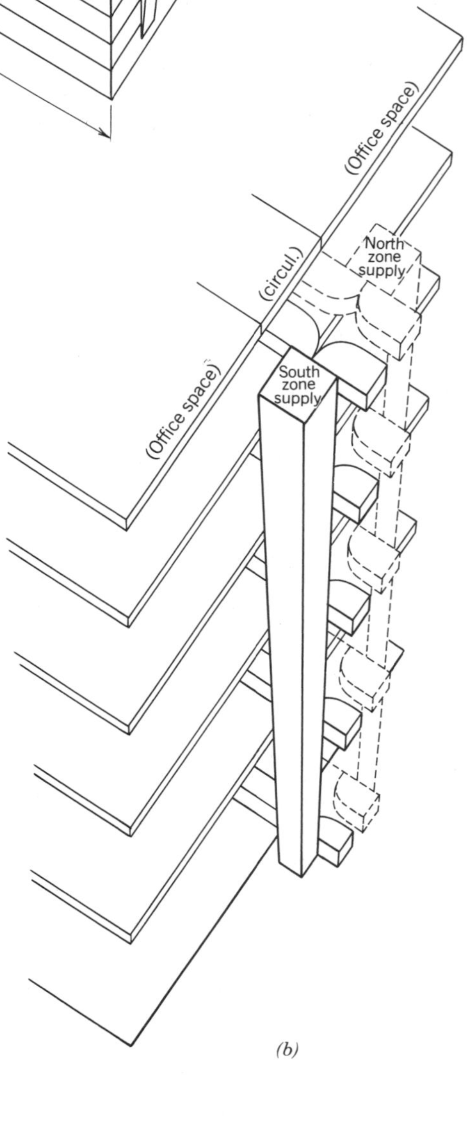

(b)

Fig. 5.22. *Axonometric views of the Denver office building discussed in Example 5.3.* (a) *Overall sizes and locations of zones and duct distribution trees.* (b) *Partial view of the supply distribution trees.*

EXAMPLE 5.3, PART B. Find the largest air ducts required for the Denver office building described in Examples 5.1 and 5.3, Part A. Establish overall building floor size as length = 5 × width, or 320 ft × 60 ft, as shown in Fig. 5.22. First, determine the quantity of air:

From Table 5.12, 6 to 20 ACH is recommended.

From Example 5.3, Part A, 6 ACH is needed for the maximum hour of night ventilation if ductwork and fan are used. So, assume that 6 ACH is adequate.

Assume a central mechanical space, on the roof. At this stage, two zones seem appropriate—one for the south-facing row of 25-ft-deep office space, and one for the north. Assume a single distribution tree for each zone (see Fig. 5.22). Return air will be collected in the corridor. For each square foot of this 12-ft floor height building, 6 ACH requires a volume of air equal to 6 ACH × 12 ft^3 = 72 cfh/ft^2 floor area. (There is no suspended ceiling, which would reduce the required air volume.) This is the volume of air that the ducts must carry.

Next, determine the recommended velocity, from Table 5.13. Since the 6 ACH is likely to also be the daytime air-change rate (Table 5.12), choose the recommended rather than the maximum velocity. (Maximum velocities produce more duct noise and require more fan energy.) Select 1200 fpm as the velocity.

The main duct size is, therefore,

$$\frac{72 \text{ cfh/ft}^2 \times 144 \times 1.05 \text{ friction allowance}}{60 \times 1200 \text{ fpm}} =$$

$$0.15 \text{ in.}^2 \text{ area duct per square foot of floor area}$$

Thus, in this two-zone, 90,000-ft² office building, the main duct area for each zone will be 45,000 × 0.15 = 6750 in.², or about 82 in. square (7 ft square). (An equally large return duct is required, as well!)

Although a flatter rectangular duct would probably be easier to integrate with the vertical circulation core, this simple square main duct is illustrated in Fig. 5.22. At each floor, ducts branch off on either side, serving 9000 ft²/2 = 4500 ft² each. As the ducts get closer to the offices, noise becomes a more pressing concern; the recommended velocity drops to about 800 fpm.

$$\frac{72 \text{ cfh/ft}^2 \times 144 \times 1.05 \text{ friction allowance}}{60 \times 800 \text{ fpm}} =$$

$$0.23 \text{ in.}^2/\text{ft}^2 \times 4500 \text{ ft}^2 = 1035 \text{ in.}^2,$$

or about 32 in. square (almost 3 ft square)

However, the 12-ft floor-to-floor height limits us to a shallower duct: (12′) − (2′ structure) − (8′ minimum corridor) = 2′ maximum duct depth. For thin rectangular ducts,

$$\frac{72 \times 144 \times 1.25}{60 \times 800} =$$

$$0.27 \text{ in.}^2/\text{ft}^2 \times 4500 \text{ ft}^2 = 1215 \text{ in.}^2$$

Select duct dimensions of 24 in. × 50 in. Note, however, that the 10-ft-wide corridor ceiling space is nearly filled by *two* 2 ft × 4 ft, 2 in. ducts, one for each zone. This first section of the horizontal duct clearly requires a higher velocity, which would reduce the size but increase the noise.

Meanwhile, the main vertical duct areas decrease as follows:

5th floor: (preceding calculation) =
6750 in.², or 82″ × 82″

4th floor: 0.15 × 36,000 =
5400 in.², or 73″ × 73″

3rd floor: 0.15 × 27,000 =
4050 in.², or 64″ × 64″

2nd floor: 0.15 × 18,000 =
2700 in.², or 52″ × 52″

1st floor: 0.15 × 9,000 =
1350 in.², or 37″ × 37″

Clearly, the use of recommended velocities is resulting in very large ducts; there is hardly room for the return air system, which has been forced to the ends of the building. If we use *maximum* velocities (1600 fpm main duct, 1300 fpm branch ducts), the main duct size becomes 1200/1600 × 0.15 = 0.086 in.²/ft². The duct sizes thus become

5th floor: 0.086 × 45,000 =
3848 in.², or 62″ × 62″

4th floor: 0.086 × 36,000 =
3078 in.², or 56″ × 56″

3rd floor: 0.086 × 27,000 =
2309 in.², or 48″ × 48″

2nd floor: 0.086 × 18,000 =
1731 in.², or 42″ × 42″

1st floor: 0.086 × 9,000 =
1770 in.², or 28″ × 28″

and the largest horizontal branch duct size becomes

$$800/1300 \times 0.27 =$$

$$0.165 \text{ in.}^2/\text{ft}^2; \quad 0.165 \times 4500$$

$$= 743 \text{ in.}^2, \text{ or } 24''$$

$$\times \text{ approximately } 30''$$

This more dimensionally (less acoustically) compatible distribution tree is shown in Fig. 5.22. Note that although higher velocities will also require more energy (to be used by the fans), less material will be needed to construct the ductwork.

Central Heating or Cooling Equipment. In the early design stages, an approximate size for the largest equipment is sometimes useful. Once the heating or cooling capacities are known, manufacturer's catalogs can be consulted for the dimensions of the heating and cooling equipment (see also Sections 6.6–6.8).

The critical decision in sizing the heating equipment is the *design temperature:* What is the lowest reasonable outdoor temperature for which a heating device can be sized if the desired interior temperature is to be maintained? These winter design temperatures are listed in Appendix A. When this is known, the next step is

$$\text{design } \Delta T = \text{inside temperature} - \text{outside design temperature}$$

In Section 5.2, the criteria for Btu/DD ft^2 were listed in Table 5.6. To convert to the required capacity of a building's heating equipment, calculate

$$\frac{\text{Btu/DD ft}^2}{24 \text{ h}} \times \Delta T \times \text{ft}^2 \text{ floor area} = \text{Btu/h heating capacity}$$

Note: For passively solar heated buildings, the backup heating unit is sometimes sized for the heat loss of the *entire* envelope and sometimes for the envelope *minus the solar wall,* just as the criteria were defined in Table 5.6. Similarly, buildings with *reliable* internal gains (lights, equipment, people, etc.) are sometimes designed with smaller heating units, because internal gains supply a constant portion of the space heating needs. For a more complete discussion, see Section 5.4.

The sizing of cooling (mechanical refrigeration) units is not so straightforward, as will become evident when detailed hourly heat gain procedures are presented (Sections 5.7, 5.8). However, a *very approximate* early estimate can be obtained from the estimated hourly gains in Table 5.10. *Warning:* This estimate is likely to be *lower* than that obtained using the peak heat gain hour, for which cooling equipment is often sized.

$$\text{sensible Btu/h cooling capacity} = [\text{approx. heat gain (Btu/h ft}^2)][\text{floor area (ft}^2)]$$

Another common unit used for sizing mechanical refrigeration is "tons" of cooling capacity, one ton being equivalent to the useful cooling effect of a ton of ice, or 12,000 Btu/h (3516 W). Therefore, the required capacity in tons is

$$\frac{\text{heat gain (Btu/h)}}{12{,}000} = \text{tons of cooling}$$

For the thermally well-designed residence of today, a rule of thumb is

$$1 \text{ ton cooling}/1000 \text{ ft}^2 \text{ of floor area}$$

For commercial buildings, see estimates in Table 6.15.

Yet another common unit used in the sizing of cooling equipment is the total (latent plus sensible) heat capacity of the machine. A quick *approximate* estimate of your building's *total* cooling load can be obtained from

$$\text{sensible heat gain (Btu/h)} \times 1.3 = \text{total (latent + sensible) Btu/h}$$

This additional 30% of latent heat represents a *typical* office or commercial building's mixture of sensible and latent heat; it does not account for food preparation or other activities that produce great amounts of moisture.

5.4 Calculating Hourly Heat Loss and Fuel Consumption

Sections 4.1 through 4.6 showed how heat is exchanged by conductance through a building's envelope (q) and through the ventilation of a building (q_v). By combining these rates of heat exchange, we can obtain a building's total hourly heat loss in winter. (To calculate the total hourly heat gain in summer, other gains must be added; this more complicated procedure is discussed in Sections 5.7 and 5.8). The total hourly heat loss of a building can be calculated under several different assumptions reflecting different purposes.

(a) Maximum Hourly Loss: Sizing Conventional Heating Equipment. The most typical use of $q + q_v$ is to determine the maximum amount of heat per hour that heaters must provide. Two important assumptions are usually made:

1. No internal heat gains (lights, people, etc.) and no solar gains are present in the building.
2. The lowest outdoor temperature ("design condition"—see Appendix A) is occurring.

These are conservative assumptions that lead to the installation of heating equipment that is rarely used to capacity. Such equipment does provide a safety margin, for those rare times when lower than design temperatures occur or when windows are inadvertently left open or other temporary and unexpected heat leaks occur in very cold weather.

Thus, to obtain design hourly heat loss, calculate

$$q_{\text{total}} = q + q_v$$

where

$$q = (\Sigma UA)\ \Delta T$$

with ΣUA the sum, for all exposed components of the building's envelope, of $U \times A$; ΔT the design condition (from Appendix A); and

$$q_v = 1.08V\ \Delta T \qquad (1200V\ \Delta T;\ \text{SI units})$$

where V is the volume in cfm (L/s) of outdoor air introduced.

(b) Maximum Hourly Heat Loss: Sizing Auxiliary Heating for Passive Solar Buildings. The one important difference between the maximum heat loss calculations for conventional and passive solar buildings is the following assumption.

> No internal heat gains (lights, people, etc.) are present, but there is sufficient stored solar energy to at least cancel out the heat losses through the south solar collection area.

Otherwise, the procedure is the same as that for conventional buildings:

$$q_{\text{total}} = q + q_v$$

where

$$q = \Sigma UA_{\text{ns}}\ \Delta T$$

$$q_v = 0.018\text{V}\ \Delta T \qquad (\Delta T \text{ is the design condition from Appendix A})$$

$$(1200\ \Delta T;\ \text{SI units})$$

ΣUA_{ns} *excludes* the solar collector area

This can lead to occasional chilly interiors, as when several days of heavily overcast skies coincide with design-condition outdoor temperatures. In some locations, therefore, designers use the more conservative, conventional procedure for sizing the auxiliary heaters for passive solar buildings.

(c) Maximum Hourly Loss: Checking Design Criteria. The calculations that produce q and q_v are also useful in reviewing a building's design. By showing where most of the heat loss is occurring, they can quickly pinpoint opportunities for energy conservation and increased comfort. If much of the building's heat loss is occurring through the large windows in one wall, for example, consider the following options.

1. Reduce window size. (Architectural and daylighting considerations may override.)
2. Add a layer of glazing, to reduce heat loss and increase the surface temperature of this large glass area. (Cost and detailing considerations may override.)
3. Add thermal shades or shutters, to dramatically reduce heat loss and increase surface temperature. (Architectural, view, cost, and detailing considerations may override.)

If much of the building's heat loss occurs through ventilation air, consider these options.

1. Reduce infiltration, by tighter construction (or reduce mechanical ventilation toward the code-required minimum).
2. Add a heat exchanger between outgoing and incoming air.

These calculations can also be used to check your building against published criteria for thermal performance (see Section 5.2). Redesign of building envelopes to meet such criteria is fairly common in the early stages of building design.

(d) Hourly Rates of Fuel Consumption. When outdoor temperatures in winter drop below the building balance point (see Section 5.5), heating systems usually begin to operate. The hourly rate of fuel consumption depends on the hourly heat loss from the interior space. If the boiler (or furnace) is selected to run continuously at the outdoor, critical winter design temperature, then it will cycle (run intermittently) at higher outdoor temperatures. The equipment, however, *is* selected on the basis of the maximum winter demand rate and therefore relates to the calculated heat loss at the design temperature.

EXAMPLE 5.6. Calculate the rates of burning for several fuels (or the rate of using electricity) to make up the hourly heat loss, under design conditions, of a mercantile store. Its maximum hourly heat loss is 159,840 Btu/h. For fuel values, refer to the data in Table 5.14.

TABLE 5.14 **Approximate Heat Values of Fuels**

NOTE: This table includes the thermal value of electricity used on-site (but not losses in fuel energy at the electrical generating plant). Approximate seasonal efficiencies of typical burner-boiler equipment are also shown.

Fuel	Heat Value: Conventional Units	Heat Value: SI Units	Typical Seasonal Efficiency, Percent
Anthracite coal	14,600 Btu/lb	33,980 kJ/kg	65–75
No. 2 oil	141,000 Btu/gal	39,300 kJ/L	70–80
Natural gas[a]	1,050 Btu/ft³	39,100 kJ/m³	70–80
Propane	2,500 Btu/ft³	93,150 kJ/m³	70–80
Electricity	3,413 Btu/kW	1 kW	95–100
Wood[b]	7,000 Btu/lb	16,290 kJ/kg	30–50

[a]Natural gas is frequently sold in "therms"; one therm = 100,000 Btu/h (SI units, 29.3 kW).
[b]At 20% moisture content.
[c]Higher efficiency furnaces are now available; see Fig. 6.46.

SOLUTION. If, for instance, coal were used, the statement would be

$$\text{lb/h} \times \text{Btu/lb heat value} \times \text{efficiency} = \text{Btu/h heat loss}$$

Transposing, we have

$$\text{lb/h} = \frac{\text{Btu/h heat loss}}{\text{Btu/lb} \times \text{efficiency}}$$

(Other efficiency statements are similar.) Applying values to this and to statements for the other fuels, we obtain the rates

$$\text{coal } \frac{159{,}840}{14{,}600 \times 0.70} = 15.6 \text{ lb/h}$$

$$\text{oil } \frac{159{,}840}{141{,}000 \times 0.75} = 1.51 \text{ gal/h}$$

$$\text{gas } \frac{159{,}840}{1052 \times 0.75} = 203 \text{ cfh}$$

$$\text{electricity } \frac{159{,}840}{3.41 \times 1.00} = 46{,}600 \text{ W (46.6 kW)}$$

The foregoing results are based on the assumption that the boiler and its piping are enclosed within the useful volume of the store, as is most usual. If they were in cold basements, or if the ducts or pipes ran through unheated space, more fuel would be used and system efficiency would decline. The rates established set the values by which the fuel-burning apparatus is selected. For instance, if oil were used, a nozzle that would inject oil at the rate of about 1½ gph should be tried.

These rates are for design (extreme) conditions and are not typical of the lower average rate of operation throughout the winter.

5.5 Calculations for Heating-Season Fuel Consumption (Conventional Buildings)

The following method of estimating the fuel used for space heating in a typical season best applies to residences and small commercial buildings that are skin load dominated and not passively solar heated beyond SSF = 10. To the extent that the combination of internal and solar gains can be predicted accurately, this method yields a reasonable estimate of annual fuel consumption for any building. For passively solar heated buildings, use instead the method given in Section 5.6.

Internal and solar gains make almost any building warmer than the outdoors during the heating season. The furnace (or other space heating device) is not needed until the outdoor temperature drops to the point at which these internal and solar gains are insufficient to heat the building by themselves: that is, when the heat lost through the building's skin and infiltration matches the heat gained through solar plus internal loads. This particular outdoor temperature is called the *balance point;* it represents the beginning of the need for space heating equipment.

To estimate the annual energy needed for a building's space heating, it is necessary to know

- The building's heat loss rate (envelope and infiltration).

- The building's internal plus solar gain rate.
- The building's balance point temperature.
- The time period during which the outside temperature falls below the building's balance point temperature (''degree days'').

(a) The Balance Point Temperature. This measurement can be expressed as

$$Q_i = UA(T_i - T_b)$$

or, frequently,

$$T_b = T_i - \frac{Q_i}{UA}$$

where T_b = balance point temperature

T_i = average interior temperature over 24 h, winter

Q_i = internal gains plus solar gains (Btu/h, or W)

UA = total heat loss rate (envelope plus infiltration) (Btu/h F, or W/°C)

To determine the total heat loss rate UA, combine the envelope (or skin) losses ΣUA (as was done in Sections 4.4 and 5.4) and the infiltration (or ventilation) losses $1.08V$ (or, in SI units, $1200V$), where V is the volume in cfm (or L/s).

The quantity Q_i cannot be determined so straightforwardly. The internal gains can be estimated as shown in Table 5.10 or calculated more precisely from known building population, lighting, and equipment data. The solar gains are elusive; each month has a different average gain. For simplicity (but less accuracy), use the average January daily insolation on a vertical surface (found in Appendix B) with this approach.

$$Q_i = \frac{\begin{bmatrix}\text{internal gains} & & \\ \text{(Btu/h while} & \times & \text{hours occupied} \\ \text{occupied)} & & \end{bmatrix}}{24\text{ h}} + \frac{\begin{bmatrix}\text{January insolation} & & \text{area (ft}^2\text{),} \\ \text{(Btu/ft}^2\text{ day average),} & \times & \text{south} \\ \text{vertical surface} & & \text{glass}\end{bmatrix}}{24\text{ h}}$$

The balance point temperature T_b can be used to do several things besides predict fuel consumption. By adding the T_b to a graph of monthly outdoor temperatures (such as in Fig. 2.7), the designer can quickly see whether heating or cooling will be the major problem for a specific building in a given climate. Also, it can be used to gain a better understanding of how zones in a building interrelate. Once the T_b is calculated for each zone, it can be determined when the entire building needs heating (outdoor temperature below any zone's T_b) or cooling (outdoor temperature above any zone's T_b), or when one zone's surplus heat can be another zone's space heating source (outdoor temperature higher than one zone's T_b, but lower than that of another). If thermal exchange between zones occurs for a major portion of the typical year, the choice of either a heating or a cooling strategy could be influenced.

(b) Degree Days. These data are published for each climate station and are calculated to various *base* temperatures (found in Appendix B). Until recently, degree days (DD) were always based on 65 F, because older, indifferently insulated buildings, with low internal gains, had a typical balance point of about 65 F. The combination of much higher levels of insulation and much more electric equipment has shoved the average building's balance point temperature downward; hence DD50, DD55, and DD60 are included with the traditional DD65 in Appendix B.

To derive heating DD for a particular climate and X base temperature DDX, each day's mean temperature (halfway between high and low) is subtracted from the base temperature; the result is the number of DDX for that day. If the mean temperature equals or exceeds the base temperature, no DDX are recorded. Then the DDX are totaled for an average year.

For example, assume that a day in Troy, New York, had a high of 60 F and a low of 34 F. The mean temperature was $(60 - 34)/2 + 34 = 47$ F.

$$65 - 47 = 18 \text{ DD65}$$

$$60 - 47 = 13 \text{ DD60}$$

$$55 - 47 = 8 \text{ DD55}$$

$$50 - 47 = 3 \text{ DD50}$$

Clearly, a building with a 65 F balance point will need more heat that day than will a building with a 50 F balance point temperature.

To convert DD conventional to DD SI, simply multiply DD conventional by 5/9:

$$\text{DDSI} = 0.56 \text{ DD conventional}$$

To obtain the ''DD balance point'' needed to estimate your particular building's heating needs, interpolate between the various base DDs as required. (If your balance point is below 50 F, get lower DD base figures from your local weather station. Do not extrapolate!)

(c) Yearly Space Heating Energy. To estimate the energy needed over an average year for a building E, calculate

$$E = \frac{(UA)(\text{DD balance point})(24 \text{ h})}{kV}$$

where E is in units of fuel consumed per year (e.g., therms of gas or kWh of electricity).

UA is the total heat loss rate, envelope + infiltration (Btu/h F, W/°C).

DD balance point is obtained as just described.

k is a factor that includes the effects of equipment efficiency at both full and potential capacity, of energy conservation devices, of energy-efficient buildings and equipment (see Table 5.15).

V is the heating value of the fuel, from Table 5.14.

TABLE 5.15 **Heating Equipment Efficiency Factors, *k***

For use in predicting annual energy consumption for space heating.

Space Heating Equipment	*k*
Electric resistance heaters within the heated space	1.0
Gas-fired boiler, fully condensing	0.9
Gas furnace: forced-air, spark-ignited, and induced-draft	0.8
Oil-fired boiler, energy-efficient model	0.65
Gas furnace: forced-air, atmospherically vented	0.60

Source: Based on *1981 Handbook of Fundamentals,* published by American Society of Heating, Refrigerating, Air Conditioning Engineers, Inc., Atlanta, GA.

A detailed list of certified furnace and boiler efficiency ratings is available from the American Gas Association (1515 Wilson Blvd., Alexandria, VA 22209).

EXAMPLE 5.7. A residence in Springfield, Illinois, has a total heat loss rate UA of 544 Btu/h F and a balance point temperature of 55 F. It will have an induced-draft, natural gas forced-air furnace, for which $k = 0.8$. For Springfield, DD55 = 3434. The average annual energy used for space heating is

$$E_{\text{therms}} = \frac{(544 \text{ Btu/h F})(3434 \text{ DD})(24 \text{ h/day})}{0.8 \times 100{,}000 \text{ Btu/therm}}$$
$$= 560 \text{ therms}$$

5.6 Passive Solar Heating Performance

Section 5.3 presented the rules of thumb for determining solar opening size, thermal mass, and the solar savings fraction (SSF). As a building design takes shape, more-detailed information becomes useful: Which passive system matches the architectural program? Which performs better thermally? For a given solar opening, what exactly is the resulting SSF? How much auxiliary fuel consumption per year must accompany that SSF? If a building overheats on sunny winter days, how hot will it get?

To this point, passive solar heating has been treated as a single approach; the rules of thumb for SSF distinguished only between systems with night insulation and those without. Important architectural differences, however, characterize the various passive solar heating approaches, as summarized in Table 5.16. On the basis of the wider architectural implications presented in Table 5.16 and the detailed sizing information found in Appendix C, the designer can select a passive solar heating approach with some confidence in both its applicability and its yearly need for auxiliary space heating.

An example of a passively solar heated building is the Visitor Center for the Antelope Valley, California, Poppy Reserve, about 85 miles northeast of Los Angeles (Fig. 5.23). Set in a remote desert where winter temperatures sometimes fall below 20 F, the building is 100% passively solar

TABLE 5.16 **Passive Solar Heating Systems Compared**

	Influence on Plan	*Heating Characteristics*
Direct gain (DG) 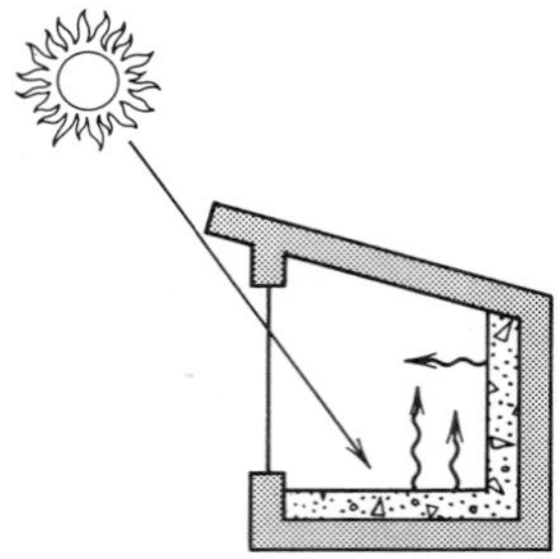	Sun can enter through south windows or skylights; open plan can allow sun and stored heat to serve entire top floor of building. Large areas of thermal mass surface should be darker colored and free of rugs, wall hangings, etc. Light-colored surfaces near glass reduce glare. Outdoor view and access to south are encouraged.	Quick to warm up in the morning, fast response to sun. Tendency to overheat at midday; large temperature swings. Much radiant loss to bare window by night; movable insulation encouraged (or triple glazing or selective film). Warmth spread throughout space along with thermal mass.
Thermal storage wall Trombe (masonry) Wall (TW) Water wall (WW) 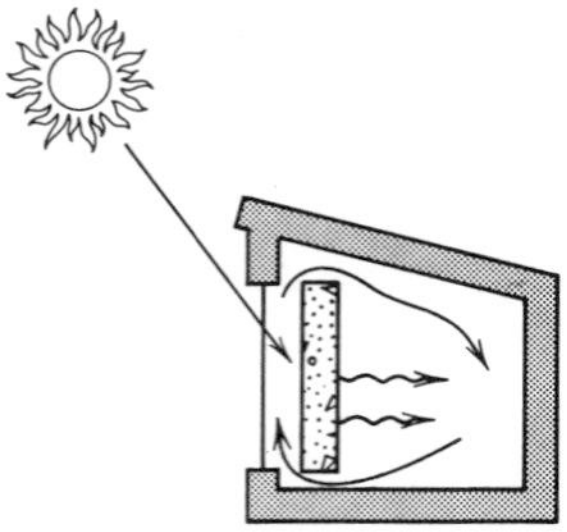	Needs to be on south wall of building. Inner wall of TW or WW should be kept clear of hangings, furniture, etc., but rest of space is unrestricted. Not much solar impact beyond about 25 ft from TW or WW. Outdoor view and access to south are discouraged.	Unvented TW, WW are slow to warm by day, slow to cool by night; small temperature swings. Most radiant heat arrives in evening; comfort is most likely near the TW or WW surface. (Vented TW behave midway between unvented TW and DG systems.)
Sunspace (SS) 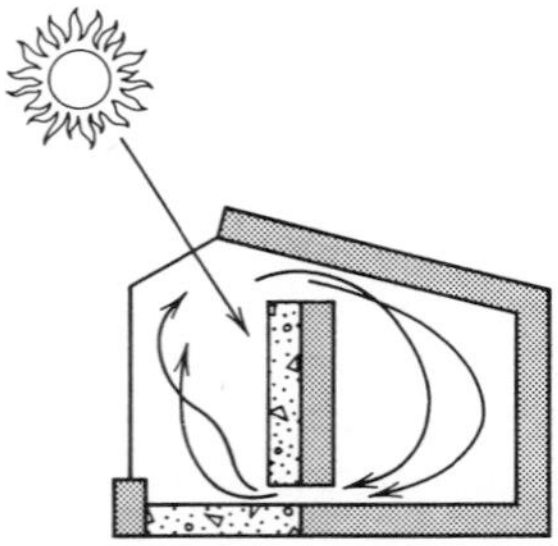	Same influences as TW and WW above; or sunspace can be insulated from building, becoming less efficient heat source for a rock bed. Floor above rock bed should be kept free of rugs. Sunspace becomes a special place with different characteristics from rest of building. Access to sunspace thereby encouraged. Access to south encouraged, view to south filtered.	Sunspace thermally like DG, but with extreme temperature swings and accentuated radiant loss by night. Movable insulation often omitted and night use curtailed. Building beyond is thermally like TW or WW, depending on its connection with sunspace. Warm floor above rock bed in months with SS surplus heat.
Roof pond 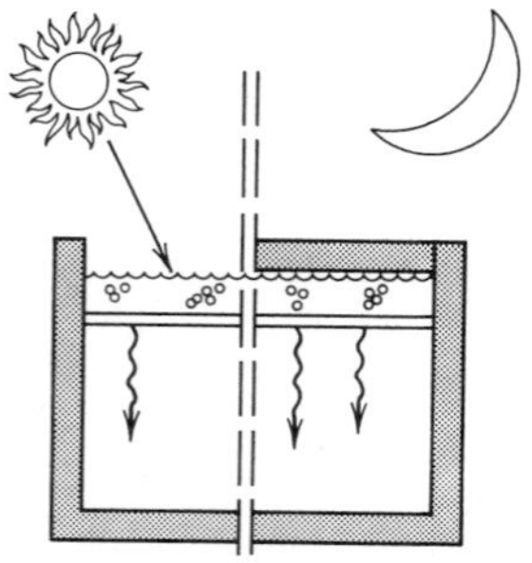	Flat or nearly flat roof is desirable. Skylight is discouraged. Plan is completely unrestricted; sidelighting and views, access to outdoors encouraged.	Low temperature swings; steady temperatures both summer and winter. Winter air stratification possible, since warmest surface is also highest surface in the space.

TABLE 5.16 **Passive Solar Heating Systems Compared (*continued*)**

	Daylighting	*Cooling*
DG	Very high DF possible, with high glare potential at lower windows. Possible conflict between light-colored surfaces for glare reduction and dark-colored surfaces for solar absorption. Summer shading greatly reduces DF. Encourages both side and top lighting.	Cross ventilation: encouraged by large windows to south. Stack ventilation: helped when clerestories are used. Night vent./thermal mass: excellent potential, much mass surface and capacity. Other closed-bldg. cooling: large south windows are big threat. Shading (and lower DF) necessary in overheating months.
TW, WW	No daylighting through TW, unless interrupted by windows (high glare potential). Diffuse light possible through translucent WW. Discourages sidelighting.	Cross ventilation: discouraged by solid TW, WW. Stack ventilation: TW or WW can itself produce stack effect, but with risk of evening overheating Night vent./thermal mass: limited interior, mass surface exposure, but lots of mass capacity. Other closed-bldg. cooling: with summer shading of TW or WW, good match with controlled cooling; thermal mass delays and reduces peak heat gains.
SS	Very high DF within sunspace; little or no daylight through common wall, except as encouraged by view and access to surface. Summer shading of sunspace reduces DF.	Cross ventilation: only to extent that common wall is penetrated for view and access. Stack ventilation: SS can become moderately effective stack. Night vent./thermal mass: both surfaces of common mass wall are available, but two spaces must be cooled. Other closed-bldg. cooling: with summer shading of SS, good match with controlled cooling (except in SS itself).
Roof pond	Discourages toplighting, encourages sidelighting, with very light color on underside of roof.	Cross ventilation: excellent potential. Stack ventilation: discouraged. Night vent./thermal mass: easily achieved with roof undersurface. Other closed-bldg. cooling: roof pond night sky cooling is often sufficient by itself, making other cooling unnecessary.

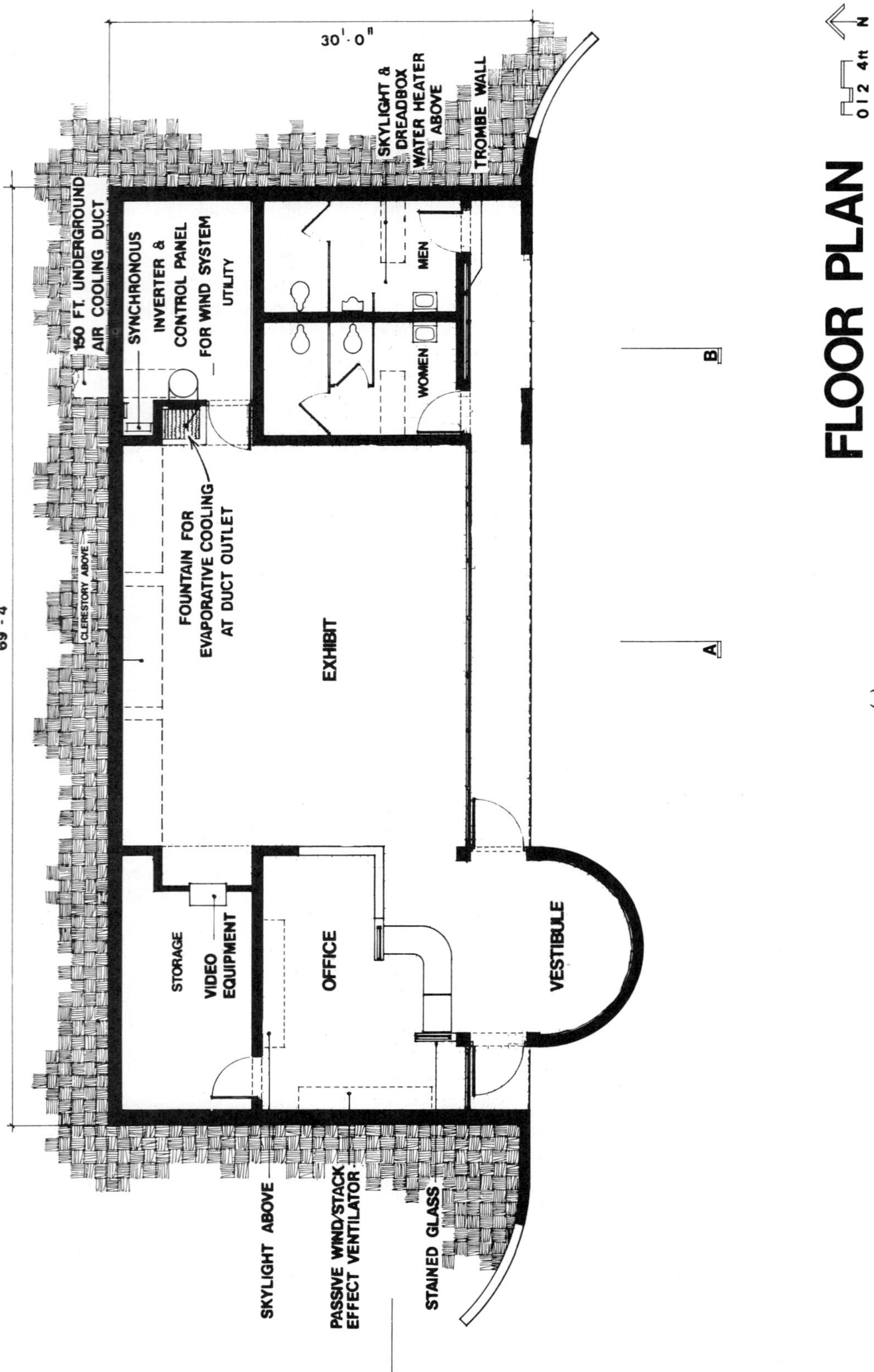

30'-0"
69'-4"
150 FT. UNDERGROUND AIR COOLING DUCT
SYNCHRONOUS INVERTER & CONTROL PANEL FOR WIND SYSTEM
UTILITY
SKYLIGHT & DREADBOX WATER HEATER ABOVE
TROMBE WALL
MEN
WOMEN
CLERESTORY ABOVE
FOUNTAIN FOR EVAPORATIVE COOLING AT DUCT OUTLET
EXHIBIT
A
B
C
STORAGE
VIDEO EQUIPMENT
OFFICE
VESTIBULE
SKYLIGHT ABOVE
PASSIVE WIND/STACK EFFECT VENTILATOR
STAINED GLASS
FLOOR PLAN
0 1 2 4ft
N

(a)

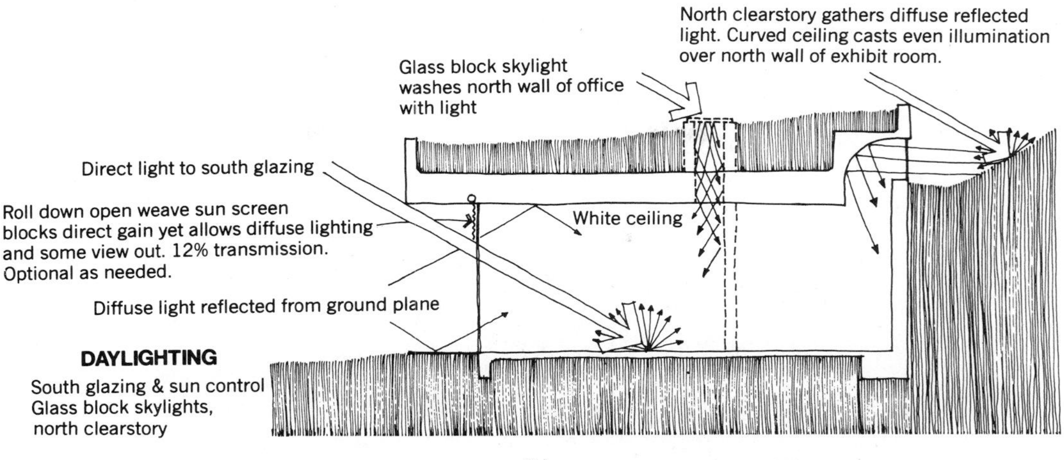
North clearstory gathers diffuse reflected light. Curved ceiling casts even illumination over north wall of exhibit room.
Glass block skylight washes north wall of office with light
Direct light to south glazing
Roll down open weave sun screen blocks direct gain yet allows diffuse lighting and some view out. 12% transmission. Optional as needed.
White ceiling
Diffuse light reflected from ground plane
DAYLIGHTING
South glazing & sun control
Glass block skylights,
north clearstory

(b)

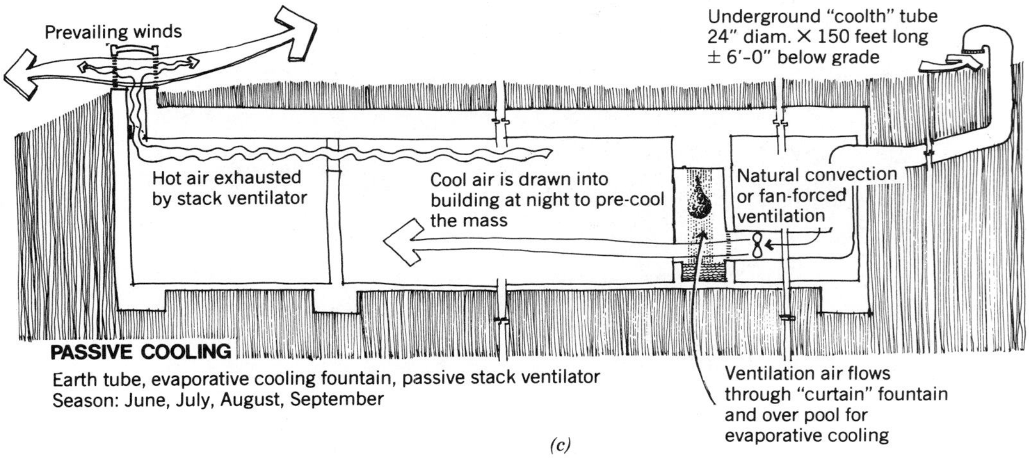
Prevailing winds
Underground "coolth" tube
24" diam. × 150 feet long
± 6'-0" below grade
Hot air exhausted by stack ventilator
Cool air is drawn into building at night to pre-cool the mass
Natural convection or fan-forced ventilation
PASSIVE COOLING
Earth tube, evaporative cooling fountain, passive stack ventilator
Season: June, July, August, September
Ventilation air flows through "curtain" fountain and over pool for evaporative cooling

(c)

Fig. 5.23. *The Visitor Center for the Antelope Valley California Poppy Reserve, the Colyer/Freeman Group, architects.* (a) *The plan shows earth sheltering for a somewhat elongated E–W axis, served by both direct-gain and Trombe wall heating strategies and by earth tube intake/stack exhaust for passive cooling.* (b) *Solar energy enters mostly through the south glass; the glass block skylights and north clerestory are primarily for even distribution of daylighting. At restrooms, a Trombe wall is used. Thermal shades on the interior protect the south glass and north clerestory on winter nights; a roll-down exterior sunscreen nearly eliminates direct sun through south glass in summer. Concrete block walls, concrete roof, and floor provide thermal mass, as useful for summer cooling as for winter heating.* (c) *In summer, the desert air first passes through an underground tube, then is evaporatively cooled within the building, in sight of visitors. This cooled air picks up the building's heat and is exhausted from the stack ventilator, assisted by a fan when necessary.* (d) *The most obvious feature of the Visitor Center, which is set into a hillside, is the 8-kW wind generator to supply display lights and video equipment, a fan, and a pump for the well. The cooling tube inlet is visible at the base of the tower.*

heated by a combination of direct-gain and Trombe wall strategies, interior thermal shades, and thermally massive ceiling, walls, and floor. An earth-sheltered, thermally massive building was chosen for its suitability to the environmentally sensitive site, as well as for its all-year thermal advantage in a desert climate. On summer days, when air temperatures frequently exceed 100 F, a combination of underground air tubes, evaporative cooling, and stack ventilation provides a modest sup-

Fig. 5.23. *(continued)*

ply of fresh air without overheating. Continuous power ventilating of the building at night draws the cool outdoor air over the thermally massive surfaces to provide added cooling capacity for the following day. The 2100-ft^2 building also is served by an 8-kW wind generator and has its own well.

The following procedure is based on Balcomb et al., the *Passive Solar Design Handbook,* Volume 3, (1983), and reprinted by permission. Portions of Volume 2 published by the U.S. Department of Energy, are also used. Volume 3 offers a much wider variety of passive systems and a wider network of location listings than can be presented in this book. Along with more "sensitivity curves," to allow prediction for nonstandard passive systems, that volume also provides a much more time-consuming and detailed method for calculating the *monthly* SSF and auxiliary energy needs; the method presented here gives annual results only. Thus, Volume 3 is an important, perhaps indispensable reference for the serious passive solar designer. (In addition, the helpful *Simplifying Guide to Volume Three,* by W. A. Shurcliff, is available.)

(a) Load Collector Ratio (LCR) Annual Performance. The method presented here, called the load collector ratio (LCR), yields the annual SSF and auxiliary energy needs for a building. This method has the following steps:

STEP 1. Choose the location and the reference passive system that most closely coincide with your building and its site. The locations listed in Appendices B and C are shown in Fig. B.1; the reference passive systems for which performance data are available (Appendix C) are summarized in Table

C.1. If your system differs significantly from the closest reference system, see Section 5.6*b*, "Variations on Reference Systems."

STEP 2. Tentatively select a size for the solar openings, balancing the rule of thumb for SSF (Table 5.8) with that for daylighting (Table 5.3).

STEP 3. Calculate the UA_{ns} for the building design—one that *excludes* the solar openings but *includes* all other envelope losses, as well as the infiltration loss. Then multiply UA_{ns} by 24 hours to obtain Btu/DD; this is called the building load coefficient (BLC).

$$\text{BLC} = 24 \times UA_{ns}$$

STEP 4. Check your building's overall loss rate against the criteria from Table 5.6:

$$\text{Btu/DD ft}^2 = \frac{\text{BLC}}{\text{floor area (ft}^2)}$$

Does your building envelope conserve energy sufficiently, or do you need more insulation (or less nonsouth glass, or less infiltration)?

STEP 5. Determine the *vertical projection* of the solar opening area A_p. (For a vertical solar opening, A_p is identical to the actual opening area; for a 45°-inclined solar opening, A_p = 0.707 actual area.)

STEP 6. Find the load collector ratio LCR, expressed in Btu/DD ft²:

$$\text{LCR} = \frac{\text{BLC}}{A_p}$$

STEP 7. For the reference systems that most closely approaches your design, consult Table C.2 (Appendix C) for the appropriate location. By interpolation, find the annual SSF that corresponds to your passive system's listed LCR. Note also the annual degree days listed in this table. (These are DD65.)

STEP 8. Finally, determine the approximate annual auxiliary heating Q required:

$$Q = (1 - \text{SSF}) \times \text{BLC} \times \text{DD}$$

Although this quick annual results method is based on the DD65 listed in Table C.2, it is possible to adjust Q to *approximately* account for higher internal gains or better-insulated envelopes. In this adjustment, use DD balance point instead of DD65 in the equation above for Q.

STEP 9. Now compare the rule-of-thumb predicted relationship between collector size and SSF to the actual one you've just calculated. If the SSF is *smaller* than you had hoped, can you decrease BLC (improve conservation) or increase collector size, or switch to another passive system with a more favorable SSF for the same LCR? If the SSF is *larger,* will you be happy with the increased fuel savings, or will you reduce the collector size, or consider another, less-efficient passive system that has some architectural advantage over the one for which you calculated SSF?

EXAMPLE 5.8. We will now take a more detailed look at the solar collecting area required for the passively solar heated building in Omaha, Nebraska, discussed in Example 5.2.

From the rules of thumb used in Example 5.2, we expected that 20% of the floor area in south glass with night insulation would yield a 51% SSF. More detailed characteristics of this building are shown in Fig. 5.24. From a program requirement of 2900 ft² of floor area, the 20% south glass area equals 0.20 × 2900 = 580 ft². For daylighting by sidelight only, DF_{av} = 0.2 (window area/floor area); if all the south glass area is available for daylighting (as with direct gain systems), DF_{av} = 0.2 × 20%, or 4%. This is adequate for office work. However, because you want to avoid dark areas near the rear walls of these spaces and to investigate Trombe walls, and so on, some north light is desirable. Choose about 3% of the floor area in north glass (say, 90 ft² glass); added DF_{av} = 0.2 × 3%, or 0.6%.

Wall insulation levels of *R*30 to *R*35 were recommended in the rule of thumb. Checking the criteria for ASHRAE 90A-1980 (Section 5.3), the maximum U_o wall for Omaha's 6601 DD65 (Appendix B) is 0.26. (Omit the south solar wall, since these criteria were developed for conventionally heated buildings.)

East, west, north opaque wall area	2210 ft²
North glass area	90 ft²

Assume night insulation on north windows; from Table 4.17, U = 0.28. Assume wall insulation *R*35; U_{wall} = 1/35, or 0.029.

THERMAL CONTROL

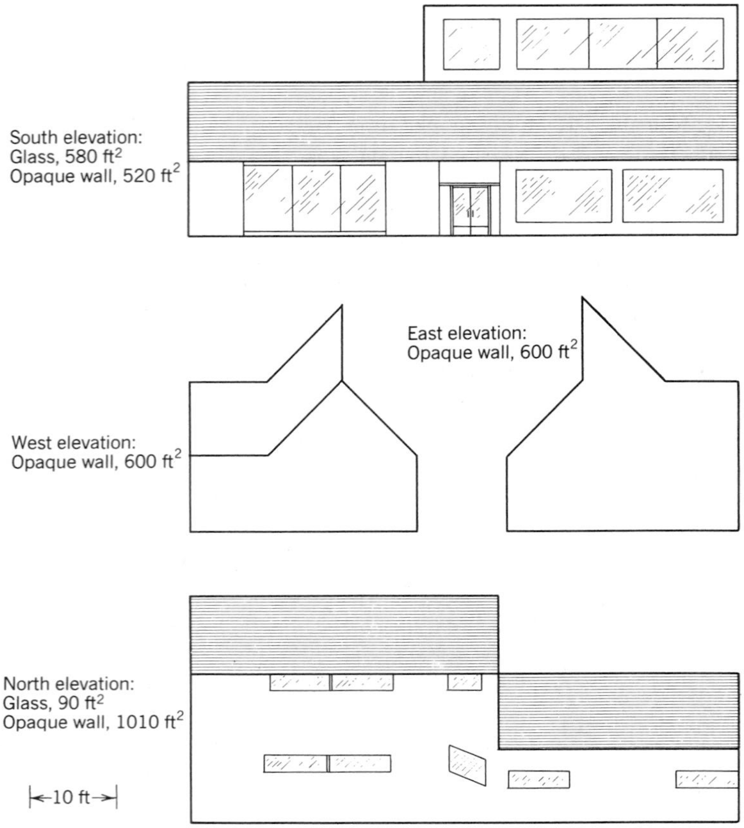

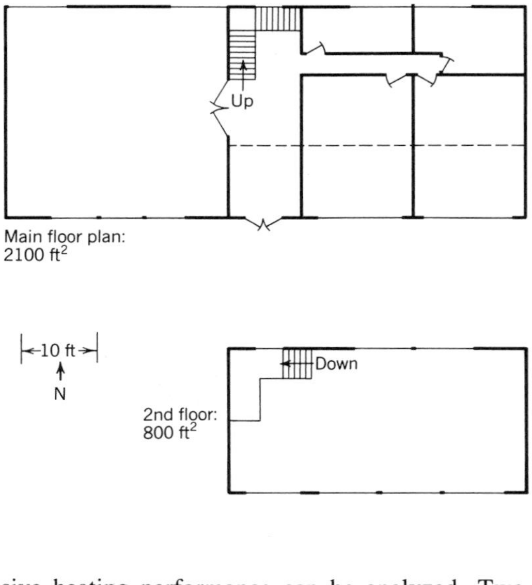

Fig. 5.24. *Plans and elevations of the Omaha office building discussed in Example 5.8.*

U_o walls =

$$= \frac{(0.029 \times 2210 \text{ ft}^2) + (0.28 \times 90 \text{ ft}^2)}{2300 \text{ ft}^2}$$

$$= \frac{64.09 + 25.2}{2300} = 0.039$$

This is well below the allowable U_o maximum of 0.26, so this $R35$ insulation will also be used in the opaque portion of south walls.

Roof insulation will be useful for coping with Omaha's summer sun as well as for winter heat conservation: select $R33$, with $U = 0.03$.

Floor slab perimeter insulation is selected as follows. From Table 4.16, for $R5.4$ insulation, at Omaha's 6601 DD, and with a metal stud wall: F_2 = about 0.53 Btu/h F ft of perimeter.

As far as infiltration is concerned, with reasonably tight construction, low population, and few entries and exits per hour you can assume ½ ACH.

Once these decisions have been made, the passive heating performance can be analyzed. Two options will be shown: one with a fixed SSF, and the other with a fixed solar glazing area.

Option A: Fix a goal of about 50% SSF and see

how the required solar opening size varies with different systems. This option stresses a rather high solar contribution and shows the architectural consequences (window size variations).

STEP 1. Investigate *each* passive system for Omaha.

STEP 2. Solar opening size will vary, but assume that it averages 580 ft² for the calculation of UA_{ns}.

STEP 3. Calculate UA_{ns}:

North, east, and west opaque walls	2210 ft² × 0.029	= 64
South opaque wall	520 ft² × 0.029	= 15
Roof	1974 ft² × 0.03	= 59
Slab perimeter	0.53 × 200 lineal feet	= 106
North glass	90 ft² × 0.28	= 25
Infiltration	0.5 ACH × 31,000 ft³ × $\frac{1.08}{60 \text{ min}}$	= 281
		550

Then BLC = 24 × 550 = 13,200 Btu/DD.

STEP 4. Calculate overall loss rate:

$$\frac{13{,}200 \text{ Btu/DD}}{2900 \text{ ft}^2 \text{ floor}} = 4.55 \text{ Btu/DD ft}^2$$

This is close enough to the Table 5.6 criterion of 4.6 Btu/DD ft² for this 6601 DD65 climate.

STEP 5. Assume all solar openings are vertical, so that A_p equals the actual area of solar opening.

STEPS 6 AND 7. For the initial assumption of 580 ft² solar opening,

$$\text{LCR} = \frac{\text{BLC}}{A_p} = \frac{13{,}200}{580} = 22.8$$

However, in this option we are looking for *variations* in solar collector area, with SSF fixed at about 50% (or 0.50). The solar collector sizes for each reference passive system that will produce an SSF of 0.50, taken from Appendix C, are listed below. Since a maximum south aperture of 580 glass + 520 wall = 1100 ft² is possible, any system exceeding this limit is shown in parentheses.

Omaha: Passive	*System*	*LCR*	*Solar Collector (Window) Area (ft²)*
Water walls:	WWA3	13	1015
	WWA6	16	825
	WWB2	17	776
	WWB4	27	489
	WWC3	36	367
Trombe walls:	TWA3	11	(1200)
	TWE3	32	413
	TWF3	9	(1467)
	TWF4	9	(1467)
	TWG3	8	(1650)
	TWG4	7	(1886)
	TWI2	13	1015
	TWI3	16	825
	TWI4	21	629
	TWJ2	21	629
Direct gain:	DGA1	(Not attainable)	
	DGA2	6	(2200)
	DGA3	18	733
	DGB2	10	(1320)
	DGC3	23	574[a]
Sunspaces:	SSA1	12	1100
	SSA2	25	528
	SSA5	8	(1650)
	SSB1	9	(1467)
	SSB3	6	(2200)
	SSB5	5	(2640)
	SSC4	15	880
	SSD1	17	776
	SSD2	32	413
	SSE1	11	(1200)

[a]Closest to rule-of-thumb prediction.

STEP 8. The approximate annual auxiliary energy required may be estimated as follows.

$$Q = (1 - \text{SSF}) \times \text{BLC} \times \text{DD}$$

where DD will be D65, since the internal loads have not yet been determined. For this building's SSF, which is 0.50, we have

$$Q = (1 - 0.50) \times 13{,}200 \times 6601 = 43{,}566{,}600 \text{ Btu annually}$$

STEP 9. For this building and climate, 16 variations of passive designs were found to be possible at an SSF of 50%. The necessary solar collector size (south glass) varied from about 370 ft^2 for WWC3 to the entire wall glazed at 1100 ft^2 for SSA1. System DGC3 came closest to the performance predicted by the rule of thumb.

Option B: Keep the solar collector area at about 580 ft^2, and see how the annual SSF varies with different systems. This option stresses architectural characteristics, and shows the energy consumption consequences.

STEP 1. For Omaha, Nebraska, investigate *each* passive system.

STEP 2. Solar opening fixed at 580 ft^2.

STEP 3. As before, $UA_{ns} = 550$; BLC = 13,200 Btu/DD.

STEP 4. As before, overall loss rate = 4.55 Btu/DD ft^2.

STEP 5. A_p = actual area of solar opening.

STEPS 6 AND 7. Again, LCR = 22.8. From Appendix C, the annual SSF for each reference passive system are listed below.

Omaha: Passive System		*SSF (at LCR = 22.8)*
Water walls:	WWA3	0.36
	WWA6	0.40
	WWB2	0.43
	WWB4	0.56
	WWC3	0.67
Trombe walls:	TWA3	0.33
	TWE3	0.62
	TWF3	0.29
	TWF4	0.29
	TWG3	0.26
	TWG4	0.22
	TWI2	0.36
	TWI3	0.40
	TWI4	0.48
	TWJ2	0.48
Direct gain:	DGA1	None
	DGA2	0.30
	DGA3	0.43
	DGB2	0.31
	DGC3	0.50[a]
Sunspaces:	SSA1	0.37
	SSA2	0.53
	SSA5	0.34
	SSB1	0.32
	SSB3	0.28
	SSB5	0.28
	SSC4	0.39
	SSD1	0.44
	SSD2	0.60
	SSE1	0.36

[a]Closest to performance predicted by rule of thumb.

STEP 8. The annual auxiliary energy required will be greatest for TWG4 (SSF = 0.22) and least for WWC3 (SSF = 0.67). For TWG4,

$$Q = (1 - 0.22) \times 13{,}200 \times 6601$$
$$= 67{,}963{,}900 \text{ Btu annually}$$

For WWC3,

$$Q = (1 - 0.67) \times 13{,}200 \times 6601$$
$$= 28{,}753{,}960 \text{ Btu annually}$$

STEP 9. Again, system DGC3 came closest to the performance predicted. All systems except DGA1 were able to make some contribution. The night-insulated systems performed quite well (near or above SSF = 0.50), as did SSA2. The best performer was WWC3, which for the same investment of 580 ft^2 required only 42% of the annual auxiliary energy needed by TWG4.

Summary: It appears that DGC3 offers the best approach for both ample daylighting and solar savings. However, a combination of DGC3 for daylighting and WWC3 for highest SSF appears highly attractive. In such a mixture, WWC3 could be used from floor level to counter height, and DGC3 used above counter height, for plentiful daylighting.

What if two or more passive systems are used in the same building? In that case, first do calculations for the entire building assuming that *one* system has *all* the solar area, and find the SSF. Next, do the calculations assuming that the second system has all the solar area, and find its SSF.

Then average the SSFs according to the relative solar areas of the two systems.

EXAMPLE 5.9. A veterinary clinic in Buffalo, New York, is using two systems: water walls for examining rooms and direct gain for the waiting/reception area. The building's characteristics are

balance point = 50 F

UA_{ns} = 356

A_p water wall =

240 ft^2 (reference system WWB4)

A_p direct gain =

150 ft^2 (reference system DGA3)

total floor area = 1900 ft^2

predicted SSF from rule of thumb =

about 36%, with night insulation

Checking the overall heat loss criteria (Table 5.6),

$$\frac{356 \times 24}{1900 \text{ ft}^2} = 4.5$$

This is less than 4.6 (for 5000–7000 DD), so it's acceptable.

First, calculate as though all A_p (390 ft^2) were system WWB4:

$$\text{BLC} = 24 \times 356 = 8544 \text{ Btu/DD}$$

$$\text{LCR} = \frac{8{,}544}{390} = 22$$

for Buffalo, SSF = 0.35, by interpolation

Next, calculate as though all A_p were system DGA3:

for LCR 22, Buffalo SSF =

0.28, by interpolation

Now calculate the average SSF, given that water wall comprises 62% of total solar opening and direct gain comprises 38%:

$$\frac{0.35(62\%) + 0.28(38\%)}{100\%} = 0.217 + 0.106 = 0.323$$

Result: The SSF is about 90% of what was predicted (0.32 as opposed to 0.36); since the glass area is already large, program requirements probably prevented further increases in it. The building already seems to conserve energy well, since it meets the heat loss criteria. So we should consider some better performing systems for Buffalo. By inspection of Appendix C (Buffalo data) system WWC3 looks better; its selective surface (on water containers) will cost more, but only single glazing is required. (Both WW systems have night insulation.)

A switch from WWB4 to WWC3, whose SSF is 0.475 at LCR 22, yields an annual SSF of

$$\frac{0.475(62\%) + 0.28(38\%)}{100\%} = 0.295 + 0.106 = 0.401$$

The approximate annual auxiliary heating energy required is

$$Q = (1 - 0.401) \times 8544 \text{ Btu/DD} \times 3322 \text{ DD50 (balance point)}$$

$$= 17 \text{ million Btu}$$

(This is equivalent to 4981 kWh, or 227 therms of natural gas burned at 75% efficiency.)

(b) Variations on Reference Systems. A particularly wide set of choices faces the designer of direct-gain and sunspace systems, in which mass distribution and glass orientation can assume thousands of different combinations. The ''sensitivity curves'' furnished in Balcomb et al. (1983) give some guidance on how a predicted SSF might vary as an actual passive system departs from a reference system.

Sensitivity curves can serve as early general design guidelines. Looking at the curves for your location, which design changes yield dramatic results, and which make little difference? The curves may also be used to adjust the SSF you found for a reference design.

(c) Internal Temperatures. Two quantities are of particular interest to passive solar designers: How much higher, compared to outdoors, will the average indoor temperature be from solar heating alone? And, how widely will this internal temperature vary (swing) on a clear winter day?

The approximate temperature difference between inside and outside on a *clear* January day, called ΔT solar, can be estimated from Fig. 5.25;

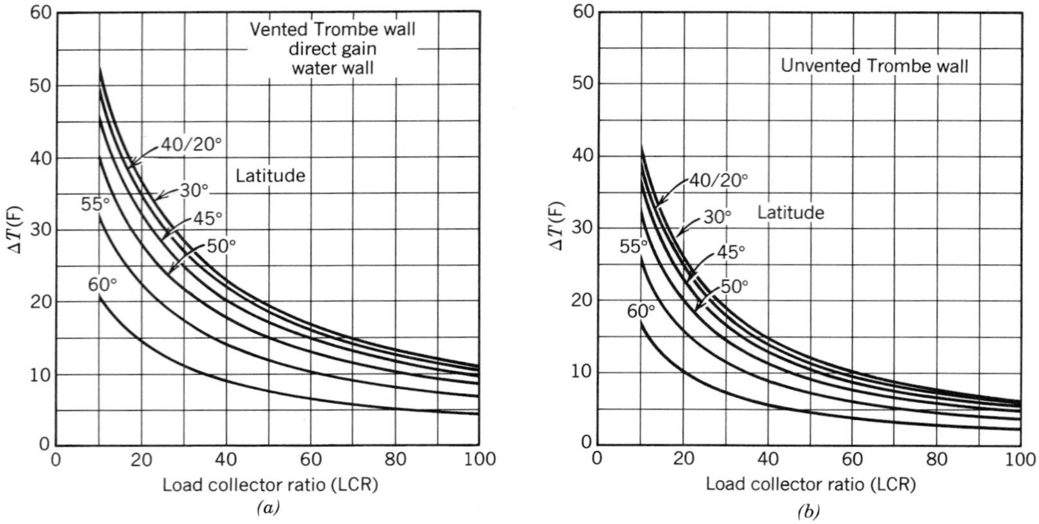

Fig. 5.25. *Graphs of "*ΔT *solar," the temperature difference to be expected between the average inside temperature and the average outside temperature on a clear January day. The curve marked "40/20°" applies to both 40° latitude and 20° latitude.* (a) ΔT *solar for direct gain, water wall, or vented Trombe wall systems.* (b) ΔT *solar for unvented Trombe wall systems. From Balcomb et al., 1980.*

it varies with latitude and with the LCR. Sunspaces are not shown in the figure. Although the temperature within the sunspace cannot be easily approximated, the ΔT solar for the room beyond the sunspace can be approximated by using Fig. 5.25*a*, if these spaces are connected by vents, or Fig. 5.25*b*, if they are connected only by a solid masonry wall.

To determine the average winter indoor temperature,

1. Find the average January ambient (outdoor) temperature, *TA*, from Appendix B.
2. Find ΔT solar for the building and its site (Fig. 5.25).
3. Find the ΔT due to internal heat sources:

$$\Delta T \text{ internal} = \frac{\text{total internal gains (Btu/day)}}{\text{BLC} + (UA_s \times 24)}$$

   where UA_s is for the solar area only. *Note:* ΔT internal averages 5 to 7F° for residences.
4. Add the quantities from the first three steps to find the average January clear-day indoor temperature.

When internal gains are high, there is less need for ΔT solar; the building is mostly "heating itself" (becoming an ILD rather than an SLD building). If the average indoor temperature is too high, smaller solar openings should be considered un-

TABLE 5.17 **Indoor Temperature Swing, ΔT Swing**

NOTE: These swings are based upon a thermal storage mass capacity of 45 Btu/ft² F; ΔT solar can be found in Fig. 5.25.

Passive Solar System	*ΔT Swing*
Direct gain: $\frac{\text{mass area}}{\text{glass area}} = 1.5$	1.11 × ΔT solar
$= 3$	0.74 × ΔT solar
$= 9$	0.37 × ΔT solar
Water wall	0.39 × ΔT solar
Trombe wall, vented for 3% of wall area	0.65 × ΔT solar
Trombe wall, unvented	0.13 × ΔT solar

Source: Balcomb et al. (1980).

less your climate is predominantly cloudy (clear days rare) in November, December, and January.

The other important comfort question is the size of *temperature swing* due to passive solar heating. Controlled by the sun and by the actions of users, rather than by a thermostat, passive solar buildings typically experience larger daily variations (swings) in indoor temperature than do conventional buildings, especially on clear days. To estimate your building's clear-day January temperature swing, or ΔT swing, see Table 5.17. The average indoor temperature determined previously will fall in the middle of this ΔT swing.

5.7 Approximate Method for Calculating Heat Gain (Cooling Load)

Unlike the calculations for winter worst-hour heat loss, which simply assume nighttime conditions and few if any internal gains, summer worst-hour heat gain calculations are very complex. The difference in air temperature between inside and outside, which was so influential in winter, is much less important in summer. Solar gains and internal gains from lights, people, and equipment *must* be included. Summer calculations are complicated further by the fact that because the hourly change in summer load can be very great, the hour at which calculations will be done must be chosen. Also, in summer the thermal mass of the building becomes influential, delaying the impact of the radiant component of heat gains from all sources. The rule-of-thumb approach represented in Table 5.10 cannot respond to hourly changes or to thermally massive construction.

A simplified heat gain procedure has been developed for residential buildings; with some risk and some judgment, it can also be used for a quick *approximation* of the conditions in commercial buildings. This simplified method was devised to be used with buildings that, like residences,

1. Are occupied and air conditioned (internal temperatures closely controlled) for 24 h/day, including weekends.
2. Derive much of their gains through the building envelope and ventilation, rather than internally.
3. Are tolerant of undersized cooling equipment, with the result that unusually hot weather means noticeably higher indoor temperatures (and that interior temperatures will vary, or "swing," during a typical summer day).

Since many commercial buildings do *not* have these characteristics, this method should not be applied if very accurate estimates are desired for such buildings. But the method is rapid, if risky. Compare the results from this procedure (Example 5.10) with those from the more exact procedure for commercial buildings (Example 5.14).

Note: This procedure is based on ASHRAE tables that consistently list SI units (with conventional units in parentheses); therefore, in Sections 5.7 and 5.8 *only* we will use the order SI (conventional).

(a) Gains Through Roof and Walls. The sensible heat gains through opaque parts of a building's envelope are calculated with the equation

$$q = U \times A \times \text{DETD}$$

where the U value sometimes differs from winter and summer conditions (e.g., Table 4.11). DETD (design equivalent temperature differences) values are listed for broad categories of construction in Table 5.18; the DETD values are based on an average indoor temperature of 23.8°C (75 F) and the outdoor conditions listed. A means of correcting DETD for other temperatures is as follows:

Where the design temperature difference [outdoor design temperature, minus 23.8°C (75 F)] is not an even increment of 2.8C° (5F°), the equivalent temperature difference should be corrected 0.5C° (1F°) for each 0.5C° (1F°) difference from the tabulated values.

For rapid approximation, however, select the DETD directly from the table for the conditions nearest to your building/climate combination. The daily temperature for your climate is listed in Appendix A.

(b) Gains Through Glass. The quick way to approximate these gains is

$$q = A \times \text{DCLF}$$

DCLF (design cooling load factors) values are listed in Table 5.19 and *include the U values* as well as the equivalent temperature differences. (These *DCLF* values do *not* correspond to the worst-hour

TABLE 5.18 **Design Equivalent Temperature Differences**

29.4°C		32.2			35.0			37.7		40.5	43.3	Outdoor Design Temperature	85 F		90			95			100		105	110
L	*M*	*L*	*M*	*H*	*L*	*M*	*H*	*M*	*H*	*H*	*H*	Daily Temperature Range[a]	*L*	*M*	*L*	*M*	*H*	*L*	*M*	*H*	*M*	*H*	*H*	*H*
												Walls and Doors												
9.7	7.5	12.5	10.3	7.5	15.3	13.1	10.3	15.8	13.1	15.8	18.6	1. Frame and veneer-on-frame	17.6	13.6	22.6	18.6	13.6	27.6	23.6	18.6	28.6	23.6	28.6	33.6
5.7	3.5	8.5	6.3	3.5	11.3	9.1	6.3	11.8	9.1	11.8	14.6	2. Masonry walls, 203.2-mm (8-in.) block or brick	10.3	6.3	15.3	11.3	6.3	20.3	16.3	11.3	21.3	16.3	21.3	26.3
5.0	2.7	7.7	5.5	2.7	10.5	8.3	5.5	11.1	8.3	11.1	13.8	3. Partitions, frame	9.0	5.0	14.0	10.0	5.0	19.0	15.0	10.0	20.0	15.0	20.0	25.0
1.4	0.0	4.2	1.9	0.0	6.9	4.7	1.9	7.5	4.7	7.5	10.3	masonry	2.5	0	7.5	3.5	0	12.5	8.5	3.5	13.5	8.5	13.5	18.5
7.7	7.5	12.5	10.3	7.5	15.3	13.1	10.3	15.8	13.1	15.8	18.6	4. Wood doors	17.6	13.6	22.6	18.6	13.6	27.6	23.6	18.6	28.6	23.6	28.6	33.6
												Ceilings and Roofs[b]												
21.1	18.8	23.8	21.6	18.8	26.6	24.4	21.6	27.2	24.4	27.2	30.0	1. Ceilings under naturally vented attic or vented flat roof—dark	38.0	34.0	43.0	39.0	34.0	48.0	44.0	39.0	49.0	44.0	49.0	54.0
16.6	14.9	19.4	17.2	14.4	22.2	20.0	17.2	22.7	20.0	22.7	25.5	—light	30.0	26.0	35.0	31.0	26.0	40.0	36.0	31.0	41.0	36.0	41.0	46.0
21.1	18.8	23.3	21.6	18.8	26.6	24.4	21.6	27.2	24.4	27.2	30.0	2. Built-up roof, no ceiling—dark	38.0	34.0	43.0	39.0	34.0	48.0	44.0	39.0	49.0	44.0	49.0	54.0
16.6	14.4	19.4	17.2	14.4	22.2	20.0	17.2	22.7	20.0	22.7	25.5	—light	30.0	26.0	35.0	31.0	26.0	40.0	36.0	31.0	41.0	36.0	41.0	46.0
5.0	2.7	7.7	5.5	2.7	10.5	8.3	5.5	11.1	8.3	11.1	13.8	3. Ceilings under unconditioned rooms	9.0	5.0	14.0	10.0	5.0	19.0	15.0	10.0	20.0	15.0	20.0	25.0
												Floors												
5.0	2.7	7.7	5.5	2.7	10.5	8.3	5.5	11.1	8.3	11.1	13.8	1. Over unconditioned rooms	9.0	5.0	14.0	10.0	5.0	19.0	15.0	10.0	20.0	15.0	20.0	25.0
0.0	0.0	0.0	0.0	0.0	0.0	0.0	0.0	0.0	0.0	0.0	0.0	2. Over basement, enclosed crawl space or concrete slab on ground	0	0	0	0	0	0	0	0	0	0	0	0
5.0	2.7	7.7	5.5	2.7	10.5	8.3	5.5	11.1	8.3	11.1	13.8	3. Over open crawl space	9.0	5.0	14.0	10.0	5.0	19.0	15.0	10.0	20.0	15.0	20.0	25.0

Source: Copyright © by the American Society of Heating, Refrigerating and Air Conditioning Engineers, Inc., Atlanta, GA. Reprinted by permission from the *ASHRAE Handbook of Fundamentals,* 1981.

[a]Daily temperature range: L (Low) calculation value: 6.7C° (12F°) Applicable range: less than 8.3C° (15F°); M (Medium) calculation value: 11.1C° (20F°) Applicable range: 8.3 to 13.8C° (15 to 25F°); H (High) calculation value: 16.7C° (30F°) Applicable range: more than 13.8C° (25F°)

[b]For roofs in shade, 18-h average = 6.1C° (11F°) temperature differential. At 32.2°C (90F) design and medium daily range, equivalent temperature differential for light-colored roof equals 6.1 + (0.71)(21.6 − 6.1) = 17.1C° [11 + (0.71)(39 − 11) = 31F°].

gains; they were obtained by averaging the hours from 5:30 A.M. to 6:30 P.M., at both 30°N and 40°N latitudes.) Again, the DCLF values were based on an inside temperature of 23.8°C (75 F) and outside temperatures as listed. (For *rapid* approximation, ignore the corrections procedure shown in note *a*.)

Glass protected by exterior shading devices that exclude all direct sun may be considered equal to the table's values for north glass protected by awnings.

(c) Gains from Outdoor Air. In residences, outdoor air usually enters by infiltration. In many other buildings, codes require the deliberate introduction of outdoor air by mechanical ventilation. Whichever way outdoor air enters your building in summer, its sensible heat gain can be calculated by either

$$q_{\text{infiltration}} = (A_{\text{exposed}})(\text{infiltration factor})$$

where A_{exposed} is the total area of exposed wall surface, including windows and doors and the infiltration factor is found from Table 5.20; or

$$q_{\text{mechanical ventilation}} = (Q)(\text{ventilation factor})$$

where Q is the volume of outdoor air (cfm or L/s; see Section 4.6), and the ventilation factor is found from Table 5.20.

(d) Gains from People. Only the sensible gains are tabulated, since a simple overall factor for latent gains is included later. The rate of heat gain from people in various activities is shown in Table 5.21; values in the ''Sensible Heat'' column are generally used. [For residences, sensible heat gain per occupant is assumed at 66 W (225 Btu/h)].

$$q_{\text{people}} =$$

(number of occupants)(sensible gain per occupant)

(e) Gains from Lights. The power supplied to electric lights (those that normally are on while cooling equipment is functioning) can be added directly to the sensible heat gain. Be sure to include ballast heat gains along with fluorescent lights; this is usually done by taking *from 1.12 to 1.2 times the total bulb wattage* of such lights (use the lower figure with energy-efficient ballasts).

(f) Gains from Equipment. In residences, a standard assumption is that 350 W (1200 Btu/h) of sensible heat gain is produced by kitchen appliances. (Other residential heat loads are assumed to be vented.) For other buildings, see the estimated range of gains given in Table 5.22.

EXAMPLE 5.10. (This example is adapted from the *ASHRAE Handbook of Fundamentals,* 1981. Reprinted by permission of the American Society of Heating, Refrigerating and Air Conditioning Engineers, Inc., Atlanta, GA.)

A one-story office building (Fig. 5.26) is located in an eastern state near 40°N latitude. The adjoining buildings on the north and west are not conditioned, and their inside air temperatures are basically equal to the outdoor air temperature at any time of day.

Roof construction. 114-mm (4.5-in.) flat roof deck of 50.8-mm (2-in.) gypsum slab [equivalent to four 12.7-mm (0.5-in.) layers of gypsum board], 50.8-mm (2-in.) preformed above-deck roof insulation, 6.4-mm (0.25-in.) asbestos board under gypsum slab, and two layers of mopped 6.8-kg (15-lb) felt vapor seal over insulation, no false ceiling [weight approximately 60 kg/m^2 (12 lb/ft^2)] [U = 0.68 W/m^2 °C (0.12 Btu/h ft^2 F)].

South wall construction. 101.6-mm (4-in.) face brick, 101.6-mm (4-in.) common brick, 9.5-mm (0.375-in.) cement plaster sand aggregate, 6.4-mm (0.25-in.) plywood panel glued to plaster [U = 2.21 W/m^2 °C (0.39 Btu/h ft^2 F)].

West wall and adjoining north party wall construction. 330-mm (13-in.) solid brick, no plaster [U = 1.42 W/m^2 °C (0.25 Btu/h ft^2 F)].

North exposed wall and east wall construction. 203.2-mm (8-in.) concrete block and 19-mm (¾-in.) plaster [U = 2.5 W/m^2 °C (0.44 Btu/h ft^2 F)].

Floor construction. 101.6-mm (4-in.) concrete on ground.

Fenestration. 0.91-m × 1.52-m (3-ft × 5-ft) window of regular plate glass with light-colored venetian blinds, not openable [U = 4.6 W/m^2 °C (0.81-Btu/h ft^2 F)].

Front doors. Two, 0.76 m × 2.13 m (2.5 × 7 ft).

Side doors. Two, 0.76 m × 2.13 m (2.5 ft × 7 ft).

Rear doors. Two, 0.76 m × 2.13 m (2.5 ft × 7 ft).

Door construction. Approximately 30% 38-mm (1.5-in.) solid pine frame and 70% 6.4-mm (0.25-

TABLE 5.19 **Design Cooling Load Factors Through Glass**

Part A. **W/m²**

Outdoor Design Temp[a]	Regular Single Glass						Regular Double Glass						Heat-Absorbing Double Glass						Clear Triple Glass		
	29.4	32.2	35.0	37.7	40.5	43.3	29.4	32.2	35.0	37.7	40.5	43.3	29.4	32.2	35.0	37.7	40.5	43.3	29.4	32.2	35.0
No Awnings or Inside Shading																					
North	72.6	85.2	97.8	110.4	123.0	138.8	59.9	66.2	75.8	82.0	88.3	94.7	37.9	44.1	53.7	60.0	66.2	72.6	53.7	60.0	63.1
NE and NW	176.6	189.3	202.0	214.6	227.1	243.0	145.1	151.4	161.0	167.2	173.6	179.9	85.1	91.4	101.0	107.2	113.6	119.9	132.6	135.7	138.9
East and west	255.5	268.1	280.8	293.4	306.0	321.9	214.6	220.9	230.3	236.7	243.0	249.2	132.6	138.9	148.2	154.6	160.9	167.2	195.7	198.8	202.6
SE and SW	220.9	233.4	246.0	258.8	271.3	287.1	186.1	192.4	202.0	208.2	214.6	220.9	110.4	116.8	126.2	132.6	138.9	145.1	167.2	173.6	176.7
South	126.2	138.9	151.4	164.0	176.7	192.4	104.1	110.4	119.9	126.2	132.6	138.9	66.0	75.8	75.8	82.0	88.3	94.7	94.7	97.9	104.1
Horiz. skylight	504.8	517.4	530.0	542.7	555.2	571.0	438.6	444.9	454.3	460.7	467.0	473.2	280.8	296.6	296.6	302.9	309.1	315.6	397.6	400.7	407.0
Draperies or Venetian Blinds																					
North	47.3	60.0	72.6	85.1	97.9	113.6	37.9	44.1	53.7	60.0	66.2	72.6	28.3	34.7	44.1	50.4	56.8	63.1	34.7	37.9	44.2
NE and NW	101.0	113.6	126.2	138.9	151.4	167.2	85.1	91.4	101.0	107.2	113.6	119.9	63.1	69.4	78.9	85.1	91.4	97.9	75.8	82.0	85.1
East and west	151.4	164.0	176.7	189.3	202.0	217.7	132.6	138.9	148.2	154.6	161.0	167.2	94.7	101.0	110.4	116.8	123.0	129.3	119.9	123.0	129.3
SE and SW	126.2	138.9	151.4	164.0	176.7	192.4	110.4	116.8	126.2	132.6	138.9	145.1	75.8	82.0	91.4	97.9	104.1	110.4	101.0	104.1	107.2
South	72.5	85.1	97.9	110.4	123.0	138.9	63.1	69.4	78.9	85.1	91.4	97.9	47.3	53.7	63.1	75.8	75.8	82.0	56.8	60.0	66.2
Roller Shades Half-Drawn																					
North	56.8	69.4	82.0	92.7	102.2	123.0	47.3	53.7	63.1	69.4	75.8	82.0	31.6	37.9	47.3	53.7	60.0	66.2	41.0	44.2	47.3
NE and NW	126.2	138.9	151.4	164.0	176.7	192.4	119.9	126.2	135.7	142.0	148.2	154.6	75.8	82.0	91.4	97.9	104.1	110.4	107.2	110.4	110.4
East and west	192.5	205.0	217.7	230.3	243.0	258.8	170.3	176.7	186.1	192.4	198.8	205.0	110.4	116.8	126.2	132.5	138.9	145.1	154.6	154.6	157.8
SE and SW	164.0	176.7	189.3	202.0	214.6	230.3	145.1	151.4	161.0	167.2	173.6	179.9	94.7	101.0	110.4	123.0	123.0	129.3	129.3	132.6	135.7
South	91.4	104.1	116.8	129.3	142.0	157.8	85.1	91.4	101.0	107.2	113.6	119.9	56.8	63.1	72.6	78.9	85.1	91.4	78.9	82.0	82.0
Awnings[b]																					
North	63.1	75.8	88.3	101.0	113.6	129.3	41.0	47.3	56.8	63.1	69.4	75.8	31.6	37.9	47.3	53.7	60.0	66.2	34.7	37.9	41.0
NE and NW	66.2	78.9	91.4	104.1	116.8	132.6	44.2	50.4	60.0	66.2	72.6	78.9	34.7	41.0	50.4	56.8	63.1	69.4	37.9	41.0	44.2
East and west	69.4	82.0	94.7	107.2	119.9	135.7	44.2	50.4	60.0	66.2	72.6	78.9	37.9	44.2	53.6	60.0	66.2	72.6	37.9	41.0	44.2
SE and SW	66.2	78.9	91.4	104.1	116.8	132.6	44.2	5.04	60.0	66.2	72.6	78.9	34.7	41.0	50.4	56.8	63.1	69.4	37.9	41.0	44.2
South	66.2	75.8	88.3	101.0	113.6	129.3	41.0	47.3	56.8	63.1	69.4	75.8	34.7	41.0	50.4	56.8	63.1	69.4	34.7	37.9	41.0

Part B. **Btu/h ft²**

Outdoor Design Temp.[a]	Regular Single Glass						Regular Double Glass						Heat-Absorbing Double Glass						Clear Triple Glass		
	85	90	95	100	105	110	85	90	95	100	105	110	85	90	95	100	105	110	85	90	95
No Awnings or Inside Shading																					
North	23	27	31	35	39	44	19	21	24	26	28	30	12	14	17	19	21	23	17	19	20
NE and NW	56	60	64	68	72	77	46	48	51	53	55	57	27	29	32	34	36	38	42	43	44
East and west	81	85	89	93	97	102	68	70	73	75	77	79	42	44	47	49	51	53	62	63	64
SE and SW	70	74	78	82	86	91	59	61	64	66	68	70	35	37	40	42	44	46	53	55	56
South	40	44	48	52	56	61	33	35	38	40	42	44	19	21	24	26	28	30	30	31	33
Horiz. skylight	160	164	168	172	176	181	139	141	144	146	148	150	89	91	94	96	98	100	126	127	129
Draperies or Venetian Blinds																					
North	15	19	23	27	31	36	12	14	17	19	21	23	9	11	14	16	18	20	11	12	14
NE and NW	32	36	40	44	48	53	27	29	32	34	36	38	20	22	25	27	29	31	24	26	27
East and west	48	52	56	60	64	69	42	44	47	49	51	53	30	32	35	37	39	41	38	39	41
SE and SW	40	44	48	52	56	61	35	37	40	42	44	46	24	26	29	31	33	35	32	33	34
South	23	27	31	35	39	44	20	22	25	27	29	31	15	17	20	22	24	26	18	19	21
Roller Shades Half-Drawn																					
North	18	22	26	30	34	39	15	17	20	22	24	26	10	12	15	17	19	21	13	14	15
NE and NW	40	44	48	52	56	61	38	40	43	45	47	49	24	26	29	31	33	35	34	35	35
East and west	61	65	69	73	77	82	54	56	59	61	63	65	35	37	40	42	44	46	49	49	50
SE and SW	52	56	60	64	68	73	46	48	51	53	55	57	30	32	35	37	39	41	41	42	43
South	29	33	37	41	45	50	27	29	32	34	36	38	18	20	23	25	27	29	25	26	26
Awnings[b]																					
North	20	24	28	32	36	41	13	15	18	20	22	24	10	12	15	17	19	21	11	12	13
NE and NW	21	25	29	33	37	42	14	16	19	21	23	25	11	13	16	18	20	22	12	13	14
East and west	22	26	30	34	38	43	14	16	19	21	23	25	12	14	17	19	21	23	12	13	14
SE and SW	21	25	29	33	37	42	14	16	19	21	23	25	11	13	16	18	20	22	12	13	14
South	21	24	28	32	36	41	13	15	18	20	22	24	11	13	16	18	20	22	11	12	13

Source: Copyright © by the American Society of Heating, Refrigerating and Air Conditioning Engineers, Inc., Atlanta, GA. Reprinted by permission from the *ASHRAE Handbook of Fundamentals*, 1981.

[a]Based upon indoor design temperature of 23.8°C (75 F) and outdoor design temperatures as indicated. Interpolate to obtain factors for outdoor design temperatures other than those given.

[b]For other external shading devices which completely shade the glass at any orientation, use the values for "Awnings, north."

TABLE 5.20 **Sensible Cooling Load Factors Due to Infiltration and Ventilation**

°C: 29.4	32.2	35.0	37.7	41.5	43.3	(Units)	Design Temperature	(Units) F:	85	90	95	100	105	110
2.2	3.5	4.7	6.0	6.9	8.2	W/m²	Infiltration, per gross exposed wall area	Btu/h ft²	0.7	1.1	1.5	1.9	2.2	2.6
6.8	9.9	13.6	16.7	19.8	23.6	W per L/s	Mechanical ventilation	Btu/h cfm	11.0	16.0	22.0	27.0	32.0	38.0

Source: Copyright © by the American Society of Heating, Refrigerating and Air Conditioning Engineers, Inc., Atlanta, GA. Reprinted by permission from the *ASHRAE Handbook of Fundamentals,* 1981.

TABLE 5.21 **Rates of Heat Gain from Occupants of Conditioned Spaces**[a]

Degree of Activity	Typical Application	Total Heat Adults, Male		Total Heat Adjusted[b]		Sensible Heat		Latent Heat	
		W	Btu/h	W	Btu/h	W	Btu/h	W	Btu/h
Seated at rest	Theater, movie	115	400	100	350	**60**	**210**	40	140
Seated, very light work writing	Offices, hotels, apts.	140	480	120	420	**65**	**230**	55	190
Seated, eating	Restaurant[c]	150	520	170	580[c]	**75**	**255**[c]	95	325[c]
Seated, light work, typing	Offices, hotels, apts.	185	640	150	510	**75**	**255**	75	255
Standing, light work or walking slowly	Retail store, bank	235	800	185	640	**90**	**315**	95	325
Light bench work	Factory	255	880	230	780	**100**	**345**	130	435
Walking, 1.3 m/s (3 mph), light machine work	Factory	305	1040	305	1040	**100**	**345**	205	695
Bowling[d]	Bowling alley	350	1200	280	960	**100**	**345**	180	615
Moderate dancing	Dance hall	400	1360	375	1280	**120**	**405**	255	875
Heavy work, heavy machine work, lifting	Factory	470	1600	470	1600	**165**	**565**	300	1035
Heavy work, athletics	Gymnasium	585	2000	525	1800	**185**	**635**	340	1165

NOTE: All values rounded to nearest 5 watts or to nearest 10 Btu/h.

Source: Copyright © by the American Society of Heating, Refrigerating and Air Conditioning Engineers, Inc., Atlanta, GA. Reprinted by permission from the *ASHRAE Handbook of Fundamentals,* 1981.

[a]Tabulated values are based on 25.5°C (78 F) room dry-bulb temperature. For 26.6°C (80 F) room dry-bulb, the total heat remains the same, but the sensible heat value should be decreased by approximately 8% and the latent heat values increased accordingly.

[b]Adjusted total heat gain is based on normal percentage of men, women, and children for the application listed, with the postulate that the gain from an adult female is 85% of that for an adult male, and that the gain from a child is 75% of that for an adult male.

[c]Adjusted total heat value for eating in a restaurant; includes 17.6 W (60 Btu/h) for food per individual [8.8 W (30 Btu/h) sensible and 8.8 W (30 Btu/h) latent].

[d]For bowling, figure one person per alley as actually bowling and all others as sitting 117 W (400 Btu/h) or standing and walking slowly 231 W (790 Btu/h).

TABLE 5.22 **Approximate Rates of Internal Sensible Heat Gain from Equipment**

NOTE: These values do *not* include gains from electric lighting or from people.

Offices	*Btu/h ft²*	*W/m²*
General offices	3 to 4	9.5 to 12.6
Purchasing and accounting	6 to 7	19 to 22
Office with computer display units	up to 15	up to 47
Computer areas		
Digital	75 to 175	237 to 552
Analog	50 to 150	158 to 475
(For equipment with all transistors, multiply values by 0.75)		
Laboratories	15 to 70	47 to 220
Restaurants,[a]		
per meal served in dining area		
Sensible	(30 Btu/h)	(8.8 W)
Latent	(30 Btu/h)	(8.8 W)

Source: ASHRAE Handbook of Fundamentals (1981), Chapter 26.

[a]Note that in Table 5.21, these gains are *already included* as heat gains from people eating in restaurants.

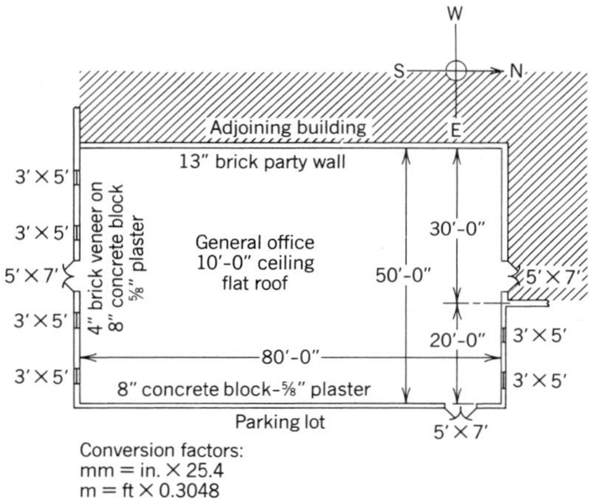

Fig. 5.26. *Plan of the office building for which heat gain calculations are shown in Examples 5.10 and 5.11. Copyright © by the American Society of Heating, Refrigerating and Air Conditioning Engineers, Inc., Atlanta, GA. Reprinted by permission from the* ASHRAE Handbook of Fundamentals, *1981.*

in.) pine panels [$U = 3.35$ W/m² °C (0.59 Btu/h ft² F)].

U values for all doors and outside walls were calculated assuming a wind speed of 3.35 m/s (7.5 mph). For party and inside walls, still air was assumed.

Outdoor design conditions. Dry-bulb temperature, 34.4°C (94 F); daily range, 11.1C° (20 F°). Wet bulb temperature, 25°C (77 F).

Indoor design conditions. Dry-bulb temperature, 25.6°C (78 F); wet-bulb temperature, 18.3°C (65 F).

Occupancy. 85 office workers from 8 A.M. to 5 P.M.

Lights. 17,500 W, fluorescent, from 8 A.M. to 6 P.M.; 4000 W, tungsten, continuous.

The conditioning equipment is located in the adjoining structure to the north.

Determine the sensible, latent, and total space cooling load at design conditions.

Note: This example is repeated for a much more detailed calculation procedure in Example 5.14.

Cooling Load Due to Heat Gain Through Building Envelope: The heat gains through roof, exposed walls, and doors shown in Table 5.23 were calculated by

$$q = U \times A \times \text{DETD}$$

for which DETD values are taken from Table 5.18. In line with the approximate (and therefore rapid) nature of this calculation method, *no corrections* were made to DETD values to adjust for either

TABLE 5.23 **Cooling Load Through Building Envelope (Example 5.10)**

Net Area (m^2)	U Value ($W/m^2\ °C$)	ΔT (C°)	DETD (C°)	Reference	DCLF	Cooling Load (W)	Section	Net Area (ft^2)	U Value ($Btu/h\ ft^2\ F$)	ΔT (F°)	DETD (F°)	Reference	DCLF	Cooling Load (Btu/h)
371.6	0.68		24.4	Table 5.18		6,170	Roof	4000	0.12		44.0	Table 5.18		21,120
37.6[a]	2.21		9.1	Table 5.18		760	South wall	405[a]	0.39		16.3	Table 5.18		2,580
71.1[a]	2.49		9.1	Table 5.18		1,610	East wall	765[a]	0.44		16.3	Table 5.18		5,490
15.8[a]	2.49		9.1	Table 5.18		360	North exposed wall	170[a]	0.44		16.3	Table 5.18		1,220
98.9[a]	1.41	11.2[b]				1,560	Party walls	1065[a]	0.25	20[b]				5,330
3.25	3.35		13.1	Table 5.18		140	Doors: S	35	0.59		23.6	Table 5.18		490
3.25	3.35	11.2[b]				120	N	35	0.59	20[b]				410
3.25	3.35		13.1	Table 5.18		140	E	35	0.59		23.6	Table 5.18		490
5.6	(4.59)[c]			Table 5.19	97.9	550	Windows: S	60	(0.81)[c]			Table 5.19	31	1,860
2.8	(4.59)[c]			Table 5.19	72.6	200	N	30	(0.81)[c]			Table 5.19	23	690
						Total: 11,610								Total: 39,680

[a]Calculated from gross wall area, less windows and doors.

[b]Design temperature difference, inside to outside.

[c]DCLF for glass *includes U* value.

outside or inside design temperatures differing from those listed, and actual gains were rounded to nearest 10. The climate's daily temperature range is medium; the roof is dark (typical of commercial buildings in air-polluted areas). Outside design temperatures of 35°C (95 F) were used. To keep things simple, the temperature difference through party walls was made equal to that on which the table was based; in this case, 35 − 23.8 = 11.2 (95 − 75 = 20).

Cooling Load Due to Heat Gain Through Glass: These heat gains were calculated by

$$q = A \times \text{DCLF}$$

for which the DCLF values (which include U value) were taken from Table 5.19. Again, *no corrections* were made, and gains were rounded to nearest 10. (*Note:* The total building envelope gains shown by this approximate method can be directly compared with results from the more exact method; compare Tables 5.23 and 5.53.)

Cooling Load Due to Heat Gain from Lights, People, and Equipment: The rate of heat gain from the lighting can be approximated simply by taking the energy input (including extra power required by ballasts for fluorescents). For fluorescent lights,

$$17{,}500 \text{ W} \times 1.2 \text{ ballast factor} = 21{,}000 \text{ W } (17{,}670 \text{ Btu/h})$$

For incandescent lights,

$$4000 \text{ W } (13{,}650 \text{ Btu/h})$$

The sensible gains from people can be determined from Table 5.21. For office work, assume 75 W (255 Btu/h):

$$85 \text{ people} \times 75 = 6380 \text{ W } (21{,}680 \text{ Btu/h})$$

Heat gains from office equipment are assumed to be small in this example. (See Table 5.10 for typical ranges of such gains.) At the common rate of 0.5 W/ft^2,

$$4000 \text{ ft}^2 \times 0.5 \text{ W/ft}^2 = 2000 \text{ W } (6280 \text{ Btu/h})$$

The sensible gains from lights, people and equipment thus total 33,380 W (113,280 Btu/h).

Cooling Load Due to Ventilation or Infiltration: Since this is a commercial building incorporating deliberate introduction of outdoor air, infiltration will be ignored. The mechanical ventilation suggested in Table 4.19 is 7 L/s (15 cfm) per person. Total mechanical ventilation is therefore 85 people × 7 L/s = 595 L/s (85 × 15 = 1275 cfm). Sensible heat gains (from Table 5.20):

$$\text{mechanical ventilation } 13.6(595 \text{ L/s}) = 8090 \text{ W}$$

$$[22.0(1275 \text{ cfm}) = 28{,}050 \text{ Btu/h}]$$

Latent Heat Gains: The approximate method estimates latent heat at 30% of sensible gains (in dry climates, 20%); the climate description suggests a typical northeastern U.S. climate, for which 30% is appropriate.

	W	*Btu/h*
Sensible gains through building envelope	11,610	39,680
Sensible gains, lights, people, equipment	33,380	113,230
Sensible gains, ventilation	8,090	28,050
Sensible total	53,080 W	181,010 Btu/h

Total gains due to sensible plus latent heat = 1.3 sensible = 69,000 W (235,310 Btu/h)

Note: In this particular example, the approximate heat gain obtained for a nonresidential structure is very close indeed to that obtained by the much more rigorous and exacting procedure outlined in Section 5.8 (overestimating the peak load by less than 1%). Note, however, that the approximate method

Underestimated the load through the building envelope—in this example, at about two-thirds of the actual worst-hour envelope heat gain.

Overestimated the load due to lights, equipment, people, and ventilation air—in this example, at about 125% of the actual worst-hour internal gain.

Gives *no* indication of how the peak cooling load changes from hour to hour.

For a quick, rough estimate of cooling equipment size for a building with only one thermal zone, the method outlined served *this example* very well.

5.8 Detailed Hourly Heat Gain (Cooling Load) Calculations

Ordinarily, the approximate method of calculating the peak cooling load (Section 5.7) is sufficient for a designer's preliminary estimates of cooling equipment size. However, much more detailed procedures are used by engineers to actually size the equipment and to assess the peak-load impact of various design options such as shading devices. This complex procedure takes account of both thermal mass and time of occupancy, and it requires a judgment as to which hour is likely to be most critical. Therefore, the more exact method of calculating cooling load (based on the *ASHRAE Handbook of Fundamentals,* 1981) can be used only if the following information is available.

1. Characteristics of the building's envelope: materials, sizes, external surface colors, and shapes.
2. Building location and orientation, as well as the extent of external shading of the building by trees or adjacent structures.
3. Outdoor design conditions (see Appendix A).
4. Indoor design conditions: DB, WB, and ventilation rate.
5. The schedule of lighting, occupancy, equipment, and any other processes that contribute to internal heat gain.
6. The time of day and the month for which to calculate cooling load. Determination of this factor may require a quick preliminary comparison between the relative impact of solar gains through roofs and glass at various orientations.
7. Thermal zoning requirements.

A summary of the entire process is shown in Fig. 5.27 (see pp. 230–231).

(a) Gains Through Roofs and Walls. The rate at which exterior heat is conducted to cooler interiors varies according to the combined and varying impact of the sun and the exterior air temperature. Heat gains through opaque surfaces such as walls and roofs involve *sol-air temperature:* the equivalent temperature of the outdoor air that, neglecting radiation, would give the same rate of heat entry into the surface as would exist with the actual combination of incident solar radiation, radiant energy exchange with the sky and other outdoor surroundings, and convective heat exchange with the outdoor air. For a sunny surface, then, sol-air temperature will be higher than simple outdoor air temperature, and it will vary from hour to hour. The heat gain situation is further complicated by both the thermal mass and the insulation of roofs and walls, which will affect the rate at which heat moves through.

Cooling load temperature differences (CLTD) have been developed to account for sol-air temperatures and weights of construction; see Tables 5.24 through 5.32. These are substituted for the more simple but familiar temperature difference (ΔT) to obtain hourly heat gain:

$$q = U \times A \times \text{CLTD}$$

Note: These CLTD values were computed for the following "standard" conditions.

Indoor air temperature: 78 F (25.5°C).

Outdoor air maximum temperature: 95 F (35°C).

Outdoor mean temperature: 85 F (29.4°C).

Outdoor daily range: 21F° (11.6C°).

Solar radiation variation typical of 40°N latitude on July 21.

The notes below the CLTD tables outline procedures for adjusting CLTDs for other conditions.

CLTD values are calculated only for more familiar roof and wall constructions. When using a construction other than those listed, consult Tables 5.24 and 5.26 to determine which listed construction is most similar in *both weight and insulation* to your construction, and use CLTDs for that listed construction.

Solid exterior doors are not specifically listed in the wall CLTD table; the *ASHRAE Handbook* presents example solutions in which solid, uninsulated wood doors are considered to be "Group F."

TABLE 5.24 **Thermal Properties and Code Numbers of Layers Used in the Calculation of CLTDs for Roof and Wall**

Thickness and Thermal Properties[a]						Description	Code	Thickness and Thermal Properties[b]					
L	K	D	SH	R	WT		Number	L	K	D	SH	R	WT
				0.059		Outside surface resistance	A0					0.333	
25.4	0.692	1858	0.233	0.036	47.2	25.4-mm (1-in.) Stucco (asbestos cement or wood siding plaster, etc)	A1	0.0833	0.4	116	0.20	0.208	9.66
101.6	1.298	2082	0.256	0.078	211.4	101.6-mm (4-in.) facebrick (dense concrete)	A2	0.3333	0.75	130	0.22	0.444	43.3
1.5	44.99	7689	0.116	0.00003	11.7	Steel siding (aluminum or other lightweight cladding)	A3	0.0050	26.0	480	0.10	0.000	2.40
				0.059		Outside surface resistance						0.333	
12.7	1.143	881	0.465			12.7-mm-(0.5-in.) slag, membrane	A4	0.0417	0.83	55	0.40		
9.5	0.190	1121	0.465			9.5-mm. (0.375-in.) felt		0.0313	0.11	70	0.40		
12.7	0.415	1249	0.302	0.031	15.9	Finish	A6	0.0417	0.24	78	0.26	0.174	3.25
10.16	1.332	2002	0.256	0.076	203.1	101.6-mm (4-in.) facebrick	A7	0.3333	0.77	125	0.22	0.433	41.6
				0.160		Air Space Resistance	B1					0.91	
25.4	0.043	32	0.233	0.585	0.8	25.4-mm (1-in.) insulation	B2	0.0833	0.025	2.0	0.2	332	0.17
50.8	0.043	32	0.233	1.176	1.6	50.8-mm (2-in.) insulation	B3	0.1667	0.025	2.0	0.2	6.68	0.33
76.2	0.043	32	0.233	1.766	2.4	76.2-mm. (3-in.) insulation	B4	0.2500	0.025	2.0	0.2	10.03	0.50
25.4	0.043	91	0.233	0.586	2.3	25.4 mm. (1-in.) insulation	B5	0.0833	0.025	5.7	0.2	3.33	0.47
50.8	0.043	91	0.233	1.176	4.6	50.8 mm. (2-in.) insulation	B6	0.1667	0.025	5.7	0.2	6.68	0.95
25.4	0.116	592	0.699	0.209	15.0	25.4-mm (1-in.) wood	B7	0.0833	0.067	37.0	0.6	1.19	3.08
62.4	0.116	592	0.699	0.525	37.6	62.5 mm (2.5-in.) wood	B8	0.2083	0.067	37.0	0.6	2.98	7.71
101.6	0.116	592	0.699	0.838	60.0	14.6 mm (4-in.) wood	B9	0.3333	0.067	37.0	0.6	4.76	2.3
50.8	0.116	592	0.699	0.421	30.2	50.8 mm (2-in.) wood	B10	0.1667	0.067	37.0	0.6	2.39	6.18
76.2	0.116	592	0.699	0.631	45.2	76.2 mm (3-in.) wood	B11	0.2500	0.067	37.0	0.6	3.58	9.25
76.2	0.043	91	0.233	1.761	6.9	76.2 mm (3-in.) insulation	B12	0.2500	0.025	5.7	0.2	10.00	1.42
101.6	0.043	91	0.233	2.346	9.3	101.6 mm (4-in.) insulation	B13	0.3333	0.025	5.7	0.2	13.33	1.90
127.0	0.043	91	0.233	2.934	11.6	127.0 mm (5-in.) insulation	B14	0.4167	0.025	5.7	0.2	16.67	2.38
152.4	0.043	91	0.233	3.520	13.9	152.4 mm (6-in.) insulation	B15	0.5000	0.025	5.7	0.2	20.00	2.85
101.6	0.571	1121	0.233	0.178	113.7	101.6 mm (4-in.) clay tile	C1	0.3333	0.33	70.0	0.2	1.01	23.3
101.6	0.381	608	0.233	0.266	62.0	101.6 mm (4-in.) l.w. concrete block	C2	0.3333	0.22	38.0	0.2	1.51	12.7
101.6	0.813	977	0.233	0.125	99.1	101.6-mm (4-in.) l.w. concrete block	C3	0.3333	0.47	61.0	0.2	0.71	20.3
101.6	0.727	1922	0.233	0.139	195.3	101.6-mm (4-in.) common brick	C4	0.3333	0.42	120.0	0.2	0.79	40.0
101.6	1.730	2242	0.233	0.059	227.5	101.6-mm (4-in.) l.w. concrete	C5	0.3333	1.00	140.0	0.2	0.333	46.6
203.2	0.571	1121	0.233	0.356	227.9	203.2-mm (8-in.) clay tile	C6	0.6667	0.33	70.0	0.2	2.02	46.7
203.2	0.571	608	0.233	0.356	124.0	203.2 mm (8-in.) l.w. concrete block	C7	0.6667	0.33	38.0	0.2	2.02	25.4
203.2	1.038	977	0.233	0.195	198.7	203.2-mm (8-in.) l.w. concrete block	C8	0.6667	0.6	61.0	0.2	1.11	40.7
203.2	0.727	1922	0.233	0.280	390.6	203.2-mm. (8-in.) common brick	C9	0.6667	0.42	120.0	0.2	1.59	80.0
203.2	1.730	2242	0.233	0.117	455.9	203.2 mm. (8-in.) l.w. concrete	C10	0.6667	1.00	140.0	0.2	0.667	93.4
304.8	1.730	2242	0.233	0.176	683.5	304.8 mm (12-in.) l.w. concrete	C11	1.0000	1.00	140.0	0.2	1.08	140.0
50.8	1.730	2242	0.233	0.029	114.2	50.8-mm (2-in.) l. w. concrete	C12	1.6667	1.00	140.0	0.2	1.167	23.4
152.4	1.730	2242	0.233	0.088	341.7	151.4-mm (6-in.) l.w. concrete	C13	0.5000	1.00	140.0	0.2	0.500	70.0
101.6	0.173	640	0.233	0.586	64.9	101.6-mm (4-in.) l.w. concrete	C14	0.3333	0.10	40.0	0.2	3.333	13.3
152.4	0.173	640	0.233	0.088	97.6	152.4-mm (6-in.) l.w. concrete	C15	0.5000	0.10	40.0	0.2	5.000	20.0
203.2	0.173	640	0.233	1.174	130.3	203.2-mm (8-in.) l.w. concrete	C16	0.6667	0.10	40.0	0.2	6.667	26.7
203.2	0.138	288	0.233	1.584	58.6	203.2-mm (8-in.) l.w. concrete block (filled insulation)	C17	0.6667	0.08	18.0	0.2	9.00	12.0
203.2	0.588	849	0.233	0.348	172.8	203.2-mm (8-in.) l.w. concrete block (filled insulation)	C18	0.6667	0.34	53.0	0.2	1.98	35.4
304.8	0.138	304	0.233	2.376	92.8	204.8-mm (12-in.) (l. w. concrete (block filled insulation)	C19	1.0000	0.08	19.0	0.2	13.5	19.0
304.8	**0.675**	**897**	**0.233**	**0.456**	**273.4**	**204.8-mm (12-in.) l.w. concrete block (filled insulation)**	**C20**	**1.0000**	**0.39**	**56.0**	**0.2**	**2.59**	**56.0**
				0.121		**Inside surface resistance**	E0					.685	
19.0	0.727	1601	0.233	0.026	30.5	**19.0-mm (0.75-in.) plaster; 19.0-mm (0.75-in.) gypsum or other similar finishing layer**	E1	0.0625	0.42	100	0.2	0.149	6.25
12.7	1.436	881	0.465	0.009	11.2	12.7-mm (0.5-in.) slag or stone	E2	0.0417	0.83	55	0.40	0.050	2.29
9.5	0.190	1121	0.465	0.050	10.7	9.5-mm (0.375-in.) felt & membrane	E3	0.0313	0.11	70	0.40	0.285	2.19
				0.176		Ceiling air space	E4					1.0	
15.9	0.061	480	0.233	0.315	9.2	Acoustic tile	E5	0.0625	0.035	30	0.20	1.786	1.88

Source: Copyright © by the American Society of Heating, Refrigerating and Air Conditioning Engineers. Inc., Atlanta, GA. Reprinted by permission from the *ASHRAE Handbook of Fundamentals,* 1981.

[a]Units: L (thickness) = mm; K = W/m C; D = kg/m^3; SH = kJ/kg °C; R = m^2 °C/W; WT = kg/m^2.

[b]Units: L (thickness) = ft; K = Btu/h ft F; D = lb/ft^3; SH = Btu/lb F; R = ft^2 F h/Btu; WT = lb/ft^2.

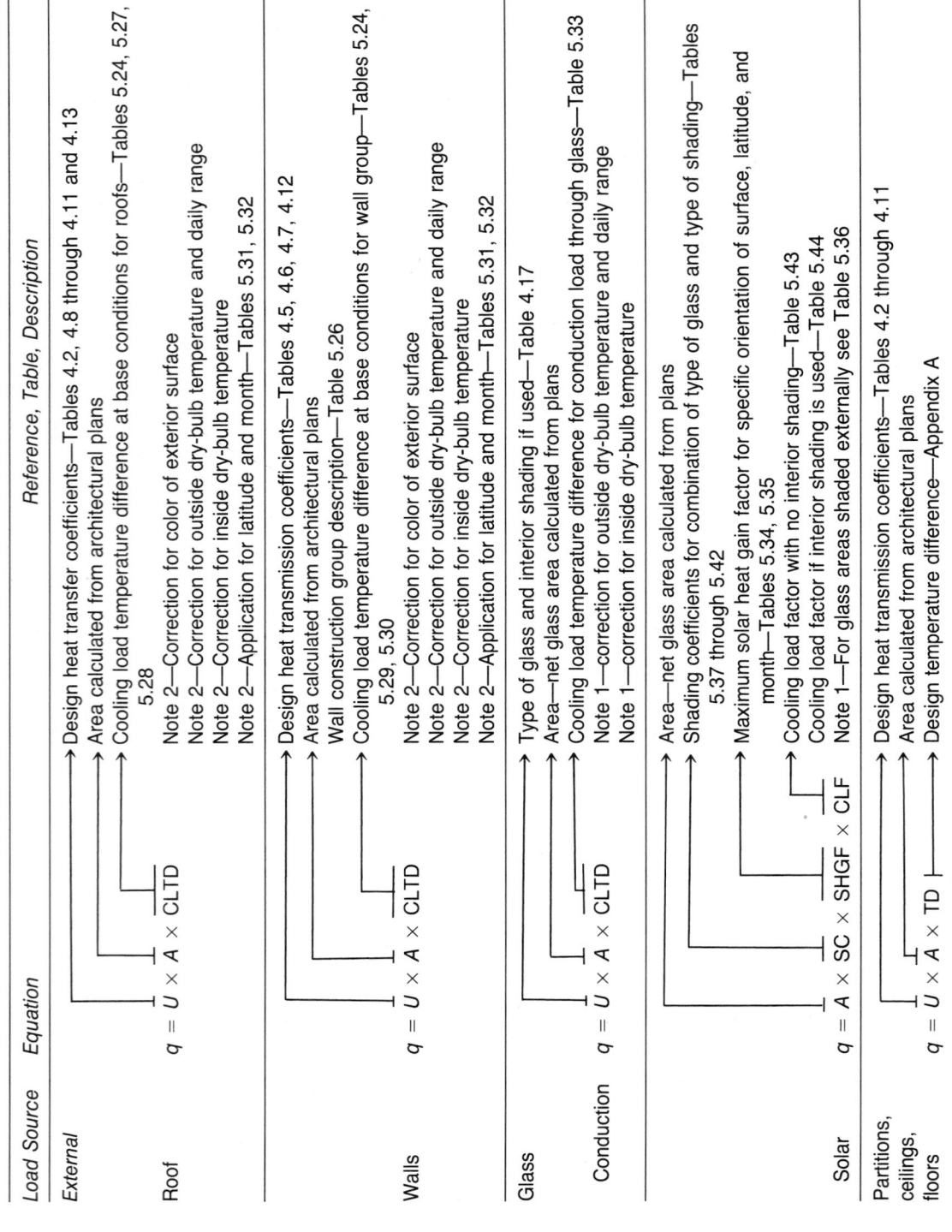

Load Source	*Equation*	*Reference, Table, Description*
External		Design heat transfer coefficients—Tables 4.2, 4.8 through 4.11 and 4.13
		Area calculated from architectural plans
		Cooling load temperature difference at base conditions for roofs—Tables 5.24, 5.27, 5.28
Roof	$q = U \times A \times \text{CLTD}$	Note 2—Correction for color of exterior surface
		Note 2—Correction for outside dry-bulb temperature and daily range
		Note 2—Correction for inside dry-bulb temperature
		Note 2—Application for latitude and month—Tables 5.31, 5.32
		Design heat transmission coefficients—Tables 4.5, 4.6, 4.7, 4.12
		Area calculated from architectural plans
		Wall construction group description—Table 5.26
		Cooling load temperature difference at base conditions for wall group—Tables 5.24, 5.29, 5.30
Walls	$q = U \times A \times \text{CLTD}$	Note 2—Correction for color of exterior surface
		Note 2—Correction for outside dry-bulb temperature and daily range
		Note 2—Correction for inside dry-bulb temperature
		Note 2—Application for latitude and month—Tables 5.31, 5.32
Glass		Type of glass and interior shading if used—Table 4.17
		Area—net glass area calculated from plans
		Cooling load temperature difference for conduction load through glass—Table 5.33
Conduction	$q = U \times A \times \text{CLTD}$	Note 1—correction for outside dry-bulb temperature and daily range
		Note 1—correction for inside dry-bulb temperature
		Area—net glass area calculated from plans
		Shading coefficients for combination of type of glass and type of shading—Tables 5.37 through 5.42
		Maximum solar heat gain factor for specific orientation of surface, latitude, and month—Tables 5.34, 5.35
		Cooling load factor with no interior shading—Table 5.43
		Cooling load factor if interior shading is used—Table 5.44
Solar	$q = A \times \text{SC} \times \text{SHGF} \times \text{CLF}$	Note 1—For glass areas shaded externally see Table 5.36
Partitions,		Design heat transmission coefficients—Tables 4.2 through 4.11
ceilings,		Area calculated from architectural plans
floors	$q = U \times A \times \text{TD}$	Design temperature difference—Appendix A

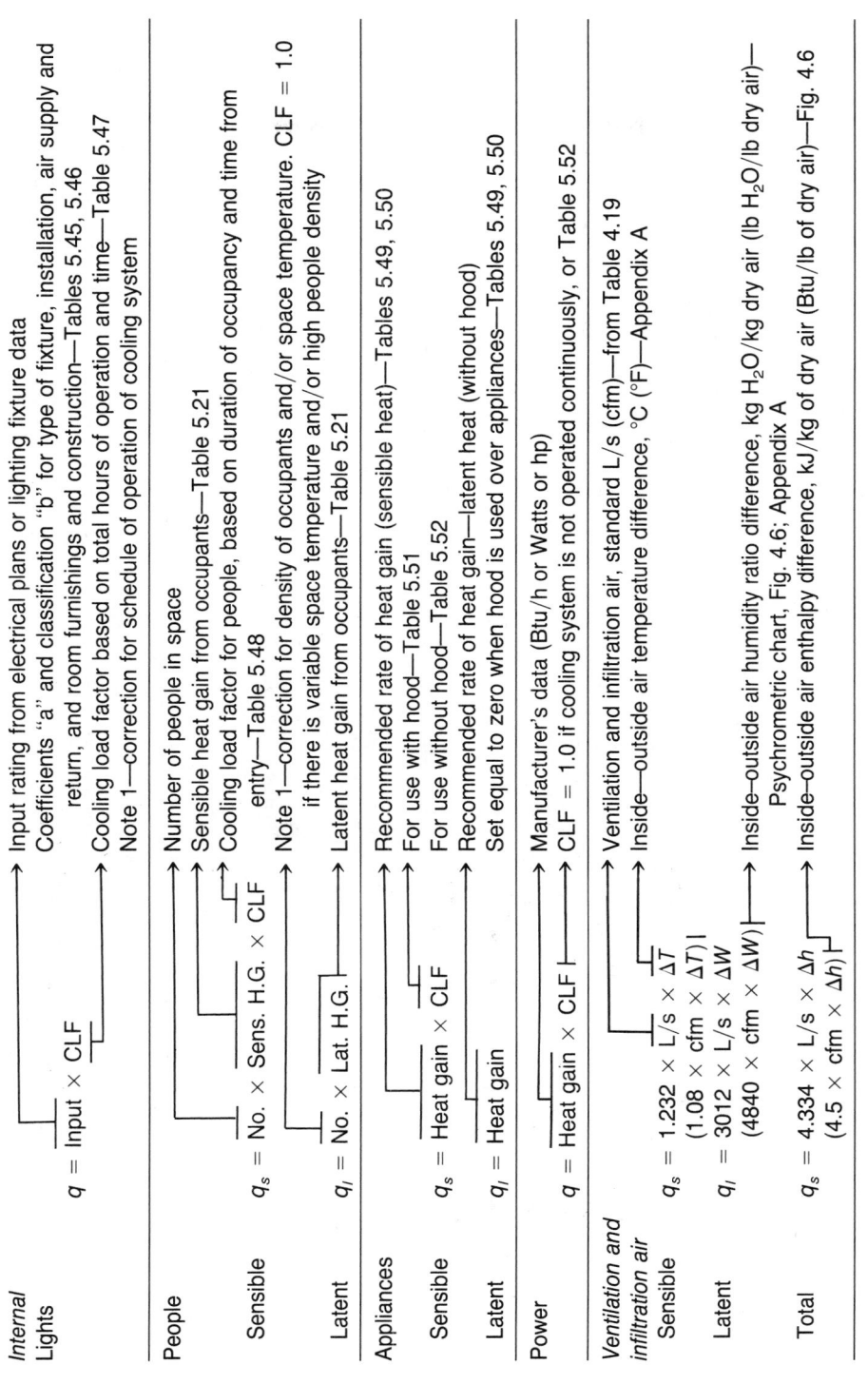

Load source	Equation	Reference sources
Internal Lights	q = Input × CLF	Input rating from electrical plans or lighting fixture data
		Coefficients "a" and classification "b" for type of fixture, installation, air supply and return, and room furnishings and construction—Tables 5.45, 5.46
		Cooling load factor based on total hours of operation and time—Table 5.47
		Note 1—correction for schedule of operation of cooling system
People		Number of people in space
		Sensible heat gain from occupants—Table 5.21
Sensible	q_s = No. × Sens. H.G. × CLF	Cooling load factor for people, based on duration of occupancy and time from entry—Table 5.48
		Note 1—correction for density of occupants and/or space temperature. CLF = 1.0 if there is variable space temperature and/or high people density
Latent	q_l = No. × Lat. H.G.	Latent heat gain from occupants—Table 5.21
Appliances		Recommended rate of heat gain (sensible heat)—Tables 5.49, 5.50
		For use with hood—Table 5.51
Sensible	q_s = Heat gain × CLF	For use without hood—Table 5.52
		Recommended rate of heat gain—latent heat (without hood)
Latent	q_l = Heat gain	Set equal to zero when hood is used over appliances—Tables 5.49, 5.50
Power		Manufacturer's data (Btu/h or Watts or hp)
	q = Heat gain × CLF	CLF = 1.0 if cooling system is not operated continuously, or Table 5.52
Ventilation and infiltration air		Ventilation and infiltration air, standard L/s (cfm)—from Table 4.19
		Inside—outside air temperature difference, °C (°F)—Appendix A
Sensible	$q_s = 1.232 \times \text{L/s} \times \Delta T$ $(1.08 \times \text{cfm} \times \Delta T)$	
Latent	$q_l = 3012 \times \text{L/s} \times \Delta W$ $(4840 \times \text{cfm} \times \Delta W)$	Inside–outside air humidity ratio difference, kg H_2O/kg dry air (lb H_2O/lb dry air)—Psychrometric chart, Fig. 4.6; Appendix A
Total	$q_s = 4.334 \times \text{L/s} \times \Delta h$ $(4.5 \times \text{cfm} \times \Delta h)$	Inside–outside air enthalpy difference, kJ/kg of dry air (Btu/lb of dry air)—Fig. 4.6

Fig. 5.27. *Guide to procedures for calculating space cooling load: load sources (also called heat gains), equations, and reference sources. Adapted from the* ASHRAE 1981 Handbook of Fundamentals, *© 1981 by the American Society of Heating, Refrigerating and Air Conditioning Engineers, Inc., Atlanta, GA.*

TABLE 5.25 **Wall Construction Group Descriptions**

Weight (kg/m²)	U Value (W/m²°C)	Group No.	Description of Construction	Weight (lb/ft²)	U value (Btu/h ft² F)	Code Numbers of Layers (See Table 5.24)
Weight (kg/m²)	***U* Value (W/m²·°C)**	**Group No.**	**Description of Construction**	**Weight (lb/ft²)**	***U* Value (Btu/h · ft²· F)**	**Code Numbers of Layers (see Table 8)**
405	2.033		101.6-mm (4-in.) Face Brick + (*Brick*)			
405	2.033	C	Air Space + 101.6-mm (4-in.) Face Brick	83	0.358	A0, A2, B1, A2, E0
439	2.356	D	101.6-mm (4-in.) Common Brick	90	0.415	A0, A2, C4, E1, E0
439	0.987-1.709	C	25.4-mm (1-in.) Insulation or Air Space + 101.6-mm (4-in.) Common Brick	90	0.174-0.301	A0, A2, C4, B1/B2, E1, E0
430	0.630	B	50.8-mm (2-in.) Insulation + 101.6-mm (4-in.) Common Brick	88	0.111	A0, A2, B3, C4, E1, E0
635	1.714	B	203.2-mm (8-in.) Common Brick	130	0.302	A0, A2, C9, E1, E0
635	0.874-1.379	A	Insulation or Air Space + 203.2-mm (8-in.) Common brick	130	0.154-0.243	A0, A2, C9, B1/B2, E1, E0
			101.6-mm (4-in.) Face Brick + (*H.W. Concrete*)			
459	1.987	C	Air Space + 50.8-mm (2-in.) Concrete	94	0.350	A0, A2, B1, C5, E1, E0
474	0.658	B	50.8-mm (2-in.) Insulation + 101.6-mm (4-in.) Concrete	97	0.116	A0, A2, B3, C5, E1, E0
698-928	0.625-0.636	A	Air Space or Insulation + 203.2-mm (8-in.) or more Concrete	143-190	0.110-0.112	A0, A2, B1, C10/11, E1, E0
			101.6-mm (4-in.) Face Brick + (*L.W. or H.W. Concrete Block*)			
303	1.811	E	101.6-mm (4-in.) Block	62	0.319	A0, A2, C2, E1, E0
303	0.868-1.397	D	Air Space or Insulation + 101.6-mm (4-in.) Block	62	0.153-0.246	A0, A2, C2, B1/B2, E1, E0
342	1.555	D	203.2-mm (8-in.) Block	70	0.274	A0, A2, C7, A6, E0
356-434	1.255-1.561	C	Air Space or 25.4-mm (1-in.) Insulation + 152.4-mm (6-in.) or 203.2-mm (8-in.) Block	73-89	0.221-0.275	A0, A2, B1, C7/C8, E1, E0
434	0.545-0.607	B	50.8-mm (2-in.) Insulation + 203.2-mm (8-in.) Block	89	0.096-0.107	A0, A2, B3, C7/C8, E1, E0
			101.6-mm (4-in.) Face Brick + (*Clay Tile*)			
347	2.163	D	101.6-mm (4-in.) Tile	71	0.381	A0, A2, C1, E1, E0
347	1.595	D	Air Space + 101.6-mm (4-in.) Tile	71	0.281	A0, A2, C1, B1, E1, E0
347	0.959	C	Insulation + 101.6-mm (4-in.) Tile	71	0.169	A0, A2, C1, B2, E1, E0
469	1.561	C	203.2-mm (8-in.) Tile	96	0.275	A0, A2, C6, E1, E0
469	0.806-1.255	B	Air Space or 25.4-mm (1-in.) Insulation + 203.2-mm (8-in.) Tile	96	0.142-0.221	A0, A2, C6, B1/B2, E1, E0
474	0.551	A	50.8-mm (2-in.) Insulation + 203.2-mm (8-in.) Tile	97	0.097	A0, A2, B3, C6, E1, E0

		H.W. Concrete Wall + (*Finish*)				
308	3.321	E	101.6-mm (4-in.) Concrete	63	0.585	A0, A1, C5, E1, E0
308	0.675-1.136	D	101.6-mm (4-in.) Concrete + 25.4-mm (1-in.) or 50.8-mm (2-in.) Insulation	63	0.119-0.200	A0, A1, C5, B2/B3, E1, E0
308	0.675	C	50.8-mm (2-in.) Insulation + 101.6-mm (4-in.) Concrete	63	0.119	A0, A1, B6, C5, E1, E0
532	2.782	C	203.2-mm (8-in.) Concrete	109	0.490	A0, A1, C10, E1, E0
537	0.653-1.061	B	203.2-mm (8-in.) Concrete + 25.4-mm (1-in.) or 50.8-mm (2-in.) Insulation	110	0.115-0.187	A0, A1, C10, B5/B6, E1, E0
537	0.653	A	50.8-mm (2-in.) Insulation + 203.2-mm (8-in.) Concrete	110	0.115	A0, A1, B3, C10, E1, E0
762	2.390	B	304.8-mm (12-in.) Concrete	156	0.421	A0, A1, C11, E1, E0
762	0.642	A	304.8-mm (12-in.) Concrete + Insulation	156	0.113	A0, C11, B6, A6, E0
		L. W. and H.W. Concrete Block + (*Finish*)				
142	0.914-1.493	F	101.6-mm (4-in.) Block + Air Space/Insulation	29	0.161-0.263	A0, A1, C2, B1/B2, E1, E0
142-181	0.596-0.647	E	50.8-mm (2-in.) Insulation + 101.6-mm (4-in.) Block	29-37	0.105-0.114	A0, A1, B3, C2/C3, E1, E0
229-249	1.669-2.282	E	203.2-mm (8-in.) Block	47-51	0.294-0.402	A0, A1, C7/C8, E1, E0
200-278	0.846-0982	D	203.2-mm (8-in.) Block + Air Space/Insulation	41-57	0.149-0.173	A0, A1, C7/C8, B1/B2, E1, E0
		Clay Tile + (*Finish*)				
190	2.379	F	101.6-mm (4-in.) Tile	39	0.419	A0, A1, C1, E1, E0
190	1.720	F	101.6-mm (4-in.) Tile + Air Space	39	0.303	A0, A1, C1, B1, E1, E0
190	0.993	E	101.6-mm (4-in.) Tile + 25.4-mm (1-in.) Insulation	39	0.175	A0, A1, C1, B2, E1, E0
195	0.625	D	50.8-mm (2-in.) Insulation + 101.6-mm (4-in.) Tile	40	0.110	A0, A1, B3, C1, E1, E0
308	1.681	D	203.2-mm (8-in.) Tile	63	0.296	A0, A1, C6, B1/B2, E1, E0
308	0.857-1.312	C	203.2-mm (8-in.) Tile + Air Space/25.4-mm (1-in.) Insulation	63	0.151-0.231	A0, A1, C6, B1/B2, E1, E0
308	0.562	B	50.8-mm (2-in.) Insulation + 203.2-mm (8-in.) Tile	63	0.099	A0, A1, B3, C6, E1, E0
		Metal Curtain Wall				
24-29	0.516-1.306	G	With/ withou t air Space + 25.4 -mm (1- in.)/50 .8-mm (2-in.) 76.2- mm (3-in.) Insulation	5-6	0.091-0.2 30	A0, A3, B5/B6/B12, A3, E0
		Frame Wall				
78	0.459-1.010	G	25.4-m m (1- in.) to 76.2- mm (3-in.) Insulation	16	0.081-0.1 78	A0, A1, B1, B2/B3/B4, E1, E0

NOTES: In the description of construction, exterior materials are listed first. Solid uninsulated doors may be considered to be group F.

Source: Copyright © by the American Society of Heating, Refrigerating and Air Conditioning Engineers, Inc., Atlanta, GA. Reprinted by permission from the *ASHRAE Handbook of Fundamentals,* 1981.

TABLE 5.26 **Roof Assembly Construction Codes**

Roof No.	Description	Code Numbers of Layers (See Table 5.24)
1.	Steel Sheet with 25.4-mm (1-in.) insulation	A0, E2, E3, B5, A3, E0
2.	25.4-mm (1-in.) wood with 25.4-mm (1-in.) insulation	A0, E2, E3, B5, B7, E0
3.	101.6-mm (4-in.) l.w. concrete	A0, E2, E3, C14, E0
4.	50.8-mm (2-in.) h.w. concrete with 25.4 (1-in.) insulation	A0, E2, E3, B5, C12, E0
5.	25.4-mm (1-in.) wood with 50.8-mm (2-in.) insulation	A0, E2, E3, B6, B7, E0
6.	152.4-mm (6-in.) l.w. concrete	A0, E2, E3, C15, E0
7.	63.5-mm (2.5-in.) wood with 25.4-mm (1-in.) insulation	A0, E2, E3, B5, B8, E0
8.	203.2-mm (8-in.) l.w. concrete	A0, E2, E3, C16, E0
9.	101.6-mm (4-in.) h.w. concrete with 25.4-mm (1-in.) insulation	A0, E2, E3, B5, C5, E0
10.	63.5-mm (2.5-in.) wood with 50.8-mm (2-in.) insulation	A0, E2, E3, B6, B8, E0
11.	Roof terrace system	A0, C12, B1, B6, E2, E3, C5, E0
12.	152.4-mm (6-in.) h.w. concrete with 25.4-mm (1-in.) insulation	A0, E2, E3, B5, C13, E0
13.	101.6-mm (4-in.) wood with 25.4-mm (1-in.) insulation	A0, E2, E3, B5, B9, E0

Source: Copyright © by the American Society of Heating, Refrigerating and Air Conditioning Engineers, Inc., Atlanta, GA. Reprinted by permission from the *ASHRAE Handbook of Fundamentals,* 1981.

TABLE 5.27 **Cooling Load Temperature Differences (CLTD) for Calculating Cooling Load from Flat Roofs (SI Units)**

Part A.

Roof No	Description of Construction	Weight, kg/m²	U value, W/m²·°C	1	2	3	4	5	6	7	8	9	10	11	12	13	14	15	16	17	18	19	20	21	22	23	24	Hour of Maximum CLTD	Minimum CLTD	Maximum CLTD	Difference CLTD
	Without Suspended Ceiling																														
1	Steel sheet with 25.4-mm (or 50.8-mm) insulation	34 (39)	1.209 (0.704)	0	−1	−2	−2	−3	−2	3	11	19	27	34	40	43	44	43	39	33	25	17	10	7	5	3	1	14	−3	44	47
2	25.4-mm wood with 25.4-mm insulation	39	0.965	3	2	0	−1	−2	−2	−1	2	8	15	22	29	35	39	41	41	39	35	29	21	15	11	8	5	16	−2	41	43
3	101.6-mm l.w concrete	88	1.209	5	3	1	0	−1	−2	−2	1	5	11	18	25	31	36	39	40	40	37	32	25	19	14	10	7	16	−2	40	42
4	50.8-mm h.w. concrete with 25.4-mm (or 50.8-mm) insulation	142	1.170 (0.693)	7	5	3	2	0	−1	0	2	6	11	17	23	28	33	36	37	37	34	30	25	20	16	12	10	16	−1	37	38
5	25.4-mm wood with 50.8-mm insulation	44	0.619	2	0	−2	−3	−4	−4	−4	−2	3	9	15	22	27	32	35	36	35	32	27	20	14	10	6	3	16	−4	36	40
6	152.4-mm l.w. concrete	117	0.897	12	10	7	5	3	2	1	0	2	4	8	13	18	24	29	33	35	36	35	32	28	24	19	16	18	0	36	36
7	63.5-mm wood with 25.4-mm insulation	63	0.738	16	13	11	9	7	6	4	3	4	5	8	11	15	19	23	27	29	31	31	30	27	25	22	19	19	3	31	28
8	203.2-mm l.w. concrete	151	0.715	20	17	14	12	10	8	6	5	4	4	5	7	11	14	18	22	25	28	30	30	29	27	25	22	20	4	30	26
9	101.6-mm h.w. concrete with 25.4-mm (or 50.8-mm) insulation	254 (254)	1.136 (0.681)	14	12	10	8	7	5	4	4	6	8	11	15	18	22	25	28	29	30	29	27	24	21	19	16	18	4	30	26
10	63.5-mm wood with 50.8-mm insulation	63	0.528	18	15	13	11	9	8	6	5	5	5	7	10	13	17	21	24	27	28	29	29	27	25	23	20	19	5	29	24
11	Roof terrace system	366	0.602	19	17	15	14	12	11	9	8	7	8	8	10	12	15	18	20	22	24	25	26	25	24	22	21	20	7	26	19
12	152.4-mm h.w. concrete with 25.4-mm (or 50.8-mm) insulation	366 (366)	(1.090) (0.664)	18	16	14	12	11	10	9	8	8	9	10	12	15	17	20	22	24	25	25	25	24	22	20	19	19	8	25	17
13	101.6-mm wood with 25.4-mm (or 50.8-mm) insulation	83 (88)	0.602 (0.443)	21	20	18	17	15	14	13	11	10	9	9	9	10	12	14	16	18	20	22	23	24	24	23	22	22	9	24	15
	With Suspended Ceiling																														
1	Steel Sheet with 25.4-mm (or 50.8-mm) insulation	44 (49)	0.761 (0.522)	1	0	−1	−2	−3	−3	0	5	13	20	28	35	40	43	43	41	37	31	23	15	10	7	5	3	15	−3	43	46
2	25.4-mm wood with 25.4-mm insulation	49	0.653	11	8	6	5	3	2	1	2	4	7	12	17	22	27	31	33	35	34	32	28	24	20	17	14	17	1	35	34
3	101.6-mm l.w. concrete	97	0.761	10	8	6	4	2	1	0	0	2	6	10	16	21	27	31	34	36	36	34	30	26	21	17	13	17	0	36	36
4	50.8-mm h.w. concrete with 25.4-mm insulation	146	0.744	16	14	13	11	10	8	7	7	8	9	11	14	17	19	22	24	25	26	26	25	23	21	20	18	18	7	26	19

(continued)

TABLE 5.27 **Cooling Load Temperature Differences (CLTD) for Calculating Cooling Load from Flat Roofs (SI Units)** (*continued*)

Roof No	Description of Construction	Weight lb/ft^2	U-value Btu/(h·ft^2·°F)	1	2	3	4	5	6	7	8	9	10	11	12	13	14	15	16	17	18	19	20	21	22	23	24	Hour of Maximum CLTD	Minimum CLTD	Maximum CLTD	Difference CLTD
	With Suspended Ceiling																														
5	25.4-mm wood with 50.8-mm insulation	49	0.471	14	11	9	7	5	4	3	3	4	6	10	14	18	23	27	30	31	32	31	29	26	22	19	16	18	3	32	30
6	152.4-mm l.w. concrete	127	0.619	18	15	13	11	9	7	6	4	4	4	6	9	12	16	20	24	27	29	30	30	28	26	23	20	20	4	30	26
7	63.5-mm wood with 25.4-mm insulation	73	0.545	19	18	16	14	13	12	10	9	8	8	9	10	12	14	17	19	21	23	24	25	24	23	22	21	20	8	25	17
8	203.2-mm l.w. concrete	161	0.528	22	20	18	16	15	13	11	10	9	8	8	8	9	11	14	16	19	21	23	25	25	25	24	23	20	8	25	17
9	101.6-mm h.w. concrete with 25.4-mm (or 50.8-mm) insulation	259 (264)	0.727 (0.511)	17	16	15	14	13	13	12	11	11	11	12	13	15	16	18	19	20	21	21	21	21	20	19	18	19	11	21	10
10	63.5-mm wood with 50.8-mm insulation	73	0.409	19	18	17	16	14	13	12	11	10	10	10	11	12	14	16	18	19	21	22	23	23	22	22	21	21	10	23	13
11	Roof terrace system	376	0.466	17	16	16	15	15	14	13	13	13	12	12	13	13	14	15	16	16	17	18	18	19	18	18	18	21	12	19	7
12	152.4-mm h.w. concrete with 25.4-mm (or 50.8-mm) insulation	376 (376)	0.710 0.499	16	16	15	15	14	13	13	12	12	12	12	13	14	15	16	17	18	18	19	19	19	18	18	18	20	12	19	7
13	101.6-mm wood with 25.4-mm (or 50.8-mm) insulation	93 (97)	0.465 (0.363)	20	19	19	18	17	16	15	14	14	13	12	12	12	12	13	14	15	16	18	19	20	20	20	20	23	12	20	8

TABLE 5.27 **Cooling Load Temperature Differences (CLTD) for Calculating Cooling Load from Flat Roofs (SI Units)** ***(continued)***

Part B.

1. *Direct Application of This Table Without Adjustments*

Values were calculated using the following conditions:

- Dark flat surface roof ("dark" for solar radiation absorption)
- Indoor temperature of 25.5°C (= T_R)
- Outdoor maximum temperature of 35°C with outdoor mean temperature of 29.4°C and an outdoor daily range of 11.6C°
- Solar radiation typical of 40°N latitude on July 21
- Outside surface resistance, R_o = 0.059 m² °C/W
- Without and with suspended ceiling, but no attic fans or return air ducts in suspended ceiling space
- Inside surface resistance, R_i = 0.121 m² °C/W

2. *Adjustments to Values*

The following equation makes adjustments for deviations of design and solar conditions from those listed in (1) above.

$$\text{CLTD}_{corr} = [(\text{CLTD} + \text{LM}) \times K + (25.5 - T_R) + (T_o - 29.4) \times f$$

where CLTD is from this table and

(a) LM is the latitude-month correction from Table 5.31 for a horizontal surface

(b) K is a color adjustment factor, applied after month-latitude adjustments are made. Credit should not be taken for a light-colored roof except where permanence of light color is established by experience, as in rural areas or where there is little smoke.
K = 1.0 if dark-colored or light-colored in an industrial area
K = 0.5 if permanently light-colored (rural area)

(c) $(25.5 - T_R)$ is the indoor design temperature correction.

(d) $(T_o - 29.4)$ is the outdoor design temperature correction, where T_o is the average outside temperature on design day.

(e) f is a factor for attic fan and/or ducts above ceiling, applied after all other adjustments have been made.
f = 1.0, no attic or ducts
f = 0.75, positive ventilation
Values were calculated without and with a suspended ceiling, but no allowances were made for positive ventilation or return ducts through the space. If ceiling is insulated and a fan is used between ceiling and roof, CLTD may be reduced by 25% (f = 0.75). Use of the suspended ceiling space for a return air plenum or with return air ducts should be analyzed separately.

3. *Roof Constructions Not Listed in Table*

The U values listed are to be used only as guides. The actual value of U as obtained from tables or as calculated for the actual roof construction should be used.

An actual roof construction not in this table would be thermally similar to a roof in the table, if it has similar mass (kg/m²) and similar heat capacity (kJ/m² °C). In such a case, use the CLTD from this table as corrected by note 2.

Example: A flat roof without a suspended ceiling has mass properties of 87.88 kg/m², U of 1.136 W/m² °C, and heat capacity of 53.9 kJ/m² °C. Use CLTD_{uncorr} from roof no. 13, to obtain CLTD_{corr}, and use the actual U value to calculate $q/A = U(\text{CLTD}_{corr}) = 1.136(\text{CLTD}_{corr})$.

4. *Additional Insulation*

For each $R7$ increase in R value due to insulation added to the roof structure, use a CLTD for a roof whose weight and heat capacity are approximately the same, but whose CLTD has a maximum value 2 h later. If this is not possible, because a roof with the longest time lag has already been selected, use an effective CLTD in the cooling load calculation equal to 16.1C°.

Source: Copyright © by the American Society of Heating, Refrigerating and Air Conditioning Engineers, Inc., Atlanta, GA. Reprinted by permission from the *ASHRAE Handbook of Fundamentals,* 1981.

TABLE 5.28 **Cooling Load Temperature Differences (CLTD) for Calculating Cooling Load from Flat Roofs (Conventional Units)**

Part A.

Roof No	Description of Construction	Weight lb/ft²	U value Btu/(h·ft²·°F)	1	2	3	4	5	6	7	8	9	10	11	12	13	14	15	16	17	18	19	20	21	22	23	24	Hour of Maximum CLTD	Minimum CLTD	Maximum CLTD	Difference CLTD
				Solar Time, hr																											
	Without Suspended Ceiling																														
1	Steel sheet with 1-in. (or 2-in.) insulation	7 (8)	0.213 (0.124)	1	−2	−3	−3	−5	−3	6	19	34	49	61	71	78	79	77	70	59	45	30	18	12	8	5	3	14	−5	79	84
2	1-in. wood with 1-in. insulation	8	0.170	6	3	0	−1	−3	−3	−2	4	14	27	39	52	62	70	74	74	70	62	51	38	28	20	14	9	16	−3	74	77
3	4-in. l.w concrete	18	0.213	9	5	2	0	−2	−3	−3	1	9	20	32	44	55	64	70	73	71	66	57	45	34	25	18	13	16	−3	73	76
4	2-in. h.w. concrete with 1-in. (or 2-in.) insulation	29	0.206 (0.122)	12	8	5	3	0	−1	−1	3	11	20	30	41	51	59	65	66	66	62	54	45	36	29	22	17	16	−1	67	68
5	1-in. wood with 2-in. insulation	9	0.109	3	0	−3	−4	−5	−7	−6	−3	5	16	27	39	49	57	63	64	62	57	48	37	26	18	11	7	16	−7	64	71
6	6-in. l.w. concrete	24	0.158	22	17	13	9	6	3	1	1	3	7	15	23	33	43	51	58	62	64	62	57	50	42	35	28	18	1	64	63
7	2.5-in. wood with 1-insulation	13	0.130	29	24	20	16	13	10	7	6	6	9	13	20	27	34	42	48	53	55	56	54	49	44	39	34	19	6	56	50
8	8-in. l.w. concrete	31	0.126	35	30	26	22	18	14	11	9	7	7	9	13	19	25	33	39	46	50	53	54	53	49	45	40	20	7	54	47
9	4-in. h.w. concrete with 1-in. (or 2-in.) insulation	52 (52)	0.200 (0.120)	25	22	18	15	12	9	8	8	10	14	20	26	33	40	46	50	53	53	52	48	43	38	34	30	18	8	53	45
10	2.5-in. wood with 2-in. insulation	13	0.093	30	26	23	19	16	13	10	9	8	9	13	17	23	29	36	41	46	49	51	50	47	43	39	35	19	8	51	43
11	Roof terrace system	75	0.106	34	31	28	25	22	19	16	14	13	13	15	18	22	26	31	36	40	44	45	46	45	43	40	37	20	13	46	33
12	6-in. h.w. concrete with 1-in. (or 2-in.) insulation	(75) 75	0.192 (0.117)	31	28	25	22	20	17	15	14	14	16	18	22	26	31	36	40	43	45	45	44	42	40	37	34	19	14	45	31
13	4-in. wood with 1-in. (or 2-in) insulation	17 (18)	0.106 (0.078)	38	36	33	30	28	25	22	20	18	17	16	17	18	21	24	28	32	36	39	41	43	43	42	40	22	16	43	27
	With Suspended Ceiling																														
1	Steel Sheet with 1-in. (or 2-in.) insulation	9 (10)	0.134 (0.092)	2	0	−2	−3	−4	−4	−1	9	23	37	50	62	71	77	78	74	67	56	42	28	18	12	8	5	15	−4	78	82
2	1-in. wood with 1-in. insulation	10	0.115	20	15	11	8	5	3	2	3	7	13	21	30	40	48	55	60	62	61	58	51	44	37	30	25	17	2	62	60
3	4-in. l.w. concrete	20	0.134	19	14	10	7	4	2	0	0	4	10	19	29	39	48	56	62	65	64	61	54	46	38	30	24	17	0	65	65
4	2-in. h.w. concrete with 1-in. insulation	30	0.131	28	25	23	20	17	15	13	13	14	16	20	25	30	35	39	43	46	47	46	44	41	38	35	32	18	13	47	34
5	1-in. wood with 2-in. insulation	10	0.083	25	20	16	13	10	7	5	5	7	12	18	25	33	41	48	53	57	57	56	52	46	40	34	29	18	5	57	52
6	6-in. l.w. concrete	26	0.109	32	28	23	19	16	13	10	8	7	8	11	16	22	29	36	42	48	52	54	54	51	47	42	37	20	7	54	47
7	2.5-in. wood with 1-in. insulation	15	0.096	34	31	29	26	23	21	18	16	15	15	16	18	21	25	30	34	38	41	43	44	44	42	40	37	21	15	44	29
8	8-in. l.w. concrete	33	0.093	39	36	33	29	26	23	20	18	15	14	14	15	17	20	25	29	34	38	42	45	46	45	44	42	21	14	46	32

9	4-in. h.w. concrete with 1-in. (or 2-in.) insulation	53 (54)	0.128 (0.090)	30	29	27	26	24	22	21	20	20	21	22	24	27	29	32	34	36	38	38	38	37	36	34	33	19	20	38	18
10	2.5-in. wood with 2-in. insulation	15	0.072	35	33	30	28	26	24	22	20	18	18	18	20	22	25	28	32	35	38	40	41	41	40	39	37	21	18	41	23
11	Roof terrace system	77	0.082	30	29	28	27	26	25	24	23	22	22	22	23	23	25	26	28	29	31	32	33	33	33	33	32	22	22	33	11
12	6-in. h.w. concrete with 1-in. (or 2-in) insulation	77 (77)	0.125 (0.088)	29	28	27	26	25	24	23	22	21	21	22	23	25	26	28	30	32	33	34	34	34	33	32	31	20	21	34	13
13	4-in. wood with 1-in (or 2-in.) insulation	19 (20)	0.082 (0.064)	35	34	33	32	31	29	27	26	24	23	22	21	22	22	24	25	27	30	32	34	35	36	37	36	23	21	37	16

Part B.

1. *Direct Application of This Table Without Adjustments*
Values were calculated using the following conditions:

- Dark flat surface roof ("dark" for solar radiation absorption)
- Indoor temperature of 78 F (= T_R)
- Outdoor maximum temperature of 95 F with outdoor mean temperature of 85 F and an outdoor daily range of 21F°
- Solar radiation typical of 40°N latitude on July 21
- Outside surface resistance, R_o = 0.333 ft² F h/Btu
- Without and with suspended ceiling, but no attic fans or return air ducts in suspended ceiling space
- Inside surface resistance, R_i = 0.685 ft² F h/Btu

2. *Adjustments to Values*

The following equation makes adjustments for deviations of design and solar conditions from those listed in (1) above.

$$\text{CLTD}_{\text{corr}} = [(\text{CLTD} + \text{LM}) \times K + (78 - T_R) + (T_o - 85)] \times f$$

where CLTD is from this table and

(a) LM is the latitude-month correction from Table 5.32 for a horizontal surface.

(b) K is a color adjustment factor, applied after month-latitude adjustments are made. Credit should not be taken for a light-colored roof except where permanence of light color is established by experience, as in rural areas or where there is little smoke.
K = 1.0 if dark-colored or light-colored in an industrial area
K = 0.5 if permanently light-colored (rural area)

(c) $(78 - T_R)$ is the indoor design temperature correction.

(d) $(T_o - 85)$ is the outdoor design temperature correction, where T_o is the average outside temperature on design day.

(e) f is a factor for attic fan and/or ducts above ceiling, applied after all other adjustments have been made.
f = 1.0 no attic or ducts
f = 0.75 positive ventilation
Values were calculated without and with a suspended ceiling, but no allowances were made for positive ventilation or return ducts through the space. If ceiling is insulated and a fan is used between ceiling and roof, CLTD may be reduced by 25% (f = 0.75). Use of the suspended ceiling space for a return air plenum or with return air ducts should be analyzed separately.

3. *Roof Constructions Not Listed in Table*

The U values listed are to be used only as guides. The actual value of U as obtained from tables or as calculated for the actual roof construction should be used.

An actual roof construction not in this table would be thermally similar to a roof in the table, if it has similar mass (lb_m/ft^2) and similar heat capactiy (Btu/ft² F). In such a case, use the CLTD from this table as corrected by note (2) above.

Example: A flat roof without a suspended ceiling has mass properties of 18.0 lb/ft², U of 0.20 Btu/h ft² F, and heat capacity of 9.5 Btu/ft² F. Use $\text{CLTD}_{\text{uncorr}}$ from roof no. 13, to obtain $\text{CLTD}_{\text{corr}}$ and use the actual U value to calculate $q/A = U\,(\text{CLTD}_{\text{corr}}) = 0.20\,(\text{CLTD}_{\text{corr}})$.

4. *Additional Insulation*

For each $R7$ increase in R value due to insulation added to the roof structure, use a CLTD for a roof whose weight and heat capacity are approximately the same, but whose CLTD has a maximum value 2 h later. If this is not possible, because a roof with longest time lag has already been selected, use an effective CLTD in cooling load calculation equal to 29 F°.

Source: Copyright © by the American Society of Heating, Refrigerating and Air Conditioning Engineers, Inc., Atlanta, GA. Reprinted by permission from the ***ASHRAE Handbook of Fundamentals***, 1981.

TABLE 5.29 **Cooling Load Temperature Differences (CLTD) for Calculating Cooling Load from Sunlit Walls (SI Units)**

Part A.

North latitude wall facing	1	2	3	4	5	6	7	8	9	10	11	12	13	14	15	16	17	18	19	20	21	22	23	24	Hour of Maximum CLTD	Minimum CLTD	Maximum CLTD	Difference CLTD
	Solar Time (h)																											
Group A Walls																												
N	8	8	8	7	7	7	7	6	6	6	6	6	6	6	6	6	6	6	7	7	7	7	8	8	2	6	8	2
NE	11	11	10	10	10	9	9	9	8	8	8	9	9	9	9	10	10	10	11	11	11	11	11	11	22	8	11	3
E	14	13	13	13	12	12	11	11	10	10	10	11	11	12	12	13	13	13	14	14	14	14	14	14	22	10	14	4
SE	13	13	13	12	12	11	11	10	10	10	10	10	10	11	11	12	12	13	13	13	13	13	13	13	22	10	13	3
S	11	11	11	11	10	10	9	9	9	8	8	8	8	8	8	8	9	9	10	10	11	11	11	11	23	8	11	3
SW	14	14	14	14	13	13	12	12	11	11	10	10	10	9	9	10	10	10	11	12	13	13	14	14	24	9	14	5
W	15	15	15	14	14	14	13	13	12	12	11	11	10	10	10	10	10	11	11	12	13	14	14	15	1	10	15	5
NW	12	12	11	11	11	11	10	10	10	9	9	8	8	8	8	8	8	8	9	9	10	11	11	11	1	8	12	4
Group B Walls																												
N	8	8	8	7	7	6	6	6	5	5	5	5	5	5	5	6	6	7	7	8	8	8	8	8	24	5	8	3
NE	11	10	10	9	9	8	7	7	7	7	8	8	9	9	10	10	11	11	11	12	12	12	11	11	21	7	12	5
E	13	13	12	11	10	10	9	8	8	9	9	10	12	13	13	14	14	15	15	15	15	15	14	14	20	8	15	7
SE	13	12	12	11	10	10	9	8	8	8	8	9	10	11	12	13	14	14	14	14	14	14	14	14	21	8	14	6
S	12	11	11	10	9	9	8	7	7	6	6	6	6	7	8	9	10	11	11	12	12	12	12	12	23	6	12	6
SW	15	15	14	13	13	12	11	10	9	9	8	8	7	7	8	9	10	11	13	14	15	15	16	16	24	7	16	9
W	16	16	15	14	14	13	12	11	10	9	9	8	8	8	8	8	9	11	12	14	15	16	16	17	24	8	17	9
NW	13	12	12	11	11	10	9	9	8	7	7	7	6	6	7	7	8	8	9	11	12	13	13	13	24	6	13	7
Group C Walls																												
N	9	8	7	7	6	5	5	4	4	4	4	4	5	5	6	6	7	8	9	9	9	10	9	9	22	4	10	6
NE	10	10	9	8	7	6	6	6	6	7	8	10	10	11	12	12	12	13	13	13	13	12	12	11	20	6	13	7
E	13	12	11	10	9	8	7	7	8	9	11	13	14	15	16	16	17	17	16	16	16	15	14	13	18	7	17	10
SE	13	12	11	10	9	8	7	6	7	7	9	10	12	14	15	16	16	16	16	16	16	15	14	13	19	6	16	10
S	12	11	10	9	8	7	6	6	5	5	5	5	6	8	9	11	12	13	14	14	14	14	13	12	20	5	14	9
SW	16	15	14	12	11	10	9	8	7	7	6	6	6	7	8	10	12	14	16	18	18	18	18	17	22	6	18	12
W	17	16	15	14	12	11	10	9	8	7	7	7	7	7	8	9	11	13	16	18	19	20	19	18	22	7	20	13
NW	14	13	12	11	10	9	8	7	6	6	5	5	6	6	6	7	9	10	12	14	15	15	15	15	22	5	15	10
Group D Walls																												
N	8	7	7	6	5	4	3	3	3	3	4	4	5	6	6	7	8	9	10	11	11	10	10	9	21	3	11	8
NE	9	8	7	6	5	5	4	4	6	8	10	11	12	13	13	13	14	14	14	13	13	12	11	10	19	4	14	10
E	11	10	8	7	6	5	5	5	7	10	13	15	17	18	18	18	18	18	17	17	16	15	13	12	16	5	18	13
SE	11	10	9	7	6	5	5	5	5	7	10	12	14	16	17	18	18	18	17	17	16	15	14	12	17	5	18	13
S	11	10	8	7	6	5	4	4	3	3	4	5	7	9	11	13	15	16	16	16	15	14	13	12	19	3	16	13
SW	15	14	12	10	9	8	6	5	5	4	4	5	5	7	9	12	15	18	20	21	21	20	19	17	21	4	21	17
W	17	15	13	12	10	9	7	6	5	5	5	5	6	6	8	10	13	17	20	22	23	22	21	19	21	5	23	18
NW	14	12	11	9	8	7	6	5	4	4	4	4	5	6	7	8	10	12	15	17	18	17	16	15	22	4	18	14
Group E Walls																												
N	7	6	5	4	3	2	2	2	3	3	4	5	6	7	8	10	10	11	12	12	11	10	9	8	20	2	12	10
NE	7	6	5	4	3	2	3	5	8	11	13	14	14	14	14	14	15	14	14	13	12	11	9	8	16	2	15	13
E	8	7	6	5	4	3	3	6	10	15	18	20	21	21	20	19	18	18	17	15	14	12	11	9	13	3	21	18
SE	8	7	6	5	4	3	3	4	7	10	14	17	19	20	20	20	19	18	17	16	14	13	11	10	15	3	20	17
S	8	7	6	5	4	3	2	2	2	3	5	7	10	14	16	18	19	18	17	16	14	13	11	10	17	2	19	17
SW	12	10	8	7	6	4	4	3	3	3	4	5	7	10	14	18	21	24	25	24	22	19	17	14	19	3	25	22
W	14	12	10	8	6	5	4	3	3	4	4	5	6	8	11	15	20	24	27	27	25	22	19	16	20	3	27	24
NW	11	9	8	6	5	4	3	3	3	3	4	5	6	7	9	11	14	18	21	21	23	18	15	13	20	3	21	18
Group F Walls																												
N	5	4	3	2	1	1	1	2	3	4	5	6	8	9	11	12	12	13	13	13	11	9	7	6	19	1	13	12
NE	5	4	3	2	1	1	3	8	13	16	17	16	16	15	15	15	15	14	13	12	10	9	7	6	11	1	17	16
E	5	4	3	2	2	1	4	9	16	21	24	25	24	22	20	19	18	17	15	13	11	10	8	7	12	1	25	24
SE	5	4	3	2	2	1	2	6	10	15	20	23	24	23	22	20	19	17	16	14	12	10	8	7	13	1	24	23
S	5	4	3	2	2	1	1	1	2	4	7	11	15	19	21	22	21	19	17	15	12	10	8	7	16	1	22	21
SW	8	6	5	4	3	2	1	1	2	3	4	6	10	14	20	24	28	30	29	25	20	16	13	10	18	1	30	29
W	9	7	5	4	3	2	2	2	2	3	4	6	8	11	16	22	27	32	33	30	24	19	15	12	19	2	33	31
NW	8	6	4	3	2	2	1	1	2	3	4	6	7	9	12	15	19	24	26	24	20	16	12	10	19	1	26	25
Group G Walls																												
N	2	1	0	0	0	1	4	5	5	7	8	10	12	13	13	14	14	15	12	8	6	5	4	3	18	0	15	15
NE	2	1	1	0	0	5	15	20	22	20	16	15	15	15	15	15	14	12	10	8	6	5	4	3	9	0	22	22
E	2	1	1	0	0	6	17	26	30	31	28	22	19	17	17	16	15	13	11	8	7	5	4	3	10	0	31	31
SE	2	1	1	0	0	3	10	18	24	27	28	27	23	20	18	16	15	13	11	8	7	6	4	3	11	0	28	28
S	2	1	1	0	0	0	1	3	7	12	17	22	25	26	24	21	17	14	11	8	7	5	4	3	14	0	26	26
SW	3	2	2	1	0	0	1	3	4	6	9	14	21	28	33	35	34	29	20	13	10	7	6	4	16	0	35	35
W	4	3	2	1	1	1	1	3	5	6	8	10	15	23	31	37	40	37	27	16	11	8	6	5	17	1	40	39
NW	3	2	1	1	0	0	1	3	4	6	8	10	12	15	20	26	31	31	23	14	10	7	5	4	18	0	31	31

Part B.

1. *Direct Application of This Table Without Adjustments*

Values were calculated using the following conditions for walls as outlined for the roof CLTD table, Table 5.27. These values may be used for all normal air conditioning estimates usually without correction (except as noted below) when the load is calculated for the hottest weather.

For totally shaded walls use the north orientation values.

2. *Adjustments to Table Values*

The following equation makes adjustments for conditions other than those listed in Note 1.

$$\text{CLTD}_{\text{corr}} = (\text{CLTD} + \text{LM}) \times K + (25.5 - T_R) + (T_o - 29.4)$$

where CLTD is from this table at the wall orientation and

(a) LM is latitude-month correction from Table 5.31

(b) K is a color adjustment factor and is applied after first making month-latitude adjustments

- $K = 1.0$ if dark colored or light in an industrial area
- $K = 0.83$ if permanently medium-colored (rural area)
- $K = 0.65$ if permanently light-colored (rural area)

Credit should not be taken for a light-colored roof except where permanence of light color is established by experience, as in rural areas or where there is little smoke. Colors: Light—Cream

Medium—Medium blue, medium green, bright red, light brown, unpainted wood, and natural color concrete

Dark—Dark blue, red, brown, and green

(c) $(25.5 - T_R)$ is indoor design temperature correction

(d) $(T_o - 29.4)$ is outdoor design temperature correction, where T_o is the average outside temperature on design day.

3. *Wall Construction Not Listed*

The U values listed are to be used only as guides. The actual value of U as obtained from tables or as calculated for the actual wall structure should be used.

An actual wall construction not listed in this table (or Table 5.26) would be thermally similar to a wall in the table, if it has similar mass (kg/m^3) and similar heat capacity (kJ/m^2 °C). In that case, use the CLTD from this table as corrected by Note 2.

4. *Additional Insulation*

For each 7 increase in R value due to insulation added to the wall structures, in Table 5.26 use the CLTD for the wall group with the next higher letter in the alphabet (e.g., C → A). When the insulation is added to the exterior of the construction rather than the interior, use the CLTD for the wall group two letters higher. If this is not possible, due to having already selected a wall in group A, use an effective CLTD in the load calculation as given in the following table.

CLTD, Uncorrected, When Vertical Wall Structure Is "Thermally" Heavier than Group A Due to Added Insulation

N	NE	E	SE	S	SW	W	NW
6.1	9.4	12.2	11.6	9.4	11.6	12.2	9.4

Source: Copyright © by the American Society of Heating, Refrigerating and Air Conditioning Engineers, Inc., Atlanta, GA. Reprinted by permission from the *ASHRAE Handbook of Fundamentals,* 1981.

TABLE 5.30 **Cooling Load Temperature Differences (CLTD) for Calculating Cooling Load from Sunlit Walls (Conventional Units)**

Part A.

North latitude wall facing	Solar Time (h) 1	2	3	4	5	6	7	8	9	10	11	12	13	14	15	16	17	18	19	20	21	22	23	24	Hour of Maximum CLTD	Minimum CLTD	Maximum CLTD	Difference CLTD
Group A Walls																												
N	14	14	14	13	13	13	12	12	11	11	10	10	10	10	10	10	11	11	12	12	13	13	14	14	2	10	14	4
NE	19	19	19	18	17	17	16	15	15	15	15	15	16	16	17	18	18	18	19	19	20	20	20	20	22	15	20	5
E	24	24	23	23	22	21	20	19	19	18	19	19	20	21	22	23	24	24	25	25	25	25	25	25	22	18	25	7
SE	24	23	23	22	21	20	20	19	18	18	18	18	18	19	20	21	22	23	23	24	24	24	24	24	22	18	24	6
S	20	20	19	19	18	18	17	16	16	15	14	14	14	14	14	15	16	17	18	19	19	20	20	20	23	14	20	6
SW	25	25	25	24	24	23	22	21	20	19	19	18	17	17	17	17	18	19	20	22	23	24	25	25	24	17	25	8
W	27	27	26	26	25	24	24	23	22	21	20	19	19	18	18	18	18	19	20	22	23	25	26	26	1	18	27	9
NW	21	21	21	20	20	19	19	18	17	16	16	15	15	14	14	14	15	15	16	17	18	19	20	21	1	14	21	7
Group B Walls																												
N	15	14	14	13	12	11	11	10	9	9	9	8	9	9	9	10	11	12	13	14	14	15	15	15	24	8	15	7
NE	19	18	17	16	15	14	13	12	12	13	14	15	16	17	18	19	19	20	20	21	21	21	20	20	21	12	21	9
E	23	22	21	20	18	17	16	15	15	15	17	19	21	22	24	25	26	26	27	27	26	26	25	24	20	15	27	12
SE	23	22	21	20	18	17	16	15	14	14	15	16	18	20	21	23	24	25	26	26	26	26	25	24	21	14	26	12
S	21	20	19	18	17	15	14	13	12	11	11	11	11	12	14	15	17	19	20	21	22	22	22	21	23	11	22	11
SW	27	26	25	24	22	21	19	18	16	15	14	14	13	13	14	15	17	20	22	25	27	28	28	28	24	13	28	15
W	29	28	27	26	24	23	21	19	18	17	16	15	14	14	14	15	17	19	22	25	27	29	29	30	24	14	30	16
NW	23	22	21	20	19	18	17	15	14	13	12	12	12	11	12	12	13	15	17	19	21	22	23	23	24	11	23	9
Group C Walls																												
N	15	14	13	12	11	10	9	8	8	7	7	8	8	9	10	12	13	14	15	16	17	17	17	16	22	7	17	10
NE	19	17	16	14	13	11	10	10	11	13	15	17	19	20	21	22	22	23	23	23	23	22	21	20	20	10	23	13
E	22	21	19	17	15	14	12	12	14	16	19	22	25	27	29	29	30	30	30	29	28	27	26	24	18	12	30	18
SE	22	21	19	17	15	14	12	12	12	13	16	19	22	24	26	28	29	29	29	29	28	27	26	24	19	12	29	17
S	21	19	18	16	15	13	12	10	9	9	9	10	11	14	17	20	22	24	25	26	25	25	24	22	20	9	26	17
SW	29	27	25	22	20	18	16	15	13	12	11	11	11	13	15	18	22	26	29	32	33	33	32	31	22	11	33	22
W	31	29	27	25	22	20	18	16	14	13	12	12	12	13	14	16	20	24	29	32	35	35	35	33	22	12	35	23
NW	25	23	21	20	18	16	14	13	11	10	10	10	10	11	12	13	15	18	22	25	27	27	27	26	22	10	27	17
Group D Walls																												
N	15	13	12	10	9	7	6	6	6	6	6	7	8	10	12	13	15	17	18	19	19	19	18	16	21	6	19	13
NE	17	15	13	11	10	8	7	8	10	14	17	20	22	23	23	24	24	25	25	24	23	22	20	18	19	7	25	18
E	19	17	15	13	11	9	8	9	12	17	22	27	30	32	33	33	32	32	31	30	28	26	24	22	16	8	33	25
SE	20	17	15	13	11	10	8	8	10	13	17	22	26	29	31	32	32	32	31	30	28	26	24	22	17	8	32	24
S	19	17	15	13	11	9	8	7	6	6	7	9	12	16	20	24	27	29	29	29	27	26	24	22	19	6	29	23
SW	28	25	22	19	16	14	12	10	9	8	8	8	10	12	16	21	27	32	36	38	38	37	34	31	21	8	38	30
W	31	27	24	21	18	15	13	11	10	9	9	9	10	11	14	18	24	30	36	40	41	40	38	34	21	9	41	32
NW	25	22	19	17	14	12	10	9	8	7	7	8	9	10	12	14	18	22	27	31	32	32	30	27	22	7	32	25
Group E Walls																												
N	12	10	8	7	5	4	3	4	5	6	7	9	11	13	15	17	19	20	21	23	20	18	16	14	20	3	22	19
NE	13	11	9	7	6	4	5	9	15	20	24	25	25	26	26	26	26	26	25	24	22	19	17	15	16	4	26	22
E	14	12	10	8	6	5	6	11	18	26	33	36	38	37	36	34	33	32	30	28	25	22	20	17	13	5	38	33
SE	15	12	10	8	7	5	5	8	12	19	25	31	35	37	37	36	34	33	31	28	26	23	20	17	15	5	37	32
S	15	12	10	8	7	5	4	3	4	5	9	13	19	24	29	32	34	33	31	29	26	23	20	17	17	3	34	31
SW	22	18	15	12	10	8	6	5	5	6	7	9	12	18	24	32	38	43	45	44	40	35	30	26	19	5	45	40
W	25	21	17	14	11	9	7	6	6	6	7	9	11	14	20	27	36	43	49	49	45	40	34	29	20	6	49	43
NW	20	17	14	11	9	7	6	5	5	5	6	8	10	13	16	20	26	32	37	38	36	32	28	24	20	5	38	33
Group F Walls																												
N	8	6	5	3	2	1	2	4	6	7	9	11	14	17	19	21	22	23	24	23	20	16	13	11	19	1	23	23
NE	9	7	5	3	2	1	5	14	23	28	30	29	28	27	27	27	27	26	24	22	19	16	13	11	11	1	30	29
E	10	7	6	4	3	2	6	17	28	38	44	45	43	39	36	34	32	30	27	24	21	17	15	12	12	2	45	43
SE	10	7	6	4	3	2	4	10	19	28	36	41	43	42	39	36	34	31	28	25	21	18	15	12	13	2	43	41
S	10	8	6	4	3	2	1	1	3	7	13	20	27	34	38	39	38	35	31	26	22	18	15	12	16	1	39	38
SW	15	11	9	6	5	3	2	2	4	5	8	11	17	26	35	44	50	53	52	45	37	28	23	18	18	2	53	48
W	17	13	10	7	5	4	3	3	4	6	8	11	14	20	28	39	49	57	60	54	43	34	27	21	19	3	60	57
NW	14	10	8	6	4	3	2	2	3	5	8	10	13	15	21	27	35	42	46	43	35	28	22	18	19	2	46	44
Group G Walls																												
N	3	2	1	0	−1	2	7	8	9	12	15	18	21	23	24	24	25	26	22	15	11	9	7	5	18	−1	26	27
NE	3	2	1	0	−1	9	27	36	39	35	30	26	26	27	27	26	25	22	18	14	11	9	7	5	9	−1	39	40
E	4	2	1	0	−1	11	31	47	54	55	50	40	33	31	30	29	27	24	19	15	12	10	8	6	10	−1	55	56
SE	4	2	1	0	−1	5	18	32	42	49	51	48	42	36	32	30	27	24	19	15	12	10	8	6	11	−1	51	52
S	4	2	1	0	−1	0	1	5	12	22	31	39	45	46	43	37	31	25	20	15	12	10	8	5	14	−1	46	47
SW	5	4	3	1	0	0	2	5	8	12	16	26	38	50	59	63	61	52	37	24	17	13	10	8	16	0	63	63
W	6	5	3	2	1	1	2	5	8	11	15	19	27	41	56	67	72	67	48	29	20	15	11	8	17	1	72	71
NW	5	3	2	1	0	0	2	5	8	11	15	18	21	27	37	47	55	55	41	25	17	13	10	7	18	0	55	55

Part B.

1. *Direct Application of the Table Without Adjustments*

Values were calculated using the same conditions for walls as outlined for the roof CLTD table, Table 5.28. These values may be used for all normal air conditioning estimates usually without correction (except as noted below) when the load is calculated for the hottest weather.

For totally shaded walls use the north orientation values.

2. *Adjustments to Table Values*

The following equation makes adjustments for conditions other than those listed in Note 1.

$$CLTD_{corr} = (CLTD + LM) \times K + (78 - T_R) + (T_o - 85)$$

where CLTD is from this Table at the wall orientation and

(a) LM is the latitude-month correction from Table 5.32.

(b) K is a color adjustment factor and is applied after first making month-latitude adjustment

K = 1.0 if dark-colored or light in an industrial area

K = 0.83 if permanently medium-colored (rural area)

K = 0.65 if permanently light-colored (rural area)

Credit should not be taken for wall color other than dark except where permanence of color is established by experience, as in rural areas or whee there is little smoke.

Colors: Light—Cream

Medium—Medium blue, medium green, bright red, light brown, unpainted wood, and natural color concrete

Dark—Dark blue, red, brown, and green

(c) $(78 - T_R)$ is indoor design temperature correction.

(d) $(T_o - 85)$ is outdoor design temperature correction, where T_o is the average outside temperature on design day.

3. *Wall Construction Not Listed*

The U values listed are to be used only as guides. The actual value of U as obtained from tables (Chapter 4) or as calculated for the actual wall structure should be used.

An actual wall construction not listed in this table (or Table 5.26) would be thermally similar to a wall in the table, if it has similar mass (lb/ft^2) and similar heat capacity (Btu/ft^2 F). In that case, use the CLTD from this table as corrected by Note 2.

4. *Additional Insulation*

For each 7 increase in R value due to insulation added to the wall structures in Table 5.26, use the CLTD for the wall group with the next higher letter in the alphabet (e.g., C → A). For example, move to a group B wall when the initial wall group is C. When the insulation is added to the exterior of the construction rather than the interior, use the CLTD for the wall group two letters higher. If this is not possible, due to having already selected a wall on Group A, use an effective CLTD in the load calculation as given in the following table.

CLTD, Uncorrected, When Vertical Wall Structure Is "Thermally" Heavier than Group A Due to Added Insulation

N	NE	E	SE	S	SW	W	NW
11	17	22	21	17	21	22	17

Source: Copyright © by the American Society of Heating, Refrigerating and Air Conditioning Engineers, Inc., Atlanta, GA. Reprinted by permission from the *ASHRAE Handbook of Fundamentals,* 1981.

TABLE 5.31 **CLTD Correction (°C)[a,b] for Latitude and Month Applied to Walls and Roofs, North Latitude[c] (SI Units)**

Lat.	*Month*	*N*	*NNE NNW*	*NE NW*	*ENE WNW*	*E W*	*ESE WSW*	*SE SW*	*SSE SSW*	*S*	*Hor.*
16	Dec.	−2.2	−3.3	−4.4	−4.4	−2.2	−0.5	2.2	5.0	7.2	−5.0
	Jan./Nov.	−2.2	−3.3	−3.8	−3.8	−2.2	−0.5	2.2	4.4	6.6	−3.8
	Feb./Oct.	−1.6	−2.7	−2.7	−2.2	−1.1	0.0	1.1	2.7	3.8	−2.2
	Mar./Sept.	−1.6	−1.6	−1.1	−1.1	−0.5	−0.5	0.0	0.0	0.0	−0.5
	Apr./Aug.	−0.5	0.0	−0.5	−0.5	−0.5	−1.6	−1.6	−2.7	−3.3	0.0
	May/Jul.	2.2	1.6	1.6	0.0	−0.5	−2.2	−2.7	−3.8	−3.8	0.0
	Jun.	3.3	2.2	2.2	0.5	−0.5	−2.2	−3.3	−4.4	−3.8	0.0
24	Dec.	−2.7	−3.8	−5.0	−5.5	−3.8	−1.6	1.6	5.0	7.2	−7.2
	Jan./Nov.	−2.2	−3.3	−4.4	−5.0	−3.3	−1.6	1.6	5.0	7.2	−6.1
	Feb./Oct.	−2.2	−2.7	−3.3	−3.3	−1.6	−0.5	1.6	3.8	5.5	−3.8
	Mar./Sept.	−1.6	−2.2	−1.6	−1.6	−0.5	−0.5	0.5	1.1	2.2	−1.6
	Apr./Aug.	−1.1	−0.5	0.0	−0.5	−0.5	−1.1	−0.5	−1.1	−1.6	0.0
	May/Jul.	0.5	1.1	1.1	0.0	0.0	−1.6	−1.6	−2.7	−3.3	0.5
	Jun.	1.6	1.6	1.6	0.5	0.0	−1.6	−2.2	−3.3	−3.3	0.5
32	Dec.	−2.7	−3.8	−5.5	−6.1	−4.4	−2.7	1.1	5.0	6.6	−9.4
	Jan./Nov.	−2.7	−3.8	−5.0	−6.1	−4.4	−2.2	1.1	5.0	6.6	−8.3
	Feb./Oct.	−2.2	−3.3	−3.8	−4.4	−2.2	−1.1	2.2	4.4	6.1	−5.5
	Mar./Sept.	−1.6	−2.2	−2.2	−2.2	−1.1	−0.5	1.6	2.7	3.8	−2.7
	Apr./Aug.	−1.1	−1.1	−0.5	−1.1	0.0	−0.5	0.0	0.5	0.5	−0.5
	May/Jul.	0.5	0.5	0.5	0.0	0.0	−0.5	−0.5	−1.6	−1.6	0.5
	Jun.	0.5	1.1	1.1	0.5	0.0	−1.1	−1.1	−2.2	−2.2	1.1
40	Dec.	−3.3	−4.4	−5.5	−7.2	−5.5	−3.8	0.0	3.8	5.5	−11.6
	Jan./Nov.	−2.7	−3.8	−5.5	−6.6	−5.0	−3.3	0.5	4.4	6.1	−10.5
	Feb./Oct.	−2.7	−3.8	−4.4	−5.0	−3.3	−1.6	1.6	4.4	6.6	−7.7
	Mar./Sept.	−2.2	−2.7	−2.7	−3.3	−1.6	0.5	2.2	3.8	5.5	−4.4
	Apr./Aug.	−1.1	−1.6	−1.1	−1.1	0.0	0.0	1.1	1.6	2.2	1.6
	May/Jul.	0.0	0.0	0.0	0.0	0.0	0.0	0.0	0.0	0.5	0.5
	Jun.	0.5	0.5	0.5	0.0	0.5	0.0	0.0	−0.5	−0.5	1.1
48	Dec.	−3.3	−4.4	−6.1	−7.7	−7.2	−5.5	−1.6	1.1	3.3	−13.8
	Jan./Nov.	−3.3	−4.4	−6.1	−7.2	−6.1	−4.4	−0.5	2.7	4.4	−13.3
	Feb./Oct.	−2.7	−3.8	−5.5	−6.1	−4.4	−2.7	0.5	4.4	6.1	−10.0
	Mar./Sept.	−2.2	−3.3	−3.3	−3.8	−2.2	−0.5	2.2	4.4	6.1	−6.1
	Apr./Aug.	−1.6	−1.6	−1.6	−1.6	−0.5	0.0	2.2	3.3	3.8	−2.7
	May/Jul.	0.0	−0.5	0.0	0.0	0.5	0.5	1.6	1.6	2.2	0.0
	Jun.	0.5	0.5	1.1	0.5	1.1	0.5	1.1	1.1	1.6	1.1
56	Dec.	−3.8	−5.0	−6.6	−8.8	−8.8	−7.7	−5.0	−2.7	−1.6	−15.5
	Jan./Nov.	−3.3	−4.4	−6.1	−8.3	−7.7	−6.6	−3.3	−0.5	1.1	−15.0
	Feb./Oct.	−3.3	−4.4	−5.5	−6.6	−5.5	−3.8	0.0	3.3	5.0	−12.2
	Mar./Sept.	−2.7	−3.3	−3.8	−4.4	−2.7	−1.1	2.2	4.4	6.6	−8.3
	Apr./Aug.	−1.6	−2.2	−2.2	−2.2	−0.5	0.5	2.7	3.8	5.0	−4.4
	May/Jul.	0.0	0.0	0.0	0.0	1.1	1.1	2.7	3.3	3.8	−1.1
	Jun.	1.1	0.5	1.1	0.5	1.6	1.6	2.2	2.7	3.3	0.5
64	Dec.	−3.8	−5.0	−6.6	−0.8	−9.4	−10.0	−8.8	−7.7	−6.6	−16.6
	Jan./Nov.	−3.8	−5.0	−6.6	−8.8	−8.8	−8.8	−7.2	−5.5	−4.4	−16.1
	Feb./Oct.	−3.3	−4.4	−6.1	−7.7	−7.2	−5.5	−2.2	0.5	2.2	−14.4
	Mar./Sept.	−2.7	−3.8	−5.0	−5.5	−3.8	−2.2	1.1	3.8	6.1	−11.1
	Apr./Aug.	−1.6	−2.2	−2.2	−2.2	−0.5	0.5	2.7	5.0	6.1	−6.1
	May/Jul.	0.5	0.0	0.5	0.0	1.6	2.2	3.3	4.4	5.5	−1.6
	Jun.	1.1	1.1	1.1	1.1	2.2	2.2	3.3	3.8	5.0	0.0

Source: Copyright © by the American Society of Heating, Refrigerating and Air Conditioning Engineers, Inc., Atlanta, GA. Reprinted by permission from the *ASHRAE Handbook of Fundamentals,* 1981.

[a]The correction is applied directly to the CLTD for a wall or roof as given in Tables 5.27 and 5.29.

[b]The CLTD correction given is *not* applicable to Table 5.33, Cooling Load Temperature Differences for Conduction Through Glass.

[c]For south latitudes, replace January through December by July through June.

TABLE 5.32 **CLTD Correction (F)[a,b] for Latitude and Month Applied to Walls and Roofs, North Latitudes[c] (Conventional Units)**

Lat.	Month	N	NNE NNW	NE NW	ENE WNW	E W	ESE WSW	SE SW	SSE SSW	S	Hor.
16	Dec.	−4	−6	−8	−8	−4	−1	4	9	13	−9
	Jan./Nov.	−4	−6	−7	−7	−4	−1	4	8	12	−7
	Feb./Oct.	−3	−5	−5	−4	−2	0	2	5	7	−4
	Mar./Sept.	−3	−3	−2	−2	−1	−1	0	0	0	−1
	Apr./Aug.	−1	0	−1	−1	−1	−3	−3	−5	−6	0
	May/Jul.	4	3	3	0	−1	−4	−5	−7	−7	0
	Jun.	6	4	4	1	−1	−4	−6	−8	−7	0
24	Dec.	−5	−7	−9	−10	−7	−3	3	9	13	−13
	Jan./Nov.	−4	−6	−8	−9	−6	−3	3	9	13	−11
	Feb./Oct.	−4	−5	−6	−6	−3	−1	3	7	10	−7
	Mar./Sept.	−3	−4	−3	−3	−1	−1	1	2	4	−3
	Apr./Aug.	−2	−1	0	−1	−1	−2	−1	−2	−3	0
	May/Jul.	1	2	2	0	0	−3	−3	−5	−6	1
	Jun.	3	3	3	1	0	−3	−4	−6	−6	1
32	Dec.	−5	−7	−10	−11	−8	−5	2	9	12	−17
	Jan./Nov.	−5	−7	−9	−11	−8	−4	2	9	12	−15
	Feb./Oct.	−4	−6	−7	−8	−4	−2	4	8	11	−10
	Mar./Sept.	−3	−4	−4	−4	−2	−1	3	5	7	−5
	Apr./Aug.	−2	−2	−1	−2	0	−1	0	1	1	−1
	May/Jul.	1	1	1	0	0	−1	−1	−3	−3	1
	Jun.	1	2	2	1	0	−2	−2	−4	−4	2
40	Dec.	−6	−8	−10	−13	−10	−7	0	7	10	−21
	Jan./Nov.	−5	−7	−10	−12	−9	−6	1	8	11	−19
	Feb./Oct.	−5	−7	−8	−9	−6	−3	3	8	12	−14
	Mar./Sept.	−4	−5	−5	−6	−3	−1	4	7	10	−8
	Apr./Aug.	−2	−3	−2	−2	0	0	2	3	4	−3
	May/Jul.	0	0	0	0	0	0	0	0	1	1
	Jun.	1	1	1	0	1	0	0	−1	−1	2
48	Dec.	−6	−8	−11	−14	−13	−10	−3	2	6	−25
	Jan./Nov.	−6	−8	−11	−13	−11	−8	−1	5	8	−24
	Feb./Oct.	−5	−7	−10	−11	−8	−5	1	8	11	−18
	Mar./Sept.	−4	−6	−6	−7	−4	−1	4	8	11	−11
	Apr./Aug.	−3	−3	−3	−3	−1	0	4	6	7	−5
	May/Jul.	0	−1	0	0	1	1	3	3	4	0
	Jun.	1	1	2	1	2	1	2	2	3	2
56	Dec.	−7	−9	−12	−16	−16	−14	−9	−5	−3	−28
	Jan./Nov.	−6	−8	−11	−15	−14	−12	−6	−1	2	−27
	Feb./Oct.	−6	−8	−10	−12	−10	−7	0	6	9	−22
	Mar./Sept.	−5	−6	−7	−8	−5	−2	4	8	12	−15
	Apr./Aug.	−3	−4	−4	−4	−1	1	5	7	9	−8
	May/Jul.	0	0	0	0	2	2	5	6	7	−2
	Jun.	2	1	2	1	3	3	4	5	6	1
64	Dec.	−7	−9	−12	−16	−17	−18	−16	−14	−12	−30
	Jan./Nov.	−7	−9	−12	−16	−16	−16	−13	−10	−8	−29
	Feb./Oct.	−6	−8	−11	−14	−13	10	−4	1	4	−26
	Mar./Sept.	−5	−7	−9	−10	−7	−4	2	7	11	−20
	Apr./Aug.	−3	−4	−4	−4	−1	1	5	9	11	−11
	May/Jul.	1	0	1	0	3	4	6	8	10	−3
	Jun.	2	2	2	2	4	4	6	7	9	0

Source: Copyright © by the American Society of Heating, Refrigerating and Air Conditioning Engineers, Inc., Atlanta, GA. Reprinted by permission from the *ASHRAE Handbook of Fundamentals,* 1981.

[a]The correction is applied directly to the CLTD for a wall or roof as given in Tables 5.28 and 5.30.

[b]The CLTD correction given is *not* applicable to Table 5.33, Cooling Load Temperature Differences for Conduction Through Glass.

[c]For south latitudes, replace January through December by July through June.

EXAMPLE 5.11. (Copyright © by the American Society of Heating, Refrigerating and Air Conditioning Engineers, Inc., Atlanta, GA. Reprinted by permission from the *ASHRAE Handbook of Fundamentals*, 1981.)

A building 9.14 m × 30.48 m × 3.04 m (30 ft × 100 ft × 10 ft) is located at 32°N latitude in an area with a design outside dry-bulb temperature of 32.2°C (90 F) and a daily range of 11.1C° (20 F°). The inside design dry-bulb temperature is 25.5°C (78 F). Determine the cooling load as a result of heat gain through the roof and south and west walls at 1200, 1400, and 1600 hours for August 21. The south wall measures 30.48 m × 3.04 m (100 ft × 10 ft); the west wall, 9.14 m × 3.04 m (30 ft × 10 ft). Further constructions are

Roof. 12.7-mm (0.5-in.) slag plus 9.5-mm (0.375-in.) felt and membrane, 50.8-mm (2-in.) heavyweight concrete, 50.8-mm (2-in.) insulation [R = 0.327(6.7)], and a suspended ceiling of acoustic tile. Assume f = 1.0; no attic or ducts; see notes to Table 5.27.

South Wall. 203.2-mm (8-in.) heavyweight concrete with 50.8-mm (2-in.) insulation [R = 1.18(6.7)] and interior finish.

West Wall. Stucco on 101.6-mm (4-in.) L.W. concrete block with 25.4-mm (1-in.) insulation [R = 0.58(3.3)] and interior finish [R = 0.026(0.15)].

SOLUTION. The U values and respective areas are as follows.

	Roof (No. 4)	South Wall (Group B)	West Wall (Group F)
U^a	0.516 (0.091)	0.653 (0.115)	0.914 (0.161)
Area^b	278.5 (3000)	92.7 (1000)	27.9 (300)

^aU = W/m² °C (Btu/h ft² F).
^{b}Area = m²(ft²).

The correction factors to be applied to the tabulated CLTDs are:

1. Correction factor for outside conditions:

average temperature =

$$32.2 - 0.5(11.1) = 26.7°\text{C (80 F)}$$

$$\text{correction} = 26.7 - 29.4 = -2.7\text{C}° \ (-5\ \text{F}°).$$

2. Correction factor for inside design dry-bulb temperature:

$$\text{correction} = (25.5 - 25.5) = 0\text{C}° \ (0\text{F}°).$$

3. Latitude-month correction (Table 5.31):

$$\text{Roof} = -0.5\ (-1)$$
$$\text{South wall} = 0.5\ (1)$$
$$\text{West wall} = 0.0\ (0)$$

4. Color correction:

$$\text{dark roof } K = 1.0$$
$$\text{medium walls } K = 0.83$$

5. Total correction (SI units):

$$\text{roof, CLTD}_{\text{corr}} = [(\text{CLTD} - 0.5) \times 1.0 + (0) + (-2.7)] \times 1.0 = \text{CLTD} - 3.2$$

$$\text{south wall, CLTD}_{\text{corr}} = (\text{CLTD} + 0.5) \times 0.83 + (0) + (-2.7) = (\text{CLTD} + 0.5) \times 0.83 - 2.7$$

$$\text{west wall, CLTD}_{\text{corr}} = (\text{CLTD} - 0.0) \times 0.83 + (0) + (-2.7) = \text{CLTD} \times 0.83 - 2.7$$

Note: CLTD corrections in English units were derived by multiplying the SI corrections by 1.8.

Roof

Time	Table 5.27 CLTD	CLTD Corrected	$q = U \times A \times$ CLTD [W (Btu/h)]
1200	14 (25)	10.8 (19.4)	1552 (5296)
1400	19 (35)	15.8 (28.4)	2272 (7753)
1600	24 (43)	20.8 (37.4)	2990 (10 210)

South Wall

Time	Table 5.29 CLTD	CLTD Corrected	$q = U \times A \times$ CLTD [W (Btu/h)]
1200	6 (11)	2.7 (4.9)	163 (564)
1400	7 (12)	3.5 (6.3)	212 (725)
1600	9 (15)	5.2 (9.4)	315 (1081)

West Wall

Time	Table 5.29 CLTD	CLTD Corrected	$q = U \times A \times$ CLTD [W (Btu/h)]
1200	6 (11)	2.3 (4.1)	58 (198)
1400	11 (20)	6.4 (11.5)	163 (555)
1600	22 (39)	15.6 (28.1)	398 (1357)

Note that during the time period from noon to 4 P.M., the cooling loads through the roof and south wall nearly doubled, whereas those through the west wall quintupled.

TABLE 5.33 **Cooling Load Temperature Differences (CLTD) for Conduction Through Glass**

Solar time (h)	1	2	3	4	5	6	7	8	9	10	11	12	13	14	15	16	17	18	19	20	21	22	23	24
CLTD																								
C°	1	0	−1	−1	−1	−1	−1	0	1	2	4	5	7	7	8	8	7	7	6	4	3	2	2	1
F°	1	0	−1	−2	−2	−2	−2	0	2	4	7	9	12	13	14	14	13	12	10	8	6	4	3	2

Corrections: The values in the table were calculated for an inside temperature of 25.5°C (78 F) and an outdoor maximum temperature of 35°C (95 F) with an outdoor daily range of 11.6C° (21F°). The table remains approximately correct for other outdoor maximums 33.8–38.8°C (93–102 F) and other outdoor daily ranges 8.9–18.9C° (16–34F°), provided the outdoor daily average temperature remains approximately 29.4°C (85 F). If the room air temperature is different from 25.5°C (78 F) and/or the outdoor daily average temperature is different from 29.4°C (85 F), the following rules apply: (a) For room air temperature less than 25.5°C (78 F), add the difference between 25.5°C (78 F) and room air temperature; if greater than 25.5°C (78 F), subtract the difference. (b) For outdoor daily average temperature less than 29.4°C (85 F), subtract the difference between 29.4°C (85 F) and the daily average temperature; if greater than 29.4°C (85 F), add the difference.

Source: Copyright © by the American Society of Heating, Refrigerating and Air Conditioning Engineers, Inc., Atlanta, GA. Reprinted by permission from the *ASHRAE Handbook of Fundamentals,* 1981.

(b) Gains from Adjacent Spaces. If adjacent spaces are at a higher temperature than the cooled space, the adjacent spaces will yield heat gains. These gains, through the walls, ceilings, or floors that separate the spaces, are calculated in the simple, familiar manner:

$$q = UA\Delta T$$

where ΔT is the temperature difference between the cooled space and the adjacent space.

Where floors are either in direct contact with the ground, or over an underground basement that is neither ventilated nor warmed, do not assume any heat gain through floors.

(c) Gains Through Glass. There are heat gains through glass by conduction from hot air outside to colder inside and by solar radiation. The CLTD for conduction through glass (Table 5.33) is used in a calculation similar to simple opaque conduction gain:

$$q = U \times A \times \text{CLTD}$$

The gains due to solar radiation are much more complex, because they must include variations in type of window, shading devices, and the incident angles and intensities of sunlight. ASHRAE has combined some of these variations into solar heat gain factors (SHGF), which are listed in Tables 5.34–5.36. (Note that these SHGF are listed only for the *maximum* hour; later in this process, they are adjusted to account for the *actual* hour of calculation.)

Further adjustment is needed for different fenestration types and shading devices; this is accomplished by the *shading coefficient* (SC), listed in Tables 5.37 through 5.42. The shading coefficient for any fenestration is:

$$\text{SC} = \frac{\text{solar heat gain of fenestration}}{\text{solar heat gain of double-strength glass}}$$

(More detailed SCs are listed in the ASHRAE *Handbook of Fundamentals*.) In addition, cooling load factors (CLF) must be used, along with SHGF, to account for delay in the impact of incoming solar radiation on the air temperature of a cooled space. These are presented for two basic cases: with interior shading, and without (in which case the thermal massiveness of the room construction becomes influential). These CLF values are listed in Tables 5.43 and 5.44.

TABLE 5.34 **Maximum Solar Heat Gain Factor (W/m^2) for Sunlit Glass, North Latitudes**[a]

16 Deg

	N	NNE/ NNW	NE/ NW	ENE/ WNW	E/ W	ESE/ WSW	SE/ SW	SSE/ SSW	S	Hor.
Jan.	95	95	174	464	663	770	792	704	628	782
Feb.	104	104	303	568	729	779	735	593	486	868
Mar.	110	167	441	647	745	741	622	435	293	918
Apr.	123	312	543	681	716	644	473	243	142	912
May	164	416	596	688	678	565	363	142	129	890
June	208	448	612	685	653	527	312	129	129	874
July	174	416	590	675	663	549	350	139	133	874
Aug.	129	316	530	659	691	644	451	233	145	890
Sep.	114	158	423	618	716	707	603	423	293	890
Oct.	104	104	300	549	704	748	710	577	473	852
Nov.	95	95	174	457	650	760	779	694	618	776
Dec.	91	91	129	416	625	760	801	735	669	738

24 Deg

	N	NNE/ NNW	NE/ NW	ENE/ WNW	E/ W	ESE/ WSW	SE/ SW	SSE/ SSW	S	Hor.
Jan.	85	85	129	404	599	757	798	760	716	675
Feb.	95	95	252	521	694	770	767	672	606	786
Mar.	107	142	391	615	738	748	675	530	432	868
Apr.	117	278	502	659	719	669	533	338	237	893
May	136	369	562	675	688	599	416	211	145	890
June	174	401	581	675	669	565	369	174	136	880
July	142	366	555	663	672	584	407	205	145	877
Aug.	120	274	492	640	694	644	511	325	227	874
Sep.	110	133	375	584	700	710	650	514	423	839
Oct.	98	98	249	502	666	748	741	653	590	770
Nov.	85	85	133	398	590	745	786	748	707	672
Dec.	82	82	91	353	568	738	779	779	748	628

32 Deg

	N (Shade)	NNE/ NNW	NE/ NW	ENE/ WNW	E/ W	ESE/ WSW	SE/ SW	SSE/ SSW	S	Hor.
Jan.	76	76	91	331	552	722	786	789	776	555
Feb.	85	85	205	470	647	764	782	732	697	685
Mar.	101	117	338	577	716	748	716	615	555	795
Apr.	114	252	461	631	716	691	590	445	363	855
May	120	350	536	656	694	628	489	312	233	874
June	139	385	555	656	675	596	439	262	189	871
July	126	350	527	643	678	612	473	303	227	861
Aug.	117	249	445	615	691	663	571	429	350	836
Sep.	104	110	325	546	678	716	688	596	540	770
Oct.	88	88	199	451	615	738	754	710	678	672
Nov.	76	76	91	325	546	710	773	776	767	552
Dec.	69	69	69	265	511	688	776	795	795	498

40 Deg

	N (Shade)	NNE/ NNW	NE/ NW	ENE/ WNW	E/ W	ESE/ WSW	SE/ SW	SSE/ SSW	S	Hor.
Jan.	63	63	53	233	486	647	760	795	801	420
Feb.	76	76	158	407	587	738	776	770	760	568
Mar.	91	91	293	533	688	751	745	681	650	704
Apr.	107	224	441	599	707	704	640	536	486	795
May	117	322	521	637	694	656	552	420	357	836
June	151	357	543	647	681	628	508	366	300	842
July	120	322	514	625	681	641	536	681	344	827
Aug.	110	224	426	584	681	675	618	536	470	779
Sep.	95	95	274	505	640	716	713	659	631	678
Oct.	79	79	154	388	568	710	751	745	738	558
Nov.	63	63	63	230	476	634	748	782	789	416
Dec.	57	57	57	189	476	593	732	786	798	357

48 Deg

	N (Shade)	NNE/ NNW	NE/ NW	ENE/ WNW	E/ W	ESE/ WSW	SE/ SW	SSE/ SSW	S	Hor.
Jan.	47	47	47	167	372	552	681	754	773	268
Feb.	63	63	114	325	530	681	764	786	789	435
Mar.	82	82	252	486	644	738	754	732	719	593
Apr.	98	192	416	568	691	710	678	612	587	713
May	110	306	498	631	290	675	606	514	473	779
June	145	347	521	644	678	650	568	467	423	795
July	117	503	492	618	675	659	590	498	461	770
Aug.	104	192	404	549	666	681	656	593	568	704
Sep.	85	41	227	454	290	704	719	704	694	574
Oct.	66	66	110	303	508	653	735	760	764	42
Nov.	47	47	492	164	363	543	669	738	757	268
Dec.	41	41	41	114	287	492	615	710	735	205

56 Deg

	N (Shade)	NNE/ NNW	NE/ NW	ENE/ WNW	E/ W	ESE/ WSW	SE/ SW	SSE/ SSW	S	Hor.
Jan.	32	32	32	66	233	398	533	612	647	126
Feb.	50	50	66	224	439	581	704	754	770	287
Mar.	69	69	205	429	584	707	751	760	760	470
Apr.	88	183	388	546	666	704	704	672	663	615
May	114	312	470	615	678	688	650	590	571	700
June	167	350	505	628	672	672	618	549	530	729
July	117	309	464	606	666	675	634	577	558	697
Aug.	95	177	375	521	640	681	678	650	640	609
Sep.	73	73	183	398	540	666	650	726	729	454
Oct.	50	50	63	215	416	555	672	722	738	287
Nov.	32	32	32	66	227	385	521	599	631	126
Dec.	22	22	22	22	148	290	426	502	540	73

64 Deg

	N (Shade)	NNE/ NNW	NE/ NW	ENE/ WNW	E/ W	ESE/ WSW	SE/ SW	SSE/ SSW	S	Hor.
Jan.	9	9	9	9	47	142	211	281	303	25
Feb.	35	35	35	136	281	454	558	637	663	142
Mar.	57	57	148	357	502	640	713	745	754	331
Apr.	79	186	357	514	634	691	710	710	707	505
May	151	306	473	596	666	694	678	653	644	606
June	196	360	511	609	672	681	656	618	609	640
July	155	303	467	587	653	678	666	637	631	606
Aug.	85	183	344	495	609	666	685	685	685	502
Sep.	60	60	136	325	467	596	672	707	716	319
Oct.	35	35	35	126	262	426	527	603	628	145
Nov.	13	13	13	13	47	139	208	274	293	25
Dec.	0	0	0	0	3	15	35	44	47	3

Source: Copyright © by the American Society of Heating, Refrigerating and Air Conditioning Engineers, Inc., Atlanta, GA. Reprinted by permission from the *ASHRAE Handbook of Fundamentals,* 1981.

[a]Solar gains through one layer of double-strength clear sheet glass (3.175-mm). See also Appendix B.

TABLE 5.35 **Maximum Solar Heat Gain Factor (Btu/h ft^2) for Sunlit Glass, North Latitudes**[a]

16 Deg	N	NNE/ NNW	NE/ NW	ENE/ WNW	E/ W	ESE/ WSW	SE/ SW	SSE/ SSW	S	Hor.
Jan.	30	30	55	147	210	244	251	223	199	248
Feb.	33	33	96	180	231	247	233	188	154	275
Mar.	35	53	140	205	239	235	197	138	93	291
Apr.	39	99	172	215	227	204	150	77	45	289
May	52	132	189	218	215	179	115	45	41	282
June	66	142	194	217	207	167	99	41	41	277
July	55	132	187	214	210	174	111	44	42	277
Aug.	41	100	168	209	219	196	143	74	46	282
Sep.	36	50	134	196	227	224	191	134	93	282
Oct.	33	33	95	174	223	237	225	183	150	270
Nov.	30	30	55	145	206	241	247	220	196	246
Dec.	29	29	41	132	198	241	254	233	212	234

24 Deg	N	NNE/ NNW	NE/ NW	ENE/ WNW	E/ W	ESE/ WSW	SE/ SW	SSE/ SSW	S	Hor.
Jan.	27	27	41	128	190	240	253	241	227	214
Feb.	30	30	80	165	220	244	243	213	192	249
Mar.	34	45	124	195	234	237	214	168	137	275
Apr.	37	88	159	209	228	212	169	107	75	283
May	43	117	178	214	218	190	132	67	46	282
June	55	127	184	214	212	179	117	55	43	279
July	45	116	176	210	213	185	129	65	46	278
Aug.	38	87	156	203	220	204	162	103	72	277
Sep.	35	42	119	185	222	225	206	163	134	266
Oct.	31	31	79	159	211	237	235	207	187	244
Nov.	27	27	42	126	187	236	249	237	224	213
Dec.	26	26	29	112	180	234	247	247	237	199

32 Deg	N (Shade)	NNE/ NNW	NE/ NW	ENE/ WNW	E/ W	ESE/ WSW	SE/ SW	SSE/ SSW	S	Hor.
Jan.	24	24	29	105	175	229	249	250	246	176
Feb.	27	27	65	149	205	242	248	232	221	217
Mar.	32	37	107	183	227	237	227	195	176	252
Apr.	36	80	146	200	227	219	187	141	115	271
May	38	111	170	208	220	199	155	99	74	277
June	44	122	176	208	214	189	139	83	60	276
July	40	111	167	204	215	194	150	96	72	273
Aug.	37	79	141	195	219	210	181	136	111	265
Sep.	33	35	103	173	215	227	218	189	171	244
Oct.	28	28	63	143	195	234	239	225	215	213
Nov.	24	24	29	103	173	225	245	246	243	175
Dec.	22	22	22	84	162	218	246	252	252	158

40 Deg	N (Shade)	NNE/ NNW	NE/ NW	ENE/ WNW	E/ W	ESE/ WSW	SE/ SW	SSE/ SSW	S	Hor.
Jan.	20	20	20	74	154	205	241	252	254	133
Feb.	24	24	50	129	186	234	246	244	241	180
Mar.	29	29	93	169	218	238	236	216	206	223
Apr.	34	71	140	190	224	223	203	170	154	252
May	37	102	165	202	220	208	175	133	113	265
June	48	113	172	205	216	199	161	116	95	267
July	38	102	163	198	216	203	170	129	109	262
Aug.	35	71	135	185	216	214	196	165	149	247
Sep.	30	30	87	160	203	227	226	209	200	215
Oct.	25	25	49	123	180	225	238	236	234	177
Nov.	20	20	20	73	151	201	237	248	250	132
Dec.	18	18	18	60	135	188	232	249	253	113

48 Deg

	N (Shade)	NNE/ NNW	NE/ NW	ENE/ WNW	E/ W	ESE/ WSW	SE/ SW	SSE/ SSW	S	Hor.
Jan.	15	15	15	53	118	175	216	239	245	85
Feb.	20	20	36	103	168	216	242	249	250	138
Mar.	26	26	80	154	204	234	239	232	228	188
Apr.	31	61	132	180	219	225	215	194	186	226
May	35	97	158	200	218	214	192	163	150	247
June	46	110	165	204	215	206	180	148	134	252
July	37	96	156	196	214	209	187	158	146	244
Aug.	33	61	128	174	211	216	208	188	180	223
Sep.	27	27	72	144	191	223	228	223	220	182
Oct.	21	21	35	96	161	207	233	241	242	136
Nov.	15	15	15	52	115	172	212	234	240	85
Dec.	13	13	13	36	91	156	195	225	233	65

56 Deg

	N (Shade)	NNE/ NNW	NE/ NW	ENE/ WNW	E/ W	ESE/ WSW	SE/ SW	SSE/ SSW	S	Hor.
Jan.	10	10	10	21	74	126	169	194	205	40
Feb.	16	16	21	71	139	184	223	239	244	91
Mar.	22	22	65	136	185	224	238	241	241	149
Apr.	28	58	123	173	211	223	223	213	210	195
May	36	99	149	195	215	218	206	187	181	222
June	53	111	160	199	213	213	196	174	168	231
July	37	98	147	192	211	214	201	183	177	221
Aug.	30	56	119	165	203	216	215	206	203	193
Sep.	23	23	58	126	171	211	227	230	231	144
Oct.	16	16	20	68	132	176	213	229	234	91
Nov.	10	10	10	21	72	122	165	190	200	40
Dec.	7	7	7	7	47	92	135	159	171	23

64 Deg

	N (Shade)	NNE/ NNW	NE/ NW	ENE/ WNW	E/ W	ESE/ WSW	SE/ SW	SSE/ SSW	S	Hor.
Jan.	3	3	3	3	15	45	67	89	96	8
Feb.	11	11	11	43	89	144	177	202	210	45
Mar.	18	18	47	113	159	203	226	236	239	105
Apr.	25	59	113	163	201	219	225	225	224	160
May	48	97	150	189	211	220	215	207	204	192
June	62	114	162	193	213	216	208	196	193	203
July	49	96	148	186	207	215	211	202	200	192
Aug.	27	58	109	157	193	211	217	217	217	159
Sep.	19	19	43	103	148	189	213	224	227	101
Oct.	11	11	11	40	83	135	167	191	199	46
Nov.	4	4	4	4	15	44	66	87	93	8
Dec.	0	0	0	0	1	5	11	14	15	1

Source: Copyright © by the American Society of Heating, Refrigerating and Air Conditioning Engineers, Inc., Atlanta, GA. Reprinted by permission from the *ASHRAE Handbook of Fundamentals*, 1981.

[a]Solar gains through one layer of double-strength clear sheet glass (0.125 in.) See also Appendix B.

TABLE 5.36 **Maximum Solar Heat Gain Factor for Externally Shaded Glass (Based on Ground Reflectance of 0.2)**

Part A. **W/m²**

Use for latitudes 0–24°
For latitudes greater than 24, use north orientation, Table 5.34.
For horizontal glass in shade, use the tabulated values for all latitudes.

	N	NNE/ NNW	NE/ NW	ENE/ WNW	E/ W	ESE/ WSW	SE/ SW	SSE/ SSW	S	(All Latit.) Hor.
Jan.	98	98	98	101	107	114	117	117	120	50
Feb.	107	107	107	110	114	117	120	120	123	50
Mar.	114	114	117	120	123	126	126	123	123	60
Apr.	126	126	130	133	133	133	129	126	126	76
May	137	139	142	145	142	136	129	126	126	88
June	142	145	148	148	145	139	129	126	126	98
July	142	142	145	148	148	142	133	129	129	98
Aug.	133	133	136	142	145	142	136	133	133	88
Sep.	117	117	120	126	129	133	133	129	129	73
Oct.	107	107	107	114	120	123	126	126	126	60
Nov.	101	101	101	101	107	114	120	120	123	54
Dec.	95	95	95	98	101	107	114	117	117	47

Part B. **Btu/h ft²**

Use for Latitudes 0–24°.
For latitudes greater than 24, use north orientation, Table 5.35.
For horizontal glass in shade, use the tabulated values for all latitudes.

	N	NNE/ NNW	NE/ NW	ENE/ WNW	E/ W	ESE/ WSW	SE/ SW	SSE/ SSW	S	(All Latit.) Hor.
Jan.	31	31	31	32	34	36	37	37	38	16
Feb.	34	34	34	35	36	37	38	38	39	16
Mar.	6	36	37	38	39	40	40	39	39	19
Apr.	40	40	41	42	42	42	41	40	40	24
May	43	44	45	46	45	43	41	40	40	28
June	45	46	47	47	46	44	41	40	40	31
July	45	45	46	47	47	45	42	41	41	31
Aug.	42	42	43	45	46	45	43	42	42	28
Sep.	37	37	38	40	41	42	42	41	41	23
Oct.	34	34	34	36	38	39	40	40	40	19
Nov.	32	32	32	32	34	36	38	38	39	17
Dec.	30	30	30	31	32	34	36	37	37	15

Source: Copyright © by the American Society of Heating, Refrigerating and Air Conditioning Engineers, Inc., Atlanta, GA. Reprinted by permission from the *ASHRAE Handbook of Fundamentals,* 1981.

TABLE 5.37 **Shading Coefficients for Single Glass and Insulating Glass**[a]

Part A. ***Single Glass***

Type of Glass	*Nominal Thickness*[b]	*Solar Trans.*[b]	*Shading Coefficient, SC*[c]
Clear	3 mm (1/8 in.)	0.86	1.00
	6 mm (1/4 in.)	0.78	0.94
	10 mm (3/8 in.)	0.72	0.90
	12 mm (1/2 in.)	0.67	0.87
Heat absorbing	3 mm (1/8 in.)	0.64	0.83
	6 mm (1/4 in.)	0.46	0.69
	10 mm (3/8 in.)	0.33	0.60
	12 mm (1/2 in.)	0.24	0.53

Part B. ***Insulating Glass***

Clear out, clear in	3 mm (1/8 in.)[d]	0.71[e]	0.88
Clear out, clear in	6 mm (1/4 in.)	0.61	0.81
Heat absorbing[f] out, clear in	6 mm (1/4 in.)	0.36	0.55

Source: Copyright © by the American Society of Heating, Refrigerating and Air Conditioning Engineers, Inc., Atlanta, GA. Reprinted by permission from the *ASHRAE Handbook of Fundamentals,* 1981.

[a]Refers to factory-fabricated units with 5-, 6-, or 12-mm (3/16-, 1/4-, or 1/2-in.) air space or to prime windows plus storm sash.

[b]Refer to manufacturer's literature for values.

[c]Based on outdoor air at 12 km/h (7.5 mph) and still air indoors.

[d]Thickness of each pane of glass, not thickness of assembled unit.

[e]Combined transmittance for assembled unit.

[f]Refers to gray-, bronze-, and green-tinted heat-absorbing float glass.

THERMAL CONTROL

TABLE 5.38 **Solar Optical Properties and Shading Coefficients of Transparent Plastic Sheeting**

Type of Plastic	*Transmittance* Visible	*Transmittance* Solar	SC
Acrylic			
Clear	0.92	0.85	0.98
Gray tint	0.16	0.27	0.52
Gray tint	0.33	0.41	0.63
Gray tint	0.45	0.55	0.74
Gray tint	0.59	0.62	0.80
Gray tint	0.76	0.74	0.89
Bronze tint	0.10	0.20	0.46
Bronze tint	0.27	0.35	0.58
Bronze tint	0.49	0.56	0.75
Bronze tint	0.61	0.62	0.80
Bronze tint	0.75	0.75	0.90
Reflective[a]	0.14	0.12	0.21
Polycarbonate			
Clear, 3-mm (1/8-in.)	0.88	0.82	0.98
Gray, 3-mm (1/8-in.)	0.50	0.57	0.74
Bronze, 3-mm (1/8-in.)	0.50	0.57	0.74

Source: Copyright © by the American Society of Heating, Refrigerating and Air Conditioning Engineers, Inc., Atlanta, GA. Reprinted by permission from the *ASHRAE Handbook of Fundamentals,* 1981.

[a]Aluminum metallized polyester film on plastic.

TABLE 5.39 **Shading Coefficients (SC)**

Part A. ***Shading Coefficients for Single Glass with Indoor Shading by Venetian Blinds or Roller Shades***

Type of Glass	*Nominal Thickness[a], [mm (in.)]*	*Solar Trans[b]*	*Venetian blinds[b]* Medium	*Venetian blinds[b]* Light	*Roller Shade, Opaque* Dark	*Roller Shade, Opaque* White	*Roller Shade, Translucent* Light
Clear	2.5 to 6 (3/32 to 1/4)	0.87 to 0.80					
Clear	6 to 12 (1/4 to 1/2)	0.80 to 0.71					
Clear pattern	3 to 12 (1/8 to 1/2)	0.87 to 0.79	0.64	0.55	0.59	0.25	0.39
Heat-absorbing pattern	3 (1/8)	—					
Tinted	5 to 5.5 (3/16, 7/32)	0.74, 0.71					
Heat-absorbing[c]	5 to 6 (3/16, 1/4)	0.46					
Heat-absorbing pattern	5 to 6 (3/16, 1/4)	—	0.57	0.53	0.45	0.30	0.36
Tinted	3 to 5.5 (1/8, 7/32)	0.59, 0.45					
Heat-absorbing or pattern	—	0.44 to 0.30	0.54	0.52	0.40	0.28	0.32
Heat-absorbing[c]	10 (3/8)	0.34					
Heat-absorbing or		0.29 to 0.15					
pattern	—	0.24	0.42	0.40	0.36	0.28	0.31
Reflective coated glass							
SC[d] = 0.30			0.25	0.23			
0.40			0.33	0.29			
0.50			0.42	0.38			
0.60			0.50	0.44			

Part B. ***Shading Coefficients for Insulating Glass[e] with Indoor Shading by Venetian Blinds or Roller Shades***

				Type of Shading				
		Solar Trans.[a]		Venetian Blinds[f]		Roller Shade		
						Opaque		Translucent
Type of Glass	Nominal Thickness, Each Light	Outer Pane	Inner Pane	Medium	Light	Dark	White	Light
Clear out	2.5, 3 mm (3/32, 1/8 in.)	0.87	0.87					
Clear in				0.57	0.51	0.60	0.25	0.37
Clear out								
Clear in	6 mm (1/4 in.)	0.80	0.80					
Heat-absorbing[c] out	6 mm (1/4 in.)	0.46	0.80	0.39	0.36	0.40	0.22	0.30
Clear in								
Reflective coated glass								
SC[d] = 0.20				0.19	0.18			
0.30				0.27	0.26			
0.40				0.34	0.33			

Part C. ***Shading Coefficients for Double Glazing with Between-Glass Shading***

		Solar Trans.[a]			Type of Shading		
					Venetian Blinds		
Type of Glass	Nominal Thickness, Each Pane	Outer Pane	Inner Pane	Description of Air Space	Light	Medium	Louvered Sun Screen
Clear out	2.5, 3 mm (3/32, 1/8 in.)	0.87	0.87	Shade in contact with glass or shade separated from glass by air space	0.33	0.36	0.43
Clear in							
Clear out	6 mm (1/4 in.)	0.80	0.80	Shade in contact with glass-voids filled with plastic	—	—	0.49
Clear in							
Heat-absorbing[c] out				Shade in contact with glass or shade separated from glass by air space	0.28	0.30	0.37
Clear in	6 mm (1/4 in.)	0.46	0.80	Shade in contact with glass-voids filled with plastic	—	—	0.41

Source: Copyright © by the American Society of Heating, Refrigerating and Air Conditioning Engineers, Inc., Atlanta, GA. Reprinted by permission from the *ASHRAE Handbook of Fundamentals,* 1981.

[a]Refer to manufacturer's literature for exact values.

[b]For vertical blinds with opaque white and beige louvres, tightly closed, SC is 0.25 and 0.29 when used with glass of 0.71 to 0.80 transmittance.

[c]Refers to gray-, bronze-, and green-tinted heat-absorbing glass.

[d]SC for glass with no shading device.

[e]Refers to factory-manufactured units with 5-, 6-, or 13-mm (3/16-, 1/4-, or 1/2-in.) air space, or to prime windows plus storm windows.

[f]For vertical blinds with opaque white or beige louvres, tightly closed, SC is approximately the same as for opaque white collar shades.

TABLE 5.40 **Shading Coefficients for Single and Insulating Glass with Draperies**

	Range of Shading Coefficients			
			Drapery Fabrics[b]	
Glazing	*Glass Trans.*	*Glass SC*[a]	*High Transmittance, Low Reflectance*	*Low Transmittance, High Reflectance*
Single Glass				
6 mm (¼ in.) clear	0.80	0.95	0.80	0.35
12 mm (½ in.) clear	0.71	0.88	0.74	0.35
6 mm (¼ in.) heat abs.	0.46	0.67	0.57	0.33
12 mm (½ in.) heat abs.	0.24	0.50	0.43	0.30
Reflective coated (see	—	0.60	0.57	0.33
manufacturer's literature for	—	0.50	0.46	0.31
exact values)	—	0.40	0.36	0.26
	—	0.30	0.25	0.20
Insulating Glass, 12-mm (½-in.) air				
space clear out and clear in	0.64	0.83	0.66	0.35
Heat abs. out and clear in	0.37	0.55	0.49	0.32
Reflective coated (see	—	0.40	0.38	0.28
manufacturers' literature for	—	0.30	0.29	0.24
exact values)	—	0.20	0.19	0.15

Source: Based on *Handbook of Fundamentals,* published by the American Society of Heating, Refrigerating and Air Conditioning Engineers, Inc., Atlanta, GA.

[a]For glass alone, with no drapery.

[b]Drapes of 100% fullness, loose hanging.

[c]See *ASHRAE Handbook of Fundamentals* (1981), Chapter 27, Table 38, for more detailed listings.

TABLE 5.41 **Shading Coefficients for Domed Skylights**

Dome	Light Diffuser (Translucent)	Curb (See Below) Height, [mm (in.)]	Width-to-Height Ratio	Shading Coefficient	U Factor
Clear	Yes	0 (0)	∞	0.61	2.6 (0.46)
τ = 0.86[a]	τ = 0.58	230 (9)	5	0.58	2.4 (0.43)
		460 (18)	2.5	0.50	2.3 (0.40)
Clear		0 (0)	∞	0.99	4.5 (0.80)
τ = 0.86	None	230 (9)	5	0.88	4.3 (0.75)
		460 (18)	2.5	0.80	4.0 (0.70)
Translucent	None	0 (0)	∞	0.57	4.5 (0.80)
τ = 0.52		460 (18)	2.5	0.46	4.0 (0.70)
Translucent	None	0 (0)	∞	0.34	4.5 (0.80)
τ = 0.27		230 (9)	5	0.30	4.3 (0.75)
		460 (18)	2.5	0.28	4.0 (0.70)

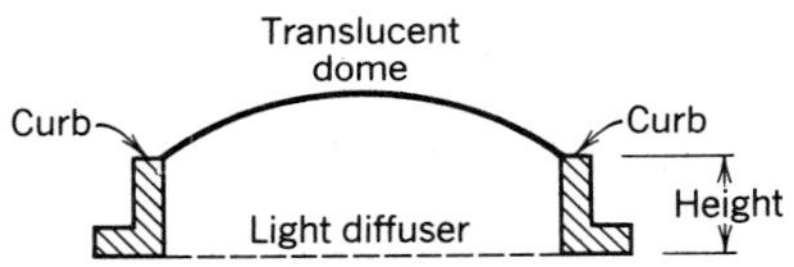

Source: Copyright © by the American Society of Heating, Refrigerating and Air Conditioning Engineers, Inc., Atlanta, GA. Reprinted by permission from the *ASHRAE Handbook of Fundamentals,* 1981.

[a]τ = transmittance.

TABLE 5.42 **Approximate Shading Coefficients of External Shading Devices**

Awnings	
Of venetian blind type, ⅔ drawn	0.43
fully drawn	0.15
Dark or medium canvas	0.25
Shading screens	0.28–0.23
Louvres, movable	0.15–0.10
Overhang: continuous, completely shading window	0.25
Dense tree casting heavy shade	0.25

Source: Adapted from Olgyay and Olgyay (1957).

TABLE 5.43 **Cooling Load Factors (CLF) for Glass Without Interior Shading, North Latitudes**

Fenes-tration Facing	Room Con-struction	Solar Time (h)																							
		1	2	3	4	5	6	7	8	9	10	11	12	13	14	15	16	17	18	19	20	21	22	23	24
N (shaded)	L	0.17	0.14	0.11	0.09	0.08	0.33	0.42	0.48	0.56	0.63	0.71	0.76	0.80	0.82	0.82	0.79	0.75	0.84	0.61	0.48	0.38	0.31	0.25	0.20
	M	0.23	0.20	0.18	0.16	0.14	0.34	0.41	0.46	0.53	0.59	0.65	0.70	0.73	0.75	0.76	0.74	0.75	0.79	0.61	0.50	0.42	0.36	0.31	0.27
	H	0.25	0.23	0.21	0.20	0.19	0.38	0.45	0.49	0.55	0.60	0.65	0.69	0.72	0.72	0.72	0.70	0.70	0.75	0.57	0.46	0.39	0.34	0.31	0.28
NNE	L	0.06	0.05	0.04	0.03	0.03	0.26	0.43	0.47	0.44	0.41	0.40	0.39	0.39	0.38	0.36	0.33	0.30	0.26	0.20	0.16	0.13	0.10	0.08	0.07
	M	0.09	0.08	0.07	0.06	0.06	0.24	0.38	0.42	0.39	0.37	0.37	0.36	0.36	0.36	0.34	0.33	0.30	0.27	0.22	0.18	0.16	0.14	0.12	0.10
	H	0.11	0.10	0.09	0.09	0.08	0.26	0.39	0.42	0.39	0.36	0.35	0.34	0.34	0.33	0.32	0.31	0.28	0.25	0.21	0.18	0.16	0.14	0.13	0.12
NE	L	0.04	0.04	0.03	0.02	0.02	0.23	0.41	0.51	0.51	0.45	0.39	0.36	0.33	0.31	0.28	0.26	0.23	0.19	0.15	0.12	0.10	0.08	0.06	0.05
	M	0.07	0.06	0.06	0.05	0.04	0.21	0.36	0.44	0.45	0.40	0.36	0.33	0.31	0.30	0.28	0.26	0.23	0.21	0.17	0.15	0.13	0.11	0.09	0.08
	H	0.09	0.08	0.08	0.07	0.07	0.23	0.37	0.44	0.44	0.39	0.34	0.31	0.29	0.27	0.26	0.24	0.22	0.20	0.17	0.14	0.13	0.12	0.11	0.10
ENE	L	0.04	0.03	0.03	0.02	0.02	0.21	0.40	0.52	0.57	0.53	0.45	0.39	0.34	0.31	0.28	0.25	0.22	0.18	0.14	0.12	0.09	0.08	0.06	0.05
	M	0.07	0.06	0.05	0.05	0.04	0.20	0.35	0.45	0.49	0.47	0.41	0.36	0.33	0.30	0.28	0.26	0.23	0.20	0.17	0.14	0.12	0.11	0.09	0.08
	H	0.09	0.09	0.08	0.07	0.07	0.22	0.36	0.46	0.49	0.45	0.38	0.33	0.30	0.27	0.25	0.23	0.21	0.19	0.16	0.14	0.13	0.12	0.11	0.10
E	L	0.04	0.03	0.03	0.02	0.02	0.19	0.37	0.51	0.57	0.57	0.50	0.42	0.37	0.32	0.29	0.25	0.22	0.19	0.15	0.12	0.10	0.08	0.06	0.05
	M	0.07	0.06	0.06	0.05	0.05	0.18	0.33	0.44	0.50	0.51	0.46	0.39	0.35	0.31	0.29	0.26	0.23	0.21	0.17	0.15	0.13	0.11	0.10	0.08
	H	0.09	0.09	0.08	0.08	0.07	0.20	0.34	0.45	0.49	0.49	0.43	0.36	0.32	0.29	0.26	0.24	0.22	0.19	0.17	0.15	0.13	0.12	0.11	0.10
ESE	L	0.05	0.04	0.03	0.03	0.02	0.17	0.34	0.49	0.58	0.61	0.57	0.48	0.41	0.36	0.32	0.28	0.24	0.20	0.16	0.13	0.10	0.09	0.07	0.06
	M	0.08	0.07	0.06	0.05	0.05	0.16	0.31	0.43	0.51	0.54	0.51	0.44	0.39	0.35	0.32	0.29	0.26	0.22	0.19	0.16	0.14	0.12	0.11	0.09
	H	0.10	0.09	0.09	0.08	0.08	0.19	0.32	0.43	0.50	0.52	0.49	0.41	0.36	0.32	0.29	0.26	0.24	0.21	0.18	0.16	0.14	0.13	0.12	0.11
SE	L	0.05	0.04	0.04	0.03	0.03	0.13	0.28	0.43	0.55	0.62	0.63	0.57	0.48	0.42	0.37	0.33	0.28	0.24	0.19	0.15	0.12	0.10	0.08	0.07
	M	0.09	0.08	0.07	0.06	0.05	0.14	0.26	0.38	0.48	0.54	0.56	0.51	0.45	0.40	0.36	0.33	0.29	0.25	0.21	0.18	0.16	0.14	0.12	0.10
	H	0.11	0.10	0.10	0.09	0.08	0.17	0.28	0.40	0.49	0.53	0.53	0.48	0.41	0.36	0.33	0.30	0.27	0.24	0.20	0.18	0.16	0.14	0.13	0.12
SSE	L	0.07	0.05	0.04	0.04	0.03	0.06	0.15	0.29	0.43	0.55	0.63	0.64	0.60	0.52	0.45	0.40	0.35	0.29	0.23	0.18	0.15	0.12	0.10	0.08
	M	0.11	0.09	0.08	0.07	0.06	0.08	0.16	0.26	0.38	0.48	0.55	0.57	0.54	0.48	0.43	0.39	0.35	0.30	0.25	0.21	0.18	0.16	0.14	0.12
	H	0.12	0.11	0.11	0.10	0.09	0.12	0.19	0.29	0.40	0.49	0.54	0.55	0.51	0.44	0.39	0.35	0.31	0.27	0.23	0.20	0.18	0.16	0.15	0.13
S	L	0.08	0.07	0.05	0.04	0.04	0.06	0.09	0.14	0.22	0.34	0.48	0.59	0.65	0.65	0.59	0.50	0.43	0.36	0.28	0.22	0.18	0.15	0.12	0.10
	M	0.12	0.11	0.09	0.08	0.07	0.08	0.11	0.14	0.21	0.31	0.42	0.52	0.57	0.58	0.53	0.47	0.41	0.36	0.29	0.25	0.21	0.18	0.16	0.14
	H	0.13	0.12	0.12	0.11	0.10	0.11	0.14	0.17	0.24	0.33	0.43	0.51	0.56	0.55	0.50	0.43	0.37	0.32	0.26	0.22	0.20	0.18	0.16	0.15
SSW	L	0.10	0.08	0.07	0.06	0.05	0.06	0.09	0.11	0.15	0.19	0.27	0.39	0.52	0.62	0.67	0.65	0.58	0.46	0.36	0.28	0.23	0.19	0.15	0.12
	M	0.14	0.12	0.11	0.09	0.08	0.09	0.11	0.13	0.15	0.18	0.25	0.35	0.46	0.55	0.59	0.59	0.53	0.44	0.35	0.30	0.25	0.22	0.19	0.16
	H	0.15	0.14	0.13	0.12	0.11	0.12	0.14	0.16	0.18	0.21	0.27	0.37	0.46	0.53	0.57	0.55	0.49	0.40	0.32	0.26	0.23	0.20	0.18	0.16

	L	0.12	0.10	0.08	0.06	0.05	0.06	0.08	0.10	0.12	0.14	0.16	0.24	0.36	0.49	0.60	0.66	0.66	0.58	0.43	0.33	0.27	0.22	0.18	0.14
SW	M	0.15	0.14	0.12	0.10	0.09	0.09	0.10	0.12	0.13	0.15	0.17	0.23	0.33	0.44	0.53	0.58	0.59	0.53	0.41	0.33	0.28	0.24	0.21	0.18
	H	0.15	0.14	0.13	0.12	0.11	0.12	0.13	0.14	0.16	0.17	0.19	0.25	0.34	0.44	0.52	0.56	0.56	0.49	0.37	0.30	0.25	0.21	0.19	0.17
	L	0.12	0.10	0.08	0.07	0.05	0.06	0.07	0.09	0.10	0.12	0.13	0.17	0.26	0.40	0.52	0.62	0.66	0.61	0.44	0.34	0.27	0.22	0.18	0.15
WSW	M	0.15	0.13	0.12	0.10	0.09	0.09	0.10	0.11	0.12	0.13	0.14	0.17	0.24	0.35	0.46	0.54	0.58	0.55	0.42	0.34	0.28	0.24	0.21	0.18
	H	0.15	0.14	0.13	0.12	0.11	0.11	0.12	0.13	0.14	0.15	0.16	0.19	0.26	0.36	0.46	0.53	0.56	0.51	0.38	0.30	0.25	0.21	0.19	0.17
	L	0.12	0.10	0.08	0.06	0.05	0.06	0.07	0.08	0.10	0.11	0.12	0.14	0.20	0.32	0.45	0.57	0.64	0.61	0.44	0.34	0.27	0.22	0.18	0.14
W	M	0.15	0.13	0.11	0.10	0.09	0.09	0.09	0.10	0.11	0.12	0.13	0.14	0.19	0.29	0.40	0.50	0.56	0.55	0.41	0.33	0.27	0.23	0.20	0.17
	H	0.14	0.13	0.12	0.11	0.10	0.11	0.12	0.13	0.14	0.14	0.15	0.16	0.21	0.30	0.40	0.49	0.54	0.52	0.38	0.30	0.24	0.21	0.18	0.16
	L	0.12	0.10	0.08	0.06	0.05	0.06	0.07	0.09	0.10	0.12	0.13	0.15	0.17	0.26	0.40	0.53	0.63	0.62	0.44	0.34	0.27	0.22	0.18	0.14
WNW	M	0.15	0.13	0.11	0.10	0.09	0.09	0.10	0.11	0.12	0.13	0.14	0.15	0.17	0.24	0.35	0.47	0.55	0.55	0.41	0.33	0.27	0.23	0.20	0.17
	H	0.14	0.13	0.12	0.11	0.10	0.11	0.12	0.13	0.14	0.15	0.16	0.17	0.18	0.25	0.36	0.46	0.53	0.52	0.38	0.30	0.24	0.20	0.18	0.16
	L	0.11	0.09	0.08	0.06	0.05	0.06	0.08	0.10	0.12	0.14	0.16	0.17	0.19	0.23	0.33	0.47	0.59	0.60	0.42	0.33	0.26	0.21	0.17	0.14
NW	M	0.14	0.12	0.11	0.09	0.08	0.09	0.10	0.11	0.13	0.14	0.16	0.17	0.18	0.21	0.30	0.42	0.51	0.54	0.39	0.32	0.26	0.22	0.19	0.16
	H	0.14	0.12	0.11	0.10	0.10	0.10	0.12	0.13	0.15	0.16	0.18	0.18	0.19	0.22	0.30	0.41	0.50	0.51	0.36	0.29	0.23	0.20	0.17	0.15
	L	0.12	0.09	0.08	0.06	0.05	0.07	0.11	0.14	0.18	0.22	0.25	0.27	0.29	0.30	0.33	0.44	0.57	0.62	0.44	0.33	0.26	0.21	0.17	0.14
NNW	M	0.15	0.13	0.11	0.10	0.09	0.10	0.12	0.15	0.18	0.21	0.23	0.26	0.27	0.28	0.31	0.39	0.51	0.56	0.41	0.33	0.27	0.23	0.20	0.17
	H	0.14	0.13	0.12	0.11	0.10	0.12	0.15	0.17	0.20	0.23	0.25	0.26	0.28	0.28	0.31	0.38	0.49	0.53	0.38	0.30	0.25	0.21	0.18	0.16
	L	0.11	0.09	0.07	0.06	0.05	0.07	0.14	0.24	0.36	0.48	0.58	0.66	0.72	0.74	0.73	0.67	0.59	0.47	0.37	0.29	0.24	0.19	0.16	0.13
Hor.	M	0.16	0.14	0.12	0.11	0.09	0.11	0.16	0.24	0.33	0.43	0.52	0.59	0.64	0.67	0.66	0.62	0.56	0.47	0.38	0.32	0.28	0.24	0.21	0.18
	H	0.17	0.16	0.15	0.14	0.13	0.15	0.20	0.28	0.36	0.45	0.52	0.59	0.62	0.64	0.62	0.58	0.51	0.42	0.35	0.29	0.26	0.23	0.21	0.19

Source: Copyright © by the American Society of Heating, Refrigerating and Air Conditioning Engineers, Inc., Atlanta, GA. Reprinted by permission from the *ASHRAE Handbook of Fundamentals,* 1981.

[a]L = light construction: frame exterior wall, 50.8-mm (2-in.) concrete floor slab, approximately 146 kg (30 lb) of material/m^2 (ft^2) of floor area.

M = medium construction: 101.6-mm (4-in.) concrete exterior wall, 101.6-mm (4-in.) concrete floor slab, approximately 341 kg (70 lb) of building material/m^2 (ft^2) of floor area.

H = heavy construction: 152.4-mm (6-in.) concrete exterior wall, 152.4-mm (6-in.) concrete floor slab, approximately 635 kg (130 lb) of building materials/m^2 (ft^2) of floor area.

TABLE 5.44 **Cooling Load Factors (CLF) for Glass with Interior Shading, North Latitudes (All Room Constructions)**

Fenestration Facing	Solar Time, h																							
	1	2	3	4	5	6	7	8	9	10	11	12	13	14	15	16	17	18	19	20	21	22	23	24
N	0.08	0.07	0.06	0.06	0.07	0.73	0.66	0.65	0.73	0.80	0.86	0.89	0.89	0.86	0.82	0.75	0.78	0.91	0.24	0.18	0.15	0.13	0.11	0.10
NNE	0.03	0.03	0.02	0.02	0.03	0.64	0.77	0.62	0.42	0.37	0.37	0.37	0.36	0.35	0.32	0.28	0.23	0.17	0.08	0.07	0.06	0.05	0.04	0.04
NE	0.03	0.02	0.02	0.02	0.02	0.56	0.76	0.74	0.58	0.37	0.29	0.27	0.26	0.24	0.22	0.20	0.16	0.12	0.06	0.05	0.04	0.04	0.03	0.03
ENE	0.03	0.02	0.02	0.02	0.02	0.52	0.76	0.80	0.71	0.52	0.31	0.26	0.24	0.22	0.20	0.18	0.15	0.11	0.06	0.05	0.04	0.04	0.03	0.03
E	0.03	0.02	0.02	0.02	0.02	0.47	0.72	0.80	0.76	0.62	0.41	0.27	0.24	0.22	0.20	0.17	0.14	0.11	0.06	0.05	0.05	0.04	0.03	0.03
ESE	0.03	0.03	0.02	0.02	0.02	0.41	0.67	0.79	0.80	0.72	0.54	0.34	0.27	0.24	0.21	0.19	0.15	0.12	0.07	0.06	0.05	0.04	0.04	0.03
SE	0.03	0.03	0.02	0.02	0.02	0.30	0.57	0.74	0.81	0.79	0.68	0.49	0.33	0.28	0.25	0.22	0.18	0.13	0.08	0.07	0.06	0.05	0.04	0.04
SSE	0.04	0.03	0.03	0.03	0.02	0.12	0.31	0.54	0.72	0.81	0.81	0.71	0.54	0.38	0.32	0.27	0.22	0.16	0.09	0.08	0.07	0.06	0.05	0.04
S	0.04	0.04	0.03	0.03	0.03	0.09	0.16	0.23	0.38	0.58	0.75	0.83	0.80	0.68	0.50	0.35	0.27	0.19	0.11	0.09	0.08	0.07	0.06	0.05
SSW	0.05	0.04	0.04	0.03	0.03	0.09	0.14	0.18	0.22	0.27	0.43	0.63	0.78	0.84	0.80	0.66	0.46	0.25	0.13	0.11	0.09	0.08	0.07	0.06
SW	0.05	0.05	0.04	0.04	0.03	0.07	0.11	0.14	0.16	0.19	0.22	0.38	0.59	0.75	0.83	0.81	0.69	0.45	0.16	0.12	0.10	0.09	0.07	0.06
WSW	0.05	0.05	0.04	0.04	0.03	0.07	0.10	0.12	0.14	0.16	0.17	0.23	0.44	0.64	0.78	0.84	0.78	0.55	0.16	0.12	0.10	0.09	0.07	0.06
W	0.05	0.05	0.04	0.04	0.03	0.06	0.09	0.11	0.13	0.15	0.16	0.17	0.31	0.53	0.72	0.82	0.81	0.61	0.16	0.12	0.10	0.08	0.07	0.06
WNW	0.05	0.05	0.04	0.03	0.03	0.07	0.10	0.12	0.14	0.16	0.17	0.18	0.22	0.43	0.65	0.80	0.84	0.66	0.16	0.12	0.10	0.08	0.07	0.06
NW	0.05	0.04	0.04	0.03	0.03	0.07	0.11	0.14	0.17	0.19	0.20	0.21	0.22	0.30	0.52	0.73	0.82	0.69	0.16	0.12	0.10	0.08	0.07	0.06
NNW	0.05	0.05	0.04	0.03	0.03	0.11	0.17	0.22	0.26	0.30	0.32	0.33	0.34	0.34	0.39	0.61	0.82	0.76	0.17	0.12	0.10	0.08	0.07	0.06
HOR.	0.06	0.05	0.04	0.04	0.03	0.12	0.27	0.44	0.59	0.72	0.81	0.85	0.85	0.81	0.71	0.58	0.42	0.25	0.14	0.12	0.10	0.08	0.07	0.06

Source: Copyright © by the American Society of Heating, Refrigerating and Air Conditioning Engineers, Inc., Atlanta, GA. Reprinted by permission from the *ASHRAE Handbook of Fundamentals*, 1981.

So, the heat gain due to solar radiation through glass is:

$$q = \text{area} \times \text{maximum SHGF} \times \text{SC} \times \text{CLF}$$

Finally, the *total* heat gain through glass includes both the conduction gains and those from solar radiation:

$$\text{window } q_{\text{total}} = (UA \times \text{CLTD}) + (A \times \text{SHGF} \times \text{SC} \times \text{CLF})$$

EXAMPLE 5.12. (Based on the *ASHRAE Handbook of Fundamentals,* 1981).

Determine the cooling load due to glass on the south and west walls of a building at noon, 2 P.M. and 4 P.M. in August. The building is located at 32°N latitude with outside design conditions of 32.2°C (90 F) dry-bulb temperature and a 11.1C° (20F°) daily range. The inside design dry-bulb temperature is 25.5°C (78 F). Assume the room construction is of medium weight. The south glass is insulating glass [6.35-mm (0.25-in.) air space] having an area of 9.29 m² (100 ft²) with no interior shading. The west glass is 5.56-mm (0.21875-in.) single gray-tinted glass with an area of 9.29 m² (100 ft²) and with light-colored venetian blinds.

SOLUTION. The data required for the calculations are as follows:

	U_2 [W/m² K (Btu/h ft² F)]	Shading Coefficient, SC	Maximum SHGF (Table 5.34) [W/m² (Btu/h ft²)]
South glass	3.5 (0.61)	0.81 (Table 5.37)	350 (111)
West glass	4.6 (0.81)	0.53 (Table 5.39)	691 (219)

The conduction heat gain component of the cooling is calculated by

$$q = U \times A \times \text{CLTD}$$

Time	CLTD (Table 5.33)	CLTD Corrected	South Glass U × A × CLTD [W(Btu/h)]	West Glass U × A × CLTD [W(Btu/h)]
1200	5 (9)	2.2 (4)	71 (244)	94 (324)
1400	7 (13)	4.2 (8)	135 (488)	179 (648)
1600	8 (14)	5.2 (9)	167 (550)	222 (739)

The correction factor applied to the above values was −2.8C° (−5F°), computed from the notes of Table 5.33. The CLTDs are rounded off, and inconsistencies in conversion between SI and English units may occur.

Outdoor daily average temperature is 32.2 − 0.5 (11.1) = 26.6°C (80 F); CLTD correction is 29.4 − 26.6 = 2.8 less (85 − 80 = 5 less).

The solar heat gain component of the cooling load is calculated by

$$q = A(\text{SC})(\text{SHG})(\text{CLF})$$

	South Glass		West Glass	
Time	CLF (Table 5.43)	A(SC)(SHG)(CLF) [W(Btu/h)]	CLF (Table 5.44)	A(SC)(SHG)(CLF) [W(Btu/h)]
1200	0.52	1386 (4733)	0.17	578 (1970)
1400	0.58	1546 (5280)	0.53	1803 (6140)
1600	0.47	1253 (4278)	0.82	2789 (9500)

The total cooling load due to heat gain through the glass is therefore

Time	South Glass [W(Btu/h)]	West Glass [W(Btu/h)]
1200	1457 (4977)	672 (2 254)
1400	1681 (5766)	1982 (6 788)
1600	1420 (4828)	3011 (10 229)

(d) Gains from Electric Lighting. Although we commonly assume that all the energy supplied to lights becomes immediate heat gain, this is often untrue. The radiant component of light raises air temperature only *after* it has been absorbed and released by materials in the space. Thus, again, cooling load factors (CLF) are necessary to account for this lag (see Fig. 5.28).

First, the types of furnishings, the air supply and return, and the type of electric lights are compared (Table 5.45); then the type of room-envelope thermal mass and the air circulation are considered (Table 5.46). Given this information, the actual CLF can be obtained from Table 5.47.

If the cooling system operates only during occupied hours (and thereby does not remove stored heat during unoccupied times), then CLF = 1.0 (and all energy for lights is assumed to be heat gain). Where lights are on continuously, CLF also equals 1.0:

$$q_{\text{lights}} = W \times \text{CLF}$$

where W represents the input watts to lights (including ballasts) (1 W = 3.41 Btu/h).

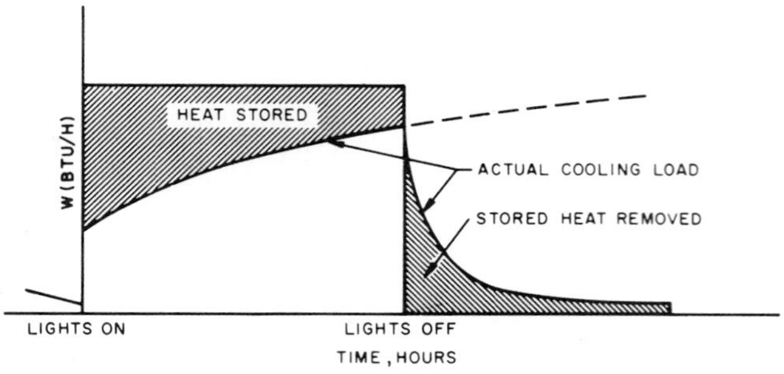

Fig. 5.28. Thermal storage effect in the cooling load caused by electric lights. Copyright © by the American Society of Heating, Refrigerating and Air Conditioning Engineers, Inc., Atlanta, GA. Reprinted by permission from the AHSRAE Handbook of Fundamentals, *1981.*

TABLE 5.45 **Design Values of *a* Coefficient: Features of Room Furnishings, Light Fixtures, and Ventilation Arrangements**

a	*Furnishings*	*Air Supply and Return*	*Type of Light Fixture*
0.45	Heavyweight, simple furnishings, no carpet	Low rate; supply and return below ceiling [$V \leq 2.5(0.5)$][a]	Recessed, not vented
0.55	Ordinary furniture, no carpet	Medium to high ventilation rate; supply and return below ceiling or through ceiling grille and space [$V \geq 2.5(0.5)$][a]	Recessed, not vented
0.65	Ordinary furniture, with or without carpet	Medium to high ventilation rate or fan coil or induction type air-conditioning terminal unit; supply through ceiling or wall diffuser; return around light fixtures and through ceiling space. [$V \geq 2.5(0.5)$][a]	Vented
0.75 or greater	Any type of furniture	Ducted returns through light fixtures	Vented or free-hanging in air stream with ducted returns

Source: Copyright © by the American Society of Heating, Refrigerating and Air Conditioning Engineers, Inc., Atlanta, GA. Reprinted by permission from the *ASHRAE Handbook of Fundamentals,* 1981.

[a]*V* is room air supply rate in L/s m^2 (cfm/ft^2) of floor area.

(e) Gains from People. This more exacting method of heat gain calculation requires a distinction between sensible and latent gains (see Table 5.21). Sensible gains include a large radiation component (70% of sensible bodily loss, assumed in this procedure)—and again, this will be soaked up by furnishings and materials before it contributes to a space's cooling load. The latent heat gains, by contrast, immediately become part of the space's cooling problem.

Table 5.21 presents the latent (l) and sensible (s) components of gains from people, and Table 5.48 lists the CLF to be applied to the sensible portion of such gains.

People: q_s = (number of occupants) (sensible Btu/h per person) (CLF)

q_l = (number of occupants) (latent Btu/h) per person)

(In SI units, use watts instead of Btu/h.)

TABLE 5.46 **The *b* Classification Values Calculated for Different Envelope Constructions and Room Air Circulation Rates**

	Floor Construction and Floor Weight in Pounds per Square Foot of Floor Area[a]				
aRoom Air Circulation and Type of Supply and Return — *mm thickness, Kg/m² mass:*	*2-in. Wooden Floor, 10 lb/ft², 50.8, 48.8*	*3-in. Concrete Floor, 40 lb/ft², 76.2, 195.3*	*6-in. Concrete Floor, 75 lb/ft², 152.4, 366.2*	*8-in. Concrete Floor, 120 lb/ft², 203.2, 585.8*	*12-in. Concrete Floor, 160 lb/ft², 304.8, 781.1*
Low ventilation rate—minimum required to handle cooling load. Supply through floor, wall, or ceiling diffuser. Ceiling space not vented.	B	B	C	D	D
Medium ventilation rate. Supply through floor, wall, or ceiling diffuser. Ceiling space not vented.	A	B	C	D	D
High room air circulation induced by primary air of induction unit or by fan coil unit. Return through 'ceiling space.	A	B	C	C	D
Very high room air circulation used to minimize room temperature gradients. Return through ceiling space.	A	A	B	C	D

Source: Cooling and Heating Load Calculation Manual, 1979, published by the American Society of Heating, Refrigerating and Air Conditioning Engineers, Inc., Atlanta, GA.

[a]This table is based on floor covered with carpet and rubber pad. For floor covered with floor tile, use letter designation in next row down with the same floor weight.

TABLE 5.47 **Cooling Load Factors (CLF) for Electric Lighting**

Part A. ***When Lights Are On for 8 Hours***

"a" Coef-ficients	"b" Class-ification	Number of hours after lights are turned on 0	1	2	3	4	5	6	7	8	9	10	11	12	13	14	15	16	17	18	19	20	21	22	23
	A	0.02	0.46	0.57	0.65	0.72	0.77	0.82	0.85	0.88	0.46	0.37	0.30	0.24	0.19	0.15	0.12	0.10	0.08	0.06	0.05	0.04	0.03	0.03	0.02
	B	0.07	0.51	0.56	0.61	0.65	0.68	0.71	0.74	0.77	0.34	0.31	0.28	0.25	0.22	0.20	0.18	0.16	0.15	0.13	0.12	0.11	0.10	0.09	0.08
0.45	C	0.11	0.55	0.58	0.60	0.63	0.65	0.67	0.69	0.71	0.28	0.26	0.25	0.23	0.22	0.20	0.19	0.18	0.17	0.16	0.15	0.14	0.13	0.12	0.12
	D	0.14	0.58	0.60	0.61	0.62	0.63	0.64	0.65	0.66	0.22	0.22	0.21	0.20	0.20	0.19	0.19	0.18	0.18	0.17	0.16	0.16	0.16	0.15	0.15
	A	0.01	0.56	0.65	0.72	0.77	0.82	0.85	0.88	0.90	0.37	0.30	0.24	0.19	0.16	0.13	0.10	0.08	0.07	0.05	0.04	0.03	0.03	0.02	0.02
	B	0.06	0.60	0.64	0.68	0.71	0.74	0.76	0.79	0.81	0.28	0.25	0.23	0.20	0.18	0.16	0.15	0.13	0.12	0.11	0.10	0.09	0.08	0.07	0.06
0.55	C	0.09	0.63	0.66	0.68	0.70	0.71	0.73	0.75	0.76	0.23	0.21	0.20	0.19	0.18	0.17	0.16	0.15	0.14	0.13	0.12	0.11	0.11	0.10	0.10
	D	0.11	0.66	0.67	0.68	0.69	0.70	0.71	0.72	0.72	0.18	0.18	0.17	0.17	0.16	0.16	0.15	0.15	0.14	0.14	0.13	0.13	0.13	0.12	0.12
	A	0.01	0.66	0.73	0.78	0.82	0.86	0.88	0.91	0.93	0.29	0.23	0.19	0.15	0.12	0.10	0.08	0.06	0.05	0.04	0.03	0.03	0.02	0.02	0.01
	B	0.04	0.69	0.72	0.75	0.77	0.80	0.82	0.84	0.85	0.22	0.19	0.18	0.16	0.14	0.13	0.12	0.10	0.09	0.08	0.08	0.07	0.06	0.06	0.05
0.65	C	0.07	0.72	0.73	0.75	0.76	0.78	0.79	0.80	0.82	0.18	0.17	0.16	0.15	0.14	0.13	0.12	0.11	0.11	0.10	0.10	0.09	0.08	0.08	0.07
	D	0.09	0.73	0.74	0.75	0.76	0.77	0.77	0.78	0.79	0.14	0.14	0.13	0.13	0.13	0.12	0.12	0.11	0.11	0.11	0.10	0.10	0.10	0.10	0.09
	A	0.01	0.76	0.80	0.84	0.87	0.90	0.92	0.93	0.95	0.21	0.17	0.13	0.11	0.09	0.07	0.06	0.05	0.04	0.03	0.02	0.02	0.02	0.01	0.01
	B	0.03	0.78	0.80	0.82	0.84	0.85	0.87	0.88	0.89	0.15	0.14	0.13	0.11	0.10	0.09	0.08	0.07	0.07	0.06	0.05	0.05	0.04	0.04	0.04
0.75	C	0.05	0.80	0.81	0.82	0.83	0.84	0.85	0.86	0.87	0.13	0.12	0.11	0.10	0.10	0.09	0.09	0.08	0.08	0.07	0.07	0.06	0.06	0.06	0.05
	D	0.06	0.81	0.82	0.82	0.83	0.83	0.84	0.84	0.85	0.10	0.10	0.10	0.09	0.09	0.09	0.08	0.08	0.08	0.08	0.07	0.07	0.07	0.07	0.07

Part B. ***When Lights Are On for 12 Hours***

"a" Coef-ficients	"b" Class-ification	Number of hours after lights are turned on 0	1	2	3	4	5	6	7	8	9	10	11	12	13	14	15	16	17	18	19	20	21	22	23
	A	0.05	0.49	0.59	0.67	0.73	0.78	0.83	0.86	0.89	0.91	0.93	0.94	0.95	0.51	0.41	0.33	0.27	0.22	0.17	0.14	0.11	0.09	0.07	0.06
	B	0.13	0.57	0.61	0.65	0.69	0.72	0.75	0.77	0.79	0.82	0.83	0.85	0.87	0.43	0.39	0.35	0.31	0.28	0.25	0.23	0.21	0.18	0.17	0.15
0.45	C	0.19	0.63	0.65	0.67	0.69	0.71	0.73	0.74	0.76	0.77	0.79	0.80	0.81	0.37	0.35	0.33	0.31	0.29	0.27	0.26	0.24	0.23	0.21	0.20
	D	0.22	0.66	0.67	0.68	0.69	0.70	0.71	0.72	0.73	0.74	0.74	0.75	0.76	0.32	0.31	0.30	0.29	0.28	0.27	0.26	0.26	0.25	0.24	0.23
	A	0.04	0.58	0.66	0.73	0.78	0.82	0.86	0.89	0.91	0.93	0.94	0.95	0.96	0.42	0.34	0.27	0.22	0.18	0.14	0.11	0.09	0.07	0.06	0.05
	B	0.11	0.65	0.68	0.72	0.74	0.77	0.79	0.81	0.83	0.85	0.86	0.88	0.89	0.35	0.32	0.28	0.26	0.23	0.21	0.19	0.17	0.15	0.14	0.12
0.55	C	0.15	0.69	0.71	0.73	0.75	0.76	0.78	0.79	0.80	0.81	0.83	0.84	0.85	0.30	0.29	0.27	0.25	0.24	0.22	0.21	0.20	0.19	0.17	0.16
	D	0.18	0.72	0.73	0.74	0.75	0.76	0.76	0.77	0.78	0.78	0.79	0.80	0.80	0.26	0.25	0.24	0.24	0.23	0.22	0.22	0.21	0.20	0.20	0.19
	A	0.03	0.67	0.74	0.79	0.83	0.86	0.89	0.91	0.93	0.94	0.95	0.96	0.97	0.33	0.26	0.21	0.17	0.14	0.11	0.09	0.07	0.06	0.05	0.04
	B	0.09	0.73	0.75	0.78	0.80	0.82	0.84	0.85	0.87	0.88	0.89	0.90	0.91	0.27	0.25	0.22	0.20	0.18	0.16	0.15	0.13	0.12	0.11	0.10
0.65	C	0.12	0.76	0.78	0.79	0.80	0.81	0.83	0.84	0.85	0.86	0.86	0.87	0.88	0.24	0.22	0.21	0.20	0.19	0.17	0.16	0.15	0.14	0.14	0.13
	D	0.14	0.79	0.79	0.80	0.80	0.81	0.82	0.82	0.83	0.83	0.84	0.84	0.85	0.20	0.20	0.19	0.18	0.18	0.17	0.17	0.16	0.16	0.15	0.15
	A	0.02	0.77	0.81	0.85	0.88	0.90	0.92	0.94	0.95	0.96	0.97	0.97	0.98	0.23	0.19	0.15	0.12	0.10	0.08	0.06	0.05	0.04	0.03	0.03
	B	0.06	0.81	0.82	0.84	0.86	0.87	0.88	0.90	0.91	0.92	0.92	0.93	0.94	0.19	0.18	0.16	0.14	0.13	0.12	0.10	0.09	0.08	0.08	0.07
0.75	C	0.09	0.83	0.84	0.85	0.86	0.87	0.88	0.88	0.89	0.90	0.90	0.91	0.91	0.17	0.16	0.15	0.14	0.13	0.12	0.12	0.11	0.10	0.10	0.09
	D	0.10	0.85	0.85	0.86	0.86	0.86	0.87	0.87	0.88	0.88	0.88	0.89	0.89	0.14	0.14	0.14	0.13	0.13	0.12	0.12	0.12	0.11	0.11	0.11

Part C. ***When Lights Are On for 16 Hours***

"a" Coef-ficients	"b" Class-ification	0	1	2	3	4	5	6	7	8	9	10	11	12	13	14	15	16	17	18	19	20	21	22	23
		Number of hours after lights are turned on																							
0.45	A	0.12	0.54	0.63	0.70	0.76	0.81	0.85	0.88	0.90	0.92	0.94	0.95	0.96	0.97	0.97	0.98	0.98	0.54	0.43	0.35	0.28	0.23	0.18	0.15
	B	0.23	0.66	0.69	0.72	0.75	0.78	0.80	0.82	0.84	0.85	0.87	0.88	0.89	0.90	0.91	0.92	0.93	0.49	0.44	0.39	0.35	0.32	0.29	0.26
	C	0.29	0.72	0.74	0.75	0.77	0.78	0.80	0.81	0.82	0.83	0.84	0.85	0.86	0.87	0.88	0.88	0.89	0.45	0.42	0.39	0.37	0.35	0.33	0.31
	D	0.31	0.75	0.76	0.77	0.77	0.78	0.79	0.79	0.80	0.81	0.81	0.82	0.82	0.83	0.83	0.84	0.84	0.40	0.39	0.37	0.36	0.35	0.34	0.33
0.55	A	0.10	0.63	0.70	0.76	0.81	0.84	0.87	0.90	0.92	0.93	0.95	0.96	0.97	0.97	0.98	0.98	0.99	0.44	0.35	0.28	0.23	0.18	0.15	0.12
	B	0.19	0.72	0.75	0.77	0.80	0.82	0.84	0.85	0.87	0.88	0.89	0.90	0.91	0.92	0.93	0.94	0.94	0.40	0.36	0.32	0.29	0.26	0.24	0.21
	C	0.24	0.77	0.79	0.80	0.81	0.82	0.83	0.84	0.85	0.86	0.87	0.88	0.88	0.89	0.90	0.90	0.91	0.37	0.34	0.32	0.30	0.29	0.27	0.25
	D	0.26	0.80	0.80	0.81	0.82	0.82	0.83	0.83	0.84	0.84	0.85	0.85	0.86	0.86	0.86	0.87	0.87	0.33	0.32	0.31	0.30	0.29	0.28	0.27
0.65	A	0.07	0.71	0.77	0.81	0.85	0.88	0.90	0.92	0.94	0.95	0.96	0.97	0.97	0.98	0.98	0.99	0.99	0.34	0.27	0.22	0.18	0.14	0.12	0.09
	B	0.15	0.78	0.81	0.82	0.84	0.86	0.87	0.88	0.90	0.91	0.92	0.92	0.93	0.94	0.94	0.95	0.96	0.31	0.28	0.25	0.23	0.20	0.18	0.16
	C	0.18	0.82	0.83	0.84	0.85	0.86	0.87	0.88	0.89	0.89	0.90	0.90	0.91	0.92	0.92	0.93	0.93	0.28	0.27	0.25	0.24	0.22	0.21	0.20
	D	0.20	0.84	0.85	0.85	0.86	0.86	0.87	0.87	0.87	0.88	0.88	0.88	0.89	0.89	0.89	0.90	0.90	0.25	0.25	0.24	0.23	0.22	0.22	0.21
0.75	A	0.05	0.79	0.83	0.87	0.89	0.91	0.93	0.94	0.95	0.96	0.97	0.98	0.98	0.98	0.99	0.99	0.99	0.24	0.20	0.16	0.13	0.10	0.08	0.07
	B	0.11	0.85	0.86	0.87	0.89	0.90	0.91	0.92	0.93	0.93	0.94	0.95	0.95	0.96	0.96	0.96	0.97	0.22	0.20	0.18	0.16	0.15	0.13	0.12
	C	0.13	0.87	0.88	0.89	0.89	0.90	0.91	0.91	0.92	0.92	0.93	0.93	0.94	0.94	0.94	0.95	0.95	0.20	0.19	0.18	0.17	0.16	0.15	0.14
	D	0.14	0.89	0.89	0.89	0.90	0.90	0.90	0.91	0.91	0.91	0.91	0.92	0.92	0.92	0.92	0.93	0.93	0.18	0.18	0.17	0.17	0.16	0.16	0.15

Source: Copyright © by the American Society of Heating, Refrigerating and Air Conditioning Engineers, Inc., Atlanta, GA. Reprinted by permission from the *ASHRAE Handbook of Fundamentals,* 1981.

TABLE 5.48 **Sensible-Heat Cooling Load Factors (CLF) for People**

Total Hours in Space	*Hours After Each Entry into Space*																							
	1	*2*	*3*	*4*	*5*	*6*	*7*	*8*	*9*	*10*	*11*	*12*	*13*	*14*	*15*	*16*	*17*	*18*	19	20	21	22	23	24
2	0.49	0.58	0.17	0.13	0.10	0.08	0.07	0.06	0.05	0.04	0.04	0.03	0.03	0.02	0.2	0.02	0.02	0.01	0.01	0.01	0.01	0.01	0.01	0.01
4	0.49	0.59	0.66	0.71	0.27	0.21	0.16	0.14	0.11	0.10	0.08	0.07	0.06	0.06	0.05	0.04	0.04	0.03	0.03	0.03	0.02	0.02	0.02	0.01
6	0.50	0.60	0.67	0.72	0.76	0.79	0.34	0.26	0.21	0.18	0.15	0.13	0.11	0.10	0.08	0.07	0.06	0.06	0.05	0.04	0.04	0.03	0.03	0.03
8	0.51	0.61	0.67	0.72	0.76	0.80	0.82	0.84	0.38	0.30	0.25	0.21	0.18	0.15	0.13	0.12	0.10	0.09	0.08	0.07	0.06	0.05	0.05	0.04
10	0.53	0.62	0.69	0.74	0.77	0.80	0.83	0.85	0.87	0.89	0.42	0.34	0.28	0.23	0.20	0.17	0.15	0.13	0.11	0.10	0.09	0.08	0.07	0.06
12	0.55	0.64	0.70	0.75	0.79	0.81	0.84	0.86	0.88	0.89	0.91	0.92	0.45	0.36	0.30	0.25	0.21	0.19	0.16	0.14	0.12	0.11	0.09	0.08
14	0.58	0.66	0.72	0.77	0.80	0.83	0.85	0.87	0.89	0.90	0.91	0.92	0.93	0.94	0.47	0.38	0.31	0.26	0.23	0.20	0.17	0.15	0.13	0.11
16	0.62	0.70	0.75	0.79	0.82	0.85	0.87	0.88	0.90	0.91	0.92	0.93	0.94	0.95	0.95	0.96	0.49	0.39	0.33	0.28	0.24	0.20	0.18	0.16
18	0.66	0.74	0.79	0.82	0.85	0.87	0.89	0.90	0.92	0.93	0.94	0.94	0.95	0.96	0.96	0.97	0.97	0.97	0.50	0.40	0.33	0.28	0.24	0.21

Source: Copyright © by the American Society of Heating, Refrigerating and Air Conditioning Engineers, Inc., Atlanta, GA. Reprinted by permission from the *ASHRAE Handbook of Fundamentals,* 1981.

For spaces in which people are packed closely together, CLF = 1.0, since there is less opportunity to radiate to walls or ceilings. Where the cooling system does not operate continuously, also use CLF = 1.0.

EXAMPLE 5.13. Determine the cooling load in a building at noon, 2 P.M., and 4 P.M. due to four people occupying an office from 9 A.M. to 5 P.M. The office temperature is 25.5°C (78 F).

SOLUTION. From Table 5.21, $q_s = 75$ W (255 Btu/h) per person and $q_l = 75$ W (255 Btu/h) per person. The people occupy the space for 8 h, and noon is 3 h after the people enter. Therefore,

Time	*CLF (Table 5.48)*	*Sensible Load No. × q_s × CLF [W(Btu/h)]*	*Latent Load No. × q_l [W(Btu/h)]*
1200	0.67	201 (683)	300 (1020)
1400	0.76	228 (775)	300 (1020)
1600	0.82	246 (836)	300 (1020)

(f) Gains from Appliances and Equipment. It is particularly difficult to predict cooling loads from cooking, which vary greatly according to appliance efficiency, menu, and chef. The *ASHRAE Handbook of Fundamentals* (1981) presents detailed data on commercial cooking appliances; a simple, more approximate method is to find the *total input rated energy* q_r (Btu/h or W) of all appliances within a few broad categories. These values are then adjusted for ''normal'' usage, as shown in Tables 5.49 and 5.50.

For other equipment, such as that found in laboratories and offices, Table 5.50 presents a range of ordinary heat gain rates, all of which are converted as sensible gains.

Cooling loads from motors located within conditioned spaces are explained in detail in the *ASHRAE Handbook of Fundamentals* (1981); as an approximation, a motor's horsepower (hp) rating can be converted directly to sensible heat gain via the formula

$$q = 745.7 \text{ hp}$$

For all the sensible heat gains, cooling load factors must again be utilized, to correspond to the delay in the impact of the radiant fraction of such gains. These CLF values are presented in Tables 5.51 and 5.52. (For all equipment not specifically vented or under hoods, use the CLF for unhooded appliances.) Latent heat gains, in contrast, immediately become a part of the space's cooling load.

(g) Gains from Outdoor Air. It is commonly assumed that the quantities of outdoor air listed in Table 4.19 are so great that when they are deliberately introduced into a space, they maintain enough outward pressure to prevent infiltration. However, to the extent that outdoor air is replaced by recycled or purified indoor air, additional infiltration is more likely to occur.

Whichever assumption is made, the total amount of outdoor air brings both sensible and latent heat gains to the space; both impact immediately, so no cooling load factors are necessary. (Actually,

TABLE 5.49 **Approximate Cooling Load Factors (CLF) for Cooking Appliances**

	Cooking Fuel Used	
	Electric and Steam	*Gas*
Appliances under exhaust hoods (or otherwise directly vented)	$q_s = 0.11q_r$ $q_l = 0.05q_r$	$q_s = 0.07q_r$ $q_l = 0.03q_r$
Appliances not vented or under hood, but located in conditioned area	$q_s = 0.33q_r$ $q_l = 0.17q_r$	$q_s = 0.20q_r$ $q_l = 0.10q_r$

q_r = total rated input energy, W (Btu/h)

q_s = sensible portion of q_r in normal usage, W (Btu/h)

q_l = latent portion of q_r in normal usage, W (Btu/h)

Source: Adapted from *1981 Handbook of Fundamentals,* published by the American Society of Heating, Refrigerating and Air Conditioning Engineers, Inc., Atlanta, GA.

ventilation air is cooled at the air conditioning equipment *before* it enters the space. Technically, it is not a part of the *space's* cooling load, but the air conditioner must still be sized to accommodate the load from ventilation air.) In a simple approximation applicable to most cooling situations,

$$\text{outdoor air } q_s = 1.08 \text{ cfm } \Delta T$$

where ΔT is the temperature difference (F°), between inside and outside air, and

$$\text{outdoor air } q_l = 4840 \text{ cfm } \Delta W$$

where ΔW is the difference in humidity ratio between inside and outside air (see the psychrometric chart shown in Fig. 4.6). In SI units,

$$\text{outdoor air } q_s = 1.232 \text{ L/s } \Delta T$$

$$\text{outdoor air } q_l = 3012 \text{ L/s } \Delta W$$

(h) Total Load on Cooling Equipment. This section has discussed cooling loads from the following sources within spaces, which may be added together (see also Fig. 5.27).

Sensible	*Latent*
Roofs and walls	
+ Glass	
+ Adjacent spaces	
+ Electric lighting	
+ People	
+ Appliances and equipment	People + Appliances
+ Outdoor air	+ Outdoor air
= Total sensible heat gain within space	= Total latent heat gain within space

These totals may be used to determine the cooling equipment necessary. However, there may be further (although often minor) heat gains from fans and motors in the air conditioning equipment, air duct leakage, and the friction of air against ductwork. In the case of smaller equipment, manufacturers often list the capacity as including the equipment's heat contribution to the space. For larger or more complicated buildings, the additional load (beyond gains within spaces) is calculated by mechanical or architectural engineers.

EXAMPLE 5.14. (Copyright © by the American Society of Heating, Refrigerating and Air Conditioning Engineers, Inc., Atlanta, GA. Adapted from the *ASHRAE Handbook of Fundamentals*, 1981.)

A one-story office building (see Fig. 5.26 and Example 5.10) is located in an eastern state near 40°N latitude. The adjoining buildings on the north and west are not conditioned, and their inside air temperatures are basically equal to the outdoor air temperature at any time of day.

Roof construction. 114-mm (4.5-in.) flat roof deck of 50.8-mm (2-in.) gypsum slab [equivalent to four 12.7-mm (0.5-in.) layers of gypsum board], 50.8-mm (2-in.) preformed above-deck roof insulation, 6.4-mm (0.25-in.) asbestos board under gypsum slab, and two layers of mopped 6.8-kg (15-lb) felt vapor seal over insulation, no false ceiling. [Weight approximately 60 kg/m² (12 lb/ft²)] [U = 0.68 W/m² °C (0.12 Btu/h ft² F)].

South wall construction. 101.6-mm (4-in.) face brick, 101.6-mm (4-in.) common brick, 9.5-mm (0.375-in.) cement plaster sand aggregate, 6.4-mm (0.25-in.) plywood panel glued to plaster. [U = 2.21 W/m² °C (0.39 Btu/h ft² F)].

West wall and adjoining north party wall construction. 330-mm (13-in.) solid brick, no plaster [U = 1.42 W/m² °C (0.25 Btu/h ft² F)].

North exposed wall and east wall construction. 203.2-mm (8-in.) concrete block and 19-mm (¾-in.) plaster [U = 2.5 W/m² °C (0.44 Btu/h ft² F)].

Floor construction. 101.6-mm (4-in.) concrete on ground.

Fenestration. 0.91-m × 1.52-m (3-ft × 5-ft) window of regular plate glass with light-colored venetian blinds, not openable [U = 4.6 W/m² °C (0.81 Btu/h ft² F)].

Front doors. Two, 0.76 m × 2.13 m (2.5 ft × 7 ft).

Side doors. Two, 0.76 m × 2.13 m (2.5 ft × 7 ft).

Rear doors. Two, 0.76 m × 2.13 m (2.5 ft × 7 ft).

Door construction. Approximately 30% 38-mm (1.5-in.) solid pine frame and 70% 6.4-mm (0.25-in.) pine panels. [U = 3.35 W/m² °C (0.59 Btu/h ft² F)].

U values for all doors and outside walls were calculated assuming a wind speed of 3.35 m/s (7.5 mph). For party and inside walls, still air was assumed.

TABLE 5.50 **Rate of Heat Gain from Miscellaneous Appliances**

Miscellaneous Data	Manufacturer's Rating		Recommended Rate of Heat Gain, (W)			Miscellaneous Data	Manufacturer's Rating		Recomended Rate of Heat Gain (Btu/h)	
	Watts	*Sensible*	*Latent*	*Total*			*Btu/h*	*Sensible*	*Latent*	*Total*
					Electrical Appliances					
Blower type	1580	675	120	785	Hair dryer	Blower type	5,400	2,300	400	2,700
Helmet type 60 heaters @ 25 W	700	550	100	650	Hair dryer	Helmet type 60 heaters @ 25 W	2,400	1,870	330	2,200
91.44-cm normal use	1500	250	50	300	Permanent wave machine	36-in. normal use	5,000	850	150	1,000
1.27-cm diameter		28		28	Neon sign, per linear meter of tube[a]	0.5-in. diameter		30		30
0.95-cm diameter		56		56		0.375-in. diameter		60		60
					Sterilizer, instrument		3,750	650	1200	1,850
	1100	190	350	540	Magnetic card typewriter		690	350	0	350
					Small copier	Running	6,000	6,000	0	6,000
Running	1760	1760	0	1760		Standby	3,000	3,000	0	3,000
Standby	880	880	0	880	Large copier	Running	12,000	12,000	0	12,000
Running	3515	3515	0	3515		Standby	6,000	6,000	0	6,000
Standby	1760	1760	0	1760						
					Gas-Burning Appliances					
					Lab burners					
1.1-cm barrel	880	495	125	620	Bunsen	0.4375-in. barrel	3,000	1,680	430	2,100
3.8-cm wide	1465	820	205	1025	Fishtail	1.5-in. wide	5,000	2,800	700	3,500
2.54-cm diameter	1760	985	245	1230	Meeker	1-in. diameter	6,000	3,360	840	4,200
Mantle type	585	530	60	590	Gas light, per burner	Mantle type	2,000	1,800	200	2,000
Continuous flame	730	265	30	295	Cigar lighter	Continuous flame	2,500	900	100	1,000

Source: Copyright © by the American Society of Heating, Refrigerating and Air Conditioning Engineers, Inc., Atlanta, GA. Reprinted by permission from the *ASHRAE Handbook of Fundamentals,* 1981.

[a]Conventional (Btu/h) values are per linear foot of tube.

TABLE 5.51 **Sensible-Heat Cooling Load Factors (CLF) for Hooded Appliances**

Total Operational Hours	Hours After Appliances Are On																							
	1	2	3	4	5	6	7	8	9	10	11	12	13	14	15	16	17	18	19	20	21	22	23	24
2	0.27	0.40	0.25	0.18	0.14	0.11	0.09	0.08	0.07	0.06	0.05	0.04	0.04	0.03	0.03	0.03	0.02	0.02	0.02	0.02	0.01	0.01	0.01	0.01
4	0.28	0.41	0.51	0.59	0.39	0.30	0.24	0.19	0.16	0.14	0.12	0.10	0.09	0.08	0.07	0.06	0.05	0.05	0.04	0.04	0.03	0.03	0.02	0.02
6	0.29	0.42	0.52	0.59	0.65	0.70	0.48	0.37	0.30	0.25	0.21	0.18	0.16	0.14	0.12	0.11	0.09	0.08	0.07	0.06	0.05	0.05	0.04	0.04
8	0.31	0.44	0.54	0.61	0.66	0.71	0.75	0.78	0.55	0.43	0.35	0.30	0.25	0.22	0.19	0.16	0.14	0.13	0.11	0.10	0.08	0.07	0.06	0.06
10	0.33	0.46	0.55	0.62	0.68	0.72	0.76	0.79	0.81	0.84	0.60	0.48	0.39	0.33	0.28	0.24	0.21	0.18	0.16	0.14	0.12	0.11	0.09	0.08
12	0.36	0.49	0.58	0.64	0.69	0.74	0.77	0.80	0.82	0.85	0.87	0.88	0.64	0.51	0.42	0.36	0.31	0.26	0.23	0.20	0.18	0.15	0.13	0.12
14	0.40	0.52	0.61	0.67	0.72	0.76	0.79	0.82	0.84	0.86	0.88	0.89	0.91	0.92	0.67	0.54	0.45	0.38	0.32	0.28	0.24	0.21	0.19	0.16
16	0.45	0.57	0.65	0.70	0.75	0.78	0.81	0.84	0.86	0.87	0.89	0.90	0.92	0.93	0.94	0.94	0.69	0.56	0.46	0.39	0.34	0.29	0.25	0.22
18	0.52	0.63	0.70	0.75	0.79	0.82	0.84	0.86	0.88	0.89	0.91	0.92	0.93	0.94	0.95	0.95	0.96	0.96	0.71	0.58	0.48	0.41	0.35	0.30

TABLE 5.52 **Sensible-Heat Cooling Load Factors (CLF) for Unhooded Appliances[a]**

Total Operational Hours	Hours After Appliances Are On																							
	1	2	3	4	5	6	7	8	9	10	11	12	13	14	15	16	17	18	19	20	21	22	23	24
2	0.56	0.64	0.15	0.11	0.08	0.07	0.06	0.05	0.04	0.04	0.03	0.03	0.02	0.02	0.02	0.02	0.01	0.01	0.01	0.01	0.01	0.01	0.01	0.01
4	0.57	0.65	0.71	0.75	0.23	0.18	0.14	0.12	0.10	0.08	0.07	0.06	0.05	0.05	0.04	0.04	0.03	0.03	0.02	0.02	0.02	0.02	0.01	0.01
6	0.57	0.65	0.71	0.76	0.79	0.82	0.29	0.22	0.18	0.15	0.13	0.11	0.10	0.08	0.07	0.06	0.06	0.05	0.04	0.04	0.03	0.03	0.03	0.02
8	0.58	0.66	0.72	0.76	0.80	0.82	0.85	0.87	0.33	0.26	0.21	0.18	0.15	0.13	0.11	0.10	0.09	0.08	0.07	0.06	0.05	0.04	0.04	0.03
10	0.60	0.68	0.73	0.77	0.81	0.83	0.85	0.87	0.89	0.90	0.36	0.29	0.24	0.20	0.17	0.15	0.13	0.11	0.10	0.08	0.07	0.07	0.06	0.05
12	0.62	0.69	0.75	0.79	0.82	0.84	0.86	0.88	0.89	0.91	0.92	0.93	0.38	0.31	0.25	0.21	0.18	0.16	0.14	0.12	0.11	0.09	0.08	0.07
14	0.64	0.71	0.76	0.80	0.83	0.85	0.87	0.89	0.90	0.92	0.93	0.93	0.94	0.95	0.40	0.32	0.27	0.23	0.19	0.17	0.15	0.13	0.11	0.10
16	0.67	0.74	0.79	0.82	0.85	0.87	0.89	0.90	0.91	0.92	0.93	0.94	0.95	0.96	0.96	0.97	0.42	0.34	0.28	0.24	0.20	0.18	0.15	0.13
18	0.71	0.78	0.82	0.85	0.87	0.89	0.90	0.92	0.93	0.94	0.94	0.95	0.96	0.96	0.97	0.97	0.97	0.98	0.43	0.35	0.29	0.24	0.21	0.18

[a]Includes ordinary office equipment.

Outdoor design conditions. Dry-bulb temperature, 34.4°C (94 F); daily range 11.1C° (20 F°); wet-bulb temperature, 25°C (77 F); $W_o = 0.0161$ kg of vapor per kilogram of dry air (lb vapor/lb dry air).

Indoor design conditions. Dry-bulb temperature, 25.6°C (78 F); wet-bulb temperature, 18.3°C (65 F); $W_i = 0.0102$ kg of vapor per kilogram of dry air (lb vapor/lb dry air).

Occupancy. 85 office workers from 8 A.M. to 5 P.M.

Lights. 17,500 W, fluorescent, from 8 A.M. to 6 P.M.; 4000 W, tungsten, continuous. The fixtures are not vented.

The conditioning equipment is located in the adjoining structure to the north.

Determine the sensible, latent, and total space cooling load at design conditions.

Note: Heat gains from office equipment have been assumed, as in Example 5.10, at 0.5 W per square foot of floor area.

SOLUTION. For this job, judgment indicates that the roof will make the greatest single contribution to the cooling load. The roof receives more sun during the summer than do walls, and most of the glass faces south (none east or west). Hence, the time of maximum cooling load will probably be the time of maximum CLTD for the roof. From Table 5.27, the CLTD for roof 5, a 25.4-mm (1-in.) wood roof with 50.8 mm (2 in.) of insulation (closest to the actual roof in general construction and U value), has a maximum value of 36 (64) at 4 P.M. but is only slightly less [35 (63)] at 3 P.M. From the variation of cooling load factors with time for glass facing south given in Table 5.44, we note that the maximum cooling load due to the windows would be only slightly more at noon than at 1 P.M. It is necessary to make a quick estimate.

Load		Solar Time 12	1	2	3	4
Roof $= U \times A \times$ CLTD	W	5,340	6,750	7,875	8,720	8,860
	Btu/h	18,240	23,040	26,880	29,760	30,240
South glass, solar $= A \times$ SC $\times$ SHG $\times$ CLF	W	1,195	1,152	980	720	504
	Btu/h	4,081	3,934	3,344	2,459	1,721
Totals	W	6,535	7,902	8,855	9,440	9,364
	Btu/h	22,321	26,974	30,223	32,219	31,961

Evaluation of how the values above were computed using the indicated equations will be explained shortly. Based on the estimate, the cooling load calculations will be made for 3 P.M. In some cases, there would be no clear-cut evidence of this nature; consequently, it would be necessary to estimate the total load for a number of hours and then select the maximum.

Cooling Load Due to Heat Gain Through the Roof and Exposed Walls: Cooling load due to heat gain through the roof, exposed walls, and doors is calculated using the equation

$$q = U \times A \times \text{CLTD}$$

where the cooling load temperature differences (CLTD) are taken from Tables 5.27 and 5.29, respectively. The building is considered to be of medium construction. The CLTDs chosen were for roofs and walls whose U values and general construction were as close to the actual components as possible. All tabular CLTD values were decreased by 0.5 (1) to correct for the outdoor daily average temperature being 0.5C° (1F°) less than 29.4°C (85 F). No other corrections were necessary. The calculations above were also performed for the doors in the exposed walls. See Table 5.53.

Cooling Load Due to Heat Gain Through Fenestration Areas: Examination of the values of maximum solar heat gain factors (Table 5.34) for south windows at 40°N latitude indicates that, of the three summer months (June, July, and August), values are the highest in August. Therefore, August is selected for calculation of the cooling load component due to heat gain through the windows.

The load component due to conduction heat gain is calculated using the equation

$$q = U \times A \times \text{CLTD}$$

where the CLTD value is taken from Table 5.33. Again, the tabular value was decreased by 0.5 (1) to correct for the outdoor daily average temperature being 28.9°C (84 F).

The U value of the glass was taken as 4.59 W/m^2 °C (0.81 Btu/h ft^2 F) from Table 4.17, for single glass with interior shading.

The load component due to solar heat gain was computed using the equation

$$q = A \times \text{SC} \times \text{SHGF} \times \text{CLF}$$

TABLE 5.53 **Cooling Load Summary for Various Components, Example 5.14**

Net Area (m^2)	U Value (W/ m^2 °C)	ΔT (C°)	CLTD (C°)	Ref. for CLTD	SC[c]	SHG[d] (W/m^2)	CLF[e]	Cooling Load (W)	Component	Net Area (ft^2)	U Value (Btu/h ft^2 F)	ΔT (F°)	CLTD (F°)	Ref for CLTD	SC[c]	SHG[d] (Btu/ h ft^2)	CLF[e]	Cooling Load (Btu/h)
371.6	0.68	—	34.5	Table 5.27 Roof #5	—	—	—	8,717	Roof	4000	0.12	—	62	Table 5.28 Roof #5	—	—	—	29,760
37.6[a]	2.21	—	10.5	Table 5.26 Group D	—	—	—	872	South wall	405[a]	0.39	—	19	Table 5.26 Group D	—	—	—	3,001
71.1[a]	2.49	—	19.5	Table 5.26 Group E	—	—	—	3,452	East wall	765[a]	0.44	—	35	Table 5.26 Group E	—	—	—	11,781
15.8[a]	2.49	—	7.5	Table 5.26 Group E	—	—	—	295	North exposed wall	170[a]	0.44	—	14	Table 5.26 Group E	—	—	—	1,047
98.9[a]	1.41	8.8[b]	—	—	—	—	—	1,241	West and north party walls	1065[a]	0.25	16[b]	—	—	—	—	—	4,260
3.25	3.35	—	20.5	Table 5.26 Group F	—	—	—	223	Doors in south wall	35	0.59	—	37	Table 5.26 Group F	—	—	—	764
3.25	3.35	8.8[b]	—	—	—	—	—	97	Doors in north wall	35	0.59	16[b]	—	—	—	—	—	330
3.25	3.35	—	19.5	Table 5.26 Group F		—	—	212	Doors in east wall	35	0.59	—	35	Table 5.26 Group F	—	—	—	723
5.6	4.59	—	7.5	Table 5.33	0.55	470	0.50	916	South windows	60	0.81	—	13	Table 5.33	0.55	149	0.50	3,090
2.8	4.59	—	7.5	Table 5.33	0.55	110	0.82	235	North windows	30	0.81	—	13	Table 5.33	0.55	35	0.82	789
								Total: 16,260 W										Total: 55,545 Btu/h

Source: Adapted from *1981 Handbook of Fundamentals*, published by the American Society of Heating, Refrigerating and Air Conditioning Engineers, Inc., Atlanta, GA.

[a]Calculated from gross wall area, less fenestration and doors.

[b]34.4°C − 25.6°C (94 F − 78 F), outdoor minus indoor at design conditions.

[c]From Table 5.39.

[d]From Tables 5.34 and 5.35.

[e]From Table 5.44.

Table 5.39 shows a shading coefficient of 0.55 for clear glass with light-colored venetian blinds. The SHGF and CLF values were taken from Tables 5.34 and 5.44, respectively.

Results of the calculation for fenestration areas are shown in Table 5.53.

Cooling Load Due to Heat Gain Through Party Walls: For the north and west party walls, cooling load was calculated using the equation

$$q = U \times A \times \Delta T$$

where the temperature difference was that which existed at 3 P.M., 8.8C° (16 F°). The results are shown in Table 5.53.

Cooling Load Due to Internal Heat Sources: For the cooling load component due to lights, the tungsten lights are calculated as given, while the added heat of the ballasts of the fluorescent lights is accounted for by the factor 1.2 times lighting watts:

$$q_{\text{tun}} = 4000 \text{ W } (13{,}650 \text{ Btu/h})$$

$$q_{\text{fl}} = 17{,}500 \times 1.20$$

$$= 21{,}000 \text{ W } (71{,}673 \text{ Btu/h})$$

From Table 5.45, an *a* coefficient of 0.55 is selected for a medium weight structure and non-vented light fixtures. From Table 5.46, a *b* classification of B is chosen. From Table 5.47, for lights on for 10 h and a load calculation done 7 h after they were turned on, CLF = 0.80, by interpolation. Therefore, the cooling load at 3 P.M. due to the fluorescent lights is 21,000 W × 0.80 = 16,800 W (57,340 Btu/h). (Because so much of the tungsten lamps' heat is radiated directly at occupants, no CLF is applied.)

For the cooling load due to people, Table 5.21 is used for seated occupants doing light office work. The CLF of 0.83 for the sensible portion, taken from Table 5.48, reflects a total of 10 h of occupancy and a load calculation 7 h after entry.

		W	*(Btu/h)*
sensible cooling load			
	85 people × 0.83 × 75	= 5,270	(17,990)
latent cooling load			
	85 × 75	= 6,350	(21,675)
total, due to people		11,620	(39,665)

For equipment, the assumption is that 0.5 W per square foot of floor area will be installed. Table 5.52 indicates that for unhooded appliances on for 10 h at 7 h after appliances are on, CLF = 0.85. The cooling load at 3 P.M. due to equipment, then, is

$$0.5 \text{ W/ft}^2 \times 4000 \text{ ft}^2 = 2000 \text{ W}$$

$$2000 \text{ W} \times 0.85 = 1700 \text{ W } (5800 \text{ Btu/h})$$

Cooling Load Due to Ventilation and Infiltration Air: From Table 4.19, the required ventilation rate is 7 L/s per person (15 cfm/person). The total necessary ventilation is then 85 × 7 = 595 L/s (1275 cfm). The total space volume is 1133 m^3 (40,000 ft^3), which therefore corresponds to

$$\frac{595 \text{ L/s} \times 3600}{1133 \times 1000}, \text{ or } \frac{1275 \text{ cfm} \times 60}{40{,}000}$$

$$= 1.91 \text{ air changes per hour}$$

The space floor area is 371.6 m^2 (4000 ft^2), 595/371.6 = 1.6 L/s m^2 (1275/4000 = 0.32 cfm/ft^2). For this example, assume the ventilation air will be introduced directly into the space and included as part of the space cooling load.

Window infiltration can be considered zero, since the windows are sealed. Calculation of door infiltration requires some judgment. Assume that outside and inside doors are frequently opened simultaneously and that door infiltration should be included. Allow 2.83 m^3 (100 ft^3) of outdoor air per person per door passage. Assume that outside doors will be used at a rate of 10 persons/h and inside doors will be used at a rate of 30 persons/h. Total infiltration will then be (40 × 2.83 × 1000)/3600 = 31 L/s (67 cfm).

The sensible and latent portions of the load component are calculated using the equations

$$q_s = 1.232 \text{ L/s } \Delta T$$

$$q_s = 1.10 \text{ cfm } \Delta T$$

$$q_l = 3012 \text{ L/s } \Delta W$$

$$q_l = 4840 \text{ cfm } \Delta W$$

where, at 3 P.M., t_0 = 34.4°C (94 F), t_i = 25.6°C (78 F), W_o = 0.0161, and W_i = 0.0102.

For ventilation 595 L/s (1275 cfm),

$$q_s = 1.232 \times 595 \times (34.44 - 25.55)$$

$$= 6575 \text{ W}$$

$$[1.10 \times 1275 \times (94 - 78)$$

$$= 22{,}440 \text{ Btu/h}]$$

$$q_l = 3012 \times 595 \times (0.0161 - 0.0102) = 10{,}664 \text{ W}$$

$$[4840 \times 1275 \times (0.0161 - 0.0102) = 36{,}410 \text{ Btu/h}]$$

For infiltration 31 L/s (67 cfm),

$$q_s = 1.232 \times 31 \times (34.44 - 25.55) = 350 \text{ W}$$

$$[1.10 \times 67 \times (94 - 78) = 1180 \text{ Btu/h}]$$

$$q_l = 3012 \times 31 \times (0.0161 - 0.0102) = 568 \text{ W}$$

$$[4840 \times 67 \times (0.0161 - 0.0102) = 1915 \text{ Btu/h}]$$

These calculations are summarized in Table 5.54. Note that in this rather typical office build-

TABLE 5.54 **Summary of Calculations for Example 5.14**

DB(°C)	*WB(°C)*	*Hum Ratio*		*DB(F)*	*WB(F)*	*Hum Ratio*
34.4	25.0	0.0161	Outdoor conditions	94	77	0.0161
25.6	18.3	0.0102	Indoor conditions	78	65	0.0102
8.8		0.0059	Difference	16		0.0059

W	*Sensible Cooling Load at 3 P.M.*	*Btu/h*
	Roof and exposed walls	
8,717	Roof	29,760
872	South wall	3,001
3,452	East wall	11,781
295	North wall	1,047
223	Doors in south wall	764
212	Doors in east wall	723
	Fenestration areas	
916	South windows	3,090
235	North windows	789
	Party Walls	
1,241	West and north wall	4,260
97	Doors in north wall	330
	Internal Sources	
4,000	Tungsten lights	13,650
16,800	Fluorescent lights	57,340
5,270	People	17,990
1,700	Equipment	5,800
	Ventilation and infiltration	
6,575	Ventilation	22,440
350	Infiltration	1,180
50,955	Total	173,945

W	*Latent Cooling Load at 3 P.M.*	*Btu/h*
6,350	*People*	21,675
10,664	*Ventilation*	36,410
568	*Infiltration*	1,915
17,582	Total	60,000
68,537	Grand Total Load	233,945

Source: Adapted from *1981 Handbook of Fundamentals*, published by the American Society of Heating, Refrigerating and Air Conditioning Engineers, Inc.

ing and climate, the worst-hour cooling load is composed of 74% sensible heat and 26% latent heat. Note also that the peak hourly gain from sensible and latent heat is

$$\frac{68{,}537}{371.6} = 185\ \text{W/m}^2 \left(\frac{233{,}945}{4000} = 58\ \text{Btu/h ft}^2\right)$$

These results differ slightly from the more quickly derived values of Example 5.10; they differ drastically from the rule-of-thumb guide to heat gain presented in Table 5.10, which would have led us to expect gains in the neighborhood of 16 Btu/h ft^2. However, the installed lighting totals 21,500 W/4000 ft^2 of floor area = 5.4 W per square foot, more than twice the current energy-conserving maximums that were assumed in the rule of thumb. Furthermore, the windows were shaded only on the inside (SC = 0.55), rather than with the more effective exterior shading that the rule of thumb assumed (SC about 0.25, awnings). If the *sensible* peak load were to be adjusted for half the electric lights and half the solar gains through windows, this building's peak would drop from 50,955 to 39,980 W, or 108 W/m^2 (34 Btu/h ft^2). Further targets for energy conservation include the roof, whose U = 0.12 Btu/h ft^2 F is considerably above the U_o maximums shown in Fig. 5.8 for commercial buildings in *any* climate.

5.9 Passive Cooling Calculation Procedures

When outdoor air is 85 F or below, it is possible to cool buildings by simple ventilation, maintaining conditions indoors that are within the "comfort zone." The equations for sizing windows and stacks for either cross ventilation or stack ventilation were presented in Section 4.6.

In the many climates in which the outdoor temperature is above 80 F for a large number of working-day hours (see Table 5.55), buildings that are closed to the hot exterior during those hours are practical. [Between 80 and 85 F, outdoor air can keep people within the comfort zone if it moves across the body fast enough (see Fig. 2.4). In areas with reliable winds, open buildings are feasible up to 85 F.] From the advent of air conditioning to recent years, the standard response of designers was to turn to mechanical cooling, utilizing the refrigeration cycle and forced air. Recently, some passive cooling alternatives have proved to be effective in climates that have clear or cool summer nights. The following procedures go beyond the quicker rules of thumb that appeared in Section 5.3 and allow the designer to better adjust a preliminary design to the program and the climate.

TABLE 5.55 **Cooling Degree Days (CDD), to Base 18.3°C (65 F), and Cooling Hours (CH) over 26.7°C (80 F)**[a]

City, State	*CDD* [°C-days (F-days)]	*CH* (h)
Albuquerque, New Mexico	747.2 (1345)	922
Albany, New York	273.2 (492)	267
Amarillo, Texas	698.9 (1258)	951
Atlanta, Georgia	816.1 (1469)	695
Birmingham, Alabama	918.9 (1654)	1020
Bismarck, North Dakota	293.3 (528)	545
Brownsville, Texas	2139.4 (3851)	2504
Boise, Idaho	416.1 (749)	715
Boston, Massachusetts	374.4 (674)	310
Buffalo, New York	222.2 (400)	157
Burlington, Vermont	204.4 (368)	232
Charleston, South Carolina	1101.7 (1983)	1084
Cheyenne, Wyoming	171.1 (308)	299
Chicago, Illinois	396.1 (713)	377
Cincinnati, Ohio	637.2 (1147)	663

TABLE 5.55 **Cooling Degree Days (CDD), to Base 18.3°C (65 F), and Cooling Hours (CH) over 26.7°C (80 F)[a] (*continued*)**

City, State	*CDD [°C-days (F-days)]*	*CH (h)*
Cleveland, Ohio	377.3 (670)	364
Columbia, Missouri	669.4 (1205)	765
Detroit, Michigan	381.7 (687)	372
Dodge City, Kansas	736.7 (1326)	986
El Paso, Texas	1083.9 (1951)	1576
Fort Worth, Texas	1388.9 (2500)	1742
Fresno, California	910.6 (1639)	1377
Great Falls, Montana	190.6 (343)	299
Houston, Texas	1525.0 (2745)	1593
Indianapolis, Indiana	501.1 (902)	542
Jackson, Mississippi	1311.7 (2361)	1490
Jacksonville, Florida	1517.2 (2731)	1509
Kansas City, Missouri	819.4 (1475)	1034
Lake Charles, Louisiana	1462.8 (2633)	1463
Los Angeles, California	198.3 (357)	73
Louisville, Kentucky	670.6 (1207)	696
Lubbock, Texas	876.1 (1557)	1234
Madison, Wisconsin	235.6 (424)	293
Medford, Oregon	253.3 (456)	558
Memphis, Tennessee	1040.0 (1872)	1272
Miami, Florida	2327.2 (4189)	2495
Minneapolis, Minnesota	496.7 (894)	607
Nashville, Tennessee	831.1 (1496)	893
New Orleans, Louisiana	1473.9 (2653)	1419
New York, New York	570.6 (1027)	377
Norfolk, Virginia	754.4 (1358)	634
Oklahoma City, Oklahoma	1045.6 (1882)	1280
Omaha, Nebraska	559.4 (1007)	738
Philadelphia, Pennsylvania	600.6 (1081)	563
Phoenix, Arizona	1852.2 (3334)	2710
Pittsburgh, Pennsylvania	406.7 (732)	390
Portland, Maine	162.2 (292)	181
Portland, Oregon	137.8 (248)	200
Raleigh, North Carolina	752.8 (1355)	750
Richmond, Virginia	686.7 (1236)	684
Sacramento, California	540.6 (973)	851
Salt Lake City, Utah	532.2 (958)	844
San Antonio, Texas	1588.9 (2860)	1816
San Diego, California	333.3 (600)	55
San Francisco, California	54.4 (98)	58
Saint Louis, Missouri	772.2 (1390)	927
Seattle–Tacoma, Washington	74. (134)	116
Tampa, Florida	1751.1 (3152)	1794
Tulsa, Oklahoma	955.0 (1719)	1142
Washington, D.C.	828.3 (1491)	939

Source: Copyright © by the American Society of Heating, Refrigerating and Air Conditioning Engineers, Inc., Atlanta, GA. Reprinted by permission from the *ASHRAE Handbook of Fundamentals,* 1981.

[a]From NOAA's test reference year (TRY) weather tapes.

(a)

As an introduction to today's passively cooled buildings, consider the Bateson Building in Sacramento, California, the first in an impressive series of state office buildings that are largely passively heated and cooled. The Bateson Building (Fig. 5.29*a–c*) is a four-story structure of 285,000 ft^2 occupied by 1200 workers. Although most of the building's needed heat is internally produced by lights, people, and equipment (Fig. 5.29*d*), direct-gain passive heating is also used. Winter sun enters some office-facing windows but comes mostly through the banks of clerestories that cover the atrium. Heat is stored in the thermally massive structure; hot air from the atrium's ceiling can be blown down to floor level through huge, bright yellow tubes, or forced into two 660-ton rock beds below the building for storage. Night insulation is not utilized because night heating is a relatively minor issue. Cooling is the major problem for such a building in Sacramento, so shading devices are plentiful. Movable exterior vertical louvres shade the south-facing clerestories; horizontal louvres shade the north-sloped glass that admits more daylight to the atrium. During the hotter months, bright orange roll-down fabric shades are automatically deployed to block direct sun in the morning on

Fig. 5.29. The Bateson Building, an office building for Sacramento's climate. (Reprinted courtesy of the Office of the State Architect, Sacramento.) A variety of approaches to envelope and equipment distinguishes this design, developed by the staff of the California State Architect's office. (a) *From the exterior the east facade (and its counterpart on the west) is brightened by roll-down fabric shades. These disappear daily, when direct sun no longer threatens to overheat the perimeter offices. (Sun control on the south facade is a simple trellis.)* (b) *The interior court is roofed to admit south sun in winter for passive solar heating; movable louvres can close out direct sun whenever overheating is a problem.* (c) *The range of heating and cooling devices, including solar collectors for domestic water heating, rock bed, and building mass to store ''coolth'' (obtained from venting with Sacramento's cool night air), and vertical canvas tubes with fans to reduce the courtyard's air stratification.* (d) *Graphs of the average annual heat losses and gains. Note the very small portion of the total that is required of the auxiliary heating and cooling equipment. Electric lights in this daylight-emphasizing example still constitute nearly half the cooling problem in the summer—and nearly half the heating source in the winter.*

(b)

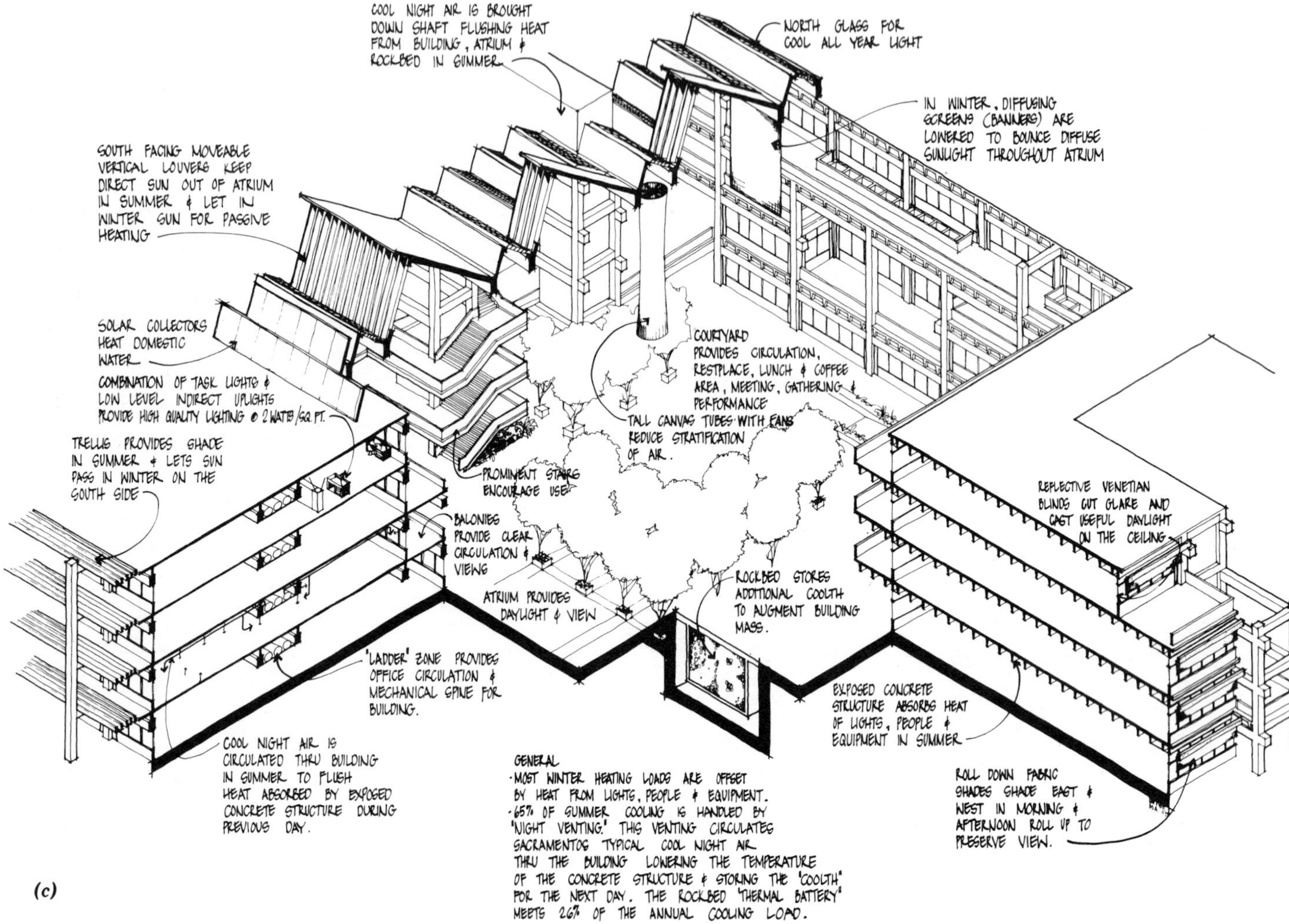

(c)

Fig. 5.29. (*continued*)

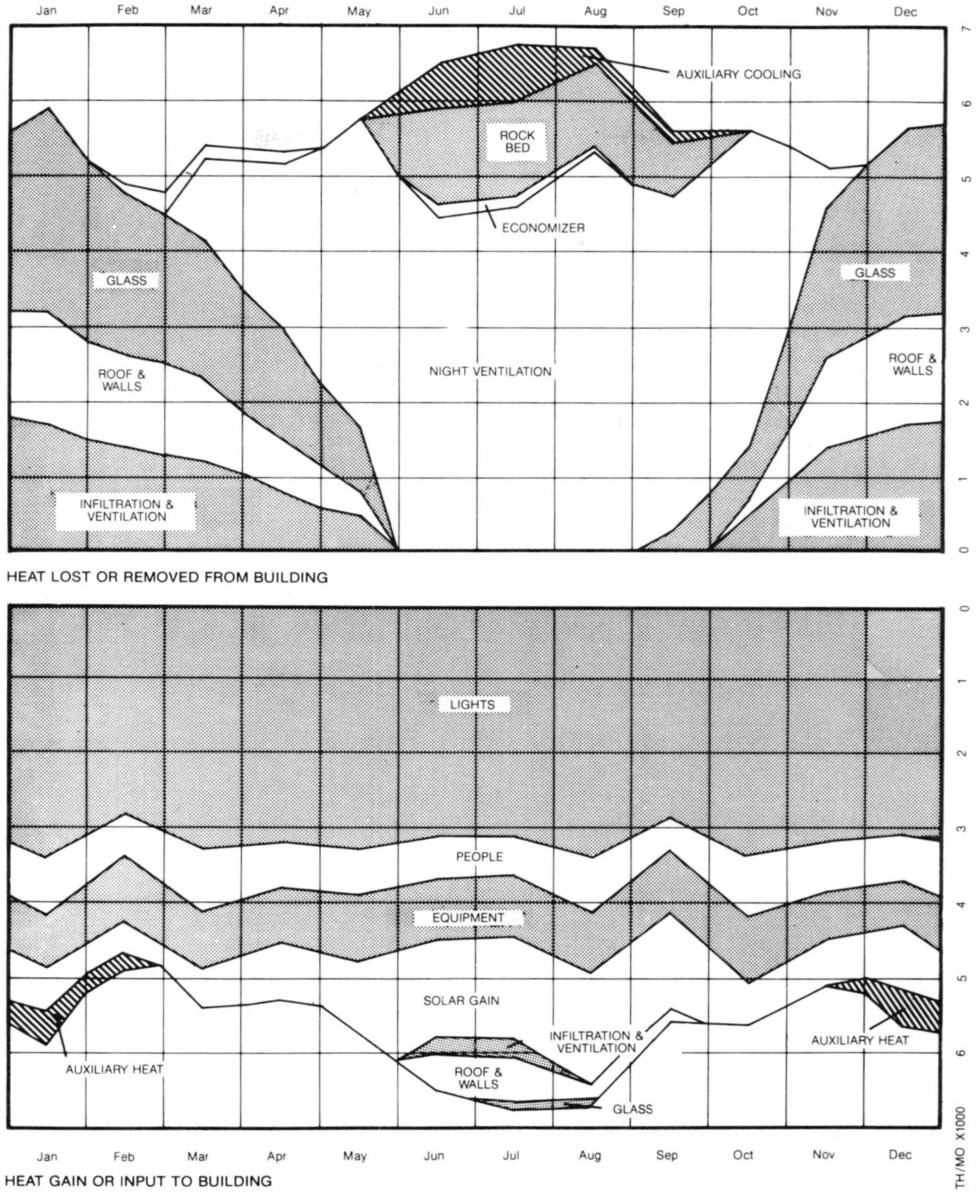

(d)

Fig. 5.29. *(continued)*

east windows and in the afternoon on west windows. From March through September, simple fixed overhangs block all south sun from offices.

Daylighting is an important design strategy: every portion of the building is within 40 ft of a window or of the daylit atrium. Each window has venetian blinds to control daylight distribution, and direct sun entering the atrium is intercepted and diffused by hanging banners before it can annoy office workers. The supplementary electric lighting still required is provided as indirect light, so that much of its heat is immediately absorbed by the exposed-concrete, thermally massive ceiling structure. Direct task lights can be used as required.

The exposed concrete structure stores the

building's heat by day; on really hot days, warm indoor air can be blown into the rock beds in exchange for their stored cool air. At night, cool outdoor air is blown throughout the building by a forced-air distribution system, capable of 6 ACH, thus removing the structure's stored heat. The Bateson Building is designed to require 19,900 Btu/ft^2 per year.

(a) Night Ventilation of Thermal Mass. Before doing these detailed calculations, be sure that (1) you have checked your summer climate against the passive cooling design strategies (Fig. 2.8) and that night ventilation of thermal mass is appropriate and (2) your building and climate have been checked for approximate performance based on rules of thumb (Fig. 5.19) for this cooling strategy. The following procedure is adapted from one developed by Karen Crowther for a workshop at the Fifth National Passive Solar Conference in Amherst, Massachusetts, 1980.

STEP 1. In Column II of Table 5.56, list the hourly outdoor temperatures for the design condition; (these may be approximated from the summer DB temperature and mean daily range given in Appendix A). This will give the worst-day performance. (To get average-day performance, list average hourly temperatures, which are available from local weather service.) You need not list temperatures above 80 F.

STEP 2. Calculate the 24-h heat gain for the building, in Btu. Find the sum of all the hourly heat gains through the envelope, the minimum ventilation while "closed," and the internal gains while operating. List on line H of Table 5.56.

TABLE 5.56 **Night Ventilation of Thermal Mass**

Calculation Procedure

A. Mass surface area (from step 3)	______ ft^2
B. Mass heat capacity (from step 4)	______ Btu/F
C. Floor area (supplementary cooling, step 5), and	______ ft^2
Total building volume	______ ft^3

(I) Hour	*(II)* Outside Air Temperature (F)	*(III)* Cooling (Btu)	*(IV)* Mass Temperature (F)
8 P.M.	______	______	______
9 P.M.	______	______	______
10 P.M.	______	______	______
11 P.M.	______	______	______
Midnight	______	______	______
1 A.M.	______	______	______
2 A.M.	______	______	______
3 A.M.	______	______	______
4 A.M.	______	______	______
5 A.M.	______	______	______
6 A.M.	______	______	______
7 A.M.	______	______	______
8 A.M.	______	______	______
9 A.M.	______	______	______

D. Total mass cooling	______ Btu	
E. Final mass temperature		______ F
F. Supplementary cooling (see Steps 5 and 11)	______ Btu	
G. Total cooling, D + F	______ Btu	
H. 24-h heat gain, from step 2	______ Btu	
I. Flow rate required for night ventilation	______ cfh	or ______ ACH

STEP 3. Find the total area of the thermal mass surface that is *exposed* (no rugs, etc.) both to the space to be cooled *and* to moving night air during the ventilation (''open'') cycle; list on line A of Table 5.56. *Note:* The larger the mass area exposed, the better the performance. This is why direct-gain solar-heated buildings often make such good candidates for night ventilation cooling. Two additional comments on mass surface: place it where people ''see'' it, so that it can readily receive their radiant heat; and keep direct sun off the mass (and out of the building) during the cooling season.

STEP 4. Find the mass heat capacity for the entire space to be cooled: mass volume × density × specific heat. Table 4.2 lists both density and specific heat for most common building materials; Table 5.57 shows a quick way to get mass heat capacities for the most common thermal mass materials. Enter this total mass heat capacity on line B of Table 5.56.

STEP 5. For ''supplementary'' cooling due to surfaces *other* than those of the principal thermal mass, list the space's *floor area*. This step should only be taken for spaces with a significant amount of roof, wall, or floor area *in addition* to the thermal mass areas counted in step 3. *Examples:* If your space has exposed concrete ceilings, walls, and floors—all counted as thermal mass—skip this step. If all your thermal mass is in the form of free-standing water containers, enter the entire floor area. If the entire ceiling is thermal mass, but the walls or the floors are not, enter half the floor area. If the entire ceiling is thermal mass, but the floor is not, and there are few or no walls (e.g., open office plan), enter one-third to one-fourth of the floor area.

STEP 6. Complete column III hour by hour, after determining the mass temperature (column IV for the preceding hour):

$$\text{cooling Btu} = \begin{matrix}\text{previous}\\ \text{hour}\\ \text{mass}\\ \text{temp.}\\ \text{(col. IV)}\end{matrix} - \begin{matrix}\text{outside}\\ \text{temp.}\\ \text{(col. II)}\end{matrix} \times \begin{matrix}\text{mass}\\ \text{surface}\\ \text{area}\\ \text{(line A)}\end{matrix}$$

For the first hour, assume that the mass temperature is 80 F.

STEP 7. Complete column IV hour by hour, after calculating the cooling Btu (column III):

$$\text{mass temp.} = \begin{matrix}\text{previous}\\ \text{hour}\\ \text{mass temp.}\\ \text{(col. IV)}\end{matrix} - \frac{\text{cooling Btu (col. III)}}{\text{mass heat capacity (line B)}}$$

STEP 8. Continue this hourly process using columns III and IV *until* the falling temperature of the mass equals the rising temperature of the outdoor air. At that point, continuing with plentiful ventilation will only rob the mass of its ''coolth''; the building therefore switches to closed mode, with minimal ventilation.

STEP 9. Add all the hourly cooling Btu values (column III) to obtain the total mass cooling (line D).

STEP 10. Note the final mass temperature, from column IV. This is likely to be at least 5 F° *above* the *lowest* air temperature of the night (column II). If the mass temperature is significantly higher, consider redesigning for more exposed mass surface area.

TABLE 5.57 **Common Mass Heat Capacities**

Conventional Units			*S.I. Units*
ft³	*× (Btu/ft³ F) = Btu/F*		*m³ × (kJ/m³K) = kJ/K*
Volume ×	(62.4)	Water	volume × (4181)
Volume ×	(22.5)	Ordinary concrete	volume × (1507)
Volume ×	(18.7)	Masonry, grout-filled	volume × (1253)
Volume ×	(15.6)	Brick	volume × (1045)

Source: Adapted from Crowther, "Night Ventilation Cooling of Mass," in Miller (1980).

STEP 11. If supplementary cooling is appropriate (see step 5), calculate it as follows.

$$\text{supplementary cooling Btu} = \left[80\text{ F} - \text{final mass temperature (line E)}\right] \times 2.25 \times \left[\text{floor area (line C)}\right]$$

The 2.25 factor assumes a modest role for the other, less thermally massive surfaces. Enter this supplementary cooling on line F.

STEP 12. Obtain total cooling by adding lines D and F; enter total on line G.

STEP 13. In step 2, you entered the 24-h heat gain for the building on line H. Compare this cooling Btu needed to the total provided (line G). If you haven't provided enough cooling, and the final mass temperature is more than 7 F° above the lowest nighttime outdoor temperature, consider redistributing the building mass over a wider surface (e.g., 3000 ft² of 4-in. slab rather than 2000 ft² of 6-in. slab) and trying again. If you don't have enough cooling, and the final mass temperature is 5 F° to 7 F° above the lowest nighttime outdoor temperature, consider providing both more mass and more surface area (e.g., 3000 ft² of 4-in. slab rather than 2000 ft² of 4-in. slab) and trying again.

STEP 14. Determine and enter on line I the approximate flow required for night ventilating air. Use the formula

$$\text{cfh} = \frac{\text{Btu/h}}{0.018\ \Delta T}$$

where cfh is the minimum required flow rate of night air, Btu/h is the cooling Btu for the hour of maximum cooling during the night (column III), and ΔT is the temperature difference between the final mass temperature (column IV) and the outdoor air (column II) for that same hour of maximum cooling. It is often useful to express this night ventilation flow rate in terms of air changes per hour (ACH).

$$\text{ACH} = \frac{\text{cfh required}}{\text{building volume (ft}^3\text{)}}$$

EXAMPLE 5.3, PART C. Find the detailed cooling performance in night ventilation of thermal mass for a typical bay of the Denver office building described in Example 5.3, Part A (also in Example 5.1 and Figs. 5.14 and 5.22). Begin by assuming a structural configuration as shown in Fig. 5.30; the space between the concrete joists is washed by night-ventilation air, distributed from supply ducts at alternate girders, and collected by return ducts at alternate girders. The bay size is 25 × 30 = 750 ft² (to include corridor).

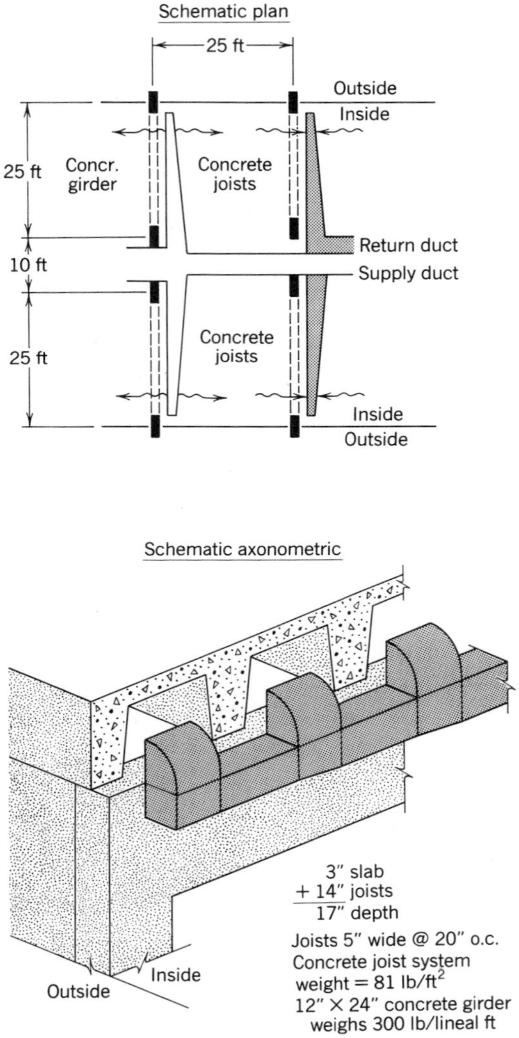

Fig. 5.30. Relationship between ductwork for night ventilation and the thermal mass of the concrete structural system, for the Denver office building discussed in Example 5.3. (See also Fig. 5.22.)

STEP 1. From Appendix A, the design temperature is 91 F, mean daily range 28. This is distributed (with the help of the sine curve equation) as shown in Table 5.58; minimum temperature assumed at 3 A.M., maximum at 3 P.M.

STEP 2. The 24-hour heat gain can be approximated from the results of Example 5.3, Part A. While the building is occupied, the heat gains approximate 12.3 Btu/h ft^2 for a typical floor (including the effects of the opaque end walls). When it is unoccupied, the gains disappear for lights, people, equipment, windows, and ventilation. However, the opaque envelope gains remain, since this envelope stores

TABLE 5.58 **Denver Office Building Performance: Cooling by Night Ventilation of Thermal Mass (for a Typical Bay, 25 ft × 30 ft)**

A. Mass surface area	1,630 ft^2
B. Mass heat capacity	10,238 Btu/F
C. Floor area (supplementary cooling)	(188 ft^2)
Total building volume (25 × 30 × 12)	9,000 ft^3

(I) Hour	*(II)* Outside Air Temperature (F)	*(III)* Cooling Btu		*(IV)* Mass Temperature (F)
8 P.M.	80	none		80
9 P.M.	77	(80 − 77)1630 (line A) =	4,890	$80 - \left(\frac{4{,}890}{10{,}238, \text{ line B}}\right) = 79.5$
10 P.M.	74	(79.5 − 74)A =	8,965	$79.5 - \frac{(8{,}965)}{B} = 78.6$
11 P.M.	71	(78.6 − 71)A =	12,390	$78.6 - \frac{(12{,}390)}{B} = 77.4$
Midnight	68	(77.4 − 68)A =	15,320	$77.4 - \frac{(15{,}320)}{B} = 75.9$
1 A.M.	66	(75.9 − 66)A =	16,135	$75.9 - \frac{(16{,}135)}{B} = 74.3$
2 A.M.	64	(74.3 − 64)A =	16,790	$74.3 - \frac{(16{,}790)}{B} = 72.7$
3 A.M.	63	(72.7 − 63)A =	15,810	$72.7 - \frac{(15{,}810)}{B} = 71.2$
4 A.M.	64	(71.2 − 64)A =	11,735	$71.2 - \frac{(11{,}735)}{B} = 70.1$
5 A.M.	66	(70.1 − 66)A =	6,685	$70.1 - \frac{(6.685)}{B} = 69.4$
6 A.M.	68	(69.4 − 68)A =	2,280	$69.4 - \frac{(2{,}280)}{B} = 69.2$
7 A.M.	71	Stop—mass temperature is below outdoor air temperature.		
8 A.M.	74			
9 A.M.	77			

D. Total mass cooling	111,000 Btu
E. Final mass temperature	69.2 F
F. Supplementary cooling	4,570 Btu
G. Total cooling	115,570 Btu
H. 24-hour heat gain	94,275 Btu
I. Flow rate required for night ventilation	about 10 ACH

daily heat and releases it gradually. Therefore, unoccupied gains total 0.7 for north/south walls and 0.3 for end walls, for a total of 1.0 Btu/h ft^2.

Occupied: 9 h × 12.3 = 110.7 Btu/day ft^2
Unoccupied: 15 h × 1.0 = 15.0 Btu/day ft^2
127.5 Btu/day ft^2 × 750 ft^2 = 94,275 Btu/day

STEP 3. Find the total area of mass. Assume that a carpet covers the floor (for comfort, appearance, and acoustic absorption). The concrete girders are exposed on the bottom and on one side; the other side is mostly covered by ductwork (supply or return occurs at each girder). Therefore, the exposed surface, per bay, is

12-in.-wide bottom + 24-in.-high side + 12-in.-wide top = 48 in. per linear foot

48 in × 25 ft length per bay = 100 ft^2

The concrete joists are 25 in. on center; in that 25 in., they expose 5 in. of joist bottom, 14 in. + 14 in. = 28 in. of their sides, and about 18 in. of the underside of concrete slab, providing a total of 51 in. of exposed concrete per 25-in. spacing:

$$\frac{51}{25} = 2.04 \text{ ft}^2 \text{ exposed concrete per ft}^2 \text{ floor area}$$

Note: The corridor ceiling is largely hidden by air ducts, and largely unavailable for cooling. However, the gains from lights, people, and equipment will be considerably less in this circulation area. For simplicity, these effects are assumed to cancel one another out.

For a typical 25 ft × 30 ft bay = 750 ft^2 floor area, the exposed mass surface is

Girders	= 100 ft^2
Slab/joists, 2.04 × 750	= 1530 ft^2
	1630 ft^2

STEP 4. Determine the mass heat capacity. The girder volume is 2 ft^3/lin. ft × 25 ft = 50 ft^3. The slab/joists weigh 81 lb/ft^2. Since structural concrete has a typical density of about 150 lb/ft^3, we can assume that the slab/joist volume is 81/150 = 0.54 ft^3 per square foot of floor, or 0.54 × 750 ft^2 = 405 ft^3.

total volume = 50 ft^3 girder + 405 ft^3 slab/joist = 455 ft^3

From Table 5.56

mass heat capacity = 455 ft^3 × 22.5 = 10,238 Btu/F

STEP 5. Little supplementary cooling will be provided, since the floor is carpeted and the ceiling is open so that it can be flushed with night air. Therefore, assume that only ¼ of the floor area contributes to cooling.

$$0.25 \times 750 = 188 \text{ ft}^2$$

STEPS 6 through 10. See Table 5.58.

STEP 11. Determine the supplementary cooling:

(80 F − 69.2 F) × 2.25 = 188 ft^2 = 4,570 Btu

STEP 12. See Table 5.58.

STEP 13. The building has more than adequate cooling capacity to meet this design condition. On an average summer day, with even lower outdoor temperatures (and lower infiltration heat gains), the building is clearly adequate for night ventilation of thermal mass.

STEP 14. The air flow required will be checked at 2 A.M., the hour of most cooling:

$$\frac{16{,}790 \text{ Btu/h}}{0.018 \times (72.7 - 64)} = 107{,}220 \text{ cfh}$$

$$\text{ACH} = \frac{107{,}220 \text{ cfh}}{9{,}000 \text{ ft}^3} = 12$$

However, since the cooling capacity exceeds the actual total stored heat gain, the actual flow rate can be approximated as

$$\frac{94{,}275 \text{ stored heat}}{115{,}570 \text{ cooling capacity}} = 0.82$$

Then, 0.82 × 12 ACH = about 10 ACH actual flow rate. Note that this total exceeds the flow predicted by the rule of thumb (6 ACH—see Example 5.3, Part A). The ducts are designed for 6 ACH at acceptable velocities for noise considerations; the noisier air flow rate of 10 ACH should not be a problem at night in an unoccupied office building.

(b) Roof Pond Cooling. Before doing the following detailed calculations be sure that (1) you have checked your summer climate against the passive cooling design strategies (Fig. 2.8) and that either the "evaporative" or the "high thermal mass" strategy is appropriate and (2) your building and climate have been checked for approximate performance based on the rules of thumb. Note, however, that the rule of thumb is based only upon summer night DB temperature. With evaporative cooling (such as a light spray onto the surface of the roof pond containers), better performance can be expected, as the following calculation procedure, based on Fleischhacker, Bentley, and Clark "A Simple Verified Methodology for Thermal Design of Roof Pond Cooled Buildings," in ASES (1982) shows. This procedure can be used to check the size and depth required for your building's roof pond. It assumes an *optimum pond depth of 4 in.*, but allows for other depths as well.

STEP 1. First, assemble the following data about your climate.

Maximum DB temperature (Appendix A).

Mean daily range (Appendix A).

Minimum DB temperature (= max DB − mean daily range).

Average maximum rh for July (from local climatological data, such as those given in Table 2.4)

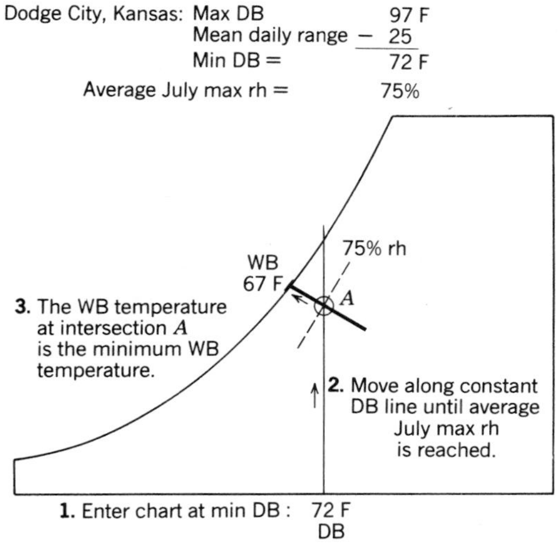

Fig. 5.31. Finding minimum WB temperature for Dodge City, Kansas.

July average temperature (TA July, in Appendix B).

From these data, determine two further characteristics for your climate: (1) minimum WB temperature; (2) average July operating hours for residential air conditioning, or N.

Minimum WB temperature can be determined from the psychrometric chart shown in Fig. 4.6. Enter the chart with minimum DB and move vertically along the constant DB line until you reach the average maximum rh for July. At that intersection, refer to the diagonal WB temperature lines, from which minimum WB temperature can be determined (see Fig. 5.31).

Average July operating hours, N, can be estimated from Fig. 5.32. Enter at July TA for your climate. Your climate's N will fall somewhere between the maximum N and minimum N lines; as a guide to estimating N for your climate compare your climate's mean daily range and rh to those of the cities shown in Fig. 5.32.

STEP 2. Calculate the peak hourly heat gain, *but exclude internal gains*. (The shorter method given in Section 5.7 is appropriate for this roof-pond-sizing procedure.)

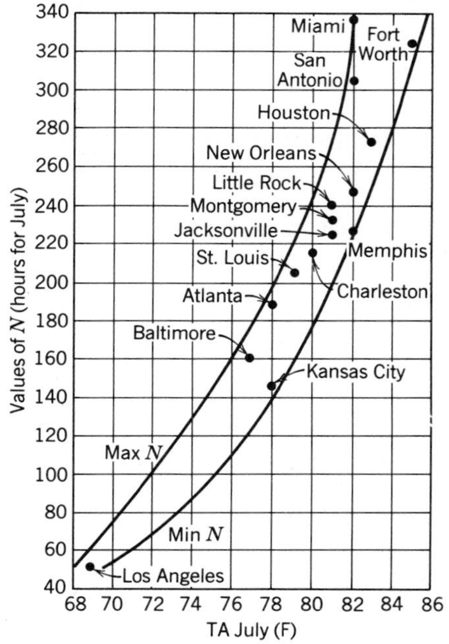

Fig. 5.32. Range of N*, average July operating hours for residential air conditioners.*

STEP 3. Approximate the daily total heat (excluding internal heat) to be stored in the roof pond, or Q_E, in Btu.

$$Q_E = \frac{\text{(Btu/h, peak hourly gain)}(N\text{, hours for July)}}{\text{31, days in July}}$$

STEP 4. Determine the rate of the internal gain, in Btu/h, while the building is occupied.

STEP 5. Approximate the daily total internal heat to be stored in the roof pond, or Q_I, in Btu.

$$Q_I = \text{hourly internal gains, in Btu/h} \times \text{daily hours of building occupation}$$

STEP 6. Calculate the daily heat gain directly to the roof pond through its insulated covers, or Q_P, in Btu. This formula assumes $R16$ insulating panels with white upper surfaces and foil faces on the under surface.

$$Q_P = 0.4(A_c)(4 \times \text{DB max, F} - \text{DB min, F} - 200)$$

where A_c is the horizontal surface area of pond in square feet.

STEP 7. Consider whether fans will be used to stir the room air below the roof pond (to help the heat exchange between the pond and the room, as well as to aid comfort) and determine the value to be used for h, the overall heat transfer coefficient.

	fpm (m/s)	0 (0)	44 (0.22)	73 (0.37)	115 (0.58)
Air velocities:					
Values of h:		1.25	1.47	1.53	1.70

STEP 8. Determine the highest comfortable internal air temperature, T_{op}, from the comfort criteria provided in Fig. 2.4.

STEP 9. Calculate the maximum allowable internal air temperature, T_{imax}, which will be higher than T_{op} by a factor F that is related to your climate's characteristics.

$$T_{imax} = T_{op} + F$$

Values of *F*

	Miami	*San Antonio*	*Phoenix*
(Max. DB, coincident WB)	(90/77)	(97/73)	(107/71)
(Mean daily range)	(15)	(19)	(27)
For 0 fpm interior air motion, F =	2.0	3.0	4.0
For 115 fpm interior air motion, F =	1.0	1.5	2.0

STEP 10. Determine the allowable temperature swing of the roof pond.

(a) Calculate the maximum pond T:

$$\text{max pond } T = T_{imax} - \frac{\left(\begin{array}{l}\text{peak total hourly heat gain,}\\ \text{including internal gains, Btu/h}\end{array}\right)}{h(A_c)}$$

where T_{imax} is from step 9, h is from Step 7, and A_c is the horizontal surface area of the pond in square feet.

(b) Calculate the minimum pond T, which depends on whether or not evaporative cooling will be used to help lower the pond's temperature and on several characteristics of the pond and the building.

1. For a *dry* pond surface,

$$\text{min pond } T_{dry} = \text{DB}_{min} + 1.5 \pm \text{corrections F° (if any)}$$

No Corrections Required	*Corrections Required*								
	Pond Depth (in.)			*Night Internal Load*			*Portion Exposed to Sky*		
For a 4-in. deep pond, 2 Btu/hr ft² night internal load, 100% exposed to sky, F° corrections:	2	6	10	0	4 Btu/h ft²	8 Btu/h ft²	⅓	½	⅔
+0	−1.5	+0.7	+1.0	+0	+0.8	+2.5	+3.3	+2.1	+1.2

2. For a *wet* pond surface,

$$\text{min pond } T_{\text{wet}} = \text{DB}_{\text{min}} - \frac{(\text{DB}_{\text{min}} - \text{WB}_{\text{min}})}{2}$$

c. Find the pond temperature swing ΔT_P:

$$\Delta T_{p\text{dry}} = \text{max pond } T - \text{min pond } T_{\text{dry}}$$

$$\Delta T_{p\text{wet}} = \text{max pond } T - \text{min pond } T_{\text{wet}}$$

(Obviously, if neither minimum pond T dry nor minimum pond T wet is lower than maximum pond T, a roof pond cannot be used!

STEP 11. Determine the required pond depth D (in inches).

$$D = \frac{(0.19)(Q_E + Q_I + Q_p)}{(\Delta T_p)(A_c)}$$

where Q_E, Q_I, and Q_P are the daily pond heat gains, from steps 3, 5, and 6; ΔT_p is from step 10; and A_c is the horizontal surface area of the pond in square feet.

Note: If D is less than 4 in., consider reducing the pond size and recalculating. If D is much more than 4 in., consider a larger pond area, more air motion, or the use of a wet pond surface to more closely approach the optimum 4-in. depth, or see optional step 12.

OPTIONAL STEP 12. Auxiliary mechanical air conditioning may offer a more economical alternative than increased pond size. The size, in "tons" of air conditioning required, is determined as follows.

a. Desired D is 4 in., optimum.

b. $\Delta T_p = \left(\dfrac{0.19}{D}\right)\dfrac{(Q_E + Q_I + Q_p)}{A_c}$

c. So,

$$\text{max pond } T = \text{min pond } T + \Delta T_p$$

d. And

$$\text{tons of AC} = \frac{\text{peak total hourly heat gain (Btu/h)} - [h(A_c)](T_{i\max} - \text{max pond } T)}{6000}$$

where the peak total hourly heat gain *includes* internal gains, h is from step 7, A_c is the horizontal surface area of the pond in square feet, $T_{i\max}$ is from step 9, and max pond T is from step 10.

EXAMPLE 5.4, PART B. Size the roof pond for the Albuquerque office building for which we predicted that a 4-in.-deep pond equal to ¾ of the building's floor area would be sufficient. Assume that the one story office is 4000 ft^2 in area. Try a 4-in.-deep pond of 3000 ft^2.

STEP 1. Albuquerque,

DB_{max} is 94 F (Appendix A).

MDR is 27F° (Appendix A).

DB min is 67 F.

July rh maximum is approximately 50% (local data).

July TA is 79 F (Appendix B).

WB_{min}, from Fig. 4.6 at the intersection of the 67 F DB line and 50% rh, is 56 F.

July operating hours N for this dry, high area should be close to the minimum for TA = 79; from Fig. 5.32, this is about 160 h.

STEP 2. Determine the peak hourly gain. Earlier in this exercise, a total gain of 15 Btu/h ft^2 was assumed as an average. Assume that 9 Btu/h ft^2 of this total represents the load from electric light, people, and equipment, and that 6 Btu/h ft^2 is due to envelope and ventilation gains. Then,

$$\text{peak hourly gains (excluding internal)} = 6 \times 4000 \text{ ft}^2 = 24{,}000 \text{ Btu/h}$$

STEP 3.

$$Q_E = \frac{24{,}000 \text{ Btu/h} \times 160 \text{ h}}{31 \text{ days}} = 123{,}870 \text{ Btu}$$

STEP 4.

$$\text{internal gains} = 9 \text{ Btu/h ft}^2 \times 4000 = 36{,}000$$

STEP 5.

$$Q_I = (36{,}000 \text{ Btu/h})(9 \text{ h operation}) = 324{,}000 \text{ Btu}$$

STEP 6.

$$Q_p = 0.4(3000 \text{ ft}^2)(4 \times 94 \text{ F} - 67 \text{ F} - 200) = 1200 \times 109 = 130{,}800 \text{ Btu}$$

THERMAL CONTROL

STEP 7. Fans will be used, at 115 fpm, for added comfort as well as for heat transfer. Therefore, $h = 1.7$.

STEP 8. From Fig. 2.4, with 115 fpm (0.58 m/s) air speed, it appears that 83 F is within the comfort zone. So $T_{op} = 83$ F.

STEP 9. $T_{imax} = T_{op} + F$. Albuquerque's conditions appear to be somewhere between those of San Antonio and those of Phoenix, so assume F to be 1.75.

$$T_{imax} = 83 + 1.75 = 84.75 \text{ F}$$

STEP 10.

$$\text{max pond T} = T_{imax} - \frac{\text{(total hourly gain)}}{h(A_c)}$$

$$= 84.75 - \frac{(24{,}000 + 36{,}000)}{1.7 \times 3000}$$

$$= 84.75 - 11.75 = 73 \text{ F}$$

(a) Assume a dry pond surface. Since this is a 4-in.-deep pond that is fully exposed to sky, with no night load, there is no correction factor.

(b) min pond T_{dry} = DB min + 1.5 = 67 + 1.5 = 68.5 F
For a wet pond,

$$\text{min pond } T_{wet} = DB_{min} - \frac{(DB_{min} - WB_{min})}{2}$$

$$= 67 - \frac{(67 - 56)}{2}$$

$$= 61.5 \text{ F}$$

(c) the pond temperature swing is therefore

$$\Delta T_p \text{ dry} = 73 - 68.5 = 4.5$$

$$(\Delta T_p \text{ wet} = 73 - 61.5 = 11.5)$$

STEP 11.

$$D_{rqd} = \frac{0.19(123{,}870 + 324{,}000 + 130{,}800)}{4.5 \times 3000}$$

$$= \frac{109{,}947}{13{,}500} = 8.1 \text{ in., for a dry pond.}$$

(For a wet pond, D_{rqd} = 3.2 in.)

It appears that a 4-in. deep pond with a wet surface of somewhat less than 3000 ft^2 would be sufficient for this building in this climate.

References

Andrews, F. (1966). *The Architect's Guide to Mechanical Systems,* Reinhold, New York.

ASES (1982). *Progress in Passive Energy Systems,* Vol. 7, Proceedings of the 7th National Passive Solar Conference, American Solar Energy Society, Inc., Boulder.

Publications of the American Society of Heating, Refrigerating and Air Conditioning Engineers, Inc., Atlanta: *Handbook of Fundamentals,* 1981, *Cooling and Heating Load Calculation Manual,* 1979.

Balcomb, J. D., Barley, D., McFarland, R., Perry, Jr., J., Wray, W., and Noll, S. (1980). *Passive Solar Design Handbook,* Vol. 2: *Passive Solar Design Analysis,* U.S. Department of Energy, Washington, D.C.

Balcomb, J. D., Jones, R. (Ed.), Kosiewicz, C., Lazarus, G., McFarland, R., and Wray, W. (1983). *Passive Solar Design Handbook,* Vol. 3, American Solar Energy Society, Inc., Boulder.

BEPS, Building Energy Performance Standards (1979). *Federal Register,* Vol. 44, no. 230, November 28. Proposed by the U.S. Department of Energy.

Bowen, A., Clark, E., and Labs, K. (1981). *Passive Cooling,* American Solar Energy Society, Boulder.

Brown, G. Z., Reynolds, J. S., and Ubbelohde, S. (1982). *Inside/Out: Design Procedures for Passive Environmental Technologies,* Wiley, New York.

Buehrer, H. (1978). *Estimating Energy Usage.* Developed for the AIA Energy Audit Seminars.

Egan, M. C. (1975). *Concepts in Thermal Comfort,* Prentice-Hall, Englewood Cliffs, NJ.

Johnson, Timothy (1981). *Solar Architecture, The Direct Gain Approach,* McGraw-Hill, New York.

Mazria, Edward (1979). *The Passive Solar Energy Book,* Rodale Press, Emmaus, PA.

Miller, H. (Ed.) (1980). *Proceedings of the Passive Cooling Workshop,* 5th National Passive Solar Conference, Amherst. Published by American Solar Energy Society, Inc., Boulder.

Millet, M. and Bedrick, J. (1980). *Manual Graphic Daylighting Design Method,* Department of Architecture, University of Washington, Seattle; Sponsored by the U.S. Department of Energy.

Olgyay, V., and Olgyay, A. (1957). *Solar Control and Shading Devices,* Princeton University Press, Princeton, NJ.

Watson, D., and Labs, K. (1983). *Climatic Design, Energy Efficient Building Principles and Practices,* McGraw-Hill, New York.

6

SYSTEMS AND EQUIPMENT FOR HEATING AND COOLING

The preceding chapters have dealt with most of the early steps in the design process, such as the choice of thermal design strategy (Chapter 2), the choice of siting (Chapter 3), and the choice of the components and initial sizing of envelopes (Chapters 4 and 5). This chapter carries the design process into the specifics of mechanical systems and equipment.

There are various ways to organize this large collection of components and systems. This chapter is organized as follows.

1. A review of the relative thermal role of building envelopes versus their internal heating/cooling equipment.
2. Some details of *envelope components* typically used to reduce the need for heating and cooling.
3. A look at the heating/cooling *system design process* begun by the architect and completed by the engineer.
4. The basics of the process by which mechanical systems are integrated into building design.
5. Some details of *internal components*; the refrigeration cycles.
6. More details of small-building heating/cooling systems.
7. The four generic HVAC system types for large buildings.
8. Larger, centralized HVAC equipment.
9. *through* 12. Details of the four generic HVAC systems.
13. *and* 14. Large-scale system opportunities.

Items 1, 3, 4, and 7 are a more theoretical introduction to systems and equipment; hence, rather than reading the chapter in order, you could read those items first, then go on to the component-and-equipment sections.

6.1 Review of the Need for Mechanical Equipment

One of the most basic functions of buildings is to provide shelter from weather. In a carefully designed building, the roofs, walls, windows, and interior surfaces alone can maintain comfortable interior temperatures for most of the year in most North American climates. With appropriate scheduling, the most uncomfortable hours within buildings can often be avoided; for example, the siesta avoids the hottest afternoon hours within stores and office buildings. Several aspects of comfort and climate, however, pose difficult challenges for ordinary building forms and materials.

A building surface primarily influences comfort through its *surface temperature*; secondarily, it slowly changes *air temperature* (as when cool air moves across a warm surface). Important as these two determinants of comfort are, they sometimes are not sufficient by themselves. Especially in cooling situations, *air motion* and *relative humidity* are significant comfort determinants; and for many indoor activities, *air quality* becomes an important issue in both heating and cooling.

Building form can work with climate to produce air motion for cooling, although the faster air speeds that extend the human comfort zone above 83 F (28°C) may be difficult to provide without mechanical assistance. Relative humidity is still controlled by mechanical (or chemical) means. Building form and materials may be able to keep spaces surprisingly cool, but without dehumidification, surfaces in many North American summer climates can become clammy and covered with mold.

It is difficult to filter air without a fan to force the air through the filtering media. Electrostatic filtering, of course, is done within mechanical equipment.

Thus, whereas the desired air and surface temperatures can often be achieved by passive means (a combination of building form, surface material, and occasional user response), the comfort determinants of air motion, relative humidity, and air quality often require mechanical devices. As the control of air properties—motion, moisture, particulate content—becomes more critical to comfort, the designer becomes more likely to respond with a sealed building, excluding outdoor air except through carefully controlled mechanical equipment intakes. In the recent past, this exclusion of outdoor air has often been accompanied by the exclusion of daylight, of view, of solar heat on cold days—in sum, by a general rejection of all aspects of the exterior environment. As designers come to terms with the role of mechanical equipment, we should also clarify the role of these devices in relation to the climate: Are they occasional modifiers, permanent interpreters, or permanent excluders of the outdoors? The changing relationships, as the year progresses, among microclimate, building design, and mechanical equipment are shown in Fig. 6.1.

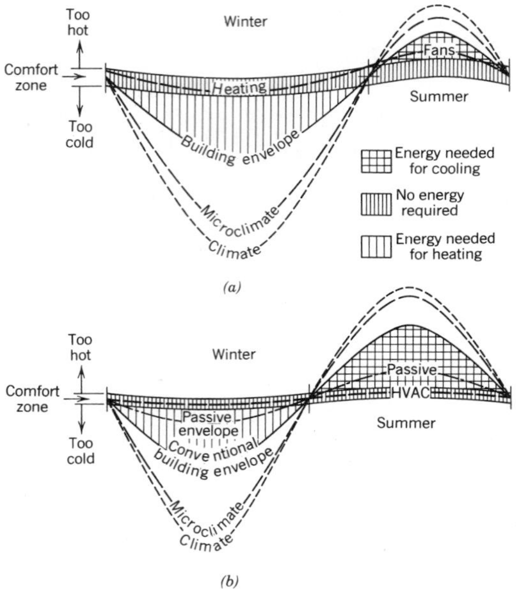

Fig. 6.1 Meeting the heating and cooling loads imposed by a "temperate" climate: by siting (using microclimate), by building design, and by utilizing energy consumed by mechanical equipment. (a) *Many of today's older buildings were designed before air conditioning; lower expectations of indoor comfort meant a wider "comfort zone" range, and conditions in the shorter summer season (aided by fans) frequently exceeded the comfort zone. Energy used for heating was frequently of far more concern than was the electricity to run the fans.* (b) *With the advent of full mechanical air conditioning, a narrower range of conditions could be maintained. Higher lighting levels provided much of the energy used for heating and extended the cooling season considerably. Energy used for cooling became the main concern. With sealed buildings, utilization of microclimate advantages (such as windier summer locations) presents less opportunity for energy savings. Passive heating and cooling, through design of the building envelope, offers considerable energy savings in both seasons.*

6.2 Passive Heating and Cooling Components

This section precedes the discussion of conventional heating and cooling equipment and its related design process, because so many "passive" components *reduce the need* for conventional thermal devices. Yet the *preliminary* design process—discussed at the beginning of Section 6.3—clearly includes the consideration of these passive components.

At first glance, the office building in Fig. 6.2 appears to be a conventional glass—perhaps curtainwall—box. But passive approaches in this Niagara Falls, New York, example nearly eliminate infiltration, provide almost equal inside surface temperatures on all four exterior walls, and provide enough daylighting to eliminate daytime electric lights for almost half the office space. In this case, it all happens at the perimeter, where two vertical planes of glass are spaced 4 ft apart; in this void are movable horizontal sunshading louvres, which also help distribute daylight evenly and provide some night insulation. For cooling, natural ventilation of this glass cavity occurs through the stack effect.

At the end of this section are two other examples in which passive design principles more thoroughly organize the interior arrangement and overall building form. Before these examples, an array of components is discussed, although we can only begin to cover the rapidly expanding range

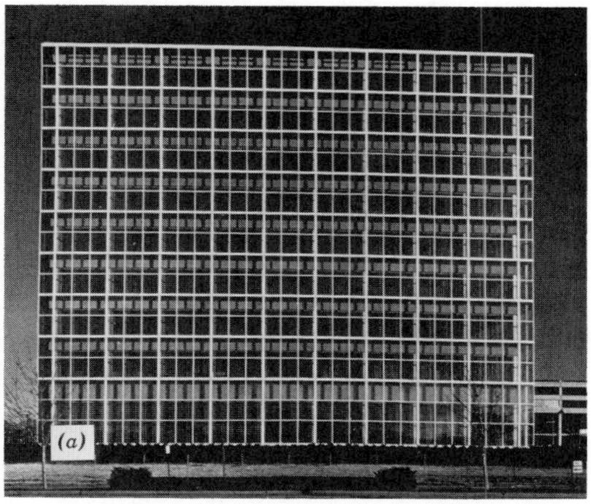

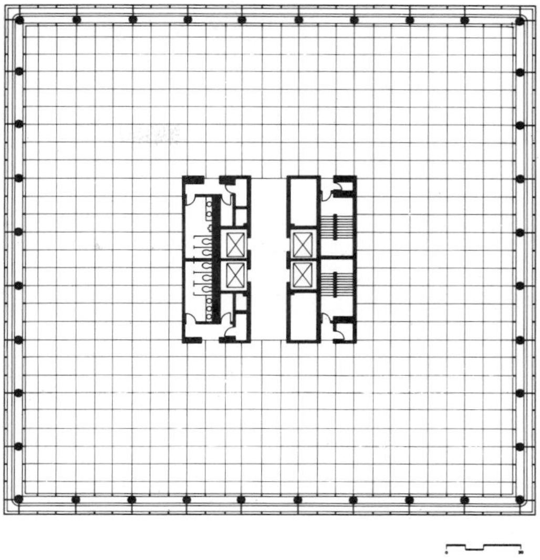

(b)

Fig. 6.2 *The Occidental Chemical Corporate Office Building, Niagara Falls, New York. Architects: Cannon Design, Inc.* (a) *Fully exposed to sun and wind despite its downtown site, the building appears as a conventional curtainwall box.* (b) *An ordinary-looking plan utilizes a central core, with suspended ceilings for horizontal service distribution.* (c) *A corner office displays the horizontal louvres, 15 ft long and 8 in. apart, painted white for best diffusion of daylight. They are motor-controlled by light sensors and can be closed on cold nights. Though all facades appear to be alike, the louvre positions will vary on sunny days.* (d) *The 4 ft. wide cavity allows for maintenance walkways and ample natural ventilation by stack effect. Internal and solar gains have made central heating unnecessary. Temperatures of the inside glass surface are nearly equal on all four sides; a perimeter heating/cooling system is therefore unnecessary, and the entire building is served by an all-air, variable-volume HVAC system.*

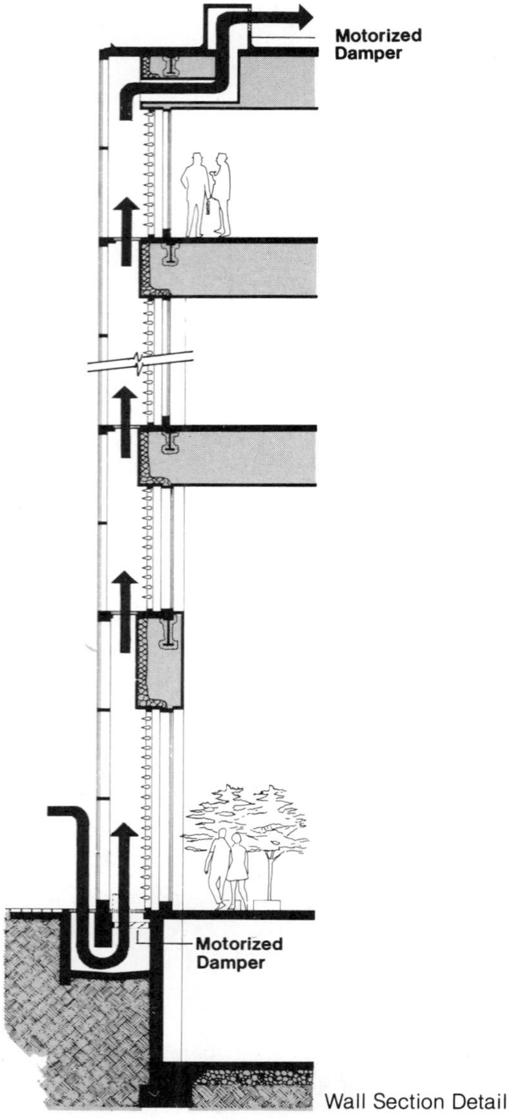

(d)

Fig. 6.2 *(continued)*

of products and processes available to the energy-conscious designer.

(a) Sunshading. Perhaps the single most important energy-related component for passively cooled buildings is the sunshade. (And since most solar-heated buildings will experience overheating in hot weather, sunshading is critical for passively heated buildings as well.) If correctly implemented, sunshading rejects nearly 80% of solar heat gains yet aids in distributing daylighting deep into buildings, where it replaces or reduces the internal gains due to electric lights.

If a building is arranged to intercept the intense rays of the sun before they pass through its glass walls instead of afterward, the air conditioning cooling load can often be cut in half. In approximate terms, external shading rejects about 80% of the fierce attack of solar energy, whereas internal shading accepts and reradiates 80% of it. Outside louvres have a chance to cool off in an occasional breeze, but inside drapes are part of a heat trap, and they constitute a system of hot-weather radiant heating that discomforts those who must work near perimeter surfaces.

Just how fierce is the sun? Compare it to the traditional heating effect of cast-iron steam radiation. Each square foot of such radiation produces 240 Btu/h. The amount of energy passing through 1 ft^2 of unshaded glass on an east wall in the morning is often evaluated at over 200 Btu/h. It can be frightening to think of an entire wall, blanketed *in summer* by a heat-producing source that is potentially almost as powerful as cast-iron radiation. The rejection of a maximum amount of this solar energy is obviously important for cooling performance. Exterior sunshades, then, represent the better initial design choice. In order to properly reject direct sun yet allow for view and daylight, many sunshades project out from the windows they protect. These exterior projections become highly visible elements of facades, and they tempt some designers into superimposing formal esthetic criteria that can be damaging to the solar control functions. A frequent example is the application of the same sunshade geometry to *all* elevations of a building. When the sunshades are fixed in position, this is usually counterproductive (see Fig. 3.12). Where they are movable, as in Fig. 6.2, this same sunshade approach is not so serious.

Fixed sunshading devices are very common, partly because they lack the moving parts and controls that can be expensive and troublesome. As explained in Section 3.4, they do not ordinarily serve very well, since in order to block the sun on any elevation in September, they must also block it in March. For many buildings in temperate climates, March is a heating-need month, whereas September is a cooling-need month. Because of this disadvantage, fixed sunscreens are not discussed further here; a procedure for evaluating their shading performance is presented in Appendix D.

Adjustable sunshading, once enormously common (it seemed as if every 1930s shop had an awning) but later considered "old-fashioned," has made a comeback with the advent of passive cooling. Some basic approaches to adjustable devices are shown in Table 6.1. Roll-down shades can be seen on Sacramento's Bateson Building (Fig. 5.29). Adjustable awnings, shown in Fig. 6.3, were also used on the buildings shown in Figs. 2.20 and 2.21 and in the Conservation Center for the Forest Society in Concord, New Hampshire (Fig. 6.20). Horizontal movable louvres enclose the Occidental Chemical offices (Fig. 6.2). Vertical movable louvres are particularly effective on east and west orientations, where they can block the sun from (for example) the southwest while maintaining a view out to the northwest.

How are movable shading devices adjusted to the desired position? Most manufacturers offer three types of controls: manual, motorized, and automatic. Manual systems are cheap and relatively trouble-free, but they require thoughtful, timely action on the part of the users of the building. So do motorized controls, but for large or heavy devices in remote places (clerestories, for example), the motorized control provides a steady hand for smooth operation. Automatic systems have the advantages of freeing the users from adjustment tasks and of taking into account the thermal needs of the building as a whole when setting the sunshade's position. With computerized controls becoming widespread for larger buildings, the added costs of automatic sunshading control can be easily incorporated into the overall cost of controls.

(b) Glazing. The thermal insulating properties of glass and plastics were detailed in Section 4.4 (see Table 4.17 in particular), and glazing's light transmission will be discussed in later chapters on lighting. A third important characteristic of glazing, for passive solar performance, is *solar transmission*. Several recently developed glazing products offer either increased solar transmittance or improved insulating performance. Either approach will aid in passive solar heating. Provided that careful attention is paid to glazing's role in daylighting year-round and that shading protection is provided in summer, good choices for all-season performance can be identified.

When cooling is the dominant concern, many designers use glazing that permanently rejects solar heat. The potential problems with reflective glazing were shown in Figs. 3.9, 3.10, and 3.11. However, heat-rejecting glass *is* appropriate when passive solar heating is *never* advantageous for a building.

Table 6.2 shows the ratios of solar transmission to heat loss for a variety of glazings used in passive solar heating applications. High ratios indicate materials that perform well. Glazings with lower ratios may have net daily heat losses; in colder climates, such glazings will require movable insulation at night, in order to admit more daytime solar heat than is lost through the glazing over 24 hours.

Recent improvements in the insulating performance of glazing commonly have followed two approaches: reduction of losses by convection and reduction of losses by radiation. *Convection losses* can be cut by the introduction of a screen of transparent cellular materials between the sheets of double glazing. This breaks up the convection currents that form in the glazing cavity, but it also interferes with the view through the window. Alternatively, the glazing cavity can be filled with heavy gas (such as argon) and then sealed.

Radiation losses can be greatly reduced by the introduction of *selective transmitter* film somewhere within the glazing cavity. As shown in Fig. 6.4, these films admit most of the incoming solar radiation in both visible and near-infrared (short) wavelengths. As objects become warmed within the room, however, they begin to emit far-infrared (long wave) radiation that is reflected back into the room by the selective film. These selective films typically are available either as separate sheets that can be inserted between sheets of glazing as a window is fabricated or as coatings that are applied to glass at the factory. As a separate sheet, a selective film could also be applied to existing windows—for instance, between storm windows and the ordinary windows they protect.

Selective films typically transmit slightly less solar energy than is transmitted by ordinary glass. Another approach to solar windows is to use a film that *increases* solar transmittance over ordinary

TABLE 6.1 **Adjustable Sunshading**

	Cooling and Daylighting Performance While Shading					
	Solar Heat Rejection	*Daylight Distribution*	*View*	*Ventilation*	*Typical Orientation*	*Heating: Night Insulation*
Roll-down shades						
Solid 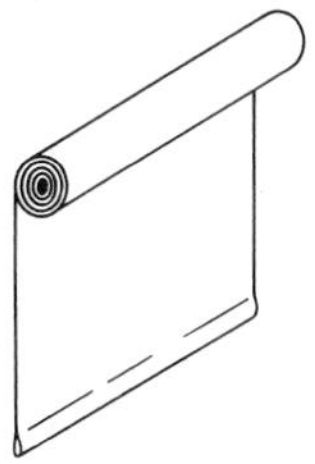	Blocks both direct and ground-reflected light. Heat builds up behind shade, unless vented.	Translucent shades diffuse sunlight evenly but may themselves become sources of glare, due to their brightness. If colored, they will color incoming diffuse light.	Blocks view.	Blocks ventilation.	East and west, where it is rolled down daily during hours of threat from direct sun.	Can be effective if infiltration above and around the shade is blocked.
Slatted 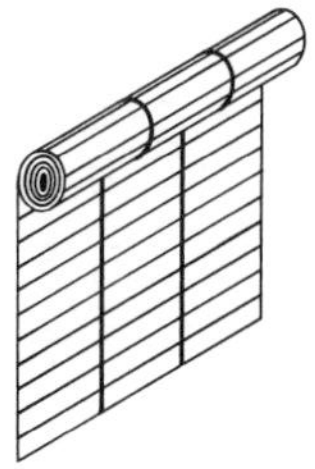	Blocks most direct sun and ground reflected light.	Thinly striped pattern of direct sun is admitted.	Heavily filters view.	Greatly reduces ventilation.		No value, unless slats can fit together to eliminate openings.

Fold-out Retractable Awning 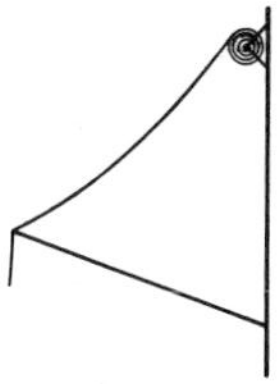	Blocks direct sun, admits ground reflected light. Heat on awning easily dissipated by breezes.	Translucent awnings diffuse sunlight evenly. (Colored translucent awnings color the diffused light.) Awnings admit ground-reflected light toward ceiling, where white surfaces diffuse it deeper into space.	Blocks sky view. Allows ground view.	Reduces ventilation somewhat, tends to direct wind upward into space.	Any orientation (on south, awnings can be deep and can be left in position for weeks at a time).	If awning folds back against window, it provides some insulation, but infiltration reduces its value.
Pivoting Louvres						
Horizontal 	Blocks direct sun, admits ground-reflected light.	Sunlight and ground-reflected light directed toward ceiling, where white surfaces diffuse the light deeper into space.	Blocks sky view. Filters ground view.	Reduces ventilation somewhat; directs wind upward into space.	South; also used on other elevations.	Closed louvres provide some insulation, but infiltration reduces its value.
Vertical 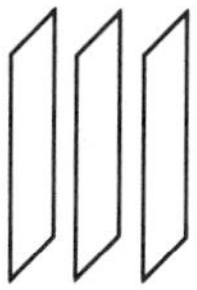	Blocks direct sun, admits some sky and ground-reflected light.	Diffuse and ground-reflected light directed sideways into space. Does not usually help with deeper daylight penetration.	Filters view.	Reduces ventilation somewhat; directs wind sideways into space.	East and west; sometimes south and north.	

Fig. 6.3 Movable awnings can become an important design element, by helping to change a building's appearance with changing sun positions. (Courtesy of Levolor-Lorentzen, Inc.)

TABLE 6.2 **Ratios of Average Solar Transmission to Heat Loss for Some Vertical Glazing Assemblies**[a]

1. Single-pane D.S. flat glass[b]	$\frac{84\%}{1.13 \text{ Btu/h F}} = 74.3$
2. Double glass (½″ air space)	$\frac{71\%}{0.58 \text{ Btu/h F}} = 121.6$
3. Triple glass (½″ air space)	$\frac{59\%}{0.36 \text{ Btu/h F}} = 164.6$
4. Quadruple glass (½″ air space)	$\frac{49\%}{0.28 \text{ Btu/h F}} = 175.0$
5. Selective (ITO) transmitter with D.S. glass spaced ½″ away[c]	$\frac{65\%}{0.29 \text{ Btu/h F}} = 224$
6. Double FRP (½″ air space)[d]	$\frac{66\%}{0.56 \text{ Btu/h F}} = 118$
7. Double 2-mil Teflon (½″ air space)[e]	$\frac{84\%}{0.61 \text{ Btu/h F}} = 138$
8. Quadruple Teflon (½″ air space)[e]	$\frac{71\%}{0.31 \text{ Btu/h F}} = 229$

Source: Johnson (1981).

[a] *U* values for glass systems are from the ASHRAE *1972 Handbook of Fundamentals.*

[b] D.S. stands for double-strength glass.

[c] ITO stands for indium tin oxide coating.

[d] FRP stands for fiberglass-reinforced plastic.

[e] Teflon™ is a very-high-solar-transmittance plastic film.

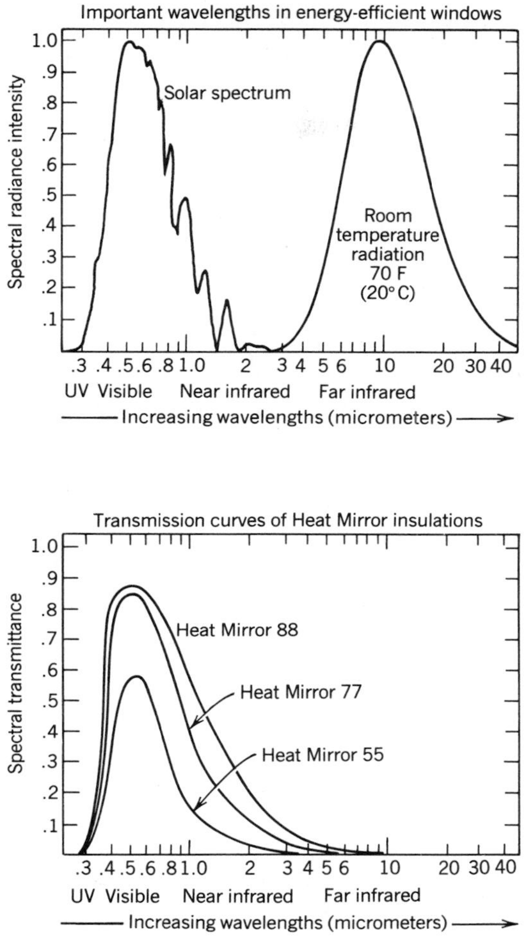

Fig. 6.4 Transmission and reflection performance for a selective transmitter film (Heat Mirror™). Incoming solar radiation (both visible and near-infrared) is mostly transmitted, whereas radiant heat from room-temperature objects is reflected and thereby kept within heated spaces. (Courtesy of Southwall Corporation.)

glass (see Fig. 6.5). When such an assembly is quadruple glazed (item 8 in Table 6.2), it essentially matches the selective film's performance in triple glazing.

The high insulating value of these window assemblies is as attractive as the performance of ordinary double-glazed windows with movable night insulation, as indicated in Fig. 6.6. Given the dependence of movable insulation on either an automated control system or user response, the selective film assemblies may be the more attractive alternative in many cases.

A related component that can combine the functions of both glazing and sunshading is the *air-extract window* diagrammed in Fig. 6.7. Developed in Scandinavia in the 1950s and now utilized in the United States as well, this is a triple-glazed window that passes room air between a typical outer double-glazed window and an inner,

Fig. 6.5 Cutaway view of window with two panes of glass and either (a) *one layer or* (b) *two layers of high-solar-transmitting film (SunGain™) suspended between.* (c) *Transmittance of film compared to ordinary glass. (Courtesy of 3M Corporation.)*

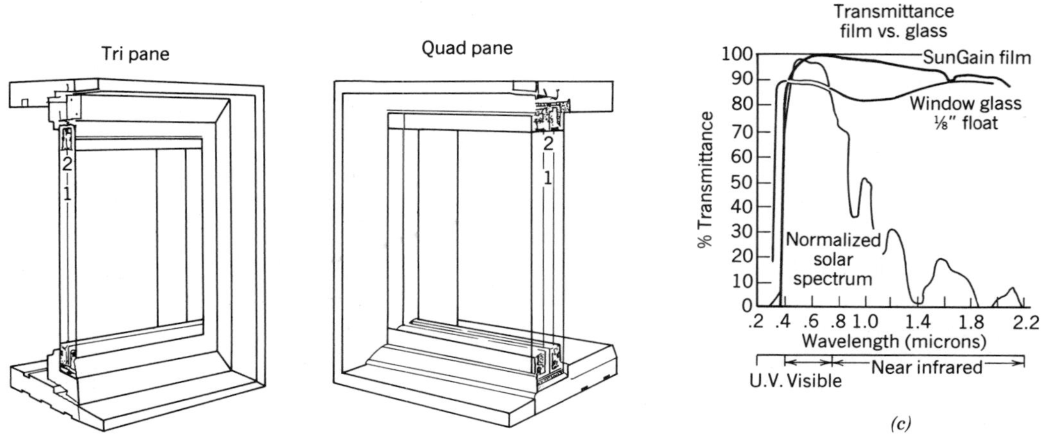

(c)

1 Clear, resilient SunGain® window film.
2 Spring-mounted spacers from which film is suspended (visible on 2 sides).

(a) (b)

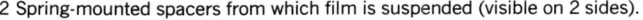

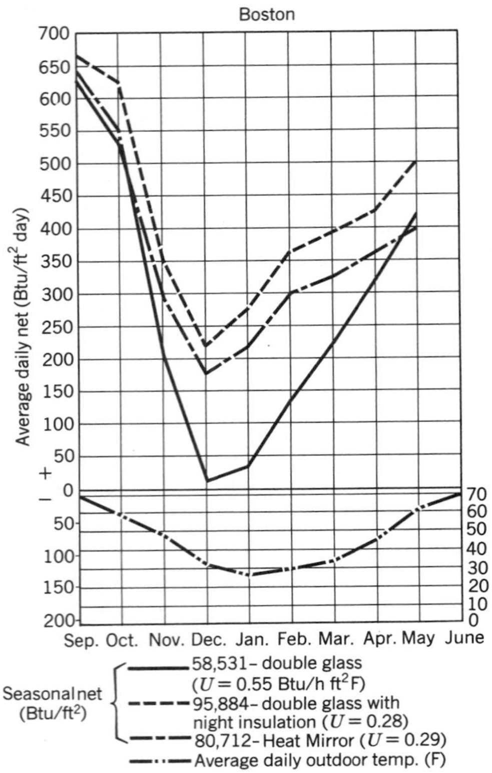

(a)

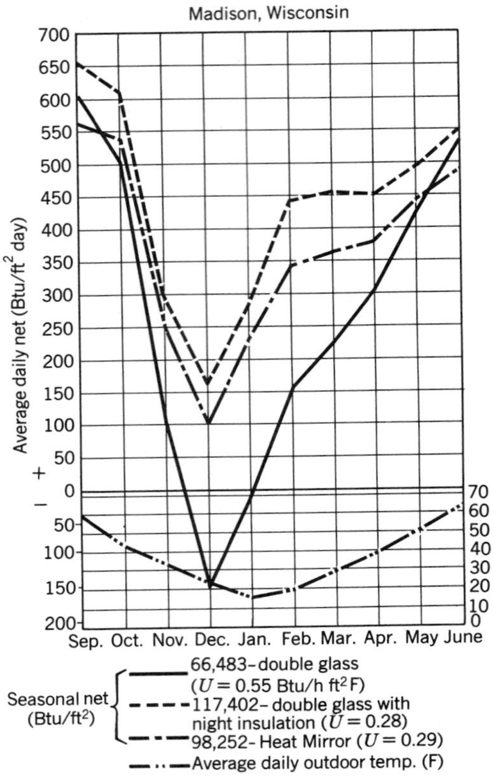

(b)

Fig. 6.6 Average daily net heating, per square foot of south window, for three glazing assemblies in three climates. (a) *In Boston, ordinary double glazing is a slight energy plus even in December.* (b) *In Madison, ordinary double glazing is a net loser for almost two months.* (c) *In Seattle, ordinary double glazing is a net loser for more than two months. (From Johnson, 1981.)*

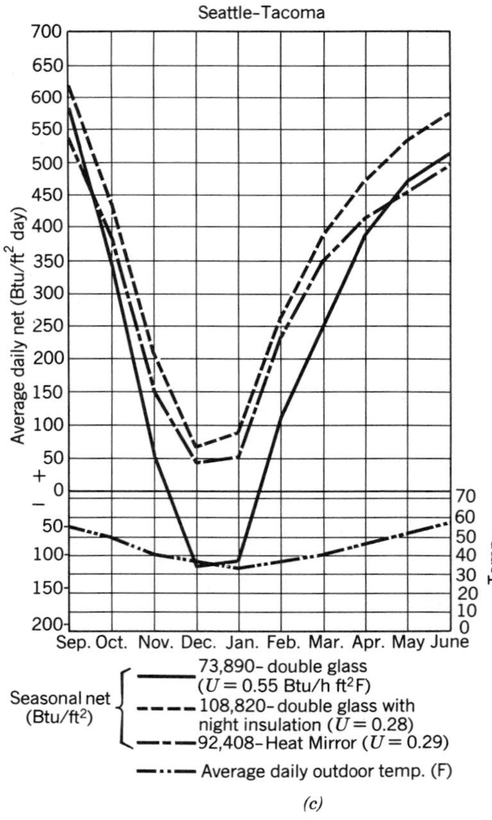

(c)

single pane. The inner pane thus is kept at very nearly the same temperature as the room air, which greatly increases comfort near windows on very cold (or very hot) days. Venetian blinds are often inserted in this cavity, where they can intercept direct sun and redirect its light toward the ceiling (see also Fig. 6.11). The solar heat intercepted by the blind is carried off by the room air to a plenum above the ceiling, where such air can be either exhausted or recirculated, and its heat content either reclaimed or rejected. The U value of these windows is dependent upon the rate of air flow between the glazing (Fig. 6.7*c*); typical flow rates are 4 to 6 cfm per foot of window width. For an example of such windows in a large U.S. building, see Fig. 6.124, which shows the Comstock Center in Pittsburgh.

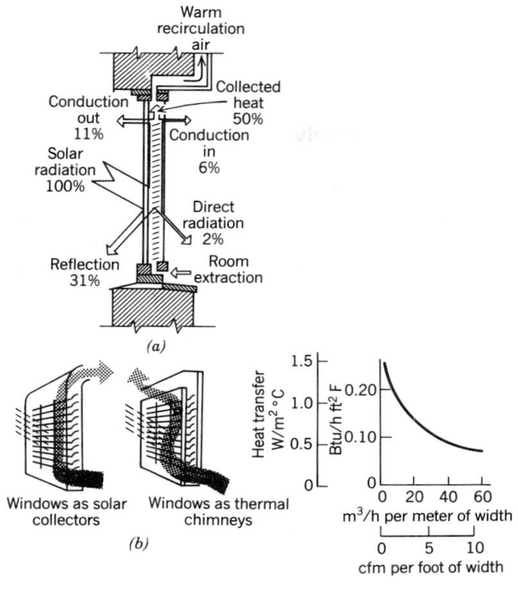

*Fig. 6.7 Air extract windows. (*a*) Cross section, showing the window as a solar collector. (*b*) Contrasts between the solar collector (winter) and the solar chimney (summer) modes. (*c*) The* U *value of these windows varies with the rate of airflow within the glazing; typical flow rate is 4 to 6 cfm/ft of window width. (From articles by D. Aitkin and O. Seppanen in* Proceedings of the Sixth National Passive Solar Conference. © *1981 by the American Solar Energy Society, Inc.)*

(c) Movable Insulation. This component is unique to passive solar heating. Usually one of the costliest components, it is also the one most likely to be made by hand and on site. It demands the most from the users of the buildings. Living in passively heated and cooled buildings is often likened to "thermal sailing" (Figs. 2.18, 2.19); movable insulation is the most visible sail. Many varieties of these thermal shades and shutters are available; they can be placed inside, within, or outside the glazing they protect. These options are summarized in Table 6.3. Typical *U* values for windows with movable insulation can be found in Table 4.17.

Along with movable insulation come some common problems. The first is control: Will the insulation be closed when it's cold and sunless and open when the sun should be admitted? As in the case of movable sunshading, the options here are manual, motorized, and automatic controls. For homes, the automatic option is certainly available but as yet is rarely installed. Another problem is the thickness needed to achieve good insulating performance. Bulky shades are awkward both in appearance and operation; thick shutters are less of a problem unless they stack when open (which rapidly increases the storage space they require). Unfortunately, some of the materials with the highest *R* value per inch emit toxic gases when burned, so their use within rooms requires costly fire-retardant covering materials. Another problem is infiltration into the air space formed between the movable insulation and the glazing, which can reduce this assembly's thermal performance. Tight seals at the top and sides of thermal shades and shutters are particularly important for stopping the convection currents that would readily bring warmed room air into the top of the air space and send chilled air out of the bottom.

Available references that discuss the details of movable insulation options include Shurcliff (1980) and Niles and Haggard (1980); the latter is particularly rich in architectural detailing examples. Figure 6.8 shows some of the many approaches to movable insulation; they can also be found in the buildings shown in Figs. 1.12, 2.18, 5.23, and 6.2.

(d) Thermal Storage. The thermal storage properties of many common building materials—including density, conductivity, and specific heat—can be found in Table 4.2. The procedures for sizing of thermal storage for passive heating and cooling were outlined in Sections 5.3 (preliminary guidelines), 5.6 (passive solar heating), and 5.9 (passive cooling). The performance of varying degrees of thermal mass can be found in Appendix C. Here, information about processes and products for thermal storage will be presented.

For *direct-gain* spaces, the thermal mass must be widely distributed around the room, so that direct sun can strike the mass surface and/or be reflected to the mass surface as soon as possible upon entering the window. Table 5.9 gave rules of thumb for mass; a common recommendation for direct-gain spaces above 50% SSF is for a thermal mass area five to seven times the area of glass (for no more than the optimum 4-in. masonry thickness). Appendix C shows how various ratios of mass to glass affect performance. There

TABLE 6.3 **Movable Window Insulation**

	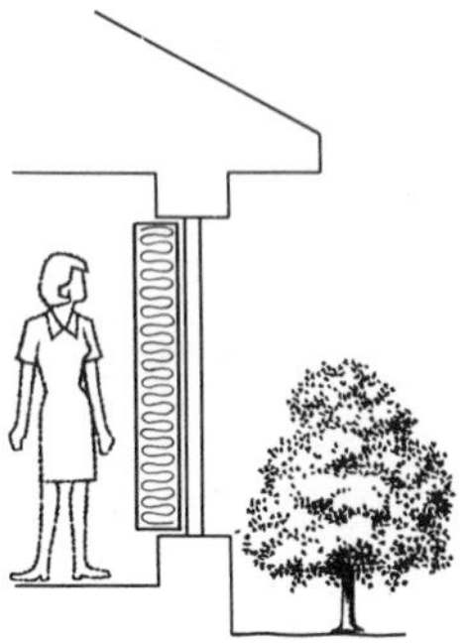 *Inside the Space*	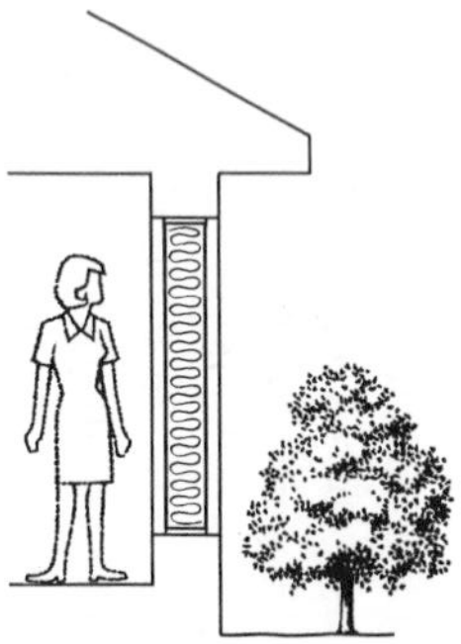 *Between Glazings (or with Trombe Wall)*	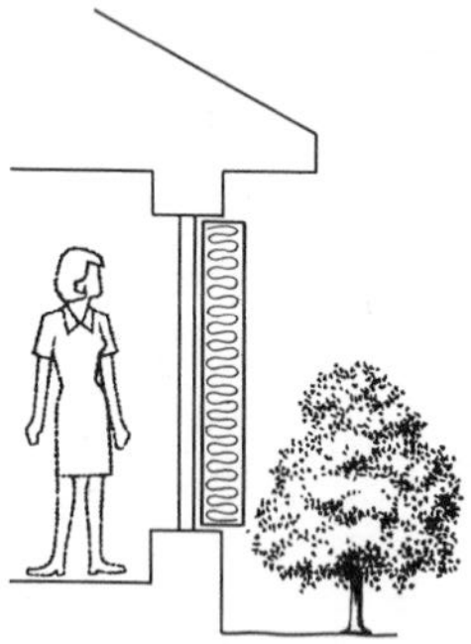 *Outside*
General opportunities	Insulation unaffected by weather or exterior abuse.	Insulation unaffected by weather and exterior or interior abuse, and stays relatively dust-free.	Insulation protects window weathering or exterior abuse.
	Window surface changes texture when insulation covers it.		Most effective position to act as sunshade.
General problems	Least effective position to act as sunshade.	Between glazings, insufficient thickness inhibits developing adequate *R* value.	Wind and sleet can inhibit moving the insulation.
	Subject to interior abuse.	Difficult maintenance access.	Must be weatherproof construction and is subject to exterior abuse.
	Can affect furniture arrangement near window.	Storage is difficult.	Can affect landscaping materials near window.
	Some foams produce toxic gases when burned.		
How insulation moves			
Blow in/out 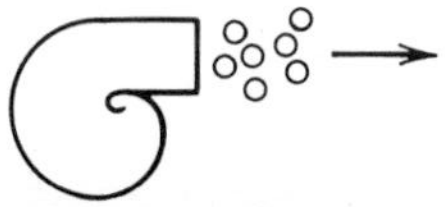	(Approach not developed.)	Bead-Wall(TM): Styrofoam beads fill glazing cavity. Vacuum-cleaner-type motor sucks beads out, blows beads in. Bead storage tanks required. Moderately poor visibility through	(Approach not developed.)

Roll up/down 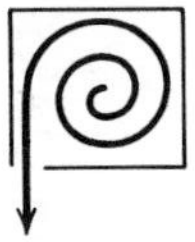 Insulation disappears in open position.	Many available products; often site-constructed. High R value usually means a bulky roller and awkward operation. Both automatic and manual operation are common. Little impact on furniture arrangement.	For Trombe walls, motorized and automated operation is common.	Widely used as sunshading and security, but seldom has high R value. Both automatic and manual operation are common.
Slide up/down/side 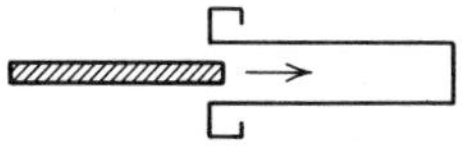 Insulation can disappear into pocket.	High R value can be obtained, if wide pocket is acceptable. Widely used, with or without pocket. Little impact on furniture arrangement. Usually manually operated.	Not frequently used.	High R value easily attained, since pocket usually omitted on exterior. Shutter therefore influences exterior appearance, either in open or closed position.
Fold up/down/side 	High R value easily obtained. Needs some clearance in front of window; can affect furniture. Usually manually operated.	Not frequently used due to insufficient clearance.	High R value easily obtained. Snow inhibits up/down operation. Some potential as reflector for added sun through glass. Needs some clear space outside window.
Swing up/down/side 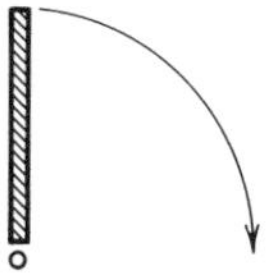	High R value easily obtained. Substantial visual impact, in any position. Needs considerable clearance in front of window, affecting furniture.	Rare: insufficient clearance.	High R value easily obtained. Snow inhibits up/down operation. Good potential as reflector for added sun through glass. Needs considerable clear space outside window.

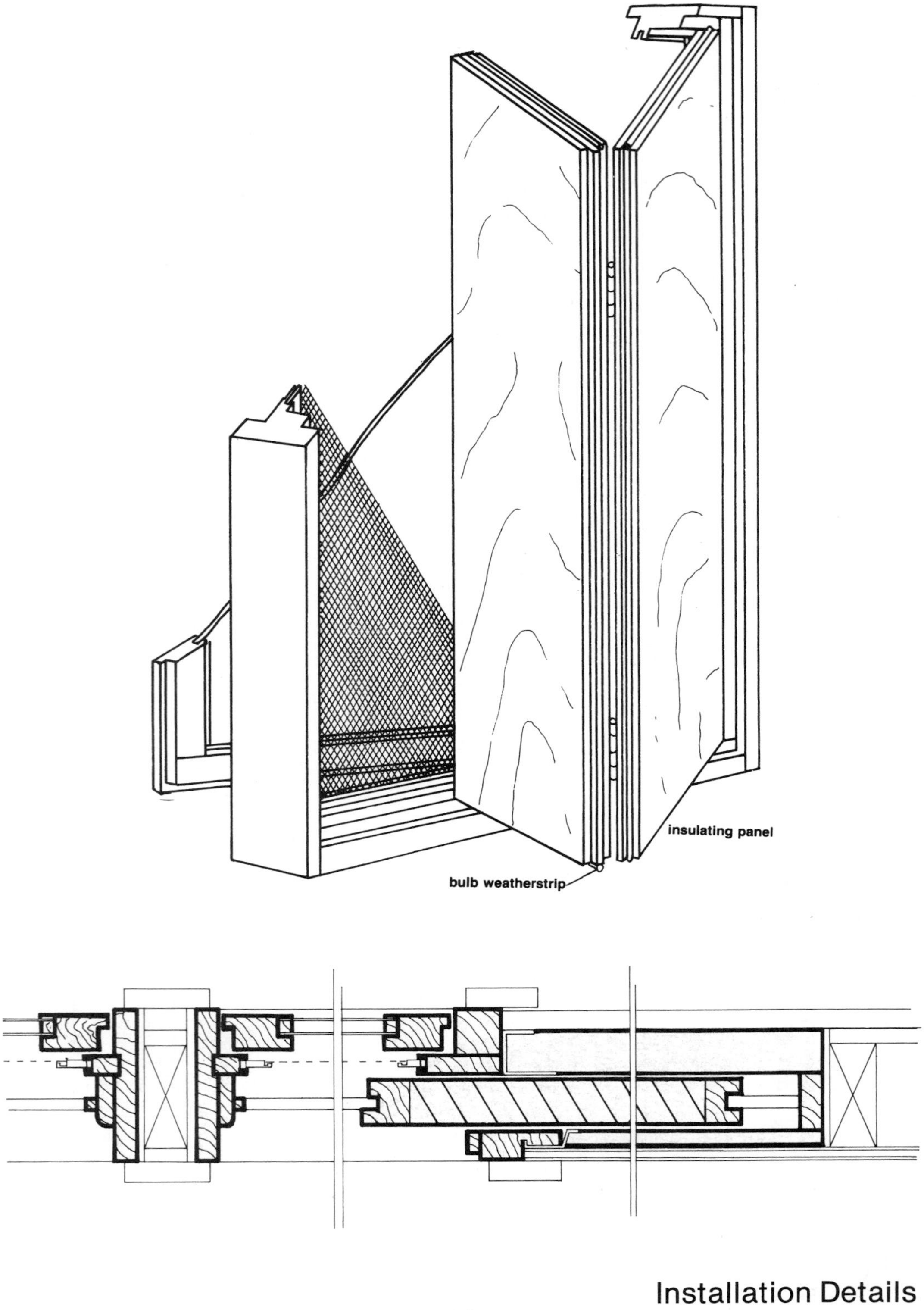
insulating panel
bulb weatherstrip
Installation Details

(a)

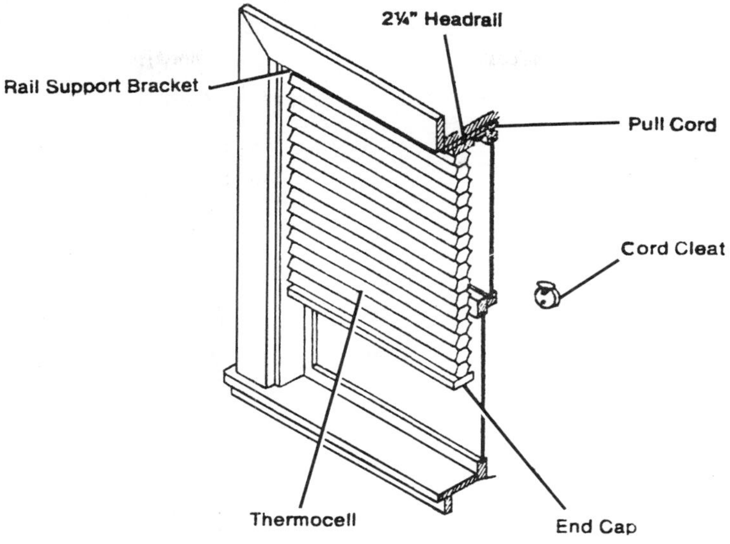

(b)

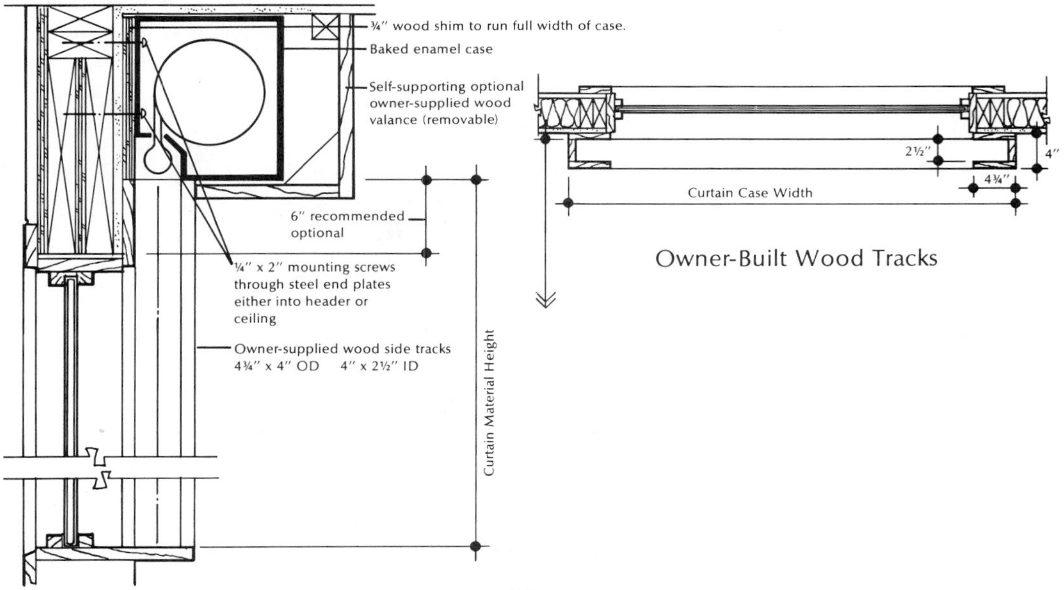

(c)

Fig. 6.8 *Variations on movable insulation.* (a) *Bi-fold and sliding solid panels (or shutters) are available up to* R*14; built-in weather stripping is important if* R*14 performance is to be approached. (Courtesy of Sunflake.)* (b) *Manual pull-down honeycomb-film shade has* R*9, in a variety of metallic finishes.* (c) *Roll-down insulating curtain is fully automatic and requires side tracks to achieve its maximum potential of* R*12. (Courtesy of Thermal Technology Corporation.)*

are many common ways to provide such mass surfaces, including brick veneer and clay tile over a bed of grout. These surfaces can be applied to frame construction.

For *thermal storage walls* (Trombe walls), previous design decisions have been made regarding size and thickness, to vent or not to vent, and movable insulation. Some detailing for vents and

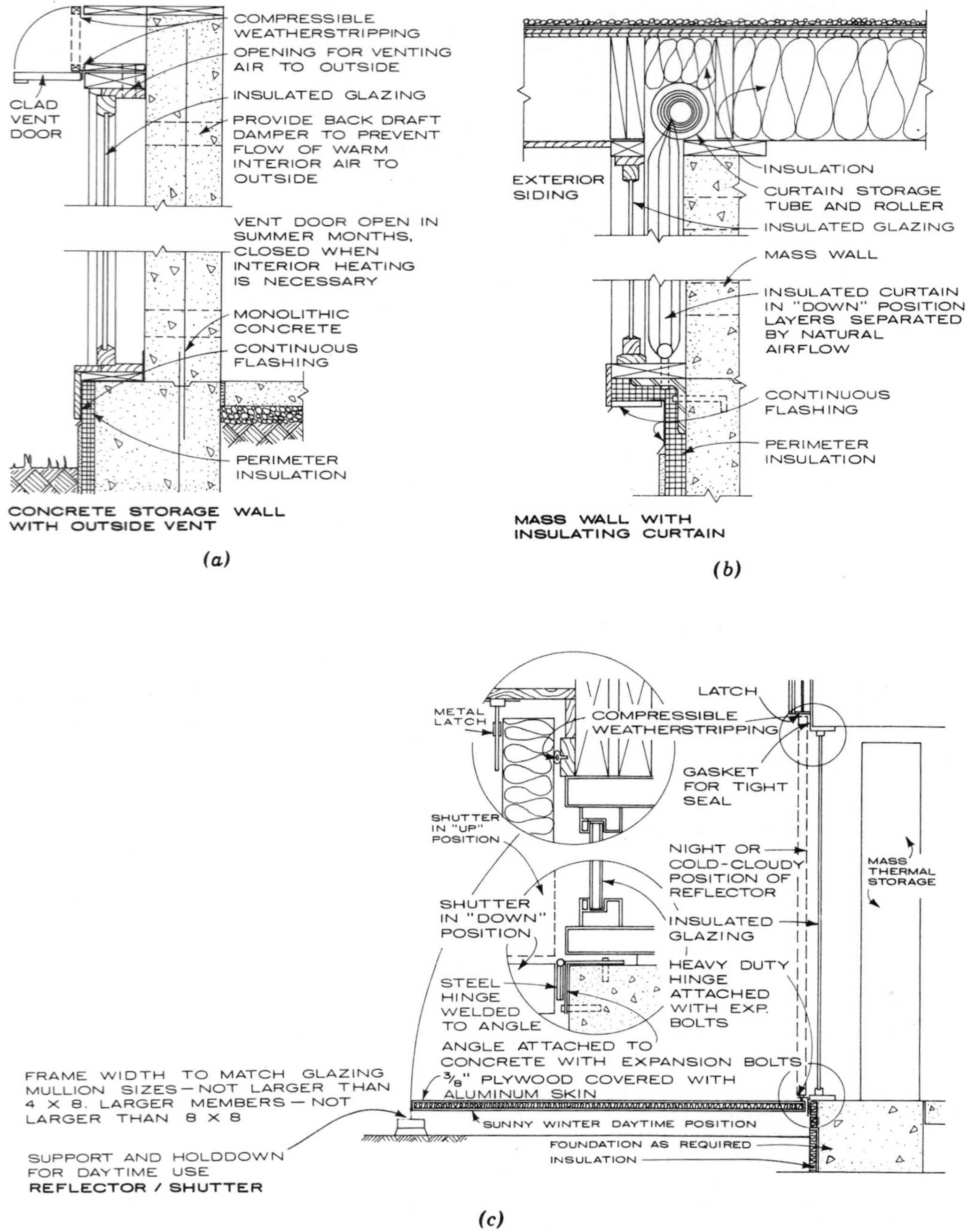

Fig. 6.9 Detailing for thermal storage walls (Trombe walls). (a) *If vented Trombe walls are used, then a top vent to the exterior can provide low-velocity air exhaust from the space served by the Trombe wall.* (b) *Roll-down movable insulation such as this usually has motorized (frequently automated) controls. The cavity between wall and glass should be generous enough to avoid abrasion on the curtain as it moves.* (c) *Fold-down exterior insulating shutters are much easier to maintain than are curtains enclosed behind glass, but they are subject to severe weathering, and snow can inhibit operation. They have the significant added benefit of acting as reflectors to augment solar collection. (From AIA, Ramsey, and Sleeper,* Architectural Graphic Standards, *7th ed., © 1981 by John Wiley & Sons. Reprinted by permission.)*

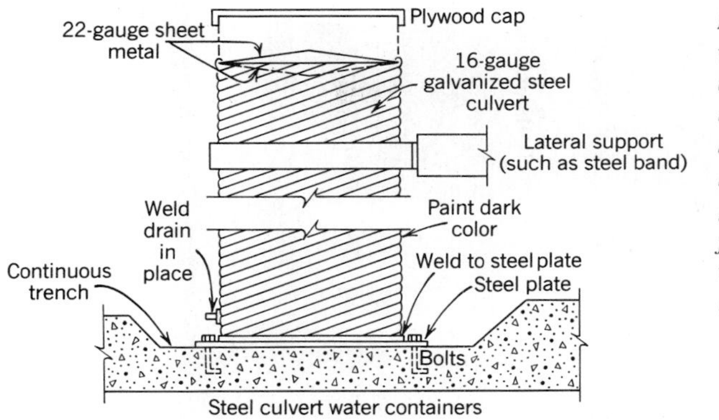

Fig. 6.10 Detailing for a steel culvert water wall. The need for a cap (to prevent evaporation), lateral support at the top (to meet earthquake resistance requirements), and a trench for containment of accidental spillage is common to all water walls, regardless of container material. (Adapted from AIA, Ramsey, and Sleeper, Architectural Graphic Standards, *7th ed., © 1981 by John Wiley & Sons.)*

Water-Storage Tubes: Typical Data

Diameter (in.)	*Height (ft)*	*Volume (gal)*	*Weight (lb) Full*	*Heat Capacity, 20F° Rise (Btu)*	*Floor Loading (lb)*	
					per lin. ft	*per ft²*
12	4	23.5	204	3,900	204	260
12	7	41	354	6,800	354	451
12	8	47	404	7,800	404	514
18	5	66	567	11,000	378	321
18	10	132	1122	22,000	748	635

movable insulation is shown in Fig. 6.9. *Water walls* (Fig. 6.10) also have been sized previously. They are commonly made of corrugated galvanized steel culverts, steel drums, or fiberglass-reinforced plastic tubes (for which manufacturers will provide suggested installation details). Within water wall containers, some air space should be provided, as water expands when it heats. A rust inhibitor should be added to the water within steel containers; algae growth in plastic tubes, encouraged by the daylight that passes through them, is usually controlled by the addition of an algicide to the water. Dyes may be added to change the color of daylight through the tubes.

Phase change materials, briefly described in Section 5.3, are available in configurations other than the flat bags of eutectic salts shown in Fig. 6.11. As flat tiles enclosing bags of salt, they can form the finished surface in any horizontal application—floors, ceilings, or counter and table tops,

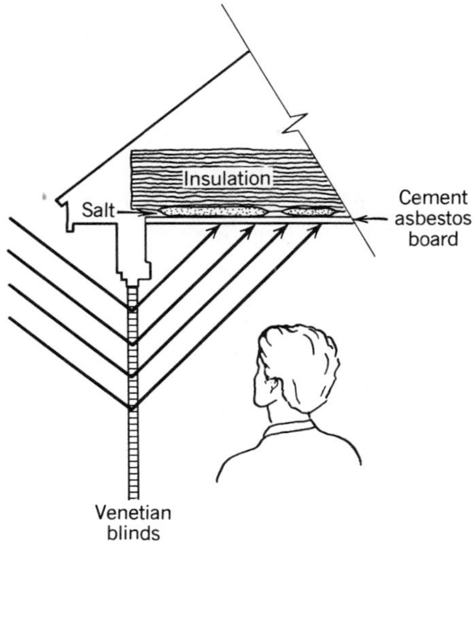

Fig. 6.11 Using bags of phase change material (salt) for thermal storage above a flat ceiling. Mirrored-surface venetian blinds reflect direct sun to the ceiling, where it strikes a heavy (at least 90 lb/ft²) board product with high conductivity. Plaster is also suitable; ordinary gypsum board is not sufficiently conductive. Bags of salts weighing 5 lb/ft² are in contact with the ceiling board and must be installed in horizontal position. They are usually formulated to melt between 70 and 75 F. (Reprinted by permission from Johnson, 1981).

Fig. 6.12 Other approaches to phase change thermal storage. (a) *The Heat Sponge™ is a cabinet that uses a small fan to circulate air through the trays of salts. (Photo by Seymour Jarmul.)* (b) *Stud wall cavities can contain Enerphase™ panels of salts, which can be used either as thermal mass in a solar wall (as shown) or as interior storage devices. A fan-forced airflow of 50 cfm per stud cavity improves their storage performance. (Courtesy of Dow Chemical Company.)*

for example. Tubes and trays of salts can be arranged as desired or purchased as a prefabricated unit (see Fig. 6.12). The function of these materials is to keep room temperatures steady; their performance in preventing overheating on a sunny winter day will also be appreciated on hot summer days, provided they are taken below the phase change or melting temperature at night. For U.S. locations with large daily temperature ranges in summer, thermal storage surfaces for passive solar heating in winter are potentially useful for night ventilation cooling in summer.

(e) Roof Ponds. Once the area and the depth of roof ponds have been determined (Sections 5.3 and 5.9), questions arise about the architectural integration of such large horizontal surfaces and their relative emphasis upon heating or cooling. Figure 6.13 shows variations in the treatment of containers and the insulating panels to match the most important thermal role of roof ponds; Fig. 6.14 shows three typical approaches to the placement of roof ponds on buildings. A valuable source of information about the container bags, sliding insulation, roof decks, and other components

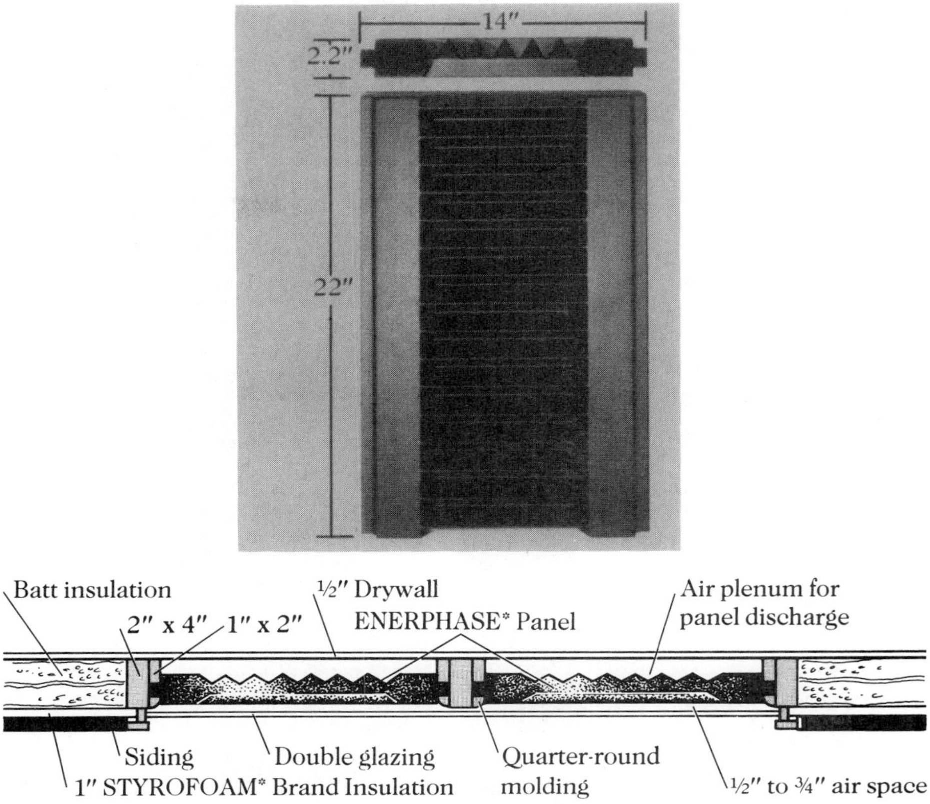

StoragePanels are combustible and should not be exposed to flame or other ignition sources.

(b)

***Fig. 6.12** (continued)*

needed for roof ponds is the *California Passive Solar Handbook* by Phillip Niles and Ken Haggard, written for the California Energy Commission in 1980. (For ordering information, write to their Solar Office, 1111 Howe Avenue, Sacramento CA 95825.) Commercially available products include the Skytherm® system, developed and pioneered by Harold R. Hay.

(f) Ventilation. Simple calculation procedures for cross ventilation and stack effect ventilation of buildings were detailed in Section 4.6. Several

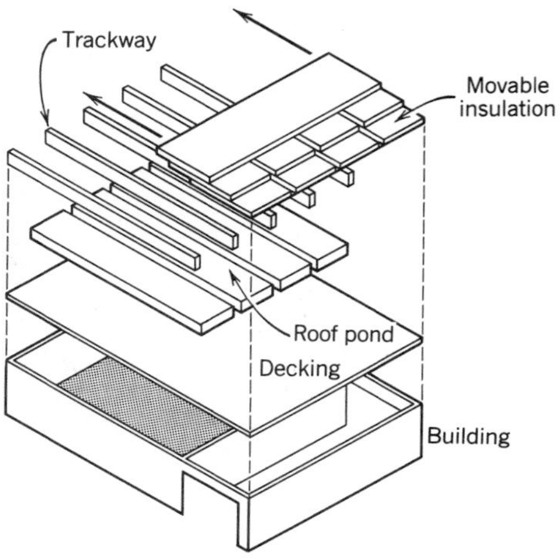

(a)

***Fig. 6.13** (a) The principal components of roof ponds.*

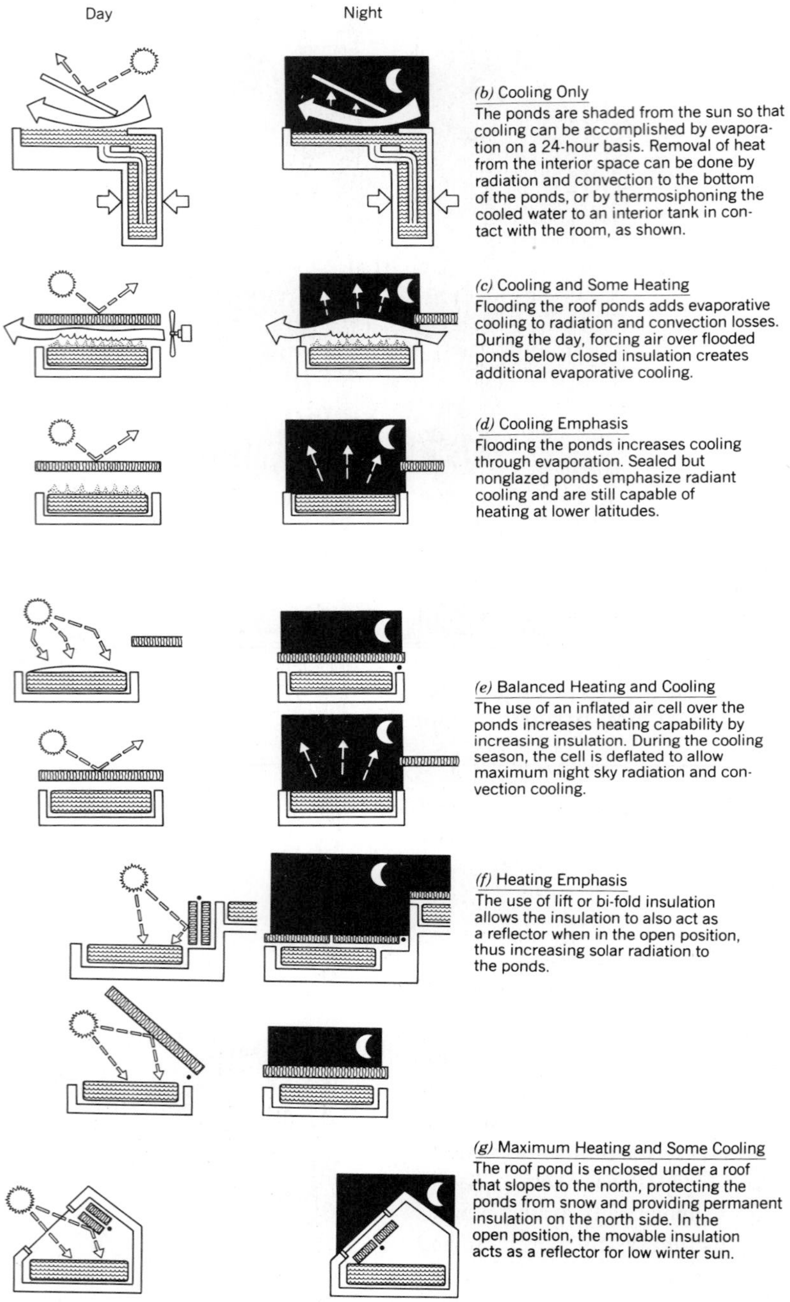
Day
Night
(b) Cooling Only
The ponds are shaded from the sun so that cooling can be accomplished by evaporation on a 24-hour basis. Removal of heat from the interior space can be done by radiation and convection to the bottom of the ponds, or by thermosiphoning the cooled water to an interior tank in contact with the room, as shown.
(c) Cooling and Some Heating
Flooding the roof ponds adds evaporative cooling to radiation and convection losses. During the day, forcing air over flooded ponds below closed insulation creates additional evaporative cooling.
(d) Cooling Emphasis
Flooding the ponds increases cooling through evaporation. Sealed but nonglazed ponds emphasize radiant cooling and are still capable of heating at lower latitudes.
(e) Balanced Heating and Cooling
The use of an inflated air cell over the ponds increases heating capability by increasing insulation. During the cooling season, the cell is deflated to allow maximum night sky radiation and convection cooling.
(f) Heating Emphasis
The use of lift or bi-fold insulation allows the insulation to also act as a reflector when in the open position, thus increasing solar radiation to the ponds.
(g) Maximum Heating and Some Cooling
The roof pond is enclosed under a roof that slopes to the north, protecting the ponds from snow and providing permanent insulation on the north side. In the open position, the movable insulation acts as a reflector for low winter sun.

***Fig. 6.13*(b) *through* (g).** *Variations on roof ponds for optimization of heating or cooling performance. (Reprinted from Niles and Haggard, 1980, by permission of the California Energy Commission.)*

devices can be used to enhance the stack effect by creating suction when wind blows across the top of the stack. The most common such device (available in chain store catalogs) is shown in Fig. 6.15*a*; its performance characteristics are listed in Table 6.4. This is probably not the most effective

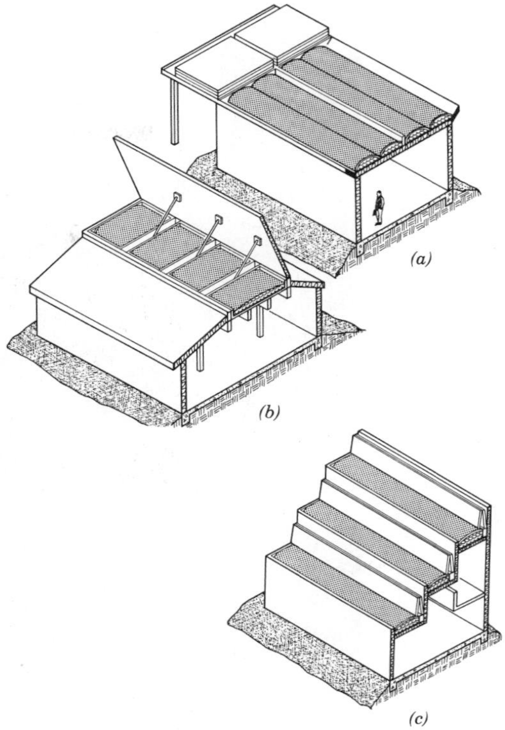

Fig. 6.14 *Three typical roof pond integrations with building form.* (a) *Insulating panels slide open to stack over exterior or untempered space. This is the most simple and common application.* (b) *Hinged insulating panels with reflective undersurfaces can act as reflectors to increase winter insulation. Snow and wind can inhibit operation.* (c) *Bi-fold insulating panels also act as reflectors, interfering less with diffuse radiation and allowing for exposure to night sky for summer cooling. (Reprinted from Niles and Haggard, 1980, by permission of the California Energy Commission.)*

topping device, however. Figure 6.16 compares the volumetric airflow results for the turbine and several other ventilators with those for a simple open stack. The tests were done in a wind tunnel at Virginia Polytechnic Institute.

Passive ventilation through windows and skylights is influenced by the position of the open window; if wind strikes the glass surface in its open position, it will be deflected. The direction of the wind approaching the window is generally unpredictable. Also, whereas for simple ventilation (without cooling) it is usually desirable to keep wind *away* from people, for cooling above the standard comfort zone, wind *across* the body is helpful (Fig. 3.20). For these reasons, a window that can be opened in a variety of positions can be useful; some examples are shown in Fig. 6.17.

Fig. 6.15 *A collection of passive heating and cooling components at the Cottage Restaurant, Cottage Grove, Oregon. Equinox Design, Inc.* (a) *Northeast view, showing wind-gravity (turbine) ventilators (see Fig. 6.16) projecting above the roof; they provide night ventilation of thermal mass for summer cooling.*

*Fig. 6.15 (continued) (*b*) Southeast view, showing the solar-heating features: clerestory, protected by interior thermal shutters; and lower south windows, protected inside by automatic roll-down thermal shades (see Fig. 6.8) and shaded outside in summer by deciduous vines. (*c*) Interior view, showing thermal mass in the concrete floor and in the many 30-gal barrels of water placed throughout the interior. Barrels on suspended rack receive winter sun through clerestory and are cooled by summer night air drawn from below floor slab and out through turbine ventilators. (Photos by G. Z. Brown.)*

(g) Heat Exchangers. As attention to the design of energy-conserving envelopes increases, so does the proportion of winter heat loss (and summer heat gain) attributable to infiltration or ventilation. (In Example 5.8, 375, or 68%, of the total UA_{ns} of 550 for an Omaha office building was attributable to infiltration.) As the tightness of construction increases, fewer air changes per hour (ACH) occur from natural infiltration; this makes forced ventilation attractive as a means of reducing indoor air pollution. If a heat exchanger is used, it is possible to maintain an adequate supply of fresh air without severe energy consumption consequences. Figure 6.18 illustrates the basic principle of the air-to-air heat exchanger that is becoming increasingly common for tightly built

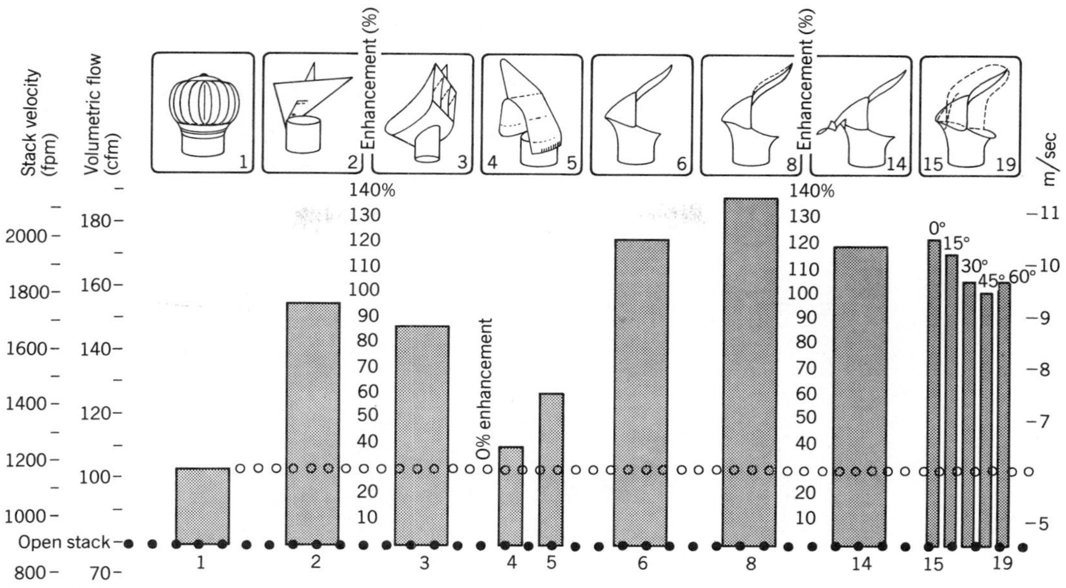

Fig. 6.16 *Performance of some passive ventilators compared to a simple open stack. (From the work of R. P. Schubert and Philip Hahn; reprinted from* Progress in Passive Solar Energy Systems, *© 1983, by permission of the publisher, the American Solar Energy Society, Inc.)*

Fig. 6.17 *Window that can open in more than one position, providing flexibility for passive cooling performance.* (a) *The window tilts, directing the incoming air toward the ceiling for simple ventilation.* (b) *The window swings in, allowing incoming air to move across people and thus enhancing warm weather cooling. (Courtesy of Three Rivers Aluminum Company, Inc.)*

TABLE 6.4 **Turbine Ventilator Performance**

NOTE: The combination of wind suction and stack effect produces the following exhaust capacities (cfm) for various throat diameters and stack heights of turbine ventilators. Recommended minimum spacing between ventilators is 20 ft.

Outdoor Wind Velocity (mph):		2			4			5			6			8			10		
Temp. Diff. (F°):		10	20	30	10	20	30	10	20	30	10	20	30	10	20	30	10	20	30
Throat Diameter	Ht. above Intake (ft.)	Exhaust Capacity (cfm)																	
6″	10	114	125	130	210	221	226	266	277	282	314	325	330	426	437	442	534	545	550
	20	122	135	144	218	231	240	274	287	296	322	335	344	434	447	456	542	555	564
	30	129	144	156	225	240	252	281	296	308	329	344	356	441	456	468	549	564	576
	40	135	152	166	231	248	262	287	304	318	335	352	366	447	464	478	555	572	586
10″	10	209	222	274	370	383	435	463	476	528	545	558	610	728	741	793	915	928	980
	20	234	269	301	395	430	462	488	523	555	570	605	637	753	788	820	940	975	1007
	30	254	301	328	415	462	489	508	555	582	590	637	664	773	820	847	960	1007	1034
	40	269	318	355	430	479	516	523	572	609	605	654	691	788	837	874	975	1024	1061
14″	10	333	383	422	558	608	647	691	741	780	804	854	893	1062	1112	1151	1324	1374	1413
	20	376	444	496	601	669	721	734	802	854	847	915	967	1105	1173	1225	1367	1435	1487
	30	413	496	560	638	721	785	771	854	918	884	967	1031	1142	1225	1289	1404	1487	1551
	40	444	539	614	669	764	839	802	897	972	915	1010	1085	1173	1268	1343	1435	1530	1605
18″	10	476	564	623	755	843	902	924	1012	1071	1071	1159	1218	1399	1487	1546	1737	1825	1884
	20	549	662	747	828	941	1026	997	1110	1195	1144	1257	1342	1472	1585	1670	1810	1923	2008
	30	611	747	853	890	1026	1132	1059	1195	1301	1206	1342	1448	1534	1670	1776	1872	2008	2114
	40	662	819	941	941	1098	1220	1110	1267	1389	1257	1414	1536	1585	1742	1864	1923	2080	2202

24″	10	716	874	978	1101	1259	1363	1327	1485	1589	1522	1680	1784	1963	2121	2225	2412	2570	2674
	20	844	1046	1196	1229	1431	1581	1455	1657	1807	1650	1852	2002	2091	2293	2443	2540	2742	2892
	30	954	1196	1384	1339	1581	1769	1565	1807	1995	1760	2002	2190	2201	2443	2631	2650	2892	3080
	40	1046	1324	1542	1431	1709	1927	1657	1935	2153	1852	2130	2348	2293	2571	2789	2742	3020	3238
30″	10	1139	1385	1545	1719	1965	2125	2070	2316	2476	2379	2625	2785	3068	3314	3474	3769	4015	4175
	20	1342	1655	1890	1922	2235	2470	2273	2586	2821	2582	2895	3130	3271	3584	3819	3972	4285	4520
	30	1514	1890	2185	2094	2470	2765	2445	2821	3116	2754	3130	3425	3443	3819	4114	4144	4520	4815
	40	1655	2090	2430	2235	2670	3010	2586	3021	3361	2895	3330	3670	3584	4019	4359	4285	4720	5060
36″	10	1613	1967	2201	2475	2829	3063	2988	3342	3576	3418	3772	4006	4414	4768	5002	5428	5782	6016
	20	1901	2354	2692	2763	3216	3554	3276	3729	4067	3706	4159	4497	4702	5155	5493	5716	6169	6507
	30	2148	2692	3115	3010	3554	3977	3523	4067	4490	3953	4497	4920	4949	5493	5916	5963	6507	6930
	40	2354	2981	3470	3216	3843	4332	3729	4356	4845	4159	4786	5275	5155	5782	6271	6169	6796	7285
42″	10	2183	2663	2998	3350	3835	4170	4047	4527	4862	4645	5125	5460	6000	6480	6815	7365	7845	8180
	20	2588	3203	3668	3760	4375	4840	4452	5067	5532	5050	5665	6130	6405	7020	7485	7770	8385	8850
	30	2928	3668	4243	4100	4840	5415	4792	5532	6107	5390	6130	6705	6745	7485	8060	8110	8850	9425
	40	3203	4058	4723	4375	5230	5895	5067	5922	6587	5665	6520	7185	7020	7875	8540	8385	9240	9905
48″	10	2868	3500	3925	4412	5044	5469	5308	5940	6365	6078	6710	7135	7843	8475	8900	9638	10270	10695
	20	3378	4185	4785	4922	5729	6329	5818	6625	7225	6588	7395	7995	8353	9160	9760	10148	10955	11555
	30	3817	4785	5535	5361	6329	7079	6257	7225	7975	7027	7995	8745	8792	9760	10510	10587	11555	12305
	40	4185	5300	6175	5729	6844	7719	6625	7740	8615	7395	8510	9385	9160	10275	11150	10955	12070	12945

Source: Reprinted courtesy of Western Ventilating Equipment, Inc., Los Angeles, California.

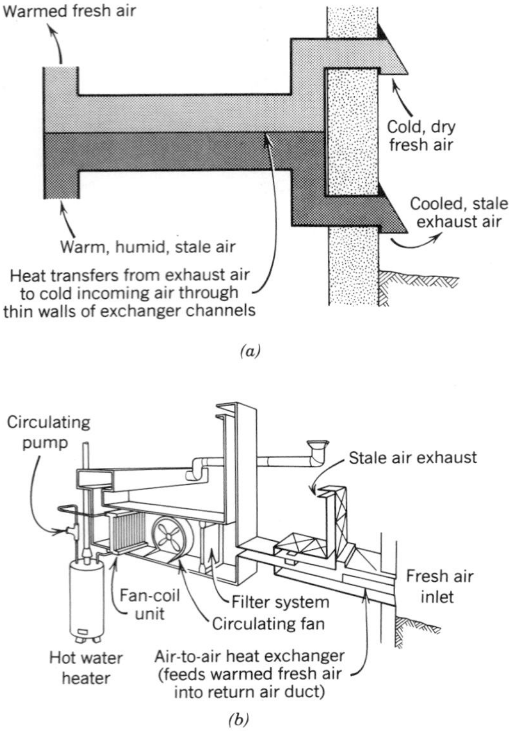

Fig. 6.18 *Air-to-air heat exchangers are particularly helpful for smaller buildings in cold climates.* (a) *The basic principle of operation.* (b) *Installation in a small hot water heating application. (Reprinted by permission of Alberta Agriculture, Home and Community Design Branch,* Low Energy Home Designs, © *1983 by Brick House Publishing Co.)*

small buildings (large-building heat exchangers, often called heat recovery devices, are illustrated in Section 6.8).

Some commercially available heat exchangers are capable of extracting 70% or more of the heat from exhaust air. For the best diffusion of incoming fresh air through a building, the heat exchanger should be incorporated at the central forced-air fan (Fig. 6.18*b*). When a central forced-air system is not available, heat exchangers can be placed at various points in the building; typically, each heat exchanger is then equipped with its own fan.

Some cautions about air-to-air heat exchangers:

1. Avoid using them on exhaust airstreams that are contaminated with grease, lint, or excessive moisture (through cooking and clothes drying in particular) since clogging, frosting, and fire hazard problems can develop.
2. In colder winter conditions, a built-in defroster, which will require energy, will be needed.
3. Keep the outdoor fresh air intake as far as possible from the exhaust air outlet, to avoid drawing indoor-contaminated air back into the building.

Indoor pollution of air was discussed in Section 4.6, along with requirements for fresh air (Table 4.19). We usually refer to the quantity of fresh air assumed in heating or cooling calculations in terms of ACH. Although the guidelines given in Table 4.19 should provide for sufficient fresh air, the general question arises, What is the minimum ACH level compatible with adequate fresh air? A frequently cited (although often disagreed with) minimum is 0.4 ACH. Table 6.5 shows that this 0.4 ACH rate of infiltration-plus-ventilation would produce *less* than 40% ''new'' air within a building after one hour; even after eight hours at this rate, not quite all the ''old'' air will be exhausted. There is, then, a difference between the fresh air input rate (ACH) and the *replacance*—the fraction of air molecules at one specified time that was *not* in the house at an earlier, reference time. This relationship, along with details of air pollutants and of heat exchanger design, is thoroughly discussed in Shurcliff (1981).

Water-to-water heat exchangers are also used to recover otherwise-wasted energy; since heat exchange is aided by steady flow, pumps are usually part of the water heat-exchange process. More information on these devices can be found in Section 9.4.

(h) Dehumidification. Passive cooling techniques such as roof ponds and night ventilation of mass (discussed in Chapters 2 and 5) are *sensible* cooling processes: that is, they work by lowering surface temperatures, which lowers air temperatures. In many hot and humid climates (and in tightly enclosed buildings that produce moisture, even in winter), *latent* cooling is also needed, to take excessive moisture from the indoor air. Either of two common processes is used for dehumidification: a chemical process of *absorption* (using desiccants) or a mechanical process of *refrigeration*. In absorption dehumidification, moisture is

TABLE 6.5 **Air Replacance Compared to Input Air Changes per Hour (ACH)**

NOTE: Mixing is considered continuous and vigorous—as would be obtained in a forced-ventilation system.

Time from Start of Run (h)	*Rate of Air Input (ACH):*	*Replacance (%)*								
		0.06	*0.12*	*0.25*	*0.5*	*1*	*2*	*4*	*8*	*16*
1/16		0.4	0.8	1.6	3.1	6.1	11.7	22.1	39.3	63.2
1/8		0.8	1.6	3.1	6.1	11.7	22.1	39.3	63.2	86.5
1/4		1.6	3.1	6.1	11.7	22.1	39.3	63.2	86.5	98.2
1/2		3.1	6.1	11.7	22.1	39.3	63.2	86.5	98.2	99.9
1		6.1	11.7	22.1	39.3	63.2	86.5	98.2	99.9	100
2		11.7	22.1	39.3	63.2	86.5	98.2	99.9	100	100
4		22.1	39.3	63.2	86.5	90.2	99.9	100	100	100
8		39.3	63.2	86.5	98.2	99.9	100	100	100	100
16		63.2	86.5	98.2	99.9	100	100	100	100	100

Source: Shurcliff (1981). Reprinted by permission.

absorbed from the air by a chemical such as calcium chloride, silica gel, or lithium chloride. When these chemicals are saturated, they must be "regenerated" (relieved of excessive moisture) before they can again serve as dehumidifiers. (One absorption-regeneration process, applied to mechanical equipment, is illustrated in Fig. 6.33.) Passive techniques such as solar drying can be used for regeneration. The principal disadvantage of chemical dehumidification is that the process *raises the temperature* of the air while it lowers the relative humidity; it is the reverse of the evaporative cooling process, as was shown in Fig. 4.8. If desiccants are used, the passive cooling capacity must be great enough to further lower the indoor air temperature after it has been raised by dehumidification. Details of several chemical dehumidification-regeneration processes can be found in Bowen, Clark, and Labs (1981).

Refrigeration dehumidification is quite common; dehumidification is usually a part of the refrigeration cooling process, which is described in Section 6.5. For smaller buildings that need only dehumidification, rather than mechanical cooling, *refrigerant dehumidifiers* are commercially available (Fig. 6.19). Their advantage is that the air

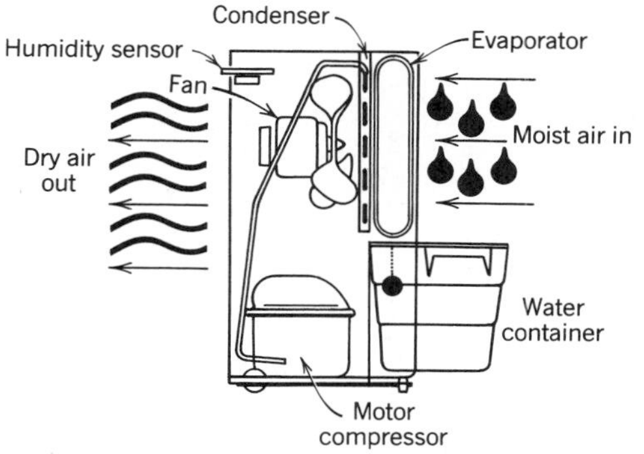

Typical Performance Data

Watts	290	450	465	630
Pints per day[a]	14	20	25	30
Unit dimensions (in.)				
Height	20½	20½	21½	21½
Width, 11¾				
Depth, 16¾				
"Average" room size served	20′ × 32′	35′ × 40′	40′ × 48′	40′ × 60′

[a]For air at 80 F, 60% rh.

Fig. 6.19 *Refrigerant dehumidifiers typically are installed as freestanding units in smaller buildings.*

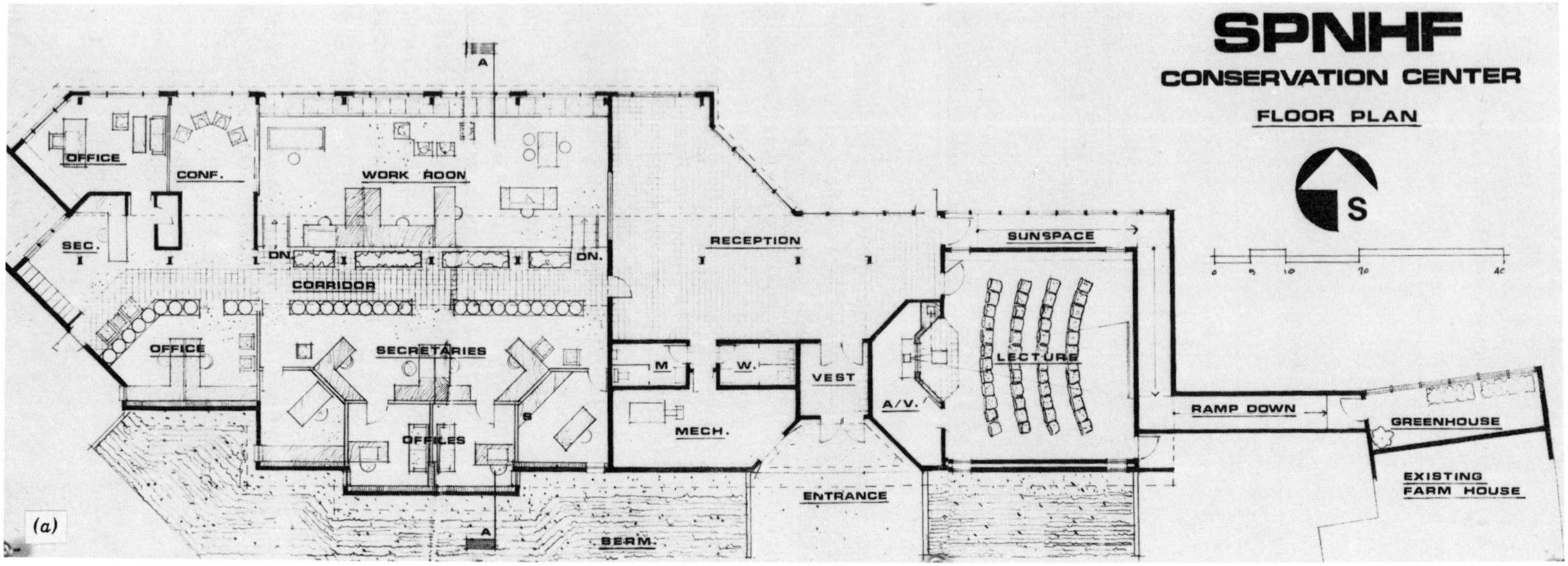

Fig. 6.20 The Conservation Center for the Society for the Protection of New Hampshire Forests, Concord, New Hampshire. Banwell White & Arnold, architects. (a) The plan is elongated on the E–W axis, to maximize south solar exposure and facilitate daylighting. Direct gain serves the reception area, workroom, and offices, and a sunspace–double-envelope combination warms the lecture room. A wood-fired boiler provides backup heat. (b) This section relates south glazing to heat storage materials: translucent water tubes, masonry walls, and phase change materials both in ceiling and windowsill positions. The circulation of hot air that collects at the clerestory is also evident. In summer, the hot air is vented and an awning will shade the clerestory. Daylighting is diffused through the translucent tubes to the spaces beyond. (c) At the workroom, mirror-finish venetian blinds reflect solar energy to the phase change materials in bags above the metal deck ceiling. (Photo by C. Stuart White, Jr.)

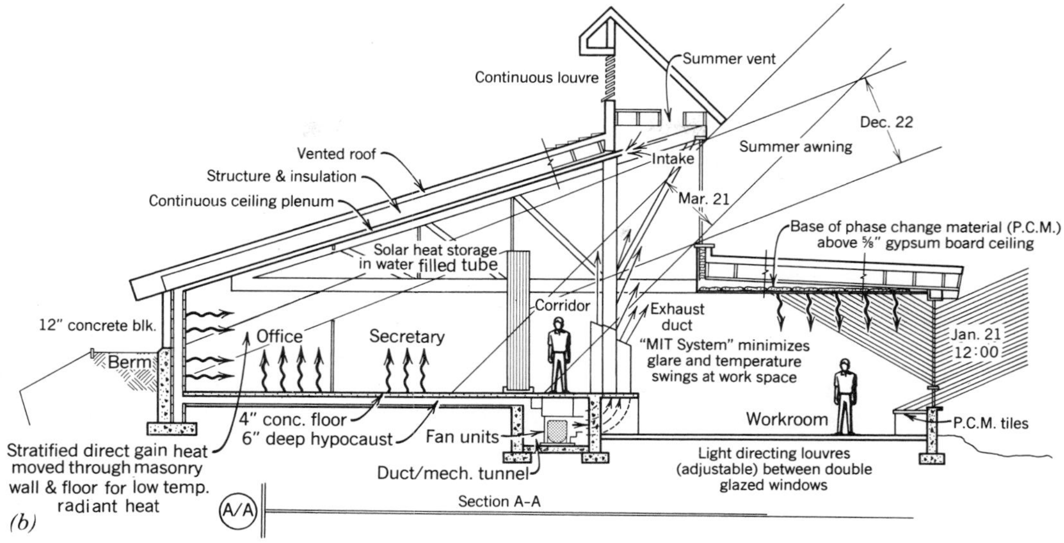

(b)

***Fig. 6.20** (continued)*

temperature remains essentially unchanged during dehumidification. However, these devices consume energy to run the refrigeration cycle, and this is added, as heat, to the space. Accumulated water must be periodically removed from the unit. Most refrigerant dehumidifiers encounter operating difficulties at air temperatures below 65 F (18°C), at which point frost forms on their cooling coils. This could cause problems in a tightly enclosed residence in winter.

(i) Passive Components Summary. A look at two passively heated and cooled projects will serve as a way of interrelating the array of components discussed above. The Conservation Center for the Society of the Protection of New Hampshire Forests, located near Concord, New Hampshire, is shown in Fig. 6.20. This 7000-ft^2 combination of office space, reception area, and lecture hall demonstrates the uses of New England forest products, as well as both direct-gain and sunspace heating strategies. Reflecting roofs and window louvres augment and direct sunlight, and water tubes, masonry walls, concrete floors, and phase change materials (which fuse at 73 F) aid in thermal storage. Warm air is circulated behind and below spaces to distribute solar heat. Daylighting, summer shading, and clerestory ventilation are the passive cooling strategies. Although night-insulating thermal shutters were eliminated (due to budget considerations) the building apparently achieves over 60% solar heating; backup heat is provided by a wood-fired boiler whose hot water is distributed to radiant baseboard panels. The total south glass area is 2100 ft^2; the envelope heat

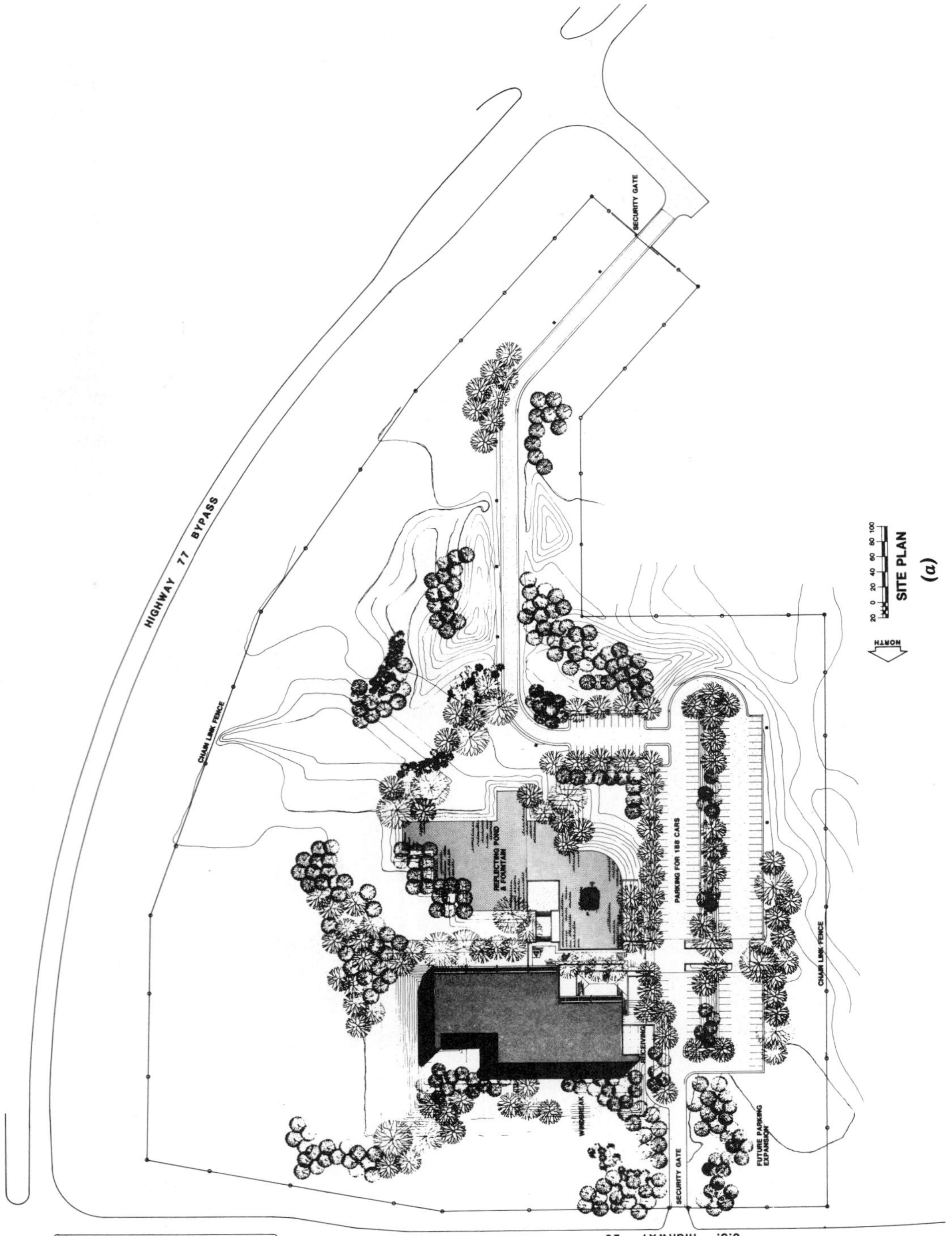
HIGHWAY 77 BYPASS
CHAIN LINK FENCE
SECURITY GATE
REFLECTING POND & FOUNTAIN
PARKING FOR 188 CARS
RECEIVING
WINDBREAK
SECURITY GATE
FUTURE PARKING EXPANSION
CHAIN LINK FENCE
U.S. HIGHWAY 20
NORTH
20 0 20 40 60 80 100

SITE PLAN
(*a*)

loss rate is 6.9 Btu/DD ft^2. The total auxiliary energy consumed is about 23,500 Btu/ft^2 per year (for lighting, heating, and office machines).

The other example is a 68,000-ft^2 insurance office building located on a suburban site in Nebraska (Fig. 6.21). Earth was scooped up to form protective berm walls on the north, east, and west; a reflecting pond on the south now fills the excavation. Landscaping materials and forms provide shelter from winter winds, control snow drifting, provide some shading, and control storm water runoff. Ventilation air is taken through 8- and 12-in. diameter earth tubes that are buried 20 ft under the north berm wall and total 3200 ft in length. In summer this air is cooled to as low as 60 F, while in winter it can reach 55 F upon entry to the building. Solar collectors shade south windows from direct sun. Collected solar heat is stored and used for building heating (by water-to-air heat pumps) in winter and rejected to the pond in summer. The

Fig. 6.21 A suburban office building in the Plains states: the Great West Casualty Company/Joe Morten & Sons, Inc., Building, South Sioux City, Nebraska. FEH Associates, architects. (a) *Site plan, with berm walls and reflecting (cooling) pond.* (b) *South elevation, showing cooling pond; solar collectors shade south windows.* (c) *A wind-protected, sun-trapping outdoor terrace. (Photos by Joel Strasser.)*

wall insulation is on the outer side, so that the building's thermal mass can help to stabilize internal temperatures.

6.3 Heating, Ventilating, and Air Conditioning (HVAC): Typical Design Processes

HVAC systems usually involve a minimum of three design stages. In the *preliminary design* phase, the most general combinations of comfort needs and climate characteristics are considered:

Activity comfort needs are listed.

Activity schedule is listed.

Site energy resources are analyzed.

Climate design strategies are listed.

Building form alternatives are considered.

Combinations of passive and active systems are considered.

One or several alternatives are sized by rule of thumb.

This level of analysis covers the processes discussed in Chapters 1–4 and Sections 5.1–5.3. For smaller buildings, this analysis is often done by the architect alone. For innovative or unusual systems in smaller buildings, and especially for larger, multiple-zone buildings, consultants such as engineers and landscape architects often are included. The team approach is particularly valuable in assessing the strengths of various design alternatives. The architect and the consultants have very different perspectives, and when mutual goals can be clearly agreed upon early in the design process, these perspectives are not only mutually supporting but can produce striking innovations whose benefits extend far beyond services to the clients of a particular building. By setting an example, the team can make available better environments for less energy for hundreds of subsequent buildings. With inspired teamwork, the distribution of HVAC services can enhance building form, as many examples in this chapter show.

By the time the *design development* phase is reached, one of the design alternatives has probably been chosen as the most promising combination of esthetic, social, and technical solutions for the program. The consulting engineer (or architect, on a smaller job) is furnished with the latest set of drawings and the program. Typically, the architectural or mechanical engineer then:

1. Establishes design conditions.
   (a) By activity, lists the range of acceptable air and surface temperatures, air motions, relative humidities, lighting levels, and background noise levels.
   (b) Establishes the schedule of operations.
2. Determines the HVAC zones, considering:
   (a) Activities.
   (b) Schedule.
   (c) Orientation.
   (d) Internal heat gains.
3. Estimates the thermal loads on each zone.
   (a) For worst winter conditions.
   (b) For worst summer conditions.
   (c) For the average condition or conditions that represent the great majority of the building's operating hours.
   (d) Frequently, an estimate of annual energy consumption is made.
4. Selects the HVAC systems. Often, several systems will be used within one large building, since orientation, activity, or scheduling differences may dictate different mechanical solutions. Especially common is one system for the all-interior zones of large buildings and another system for the perimeter zones.
5. Identifies the HVAC components and their locations:
   (a) Mechanical rooms.
   (b) Distribution trees—vertical chases, horizontal runs.
   (c) Typical in-space components, such as underwindow fan-coil units, air grilles, and so on.
6. Sizes the components.
7. Lays out the system. At this stage, conflicts with other systems (structure, plumbing, fire safety, circulation, etc.) are most likely to become evident. Since insufficient vertical clearance is one of the most common building coordination problems with HVAC systems (especially all-air systems—see Sections 6.7 and 6.9), the layouts must include sections as well as plans. Opportunities for integration with other systems also become more apparent at this stage: air ducts can

also help distribute daylighting, act as sunshading devices, or fulfill other functions.

After the architect and the other consultants hold conferences in which HVAC system layout drawings are compared to those for other primary systems (structure, plumbing, electrical, etc.), *design finalizing* occurs. At this final stage, the HVAC system designer verifies the match between the loads on each component and the component's capacity to meet the load. Final layout drawings then are completed.

6.4 HVAC and Building Organization

By this time, many decisions about a building design have been made: design strategies appropriate to the climate and the building's activities have been identified, and the basic siting and overall form of the building have been determined from daylighting and thermal considerations, among others. This section begins by considering the internal, yet broad issues of zoning and system choice and ends with a discussion of the more detailed consequences of system choice. A general guide to estimation of a building's thermal zone requirements was developed in Section 5.1, which discussed the importance of differences in function, schedule, and orientation.

(a) Zoning. The minimum number of thermal zones for a conventionally designed multipurpose building is shown in Fig. 6.22; more than these 16 zones could be produced by differences in scheduling within a zone, such as between "offices" and "stores." As is true of the other occupied floors, "apartments" have a minimum of five zones (based on orientation); however, the emphasis on individual controls—and the variation in usage patterns—often produces as many zones as there are apartments. When the details of zoning are added to the other preliminary design decisions, the details of HVAC systems can be considered.

(b) System Anatomy. Table 6.6 describes the basic organization of any HVAC system. Three kinds of common task (heating, cooling, ventilating) are done by production components; usually, they require distribution and delivery components.

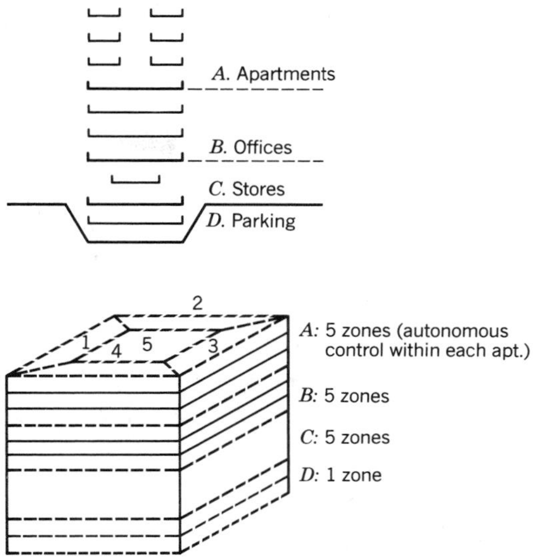

Fig. 6.22 The minimum number of thermal zones for a conventionally designed, multipurpose building of medium size and height. Scheduling and/or internal load differences within a zone could require division into additional zones.

Intake supplies and exhaust by-products accompany each task.

(c) Early Choices. Although the eventual choice of HVAC system should follow an analysis of the zone's needs, some early concepts underlie system choices. The first we will consider is the question of *central versus local* systems.

Central systems require one or several large mechanical spaces (often in basements and/or on roofs), sizable distribution trees, and complex control systems. The noise, heat, and other characteristics of such mechanical rooms can be controlled fairly easily, since the machinery is concentrated at a few locations. Similarly, maintenance is easy to perform, although breakdowns in central equipment can paralyze the entire building. Air quality can easily be controlled by location of the air intakes high above the pollution at street level and by regular maintenance of the centralized air filtering equipment. Longer equipment life can be expected, with regular maintenance. Energy conservation can be served by the recovery of one machine's heat by-product for a nearby machine's heat input. Although there are many ways to provide for the differing thermal needs of the many

THERMAL CONTROL

TABLE 6.6 **Basic HVAC Systems: Tasks and Components**

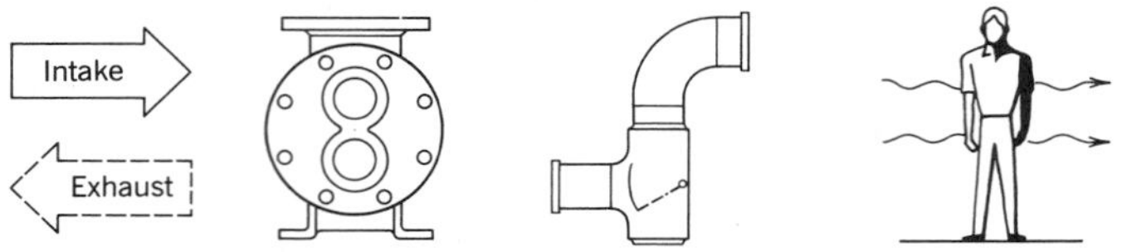

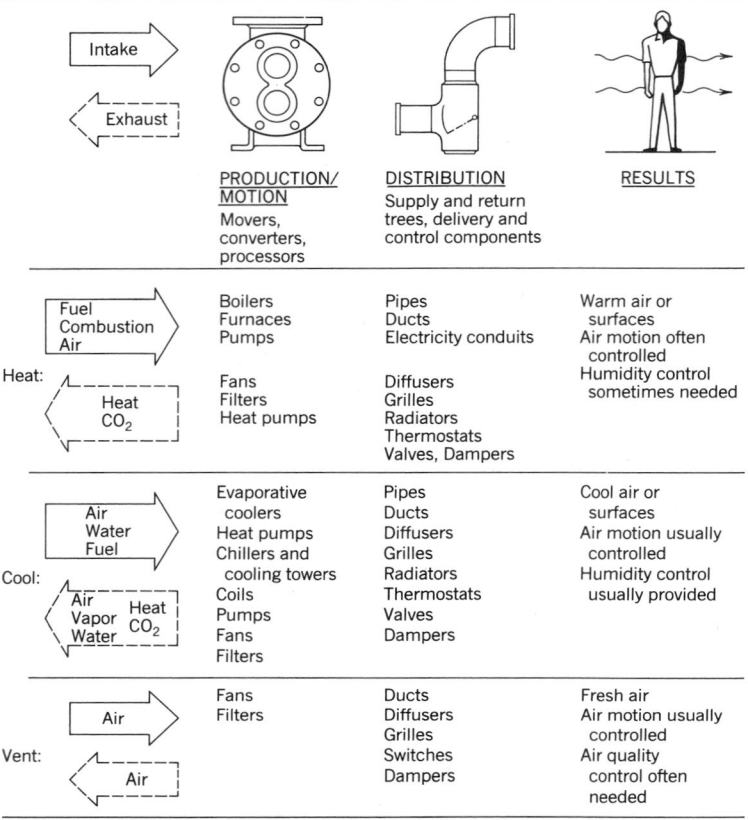

	PRODUCTION/MOTION Movers, converters, processors	DISTRIBUTION Supply and return trees, delivery and control components	RESULTS
Heat: Fuel, Combustion, Air (in) / Heat, CO_2 (out)	Boilers Furnaces Pumps Fans Filters Heat pumps	Pipes Ducts Electricity conduits Diffusers Grilles Radiators Thermostats Valves, Dampers	Warm air or surfaces Air motion often controlled Humidity control sometimes needed
Cool: Air, Water, Fuel (in) / Air, Vapor, Water, Heat, CO_2 (out)	Evaporative coolers Heat pumps Chillers and cooling towers Coils Pumps Fans Filters	Pipes Ducts Diffusers Grilles Radiators Thermostats Valves Dampers	Cool air or surfaces Air motion usually controlled Humidity control usually provided
Vent: Air (in) / Air (out)	Fans Filters	Ducts Diffusers Grilles Switches Dampers	Fresh air Air motion usually controlled Air quality control often needed

Source: Class notes developed by G. Z. Brown, University of Oregon.

zones served by central systems, one important drawback of central systems is a difference in zone scheduling: when the entire system must be activated to serve one zone (such as computer operations in an office building on a weekend), energy is wasted.

Local systems therefore become increasingly attractive as scheduling differences multiply. Also, pronounced differences in other factors—function (with resulting comfort expectations) or placement within the building, for example—can lead to the choice of local systems. Large and centralized equipment spaces are not required with local systems; rather, production equipment is distributed throughout the building (or over the roof of low-rise structures). Dispersal of equipment minimizes the size of distribution trees and greatly simplifies control systems. Moreover, system breakdowns affect only small portions of the building. However, noise and other by-products of multiple machines pose numerous potential threats to occupied spaces, and maintenance is demanding, because access to so many locations is often not easy. Then, too, air quality depends on the regular cleaning of many filters scattered over the building, often within occupied spaces. The potential for energy conservation seems promising because heating or cooling is produced only as locally needed, but there is little chance to use one zone's waste heat as another's needed source.

Another basic question is that of *uniformity versus diversity* in the interior environments of buildings. This question encompasses not only thermal experiences, but visual and acoustical ones as well.

The advantages of uniformity are most evident

in a rapidity of design and construction that, through mass production and speed, often brings lower first costs. As mentioned in Chapter 1, the flexibility in office arrangements that are accompanied by uniform ceiling heights, light placement, grille locations, and so on, can extend a building's usable life span. However, uniformity is not always attractive to users, and diversity is often encouraged at a more personal level—with office furnishings, for example. A more thorough approach to diversity can provide stimulus to the user who spends many hours away from the variability of the exterior climate.

If offices must be uniform in ceiling lighting, air handling, and size, the corridors that connect them and the lounges, or other supporting service spaces, can deliberately be made different. Diversity requires a complete and detailed design of places; it gives the builder a more complex and interesting task; and it can provide orientation and interest to the users. The attractiveness of diversity is evident in most collections of retail shops, in which light and sound—and sometimes heat and aroma—are used to distinguish one shop from the next.

Diversity in the thermal conditions to be maintained, such as warmer offices and cooler circulation spaces in the winter, can be used to enhance the comfort of the office users. Designers have long recognized that a space can be made to seem brighter and higher if it is preceded by a dark, low transition space. Thermal comfort impressions can be manipulated similarly. Less-than-comfortable conditions in circulation spaces or other less critical zones not only make the critical spaces seem more comfortable by contrast, but also save significant amounts of energy over the life of a building. Furthermore, such conditions can make passive strategies more attractive.

A large-scale demonstration of diversity in thermal zones is shown in Fig. 6.23. Passive solar heating can make a significant contribution, even through a shallow-sloped, single-glazed cover in cloudy Glasgow, Scotland, largely because the mall area and leisure areas are allowed a much wider thermal range than would be permitted in stores and offices. The overcast skies are quite suitable for daylight, and the addition of summer sunshading makes natural ventilation (through stack effect, assisted by fans) possible during the cool summers. U.S. Pacific Northwest climate conditions are similar.

(d) Comparing Systems and Zones. In the process of selecting systems from the wide variety available, it is helpful to consider the match between the zones' characteristics and those of various systems. Among the considerations are zone placement (near to or away from the building skin), the zone's thermal loads, the comfort determinants based on the zone's activities, the space available for system components within the zone, and the life-cycle costs of various system alternatives.

Zone placement will sometimes preclude local systems, which depend on easy access to outdoor air both for fresh air and for a heat source or sink. Local systems for interior (away-from-skin) zones are awkward. Relationships between zone placement and building forms are shown in Fig. 6.24.

The thermal loads on each zone will determine the extent to which heating or cooling is the dominant problem—which, in turn, can influence the choice of system. A zone with little cooling load and low moisture production may be well served by a simple system of fresh air plus heating, with no humidity control. Zones that require cooling will usually also require more complete control of air motion and relative humidity. Although it is risky to generalize about which comfort determinants are most important (given differences between activities and between individuals), it can generally be assumed that comfort and thermal tasks are related.

	For Heating of Spaces	*For Cooling of Spaces*
More important	Surface, air temperatures	Air motion
↓		Relative humidity
	Air motion	Air, surface temperatures
	Relative humidity	
Less important	Air quality	Air quality

Thus, the choice of systems can be based partly on whether the system provides good control of the more important comfort determinants.

Another important factor in system choice is the amount of space the system requires. In some cases, it is easy to provide small equipment rooms at regular intervals throughout a building, such that little or nothing in the way of a distribution

THERMAL CONTROL

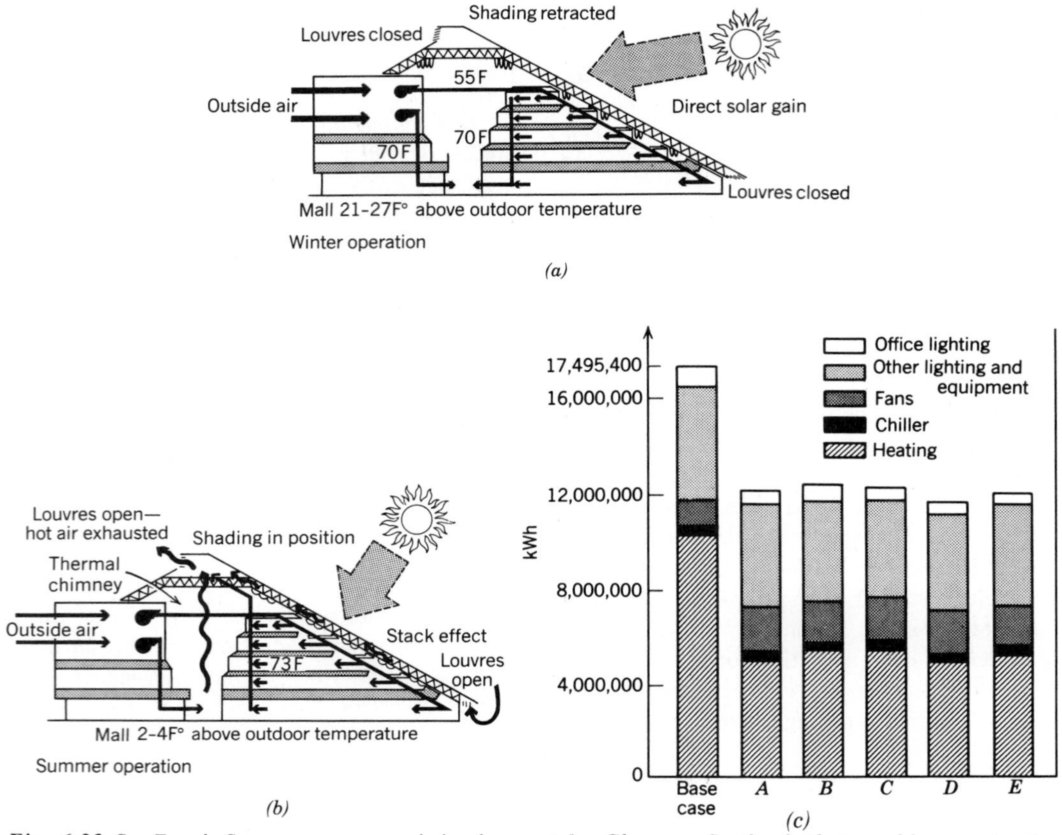

Fig. 6.23 St. Enoch Square, a proposed development for Glasgow, Scotland, that would use extensive passive solar heating, daylighting, and natural ventilation. Reiach & Hall and GMW Partnership, architects (joint venture); Cosentini Associates, energy consultants; Princeton Energy Group, daylighting consultants. (a) *Schematic section showing winter operation; the mall temperature varies around 63 F during operating hours, while offices are kept near 70 F.* (b) *Schematic section showing summer operation; the mall temperature swings from 68 to 74 F during operating hours.* (c) *Estimates of annual energy consumption, for a conventional-design "base case" and several alternative configurations of the proposed complex. Note the significantly lower heating energy requirements, resulting in part from the lower winter temperatures allowed for less-critical zones such as the mall and the leisure areas in configurations A–E.*

tree will be required. In other cases, a network of distribution trees and central, large equipment spaces are easier to accommodate. These central systems typically fall into one of three classifications:

All-air (the largest distribution trees).

Air & water.

All-water (the smallest distribution trees, with local control of fresh air).

The details of these systems can be found in Sections 6.7–6.11, along with typical applications and space requirements. For now, a preliminary screening is necessary, to match zones and systems. A simplified procedure is shown in Table 6.7, in which preliminary system choices are made for a building such as the multipurpose structure shown in Fig. 6.22. In this process, the 16 zones first apparent are translated into three local systems and three central systems: one all-air, one air & water, and one all-water.

The Fox Plaza Building in San Francisco, which illustrates many of the matches between systems and zones, is shown in Fig. 6.25. This project

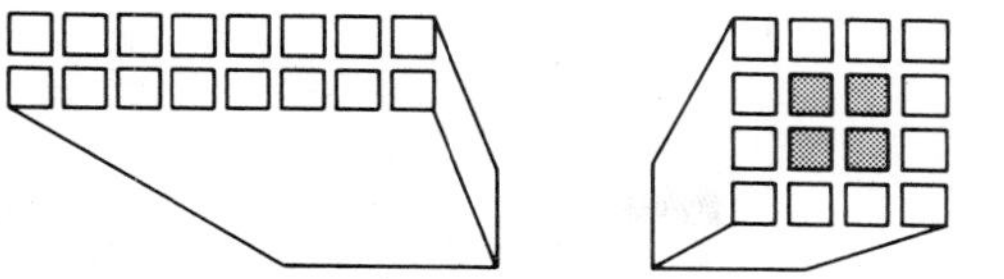

TASKS:	BUILDING FORMS:	
Vent each space	• Directly, through skin	• Indirectly, requiring distribution
Heat each space	• Directly, through solar • Combustion air directly available, as is exhaust of gases • Through-skin heat pump	(Not frequently required) • Directly, with electric resistance • Indirectly, requiring distribution
Cool each space	• Directly, through ventilation • Through-skin heat pump	• Indirectly, requiring distribution

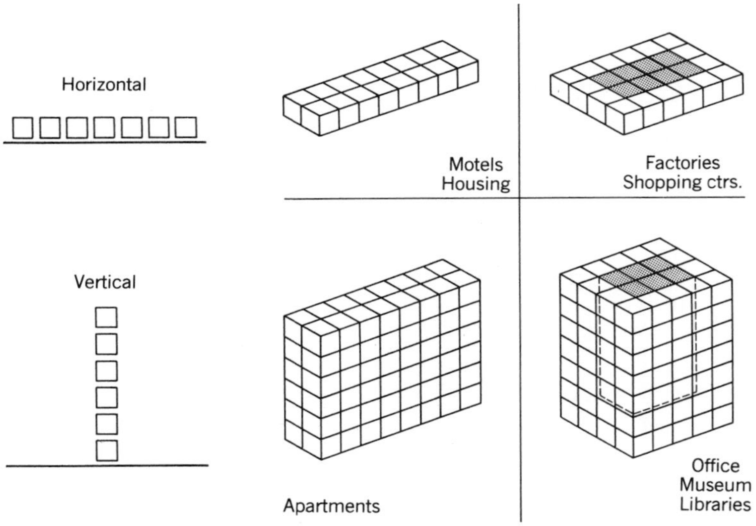

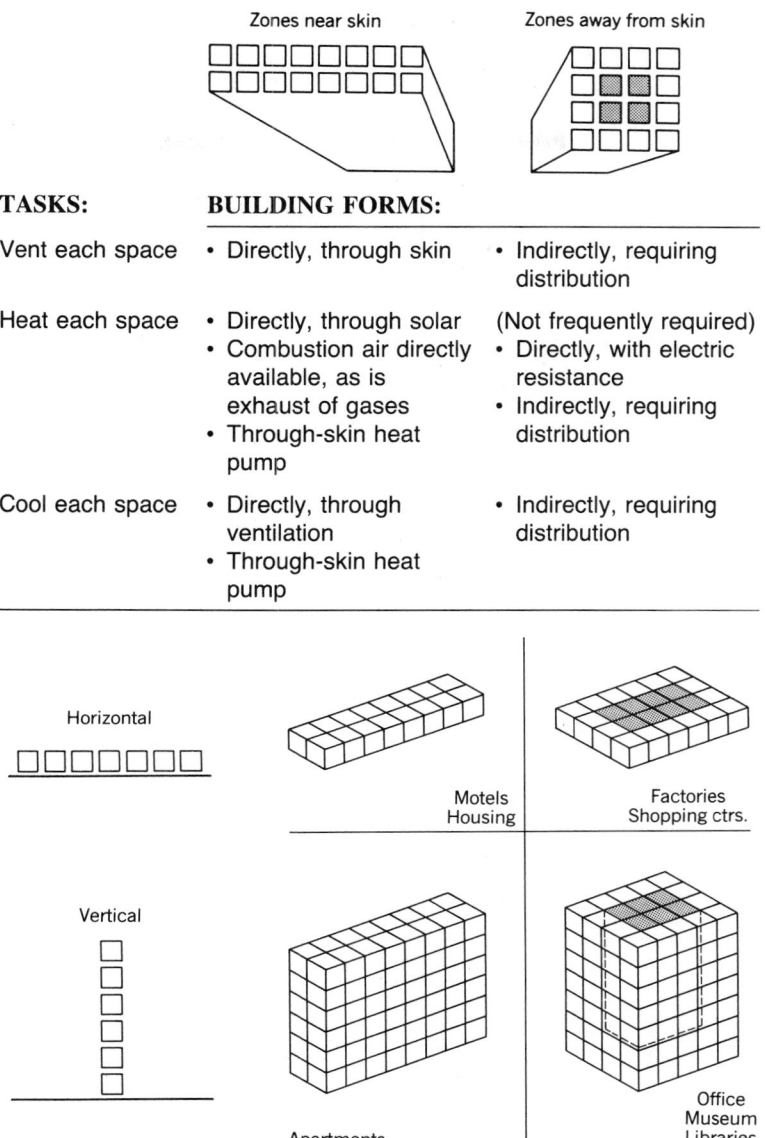

Fig. 6.24 *Zone placement and building form are related to heating, cooling, and ventilating tasks; some applications take on typical building forms. (From class notes developed by G. Z. Brown, University of Oregon.)*

includes four major building types in one structure:

1. Underground garage for the storage of cars.
2. Commercial center at ground level, including a bank, a women's specialty store, and other commercial establishments.
3. Ten floors of offices.
4. Sixteen floors of apartments.

The mechanical level is located between the office portion of the building and the apartments above. In this way, the main runs of all the equipment—heating, air conditioning, electrical, etc.—are directed both upward and downward, in the shortest possible distance. The spatial requirements of of-

TABLE 6.7 **Procedure for Matching Zones and Systems**

Capsule Description

A multipurpose building (similar to that shown in Fig. 6.22) is situated in a cold winter–mild summer climate.

Apartments are on upper floors, surrounding an open-air central court; they have adequate daylight and cross ventilation. Floor heights are quite low.

Offices are rented to various tenants. Exterior offices have high ceilings, to facilitate daylighting, and therefore have low internal gains but restricted clearance for horizontal ducts. Interior offices have lower ceilings; vertical chase space is limited, since it reduces rentable area.

Shops are located around the perimeter on the high-ceilinged ground floor; some smaller shops are located on the mezzanine and ground floors in the interior zone. Space for vertical chases is severely limited on these highest rental floors.

The parking area is below grade, surrounded by air and light wells. Floor heights are very limited, to reduce ramp length.

Activities		*Offices*			*Shops*			
(Program)	*Apartments*	*Computer Center*					*Restaurant*	*Parking*
Schedule	24 h	24 h	9 hr		*9 h*		*12 h*	
Placement	Exterior	Exterior for access	Exterior	Interior	Interior	Exterior	Exterior for access	Entire floor
Internal gains	Vary	High	Low	Medium	High	Medium	High plus moisture	Low with exhaust gas
Dominant HVAC task(s)	Heat Ventilate	Cool	Heat/cool	Cool	Cool	Cool	Cool	Ventilate
HVAC space available								
Vert. (in plan)	Medium	Medium	Medium	Tight	Tight	Tight	Tight	Medium
Horiz. (in section)	Tight	Medium	Tight	Medium	Tight	Ample	Ample	Tight
System choices								
Local		*A*					*B*	*C*
Central								
All-air			*(D)*	*D*	*D*	*(D)*		
Air & water			*E*			*E*		
All-water	*F*							

Summary

- *A.* The computer center's unique schedule and rate of internal gain usually requires a separate system, equipped with humidity and air quality controls to protect the equipment. Some heat recovery for use in *E* and *F* seems advisable.
- *B.* The restaurant's special problems of heat, moisture, and aroma, as well as its schedule, require a separate system.
- *C.* The parking area needs only plenty of fresh air; it requires no tie with the other zones at all.
- *D.* The always-cooling loads of interior zones are best served by all-air systems offering control of humidity and air quality. However, vertical chase size is tight, and high-velocity distribution may be required. (The exterior zones could also be served by all-air. But the need for heating, plus the likelihood of fresh air infiltration and the tight clearance for horizontal ductwork, all suggest that the system for exterior zones be separated from the system for interior zones.)
- *E.* Quick changes from heating to cooling are best handled by water; some central air-quality control is offered by air & water systems.
- *F.* A central all-water system offers energy conservation advantages, recovering waste heat from system *D* (and potentially from *A*). Fresh air is easily and cheaply handled on a local basis, which also provides cooling.

Mechanical Space

Probably best located on the top office floor, or on a floor of its own between offices and apartments. Distribution tree sizes will thereby be minimized on the high-rent ground floor.

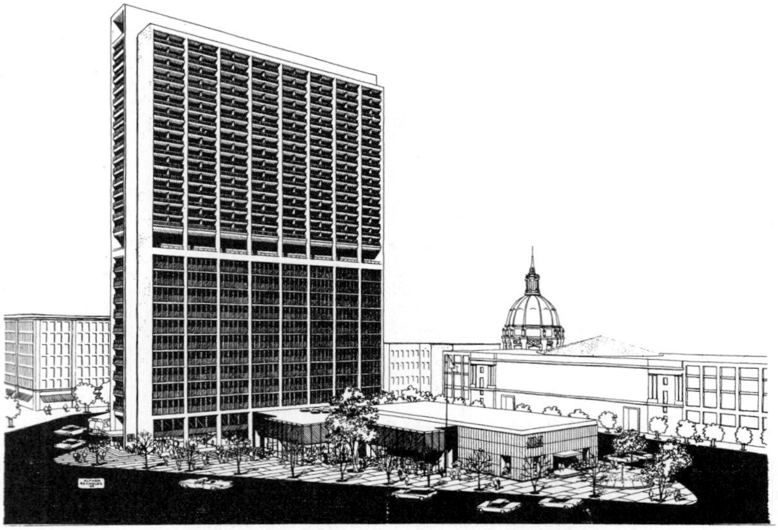

(a)

Fig. 6.25 (a) *The Fox Plaza Building, San Francisco. Victor Gruen Associates, Inc., architects and engineers. Top mechanical story encloses cooling towers and shelters domestic hot water generator-storage units for apartments. Upper stories consist of 16 floors of apartments. Intermediate (13th-floor), mechanical story is the location of water chillers, pumps for circulation of chilled water, and cooling tower water. Hot and cold high-velocity air ducts for downfeed to lower story conditioning originate there. It is also the location of steam boilers and converters (steam to hot water), for fan-coil units in residential stories above and similar converters for hot water for coils in lower story office air units. Also on the 13th floor, adjacent is the domestic hot water generator-storage receiver for the offices below. Below the 13th floor, climate-making center are 10 stories of offices and 2 stories of commercial area. (Courtesy of* Progressive Architecture.) (b) *The Fox Plaza Building in construction. Clearly shown are the air-handling units on the 13th floor. Also visible are downfeed ducts that supply high-velocity hot and cold air from these units to the office stories below. Other ducts return the air at normal velocities to the air-handling units for reconditioning. Upper residential stories in this equable climate of San Francisco are supplied with heating but not cooling. Office cooling deals with people-concentration and lights. (Photo by Morley Baer.)*

fices and the spatial requirements of apartments are quite different; thus, the total design layout of each of the two uses differs—floor-to-floor heights, window treatment, and the heating, cooling, electrical, elevators, and other services are all different. The placement of the mechanical level between the offices and the apartments not only facilitates an efficient operation but also provides for a definite visual separation between the two functions.

Quite unusual is the placement of the steam boilers on the 13th floor instead of in the conventional basement location. Adjunct to the general 13th-floor center, only a small amount of auxiliary equipment is located on the roof and in a small portion of the garage. Residential areas have hot-water heating, offices have dual-duct, high-velocity heating/cooling, and commercial tenants are supplied with hot and chilled water for individual climate control requirements.

(e) Central Equipment Location. The Fox Plaza Building exemplifies an intermediate location for the heating and cooling production equipment—one that separates floors of apartments from floors of offices. Other typical locations for central equipment are in the basement (where noise is most easily isolated, utilities are easily accessed, and machine weight is little problem) and on the roof, where access to air as a sink for reject heat is easiest of all and headroom is unlimited. Very tall buildings may require several intermediate mechanical floors. Examples of these approaches are found throughout the rest of the chapter.

Fig. 6.25 (continued)

These equipment spaces have special environmental needs: lots of fresh air, strong support, and high headroom. The equipment can generate considerable heat, moisture, air motion, noise, and vibration—to the potential annoyance of nearby floors (or neighboring buildings). As shown in the Fox Plaza example, the equipment can be strongly expressive of building services and can serve a useful demarcation role between vertical layers of high-rise buildings.

(f) Distribution Trees. Section 5.1 raised the preliminary questions about distribution trees: how many, what kind (air or water), and where to place them within buildings. Before carrying these questions further, we must consider two other basic concepts: the extent to which the distribution system should be architecturally expressed or concealed, and the extent to which the HVAC system might be integrated with structural elements.

Concealment and Exposure. The pipes, ducts, and conduits that take the necessary resources to and from the interior are often carried within a network of spaces unseen by anyone except builders and repair people. The advantages of concealment include less water and air noise, fewer surfaces requiring cleaning, less care necessary in construction (leaks, not looks, are important), and more control over the appearance of the interior ceiling and wall surfaces. Although maintenance access to such hidden supply lines is more difficult, a variety of readily removable covers is available, particularly in suspended ceilings.

On the other hand, the exposure of these supply networks provides an honest and direct source of visual (and occasionally acoustical) interest. Exposure in corridors and service areas and concealment in offices constitute an approach used in many office buildings. Flexibility is usually encouraged by exposure; changes can be easily made when there is no need for neatly cut holes in concealing surfaces. However, flexibility from movable partitions requires constant ceiling heights—a feature of the suspended-ceiling approach.

One of the more spectacular examples of exposed mechanical (and structural) systems is shown in Figure 6.26—the result of a design competition for a museum of modern art, reference library, center for industrial design, center for music and acoustic research, and supporting services, in downtown Paris.

When users are invited to play an active role in adjusting conditions inside, exposure of the switches they manipulate is helpful. Visible mechanisms not only remind users of their opportunities, but also encourage user interaction. In this way, adjustments are sometimes discovered that the designer had not anticipated.

Fig. 6.26 *Centre Georges Pompidou, Paris. Some views of the mechanical support systems. The design competition for this complex was announced in 1971, but subsequent lawsuits and budget cuts delayed its opening until January 1977. Piano + Rogers architects. (Photo by John Tingley.)*

Mechanical-Structural Integration or Separation. The similarity of these two technical support systems—structures and environmental controls—has intrigued designers ever since mechanical systems began to require substantial volume for distribution, as in air-duct systems. As the complexity and size of the mechanical distribution systems was *increasing* with technological development (typically, more air is required to cool a space than simply to heat it), the increased strength of materials was *reducing* the size of the structural system. The ''uncluttered'' floor areas between the more widely spaced columns became desirable for flexibility in spatial layout. With the mechanical systems at or within these columns, floor areas remained clear, thus giving mechanical-structural integration further impetus. With the new expectations for cooling, the refrigeration cycle's cooling tower often moved to the roof, taking the air-handling machinery with it. This further en-

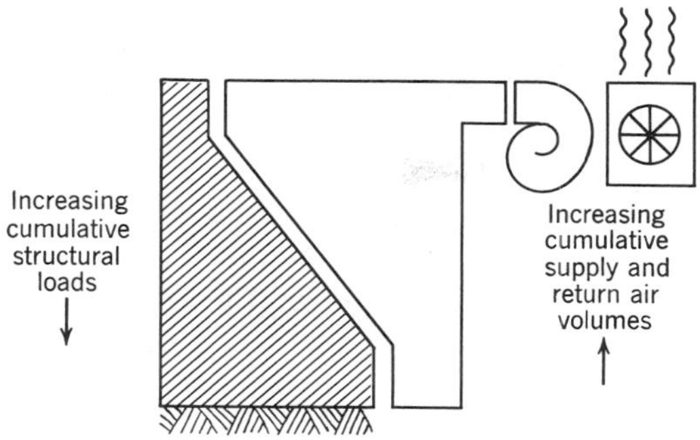

Fig. 6.27 Technical support system. With rooftop air handling, the total air duct size decreases toward the ground. Conversely, the total structural load increases toward the ground.

couraged the merging of systems, for one system was growing wider as the other diminished (see Fig. 6.27). Thus a fixed-column cross section, consisting mostly of the structural column at the base and the air duct at the top, became possible. (One of the responses to this opportunity, expressed as a dual-duct supply and return system, is shown in Fig. 6.110, page 421.)

Yet the functions of these systems differ widely: compared to the dynamic on-off air, water, and electrical distribution systems, the structural system is static—gravity never ceases. The moving parts in mechanical systems need maintenance far more frequently than the connections of structural components. Changes in occupancy can mean enormous changes in mechanical systems, requiring entirely different equipment; structural changes of such magnitude usually occur only at demolition. Mechanical systems can invite user adjustment; structural systems rarely do.

Thus, while it is possible to wrap the mechanical systems in a structural envelope, it is of questionable long-term value, given the differing life spans and characteristics of these systems. The probability of future change suggests that the *mechanical* system be the exposed one, despite the appeal to many designers of the structural system's cleaner lines.

Distribution Tree Placement Options. These options are summarized in Fig. 6.28. Vertical placement options are important because they affect floor space, influencing the flexibility of spatial layout and the availability of usable (or rentable) floor space. Horizontal placement options affect ceiling height—a particular issue in daylighting design, and sometimes a critical factor when overall height limits are imposed yet a maximum of usable floor space is desired. (In Washington, D.C., for instance, no building can rise higher than the Capitol.) Both vertical and horizontal distribution at the edges can have dramatic impact on building appearance.

As the popularity of the glass facade of the modern office building developed, pipes and ducts were gradually cleared away from the building exterior. In earlier years, steam and condensate risers at the exterior columns had inevitably given way to risers in the core, near the elevators. Often hot *and* chilled water risers branched laterally in the space below each floor and were turned up to fan-coil units below the glass. Systems of high-pressure, high-velocity, conditioned air increased in popularity, also taking the form of risers in the central core, with supply and return runouts in the space above the ceiling at each story. The concomitant architectural form was a horizontal glass strip with uniform, pipeless columns set back from the building line. Between these facade strips of glass were bands of opaque glass, metal, or masonry; behind these lay the floor structure, with horizontal pipes and ducts below and sometimes a sill-high space above. There was a strong horizontal accent. Because heat gain and heat loss are maximum at the building perimeter, as a result of the effects of the sun in summer and low temperature in winter, the pipe and duct runouts continued at large dimension from the core to the outside curtain of the building. As the intricacy of the mechanical services increased, the depth of the structure-plus-equipment layer attained about one-third of the total, overall floor-to-floor height. High-velocity air systems with central units began to replace the formerly popular systems that used air-handling rooms at each story. This development stressed the importance of the vertical riser. It began to be difficult to get branches out of the central core past elevators, stairs, and other risers.

Much needed was relief for the central core through the relocation of risers at the perimeter (where they are most needed) and a reduction in the length—and consequently in the size—of pipe and duct branches. Not least in importance was the avoidance of bulky ceiling crossovers, which was effected by moving large assemblies of building "arteries and veins" to the surfaces they serve.

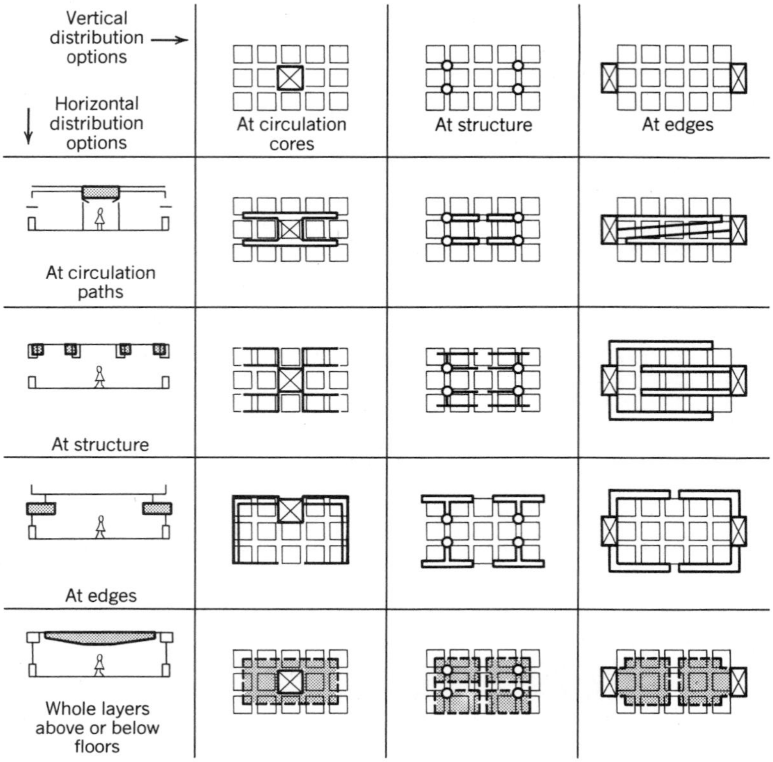

Fig. 6.28 Distribution tree placement options, both vertical (impact on plan) and horizontal (impact on section). (Based on class notes developed by G. Z. Brown, University of Oregon.)

It was logical to place at the perimeter the parts of the system that deal with the effects of sun, shade, and temperature change in the several perimeter zones, leaving at the core a separate network to handle the more stable interior areas. The disadvantages of perimeter distribution include (usually) higher construction costs and an environment that is more thermally hostile, due to the extremes of outdoor temperature.

Vertical distribution within *internal circulation cores* is very common (see Fig. 6.29*b*), as it leaves a maximum of plan flexibility for the rest of each floor and does not disturb the prized floor areas nearest windows. However, one centralized vertical distribution trunk will require large horizontal branches near the core, so with this choice early thought must be given to the horizontal placement options. Vertical distribution integrated with *structure* has some intriguing possibilities (see Fig. 6.31) where the structure-HVAC integration concept is suitable. Because multiple HVAC trees are implied (since there are multiple columns with which they are integrated), the horizontal branches tend to be small. However, these branches often join the vertical trunk at the same place that critical column-to-girder structural connections need to be made; interference is common and can be costly to solve. Vertical distribution at the *edges* is potentially dramatic in form but costly to enclose (if outside) or wasteful of prime floor space (if inside).

Horizontal distribution above *corridors* is very common, since reduced headroom here is more acceptable than in the main activity areas. Furthermore, corridors tend to be away from windows, so their lower ceilings don't interfere with daylight penetration. Since corridors connect nearly all spaces, horizontal service distribution to such spaces is also provided. Furthermore, exposure of these services above corridors can heighten the contrast between such serving spaces and the uncluttered, higher ceiling offices that are served.

Horizontal distribution at the *structure* is sometimes chosen, particularly where U-shaped beams or box beams provide ready channels for HVAC distribution. However, the penetration of horizontal structural members by these continuous service runs must be coordinated. Horizontal distribution at the *edges* can be integrated usefully with sunshading devices and light shelves; it can also act as a spandrel element that contrasts with the window strips. Horizontal distribution within *whole layers* above or below floors is often utilized, particularly for return-air plenums (contrasted with supply air carried within ducts).

An impressive example of the typical combination of vertical distribution at circulation cores and horizontal distribution in whole layers is provided by the World Trade Center Towers, shown in Fig. 6.29. In these towers the tenant space in typical floors is very wide, in contrast to the usual 20- to 30-ft column spacing seen in typical plans. In achieving such long spans between core and perimeter, the need for greater structural strength at core and perimeter is essential. The multicolumn exterior wall panels are especially important in resisting vertical, lateral, and other forces in this tall and unusual building. For this reason, vertical air distribution at exerior walls would not be appropriate, so vertical air ducts are located in the core.

The horizontal ducts and pipes run through a layer formed by the truss-joist construction (Fig. 6.29*c*), from core to perimeter, where induction units (see Section 6.10) supply cool or warm air as needed. Also, supply ducts from the core to the interior zones utilize this layer, which further serves as a return-air plenum. Thus, an efficient minimum 12-ft story height, floor-to-floor, is maintained with an adequate floor-to-ceiling height in occupied spaces.

Two-story mechanical floors in the towers have wider exterior columns, due to a larger unbraced height, than do typical stories. Behind their unglazed exterior walls is a 4-ft open-air perimeter walk. On the inner side is a "breathing" wall of louvres, permitting intake and exhaust air for the mechanical equipment.

Another example of supply at the edge for both vertical and horizontal distribution can be found in the International Building in San Francisco (Fig. 6.30). Here the vertical shafts are prominently exposed at the corners; these shafts carry supply and return ducts serving the four perimeter air conditioning zones. Air-handling equipment and a 750-ton refrigeration plant are located on the floors just

Fig. 6.29 *(overleaf) The World Trade Center Towers, New York City. The Port Authority of New York and New Jersey, owner. Minoru Yamasaki & Associates, architects. Emery Roth & Sons, architects. Skilling, Helle, Christiansen & Robertson, structural engineers. Jaros, Baum & Bolles, mechanical engineers. Joseph R. Loring & Associates, electrical engineers.* (a) *In each of the twin towers, the mechanical floors, of which there are four above grade and one below, are each two stories high (24–28 ft, finish floor to finish floor). At the intermediate level of each of these two-story stations, there is no structural floor outside the core. This omission affords a two-story height for equipment and for horizontal runs of ducts and pipe. The planned volume of the space required for equipment is a vital consideration in the early design stages. The mechanical stories above grade total 105.7 ft in height. Of the 1350-ft full height of each tower, this comprises 105.7/1350.0 = 0.078, or 7.8%. The subgrade mechanical story, vertical distribution in the core, and other miscellaneous mechanical space increase this percentage considerably. From intermediate mechanical floors, such as the 41st, the distribution of air conditioning services is both up and down; from the 41st down to the 25th and up to the 57th—16 floors in each direction.* (b) *Plan of typical floor. Deck spans as great as 36 and 60 ft are quite unusual. Great flexibility of later planning in tenant spaces is achieved. The structural floor comprises large, preassembled panels consisting of custom-designed truss-joists carrying steel decking. The joists are in pairs. The pairs are spaced 6 ft, 8 in. apart on centers. This spacing affords adequate width for the air ducts that run between them and connect the vertical ducts in the core with the system of air delivery units in the ceiling of the tenant space.* (c) *The open nature of the truss-joist construction, somewhat unusual in tall buildings, permits horizontal ducts emanating from the vertical risers in the core to reach the perimeter. The space between joists and the triangular openings within the joist make this possible. A similar arrangement, also within the joist depth, serves a distribution duct system in the* tenant *areas. There, rectangular and flexible round ducts form a grid that delivers air down through lighting fixtures in the ceiling. Return air is taken back through plenums and the core return ducts to mechanical floors.*

108 to 110
14'–0"
+ 11'–8"
25'–8"
Floors
Story height, fin. floor to fin. floor.
33 Stories
Spacing
75 to 77
Mechanical floors
14'–0"
+ 14'–0"
28'–0"
34 Stories
41 to 43
14'–0"
+ 14'–0"
28'–0"
34 Stories
14'–0"
+ 10'–0"
24'–0"
7 to 9
12
Subgrade
–3 to –5
11'–0"
+ 11'–0"
22'–0"

(a)

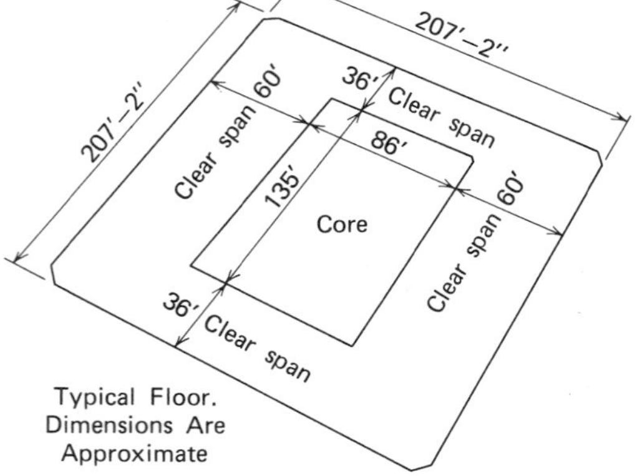
207'–2"
207'–2"
60'
36'
Clear span
Clear span
86'
135'
Core
60'
Clear span
36'
Clear span
Typical Floor. Dimensions Are Approximate

(b)

(c)

Fig. 6.30 The International Building, San Francisco. Anshen and Allen, architects; Eagelson, Engineers (Charles Krieger, E.E.), mechanical designers. (Courtesy of Progressive Architecture.*) (a) One of the four corner main ducts. (b) In this ingenious scheme, the major supply arteries for conditioned air are located in alternate corners. In diagonally opposite locations, supply risers are placed in large square enclosures. Although nonstructural, each enclosure is emphasized as a distinct vertical design element. Each encloses both hot- and cold-air ducts, which supply two separately controlled orientations on all 21 stories. Conditioned air originates at an intermediate floor, the third. In the opposite two corners, similar ducts return much of the air to the equipment story. The balance is returned through duct risers in the core.*

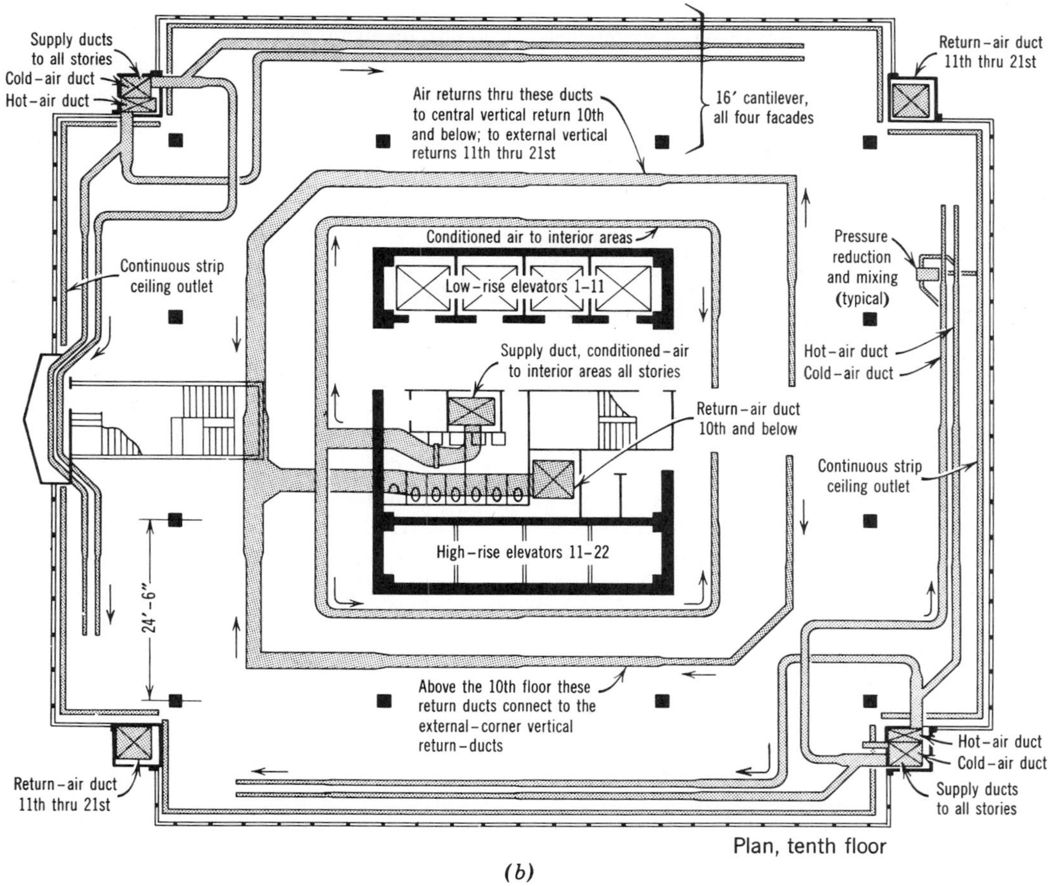

Plan, tenth floor

(b)

below the terrace level (those least desirable for renting). Each corner duct branches to serve two zones, separately controlled. Pressure reduction and blending are done by equipment in the hung ceiling, and from these points air flows to strip-grille diffusers directly above the glass on the four sides of the building. Local controls assure comfort to personnel in each area.

Interior zones on each floor are supplied by a riser duct in the building core, which branches at each floor to a loop just outside the line of elevators. The loop serves ceiling diffusers.

Between the perimeter loop and the interior loop just described, a return loop collects air for return to the central station (second and third floors). These return loops on the 11th to the 21st floors are picked up by external return risers on alternate exterior corners. From the 10th floor on down, the loops are picked up (as shown in Fig. 6.30) by an interior return riser that extends down through the core in front of the blank faces of the high-rise elevators. To provide a clear space between the elevator banks on the main floor (4th or terrace), the two core ducts-risers are offset at the ceiling of that story.

In summary, perimeter air for all stories is supplied through corner ducts. Central air for all stories is supplied through a core duct. All return air above the 10th floor is carried down through the return ducts at the *other* two corners. Return air from the 10th floor and below is carried down by a return duct in the core.

A final example of vertical distribution at the edge, this time integrated with large hollow structural columns, is provided by the Chemistry Building at the University of California, Berkeley (Fig. 6.31). Because of the large number of distribution trees necessary for laboratory buildings, 16 hollow exterior reinforced-concrete structural columns were used to enclose pipes and ducts connected to short lateral branches, thereby minimizing crossovers and avoiding ''spaghetti'' patterns on the ceilings.

These are truly multipurpose columns. Together with the slimmer interior columns, they support the building. In the late afternoon, when the balconies (between adjacent columns) have ceased to exclude all direct sunlight, the deep exterior columns take over as totally effective sunshades. Finally, they enclose exhaust ducts from all exterior (and some interior) fume hoods (*FH* in Fig. 6.31). These ducts are exhausted by fans behind the masonry grill at the roof level. Four of the exterior columns are *pipe-station columns*; of the two shown in Fig. 6.31, one is detailed. The piping services include domestic hot and cold water, industrial hot and cold water, distilled water, demineralized water, natural gas, vacuum, compressed air, hydrogen sulfide, nitrogen, oxygen, steam, acid waste, vent pipes, and roof leader pipes. There is access to all exterior columns at all floors.

This is a once-through system; no air is recirculated. The pressure in the halls is greater than that in the rooms, preventing the flow of air or gases back out into the corridors.

Summary: All fresh air is brought in at ground level. It is partially conditioned and delivered to rooms by means of a central riser and ceiling ducts in corridors. After reheat and delivery to rooms, it is exhausted through fume hoods to ducts in hollow exterior columns; and in the case of interior rooms, to a central group of exhaust ducts. All exhaust fans are on the roof. Four exterior columns are also *pipe station* columns.

Further treatment of the relationships among HVAC systems, their distribution trees, and buildings is given in the examples that accompany the more detailed descriptions of large-building HVAC systems in Sections 6.7–6.11.

6.5 Refrigeration Cycles

Unlike the slow and diffuse heat transfer processes that characterize passive heating and cooling approaches, mechanical equipment can rapidly concentrate heating or cooling, on demand. The refrigeration cycle is a particularly useful mechanical process in heating as well as cooling applications. The two types of heat transfer process commonly used in mechanical equipment for buildings are the *compressive* and the *absorption* refrigeration cycles. (Equipment that utilizes these cycles is discussed in Sections 6.6 and 6.8.)

(a) Compressive Refrigeration. As applied in Fig. 6.32, the compressive refrigeration cycle is a scheme for transferring heat from one circulated water system (chilled water) to another (condenser water). The means for doing this is the

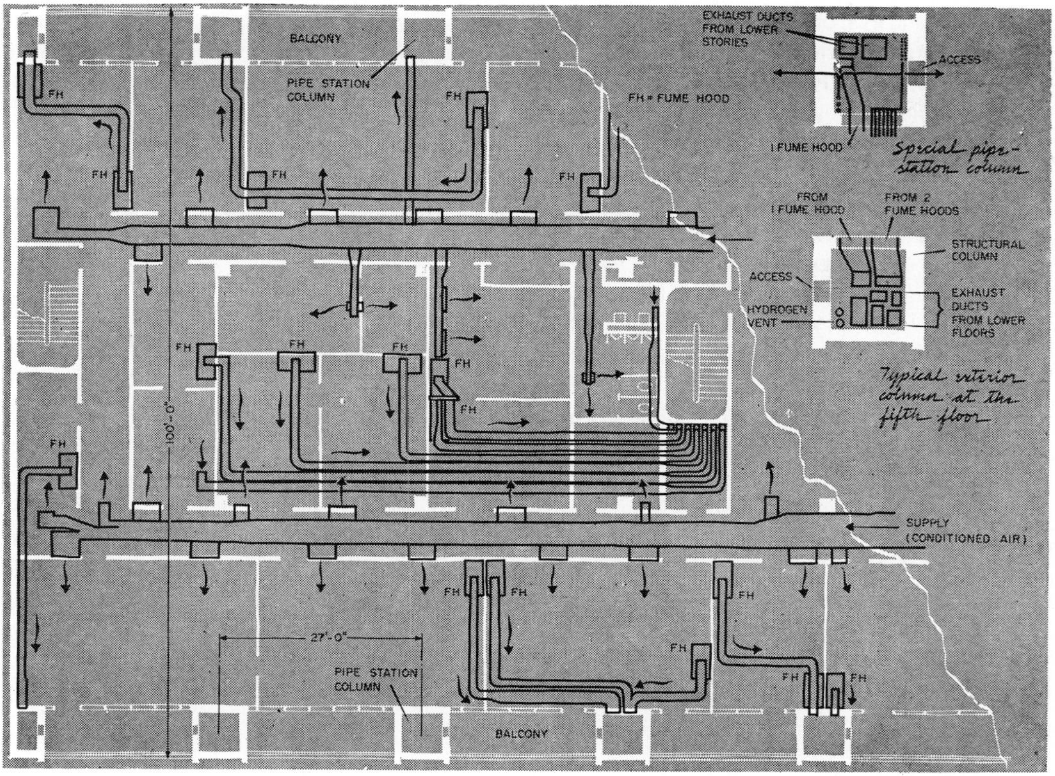

(a)

Fig. 6.31 Chemistry Building, Unit 1, University of California at Berkeley. Anshen & Allen, architects. Mechanical design by Bayha, Weir & Finato. (a) *Preliminary rendering and layout of supply and exhaust air shown in four of seven bays at typical floor, the fifth. The burden of exhaust from fume hoods is relieved by routing those located in outside rooms to exterior hollow structural columns. Hoods in* interior *rooms are ducted to exhaust stacks adjacent to the stairway. Hollow columns provide distribution space for piping. Lateral crossover of mechanical services at ceilings is minimized.* (b) *Behind the decorative masonry screen at the roof edge are exhaust fans that pick up the vertical ducts in the hollow columns to relieve the buildup of gases produced by a multitude of chemical research projects. At this hour the sun is intercepted by masonry walls that form an effective barrier against heat transmission to the indoors. This relieves a cooling system that must cope with the burden of people, lights, and heat-producing research.*

(b)

liquification and evaporation of a refrigerant, such as Freon, during which processes it gives off and takes on heat, respectively. The heat that it gives off must be disposed of (except in the heat pump), but the heat that it acquires is drawn out of the circulated water known as the chilled water, which is the medium for subsequent cooling processes.

Freon, a gas at normal temperatures and pressures, must be compressed and liquefied to be of service later as a heat absorber. To be liquefied (see Fig. 6.32), Freon must first be compressed to a high-pressure vapor; then, by means of cool water, latent heat is extracted from the Freon, which condenses it to a liquid. This product, high-pressure liquid Freon, is a potential heat absorber since, when it is released through an expansion valve, it springs back mechanically to gaseous form. In this change of state it must take on latent heat, by drawing heat out of the circulated water of the chilled water system. It may be said that the refrigeration cycle pumps the heat out of the chilled water system into the condenser water system. Indeed, by special (reverse cycle) arrangements of the water systems, a *heat pump* is the result. (The compressive refrigeration cycle can be used to transfer heat between almost any media; whereas Fig. 6.32 illustrates a water-water cycle, Fig. 6.55 shows the heat pump in both air-air and water-air applications.)

(b) Absorption Refrigeration Cycle. This process is illustrated in Fig. 6.33. In order to remove heat from chilled water, it uses still more heat in regenerating the salt solution and is typically somewhat less efficient than the simple compressive cycle. However, this heat for regeneration may be provided by solar energy, or by relatively high-temperature waste heat from another source. Because the high-grade energy (electricity) needed to run a compressor with is replaced by the lower grade heat needed to run the generator, the absorption cycle can enjoy an energy advantage over the compressive cycle.

6.6 Heating and Cooling Systems for Small Buildings

Smaller buildings are typically skin load dominated: that is, for them the climate dictates whether heating or cooling is the major concern. In some climates, only heating systems are needed; the building can "keep itself cool" during hot weather without mechanical assistance. In other climates, only cooling is needed. In still others, both heating and cooling are required. In this section, heating-only mechanical systems will be treated first, then heating-and-cooling systems, then cooling-only systems.

(a) Heating-Only Systems, For Rooms. After the sun, the most ancient method of heating is the radiant effect of fire. With each step from campfire to fireplace to wood stove, more of the fuel's heat was captured for the room rather than wasted to the outdoors (Fig. 6.34). Although many people enjoy the sight, sound, and smell of the open fireplace, the tightly enclosed wood stove with catalytic combustor is a substantially more efficient and less polluting approach to heating.

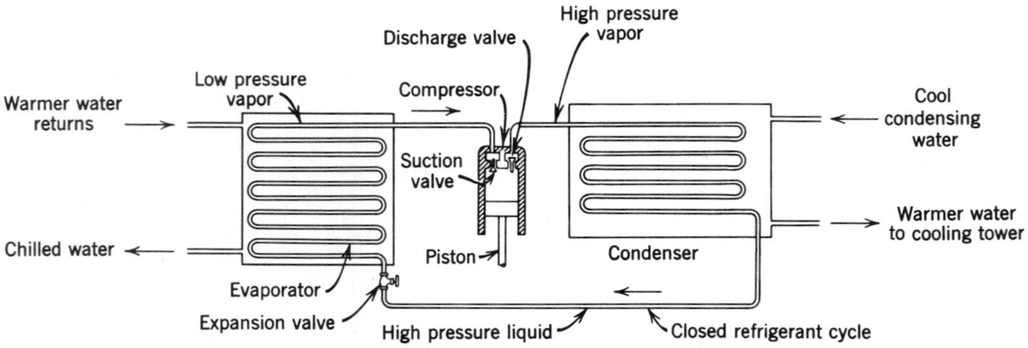

Fig. 6.32 *Schematic arrangement of a compressive refrigeration cycle.*

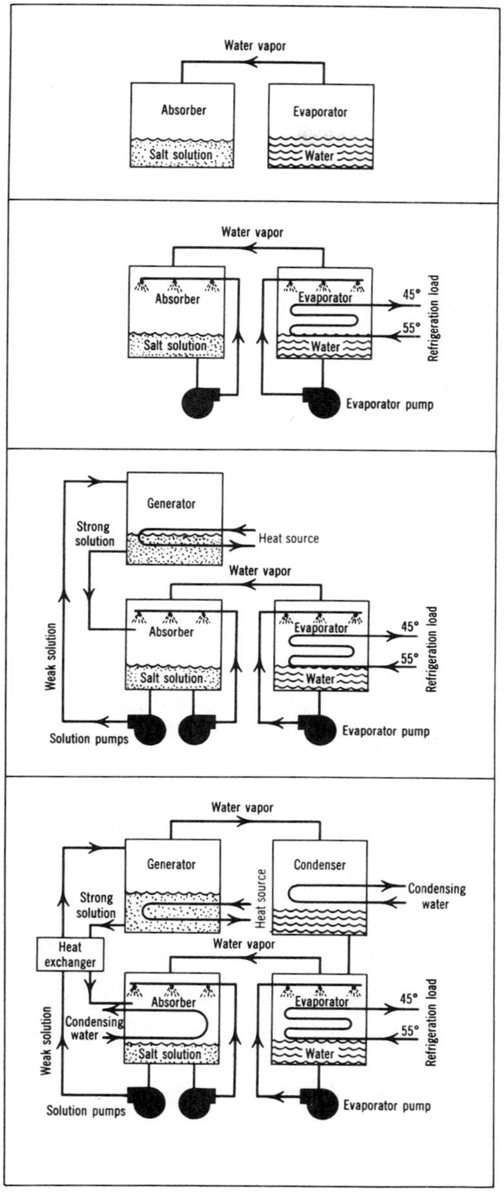

1 Evaporator and absorber

Consider two connected, closed tanks with a salt solution (lithium bromide) in one and water in the other. Just as common table salt absorbs water on a damp day, the salt solution in the absorber soaks up some of the water in the evaporator. The water remaining is thereby cooled by evaporation.

2 Evaporator coil and pump added

This refrigeration effect is utilized by putting a coil in the evaporator tank. Water from this tank is pumped to a spray header which wets the coil. The spray's evaporation chills water in the coil as it circulates to the refrigeration load. Solution pumped to spray in absorber raises efficiency.

3 Solution pumps and generator added

In an actual operating cycle, the salt solution is continuously absorbing water vapor. To keep the salt solution at proper concentration, part of it is pumped directly to a generator where excess water vapor is boiled off. The reconcentrated salt solution is returned to the absorber tank where it mixes with the solution sprayed to absorber in step 2.

4 Condenser and heat exchanger added

Water vapor boiled off from the weak solution is condensed and returned to the evaporator. A heat exchanger uses the hot, concentrated salt solution leaving the generator to preheat the cooler, weak solution coming from the absorber. Finally, condensing water circulating through the absorber and condenser coils removes the waste heat.

Fig. 6.33 *Absorption-refrigeration cycle. Connected systems are a heat source (such as steam, solar-heated water, or waste heat from an industrial process), condensing water, and chilled water (refrigeration load). (Courtesy of the Carrier Corporation.)*

Wood stoves are available in a wide variety of styles and are made of several materials. The sizes of such stoves are often difficult to determine. Manufacturers rarely specify the Btu/h output, which depends on the density, moisture content, and burn time of the wood fuel. Wood that has been split, loosely stacked, and covered from rain for at least 6 months should achieve a moisture content of about 20% by weight. The following sizing procedure assumes no more than this 20%

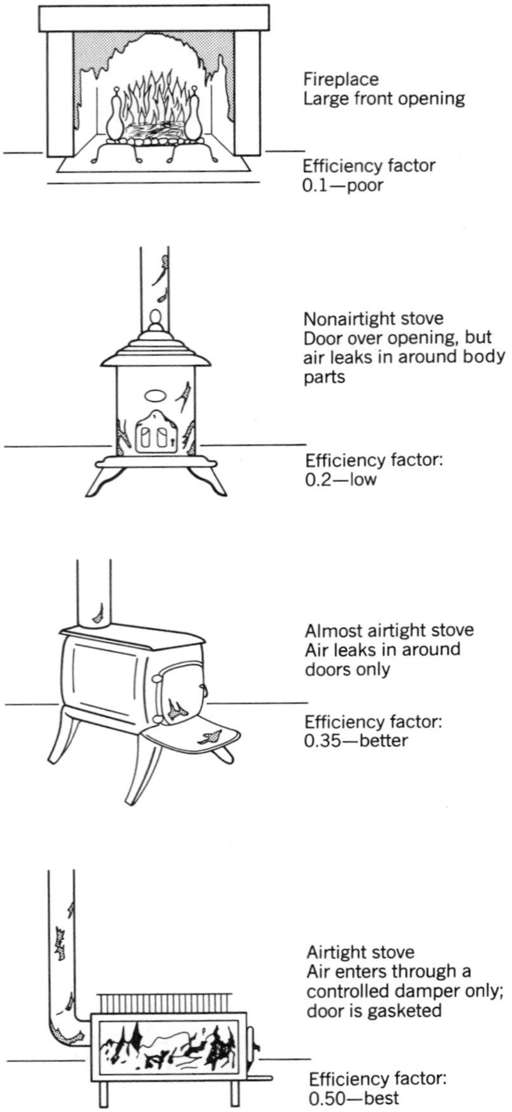

Fig. 6.34 *The efficiency of wood-burning devices has improved substantially. (Copyright © 1978 by* Alternative Sources of Energy. *Issue #35; 107 S. Central Ave., Milaca, MN 56353; Reprinted by permission.)*

moisture content. For more details, see *Alternative Sources of Energy Magazine* #35 (1978).

The formula for the hourly heat output to a room from a wood stove is

$$\text{Btu/h} = \frac{(V)(E)(D)(7000)}{T}$$

where V = useful (loadable) volume of the stove (ft^3)

TABLE 6.8 **Approximate Average Wood Density**

Type	*Density (lb/ft^3)[a]*
Shagbark hickory	40.5
White oak	37.4
Red oak	36.2
Beech	36.2
Sugar maple	34.9
Yellow birch	34.3
White ash	33.7
Black walnut	31.2
American elm	28.7
Spruce	25.6
Hemlock	23.7
Aspen	23.1
Red cedar	18.7
White pine	17.5

Source: American Sources of Energy Magazine, #35, © 1978. Reprinted by permission.

[a]These values are approximate; they vary a great deal.

E = percent efficiency, expressed as a decimal (<1.0); see Fig. 6.34

D = density of the wood fuel (Table 6.8)

T = burn time (h) for a complete load of firewood; usually assumed at 8 h minimum

7000 = Btu/lb of firewood, 20% moisture content

(''Bone-dry'' wood can be assumed to have 8600 Btu/lb.) Note that a *drop* in burn time *increases* the Btu/h output; it is evident that when the air supply is increased to the stove, the fire burns hotter, consuming the wood more quickly. To meet the design heat loss (worst condition) for a room, burn times of 8 to 10 h should be assumed. Stoves rarely need relighting with a 10-h burn time.

Wood stoves are frequently used as the sole mechanical heat source for an entire building, such as a residence or a small commercial building that is passively solar heated. Since radiant heat is the dominant form of heat output, the areas that ''see'' the stove will get most of the benefit. However, ''circulating'' stoves convert a larger portion of their heat to convected heat, which produces a layer of hot air at ceiling level. By providing a path between rooms at the ceiling, this hot air will slowly spread throughout a building; it also easily finds its way upstairs, since warm air rises. The more thermally massive the ceiling construction, the longer it will store and reradiate the heat from this warm air mass.

The flue leading from a wood stove carries very hot gases, which are a potential source of heat (and pollution). The flue can be exposed to a space, making its radiant heat available, or simple heat exchangers can be constructed (such as the preheating of domestic hot water). More elaborate heat recovery devices—for boiler flue heat recovery—are discussed in Section 6.8.

Catalytic combustors, a recent development, reduce the air pollution from wood burning. These devices are honeycomb-shaped, chemically treated discs, as much as 6 in. in diameter and 3 in. thick. They are either inserted in the flue or built into the stove itself. When wood smoke passes through the combustor, it reacts with the chemical and ignites at a much lower temperature; this causes gases to burn that otherwise would have gone up the flue. The result is more heat produced, less creosote buildup in the flue, and fewer pollutants in the atmosphere. Like the catalytic converters in autos, these devices impose limits on the fuel: plastic, colored newsprint, metallic substances, and sulfur are ruinous to combustors, which means that the stove must be used as a wood burner, not a trash incinerator.

Wood stoves have a larger impact on building design than most other heating devices have. Either noncombustible materials must be placed below and around them, or minimum clearance to ordinary combustible building materials must be provided. Furniture arrangements and circulation paths must be designed with the very hot stove surfaces in mind. Hot spots occur near the stove; cold spots occur whenever visual access to the stove is blocked. Thermally massive materials near the stove are advantageous in leveling the large temperature swings that can accompany the on-off cycle of the stove; this affinity for thermal mass has made the wood stove a popular choice for backup heat in passively solar heated buildings. Finally, the amount of space required for wood storage should not be overlooked; recall the impact of the wood storage space on the house shown in Fig. 1.10. A covered, well-ventilated, easily accessible, and quite large space is optimum.

Electric resistance heaters carry the disadvantage of using high-grade energy to do a low-grade task, as shown in Fig. 1.20. Their advantages, however, are impressive: low first cost and individual thermostatic control that can easily be used to make each room a separate heating zone. Thus, the energy wasted at the electricity-generating plant (usually 60 to 70%) can be partially recovered at the building, where unused rooms can remain unheated. A few of the many types of electric resistance heaters are shown in Fig. 6.35. As in the case of wood stoves, surfaces can sometimes reach high temperatures, requiring care in the location of heaters relative to furniture placement and traffic flow. Electric heaters are sized by their capacity, in kilowatts (1 kW = 3413 Btu/h).

Ceilings (or floors) can be constructed to include electric resistance wiring (Fig. 6.36). The primary disadvantage of ceiling heat is that hot air stratifies just below the ceiling, so that air motion is discouraged.

(b) Heating-Only Systems: Hot Water. The many remaining choices for whole-building heating systems will be discussed in the order (1) hot water and (2) forced air. Such systems will include a fuel, a heat source, a "mover" (such as pump or fan), a distribution system, a heat exchanger or terminal within the space, and a control system.

Hot water and steam boilers are rated according to capacity by the Hydronics Institute. The ratings also distinguish between cast iron and steel boilers: cast iron boilers have IBR ratings and steel boilers have SBI ratings. The initials refer, respectively, to the Institute of Boiler and Radiator Manufacturers and the Steel Boiler Institute, which combined to form the Hydronics Institute. The ratings are expressed in Btu/h. Steam boilers, are, additionally, evaluated in square feet of radiation, an older classification (1 sq ft of radiation = 240 Btu/h).

When interpreting the ratings, be careful to select a boiler whose net (useful) rating matches the calculated critical heat loss of the house or building.

Boilers and their accessories comprise a wide inventory. A few selected types are discussed here.

1. *Oil-fired steel boiler*. A refractory chamber receives the hot flame of the oil fire. Combustion continues within the chamber and the fire tubes. Smoke leaves through the breeching at the rear. Water, *outside* the chamber, receives the heat gen-

NATURAL CONVECTION UNITS

NATURAL CONVECTION UNITS

Heating units for wall mounting, recessed placement or surface placement are made with elements of incandescent bare wire or lower temperature bare wire or sheathed elements. An inner liner or reflector is usually placed between elements so that part of the heat is distributed by convection and part by radiation. Electric convectors should be located so that air movement across the elements is not impeded. Small units with ratings up to 1650 W operate at 120 V. Higher wattage units are made for 208 or higher voltages and require heavy duty receptacles.

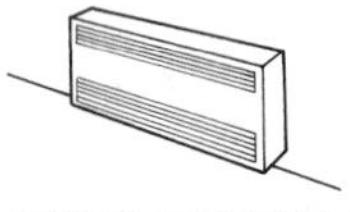

L 24″ to 96″
D 3″ to 8″
H 11″ to 32″
CAP 1000 W to 4000 W

CABINET CONVECTOR (Surface Mounted or Recessed)

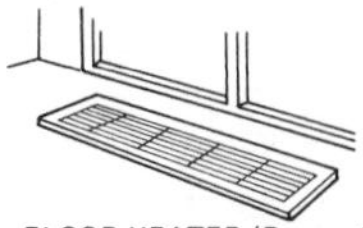

L 14″ to 108″
W 5″ to 8″
H 8″ to 11″
CAP 300 W to 2000 W

FLOOR HEATER (Recessed)

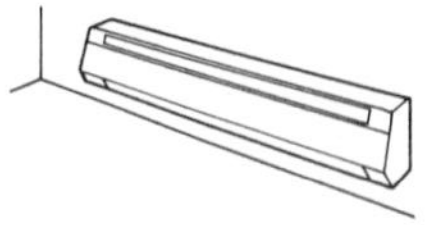

L 24″ to 120″
D 2″ to 8″
H 4″ to 12″
CAP 300 W to 4000 W

BASEBOARD HEATER (Wall Mounted)

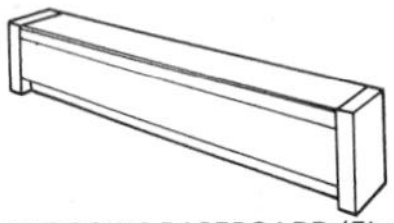

L 23″ to 107″
D 3″ to 6″
H 9″ to 12″
CAP 300 W to 2000 W

HYDRONIC BASEBOARD (Floor Mounted)

NATURAL CONVECTION UNITS

RADIANT HEATING UNITS

RADIANT HEATING

Heat is produced by a current that flows in a high resistance wire or ribbon and is then transferred by radiation to a heat absorbing body. Manufacturer's recommendations for clearance between a radiant fixture and combustible materials or occupants should be followed.

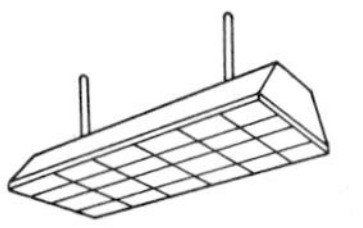

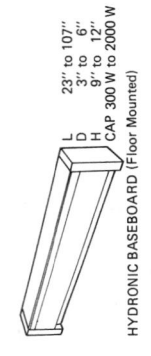

L 14″ to 86″
W 4″ to 12″
H 3″ to 16″
CAP 500 W to 7000 W

INFRARED HEATER (Pendant Mounted) Circular heat lamp is available

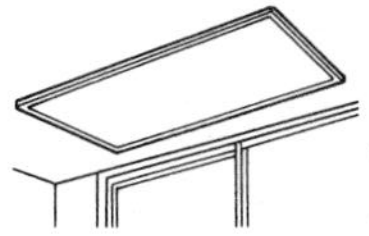

L 48″ to 144″
W 24″ to 48″
D 1″
CAP 500 W to 1000 W

RADIANT HEAT PANEL (Surface Mounted or Recessed) Decorative murals are available

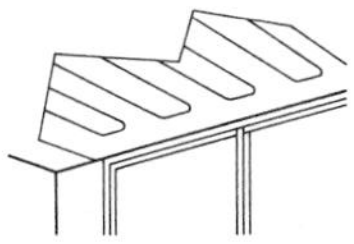

Dimensions and capacity vary with coverage

RADIANT CEILING WITH EMBEDDED CONDUCTORS

RADIANT HEATING UNITS

FORCED AIR UNITS

FORCED AIR UNITS

Unit ventilators and heaters combine common convective heating with controlled natural ventilation.

Unit ventilators are most often mounted on an outside wall for air intake and at windowsills to prevent the down draft of cold air.

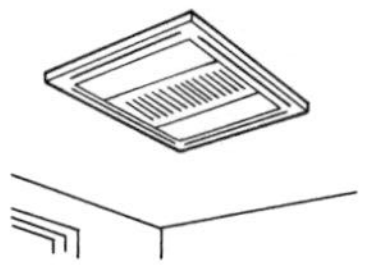

L 10″ to 14″
W 8″ to 14″
D 4″ to 8″
CAP 500 W to 1500 W

CEILING HEATER (Recessed) Circular unit with light is available

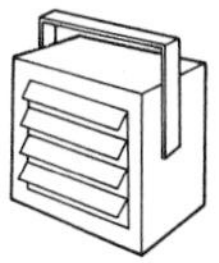

W 12″ to 52″
D 6″ to 22″
H 12″ to 26″
CAP 1.5 KW to 50 KW

UNIT HEATER (Bracket Mounted)

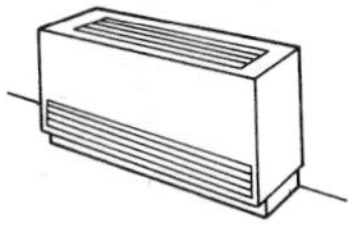

L 48″ to 104″
D 11″ to 26″
H 26″ to 32″
CAP 1 KW to 36 KW

UNIT VENTILATOR (Surface Mounted or Recessed)

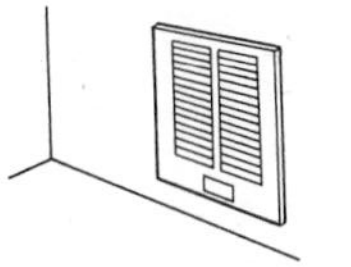

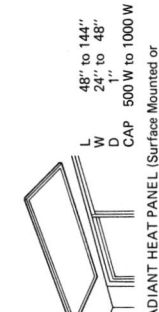

W 10″ to 18″
D 2″ to 6″
H 9″ to 24″
CAP 750 W to 4000 W

WALL HEATER (Recessed)

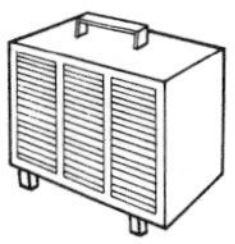

W 10″ to 72″
D 2″ to 12″
H 7″ to 24″
CAP 500 W to 5000 W

PORTABLE HEATER

FORCED AIR UNITS

Fig. 6.35 *Varieties of electric resistance heating units. (From AIA, Ransey and Sleeper* Architectural Graphic Standards, *7th ed., © 1981 by John Wiley & Sons. Reprinted by permission.)*

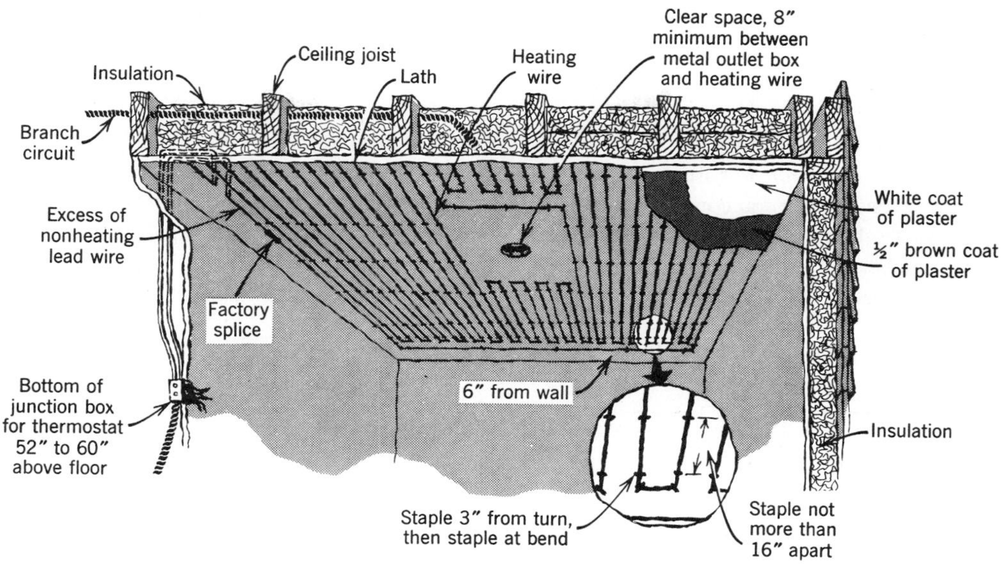

Fig. 6.36 *Typical installation of radiant-heating cable ready for completion of plaster ceiling. (Courtesy of General Electric Company.)*

erated in the combustion chamber. If a domestic hot water coil is connected for use, a larger capacity boiler is selected. An aquastat (water thermostat) turns on the burner whenever the boiler water cools off, thereby maintaining a reservoir of hot water ready for heating the building.

2. *Gas-fired cast iron hot water boiler* (Fig. 6.37). Cast iron sections contain water that is heated by hot gases rising through these sections. Output is related to the *number* of sections. Additional heat is gained from a heat extractor in the flue. With induced draft combustion and intermittent

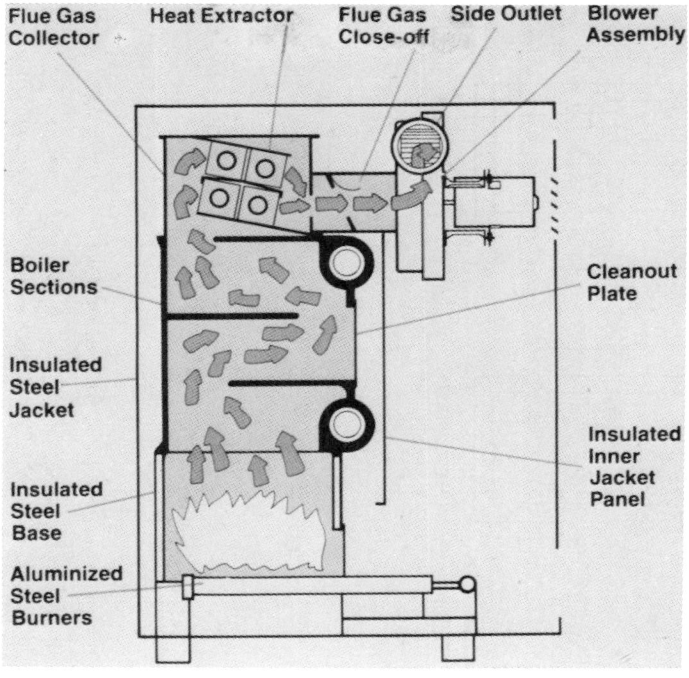

Fig. 6.37 *Gas-fired cast iron sectional boiler for hot water heating. Very high operating efficiency is possible; no chimney is required, as lower temperature exhaust gases can be vented through wall to exterior.*

electronic ignition instead of a pilot light, 87% seasonal operating efficiency is attainable. The American Gas Association (AGA) sets standards for gas-fired equipment.

3. *Oil-fired, cast iron hot water boiler*. Primary and secondary air for combustion may be regulated at the burner unit. Flame enters the refractory chamber and continues around the outside of the water-filled cast iron sections.

Hot water heating circuits come in four principal arrangements. Figure 6.38*a* shows the series loop system, usually run at the building's perimeter. The water flows to and *through* each baseboard or fintube in turn. Obviously, the water at the end of the circuit is a little cooler, but since in all hot water systems the water temperature drop seldom exceeds 20F°, the *average* temperature can usually be used to select the baseboard or other elements. Valves at each heating element are not

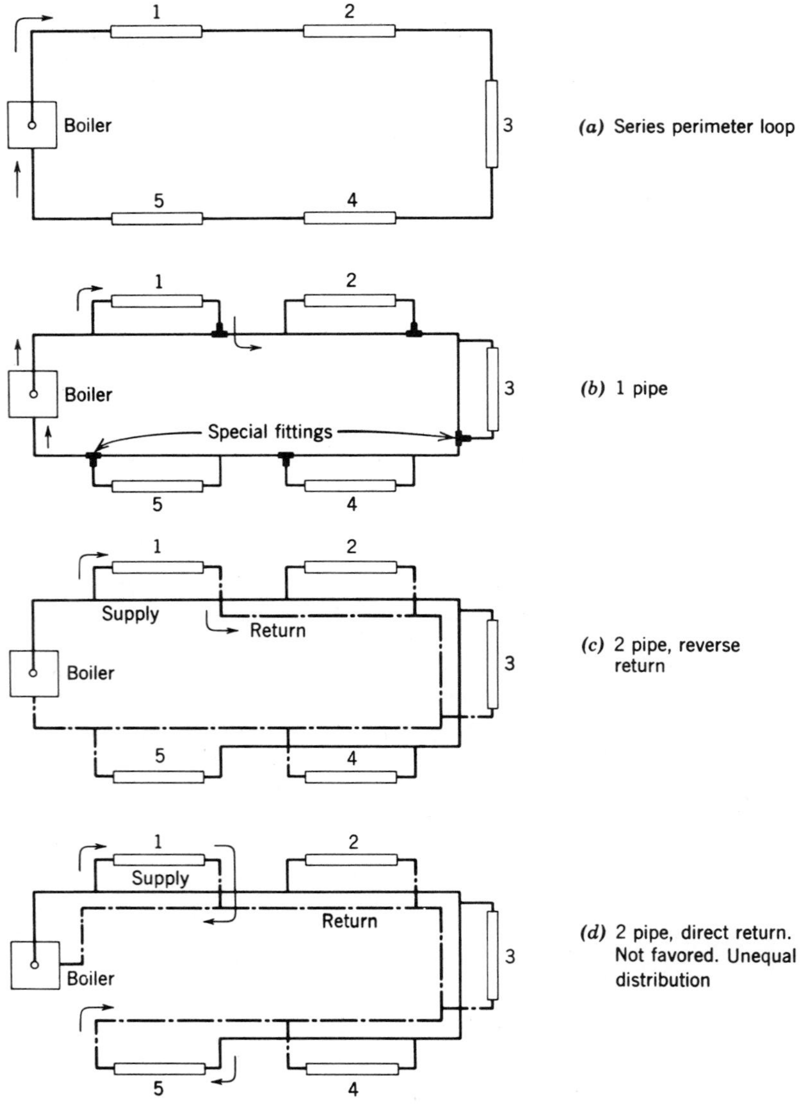

Fig. 6.38 *Plan views showing principles of piping for water distribution to heating elements (baseboard convectors shown here) in hot-water-heating systems. For simplicity, controls are not shown.*

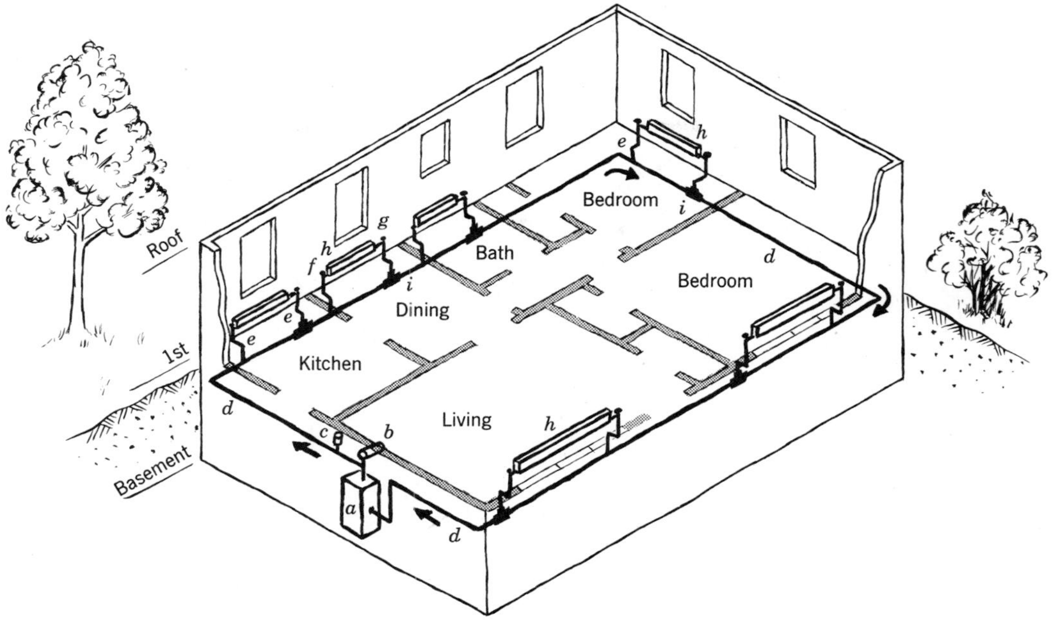

Fig. 6.39 Hot water, one-pipe system—type b *of Fig. 6.44.* (a) *Boiler.* (b) *Compression tank.* (c) *Circulator (pump).* (d) *Hot water main.* (e) *Runout (branch).* (f) *Control valve.* (g) *Air vent.* (h) *Baseboard heating unit.* (i) *Special return fitting.*

possible, since any valve would shut off the entire loop. Adjustment is by a damper at each baseboard, which reduces the natural convection of air over the fins. This is a "one-zone" system—all elements on, or off, together. There is no general rule about the maximum allowable length of water circuit, but for long runs, the pipe size can be increased or *several* loops used in parallel.

The one-pipe system shown in Figs. 6.38*b* and 6.39 is a very popular choice. Special fittings act to divert part of the flow into each baseboard. A valve may be used at each one to allow for reduced heat or for a complete shutoff, to conserve energy—an advantage that the loop system does not provide. The one-pipe system uses a little more piping and thus is not as economical to install as the loop system, in which piping is minimal.

The two-pipe reverse-return shown in Fig. 6.38*c* may be considered the classic piping arrangement. Water nearly at boiler temperature is supplied to each baseboard without being cooled by passing through a previous baseboard or accepting the cooler return water. Equal friction, resulting in equal flow, is achieved through all baseboards (nos. 1 to 5) by *reversing* the return instead of running it directly back to the boiler. This equality is effected by equal lengths of water flow through any baseboard together with its lengths of supply-and-return main. More pipe is required for this system than for the systems shown in Fig. 6.38*a* or 6.38*b*.

Figure 6.38*d* shows an arrangement that is not favored because the path of water through baseboard no. 1 is much shorter than that through the others, and especially no. 5. Baseboard no. 5 could easily be undesirably cool, since it is short-circuited by the others.

Air vents and water drains are part of the distribution system. Except for the necessary air cushion in the upper part of the compression tank above the boiler, air must not be allowed to accumulate at high points in the piping or at the convector branches. Air vents relieve these possible air pockets, which would otherwise make the system air-bound and inoperative.

If a system is to be drained and left idle in a cold house, water trapped in low points could freeze and burst the tubing or fittings. Operable drain valves must be provided at such locations and, of course, at the bottom of the boiler, as shown in Fig. 6.40.

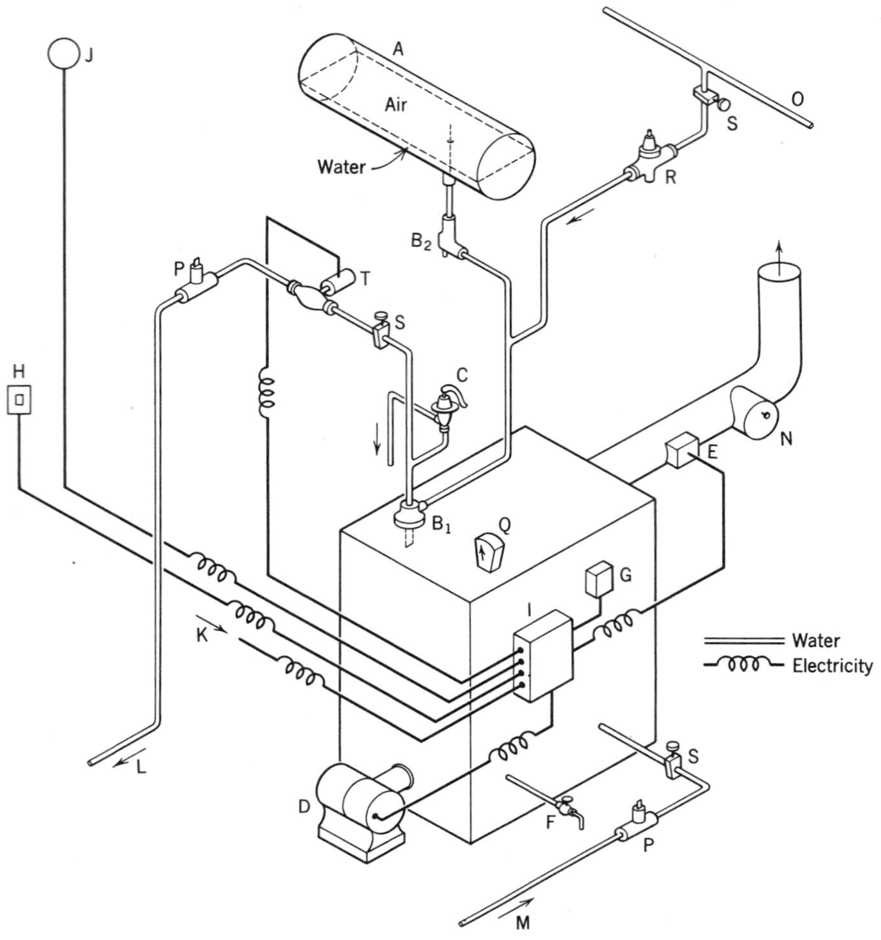

Fig. 6.40 Hydronic and electrical controls; an oil-fired boiler for heating by hot water.

(A) *Compression tank.* Accommodates the expansion of the water in the system.
(B) *Air control fittings.* Vent out unwanted air in the boiler and maintain the level in the compression tank.
(C) *Pressure relief valve.* Usually set for 30 psi. Initial cold pressure about 12 psi. Relieves excessive system pressure.
(D) *Oil burner.* Responds to aquastat or thermostat.
(E) *Stack temperature control.* Senses stack temperature and stops oil injection if ignition has not occurred.
(F) *Drain valve.* At low point in the water system.
(G) *Aquastat.* Maintains temperature of boiler water by starting the oil burner when temperature of water drops below the aquastat's setting. Sometimes set at about 180 F.
(H) *Remote switch.* At a safe distance from the boiler so that the plant can be turned off in case of trouble during which the boiler cannot be approached.
(I) *Junction box and relays.* General control center.
(J) *Thermostat.* When the room temperature drops below its setting, it turns on both the oil fire and the circulating pump.
(K) *Electrical power source.* Operates from a separate individual circuit at the power panel.
(L) *Hot water supply.* Copper tubing to convectors or baseboards.
(M) *Hot water return.* Copper tubing from convectors or baseboards.
(N) *Draft adjuster.* Regulates the draft (combustion air) over the fire.
(O) *House cold water main.* From which water is fed automatically into boiler.
(P) *Flow control valves.* Prevent casual flow of water by gravity when the circulator is not running.
(Q) *Temperature pressure gauge.* Indicates water temperature and pressure. Sometimes supplemented by immersion thermometers in supply and return mains.
(R) *Pressure-reducing valve.* Admits water into the system when the pressure there drops below about 12 psi. Has a built-in check valve to prevent backflow of boiler water into the water main.
(S) *Shutoff valves.* Normally open. Can be closed to isolate the system and permit servicing of components.
(T) *Circulator.* Centrifugal circulating pump that moves the water through the tubing and heating elements.

Hydronic and electrical controls allow automatic operation, described in Fig. 6.40. Makeup water is added as required, the air level in the tank is regulated by the air control fittings, and the circulator and burner operate as controlled by the aquastat and thermostat. If air vents in the *piping* are not automatic, they will require periodic manual "bleeding" of unwanted air.

Circulating pumps are used to overcome the friction-of-flow in the piping and fittings and to deliver water at a rate sufficient to offset the hourly heat loss of the house or building.

(c) Hydronic Heating Design. The calculations for the sizing of a water distribution system are presented in Section 9.7; they are based upon the required flow and the friction in the piping. (For domestic water supply, another factor is the vertical distance through which the water must be raised. In these closed-loop heating systems, however, the weight of the cooler water falling back to the boiler essentially counterbalances the weight of the hot water being raised. Furthermore, gravity is helping the hot—lighter—water rise, and the cooler water fall.)

The key to pipe sizing is the overall required flow rate. Ordinarily, the temperature drop that occurs as the hot supply water gives up heat to the space (through the convector) is about 20F°. Because the entire building's design heat loss is to be overcome by this system,

$$\text{overall flow rate, gpm} = \frac{\text{design heat loss, Btu/h}}{20\text{F}^\circ \times 60\text{ min/h} \times 8\text{ lb/gal}} = \frac{\text{design heat loss}}{9600}$$

Then, by using Section 9.7, we can account for friction through piping and fittings and size the main supply and return pipes. The same procedure can be applied to branches, proportioned to the heat they must deliver.

The critical choice, however, is not pipe size: it is relatively easy to distribute pipes within wall and floor/ceiling construction. Rather, the critical choice is *hot-water supply temperature:* the higher this temperature, the smaller the convector units that discharge the heat to each space. However, higher temperatures endanger occupants, who may suffer skin burns if they touch exposed parts of the convectors or distribution tree. Higher temperatures also can lead to steam within the boiler/distribution tree. These systems are not designed to accommodate steam, and serious injury can sometimes result. A safer choice of average water distribution temperature is 170 to 180 F, even though these temperatures result in larger convectors.

Baseboard convector selection is then made from manufacturer's data tables, such as the one shown in Table 6.9.

EXAMPLE 6.1. Select a Heatrim R-750 baseboard convector (Table 6.9) for a 20-ft-long living room wall. From heat loss calculation, the living room requires 9000 Btu/h. The average water temperature is 180 F, and water flow is 500 lb/h (equivalent to about 1 gpm).

SOLUTION. From the 180 F column of the left section of Table 6.9, choose a 16-ft-long baseboard convector for its capacity of 8960 Btu/h. It would fit in the 20-ft length of the wall, and 8960 Btu/h is close enough to the design figure (i.e., 9000 Btu/h).

Zoning is relatively easy to accomplish with hydronic systems, as shown in Fig. 6.41. The installation shown in the figure is made up of three separately heated areas—the first, second, and third floors. Each can be heated to different temperatures as called for by thermostats in each separate apartment. For example, if the thermostat serving the second floor (zone *B*) calls for heat, it turns on pump *B*. Flow-control valves *B* open, admitting hot water from the boiler header to main *B*. Flow-control valves *A* and *C* remain closed, preventing flow in mains *A* and *C*. Any or all of the zones may operate at one time. The boiler keeps a supply of hot water continually ready to supply any zone on demand. This is achieved by an aquastat immersed in the boiler water. When the boiler water drops below the prescribed temperature, it turns on the firing device, such as an oil burner or a gas burner, which brings the water up to temperature. If an overhead main supplies downfeed, as in the first floor of this installation, special downfeed supply and return fittings are necessary. For the second- and third-floor zones, one special return tee is sufficient. If the designer also elects to use a special upfeed *supply* tee of the venturi type, higher outputs of the convectors will result.

(d) Heating Equipment Combustion and Fuel Storage. As fuels burn to produce heat, they require oxygen to support the combustion. Since oxygen constitutes only about one-fifth of the volume of air, reasonably large rates of air flow are required. The air should be drawn in from outdoors at a position close to the fuel burner or (preferably) led to this location by a duct. For residences and other small buildings a louver *about*

TABLE 6.9 **Manufacturer's Data for Hot Water Baseboard Convectors**

Approved IBR water ratings[a]; model R-750 Heatrim—capacities Btu/h ft.

Number of Lineal Feet[b]	Water Flow Rate 500 lb/h Average Water Temperature (F)					
	170	180	190	200	210	220
1	500	560	630	690	760	820
2	1000	1120	1260	1380	1520	1640
3	1500	1680	1890	2070	2280	2460
4	2000	2240	2520	2760	3040	3280
5	2500	2800	3150	3450	3800	4100
6	3000	3360	3780	4140	4560	4920
7	3500	3920	4410	4830	5320	5740
8	4000	4480	5040	5520	6080	6560
9	4500	5040	5670	6210	6840	7380
10	5000	5600	6300	6900	7600	8200
11	5500	6160	6930	7590	8360	9020
12	6000	6720	7560	8280	9120	9840
13	6500	7280	8190	8970	9880	10660
14	7000	7840	8820	9660	10640	11480
15	7500	8400	9450	10350	11400	12300
16	8000	8960	10080	11040	12160	13120
17	8500	9520	10710	11730	12920	13940
18	9000	10080	11340	12420	13680	14760
19	9500	10640	11970	13110	14440	15580
20	10000	11200	12600	13800	15200	16400
21	10500	11760	13230	14490	15960	17220
22	11000	12320	13860	15180	16720	18040
23	11500	12880	14490	15870	17480	18860
24	12000	13440	15120	16560	18240	19680
25	12500	14000	15750	17250	19000	20500
26	13000	14560	16380	17940	19760	21320
27	13500	15120	17010	18630	20520	22140
28	14000	15680	17640	19320	21280	22960
29	14500	16240	18270	20010	22040	22780
30	15000	16800	18900	20700	22800	24600

Source: Courtesy of Burnham Corporation.

[a]Approved IBR water ratings shown above for American-Standard Heatrim Panels (with Model E-750 element) are based on a water flow of 500 lb/h with a pressure drop of 0.047 in. of water per lineal foot and a water flow of 2000 lb/h with a pressure drop of 0.530 in. of water per lineal foot. As allowed by the Institute of Boiler and Radiator Manufacturers (IBR) Testing and Rating Code for Baseboard Type of Radiation, 15% is added to water heat capacity.

The use of IBR ratings at water flow rates of 2000 lb/h is limited to installations *twice the cross-sectional area of the flue* should prove satisfactory. It should be arranged to remain open at all times. This combustion air should *not* be drawn from the general building space—it is a waste of energy, and contemporary ''tight'' construction inhibits such airflow. A dangerous condition is created whenever stack flow is restricted.

The most important combustible element in the chemical makeup of fuels is carbon. It may be burned well or poorly. When burned poorly, it can cause great energy losses and sooty operation. For success, much depends upon the proper selection of well-designed boilers, furnaces, and burners. Adjustments of primary and secondary air rates of

Water Flow Rate 2000 lb/h Average Water Temperature (F)					
170	*180*	*190*	*200*	*210*	*220*
530	590	670	730	800	870
1060	1180	1340	1460	1600	1740
1590	1770	2010	2190	2400	2610
2120	2360	2680	2920	3200	3480
2650	2950	3350	3650	4000	4350
3180	3540	4020	4380	4800	5220
3710	4130	4690	5110	5600	6090
4240	4720	5360	5840	6400	6960
4770	5310	6030	6570	7200	7830
5300	5900	6700	7300	8000	8700
5830	6490	7370	8030	8800	9570
6360	7080	8040	8760	9600	10440
6890	7670	8710	9490	10400	11310
7420	8260	9830	10220	11200	12180
7950	8850	10050	10950	12000	13050
8480	9440	10720	11680	12800	13920
9010	10030	11390	12410	13600	14790
9540	10620	12060	13140	14400	15660
10070	11210	12730	13870	15200	16530
10600	11800	13400	14600	16000	17400
11130	12390	14070	15330	16800	18270
11660	12980	14740	16060	17600	19140
12190	13570	15410	16790	18400	20010
12720	14160	16080	17520	19200	20880
13250	14750	16750	18250	20000	21750
13780	15340	17420	18980	20800	22620
14310	15930	18090	19710	21600	23490
14840	16520	18760	20440	22400	24360
15370	17110	19430	21170	23200	25230
15900	17700	20100	21900	24000	26100

where the water flow rate through the baseboard unit is equal to or greater than 2000 lb/h. Where the water flow rate through the baseboard is not known, the IBR rating at the standard water flow rate of 500 lb/h must be used.

[b]These ratings are based on active (finned) Heatrim lengths. Difference between active length and total length of the standard Heatrim heating elements is $2\frac{15}{32}$ in. Elements are unpainted.

Nonferrous fins on model E-750 elements measure $2\frac{1}{8} \times 2\frac{1}{8} \times 0.008$ in., spaced 52 fins per foot.

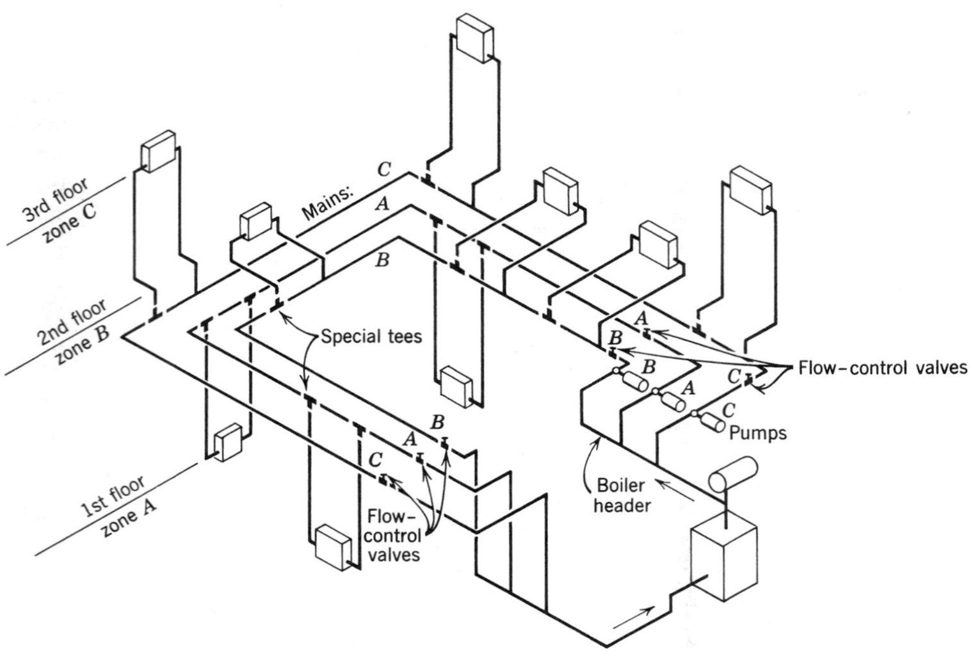

(a)

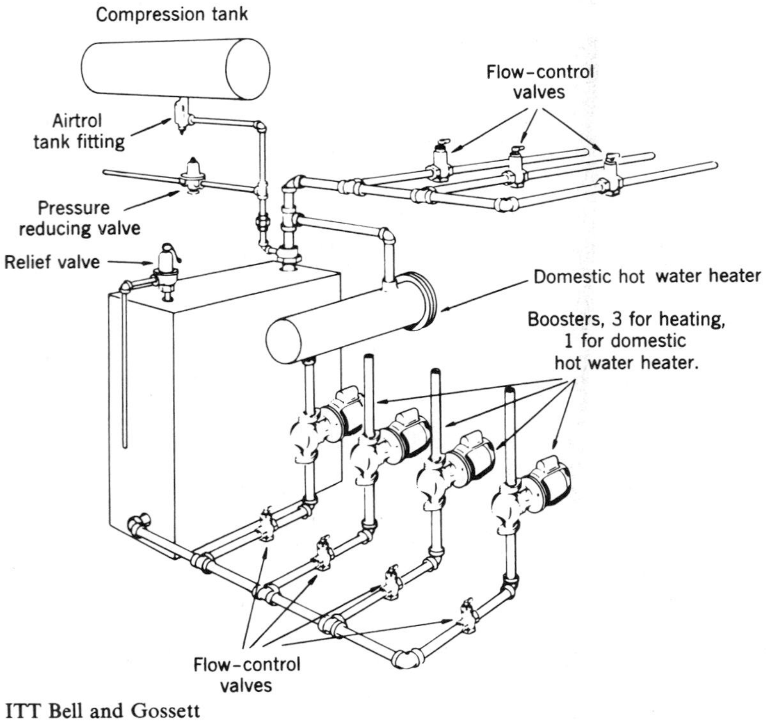

ITT Bell and Gossett

(b)

Fig. 6.41 *Three-zone, multicircuit, one-pipe system.* (a) *Each convector has connections to one pipe.* (b) *Boiler, piping, and water controls suitable for this three-zone, one-pipe system. Each one-pipe circuit should be provided with two flow-control valves and a circulator on the supply or return pipe. The terms* booster, pump, *and* circulator *are interchangeable.*

flow and of draft (flow of air and gases through the boiler) are important responsibilities of the engineer and the heating contractor. Carbon may burn to carbon monoxide (CO) or (more completely) to carbon dioxide (CO_2) with greater heat production. Flue gases should be analyzed and the percentage of carbon dioxide measured. The best economy and the cleanest and most efficient combustion occur when the CO_2 content of the flue gases most nearly approaches the values given in Table 6.10 for maximum theoretical stoichiometric percent CO_2 (see also IBR standards, Table 6.11). The architect should require that these tests be made with the adjustments necessary to gain the most efficient performance. Flue gas temperatures should be taken. Temperatures lower than 600 F indicate that heat is properly retained in the boiler instead of escaping up the chimney (see the note on *efficiency* that follows the numerical tabulation in Table 6.11).

It is important that chimneys, which carry high-temperature flue gases, be safely isolated from combustible construction, to prevent the possibility of fire. Conventional standards for houses call for a terra cotta flue lining surrounded by 8 in. of brick, with an additional 2 in. of space between the brick and any wood. The space is usually filled with incombustible mineral wool. The size of flue will be dictated by the specification for the boiler or furnace selected for use. Flue height (see Fig. 6.42*a*) had traditionally been 35 to 40 ft. The function of providing a draft, for which chimney height was an important consideration, is no longer as necessary as it previously was, because fans are used now. For example, oil is injected under pressure, accompanied by air, and forced in by a fan. Often a draft adjuster in the breeching (smoke pipe) that carries the flue gases to the chimney is arranged to open slightly to *reduce* the normal stack draft. If increased draft should ever be required, an induced draft fan that puts a suction on the flue side of the fire is usually chosen instead of greater stack height. Draft hoods above gas burners prevent downdraft from blowing out the flames.

Prefabricated chimneys (see Fig. 6.42*b*) are replacing with increasing frequency the bulkier and heavier field-built masonry. They offer a number of advantages and may be easily supported on a normal structure.

The storage space to be allowed for fuel oil depends on the proximity of the supplier and the space available at the building. For oil, when more than 275 gal is stored it is common practice to use

TABLE 6.10 **Approximate Maximum Theoretical (Stoichiometric) CO_2 Values, and CO_2 Values for Various Fuels with Different Percentages of Excess Air**

Type of Fuel	*Stoichiometric Percent CO_2*	*Percent CO_2 at Given Excess Air Values* 20%	40%	60%
Coke	21.0	17.5	15.0	13.0
Anthracite	20.2	16.8	14.4	12.6
Bituminous coal	18.2	15.1	12.9	11.3
Nos. 1 and 2 fuel oil	15.0	12.3	10.5	9.1
No. 6 fuel oil	16.5	13.6	11.6	10.1
Natural gas	12.1	9.9	8.4	7.3
Carbureted water gas	17.2	14.2	12.1	10.6
Coke oven gas	11.2	9.2	7.8	6.8
Mixed gas (natural and carbureted water gas)	15.3	12.5	10.5	9.1
Propane gas (commercial)	13.9	11.4	9.6	8.4
Butane gas (commercial)	14.1	11.6	9.8	8.5

Source: Copyright © by the American Society of Heating, Refrigeration and Air Conditioning Engineers, Inc., Atlanta, GA. Reprinted by permission from the *ASHRAE Handbook of Fundamentals,* 1967.

TABLE 6.11 **Conditions Set by the Institute of Boiler and Radiator Manufacturers for the Rating of Boilers**

	Type Oil	*Minimum Overall Efficiency*	*CO_2*	*Maximum Flue Gas Temperature*	*Maximum Smoke Reading*
Natural draft or induced-draft–bare boilers or boiler-burner units	Light	70%	10%	600 F	#2
	Heavy	75%	12¼%	600 F	#4
Natural draft or induced-draft–boiler-burner units only	Light	70%	12¼%	600 F	#2
Forced draft–boiler-burner units only	Light	75%	10%	600 F	#2
	Light	75%	12¼%	600 F	#2
	Heavy	75%	12¼%	600 F	#4

Efficiency: The overall efficiency of a boiler must be not less than that specified. This requirement has served to improve the performance of boilers since the establishment of the IBR Testing and Rating Code.

Carbon dioxide (CO_2) in the flue gas: During oil-fired tests, the oil burner must be set to produce the specific percentage of carbon dioxide in the flue gas. This requirement was established to allow all boilers to be rated on a comparable basis and to prevent anyone from obtaining ratings that are unrealistic by testing the boiler with an exceptional oil boiler under ideal conditions at a combustion efficiency that could not be reproduced in the field.

Flue gas temperature: The flue gas temperature limitation insures the user of safe and economical operation. The temperature is measured in a thoroughly insulated stack.

Draft loss through boiler (oil-fired): The draft loss through the boiler, or the difference between the draft in the stack and the draft in the combustion chamber, must not exceed a specified value. This requirement is included because excessive draft losses invariably lead to trouble when boilers are connected to poor chimneys or sited in poor draft areas.

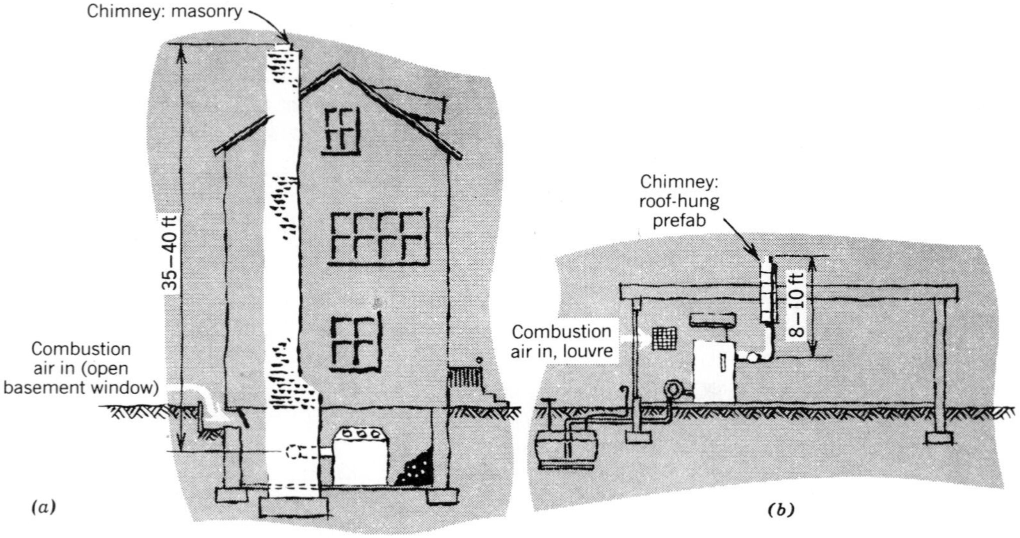

Fig. 6.42 Controlled draft in burners (b) *has eliminated the need for 40-ft chimneys* (a)*. Check with your engineer about minimum height.*

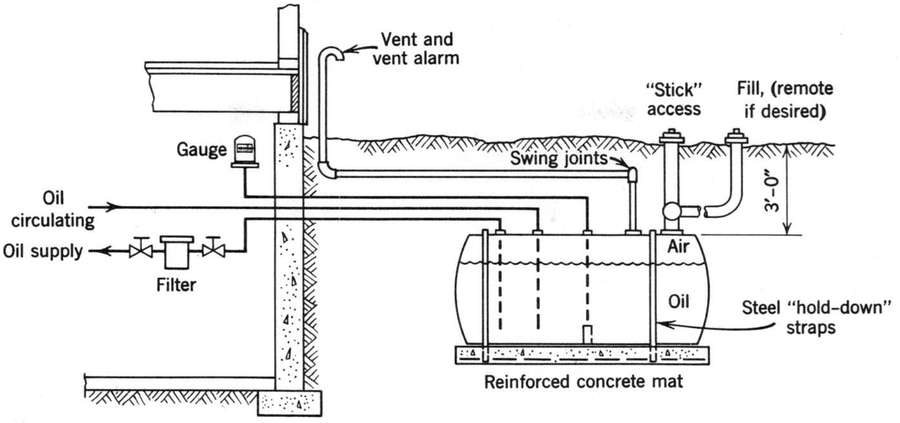

Fig. 6.43 *Details of fuel oil storage tank.*

an outside tank buried in the ground (see Fig. 6.43). The tank is often set on a concrete slab and strapped down to the slab. This prevents the tank from sinking when full or rising in flotation buoyancy that might be caused by adjacent groundwater when the tank is empty. The tank, usually made of steel, receives two coats of asphalt emulsion, to inhibit rust. Tubing for the gauge and for the supply and circulating lines are made of copper, and the fill and vent lines of wrought iron with swing joints to accommodate possible slight settlement of the tank. Oil deliveries are often made on the basis of the degree days elapsed since the last fill-up of the tank. Thus, the customer is relieved of the chore of checking the gauge and ordering periodically.

(e) Warm Air Heating Systems. These system began to supersede the open fireplace in about 1900. Originally, an iron furance that stood in the middle of the basement was hand-fired by coal. Surrounding it was a sheet metal enclosure. An opening in its side near the bottom admitted cool air that gravitated to the basement. A short duct from the top of the enclosure delivered the warm air by gravity to a large grille in the middle of the floor of the parlor. Other rooms, including those in upper stories, shared a little of this warmth when doors were left open.

Very gradual changes had culminated by mid-century in systems essentially like the ones described in Fig. 6.44. The improvements included

Automatic firing of oil or gas.

Operational and safety controls.

Ducted air to and from each room.

Blowers to replace gravity.

Filters.

Adjustable registers.

By the 1960s it became apparent that the basement was beginning to disappear, as subslab perimeter systems became popular for basementless houses (see Fig. 6.45). In general, the features above were retained and air was delivered upward across glass, to be taken back at high-return grilles.

As electricity began to replace oil and gas, former necessities such as combustion, chimneys, and fuel storage became nonessential. Horizontal electric furnaces began to appear in shallow attics or above furred ceilings. Air was delivered down from ceilings to warm exterior glass and taken back through door-grilles and open plenum space.

A new energy-saving trend appeared. The heat source (heat pump) was located centrally and fully *within* the insulated volume of the house. No stray heat escape from the unit was possible. Also, because of very small windows and double glazing, short ducts could deliver warm air from the inner side of each room, since warming of the double glass was less essential. Air returned to the unit through open grilles in doors and at the heat pump enclosure.

Today, solar energy can be used as a supplementary source of warm air (Fig. 6.44*b*).

Comfort is one of warm air heating's advantages. The motion of air in the space helps to assure uniform conditions and reasonably equal

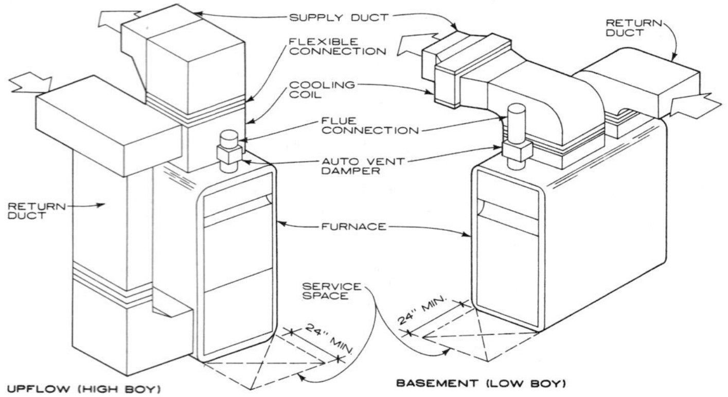

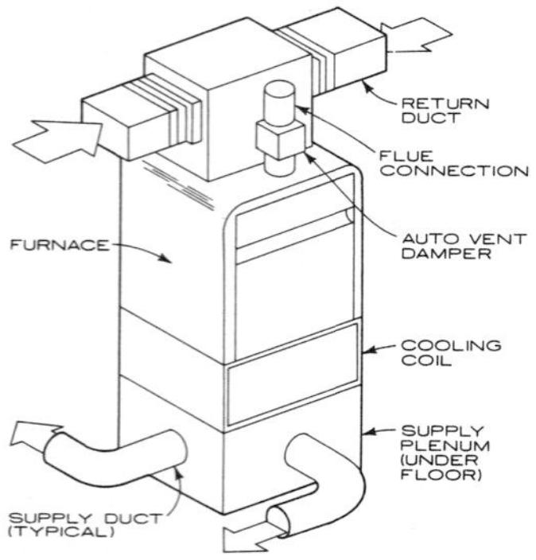

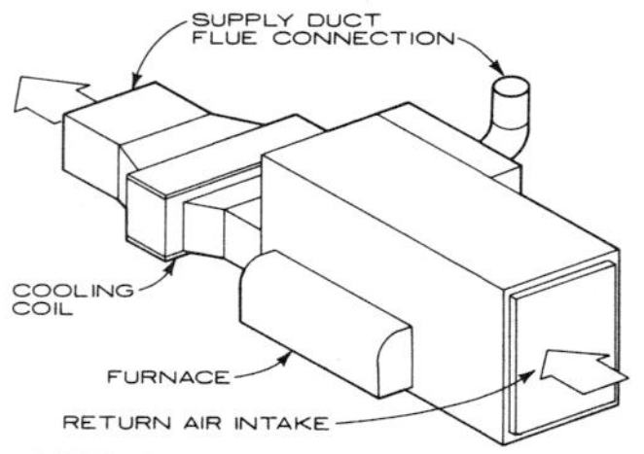

WARM AIR FURNACES

FLOOR AREA REQUIRED BY WARM AIR FURNACE

OUTPUT CAPACITY (BTU/HR)	FURNACE FLOOR AREA (SQ FT)*
Up to 52,000	2.4
52,000-84,000	4.2
84,000-120,000	6.6
120,000-200,000	13.1

*Based on net floor area occupied by the upflow or downflow furnace. Low boy unit requires 50% more floor area. Space for combustion air should be added as required by local codes. Adequate space should be provided for service.

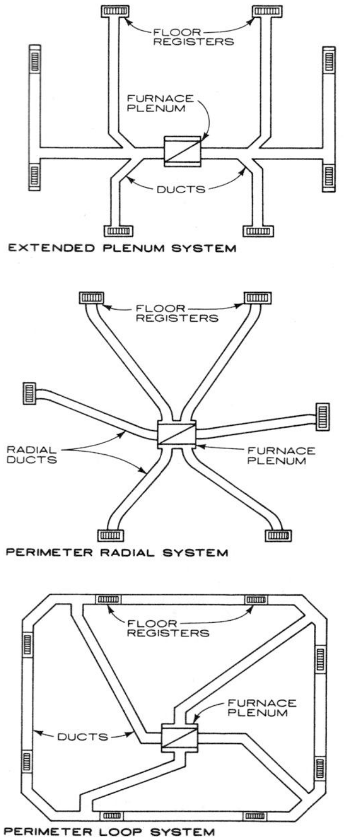

DUCT SYSTEMS

(a)

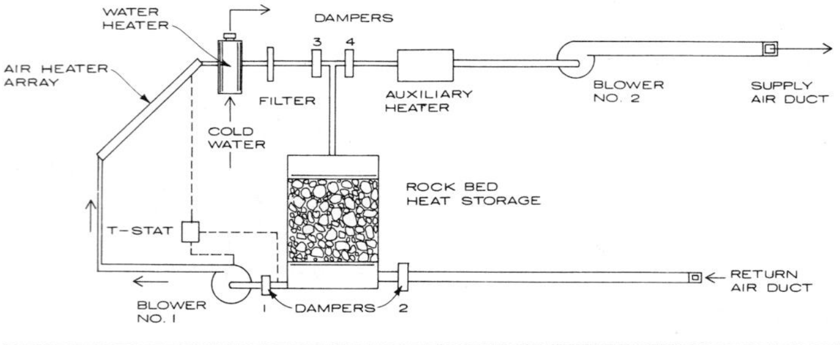

SOLAR AIR HEATER SYSTEM WITH ROCK BED STORAGE

(b)

Fig. 6.44 (a) *Typical furnace types and duct distribution arrangements.* (b) *Solar collectors used as a source of warm air. (From AIA, Ramsey and Sleeper,* Architectural Graphic Standards, *7th ed., © 1981 by John Wiley & Sons. Reprinted by permission.)*

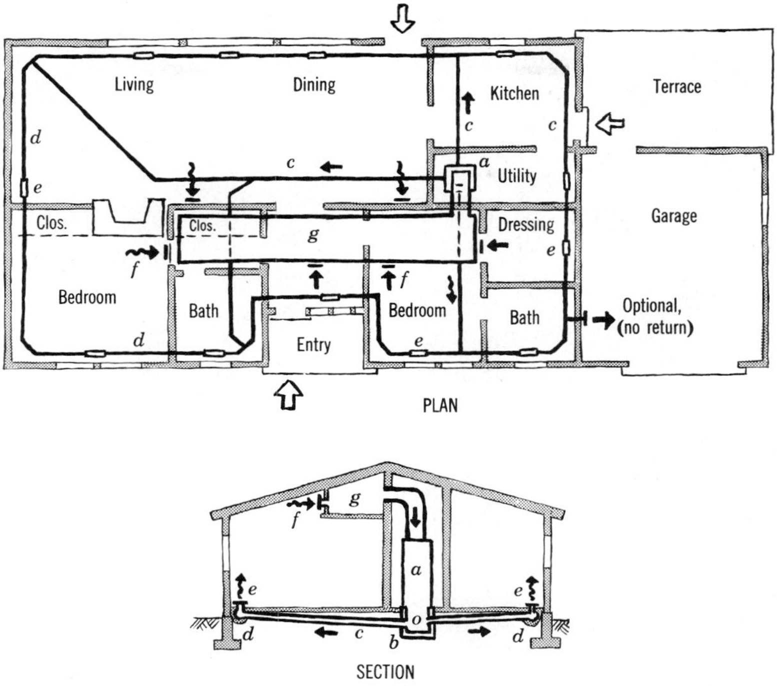

Fig. 6.45 Forced-warm-air, perimeter loop system, adaptable for cooling. No returns from kitchen, baths, or garage. (a) *Downflow air furnace* (b) *Supply plenum.* (c) *Eight-inch (plus) subslab supply ducts (encased in concrete).* (d) *Eight-inch perimeter duct (encased in concrete).* (e) *Floor register, adjustable for direction and flow rate (Fig. 6.51)* (f) *Return grille* (g) *Return plenum.*

temperatures in all parts of a building. It is possible to clean both the recirculated air and the outdoor air by means of filters and other special air-cleaning equipment. Air may be circulated in non-heating seasons. Fresh air may be introduced to reduce odors and to make up the air exhausted by fans in kitchens, laundries, and bathrooms. Central cooling can be incorporated or introduced if ducts are designed originally to do so. Cooling sometimes calls for greater rates of air circulation. Humidification can be achieved by a humidifier in the air stream, and, if cooling is included in the design, dehumidification can be accomplished in summer. For both heating and cooling, a good arrangement is to place the supply registers in the floor, below areas of glass. This is important for winter operation. With adequate attention to supply register placement, return grilles can be located so as to minimize return air ductwork. High return grilles pick up the warmer air for recooling at the equipment. In many systems, air circulates at all times and is warmed or cooled as required.

Planning for warm air systems begins with the attempt to locate the furnace reasonably close to the center of the building. After the system is designed, a furnace must be selected. It should be capable of burning fuel at a rate suitable to make up the building's hourly heat loss. The rate of air delivery must be correct to transmit this heat at the air temperature *rise* that is planned. Finally, the motor and blower must be powerful enough to overcome the friction of air against metal in both the supply and return duct systems, as well as the friction of air flowing through the furnace, filters, registers, and grilles (see Fig. 6.50). Minor adjustments can be made at the furnace to adapt to the demands of the system and the building.

Some of the system components are discussed below.

Furnaces have become much more efficient in recent years, thanks to forced-draft chimneys and heat exchangers, as shown in Fig. 6.46. Seasonal efficiencies of up to 95% are possible, in contrast to about 62% for older furnaces. Fig. 6.47 shows

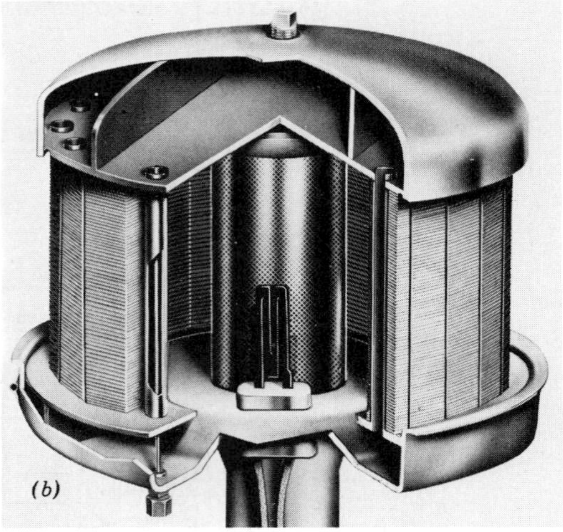

Fig. 6.46 *Furnaces with greatly increased operating efficiencies are now available.* (a) *An Amana gas furnace.* (b) *The small, high-efficiency heat exchanger utilized by the furnace which recovers heat from exhaust gases. (Courtesy of Amana Refrigeration, Inc.)*

Fig. 6.47 *Conventional warm air furnace and ducts. This system, with supply registers in the floor under glass and interior, high-return registers, is suitable for heating or cooling. Furnace and ducts are in the basement of this basement-and-one-story house. In two-story houses, supply and return registers should be in the same relative positions in each story.*

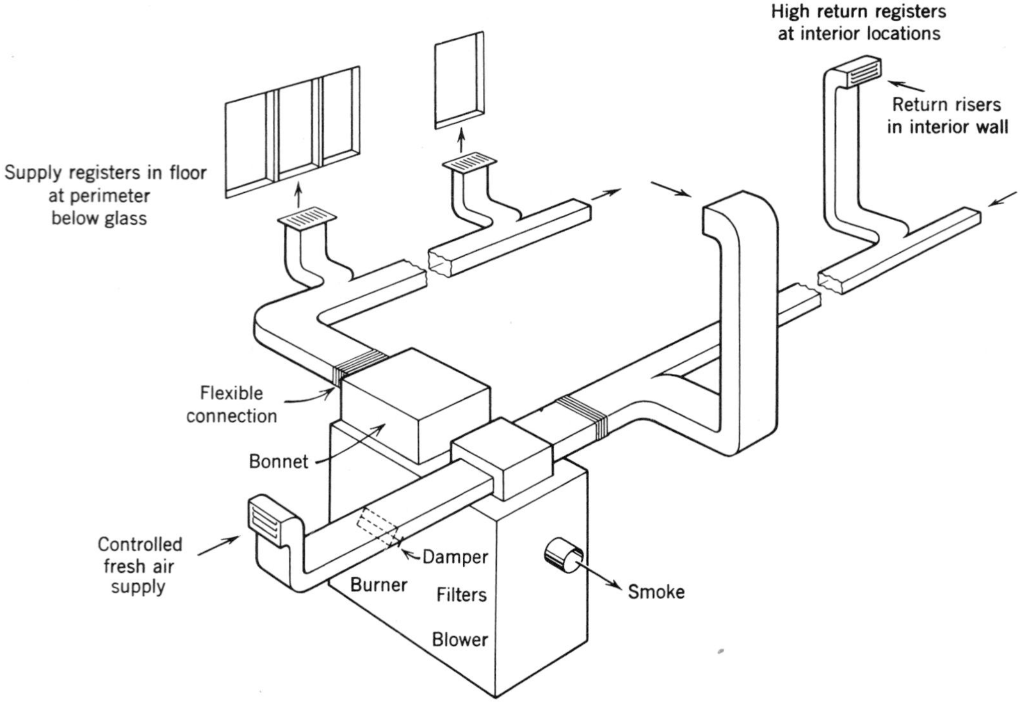

the relationship between a furnace, the duct distribution tree, and some elements of the spaces they serve.

Ducts are constructed of sheet metal or glass fiber and are either round or rectangular. Ductwork will conduct noise unless these suggestions are followed:

Do not place the blower too close to a return grille.

Select quiet motors and cushioned mountings.

Do not permit connection or contact of conduits or water piping with the blower housing.

Use rubberized canvas flexible connection between bonnet and ductwork.

Ducts also can be lined with sound-absorbing material to further discourage noise transfer.

Duct sizes may be selected on the basis of permissible air velocity in the duct.

EXAMPLE 6.2. The main duct in the low-velocity, warm air system of a residence delivers 1600 cfm. Select a size for this duct.

SOLUTION. Table 5.13 indicates that 800 fpm would be an acceptable velocity. The area of the duct in square inches would be

$$\frac{1600 \text{ cfm} \times 144 \text{ (in.}^2/\text{ft}^2)}{800 \text{ fpm}} = 288 \text{ in.}^2$$

A 20 × 14-in. (280-in.^{2}) duct is acceptable and will operate without undue noise or friction. (See the friction allowance approximations given in Example 5.3, Part B.)

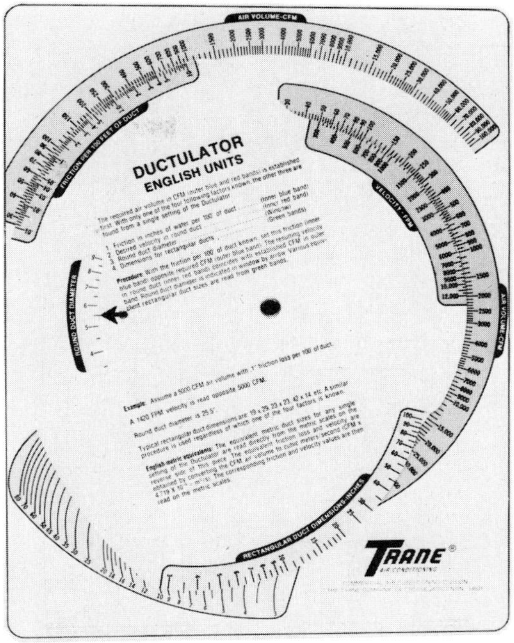

Fig. 6.48 The "Ductulator" is a duct-sizing device available from the Trane Company. The designer selects any two given factors (e.g., friction and airflow volume) and the device yields all the other factors (e.g., air velocity, diameter of the round duct required, or combinations of rectangular-duct cross sections required).

Figure 6.48 illustrates a device that simplifies duct sizing. At a glance, it will show many duct cross-sectional configurations that will satisfy the combined requirements of friction, airflow, and air velocity.

Dampers will be necessary to balance the system and adjust it to the desires of the occupants (see Fig. 6.49). Splitter dampers are used where branch ducts leave the larger trunk ducts. The flow of each riser can be controlled by an adjustable damper in the basement at the foot of the riser. Labels should indicate the rooms served. Some codes require dampers of fire-resistant material actuated by fusible links, to prevent the possible spread of fire through a duct system (see Chapter 13). Figure 6.49*d* shows how turning vanes can be used to assist airflow at sharp turns in ductwork. Such assistance reduces friction within the ductwork, thus reducing the total static head (Fig. 6.50) against which the supply fan must work.

Supply registers (Fig. 6.51) should be equipped with dampers, and their vanes should be arranged to disperse the air and to reduce its velocity as soon as possible after it enters the room. This is commonly done by providing vanes that divert the air, half to the right and half to the left. When a supply register is in the corner of a room, it is best if the vanes deflect all the air in one direction, away from the corner. Return grilles are of the slotted type in walls and of the grid type in floors. All registers and grilles should be made tight at the duct connection. See Table 6.12 for selection of registers based on output and recommended face velocity.

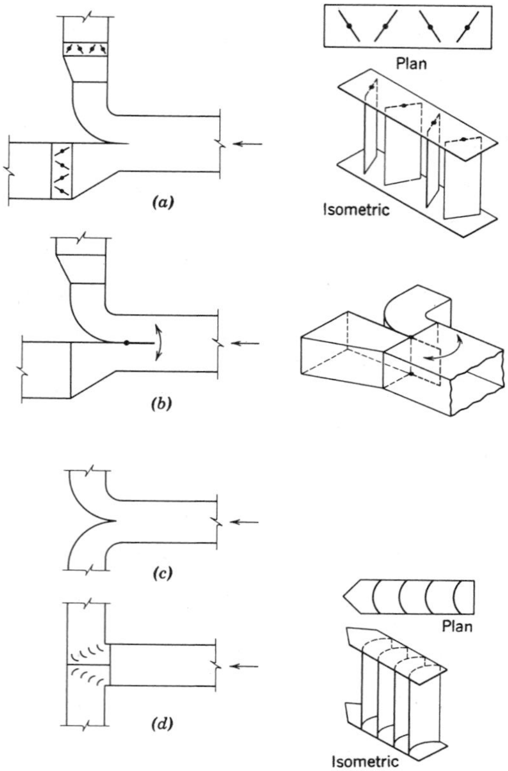

Fig. 6.49 *Air controls in ducts.* (a) *Air adjustment by opposed-blade dampers.* (b) *Air adjustment by splitter damper.* (c) *Conventional turns in ducts.* (d) *Right-angle turns with turning vanes—a more compact method.*

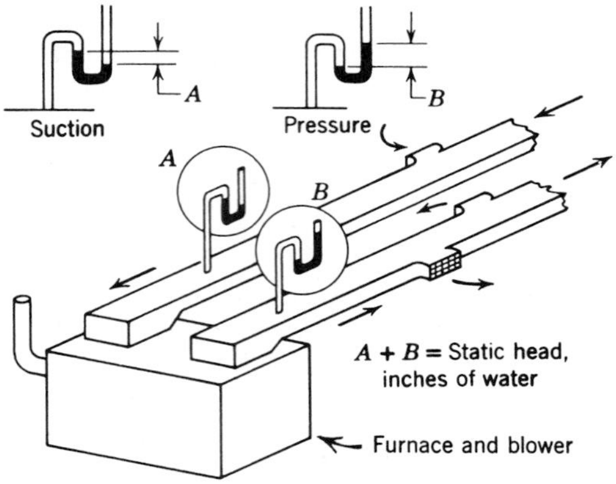

Fig. 6.50 *The static head is the pressure, measured in inches of water, available to overcome friction in the entire system.*

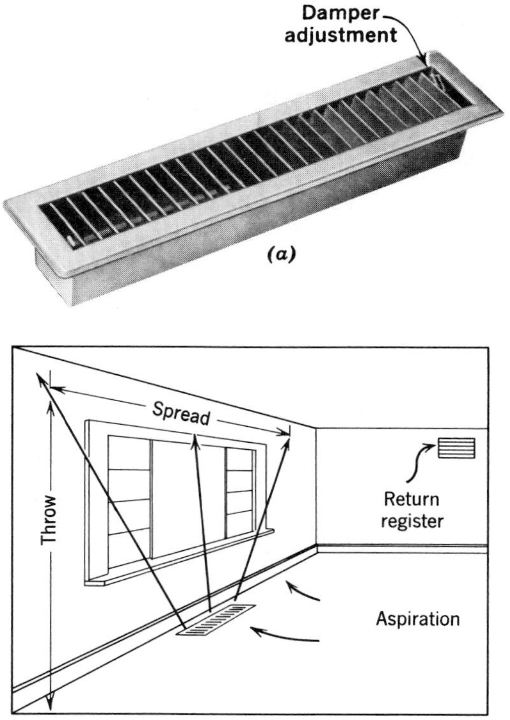

Fig. 6.51 *Floor registers and their action.* (a) *A 2¼ × 12-in. floor register (diffuser)—one of many sizes and shapes. It has diverting vanes for "spread" and an adjustable damper. See Table 6.12 for characteristics.* (b) *Concept of spread and throw. By aspiration (suction), cooler room air is induced to join the stream of warm air, resulting in a bland and pleasant airstream that crosses the room.*

Controls: The burner is started and stopped by a thermostat, which is placed in or near the living room at a thermally stable location that is protected from cold drafts, direct sunlight, and the warming effects of nearby warm air registers. A cut-in temperature of between 80 and 95 F is selected for the fan switch in the furnace bonnet. After the burner starts, the fan switch turns on the blower when the furnace air reaches the selected cut-in temperature. Burner and blower then continue to run while heat is needed. When the burner turns off, the blower continues to run until the temperature in the furnace drops to a level a little below the cut-in temperature of the fan switch. If, during operation, the temperature unexpectedly exceeds 200 F, a high-limit switch turns off the burner in the interest of comfort and safety. As in all automatically fired heating units, a stack tem-

TABLE 6.12 **Forced-Air Registers**

*Part A. **Characteristics of 2¼ × 12–in. Floor Register***

Heating (Btu/h)	3045	4565	6090	7610	9515	11,415	13,320	15,220
Cooling (Btu/h)	855	1280	1710	2135	2670	3,200	3,735	4,270
Cfm	40	60	80	100	125	150	175	200
TP loss	0.009	0.015	0.027	0.037	0.050	0.080	0.105	0.134
Vertical throw (ft)	3	4	5	6	8	10	12	14
Vertical spread (ft)	6	8	10	11	14	17	22	25
Face velocity (fpm)	280	420	565	705	880	1,050	1,230	1,400

*Part B. **Recommended Delivery Face Velocities for Various Applications (Registers)***

Application	*Recommended Velocities (fpm)*
Broadcasting studios	500
Residences	500 to 750
Apartments	500 to 750
Churches	500 to 750
Hotel bedrooms	500 to 750
Legitimate theaters	500 to 1000
Private offices, acoustically treated	500 to 1000
Motion picture theaters	1000 to 1250
Private offices, not treated	1000 to 1250
General offices	1250 to 1500
Stores, upper floors	1500
Stores, main floors	1500
Industrial buildings	1500 to 2000

Source: The *Catalog* of the Lima Register Company.

NOTE: The sound caused by an air outlet in operation varies in direct proportion to the velocity of the air passing through it. The air velocity can be controlled by selecting outlets of proper sizes. The outlet velocities recommended above are within safe sound limits for most applications.

NOTE: For residences (500–750 fpm), any of four velocities of Part A qualify (280, 420, 565, 705 fpm). For stores (1500 fpm), a higher velocity (1400 fpm) would be permissible.

perature control in the breeching cuts off the fuel if ignition fails.

(f) Heating and Cooling Systems. Mechanical cooling is a much more recent development than heating; in its early days, it was widely adopted as a retrofit to existing heating systems. The first two approaches to mechanical cooling reflect this early attitude.

Cooling coils added to warm air furnaces, which are commonplace, utilize a rather simple arrangement of the refrigeration cycle. Figure 6.32 illustrates the circuit of a refrigerant in compression, condensing, and evaporation, in which the condenser heat is carried away by water and the evaporation process draws heat out of water in another circuit to produce *chilled* water. Thus, the heat is *moved* to a heat rejection location outdoors. Figure 6.52 is a schematic diagram of an air-to-air (in distinction to a water-to-water) refrigeration device. Air instead of water can be used to cool the condenser, and indoor air can be cooled directly by being passed over the evaporator coil in which the refrigerant is expanding from a liquid to a gas. Thus, heat is moved from the indoor air to the outdoor air by the step-up action or heat-pumping nature of the refrigeration cycle. When indoor air is cooled directly in this manner, by the expanding refrigerant, the process is usually known as *direct expansion*. The cooling coils therefore are often

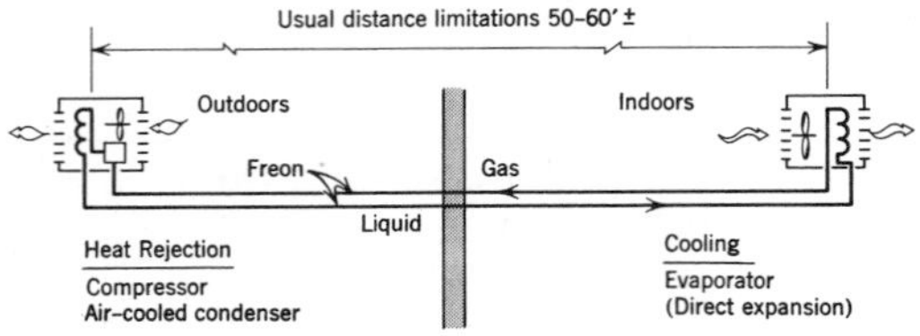

(a)

Fig. 6.52 *(*a*) Schematic diagram of an air-cooled conditioner combining remote outdoor heat rejection and indoor cooling coil suitable for placement in central airstream. (Courtesy of* Progressive Architecture.*) (*b*) Cooling/heating air-handling unit. Removal of the front panel reveals upflow circuit. At lower left is air intake and filter; adjacent are fan and motor. Above these components are gas-burning elements. At the top is a direct expansion cooling coil. (Courtesy of American Furnace Division, Singer Company.)*

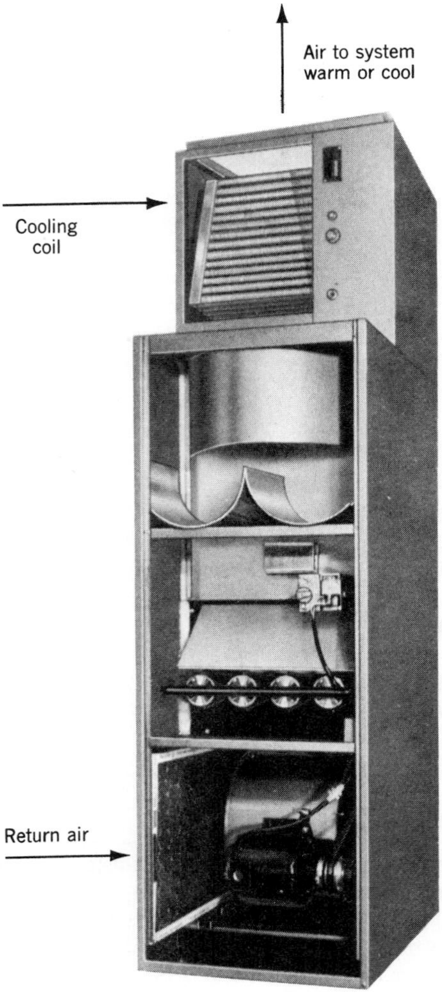

(b)

referred to as "DX coils." Figure 6.53 shows another popular arrangement, in which the airflow through the furnace is horizontal.

Meanwhile, the compressor-condenser unit is placed outdoors, on a concrete slab or on the roof. The unit creates a noisy, hot microclimate in summer—an influence on both site and building planning.

Cooling by air/heating hydronically is the approach taken by, for example, Levitt & Sons in Levittown developments. This system combines a perimeter hot water heating pipe with an overhead air-handling system. A boiler with a tankless coil supplies domestic hot water. The heat output supplies both the perimeter loop and a coil in the air-handling unit of the duct system. The total heating load is met by the combination of radiant heat generated by the perimeter loop and heated air from the overhead air-handling system.

The perimeter loop consists only of ½- or ¾-in. copper tubing embedded 4 in. below the top of the floor slab to kill the cold slab effect. It has the capacity to maintain a 35F° differential between the inside and outside temperatures at the perimeter.

The air-handling unit and overhead duct system, incorporating supply outlets in each room and central return, is used throughout the year. Its cooling coil is connected to an adjacent outdoor condensing unit (see Fig. 6.54).

Standardization was one of the goals the Levitt organization aimed at with this approach to heating and cooling. Mechanical engineer John Liebl, the designer of the system, says that its major advantage is the fact that it provides year-round comfort that works with all types of slab, on any terrain, using any fuel. This standardization simplifies design and construction costs.

An important advantage of the system from a performance viewpoint is that problems of short cycling are minimized, as part of the heating load is carried by the loop and air discharge can be

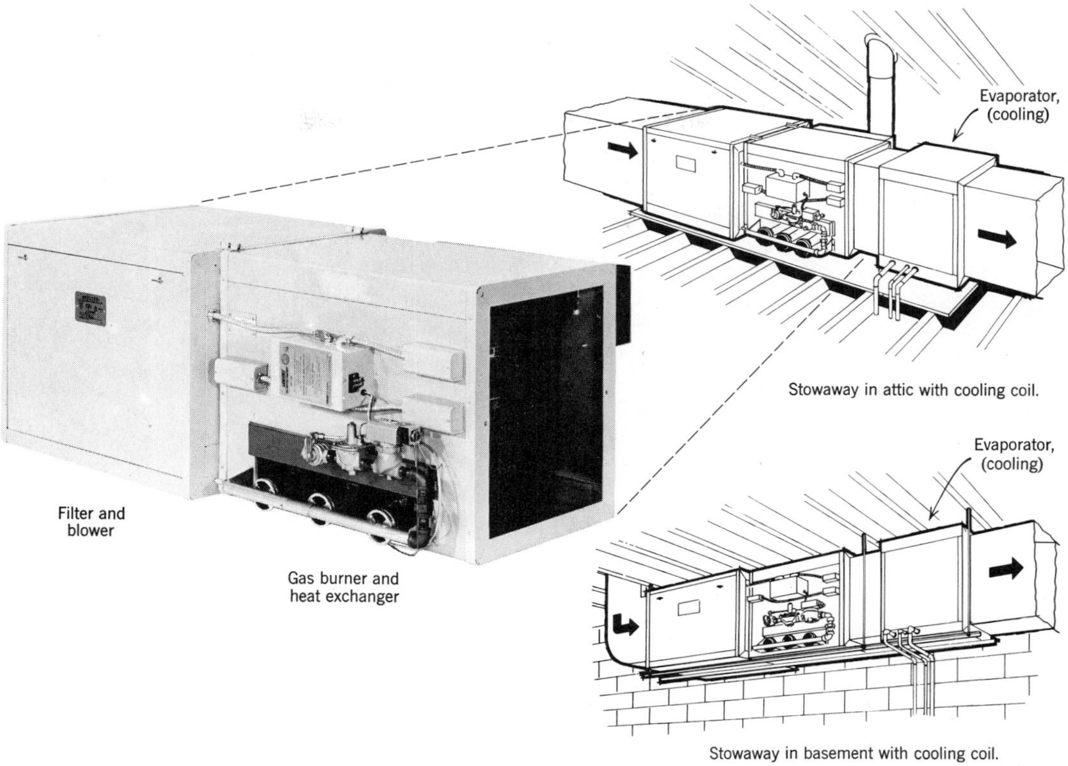

Fig. 6.53 Compact, horizontal flow combinations for heating and cooling by air. The small pipe between the two refrigerant *pipes of the cooling unit is a water drain that carries away the condensed moisture from the recirculated and outdoor air. Heating unit will require gas and flue connections. Refrigerant pipes connect to an outdoor compressor-condensor unit.*

kept at about 120 F, or 20F° lower than with a conventionally ducted system.

With respect to comfort, it is claimed not only that the radiant effect eliminates any cold slab feeling, but also that the temperature variation from floor to ceiling in test houses has not exceeded 4F°.

The *heat pump* is a single device that uses the refrigeration cycle to both heat *and* cool, thus eliminating the distinction between furnace (or boiler) and DX cooling coils. As shown in Fig. 6.55, heat is "pumped" from indoors to outdoors in summer (Fig. 6.55*a*) and from outdoors to indoors in winter (Fig. 6.55*b*). Heat pumps can transfer heat air–air, air–water, and water–water. The most common application for smaller buildings is the air–air heat pump, shown in Figs. 6.55 and 6.56. Although most of today's heat pumps utilize electricity to drive the compressive refrigeration cycle, there is a growing trend toward absorption cycle heat pumps that utilize natural gas or solar energy. Solar-driven refrigeration is a particularly elegant blend of energy source and task—the hotter the sun, the higher the cooling capacity.

One of the primary attractions of the heat pump is that in its heating mode it can give more energy than it receives (electrically). Although energy (usually electricity) is required to run the cycle, the pump draws "free" heat from a source such as outdoor air. The total heat delivered to the building is more than the heat (electricity) required to run the cycle. The measure of this heat advantage is called the *coefficient of performance,* or COP, defined as

$$\text{COP} = \frac{\text{heat delivered to space}}{\text{necessary work input}}$$

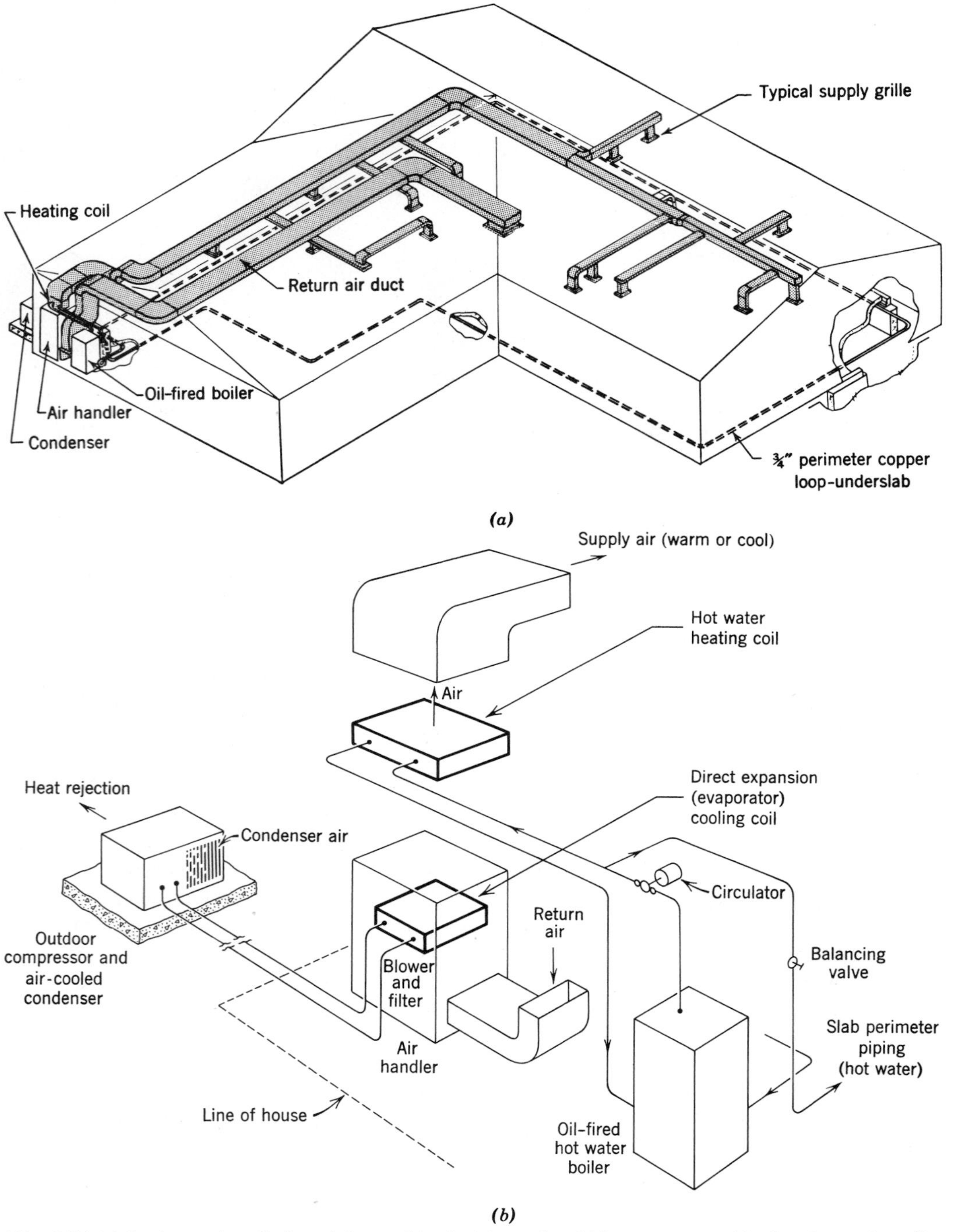

Fig. 6.54 *(*a*) Perimeter loop below slab, combined with overhead air system, provides heating and cooling. (Courtesy of Levitt & Sons. Design by mechanical engineer John Liebl. Reprinted from* Air Conditioning, Heating and Ventilating.*)* (b) *Expanded and partial view of the heating and cooling system, illustrating schematically the circulation of water, air, and refrigerant. Heating or cooling coils operate, as called for, to warm or cool the circulated air. This system combines air and water as thermal media with hydronic warming of the slab perimeter.*

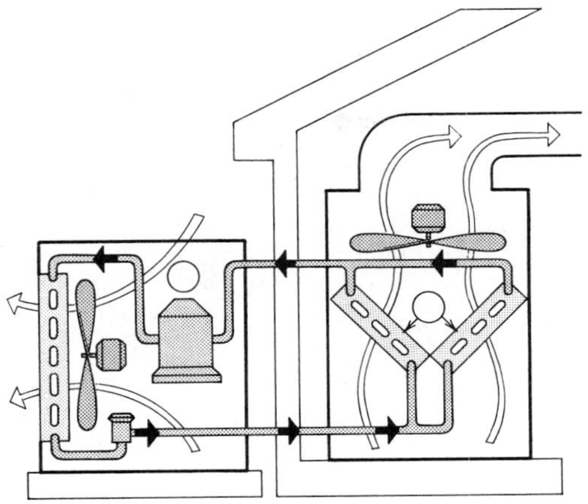

(a)

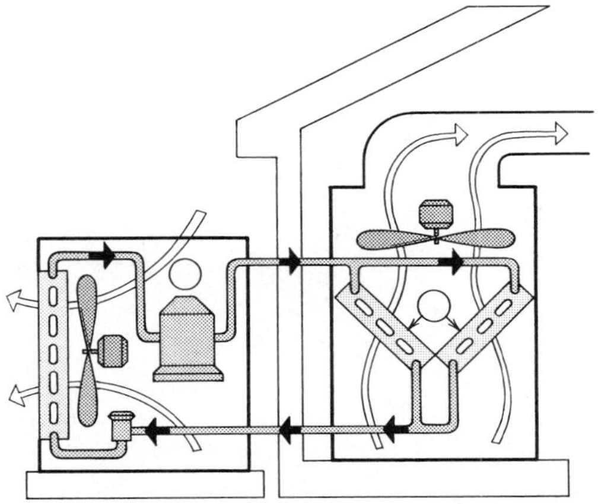

(b)

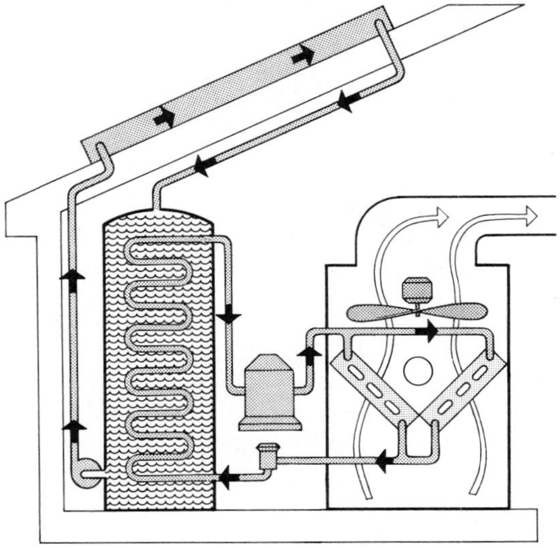

(c)

Fig. 6.55 *Applications of the compressive refrigeration cycle: as a simple heat pump, providing cooled air* (a) *or heated air* (b)*; and as* (c) *a device that increases solar collector efficiency by allowing the solar storage tank to operate at low temperatures (thus making it easily heated by the sun, even on cold days). To get usefully high temperatures from the low-temperature solar storage tank, the refrigeration cycle is utilized. (Reprinted with permission from* Popular Science *© 1978 by Times Mirror Magazines, Inc.)*

In typical space-heating applications, a seasonal COP of 2 or more is common in mild-winter areas. The energy advantage of heat pumps that utilize electricity vis-à-vis simple electric-resistance heating coils, is shown in Fig. 6.57, in which a COP of $71/27 = 2.6$ is demonstrated.

However, as might be expected from a device that draws heat from winter outdoor air, there are limitations to its heating performance. As outdoor temperatures approach 32 F, the COP drops and the outdoor coil tends to ice over. Built-in electric resistance coils must then be used; this, of course, ends the efficiency advantage that made the heat pump attractive. (See Fig. 6.58 for a demonstration of falling performance with falling temperatures.) Because of this characteristic, air–air heat pumps are less frequently used in cold-winter climates. They also generally make poor backup choices for passively solar heated buildings, since backup sources are typically needed only in the coldest weather. Heat pumps that pump heat from *water* sources, such as wells or solar-heated storage tanks, or from the *ground,* are much more dependable cold weather performers. Heat pumps have a high initial cost, and they have shown a relatively high frequency of compressor failure. Noise from the compressor may affect site planning, especially for residences. Section 6.11 provides numerous examples of multiple heat pumps for zoning within larger buildings, as well as systems that allow one pump's reject heat to become another's needed heat source.

Solar-assisted heat pumps help to overcome the heat pump's winter disadvantage. For typical air–air systems, there are four stages to winter heating (see Fig. 6.59):

Stage one: outdoor temperatures above 47 F. The solar collector-storage system is the sole

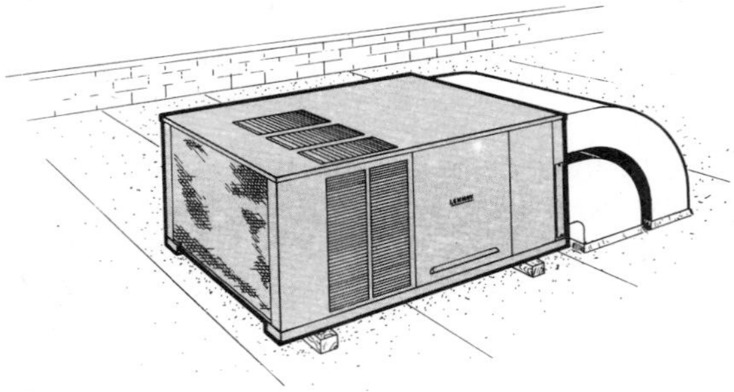

Rooftop installation

(a)

Unit on slab at grade level

(b)

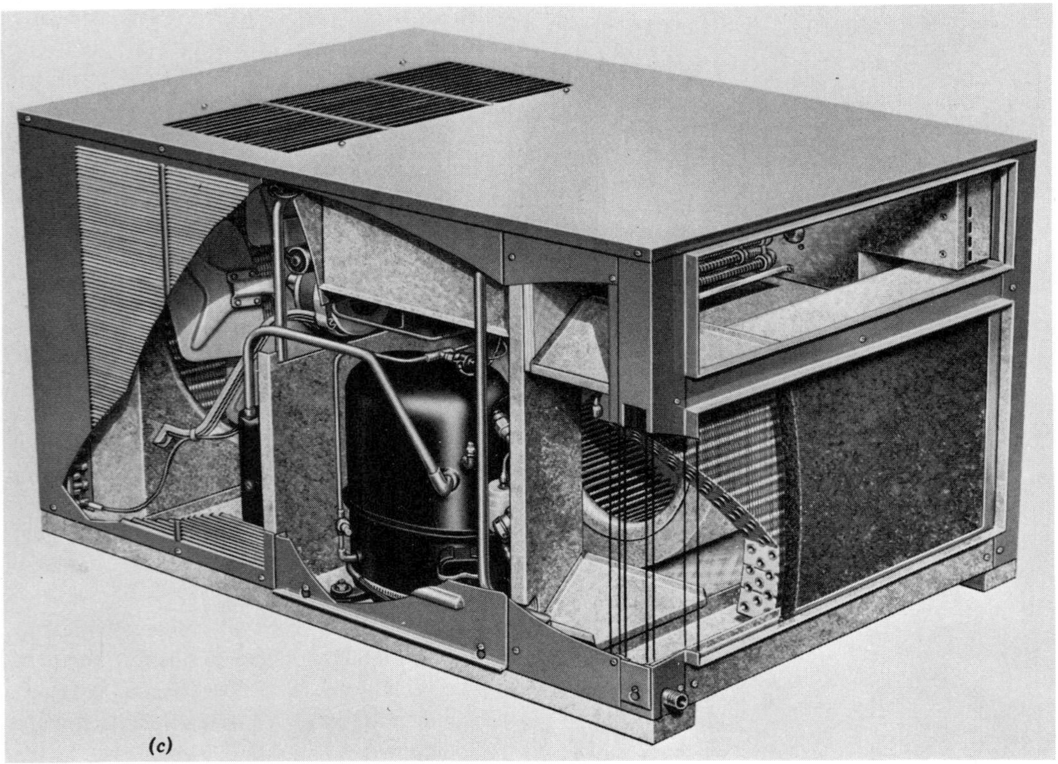

(c)

Fig. 6.56 (a) *Series single-package heat pump, horizontal, CHP9, installed on a roof: 23,000–56,000 Btu/h cooling capacity, 25,000–63,000 Btu/h heating capacity, optional electric heat. (Courtesy of Lennox Industries, Inc.)* (b) *The same heat pump, installed adjacent to a wall.* (c) Cutaway, showing the indoor air circuit (right end): *Return air from the house is drawn into the unit through the (lower) duct opening and is passed over the filter and coil. After being heated (or cooled) by the coil, the conditioned air is delivered by the cylindrical blower to the house duct system through the upper opening.* Outdoor air circuit (left end): *Outdoor air is drawn into the left compartment through the grilles on the sides and top of the cabinet. The large fan then blows this outdoor air across the coil in the left end of the cabinet. Refrigerant compressor is seen at the center of the unit. (Courtesy of Lennox Industries, Inc.)*

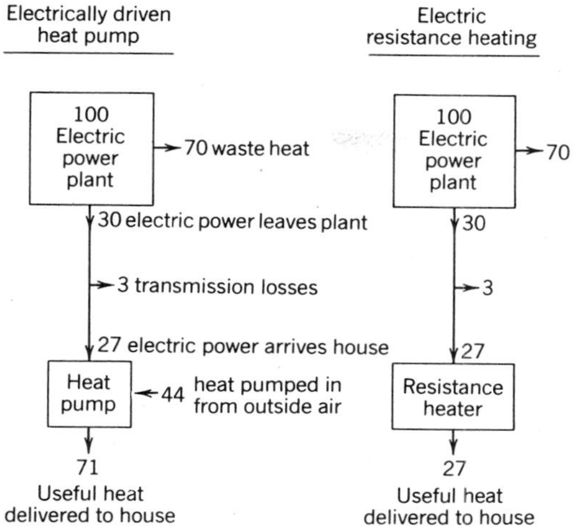

Fig. 6.57 *Efficiency comparisons between simple electric resistance heating and the electrically driven heat pump under optimum operating conditions. See Fig. 1.20 for natural gas furnace comparison. (Reprinted by permission from Fisher, 1974.)*

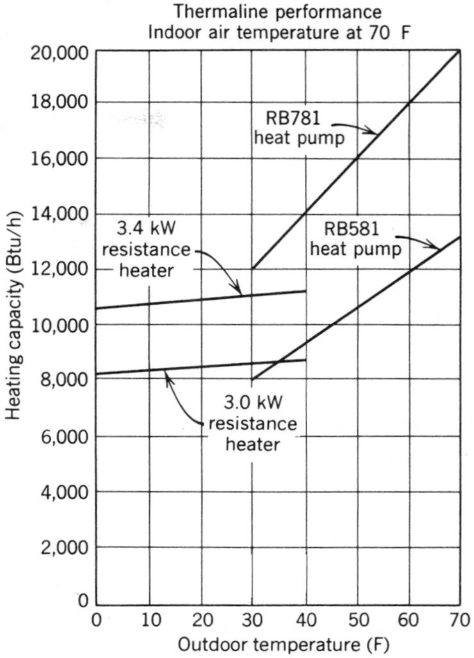

Fig. 6.58 *Typical air–air heat pump operating characteristics.*

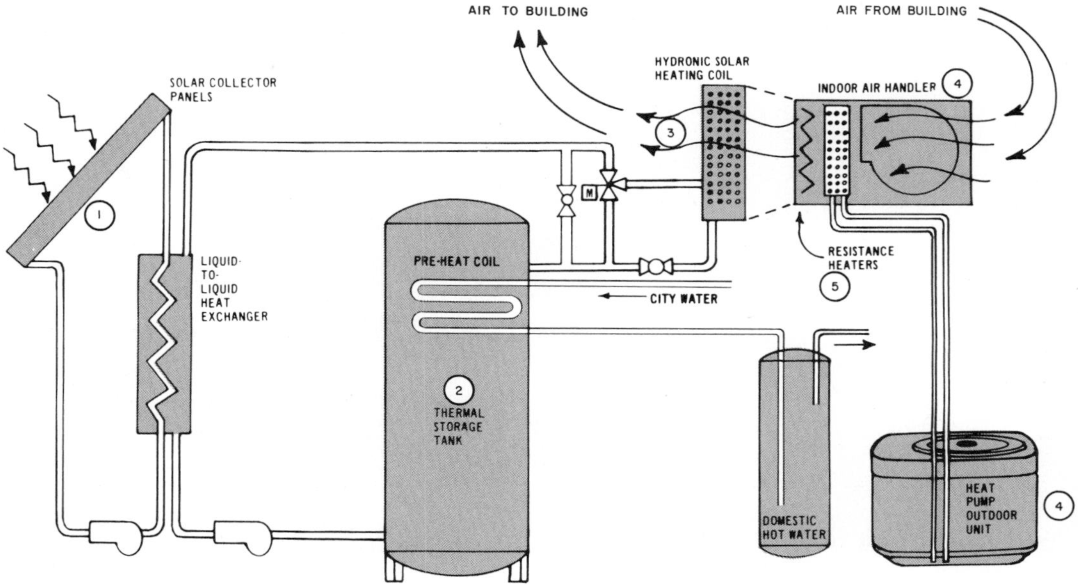

Fig. 6.59 *Components and assembly of the Climatrol Solartrol System. The numbers 1 to 5 identify equipment described in Section 6.6d. (Courtesy of Fedders Corporation and Mueller Climatrol Corporation; reprinted by permission from ''Airtemp Application Guide for Contractors, Builders.'')*

heat source. Only the blower of the air handler and the two small water circulators are in operation. Solar-heated water from the collectors (1) or the storage tank (2) is pumped to the hydronic coil (3), where it heats indoor air.

Stage two: outdoor temperature below 47 F but above 28 to 35 F. The heat pump (4) is the sole heat source. As temperatures fall, the declining output of the heat pump eventually matches (balances) the increasing heat loss of the house—usually between 28 and 35 F.

Stage three: outdoor temperature below 28 F to 35 F but above design minimums. Solar and heat pump work together. Solar-heated water is pumped from either collector (1) or storage tank (2) into the hydronic coil (3), over which passes indoor air that has been somewhat preheated by the heat pump (4).

Stage four: near winter minimum temperatures (a small percentage of the heating season). Electric resistance heaters (5) take over the main heating task, assisted by either the stored solar heat or the heat pump.

Water–air heat pumps (shown in Fig. 6.55) and solar collectors (Fig. 6.60) make an effective team. As these rather high-COP heat pumps remove heat from the solar storage tank (to deliver it to the indoor air), the resulting lower temperature of the solar-heated water *increases* the solar collector's performance. Assume that on a cold, partly sunny day, the collector being fed water from the tank at 90 F is able to raise its temperature to 94 F.

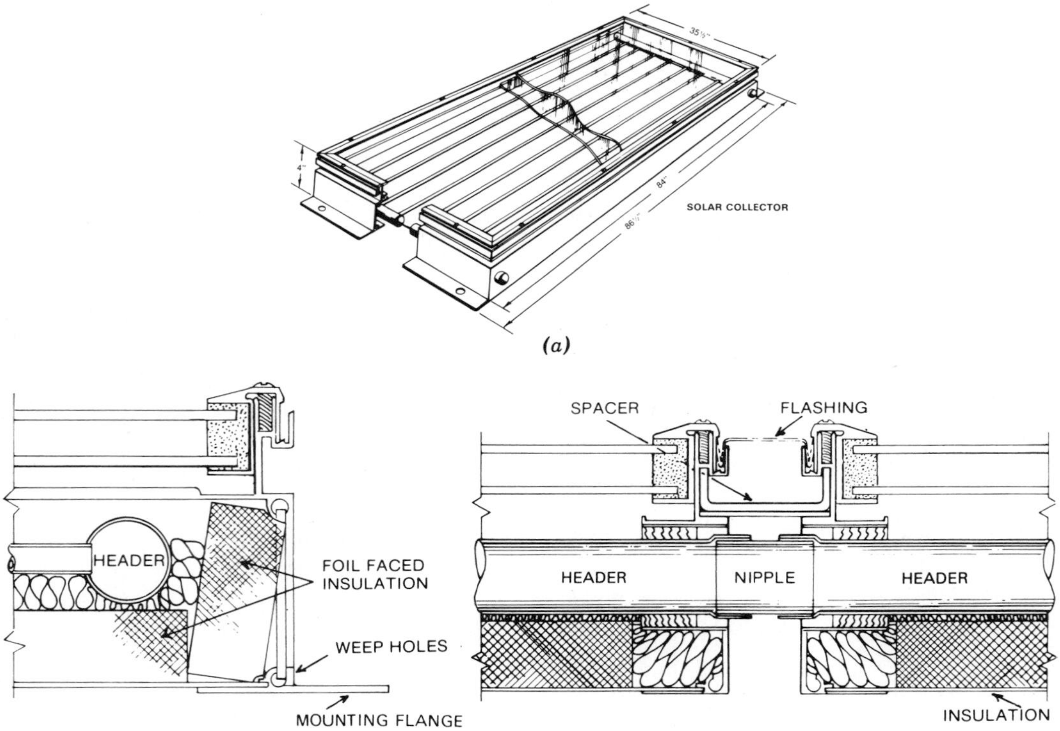

Cross section showing Mounting Flange on Top and Bottom Ends of Collectors.

(b)

Cross Section Showing Connection of Internal Manifolds of Two Adjoining Collectors.

(c)

Fig. 6.60 (a) *Cutaway section of a typical solar collector.* (b) *Cross section of header.* (c) *Method of connecting manifolds of adjacent solar collectors.*

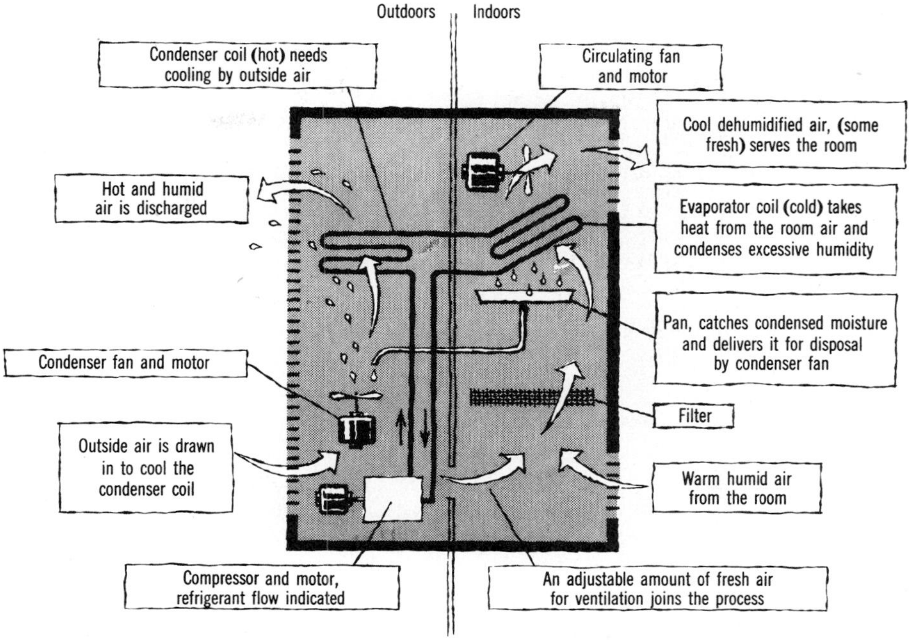

Fig. 6.61 *Schematic diagram of the operation of a through-the-wall, air-to-air conditioning (cooling) unit. Direct heat exchange occurs between air and the process of evaporation or condensation of the refrigerant. The unit is quite self-contained, requiring only access to outdoor air and an electrical connection that powers the motors of two fans and a compressor. The usual capacity is about 1–2 tons of refrigeration (12,000–24,000 Btu/h).*

This improvement is slight because of the rather high heat loss that a 94 F collector experiences when surrounded by cold air. If, however, the collector were to be fed 60 F water, its heat loss would be greatly reduced. The heat that the collector *doesn't* lose to the cold air can be invested in the 60 F water, which will leave at a considerably higher temperature than 64 F. Thus, more solar energy is collected, and more is available for transfer to the building via the heat pump.

(g) Cooling-Only Systems. The device shown in Fig. 6.61 is perhaps the most commonly seen piece of mechanical equipment in the United States. Perched in windows in full view of passersby, these window-box air conditioners noisily remind us that most of our buildings still are *not* centrally mechanically cooled. Mechanical cooling was considered a luxury until well after World War II. Built-in, through-wall air conditioners are still very popular; they offer a low-first-cost way to provide separate zones for individual apartments, motel rooms, and so on. In noisy cities, the drone of these units masks street noise for the interior, thus actually helping to promote relaxation. Unfortunately, such scattered units rarely afford the chance to conserve energy through the exchange of waste heat or the higher efficiencies that can accompany larger equipment. But if turned on only when cooling is needed (i.e., when people are present), they can provide substantial savings over the larger always-on systems.

Evaporative coolers (also affectionately termed "swamp coolers" and "desert coolers") are familiar devices in hot, arid climates (Fig. 6.62). (They are also used in other climates for special, high-heat applications such as restaurant kitchens.) They require a small amount of electricity, to run a fan, and some water, to increase the relative humidity of the air they supply to the building. As explained in Section 4.7, the net effect of this device is *no total change* in heat content (enthalpy) of the indoor air; its DB temperature is *lowered,* but there is an *increase* in relative hu-

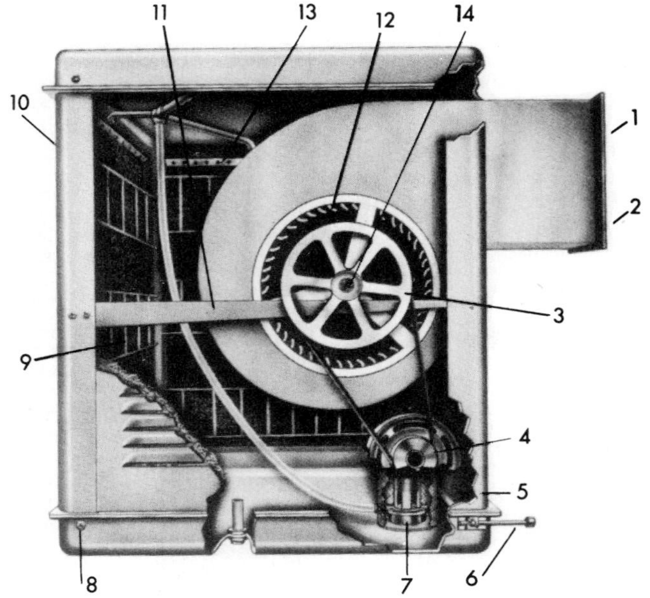

1	Control for air volume	8	Bolts
2	Switches	9	Snap—shut type pad frames
3	Blower pulley (quiet action)	10	Blower mount support
4	Motor	11	Blower wheel
5	Weather resistant exterior finish	12	Even—drip water trough
6	Supports	13	Bronze bearings
7	Water pump, plastic impeller	14	Blower shaft

(a)

Fig. 6.62 (a) *Evaporative cooler with side removed to show operation and components. Pump (lower right) circulates water from the bottom pan to three horizontal perforated tubes at the top that drip to moisten the pads. Water makeup is automatic. Humidification can, by choice, be discontinued and the blower used for air delivery only. (Courtesy of Champion Cooler Corporation.)* (b) *Cooler on the roof of a residence in Palm Springs, California.* (c) *Model numbers and data, used in Example 6.3.*

MODEL	CFM	H. P.	BLOWER WHL.		Wt.	CABINET DIMENSIONS			Lower Opening to Cooler Bottom
			Diam.	Width		Ht.	Width	Depth	
FD30H-2	3000	⅓	12	12	148	34	28	28	19
FD42H-2	4200	⅓	16	16	173	40	34	28	25
FD47H-2	4700	½	16	16	178	40	34	28	25

midity. *People* feel cooler, although no change in total heat has occurred. However, as the air passes through the space, some "actual" cooling occurs, as explained below.

The typical evaporative cooler shown in Fig. 6.62 needs full access to outdoor air and is thus often set on the roof; through-the-wall units are also available. Great quantities of dry, hot outdoor air are blown through pads kept moist by recirculated and make-up water. The "cooled" air is then delivered to the indoor space. The effect of the gently moving cool air is to cool the body and, additionally, to produce further cooling by evaporation of body moisture. The thermal-evaporative cycle of the cooler is shown in Fig. 6.63; the process follows a constant WB line (points *A* to *B*). Outdoor air at 105 F and 10% relative humidity can be considered as too *unbalanced* a condition to provide comfort for humans. When outdoor air is as dry as this, the adiabatic process of the air cooler results in indoor air that is quite the same as that provided by refrigerated cooling. The indoor supply condition of 78 F and 50% relative humidity thus produced is the same as those usually chosen as design standards for refrigerated cooling. As the supply air draws heat from the space, it will increase in temperature (points *B* to *C*). The sensible cooling provided by such air is found by the familiar ventilation formula:

$$q_v = (\text{cfm})(1.08)(\Delta T)$$

where ΔT is the difference, within the space, between supply and exhaust temperatures.

Outdoor conditions in about half the United States (see Fig. 6.64) lend themselves well to the use of the cooler. In regions approaching the East and West coasts, humid conditions increase. There, more frequent air changes should be as recommended in Fig. 6.64. In zone 4 (very humid), evaporative coolers are hard pressed to provide comfort.

Air introduced into the indoor space must be exhausted for the system to operate properly. By selecting the room through which the air is exhausted, one can route cool air as desired in any chosen path from unit to relief opening.

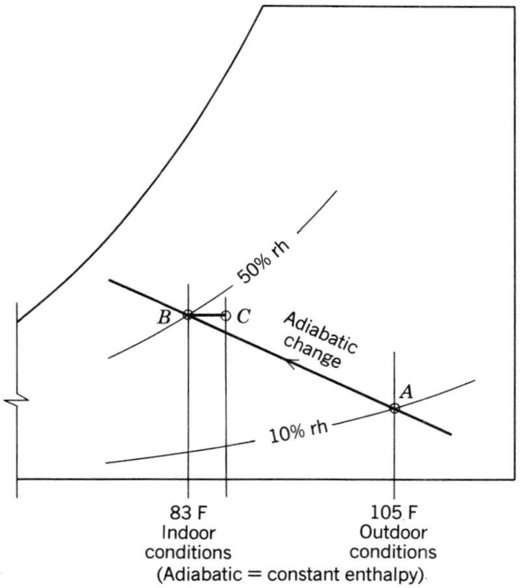

Fig. 6.63 Once-through cycle of outdoor air through an evaporative cooler. Values are transcribed from the psychrometric chart in Fig. 4.6 (page 144). Hot-dry air can be humidified adiabatically (from point A *to point* B*) to indoor conditions that fall within, or close to, the optimum comfort envelope shown in Figure 2.2, without the high-energy-consuming refrigeration cycle. As the humidified air moves through the space (point* B *to point* C*) it removes some heat.*

EXAMPLE 6.3. A one-story ranch-type house in San Diego is to be provided with evaporative coolers. The open planning of the house is such that three coolers in through-wall locations can serve all areas. From the data given in Fig. 6.64 and for *normal* conditions, calculate the cfm required for each unit and select a suitable FD unit from Fig. 6.62.

floor area of the house = 2000 ft^2

ceiling height = 8 ft

regional location = zone 2

desired number of units (zoning) = 3

SOLUTION

volume of the house: 2000 × 8 = 16,000 ft^3

total cfm rate, entire house → 16,000 ÷ 2 min = 8000 cfm

cfm per unit: → 8000 ÷ 3 units = 2700 cfm

Climate Control Chart

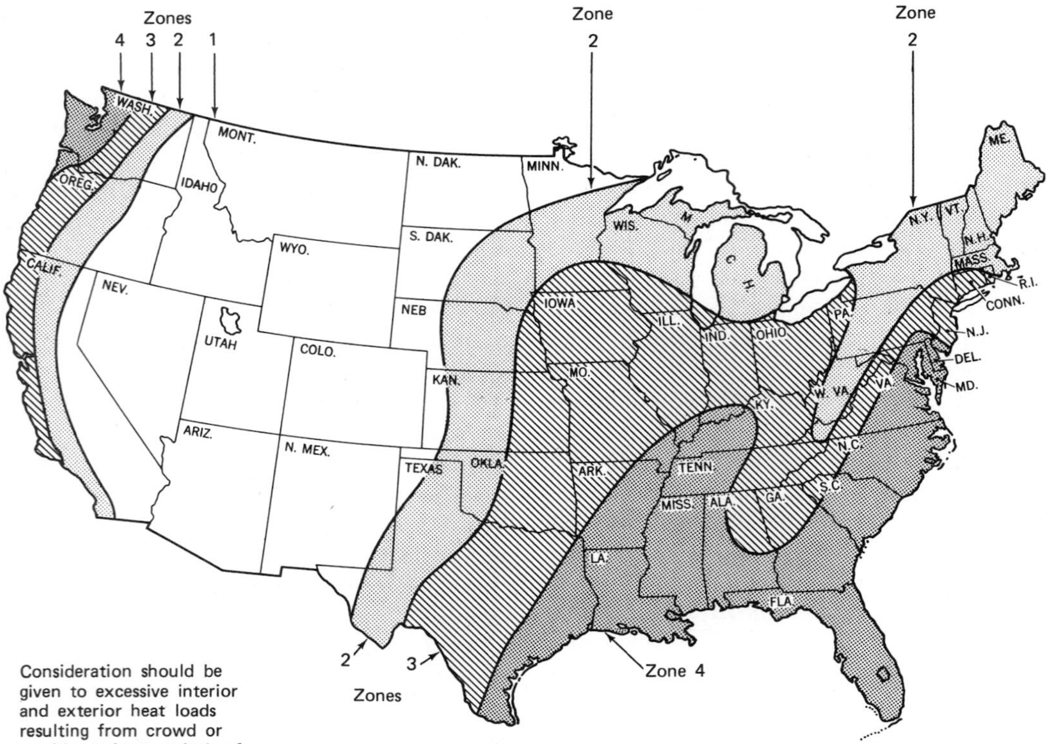

Consideration should be given to excessive interior and exterior heat loads resulting from crowd or machinery heat or lack of sun protection such as poor insulation.

Example: Area 19′ × 41′ × 10′. Total area: 7,790 cu. ft. 7,790 cubic feet divided by 2 (zone) totals 3,895 cu ft. Size Evaporative Cooler needed: 4,000 CFM.

1. Determine the cubical content of the area to be cooled by multiplying the width times length times height.
2. Check map above and find the zone in which you are located. Divide the cubical contents of the area to be cooled by the number of air changes indicated in the air change table.

Table of Air Changes

Zone		Normal	Excessive
1	change air each	3 minutes	2 minutes
2	change air each	2 minutes	$1\frac{1}{2}$ minutes
3	change air each	$1\frac{1}{2}$ minutes	1 minute
4	change air each	1 minute	$\frac{2}{3}$ minute

Fig. 6.64 *Manufacturer's recommendations for the selection of units with respect to regional zone. Air changes per minute and cfm ratings are established. (Courtesy of Champion Cooler Corporation.)*

unit selected (Fig. 6.62): → FD 30H-2

actual cfm: → 3000 × 3 units = 9000 cfm

Note that 9000 cfm of air will remove heat, as the air within the space rises in temperature. If the air rises 5F° from supply to exhaust,

$$\text{sensible cooling} = 9000 \text{ cfm } (1.08)(5\text{F}^\circ) = 48{,}600 \text{ Btu/h}$$

With such high flow rates, supply openings should be sized to avoid excessive air velocity. Assuming a maximum face velocity of 750 fpm (Table 6.12), the supply register's "free area" size for each unit is

$$\frac{3000 \text{ cfm}}{750 \text{ fpm}} = 4 \text{ ft}^2$$

This area must be increased to allow for the ob-

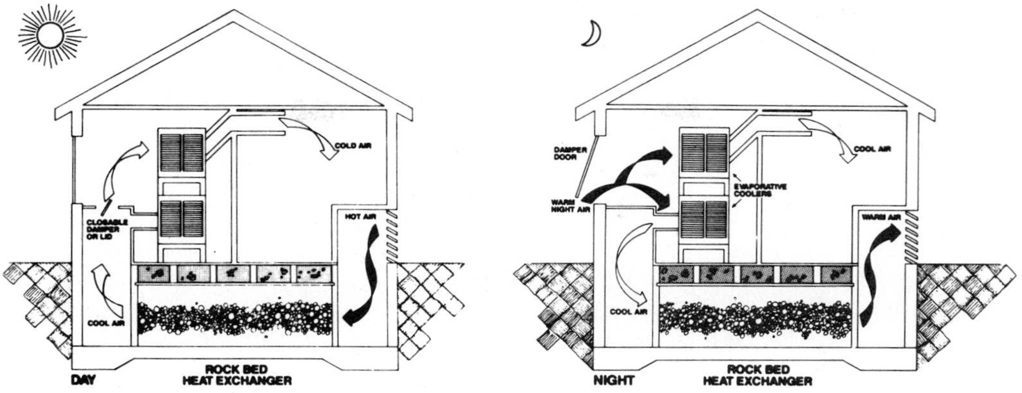

Fig. 6.65 Direct-indirect evaporative cooling, utilizing a rock bed for storage and heat exchange. (Reprinted by permission of the Environmental Research Laboratory, University of Arizona.)

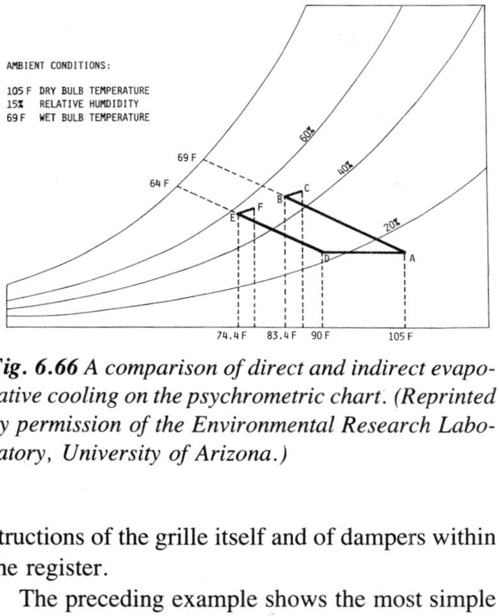

Fig. 6.66 A comparison of direct and indirect evaporative cooling on the psychrometric chart. (Reprinted by permission of the Environmental Research Laboratory, University of Arizona.)

structions of the grille itself and of dampers within the register.

The preceding example shows the most simple application of evaporative cooling, known as the *direct* process (once-treated outdoor air directly introduced to the space). Unfortunately, some areas have such hot summer daytime conditions that such simple evaporative approaches cannot produce comfort indoors. Recently, *direct and indirect* processes have been combined to achieve more "real" cooling and better indoor comfort conditions. One of several such approaches is shown in Figs. 6.65 and 6.66. Warm, rather dry night outside air is evaporatively cooled and fed into a rock bed. The air's temperature is low enough to cool the rock bed, and its relative humidity is moderate. (At the same time, the house is directly evaporatively cooled by a second cooling unit.) Figure 6.66 traces the process by day. Extremely hot, dry outdoor air (*A*) is drawn into the rock bed, where it is cooled by contact (*D*). It can then be passed through an evaporative cooler, to achieve a better combination of rh and DB temperature (*E*). After picking up both sensible and latent heat, the air is exhausted (at approximately temperature *F*).

By comparison, simple direct evaporative cooling by day would have produced indoor supply air too hot and humid for comfort (*B*). Further information on such processes is available from John Peck and Helen Kessler at the University of Arizona Environmental Research Laboratory (Tucson International Airport, Tucson AZ 85706).

6.7 Heating, Ventilating, Air Conditioning (HVAC) System Types

Large buildings have so many thermal zones, and there are so many ways to move heat from one place to another, that hundreds of HVAC systems have been devised. A few of the most typical are introduced in this section; the following section will treat in detail the major components of HVAC production and delivery. Finally, some common variations on each of the four main system classifications will be presented.

One way to classify HVAC systems is by the media used to transfer heat. Although thousands of liquids and gases can be used as carriers of heat, the three most common in building applications are air, water, and refrigerant. Traditionally, there are four main system classifications:

Air-air systems.

Air & water systems.

All-water systems.

Direct refrigerant systems.

In the first three cases, the heating/cooling production equipment typically is located centrally in a large building, often rather far from the thermal zones it serves. Distribution tree size and placement thus become important issues when those systems are selected. In direct refrigerant systems, the heating/cooling machine usually is located adjacent to the zone(s) it serves; thus, the machine's environment—the microclimate it creates, and its needs—relative to the zone's environment becomes an important consideration.

(a) All-Air Systems. The variations on all-air systems are shown in Fig. 6.67. Since air is the only heat transfer media used between the mechanical room (central station) and the zones it serves, and since air holds much less heat per unit volume than water, the distribution trees for this class are quite large. For comfort, however, these systems are, overall, the best. The quantities of air moved through the central station(s) are heated or cooled, humidity-controlled, filtered, and freshened with outdoor air—all under controlled conditions. Within the zones, supply registers and return grilles allow a well-planned stream of conditioned air to thoroughly permeate all work areas. More details on this HVAC class will be found in Section 6.9.

Single-Duct, Variable-Volume (VAV) Systems (Fig. 6.67a). Rapidly becoming the most popular of this class for large buildings, the single duct requires less building volume for distribution, and the variation of air *volume* flow rate (rather than of air temperature) saves energy relative to the single duct with reheat (Fig. 6.67*c*). Depending on outdoor conditions and prevailing indoor needs, the central station supplies at normal velocity either a heated or a cooled stream of air. Automatic volume controls (linked to each zone's thermostat) adjust the volume admitted to that zone, within an air terminal diffuser (often located above a suspended ceiling). When the central station is supplying cold air, a zone that needs more cooling will get more air; an unoccupied room with no internal gains, or a space with heat loss through an exterior wall, will get less air. Clearly, such a system is well suited to serve the interior, always-hot zones of internal-load-dominated buildings. Less clear is its suitability for the perimeter zones of buildings in cold, cloudy conditions.

Dual-Conduit Systems (Fig. 6.67b). In these systems, the space-consuming disadvantages of two distribution trees are partially overcome by high velocity (reducing duct size, but increasing noise and friction). The *primary* air supply deals with heat gains or losses through the building's skin; accordingly, its temperature is variable but its volume flow rate is constant. The *secondary* air supply is as in the VAV constant system—predetermined temperature, but variable volume. The two streams are mixed to order for each zone.

Single Duct with Reheat (Fig. 6.67c). This system (along with VAV) has the smallest distribution tree of this class, because at each zone the only object added to the duct is a small reheat coil (heat provided by steam, hot water, or electric resistance). (Technically, this could be also called an air & water system.) The central station provides a single stream of cold air that must be cold enough to meet the maximum cooling demand of any one zone. All other zones *reheat* this air as needed. In cold weather, outdoor air at temperatures as low as 38 F can be used; the colder this single central airstream, the less air need be circulated (and the smaller the ducts). For large buildings in most U.S. climates, however, the central airstream must be *cooled* most of the time; then, more energy must be spent to reheat the airstream at most zones. These systems thus are notorious for energy wastage, although careful engineering can make them attractive for some climates.

Multizone Systems (Fig. 6.67d). Because each zone has an individual, centrally conditioned airstream, the total distribution tree volume grows to astonishing size with only a few zones. The central station produces both warm and cool air-

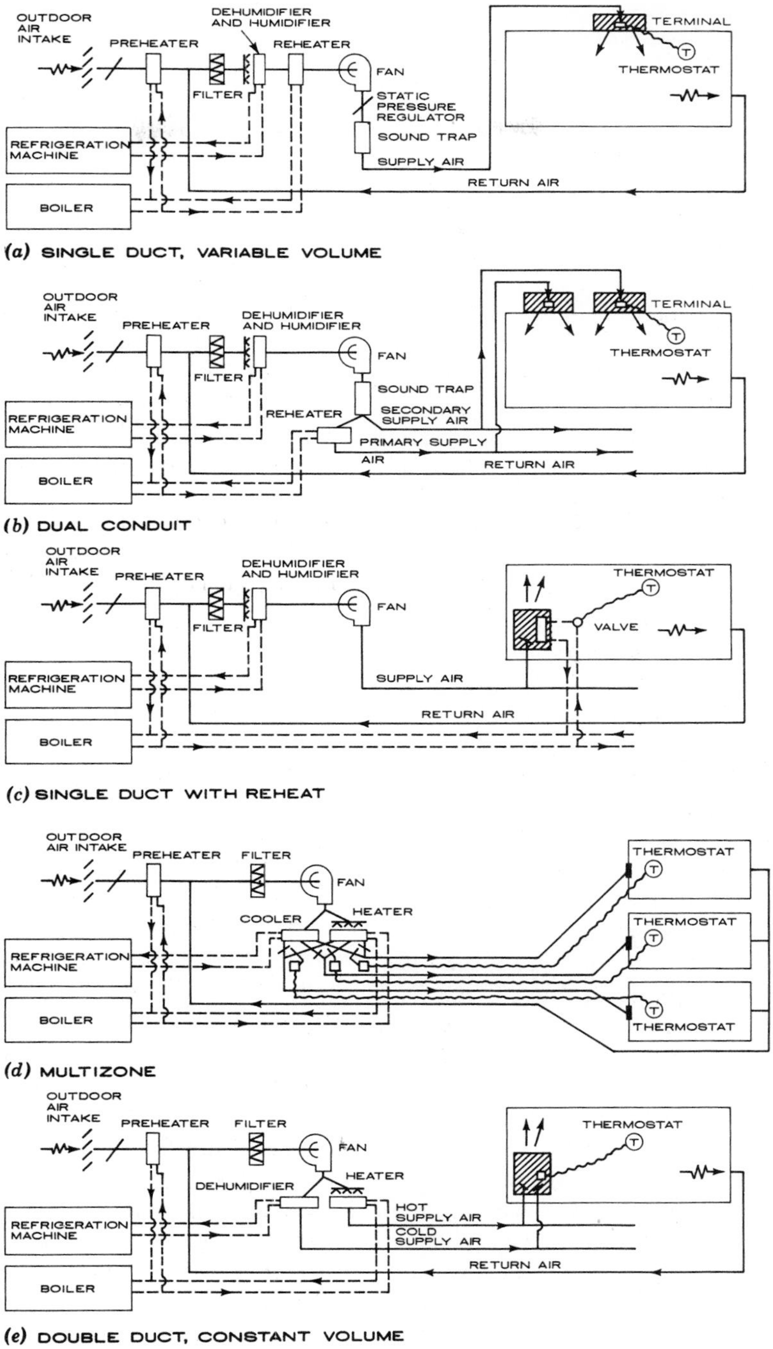

Fig. 6.67 *All-air HVAC system variations. Typical of these systems are large distribution trees, along with complete conditioning of the air. (Reprinted by permission from AIA, Ramsey, and Sleeper,* Architectural Graphic Standards, *7th ed., © 1981 by John Wiley & Sons.)*

streams, which are mixed at the central location to suit each zone. These systems are more likely to be found on medium-sized buildings, or on larger buildings in which smaller central stations are located on each floor. The single return airstream collects air from all zones (as is the case for the other systems in this class).

Double-Duct, Constant-Volume Systems (Fig. 6.67e). Two complete distribution trees are required: at the height of summer the cooling airstream does all the work, whereas in the coldest winter conditions the heating airstream carries the load. Most of the time, air from these two streams is mixed to order at each zone's air terminals. Because both temperature and volume can be controlled, this sytem offers better comfort under reduced load conditions than does the single-duct, variable-volume system. (Example: an only partially occupied room.) However, it is much more expensive to install, consumes much building volume for the two distribution trees, and is usually more energy consuming than the single-duct, variable-volume system that has largely replaced it.

(b) Air & Water Systems. Several variations on air & water systems are shown in Fig. 6.68. Most of the heating and cooling of each zone is accomplished via the water distribution tree, which is much smaller than that needed by air. For air quality—filtering, humidity, freshness—a small, centrally conditioned airstream, equal to the total fresh air required, is provided. Thus, several distribution trees are involved, yet the total space they require is almost always less than that required by all-air systems.

Exhaust air may be gathered in a return air duct system, making heat recovery possible. Or (a cheaper alternative) air can be exhausted locally, to avoid the construction of yet another distribution tree. If the water distribution provides either heating *or* cooling only, it is called a two-pipe system (shown throughout Fig. 6.68). If it provides simultaneous heating *and* cooling, it is either a three- or a four-pipe system, or yet another heating and cooling variation utilizing heat pumps (see Fig. 6.69). This class of system frequently serves the perimeter zones of large buildings, whereas all-air systems (commonly, single-duct, variable-volume) are used for the interior zones. More details on this HVAC class can be found in Section 6.10.

Induction Systems (Fig. 6.68a). This familiar system's air terminal may be found below windows throughout the United States. A high-velocity (and high-pressure), constant-volume fresh air supply is brought to each terminal, where it is forced through an opening in such a way that air already within the room (''bypass,'' or secondary, air) is *induced* to join the incoming jet of air. A fairly thorough circulation of room air is thus accomplished with only a little centrally treated air. Air then passes over finned tubes for heating or cooling. Thermostats control the unit's output by controlling either the flow of the water or the flow of secondary air.

Fan-Coil with Supplementary Air (Fig. 6.68b). Another familiar piece of below-window equipment is the fan coil, which moves the room air as it provides either heating or cooling. Centrally conditioned, tempered fresh air is brought to the space in a constant-volume stream; the fan moves both fresh and room air across a coil that either heats or cools the air, as required.

Radiant Panels with Supplementary Air (Fig. 6.68c). Either ceiling or wall panels contain the heated or cooled water, to provide a large surface for radiant heat exchange. Centrally conditioned, tempered fresh air is brought to the space in a constant-volume stream. The ''piece of equipment'' within the space is replaced by a large surface, which must be kept clear of obstructions to radiant heat exchange.

(c) All-Water Systems. The more simple-appearing all-water systems are shown in Fig. 6.69. These systems only heat and cool; air quality is dealt with elsewhere—either locally, by means of infiltration or windows; or by a separate fresh air supply system; or simply by fresh air from an adjacent system, such as a ventilated interior zone. A fan-coil terminal is again employed, so that air motion occurs along with heating or cooling. (Sometimes the fan-coil unit is located against the exterior wall, so that fresh air may be brought in and mixed with the room air through the fan.) Both baseboard and valence (above-window) units

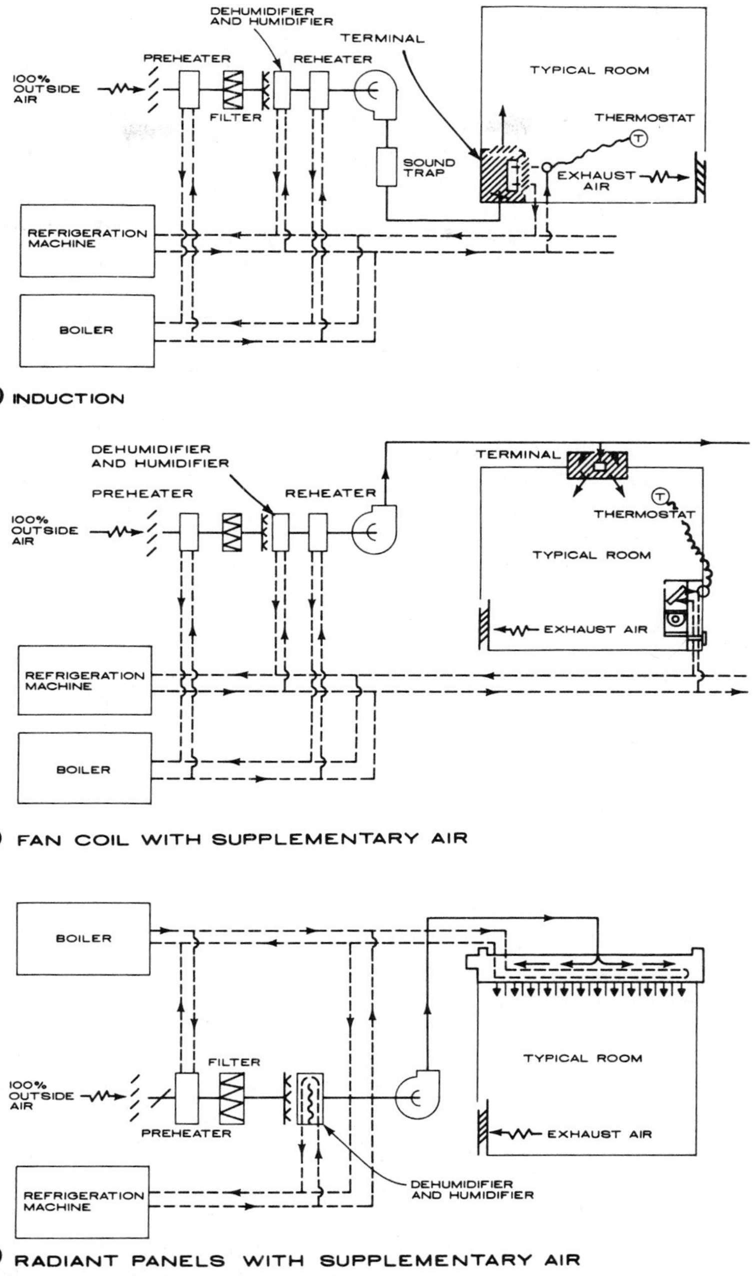

Fig. 6.68 *Air & water HVAC system variations. Although these systems are characterized by the use of several distribution trees, the volume they require is usually less than that required by all-air systems, and they condition the air nearly as thoroughly. (Reprinted by permission from AIA, Ramsey, and Sleeper,* Architectural Graphic Standards, *7th ed., © 1981 by John Wiley & Sons.)*

are also commonly available. More details on this HVAC class can be found in Section 6.11.

Two-pipe water distribution systems are shown throughout Fig. 6.68. They provide either heating or cooling. One pipe is for supply, the other for return.

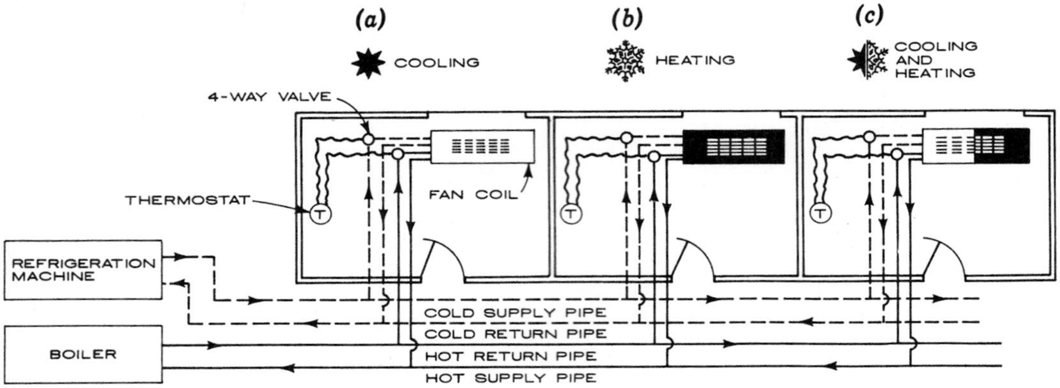

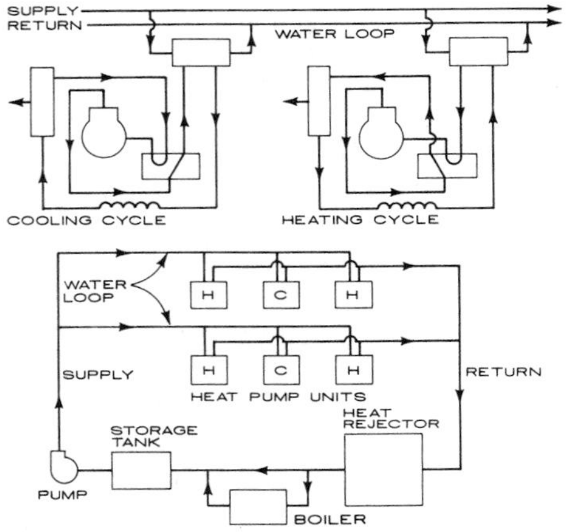

(d)

*Fig. 6.69 All-water HVAC systems. (*a, b, c*) Four-pipe distribution trees, which need smaller volumes than do those for air systems; however, less-thorough conditioning of the air is provided. (*d*) The water loop heat pump. This unit employs water–air units, allowing energy savings during periods of simultaneous heating and cooling. The heat pumps are usually acoustically isolated from the spaces they serve. (Reprinted by permission from AIA, Ramsey, and Sleeper,* Architectural Graphic Standards, *7th ed., © 1981 by John Wiley & Sons.)*

Three-pipe systems are a first-cost-saving but energy-wasting alternative to four-pipe systems. They allow simultaneous heating and cooling (two supply pipes) but mix the very warm hot return and the mildly warm cold return in only one return pipe.

Four-pipe systems are shown in Fig. 6.69*a, b, c*. They allow quick changeover heating and cooling, utilizing two supply and two return pipes. Where simultaneous cooling and heating is required within the same fan-coil unit (Fig. 6.69*c*) a split or double coil is used.

A variation on the two-pipe distribution system is the *water loop heat pump,* shown in Fig. 6.69*d*. Heat pumps (water–air) either draw heat from the loop (heating mode) or discharge heat to it (cooling mode). The loop's temperature ranges between 60 and 90 F; in hot weather, a central cooling tower disposes of the loop's excess heat, while in cold weather a central boiler adds the loop's needed heat. Since individual heat pumps are used, this system is closely related to direct refrigerant systems.

(d) Direct Refrigerant Systems (Fig. 6.70). These systems nearly eliminate the distribution trees of air or water, relying instead on a heating/cooling device adjacent to or within the space to be served. The majority of such systems are air–air devices that can be located either on rooftops or on exterior walls—wherever a plentiful supply of outdoor air can be assured. In a *single-package* system (Fig. 6.70*a*), only one piece of equipment is involved. A single-package air–air heat pump, for example, moves heat between an outdoor airstream and an indoor airstream; although kept separate, both streams pass through a *single* outdoor unit. A system with both outside and inside components is called a *split system* (Fig. 6.70*b*). A split-system air–air heat pump moves heat between the outdoor unit (which also contains the

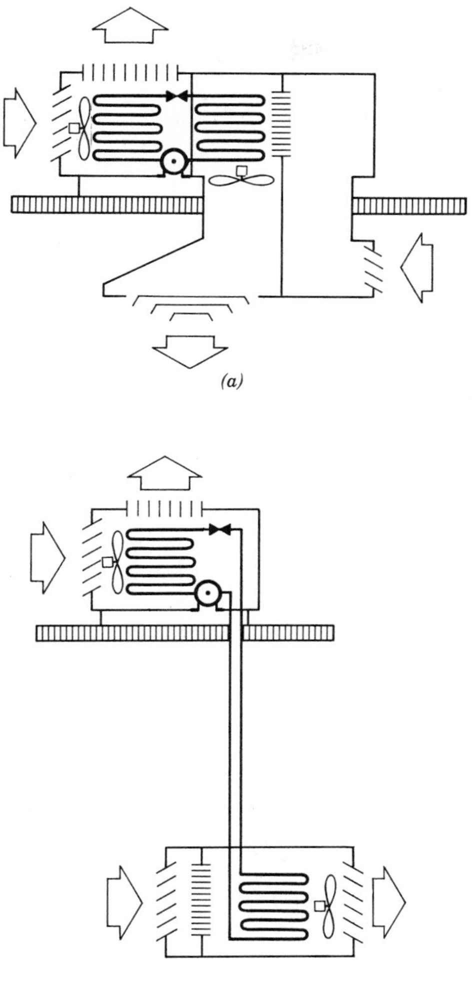

(a) (b)

Fig. 6.70 Typical arrangements for direct-refrigerant HVAC systems. In addition to the roof-mounted units shown, the equipment may also be through-the-wall. (a) *Single package.* (b) *Split system.*

compressor), through which outdoor air passes, and the indoor unit (which usually contains back-up heating coils) for the treatment and circulation of indoor air. Further details and examples are presented in Section 6.12.

(e) Typical Applications. Although each of these HVAC systems has been used in hundreds of different applications, the most common match between system and function is shown in Table 6.13. Note that several systems do not appear in this table: dual-duct, dual-conduit, four-pipe induction, and radiant panel with supplementary air. These systems are usually more expensive to install and/or operate, and thus less frequently used.

(f) Approximate Space Requirements. A general rule of thumb for the space taken by mechanical and electrical equipment was presented in Table 5.11. A more detailed idea of space requirements can now be presented in Table 6.14. The percentage of the building's *gross floor area* is shown for equipment rooms, for vertical ductwork, and for the fan-coil units and other in-the-space equipment that typically occupy part of a building's floor area. For other parts of an HVAC system, the percentage of the building's *gross cubage* (volume) is indicated: horizontal ducts, air terminals, and outlets that typically are placed overhead and therefore do not require floor area allocations.

(g) Approximate Air Quantities. Especially for all-air HVAC systems, it is useful to get an early idea of the *total quantity* of air to be moved. Duct distribution trees often strongly influence floor-to-floor heights and the sizes of central core floor areas. Therefore, Table 6.15 presents some guidelines for the approximation of these air quantities and of the refrigeration equipment necessary to treat them. It is important to remember that these figures assume *all-air systems* (except as noted) and *normal percentage outdoor air* (ventilation rates).

6.8 HVAC Central Equipment for Large Buildings

The many HVAC systems that are included in the categories of all-air, air & water, and all-water have in common a dependence on *central equipment* for the generation of heating and cooling, and/or air quality control. Figure 6.71 shows the basic relationships between some of the major pieces of central equipment and the spaces they serve. This section offers a general guide to some central equipment options and sizes.

TABLE 6.13. **Typical HVAC System Applications**

NOTE: The more expensive systems within each classification—such as dual-duct and dual-conduit (for all-air) and four-pipe induction and radiant panel with supplementary air (for air & water)—may be substituted for those shown below.

	Individual Room or Zone Unit Systems				Central Station Apparatus Systems						
	DX Self-Contained		All-Water		All-Air				Air & Water		
			Room Fan-Coil		Single Air Stream				Primary Air Systems		
							Reheat				
Applications	Room 0.3–2 Tons	Area 2 Tons and Over	Recir. Air	With Outdoor Air	Variable Volume[a]	Bypass	At Terminal	Zone In Duct	Multizone Single Duct	Secondary Water H-V H-P Induction	Room Fan-Coil With O.A.
Single-purpose occupancies											
Residential:											
Medium	×										
Large		×	×						×		
Variety and specialty shops		×			×						
Restaurants											
Medium		×			×			×			
Large		×			×	×	×	×	×		
Bowling Alleys		×			×	×					
Radio and TV studios:											
Small		×			×	×		×	×		
Large		×			×	×		×	×		
Country clubs		×			×	×		×	×		
Funeral homes		×			×				×		

Beauty salons	×	×			×						
Barber shops	×	×			×						
Churches		×			×	×			×		
Theaters					×	×					
Auditoriums					×	×					
Dance and roller skating pavilions		×			×	×					
Factories (comfort)		×				×		×			
Multipurpose occupancies											
Office buildings					×				×	×	
Hotels, dormitories			×	×	×					×	×
Motels			×								
Apartment buildings				×						×	×
Hospitals				×	×			×		×	
Schools and colleges				×	×	×	×	×			
Museums					×			×	×		
Libraries:											
Standard		×			×	×		×	×		
Rare books		×			×			×			
Department stores					×	×					
Shopping centers		×			×	×			×		
Laboratories:											
Small		×			×			×	×		
Large					×		×			×	
Marine							×			×	

Source: Reprinted by permission from *The ABC's of Air Conditioning,* © 1975, Carrier Corporation.

[a]Conventional air outlets are not satisfactory when the variation in air quantity (volume) exceeds 20%, in which case self-controlled linear diffusers are used.

TABLE 6.14 **Typical HVAC System Space Requirements**

For equipment rooms, use the percentage of the entire building's gross area.
For distribution trees, use the percentage of the total served space's gross area (or cubage).

	Percent of Gross Area (or Cubage)					
	Equipment Room		*Distribution Tree*			
			Ductwork Distribution			
System	*Air Handling %*	*Refrig.*	*Riser*	*Horizontal*	*Piping*	*Outlets and Terminals*
Low-velocity conventional[a]	2.2–3.5	0.2–1.0	[b]	0.7–0.9	—	0.07–0.08[c]
High-velocity conventional	2.0–3.3	↓	↓	0.4–0.5	—	0.2–0.4[c]
Terminal reheat (hot water)	2.0–3.3	↓	↓	↓	0.03–0.04[c]	0.4–0.5[c]
Terminal reheat (elect.)	2.0–3.3	↓	↓	↓	—	0.4–0.5[c]
Variable volume		↓	↓	0.1–0.2	—	0.8–0.9[c]
Multizone unit		↓	↓	0.7–0.9	—	0.07–0.08[c]
Double-duct	2.2–3.5	↓	↓	0.6–0.8	—	0.4–0.5[c]
Dual-conduit	2.4–3.4	↓	↓	0.3–0.4	—	0.8–0.9[c]
All-air induction	2.0–3.3	0.2–1.0	↓	0.4–0.5	0.1–0.2	1.5–2.0
Air & water induction		↓				2.0–2.5
2-pipe	0.5–1.5	↓	0.25–0.35[d]		—	2.5–3.0
4-pipe	0.5–1.5	↓	0.3–0.4[d]		—	1.0–1.5
Fan-coil unit		↓		—	0.1–0.2	
2-pipe	—	↓	—	—	0.25–0.3	
4-pipe	—	↓	—			

Source: Reprinted by permission from *The ABC's of Air Conditioning,* © 1975, Carrier Corporation.

[a]"Conventional" refers to simple single-zone systems.

[b]Included in the equipment room area.

[c]Percent of gross cubage (volume).

[d]Includes space for pipe risers.

TABLE 6.15 **All-Air HVAC Systems: Approximate Air Quantities and Refrigeration Sizes**

	Internal Gains						*Approximate Requirements*											
	Occupancy (ft^2/person)			*Lights (W/ft^2)*			*Refrigeration (ft^2/ton^a)*			*Air Quantities (cfm/ft^2)*								
										East–South–West			*North*			*Internal*		
Applications	*Low*	*Av.*	*High*	*Low*	*Av.*	*High*	*Low*	*Av.*	*High*	*Low*	*Av.*	*High*	*Low*	*Av.*	*High*	*Low*	*Av.*	*High*
Apartment, high-rise	325	175	100	1.0	2.0	4.0	450	400	350	0.8	1.2	1.7	0.5	0.8	1.3	—	—	—
Auditoriums, churches, theaters	15	11	6	1.0	2.0	3.0	400	250	90	—	—	—	—	—	—	1.0	2.0	3.0
Educational facilities	30	25	20	2.0	4.0	6.0	240	185	150	1.0	1.6	2.2	0.9	1.3	2.0	0.8	1.2	1.9
Factories																		
Assembly areas	50	35	25	3.0[b]	4.5[b]	6.0[b]	240	150	90	—	—	—	—	—	—	2.0	3.6	5.5
Light manufacturing	200	150	100	9.0[b]	10.0[b]	12.0[b]	200	150	100	—	—	—	—	—	—	1.6	2.5	3.8
Heavy manufacturing[c]	300	250	200	15.0[b]	45.0[b]	60.0[b]	100	80	60	—	—	—	—	—	—	2.5	4.0	6.5
Hospitals																		
Patient rooms[d]	75	50	25	1.0	1.5	2.0	275	220	165	0.33	0.55	0.67	0.33	0.50	0.67	—	—	—
Public areas	100	80	50	1.0	1.5	2.0	175	140	110	1.0	1.25	1.45	1.0	1.1	1.2	0.95	1.0	1.1
Hotels, motels, dormitories	200	150	100	1.0	2.0	3.0	350	300	220	1.0	1.40	1.5	0.9	1.2	1.4	—	—	—
Libraries and museums	80	60	40	1.0	1.5	3.0	340	280	200	1.0	1.6	2.1	0.9	1.1	1.3	0.9	1.0	1.1
Office buildings[d]	130	110	80	4.0	6.0[b]	9.0[b]	360	280	190	0.25	0.5	0.9	0.25	0.5	0.8	0.8	1.1	1.8
Private offices[d]	150	125	100	2.0	5.8	8.0	—	—	—	0.25	0.5	0.9	0.25	0.5	0.8	—	—	—
Stenographic	100	85	70	5.0[b]	7.5[b]	10.0[b]	—	—	—	—	—	—	—	—	—	0.9	1.3	2.0
Residential																		
Large	600	400	200	1.0	2.0	4.0	600	500	380	0.8	1.2	1.6	0.5	0.8	1.3	—	—	—
Medium	600	360	200	0.7	1.5	3.0	700	550	400	0.7	1.1	1.4	0.5	0.7	1.2	—	—	—
Restaurants																		
Large	17	15	13	1.5	1.7	2.0	135	100	80	1.8	2.4	3.7	1.2	1.6	2.1	0.9	1.1	1.4
Medium							150	120	100	1.5	2.0	3.0	1.1	1.4	1.8	0.9	1.0	1.3

TABLE 6.15 **All-Air HVAC Systems: Approximate Air Quantities and Refrigeration Sizes (*continued*)**

	Internal Gains						*Approximate Requirements*											
	Occupancy (ft^2/person)			*Lights (W/ft^2)*			*Refrigeration (ft^2/ton^a)*			*Air Quantities (cfm/ft^2)*								
										East–South–West			*North*			*Internal*		
Applications	*Low*	*Av.*	*High*	*Low*	*Av.*	*High*	*Low*	*Av.*	*High*	*Low*	*Av.*	*High*	*Low*	*Av.*	*High*	*Low*	*Av.*	*High*
Shopping Centers																		
Beauty and barber shops	45	40	25	3.0[b]	5.0[b]	9.0[b]	240	160	105	1.5	2.6	4.2	1.1	1.7	2.6	0.9	1.3	2.0
Department stores																		
Basement	30	25	20	2.0	3.0	4.0	340	285	225	—	—	—	—	—	—	0.7	1.0	1.2
Main floor	45	25	16	3.5	6.0[b]	9.0[b]	350	245	150	—	—	—	—	—	—	0.9	1.4	2.0
Upper floors	75	55	40	2.0	2.5	3.5[b]	400	340	280	—	—	—	—	—	—	0.8	1.0	1.2
Dress shops	50	40	30	1.0	2.0	4.0	345	280	185	0.9	1.2	1.6	0.7	1.0	1.4	0.6	0.8	1.1
Drug stores	35	23	17	1.0	2.0	3.0	180	135	110	1.8	2.3	3.0	1.0	1.4	1.8	0.7	1.0	1.3
Variety stores	35	25	15	1.5	3.0	5.0	345	220	120	0.7	1.4	2.0	0.6	1.2	1.6	0.5	0.9	1.1
Hat shops	50	43	30	1.0	2.0	3.0	315	270	185	1.0	1.3	1.9	0.7	1.0	1.5	0.6	0.8	1.2
Shoe stores	50	30	20	1.0	2.0	3.0	300	220	150	1.2	1.6	2.1	1.0	1.4	1.8	0.8	1.0	1.2
Malls	100	75	50	1.0	1.5	2.0	365	230	160	—	—	—	—	—	—	1.1	1.8	2.5
Refrigeration for central heating and cooling plant																		
Urban districts							475	380	285									
College campuses							400	320	240									
Commercial centers							330	265	200									
Residential centers							625	500	375									

Source: Reprinted by permission from *The ABC's of Air Conditioning,* © 1975, Carrier Corporation.

[a]Refrigeration loads are for entire application.

[b]Includes other loads that are ordinarily expressed in watts per square foot.

[c]Here, air quantities assume that supplementary means are used to remove excessive heat.

[d]Induction (air & water) systems are assumed for hospital patient rooms and for the perimeter zones of office buildings.

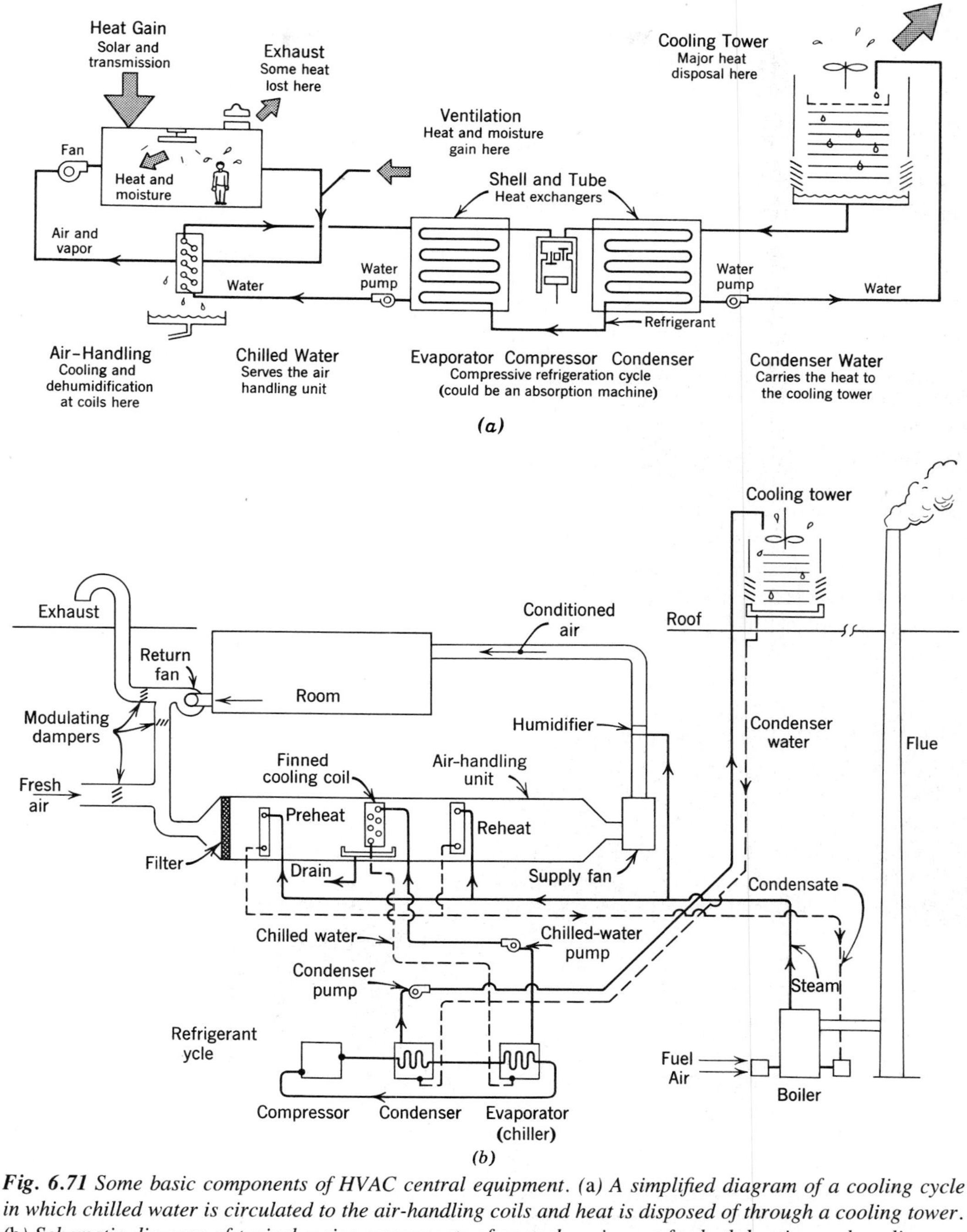

***Fig. 6.71** Some basic components of HVAC central equipment.* (a) *A simplified diagram of a cooling cycle in which chilled water is circulated to the air-handling coils and heat is disposed of through a cooling tower.* (b) *Schematic diagram of typical major components of central equipment for both heating and cooling.*

(a) Boilers. These devices produce the heat for the recirculating hot water system used for building heating. The type of boiler selected depends on the size of the heating load, the heating fuels available, the desired efficiency of operation, and whether single or modular boilers are to be installed. Boiler sizes are commonly stated either in Btu/h of net output or in (gross) horsepower, where

$$\text{boiler horsepower} = \frac{\text{heating load (Btu/h)}}{\text{\% efficiency of boiler} \times 2247 \text{ Btu/h per horsepower}}$$

Efficiency depends partly on the number of passes that the hot gases make through the water—the more passes, the higher the efficiency. It also depends on burner efficiency and on regular maintenance. Finally, efficiency is best when the equipment is operating near its capacity. Table 5.14 can be used as a guide to the expected efficiency of boilers, according to fuel source. Figure 6.72 compares typical boiler types, including two- and three-pass boilers. Note that in ''water tube'' boilers (not shown), the water to be heated is taken through tubes that are surrounded by the boiler's fire. In ''fire tube'' boilers, the hot gases of the fire are taken in tubes that are surrounded by the water to be heated. In ''wet-back'' boilers, the boiler fire is always surrounded by water, beginning at the burner nozzle.

Fossil-fuel-burning boilers need flues for exhaust gases, fresh air for combustion, and required air pollution control equipment. The exhaust gas is usually first taken *horizontally* from the boiler; this horizontal enclosure, or flue, is called the breeching. The *vertical* flue section is called the stack. Guidelines for sizes and arrangements of breeching and stacks are shown in Fig. 6.73. Local codes determine the quantity of air required for combustion; local air pollution authorities set pollution control requirements. As a general rule, combustion air can be supplied in a duct to the boiler, at an average velocity of 1000 fpm. The duct should be large enough to carry at least 2 cfm per boiler horsepower. Furthermore, ventilation air to the boiler room should be provided; preferably, the inlet and outlet should be on opposite sides of the room. Minimum sizes: enough for 2 cfm per boiler horsepower, at a velocity of about 500 fpm.

Space requirements for boilers are summarized in Fig. 6.74, in which multiple boilers are shown. Note that clear space within the room must be provided, so that the tubes of the boiler can be pulled when they must be replaced.

Several types of single boilers are discussed below. The final boiler type discussed, the modular boiler, is preferred for energy conservation.

1. *High-output, package-type steel boiler*. For large buildings that use steam as a primary heating medium, one or several such boilers may be used. Direct use of steam can be seen in Fig. 6.71, supplying preheat and reheat coils and also a humidifying unit. The relative lightness of this boiler type, compared to the older styles with ponderous masonry bases (boiler settings), makes it suitable for use on upper floors of tall buildings. Figure 6.25 shows two such boilers on the 13th floor of the Fox Plaza Building.

2. *Converter, steam to hot water (Fig. 6.75)*. When, in a building that uses primary steam boilers, secondary circuits that use hot water for heating are required, a converter is used. It is considered a heat exchanger. Figure 6.25 shows downfeed steam supply for the two boilers on the 13th floor to two such converters, one for hot water heating in the apartments and one below the garage ceiling for hot water heating in the commercial area. The secondary (hot water) circuits are not detailed in that illustration. A converter may also be used to transfer heat from steam to *domestic* (service) water. Converters are frequently used where central steam supply systems are available, as in large-city downtown areas. The easier, quieter distribution of heat by hot water has largely replaced steam heating distribution trees.

3. *Electric boilers (Fig. 6.76)*. Where electric rates are competitive with those of fossil fuels, electric boilers are sometimes used. Both hot water and steam electric boilers are available. The advantage of electric boilers is the elimination of combustion air, the flue, and air pollution at the building. The disadvantages are the use of a high-grade energy source for a relatively low-grade task and the pollution impact at the electric generating plant.

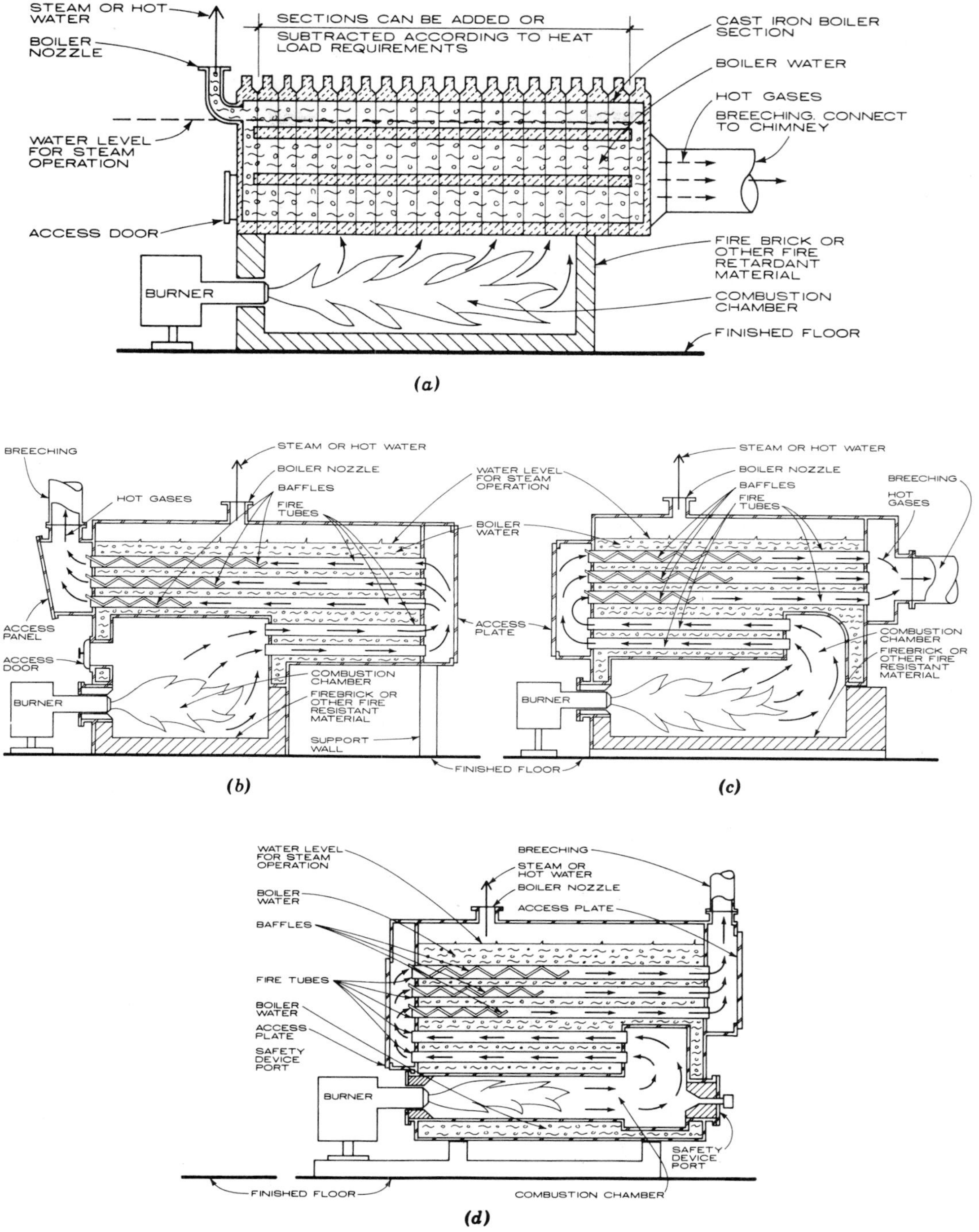

Fig. 6.72 *Comparisons of boiler types.* (a) *Cast iron sectional type.* (b) *Two-pass fire tube.* (c) *Three-pass fire tube.* (d) *Three-pass wet-back scotch marine. (Reprinted by permission from AIA, Ramsey, and Sleeper,* Architectural Graphic Standards, *7th ed., © 1981 by John Wiley & Sons.)*

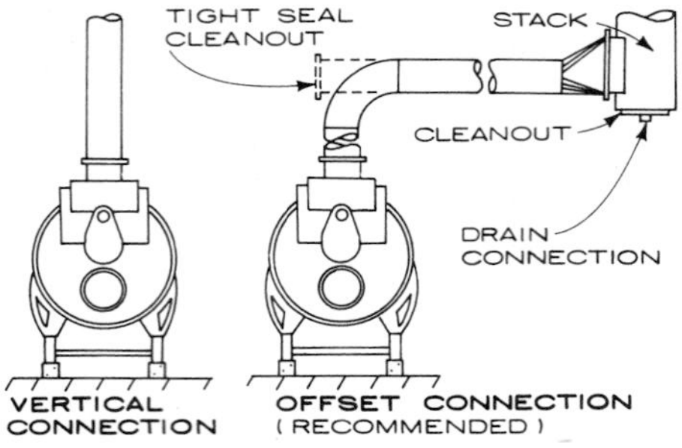

STACK DIAMETER—SINGLE BOILER VENT OR STACK

BOILER HORSE-POWER	STACK DIAMETER (IN.)	A (IN.)	B (IN.)	C (IN.)
15-20	6	15	15	12
25-40	8	20	20	16
50-60	10	25	25	20
70-100	12	30	30	24
125-200	16	40	40	32
250-350	20	50	50	40
400-800	24	60	60	48

STACK DIAMETER—MULTIPLE BOILERS: COMMON BREECHING AND STACK

BOILER HORSE-POWER	MINIMUM STACK DIAMETER (IN.)					
	NUMBER OF BOILERS					
	2		3		4	
	100 FT	200 FT	100 FT	200 FT	100 FT	200 FT
25-40	11	12	13	14	14	16
50-60	13	14	15	16	17	18
70-100	16	17	19	20	21	23
125-200	21	22	24	26	28	30
250-350	26	28	32	34	34	40
400-600	32	34	38	40	42	46

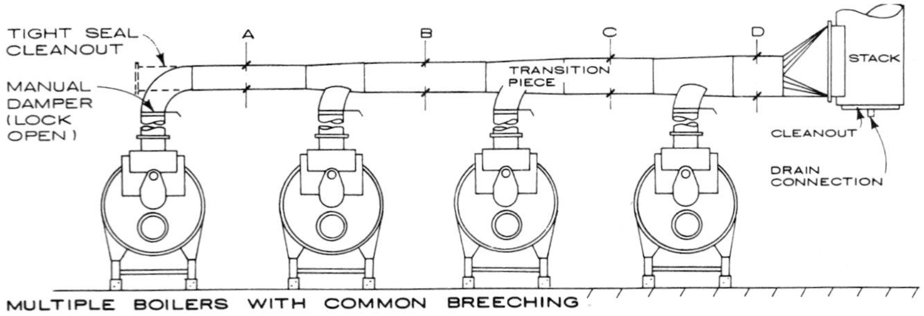

BREECHING DIAMETER—SINGLE AND MULTIPLE BOILERS

BOILER HORSE-POWER	MINIMUM BREECHING DIAMETER (IN. OD)			
	A (IN.) 1 BOILER	B (IN.) 2 BOILERS	C (IN.) 3 BOILERS	D (IN.) 4 BOILERS
15-20	6	8	9	9
25-40	8	10	11	12
50-60	10	12	14	15
70-100	12	15	17	18
125-200	16	20	22	24
250-350	20	25	28	30
400-600	24	30	33	36
700-800	24	34	38	42

Note: Stack diameter should be larger than breeching diameter.

***Fig. 6.73** Breeching and stack guidelines for fossil-fuel-fired boilers. (Reprinted by permission from AIA, Ramsey, and Sleeper,* Architectural Graphic Standards, *7th ed., © 1981 by John Wiley & Sons.)*

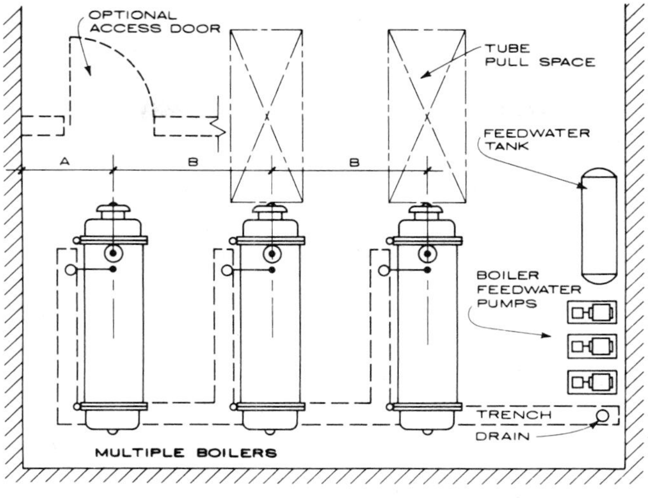

BOILER ROOM SPACE REQUIREMENTS

BOILER HP	15-40	50-100	125-200	250-350	400-800
Dimension A	5′-9″	6′-6″	6′-10″	7′-9″	8′-6″
Dimension B	7′-5″	8′-9″	9′-7″	11′-9″	14′-3″

Fig. 6.74 *Boiler room space requirements. Dimension* A *includes an aisle of 3 ft, 6 in. between the boiler and the wall. Dimension* B *between boilers, includes an aisle of at least 3 ft, 6 in. (up to 5 ft for the largest boilers). (Reprinted by permission from AIA, Ramsey, and Sleeper,* Architectural Graphic Standards, *7th ed., © 1981 by John Wiley & Sons.)*

4. *Modular boilers (Fig. 6.77).* The primary advantage of modular boilers is efficiency. Boilers achieve maximum efficiency when they are operated continuously, at their full rated fuel input. The single boilers discussed above operate this way only under outside design conditions, which by definition occur, at most, during 5% of a normal winter. In a modular boiler design, each section is run independently, Therefore, only one section need be fired for the mildest heating needs; as the weather gets colder, more sections are gradually added. Because each section operates continuously at full rated fuel input, efficiency is greatly increased (see Fig. 6.78). Each module, being rather small, requires little time to reach useful temperature and (unlike the larger, single boilers) does not waste a lot of heat as it cools down. Thus, modular boilers usually produce a 15 to 20% fuel savings for the heating season, relative to single boilers. Their other advantages include ease of maintenance (one module can be cleaned while others carry the heating load) and small size (allowing easy installation and replacement in existing buildings).

(b) Chillers. These devices remove the heat gathered by the recirculating chilled water system as it cools the building. The selection of chillers depends largely on the fuel source and the total cooling load. Chillers include absorption, or centrifugal, and reciprocating machines. Where central steam is available, or high-temperature water (from solar collectors, as waste heat from an industrial process, etc.), the *absorption chiller* (Fig. 6.79) is attractive. This device uses the absorptive refrigeration cycle (explained in Fig. 6.33), requiring approximately 18 lb of steam at 14 lb of pressure to produce a ton (12,000 Btu/h) of cooling. In general, absorption equipment is less efficient than compressive refrigeration cycle equipment, although a "cheap" or even "free" heat source to power the cycle can rapidly overcome efficiency disadvantages. Absorption machines have fewer moving parts (and therefore require less

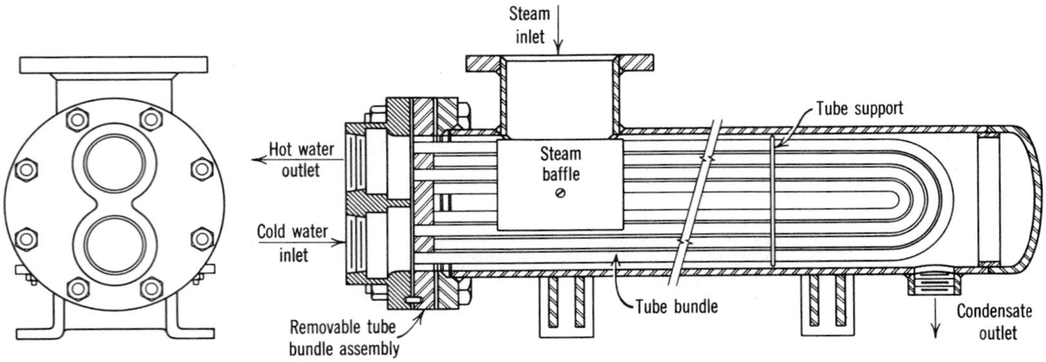

(*a*) Section illustrating the principle of heat transfer from steam to water.

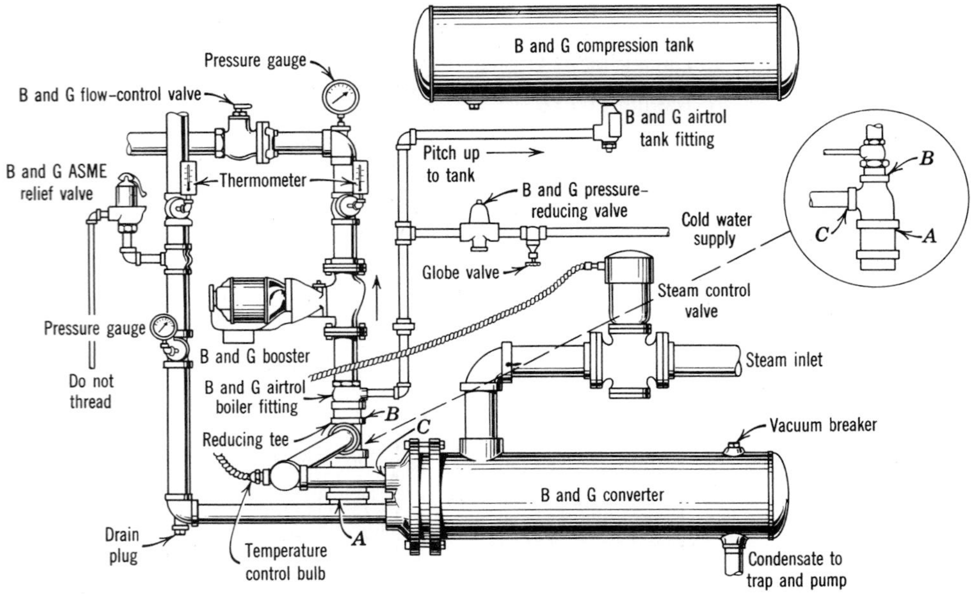

(*b*) A converter connected to steam supply and equipped with all devices necessary for a complete hot water heating system.

Fig. 6.75 *Conversion unit that transfers heat from steam to hot water. (Courtesy of ITT Bell and Gossett.)*

maintenance) and are generally quieter than compressive cycle equipment.

The compressive refrigeration cycle (explained in Fig. 6.32) is used in the other two types of chillers. Larger units of this type are *centrifugal chillers* (Fig. 6.80), whose compressors either can be driven by an electric motor or can utilize a turbine driven by steam or gas. (When a steam-driven turbine is used, the exhaust steam is often used to run an auxiliary absorption cycle machine. These two devices make an efficient combination, and the steam plant that supplied them in summer can supply heating in winter.) Centrifugal chillers usually require about 1 hp/ton (0.75 kW, or 10 ft^3 gas, or about 15 lb of steam). These large chillers usually require a cooling tower.

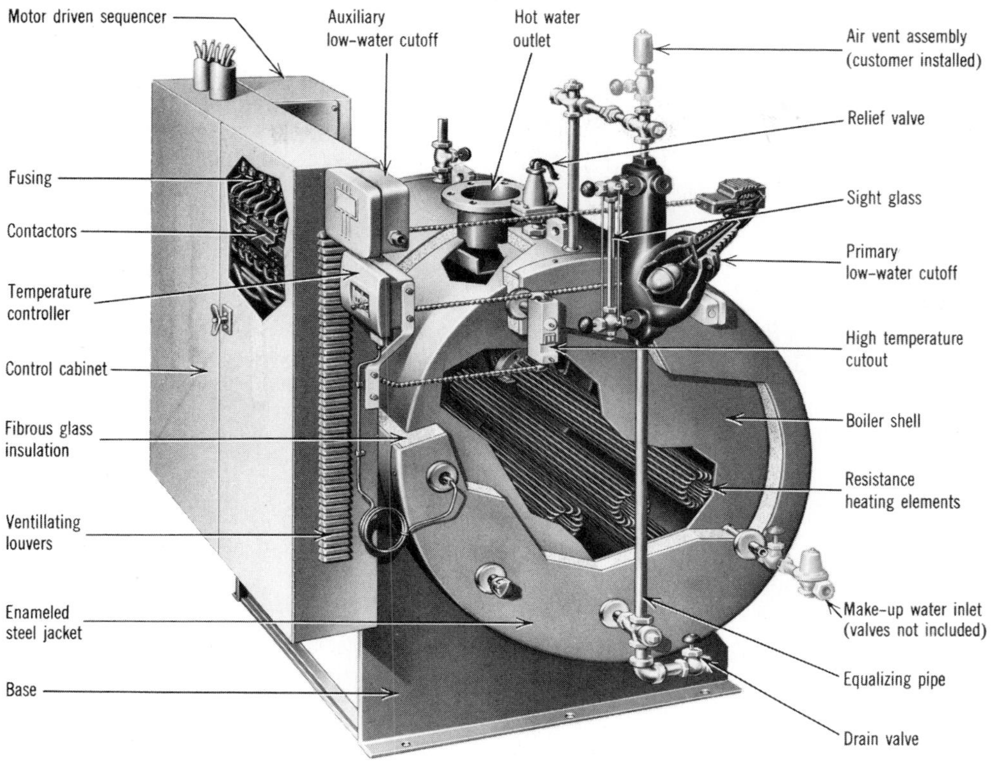

Fig. 6.76 *Electric hot water boiler, available in ratings from 180 to 1800 kW (614,160 to 6,141,600 Btu/h). Similar boilers with appropriate fittings are available for steam systems.*

Smaller compressive-cycle machines are called *reciprocating chillers* (Fig. 6.81). Usually electrically driven, they are often combined with an air-cooled heat rejection process, rather than a cooling tower. This makes them a closer relative of the smaller direct refrigerant machines introduced in Section 6.7.

Chilled water is usually supplied at between 40 and 55 F. When the chilled water is supplied cold and returns much warmer, the large rise in temperature reduces the initial size (cost) of equipment and increases the efficiency (thereby reducing operating cost as well). Water treatment may be needed for chilled water, to control corrosion or scaling.

Typical cooling capacities and space requirements of chillers are shown in Fig. 6.82. Each refrigeration machine in this illustration requires two pumps—one for the chilled water (to cool the building) and one for condenser water (to deal with reject heat). Typically, space is provided for future chiller additions, which may be required by building expansion and/or by higher internal gains from as-yet-uninstalled equipment, such as computer terminals within offices.

(c) Condensing Water Equipment. With chillers, there must be a way to reject the heat that is removed from the recirculating chilled water system. Reject heat is taken care of by the condensing water system, which serves the condensing process within refrigeration cycles. For larger buildings, the condensing water requirement is most likely to be met by a *cooling tower*.

The cooling tower's place within the overall equipment layout was shown in Fig. 6.71*b*; a more

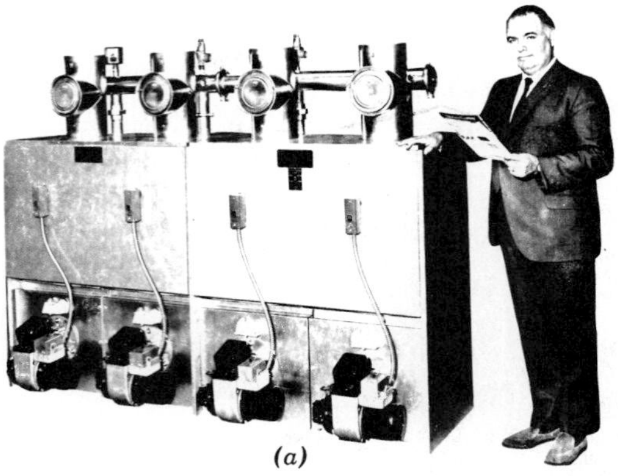

(a)

Fig. 6.77 Modular boilers. (a) A bank of four modules—total input, 1.5 million Btu/h. (b) Details of one module. (c) Schematic of flow conditions in mild weather, with only one module in operation.

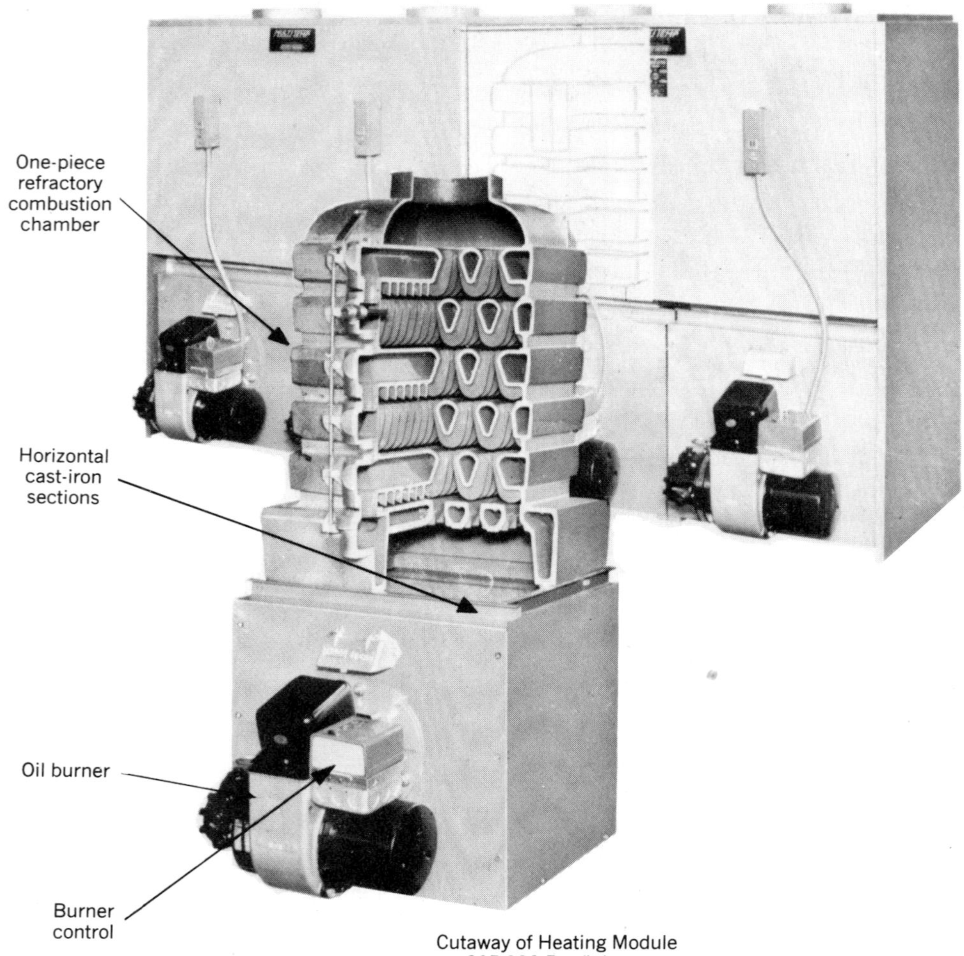

Cutaway of Heating Module
385,000 Btu/h input
Factory assembled with burners and all oil controls
20″ × 32″ × 48½″ high

Shipping weight:	644 lb	Floor loading:	179 lb/ft.^{2}
Water content:	7.25 gal	Pressure drop ($\Delta T = 20F°$):	1.7′ WC
Fire side surface:	49 ft.^{2}	Water side surface:	42 ft.^{2}
Horsepower:	9.1 hp	Pressure rating:	100 psi ASME

(b)

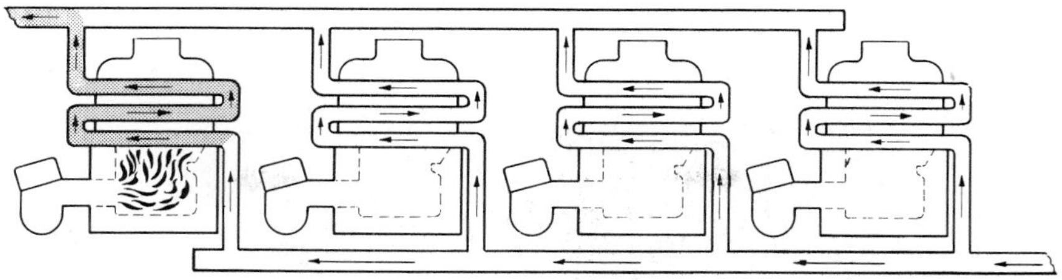

SCHEMATIC FLOW DIAGRAM
For 4-step 1,540,000 Btu/h Input Multi-Temp (MO-1540)

(c)

Fig. 6.77 *(continued)*

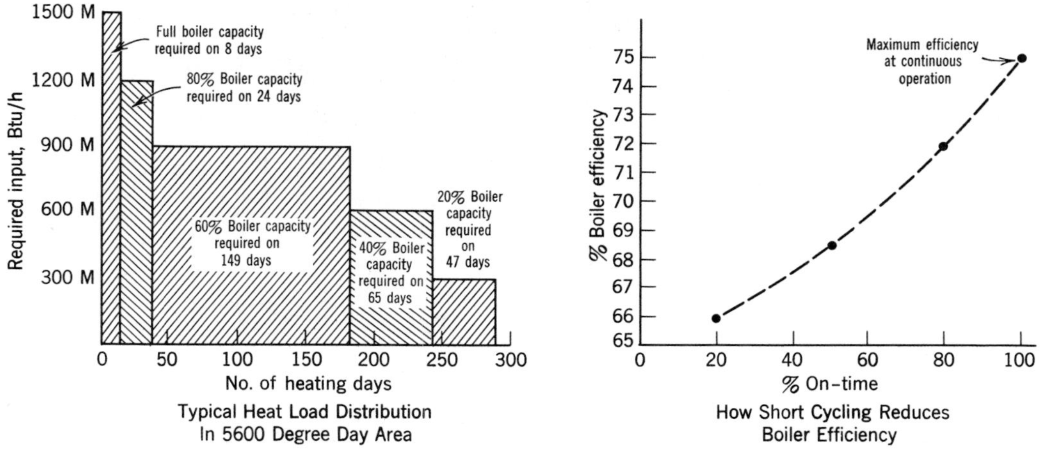

Fig. 6.78 *One large boiler versus many small ones. Whether in mild weather with few modules working, in cooler weather with several working, or in extreme winter weather with all working, the modules actually operating contribute with minimal short-cycling (right-hand diagram). Short-cycling of a single large boiler could drop the general efficiency to the lower levels of the 66–75% efficiency range.*

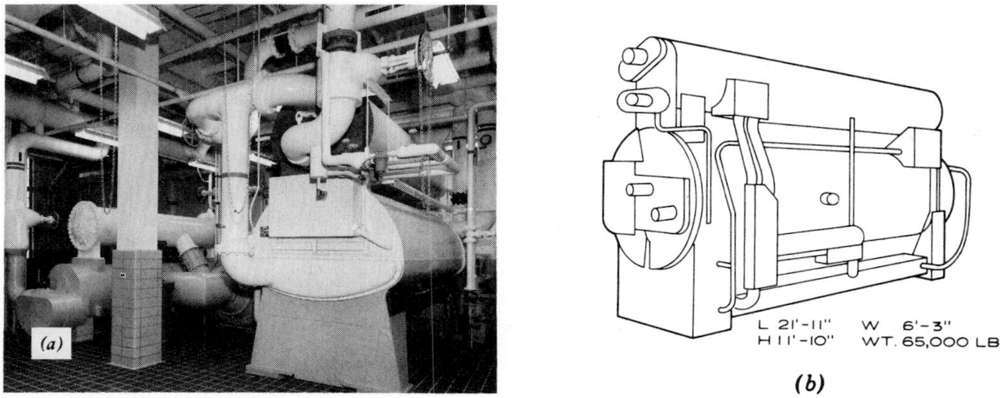

Fig. 6.79 (a) *An absorption chiller, used for producing chilled water. The Carrier Corporation. (Courtesy of Ingersoll-Rand.)* (b) *Two-stage absorption chiller utilizing steam, producing from 200 to 600 tons of cooling. (Reprinted by permission from AIA, Ramsey, and Sleeper,* Architectural Graphic Standards, *7th ed. © 1981 by John Wiley & Sons.)*

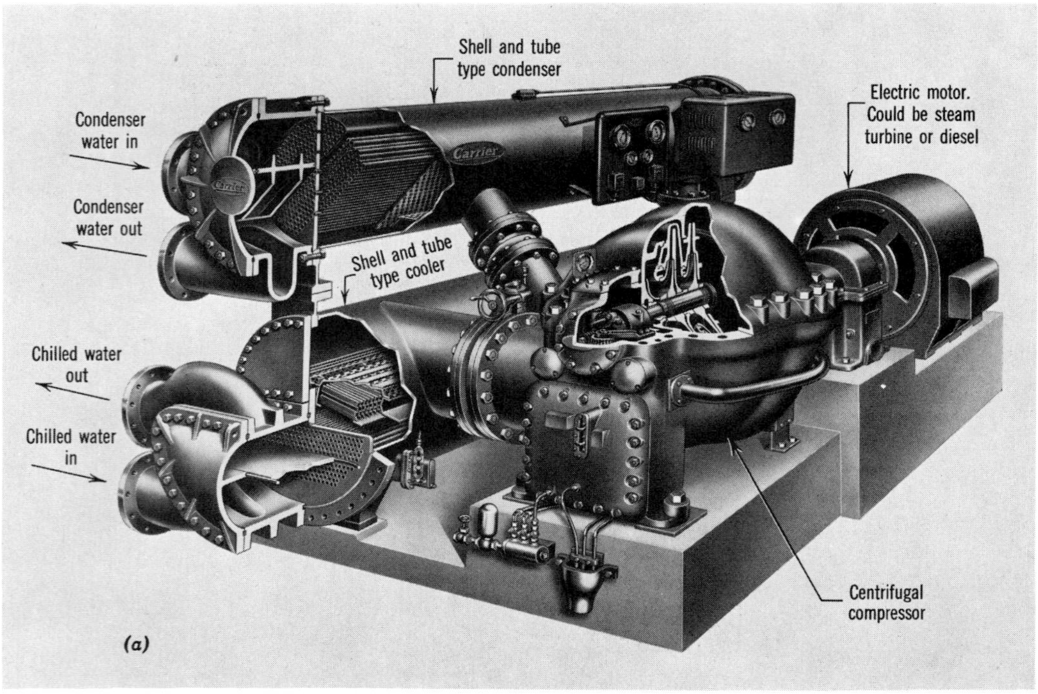

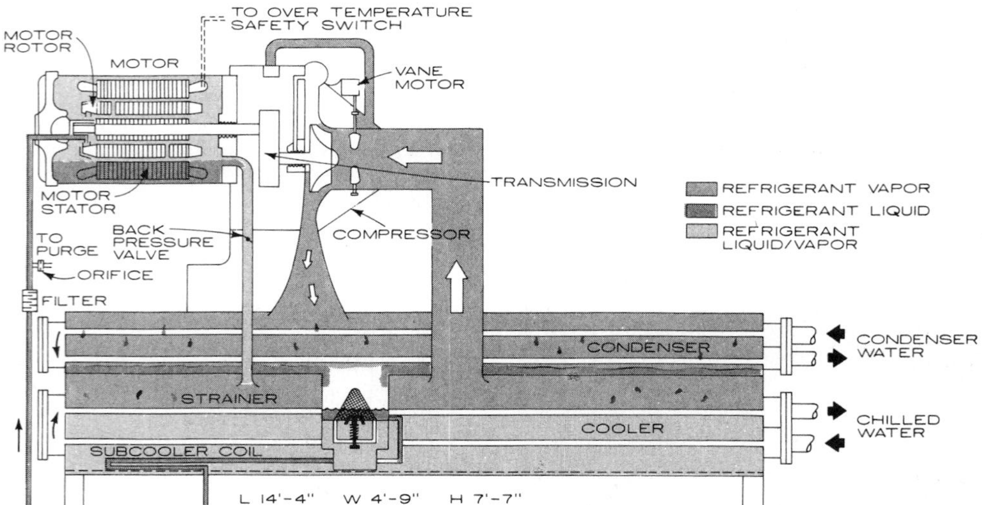

(b)

Fig. 6.80 (a) *A centrifugal chiller—a machine of large capacity that uses the compressive refrigeration cycle. (Courtesy of The Carrier Corporation.)* (b) *Centrifugal chiller with a flooded cooler and condenser within a single outer shell. This low-pressure unit typically produces 100–400 tons of cooling. (Reprinted by permission from AIA, Ramsey, and Sleeper,* Architectural Graphic Standards, *7th ed., © 1981 by John Wiley & Sons.)*

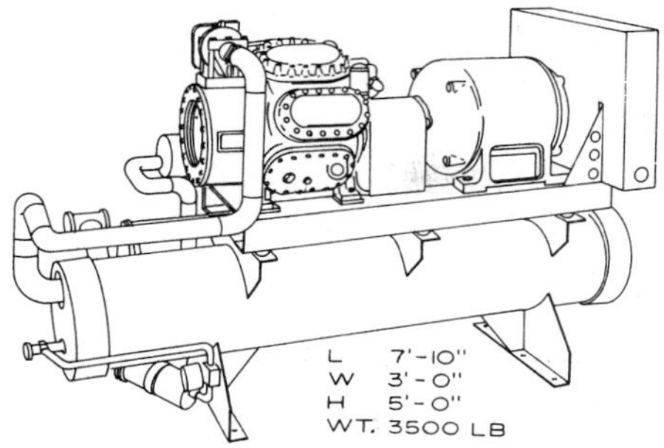

Fig. 6.81 *A reciprocating chiller—a small-capacity machine that uses the compressive refrigeration cycle. Typical at sizes of less than 200 tons of cooling. (Reprinted by permission from AIA, Ramsey, and Sleeper,* Architectural Graphic Standards, *7th ed.,* © *1981 by John Wiley & Sons.)*

detailed guide to sizes and types is given in Fig. 6.83. The object is to maximize the surface area contact between outdoor air and the heat condensing water. In crossflow towers, fans move air horizontally through water droplets and wet layers of fill (or packing), while in counterflow towers, fans move the air up as the water moves down.

Cooling towers create a special—and usually unpleasant—microclimate. They demand huge quantities of outdoor air, which they make considerably more humid. In cold weather, they can produce fog. They are typically very noisy—a natural consequence of forced-air motion. The condensing water they evaporate (between 1.6 and 2 gph per ton of refrigeration) must be replaced, which is done automatically. The steady evaporation and exposure to the outdoors under hot and humid conditions spells trouble for the condensing water: controls for scaling, corrosion, and algae growth are especially important.

Although it is tempting to try to block the noise of cooling towers with solid barriers, it is critical that noise control not interfere with air circulation. The manufacturer's recommended clearances to solid objects near cooling towers must be consulted before a tower is enclosed in any way. The roof is a favorite location for cooling towers; at least nine such devices can be seen on the roof of the Chase Manhattan Building, shown in Fig. 9.32 (p. 547).

When fouling of the condensing water system cannot be tolerated, an alternative approach, called the *closed-circuit evaporative cooler,* is taken. Its schematic operation is described in Fig. 6.84. While the condenser water is protected within an always-closed loop, a separate body of water is recirculated through the cooler, with steady evaporation and attendant problems.

(d) Energy Conservation Equipment. One big advantage of central equipment rooms is the opportunity they present for energy conservation. Regular maintenance is simplified when all the equipment lives in a generous space kept at optimum conditions; with regular maintenance comes increased efficiency of operation. Another conservation opportunity is that of heat transfer between various machines, or between distribution trees, where one's waste meets another's need.

Within equipment, heat transfer can occur in a *boiler flue economizer,* through which the hot gases in a boiler's stack are passed for use in preheating of the incoming boiler water (Fig. 6.85). For cooling equipment, *dual-condenser chillers* (Fig. 6.86) can choose whether to reject their heat to a cooling tower (via the heat rejection condenser) or to building heating (via the heat recovery condenser).

There are numerous methods for heat transfer between distribution trees—especially between building exhaust air and fresh air. When these two airstreams are rather far apart, a set of *runaround coils* (Fig. 6.87) can be used. This circulating heat-transfer fluid usually contains antifreeze; it provides simple sensible heat transfer, with no restrictions on exhaust and intake location. No contamination of intake air by exhaust air is likely. The efficiency of such coils runs between 50 and 70%, and they are available in modular sizes up to 20,000 cfm.

The *heat pipe* (Fig. 6.88) also transfers sensible heat, but the airstreams must be adjacent. Within the heat pipe, a charge of refrigerant spends its life alternately evaporating, condensing, and migrating by capillary action through the porous wick. Since the only thing that moves is refrigerant, and it is self-contained, the ideal of no maintenance and long life is obtained. Efficiency is from 50 to 70%; modular sizes are available to 54 in. × 138 in. × 8 rows deep.

Plan

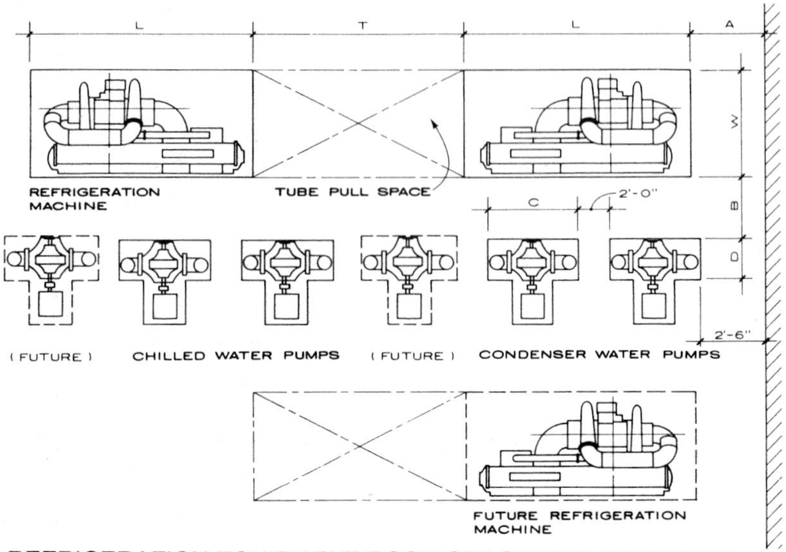

Fig. 6.82 *Chiller room space requirements. Each refrigeration machine is served by two pumps (chilled water and condenser water). (Reprinted by permission from AIA, Ramsey, and Sleeper,* Architectural Graphic Standards, *7th ed., © 1981 by John Wiley & Sons.)*

REFRIGERATION EQUIPMENT ROOM SPACE REQUIREMENTS

EQUIPMENT (TONS)	DIMENSIONS								MINIMUM ROOM HEIGHT
	L	W	HEIGHT	T	A	B	C	D	
RECIPROCATING MACHINES									
Up to 50	9′-0″	3′-0″	4′-0″	6′-0″	2′-6″	3′-6″	4′-0″	3′-0″	9′-0″
50 to 100	9′-6″	3′-0″	4′-6″	6′-0″	2′-6″	3′-6″	4′-0″	3′-6″	9′-0″
CENTRIFUGAL MACHINES									
Up to 120	14′-6″	6′-0″	6′-6″	14′-0″	2′-6″	5′-6″	4′-6″	4′-0″	10′-0″
120 to 225	15′-0″	6′-0″	6′-0″	14′-0″	2′-6″	5′-6″	4′-6″	4′-0″	10′-0″
225 to 350	15′-0″	6′-6″	7′-0″	14′-0″	2′-6″	5′-6″	5′-0″	5′-0″	11′-0″
350 to 550	15′-0″	8′-0″	7′-0″	14′-0″	2′-6″	5′-6″	6′-0″	5′-6″	11′-0″
550 to 750	15′-6″	11′-0″	8′-6″	14′-6″	2′-6″	5′-6″	6′-0″	5′-6″	12′-0″
750 to 1500	18′-6″	13′-6″	10′-0″	16′-0″	2′-6″	5′-6″	7′-6″	6′-0″	14′-0″
ABSORPTION MACHINES									
Up to 200	18′-0″	4′-0″	8′-0″	17′-0″	2′-6″	3′-6″	4′-6″	4′-0″	11′-0″
200 to 450	20′-0″	5′-6″	10′-0″	19′-0″	2′-6″	3′-6″	5′-0″	5′-0″	13′-0″
450 to 550	24′-0″	5′-6″	10′-0″	23′-0″	2′-6″	3′-6″	6′-0″	5′-6″	13′-0″
550 to 750	24′-0″	6′-6″	12′-0″	23′-0″	2′-6″	3′-6″	6′-0″	5′-6″	15′-0″
750 to 1000	27′-0″	7′-6″	12′-6″	26′-0″	2′-6″	3′-6″	7′-0″	6′-0″	16′-0″

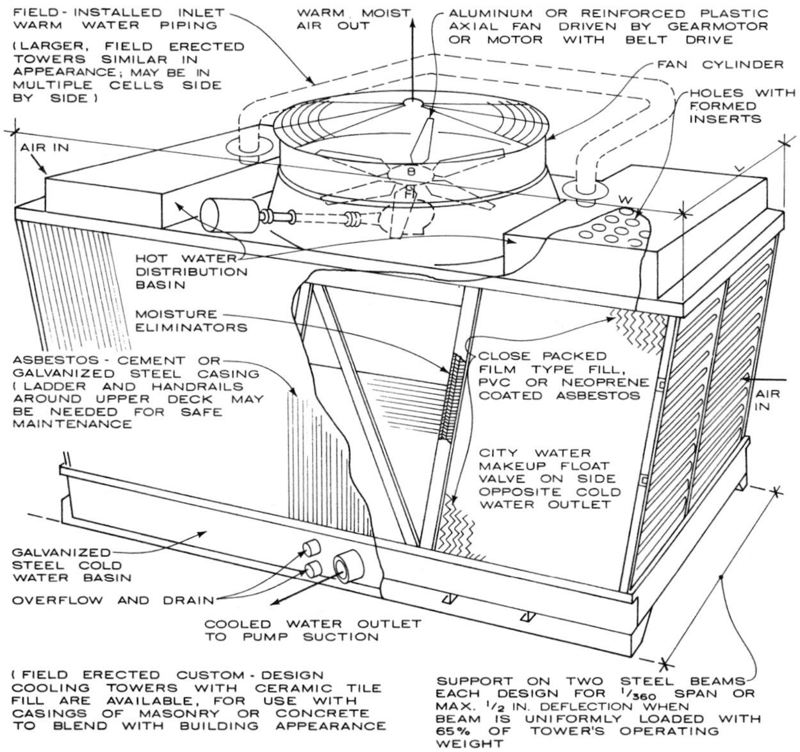

(a)

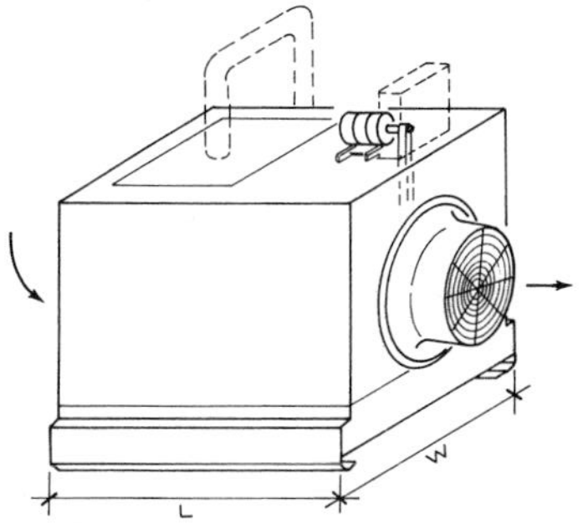

SMALL CROSSFLOW INDUCED DRAFT PACKAGE COOLING TOWER

TONS 3GPM/ TON 95-85-78	OVERALL DIMENSIONS (IN.)			OPERATING WEIGHT (LB)	MOTOR (HP)
	L	W	HT.		
5	62	34	49	750	1/3
25	70	48	78	1,200	3/4
50	82	76	92	2,400	1 1/2
100	100	95	100	5,600	5
200	83	164	117	8,300	10
500	113	232	160	19,300	25

(b)

NOTE
AVAILABLE IN SINGLE MODULES AS SKETCHED, OR BACK-TO-BACK DOUBLE INLET OR END-TO-END

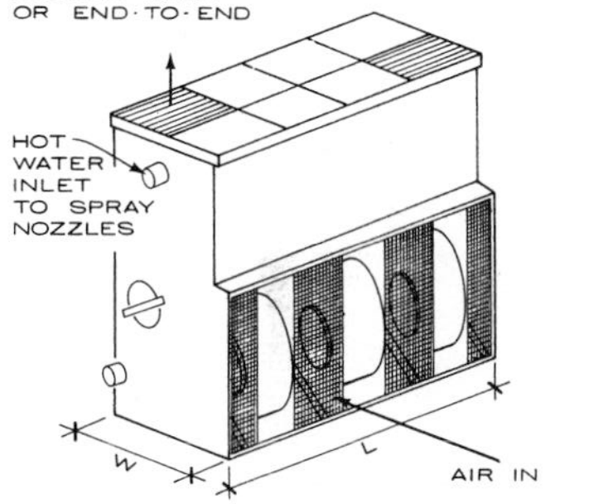

COUNTERFLOW FORCED DRAFT PACKAGE COOLING TOWER

TONS 3GPM/ TON 95-85-78	OVERALL DIMENSIONS (IN.)			OPERATING WEIGHT (LB)	MOTOR (HP)
	L	W	HT.		
20	45	48	78	950	1 1/2
50	81	48	78	1,500	5
200	156	63	161	5,600	25
400	156	115	198	12,000	50
800	300	115	198	24,000	Two-50
1600	448	115	198	47,000	Four-50

(c)

Fig. 6.83 *Cooling towers that serve the condensing water system for large buildings.* (a) *Cutaway view of a large-capacity (200–500 tons) crossflow induced-draft package cooling tower.* (b) *Size ranges for crossflow induced-draft package cooling towers.* (c) *Size ranges for counterflow forced-draft package cooling towers.* Note: *"3 gpm/ton 95-85-78" refers to the cooling capacity in tons, with condensing water flow at 3 gpm per ton of cooling, the condensing water entering the tower at 95 F and leaving at 85 F, and with outside air at no more than 78 F WB. (Reprinted by permission from AIA, Ramsey, and Sleeper,* Architectural Graphic Standards, *7th ed., © 1981 by John Wiley & Sons.)*

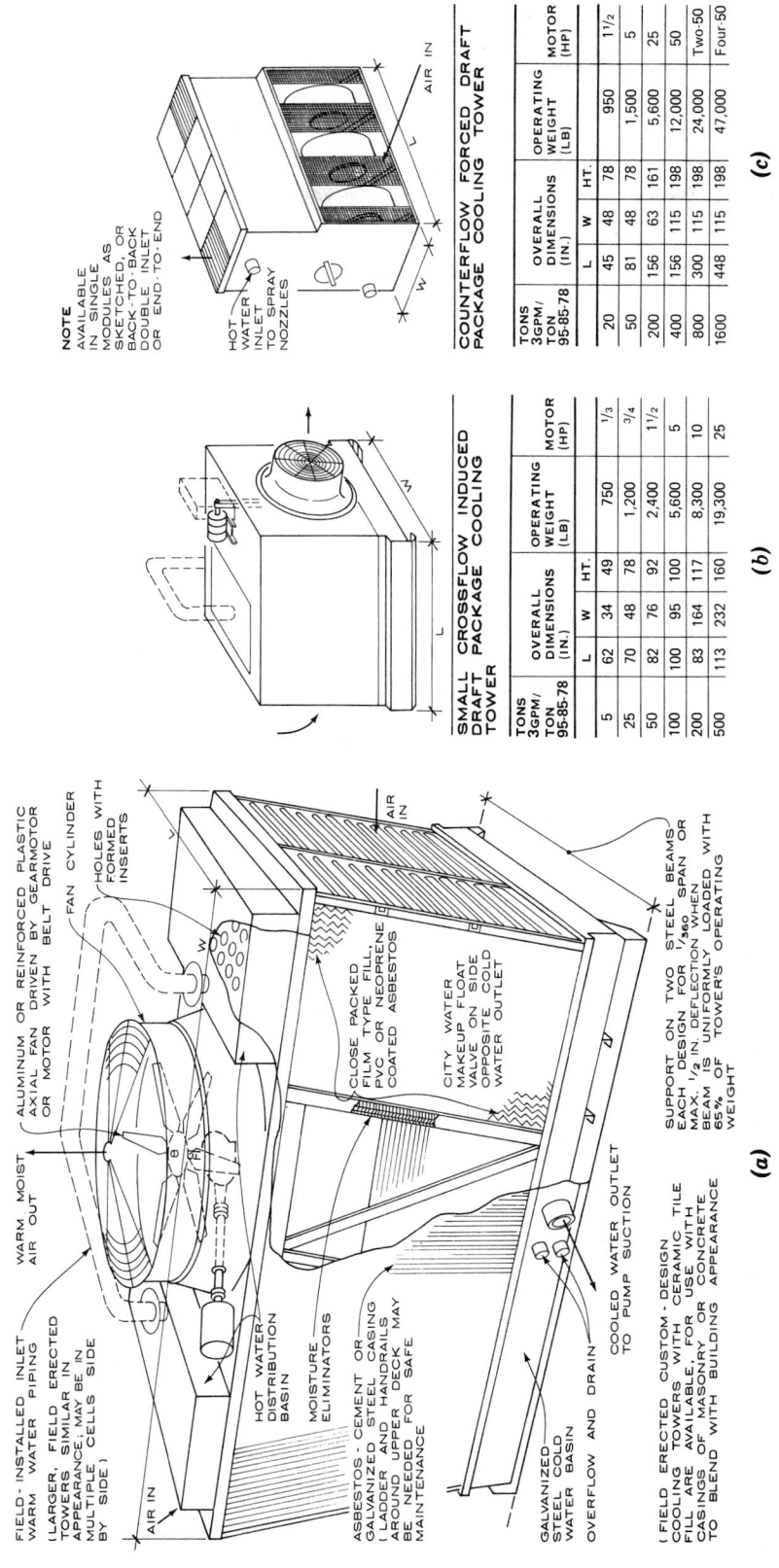

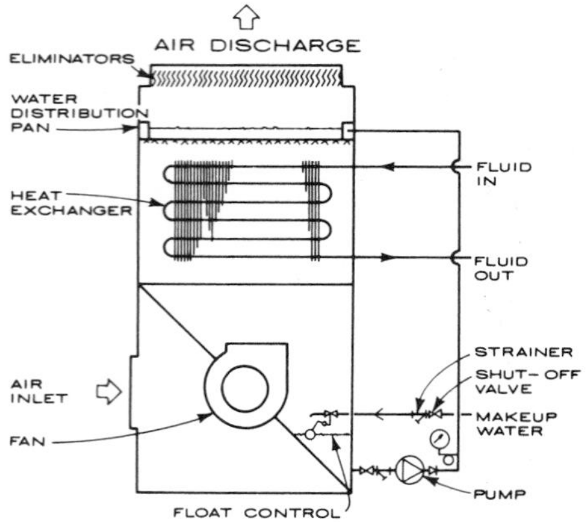

Fig. 6.84 Closed-circuit evaporative coolers, which cool the condensing water system while protecting it from contact with outside air. A self-contained water system is circulated through the evaporative cooler; steady evaporation is replaced by makeup water. (Reprinted by permission from AIA, Ramsey, and Sleeper, Architectural Graphic Standards, *7th ed., © 1981 by John Wiley & Sons.)*

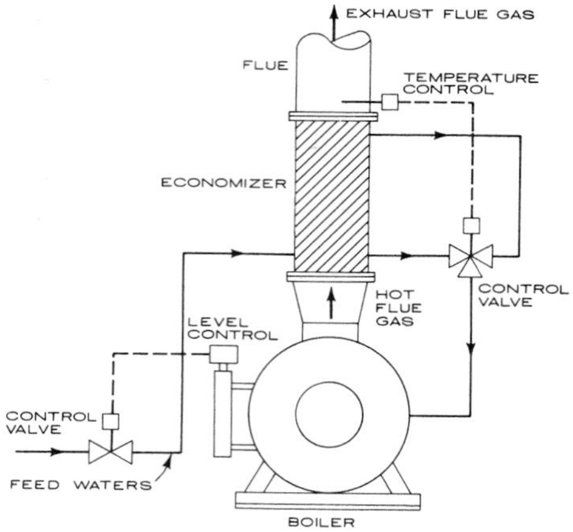

Fig. 6.85 Heat recovery for boilers. Flue gas entering at 500 F leaves the "economizer" at 325 F (a temperature still high enough to prevent condensation in the stack). The heat recovered here is added to incoming boiler water, raising its temperature from 200 to 248 F. (Reprinted by permission from AIA, Ramsey, and Sleeper, Architectural Graphic Standards, *7th ed., © 1981 by John Wiley & Sons.)*

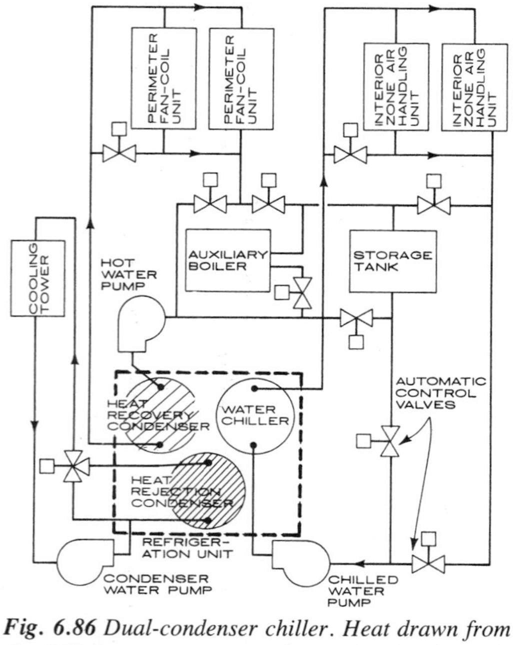

Fig. 6.86 Dual-condenser chiller. Heat drawn from the chilled water system is either rejected to the cooling tower or recovered for use in building heating. (Reprinted by permission from AIA, Ramsey, and Sleeper, Architectural Graphic Standards, *7th ed., © 1981 by John Wiley & Sons.)*

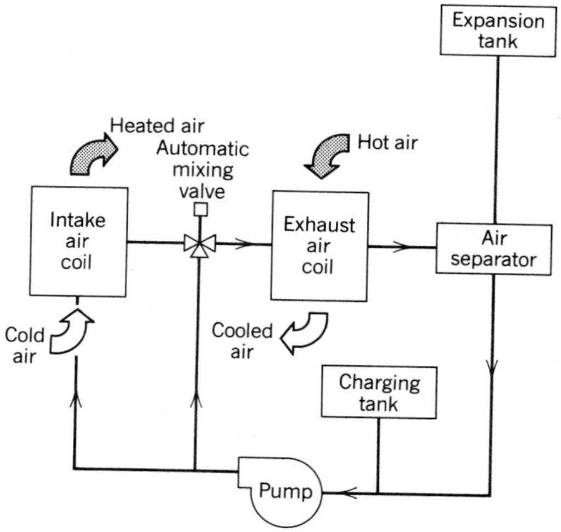

Fig. 6.87 Runaround coils for heat transfer between fresh intake air and stale exhaust air, used where airstreams are in separate locations. (Reprinted by permission from AIA, Ramsey, and Sleeper, Architectural Graphic Standards, *7th ed., © 1981 by John Wiley & Sons.*

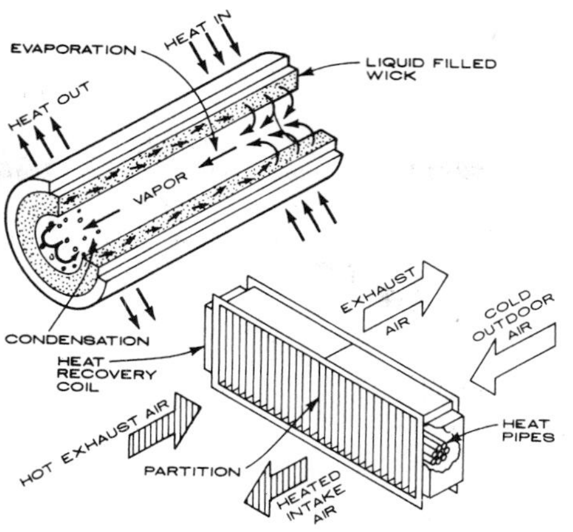

Fig. 6.88 Heat pipe. These self-contained, no-moving-part devices have many applications. Shown here is sensible heat transfer between adjacent fresh-intake and stale exhaust airstreams. (Reprinted by permission from AIA, Ramsey, and Sleeper. Architectural Graphic Standards, *7th ed., © 1981 by John Wiley & Sons.)*

Energy transfer wheels (Fig. 6.89) go further than the two preceding devices, in that they transfer latent as well as sensible heat. In winter, they recover both sensible and latent heat from exhaust air; in summer, they can both cool and dehumidify the incoming fresh air. Seals and laminar flow of air through the wheels prevent mixing of exhaust air and incoming air. A further precaution in the process purges each sector of the wheel briefly, using fresh air to blow away any unpleasant residual effects of the exhaust air on the wheel surfaces. Carryover of exhaust air qualities, except those of heat and moisture, is between 4 and 8% without purging, and less than 1% with purging. Efficiency is from 70 to 80%, and available sizes range up to 144 in. in diameter.

A final conservation opportunity is offered by the *economizer cycle* (Fig. 6.90), which uses cool outdoor air, as available, to ease the burden on a refrigeration cycle as it cools the recirculated indoor air. The economizer cycle can thus be thought of as a central mechanical substitute for the open window; when it is cool enough (about 60 F or below), 100% outside air can be provided. Relative to open windows, this cycle has several advantages: energy-optimizing automatic thermal control, filtering of the fresh air, tempering of the cool outdoor air to avoid unpleasant drafts, and an orderly diffusion of fresh air throughout the building. Its disadvantages are the loss of personal control that windows offer and little awareness of exterior-interior interaction. Economizer cycles are available as options on most direct refrigerant machines (such as single-package rooftop units) and are typically installed for large-building central air supply systems. Buildings with high internal gains (internal load dominated) are particularly good targets for economizer cycles, since they need cooling even when the outside temperature is chilly.

(e) Central Solar Energy Applications. In large buildings, mechanical equipment for solar energy is most commonly used for domestic (service) water heating and/or for powering the absorption refrigeration cycle. (Solar energy can also be passively collected in perimeter zones facing the sun and redistributed to the rest of the building by the central HVAC system.) Solar water heating will be discussed in Chapter 9. Solar cooling, using moderately high-temperature (175 to 195 F) water provided by collectors, is shown in Fig. 6.91. Although the efficiency of such a moderate-temperature absorption unit is rather low (COP 0.5 to 0.7), the cost of the solar heat supplied is usually less (particularly in summer!) than that of the alternative—using electricity to run higher efficiency compressive refrigeration cycles. Simple flat-plate solar collectors can easily deliver 160 F water under summer conditions in most U.S. locations.

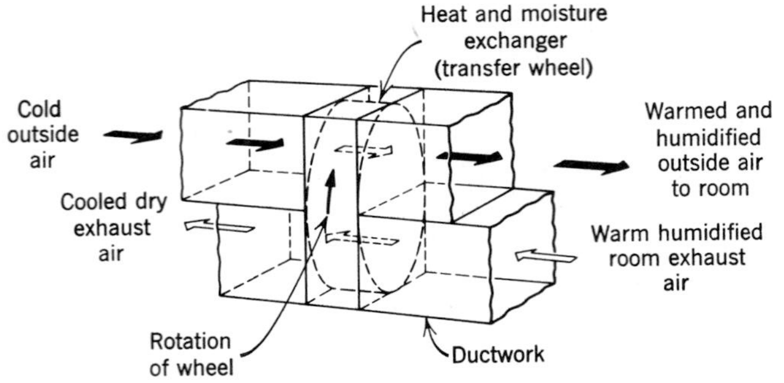

(a) Winter Operation

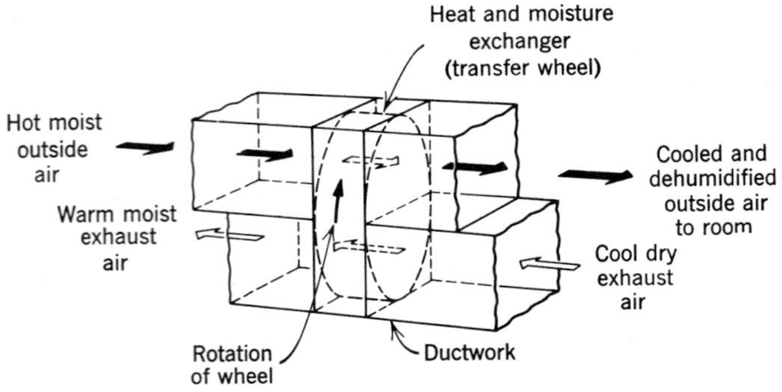

(b) Summer Operation

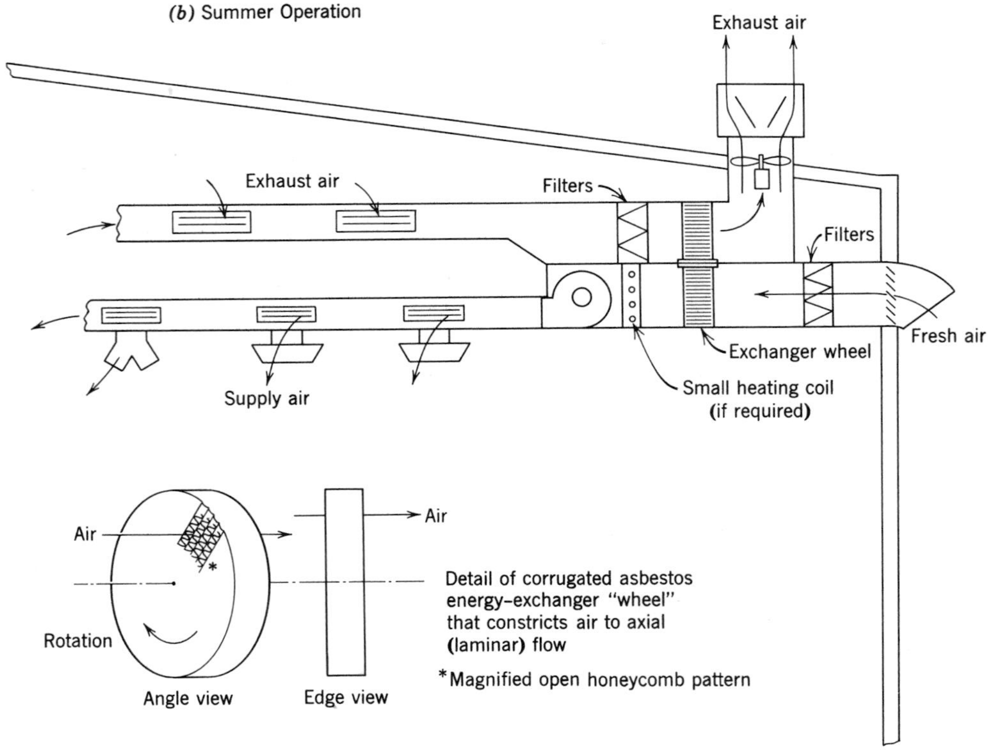

(c)

Fig. 6.89 *Energy transfer wheels. The wheel surface is impregnated with lithium chloride, which absorbs moisture and transfers it to the other airstream* (a,b) *The wheel delivers moist air in winter and dry air in summer.* (c) *Cross section through the wheel and the two airstreams it serves. Exhaust air may be filtered to keep the wheel clean.* (d) *Multiple-unit installation. In this illustration, room exhaust air passes through the upper chambers and incoming fresh air passes through the lower chambers. Wheels rotate at 8–10 rpm.*

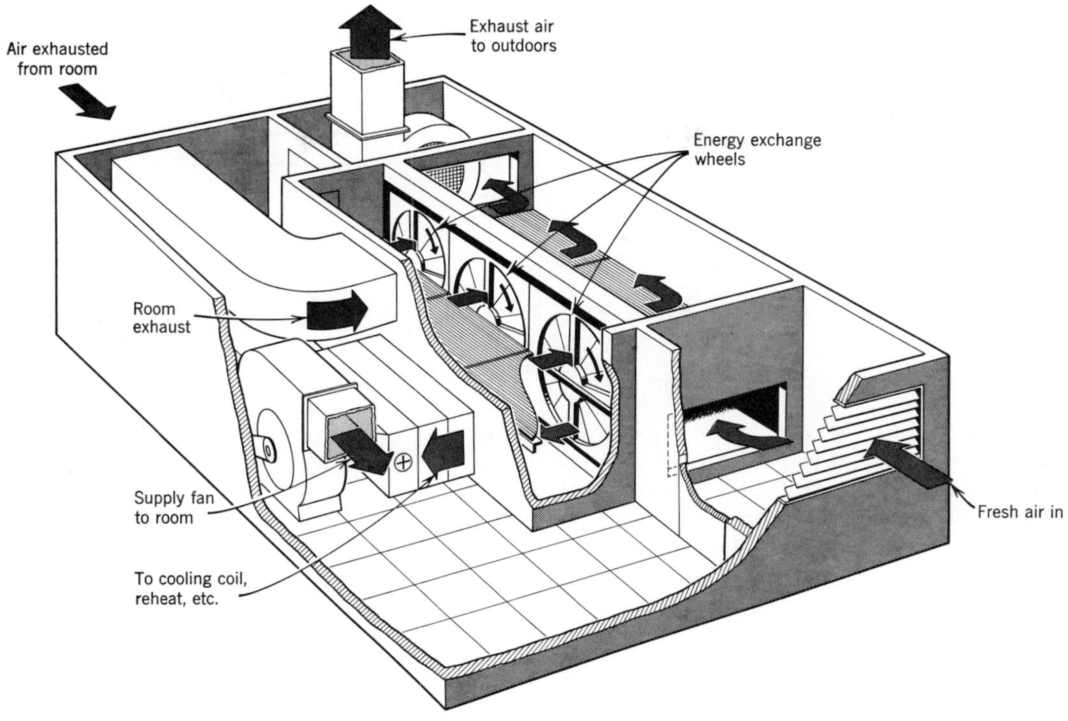

(d)

Fig. 6.89 (continued)

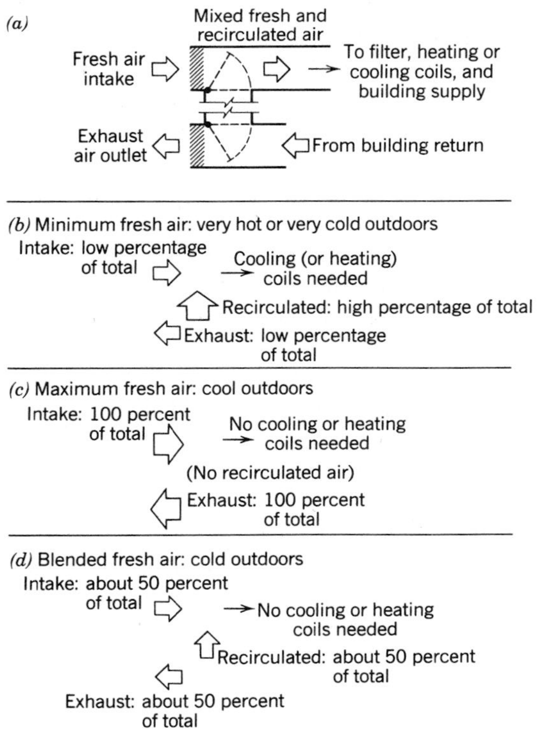

Fig. 6.90 *Economizer cycle.* (a) *The basic relationships between supply and return air: fresh, exhaust, and recirculated air.* (b) *When outside air is hot (or very cold), the economizer cycle is inactive.* (c) *When outside air is cool, it can completely replace recirculated air, making mechanical cooling inactive.* (d) *As outside air gets colder, it can be blended with recirculated air, to delay the need for the heating coil.*

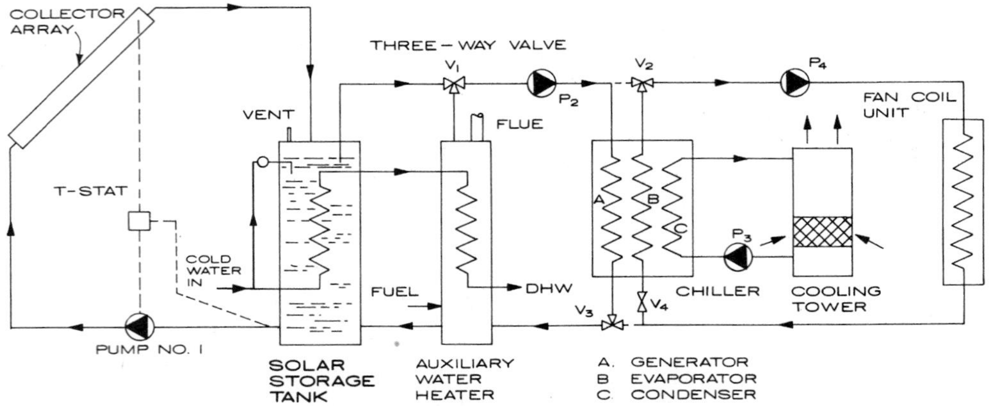

Fig. 6.91 *Solar-heated water for the absorption refrigeration cycle. Moderately hot water from the solar storage tank is supplied to the generator of the absorption cycle (explained in Fig. 6.33). When building heating is needed, solar-heated water (as available) can be supplied directly to the fan-coil unit, by reversing valves V_2 and V_3. (Reprinted by permission from AIA, Ramsey, and Sleeper,* Architectural Graphic Standards, *7th ed., © 1981 by John Wiley & Sons.)*

Parabolic concentrating collectors can deliver higher temperature water under clear sky conditions, producing somewhat higher efficiency in absorption chillers. A schematic of the parabolic collector-absorption chiller system for an office building at Disney World in Orlando, Florida, is shown in Fig. 6.92.

(f) Energy Storage. We commonly experience daily changes from warmer to colder conditions, even in summer. Central storage equipment for large buildings can take advantage of this cycle to increase operating efficiency and save energy.

One common approach to storage is the use of *water storage tanks,* such as those shown in Fig. 6.93. On typical winter days, the total internal heat generated by a large building is somewhat greater than its total need for heating at the perimeter zones. Instead of being thrown away as exhaust air, this surplus heat is captured and stored in large water tanks, from which it can be withdrawn and used on cold winter nights and weekends. In the summer, chillers can work at night, when efficiency is high because cool outdoor air helps the refrigeration cycle reject its heat. By storing the "coolth" produced, less work need be done by chillers during the next day's peak, when electric rates are highest and operating efficiency lowest. Daily cycles can also be used in rock bed storage, as we saw in Sacramento's Bateson Building (see Fig. 5.29).

Another storage approach is offered by *ice storage tanks,* as shown in Fig. 6.94. Because they take advantage of the latent heat of fusion (143.5 Btu/lb of water at 32 F), such units can store far more energy in a given-sized tank than can water (1 Btu/lb of water per °F of temperature change). A comparison of the relative sizes of storage tanks needed for an 18-story office building is shown in Table 6.16. Also, with ice storage there is less undesired mixing of hot and cold water within the tank.

Ice is usually made as a layer around a pipe that carries refrigerant in a closed circuit through the tank. Control of the thickness of this layer is important, because if too little ice is made at night, too little cooling will be available the following day. Considerable savings in electricity costs can be realized by this off-peak (nighttime) refrigeration operation. For a typical ice storage installation, see the Iowa Public Service Building (Fig. 6.117).

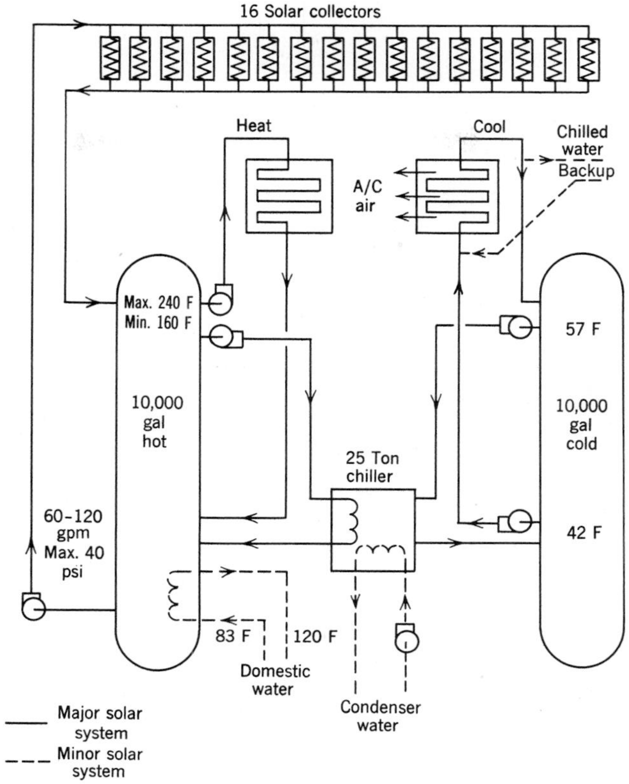

Application: Cooling, heating, hot water hydronic system.
Collector Type: Fixed horizontal parabolic trough mirrored roof with moving absorber.
Collector Manufacturer: AAI Corporation.

Collector Area: 3840 sq ft (356.75 m^2).
Storage Capacity: 10,000-gal, hot water steel tank (37,854 liters) plus 10,000-gal, chilled water steel tank (37,854 liters).
Total Btus: 400 x 10^6 Btu/yr.

(a)

Fig. 6.92 *Solar-powered absorption cooling for an office building at Disney World, Orlando, Florida. (a) Schematic diagram of the system. For clarity, the piping at the 16 collector tubes is shown in simplified form. Actually the water passes through each collector tube twice before returning to the hot water storage tank. (b) The stationary mirrored panels concentrate the sun's heat on the absorber, which is an aluminum extrusion. Above this is swaged a U-shaped copper tube.*

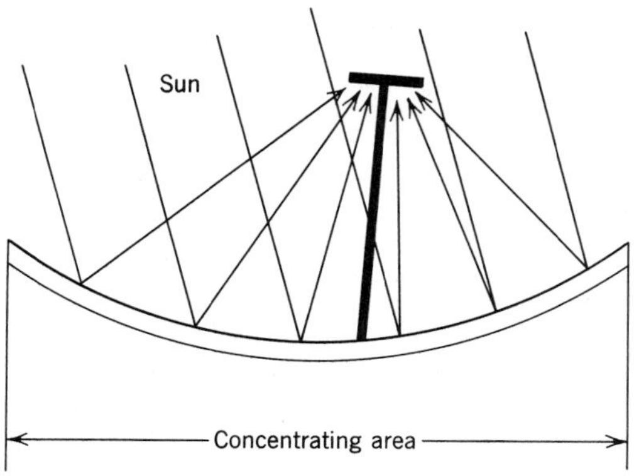

(b)

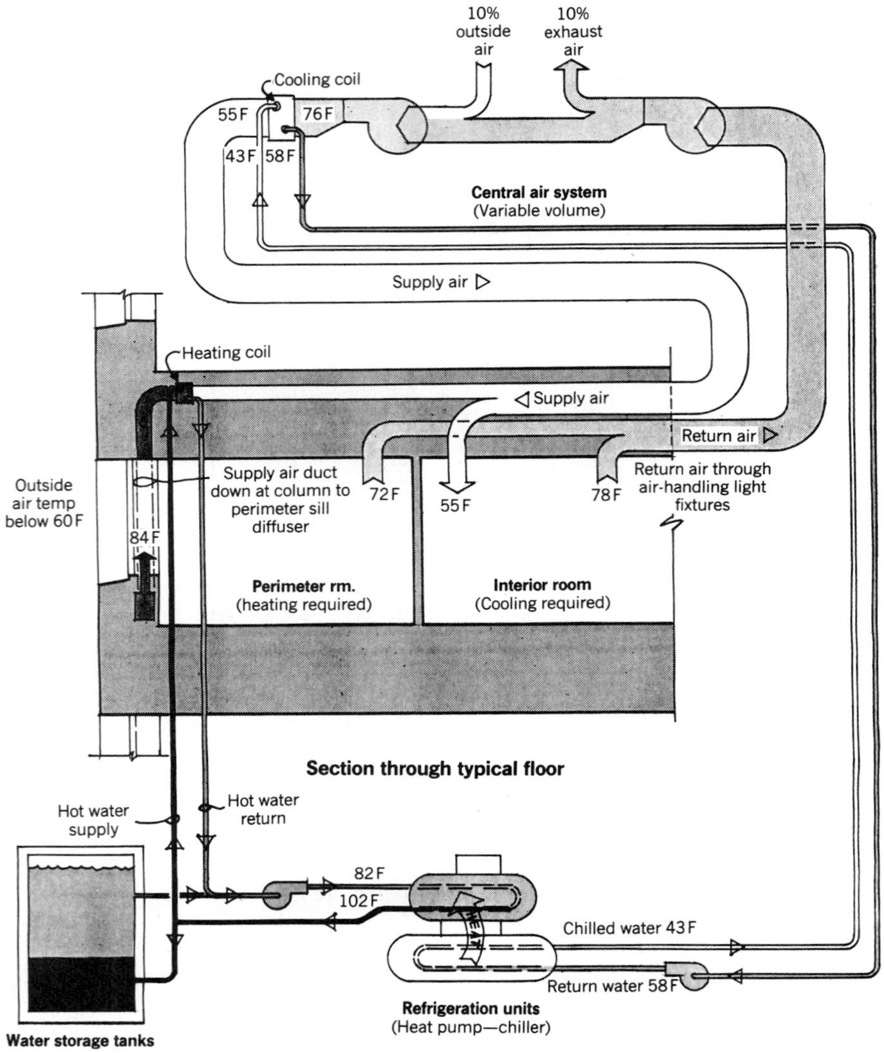

(a)

Fig. 6.93 Water tank heat storage. The 870,000-ft² transportation (Park Plaza, Boston) office building, population 2000, uses a three-compartment insulated concrete tank storing 750,000 gal of water. (a) *At outside air 40 to 50 F, the surplus internal heat is stored in the tanks, rather than rejected as exhaust air.* (b) *By the time outside air is about 50 F, the tanks are fully charged; up to outside 60 F, the economizer cycle (see Fig. 6.90) provides energy-conserving cooling.* (c) *At about 60 F outside, chillers must operate; but working all night when outside air is cooler, enables them to work less by day. Their night production is stored as cold water, available to help out at the following day's peak. Smaller machines and more efficient operation are the result. (Courtesy of Shooshanian Engineering Associates, Inc.)*

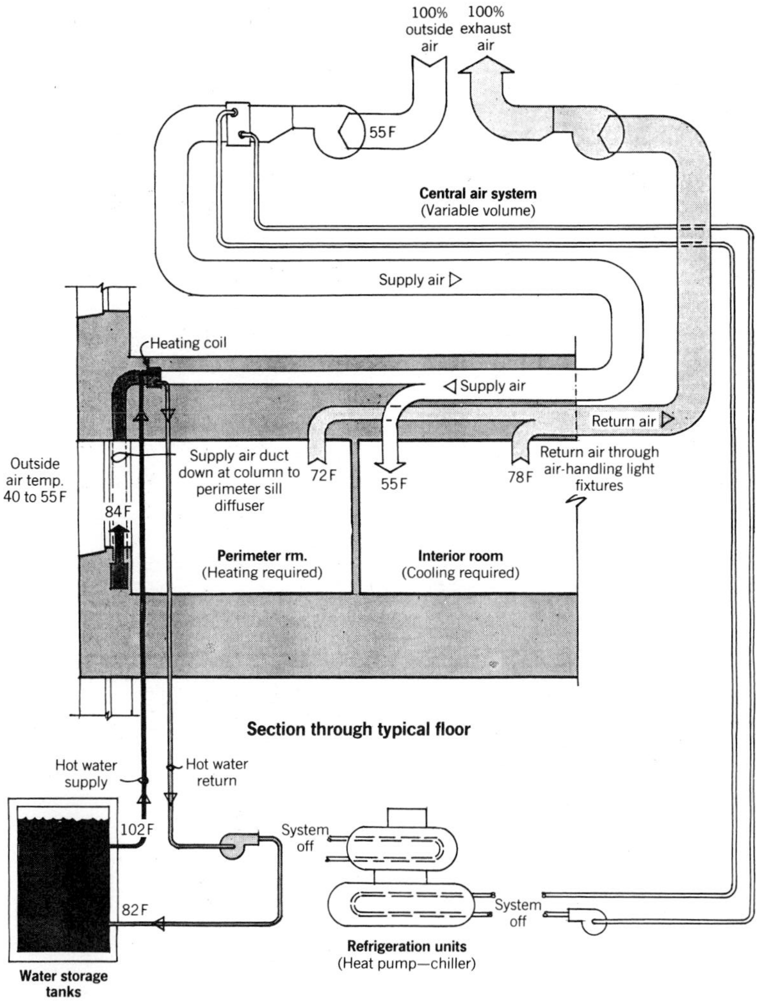

(b)

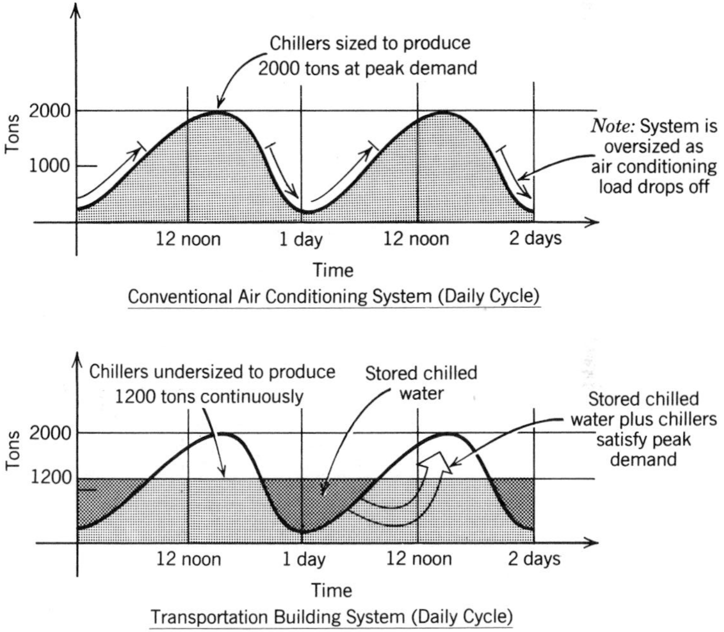

Note: Charging of storage tanks with cold water during off hours provides added cooling media for peak air conditioning needs.

(c)

Fig. 6.93 *(continued)*

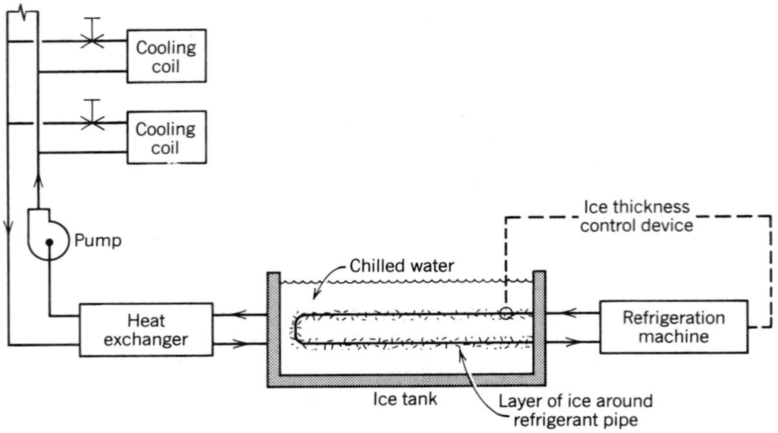

Fig. 6.94 Ice tank heat storage. Chilled water in the ice tank forms a layer of ice around the refrigerant pipe (supplied by the refrigeration machine). As chilled water at about 35 F is removed from the tank, it enters a heat exchanger, so that its temperature is closer to 45 F for distribution to cooling coils throughout the building. (Based on diagrams from Specifying Engineer, *January 1983.)*

TABLE 6.16 **Cooling Storage Comparison**

Eighteen-story office building: 375,000 gross ft^2, 800 tons peak load, 6250-ton h/day at design, 75 million Btu storage needed.

	Water Storage	*Ice Storage*
Btu/lb[a]	15	164
Pounds storage	5×10^6	0.457×10^6
Gallons	599,500	54,800
Storage efficiency	0.90	1.0
Percent ice		66%
Net gallons	666,000	83,030
Cubic feet	89,046	11,101
Size of tank approx. 8 ft deep	80- by 150-ft	30- by 50-ft
Storage ratio, water to ice	8 : 1	

Source: Reprinted by permission from *Specifying Engineer,* January 1983.

[a]Btu/lb based on ΔT of 15F° for water storage and on ΔT of 20.5F° in the melted ice water (added to 143.5 Btu/lb at fusion).

(g) Air-Handling Equipment. A given HVAC system may have many variations. In some variations, all the air is passed through one central equipment room. In others, air handling may be done in many separate, smaller rooms, whereas central heating and cooling will require only one equipment room. The guidelines for air-handling equipment for either case are shown in Fig. 6.95. Total air quantities may be estimated from Table 6.15 or obtained more precisely from cooling load calculations.

Air filtration and odor removal comprise an important part of the process; characteristics of various filters are shown in Table 6.17. Filters are selected on the basis of the degree of air cleaning desired, first cost, and ease of maintenance.

Air filters (Fig. 6.96) include both disposable and renewable, or cleanable, types, in either dry

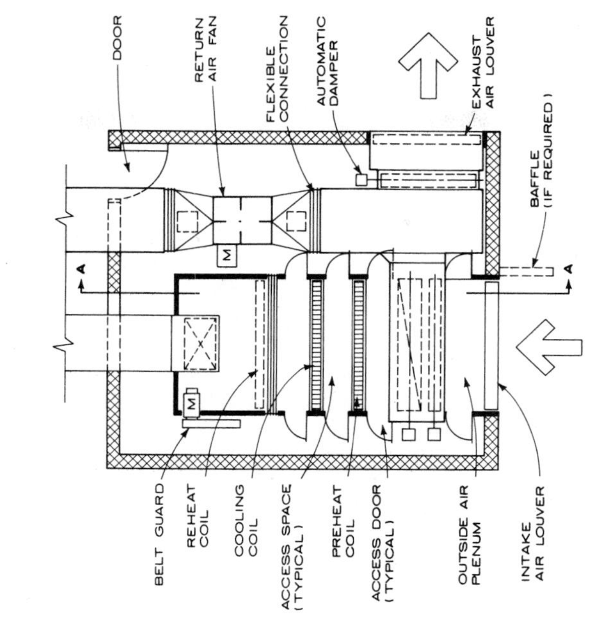

EQUIPMENT ROOM PLAN

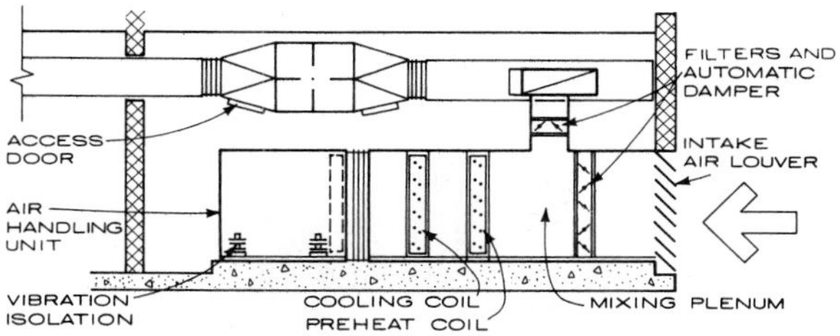

EQUIPMENT ROOM SECTION A–A

EQUIPMENT ROOM SPACE REQUIREMENTS

	APPROXIMATE OVERALL DIMENSION OF SUPPLY AIR UNITS			RECOMMENDED ROOM DIMENSIONS		
CFM RANGE	W	H	L	W	H	L
1,000– 1,800	4′-9″	2′-9″	14′-9″	12′-6″	9′-0″	18′-9″
1,801– 3,000	5′-0″	3′-6″	16′-0″	13′-9″	9′-0″	20′-0″
3,001– 4,000	6′-9″	4′-6″	16′-0″	17′-6″	9′-0″	20′-0″
4,001– 6,000	7′-6″	4′-6″	16′-9″	18′-0″	9′-0″	20′-9″
6,001– 7,000	7′-6″	4′-9″	18′-3″	18′-6″	9′-6″	22′-3″
7,001– 9,000	8′-0″	5′-0″	18′-9″	19′-0″	10′-0″	22′-9″
9,001–12,000	10′-0″	5′-6″	21′-0″	23′-0″	11′-0″	25′-0″
12,001–16,000	10′-3″	6′-0″	22′-0″	23′-6″	12′-6″	26′-0″
16,001–19,000	10′-6″	6′-6″	23′-9″	24′-0″	13′-0″	27′-9″
19,001–22,000	11′-9″	7′-3″	25′-0″	26′-9″	15′-0″	29′-0″
22,001–27,000	11′-9″	8′-6″	26′-0″	27′-0″	16′-0″	30′-0″
27,001–32,000	13′-0″	9′-9″	27′-9″	29′-0″	18′-0″	31′-9″

Fig. 6.95 *Air-handling equipment guidelines. Especially important are the separation of intake and exhaust locations, the isolation of noise and vibration from the fans, and access for maintenance of fans, filters, and coils. The return air fan and automatic damper assist in the operation of the economizer cycle (explained in Fig. 6.90).* Note: *Where higher spaces are available, air handlers may be vertical; this will shorten the length of the unit by between 2 ft and 3 ft 6 in. (Reprinted by permission from AIA, Ramsey, and Sleeper,* Architectural Graphic Standards, *7th ed., © 1981 by John Wiley & Sons.)*

TABLE 6.17 **Air Filter Characteristics**

Media and Type	*Percent Efficiency Range*		*Dust-Holding Capacity*	*Airflow Resistance (In. Water)[a]*
	Atmospheric Dust	*Small Particles*		
Dry panel, throwaway	15–30	NA	Excellent	0.1–0.5
Viscous panel, throwaway	20–35	NA	Good	0.1–0.5
Dry panel, cleanable	15–20	NA	Superior	0.08–0.5
Viscous panel, cleanable	15–25	NA	Superior	0.08–0.5
Mat panel, renewable	10–90	0–60	Good to superior	0.15–1.0
Roll mat, renewable	10–90	0–55	Good to superior	0.15–0.65
Roll oil bath	15–25	NA	Superior	0.3–0.5
Close pleat mat panel	NA	85–95	Varies	0.4–1.0
High-efficiency particulate	NA	95–99.9	Varies	1.0–3.0
Membrane	NA	to 100	NA	NA
Electrostatic with mat	80–98	NA	Varies	0.15–1.25

Source: Reprinted by permission from AIA, Ramsey, and Sleeper, *Architectural Graphic Standards,* 7th ed., © 1981 by John Wiley & Sons.

[a]Higher airflow resistance values will require increased fan energy.

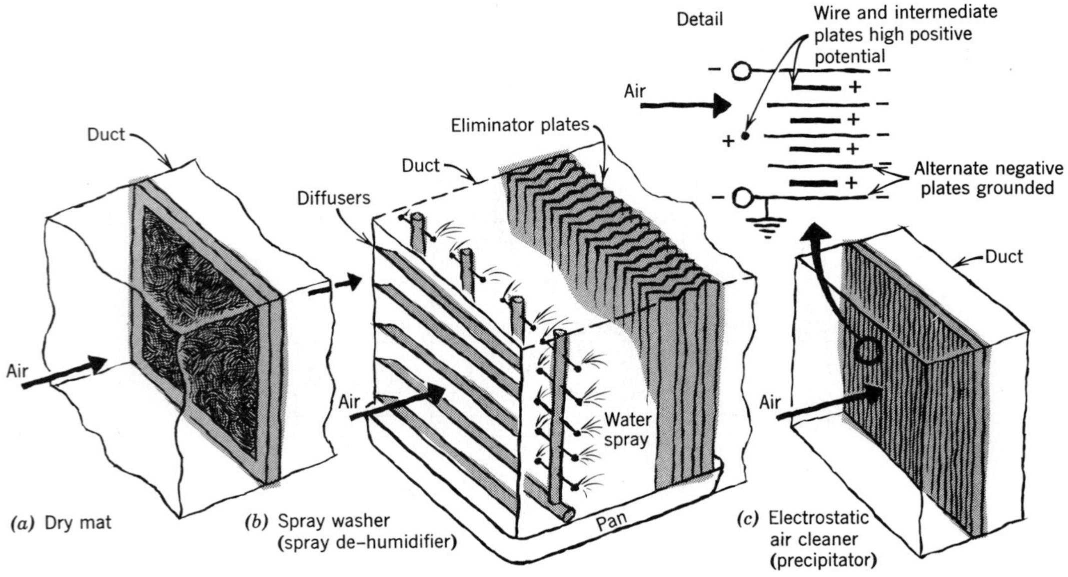

Fig. 6.96 *Air filter types. (Not shown are pleated and roll types.)*

or viscous (sticky) media. Electrostatic filters, another type, are particularly effective with atmospheric dust, including pollen. Sometimes a cheaper "prefilter" is used, to extend the life of a more expensive, higher efficiency filter downstream. Although filters may be used anywhere within air distribution trees, they commonly are placed where the air enters heating, cooling, or humidifying equipment. At such locations, they protect the equipment coils and promote higher efficiency operation.

Odor control is a problem not easily solved by filters, although air washer and carbon (activated charcoal) filters are used to help with odors. Controlling the odor at its source is the best approach; when that is impractical, a quick, thorough exhausting of odorous air is usually helpful. Less frequently used are ozone treatments and "perfume" (aerosol masking).

(h) Controls. Almost all the above-discussed devices for heating, cooling, heat transfer, and so on depend on controls. Although the most obvious control function is to maintain desired comfort conditions, controls also increase fuel economy, by promoting optimum operation, and act as safety devices, limiting or overriding mechanical equipment. They also eliminate human error; controls fall asleep only during a power failure.

Although precise control of temperature and humidity everywhere in a building is a tempting thought, controls can usually maintain only a *range* of conditions, not a set point. This range (within which neither heating nor cooling is called for) is called the "deadband." Temperature variations occur vertically within a space—the higher the space, the greater the variation. Variations between horizontal positions within a zone are highly likely, especially where one or more walls are exterior and where different rooms share a zone controlled by a single thermostat. Variations in time also occur: a building will always be warmer at the moment the furnace fan turns off than at the moment the fan turns on.

Individual controls can be classified as follows: *controllers,* which measure, analyze, and initiate action; *actuators,* which are the controller's slaves, and in turn become the bosses of pieces of equipment; *limit and safety controls,* which may function only infrequently, preventing damage to equipment or buildings; and *accessories,* a miscellaneous collection.

Systems of controls can be classified in various ways. A common distinction is by power source: electric; pneumatic (in which compressed air is the motivating force); and self-contained, including "passive" controls such as those motivated by thermal expansion of liquids or metals. Another way to classify control systems is by the motion of the controller equipment: *two-position* systems are of the simple on-off type; *multiposition* systems have several varieties of "on" position, commonly used for separate operation of more than one machine; *floating* controls can assume any position in the range between minimum and maximum; *central logic control* systems can be programmed to integrate the many aspects of building control into one decision-making unit.

Some control diagrams for common HVAC applications are shown in Fig. 6.97. Single-duct, variable-air-volume systems (Fig. 6.97*a*) are described in Section 6.9; with the constantly varying flow rate, the fan must be regulated so as to maintain the minimum pressure (and therefore, flow) needed at the most demanding outlet. This outlet may be either the one most remote from the fan or the one needing the greatest flow (because it has the highest gain) at the moment. Economizer cycles (Fig. 6.97*b*) compare outside to inside positions, and vary the proportion of fresh (outdoor) air to return air, to provide "free" cooling (see Fig. 6.90). Solar heating systems (Figs. 6.97*c* and 6.97*d*) compare solar collector temperatures to those of the storage tank, to regulate the pumped water to the collectors.

Many examples of large buildings with central logic control have already appeared in this text—notably, the Bateson Building in Sacramento (Fig. 5.29), in which the decision as to whether to lower the east–west exterior sunshades depends partly on outdoor temperature; and the Occidental Chemical Building in Niagara Falls (Fig. 6.2), in which louvre positions on each of the four elevations are adjusted for a combination of sun position, sky condition (such as overcast), and outdoor temperature. Central logic control allows for a variety of switches that not only improve a building's performance, but also can enliven its appearance.

6.9 All-Air HVAC Systems

This large and complex family of systems was introduced in Section 6.7. All-air HVAC systems require large distribution trees but can promise comfortable results. With all-air systems, air quality can be manipulated through pressure control: *negative pressures* in odorous or excessively humid locations (kitchens, toilet rooms, pet shops in shopping centers, etc.), *positive pressures* in shopping malls, corridors of apartment houses, stair towers, and so on. The difference in pressure sets up an overall direction of air flow that helps prevent the spread of odorous or otherwise contaminated air and can even help manage smoke in a fire (see Chapter 13). Positive pressures in con-

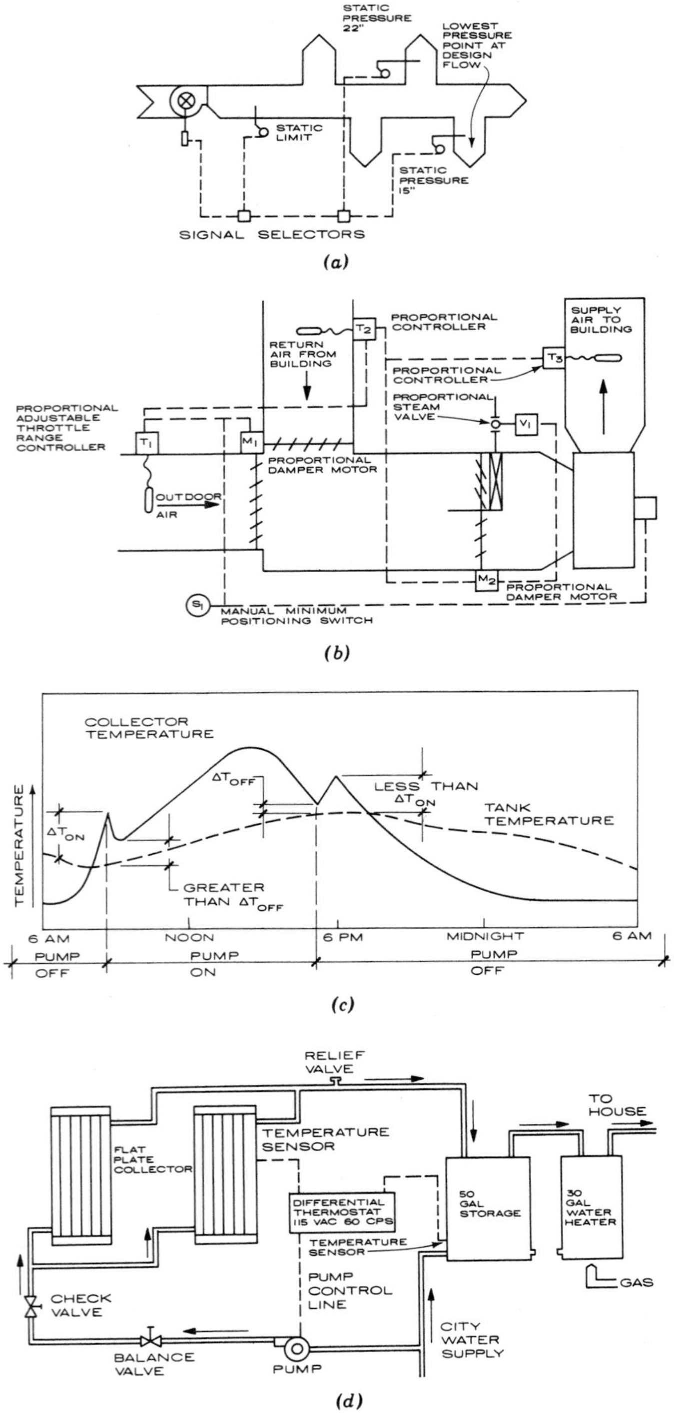

Fig. 6.97 *Control diagram for some common HVAC applications.* (a) *Single air duct with variable air volume (VAV).* (b) *Economizer cycle.* (c) *Solar water heating: the operating cycle.* (d) *Sensors within the solar water heating system. (Reprinted by permission from AIA, Ramsey, and Sleeper,* Architectural Graphic Standards, *7th ed., © 1981 by John Wiley & Sons.)*

necting spaces help each adjacent space keep its air to itself. All-air systems also offer an opportunity for increased electric lighting efficiency: because so much air is being moved, many supply air outlets and return air inlets are needed in each space. Return air can be channeled through lighting fixtures (luminaires), to serve two useful purposes: first, to lower the temperature of the air around the fluorescent tubes and thus (with most types of fluorescents) increase their light output (see Chapter 19); second, to whisk away much of the light's heat output *before* it can influence the temperature of the space. This heat is immediately available for use elsewhere, or for rejection to the outside, instead of increasing the need for cooling in the space. When the lighting heat is reused, the system is called a "heat-of-light" system.

(a) Accessories. Air diffusers and grilles (Fig. 6.98) are common to all these systems. The most popular locations for these accessories are ceilings, which are uncluttered by furniture and relatively unaffected by minor rearrangements in partitions. Also, the entire space above a suspended ceiling can be used as one huge return air duct; this arrangement is called plenum return. In a thermal zone where cooling is the prevalent condition, cool air introduced at the ceiling is ideal; being heavier, it sinks naturally upon entry. However, low wall or floor return air grilles will assure a better circulation of air. In a thermal zone where heating is the prevalent condition, the reverse locations are preferred—supply low, return high. The final size of the supply registers depends heavily on the face velocity and the throw pattern of the air (see again Fig. 6.51 and Table 6.12), which are given in manufacturers' catalogs.

Where all-air systems are used throughout larger buildings, another accessory is the *draft barrier,* which can alleviate minor and temporary cold spots in an otherwise completely internal-load-dominated building. Large exterior glass areas cool the adjacent interior air in winter. The vertical layer of cool, denser air thus created then drops to the floor and blankets it like a chilly carpet. Unless one has witnessed tests using smoke and recording thermometers, the speed and resulting discomfort of this phenomenon are seldom fully comprehended.

Under the worst conditions, this cold layer must cross the space and reach a thermostat on an interior wall before a below-the-glass heating element takes over. When it does, the problem is a dual one—to reverse the downslip of air *and* to warm the space. The heating element is seldom properly adapted to meet both challenges correctly.

Admittedly, the use of a modulated heating medium—water or air—will provide a more continuous operation to maintain a warm upflow of air at glass, yet during even brief intervals between these heating periods, the cold air slides quickly past the nonoperative heaters. An electric resistance draft barrier for under-window application is shown in Fig. 6.99.

(b) Single-Zone Systems (Fig. 6.100). Smaller buildings are often served by this least-complicated all-air example (see examples in Section 6.6). Where an entire building is essentially one thermal zone, uncomplicated enough to be served by only one thermostat, choice of this system offers very low first cost. A closely related system is the multizone (see below), which is a combination of single zones.

(c) Single-Duct, Variable-Air Volume (VAV) Systems (Fig. 6.101). As described in Section 6.7, this system is an energy-saving option ideally suited to the internal zones of large buildings, where cooling is always needed. But it can also be adapted to serve the entire building, with important savings over the constant-volume (CV) systems described below (**d** through **f**). The air-handling units (fans, etc.) for each thermal zone are sized to meet the peak demand on that zone. In most CV systems, the fans always run at this peak condition speed, even though peaks are often of rather short duration. In VAV systems, fans only run at peak speed during peak hours. This obviously saves considerable energy needed to run fans. When the building uses only one or a few central fans, a VAV system will require smaller fans, because the fans need meet only one peak condition at a time and do not have to be sized for all zones' peak flows. The variation in demands on the fan can be met either by selecting variable-pitch blades or (less expensively) by varying the speed of the fan.

Where a VAV system is used, provision must be made for at least code-minimum fresh air levels, and noise may become a problem. Although VAV systems are typically less noisy than CV systems (because less air is moving), the air motion noise *varies* with the volume, and variable noise sources are inherently more noticeable than are steady ones.

Within (or near) the spaces it serves, the VAV system typically needs a mixing box or terminal; this is often placed above a suspended ceiling. Of

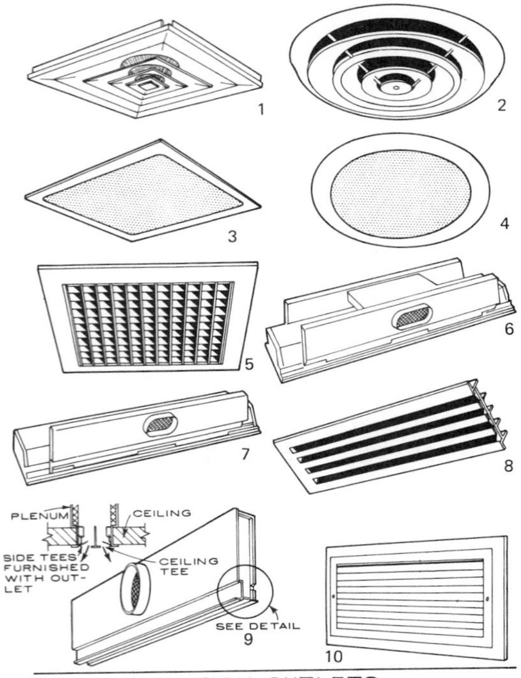

AIR DISTRIBUTION OUTLETS

KEY

1 RECTANGULAR LOUVERED FACE DIFFUSER: Available in 1, 2, 3, or 4-way pattern, steel or aluminum. Flanged overlap frame or inserted in 2 X 2 ft or 2 X 4 ft baked enamel steel panel to fit tile modules of lay-in ceilings. Supply or return.

2 ROUND LOUVERED FACE DIFFUSER: Normal 360° air pattern with blank-off plate for other air patterns. Surface mounting for all type ceilings. Normally of steel with baked enamel finish. Supply or return.

3 RECTANGULAR PERFORATED FACE DIFFUSER: Available in 1, 2, 3, or 4-way pattern, steel or aluminum. Flanged overlap frame or 2 X 2 ft and 2 X 4 ft for replacing tile of lay-in ceiling can be used for supply or return air.

4 ROUND PERFORATED FACE DIFFUSER: Normal 360° air pattern with blank-off plate for other air patterns. Steel or aluminum. Flanged overlap frame for all type ceilings. Can be used for supply or return air.

5 LATTICE TYPE RETURN: All aluminum square grid type return grille for ceiling installation with flanged overlap frame or of correct size to replace tile.

6 SADDLE TYPE LUMINAIRE AIR BOOT: Provides air supply from both sides of standard size luminaires. Maximum air delivery (total both sides) approximately 150 to 170 cfm for 4 ft long luminaire.

7 SINGLE SIDE TYPE LUMINAIRE AIR BOOT: Provides air supply from one side of standard size luminaires. Maximum air delivery approximately 75 cfm for 4 ft long luminaire.

8 LINEAR DIFFUSER: Extruded aluminum, anodized, duranodic, or special finishes, one way or opposite direction or vertical down air pattern. Any length with one to eight slots. Can be used for supply or return and for ceiling, sidewall, or cabinet top application.

9 INTEGRATED PLENUM TYPE OUTLET FOR "T" BAR CEILINGS: Slot type outlet, one way or two way opposite direction air pattern. Available in 24, 36, 48, and 60 in. lengths. Replaces or integrates with "T" bar. Approximately 150 to 175 cfm for 4 ft long, two-slot unit.

10 SIDEWALL OR DUCT MOUNTED REGISTER: Steel or aluminum for supply or return. Adjustable horizontal and vertical deflection. Plaster frame available. Suitable for long throw and high air volume.

(a)

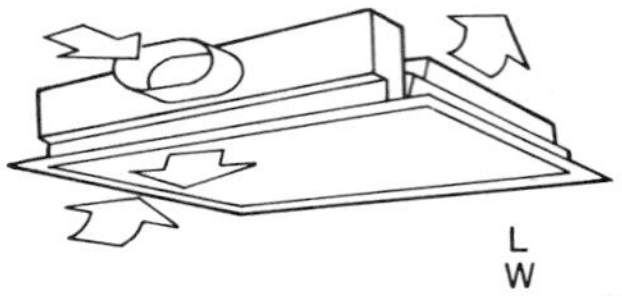

L 24" to 72"
W 24" to 72"
Capacity varies with air velocity

(b)

Fig. 6.98 *Common air distribution outlets.* (a) *Air diffusers of types 1–5 are for typical ceiling application; types 6 and 7 integrate with luminaires; types 8 and 10 may be used in other-than-ceiling applications; type 9 is a nearly "invisible" hung-ceiling adaptation.* (b) *Typical heat-of-light luminaires that integrate both supply and return air. (Reprinted by permission from AIA, Ramsey, and Sleeper,* Architectural Graphic Standards, *7th ed., © 1981 by John Wiley & Sons.)* (c) *Integral electric lighting, air distribution, and sound absorption, in an application at the* (d) *American Republic Insurance Building, Des Moines, Iowa; Skidmore Owings & Merrill, architects.*

the several terminal types available, the standard and most simple one is shown in Fig. 6.102. This terminal serves not only to vary the quantity of air, but to both attenuate the noise and reduce the velocity of air from the main trunk of the distribution tree. High velocity is commonly used in main ducts, because it reduces the size of these critical, large portions of the tree.

There are several variations on the basic VAV system, some of which respond to the problem of minimum airflows for rooms with little thermal load or to the desire to serve both interior and perimeter areas with the same HVAC system.

To maintain minimum fresh air, VAV terminals are often set so that they cannot be entirely closed off—a provision that ensures some outdoor air at all times. If this fresh air minimum, entering at low velocity, does not provide the desired air motion and mixing within the room, VAV terminals can be equipped with fans, which are activated as needed with decreasing incoming air volume. These self-contained fans mix room air with incoming air to provide an airstream of the right temperature and velocity to maintain comfort.

A more complicated approach to VAV is required when simultaneous heating and cooling are needed. This is particularly true at perimeter zones, which can generate sizable heating needs while

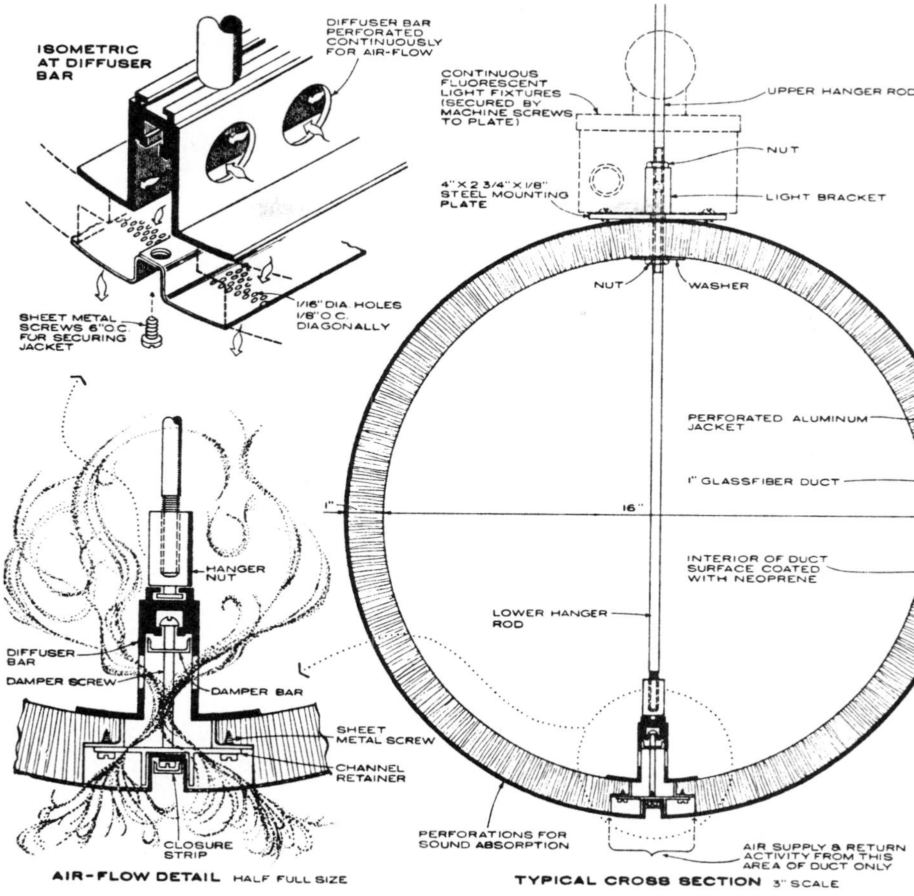
ISOMETRIC AT DIFFUSER BAR
DIFFUSER BAR PERFORATED CONTINUOUSLY FOR AIR-FLOW
SHEET METAL SCREWS 6" O.C. FOR SECURING JACKET
1/16" DIA. HOLES 1/8" O.C. DIAGONALLY
CONTINUOUS FLUORESCENT LIGHT FIXTURES (SECURED BY MACHINE SCREWS TO PLATE)
UPPER HANGER ROD
NUT
4" X 2 3/4" X 1/8" STEEL MOUNTING PLATE
LIGHT BRACKET
NUT
WASHER
PERFORATED ALUMINUM JACKET
1" GLASSFIBER DUCT
1"
16"
INTERIOR OF DUCT SURFACE COATED WITH NEOPRENE
LOWER HANGER ROD
HANGER NUT
DIFFUSER BAR
DAMPER SCREW
DAMPER BAR
SHEET METAL SCREW
CHANNEL RETAINER
CLOSURE STRIP
PERFORATIONS FOR SOUND ABSORPTION
AIR SUPPLY & RETURN ACTIVITY FROM THIS AREA OF DUCT ONLY
AIR-FLOW DETAIL HALF FULL SIZE
TYPICAL CROSS SECTION 3" SCALE

(c)

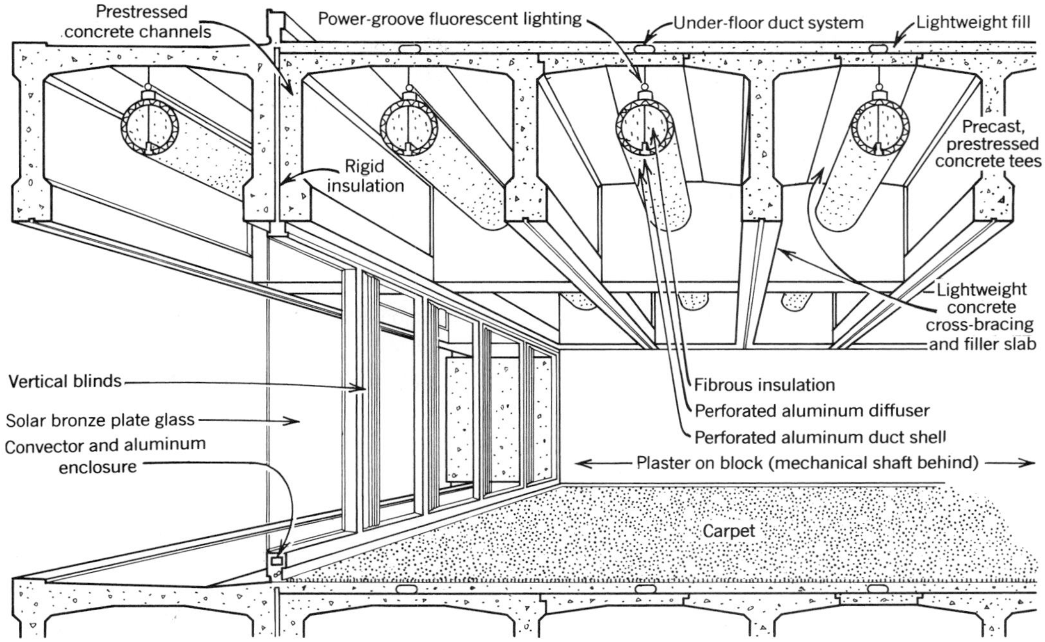
Prestressed concrete channels
Power-groove fluorescent lighting
Under-floor duct system
Lightweight fill
Precast, prestressed concrete tees
Rigid insulation
Lightweight concrete cross-bracing and filler slab
Vertical blinds
Solar bronze plate glass
Convector and aluminum enclosure
Fibrous insulation
Perforated aluminum diffuser
Perforated aluminum duct shell
Plaster on block (mechanical shaft behind)
Carpet

(d)

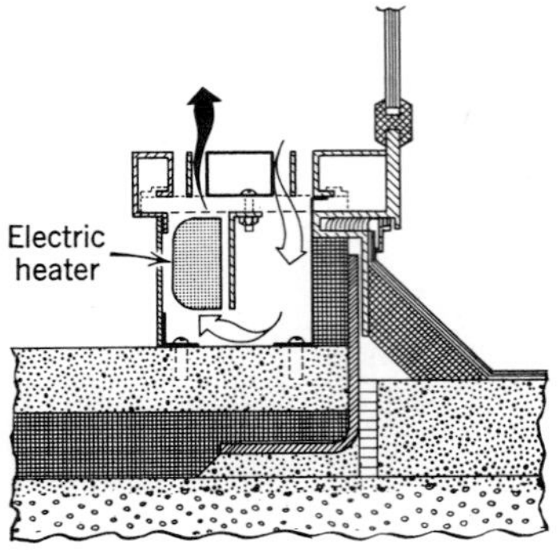

Fig. 6.99 Electric resistance draft barrier, for use below windows in buildings that generally require cooling.

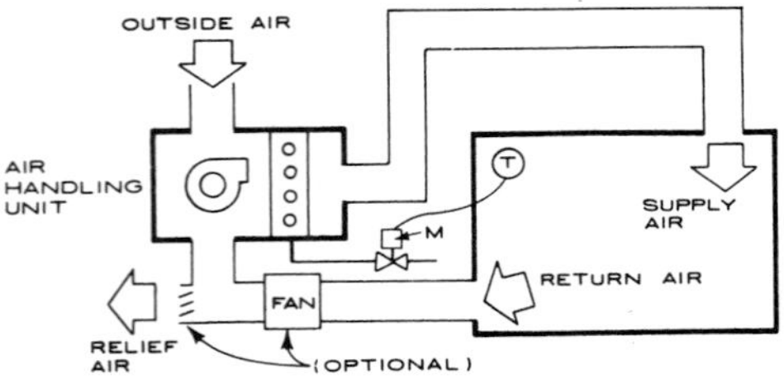

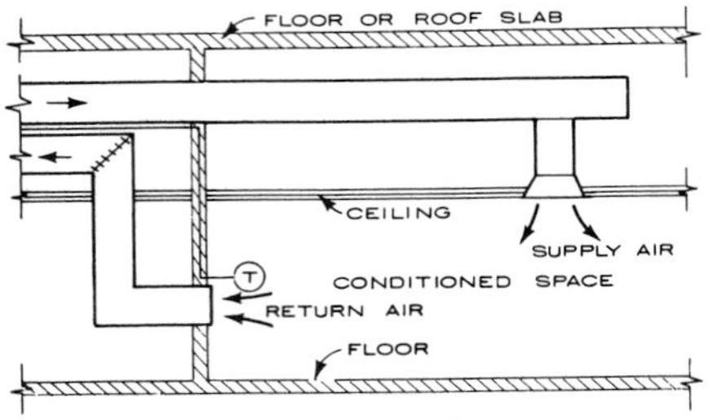

Fig. 6.100 *Single-zone system. (Reprinted by permission from AIA, Ramsey, and Sleeper,* Architectural Graphic Standards, *7th ed., © 1981 by John Wiley & Sons.)*

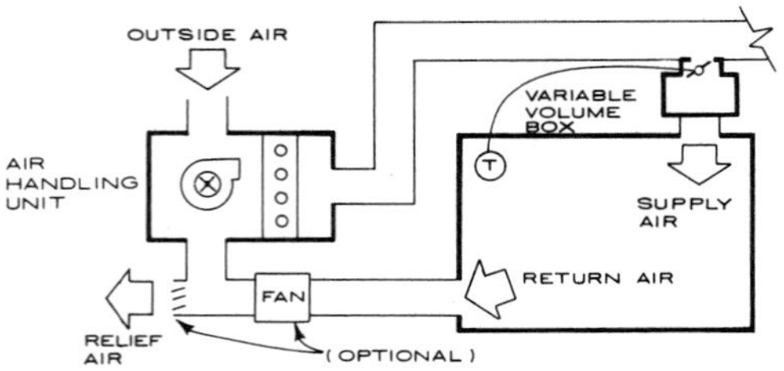

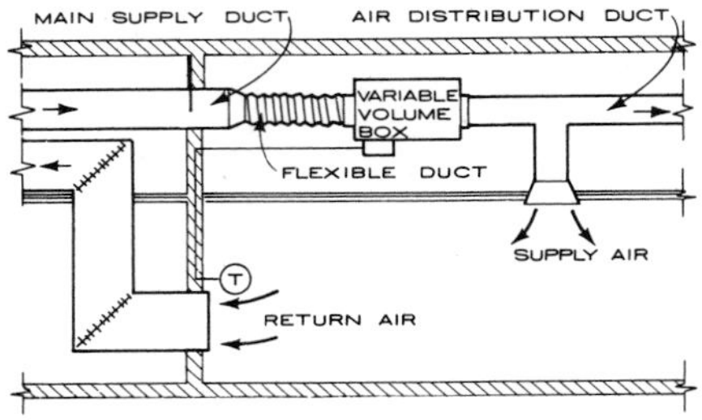

Fig. 6.101 *Single-duct, variable-air-volume (VAV) system. (Reprinted by permission from AIA, Ramsey, and Sleeper,* Architectural Graphic Standards, *7th ed., © 1981 by John Wiley & Sons.)*

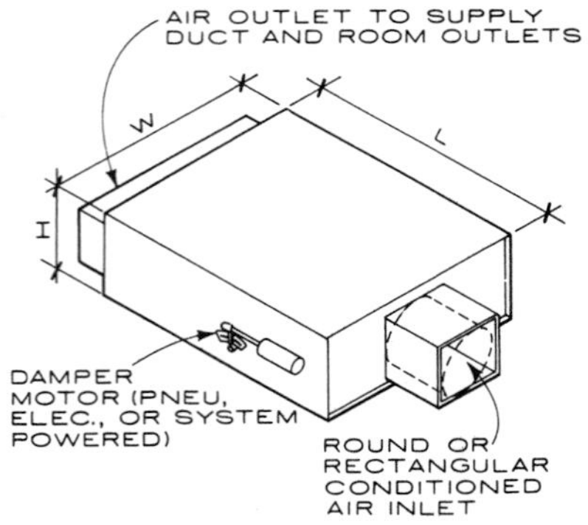

RANGE OF DIMENSIONS

CFM	HEIGHT	LENGTH	WIDTH
400	8″- 9″	24″-39″	14″-30″
800	10″-11″	24″-53″	18″-42″
1600	14″	30″-48″	22″-44″
2400	16″	42″-60″	26″-54″
3200	18″	42″-67″	33″-54″

Fig. 6.102 *The most simple type of VAV terminal which pinches back the volume of incoming air as thermal loads decrease. (Reprinted by permission from AIA, Ramsey, and Sleeper,* Architectural Graphic Standards, *7th ed., © 1981 by John Wiley & Sons.)*

the rest of the building needs cooling. One approach is to utilize an induction-type VAV terminal, with which air heated by electric lights is induced to join the incoming cool airstream. Greater heating needs often require the use of reheat terminals supplied by a circulating hot water system or by electric resistance heating. In this reheat application (more energy efficient than the standard—

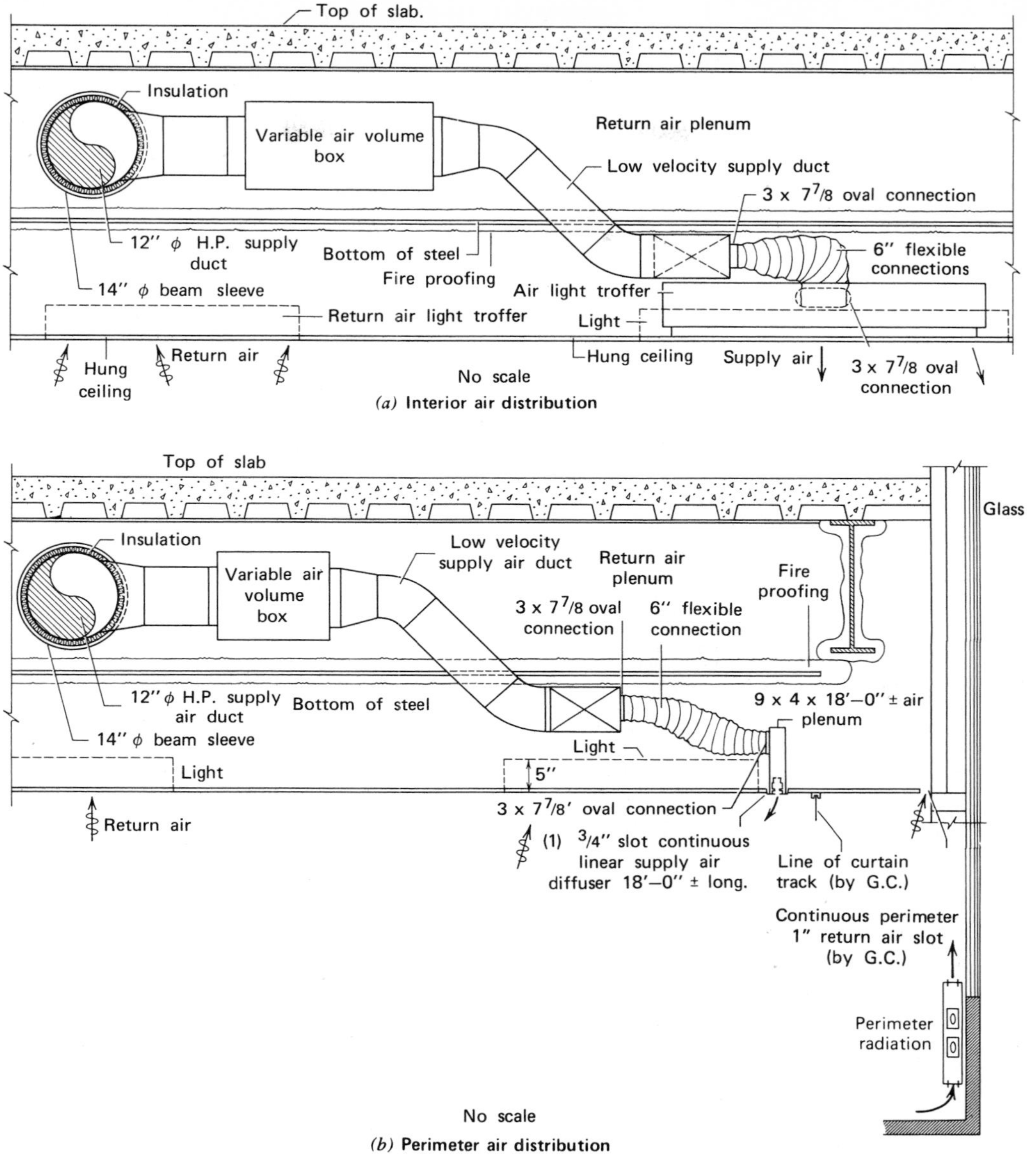

Fig. 6.103 *VAV in Denver's Anaconda Tower. Each VAV box (terminal) is controlled by a wall thermostat.* (a) *Interior zones use VAV boxes that induce low-pressure return air in the plenum to join the incoming high-pressure fresh air, for delivery to the space. Light troffers are used to admit return air to the plenum, as well as to supply air.* (b) *Perimeter zones receive only supply air, delivered through a continuous 3/4-in. slot that skirts the building perimeter inside the curtain track. A hot water perimeter radiation system works on cold days. (Anaconda Tower: Skidmore Owings & Merrill, architects; Flack & Kurtz, engineers.)*

CV—reheat system described in **d,** because a much smaller volume of air is first cooled, then reheated), the water or electric coils can be incorporated either in the VAV terminal or in the ductwork between the terminal and the space it serves.

Another approach is to provide perimeter radiation via a water system, usually placed under the window while the VAV supply is in the ceiling above (see Fig. 6.103).

A final option for VAV at perimeters is the *dual-conduit* approach, diagrammed in Fig. 6.67. The VAV conduit works year-round, providing constant-temperature (cool) fresh air. The CV conduit carries hot air in very cold weather and cool air (supplementing the VAV conduit) in very hot weather or in the case of high solar gains.

An example of a VAV system used in a retrofit is shown in Fig. 6.104. The world headquarters of the Manufacturers Hanover Trust Company is a 52-story, 1.5-million-ft² building on New York's Park Avenue. Built over 20 years ago with a perimeter induction HVAC system and an interior double-duct system, it was recently refurbished at a cost of some $60 million. Engineers Syska and Hennessy retained the perimeter induction units but replaced the double-duct interior system with VAV. With the aid of new air-handling equipment, rehabilitated chillers, and a more efficient lighting system (from 4 W/ft² to less than 2 W/ft²), the annual energy budget is expected to drop from 123,000 Btu/ft² per year to between 80,000 and 90,000 Btu/ft² per year. Window glass constitutes almost 70% of the building's skin area, which helps explain such high energy consumption.

A final example is a floor-by-floor VAV system (Fig. 6.105) in a 1-million-ft² medium-use (28-floor) Chicago office building designed (and occupied in part) by Skidmore Owings & Merrill. This lower, wider approach to office buildings utilizes three ''stacked'' atriums to relieve the monotony of the wide interior floors. Another result is lower structural and energy costs per square foot

Fig. 6.104 A retrofit application in which a new interior VAV system replaces a double-duct system. (a) Exterior of the world headquarters of the Manufacturers' Hanover Trust Company, New York City. Syska & Hennessy, engineers. (b) Detail of the retrofit energy-efficient electric lighting system. The new waffle grid luminous ceiling replaces the original flat solid vinyl panels, thus increasing lighting efficiency, diminishing acoustic problems, and conforming to current New York City code.

Fig. 6.105 Floor-by-floor VAV for a Chicago office building. (a) Exterior view of 33 West Monroe (photo by Merrick, Hedrick-Blessing.) (b) Section perspective showing "stacked atriums." Skidmore Owings & Merrill, architects and engineers.

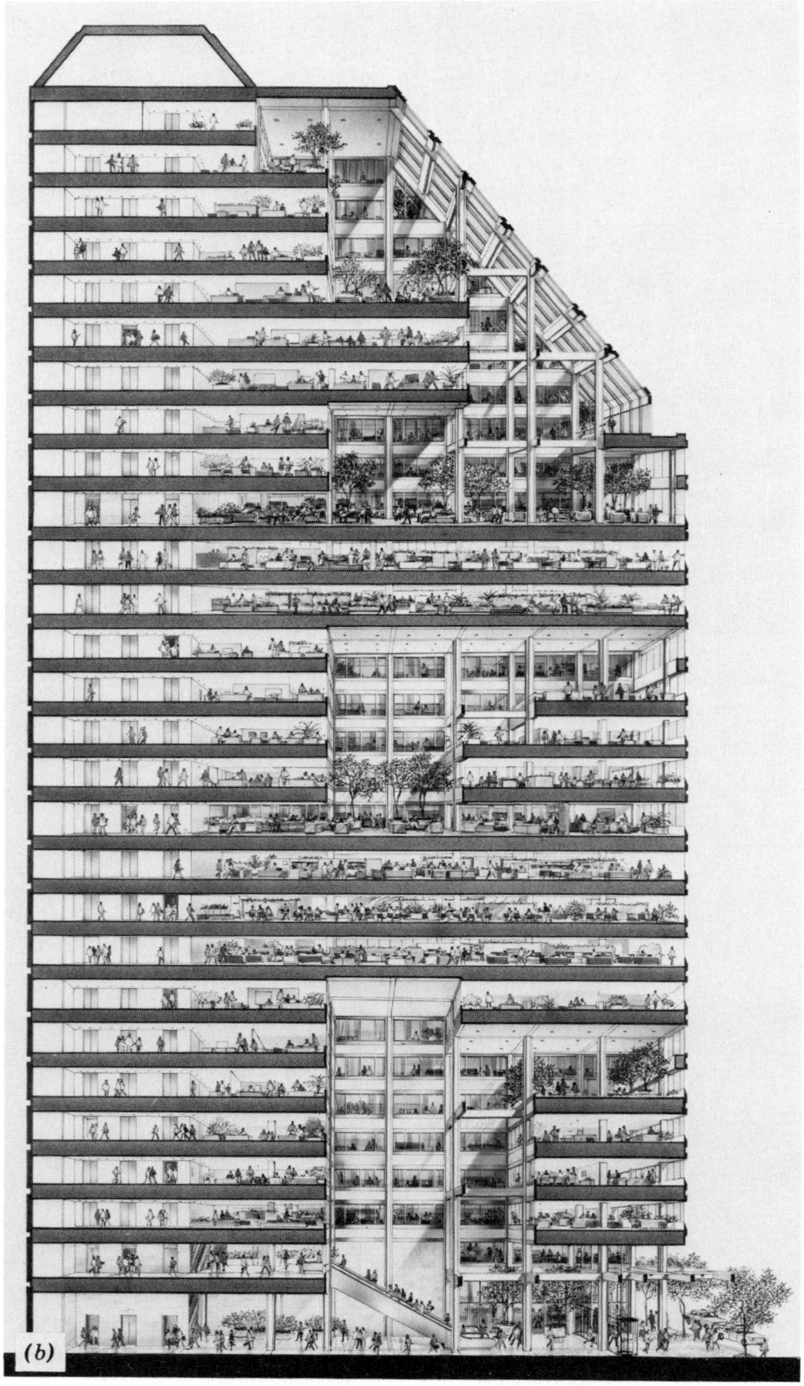

Fig. 6.105 (continued)

relative to conventional high-rise structures. The cubelike shape of the building exposes less skin area (38% of which is in insulating glass) to Chicago's cold winters; electric lighting at about 1.8 W/ft^2 holds down internal gains. To accommodate the differing schedules and comfort needs of a variety of tenants, each floor is provided with two VAV supply fans that can be operated nights and weekends, independently of the rest of the building. The mechanical core has one exterior wall (on an alley) to facilitate fresh air intake/stale air exhaust. The perimeter heating system is electric resistance fin radiation; an economizer cycle provides cooling with outdoor air below 55 F. The annual energy budget is expected to be between 49,000 and 54,000 Btu/ft^2 per year.

(d) Single Duct with Reheat (Fig. 6.106). Formerly a widespread system and usually supplied at constant volume (CV), single-duct reheat systems now are most often used when a constant air volume is important (as in hospitals and laboratories with large quantities of exhaust air) and the energy consumed in the reheating of cooled air is not a primary concern. For high-pressure and high-velocity main ducts, constant-volume reheat boxes (Fig. 6.106*b*) are used to control the noise, temperature, pressure, and velocity of the supply air. For more simple all-low-velocity systems, simple duct insert heaters (Fig. 6.106*c*) can be used. These are sized to fit the duct, which needs to be enlarged only slightly to accommodate them.

It is quite common for several HVAC systems to serve one building, as illustrated by a junior high school in Pendleton, Oregon (Fig. 6.107). This 1000-student facility has 16,400 ft^2 of solar

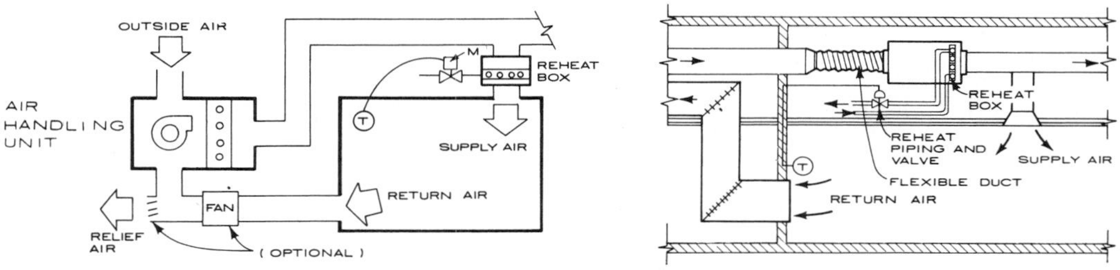

(a)

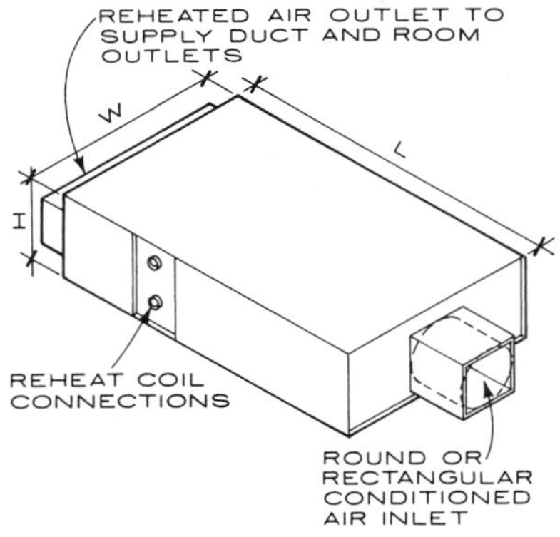

RANGE OF DIMENSIONS

CFM	HEIGHT	LENGTH	WIDTH
200	9″-11″	30″-50″	16″-22″
400	9″-11″	30″-51″	18″-30″
800	9″-11″	30″-51″	22″-42″
1600	14″-16″	48″-51″	40″-44″
2400	16″-18″	60″-55″	40″-54″
3200	16″-18″	60″-55″	16″-66″
5000	20″-18″	60″-55″	20″-80″

(b)

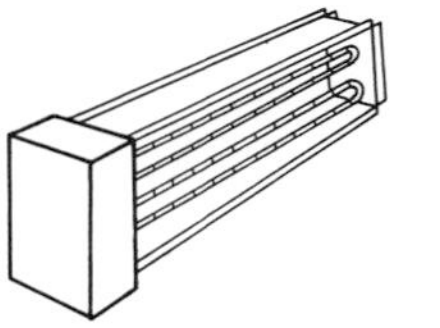

(c)

Fig. 6.106 *Single-duct with reheat system.* (a) *Diagrams of system and components.* (b) *Reheat box, or terminal, where velocity and pressure are reduced.* (c) *Simple duct heater. (Reprinted by permission from AIA, Ramsey, and Sleeper,* Architectural Graphic Standards, *7th ed., © 1981 by John Wiley & Sons.)*

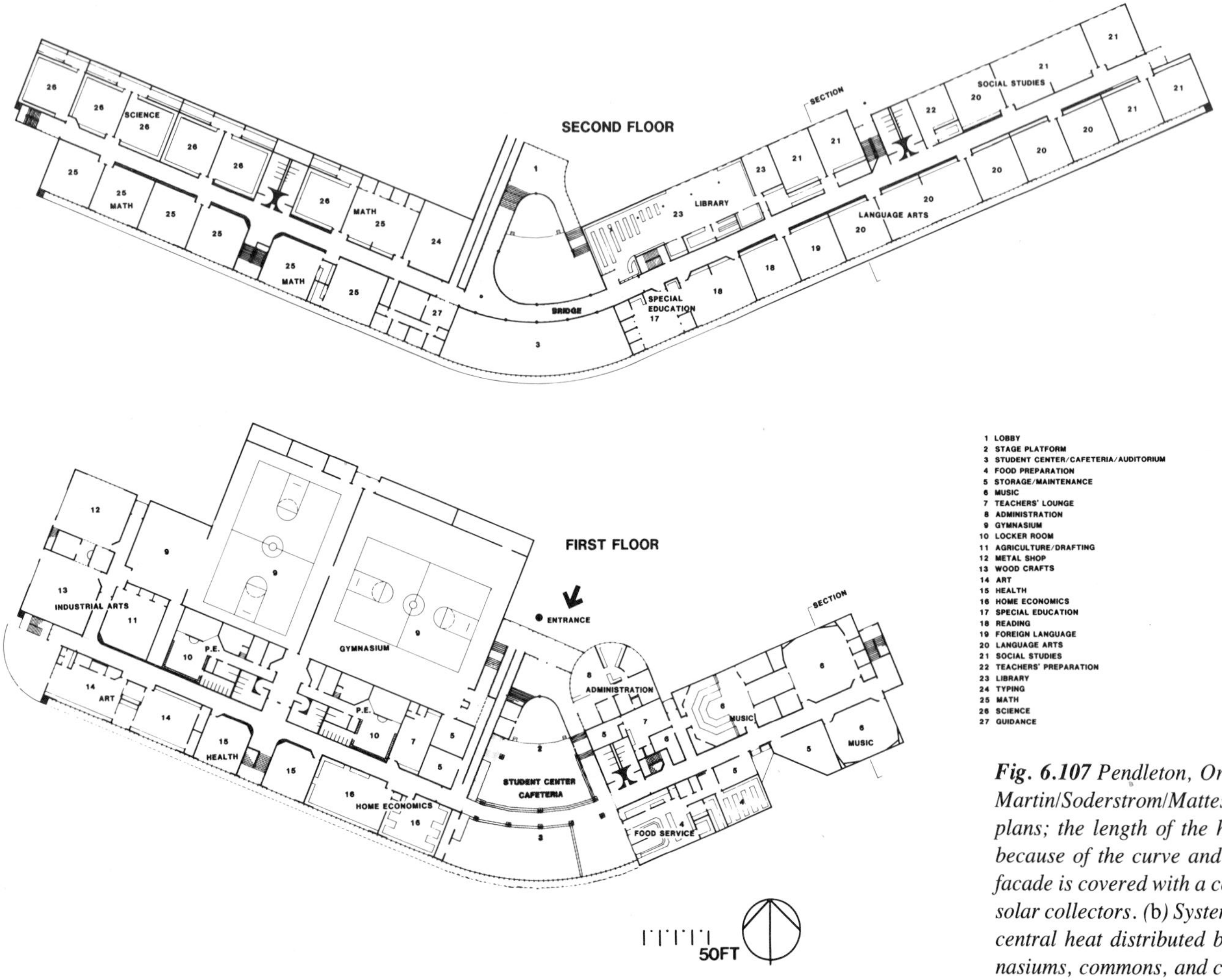

(a)

Fig. 6.107 *Pendleton, Oregon, Junior High School. Martin/Soderstrom/Matteson, architects.* (a) *Floor plans; the length of the hallways is less formidable because of the curve and the skylighting. The south facade is covered with a combination of windows and solar collectors.* (b) *System schematic, showing solar central heat distributed by CV single zone for gymnasiums, commons, and cafeteria and VAV for classroom and small zone spaces.*

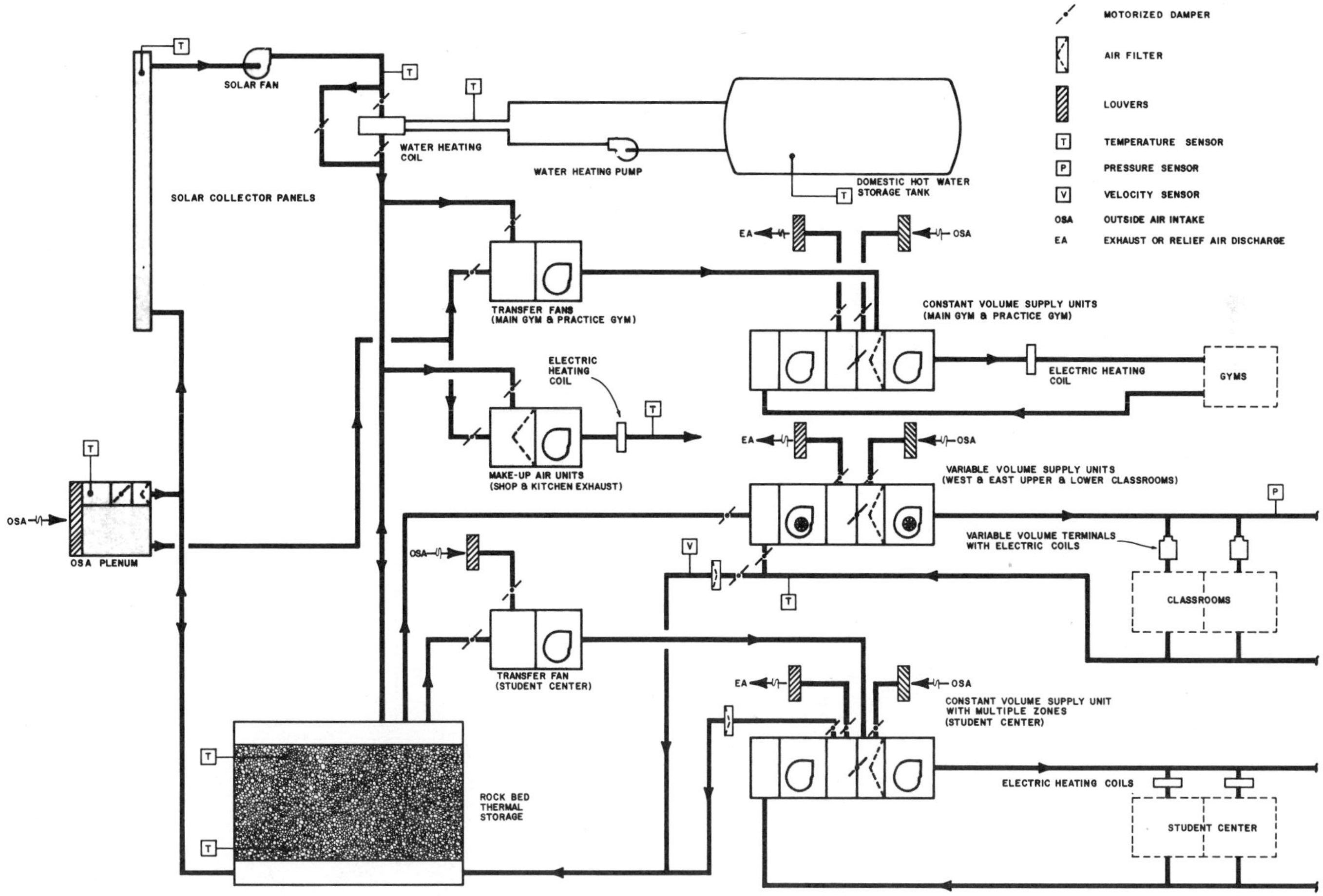
MOTORIZED DAMPER
AIR FILTER
LOUVERS
TEMPERATURE SENSOR
PRESSURE SENSOR
VELOCITY SENSOR
OSA OUTSIDE AIR INTAKE
EA EXHAUST OR RELIEF AIR DISCHARGE
SOLAR FAN
WATER HEATING COIL
WATER HEATING PUMP
DOMESTIC HOT WATER STORAGE TANK
SOLAR COLLECTOR PANELS
TRANSFER FANS (MAIN GYM & PRACTICE GYM)
CONSTANT VOLUME SUPPLY UNITS (MAIN GYM & PRACTICE GYM)
ELECTRIC HEATING COIL
GYMS
ELECTRIC HEATING COIL
MAKE-UP AIR UNITS (SHOP & KITCHEN EXHAUST)
VARIABLE VOLUME SUPPLY UNITS (WEST & EAST UPPER & LOWER CLASSROOMS)
VARIABLE VOLUME TERMINALS WITH ELECTRIC COILS
CLASSROOMS
OSA PLENUM
TRANSFER FAN (STUDENT CENTER)
CONSTANT VOLUME SUPPLY UNIT WITH MULTIPLE ZONES (STUDENT CENTER)
ELECTRIC HEATING COILS
STUDENT CENTER
ROCK BED THERMAL STORAGE

SYSTEM SCHEMATIC

(b)

collectors on the south facade; the warm air is stored in two large rock beds when not immediately needed. A large domestic hot water tank also can act as backup for the solar collectors. This centrally heated air is distributed in three separate HVAC systems: a simple one-zone CV system for the gyms; a variable-volume system for the classrooms, supplemented when necessary by electric heating coils; and a CV with reheat system for the student center.

(e) Multizone Systems (Fig. 6.108). A multizone system is a collection of single-zone systems served by a single supply fan; such systems rarely exceed eight zones per air-handling unit. Simultaneous heating of some zones and cooling of others is possible, but leakage between zones at the decks of hot and cold coils is common. Return air from all zones is mixed within one return duct. One multizone system per floor of medium- to high-rise buildings is an increasingly common application.

(f) Double-Duct Systems (Fig. 6.109). Still considered the ''Cadillac'' of HVAC systems, because of its superior comfort control, the double-duct system is more rarely seen now, due to increased energy costs and the size of such systems. Building volumes needed for a double-duct system's three (two supply, one return) full-sized air distribution trees are harder to justify, given that VAV systems can provide acceptable comfort for most of the spaces.

A most expressive example of a double-duct system serving a building's perimeter zone is the Blue Cross–Blue Shield office building in Boston (Fig. 6.110). Here, each component of the all-air distribution tree can be seen on the facades. The closely spaced and strongly emphasized vertical elements are both structural and utilitarian. Two out of every three verticals are structural, and *all* the verticals enclose ducts. Each pair (one hot, one cold) of high-velocity, round air ducts at the corresponding structural columns constitutes a vertical air supply system. At each floor they feed an attenuation and mixing chamber. At these locations, temperature is controlled as desired. The return ducts between pairs of hot-cold supplies complete the parallel, equally spaced pattern.

The mixing boxes (terminals) of double-duct systems are similar to those of other all-air systems (see Fig. 6.111). Although most double-duct

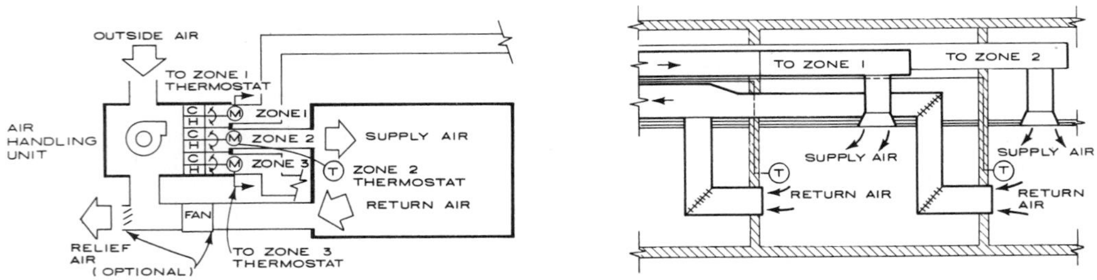

Fig. 6.108 Multizone system. (Reprinted by permission from AIA, Ramsey, and Sleeper, Architectural Graphic Standards, *7th ed., © 1981 by John Wiley & Sons.)*

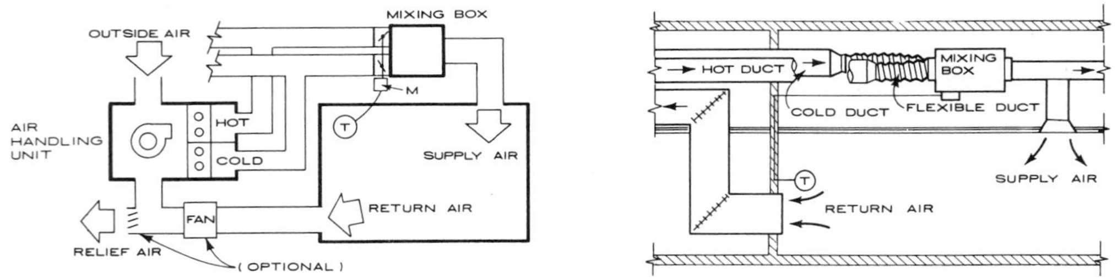

Fig. 6.109 Double-duct system. (Reprinted by permission from AIA, Ramsey, and Sleeper, Architectural Graphic Standards, *7th ed., © 1981 by John Wiley & Sons.)*

THERMAL CONTROL

Fig. 6.110 Blue Cross–Blue Shield Building, Boston. Anderson, Beckwith and Haible and Paul Rudolph, associated architects: Stressenger, Adams, Maguire, and Reidy, mechanical and electrical engineers. The two-story, ***Y****-shaped forms are structural columns that divide at mezzanine level and continue to rise in pairs to form the exterior skeleton frame. Hollow channels on the exterior of each pair enclose, individually, a hot air supply duct and a cold air supply duct. These round, high-velocity ducts join for mixing and velocity reduction in attenuation boxes, located between columns at each floor. Conditioned air is discharged upward from a window sill grille above the box. Mullions between each pair of structural columns originate at the second-floor level and extend to the mechanical story at the roof. Each mullion encloses a return air duct, which draws air through grilles in the sills of the two adjacent windows on each story. Thus the air is delivered at the exterior, accomplishes its mission at that surface, and returns in the same vertical plane to the suction side of fans on the roof.*

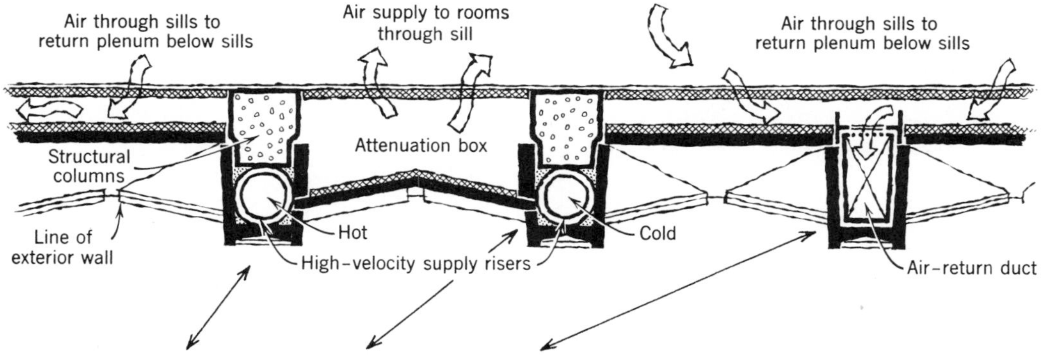

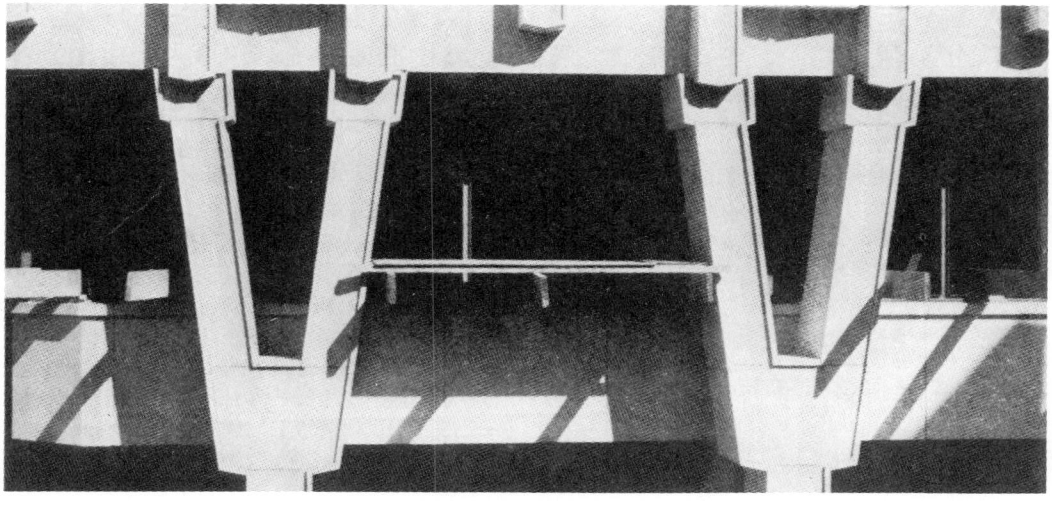

(a)

Fig. 6.111 High-velocity double-duct unit, providing terminal mixing and attenuation (pressure and sound reduction). (a) Pneumatically controlled from a thermostat, the unit blends and delivers air at a selected temperature. These constant-volume units provide accurate, constant delivery at each outlet of the system, even though the pressure in the hot and cold ducts may vary widely (Courtesy of Anemostat.) (b) Typical mixing box dimensions. (Reprinted by permission from AIA, Ramsey, and Sleeper, Architectural Graphic Standards, *7th ed., © 1981 by John Wiley & Sons.)*

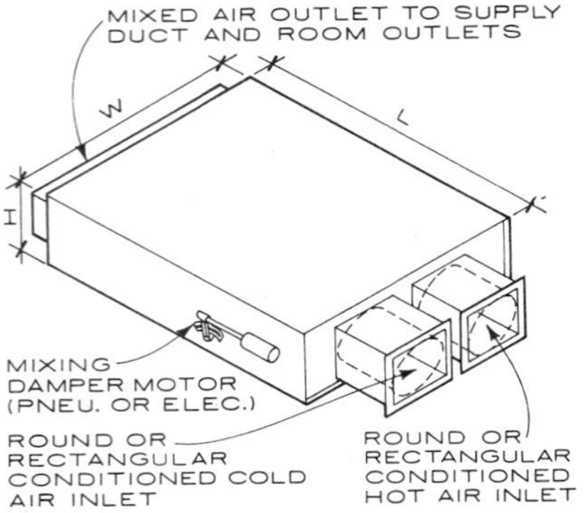

High, medium, or low velocity systems. Inlet pressure 1/4 to 1 1/2 in. W.C. Capacity range from 150 to 2000 cfm per box (low velocity) to 5000 cfm (high velocity). Box serves as converter from high to low velocity air system, noise attenuator, and control device by mixing hot and cold air streams.

RANGE OF DIMENSIONS

CFM	HEIGHT	LENGTH	WIDTH
400	6″-10″	40″-51″	30″-19″
800	8″-11″	50″-51″	42″-24″
1600	12″-14″	48″-51″	44″-40″
2400	14″-18″	60″-55″	54″-44″
3200	14″-18″	60″-55″	54″-44″
5000	16″-18″	60″-55″	54″-66″

(b)

systems are CV, they can be VAV when the reduction in airflow is no more than 50% below the maximum.

6.10 Air & Water Systems

These systems, introduced in Fig. 6.68 and Section 6.7, have the design complexity—and first cost—of supply-and-return distribution trees for both water and air. This disadvantage is offset by the space-saving advantages of water trees and the superior comfort characteristics offered by air. Since only fresh air is centrally treated and distributed, only an equal quantity of exhaust air need be ''returned.'' Because this exhaust air is not recirculated, these systems are attractive for hospitals and other buildings in which the mixing of air between zones is undesirable.

(a) Induction (Fig. 6.112). Centrally conditioned fresh air is supplied (at either high or medium pressures and velocities) to each induction terminal. Each terminal then mixes 20 to 40% incoming fresh air with 80 to 60% room air, passing it all over finned tubes for heating or cooling and circulating this mixture of air to the space.

In two-pipe systems, either hot or cold water—not both—is available to temper this air mixture ordered by the thermostat linked to each terminal. In four-pipe systems, the availability of both hot and chilled water makes it possible to switch instantly from heating to cooling, for excellent thermal control.

The induction terminals typically are located either above a space, as in Fig. 6.112, or below perimeter windows (Figs. 6.113, 6.114). An unusually low-profile below-window induction unit

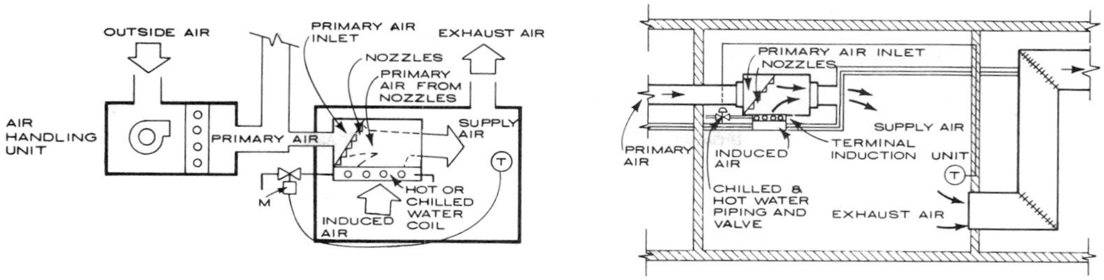

Fig. 6.112 *Induction system. Terminal induction units also are frequently located below windows. (Reprinted by permission from AIA, Ramsey, and Sleeper,* Architectural Graphic Standards, *7th ed., © 1981 by John Wiley & Sons.)*

can be seen in the Time–Life Chicago Subscription Office building (Fig. 6.115), where this perimeter system serves the outer 15-ft band of open-office work stations. The open-plan spaces are 30 ft wide, so the inner 15-ft band is served by a central system in the core, where all the return air is also gathered. Another example of perimeter high-velocity induction units is provided by the World Trade Center (Fig. 6.29), where below-window units are served by high-velocity air and by water distributed from the central core of the building.

(b) Fan Coil With Supplementary Air (Fig. 6.116). This system is closely related to the preceding induction system; however, this system uses a fan at each unit, rather than high-velocity primary air, to move room air through the unit. Figure 6.117 shows a building that includes several variations on fan coils. The Iowa Public Service Building is an innovative 167,000-ft^2 utility office building that utilizes solar collectors and ice storage, supplemented by small backup boilers. Six ice-making machines serve a 75,000-gal ice storage pit with 90 million Btu capacity. The entire system, including building security and fire alarms, is controlled by computer. Heat exchange opportunities include reject heat from the ice-making machines, heat from the central toilet exhaust air, and heat from return air taken through luminaires. Solar collectors on the roof preheat the ventilation air, which is fed into the ceiling plenum at each floor. The fan-coil units then draw from this fresh air supply.

(c) Radiant Panels with Supplementary Air (Fig. 6.118). Large areas of radiant surface can be used to offset large losses of bodily radiant heat, as in the case of large areas of cold glass on a winter day, or when users are both scantily clothed and sedentary. In summer, such panels can help offset radiant gain from electric lights or large glass areas. The ceiling is often favored for the panel location, because it is uncluttered by the furniture, tackboards, and other items that cover floors and walls. However, floors are sometimes used, as in the Bleshman Regional Day School (Fig. 6.119). The designers of this school for the handicapped recognized that the majority of its users would spend much of their time quite close to the floor, and that the colder air near the floor could be uncomfortable, especially in the New Jersey winter. The entire floor is warmed by the supply (ventilation) air in winter, which enters just below windows to counteract the downdraft off cold glass. In summer, the cool supply air first cools the floor, then cools the air in front of the warm glass. The concrete cellular "air floor" provides a thermal mass that helps maintain steady temperatures. Heated or cooled air is provided by rooftop air–air heat pumps, one of which is provided for each cluster of three to six classrooms. This HVAC example is therefore related to the direct refrigerant family, discussed in Section 6.12.

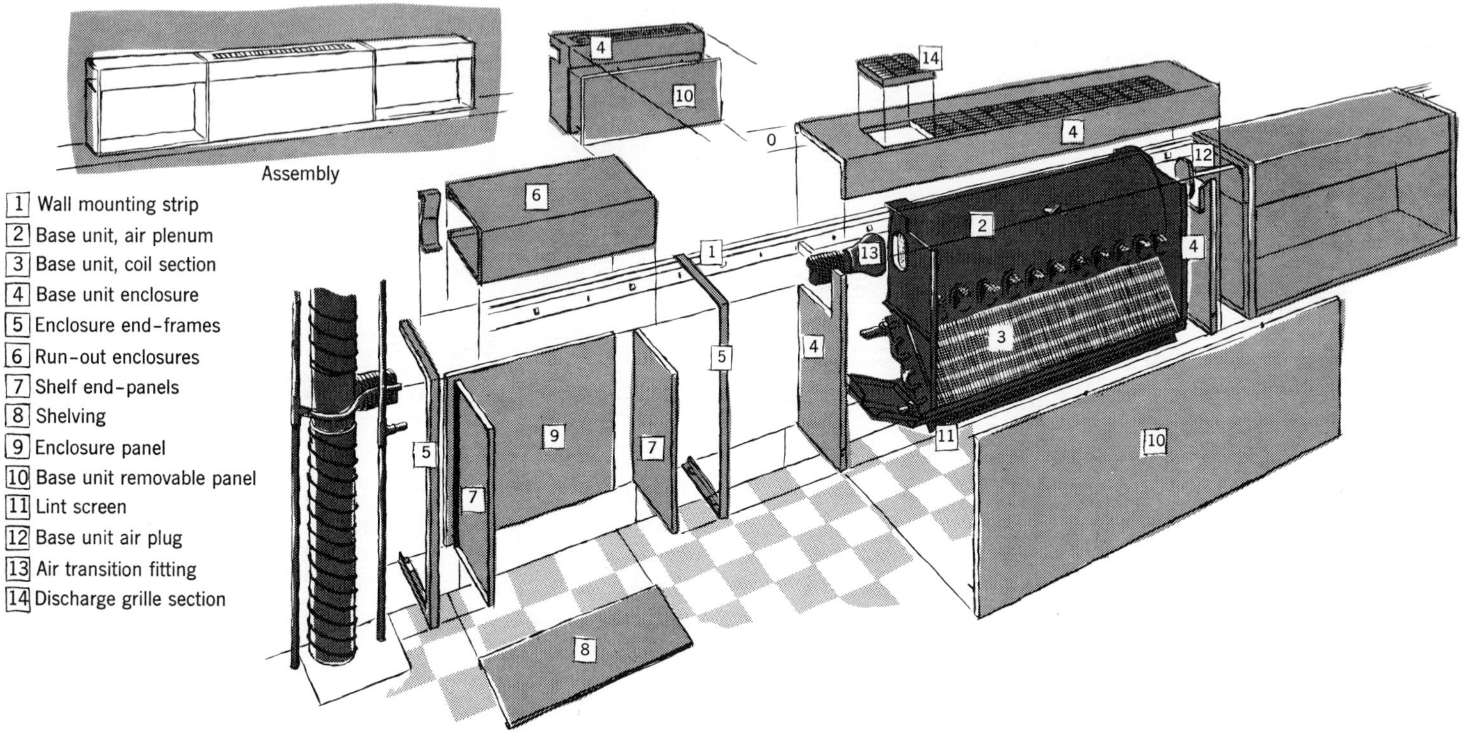

Fig. 6.113 *High-velocity induction system. Conditioned outdoor air for ventilation and to induce circulation of room air is brought in through a single high-velocity duct. It is attenuated and silenced in the chamber. (2) and then, through jets in the front of this plenum, it induces flow of room air, which is heated or cooled at finned coil (3). The lint screen (11) requires periodic maintenance for proper air flow. (Courtesy of Carrier Corporation.)*

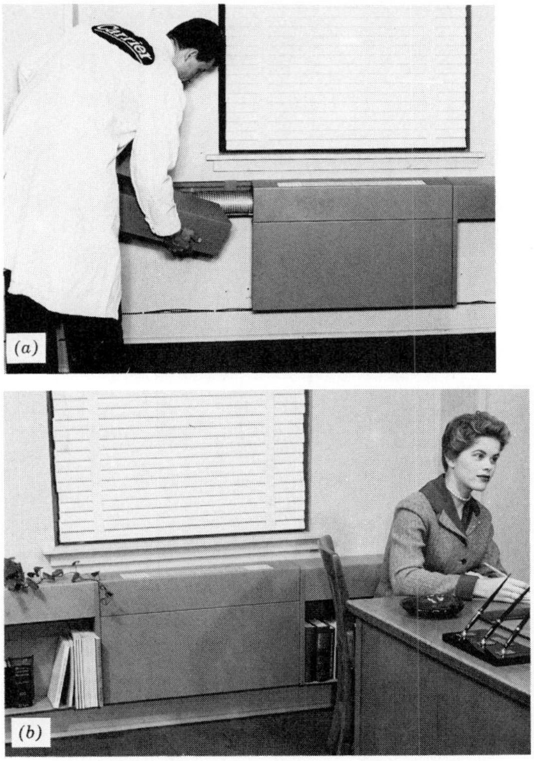

Fig. 6.114 High-velocity induction unit in place. (a) Installing the cover over the lateral branch of the duct that carries ventilation air to the unit. (b) Induction unit, as part of a bookshelf arrangement.

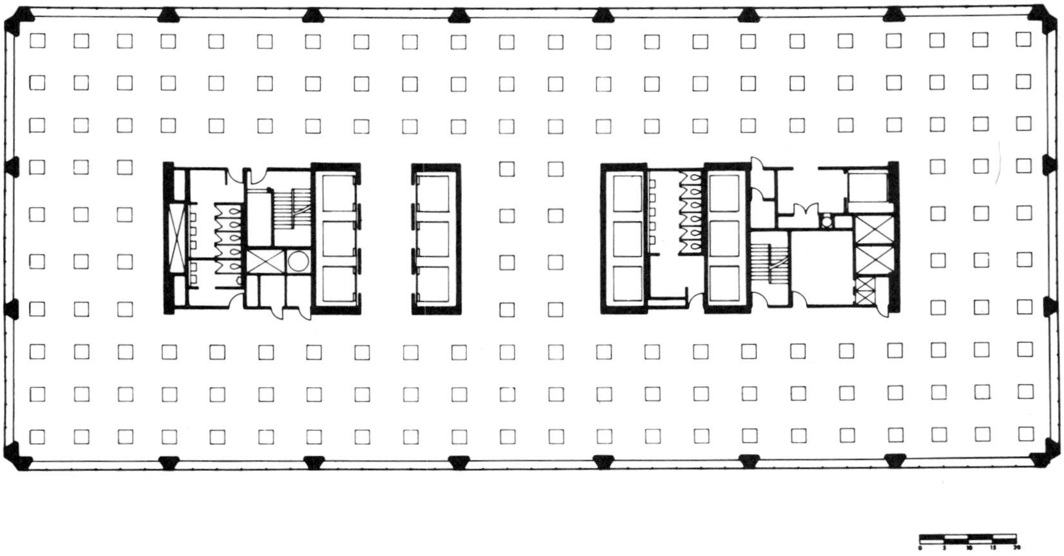

(b)

Fig. 6.115 Induction units in the Time–Life Chicago Subscription Office building. Harry Weese and Associates, architects. (a) Exterior. (Photo, Daniel Bartush.) (b) Plan; all work stations are within 30 ft of a window.

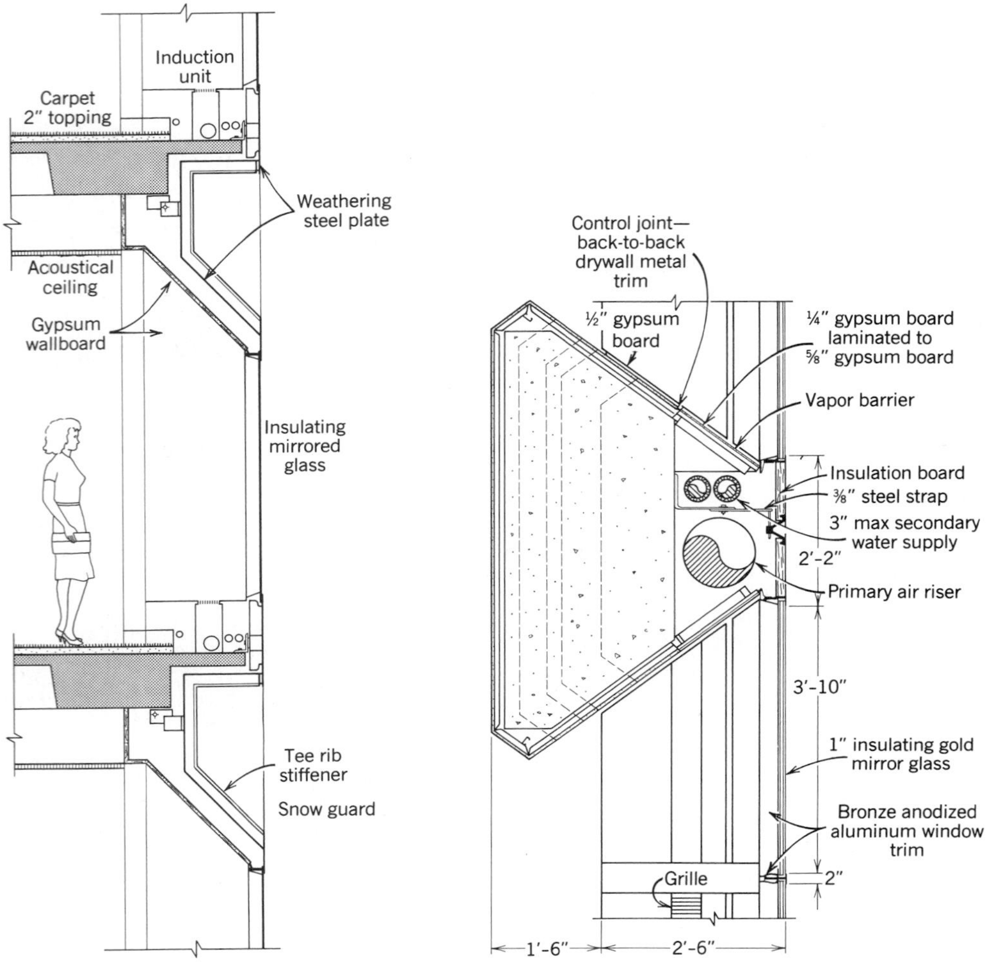

(c) *(d)*

Fig. 6.115 *(continued) (c) Section illustrating the unusually low-height induction unit. (d) Plan at exterior columns (30 ft apart), showing the primary air-supply and water-distribution trees.*

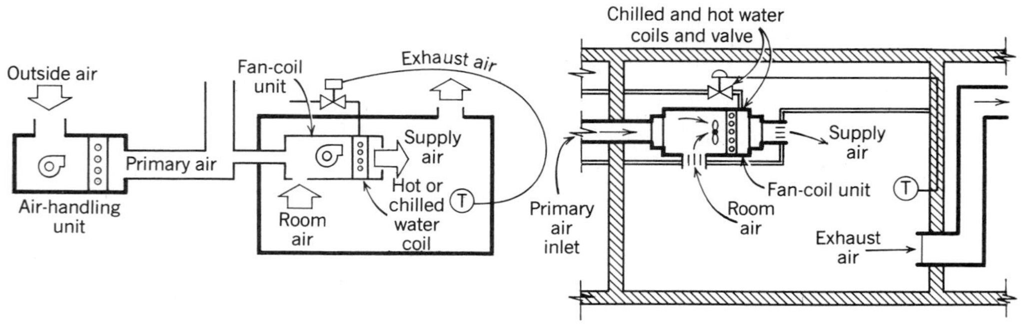

Fig. 6.116 *Fan-coil unit with supplementary air. These also are frequently located below perimeter windows.*

Fig. 6.117 *Iowa Public Service Headquarters, Sioux City, Iowa. Joint-venture architects: Rossetti Associates and Foss, Engelstad, Heil Associates.* (a) *Southeast elevation, with roof garden at the top floor.*(b) *Interior of the daylit atrium. (Photos by Balthazar Korab.)*

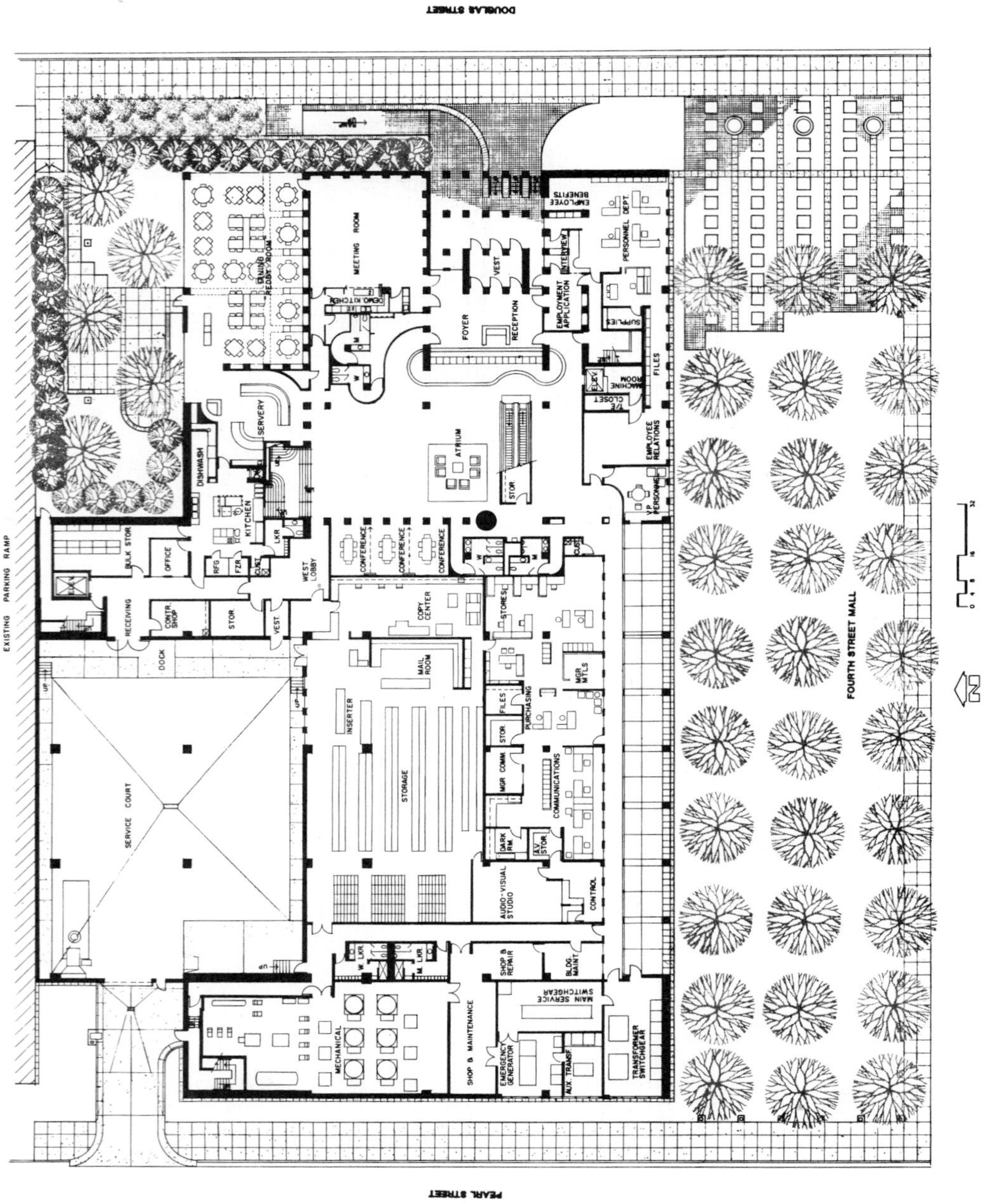

FIRST FLOOR PLAN
(c)

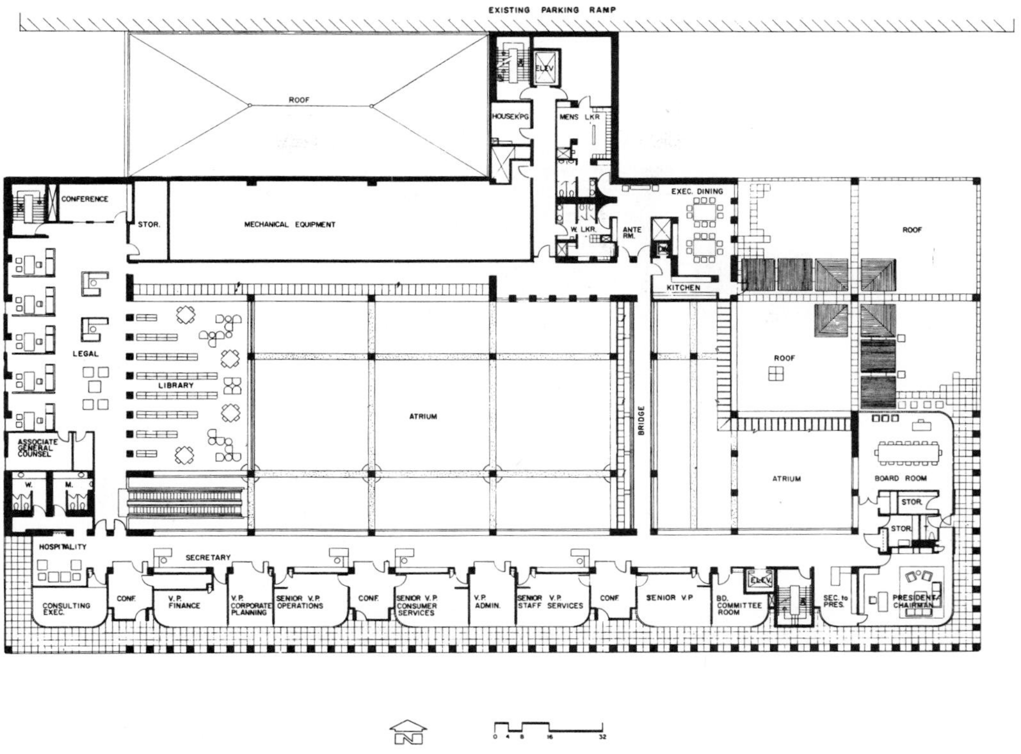

FIFTH FLOOR PLAN

(d)

Fig. 6.117 *(continued)* (c) *Ground floor plan; six ice machines sit in mechanical room.* (d) *Top (fifth) floor plan; note the proportion given over to daylighting of lower floors.*

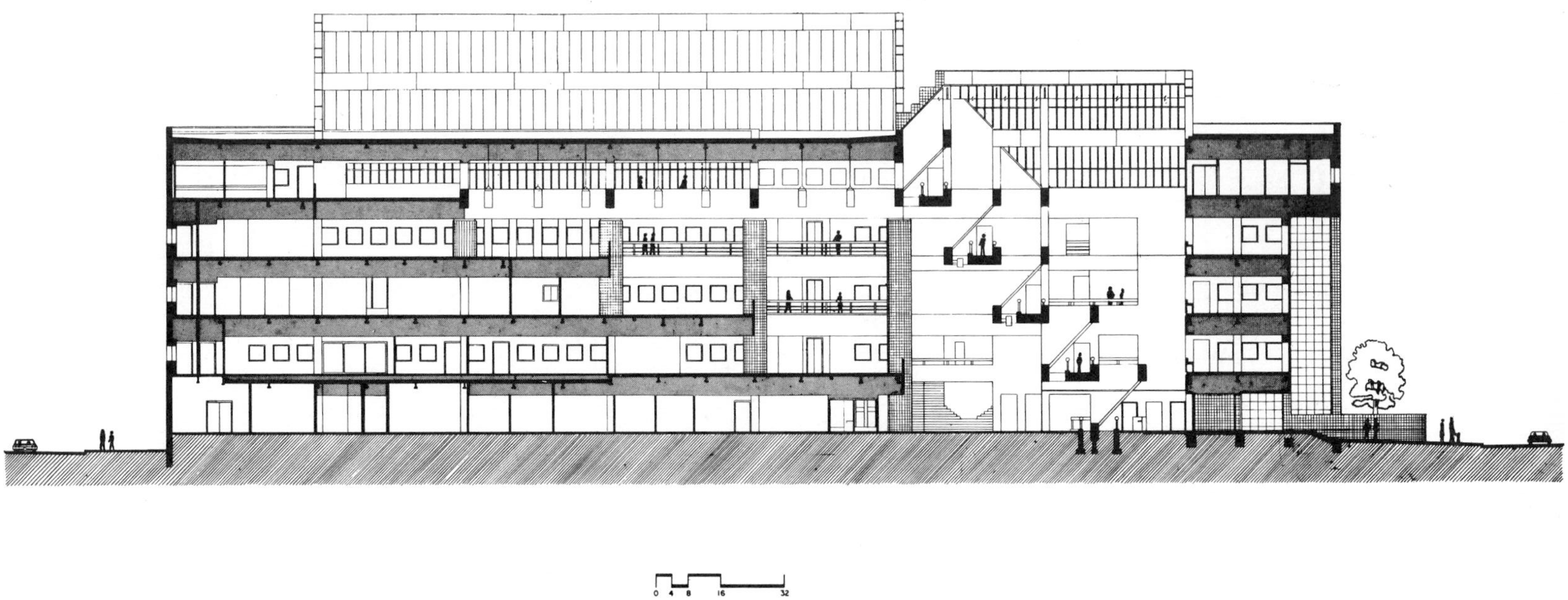

WEST - EAST SECTION LOOKING NORTH

(e)

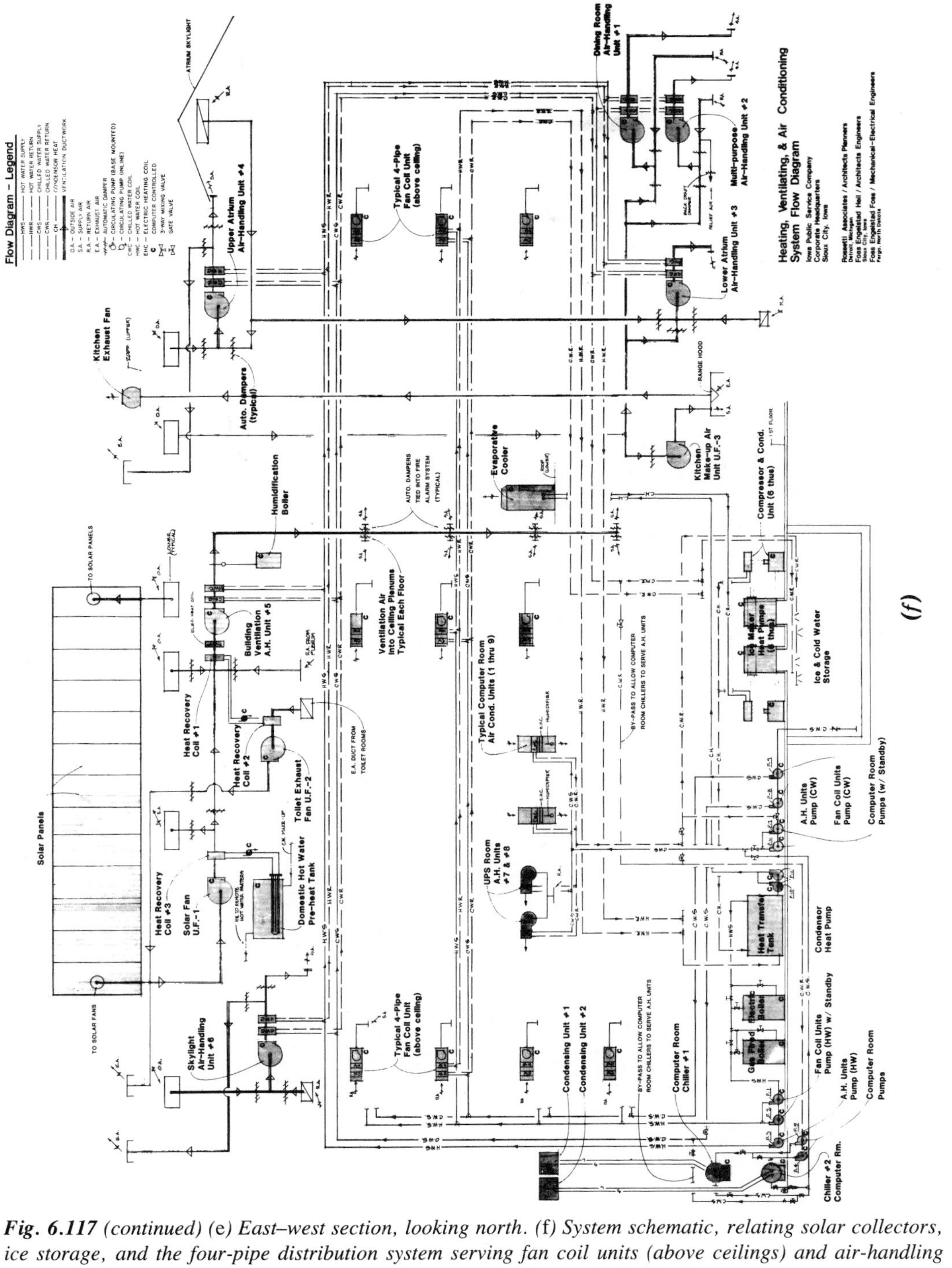

Fig. 6.117 (continued) (e) *East–west section, looking north.* (f) *System schematic, relating solar collectors, ice storage, and the four-pipe distribution system serving fan coil units (above ceilings) and air-handling units in many zones. Heat recovery units are also shown. (Courtesy of FEH Associates.)*

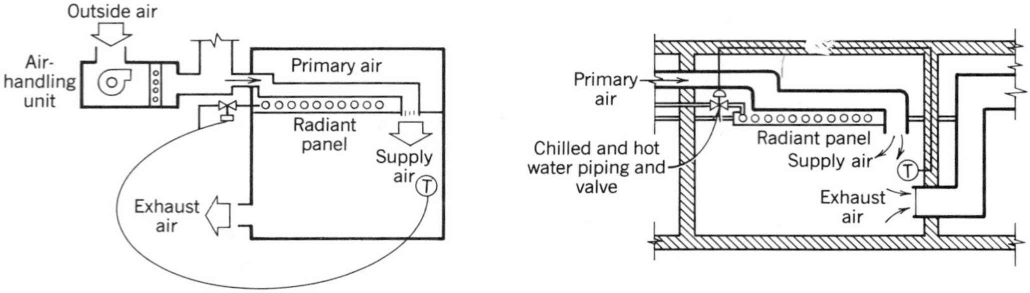

Fig. 6.118 Radiant panel with supplementary air. Walls or floors are also locations for these large-area panels.

Fig. 6.119 The Bleshman Regional Day School (for the multiple-handicapped), Paramus, New Jersey. Rothe-Johnson Associates, architects. (a) *Lobby. (Photo by Otto Baitz.)*

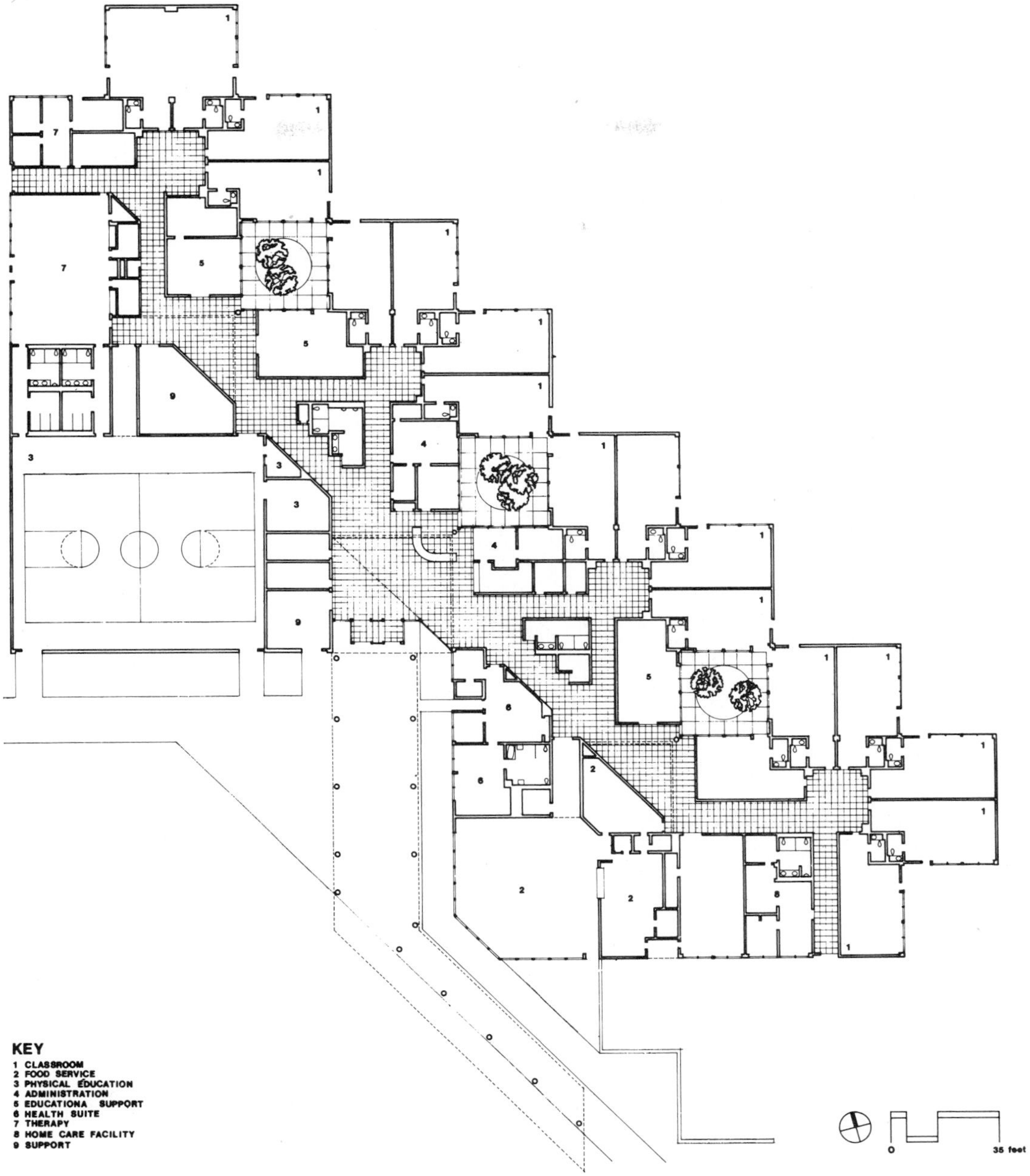

(*b*)

Fig. 6.119 *(continued) (b) Plan, showing courtyards that bring daylight between clusters of classrooms and into central circulation spine.*

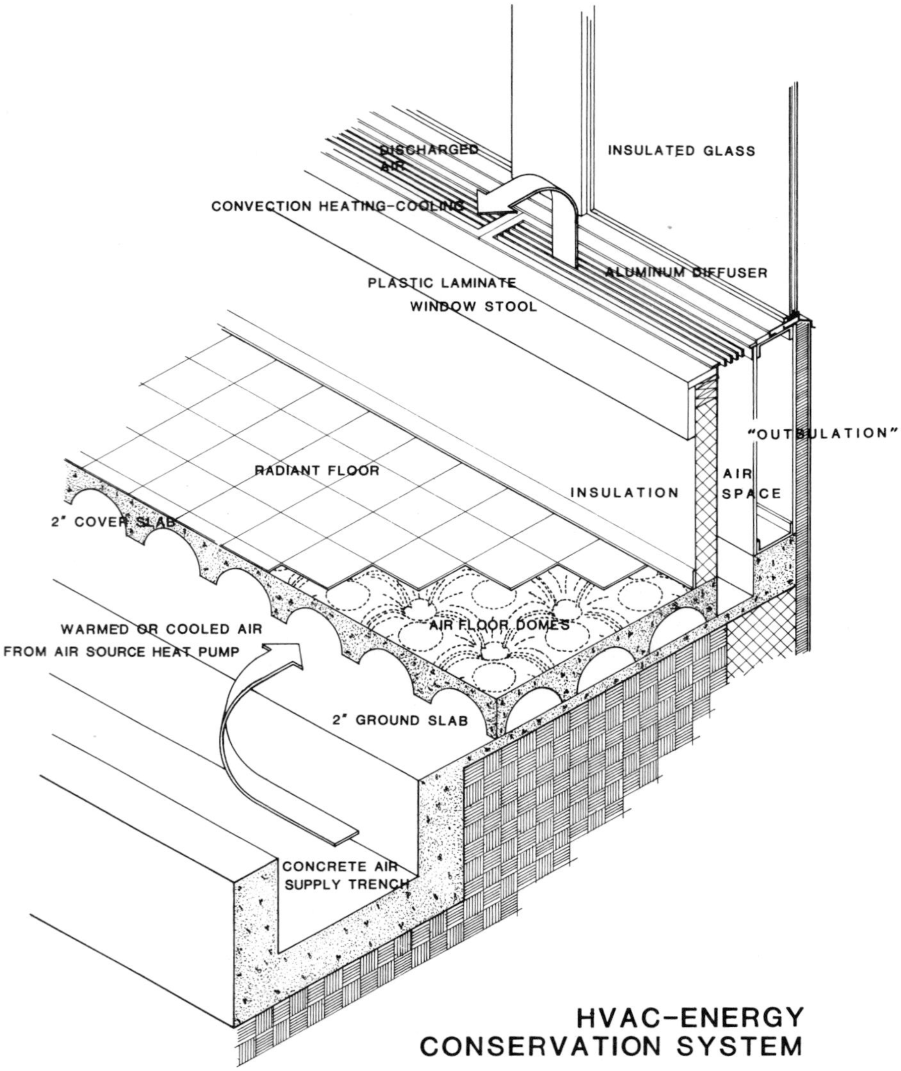

(c)

Fig. 6.119 *(continued)* (c) *Detail of the cellular concrete ''air floor'' and supply air diffuser below windows.*

6.11 All-Water Systems

These systems (see Fig. 6.69 and Section 6.7) typically deal only with temperature control; air quality is left to separate systems. In its most simple application, the *two-pipe heating-only radiator* (discussed at length in Section 6.6) is getting new exposure with some colorful and pleasing products (Fig. 6.120). The Mayer Art Center at Phillips Exeter Academy in New Hampshire (Fig. 6.121), features new exposed radiators in some older buildings. These radiators are based on simple components (typically 2¾ in. wide) that can be combined in many heights and widths, inviting the designer to feature them rather than to hide them in metal cabinets.

An especially familiar all-water application is the simple *fan-coil unit* (Fig. 6.122). These units, which may be found above ceilings, below windows, or in corners, simply control the tempera-

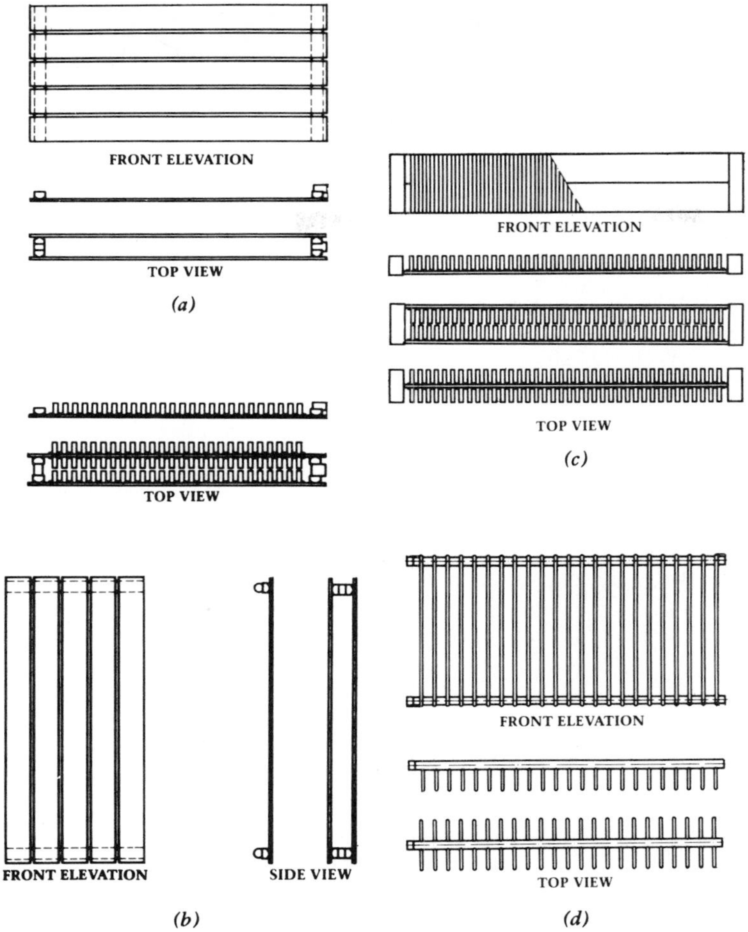

Fig. 6.120 *Hot water radiators, available in bright colors and based on simple components. (a) Flat panels, 2¾ in wide, are primarily effective in radiant heat transfer. (b) Finned panels, also 2¾ in. wide, add convective transfer. (c) Thicker, toothlike fins are particularly strong convectors. (d) Traditional ''radiators'' are formed from repeating flat tubes on edges, joined at either end with another flat tube that becomes a header. (Courtesy of Runtal/North American Energy Systems.)*

Fig. 6.121 *Exposed hot water radiators at the Mayer Art Center, Phillips Exeter Academy, Exeter, New Hampshire. Amsler Hagenah MacLean, architects. (a) Under-window panel radiator. (Photo by Alex Beatty.) (b) Radiator formed by repeating flat tubes. (Courtesy of Runtal/North American Energy Systems.)*

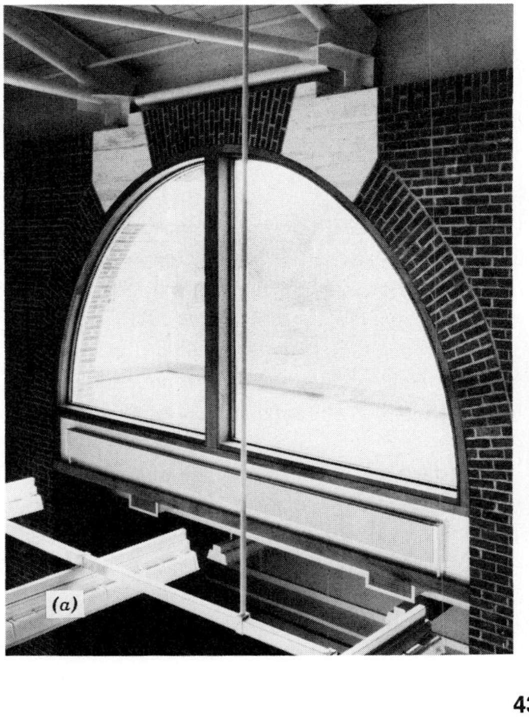

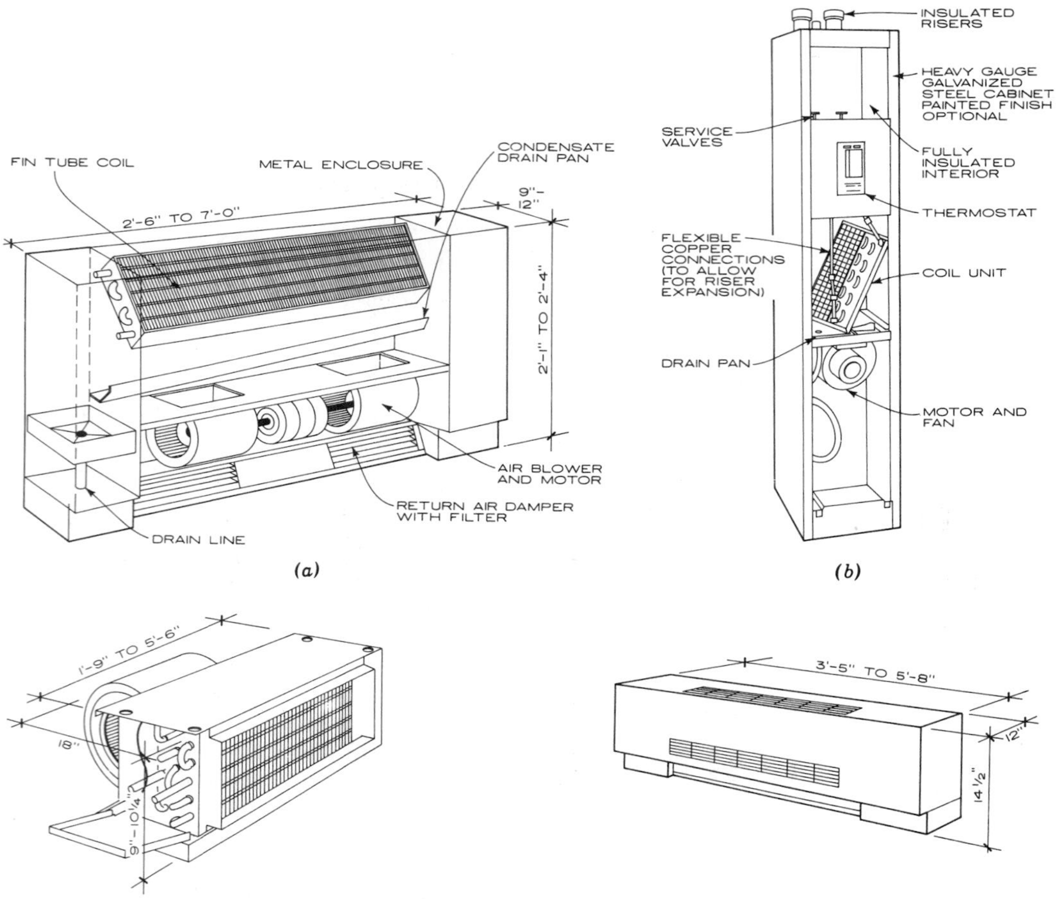

Fig. 6.122 Simple fan-coil units, without fresh air. (a) *Standard below-window unit.* (b) *''High-rise'' unit, for corners or cabinet locations.* (c) *Above-ceiling unit.* (d) *Low-profile unit (such as used in Time–Life Building, Fig. 6.115). (Reprinted by permission from AIA, Ramsey, and Sleeper,* Architectural Graphic Standards, *7th ed., © 1981 by John Wiley & Sons.)*

ture (and to an extent, the relative humidity) of the air already in the room, which is blown through the coils. Because water is often condensed from the room air when cooling is in progress, a drain line is required. Exterior air intake grilles can easily be added when fan-coil units are located below windows, to allow the local provision and tempering of ventilation air.

All-water perimeter systems with local fresh air can also take the simple form of *operable windows with hot water finned-tube radiation*. An expressive variation on this approach appears in the reading rooms and staff workrooms for the Seeley G. Mudd Library at Yale University (Fig. 6.123). Limestone spandrels are curved in to allow fresh air to enter these smaller perimeter rooms just below the windows, where hot water finned-tube radiators are available when needed. The incoming fresh air replaces exhaust air, which flows out the upper operable windows. The remainder of the building has conventional forced-air heating and cooling.

The last example of an all-water system involves the *water–air heat pump loop*. In the 175,000-ft^2 Comstock Center in Pittsburgh (Fig. 6.124), there are six to eight small heat pumps on each of 10 floors, located above the suspended ceiling. Their connecting loop doubles as the

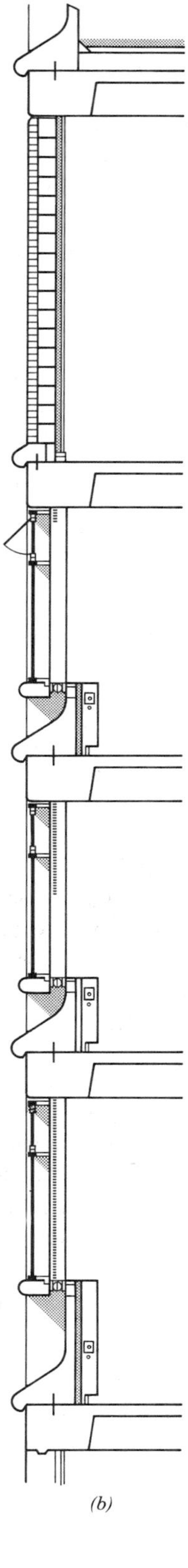

Fig. 6.123 Local fresh air and finned-tube radiation at the perimeter. The Seeley G. Mudd Library at Yale University; Roth and Moore, architects. (a) *Exterior, showing curved limestone spandrels, along which flows incoming fresh air. (Photo © Steve Rosenthal.)* (b) *Section, showing the fresh air intake, finned-tube radiation, and upper operable sash for exhaust air. This system is used for the smaller reading and staff workrooms at the perimeter (at right end of plan.)* (c) *Plan.*

building's wet-pipe sprinkler system supply. (This is possible because the heat pumps keep the loop between 65 and 85 F.) Since neither hot nor chilled water-supply and -return distribution trees are needed, there is a substantial first-cost saving, and relatively little building volume is consumed. A 23,000-gal water storage tank allows daytime heat, rejected to the loop by the heat pumps, to be stored and recalled for nighttime heating. When necessary, a small boiler will maintain the 65 F minimum temperature in the water loop.

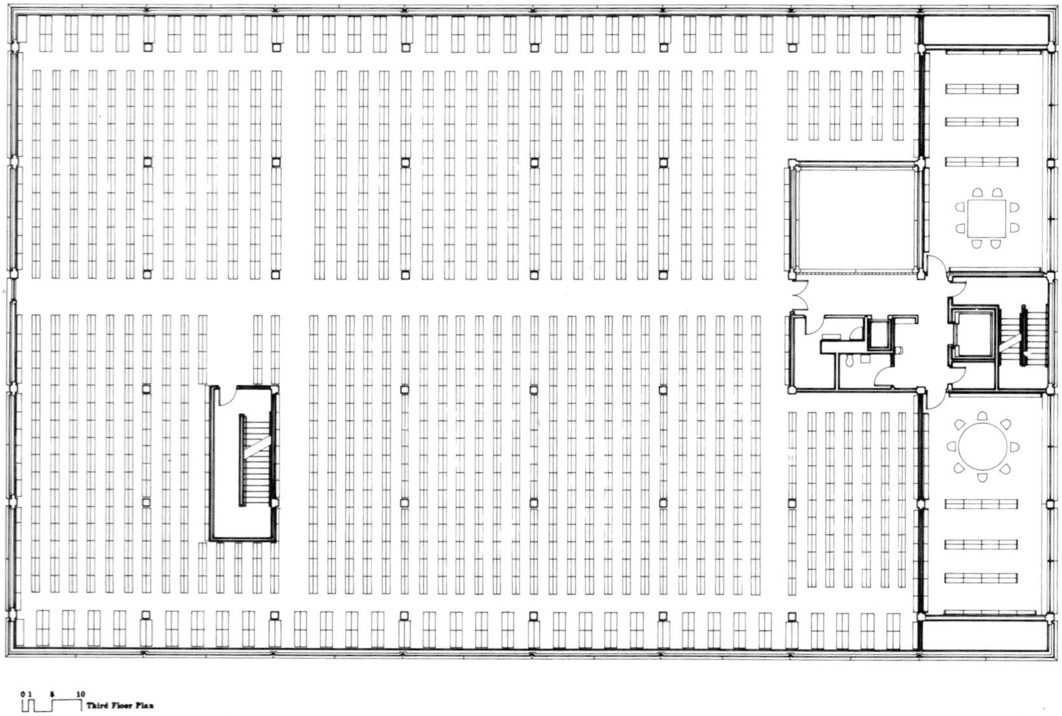

(c)

Fig. 6.123 (continued)

Another feature of the Comstock Center is its use of *extract-air windows* to control infiltration and to moderate the perimeter zone's temperatures. In these triple-glazed windows, return air from the room is drawn up between the outer double-glazed window and an inner single sheet of glass. This air warms the inner pane in winter, substantially increasing radiant comfort and eliminating the need for a separate perimeter heating system. In summer, the same procedure cools the inner glass. After the return air arrives in the ceiling plenum, it is mixed with fresh air, then tempered and recirculated by the heat pumps.

The Comstock Center also utilizes a large daylighting atrium, whose temperature control is provided largely by exhaust air from the offices; the stack effect is utilized to provide natural ventilation on its west face in summer conditions.

The proliferation of individual air-handling units illustrated in this example, linked only by a common water loop, is a characteristic shared by the next HVAC family—direct refrigerant systems.

6.12 Direct Refrigerant (Incremental) Systems

The most distinguishing feature of this approach to HVAC systems is decentralization. As shown in Fig. 6.70 and Section 6.7, these systems replace one or more central machines and associated distribution trees, with many dispersed machines requiring little, if anything, in the way of a distribution system. An analogy might be made between the first factories, which relied on a central power source (such as a water wheel) and distributed the motive power from that source to every machine within the factory via an elaborate system of belts and pulleys. Today, individually powered machines are far more common. And in buildings, dispersed machines (such as were just shown in the Comstock Center) are becoming more common as the savings (from elimination of distribution trees) and the convenience (of separate machines for separate zones) become increasingly attractive. Since these machines are easily added

GROUND FLOOR PLAN

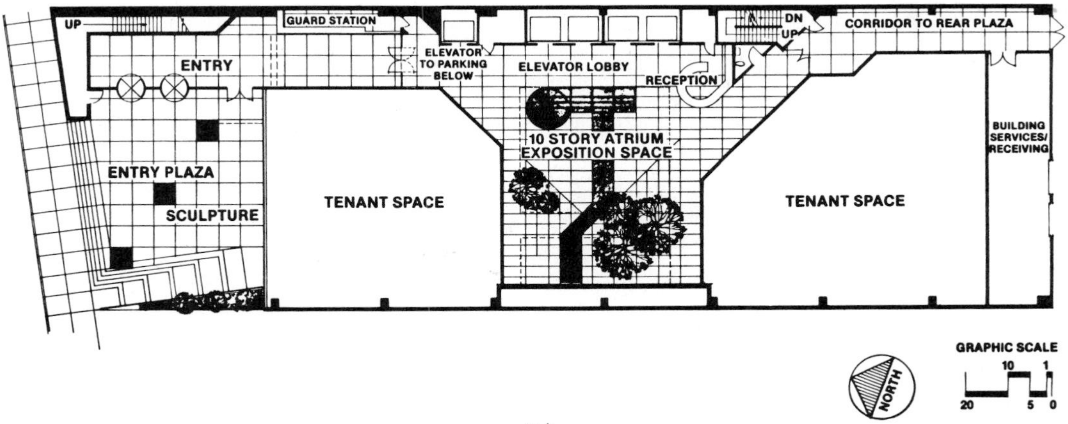

(b)

Fig. 6.124 The Comstock Center, Pittsburgh. Burt Hill Kosar Rittelmann Associates, architects. (a) View of northwest corner. (b) Ground floor plan. (c) Water loop heat pumps, combined with sprinkler system supply on each floor. Also called "tri-water system." (d) Extract-air windows control infiltration and moderate perimeter zone temperatures. Fresh air is ducted to the plenum, where it mixes with return air before being treated and recirculated by the heat pumps. (See also Fig. 6.7) (e) The central daylighting atrium is tempered by exhaust air from the offices; exhaust air then leaves via the stack effect. (f) The stack effect also controls summer overheating.

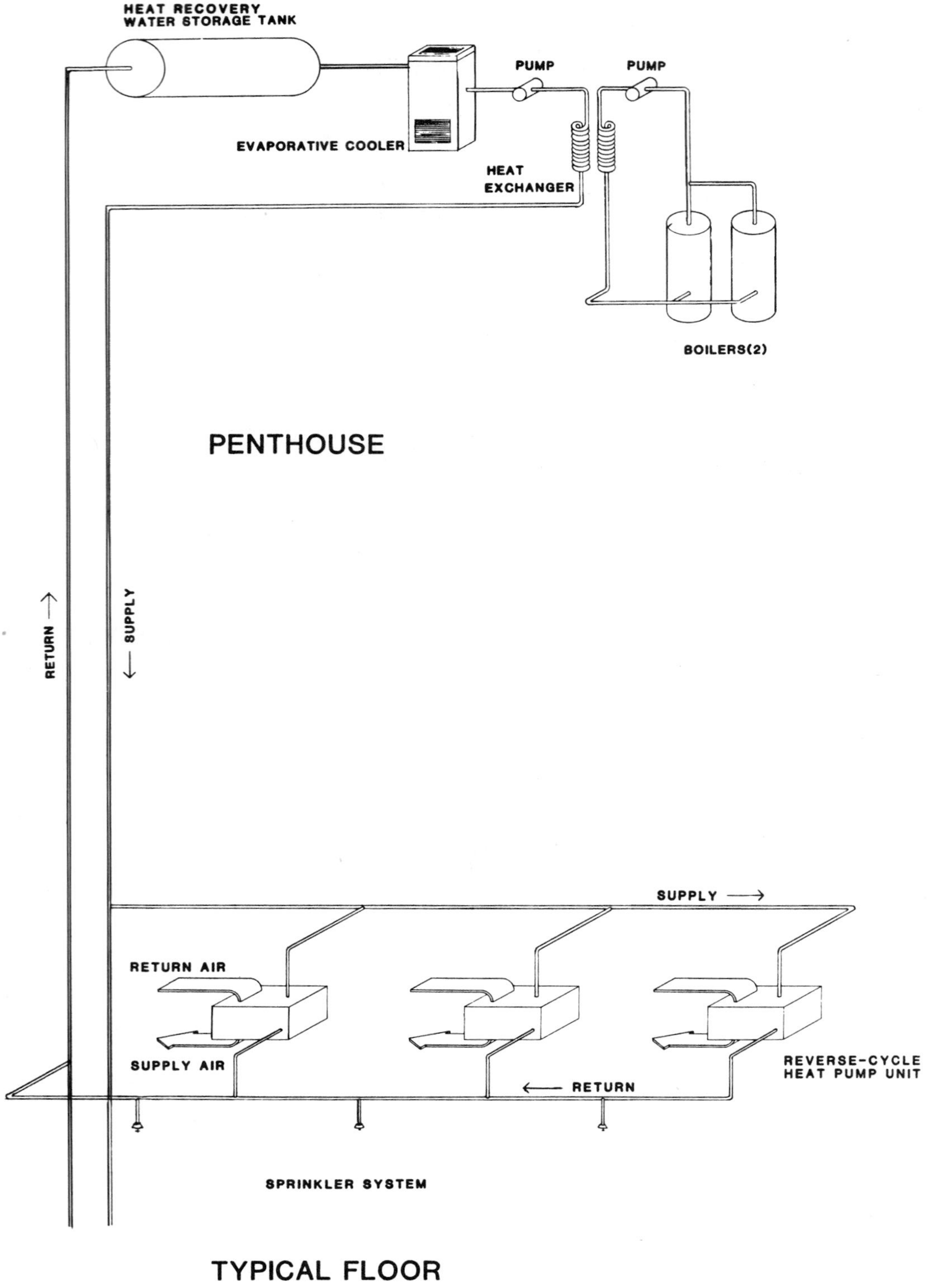
HEAT RECOVERY
WATER STORAGE TANK
PUMP
PUMP
EVAPORATIVE COOLER
HEAT
EXCHANGER
BOILERS(2)
PENTHOUSE
RETURN →
← SUPPLY
SUPPLY →
RETURN AIR
SUPPLY AIR
← RETURN
REVERSE-CYCLE
HEAT PUMP UNIT
SPRINKLER SYSTEM
TYPICAL FLOOR

(c)

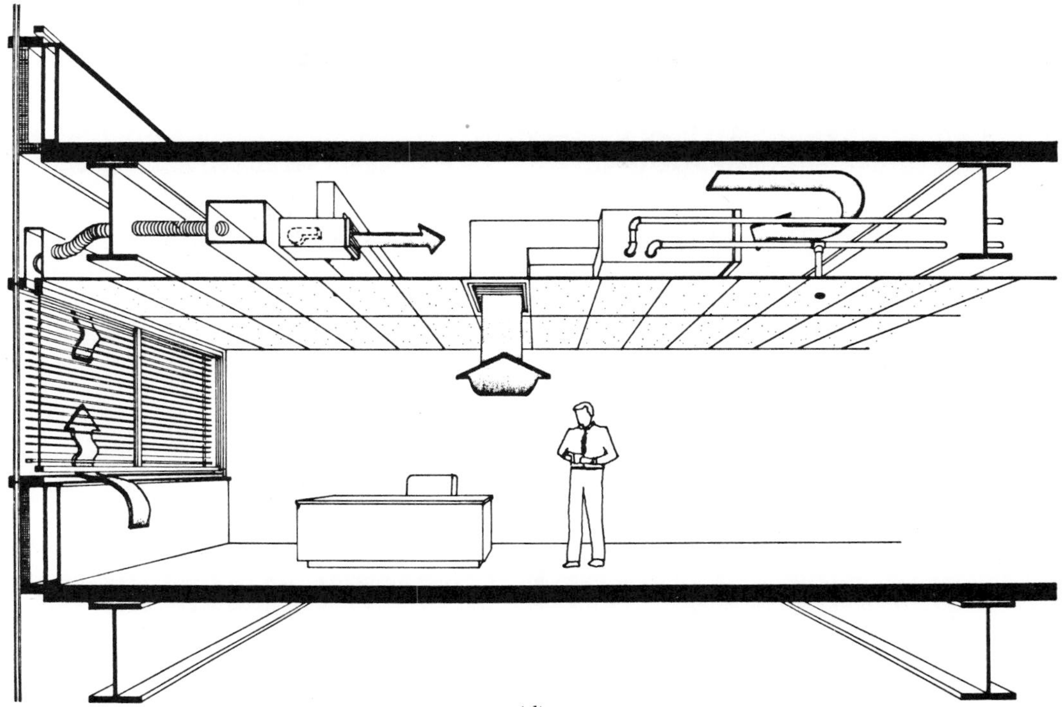

(d)

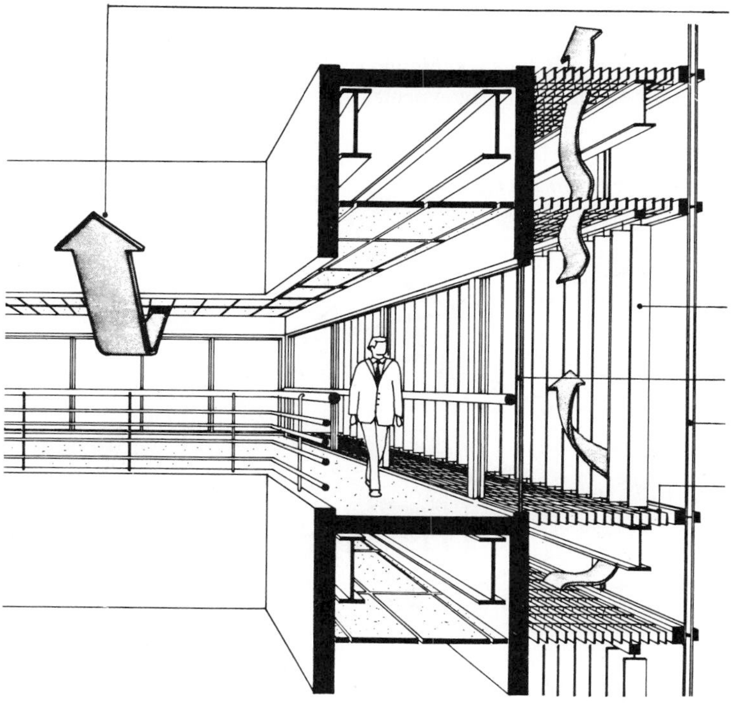

(e)

Fig. 6.124 *(continued)*

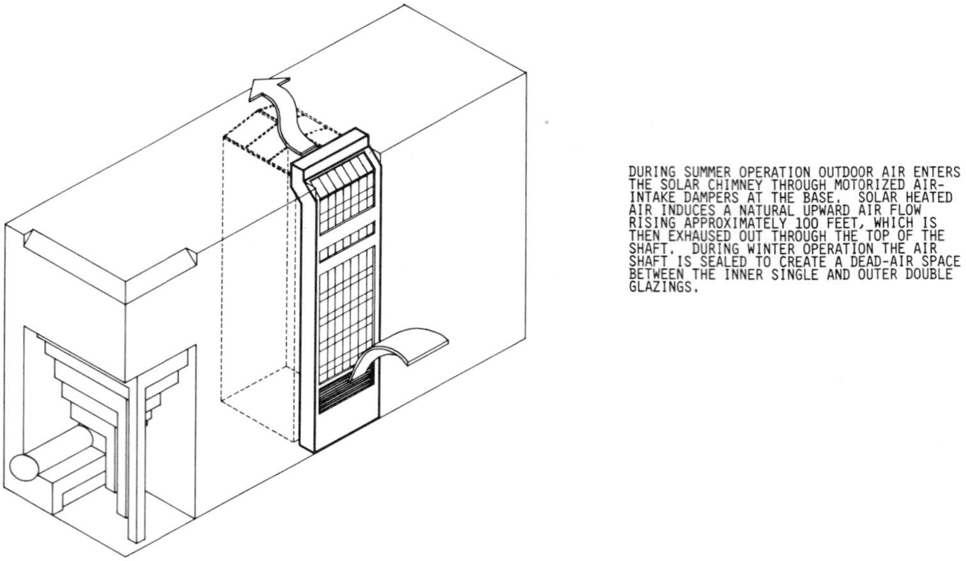
DURING SUMMER OPERATION OUTDOOR AIR ENTERS THE SOLAR CHIMNEY THROUGH MOTORIZED AIR-INTAKE DAMPERS AT THE BASE. SOLAR HEATED AIR INDUCES A NATURAL UPWARD AIR FLOW RISING APPROXIMATELY 100 FEET, WHICH IS THEN EXHAUSTED OUT THROUGH THE TOP OF THE SHAFT. DURING WINTER OPERATION THE AIR SHAFT IS SEALED TO CREATE A DEAD-AIR SPACE BETWEEN THE INNER SINGLE AND OUTER DOUBLE GLAZINGS.

(f)

Fig. 6.124 *(continued)*

as zones proliferate, they are often called incremental units.

One of the simplest of these dispersed approaches is the *makeup air* approach, shown in Fig. 6.125. Especially common in factories or laboratory buildings with high exhaust air requirements, these simple devices either heat or cool the incoming fresh air that replaces air being exhausted. They often supplement the building's main heating/cooling system, which deals with heat gains/losses through the building's skin.

Incremental (or direct refrigerant) units are

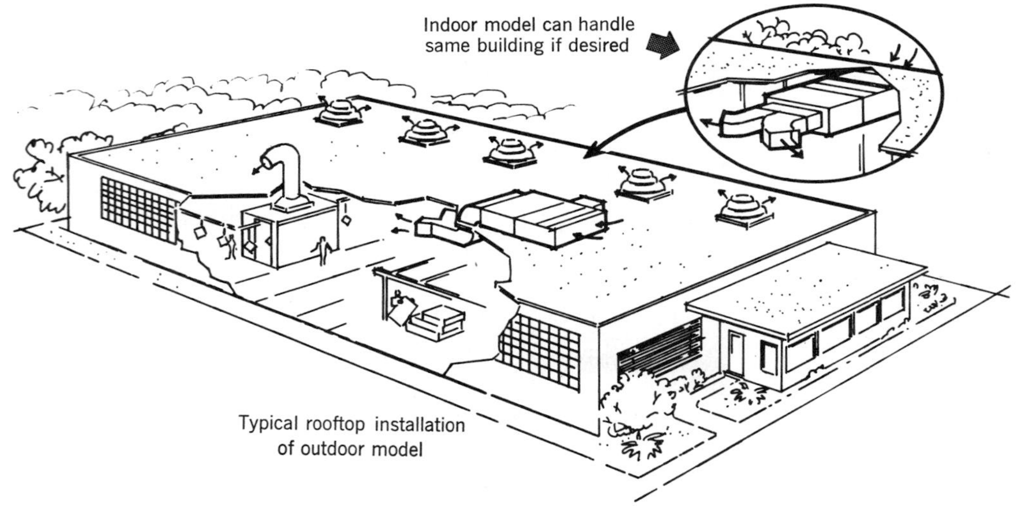
Indoor model can handle same building if desired
Typical rooftop installation of outdoor model

Fig. 6.125 *Makeup air unit heaters/coolers, distributed on a factory roof. These units prevent negative indoor air pressures by providing heated or cooled outdoor air to replace air being exhausted. Often, the building will have an additional (main) heating/cooling system, for overcoming heat losses/gains through the building's skin.*

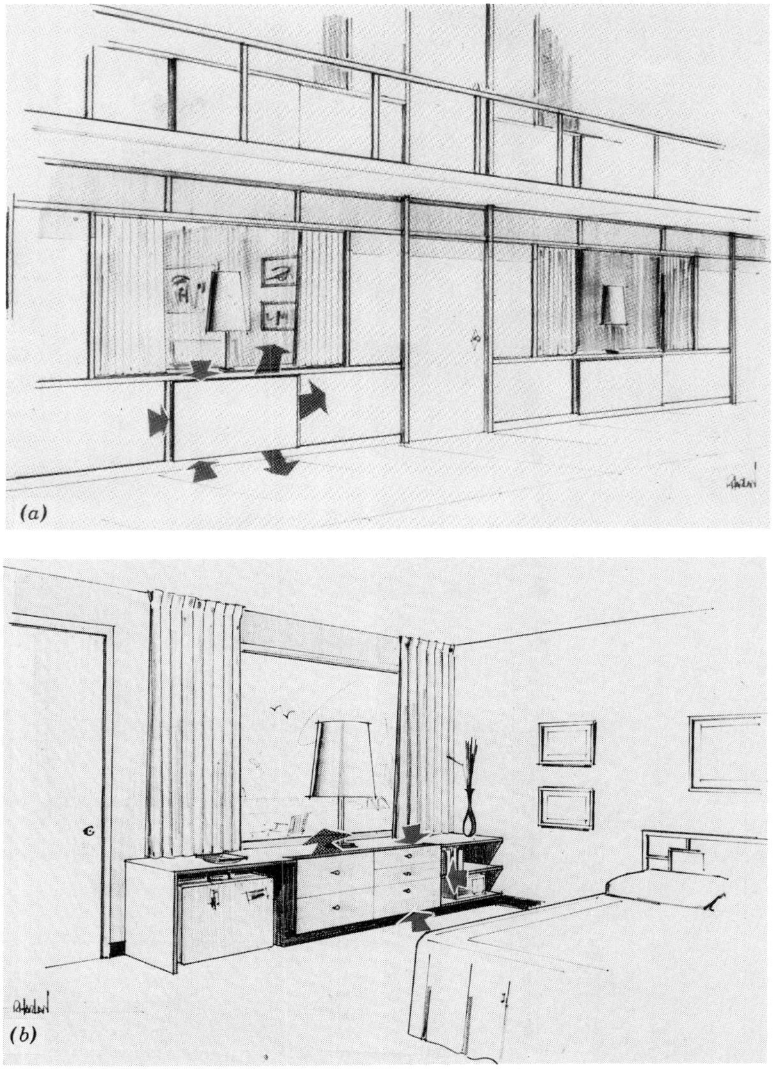

***Fig. 6.126** (a) In this application of the heat pump in a motel, a panel set forward of the exterior wall line allows the pump to inhale and exhale outdoor air around the panel edges. In summer it discharges warm air; in winter, cool air. (b) Interior cabinet detail and depth serve the heat pump. Room air is taken back as shown and discharged (after cooling or heating) upward across the glass surface. (Courtesy of General Electric Company.)*

especially common in building types distinguished by all-perimeter spaces with varying orientations and numerous thermal zones. Motels are a prime example. In Fig. 6.126, separate air–air heat pumps serve each motel room; at best, their constant noise helps mask the intermittent sounds from the adjacent parking lot/circulation space. Opportunities for heat exchange between these heat pumps are scarce; if a central water loop were substituted for outdoor air as the heat source/sink, energy costs would go down (although first cost would rise).

Figure 6.127 shows another incremental air conditioner. Note that the compressor is placed on the outdoor side of the baffle, which helps to muffle the indoor sound that it makes. These units have proven quite popular for apartment houses and motels and are being adapted for use in some office buildings. They are advantageous for sev-

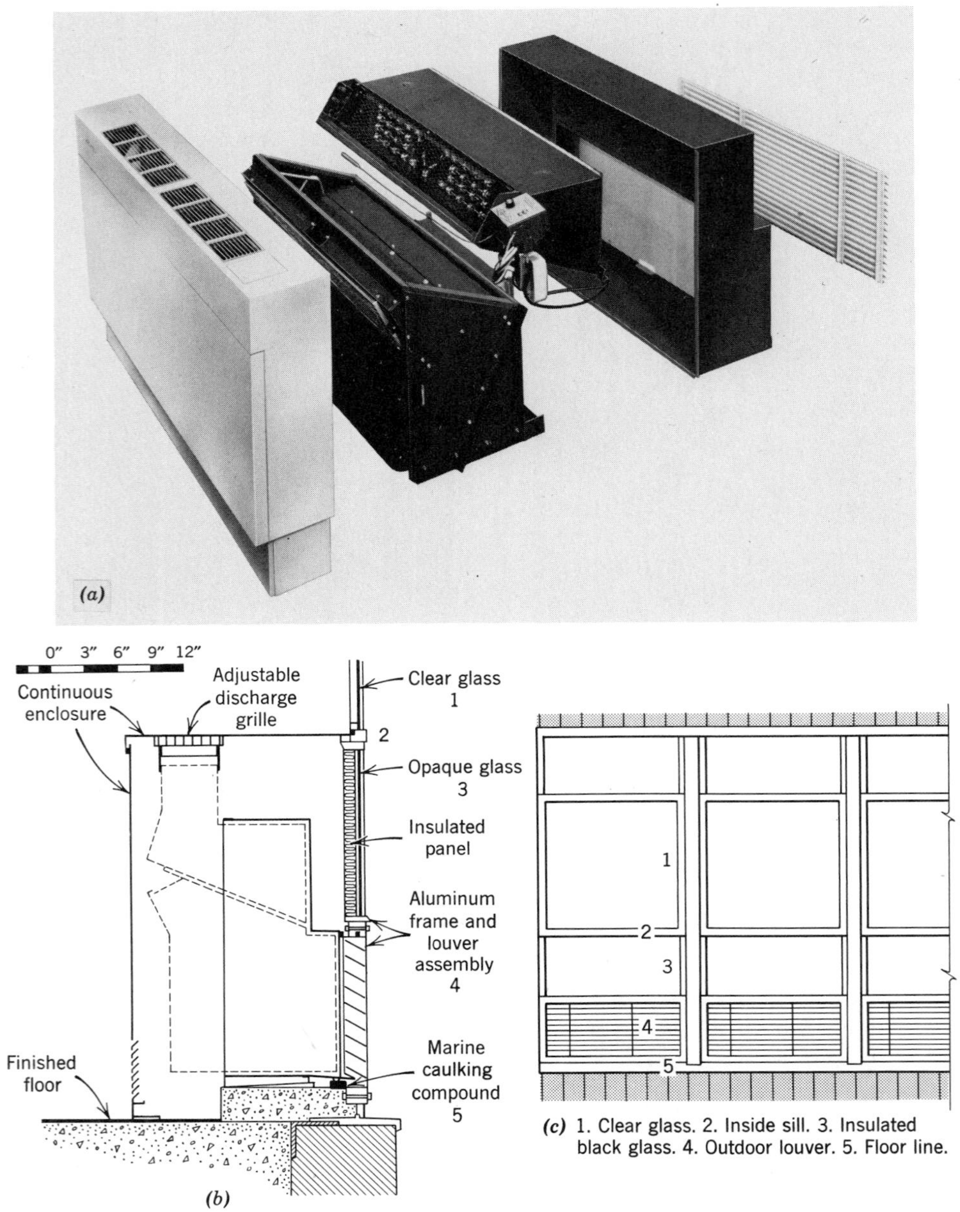

Fig. 6.127 *Incremental direct refrigerant unit.* (a) *Exploded view of the unit: left to right—room cabinet, electric-cooling chassis, room air circulation and electric-resistance heating element, wall box, and outside louver.* (b) *Section.* (c) *Facade treatment, Travelers Insurance Building, Boston. Kahn and Jacobs, architects (Courtesy of Climate Control Division, the Singer Company.)*

eral reasons. Control is in the hands of the occupant or tenant, relieving the management of complaints and making every room an individual zone. If a unit needs servicing, it is easy to remove the defective element and insert another. Cooling towers, central chillers, pumps, and piping for chilled and condenser water are all avoided, saving space and making unnecessary the services of a resident operating engineer. A disadvantage is that a compressor so close to the occupied space will always create some sounds, whereas a remote central chiller would be inaudible in the space to be conditioned. Also, large outdoor grilles sometimes present problems in architectural design, although a repeating element such as three-dimensional grilles (and their shadow patterns cast by the louvres) can provide welcome relief in an otherwise flat facade.

Incremental units easily accommodate the varying schedules of different tenants. Some years ago, the Remington Corporation developed a "triple overriding dual control" (TODC) system for the numerous incremental units in an office building in Syracuse, New York. The building owner establishes a master schedule—from 7:00 A.M. to 5:30 P.M., for example—during which all incremental units are turned on automatically. At 5:30 P.M., however, when all units are turned off, they are all immediately reset by an electric impulse over the regular power wiring. No additional electric or pneumatic controls are required for this—an economy in installation cost. When units are thus reset through the special TODC panel within each unit, a single tenant may turn on his or her conditioner for full operation by pressing a button. This will be the *only* conditioner operating in the building, unless others are similarly activated. At 9:00 P.M. and 12:00 midnight, impulses again turn off and reset all units then operating. The individual occupants may again press their buttons to continue the service, but if everyone has gone, the entire system shuts down. The occupant is only slightly inconvenienced by the need to push a button at three-hour intervals during an evening or on a weekend or holiday.

Another common application for incremental units is in schools, such as the Bleshman Regional School (see Fig. 6.119). In Fig. 6.128, a passively solar heated elementary school in New York combines several rooms with similar orientations and activities into a zone served by electric heating units above the ceilings. Fresh air is ducted to each unit; mechanical cooling is not provided.

Direct refrigerant (incremental) units are commonly located on roofs, where they have unlimited access to outdoor air and where their noise is less likely to annoy—provided they are sufficiently isolated from the building's structure. This approach is shown in the daylighted, passively solar heated Mount Airy, North Carolina, library (Fig. 6.129). This 14,000-ft^2 building also has a solar preheating system for its hot water. The five heat pumps are air–air and utilize economizer cycles. The average annual building energy consumption has been monitored at about 17,000 Btu/ft^2—approximately one-third of nearby similar-function buildings.

The final example of a direct refrigerant system is provided by an innovative tent structure over a San Francisco department store (Fig. 6.130). Two layers of fiberglass, Teflon-coated fabric, separated by an average of 12 in., are supported by a network of cables hung from eight masts. The tent roof covers about 70,000 ft^2 of sales floor; its 7% translucency to sunlight provides 450 to 550 fc of daylight. This greatly reduces the need for electric lighting, although some, clipped onto exposed fire sprinkler pipes, is still used as accent lighting. About 3.5 W/ft^2 of solar gain, mostly in the form of this diffused daylight, penetrates the tent cover. When this solar gain is combined with heat gains from people and electric accent lights, a cooling load is always generated in San Francisco's mild

Fig. 6.128 The Vincent Smith School, Port Washington, New York. Budd Mogensen, AIA, architect. South elevation, construction phase. Passive solar heat in winter serves the four classrooms and (through clerestory glass) also warms the student commons area. Natural daylighting is also characteristic for these spaces, and for the two north classrooms.

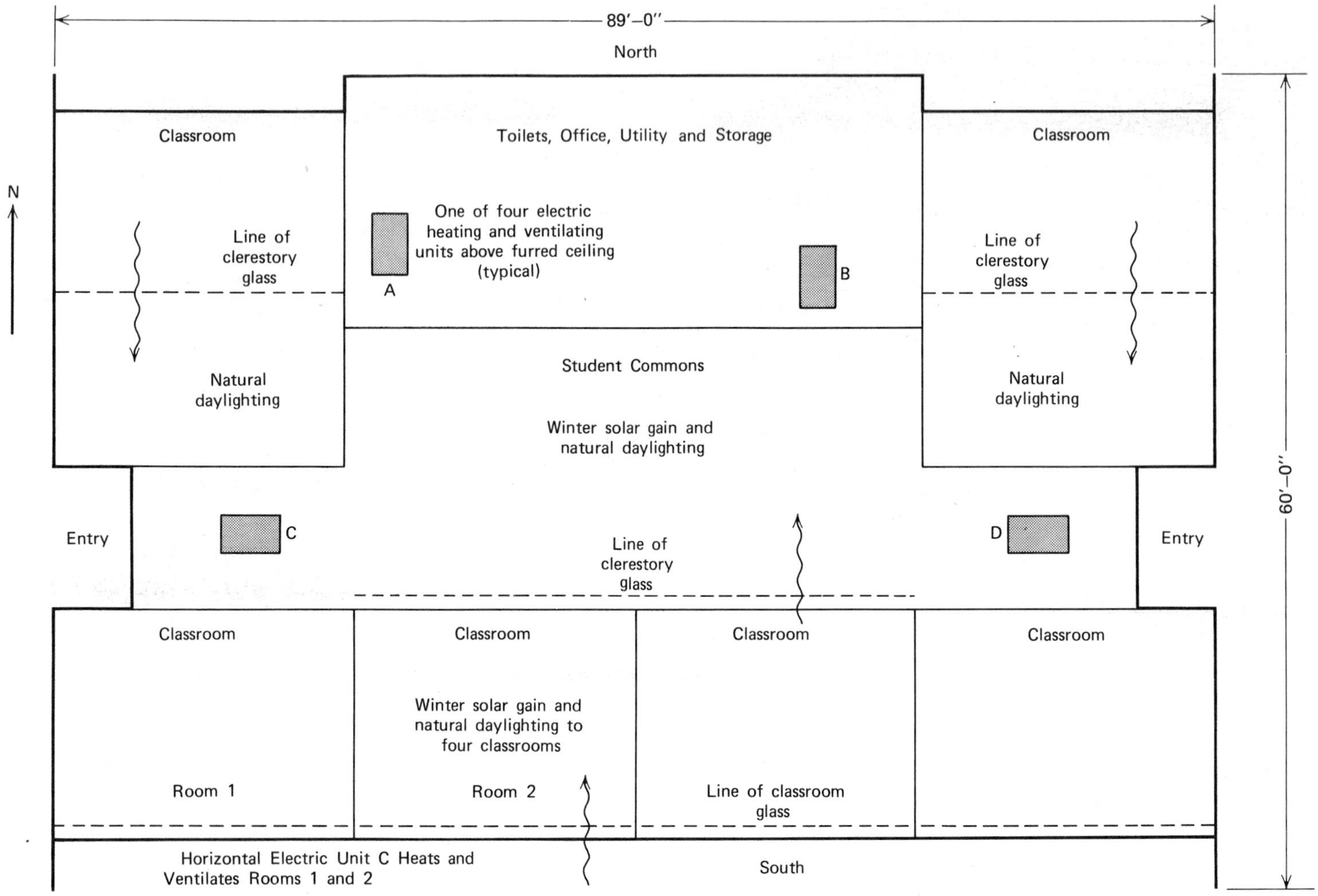

(b)

Fig. 6.128 *(continued)* (b) *Line drawing depicting space allocations. The four south classrooms and the student commons receive solar warmth and natural daylighting. The two north classrooms are lighted from above by clerestory fenestration in addition to their first-story glass as shown in* (a). *Shown also are the four climate control units, each of which adds some fresh air to the circulated warm air.*

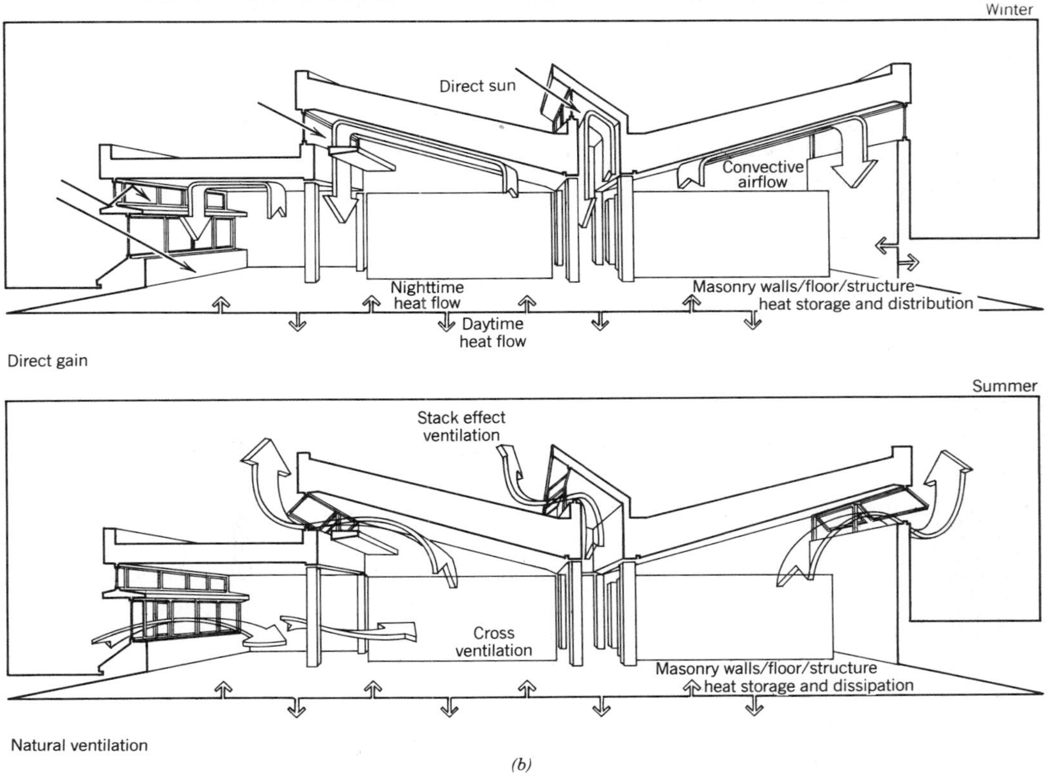

Fig. 6.129 *Mount Airy Library, Mount Airy, North Carolina. Mazria/Schiff & Associates, architects.* (a) *View from southwest. (Photo by Gordon H. Schenck, Jr.)* (b) *Thermal south–north sections showing natural flow of air and winter heating conditions. Incremental heat pumps are set on the flat-roof sections (left).* (c) *Plan.*

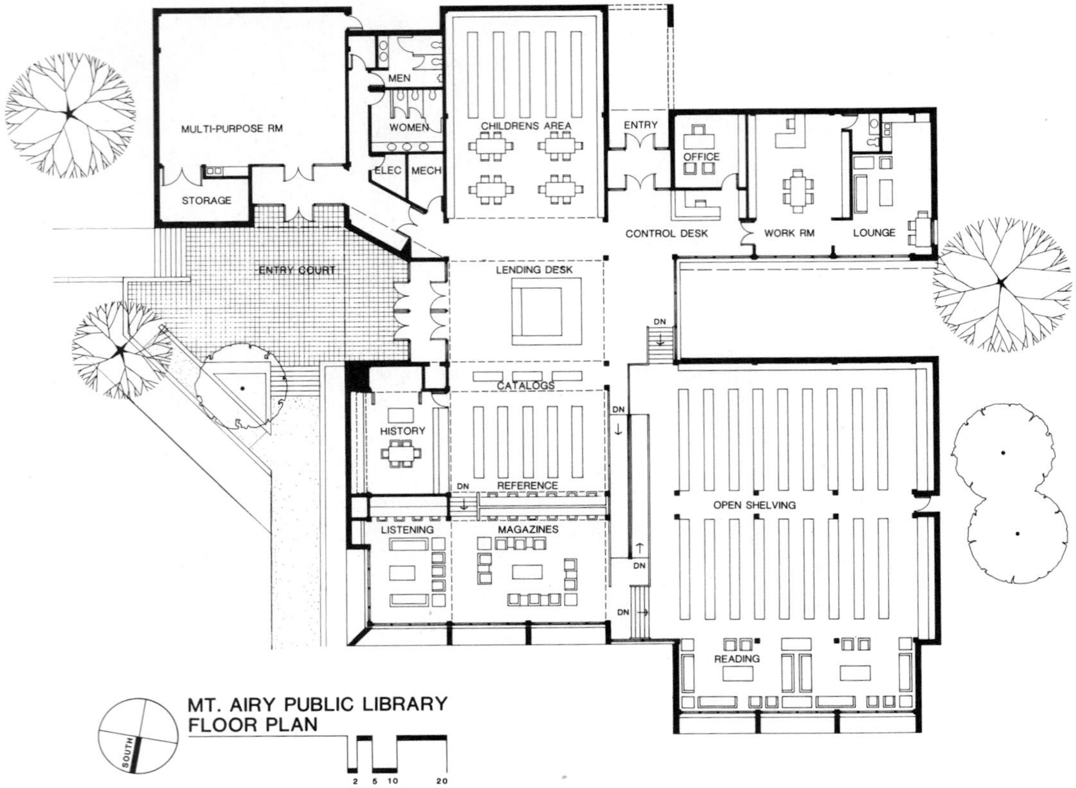

(c)

Fig. 6.129 (continued)

climate, which rarely falls below 45 F (7°C). The roof's relatively high *U* value is thus advantageous in helping lose heat. Since San Francisco overheats even more rarely, such a low resistance to heat flow is not seriously disadvantageous in summer.

To remove this heat, four sets of direct refrigerant equipment are provided at the tent perimeter, on grade. These feed into a perimeter plenum, from which the entire store is supplied with cooled air. The exhaust air is collected at the center and returned to help with the task of cooling. This is possible because *indirect evaporative cooling* (also called sensible evaporative refrigeration) units, rather than conventional compressive refrigeration equipment, are used to cool air (see Fig. 6.130*b*). In this application, two units work in tandem. Air to be cooled (supply air) enters the heat exchanger of the first unit, which is cooled by evaporative cooling of outside air. As explained in Section 6.6, this process depends on a WB temperature well under the DB temperature, a condition found throughout the West for most of the year. During peak temperature periods, the supply air is only somewhat cooled by this process; sufficiently low temperatures for use on the interior are obtained by passing it through the second unit's heat exchanger. This second unit is cooled by evaporative cooling of the *exhaust* air from the store, which is cooler than outside air under summer conditions. Thus, the exhaust air does some work beyond the direct cooling of the tent's interior. This two-stage, indirect evaporative cooling process allows the supply air to be cooled without the relative humidity being raised, as would be the case if direct evaporative cooling were used. (Indirect evaporative cooling was also shown in Figs. 6.65 and 6.66.)

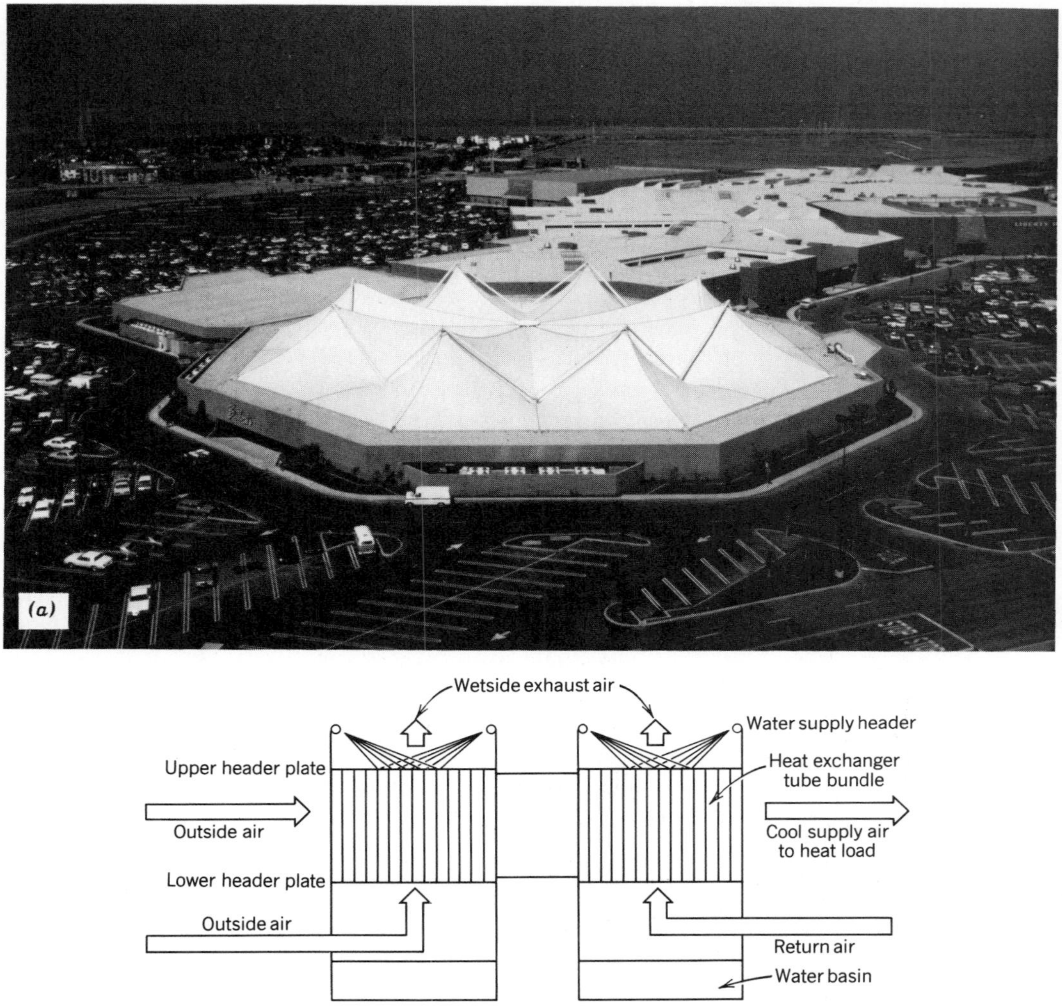

(b)

Fig. 6.130 *Bullock's Department Store at Fashion Island, San Mateo, California. Environmental Planning and Research, Inc., architects; Giampaolo and Associates, Inc., mechanical and electrical engineers.* (a) *The eight-masted white fabric roof, highly reflective to ward off solar heat, transmits some 7%, to provide ample diffuse daylight for sales areas. (Photo by Steve Proehl.)* (b) *Cooling is provided by a two-stage, indirect evaporative cooling system, which uses much less energy than does conventional compressive refrigeration.*

6.13 District Heating and Cooling

Often, large projects made up of many large buildings are well served by *one* central station heating/cooling plant. The familiar "economies of scale" apply here; very large, efficient, and well-maintained boilers and chillers encourage energy recovery through heat exchange and remove the noise and other nuisances associated with heating and cooling from the other buildings. This approach is called district heating and cooling.

(a) High-Temperature Water and Chilled Water for Airports. Although long-distance steam distribution has been used for almost a century, the development of the *high-temperature water* (HTW) principle is a product of recent decades. Offering many advantages (though steam is still frequently chosen for city distribution), circulated high-temperature, high-pressure hot water in closed systems has found great favor in Air Force bases and airports, and for groups of buildings such as hospital complexes and college campuses. Water will not flash into steam if kept at sufficiently high pressure. It may then be circulated by pumps through supply and return mains and through branches to heat exchangers, which operate conventional low-pressure hot water systems, generate steam, and perform numerous other thermal tasks. Pressures are on the order of 400 psig (pounds per square inch, gauge) and temperatures are about 300 F. During its circuit the water will sometimes lose about 150F° and 60 psig in pressure. A section shown in Fig. 6.131 illustrates a common arrangement.

High-temperature water has a number of advantages over steam for certain installations. It is a two-pipe system, and the temperature drop in the *supply* main is often as little as 10F°. With reasonably high water velocities, mains can be reduced to almost half the size of those required for steam distribution. Simplicity results from the lack of need for steam traps and pressure-reducing valves. The pipes need not pitch to low points, as in the case of steam, but may follow the contours of the ground. Although installation costs are greater, operational costs are less than for steam. Feed water treatment is negligible and corrosion is minimal. Underground problems of expansion and insulation are the same as in other subterranean systems. Large sweep-type loops accommodate expansion between fixed points, and underground piping is embedded in special, thermally efficient insulative fill.

(b) John F. Kennedy International Airport. During planning of the heating and cooling for complete air conditioning in the principal buildings of the Kennedy International (originally Idlewild) Airport, the initial concept was to have the facilities located in utility space in each of the buildings. However, the advantages of centralizing the basic equipment for both heating and cooling soon became apparent. Under the guidance of Charles Broder, mechanical engineer of the Port Authority of New York and New Jersey, planning moved in this direction. A central heating/cooling plant was designed by the architects Skidmore Owings & Merrill and engineered by Seelye, Stevenson, Value, and Knecht. In the selected scheme (see Fig. 6.132), all facilities except air handling and ducts were assembled in the central plant. The elimination of a multitude of stacks, boilers, fuel storage, water chillers, and cooling towers at each building was considered to be a great step forward in the release of valuable space and for cleanliness, control, and architectural freedom. Moreover, it proved to be economically advantageous.

Four La Mont-type International Boiler Works Company high-temperature water boilers supply the heating needs and also serve the Carrier absorption-refrigeration machines. These were specially adapted to use this high-temperature water instead of steam. Thus the boilers are active throughout the year, burning gas or oil. The latter fuel is stored in a 210,000-gal spheroid tank ad-

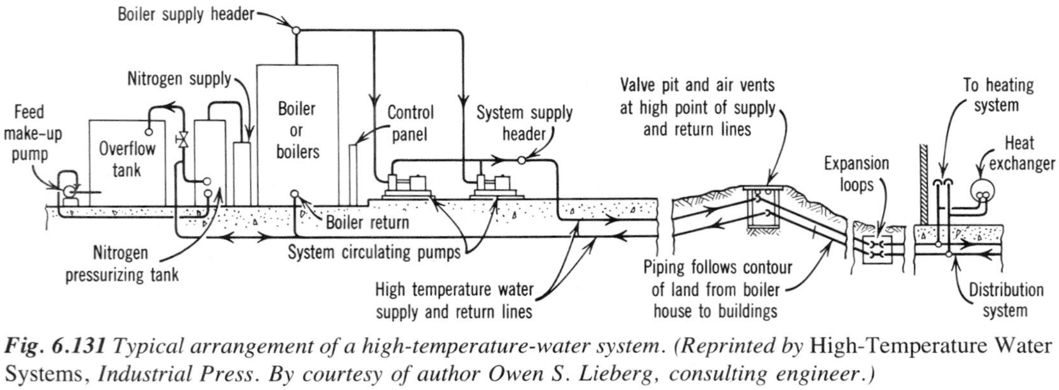

Fig. 6.131 Typical arrangement of a high-temperature-water system. (Reprinted by High-Temperature Water Systems, *Industrial Press. By courtesy of author Owen S. Lieberg, consulting engineer.)*

Fig. 6.132 (a) *John F. Kennedy International Airport. The central heating and cooling plant is to the left of the central pool. It is identified as an L-shaped building with a spherical oil-storage tank in its foreground. (Photos courtesy of Port Authority of New York and New Jersey.)*

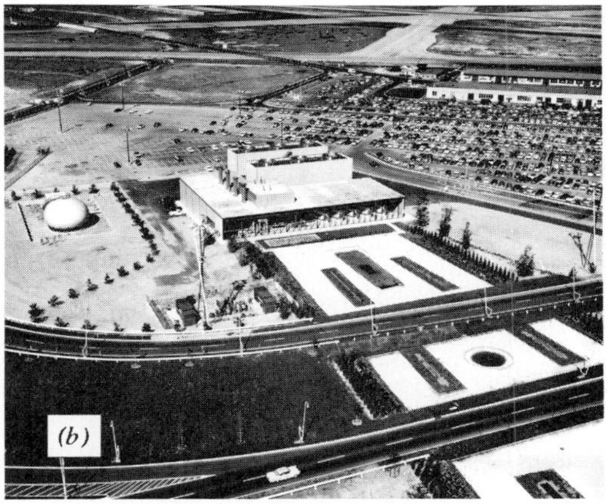

(b) *Heating and cooling plant, initial stage. At left, spherical oil storage tank. Four stacks serve the high-temperature water boilers. Five of the nine absorption refrigeration machines appear through glass at the right. Six cells identify the two cooling tower structures in the rear. Skidmore Owings & Merrill, architects. Seelye, Stevenson, Value, and Knecht, engineers.*

(c) *Night view of the refrigeration-absorption machines in the heating-cooling plant. Distinctive colors identify the piping connections to the machines. Chilled water, condenser water and, in this case, high-temperature water instead of steam, comprise the three circuits connected to each of the nine machines.*

jacent to the plant. Boilers, chillers, and cooling towers are all in close proximity.

Statistics for the initial stage of the plant are as follows.

1. *Hot water:* 160,000,000 Btu/h, 1,140,000 lb/h pump capacity, 160-ft head, 380 F supply, 240 F return.
2. *Chilled water:* 6210 tons, 16,800 gpm pump capacity, 150-ft head, 55 to 45 F cooling range.

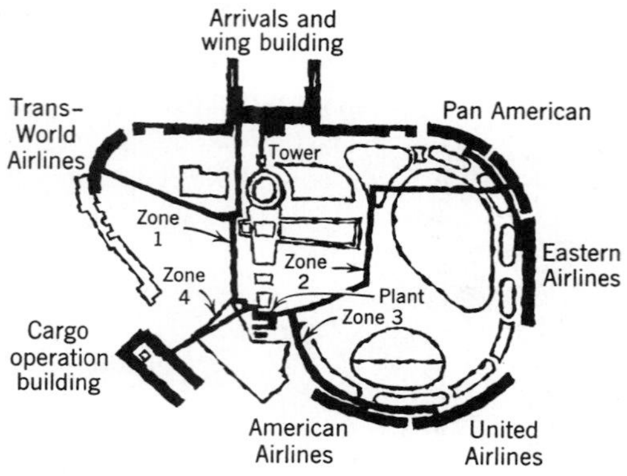

(d)

Figure *(continued)*
(d) *Control terminal area, showing location of heating and cooling plant and the four zones of distribution for high-temperature water and chilled water. (Reprinted by permission from "Heating and Air Conditioning a Civilian Airport," by Charles Broder, M.E., in* High Temperature Water, *a Symposium bulletin published by the American Society of Heating, Refrigerating and Air Conditioning Engineers, Inc.)*

3. *Condenser water:* 24,600 gpm pump capacity, cooling tower capacity for rated flow from 102.4 to 85 F with WB at 78 F.

At the ends of the four zones for water distribution—namely, the various buildings of the airport group—chilled water and hot water are used in the coils of air-handling units from which conditioned air is circulated for complete climate control. Automatic electronic control and monitoring equipment was used. Checked periodically, this equipment is all but *completely* automatic in its regulation of the temperature and humidity throughout the building. (The initial stage as described here has been subject to expansion as air travel has increased.)

6.14 Total Energy and Cogeneration

In the course of these first six chapters, electricity has repeatedly been called a "high-grade" source, in reference to the high temperatures needed to

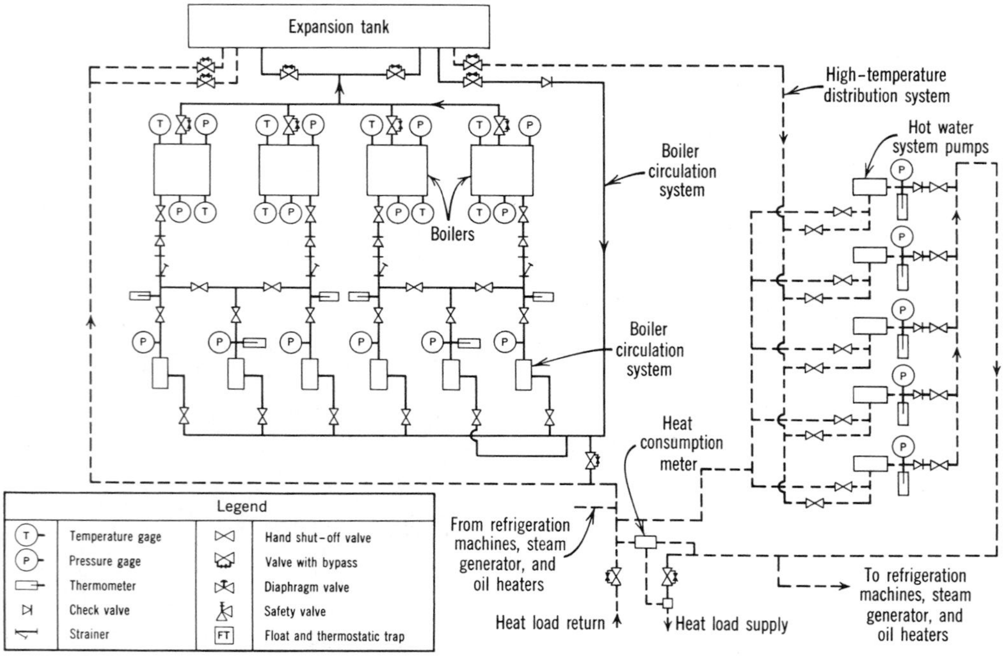

(e)

(e) *Piping diagram for high-temperature water system, showing circulation through boilers and expansion tank and connection to heating load and refrigeration load. (From "Space Heating at Idlewild" by Charles Broder, M.E., in* Industry Power.*)*

produce electricity by conventional (fossil fuel) means and to the large amount of waste heat (often twice the fraction of electricity) produced in the process. Total energy (also called cogeneration, especially when an industrial process is involved) is an attempt to recover some of the otherwise-wasted, lower grade heat that accompanies the generation of electricity by steam turbines.

(a) Electrical Power Generation at the Site. It has been found that *where conditions are favorable,* electricity for power and light can be generated economically by a system that also supplies the building with heating in winter and cooling in summer. Such a system utilizes a fuel such as gas or oil and is often independent of the local electric utility company. Experience has indicated that although installation costs are greater than for the more conventional systems that use separate services of electricity and fuel, the savings in annual operating cost can sometimes pay for the excess installation cost in only a few years. Operational savings continue thereafter. The success of this method, which developed largely in the 1960s, is evident by its use in hundreds of commercial and industrial buildings and in many schools.

Total energy is particularly attractive in district heating/cooling plants. This strongly suggests the need for expert engineering analysis of conditions before selection of the most appropriate method of providing the thermal and power services in a building. The conditions include type and size of building, its geographic location, occupancy characteristics, relative costs of oil, gas, and electricity, and whether cooling as well as heating is essential.

(b) Early On-Site Power Generation. Before 1900 and in the early years of this century, nearly all large buildings and groups of buildings were supplied with direct current generated on or near the premises. The motive power was usually in the form of steam-driven reciprocating engines with belt connections to direct-current generators. There were very natural reasons for this local operation. Direct current cannot be transformed to different voltages and must be generated and distributed at the voltage used in the building. At these relatively low voltages, power loss in the distribution system is great, and distance adds greatly to the loss. With the development and use of alternating-current machines, utility companies were able to establish central power stations from which electricity could be transmitted great distances to the user at high voltage. There it was transformed down to domestic voltages for use. Since, at high voltage, power losses are very low, this system became universal. During the 1920s and 1930s, owners removed their private power generators and accepted utility service, with its savings in operating expense.

(c) How Total Energy Developed. There are valid reasons for the return of power generation to the site, and they relate to the technical advances of recent times. In earlier installations, in which steam, produced from coal-burning boilers, was used, there was little or no energy salvage from the steam, which was mostly wasted. The newer fuels as used directly in reciprocating engines or turbines have residual heat value that can be recovered for purposes of heating or cooling. The techniques for this recovery have been perfected by experience with nuclear power, rocketry, aviation, and other recent developments. For total energy to be successful, there has to be a reasonably steady demand in the building for the power generated and also for the heat recovered. It is easily understood that many buildings have this need. Lighting and the demand for power by computers, electrical business machines, electrical cooking, and the great multitude of electrically powered devices often make the call for power nearly constant throughout the year. Similarly, the exhaust heat recovery from the engines or turbines that power the generators is in demand for either heating or cooling at any time of the year. Cooling, formerly something of a luxury, is now frequently essential. It also fills the gap of energy demand that formerly existed in summer, when heating was not required.

(d) How It Works. Figure 6.133 shows the two principal systems for total energy, reprinted here through the courtesy of Educational Facilities Laboratories. In both systems—one using a turbine and the other a reciprocating engine to operate the generator that supplies electric power—heat is reclaimed to produce steam or hot water. The steam or hot water is then used for heating

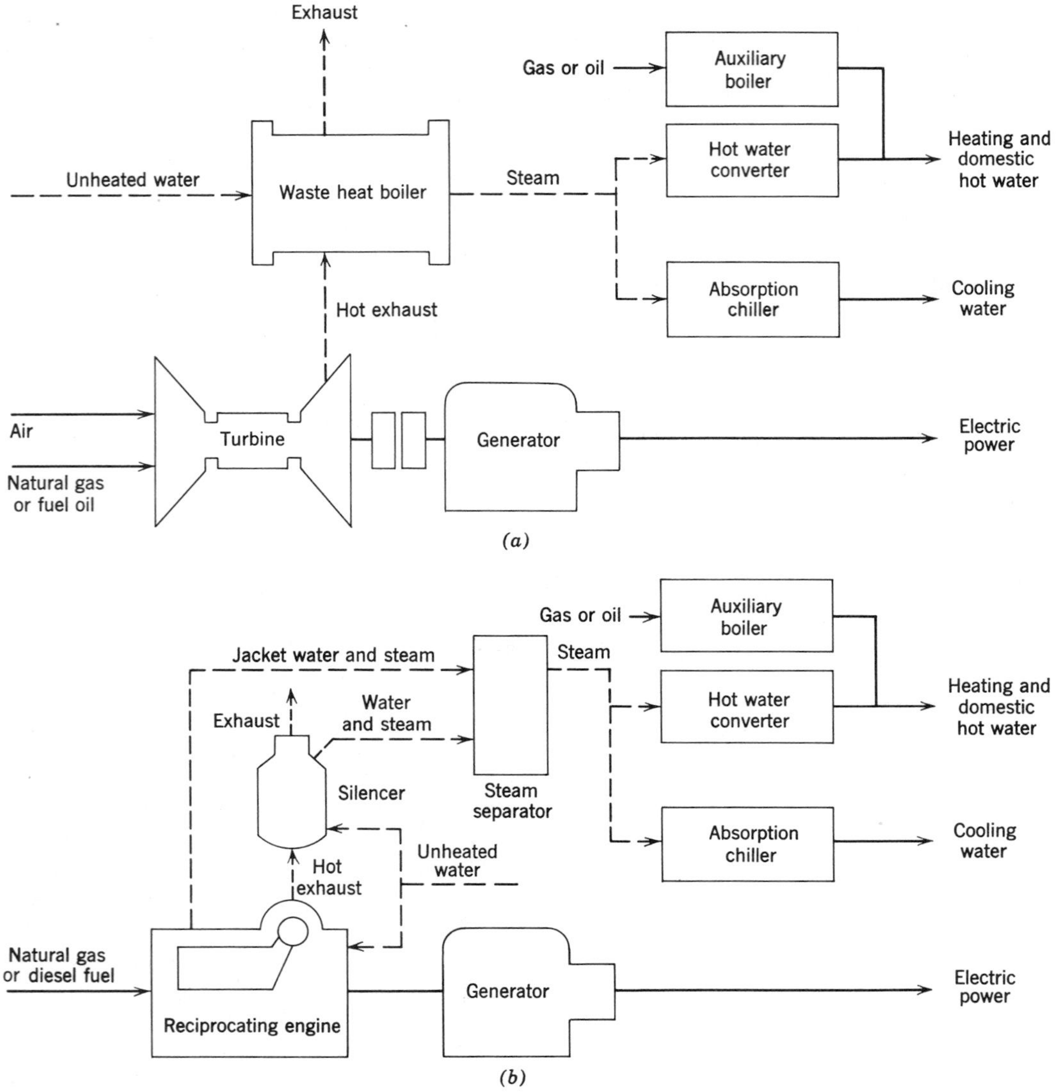

Fig. 6.133 Total energy systems. (a) *Using turbine.* (b) *Using reciprocating engine. Reprinted by permission from* Total Energy, *Educational Facilities Laboratories.*

or, by use of an absorption chiller, for cooling. When a turbine is used, the fuels are natural gas or fuel oil. The heat is recovered by passing the hot turbine exhaust through a waste heat boiler to produce steam. Fuel for a reciprocating engine is natural gas or diesel fuel. Both the jacket cooling water and the hot engine exhaust are passed through heat exchangers that utilize the heat to produce steam or hot water for heating or cooling, as already described. In both systems, an auxiliary boiler, fired directly by gas or oil, stands ready to help maintain a balance in the system.

The unique feature of total energy is the use of the recovered heat. This is essentially free energy, since it would otherwise have to be purchased and paid for separately in the form of electricity or other fuels.

(e) Total Energy for Schools. Many recent changes in school buildings and in school programs prompted the Education Facilities Laboratories to support a study of the feasibility of total energy for schools, methods of determining feasibility in specific instances, and guidelines for school administrators and engineers in developing solutions.

The changing demands are numerous. They include large energy requirements that make the school of two decades ago seem quite obsolete. In the modern school, lighting has been upgraded and cooling has largely become as high a priority as heating. Electronic teaching aids add to these energy demands. Evening classes, multifunction buildings, and summer programs put many schools into a new category.

The study was conducted as a research project by Fred S. Dubin Associates, consulting engineers. Norman Kurtz, P.E., was the research director and prepared the original report. EFL's interest in sponsoring this work stemmed not only from the foregoing school demands but also from the rapidly increasing costs of mechanical and electrical systems in schools and, with regard to total energy, the less objective and inescapably conflicting commercial interests of electric utility companies and the manufacturers of electric generators. The report included a consideration of college buildings as well as of schools.

The foregoing conditions and criteria that evaluate schools and college buildings, as possibly suitable for the use of total energy, can also be applied to other building types. These can include commercial and industrial buildings as well as offices and motels.

(f) An Example of Suitability in a Motel. As we have seen, total energy is appropriate under special conditions that relate strongly to the type of building, its occupancy, and its use. One suitable and successful application can be seen at the Midtown Holiday Inn at Montgomery, Alabama. Its urban location makes it a center for public gatherings, functions, entertainment, and meetings. Unlike motels in which residential occupancy and the services to care for residents are the principal concern, this facility combines those requirements with a very large and diverse occupancy by visitors. The result is that the need for electric power generation and its by-products, heating and cooling, is quite constant. The motel is served for its electric requirements by two gas engine-generator sets. Heat recovered from the engines is used for heating, air conditioning, and domestic hot water. The several areas at the motel occupy the following amounts of space (see Fig. 6.134).

174 guest rooms	65,000
Commercial area	13,000
Rooftop supper club	6,000
Total	84,000 ft^2

The population with which the air conditioning system must cope on many special occasions is

Meeting room	425
Dining room	94
Coffee shop	82
Bar	60
Small meeting rooms, 4 @ 20	80
Total	761 people
Food service	2000 meals per day

The impressive problem of keeping these visitors cool or warm, and supplied with food, hot water, and proper illumination, is handled by one fuel—gas. It is used in two 450-hp, engine-generator sets with 300-kW generators operating at 480 V (see Fig. 6.134*a*). Recovered heat from the engines is used to produce steam. Two steam heat exchangers are used to generate hot water for heating and for domestic needs. An additional heat exchanger heats the swimming pool when necessary. A gas-fired boiler stands by as an auxiliary to supply additional steam for peak loads. Cooling is provided by an absorption machine with a gas-engine-driven compressor as a standby. The principal equipment consists of

Two Caterpillar G-379 engine-generator sets, 450 hp, 480 V.

Auxiliary gas-fired steam boiler, 5,500,000 Btu/h.

Three heat exchangers for heating, hot water, and swimming-pool heating.

Absorption machine for cooling: York EK-13, 125 tons.

Standby for cooling: gas-engine-driven compressor, 100 tons.

Fig. 6.134 *Midtown Holiday Inn, Montgomery, Alabama.* (a) *Interior view of equipment room of the natural gas total energy system.* (b) *Flow diagram for a typical system recovering both jacket water heat and exhaust heat.* (c) *Exterior view of equipment room, housing entire total energy plant. Visible elements are louver for ventilation and combustion air, exhausts, and cooling tower on top.*

Cooling tower on roof, 350 tons capacity.

Air-handling units: for large meeting room, three at 4000 cfm each (15% fresh air); for dining room, 4000 cfm plus exhaust fan; for kitchen, 3600-cfm exhaust fan.

Although not related to the use of total energy, the air-handling figures indicate the problem of dealing with large and instant crowd gatherings. Moving cooled and dehumidified air into, and the vitiated air out of, these spaces is a task that not only involves the need to provide comfort conditions, but also requires control of air speed and sound.

Resident guests control their own environment by means of three-pipe water systems serving fan-coil units in each room. Both hot and chilled water are available when needed, and the third pipe carries away the return water. Fan-coil unit blowers in guest rooms operate at 200 or 400 cfm, depending on room size. In both the guest rooms and the public spaces, heating and cooling are available simultaneously when outside temperatures are between 45 and 65 F, during which time both chilled and hot water are circulated. When outside temperature is over 65 F, only chilled water is circulated. When outside temperature is below 45 F, only hot water is circulated.

The mechanical room that houses the principal components of the system is a separate and isolated building seen in full view in Fig. 6.134*c*. With the advice of acoustical consultants, the escaping sound has been reduced to a whisper.

Either of the generators at the Midtown Holi-

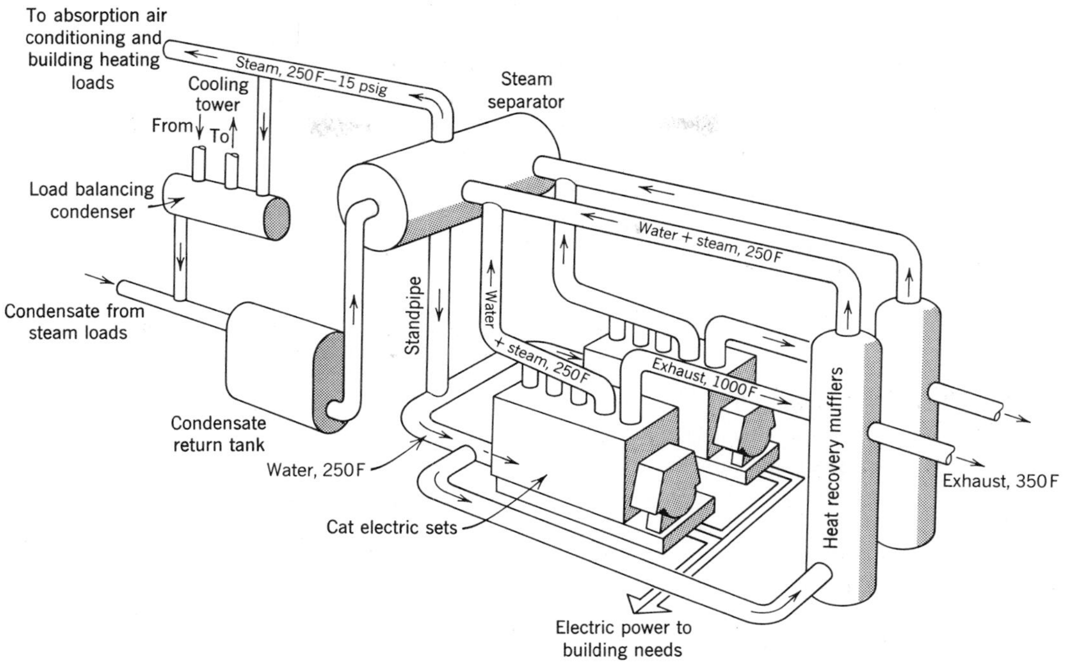

(b)

(c)

Fig. 6.134 *(continued)*

day Inn can carry the entire load of the building, permitting alternate operation of the generators and the assurance of one as a standby in case of mechanical difficulty. As for the dependability of gas service, the inn experienced only one 20-min outage in its first year of operation.

The net additional installation cost for the total energy system over the originally designed conventional system was $102,400. The net annual savings in operation from the use of total energy instead of a conventional system was $13,162 for a test period of one year.

(g) System Components. The differences in the equipment needed for a total energy system and a conventional one are generally limited to the equipment room. No change is needed in the piping for chilled and hot water, the fan-coil units, the absorption machine, or the air-handling units. The manufacturer's engine-generator set that was used at the Midtown Holiday Inn is pictured in Fig. 6.134*a*, together with its heat recovery equipment and control panel. Figure 6.134*b* shows in schematic form the piping and a few additional devices commonly used. They comprise a steam separator, load-balancing condenser, and condensate return tank. After the steam has passed through the heat exchangers to produce hot water or through the absorption chiller for chilled water, the rest of the distribution system is as it would normally be in any design suitable for the specific building.

Although the terms just mentioned are typical, it must be understood that there are many reasonable or necessary variations within the design of an efficient total energy system.

References

American Institute of Architects, Ramsey, C. G., and Sleeper, H. R. (1981). *Architectural Graphic Standards*, 7th ed., Wiley, New York.

American Society of Heating, Refrigerating and Air Conditioning Engineers, Inc., (1981). *Handbook of Fundamentals*, ASHRAE, Atlanta.

Bowen, A., Clark, E., and Labs, K. (1981). *Passive Cooling*, American Solar Energy Society, Boulder.

Fisher, J. (1974). *Energy Crisis in Perspective*, New York.

Johnson, Timothy (1981). *Solar Architecture: the Direct Gain Approach*, McGraw-Hill, New York.

Niles, P., and Haggard, K. (1980). *California Passive Solar Handbook*, California Energy Commission, Sacramento.

Shurcliff, W. A. (1980). *Thermal Shutters and Shades*, Brick House, Andover, MA.

Shurcliff, W. A. (1981). *Air-to-Air Heat Exchanges for Houses*, Brick House, Andover, MA.

PART III
WATER AND WASTE

Almost every building designed today in the United States is supplied with potable (drinkable) water. In the vast majority of buildings, most of this clean water is used to carry away organic waste. The result is a wide range of design impacts covering everything from the detailed arrangement of bathroom fixtures and interior surfaces to the overall plans of very large and complex water and sewage treatment facilities.

This section of the book presents the following topics: basic planning information (Chapter 7); rainwater (Chapter 8); water supply (Chapter 9); organic waste disposal, both water and waterless (Chapter 10); the particular design issues of the bathroom (Chapter 11); and solid waste design issues (Chapter 12). Fire protection, which requires the largest amount of water supply piping within a building, is covered in Chapter 13.

7

WATER AND BASIC DESIGN

7.1 Water in Architecture

Throughout history, in nearly all climates and cultures, the designer's major concern about water was how to keep it *out* of a building. Only within the last 100 years has water supply *within* a building become commonplace in industrialized countries. In the rest of the world today, running water is still not available within most buildings. Water's potential contributions to life-style and architecture are as numerous and varied as are appropriate design responses to the supply, usage, and return of such a versatile commodity.

(a) Nourishment. Much of the human body is water, the most abundant chemical in our bodies as well as in our diet. The amount of really "pure" (potable) water that we need for drinking and cooking is very small—only about 3 g/cd (gallons per capita per day) (11.4 L/cd) in the United States. The most common supply system throughout history has been the central municipal fountain or well, whose technical importance to the community has often been emphasized by the esthetics of both the fountain's overall sculptural composition and the elegance of detail in water spouts, basins, and other elements. The social importance is evident from its location in the central plaza, from which site small amounts of water are carried daily by townspeople to homes or workplaces.

As daily potable water became available within buildings, water-related social opportunities diminished. However, employees still converse around the drinking fountain—for example, when it is located in a place that invites lingering or relaxing.

(b) Cleansing and Hygiene. Water is a nearly ideal medium for the dissolution and transport of organic waste, and its high heat-storage capacity makes the attainment of comfortable temperatures for bathing easy. Much larger quantities of water are used for cleaning than for nourishment: in the average U.S. home, about 14 g/cd (53 L/cd) is used for clothes washing and dishwashing and another 21 g/cd (79.5 L/cd) is used for bathing and personal hygiene.

In the past, water for cleaning was carried to the home less frequently; the Saturday night (only) bath was typical well into the 20th century in this country. Bathing vessels were usually portable and sometimes were combined with other pieces of furniture (a couch that sat over a tub, a metal tub that folded up inside a tall wooden cabinet, etc.) Thus, a bathplace rather than a bathroom was the common design response.

Although the physical constraints of water carrying were important design influences on bathing, cultural attitudes were at least as strong an influence throughout history. Perhaps the most startling contrast is between the medieval concept of bathing as an almost sinful indulgence and the earlier Roman attitude toward public baths as the social centers of cities.

Now, bathing facilities are commonly designed to be used on a very personal scale, in privacy. There also are welcome opportunities to design more-social bathing places, such as swimming pools, bath houses, and hot tubs. The characteristics of the water supply (spouts, jets, cascades) can be matched with those of the water body (mirror-smooth, gently flowing, rippled, rolling, foaming) to obtain the desired atmosphere.

(c) Ceremonial Uses. Largely through its associations with cleaning, water acquired a ceremonial significance that remains particularly evident in religious services. Examples of the ceremonial use of water include vessels containing holy water at entrances to Catholic churches, pools in the forecourts of mosques, and full-immersion baptismal fonts at the altars of some Protestant churches. The opportunities for esthetic expression are particularly rich in these ceremonial applications.

(d) Transportational Uses. In stark contrast to its uses in nourishing, cleansing, and celebrating, water is used in our buildings principally to transport organic waste. There is perhaps no more flagrant example of a mismatch in architecture than the high-grade resource of pure water being used for the low-grade task of carrying away a cigarette butt. The typical U.S. home uses 32 g/cd (121 L/cd) just to flush toilets.

In the past, table scraps were commonly fed to animals or composted, and human waste thrown out of windows (accompanied by warning cries) or deposited in holes below outhouses. Organic waste disposal was thus dependent on either portable vessels or special structures set apart from the typical building.

As water supplies were developed, water's advantages over the foul smell and inconvenience of these methods became irresistible. A typical sequence of events unfolded on Manhattan Island. In the 1700s, Manhattan was farm country that, like all other areas that later developed into large cities, had minimal water needs. Potable water was available in shallow wells and from some springs and streams. These sources were largely unaffected by the minor ground pollution from widely separated dry-pit privies (outhouses) that received human wastes. Paved city streets appeared in the 1800s, at which time the natural streams were enclosed in pipes, called *storm sewers;* these pipes led the rainfall to the many waterways surrounding the island. Thus far, nature had not been seriously hurt. Then, in the later 1800s, flush toilets appeared. It seemed natural to connect the toilets to the already established "storm sewers" and to rename the pipeways "combined" sewers, which now carried both storm water and so-called "sanitary" (really, polluted) drainage to the rivers. Fast-flowing rivers are natural sewage treatment plants, and, surprisingly, for many decades they did a fair job of keeping pollution reasonably in check. With the prospect of future sewage treatment plants, separate "sanitary" sewers were built. Also, there were some remaining (and some newly built) storm sewers that did not carry the wastes from toilets.

In cities where this confused pattern of sewer systems still exists (this includes most larger and older cities), it is now extremely difficult and expensive to sort out and reroute sewers so that *only* sanitary drainage goes to treatment plants and *all* storm drainage goes to waterways or into the ground. It seems particularly ironic that in most U.S. locations the rainwater that falls on a home's roof is adequate in both quality and quantity to supply a family's cleansing needs [21 + 14 = 35 g/cd (132 L/cd)]. In Chapters 8, 9, and 10, these possibilities for rainwater will be developed further.

As the human waste disposal place became a room within the building, the design issues grew more complex. Physically, there was the need for running water, and for large-diameter pipes that sloped downward continuously from the toilet to a sewer or septic tank. As sewer gas became a recognized problem, an elaborate system of traps and vents became necessary. Again, cultural attitudes were also influential: How private an activity was this elimination? To what extent could one plumbing fixture accommodate both body cleansing and waste elimination? If males insisted on standing rather than sitting while urinating, how to devise a toilet that would also properly accommodate defecation, which requires a low seat? Some designer responses to these questions can be found in Chapter 11.

(e) Cooling. Water has a remarkable cooling potential: it stores heat readily, removes large quantities of heat when it evaporates, and vaporizes readily at temperatures commonly found at the human skin surface. In hot-dry climates, designers can place water surfaces (or sprays) upwind from the place to be cooled, or resort to the evaporative coolers discussed in Sections 2.4, 4.7, and 6.6. Cooling towers (Section 6.8) are familiar components of large-building cooling systems.

Because all of us have experienced the physical cooling of the skin by water, we all carry psychological associations between water and cooling that can enhance our comfort on hot days. The sight of sunlight reflected on a water surface, with its characteristic "dancing" quality, connotes coolness, as does the sound of running or splashing water. Thus, even when water does not physically cool people, it can make an important psychological contribution to human comfort (Fig. 7.1).

(f) Ornamental Uses. In almost any landscaping application, indoors or out, water be-

Fig. 7.1 Water as an esthetic feature of a courtyard in the hot-arid part of Mexico. The sound and sight of running water add to a psychological impression of coolness.

comes a center of interest. Our association of water with nourishing, cleansing, and cooling make water a very powerful design element—a fact recognized by landscape designers throughout history. In arid regions, water is often used sparingly, in small, tightly controlled channels and at lower flow rates. The gardens of Islamic architecture are especially effective demonstrations of such design restraint. Where water is more plentiful, it has been used lavishly, as at the Villa d'Este in Tivoli, outside Rome.

Especially useful design characteristics of water include its *reflectivity,* which sets it apart from most plant and ground materials in a garden; its *liquidity,* which creates unique sounds wherever it is moved; and its *life-sustaining* potential, which allows the addition of both water plants and animals to a garden.

(g) Protective Uses. Every designer dreads water's ability to penetrate a roof and damage a building and its contents. But we all depend on water as the best fire protection available in most buildings. The vast quantities of water potentially required for firefighting must be delivered quickly; the result is pipes of enormous diameter regulated by very large valves. Because this system's distribution tree must be immediately obvious to firefighters, some degree of exposure is prudent. Despite its size and guarantee of at least partial exposure in public places, a fire protection water supply system is rarely treated as a visually integral design element. This mismatch of potential and actuality will be discussed further in Chapter 13.

7.2 The Hydrologic Cycle

There is a finite quantity of water in the earth and its atmosphere. The process whereby this water constantly circulates, powered by about one-fourth of earth's solar energy, is called the hydrologic cycle (Fig. 7.2). More than 99% of this water is "inaccessible"—either because it is salt water or because it is frozen in glaciers on polar ice caps. The most accessible sources of water are precipitation and runoff.

Precipitation has the advantage of relative purity, although acid rain is a growing threat in many parts of the world, including much of the United States and Canada. Like solar energy, precipitation is a very large but very thinly spread resource; its capture is therefore likely to take place on an individual basis. Until we experience a "water crisis" similar to the "energy crisis" of the 1970s, rainwater capture will remain a mostly untapped fresh water source in the United States.

Runoff enjoys the advantage of a concentrated flow of water, which permits easy capture of large quantities. Its most serious disadvantage is the possibility of pollution—organic, chemical, and radioactive—depending on what is upstream from the point of capture. In some regions of North America, river water is reused 50 times on its way to the ocean. Further discussion of water sources and treatments occurs in Chapters 8 and 9.

That part of daily precipitation which neither evaporates nor joins the runoff becomes part of *soil moisture*. Much soil moisture is used by growing plants and is soon transpired (evaporated) by

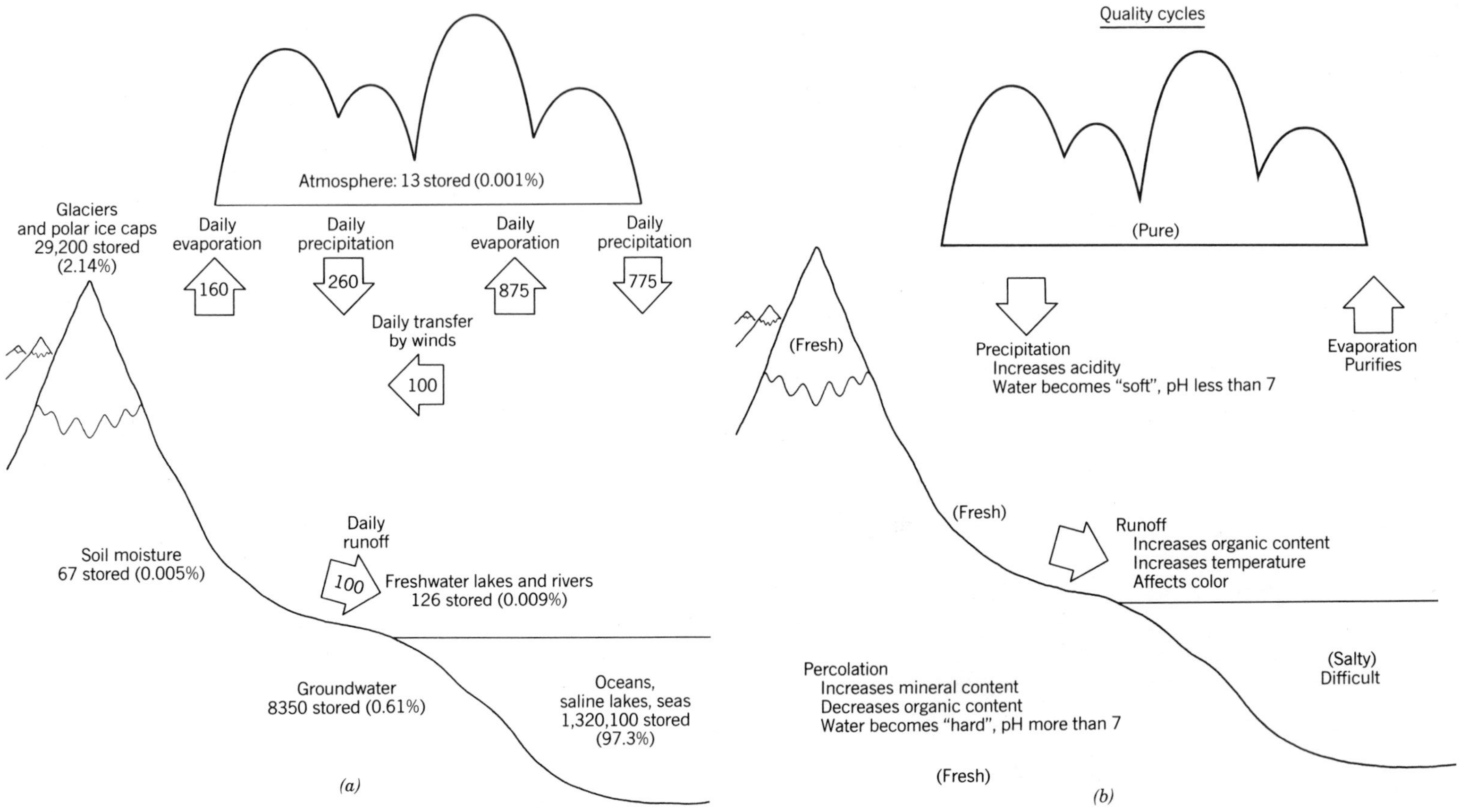

Fig. 7.2 *(a) The hydrologic cycle. The figures given are in cubic kilometers. ''Evaporation'' includes transpiration from plants as well as evaporation from surfaces. ''Precipitation'' can be rain, hail, sleet, or snow. The vast majority of stored water is in the ocean. (b) Quality of water at various stages within this cycle. See also Chapter 9 for quality and treatment considerations.*

the plant to the atmosphere. As water works downward below the root zone of plants, it eventually reaches a zone of saturation, where all voids in the earth's material are filled with water. This zone of saturation is called *groundwater*; the upper surface of groundwater is called the water table. Wells are commonly sunk to a point well below the water table, so that the latter's seasonal fluctuations will not interrupt the well's access to groundwater.

7.3 Basic Planning

After considering the relationship between a building and the roles that water plays within it, the designer must do basic sizing—of the quantities of water needed, of the areas in which water will be used, and of the areas and equipment associated with water's return to the hydrologic cycle.

But before discussing basic planning for the amounts of water used daily within buildings and for the treatment of that water, we should note that water is also an important component of building construction. The production of 1 ton (907 kg) of bricks requires 580 gal (2200 L) of water; 1 ton of steel requires 43,600 (165,000 L); 1 ton of plastic requires 348,750 gal (1.32 million L). Water is one of the main ingredients of concrete; a typical 94-lb (42.6 kg) bag of cement requires about 6 gal (23 L) of water.

Another way in which buildings contribute to water consumption is through electricity consumption. Most power plants require very large quantities of water, which they quickly return to the hydrologic cycle warmer in temperature (and perhaps as vapor). A large nuclear power plant that utilizes a cooling tower can evaporate daily the approximate equivalent of about 9 mi^2 (23 km^2) of forest.

(a) Water Supply. The task of rough sizing for water needs is complicated by the conflict between *current practice* and *conservation*. Current practice tends toward the use of large amounts of water for very low-grade tasks. Conservation reserves high-quality water for high-grade tasks and emphasizes recycling, as well as diminished overall usage, of water.

The supply of water often is first estimated in terms of gallons per capita per day (liters per capita per day). Table 7.1 shows some common terms used in water supply. Typical quantities were matched with nourishing, cleansing, and other usages in the United States in Section 7.1. To gain an appreciation of how the rate of urban water

TABLE 7.1 **Some Water Measurement Terms**

Quantity	
1 gal	= 8.33 lb
	= 231 in.^{3}
	= 0.134 ft^3
	= 3.785 L
1 liter	= 0.263 gal
1 cubic foot	= 7.48 gal
	= 6.24 lb
1 ton	= 240 gal
1 acre-foot (a-f)	= 325,851 gal
	= 43,560 ft^3
1 million gallons	= 3.07 a-f
Flow	
1 cfs	= 1.98 a-f/day
1 a-f/yr	= water supply for five for one year at 180 gal per person per day
1000 gpm	= 2.23 cfs
	= 4.42 a-f/day
1 million gpd	= 694.4 gpm
	= 1.55 cfs
	= 1,120 a-f/yr
Cost	
10¢ per 1000 gal	= 7.48¢ per 100 ft^3
	= \$32.59 per a-f
10¢ per 100 ft^3	= \$43.56 per a-f
	= \$13.40 per 1000 gal
10¢ per ton	= \$136 per a-f
Leaks	
Slow drip	= 170 gpd
	= 62,050 gal/yr
⅛″ (3.2-mm)	= 3600 gpd
diameter stream	= 4 a-f/yr

Source: Milne (1976).

usage has changed over time, consider the following figures (from Milne, 1976).

Imperial Rome, *c.*	38 g/cd (144 L/cd)
London, 1912	40 g/cd (151 L/cd)
American cities just before World War II	115 g/cd (435 L/cd)
Los Angeles, mid-1970s	182 g/cd (689 L/cd)

Although the historical trend has clearly been toward higher per capita use of water, the recent emphasis on conservation has resulted in significant changes in this pattern. In Section 10.1, the influences of various conservation practices will be considered in more detail. Table 7.2 can be used to estimate the daily indoor usage of water in various facilities, if *current practices* are antic-

TABLE 7.2 **Planning Guide for Water Supply**

Building Usage	*Per Capita (As Listed) Daily Usage* Gallons	*(Liters)*
Airports (per passenger)	3–5	(11–19)
Apartments, multiple family (per resident)	60	(227)
Bath houses (per bather)	10	(38)
Camps		
Construction, semipermanent (per worker)	50	(189)
Days with no meals served (per camper)	15	(57)
Luxury (per camper)	100–150	(378–568)
Resorts, day and night, with limited plumbing (per camper)	50	(189)
Tourist, with central bath and toilet facilities (per person)	35	(132)
Cottages with seasonal occupancy (per resident)	50	(189)
Courts, tourist, with individual bath units (per person)	50	(189)
Clubs		
Country (per resident member)	100	(378)
Country (per nonresident member present)	25	(95)
Dwellings		
Boardinghouses (per boarder)	50	(189)
Additional kitchen requirements for nonresident boarders	10	(38)
Luxury (per person)	100–150	(378–568)
Multiple-family apartments (per resident)	40	(151)
Rooming houses (per resident)	60	(227)
Single family (per resident)	50–75	(189–284)
Estates (per resident)	100–150	(378–568)
Factories (gallons per person per shift)	15–35	(57–132)
Highway rest area (per person)	5	(19)
Hotels with private baths (two persons per room)	60	(227)
Hotels with private baths (per person)	50	(189)
Institutions other than hospitals (per person)	75–125	(284–473)
Hospitals (per bed)	250–400	(946–1514)
Laundries, self-service (gallons per washing—i.e., per customer)	50	(189)
Livestock (per animal)		
Cattle (drinking)	12	(45)
Dairy (drinking and servicing)	35	(132)
Goat (drinking)	2	(8)
Hog (drinking)	4	(15)

ipated. For very rough approximations of *conservation* effects:

- Reduce Table 7.2 values by 25%, assuming simple conservation measures such as flow controls.
- Reduce Table 7.2 values by 50%, assuming partial recycling.

The more water used for flushing toilets within the building types listed in Table 7.2, the greater the potential for savings through conservation.

In urban areas, public water mains usually provide the necessary quantities of water at the pressures and rates of flow required to operate typical plumbing fixtures. For isolated systems, or those independent of public networks, water supply can

TABLE 7.2 **Planning Guide for Water Supply (*continued*)**

Building Usage	Per Capita (As Listed) Daily Usage	
	Gallons	*(Liters)*
Horse (drinking)	12	(45)
Mule (drinking)	12	(45)
Sheep (drinking)	2	(8)
Steer (drinking)	12	(45)
Motels with bath, toilet, and kitchen facilities (per bed space)	50	(189)
With bed and toilet (per bed space)	40	(151)
Parks		
Overnight, with flush toilets (per camper)	25	(95)
Trailer, with individual bath units, no sewer connection (per trailer)	25	(95)
Trailer, with individual baths, connected to sewer (per person)	50	(189)
Picnic		
With bathhouses, showers, and flush toilets (per picnicker)	20	(76)
With toilet facilities only (gallons per picnicker)	10	(38)
Poultry		
Chickens (per 100)	5–10	(19–38)
Turkeys (per 100)	10–18	(38–68)
Restaurants with toilet facilities (per patron)	7–10	(26–38)
Without toilet facilities (per patron)	2½–3	(9–11)
With bar/cocktail lounge (additional quantity per patron)	2	(8)
Schools		
Boarding (per pupil)	75–100	(284–378)
Day, with cafeteria, gymnasium, and showers (per pupil)	25	(95)
Day with cafeteria but no gymnasiums or showers (per pupil)	20	(76)
Day without cafeteria, gymnasiums, or showers (per pupil)	15	(57)
Service stations (per vehicle)	10	(38)
Stores (per toilet room)	400	(1514)
Swimming pools (per swimmer)	10	(38)
Theaters		
Drive-in (per car space)	5	(19)
Movie (per auditorium seat)	5	(19)
Workers		
Construction (per person per shift)	50	(189)
Day (school or office, per person per shift)	15	(57)

Source: Manual of Individual Water Supply Systems (1975).

come from individual sources: wells, springs, cisterns, lakes, and so forth. For the minimum pressures and flow rates necessary from these sources (and/or the storage vessels associated with them), see Table 9.14.

(b) Cisterns. Where rainwater is to be utilized, a rough approximation of catchment area and cistern storage volume is initially needed. This procedure is detailed in Chapter 8. For now,

1. From Table 7.2, find the quantity of rainwater to be used daily:

$$\text{g/cd} \times \text{population} = \text{gpd}$$

$$(\text{L/cd} \times \text{population} = \text{L/d})$$

2. Convert this quantity to the yearly need for water:

$$\text{gpd} \times 365 \text{ days} = \text{gal/yr}$$

$$(\text{L/d} \times 365 \text{ days} = \text{L/yr})$$

3. Assume, conservatively, that a "dry" year will have two-thirds the precipitation of an average year; this measurement is the "design precipitation." (Average annual precipitation is available from NOAA annual summaries.)

$$\text{Average annual precipitation} \times 2/3 = \text{design precipitation}$$

4. From Fig. 7.3, determine the catchment area required.
5. Roughly size the cistern (storage) capacity by finding the longest dry period (in days of negligible rainfall, from NOAA local climatological data):

$$\text{cistern capacity} = \text{gpd} \times \text{days of dry period}$$

6. Convert capacity to volume by the formula

$$1 \text{ ft}^3 \text{ stores } 7.48 \text{ gal of water}$$

$$(1 \text{ m}^3 \text{ stores } 1000 \text{ L water})$$

EXAMPLE 7.1. A 20,000-ft^2 one-story factory near Salem, Oregon, will use roof-collected rainwater to flush its toilets. Water-conserving toilets using 3 gal/flush and serving 20 workers are to be used. Table 7.2 shows that factory workers use between 15 and 35 g/cd. Since usage at this factory will be low—no showers, for example—assume 15 g/cd:

$$15 \text{ g/cd} \times 20 \text{ workers} = 300 \text{ gpd, for } \textit{all} \text{ usages}$$

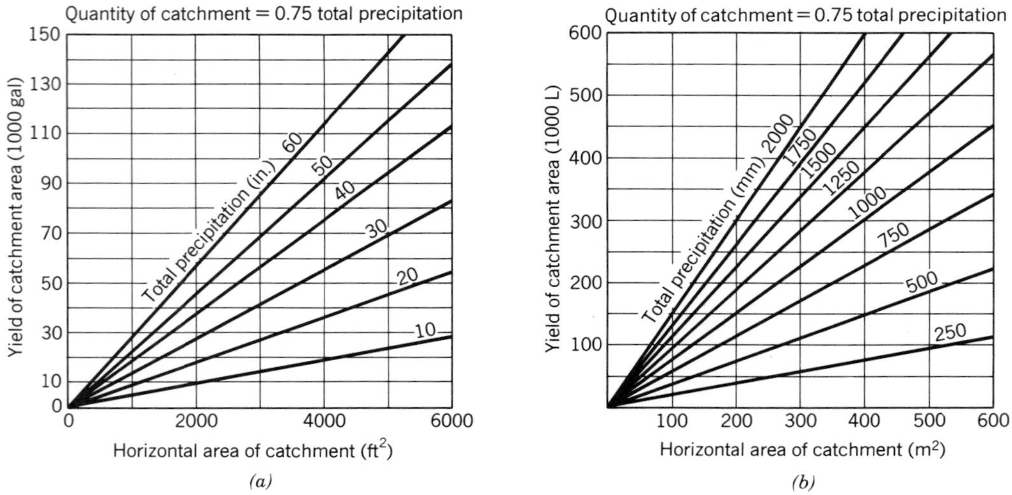

Fig. 7.3 Yields of rainfall catchment areas (roofs), in terms of total precipitation. In these graphs, 75% of the total precipitation is assumed to be catchable; the remainder is lost to evaporation or spillage. (a) *Conventional units, from the U.S. Environmental Protection Agency's* Manual of Individual Water Supply Systems, *1975.* (b) *SI units (1 m^3 = 1000 L).*

Since low-flush toilets are to be used, reduce this figure by 25%:

$$0.75 \times 300 = 225 \text{ gpd}$$

Toilets will probably account for most of this 225 gpd: for example, at three flushes per day per worker,

$$3 \text{ flushes/day} \times 3 \text{ gal/flush} \times 20 \text{ workers} = 180 \text{ gpd}$$

Assume, then, that up to 200 gpd of rainwater will be utilized.

Catchment Area: Salem's average annual rainfall is 41 in. Design rainfall is ⅔ × 41 = 27.3 in., in a "dry" year; yearly need is 200 gpd × 365 days = 73,000 gal. The combination of 73,000 gal and 27.3 in. is off the chart in Fig. 7.3*a*, so divide 73,000 gal by 2, to obtain 36,500 gal, and then double the resulting catchment area. At 27.3 in. of precipitation, 36,500 gal will be caught by 2800 ft^2. The catchment area for 73,000 gal is, therefore, 2 × 2800 = 5600 ft^2. (So about 28% of the 20,000-ft^2 factory's roof area will suffice for a catchment area.)

Cistern Capacity: Salem normally has very dry summers; average monthly rainfall is as follows:

May	2.1 in.
June	1.4 in.
July	0.4 in.
August	0.6 in.
September	1.5 in.
October	4.0 in.

The dry period, then, runs from mid-June to mid-September—about 90 days. Thus,

$$\text{capacity} = 200 \text{ gpd} \times 90 \text{ days} = 18{,}000 \text{ gal}$$

$$\text{volume} = \frac{18{,}000 \text{ gal}}{7.48 \text{ gal/ft}^3} = 2406 \text{ ft}^3$$

(For example, 5-ft deep × 22-ft square = 2420 ft^3.) A more detailed sizing procedure for this building's cistern is presented in Chapter 8.

(c) Required Facilities. Another important early design question is that of how many plumbing fixtures should be provided. Table 7.3 lists the minimum plumbing facilities required for various types and sizes of building occupancies. (Note that local requirements may differ somewhat from this particular guide, which is taken from the 1985 Uniform Plumbing Code. There are many plumbing codes in use in North America, and they sometimes disagree.) Since these are considered "minimal" requirements, more generous provisions are sometimes appropriate. Chapter 11 discusses the design of the spaces in which these facilities are used.

(d) Sewage. Where public sewers are available (as in most urban areas), the designer usually is not concerned with estimating the flow of sewage. However, where private or on-site sewage treatment is required, or where public sewage treatment facilities are overtaxed, total sewage flow is an early design concern. Table 7.4 lists these sewage flows, by type of occupancy, again in g/cd (L/cd). Note that sewage flow may differ from supply flow, especially where supply water is used for irrigation or in evaporative processes.

Once the daily flow of sewage is established, some early guidelines are needed for determining the suitability of and the area required by some treatment processes. One of the most common reasons for the rejection of a potential building site is a lack of suitability for sewage disposal. A geologic analysis of structural and sewage disposal potential is one of the first documents needed by the designer. Chapter 10 details the sizing of some common treatment methods. For this earlier stage, the following rules of thumb can be useful:

Septic Tank Drainfields. In conventional units (minimum 750-ft^2 area for any system):

- For shallow trenches in poorly draining soil: drainfield area = total sewage flow in gpd × 3.6 ft^2/gal.
- For deep trenches in well-draining soil: drainage area = total sewage flow in gpd × 0.4 ft^2/gal.

In S.I. units (minimum 70 m^2):

- For shallow trenches, poor soil: L/day × 0.087 m^2/L
- For deep trenches, good soil: L/day × 0.01 m^2/L

(These guidelines allow for an expansion area equal to the original size of the drainage field, in case of field failure.)

TABLE 7.3 **Minimum Plumbing Facilities**[a]

Type of Building or Occupancy[b]	*Water Closets (Fixtures per Person)*	*Urinals*[c] *(Fixtures per Male)*	*Lavatories (Fixtures per Person)*	*Bathtubs or Showers (Fixtures per Person)*	*Drinking Fountains*[d,e] *(Fixtures per Person)*
Assembly places (theaters, auditoriums, convention halls, etc.)—for permanent employee use	*Male* / *Female* 1:1–15 / 1:1–15 2:16–35 / 2:16–35 3:36–55 / 3:36–55 Over 55, add 1 fixture for each additional 40 persons	1 per 50	*Male* / *Female* 1 per 40 / 1 per 40	—	—
Assembly places (theaters, auditoriums, convention halls, etc.)—for public use	*Male* / *Female* 1:1–100 / 3:1–100 2:101–200 / 6:101–200 3:201–400 / 8:201–400 Over 400, add 1 fixture for each additional 500 males and 2 for each 300 females	1:1–100 2:101–200 3:201–400 4:401–600 Over 600, add 1 fixture for each additional 300 males	*Male* / *Female* 1:1–200 / 1:1–200 2:201–400 / 2:201–400 3:401–750 / 3:401–750 Over 750, add 1 fixture for each additional 500 persons	—	1 per 75[f]
Dormitories[g]—school or labor	*Male* / *Female* 1 per 10 / 1 per 8 Add 1 fixture for each additional 25 males (over 10) and 1 for each additional 20 females (over 8)	1 per 25 Over 150, add 1 fixture for each additional 50 males	*Male* / *Female* 1 per 12 / 1 per 12 Over 12, add 1 fixture for each additional 20 males and 1 for each 15 females	1 per 8 For females, add 1 bathtub per 30; over 150, add 1 per 20	1 per 75[f]
Dormitories—for staff use	*Male* / *Female* 1:1–15 / 1:1–15 2:16–35 / 2:16–35 3:36–55 / 3:36–55 Over 55, add 1 fixture for each additional 40 persons	1 per 50	*Male* / *Female* 1 per 40 / 1 per 40	—	—

Dwellings[h] Single Multiple or apartment house	 1 per dwelling 1 per dwelling or apartment unit	 —	 1 per dwelling 1 per dwelling or apartment unit	 1 per dwelling 1 per dwelling or apartment unit	 —
Hospitals Waiting room For employee use	 1 per room *Male* / *Female* 1:1–15 / 1:1–15 2:16–35 / 2:16–35 3:36–55 / 3:36–55 Over 55, add 1 fixture for each additional 40 persons	 — 1 per 50	 1 per room *Male* / *Female* 1 per 40 / 1 per 40	 — —	 1 per 75[f]
Hospitals Individual room Ward room	 1 per room 1 per 8 patients	 — —	 1 per room 1 per 10 patients	 1 per room 1 per 20 patients	 — 1 per 75[f]
Industrial[i] warehouses, workshops, foundries and similar establishments (for employee use)	*Male* / *Female* 1:1–10 / 1:1–10 2:11–25 / 2:11–25 3:26–50 / 3:26–50 4:51–75 / 4:51–75 5:76–100 / 5:76–100 Over 100, add 1 fixture for each additional 30 persons		Up to 100, 1 per 10 persons Over 100, 1 per 15 persons[j,k]	1 shower for each 15 persons exposed to excessive heat or to skin contamination with poisonous, infectious, or irritating material.	1 per 75[f]
Institutional (other than hospitals or penal institutions), on each occupied floor	*Male* / *Female* 1 per 25 / 1 per 20	1 per 50	*Male* / *Female* 1 per 10 / 1 per 10	1 per 8	1 per 75[f]
Institutional (other than hospitals or penal institutions), on each occupied floor—for employee use	*Male* / *Female* 1:1–15 / 1:1–15 2:16–35 / 2:16–35 3:36–55 / 3:36–55 Over 55, add 1 fixture for each additional 40 persons	1 per 50	*Male* / *Female* 1 per 40 / 1 per 40	—	—

(continued)

TABLE 7.3 **Minimum Plumbing Facilities**[a] **(*continued*)**

Type of Building or Occupancy[b]	*Water Closets (Fixtures per Person)*		*Urinals*[c] *(Fixtures per Male)*	*Lavatories (Fixtures per Person)*		*Bathtubs or Showers (Fixtures per Person)*	*Drinking Fountains*[d,e] *(Fixtures per Person)*
Office or public building	*Male* 1:1–15 2:16–35 3:36–55 4:56–80 5:81–110 6:111–150	*Female* 1:1–15 2:16–35 3:36–55 4:56–80 5:81–110 6:111–150		*Male* 1:1–15 2:16–35 3:36–60 4:61–90 5:91–125	*Female* 1:1–15 2:16–35 3:36–60 4:61–90 5:91–125	—	1 per 75[f]
	Over 150, add 1 fixture for each additional 40 persons			Over 125, add 1 fixture for each additional 45 persons			
Office or public building—for employee use	*Male* 1:1–15 2:16–35 3:36–55	*Female* 1:1–15 2:16–35 3:36–55	1 per 50	*Male* 1per 40	*Female* 1 per 40	—	—
	Over 55, add 1 fixture for each additional 40 persons						
Penal institutions—for employee use	*Male* 1:1–15 2:16–35 3:36–55	*Female* 1:1–15 2:16–35 3:36–55	1 per 50	*Male* 1 per 40	*Female* 1 per 40	— —	1 per 75[f]
	Over 55, add 1 fixture for each additional 40 persons						
Penal institutions—for prisoner use							
Cell	1 per cell		—	1 per cell		—	1 per cell block floor
Exercise room	1 per exercise room		1 per exercise room	1 per exercise room		—	1 per exercise room

Restaurants, pubs and lounges[l]	*Male* 1:1–50 2:51–150 3:151–300 Over 300, add 1 fixture for each additional 200 persons	*Female* 1:1–50 2:51–150 3:151–300	1:1–150 Over 150, add 1 fixture for each additional 150 males	*Male* 1:1–150 2:151–200 3:201–400 Over 400, add 1 fixture for each additional 400 persons	*Female* 1:1–150 2:151–200 3:201–400		
Schools—for staff use	*Male* 1:1–15 2:16–35 3:36–55 Over 55, add 1 fixture for each additional 40 persons	*Female* 1:1–15 2:16–35 3:36–55	1 per 50	*Male* 1 per 40	*Female* 1 per 40	—	—
Schools[m]—for student use Nursery	*Male* 1:1–20 2:21–50 Over 50, add 1 fixture for each additional 50 persons	*Female* 1:1–20 2:21–50	—	*Male* 1:1–25 2:26–50 Over 50, add 1 fixture for each additional 50 persons	*Female* 1:1–25 2:26–50	—	1 per 75[f]
Elementary	*Male* 1 per 30	*Female* 1 per 25	1 per 75	*Male* 1 per 35	*Female* 1 per 35	—	1 per 75[f]
Secondary	*Male* 1 per 40	*Female* 1 per 30	1 per 35	*Male* 1 per 40	*Female* 1 per 40	—	1 per 75[f]
Others (colleges, universities, adult centers, etc.)	*Male* 1 per 40	*Female* 1 per 30	1 per 35	*Male* 1 per 40	*Female* 1 per 40	—	1 per 75[f]
Worship places—educational and activities unit	*Male* 1 per 250	*Female* 1 per 125	1 per 250	1 per toilet room		—	1 per 75[f]
Worship places—principal assembly place	*Male* 1 per 300	*Female* 1 per 150	1 per 300	1 per toilet room		—	1 per 75[f]

(continued)

TABLE 7.3 **Minimum Plumbing Facilities**[a] ***(continued)***

Whenever urinals are provided, one (1) water closet less than the number specified may be provided for each urinal installed, except that the number of water closets in such cases shall not be reduced to less than two-thirds (⅔) of the minimum specified.

Source: Uniform Plumbing Code, copyright © 1985. Printed by permission of the International Association of Plumbing and Mechanical Officials.

[a]The figures shown are based upon one (1) fixture being the minimum required for the number of persons indicated or any fraction thereof.

[b]Building categories not shown on this table shall be considered separately by the Administrative Authority.

[c]In applying this schedule of facilities, consideration must be given to the accessibility of the fixtures. Conformity purely on a numerical basis may not result in an installation suited to the needs of the individual establishment. For example, schools should be provided with toilet facilities on each floor having classrooms. Temporary workingmen facilities: one (1) water closet and one (1) urinal for each thirty (30) workmen.

a. Surrounding materials: wall and floor space to a point two (2) feet (0.6 m) in front of urinal lip and four (4) feet (1.2 m) above the floor, and at least two (2) feet (0.6 m) to each side of the urinal shall be lined with nonabsorbent material.

b. Trough urinals are prohibited.

[d]Drinking fountains shall not be installed in toilet rooms.

[e]There shall be a minimum of one (1) drinking fountain per occupied floor in schools, theaters, auditoriums, dormitories, offices or public buildings.

[f]Where food is consumed indoors, water stations may be substituted for drinking fountains. Theaters, auditoriums, dormitories, offices, or public buildings for use by more than six (6) persons shall have one (1) drinking fountain for the first seventy-five (75) persons and one (1) additional fountain for each one hundred and fifty (150) persons thereafter.

[g]Laundry trays: one (1) for each fifty (50) persons. Slop sinks: one (1) for each hundred (100) persons.

[h]Laundry trays: one (1) laundry tray or one (1) automatic washer standpipe for each dwelling unit or two (2) laundry trays or two (2) automatic washer standpipes, or combination thereof, for each ten (10) apartments. Kitchen sinks: one (1) for each dwelling or apartment unit.

[i]As required by ANSI Z4.1–1968, Sanitation in Places of Employment.

[j]Where there is exposure to skin contamination with poisonous, infectious, or irritating materials, provide one (1) lavatory for each five (5) persons.

[k]Twenty-four (24) lineal inches (609.6 mm) of wash sink or eighteen (18) inches (457.2 mm) of a circular basin, when provided with water outlets for such space, shall be considered equivalent to one (1) lavatory.

[l]A restaurant is defined as a business that sells food to be consumed on the premises.

a. The number of occupants for a drive-in restaurant shall be considered as equal to the number of parking stalls.

b. Employee toilet facilities are not to be included in the above restaurant requirements. Hand washing facilities must be available in the kitchen for employees.

[m]This schedule has been adopted by the National Council on Schoolhouse Construction.

TABLE 7.4 **Estimated Sewage Flow Rates**

Type of Occupancy	*Unit Gallons (Liters) per Day*
Airports	15 (56.8) per employee
	5 (18.9) per passenger
Auto washers	Check with equipment manufacturer
Bowling alleys (snack bar only)	75 (283.9) per lane
Camps, campgrounds with central comfort station	35 (132.5) per person
Campgrounds with flush toilets, no showers	25 (94.6) per person
Day camps (no meals served)	15 (56.8) per person
Summer and seasonal	50 (189.3) per person
Churches (sanctuary)	5 (18.9) per seat
With kitchen waste	7 (26.5) per seat
Dance halls	5 (18.9) per person
Factories, no showers	25 (94.6) per employee
With showers	35 (132.5) per employee
Cafeteria, add	5 (18.9) per employee
Hospitals	250 (946.3) per bed
Kitchen waste only	25 (94.6) per bed
Laundry waste only	40 (151.4) per bed
Hotels (no kitchen waste)	60 (227.1) per bed (2 person)
Institutions (resident)	75 (283.9) per person
Nursing home	125 (473.1) per person
Rest home	125 (473.1) per person
Laundries, self-service	50 (189.3) per wash cycle
Commercial	Per manufacturer's specifications
Motel	50 (189.3) per bed space
With kitchen	60 (227.1) per bed space
Offices	20 (75.7) per employee
Parks, mobile homes	250 (946.3) per space
Picnic parks (toilets only)	20 (75.7) per parking space
Recreational vehicles, without water hook-up	75 (283.9) per space
With water and sewer hook-up	100 (378.5) per space
Residences	75 (283.9) per person
Restaurants/cafeterias	20 (75.7) per employee
Toilet	7 (26.5) per customer
Kitchen waste	6 (22.7) per meal
Add for garbage disposal	1 (3.8) per meal
Add for cocktail lounge	2 (7.6) per customer
Kitchen waste—disposable service	2 (7.6) per meal
Schools—staff and office	20 (75.7) per person
Elementary students	15 (56.8) per person
Intermediate and high	20 (75.7) per student
With gym and showers, add	5 (75.7) per student
With cafeteria, add	3 (11.4) per student
Boarding, total waste	100 (378.5) per person
Service station, toilets	1000 (3785) for 1st bay
	500 (1892.5) for each additional bay
Stores	20 (75.7) per employee
Public restrooms, add (per unit of floor space)	1 (3.8) per 10 ft^2 (.9 m^2)
Swimming pools, public	10 (37.9) per person
Theaters, auditoriums	5 (18.9) per seat
Drive-in	10 (37.9) per space

Source: Uniform Plumbing Code, copyright © 1985. Printed by permission of the International Association of Plumbing and Mechanical Officials.

Mounds. These are built-up leaching fields on top of an existing grade (see Fig. 10.16). For a single-family dwelling, allow for a 4-ft-high mound whose bottom area is a square 44 ft (13.4 m) on each side and whose sides slope at a 1:3 vertical-to-horizontal ratio.

Package Sewage Plant Drainfields. In these, sewage is treated to a much greater extent than in septic tanks, and effluent is filtered:

- In poorly draining soil: total sewage flow in gpd × 0.49 ft^2/gal (L/day × 0.012 m^2/L).
- In well-draining soil: total sewage flow in gpd × 0.23 ft^2/gal (L/day × 0.006 m^2/L).

Sewage Lagoons. These consist of two open treatment ponds (primary and secondary) and are sized on the basis of pounds of biological oxygen demand (BOD) rather than gallons of sewage flow. A typical assumption for estimating BOD is

- 0.2 lb BOD/person for ''ordinary'' domestic sewage.
- 0.3 lb BOD/person where garbage grinders or other devices contribute added organic material to domestic sewage.

The *total* acreage needed for the two ponds can be estimated as

- 20 lb BOD/a for the (colder) northern United States.
- 35 lb BOD/a for the (drier, warmer) southern United States.

(The primary pond is usually sized for 50 lb BOD/a.)

References

Milne, M. (1976). *Residential Water Conservation,* U.S. Office of Water Research and Technology, Department of Commerce, NTIS.

U.S. Environmental Protection Agency (1975). *Manual of Individual Water Supply Systems,* U.S. EPA, Washington, D.C.

8

RAINWATER

There is a striking similarity between rainwater and solar energy. Both are essential for agriculture, and both have been well understood and utilized by farmers since agrarian societies first emerged. Both can be very beneficial to architecture and were utilized as needed by anonymous builders for centuries. Both fell out of favor with designers as plentiful supplies of pure, centrally treated water and concentrated, centrally controlled fuels became commonplace. Both are thinly yet relatively evenly spread over the world's population, so they are at least seasonally available to help meet nearly every building's needs for water and heat. Yet both are difficult to utilize in industrialized societies, because they require *individual* expenditures.

Consider the typical public water and electric utility in a U.S. city. It can raise the funds to build large water treatment plants, electricity generating plants, and the network of pipes and wires that bring these commodities to every building. The utility's costs, including interest on its construction debts, will be passed on to its consumers, along with a margin of profit that is usually controlled by state governments. Thus, our society has a well-established method for encouraging central suppliers of water and heat.

Now consider the individual building owner. To build a cistern and a solar-heated building, she must borrow money at an interest rate higher than that which the utility pays; and both options cost more initially than simple connections to the utility's pipes and lines. Even though she is willing to flush her toilets with rainwater rather than with chlorinated and filtered potable water from the utility, and even though she is willing to heat with lower-grade solar energy than with higher-grade electricity, she must pay a substantial first-cost penalty—with interest—to do so.

The overall public good could be well served by a mixture of public networks of pure water and electricity and individual cisterns and solar applications. The environmental benefits would be substantial—less water withdrawn from rivers, lakes, and underground aquifers; less stormwater discharged to pollute rivers; less fuel used to generate electricity; less environmental damage from power plants. Yet we continue to economically discourage the individual who uses the rain and the sun.

8.1 Collection and Storage

In terms of both quality and quantity, rainwater is an attractive alternative. Figure 7.2*b* showed that rainwater is near to the purest state in the hydrologic cycle. More recently, it is true, air pollution began to threaten the quality of rainwater in some areas, as ''acid rain'' has become widespread in the northeastern section of the North American continent and in Europe. In some particularly air-polluted locations, lead poses a threat to rainwater quality. Also, on any catchment surface, dust and bird droppings are common pollutants that must be considered. Other factors that bear on rainwater quality are roofing materials and the form of the roof. The appropriate health authority should be consulted for a list of roofing materials that will have no toxic effects on rainwater. Steeper roofs are scoured by winds and thus collect less dust and give cleaner runoffs. Devices to discourage roosting birds are strongly recommended, as are periodic checks of cistern water for bacteria. Fungicides (for moss control) should be scrupulously avoided, as should roofing paints containing lead. For these reasons, urban rainwater commonly is not used for drinking and cooking. (Bottled water or on-site water distilling can supply potable water.) For the typical residence, however, that still leaves about 95% of indoor water usage that could be provided by rainwater. And in those cities that have very ''hard'' public water supplies, rainwater's ''soft'' characteristics make it particularly attractive.

The quantity of rainwater available in most U.S. locations could meet a high percentage of typical home or business needs. Milne (1976) pointed out that the 42 in. of rain that falls annually on the streets and roofs of Manhattan Island could, if collected and stored, provide 148% of the residential needs of the 1.7 million inhabitants. For the typical U.S. suburban house (roof area of 1500 ft^2), the annual catchment can be estimated by combining the rainfall quantities (Fig. 8.1) with the resulting catchment yield (Fig. 7.3). Even at a "dry" rate of only 20 in. annually, a 1500-ft^2 roof would yield about 12,000 gal, or 33 gpd—nearly enough to meet the clothes cleaning and dishwashing needs of the family. In SI units, a 140-m^2 roof with annual 500 mm precipitation will yield 50,000 L, or 137 L/d.)

Unfiltered rainwater seems particularly well-suited to the irrigation of small lawns or gardens, both because it lacks additives unneeded by plants, such as chlorine or sodium fluoride, and because it can reduce the user's demand on the public water supply on the hottest summer days. Cisterns located above the irrigated area have the advantage of replacing a pump with simple gravity flow. The rainbarrel at the bottom of a downspout was an example of this approach—though one of limited capacity.

In many of the world's drier areas, smaller cisterns within the home are common. Such cisterns, which can be fed both by rainwater and by the public supply, are frequently used for all domestic purposes, including drinking. The presence of such a large water volume also can be advantageous in the event of fire. Sometimes these storage cisterns are required because the public supply is diminished or cut off at peak-usage hours, due to insufficient water main capacity. Figure 8.2 shows a cistern in the typical outdoor location, along with various options by which pollution from dirty roofs can be minimized.

***Fig. 8.2** (opposite) A section through a typical outdoor cistern.* (a) *A "roof washer" gets the dirtiest, first runoff from the roof. It can later be emptied either by opening the faucet wide or by leaving the faucet slightly open so that it will slowly drain.* (b) *In place of the "roof washer," a sand filter may be used. The "flapper valve" is then used rarely, perhaps to divert the first rainfall after a prolonged dry spell. After this, the valve is left in a position to divert all rainwater to the sand filter. From the U.S. Environmental Protection Agency's* Manual of Individual Water Supply Systems, *1975.* (c) *Another roof-washing option is a "tipping valve," which dumps the first* X *gallons of each rainfall. After each rainfall, the valve must be manually (or spring) reset to "dry" position if it is to again intercept dirty water. In an extended wet period, it would probably be left in the "wet" position.*

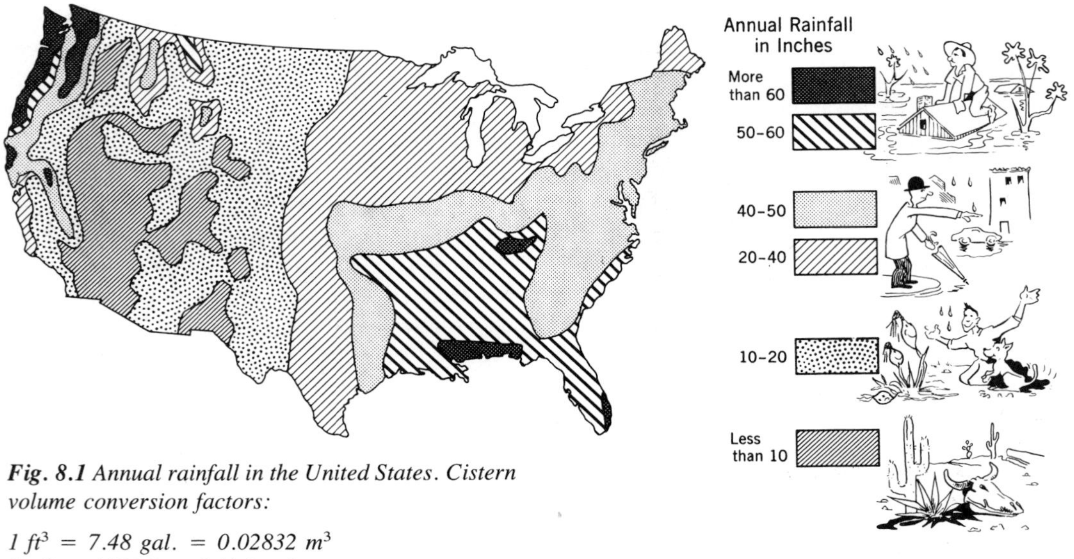

***Fig. 8.1** Annual rainfall in the United States. Cistern volume conversion factors:*

1 ft^3 *= 7.48 gal. = 0.02832* m^3
1 m^3 *= 1000 L = 264.2 gal.*

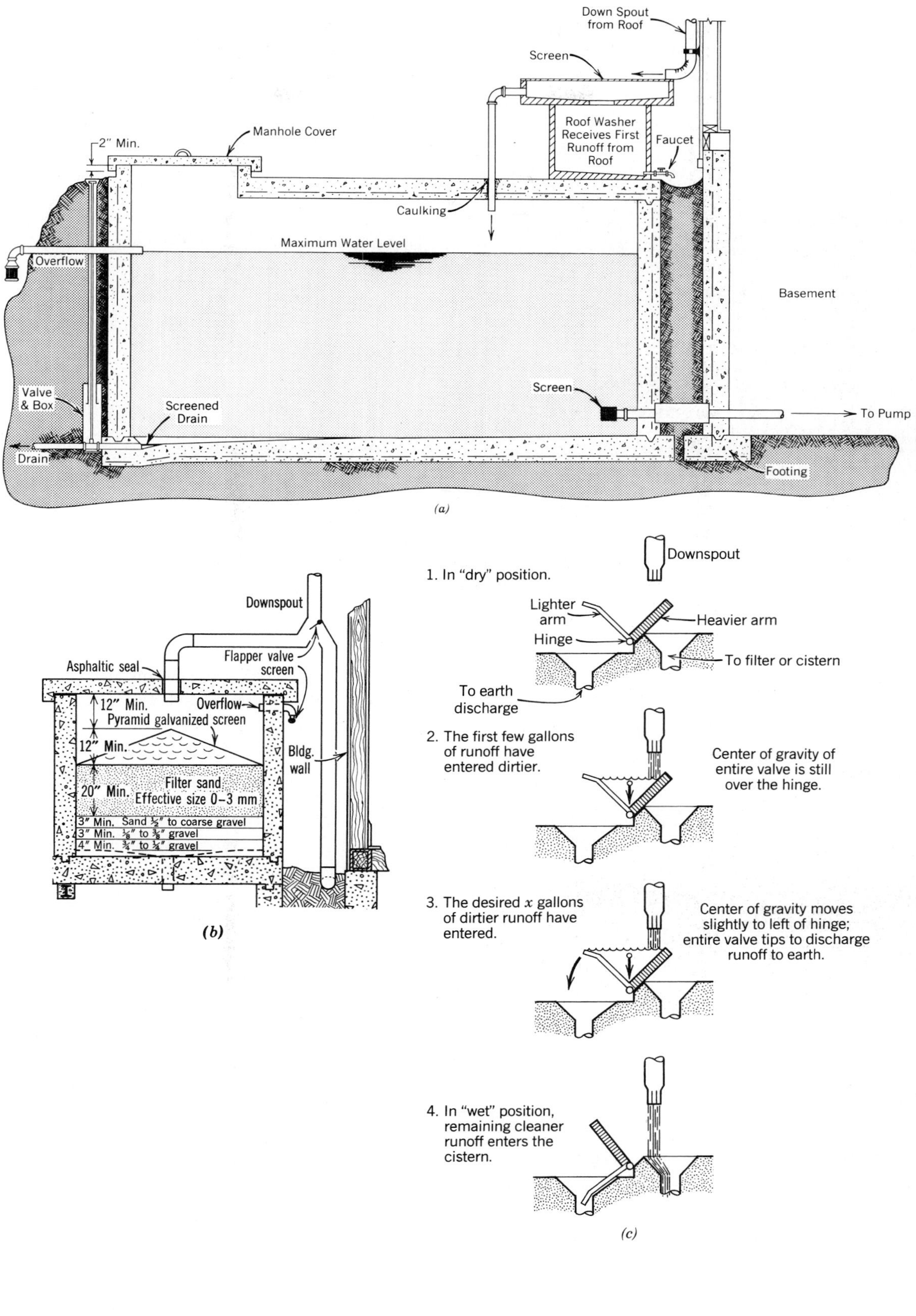

Down Spout from Roof
Screen
Roof Washer Receives First Runoff from Roof
Faucet
2" Min.
Manhole Cover
Caulking
Maximum Water Level
Overflow
Basement
Valve & Box
Screened Drain
Screen
To Pump
Drain
Footing
(a)
Downspout
Asphaltic seal
Flapper valve screen
12" Min.
Overflow
Pyramid galvanized screen
12" Min.
Bldg. wall
Filter sand
20" Min. Effective size 0–3 mm
3" Min. Sand ½" to coarse gravel
3" Min. ⅛" to ⅜" gravel
4" Min. ¾" to ¼" gravel
(b)
1. In "dry" position.
Downspout
Lighter arm
Heavier arm
Hinge
To filter or cistern
To earth discharge
2. The first few gallons of runoff have entered dirtier.
Center of gravity of entire valve is still over the hinge.
3. The desired x gallons of dirtier runoff have entered.
Center of gravity moves slightly to left of hinge; entire valve tips to discharge runoff to earth.
4. In "wet" position, remaining cleaner runoff enters the cistern.
(c)

Fig. 8.3 The country home of architect John Andrews, Eugowra, NSW, Australia. (a) *View from the southwest, showing corner cisterns, central daylight/ventilation barrel vault, and "energy tower" with water storage tank.* (b) *The view from the south, showing shading pergola to be covered by vines.* (c) *Floor plan, showing fireplace at centers and cisterns at all corners. Insect screens protect the openings to summer breezes.* (d) *Section. Courtesy of John Andrews International Pty. Ltd.*

When cisterns are taken seriously as water storage devices, their function and size can become strong design-form determinants. The country home of architect John Andrews in the dry ranchland of New South Wales, Australia, offers a particularly striking example of cisterns and form (Fig. 8.3). The corner rainwater collectors also help deflect cooling summer breezes into the house, along the diagonal walls at each corner. The central skylight then vents the breezes, along with the house's heat. From the corner cisterns, rainwater is pumped up to the central tower's storage tank. It then can be heated by solar collectors (to be installed on the sloping top of the tower), or used to supply the house's plumbing fixtures, or even used to sprinkle the metal roof surface, whose surface temperature could quickly be lowered by evaporation. Unevaporated roof water simply runs back into the cisterns. As a final design-with-water gesture, the shower (off bedroom 1) has been made into a true celebration place for cleansing and refreshing.

In this passively solar heated home, the living areas are placed on the elongated north side—the warmer side in the cold season of this Southern Hemisphere house. Passive cooling is aided by pergolas on the west and south, which are covered with vines for hot season shading of windows.

Sizing. In Chapter 7, procedures were given for rough sizing of both the catchment area and the storage capacity for cisterns. When rainwater is to be a primary, as opposed to merely a supplementary source, a closer look must be taken at rainfall deposits and user withdrawals from a cistern.

This procedure depends upon the monthly average rainfall (from NOAA Local Climatological Data), the monthly water usage, and the catchment area yield (from Fig. 7.3).

EXAMPLE 8.1. Take a closer look at the cistern that was approximately sized in Example 7.1. For this cistern, to be used for toilet flushing in a factory near Salem, Oregon, daily usage was estimated at 200 gal and catchment area at 5600 ft^2. The cistern capacity was estimated at 18,000 gal.

(b)

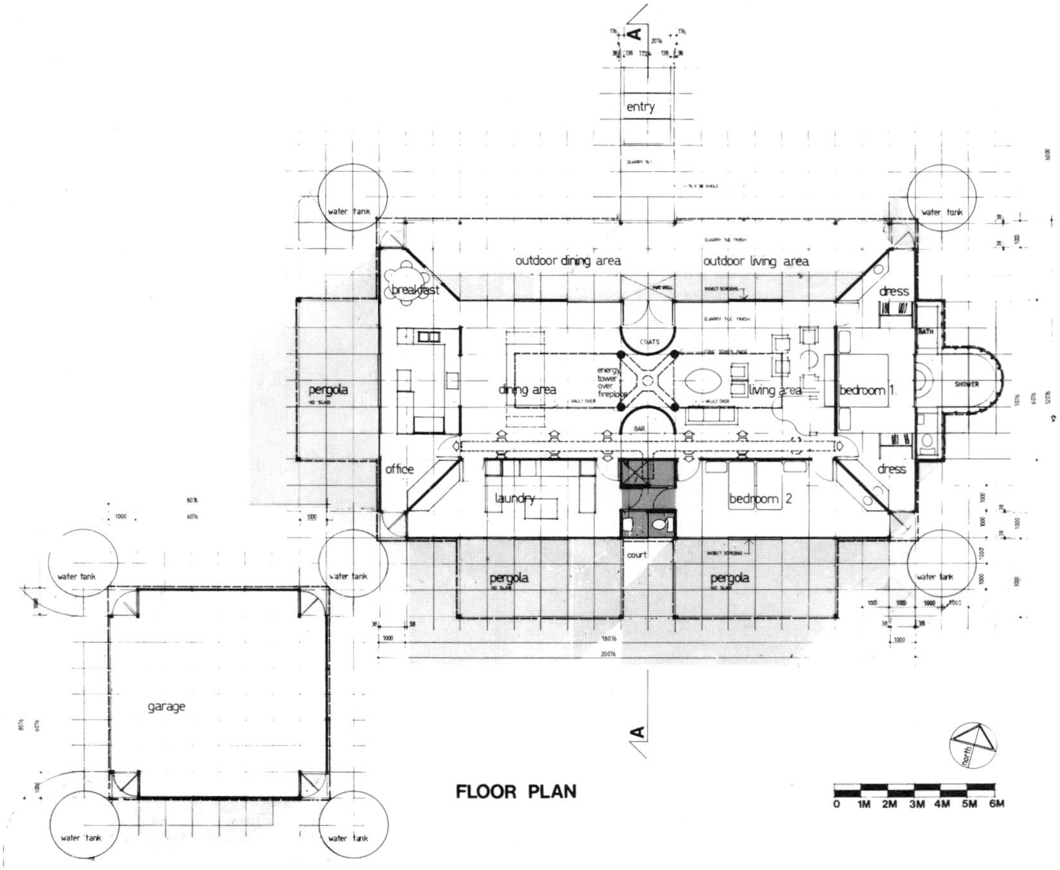
entry
water tank
outdoor dining area
outdoor living area
breakfast
dress
pergola
dining area
energy tower over fireplace
living area
bedroom 1.
office
laundry
bedroom 2
dress
court
pergola
pergola
garage
FLOOR PLAN
north
0 1M 2M 3M 4M 5M 6M

(c)

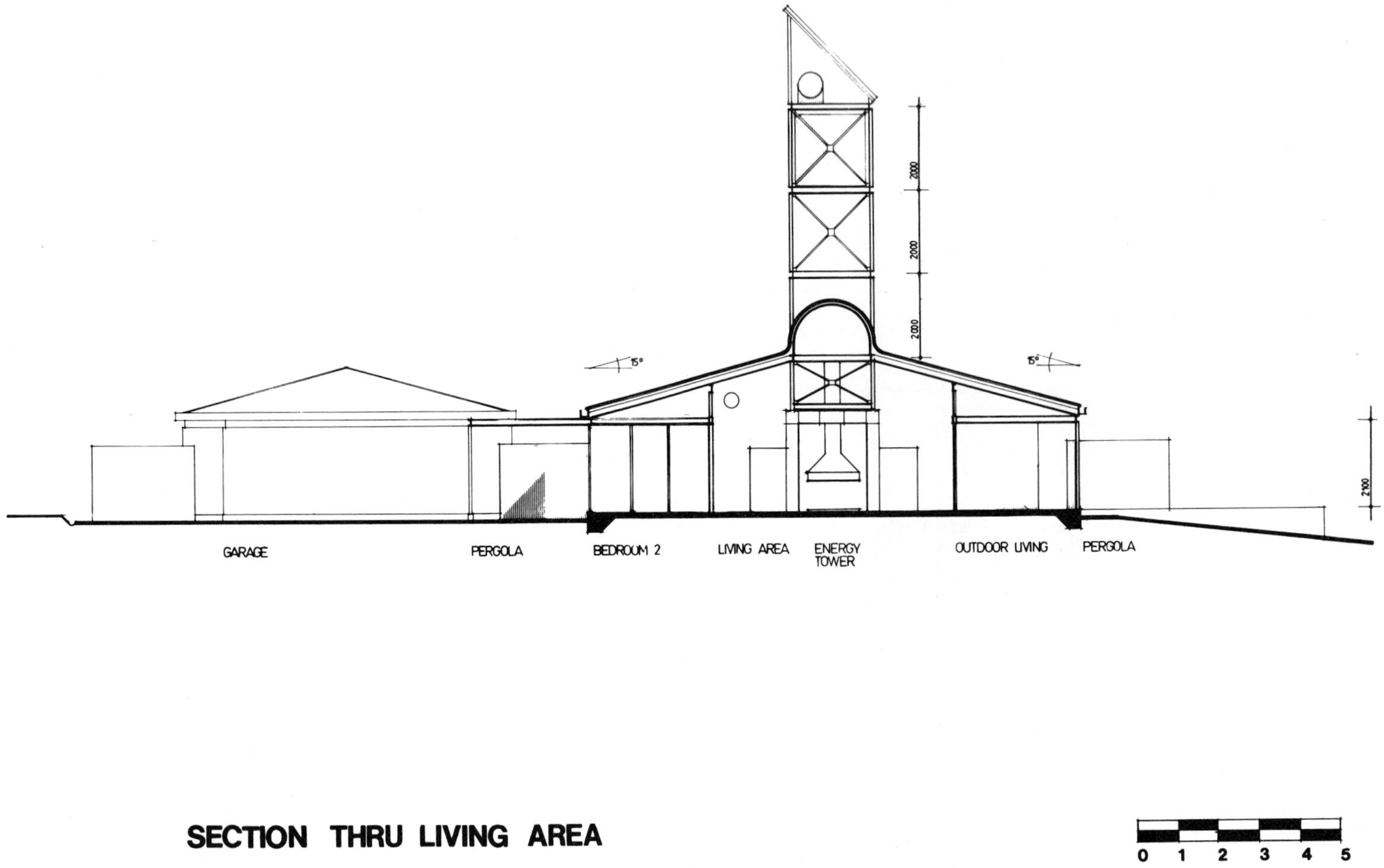

SECTION THRU LIVING AREA

(d)

Fig. 8.3 (*continued*)

Begin the process in the midst of the wettest months.

From Table 8.1, the following conclusions can be drawn.

1. For an 18,000-gal cistern, when the end-of-December cumulative capacity is added to January's surplus, the cistern will be at capacity from November through April.
2. With no allowances for abnormally dry months, we could reduce the cistern size by about 3000 gal (the size of the smallest cumulative capacity, in September).
3. A larger cistern—one that could utilize everything from the catchment area—could be built. Its approximate size would be the year-end surplus of 38,510 gal plus the maximum spring monthly cumulative capacity of 25,410 gal, in April.
4. Or the surplus could be devoted to additional usage of rainwater, beyond mere toilet flushing, from November through April.

8.2 Rainwater and Site Planning

Prior to the spread to rural areas of buildings, streets, roads, and paved parking lots, water from rainfall and melting snow found its own way to natural destinations. Surface flow to creeks, streams, and rivers accounted for part of this drainage. Underground flow aided the general runoff. Outcropping of flowing groundwater created springs and artesian wells. Low, dished areas formed lakes that in turn overflowed to outlet streams. Flat areas sometimes developed into swamps or marshes.

At a time when there was a choice of locations for towns and villages, sites next to rivers were usually chosen. The waterways provided transportation and water was supplied from the river or from adjacent wells. As streets and roads were built, slopes could be arranged whereby the rain falling on these areas and flowing onto them from roofs of buildings could run to the river. At inte-

TABLE 8.1 **Rainfall Cistern Sizing Procedure**

I		*II*	*III*	*IV*	*V*	*VI*
Month and Rainfall[a] *(in.)*		*Catchment Yield*[b] *(gal)*	*Usage*[c] *(gal)*	*Net*[d] *(gal)*	*Cumulative Capacity*[e] *(gal)*	*Cumulative Capacity Adjusted for Actual Size*[f] *(gal)*
January	6.9	18,630	6,200	12,430	12,430	12,430
February	4.8	12,960	5,600	7,360	19,790[g]	18,000
March	4.3	11,610	6,200	5,410	25,200[g]	18,000
April	2.3	6,210	6,000	210	25,410[g]	18,000
May	2.1	5,670	6,200	−530	24,880[g]	17,470
June	1.4	3,780	6,000	−2,220	22,660[g]	15,250
July	0.4	1,080	6,200	−5,120	17,540	10,130
August	0.6	1,620	6,200	−4,580	12,960	5,550
September	1.5	4,050	6,000	−1,950	11,010	3,600
October	4.0	10,800	6,200	4,600	15,610	8,200
November	6.1	16,470	6,000	10,470	26,080[g]	18,000
December	6.9	18,630	6,200	12,430	38,510[g]	18,000

Source: Based upon procedure from Brown, Reynolds, Ubbelohde, *InsideOut: Design Procedures for Passive Environmental Technologies,* copyright © 1982 by John Wiley & Sons.

[a]NOAA data for Salem, Oregon.

[b]From Fig. 7.3, according to which 10 in. precipitation yields about 27,000 gallons for this 5,600-ft^2 catchment area.

[c]The factory uses 200 gpd times days in each month.

[d]Col. II minus col. III.

[e]Values in col. IV, added month by month.

[f]Set at 18,000 gallons for the factor in Example 7.1.

[g]The cumulative capacity exceeds the actual storage capacity of 18,000 gallons.

WATER AND WASTE

rior parts of the country, high ground was favored for building sites and growing communities. Obviously, swampy or marshy ground would not be chosen, but it did provide terminal locations for the stormwater that ran off the high ground. In the course of this natural flow, much of the water was drawn by evaporation to the clouds. The rest, conforming to topographic river basins, continued along to seek its way to the sea.

Unfortunately, as building increased, desirable locations grew scarce. The possibility of selecting high, dry ground diminished. Great areas, formerly low and marshy, were filled in and buildings constructed, often on piles. From such locations stormwater could not be disposed of by drainage to some adjacent lower area, or even recharged to the earth through dry wells. Moreover, extensive grids of paved streets and sidewalks in these level developments caught and held the water, resulting in considerable "ponding." Storm sewers had to be built and the water transported great distances, often having to be lifted at intermediate pumping stations before reaching its destination, which might be a remote river.

This emphasis on the removal of stormwater has led to an expensive and elaborate system based on quick "disposal" of rainfall. By decreasing the time between precipitation and runoff, quick disposal increases the peak flows within such systems, thereby increasing flooding of rivers during storms but reducing the rivers' flow between storms.

The overloading of storm sewers not only causes minor flooding, but also can influence building design. The designers of the New Orleans Convention and Exhibition Center, a new building with 610,000 ft^2 of roof area, spared the city's storm sewers from the impact of a 14-acre runoff by carrying the rainwater over the roofs of adjoining wharfs, and thus discharging it directly into the Mississippi River.

As urban storm sewers reach capacity and suburban groundwater levels drop, designers have begun to emphasize "stormwater infiltration" (or recharge of groundwater), rather than "quick runoff." Three design strategies for encouraging such recharge have emerged: roofs that will retain water and slowly release it, porous pavement, and on-site infiltration of runoff.

(a) Roof Retention. If stormwater is to be sent to storm sewers or to soak into the ground, a *slow flow* from roofs will help by diminishing peak flows in sewers and giving soaked soil a longer time to absorb still more runoff. Nearly flat roofs with specially designed drains (see Fig. 8.13*c*) permit slower discharge yet eventually drain completely dry (to discourage mosquito breeding, etc.). (For cisterns, however, sloped roofs should be used, as they stay cleaner than flat roofs.) This temporary pond on top of a building will clearly add to structural requirements. Another problem could be posed by high winds blowing sheets of water onto people below. In summer, however, a flooded roof can provide the significant thermal advantage of greatly lowering daytime roof surface temperatures. Rainfall-retaining roofs are now required in some urban areas with overtaxed storm sewers.

(b) Porous Pavement. Where groundwater is in short supply, new building sites now are often required to retain rainfall on-site. To accomplish this, many builders use either porous concrete or "incremental" paving (alternation of paving materials with grass or ground-cover plants) for parking lots and roadways.

Porous concrete has been used for many years in building construction as a low-strength, high-porosity material that has some insulating properties (R5 for 10-in. thickness). Patented porous concrete pavement now in use in Florida has a strength of over 3800 psi and a permeability of 2.3 gallons of water per minute per square foot. (A 2500 psi mix has a permeability of 18.5 gallons per minute per square foot.) In cold weather areas, the freeze-thaw cycle could be destructive to porous concrete.

"Incremental" paving, in which the many joints allow water to pass through, offers another possibility. So does the handsome, if more expensive, approach of alternating concrete paving with grass or ground cover, as shown in Fig. 8.4.

(c) Site Design for Recharging. This tactic is especially advisable in suburban-density developments in drier climates with absorptive soil (sand, gravel, etc.). The first option is at each house; design of the entire subdivision can also be influenced.

The simplest design approach is the gutterless sloped roof illustrated in Fig. 8.5, which is appli-

Fig. 8.4 An Oregon application of "Grasscrete," a system in which concrete blocks are alternated with tufts of grass. Unthank Seder Poticha, Architects. Photo by Jane Lidz.

cable to one-story, basementless homes with wide, overhanging roofs. A gravel-filled trench skirting the perimeter directly below the edge of the eaves catches the water flowing off the roof.

Some designers do not like the appearance of conventional gutters and leaders. There are many ingenious ways to avoid or modify their use and yet provide proper drainage. In many cases, however, gutters and leaders will be required, either to collect rainwater for cisterns or to control conditions around the perimeter of the house. Several options for stormwater recharge can be used with the gutter-leader combination.

A splash pan at the foot of each leader (Fig. 8.6*a*) offers the simplest method. It will lead the water a few feet from the house but will accommodate only a relatively low rate of flow. A gravel-filled pipe is somewhat more effective (Fig. 8.6*b*). When the soil is not very permeable (as, for instance, with clay), it is best to use a dry well with extended area and many perforations through which the water can be discharged to the ground.

Footing drains are often used to collect and lead away groundwater that accumulates around the foundations. This reduces the likelihood of basement wall leakage. These drains are most necessary when higher ground near the buildings increases the flow of groundwater against underground walls. Figure 8.7 illustrates this and also shows how stormwater from drains and roofs may be led to a surface absorption area of rock and gravel beyond a head wall where the general storm drain outcrops. This method can be chosen where there is sufficient property area and slope. It has the advantages of easy maintenance and service. Also, one can observe whether or not it is functioning correctly.

In new suburban developments, for which there are no storm sewers, recharge basins are some-

Fig. 8.5 Gutters and leaders are not always essential, provided doorways, walls, foundations, and landscaped lawns are not subjected to rain concentrations.

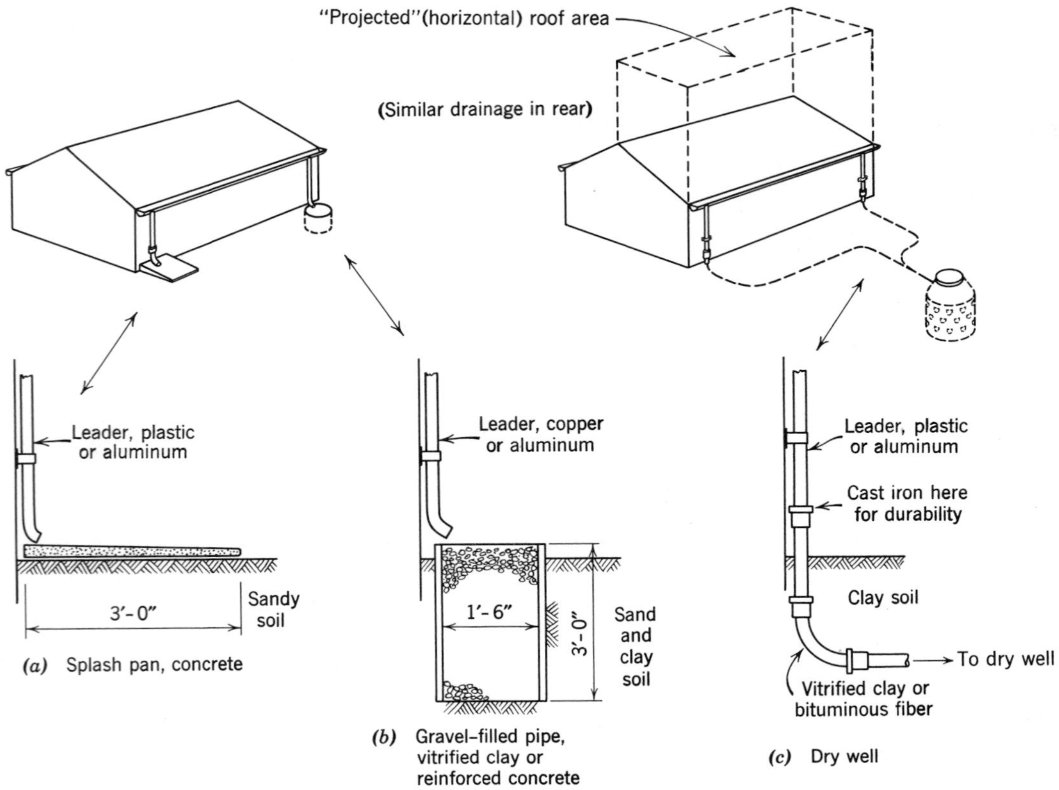

Fig. 8.6 Roof drainage for houses. Method (a) *is suitable for low rates of flow introduced into very pervious soil. When denser soil is encountered,* (b) *is used to get the water into the ground and thus avoid surface erosion. For heavy flow or to lead the water further from the structure,* (c) *may be used with one or several dry wells.*

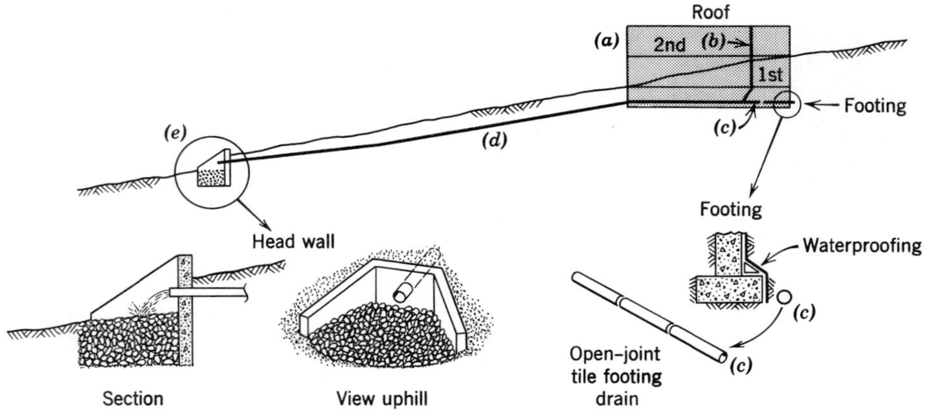

Fig. 8.7 Disposal of stormwater on the site but remote from the house or building. When a wall is against a hill, it is usually subjected to pressure of groundwater during storms. Open-joint clay, plastic or fiber tile accepts this water and carries it away. (a) *Position of building.* (b) *Roof drainage.* (c) *Footing drains.* (d) *Tight-joint clay tile or bituminous fiber pipe.* (e) *Flow through stone and gravel returns the water to the earth. Head wall is appropriate in lieu of a dry well if the site permits.*

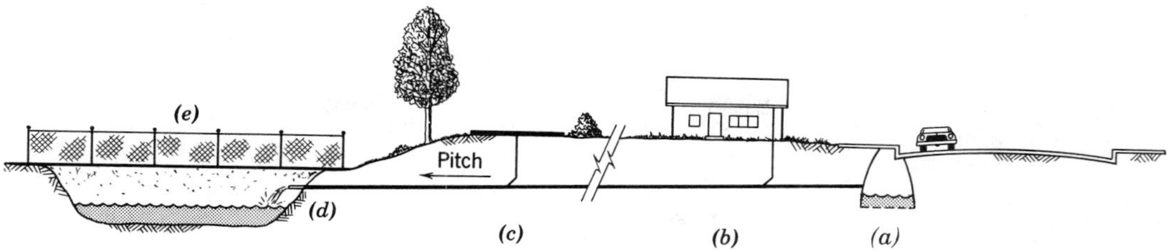

Fig. 8.8 Recharge basins in suburban communities. When topography, groundwater level, and porosity of the soil permit, developers are sometimes required to install systems that collect stormwater and carry it from (a) *catch basins at street curbs,* (b) *the roofs of all houses, and* (c) *paved areas to a recharge basin* (d) *that receives the water and returns it to the ground. For the safety of children, a fence* (e) *must be used to prevent unauthorized access to the basin.*

times required to deliver stormwater to the ground. Water from numerous roofs, paved areas, and curb catch basins is collected and piped to an open, unpaved pit, where it sinks into the earth. This method is not recommended in areas of dense, impervious clay soil (see Fig. 8.8). A particularly effective example of this approach is offered by the community of passively solar heated residences known as Village Homes, outside Davis, California. This area receives only about 20 in. of rain annually, so the recharge of groundwater seemed to this garden-oriented community to be a more attractive option than loss of the rainwater to the storm sewer. Stormwater flows from leaders to dry, rockbed channels, along which are gardens and bicycle paths (Figs. 8.9 and 8.10). Occasion-

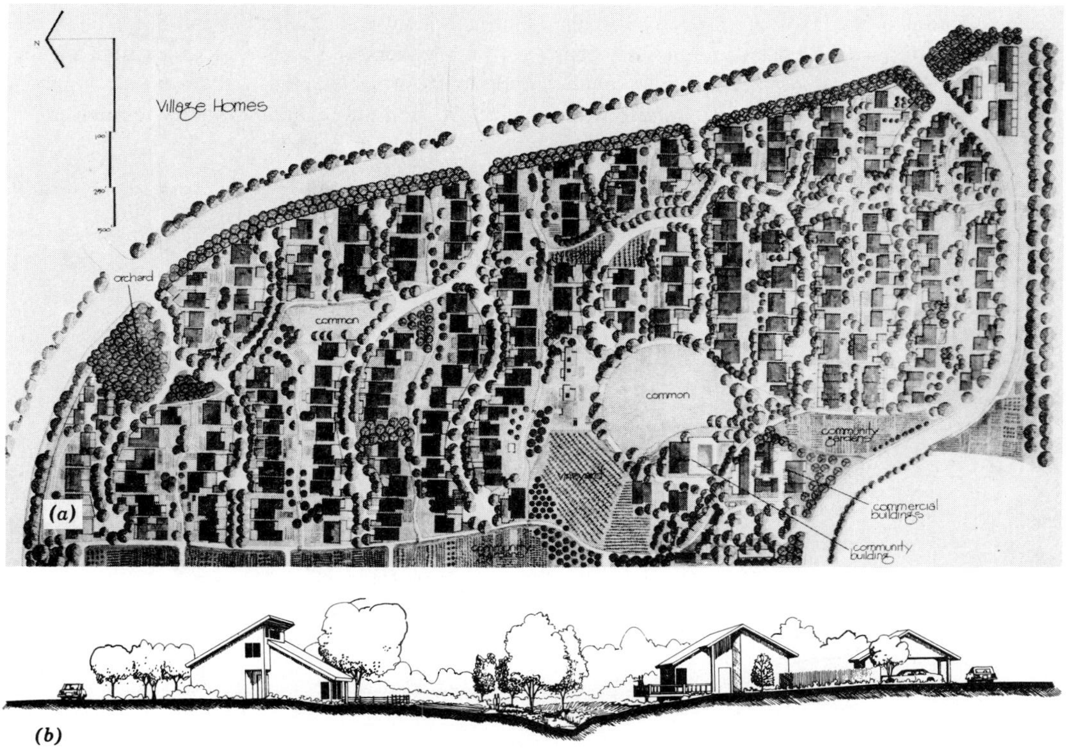

Fig. 8.9 Village Homes, a California subdivision of solar homes and stormwater recharge areas. (a) *Plan shows the emphasis on bicycle paths, narrow streets, and widespread community-maintained garden space through which the recharge streambeds are led.* (b) *Site section explains the gradual drainage from leaders at houses to recharge stream. Reprinted by permission from Corbett, M.,* A Better Place to Live, *copyright © 1983 by the Rodale Press.*

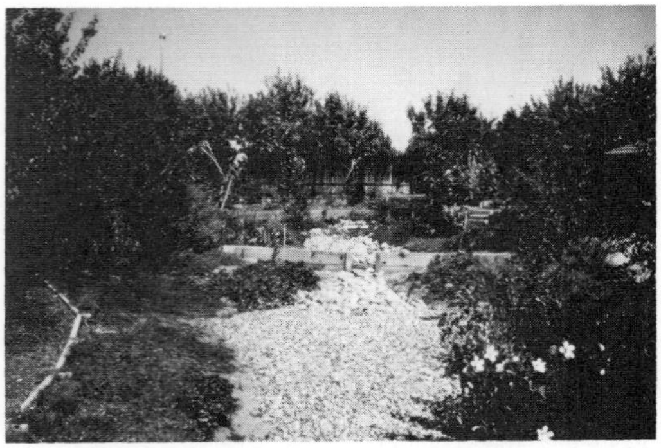

Fig. 8.10 The open drainage ways and pedestrian paths of Village Homes. Photo by Alan Butler.

ally, small dams across these channels create temporary holding ponds, in case the runoff has not yet soaked through the channel bottom. In the event of extraordinarily heavy rainfall, an inlet to the public storm sewer is available beyond the final holding pond; this inlet is needed approximately once every five years.

The planning of the landscape around buildings may closely follow such considerations as irrigation. The ''hydrozone'' concept of landscape planning is shown in Fig. 8.11. To minimize water consumption, exotic plant species are kept to a minimum and located near the house, where storm runoff and irrigating water is readily available. Native and adapted plantings, which can survive on normal rainfall, are utilized elsewhere. (Systems for irrigation are discussed in Section 9.6.)

8.3 Components

The first stormwater system design decision to be made involves the establishment of ''watersheds'' on a building's roof. To what edges, or at what points, will runoff be directed? To what depths will it accumulate before it leaves the roof? Since the answers to these questions depend on the intensity of storms as well as on the roof's geometry, it is necessary to find the maximum hourly rainfall for each location. This figure is available from local building code officials, or from Fig. 8.12.

Gutters and *leaders* (downspouts) can be sized through the use of Tables 8.2 and 8.3. The sizing of both gutters and leaders depends both on the horizontal projected area of the roof, as shown in Fig. 8.6, and the maximum hourly rainfall.

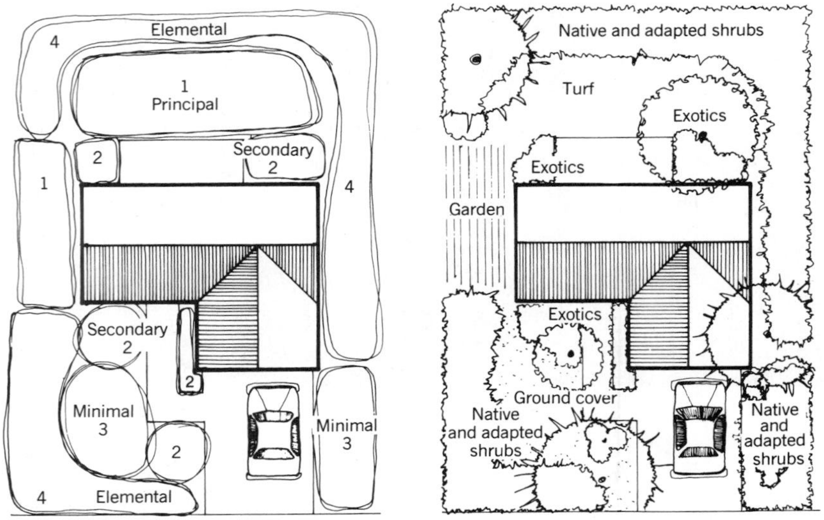

(a)

Fig. 8.11 The Hydrozone concept of landscape planning restricts exotic planting to special, easily watered areas. Native and adapted plantings are used elsewhere. (a) *Hydrozonics on a typical suburban lot.* (b) *Hydrozonics in a town-house subdivision.* (c) *Hydrozonics applied to more radical planning for water conservation. Reprinted by permission from* Energy Conserving Site Design, *E. G. McPherson, ed., copyright © 1984 by the American Society of Landscape Architects.*

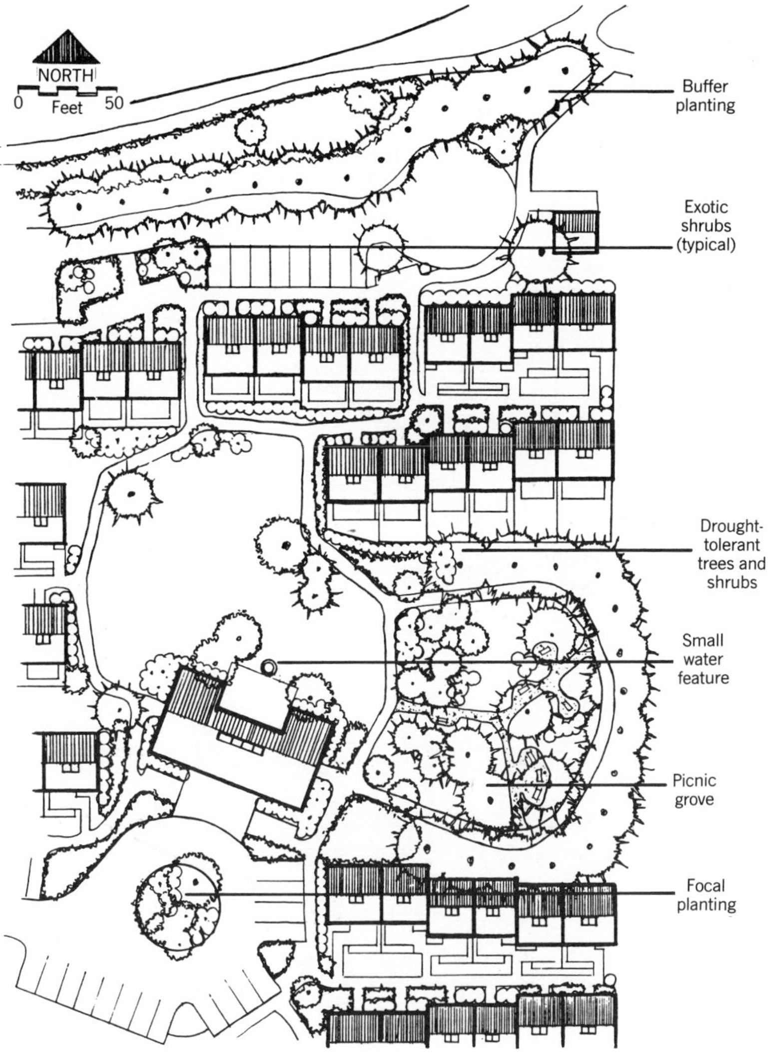
NORTH
0
Feet
50
Buffer planting
Exotic shrubs (typical)
Drought-tolerant trees and shrubs
Small water feature
Picnic grove
Focal planting

(b)

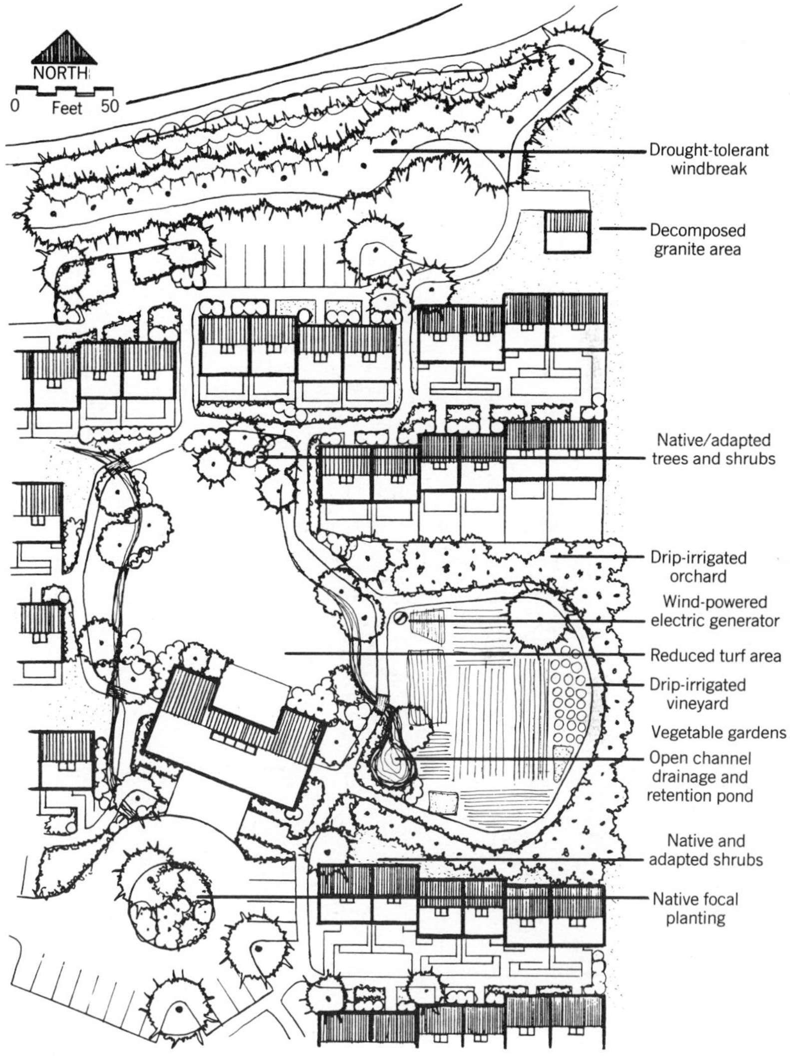
NORTH
0
Feet
50
Drought-tolerant windbreak
Decomposed granite area
Native/adapted trees and shrubs
Drip-irrigated orchard
Wind-powered electric generator
Reduced turf area
Drip-irrigated vineyard
Vegetable gardens
Open channel drainage and retention pond
Native and adapted shrubs
Native focal planting

(c)

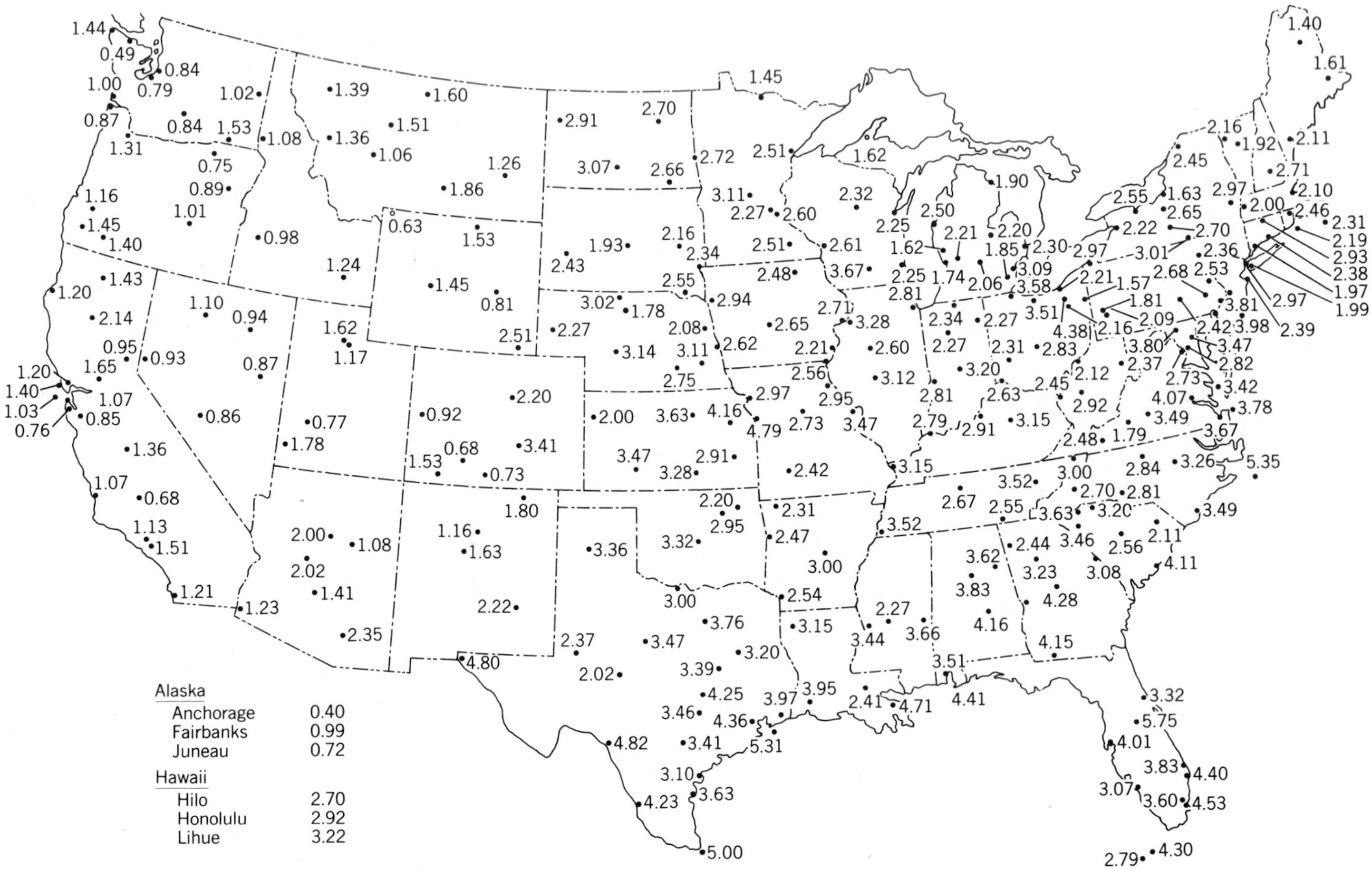

Fig. 8.12 *Maximum recorded hourly rainfall, in inches. (Latest data, 1961.) U.S. Weather Bureau Technical Paper #2, last issued in 1963.*

TABLE 8.2 **Gutter Sizes**

*Part A. **Conventional Units: Maximum Horizontal Projected Roof Areas, Square Feet***

Diameter of Gutter 1/16" Slope	Maximum Rainfall (in./hr)				
	2	3	4	5	6
3	340	226	170	136	113
4	720	480	360	288	240
5	1250	834	625	500	416
6	1920	—	960	768	640
7	2760	1840	1380	1100	918
8	3980	2655	1990	1590	1325
10	7200	4800	3600	2880	2400
Diameter of Gutter 1/8" Slope	Maximum Rainfall (in./hr)				
	2	3	4	5	6
3	480	320	240	192	160
4	1020	681	510	408	340
5	1760	1172	880	704	587
6	2720	1815	1360	1085	905
7	3900	2600	1950	1560	1300
8	5600	3740	2800	2240	1870
10	10200	6800	5100	4080	3400
Diameter of Gutter 1/4" Slope	Maximum Rainfall (in./hr)				
	2	3	4	5	6
3	680	454	340	272	226
4	1440	960	720	576	480
5	2500	1668	1250	1000	834
6	3840	2560	1920	1536	1280
7	5520	3680	2760	2205	1840
8	7960	5310	3980	3180	2655
10	14400	9600	7200	5750	4800
Diameter of Gutter 1/2" Slope	Maximum Rainfall (in./hr)				
	2	3	4	5	6
3	960	640	480	384	320
4	2040	1360	1020	816	680
5	3540	2360	1770	1415	1180
6	5540	3695	2770	2220	1850
7	7800	5200	3900	3120	2600
8	11200	7460	5600	4480	3730
10	20000	13330	10000	8000	6660

(continued)

TABLE 8.2 **Gutter Sizes (*continued*)**

Part B. ***SI Units: Maximum Horizontal Projected Roof Areas, Square Meters***

Diameter of Gutter 5.2 mm/m Slope	Maximum Rainfall (mm/hr)				
	50.8	76.2	101.6	127	152.4
76.2	31.6	21	15.8	12.6	10.5
101.6	66.9	44.6	33.4	26.8	22.3
127	116.1	77.5	58.1	46.5	38.7
152.4	178.4	—	89.2	71.4	59.5
177.8	256.4	170.9	128.2	102.2	85.3
203.2	369.7	246.7	184.9	147.7	123.1
254	668.9	445.9	334.4	267.6	223

Diameter of Gutter 10.4 mm/m Slope	Maximum Rainfall (mm/hr)				
	50.8	76.2	101.6	127	152.4
76.2	44.6	29.7	22.3	17.8	14.9
101.6	94.8	63.3	47.4	37.9	31.6
127	163.5	108.9	81.8	65.4	54.5
152.4	252.7	168.6	126.3	100.8	84.1
177.8	362.3	241.5	181.2	144.9	120.8
203.2	520.2	347.5	260.1	208.1	173.7
254	947.6	631.7	473.8	379	315.9

Diameter of Gutter 20.9 mm/m Slope	Maximum Rainfall (mm/hr)				
	50.8	76.2	101.6	127	152.4
76.2	63.2	42.2	31.6	25.3	21
101.6	133.8	89.2	66.9	53.5	44.6
127	232.3	155	116.1	92.9	77.5
152.4	356.7	237.8	178.4	142.7	118.9
177.8	512.8	341.9	256.4	204.9	170.9
203.2	739.5	493.3	369.7	295.4	246.7
254	133.8	891.8	668.9	534.2	445.9

Diameter of Gutter 41.7 mm/m Slope	Maximum Rainfall (mm/hr)				
	50.8	76.2	101.6	127	152.4
76.2	89.2	59.5	44.6	35.7	29.7
101.6	189.5	126.3	94.8	75.8	63.2
127	328.9	219.2	164.4	131.5	109.6
152.4	514.7	343.3	257.3	206.2	171.9
177.8	724.6	483.1	362.3	289.9	241.4
203.2	1040.5	693	520.2	416.2	346.5
254	1858	1238.4	929	743.2	618.7

NOTE: Round, square, or rectangular gutters, leaders, or pipe may be used. They are considered equivalent when enclosing a scribed circle equivalent to the diameter listed above.

Source: Reprinted by permission from *Uniform Plumbing Code,* 17th ed., copyright © 1985 by the International Association of Plumbing and Mechanical Officials.

TABLE 8.3 **Roof Drain and Leader Sizes**

Part A. ***Conventional Units: Maximum Horizontal Projected Roof Area, (ft²)***

Max. Hourly Rainfall (in.)	Size of Drain or Leader (in.)					
	2	3	4	5	6	8
1	2880	8800	18400	34600	54000	116000
2	1440	4400	9200	17300	27000	58000
3	960	2930	6130	11530	17995	38660
4	720	2200	4600	8650	13500	29000
5	575	1760	3680	6920	10800	23200
6	480	1470	3070	5765	9000	19315
7	410	1260	2630	4945	7715	16570
8	360	1100	2300	4325	6750	14500
9	320	980	2045	3845	6000	12890
10	290	880	1840	3460	5400	11600
11	260	800	1675	3145	4910	10545
12	240	730	1530	2880	4500	9660

Part B. ***SI Units: Maximum Horizontal Projected Roof Area, Square Meters***

Max. Hourly Rainfall (mm)	Size of Drain or Leader (mm)					
	50.8	76.2	101.6	127	152.4	203.2
25.4	267.6	817.5	1709.4	3214.3	5016.6	10776.4
50.8	133.8	408.8	854.7	1607.2	2508.3	5388.2
76.2	89.2	272.2	569.5	1071.1	1671.7	3591.5
101.6	66.9	204.4	427.3	803.6	1254.2	2694.1
127	53.4	163.5	341.8	642.9	1003.3	2155.3
152.4	44.6	136.6	285.2	535.6	836.1	1794.4
177.8	38.1	117.1	244.3	459.4	716.7	1539.4
203.2	33.4	102.2	213.7	401.8	627.1	1347.1
228.6	29.7	91	190	357.2	557.4	1197.5
254	26.9	81.8	170.9	321.4	501.7	1077.6
279.4	24.2	74.3	155.6	292.2	456.1	979.6
304.8	22.3	67.8	142.1	267.6	418.1	897.4

NOTE: Round, square, or rectangular gutters, leaders, or pipe may be used. They are considered equivalent when enclosing a scribed circle equivalent to the diameter listed above.

Source: Reprinted by permission from *Uniform Plumbing Code,* 17th ed., copyright © 1985 by the International Association of Plumbing and Mechanical Officials.

WATER AND WASTE

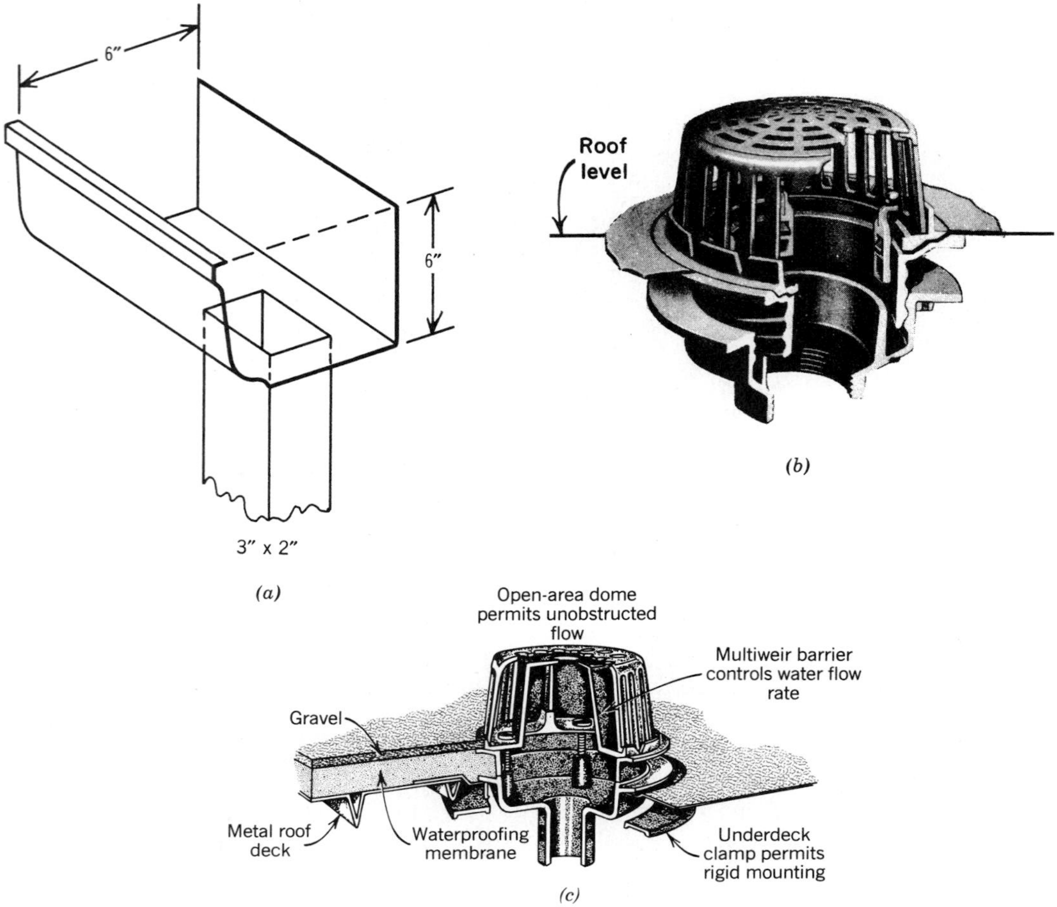

Fig. 8.13 Storm drainage components. (a) *Conventional gutter and leader for houses; sizes vary by manufacturer.* (b) *Ordinary roof drain (Josam Manufacturing Co.).* (c) *Roof drain for controlled flow. From* Specifying Engineer *magazine, Nov. 1982.*

EXAMPLE 8.2. Select a gutter and two leaders for the front half of a house, as shown in Fig. 8.6. Rainfall rate is 4 in./h. Projected roof area is 700 ft^2, and the slope of the gutter is 1⁄16 in. in 1 ft of length.

SOLUTION. From Table 8.2, choose a semicircular gutter with a 6-in. diameter. (Note that if a steeper gutter slope were designed, then at ½-in. slope per foot, only a 4-in.-diameter gutter would be required.)

Since two leaders will be used, each will drain 350 ft^2. Table 8.3 shows that a 2-in. leader can be used. For this gutter-leader combination, specify the detail of Fig. 8.13*a*.

Storm gutters and leaders can have an important impact on a building's appearance (Fig. 8.14). Alternatively, leaders can be set within buildings and gutters can be built into a roof's surface, to minimize the visual impact.

Fig. 8.14 *Hayward Field West Grandstand at the Eugene Campus of the University of Oregon, built in 1976. The Amundson Associates, Architects, Springfield, Oregon. This is the track-and-field facility for the university.* (a) *In addition to the architectural and esthetic reasons for sloping the roof to the rear of this attractive and functional structure, the design allows for the collection of stormwater. The water collects at a horizontal gutter at the rear roof edge and drains through six leaders.* (b) *Portion of the roof gutter and four of the six leaders. Storm drainage from all downspouts is collected below ground and led to the street, where it empties into an 18-in. storm sewer.*

Where routing of stormwater inside a building is preferable, drains and leaders can be sized from Table 8.3, and *horizontal piping* can be sized from Table 8.4. Care should be taken to insulate such lines; cold rainwater inside pipes can cause condensation to form on the outside, sometimes resulting in staining and other water damage.

TABLE 8.4 **Horizontal Rainwater Piping Sizes**

Part A. ***Conventional Units: Maximum Horizontal Projected Roof Area, Square Feet***

Size of Pipe (in.), ⅛″ Slope	Maximum Rainfall (in./hr)				
	2	3	4	5	6
3	1644	1096	822	657	548
4	3760	2506	1880	1504	1253
5	6680	4453	3340	2672	2227
6	10700	7133	5350	4280	3566
8	23000	15330	11500	9200	7600
10	41400	27600	20700	16580	13800
12	66600	44400	33300	26650	22200
15	109000	72800	59500	47600	39650
Size of Pipe (in.), ¼″ Slope	**Maximum Rainfall (in./hr)**				
	2	3	4	5	6
3	2320	1546	1160	928	773
4	5300	3533	2650	2120	1766
5	9440	6293	4720	3776	3146
6	15100	10066	7550	6040	5033
8	32600	21733	16300	13040	10866
10	58400	38950	29200	23350	19450
12	94000	62600	47000	37600	31350
15	168000	112000	84000	67250	56000
Size of Pipe (in.), ½″ Slope	**Maximum Rainfall (in./hr)**				
	2	3	4	5	6
3	3288	2295	1644	1310	1096
4	7520	5010	3760	3010	2500
5	13360	8900	6680	5320	4450
6	21400	13700	10700	8580	7140
8	46000	30650	23000	18400	15320
10	82800	55200	41400	33150	27600
12	133200	88800	66600	53200	44400
15	238000	158800	119000	95300	79250

TABLE 8.4 **Horizontal Rainwater Piping Sizes (*continued*)**

Part B. ***SI Units: Maximum Horizontal Projected Roof Area, Square Meters***

Size of Pipe (mm), 10.4 mm/m Slope	*Maximum Rainfall (mm/hr)*				
	50.8	*76.2*	*101.6*	*127*	*152.4*
76.2	152.7	101.8	76.4	61	50.9
101.6	349.3	232.8	174.7	139.7	116.4
127	620.6	413.7	310.3	248.2	206.9
152.4	994	662.7	497	397.6	331.3
203.2	2136.7	1424.2	1068.4	854.7	706
254	3846.1	2564	1923	1540.3	1282
279.4	6187.1	4124.8	3093.6	2475.8	2062.4
381	10126.1	6763.1	5527.6	4422	3683.5
Size of Pipe (mm), 20.9 mm/m Slope	*Maximum Rainfall (mm/hr)*				
	50.8	*76.2*	*101.6*	*127*	*152.4*
76.2	215.5	143.6	107.8	86.2	71.8
101.6	492.4	328.2	246.2	197	164.1
127	877	584.1	438.5	350.8	292.3
152.4	1402.8	935.1	701.4	561.1	467.6
203.2	3028.5	2019	1514.3	1211.4	1009.5
254	5425.4	3618.5	2712.7	2169.2	1806.9
304.8	8732.6	5815.5	4366.3	3493	2912.4
381	15607.2	10404.8	7803.6	6247.5	5202.4
Size of Pipe (mm), 41.7 mm/m Slope	*Maximum Rainfall (mm/hr)*				
	50.8	*76.2*	*101.6*	*127*	*152.4*
76.2	305.5	213.2	152.7	121.7	101.8
101.6	698.6	465.4	349.3	279.6	232.3
127	1241.1	826.8	620.6	494.2	413.4
152.4	1988.1	1272.3	994	797.1	663.3
203.2	4274.4	2847.4	2136.7	1709.4	1423.2
254	7692.1	5128.1	3846.1	3079.6	2564
304.8	12374.3	8249.5	6187.1	4942.3	4124.8
381	22110.2	14752.5	11055.1	8853.4	7362.3

Source: Reprinted by permission from *Uniform Plumbing Code,* 17th ed., copyright © 1985 by the International Association of Plumbing and Mechanical Officials.

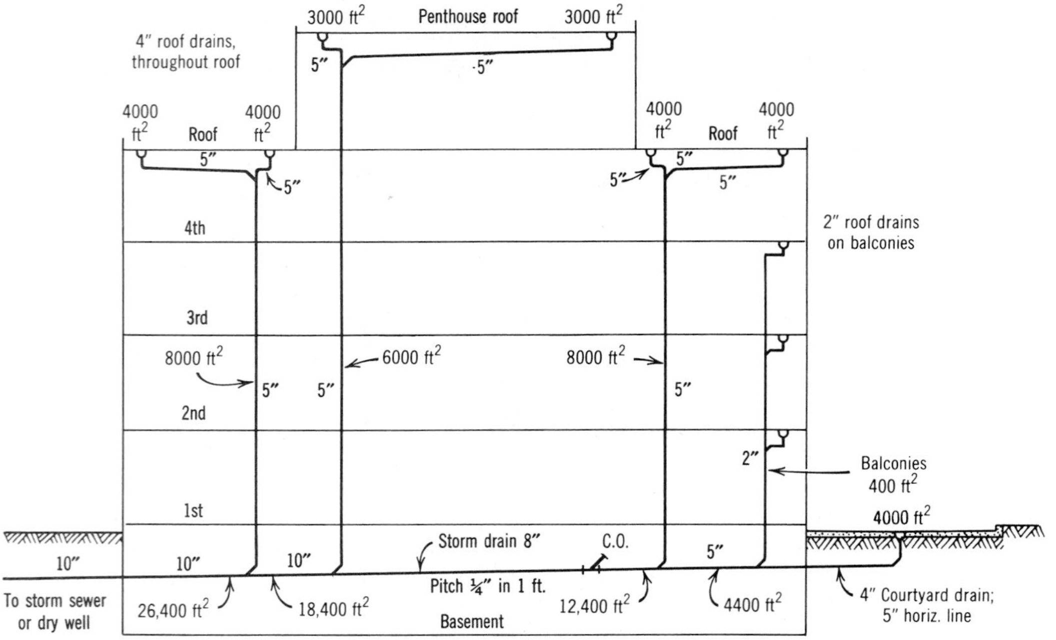

Fig. 8.15 (Example 8.3.) Separate storm drainage. Areas drained and corresponding sizes of vertical leaders and horizontal drains are from Tables 8.3 and 8.4. Storm drain piping within a building needs insulative covering with a vapor barrier on the outside. This prevents condensation (sweating) on the pipes when in winter, warm, moisture-laden air in the building could otherwise reach the pipe surface (which would be cold from carrying icy water), condense there, and lead to wet, dripping conditions on the pipes. Each roof has two drains, in case one is temporarily blocked.

EXAMPLE 8.3. Select sizes for vertical conductors and horizontal storm drains for the building shown in Fig. 8.15. Roof, balcony, and courtyard areas are as shown, rainfall rate is 4 in./h (table value), and the pitch of horizontal drains is ¼-in. slope in 1 ft of run.

SOLUTION. Sizes selected and shown in Fig. 8.15 may be verified in Tables 8.3 and 8.4.

References

Milne, M. (1976). *Residential Water Conservation*, U.S. Office of Water Research and Technology, Department of Commerce, NTIS.

9

WATER SUPPLY

With water, one of the designer's first concerns is to match the quality of the water to the task it performs. As this becomes a more serious design issue, designers will begin to provide for the recycling of water within and around buildings, as well as to specify plumbing fixtures that use less water. Table 9.1 shows typical relationships between quality and usage. Water recycling and conservation are discussed in detail at the beginning of Chapter 10. This chapter will deal primarily with *potable* (drinkable) water: first with issues of water quality, and then with the matter of assuring an adequate supply of water throughout the building.

9.1 Water Quality

A summary of water quality at the various stages of the hydrologic cycle was shown in Fig. 7.2. As precipitation, water contains few impurities: almost no bacterial content is present, and only small amounts of minerals and gases can be expected. To collect this nearly pure water, surfaces are needed—and at this point, foreign substances can readily contaminate the water. These pollutants can affect water's physical (mostly organic), chemical (mostly inorganic), biological, or radiological characteristics. Both surface water and groundwater are subject to pollution.

(a) Physical Characteristics. Some of the most noticeable alterations of water quality fall within this category. Water from surface sources (roof runoff, streams, rivers, lakes, ponds) is particularly subject to physical pollutants.

Turbidity is easy to see, and thus a likely source of dissatisfaction for the would-be consumer. It is caused by the presence of suspended material such as clay, silt, other inorganic material, plankton, or finely divided organic material. Even those materials that do not adversely affect health are usually esthetically objectionable.

Color, another visible alteration, is often caused by dissolved organic matter, as from decaying

TABLE 9.1 **Water Use and Quality in Buildings**

Use	*Desired Quality*
A. Consumed	
1. Drinking and cooking	Potable
2. Bathing	Potable
3. Laundering	Soft
4. Irrigation and watering of livestock	Unpolluted
5. Industrial processes	As required
6. Vapor to increase the relative humidity of air	
B. Circulated	
1. Hot water for heating	Note: Make-up water should be soft or neutral and, for swimming, potable
2. Chilled water for cooling	
3. Condenser cooling water	
4. Swimming pool water	
5. Steam for heating, later condensed	
C. Generally static	
1. Water stored for fire protection	No special requirement
2. Water in fire standpipes	
3. Water in sprinkler piping	
D. Controlled	
1. Vapor condensed to reduce relative humidity of air	

NOTE: For water uses in Section A, above, flow is often continuous. Section B comprises uses for which flow other than circulation is intermittent or at a relatively low rate, the water added to the systems being known as "make-up water." Items C2 and C3 call for piping to provide adequate though infrequent flow in emergencies. Obviously, item D1 relates only to moisture condensed out of the air and involves no supply.

vegetation. Some inorganic materials also color water, as do growths of microorganisms. Like turbidity, such color changes usually do not threaten health but often are psychologically undesirable.

Taste and *odor* can be caused by organic compounds, inorganic salts, or dissolved gases. This condition can be treated only after a chemical analysis has identified which source is responsible.

Temperature is another characteristic of psychological importance—we expect drinking water to be cool. In general, water supplied between 50–60 F (10°–16°C) is preferred.

Foamability is usually caused by concentrations of detergents. The foam itself does not pose a serious health threat, but it may indicate that other, more dangerous pollutants associated with domestic waste are also present. Because of increased foaming in water in the 1960s, today's detergents must use biodegradable LAS (linear alkylate sulfonate), which biodegrades rapidly—except in the absence of oxygen. Since this lack of oxygen is characteristic of some septic tank drainage fields, foam in drinking water should be promptly investigated.

TABLE 9.2 **Example of Chemical Analysis**

Quality		*Parts per Million (ppm)*[a]
Total hardness	as $CaCO_3$	30
Calcium hardness	as $CaCO_3$	20
Alkalinity (Methyl Orange)	as $CaCO_3$	27
Alkalinity (Phenolpntalein)	as $CaCO_3$	0
Free Carbon Dioxide	as CO_2	13.5
Chlorides	as Cl	6
Sulfates	as SO_4	4
Silica	as SiO_2	19
Phosphates—normal	as PO_4	0
Phosphates—total	as PO_4	0.5
Iron—total	as Fe	1.6
Total dissolved solids		66
Turbidity or sediment		present

Source: A report by Olin Water Service, for a private well in Virginia.

[a]Note that "ppm" and "mg/L" are essentially equivalent terms for concentration in water.

(b) Chemical Characteristics. Groundwater is particularly subject to chemical alteration, because as it moves downward from the surface it slowly dissolves some minerals contained in rocks and soils. A chemical analysis (such as that given in Table 9.2) is usually necessary for individual water supply sources. These analyses will indicate (1) the possible presence of harmful or objectionable substances, (2) the potential for corrosion within the water supply system, and (3) the tendency for the water to stain fixtures and clothing. Concentrations are expressed in mg/L (milligrams per liter), which is essentially equivalent to ppm (parts per million).

Some general terms commonly used to describe chemical characteristics of water are defined as follows.

Alkalinity is caused by bicarbonate, carbonate, or hydroxide components. Testing for these components of water's alkalinity is a key to determining which treatments to use.

Hardness. This is a relative term (see Fig. 9.1). "Hard" water inhibits the cleaning action of soaps and detergents, and it deposits scale on the inside of hot water pipes and cooking utensils, thus wasting heating fuel and making utensils unusable. Hardness, which is caused by calcium and magnesium salts, can be classified as temporary (carbonate) or permanent (noncarbonate). Temporary hardness is largely removed when the water is heated—it forms the scale just described. Permanent hardness cannot be removed by simple heating (see Section 9.2*g*).

pH. This is a measure of the water's hydrogen ion concentration, as well as its relative acidity or alkalinity (see Fig. 9.1). A pH of 7 is neutral. Measurements below 7 indicate increasing acidity (and corrosiveness); water in its natural state can have a pH as low as 5.5, with 0 the ultimate acidity. Measurements higher than 7 indicate increasing alkalinity; a pH as high as 9 can be found in water in its natural state, with 14 the ultimate alkalinity. The pH value is the starting point for determining treatments for corrosion control, chemical dosages, and disinfection.

Chemical additions to water supplies most commonly include the following elements.

Toxic substances are occasionally present in water supplies. Local health authorities can pro-

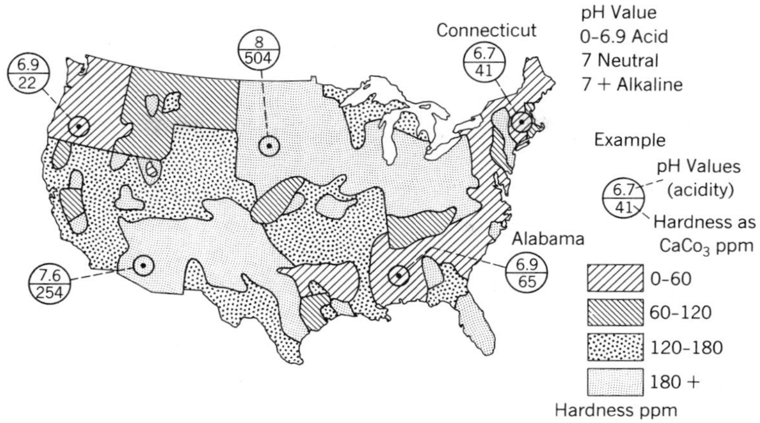

Fig. 9.1 Approximate groundwater chemical characteristics across the United States. Treatment may be needed when pH is less than 7.0 (acidity results in corrosion) or when hardness as $CaCo_3$ exceeds 65 ppm (essentially, ppm and mg/L are identical measures). Courtesy of Progressive Architecture.

vide information about acceptable concentrations of such substances as arsenic (As), barium (Ba), cadmium (Cd), chromium (Cr^{+6}), cyanides (CN), fluoride (F), lead (Pb), selenium (Se), and silver (Ag). Although limited amounts of *fluoride* are frequently added to water supplies to help prevent tooth decay, fluorides in excess of such ''optimum concentrations'' can produce mottling of teeth. *Lead* poses a dangerous threat, even in relatively small amounts, because it is a cumulative poison. Lead in water usually comes from lead piping (in older buildings or cities) or corrosive water on lead-painted roofs. A maximum recommended concentration is 0.05 mg/L.

Unfortunately, the list of toxic chemicals is expanding, and as chemical waste dumps have been abandoned or mismanaged, groundwater has become contaminated. The U.S. Environmental Protection Agency has estimated that 75% of both active and abandoned chemical waste dumps—some 51,000 in all—are leaking. In addition to the following list of inorganic chemicals, we are becoming aware of many new organic chemicals as well, some of which are suspected of causing 5 to 20% of U.S. cancers. The threat to our groundwater supplies is illustrated in Fig. 9.2. Once polluted, aquifers are extremely difficult to clean.

Chlorides can enter water as it passes through

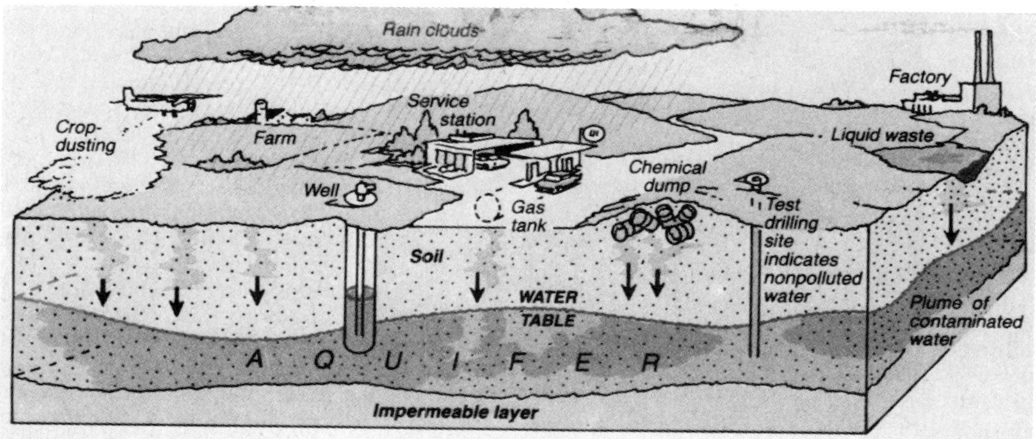

Fig. 9.2 How groundwater becomes contaminated. The ''plume'' formed by contaminants can often go undetected. Copyright © 1982 by Newsweek, Inc., all rights reserved. Reprinted by permission.

geological deposits formed by marine sediment, or because of pollution from sea water, brine, or industrial or domestic wastes. A noticeable taste will result from chloride in excess of 250 mg/L.

Copper can enter water from natural copper deposits or from copper piping that contains corrosive water. Concentrations of copper in excess of 1.0 mg/L can result in an undesirable taste.

Iron is frequently present in groundwater. Corrosive water in iron pipes will also add iron to water. At concentrations above 0.3 mg/L, iron can lend a brownish color in washed clothes and can affect the taste of the water.

Manganese can both pose a physiological threat (it is a natural laxative) and produce color and taste effects similar to those produced by iron. The recommended limit is 0.05 mg/L.

Nitrates in high concentrations pose a threat to infants, in whom it can cause "blue baby" disease. In shallow wells, nitrate concentrations can indicate seepage from deposits of livestock manure.

Pesticides, a growing threat to water supplies, are particularly common in wells near homes that have been treated for termite control. Avoid using pesticides near wells.

Sodium is primarily dangerous for people with heart, kidney, or circulatory ailments. For a low-sodium diet, the sodium in drinking water should not exceed 20 mg/L. Salts spread on roadways for ice control can leach into the soil and enter groundwater. Note that some "water softeners" (discussed below, in Section 9.2*g*) can raise sodium concentrations in water.

Sulfates, which have laxative effects, can enter groundwater from natural deposits of Epsom salts (magnesium sulfate) or Glauber's salt (sodium sulfate). Concentrations should not exceed 250 mg/L.

Zinc sometimes enters groundwater in areas where it is found in abundance. Although not a health threat, it can cause an undesirable taste at concentrations above 5 mg/L.

(c) Biological Characteristics. Potable water should be kept as free as possible from disease-producing organisms—bacteria, protozoa, and viruses. These organisms are not easily identified; a thorough biological water test is complex and time consuming. For this reason, the standard test is for *one* kind of bacteria—the coliform group (*E. coli*), which is always present in the fecal wastes of humans (as well as those of many animals and birds) and which outnumbers all other disease-producing organisms in water. The recommended maximum concentration for coliform bacteria is one organism per 100 mL (about ½ cup) water.

For biological activity to be kept to a minimum in drinking water, a water source should be chosen that does not normally support much plant or animal life; hence the popularity of groundwater, rather than surface water, as a source. In addition, the supply should be protected from subsequent biological contamination. Where cities depend on small lakes for water, humans are frequently excluded from the watersheds. Organic fertilizers and nutrient minerals should also be kept out of the water supply, to further discourage biological activity. For the same reason, stored water should be kept dark and at low temperatures. Finally, organisms (or their by-products) are commonly destroyed at treatment facilities.

(d) Radiological Characteristics. The mining of radioactive materials and the use of such materials in industry and power plants have produced radiological pollution in some water supplies. Since radiological effects are cumulative, concentrations of radioactive materials should be low indeed. "Safe" minimum concentrations have continually been revised downward for other radiation exposures; consult the local public health service for current recommendations.

9.2 Water Treatment

In the preceding section, water pollution was broken down into physical, chemical, biological, and radiological categories. The various forms of treatment for such pollutants do not necessarily fall into the same categories, as one treatment may be effective for several different polluted conditions. A general look at common domestic water quality problems is provided in Table 9.3. The treatment processes available range from the mild (sedimentation) to the extreme (distillation).

(a) Sedimentation. This process removes suspended matter from water simply by allowing time

TABLE 9.3 **Common Water Quality Problems and Treatment in Small Systems**

Item	*Cause*	*Bad Effect*	*Correction*
Hardness	Calcium and magnesium salts from underground flow	Clogging of pipes by scale, burning out of boilers, and impaired laundering and food preparation	Ion-exchanger (Zeolite process)
Corrosion	Acidity, entrained oxygen and carbon dioxide (low pH)	Closing of iron pipe by rust, leaking connections, destruction of brass pipe	Raising the alkaline content (Neutralizer)
Biological pollution	Contamination by organic matter or sewage	Disease	Chlorination by sodium hypochlorite or chlorine gas
Color	Iron and manganese	Discoloration of fixtures and laundry	Chlorination and fine filtration
Taste and odor[a]	Organic matter	Unpleasantness	Filtration through activated carbon (Purifier); aeration
Turbidity[a]	Silt or suspended matter picked up in surface or near-surface flow	Unpleasantness	Filtration

[a]These problems are not common in private systems that use deep wells.

and the inactivity of the water to do the work of settling out heavier suspended particles. Simple basins, ponds, or tanks constructed for this purpose are large enough to retain the water for at least 24 hours and are equipped with baffles to slow the water flow. To clean out the sediment, water usually is diverted to an identical second basin while the first is being cleaned.

(b) Coagulation (Flocculation). This process also removes suspended matter, along with some coloration. A chemical such as alum (hydrated aluminum sulfate) is added to turbulent water. The water is then held in a quiet condition, in which the suspended particles will combine with the alum to form floc. These heavy particles then settle out, in a process similar to sedimentation. Some adjustment of the pH may be necessary.

(c) Aeration. This simple process can improve the taste and color of water, help remove iron and manganese, and sometimes decrease corrosiveness. In aeration, as much of the water's surface as possible is exposed to air. The methods used are rich in esthetic possibilities—the spraying of water into air, the fall of a turbulent stream of water over a spillway, and others. To guard against contamination, these processes are often enclosed; if exposed, they must be kept clean. For aeration within tanks, water is passed through a series of perforated plates, in streams or droplets.

Aeration improves the flat taste of distilled water and cistern water, by adding oxygen. It also oxidizes iron or manganese, which then can easily be removed by filtration, and removes odors caused by hydrogen sulfide and algae. Finally, it can reduce corrosiveness caused by carbon dioxide and other gases, although the increase in oxygen content will partially offset this process.

(d) Filtration. This very common treatment can remove suspended particles, some bacteria, and color or taste. The more common approaches are listed below, beginning with filtering to remove suspended particles then moving on to more specialized applications for the removal of iron and/or manganese, tastes, and odors.

Slow sand filters (see Fig. 8.2, for rainwater application) are low-maintenance, easily constructed devices that should be cleaned as often as the turbidity of the water demands—from once a day to perhaps once a month. They are cleaned by removal and replacement of the top 1 in. of sand, which is then either washed for reuse or discarded. The approximate rate of flow is between 60 and 180 gallons per day per square foot (2450 to 7350 liters per day per square meter) of filter bed surface. Overall thickness is usually 30–48 in. of sand over 6–8 in. of gravel (760 to 1220 mm of sand over 150 to 200 mm of gravel).

Pressure sand filters require controls and the attention of an operator, and thus are rarely used for individual water systems.

Diatomaceous earth filters can be of either the vacuum or the pressure type. They require periodic attention to remain effective.

Porous stone, ceramic, or unglazed porcelain filters (also called Pasteur filters) are usually made in small sizes so that they can be attached to water faucets. They are used widely in some countries, such as Mexico, but poor maintenance or hairline cracks often lead to bacterial infiltration, complicating the filtration process. A more positive approach to the disinfection of drinking water thus is desirable.

Chlorination/fine filtration is a combined process that removes iron and/or manganese from water. The chlorine chemically oxidizes the iron or manganese, which forms a precipitate. Chlorine also kills iron bacteria (which can form a slimy mass when present) and disinfects. Then, the fine filter removes the precipitated iron or manganese. Since water containing iron or manganese poses a threat to plumbing and other water treatment processes, this procedure should be the first one applied to water supplies that suffer from excess iron and/or manganese.

Alternative treatments for iron and/or manganese are aeration followed by filtration and treatment with potassium permanganate followed by filtration.

Activated carbon filters are particularly effective for removing tastes and odors. The water is passed through granular carbon, which attracts large quantities of dissolved gases, soluble organics, and fine solids.

Reverse osmosis is sometimes used to reduce the mineral content in water. An inert, semipermeable membrane has higher-pressure supply water on one side; as the pressure slowly forces water through this filtering membrane, most of the minerals (dissolved solids) are removed. Dissolved chemicals, however, remain. The principal disadvantage of this process is its enormous appetite for water; only a portion of the water supplied to a reverse-osmosis filtering unit is delivered to the user. The remainder must be used to flush the semipermeable membrane, so that mineral buildup is avoided.

Commercial reverse-osmosis units are available in sizes ranging from a 1 gpm (3.9 L/m) water delivery rate [using two membranes and a 3-hp motor, requiring 4.5 gpm (17 L/m) feedwater, for a 22% recovery rate] to a water delivery rate of 12.5 gpm (47.3 L/m) [using 12 membranes and a 15-hp motor, requiring 19.2 gpm (72.8 L/m), for a 65% recovery rate).

(e) Disinfection. This is the most important health-related water treatment. Although chlorination has become the standard approach to removing harmful organisms from water, there are alternatives: ultraviolet light (unsuitable for water with high turbidity, since the light cannot easily penetrate), bromine, iodine, ozone, and heat treatment, among others. Chlorine, however, continues to disinfect after the initial application. It is this continuing protection that has made it so universally relied on, despite dangers such as that posed by deadly chlorine gas. Although chlorine affects the taste and odor of water, it is also effective in removing less-desirable tastes and odors.

Factors that affect chlorine's ability to disinfect include

1. *Chlorine concentration*. The higher the concentration, the faster and more complete the rate of disinfection.
2. *Contact time*. The longer the chlorine contacts the organisms in water, the more complete the disinfection. At a minimum, 0.4 mg/L of chlorine should contact water 30 minutes before use.
3. *Water temperature*. The higher the temperature during contact, the more complete the disinfection.
4. *pH*. The lower the pH, the more effective the disinfection.

Chlorine can be added automatically to water supplies by *hypochlorinators*. These devices automatically pump (or inject) into water a chlorine solution formed from powder or tablets or from a prepared solution. They are usually no larger than the pumps used in small water systems (see Fig. 9.9). Some hypochlorinators are specially designed for low and fluctuating water pressures, or for use where electricity is not available.

(f) Corrosion Control. It is important to control corrosion in order both to keep water systems operating freely and to prevent corrosive water from increasing the concentration of hazardous materials (as from copper pipes). Corrosion is a slow degradation of a metal by a flow of electric current from the metal to its surroundings. The factors involved in corrosion control are:

1. *Acidity*. The more acid (low pH), the more corrosive the water.
2. *Conductivity*. As dissolved mineral salts increase the water's conductivity, they encourage the flow of the electrical current of corrosion.
3. *Oxygen content*. Dissolved oxygen destroys the thin protective hydrogen film on immersed metals, thus promoting corrosion.
4. *Carbon dioxide content*. Carbon dioxide forms carbonic acid, which attacks metal surfaces.
5. *Water temperature*. Increased temperature increases corrosion.

The products of corrosion often contribute to scale formation. Scale then lines surfaces, eventually clogging openings.

Acid neutralizers can be installed on water supplies with low pH; their function is often combined with those of hypochlorinators. Typically, neutralizing solutions are mixtures of soda ash and water.

Alternatives to corrosion control include feeding into the water small amounts of film-forming materials (polyphosphates or silicates), installing dielectric or insulating unions (to avoid complications from dissimilar pipe metals), and avoiding metal piping and fixtures altogether.

(g) Softening. Water "hardness" is caused primarily by calcium and magnesium deposits; when they are removed, water will be "soft." Where water hardness is mild enough to affect only laundering, cisterns may be used to collect "soft" rainwater to use for clothes washing. Where water hardness produces scale in pipes and within water heating appliances, and cisterns are not feasible, water softening equipment is used.

This equipment commonly uses zeolite, in an ion-exchange process. Water containing calcium and/or magnesium is passed through a bed of zeolite, which exchanges its sodium ions for those of calcium and/or magnesium. The water thus loses hardness, but it gains sodium content—an undesirable development for people on low-sodium diets. Periodically, as the zeolite loses its sodium ions, it must be regenerated. This is done automatically within the softening equipment, in which a brine solution is passed through the bed of zeolite. Alternatives to zeolite as the ion-exchange material include glauconite (or greensand) and a combination of resins [precipitated synthetic, organic (carbonaceous), and synthetic].

(h) Nuisance Control. Some organisms may not be injurious to health but can multiply so rapidly that piping or filters become clogged or the water's appearance, odor, and taste are affected. This process is most common in surface water sources, and it is within surface reservoirs that these treatments are most often applied. Algae growths, the most prevalent nuisance, can usually be controlled by applying copper sulfate (blue stone or blue vitriol) to the water body. Sudden and massive algae kills can have adverse impacts on other life forms within the water, because the decomposing algae rob the water of oxygen. As a further precaution, stored water should be shielded from sunlight whenever possible.

(i) Fluoridation. A heated controversy has developed over the addition of fluoride to drinking water. The advantage of fluoridation is that children who drink fluoridated water have lower rates of tooth decay. And since everyone drinks water, all children benefit, not just those who can afford fluoride pills. Its disadvantages are that only children need the fluoride, not adults, and that in amounts above those used in water treatment, fluoride is toxic and can cause mottled teeth. Opponents of fluoridation suggest that because sugar is a leading cause of tooth decay in children, sugar, rather than water, should be fluoridated.

Small water systems can be equipped with fluoridation units. However, fluoride levels in the water supply must be carefully monitored.

Fig. 9.3 A solar still can be used to provide a small daily quantity of pure water; this installation serves a laboratory. Courtesy of Horace McCracken, Sunwater Solar Stills.

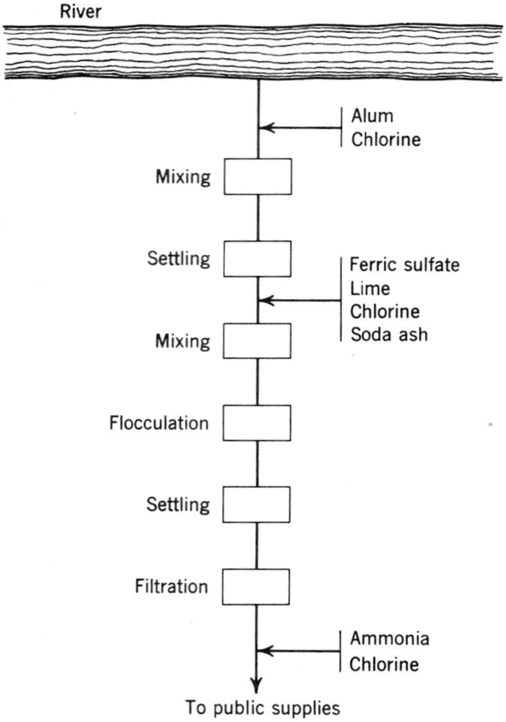

Fig. 9.4 When a city's water is taken from surface sources, such as rivers, multistage treatment is usually necessary. From Water Quality and Treatment, *2nd ed., American Water Works Association, Inc.*

(j) Distillation. Where water pollution is extreme, as in the case of sea (salt) water, distillation may be the best treatment. Water is heated, to encourage evaporation. As the water turns to vapor, virtually all pollutants are left behind. When this vapor encounters a cooler surface, it condenses, and pure water (although somewhat flat in taste) can be collected from this surface.

Any heat source can be used in the distilling of water; solar stills (Fig. 9.3) are gaining in popularity because the energy used is free. In semiarid, rather sunny climates, a solar still should produce about 1⁄12 gallon per square foot of collector surface area (4 liters per square meter per day). This slow rate of production suggests that only the water used directly for drinking is usually feasible for distillation. Another factor to consider is that the cleaning of the still is generally accomplished by flushing it with twice as much water as was delivered. If not excessively brackish, this flush water could be used for irrigation or other nonpotable-quality tasks.

9.3 Water Sources

This section focuses on the equipment used to capture and store groundwater from wells. Other water sources are less often used for smaller systems, either because they require much more extensive treatment (surface water from lakes or rivers) or because they provide water intermittently (cisterns). The sequence of treatment for a city's water taken from a river is shown in Fig. 9.4; this multistage treatment is inappropriate for small water systems that receive only occasional maintenance. Cisterns were discussed in Chapter 8. Another increasingly important source of city water, the treated effluent of sewage treatment plants, is discussed in Chapter 10.

(a) Wells. Farms and remote housing developments usually have private water systems. In rural and suburban areas where the progress of building is faster than the development of municipal water supplies, private sources must be sought. The elaborate treatment necessary for the use of surface waters or those in dug wells of the old type makes the use of driven or drilled wells preferable. Water from these sources usually has at least the advantages of purity, coolness, and freedom from turbidity, odor, and unpleasant taste—

any of which *may* be encountered in addition to either acidity or hardness.

Bored wells, which are dug with earth augers, are usually less than 100 ft deep. They are used when the earth to be bored through is boulder-free and will not cave in. The diameter range is 2 to 30 in. The bored well is then cased with metal, vitrified tile, or concrete.

Driven wells are the simplest and, usually, the least expensive type. A steel drive-well point is fitted on the end of pipe sections and driven into the earth. The drive point is usually 1¼ to 2 in. (32 to 51 mm) in diameter. The materials and design of drive-well points vary according to the expected characteristics of the earth in which the well is driven. First, a pilot hole is bored (frequently, with a simple hand auger), and the drive-well point and pipe sections lowered into it. Then the well is driven to a point well below the water table.

The construction of *jetted wells* requires a source of water and a pressure pump. A washing well point is supplied with water under pressure; this loosens the earth and allows the point and pipe to penetrate.

Drilled wells require more elaborate equipment of several types, depending on the geology of the site. The percussion (or cable-tool) method involves the raising and dropping of a heavy drill bit and stem. Having thus been pulverized, the earth being drilled is mixed with water to form a slurry, which is periodically removed. As drilling proceeds, a casing is also lowered (except when drilling through rock).

Rotary drilling methods (either hydraulic or pneumatic) utilize a cutting bit at the lower end of the drill pipe; a drilling fluid (or pressurized air) is constantly pumped to the cutting bit to aid in the removal of particles of earth, which are then brought to the surface. After the drill pipe is withdrawn, a casing is lowered into position.

Another method is the down-the-hole pneumatic (air) hammer, which combines the percussion effect with the rotary drill bit.

Local well drillers, who usually use the method most suited to the prevailing geology of the region, can offer useful advice about well construction methods. When clients plan to build in a remote location, the architect and engineer will want to advise them about water problems. Quality-corrective measures can always be taken and pumping equipment purchased, but the amount of water that can be obtained from the ground and the depth and cost of wells are all-important considerations. There are some problem areas where wells several hundred feet in depth will yield as little as 5 gpm, or nothing. The cost of drilling a number of exploratory wells may sometimes be excessive. Unfortunately, when such difficulties occur, there is often no easy solution. Conferences with neighboring owners, state and federal geologists, and local well drillers are all helpful. Many regions, of course, yield plentiful water, but the cost of the probable depth of the well should be considered.

A low-yield well can be combined with storage tanks, so that the pump can run all night, slowly filling the tanks to meet the next day's demands.

(b) Pumps. Three common types of pumps used in well water supply are the positive displacement, the centrifugal, and the jet. Their variations and characteristics are summarized in Table 9.4.

Positive Displacement Pumps. There are two principal types of positive displacement pump. In a *reciprocating pump,* a plunger moves back and forth within a cylinder equipped with check valves. The cylinder is best located near or below the groundwater level. Water enters the cylinder through an initial check valve (which allows flow in only one direction). As the plunger moves toward this check valve, the water is forced through the second check valve, located within the plunger itself. Then, as the piston returns to its original position, the water is forced upward toward the surface.

A *rotary pump* has a helical or spiral rotor—a turning vertical shaft within a rubber sleeve. As the rotor turns, it traps water between it and the sleeve, thus forcing the water to the upper end of the rotor.

Centrifugal Pumps. This type of pump contains an impeller mounted on a rotating shaft. The rotating impeller increases the water's velocity while forcing the water into a casing, thus converting the water's velocity into higher pressure. Each impeller and casing is called a stage; many stages can be combined in a multistage pump. The number of stages depends upon the pressure needed to

TABLE 9.4 **Pumps for Water Supply**

Type of Pump	Practical Suction Lift[a]	Usual Well-Pumping Depth	Usual Pressure Heads	Advantages	Disadvantages	Remarks
Positive Displacement						
1. Reciprocating						
(a) Shallow well **(b)** Deep well	*22–25 ft* *22–25 ft*	*22–25 ft* *Up to 600 ft*	*100–200 ft* *Up to 600 ft above cylinder*	1. Positive action. 2. Discharge against variable heads. 3. Pumps water containing sand and silt. 4. Especially adapted to low capacity and high lifts.	1. Pulsating discharge. 2. Subject to vibration and noise. 3. Maintenance cost may be high. 4. May cause destructive pressure if operated against closed valve.	1. Best suited for capacities of 5–25 gpm against moderate to high heads. 2. Adaptable to hand operation. 3. Can be installed in very small diameter wells (2″ casing). 4. Pump must be set directly over well (deep well only).
2. Rotary						
(a) Shallow well (gear type)	*22 ft*	*22 ft*	*50–250 ft*	1. Positive action 2. Discharge constant under variable heads. 3. Efficient operation.	1. Subject to rapid wear if water contains sand or silt. 2. Wear of gears reduces efficiency.	
(b) Deep well (helical rotary type)	Usually submerged	50–500 ft	100–500 ft	1. Same as shallow well rotary. 2. Only one moving pump device in well.	1. Same as shallow well rotary except no gear wear.	1. A cutless rubber stator increases life of pump. Flexible drive coupling has been weak point in pump. Best adapted for low capacity and high heads.
Centrifugal						
1. Shallow well						
(a) Straight centrifugal (single stage)	*20 ft max.*	*10–20 ft*	*100–150 ft*	1. Smooth, even flow. 2. Pumps water containing sand and silt. 3. Pressure on system is even and free from shock. 4. Low-starting torque. 5. Usually reliable and good service life.	1. Loses prime easily. 2. Efficiency depends on operating under design heads and speed.	1. Very efficient pump for capacities above 60 gpm and heads up to about 150 ft.

(b) Regenerative vane turbine type (single impeller)	*28 ft max.*	*28 ft*	*100–200 ft*	1. Same as straight centrifugal except not suitable for pumping water containing sand or silt. 2. They are self-priming.	1. Same as straight centrifugal except maintains priming easily.	1. Reduction in pressure with increased capacity not as severe as straight centrifugal.
2. Deep well						
(a) Vertical line shaft turbine (multistage)	*Impellers submerged*	*50–300 ft*	*100–800 ft*	1. Same as shallow well turbine. 2. All electrical components are accessible, above ground.	1. Efficiency depends on operating under design head and speed. 2. Requires straight well large enough for turbine bowls and housing. 3. Lubrication and alignment of shaft critical. 4. Abrasion from sand.	
(b) Submersible turbine (multistage)	*Pump and motor submerged*	*50–400 ft*	*50–400 ft*	1. Same as shallow well turbine. 2. Easy to frost-proof installation. 3. Short pump shaft to motor. 4. Quiet operation. 5. Well straightness not critical.	1. Repair to motor or pump requires pulling from well. 2. Scaling of electrical equipment from water vapor critical. 3. Abrasion from sand.	1. 3500 RPM models, while popular because of smaller diameters or greater capacities, are more vulnerable to wear and failure from sand and other causes.
Jet						
1. Shallow well	*15–20 ft below ejector*	*Up to 15–20 ft below ejector*	*80–150 ft*	1. High capacity at low heads. 2. Simple in operation. 3. Does not have to be installed over the well. 4. No moving parts in the well.	1. Capacity reduces as lift increases. 2. Air in suction or return line will stop pumping.	
2. Deep well	*15–20 ft below ejector*	*25–120 ft 200 ft max.*	*80–150 ft*	1. Same as shallow well jet. 2. Well straightness not critical.	1. Same as shallow well jet. 2. Lower efficiency, especially at greater lifts.	1. The amount of water returned to ejector increases with increased lift—50% of total water pumped at 50-ft lift and 75% at 100-ft lift.

Source: U.S. Environmental Protection Agency, *Manual of Individual Water Supply Systems,* 1975.

[a]Practical suction lift at sea level. Reduce lift 1 ft for each 1000 ft above sea level.

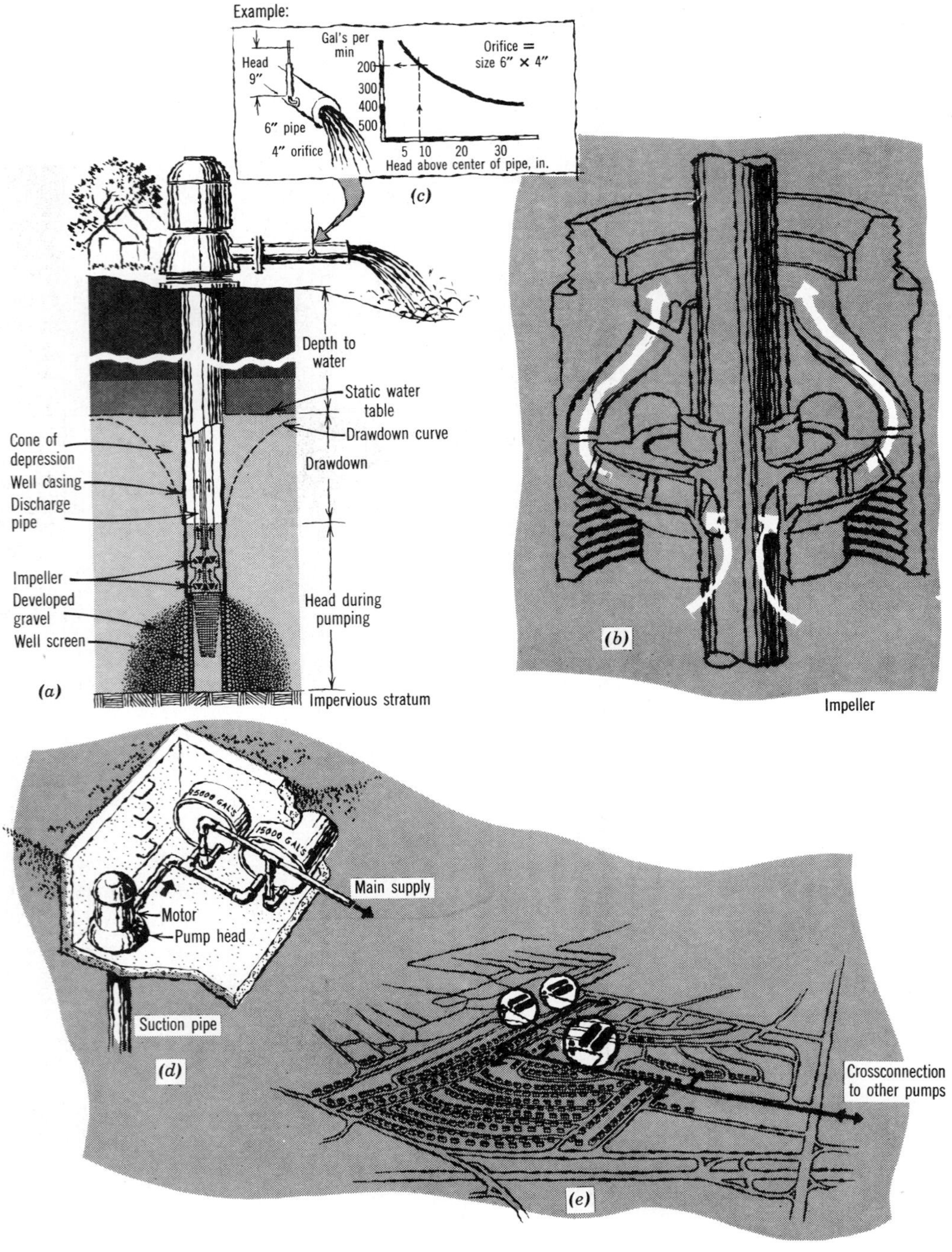

Fig. 9.5 (a) *A turbine well-pump.* (b) *Its operation.* (c) *Measurement of its capacity.* (d, e) *Its use in supplying a small community with groundwater.* (f) *A Jacuzzi multistage lineshaft turbine well-pump. (Jacuzzi Bros., Inc.) Capacities of turbine pumps range from 50 to 16,000 gpm (gallons per minute). By permission of* Progressive Architecture.

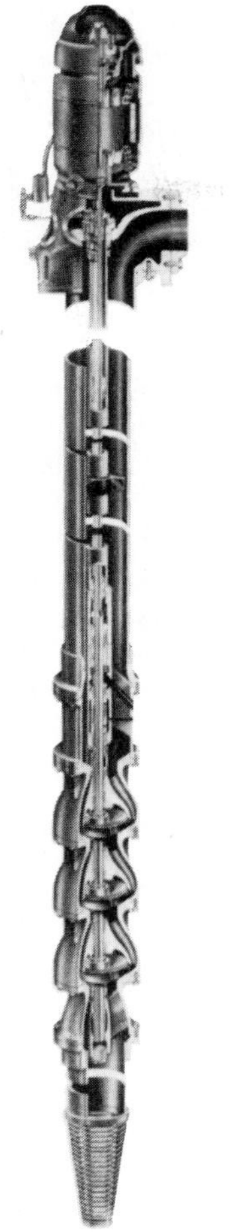

(f)

Fig. 9.5 (continued)

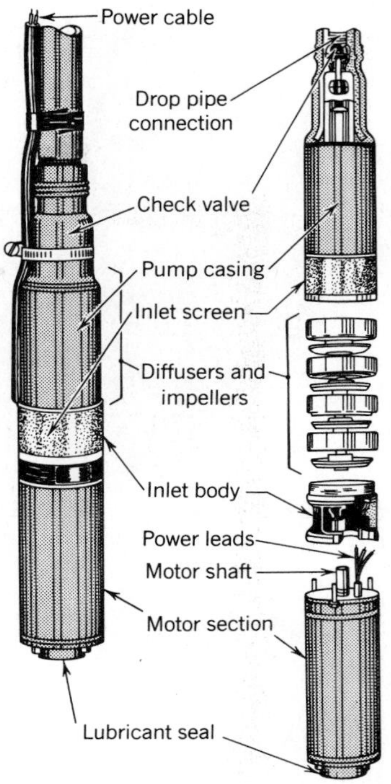

Fig. 9.6 *A submersible pump (centrifugal type), in exploded view. From the U.S. Environmental Protection Agency's* Manual of Individual Water Supply Systems, *1975.*

operate the water supply system, as well as the height to which the water must be raised. The most common centrifugal pumps are those used in deep wells.

The *turbine pump* has a vertical turbine located below groundwater levels and a driving motor located higher up, usually over the well casing at grade. A long shaft is thus required between the motor and the turbine. Substantial head clearances for this shaft's removal can be required. Figure 9.5 shows a turbine pump installation for a small community on Long Island, New York. The water is taken from the ground by multistage turbine pumps at depths of several hundred feet. The water is delivered to submerged hydropneumatic tanks at a pressure of about 80 psi. As water is demanded in the houses, the air under pressure in the upper part of the tanks forces it through the mains.

Submersible pumps are designed so that the motor can be submerged along with the turbine (Fig. 9.6). The lengthy shaft is thus eliminated.

Jet (or Ejector) Pumps. In a jet pump a venturi tube is added to the centrifugal pump. A portion of the water that is discharged from a centrifugal pump at the wellhead is forced down to a nozzle and venturi tube (Fig. 9.7). The lower pressure within the venturi tube induces well water to flow in, and the velocity of the water from the nozzle pushes it upward toward the centrifugal pump, which can then more easily lift it by suction.

(a)

Fig. 9.7 Details of the deep-well jet pump. (a) *Photograph of a multistage jet pump housing and equipment. At the top is the on/off electrical switch activated by pressure settings. It controls the direct-connected electric motor at the left. Impellers are enclosed in the pump housing at the right. Circulating connections to and from the jet can be seen at the right, the pump discharge at the top. Note the pressure gauge readings up to 100 psi, within which range the pump can be set to operate.* (b) *Well casing and circulating lines. Jet element can be seen at the bottom of the left-hand (larger) pipe.* (c) *Cutaway section of the pump.* (d) *Pumping capacity gph under various conditions. Note the two pressure ranges, 20 to 50 psi and 30 to 60 psi. Jacuzzi Bros., Inc.* (e) *Jet-type (also known as venturi or ejector) deep-well pump and storage tank for a house or small building (for well lifts greater than 25 ft). Reduced pressure at* (a), *the jet nozzle, induces flow of groundwater into the circulated flow.*

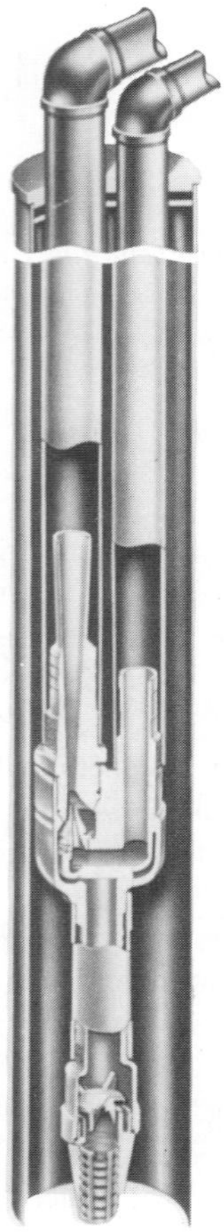

(b)

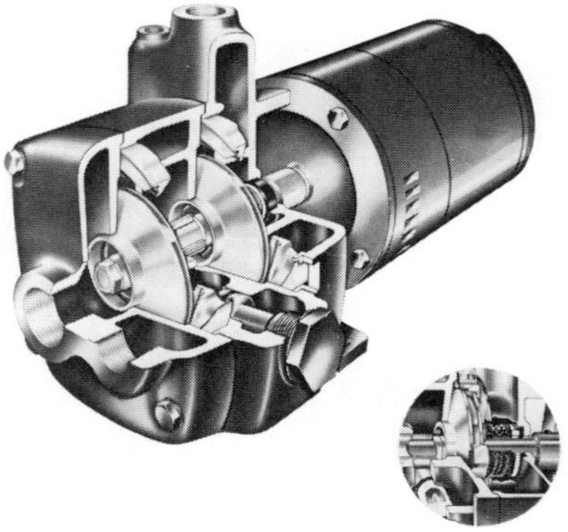

(c)

CHOOSE THE CORRECT JH FROM THESE CHARTS
DEEP WELL (Down to 120 feet)

The JH in the ½, ¾, 1 and 1½ horsepower rating, matched with the appropriate injector and pipe sizes, will pump this amount of water at the indicated lift or depth to the water in the well:

If suction lift or depth to water is	JH ½ HORSEPOWER	JH ¾ HORSEPOWER	JH 1 HORSEPOWER	JH 1½ HORSEPOWER
	Produces these gallons per hour between 20-50 lbs. discharge pressure		Produces these gallons per hour between 30-60 lbs. discharge pressure	
30 feet	795 G.P.H.	990 G.P.H.	1140 G.P.H.	1620 G.P.H.
40	680	875	1000	1470
50	575	735	875	1300
60	445	630	745	1200
70	360	495	620	920
80	310	385	530	820
90	255	315	435	700
100	220	275	340	590
110	195	240	295	540
120		205	250	480

(d)

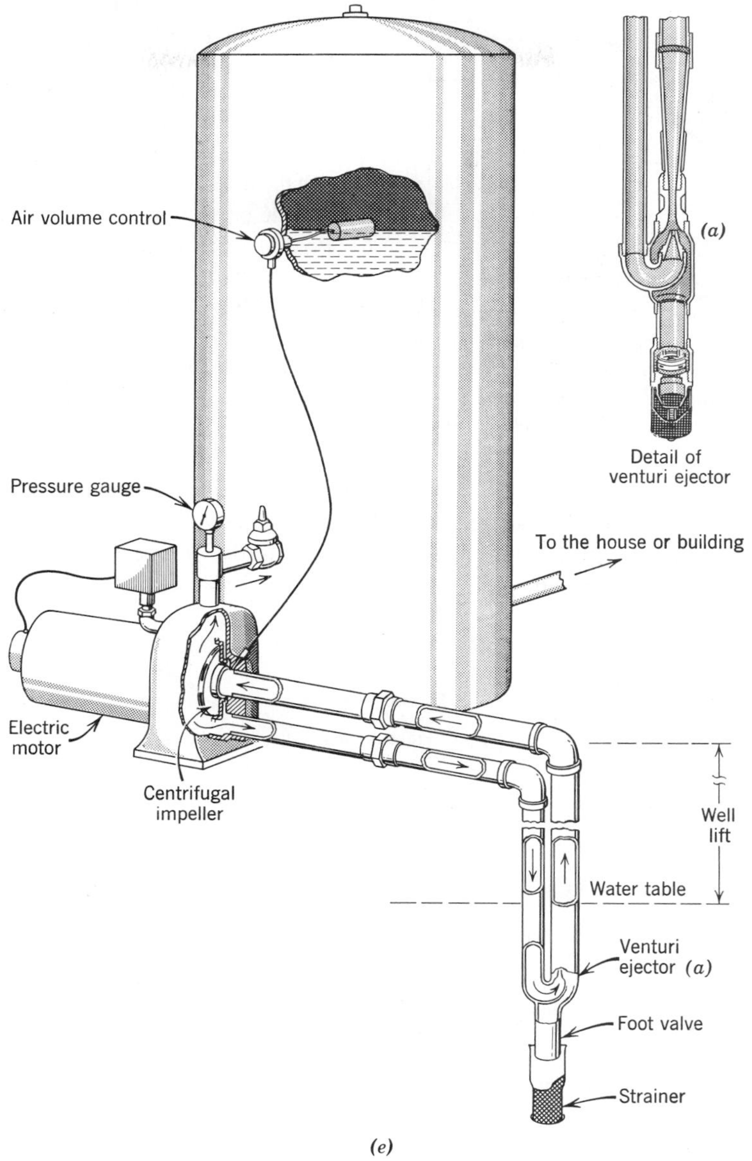

Fig. 9.7 (continued)

Pump Selection. The type of pump selected depends on many factors, including the rate of yield of the well, the daily flow (and maximum instantaneous flow rate) needed by the users, the size of storage or pressure tank used, and the total operating pressure against which the pump works (including the height to which water must be raised within the well). First cost, maintenance, and reliability are also factors, as is the energy used by the pump. In cold climates, the pump and the water supply system must be protected from freezing.

Of these factors, the two critical determinants are the flow rate (gallons per minute or per hour to be delivered) and the total pressure (or "head"). The flow rate depends on the number of fixtures to be served (Fig. 9.8), and the total pressure (Fig.

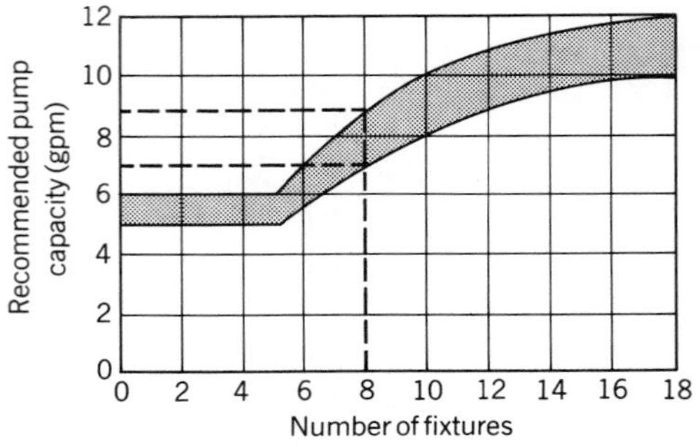

Fig. 9.8 The relationship between the recommended flow-rate capacity of a pump and the number of domestic plumbing fixtures it supplies. See details in Section 9.6. From the U.S. Environmental Protection Agency's Manual of Individual Water Supply Systems, *1976.*

9.9) includes the suction lift, static head, and friction loss plus pressure head. This relationship will be explained in detail in Section 9.6.

Storage tanks are frequently used both to maintain a constant pressure on a pump-supplied water system and to allow for temporary peaks in water supply rates that exceed the capacity of the pump. Pressure tanks are most frequently used. Elevated tanks offer one alternative, cisterns another—although the latter usually are not located high enough to provide pressure to the supply system.

Pressure tanks are often housed in outbuildings, along with the pump and any water treatment equipment (Fig. 9.9). The temperature of the outbuilding must be kept above freezing, and its roof and walls should be removable, to allow for replacement of parts over time. One type of pressure-storage tank is shown in Fig. 9.10.

The capacity of pressure tanks usually is small in comparison to the daily *total* water consumption; they provide short-term responses to peak flow demands. As a general rule, the pressure tank should be sized to deliver about 10 times the pump's

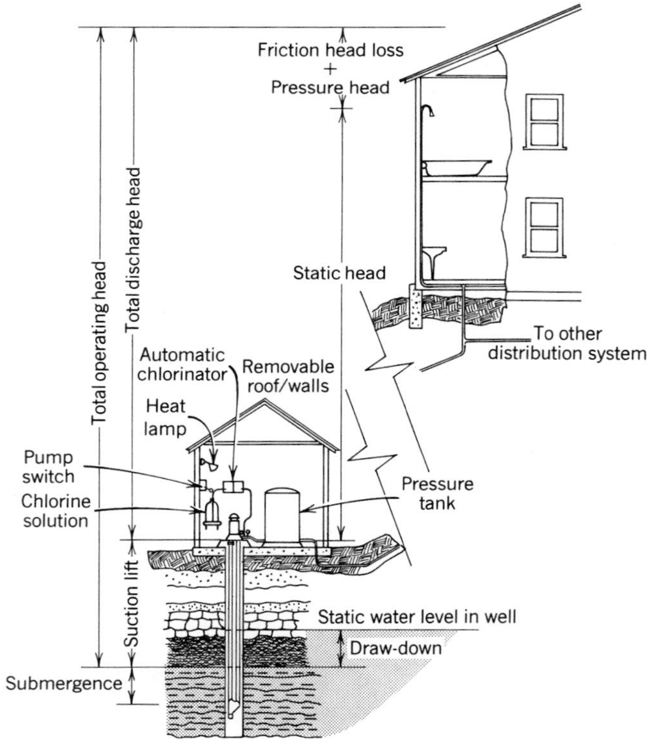

Fig. 9.9 The components of the total operating pressure (or head), a crit·cal determinant of pump size. (For details, see Section 9.6.) The pumphouse is usually a separate structu e with water treatment and storage components. From the U.S. Environmental Protection Agency's Manual of Individual Water Supply Systems, *1976.*

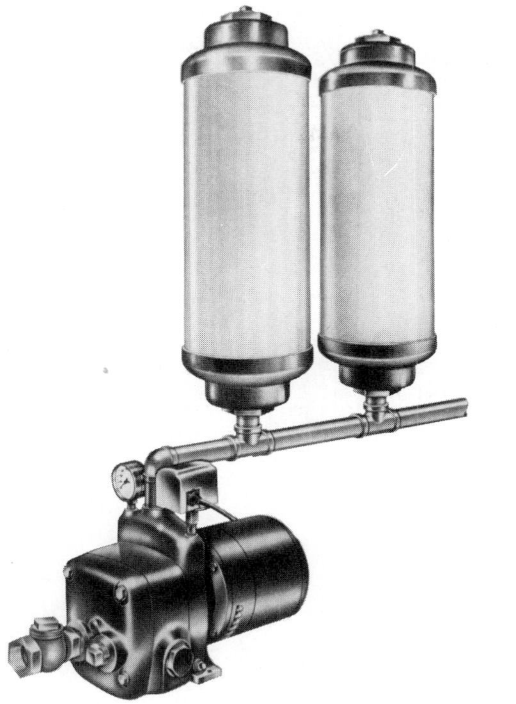

Fig. 9.10 Small pressure-storage tanks are installed primarily to keep a water supply system at constant pressure. These "Hydrocel" models are small (8½ in. in diameter, 27 in. long), so they can be installed almost anywhere along the supply system. Courtesy of Jacuzzi Bros., Inc.

capacity in gpm (L/m). For a typical residence, allow 10–15 gal (38–57 L) tank capacity per person served.

For larger installations, the size of a pressure storage tank can be calculated by

$$Q = \frac{Qm}{1 - \dfrac{P_1}{P_2}}$$

where Q = tank volume (gal)
Qm = 15 minutes of storage at peak usage rate (gal)
P_1 = minimum allowable operating pressure (psi) plus atmospheric pressure (14.7 psi)
P_2 = maximum allowable operating pressure (psi) plus atmospheric pressure (14.7 psi)

The ranges of allowable pressures are discussed in Section 9.6. (An alternative tank-sizing procedure is shown in Table 9.5.)

EXAMPLE 9.1. An office building in a remote location has a water supply system served by a pump and well. The peak demand is 50 gpm. The fixtures will operate at a minimum of 50 psi; a maximum of 70 psi should not be exceeded. Therefore,

$$Qm = 15 \text{ min} \times 50 \text{ gpm} = 750 \text{ gal}$$

and

$$Q = \frac{Qm}{1 - \dfrac{P_1}{P_2}} = \frac{750}{1 - \left[\dfrac{50 + 14.7}{70 + 14.7}\right]} = \frac{750}{1 - 0.76} = 3125 \text{ gal}$$

TABLE 9.5 **Calculations to Establish Capacity of Water Storage Tanks at Kinloch (Fig. 9.11)**

1. Potable domestic water demand per day	30,000 gpd
2. Assumed hours of use	7 A.M. to 9 P.M. 14 hours
3. Assumed hours of virtual nonuse	9 P.M. to 7 A.M. 10 hours
4. Well-yield rate per hour	25 gpm × 60 min = 1500 gph
5. Total well yield in 14 hours	15000 gph × 14 hr = 21,000 gal
6. Total water needed in 14 hours	30,000 gal
7. Well yield in 14 hours	21,000 gal (see 5 above)
8. Net-tank capacity required, minimum	30,000 − 21,000 = 9000 gal
9. Net-tank capacity as designed	
20,000 × 0.80 (80% full) = 16,000 (2 tanks at 10,000 gal each)	
2000 × 0.70 (70% full) = 1,400 (1 tank at 2,000 gal)	
17,400 gal, O.K., greater than 9000	
10. Well operation at night to restore 9000 gal to tanks	9000 ÷ 1500 = 6 hours of the 10 night hours available.

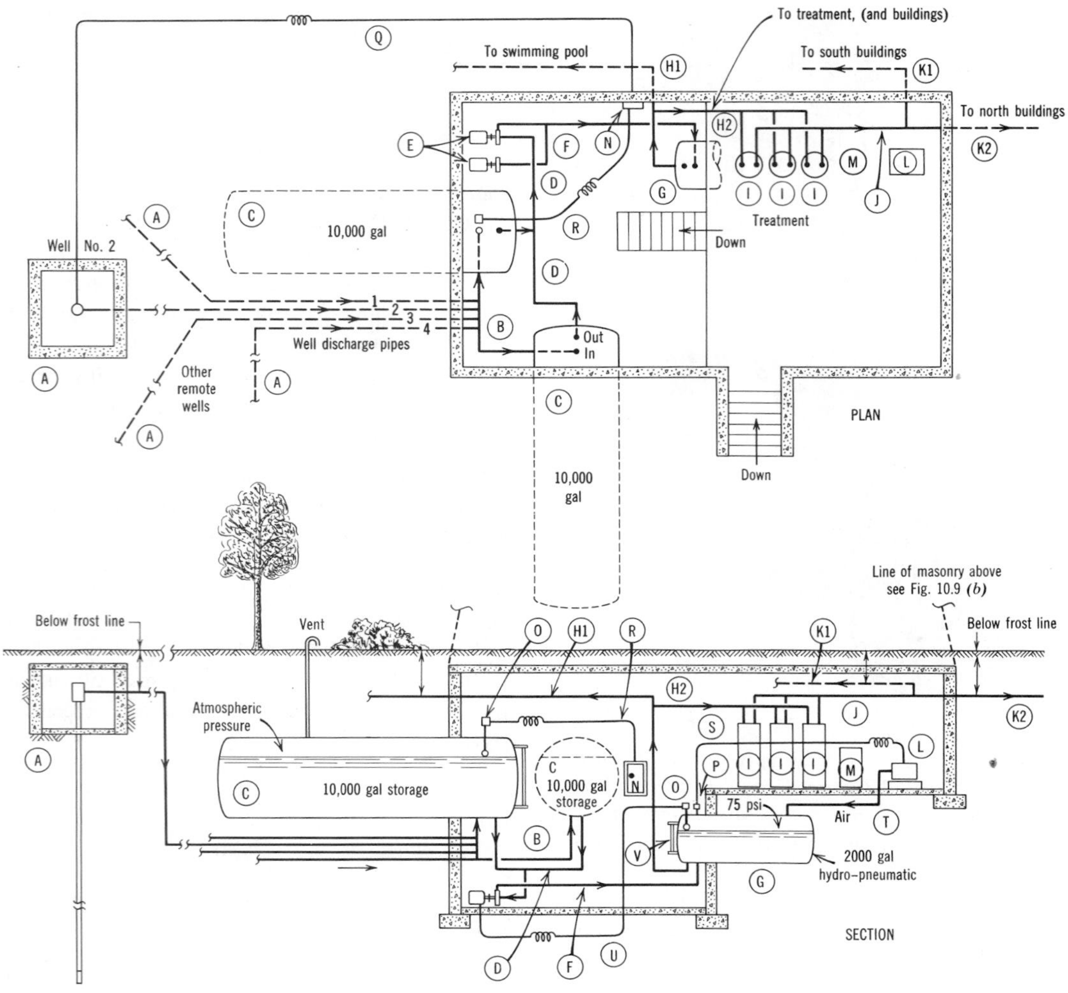

(a)

Legend

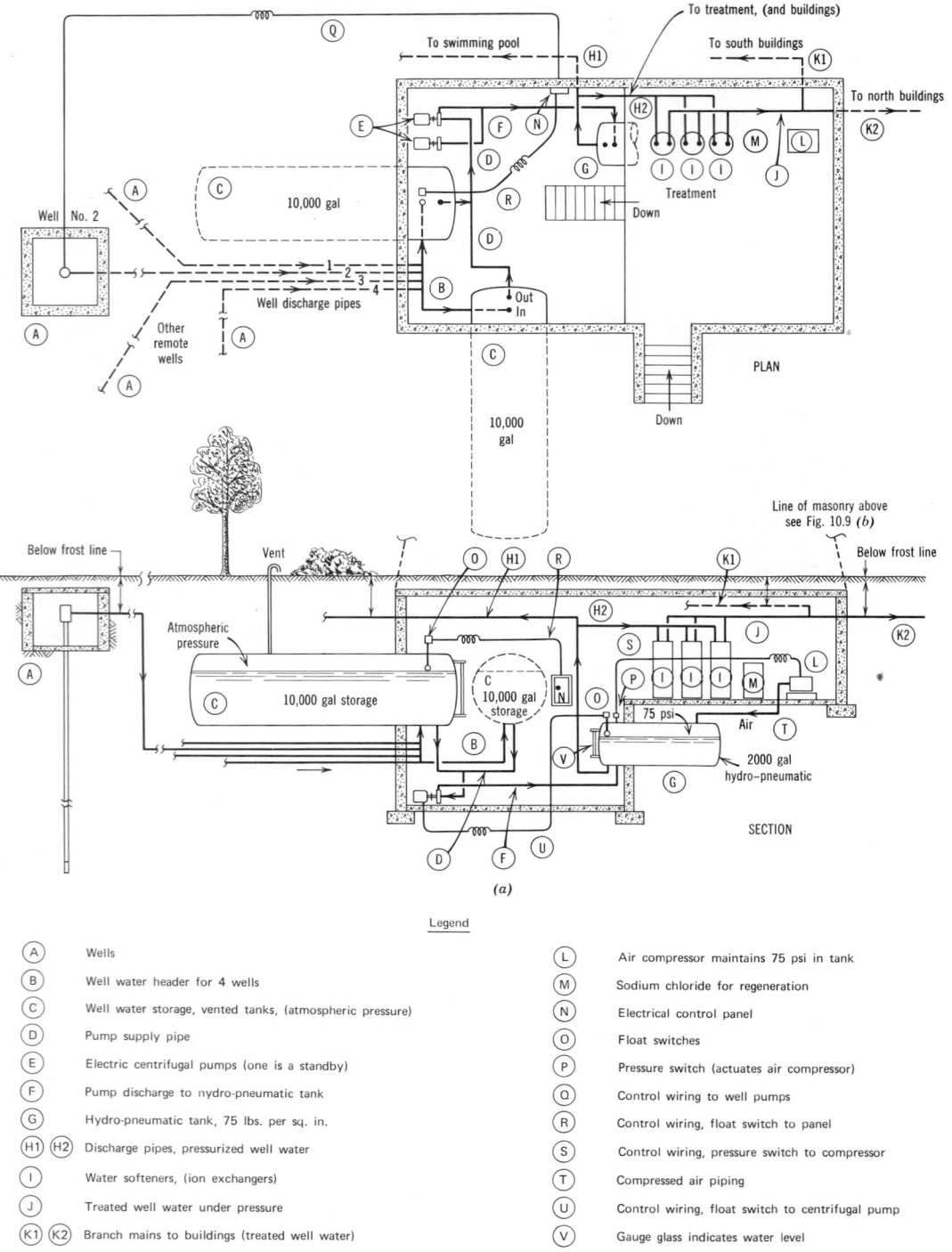

(A)	Wells	(L)	Air compressor maintains 75 psi in tank
(B)	Well water header for 4 wells	(M)	Sodium chloride for regeneration
(C)	Well water storage, vented tanks, (atmospheric pressure)	(N)	Electrical control panel
(D)	Pump supply pipe	(O)	Float switches
(E)	Electric centrifugal pumps (one is a standby)	(P)	Pressure switch (actuates air compressor)
(F)	Pump discharge to nydro-pneumatic tank	(Q)	Control wiring to well pumps
(G)	Hydro-pneumatic tank, 75 lbs. per sq. in.	(R)	Control wiring, float switch to panel
(H1) (H2)	Discharge pipes, pressurized well water	(S)	Control wiring, pressure switch to compressor
(I)	Water softeners, (ion exchangers)	(T)	Compressed air piping
(J)	Treated well water under pressure	(U)	Control wiring, float switch to centrifugal pump
(K1) (K2)	Branch mains to buildings (treated well water)	(V)	Gauge glass indicates water level

This schematic diagram shows system components and connections. For clarity, details (valves, drains, check-valves etc.) are not shown. All control wiring, here simply indicated, operates through control panel (N). See text for operation.

Fig. 9.11 (a) *Water control and distribution center for the Kinloch Estate. (Bentel & Bentel, Architects, FAIA.) The plant is located in a hillside and forms a story below the bathing and dressing pavilion.*

The capacity of elevated tanks usually is equal to at least two days of average water usage. For firefighting or other special requirements, the capacity may have to be even greater.

A summary of the quality, treatment, and supply issues raised thus far is provided in the example of a rural estate shown in Fig. 9.11 and Table 9.5. On this large estate, water was required for an estimated demand of 30,000 gpd. This was for domestic use only, the irrigation needs being solved by a separate installation pumping from a lake. Wells were dug for the domestic supply. It was quickly evident that despite the great depths of the wells drilled, the available flow would be small. Four wells yielded a total rate of only 25 gpm.

Calculations for the amount of water to be stored to supplement this meager supply are shown in Table 9.5. The table shows that during the 14 daytime hours under conditions of peak demand, the well pumps would run continuously. Concurrently, an additional 9000 gal would be drawn from the tanks. At night, the well pumps would run for six hours to restore the 9000 gal drawn from the tanks during the day.

The operation of the system is illustrated in the diagram and notes of Fig. 9.11*a*. The pressure of 75 psi in the hydropneumatic tank is sufficient to raise the water to the greatest height in the distribution system, overcome friction in the piping, and leave a residual pressure available at each fixture of about 10 to 15 psi, as needed. Excessive pressure can always be moderated by a valve in the branch supply of the fixture.

Pressure in this tank is assured by the operation of the air compressor actuated by the pressure switch. The float switch starts the centrifugal pump, to deliver water from the storage tanks as needed. The storage tanks, piped together and acting as a single reservoir, are fed by the four wells. The wells operate in unison, singly, or in groups, depending on the level in the storage tanks. If the level drops rapidly, all well pumps can run. Although they are arranged to operate all day when needed, in periods of minimum demand only one or two may be called upon. Each has its own power supply, but controls emanate from panel *N*.

Swimming pools should be supplied with pure, potable water. Biological tests showed that this well water was safe. Thus, the fill-line to the swimming pool is connected from *this* control center rather than from the supply of lake water used for irrigation. Since swimming pool water is separately recirculated and provided with its own purification treatment at a location adjacent to the pool, this supply line for make-up water to the pool would have infrequent and off-hour use. Water for domestic use in the buildings is passed through one of three ion-exchangers to provide softening and to make the water more suitable for washing, bathing, and cooking. Periodically, the calcium precipitate can be flushed out and the tanks regenerated by sodium chloride.

Fig. 9.11 (b) *One of two pavilions adjacent to the swimming pool that serves residents and guests at Kinloch. Below this pavilion is a separate story, with its own downhill access, that houses the control center. Exterior tubing to and from the center is all below grade.*

WATER AND WASTE

9.4 Hot Water Systems and Equipment

There are many ways to provide the "domestic" or "service" hot water needs within a building—the hot water used *not* for space heating, but for bathing, clothes washing, dishwashing and other related functions. Whereas much of the world either heats such water on a cookstove or does without it, North Americans enjoy an array of choices for DHW (domestic hot water) supply systems. In this section, rough supply estimates are considered first, then basic design choices, conventional heater tanks, solar water heaters, and, finally, energy recovery heaters.

(a) Estimating the Demand. One of the more familiar applications for DHW is in the home. In

TABLE 9.6 **Domestic Hot Water Consumption—Residences**

	Gallons (Liters) per Use Hot Water Required	
	14-lb (6.4-kg) Machine	*18-lb (8.2-kg) Machine*
Clothes Washing Machine		
Hot wash/hot rinse	38 gal (144 L)	48 gal (182 L)
Hot wash/warm rinse	28 gal (106 L)	36 gal (136 L)
Hot wash/cold rinse	19 gal (72 L)	24 gal (91 L)
Warm wash/cold rinse	10 gal (38 L)	12 gal (45 L)
Dishwashing	*Small*	*Large*
Dishwashing machine	10 gal (38 L)	15 gal (57 L)
Sink washing	4–8 gal (15–30 L)	
Personal Hygiene		
Tub bathing	12–30 gal (45–134 L)	
Wet shaving/hair washing	2–4 gal (8–15 L)	
Showering	2–6 gpm (13–38 L/s)	

Source: Reprinted by permission from Russell Plante, *Solar Domestic Hot Water,* copyright © 1983 by John Wiley & Sons.

the United States, it is often assumed that in a two-person family, each person will use 20 gal (76 L) of hot water per day, and that each additional family member will use another 15 gal (57 L) daily. Thus, a family of four would require 70 gal (266 L) of hot water daily. To gain a better understanding of this quantity, see Table 9.6.

Estimation of the demand for commercial and institutional buildings is not so simple, as rules of thumb are less reliable. For larger buildings, there is a trade-off between quick recovery (high heating capacity) and storage size; big tanks have small heaters, and vice versa. The details of both residential and commercial water heater sizing will be explored in Section 9.4*c*.

An important variable is the temperature at which hot water is to be provided. Table 9.7 shows that although water becomes uncomfortably hot to the touch above 110 F (43.3°C), much higher temperatures are often used for some commercial

TABLE 9.7 **Temperatures at Point of Use, DHW**

Use	*Temp. F*	*(°C)*
Lavatory		
Hand washing	105	(40.6)
Shaving	115	(46.1)
Showers and tubs	110	(43.3)
Therapeutic baths	95	(35.0)
Commercial and institutional dishwashing		
Wash	140	(60.0)
Sanitizing rinse	180	(82.2)
Commercial and institutional laundry	180	(82.2)
Residential dishwashing and laundry	140	(60.0)
Surgical scrubbing	110	(43.3)

Source: Copyright © by the American Society of Heating, Refrigerating and Air Conditioning Engineers, Inc., Atlanta, GA. Reprinted by permission from the *ASHRAE Systems Handbook,* 1984.

processes. When determining the temperature at which DHW is to be supplied, consider the following factors.

High temperatures:

- Allow the installation of smaller storage tanks (less hot water is mixed with cold to achieve final usage temperature at the shower, sink, or lavatory)—but require larger heating units.
- Can be achieved at the point of use by heaters built into equipment such as dishwashers.
- Can cause scale to form on heating coils and within piping [above 140 F (60°C) in areas of ''hard'' water quality].
- May be required by code for some applications.

Lower temperatures:

- Are less likely to cause burns, but may not achieve desired sanitation.
- Mean less energy consumed, because storage and pipe heat losses are lower.
- Allow the installation of smaller heating units—but require larger storage tanks.
- Make possible the use of lower-grade heat sources for DHW, such as solar energy or waste heat recovery.

Figure 1.14 showed that 4% of annual U.S. energy usage is for DHW, which requires the lowest-grade heat source of all the common usages.

(b) Basic System Choices. It is as difficult to categorize the array of available DHW systems as it was to categorize the many HVAC systems in Chapters 5 and 6. The most important variables are heat source, method of heat transfer to the water, central versus distributed heater/storage equipment, and related distribution trees.

Heat Sources. These include the familiar concentrated-energy, high-grade sources such as natural gas and electricity. Oil- and coal-fired boilers are frequently equipped with DHW coils as well. Buildings served by steam use steam as a DHW heat source. Because of the relatively low-grade final temperatures needed, DHW can also be readily provided by wood-burning equipment, incinerators, solar energy equipment, heat pumps, and heat recovery devices (as in commercial ice-making machines whose discharged heat is contributed to DHW).

Heating Methods. There are two basic methods:

1. *Direct* heating brings water against directly heated surfaces: electric-resistance elements within tanks, or other electrically warmed surfaces, or surfaces directly exposed to fire or hot gases.

2. *Indirect* heating can be accomplished in several ways. Coils containing steam or fluids can be submerged within water tanks or (Fig. 9.12) set within boilers, whose primary function usually is to provide space or industrial process heating. Alternatively, the coils containing DHW can be placed outside the boiler but within a casing containing steam, hot exhaust gases, or very hot water.

Both direct or indirect methods can be utilized in a variety of equipment:

Storage tank water heaters, the type most commonly used for residential and small commercial purposes (see Section 9.4*c*).

Circulating storage water heaters, in which the water is first heated by a coil, then circulated through the storage tank (as in some solar heaters—see Section 9.4*d*).

Tankless heaters, in which the water is very quickly raised to a desired temperature within a heating coil and immediately sent to the point of usage.

Tankless water heaters are available in larger sizes for central hot water systems (Fig. 9.12) or in smaller sizes for remote plumbing fixtures that occasionally need hot water or for isolated bathrooms, laundry rooms, and so forth. Decentralized tankless units can be as small as ''instant hot water taps'' for kitchen or bar sinks—electric-resistance heaters capable of generating perhaps ½ gph at up to 200 F (2 L/h at up to 93°C). Bathroom groups can require either electric resistance or gas-fired units (Fig. 9.13) producing up to 5 gpm (0.3 L/s) with a temperature rise of as much as 40F° (22C°). Because tankless heaters often consist of fairly long coils through which the water passes as it is heated, they may add considerable friction to the total for which the supply system is designed (see Section 9.7).

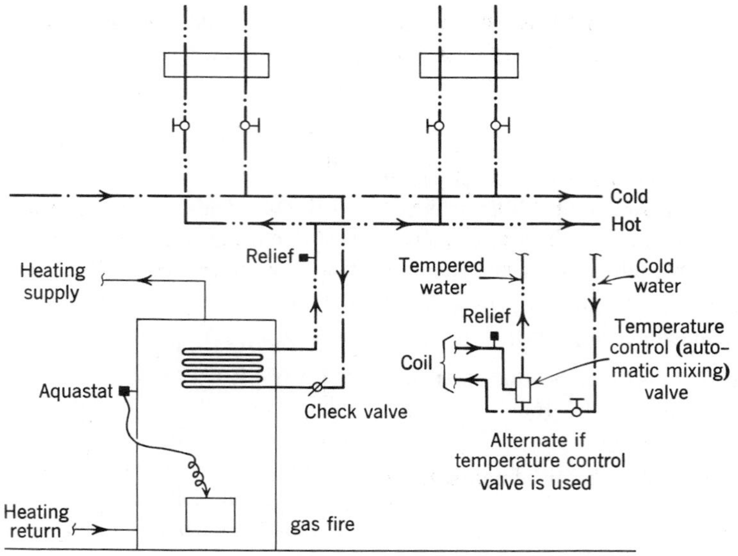

(a)

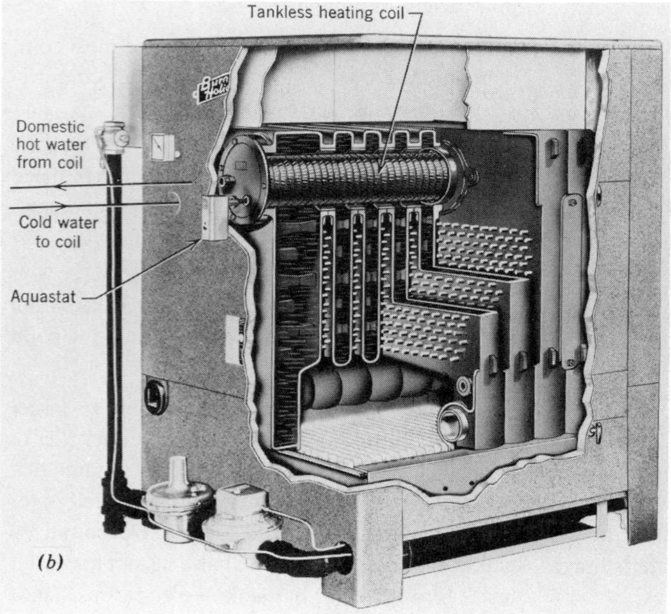

(b)

Fig. 9.12 (a) *An example of indirect, tankless heater for DHW, utilizing a boiler that provides hot water for space heating. Since no storage is used, the point of use for DHW should be very close to the boiler.* (b) *Internal tankless heating coil for domestic hot water immersed in the jacket water of a gas-fired hot water heating boiler. Approximate capacity range 3 to 15 gpm at 100F° rise.* (c) *External-type, tankless heater for domestic hot water. Boiler water is piped to the unit and circulates by gravity, transferring heat to the coil. Approximate capacity range 3 to 15 gpm at 100F° rise.* NOTE: *Because it can be highly inefficient to provide summertime DHW with a boiler designed for wintertime space-heating loads, codes may alter or prohibit this arrangement.*

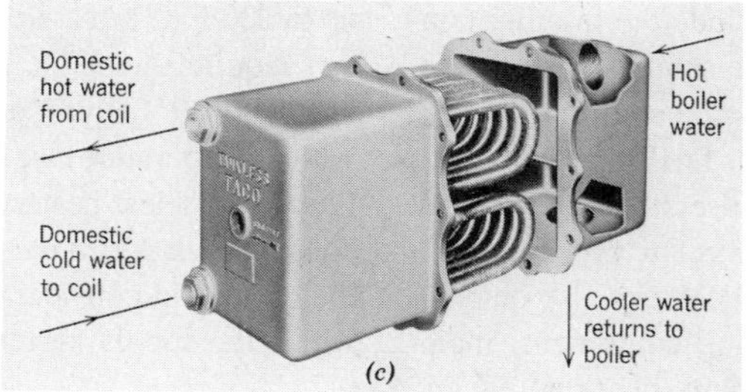

(c)

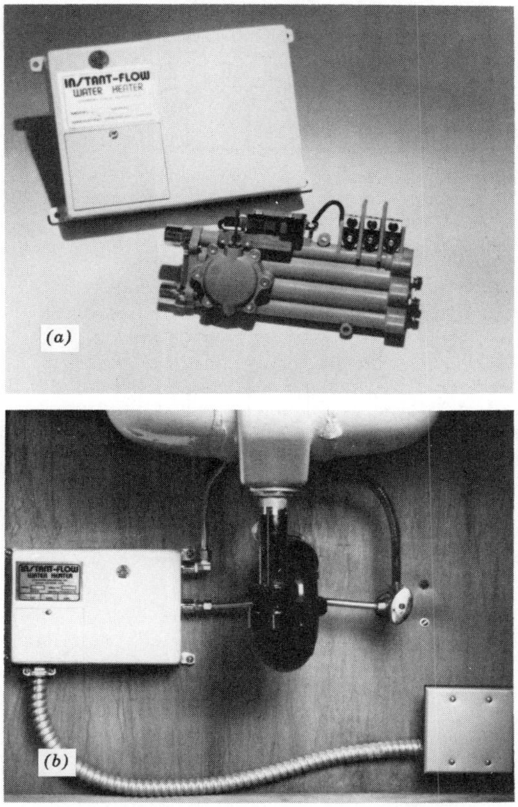

Fig. 9.13 *Instantaneous or tankless water heater.* (a) *In this "instant-flow" water heater, a series of coils heats water as it flows through.* (b) *Installation below a typical lavatory. A wide range of sizes is available; the 4.6 kW unit raises water by 31F° at 1 gpm; the 9 kW unit raises water by 61F° at 1 gpm. Courtesy of Chronomite Laboratories, Inc.*

Central versus Distributed Equipment. This choice can be particularly complex. In the United States a central water-heater storage tank is standard equipment in homes and small stores. But the growing use of solar water heating, along with awareness of the heat losses from water-heater storage tanks, has created new interest in *combinations* of central and distributed DHW.

Consider a larger residential DHW application such as that shown in Fig. 9.14. Here, two areas in which hot water is used are separated by some 50 ft. If the simple, centralized water-heater storage tank is used (Fig. 9.14*a*), it would probably be placed nearest the maximum-use fixtures—dishwashers and clothes washers. (Also, floor space for the water-heater storage tank is usually more plentiful there.) However, each time hot water is needed in the remote bathroom serving the bedrooms, the previously heated water in at least 50 ft of supply pipe—which, despite insulation, will almost certainly have cooled—must be drained off before hot water finally arrives. (A ¾-in. diameter pipe, for example, will hold 4.6 gal in 50 ft of pipe. If the water heater is set at an energy-conserving level of 120 F, and the incoming city water is at 50 F, the 4.6 gal of wasted water will also waste the 2682 Btu invested when it was heated.)

Another way to conserve water is with a recirculating hot water system (Fig. 9.14*b*). The primary disadvantages of this system are the increased heat loss through the hot water pipe (now kept at 120 F for 24 hours per day, rather than for just a few minutes) and the energy required to run the circulating pump (again for 24 hours daily).

A decentralizing option (Fig. 9.14*c*) is to install two water-heater storage tanks, one for each group. First cost probably will be greater, and more floor space required. The daily waste heat from *two* water heaters is clearly a disadvantage that will at least partially offset the energy saved by eliminating the 50-ft run of pipe. (Further, in warm weather this waste heat will add to discomfort within the house.) A decentralized-centralized mix (Fig. 9.14*d*) will save energy, water, and floor space, but it will almost certainly be costlier to install. A central solar water heater—either passive (shown) or active—brings the water up to warm (winter) or very hot (summer). At each fixture group, a tankless heating coil instantaneously heats the water, as needed. Almost no water is wasted, and the energy lost in the 50 ft of pipe will be solar energy, not purchased energy from natural gas or electricity.

Another option would be to omit the solar water heater and simply install two tankless heaters. The primary disadvantage of this setup would be the need for larger-capacity heaters, whose instantaneous demand for energy could lead to electric cost penalties in areas where peak-load rates are high.

Distribution Trees. It follows from the preceding discussion that the choice of a distribution tree associated with central water heaters must take into account the gradual cooling of water

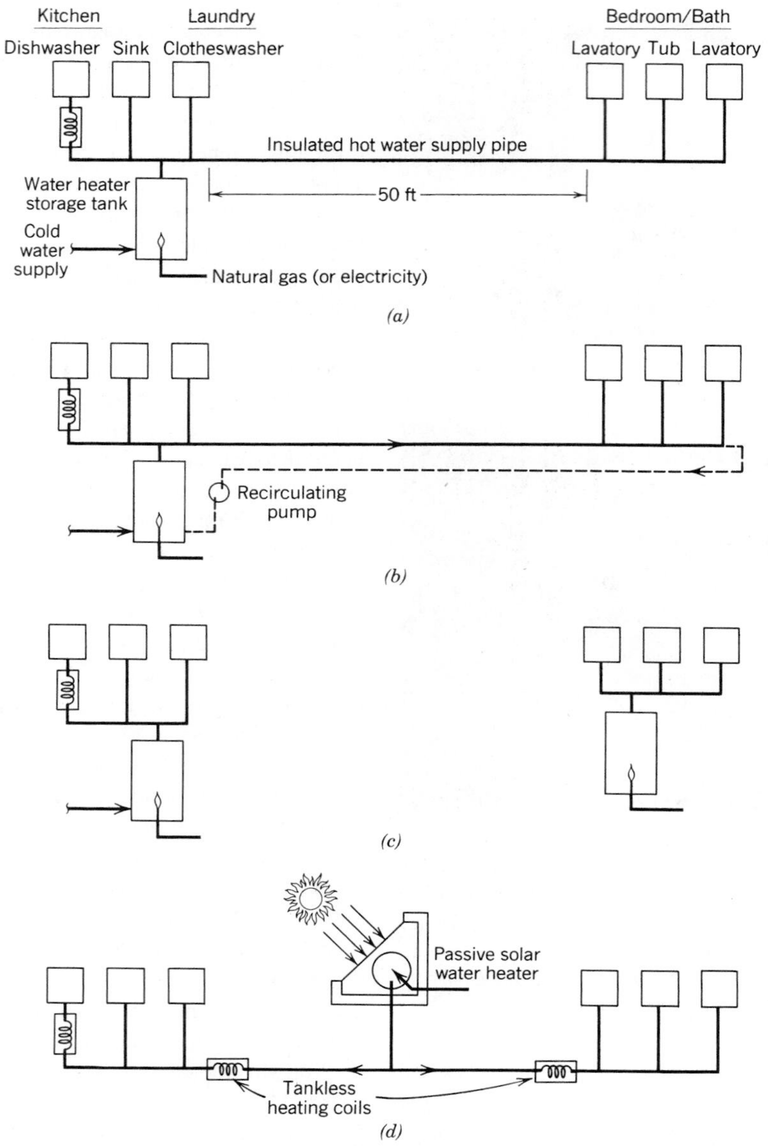

***Fig. 9.14** Options for DHW in a larger residence.* (a) *Typical centralized water-heater storage tank; water and energy is wasted in the 50-ft. run.* (b) *Recirculating pump added; this saves water but requires continuous energy to operate pump and make up the pipe's heat losses.* (c) *Decentralized approach using two water-heater storage tanks; this roughly doubles the energy lost from storage but eliminates water and energy loss from the 50-ft run.* (d) *Central solar water heater/decentralized tankless heating coil combination; saves on everything except initial cost.*

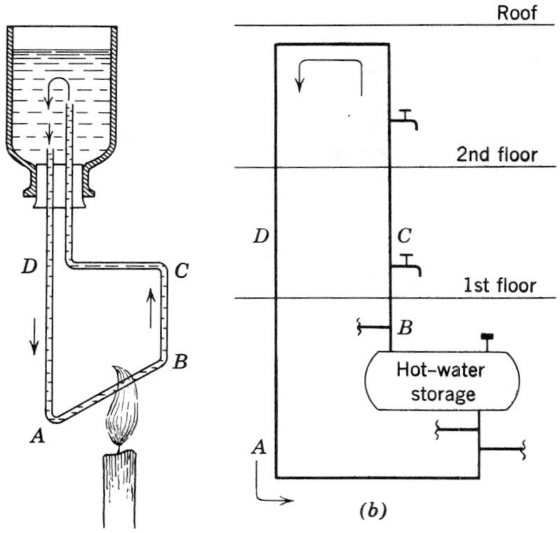

Fig. 9.15 (a) *Principle of hot water circulation by gravity (thermosiphon).* (b) *Its application to hot water service. During periods of no demand there is sufficient incidental cooling between* C *and* D *so that dense (less warm) water at* A *forces the lighter (hot) water at* B *to rise for speedy availability at each faucet.*

within the distribution system, once it has left the central heater. If the heat losses associated with constantly-circulating hot water (looped trees) are preferable to the heat and water wastage associated with simple, single hot water distribution trees, such loop systems can be achieved in either of two ways.

Thermosiphon hot water circulation depends on the fact that water expands and becomes lighter when heated, as may be seen in Fig. 9.15*a*. If heat is applied to the lower loop of a glass tube, both ends of which have been inserted in an inverted bottle containing water, the water will move from *A* to *B* and will rise through the tube *BC* into the bottle. It here becomes cooled and drops through the tube *DA* to *A*, is again heated, and rises in the tube *BC*—thus completing the circulation. Since the movement depends on the difference in weight between the two columns of water, the velocity and consequent efficiency of the circulating system increase with the temperature of the water and the height of the circuit. Hot water supply systems therefore usually consist of a heater with a storage tank, piping to carry the heated water to the farthest fixture, and a continuation of this piping to return the unused cooled water back to the heater. A constant circulation is thereby maintained, and hot water may be drawn at once from a fixture without first draining off through the faucet the cooled water that would be standing in the supply pipe if there were no return conduit for its escape. Because heat increases the corrosive action of water on metals, copper tubing is usually the best choice for use in hot water and hot water circulating systems. Thermosiphon circulating systems are particularly effective in multistory buildings, because of the beneficial effect of increased circuit height on circulation efficiency.

Forced circulation of hot water is often utilized where a height advantage is unavailable. Low, long, rambling buildings, such as some large one-story residences, schools, and factories, lack the height needed to set up good hot water circulation by gravity. Also, the flow is diminished by friction in long pipe runs. For such buildings the forced-circulation scheme shown in Fig. 9.16*a* offers one option. Three independent aquastats—devices that create an electrical signal impulse when water temperature drops—control this system. Aquastats *A*, *B*, and *C*, respectively, sense the temperatures of the water in the heater, the tank, and the end of the circulation-return main. As needed, they turn on the oil or gas burner, the tank-circulating pump, and the system-circulating pump. Fixtures remote from the tank are as close to hot water as the length of their hot water runout pipes. Water is usually available at full temperature in 5 to 10 s. Trial aquastat settings in °F could be *A* 180, *B* 160, *C* 120 (°C: *A* 82, *B* 71, *C* 49).

An energy-saving computer control is available for hotels, motels, apartments, and larger commercial buildings. This device (Fig. 9.16*b*) *varies the supply temperature* of hot water, so that the hottest water is supplied at the busiest hours. In hours of low usage, much lower supply temperatures mean that more ''hot'' water will be mixed with less ''cold'' water at showers, lavatories, and sinks. This increased ''hot'' water quantity poses no problem at off-peak hours, and the lower temperature means significant decreases in heat loss from the recirculating supply water system. A typical payback period is one year. The device stores

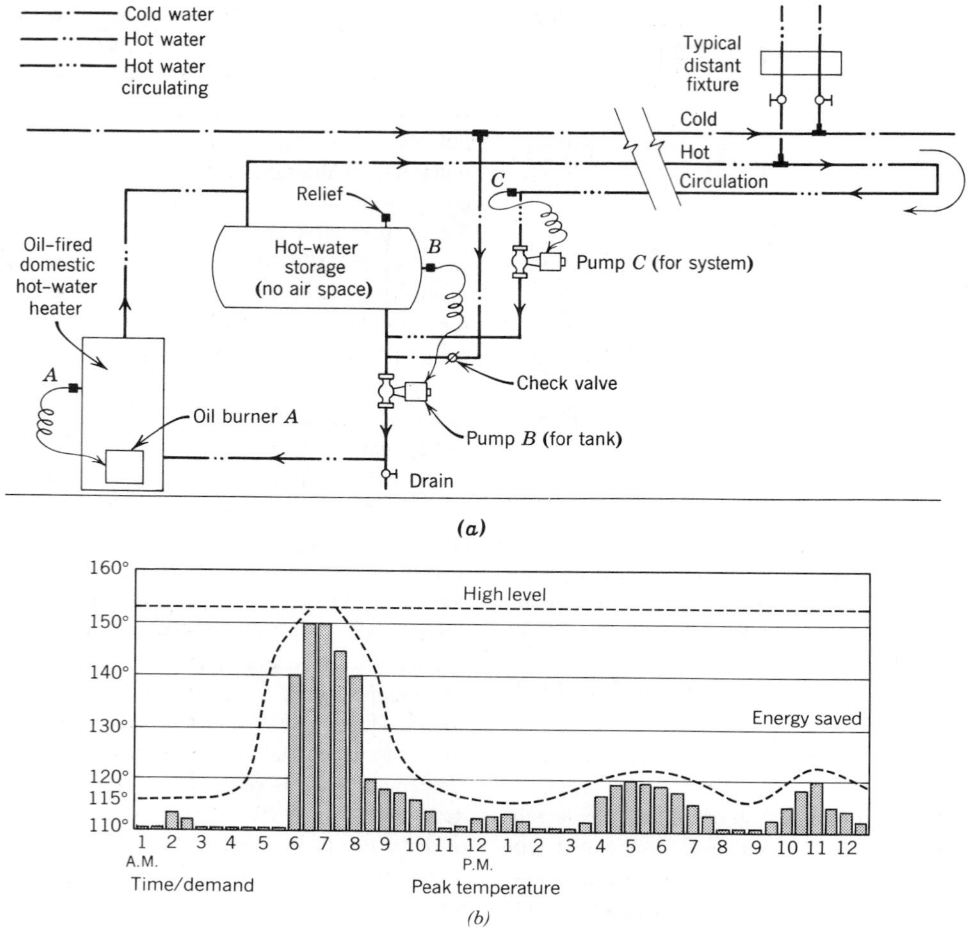

Fig. 9.16 (a) *Forced circulation of DHW, for a long, low building. (Note that the use of space-heating boilers for DHW may cause system inefficiencies in warm weather.* (b) *Energy savings possible when the supply temperature of hot water is varied. Heat losses from supply pipes are greatly reduced for most of the typical day. A computer controls the supply temperature. Courtesy of Fluidmaster, Inc.*

a memory (adjusted weekly) of the typical daily patterns of usage, and varies the supply temperature accordingly.

(c) Conventional Heat Sources. One of the most common appliances used in the United States is the water-heater storage tank, an energy-conserving model of which is shown in Fig. 9.17.

For residences, the capacity of hot water heating/storage equipment can be taken from the values in Table 9.8. Note that for the familiar tank-type direct water heaters (gas, electric, or oil), there is a stated relationship among storage size, rate of hourly heat input, rate of draw (demand) over one hour, and recovery rate. Example 9.2 uses Table 9.8 for the sizing of a residential water heater.

EXAMPLE 9.2. Select a natural gas water heater for a four-bedroom house with 2½ baths (two full baths and one powder room). The *minimum* requirements of HUD-FHA are shown in Table 9.9,

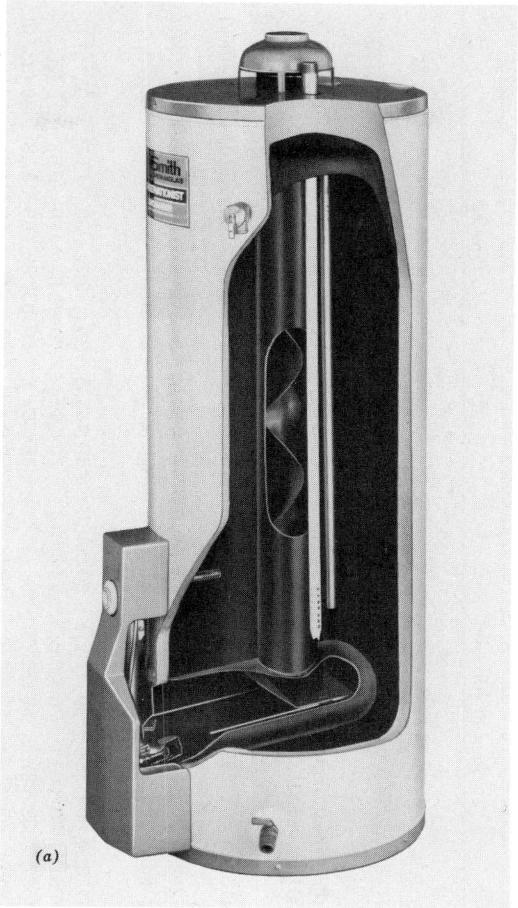

(a)

		All Dimensions in Inches				Natural & Propane Gas		
Model No.	Gal. Cap.	Vent Size	Height Plus Diverter	Height Less Diverter	Diameter	BTU Input Per Hr.	Gal./Hr.* Recovery 90° Rise	Approx. Ship Wt. (Lbs.)
PGCS-40	40	3	54	50 7/16	18 1/4	32,000	35.3	127
PGCS-50	50	3	57 1/4	53 11/16	20 3/16	40,000	43.1	156
PGCS-65	65	4	59	55 1/2	21 3/4	50,000	53.9	192

*Recovery capacities are based on DOE Method of Test (90° Rise).

Note: To compensate for the effects of high altitude areas above 2,000 feet, recovery capacity should be reduced approximately 4% for each 1,000 feet above sea level.

(b)

Fig. 9.17 (a) *Water-heater storage tanks, with such energy-conserving features as a gradually restricted flue outlet to maximize thermal efficiency and a combustion chamber that prevents heat from escaping directly up the flue (thus preventing continuing heat loss from the stack effect).* (b) *Numerical characteristics for such water heater storage tanks. Courtesy of the A. O. Smith Corporation.*

TABLE 9.8 **Minimum Residential Water Heater Capacities**

Number of Baths	*1 to 1.5*			*2 to 2.5*				*3 to 3.5*			
Number of Bedrooms	*1*	*2*	*3*	*2*	*3*	*4*	*5*	*3*	*4*	*5*	*6*
Gas[a]											
Storage, gal (L)	20 (75.8)	30 (113.7)	30 (113.7)	30 (113.7)	40 (151.6)	40 (151.6)	50 (189.5)	40 (151.6)	50 (189.5)	50 (189.5)	50 (189.5)
1000 Btu/h (1 kW) input	27 (7.9)	36 (10.5)	36 (10.5)	36 (10.5)	36 (10.5)	38 (11.1)	47 (13.8)	38 (11.1)	38 (11.1)	47 (13.8)	50 (14.6)
1-h draw, gal (L)	43 (163.0)	60 (227.4)	60 (227.4)	60 (227.4)	70 (265.3)	72 (272.9)	90 (341.1)	72 (272.9)	82 (310.8)	90 (341.1)	92 (348.7)
Recovery, gph (mL/s)	23 (24.1)	30 (31.5)	30 (31.5)	30 (31.5)	30 (31.5)	32 (35.6)	40 (42.0)	32 (33.6)	32 (33.6)	40 (42.0)	42 (44.1)
Electric[a]											
Storage, gal (L)	20 (75.8)	30 (113.7)	40 (151.6)	40 (151.6)	50 (189.5)	50 (189.5)	66 (250.1)	50 (189.5)	66 (250.1)	66 (250.1)	80 (303.2)
kW input	2.5	3.5	4.5	4.5	5.5	5.5	5.5	5.5	5.5	5.5	5.5
1-h draw, gal (L)	30 (113.7)	44 (166.8)	58 (219.8)	58 (219.8)	72 (272.9)	72 (272.9)	88 (333.5)	72 (272.9)	88 (333.5)	88 (333.5)	102 (386.6)
Recovery, gph (mL/s)	10 (10.5)	14 (14.7)	18 (18.9)	18 (18.9)	22 (23.1)	22 (23.1)	22 (23.1)	22 (23.1)	22 (23.1)	22 (23.1)	22 (23.1)
Oil[a]											
Storage, gal (L)	30 (113.7)	30 (113.7)	30 (113.7)	30 (113.7)	30 (113.7)	30 (113.7)	30 (113.7)	30 (113.7)	30 (113.7)	30 (113.7)	40 (151.6)
1000 Btu/h (1 kW) input	70 (20.5)	70 (20.5)	70 (20.5)	70 (20.5)	70 (20.5)	70 (20.5)	70 (20.5)	70 (20.5)	70 (20.5)	70 (20.5)	70 (20.5)
1-h draw, gal (L)	89 (337.3)	89 (337.3)	89 (337.3)	89 (337.3)	89 (337.3)	89 (337.3)	89 (337.3)	89 (337.3)	89 (337.3)	89 (337.3)	89 (337.3)
Recovery, gph (mL/s)	59 (61.9)	59 (61.9)	59 (61.9)	59 (61.9)	59 (61.9)	59 (61.9)	59 (61.9)	59 (61.9)	59 (61.9)	59 (61.9)	59 (61.9)

Tank-type indirect[b,c]										
1-W-H rated draw, gal (L) in 3-h, 100 F° (55.6 C°) rise	40 (151.6)	40 (151.6)		66 (250.1)	66 (250.1)[d]	66 (250.1)	66 (250.1)	66 (250.1)	66 (250.1)	66 (250.1)
Manufacturer-rated draw, gal (L) in 3-h, 100 F° (55.6 C°) rise	49 (185.7)	49 (185.7)		75 (284.2)	75 (284.2)[d]	75 (284.2)	75 (284.2)	75 (284.2)	75 (284.2)	75 (284.2)
Tank capacity, gal (L)	66 (250.1)	66 (250.1)		66 (250.1)	66 (250.1)[d]	82 (310.8)	66 (250.1)	82 (310.8)	82 (310.8)	82 (310.8)
Tankless-type indirect[c,e]										
1-W-H-rated, gpm (L/s) 100 F° (55.6C°) rise	2.75 (0.17)	2.75 (0.17)		3.25 (0.205)	3.25 (0.205)[d]	3.75 (0.205)	3.25 (0.205)	3.75 (0.205)	3.75 (0.205)	3.75 (0.205)
Manufacturer-rated draw, gal (L) in 5 min, 100 F° (55.6C°) rise	15 (56.8)	15 (56.8)		25 (94.7)	25 (94.7)[d]	35 (132.6)	25 (94.7)	35 (132.6)	35 (132.6)	35 (132.6)

Source: U.S. HUD-FHA, for one- and two-family living units.

[a]Storage capacity, input, and recovery requirements indicated in the table are typical and may vary with each individual manufacturer. Any combination of these requirements to produce the stated 1-h draw will be satisfactory.

[b]Boiler-connected water heater capacities [180 F (82.2°C) boiler water, internal or external connection].

[c]Heater capacities and inputs are minimum allowable. Variations in tank size are permitted when recovery is based on 4 gph/kW (4.2 mL/s · kW) @ 100 F° (55.6C°) rise for electrical. A.G.A. recovery ratings for gas heaters, and IBR ratings for steam and hot water heaters.

[d]Also for 1 to 1.5 baths and 4 bedrooms for indirect water heaters.

[e]Boiler-connected heater capacities [200 F (93.3°C) boiler water, internal or external connection].

TABLE 9.9 **Procedure for Selecting a Residential Water Heater Storage Tank**

	Column A HUD-FHA Minimum (Table 9.8)	Column B Conservationist Water Heater, A.O. Smith (Figure 9.17) PGCS-40	Column C PGCS-50
Storage (gal)	40	40	50
1000 Btu input	38	32.0	40
1-hr draw (gal)[a]	72	75.3	93.1
Recovery (gph)	32	35.3	43.1

[a]1-h draw = tank capacity + 1-h recovery.

column *A*. From Fig. 9.17, we select two trial units (40 gal and 50 gal) that *might* comply; see columns *B* and *C*.

SOLUTION. PCGS-50 is selected. (However, PGCS-40 is acceptable in all respects except that of Btu input.)

Small shops and office buildings can be treated like residences for hot water sizing. For larger buildings, the trade-off between heaters and storage tanks should be explored. To begin this process, consult Table 9.10, which shows the maximum hourly and daily demands and contrasts them to the average day (which should be used to estimate monthly energy consumption).

The relationship between heater size and tank size is graphed in Fig. 9.18. The advantage of a large heater is its smaller tank, which consumes less floor space and volume, weighs less, and probably has a lower first cost. The advantage of a large tank is that its smaller heater tends to work steadily rather than in spurts; this lower, steadier demand for energy will lead to lower utility rates in the case of electric heating, and thus to money savings over the life of the system. When the fuel supply is solar energy or waste heat, the larger tank again is usually better suited to the characteristics of the fuel supply.

EXAMPLE 9.3. (Adapted from the *ASHRAE Systems Handbook,* 1980). A women's dormitory housing 300 students, with a cafeteria serving 300 meals within one hour, is to be built. Find the required storage size for two conditions: (*a*) assuming a minimum recovery rate for both dorm and cafeteria; and (*b*) assuming a dorm recovery rate of 2.5 gph (2.6 mL/s), which is half the maximum hourly value given in Table 9.10, and a cafeteria recovery rate of 1.0 gph (1.1 mL/s), which is two-thirds the maximum hourly value given in Table 9.10.

SOLUTION.

Minimum Recovery: From Fig. 9.18*a*, the minimum recovery rate for women's dormitories is 1.1 gph. For 300 students,

$$300 \times 1.1 = 330 \text{ gph recovery}$$

At this rate, again from Fig. 9.18*a*, the minimum usable storage capacity is 12 gal/student. Assume that 70% of the total size is this "usable" capacity. This means that after 70% of the stored hot water is withdrawn, the remaining water has been cooled (by incoming water) to an unusably low temperature. Storage size must be increased by

$$\frac{100\%}{70\%} = 1.43$$

Thus,

$$12 \times 300 \times 1.43 = 5150 \text{ gal storage}$$

From Fig. 9.18*e*, the minimum recovery rate for the cafeteria (serving full meals, type *A*) is 0.45 gph. For 300 meals,

$$300 \times 0.45 = 135 \text{ gph recovery}$$

TABLE 9.10 **Domestic Hot Water, Commercial, Institutional**

Type of Building	Maximum Hour	Maximum Day	Average Day
Men's dormitories	3.8 gal (14.4 L)/student	22.0 gal (83.4 L)/student	13.1 gal (49.7 L)/student
Women's dormitories	5.0 gal (19 L)/student	26.5 gal (100.4 L)/student	12.3 gal (46.6 L)/student
Motels: no. of units[a]			
20 or less	6.0 gal (22.7 L)/unit	35.0 gal (132.6 L)/unit	20.0 gal (75.8 L)/unit
60	5.0 gal (19.7 L)/unit	25.0 gal (94.8 L)/unit	14.0 gal (53.1 L)/unit
100 or more	4.0 gal (15.2 L)/unit	15.0 gal (56.8 L)/unit	10.0 gal (37.9 L)/unit
Nursing homes	4.5 gal (17.1 L)/bed	30.0 (113.7 L)/bed	18.4 gal (69.7 L)/bed
Office buildings	0.4 gal (1.52 L)/person	2.0 gal (7.6 L)/person	1.0 gal (3.79 L)/person
Food service establishments:			
Type A—full meal restaurants and cafeterias	1.5 gal (5.7 L)/max meals/h	11.0 gal (41.7 L)/max meals/h	2.4 gal (9.1 L)/avg meals/day[b]
Type B—drive-ins, grilles, luncheonettes, sandwich and snack shops	0.7 gal (2.6 L)/max meals/h	6.0 gal (22.7 L)/max meals/h	0.7 gal (2.6 L)/avg meals/day[b]
Apartment houses: no. of apartments			
20 or less	12.0 gal (45.5 L)/apt.	80.0 gal (303.2 L)/apt.	42.0 gal (159.2 L)/apt.
50	10.0 gal (37.9 L)/apt.	73.0 gal (276.7 L)/apt.	40.0 gal (151.6 L)/apt.
75	8.5 gal (32.2 L)/apt.	66.0 gal (250 L)/apt.	38.0 gal (144 L)/apt.
100	7.0 gal (26.5 L)/apt.	60.0 gal (227.4 L)/apt.	37.0 gal (140.2 L)/apt.
200 or more	5.0 gal (19 L)	50.0 gal (195 L)/apt.	35.0 gal (132.7 L)/apt.
Elementary schools	0.6 gal (2.3 L)/student	1.5 gal (5.7 L)/student	0.6 gal (2.3 L)/student[b]
Junior and senior high schools	1.0 gal (3.8 L)/student	3.6 gal (13.6 L)/student	1.8 gal (6.8 L)/student[b]

[a]Interpolate for intermediate values.
[b]Per day of operation.

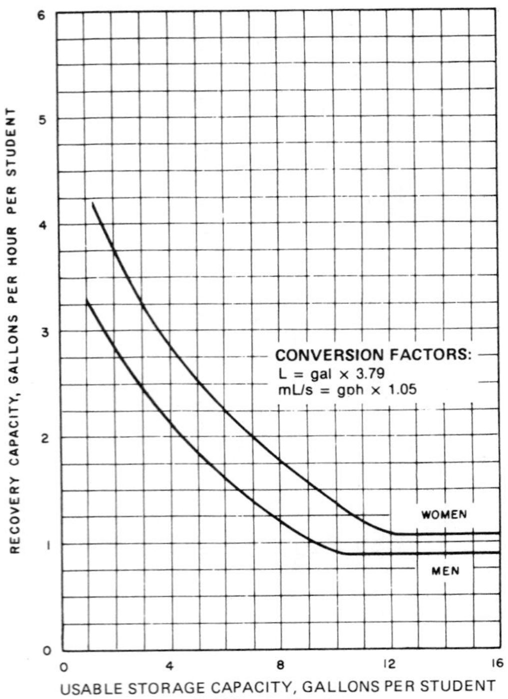

(*a*) Dormitories

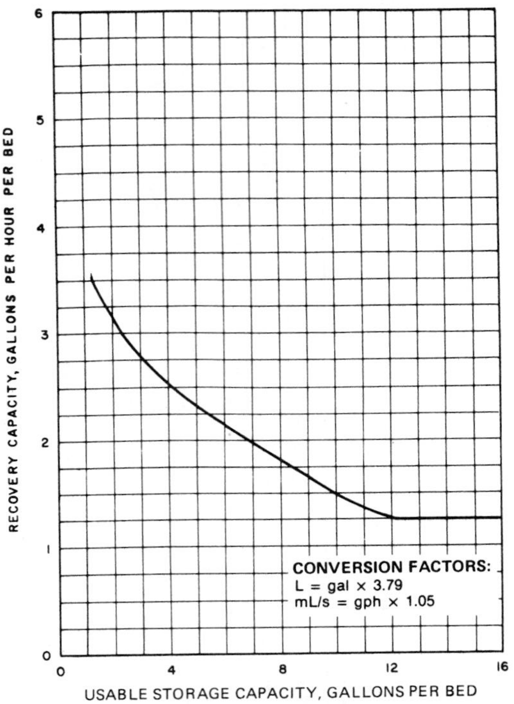

(*c*) Nursing Homes

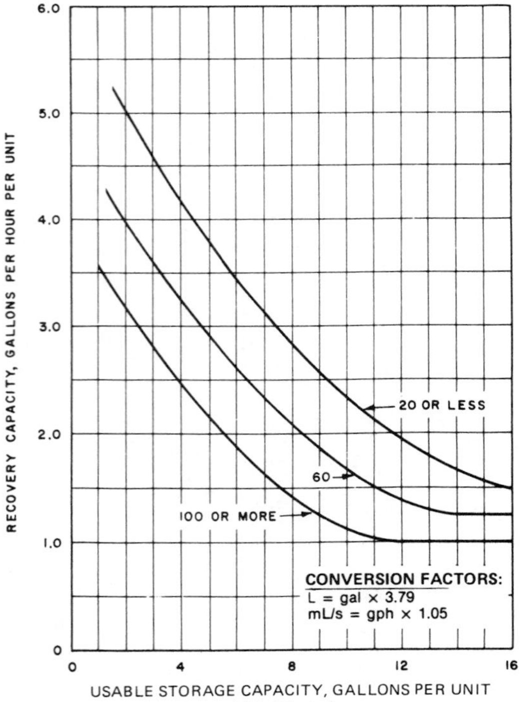

(*b*) Motels

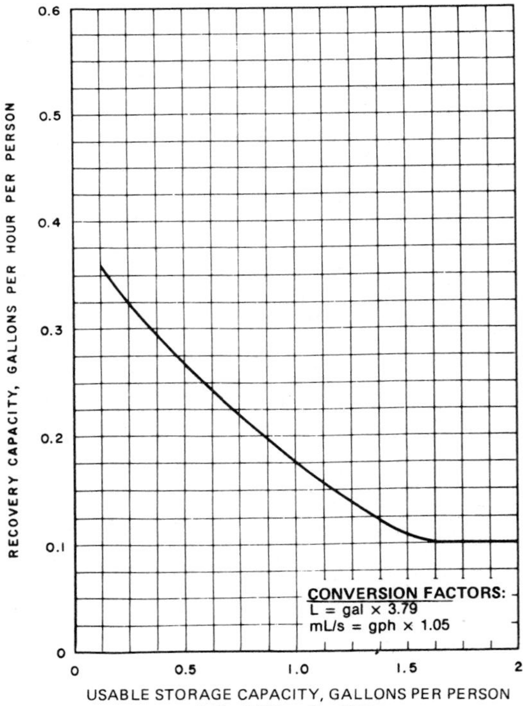

(*d*) Office Buildings

Fig. 9.18 *Domestic hot water sizing: the tradeoff between recovery (heater) capacity and storage capacity. "Usable" storage capacity is usually considered to be between 60% and 80% of the total tank capacity. Copyright © by the American Society of Heating, Refrigerating and Air Conditioning Engineers, Inc., Atlanta, GA. Reprinted by permission from the* ASHRAE Systems Handbook, *1984.*

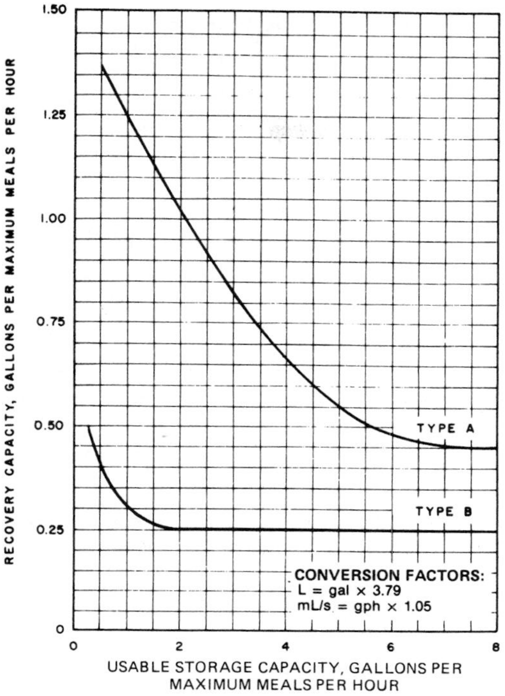

(*e*) Food Service

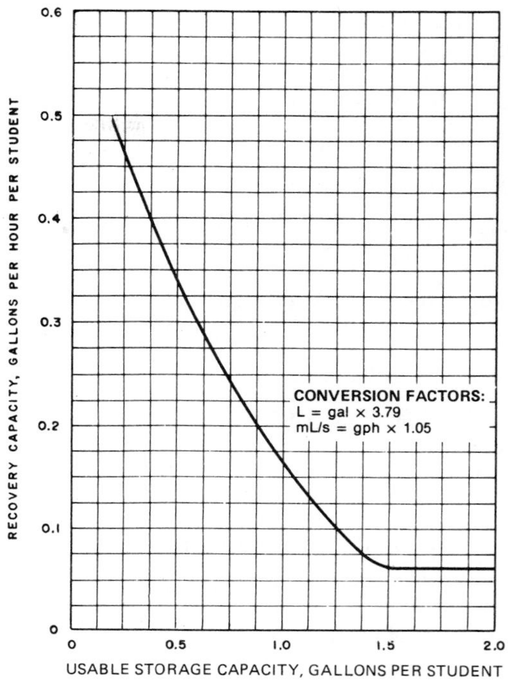

(*g*) Elementary Schools

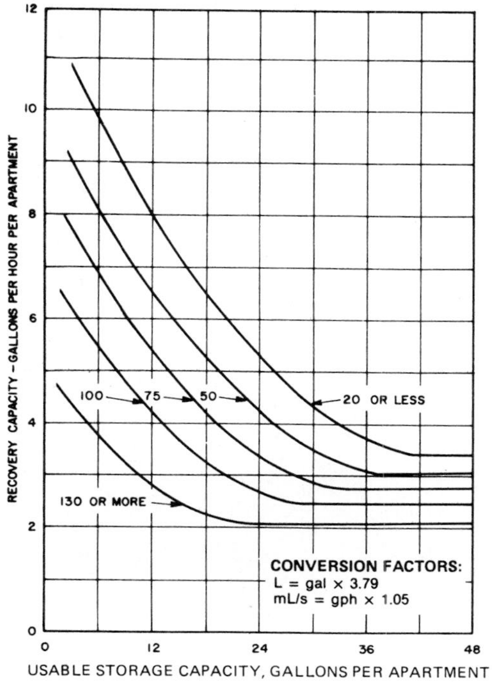

(*f*) Apartments

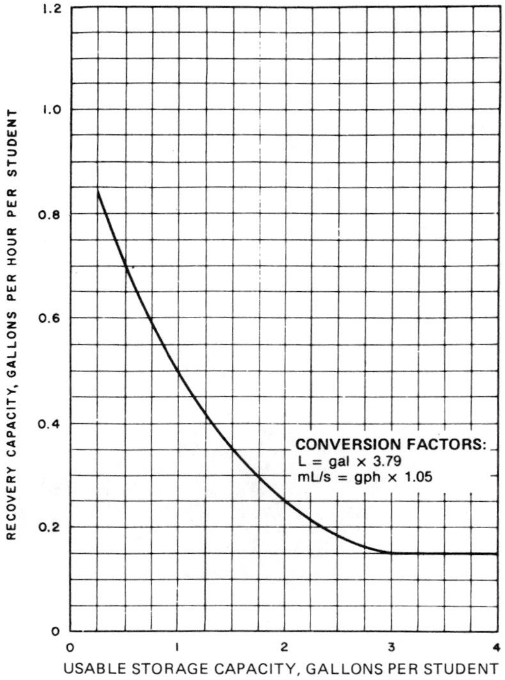

(*h*) High Schools

At this rate, the minimum usable storage capacity is 7 gal/meal. Thus,

$$300 \times 7 \times 1.43 = 3000 \text{ gal storage}$$

Combining these requirements for dorm and cafeteria,

$$\text{recovery} = 330 + 135 = 465 \text{ gph}$$
$$\text{storage} = 5150 + 3000 = 8150 \text{ gal}$$

Faster Recovery: At the specified dorm recovery rate of 2.5 gph,

$$300 \times 2.5 = 750 \text{ gph}$$

and the minimum usable storage required is 5 gal/student. Thus,

$$300 \times 5 \times 1.43 = 2150 \text{ gal}$$

At the specified cafeteria recovery rate of 1.0 gph,

$$300 \times 1.0 = 300 \text{ gph}$$

and the minimum usable storage required is 2 gal/meal. Thus,

$$300 \times 2.0 \times 1.43 = 860 \text{ gal}$$

Combining these requirements for dorm and cafeteria,

$$\text{recovery} = 750 + 300 = 1050 \text{ gph}$$
$$\text{storage} = 2150 + 860 = 3010 \text{ gal}$$

Note: This sizing procedure does not account for "system losses": that is, heat lost from the storage tanks and from the hot water piping. Recovery capacities are usually increased because of these losses, which are simple to calculate when the U value of the tank and pipe insulation is known: $\text{Btu/h} = U \times A \times \Delta T$.

For this example, an increase in heater size of 225% for faster recovery allows the size of tank to be reduced to only 37% of original size. For these larger DHW applications, central steam is often used as a heat source. Figure 9.19 shows an indirect storage tank system. Such systems may cost more than tankless systems, but they are usually cheaper to operate, both because the peak prices for the instantaneous fuel demands of tankless heaters are avoided and because low-grade steam or waste high-temperature water sources can be utilized.

In any system that heats water under pressure, safety precautions are necessary. To minimize the dangers of excessive pressures (which can damage the system) and of superheated water (which can severely injure people), most codes require *pressure and temperature (P/T) relief valves* to be installed on top of all water heaters. Since these valves are designed to release hot water whenever dangerous pressure or temperature levels are reached, they should be attached to a length of pipe that will conduct the released water to a drain.

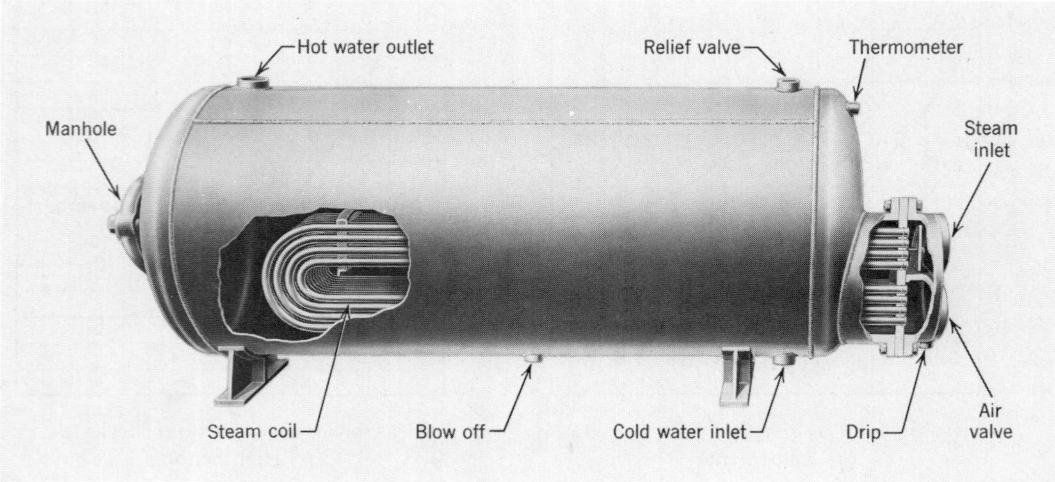

Fig. 9.19 *Storage tank heater and storage for domestic hot water for large-demand applications. Steam coil submerged in tank (see Fig. 9.33). Capacities 100 to 10,000 gph, varying by length of coil, for 140F° (40 to 180 F) temperature rise.*

(d) Solar Water Heating. Solar energy is most attractive to a designer when it will do most of its work during the time that it is most available. On a seasonal basis (Fig. 9.20), solar energy is especially attractive for outdoor swimming pool heating, as it will be used only in sunny, warm months. Throughout the United States, solar energy can easily meet most of the summertime demand for DHW as well. In the northern part of the country, a much smaller solar contribution can be expected in winter. An especially difficult problem is the winter space-heat mismatch between greatest need and least supply. For both solar DHW and space-heating systems, winter brings the added complication of the need to protect against freezing.

Again, there are many ways in which solar energy can be used to heat water. Solar water heating systems are usually classified by the means of fluid circulation (''passive'' or ''active''), the means by which heat from the collector piping is transferred to the DHW itself (''direct'' or ''indirect,'' or closed loop), and the means of protection against freezing.

Passive systems rely on gravity for circulation, as was explained in Section 9.4*b*. Hence, the storage tank usually must be placed above the collector and the number of bends in the collector's supply and return piping minimized, to minimize friction. Heavy storage tanks located high in a building can cause structural problems. The advantages of the passive approach include lower cost of components, high mechanical reliability (no pump, etc.), and low operational costs.

Active systems use pumps to force the fluid to the collector. Although this setup allows the collector and storage to be located anywhere that is convenient for the designer, it introduces the complications of mechanical breakdown, increased maintenance, and the high cost of the energy needed to run the pump.

Direct systems utilize only one fluid: the water to be heated for use in the building is circulated through the solar collector. Such a system has the advantages of simplicity and efficiency, as it does not require a separate loop of fluid and the attendant piping complications and inefficiencies of heat exchange.

Indirect systems use a closed loop containing a fluid that circulates through the collector and storage tank. The fluid is not mixed with the DHW itself; rather, heat is passed from one fluid to the other through a heat exchanger. One advantage of this system is that it allows nonfreezing solutions to be used in the collector loop. It also allows collectors to be operated at low pressures, rather than at the high pressures typical of urban public water systems. The choice of fluid is determined by its freezing and boiling points, its specific heat, and its level of toxicity.

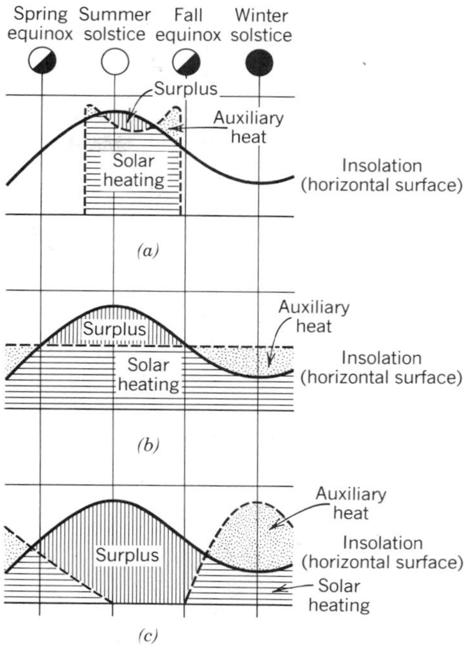

Fig. 9.20 *Comparison of the pattern of supply of solar energy (on a horizontal surface) to various patterns of heating needs.* (a) *Swimming pool heating.* (b) *DHW.* (c) *Space heating.*

As Table 9.11 shows, these design elements (passive and active, direct and indirect) are combined in a number of typical solar DHW systems. Brief descriptions of some of the more common systems follow.

1. *Batch systems*. In these, the simplest of all such systems, a black-painted storage tank is exposed to the sun within a glazed collector box (Fig. 9.21). The price of simplicity is the inefficiency of exposing a tank of hot water to the cold nighttime conditions of the collector. Movable insulation can be used to reduce this problem. These heaters are also referred to as ''breadbox'' water

TABLE 9.11 **Solar Domestic Hot Water (DHW) Systems**

Type	*Main Features*	*Advantages*	*Disadvantages*
Batch system	Batch tank inside collector box. Potable water within collector/tank.	No external power. Few components. Collector/tank at any location.	Seasonal; dependent on freezing locations. Heat loss at night from storage.
Thermosiphon system[a]	Flat plate liquid collectors. Normally open loop but no pump or external power (passive DHW system). Storage tank higher than collector.	No external power. Few components. High performance.	Seasonal; dependent on freezing locations (if water is collector fluid). Needs structural support for high storage tank.
Closed-loop freeze-resistant system[a]	Flat plate liquid collectors. Closed loop of piping from collectors to storage tank. Uses external energy (circulator and differential controller). Uses nonfreezing collector fluid. Pressurized stonelined storage tank.	Can be used in coldest climates. More and better-established competition. Circulator; small consumption of external energy. High performance.	Liquid; service/maintenance required. More components required.
Drain-back system[a]	Flat plate liquid collectors. Water is collector fluid (open loop). Potable water circulates through the heat exchanger in the storage tank (not through the collectors). Large heat exchanger. Pitched headers.	Can be used in the coldest climates. No antifreeze used. Most simple of active flat plate systems (no valves).	Larger pump; larger consumption of external energy. System must drain thoroughly. Use of corrosion inhibitor recommended.

Drain-down system[a]	Flat plate liquid collectors. Potable water circulated through collectors. Line pressure feeds collectors (open loop). Has automatic drainage valves. Pitched headers.	No heat exchanger or extra storage tank needed. High performance.	In some instances a larger pump; larger consumption of external energy. System must drain thoroughly. No corrosion inhibitor possible. Freeze danger with valve failure.
Air-to-liquid system[a]	Flat plate air collectors. Air-to-water heat exchanger. Ductwork and blower. Pipes and circulator. Larger collector area than liquid system.	Won't freeze (dependent on exchanger location). Air leaks won't cause damage. Integrates well with space heating.	Hard to detect leaks. More space required for ducts. Blower and circulator required. More carpentry involved. Less efficient than other systems.
Phase-change system[a]	Flat plate liquid collectors. Freon 114 or R12. Storage tank higher than collectors (passive type). Closed-loop refrigerant-grade piping from collectors to storage tank (passive type) or to condenser (subambient type).	No external power (passive type). Can be mounted at any location (subambient type).	Very hard to detect leaks. Special equipment to install. More components required (subambient type).

[a]Reprinted by permission from Russell H. Plante, *Solar Domestic Hot Water: A Practical Guide,* Copyright © 1983 by John Wiley & Sons.

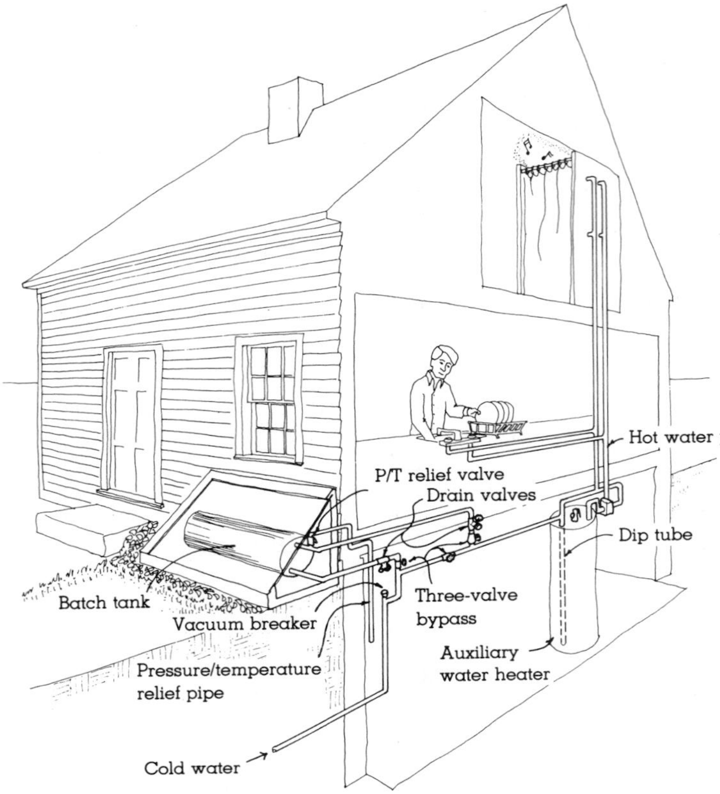

Fig. 9.21 *Batch solar water-heating system. Reprinted by permission from Daniel K. Reif,* Passive Solar Water Heaters, *copyright © 1983 by Brick House Publishing Company, Inc.*

heaters and ''integral passive solar water heaters.'' The batch system was used at the Antelope Valley California Poppy Reserve (Fig. 5.23), as shown in Fig. 9.22. Because of their simplicity, they are favored by do-it-yourself builders, especially in mild-winter climates.

2. *Thermosiphon systems*. The sun acts as both the ''pump'' and the heat source for these systems (Fig. 9.23). With no moving parts, maintenance needs are low. The collector, however, must be lower than the tank, and piping must be kept as simple as possible. Since the coldest water remains in the lower collector at night, the hot water in the upper tank is not threatened with undue heat loss. However, freezing conditions pose a severe threat to the collector. Accordingly, indirect (closed-loop) systems containing a nonfreezing fluid are frequently used; *phase-change systems* are fast becoming the cold-winter option for this type of passive system.

3. *Closed-loop freeze-resistant systems*. In addition to being used in thermosiphoning systems, these are commonly used in active systems (Fig. 9.24). A small pump circulates nonfreezing fluid to the collector when there is sufficient sun. This process is governed by a ''differential controller,'' a device that compares the temperature of the stored hot water to that of the collector. The price of this freeze protection is the inefficiency of the heat exchanger, used between the collectors' fluid and the DHW itself. Some codes require a *double* wall between any toxic nonfreezing fluid and the DHW, which further reduces efficiency.

4. *Drain-back systems* (Fig. 9.25*a*). Although these systems use water as the fluid pumped from tank to collector, the water is not the DHW itself. Instead, the DHW passes through a heat exchanger in the solar storage tank. With the DHW kept out of the collector, the solar collector can operate at lower water pressure. This, however,

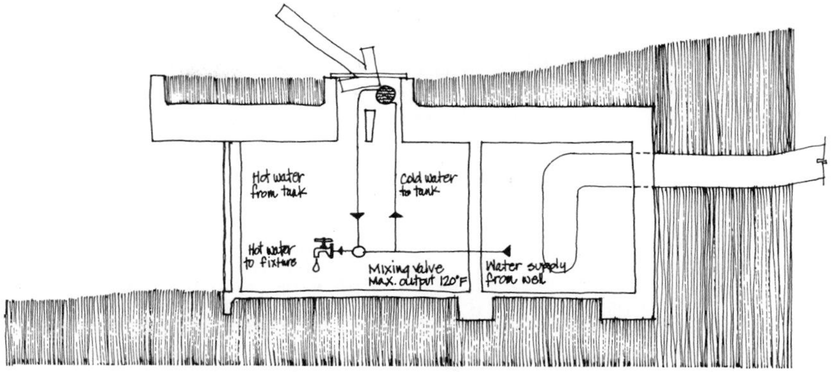

Fig. 9.22 A batch solar water heater supplies the hot water for the Antelope Valley California Poppy Reserve (also described in Fig. 5.23). Courtesy of the Colyer/Freeman Group, Architects.

requires a large heat exchanger, with attendant inefficiency. It also requires care in the design and installation of piping between collector and tank, so that the collector will drain thoroughly. When the controller senses that no solar energy can be gathered, it cuts off the pump, and water drains back into the tank. Therefore, the collector will be filled with air, not water, for all nighttime and cloudy-cold daytime hours. This suggests that corrosion inhibitors should be added to the collector/tank's water, since the piping is frequently exposed to air.

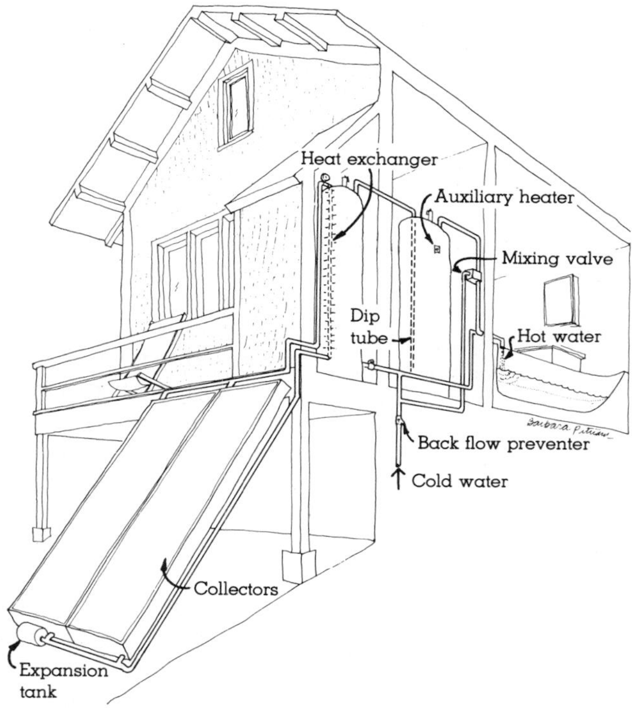

Fig. 9.23 Thermosiphon solar water-heating system. Reprinted by permission from Daniel K. Reif, Passive Solar Water Heaters, *copyright © 1983 by Brick House Publishing Company, Inc.*

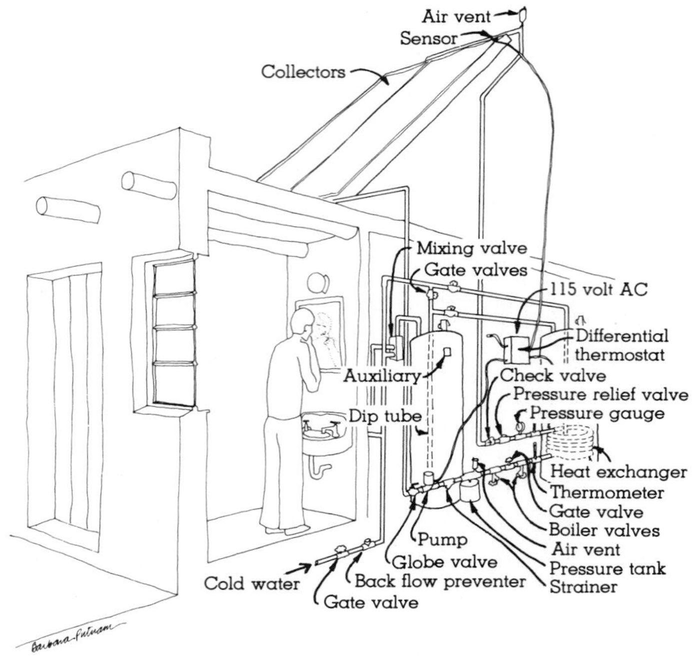

Fig. 9.24 *Closed-loop freeze-resistant solar water heating system. Reprinted by permission from Daniel K. Reif,* Passive Solar Water Heaters, *copyright © 1983 by Brick House Publishing Company, Inc.*

5. *Drain-down systems* (Fig. 9.25*b*). These are the only active systems that do not utilize heat exchangers; DHW is circulated directly through the collector. Both higher pressure and higher efficiency result. In this system, the collector is usually filled with water that moves only when the differential controller activates the pump. Whenever the outside temperature drops near freezing, the controller activates solenoid valves and the water in the collector is drained down (dumped). In cold-winter areas, this process could result in several gallons of water wasted per winter day. Although the lack of a heat exchanger is attractive from the standpoint of cost savings and thermal efficiency, the set of electrically controlled solenoid valves is not foolproof. When malfunctions occur in pressurized systems, the loss of water can be enormous, and there is great potential for water damage to the building.

6. *Air-to-liquid systems*. These systems use air collectors and rock storage beds; a heat exchanger transfers heat from the collector's air to the DHW. Much lower efficiencies result, and the ductwork is much more space consuming than are pipes to water collectors. Worse yet, leaks in air collectors or storage beds are very difficult to detect. However, air collectors aren't damaged by freezing.

7. *Phase-change systems* (Fig. 9.26). In any of the above-discussed systems, in which DHW is kept out of the collector, the fluid within the collector not only can be freeze resisting, but also can take advantage of latent heat (the considerable amount of heat stored when this fluid vaporizes and released when the fluid condenses). The primary disadvantages are the first cost, the difficulty in detecting leaks, and the resulting threat of some refrigerant fluids to the environment (such as the ozone layer).

Passive approaches to phase-change materials were outlined in the discussion of thermosiphoning systems. *Subambient* approaches utilize a heat pump and can glean heat even from cold-cloudy ("subambient") conditions. They thus represent

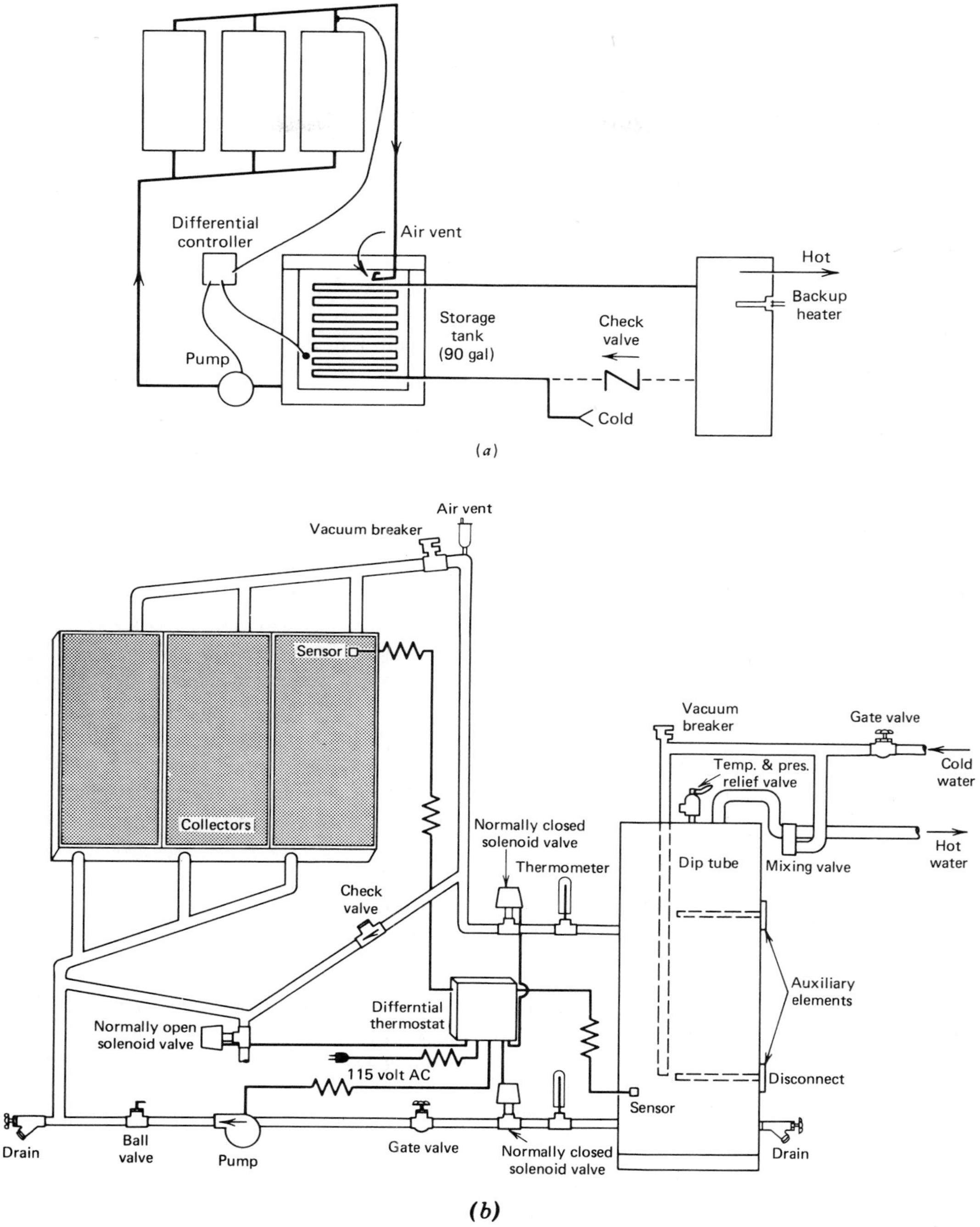

Fig. 9.25 *Comparison of* (a) *drain-back solar water-heating system with* (b) *drain-down system. The* drain-back *system's performance can be increased by the addition of the check valve (dotted lines), this allows the colder water at the bottom of the storage tank to circulate into the heat exchanger. The* drain-down *system is best used in mild-winter climates with infrequent freezing temperatures, since the water in the collector is dumped with each freezing threat. Reprinted by permission from Russell Plante,* Solar Domestic Hot Water, *copyright © 1983 by John Wiley & Sons.*

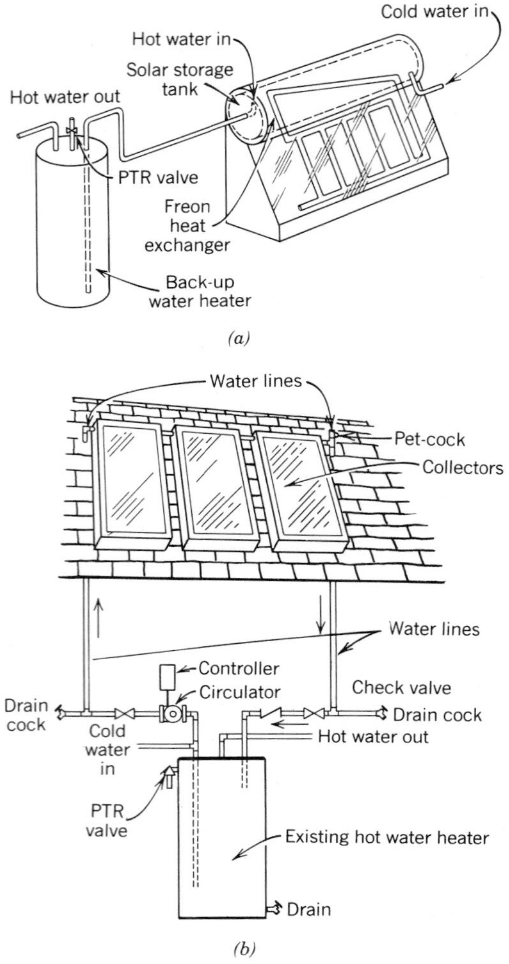

Fig. 9.26 *Various approaches to phase-change solar water heating.* (a) *A smaller solar-heated water tank can be placed at the top of the collector. The circulation of the phase change fluid within the collector is by thermosiphon. The exposed storage can cause problems in cold-winter areas.* (b) *Small heat exchangers can be built into the tops of thermosiphoning collectors; DHW is drained down when the threat of freezing arises. Reprinted by permission. Copyright © by* Solar Age *magazine, November 1983.*

a form of solar-assisted heat pump, closely related to those discussed below.

The *sizing* of solar water heaters usually begins with rules of thumb. For batch heaters, the sizing rule of thumb (Fig. 9.27) is

from 0.45 to 0.65 ft² glazing per gallon of water stored

For the flat plate solar collectors used in all the other solar DHW systems listed above, the rules of thumb are

12 to 25 ft² collector/person (residential)

optimum tilt (up from horizontal) equal to latitude (or less)

1 to 1.5 gal of storage per ft² collector area

This collector-sizing rule of thumb is for *residential* hot water usage, including cooking and bathing. For warmer climates with ample insolation, use the lower figure; this will supply about half of the hot water on an annual average basis. (For *nonresidential* DHW, adjust the rule of thumb by comparing gallons of hot water per person per day used in residences to those used in the nonresidential function for which you are designing—see Section 7.3.)

A more detailed sizing procedure would consider the collector's expected heat contribution for some typical months. Such a procedure requires both weather data and assumptions about water temperatures and other values. As a result, detailed answers are best obtained from computer programs, such as those listed in Appendix H. For those interested in something more than rules of thumb but less than computer programs, however, the following approach to collector sizing (adapted from Brown, Reynolds, and Ubbelohde, *Inside Out: Design Procedures for Passive Environmental Technologies,* copyright © 1982 by John Wiley & Sons) is provided.

STEP 1. *Select the collector tilt angle.* This can be done by consulting Tables B.11–B.14, which present clear-day insolation values for various tilt angles. Appendix B, Table B.15, presents average insolation values on horizontal surfaces, which should also be checked. The tilt angle selected should be close to the optimum angle for the best month for total insolation.

STEP 2. *Check the collector efficiency.* This should be done at least twice: for the best insolation month and for the worst month. This step requires the following data:

Hourly insolation on tilted surface (Appendix B, Table B.11–B.14).

Outdoor average temperature (Appendix B, Table B.15).

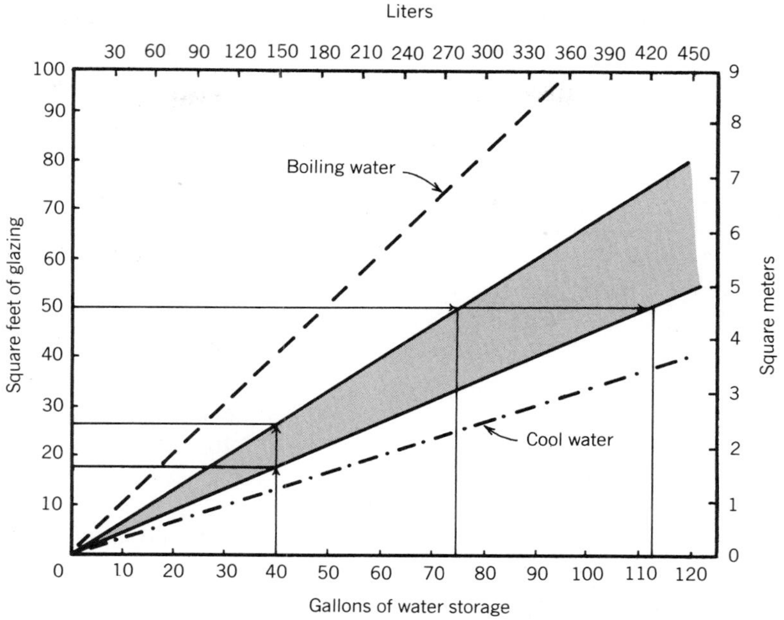

Fig. 9.27 *Rule-of-thumb guide to the sizing of batch solar water heaters. In this example, a 40-gal storage tank should be contained within a collector that has between 18 and 27 ft² of south-facing glazing. Or a 50-ft² collector box should contain a tank of between 75 and 113 gal. Adapted from Daniel K. Reif,* Passive Solar Water Heaters, *copyright © 1983 by Brick House Publishing Co. Reprinted by permission.*

The hourly insolation values on tilted surfaces in Tables B.11–B.14, are for *clear* days. For most locations, this should be adjusted for *average* conditions, which are shown for vertical (south) and horizontal surfaces in Appendix B, Table B.15. Since for most of North America the optimum yearly DHW tilt angle will be closer to horizontal than to vertical, a simple correction to insolation must be made:

$$\text{Hourly average day insolation on tilted surface} = \left[\begin{array}{l}\text{hourly clear-day}\\ \text{insolation on tilted}\\ \text{surface}\end{array}\right]\left[\frac{\text{average-day total horizontal insolation}}{\text{clear-day total horizontal insolation}}\right]$$

This step also requires assumptions about the input temperature of the water supplied to the collector, or T_i. In summer, this can be quite high—even higher than the thermostat set-point temperature of the hot water tank. In winter, it is likely to be lower, by perhaps 10 to 20F°, than the thermostat set-point temperature. Finally, this step requires a choice of collector type, so that efficiency can be determined. Figure 9.28 shows the efficiency curves for a variety of flat plate collectors. Note that the simplest of all, the unglazed flat black collectors, have the highest efficiency of all collectors at combinations of very low $T_i - T_a$ and very high I_o. These conditions are typical of summer days, on which the water is to be heated only a few degrees. The fact that this corresponds to swimming pool applications is one reason why such collectors are so popular for that purpose. Low cost is another reason. For the opposite conditions—those of winter space heating—more-elaborate collectors are appropriate. Evacuated-tube collectors have the advantage of nearly eliminating collector heat loss by convection. The Fresnel lens and/or tracking-concentrating collectors increase the incoming solar energy per unit area, while heat loss increases only slightly (due to higher ΔT). Both these approaches are expensive, and tracking collectors require added maintenance for the additional motion. Selective surfaces represent a simple, relatively cheap improvement over flat black collectors; they absorb solar energy just as well but emit radiant

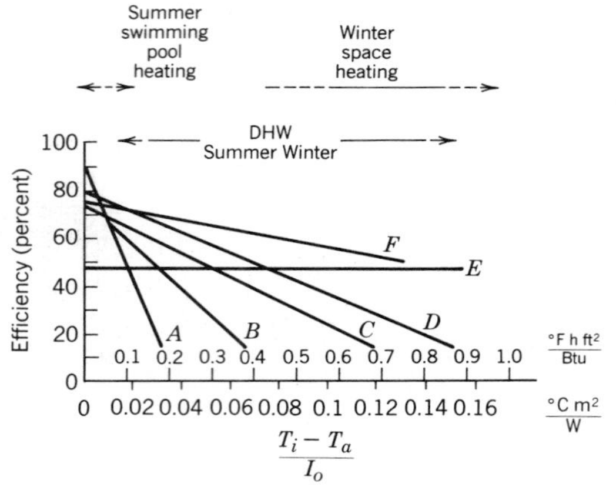

***Fig. 9.28** Collector efficiencies depend on insolation, water temperature, and ambient (air) temperature. Collector types are*

A: *unglazed, flat plate, flat black.*

B: *single-glazed, flat plate, flat black.*

C: *double-glazed, flat plate, flat black.*

D: *single-glazed, flat plate, black chrome selective surface.*

E: *single-glazed, evaporated tube, concentric selective absorber (no rear reflector).*

F: *linear Fresnel lens, tracking-concentrating, black chrome selective surface.*

Based on the U.S. Dept. of Housing and Urban Development's Intermediate Minimum Property Standards Supplement, *1977.*

energy at a vastly lower rate, thus cutting collector heat loss by radiation.

STEP 3. *Approximate system efficiency.* Once the hot water leaves the collector, it must be led back to storage. In indirect systems, it must go through a heat exchanger. The tank will lose some heat; heat losses also occur as the cooled water is led back to the collector. Another ''loss'' to be considered is noncollection time. Estimates of collector performance are based on total daily insolation, yet the first few and last few hours of daylight generally will *not* bring enough insolation to warm the collector. Therefore, some reduction should be made in daily insolation totals when performance is estimated. All these factors can be accounted for by assuming a lower system efficiency. A perhaps optimistic assumption is

$$\text{system efficiency} = 0.8\ (\text{collector efficiency})$$

STEP 4. *Determine the total heat needed for DHW.* Estimates of the quantity of hot water needed were made in Sections 9.4*a* and 9.4*c*. Once the total gallons (or liters) per day are known, and the desired storage temperature determined, the heat needed for the daily supply of DHW is easy to calculate. In conventional units,

$$Q = 8.33\ (\text{gpd})\ (T_s - T_g)$$

where Q = daily heat need, in Btu
8.33 = weight of water, pounds per gallon $\times\ 1^{\text{Btu}}/\text{lb F}$
T_s = storage-temperature, F (see Table 9.7)
T_g = groundwater temperature, F (see Fig. 5.20, well-water temperatures)

In SI units,

$$Q = 1.16\ (\text{L/d})\ (T_s - T_g)$$

where Q = daily heat need, in W-h
1.16 = W-h/L °C
T_s = storage temperature, °C
T_g = groundwater temperature, °C

STEP 5. *Determine the desired percentage of total heat to be solar, for DHW.* There are several approaches to this problem. One common strategy is to provide 100% of the hot water needs in the best insolation month. A closely related strategy is to choose the *yearly* percent solar desired, then provide that percentage in the ''average'' month (such as March). Whatever strategy is chosen, a month and its percent solar must be identified.

STEP 6. *Size the collector.* This is done by combining the preceding steps, as follows.

$$\text{collector area} = \frac{\text{daily heat need} \times \text{percent solar desired}}{\text{daily insolation} \times \text{system efficiency}}$$

This may be done for several months, as a check on optimum collector size.

EXAMPLE 9.4. Determine the approximate size of the solar collectors needed to serve the women's dormitory described in Example 9.3. Assume the location to be Springfield, Illinois.

STEP 1. *Tilt angle.* Refer to Appendix B (Table B.15). Springfield's latitude is 39.8°N. In Table B.15, average insolation is given for July. Use this month as the ''best'' month.

From Table B.12, the optimum tilt angle for July at 40°N latitude is horizontal. However, at 30° (latitude − 10°), much better performance will be obtained in "average" months, such as March and September. Choose a 30° tilt angle.

STEP 2. *Collector efficiency*. For the best month (July), the best hourly clear-day insolation is 307 Btu/h ft^2, at 30° tilt, in the hour centered at noon. To adjust this for *average* hourly value, calculate

$$307 \times \frac{2058 \text{ Btu/day average horizontal (Table B.15)}}{2534 \text{ Btu/day clear horizontal (Table B12)}} = 249 \text{ Btu/h}$$

Outdoor average temperature is approximately obtainable from Table B.15; TA for July is 76 F. This is quite conservative, since it is the average daily temperature; the temperature at noon should be higher. From Appendix A, the mean daily range for Springfield is 21F°. Thus, 76 + 21/2 = about 86 F. The thermostat set-point temperature of the water storage tank is assumed at 115 F. This allows adequately hot water for most dormitory usages, although additional added heat will be required for the cafeteria's dishwashing.

Assume, since it's summer, that T_i is at the set-point of 115 F. (In winter, it would be lower). Therefore, the quantity $(T_i - T_a)/I_o$ can be calculated at

$$\frac{115 \text{ F} - 86 \text{ F}}{249 \text{ Btu/h}} = 0.116$$

Enter Fig. 9.28 with this number, and assume a single-glazed, flat plate selective surface collector (type D). The efficiency will be about 70% at these conditions.

STEP 3. *System efficiency*. Use the rule of thumb

$$0.8 \text{ (70\% collector efficiency)} = 56\%$$

STEP 4. *Total heat needed*. The average usage in gpd can be estimated from Table 9.10. For women's dormitories, the rate is 12.3 gal/student × 300 students = 3690 gal. For the cafeteria, the average rate (type A) is 2.4 gal/meal = 2.4 × 300 = 720. Total gallons needed: 3690 + 720 = 4410.

Groundwater temperature for Springfield (from Fig. 5.20) is about 56 F. Assume that solar energy will be used to raise it to 115 F.

$$Q = 8.33 \text{ (4410 gpd) } (115 - 56) = 2{,}167{,}380 \text{ Btu}$$

STEP 5. *Desired percent solar*. For this best hour in July, assume 100% solar contribution.

STEP 6. *Collector size*. Average daily insolation can be obtained from Appendix B, as before:

clear day total, 30° tilt, July: 2409 (Table B.12)

clear day total, horizontal, July: 2534 (Table B.12)

average day total, horizontal, July: 2058 (Table B.15)

average July daily insolation, at 30° tilt =

$$\frac{2409 \times 2058}{2534} = 1956$$

collector area =

$$\frac{2{,}167{,}380 \text{ Btu} \times 100\% \text{ solar}}{1956 \text{ Btu/ft}^2 \times 56\% \text{ system efficiency}} = 1979 \text{ ft}^2$$

This size should be checked against average-month performance and adjusted as desired. Note that this is a ratio of about 1979/300 = 6.6 ft^2 of collector per student—lower than the rule of thumb for typical residential DHW systems.

Note also that at the rule of thumb rate of 1 to 1.5 gal storage per ft^2 collector, a tank size of 2000 to 3000 gal is indicated. In Example 9.3, the minimum recovery rate needed a storage tank of 8150 gallons, and the faster recovery rate needed 3010 gallons. Either seems sufficient for this size of collector.

Swimming pool heating is an especially attractive application for solar energy. A common rule of thumb for collector sizing is

collector area = 0.5 pool surface area

For summer ambient temperature operations, unglazed collectors are both the best performing and the least expensive, as explained for Fig. 9.28. Another important consideration is a pool cover, which will not only conserve the pool's heat, but also conserve water lost by evaporation. On very hot-dry days, water losses can reach 100 gpd from a 20 × 40–ft pool.

*(e) **Heat Pump Water Heaters.*** The heat pump's use of the compressive refrigeration cycle was explained in Section 6.6 (see Figs. 6.55–6.59). Air-to-water heat pumps are also used for DHW, as shown in Fig. 9.29. Since these devices remove heat from the air, they are usually installed either in normally overheated spaces (such as restaurant kitchens) or in unheated spaces such as garages or basements. The spaces that contain these units will be cooled and dehumidified by them. Heat pumps require only a little more space than that required by the simple hot-water storage tanks they serve. Some noise is created by the compressor and the fan that moves air across the evaporator.

Heat rejected from any refrigeration unit can be used for heating water, via the heat pump. Applications include ice making machines, refrigerated display units in grocery stores, walk-in freezers, and many others. Whenever constant refrigeration loads are present, there is the opportunity to utilize waste heat.

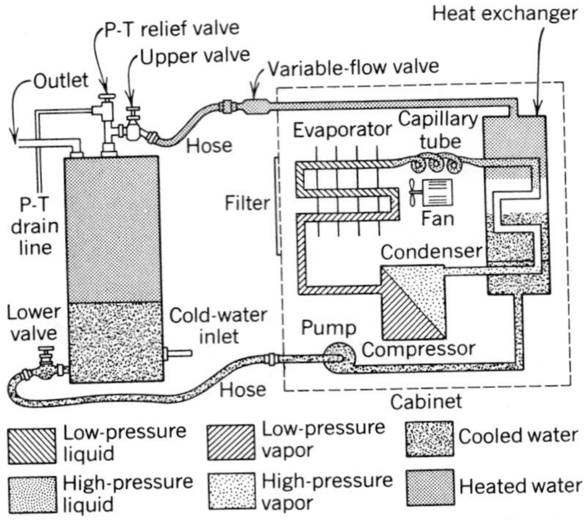

Fig. 9.29 Heat pump (air–water) used for heating DHW. Reprinted by permission from Specifying Engineer *magazine, Oct. 1983. Copyright © 1983 by Cahners Publishing Co.*

9.5 Water Distribution

This section looks at ways to supply water throughout buildings at pressures sufficient to operate plumbing fixtures. Smaller buildings may be served simply by the pressure available in water mains (or pressure tanks fed by pumped wells). This is called upfeed distribution, because the water rises directly from mains to the plumbing fixtures. For taller buildings, two other options are available: pumped upfeed (in which pumps supply the additional pressure needed) and downfeed (in which pumps raise the water to storage tanks at the top of the building, and water then drops down to the plumbing fixtures).

In cities with municipal water supply systems, water is distributed through street mains at pressures varying from about 50 psi at the main to about 70 psi. For low-rise buildings of two or three stories, these pressures are adequate to act against the static pressure of water standing in the vertical piping, overcome the frictional resistance of water flow in the pipes, and still deliver water at the pressure required to operate plumbing fixtures. The flow-pressure required at the fixtures varies from 5 to 20 psi, depending on the type of fixture—for example, the basin faucet, showerhead and faucet, or water closet fixture. (Table 9.14, in Section 9.7, lists the minimum flow rates and pressures for typical plumbing fixtures).

*(a) **Static Pressure.*** The pressure exerted at the bottom of a stationary ''head'' of water is related directly to its height. One cubic foot of water weighs 62.4 lb. Consider a ''cube'' of water 1 ft square and 1 ft high. Its weight (62.4 lb) rests on a bottom area of 1 ft^2 (144 in.^{2}). The *static* pressure at the bottom is 62.4/144 = 0.433 psi. Reciprocally, 1 psi of pressure will *sustain* a static (stationary) column of water 1/0.433 = 2.3 ft in height. When fixture pressure and pressure lost in friction-of-flow in pipes are considered, the problem becomes more complex. Example 9.6 in Section 9.7 illustrates this problem in the calculation of pipe size. For upfeed and downfeed distribution, the relationship of heights and static pressure is one controlling factor.

*(b) **Upfeed Distribution.*** In small, low buildings of moderate water demand, it is seldom difficult to achieve the proper flow-pressure at fixtures by the use of an upfeed system. Pressure at the fixtures is usually more than required. When

this causes an inconvenient splash, as at lavatory basins, a flow restrictor can be used in the faucet outlet.

Various parts of the typical upfeed system shown in Fig. 9.30 will now be considered, beginning at the point where supply water enters the building. In cold climates, water in the service entry pipe (or tube) must not freeze. The pipe must therefore be below the *frost line* of frozen ground. This could vary from 0 to 7 ft, depending on geographical location. The onset of winter in such climates requires the closing and draining of pipes supplying the hose bibbs (and other external piping) by means of a *stop-and-waste* valve. Houses left *unheated* in cold-winter weather must be entirely free of water that could freeze and burst the pipes. Note the drain valve at every low point in the system. House shut-off controls are usually located at the main, at the curb, and within the house.

Meters have recently taken on a new role: along with measuring the water quantity for which the occupant is to be charged, they now serve a restrictive function. During water shortages, they can be used to indicate water use in excess of established limits, beyond which limit fines are imposed and in some cases water supply reduced by valves controlled by the water company.

Treatment is most often performed to reduce hardness that could clog piping and equipment, or to neutralize acidity—a source of corrosion. During the short periods when the treatment tanks are valved off for backwashing or other servicing, the *bypass shut-off valve* is opened.

From this point on, the water continues under pressure to

1. Supply make-up water to the house-heating boiler, as required.
2. Supply water to and pressurize the cold water mains and branches, including the garden hose bibbs.
3. Supply water to and pressurize the domestic

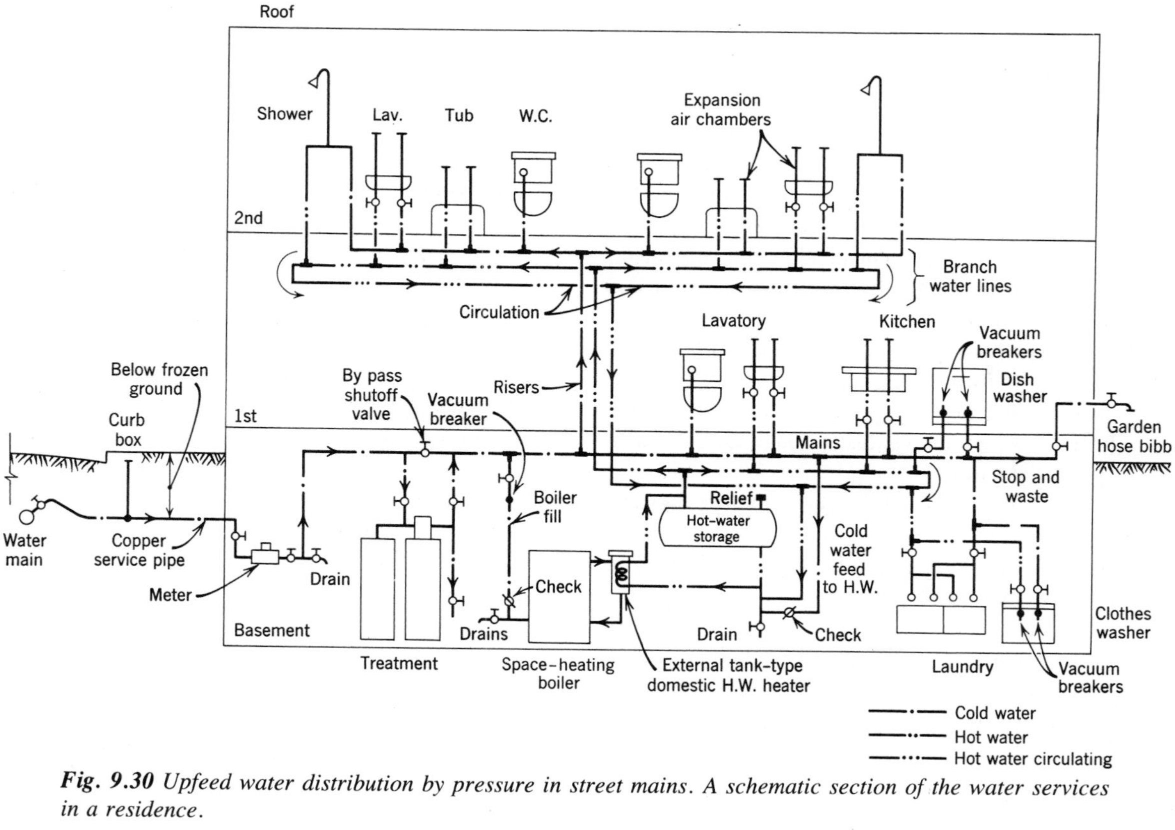

Fig. 9.30 *Upfeed water distribution by pressure in street mains. A schematic section of the water services in a residence.*

hot water system through the hot water heater, the hot water storage tank, and the mains, branches, and circulating lines.

The air-filled expansion chambers on cold water *runouts* absorb and reduce the shock of so-called "water hammer" when faucets are shut off. On hot water runouts they perform the same function, *plus* they allow for the expansion of the hot water as it increases in volume with increasing temperature. Vacuum breakers prevent back-flow of polluted water into pipes carrying potable (hot *or* cold) water. Water from all fixtures and appliances, such as dishwashers, clothes washers, and boilers, is thus isolated.

In the legend for Fig. 9.30, the word *schematic* is significant. The entire diagram shows all parts lying in a two-dimensional *plane*. Obviously, in a real house a three-dimensional system would exist. For instance, economy of piping would suggest that, if possible, kitchen and lavatory, as well as the two upstairs bathrooms, be placed back-to-back.

(c) Principles of Downfeed Distribution. Water pumped directly from the street main (or from a basement "suction tank" filled by gravity from the main) is lifted to a roof-storage tank. In cold climates, the tank water in such exposed locations is kept at temperatures above freezing by heating coils in the tank. The fact that water pressure increases with distance below the tank water level is clearly shown by the construction of the tank (see Fig. 9.31). The iron rings, tensioned by adjustable threaded clamps, become more closely spaced toward the bottom of the tank, where the greater water pressure makes it increasingly difficult to restrain the vertical wooden staves of this cylindrical barrel.

Tanks like these add interest, though perhaps not beauty, to a building's silhouette. A "fence" enclosing this tank would have to be about 24 ft high. Architecturally, this rectangular lump on the roof would not be much of an improvement. Yet in the century following the appearance of such structures as the Goodwill Building, our technology has become more complex. Presently, for most high-rise buildings, including many 60 stories and more in height, an entire rooftop crowded with equipment and technical facilities is needed to serve

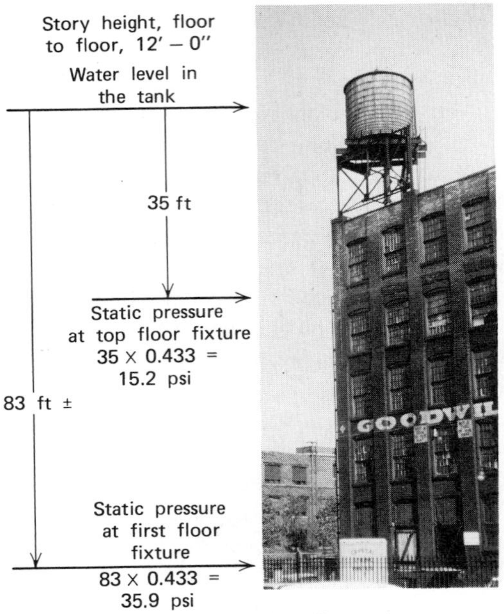

Fig. 9.31 *High-rise building of the 1870s, with downfeed distribution. Although upfeed distribution from a street main had been attempted in buildings higher than two or three stories, it was evident then (as it is now) that street main pressures that may drop below the usual 50 psi, together with heavy use, would result in very low pressure at fixtures in upper stories. Therefore, water is pumped from the main to elevated wooden roof tanks high enough to assure reasonable pressure at the top story and ample pressure at the bottom of the downfeed run.*

the stories below, or the uppermost *zone*. The items could include

- Water storage tanks.
- Two-story penthouses over elevator banks.
- Chimneys.
- Numerous plumbing vents.
- Exhaust blowers.
- Air conditioning cooling towers.
- Cantilevered rolling rig to support scaffold for exterior window-washing.
- Perimeter track for window washing rig.

Many of these units can be seen in the photographs of the roof of the Chase Manhattan Building shown in Fig. 9.32.

Fig. 9.32 (a) *High-rise building of the 1970s, with downfeed distribution. Uniquely, the elevated wooden outdoor water storage tank has not changed very much in 100 years (see Fig. 9.31). Its function is the same, and the tank's water level remains several stories above the top floor plumbing fixtures. As part of what might be called "Rooftop City," the tank has joined the contemporary equipment complex, surrounded by the visual barrier of a louvered "skirt."* (b) *The Chase Manhattan Building in New York City. Skidmore, Owings and Merrill, Architects. A helicopter view. Photo by Sky Service.*

Thus, since the 1960s, a tall building commonly has a *band* or *screen* two stories (or more) high above the structural roof. It might be said that it all began with the need for elevated water-storage tanks.

(d) Tall Building Downfeed Distribution. Figure 9.33*a* shows a medium-rise building, in which one elevated tank can serve all the lower floors. For taller buildings, it is advisable to separate groups of floors into *zones* with a maximum height (for plumbing pressure limits) of about 150 ft. This zone-height limitation is based on the height-to-static pressure relationship. At the top of the zone (about 35 ft below the storage tank), the minimum desirable pressure is probably at least 15 psi. At the bottom of the zone, the maximum desirable pressure is perhaps 80 psi; above this pressure, damage to fixtures might occur.

$$80 \text{ psi} - 15 \text{ psi} = 65 \text{ psi difference}$$

$$65 \text{ psi} \times 2.3 \text{ ft/psi} = \text{about } 150 \text{ ft}$$

With pressure-reducing valves at lower floors, however, these zones can be much higher, as shown in Fig. 9.33*b*.

Consider the system described in Fig. 9.33*a*, beginning with the elevated tank. The lower part of the tank often serves as a reserve space to hold a supply of water for a fire hose system. In this case, only the water in the upper part is available for use as domestic (service) water. The amount stored must be enough to supplement what the pump will deliver during the several daily hours of high demand that occur in most buildings. The pump then continues, often for several hours, to replenish the house supply that had become partially depleted during the busy period. The suction tank is a buffer between the system and the street mains. It usually holds enough reserve to allow the pumps to make up the periodic depletion in the house tank. It refills automatically by gravity flow from the street main that, consequently, will not suffer as much of a drop in pressure as it would if it were connected directly to the suction side of

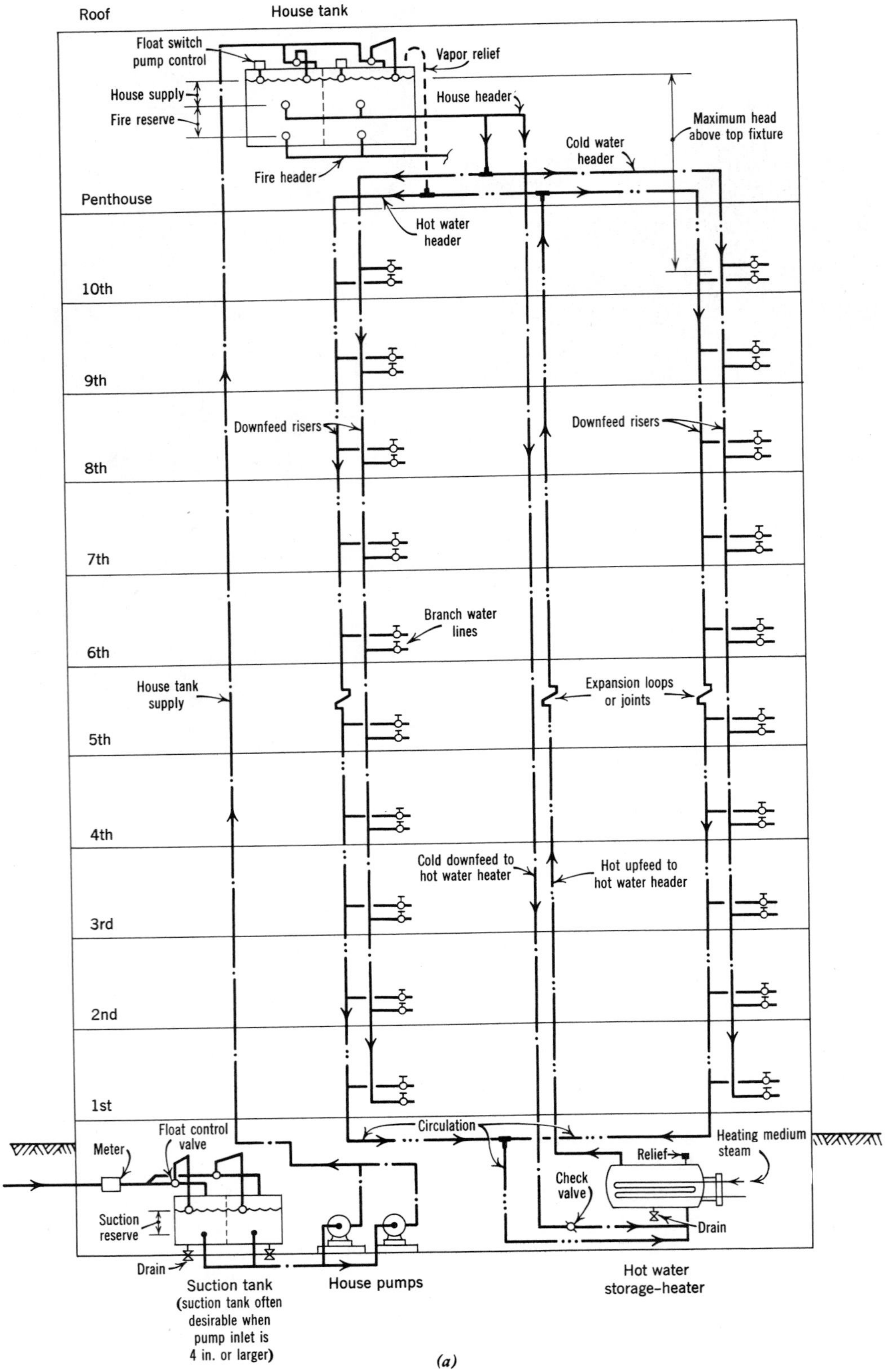

(a)

Fig. 9.33 (a) *Downfeed water distribution, a schematic section. Part of the water services for a 10-story building. Hot water circulation moves through the hot upfeed in two directions at the hot water header and*

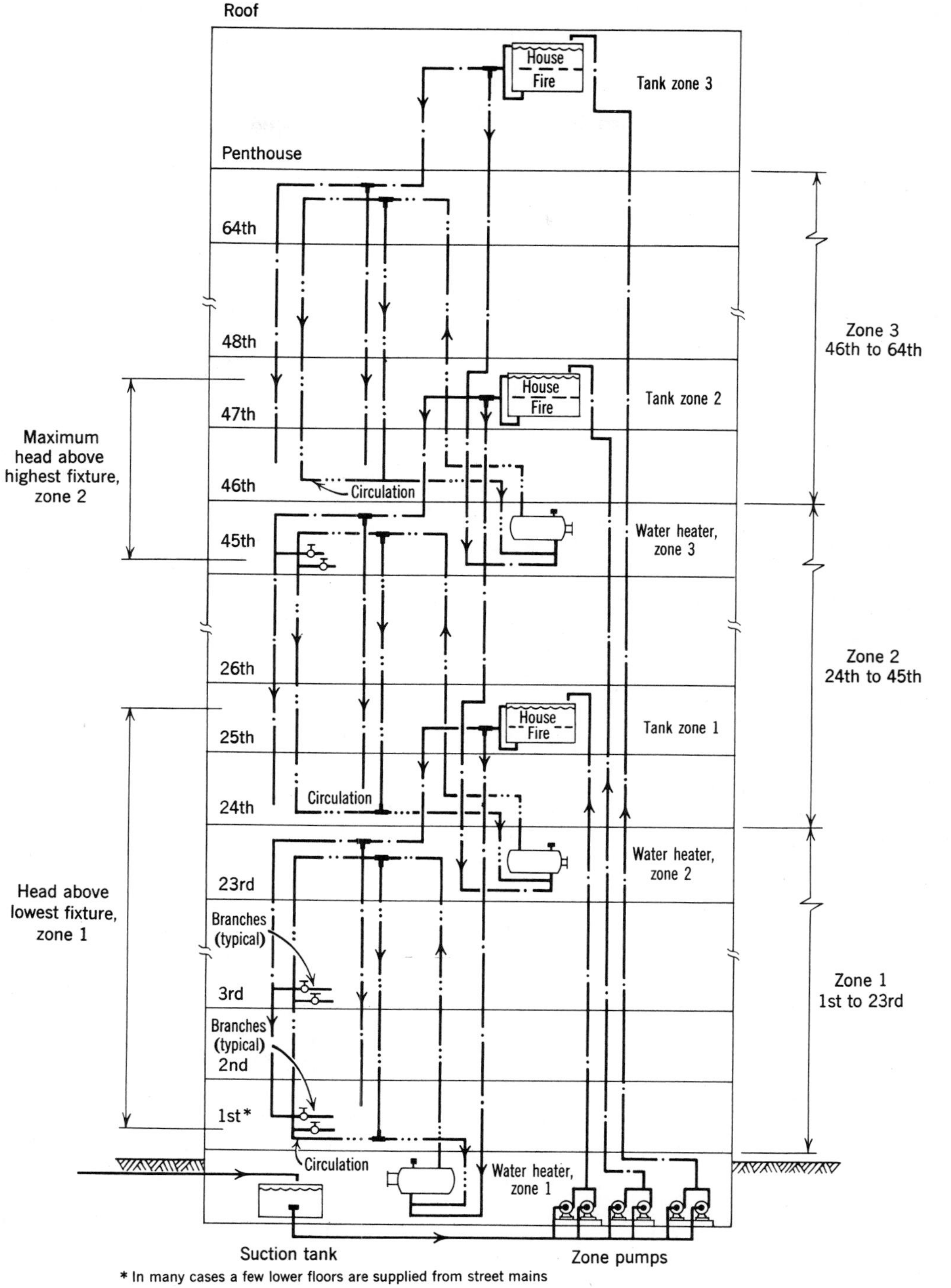

(b)

down to the tank through the two downfeed hot water risers. For details of one type of steel house tank, a typical centrifugal house pump, expansion joints, and expansion in hot water riser, see Figure 9.34. (b) *Downfeed water distribution, a schematic section. Part of the water services for a zoned building. Zone tanks include a fire reserve but standpipes are omitted from* this *drawing. For detail of steam-type domestic water heater, see Figure 9.19.*

the house pumps. Neighboring water users are protected from the adverse effects of sudden demands within adjacent large buildings. See Fig. 9.34.

House tanks and suction tanks are sometimes made of steel plate and divided vertically in half, each half having identical piping and controls. Hence, one half of the tank can be cleaned out at a time during hours of low demand without shutting down the entire system. One full-capacity pump is supplemented by an equivalent standby pump for alternate use. Since there is no suction lift below the pump or any fixture pressure at the top of the delivery leg (house tank supply), the head against which the pump works is the sum of the distance from the suction tank water level to the top of the house tank and the feet of head equivalent to the friction loss in the tank supply pipe. For this kind of service the vertical piping is on the order of 3 or 4 in. in diameter for large buildings. Sizes are established by a formal engineering design.

The house supply is fed by a short pipe from the house header to the cold water header that circles the top story and connects to many downfeed cold water risers. For simplicity, Fig. 9.33*a*

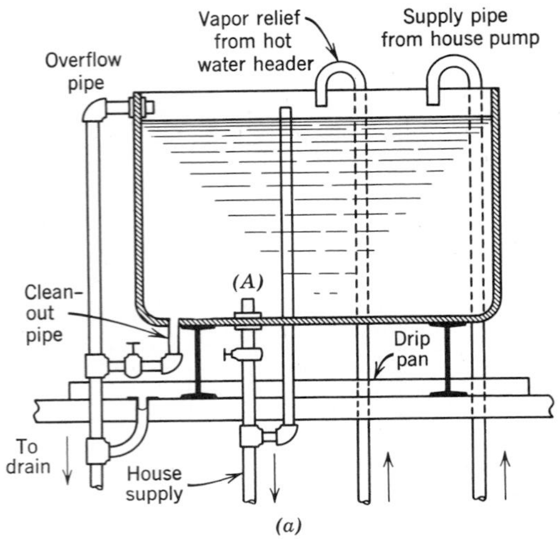

Fig. 9.34 (a) *House tank in elevated position for downfeed by gravity. Sediment in tank is drawn off through clean-out pipe and is prevented from entering house supply by pipe projection at* A. *Water for fire reserve could be provided by additional piping or by a separate tank.* (b) *Centrifugal house pump, commonly used to fill elevated house tank in top story or at intermediate mechanical floor. Electric motor responds to float switch pump control at the tank.*

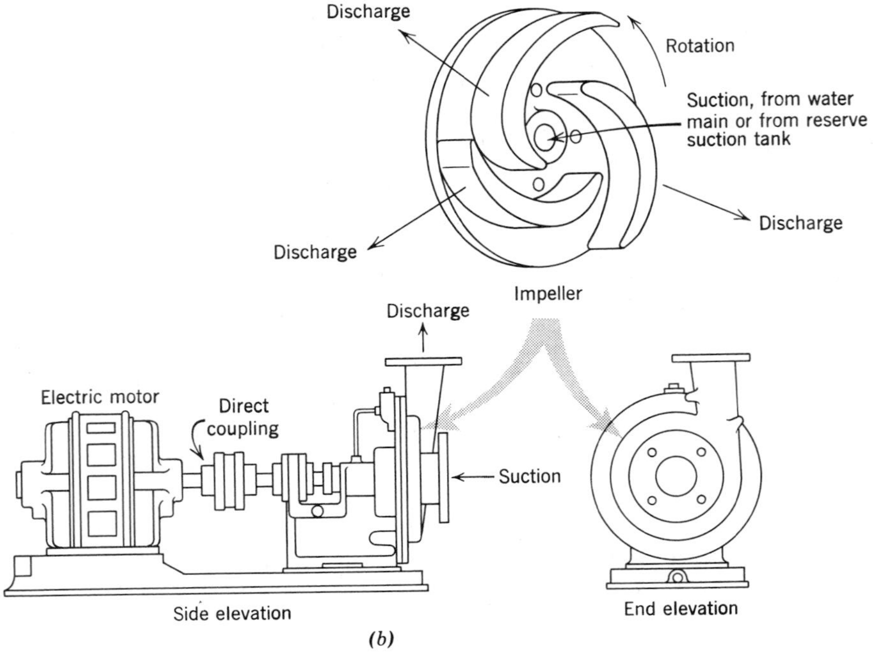

shows only two risers and also omits many valves and controls. Figure 9.33*b* is even more simplified. The hot water circuiting originates as cold water at the house tank header and takes quite a long route. Descending to the bottom of the hot water heater, it rises to seek its own level at the hot water header, becoming available there for hot water downfeed on demand. All of this occurs as flow below the general pressurizing effect of the house tank. In effect, when there is a cold- and hot-water demand on a story near the top of the building, the cold water makes a short trip down to the faucet, while the hot water goes through three vertical pipes instead of one.

The general scheme just described, with tank above and heater below, is used in multiple forms for very tall buildings. The zones are quite independent; the only common service is that provided by the general suction tank. By this zoning method, problems of pipe expansion, excessive pipe sizes, and high pressures in lower stories are minimized. Commonly, 2½ stories, or about 35 ft, comprise the minimum pressure head above the top fixture served by any zone tank. The static pressure created at the fixture is thus $35 \times 0.433 = 15$ psi. If, during flow, not too much pressure is lost in friction, flushometers (flush valves) can be placed at this level, though flush tanks, because of their lower fixture-pressure demand, most often be accepted. Quite the opposite problem occurs at the bottom of the zones, where excessive pressures must be reduced at the fixtures. In zone I of Fig. 9.33*b*, first-floor fixtures are below a head of 24½ stories, or about 149 psi of static pressure. It is obvious that pressure-reducing valves must be used, and that fixture control valves must be throttled.

Pipe and Tube Expansion. The range of temperature from the normal indoor air temperature of about 70 F to that of service hot water (which often exceeds 160 F) is an index of the expansion of pipes and water as their temperatures rise from shut-down status (70 F) to operating status (160 F). (See Table 9.12.) The longitudinal elongation of pipe, though negligibly small in houses, can be appreciable in a tall building. Two methods of allowing freedom for this longitudinal motion in long runs of expanding hot water piping are shown in Fig. 9.35*a*. The use of these devices precludes the buildup of excessive stresses in the metal of the pipes and the tendency of the pipes to buckle laterally.

TABLE 9.12 **Thermal Expansion of Pipe and Tubing**

Elongation in in./100 ft of pipe or tube for various increases in temperature.

Increase in Temperature (F)	*Steel Pipe*	*Copper Tubing*
20	0.149	0.222
40	0.299	0.444
60	0.449	0.668
80	0.601	0.893
100	0.755	1.119
120	0.909	1.346
140	1.066	1.575
160	1.224	1.805
180	1.384	2.035
200	1.545	2.268

NOTE: Special care must be taken to adapt for the expansion of *plastic* pipe, the elongation of which is about five times that of copper tubing under the same conditions.

EXAMPLE 9.5. A 20-story zone in a tall building has a height of 280 ft. What will be the increase in length of a copper tube carrying "service hot water" (domestic hot water), when its temperature increases from 70 to 160 F?

SOLUTION. The difference in temperature is 90F°.

Elongation per 100 ft for 90F° increase is 1.01 in. (Refer to Table 9.12). Thus,

$$2.8 \times 1.01 = 2.82 \text{ in.}$$

There are a number of ways of providing for this expansion. The one shown in Fig. 9.35*b* consists of accepting the motion at two locations, which would make the expansion in each case 1.41 in. Equidistant anchorage to fix the tubing is provided at the bottom, the 10th floor, and the 20th floor. The support of the vertical riser at floors other than those at which it is anchored could consist of clamps of the type illustrated in Fig. 9.40, supported on springs.

(e) Pumped Upfeed Distribution. This distribution system (see Fig. 9.36) is for medium-

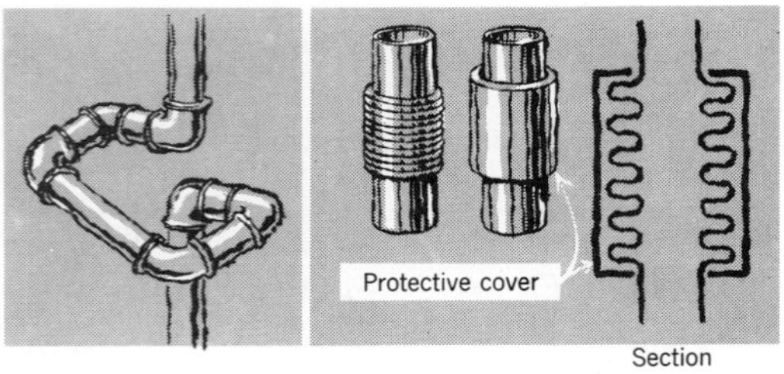

Fig. 9.35 (a) *Accommodation for the expansion of hot water piping or tubing. Left: Expansion joint of pipe and fittings. Right: A manufactured product.* (b) *(Example 9.5) Suggested scheme for location of the points of anchorage and expansion for service hot water tubing in a 20-story zone.*

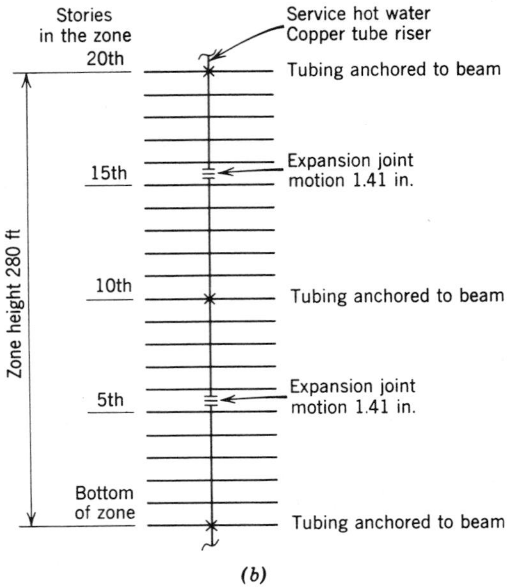

size buildings—those too tall to rely on street main pressure but low enough to do without heavy storage tanks on the roof. This sophisticated equipment can deliver water at rates varying from what is needed for two or three faucets to full building demand, while maintaining at each outlet a pressure within 2 psi of the design pressure for that outlet.

The installation shown in Fig. 9.36*a* and *b* is a triplex pump group. According to demand, one, two, or three pumps will operate. Since each pump is of the variable-speed type, virtually an infinite

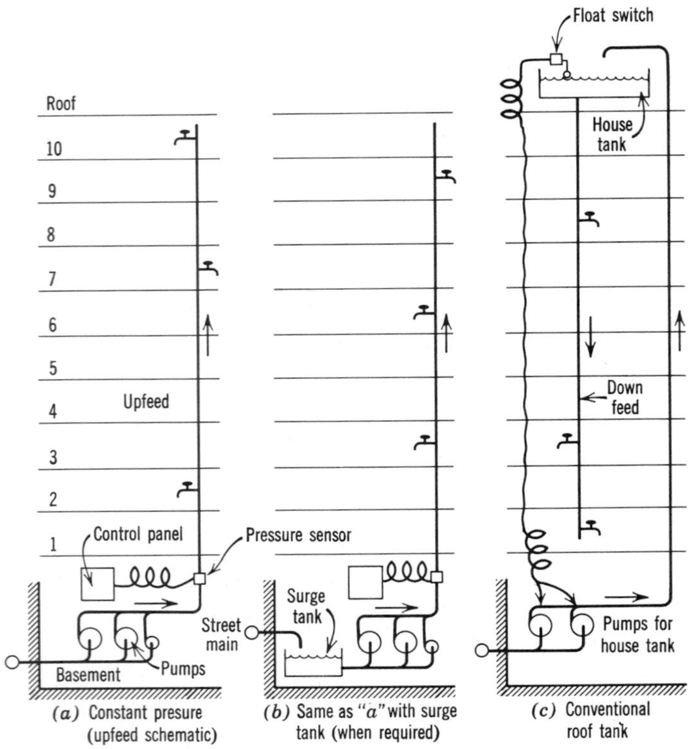

Fig. 9.36 Constant pressure upfeed pumping (a) *and* (b), *compared to gravity downfeed from house tank on roof* (c). *By permission of* Progressive Architecture.

number of delivery rates can be achieved within the zero to maximum design rate. The pumps operate in sequence. When a very small rate of demand occurs, the smallest or "jockey" pump starts in response to a low voltage impressed on its motor. This and all other operations are triggered and adjusted by the pressure sensor at the base of the riser. The jockey pump continues to run until it has reached its maximum delivery rate, at which time the first of the larger pumps cuts in, joined by the third when required. Sequential operation of the three pumps, each increasing in delivery as called for by the sensor, meets the requirement for increasing supply at nearly constant pressure.

It would appear that the second of the two larger pumps would run less frequently and thus get the least wear. However, wear is equalized by the assigning "lead" and "lag" positions. For a period of 24 hours, one of the large pumps holds the lead position and starts after the jockey pump, giving the other large pump a smaller burden. The next day the rested pump takes over the more active assignment. All of this occurs automatically.

At full operation this triplex unit can put a suction demand on the street main that could seriously reduce the available water pressure in the neighborhood (Fig. 9.36*a*). Therefore, city officials sometimes require that the system feed on a surge tank, filled by casual flow from the street main, independent of the building requirements (Fig. 9.36*b*). This requirement is often made when the design indicates a maximum building demand in excess of 400 gpm.

Obvious advantages of upfeed pumping are the elimination of the house tank and the heavy structure that transmits its weight down to the footings and, of course, the elimination of the necessary periodic cleaning of the tank. A shortcoming is the lack of reserve storage, which could cause a serious problem during an electrical power failure. However, minimum flow during this kind of emergency can be arranged if a diesel or other independent standby motor is available to drive one of the pumps.

9.6 Piping, Tubing, Fittings, and Controls

(a) Piping, Tubing, and Fittings. The conveyance of water through buildings to locations of use implies the design of a system of piping or tubing that efficiently fulfills its purpose, is easily maintained, and interferes as little as possible with the interior architectural form. It may be assumed that except in basements, in utility rooms, and at points of access to controls, the system will be concealed. Stud-and-joint construction provides space for concealment, but in fireproof buildings, vertical and horizontal furred spaces must often be provided.

The corrosive effects of water and the resistance of metals to corrosion are usually matters for chemists and metallurgists to deal with. In general, however, public or private treatment should be provided to correct corrosive qualities. When this is done, it is theoretically suitable to use a cheaper piping material—steel. Yet prudence suggests that a better material be selected. In the nonferrous group, red brass and copper tubing are effective in corrosion resistance. Copper tubing is a very popular choice. It is less expensive than brass, assembles more easily, and is not subject to dezincification (attack by acids on the zinc in brass). For use in handling aggressive waters, plastic is often a good choice. Like copper, it is light in weight and assembles with great ease.

For ferrous pipes and "iron pipe size" brass, threaded connections are used. The external, tapered thread on the pipe is covered with pipe compound and screws tight against the internal tapered thread of the coupling or other fitting. The solder-joint connection in copper depends on capillary attraction that draws the solder into a cylinder of clearance between the mating surfaces of tube and fitting. This occurs after the surfaces are polished and fluxed and the parts placed in final position. They are then heated, and molten solder is applied to the circular opening where the fitting-edge surrounds the tube, with a small clearance. It is then drawn into the cylindrical connection. Solders are tin-lead or tin-antimony alloys. This kind of joint permits the advantageous setting up of an entire tubing assembly without turning the parts (as in threaded installations), and before the soldering commences. For the same strength, copper tubing may have thinner walls because no threads need to be cut into it. Its smooth interior surface offers less friction to flowing water. While threaded- and solder-joint connections are the most common in small work, there are many other types. Ferrous pipes in the larger sizes are often welded or connected by bolted flanges (see Figs. 9.37 and 9.38 and Table 9.13).

(b) Plastic Pipe. Most of the plastic pipes and fittings now produced are synthetic resins derived from such materials as coal and petroleum.

Rapid increase in the development, acceptance, and use of plastics for water piping, fittings, and, indeed, drainage systems (see Chapter 10) suggests a separate discussion of this *family* of materials.

1. *Selection of Material.* The chemistry of plastics is quite intricate. The material can appear in a great variety of forms, a few of which, especially suitable for water piping, are listed in Table 9.13. The specifier is well advised to seek information from official and established sources.

2. *Sanitation.* Early concern about the possibility of adding odor, taste, or toxicity to potable water has long been dispelled. The National Sanitation Foundation tests and certifies plastic pipe. Their NSF seal must appear on pipes that are to carry potable water.

3. *History.* Almost 100 years ago the first plastic material, celluloid, appeared. Another product, Bakelite, was developed in 1905. Used for water piping, polyethylene has given successful service for more than 30 years. Intensive research and standardization beginning about 1945 has established many plastic materials as appropriate for plumbing.

4. *Code Acceptance.* The plumbing industry is closely governed by regulations referred to as *codes.* These codes usually govern which materials are acceptable to the local jurisdiction. Codes written by associations of plumbing officials or plumbing contractors are called *model codes.* The following model codes approve the use of plastics piping for all or a portion of the plumbing system.

- BOCA Basic Plumbing Code (Building Officials and Code Administrators International).
- National Standard Plumbing Code (National Association of Plumbing, Heating, Cooling Contractors).
- Southern Standard Plumbing Code (Southern Building Code Congress).

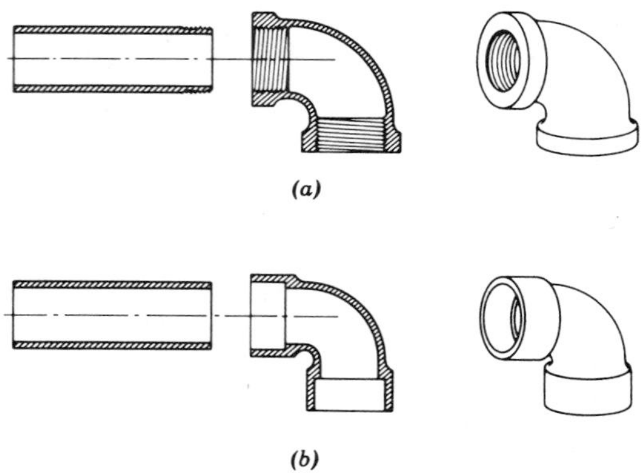

Fig. 9.37 Methods of connecting pipes and fittings, and tubes and fittings. (a) *Threaded: for ferrous pipe and fittings and for "iron pipe size" (IPS) brass.* (**b**) *Soldered: for copper tubing and fittings. A sliding fit similar to that of* (b) *is used for the solvent weld of plastic connections.*

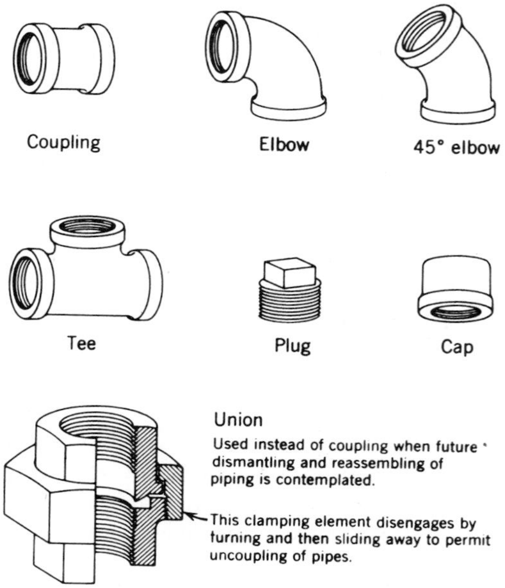

Fig. 9.38 Examples of threaded *pipe fittings for ferrous or brass pipe. A few of the many fittings used in water piping. These and all common fittings are also available for solder-joint (copper) or solvent-weld (plastic) connections, and usually for transition from one system or material to another.*

TABLE 9.13 **Pipe and Tubing for Water Services**

Part A. ***Characteristics of Materials***

Kind of Pipe	*Material or Manufacture*	*Connections*	*Qualities*	*Notes*
Steel	Butt welded to 2 in. diameter, seamless, large sizes	Threaded	Basic	Should be used only when water is not corrosive
Brass, red	85% copper 15% zinc	Threaded, "IPS" (iron pipe size)	Corrosion-resistant	Bulky, because of the need for threading
Copper tube, type "K"	Seamless, hard or soft temper	Soldered fittings	Corrosion-resistant and easy to fabricate	Thinner-walled than brass; easy to put together and dismantle
Copper tube, type "L"	Seamless, thinner walls than type "K," hard or soft temper	Soldered fittings	Corrosion-resistant and easy to fabricate	Thinner-walled than brass; easy to put together and dismantle
Plastic[a]	See Part B	Solvent cement weld[b]	Very easy to fabricate	Not subject to electrolytic corrosion
Nickel, silver and chrome	Copper, nickel, and zinc, steel, and chromium	Threaded	Corrosion-resistant	Special applications
Galvanized steel	Zinc-coated steel	Threaded	Moderately corrosion-resistant	Suitable for mildly acid waters

Part B. ***Plastic Piping***

Symbol	*Material*	*Cold Water*	*Hot Water*
PE	Polyethylene	√	
ABS	Acrilylonitrile-Butadiene Styrene	√	
PVC	Polyvinyl Chloride	√	
PVDC	Polyvinyl Dichloride	√	√[c]

Source: Courtesy of the Plastics Pipe Institute.

[a]Upper limit of temperature, hot water, 180 F.

[b]For ABS and PVC (Part B).

[c]Developed for this special use. Most plastic materials not currently approved for hot water piping.

- Uniform Plumbing Code (International Association of Plumbing and Mechanical Officials).

5. *Effects of Temperature and Vibration*. Most of the materials used for piping are thermoplastics and have the quality of repeated softening under the application of heat. PVDC material can carry water at 180 F, but plastic pipe should not be subjected to temperatures higher than this. Expansion (see note to Table 9.12) is great and affects the piping design. Quite shockproof, plastic is used in the plumbing systems of about 80% of mobile homes.

6. *Information Source*. The Plastics Pipe Institute, a division of the Society of the Plastics Industry, will furnish objective information about plastic pipe.

(c) Valves and Controls. A good system utilizes many valves. It is usually desirable to valve every riser, the branches that serve bathrooms or kitchens, and the runouts to individual fixtures. This setup facilitates repairs at any location with a minimum of shutdown within the system. Treatment equipment will have a valved bypass (see Fig. 9.30). Pumps and other devices that may need repair should be disconnectable by unions (Fig. 9.38) after valves are closed.

The gate valve (Fig. 9.39*a*), with its retractable leaf machined to seal tightly against two sloping metal surfaces when closed, offers the least resistance to water flow when open. It is usually chosen for locations where it is left completely open most of the time. The compression-type globe valve (Fig. 9.39*b*) is usually used for the closing or throttling of flow near a point of occasional use. Faucets are usually of the compression type, as are drain valves or hose connections. They are similar to the angle valve (Fig. 9.39*d*). When it is necessary to prevent flow in a direction opposite to that which is planned, a check valve (Fig. 9.39*c*) is introduced. The hinged leaf swings to permit flow in the direction of the arrow but closes against flow in the other direction.

(d) Pipe Supports. If a piping system of conventional dimensions were to stand alone, without a building to rest on, it would quickly collapse. Quite heavy because of its metallic nature and water content, it needs closely spaced supports (Fig. 9.40). Vertical runs of 1-in. piping should be supported at every story, but larger sizes may extend for two stories. Horizontal pipes should be supported at intervals of 10 ft. Closer spacing (6 to 8 ft) is preferred for sizes of ½ in. and smaller. Horizontal copper tubing should always be supported at closer spacing then steel. Adequate positioning of horizontal runs is important to assure correct pitch and drainage. Hangers are adjustable for this purpose.

(e) Shock and Hot Water Expansion. Water systems can be noisy. When faucets are shut off abruptly, the force exerted by the decelerated

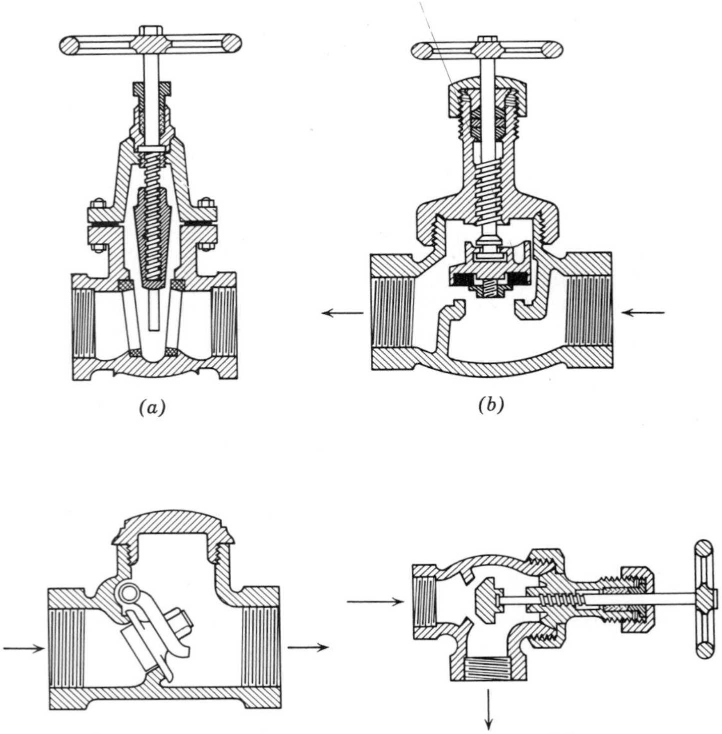

(*a*) (*b*) (*c*) (*d*)

Fig. 9.39 *Typical valves for water systems.* (a) *Gate valve.* (b) *Globe valve.* (c) *Check valve.* (d) *Angle valve.*

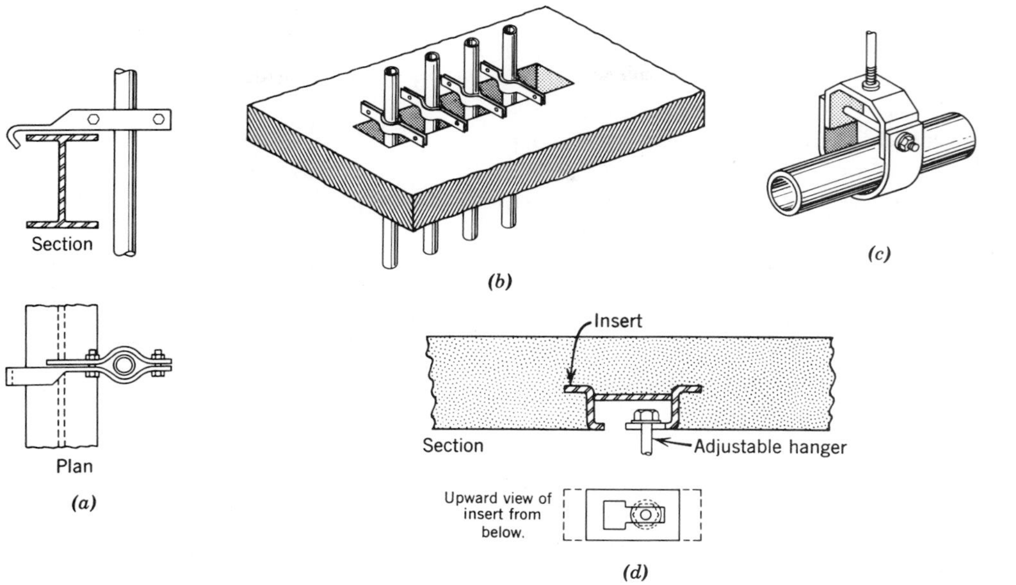

Fig. 9.40 *Pipe supports.* (a) *Vertical riser supported at steel beam.* (b) *Vertical riser group supported at slot in concrete slab.* (c) *Horizontal pipe hung from slab above by adjustable-length clevis hanger.* (d) *Typical metal insert in soffit of concrete to receive hanger rod.*

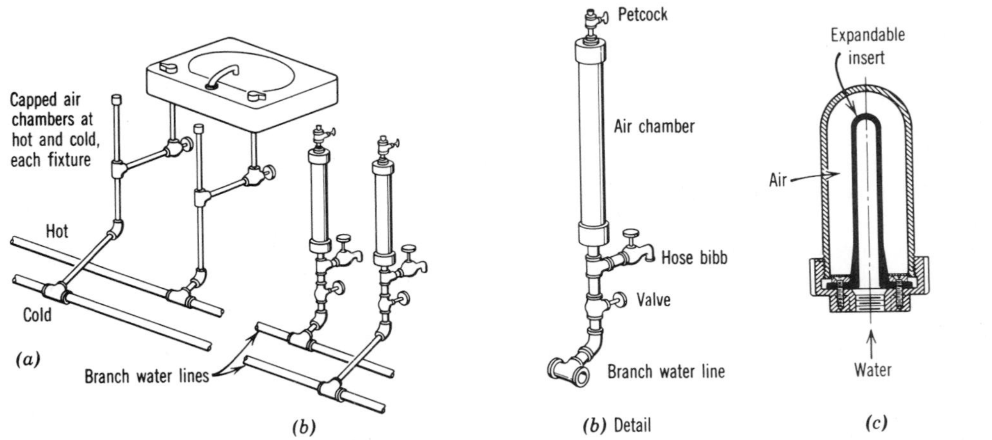

Fig. 9.41 *Shock relief and expansion chambers. Air chambers cushion the shock of the water hammer when the fixture faucets are shut off abruptly. They also permit hot water to expand, instead of periodically forcing open the hot-water-emergency pressure relief valve at the heater or tank.* (a) *Capped air chambers at each supply pipe of each fixture.* (b) *Rechargeable air chambers on hot and cold branch water lines. (Individual fixture chambers are omitted when these are used.)* (c) *Special shock absorber.*

flowing water shakes and rattles the pipes. This phenomenon is called water hammer. Lengths of vertical pipe about 2 ft long at the fixture branches (Fig. 9.41*a*) will usually solve this problem. They trap a certain amount of air, which absorbs the impact of the water with some resilience. A somewhat better and a more controllable device is a "rechargeable air chamber" (Fig. 9.41*b*). By closing the valve and draining the water through the hose bibb while the petcock above is open to admit air, the chamber may be refilled with air. Closing the petcock and hose bibb and opening the valve completes the service operation and reconnects the device with the water system. Rechargeable chambers are used on branch lines adjacent to groups of fixtures. Access for service must be provided. When this method is chosen, the smaller pipe extensions (Fig. 9.41*a*) are usually omitted. Perhaps the best method is the use of a special shock absorber (Fig. 9.41*c*).

Air cushions also protect the relief valve against frequent operation, with the resultant leakage of hot water, as the hot water periodically expands and contracts in closed systems.

(f) Condensation, or "Sweating." The moisture that is always present in air often condenses on the exterior surface of cold pipes. Dropping off the pipes, it creates an unpleasantly wet condition and disfigures finished surfaces. Groundwater in some parts of the United States is 50 F and colder. A pipe carrying such water might have a surface temperature of about 60 F. The psychrometric chart (page 144) indicates that at a summer air temperature of 85 F, condensation will occur on this pipe when the relative humidity exceeds 40%. All cold water piping and fittings should be covered. Glass fiber ½ or 1 in. thick is commonly chosen for this purpose. A tight vapor barrier on the exterior surface of the covering prevents the moisture-laden air from penetrating the insulation to reach the colder surface. The insulation provides another advantage of equal importance: it retards heat flow from the warmer air to the water, thus preventing the water from becoming disagreeably warm.

(g) Heat Conservation. Pipes carrying domestic hot water should be insulated to conserve the fuel used to heat the water and to assure a correct water temperature at the point of use. Minimal covering is a ½-in. thick glass fiber. Parallel hot and cold water piping, even though insulated, should be separated by 6 in. or more, to prevent heat interchange.

Storage tanks and water heaters are usually manufactured with integral insulation. Older devices, however, may have less insulation than today's energy prices warrant. As a result, many of the older water heaters are retrofitted with added insulation, as shown in Fig. 9.42. Some electric utilities find the savings from conservation so attractive (relative to the cost of building new generating facilities) that they provide water-heater wrapping as a free service to customers.

Fig. 9.42 *Retrofitting insulation on existing, minimally-insulated water-heater storage tanks.*

(h) Irrigation. The highest priority for landscape watering is to design for optimum rainfall retention (see Figs. 8.4 through 8.10). In the past, provision for the watering of the landscape around buildings has frequently been limited to the installation of sill cocks (hose bibbs) at building exteriors. In many areas of North America, half the residential water usage is for outdoor purposes. With increasing demands on a finite water supply, water-conserving irrigation equipment has become available.

Lawn sprinklers are relatively inefficient irrigating devices, as much of their water is lost to evaporation and to runoff. It is commonly estimated that lawn sprinkling will provide ½ in. of water per hour per square foot of lawn (or 0.3 gph per square foot).

Landscape sprinkling is least efficient during the daytime, when hot sun and dry air combine to increase the rate of evaporation. However, nighttime sprinkling is rarely convenient for building custodians or homeowners. One solution to this problem is the use of *sprinkler timing devices* (Fig. 9.43*a*), which control electronic valves and a network of underground supply pipes and sprinkler heads that are permanently installed within landscaped areas. These timing devices typically in-

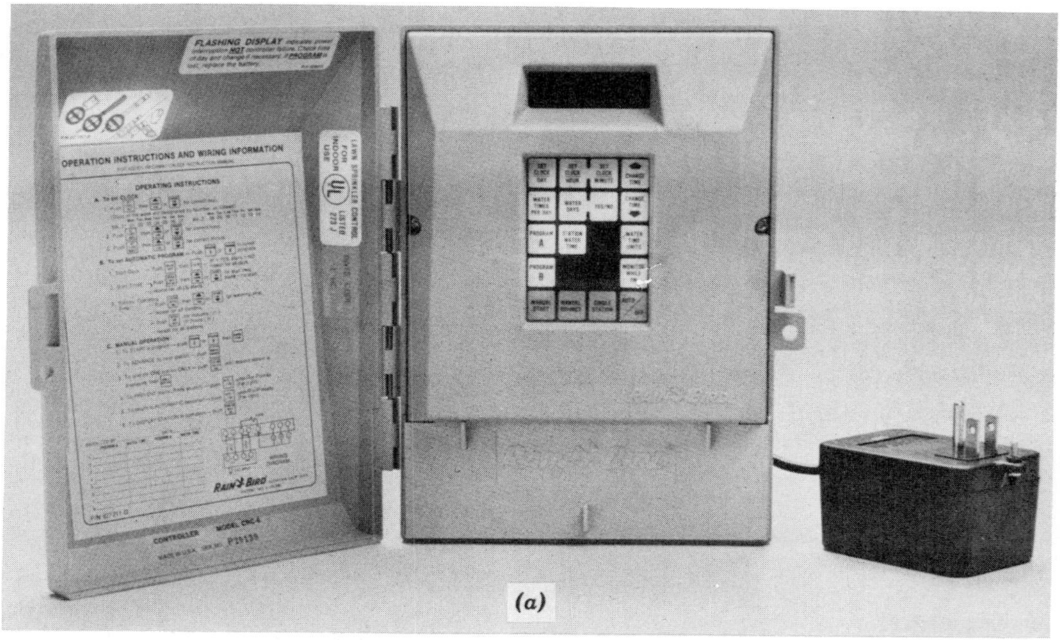

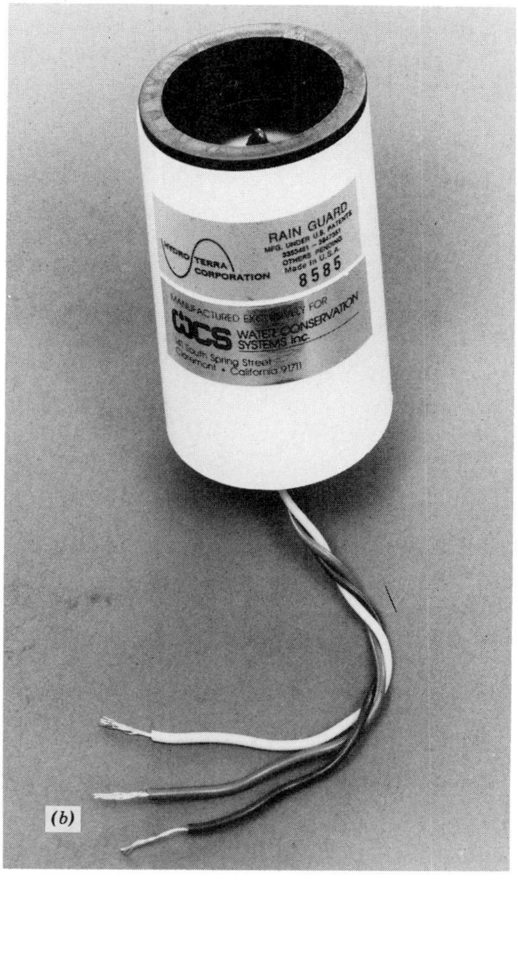

Fig. 9.43 *Components for automatic landscape sprinkling systems.* (a) *Programmable controllers feature multiple stations, up to 14-day sequences, and memory provisions in case of power failure. Courtesy of Rain Bird Sales, Inc.* (b) *A solid-state rain sensor automatically prevents needless watering during rainfall.*

clude controls governing the length of the watering cycle, the time at which the cycle begins (usually before sunrise, when relative humidity is highest), and the number of days between cycles. A "rainy day switch" is often included, so that the irrigation can be discontinued during rainy weather. Rain sensors can be used to shut off irrigation automatically (Fig. 9.43*b*). Tighter control over the water and plant relationship can be obtained with *tensiometers* (Fig. 9.43*c*), which monitor the moisture content of the soil at the depth of the plants' root zone. These can be installed so as to override the automatic timing device, thus watering more—or less—frequently, depending on the plants' needs. Instead of sprinklers, *bubblers* (Fig. 9.43*d*) can be installed, with very low flow rates and less evaporative water losses.

Drip irrigation takes a very different approach to that of the flooding method typical of sprinklers. From a network of plastic tubes, either just

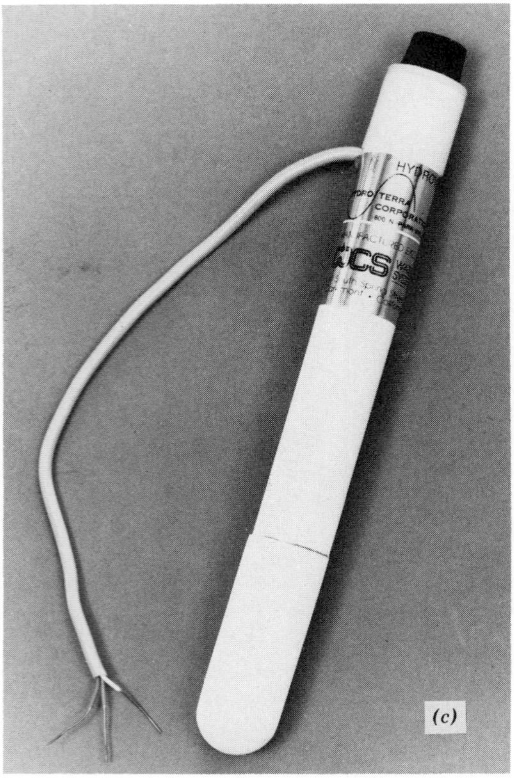

Fig. 9.43 *(continued)* (c) *A tensometer controls irrigation by monitoring the moisture content of the soil. Courtesy of Water Conservation Systems, Inc.* (d) *Bubblers are low-flow substitutes for sprinklers—a step toward drip irrigation. These can be pressure-compensating, to permit constant flow. Courtesy of Rain Bird Sales, Inc.*

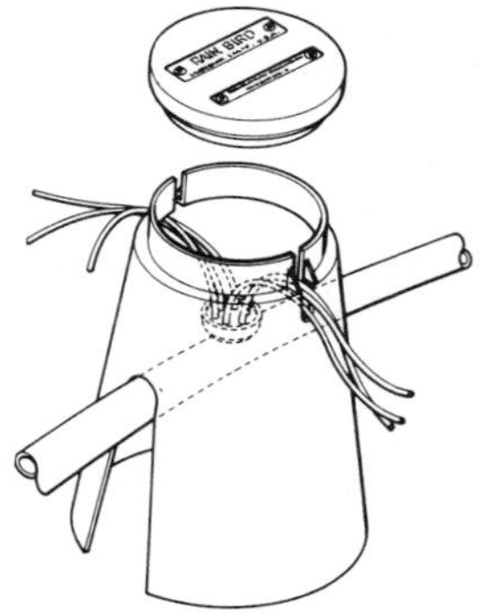

(b)

Fig. 9.44 *Equipment for drip irrigation.* (a) *On low-pressure lines, emitters are installed—one for each group of plants to be watered. The lines can be laid on the ground surface, or just under the surface.* (b) *Simple emitter boxes allow easy access. Courtesy of Rain Bird Sales, Inc.*

underground or on the surface, *emitters* (Fig. 9.44) slowly and steadily drip water onto the ground surface at each needy plant. Most of this water soaks into the soil at a rate that is better for most plants than the sudden, short flooding of intermittent sprinkling. Two requirements are especially important for drip systems: the water must not contain materials that can clog the small holes of the emitters, and the pressure must be low. Pressure-reducing valves at the source of the drop system are advisable; if necessary, filters should also be installed there.

Drip irrigation is not a universal solution to landscape watering; it is best for individual plants such as shrubs and small trees, but is difficult to apply to large lawns. Where appropriate, it can achieve significant water conservation compared to sprinklers.

9.7 Sizing of Water Pipes

There must be sufficient pressure at fixtures to assure the user of a prompt and generous flow of water. Municipal ordinances often state that the flow must be adequate to keep the fixtures clean and sanitary. The convenience of the user and the objectives of sanitation are consistent with each other and have resulted in prescribed pressures that must be maintained at the various fixtures to assure the proper flow rates. These pressures and flows are listed in Table 9.14.

Fixture pressures vary from 5 to 15 psi for fixtures other than hose bibbs. Since the pressure in street mains is usually about 50 psi, it is possible to assure the correct fixture pressure, provided that the water does not have to be lifted to too great a height and that too much pressure is not lost by friction in distribution piping that is too long in *developed length* (actual distance of water flow), or that interposes too many fittings such as elbows and tees, or is too small in diameter.

The pressure components and their total in an upfeed system actuated by street main pressure are as follows.

Proper fixture flow pressure	A
Pressure lost because of height	B
Pressure lost by friction in piping	C
Pressure lost by flow through meter	D
Total street main pressure	E

In a design, items A, B, and E are known and are reasonably constant. The value of A can be

TABLE 9.14 **Minimum Flow and Pressure Required by Typical Plumbing Fixtures**

	Flow Pressure,[a] *Pounds per Square Inch (psi)*	*Flow Rate, Gallons per Minute (gpm)*
Ordinary basin faucet	8	2.0
Self-closing basin faucet	8	2.5
Sink faucet, ⅜ inch	8	4.5
Sink faucet, ½ inch	8	4.5
Bathtub faucet	8	6.0
Laundry tub faucet, ½ inch	8	5.0
Shower	8	5.0
Ball-cock for closet	8	3.0
Flush valve for closet	15	15–40[b]
Flushometer valve for urinal	15	15.0
Garden hose (50 ft., ¾-inch sill cock)	30	5.0
Garden hose (50 ft., ⅝-inch outlet)	15	3.33
Drinking fountains	15	0.75
Fire hose 1½ inches, ½-inch nozzle	30	40.0

Source: U.S. Environmental Protection Agency, *Manual of Individual Water Supply Systems,* 1975.

[a]Flow pressure is the pressure in the supply near the faucet or water outlet while faucet or water outlet is wide open and flowing.

[b]Wide range due to variation in design and type of closet flush valves.

found in Table 9.14. Street main pressure, E, is a characteristic of the local water supply. Item B, the pressure lost due to height, can be found by multiplying the height in feet by 0.433 (see the discussion of static head, Section 9.5*a*). Item D, the pressure lost in flow through the water meter, depends on flow (in gpm) and pipe size (see Fig. 9.45), neither of which is yet known. Therefore, the value of Item D is *estimated*. (For residences and small commercial buildings, it rarely exceeds 2 in.) Later, it must be checked and a recalculation made if necessary.

The selection of a pipe size is facilitated by Fig. 9.46. Pipe diameter is determined by the point of intersection of a horizontal line representing flow in gpm and a vertical line expressing friction loss in psi/100 ft of pipe length. To select a pipe size, one needs to know the probable flow and the *unit*-friction loss in the pipe and fittings. Also, the noise created by water flow must be considered. Above 10 fps is usually too noisy; above 6 fps may be too noisy in acoustical-critical locations.

Flow can be found by assigning the fixture units listed in Table 9.15, the sum of which is an index of the demand flow that can be found in Fig. 9.47. These curves, based on experience, indicate that flow does *not* increase in direct proportion to an increase in fixture units. In larger installations, there is less likelihood that many fixtures will be operating concurrently.

To establish the desired friction loss, divide value C (pressure lost by friction in piping) by the *total equivalent length* (TEL) of the piping. This length is the sum of the *developed length* (DL) (total linear distance of water travel) and the length equivalent to the fittings. For instance, Table 9.16 shows that a 90° ell causes a friction loss equivalent to that of 3 ft of pipe in a 1-in.-diameter pipe run. Obviously the number and style of fittings must be estimated, and the *size* of fittings assumed. This is a puzzling but common engineering procedure that sometimes requires several recalculations.

EXAMPLE 9.6. Using the following data—some of which have been arrived at by the assumptions referred to above—find the proper size for a metered water supply main.

Street main pressure (minimum)	50 psi
Height, topmost fixture above main	30 ft
Topmost fixture type	Water closet with flush valve
Fixture units in the system	85

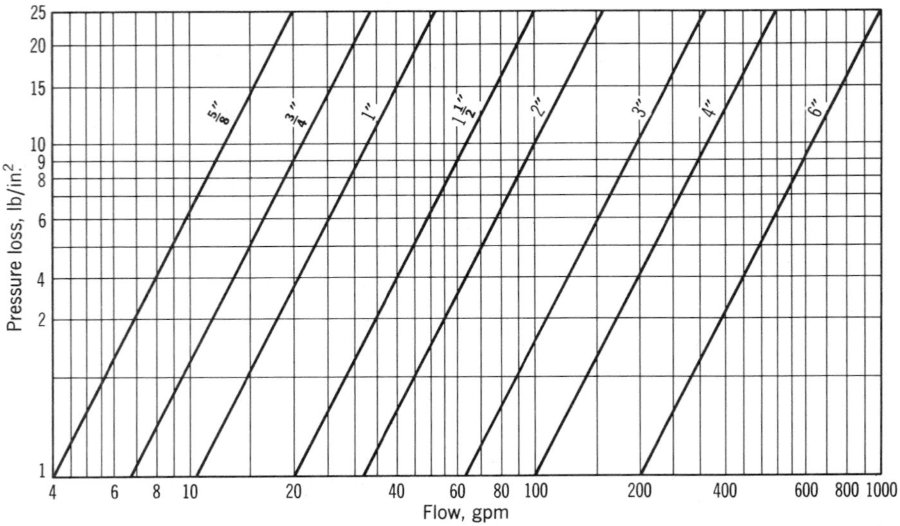

Fig. 9.45 Pressure losses in water meters. From the ASHRAE Heating, Ventilating, Air Conditioning Guide, *31st ed. Reprinted by permission.*

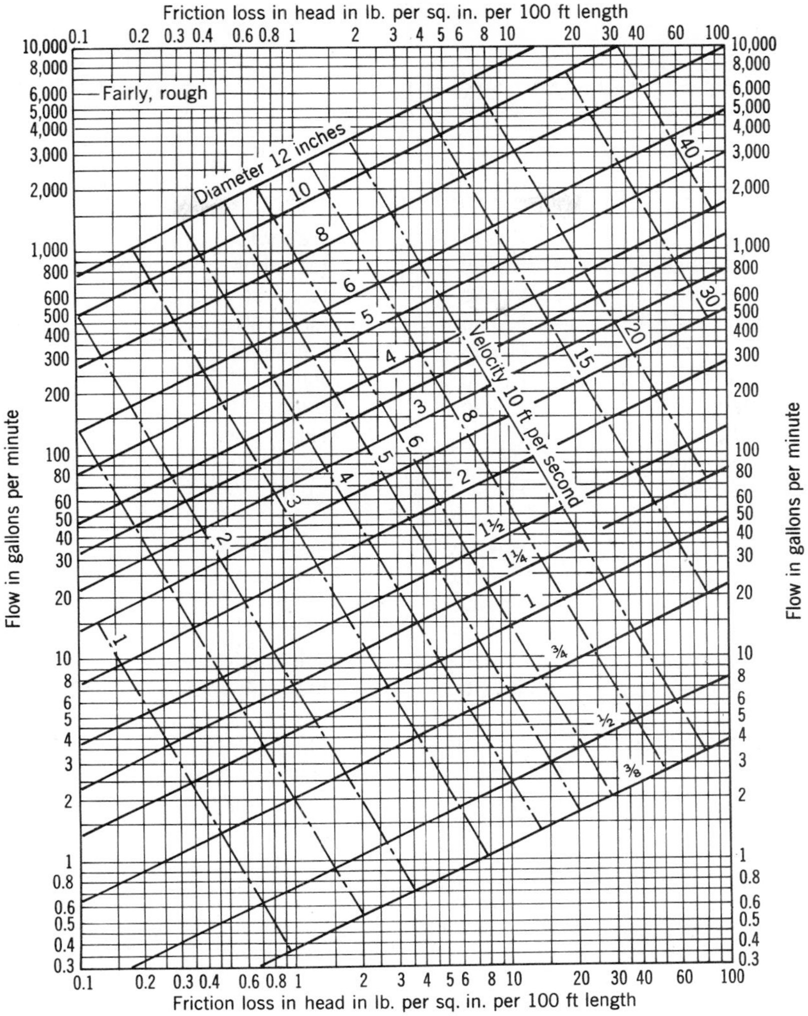

Fig. 9.46 *Flow chart for typical (fairly rough) pipe. Velocity is shown, as an aid in noise control: above 10 fps, moving water can be heard within pipes. Copyright © by the American Society of Heating, Refrigerating and Air Conditioning Engineers, Inc., Atlanta, GA. Reprinted by permission from* ASHRAE Fundamentals, *1972*.

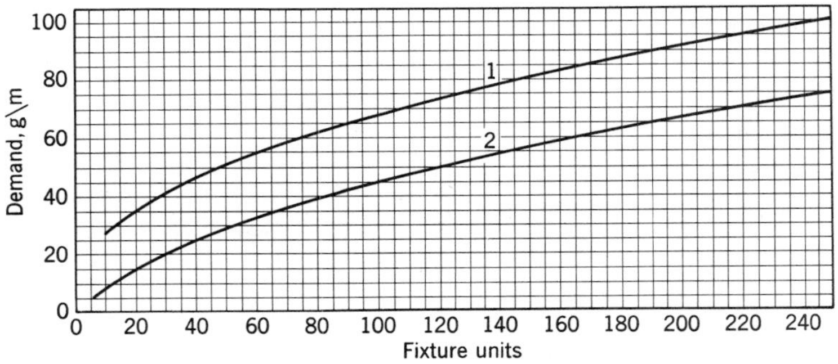

Fig. 9.47 *Estimate curves for demand load. Curve no. 1 is for a system of predominantly flush valves. Curve no. 2 is for a system of predominantly flush tanks. Copyright © by the American Society of Heating, Refrigerating and Air Conditioning Engineers, Inc., Atlanta, GA. Reprinted by permission from* ASHRAE Handbook of Fundamentals, *1972*.

TABLE 9.15 **Demand Weights of Fixtures in Supply Fixture Units**[a]

Fixture or Group[b]	*Occupancy*	*Type of Supply Control*	*Weight in Fixture Units*[c]
Water closet	Public	Flush valve	10
Water closet	Public	Flush tank	5
Pedestal urinal	Public	Flush valve	10
Stall or wall urinal	Public	Flush valve	5
Stall or wall urinal	Public	Flush tank	3
Lavatory	Public	Faucet	2
Bathtub	Public	Faucet	4
Shower head	Public	Mixing valve	4
Service sink	Office, etc.	Faucet	3
Kitchen sink	Hotel or restaurant	Faucet	4
Water closet	Private	Flush valve	6
Water closet	Private	Flush tank	3
Lavatory	Private	Faucet	1
Bathtub	Private	Faucet	2
Shower head	Private	Mixing valve	2
Bathroom group	Private	Flush valve for closet	8
Bathroom group	Private	Flush tank for closet	6
Separate shower	Private	Mixing valve	2
Kitchen sink	Private	Faucet	2
Laundry trays (1–3)	Private	Faucet	3
Combination fixture	Private	Faucet	3

Source: NBS Report BMS79, *Water-Distributing Systems for Buildings.* Copyright © by the American Society of Heating, Refrigerating and Air Conditioning Engineers, Inc., Atlanta, GA. Reprinted by permission from *ASHRAE Handbook of Fundamentals,* 1972.

[a]For supply outlets likely to impose continuous demands, estimate continuous supply separately, and add to total demand for fixtures.

[b]For fixtures not listed, weights may be assumed by comparing the fixture to a listed one using water in similar quantities and at similar rates.

[c]The given weights are for total demand. For fixtures with both hot and cold water supplies, the weights for maximum separate demands may be taken as ¾ the listed demand for the supply.

Developed length (DL) of the piping to the highest and most remote fixture	100 ft
Pipe length equivalent to fittings (commonly estimated at 50% of the DL)	50 ft
System uses predominantly	Flush valves

SOLUTION. From the minimum street main pressure, subtract the sum of the fixture pressure, the static head, and the pressure lost in the meter. This sum is

	psi
A—fixture pressure (Table 9.14)	15
B—static head 30 × 0.433	13
D—pressure loss in meter (estimated, Fig. 9.45)	8
	36
E	50.0
Subtract (A + B + D)	−36.0
	14.0

TABLE 9.16 **Allowance in Equivalent Length of Pipe for Friction Loss in Valves and Threaded Fringes**[a]

Diameter of Fitting inches	*Equivalent Length of Pipe for Various Fittings* 90-Deg Standard Ell, feet	45-Deg Standard Ell, feet	90-Deg Side Tee, feet	Coupling or Straight Run of Tee, feet	Gate Valve, feet	Globe Valve, feet	Angle Valve, feet
$\frac{3}{8}$	1	0.6	1.5	0.3	0.2	8	4
$\frac{1}{2}$	2	1.2	3	0.6	0.4	15	8
$\frac{3}{4}$	2.5	1.5	4	0.8	0.5	20	12
1	3	1.8	5	0.9	0.6	25	15
$1\frac{1}{4}$	4	2.4	6	1.2	0.8	35	18
$1\frac{1}{2}$	5	3	7	1.5	1.0	45	22
2	7	4	10	2	1.3	55	28
$2\frac{1}{2}$	8	5	12	2.5	1.6	65	34
3	10	6	15	3	2	80	40
$3\frac{1}{2}$	12	7	18	3.6	2.4	100	50
4	14	8	21	4.0	2.7	125	55
5	17	10	25	5	3.3	140	70
6	20	12	30	6	4	165	80

[a]From NBS Report BMS66 *Plumbing Manual.*

The pressure lost in 100 ft (DL) of piping plus the 50 ft of piping equivalent to the pressure lost by friction in the fittings therefore totals 14 psi. Total equivalent length (TEL) is 150 ft. This procedure assures 15 psi at the critical fixture. The unit-friction loss, psi/100 ft of pipe, will be

$$14 \text{ psi} \times 100/150 \text{ TEL} = 9.33 \text{ psi}/100 \text{ ft}$$

From Fig. 9.47, curve 1, a flush-valve system with 85 fixture units will have a probable flow (demand) of 64 gpm. Given this information, enter Fig. 9.46 horizontally at 64 gpm and vertically at 9.3 psi/100 ft. At the intersection of these lines, the flow rate and velocity are determined. Between 1½-in.- and 2-in.-diameter pipe,

$$\text{velocity} = 8 \text{ fps (less than 10, so OK)}$$

Therefore, a 2-in.-diameter supply pipe will be chosen, with a 2-in. meter.

Now find the actual pressure loss in the 2-in. meter for a flow of 64 gpm. Figure 9.45 shows that this is 4 psi. Since this is *less* than the 8 psi estimated, the pressure at the *fixture* will be slightly higher than planned. When a final system layout is made, the fittings are tabulated and the length in piping equivalent to fittings is found. If this differs greatly from the 50 ft estimated in Example 9.6, a recalculation is made.

References

Milne, M. (1976). *Residential Water Conservation,* U.S. Dept. of Commerce, NTIS.

U.S. Environmental Protection Agency (1975). *Manual of Individual Water Supply Systems.*

10

WATER AND WASTE

The waste of resources inherent in the use of much of our potable water to flush toilets was mentioned in Chapter 7. Here we will examine more closely this conventional approach to bodily waste removal, along with alternatives that use less water—or no water at all. For typical U.S. residences, the potential impact of such alternatives on water usage and treatment is shown in Fig. 10.1.

Almost every plumbing fixture within buildings is provided with both water supply and waste pipes. Since the toilet is usually the largest user of water (Fig. 10.1), as well as one of the worst polluters, this chapter's comparison of conventional, conserving, partial recycling, and full recycling/waterless alternatives is focused on toilets. Other water-conserving alternatives to conventional fixtures can be found in Chapter 11.

10.1 Toilets, Water, and Waste

(a) The Conventional Toilet. In the common toilet, or water closet (WC), a sudden deluge of water removes human waste, and simultaneously helps cleanse the toilet. Fast-moving water requires pressure and makes noise. The older flush-tank toilet (Fig. 10.2*a*) stores a smaller quantity of water (about 2.5 gal, or 9.5 L) at sufficient height above the toilet bowl to achieve fast flow. Although it uses minimal amounts of water, it is noisy and its elevated tank has a maximum visual impact. The more common (in North America) version is the two-piece toilet with the tank bolted to the bowl (Fig. 10.2*b*). With much less pressure available, this toilet requires 5 to 7 gal (19 to 26 L) per flush, but it is quieter. Newer "shallow

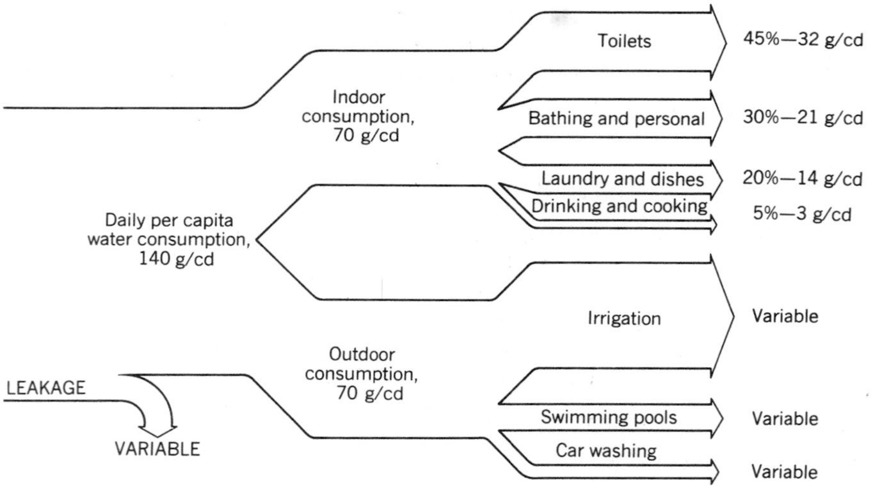

(*a*)

Fig. 10.1 *Opportunities for water conservation for the typical U.S. family.* (a) *At present the average U.S. residential usage is about 140 g/cd—all of it potable.* (b) *With attention to recycling and the matching of water quality to usage, potable water usage could be cut to 105 g/cd—a 25% reduction.* (c) *With further on-site treatment and recycling, potable water usage drops to 73 g/cd—a 40% reduction—and the need for public sewage treatment (for residences) disappears. From Milne (1976).*

trap'' models reduce the quantity needed to 3.5 gal. (13.2 L). The maintenance problem posed by the seam between tank and toilet is eliminated in the one-piece toilet (Fig. 10.2*c*). The cost of its low profile is an even greater need for water: 6 to 8 gal (23 to 30 L) per flush.

The fourth common alternative is the flush-valve toilet, which depends on the building's water pressure rather than its own stored water. A noisy system requiring very high flow rates, it is seldom found in residences. However, it requires as little as 3 gal (11.4 L) per flush.

Architects could easily design toilets that require less water per flush and yet have very little impact on the user. With effort on the part of both users and architects, water can be recycled so as to supply conventional toilets with lower-quality water.

(b) Water-Conserving Toilets. In recent years, most manufacturers have produced toilets that use well under the ''five-gallon flush'' the above models typically use. More details on conventional toilets can be found in Section 11.2 and in Fig. 11.9.

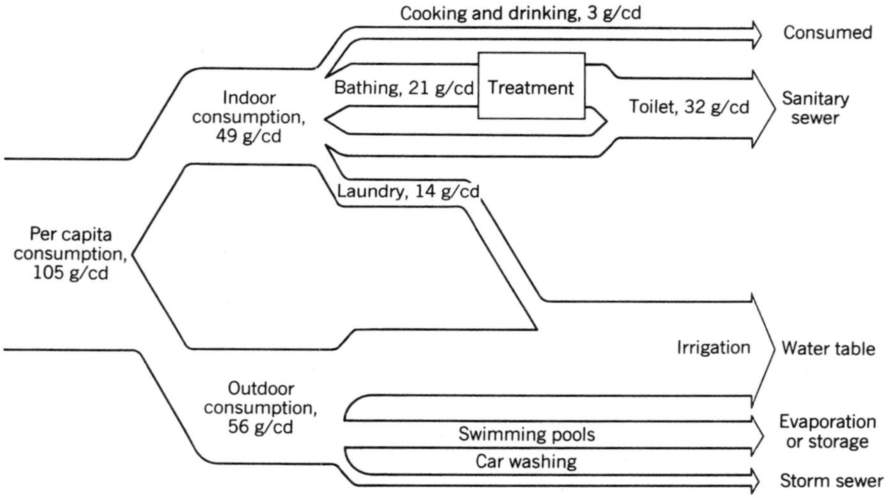

(b)

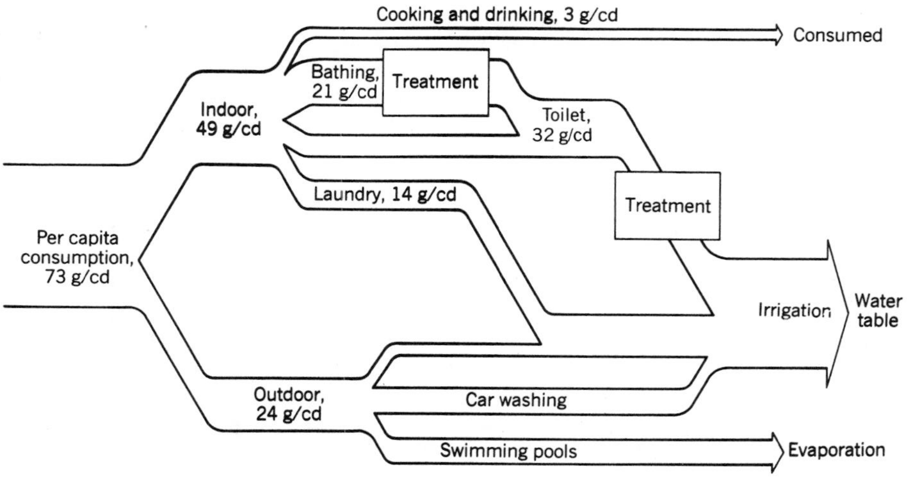

(c)

Fig. 10.1 *(continued)*

Fig. 10-2 *Conventional flush tank toilets.* (a) *Two-piece flush tank toilet with elevated, wall-mounted tank and pull chain.* (b) *Two-piece flush tank toilet with tank bolted to bowl ("close coupled").* (c) *One-piece flush tank toilet with integral tank. From Milne, (1976).*

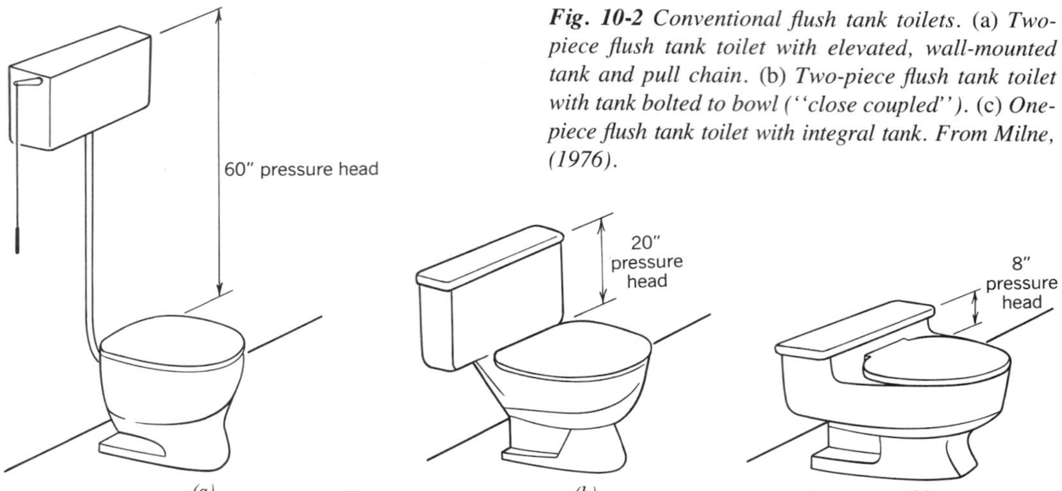

(a) *(b)* *(c)*

In one new approach to the conventional flush toilet, shown in Fig. 10.3, the innovations include a user-controlled quantity of flush water and a straight drop-flapper valve combination that replaces the conventional trap. Since the user controls the duration of the flush, water consumption varies. The manufacturers estimate an average of 1 qt (0.95 L) per flush.

Dramatic water savings are possible when *air*

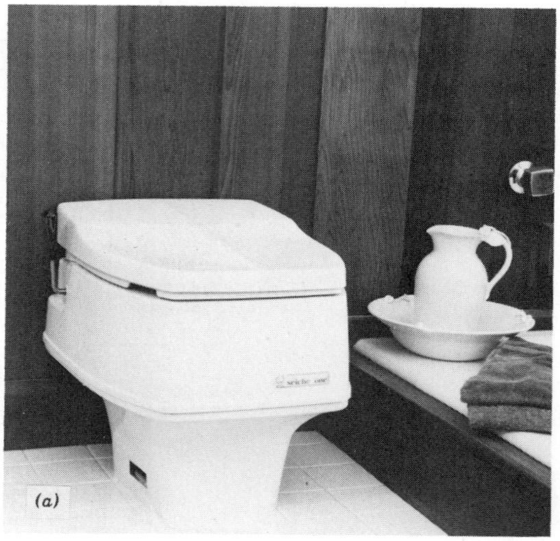

Fig. 10.3 (a) *The user-controlled Seiche One flush toilet averages about 1 quart (0.95 liter) per flush.* (b) *The conventional trap is replaced by a "flapper" valve and a straight drop toward the sewer line. Courtesy of Patrick Creek Corporation.*

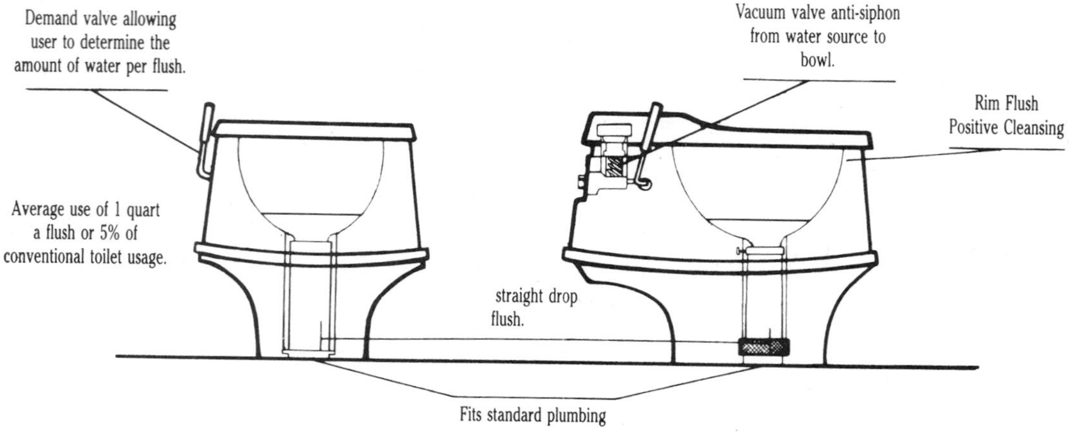

(b)

pressure is teamed with water to provide flushing. The Microphor flush toilet (Fig. 10.4*a*) uses compressed air and 2 qt (1.9 L) of water per flush in a two-chambered toilet. When the flush lever is pressed, the water and waste in the bowl (upper chamber) is deposited into the secondary (lower) chamber, and 2 qt of water (direct from the supply pipe) washes down the sides of the bowl to await the next user. The now-closed secondary chamber then is pressurized with compressed air, and its contents deposited into the conventional sewer line.

Air compressors are needed, along with compressed air lines. A small compressor (¼ to ½ hp) with accompanying air tank will operate up to three toilets. Although the compressor is noisy, the toilets themselves are no more noisy than conventional toilets; they use the same plumbing lines. With such low flows, the designer may choose to increase the slope of waste lines; however, the 2-qt quantity has proven sufficient to carry the waste in existing installations.

In contrast, the Envirovac flush toilet (Fig. 10.4*d*) uses a vacuum and 1.5 qt (1.4 L) per flush. Because a central sewage tank (kept under vacuum by a pump) must be used, the sewer line from the toilet to this tank may run horizontally without a slope, or even run vertically upward from the toilet. This can have significant architectural advantages when vertical clearance is tight or toilets must be located below the level of the public sewer, as in marinas or basements. In tall buildings, significant savings in power for water pumping are achieved. Further, the central sewage holding tank can be flushed into the sewer at off-peak times—a benefit to the treatment plant. These water-saving, architectural, and water treatment advantages must be weighed against the cost of the toilets and the tank/pump combinations, the space required for the tank/pump, the power used by the pump, and the possibility that a power failure will halt system operation.

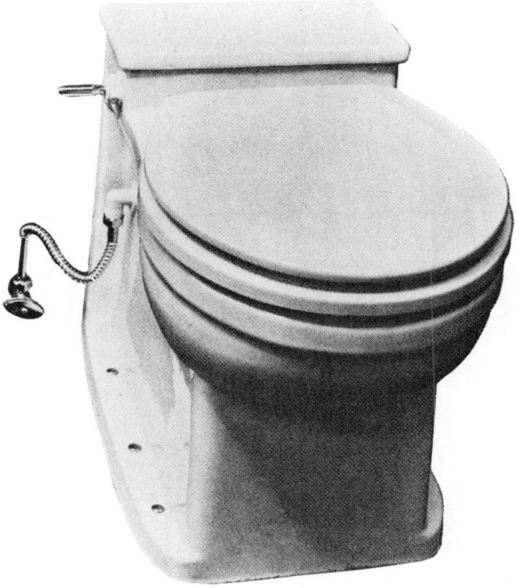

(a)

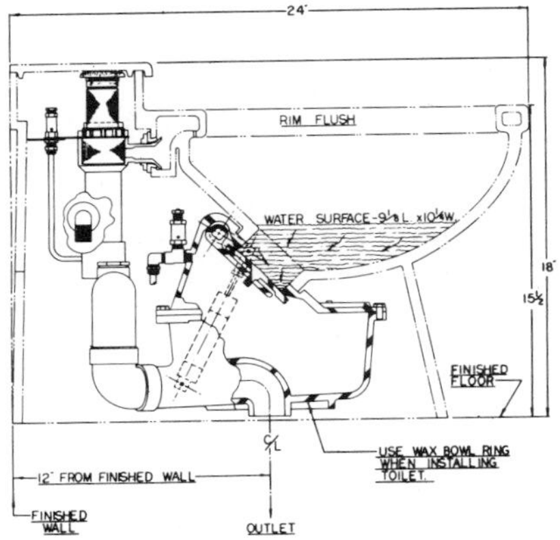

(b)

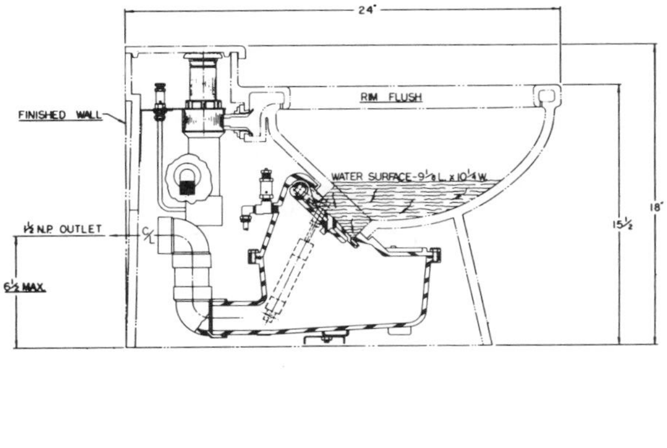

(c)

Fig. 10.4 (a) *Compressed air is combined with a 2-quart flush in the two-chambered Microphor toilet.* (b) *Section through the floor-discharge model.* (c) *Section through the wall-discharge model. Courtesy of the Microphor Company.* (d) *The Envirovac*

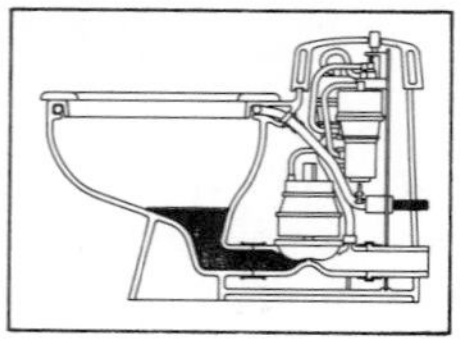

(d)

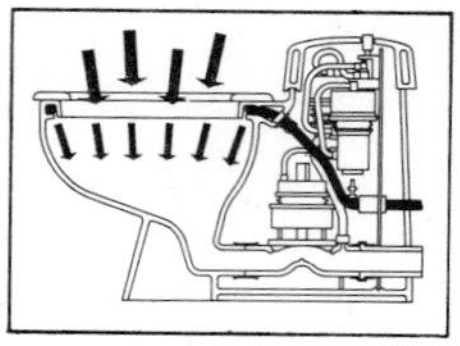

(e)

Fig. ***10.4*** *(continued)* *system utilizes a vacuum and a 3-pt flush. Less than three pints of water remain in the bowl until toilet is used.* (e) *When pushbutton flush is pressed, the discharge valve opens for about three seconds; a rush of air into the evacuated piping carries along sewage, odor, and airborne bacteria. Simultaneously, the washdown water flow begins. The flow continues for four seconds after the discharge valve closes, so that most of the 3-pt flush remains in the bowl.* (f) *Toilets are directly connected to a central holding tank, kept under a vacuum by a pump.* (g) *Central tank/pump combinations can serve one house or a large building.* (h) *The separation of graywater and blackwater is another water-conserving opportunity. Courtesy of Envirovac, Inc.*

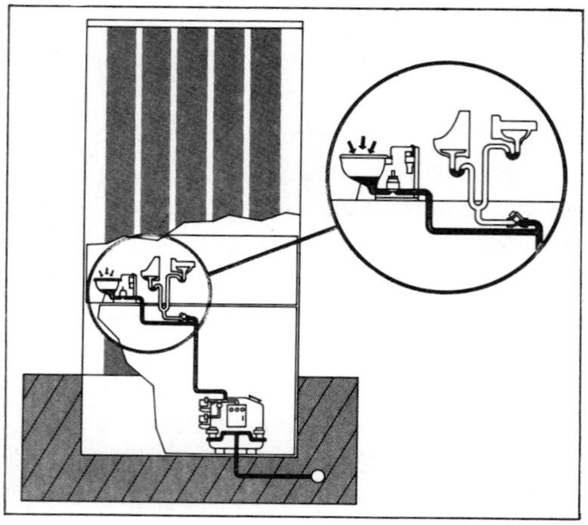

(f)

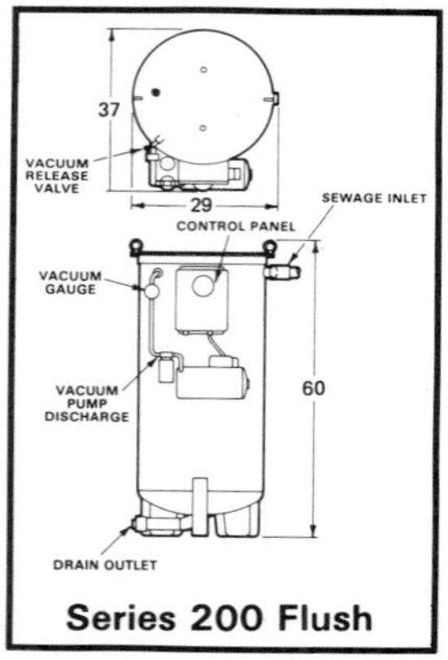

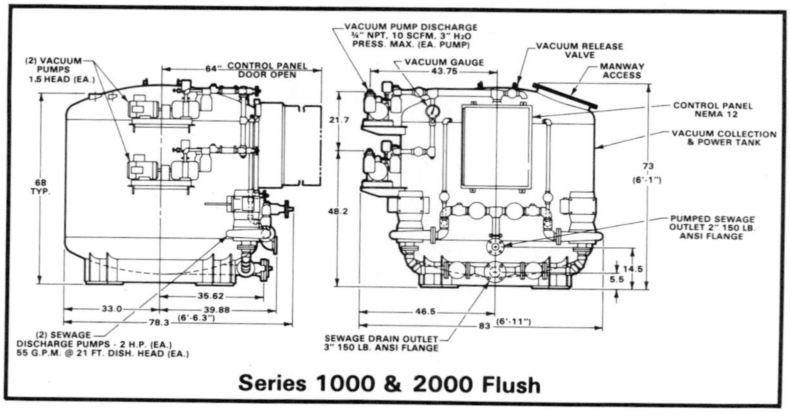

(g)

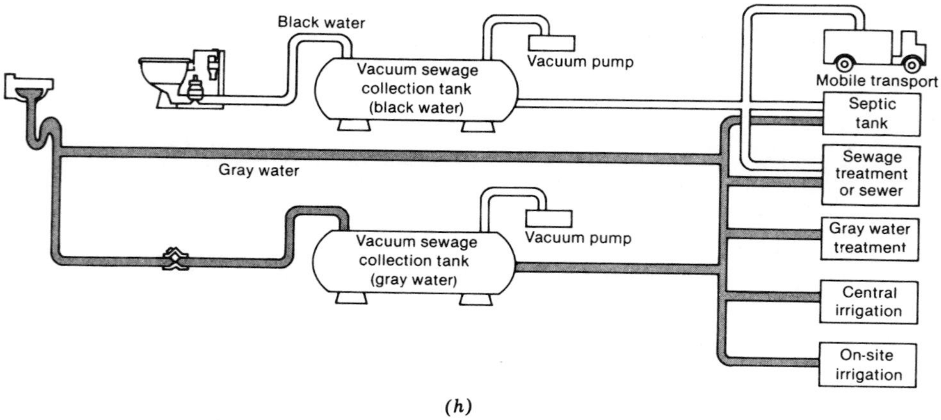

(h)

Vacuum systems can be installed for groups of buildings, such as subdivisions. Small pipe sizes, freedom from the need for continuous-sloping sewer lines, and conservation advantages for both supply and treatment are the benefits. Additionally, ''graywater'' (from kitchen, laundry, and bathing fixtures) can be kept separate from toilet water (''blackwater''), and thus made easier to recycle after moderate treatment.

(c) Recycling of Water. These chapters on water and waste illustrate four common ''grades'' of water in buildings:

Potable water (usually treated, suitable for drinking).

Rainwater.

Graywater (waste water not from toilets or urinals).

Blackwater (water containing toilet or urinal waste).

Many uses of water in and around buildings could be met satisfactorily with rainwater: bathing and laundry, irrigation, and toilet flushing, for example. Opportunities for the reuse of graywater and blackwater are listed in Table 10.1. In gen-

TABLE 10.1 **Potential for Domestic Water Recycling**

Original Use	Reuse: Toilet	Irrigation	Sprinkler	Kitchen Sink	Carwash	Laundry	Pool	Shower/Tub	Bathroom Sink	Dishwasher	Drinking	Cooking
1. Toilet[a]	2	–	–	–	–	–	–	–	–	–	–	–
2. Irrigation*[b]	1	1	1	–	1	–	–	–	–	–	–	–
3. Sprinkler*[c]	1	1	1	–	1	–	–	–	–	–	–	–
4. Kitchen sink with grinder	1	0	1	–	–	–	–	–	–	–	–	–
5. Carwash*	1	0°	1°	–	1	–	–	–	–	–	–	–
6. Laundry[d]	1	0°	1°	–	1	1	–	–	–	–	–	–
7. Pool (chlorinated)	1	–	–	–	1	1	2	–	–	–	–	–
8. Shower/tub	1	0°	1°	–	1	1	–	1	–	–	–	–
9. Bathroom sink[e]	1	0°	1°	–	–	–	–	–	–	–	–	–
10. Dishwasher	1	0°	1°	0	1	–	–	–	–	0	–	–
11. Drinking*	1	0	1	0	1	–	–	–	–	–	–	–
12. Cooking	1	0	1	0	1	–	–	–	–	0	0	0

Legend

0 Reusable directly, without treatment.

1 Reusable with settling and/or filtering (primary treatment).

2 Reusable with settling, filtering, and chemical treatment—usually chlorination (secondary treatment).

– Not reusable.

Source: Milne (1976).

*Very difficult to collect.

°Special soaps required.

[a]Small valves and underwater moving parts—clogging problem.

[b]Large orifice: unpressurized open hose or channel.

[c]Small orifice: pressurized.

[d]Assumes no diapers with fecal matter.

[e]Shaving and brushing teeth.

eral, graywater must undergo treatment before reuse because it is likely to contain soap, hair, grease from the kitchen, and occasionally human waste (soiled clothes in the laundry). Considering these contaminants, "light graywater" might describe water from lavatories, tubs, and showers, and "dark graywater" that from washing machines, kitchen sinks, and dishwashers. More extensive treatment of blackwater is required, given the higher concentration of human waste in it.

The collection of these separate types of water poses a problem. In the past, rainwater, graywater, and blackwater were mixed together, which severely overloaded sewage treatment facilities in rainy weather. Today, we usually mix graywater and blackwater. There are precedents for keeping them separate, however (Fig. 10.5). The potential for graywater reuse with separate waste collection systems is shown in Fig. 10.1*b*: bathing's 21 g/cd (gallons per capita per day) can meet much of the conventional toilet's 32 g/cd, and laundry's 14 g/cd can help with irrigation. Another potential benefit of separate graywater collection is that the water's heat content can be partially reclaimed (Table 10.2).

Fig. 10.5 *(opposite) Residence of the 27th Vicar of the Parish Church at Bibury (established* A.D. *1086) in the English Cotswolds.* (a) *It is evident that the plumbing was added after the construction of the residence—perhaps several hundred years later. In England the drainage system of an older building often appears on the outside of the building. It seemed inappropriate to ask the Vicar's permission to inspect his indoor facilities, so we have taken the liberty of assigning probable uses and examining this conveniently visual "printed circuit" of drains. Note that: 1. Offsets of the vertical piping are made with "easy bends" and are pitched down in the direction of the flow. 2. Vents are located at high outdoor points. 3. Stacks carrying so-called blackwater are of larger pipe size than those carrying graywater.* (b) *In this very sparsely settled region of the Cotswolds, the dispersal of stormwater to the ground and the private septic tank treatment of blackwater plus graywater can be satisfactory. In planning new systems the possible separate treatment of B, G, and S waters is becoming an important consideration.*

Various stages of treatment are appropriate for different kinds of graywater (Fig. 10.6). Kitchen sink waste contains grease, which should be trapped (and periodically removed) before it can clog the filters and heat exchangers that serve graywater recycling systems. Similarly, lavatory showers and laundry waste contains lint and hair that must be intercepted quickly. Devices that do so, which are often called *interceptors,* are described in Fig. 10.39.

Various graywater-filtering devices are now available, including that shown in Fig. 10.7, whose effluent (outflowing water) could be used for either toilet flushing or irrigation.

In the blackwater-recycling toilet systems currently available, treatment is more extensive and first costs much higher. Although water is reused, the sewage sludge must be removed periodically from the treatment tank.

(d) Waterless Toilets. The array of waterless alternatives include toilets in which chemicals or oil are substituted for water. These devices are commonly found in airplanes, vehicles, and boats, as well as in remote and environmentally sensitive areas. The chemicals must be frequently recharged, and the waste products removed. Other waterless toilets temporarily treat the waste awaiting discharge to a sewer by freezing it, burning it, or otherwise packaging it so as to remain unoffensive. Obviously, such devices can become energy-intensive solutions to waterless waste disposal. Also, they are still dependent upon public sewer systems.

Composting toilets offer a self-contained and much less energy-intensive strategy for waterless toilets. Of the two systems presented here, the Mullbänk has less architectural impact but requires more energy. Both systems rely upon *aerobic* digestion of waste (i.e., that which occurs in the presence of oxygen). Aerobic systems usually are essentially odor-free, and the exhaust air is rich in CO_2 and water vapor. In contrast, *anaerobic* decomposition (that which occurs without oxygen) is malodorous and produces methane gas as an important by-product.

In the *Mullbänk toilet* (Fig. 10.8), heat and controlled humidity rapidly decompose human waste within a polystyrene housing. Requirements are energy input and forced ventilation, as well as periodic waste removal. Operation is continuous, uniform, and usually odor-free. This toilet is primarily for human waste, although organic kitchen

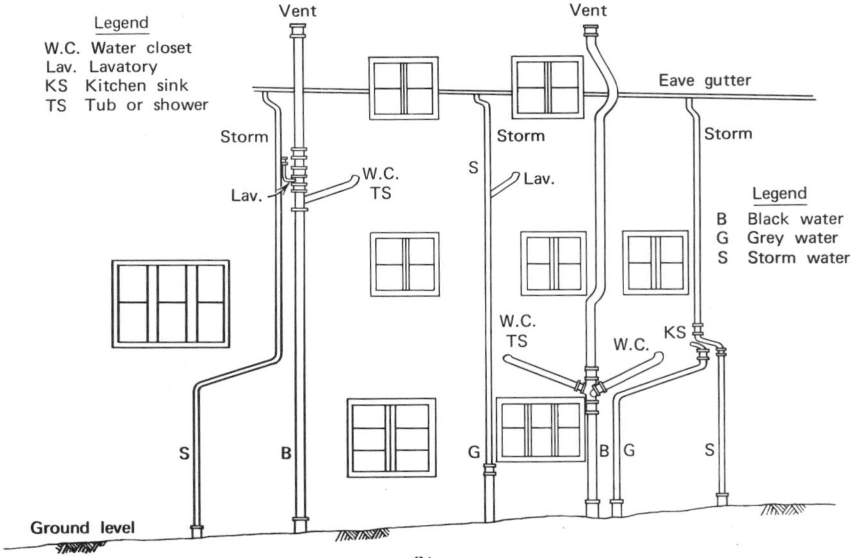

(b)

TABLE 10.2 **Heat Recovery from Graywater**

Source	*Volume/Use, Gal (L)*	*Flow Rate, Gal (L)*	*F*	*(°C)*	*Quality*	*Coincident*[a]
Kitchen sink	Up to 5 (1–20)	2.5 (10)/min	85	(30)	Poor	No
Lavatory	Up to 1.5 (1–5)	1.25 (5)/min	85	(30)	Fair	No
Bath	32 (120)	5 (20)/min	100	(37)	Good	No
Shower	15 (50)	2 (7)/min	100	(37)	Good	Yes
Washer	40 (150)	7 (25)/min	60	(15)	Moot	No

Source: Glenn Nelson, "Graywater Heat Recovery," © *Solar Age,* August 1981. Reprinted by permission.

[a]Coincident flow of warm waste water and input water to heater. If "no," a holding tank is needed for heat exchange. If "yes," a countercurrent heat exchanger can be built into the drain line. (See Fig. 10.6.)

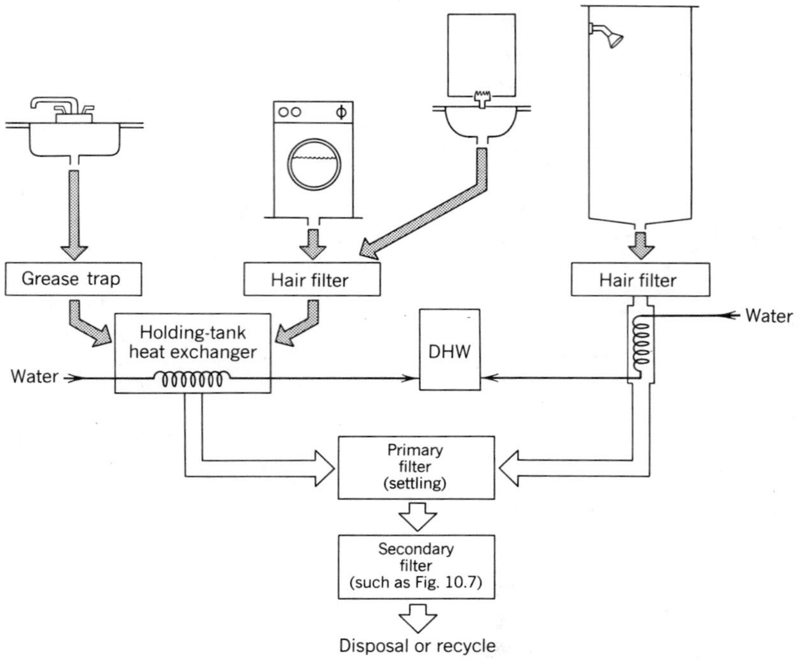

Fig. 10.6 *Sequence of water treatment and heat reclamation for domestic graywater.*

and paper refuse in limited quantities are also acceptable. The waste product is a fertilizer/soil amendment, suitable for gardens.

The *Clivus Multrum System* (Fig. 10.9) represents a slower and less energy-intensive approach. It has a much larger decomposition chamber, which must be below the toilet and the kitchen, from which it readily accepts organic waste. It must also be accessible to remove the humus—from 3 to 10 gal of soil per person per year.

The relatively large chamber (about the size of a Volkswagen ''bug'') and its low position/access requirement can have a significant impact on design. Also of significance is the device's appetite for fresh, and preferably warm, air. The more warm air, the speedier the process of aerobic composition and, therefore, the less odor. In winter, this air could cause increased heat losses by infiltration in a building. The natural ventilation due to the stack effect (Section 4.6) is at least 15 cfm (7 L/s) in the Clivus Multrum. A ventilating fan in the stack will increase this rate significantly. On the other hand, if outdoor air is brought directly to the chamber, it could be too cold for proper decomposition. Solar heating of such input air offers one alternative.

10.2 Waste Treatment

In addition to the self-contained composting toilets just described, there are numerous opportunities for the treatment of human and domestic wastes from buildings. In this section, individual building-scale systems will be considered first, then clusters of buildings, and finally municipal treatment plants.

(a) Individual Systems. The great majority of individual sewage treatment systems in North America use the *septic tank* (Fig. 10.10) as a *primary* treatment, where the settling of solids and anaerobic digestion take place. Subsequently, the effluent receives *secondary* treatment, which usually consists of a filtering process. Four common filtration systems are seepage pits, drainfields, mounds, and sand filters. Occasionally, a *tertiary* treatment (usually disinfection with chlorine) must be used.

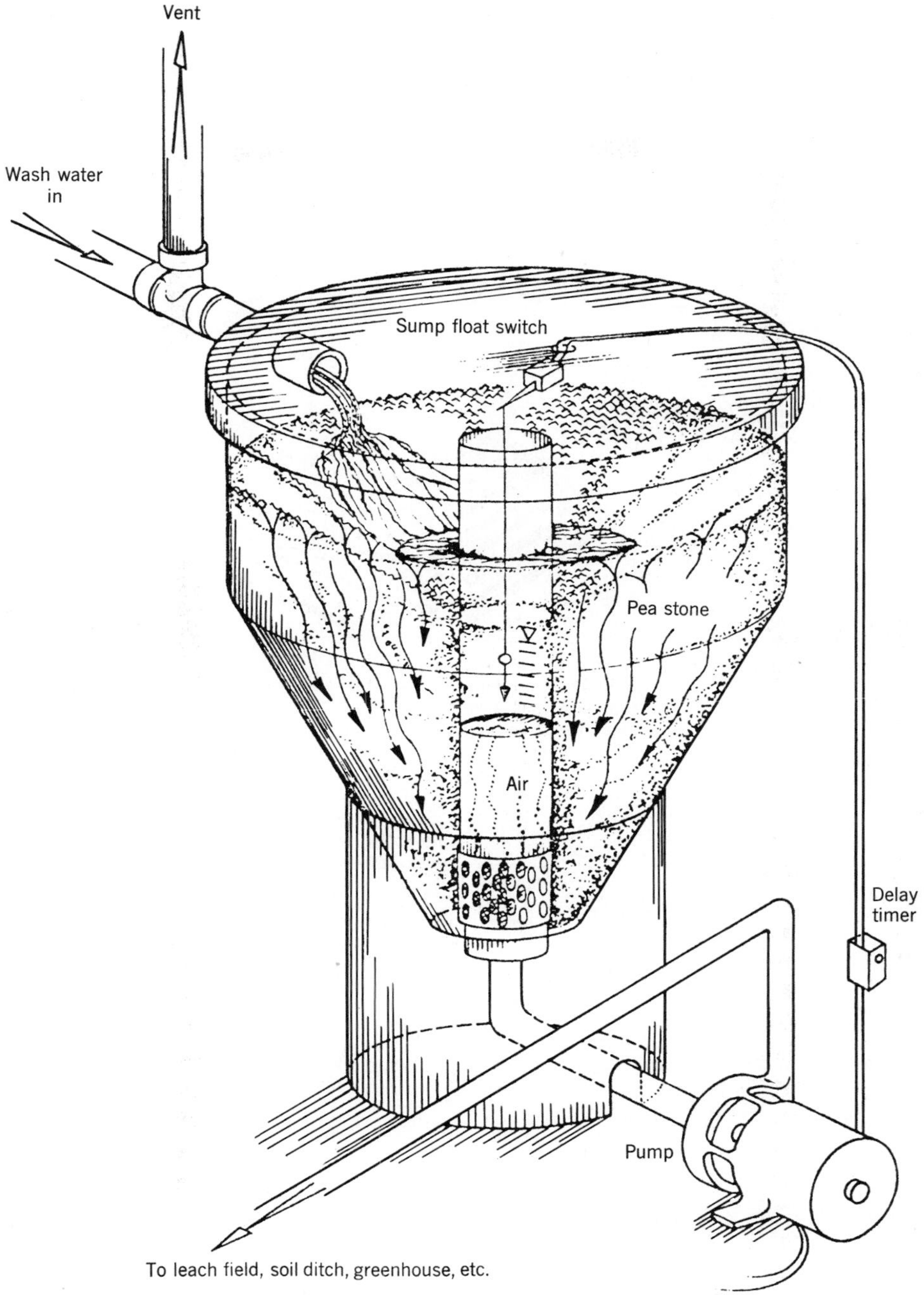

Fig. 10.7 *Graywater filtering: the Clivus Multrum Washwater Treatment System, for water from sink, lavatory, laundry, tub, and shower. Because of aerobic conditions within this filter (see discussion below), the effluent is virtually odor-free. Courtesy of Clivus Multrum USA, Inc.*

(a)

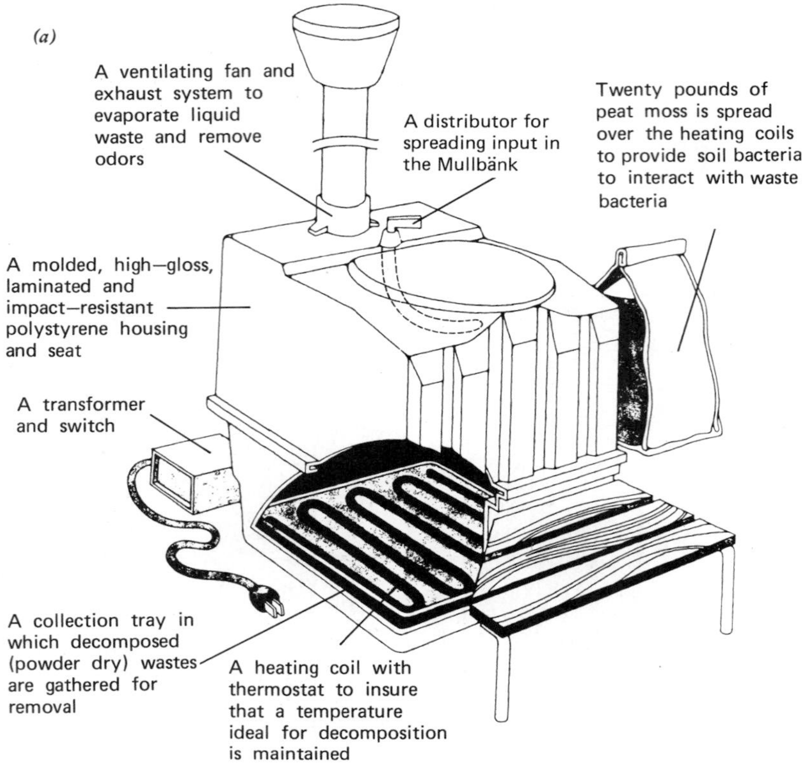
A ventilating fan and exhaust system to evaporate liquid waste and remove odors
A distributor for spreading input in the Mullbänk
Twenty pounds of peat moss is spread over the heating coils to provide soil bacteria to interact with waste bacteria
A molded, high–gloss, laminated and impact–resistant polystyrene housing and seat
A transformer and switch
A collection tray in which decomposed (powder dry) wastes are gathered for removal
A heating coil with thermostat to insure that a temperature ideal for decomposition is maintained

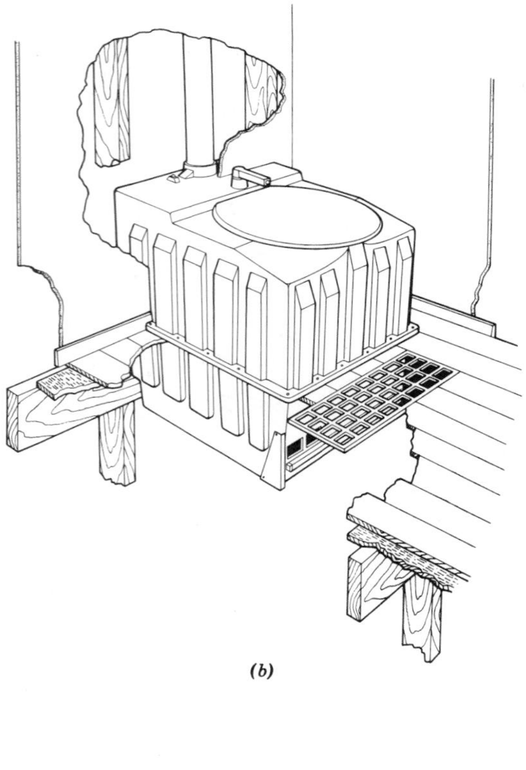

(b)

Fig. 10.8 *The Mullbänk composting toilet.* (a) *The user steps up before sitting down on the self-contained in-bathroom unit.* (b) *Alternatively, it can be framed in at seat height, providing that access and ventilation are maintained for the bottom of the unit. Energy is required for the heating coils and for the ventilating fan. The dry wastes must be removed periodically. Courtesy of Recreation Ecology Conservation of the U.S., Inc., Milwaukee, WI.*

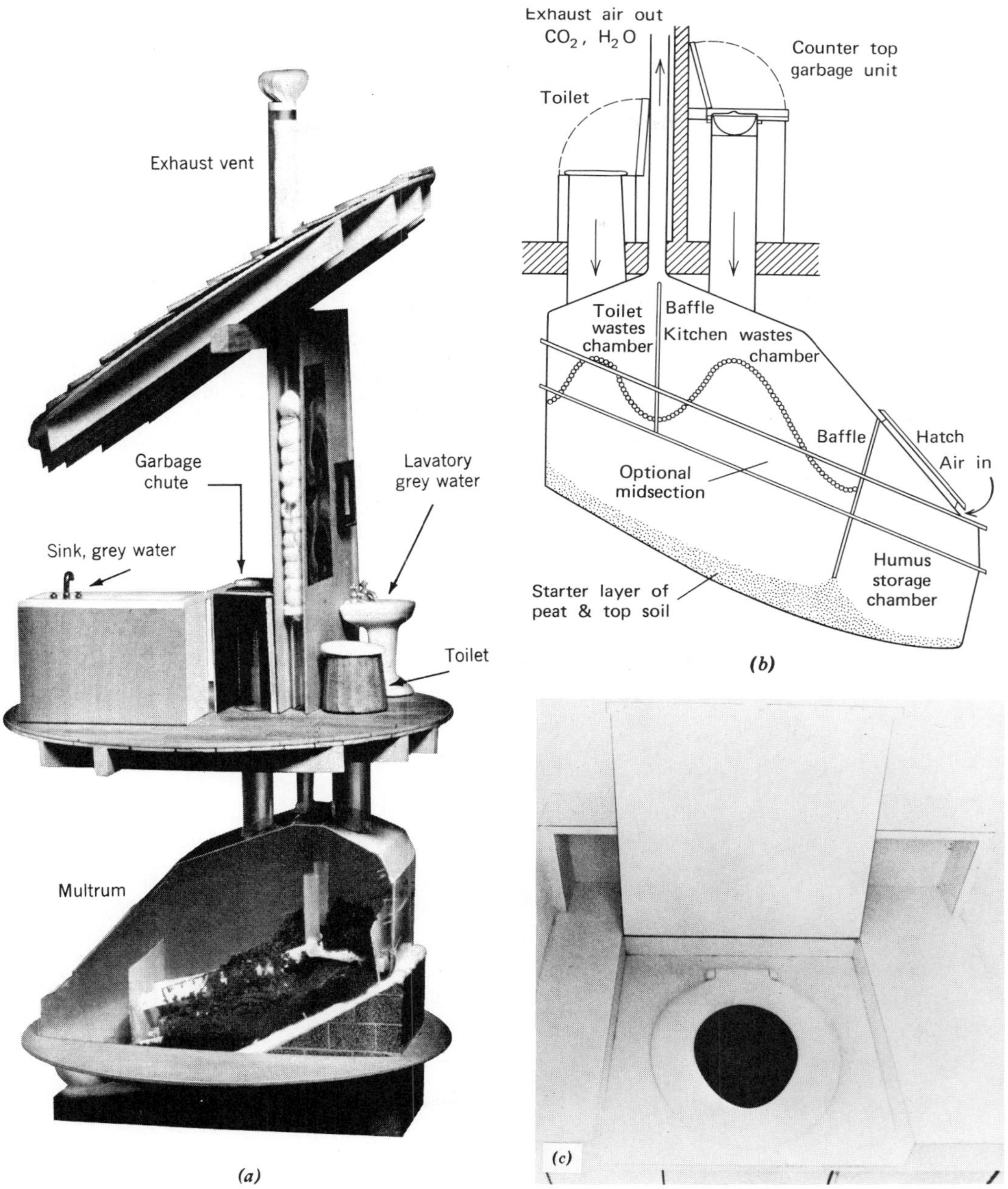

Fig. 10.9 *The Cilvus Multrum System, with composting of both kitchen and toilet organic wastes.* (a) *The lower composting chamber requires 4′ × 8′ × 7′ floor space (1.2 × 2.4 × 2.1 m), a generous supply of air (warm air speeds decomposition), and access for periodic removal of the garden-ready humus.* (b) *It must be arranged so that its upper end receives toilet wastes; as the bottom slopes downward, kitchen wastes are deposited on top of the decomposing toilet wastes. The vent stack assures continuous airflow, preventing odors from entering bathroom or kitchen.* (c) *The toilet seat and cover must be kept closed when not in use, so as to keep air flowing through the composting chamber and out the stack. The bowl (containing no water) swings open when the seat is occupied, exposing the composting chamber below. Courtesy of Clivus Multrum USA, Inc.*

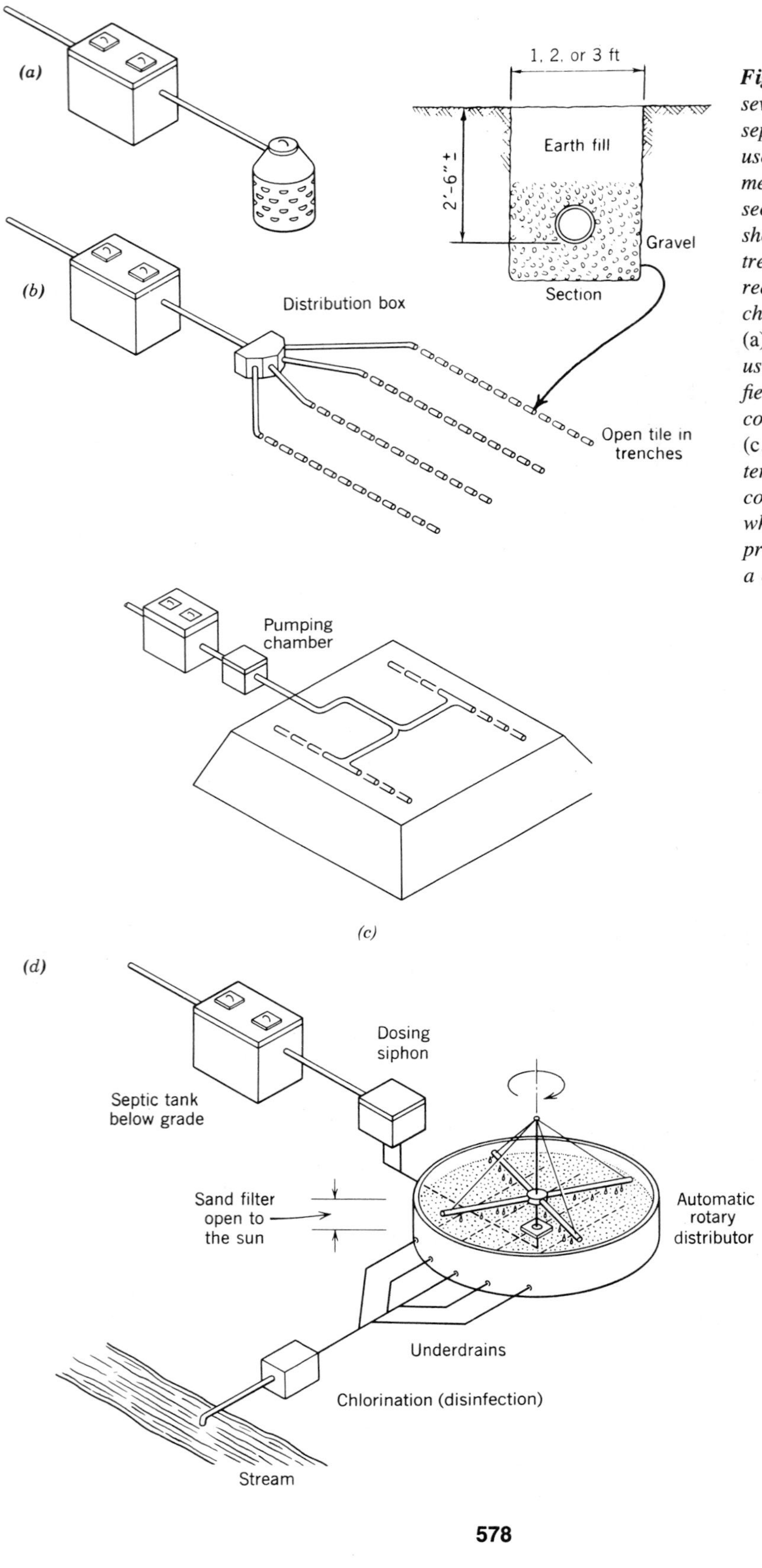

Fig. 10.10. *In individual sewage treatment systems, septic tanks are commonly used for primary treatment. Four options for secondary treatment are shown here. Tertiary treatment usually is only required for effluent discharge into waterways.* (a) *Seepage pits are not usually used.* (b) *Drainfields constitute the most commonly used options.* (c,d) *Mounds and sand filters are more expensive to construct and are used where high water tables preclude the use of options a or b.*

Septic tanks (Fig. 10.11) are commonly constructed of precast concrete. The sewage enters the first chamber, where solids sink to the bottom as sludge, and scum forms on the surface. Anaerobic decomposition proceeds. The liquid moves through the submerged opening in the middle of the tank to the second chamber, where finer solids continue to sink, and less scum forms on the surface. Finally, the effluent, about 70% purified, leaves the septic tank for secondary treatment.

The anaerobic decomposition process is so thorough that the sludge needs only occasional removal—once in several years is about average for residences. The longer that the sewage stays in the septic tank, the less polluted the effluent. This is why water conservation measures are so welcome with septic tank systems; the less the flow, the longer the water remains in the tank.

Most systems with septic tanks will eventually experience "failure," usually due to a breakdown in the secondary treatment rather than in the septic tank. With periodic removal of sludge, the septic tank itself is a reliable and simple primary treatment device. However, some types of domestic waste can disrupt the anaerobic process within the tank: notably, towels, rags, insecticides, excessive paper, and hazardous materials.

Septic tank sizes are commonly based on code requirements that consider the number of bedrooms in residences (Table 10.3) or the number

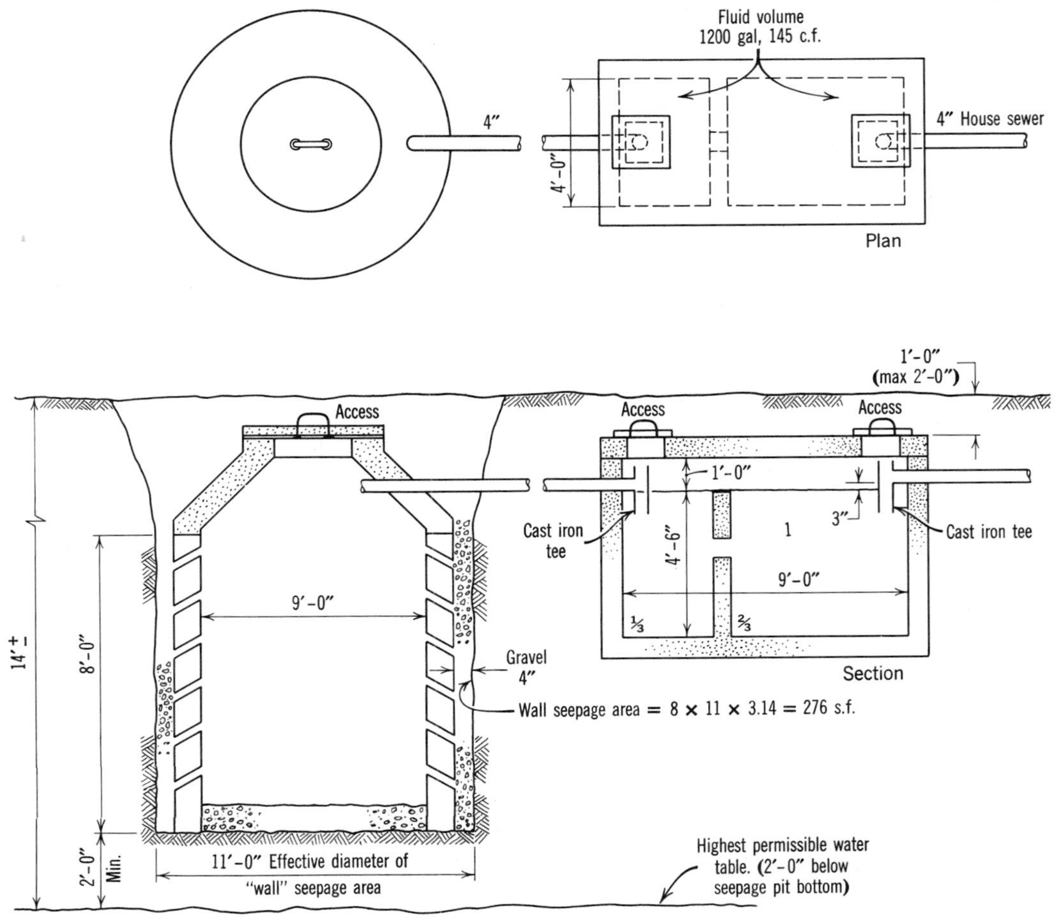

Fig. 10.11 *(Example 10.1.) Plan and section of septic tank and seepage pit for four-bedroom house. Pit is suitable when the earth is absorbent and the water table low (below pit-bottom). Drawing is not to scale.*

TABLE 10.3 **Septic Tank Capacity**[a]

Single-Family Dwellings—Number of Bedrooms	*Multiple Dwelling Units or Apartments—One Bedroom Each*	*Other Uses: Maximum Fixture Units Served*[b,c]	*Minimum Septic Tank Capacity, Gal (L)*[c]
1 or 2		15	750 (2838)
3		20	1000 (3785)
4	2 units	25	1200 (4542)
5 or 6	3	33	1500 (5677.5)
	4	45	2000 (7570)
	5	55	2250 (8516.3)
	6	60	2500 (9462.5)
	7	70	2750 (10408.8)
	8	80	3000 (11355)
	9	90	3250 (12301.3)
	10	100	3500 (13247.5)

Extra bedroom: 150 gal (567.8 L) each.
Extra dwelling units over 10: 250 gal (946.3 L) each.
Extra fixture units over 100: 25 gal (94.6 L) per fixture unit.

Source: Reprinted by permission from the Uniform Plumbing Code, copyright © 1982 by the International Association of Plumbing and Mechanical Officials.

[a]Septic tank sizes in this table include sludge storage capacity and the connection disposal of domestic food waste units without further volume increase.

[b]See Table 10.9.

[c]For larger or nonresidential installations in which sewage flow rate is known, size the septic tank as follows.

1. Flow up to 1500 gpd (5677.5 L/d):

$$\text{flow} \times 1.5 = \text{septic tank capacity}$$

2. Flow over 1500 gpd (5677.5 L/d)

$$(\text{flow} \times 0.75) + 1125 = \text{septic tank capacity in gallons}$$

$$[(\text{flow} \times 0.75) + 4258 = \text{liters}]$$

of waste fixture units served (see Table 10.9). (As shown in Table 10.4, maximum size may also be related to the soil condition.) Sewage flow rates are also considered. Oversized septic tanks are more expensive to install, but they release cleaner effluent and will prolong the life of the secondary treatment process.

The size of the secondary treatment system is usually based on the expected total flow over a 24-hour period. (This can be estimated from Table 7.4.)

Aerobic treatment units (Fig. 10.12) represent an increasingly popular alternative to septic tanks for primary sewage treatment. They depend upon air bubbled through the sewage to achieve aerobic digestion, which is faster than anaerobic digestion; hence, they can be smaller in size than septic tanks.

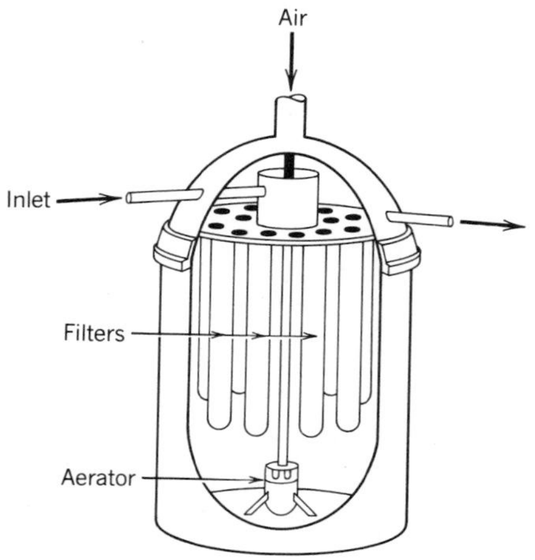

Fig. 10.12 *Aerobic treatment unit, an alternative to the simple anaerobic septic tank. From Milne (1976).*

They are energy-intensive and may require more maintenance than will the anaerobic tank. A secondary treatment process is required, as well. The effluent typically is less polluted than is that of septic tanks.

The sewage first enters an aeration chamber, where it is kept in turmoil so that air can continue to percolate through it. The air is forced by an air compressor; the sewage can be stirred by a variety of devices, depending upon the manufacturer. A second chamber is then used, to allow remaining solids to settle and to be filtered out before the effluent leaves the secondary treatment process.

Seepage pits (Figs. 10.11 and 10.13) are *secondary* treatment facilities that are only appropriate in very porous soil, where the water table is at least 2 ft (0.6 m) below the bottom of the pit. (They should not be used as the only treatment process.) Since these pits are commonly 10 to 15 ft below the earth surface, very low water table may be required. Another common usage of precast seepage pits is for "dry wells" that receive runoff from paved areas during rainstorms.

Seepage pits are sized by the square footage of wall area exposed to the earth—the "leaching area," as listed in Table 10.4. Placement of them relative to buildings, water sources, waterways, and property lines is strictly controlled—see Table 10.5 and Fig. 10.14.

Fig. 10.13 *Precast elements used for recharge to the soil of the partially purified effluent of septic tanks or separately collected stormwater. In the former use, they are known as seepage pits; in the latter, dry wells. They are often made of prestressed concrete of high strength (4000 psi ultimate). Here, shown in the manufacturer's yard, are, right to left, two perforated rings with conical top, cylindrical extender and concrete cover for access, alternate perforated cone, and typical extenders for greater depth. Rings are usually available in 8- and 10-ft diameters and in heights of 3, 4, and 5 ft.*

TABLE 10.4 **Septic Tank and Leaching Area Design Criteria for Five Typical Soils**

Type of Soil	Required ft² of Leaching Area/100 Gal (m²/L)	Maximum Absorption Capacity, Gal/ft² of Leaching Area for a 24-h Period (L/m²)	Maximum Septic Tank Size Allowable: Gallons	Maximum Septic Tank Size Allowable: Liters
1. Coarse sand or gravel	20 (0.005)	5 (203.7)	7500	(28387.5)
2. Fine sand	25 (0.006)	4 (162.9)	7500	(28387.5)
3. Sandy loam or sandy clay	40 (0.010)	2.5 (101.9)	5000	(18925)
4. Clay with considerable sand or gravel	90 (0.022)	1.10 (44.8)	3500	(13247.5)
5. Clay with small amount of sand or gravel	120 (0.029)	0.83 (33.8)	3000	(11355)

Source: Reprinted by permission from the *Uniform Plumbing Code,* copyright © 1982 by the International Association of Plumbing and Mechanical Officials.

TABLE 10.5 **Location of Sewage Disposal Systems**

Minimum Horizontal Distance Clear Required From:	*Building Sewer*		*Septic Tank*		*Disposal Field*		*Seepage Pit or Cesspool*	
Buildings or structures[a]	2 ft	(0.6 m)	5 ft	(1.5 m)	8 ft	(2.4 m)	8 ft	(2.4 m)
Property line adjoining private property	Clear		5 ft	(1.5 m)	5 ft	(1.5 m)	8 ft	(2.4 m)
Water supply wells	50 ft[b]	(15.2 m)	50 ft	(15.2 m)	100 ft	(30.5 m)	150 ft	(45.7 m)
Streams	50 ft	(15.2 m)	50 ft	(15.2 m)	50 ft	(15.2 m)	100 ft	(30.5 m)
Trees	—		10 ft	(3 m)	—		10 ft	(3 m)
Seepage pits or cesspools	—		5 ft	(1.5 m)	5 ft	(1.5 m)	12 ft	(3.7 m)
Disposal field	—		5 ft	(1.5 m)	4 ft[d]	(1.2 m)	5 ft	(1.5 m)
On-site domestic water service line	1 ft	(0.3 m)	5 ft	(1.5 m)	5 ft	(1.5 m)	5 ft	(1.5 m)
Distribution box	—		—		5 ft	(1.5 m)	5 ft	(1.5 m)
Pressure public water main	10 ft[c]	(3 m)	10 ft	(3 m)	10 ft	(3 m)	10 ft	(3 m)

NOTE: When disposal fields and/or seepage pits are installed in sloping ground, the minimum horizontal distance between any part of the leaching system and ground surface shall be fifteen (15) feet (4.6 m).

Source: Reprinted by permission from the *Uniform Plumbing Code,* copyright © 1982 by the International Association of Plumbing and Mechanical Officials.

[a]Including porches and steps, whether covered or uncovered, breezeways, roofed porte-cocheres, roofed patios, car ports, covered walks, covered driveways and similar structures or appurtenances.

[b]All drainage piping shall clear domestic water supply wells by at least fifty (50) feet (15.2 m). This distance may be reduced to not less than twenty-five (25) feet (7.6 m) when the drainage piping is constructed of materials approved for use within a building.

[c]For parallel construction—For crossings, approval by the Health Department shall be required.

[d]Plus two (2) feet (.6 m) for each additional foot (.3 m) of depth in excess of one (1) foot (.3 m) below the bottom of the drain line.

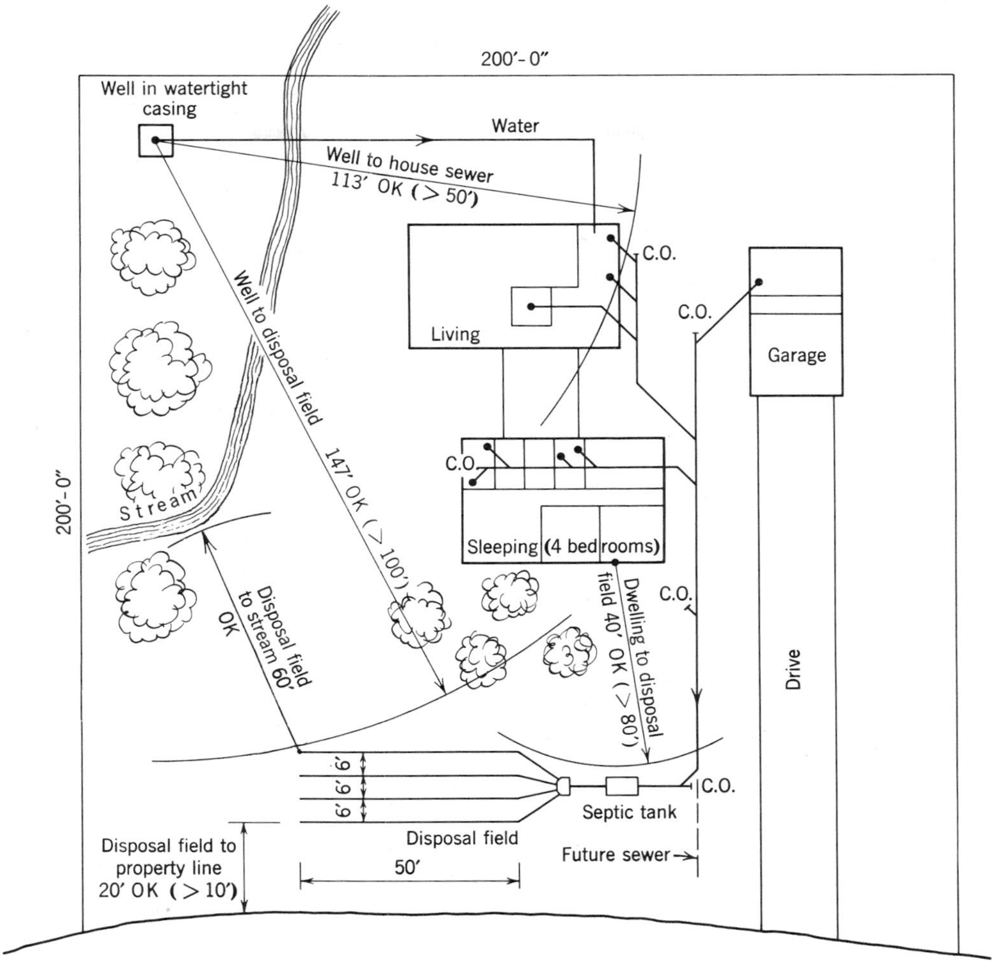

Fig. 10.14 *Checking for required clearances on a suburban lot. Although the 200-ft² lot is more than the minimum required, space should be left for a backup disposal field (of equal size) in case of failure of the original field. A seepage pit adjacent to this septic tank would barely meet the clearance criteria. With a stream nearby, the water table is probably too high for such a pit (see Table 10.5).*

EXAMPLE 10.1 (Fig. 10.11). Design a septic tank seepage pit system for a suburban residence under the following conditions.

Bedrooms	4
Occupants	8
Soil	Sandy loam
Depth to water table	14 ft

SOLUTION.

Daily flow (Table 7.4) =

$$75 \text{ gpd} \times 8 \text{ people} = 600 \text{ gpd}$$

Septic tank capacity (Table 10.3) = 1200 gal

Fluid volume of tank =

$$1200 \div 7.48 \text{ gal/ft}^3 = 160 \text{ cu ft}$$

Pit surface required (Table 10.4) =

$$40 \times 600/100 = 240 \text{ ft}^2$$

Establish dimensions for the septic tank. Dimensions from the chosen tank (Fig. 10.11) are

$$4.5 \times 4.0 \times 9.0 = 162 \text{ cu ft volume}$$

OK, 162 > 160 (required)

Check the cylindrical effective seepage area of the selected pit (Fig. 10.11):

$$8 \times 11 \times 3.14 = 276 \text{ ft}^2$$

$$\text{OK, } 276 > 240 \text{ (required).}$$

Tile drainage (disposal) fields (Fig. 10.15) are very commonly used as a secondary treatment method, since they are relatively inexpensive to build and do not require a water table so deep—or soil so permeable—as seepage pits require. For a tile field, drains of square-edge agricultural tile, 4 in. or more in diameter and placed in shallow trenches, are covered with gravel. The ends of the tiles are separated by ¼-in. openings. The effluent runs out of these spaces and stands in the interstices of the gravel until it seeps into the earth. In effect, the gravel provides spaces that act as a dry well to receive the fluids and accommodate them until they slowly sink into the ground.

Disposal fields are located according to regulations such as those given in Table 10.5 and sized in relation to total sewage flow and septic tank size (see Tables 10.4 and 10.6). Although Table 10.4 shows the maximum absorption capacities for soils based on gal/ft² over a 24-hour period, a common alternative is to use *percolation tests*. For such tests, a "test pit" is dug and water poured into it. The number of minutes it takes for the water level to drop 1 in. is then recorded. Some codes present design sizes based on percolation test data. In areas with poor soil or occasional flooding, codes often require that an area equal to the drainage field's required size be set aside for use in the event of failure of the original field.

The rules of thumb for sizing disposal fields were presented in Section 7.3. The trench width, depth, and spacing shown in Fig. 10.15 are typical, but other combinations can be used, provided that they meet the requirements listed in Table 10.6.

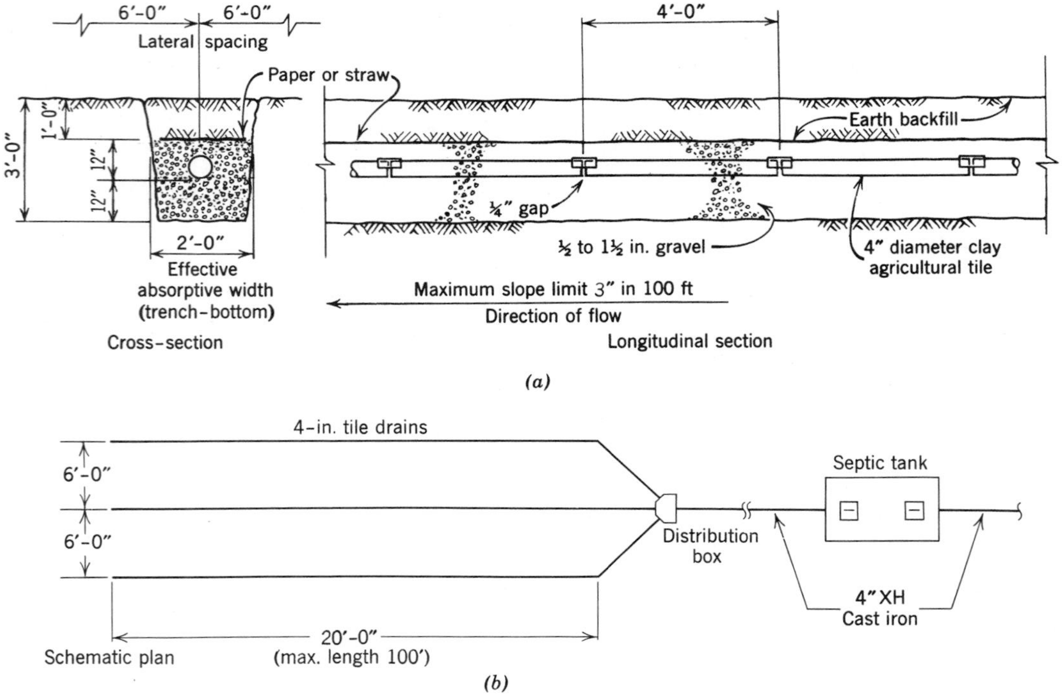

Fig. 10.15 (Example 10.2.) Tile drainfield for a four-bedroom, eight-person house. Although the drawings are not to scale, the dimensions would indicate a required area of about 20 × 70 ft on the lot. When it is considered that it is best not to have the elements run below walks, drives, or other paved areas, sewage treatment on a small lot demands considerable space. (a) *Transverse and longitudinal sections.* (b) *Schematic plan.*

TABLE 10.6 **Disposal Field Trenches**

Part A. ***Dimensions***

	Minimum	*Maximum*
Length of drain line(s)	—	100 ft (30.5 m)
Bottom width of trench	18 in. (457.2 mm)	36 in. (914.4 mm)
Spacing of lines, o.c.[a]	6 ft. (1.8 m)	—
Depth of earth cover over lines	12 in. (304.8 mm) [NOTE: 18 in. (457.2 mm) preferred]	—
Grade of lines	Level	3 in./100 ft (25 mm/m)
Filter material		
Over drain lines	2 in. (50.8 mm)	—
Under drain lines[a]	12 in. (304.8 mm)	—[d]

Part B. ***Leaching Areas***

Trench bottom[b]: minimum 150 ft^2 (14 m^2) per system
Trench side wall: minimum[c] 2 ft^2/ft
maximum[d] 6 ft^2/ft

Source: Reprinted by permission from the *Uniform Plumbing Code,* copyright © by the International Association of Plumbing and Mechanical Officials.

[a]Minimum spacing of drain lines: 4 ft (1.2 m) plus 2 ft (0.6 m) for *each* additional foot (0.3 m) of depth *beyond* 1 ft (0.3 m) below the bottom of the drain line.

[b]Exclusive of rock, clay, or other impervious formations.

[c]Based on the minimum 12-in. trench depth below drain tile.

[d]Maximum of 36-in. trench depth below drain line can be counted when calculating required absorption area.

EXAMPLE 10.2 (Fig. 10.15). Design a tile drainfield for the conditions given in Example 10.1.

SOLUTION. From the rule of thumb (Section 7.3), between 0.4 ft^2 and 3.6 ft^2 of drainfield area will be required per gallon of sewage; this will also allow for a second, backup drainfield.

$$600 \text{ gpd} \times 0.4 = 240 \text{ ft}^2$$
$$\times 3.6 = 2160 \text{ ft}^2$$

Since this is moderately well-draining soil, the total area will probably be about 1200 ft^2: 600 ft^2 of actual field and 600 ft^2 of backup area.

The effective absorption area of the typical trench depth and spacing shown in Fig. 10.15 is

trench width 2.0 ft^2/ft

trench sides <u>2.0</u> ft^2/ft (12″ each side)

4.0 ft^2/ft

From Table 10.4, the required square footage of leaching area for sandy loam is 40 ft/100 gal:

$$40 \text{ ft}^2/100 \text{ gal} \times 600 \text{ gpd} = 240 \text{ ft}^2$$

Since the trench has 4.0 ft^2/ft, the total trench length is

$$240 \text{ ft}^2/4.0 \text{ ft}^2/\text{ft} = 60 \text{ ft}$$

A three-line disposal field is selected, so 60 ft/3 = 20 ft per line. Given the required clearances to the edges of the disposal field (Table 10.5) of 5 ft to property lines, the disposal field's area can be calculated as:

width: 5 ft + 12 ft (two spaces of 6 ft each) + 5 ft = 22 ft

length: 5 ft + 20 ft line + 5 ft = 30 ft

The area, then, is 22 × 30 = 660 ft^2 (about equal to the rule of thumb approximation).

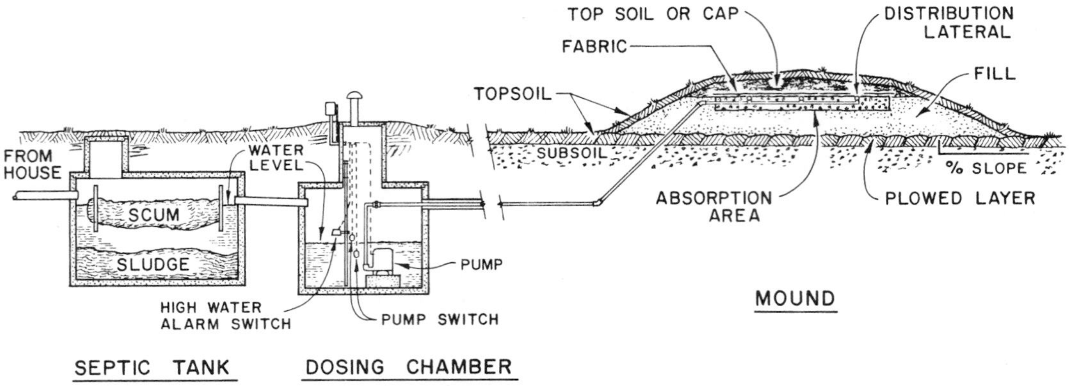

Fig. 10.16 Mounds with leaching beds offer an option when the water table is high. They system serves a two- or three-bedroom home. From Converse (1978).

Mounds with leaching beds (Fig. 10.16) represent a newer solution in the United States, and thus are likely to require special approval. The guidelines for leaching-bed sizing are essentially similar to those for drainage tile disposal fields. The absorption area for leaching beds, however, must be *50% greater* than that required for trenches. The bottom of the leaching bed generally must be at least 5 ft (1.5 m) above the water table, although in water-scarce areas, officials may reduce this requirement.

(b) Treatment Plants for Several Buildings. A variety of ''package'' treatment plants can be adapted for institutions or other groups of buildings such as subdivisions or shopping centers. One typical process is shown in simplified form in Fig. 10.17. Another example of the complexity of and the area required for this size of system is described here (Figs. 10.18–10.20).

A method of sewage treatment known as the Pasveer Oxidation Stream, which employs a principle similar to that followed by a natural stream, has been adopted at New York Institute of Technology. Serving the 450-acre campus at Old Westbury, Long Island, New York, this system provides an on-campus sewage treatment, which returns the purified effluent to the ground through 48 leaching wells located under the athletic field. The groundwater, thus restored, provides a contributing source of water for 400-ft-deep wells, distantly located, that furnish part of the water supply for the campus buildings (see Fig. 10.18).

During the development of the campus in the 1960s and 1970s, studies were made of the sewage problem that would develop with the growth of the college. At the beginning, there were a few small buildings, later converted for administration offices and classrooms. These were, and still are, served by septic tanks and leaching fields. Presently, they constitute less than 20% of the sewage flow of the expanded building complex and may in future years be connected to the campus plant. There was no public sewer near the campus, and the health authorities ruled out the use of septic tanks for the numerous additional buildings that were contemplated.

The oxidation stream process applied here was developed by the Netherlands Research Institute for Public Health Engineering (T.N.) and is now in operation in many U.S. locations as well as in Europe. It is considered to be a modified form of the *activated sludge* process, with aerobic digestion and periodic sludge removal.

An important feature of the system is the mechanical aerator (Fig. 10.19), which keeps the stream of sewage moving and provides the oxidation necessary for aerobic digestion. In this design, sludge-drying beds are placed on an island, surrounded by the continuously moving oxidation stream. Another feature of the system is its low profile (Fig. 10.20), readily screened by trees. The

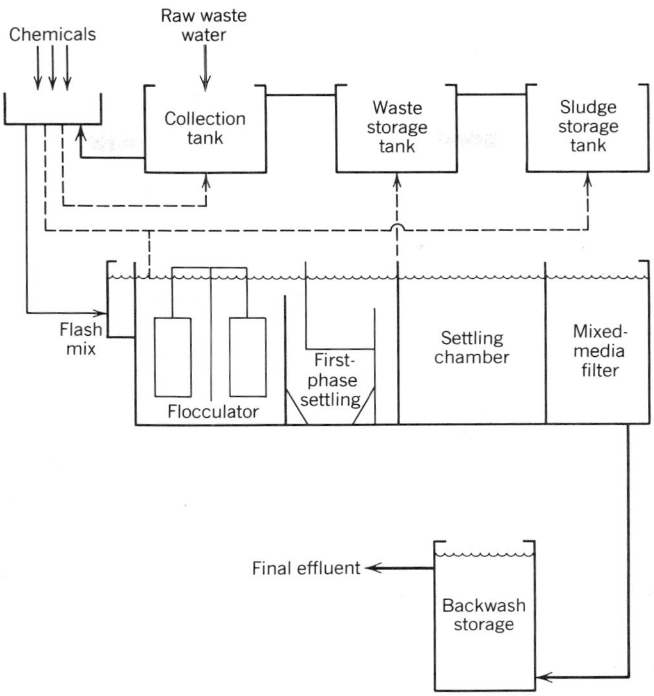

Fig. 10.17 Sewage treatment for groups of buildings. This is a combined physical-chemical treatment process: settling, anaerobic decomposition, filtering and chemical treatment from a primary-secondary-tertiary combination that results in a nearly-pure effluent suitable for release to water bodies. From Milne (1976).

aerobic digestion process is not malodorous. The sounds are produced by the water-wheel action of the mechanical aerators. No air compressor is required.

The plant has a full-time accredited operator; there are also two assistants and one relief operator. This design provides for a population of 4330, with a 340,000-gpd flow. It is presently used at about one-quarter of that capacity. The sludge has been removed from the drying beds twice in the first five years of operation.

(c) Large-Scale Sewage Treatment. Two examples of community-wide sewage treatment are presented here—one for a small, water-scarce community near San Diego, California, and one for the land-scarce conditions of New York City.

The *Padre Dam Municipal Water District* (Figs. 10.21–10.23) took brave steps toward water conservation and recycling during the 1960s and 1970s. According to Mr. Gary Stevens, the plant superintendent, the decision to build the Santee Water Reclamation Plant evolved as follows. The Padre Dam Municipal Water District, Mr. Stevens noted, is a region where rainfall is less than 15 in. per year. It has no local water supplies available and water has to be imported 300 miles from the Colorado River. At the same time it was contemplated that the waste water resulting from this flow might have to be thrown away by discharging it into the Pacific Ocean. Adequate sewage treatment could process this fluid waste for secondary uses such as irrigation and recharging of groundwater. Considering further that many municipal supplies for potable water were taken from rivers often heavily polluted with sewage [see Fig. 9.4] and, after treatment, pumped into the domestic water mains, [district] officials decided to attempt to perfect a system that would make use of purified waste water for many secondary uses.

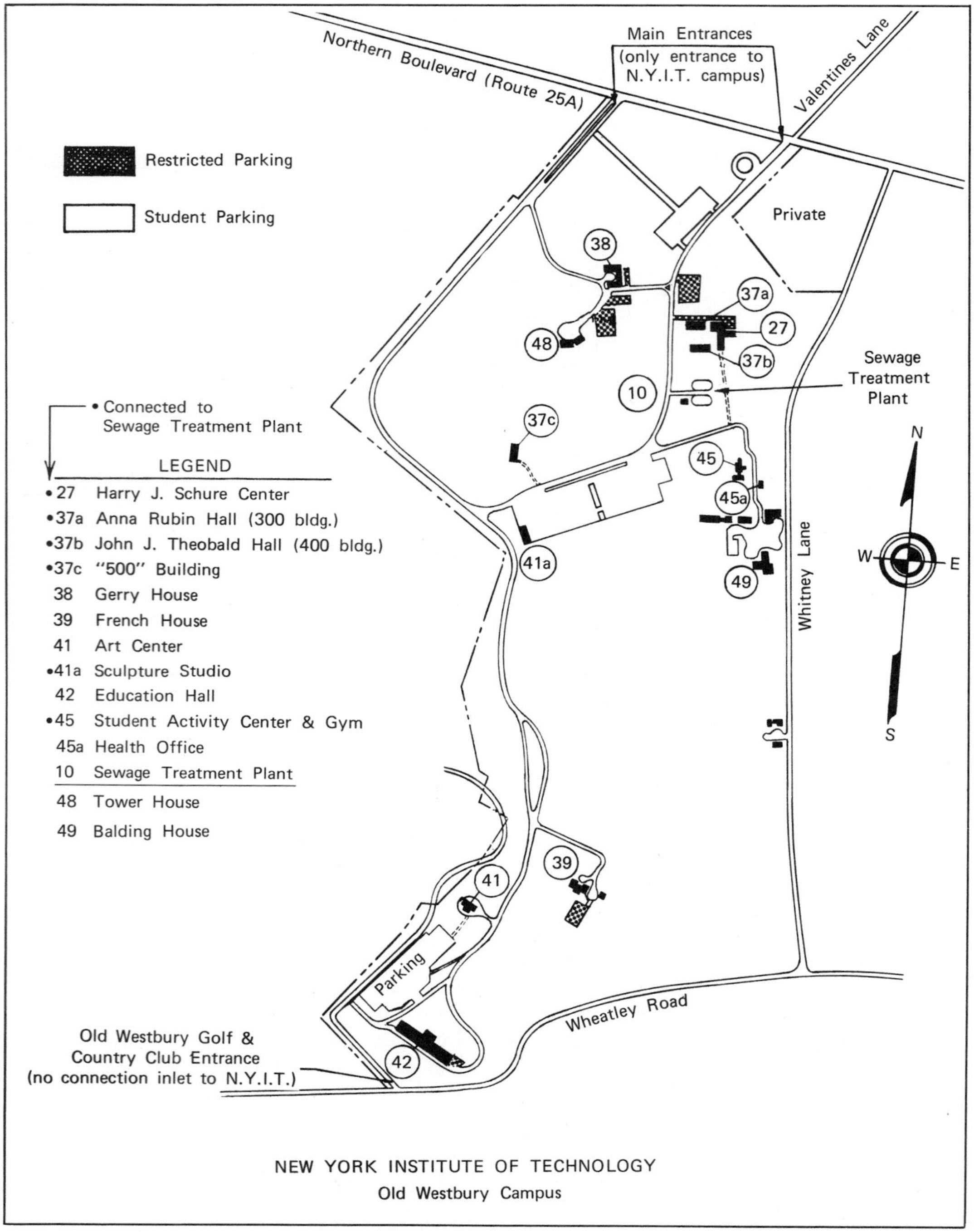
Northern Boulevard (Route 25A)
Main Entrances (only entrance to N.Y.I.T. campus)
Valentines Lane
Restricted Parking
Student Parking
Private
Sewage Treatment Plant
Whitney Lane
N
S
W
E
• Connected to Sewage Treatment Plant
LEGEND
•27 Harry J. Schure Center
•37a Anna Rubin Hall (300 bldg.)
•37b John J. Theobald Hall (400 bldg.)
•37c "500" Building
38 Gerry House
39 French House
41 Art Center
•41a Sculpture Studio
42 Education Hall
•45 Student Activity Center & Gym
45a Health Office
10 Sewage Treatment Plant
48 Tower House
49 Balding House
38
48
37a
27
37b
10
37c
45
45a
41a
49
39
41
42
Parking
Wheatley Road
Old Westbury Golf & Country Club Entrance (no connection inlet to N.Y.I.T.)
NEW YORK INSTITUTE OF TECHNOLOGY
Old Westbury Campus

(a)

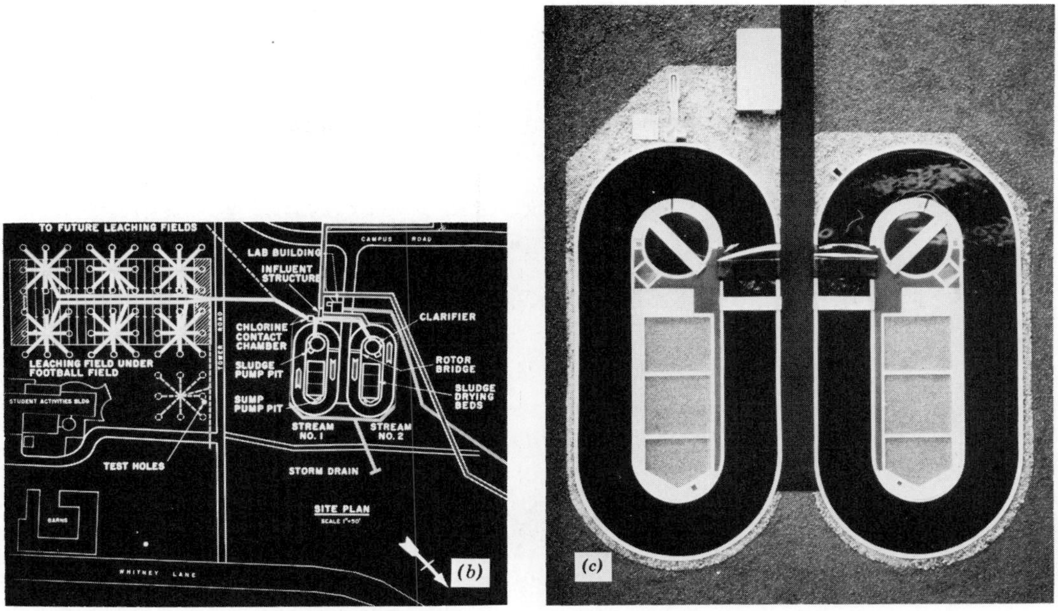

Fig. 10.18 *Sewage treatment plant, New York Institute of Technology (NYIT). The decision to locate the treatment plant at a position central to all the buildings it serves was a very appropriate choice for efficiency.* (a) *(opposite) Surrounded by heavily wooded areas (see Fig. 10.20), it is not a prominent landmark, though it offers a pleasant appearance to the occasional viewer. Although only half of the campus buildings are presently connected to the treatment plant, they are the largest ones and are responsible for 80% of the total campus sewage flow. A frequently used walkway (dotted lines on the map) connects the north academic center with the student activity and athletic center. This path passes just east of the woods that surround the plant. At certain points the plant can be seen through the trees, but no odor betrays its presence.* (b) *Layout of the treatment plant.* (c) *Plan of the two streams. Each of these units includes stream, rotor, clarifier with adjacent sludge pump pit, and three drying beds. Because of odorless operation, the proximity of the plant to the Student Activities Building poses no problem. Courtesy of Bogen Jenal, Engineers, P.C.*

Fig. 10.19 (a) *Rotor with plastic "greenhouse" cover.* (b) *The rotor induces oxidation in this oval, circulating stream.*

Fig. 10.20 *The NYIT sewage treatment plant at completion.* (a) *The only projections above ground level are the office/laboratory and the plastic covers over the rotors. Of the two lagoons, only the one in the foreground has been used, and only to a small percentage of its design capacity. The unused balance of its rating and the entire capacity of the second lagoon are available for future campus expansion.* (b) *The oxidation stream does not appear turbulent here, but it becomes so under the action of the mechanical aerator (rotor).*

Fig. 10.21 *Santee Water Reclamation Plant and Santee Park and Recreational Facilities, Padre Dam Municipal Water District.* (a) *Aerial photograph of the Santee Plant, including the seven recreational lakes. Note the location of the San Diego River, at the bottom of the photograph. The water reclamation plant and the stabilization ponds do not appear in this aerial picture.* (b) *Raw sewage from the community of Santee enters the treatment plant, which is located at the top of this diagram. The process then proceeds southward*

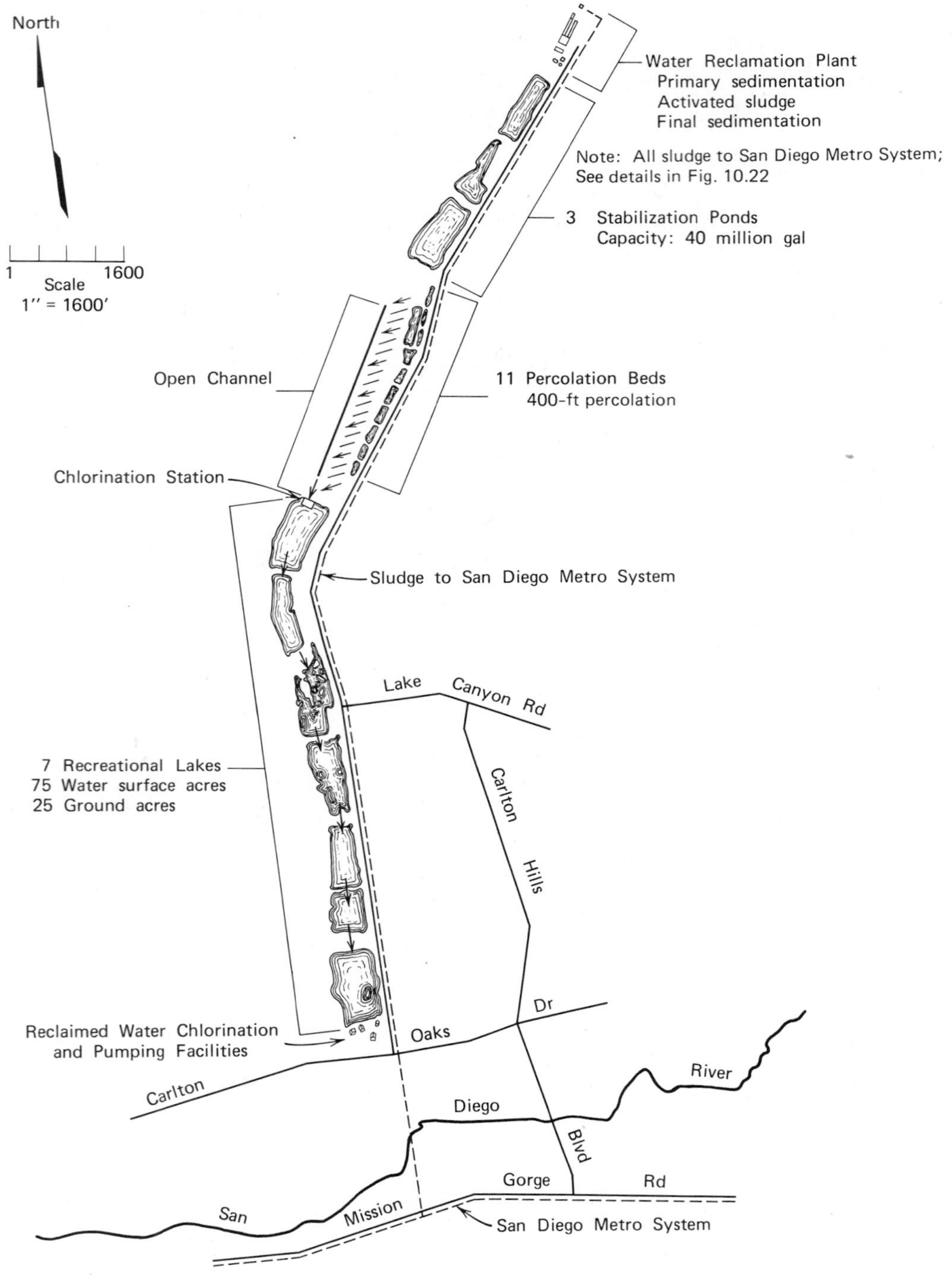

(b)

Fig. 10.21 *(continued)*
to the point where reclaimed water is pumped to irrigation, or recharges ground water. Sludge does not enter the San Diego River, but is pumped to the San Diego Metro System. For details of the system see Figure 10.23.

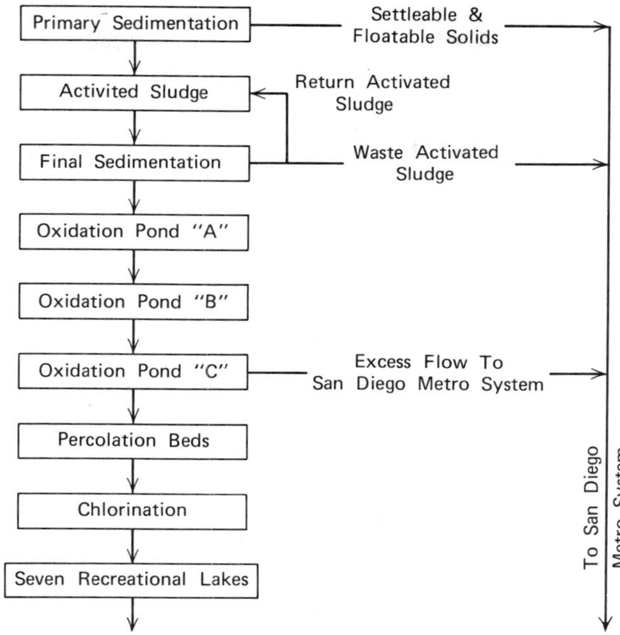

Fig. 10.22 *Santee Water Reclamation System, Padre Dam Municipal Water District. This "laundry" process cleans up and makes reusable the valuable water that constitutes the major part of raw sewage.*

Contrary to common opinion that such a concept might not be acceptable to the public, an effective public relations program concurrent with the technological advances has won over the citizens of the district.

The project involved building a sewage treatment plant and utilizing seven pits left over from a prior unrelated operation consisting of surface mining of sand and gravel. After partial purification of the sewage at the plant by the first two stages of the conventional activated sludge process, the effluent is discharged to form lakes adjacent to the plant. This provides tertiary (oxidation) treatment after which the purer effluent is pumped to a filter area at the north end of the complex where it is further purified by flow through sand and gravel. Chlorination is administered at one point. The water then flows through a series of seven lakes Nos. 7–6–5–4–3–2–1, in that order [see Fig. 10.21].

The appearance of new lakes in this semiarid region created an initial interest that was rapidly augmented by a very clever program. At first the lakes were fenced in. Then the fences were removed. The seven lakes were next made available for boating. They were stocked with fish and careful studies were made which indicated that the fish were healthy and flourishing. Fishing was permitted, but for a while all fish had to be returned to the water. Anglers were later permitted to keep and consume the fish. Finally swimming was permitted, a use which connotes pure water that is in no way harmful to health. The overflow from the last lake, No. 1, discharges into Sycamore Canyon Creek and is used for irrigation of a golf course and the recharging of groundwater.

The *Ward's Island Sewage Treatment Works* (Fig. 10.24) in New York City provides a vastly larger example. Handling over 200 million gpd, it serves large parts of two city boroughs, Manhattan and the Bronx. Along the shorelines, "intercepting" sewers were built to head off the multitude of street sewers previously emptying into

Santee Water Reclamation Plant

1 – Available but not in use
2 – Not available
3 – In use

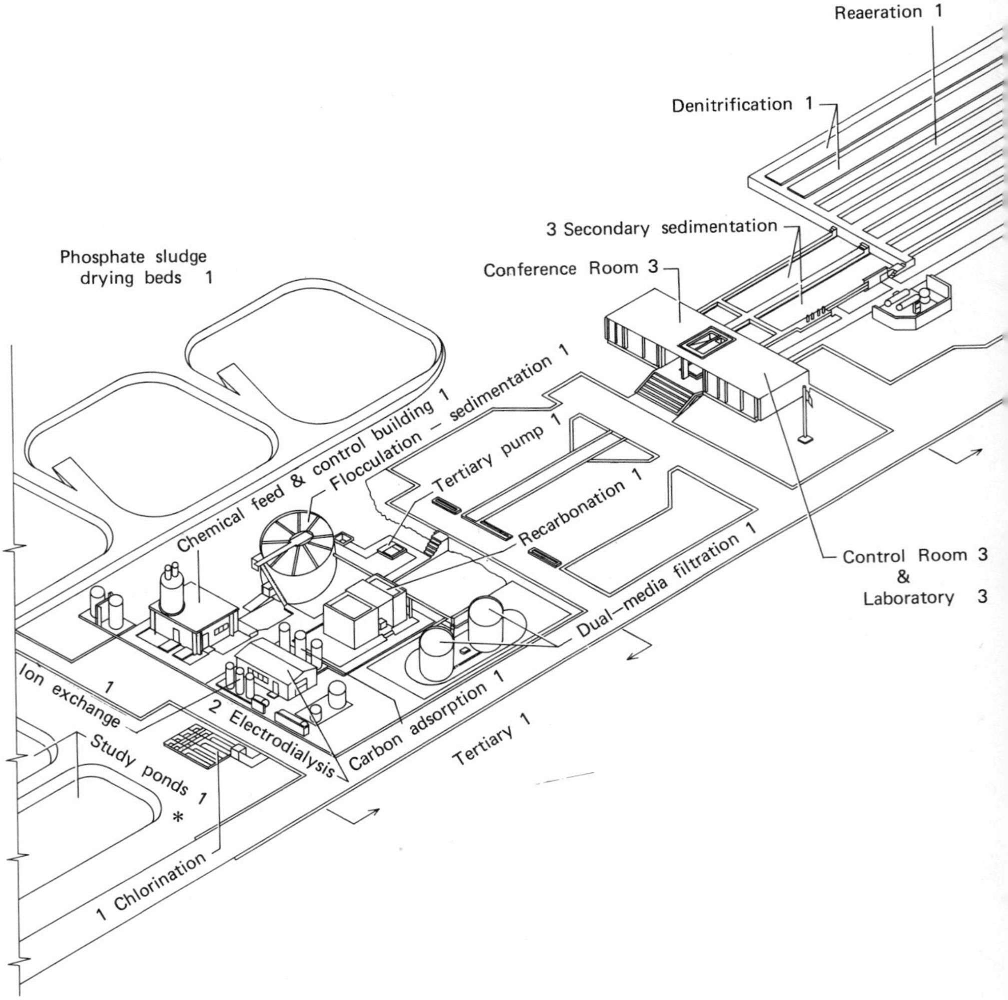

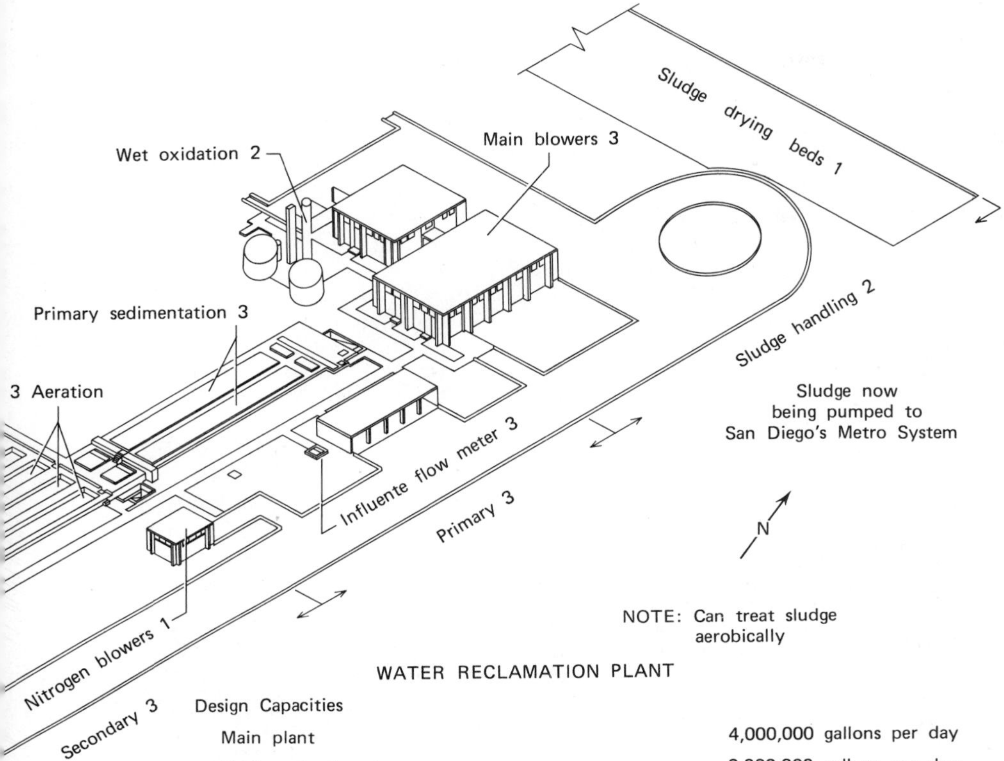

WATER RECLAMATION PLANT

Design Capacities	
Main plant	4,000,000 gallons per day
Tertiary treatment	2,000,000 gallons per day
* Potable water study ponds	100,000 gallons per day
Total Construction Cost	$3,825,000
U.S. Government Construction Grants	$1,497,000
Joint Demonstration Research Project	$511,000
Santee County Water District	
Federal Water Quality Administration	

Fig. 10.23. *Details of the reclamation plant. The item "Study Ponds," marked with an asterisk, indicates the continuing study and preparation for future distribution of drinking water that is reclaimed from sewage. The plant is not burdened by stormwater, which is recharged to the ground where it occurs locally in the community.*

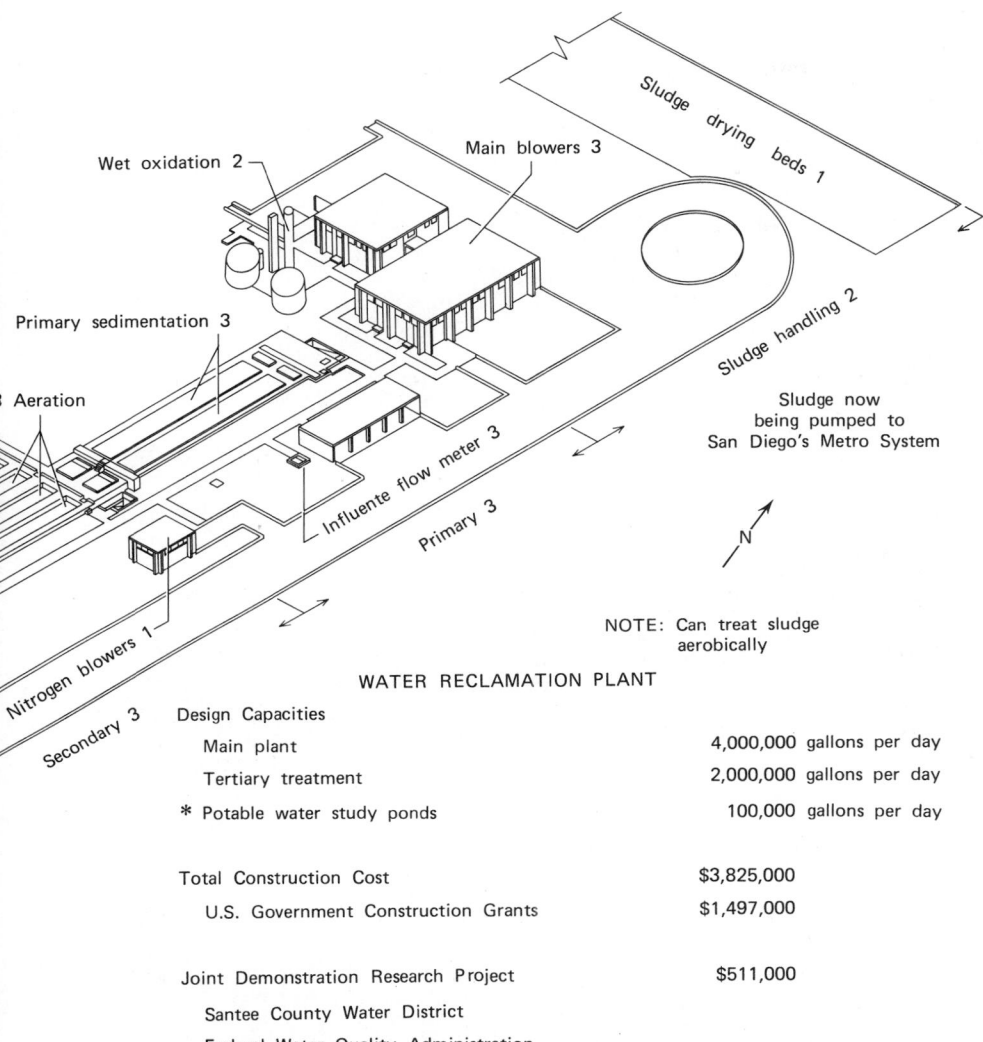

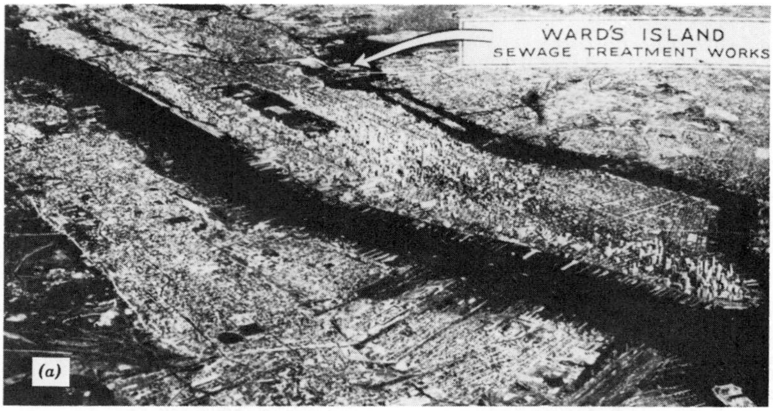

Fig. 10.24 *The Ward's Island Sewage Treatment Works, New York City.* (a) *In the center of the aerial photograph is Manhattan Island, identified by Central Park. The island is flanked in the foreground by the Hudson River and, on its other side, by the East River, identified by Roosevelt Island. To the north (upper left), Manhattan is separated from the borough of The Bronx by the Harlem River.* (b) *Activated sludge process.* (A) *Manhattan grit chamber.* (B) *Manhattan sewage tunnel.* (C) *Bronx sewage tunnel.* (D) *Laboratory and administration.* (E) *Power plant.* (F) *Pump and blower building.* (G) *Preliminary settling chambers.* (H) *Aeration chambers.* (I) *Final settling chambers.* (J) *Sludge storage building.* (K) *and* (L) *95% pure water discharge.* (M) *Dock for sludge boats.*

the waters surrounding the city. They lead the sewage through deep under-river tunnels to the island plant.

Ward's Island uses the activated sludge process—one of the more efficient methods of treatment. Compressed air is bubbled through the sludge so that aerobic digestion can occur. On a much larger scale, it is the same process as described in Fig. 10.12. Grit settles out at chambers in Manhattan and the Bronx. Together with other intercepted solids, such as pieces of wood, it is trucked to outlying districts, where it is dumped as "garbage fill" along with garbage from other sources. In the preliminary settling tanks, other heavy solids drop during an hour's pause of the sewage flow. The remaining, highly polluted fluid flows through the aeration tanks for a 3-hour trip in the presence of a biologically active culture. As a result of bacterial action, accelerated by air pumped in along the way, digestion takes place and the fluid can be made 95% pure before it is discharged at two points to the river. The resulting floc, digested and much purified, is collected in the final settling tanks. The practice of dumping this floc at sea was stopped in 1981.

10.3 Principles of Drainage

Early in the history of indoor plumbing, waste drainage was a simple matter: a pipe containing the waste water led to a sewer (Fig. 10.25*a*). Before long, the noxious gases that were created by the anaerobic conditions in the sewer became a threat to the health of those indoors. Thus, the *trap* was invented (Fig. 10.25*b*) to block the pipe so that gases could not pass. However, as moving water filled the pipe downstream from the traps, the mass of water acted as a plunger, creating higher pressures in front of it and negative pressures behind. The positive pressures might force sewer gas through the water in other traps; worse, the negative pressures could suck (or siphon) the water from the trap, leaving it open to gas pas-

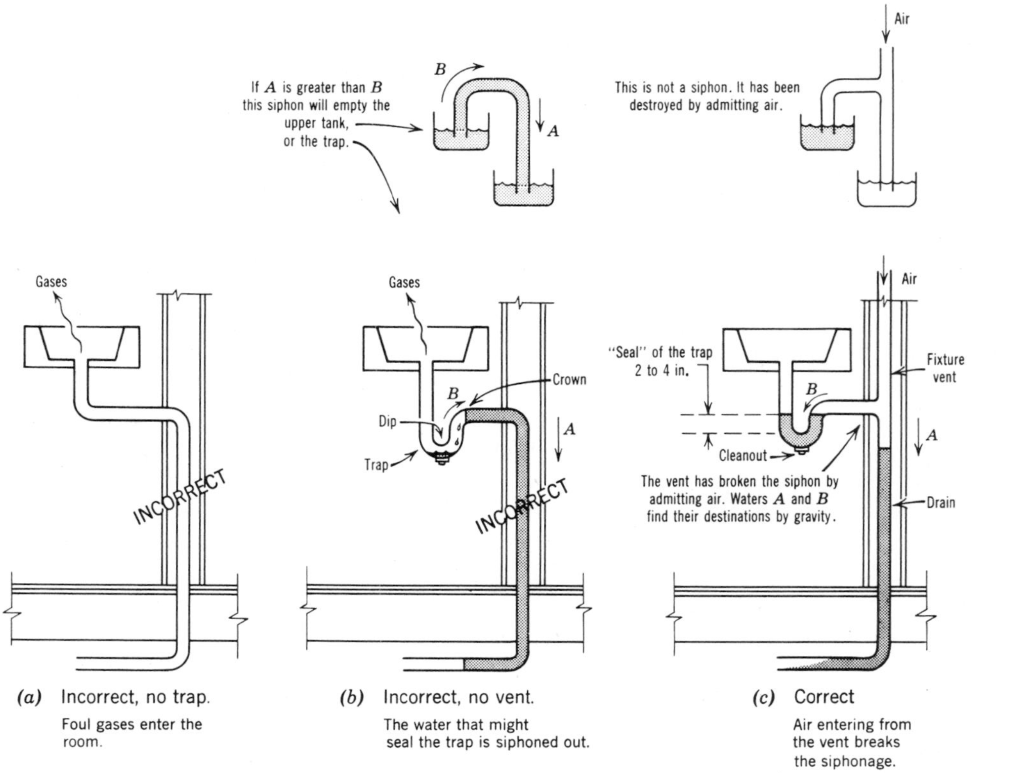

(*a*) Incorrect, no trap.
Foul gases enter the room.

(*b*) Incorrect, no vent.
The water that might seal the trap is siphoned out.

(*c*) Correct
Air entering from the vent breaks the siphonage.

***Fig. 10.25** The function of a trap and one of the several functions of a vent (preventing siphonage).*

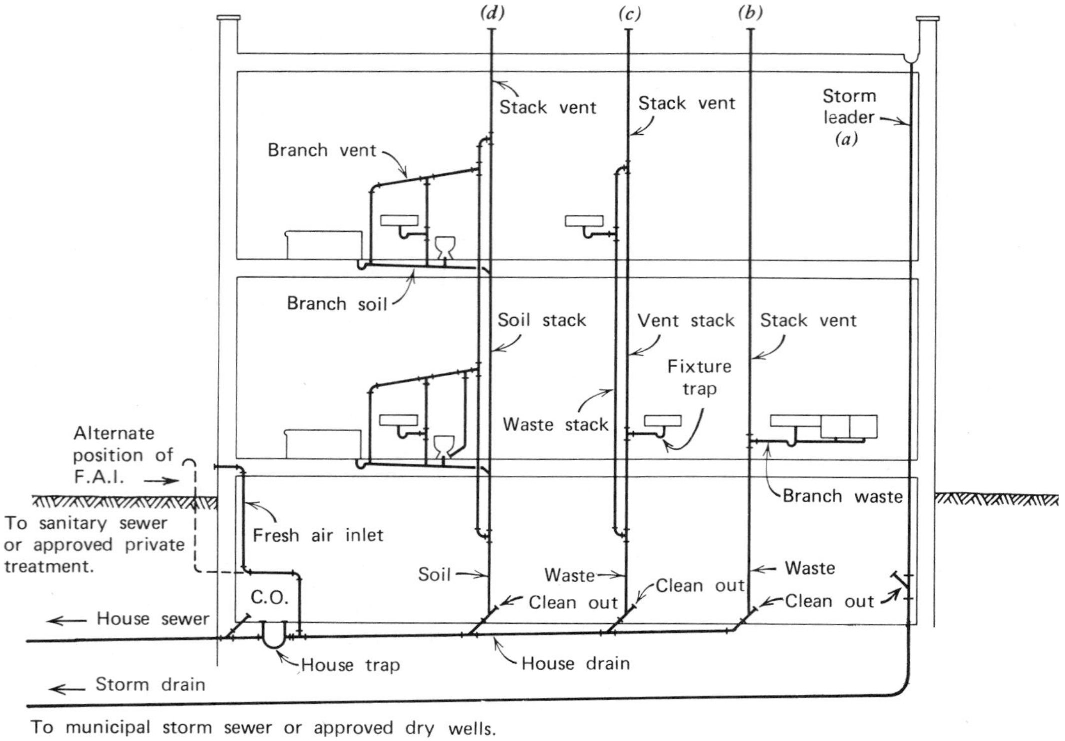

Fig. 10.26 A typical sanitary drainage system, which separates stormwater from sewage. This installation combines blackwater and graywater, although their separation would encourage graywater recycling. The housetrap is optional under some plumbing codes, and illegal under others (since sanitary sewers could be vented at street level through the "fresh air inlet").

sage. A way to deal with these pressures was through the installation of *vents,* so that the suction would draw air down through vents rather than water from the traps (Fig. 10.25*c*). A typical arrangement of fixtures, traps, and vents is shown in Fig. 10.26.

(a) Traps. The only separation between the unpleasant and dangerously unhealthy gases in a sanitary drainage system and the air breathed by room occupants is the water caught in the fixture trap after each discharge from a fixture. Sufficient water must flow, especially in water closets, to keep this residual water clean. Traps are made of steel, cast iron, copper, plastic, or brass—except those in water closets and urinals, which are often made of vitreous china cast integrally with the fixture. The deeper the seal, the more resistance to siphonage but the greater the fouling area; therefore, a minimum depth of 2 in. and a maximum depth of 4 in. are common standards. All traps should be self-cleaning: that is, capable of being completely flushed each time the trap operates, so that no sediment will remain inside to decompose.

There are a few exceptions to the rule that each fixture should have its own trap. Common exceptions include two laundry trays and a kitchen sink connected to a single trap, not more than three laundry trays using one trap, and three lavatories on a single trap. In the case of the laundry trays and sink, the sink is equipped with the trap and set nearest to the stack. (See stack *b* in Fig. 10.26).

Traps are usually placed within 2 ft of the fixture and should be accessible for cleaning through a bottom opening that is otherwise closed by a plug. Overflow pipes from fixtures are connected into the inlet side of the trap. In long runs of horizontal pipe, so-called ''running traps'' are used only near the drains of floors, areas, or yards and should be provided with hand-hole cleanouts. ''Island'' sinks pose a special problem when the vent line cannot lead upward from such an exposed location. The sink's waste line can be taken to a distant sump, which is then itself trapped and vented. (See Fig. 10.28*c* for a similar process.)

When fixtures are used very infrequently, the water in traps can evaporate into the air, breaking the seal of the trap. In contemplating the possible frequency of use, this fact should be kept in mind by the designer. Unoccupied residences (such as weekend or vacation homes) are likely candidates for sewer gas penetration through traps emptied by evaporation. Otherwise, evaporation to a dangerous degree rarely occurs, except in the case of floor drains. Trapped drains of this type, employed to carry away the water used in washing floors or drained from heating equipment, may often lose the water seal between infrequent operations. Many authorities are reluctant to approve floor drains connected to the building's sewer, requiring instead that they be separately connected to a dry well. In either case the use of a special hose bibb, affording a source of water directly above the drain, is a wise precaution. It can easily be used to manually refill the trap of the drain, or, overflow lines from lavatories can lead to the floor drain trap.

(b) Vents. For the admission of air and the discharge of gases, soil, and waste, stacks are extended through roofs and a system of air vents, largely paralleling the drainage system, is provided. As in the case of drainage stacks, the ventilating stacks extend through the roof or vent through the drainage stack. The functions of venting are often misunderstood. It is true, of course, that one important purpose is to ventilate the system by allowing air from the fresh-air inlet (or from the sewer, if there is no house trap or fresh-air inlet) to rise through the system and carry away offensive gases. This provides some purification for the piping. However, several other purposes are served by the vent piping. The introduction of air near the fixture (and in the case of circuit vents, at the branch soil line) breaks the possible siphonage of water out of the trap. Under other circumstances—namely, when drainage fluids descend to a fixture group through the soil stack—the foul gases would bubble through the trap-seals of that group. The vent system provides a local escape for these gases. Comprehensive experiments have shown that circuit venting—which permits air and gases to pass in and out of the soil or waste branch instead of at each fixture, as in the case of continuous venting (individual fixture venting)—can fully prevent the siphonage of trap-seals or their penetration by gases (see Fig. 10.27).

(c) Air Gaps and Vacuum Breakers. Every plumbing fixture is supplied with pure water at one point; most discharge contaminated fluids at another. The proximity of sewage to potable water at fixtures is inescapable. It is possible that sewage could accidentally be siphoned into a pipe carrying potable water. Consider an improperly placed faucet whose outlet is below the rim of a fixture. If the fixture overflow is plugged and the fixture bowl full, the faucet can easily project into the foul drainage water. If, in this circumstance, the water piping is drained while the faucet is open, contaminated water could be drawn by suction into the water piping.

In water closets served by flushometers (flush valves), the water supply unavoidably enters the bowl below the rim. A vacuum breaker placed in the flushometer closes with water pressure but opens to admit air if there is suction in the water pipe. This prevents siphonage in much the same way that a vent prevents trap siphonage (see Fig. 10.28). The use of vacuum breakers at dishwashers and clothes washers is diagrammed in Figure 9.30. These are especially important locations, since in both these appliances pumps force the waste water into the drainline.

10.4 Piping, Fittings, and Accessories

(a) Piping and Fittings. The principal materials used for soil and waste piping and for venting are cast iron, copper, and plastic. Galvanized steel is sometimes chosen for vents and for tall stacks in high-rise structures (see Fig. 10.29). Sometimes, different materials are used in the same system. Where dissimilar metals are connected, dielectric unions are used to prevent corrosion due to electrolysis.

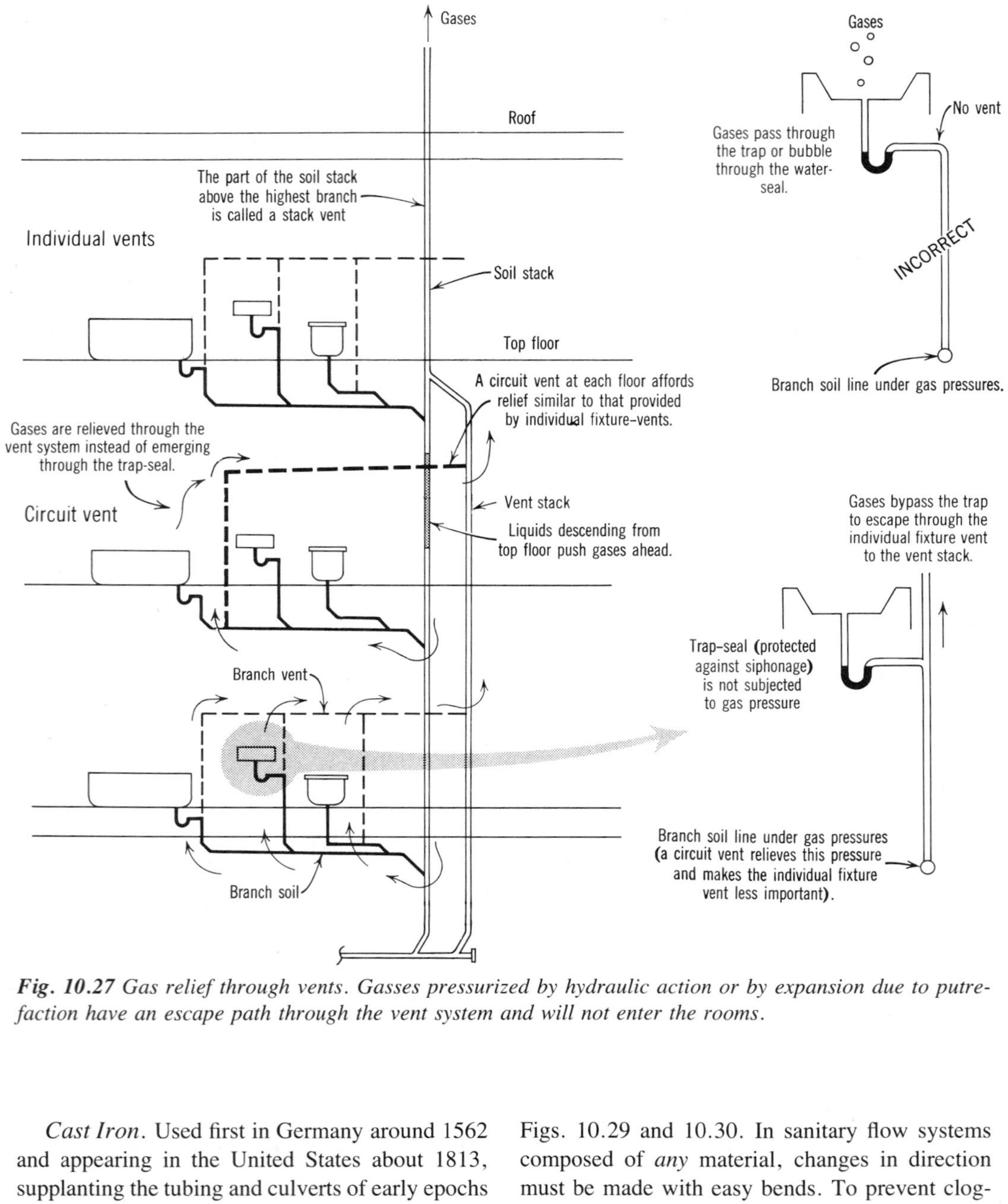

Fig. 10.27 *Gas relief through vents. Gasses pressurized by hydraulic action or by expansion due to putrefaction have an escape path through the vent system and will not enter the rooms.*

Cast Iron. Used first in Germany around 1562 and appearing in the United States about 1813, supplanting the tubing and culverts of early epochs that employed clay, lead, bronze, and wood, cast iron was the earliest of the modern materials used for piping. Its durability and resistance to corrosion makes it eminently suitable for the components of sanitary drainage systems. It is appropriate for a wide range of uses, from small residential work to the stacks and branches of tall buildings.

Typical fittings for sanitary drainage appear in Figs. 10.29 and 10.30. In sanitary flow systems composed of *any* material, changes in direction must be made with easy bends. To prevent clogging or fouling by the solid materials in the piping, right-angle connections are not used. Thus, the choices in Fig. 10.29 would be for ⅛ bend plus a 45° **Y**, or a ¼ bend long sweep. The top connection of the 90 deg **T**, in the position shown, would connect *only* to a vent.

The three cast-iron soil pipe joints shown in Fig. 10.31 are semirigid, watertight, and gastight

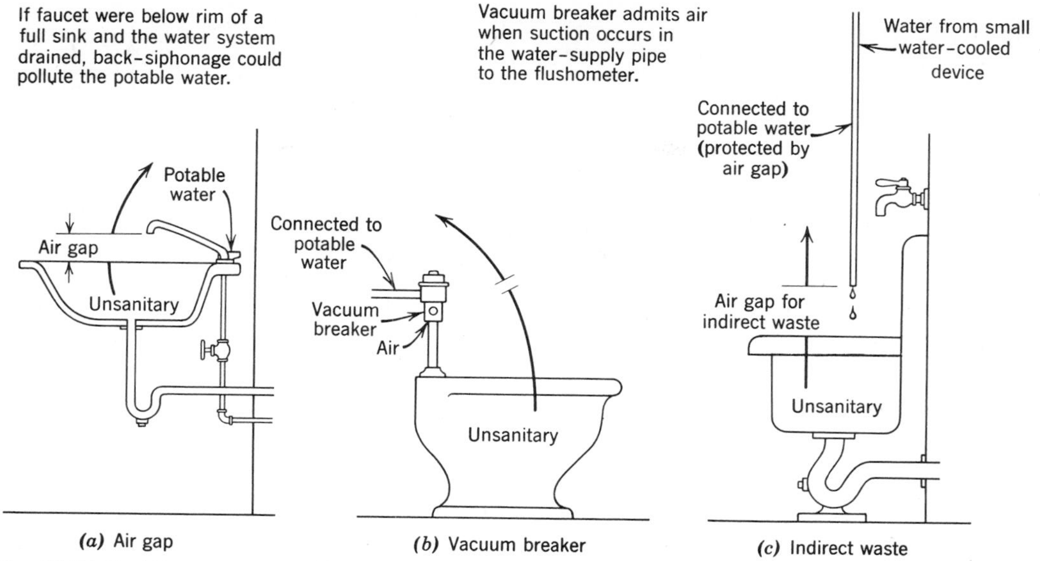

Fig. 10.28 Backflow preventers. Unsanitary fluid wastes cannot be siphoned into the potable water piping.

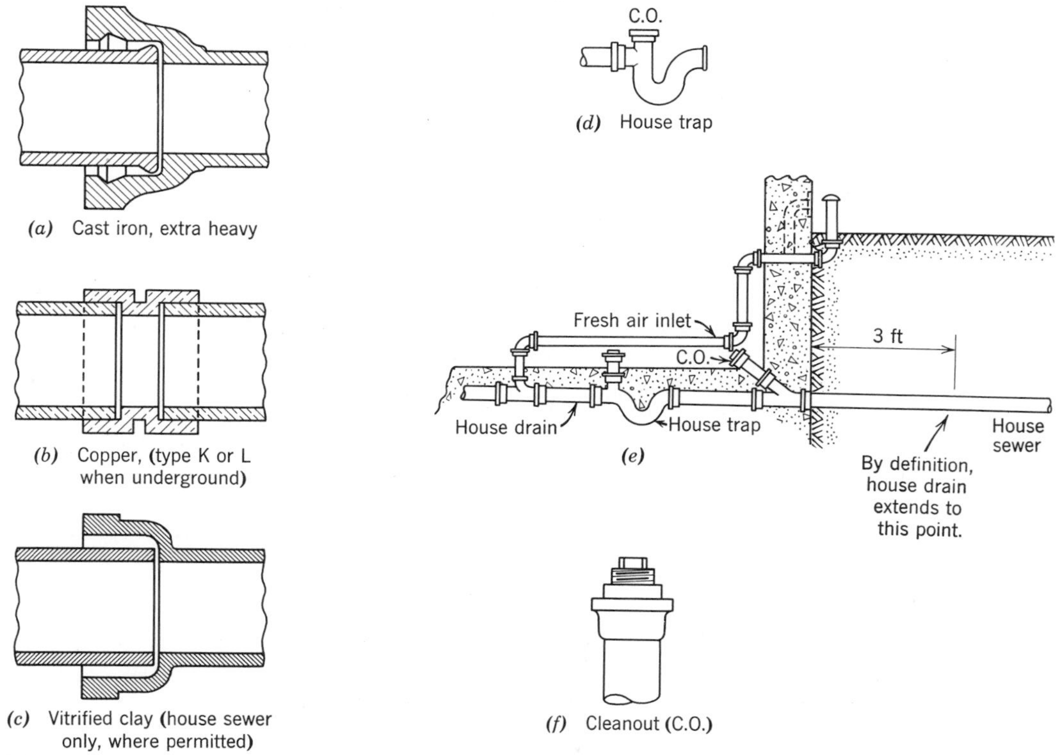

***Fig. 10.29** Piping and fittings.* (a) *Connection of cast iron piping.* (b) *Coupling to connect copper tubing.* (c) *Connection of vitrified clay piping.* (d) *Detail of house trap fitting.* (e) *House drain, house trap with cleanouts and vent (fresh air inlet), and house sewer.* (f) *Cleanout showing removable threaded plug. For large buildings the terms building drain, building sewer, and so on supplant "house." The inclusion or omission of the house trap depends on the local code.* NOTE: *Plastic connection is similar to (b).*

Fig. 10.30 Cast iron fittings: principal types and method of flashing at roofs.

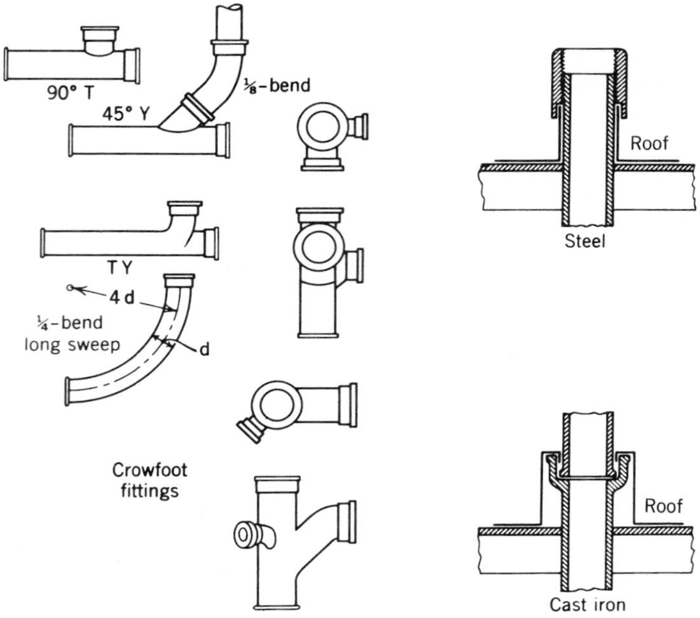

Fig. 10.31 The various joints used to connect cast-iron soil pipe and fittings. Courtesy of Cast Iron Soil Pipe Institute.

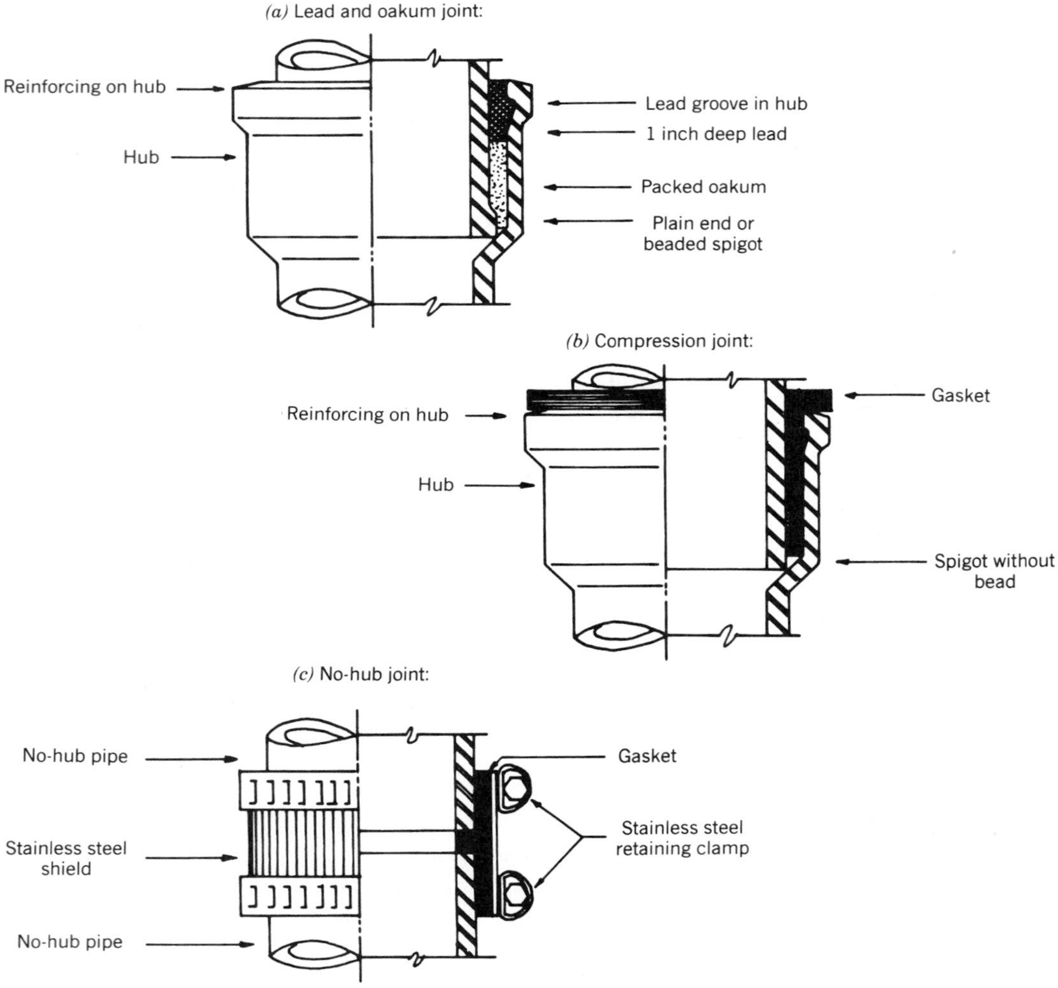

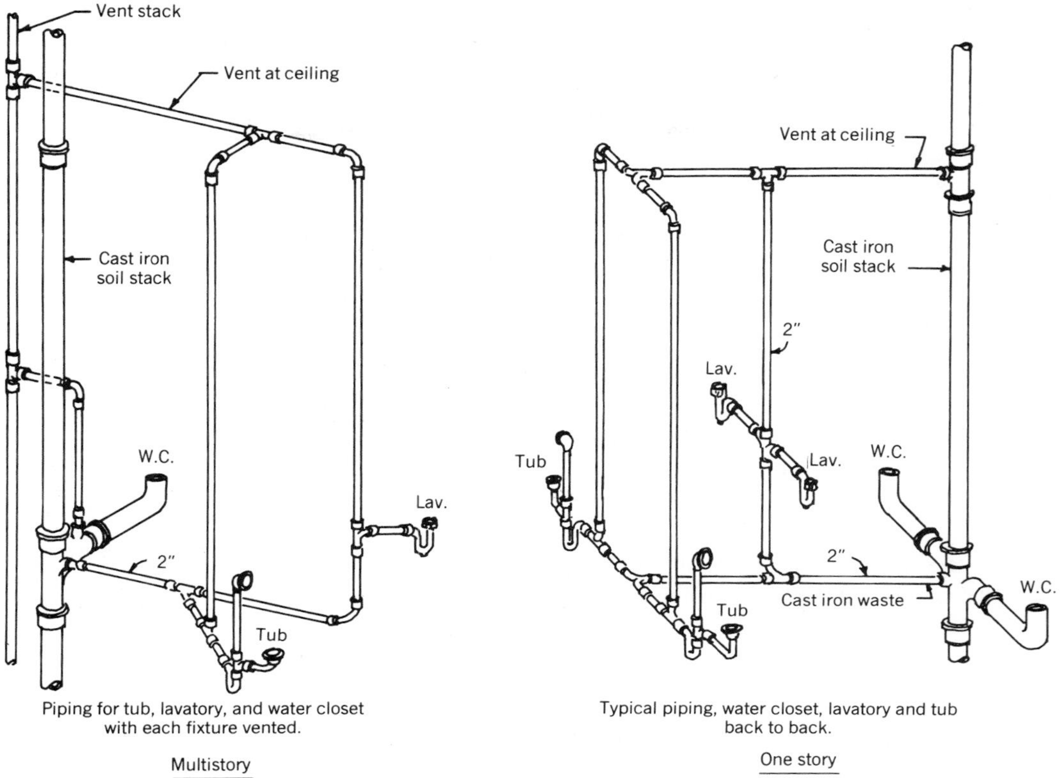

Fig. 10.32 Two typical piping arrangements for water closet, lavatory, and tub. Courtesy of Cast Iron Soil Pipe Institute.

connections of two or more pieces of pipe or fittings in a sanitary system. A special characteristic common to types *b* and *c* is that they provide a quieter plumbing system and slightly more flexible joints. The use of cast-iron soil pipe and fittings in bathroom groups can be seen in Fig. 10.32.

Copper Tube and Fittings for DWV. There are several tube classifications for the copper products used in *plumbing* systems: K, L, and M are the choices for *water* systems, and DWV for *drainage, waste,* and *vent* installations (as the initials indicate). Connections between copper tubing and its couplings or fittings are made by a sliding fit (see Fig. 10.29*b*). Between the mating surfaces is a cylindrical capillary space filled with solder. The process of making the joint is a simple one, utilizing a ''flux,'' heat, and solder. Properly made, the joint will be airtight and capable of withstanding high pressure. (High-temperature solder should be used for hot water lines, such as those to solar collectors.) To undo the joint for repair or renovation, it is simply reheated until the solder melts. Like cast iron, copper has a history of use in ancient installations. Updated and highly developed in recent decades, it is now in widespread use.

Plastic Materials for DWV. Along with copper and cast iron, plastics are very suitable for sanitary drainage systems. Plastics comprise a *family* of materials. Table 10.7 lists the three kinds of plastics most suitable for drainage, waste, and vent. One of these materials, acrylonitrile-butadiene-styrene (ABS), is identified and further evaluated by the labeling shown in Fig. 10.33. One of several steps used in making a ''solvent-weld'' connection is seen in Fig. 10.33, as is a method of support in wood frame construction.

TABLE 10.7 **Suitable Choices of Material for Plastic Piping in DWV (Drainage, Waste, and Vent) and Sewer Systems**

Symbol	*Material*	*DWV*	*Sewer*
ABS	Acrilylonitrile-Butadiene Styrene	√	√
PVC	Polyvinyl Chloride	√	√
SRP	Styrene Rubber Plastic		√

Source: Reprinted by permission of *Progressive Architecture.*

Fig. 10.34 shows assembled bathroom piping in place, and Fig. 10.35 illustrates the lightness of the material and its adaptability to prefabrication.

(b) Accessories. Among the many special devices that can form part of a plumbing system, a few are described here, including floor drains, backwater valves, ejectors, and interceptors.

Floor Drains. When floors in buildings must be washed down after such operations as food preparation and cooking, floor drains are usually necessary. Since they are sometimes connected to sanitary drainage systems and, in long periods of disuse, might lose their trap-seals by evaporation, special precautions are necessary to preserve the trap-seal and avoid odors and unsanitary conditions in the room (see Fig. 10.36).

Backwater Valves. These devices (Fig. 10.37) are sometimes used when plumbing fixtures are installed at low elevations, such as in basements, or in other locations that are near the level of the sewer. They cannot be used to protect the entire plumbing system, and should only be used when

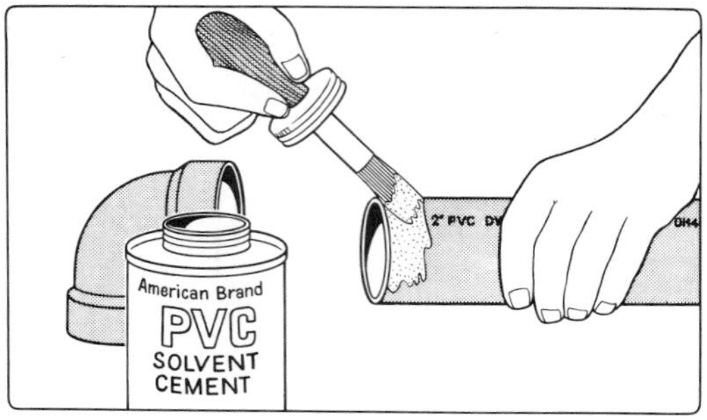

(a)

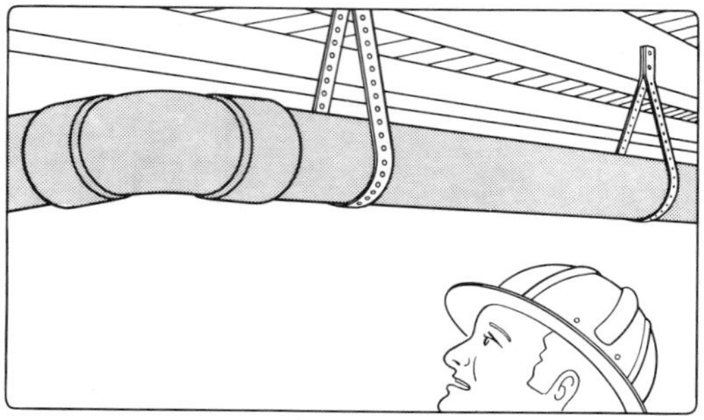

(b)

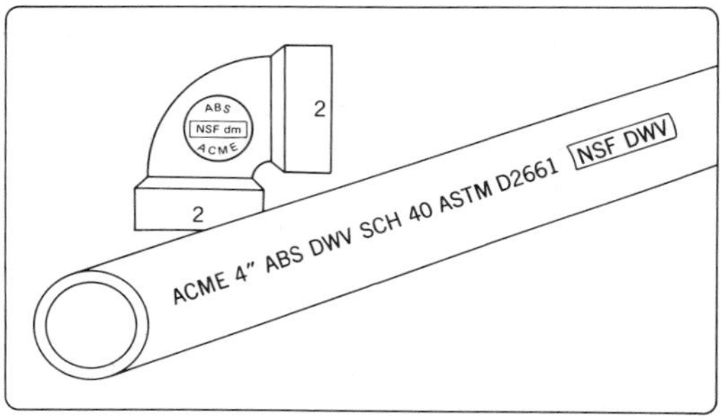

(c)

ACME	The name of the manufacturer.
4 in.	Diameter of the pipe.
ABS	Acrylonitrile-Butadiene-Styrene, the material.
DWV	Suitable for drainage waste and vent.
SCH 40	Schedule 40. This identifies the wall thickness of the pipe.
ASTM D2661	"Standards Number" assigned by the American Society for Testing Materials.
NSF DWV	Tested by the National Sanitation Foundation Testing Laboratory. The pipe meets or exceeds the current standards for sanitary service.

Fig. 10.33 *Details in the use of plastic pipe.* (a) *One of the steps in making a "solvent weld" of a plastic pipe to a plastic fitting.* (b) *In wood frame construction, plastic pipe assemblies can be supported by metal straps nailed to the wood joists. Flexibility of the plastic material suggests that the supports be more closely spaced than in the case of metal piping.* (c) *Typical identification symbols on plastic pipe. Courtesy of Plastic Pipe Institute.*

Fig. 10.34 *Plastic piping (solid lines) for water service, gas service, hot and cold water lines, and for drainage, waste, and vent. Gas service below grade can be PE, PB, or PVC. Courtesy of Plastic Pipe Institute.*

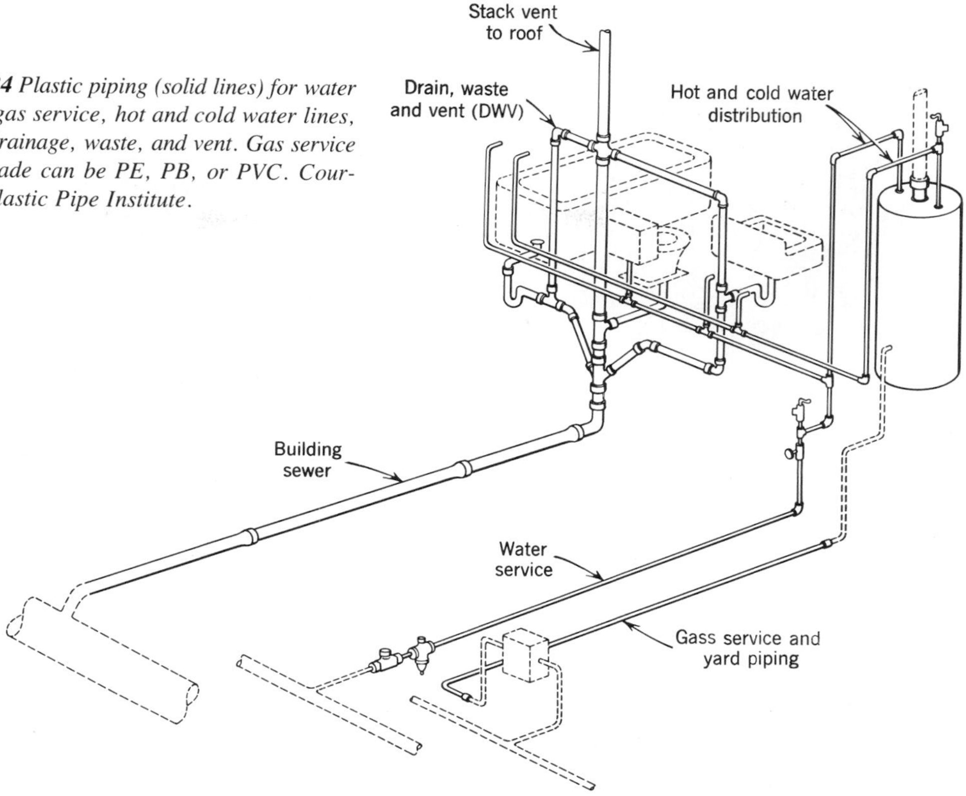

Fig. 10.35 *Plastics lend themselves to preassembly of sections of DWV piping. Materials used for DWV are polyvinyl chloride (PVC), either tan or white, and acrylonitrile-butadiene-styrene (ABS), black. The architect or engineer should check structure that may be cut to accommodate piping. Notice in this picture that the notched joists are deeper and more closely spaced than other floor beams. Courtesy of Plastics Pipe Institute. Photo by Richards Studio.*

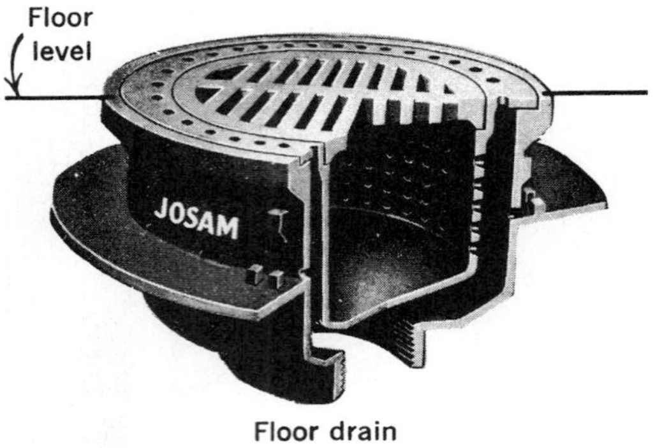

Fig. 10.36 Floor drain. Josam Manufacturing Company.

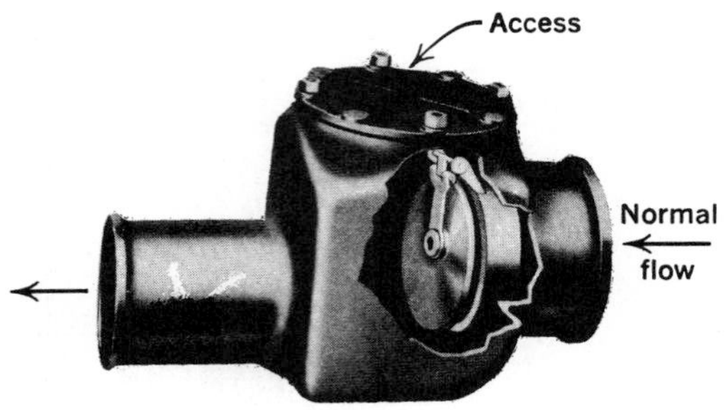

Fig. 10.37 Backwater valve.

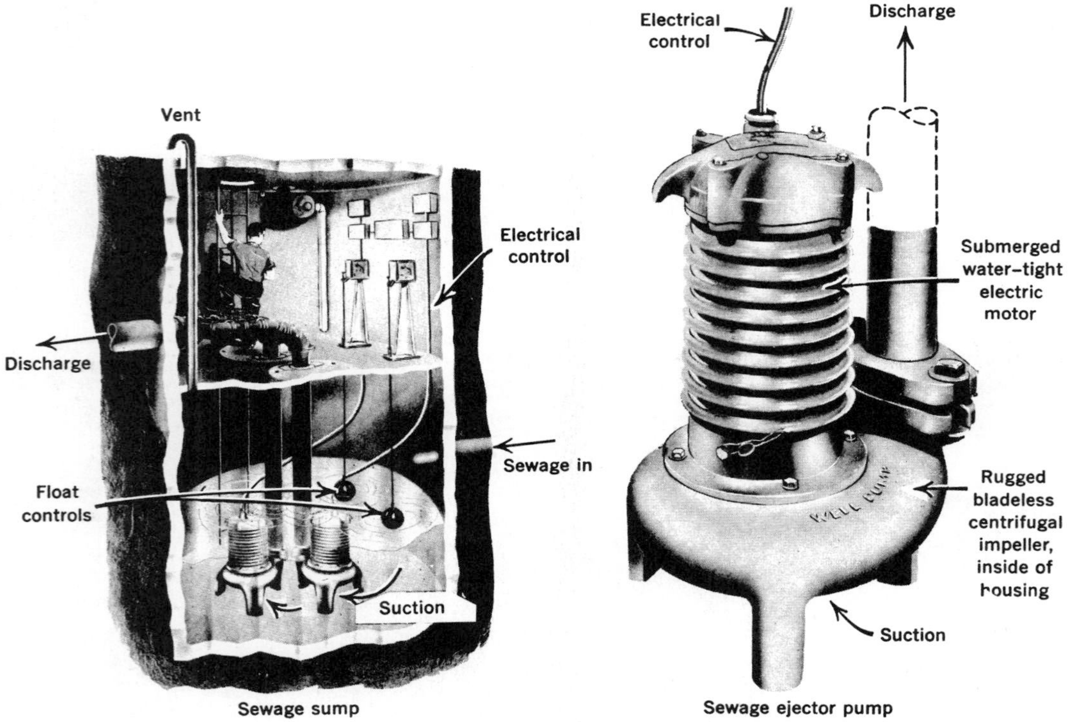

Fig. 10.38 Sump and ejector: a submersible-type centrifugal pump for raising sewage to a higher level. This principle, shown here for an outdoor subgrade sewer installation, may be used in basement applications within buildings. Venting must be carried to roof. Weil Pump Company.

necessary. They must be accessible for maintenance. An alternative to the use of backwater valves is the sewage sump.

Sewage Sumps and Ejectors. Whenever subsoil drainage, fixtures, or other equipment are situated below the level of public sewers, a sump pit or receptacle must be installed. Into this pit the drainage from the low fixtures may flow by gravity, and from it the contents are then lifted up into the building sewer. Sewage ejectors may be motor-driven centrifugal pumps (see Fig. 10.38) or they may be operated by compressed air. The latter have no revolving parts within the receptacle. An air compressor is started when the float within the sump reaches a certain level, and air at a pressure greater than 0.433 psi for each foot of lift is delivered into the space above the liquid. The air pressure closes the inlet and opens the outlet check valves, expelling the contents of the sump and elevating it to the sewer.

Interceptors. Sanitary drainage installations ultimately discharge their waste matter into private or public sewage treatment plants that attempt to digest or cope with anything that may come through the pipes. Public plants are somewhat better equipped to handle this problem than private installations. Of course, it is true that from any plumbing fixture to the end of the disposal process, all parts of systems should be openable through cleanouts and other points of access, to relieve clogging that will often occur in the piping as well as in the septic tank or public disposal plant. Since it is quite impossible to control human judgment about what should or should not be discarded into the plumbing drains, trouble can usually be expected. This problem can be reduced somewhat by devices known as interceptors, which catch foreign matter before it travels too far into the system.

It is evident that interceptors require periodic servicing. Interceptors for as many as 25 different kinds of extraneous material are listed by some manufacturers. They include devices to catch hair, grease, plaster, lubricating oil, glass grindings, or troublesome unwanted material from many industrial processes. One of the few interceptors that is sometimes needed in homes, and more often in institutional kitchens, is the grease interceptor, or *grease trap*.

As the waste is passed from a kitchen sink through the circuitous path of the grease interceptor, the grease floats to the top, where it is trapped between baffles, while the more fluid wastes pass through at a lower level. There are special reasons for removing grease; for one thing, it congeals within piping and thus physically retards the sewage digestion process (see Figs. 10.39*a* and 10.39*b*).

(c) Examples of Installations. In residential work, the piping assemblies may often be viewed

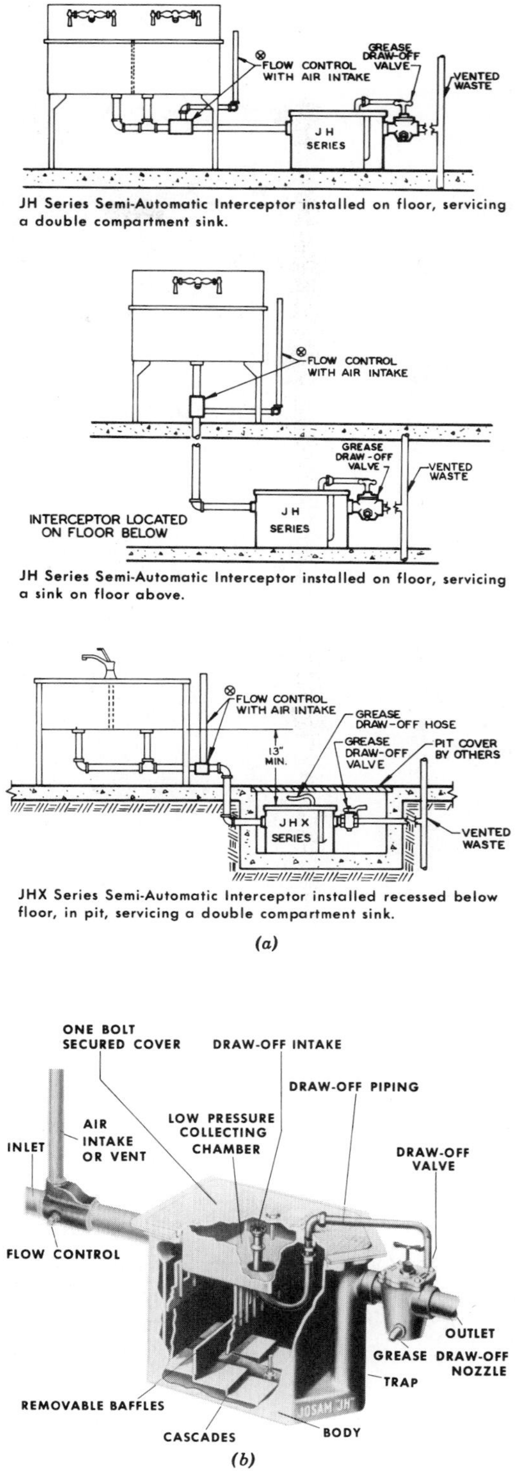

Fig. 10.39 (a) *One type of institutional interceptor, a grease trap. Choice of three locations—adjacent to sink, on floor below, or in a pit.* (b) *Cutaway view with identification of component parts. Josam Manufacturing Company.*

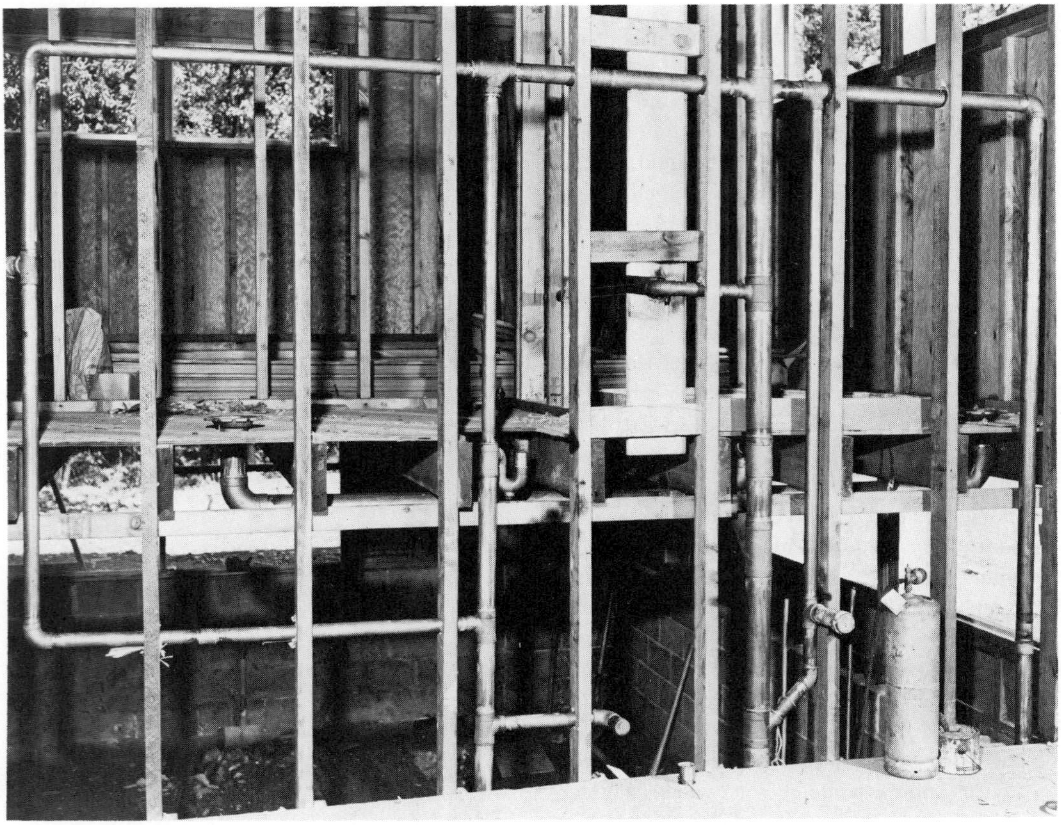

Fig. 10.40 *Drainage vent pipe in a frame residence. This installation in copper "drainage waste and vent tubing" (DWV) serves two bathrooms at the upper level behind the 6-in. stud partition and a kitchen sink and laundry tray at the lower level, which are on this side of the partition. In the bathrooms the roughing serves (from left to right) a lavatory, water closet, and bathtub and a lavatory, shower, and water closet. Bathtub and shower traps can usually be accommodated within the joist depth. The bend below the water closets, however, often leads to a horizontal branch exposed or furred-in below the joists. Some codes permit this branch from a water closet to be up to 6 to 10 ft long before it joins a vent. Courtesy of Copper Development Association.*

as a "flag," as can be seen in Fig. 10.40. The mast is the soil stack, the horizontal top of the flag is the branch vent, the bottom is the soil or waste branch, and the outer edge is the vertical pipe of the last fixture. In frame construction, the flag usually fits into a 6-in. partition. Fixture branches project from the surface of the flag. There is considerable advantage in "back-to-back" planning of baths and kitchens; this allows the piping assembly to pick up the drainage of fixtures on both sides of it. When all the fixtures are on nearly the same level, as in Fig. 10.40, it is unnecessary to have a separate vent stack standing beside the soil stack, as is often the case in multistory construction. In this illustration, and generally in one-story construction, the upper part of the soil stack forms a vent called a stack vent, to which the branch vents connect. A separate major vertical vent would be called a vent stack.

Multistory construction, especially in office buildings, must be flexible and free of random partitions that would interfere in the periodic replanning of interior spaces and the relocating of dividing partitions. The use of "cores" has offered a solution to this problem. Risers of the various systems are grouped in planes that coincide

Fig. 10.41 Risers in a fireproof, multistory building. Pipes, tubes, conduits, and ducts virtually enclose toilet rooms and utility spaces. Ventilation ducts and a master 5-in. copper hot-water riser are just left of center. Soil and vent stacks with hot and cold water supplies, all of copper, are to the right of this group. At the left in lighter tone is the galvanized steel feeder conduit and distribution circuit conduits, of the same metal, for a local electrical control panel box. Note that some pipes and tubes are supported at this floor by bolted clamps. After testing and before pipes are enclosed, covering will be completed. Copper Development Association.

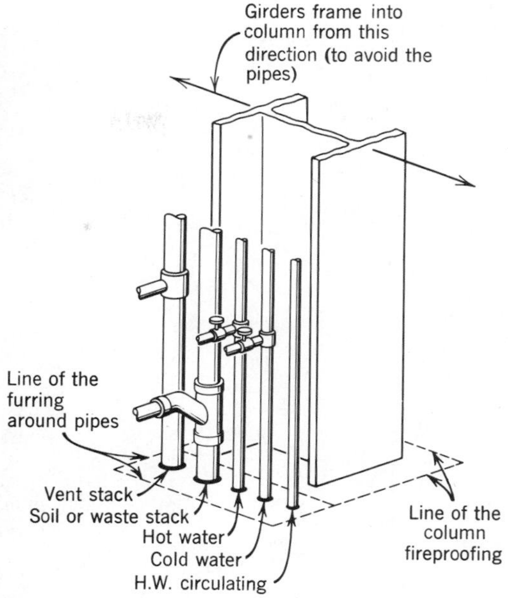

Fig. 10.42 Piping at a "wet" column. In large office buildings, there are usually several of these, remote from the core and out in the general office area.

with permanent partitions of block masonry. This island of fixed construction is often placed at the central section of the building, freeing the surrounding areas for access to daylight. A hole in the floor for each pipe is often chosen in preference to a slot or shaft. This method interferes less with the floor construction (Fig. 10.41).

Offices often need a single lavatory or a complete toilet room for executives at locations away from the central core of the building. "Wet" columns with a full complement of plumbing pipes make this possible. If the pipes are to accompany a column in a steel building, structural coordination must be sought early in the planning if the pipes are to clear the structural framing of the floor (Fig. 10.42).

In some installations, the branch soil and waste piping perforates a floor and crosses below the slab to join the stack. Tubing has been developed, however, that sits above the structural slab, obviating the need for hung ceilings below (see Fig. 10.43). A lightweight concrete fill is cast to cover the tubing, raising the floor by 5 or 6 in. This can create a raised floor in the toilet room—not the best planning—so the higher floor level is usually carried throughout the floor of the entire story, forming a convenient space into which the electrical conduit can be placed at a time later than would have been required if it were to be placed in the structural slab. This affords some freedom in construction, because conduits in the structural slab conflict with reinforcement, and, furthermore, they must be planned and placed earlier.

WATER AND WASTE

Fig. 10.43 *Horizontal waste piping* above *the structural slab, in an example of plumbing roughing for two lavatory rooms in a fireproof office building. A lavatory and water closet in each room are served by soil and waste branches below and vent branches above. Hot and cold water tubing with air chambers can be seen. Although the extensions of the water tubing above the two flushometer connections appear to connect into the horizontal vent branches, they actually do not; they are capped and merely touch the bottoms of the vent branches. Note that soil branches are above the structural slab. A fill of 5 or 6 in. will be necessary to cover the tubing. All vertical tubing will be within the masonry block used to enclose the cubicles. Copper Development Association.*

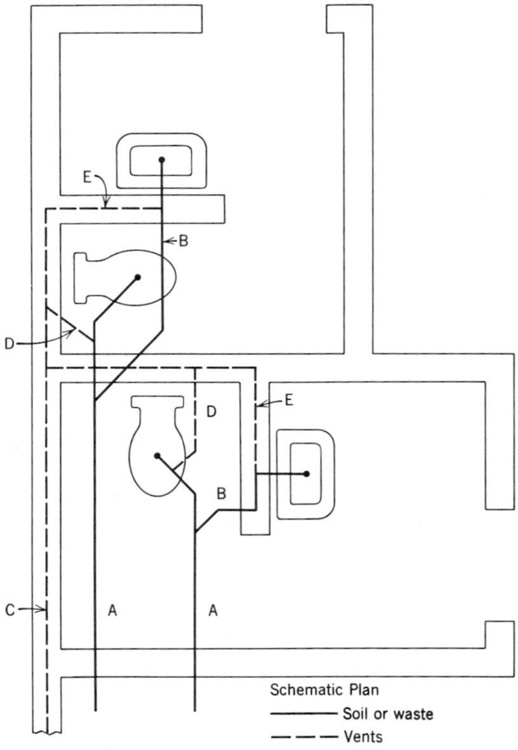

10.5 Design of Residential Waste Piping

The sanitary drainage requirements for a residence, plans of which are shown in Fig. 10.44, provide the basis for an illustrative example.

EXAMPLE **10.3.** Design, lay out, and size the piping for the sanitary drainage system for the house shown in Fig. 10.44.

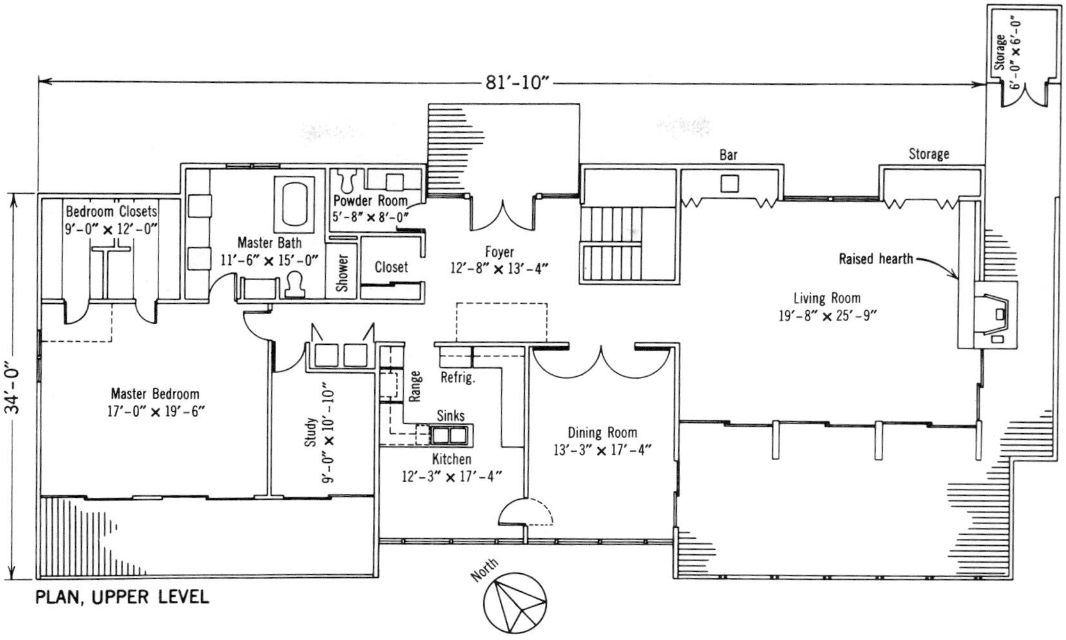

PLAN, UPPER LEVEL

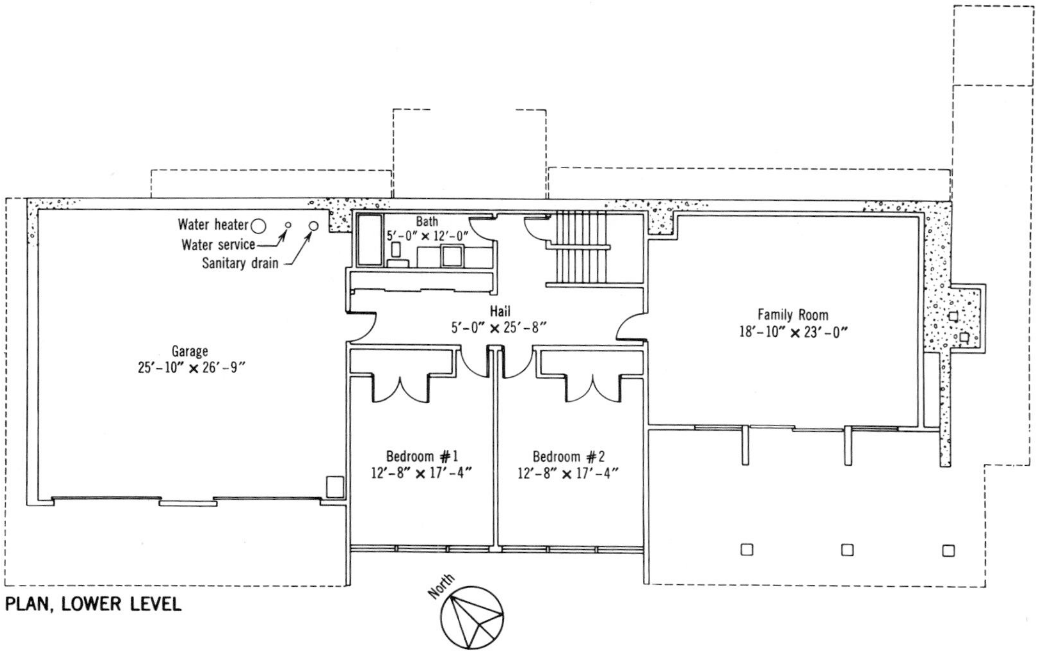

PLAN, LOWER LEVEL

Fig. 10.44 *(Example 10.3.) House on Long Island, Budd Mogensen, Architect and Planner. Floor plans to be used in the solution of the example.*

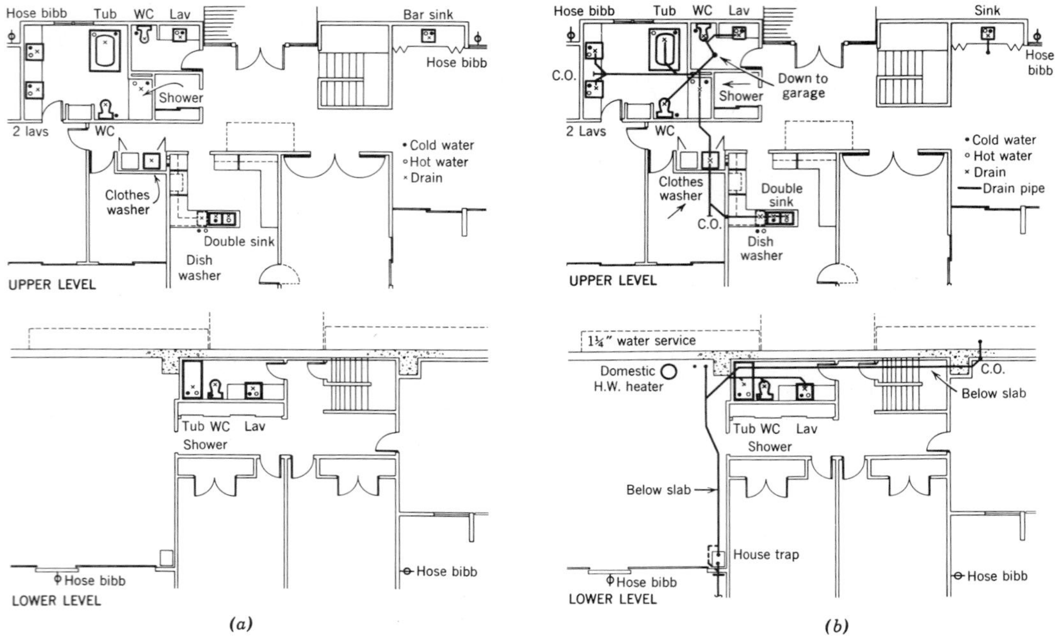

Fig. 10.45 *(Example 10.3.)* (a) *Plumbing requirements, water supply, and partial sanitary drainage.* (b) *Sanitary drainage plan.*

SOLUTION. The first step is to identify the locations where hot and cold water is needed at fixtures and where soil or waste drains must be provided. Figure 10.45*a* illustrates how this is done. A plan layout for the drains in both levels follows (see Fig. 10.45*b*).

The task of fitting the two plumbing ''distribution trees'' (supply and waste) into available horizontal and vertical chases can be difficult. This is especially true of the waste system, which has larger pipes, because it is not under pressure, and must carry solids as well as liquid. To compound the problem, because of gravity flow requirements these larger pipes must slope continuously downward, from fixtures to the sewer.

Next comes the ''plumbing section'' (Fig. 10.46). The local administrative authority usually requires this to be submitted for approval. Sizes of all piping are determined from Tables 10.8 through 10.12. Drainage fixture units (d.f.u.) for the system that we are designing are summarized in Table 10.13 from data given in Table 10.9. Runouts from *individual* fixtures should be the same size as the fixture trap. Branch runouts from *groups* of fixtures are sized by the drainage fixture units of the group. Although Table 10.11 permits a 3-in. branch for not more than two water closets, it is common practice to use a 4-in. runout for *every* water closet and not less than a 4-in. branch for every group of water closets. Since Table 10.11 allows 160 fixture units for a 4-in. branch, this size is acceptable for any branch to which a water closet is connected and, as can be seen, for any *stack*. The house drain cannot be less than 4 in., and because (at a ¼-in. fall per ft) a 4-in. house drain (Table 10.10) will carry 216 d.f.u., that size is chosen and is more than adequate for our 28 d.f.u. system (see Table 10.13). A water closet

Fig. 10.46 *(Example 10.3.) Plumbing section. When every fixture is vented individually, as in this example, the method is known as* continuous venting. *In larger systems, batteries of fixtures may be vented by a loop or circuit vent. This reduces the piping of the vent system (see Fig. 10.47).*

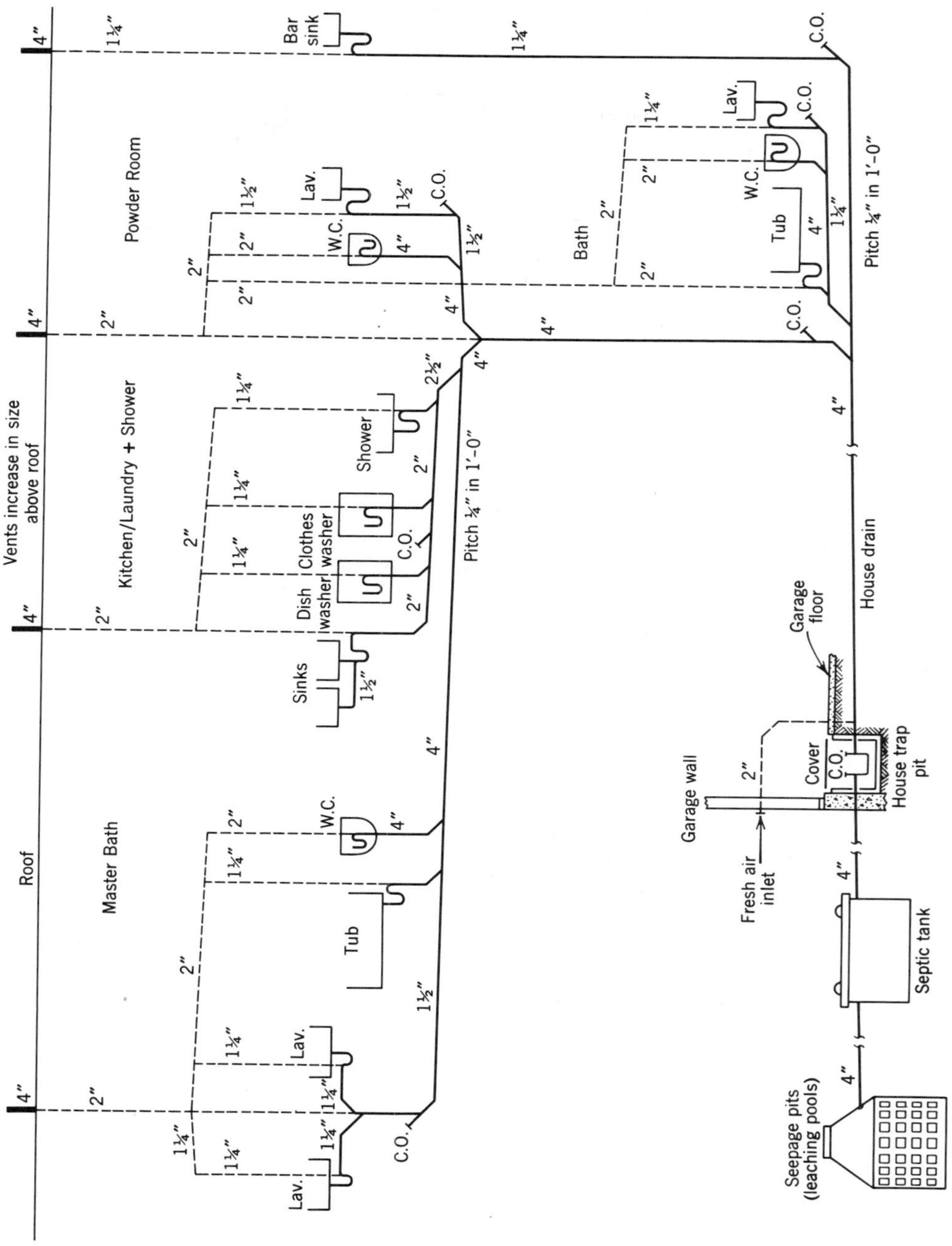
Roof
Vents increase in size above roof
Master Bath
Kitchen/Laundry + Shower
Powder Room
Lav.
Tub
W.C.
Sinks
Dish washer
Clothes washer
Shower
Bar sink
Bath
C.O.
Pitch ¼" in 1'-0"
House drain
Garage floor
Garage wall
Fresh air inlet
Cover
House trap pit
Septic tank
Seepage pits (leaching pools)

TABLE 10.8 **Size of Nonintegral Traps for Different-Type Plumbing Fixtures**

Plumbing Fixture	*Trap Size in Inches*
Bathtub (with or without overhead shower)	$1\frac{1}{2}$
Bidet	$1\frac{1}{4}$
Combination sink and wash (laundry) tray	$1\frac{1}{2}$
Combination sink and wash (laundry) tray with food waste grinder unit	$1\frac{1}{2}$[a]
Combination kitchen sink, domestic, dishwasher, and food waste grinder	2
Dental unit or cuspidor	$1\frac{1}{4}$
Dental lavatory	$1\frac{1}{4}$
Drinking fountain	$1\frac{1}{4}$
Dishwasher, commercial	2
Dishwasher, domestic (nonintegral trap)	$1\frac{1}{2}$
Floor drain	2
Food waste grinder—commercial use	2
Food waste grinder—domestic use	$1\frac{1}{2}$
Kitchen sink, domestic, with food waste grinder unit	$1\frac{1}{2}$
Kitchen sink, domestic	$1\frac{1}{2}$
Kitchen sink, domestic, with dishwasher	$1\frac{1}{2}$
Lavatory, common	$1\frac{1}{4}$
Lavatory (barber shop, beauty parlor or surgeon's)	$1\frac{1}{2}$
Lavatory, multiple type (wash fountain or wash sink)	$1\frac{1}{2}$
Laundry tray (1 or 2 compartments)	$1\frac{1}{2}$
Shower stall or drain	2
Sink (surgeon's)	$1\frac{1}{2}$
Sink (flushing rim type, flush valve supplied)	3
Sink (service type with floor outlet trap standard)	3
Sink (service trap with P trap)	2
Sink, commercial (pot, scullery, or similar type)	2
Sink, commercial (with food grinder unit)	2

Source: National Standard Plumbing Code.

[a]Separate traps required for wash tray for sink compartment with food waste grinder unit.

TABLE 10.9 **Drainage Fixture Unit Values for Various Plumbing Fixtures**

Type of Fixture or Group of Fixtures	*Drainage Fixture Unit Value (d.f.u.)*
Automatic clothes washer (2-in. standpipe)	3
Bathroom group consisting of a water closet, lavatory, and bathtub or shower stall:	
Flushometer valve closet	8
Tank type closet	6
Bathtub[a] (with or without overhead shower)	2
Bidet	1
Clinic Sink	6
Combination sink-and-tray with food waste grinder	4
Combination sink-and-tray with one $1\frac{1}{2}$-in. trap	2
Combination sink-and-tray with separate $1\frac{1}{2}$-in. traps	3
Dental unit or cuspidor	1
Dental lavatory	1
Drinking fountain	$\frac{1}{2}$
Dishwasher, domestic	2
Floor drains with 2-in. waste	3
Kitchen sink, domestic, with one $1\frac{1}{2}$-in. trap	2
Kitchen sink, domestic, with food waste grinder	2
Kitchen sink, domestic, with food waste grinder and dishwasher 2-in. trap	3
Kitchen sink, domestic, with dishwasher $1\frac{1}{2}$-in. trap	3
Lavatory with $1\frac{1}{4}$-in. waste	1
Laundry tray (1 or 2 compartments)	2
Shower stall, domestic	2
Showers (group) per head	2
Sinks:	
Surgeon's	3
Flushing rim (with valve)	6
Service (trap standard)	3
Service (P trap)	2
Pot, scullery, etc.	4
Urinal, pedestal, syphon jet blowout	6
Urinal, wall lip	4
Urinal, stall, washout	4
Urinal trough (each 6-ft section)	2
Wash sink (circular or multiple) each set of faucets	2
Water closet, tank-operated	4
Water closet, valve-operated	6
Fixtures not listed above:	
Trap size $1\frac{1}{4}$ in. or less	1
Trap size $1\frac{1}{2}$ in.	2
Trap size 2 in.	3
Trap size $2\frac{1}{2}$ in.	4
Trap size 3 in.	5
Trap size 4 in.	6

Source: National Standard Plumbing Code.

[a]A shower head over a bathtub does not increase the fixture unit value.

TABLE 10.10 **Building Drains and Sewers**[a]

Diameter of Pipe (in.)	Maximum Number of Fixture Units That May Be Connected to Any Portion of the Building Drain or the Building Sewer Including Drain Branches			
	Fall per Foot			
	1/16 in.	1/8 in.	1/4 in.	1/2 in.
2			21	26
2½			24	31
3		36[b]	42[b]	50[b]
4		180	216	250
5		390	480	575
6		700	840	1,000
8	1,400	1,600	1,920	2,300
10	2,500	2,900	3,500	4,200
12	2,900	4,600	5,600	6,700
15	7,000	8,300	10,000	12,000

Source: National Standard Plumbing Code.

[a]On-site sewers that serve more than one building may be sized according to the current standards and specifications of the Administrative Authority for public sewers.

[b]Not more than two water closets or two bathroom groups.

TABLE 10.11 **Horizontal Fixture Branches and Stacks**

Diameter of Pipe (in.)	Maximum Number of Fixture Units that May Be Connected to:			
	Any Horizontal Fixture Branch[a]	Stack Sizing for 3 Stories in Height or 3 Intervals	Stack Sizing for More than 3 Stories in Height	
			Total for Stack	Total at 1 Story or 1 Branch Interval
1½	3	4	8	2
2	6	10	24	6
2½	12	20	42	9
3	20[b]	48[b]	72[b]	20[b]
4	160	240	500	90
5	360	540	1,100	200
6	620	960	1,900	350
8	1,400	2,200	3,600	600
10	2,500	3,800	5,600	1,000
12	3,900	6,000	8,400	1,500
15	7,000			

NOTE: Stacks shall be sized according to the total accumulated connected load at each story or branch interval and may be reduced in size as this load decreases to a minimum diameter of ½ of the largest size required.

Source: National Standard Plumbing Code.

[a]Does not include branches of the building drain.

[b]Not more than two water closets or bathroom groups within each branch interval or more than six water closets or bathroom groups on the stack.

TABLE 10.12 **Size and Length of Vents**

Size of Soil or Waste Stack	Fixture Units Connected	Diameter of Vent Required (in.)								
		1¼	1½	2	2½	3	4	5	6	8
		Maximum Length of Vent (ft)								
Inches										
1½	8	50	150							
1½	10	30	100							
2	12	30	75	200						
2	20	26	50	150						
2½	42		30	100	300					
3	10		30	100	100	600				
3	30			60	200	500				
3	60			50	80	400				
4	100			35	100	260	1000			
4	200			30	90	250	900			
4	500			20	70	180	700			
5	200				35	80	350	1000		
5	500				30	70	300	900		
5	1100				20	50	200	700		
6	350				25	50	200	400	1300	
6	620				15	30	125	300	1100	
6	960					24	100	250	1000	
6	1900					20	70	200	700	
8	600						50	150	500	1300
8	1400						40	100	400	1200
8	2200						30	80	350	1100
8	3600						25	60	250	800
10	1000							75	125	1000
10	2500							50	100	500
10	3800							30	80	350
10	5600							25	60	250

Source: National Standard Plumbing Code.

TABLE 10.13 **Drainage Fixture Units (Example 10.3)**

	Units
Bar sink	2
Kitchen sink and dishwasher	3
Lavatory	1
Water closet	4
Clothes washer	3
Master bath, lavatory, water closet, tub	6
Extra lavatory	1
Shower	2
Lower bath, lavatory, water closet, tub and shower	6
	28
Values are from Table 10.9	
Hose bibb drainage to ground	
Roof drainage to dry wells	

must have a 2-in. vent. Table 10.12 shows that a 2-in. vent, for vent-lengths not exceeding 150 ft, will serve 20 fixture units. This size is acceptable for branch vents and stack vents. For individual fixture vents, the vent size is usually the same as the size of the fixture runout. Under most codes, vertical vents that penetrate the roof must increase to a 4-in. size to prevent blocking by icing in freezing weather.

It will be evident to the reader that in residential applications and in other relatively small buildings certain fairly standard minimum sizes, such as a 4-in. soil stack and 2-in. vent, are usually adequate. Tables 10.8 to 10.12 are especially useful in the design of drainage systems for larger buildings.

10.6 Design of Larger-Building Waste Piping

EXAMPLE 10.4. Select sizes for drainage and vent piping for the plumbing in an office building for which the fixtures are shown in Fig. 10.47.

SOLUTION. Individual fixture branches shall not be less than the size indicated in Table 10.8 for the minimum size of trap for each fixture. Selected fixture units from Table 10.9 are applied to each section of the piping and totaled for each branch and stack, and for the building drain and the building sewer. An example of a fixture-unit summary and sample sizes of individual branches that connect into a typical branch of the men's toilet group on any floor are shown in the following table.

Features	*Units per Fixture*	*Total Fixture Units*	*Diameter, Fixture Branch (in.)*
1 service sink	3	3	3
3 lavatories	1	3	1½
3 urinals, washout	4	12	2
3 water closets, valve operated	6	18	4
Total fixture units, men's toilet branch		36	

Reference to Table 10.11 indicates that a 3-in. horizontal fixture branch is inadequate for the above group because it will handle only 20 fixture units and not more than two water closets. Therefore, a 4-in. pipe is selected. Its capacity of 160 fixture units will be more than enough for the 36 needed here. The same table shows that the soil stack can be 4 in. in diameter (it is run thus for its entire height). Its capacity of 90 fixture units per story is sufficient for the 64 that connect in at each T-Y connection.

According to Table 10.10, the building drain and the building sewer at their pitch of ¼ in. in 1 ft should be 5 in. in diameter. Their capacity for 480 fixture units exceeds the 350½ placed upon them. The vent stack at a 70 ft length and 338 fixture units would be 2½ in., but 3 in. is a better choice. This is increased to 4 in. as it passes through the roof.

Although opinion may vary about the relative merits of continuous or circuit venting, either system, properly designed, will effectively prevent the siphoning of traps or relieve air pressures that could cause foul gases to bubble through the traps into the occupied space. Another system, especially suitable for high-rise buildings, eliminates the vent stack completely with equal effectiveness.

(a) The Sovent System. This essentially ventless system changes the nature of the effluent (discharge of wastes and sewage from the fixtures) instead of coping with the pressures and suctions that normal effluent would cause (see Figs. 10.48 and 10.49).

Fig. 10.47 (opposite) *(Example 10.4.) Plumbing section of an office building in general conformity with National Standard Plumbing Code. Circuit vents serve branch soil lines. The house trap and fresh-air inlet are omitted from building drain. Some codes require continuous venting (see Fig. 10.46). Note that* (a) *Relief vent not required on top floor.* (b) *Men's and women's toilets on 3rd floor are typical and would be repeated on 1st, 2nd, 4th, and 5th.* (c) *Drinking fountains on 5th and 4th floors would be repeated on 1st, 2nd, and 3rd.*

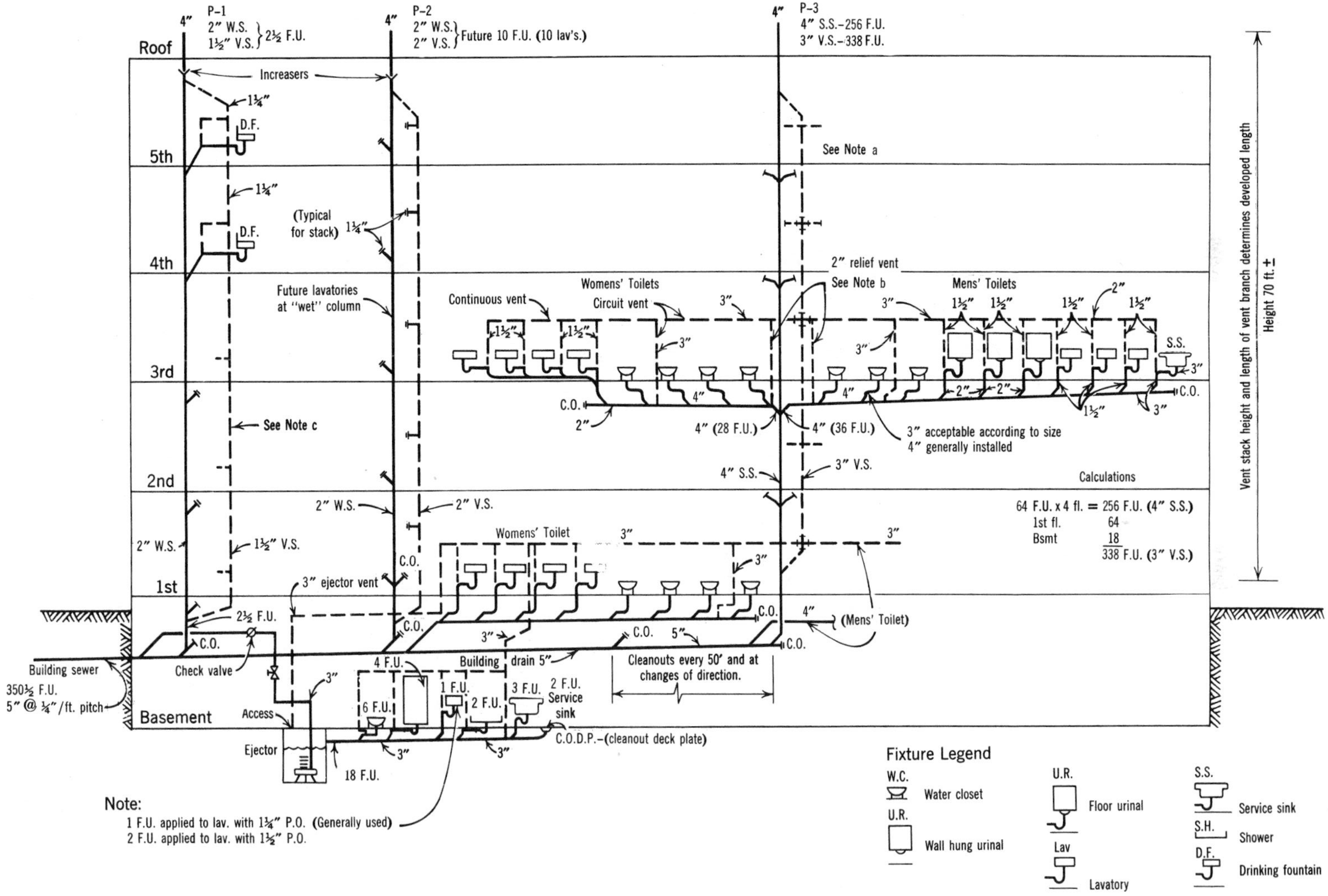

Note:
1 F.U. applied to lav. with 1¼" P.O. (Generally used)
2 F.U. applied to lav. with 1½" P.O.

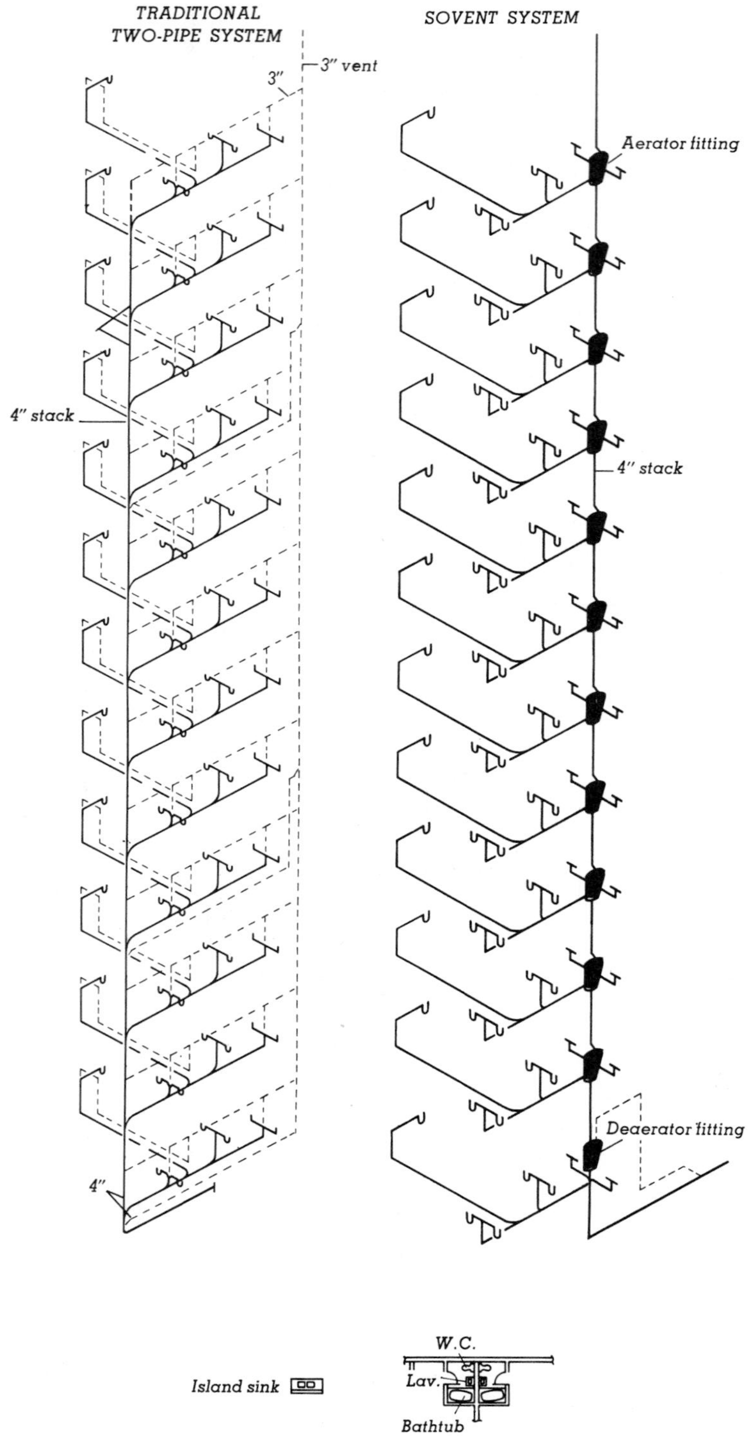

Fig. 10.48 Cost-saving potential can be seen by the choice of the Sovent system over a two-pipe system for this 12-story stack serving an apartment grouping. Courtesy of the Copper Development Association.

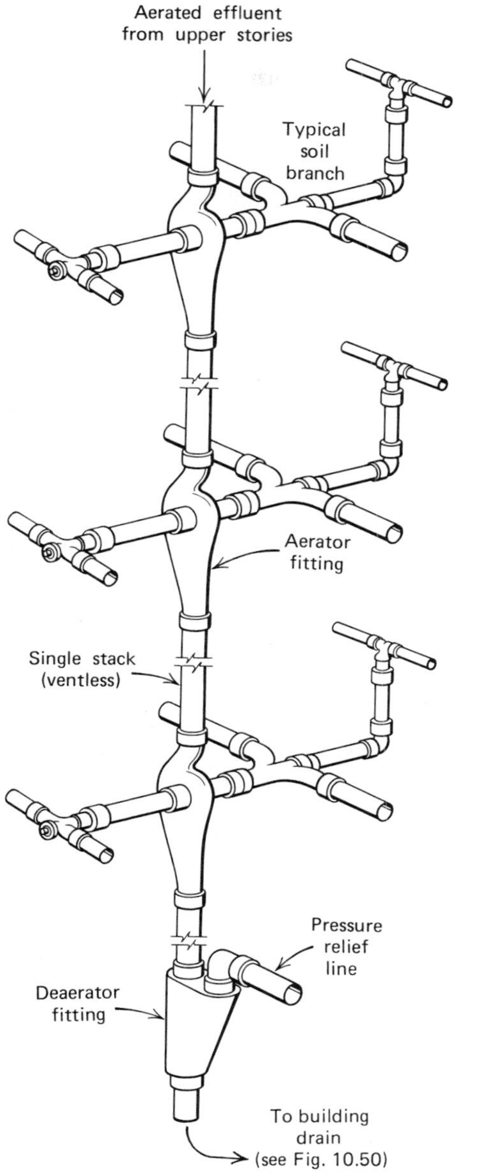

Fig. 10.49 The Sovent drainage stack consists of aerator fittings that join the horizontal branches to the stack at each floor level, and a deaerator fitting at the bottom of the stack. The stack is open to the atmosphere above the roof at the top. Courtesy of Copper Development Association.

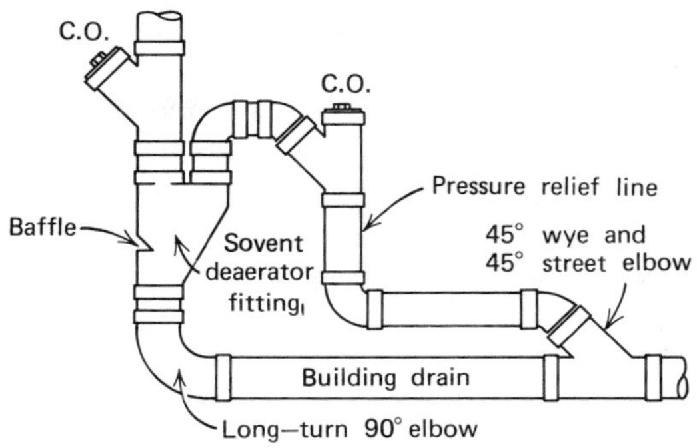

Fig. 10.50 The deaerator consists of an air separation chamber with an internal nosepiece, a stack inlet, a pressure relief outlet at the top, and a stack outlet at the bottom. The deaerator fitting at the bottom of the stack functions in combination with the aerator fittings above to make the single stack self-venting. The deaerator is designed to overcome the tendency for the falling waste to build up excessive back pressure at the bottom of the stack when the flow decelerates at the bend into the horizontal drain. Courtesy of Copper Development Association.

The "plunger" effect of a descending "slug" of water/waste within pipes was described in Section 10.3. If the effectiveness of the "plunger" can be reduced, the negative and positive pressures created by it will be also reduced. If their values can be brought down below the holding power of the several inches of water in the trap, no vents will be necessary. In the single-stack *Sovent* system illustrated in Fig. 10.49, this is done by dealing with the normal liquid effluent at each floor. Aeration there produces a *foam* that lacks the stack-filling tendency of the liquid effluent. Thus, through the creation of a *soft* plunger, pressure variations in the single stack are minimized.

Tests have shown that the positive and negative pressures produced by *normal* liquid effluent during its descent and relieved by the vent piping are often about 5 to 12 in. water gauge. Obviously, if the vents were not provided, the 2 to 4 in. of water seal in the traps would be vulnerable to penetration by gases from pipes under positive pressure or siphonage of water seals into pipes that may be under negative pressure.

Figures 10.48, 10.49, and 10.50 illustrate the components and the action of the Sovent system. Effluent, already aerated and descending from upper stories, is diverted in the stack at each lower story. The aerator fitting there affords a passage for this diverted flow and also an air space into

WATER AND WASTE

which the effluent from the local branch soil or waste can drop. Here it spatters, mixing with the air to form a rarified mixture of air and liquid. Tests show that this mixture does not produce pressures, positive or negative, of more than 1 in. water gauge. Thus, a trap-seal of 2 in. or more is safe against siphonage or penetration.

At the foot of the single stack the aerated effluent is compacted—a process aided by a baffle in the path of the flow in the deaerator fitting (see Fig. 10.50). If not relieved, air piling up at this point could cause pressures in the stack at the first floor. An air-discharge pipe provides this relief of air from the deaerator fitting to the upper part of the building drain, above the liquid flow.

The Sovent system was invented by Fritz Sommer of Switzerland, who tested it in a 10-story drainage test tower. Since its introduction in 1962, it has been installed and used in hundreds of buildings in Europe and Africa. Canada used the Sovent method in the Habitat apartments at the 1967 Montreal Expo. It was first used in the United States in 1968 at the Uniment Apartments in Richmond, California; there it was tested and approved before going into service. Thus Sovent was first granted U.S. code acceptance in 1968 in Richmond, California. Following this success, its code acceptance grew rapidly during the early 1970s. Today, Sovent is accepted by the major model codes used nationwide.

References

Converse, J., et al. (1978). *Design and Construction Manual for Mounds . . .,* Small Scale Waste Management Project, University of Wisconsin Extension, Madison, Wis.

Milne, M. (1976). *Residential Water Conservation,* U.S. Office of Water Research and Technology, Department of Commerce, NTIS.

Stoner, C. Ed. (1977). *Goodbye to the Flush Toilet,* Rodale Press, Emmaus, Penn.

11
BATHROOM DESIGN

More than any other familiar space, the bathroom reflects the interrelation of those factors—comfort, heating, cooling, ventilation, moisture control, water supply, and waste—that have provided the basis of the first 10 chapters. It also furnishes an excellent model through which to illustrate lighting considerations, sound isolation, and the special needs of the handicapped.

One of the more challenging aspects of bathroom design is the frequent conflict between physiological design criteria and such psychological influences as cultural attitudes about bathroom activities. The physiological criteria change as slowly as evolution. The cultural attitudes can change very rapidly, even within a generation. The bathroom supports two primary human activities: cleansing and elimination. Today's common attitudes about these closely linked activities are that one is "clean," the other "dirty." One might be discussed (though rarely) among friends; the other is simply not fit for conversation. This conflict in attitudes influences the design of our plumbing fixtures. The toilet is the fixture used for elimination; it is the logical place to provide for the cleansing of the perineal zone that should immediately follow elimination. (This is especially true of public toilets, where stalls close off toilets from lavatories.) Yet very few toilets presently incorporate the cleansing feature: the mixture of cleansing and elimination in the same fixture too often seems abhorrent.

Another reason to single out the bathroom for special attention is its role in the conservation of energy and water. Chapters 7 and 10 established the toilet as the key to significant water conservation in residences. Bathing—in particular, showering—represents another important target, one in which savings in water also mean savings in the energy used to heat water.

In discussing these topics, we will look first at the influences of physiology, psychology, and conservation on typical plumbing fixtures. Then we will consider how to put them together in various combinations, in spaces that are comfortable, pleasantly lighted, private, and fully accessible. Lastly, we will address some aspects of construction—the roughing in and maintenance of the fixtures.

11.1 Physiology, Psychology, and Fixtures

The brief summary presented in this section is based largely on a particularly revealing (and entertaining) study of the struggle between physiological and cultural criteria in fixture and bathroom design—Alexander Kira's *The Bathroom: Criteria for Design,* an expanded edition of which was published in 1976 by Viking Press.

(a) Cleansing. A key issue here is the contrast betwen running water and a standing water body. *Lavatories* are used primarily for the cleansing of hands, face, and teeth—activities done quickly with running water that is wasted directly, rather than being collected. Most lavatories are designed as collection bowls—perhaps a reflection of the days when washbasins of standing water, drawn and heated elsewhere, were brought to the bathing place. Most lavatories also have fittings that project out over the sink. Although these fittings have slowly evolved into very sleekly designed objects, most of them still dump running water directly into the drain, are hard to use as drinking fountains, and can be wounding to those who try to wash hair in the lavatory. In considering the prevalence of running as opposed to standing water, the role that running water could play in keeping lavatory surfaces clean, and the need for a drinking fountain when teeth are brushed, Kira proposed a quite different lavatory design (Fig. 11.1). For public toilet rooms, used almost exclusively for hand washing, the lavatory can be kept very simple (Fig. 11.2).

Whole-body cleansing more clearly illustrates

WATER AND WASTE

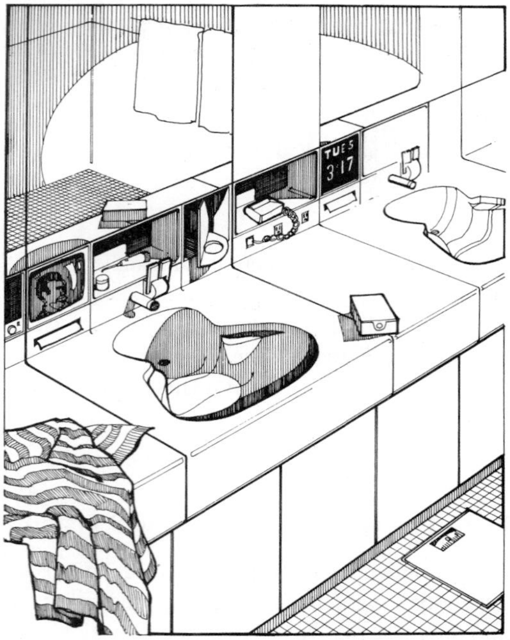

Figure 11.1 *A proposed lavatory that exploits running water. The water issues from a fountainlike stream, for ease in drinking and in hair and face washing. The stream strikes the lavatory bowl in such a way as to minimize splash outside the bowl, yet sets up a self-cleaning swirling action. A small repository at the back, over the drain, can serve for standing water when desired. From The Bathroom: New and Expanded Edition, copyright 1966, 1976 by Alexander Kira. By permission of Bantam Books, Inc. All rights reserved.*

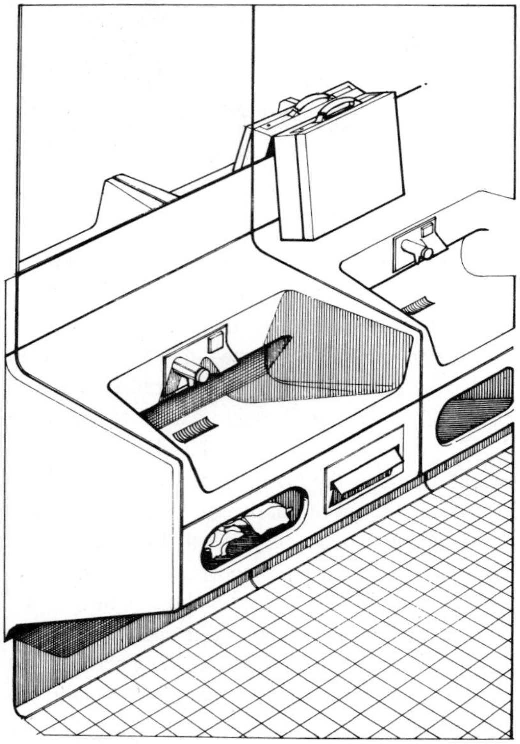

Fig. 11.2 *A proposed public washroom lavatory in which hand washing is the main concern. A single source of tempered water is provided, again with an eye toward self-cleaning of the fixture. Facilities for soap and for hand drying are built directly into the lavatory. From The Bathroom: New and Expanded Edition, copyright © 1966, 1976 by Alexander Kira. By permission of Bantam Books, Inc. All rights reserved.*

the running/standing water contrast. In cleansing, the ordinary sequence is *wet, soap, scrub, and rinse*. Standing water is good for wetting, and very good for soaping/scrubbing; running water is superior for rinsing. The psychological aspect enters here in common attitudes toward tubs and showers. Tubs are often seen as places to relax in, to spend more time in, perhaps to read in. Showers are viewed as quicker, "no-nonsense," stand-up places. Yet each could benefit from some features of the other.

Tubs should be designed so that the reclining body is supported at the back; this requires a contoured surface (Fig. 11.3) rather than the ordinary straight-line design. It also requires braces for one's feet, since the body will otherwise tend to float up and way from such a backrest. These tubs can be designed to accommodate persons of various leg-lengths, as shown in Fig. 11.3*b*. There must also be a seat, to allow most of the body a chance to be out of the water and to facilitate safe entry/exit from the tub. Especially needed is a hand-held shower for the final rinse; soapy standing water leaves a scummy film on both people and fixtures.

Showers may seem very "efficient," but cleaning would be more thorough and safe if the bather could turn off the water and sit for at least part of the soap/scrub activity, especially for the lower legs and feet. Showers with integral seats (Fig. 11.4) are now common.

Another consideration is the location of water controls (fittings). The user should be able to reach

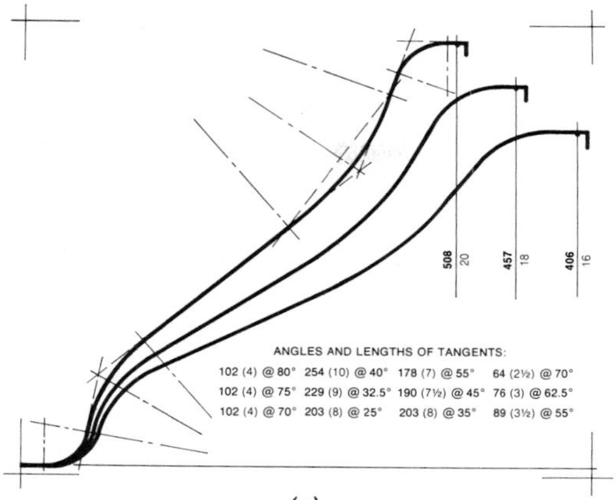

(a)

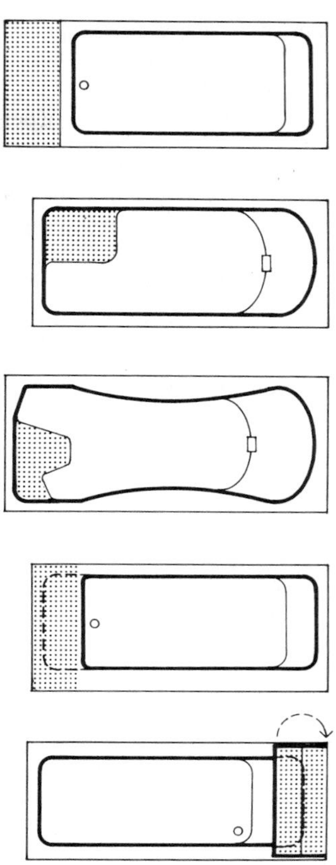

(b)

Fig. 11.3 *Some considerations for tub design, to facilitate relaxation during whole body cleansing.* (a) *A contoured backrest allows comfortable reclining, with support to both the lower back and shoulder. Curve #2, with a median angle of 32.5°, promises comfort for most users.* (b) *Various approaches to tub plans, allowing long- and short-legged persons to brace their bodies against the contoured backrest. Raised seats are also provided, to facilitate entry/exit. From The Bathroom: New and Expanded Edition, copyright © 1966, 1976 by Alexander Kira. By permission of Bantam Books, Inc. All rights reserved.*

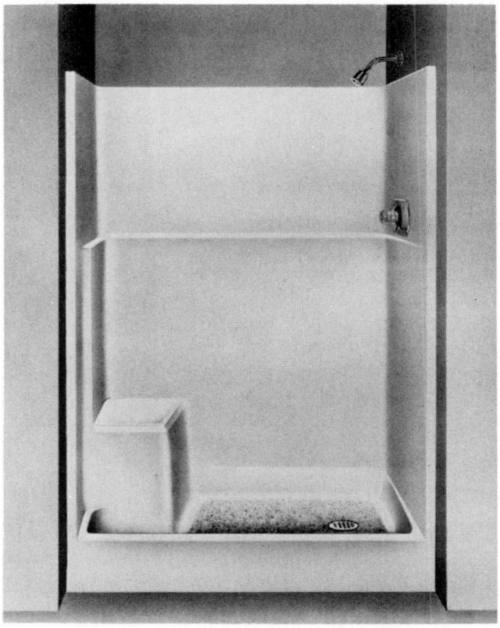

Fig. 11.4 *Many prefabricated showers now include seats, to encourage a safer posture while soaping and scrubbing the feet and legs.*

them easily from outside the tub or shower without wetting his or her arm, but also able to manipulate them from within the tub/shower even if temporarily blinded by soap.

(b) Elimination. Two conflicts arise with toilets: the already-mentioned difficulty of combining cleansing with elimination, and the conflict over the height of the toilet. A *lower toilet* is definitely of benefit to the average person, who will achieve far better bowel evacuation in a full squatting position. If combined with a toilet seat contoured to best support the body during defecation, the proposed toilet in Fig. 11.5*a* would be physiologically superior to most of today's fixtures. There are problems with low height. The typical male stands while urinating, and the lower toilet presents a more difficult target. This can have serious maintenance consequences unless males can be induced to sit or a separate urinal is provided—an unlikely option for residential bathrooms. (In our culture, urinals are "institutional"; they are also expensive.) Another problem with the lower toilet is that the elderly and some handicapped people will have difficulty getting on and off the seat (see also Fig. 11.17). Yet another problem arises from the fact that toilets are considered to be seats (for reading, toenail clipping, etc.). The low, squat-inducing toilet is decidedly not at a comfortable chair height.

As a result, a *higher toilet* with otherwise similar features is proposed (Fig. 11.5*b*). Note that both of these toilets feature openings whose shape differs from that of the conventional oval. This hourglass-shaped opening provides proper support for the body during defecation, and it allows a more generous opening, front and back, for proper perineal cleansing by hand. Another change is that there is no separate toilet seat; instead, these fixtures incorporate electric resistance heaters that warm the small portion of the toilet that contacts the body. Thus, the toilet becomes a consumer of

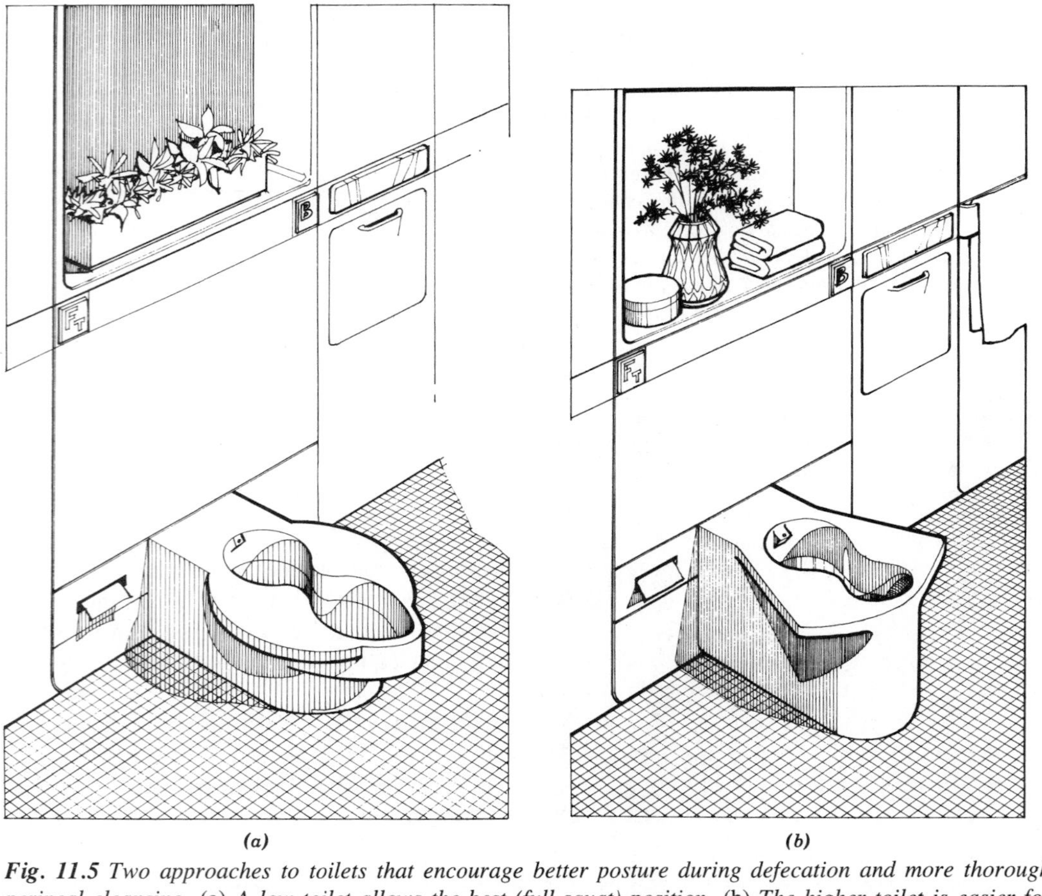

(a) *(b)*

Fig. 11.5 *Two approaches to toilets that encourage better posture during defecation and more thorough perineal cleansing.* (a) *A low toilet allows the best (full-squat) position.* (b) *The higher toilet is easier for the elderly and handicapped and better intercepts the urine from a standing male (although a separate foldout urinal is provided; see Fig. 11.8). Both toilets eliminate the separate toilet seat and its maintenance problems; both use small amounts of energy to warm the toilet at its point of contact with the body and to heat the water for perineal cleansing. Such cleansing is also the reason for larger front and back openings. The pushbuttons are labeled F_T for the flush, and B for the bidet (cleansing) function. From The Bathroom: New and Expanded Edition, copyright © 1966, 1976 by Alexander Kira. By permission of Bantam Books, Inc. All rights reserved.*

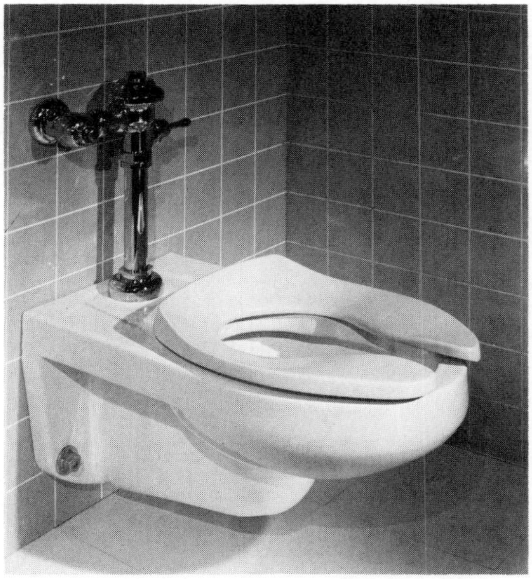

Fig. 11.6 A toilet seat for the coventional WC, which encourages better posture for defecation and somewhat better access for perineal cleansing. Courtesy of American Standard, Inc.

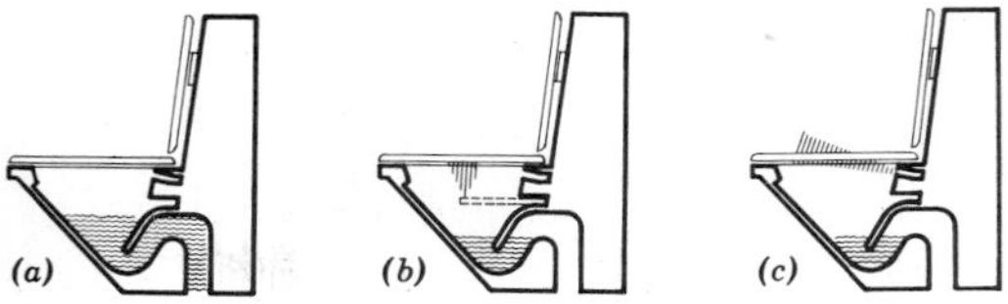

Fig. 11.7 Provisions for perineal cleansing within toilets. (a) *Section through toilet as it usually appears.* (b) *At the initiation of perineal cleansing, a pipe extends and emits a controlled spray of warmed water.* (c) *After pipe is extracted, a warm air jet issues from just below the seat. From The Bathroom: New and Expanded Edition, copyright © 1966, 1976 by Alexander Kira. By permission of Bantam Books, Inc. All rights reserved.*

energy—which is also used to heat the water that is provided for perineal cleansing. For those who insist on a seat, or for ordinary posture on conventional toilets, a physiologically sound contoured toilet seat is available (Fig. 11.6).

It is especially important that perineal cleansing be encouraged by the designers of bathrooms. The typical dry toilet tissue provided usually is not adequate for this task. In small, private bathrooms this problem is easily solved by placement of the toilet adjacent to the lavatory (or tub), where toilet tissue can be wetted with clean water. In public toilets separated by stalls, perineal cleansing requires either a clean water source built into the toilet, or a separate *bidet* within the stall—a highly unlikely provision, given the extra floor space and the cost of the bidet. Again, cultural issues arise: bidets are quite common in the private bathrooms of Europe, but a rarity in North America. Two common alternatives are (1) to build a cleansing source into the toilet itself (Fig. 11.7) or (2) to construct a toilet seat that provides for such cleansing. Several American manufacturers offer such seats, which can be easily adapted to existing toilets.

The *urinal* is an answer to the problem posed by lower toilets to males who stand while urinating. Although urinals are culturally acceptable in public toilet rooms, they have never achieved a place in private bathrooms. One solution proposed by Alexander Kira is to build the home urinal into the wall (Fig. 11.8*a*); it could be pulled out for use, at the optimum mounting height, and at an optimum receiving shape to eliminate backsplash onto the user. When pushed back into the wall, it would flush automatically. A less-satisfactory solution would be to modify the toilet by raising its back surface (Fig. 11.8*b*).

11.2 Fixtures and Conservation

Several of the important options for conservation have included the substitution of rainwater or graywater for potable water in fixtures such as toilets (Chapters 8 and 10) or the elimination of the use of water in toilets (Section 10.1). We now will take a closer look at a third conservation option: the use of less water in conventional fixtures.

(a) Toilets. The more dramatic water-saving opportunities in flush toilets were presented in Section 10.1: small quantities of flush water aided by compressed air (Fig. 10.3) or by a vacuum–sewage tank combination (Fig. 10.4). Unfortunately, each of these alternatives requires the complicating factors of an air system and its equipment. The four common flush toilets are compared in Fig. 11.9. Refer also to Fig. 10.2,

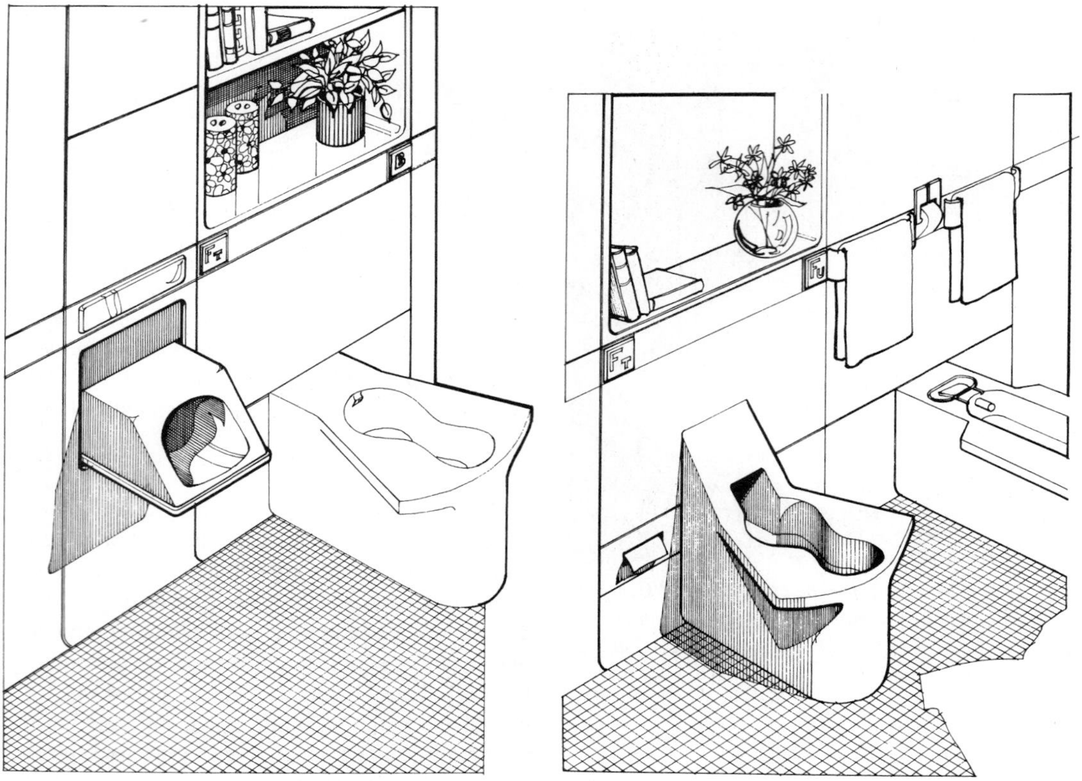

(a) (b)

Fig. 11.8 Designs to accommodate male urination. (a) *A built-in, tilt-out urinal for the home, visible only when in use.* (b) *A higher back for the toilet. A bidet for perineal cleansing is shown to the right of the toilet. From The Bathroom: New and Expanded Edition, copyright © 1966, 1976 by Alexander Kira. By permission of Bantam Books, Inc. All rights reserved.*

for the characteristics of one- and two-piece toilets, flush tanks, and flush valves. Some states place limits on the number of gallons per flush for water closets.

Washdown toilets are no longer made in the U.S., although many are still in use. This is the noisiest toilet, and the most likely to become plugged, since it has the smallest-diameter trap. It was usually found in the two-piece flush tank toilet; with an elevated tank, the flush required was only about 2.5 gal (9.5 L).

Siphon jet toilets are in widespread use in North America, particularly in residences. A small priming jet hurries the bowl's contents along into the trap, and hastens the siphon action. With the elevated tank (two-piece) toilet, this process requires a flush of about 3.75 gal (14.2 L). With the more common close-coupled two-piece toilet, it requires a flush of from 5 to 7 gal (19 to 26.5 L). Siphon jet toilets are sometimes equipped with flush valves, which use less water (see the discussion of ''Blowout,'' below).

Siphon vortex toilets are especially suitable for low-velocity water (often as a result of low pressure). They are therefore also the quietest, making them a favorite wherever bathrooms are adjacent to sleeping areas or other acoustically sensitive spaces. The water enters the bowl off-center in such a way as to form a vortex; this swirling action cleans the sides of the bowl and the trap, helping the siphon action in emptying both bowl and trap. The newer one-piece flush tank toilets usually have the siphon vortex flushing action, which typically requires 6 to 8 gal (22.7 to 30.3 L).

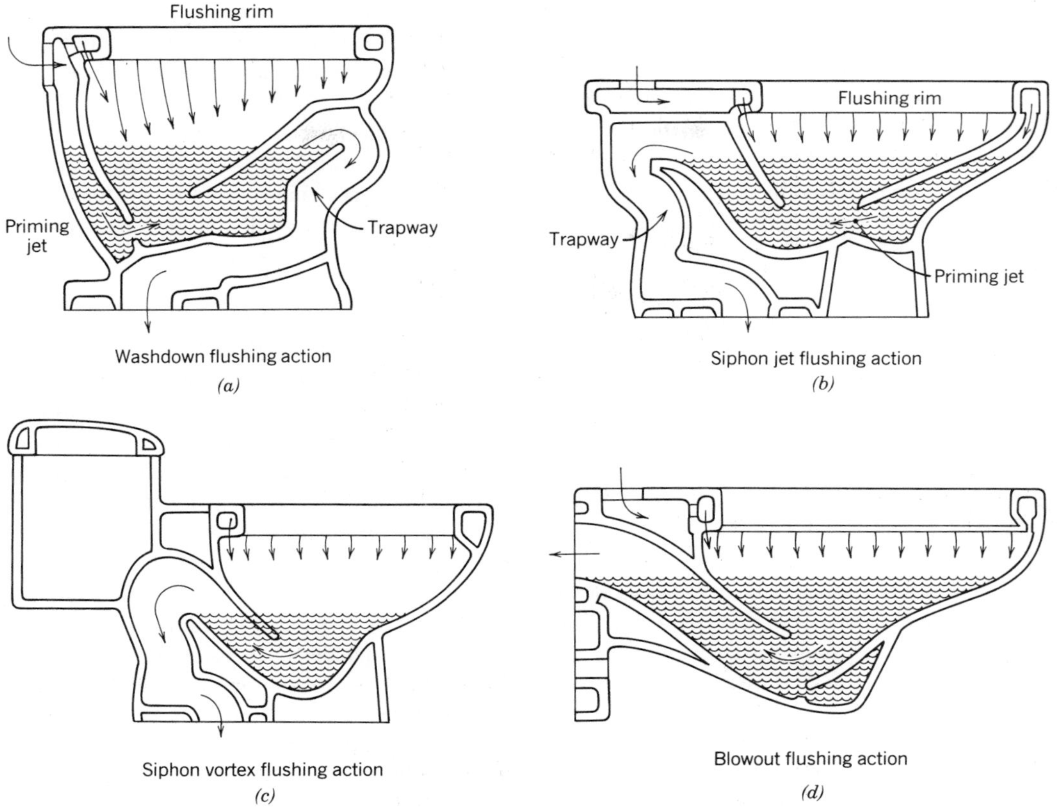

Fig. 11.9 Four common flushing actions that are built into toilets. Newer flush toilets typically use either type b or type c; flush valve toilets use either type b or type d. (See also Fig. 10.2.) From Milne (1976).

Blowout toilets combine very-high-velocity water and a simple trap to offer a noisy but very-low-maintenance toilet dependent on flush valves rather than tanks. They are very common in commercial and institutional toilet rooms, where large water supply lines and high pressures are available. The high velocity of the water lowers the quantity required from 3 to 4 gal (11.4 to 15.1 L) per flush.

(b) Flushing Controls. Water conservation is encouraged by the *dual cycle* toilet, whose flushing mechanism allows a choice of fewer gallons for liquid wastes, more gallons for solid wastes. More common outside the United States, this simple mechanism's handle is pushed up for liquid flushing and down for solids.

Another, newer development is the *automatic flush,* triggered by radiant heat from the pressure of a body at the fixture or by light reflected off the user and back to the control. This ''touchless'' approach seems to promise more for hygiene than for water conservation, although it does prevent the flush valve from being held open too long by a careless user.

In general, plumbing fixtures will emit less water as the supply water pressure is reduced. For this reason, *pressure-reducing valves* are becoming popular as water conservation devices (when they would not otherwise be required to protect fixtures from overpressure). Installed on the supply line to a building, they can save water throughout the structure.

(c) Shower Heads. These devices have been notorious for encouraging prodigal water usage; typical flow rates of 6 gpm (0.4 L/s) and maximum rates of 12 gpm (0.7 L/s) once were common. Even in a ''short'' (five-minute) shower, this rate of use could consume as much as 60 gal (227 L) of water, much of it heated. Many codes now require a limitation on shower-head flow; a fixture of 2.5 gpm (0.2 L/s) is common. These flows can be designed into the shower head, or they can be achieved by cheap, simple flow restricters in retrofit applications. Some utilities distribute flow restrictors free of charge. Most bathers notice no difference, either in enjoyment or in cleansing, when using restricted-flow shower heads.

(d) Lavatory Faucets. At full flow, lavatory faucets typically deliver 4 to 5 gpm (0.25 to 0.3 L/s). Newer, low-flow faucets utilize a variety of devices to function as well (or better) with less water. Such devices include aerators (which add air bubbles to the stream, making it splash less and appear larger), flow restrictors, and mixing valves to control temperature. The lower flows achieved range from ½ to 2½ gpm (0.03 to 0.16 L/s). Another promising development is the foot-operated faucet, which frees the hands from having to control water flow, thus saving a few seconds of flow during each lavatory usage. These devices could be particularly helpful at kitchen sinks, where extensive washing of objects takes place.

(e) Appliances. Dishwashers, washing machines and other appliances are big users of water—and of the energy needed to heat it. Dishwashers use 12 to 18 gal (45 to 68 L) per cycle, heated well beyond the 120 F (48°C) typical of household hot water supply. Some models now allow shorter cycles, which can cut use to perhaps 7 gal (26.5 L).

Clothes washing machines use from 40 to 55 gal (151 to 208 L) for full-size loads. In the past, ''suds saver'' features allowed soapy, hot wash water to be reused. Many newer washers allow for a wider selection of water quantities and temperatures—a feature that can save considerable water and energy.

11.3 The Space Itself

After plumbing fixtures designed for the human physique and for water conservation have been chosen, they are placed in a space that has several unusual environmental needs. In this section, the first topic addressed is accessibility and privacy for the room and its fixtures; the emphasis here will be on the special problems of the handicapped. The next topic is the public bathroom, as typically designed into ''cores'' of multistory buildings. Finally, we will consider some special environmental control systems: thermal, luminous, and sonic.

(a) Accessibility and Privacy. One of the most obvious requirements for bathroom design is that a person using any fixture within the bathroom should not be visible to someone outside the room. In private bathrooms, a simple closed door usually provides the solution. In public toilet rooms, however, entrances should be arranged so as to break the line of sight from hallways. The approaches shown in Fig. 11.10 show the amount of floor space required to balance privacy with adequate clearance for handicapped users in wheelchairs. (Much of the information in this section is reproduced with permission from *American National Standard A117.1-1980,* ''specifications for making buildings and facilities accessible to and usable by physically handicapped people,'' copyright © 1980 by the American National Standards Institute (ANSI). Copies of this standard may be purchased from ANSI, 1430 Broadway, New York, NY 10018).

Another accessibility issue is that of room for wheelchair maneuvering within such tight spaces as toilet rooms. The minimum clear floor area should be a circle 5 ft (1525 mm) in diameter. Minimum clearances for wheelchair maneuvering are shown in Fig. 11.11. Provisions for people in wheelchairs also influence the heights at which items such as light switches, electrical receptacles, paper towels, and water controls should be installed. The height limitations shown in Fig. 11.12 suggest that, in general, all objects to be reached by hand in toilet rooms be placed more than 15 in. (380 mm) and less than 48 in. (1220 mm) above the floor.

Floor space requirements and mounting-height limitations also apply to the various plumbing fixtures in bathrooms, at least one of which (in each case) should be accessible to people in wheelchairs in most buildings. These requirements are

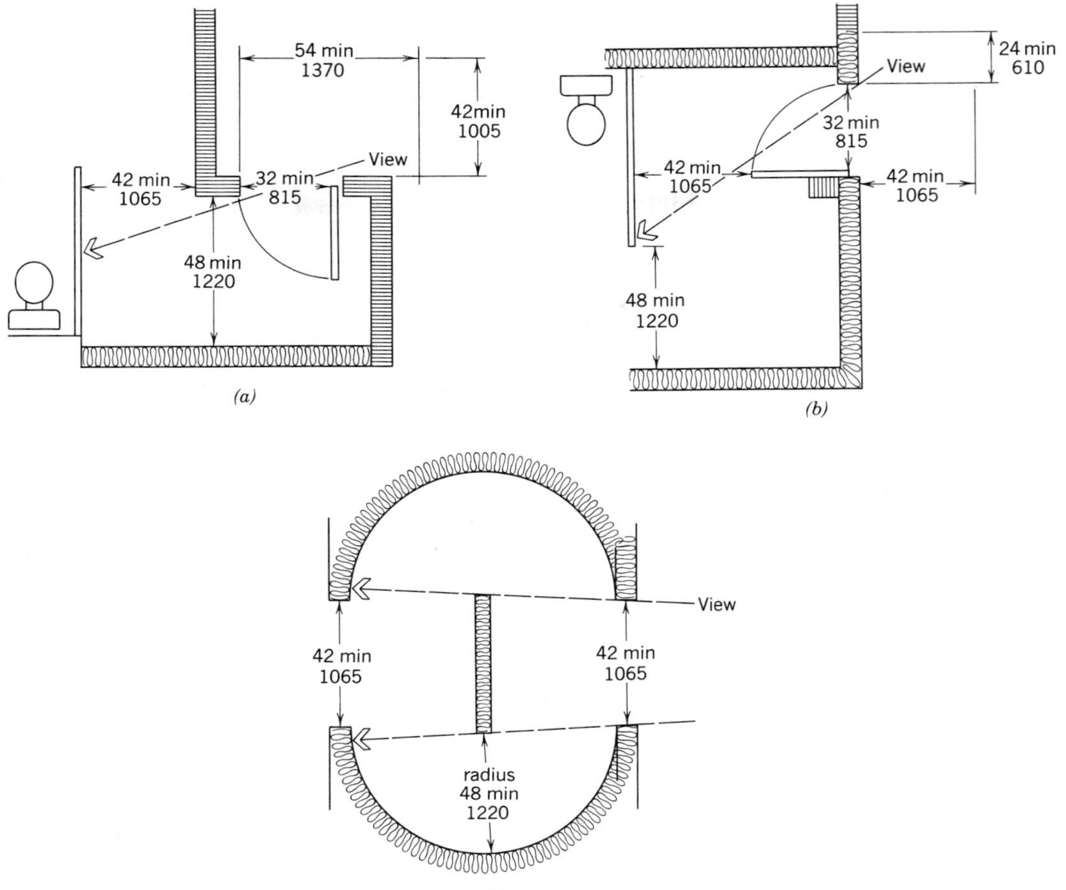

(c)

Fig. 11.10 *Entries to public toilet rooms should allow both visual privacy for those inside and easy access for the handicapped. All doorways must be at least 32 in (815 mm) in width and be clear of any obstruction, including protruding hardware. Designs a and b require the opening of doors, whereas design c allows rapid, easy passage of people and air—and sound.*

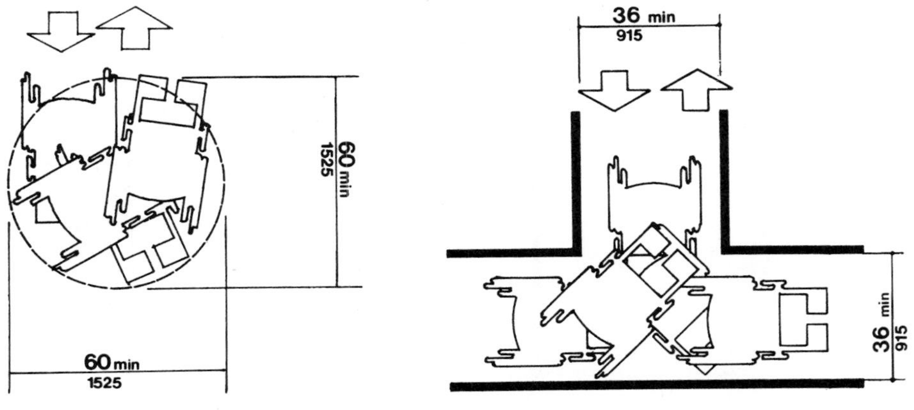

(a) **(b)**

Fig. 11.11 *Wheelchair turning space, minimum requirements.* (a) *60 in. (1525 mm) diameter space.* (b) *T-shaped for 180° turns. Reprinted by permission from American National Standard A117.1-1980, copyright © 1980 by the American National Standards Institute (ANSI).*

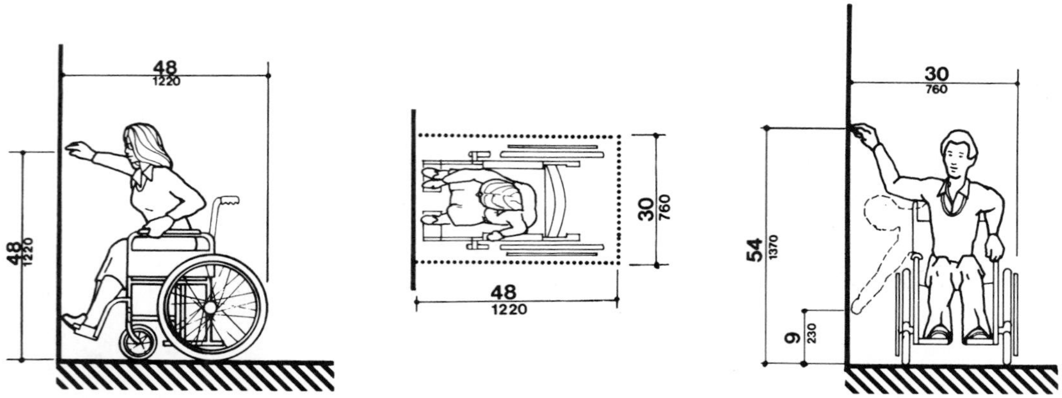

Fig. 11.12 Forward-reach and side-reach limits for people in wheelchairs. Reprinted by permission from American National Standard A117.1-1980, copyright © 1980 by the American National Standards Institute.

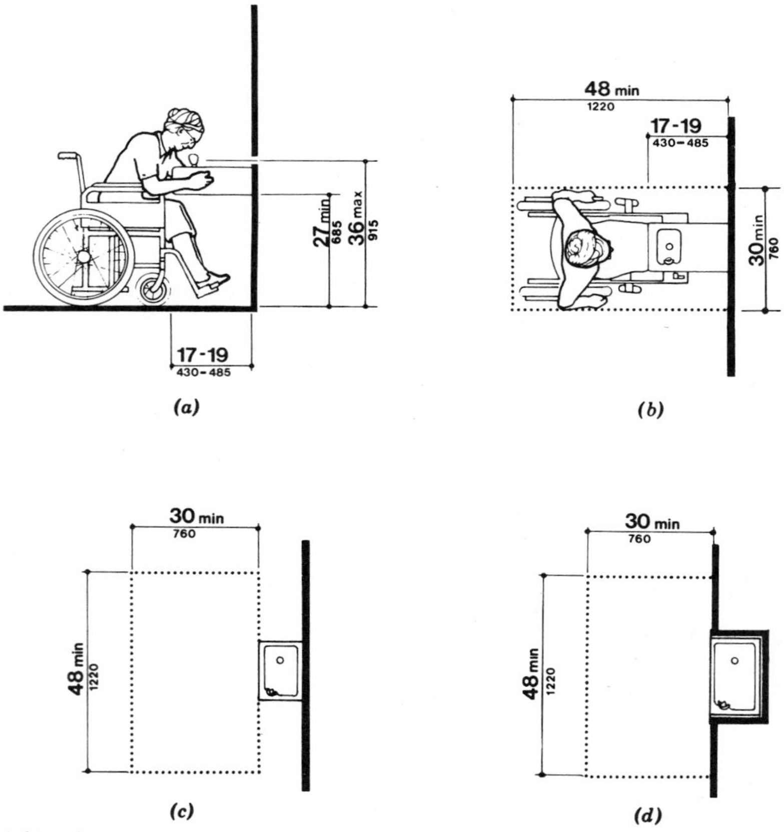

Fig. 11.13 Drinking fountain installation dimensions. (a) *Spout height and knee clearance.* (b) *Clear floor space.* (c) *Free-standing fountain or cooler.* (d) *Built-in fountain or cooler. Reprinted by permission from American National Standard A117.1-1980, copyright © 1980 by the American National Standards Institute.*

shown for drinking fountains (which often must be located outside toilet rooms) in Fig. 11.13, lavatories in Fig. 11.14, bathtubs in Fig. 11.15, and shower stalls in Fig. 11.16. The grab bars shown should have a diameter (or width) of 1¼ in. to 1½ in. (32 to 38 mm), with a clear space to the wall of 1½ in. (38 mm). They should be capable of resisting loads up to 250 lb-ft (1112 N), and not rotate within their fittings.

Toilets (water closets) have particularly large floor space requirements, as shown in Fig. 11.17. There is a conflict between the lower mounting

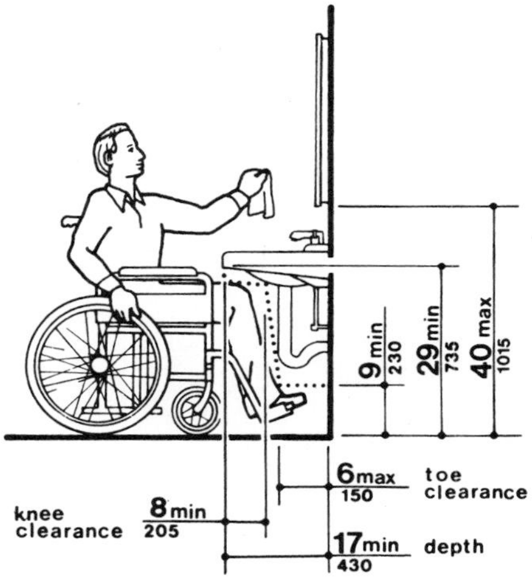

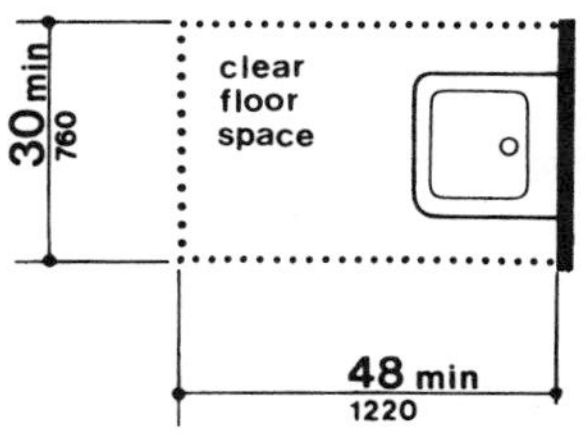

Fig. 11.14 Clearances for lavatories. Hot water and drain pipes below lavatories must be insulated or otherwise covered. Mirrors should be mounted so that the lower edge is no more than 40 in. (1015 mm) above the floor. Reprinted by permission from American National Standard A117.1-1980, copyright © 1980, by the American National Standards Institute. NOTE: *Standard height to lavatory counter is 31 in. (787 mm) (from Architectural Graphic Standards, 7th ed.).*

height needed for proper defecation, the conventional height of 15 in. (380 mm), and the recommended height for handicapped users of 17 to 19 in. (430–485 mm). The doors on stalls designated for wheelchair access should swing out rather than in. Provision for a wheelchair sitting beside the outswinging door must also be made (refer again to Fig. 11.10 for door clearances). No doors should swing into the clearance space required for any fixture.

Urinals for handicapped users should either be of the floor-mounted stall-type or, if wall-hung, have an elongated rim at a maximum of 17 in. (430 mm) above the floor. A clear floor space of 30 × 48 in. (760 × 1220 mm) is required in front of the urinal, and its flush controls should be no more than 44 in. (1120 mm) above the floor.

(b) Public Toilet Rooms. These spaces are especially challenging, because of several conflicting design criteria. They should be in easily accessible places, yet be as acoustically and visually private as possible. Unfortunately, they can easily become security problem areas in some buildings and cities. They generate no rental income (in contrast to office or commercial space around them), yet they cost significantly more per square foot to construct and require more-intensive maintenance. They need high rates of ventilation and could benefit from the psychologically purifying effects of plentiful sunlight—they even have high thermal mass for solar heat storage—yet they are usually placed in the innermost, lowest-rent zones of buildings (such as central service cores), far from wind or sun's benefits.

Further discussion of service cores and their related spaces can be found in Chapter 12. At this point, some design considerations for public toilet rooms can be listed.

Location should usually be central without being "featured." They are necessary public services; people expect to find toilet rooms near reception areas, waiting rooms, eating places, elevators or stairs, coat or package-checking services, public telephones, drinking fountains, and major entrances.

Entries must balance accessibility and privacy (as in Fig. 11.10). Where possible, men's and women's toilet rooms should be located next to

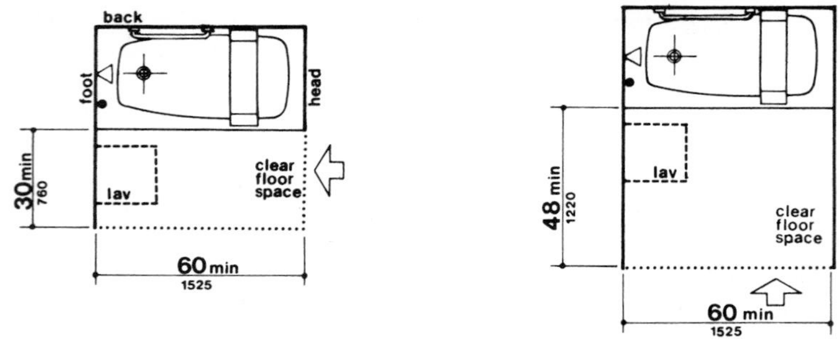

(a)

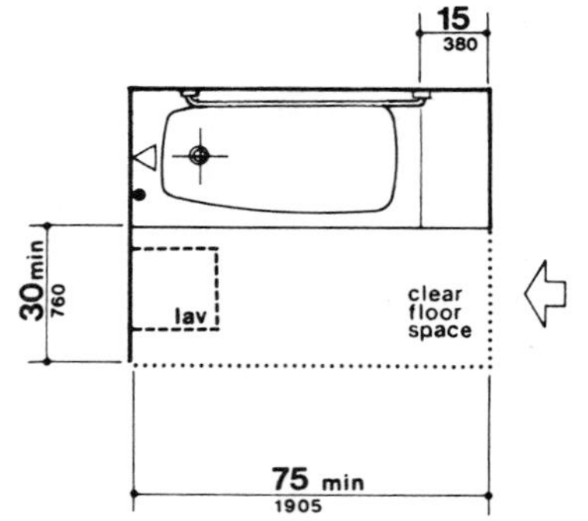

SYMBOL KEY:
● Shower controls
◁ Shower head
⊕ Drain

(b)

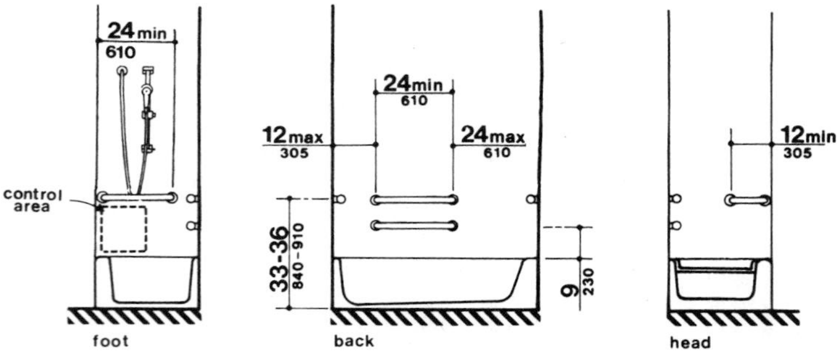

(c)

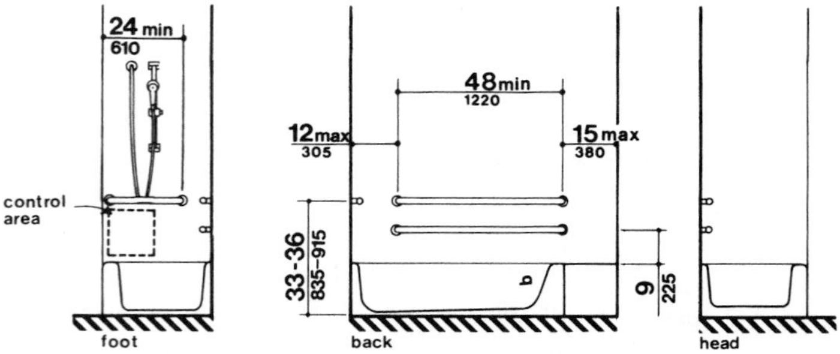

(d)

Fig. 11.15 *Clearances and grab bars for bathtubs;* (a) *and* (c) *are with seat in tub, whereas* (b) *and* (d) *are with seat at head of tub. Reprinted by permission from American National Standard A117.1-1980, copyright © 1980 by the American National Standards Institute.*

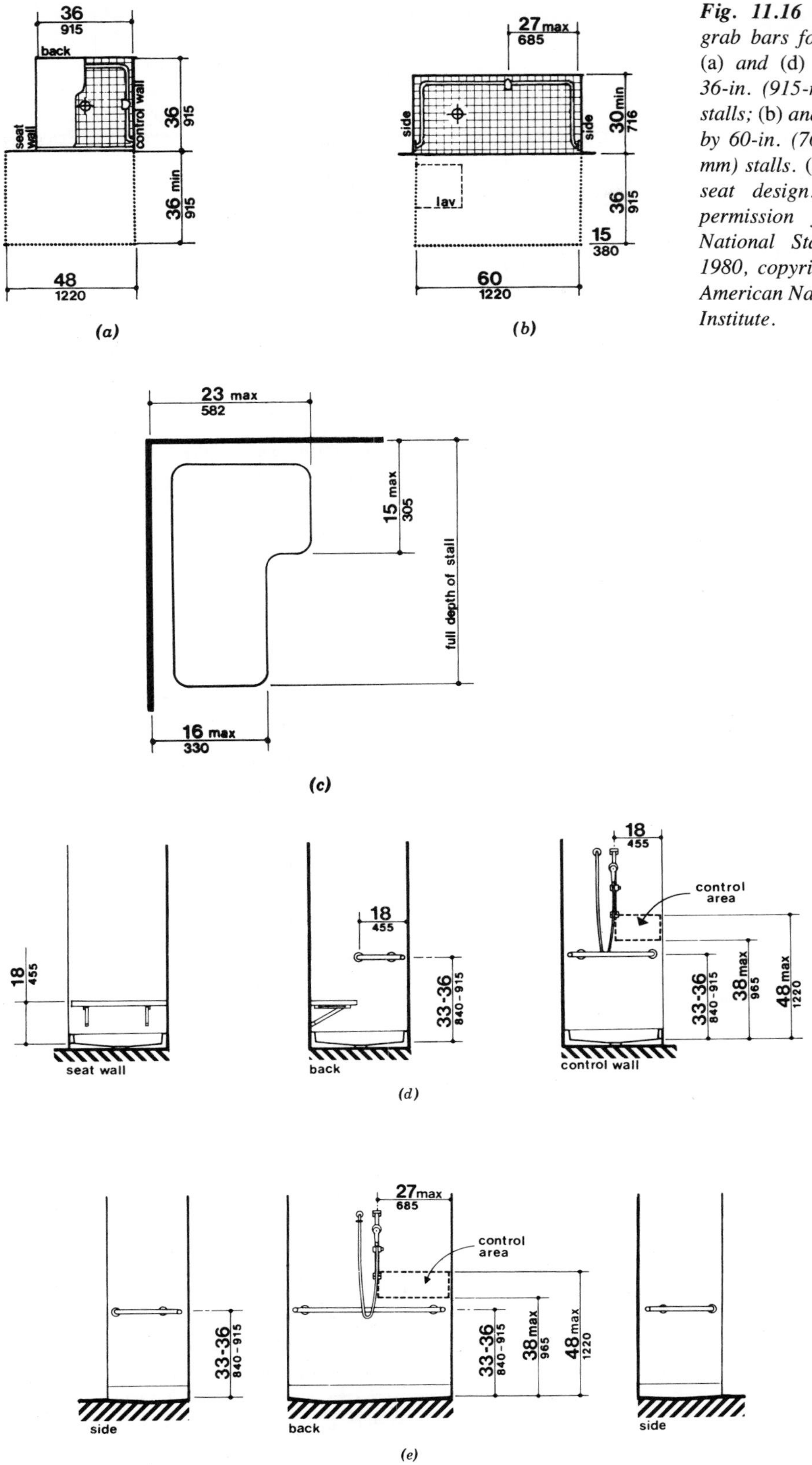

Fig. 11.16 *Clearances and grab bars for shower stalls.* (a) *and* (d) *show 36-in. by 36-in. (915-mm by 915-mm) stalls;* (b) *and* (e) *show 30-in. by 60-in. (760-mm by 1525-mm) stalls.* (c) *shows shower seat design. Reprinted by permission from American National Standard A117.1-1980, copyright © 1980, the American National Standards Institute.*

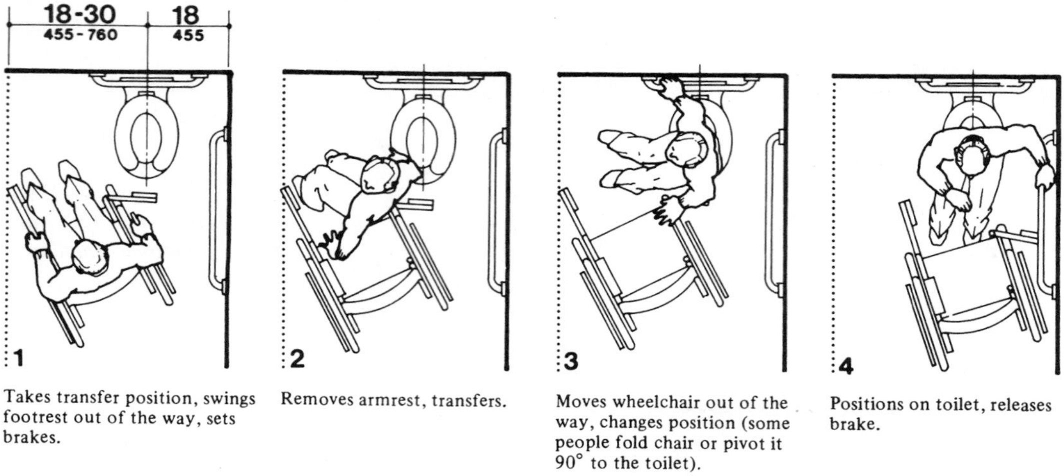
18-30
455-760
18
455
1
Takes transfer position, swings footrest out of the way, sets brakes.
2
Removes armrest, transfers.
3
Moves wheelchair out of the way, changes position (some people fold chair or pivot it 90° to the toilet).
4
Positions on toilet, releases brake.

(a)

Diagonal Approach

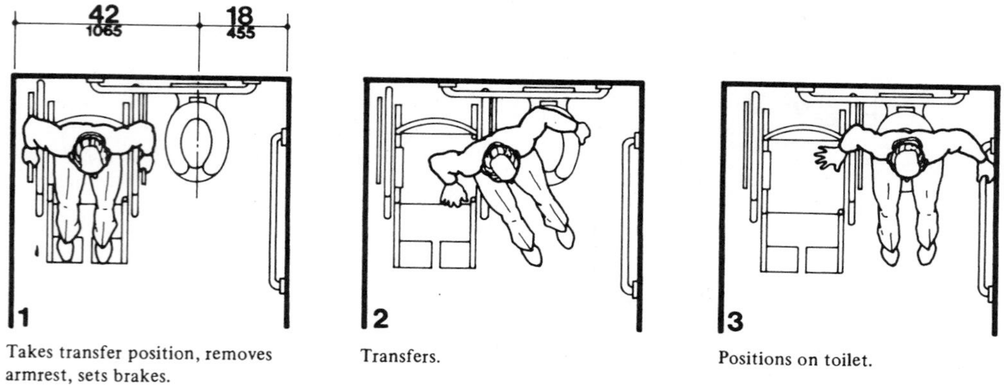
42
1065
18
455
1
Takes transfer position, removes armrest, sets brakes.
2
Transfers.
3
Positions on toilet.

(b)

Side Approach

Fig. 11.17 *Clearances and grab bars for toilets. Two typical approaches* (a, b) *necessitate the clearances of the stalls shown* (c, d, e, f), *as well as arrangements for toilets not within stalls* (g, h). *Reprinted by permission from American National Standard A117.1-1980, copyright ©1980 by the American National Standards Institute.*

each other and both entries should be visible at a glance.

Fixtures are almost always grouped together for economy in both installation costs and floor area. A simple, clean counter of lavatories can be visually appealing; large mirrors add to a sense of spaciousness in an otherwise crowded space. A row of toilet stalls, by contrast, rarely is visually inviting, and the separation of lavatories from toilets raises physiological problems, as discussed earlier. It is unfortunate that this stall, the most private place in public buildings, is usually the worst lighted, most crowded, worst smelling, and most poorly performing of all spaces. Some im-

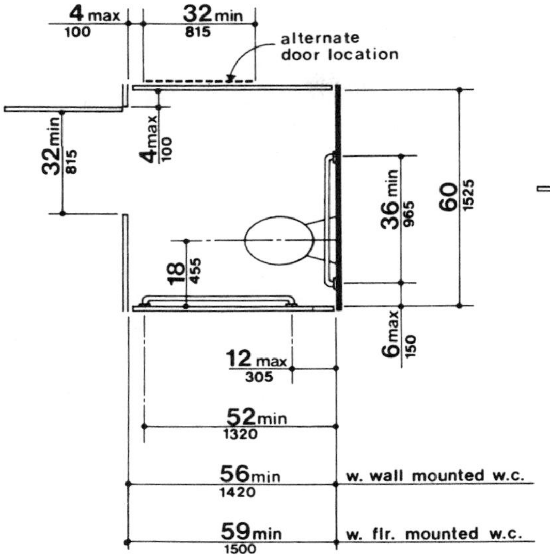

(c)
Standard Stall

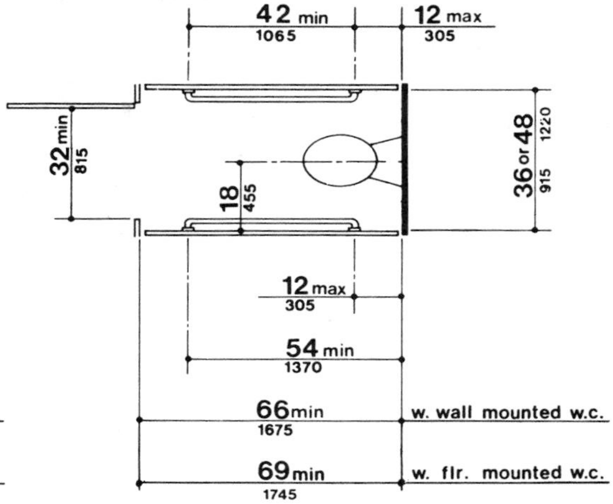

(d)
Alternative Stall

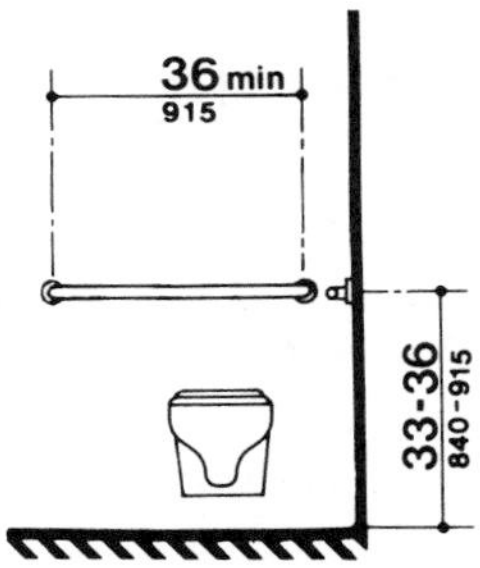

(e)
Rear Wall of Standard Stall

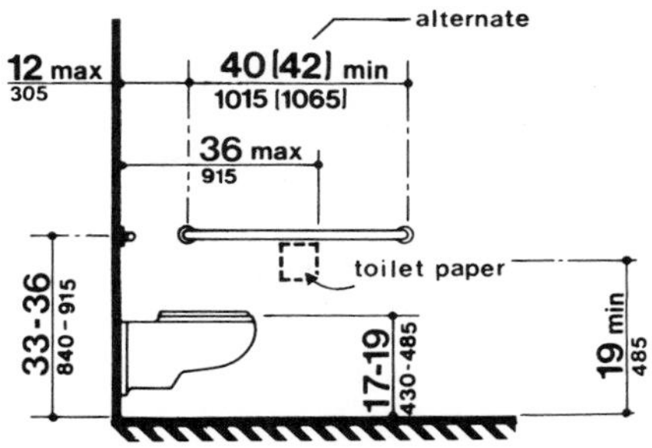

(f)
Side Walls

Fig. 11.17 (continued)

provements for such toilet stalls are suggested in Fig. 11.18.

The required number of fixtures in such rooms is determined by codes, as was shown in Table 7.3. Although requirements differ by function, in general there will be somewhat fewer lavatories than urinals or water closets. Lavatories are usually closest to the toilet room entry, since some users need only cleansing facilities. Preferably, users of toilets/urinals will be brought next to lavatories before they leave.

Lighting can be of relatively low intensity compared to that required for work areas. Task lighting in toilet rooms should be concentrated on lavatories, urinals, and toilets. Other areas can be much less brightly lit. In keeping with the "restroom" concept of a refreshing space, designers might consider deliberately changing the lighting approach within toilet rooms from that of the surrounding work areas. If one is uniformly and diffused lighted, the other might consist of pools of bright light within a darker space. If one is primarily electrically lighted, the other might be dominated by daylight. These deliberate contrasts can be extended to materials, ceiling heights, reverberance, temperature/air motion, and so on.

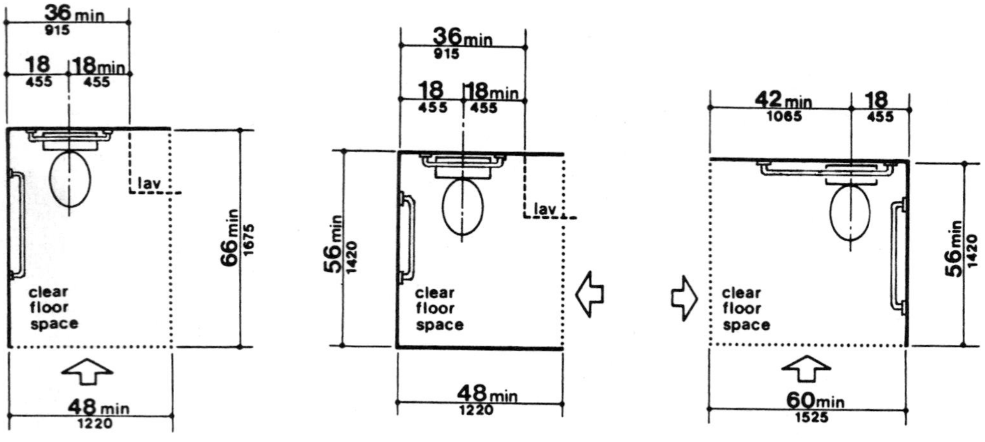

(g)
Clear Floor Space at Water Closets Not in Stalls

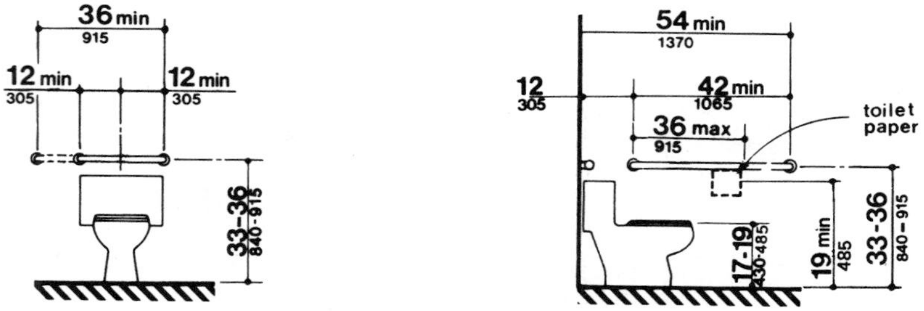

Back Wall

Side Wall

(h)
Grab bars

Fig. 11.17 (continued)

Ventilation is of particular importance. The primary things to remember are to keep the toilet downstream in the airflow and not to recycle air from toilet rooms. Accordingly, the air pressure within toilet rooms should be kept slightly below that of surrounding space, so that airflow is always *into* the toilet room, and exhaust vent intakes should be as close as possible to the toilet itself, yet kept above it to minimize maintenance problems and thermal discomfort due to drafts.

Although fresh air is sometimes brought directly to toilet rooms, a more common procedure is to supply slightly more air to surrounding office or commercial spaces than is returned; the resulting surplus is available to be drawn into toilet rooms, then exhausted.

Acoustic privacy is usually achieved by the placement of toilet rooms within service cores. However, ventilating openings and doorless-but-screened access ways present ready paths for sound. Some users of toilets respond to lack of acoustic privacy by repeatedly flushing the toilet to mask the sounds associated with elimination. This is a noisy and very wasteful practice, which could be minimized if the designer provides for acoustic privacy. Isolated toilet stalls usually are not practical—maintenance is complicated, and walls are far more expensive than simple visual-screen partitions. Instead, a steady masking sound is often provided. A higher-sound-level ventilating system is frequently chosen, since noisy fans are often chepater than the quiet ones, and they reassure the

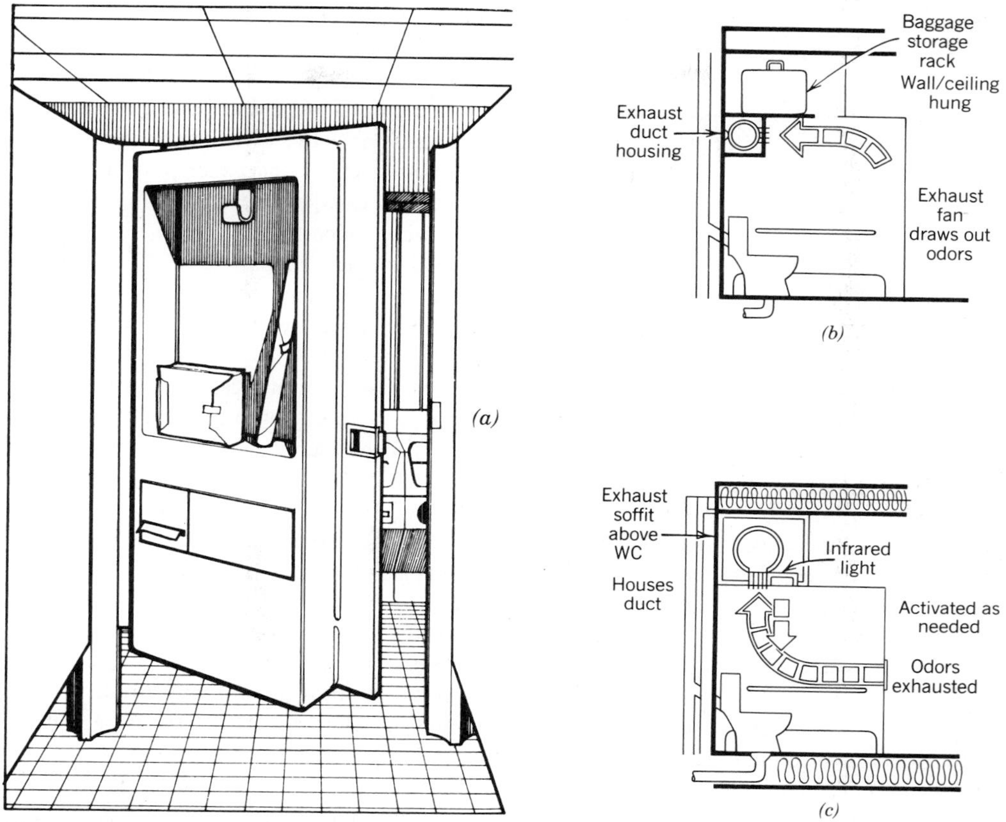

Fig. 11.18 Improving the public toilet stall. (a) *Stall doors could be designed to house small articles and outer clothing carried by the user. From The Bathroom: New and Expanded Edition, copyright © 1966, 1976 by Alexander Kira. By permission of Bantam Books, Inc. All rights reserved.* (b) *For transportation terminals, added provision for luggage, without an increase in floor area requirements, is desirable.* (c) *Exhaust ventilation directly above toilets is desirable. In cold weather, small radiant heat lamps with timer switches could provide task heating and lighting in an otherwise darker, colder environment. From a study by Michael Bush, University of Oregon.*

user that polluted air is constantly being removed from the space.

(c) The Private Bathroom. Individual bathrooms within large buildings, such as those for hotel rooms, apartments, and hospital patients' rooms, commonly are subject to similar location constraints as the larger public toilet rooms: that is, a location away from the more desirable perimeter zone, with its daylight and fresh air. Accordingly, the ventilation strategy for such individual bathrooms is similar—an exhaust fan, drawing air from the spaces near the bathroom. Individual residences can often provide a happier environment for the bathroom, with daylight and fresh air available.

One of the first design considerations for private bathrooms is recognition that three distinct categories of activity are involved: partial cleansing/grooming, full body cleansing, and elimination. A typical private bathroom is the *three-fixture type:* lavatory, toilet, and tub/shower, arranged within one relatively small room [35 ft^2 (3.3 m^2) is minimal]. Advantages include a potentially efficient use of floor space and the convenience of lavatory/toilet adjacency for perineal cleansing. The great disadvantage is that the room is designed for one user at a time, so that others are deprived of access to fixtures for elimination or partial cleansing when one person is engaged in full-body cleansing.

As a result, there has been a trend toward *compartmented* bathroom, in which the lavatory may be located in hallways, bedrooms, or small alcoves. This less-modest environment is usually suitable for the partial cleansing/grooming activities associated with lavatories. Also, lavatories are less expensive fixtures, require less floor area, and are less likely to do great damage if they malfunction. Most commonly, the toilet and tub/shower are then grouped in a more isolated, private space. Where many users of one such compartmented bathroom are expected, it is advisable also to separate the toilet (along with a smaller lavatory) from the tub/shower room.

Another typical approach to private bathrooms is the *guest bath,* in which a lavatory, a toilet, and a shower stall are included. The substitution of a shower stall for a tub cuts down the space needed to about 30 ft^2 (2.8 m^2). The guest bath is often the second bathroom in a residence, and sometimes is physically remote from the main one. A more limited and very common private bathroom is the *half-bath,* containing only a lavatory and a toilet. This room can be less than 25 ft^2 (2.2 m^2), and it is frequently found crouched below a stairway on the main floor of houses with upstairs bedrooms and main bath(s). The tight quarters and steeply sloped ceilings involved in such an arrangement offer a great opportunity for the use of mirrors as space expanders and light reflectors, as well as for aids to grooming.

Fixtures for private bathrooms commonly are more visually elegant, and somewhat more demanding of maintenance, than are their counterparts in public toilet rooms. The most dramatic difference is the presence of a tub, which can be large or small, raised or sunken, or built as a skylit greenhouse-like appendage to the bathroom (see Fig. 8.3, showing Andrews farmstead house). Special attention can easily be paid to accommodations such as pullout stepstools for small children at lavatories, and especially to storage—towels and soaps, at tubs/showers, grooming aids and medicines at the lavatory. It is common for as much design attention to be paid to these accessories in private bathrooms as to the selection and placement of the fixtures themselves.

Lighting affords a special opportunity at the lavatory, because of the availability of natural light and the tradition of electric lights for grooming. There are several important points to remember here: that the objective is to light the human face, not the mirror surface; that excessively bright lights near the eyes can cause glare; and that the background seen behind the face in the mirror should be no lighter than the face itself. For these reasons, direct sun on the face at lavatories is usually avoided, as are windows directly opposite the mirror (unless they open into a low-daylight scene). Some possibilities for natural and electric lighting for lavatories are shown in Fig. 11.19.

Thermal considerations are likely to be dominated by the ventilation needs, which are higher than those for any other residential space except for kitchens. If residences were to be designed as superinsulated buildings, a constant ventilating fan could be installed, pulling from the bathroom that amount of minimum fresh air required for the rest of the entire house. (However, if forced-air heat exchangers are provided for such residences, the bathroom fan should operate only as needed.) In hotels, this is a common way to assure fresh air to each room, and some multistory apartment houses also assure ventilation in this manner. Since most climates are not severe enough to warrant such tight regulation of outdoor air year-round, residential bathrooms are often ventilated, either naturally or with fans that are used only as needed.

Heating is the next most significant need, since a body that is naked and wet is one that is most susceptible to heat loss through both radiation and evaporation. Solar energy has much to offer the bathroom: strong radiant effects; the potential for some overheating, which may be appropriate in this more-needy space; and a psychological association between direct beams of sun and purification of the surfaces that it strikes. The bathroom contains fixtures and surfaces that are typically more thermally massive than those of most other spaces in houses. However, night insulation should cover windows and especially skylights, since radiant losses through unprotected glass are so great, and since the more humid air of the bathroom will readily condense on cold surfaces such as skylights exposed to the night sky.

Task heating is very popular and appropriate for the bathroom; radiant heat lamps provide the

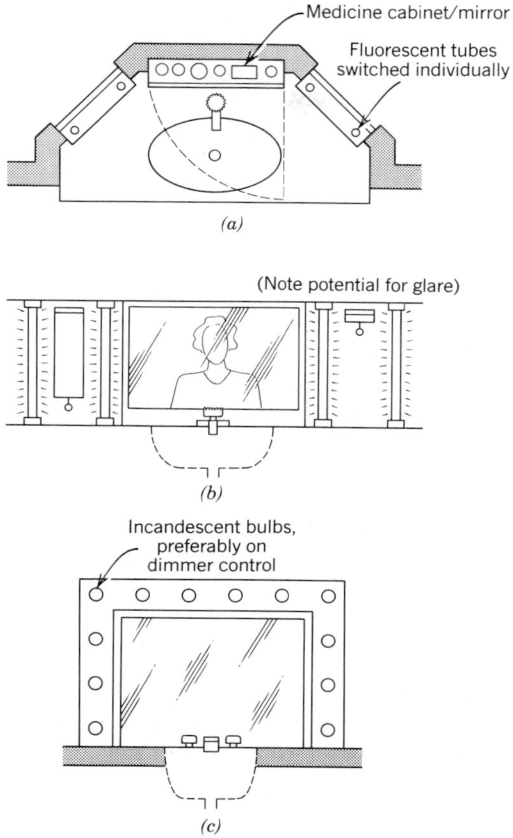

***Fig. 11.19** Lighting for lavatories.* (a) *A variation on a bay window approach, with daylight and electric light sources on either side of a central mirror.* (b) *For modeling across the face, there must be more light from one side than from another. This can be achieved with switches for each electric lamp, or by daylight controls in each window.* (c) *The traditional "theatrical" lighting has a high potential for direct glare. The larger the mirror, the farther the row of lamps from the face's image, and the less the glare problem. Less glare would be produced by a large, low-brightness overhead light source, with light-colored diffusing surfaces on walls and counters. From a study by Michael Bush, University of Oregon.*

added heat for the body fresh from tub or shower, without overheating the entire bathroom. Timer switches are frequently used with these heat lamps. If a forced-air heating system is used, an outlet near the lavatory is welcome for early morning grooming in winter. Air grilles should be installed on ceilings or on vertical surfaces in rooms that contain water, such as bathrooms and kitchens, since floors are frequently mopped and water spills are common. The toe space for cabinets beneath kitchen and bathroom sinks is a popular location for these supply air grilles.

When it is required, cooling is often linked with ventilation; the more cooling, the greater the ventilation rate. In very hot conditions, this is counterproductive, as it brings too much hot air into the house. A means of providing cool vertical surfaces (from night-ventilated thermal mass, for example), and a fan to stir the humid air, are a promising combination for hot-day comfort.

Some combinations of lighting and thermal considerations in bathrooms are shown in Fig. 11.20.

Acoustic considerations comprise another important aspect of bathroom design. Because of their nonabsorbent surfaces, bathrooms are frequently the most reverberant spaces in the house. They encourage the singer or the whistler, and may even serve as practice rooms for the family musician. But they can also intimidate a person who prefers to be quiet, such as a guest using the toilet while the rest of the dinner party sits quietly in the adjacent room.

Two common design responses are isolation and masking sound. Bathrooms can be separated from acoustically sensitive spaces by closets or hallways. With careful attention to sufficiently massive construction and/or other construction details, the doors, walls, ceilings, and floors of bathrooms can be constructed so as to reduce the passage of sound to acceptably low levels. (These considerations are discussed in Chapters 26 and 27). Sound isolation also depends on attention to detail: no cracks around the bathroom door, no back-to-back electrical outlets between bathrooms and adjacent spaces, no air grilles into ducts that have other grilles nearby, and no other open window near the bathroom window. If such goals are unattainable, or if additional acoustic security is desired, masking sound can be added as needed within the bathroom. All too often, the masking noise is provided by repeated flushings of the toilet (itself an embarrassing sound for some) or the running of water in the lavatory. But a noisy ventilating fan would serve the purpose, as would a music source, such as a radio. An elegant, expen-

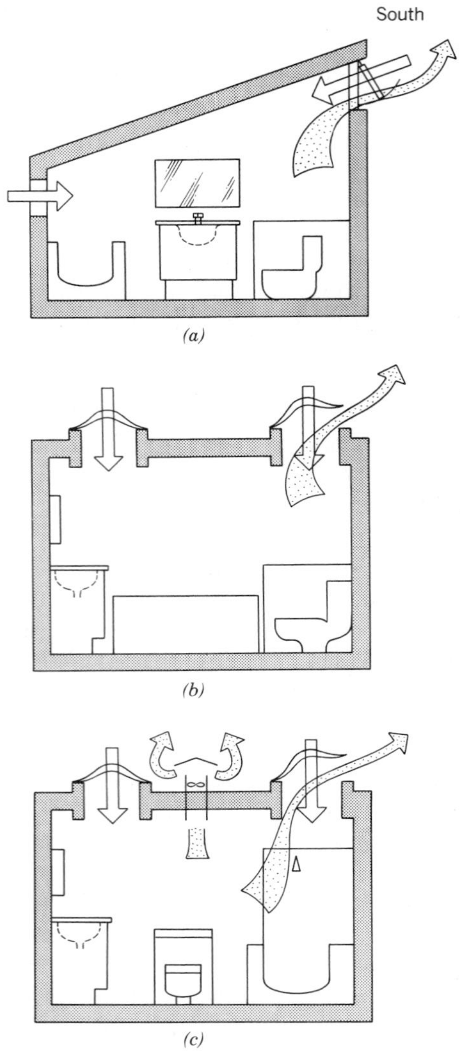

Fig. 11.20 Light and air considerations in bathrooms. (a) *An all-sidelight approach, in which south sun enters through clerestory and strikes the tub and floor. The lavatory is daylit from either side. Air past the toilet is drawn directly up and out of the clerestory, helped by the stack effect.* (b) *In this two-skylight approach, most light is given the lavatory and the toilet. An operable skylight over the toilet adds ventilation where it is needed. Without night insulation, these skylights could be the source of annoying droplets of condensation in winter. Horizontal skylights pose a problem for summer solar gains, however.* (c) *This two-skylight variation gives primary light and ventilation to the tub rather than to the toilet. Powered ventilating fans can assist when natural ''stack'' ventilation is insufficient. From a study by Michael Bush, University of Oregon.*

sive touch would be a recirculating fountain gracing one corner; it could provide splashing sounds only when the light or fan is on.

The final consideration in bathroom acoustics is the sound of water moving within pipes. Where such pipes are exposed or barely concealed within, or firmly attached to, thin walls, perceptible sounds are highly likely. Although many of us may have experienced this sudden acoustic signal that a bathroom is in use, few have experienced it quite like the owner of a house on Long Island who was cited by Reyner Banham (1969). Designed at the height of the modernist expression of mechanical services, this house featured a dining table cantilevered from the exposed waste stack serving the bathroom above! Where water noise is undesirable, pipes should be wrapped, resiliently mounted, and/or located in a less acoustically critical wall.

11.4 Installation and Maintenance

The typical bathroom is a collection of individual fixtures, with fittings and accessories, combined with some custom cabinetry or shelving. The designer faces a wide range of styles offered by various manufacturers; the final array of components can be assembled from many different sources and chosen to create the most special effect or to meet the lowest budget. The preceding chapters have discussed fixture selection based on criteria such as physiology and water and energy consumption. To these must be added esthetics, durability and ease of cleaning, compatibility between fitting and fixture, and cost.

There are prefabricated bathrooms, in which one manufacturer assembles the piping for preselected fixtures (Fig. 11.21). Going further, there are a few examples of entirely one-piece bathrooms, which incorporate the maintenance advantage of having no seams between fixtures, walls, and floors. This, of course, makes fixture replacement expensive and difficult.

For detail designers however, the bathroom presents the demanding task of getting the right-sized pipes to appear in the right place. Figure 11.22 furnishes an example from a manufacturer's catalog, along with the information needed both for working drawings and for ''roughing in''—the process of getting all pipes installed, capped, and

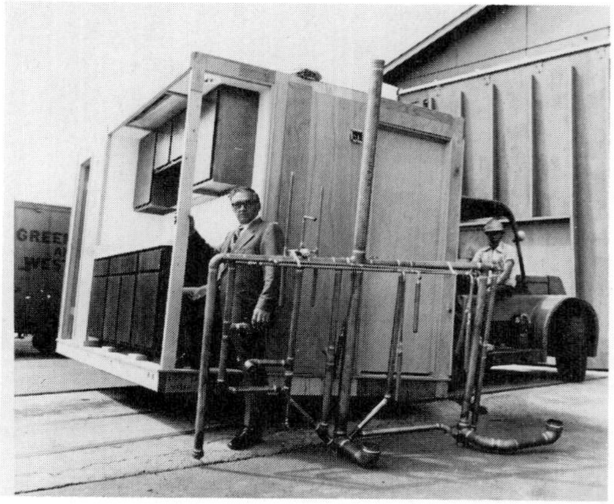

Fig. 11.21 *A preassembled plumbing "tree" for manufactured housing, along with the prefabricated kitchen/bath in which it is installed. Courtesy of Wausau Homes, Inc., and the Copper Development Association.*

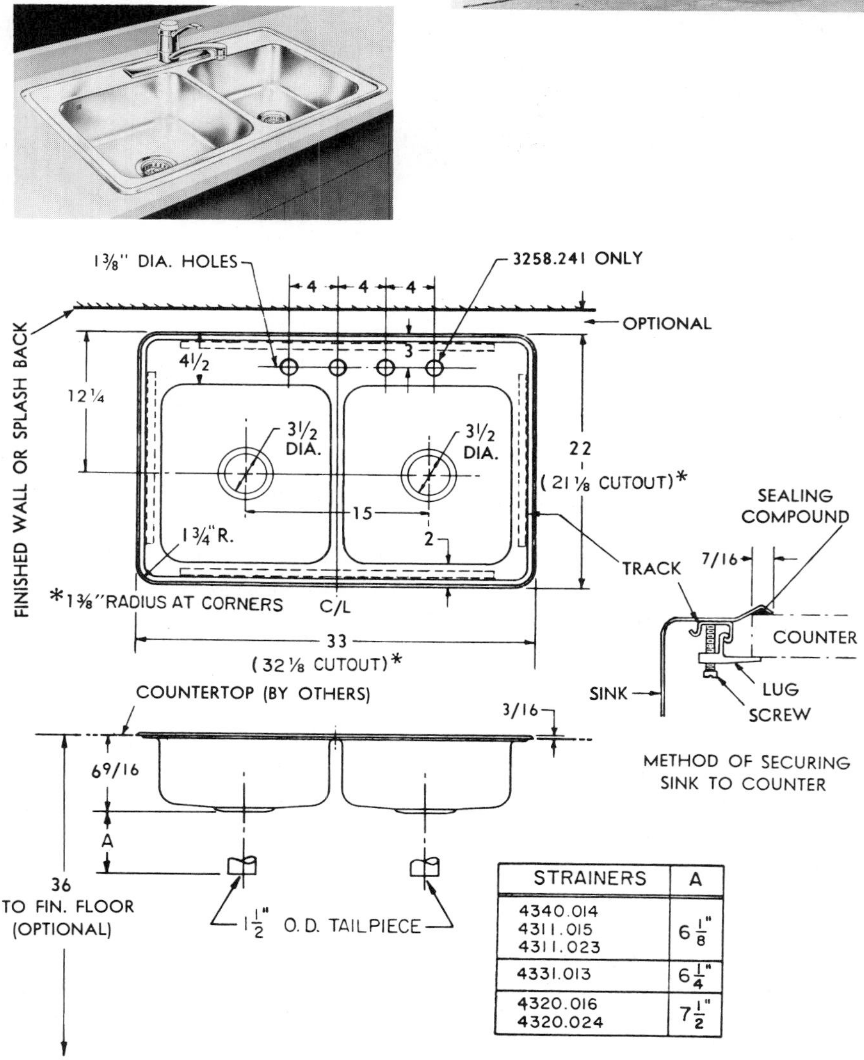

Fig. 11.22 *A typical catalog description of a stainless steel sink. Included are fixture dimensions and roughing dimensions. Courtesy of American Standard, Inc.*

Fig. 11.23 Roughing in place for four lavatories in the washroom of a school. At this stage, the waste branches have been capped and the system tested against possible leakage. The waste branches are made of cast iron, the vents of galvanized steel, and the water lines of copper with soldered fittings. Roughing dimensions have been followed. Vertical capped expansion and shock tubes serve as extensions of the entering water pipes that serve the hot and cold water branches.

pressure-tested before the actual fixtures are installed.

An example of roughing in for an office building was shown in Fig. 10.43. The roughing in of supply and waste piping for school lavatories is given in Fig. 11.23. Institutions such as schools have extensive requirements for durability and ease of maintenance. The fixtures are made of such resistant materials as stainless steel, chrome-plated cast brass, precast stone or terrazzo, or high-impact fiberglass. The fixture controls are designed to withstand heavy use—or misuse—and the fixtures are securely tied into the structure with concealed mounting hardware designed to resist extraordinary forces. (Some schools even move the lavatories into the hallway for better visual control of at least a part of the restroom facilities.)

In prisons, extreme measures are taken to prevent plumbing fixtures from becoming weapons. Heavy-gauge stainless steel fixtures with nonremovable fittings are provided, at very high cost for both the fixture and its tamper-proof installation.

For the more ordinary bathroom, the two maintenance questions of greatest concern are (1) how easy is cleaning around and within the fixture, and (2) how accessible are those parts of the fixture most likely to need repair or replacement? Ease of cleaning is often determined more by the space around the fixture than by the design of the fixture itself. This is particularly true of toilets, where a generous space of open floor to either side makes maintenance easy and therefore likely to be more frequent. Access to fixture parts may be more difficult. An access panel in the wall of the room behind the fixtures is often provided, encouraging speedy repair and replacement for fixtures at tubs, showers, and lavatories. Ideally, accessibility to all plumbing lines—whether via access panels in walls, trenches in concrete floors, exposed basement ceilings, or adequately deep crawl spaces—should be provided. The bathroom is likely to undergo thorough ''remodeling,'' including fixture replacement, as styles change and water/energy conservation becomes more important.

References

ANSI (1980). *American National Standard A117.1-1980, Specifications for Making Buildings and Facilities Accessible to and Useable by Physically Handicapped People,* American National Standards Institute, New York.

Banham, R. (1969). *The Architecture of the Well Tempered Environment,* Architectural Press, London.

Kira, A. (1966, 1976). *The Bathroom: New and Expanded Edition,* Bantam Books, New York.

Milne, M. (1976). *Residential Water Conservation,* U.S. Department of Commerce, NTIS.

12

SOLID WASTE

For some designers, the last building distribution system considered is the one involving the bulkiest items: the flow of supplies in and solid waste out. Since this system usually is not seen as consuming building energy or requiring specialized equipment, it ordinarily becomes a lower-priority system in the design process. Yet provisions for delivery of supplies, and especially for the collection and storage of solid wastes, can be more space-consuming than water/waste systems, can present a fire danger, and can create severe local environmental problems. The separation of solid waste for resource recovery involves significant energy consequences. Finally, mechanical equipment associated with solid waste is now more commonly installed.

12.1 Waste and Resources

In the past 30 years or so, there has been a marked increase in the amount of packaging material used for consumer products. Where shoppers once refilled reusable containers for bulk supplies at the market, for example, they now buy food in bags or cans that are discarded after use. This trend has increased spatial needs in the store, where shelves of cans are required instead of a bin of bulk products, and in the home, where such packaging soon turns into waste and must be stored until garbage collection day. Energy is required to make the boxes, bags, cans, and other containers; to transport them; and to collect them as trash. Devices such as trash compactors add both space and energy requirements to the process. Landfills for garbage disposal fill more rapidly as solid-waste flows increase.

There may not be much that building designers can do about these increased packaging trends. But the solid wastes for buildings do contain important resources, and a designer can help society to recover those resources, rather than to bury them in landfill or in the ocean. Further, the buildings themselves can be designed for materials recovery upon remodel or demolition, as was discussed in Section 1.7.

The resource within solid waste can be divided into the high-grade resources represented by recyclable materials and the low-grade resource of heat obtainable from the burning (incineration) of combustible solid wastes.

(a) High-Grade Resources. These include metals such as aluminum and steel, paper and paperboard, and some plastics. Glass is especially suitable for recycling when it can be reused after simple washing, as in the case of beer and soft drink bottles. ''Returnable'' bottles and cans have measurably decreased roadside litter in states that now require such containers. Newspaper is so readily stored and recycled that in many communities, charitable groups or service organizations recover newspapers for profit. Recycled paperboard can easily save 50% of the energy that would be required for pulp from virgin material.

The recovery of aluminum saves 96% of the energy necessary to produce it originally. Since the aluminum production process is dependent upon electricity, the recycling of aluminum is even more attractive. Energy conservation in aluminum production has been developed to the point that only 82% as much energy was needed to make a pound of aluminum in 1982 as in 1972. More than a third of that reduction was attributable to aluminum recycling programs; the remainder came from improvements in the production process. And as for steel, recycling can produce a 52% energy savings, compared to the use of virgin material.

Plastics are more difficult to recycle. Currently, USFDA regulations prevent the use of recycled plastics in food-related items. However, the plastics from items such as soft drink containers, margarine tubs and lids, and milk jugs can be reprocessed into plastic pellets, which are cheaper than virgin plastic pellets. These pellets are made into nonfood items—toys, building products, sports products, and other things.

(b) Low-Grade Resources. These resources include materials for which recycling is impractical, but which are combustible. Although to recover heat from such materials obviously seems better than to waste them altogether, the resulting additional problems of air pollution (and ash disposal) must be minimized. Low-grade resources include gaseous wastes, liquid and semiliquid wastes, and solid wastes. Industrial and commercial processes can generate wastes of all types, including some with very high heat content and some that are very toxic. In buildings, however, solid wastes are the ones most frequently subject to burning.

Some typical waste products are classified in Table 12.1; the higher the classification number, the more difficult the task of incineration. The heating values of these materials may be compared with those of the typical fuels listed in Table 5.14; this comparison shows that the most highly combustible grade of common solid waste has, per pound, better than one-half the heating value of

TABLE 12.1 **Types, Composition, Heating Values of Wastes**

Type	*Typical Mixture*	*Incombustible Solids (%)*	*Moisture (%)*	*Heating Value (Btu/lb)*
0	Trash, a mixture of highly combustible waste such as paper, cardboard, cartons, wood boxes, and combustible floor sweepings, from commercial and industrial activities. The mixtures contain up to 10% by weight of plastic bags, coated paper, laminated paper, treated corrugated cardboard, oily rags, and plastic or rubber scraps.	5	10	8500
1	Rubbish, a mixture of combustible waste such as paper, cardboard cartons, wood scrap, foliage, and combustible floor sweepings, from domestic, commercial, and industrial activities. The mixture contains up to 20% by weight of restaurant or cafeteria waste, but contains little or no treated papers, plastic, or rubber wastes.	10	25	6500
2	Refuse, consisting of an approximately even mixture of rubbish and garbage by weight. (Common to apartment and residential occupancy)	7	50	4300
3	Garbage, consisting of animal and vegetable wastes from restaurants, cafeterias, hotels, hospitals, markets, and like installations.	5	70	2500
4	Human and animal remains, consisting of carcasses, organs, and solid organic wastes from hospitals, laboratories, abattoirs, animal pounds, and similar sources.	5	85	1000
5	By-product waste—gaseous, liquid, or semi-liquid—such as tar, paints, solvents, sludge, fumes, etc., from industrial operations.	—	—	[a]
6	Solid by-product waste, such as rubber, plastics, wood waste, etc., from industrial operations.	—	—	[a]

Source: Incinerator Institute of America.

[a]Btu values must be determined by the individual materials incinerated.

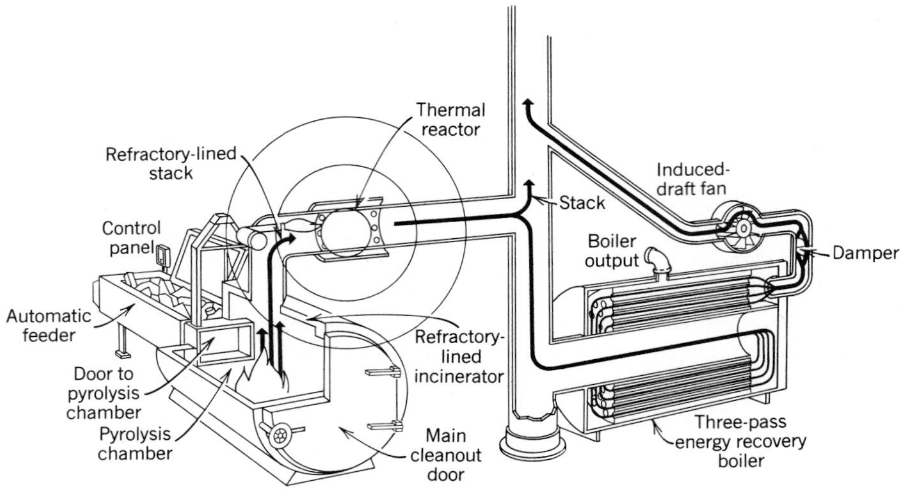

Fig. 12.1 A pyrolytic incinerator controls the amount of combustion air to its firebed, producing temperatures of about 1200 F. At such temperatures in an oxygen-lean atmosphere, the waste thermally degrades, giving off combustion gases. These gases flow to a thermal reactor, where they are ignited at even higher (1800–2000 F) temperatures. The resulting flue gas is mostly carbon dioxide and water vapor, and thus meets most pollution standards. Courtesy of the Kelley Company.

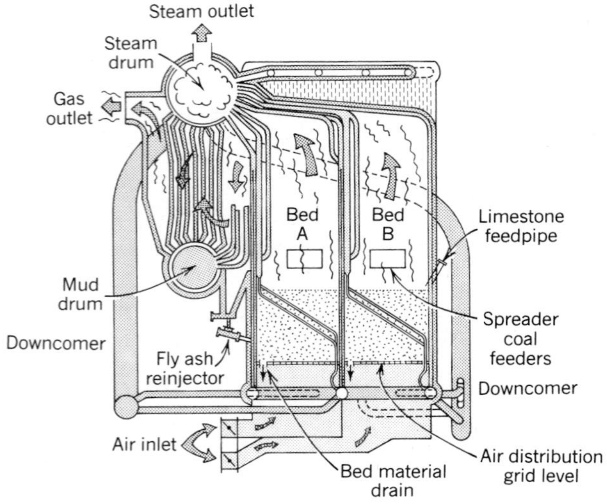

Fig. 12.2 A "universal" burning device for solid, gaseous, or liquid wastes is the fluidized-bed steam generator. The fluidized bed is composed of selected materials that react chemically with the fuel or the flue gas; air is injected from below to keep the "bed" in constant motion. The high temperature and chemical reaction produce nearly complete combustion of a wide range of fuel types. This reduces or eliminates the need for air pollution control equipment on the flue. Two (or more) beds allow the burning of smaller amounts of wastes at off-peak times, while maintaining high efficiency. Reprinted from Specifying Engineer, November 1982. Copyright © 1982 by the Cahners Publishing Co.

anthracite coal. Air pollution regulations have severely restricted simple trash burning as a means of solid waste disposal; incinerators now must meet increasingly strict regulations. Consequently, up-to-date incinerators are of the controlled-air type or the pyrolitic type (Fig. 12.1) and are designed to recover the energy from burning wastes. A newer "fluidized bed" steam generator for large, professionally maintained buildings, is shown in Fig. 12.2; it incorporates a burning process that greatly reduces air pollution.

Where large quantities of mixed trash, garbage, and other refuse are collected, special resource-recovery plants can be built to recover materials, produce useful steam for electricity generation, and reduce the flow of waste to landfill

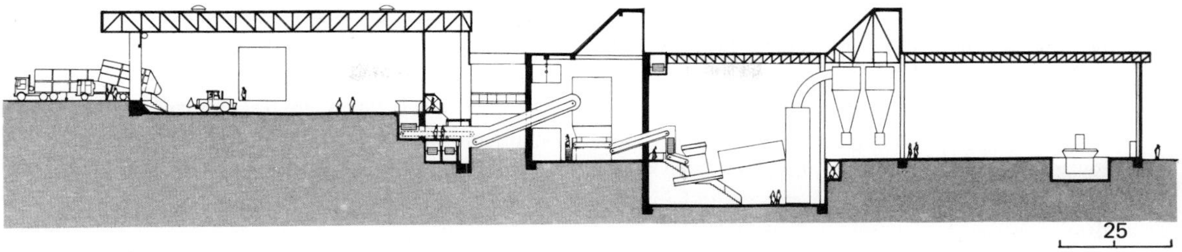

(a)

(b)

Fig. 12.3 Solid-waste resource recovery project. The flow of initially unseparated garbage through a proposed facility is shown in (a). *The rendering* (b) *shows the recovery plant in the foreground, the electric generating station beyond. Cone-shaped storage bins hold shredded refuse awaiting conversion to steam for the turbines. Clean Communities Corporation, Haverhill, Mass.; Camp Dresser & McKee, Inc., Engineers. Reprinted from Architectural Record, June 1975. Copyright © 1975 by McGraw-Hill, Inc. All rights reserved.*

(Fig. 12.3). In this energy-intensive process, the mixed garbage is shredded and blown through large ''air classifiers'' that separate the organic (burnable) wastes from metals and glass. The burnable wastes are then used, under controlled combustion, to generate electricity. The metals are further separated magnetically into ferrous and nonferrous classes; these and the glass are then recycled.

12.2 Resource Recovery: Central or Local?

In general, the more thoroughly mixed are the different types of solid waste, the harder it is to recover their high- and low-grade resources. From an energy conservation viewpoint, solid wastes should be kept as separate as possible; glass bottles should be washed and reused rather than broken and recycled, and unrecyclable but burnable solid wastes should be kept clean and dry until they can be burned. The earlier that different types of metals are separated, for example, the less energy spent later on to separate aluminum from steel, and so forth. Organic food wastes could be composted for use on site, rather than ground up and added to the load on the sewage treatment system.

However, there are two disadvantages to "local" solid waste separation (at the point of their discard)—one cultural, one physical. Keeping solid wastes separate requires somewhat more effort and time on the part of the consumer. Rather than dumping everything in a garbage can, the resource-conscious consumer will wash metal cans, remove both ends and flatten them, then deposit them in a container reserved for ferrous metal. Newspaper and box cardboard are stacked separately; returnable bottles and cans are kept apart from glass; compostable kitchen wastes are kept separate from garbage. This disadvantage of slightly more work and time for waste separation is compounded by the physical disadvantage of the floor or cabinet space taken up by so many separate containers. In communities fortunate enough to have recycling-oriented garbage service, these containers can be carried out and lined up to await garbage collection. The garbage trucks used for this purpose are complex assemblages of various bins, rather than simple massive caverns. In many communities, however, each pile of recyclable materials must be taken to a different point—a disadvantage in terms of both the time and energy used in transportation.

The characteristics of local waste separation, then, are increased consumer time and building space requirements. Central waste separation is characterized by energy-intensive (and noisy) industrial processes. The next two sections offer a detailed look at the consequences of local waste separation on some common building types.

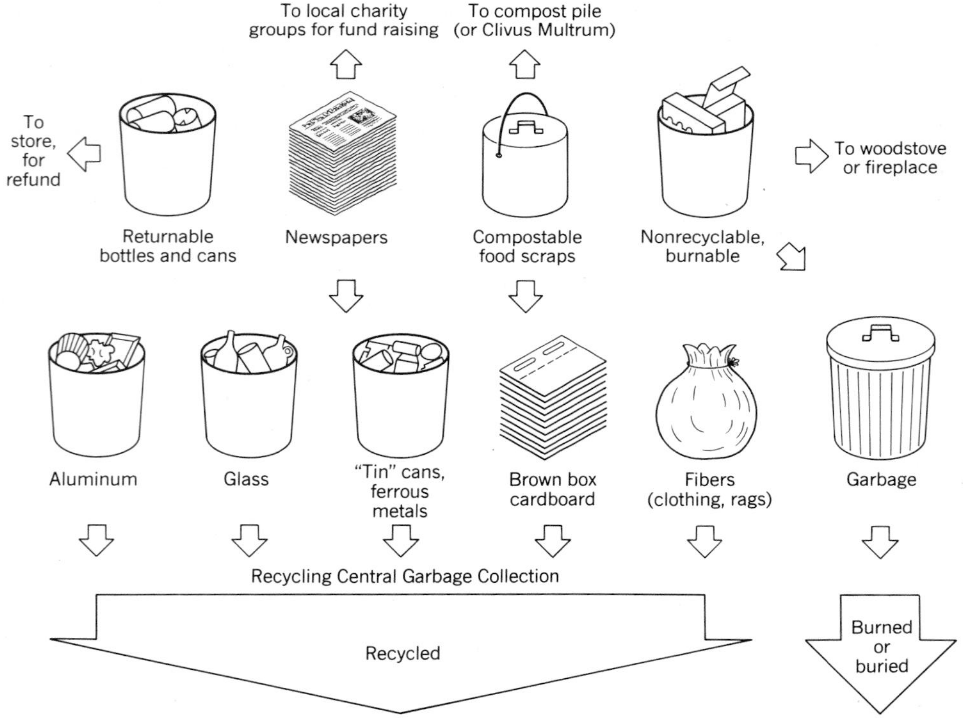

Fig. 12.4 Local separation of solid wastes can require many containers. In communities that offer recycling garbage collection service, the consumer can recycle without transporting individual containers to many different collection points. Some recycling services ask for glass to be separated into green, brown, and clear; some offer to recycle white paper.

12.3 Solid Waste in Small Buildings

The choice between separation of waste or the mixing of garbage in one can is most often enjoyed by the occupants of small buildings. Where food preparation is involved, as in residences or restaurants, the collection of special containers can reach surprising complexity (Fig. 12.4).

As the point of origin for so many types of solid waste, the kitchen is often the location of waste separation and storage. There is an inherent conflict, however, between the kitchen's frequently hot and humid environment and the need for a cool, dry, well-aired place for solid wastes. This suggests that a space—pantry, airlock entry, cabinet, closet—be provided that opens to the kitchen on one side and to the outside on the other (Fig. 12.5). Both the daily deposits to solid waste and the weekly waste removal are made easy, and the near-outdoor conditions are better for the waste in most U.S. climates. It is also important that cleaning of the waste storage area be easy.

Two mechanical devices increasingly found in homes and apartments are garbage compactors and garbage disposers, which usually are installed directly beneath the drain of the kitchen sink.

The *garbage compactor* allows a much less bulky storage arrangement. Used selectively, it can compact several of the stacks of items shown in Fig. 12.4, such as nonrecyclable burnable items, aluminum for recycling, ferrous metals, and box cardboard. Used indiscriminately, it can make the central process of garbage separation more difficult by crushing dissimilar items together. For a recycling-conscious household, it is questionable whether the compactor will save much more storage volume than it takes for itself.

The *garbage disposer* (Fig. 12.6) grinds up organic food scraps and sends them on through the

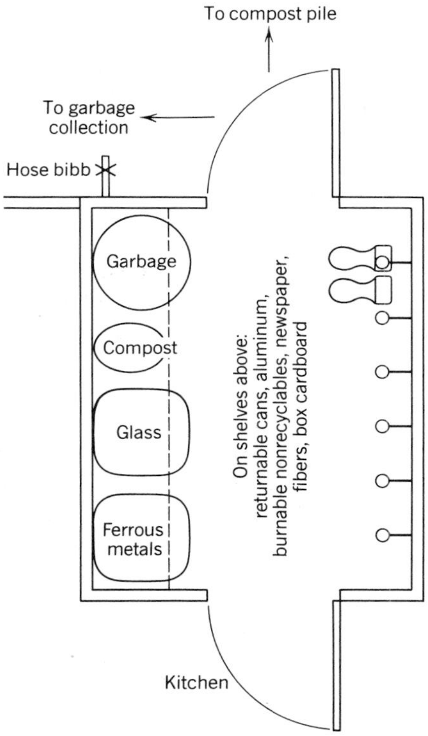

__Fig. 12.5__ An entry vestibule (or airlock) to the kitchen can serve as a depository for solid wastes, as well as for coats, boots, and other items. Heavy, dirtier items belong on the floor, and many recyclables can be readily and conveniently stored on shelves. An outdoor hose bibb is useful for washing out soiled compost containers and garbage cans.

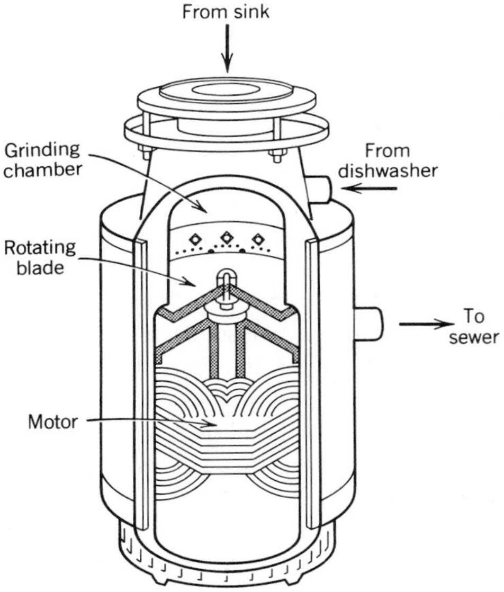

__Fig. 12.6__ The garbage disposer (or grinder) diverts solid waste (food scraps) from landfills to sewage treatment plants. Occupying less than a cubic foot, it saves garbage storage space and some of the users' time, but it requires more energy and much more water than treatment of the food scraps as garbage. Composting is another alternative. From Milne (1976).

sewer. This device is a boon to central garbage collection, since it lightens the weight of the garbage can and adds less moisture to the garbage (which thus can be burned more efficiently). However, garbage disposer units require both water and energy. Water must be kept running during the grinding process, to coagulate grease for chopping, to wash the blades and keep them free of debris, and to cool the grinder's motor. In all, from 2 to 4 gal (7.5 to 15 L) are required for one minute of operation. Because more water and solid waste is deposited in the sewer system, moreover, the central sewage treatment plant requires more energy to operate.

The alternative is the *compost pile,* familiar to most home gardeners as a source of excellent soil conditioners. Urban opportunities for composting seem very limited; yet raised growing beds on a balcony or window flower boxes are logical recipients of a family's compost. The problem may not be so much what to do with the final product (humus), as where to locate the compost pile.

The outdoor compost pile has several characteristics that challenge the designer. At its best, it is a frequently turned, quite warm, damp, well-aired source of rich humus (and red worms) for gardens; odors are noticeable only while the pile is turned. At its worst, it is a source of unpleasant odors and a breeding place for vermin. Where odors are not objectionable, the heat generated in a frequently fed and tended compost pile would be welcome against the exterior walls of residences. Clearly, these walls must have nonorganic exterior materials!

As groups of residences are combined into large apartment complexes, the solid waste systems can make more significant demands on the designer. Where central storage compounds are provided, garbage cans should be fenced to ward off dogs and other marauders. Different bins for recyclable materials might be provided. A central compost pile could provide humus for the landscaping of the complex. An incinerator is frequently installed to recover heat from nonrecyclable burnable wastes. This combination of space and equipment has special environmental needs, including garbage truck access, noise control, and location of both incinerator stack and compost pile with respect to prevailing winds.

12.4. Solid Waste in Large Buildings

As buildings become larger, it becomes more likely that solid wastes will be handled several times in the storage and collection process. This may inhibit the separated-waste approach, since not only the employees who generate it, but also the custodians who collect and store the waste, must understand what items go into which bins. However, larger buildings tend to generate large and concentrated kinds of wastes, which makes recycling more attractive.

Consider the multistory office building. The office floor operations are likely to waste large quantities of white paper and smaller quantities of newspaper, box cardboard, and unrecyclable burnable trash (including floor sweepings). Much smaller quantities of food scraps (coffee grounds), metals, and glass are also generated. Given the high cost of rental space, the pressures are very high for a simple mixed-garbage can (rather than multiple separate bins). However, a building that utilizes an incinerator for heat recovery might provide for three waste categories: white paper for recycling, burnable trash, and garbage (glass, metal, food scraps).

(a) The Collection Process. In larger buildings, the collection of solid waste typically involves a three-stage process (Fig. 12.7). The first stage is the generation of the waste itself, by employees who might be provided with one wastebasket apiece. If each employee is expected to separate wastes, this suggests a redesigned receptacle in which the most frequently used basket is on top. At the typical desk in an office building, white paper would thus be the easiest to deposit. In lower baskets, burnable trash would go on one side, garbage on the other. Waste separation at the point of origin thus may not require more floor area.

The second stage begins as custodians disconnect these individual baskets, dump them into separate bins on a collection cart, and reconnect the individual empty baskets for the next day's deposits. At various special-purpose stations, special wastebaskets can be supplied: white paper at the copying machine, garbage at the employee lounge, burnable trash at the shipping/receiving station.

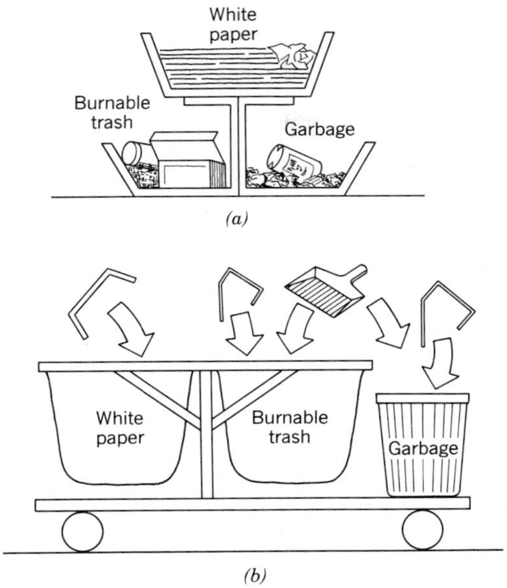

Fig. 12.7 Three-stage collection process, with three category waste separation, for office buildings. (a) *At each work station, a three-compartment waste receptacle is provided. White paper is the predominant waste product, so its compartment is the easiest to use.* (b) *Custodians begin the second stage by collecting waste in separate bins. Floor sweepings are added to the burnable trash.* (c) *At the end of the second stage, the three categories of wastes are deposited in separate chutes. Service sinks, paper shredders, and other maintenance items can be incorporated in such a service closet.* (d) *The third stage begins with compactors at the base of each chute. White paper is compacted, baled, and stored; burnable trash is shredded and conveyed; garbage is compacted, baled, and stored.* (e) *At the end of the third stage, white paper is collected for recycling, burnable trash is incinerated and its heat recovered, and garbage is collected for separation at a central plant.*

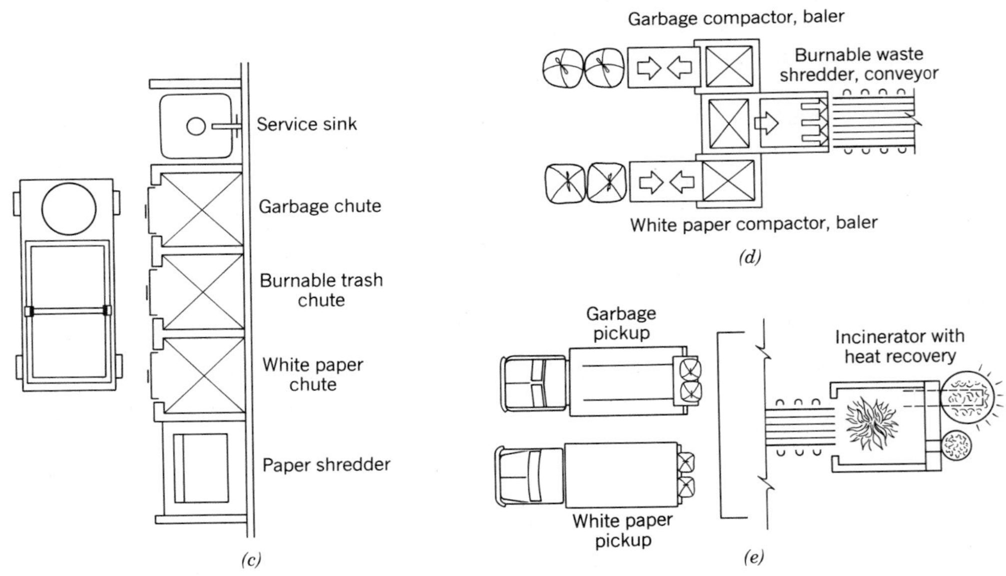

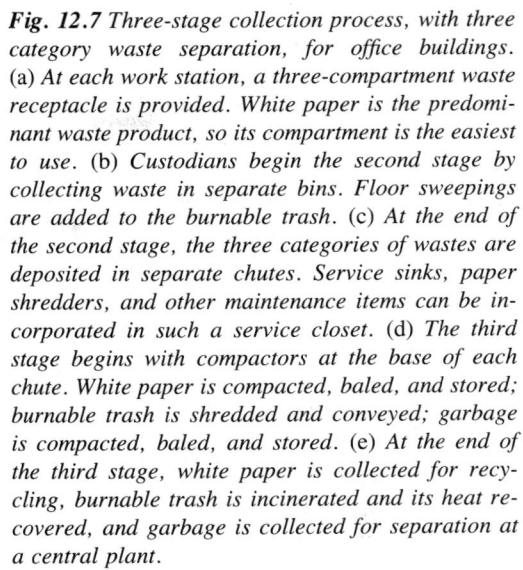

Floor sweepings are added to burnable trash (or garbage, as is appropriate). When the cart is full, it is wheeled to the service closet, which is probably located within the core of the building. Here would be a chute for each of the three categories of waste, along with a service sink to wash out the garbage bin (and perhaps a paper shredder to be used selectively by employees). Where chutes are impractical, cans of waste storage are closeted near service elevators.

The third stage begins at the bottom of the chutes, where white paper is compacted and baled, burnable trash shredded and conveyed toward the incinerator, and garbage compacted and bagged. In the storage space needed for these bags and bales, cool, dry, and fresh air is desirable. The compactors and shredders are noisy, and must be vibration-isolated from the floor. At the end of the third stage, a truck or van from the recycling center collects bales of white paper, a garbage truck collects garbage bags, and the heat-recovery incinerator turns burnable trash into much heat and a little ash.

Where incinerators are impractical because of first-cost/space constraints or air quality considerations, the burnable trash can be combined with garbage throughout all three stages. In cities with resource recovery stations (see Fig. 12.3), the end result is similar, although more energy is consumed in the process.

(b) The Service Core. Medium- and high-rise buildings usually concentrate many building services within a "core," from which services can be distributed as needed throughout each floor. The core typically contains stairs; passenger and service elevators; toilet rooms; service closets; mechanical, plumbing, and electrical chases; electrical/telephone closets with local switching capabilities; fire protection equipment; and supply closets. The size of such cores varies widely; for example, the taller the building, the more elevators.

Often, these service cores are identical in plan from one floor to the next. Alternatively, the vertical services can be identical (stairs, elevators, chases) and the arrangement of other elements (toilet rooms, supply closets) allowed to vary. Whether or not minor floor-plan variations occur, the cores usually depart radically from the typical plan—and ceiling height—at both roof and ground floor (Fig. 12.8). (Middle interruptions also occur where intermediate mechanical floors are provided.) Depending upon the type of elevator machinery selected, the machine room ceiling height above the top floor served might be three times the ordinary floor height (see Chapter 23). On the delivery/loading floor, large trucks must be accommodated. For chillers, boilers, incinerators, and fans, higher ceilings may be required to accommodate the flues, ducts, and pipes to which they must be connected; see Section 6.8 for the dimensions required. Figure 12.9 offers an example of the way in which a building's service core can expand to nearly fill the ground floor.

A service core can be related to the remaining service floor area in any of several ways (Fig. 12.10). A single central core is a familiar arrangement (Occidental Chemical Offices, Fig. 6.2). Frequent variations include a core at the edge (Seeley Mudd Library, Fig. 6.123; Comstock Center, Fig. 6.124) or one that is detached, two cores symmetrically placed (Time-Life Building, Fig. 6.115), corner cores, or core services dispersed somewhat randomly (Bateson Building, Fig. 5.29; Iowa Public Service Building, Fig. 6.117).

Each core location has advantages and drawbacks. From the viewpoint of rentable space, important factors are the flexibility of the served (rental) floor area; the exposure of rental area to the perimeter, and thus to light, air, and view; and the extent to which ground floor high-rent space gets commercial exposure. For fire safety and occupant convenience, the distance of travel from the furthest rental floor area to the service core and the clarity of circulation within the rental area, are important. For the environment within the core—toilet rooms, stairs, elevator waiting areas—access to light, air, and view are desirable. For the mechanical services, ease of core area expansion on the roof, at the loading dock, and in the basement are important, as is the length of travel for ducts and other such elements within each floor. These factors are compared for various types of core designs in Table 12.2.

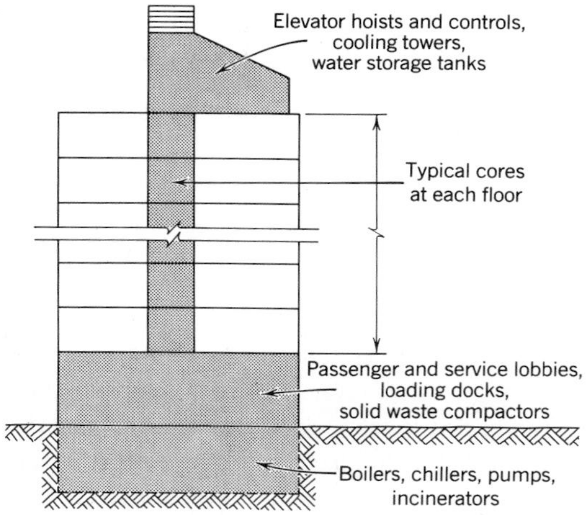

Fig. 12.8 *Schematic section through a core of a multistory building. At the top and bottom, floor space and ceiling height requirements increase. The consequences on the ground (or delivery/collection) floor are particularly great.*

Fig. 12.9 *Pan Am Building, New York City. Emery Roth and Sons, architects; Walter Gropius, Pietro Belluschi, associated architects. The distribution principles of the building in Figure 9.33 can result in the "mechanical stories" as used in the Pan Am Building. Stories of this type may also contain air conditioning compressors, water chillers, and blowers.* (b) *Partial model, first floor. Typical of planning in many high-rise buildings, this is one of the early models used in developing space concepts for the Pan Am Building. Equipment risers have a way of being vertical with no (or few) offsets. The first floor and typical upper floor set the pattern. Space allocations must be made for sanitary and storm drainage, elevators, ducts, electrical and telephone risers, water risers, and fire stairs. Equipment planners must aid in establishing rooftop areas, mechanical stories, and the general building horizontal cross section. Courtesy of Emery Roth & Sons, Architects.*

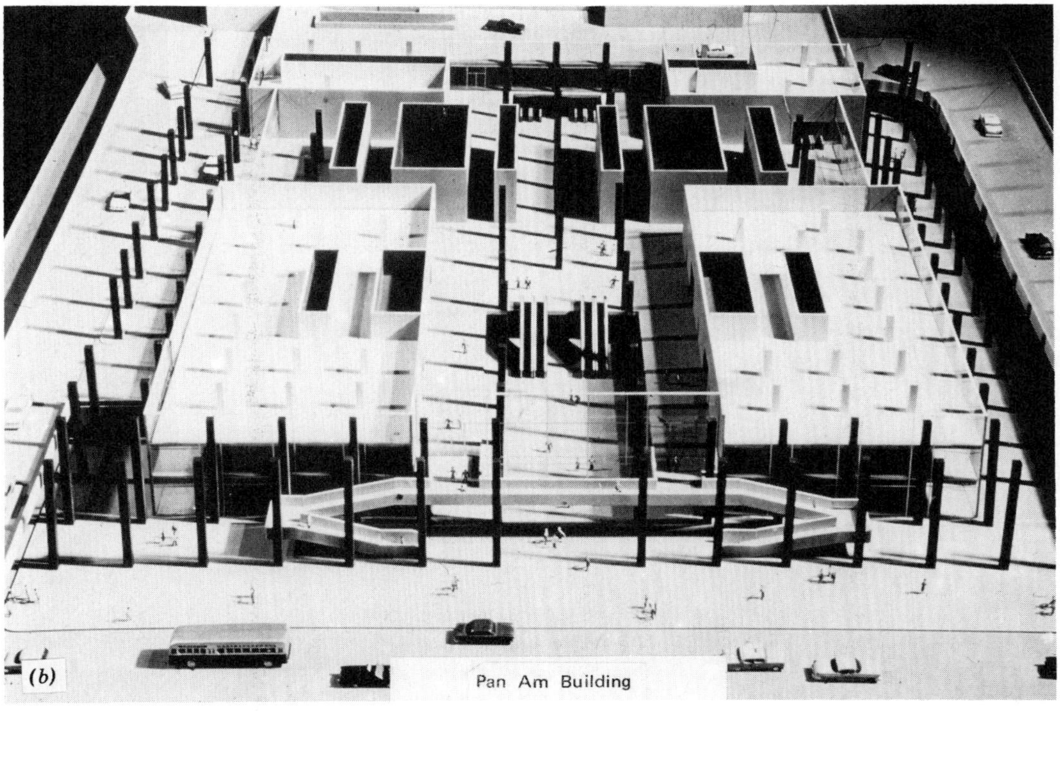

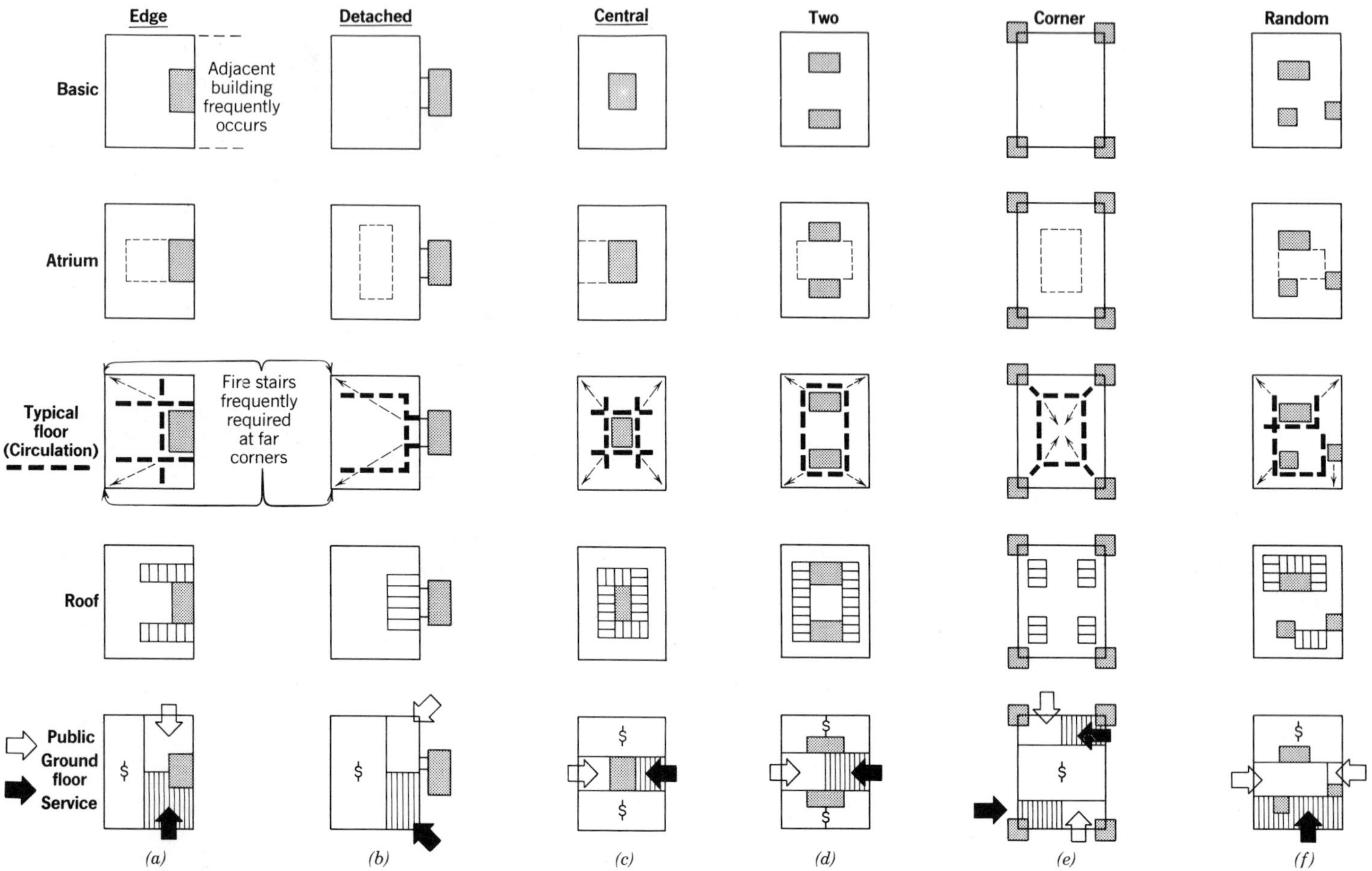

Fig. 12.10 *Comparison of core arrangements and building plans.* (a,b) *The edge and the detached cores give great flexibility of rental floor area, with light and view for core spaces.* (c) *The central core expands readily at the roof and the ground floor and has clear circulation and fairly flexible rental space.* (d) *Two-core is a popular, workable arrangement.* (e) *The corner cores give great flexibility to rental floors but are difficult at roof and ground floor.* (f) *Random cores generally occur in low-rise buildings, in which the benefits of repetitious plans are minimal.*

TABLE 12.2 **Building Characteristics and Core Placement**

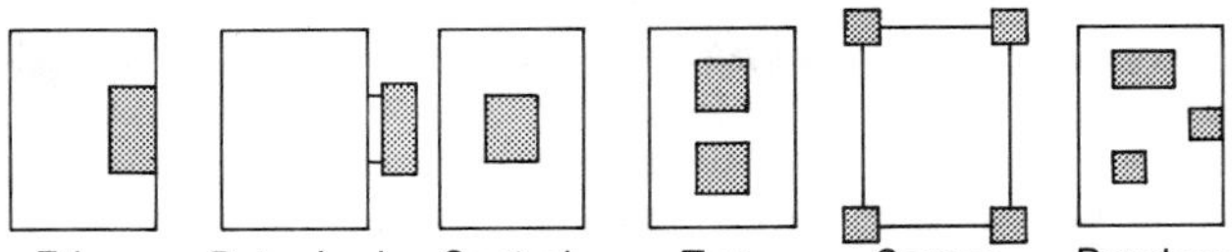

Core Designs	Edge	Detached	Central	Two	Corner	Random
Flexibility of typical rental areas	Good	Excellent	Average	Fair	Good	Poor
Perimeter for rental areas	Fair	Average	Excellent	Excellent	Poor	Good
Ground floor high rent area	Average	Excellent	Average	Fair	Good	Poor
Distance of travel from core, typical	Fair	Poor	Good	Excellent	Average	Average
Clarity of circulation, typical	Average	Fair	Good	Excellent	Average	Poor
Daylight, view for core spaces	Good	Excellent	Poor	Poor	Excellent	Fair
Core expansion at roof	Average	Poor	Excellent	Good	Fair	Average
Core expansion at ground floor	Average	Fair	Good	Excellent	Poor	Average

Excellent Good Average Fair Poor

12.5 Equipment for the Handling of Solid Waste

Where large amounts of solid waste are generated, and little space is available in which to store it, various types of equipment can be used to change the volume or the composition of the waste, so as to facilitate its transportation and storage. Incinerators were discussed in Section 12.1; other approaches are considered here. The first step is to determine the probable daily flow of solid waste, as is done in Table 12.3. When the extent of the problem is known, equipment can be selected.

(a) Compactors (Fig. 12.11). Of the wide variety of compacting devices, most are able to reduce the volume of solid waste to as little as 10% of the original volume. Among the many choices to be made are vertical versus horizontal compaction, automatic chute fed versus manual freestanding, whether wastes are to be bagged or baled, and the final size of each unit of compacted waste. Compactors can be noisy; they can also be prone to fires as heat is generated in the compaction process. Many compactors have built-in sprays for both fire control and disinfecting. Access to wash water and a floor drain are highly desirable.

(b) Pulping Systems (Fig. 12.12). Rather than extend space-consuming chutes up through many floors, these systems grind waste into pulp in the presence of water, and thus make a readily transportable slurry. At the loading docks, this slurry enters a water press, where about 90% of the water is squeezed out, reducing the volume to about one-fifth that of the original wastes. This water is reused and replenished as required. Such pulping systems are used not only for general refuse, as illustrated, but also for the destruction of documents and for food service wastes. Pulping systems have limitations; they should not be used to handle metal or plastics, which are recoverable. Pulping systems are replacing incinerators in urban areas; their advantages in reduced air pollution must be weighed against the possible energy recovery from incineration.

TABLE 12.3 **Approximate Daily Flows of Solid Waste**

Classification	*Building Type*	*Amount of Waste*
Industrial buildings	Factories	Survey required
	Warehouses	1½ lb per 100 ft²/day
Commercial buildings	Office buildings	1 lb/100 ft²/day
	Department stores	3 lb/100 ft²/day
	Shopping centers	Survey required
	Supermarkets	7 lb/100 ft²/day
	Restaurants	2 lb/meal
	Drug stores	3 lb/100 ft²/day
	Banks	Survey required
	Cafeterias	½–¾ lb/meal
	Clubs	1½ lb/meal
Residential buildings[a]	Private homes	3 lb basic + 1 lb/bedroom
	Apartment buildings	3 lb/bedroom/day
Schools	Grade, high, university	6 lb/room & ¼ lb/pupil/day
Institutions	Hospitals	20 lb/bed/day & 2 lb/meal
	Nursing homes	4 lb/person/day
Hotels	First class	2 lb/room & 2 lb/meal/day
	Medium class	1½ lb/room & 2 lb/meal/day
	Motels	2 lb/room/day
	Trailer camps	6–10 lb/trailer/day

Source: Courtesy of International Dynetics Corporation.

[a]Residential wastes are typically 130–150 lb/yd³, with a moisture content of 10–15%. On a per capita basis, it can average as high as 2.7 lb per person per day.

Fig. 12.11 A refuse compactor suitable for apartment house or office building service. This will compact 54 yd³ of refuse per hour, packing a front- or rear-load commercial container. Overall size is 32 in. wide, 66 in. long (without container), and 67 in. high; minimum room size is 6 ft wide by 13 ft long. Courtesy of General Dynetics Corporation.

(c) Vacuum Systems. The primary advantage of the pulping system is that it reduces the building volume consumed by waste chutes or storage cans on each floor. A similar approach might be a grinder-plus-evacuated-tube system, similar to the vacuum sewage systems described in Section 10.1 (Fig. 10.4). Such a system would use less water, unless the recirculating slurry lines were omitted. Vacuum conveying systems are now used in several industrial applications, such as ash and woodchip transporting. Vacuum systems that use air, not water, as the transport medium are frequently used for linens in hotels and for trash. Separate vacuum systems for trash and linens are desirable, both to lessen the soiling of the linens and to allow them to be delivered to a destination different from that for the trash.

The advantages of pressurized systems that use pulping or vacuum is that lines can be small and the contents can be moved horizontally or even upward. This allows far greater flexibility in design than is possible with gravity systems.

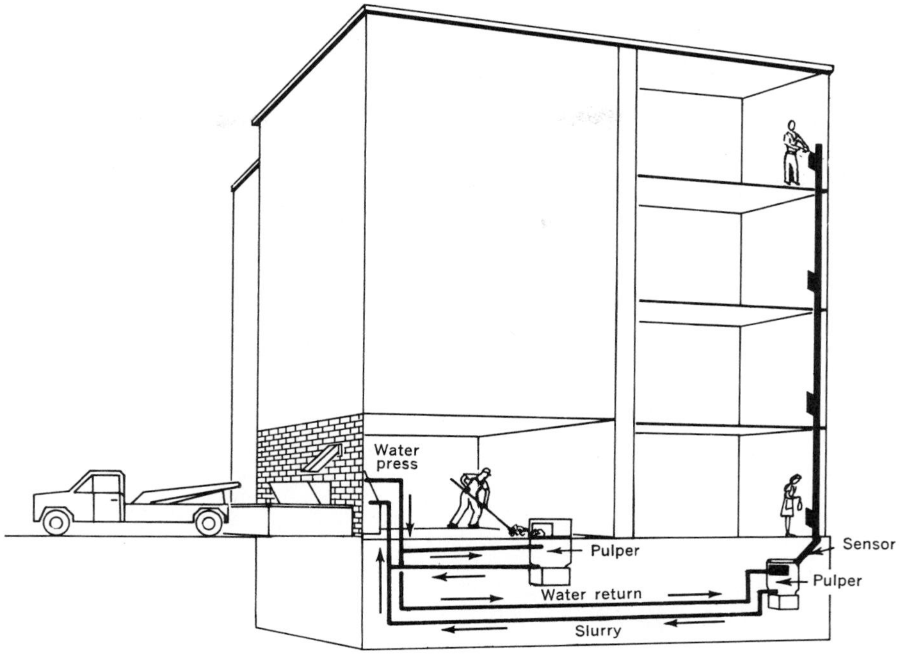

Fig. 12.12 A pulping installation for apartment refuse. The pulper is located to deal directly with solid waste; the slurry then transports the waste to the press at the loading dock. The pulper can operate manually or automatically by means of a sensor as shown. One water press can serve two or more pulping units. Wascon Systems, Inc.

(d) Summary. There are many options to be considered within the present waste-handling process of grind, crush, burn, and bury. As earth's nonrenewable resources are stripped away, the recovery of materials from solid waste becomes increasingly attractive. The situation is similar to energy-and-design issues: after decades of energy-intensive building trends, designers are now giving space to daylighting atriums and allowing surfaces to act as thermal mass. A modest increase in floor space and equipment to encourage resource recovery from solid waste seems an increasingly worthwhile design investment.

Reference

Milne, M. (1976). *Residential Water Consumption,* U.S. Department of Commerce, NTIS.

PART IV
FIRE PROTECTION

One of the most challenging aspects of building design is that of resistance to fire. Designers find it easy to consider how their buildings can be made bearably cool on a stiflingly hot day or lighted and warm on a bitterly cold night. It is less pleasant to imagine a building burning, and to consider how first its occupants, then its contents—including the building itself—can be saved.

To complicate this process, there are inherent conflicts between some of the optimum features for fire resistance and some of those optimum features for passive strategies such as daylighting or passive cooling, or for forced-air systems. There are even potential design conflicts between systems for the safe evacuation of people and systems for the suppression of fire, since both people and fire thrive on oxygen.

This discussion of these topics in fire protection will begin with basic design considerations for fire resistance. Smoke management (for safe evacuation and for limited smoke damage) is next, followed by fire suppression systems such as sprinklers and non-water-based approaches. Lightning protection is then discussed, along with the many fire detection and alarm systems that are keyed to the four stages of the "typical" building fire.

Throughout Part IV, the influence of the National Fire Protection Association (NFPA) will be apparent. An especially useful source of information for diagnosis is the periodically updated *Fire Protection Handbook,* from which many illustrations have been reprinted in our text. Publications can be ordered from the NFPA, Batterymarch Park, Quincy, MA 02269.

13

FIRE PROTECTION

13.1. Design for Fire Resistance

Fire is a special kind of oxidation known as *combustion*. Oxidation, which has been discussed in the previous chapters in terms of rust within water supply equipment and aerobic digestion in waste disposal systems, is a process in which molecules of a fuel are combined with molecules of oxygen, producing a mixture of gases and energy. When this occurs rapidly, as in a fire, energy is released as heat and light, and some gases become visible as smoke.

Fire has a triangle of needs: fuel, high temperature, and oxygen. If deprived of any of these needs, building fires will be extinguished. In general, this triangle's influence on building design is as follows. The *fuel* is the building's structure and contents; the designer controls the choice of structural and finish materials, but rarely the final contents. The *temperatures* achieved in fires are well beyond the ability of building cooling systems to control, so special water systems (in the form of sprinklers) are often installed to deprive fire of the high temperatures it needs. *Oxygen* may be denied to a fire partly by limitations on ventilation, but these can have serious safety consequences. Another design response is to install fire suppression systems that either cover the fuel (foam, dry chemicals) or displace oxygen with another gas—for example, carbon dioxide or halogenated agents such as Halon 1301. The later agent inhibits the chemical action of the flame itself.

(a) Sources of Ignition. Buildings commonly contain three basic sources of ignition: chemical, electrical, and mechanical. Much more rarely, nuclear sources are present. In *chemical* combustion, most commonly known as "spontaneous combustion," some chemicals reach ignition at ordinary temperatures within buildings. Chemical combustion depends on the rate of heat generation (related to the degree of saturation of combustible products by the chemicals involved), the air supply (enough to supply oxygen but not enough to lower the temperature), and the insulation provided by the immediate surroundings (again—the more insulation, the easier the attainment of combustion temperatures).

Electrical heat energy is most commonly supplied by resistance heating, a familiar process in many appliances and in space-heating equipment. Less common are electric ignition by induction, dialectic process, arcing, and static electricity. An infrequent but enormously destructive electrical source is lightning.

Mechanical heat energy is produced by friction (including sparks), by overheating of machinery, and occasionally by the heat of compression.

(b) Products of Combustion (Fig. 13.1). Everyone has experienced to some degree the dangers of the thermal products of combustion—flame and heat. These visible and tactile elements of fire can cause burns, dehydration, heat exhaustion, and fluid blockage of the respiratory tract, but they are responsible for only about one-quarter of the deaths resulting from building fires. The bulk of such deaths are caused by the nonthermal products—

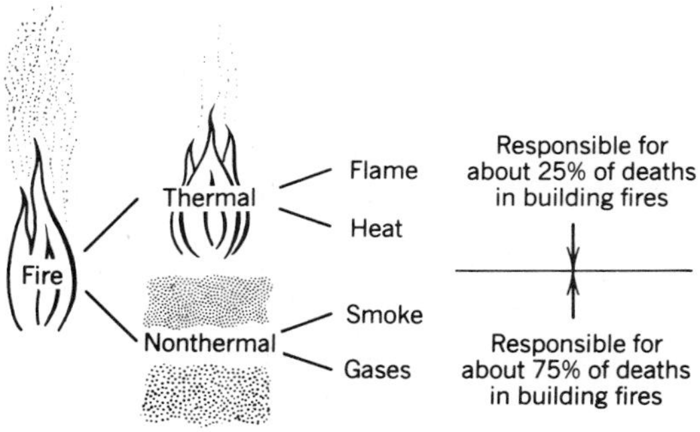

Fig. 13.1 *Although fire's thermal products—flame and heat—make strong visible and tactile impressions, fire's nonthermal products—smoke and gases—pose the greater threat to life.*

smoke and gases. Smoke can usually be seen and smelled. Made up of droplets of flammable tars and small particles of carbon suspended in gases, it irritates the eyes and nasal passages, sometimes blinding and/or choking a person. Gases are especially dangerous because, without smoke, they are so often difficult to detect. Some gases are directly toxic, but all are dangerous because they displace oxygen. Many gases of combustion are discussed as air pollutants in Fig. 3.16. Common gases released in building fires include carbon monoxide and carbon dioxide. *Carbon monoxide* is not the most toxic product of combustion, but it is the most abundant. It is produced when insufficient oxygen is available to completely oxidize the burning material. Since *carbon dioxide* is produced in large quantities, it rapidly overstimulates breathing and thus causes the lungs to swell. Other dangerous and common building-fire gases include hydrogen sulfide, sulfur dioxide, ammonia, cyanide, and hydrogen chloride (from burning polyvinyl chloride).

In indoor fires oxygen commonly becomes insufficient, since the fire consumes it so rapidly. The normal concentration of oxygen in air is about 21%. At 15%, muscular skill is diminished; at 14% down to 10%, people remain conscious but judgment is faulty and fatigue is rapid; at 10% down to 6%, collapse occurs, but revival is possible when increased oxygen is supplied. The technique of starving the fire of oxygen can therefore pose a threat to humans, both by increasing the chances of carbon monoxide production and by depriving people of oxygen.

(c) Objectives in Fire Safety. Many older buildings were designed at a time when building fires usually resulted in the loss of several adjacent structures and, hence, the primary objective of fire fighting was to limit the conflagration to as few city blocks as possible. With the increased use of fire-resistant construction and code control of site and building planning, fires typically were confined to but one building at a time. As fire suppression systems came into common usage within buildings, it came to be expected that fires could be confined to one or two floors within a building. And now that automatic detection/suppression systems are technically advanced, fires can usually be confined to only one room, or to even smaller areas.

Three common objectives of building fire safety, in order of usual importance, are

- Protection of life.
- Protection of property.
- Continuity of operation.

Many of the elements of building fire safety are covered by building codes, but it is important to remember that codes specify the *minimum* acceptable protection. Designers can go much further than the codes require to enhance fire safety. It is also important to realize that codes *cannot* cover all aspects of fire safety, as there are too many variables to building design. The inadequacy of attitudes such as "leave it to the code and the sprinkler system" can be seen in the list of common design deficiencies given in Tables 13.1 and 13.2, all of which can contribute to the spread of smoke and flames through—and between—buildings.

(d) Fire Safety and Other Environmental Control Systems. In some instances, the optimum design for fire safety will resemble the optimum design for lighting, thermal, acoustic, and water systems. One such matching characteristic is a high degree of thermal mass, which is useful for passive heating and cooling systems, for acoustic isolation of airborne sound, and for fire barriers (most thermally massive materials will not burn easily). Another such characteristic is the elevated water storage tank, which provides both adequate water pressure for plumbing fixtures and water for firefighting in the first few minutes of a fire, before firefighters arrive. Finally, solid (noncombustible) overhangs over windows not only provide sunshading, but also discourage the vertical spread of fire over the building face and can serve as emergency exterior places of refuge.

In many other instances, there are potential design conflicts between building performance under ordinary conditions and building performance in the extraordinary event of a fire. Daylighting and natural ventilation are best served by high ceilings and low partitions, which encourage light and air from the perimeter to pervade building interiors. Fire and smoke can easily spread through such "open plans." (On the other hand, fire can build up much more rapidly in small, enclosed rooms, which retain heat.) Forced-air systems that can

TABLE 13.1 **Design Deficiencies That Can Contribute to Fire Spread Through a Building**

Lack of or inadequate vertical and/or horizontal fire separations.
Unprotected or inadequately protected floor and wall openings for stairs, doors, elevators, escalators, dumbwaiters, ducts, conveyors, chutes, pipes, and windows.
Concealed spaces in walls and above ceilings without adequate fire-stopping or fire divisions.
Combustible interior finish, including combustible protective coatings and insulation.
Combustible structural members (beams, girders, and joists) framed into firewalls.
Improper anchorage of structural members in masonry-bearing walls.
Explosion or pressure damage to the building due to lack of or inadequate explosion venting where required.
Damage to unprotected framing resulting in weakening or destruction of floors and walls used as fire barriers.
Lack of means to ventilate fire gasses.

Source: Reprinted with permission from *Fire Protection Handbook,* 15th ed., copyright © 1981 by the National Fire Protection Association, Quincy, MA.

TABLE 13.2 **Design Deficiencies That Can Contribute to Fire Spread From One Building to Another**

Lack of or inadequate fire division walls between adjoining buildings.
Unprotected or inadequately protected openings in fire division walls between adjoining buildings or in firewalls between detached buildings.
Exterior walls lacking adequate fire resistance.
Inadequate separation distance.
Combustible roofs, roof coverings, roof structures, overhanging eaves, trim, and so forth.
Lack of protection at openings to passageways, pipe tunnels, conveyors, ducts, and so forth between detached buildings.
Explosion or pressure damage to adjoining or detached buildings.
Collapse of exterior walls.

Source: Reprinted with permission from *Fire Protection Handbook,* 15th ed., copyright © 1981, National Fire Protection Association, Quincy, MA.

heat, ventilate, and cool are also potential pathways for smoke and fire; this hazard is especially serious when the systems penetrate floors, since vertically spreading fires are harder to fight than are horizontally spreading ones. (On the other hand, carefully designed forced-air systems can aid in smoke management.) "Windowless" buildings that rely on electric light in place of daylight are especially dangerous in fires, because firefighters cannot easily evacuate occupants or gain access to the building. Sunscreens that completely cover windows are disadvantageous for the same reason; nonoperable windows, while considered advantageous for tightly controlled air conditioning, must be broken for fire evacuation/access. Many excellent insulating materials will burn readily, and some will give off toxic gases. Some of the loveliest interior finishes are both flammable and deadly sources of gases in a fire.

This conflict, between interior comfort throughout a building's life and safety at the moment of its impending death by fire, is the principal reason why codes cannot alone assure fire safety. The dilemma of ordinary versus extraordinary performance is one that must be faced by the designer, owner, and occupants of buildings.

(e) Protection of Life. For most buildings, a reasonable goal is the evacuation of all occupants in the time interval between the detection of a fire and the arrival of the firefighters. Designers can provide clearly defined pathways to exits ("exit access") that can be kept relatively clear of smoke (Fig. 13.2). The absolute minimum width of an exit access is 28 in. (356 mm). "Exits" can take a variety of forms. Vertical exits (Figs. 13.3, 13.4) include smokeproof towers, exterior and interior stairs and ramps, and escalators that meet specific requirements. *Vertical exits do not include elevators;* they are too easily stalled or, worse, opened at the floor of a fire by malfunctioning signal equipment. Exits in the horizontal plane include doors leading directly to the outside, two-hour fire-enclosed hallways, and moving walks. Special ''horizontal exits'' are provided by internal fire-walls penetrated by two fire doors—one swinging open in each direction. ''Exit discharge'' is the actual point at which exit ways open to the outside.

Distances to and the sizes of exits are controlled by code minimums (Tables 13.3 and 13.4). When automatic fire suppression systems such as sprinklers are used, the allowable distances to exits are increased. Designers should remember, however, that at least 30% of building fire deaths result

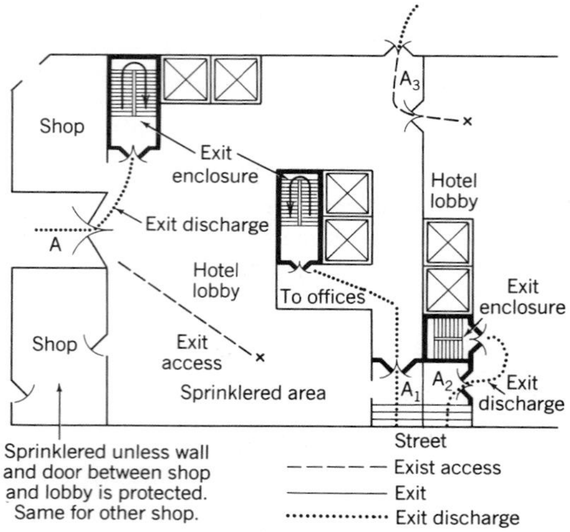

Fig. 13.2 *Exit access, exit, exit enclosure, and exit discharge on the first floor of a multistory building. Reprinted with permission from* Fire Protection Handbook, *15th ed., copyright © 1981 by the National Fire Protection Association, Quincy, MA.*

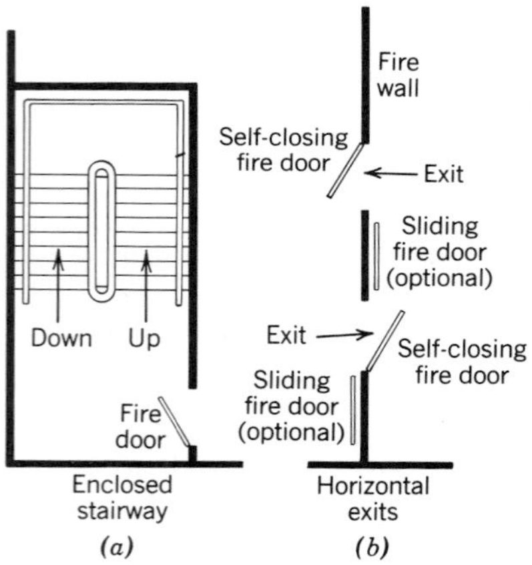

Fig. 13.3 *Plan views of exits.* (a) *Enclosed stairway allows occupants on any floor above a fire to escape.* (b) *Horizontal exit through an interior firewall provides a quick refuge and lessens the need of a hasty flight down the stairs. Two wall openings are needed to facilitate exit in either direction. The swinging doors are self-closing so as to discourage spreading smoke. Horizontal sliding fire doors, provided for safeguarding property, are normally open but close automatically in case of fire. Reprinted with permission from* Fire Protection Handbook, *15th ed., copyright © 1981 by the National Fire Protection Association, Quincy, MA.*

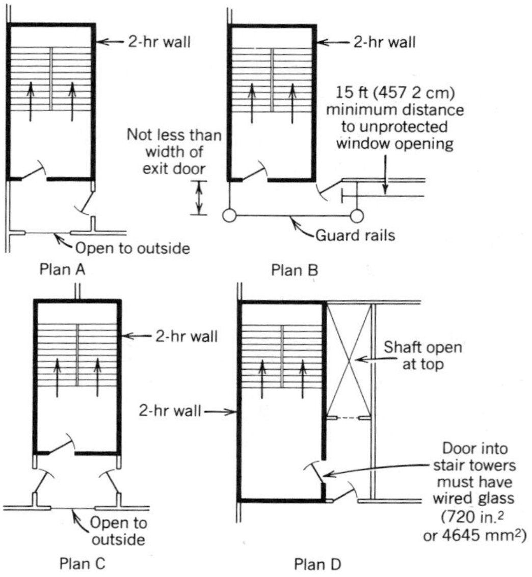

Fig. 13.4 *Four variations of smokeproof towers. Plan A has a vestibule opening from a corridor. Plan B shows an entrance by way of an outside balcony. Plan C could provide a stair tower entrance common to two buildings. In Plan D, smoke and gasses entering the vestibule would be exhausted by natural or induced draft in the open air shaft. In each case a double entrance to the stair tower with at least one side open or vented is characteristic of this type of construction. Pressurization of the stair tower in the event of fire provides an attractive alternative for tall buildings and is a means of eliminating the entrance vestibule. Reprinted with permission from* Fire Protection Handbook, *15th ed., copyright © 1981 by the National Fire Protection Association, Quincy, MA.*

TABLE 13.3 **Summary of Life Safety Code Provisions for Travel Distances to Exits**

Occupancy	*Dead-End[a] Limit, Ft (m)*	*Travel Limit to an Exit, ft (m)*	
		Unsprinklered	*Sprinklered*
Places of Assembly	20[b] (6.1)	150 (45.7)	200 (61.0)
Educational	20 (6.1)	150 (45.7)	200 (61.0)
Open plan	N.R.[c]	150 (45.7)	200 (61.0)
Flexible plan	N.R.	150 (45.7)	200 (61.0)
Health Care			
New	30 (9.1)	100 (30.5)	150 (45.7)
Existing	N.R.	100 (30.5)	150 (45.7)
Residential			
Hotels	35 (10.7)	100 (30.5)	150 (45.7)
Apartments	35 (10.7)	100 (30.5)	150 (45.7)
Dormitories	0	100 (30.5)	150 (45.7)
Lodging or rooming houses, 1- and 2-family dwellings	N.R.	N.R.	N.R.
Mercantile			
Class A, B, and C	50 (15.2)	100 (30.5)	150 (45.7)
Open Air	0	N.R.	N.R.
Covered Mall	50 (15.2)	200 (61.0)	300 (91.4)
Business	50 (15.2)	200 (61.0)	300 (91.4)
Industrial			
General, and special purpose	50 (15.2)	100 (30.5)	150[d] (45.7)
High hazard	0	75 (22.9)	75 (22.9)
Open structures	N.R.	N.R.	N.R.
Storage			
Low	N.R.	N.R.	N.R.
Ordinary hazard	N.R.	200 (61.0)	400 (122.0)
High hazard	0	75 (22.9)	100 (30.5)
Open parking garages	50 (15.2)	200 (61.0)	300 (91.4)
Enclosed parking garages	50 (15.2)	150 (45.7)	200 (61.0)
Aircraft hangars, ground floor	20 (6.1)	Varies[e]	Varies[e]
Aircraft hangars, mezzanine floor	N.R.	75 (22.9)	75 (22.9)
Grain elevators	N.R.	N.R.	N.R.
Miscellaneous occupancies	N.R.	100 (30.5)	150 (45.7)

Source: Reprinted with permission from *Fire Protection Handbook,* 15th ed., copyright © 1981 by the National Fire Protection Association, Quincy, MA.

[a]Dead-end is an extension of a corridor or aisle beyond an exit (or exit access) that forms a pocket in which occupants may be trapped.

[b]In aisles.

[c]No requirement or not applicable.

[d]A special exception is made for one-story, sprinklered, industrial occupancies.

[e]See Paragraph 15–4.2 of Life Safety Code for special requirements.

from fire cutting off the paths to exits. Another point to consider is that the building population as estimated for fire safety is usually much greater than the population for which HVAC, water, or elevator service is designed. The stairs must be designed so as to allow those already within the stairwell to continue down without interference from access doors on any floor. The minimum dimensions of stairs are specified by codes (Table 13.5). Stairs with direct access to outdoor air at each floor—so-called smokeproof towers—are the safest kind.

TABLE 13.4 **Summary of Life Safety Code Provisions for Occupant Load and Capacity of Exits**

Occupancy	Occupant[a] Load: Ft² (m²) per Person	Capacity of Exits, Number of Persons per Units of Exit Width[b]: Doors[c] Outside	Horizontal Exit	Ramp Class A	Ramp Class B	Escalator	Stairs
Places of Assembly	15 Net (1.4)	100	100	100	75	75	75
Areas of concentrated use without fixed seating	7 Net (0.7)						
Standing space	3 Net (0.3)						
Educational		100	100	100	60		60
Classroom area	20 Net (1.9)						
Shops and vocational	50 Net (4.6)						
Day Nurseries with sleeping facilities	35 Net (3.3)						
Health Care		30	30	30	30		22
Sleeping departments	120 Gross (11.1)						
Inpatient departments	240 Gross (22.3)						
Residential	200 Gross (18.6)	100	100	100	75	75	75
Mercantile		100	100			60	60
Street floor and sales basement	30 Gross (2.8)						
Other floors	60 Gross (5.6)						
Storage-shipping	300 Gross (27.9)						
Office areas	100 Gross (9.3)						
Business	100 Gross (9.3)	100	100	100	60	60	60
Industrial	100 Gross (9.3)	100	100	100	60	60	60
Detention and Correctional occupancies	120 Gross (11.1)	100	100	100	100[d] 60[e]		

Source: Reprinted with permission from *Fire Protection Handbook,* 15th ed., copyright © 1981 by the National Fire Protection Association, Quincy, MA.

[a]When actual maximum population count can be determined from plans, it can be substituted for these figures.

[b]An exit width is defined as 22 in. (559 mm).

[c]Not more than three risers or 21 in. (533 mm) above or below grade.

[d]100 (down).

[e]60 (up).

TABLE 13.5 **Requirements for Exit Stairs**

	Conventional Units			SI Units		
Stair Requirements	*New Buildings*	*Class A*[a]	*Class B*[a]	*New Buildings*	*Class A*[a]	*Class B*[a]
Width, minimum	44 in.[b]	44 in.	44 in.[b]	1118 mm	1118 mm	1118[b] mm
Width inside hand rails, minimum	37 in.	37 in.	37 in.	940 mm	940 mm	940 mm
Tread, without nosing, minimum	11 in.	10 in.	9 in.	279 mm	254 mm	229 mm
Riser height, maximum	7 in.	7½ in.	8 in.	178 mm	191 mm	203 mm
Landing height, maximum	12 ft.	8 ft.	12 ft.	3.7 m	2.4 m	3.7 m
Maximum landing dimension in travel direction	44 in.	44 in.	44 in.	1118 mm	1118 mm	1118 mm

Source: Reprinted with permission from *Fire Protection Handbook,* 15th ed., copyright © 1981 by the National Fire Protection Association, Quincy, MA.

[a]Classes A and B are defined as stairs having these dimensional characteristics in *existing* buildings.

[b]36 in. (914 mm) where total occupancy of all floors served by stairway is less than 50.

High-rise buildings present much more difficult problems, since firefighting equipment can ordinarily reach no higher than seven floors (about 90 ft) and since, typically, only two exit stairways are provided. A fire stair must allow firefighters to move up while occupants are moving down. With a typical exit stair, a 15-story building housing 60 persons per floor per stair can be evacuated in about nine minutes. However, with the same stair in a 50-story building housing 240 persons per floor per stair, evacuation will take at least 2 hours, 11 minutes! When doors are held open at each floor to admit occupants fleeing the fire, smoke can readily enter the stairwell. Because of the impracticality of rapid evacuation, larger buildings are required to provide "refuge areas" where smoke penetration is less likely. These details on smoke management are presented in the next section.

(f) Property Protection. One of the earliest design concerns in this category is that the site should permit access for firefighting equipment. Ideally, fire trucks should be able to pull alongside each exterior wall. If accessibility is limited by adjacent buildings, and roadways alongside buildings and other measures are impractical, more reliance must be placed on internal fire suppression systems. Another factor is the amount of time it will ordinarily take for firefighters to reach a site. In congested urban areas, or remote rural ones, time that elapses between the alarm and the arrival of firefighters can permit a fire to grow to unstoppable proportions. In these cases, emphasis again shifts to internal fire suppression systems.

Another design concern is for adequate water to fight the fire. Reliance on city water mains is not always a good solution. Elevated tanks on buildings can help in the early moments of firefighting, but their capacity is soon exhausted. Some buildings therefore rely on lakes or enclosed reservoirs for firefighting supply (see the Morton Building, Fig. 6.21, and the discussion in Section 13.3).

Exposure protection is becoming common in areas where highly flammable surroundings pose a serious threat of fires originating *outside* a building. Candidates for exposure protection include buildings surrounded by older wooden buildings, bordered by lumberyards or other commercial activities that utilize highly flammable materials, or even bordered by open fields of dry grass or brush. Exposure protection guards against heat transfer by radiation and convective currents and against direct fire transfer via flying embers. Exposure protection begins with the use of nonflammable materials for the building's exterior. Erecting firewalls between the building and a fire-threatening neighbor is a more drastic, but sometimes necessary, step. Exposure protection sometimes includes external water sprinkler systems, in which a sprayhead is placed over (or under) each opening such as windows or doors. Sprinkler systems that soak the roof can play a cooling role on summer days, as well as an exposure protection role. Exterior doors can be chosen for their fire-delaying characteristics; fire-rated doors are shown in Table 13.6. Windows can also be protected by shutters with similar fire ratings that are designed to close automatically at high temperatures.

Compartmentation has become increasingly important as buildings have become lightweight structures incorporating decreased fire resistance and open floor areas that encourage the spread of fires. Building codes establish the maximum floor areas permissible for various constructions and occupancies. If a building's floor area exceeds such limits, it must be subdivided by firewalls into areas that fall within the code limitations. Openings in the firewalls must be protected by fire doors (Table 13.6) and fire dampers in forced-air systems (Fig. 13.5). As important as compartmentation is in buildings with large floor areas, however, the vertical spread of fire poses a more serious problem. For this reason, compartmentation requirements around vertical openings are often especially strict.

Concealed spaces are found in many contemporary buildings, especially over suspended ceilings but also behind walls, within pipe chases, in attics, and in other places. All such spaces can offer paths for the spread of fire. The designer can utilize noncombustible materials, insofar as possible, in such spaces. Another important step is to include automatic fire detection and suppression systems in these uninhabited spaces—and often, the use of oxygen-deprivation approaches to fire suppression. Another design response is compartmentation; using firestops (or firewalls) to break up otherwise continuous concealed spaces. Most codes, for example, require firestopping around each 1000 ft^2 (93 m^2) of suspended ceiling area, and every 2000 ft^2 (186 m^2) of attic floor area.

Structural protection, another important requirement, allows a building to continue to stand during a fire and enables the building to be salvaged rather than demolished after the fire. Codes require various protective layers for structural materials. In order of importance, it is usually most critical to protect columns, girders, beams, and, finally, the floor slab.

(g) Continuity of Operations. For most building functions, it is desirable to minimize the disruption of operation that fire will cause. Design strategies to encourage continuity of operations include special fire alarm/suppression systems for especially critical operations areas (control rooms, for example), the design of HVAC systems to allow for 100% outside air (to aid in purging a building of smoke after the fire is out), and provision for the speedy removal of the water dumped on a fire from a sprinkler system. The details of the wall scuppers and floor drains for such water re-

TABLE 13.6 **Fire-Rated Doors**[a]

Class	*Rating*	*Location*
A	3 h	Between buildings, or between separate fire areas in a single building
B	1 to 1½ h	Around vertical openings within buildings, such as stairs, elevator shafts, or mechanical chases
C	¾ h	In partitions between corridors and rooms
D	1½ h	Exterior walls with severe fire exposure from outside the building
E	¾ h	Exterior walls with light or moderate fire exposure from outside the building

Source: Excerpted from *Fire Protection Handbook,* NFPA (1976).

[a]Tight-fitting doors will help stop the spread of smoke.

moval can be found in Section 13.3. The floors in sprinkler-served buildings should be waterproof; waterproofing should also be carried up walls, columns, pipes, and other elements to a height of 4 to 6 in. (102 to 152 mm) above the floor. These provisions will help to minimize the water damage associated with fires.

13.2 Smoke Management

As it has become increasingly evident that smoke kills more people in building fires than either heat or structural collapse, it has also become more common to design for smoke management as well as for resistance to fire and fire suppression. The

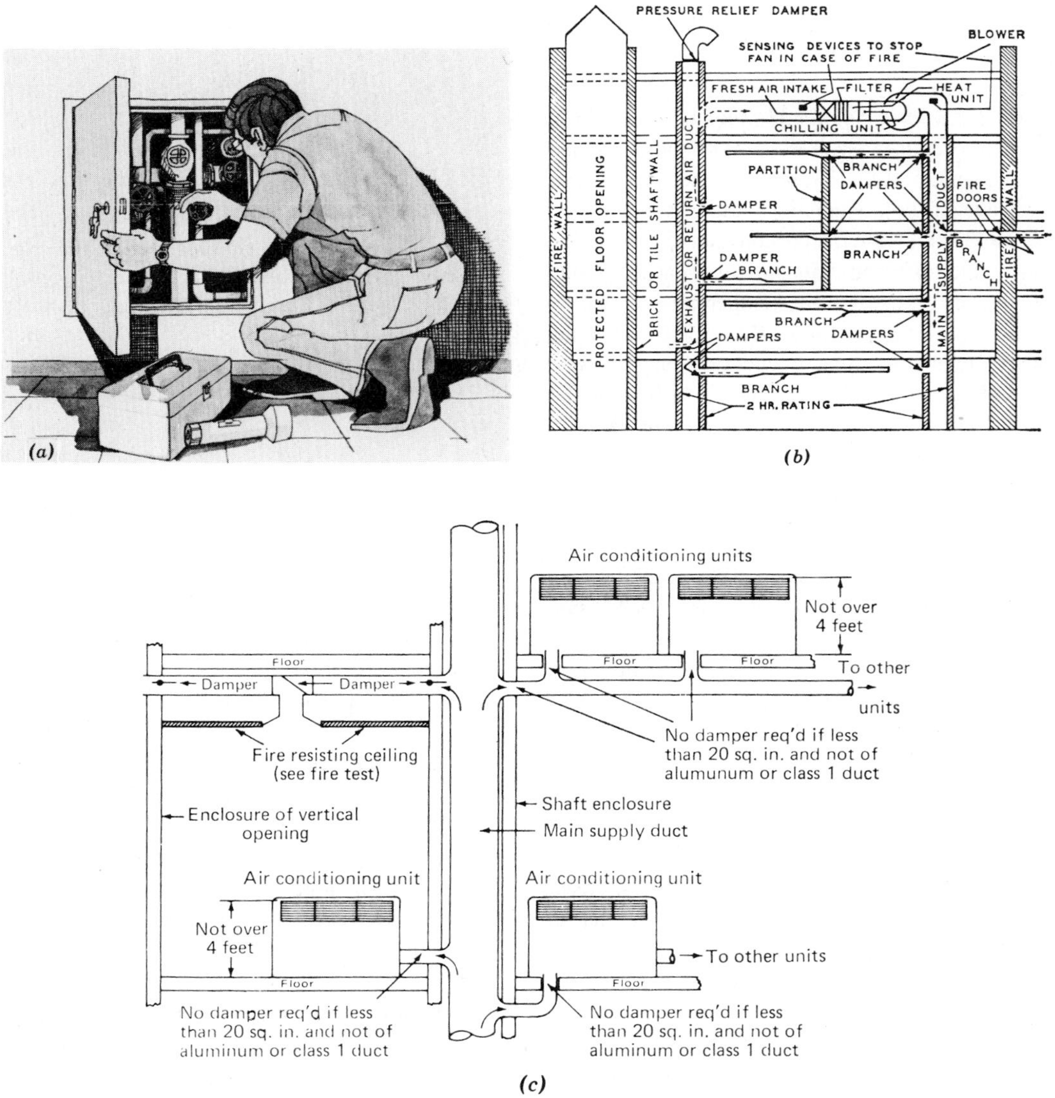

Fig. 13.5 Openings in firewalls are protected by various devices. (a) *Access doors to controls and other elements within firewalls can be of firerated construction.* (b) *Locations of fire dampers in a typical forced-air system—at firewalls, partitions, and shaft enclosures.* (c) *Detail of a forced-air installation, showing required and optional locations of fire dampers. Parts* (b) *and* (c) *are reproduced with permission from* Fire Protection Handbook, *15th ed., copyright © 1981 by the National Fire Protection Association, Quincy, MA.*

objectives in smoke management are the same as those for fire resistance: to reduce deaths and property damage due to smoke, and to provide for continuity of operations with minimal smoke interference. Smoke management systems have been used for several generations; scientific laboratory buildings equipped with fume hoods (such as the Berkeley Chemistry Bldg., Fig. 6.31) offer a familiar example. Several options for smoke management are now available for ordinary building design.

(a) Factors in Smoke Management. The heat of a fire produces air pressure and buoyancy that aid the spread of smoke well beyond the scene of the fire itself. In low buildings, the smoke is spread primarily by heat-induced convective air motion and by the differential air pressures caused by the expansion of gases as the temperature increases. In tall buildings, the stack effect complicates this pattern of smoke spread, encouraging the rapid vertical rise of heated air within vertical shafts. Wind forces from outside are also more likely to be a factor in tall buildings. (Wind velocities are usually higher with increasing distance from the earth's surface). Forced-air systems, so common in high-rise buildings, can also contribute to the spread of smoke. The interactions among fire, stack effect, wind, building geometry, and HVAC systems are complex and not as yet completely understood. The ASHRAE publication *Design of Smoke Control Systems for Buildings* is a good source of more information.

(b) Confinement. The most passive design response to smoke is to try to confine it to the fire area itself. Another confinement technique is to exclude smoke from specially protected areas, called refuges. Compartmentation is important in these approaches: where firewalls cannot be used, special smoke barriers (often called curtain boards) are suspended from the ceiling (Fig. 13.6) in an effort to trap the initial layer of hot air and smoke, and therefore to set off the fire detection/suppression system more quickly. As useful as these partial barriers can be in the fire's early stages, they quickly lose effectiveness as the smoke layer thickens or as air pressures force the smoke below the barrier. Even in firewalls with fire door protection, small cracks around such openings provide smoke paths, aided by the air pressures and velocities that fires can quickly produce. Therefore, the typical ''barrier'' is not very resistant to spreading smoke, and smoke control based solely on confinement must be closely linked to an effective early detection/suppression system, such as a sprinkler system.

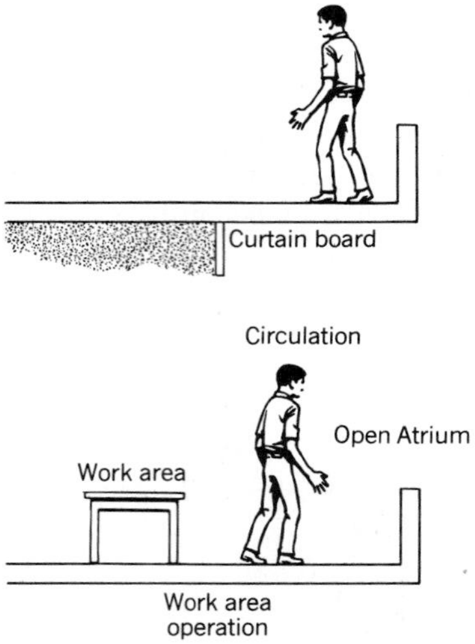

Fig. 13.6 *Smoke barriers, often called curtain boards, are useful in a fire's early stages. By confining the initial layer of heated air and smoke produced by the fire, they help to slow the spread of smoke while making more likely the early detection and suppression of the fire.*

(c) Dilution. For a limited time in a fire's early stages, the dilution of smoke with 100% outdoor air (provided by the HVAC system) may make conditions bearable during occupant evacuation. However, such large quantities of fresh air are needed in so short a time that dilution alone is rarely sufficient for smoke control, particularly when the smoke contains toxic fumes. When the dilution system is combined with both confinement and early detection/suppression systems, it becomes more attractive. Dilution systems that dump large quantities of outdoor air into refuge areas can create such high pressures within the refuge that smoke cannot enter through cracks around doors. However, when such doors are opened (as when more people enter the refuge), dilution systems that rely on the conventional

HVAC system can rarely provide sufficient pressure to exclude smoke through the open doorway.

(d) Exhaust. Special exhaust systems that function only in a fire are becoming more common. These systems employ both air velocity and air pressure to help control smoke. Although they involve a greater initial expense, because they require special fans and special smoke-exhaust shafts, exhaust systems have several advantages over simple confinement or dilution approaches.

1. They can remove toxic gases from refuge areas (particularly when there is a dilution or outdoor air supply to such areas).
2. They help firefighters by improving the air quality in the vicinity of the fire itself.
3. They can help control the direction that a fire takes, by creating air currents that a fire will follow.
4. They remove the unburned but combustible gases from a fire before the latter can cause a ''backdraft,'' or ''flashover'' (smoke-explosion).
5. By creating higher air pressures in refuge areas and lower pressures in the fire zone, they keep smoke out of refuge areas, even when doors are temporarily open.
6. With them, the tall-building phenomenon of stack effect, complicated by buoyancy and wind, is less likely to overcome the smoke management system.

Two systems within the building must be closely coordinated with smoke exhaust systems: the HVAC system and the fire detection/suppression (usually sprinkler) system. As the fire detection system activates the smoke exhaust fans, it must also override the conventional HVAC system operation, usually by switching to 100% outside supply air and simultaneously blocking the return duct system. This process pressurizes each HVAC zone, to form a barrier against the smoke (Fig. 13.7), and also keeps smoke out of the return duct system. If the HVAC system is VAV, all supply control valves (dampers) must be moved to full open position. If additional protection is needed for refuge areas, additional air supply to such areas can be provided, perhaps by a separate (smoke dilution) system.

Sprinkler systems can hamper the functioning of smoke exhaust systems, both by creating a curtain of water that inhibits the movement of smoke and by cooling the smoke and thus reducing its buoyancy. The buoyancy of smoke is the factor relied on by smoke exhaust shafts, whose intakes

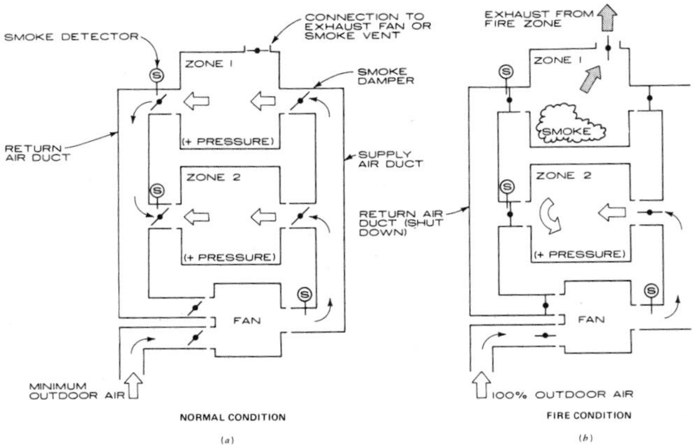

Fig. 13.7 *Smoke management by a smoke exhaust system.* (a) *The conventional HVAC system is shown operating at normal, minimum outdoor air supply conditions.* (b) *When smoke detectors trigger the smoke exhaust system, special fans begin to exhaust smoke, through a special duct, from the affected area only. Simultaneously, the conventional system undergoes two changes: All supply air becomes 100% outside air (for smoke dilution) and all return openings are closed off (so as to pressurize all nonfire areas, to keep smoke out of them and out of the return duct system). From AIA; Ramsey and Sleeper,* Architectural Graphic Standards, *7th ed., copyright © 1981 by John Wiley & Sons. Reprinted by permission.*

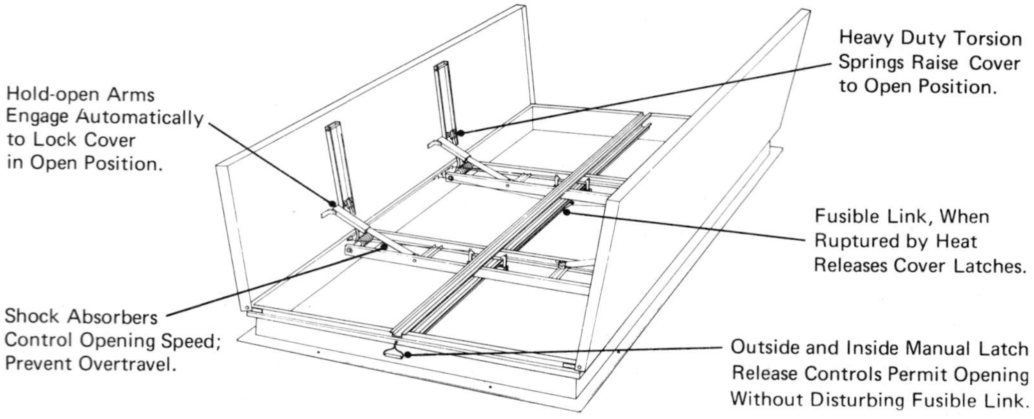
Hold-open Arms
Engage Automatically
to Lock Cover
in Open Position.
Heavy Duty Torsion
Springs Raise Cover
to Open Position.
Fusible Link, When
Ruptured by Heat
Releases Cover Latches.
Shock Absorbers
Control Opening Speed;
Prevent Overtravel.
Outside and Inside Manual Latch
Release Controls Permit Opening
Without Disturbing Fusible Link.

Fig. 13.8 *Automatic heat-and-smoke roof hatch ventilators, primarily for one-story commercial and industrial buildings. Courtesy of Inryco, Inc.*

are located at ceiling level within each zone. Although these two systems thus appear to work at cross purposes, the suppression of the fire is obviously as important a goal as smoke management. Sprinkler systems suppress fires so quickly that the size of the accompanying smoke exhaust systems can generally be reduced. When the fire suppression system relies on oxygen displacement (such as carbon dioxide or halogenated agents), smoke exhaust systems clearly pose a threat; special attention is required if both systems are to be used.

Automatic heat-and-smoke ventilating hatches (without fans) are often installed (Fig. 13.8) in smaller buildings. These hatches open individually as heat or smoke from the fire trigger their control devices. As a result, indoor conditions near the fire are improved for firefighters, and firefighters on the roof are quickly alerted to the location of the fire indoors.

In summary, designers can provide several components that aid in the exhaust of smoke and the provision of access and ventilation for firefighters for almost any building: emergency controls on HVAC systems, operable windows and skylights (to facilitate ventilation and occupant escape), and smoke-and-heat hatches in roofs.

13.3 Water for Fire Suppression

The most popular medium for building fire suppression is water, which is readily available and relatively low in cost. Water cools, smothers, emulsifies, and dilutes. As it turns to vapor, it removes 970 Btu/lb (2256 kJ/kg) at atmospheric pressure and its volume increases 1700 times—a process that helps push away the oxygen needed by the fire. Water has several disadvantages, which sometimes preclude its use for fire suppression: it damages most contents of buildings, including interior surfaces; it conducts electricity readily as a stream (less readily as a spray); many flammable oils will float on the water's surface while continuing to burn; and as water vaporizes rapidly as steam, it can harm firefighters. Where these factors are major considerations, other suppression media can be considered (see Section 13.4).

Fire-suppressing systems are commonly combined with smoke management systems; Fig. 13.9 shows a typical combination of provisions in one-story buildings. Taller buildings, and especially those with multistory interior spaces (atriums), present a special challenge (Fig. 13.10). A typical approach is to construct an approximately 6-ft (1.8 m) = deep curtain board (or fire spandrel) at the opening to the atrium on each floor. Smoke detectors, motorized dampers, and sprinklers can be combined in various ways. At the lobby level, the atrium floor level, and at any office floors open to the atrium, sprinklers at about 6-ft (1.8-m) o.c. will provide a water curtain at the edge of the atrium opening. Where glazing is used at the atrium's perimeter, the glazing frames sometimes can be designed to allow for considerable thermal expansion, and sprinklers can provide a water cur-

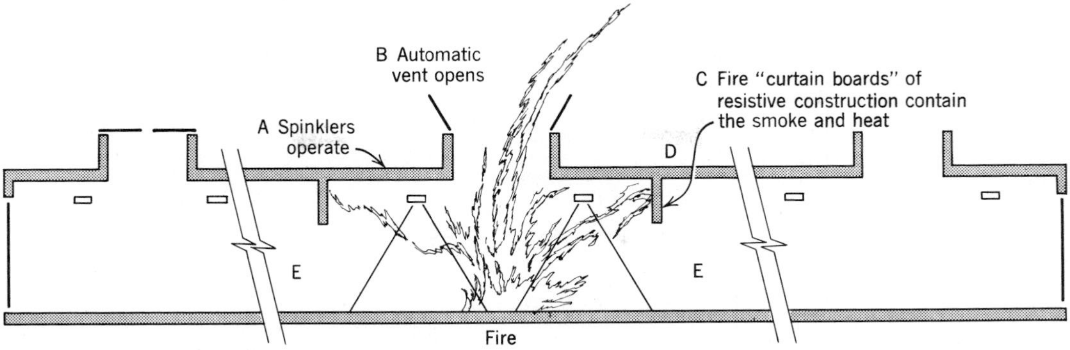

Fig. 13.9 Fire protection for large-area one-story buildings, showing the fire suppression (sprinkler) system, (a) *and the smoke management system,* (b *and* c). *Locations D and E are relatively safe positions for firefighters.*

tain on the office side of the glazing. Where balcony corridors adjoin the atrium, two sprinkler water curtains can be provided, one on each side of the partitions between corridor and office space. The smoke exhaust system often can provide six ACH (air changes per hour) for all spaces that open onto the atriums.

Sprinkler systems are widely relied on as proven, automatic fire-suppressors. However, designers must remember that the use of sprinklers does *not* give one license to ignore fire code limitations, even though many codes are more lenient for sprinklered buildings. In addition, provision must be made for adequate water supply, standby power for water pumping, and water drainage during and after the fire. Sprinklers require very large supply

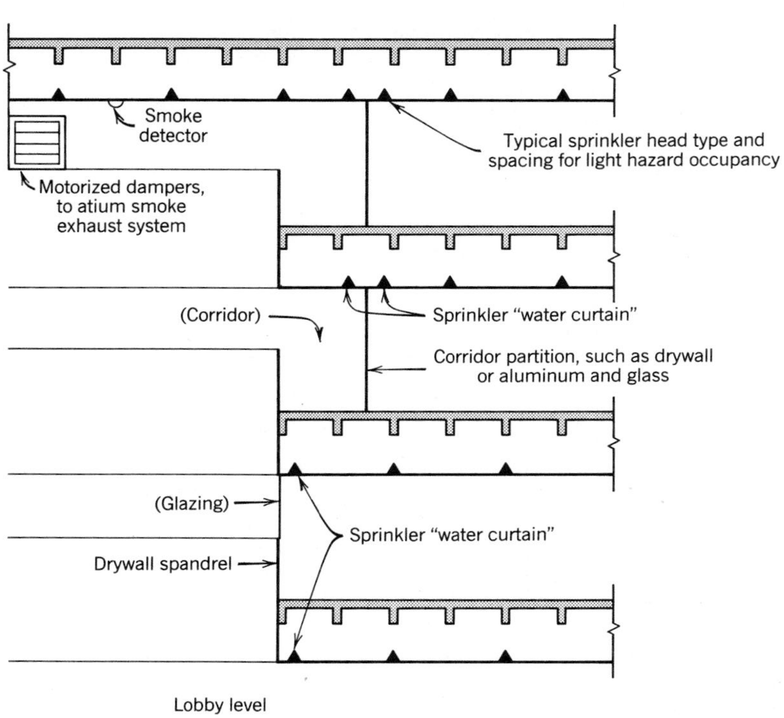

Fig. 13.10 Fire protection in an atrium-type office building: section showing the detection/suppression system and provisions for smoke management.

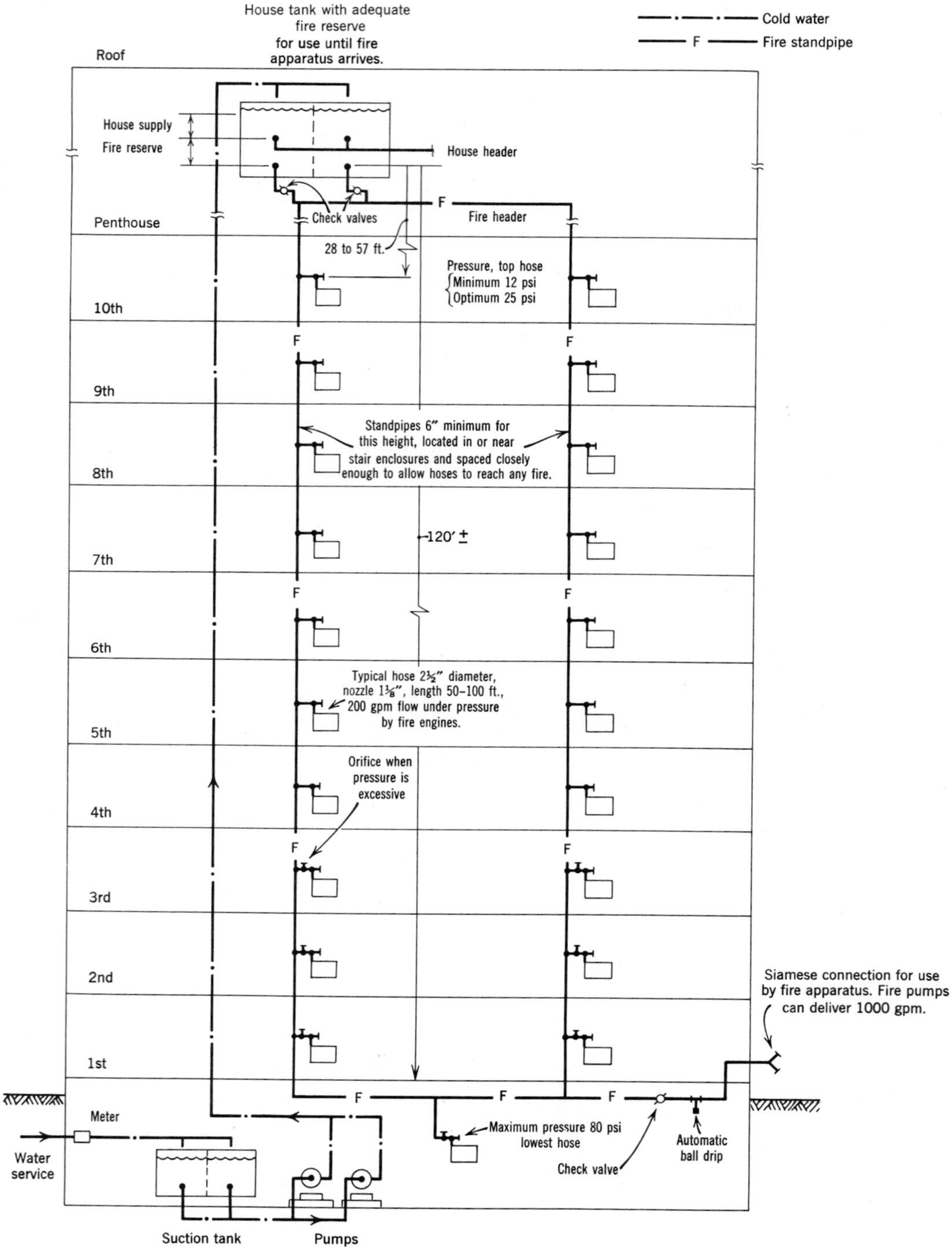

Fig. 13.11 Schematic of a standpipe and hose system, which can sometimes feed sprinkler systems as well. Water is supplied initially by the roof tank's gravity downfeed, then by firefighters, through the siamese connection. Hoses at each floor are used first by building personnel, then by firefighters.

pipes and valves, and sometimes fire pumps—all of which are frequently considered unsightly. On the other hand, sprinklers afford opportunities for integration with energy-conserving HVAC systems (see the Comstock Center, Fig. 6.124) and with display lighting (see Bullock's department store, Fig. 6.130). Also, the cost of sprinkler system installation can usually be recovered rather quickly through reductions in fire insurance premiums.

(a) Standpipes and Hoses. Standpipes and hoses with a separate water reserve or upfeed pumping are valuable in *any* building but become essential in tall buildings whose upper floors cannot be reached by firefighting equipment located at ground level. Figure 13.11 shows such a system, which is designed to be used by building personnel until the fire engines arrive, and thereafter by the firefighters. It is not practical to store enough water on the roof for a protracted firefighting period, and it is usually assumed that a half hour's supply will be more than enough to provide for the short period it takes the fire engines to arrive. When the system is used by the fire department, its pumps are attached to the street siamese (Fig. 13.12) to deliver water from street hydrants or the building's "secondary source." The check valve closest to the siamese in use opens and the check valves at the tank close, to prevent the water from rising uselessly in the tank. After the engines are disconnected from the siamese, the water between the siamese and the adjacent check valve drains out through the ball drip so that it does not freeze.

An overhead tank is considered a most dependable source, but the height required can be architecturally undesirable. In such a case, upfeed fire pumps operating automatically to deliver water to higher stories from lower suction reserve tanks

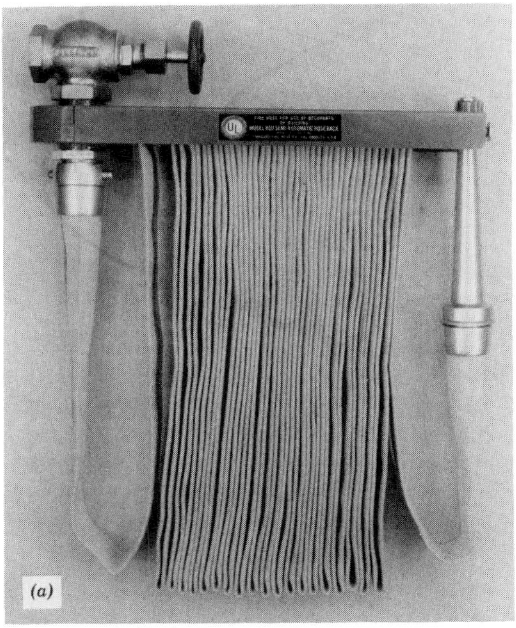

Fig. 13.12 *Components for standpipe and hose system.* (a) *Standard hose rack.* (b) *Hose rack and fire extinguisher in cabinet with glass door.* (c) *Siamese connection through which fire department pumping equipment supplies water to standpipe system. The similar sprinkler siamese is marked "sprinkler." Color codes sometimes differentiate the two.*

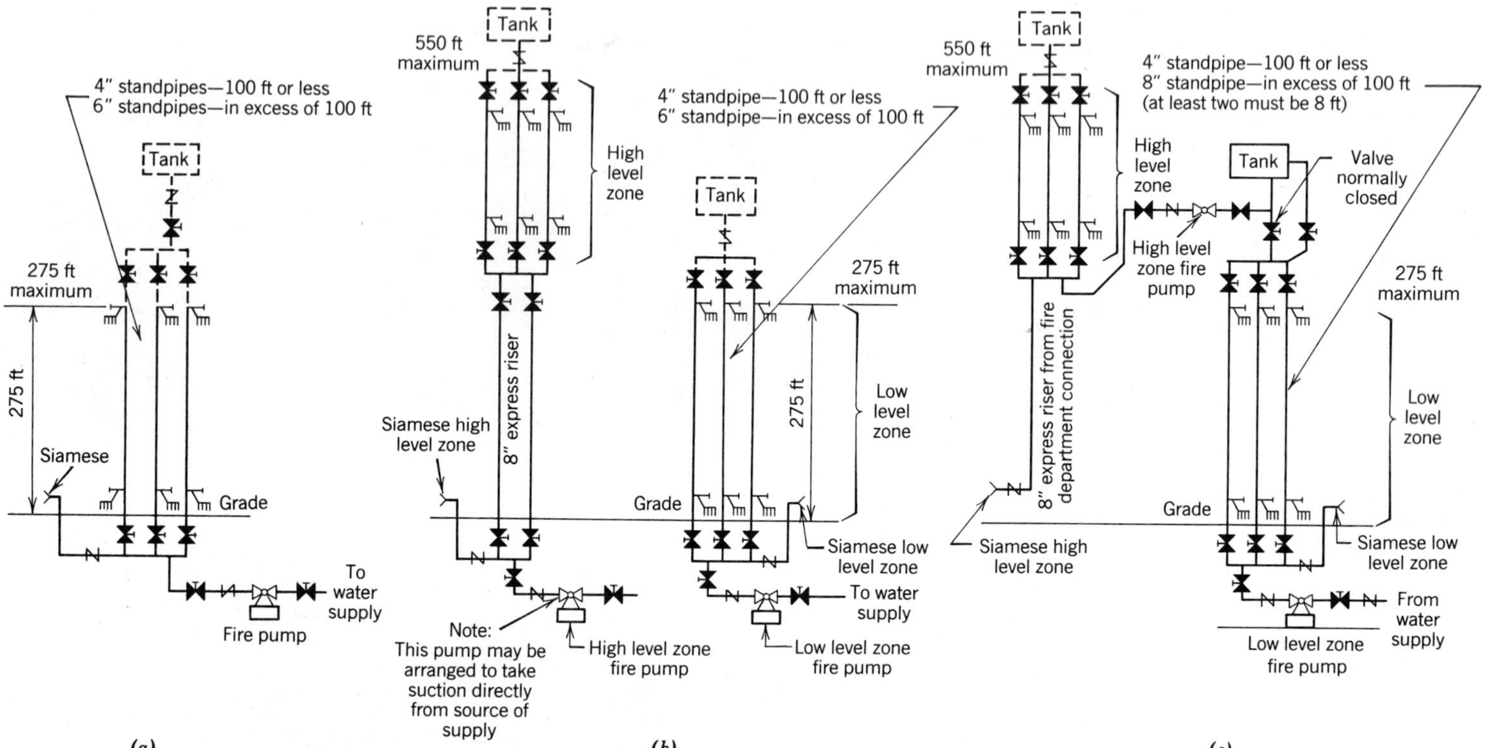

Fig. 13.13 *Zoning for standpipes in tall buildings.* (a) *A typical one-zone system will require a 1000-gpm (63.1-L/s) firepump.* (b) *An independent two-zone system for high-rise buildings, requiring two 1000-gpm (63.1-L/s) fire pumps; the first pump provides high-pressure suction to the second, as well as supplying the lower zone.* (c) *An alternate two-zone high-rise system features an elevated tank, from which the upper zone's fire pump can draw; both pumps must be 1000-gpm (63.1-L/s) capacity. Reprinted with permission from* Fire Protection Handbook, *15th ed., copyright © 1981 by the National Fire Protection Association, Quincy, MA.*

may be used. Another option in this case is a pneumatic tank that delivers water by the power of the air that is compressed in the upper portion of the tank.

A *fire* standpipe zone will usually coincide with the service water zone (see Fig. 13.13, and also Fig. 9.33). Fire standpipes and their hoses (see Fig. 13.12) are usually located at or near fire stairs from which personnel or firefighters can approach a fire (see Fig. 13.17).

(b) Sprinklers. Automatic sprinkler systems consist of a horizontal pattern of pipes placed near the ceilings of industrial buildings, warehouses, stores, theaters, offices, homes, and other structures in which a fire hazard requires their use. These pipes are provided with outlets and sprinkler heads constructed such that abnormally high temperatures will cause them to open automatically and emit a series of fine water sprays.

The temperature at which a sprinkler is triggered is usually specified; it should be at least 25F° (13.9C°) higher than the maximum ceiling temperature ordinarily expected. "Ordinary" sprinkler heads operate at between 135 and 170F. A sprinkler's heat-sensitive controls may be made of metal alloys that fuse, organic materials that soften, or organic liquids that expand and rupture their glass enclosure (Fig. 13.14). The sprinkler heads are usually upright, pendent, or sidewall types. Upright heads sit on top of the exposed supply piping. Pendent heads hang below piping, which thus can be concealed above suspended ceilings (Figs. 13.14*a* and 13.14*b*). The pendent heads themselves can be nearly concealed; low-profile, recessed, and concealed pendent heads are available (Fig. 13.4*c*). Sidewall sprinklers (Fig. 13.4*d*) usually located adjacent to one wall of smaller rooms—hotels, apartments, etc.—and throw a spray of water entirely across such rooms. Therefore only one sidewall sprinkler head per small room usually is used. A welcome development is the *flow control* sprinkler, which *closes* automatically once the ceiling temperatures are sufficiently reduced. Ordinary sprinklers, once activated, will continue to operate until a main valve is manually closed; this can waste large quantities of water and result in extensive water damage. The flow control sprinkler shown in Fig. 13.14*e* will emit water when ambient temperatures reach 165 F (74°C)—another model operates at 145 F (62.8°C)—and close again when ambient drops below 95 F (35°C). These sprinklers are more visible, since both the bottom and side elements must be exposed within a space.

All types of sprinkler heads must be replaced after use.

Sprinkler systems are of four common types.

1. *Wet-pipe* (most common), in which water is always standing under pressure in all pipes and mains.
2. *Dry-pipe* (used in unheated buildings), in which pipes are filled with compressed air (or nitrogen) until the opening of a sprinkler head permits water flow.
3. *Preaction,* which is very similar to dry pipe except that water is admitted to the pipes *before* any sprinkler head has opened, and which thus sounds a very early alarm.
4. *Deluge,* in which *all* sprinkler heads go off at once.

In the wet-pipe system, sprinklers in the area affected are opened by sensitive elements within the sprinkler heads themselves. In the dry-pipe systems, remote valves may be actuated by sensitive elements to admit water to sprinkler heads. Dry-pipe systems also require air compressors; a heated main control valve housing; and the pitching of all piping, to allow thorough drainage after usage. Preaction systems are especially popular where the building's contents are subject to water damage—computer rooms, retail stores, and so forth—since the early alarm provided by water filling the piping often permits the fire to be found and extinguished manually, before any sprinklers open. Deluge systems are used where extremely rapid fire spread is expected—in aircraft hangars, for example, or other places where flammable liquid fires may break out.

The spacing of sprinkler heads is a complex matter, and sprinkler systems are usually designed by professionals working for sprinkler manufacturers. However, some guidelines for preliminary sprinkler location can be included here. The first consideration is the degree of hazard faced by the occupants, as listed in Table 13.7. Once this is

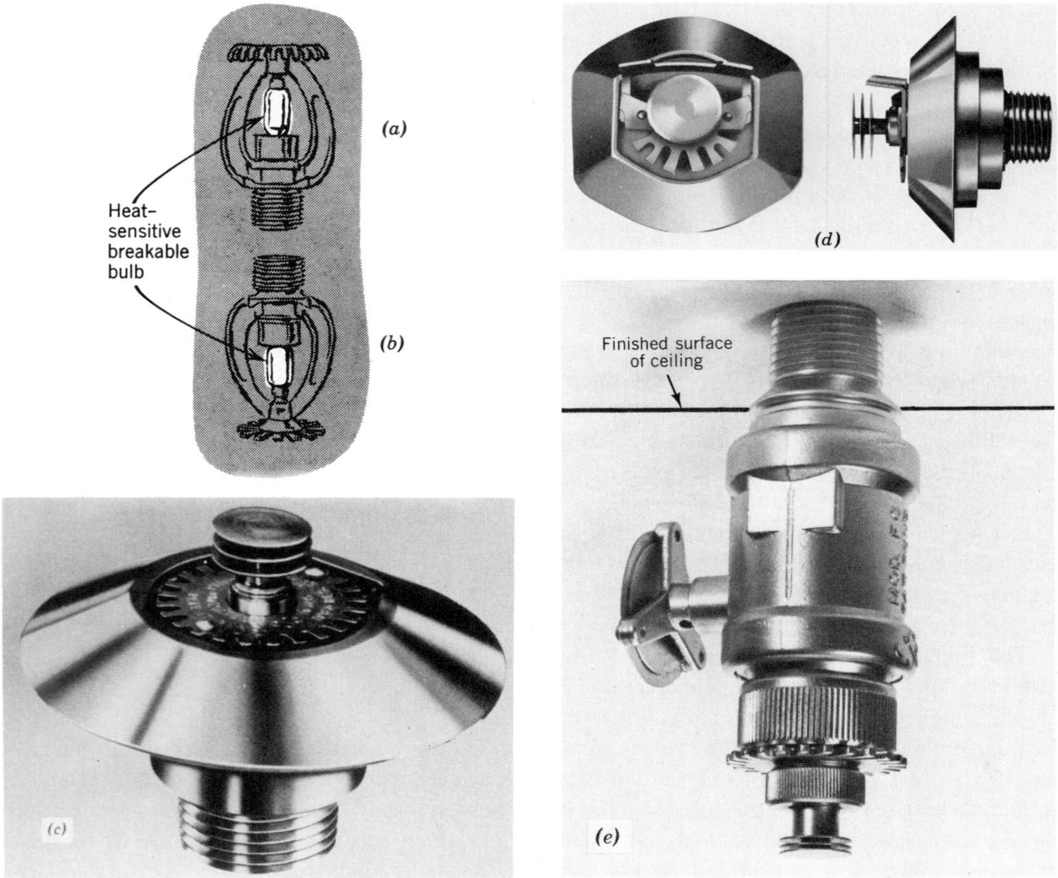

Fig. 13.14 Sprinkler heads. (a) *Upright; sits above exposed piping, just below the structural deck—where the hottest gases will first accumulate.* (b) *Pendant projecting through a suspended ceiling in most installations. Both* (a) *and* (b) *use the quartzoid bulb, a transparent bulb with a colored liquid that ruptures at a preset temperature to release the water stream; heads can be plated, polished or colored—but never painted.* (c) *Pendant, styled for a more contemporary appearance.* (d) *A sidewall, front and side views.* (e) *Flow control pendant, which must be exposed within the space it serves. (Illustrations* (c) (d), *and* (e) *courtesy of Central Sprinkler Corporation.)*

known, sprinklers and pipes can be approximately located with the aid of Table 13.8. Within each space, sprinklers should be located so as to detect a fire readily and to discharge water over the greatest area (considering obstacles such as beams, partial height partitions, etc.).

Piping for sprinkler supply can be hydraulically designed, as shown in Section 9.7. A complicating factor is the expectation that only a small percentage of the sprinklers will actually open; over 50% of the fires studied over a 49-year period were extinguished by two or fewer sprinklers. A detailed sizing procedure would consider both the available pressure at the highest sprinkler and the expected flow rate, which can vary from 500 to 5000 gpm (31.5 to 315 L/s). The sprinklers' actual performance is then determined by

$$Q = K\sqrt{p}$$

where Q = flow rate in gpm
K = K factor, published for each sprinkler head by U.S. manufacturers (see Table 13.8, Part A)
P = pressure in psi

TABLE 13.7 **Relative Fire Hazard for Various Occupancies, as Related to Sprinkler Installations**

Classification of Occupancies

Occupancy classifications for this standard relate to sprinkler installations and their water supplies only. They are not intended to be a general classification of occupancy hazards.

Light-Hazard Occupancies

Occupancies or portions of other occupancies where the quantity and/or combustibility of contents is low, and fires with relatively low rates of heat release are expected.

Light hazard occupancies include

Churches
Clubs
Educational
Hospitals
Institutional
Libraries, except large stack rooms
Museums
Nursing or convalescent homes
Office, including data processing
Residential
Restaurant seating areas
Theaters and auditoriums excluding stages and prosceniums
Unused attics

Ordinary-Hazard Occupancies

Group 1—occupancies or portions of other occupancies where combustibility is low, quantity of combustibles is moderate, stock piles of combustibles do not exceed 8 ft (2.4 m), and fires with moderate rates of heat release are expected.

Group 1 ordinary-hazard occupancies include

Automobile parking garages
Bakeries
Beverage manufacturing
Canneries
Dairy products manufacturing and processing
Electronic plants
Glass and glass products manufacturing
Laundries
Restaurant service areas

Group 2—occupancies or portions of other occupancies where quantity and combustibility of contents is moderate, stock piles do not exceed 12 ft (3.7 m), and fires with moderate rate of heat release are expected.

Group 2 ordinary hazard occupancies include

Cereal mills
Chemical plants—ordinary
Cold storage warehouses
Confectionery products
Distilleries
Leather goods manufacturing
Libraries—large stack room areas
Machine shops
Metal working
Mercantiles
Printing and publishing
Textile manufacturing
Tobacco products manufacturing
Wood product assembly

Group 3—occupancies or portions of other occupancies where quantity and/or combustibility of contents is high and fires of high rate of heat release are expected.

Group 3 ordinary hazard occupancies include

Feed mills
Paper and pulp mills
Paper process plants
Piers and wharves
Repair garages
Tire manufacturing
Warehouses (having moderate to higher combustibility of content, such as paper, household furniture, paint, general storage, whiskey, etc.)
Wood machining

TABLE 13.7 **Relative Fire Hazard for Various Occupancies, as Related to Sprinkler Installations (*continued*)**

Extra-Hazard Occupancies

Occupancies or portions of other occupancies where quantity and combustibility of contents is very high and flammable liquids, dust, lint, or other materials are present, introducing the probability of rapidly developing fires with high rates of heat release.

Group 1—occupancies with little or no flammable or combustible liquids, such as

Combustible hydraulic fluid use areas
Die casting
Metal extruding
Plywood and particle board manufacturing
Printing (using inks with below 100F [37.8°C] flash points)
Rubber reclaiming, compounding, drying, milling, vulcanizing
Saw mills
Textile picking, opening, blending, garnetting, carding, combining of cotton, synthetics, wood shoddy or burlap
Upholstering with plastic foams

Group 2—occupancies with moderate to substantial amounts of flammable or combustible liquids, or where shielding of combustibles is extensive, such as

Asphalt saturating
Flammable liquids spraying
Flow coating
Mobile home or modular building assemblies (where finished enclosure is present and has combustible interiors)
Open oil quenching
Solvent cleaning
Varnish and paint dipping

Source: Reprinted with permission from NFPA 13-1983, *Standards for the Installation of Sprinkler Systems,* copyright © 1983 by the National Fire Protection Association, MA.

TABLE 13.8 **Sprinkler Design Guidelines**

Part A. ***Sprinkler Heads***
Discharge Characteristics

Nominal Orifice Size (in.)	*Orifice Type*	K *Factor*[a]	*Percent of Nominal ½-in. Discharge*
¼	Small	1.3–1.5	25
5/16	Small	1.8–2.0	33.3
⅜	Small	2.6–2.9	50
7/16	Small	4.0–4.4	75
½	Standard	5.3–5.8	100
17/32	Large	7.4–8.2	140

Temperature Classifications

Max. Ceiling Temp.		*Temperature Rating*				
F	*°C*	*F*	*°C*	*Temperature Classification*	*Color Code*	*Glass Bulb Colors*
100	38	135 to 170	57 to 77	Ordinary	Uncolored	Orange or red
150	66	175 to 225	79 to 107	Intermediate	White	Yellow or green
225	107	250 to 300	121 to 149	High	Blue	Blue
300	149	325 to 375	163 to 191	Extra high	Red	Purple
375	191	400 to 475	204 to 246	Very extra high	Green	Black
475	246	500 to 575	260 to 302	Ultrahigh	Orange	Black
625	329	650	343	Ultrahigh	Orange	Black

(continued)

TABLE 13.8 **Sprinkler Design Guidelines** (*Continued*)

Part B. ***Sprinkler Spacing***

Maximum Coverage per Sprinkler:

Light hazard — 200 ft^2 smooth ceiling and beam-and-girder construction (225 ft^2 for hydraulically designed systems)
130 ft^2 open wood joist
168 ft^2 all other types of construction

Ordinary hazard— 130 ft^2 all types of construction except
100 ft^2 high-piled storage

Extra hazard — 90 ft^2 all types of construction
(100 ft^2 for hydraulically designed systems)

Direction of Lines: Either direction to facilitate hanging except:
Across beams for beams on girders 3 ft to 7½ ft on centers
Across joists for wood joists (open or sheathed) and bar joists (through or under)

Maximum Spacing Between Lines and Sprinklers:

Light and ordinary hazard — 15 ft except 12 ft for high-piled storage
Extra hazard — 12 ft

Part C. ***Sprinkler Pipe Sizes***[b]
Light Hazard

Number of Sprinklers at Ceiling				*Number of Sprinklers Above and Below Ceiling*			
Steel		*Copper*		*Steel*		*Copper*	
1 in. pipe	2 sprinklers	1 in. tube	2 sprinklers	1 in.	2 sprinklers	1 in.	2 sprinklers
1¼ in. pipe	3 sprinklers	1¼ in. tube	3 sprinklers	1¼ in.	4 sprinklers	1¼ in.	4 sprinklers
1½ in. pipe	5 sprinklers	1½ in. tube	5 sprinklers	1½ in.	7 sprinklers	1½ in.	7 sprinklers
2 in. pipe	10 sprinklers	2 in. tube	12 sprinklers	2 in.	15 sprinklers	2 in.	18 sprinklers
2½ in. pipe	30 sprinklers	2½ in. tube	40 sprinklers	2½ in.	50 sprinklers	2½ in.	65 sprinklers
3 in. pipe	60 sprinklers	3 in. tube	65 sprinklers				
3½ in. pipe	100 sprinklers	3½ in. tube	115 sprinklers				
4 in. pipe	See Note *c*	4 in. tube	See Note *c*				

Ordinary Hazard

Number of Sprinklers at Ceiling				Number of Sprinklers Above and Below Ceiling			
Steel		*Copper*		*Steel*		*Copper*	
1 in. pipe	2 sprinklers	1 in. tube	2 sprinklers	1 in.	2 sprinklers	1 in.	2 sprinklers
1¼ in. pipe	3 sprinklers	1¼ in. tube	3 sprinklers	1¼ in.	4 sprinklers	1¼ in.	4 sprinklers
1½ in. pipe	5 sprinklers	1½ in. tube	5 sprinklers	1½ in.	7 sprinklers	1½ in.	7 sprinklers
2 in. pipe	10 sprinklers	2 in. tube	12 sprinklers	2 in.	15 sprinklers	2 in.	18 sprinklers
2½ in. pipe	20 sprinklers	2½ in. tube	25 sprinklers	2½ in.	30 sprinklers	2½ in.	40 sprinklers
3 in. pipe	40 sprinklers	3 in. tube	45 sprinklers	3 in.	60 sprinklers	3 in.	65 sprinklers
3½ in. pipe	65 sprinklers	3½ in. tube	75 sprinklers				
4 in. pipe	100 sprinklers	4 in. tube	115 sprinklers				
5 in. pipe	160 sprinklers	5 in. tube	180 sprinklers				
6 in. pipe	275 sprinklers	6 in. tube	300 sprinklers				
8 in. pipe	See Note *c*	8 in. tube	See Note *c*				

Extra Hazard

Number of Sprinklers at Ceiling			
Steel		*Copper*	
1 in. pipe	1 sprinkler	1 in. tube	1 sprinkler
1¼ in. pipe	2 sprinklers	1¼ in. tube	2 sprinklers
1½ in. pipe	5 sprinklers	1½ in. tube	5 sprinklers
2 in. pipe	8 sprinklers	2 in. tube	8 sprinklers
2½ in. pipe	15 sprinklers	2½ in. tube	20 sprinklers
3 in. pipe	27 sprinklers	3 in. tube	30 sprinklers
3½ in. pipe	40 sprinklers	3½ in. tube	45 sprinklers
4 in. pipe	55 sprinklers	4 in. tube	65 sprinklers
5 in. pipe	90 sprinklers	5 in. tube	100 sprinklers
6 in. pipe	150 sprinklers	6 in. tube	170 sprinklers
8 in. pipe	See Note *c*	8 in. tube	See Note *c*

Source: Reprinted with permission from NFPA 13-1983, Standard for the Installation of Sprinkler Systems, copyright © 1983, National Fire Protection Association, Quincy, Massachusetts 02269. This reprinted material is not the complete and official position of the NFPA on the referenced subject, which is represented only by the standard in its entirety.

[a]The *K* factor is explained on page 680.

[b]These sizes do not apply to systems that are hydraulically designed.

[c]See NFPA 13-1983 for special requirements.

Required flow rates are determined by the "density" of the fire hazard, which ranges from 0.1 gpm/ft^2 for light hazard to 0.5 gpm/ft^2 for extra hazard. In the preliminary design, however, pipe sizing may be approximated from the data given in Table 13.8.

Special installation requirements for sprinkler systems include: (1) at least one fire department connection on each frontage; (2) a master alarm valve control for all water supplies other than fire department connections; (3) special firewalls between protected areas and unprotected areas; and (4) sloping waterproof floors with drains or scuppers to carry away waste water.

When gravity tanks are used with sprinkler systems, they should provide an adequate reserve for this purpose and, in any case, enough to operate 25% of the sprinkler heads for 20 minutes. As in the case of standpipe and hose systems, this provision gives the fire company a chance to arrive and take over.

A typical automatic sprinkler system is shown in Fig. 13.15. The building, a printing and publishing plant, is in the category of "Ordinary Hazard, Group 2" (see Table 13.7). The sprinkler design results in a nozzle spacing such that one nozzle (sprinkler head) takes care of 125 ft^2 11.6 m^2) of floor area. Standards of the National

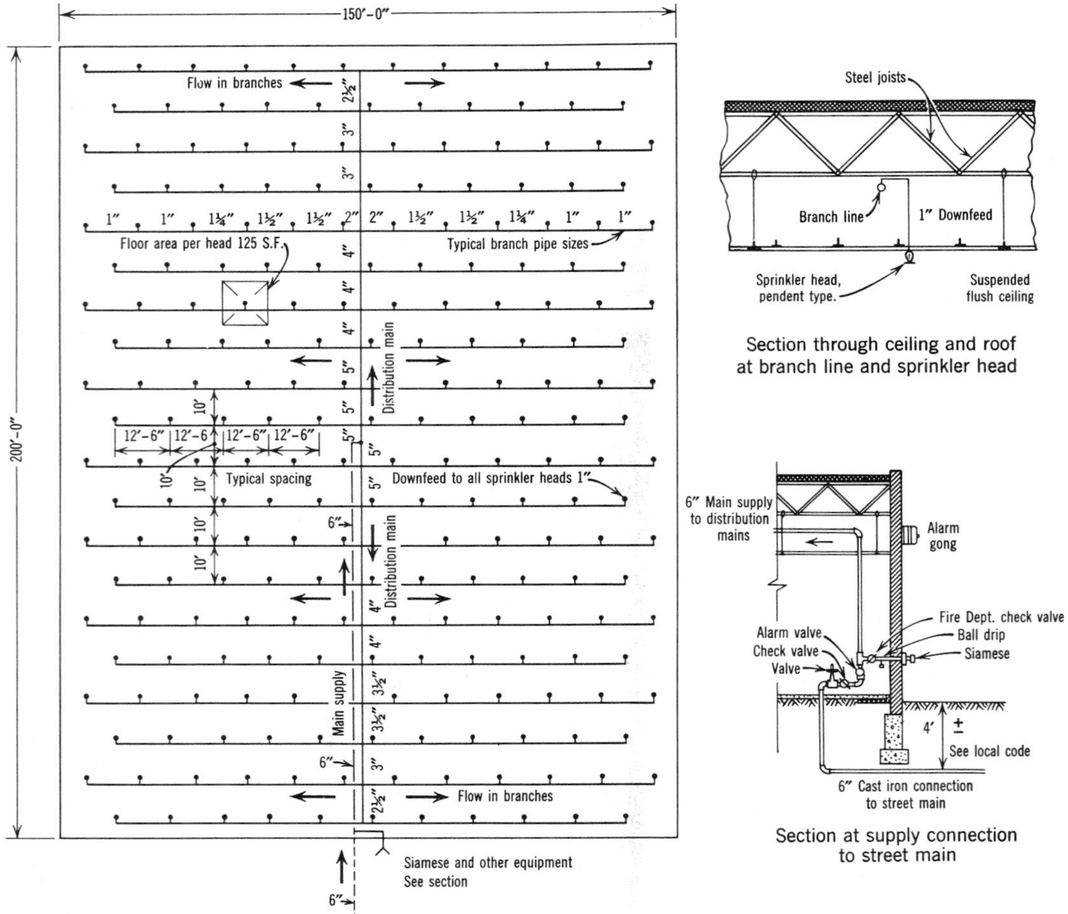

Fig. 13.15 *Plan of sprinklered industrial building, ordinary hazard occupancy, 125 ft^2 (11.6 m^2) per sprinkler head. Sprinklers (and standpipes) may use water from street mains when pressure is adequate. Either system may use pneumatic or gravity tanks. When the latter is used to supply both systems, an independent sprinkler reserve occupies the bottom, and the fire standpipe supply occupies the top. Auxiliary fire-engine feed by siamese should be provided in all cases. Auxiliary sources and standby pressurization may be required if street adequacy is questionable.*

Fire Protection Association (NFPA) are constantly being reviewed and updated. Designers are referred to the NFPA no. 13, *Installation of Sprinkler Systems,* which lists the requirements (briefly referred to here), as well as numerous other regulations.

An alarm gong mounted on the outside of the building (see Fig. 13.15) warns of water flow through the alarm valve upon the actuating of a sprinkler head. This warning gives the building personnel an opportunity to make additional fire-fighting arrangements that can minimize loss and speed the termination of the fire; in this way, the sprinklers can be turned off, to prevent excess water damage to building contents after the fire is out. Sprinkler alarms commonly are also connected to private regional supervisory offices that communicate promptly with municipal fire departments upon the receipt of a signal. Siamese connections permit fire engines to pump into the sprinkler system in a manner similar to that used for standpipe systems.

All public buildings, and other buildings as required, should be provided with fire detection and alarm systems that indicate, in the custodian's office, the location of the fire.

In future sprinkler systems, different types of sprays may be ejected from a single sprinkler head: one spray of larger droplets to penetrate the fire plume and thereby cool the burning surfaces, as well as adjacent surfaces; and another, finer spray to cool the ceiling itself.

(c) Upfeed Pumping. In tall buildings, sprinklers can be supplied with water from elevated storage tanks. These tanks, rising above large buildings throughout the country, have become a traditional part of the American architectural scene. Now their use appears to be diminishing. Architects—such as Skidmore, Owings & Merrill, in their design for General Mills's central office building in suburban Minneapolis—are turning to automatically controlled pumping systems that obviate the need for elevated storage (Fig. 13.16).

The purpose of the familiar water tower has always been fourfold: to supply a constant pressure on the distribution lines, to store sufficient water to balance supply and demand, to prevent excessive starting and stopping of the pump, and to provide a dependable fire reserve. The last fac-

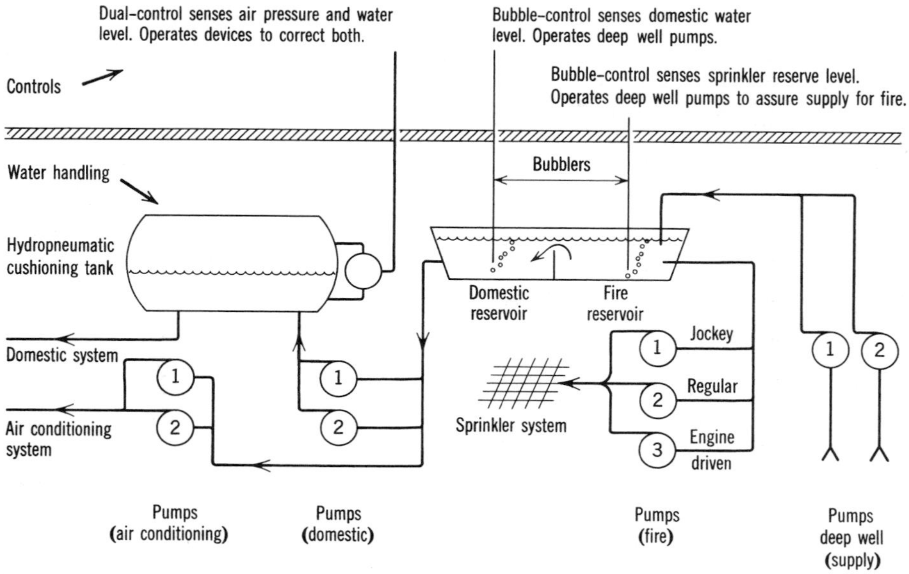

Fig. 13.16 Diagram of upfeed pumping system for supply of sprinklers and building demands for domestic water. In the General Mills Building, referred to in Section 13.3, a large subsurface concrete water tank provides the secondary water source that fire insurance carriers look for. Thus, the wells and their pumps constitute a primary water source. If the well yield fails, the concrete tank supply can be tapped. An additional resource provides engine-driven pumping that operates in the event of a power outage at an electric utility company. Automatic controls and piping for the storage tank reserve system are not shown in this illustration. Courtesy of Progressive Architecture.

tor has been of critical significance in the calculation of fire-insurance rates.

The principal objections to the use of tanks have been their unsightliness, the ever-increasing cost of steel and steel-construction work, the problem of freezing, and—in the case of large buildings—the tanks' tremendous weight. For the General Mills building, the fire underwriters possibly would have required at least 50,000 gal (189,250 L) of residual water in the tank for emergency purposes, which would mean about 100,000 gal (378,500 L) of elevated storage. The alternative chosen was a reinforced-concrete structure placed underground to one side of the building and covered with from 3 to 5 ft (0.9 to 1.5 m) of earth. Small vents rising from this reservoir blend in with the lawn and landscaped shrubbery above.

A comparison of costs would be difficult because of the number of factors involved. Generally, however, the savings in steel tended to make the overall cost comparable to that of an elevated tank of the same capacity. The savings were made possible chiefly by refined automatic-pump control. The underground reservoir eliminates the problem of appearance and weight, but the other usual advantages of an elevated tank—reliability in case of fire, a minimum starting and stopping of motors, and the maintenance of pressure while balancing supply and demand—must be equaled in the automatic circuitry. This is not so simple as it might at first seem. Factors such as fluctuations of demand, friction within the pipes, elevations, starting surges from the pumps, and pressure-flow characteristics of the pumps themselves must be met. Various combinations of these problems undoubtedly account to a great extent for the continued use of elevated tanks. Yet these are problems that can be handled efficiently on the drafting board of a controls engineer, rather than in steel mills and contractors' offices, where the expense is much greater. Hence, the trend is toward the use of more-sophisticated pump control.

It can be seen from Fig. 13.16 that a continuous flow from the deep-well pumps, through both domestic and fire reservoirs, prevents the water from becoming stale and rancid. The fire reservoir is given the necessary priority over the domestic reservoir by means of a simple weir. Even if the domestic reservoir were completely empty, the fire reservoir would remain full. Pressure for the sprinkler is supplied by a small, 20-gpm (1.26 L/s) jockey pump. Signals from the sprinkler system bring in a 750-gpm (47.3 L/s) main pump. If this should fail, a diesel-engine-driven pump of equal capacity automatically takes over.

The circuitry of two 1000-gpm (63.1 L/s) deep-well pumps and two 200-gpm (12.6 L/s) domestic pumps was designed by engineers in the office of Skidmore, Owings & Merrill. Design of the controls was made by the Automatic Control Company of St. Paul. Three sensing units govern the operation of the pumps, bubble-control units in each of the two reservoirs, and a dual-control unit that regulates supply for the pressure tank. All three elements are connected to a large central cabinet located in an underground pump room next to the reservoir. Pumps are controlled through the cabinet.

The bubble control uses a small air compressor within the central cabinet to send a flow of air through a ¼-in. (6.4-mm) tube to the reservoir. Back pressure on this flow, which varies with the level in the reservoir, operates pressure switches within the cabinet.

The hydropneumatic tank is used not for water storage, as is sometimes mistakenly supposed, but to store air under pressure that will balance out surge from the two domestic pumps and reduce the number of times the pumps must be started and stopped. This is a hybrid of the closed system in which several pumps are sequenced automatically to supply even pressure. Its advantage is that only two pumps are used.

One of the disadvantages in the past has been the difficulty of maintaining the correct ratio of 60% air to 40% water. Tanks supplied with water from deep wells become air-bound as water stored in them gives up its absorbed air. The dual-control installation in the General Mills system eliminates the need for manual adjustment of this ratio by employing two sensing devices within a single control. A drop in air pressure in the tank sends signals to start the pump. A rise in water level sends signals to stop it.

The signals from the dual-control and the bubble-control units are processed in the central cabinet for correct time delay through motor-driven relays. The central control system also either alternates the pumps, to ensure even wear, or runs them together if the demand requires. In the event of low suction, it shuts the pumps off to prevent motor damage. At the same time, it sends an alarm to the office of the maintenance engineer, indicating the location of the trouble.

Normally, the system automatically satisfies the heavy demands of air conditioning, fire control, and domestic water supply in this modern, rurally isolated office building.

Upfeed pumping for an urban high-rise building is exemplified in the Transamerica Building (Fig. 13.17) in San Francisco. Two fire pumps can deliver 750 gpm at 275 psi (47.3 L/s at 20 kg/cm^2) discharge pressure; they draw from city mains at 50 psi (3.65 kg/cm^2) or from a 5000-gal (18,925-L) closed tank in the basement. These pumps feed into two 6-in.-diameter (152-mm) "express" risers, one in each stair tower, that rise the full height of this 48-story office building. These risers serve both the sprinkler system and fire department hose lines. In each of three 16-story zones, "local" 6-in.-diameter risers branch off to feed a 2-in. (50.8-mm) looped main at each floor.

The sprinkler piping is carried above suspended ceilings; pendent sprinkler locations are coordinated with the modular grid that also locates partitions, air diffusers, and utility jacks. There are provisions for moving sprinkler heads as tenants change. At regular intervals, tees have been provided with one outlet stubbed and capped for future use. Typical office spaces are served by fully recessed, ½-in.-orifice, 165 F (12.7 mm, 74°C) pendent sprinklers. Exposed pendent sprinklers are used in toilet rooms and service areas.

(d) Provisions for Water Drainage. Sprinkler heads can release a great deal of water, most of which will remain unvaporized and quickly collect at floor level. In addition to waterproofing the floors and at lower walls, columns, and other elements, provision should be made, where possible, for gravity drainage of water. Table 13.9 provides guidelines for determining the number of 4-in. (101.6-mm) exterior wall scuppers (or floor drains) that should be provided (see also Fig. 13.18).

Fig. 13.17 *The Transamerica Building, San Francisco, California; William Pereira & Associates, Los Angeles, California, architects.* (a) *The pyramid form provides rentable floor areas ranging from 22,000 to 3000 ft^2 (2044 to 279 m^2).* (b) *Sixth-floor sprinkler plan, showing two risers, looped feed main, and branch lines. Courtesy of Copper Development Association.*

FIRE PROTECTION

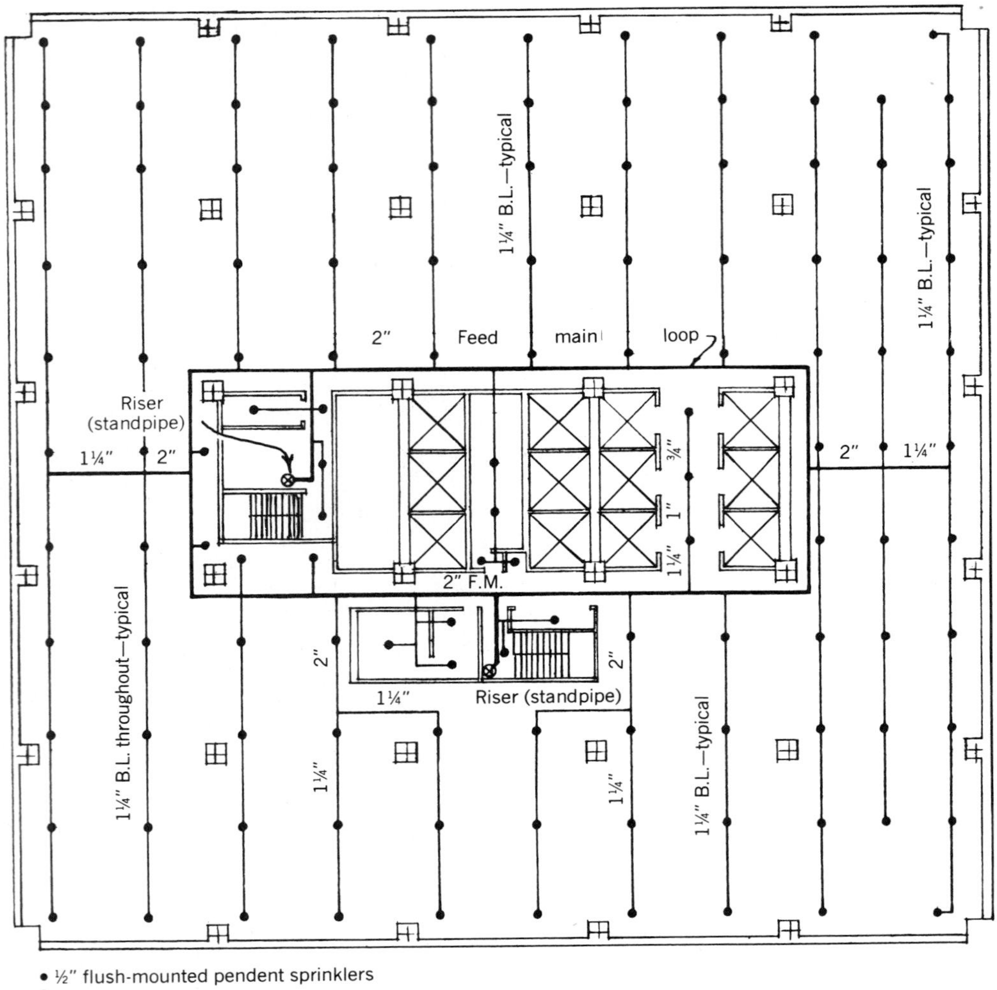

(b)

Fig 13.17 (continued)

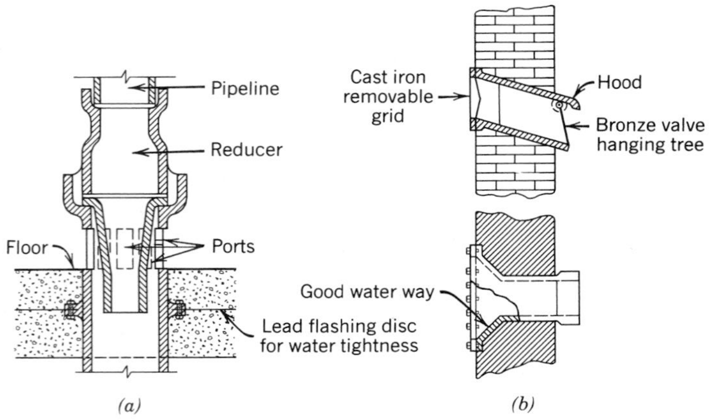

Fig. 13.18 *Water drainage for sprinklered buildings.* (a) *Floor drains, shown in a multistory application.* (b) *Wall scuppers. Reprinted with permission from* Fire Protection Handbook, *15th ed., copyright © 1981 by the National Fire Protection Association, Quincy, MA.*

TABLE 13.9 **Drains and Scuppers for Sprinklered Buildings**

Part A. ***Flow Rate of Discharge for Typical 4-in.² (101.6-mm²) Scuppers and 4-in. (101.6-mm)-Diameter Floor Drain Outlets***

Depth of Water		Discharge Flow Rate	
in.	*(mm)*	*gpm*	*(L/s)*
1	(25.4)	33	(2.08)
2	(50.8)	71	(4.48)
3	(76.2)	132	(8.33)
4	(101.6)	188	(11.86)
5	(127.0)	218	(13.75)
6	(152.0)	145	(15.45)

Part B. ***Number of 4-in. (101.6-mm) Scuppers or Drains Required***

Number of Scuppers or Drains	*Floor Area*
2	500 ft² (46.45 m²) or less
3	750 ft² (69.7 m²)
4	1000 ft² (91.9 m²)

Additional Scuppers for Larger Areas

For extra-hazard occupancies or floors of questionable watertightness, or where contents are especially subject to water damage:

1 per additional 500 ft (46.45 m²)

For moderate hazard occupancies, with watertight floors:

1 per additional 1000 ft² (92.9 m²)

For ordinary hazard occupancies with strictly watertight floors:

1 per additional 2000 ft² (185.8 m²)

Source: Reprinted by permission from *Fire Protection Handbook,* 15th ed., copyright © 1981 by the National Fire Protection Assocation, Quincy, MA.

Scuppers are preferable, since floor drains are more easily clogged by debris. Scuppers are provided with hoods that protect against control of infiltration, birds, or insects.

13.4 Other Fire Suppression Methods

When water poses almost as much of a threat as fire to a structure or its contents, a variety of other fire suppression methods are available. The most passive of such measures are intumescent materials, which expand rapidly as they are touched by fire. This process creates air pockets that insulate a surface from the fire or swell a material until it blocks openings through which fire (or smoke) could have passed. Intumescent paints, caulks, and putties are available. Some intumescent materials come in ½-in. (6.35-mm)-thick sheets, with various facing materials.

(a) Portable Fire Extinguishers. Most fires can be extinguished at an early stage with these common devices. Although some types of fire extinguishers do contain water or water mixtures, others contain dry chemicals and/or gases appropriate for various fire applications. A typical portable extinguisher was shown in Fig. 13.12*b*. These devices should be located in conspicuous places along ordinary paths of egress. Portable fire extinguishers are labeled as follows.

Class 1A to 40A: The numerals refer to the relative extinguishing potential (40A will extinguish 40 times as much as 1A). "A" refers to the contents: water or water-based agents, or multipurpose chemical agents. For use on ordinary combustibles such as wood, trash, paper, and textiles.

Class 5B to 40B: The numerals refer to the approximate square footage of deep-layer liquid fire that an inexperienced operator can extinguish. "B" refers to the contents: smothering or flame-interrupting chemicals such as CO_2, sodium, and potassium bicarbonate base dry chemicals, foam, or halogenated agents. For use on fires in liquid petroleum products or flammable liquids such as paint and solvents.

Class C: Contents are non-electrically-conducting, such as CO_2, sodium, and potassium bicarbonate base dry chemicals, or halogenated agents. For use on fires on or near electrical equipment.

Class A:B:C: "Multipurpose" dry chemical extinguishers filled primarily with ammonium phosphate. Although indicated for all three classes of use, ammonium phosphate is not ideal for electrical fires, because it leaves an especially hard residue.

Class D: Contents are dry powders, such as graphite or sodium chloride. For use on a variety of combustible metals; the specific combustible metal for which the extinguisher is designed is printed on the extinguisher's nameplate.

For the details of portable extinguisher types and location requirements, see NFPA no. 10, *Portable Fire Extinguishers,* available from the National Fire Protection Association.

(b) Carbon Dioxide (CO_2). This gas smothers a fire by displacing oxygen; therefore, CO_2 is usually used in tightly confined spaces that are free of people or animals—for example, display cases, mechanical or electrical chases, and unventilated areas above suspended ceilings. CO_2 is stored, under great pressure, as a liquid; when it is released as a gas, it absorbs about 120 Btu/lb, to provide cooling as well as smothering action. CO_2 is noncombustible and will not react with most substances. It spreads from discharge as a gas, but will stratify over time.

The main problem with CO_2 is that it must be used at concentrations of from 21% to 62% of that of the air, depending on the fire's fuel. However, at a CO_2 concentration of 9%, loss of consciousness will occur after a few minutes. This may allow enough time for an awake occupant to escape, but it doesn't help the firefighters' (or trapped persons') environment. Another potential problem is that after the initial smothering and subsequent dissipation of the CO_2, smoldering embers might reignite.

A typical CO_2 automatic detection/suppression system will have components similar to those of the Halon 1301 system shown later in Fig. 13.20.

(c) Foams. Because foams (masses of gas-filled bubbles) are lighter than water and flammable liquids, they float on the surfaces of burning liquids, smothering and cooling the fire. Foams can be designed to inundate a surface or to fill cavities; they can be thin and rapidly spread, or thick, tough, and heat-resistant. Both air-foaming and chemical-foaming methods can be employed. Because foam is so effective on flammable liquid fires, it is especially popular in airplane hangars (Fig. 13.19).

In Great Britain in the 1950s, a new high-expansion foam was introduced for firefighting in mines. Its introduction into the United States by private developers in collaboration with the U.S. Bureau of Mines proved its effectiveness against fires of a similar nature in this country. Years of research and improvement established the ideal foaming agent and led to a wide variety of uses in many other fields of firefighting. The patents were acquired in 1963 by Walter Kidde & Company. Among the many unique features of this system, which has been brought to a high degree of perfection, is the expansion rate of the foaming process. One thousand gallons of bubbles are produced from 1 gal (3.785 L) of water. This 1000-to-1 rate is most dramatic when compared with the 10-to-1 rate of the familiar protein foam that firefighters have long used effectively, especially on oil fires.

Albert Kahn Associated Architects & Engineers, Inc., designed and installed a system of fire protection, using this method, in the North Central Airlines Hangar Building at the Metropolitan Airport in Detroit.

The foam-generating equipment in this installation can fill the 38,400-ft^2 (3567-m^2) hangar with 1,400,000 ft^3 (39,648 m^3) of foam to a height of 36 ft (10.8 m) in less than 12 min. Automatic devices, upon sensing abnormal heat increase, operate the foam generators, open roof vents, start the smoke control exhaust fans in the vents and transmit a fire alarm signal to the Airport Fire Department. Foam discharge is delayed 30 seconds while evacuation sirens sound, to permit occupants to leave the fire area. Manual "override" controls allow personnel to start the system in the event of failure of the automatic controls, or to stop it if the fire is small and controllable by other methods.

The high-expansion foam is created by wetting a nylon net with a mixture of water and a special detergent soap concentrate. A large blower directs an air current through the net, producing an avalanche of foam.

Suds blanket the fire, attacking it in several ways. The water in the suds converts to steam, absorbing the heat of the fire. The expansion of the foam into steam reduces the oxygen content to about 7%, which is insufficient to support active combustion. A cooling effect is achieved by the

Fig. 13.19 Detergent foam discharged from four units, two of which are seen in the illustration, after 3 min of operation. This illustration at the North Central Airlines hanger building at Detroit Metropolitan Airport smothers fire and prevents its spread, and will not harm airplanes or machinery. Firefighters can breathe inside the foam. The system can handle combustible liquid fires. Note also the grid of upright sprinklers directly below the roof surface.

wetting action of the breaking bubbles. The movement of air currents toward the fire to replace the rising hot gases draws the foam to the center of the fire. There it blocks the air flow and cuts off the supply of oxygen.

The fire, thus contained and diminished, can be approached by firefighters for further control. Personnel advancing through the cooler sections of the foam are safe because the foam in these areas is 99% air and can support human life.

Aircraft are not harmed by the foam. Delicate machinery that might be injured by high velocity streams of water is undamaged and left quite clean when the foam is rinsed away. After all, it *is* a soap.

The structure—in this case, an open steel frame with a metal roof deck—is protected from excessive temperatures that might weaken it and cause it to collapse.

Perhaps the most serious potential disadvantage of foam is that the air turbulence created by a fire can, if violent enough, divert lightweight foam from its target.

(d) Halogenated Agents. Halogenated hydrocarbons, known commonly as "halons," are gases (stored as liquids) in which one or more hydrogen atoms have been replaced by halogen atoms. Although the original hydrocarbons are often highly flammable gases, the substitution of halogen not only makes the gas itself inflammable, but also gives it flame-extinguishing capabilities. The exact process in which halons extinguish fires is not known; apparently, unlike water or CO_2, they work primarily by inhibiting flames chemically rather than physically.

The most widely used of these agents is *Halon 1301;* the name signifies that its molecule contains one carbon atom, three fluorine atoms, no chlorine atoms, and one bromine atom. (When a fifth number is listed for halons, it refers to the number of iodine atoms.) Hydrogen atoms are not accounted for.

The advantages of Halon 1301 are that it can be released with relative safety in a flooding system in areas such as computer rooms, quickly extinguishing a fire with no harm to contents and—as yet—little demonstrated harm to people. It is also lightweight and space-saving relative to other fire suppressants. Halon 1301 is used on all commercial aircraft and in many special building applications, such as computer rooms, museums and libraries, telephone exchanges, and kitchens. Accordingly, it is effective against Classes A, B, and C fires. It is commonly used in portable fire extinguishers.

Halon 1301 does have limitations: It will not extinguish burning metals or self-oxidizing materials such as gunpowder. For deep-seated Class A fires, a higher concentration over a longer ''soaking time'' is required, necessitating much larger storage vessels and spaces that are relatively airtight. Halon is heavier than air, and it will slowly stratify in a steadily dropping layer as fresh air is introduced; also, it is more expensive than CO_2 or water. Finally, Halon 1301 at high concentrations will cause dizziness, impaired coordination, and reduced mental acuity. These are temporary effects, however; no residual toxic effects have yet been demonstrated. Ordinarily, Halon 1301 concentrations of 4% to 6% are sufficient to extinguish a fire. At concentrations below 7%, there is virtually no effect on humans. At concentrations of from 7% to 10%, the above-listed symptoms occur; at those above 10%, the toxic effects are potentially serious. When it is exposed to flame or very high surface temperatures, Halon 1301 decomposes to gases that are more dangerous; these have a sharp and acrid odor, which encourages occupants to leave but leaves a poor environment for firefighters. Halon 1301 is stored under high pressure in cylinders that should be kept from heat exposure.

Halon 1301, unlike water, is really more a protector of building *contents* than of the building *structure*. It leaves no sticky residue, such as most dry chemicals produce, and is therefore a popular choice where a clean fire-suppressing agent is required, people are present, and there are objects or processes of high value. Occasionally, it is used where water availability is low, or where space cannot be found for systems that use other, bulkier fire suppressants. A simple diagram of a Halon 1301 installation is shown in Fig. 13.20.

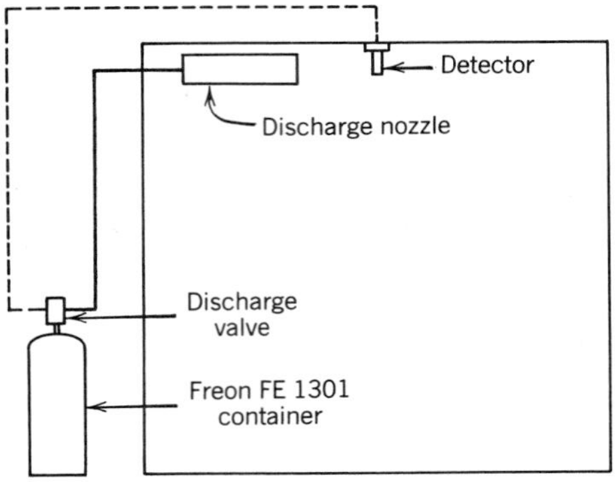

Fig. 13.20 *Schematic of a Halon 1301 automatic fire detection/suppression system that provides total flooding of a space. Reprinted by permission from Jensen (ed.),* Fire Protection for the Design Professional, *copyright © 1975 by Van Nostrand Reinhold Company, Inc.*

Many other potential halogenated agents can be used in fire suppression. The prevalence of Halon 1301 seems attributable to its especially low corrosive effects on common building metals or plastics. For a more complete listing of halons, see the NFPA's *Fire Protection Handbook,* as well as specific NFPA publications on halon systems.

13.5 Lightning Protection

All the electrically conductive metal in a building represents a potential path for the unimaginable force of a lightning bolt. It is important to consider intertying, or bonding, all metal systems to the designed lightning conducting system, to achieve full protection.

Lightning is nature's most destructive force: the average lightning discharge is estimated at 200 million v, 30,000 amp, and courses through a grounded object in less than a thousandth of a second. ''Cold'' lightning bolts have ample current, voltage, and duration to shatter and kill, but not to ignite combustibles. ''Hot'' bolts will ignite combustibles as well. Nearly half of the annual casualties from lightning occur in unprotected buildings.

Any lightning protection should consist of at least two *air terminals;* at least two *ground terminals,* spaced as far apart as possible and terminating in always-moist soil; and the *conductors* that connect air and ground terminals. The conductors may be exposed on the building exterior or concealed within a structure.

All metallic objects on the roof should be bonded to a looped conductor that joins the air terminals. These air terminals commonly are placed at both roof edges and ridges, at intervals of about 20 ft (6 m). The conductors are made usually of copper and are best housed in a plastic pipe conduit. Ground conductors, often called counterpoise conductors, are also installed to connect the earth terminals.

In reinforced-concrete buildings, column steel reinforcing can be used for lightning conduction *if* the column steel is welded, rather than merely tied. Precast-concrete buildings present special problems, as reinforcing steel within each panel must be connected to the eventual lightning conductor. Metal balcony railings present similar problems. In steel buildings, the columns relatively easily become lightning conductors; care must be taken to adequately bond both tops and foundations of such columns to the air and ground terminals, respectively.

Lightning protection systems are particularly important for very tall buildings. Figure 13.21 shows the schematic diagram for the John Hancock Center in Chicago, a structural steel building

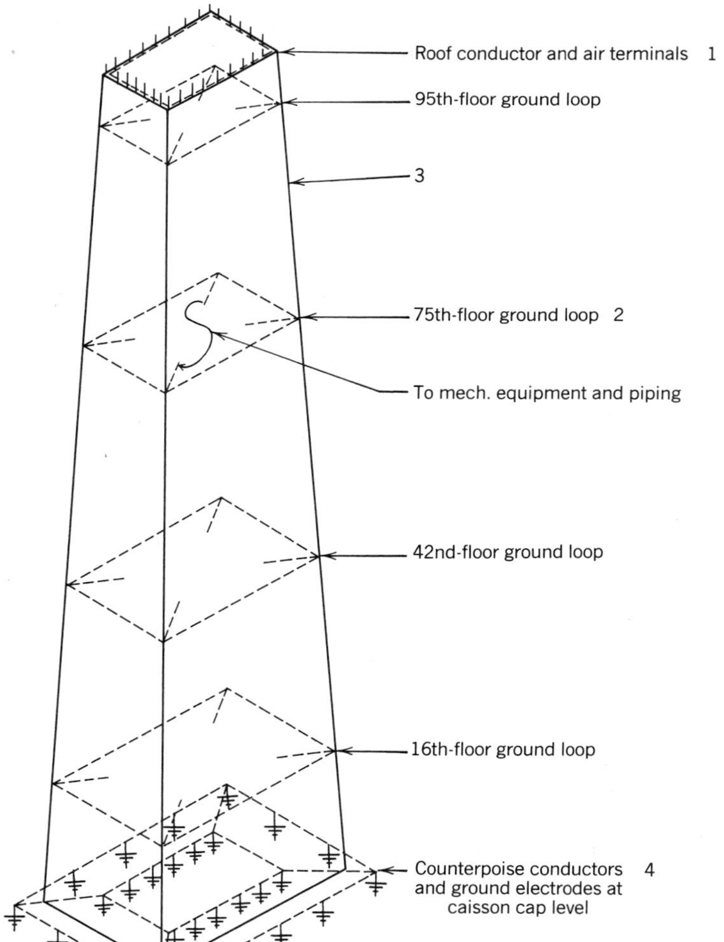

Fig. 13.21 *Lightning protection for the John Hancock Center, Chicago. Air terminals with roof conductors* (1) *are bonded to building steel and looped to assure that a single wiring break will not fault the system. Intermediate-floor ground loops* (2) *with connections to building mechanical equipment and piping protect against side-effect flashover to people or equipment. Building steel* (3), *properly bonded together, is used as a down conductor. Counterpoise conductors* (4) *cross-connect ground electrodes with ground conductors located below minimum groundwater level, to assure adequate and uniform dissipation of the charge. Reprinted by permission from Jensen (ed.)* Fire Protection for the Design Professional, *copyright © 1975 by Van Nostrand Reinhold Company, Inc.*

of unusual height. Intermediate loops at the mechanical equipment floors help to overcome the problems posed by the poorer conductivity of metal piping systems, which can become additional high-impedance paths for lightning, with resultant destructive effects. The intermediate loops are connected to each outside column, as well as to the main electrical, plumbing, air conditioning, and fire protection risers. These loops are also connected to both the roof conductor and ground counterpoise conductor.

Lightning protection codes are available both from Underwriters Laboratories and from the NFPA. They deal with the specifics of location, of sizing, and of the interconnection of lightning protection systems with all other potential conductors within buildings.

FIRE ALARM SYSTEMS

13.6 General Considerations

A fire alarm system serves primarily to protect life and secondarily to prevent property loss. Since buildings vary in occupancy, flammability, type of construction, and value, the fire alarm system must be tailored to the needs of a specific facility. In schools for instance, where the paramount consideration is rapid, orderly evacuation, the type of system used will generally be the same although the means of automatic detection may vary with construction type, building height, furnishings, and staffing. The fire alarm is part of the overall fire *protection* system design of the building. In particular, it overlaps with the design of safe egress and smoke/fire control in matters such as fan control and smoke venting, smoke door closers, rolling shutters, elevator control override, and the like. These automatic functions are initiated by operation of the alarm system but are designed in accordance with the overall fire protection plan.

As with other alarm systems, a fire alarm system has three basic parts: signal initiation, signal processing, and alarm indication. The signal initiation can be manual (pull stations or telephones) or automatic (fire and smoke detectors and/or waterflow switches). The alarm signal is processed by some sort of control equipment, which in turn activates audible and visible alarms and, in some cases, alerts a central fire station or municipal authorities.

13.7 Fire Codes, Authorities, and Standards

There are probably more codes and standards in the area of fire protection than in any other area, with the possible exception of structural standards. Although these codes are devoted primarily to fire protection and life safety, they also govern the type of fire alarm system to be installed. The following codes bear directly on fire alarm system arrangements.

NFPA (National Fire Protection Association) Life Safety Code 101—deals with municipal and auxiliary connections; auxiliary controls for elevators, door release, and so forth; type and location of alarm devices; circuit arrangement; alarm types and signals.

NFPA 72A Local Protective Signaling Systems.

NFPA 72B Auxiliary Protective Signaling Systems.

NFPA 72C Remote Station Protective Signaling Systems.

NFPA 72D Proprietary Protective Signaling Systems.

NFPA 72E Automatic Fire Detectors.

NFPA 74 Household Fire Warning Equipment.

SBC (Standard Building Code), published by the Southern Building Code Congress—deals with auxiliary system controls such as door release, elevator capture, and so forth. In addition, SBC has a special section on high-rise buildings, covering fire communication, elevator control, and smoke alarms.

BOCA (Building Officials and Code Administrators International) Basic Fire Prevention Codes—cover smoke detection, plus requirements for high-rise buildings.

HUD (U.S. Dept. of Housing and Urban Development)—covers requirements for residential buildings and care-type facilities. FHA minimum property standards are included here.

In addition, several insurance groups (e.g., Factory Mutual) issue standards that may apply to a specific facility, and there are local codes, standards, and fire marshal regulations. As with other aspects of construction—and perhaps more improtantly than with most—the architect/designer must ascertain which regulations have jurisdiction before proceeding with the design. The specific recommendations given in Sections 13.16 through 13.20 are representative of present good practice. Actual design must be based on and meet the requirements of *current* NPFA standards plus other codes having jurisdiction in the particular locale of the project.

13.8 Fire Alarm Definitions and Terms

The following list can serve as a reference to terminology that the reader will encounter in fire alarm work.

Automatic system. A system in which an alarm-initiating device operates automatically to transmit or sound an alarm signal.

Auxiliary fire alarm system. See Section 13.13 and Fig. 13.34*b*.

Breakglass. Refers to a false-alarm deterrent available in fire alarm stations; a glass rod is placed across the pull-lever that breaks easily when the lever is pulled.

Coded alarm signal. An alarm signal that represents a 1-, 2-, 3-, or 4-digit number indicative of the location of the fire alarm station operated.

Coded system. One in which not less than three rounds of coded alarm signals are transmitted, after which the fire alarm system may be manually or automatically silenced.

Common code. See *Dual-coded system, Master-coded system, Noncoded system, Selective-coded system, Zone-coded system*.

Continuous ringing. Refers to a continuous alarm. In coded systems so arranged, it refers to the signal that sounds after the completion of the normal number of rounds (usually four) of identifying coded alarm signal.

Control unit (fire alarm panel). The controls, relays, switches, and associated circuits necessary to (1) furnish power to a fire alarm system, (2) receive signals from alarm-initiating devices and transmit them to indicating devices and accessory equipment, and (3) electrically supervise the system circuitry.

Dual-coded system. See Section 13.15*d*.

Local fire alarm system. See Section 13.13 and Fig. 13.34*a*.

Local noninterfering coded station. A fire alarm station that, once actuated, will transmit not less than four rounds of coded alarm signals and cannot be interfered with by any subsequent actuation of that station until it has transmitted its complete signal.

Manual system. One in which the alarm initiating device is operated manually to transmit or sound an alarm signal.

Master coded system. See Section 13.15*b*.

Noncoded system. See Section 13.15*a*.

Presignal system. See Section 13.15*f*.

Proprietary fire alarm system. See Section 13.13*d* and Fig. 13.34*d*.

Remote-station fire alarm system. See Section 13.13*c* and Fig. 13.34*c*.

Selective-coded system. See Section 13.15*e*.

Station, fire alarm. A manually operated alarm-initiating device; may be equipped to generate a continuous signal (noncoded station) or a series of coded pulses (coded station).

Supervised system. A system in which a break or ground in the wiring, that will prevent the transmission of an alarm signal will actuate a trouble signal.

Trouble signal. A signal indicating trouble of any nature, such as a circuit break or ground, occurring in the device or wiring associated with a fire alarm system.

Zone-coded system. See Section 13.15*c*.

13.9 Principles of Automatic Fire Detection

Fire authorities agree that *most* fires pass through four stages, the last of which is the visible, flaming fire. These stages are shown in Fig. 13.22 as a function of duration and degree of hazard. Also shown are the type of automatic detector recommended.

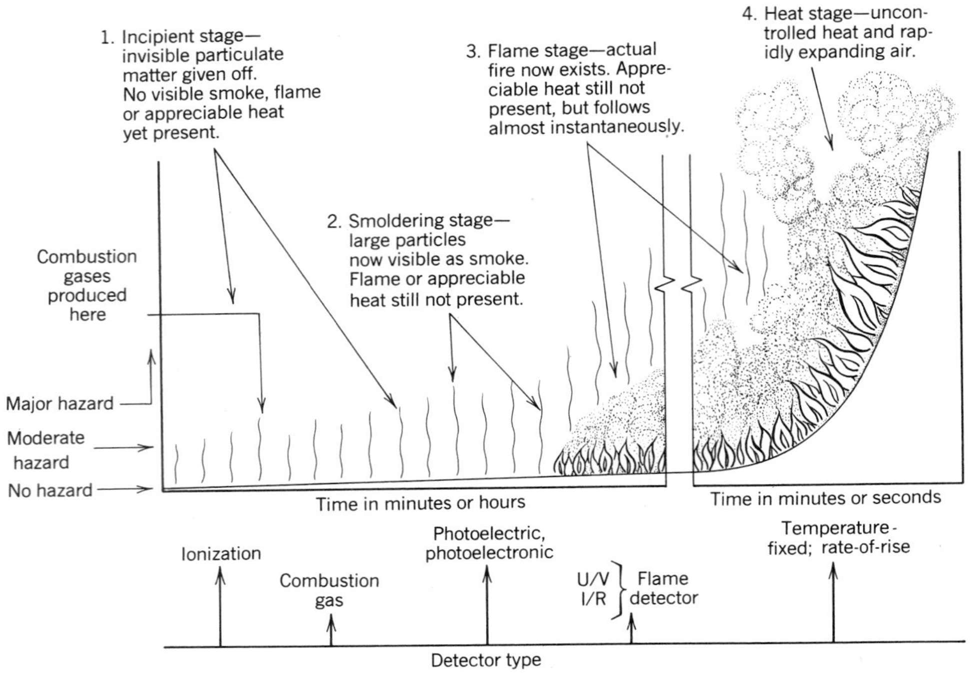

Fig. 13.22 Four stages of a fire. Early detection of each stage requires a different type of detector that will respond to the fire's particular characteristics at that stage.

(a) Incipient Stage. In this stage the combustion products comprise a significant quantity of microscopic particles (0.01–1.0 micron), which are best detected by *ionization-type* detectors (see Fig. 13.23). These detectors contain a small amount of radioactive material that serves to ionize the air between two charged surfaces, causing a current to flow. Combustion particles entering the detector

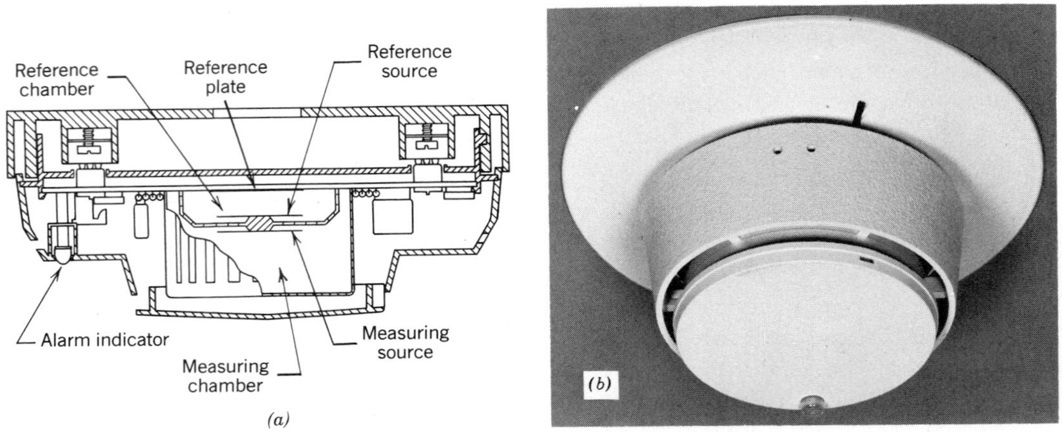

Fig. 13.23 (a) *Section through a dual-chamber ionization-type detector. The use of a reference chamber that is not exposed to combustion products reduces the influence of temperature and humidity, permitting sensitivity to be increased without triggering false alarms.* (b) *Photograph of a modern dual-chamber detector, with adjustable sensitivity and integral supervisory LED. Illustrated unit is 6 in. in diam. and 3 in. high. Courtesy of FENWAL, Division of Kidde. Ionization detectors are classified as early-warning units.*

chamber reduce air ion mobility, thus reducing current flow and increasing voltage. These changes are sensed and the alarm is set off. Response time of this type of detector depends on how rapidly the combustion particles can reach the detector—a factor that varies with room air currents and the type of material "burning." Once the particles reach the detector, response is instantaneous. Therefore, ionization detectors are best applied indoors, in spaces with stagnant air or low air velocity (below 50 fpm) and in which little visible smoke (large particles) is expected. (Some manufacturers make a special unit whose sensitivity *increases* with air velocity, for application in spaces with air velocity up to 500 fpm).

Ionization detectors should not be installed on warm or hot ceilings, or in any other location where hot air concentrates, since it will prevent the combustion particles from reaching the detectors. As a corollary, ionization detector sensitivity is higher at low ambient temperature. Since particles tend to agglomerate after leaving the combustion area, because of mutual attraction, ionization detectors are most sensitive in low ceiling spaces. Vapor and fumes from manufacturing processes interfere with detection and may cause false alarms. Also, since dust settling in the ionization chambers reduces sensitivity, these units need periodic maintenance. Coverage of detectors varies from 150 to 750 ft^2 per unit, depending on the unit used, the type of combustible material in the space, and ambient conditions. Once the fire has passed the incipient stage and becomes smokey, or where even in the incipient stage the combustion products are the large particles typical of visible smoke, the ionization detector loses sensitivity, and photoelectric detectors are recommended. Both ionization detectors and photoelectric detectors are classified as early-warning types.

(b) Smoldering Stage. Refer again to Fig. 13.22. The smoldering stage of a fire is characterized by particles up to 10 microns in size. Such particles, although small, are visible to the naked eye as smoke and are best detected by photometric means. The simplest type of *photoelectric smoke detector* operates on the principle of beam obscuration, as shown in Fig. 13.24. A beam of light is directed onto a photocell and a steady-state, no-smoke, circuit condition is established. The presence of smoke in sufficient concentration will partially obscure the beam, changing current flow in the photocell circuit and setting off an alarm response. Sensitivity is typically set at 1 to 2% obscuration per foot. Dust, dirt, and fumes from industrial processes will also cause obscuration, and light reduction caused by lamp aging will reduce current flow, resulting in false alarms. Thus, considerable maintenance is required to maintain sensitivity in this design, and it is therefore not recommended for extended unattended service.

A second design, usually referred to as a "photoelectronic," scattered-light, or "Tyndall-effect" detector, is illustrated in Fig. 13.25. In this design a beam of light is directed at a supervisory photocell, which serves to indicate that the unit is operating properly. The alarm photocell is shielded and normally receives no light. When smoke enters the unit, light is reflected from (scattered by) the smoke particles, and strikes the alarm

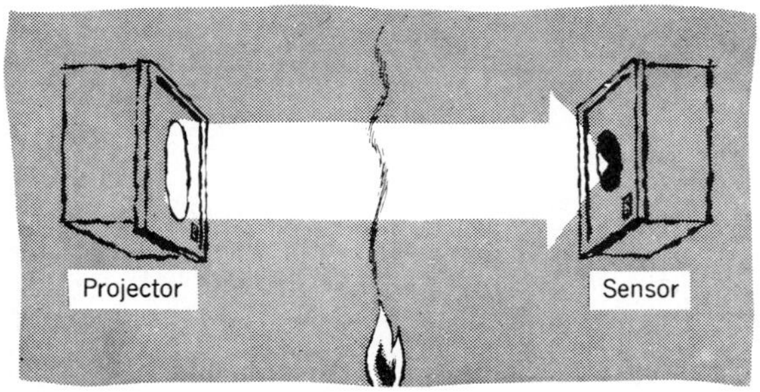

Fig. 13.24 *Principle of smoke detection by beam obscuration. This design is used in two ways: as illustrated, with a projector in one location and a photocell at a remote location; or with the entire assembly in a single small housing, with light source and receiver on opposite sides of a 2 to 3-in.-wide smoke chamber.*

FIRE PROTECTION

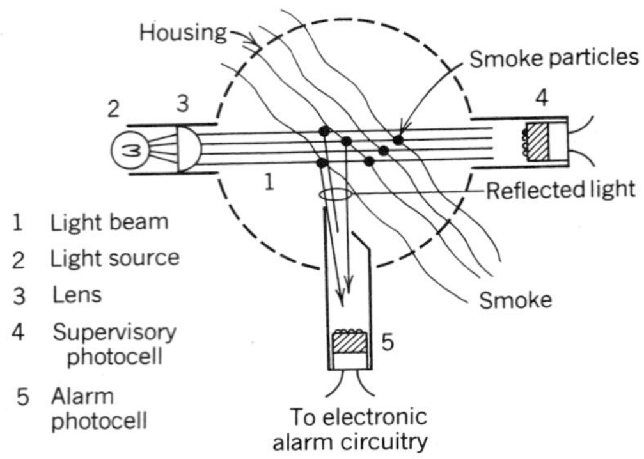

(a)

(b)

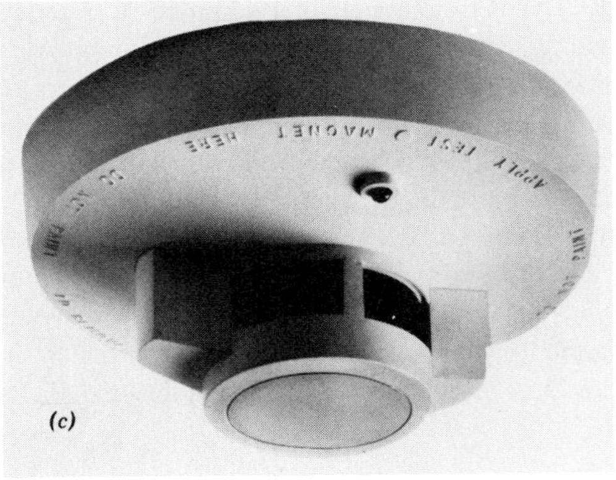
(c)

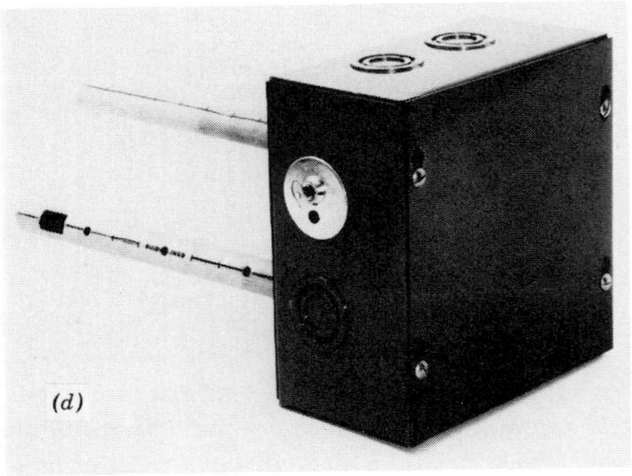
(d)

Fig. 13.25 (a) *Principle of operation of a scattered light photoelectronic spot smoke detector. A beam of light* (1) *from a source* (2) *is focused by a lens* (3) *on a supervisory cell* (4). *Smoke entering the unit reflects the light onto the alarm cell* (5), *changing its electrical characteristics and setting off an alarm. Photoelectric units with sensitivities up to 2% per foot are usually classified as early-warning units.* (b) *A typical "scattered light" spot smoke detector with integral temperature detector. This unit also contains relay contacts for operation of door releases and other auxiliary functions. It is approximately 7 in. in diameter and 2 in. high.* (c) *Rate-compensated fixed-sensitivity photoelectronic smoke detector intended for open-air use. It will alarm at 1.5% per foot obscuration or when obscuration increases at an abnormally rapid rate (usually 0.1% per foot per minute). The former detects smoldering fires and the latter fast-burning, smokey fires. Courtesy of Pyrotector, Inc.* (d) *Unit similar in design to (c), but designed to be mounted on an air duct. The sampling tubes extend inside the duct into the air stream. The unit will function with air velocities of 500–5000 fpm. These devices are principally intended to provide for shut-down and smoke damper control, in order to prevent smoke dispersal throughout the building via the air distribution system. The illustrated unit is 8 in. × 8 in. × 4 in. Courtesy of Pyrotector, Inc.*

cell. This changes the cell's resistance and the resultant signal is amplified electronically, causing an alarm. An alternate design depends on refraction of light to set off the alarm. These designs are not sensitive to normal dust and dirt accumulation or to light source depreciation, and high sensitivity can be maintained without continual maintenance. Thus, they normally are found in commercial use and good-quality residential applications. Sensitivity is usually set at 1% obscuration per foot.

Photoelectric smoke detectors of all designs will detect particles from about 0.2 microns (0.2×10^{-3} mm) to about 1000 microns (1.0 mm). Thus they are useful not only for smoldering fires but also for the smokey fires that characterize the burning of certain plastics and chemicals. Also, particle agglomeration, which increases with the distance the smoke travels, does not reduce their sensitivity, as it does with ionization types.

Maximum recommended spacing for photoelectric detectors, as for other types, is given by Underwriters Laboratories and Factory Mutual Lab standards. Closer spacing is often mandated by the particular application or by the structure's characteristics. Manufacturers' recommendations should be obtained for all installations. In order to provide early-warning detection of a wider range of combustion products than is possible with either the photoelectronic- or ionization-type detectors individually, several manufacturers produce a unit that combines a multichamber ionization detector with a photoelectronic detector. Table 13.10 gives comparative detection characteristics of these three detector types.

Several other sophisticated detectors are available for early-warning detection in the incipient and smoldering fire stages. Probably the most sensitive unit is based on the operating principle of a *Wilson cloud chamber*. When microscopic particles, such as those produced by the early stages of fire, are introduced into a saturated atmosphere (cloud chamber), they act as nuclei around which water vapor condenses to form visible droplets (Fig. 13.26). The detector operates by sampling air from the protected space and setting off an alarm when the cloud chamber indicates the presence of particulate matter. It is set after exposure to ambient conditions, in order to prevent false alarms. Its disadvantages are the need for piping and a high price in small installations. For large installations (30 or more detection points), the price is competitive with those of comparable systems. This

FIRE PROTECTION

TABLE 13.10 **Detector Response Characteristics**

This chart lists the response characteristics of different types of smoke detectors to various smoldering and burning materials. This information should be used only as a general guideline. Actual detector performance of different manufacturers and in-field conditions may vary.

Material	*Ionization*	*Photoelectronic*	*Combination Ionization/ Photoelectronic*
Paper	Good	Fair	Excellent
Wood (Smoldering)	Fair	Good	Excellent
Wood (Flaming)	Very Good	Good	Excellent
Fabric	Fair	Good	Very Good
Upholstery	Fair	Good	Very Good
Polystyrene Foam	Very Good	Very Good	Very Good
Paint	Very Good	Good	Excellent
Oil	Good	Fair	Good
Polyethylene	Good	Good	Good
Isobutyl Rubber (Wire Insulation)	Good	Fair	Good
Phenol Formaldehyde (P.C. Boards)	*Fair	Good	Excellent
P.V.C.	Fair	Good	Good

Source: Courtesy of Gamewell.

*Ionization detectors are not recommended for detecting smoldering P.C. Board type fires. An increased number (decreased spacing) of optical or combination detectors is recommended.

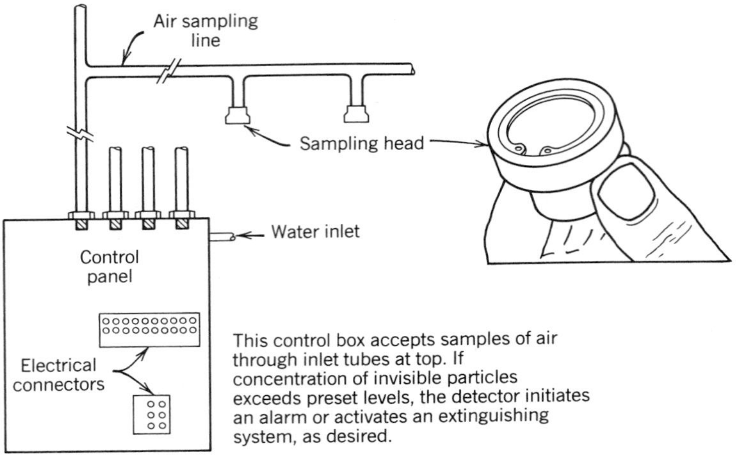

Fig. 13.26 *Cloud chamber fire detector. Courtesy of Environment One.*

detector is particularly applicable to high-value installations, such as museums, because of its very high sensitivity.

Two other ''high tech'' detector designs react to the gases produced immediately after the incipient stage of a fire (see Fig. 13.22). A *laser-beam detector* measures the change in the index of refraction of air caused by combustion gases, heat, and smoke, and alarms when the change reaches a preset level. This unit, though sensitive, is expensive, not widely accepted, and heavily dependent on air currents, which limits its application. Another type, available in several designs, measures the concentration of combustion gases with the aid of either a semiconductor or a catalytic element. These units are cheaper and simpler than the laser designs, but presently have reliability and acceptability problems.

(c) Flame Stage. As noted in Fig. 13.22, the appearance of flame is followed almost instantaneously by heat buildup and the rapid spread of flame, with a concommitant large increase in hazard. Detection of flame is no longer ''early-warning,'' and the prime requirement for a detector at this stage is speed. It is self-evident that the actions taken as a result of a flame detection alarm, such as fire suppression and/or evacuation, must also be very rapid. This is not the case with early-warning equipment alarms, with which, at the very least, several minutes normally are available to investigate the source of the alarm, *if desired,* before sounding the general alarm.

Flame detectors are of two types: those that detect ultraviolet radiation and those that detect infrared radiation. Both types of radiation are present at the beginning of the flame stage.

Ultraviolet (UV) radiation detectors operate by detecting the UV radiation produced by flames, which is typically in the 170- to 290-nm (nanometer) range. (The entire UV range is 40–390 nm). Hydrocarbon (organic material) fires in particular produce strong radiation in this range. A quartz filter in front of the UV sensor tube limits the detector's reception range to between 200 and 250 nm, which is below the visual range. This desensitizes it to sunlight and to incandescent and fluorescent sources. The detector need not ''see'' the flame directly; it can also detect UV radiation reflected from walls, ceilings, and the like. Obviously, the more direct the path, the stronger the radiation and the more rapidly the detector will respond. Since many devices—in particular, electric welders—produce UV radiation in sufficient quantity to activate the sensor, flame detectors are usually programmed to respond only to flickering sources at the usual flame flicker rate of 5 to 30 times per second.

In sum, UV detectors are very sensitive, react in seconds, and will respond to most types of fires. They are best applied in highly flammable or explosive storage and work areas, either open or

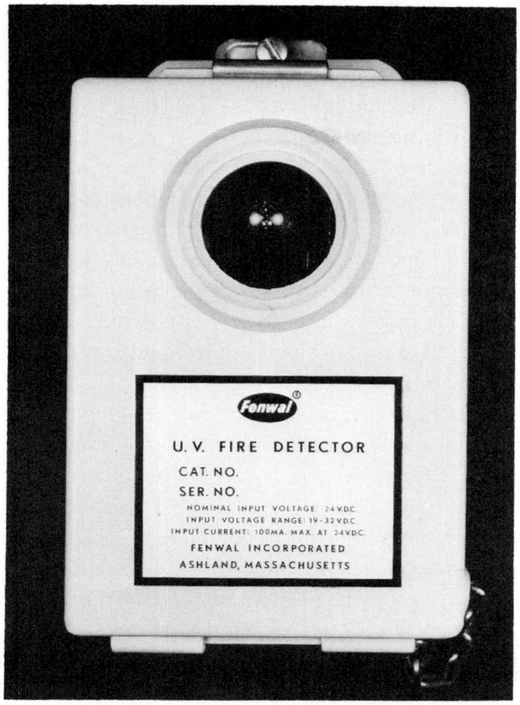

Fig. 13.27 *Typical ultraviolet flame detector, non-explosion-proof design. The quartz UV sensing tube visible behind the cabinet window has a 90° cone of vision. The illustrated unit is approximately 8 in. (high) × 4 in. × 4 in. Courtesy of Fenwal, Division of Kidde.*

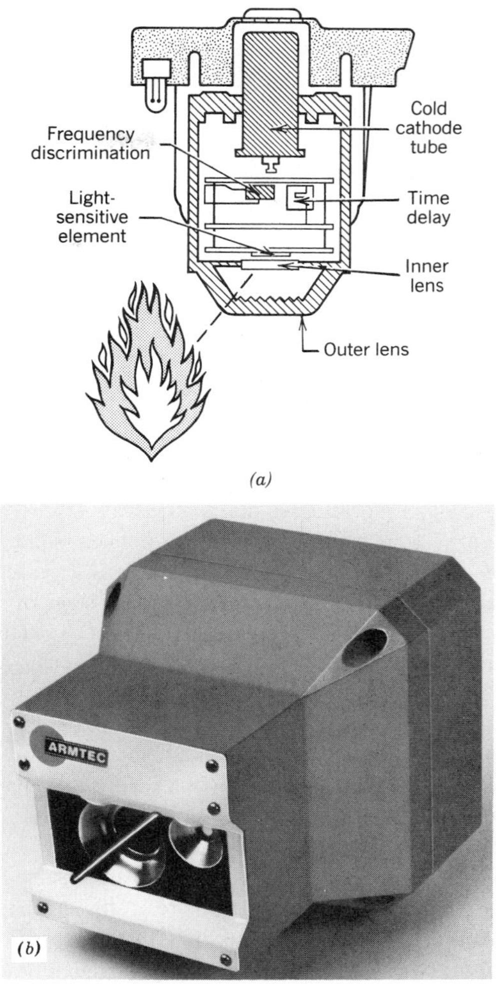

Fig. 13.28 (a) *Cross section through an IR flame detector. Courtesy of Pyrotronics.* (b) *Spot fire detection unit containing* both *UV and IR detectors. This unit, which is explosion-proof, is about 5½ in. square and 7 in. high and contains electronic alarm circuitry. Courtesy of Armtec Industries.*

closed. They are approved by major fire agencies. A typical unit is shown in Fig. 13.27.

Infrared radiation detectors are sensitive to radiation in the infrared region, between 650 and 850 nm—that is, at the edge of the visible light band. They differ from ambient heat detectors in that they respond to radiant energy and are essentially optical detectors. Like the UV detectors, IR units are very sensitive, react in seconds, and must be programmed for flicker response to avoid false alarms, which can occur if they are exposed to other sources of IR, such as sunlight. Because of interference problems, IR units are normally applied to enclosed spaces such as sealed storage vaults and the like. A typical unit is shown in Fig. 13.28.

(d) Heat Stage. Again referring to Fig. 13.22, the heat stage is the last and most hazardous stage, since by this time the fire is burning openly and producing great heat, incandescent air, and smoke. Spread of the fire depends on fuel, air currents, and the construction of the space in which the fire is burning. Detectors intended for use at this stage respond to heat and are referred to as heat-acuated, thermal, thermostatic, or simply temperature detectors. They act much like the fusible link in a sprinkler head. Effective application is restricted to locations where the subsequent alarm will permit adequate countermeasures to be taken in time to prevent injury and minimize loss. Keep in mind that the heat stage *follows* the smoke stage, and that smoke, not heat, is responsible for most casualties in fires.

Heat detectors are made in two designs: spot units, which are mounted in the center of the area to be protected; and linear units, which sense heat along their entire length. Both types respond to the high ambient temperatures caused by hot air convection from a fire. The linear type will also sense the overheating of an object or surface with which it is in contact, without the presence of fire.

Spot heat detectors are of two types: fixed-temperature units and rate-of-rise units. In the former, a set of contacts operates when a preset (nonadjustable) temperature is reached—usually 135F or 185F. The rate-of-rise type operates when the rate of ambient temperature change exceeds a predetermined amount, indicative of the heat stage of a fire. The rate-of-rise unit is normally combined with a fixed-temperature unit in a single housing. The fixed-temperature unit is available in either a one-time nonrenewable design that utilizes a low-melting-point alloy plug or an automatic resetting unit similar to a thermostat (hence the nomenclature). For most applications, the resettable unit is preferred. Two typical units are illustrated in Fig. 13.29. Spot units are best applied in areas subject to rapid-temperature-rise fires, (e.g., basements; see Fig. 13.29*d*) that are separated from occupied areas, so that even at the heat stage, enough time remains for evacuation and fire suppression measures. Obviously, some property loss will occur, and this factor must be weighed against the higher cost of an early-warning system and the degree of hazard involved.

Linear (continuous-line) heat detectors are also of two major types. One type uses a pair of steel wires under tension, held apart by thermoplastic insulation (Fig. 13.30*a*). When exposed to its rated alarm temperature, the insulation melts and the wires come into contact, changing the current flow in the circuit to which they are connected and setting off an alarm. This type is available in a range of alarm temperatures (68°C/155F, 88°C/190F, 138°C/280F) and can be used in open- or closed-circuit configurations (see Section 13.14) in lengths up to 2500 ft. Meters are available that will indicate the exact point of a fault, so that in long runs where overheating rather than fire is detected, countermeasures can be taken rapidly, without the delay caused by having to seek out the fault point. The major disadvantage to this type of detector is that it is nonrenewable; once having faulted, the melted section must be cut out and replaced.

The second type of linear detector (Fig. 13.30*b*) is basically a linear thermistor—that is, a device whose electrical resistance varies with temperature. The resistance is monitored at the control panel, and small changes noted. Depending on the thermistor material selected the detection range can be set anywhere from 70F (21°C) to 1200F (650°C). This type of linear detector is more sensitive, rapid, and expensive than the preceding design. When linear detectors are set to detect surface tempera-

Fig. 13.29 (opposite) *Spot-type heat detectors.* (a) *Fusible plug melts out at predetermined temperature, opening (or closing) the circuit and causing an alarm. Unit is indicating and nonrenewable.* (b) *Rate-of-rise unit is made of an air chamber with a restricted bleed valve. Rapid temperature rise causes the bellows to expand before air is lost by bleeding, thereby setting off the alarm. The illustrated unit is combined with fixed-temperature unit, similar to* (c). (c) *Bimetallic unit action is similar to that of a thermostat, and is self-restoring.* (d) *Combination rate-of-rise and fixed-temperature unit installed in a wood-frame house basement adjacent to the furnace. See also Fig. 13.25*b, *which shows a photoelectronic detector with an integral, fixed-temperature detector. Such combined units cover a wider range of hazards than can either single unit.* (e) *Commercial-grade surface-mount temperature detector of the self-resetting variety. This unit will alarm both at a preset fixed temperature of 135F or 200F and, regardless of temperature, when a rapid change in temperature occurs, as at the beginning of the heat stage of a fire. It operates on the principle of themal lag: prevention of heat equalization on fast ambient temperature changes sets off alarm. A separate contained element will alarm at a fixed temperature. The illustrated unit is 6 in. high overall, mounts on a standard 4-in. box, and is equipped with an indicating light. Courtesy of Pyrotronics.*

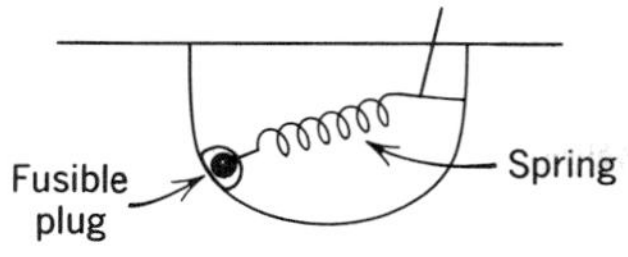

(a)

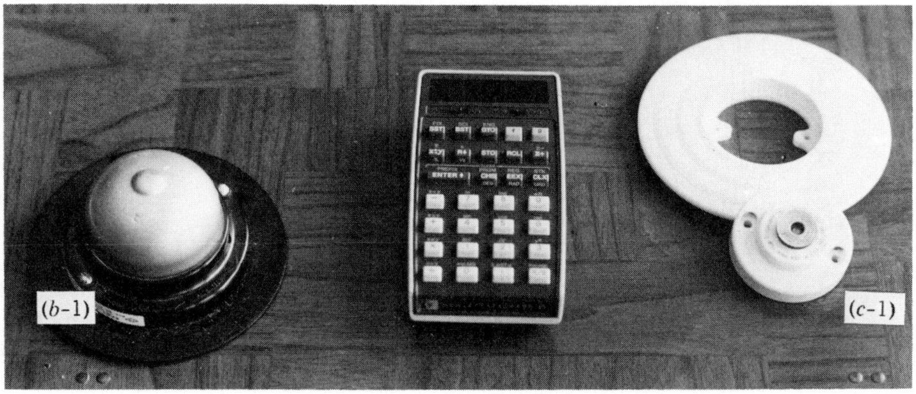

(*b*-1) (*c*-1)

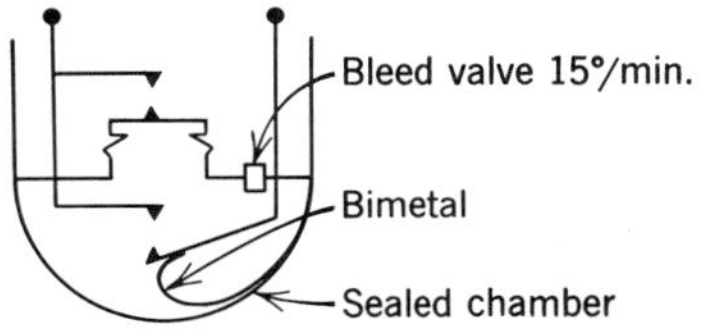

Rate of rise, fixed temp. (auto-reset)

(*b*-2)

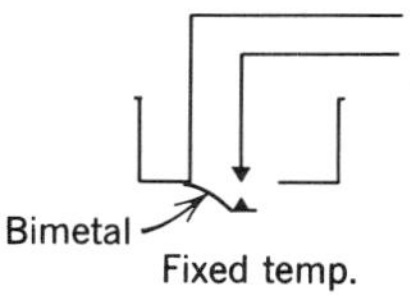

(*c*-2)

(*d*)

(*e*)

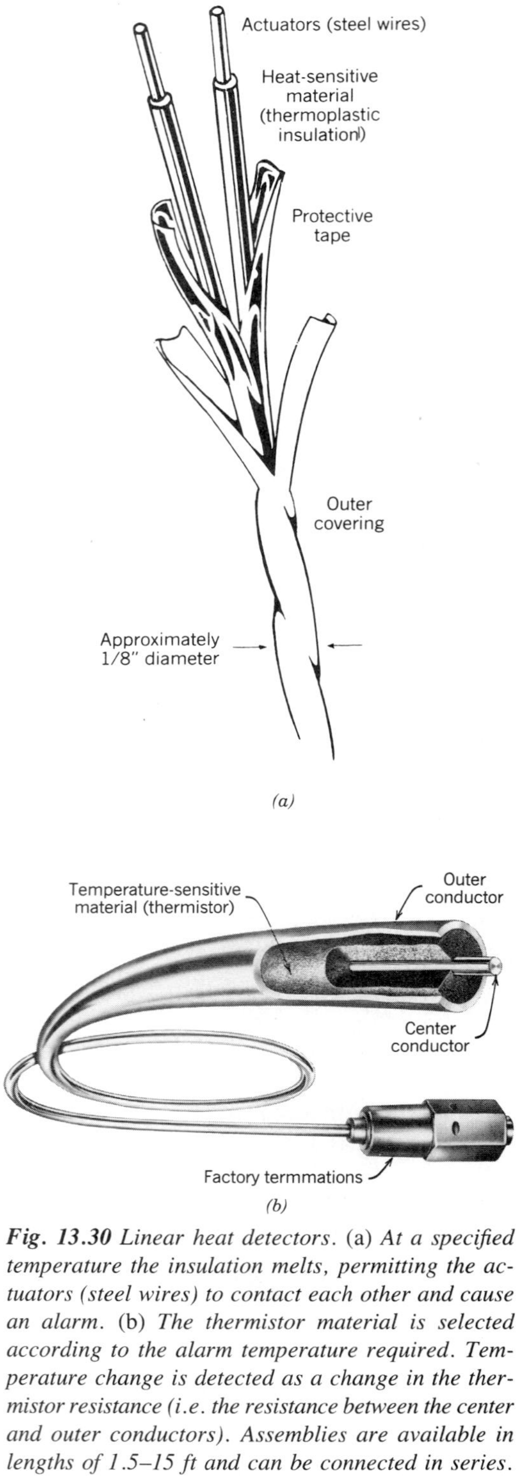

Fig. 13.30 Linear heat detectors. (a) *At a specified temperature the insulation melts, permitting the actuators (steel wires) to contact each other and cause an alarm.* (b) *The thermistor material is selected according to the alarm temperature required. Temperature change is detected as a change in the thermistor resistance (i.e. the resistance between the center and outer conductors). Assemblies are available in lengths of 1.5–15 ft and can be connected in series. Courtesy of Protectowire* (a) *and Fenwal, Division of Kidde* (b).

ture rather than air temperature, they act more like early-warning devices. They are extensively used in this fashion to continuously monitor the temperature of such devices as transformers, cable trays, generators, and all sorts of hazardous equipment.

13.10 Detector Application Guidelines

Smoke detectors of all types (ionization, photo) are subject to false alarms activated by particulate matter in the air. Detectors are by nature threshold devices; raising sensitivity increases false alarming, decreasing sensitivity shortens the crucial early-warning period. Because of this problem, most units are adjustable over a limited range and must be set in the field for optimum performance. Smoke detectors should not be placed directly in an airstream, or near the discharge from ducts or registers. However, they can be placed near the entry to return air ducts.

Smoke detectors installed in air ducts are *not* substitutes for area coverage. Because of dilution, they will detect only a very heavy concentration of smoke in the occupied area; hence, they will not function as early-warning devices. Their purpose in ducts is normally to shut down the ventilation system in order to prevent smoke from circulating through a building, in the event that area detectors do not perform this function. Another reason for not using duct detectors for area protection is that ventilation systems are shut down for extended periods (nights, weekends)—which obviously disables the duct detectors during these periods. See NFPA Standard 90 for the application of smoke detectors to air conditioning and ventilating systems.

Heat detectors that react to ambient temperature are particularly sensitive to ceiling height and air movement. Rate-of-rise temperature detectors should not be employed where rapid temperature changes occur normally, as near an open furnace or in areas exposed to the sun. Where fires can occur within an enclosure such as switchgear, a linear detector in contact with the enclosure or a smoke detector *inside* the enclosure is indicated. Where fires of various types can occur, several types of detection can be applied simultaneously, on the same circuits, with indication or annunciation to indicate the alarming device.

When the manufacturer indicates a recommended area coverage for a detector, a square or a circumscribed circle is intended. Rectangular coverage will yield a smaller area, since the maximum safe distance between detector and fire is established by the diagonal of the square-area coverage, which is the diameter of the circumscribed circular coverage. For data on spacing, refer to NFPA Standard 72F and the UL and Factory Mutual standards cited earlier. The final decision on the type and placement of detectors is best made with the aid of an expert whose objectivity is unquestioned.

Once the hazard has been detected, a signal is transmitted to a control center and appropriate action taken. In today's microprocessor world, the method of transmitting and processing the signal, is changing rapidly. In conventional (at least for the present) systems, the alarm signal is transmitted over dedicated conductors (''hard wiring'') to a control panel consisting basically of electromechanical relays. These in turn close audible device circuits, illuminate annunciator panels, control fans and door-releases, and so on, all via dedicated wiring. This arrangement has the advantages of reliability and simplicity. As the system grows, however, the wiring becomes heavy, complex, and expensive, relay panels become large and bulky, and changes become difficult. Also, troubleshooting of faults becomes time-consuming.

The newer microprocessor-controlled systems perform the same functions but use single sets of wires to receive and transmit digital data, and so in *large* systems become economically competitive. This type of system has a distinct advantage in retrofit work, in which hard wiring costs are very high; hence the higher electronic equipment cost can be offset by lower wiring cost. Also, the electronic system is simple to operate and maintain, can readily be extended, is easily incorporated into an overall security system (thus reducing overall costs in large facilities), and is capable of many functions that the conventional system cannot perform, such as remote control and monitoring, and timed or periodic function changes with no system disruption. Where devices such as coded stations are described, below, the description will be of conventional units, since the electronics used to perform the same function in the newer systems are not within our scope.

13.11 Manual Stations

In contrast to automatic detectors, the manual station is operated by hand. Manual stations serve to spread the alarm that has already been detected by other means, either human or automatic (see Fig. 13.31). Manual stations are either *coded* or *noncoded*; see Section 13.15 for an explanation of coding. If identification of the exact manual noncoded station operated is desirable, an annunciation panel can be added to the system; this is equivalent to using each station as a noncoded indicating zone. Because of wiring costs, annunciated systems become expensive, and beyond 10 stations, coding should be considered (see Fig. 13.32).

Fig. 13.31 (a) *A small manual fire alarm station with a break-glass rod and a single set of contacts, applicable to noncoded evacuation-type systems. Other types use a lock-open design that can only be reset by a key; this avoids the necessity of replacing a glass rod.* (b) *If desired, the station can be enclosed in a cabinet that is easily opened and* well marked.

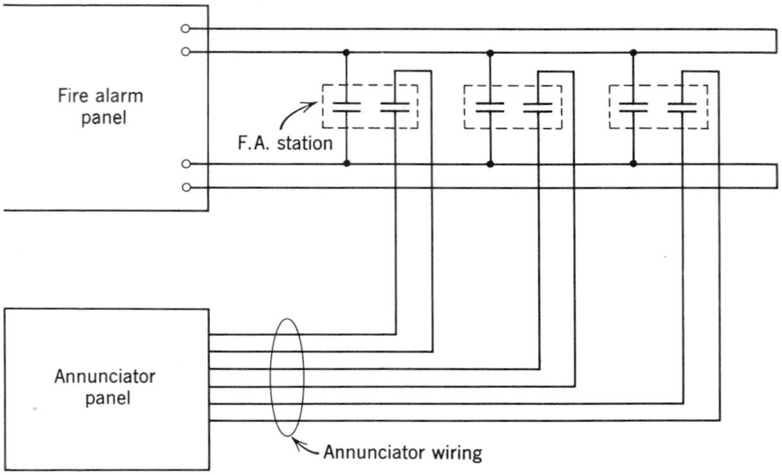

Fig. 13.32 Wiring of non-coded fire alarm manual stations. An additional set of contacts in each station provides annunciation for that station. Note that the pair of annunciation wires required for each station rapidly increases the cost of such a system. Alternatively, electronic transmitters can be used on a single set of wires, thus reducing wiring costs but increasing equipment cost.

When the system design is such that immediate aural identification of the operated station is necessary, a coded station is used. The station code is received at the control panel, processed, and then transmitted audibly on the system gongs. Not less than three rounds of code, and normally four rounds, are transmitted.

The code usually comprises three or four digits, for example, 2-3-2 with a pause between the ringing groups and a longer pause between the rounds. The first number may identify the building floor, the second digit the wing, and the third digit the individual station. Establishment of codes is left to the user. Manual stations are placed in the normal path of egress from a building, so that an alarm may be turned in by a person as he or she exits. It is *imperative,* therefore, that stations be well marked and easily found. Architects who place fire alarm stations in nooks and corners and in camouflaged cabinets because they spoil the decor of the lobby are defeating the purpose of the system. Similarly, placement of bells *inside* hung ceilings because they are unattractive is not only foolish but dangerous, and should *never* be done, regardless of the circumstances. Loss of property and even of life may result from such ill-conceived aesthetic considerations.

13.12 Sprinkler Alarms

Water flow switches are placed in sprinkler pipelines and operate when a sprinkler head goes off (see Fig. 13.33). In electrical terms, a water flow switch is a set of contacts, similar to a temperature detector. It can be used to trip a coded transmitter, setting off a sprinkler code; to show up on a sprin-

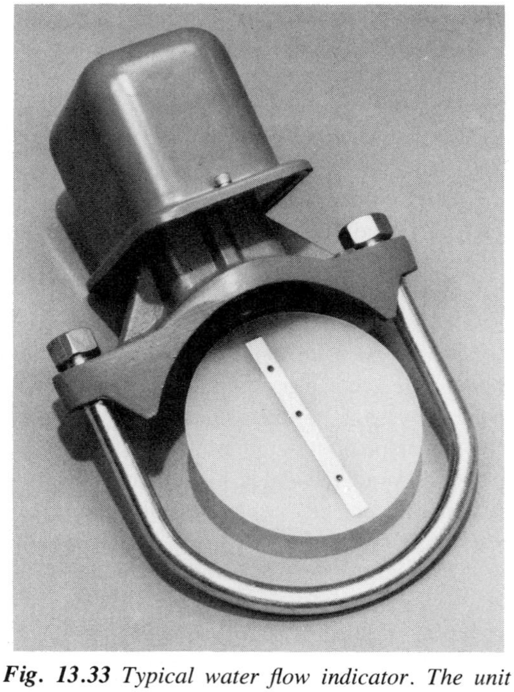

Fig. 13.33 Typical water flow indicator. The unit bolts onto a sprinkler pipe with the paddle inside *the pipe. Any water motion deflects the paddle, causing a signal to be transmitted from the microswitch mounted in the box on top of the pipe. Courtesy of Notifier Company.*

kler annunciator board, called a sprinkler alarm panel; or to act as a zone in a noncoded system. Wiring of water flow switches or of switches activated by other extinguishing systems, is the same as that for a manual station.

13.13 Types of Fire Alarm Systems

Fire alarm systems are classified according to function and application and are covered in separate NPFA standards.

(a) Local Protective Signaling System. See NFPA Standard 72A and Fig. 13.34*a*. This arrangement, as the name indicates, is intended to sound an alarm only in the protected premises. Any action must be taken locally, either manually or automatically. Thus, notification to the fire department must be manual, although fire suppression systems can be set into operation automatically. This arrangement is applicable to private residences, both single and multioccupancy, and to other private, relatively small facilities. When the building is unoccupied, notification to the fire department can come only incidentally, perhaps as a result of notification from a passerby.

(b) Auxiliary Protective Signaling System. See NFPA Standard 72B and Fig. 13.34*b*. This is simply a local system equipped with a direct connection to a municipal fire alarm box. The received alarm signal is identical to that resulting from a manual alarm at that city box. Since the fire department is aware of all direct connections, arriving firemen would check the protected premises. This type of system is usually applied to public buildings such as schools, governmental offices, museums, and the like.

(c) Remote-Station Protective Signaling System. See NFPA Standard 72C, and Fig. 13.34*c*. This system is similar to the auxiliary system, except that the alarm signal is transmitted via a leased telephone line to a remote location (a fire or police facility or a telephone answering service) that is manned 24 hours a day. The notice of alarm is then telephoned to the fire department. This arrangement is used in private buildings, such as stores and offices, that are unoccupied for considerable periods and for which reliance on passersby to turn in a fire signal is unacceptable. According to the NFPA definition, an audible alarm circuit extended from a local system to a nearby building—as, for instance, from a store to a nearby

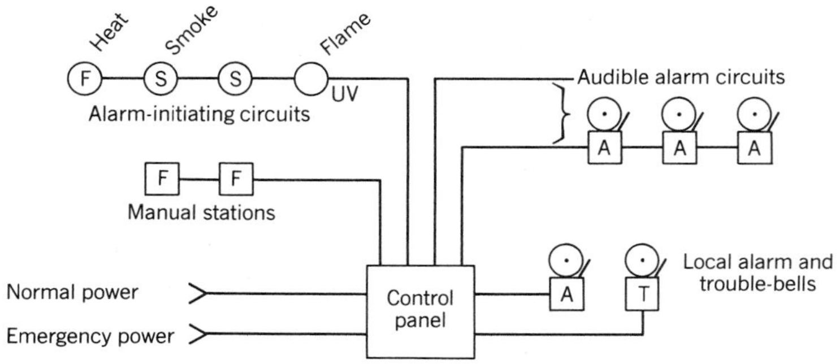

(a)

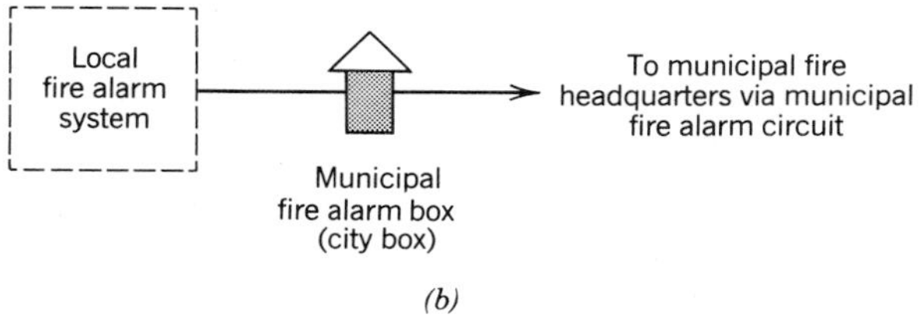

(b)

Fig. 13.34 *Fire alarm system arrangements.* (a) *Typical local system (NFPA 72a).* (b) *Typical auxiliary system (NFPA 72B).*

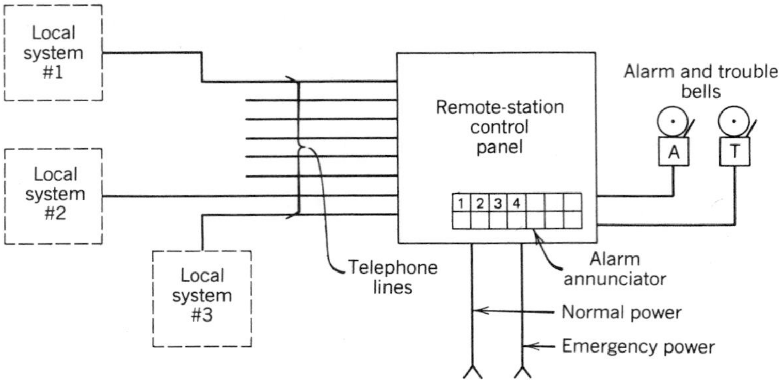

(c)

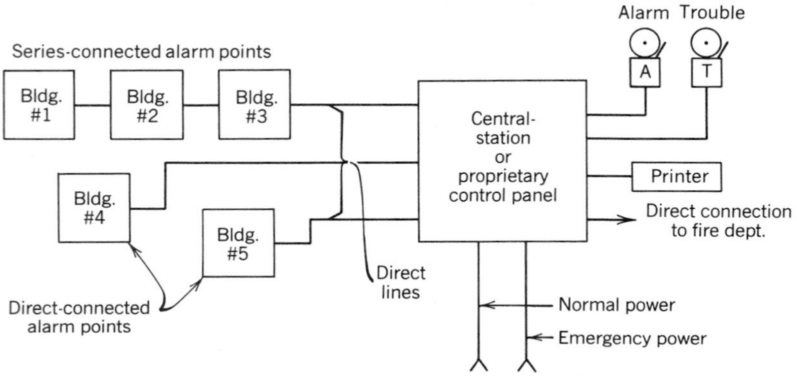

(d)

Fig. 13.34 (continued) (c) *Typical remote-station system (NFPA 72C).* (d) *Typical proprietary or central-station system (NFPA 72D).*

residence—does *not* constitute a remote-station system, unless all the requirements of Standard 72C are met.

(d) Proprietary Protective Signaling Systems. See NFPA 72D and Fig. 13.34*d*. This system, which is applicable to large, multibuilding facilities such as universities, manufacturing facilities, and the like, utilizes a dedicated central supervisory station to receive signals from all buildings. In a proprietary system, this station is on the site and is manned by persons associated with the facility. A common arrangement is for the station to be located in a guardhouse or similar supervisory location, from which point alarms are sent manually to fire stations. The central station system differs from the proprietary only in that the supervisory station is manned by persons unconnected with the facility, normally for a fee. The location may be on-site or off-site.

Information on the exact location or zone within each building at which the alarm occurred, is multiplexed or digitalized and transmitted over a single pair of wires to the central supervisory location. In older systems a separate pair of wires was required for each location. The central location has an audible alarm, some sort of visual display that indicates the alarm location, and a printer that makes a permanent record of each alarm. As has been mentioned, these central supervisory locations are frequently multipurpose, covering all aspects of facility security plus, frequently, control functions such as energy management.

13.14 Circuit Design

A system that is normally deenergized and carries no current except when functioning is called an open-circuit system (see Fig. 13.35*a*). Such a sys-

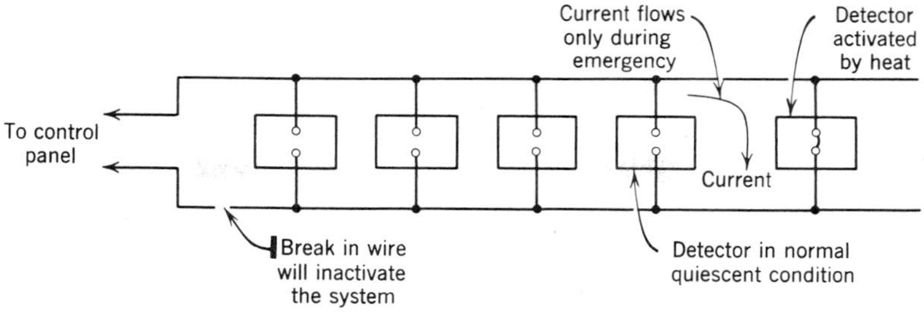

(a) Wiring of an open-circuit fire-alarm system

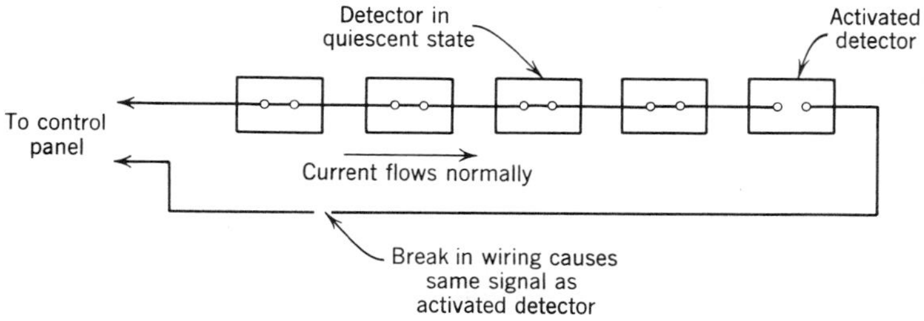

(b) Wiring of a closed-circuit fire-alarm system

Fig. 13.35 *Wiring of open and closed-circuit unsupervised systems, showing the inherent advantage of closed-circuit wiring.*

tem is the simplest and most economical type but has the disadvantage of not indicating a broken wire or other malfunction that will render it inoperative.

Figure 13.35*b* shows a closed-circuit system. This arrangement will set off the alarm bells in the event of trouble in the equipment, but since this type of "false alarm" is to an extent undesirable, a further refinement can be added in the form of a trouble bell and/or light that will indicate an equipment failure to the occupants without ringing the fire alarm bells. This feature is known as *supervision,* and such a system is known as a supervised system. Supervised systems can utilize open- or closed-circuit devices, depending upon circuit arrangement. Furthermore, with special wiring and circuit design, the system can be arranged so that a single break or ground in the wiring to the devices will not prevent the operation of the system.

As a system becomes larger, a desirable feature to incorporate in the circuitry arrangement is the grouping of the devices into zones. Each zone is covered by a separate section or module in the control panel, or is annunciated, permitting extremely rapid identification of an alarm signal.

13.15 System Coding

(a) Noncoded Systems. Noncoded systems are *continuous ringing* evacuation types using manual and automatic alarm initiation. If desired, the devices can be zoned and, if the system is sufficiently large, annunciation can be provided. Audible devices are continuous ringing vibrating bells and horns (see Fig. 13.36*a*).

(b) Master-Coded Systems. This system, also called common-coded and fixed-coded, generates four rounds of code when any signal device operates. It utilizes a single code transmitter at the panel. Normally, the system stops after four rounds of code, although it can readily be arranged to sound continuously thereafter. When the code is set to ring the bells at an even 108 strokes per minute, it is known as "march time," because of the rhythmic cadence. This beat aids in the rapid

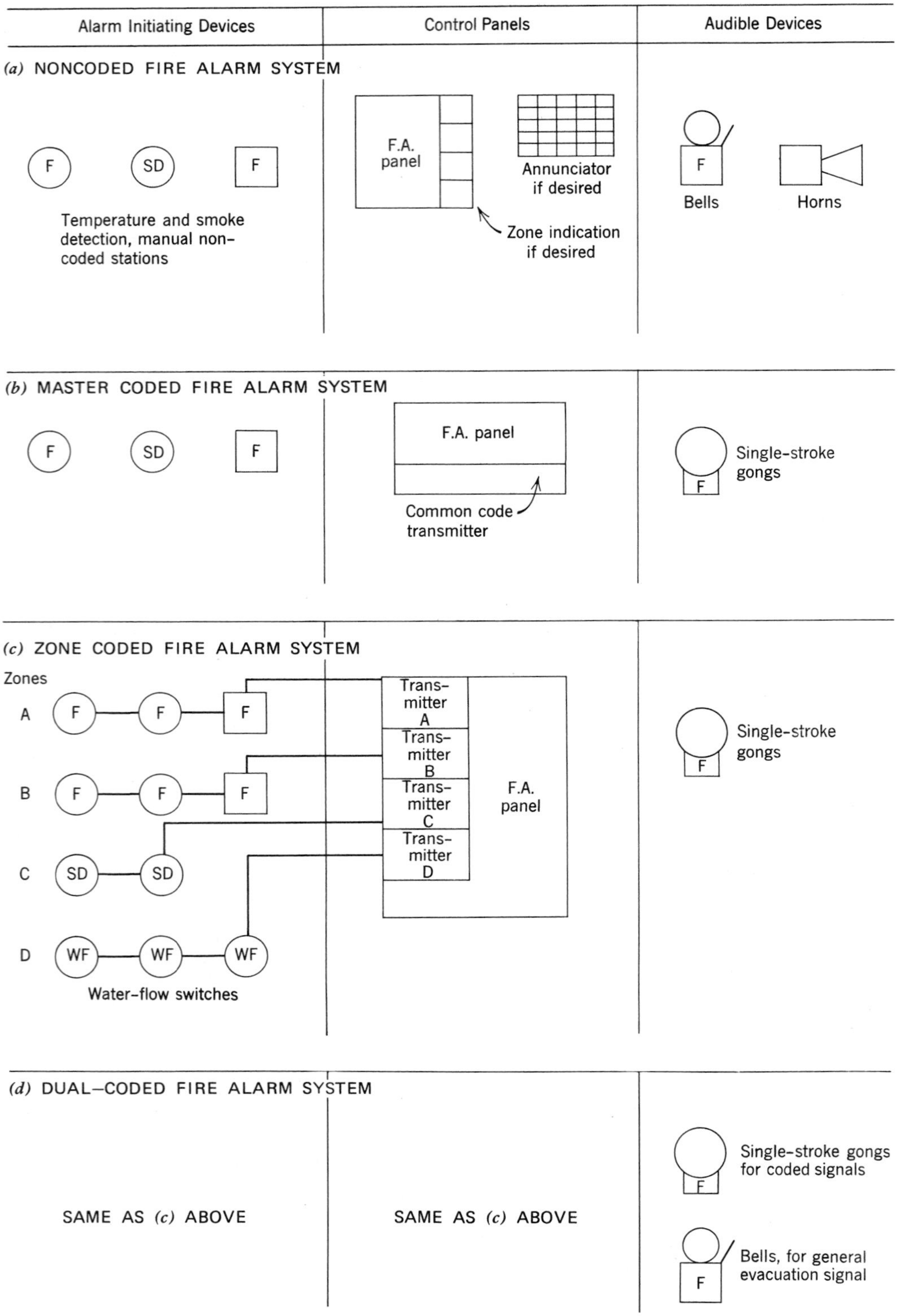
Alarm Initiating Devices
Control Panels
Audible Devices
(a) NONCODED FIRE ALARM SYSTEM
F
SD
F
Temperature and smoke detection, manual non-coded stations
F.A. panel
Annunciator if desired
Zone indication if desired
F
Bells
Horns
(b) MASTER CODED FIRE ALARM SYSTEM
F
SD
F
F.A. panel
Common code transmitter
F
Single-stroke gongs
(c) ZONE CODED FIRE ALARM SYSTEM
Zones
A
B
C
D
F
SD
WF
Water-flow switches
Trans-mitter A
Trans-mitter B
Trans-mitter C
Trans-mitter D
F.A. panel
Single-stroke gongs
(d) DUAL–CODED FIRE ALARM SYSTEM
SAME AS (c) ABOVE
SAME AS (c) ABOVE
Single-stroke gongs for coded signals
Bells, for general evacuation signal

Fig. 13.36 *Fire alarm system coding arrangements.*

Alarm Initiating Devices	Control Panels	Audible Devices

(e) SELECTIVE CODED FIRE ALARM SYSTEM.

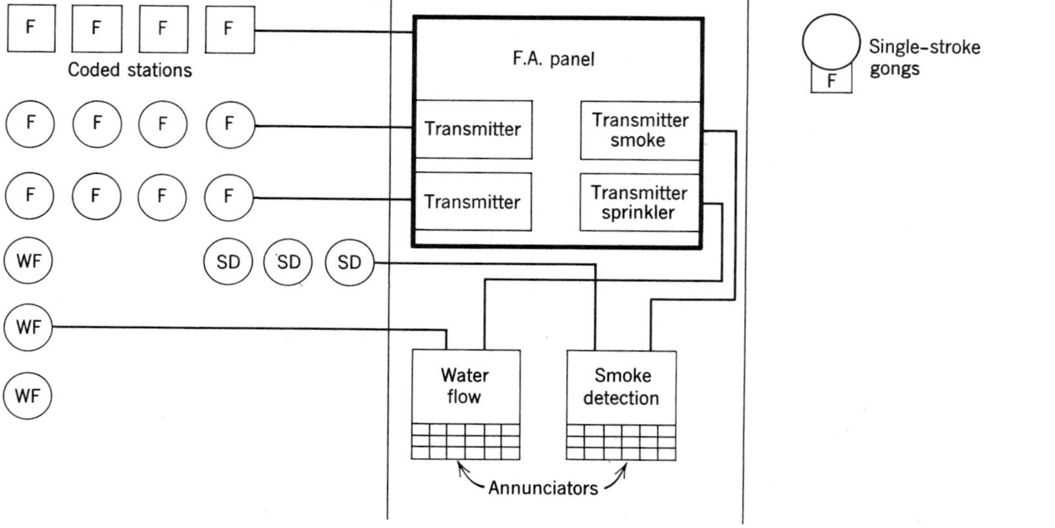

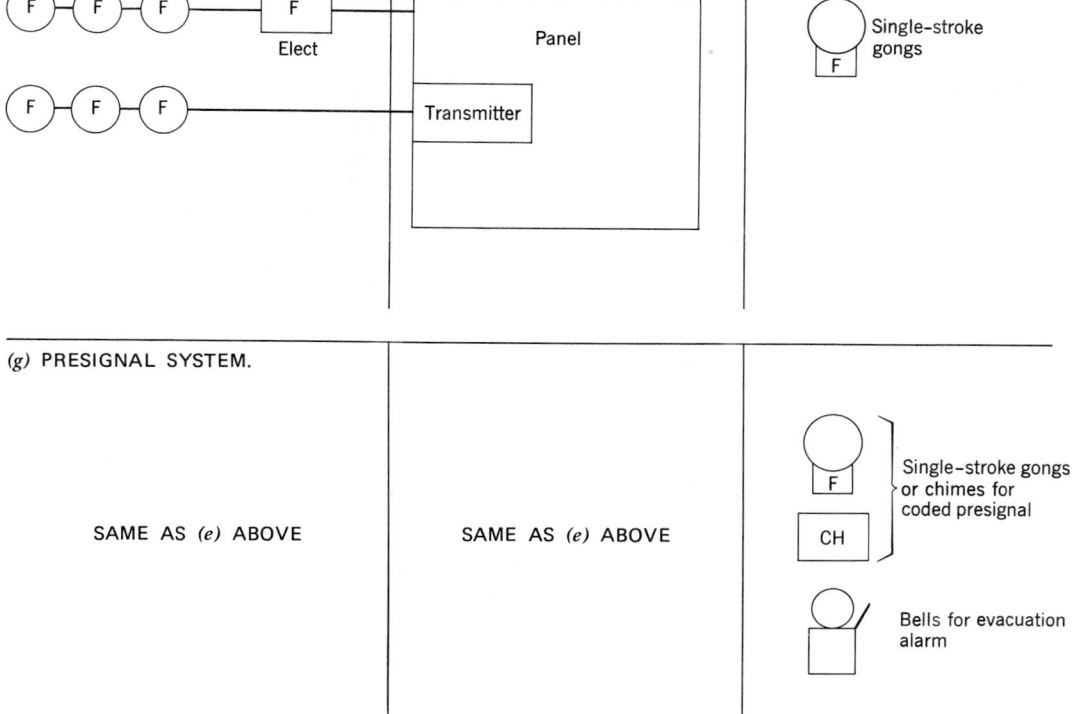

(f) SELECTIVE CODED FIRE ALARM SYSTEM WITH ELECTRICALLY OPERATED CODED STATIONS THAT ACT AS SUBMASTERS TO DETECTOR CIRCUITS. DETECTORS TRIPPING ELECTRICALLY OPERATED STATION SOUND ITS CODE.

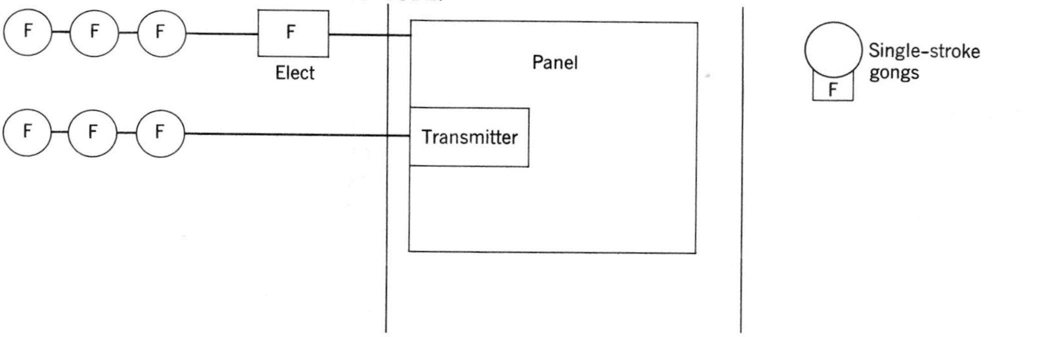

(g) PRESIGNAL SYSTEM.

SAME AS *(e)* ABOVE

SAME AS *(e)* ABOVE

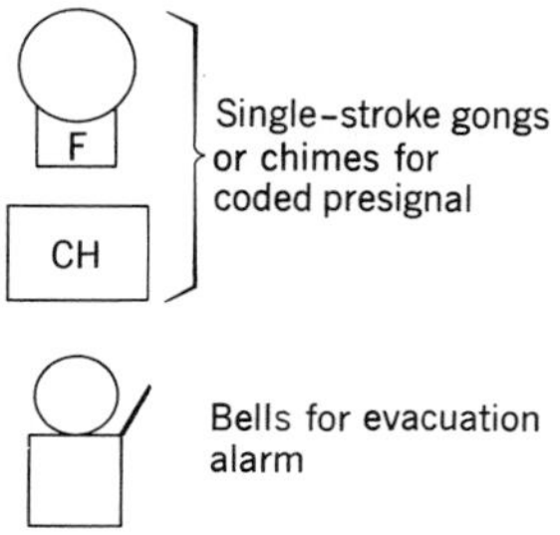

panic-free evacuation of a building and therefore is frequently used in schools (see Fig. 13.36*b*).

(c) Zone-Coded Systems. Identification of the alarmed zone in a system can be accomplished with zone lights, an annunciator, or by coding. In the first two cases, it is necessary to *go to the panel or annunciator* to determine the location of the operated device, which entails a possibly critical delay. All coded systems obviate this necessity by sounding the code on all the gongs in the building, thus immediately identifying the station and permitting the building staff to quickly investigate the cause of the alarm and take appropriate measures.

Therefore, if a *coded* system is desired, but by zone rather than by device (and this is less expensive by far), *noncoded* manual stations are used, along with automatic detectors, grouped by circuit into zones. These trip zone transmitters *in the panel,* which in turn ring the zone's code on the single stroke gongs or chimes (see Fig. 13.36*c*). As with all coded systems, four rounds of coded signal are sounded, after which the system is silenced. In all coded systems it is advisable to include a device that records in plain English all alarms, including time of receipt and the code sounded.

(d) Dual-Coded Systems. This arrangement is a combination of noncoded and zone-coded systems. When an alarm device operates, it initiates two separate functions—an identifying coded alarm and a continuous ringing evacuation alarm. The alarms are sounded simultaneously; the coded alarm in the building's maintenance office and the evacuation alarm on separate audible devices throughout the building. A requisite to the application of this system is a continuously manned office in which the coded identifying signal can be received and acted upon (see Fig. 13.36*d*).

(e) Selective-Coded Systems. These are fully coded systems in which all manual devices are coded and all automatic devices are arranged to trip code transmitters at the panel. Each manual station can be immediately identified by its distinctive code. Automatic devices may be grouped in any fashion desired, and annunciated if desired. The combinations and circuitry are entirely in the hands of the designer. In large systems, which fully selective-coded systems usually are, sprinkler transmitters and smoke detectors operate as integral subsystems of the main fire alarm panel (see Figs. 13.36*e* and 13.36*f*).

(f) Presignaling. When it is desired to alert only key personnel, a system called presignaling is used. Small bells or chimes are activated only at their work locations. Since these systems are always selectively coded, the personnel alerted can immediately investigate and, if necessary, manually turn in a general alarm by key operation of a station (see Fig. 13.36*g*). Because of the delay involved, this type of system is used only in buildings where evacuation is difficult and sufficient staff is available to immediately investigate the cause of an alarm.

The specific building system descriptions that follow are based on good present practice. Actual design must meet the requirements of the latest-edition NFPA standards, plus other codes having jurisdiction.

13.16 Residential Fire Alarms

Refer to NFPA 74, *Household Fire Warning Equipment,* and NFPA 101, Chapter 22, for detailed requirements. The system should provide sufficient time for the evacuation of the residents and for appropriate countermeasures to be initiated. The elements of the system are the various alarm-initiating devices, the wiring and control panel, and the audible alarm devices. The most basic system requires single-station (self-contained) *approved* smoke detectors with integral alarm in each sleeping area, at the top of stairs leading to occupied areas, and on each level of the structure (see Fig. 13.37). Some codes permit smoke detectors in existing construction to be battery powered provided the battery is monitored. In new construction, units must be powered by house current to avoid the common problem of dead batteries. The audible device is a bell in or near the sleeping rooms plus an exterior, weatherproof bell to alert neighbors.

An improved system would include the following:

1. Approved smoke detectors in each sleeping area and at the head of each stair, with at least

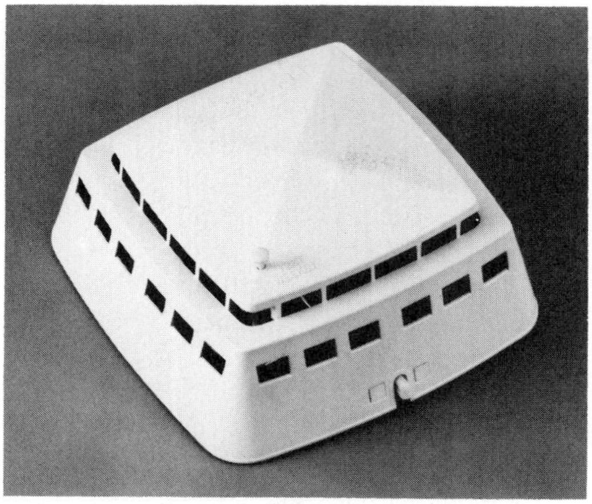

Fig. 13.37 *Typical single-station integral smoke detector. This unit contains a dual-chamber smoke detector, a battery with test switch, and a horn. The unit measures just under 6 in. square and 2 in. deep. Courtesy of Statitrol.*

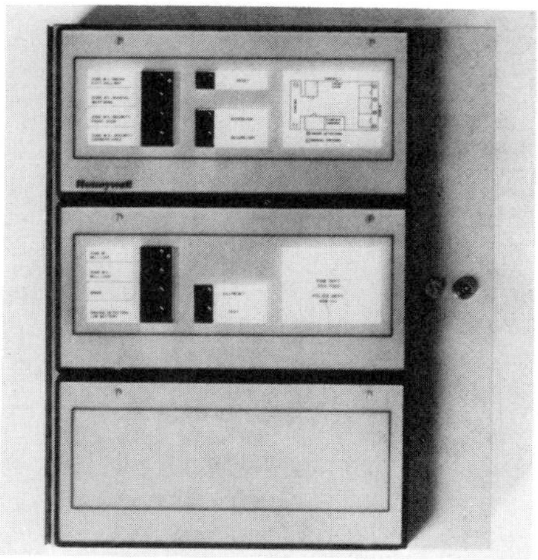

Fig. 13.38 *A modern control panel annunciator unit shows the zone and the specific device that operated. On a plan of the house, a light indicates the location of the tripped device. Such an annunciator can show all the devices and distinguish between types by use of different color lights. Reproduced courtesy of Honeywell Commercial Division.*

one on every level. Combined smoke/heat detectors in boiler room, kitchen, garage. See Fig. 13.25. Heat detectors in the attic, kitchen and boiler room are set at 185F because of high ambient temperatures. Other units are set at 135F.

2. All alarm-initiating devices connected on closed, supervised circuits.

3. Central panel annunciated to show alarmed devices (see Fig. 13.38) and arranged to shut off oil and gas lines and the attic fan (to prevent spread of smoke), and to turn on lights both inside and out. Also, an automatic dialer to ring a neighbor's phone and give a distinctive alarm sound when the phone is answered.

4. Backup power for the system—usually a large dry-cell, although a storage battery with trickle charger is better. Power supply should be supervised.

This arrangement should give the occupants sufficient time to evacuate the house and to minimize property loss.

13.17 Multiple-Dwelling Fire Alarm Systems

See NFPA 101, Chapters 18 and 19. In addition to normal requirements, NFPA 101 lists many special requirements for tall multistory buildings, buildings intended for elderly residents, and structures with fire suppression systems.

(a) Apartment Houses. The basic protection scheme requires an approved single-station smoke/heat detector in each apartment sleeping area, powered by house current. Battery units may not be used since, unlike homeowners, apartment dwellers rely on the building management for all building services, and periodic battery checks and replacement might not be carried out. Large buildings (see Code) require, in addition, a manual system with corridor and lobby manual stations, as well as loud bells in the corridors to signal general evacuation. The system is supervised and noncoded; the control panel is located in an attended location, such as the building superintendent's office. In tall multistory buildings, an annunciator is required, to indicate the floor on which the manual station was operated. A duplicate annunciator in the lobby will direct firefighters to the proper floor. Improvements in the basic protection scheme could include the following:

1. Annunciation of all the single-station apartment detectors, with a duplicate annunciator in the lobby (see Fig. 13.39). Arrangement for the apartment circuit to give local alarm on smoke detection and general alarm with annunciation when a heat detector adjacent to the smoke detector operates; this will minimize the nuisance of general evacuation on false alarms. A light over the apartment door, activated by the apartment alarm, to assist firefighters if full annunciation is not provided.

2. In large apartments, approved detectors in kitchens and in halls outside sleeping areas. All detectors in an apartment connected on a single alarm circuit.

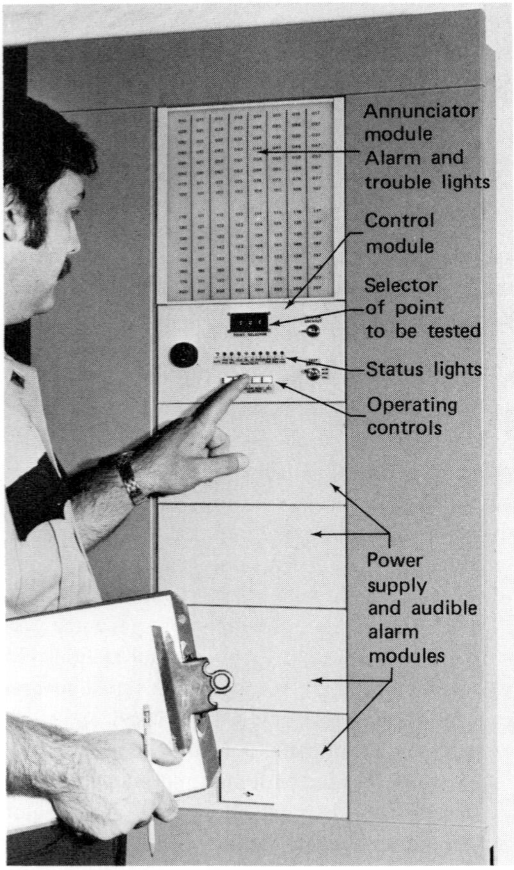

Fig. 13.39 *Control panel for a noncoded general-evacuation alarm-type system with full annunciation. Applicable to schools, multiple dwellings, industrial and commercial buildings. Courtesy of Federal Signal Corp., Autocall Division.*

3. Smoke detection in corridors, service and utility spaces, and storage areas, with anunciation.

4. Standby power to all fire-alarm system circuits. In all installations an auxiliary system is desirable, and in some local codes, one is required. See Section 13.13, and Fig. 13.34*b*.

Dormitories. See NFPA 101 Chapter 16. Dormitories are classed as apartments or hotels for Code purposes, with modification as given there.

Basic protection is provided by a supervised, zoned, noncoded system with continuously ringing bells. Manual noncoded stations are placed at each level in fire stairs and in lobbies. Automatic detectors connected in groups are installed in storage, mechanical equipment, and other unoccupied areas. Smoke detection on each floor is required using single station units in rooms or corridor stations connected to the central fire-alarm system. Bells must be placed in corridors, study rooms, and alcoves and should be of sufficient volume to waken even the soundest sleeper. Since dormitories are designed with soundproofing in mind, to allow for ideal study and sleeping conditions, bells and horns of high sound intensity must be selected. The control panel is generally placed in the mechanical equipment room. If annunciation is used to assist firefighters in locating the cause of trouble, a duplicate annunciator should be placed in the *entrance lobby*. Since most dormitories are built under the aegis of some public agency, the applicable fire codes should be consulted in all cases. An auxiliary system connection is frequently required. An improved system would include the following.

1. Corridor smoke detection.
2. Single-station smoke detectors in each sleeping room, connected into the corridor circuit. Operation of any detector will sound a general evacuation alarm.
3. Annunciation of individual detectors.
4. Standby power for all fire alarm system circuits.
5. In very large buildings, a voice communication system between the lobby and upper floors, and elevator control. (See Section 13.19, on high-rise building systems.)

(c) Hotels. See NFPA 101 Chapter 16 for all requirements.

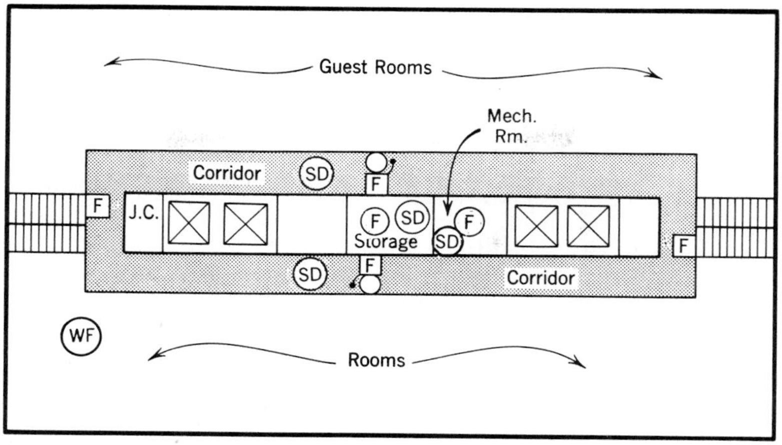

Fig. 13.40 Typical hotel floor plan. Bells are placed so that an alarm will be audible in all parts of the building. Manual stations are normally installed at all points of egress, such as stairwells and main floor exits. A typical location for sprinkler alarm transmitter would be in a janitor's closet (J.C.) or electric closet. Automatic smoke temperature detectors are properly located in storage and mechanical spaces.

A hotel of any magnitude would utilize a supervised dual-coded system with automatic stations in storerooms, kitchen, boiler room, and other unsupervised areas. A presignal system may be used where authorities permit it. Sprinkler transmitters and annunciation are common, as are smoke detectors in the ducts, activating an engineered smoke control system whose primary purpose is to prevent the spread of smoke through the building. Smoke detection in corridors is also normally required. Presignal or coded bells are installed in the office of the building engineer as well as in that of the hotel managers. The fire alarm panel is most often placed in the mechanical equipment area. An auxiliary circuit, usually required by local codes, trips a city fire alarm box or, in large cities, connects to a fire supervision company, which in turn notifies the city fire department. All such connections are rigidly controlled by city ordinances. Figure 13.40 shows a typical hotel floor plan.

An improved system would include the following.

1. Single-station smoke detectors in each room or suite, connected to the central system, and annunciated. In lieu of individual floor annunciation, over-the-door lights and floor annunciation can be used.
2. A voice communication system in high-rise buildings.

13.18 School Fire Alarm Systems

See NFPA 101 Chapter 10 for detailed requirements.

Although personal safety and prevention of property loss combine to form the purpose of all fire alarm systems, the former consideration far outweighs the latter in the instance of school buildings, particularly of the elementary grades. For this reason a general schoolwide alarm causing an immediate evacuation of the premises is the primary requirement of the system. Consideration also must be given to maintaining the uniqueness of the sound of the fire alarm gongs to allow no possibility of confusion with the program gongs, where the latter are used.

The system employed almost universally calls for a closed-circuit, supervised arrangement, noncoded or master-coded in the case of smaller schools and dual-coded in the instance of large or multibuilding institutions. For a multiple-building school, the circuitry is generally arranged to sound an evacuation alarm in the affected building only. The signal also is transmitted to administrative areas in other buildings, and almost always, via auxiliary circuit, to the municipal fire department.

Since regular fire alarm drills are mandatory in all schools, the system circuitry must be arranged to allow for this type of testing. As with other systems, manual stations are placed on each floor at exit points, such as stairways, with automatic

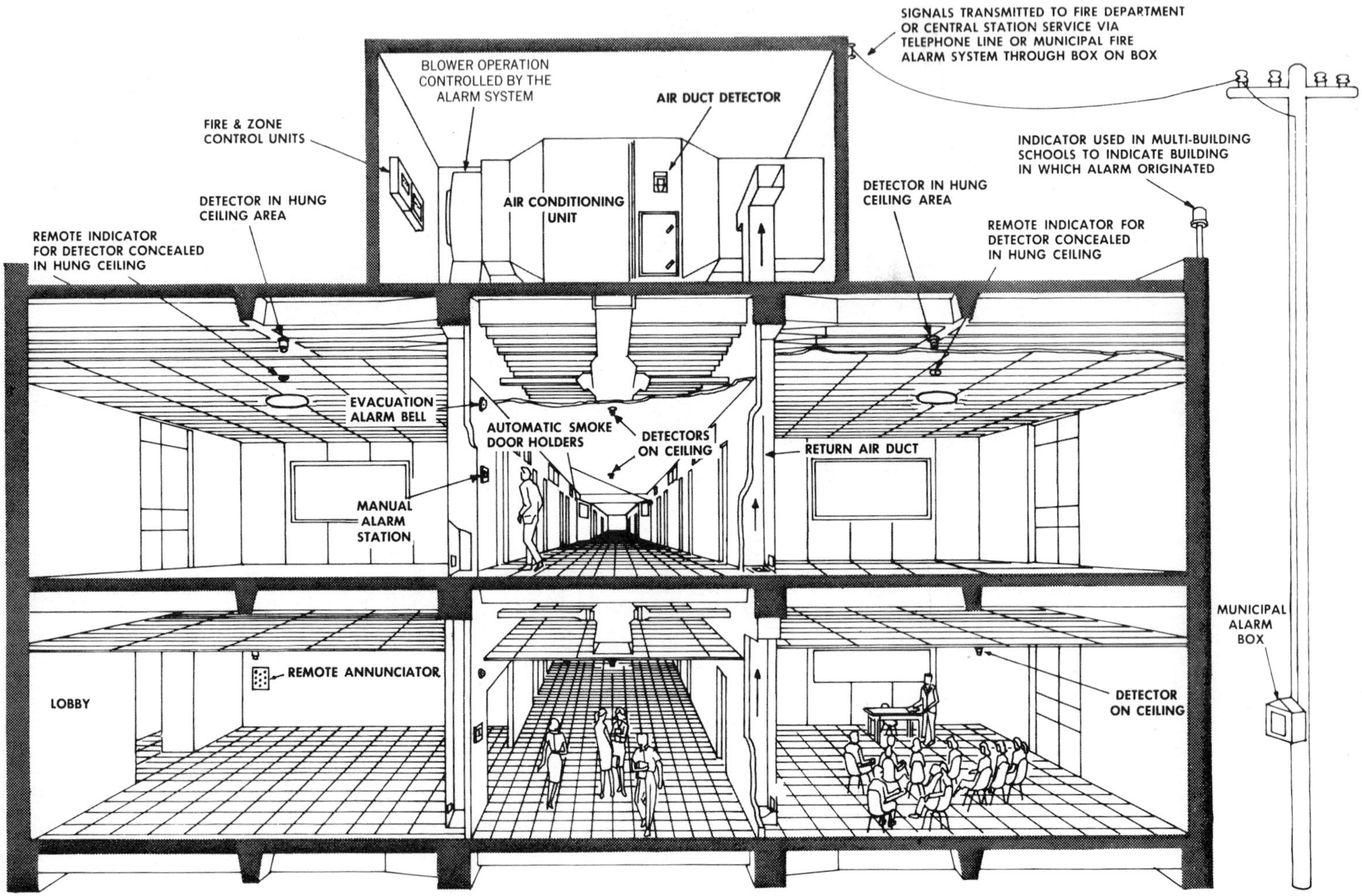

Fig. 13.41 Typical system affording good protection in a school building. Courtesy of Pyrotronics.

stations in the boiler room, kitchen, some laboratories, shop classrooms, and selected storage areas (see Fig. 13.41). It is also advisable to connect any sprinkler flow switches to the alarm system to effect building evacuation while utilizing, in larger schools, a central sprinkler annunciator panel, which will indicate the particular water flow switch involved. This accomplishes the same purpose as coding of fire alarm stations.

Most states have adopted statutes requiring the application of automatic detection in addition to the previously universal requirement for manual systems. In an unoccupied building this will reduce property loss, and in an occupied building it may give the short alarm margin time required to avoid serious injury.

A summary of basic system type recommendation for all but high-rise buildings is given in Table 13.11.

13.19 Office Building Fire Alarm Systems

See NFPA 101, Chapter 26.

(a) Small- to Medium-Size Buildings. Since these buildings are unoccupied for considerable periods of time, fire alarm systems in office buildings normally utilize automatic detection equipment connected to a supervised selectively coded system. Temperature detectors of the combination thermal and rate-of-rise type, plus smoke detectors are located in critical areas unless sprinklers are required. Smoke detection in return-air ducts is often arranged to shut down all building supply and exhaust fans in order to avoid both feeding the fire and spreading smoke. A fan restart panel at an accessible location near the lobby should then be provided for selective restart of the *exhaust* fans only. Since this arrangement of fan shutdown and restart is not universal, it is well to consult local fire authorities and to follow their recommendations. Coded manual stations are required at egress points, plus a lobby station. Additional circuits can be arranged to release smoke control doors.

Water flow switches on the sprinkler system and smoke detectors in their system should run to individual annunciators, which in turn connect to the fire alarm control panel. This panel then may be auxiliarized for outside supervision if the building is unattended for periods of time. Where permitted by codes, presignal systems may be used in fully attended buildings to avoid unnecessary building evacuation and the risk of confusion and even panic, since building tenants often, and transient occupants always, are unfamiliar with the alarms (see Table 13.11). This is particularly important in public-type structures, where most of the occupants are transient. To avoid the possibility of the presignal going uninvestigated, a timing arrangement is possible that will turn in a general alarm if the alarm device causing the signal is not reset within a predetermined time.

(b) Large and High-Rise Office Buildings. Sad experience has demonstrated that high-rise buildings, once thought to be "fireproof," are emphatically not so. Indeed, due to their size, they have particularly severe fire protection problems, one of which is reliable communications during fire emergencies. As a result, a fire-alarm-and-communications system has appeared that is specifically intended for high-rise building use. In addition to the usual fire alarm system functions, this type of system can provide, from the lobby fire command post,

1. Two-way communication with at least one station per floor, all mechanical equipment rooms, elevator machine rooms and air-handling (fan) rooms.
2. Control of alarm signals and sensors.
3. Visual display of alarms, by floor location.
4. Selective and group control of all audible devices in the building.
5. Communcation with the fire department or central station.
6. Control of all air-handling units in the building and all smoke control devices.
7. Elevator capture and emergency control.
8. Public address throughout the building.
9. Testing and supervision of all circuits and devices.

The exact equipment supplied depends on the building and the local fire code. The outstanding characteristics of this system are the communications system, the visual display of alarm locations, and the remote control of air-handling equipment. The unit shown in Fig. 13.42 provides complete

TABLE 13.11 **Fire Alarm System Recommendations**

Before establishing design, consult specific NFPA and other jurisdictional code requirements.

	Schools[b]	*Hospitals & Nursing Homes*[b]	*Colleges*[b] *(For Dorms, see Multiple Dwellings)*	*Industrial*[d]	*Commercial*	*Libraries & Record Storage*[b]	*Multiple Dwellings*[b] *(Dorms, Hotels, Motels, Apts.)*
Single-story building under 20,000 sq ft	Noncoded, annunciated general alarm system[a]	Coded, annunciated general alarm system	Noncoded, annunciated general alarm system	Noncoded, annunciated general alarm system	Noncoded, annunciated general alarm system[a]	Noncoded, annunciated general alarm system	Noncoded, annunciated general alarm system
Single-story building over 20,000 sq ft	Dual-coded, annunciated general alarm system[a]	Dual-coded, annunciated general alarm system	Coded, annunciated general alarm system	Coded, annunciated general alarm system	Coded, annunciated presignal alarm system[c]	Coded, annunciated general alarm system	Coded, annunciated presignal alarm system[c]
Multistory building	Dual-coded, annunciated general alarm system[a]	Dual-coded, annunciated general alarm system	Coded, annunciated general alarm system	Coded, annunciated general alarm system	Coded, annunciated presignal alarm system[c]	Coded, annunciated general alarm system	Coded, annunciated presignal alarm system[c]

NOTE: Based on recommendations of Autocall, Federal Signal, with emendations by the author.

[a]A common-coded, annunciated general alarm system may be desirable in these occupancies.

[b]Complete smoke detection of area or entire building is recommended for these applications.

[c]The use of presignal system (where codes permit) presupposes that trained personnel are awake and on duty at one location 24 hours a day. If this is not the case, use noncoded, annunciated general alarm in multiple dwellings and coded, annunciated general alarms in commercial buildings.

[d]Special attention should be given to the selection of signals because of the high ambient noise usually present in these areas.

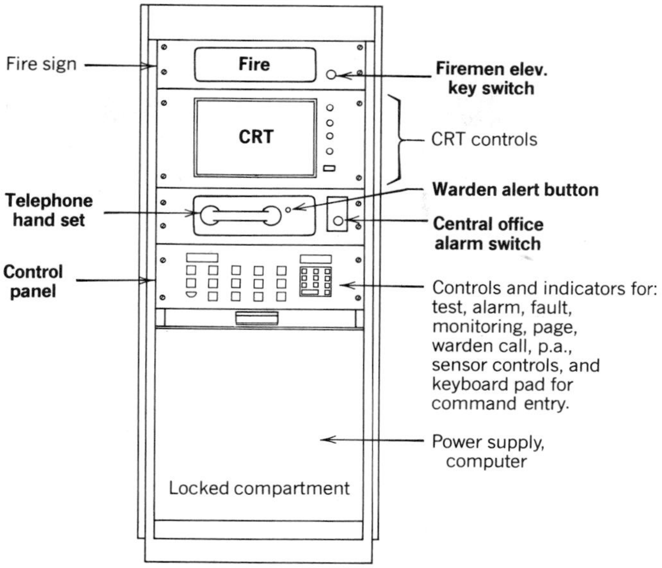

Fig. 13.42 Front view of a computerized data acquisition and control center, which provides security system supervision and environmental monitoring, in addition to high-rise fire system control. The telephone handset provides communication with the fire stations on all floors. The condition of any device can be displayed on the CRT. All data is run via a single 11-conductor, multiplex cable. In addition to fire alarm communication, the unit provides elevator capture and control, control of all fans, vents, dampers, and smoke doors, and supervision of all fire devices and circuits. Other functions include control and monitoring of up to 2000 "addresses" in the building, including other building security systems. This unit is 2 ft square and 5 ft high. Courtesy of Multiplex Electrical Services.

building system supervision and control in addition to a very sophisticated fire alarm and control system.

13.20 Industrial Building Fire Alarm Systems

Industrial building systems are normally selective-coded and fully supervised (refer to Table 13.11). Presignaling is utilized in structures where for any reason an evacuation alarm is undesirable. In addition to manual stations at points of egress, the following devices may be used.

1. Temperature and smoke detectors in storage areas and laboratories.
2. Smoke and flame detectors in record rooms and continuous process laboratories.
3. Water flow switches on all sprinklers.

The annunciators, control panel, and alarm register are best placed in the guardroom. If none is available, an auxiliary or remote-station circuit should be added to allow remote monitoring.

Because of the high ambient noise level in many plants, horns are substituted in such areas for bells and gongs, which might be inaudible. (See Figs. 22.11 through 22.13.) If the building is sufficiently large, a computer-microprocessor-controlled system of the type shown in Fig. 13.42 (and Fig. 22.11) can be used. As there, the unit can provide other control/monitoring functions as required.

In summary, the specific occupancy recommendations given in Sections 13.16 through 13.20 are representative of good practice at the present time. However, codes are constantly being changed in this particularly sensitive field, and they vary with locale. In actual design situations, all codes and regulations having jurisdiction must be complied with.

PART V

ELECTRICITY

Electricity is the most prevalent form of energy in a modern building. It supplies not only electric outlets and electric lighting, but also the motive power for ventilation, heating, and cooling equipment; traction power for elevators and material transport; and power for all signal and communications equipment. An electric power failure can paralyze a facility. If properly designed, the facility will return to partial functioning by virtue of emergency equipment that will furnish some part of the facility's electricity needs for a limited time.

Given this complete dependence on electric power for normal functioning, it is apparent that designers must be familiar with normal electrical systems. Chapter 14 reviews basic relationships in electric circuits with emphasis on power, energy, energy costs, and methods of energy managements and electric load control. Chapter 15 introduces the reader to the concept and basis of electric equipment ratings and capacity, and continues with a description of modern wiring systems and their components. Chapter 16, essentially a continuation of the previous chapter, describes service, utilization and emergency/standby power equipment. Also discussed are energy conservation considerations and economic factors. Chapter 17 draws on information given in the three preceding chapters to demonstrate straightforward design methods for building electrical systems.

14

PRINCIPLES OF ELECTRICITY

Historically, usable energy was most often produced by burning a natural fuel such as coal or oil. The resultant heat energy was used directly as heat and light or converted by machines into other desired forms of energy such as motion. Only within the last century, however, has this heat in turn been used to create another form of usable energy—electricity. Even the recent partial substitution of nuclear for fossil fuels has affected only the heat production portion of this process. Beyond that point, the heat is utilized in the same manner to drive generators that produce electricity. It is well to remember, therefore, that in terms of natural resources electricity is an expensive form of energy, since the efficiency of heat-to-electricity conversion, on a commercial scale, rarely exceeds 40%.

14.1 Electric Energy

Electricity constitutes a form of energy itself that occurs naturally only in unusable forms such as lightning and other static discharges or in the natural galvanic cells, which cause corrosion. The primary problem in the utilization of electric energy is that, unlike fuels or even heat, it cannot be stored and therefore must be generated and utilized in the same instant. This requires an entirely different concept of utilization than, for example, a heating system with its burner, piping, and associated equipment.

The bulk of electric energy utilized today is in the form of alternating current (a-c), produced by a-c generators, commonly called alternators. Direct-current (d-c) generators are utilized for special applications requiring large quantities of d-c. In the building field such a requirement is found in elevator work. Smaller quantities of d-c, furnished either by batteries or by rectifiers, are utilized for telephone and signal equipment, controls, and other specialized uses.

14.2 Unit of Electric Current: The Ampere

Electricity flowing in a conductor is called current, or amperage, and is abbreviated *amp, amps,* or simply ''A.'' When current is used in an equation, it is usually represented by the letter I or i.

It is convenient to establish an analogy between electric systems and mechanical systems as an aid to comprehension. Current is a measure of flow and, as such, would correspond to water flow in a hydraulic system (see Fig. 14.1). The correspondence is not complete, however, since in the hydraulic system the velocity of water flow varies, whereas in the electric system the velocity of propagation is constant and may be considered instantaneous; hence the need to utilize the electric energy the instant it is produced.

14.3 Unit of Electric Potential: The Volt

The electron movement and its concomitant energy, which constitutes electricity, is caused by

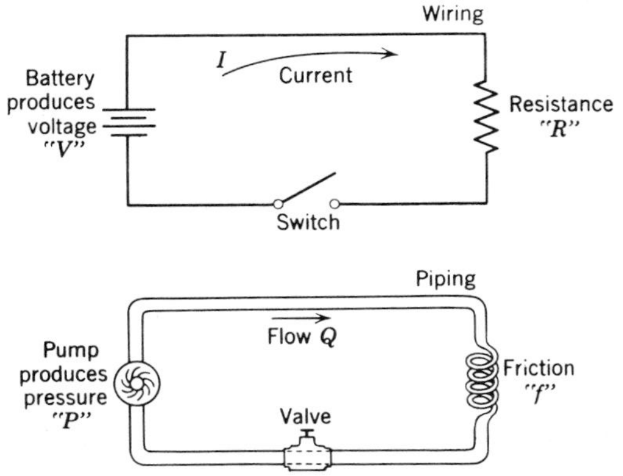

Fig. 14.1 Electric-hydraulic analogy. The circuits show that voltage is analogous to pressure, current to flow, friction to resistance, wire to piping, and switches to valves.

creating a higher positive electric charge at one point on a conductor than exists at another point on that same conductor. This difference in charge can be created in a number of ways. The oldest and simplest method is by electrochemical action, as in the battery. In the ordinary dry cell, or in a storage battery, chemical action causes positive charges (+) to collect on the positive terminal and electrons or negative charges (−) to collect on the negative terminal. Assume for the moment that nothing is connected to the battery terminals. There is a tendency to flow between the electrified particles concentrated at the positive and negative terminals. *Potential difference* or *voltage* is the name given to this force. This force is analogous to pressure in a hydraulic or pneumatic system. Just as the pressure produced by a pump or blower causes water or air to flow in a connecting pipe, so too the potential (voltage) produced by a battery (or generator) causes current to flow when the terminals between which a voltage exists are connected by a conductor. The higher the voltage (pressure), the higher the current (flow) for a given resistance (friction) (see Fig. 14.2). Other means of producing voltage, both direct (d-c) and alternating (a-c), are discussed in Section 14.8. The unit of voltage is the volt, abbreviated "V."

14.4 Unit of Electric Resistance: The Ohm

The flow of fluid in a hydraulic system is resisted by friction; the flow of current in an electric circuit

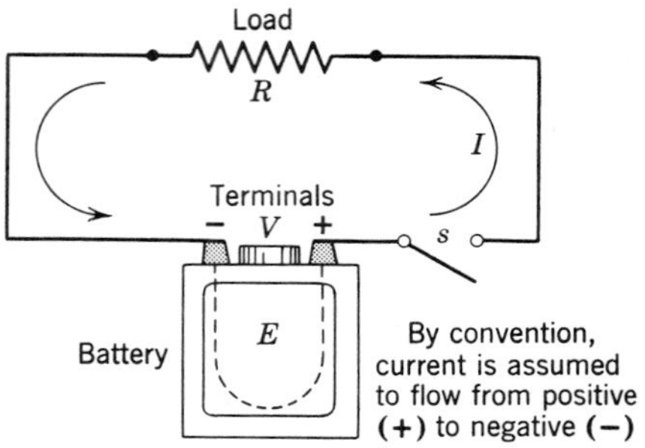

Fig. 14.2 Current flows in the electric circuit as a result of the voltage (potential difference) V *that exists between the terminals of the battery. By convention, current is shown from positive (+) to negative (−) in the circuit (and from − to + inside the battery).*

is resisted by *resistance,* which is the electrical term for friction. In a d-c circuit this unit is called resistance and is abbreviated R; in an a-c circuit it is called impedance and is abbreviated Z. The unit of measurement is the *ohm*. (It is interesting to note that the scientific convention of naming units after persons whose work is closely related to the field is here too followed. Thus the ampere, volt, and ohm are derived from André Ampere, Alessandro Volta, and Georg Ohm).

Materials display different resistance to the flow of electric current. Metals generally have the least resistance and are therefore called conductors. The best conductors are the precious metals—silver, gold, and platinum—with copper and aluminum only slightly inferior. Conversely, materials that resist the flow of current are called insulators. Glass, mica, rubber, oil, distilled water, porcelain, and certain synthetics such as phenolic compounds exhibit this insulating property and are therefore used to insulate electric conductors. Common examples are rubber and plastic wire coverings, porcelain lamp sockets, and oil-immersed switches.

14.5 Ohm's Law

The current I that will flow in a d-c circuit is directly proportional to the voltage V and inversely proportional to the resistance R of the circuit. Expressed as an equation, we have the basic form of Ohm's law:

$$I = \frac{V}{R} \tag{14.1}$$

In a-c circuits, the same relation holds true except that instead of d-c resistance we use the a-c impedance. Ohm's law is frequently written in another form, that is

$$V = IR \tag{14.2}$$

which states the mathematical relationship that volts = currents × resistance. This form has no logical basis; therefore, we recommend that the reader remember the form of equation 14.1, which clearly states the physical situation; that is, as a result of voltage V, a current I is produced that is proportional to the electric pressure V, and inversely proportional to the electric friction R.

An example will illustrate the application of

Ohm's law. The example chosen is applicable to both a-c and d-c, since the device is purely resistive. This will be more fully explained in the subsequent discussion on alternating current.

EXAMPLE 14.1. An incandescent lamp having a hot resistance of 66 ohms is put into a socket that is connected to a 115-V supply. What current flows through the lamp?

SOLUTION

$$I = \frac{V}{R} = \frac{115}{66} = 1.74 \text{ A}$$

(These figures correspond to a normal 200-W general service incandescent lamp.)

We mentioned hot resistance in the above example since some materials exhibit higher resistance when hot than when cold. A typical example of this is the tungsten-filament lamp that when first turned on takes, for a fraction of a second, 10 to 15 times the current that flows when the filament is hot.

14.6 Circuit Arrangements

The two basic electric circuit arrangements are *series* and *parallel*. These concepts are the same for both d-c and a-c. As above, we will use purely resistive circuits so that circuit calculations will be applicable to both d-c and a-c. In other than purely resistive circuits, a-c circuit calculations are different, and much more complicated, than their d-c counterparts.

(a) Series Circuits. In a series arrangement the elements are connected one after the other—in series. Thus, resistances and voltages add. This is indicated graphically in Fig. 14.3*a*. An electric circuit may be defined as a complete conducting path that carries current from a source of electricity to and through some electrical device (or load) and back to the source. A current can never flow unless there is a complete (closed) circuit. It should be obvious that because of the arrangement of the components, *the current is the same in all parts of the circuit*. A somewhat more complicated circuit is shown in Fig. 14.3*b* and is analyzed in Example 14.2.

It is customary to refer to connection points on such wiring diagrams by letters as *a, b, c, d,* and so on. The battery voltage may then be called V_{ab} = 12 V; the voltage across the load resistance, V_{cd} = 11.5 V; the resistance of the two wires $r_{bc} + r_{da}$ = 0.04 ohm. The positive and negative terminals of the battery are shown.

EXAMPLE 14.2. The battery in Fig. 14.3*b* is rated at 12 V, the line resistance (both wires) is 0.04 ohm, the battery internal resistance is 0.01 ohm,

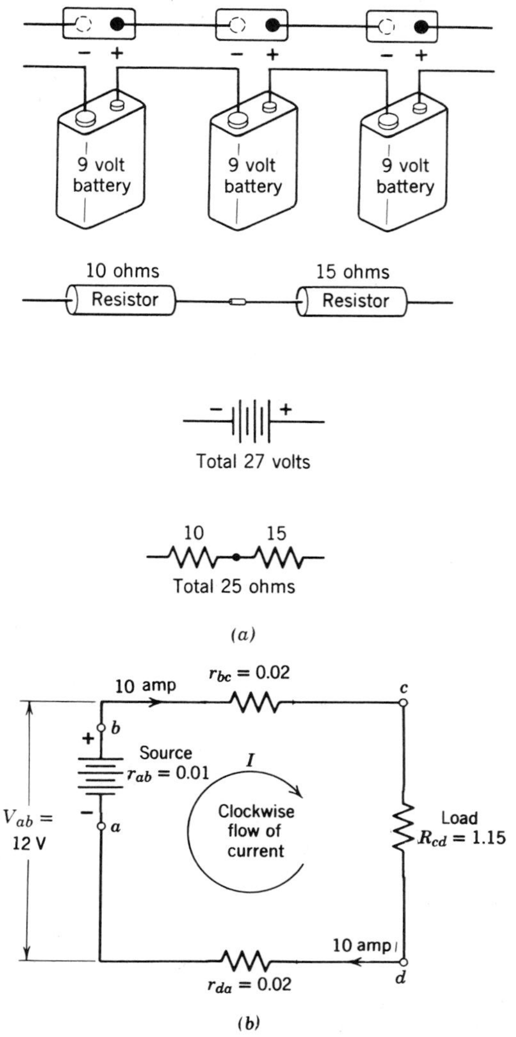

Fig. 14.3 (a) *Physical and graphic representation of series connection of batteries and resistors.* (b) *A simple series circuit.*

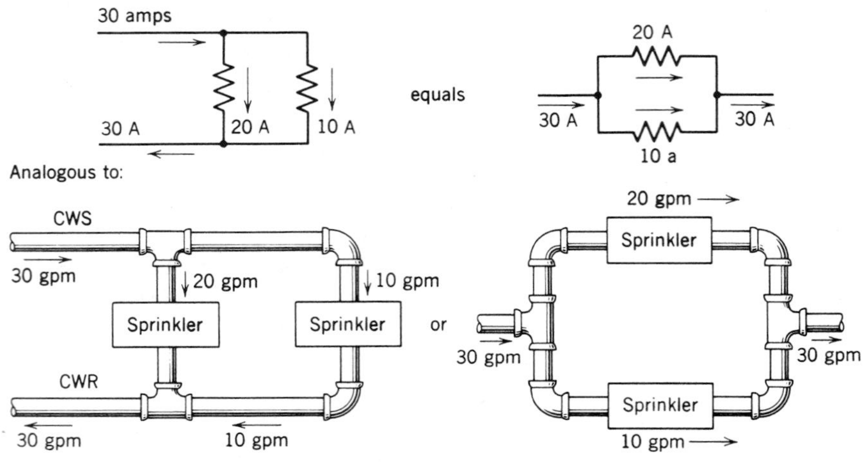

(a)

and the load resistance is 1.15 ohms. Determine (a) current flowing in the circuit, (b) the voltage across the load (V_{cd}).

SOLUTION

(a) The current flowing is

$$I = \frac{V}{R} = \frac{V_{ab}}{r_{ab} + r_{bc} + r_{cd} + r_{da}}$$

$$= \frac{12}{0.01 + 0.02 + 1.15 + 0.02} = \frac{12}{1.2}$$

$$= 10 \text{ A}$$

(b) The voltage drop across the load is

$$V_{cd} = I \times R_{cd} = 10 \times 1.15 = 11.5 \text{ V}$$

Series circuits have very limited application in building wiring.

(b) Parallel Circuits. When two or more branches or loads in a circuit are connected between the same two points, they are said to be connected in *parallel* or *multiple*. Such an arrangement and its hydraulic equivalent are shown in Fig. 14.4*a*. From the circuit of Fig. 14.4*b* it should be apparent that multiple loads are across the same voltage and, in effect, constitute separate circuits. From this we conclude that in the multiple arrangement, the total current in the circuit is the sum of the individual currents flowing in the branches, that is

$$I_T = I_1 + I_2 + I_3$$

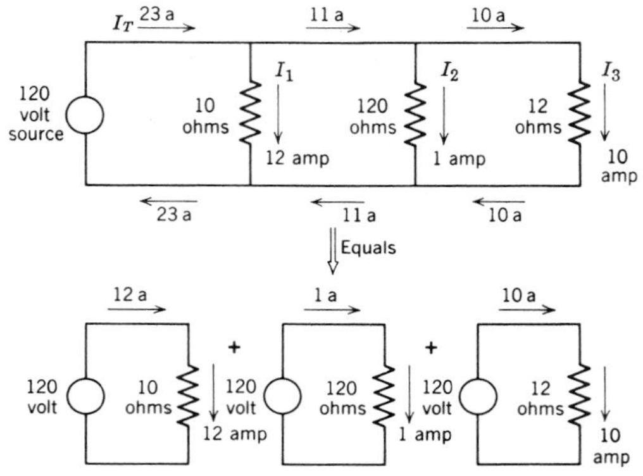

(b)

Fig. 14.4 (a) *In a parallel connection the flow divides between the branches, but the pressure is the same in each branch.* (b) *Note that loads connected in parallel are equivalent to separate circuits superimposed into a single connection. Each load acts as an independent circuit unrelated to, and unaffected by, the other circuits.*

Refer to Fig. 14.4*b*. Notice from the numbers shown there that the total current flowing in the circuit is the sum of all the branches, but that the current in each branch is the result of a separate Ohm's law calculation. Thus, in the 10-ohm load a 12-A current flows, and so forth.

The parallel connection is the standard arrangement in all building wiring. A typical lighting and receptacle arrangement for a large room is shown in Fig. 14.5. Here the lights constitute one parallel grouping, and the convenience wall outlets con-

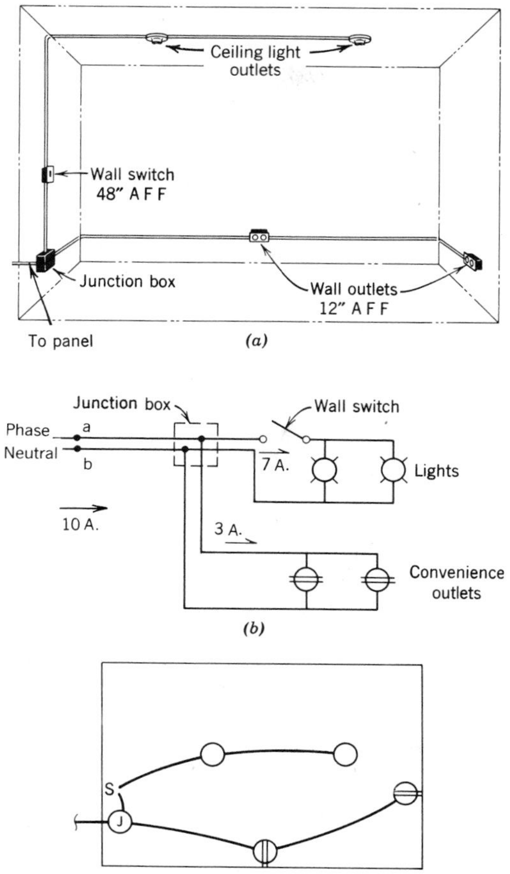

Fig. 14.5 Parallel groupings of ceiling light outlets and wall outlets are in turn connected in parallel to each other. A circuit is shown (a) *pictorially,* (b) *schematically, and* (c) *as on an architectural plan.*

stitute a second parallel grouping. The fundamental principle to remember is that loads in parallel are additive for current, and that each has the same voltage imposed.

One additional point is important to appreciate. Remember, from Ohm's law, that current is inversely proportional to resistance. Thus as resistance drops, current rises. Now look at the circuit of Fig. 14.5. Normally that circuit will carry 10 amperes. But if by some mischance, such as deterioration of the wiring insulation, a connection appears between points *a* and *b*, the circuit is *shortened* so that there is no resistance in the circuit. The current rises instantly to a very high level, and we have a *short circuit*. If the circuit is properly protected, the fuse or circuit breaker will open, and the circuit will be de-energized. If not, excessive current will probably start a fire.

14.7 Direct Current and Alternating Current (d-c and a-c)

Whenever the flow of electric current takes place at a constant time rate, practically unvarying and in the same direction around the circuit, it is called d-c. The d-c voltages of 1.5 V positive polarity, and 1.0 V negative polarity are shown in Fig. 14.6*a*.

Whenever the flow of current is periodically varying in time and in direction, as indicated by the symmetrical positive and negative loops or sine

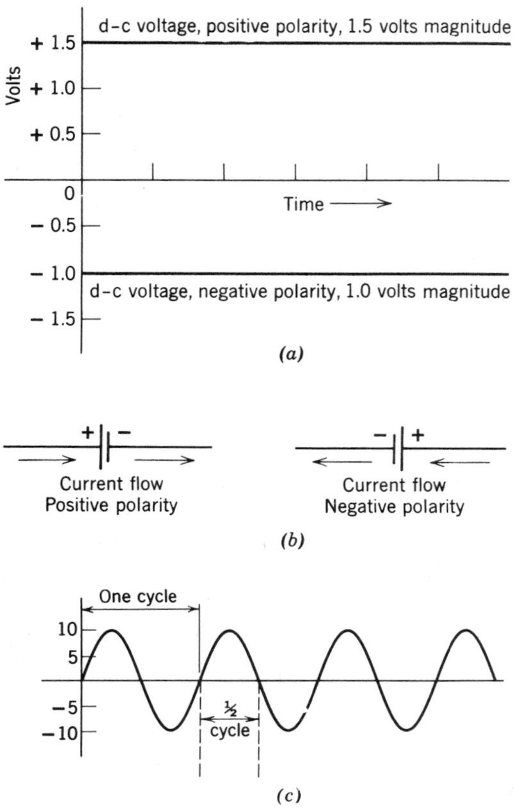

Fig. 14.6 (a) *Graphic representation of d-c voltages with positive and negative polarity.* (b) *Circuit symbol representation of a battery source. The longer bar is positive.* (c) *Alternating current (a-c).*

waves in Fig. 14.6*c*, it is called an alternating current. The distance along the time axis spanned by a positive and a negative a-c loop is called one cycle. The number of such cycles occurring in one second is known as the *frequency* of the a-c current. Modern a-c systems in the United States operate at a frequency of 60 cycles per second, or 60 *hertz* (after H. Hertz). This means that current at 60 hertz (abbreviated Hz) is delivered to the consumer.

The a-c circuits differ from d-c circuits in a number of important respects and, since normal current supply is 60 Hz a-c, it is important to understand a-c terminology and usage. Instead of resistance the corresponding parameter in an a-c circuit is impedance, which is also measured in ohms. Depending on the device, it can be markedly different from the d-c resistance. For an a-c circuit, Ohm's law is

$$I = \frac{V}{Z} \tag{14.3}$$

where Z is the symbol for impedance. We will not go into a-c circuit calculations primarily because such calculations are not especially useful to the reader. What *is* useful and important is an understanding of power and energy in both d-c and a-c circuits. This is discussed in Section 14.9.

14.8 Electric Power Generation

(a) Direct Current. With respect to generation of large amounts of power, photoelectric, piezoelectric, and thermoelectric effects including solar cells can be ignored, leaving the battery and the d-c generator as the sources of d-c current. Since the d-c generator is in reality an a-c generator with a device (commutator) attached that rectifies the a-c to d-c, the battery is still the only major direct source of direct current. (There are some special types of generators that produce direct current *directly,* but their use to date has been extremely limited.) A discussion of the application of batteries for emergency and standby power supply can be found in Section 16.29.

Another source of d-c power is rectification of a-c that can be accomplished on any desired scale to provide as much d-c power as there is available a-c power. The principal application of d-c in buildings is for elevator power. Small amounts of d-c are also used for controls and telephones.

(b) Alternating Current. Alternating current is produced commercially by an a-c generator, generally called an alternator. The prime mover may be any type of engine or turbine. The process by which electricity is produced is illustrated in Fig. 14.7. It is based on the fundamental discovery in 1831 by Michael Faraday of the principle of electromagnetic induction. Put briefly, this principle states that when an electrical conductor is moved in a magnetic field, a voltage is induced

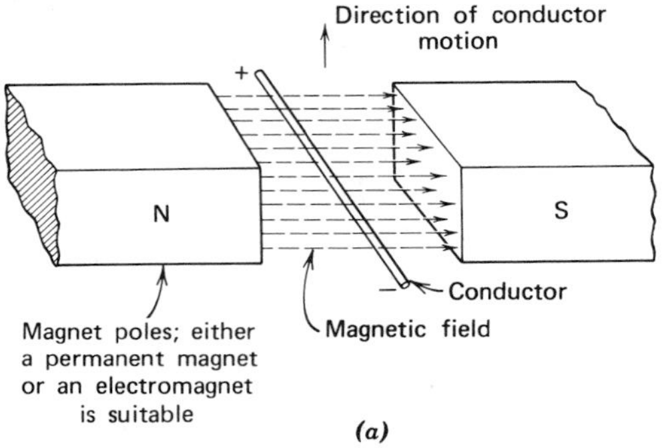

(a)

Fig. 14.7 (a) *The action fundamental to all generators is illustrated here. When a conductor of electricity moves through a magnetic field, a voltage is produced in the conductor, with polarity as shown.*

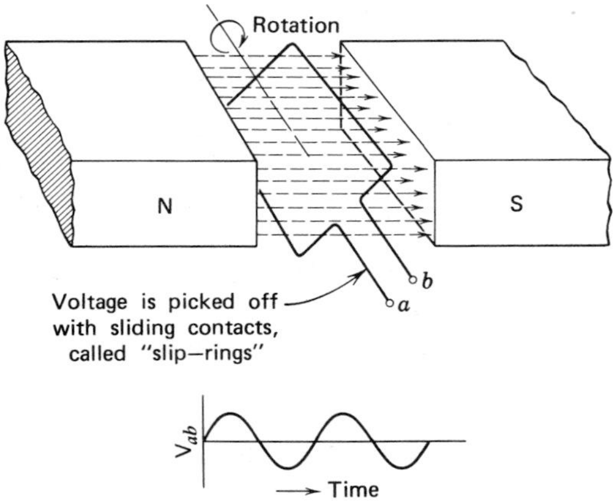

(b)

Fig. 14.7 (b) *Rotating a coil in a magnetic field produces an alternating sinusoidal voltage at terminals* a b *because of the alternating polarity (see Fig. 14.7*a*).*

in it (see Fig. 14.7*a*). The direction of movement determines the polarity of the induced voltage, as shown. If the conductor is formed into a coil and rotated in the magnetic field, a voltage of alternating polarity is produced, that is, alternating current. It does not matter whether the conductor or the magnet moves; the motion of the conductor and the field with respect to each other produces the voltage (see Fig. 14.7*b*). It is only one step (of development) from this rudimentary a-c generator to the large, powerful alternator that produces a-c in a modern power plant. The frequency of the voltage generated is a function of the machine design (number of poles) and the speed at which it is driven. Normal generator frequency in the United States is 60 Hz, whereas in Europe and the Mid-East it is 50 Hz.

14.9 Power and Energy

It is important, indeed imperative, that the distinction between power and energy be clearly understood, since all too frequently the terms are incorrectly used interchangeably. *Energy* is the technical term for the more common expression—work. *Power* is the rate at which energy is used. In terms of power, energy is the product of power and time, that is

$$\text{energy or work} = \text{power} \times \text{time} \qquad (14.4)$$

In practical terms, energy is synonymous with fuel and therefore also cost. Thus energy can be expressed as barrels (tons) of oil, cubic feet (cubic meters) of gas, tons of coal, kilowatt hours of electricity usage, and dollars of operating cost. The concept of energy efficiency of structures can be stated in terms of annual usage of oil, gas, and electricity or alternatively in terms of dollars of total fuel cost. In technical terms, energy is expressed in units of Btu (calories), foot-pounds (joules), and kilowatt-hours.

Power is the rate at which energy is used or, alternatively, the rate at which work is done, since energy and work are synonymous. The term *power* implies continuity, that is, the use of energy at a particular rate, over a given, generally considerable, span of time. The concept of power necessarily involves the factor of time since it is, as stated above, the *rate* at which work is done. Thus multiplying power by time yields energy. Typical units of power in the English system are horsepower, Btu per hour, watt, and kilowatt. In the metric or SI system the corresponding units are joule per second, calorie per second, watt, and kilowatt. In physical terms, power is also the rate at which fuel (energy) is used. Thus power can also be expressed as gallons (liters) of oil per hour, cubic feet (cubic meters) of gas per minute, and tons of coal per day.

14.10 Power in Electric Circuits

The unit of electric power is the watt (W). A larger unit of 1000 watts is the kilowatt (kW). The power input in watts to any electrical device having a resistance R and in which the current is I is given by the equation

$$W = I^2R = I(IR) \qquad (14.5)$$

where W is wattage. This is true for both d-c and a-c circuits. However, since the resistance of an item is generally not known, but the circuit voltage and current *are* known, it would be preferable to be able to calculate power using these two quantities. This can be done, but is different for d-c and a-c.

In d-c circuits

$$\text{by Ohm's law } V = IR$$

$$\text{and, since } W = I(IR),$$

$$W = VI \qquad (14.6)$$

where W is in watts, R in ohms, I in amperes, and V in volts.

In a-c circuits impedance is comprised of resistance and reactance (a-c resistance of inductance and capacitance) and causes a phase difference between voltage and current. This phase difference is represented by an angle, the cosine of which is called the power factor, abbreviated "*pf*." This quantity is extremely important in that it enables us to calculate power in an a-c circuit. The a-c power equation is similar to that for d-c (see equation 14.6) with the addition of this special a-c term of power factor, that is

$$W = VI \times \text{pf} \qquad (14.7)$$

If power factor is not used in the equation, the

product of voltage and current gives a quantity known as volt-amperes. In a purely resistive circuit, such as one with only electric heating elements, impedance equals resistance, power factor equals 1.0, and wattage equals volt-amperage. A few examples here should make applications of these equations clear.

EXAMPLE 14.3. Referring back to Example 14.1 in Section 14.5, calculate the power drawn using equations 14.5, 14.6, and 14.7.

SOLUTION. Since an incandescent lamp is purely resistive and therefore has unity (1.0) power factor, it does not matter whether the circuit is a-c or d-c.

From Example 14.1, $R = 66$ ohms, $I = 1.74$ A, and $V = 115$ V

1. In a d-c circuit, we would use equation 14.6

$$W = VI = 115(1.74) = 200 \text{ W}$$

2. In an a-c circuit, we would use equation 14.7

$$W = VI \times \text{pf} = 115 \times 1.74 \times 1.0 = 200 \text{ W}$$

3. In either a d-c or an a-c circuit, we can use equation 14.5

$$W = I^2R = (1.74)^2 \times 66 = 200 \text{ W}$$

EXAMPLE 14.4. Using the data given in Example 14.2 and Fig. 14.3*b*, determine (a) the power lost in the wiring and (b) the power input to the load.

SOLUTION

(a) The total line loss is

$$W = I^2R = I^2(r_{bc} + r_{da}) = (10)^2 \times 0.04 = 4 \text{ W}$$

(b) The power input to the load is

$$W = I^2R = I^2R_{cd} = (10)^2 \times 1.15 = 115 \text{ W or } 0.115 \text{ kW}$$

Alternatively, we can find this power by multiplying voltage and current. The voltage on the load is

$$IR = 10 \times 1.15 = 11.5 \text{ V}$$

and

$$W = VI = 11.5 \times 10 = 115 \text{ W} = 0.115 \text{ kW}$$

EXAMPLE 14.5. Refer to Fig. 14.5. Assume a 150-W incandescent lamp at each ceiling outlet. Also assume the load connected to one convenience outlet to be a 10-A hair dryer and blower, with a power factor of 0.80. Calculate the current and power in the two branches of the circuit, and the total circuit current, assuming a 120-V a-c source.

SOLUTION

(a) In the circuit branch feeding the lights we have (since power factor is unity)

$$\text{Power} = VI$$

$$300 \text{ W} = 120 \text{ V} \times I$$

$$I = \frac{300}{120} = 2.5 \text{ A}$$

(b) In the second branch we have a 10-A, 0.8 pf load.

Power in watts = volts × amperes × power factor

$$W = 120 \times 10 \times 0.8 = 960 \text{ W}$$

But the circuit volt-amperes are

$$\text{V-A} = 120 \times 10 = 1200 \text{ V-A}$$

This latter figure is important when sizing equipment.

(c) To calculate the total current flowing from the panel to both branches of the circuit, we must combine a purely resistive current (lamp circuit) with a reactive one (dryer circuit). The exact value of current is the square root of the sum of the squares of the two branch currents. However, as an approximation, the currents may simply be added arithmetically. This yields a result that is higher than actual and is, therefore, on the safe side when sizing equipment. Hence,

approximate total current = $2.5 + 10 = 12.5$ A

Actual current is 12.1 A; our error in approximating is 3.2%, which is acceptable for most uses.

One further example at this point will demonstrate the importance of power factor in normal situations.

EXAMPLE 14.6 The nameplate of a motor shows

the following data: 3 hp, 240 V, a-c, 17 A. Assume an efficiency of 90%. Calculate the motor (and, therefore, circuit) power factor.

SOLUTION

$$1 \text{ hp} = 746 \text{ W}$$

Therefore

$$3 \text{ hp} = 3 \times 746 = 2238 \text{ W output}$$

$$\text{efficiency} = \frac{\text{output}}{\text{input}}$$

so

$$\text{power input} = \frac{2238 \text{ W}}{0.9} = 2487 \text{ W}$$

But for a-c,

$$\text{power} = \text{volts} \times \text{amperes} \times \text{power factor}$$

so

$$2487 = 240 \text{ V} \times 17 \text{ A} \times \text{power factor}$$

and

$$\text{power factor} = \frac{2487}{240 \times 17} = 0.61$$

Note the large difference between volt-amperes and watts.

$$VI = 240 \times 17 = 4080 \text{ V-A}$$

$$P = \text{as above} = 2487 \text{ W}$$

Where P designates power.

14.11 Energy in Electric Circuits

Since energy = power × time, the amount of energy used is directly proportional to the power of the system and to the length of time it is in operation. Since power is expressed in watts or kilowatts, and time in hours (seconds and minutes are too small for our use), we have for units of energy: watt-hours (Wh) or kilowatt-hours (kWh). Obviously, one watt-hour equals one watt of power in use for one hour, and one kilowatt-hour equals one kilowatt in use for one hour.

EXAMPLE 14.7

(a) Find the daily energy consumption of the appliances listed if they are used daily for the amount of time shown.

Toaster (1340 W)	15 min
Percolator (500 W)	2 h
Fryer (1560 W)	½ h
Iron (1400 W)	½ h

(b) If the average cost of energy is $0.06 per kilowatt-hour, find the daily operating cost.

SOLUTION

(a) Toaster $1340 \text{ W} = 1.34 \text{ kW} \times \frac{1}{4} \text{ h} = 0.335 \text{ kWh}$

Percolator $500 \text{ W} = 0.5 \text{ kW} \times 2 \text{ h} = 1.00 \text{ kWh}$

Fryer $1560 \text{ W} = 1.56 \text{ kW} \times \frac{1}{2} \text{ h} = 0.78 \text{ kWh}$

Iron $1400 \text{ W} = 1.4 \text{ kW} \times \frac{1}{2} \text{ h} = 0.70 \text{ kWh}$

Total 2.815 kWh

(b) The cost is

$$2.815 \text{ kWh} \times \$0.06/\text{kWh} = \$0.1689$$

or approximately 17 cents

Clearly the power being used at any specific time during the day by a residential household varies considerably. If we were to graph the power in use for a typical American household during a normal weekday, the plot might look something like that in Fig. 14.8. The average power demand of the household is obviously much lower than the maximum. The ratio between the two is called the overall *load factor* and runs between 20% and 30% for a typical household. The energy used by this household for the 24-h period shown is represented by the *area* under the curve of Fig. 14.8. This can be determined by integration only, since it varies continuously. That this is exactly what a kilowatt-hour meter does will be explained in Section 14.15, which deals with electrical measurements.

EXAMPLE 14.8. It has been estimated that the average power demand of an American household is 1.2 kW. Calculate the monthly electric bill of such a household, assuming a flat rate of $0.06 per kilowatt-hour.

SOLUTION

Monthly energy consumption

$$= 1.2 \text{ kW} \times \frac{24 \text{ h}}{\text{day}} \times \frac{30 \text{ days}}{\text{month}} = 864 \text{ kWh}$$

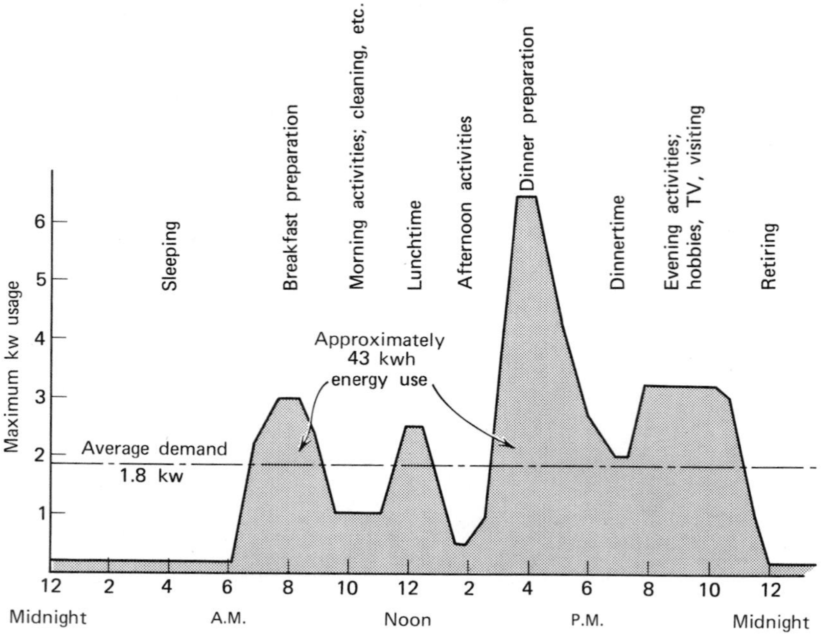

Fig. 14.8 Hypothetical graph of power usage for a typical American household. Electric cooking is assumed. Area under the curve represents energy usage. Maximum kW demand (vertical axis) is based on a 15-min integrated demand, thus eliminating spikes in demand, such as those caused by starting a refrigeration (air-conditioning) compressor. This curve has a 24-h use of approximately 43 kW-h, giving an average demand of 1.8 kW and a load factor of 27.5%.

Electric power bill

$$= 864 \text{ kWh} \times \$0.06/\text{kWh} = \$51.84$$

We stated in Example 14.8 above that the bill was based on a ''flat'' rate of 6 cents per kilowatt-hour. In actual fact the rate structure of most utilities is not so constructed. Generally, the tariff *decreases* with increasing use, thus encouraging larger use of electric power. This type of tariff was designed in the halcyon days of low-cost fuel, when the utility's costs were in large measure those of transmission, distribution, and administration. That this situation has changed is well known. Residential electric rates have been increased but, with few exceptions, have not been restructured to encourage either efficient use of electric energy or reduction in energy usage. One technique, long standard in industrial and commercial user tariffs but almost never applied to residential users, is the levying of a charge for power (kW) in addition to the normal energy (kWh) charge. This *demand charge* is primarily useful in encouraging users to reduce their peak loads. In so doing, energy use is also reduced somewhat.

14.12 Electric Demand Charges

As stated, electric utility companies normally levy a kW demand charge on all but individual residential customers. Varying with the individual utility company involved, this monthly charge runs between \$2 and \$10 per kilowatt of maximum average demand in any demand interval for that month. Demand intervals vary, being either 15 or 30 min (see Fig. 14.9).

Most utilities use a ''sliding window'' interval timing technique that starts a new interval every minute and updates the maximum interval demand accordingly. This enables them to find and bill for the maximum electric power demand in any 15- (or 30-) min period in the month. Some companies also include a ''ratchet'' clause that levies a demand charge for a number of months based on the maximum demand in any single month. This pe-

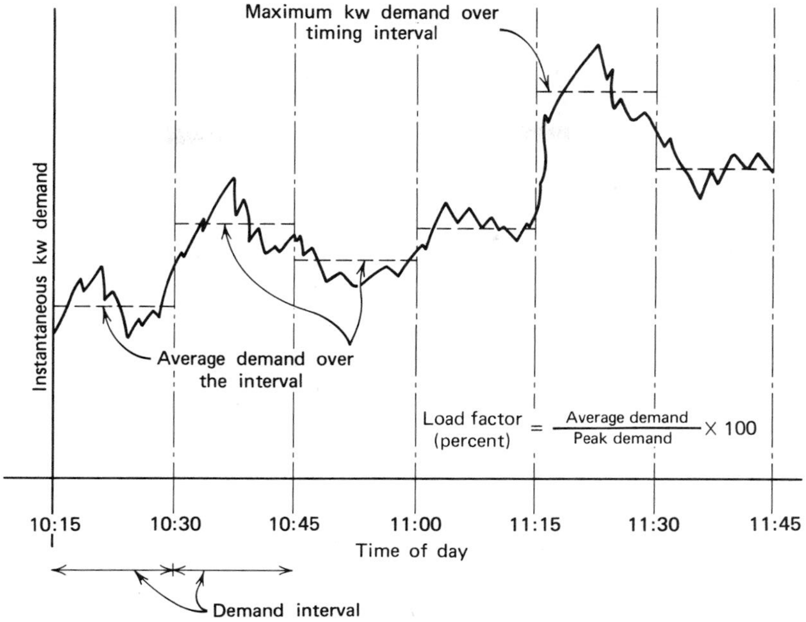

Fig. 14.9 *Typical instantaneous load curve for a facility. The utility demand meter records the average demand in each period (here 15 min). The maximum interval demand—in this case between 11:15 and 11:30—is used as the basis for monthly billing. A high load factor (utilization factor) indicates that little can be gained by demand control; a low load factor indicates the reverse.*

nalizes users with seasonal highs, that is, users with a low *yearly* load factor. The load factor is a measure of uniformity of power demand; a low load factor indicates short-time demand peaks for which the user is heavily charged. The reasoning behind the imposition of a demand charge and the significance of load factors can best be demonstrated by an example.

Assume that a pottery manufacturer, whose average 8-h daily load is 20 kW for lighting, pottery wheels, and the like, operates two 50-kW electric kilns twice a month for a 4-h period each time. Further assume an energy charge of $.05 per kilowatt-hour. The total monthly *energy* bill for the operation of the two kilns would be

$$\text{Cost} = 2 \times 50 \text{ kW} \times 8 \text{ h} \times \frac{\$.05}{\text{kWh}} = \$40$$

Thus, were it not for the demand charge, the utility company (which is required by law to supply maximum customer demand) would have to provide and maintain 100 kW of generation, transmission, and distribution facilities in return for a payment that is the equivalent of 1.11 kW of average continuous load; that is

$$\text{equivalent continuous load} = \frac{\$40}{720 \text{ hr/month} \times .05} = 1.11 \text{ kW}$$

This user's load factor can be calculated readily. By definition,

$$\text{load factor} = \frac{\text{average power demand}}{\text{maximum power demand}} \tag{14.8}$$

But, since for a given time interval the average power demand is equal to the energy used divided by the time involved, that is

$$\text{average power demand} = \frac{\text{kWh energy use}}{\text{hours of use}} \tag{14.9}$$

we have the general expression

$$\text{load factor (LF)} = \frac{\text{kWh energy use}}{\text{maximum demand} \times \text{hours use}} \tag{14.10}$$

For the case under consideration, the *monthly* load factor is

$$\mathrm{LF} = \frac{23 \text{ days} \times 8 \text{ h} \times 20 \text{ kW} + 2 \text{ days} \times 4 \text{ h} \times 100 \text{ kW}}{120 \text{ kW} \times 720 \text{ h/month}}$$

$$= \frac{3680 + 800 \text{ kWh/month}}{86400 \text{ kWh/month}}$$

$$= .052 \text{ or } 5.2\%$$

This is obviously a very poor load factor, which results in a high demand charge. Assuming a \$5.00 per kW demand tariff, this pottery manufacturer would be billed, monthly, an additional:

$$\text{demand charge} = 120 \text{ kW} \times \$5.00 = \$600$$

Since the energy bill is only

$$\text{energy cost} = 4480 \text{ kWh} \times 0.05 = \$224.00$$

this manufacturer is paying heavily for highly peaked power use.

Although the illustration selected is somewhat extreme in its type of power use, it is not uncommon to find demand charges of the same order of magnitude as energy charges. Obviously, it is impossible to eliminate demand charges entirely, but it is certainly possible and frequently very simple to reduce them. Such a step is in the interest of the user, the utility, and the public at large: the user—for simple economic reasons; the utility—to permit more efficient use of its facilities; and the general public—by avoidance of unnecessary power plant construction and concomitant inefficient use of fuel, and overall reduction in fuel use. This last item is a secondary benefit of demand control.

The next two sections will discuss two different, but frequently confused, electric system control functions. The first is electric demand control, whose primary function is to reduce electric *power* demand. This reduces demand changes and incidentally, secondarily, reduces energy consumption. The second is *energy* management, which is primarily concerned with reduction of all types of energy use, including electric. Electric demand control is frequently included as one part of an overall energy management system.

14.13 Electric Demand Control

Electric demand control methods vary greatly in complexity and in degree of automation, but all basically perform the same task—efficient utilization of available energy to produce a high load factor resulting in a lowering of demand charges. An ancillary, but important, benefit is the maximum utilization of building electrical power equipment, which normally runs underloaded. This results in smaller equipment, lower first cost, and less space utilization.

A further advantage to the user accrues from the fact that some utilities are offering their customers rebates that cover up to 40% of the cost of demand control and energy management equipment. This bonus serves to reduce by that amount the length of the payback period. (See Appendix E for a discussion of ''payback period'' calculation.) The reasoning of the utility is simple: it presently costs \$1500± per kilowatt of generating capacity for new power plant construction, whereas demand and energy management equipment costs only \$80± per kW demand reduction. Since a utility, by terms of its franchise, must supply all the power demanded, it is very much in the interest of any utility that is generating near capacity to reduce demand.

The basic technique of demand control is a simple one; electric loads are disconnected and reconnected in such a fashion that demand peaks are leveled off and load factor thereby improved. The extent to which a facility's electric loads can tolerate this type of switching is an indication of the potential effectiveness of a demand controller. An installation with a large uninterruptible load, such as computers or office lighting, will benefit minimally from demand control. Most industrial and commercial installations, however, contain a large percentage of interruptible loads (interruptions may be very short) and demand control systems frequently accomplish a 15 to 20% reduction in electric bills with a resultant short payback period on equipment investment.

The proliferation of demand control equipment has also produced a proliferation of nomenclature, including load shedding control, automated load control, peak demand control, and computerized load control. The last term refers to the type of equipment used to accomplish the intended control function. Descriptions that include the term ''energy management'' are inaccurate. Those devices are primarily concerned with control of *energy,* and are discussed in the next section. Demand control devices are intended to control *power,* with energy savings being an important, but secondary, benefit. We will use the expression ''electric demand control'' for simplicity and avoid knowingly using a term that refers to a particular manufacturer's equipment.

(a) Level 1—Load Scheduling and Duty-Cycle Control. This level is the simplest and most obvious, and it is applicable to all types of facilities. The installation's electric loads are analyzed and then scheduled to restrict demand. Thus large loads can be shifted to off-peak hours and controlled to avoid coincident operation. The user can also take advantage of special night and weekend utility rates for loads that do not require immediate operation, such as battery charging and transfer pumping. Control can be entirely manual or automated by use of a duty-cycle controller. This device is essentially a programmable time switch (see Section 16.15) with switching for a number of circuits, or loads. Typical applications of this device are control of HVAC loads, lighting loads, and process loads in small commercial, institutional, and industrial buildings.

The usefulness of these units lies in

1. Eliminating energy waste by shutting down units when not required.
2. Automatic control such as preheat and precool, which results in lower power and energy levels.

These devices are not, strictly speaking, demand controllers in that no cognizance is taken of the actual continuous electric loads. Instead, the devices operate on a preset duty cycle relying entirely on a prior analysis of the loads. Although such an analysis is a necessary first step in all levels of electric load control, its efficacy is limited since many of the loads are automatically controlled by pressure switches, thermostats, and the like. These loads cannot be scheduled with this type of controller and coincident operation cannot be prevented.

(b) Level 2—Demand Metering Alarm. If in conjunction with a duty-cycle controller some type of continuous demand metering is installed that will go into alarm when a predetermined demand level is exceeded, a basic load control system will have been established. The load analysis discussed above would have to be extended to determine load priorities so that when the preset maximum demand load is exceeded and the alarm sounds, loads can be shed (disconnected) manually in a predetermined order of priority and, subsequently, reconnected also in order of priority. This type of control is practical only for a limited size installation inasmuch as most of the load switching activity is manual. A unit typical of this type is illustrated in Fig. 14.10.

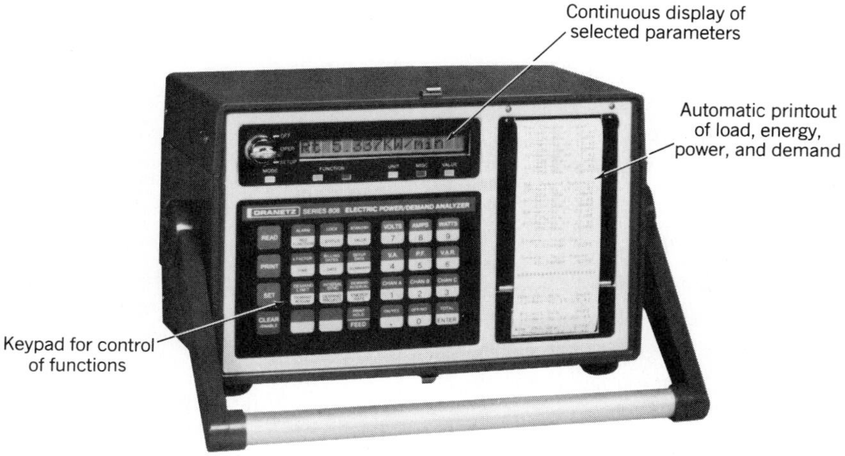

Fig. 14.10 This portable unit combines a metering device, a printer, and alarm capability. The unit accepts utility load data (pulse or conventional), automatically prints maximum demands per demand interval, day, billing period, or other desired period and can be readily arranged to initiate an alarm signal when a preset power demand or rate of energy consumption *is exceeded. The unit's portability permits analyzing separate parts of an electric power system. The unit is also arranged to display and print current, voltage, power factor, power, and energy use. Courtesy of Dranetz Technologies.*

ELECTRICITY

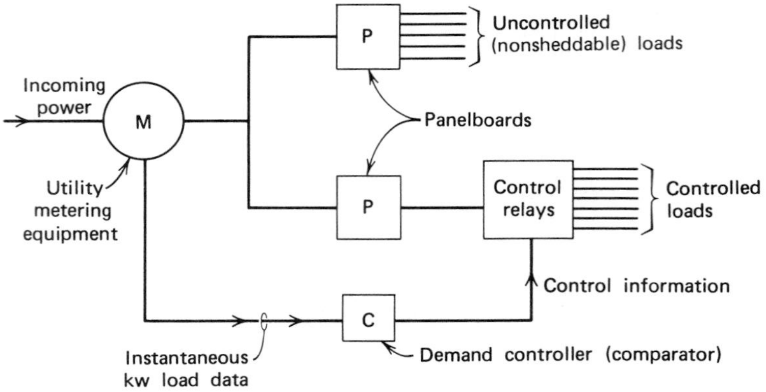

Fig. 14.11 (a) *Block diagram of a system of automatic electric power control. The demand controller receives instantaneous load data from the metering equipment, compares it to preset limits, and disconnects and reconnects controllable loads automatically to keep kW demand (load) within these limits.*

(c) Level 3—Automatic Instantaneous Demand Control. This type of control (also called "rate control") is, in effect, an automated version of the level-2 system described above. The unit accepts instantaneous kW load information from the utility system, compares this information to the preset kW limit (rate control), and acts automatically to disconnect and reconnect loads as required. These units *do not* recognize the utility's metering interval but act continuously on the basis of load comparison data. For this reason, these units are also referred to as load comparator controllers. Examination of Fig. 14.11 will make the unit's operation clear.

The first step in setting up this system is to separate the controllable ("sheddable") loads from those that must remain uninterrupted. Depending on the type of facility, the two lists that follow are typical.

Sheddable Electric Loads

Nonessential lighting	Domestic hot water heating
Ventilation fans	Sewage ejectors with appropriate level controls
Space heating	Transfer pumps
Comfort cooling	Electric snow melting
Noncritical batch process equipment	Any device with flywheel effect
Electric boilers	

Nonsheddable Electric Loads

Essential lighting	Process equipment
Elevators	Material handling equipment
Refrigeration compressors	Office machinery

The nonsheddable loads are fed directly from the power line. The sheddable loads are fed through a panel of control relays that respond to on/off instructions from the demand controller. Note that the controller acts to reduce maximum loads (peaks) and fill in low points (valleys). Although theoretically the energy use with or without the controller is the same, in actual practice energy savings of 15% and more are common. Some units have special provisions to overcome some of the limitations inherent in rate control systems. These are:

1. Excessive cycling
2. Excessive off-time
3. Inability to readily adapt to varying load patterns, resulting from variable production schedules, time schedules, changes in weather, and so on.

As a result of these limitations, this system is most useful in applications where operating modes do not change frequently and the facility is not very large. Thus stores, supermarkets, warehouses, small industrial facilities, and commercial installations are well served with this level system

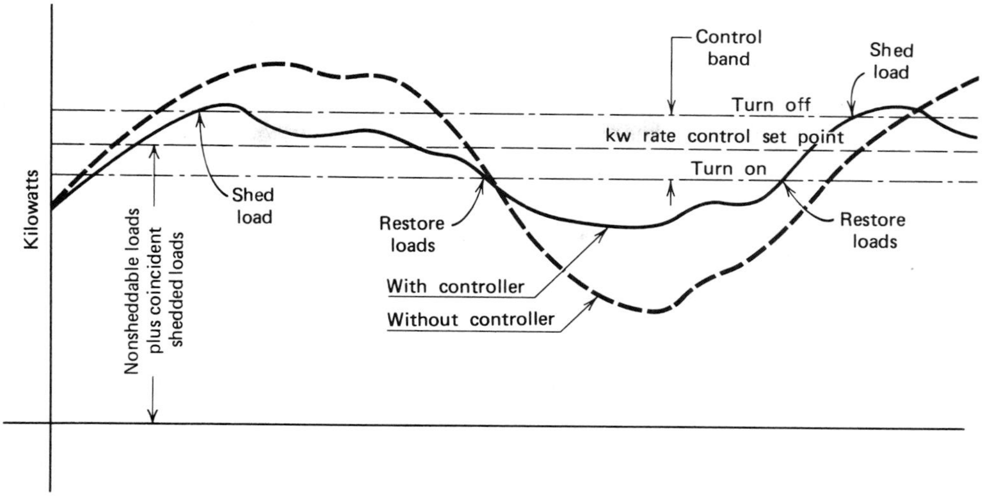

Fig. 14.11 (b) *Action of the demand controller is graphically illustrated. See text, Section 14.13.*

if they have at least 20% sheddable loads and their connected electric load is at least 150 kilovolt-amperes (kVA).

(d) Level 4—Ideal Curve Control. This controller operates by comparing the actual rate of *energy* usage to the ideal rate, and controls kW demand by controlling the total *energy* used within a metering interval. A constant rate of energy use over a demand interval would show as the set of repeated straight lines in Fig. 14.12*a*. The utility company determines the demand over the demand interval by integrating the kilowatt-hour energy over the interval and dividing by the interval time. Thus, the user is actually given a block of energy (kWh) that can be utilized at any desired rate, not necessarily at the constant rate of Fig. 14.12*a*. The desirable rate of energy use is shown as the "ideal curve" on the typical usage curve shown in Fig. 14.12*b*.

Shed points can be preprogrammed for each load independently according to a predetermined priority, and priorities can be readily adjusted and rescheduled. Loads are normally shed only toward the end of the interval when the permissible energy total is approached and all loads are restored

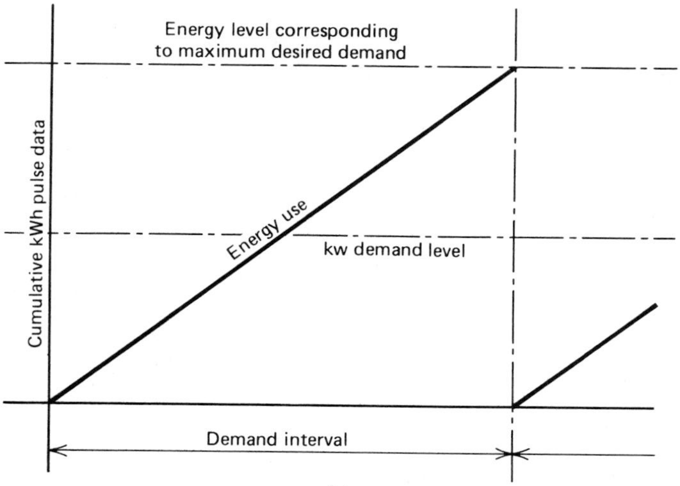

(a)

Fig. 14.12 (a) *Graph of cumulative energy use over a demand interval, corresponding to a constant kW demand. (Energy use is the time integral of power; i.e., kW-h = kW × time).*

ELECTRICITY

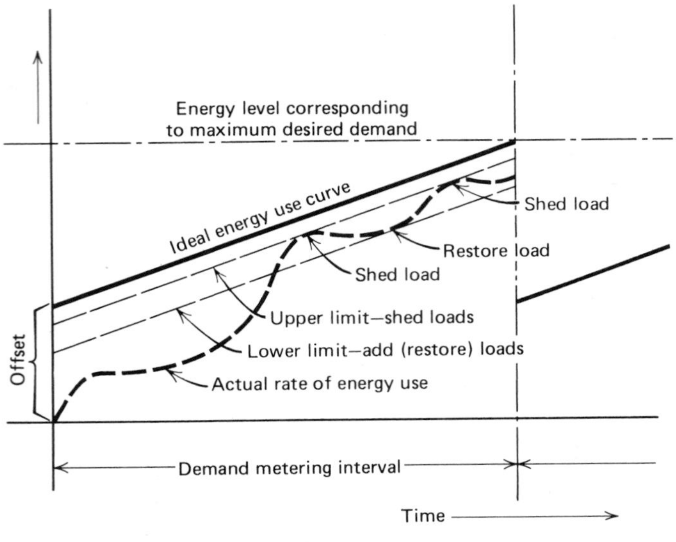

Fig. 14.12 (b) *Graph of action of an ideal rate controller. Note that toward the latter part of the period the pattern has been established, and the actual rate of energy use corresponds closely to the ideal.*

at the beginning of each interval. Thus during each interval loads are off for only a few minutes at most. Controllers operating on the "ideal curve" principle are considerably more flexible than the kilowatt rate controller described in the level-3 system above and are applicable to facilities of widely divergent load size, but with at least a 300-kW connected load. As with other controllers, the principal savings will be in demand charges, but almost always with considerable economies in energy billings.

An interesting application of these units is to utilize them to operate a standby power plant to supply peak power demands in lieu of shedding loads. In such a system, maximum demand could be set lower than otherwise would be possible, and power peaks supplied by starting and loading the standby emergency power system. An auxiliary benefit of such an arrangement would be providing necessary exercise to the standby generator unit. Such an arrangement would require the approval of the utility, since many utilities prohibit use of standby generation except under emergency conditions.

(e) Level 5—Forecasting Systems. These systems are by far the most sophisticated, the most expensive, and the most effective. They are best applied to large structures where the number of loads, load patterns, and complexity of operation precludes the use of the preceding systems. As a result of the large amount of load data, these systems frequently are installed as part of the computerized central control facilities in large facilities as noted above. Details of operation are too complex for our needs, but the basic operation can be described. These units operate by continuously forecasting the amount of energy remaining in the demand interval, based on kilowatt-hour pulse data received. They then examine the status and priority of each of the connected loads and decide on a course of action. Loads that in other systems are classified noncontrollable are in this system controlled, because of the accuracy and rapidity of the control function. A pneumatic compressor, for instance, that supplies process air might in other level control systems be classified as noncontrollable, despite the fact that it has long off periods. With computer control, such a load would be classified as "delayable" or "inhibited," since a 30-sec or 1-min delay in activation after the pressure switch closes its contact will normally be acceptable.

The advantage of these systems is that if programmed properly they can make small, accurate load changes throughout the interval, resulting in minimum load cycling and maximum efficiency.

14.14 Energy-Management Systems/Building Automation

The field of building automation has burgeoned in recent years to the extent that some sort of automated control system is standard in new construction, and is becoming so, via retrofit, in existing buildings. The problem that arises in understanding the application of this equipment lies in the fact that essentially the same item of equipment can be used for all types of load control but, depending upon the application, is variously called a lighting controller, HVAC controller, energy-management device, building programmable controller, and so on, almost ad infinitum. As explained in detail in the discussion of lighting control (Sections 21.2 to 21.4), the equipment is one of two types: preprogrammed only, or programmed with feedback and control logic. The former are simply programmable timers, the latter programmable controllers (see Fig. 21.6). Energy-management devices are generally programmable controllers.

Energy management is accomplished by controlling the functioning of most of the energy-consuming devices in the facility. These functions include duty cycling and load shedding and encompass HVAC, lighting, chiller control, and process equipment. Peak demand control and other energy functions may be included in this control device or may be accomplished by separate, dedicated devices, which may be connected to the central controller. In addition, the central programmable controller can be expanded to include signaling functions not primarily concerned with energy, such as security, fire safety, alarms of all types, communications, data logging, and so on. These are discussed in Chapter 22. The control signals, as explained in the sections on lighting control, may either be run on control wiring or else be high-frequency signals run on the existing power wiring. This latter is obviously particularly appropriate to retrofit work. Two typical energy-management system controllers are shown in Fig. 14.13.

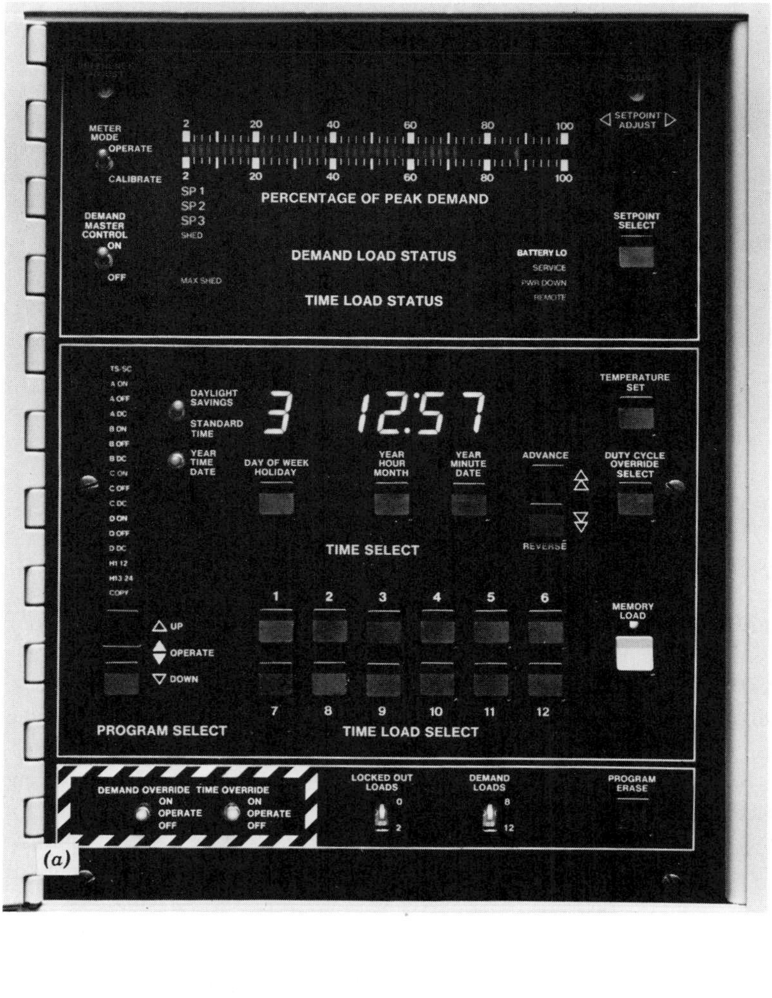

Fig. 14.13 (a) *Control panel of a medium-size energy-management controller. This device provides time-of-day scheduling, duty cycling, and demand limiting, via 24 output channels. It connects to a large number of remote sensors, receivers, and local override devices. Controlled loads include HVAC equipment, furnaces, water heaters, boilers, ovens, lighting, process equipment, and the like. Typical applications are multistory office buildings, department stores, and medium-size manufacturing facilities. The unit is suitable for use either with hard wiring or power line carrier (PLC) output for impressing on existing a-c lines. Unit dimensions are 20″ × 16″ × 7″ deep. Courtesy of Aegis Energy Systems.*

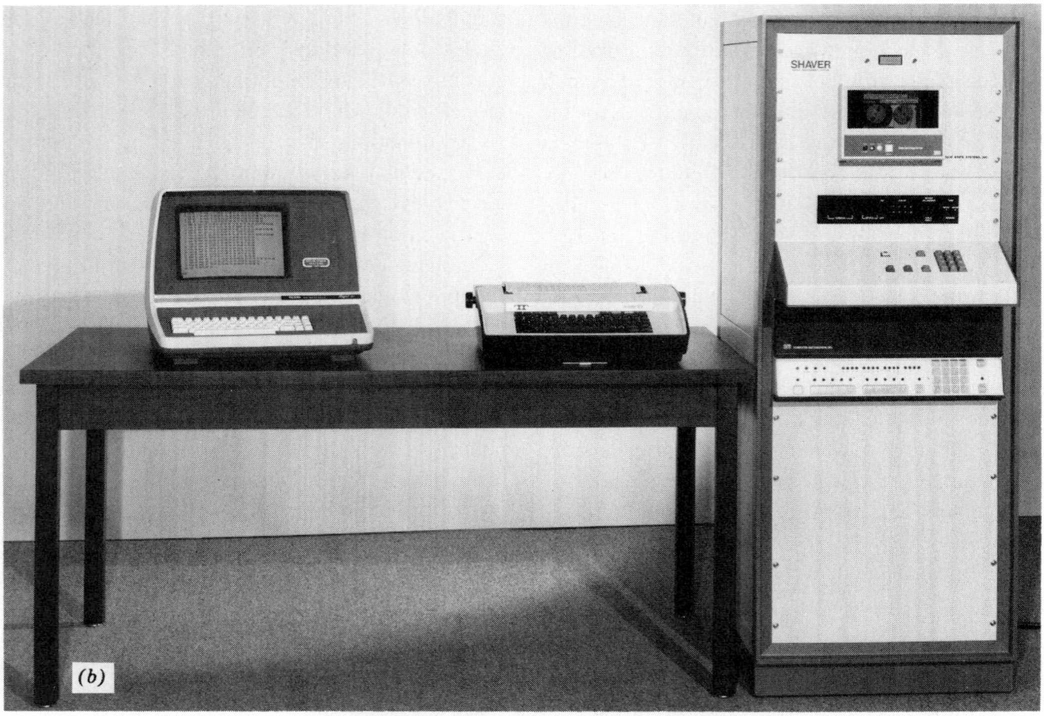

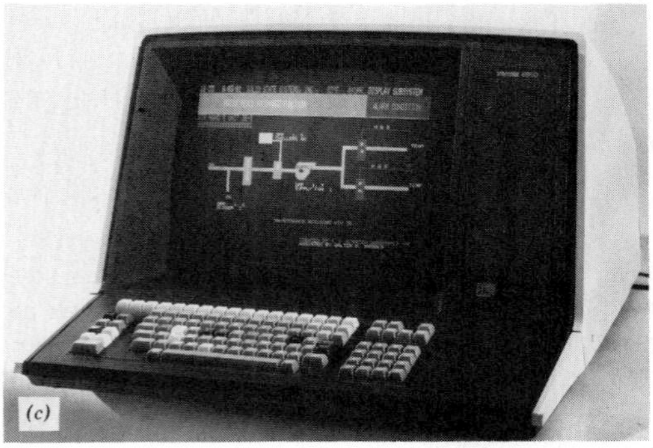

Fig. 14.13 (b) *Large, computerized, energy-management unit that combines control of HVAC, lighting, peak demand, and energy consumption. Proper programming permits optimized systems control, which leads to minimal energy use. The unit can also provide alarm monitoring, processing, and preventive-maintenance functions. The video display unit can display monitoring functions as in* (b), *systems diagrams as in* (c), *or any other desired command, feedback, or calculated function. Courtesy of Solid State Systems, Inc.*

14.15 Electrical Measurements

In the preceding sections we have explained the fundamental electric quantities of voltage and current and have defined the units involved as volts and amperes, respectively. As is true of all other physical quantities that are to be used in practical application, the need existed for a simple means of measuring these quantities. This need was met by the development of the meter movement illustrated in Fig. 14.14. Everyone at one time or another has felt the repulsion between like poles of two magnets held close together and, conversely, the attraction between opposite poles. This principle is used in the basic meter movement. It causes a deflection of the pointer as a result of the repulsion between the field of a permanent magnet and an electromagnet. The electromagnet is formed when current flows in the coil, and its strength is proportional to the amount of current flowing. Thus, a strong current causes a larger deflection of the needle and, therefore, a higher reading on the dial. A spring (see Fig. 14.14*b*) provides restraining torque on the pointer. To make this very sensitive basic unit usable for large currents (it is intrinsically a microammeter, sensitive to millionths of

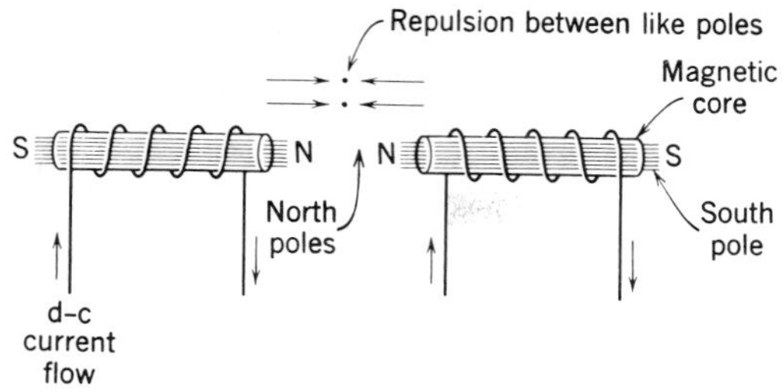

Fig. 14.14 (a) *Diagram showing basic electromagnetic principle and interaction between electromagnets. Any iron core becomes an electromagnet when current flows in a coil wound around it, as shown.*

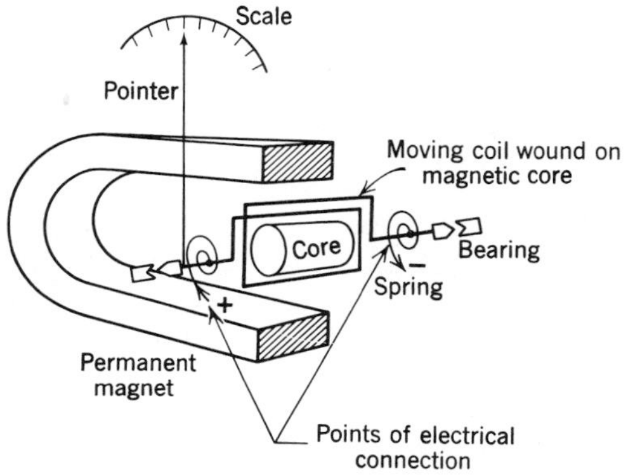

Fig. 14.14 (b) *The principle of the electromagnet is used in this basic meter movement. Current flowing in the movable coil forms an electromagnet whose field interacts (see* a*) with the permanent magnet's field, causing a deflection proportional to the current flow. Courtesy of Westinghouse, Relay-Instrument Division.*

an ampere), we simply divert, or shunt away, most of the current, allowing only a few microamperes to actually flow in the meter coil.

To use the same unit as a voltmeter, we put a large resistance (multiplier) in series with the meter, again limiting the current flowing to a few microamperes. The scale is then calibrated in volts. All d-c meters are made in this fashion. Most a-c meters operate on basically the same principle, except that instead of a permanent magnet an electromagnet is used. Thus when the polarity reverses, the deflecting force remains in the same direction. A d-c meter connected to an a-c circuit simply will not read, since inertia prevents the needle from bouncing up and down 60 times a second. Digital meters operate on an entirely different principle.

The measurement of current and voltage in practical application is generally not as important as the measurement of power and energy, as the preceding section has made abundantly clear. To measure power, we take advantage of the fact learned earlier that power is proportional to the product of the voltage and current in the circuit. Although actual construction is complex, the theory of operation of a wattmeter is simple. The meter has two coils; a current coil that is similar in connection to an ammeter, and a voltage coil that is similar in connection to a voltmeter. By means of the physical coil arrangement, the meter deflection is proportional to the product of the two, and therefore to the circuit power. The meter can be calibrated as desired, depending on the size of the shunts and multipliers. To measure energy, the factor of time must be introduced, since as we know

$$\text{energy} = \text{power} \times \text{time}$$

D-c energy meters are available but are not of general interest because of the rarity of d-c power. A-c watt-hour meters are basically small motors, whose speed is proportional to the power being used. The number of rotations is counted on the dials, which are calibrated directly in kilowatt-hours. A diagram of the basic construction of an a-c kilowatt-hour meter is shown in Fig. 14.15. As can be seen, the kilowatt-hour energy consumption and the maximum interval kW demand can be read directly from the dials. (The illustrated unit is of the standard, not energy pulse, type.) If the numbers involved are too large or a meter is used with current transformers between it and the line, or for calibration reasons, a multiplying factor is required to arrive at the proper kilowatt-hour consumption. This number is written directly on the meter nameplate, and we multiply the meter reading by it to get the actual kilowatt-hours. In

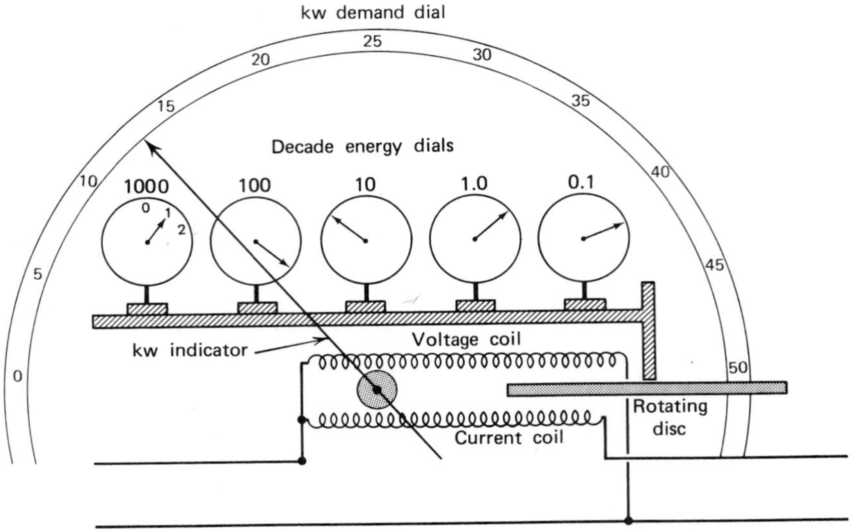

Fig. 14.15 Typical induction-type kilowatt-hour meter with kW demand dial. Decade dials register total disc revolutions that are proportional to energy. Disc speed is proportional to power. Note that the current coil is in series with the load, whereas the voltage coil is in parallel.

the absence of such a number, it can be assumed that the meter reads directly in kilowatt-hours.

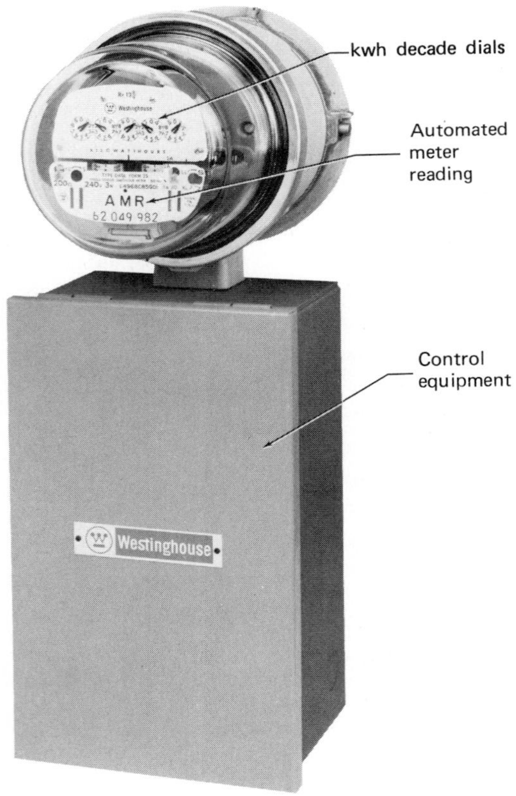

A special type of kilowatt-hour meter is illustrated in Fig. 14.16 which, although its face is similar to a standard kilowatt-hour meter, is in fact much more sophisticated. This instrument is equipped with an electro-optical automatic meter reading system that is activated from a remote location. The meter data are transmitted electrically to a data-processing center where it may be used by the utility to prepare subscribers' bills, to prepare customer load profiles, and to study, in combination with other such data, area load patterns, equipment loading, and so on. The customer can use instantaneous data to control loads as explained at length in the preceding section. Of course, the most obvious characteristic of this equipment is the elimination of the traditional (and very expensive) meter reader.

***Fig. 14.16** The illustrated encoding register (meter) for automated meter reading is mounted on a control cabinet that contains electronic equipment capable of performing a number of control functions. These include supply of pulse data for demand metering (see Section 14.13), automated metering functions including meter checking and data transmission to a central point, control of selected load, and related functions. Courtesy of Westinghouse Electric Corp., Electronics, Measurement & Control Business Unit, Raleigh, N.C.*

15

ELECTRICAL SYSTEMS AND MATERIALS: WIRING

The major components of a building's electrical power system can be arranged in three major categories; wiring, power-handling equipment, and control and utilization equipment. In the first category, we include conductors and raceways of all types; in the second, transformers, switchboards, panelboards, large switches, and circuit breakers; and in the last, actual utilization equipment such as lighting, motors, controls, and wiring devices. After some discussion that is applicable to all electrical materials, this chapter will discuss in detail the items in the first of these three categories, that is, the wiring system. Chapter 16 will cover most of the remaining items in the other two categories with the exception of lighting equipment, which will be discussed in the lighting section. Signal equipment including telephones, intercom, alarm, and control will be covered separately.

15.1 System Components

Referring to Fig. 15.1 note that the power system proceeds from the service point to the utilization point in a series of steps of decreasing circuit capacity. Although a normal single-line diagram does not differentiate by line weight between heavy and light (large and small) conductors (the heavier the conductor the greater the amount of power being carried), the single-line diagram of Fig. 15.1 does, in order to show relative power levels in the system. This "size" differentiation is more clearly

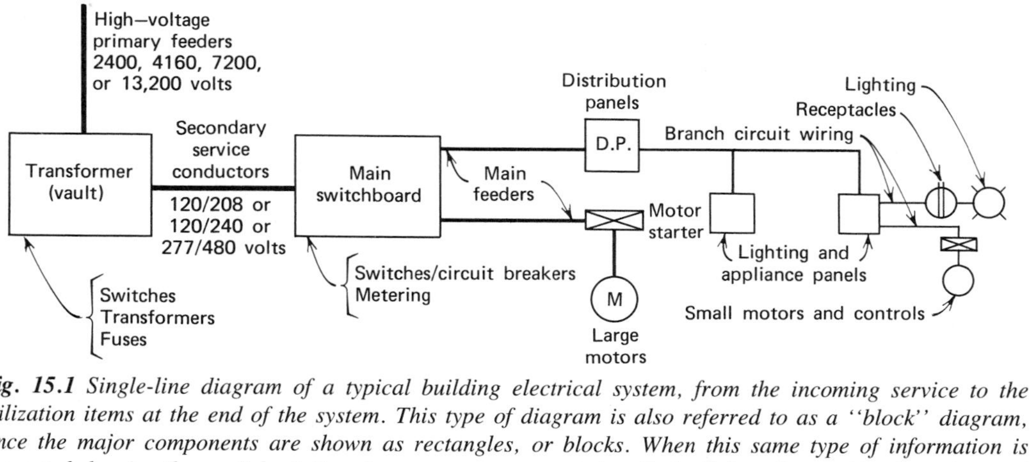

Fig. 15.1 Single-line diagram of a typical building electrical system, from the incoming service to the utilization items at the end of the system. This type of diagram is also referred to as a "block" diagram, since the major components are shown as rectangles, or blocks. When this same type of information is presented showing the spatial relations between components, it is called a "riser diagram"; when electrical symbols are used in lieu of the blocks, it is called a "one-line" or a "single-line" diagram. Were the connecting conductors between the major system components drawn to reflect size and thereby power-handling capacity, the system would appear as shown here.

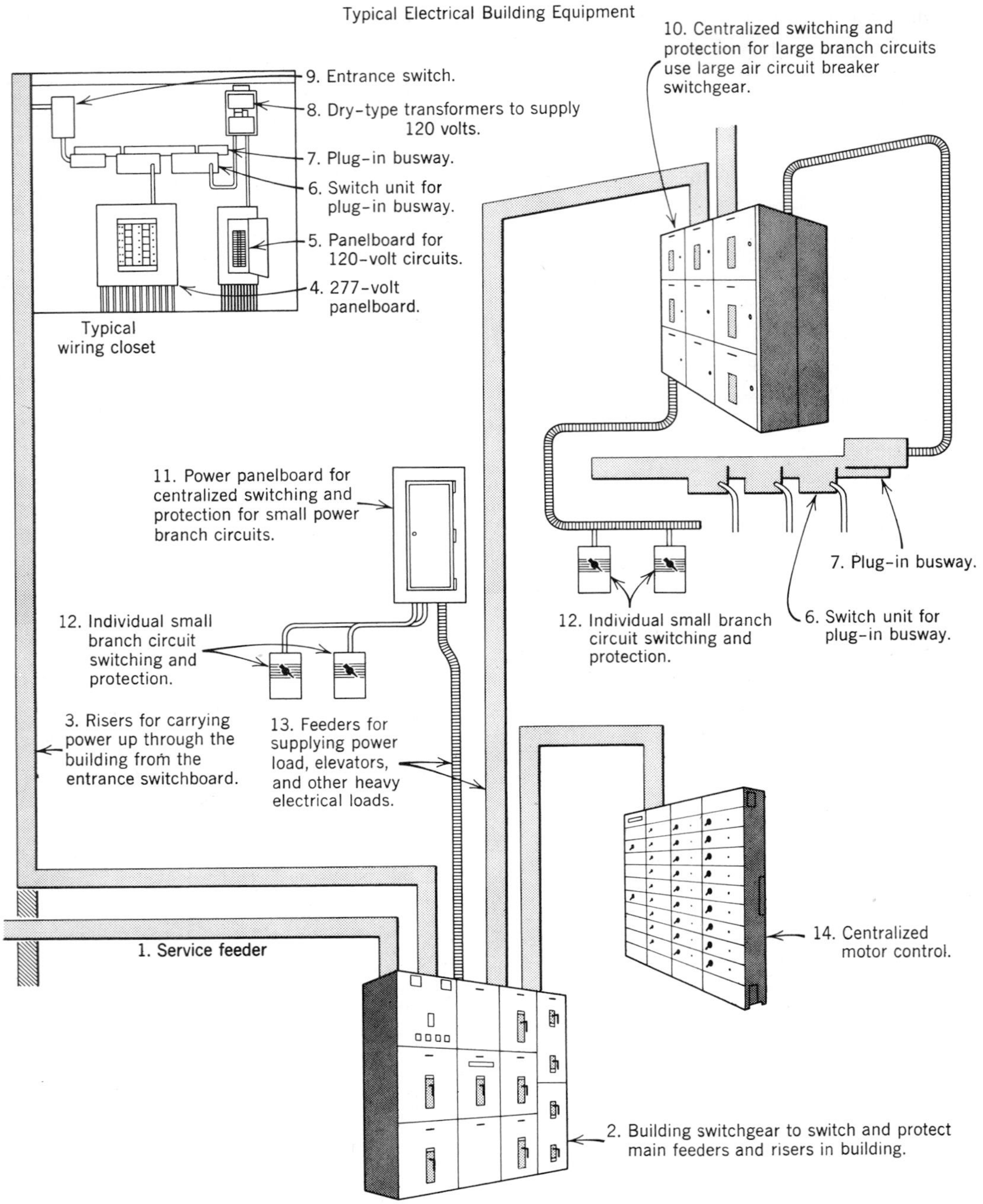

Fig. 15.2 A building's electrical system shown pictorially, with the power capacity being indicated, more or less, by the size of the conductors. Note that this diagram does not extend beyond the local panelboard. Courtesy of General Electric.

shown in Fig. 15.2 which is a pictorial representation of a system similar to that of Fig. 15.1, but in somewhat greater detail and omitting items beyond the panel board.

15.2 National Electrical Code

The National Electrical Code (NEC) of the National Fire Protection Association (NFPA) defines the fundamental safety measures that must be followed in the selection, construction, and installation of all electrical equipment. This code is used by all inspectors, electrical designers, engineers, contractors, and the operating personnel charged with the responsibility for safe operation. Having been incorporated into OSHA (Occupational Safety and Health Act) it has, in effect, the force of law. The reader of this book should obtain for personal use the latest edition of the NEC from the NFPA at 80 Batterymarch Street, Boston, Massachusetts 02110. Frequent references will be made to this code.

In addition to the National Code, many large cities such as New York, Boston, and Washington, D.C., have their own electrical codes that, though similar to the NEC, contain numerous special requirements.

In order to assure a minimum standard of intrinsic electrical safety for electrical equipment, a single agency was needed to establish standards for, and to actually test and inspect, electrical equipment. Such an organization is the Underwriters Laboratories, Incorporated, which publishes lists of inspected and approved electrical equipment. These listings are universally accepted, and many local codes state that only electrical materials bearing the Underwriters Laboratories (UL) label (of approval) will be acceptable.

15.3 Economics of Material Selection

The selection of electrical materials involves not only choosing a material or assembly that is functionally adequate and, where necessary, visually satisfactory, but also the consideration of economic factors. This is necessary since, in most instances, there is available a multiplicity of equipment that will fulfill the construction need. In such cases, economic factors often decide the issue. The decision is relatively simple when the various materials differ only slightly from each other and a straightforward first-cost comparison is all that is required. Often, however, the choice is not so simple, since the materials may vary considerably in characteristics other than functional suitability, and a more detailed cost study is required. As the reader is no doubt aware, such economic analyses are frequently of greater importance in comparisons between various HVAC systems. This is less so when dealing with electrical systems, since from an energy point of view electrical systems are more passive than HVAC systems, and it is energy costs that are often the decisive factor in economic analyses. We are speaking, of course, of life-cycle equipment costs (over the life of the *structure*) expresseed in *present-value* dollars, or annual owning and operating costs, including equipment amortization costs. The type of analysis used depends on the situation. However, such comparisons are useful only when both the initial cost and the operating costs will be borne by the same individual, that is, an owner-operator. Obviously, in the case of a speculative building venture only first cost is considered.

It is often quite difficult to perform such cost analyses since accurate data on life and maintenance costs for electrical equipment may not be readily available. Still, even if a formal analysis cannot be done, the principle involved must be considered in the selection of all electrical material, bearing in mind its particular importance in the case of energy-consuming equipment such as lighting and motors. Detailed economic discussions on these items will be found with the related technical discussion.

15.4 Energy Consideration

As mentioned above, energy costs are a major factor in economic analysis. However, energy considerations are at least as important, in and of themselves. This aspect will be examined in detail in the discussions both of building energy budgets and of their components, including lighting, elevators, and electric motors in all other systems.

15.5 Electrical Equipment Ratings

All electrical equipment is rated for the normal service it is intended to perform. These ratings may be in voltage, current, duty, horsepower, kW, kVA, temperature, enclosure, and so on. Ratings related to the specific equipment will be discussed in the sections below. The ratings that are specifically and characteristically electrical are those of voltage and current.

(a) Voltage. The voltage rating of an item of electrical equipment is the maximum voltage that can safely be applied to the unit continuously. It frequently, but not always, corresponds to the voltage applied in normal use. Thus, an ordinary wall electrical receptacle is rated at 250 V maximum, though in normal use only 120 V are applied to it. The rating is determined by the type and quantity of insulation used and the physical spacing between electrically energized parts.

(b) Current. The current rating of an item of electrical equipment is determined by the maximum operating temperature at which its components can operate properly continuously. That in turn depends on the type of insulation used. As a case in point, consider an electric motor. The current flowing in the motor windings causes a power loss (I^2R), which generates heat. If the windings are insulated with varnished cotton braid, with a maximum safe operating temperature of 65° C, the maximum permissible current (to which the horsepower rating of the motor is directly related) is that current which will produce this operating temperature. If these same windings are insulated with a silicone or glass compound with a maximum operating temperature of 150° C, obviously more current can be safely carried and the horsepower rating is consequently larger. Thus we see that, although a motor is rated in horsepower (or kW where SI units are used), a transformer is rated in kVA and a cable, as will be discussed below, is rated in amperes, the actual criterion on which all these ratings are based is maximum permissible operating temperature.

15.6 Interior Wiring Systems

At this point it will be helpful to survey the different types of interior wiring systems before commencing a discussion of components. When the primary purpose of the system is to distribute electrical energy, it is referred to as an electrical power system; when the purpose is to transmit information, it is referred to as an electrical signal system. In this chapter we will deal with electrical power systems.

Due to the nature of electricity, its distribution within a structure poses basically a single problem: how to construct a distribution system that will *safely* provide the energy required at the location required. The safety consideration is all-important, since even the smallest interior system is connected to the utility's powerful network and the potential for damage, injury, and fire is always present. The solution to this problem is to isolate the electrical conductors from the structure except at those specific points, such as wall receptacles, where contact is desired. This isolation is generally accomplished by insulating the conductors and placing them in protective raceways. The principal types of interior wiring systems in use today are exposed insulated cables, insulated cables in open raceways, and insulated conductors in closed raceways.

(a) Exposed Insulated Cables. In this category would be included (using the NEC nomenclature) NM (''Romex'') and AC (''BX'') (see Sections 15.11 and 15.12).

Also included are other types where the cable construction itself provides the necessary electrical insulation and mechanical protection.

(b) Insulated Cables in Open Raceways (Trays). This system is specifically intended for industrial application, and it relies upon both the cable and the tray for safety.

(c) Insulated Conductors in Closed Raceways. This system is the most general type and is applicable to all types of facilities. In general, the raceway is installed first and the wiring pulled in or laid in later. The raceways themselves may be

1. Buried in the structure; for example, conduit in the floor slab or underfloor duct (see Sections 15.18, 15.19, 15.24, and 15.25).
2. Attached to the structure; for example, all types of surface raceways, including con-

duit and wireways suspended above hung ceilings (see Sections 15.23 and 15.30).

3. Part of the structure; for example, cellular concrete and cellular metal floors (see Sections 15.26 and 15.27).

(d) Combined Conductor and Enclosure. This category is intended to cover all types of factory-prepared and factory-constructed integral assemblies of conductor and enclosure. Included here are flat cable intended for under-carpet installation (Section 15.29), flat cable assemblies and lighting track (Section 15.16), manufactured wiring systems (Section 15.30), and all types of busway, busduct, and cablebus (Section 15.15).

15.7 Conductors

Electrical conductors (wiring) are the means by which the current is conducted through the electrical system, corresponding to the piping in the hydraulic analogy. By convention, a single insulated conductor No. 6 AWG (American Wire Gauge) or larger, or several conductors of any size assembled into a single unit, are referred to as cable. Single conductors No. 8 AWG and smaller are called wire.

The standard of the American wire and cable industry for round cross-section conductors is the American Wire Gauge. All wire sizes up to No. 0000 (also written No. 4/0) are expressed in AWG. The AWG numbers run in *reverse* order to the size of the wire, that is, the smaller the AWG number, the larger the size. Thus No. 10 is a heavier wire than No. 12 and lighter (thinner) than No. 8. The No. 4/0 size is the largest AWG designation, beyond which a different designation called MCM (thousnd circular mil) is used. In this designation, wire diameter *increases* with number; thus, 500 MCM is a heavier wire (double the area) than 250 MCM.

Outside of the United States, where the metric system is in general use, conductor sizes are given simply as the diameter in millimeters (mm). Table 15.1 gives dimensional and stranding data for the commonest wire sizes and includes the millimeter equivalent of each size. This will prove useful in interfacing American gauges and metric sizes.

Table 15.1 *Physical Properties of Bare Conductors*

Size (AWG or MCM)	*Area (Circular Mils)*	*Diameter (Inches)*		*Diameter (Millimeters)*		*d-c Resistance Ohms/1000 ft at 77° F, 25° C (Bare Copper)*
		Solid	*Stranded*	*Solid*	*Stranded*	
16	2580	0.0508	—	1.29	—	4.10
14	4109	0.0641	—	1.63	—	2.57
12	6530	0.0808	—	2.05	—	1.62
10	10,380	0.1019	—	2.59	—	1.02
8	16,510	0.1285	—	3.26	—	0.64
6	26,240	0.162	0.184	4.11	4.67	0.41
4	41,740	0.204	0.232	5.18	5.89	0.26
2	66,360	0.258	0.292	6.55	7.42	0.16
1	83,690	0.289	0.332	7.34	8.43	0.13
0 (1/0)	105,600	0.325	0.373	8.26	9.47	0.10
00 (2/0)	133,100	0.365	0.418	9.27	10.62	0.081
000 (3/0)	167,800	0.410	0.470	10.41	11.94	0.064
0000 (4/0)	211,600	0.460	0.528	11.68	13.41	0.051
250 MCM	250,000	0.500	0.575	12.70	14.61	0.043
300 MCM	300,000	0.548	0.630	13.92	16.00	0.036
400 MCM	400,000	0.632	0.728	16.05	18.49	0.027
500 MCM	500,000	0.707	0.813	19.56	20.65	0.022

Source: Extracted from the National Electrical Code, except for millimeter dimensions.

ELECTRICITY

15.8 Conductor Ampacity

Conductor current-carrying capacity or *ampacity*, is determined as explained above, by the maximum operating insulation temperature. Heat generated as a result of the current flowing is dissipated into the environment. Thus, for a given environment (open-air or enclosed), ampacity increases with increasing conductor size and maximum permissible insulation temperature. These facts are clearly shown in Table 15.2*a*. If more than three conductors are placed in a conduit, the increase in temperature requires that the conductors be derated by the amount shown in Table 15.2*b*.

Since heat dissipation from a conductor in free air is much greater than from the same conductor enclosed in conduit or direct buried, the corresponding allowable ampacity is also greater. Conversely, if the ambient temperature around the conductor is above 30°C, on which all the ampacity tables are based, the permissible ampacity must be reduced.

Ampacity tables for conductors in free air and cable types not shown in Table 15.2*a* and derating factors for high ambient temperature are all found in the NEC. Typical ambient temperatures are given in Table 15.3.

TABLE 15.2*a* **Allowable Ampacities of Enclosed Insulated Copper Conductors (not more than three conductors in raceway or directly buried in earth) Based on Ambient Temperature of 30°C (86 F)**

	Temperature Rating of Conductor[a]		
	60°C (140 F)	*75°C (167 F)*	*90°C (194 F)*
Size *AWG MCM*	*Type UF*	*Types RHW, THW, THWN, XHHW*[b]	*Types THHN, XHHW*[b]
14[b]	15	15	15
12[b]	20	20	20
10[b]	30	30	30
8	40	50	55
6	55	65	75
4	70	85	95
3	80	100	110
2	95	115	130
1	110	130	150
0	125	150	170
00	145	175	195
000	165	200	225
0000	195	230	260
250	215	255	290
300	240	285	320
350	260	310	350
400	280	335	380
500	320	380	430

Source: Extracted from the National Electrical Code.

[a]These ampacities relate only to conductors described in Table 15.4.

[b]For dry locations use 90°C rating; for wet locations use 75°C rating.

15.9 Conductor Insulation and Jackets

Most conductors are covered with some type of insulation that provides electrical isolation and some physical protection. Additional physical shielding, where necessary, is provided by a jacket over the insulation. Insulation is rated by voltage. Ordinary building wiring is rated for 300 V or 600 V. The common types of building wire insulation are listed in Table 15.4 with the associated trade names, code letters, maximum temperatures, and special provisions.

15.10 Copper and Aluminum Conductors

Aluminum has an inherent weight advantage over copper, with concomitant lower installation costs. Economy usually lies with copper in small- and

TABLE 15.2b **Current-Carrying Capacity Derating Factors**

Number of Conductors[a] *in Raceway*	*Derating Factor*
4 to 6	0.80
7 to 24	0.70
25 to 42	0.60
43 and above	0.50

Source: Extracted from the National Electrical Code.

[a]Neutral conductors are not counted.

TABLE 15.3 **Typical Ambient Temperatures**

Location	*Temperature*	*Minimum Rating of Required Conductor Insulation*
Well ventilated, normally heated buildings	30° C (86° F)	(See note below)
Buildings with such major heat sources as power stations or industrial processes	40° C (104° F)	75° C (167° F)
Poorly ventilated spaces such as attics	45° C (113° F)	
Furnaces and boiler rooms (min.)	40° C (104° F)	75° C (167° F)
(max.)	60° C (140° F)	90° C (194° F)
Outdoors in shade in air	40° C (104° F)	75° C (167° F)
In thermal insulation	45° C (113° F)	75° C (167° F)
Direct solar exposure	45° C (113° F)	75° C (167° F)
Places above 60° C (140° F)		110° C (230° F)

NOTE: 60°C for up to and including No. 8 AWG copper and 75°C for over No. 8 AWG copper.

TABLE 15.4 **Characteristics of Selected Insulated Conductors for General Wiring**

Trade Name	*Type Letter*	*Maximum Operating Temperature*	*Application Provisions*
Moisture- and heat-resistance rubber	RHW	75°C 167 F	Dry and wet locations
Heat-resistant thermoplastic	THHN	90°C 194 F	Dry locations
Moisture- and heat-resistant thermoplastic	THW	75°C 167 F	Dry and wet locations
Moisture- and heat-resistant thermoplastic	THWN	75°C 167 F	Dry and wet locations
Moisture- and heat-resistant cross-linked thermosetting polyethylene	XHHW	90°C 194 F	Dry locations
		75°C 167 F	Wet locations
Underground feeder; moisture resistant	UF	60°C 140 F	Underground, including direct burial

Source: Extracted from the NEC.

medium-size cable, since weight is not a problem and the smaller conduit required for the smaller copper conductors generally makes them cheaper. In the larger cable sizes, the aluminum weight advantage offsets the economy of smaller copper size and smaller conduit, and generally proves less expensive, particularly in areas of high labor cost such as urban areas.

Aluminum and copper both exhibit the low electrical resistivity necessary for a good electrical conductor. There are, however, difficulties inherent in splicing and terminating aluminum. These difficulties—which can be overcome with the use of proper equipment, techniques, and workmanship—stem from aluminum's cold-flow characteristic when under pressure (causing joints to loosen) and aluminum's oxide. This oxide, which forms within minutes on any exposed aluminum surface, is an adhesive, poorly conductive film that must be removed and prevented from reforming if a successful, long-life joint or termination is to be effected. If this is not done, the oxide causes a high-resistance joint with consequent excessive heat generation and possible incendiary effects. Furthermore, when used in branch circuits, even if properly installed initially, aluminum can create problems when wiring devices are replaced by unskilled homeowners.

As a result of a number of unfortunate incidents, some localities in the United States have banned the use of aluminum wire in branch circuitry. Heavy feeders are normally installed by experienced and skilled contractors, and the risk of a poor joint is minimized. We recommend that the use of aluminum conductors be restricted to sizes not smaller than No. 4 AWG, and that installation be permitted *only* by contractors who certify expertise in the specialized techniques involved. Also, local codes and the electrical inspectors should be consulted. All references in this text, including all tables and illustrations, are to copper conductors. The following sections are devoted to a brief description of the principal building wire types.

15.11 Flexible Armored Cable

Among the most common types of cable run without raceways is the NEC type AC armored cable,

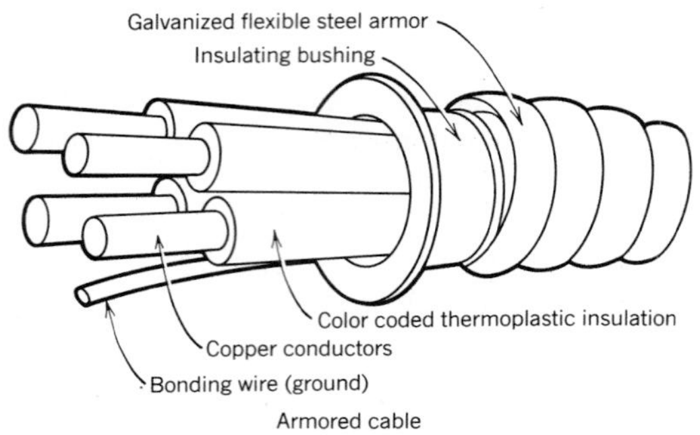

Fig. 15.3 *Type AC flexible armored cable (BX). The bushing is installed to protect the wires from the sharp metal edges of the cut armor. Courtesy of AFC/NorTek.*

commonly known in the smaller sizes by the trade name "BX." It is an assembly of insulated wires, bound together and wrapped with a spiral-wound interlocking strip of steel tape (see Fig. 15.3). The cable is installed by simple U-clamps or staples holding it against beams, walls, and so on. This type of installation is frequently used in residences and in the rewiring of existing buildings. Use of type AC cable is generally restricted to dry locations. For specific application details see NEC Article 333, "Armored Cable."

15.12 Nonmetallic Sheathed Cable (Romex)

In application, NEC types NM and NMC, also known by the trade name "Romex," are restricted to small buildings, that is, residential and other structures not exceeding three floors above grade (see Fig. 15.4). The plastic outer jacket, unlike the armor on type AC, makes type NM easier to handle but more vulnerable to physical damage. For application details and restrictions see NEC Article 336, "Non-metallic Sheathed Cable." Typical installation technique is shown in Figure 15.5.

15.13 Conductors for General Wiring

Under this heading (Article 310) the NEC lists the wire types that are generally used and installed in raceways and referred to by the term "building wire." The most common types are listed in Table

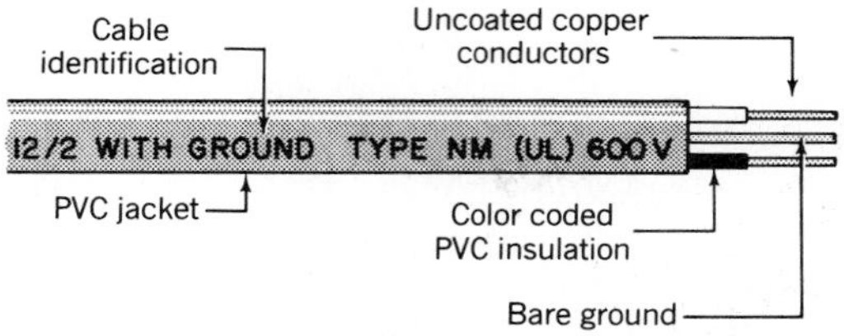

Fig. 15.4 Construction of typical NEC type NM cable. The cable is a two-conductor, No. 12 AWG with ground, insulated for 600 V. Also normally shown are the manufacturer, cable trade name, and the letters (UL), which indicate listing of this product by the Underwriters Laboratories, Inc. The ground wire may be bare or covered, and the entire cable may be obtained flat (illustrated), oval, or round.

15.4. These wires consist of a copper conductor covered with insulation, and in some instances with a jacket (see Fig. 15.6).

15.14 Special Cable Types

Although most building wiring is accomplished with plastic-insulated 300-V and 600-V conductors of the types described in the preceding sections, applications often require the use of special cables. These include high-voltage cables, armored cables, corrosion-resistant jacketed cables, underground cables, and so on. The reader is referred to manufacturers' catalogs and the NEC for construction and application details. Service entrance cables and installation are discussed in Sections 16.1 to 16.4, which cover electric service.

15.15 Busway/Busduct/Cablebus

A busway (busduct) is an assembly of copper or aluminum bars, in a rigid metallic housing. Its use is almost always preferable, from an economic viewpoint, in two instances: when it is necessary to carry large amounts of current (power), and when it is necessary to tap onto an electrical power conductor at frequent intervals along its length.

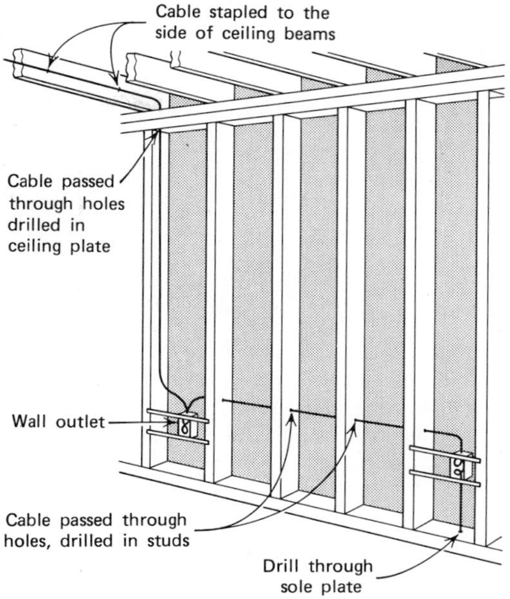

Fig. 15.5 Typical wiring technique using types NM (Romex) or AC (BX), in wood stud-type construction. With metal stud construction the cables are passed through precut openings in lieu of field-drilled holes.

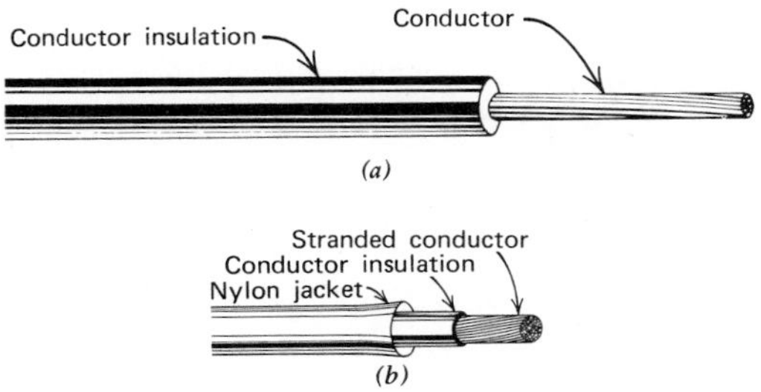

Fig. 15.6 (a) *Typical construction of unjacketed building wire such as type THW; see Table 15.4. Conductors normally are solid through No. 8 AWG, and stranded in sizes No. 6 AWG and larger; see Table 15.1.* (b) *The illustrated construction is typical for any nylon-jacketed cable such as THWN or THHN. (The first three letters indicate the type of insulation, and the final N indicates the nylon jacket). Illustration* (b) *Courtesy of ITT, Royal Electric Division.*

1. The alternative in the case of heavy current is to use paralleled sets of conductors or a single large conductor. Both of these alternatives are expensive and the latter becomes increasingly inefficient as cable size increases, because large round conductors require more cross section per ampere than small ones. This is not the case with the flat

Fig. 15.7 A sectional view of this type of busduct shows the tight assembly of insulated conductors within a metal housing. This design, unlike the ventilated type, can be mounted in any position, since heat dissipation is by conduction and radiation. The eight sets of cable shown have the same current-carrying capacity as the busduct shown. Reproduced by permission of The Square D Co.

conductors (busbars) used in busduct (see Fig. 15.7).

2. Where many power tap-offs are required, costs become very high because of the large amount of expensive hand labor involved, since a connection must be made to each conductor in the run. The preferable alternative is to use ''plug-in'' busway or busduct, to which connection can be made simply and rapidly with a plug-in device, in similar fashion to the insertion of a common plug into a wall receptacle. An additional advantage accrues in that connection and disconnection is simply a matter of inserting or withdrawing the plug-in device, whereas cable taps are permanent connections (see Fig. 15.8).

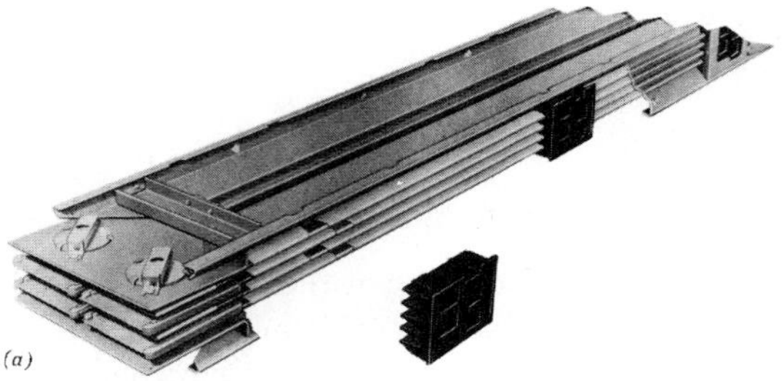

(a)

Fig. 15.8 Construction of plug-in busduct. Plug-ins are spaced evenly on alternate sides to facilitate connection of plug-in breakers, switches, transformers, or cable taps. Housing is of sheet steel with openings for ventilation. Cover plate is not shown. Reproduced by permission of The Square D Co.

Busduct is specified by type, material, number of buses, current capacity, and voltage: aluminum feeder busduct, 4 wire, 1000 A, 600 V; or copper plug-in busway, 100 A, 3 wire, 600 V. Feeder busduct (no plug-in capability) is available in ratings from 400 to 4000 A. Plug-in busway is available from 30 A for lighting or light machinery circuits (see Fig. 21.39, page 1105) to 3000 A. A wide variety of fittings and joints are available for all busways to permit easy installation (see Fig. 15.9). Designs are available for indoor and outdoor application.

Cablebus is similar to ventilated busduct, except that it uses insulated cables instead of busbars. These cables are rigidly mounted in an open space-frame. The advantage of this construction is that it carries the ampacity rating of its cables *in free air,* which is much higher than the conduit rating, thus giving a high amperes-per-dollar first-cost figure. Its principal disadvantage is bulkiness.

As an example of the type of study that should be made when considering an item as relatively simple as an electrical feeder, refer to Table 15.5, which shows the results of such a study in relative cost terms. Note that when considering first cost alone, the advantage lies in cablebus; adding the energy-loss consideration shifts the advantage to cable tray and interlocked armor cables. No general conclusion should be drawn from Table 15.5 regarding costs. A change in feeder length, num-

Fig. 15.9 A typical installation of compact design busduct. Note that the individual busducts are supported by channels hung from the ceiling and that the same hangers support more than one level of bus. Turns are easily made in the same plane (horizontal angle) and in two planes (vertical to horizontal). Reproduced by permission of The Square D Co.

ber of taps, hours of operation, energy rates, or any of the other factors can shift the advantage to a different system. The point of our study is to demonstrate clearly that life-cycle costs and first costs often yield entirely different results and, therefore, that this type of study is required before an intelligent decision can be made. (Life-cycle cost is taken to mean the present value of all costs over the installation's life cycle—in this case, 20 years for the system.)

Two additional items are worthy of note.

1. The very factors that yield lower first cost operate to yield higher operating cost. The smaller copper sizes in busduct and cablebus, permitted by high-temperature insulation and good ventilation, cause increased power loss because of their higher resistivity.
2. If the heat loss from the busduct or cablebus can be used to advantage, the related energy cost can be credited, instead of being considered a total loss, and life-cycle costs can be changed considerably. Conversely, it can also affect the building cooling load.

15.16 Flat Cable Assemblies and Lighting Track

Two special construction assemblies that act as light-duty (branch circuit) plug-in busways are widely used. The first—referred to in NEC Article 363 as "Flat Cable Assemblies"—is field constructed; the second, known in the industry as lighting track, is factory prepared and field mounted. Unlike the heavier plug-in busways, both are specifically intended to feed utilization equipment directly, which, as one of the names shows, is generally lighting equipment.

(a) Flat Cable Assemblies. A specially designed cable consisting of two, three, or four conductors, No. 10 AWG, is field installed in a rigidly mounted standard 1⅝ in. square structural channel. Power tap devices, installed where required, puncture the insulation of one of the phase conductors and the neutral. Electrical connection is then made to the pigtail wires that extend from the tap devices. This connection can extend directly to the device or to an outlet box with a receptacle, which then acts as a disconnecting means for the electric device being served. In this fashion lights, small motors, unit heaters, and other single-phase, light-duty devices can be served without the necessity of "hard" (conduit and cable) wiring. Figure 15.10 illustrates this equipment.

(b) Lighting Track. This is a factory-assembled channel with conductors for one to four circuits *permanently* installed in the track. Power is taken from the track by special tap-off devices that contact the track's electrified conductors and carry the power to the attached lighting fixture. The tracks are generally rated at 20 A, and, unlike FC cable assemblies, they are restricted to 120 V. A typical design is shown in Figure 15.11. An application of track lighting is shown in Fig. 21.18, page 1089.

15.17 Cable Tray

This system, which is covered in NEC Article 318, is simply a continuous open support for approved cables. When used as a general wiring system, the cables must be self-protected. The advantages of this system are free-air-rated cables, easy instal-

TABLE 15.5 **Life-Cycle Relative Cost Comparison of 2000-A, 208-V[a] Feeder Installation**

Feeder System Description[a,b] *1*	*Material Cost* *2*	*Labor Cost*[c] *3*	*Total First Cost* *4*	*Power*[d,e] *Loss per 100 ft (Kilowatt)* *5*	*Annual Energy Loss*[f] *(Kilowatt-hour)* *6*	*Annual Energy Cost*[g] *7*	*Life-Cycle Energy Cost*[h] *8*	*Total Life-Cycle Cost*[i] *9*
Cablebus: using 4 sets of 350 MCM, XHHW copper	0.740	0.260	1.00	5.91	25,547	0.122	1.542	2.542 138%
Wire and conduit: 4 sets of 750 MCM, XHHW copper, 3½ in. rigid conduit	0.653	0.477	1.13	2.90	12,536	0.060	0.757	1.87 102%
Busduct: 2000 A, 600 V 3ϕ, 3 W, copper	1.232	0.198	1.43	5.60	24,207	0.116	1.461	2.891 157%
Cable tray: aluminum ladder type, with 4 sets of 750 MCM armored cable	0.750	0.340	1.09	2.90	12,536	0.060	0.757	1.847 100%

[a] Equipment rating is 600 V.

[b] Data in columns 1 to 4 and the illustrations are derived from a study published by Husky Products, Inc., and is used with permission.

[c] Overall labor rate used includes both journeymen and foremen, and overhead costs.

[d] Data in columns 5 to 9 developed by the author.

[e] Based on published resistivity data for cable and bus—assuming 80% demand (1600 A) and all conductors in the system equally loaded.

[f] Based on 80% demand, 12 hours per day, 360 days per year.

[g] Using $0.07 per kilowatt-hour as the combined net rate, including demand charges.

[h] Using 20-year life cycle, 8% fixed capital cost, and 3% annual escalation in energy cost.

[i] Sum of columns 4 and 8.

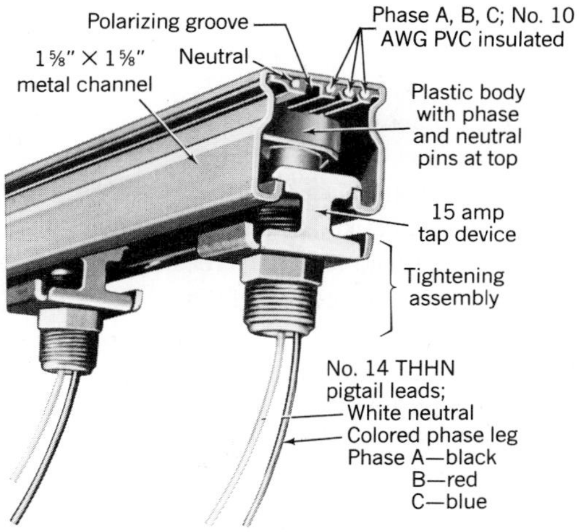

Fig. 15.10 Flat cable assembly installation. The unit illustrated is a four-conductor, 30-ampere, 600-V FC cable. Taps may be phase to neutral or phase to phase, that is, 120 V, 277 V, or 480 V. The tap can feed directly to a device or can energize a receptacle in an outlet box. A lighting fixture can be hung from the FC cable channel with a fixture hanger and hook. If the tap is removed, the pin holes in the PVC insulation "heal," maintaining the integrity of the insulation. Courtesy of Chan-L-Wire®/Wiremold Co.

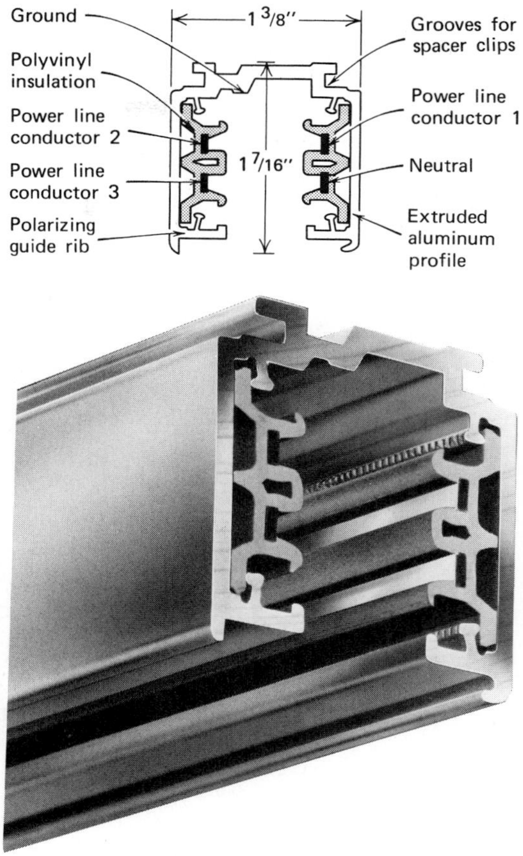

Fig. 15.11 Lighting track. The electrified conductors are permanently installed in the aluminum track, which is grounded for safety. Tracks and insert plugs are available in single-circuit design and multiple-circuit design (illustrated). Accessory fittings permit taps, joints, feeds, and connections at any point on the track. Courtesy of Nutone Division, Scovill.

lation and maintenance, and relatively low cost. The disadvantages are bulkiness and the required accessibility.

15.18 Closed Raceways

The following sections deal with closed wiring raceways, which will complete our discussion of raceways. We will not go into the details of construction and application, because of space limitation, and because those data are readily available from manufacturers and the applicable NEC articles. We will, however, provide enough material for the reader to become familiar with the types of raceways, their common applications and limitations, and, where applicable, comparative characteristics.

15.19 Steel Conduit

The purpose of conduit is to

1. Protect the enclosed wiring from mechanical injury and corrosion.
2. Provide a grounded metal enclosure for the wiring in order to avoid shock hazard.
3. Provide a system ground path.
4. Protect surroundings against fire hazard as a result of overheating or arcing of the enclosed conductors.
5. Support the conductors.

For these reasons the NEC generally requires that all wiring be enclosed in a rigid metallic conduit. Metal electrical raceways and associated fittings must be corrosion resistant. To this end, steel conduit is manufactured in several ways, among which are hot-dip galvanized, sherardized (coated with zinc dust), enameled, and plastic covered.

There are three types of steel conduit that differ basically only in wall thickness. They are, in order of decreasing weight

1. Heavy-wall steel conduit, also referred to simply as "rigid steel conduit"; it is covered in NEC Article 346.
2. Intermediate metal conduit, usually referred to as IMC; this is covered in NEC Article 345.
3. Electric metallic tubing, normally known as EMT or thin-wall conduit; this is covered in NEC Article 348.

The differences are clearly shown in Tables 15.6 and 15.7. Several types of heavy-wall conduit plus EMT are shown in Fig. 15.12. The equivalent millimeter (mm) sizes of conduits are given in Table 15.8.

Rigid conduit and IMC use the same fittings and are threaded alike. As a result of its thin wall, EMT does not lend itself to threading; instead, it uses set-screw and pressure fittings. The thinner walls of EMT and IMC yield a larger ID (inside diameter) and, therefore, easier wire pulling. This combination of lower weight and easier wire pulling gives EMT and IMC a distinct labor cost advantage over rigid conduit, which is enhanced further in jobs with a great deal of field bending and

TABLE 15.6 **Comparison of Steel Conduit Diameters**

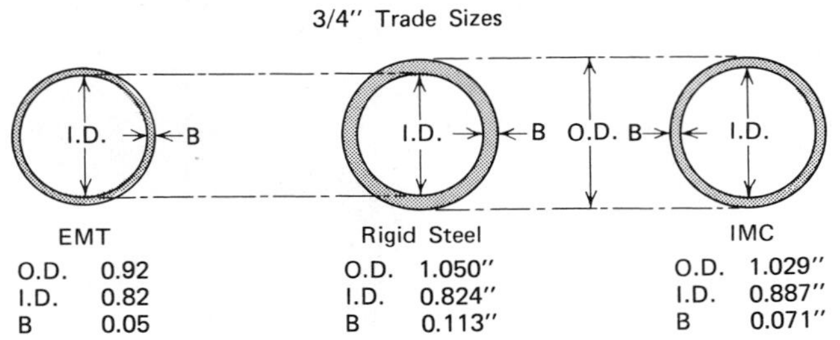

TABLE 15.7 **Comparative Dimensions and Weights of Metallic Conduit**

Nominal or Trade Size	Outside Diameter (Inches)				Weight per 10-ft Length (Pounds)[a]			
	RS[b]	IMC[c]	EMT[d]	AL[e]	RS	IMC	EMT	AL
½	0.84	0.82	0.71	0.84	7.9	5.7	2.9	2.7
¾	1.05	1.03	0.92	1.05	10.5	7.8	4.4	3.6
1	1.32	1.29	1.16	1.32	15.3	11.2	6.4	5.3
1¼	1.66	1.64	1.51	1.66	20.1	14.4	9.5	7.0
1½	1.90	1.88	1.74	1.90	24.9	17.6	11.0	8.6
2	2.38	2.36	2.20	2.38	33.2	23.5	14.0	11.6
2½	2.88	2.86	2.88	2.88	52.7	39.3	20.5	18.3
3	3.50	3.48	3.50	3.50	68.3	48.3	25.0	23.9
3½	4.00	3.97	4.00	4.00	83.1	56.1	32.5	28.8
4	4.50	4.47	4.50	—	97.2	62.5	37.0	—

Source: All data courtesy of Allied Tube & Conduit Corp.

[a]Standard length including one coupling.

[b]Standard heavy-wall rigid steel conduit.

[c]Intermediate-weight steel conduit.

[d]Electric metallic tubing.

[e]Aluminum.

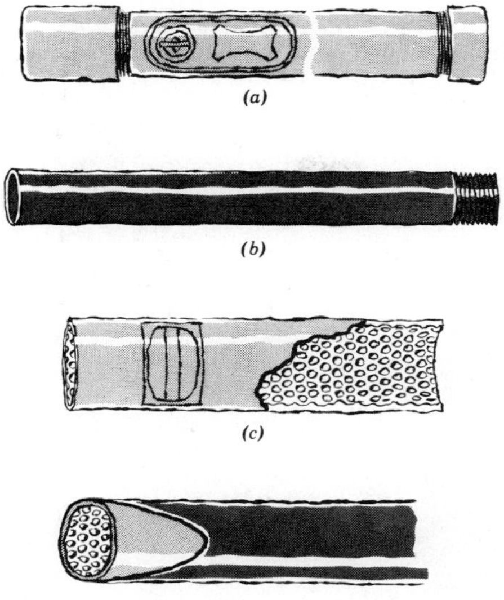

Fig. 15.12 Steel conduits: (a) *Galvanized, heavy wall, rigid;* (b) *black enameled;* (c) *EMT thin wall;* (d) *plastic-coated conduit, for use in highly corrosive atmospheres.*

handling of conduit. Both, however, have application restrictions, which are detailed in the NEC.

Generally, no conduit smaller than ½ in., nominal trade diameter, is used. Ordinary steel pipe may not be used for electric purposes, and all electric steel conduit is distinctively marked as such.

When steel conduit is installed in direct contact with the earth, it is advisable to use hot-dip galvanized type and to coat the joints with asphaltum. If the earth is very wet, the complete conduit system should be coated with an asphalt compound. Conduit is fastened to the structure in much the same way as pipe; with pipe straps and clamps. Vertical load at floor opening is taken with special support clamps. Trapeze mounting is common for conduit banks hung from the ceiling, as in Fig. 15.13.

The selection of conduit size depends on the number and diameter of the wires that may be drawn into the conduit without injuring the wire. The number and radius of bends in the conduit, as well as its total length, affect the degree of abrasion to the wiring insulation. No wires should be installed until the conduit system has been inspected and approved.

TABLE 15.8 **Metric Equivalents of American Conduit Sizes**

Conduit Size (Inches)	*Dimensions (Millimeters)*	
	OD	*ID*
$\frac{3}{8}$	16.5	9.5
$\frac{1}{2}$	20.5	12.7
$\frac{5}{8}$	23.0	15.9
$\frac{3}{4}$	26.5	19.0
1	33.0	25.4
$1\frac{1}{4}$	42.0	31.75
$1\frac{1}{2}$	48.0	38.1
$1\frac{3}{4}$	52.0	44.5
2	59.7	50.3
$2\frac{1}{2}$	76.0	63.5
3	89.0	76.2
$3\frac{1}{2}$	101.5	88.9
4	114.0	101.6

For structural reasons, conduits in concrete slabs are run close to the bottom surface (in the portion of the slab in tension) or near the central portion. If a great number of conduits must be embedded, it may be necessary to increase the slab thickness. In many instances, the structural slab is covered with a concrete topping, in which conduit may be installed without affecting slab integrity. In all cases, local building codes should be consulted for limitations on embedded conduits. In any event the top of any conduit shall be at least ¾ in. below the finished floor surface in order to prevent cracking. When heavy trucking is expected, this allowance should be increased to 1½ in. minimum.

In general, the following rules should be observed and included in all specifications for conduit work in concrete slabs:

1. Conduits shall have an OD no greater than one-third of the slab thickness as measured at its thinnest point.
2. Conduits running parallel to each other shall be spaced not less than three times the OD of the largest conduit center-to-center.
3. Conduits running parallel to beam axis shall not run above beams.
4. Conduit crossings shall be as near to a right angle as possible.
5. Minimum cover over conduits shall be ¾ in.

Fig. 15.13 Typical overhead conduit bank installation. Note that due to field conditions the insert (a) *for hanger rods was inadequate, and an additional insert* (b) *was added. This conduit bank uses EMT, which has a pipe wall thickness of approximately one-third that of heavy-wall rigid conduit. The resulting weight difference in a large bank such as this is very pronounced. EMT is suitable here since the overhead location protects the pipe from severe physical abuse. EMT joints are made with set-screw fittings* (c). *Note how individual conduits are fixed by clamps to the trapeze channel* (d). *Courtesy of Republic Steel Corp.*

15.20 Aluminum Conduit

The use of aluminum conduit has increased greatly in recent years because of the weight advantage that aluminum has over steel, being even lighter than EMT. The savings in labor cost more than offsets the additional cost of the material itself. In addition, aluminum has better corrosion resistance in most atmospheres; it is nonmagnetic, giving lower voltage drop; it is nonsparking; and, generally, it does not require painting.

Its major drawback is its deleterious effect on many types of concrete, causing spalling and cracking when embedded. Although manufacturers can demonstrate many cases of embedding in concrete without harmful effect, it is a procedure that should be avoided unless the concrete additives are rigidly controlled. It is also inadvisable to bury aluminum in earth, with or without asphalt or other coating, because of the rapid corrosion often encountered. Other difficulties frequently encountered are freezing of threaded joints, because of thread deformation, and difficulty in obtaining electrical contact with grounding straps. With the exceptions noted above, aluminum conduit may be used in all locations where steel conduit is used.

15.21 Flexible Metal Conduit

This type of conduit construction—which consists of an empty spirally wound interlocked armor raceway—is known to the trade as "Greenfield" and is covered in NEC Article 350. It is used principally for motor connections or other locations where vibration is present, where movement is encountered, or where physical obstructions make its use necessary. The acoustic and vibration isolation provided by flexible conduit is one of its most important applications. It should always be used in connections to motors, transformers, ballasts, and the like. A typical application is for wiring inside metal partitions. When covered with

Fig. 15.14 *This is a particularly good application of liquid-tight flexible conduit since it provides weatherproofing and acoustical isolation of the noise-producing transformer. Courtesy of Electri-Flex Company.*

a liquid-tight plastic jacket, it is suitable for use in wet locations (see Fig. 15.14). In this construction it is most often known by the trade name ''Sealtite.''

15.22 Nonmetallic Conduit

A separate classification of rigid conduit (NEC Article 347) covers raceways that are formed from such materials as fiber, asbestos-cement, soapstone, rigid polyvinyl chloride (PVC), and high-density polyethylene.

For use above ground, this conduit must be flame retardant, tough, and resistant to heat distortion, sunlight, and low-temperature effects. For use underground the last two requirements are waived. Generally, nonmetallic conduit may be used without restriction in nonhazardous areas, within the physical limitations of the material involved. Thus plastic conduit has a temperature limitation, asbestos-cement has considerable physical strength limitations, and so on. As a result of these limitations, PVC conduit is the material of choice for indoor exposed use and asbestos-cement, fiber, and PVC plastic for outdoor and underground use. A separate ground wire *must* be provided, since the ground provided by a metallic conduit is absent.

15.23 Metal Surface Raceways

These raceways are covered in NEC Article 352. Surface metal raceways and multi-outlet assemblies may be utilized only in dry, nonhazardous, noncorrosive locations and may generally contain only wiring operating below 300 V. Such raceways are normally installed in exposed condition and in places not subject to physical injury.

The principal applications of surface metal raceways are:

1. Where the architecture does not permit recessing (see Fig. 15.15).
2. Where economy in construction weighs very heavily in favor of surface raceways (see Fig. 15.16).
3. Where outlets are required at frequent intervals, and where rewiring is required or anticipated (see Fig. 15.17).
4. Where access to equipment in the raceways is required (see Fig. 15.18).
5. Where one is rewiring existing installations to avoid the extensive and expensive cutting and patching required to ''bury'' a raceway (see Fig. 15.19).

15.24 Floor Raceways

The NEC recognizes three types of floor raceways:

Underfloor raceways—Article 354

Cellular metal floor raceways—Article 356

Cellular concrete floor raceways—Article 358

All three types are applicable to all types of structures and none may be used in corrosive or hazardous areas. The fundamental difference between them is that underfloor raceways are added on to the structure, whereas cellular floor raceways are part of the structure itself—and therefore have a pronounced effect on the building's architecture.

Fig. 15.15 *The exposed wood members require the use of an unobtrusive surface raceway. A small flat raceway feeds receptacle outlets into which the elaborate hanging fixtures are plugged. Courtesy of Wiremold Co.*

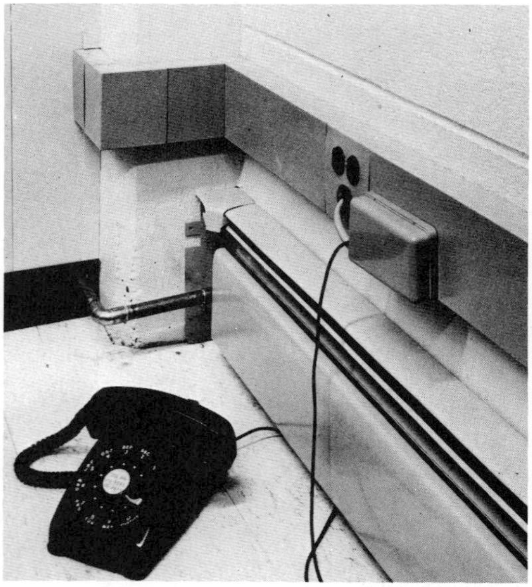

Fig. 15.16 *A split raceway under a single cover handles both power and telephone conductors. By placing it above the heating unit as shown, or below a fan-coil unit, no obstacle is created and no room space is lost. Multisection units are available for running power and signal conductors; where used, the wiring must be segregated and the compartments color coded and maintained in the same relative position throughout the installation. Courtesy of Wiremold Co.*

Fig. 15.17 (below) *Surface outlet finds ready application in areas where requirements change frequently and are usually heavy, as on test benches. Courtesy of Wiremold Co.*

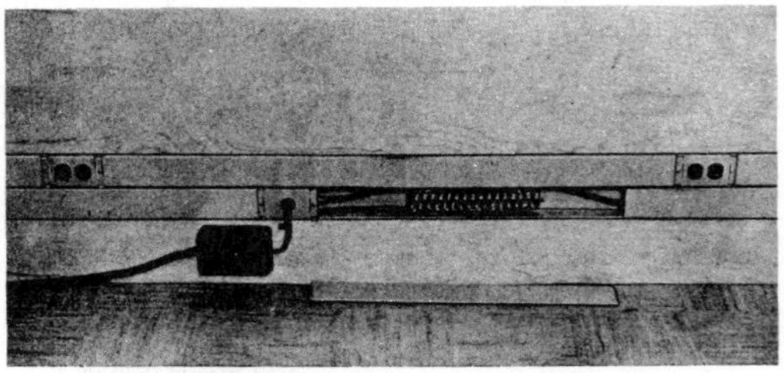

Fig. 15.18 *Parallel power and telephone raceways with separate covers are shown. Note easy access to telephone terminal board mounted inside the telephone raceway. In a conduit job, such a terminal board would require the installation of a separate cabinet. Where desired, a single cover can be installed as in Figure 15.16. Courtesy of Wiremold Co.*

Fig. 15.19 Although surface raceways may be unsightly in some installations, they are entirely unobjectionable in areas with exposed piping, as in this school rewiring job. Courtesy of Wiremold Co.

15.25 Underfloor Duct

These raceways may be installed beneath or flush with the floor. They find their widest application in office spaces, since their use permits placement of power and signal outlets immediately under desks and other furniture, regardless of furniture layout. Where such underfloor raceways are not employed, and it is desired to place an outlet *on the floor,* one of the following methods is necessary:

1. Channel the floor and install a conduit in the chase, connecting it to the nearest wall outlet. Patch the chased portion.
2. Drill through the floor and run a conduit on the ceiling below to an outlet below. When using this technique, special ''poke-through'' fittings with adequate fire rating are required to restore the integrity of the slab.
3. Drill through the floor twice and connect the new outlet to an existing outlet via a conduit on the ceiling below. This, like (2), is expensive and disturbs the occupant below.
4. Install a surface floor raceway.

Since all of these have obvious major disadvantages, underfloor (UF) duct systems were widely employed until the introduction of what may be called over-the-ceiling ducts (in apposition to under-the-floor ducts) and flat cable under-carpet wiring. These systems are discussed in Sections 15.29 and 15.30. The reason that alternate systems were developed is economic—underfloor duct systems are expensive and frequently underutilized, giving an unsatisfactory return on investment. However, before discussing the relative merits of systems, an understanding of what UF duct systems are, and how they are assembled and utilized, is in order.

Underfloor duct systems are available in two basic designs—single level and two level. In a single-level system all of the system components are on the same level—the feeder ducts from the panels, the distribution ducts with inserts for floor outlets, and, most important, the junction boxes. As the number of parallel distribution ducts grows with the size of the open floor area and the density of the furniture layout, the need for more feeder ducts and consequently larger and more complex junction boxes also grows. A modern office floor layout invariably requires a triple duct system (power, telephone, and signal). Since power and telephone wiring must be separated by metal barriers, the junction boxes become complex, and consequently large and deep. The depth of these boxes obviously controls the amount of concrete fill required. A simple single-level system requires a minimum concrete fill of 2½ in., while a complex one can easily require an inch more. With such a concrete fill requirement, a two-level system with its great flexibility becomes desirable.

Refer now to Figs. 15.20 and 15.21, which illustrate a two-level system. Note that the fundamental difference is that the feeder and distribution ducts are on different levels, thus eliminating the necessity for complex junction boxes; this gives the system unlimited feeder capacity and thereby obviates the necessity for supplemental conduit feeds, as in the single-level system. The drawback of this system is the conduit fill required. A *minimum* of 3⅝ in. is required, though additional slab thickness throughout can be avoided by depressing part of the slab to accommodate the feeder ducts, as shown in Fig. 15.22.

In a single-level system the ducts are arranged in a grid pattern, with spacing selected to provide desk coverage. Thus, if desks are 48 in. wide with a 4-ft aisle between, an 8-ft lateral duct spacing would provide adequate coverage. Since feeder ducts—particularly for telephone service—rapidly become filled, cross-connecting distribution and feeder ducts must be supplied. Also, the location of telephone closets has a pronounced effect on the pattern and spacing selected. With a two-level system, distribution ducts would be run longitu-

ELECTRICITY

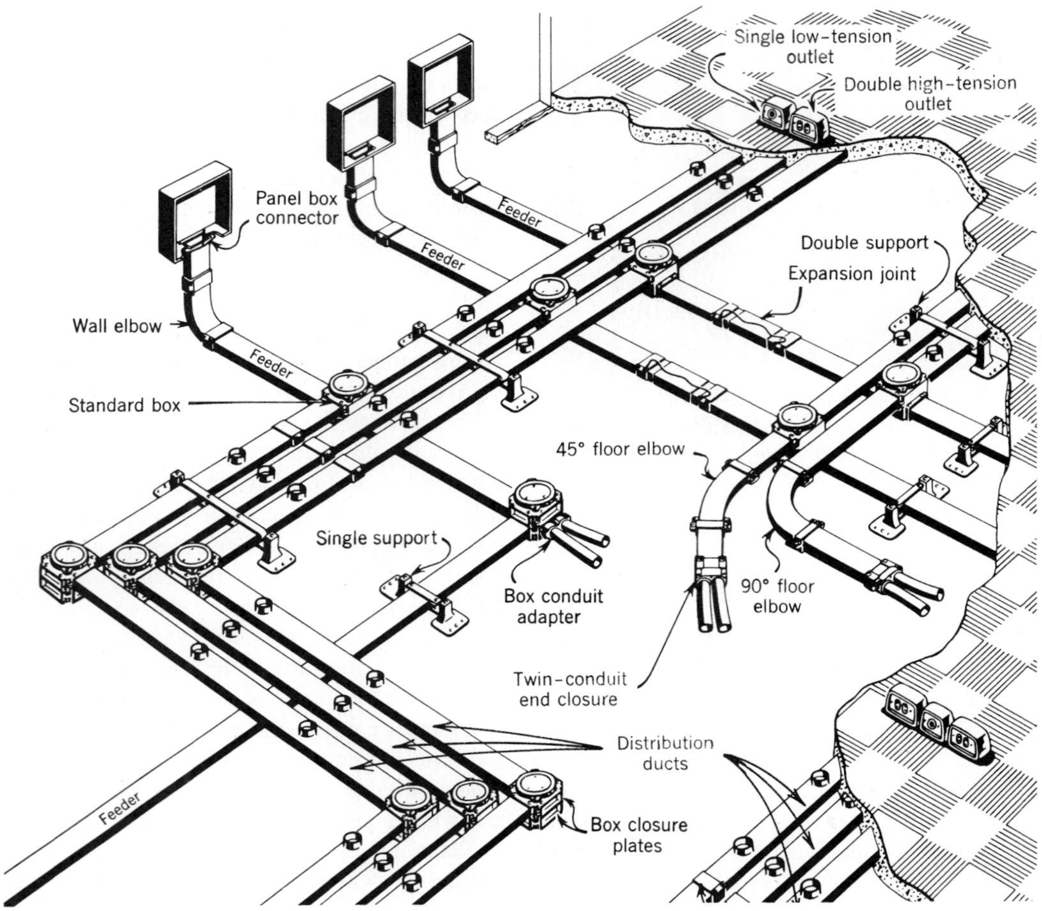

Fig. 15.20 *Two-level underfloor duct system, using standard size duct only. Feeder ducts are run below the distribution (outlet) ducts, resulting in simple standard junction boxes (see Fig. 15.21), unlimited feeder capacity, and complete separation of power and signal wiring.* Source: *General Electric—Wire and Cable Dept.*

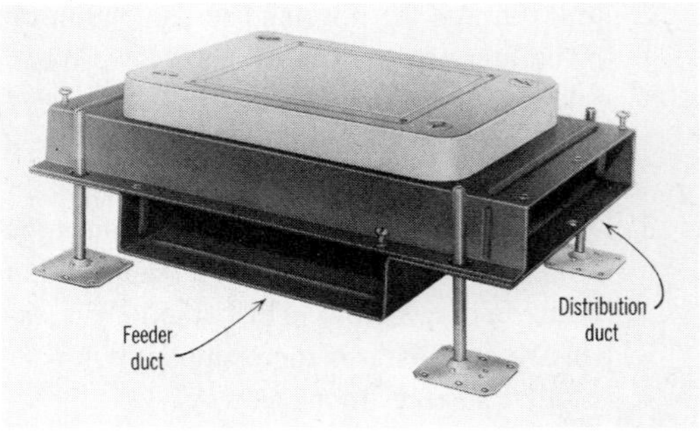

Fig. 15.21 *A typical two-level junction box demonstrates the simplicity of the two-level system. Courtesy of The Square D Co.*

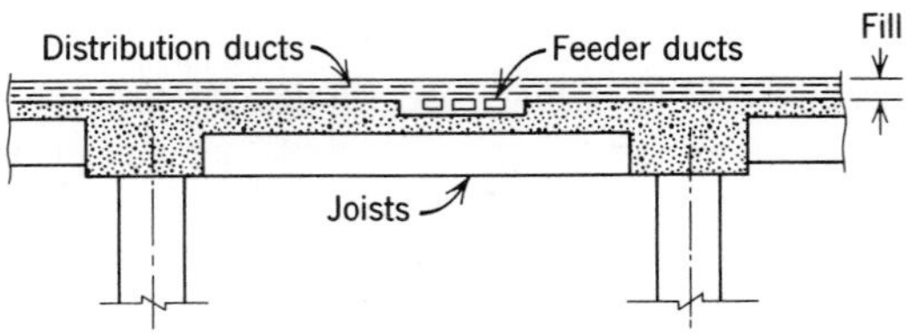

Fig. 15.22 *Setting a two-level underfloor duct system. To avoid thickening fill, a depression in the slab can accept feeder ducts. Ducts would be run near the bay center to avoid the negative steel of joists, near columns.*

dinally at 6- to 8-ft spacing, with lateral feeder ducts as necessary.

Underfloor ducts may be cast into the structural slab in lieu of being in fill or topping, but the slab must be designed to accommodate them. The use of a fill or topping on the structural slab for underfloor duct has these advantages:

1. Ducts can be run in any direction, without conflict to structural elements.
2. Finishing is simplified.
3. Coordination is simplified.
4. Formwork and construction sequence are simplified.

The disadvantages are:

1. Additional concrete increases costs directly by increasing weight. This is particularly expensive in seismic designs.
2. Height of building may be increased.

In retrofit jobs where underfloor duct is decided upon rather than one of the other floor or ceiling raceway systems, the ducts will obviously be placed in a new floor fill.

In conclusion, some general comments on the application of underfloor duct systems are in order. Underfloor duct systems are *expensive*. They can add 50% to the building's electric system cost, without consideration of the construction costs involved. To justify their use, therefore, the building should meet these criteria:

1. Open floor areas, with a requirement for outlets at locations removed from walls and partitions.
2. Under-carpet wiring system inapplicable.
3. Outlets from ceiling systems unacceptable.
4. Frequent rearrangement of furniture and other items requiring electrical and signal service.

The facilities that may meet these criteria are prestige office buildings, museums, galleries and other display-case spaces, high-cost merchandising areas, and selected areas in industrial facilities. Bear in mind that even in high-cost office construction, underfloor duct systems are difficult to justify economically unless furniture layout will change. For fixed layouts, conduit-fed floor boxes are the economical choice. In doubtful cases, alternate arrangements can be planned and an intelligent choice made after costs and impact on the building structure are studied.

15.26 Cellular Metal Floor Raceway

The underfloor duct system described above is best applied to known furniture layouts and to rectilinear arrangements. Random arrangements, such as those found in office landscaping, require a fully accessible floor—if indeed the floor is to be used for electrification. This may be provided by a cellular (metal) floor that is an integrated structural/electrical system. The floor can be fully or partially electrified. A floor designed with two or three electrified cells adjacent to several cells of structural floor, as shown in Fig. 15.23, will give sufficient coverage for most purposes. One of the many structural element designs available is shown in that figure.

The cellular floor is part of the structural system and is designed accordingly. The electrical wiring is fed into the cells from header ducts and/or trenches that run perpendicular to the floor cells and constitute a system of underfloor ducts in themselves. The header ducts in turn are fed from lighting panels and signal and telephone cabinets, in much the same manner as normal underfloor ducts.

Three types of wiring systems generally run in separate floor cells and header ducts: general lighting and appliances, computer wiring, and telephone and signal systems. The latter two may be combined in a single cell only if the signal system voltage and power level are low. A complete range of outlets and fittings is available.

15.27 Precast Cellular Concrete Floor Raceways

This structural concrete system is similar to a cellular metal floor in application and has the same advantages: large capacity, versatility in that each cell is a potential raceway, and flexibility in outlet placement and movement. Here too, as with the metal cell constructions, first cost is higher than that of a standard underfloor duct installation although life-cycle cost is frequently lower, depending on space use (see Fig. 15.24). A cell is

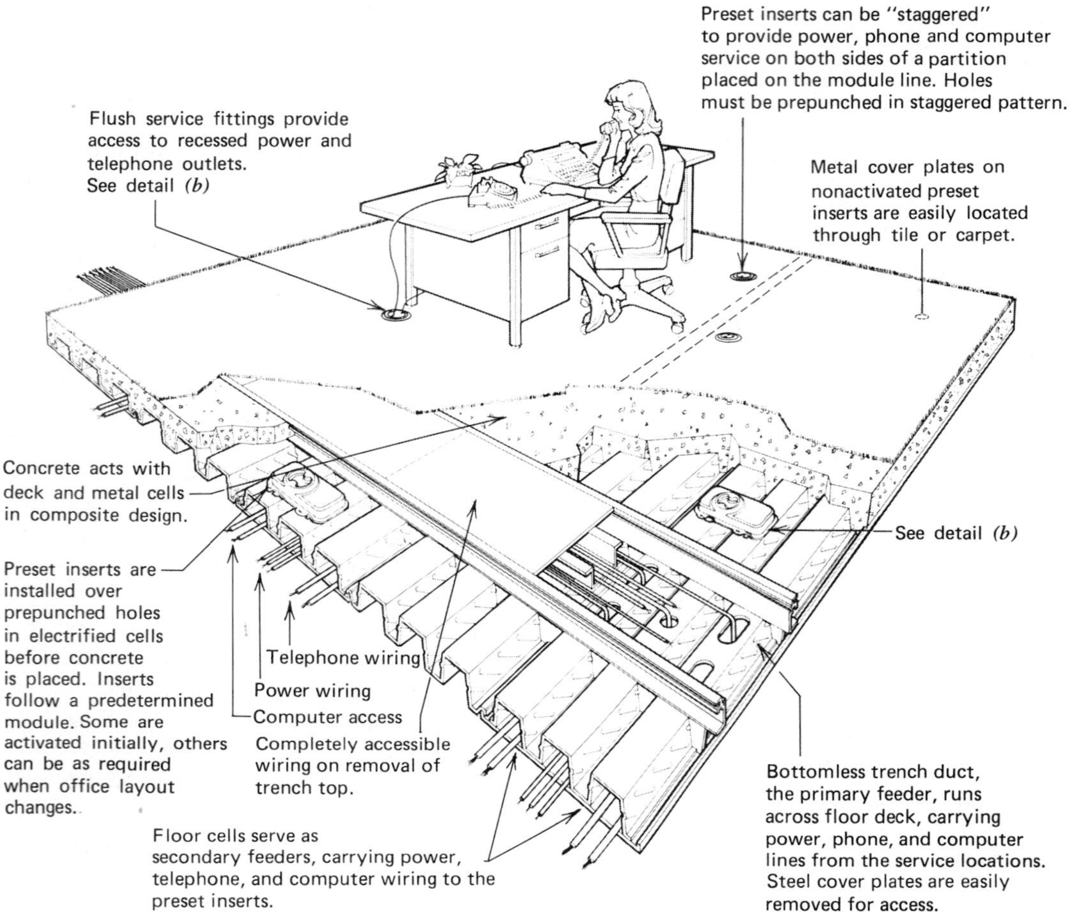
Preset inserts can be "staggered" to provide power, phone and computer service on both sides of a partition placed on the module line. Holes must be prepunched in staggered pattern.
Flush service fittings provide access to recessed power and telephone outlets. See detail (b)
Metal cover plates on nonactivated preset inserts are easily located through tile or carpet.
Concrete acts with deck and metal cells in composite design.
See detail (b)
Preset inserts are installed over prepunched holes in electrified cells before concrete is placed. Inserts follow a predetermined module. Some are activated initially, others can be as required when office layout changes.
Telephone wiring
Power wiring
Computer access
Completely accessible wiring on removal of trench top.
Bottomless trench duct, the primary feeder, runs across floor deck, carrying power, phone, and computer lines from the service locations. Steel cover plates are easily removed for access.
Floor cells serve as secondary feeders, carrying power, telephone, and computer wiring to the preset inserts.

(a)

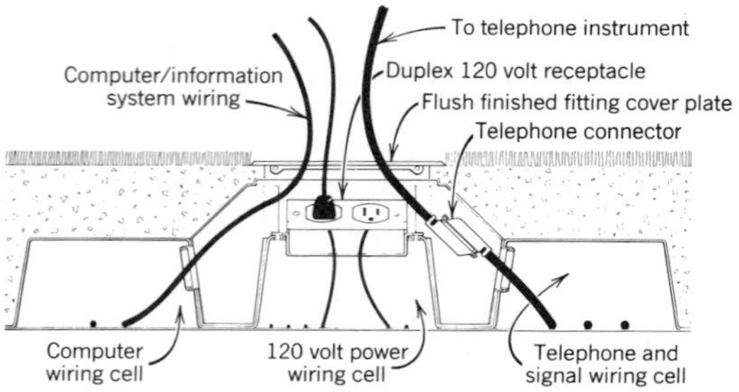

(b)

Fig. 15.23 (a) *One of many designs for electrified cellular floor. The floor cells are available in many designs depending primarily on the structural requirements. The trench that straddles the cells provides the electrical feeds through precut holes in the cells. The trench itself is completely accessible from the top and, when opened, exposes all the wiring and the cells below. Alternatively, the cells can be fed by a header duct that is accessible only through handholes.* (b) *Activated preset insert. Note that the insert straddles the center (power) cell and provides access to the two adjoining low-tension wiring cells. Power and signal wiring are completely separated at all times by metal barriers. If desired, a standard surface fitting can be mounted on the floor in lieu of the flush plate shown. When an insert is to be deactivated, the flush cover plate is simply replaced with a blank plate. Courtesy of INRYCO, Inc.*

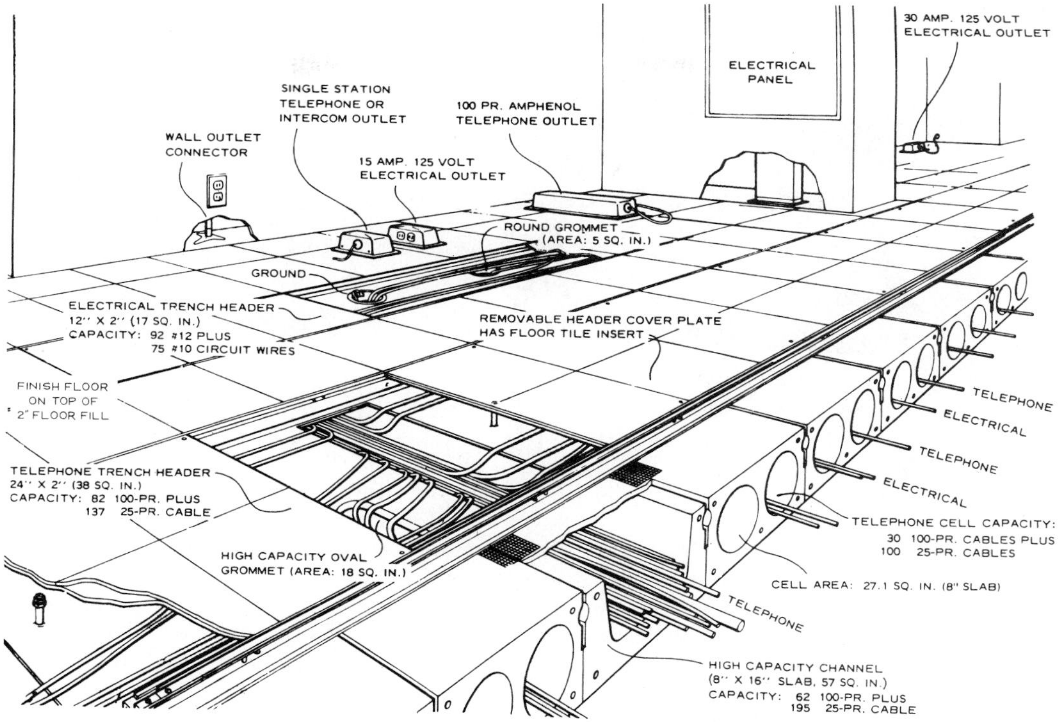

A HIGH-CAPACITY CONCRETE CELLULAR SYSTEM EMPLOYING TRENCH HEADER DUCTS

***Fig. 15.24** Precast cellular concrete slabs are the masonry equivalent of cellular metal construction. Note that cells marked "telephone" can alternatively be used for computer access wiring, as in Fig. 15.23. Courtesy of Flexicore Co. Inc.*

defined in NEC Article 358 as a "single, enclosed, tubular space in a floor made of precast cellular concrete slabs, the direction of the cell being parallel to the direction of the floor member." Feed for these cells is provided, as with metal cellular floor construction, by header ducts. Although header ducts are normally installed in concrete fill above the hollow core structural slab, a header arrangement with feed from the ceiling below is also entirely practical. Like the metallic cellular floor, the cells can be used for air distribution and even for piping, although these items are generally installed in a hung ceiling.

15.28 Full Access Floor

This construction, frequently used in retrofit jobs, provides instant and complete access to an underfloor plenum. The system was originally developed for data processing areas that have a requirement for large, fully accessible cable spaces and for large quantities of conditioned air. The solution to both of these requirements is an infinite access floor consisting of lightweight die-cast aluminum panels supported on a network of adjustable steel or aluminum pedestals (see Fig. 15.25). Panels are available from 18 × 18 in. to 36 × 36 in., and floor depth is normally 12 to 24 in., although taller pedestals are available. The construction is usually fireproof.

15.29 Under-Carpet Wiring System

This wiring system, which is covered in NEC Article 328, is a successful attempt to provide a floor-level branch-circuit wiring system with its inherent advantages, but without the high cost of an underfloor duct system. Basically the system com-

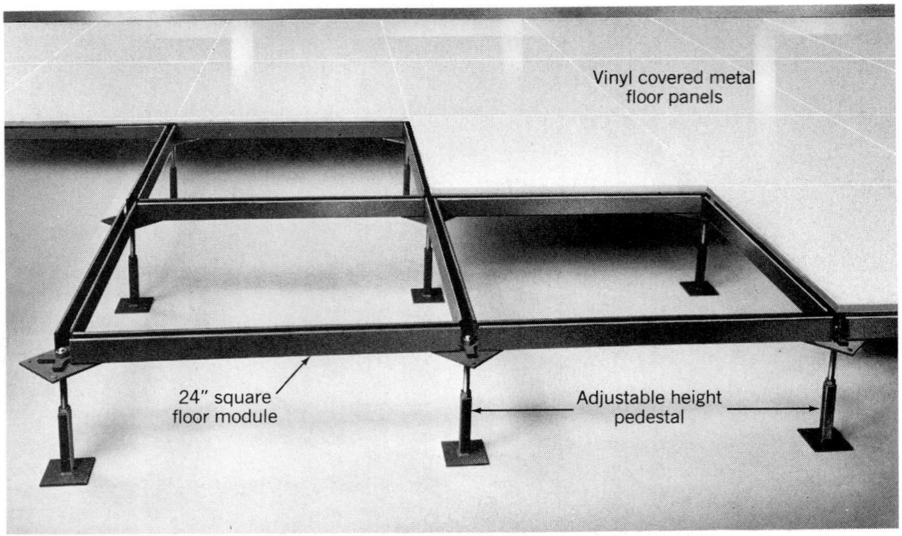

Fig. 15.25 Photo of a typical raised floor system. The illustrated construction (Met-L-Strut by Unistrut) is a 24 in. square module, with vinyl-covered metal floor panels and adjustable height floor pedestals, maximum 18 in. Other systems are available with 18 in. square floor panels and floor cavity height to 36 in. Floor cavity is often utilized as an air plenum in computer rooms, in addition to being an infinite-access wiring space. When so used, wiring must be suitable, per NEC. Courtesy of UNISTRUT Building systems.

prises a factory assembled flat cable (NEC type FCC), approved for installation *only* under carpet squares, and the concomitant accessories necessary for connection to 120-V power outlets. The cable itself consists of three or more flat copper conductors, placed edge to edge and enclosed in an insulating material (see Fig. 15.26). The entire assembly is covered with a grounded metal shield, for physical protection (Fig. 15.27).

All manufacturers also offer a complete line of boxless junction fittings, connectors, adapters, and receptacles. The cable, when properly installed on a hard flat surface, is approxmately 0.03 in. high, and this is essentially undetectable when covered with carpet. Since the carpet can readily be removed in sections, the entire system can be repositioned to meet changing furniture layout requirements, with a minimum of disruption and no structural work. The cable is designed to carry normal physical loads such as office traffic and furniture placement without affecting its electrical performance. Figure 15.27 shows typical installation. Similarly constructed multiconductor cable is available for telephone, low-tension signal, and computer access to satisfy the needs of a modern office installation (see Fig. 15.26*b*).

Because of its low profile, lack of structural impact, flexiblity, and relatively low cost, the FCC under-carpet wiring system is ideal for retrofit work and for new installations where low cost is essential and ceiling systems (Section 15.30) are undesirable. Its major drawback is the code requirement for carpet squares as the floor finish.

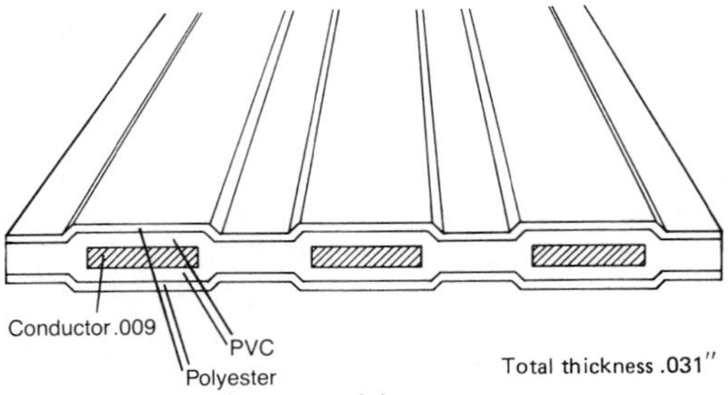

(a)

Fig. 15.26 (a) *Schematic section through one design of NEC type FCC under-carpet cable. The illustrated copper conductors are equivalent to No. 12 AWG. The PVC acts as insulation, and the polyester as both insulation and physical protection. Other designs utilize a metallic bottom shield for physical protection.*

Fig. 15.26 (b) *Multiconductor low-tension communications/data under-carpet cable comes in specific factory-prepared lengths because of the requirement for multipin end connectors, as shown. These connectors cannot be readily field installed. Courtesy of Wiremold Co.* (a), *and Thomas and Betts* (b).

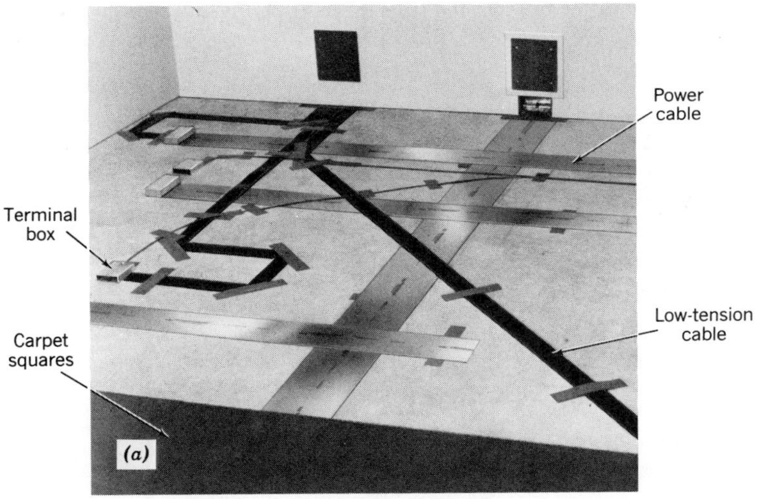

Fig. 15.27 (a) *Flexibility and absence of structural impact are the outstanding advantages of under-carpet-square wiring. Power, telephone, and data pedestals can be located wherever desired. Low-tension wiring is installed over the power cable and simply taped in place. Note the terminal boxes at the cable ends, on which floor outlets* (b) *will be mounted.* (b) *Typical 120-V duplex receptacle outlet connected to the floor wiring system. Note the metallic shield over the cable, and the simplicity of installation: the entire assembly is simply taped to the slab. Courtesy of Thomas and Betts* (a), *and Wiremold Co.* (b).

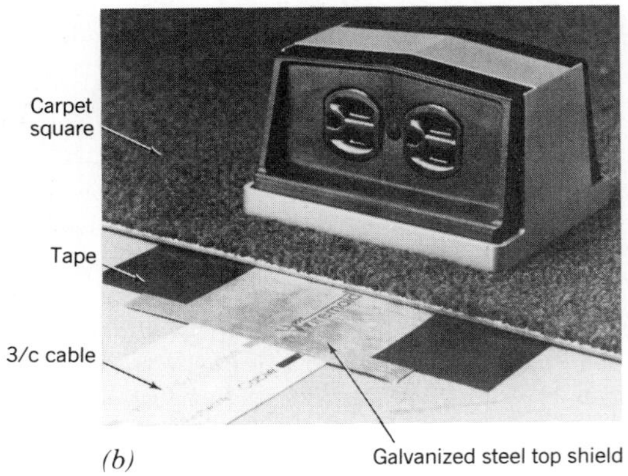

15.30 Ceiling Raceways and Manufactured Wiring Systems

The need for electrical flexibility in facilities with limited budgets coupled with the high cost of underfloor electrical raceway systems encouraged the development of equivalent over-the-ceiling systems. These systems are actually more flexible than their underfloor counterparts, since they energize lighting as well as provide power and telephone facilities; furthermore, they permit very rapid changes in layouts at low cost. This last characteristic is particularly desirable in stores where frequent display changes necessitate corresponding electrical facility changes. In addition to the extreme flexibility of this electrical layout made possible by the ceiling raceway system, it has the additional advantage that it itself, not being cast in concrete like its underfloor counterpart, can be altered at will. Thus, not only layout changes (as mentioned above) but also changes in the utilization of existing spaces can readily be accommodated. This is a particularly important characteristic in merchandising and educational facilities where spaces originally planned for one use have their application completely changed during the course of the building's life.

The systems vary among manufacturers but are essentially the same, and similar to underfloor systems. The entire system is shown clearly in Fig. 15.28. Header ducts (wireways) connect to electrical panels and telephone cabinets in the power and telephone closets, respectively. Telephone headers are normally of larger size than the power header and can carry other low-voltage, low-power signal equipment as well. Distribution ducts (laterals) tap onto the headers. These laterals may act as subdistribution raceways, feed lighting fixtures and telephone and power outlets directly, or do both.

The standard method for extending ceiling-level wiring to floor- or desk-level telephone and power outlets is my means of a vertical two-section raceway fed from the top (see Fig. 15.29). These poles or posts are prewired, contain several power outlets and a telephone connection, and are simply

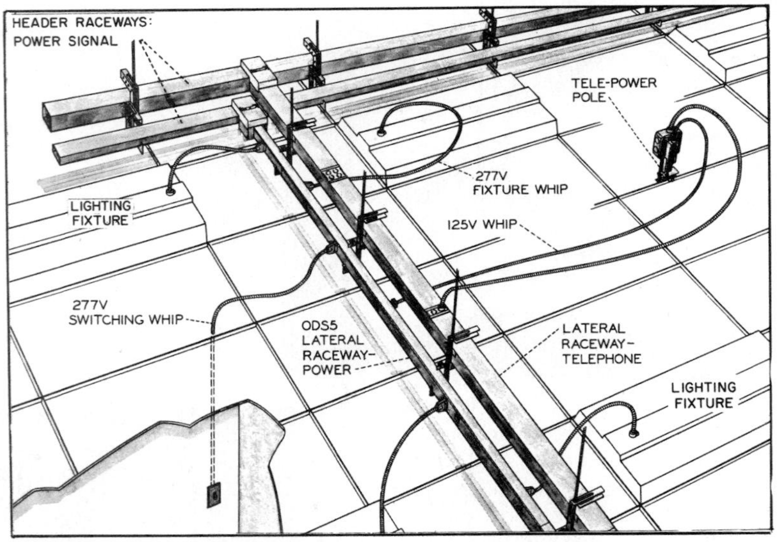

Fig. 15.28 Overhead raceway system fed from a central service core. The system is a conventional "tree" system; that is, the main feed at the panel subfeeds to headers that in turn subfeed to distribution ducts, which in turn subfeed to the utilization devices, which here are lighting fixture and ceiling-to-floor poles (Tele-Power® poles). Depending on the size and layout of the system, laterals may serve as subdistribution only or as subdistribution plus feed to utilization devices. The "whips" shown are prewired cord sets that simply plug into the power raceway to feed lighting, switches, and receptacles in Tele-Power poles. Low-tension and communications raceways are provided with takeoff points for rapid connection. All numbers and nomenclature in this and subsequent related figures refer to Wiremold® equipment. Courtesy of Wiremold Co.

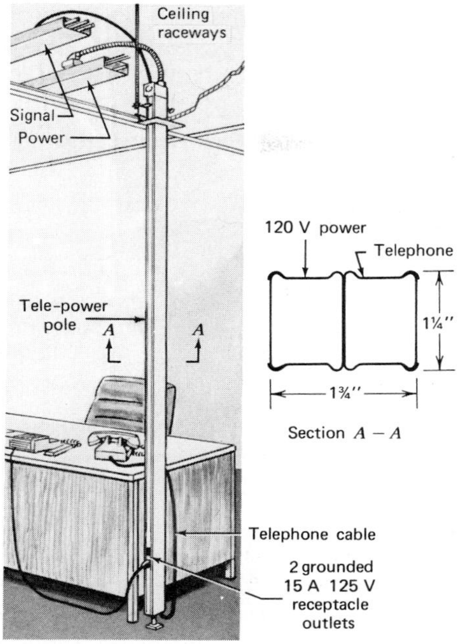

Fig. 15.29 Detail of one design (of many) of telephone-power pole. Other designs have different dimensions, outlets, bases, and colors. Some designs have a third compartment for data-system cables. In this unit, the power compartment is prewired with two single 15-a, 125-V outlets. It is fed with a flexible armored cable terminating in a special polarized plug that plugs in to the ceiling power raceway. The communications compartment is generally unwired. Pole lengths are available to match ceiling height requirement. Courtesy of Wiremold Co.

Fig. 15.30 A service pole in this consulting engineer's office feeds a portable light-table and other electrical accessories. The communication compartment has been left unwired. Courtesy of Wiremold Co.

and easily installed at any desired location. The end result in a hung ceiling office area (Fig. 15.30) or an exposed slab area (Fig. 15.31) is certainly less elegant than a floor-level wiring system, but for most users it is satisfactory, and its low cost as compared to any type of floor-duct system is a prime redeeming feature.

When power feeds to fixtures and receptacles are made with "hard" wiring, considerable field labor is involved with corresponding high cost. Furthermore, the permanent nature of such wiring lessens the inherent flexibility of the raceway sys-

Fig. 15.31 In this economical installation the absence of a hung ceiling is obscured by painting the ceiling, fixture bodies, and raceways black. Courtesy of Wiremold Co.

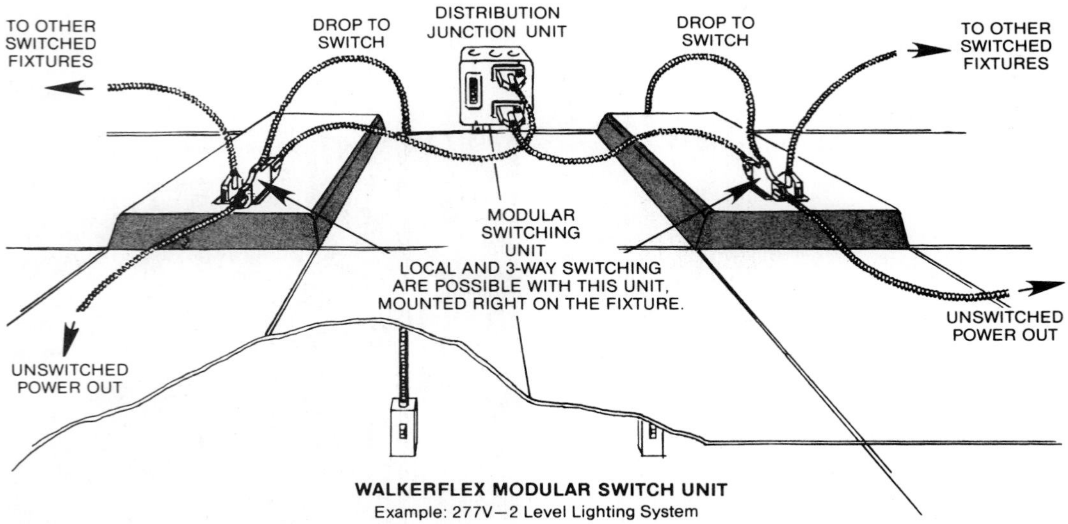

Fig. 15.32 In addition to ease of wiring (and unwiring), the modular factory-prepared wiring units permit flexible switching arrangements. In the drawing, the distribution junction unit is fed from power wiring in a ceiling raceway (not shown), and it in turn supplies lighting fixtures by simple, rapid, plug-in connection. The prefab switching units mounted on the fixtures replace the junction box of the traditional hard-wired system. Note that power poles (unswitched) can be supplied via these junction units. Courtesy of Walker, Div. of Butler Mfg. Co.

tem. To solve both of these problems a number of manufacturers have developed a line of modular branch-circuit wiring elements. These, covered in NEC Article 604, consist of jacketed or armored cable sets terminating in polarized plugs. The polarization prevents accidental interconnection of low voltage, 120-V and 277-V systems. Ceiling raceways are equipped with matching receptacles, and connection to fixtures, poles, and other devices becomes a simple matter of insertion of plugs (see Fig. 15.32).

Disconnection is an equally simple action, resulting in a system of extreme flexibility. The additional cost of the manufactured wiring elements is frequently offset by labor savings even on initial installation and certainly after one or two field changes. Cable sets are available for power (120 V and 277 V), telephone, and all types of low-voltage signal equipment. The cables are approved for use in conditioned air plenums and hung ceilings.

To take full advantage of the potential labor-cost savings inherent in the system, field labor must be minimized. This is accomplished by factory equipping all utilization equipment, including lighting fixtures, with appropriate plug-in connectors.

15.31 Boxes and Cabinets

In this category are included pull boxes, splice boxes, and outlet boxes. Splice boxes, as the name suggests, are placed in raceway runs at points where splices or taps must be made; the NEC prohibits having splices inside conduits. (Splices are permitted in wireways and troughs with removable covers.) Pull boxes are placed in conduit runs where it is necessary to interrupt the raceway for a wire-pulling point. As explained in Section 15.19, this depends on the pulling friction in the system. The size of pull boxes depends on the number and size of incoming conduits, the direction in which conduits leave, and whether or not splices will be made in the box. Minimum sizes based on the above data are specified in the NEC. When a box is equipped with a hinged door(s) and contains some equipment other than wiring, such as a terminal board, it is referred to as a cabinet. All boxes must be equipped with tightly fitting, removable covers.

16

ELECTRICAL SYSTEMS AND MATERIALS: SERVICE AND UTILIZATION

Having discussed in the preceding chapter conductors and raceways, the first of three major categories of electrical equipment, we turn now to the two remaining categories: power-handling and utilization equipment. The first step is to examine the means by which electric service is brought into a structure.

16.1 Electric Service

Public utility franchises require only that service be made available at the private property line. Thus service is normally tapped onto the utility lines at a mutually agreeable point at or beyond the property line. The service tap may be a connection on a pole with an *overhead service drop* or an *underground service lateral* to the building, or a connection to an underground utility line with a service lateral to the building. All electrical construction work on private property is normally at the owner's expense.

Under certain conditions the owner can influence the type of construction utilized by the electric utility company in conveying the electric service to the site. This is often the case in large tract developments and in places where owners are willing to share some of the cost of better grade construction. Also, in many areas, the utilities themselves have instituted "beautification" programs in an effort to decrease the objectionable appearance of much of their equipment. Service from the utility line to the building may be run overhead (OH) or underground (UG) depending on the following conditions:

1. Length of service run.
2. Type of terrain.
3. Budget limitations.
4. Utility company voltage.
5. Site and nature of electric load.
6. Importance of appearance.
7. Local practices and ordinances.
8. Maintenance and service continuity.
9. Weather conditions.

16.2 Overhead Service

The principal advantage of overhead electric lines is low cost. Depending on terrain and other factors, the cost of overhead as compared to underground installation has been in the range of 10 to 50% (the latter when compared to direct burial cable installation). This accounts for the majority of installations being overhead. In recent years special techniques in underground installation have lowered that cost, making it a feasible alternative. Where the service run is several hundred feet or more, voltages higher than utilization level may be involved. This weighs heavily in favor of overhead lines, particularly with voltages exceeding 5000 V. Similarly, when terrain is rocky and the electrical load is heavy, the cost of an underground installation rises sharply. Since overhead lines are easily maintained and repaired, and faults easily located, service continuity with overhead lines is usually quite good. In areas where there are extreme weather conditions, called heavy loading areas, where combinations of snow, wind, and ice increase the possibility of outages, underground lines are preferred if service interruptions involve hardship or financial loss.

Overhead cables are of several types: bare, weatherproof, or preassembled aerial cable. Bare

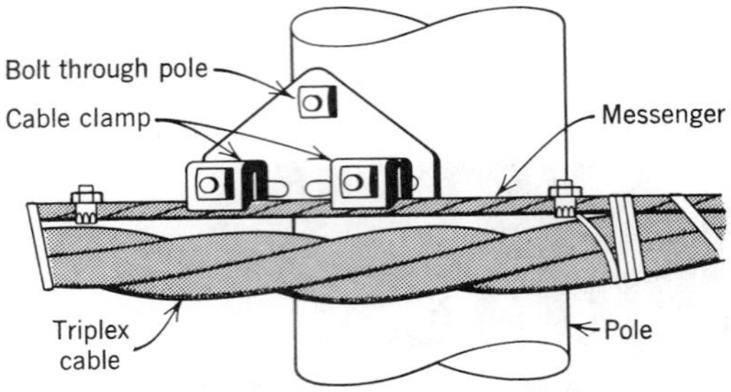

Fig. 16.1 Preassembled aerial messenger cables are carried by steel messenger cables clamped to the poles.

copper cable supported on porcelain or glass insulators on crossarms are normally used for high voltage (2.4 kV and higher) lines. Secondary circuits at 600 V and below are generally run on porcelain spool secondary racks with 1/c weatherproof cable as the conductor. Preassembled aerial cable consists of three or four insulated cables wrapped together with a metallic tape and suspended by hooks from the poles. This type of construction may be used up to 15 kV. It often proves to be more economical than crossarm or rack installation (see Fig. 16.1).

A typical detail of an overhead electric service entrance to a multiple residence is given in Fig. 16.2.

16.3 Underground Service

The advantages of underground electric service are attractiveness (lack of overhead visual clutter), service reliability, and long life. The principal disadvantage is high cost. To overcome this, utilities frequently use direct burial techniques which, by eliminating the raceway, reduce costs considerably. Since direct buried cable cannot be pulled out if it faults, as would be the case with a raceway-installed cable, it is recommended that the decision of which technique is to be used be based on the consideration of these data:

1. Cost premium for underground raceway installation, including handholes if required (see Section 16.4).
2. Record of outages for direct burial installation.
3. Cost and availability of repair service (utilities frequently will repair customer-owned underground service laterals, *for a fee*).
4. Impact of electric service outage in terms of time delays, inconvenience, necessity to dig up lawns and paved areas, and cost impact in the case of a commercial facility.

16.4 Underground Wiring

Exterior installations are generally in connection with service—either directly from the utility or a subfeed between buildings. The methods available for underground wiring are:

1. Direct burial; see Fig. 16.3.
2. Installation in Type II, direct burial duct; see Fig. 16.4.
3. Installation in Type I, concrete encased duct, see Fig. 16.4.

The first alternative offers low cost and ease of installation, with the disadvantage stated above. The second offers median cost but little strength; therefore, only installations on undisturbed earth and/or under light paving are recommended. The last (3) offers high strength and permanence, but at the highest price of the three.

Nonmetallic duct (conduit) intended for underground electrical use is commercially available in two wall thicknesses. NEMA (National Electrical Manufacturers Association) Type II with a heavy wall provides the physical protection required and is suitable for direct burial installation with no concrete encasement. Type I is manufactured with a thinner wall and is intended for encasement in a minimum of 2 in. of concrete. Common trade names for asbestos-cement and fiber ducts are ''Transite'' and ''Orangeburg.'' Plastic conduit is referred to as PVC, or simply as plastic. Nonmetallic conduit is most frequently used without concrete encasement for low-voltage and signal wiring and with encasement for high-voltage wiring. It offers several advantages over steel conduit

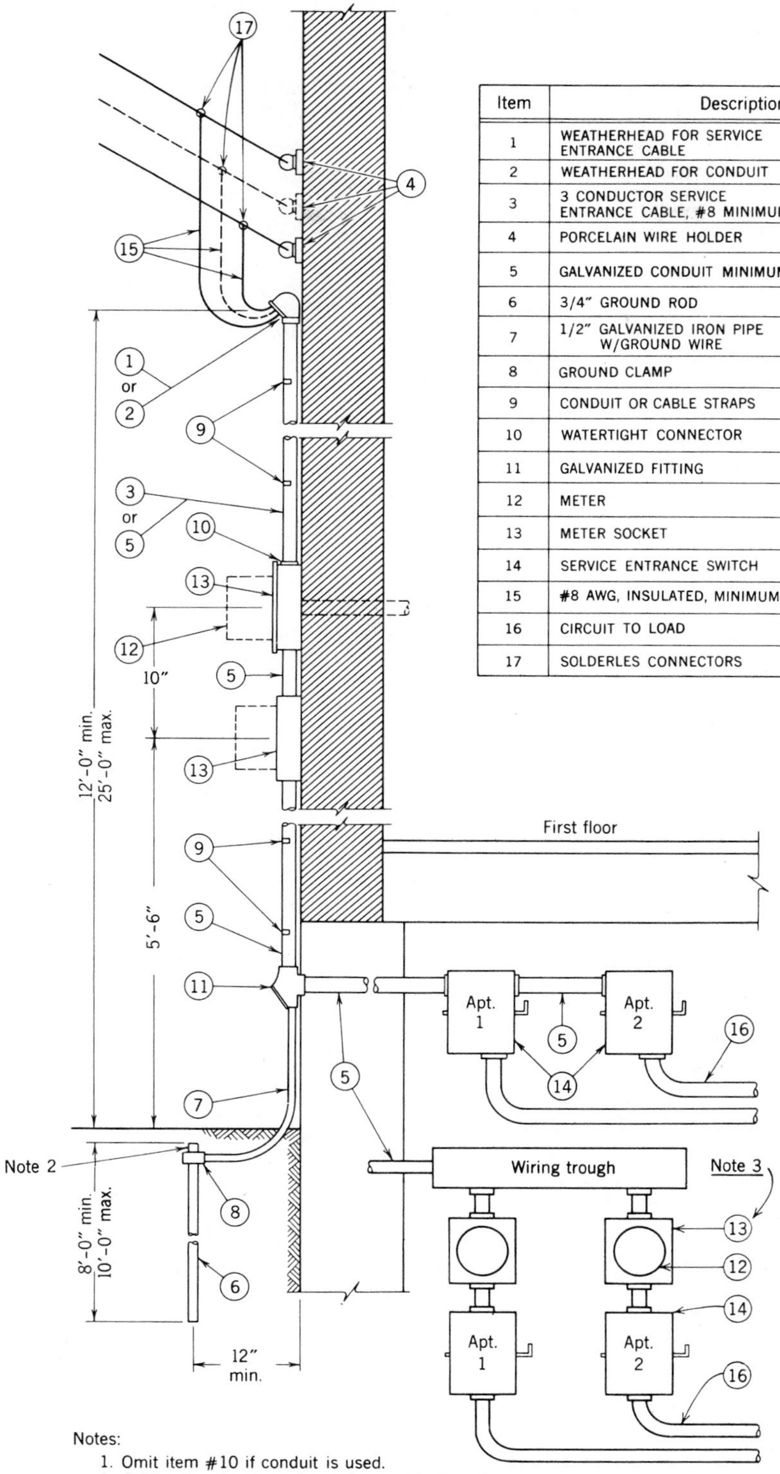

Item	Description
1	WEATHERHEAD FOR SERVICE ENTRANCE CABLE
2	WEATHERHEAD FOR CONDUIT
3	3 CONDUCTOR SERVICE ENTRANCE CABLE, #8 MINIMUM
4	PORCELAIN WIRE HOLDER
5	GALVANIZED CONDUIT MINIMUM 1"
6	3/4" GROUND ROD
7	1/2" GALVANIZED IRON PIPE W/GROUND WIRE
8	GROUND CLAMP
9	CONDUIT OR CABLE STRAPS
10	WATERTIGHT CONNECTOR
11	GALVANIZED FITTING
12	METER
13	METER SOCKET
14	SERVICE ENTRANCE SWITCH
15	#8 AWG, INSULATED, MINIMUM
16	CIRCUIT TO LOAD
17	SOLDERLES CONNECTORS

Notes:

1. Omit item #10 if conduit is used.
2. Cold water pipe ground may be used in lieu of ground rod.
3. Meters may alternatively be placed inside the building.

Fig. 16.2 *Detail of typical overhead electric service to a multiple residence. Note that meters are usually placed on the exterior of the buildings. If that is objectionable, they can alternatively be installed inside, provided access is available to the utility's meter readers.*

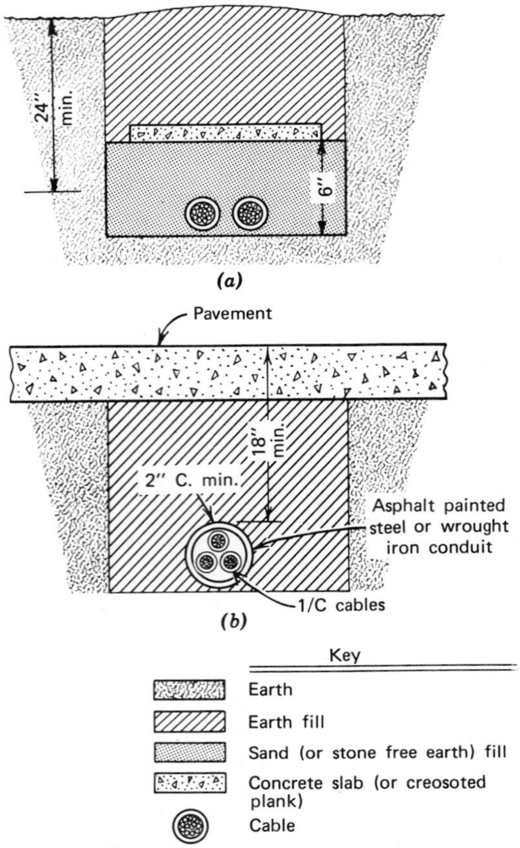

Fig. 16.3 (a) *Technique for installation of direct-burial cable.* (b) *Under highways, streets, and other high-load areas, cable should be installed in metal conduit.*

for underground work, such as lower cost and freedom from corrosion.

When underground electric wiring is duct installed and the run extends over several hundred feet (the exact distance depending on pulling tension), a pulling handhole or manhole is necessary. Handholes are used for low-voltage power and signal cables, and runs with a small number of cables. Manholes are used for high-voltage cables and where large duct banks must be accommodated. Precast handholes and manholes are readily available in most standard sizes and are usually cheaper than field-formed and poured units.

Cable used in underground wiring must be spe-

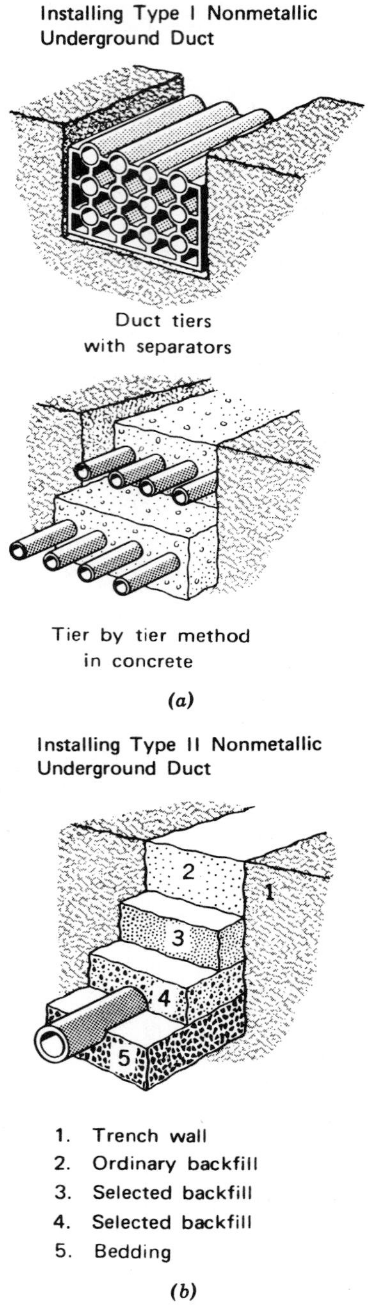

Fig. 16.4 Underground duct installation. (a) *Concrete duct bank should have minimum 6-in. earth cover.* (b) *Heavy wall duct should be buried at least 24 in. in ordinary traffic areas and 36 in. where subject to heavy traffic. Each layer is about 8 in. thick.*

cifically manufactured and approved for that purpose. Type SE is the basic service entrance cable, constructed with moisture- and flame-resistant covering. When provided with moisture proofing for underground use, the designation is SE type U, or simply USE. Underground cable for other than service runs is classified type UF (underground feeder).

16.5 Service Equipment

Referring back to Fig. 15.1, the reader will note that interposed between the high-voltage incoming utility lines and the secondary service conductors is a block labeled "transformer." This item is required whenever the building voltage is different from the utility voltage. It may be pole or pad mounted outside the building, or installed in a room or vault inside the building, as will be discussed below.

16.6 Transformers

A transformer is a device that changes or *transforms* alternating current of one voltage to alternating current of another voltage. Transformers cannot be used on d-c. A transformer would typically be used to step down an incoming 4160 V service to 480 V for distribution within a building; another transformer would be used in a local electric closet to step down the 480 V to 120 V for use on receptacle circuits. It is well to remember that ordinarily 120, 208, 240, 277, and 480 V are called secondary voltages, and 2400, 4160, 7200, 12,470, and 13,200 V are primary voltages.

Transformers are available in single-phase or three-phase construction (see Section 17.3 for an explanation of these terms). Transformer power capacity is rated in kVA (kilovolt-amperes). For a single-phase unit, this figure is the product of the full load current and the voltage. Since the voltages are different on primary and secondary, so are the currents—because the kVA remains constant. Thus a 100 kVA 2400/120 V transformer will carry at full load:

$$\text{primary current} = \frac{100{,}000\ \text{V-A}}{2{,}400\ \text{V}} = 41.6\ \text{A}$$

$$\text{secondary current} = \frac{100{,}000\ \text{V-A}}{120\ \text{V}} = 832\ \text{A}$$

Cooling medium is a transformer characteristic of primary importance. Transformers are either dry (air cooled) or liquid filled. The choice depends upon required electrical characteristics, proposed physical location, and cost factors. Although detailed considerations are beyond the scope of this book, some selection criteria are presented in the following sections. In general, units rated above 5 kV are liquid filled and units in the 600-V class are dry. Units installed indoors, except in vaults, are normally dry type, and are intended for general purpose light and power circuits (see Table 16.1). Load center transformers are installed in unit substations, both indoor and outdoor. Distribution transformers are mounted on a pole or on a concrete pad outdoors. Substation transformers are large and are always concrete-pad mounted.

The cheapest liquid transformer coolant with good electrical characteristics is mineral oil. However, because of its flammability, its application is limited. Nonflammable liquid coolants, called "askarels," generally containing polychlorinated biphenyl (PCB) were at one time very widely used. However, since 1979 PCB use in new equipment has been banned by Federal regulation due to adverse ecological impact. Because of this ban, a number of other liquid coolants with low flammability and good electrical and physical characteristics have been developed. They are all more expensive than mineral oil (see Table 16.5).

The insulation class of a transformer affects its permissible temperature rise, operating temperature, physical size, electrical power losses, overload capacity, and life.

The physical size of a transformer of given kVA rating and voltage depends on the type of insulation used. In order of decreasing physical size and increasing operating temperature we have, for dry transformers, 105°C, 150°C, 185°C, and 220°C systems that represent organic, inorganic, and silicone insulating materials, respectively (see Table 16.2).

The dimensional data in Table 16.3 vary considerably from one manufacturer to another, and therefore the table is useful only for a concept of bulk volume and weight.

TABLE 16.1 **Typical Transformer Data**

Transformer Type and Max. Capacity[a] of 3-Phase Bank	Cooling and Insulating Medium	Voltage[b] Primary		Voltage[b] Secondary	
General purpose, dry type	Air	120, 208, 240 480, 600		120, 208, 240 480, 600	
Load center	Air	2400	4160	120	208
	Silicone[c]	7200	11000	240	480
2000 kV-A	Oil	13200	13800	600	
Distribution	Silicone	2400	4160	120	208
	Oil	7200	13200	240	480
750 kV-A		13800		600	
Substation	Oil	2400	4160	480	600
		7200	12470	2400	4160
above		13200			
500 kV-A		22000	34000		

[a]Larger ratings are available on special order.
[b]These are nominal voltages.
[c]This is one type of nonflammable liquid.

Although 220°C system insulation transformers can withstand 150°C rise, many users specify 220°C system insulation with 115°C rise or even 80°C rise; that is, a better grade of insulation is used with an underrated transformer. The reason for this is threefold:

1. Longer life.
2. Higher overload capacity.
3. Lower operating cost.

A 220°C system transformer operated at full load *continuously* (an unusual situation) has a shortened life—estimated by some experts on the basis of accelerated aging tests to be between 3 and 10 years. The same class insulation transformer (220°C system) rated at 80°C has a life expectancy of over 100 years. With respect to operating cost, the same situations obtain here as with the high-temperature insulation discussed in Section 15.15 and studied in Table 15.5. The truism that one gets nothing for nothing can be expressed here as in return for the smaller, lighter, cheaper, and hotter transformer (220°C system), we have higher losses and correspondingly higher operating cost, and shorter life.

Refer to Table 16.4, which compares two dry-

TABLE 16.2 **Air-Cooled Transformer Electrical Insulation Temperature Ratings (Based on 40°C Ambient)**

Insulation Class (System)[a]	Insulation Type	Average Conductor Temperature Rise	Ambient Temperature	Hot-Spot Temperature Gradient	Total Maximum Temperature
105°C	Organic (A)	55°C	40°C	10°C	105°C
150°C	Mica, glass (B)	80°C	40°C	30°C	150°C
185°C	Asbestos (F)	115°C	40°C	30°C	185°C
220°C	Silicones (H)	150°C	40°C	30°C	220°C

[a]The modern terminology for insulation class uses "system" in lieu of "class."

TABLE 16.3 **Typical Dry-Type Transformer, Dimensions and Weights**

kV-A Output Continuous 80°C Rise	Approx. Dimensions (Inches)			Approx. Weight (Pounds)
	Height	Width	Depth	
	Primary 2400/4160 Y V Secondary 120/240 V			
Single-Phase				
3	15¼	8⅝	7¾	97
5	15⅝	11¼	9⅞	135
10	19⅛	11¼	9⅞	235
15	30⅛	22⅜	18⅛	325
25	34⅛	25⅞	20⅛	375
50	36⅛	30⅛	22⅛	600
100	40⅛	36⅛	25⅜	1100
	Primary 480 V Secondary 208Y/120 V			
Three-Phase				
45	34⅛	33¼	20⅜	580
75	36⅛	41⅛	22⅜	820
150	40⅛	50⅞	25¾	1500
225	58¾	62¼	35	2630

type transformers, type H; one is designed for 80°C rise and the other for 150°C rise. The operating cycle is chosen to be representative of commercial use. With such a cycle, even the 150°C rise unit will probably last 30 years—the specified life cycle. The current difference in first cost between the two units is less than $2000. Table 16.4 tells us that the lower energy waste of the 80°C unit will repay the first-cost differential in about 4 years. Beyond that the advantage is entirely on the side of the 80°C unit. Also, with a total permissible hot-spot temperature of 220°C (428°F), the location of the 150°C rise unit must be very carefully chosen, since it can create a serious heat generation and radiation problem.

In summary, then, a transformer is specified by type, phase, voltages, kVA rating, sound level, and insulation class. Thus, 112.5 kVA, three-phase, 480/120-208 V, air-cooled, indoor, dry-type transformer with 220°C insulation system and 115°C rise, 45 db (decibel) maximum sound level is an adequate transformer description. Sound ratings of transformers, as well as installation techniques and acoustical treatment, are discussed in Part Nine, "Acoustics."

ELECTRICITY

TABLE 16.4 **Transformer Operating-Cost Comparison, 750 kVA, Dry Types**

750 kVA Dry-Type Transformer Type H Insulated	Efficiency at Stated Load[a] (Percent)	Loss[b] (kW)	Annual Energy: Loss: kWh[c] Cost: $[d]	Life-Cycle Energy Cost[e]
80°C rise	98.8	9.0		
	99.0	5.18		
	98.9	3.71		
	98.3	2.86	35,625	$38,982
	—	2.2	$2,494	
150°C rise	98.4	12.0		
	98.7	6.45		
	98.5	4.46		
	97.8	3.09	42,786	$46,812
	—	2.2	$2,996	

[a]Efficiency figures are averaged from manufacturers' published data.

[b]Loss at 0 load is core loss of 2200 W.

[c]Operating cycle, representative of a typical commercial building is full load—2 h, ¾ load—8 h, ½ load—4 h, ¼ load—2 h, 0 load—8 h.

[d]Based on electric energy cost of $0.07 per kWh, including demand.

[e]Using 30-year life, 8% fixed capital cost, 3% annual escalation.

16.7 Transformers Outdoors

A service transformer bank is necessary, as explained above, when the facility utilization voltage is different from the utility voltage (see Fig. 16.5*a*). The designer occasionally opts for a step-up, step-down arrangement when the service run is so long that the conductor cost when run at a low voltage would be excessive because of large-sized cables. In such instances, the cost of the double transformer installation must be more than offset by the savings in feeder cost to be economically justifiable (see Fig. 16.5*b*).

The advantages of an outdoor transformer installation are:

1. No building space required.
2. Reduced noise problem within building.
3. Lower cost.
4. Ease of maintenance and replacement.
5. No interior heat problem.
6. Opportunity to use low-cost, long-life oil-filled units.

On the other hand, it is frequently easier to find space indoors (preferably in a basement) than to find a suitable exterior spot; noise may be more disturbing from the available exterior spaces, such as courtyards, than from a basement. Costs can run high if long secondary runs are required; heat can often be handled by louvers or areaways adjoining a basement or even made use of profitably if the transformer load is fairly constant. Also, since exposure to direct sunlight decreases the unit rating by increasing its temperature, a shaded spot may be difficult to find. Furthermore, an exterior transformer, except where pole mounted, is a questionable choice in an area with a high incidence of vandalism, regardless of sturdiness of construction. Finally, appearance of an exterior unit may be objectionable. This latter point has received much attention from manufacturers, and numerous designs have been developed that minimize the appearance problem (see Fig. 16.6). The most popular type of exterior transformer installation is the pad mount. It has all of the advantages listed above in addition to extreme simplicity of installation—it is simply set on a concrete pad. Consult manufacturers' catalogs for dimensional data.

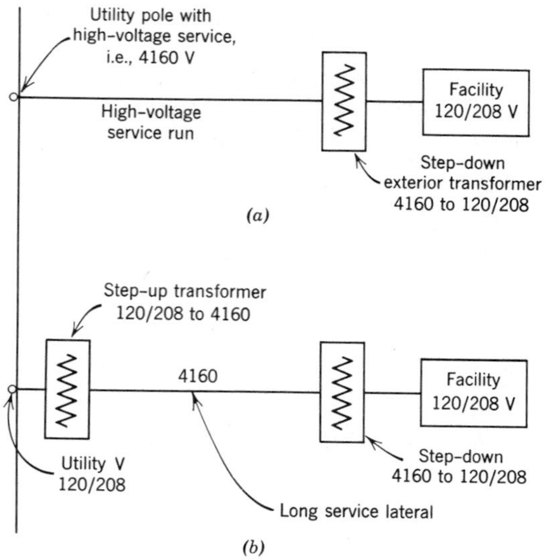

Fig. 16.5 *Service transformer arrangements.* (a) *High-voltage service with step-down service transformer at the facility, and* (b) *low-voltage service, with transformation at both ends of a long service run.*

16.8 Transformers Indoors; Heat Loss

When an indoor transformer installation is indicated, special consideration must be given to the transformer's heat-generating properties. Between 1 and 1½% of the transformer's rating, depending on type, is converted to heat at full load. Table 16.4 shows that for a 750 kVA, dry-type, 150°C rise unit, 12 kW or 41,000 Btu/h of heat loss is generated at full load! Losses are lower for 80°C rise units, as indicated. Liquid-filled units have approximately the same losses as 80°C rise dry units. Unless the heat can be used, sufficient ventilation—either natural or forced—must be provided to keep the ambient temperature from exceeding 40°C.

Ventilation by natural convection is most desirable, necessitating the location of the transformer room on an exterior wall (with an areaway if the room is located below grade). The size of the *free* area required for a ventilation opening is 3 sq in./kVA of capacity, plus an additional 1 sq in./kVA for switchgear losses, if any. If a louver is used in the openings, the size of the opening must usually be doubled (total: 8 sq in./kVA), since most louvers have a 50% free area. For good convection, it is desirable to divide the louvered

Fig. 16.6 (a) *Pad-mounted exterior transformers are neat, compact, and, if sited properly, unobtrusive. Large units can be partially screened by shrubbery.* (b) *When the size is such that visual screening becomes a problem, consideration should be given to a structural screen such as a decorative brick wall, which also provides some much needed sound screening. Courtesy of General Electric.*

areas in half, placing one-half near the ceiling, and the remaining half near the floor. To provide for equipment removal, if the areaway is large enough, it may be useful to add louver area between the upper and lower openings and make the full louver removable. A bird screen is also desirable.

Since outside air temperature will vary, it is advisable to use a temperature-controlled, adjustable louver. In severe cold climates, heat loss from the electrical equipment may not be sufficient in the winter to warm the room. In such instances, a unit heater should be installed in the room with a thermostat set at 55 F.

16.9 Transformers Indoors; Selection

When transformers are installed indoors, they are subject to stringent NEC regulations that are designed to make the installation intrinsically safe. These regulations are detailed in NEC Article 450 and no purpuse is served in duplicating them here. Instead, we wish to present the reader with the essential considerations involved.

1. Oil-filled transformers present a fire hazard indoors, because flammable oil can spread from a tank leak or rupture. To prevent this, oil-filled transformers must be installed in a fire-resistant vault, the construction of which involves a heavy cost. Advantages offsetting this cost are the oil-filled transformer's small size, low weight, low first cost, low losses, long life, excellent electrical characteristics, low noise level, and high overload capacity. Despite these, the vault requirement has had the effect of restricting oil-filled units to industrial facilities and other structures where electrical considerations require its use.

2. Nonflammable liquid-filled units have most of the advantages of oil-filled units and do *not* require a vault unless voltage is very high. They do, however, require a sump or catch basin of sufficient capacity for all of the contained liquid. This and a relatively high first cost are the negative aspects.

3. Dry-type units are the transformers of choice in the majority of indoor installations, despite shorter life, higher losses, high noise level, greater weight, and larger size than the liquid-filled units.

TABLE 16.5 **Relative Installation Costs for a 300 to 1000 kVA Transformer**

Transformer Type	*Temperature Rise*[a] *(°C)*	*Relative Transformer Cost*	*Construction Cost*	*Total Relative First Cost*
Oil filled	80	1.00	50–100[b]	1.50–2.00
Silicone, inter alia	80	1.40	20–40[c]	1.60–1.80
Dry, ventilated	80	1.65	—	1.65
	80	1.50	—	1.50
	150	1.35	—	1.35
	150	1.20	—	1.20

[a]Transformers of equal temperature rise have approximately equal life and equal losses.

[b]Cost of vault depends on local labor costs and size of transformer. The relative cost decreases with increasing transformer size.

[c]Cost of catch basin. As in the case of a vault, relative cost depends on labor rates and transformer size.

The principal advantage is ease of installation and almost unrestricted choice of location. As explained above, by using an underrated transformer (220°C system, 80°C rise), losses can be reduced and life extended. Also, for a price premium, noise level can be reduced. When a dry-type transformer is equipped and modified so that its characteristics are equal to that of liquid-filled units, its installed and owning costs are in the same range as the liquid units. Still, most users find the lack of restrictions on placement a sufficient reason for its choice. Table 16.5 gives a comparison of installed costs for these three classes of transformers.

16.10 Transformer Vaults

A transformer vault is basically a fire-rated enclosure, provided because of the possibility of transformer case rupture and an oil fire. However, this must not be construed as implying that transformers are hazardous or delicate devices prone to faults. On the contrary, transformers are extremely tough, sturdy, long-lived, capable of sustaining large and prolonged overloads, and indeed among the most reliable elements of an electrical system. However, faults do occur, and an oil-filled transformer is a potential fire hazard.

Transformer vaults should be located, if practicable, where they can be ventilated to the outside air without flues or ducts. The combined net area of all ventilating openings (gross area less screens, louvers, etc.) should be as explained in the preceding section, not less than 3 sq in./kVA of transformer, but in no case less than 1 sq ft. Further ventilation recommendations plus details of enclosure construction materials, fire rating, door and sill details, and other relevant and important information are provided by NEC Article 450.

16.11 Service Equipment Arrangements and Metering

To summarize the preceding discussion and to proceed with our study of electric power-handling equipment, refer to Fig. 16.7. Electric service connection will be overhead or underground. Where building voltage is different from utility voltage, a transformer is required. This transformer may be pole or pad mounted outside the building, or installed in a room or vault inside or outside the building.

Metering must be provided at either the utility or the facility voltage, and at either the service point or inside the building. The choice is generally left to the owner with the understanding that inside meters must be readily accessible to utility personnel. If high-voltage service is purchased, then the transformers and all equipment beyond

Fig. 16.7 *Typical electrical plan and details of an underground low-voltage service (600 V class) to an industrial facility. Note that the portion of UG service beneath the building must be concrete encased. A service transformer (if required) could be mounted on the utility pole, at its base, or on a pad at the property line.*

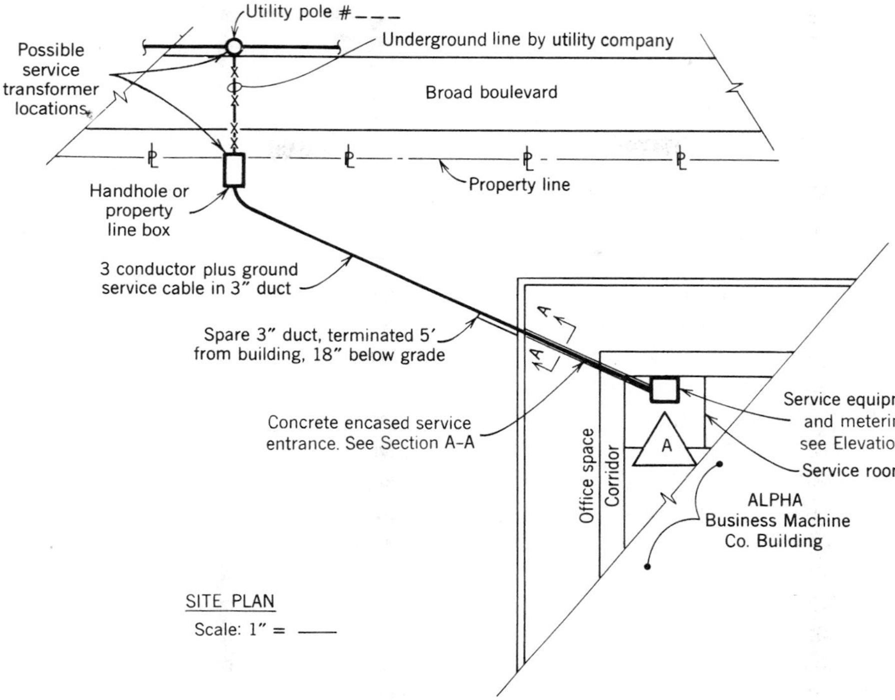

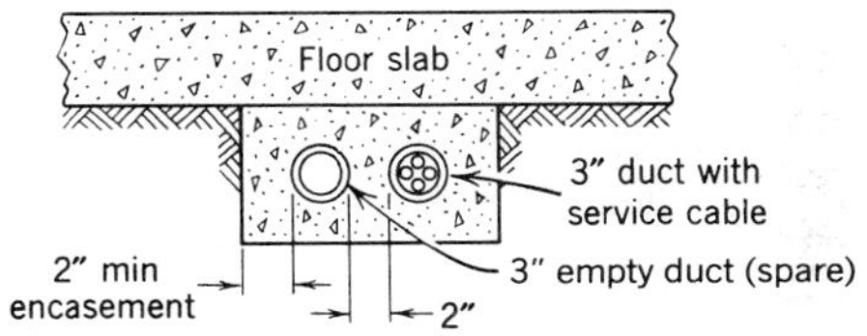

SECTION A–A
NO SCALE

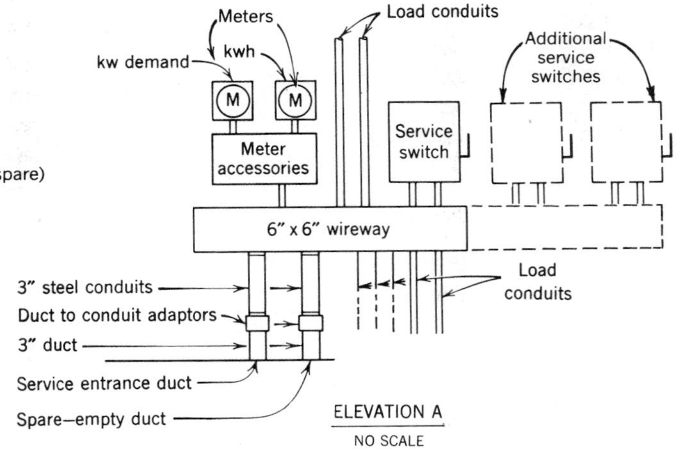

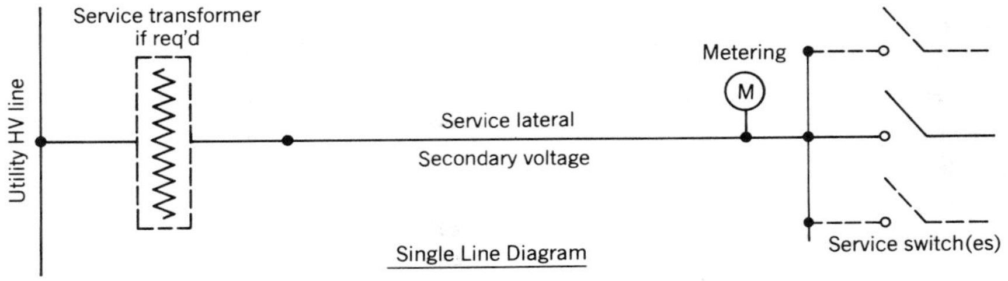

the service connection must be furnished by the owner. Conversely, if low voltage is purchased, all equipment necessary to provide low voltage is furnished by the utility. Obviously, the electric service rates for low-voltage service are *higher* than for high-voltage service in order to compensate the utility for the cost of providing and maintaining the step-down transformer and associated equipment. Therefore, it is often advisable for the owner of a facility (other than a small, residential facility) to investigate the economics of purchasing power at high voltage. Since many owners are not equipped to maintain high-voltage equipment, arrangements can sometimes be made to pay the utility to provide and maintain transformers, while taking the cost advantage of high-voltage service.

For a single-use building or a building where electric energy is included in the rental charge, only a single meter is necessary. Provision for such metering may be made in the main switchboard, or the meter may be independently mounted. In both cases, the meter is furnished and installed by the utility company. Where submetering is required, such as in apartment houses, banks of meter sockets are installed to accommodate the multiple meters. Federal regulations forbid master metering in new multiple-dwelling constructions, because it encourages energy waste. A low-voltage underground service detail as it would appear on a set of contract drawings, including relevant details, is given in Fig. 16.7. Note that here the service switch and meters are separately mounted. Meters are always installed *electrically* ahead of the service switch so that they cannot be disconnected.

16.12 Service Switch

The purpose of the electric service switch is to disconnect all of the electric service in the building except emergency equipment. Thus, in the event of fire, no electrical hazard will face fire fighters. It is therefore obvious that this disconnecting apparatus must be located at a readily accessible spot near the point at which the service conductors enter the building. If such a location is not feasible, service conductors may be run in concrete encasement under the building and will be considered ''outside the building'' up to the point at which they emerge from the floor in the building (see Fig. 16.7). At that point, the service switch must be installed. The service switch or, more accurately, the ''service disconnecting means,'' may comprise one to six properly rated switches. These are frequently assembled into a switchboard. Before discussing switchboards, a description of switches, circuit breakers, and fuses, the components out of which switchboards are constructed, is in order.

16.13 Switches

An electrical switch is a device intended for on/off control of an electrical circuit and is rated by current and voltage, duty, poles and throw, fusibility, and enclosure. The current rating is the amount of current that the switch can carry continuously and interrupt safely. Certain light-duty switches are not intended to interrupt rated current and are clearly so labeled. Switches intended for motor control are also rated in horsepower. The voltage rating of a switch is, as for other electrical equipment, by voltage class. Thus a switch is rated 250 V, 600 V, or 5 kV as required. Switches intended for normal use in light and power circuits are called general-use safety switches and are rated ND for *normal duty*. Switches intended for frequent interrupting are rated HD for *heavy duty*. Conversely, switches intended to be opened under load only occasionally, such as service switches, are rated LD for *light duty*.

An examination of Fig. 16.8 should clarify what is meant by the number of poles and throws of a switch. Unless otherwise noted, a switch is assumed to be single throw. Since the NEC states generally that the grounded neutral conductor of a circuit should not be broken, most switches carry the neutral through unbroken, by means of a solid link within the switch. This gives rise to the term *solid neutral* (SN) *switch*. Switches are available in 1, 2, 3, 4, and 5-pole construction. Poles are indicated by a ''P''; thus 3-pole is written ''3P,'' and so on.

A switch may be constructed with or without provision for fusing. If provided, the switch is fusible; if not, the switch is nonfusible. All separately enclosed switches must be in an appropriate cabinet. The National Electrical Manufacturers

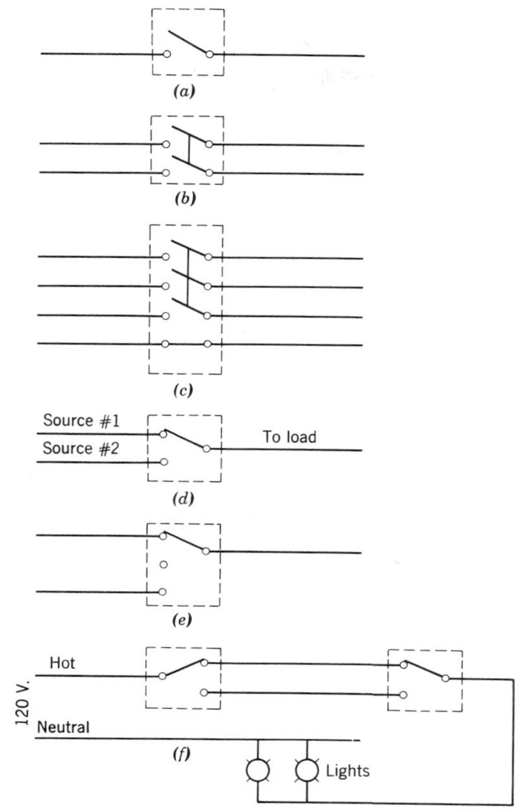

(a) Single-pole single-throw switch.

(b) Two-pole single-throw switch.

(c) Three-pole and solid-neutral (3P and SN) switch.

(d) Single-pole double-throw switch (also called, in small sizes, a 3-way switch).

(e) Single-pole double-throw switch with center "off" position (in control work called a hand-off-automatic switch).

(f) Use of two single-pole double-throw (3-way) switches for switching of a lighting circuit from two locations.

Fig. 16.8 *Typical switch configurations. Note that switches are always shown open.*

Association has standardized the nomenclature and application of enclosures for all electrical control equipment, of which switches are only one item. These are detailed in Section 16.15. Summarizing the above, then, we could adequately describe a switch thus: switch, HD, 3P & SN, 200A/150AF (fuse), 600 V, in NEMA 12 enclosure.

16.14 Contactors

A contactor is a switch. Instead of a handle-operated, movable blade, a contactor uses *contact* blocks of silver-coated copper, which are forced together to make (close) or are separated to break (open) the circuit. The common wall light switch is a small mechanically operated contactor. A relay is a small electrically operated contactor. Most contactors are operated by means of an electromagnet that causes the contacts to close (or open). Contactor terminology is somewhat different from that of a switch. Its condition when deenergized is its *normal* state. Thus, a contactor whose contacts are open when the coil is not energized is *normally open* (NO). One with normally *closed* contacts is NC. Units intended for motor control are called motor starters and are discussed in Section 16.22. Current, voltage, and number of poles have the same significance for contactors and relays as for switches.

The great advantage of contactors over switches is their facility for remote control. Switches must be manually thrown—or at best with a motor. However, the magnetic contactor is inherently a remotely controlled device, making it ideal for a myriad of control functions. They are controlled by push-button or automatic devices such as timers, float switches, thermostats, pressure switches, and so on. Since control can be both remote and automatic, the application of contactors is universal in control of lighting, heating, air conditioning, motors, and the like.

16.15 Special Switches

Many special types of switches are available. Most of these types are beyond the scope of this book, except for the following, which we will discuss briefly.

(a) Remote-Control (RC) Switches. A contactor that latches mechanically after being operated is known as a remote-control switch. These devices are useful in lighting control. A discussion of low-voltage lighting control using RC switches is found in Section 16.26.

(b) Automatic Transfer Switch. This device, which is an essential part of all standby power arrangements, is basically a double throw switch—generally 3-pole—so arranged that on failure of normal service it *automatically transfers* to the emergency service. When normal service is restored, it automatically retransfers to it. The control devices are voltage sensors that sense the condition of the service and operate the switch accordingly. Auxiliary devices can be built on to the basic switch, the most common of which are emergency generator starting equipment. Figure 16.9 illustrates a typical unit. For special application, solid-state switches are available.

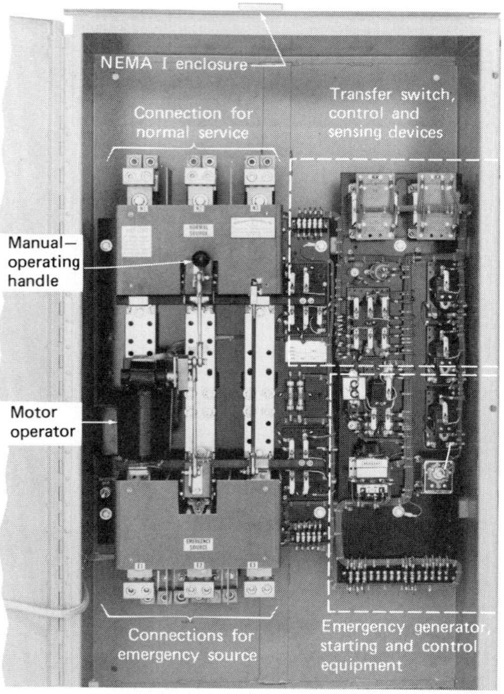

Fig. 16.9 *Automatic transfer switch in a NEMA 1, wall-mounted enclosure. This unit is rated 400 A, 3-pole, 600 V, a-c, which corresponds to approximately 150 kVA at 208 V, 3-phase. The cabinet is 58 in. high, 32 in. wide, and 18 in. deep (147.3 × 81.3 × 43.2 cm) and the entire unit weighs 354 lb (160 kg). Courtesy of Russelectric, Inc.*

(c) Time Controlled Switches. In this category are included all switches whose operation is time based. The timer can be either the familiar electromechanical device consisting of a low-speed miniature drive motor to which some type of contact-making device is physically connected, or it can be a solid-state electronic timer, which in turn controls either a relay or a solid-state switch. The latter arrangement is the basis of all modern programmable time controls, which find wide application in lighting control (Section 21.3), energy management and automated building control (Section 14.14), and institutional clock and program systems (Section 22.13).

Motor-operated timers depend for their accuracy on power line frequency, have moving parts that wear, must be reset after power outages (some units have spring-wound reserve power motors), and become cumbersome and expensive with increase in the number of controlled events. Electronic units are small, cool, quiet, independent of line frequency for accuracy (in many designs), carry through outages when standby-battery equipped, and have virtually unlimited event-control capacity. They will undoubtedly completely replace the older design except for the simplest applications. See Fig. 16.10 and Figs. 21.3 and 21.4, page 1074.

16.16 Equipment Enclosures

Proper NEMA nomenclature, descriptions, and applications for the more common enclosures are found in Table 16.6. It is important to note that there is no enclosure described as WP or weatherproof. Equipment intended for outdoor use should be specified:

1. In a Type 3R enclosure to protect against rain (see Fig. 16.11).
2. In a Type 3S enclosure to protect against wind-driven rain and sleet.

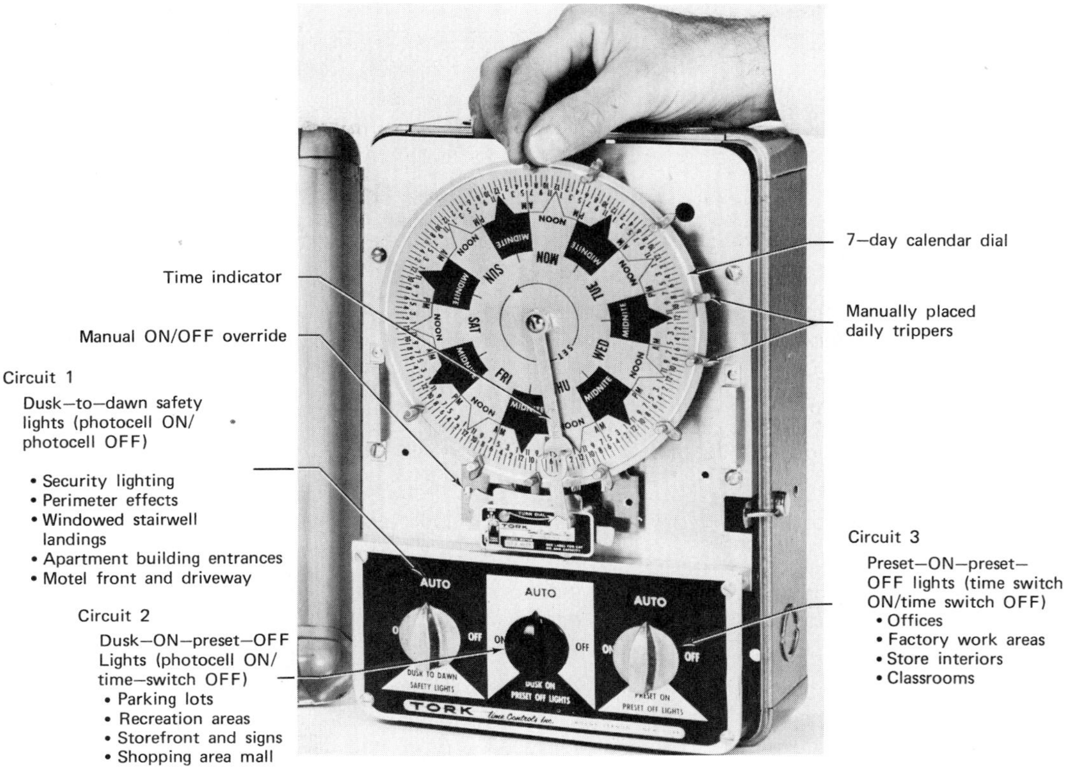

(a)

Fig. 16.10 Time switch (a) *is arranged for three different types of control functions that are applicable as indicated, to different types of load. This unit's dial is a seven-day calendar type that permits setting a different schedule for every day of the week. The dial is not astronomical (solar), since all the turn-on functions are either photocell or time-of-the-day controlled. Solar dials are useful when photocell control is not available, since sunrise and sunset times vary with the season.* (b) *Illustrates a microprocessor based seven-day time switch that controls eight circuits, up to 500 events per week, and provides holiday operation, and manual override. It is slightly larger than switch* (a) *and incomparably more flexible. Courtesy of Tork.*

3. In a Type 4 enclosure to protect against the above plus splashing and condensation.

Note also that the Type 12 industrial enclosure is similar to Type 1, except that it is gasketed for dust and drip resistance and therefore is well applied in *all* "dirty" indoor environments, including commercial and institutional spaces.

TABLE 16.6 **Control Equipment Enclosures**

NEMA Designation: Type	Description	Application
1	General purpose	Dry, indoor use
2	Drip-proof	Indoor, subject to dripping
3	Dust-tight, rain-tight, and sleet-resistant	Indoor/outdoor, where subject to windblown dust and water
3R	Rain-proof and sleet-resistant	Outdoor, subject to falling rain, snow, and sleet
3S	Dust-tight, rain-tight, and sleet-proof	Outdoor, subject to windblown water, dust, and sleet; most severe exterior duty
4	Watertight and dust-tight	Indoor/outdoor, subject to water from all directions; not sleet-proof
7–9	Hazardous	Differing in application by class and group of hazardous use; see NEC
12	Industrial use, dust-tight and drip-tight	Indoor only, general use, industrial and other "dirty" environments

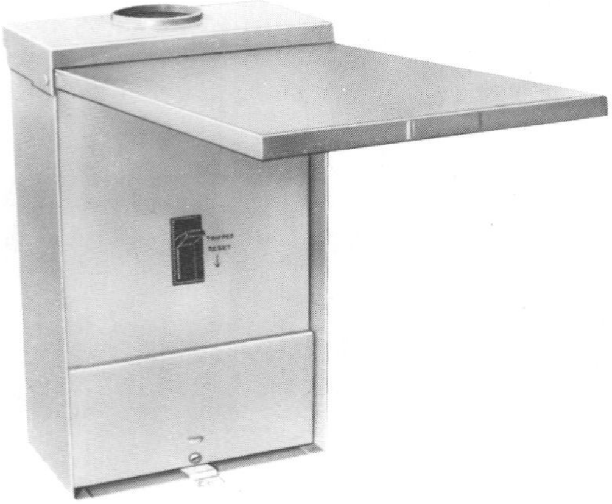

Raintight NEMA 3R circuit breaker enclosure.

Fig. 16.11 *Type 3R outdoor enclosure. This is the type usually intended when "weatherproof" is specified. Courtesy of General Electric.*

16.17 Circuit-Protective Devices

In order to protect insulation, wiring, switches, and other apparatus from overload and short-circuit currents, it is necessary to provide automatic means for opening the circuit. The two most common devices employed to fulfill this function are the fuse and the circuit breaker (c/b).

(a) Fuses. The fuse is a simple device consisting of a *fusible* link or wire of low melting temperature that when enclosed in an insulating fiber tube is called a cartridge fuse, and when in a porcelain cup is known as a plug fuse. Figure 16.12 shows common types of fuses. Plug fuses, such as those normally in residential use, are rated 5 to 30 A, 150 V max to ground. Cartridge fuses of various designs are made up to 6000 A and 600 V.

(b) Circuit Breakers. A circuit breaker is an electromechanical device that performs the same protective function as a fuse and, in addition, acts as a switch. Thus it can be used in lieu of a switch-and-fuse combination to both protect and disconnect a circuit. Most circuit breakers are equipped with both thermal and magnetic trips. The thermal trip acts on overload, whereas the magnetic trip acts on short circuit. Both the thermal and the magnetic action have inverse time characteristics, that is, the heavier the overload, the faster the trip action.

Air circuit breakers are available in two types: the molded-case breaker and the "large air breaker." Molded case breakers consist of a complete mechanism encased in a molded phenolic case. A light-duty molded case 50-A frame, plug-

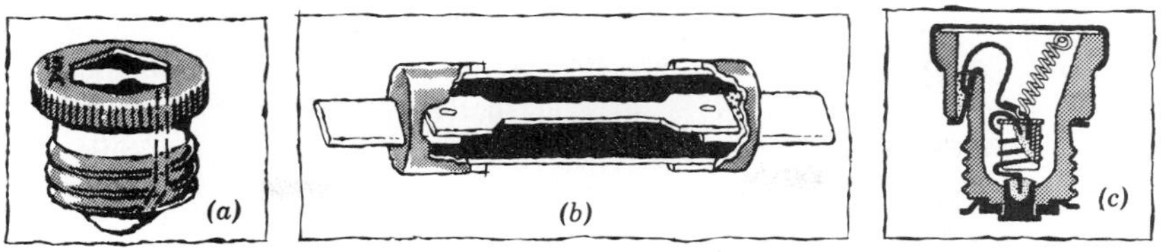

Fig. 16.12 *Standard types of fuses are* (a) *common household plug fuse;* (b) *single-element, knife-blade cartridge fuse;* (c) *dual-element, time-delay fuse with Edison base. Since fuses are inherently very fast-acting devices, time delay must be built into a fuse to prevent "blowing" on short-time overloads, such as those caused by motor starting. A dual-element fuse, such as shown in* (c) *allows the heat generated by temporary overloads to be dissipated in the large center metal element, preventing fuse blowing. If the overload reaches dangerous proportions, the metal will melt, releasing the spring and opening the circuit. The time to clear (blow) for fuses is inversely proportional to the amount of current. Courtesy of Bussman Division, McGraw Edison Co.*

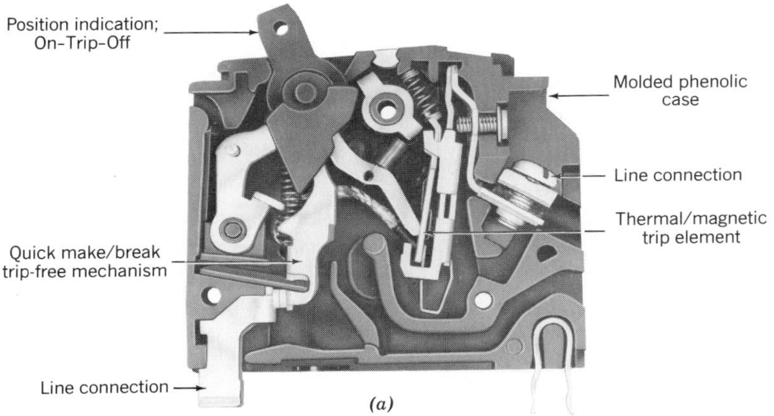

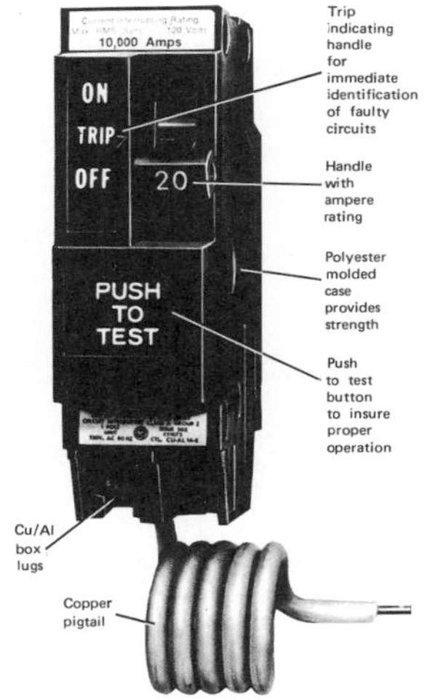

Fig. 16.13 *Molded-case circuit breakers.* (a) *Illustrates a conventional single-pole, 50A frame plug-in type c/b. Courtesy of the Square D Co.* (b) *A unit that will provide ground-fault protection in addition to functioning as an ordinary circuit breaker. This unit is designed to fit into the same physical space as a conventional molded-case circuit breaker and therefore can be used as a replacement in an existing panelboard. The units are referred to as GFCI (ground fault circuit interrupters) or GFI. Courtesy of General Electric.*

in circuit breaker, and a special type of circuit breaker designed to protect against ground faults in addition to overloads and short circuits are illustrated in Fig. 16.13. The large air circuit breaker is a more complicated and widely adjustable mechanism and can be used in applications that preclude use of molded case breakers. All breakers can be equipped with remote trip and auxiliary contacts, and all good breakers have trip-indicating handles and are "trip-free" (i.e., will trip out harmlessly if closed in on a short-circuited line). Low-voltage (600-V class) circuit breakers are available in sizes from 50 A to 2500 A and 1 to

3 poles. Characteristics vary widely and are beyond the scope of this work.

(c) Fuses and (versus) Circuit Breakers. Although both are circuit-protective devices, their characteristics differ markedly. Characteristics are tabulated below for ease of comparison.

Fuses—Switch and Fuse Combination

Advantages	*Disadvantages*
Simple and foolproof	Single pole only
Constant characteristics	Requires switch
Initial economy	Necessity for storage of replacements
Very high I.C. (interrupting capacity)	Self-destructive (one-time operation)
No maintenance	Nonadjustable
Instantaneous, energy limiting	Nonindicating
	No electric or remote control
	Not trip free

Circuit Breakers

Advantages	*Disadvantages*
Usable as switch	Low to medium I.C.
Multipole	Periodic maintenance
No replacement storage	High initial cost
Resettable	Complex construction
Indicates trip	Aging
Trip free	
Remote control	
Adjustable	
Auxiliary contacts	

The above lists demonstrate that there is no all-inclusive answer to the oft-posed question ''Which are preferable—fuses or breakers?'' The answer depends on the specific application involved and is often based on highly technical factors, beyond this book's scope. Generally, breakers are used for all lighting and appliance panels.

16.18 Switchboards and Switchgear

Switchboards and switchgear are freestanding assemblies of switches, fuses, and/or circuit breakers, which normally provide switching and feeder protection to a number of circuits connected to a main source. A switchboard may be represented in a single-line diagram as in Fig. 16.14. It serves in an electrical system to distribute, with adequate protection, large bulk power into smaller ''packages.'' Thus, by hydraulic analogy, the main buswork of the switchboard is equivalent to a main header, the switches to on/off valves, the fuses to flow-limiting devices, and the feeders to subheaders connected to the main header. Modern switchboards (Figs. 16.15 and 16.16) are all dead-front; that is, they have all circuit breakers, switches, fuses, and live parts completely enclosed in a metal structure. The operator controls all devices by means of insulated handles in the front panel. When a switchboard has circuit breakers equipped with bayonet-type contacts and mounted in a movable drawer (like the drawers of a standard letter file), they are described as the ''drawout'' type. This draw-out arrangement facilitates emergency replacements, inspection, and repairs.

There is no clear distinction made between the terms ''switchboard'' and ''switchgear,''

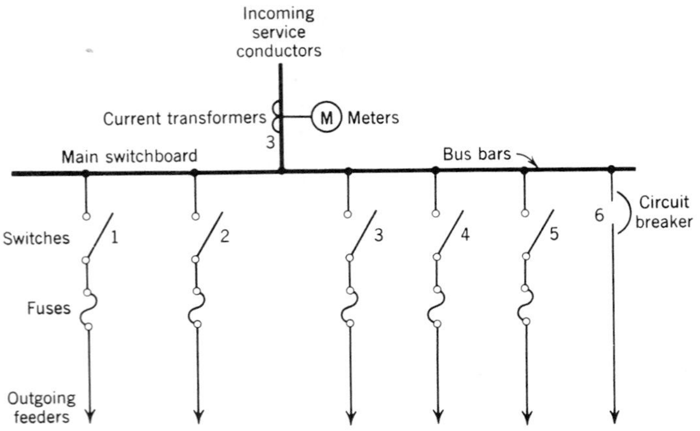

Fig. 16.14 *Typical switchboard. Switches are normally shown in the open position. Switches must be on the line (supply) side of fuses. Each line in a single-line diagram represents a 3-phase circuit. If circuit breakers were used, the entire board would be composed of units as illustrated in circuit 6.*

Fig. 16.15 *Low-voltage switchboards are available in various designs. Illustrated is a NEMA 1, panel-type, free-standing, front-accessible unit with large air main circuit breaker and molded-case branch circuit breakers. Alternatively, fused switches could be used. The compartment below the main breaker can be used for metering, current transformers, or other accessories. The section across the top is for wiring between sections. Switchboards constructed with a compact central bus in each section to which the section's circuit breakers and switches are connected are similar to panelboards and are described as "panel-type." Standard switchboard bus construction has main horizontal buses feeding all switchboard sections and vertical buses tapped from them feeding the units in each section. Standard units are from 14 to 20 in. deep and 90 in. high. Courtesy of The Square D Co.*

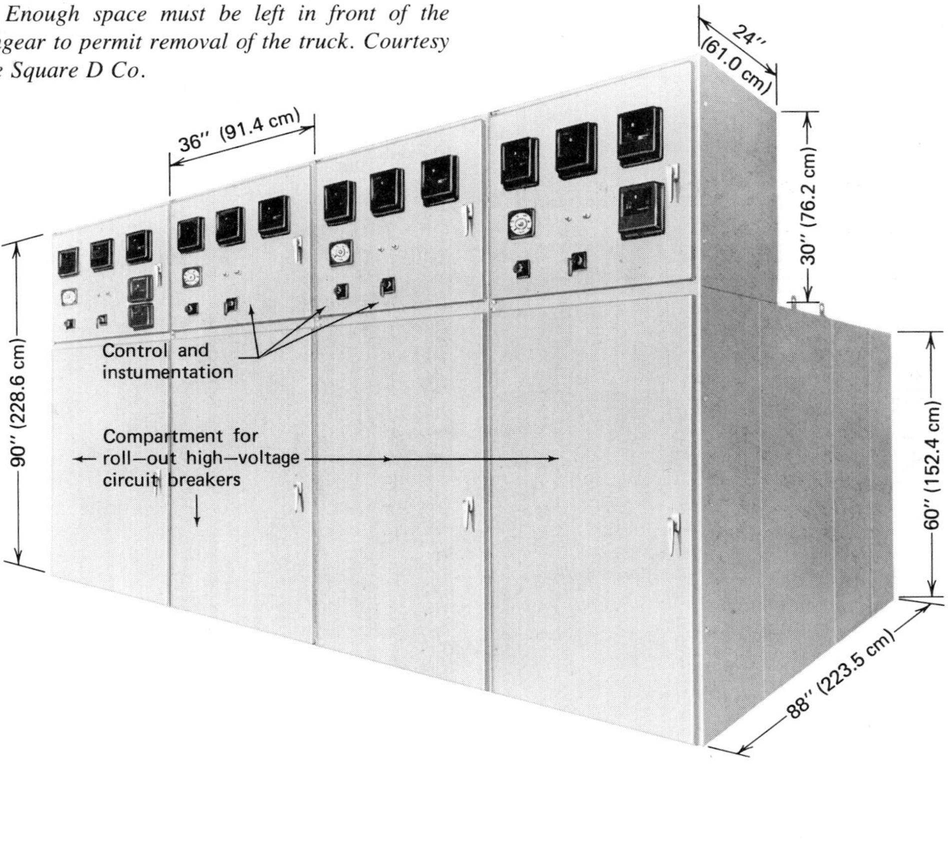

Fig. 16.16 *Typical line-up of high-voltage (15 kV) metal-clad switchgear. Length and width dimensions are typical, but vary with the item of equipment and with the manufacturer. A height of 90 in. (228.6 cm) is an industry standard. Each circuit breaker is mounted on a rolling "truck" to facilitate removal, and plugs into stationary connectors in the switchgear. Enough space must be left in front of the switchgear to permit removal of the truck. Courtesy of The Square D Co.*

although often high-voltage equipment (above 600 V) is referred to as switchgear, as is equipment that comprises individual units rather than an assembly. When molded case circuit breakers are utilized in a switchboard it is often known as a building-type switchboard. Space requirements for various types of boards are shown in Fig. 16.17. Main metal-clad switchgear for commercial, industrial, and public buildings is almost invariably located in the basement, and housed in separate well-ventilated electrical switchgear rooms. The designer must provide adequate lifting hooks, exits, hallways, and hatches for the entrance and exit of the equipment of largest dimensions and weight to be moved. Therefore, the specifications for switchgear should state the maximum number and overall maximum dimensions of sections to be bolted together as one portable section. Two, three, or four sections form the usual practical section. These sections may vary in length from about 8 to 12 ft. Smaller subdistribution switchboards require no special room enclosure, except to bar tampering or vandalism. In such instances a wire screen enclosure plus a large DANGER—HIGH VOLTAGE sign is usually adequate.

When switchgear is to be installed outdoors, three methods may be employed: build a small house to enclose normal indoor gear, utilize weatherproof outdoor gear, or utilize switchgear, which is built into its own metal house. Integral housings are equipped with heat and light and often prove the most economical choice.

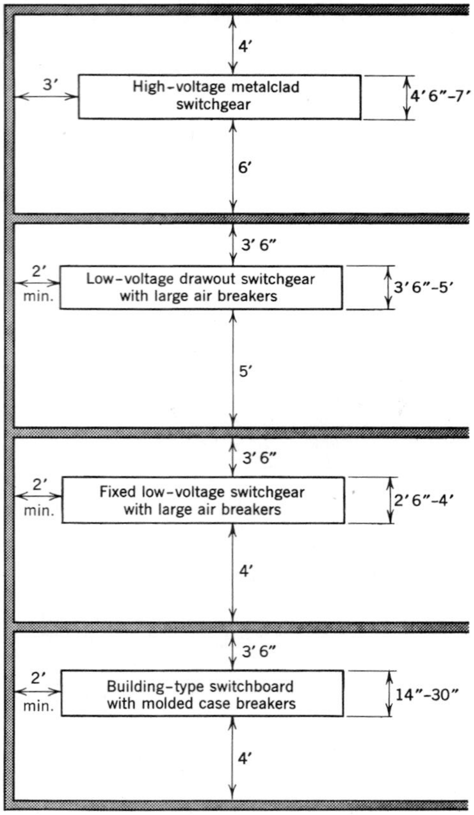

Fig. 16.17 Minimum switchgear space requirements. Clearance to ceiling should not be less than 3 ft. Each room should have two doors when switchgear is connected to high-capacity systems.

16.19 Unit Substations (Transformer Load Centers)

An assembly of primary switch-and-fuse or breaker, step-down transformer, meters, controls, buswork, and secondary switchgear is called a unit substation, or a load-center substation. It is available for indoor or outdoor use, to supply power from a primary voltage line to any large facility. The location in the building of the unit substation is governed by the type of transformer utilized, as explained above in the discussion on indoor transformer installations. For this reason, almost all indoor unit substations utilize dry-type (air-filled) transformers. Unit substations are utilized to effect the economies inherent in prefabricated construction with coordinated components. A basement location is most often selected, with ventilation requirements as detailed above. Access should be restricted to authorized persons only. A typical unit is illustrated in Fig. 16.18.

16.20 Panelboards

A panel, or panelboard, serves basically the same function as a switchboard, except on a smaller scale; that is, it accepts a relatively large block of power and distributes it in smaller blocks. Like the switchboard, it comprises main buses to which are connected circuit-protective devices (breakers or fuses), which feed smaller circuits. The panel-

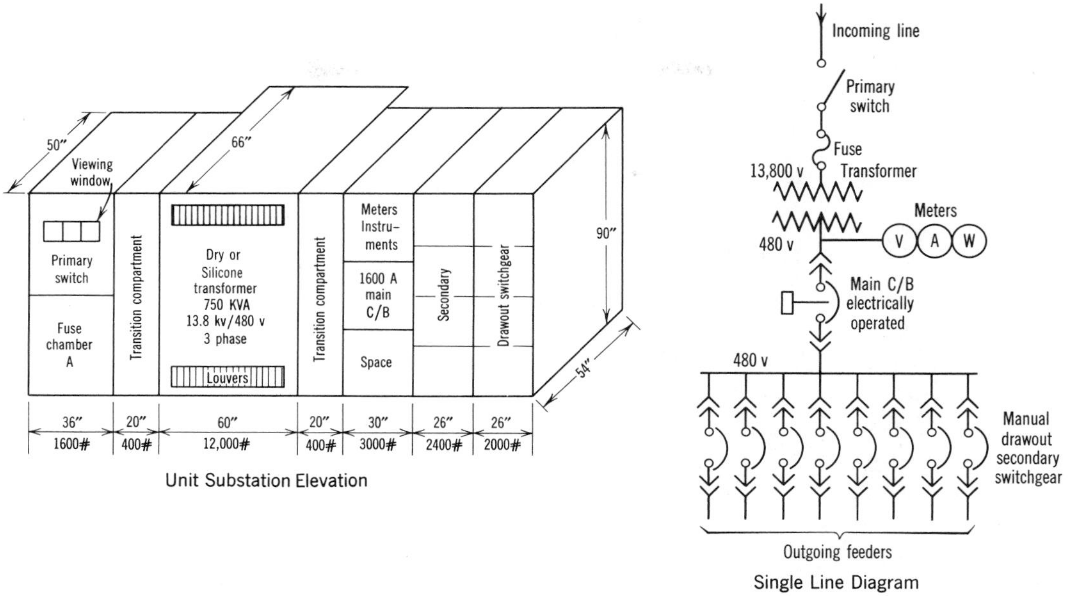

Fig. 16.18 Approximate sizes and weights of a typical large single-ended unit substation. Such a unit would supply a building with a maximum demand of 750 kVA. The incoming 13,800 V cables enter cubicle A and connect to the switch and fuses. The load side of the fuses connects to the transformer, which in turn connects to the secondary switchgear. This main secondary switchgear in turn feeds various switchboards and panelboards distributed through the building.

board level of the system is usually the final distribution point, feeding out to the branch circuits that contain the electrical utilization apparatus and devices, such as lighting, motors, and so on.

The panel components—that is, the buses, breakers, and so forth—are mounted on an insulating board that in turn is mounted inside an enclosing cabinet (see Fig. 16.19*a*). The line terminal of each circuit-protective device (breaker or fused switch) is connected to the busbars of the panelboard. The load terminal of the device then feeds the outgoing branch circuit. This is shown schematically in Fig. 16.19*b*. The busbars of the panelboard are energized by a feeder from some switchboard or load center.

Panelboards are described and specified by type, bus arrangement, branch breakers, main breaker, voltage, and mounting—though not necessarily in that order. A typical description might be: Lighting and Appliance Panel, 3-phase, 4-wire; 200A mains; main c/b, 225A frame, 150A trip. Branch breakers—all 100A frame; 8 ea. SP-20A, 4 ea. 2P-20A, 4 ea. 3P-20A; flush with hinged locked door.

16.21 Electric Motors

Motors are very frequently supplied as adjuncts to specified driven equipment, such as fans, blowers, and so on, within the constraints of the voltage and enclosure. The actual choice of motor is left to the driven-equipment supplier, it being maintained that he is best qualified to select a motor that will optimally match the driven-equipment requirements and so supply a working whole for whose proper operation he is responsible. The supplier, however, is frequently guided primarily by the price motive. The specifier therefore should be sufficiently knowledgeable so that, within the stated constraints of applicability to the load, he can specify the particular motor desired. The following paragraphs are written with that purpose in mind and therefore concentrate on application data.

(a) Direct-Current Motors. As a result of the high cost and relative rarity of direct current, these motors are used only where continuous fine speed control is required, as in the case of elevator drives.

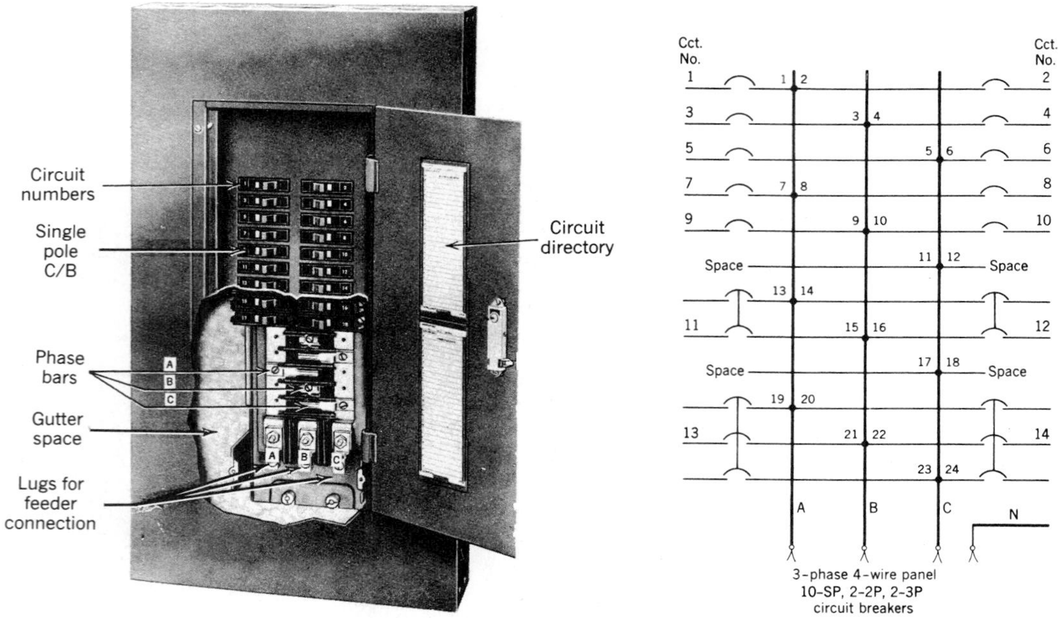

(a) (b)

*Fig. 16.19 Panelboards may be of the circuit breaker or fuse type. The illustrated panels contain 1-, 2-, and 3-pole branch circuits. Panels are provided with a minimum 4-in. gutter space, to allow routing of circuit wiring and feed-through conductors. Lighting panels average 5 in. deep- and 16 to 20 in. wide; power panels are 6 in. deep and 20 to 30 in. wide. A typical schematic diagram for a panel is shown in (*b*). Courtesy of The Square D Co.*

(b) Alternating-Current Motors. These motors fall into three general classifications: polyphase induction motors, polyphase synchronous motors, and single-phase motors. Within these categories there are further subdivisions. Of these many types, the vast majority are squirrel-cage induction machines; therefore, this type will be studied in some detail.

(c) Squirrel-Cage Induction Motors. This motor type owes its interesting name to an early design in which the rotor consisted of a group of bars welded together into a cylindrical cage-type shape. The design is basically unchanged today except for refinements. Squirrel-cage motors are manufactured in four different NEMA designs to meet different application requirements. Of these the most common are:

Type B—standard design, high efficiency and power factor, normal torque. Applicable to fans, blowers, and pumps

Type C—high starting torque, fair efficiency and power factor. Applicable to compressors, conveyors, and other devices that start under load.

A typical motor nameplate is shown in Fig. 16.20.

Items on this nameplate that are not self-evident are:

1. *Type*—this is the manufacturer's designation and indicates primarily the enclosure. Common enclosures are open drip proof, totally enclosed, fan cooled, and weather-protected.
2. *Duty*—continuous or intermittent.
3. *Service factor*—permissible overload, generally 15%.
4. *kVA code*—indicates by a letter the maximum starting current per horsepower. This is useful in selecting motor protective devices.
5. *Frame*—a NEMA standard number that indicates the motor's physical dimensions.

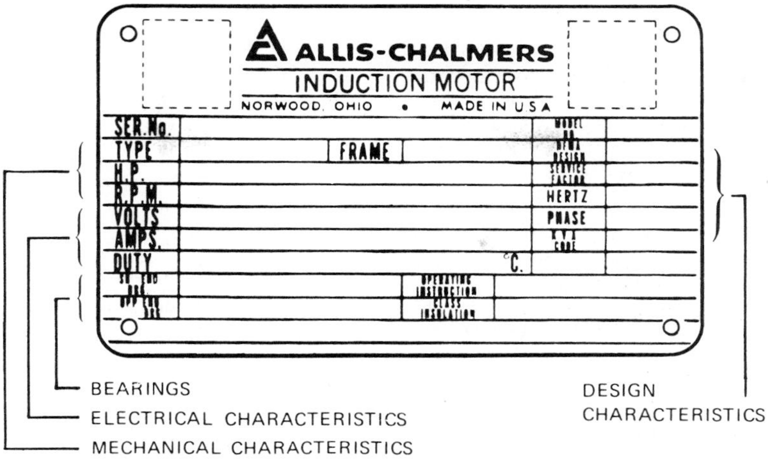

Fig. 16.20 *Typical motor nameplate. (For explanation of terms, see text.) Courtesy of Allis-Chalmers Corp.*

6. *Motor voltage*—the standard motor voltages are 208, 230/460, and 575 V. Induction motors will generally operate satisfactorily ±10% voltage. Only 208-V motors should be used on 208-V systems, since actual line voltage may be as low as 200 V. Using a 230-V motor will result in sharply reduced torque and increased temperature rise, and poor overload capacity.

(d) Electric Motor Energy Considerations. Data compiled by the Federal Energy Administration indicate that more than one-half of all the electric energy generated in the United States is used to power electric motors. Thus an easily attainable overall decrease in motor circuit power of only 1% would result in an energy savings of 3+ billion kWh or 5+ million barrels of oil annually.

Energy losses associated with electric motors stem from two sources:

1. Losses in the motor itself, as indicated by the motor's efficiency.
2. Line losses caused by the motor's power factor.

Let us consider the latter item first.

The power factor (pf) of a device is an indication of the amount of reactive current that the device requires—the lower the power factor, the larger the reactive current. Reactive current does not draw power and does not therefore show up on the kilowatt-hour meter or the kilowatt-demand meter. It *does,* however, cause I^2R line (power) losses throughout the distribution system, and it does tie up generating capacity with no financial return to the utility. Recognizing this, utility tariffs almost invariably contain a low-power-factor penalty clause. This penalty revenue compensates the utility but does nothing to diminish the energy wasted in line losses. It is specifically this energy waste that a number of proposed and enacted, state and federal regulations aim to reduce. They generally state that utilization equipment rated greater than 1000 W and lighting equipment greater than 15 W, with an inductive reactance load component, shall have a power factor of not less than 85% under rated load conditions.

To gain an appreciation of the magnitude of losses involved, consider a small industrial facility with a continuous load of 100 kW at 60% power factor. This low figure is readily realized with low power factor fluorescent lighting, and single-phase motors whose power factor is inherently low (of the order of 40 to 50%).

Consider now the effect of improving the power factor to 85%:

	60% pf	*85% pf*
Load	100 kW	100 kW
kVA (kW/pf)	166.67	117.65
Current (3ϕ, 208 V)	462 A	326 A

Losses are proportional to I^2. Assuming a 5% loss in the utility lines and another 5% loss in the building distribution system, the losses at the improved power factor are reduced to:

$$\text{loss at 85\% pf} = 10\ \text{kW}\left(\frac{326}{462}\right)^2 = 5\ \text{kW}$$

That is, the loss is cut in half. This reduction is far in excess of the minimal 1% referred to above. Assuming an 8-h day, 5-day week, this amounts to 10,400 kWh/year, which is approximately twice an average residence's annual usage. Monetarily, the improvement is no gain to the utility, which loses its surcharge. For the industrial facility, the outlay needed for capacitors and other equipment (see below) to effect this improvement will generally pay for itself within a very few years, depending on local energy rates, electrical installation costs, and other factors. For the economy, however, the energy savings is pronounced.

Most motor manufacturers now produce a line of high power factor, high-efficiency motors. Using these in lieu of standard motors with power factor correction equipment will frequently reduce the pay-back period (refer to Fig. 16.21). Note the pronounced negative effect on motor efficiency and power factor that results from operation at partial load. To avoid this, the designer should match the motor as closely as possible to the load requirement and avoid the oversizing so common in practice. This will do much to improve power factor and efficiency and concomitantly to conserve energy. Where this is impractical because of variable loads, the designer has two choices:

1. Use a standard motor, with an electronic control device that matches input to load. These devices, known as power factor controllers (*not* capacitors) act to reduce input when the motor is only partially loaded, thus reducing electrical and mechanical losses and increasing power factor.
2. Use an energy-efficient motor that has a better load versus efficiency and power factor curve than a standard machine.

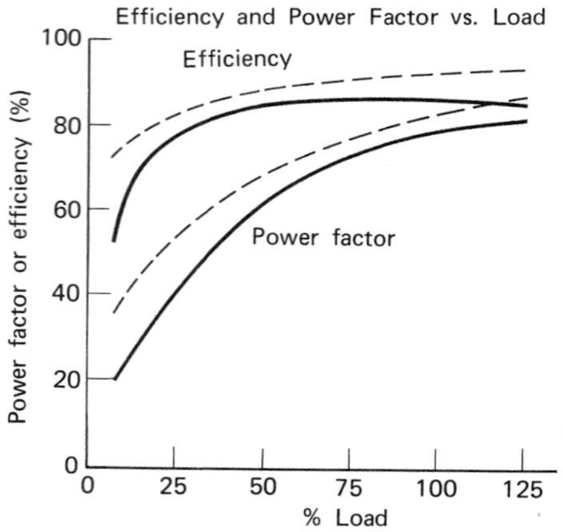

Fig. 16.21 A comparison of the efficiency and power factor characteristics of standard (solid line) and energy-efficient (dashed line) motors. Note that not only is full load performance improved, but perhaps more importantly, the partial load characteristic is much improved. (The data shown are based on industry average and not any specific manufacturer.)

With respect to motor efficiency itself, unlike power factor, any improvement is reflected in the facility's electric energy bill, in addition to effecting a savings in fuel. Unlike pf, efficiency cannot be improved by the addition of a device external to the motor, since it is inherent in the motor design.

Thus, for an *existing* facility a careful cost analysis would have to be made to determine whether replacing existing machines with newer, more expensive, high-efficiency machines is justifiable. For a new facility, the pay-back time for the cost differential of the high-efficiency machines is short. An added advantage of these machines is that the lower losses result in lower temperature operation and therefore extended life. A life-cycle cost analysis of the type given in Tables 15.5 and 16.4 should be performed before purchase of motors. The results will probably indicate a distinct advantage to the high-efficiency designs. A typical analysis of a 5-hp machine is given in Table 16.7, based on the published manufacturer's data.

16.22 Motor Control

Since d-c controllers are highly specialized and infrequently encountered, we will confine our remarks to a-c motor control. A conventional a-c motor controller is basically a contactor (see Section 16.14) designed to handle the heavy inrush currents encountered in an a-c motor starting. Its function is twofold: to start and stop the motor and to protect the machine from overload. These

TABLE 16.7 **Life-Cycle Cost Comparison for 5-hp Motor**

Motor Type	*Conventional Motor*	*High-Efficiency Motor*
Efficiency at full load	83%	87%
Power output	3730 W	3730 W
Power input	4.494 kW	4.287 kW
kwH per year (12 h/day, 5 day/wk, 52 wk/yr)	14,021	13,375
Annual kWh differential	646 kWh	
20-yr kW differential	12,920 kWh	
Annual energy cost at $.07 per kilowatt-hour	$981	$936
Initial motor cost	$225	$275
Motor life	10 yr	20 yr
Motor amortization cost	$16	$6
Total annual cost	$997	$939
Motor cost pay-back period	10 months	
20-yr life-cycle energy cost (8% capital cost, 3% annual price escalation)	$15,339	$14,635
Total 20-yr life-cycle cost	$15,668	$14,910

two separate and distinct functions are accomplished by combining a set of contacts for on/off control with a set of thermal overload elements for overload protection, in a single unit. When the contacts are operated by hand, the controller is called a manual starter; when the contacts are operated by a magnetic coil controlled by push buttons, thermostats, or other devices, the unit is known as a magnetic controller or simply and more commonly—a starter.

Recently electronic motor controllers have appeared on the market that not only provide the starting and protective functions of a conventional starter, but can also provide "soft" starts. Additional advantages of these units are reduced size and weight, long life, more sophisticated motor protection, and additional operating functions such as jogging and reversal. Disadvantages are higher cost and possible radio frequency noise problems. Motors 1 hp or less are generally controlled by a manual switch that contains an overload protection device. It is advisable to utilize such a device for all fractional horsepower motors.

Starters are available in various sizes, voltages, and NEMA enclosures. Table 16.8 gives typical electrical and physical data for full-voltage con-

TABLE 16.8 **Rating and Approximate Dimensions of a-c Full-Voltage Conventional Single-Speed Motor Controllers, 3-Phase Combination Circuit Breaker Type[a]**

NEMA Size[b] Designation	*Maximum Horsepower[b]*	*Width*	*Height*	*Depth*
			Inches	
0	3	10	24	7
1	7½	10	24	7
2	15	10	24	7
3	30	20	24	9
4	50	20	48	9
5	100	20	56	11

[a]All starters are housed in a NEMA 1 indoor ventilated enclosure.

[b]Maximum hp that can be controlled at 208–230 V. Generally, when operating at 460 V, a starter one size smaller can be used.

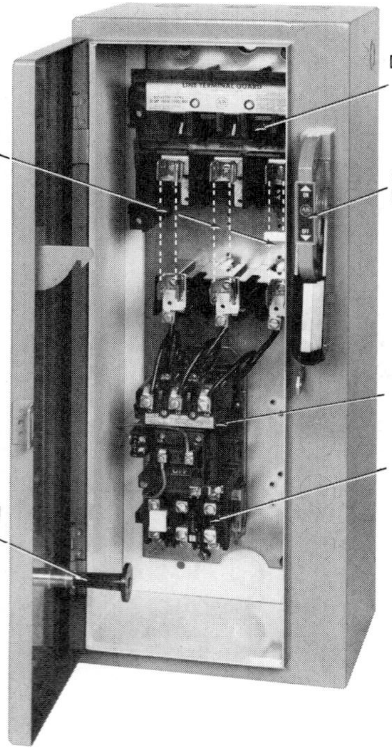

Fig. 16.22 Interior of a combination fused-switch type, across-the-line motor controller. Note that the unit is essentially a switch and a starter wired together and installed in a single cabinet. Courtesy of Allen-Bradley Co.

ventional starters. Most starters are of the full-voltage across-the-line type; that is, the contacts place the motor directly onto the line and the motor starts up immediately. When such a procedure is undesirable because of voltage dip and flicker caused by the large inrush current or because of utility company limitations, a reduced voltage starter, sometimes called a compensator, is used. These units apply reduced voltage to the motor and thus reduce starting inrush current. Every motor controller is required by the NEC to have a disconnecting means within sight of the controller. Where convenient, this disconnect switch may be combined with the starter into a single unit, known as a combination starter. A circuit breaker or fused switch is often used in such an arrangement, which then constitutes the branch circuit protection and disconnecting means (see Fig. 16.22).

When starters are to be assembled for a group of motors, the motor starters, disconnect switches, motor controls, and indicating devices may be combined into a single large assembly for convenience and economy. Such an assembly is called a motor control center. Two typical motor control center construction types are shown in Fig. 16.23. A typical, brief description of a conventional motor controller would be similar to the following:

> Combination circuit-breaker type, across-the-line motor controller, NEMA size 2, 3 O.L. elements, 208 V, in a NEMA 1 enclosure. Starter shall contain integral on/off push buttons.

16.23 Wiring Devices

The general term "wiring devices" includes all devices that are normally installed in wall outlet boxes, including receptacles, switches, dimmers, and pilot lights. Attachment plugs, also called "caps," and wall plates are also included in any discussion of wiring devices. Devices are classified (by manufacturers) in descending order of quality, as premium or specification grade, intermediate grade, and standard or economy grade. These classifications are qualitative only, and one company's specification grade may only equal another's intermediate grade. To assure a certain quality level, NEMA and/or UL standards must be specified. In application, specification grade equipment is usually used in industrial and good commercial construction, intermediate grade in most educational and good residential buildings, and standard grade in low-cost construction of all types. The grade of wiring devices should, as with all electrical equipment, be consistent with the quality of construction in the entire facility.

16.24 Receptacles

In addition to the descriptive data specified above, receptacles are identified by the number of poles and wires, and whether or not the device is designed for connection of a separate grounding wire. The grounding pole *is not counted* in the number of poles (see Fig. 16.24) but is counted in "wires." The grounding pole is connected to the green ground wire where this is run, or to the conduit system where a ground wire is not run. The equipment grounding pole must not be confused with the system ground (neutral), nor must the wiring for the two be interchanged (see Section 17.3). In a typical application, a receptacle for an electric dryer

(a) (b)

Fig. 16.23 *A typical 600-V motor control center is shown in* (a) *and* (b). *Back-to-back construction* (b) *is space saving and adds only 5 in. to the basic 15-in. depth. All units are normally 90 in. high and 20 in. wide per section, although MCCs with especially large components are deeper and wider. Courtesy of The Square D Co.*

with 4800 W, 208 V heating element, and ⅙ hp, 115 V motor would be NEMA 14-50R (see Fig. 16.24). The motor would connect across W and X, the heater across X and Y, and the appliance case to G.

Receptacles must be of the grounding type where installed on standard 15- or 20-A branch circuits. Receptacles connected to different voltages, frequencies, or current type (a-c or d-c) on the same premises must be polarized so that attachment plugs are not interchangeable. Figures 16.24 and 16.25 show some of the standard receptacle configurations and their ratings.

Receptacles are regularly available from 10 to 400 A, 2 to 4 poles, and 125 to 600 V. In addition, special types are made such as locking types and miniature (interchangeable) types. Many specific usage units are available such as range receptacles, hazardous area types, and combination receptacle and switch assemblies. A line of very high-grade devices is manufactured under the name "hospital-grade" devices, specifically intended for this exacting use. Special attention is given in manufacture to grounding provisions, firmness of contact, and long-life service. These units are marked with a distinguishing symbol (usually a green dot) to denote this grade. In addition to these types, several manufacturers produce a receptacle with built-in ground-fault protection, as illustrated in Figure 16.26. For a discussion of these GFI (ground-fault interrupter) receptacles, see Section 17.4. All receptacles other than the normal 15/20 A, 3-wire, parallel-slot type should be specified to be furnished with the required number of matching caps (plugs).

Receptacles are normally mounted vertically

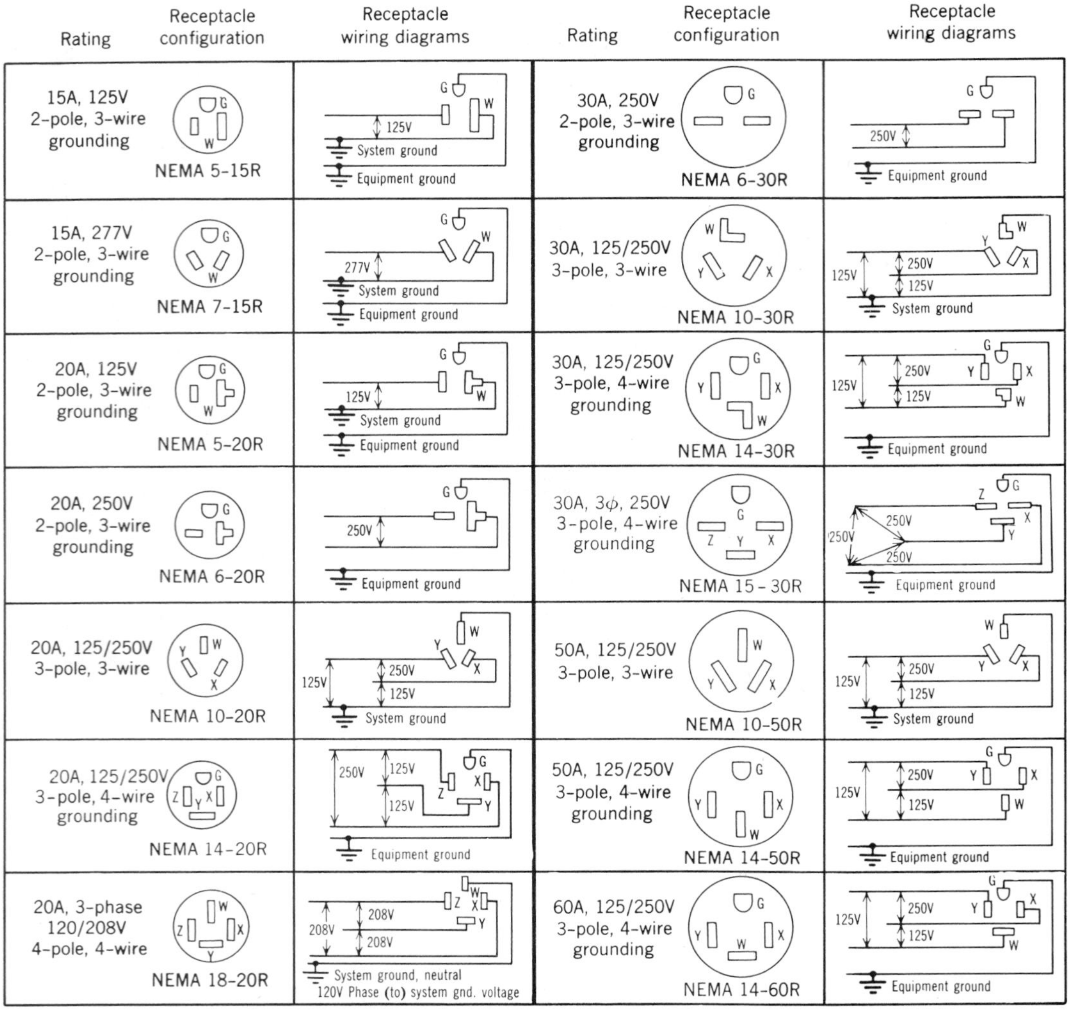

Note: "W" denotes white system ground or neutral
"G" denotes green equipment ground.

Fig. 16.24 Receptacle configuration chart of selected common general-purpose, nonlocking devices with related NEMA designations.

between 12 and 18 in. from the floor, except that in shops, labs, and other areas, where tables are used against the walls, 42 in. is the usual mounting height. A typical receptacle specification would be: Receptacle, duplex, 2-pole, 3-wire, grounding type, 20 A, 250 V, specification grade, for indoor use.

16.25 Switch Devices

By NEC definition, switches up to 30 A that can be outlet-box mounted fall into this category. Generally "a-c only" switches are preferable to the a-c/d-c type because of better construction. The usual a-c switch rating is 15, 20, or 30 A at 120 or 120/277 V. Normal constructions are single-pole, 2-pole, 3-way, 4-way, momentary-contact, 2-circuit, maintained-contact SPDT and DPDT. Operating handles are toggle-type, key, push, rocker, rotary, and tap-plate types. The mercury and a-c quiet types are relatively noiseless; the toggle, tumbler, and a-c/d-c types are generally not. "Interchangeable" devices are miniature types, fully rated, which mount three to a strap in a normal outlet box. (Other devices available in this

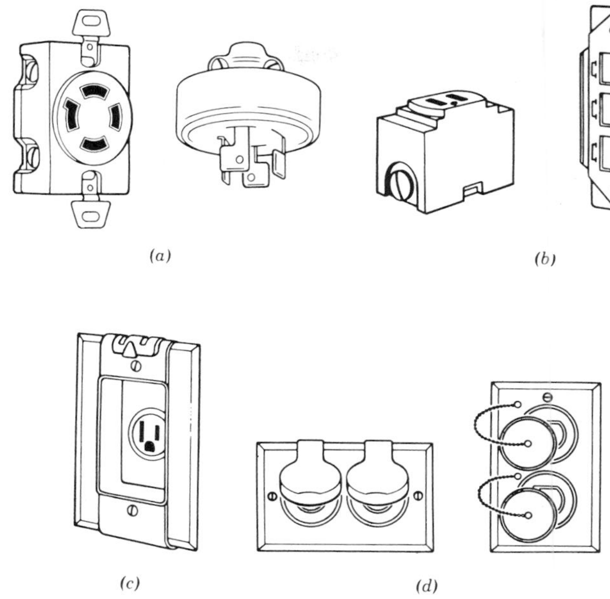

Fig. 16.25 A few receptacle and cap types. (a) *3-pole, 4-wire locking receptacle, 20 A, 125/250 V, with matching cap;* (b) *miniature (interchangeable) device with mounting strap;* (c) *clock outlet with hangar plate;* (d) *outdoor weatherproof receptacles. See Fig. 16.24 for configuration of common receptacles.*

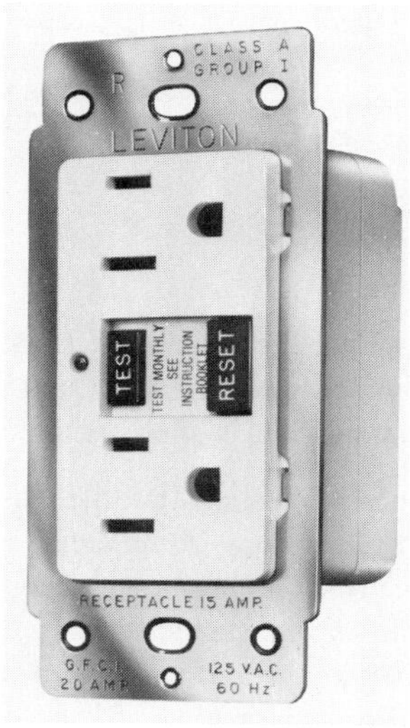

Fig. 16.26 Where it is desired to provide ground-fault protection at a single outlet, utilize a permanent, outlet-box installed unit as illustrated. Courtesy of Leviton Mfg. Co.

miniature construction are pilot lights, push buttons, and receptacles.) A typical switch specification would be: Switch, single-pole, a-c, quiet type, specification grade, 15 A, 125 V, with press handle lighted when OFF, suitable for back or side wiring.

A switch incorporating a solid-state rectifier is readily available, which will give high/off/low control for *incandescent lamps,* within its wattage rating. These devices cost very little more than an ordinary switch and serve the purpose for which a more expensive dimmer is often used. Typical applications are areas such as dining rooms, classrooms, and assembly rooms, where a lower illumination level is often acceptable and always desirable as an energy-conserving measure. In high-security areas, where the easily defeated normal key switch is inadequate, a tumbler lock controlled unit can be used. Loads that can be timed-out such as bathroom heaters and ventilating fans, can be controlled by a spring-wound timer as illustrated in Fig. 16.27. (A solid-state switch with preset time delay is shown in Fig. 21.3, page 1074.) Other common switches are also illustrated in Fig. 16.27.

Thanks to miniature electronics, a programmable switch is available that fits into a wall outlet box in lieu of an ordinary switch. The unit acts as a solid-state 15-A switch, and can be readily programmed to switch the controlled circuit or device at preset times.

Lighting control, including dimming, is covered in Sections 21.2 and 21.3.

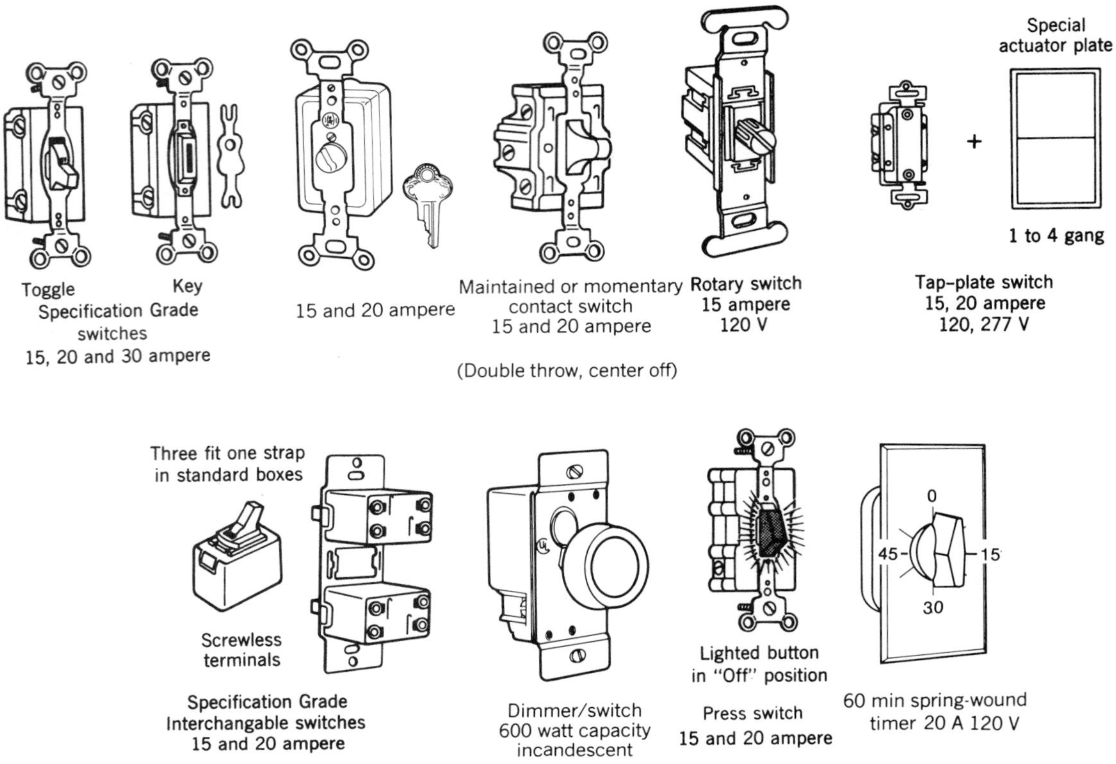

Fig. 16.27 Typical branch circuit switch types.

16.26 Low-Voltage Control Switching

The switches illustrated above are all full-voltage types; that is, they are placed directly into the line and interrupt the load circuit. A system of switching that uses light-duty, low-voltage switches to control relays, which in turn control the load circuits, is shown in Figs. 16.28 and 16.29. The system is variously called low-voltage control, low-voltage switching, and remote control switching. It is entirely electromechanical, in contrast to the solid-state electronic programmable switch/control. The latter, as applied to lighting control, is discussed in Section 21.3. The advantages of this electromechanical system over full-voltage switching are:

1. It permits local and remote control of loads, thus permitting centralized control points and greatly increased flexibility.
2. Since switch legs are of low voltage (24 V), control wires without conduit may be used, greatly reducing costs.
3. The full voltage circuitry is shortened, being run only between the loads and relays.
4. Alteration work is simple and economical.

Residential applications put the master control in the master bedroom. Commercial applications would typically be in hospitals, schools, and industrial facilities where centralization of control is desirable, without the inconvenience of panel switching. The combination of local *and* centralized control is of great importance in energy conservation, since it permits loads to be controlled without the necessity of physically going to each switch point, while at the same time retaining the local control so necessary for energy economy.

16.27 Outlet and Device Boxes

These boxes are generally of galvanized stamped sheet metal. The most common sizes are the 4-in. square and 4-in. octagonal boxes used for fixtures, junctions, and devices, and the 4 × 2⅛ in. box

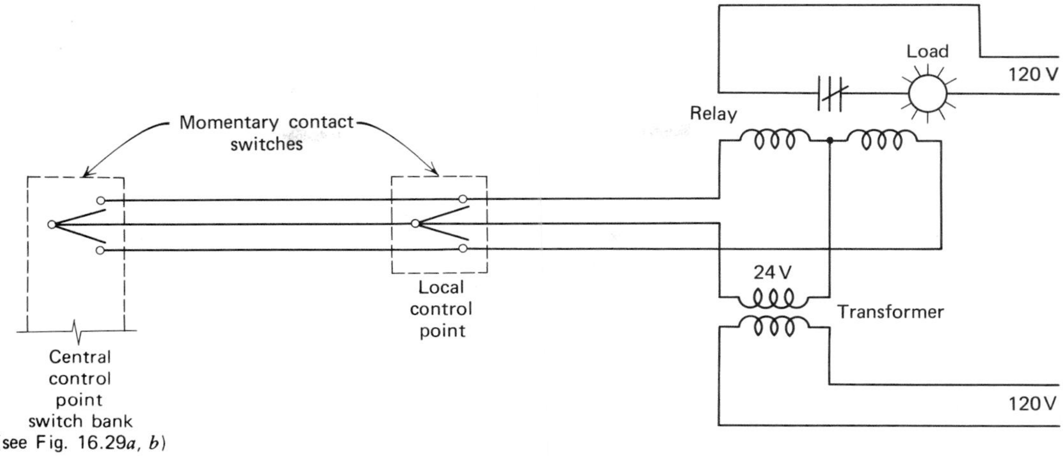

Fig. 16.28 *Low-voltage switching control. Multipoint control and central control are illustrated. The diagram shows the relays located at the load. Central relay cabinets are used in dense load areas.*

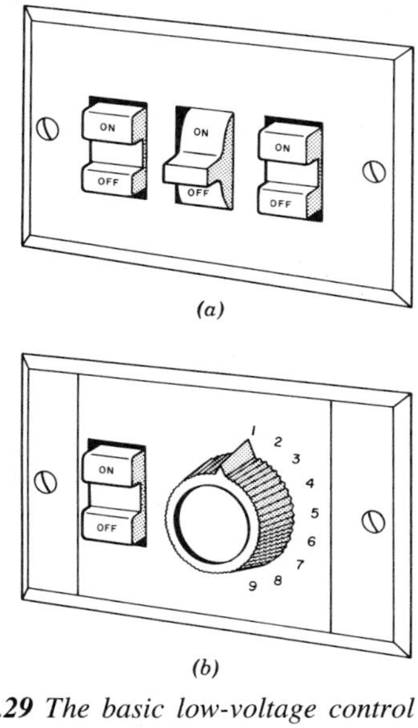

Fig. 16.29 *The basic low-voltage control device is a rocker-type illuminated momentary contact switch, with both contacts open. Switches can be ganged* (a) *or arranged with a rotary selector* (b).

Fig. 16.30 *A cleverly designed telescoping floor box locks in the "up" position during use and rests flush in the floor during periods of inactivity and when floors are cleaned. Courtesy of Maxicom Corp.*

used for single devices where no splicing is required. Box depths vary from 1½ to 3 in. Nonmetallic boxes may be used with NM and NMC cable and with nonmetallic conduit installations. In wet locations and for outdoor work, cast-iron or cast-aluminum boxes are recommended.

Floor boxes for floor outlets are usually of cast-metal construction and are installed directly in the floor slab. An interesting design is shown in Fig. 16.30, which alleviates the severe problem of outlet fouling during wet cleaning operations.

16.28 Lightning Protection Systems

The subject of lightning protection is a complex one, and the design of an adequate system is best left to specialists. Some of the initial considerations, NFPA requirements, and data on available materials will be discussed briefly below.

1. The relevant standards are NFPA Bulletin No. 78, *Lightning Code* and Underwriters Laboratories Standard UL 96A. *Master Labeled Lightning Protection Systems*.

2. The decision of whether or not to protect a structure depends on an evaluation of these factors:

(a) Frequency and severity of thunder storms.
(b) Value and nature of building and contents.
(c) Hazard to building occupants.
(d) Building exposure. Buildings in open and exposed areas are more susceptible to lightning than urban buildings.
(e) Indirect effects. For example, loss of a water tower will seriously affect fire prevention and other services.

3. If a decision is reached to protect a building, it should be done completely and properly, with UL label equipment (Label A and B) and UL approved installation (Label C). A partially protected building is in reality an improperly protected building, which may well be worse than one with no protection at all.

4. The basic principle in lightning protection is to provide a metallic path to ground for the lightning stroke, since *there is no known method of protection that will prevent the occurrence of a lightning stroke*. This will prevent the stroke from passing through the nonconducting portions of a building. Such passage is accompanied by great heat and mechanical forces (see Fig. 16.31).

5. Metal within a few feet of lightning conductors should be bonded to them to avoid arcing. Reinforcing bars in concrete construction should be electrically bonded and connected to the lightning conductors. If this is not done, a lightning stroke may cause severe damage due to the insulating gaps between the rods.

6. A lightning arrester should be placed on all aerial service conductors, whether high or low voltage.

7. Electrical and electronic equipment, which is sensitive to voltage surges such as those accompanying lightning strokes, should be individually protected by surge arresters. A typical small unit of this type is illustrated in Fig. 16.32. These units also protect equipment against voltage surges, transients, and spikes caused by phenomena other than lightning.

16.29 Emergency/Standby Power Equipment

The NEC makes a clear distinction between Emergency Systems and Standby Systems, covering the

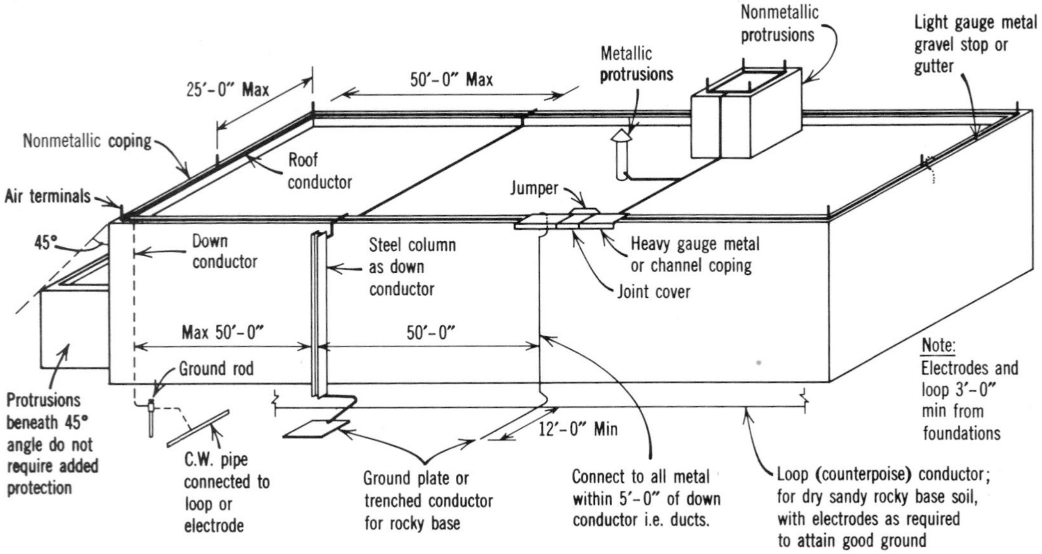

Fig. 16.31 *Diagram of a typical lightning-protection system.*

Fig. 16.32 *Small (2½ in. sq × 1⅜ in. deep) lightning arrester/surge protector useful in protecting electrical equipment from the destructiveness of electrical voltage surges. It allows the surge to bypass the protected equipment. Courtesy of Approved Lightning Protection Co.*

former in Article 700 and the latter in Article 701 (and 702). The equipment, circuitry, and arrangement of both is similar; the purpose is somewhat different.

Emergency systems are intended to supply electric power to equipment essential for *safety* to human life, upon interruption of normal supply. Included in this classification would be illumination in areas of assembly, to permit safe exiting and panic prevention, and such other vital functions as fire detection and alarm systems, elevators, fire pumps, public address and communication, and orderly shut-down or maintenance of hazardous processes.

Standby systems are divided into two categories: those legally required (Article 701) and optional systems (Article 702). The former are intended to power processes and systems whose stoppage might create hazards or hamper fire-fighting operations. This classification is broader than the emergency system one and could include HVAC systems, water supply equipment, and industrial processes. It is still intended primarily as a safety measure, however. *Optional* standby systems can cover any or all loads in a facility at the discretion of the owner, and are normally intended to protect property and prevent financial loss, in the event of a normal service interruption. A typical application would be a critical industrial process or an ongoing research project.

Health care facilities are covered by a separate set of regulations: NEC Article 517 and NFPA standard 76A, both of which are referenced as legally binding in the vast majority of jurisdictional codes. It is important to note that for both emergency and legally required standby systems, the *provision* of the system must be mandated by the authority having jurisdiction over construction, whether it be a local, state, or federal agency, or a combination of these. The NEC dictates how the system is to be designed and constructed; its existence depends on another authority. A case in point is the fundamental item of exit lights. These are mandated by the NFPA Life Safety Code 101, and Subpart E of OSHA regulations. They are not required by the NEC and reference to NEC will not assure their provision. The Code(s) having authority and jurisdiction must specifically require an emergency and/or standby system by that nomenclature, in order for NEC provisions to apply. The exact items of equipment to be powered are selected by the designer, keeping in mind the specific and general requirements of the code having jurisdiction. The majority of codes make the provisions of the NEC and the relevant NFPA standards legally binding. Most codes require emergency systems; far fewer require standby systems, and then only for essential water and water-treatment systems and a few other essential uses. The designer must thoroughly investigate the matter of jurisdictional codes before proceeding with consideration of emergency electric power systems and equipment.

System design arrangements are discussed in Section 17.22. Emergency lighting equipment and system design is covered in Section 21.29. System equipment, which falls into two principal categories, that is, generator and battery installations, is discussed below. Optional standby systems must use a fueled prime-mover; batteries are not permitted there by the NEC (see Table 16.9).

(a) Engine-Generator Sets. An engine-generator set installation comprises basically three components: the fuel system including storage, if necessary; the set itself plus exhaust facilities; and the space housing the unit (see Fig. 16.33). The principal advantages of the engine-generator set are unlimited kV-A capacity, duration of power limited only by size of fuel tank, and, if properly maintained, indefinite life. Disadvantages are noise, vibration, nuisance of exhaust piping, need for constant maintenance and regular testing, and dif-

TABLE 16.9 **Comparative Characteristics of Engine-Generator Sets and Storage Batteries**

Characteristics	*Generator Set*	*Battery*
Capacity	Unlimited	Limited
Type of power	a-c or d-c	d-c[a]
Life	Unlimited	Limited
Relative size	Small	Large
Relative initial cost:		
(1) Below 50 kva	High	Medium
(2) Above 50 kva	Medium	High
Space Required	Medium	Large
Ventilation	Large	Small
Maintenance Cost	Medium	Low
Noise	Great	None
Vibration	Some	None
Fuel Storage	Yes[b]	None

[a]Alternating current can be provided, at high cost, by a static inverter.

[b]With natural gas engines, fuel storage is eliminated.

ficulties with fuel storage. Gasoline can be stored for only a year at most, and subsequent disposal is difficult. Diesel fuel keeps somewhat longer, but disposal is also difficult. Use of gas for a diesel or gas engine obviates the fuel storage problem but poses the alternate problem of availability of gas service during emergencies.

(b) Battery Equipment. Storage batteries are often used to supply limited amounts of emergency power, primarily for lighting. Such units are mounted in individual cabinets or in racks for larger installations and are always provided with automatic charging equipment (see Fig. 16.34).

Fig. 16.33 Typical engine-generator unit and installation technique. Illustrated is a radiator-cooled diesel installation. Courtesy of ONAN Corp.

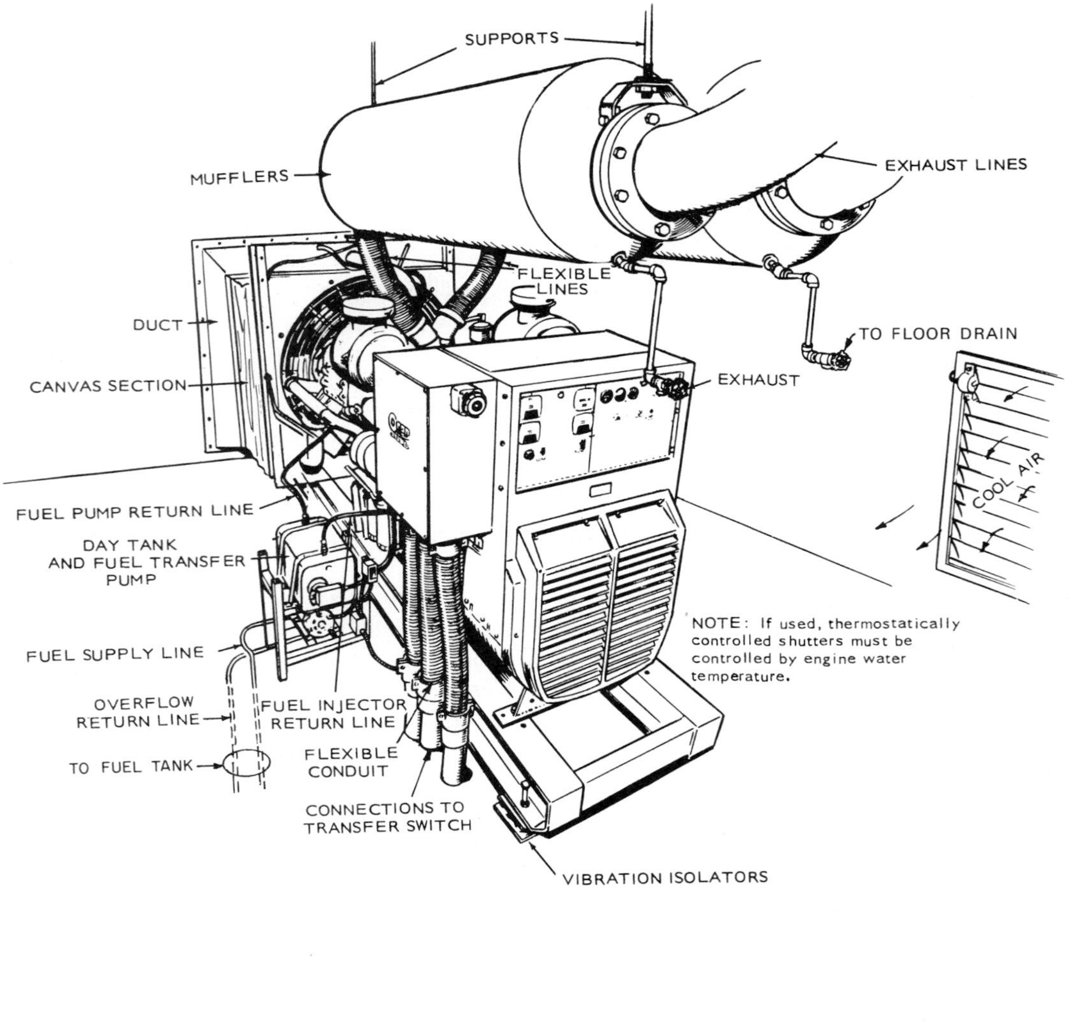

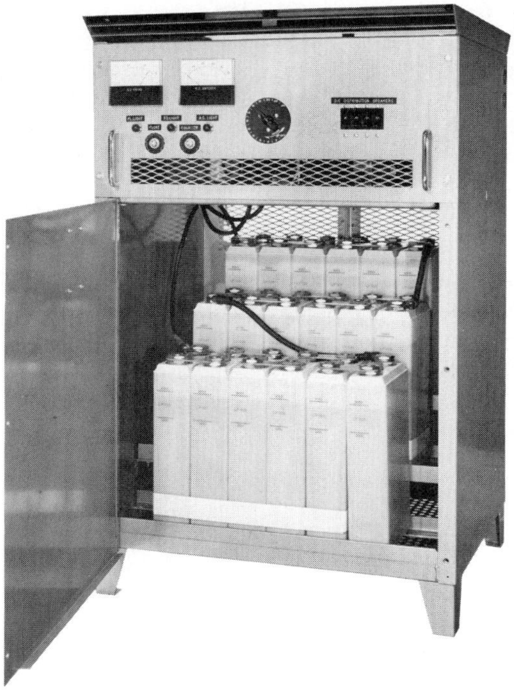

Fig. 16.34 The illustrated package unit consists of a nickel cadmium battery pack and an automatic battery charger. This unit, which is 44 in. high, 25 in. wide, and 22 in. deep, can supply 50 A at 24 V d-c. When equipped with a d-c to a-c inverter, a similar unit will supply approximately 1000 W of regulated, sinusoidal a-c power at any desired voltage. Duration of the current flow depends on the size of the battery selected. Courtesy of LaMarche Mfg. Co.

Large installations of batteries and static inverters called UPS (uninterrupted power supply) systems are commonly found in large computer installations. Such installations are complex, expensive, and designed to meet the specific requirements of the installation.

Batteries need not be installed in separate battery rooms provided that live parts are guarded and sufficient ventilation is provided to prevent gas accumulation. Battery types are undergoing constant development. At this writing the types principally in use are lead acid, nickel cadmium, lead-antimony, and lead-calcium cells. The choice depends on the application and is best left to a specialist. Batteries have the distinct advantage that they can be installed either in a central system with distribution by feeder of the battery power throughout the facility, or they can be installed in small package units around the building. Central systems are 24 to 125 V, d-c and normally feed nonfluorescent emergency lighting only. Individual packs are used often to supply a-c power via built-in inverters. The great disadvantage of battery systems is limited duration of power. The NEC requires that batteries maintain loads for a 1½ h *minimum,* but larger battery capacity is normally installed.

16.30 System Inspection

Each electric wiring system is inspected at least twice by the local inspection authorities; once after raceways (roughing) have been installed and before the wiring and closing in of walls, and once after the entire job is complete. The purpose of these inspections is to determine whether design, material, and installation techniques are meeting the national and local code requirements. Excellence of installation is the responsibility of the contractor. The designer, however, must be completely familiar with installation work and the equipment's physical characteristics in order to properly design an electrical system. Such a design will not present the contractor with unwarranted difficulties. The designer must also be very wary of equipment substitutions by the contractor who, having submitted a bid on the basis of plans and specifications, should be required to supply the specified equipment.

17

ELECTRIC WIRING DESIGN

In wiring design, as in other design, there are numerous possible solutions to each problem—some good, some fair, and some poor. Experience guides the designer to a solution that best suits the job, since it is his or her responsibility to establish the most economical design within the framework of the design criteria. This chapter opens with a discussion of these criteria and continues on to actual wiring design.

17.1 General Considerations

(a) Flexibility. Every wiring system should incorporate sufficient flexibility in branch circuitry, feeders, and panels to accommodate all probable patterns, arrangements, and locations of electric loads. The degree of flexibility to be incorporated depends in large measure on the type of facility. Thus laboratories, research facilities, and small educational buildings require a great deal more flexibility than residential, office, and fixed-purpose industrial installations. As part of the design for flexibility, provision for expansion must be provided as experience has demonstrated that most facilities will grow, both physically and in electrical demand. Overdesign, however, is as bad as underdesign, being wasteful of money and resources both initially and in operation.

(b) Reliability. The reliability of the electrical power within a facility is determined by two factors: the utility's service and the building's electrical system. The service record of the utility should be studied along with the economic impact of a power outage to determine whether, and to what extent, standby power equipment is justified (see Sections 16.29 and 17.22). Emergency equipment required for the safety of a building's occupants is determined by local, state, and national building codes. Beyond the service point, the reliability of power is entirely dependent on the wiring system. Here, too, economic studies must be made to determine the quality of equipment and the amount of redundancy (duplicate equipment) to be installed. The subject of reliability is a complex one, and we can state only a few general principles here.

1. The reliability of an electric system is only as good as that of its weakest element. Therefore, it may be necessary to provide redundancy at anticipated weak points in the system.
2. The electrical service and the building's distribution system act together. An extremely reliable (and expensive) service is of little use if the power cannot reach the desired points.
3. Critical loads within the facility should be pinpointed to determine how best to reliably serve them, that is, by establishing reliable power paths to them or by furnishing individual standby power packages for them. The latter course is often chosen for healthcare and other critical loads.

(c) Safety. Although rigid adherence to the requirement of the NEC and other applicable NFPA codes will assure an initially safe electrical installation, the designer must be constantly alert to such factors as electric hazards caused by misuse or abuse of equipment or by equipment failure. Also, attention to the size of equipment used will eliminate the oft-encountered physical hazard caused by obstruction of access spaces, passages, closets, and walls with electric equipment. Finally, lightning protection can be subsumed under the heading of safety; this topic is discussed in Section 16.28.

(d) Economic Factors. This item can readily be divided into two frequently interrelated items: first cost and operating cost. All other factors being equal, the first cost depends in large measure on whether the constructor is interested in minimum first cost or minimum owning cost. We have demonstrated that these two costs frequently stand in inverse relationship to one another (exceptions are mentioned below). Low first-cost equipment generally results in higher energy cost, higher maintenance cost, and shorter life. The decision, however, is not purely an economic one, inasmuch as the electrical energy cost factor in the operating-cost equation is directly related to energy consumption, with one exception. That exception is the utility's demand charge, which is discussed in Section 14.12. Means for minimizing this cost are covered in detail in Section 14.13. Here there occurs a coincidence of reduction to both first cost and operating costs. Load-leveling equipment permits the electrical distribution system to be sized without consideration of coincident load peaks, thus resulting in smaller equipment, operating more efficiently—near its full-load capacity. All other reductions in electric energy cost flow directly from the corresponding reduction in energy consumption.

(e) Energy Considerations. This factor is a complex one involving considerations of energy codes and budgets, energy conservation techniques (see Section 17.5), and energy control.

Buildings constructed with governmental participation may be subject to energy budget limitations expressed in Btu/square foot/year. Although the lion's share of this budget is taken by heating/cooling and lighting systems, the electrical distribution system will also be subject to conformity to stated codes. Important among these is ASHRAE Standard 90-80. Detailed energy conservation considerations are discussed in detail in Section 17.5. A discussion of the IES Recommended Procedure for Lighting Power Limit Determination is found in Section 20.5.

(f) Space Allocations. The general impression that electrical equipment is small and easily concealed is accurate only for wire and conduit. Panels, motor control centers, busduct, distribution centers, switchboards, transformers, and so on can be large, bulky, noisy, and highly sensitive to tampering and vandalism. Thus, space allocations must be concerned with maintenance ease, ventilation, expandability, centrality (to limit length of runs), limitation of access, and noise, in addition to the basic item of space adequacy.

(g) Special Considerations. These depend on the specialized nature of certain facilities and may include items such as security, central and/or remote controls, interconnection with other facilities, and the like.

17.2 Load Estimating

When initiating the wiring design of a building, it is important to be able to estimate the total building load in order to plan such spaces as transformer rooms, chases, and closets. This information is also required by the local power company well in advance of the start of construction. An exact load total can be made only after completing the design but, since this is often several months later, a good preliminary estimate is required. Such an estimate can be made from the figures given in Table 17.1. These figures are average. When it appears that the building will have heavier or lighter loads because of lighting levels or other factors, the figures should be modified accordingly. The figures having been established in the individual categories, they should be added together without application of demand or diversity factors in order to obtain the maximum demand load for which the building service equipment must be sized, in the absence of electric load-control (load-leveling) equipment. At this point an analysis must be made to determine the feasibility of incorporating such equipment into the facility (see Section 14.13). Input to this study includes the utility's complete rate schedule, including all penalty clauses, a detailed analysis of the building's equipment load patterns, and any external constraints such as maximum loads imposed by power and energy budgets.

Equipment load patterns must be carefully analyzed because they determine a load's "sheddability." Thus, for kitchen equipment, load interruption may be undesirable, but shifting of cooking time by a half hour is entirely feasible. On the other hand, for HVAC equipment, building thermal inertia and "stretching" maximum and minimum temperatures and humidities permit considerable load control without adverse effects. Also,

TABLE 17.1 **Electric Load Estimating**[a]

I	II	III	IV	V	VI
	Volt-Amperes per Square Foot[b]				*Ten Year Percent Load Growth*
			Air Conditioning[g]		
Type of Occupancy	*Lighting*[c,d]	*Misc. Power*[e]	*Electric*	*Nonelectric*	
Auditorium					
General	0.7–2.0	0	12–20	5–8	20–40
Stage	20–40	0.5			
Art gallery	2–5	0.5	5–7	2.0–3.2	20–40
Bank	2–4	3.0	5–7	2.0–3.2	30–50
Cafeteria	2.5–4.5	0.5	6–10	2.5–4.5	20–40
Church & synagogue	1.0–3.0	0.5	5–7	2.0–3.2	10–30
Computer area	1.5–2.5	1.5[f]	12–20	5–8	50–200
Department store					
Basement	3–5	1.5			
Main floor	2.5–4.5	1.0	5–7	2.0–3.2	50–100
Upper floor	2–4	0.5			
Dwelling 0–3000 ft^2	3.0	0.5	—	—	
(not hotel) 3000–120,000	2.0	0.25	—	—	50–100
above 120,000	1.0	0.15	—	—	
Garage (commercial)	0.5	0.15	—	—	10–30
Hospital	2–3	1.0	5–7	2.0–3.2	40–80
Hotel					
Lobby	1–3	0.5	5–8	2.0–3.5	30–60
Rooms (no cooking)	1.0–2.5	0.5	3–5	1.5–2.5	
Industrial loft building	1.0–2.0	1.0	—	—	50–100
Laboratories	3–4	5–20	6–10	2.5–4.5	100–300
Library	1.5–3.5	0.5	5–7	2.2–3.2	30–40
Medical center	2–4	1.5	4–7	1.5–3.2	50–80
Motel	1.0–2.5	0.5	—	—	30–60
Office building	1.5–3.5	1.5	4–7	1.5–3.2	40–80
Restaurant	1.5–2.5	0.25	6–10	2.5–4.5	20–40
School	1.5–3.5	1.5	3.5–5.0	1.5–2.2	50–80
Shops					
Barber & beauty	3–5	1.0	5–9	2–4	
Dress	2–5	0.5			
Drug	2–3	0.5	4–7	1.5–3.2	40–80
Five & ten	2–3	0.5			
Hat, shoe, specialty	2–3	0.5			
Warehouse (storage)	0.25–1.0	0.25	—	—	10–30
In the above except single dwellings:					
Halls, closets, corridors,	0.5	—	—	—	
storage spaces	0.25	—	—	—	

[a]Figures assume energy-conservation techniques applied.
[b]These figures do not include allowance for future loads.
[c]The figures given in Article 220 of the NEC are minimum figures for calculation of electric feeder sizes.
[d]See Section 20.5 for a detailed analysis of the IES lighting power limit determination.
[e]These figures are based on experience and must be applied judiciously.
[f]This figure does not include the power used by the computer.
[g]Includes the loads of air-handling equipment and pumps.

as explained in Section 14.13, the degree and duration of load shedding is a function of the type of control equipment utilized. It is well to repeat here what is stated there—load control affects maximum demand, with only minor effect on total energy consumption.

The external constraints referred to are the energy budgets recommended or required by codes, legislation, and funding bodies. After the above load control analysis is completed, or simultaneously with it, a building energy consumption analysis must be performed. This may be done manually, although the numerous computer programs available are considerably more accurate. The results of this analysis will indicate whether the target electrical energy budget is being exceeded. If so, loads will have to be modified by reconsideration of projected systems and system criteria, by incorporating energy-conservation devices and techniques into the electrical system and by drawing up energy-use guidelines that will be applied when the building is occupied. Since this last item depends for its success on the day-to-day voluntary actions of the building's occupants, it should not be considered a major conservation source. Conservation measures are covered in Section 17.5.

The electrical loads in any facility can be categorized as follows:

1. Lighting.
2. Miscellaneous power, which includes convenience outlets and small motors.
3. Heating, ventilating, and air conditioning (HVAC).
4. Plumbing or sanitary equipment.
5. Vertical transportation equipment.
6. Kitchen equipment.
7. Special equipment.

Category (1) is self-explanatory and is covered by column II of Table 17.1.

Category (2), column III, includes, in addition to receptacles and small motors, such items as small business machines, plug-in heaters, water fountains, and so on.

Category (3), column IV, includes all loads imposed by the HVAC equipment. Included therefore are fuel pumps, boiler motors, condensate pumps, all heaters, blower motors, exhaust fans, and so forth. Also included in column IV, for air conditioning loads, is refrigeration compressors. This item is omitted in column V since the air conditioning utilizes absorption machines that do not use electricity for primary power. When air conditioning is not anticipated, the HVAC load is still appreciable because of heating and ventilating (H and V) requirements. A rough estimate for this load would be ⅔ or 65% of the loads in column V.

Category (4) includes all loads associated with the water and sanitary system, including house water pumps, air compressors and vacuum pumps, sump pumps and ejectors, well pumps and fire pumps, water heaters and pneumatic tubes, plus such special items as display fountain pumps. Since these loads vary widely with local conditions and with facility design as much as type of facility, an estimate cannot be made on a volt-amperes per square foot basis, by type of building. For this reason, no figure is included in Table 17.1. If actual data cannot be used, it is helpful to remember that plumbing loads are relatively small, rarely exceeding 20% of the HVAC load, though, for the most part, they are unrelated to it.

Category (5), vertical transportation, is also obviously unrelated to square footage and therefore cannot be tabulated in Table 17.1. These loads, including elevators, moving stairs, and dumbwaiters, can be estimated from the data given in the vertical transportation chapters of this book.

Category (6), kitchen equipment, is also not included in Table 17.1, though obviously present in all restaurants, in most hospitals, and in some office and religious-use buildings. The reason for this omission is that the primary power for the major load, the cooking equipment, may be either gas or electric. Other large energy-use equipment such as dishwashers can be electric, gas, or steam fed. Furthermore, no correlation can be made between facility type, area, and load, even if electrically powered, since population and schedule are also major factors. When kitchens are planned, a kitchen consultant or other experienced planner can usually supply an estimate of the electric power requirement, which can then be added to the other figures.

Category (7), special equipment, is so variegated that no figures can be listed. Under this title is subsumed such items as laboratory equipment, shop loads, display area loads, floodlighting, canopy heaters, display window lighting, and so on.

TABLE 17.2 **Nominal Service Size in Amperes**

Nominal service sizes are 100 A, 150 A, 200 A, 400 A, 600 A.

	Area in Square Feet				
Facility	*1000*	*2000*	*5000*	*10000*	*Remarks*
Single-phase, 120/240 v, 3-wire					
Residence	100A	100A	150A	—	Minimum 100A
Store[a]	100A	150A	—	—	
School	100A	100A	150A	—	
Church[a]	100A	150A	—	—	
3-phase, 120/208 v, 4-wire					
Apartment House	—	—	100A	150A	
Hospital[a]	—	—	200A	400A	
Office[a]	—	100A	400A	600A	
Store[a]	—	100A	400A	600A	
School	—	100A	100A	200A	

[a]Fully air conditioned using electric-driven compressors. Based on figures in Table 17.1.

This load data must be gathered for individual items of equipment and added to the foregoing totals.

Table 17.2 gives a tabulation of service entrance sizes in amperes, based on single- and three-phase service for typical occupancies. These figures are for quick estimate purposes and should be adjusted after the design is completed.

17.3 System Voltage

There are several systems of voltage commonly available in the United States and Canada. The simplest of these is:

1. 120 V, single-phase, 2-wire (see Fig. 17.1*a*). This is used for the smallest of facilities such as small residences, out-buildings, and isolated small loads up to 6 kVA. Load is calculated by multiplying current and voltage. For 60-A service, which this type of service is normally limited to, no more than 50 A are usually drawn. Thus

$$VI = 120 \times 50 = 6000 \text{ V-A} = 6 \text{ kVA}$$

The nominal system voltage is 120 V, although it is also referred to as 110 V and 115 V (see Fig. 17.1*a*).

For somewhat heavier loads, the system normally used is:

2. 120/240 V, single-phase, 3-wire (see Fig. 17.1*b*). The Code requires that all residences with five or more 2-wire circuits have a minimum of 100 A, 3-wire service. Service disconnect for 100-A service would be a 100 A, 2-pole, and solid neutral switch, fused at no more than 80% of rating, or 80 A. This is usually written 100A, 2P & SN, 80AF. This service is used principally for residences, small stores, and other occupancies where the load does not exceed 80 A or 19.2 kVA.

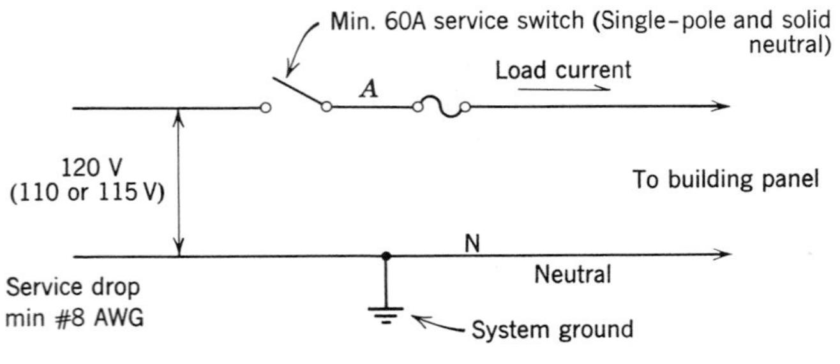

Fig. 17.1 (a) *120 V, single-phase, 2-wire service. This is also the arrangement of the usual branch circuit.*

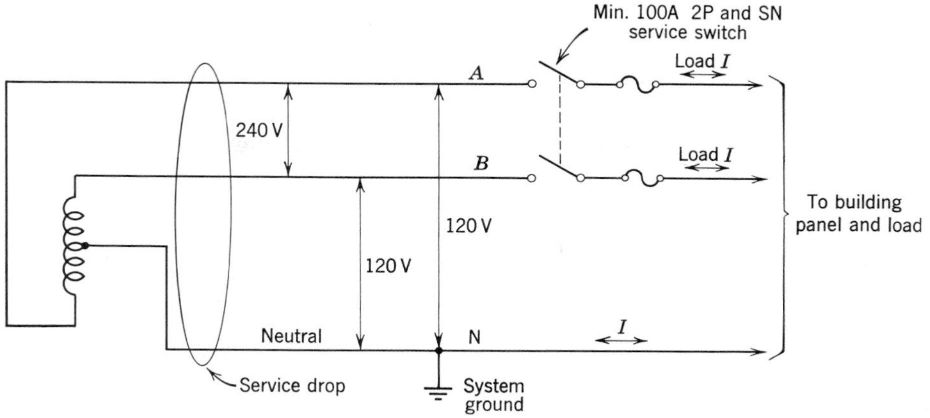

(*b*-1)

Fig. 17.1 (b-*1*) *120/240 V, single-phase, 3-wire service. The single-phase transformer is center-tapped to establish a neutral. The neutral connection is always grounded.*

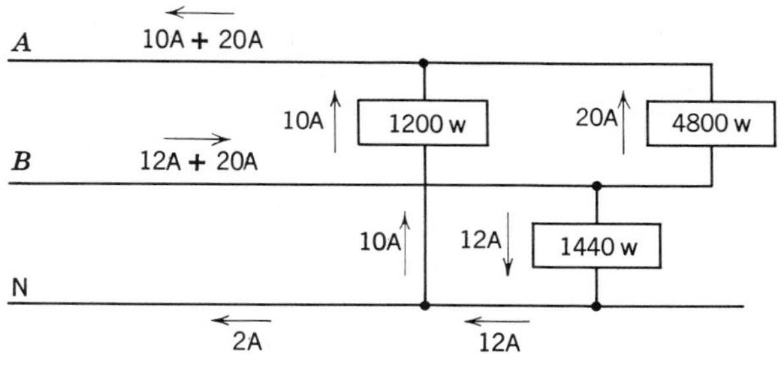

(*b*-2)

Fig. 17.1 (b-2) *Note that the neutral carries the difference in current between the A and B legs and therefore a maximum that is equal to the current in one of the legs (when the other is zero).*

Load is calculated thus:

$$\text{kVA} = \frac{V \times I}{1000} = \frac{240 \times 80}{1000} = 19.2$$

See Fig. 17.1*b*-1. Although it may appear otherwise, the neutral carries no more than full-load current. Note that each "hot leg" of the 3-wire system carries line current. Thus, total load can also be calculated:

$$\text{Load kVA} = \text{twice load on each line}$$

Assuming a balanced 80-A load

$$\begin{aligned}\text{Total kVA} &= 2 \times 80\text{ A} \times 120\text{ V}\\ &= 2 \times 9{,}600\text{ V-A} = 19{,}200\text{ V-A}\\ &= 19.2\text{ kVA}\end{aligned}$$

If load is unbalanced, with say 30 A in one line and 50 A in the second, total load is

$$\begin{aligned}120 \times 30 + 120 \times 50 &= 3600 + 6000\\ &= 9600\text{ V-A} = 9.6\text{ kVA}\end{aligned}$$

Depending on the rating of the service transformer, system voltages can be 120/240, 115/230, or 110/220, although 120/240 is the accepted industry standard. Loads that are 120 V cause a current in only one line. Loads that are 240 V, such as a clothes dryer, cause current in both lines. For example, to find the line currents caused by the three loads shown on Fig. 17.1*b*-2.

120 V, 1200 W iron on line *A*

120 V, 1440 W hair dryer on line *B*

240 V, 4800 W dryer on lines *A* and *B*

we calculate

$$I_A = \frac{1200}{120} = 10\text{ A}$$

$$I_B = \frac{1440}{120} = 12\text{ A}$$

$$I_{AB} = \frac{4800}{240} = 20\text{ A}$$

Note that the neutral only carries the unbalance of 2 A.

Total current in $A = 20 + 10 = 30$ A
Total current in $B = 20 + 12 = 32$ A
Total load $= 120(30) + 120(32)$
$= 3600 + 3840 = 7440$ W
Loads are $1200 + 1440 + 4800 = 7440$ W

The 120/240 V, single-phase system comes from a center-tapped single transformer. When two hot legs and a neutral of a 3-phase system are taken, a similar system is obtained with a rating of:

3. 120/208 V, single-phase, 3-wire (see Fig. 17.1*c*). This system is really part of a 3-phase system and is most often found *within* a building with 3-phase distribution, and is used to serve a small load that does not require 3-phase, 4-wire. Calculation of loads and line currents is considerably more complex than in (2) above because of the 120° phase displacement between phases *A* and *B*. Here, as before, the neutral will carry no more than line current whether the system is balanced or not.

When feeding large facilities, a 3-phase, 4-wire system is normally employed. This is usually:

4. 120/208 V 3-phase, 4-wire (see Fig. 17.1*d*). This system is the most widely used 3-phase arrangement and is applicable to all facilities except the very largest ones. In those, lengths of feeders and sizes of loads become so great that a higher voltage is required. In this system, 120-V loads such as lighting, small machines, receptacles, and so on, are fed at 120 V by connection between each phase leg (see Fig. 17.1*d*-2) and neutral. Motors larger than ½ hp and all 3-phase loads are fed at 208 V by connection between the 3-phase legs. Single-phase, 208 V loads such as heaters are accommodated as in (3) above, by connection between two phase legs. Such loads are often referred to as "2-pole" loads, alluding to the 2-pole current breakers used to feed them. Although this system is nominally 120/208 V, it is also used as 115/200. However, this is inadvisable because of nominal motor voltage, which will be discussed below.

Where buildings are large, either horizontally or vertically, lighting is principally fluorescent, and the 120-V load does not exceed one-third of the total load, a more economical system of voltages is available. This system is:

5. 277/480 V, 3-phase, 4-wire (see Fig. 17.1*e*). It utilizes 277-V fluorescent lighting, 480-V machinery, and small (3 to 25 kVA) dry-type closet-installed transformers to step down from 480 to 120 V for supplying receptacles and other 120-V loads. This system is ideally suited to multistory office buildings and large single-level or multi-level industrial buildings. Savings are generated by the smaller feeder and conduit sizes and smaller switchgear, which more than offset the additional cost of step-down transformers for the 120-V load.

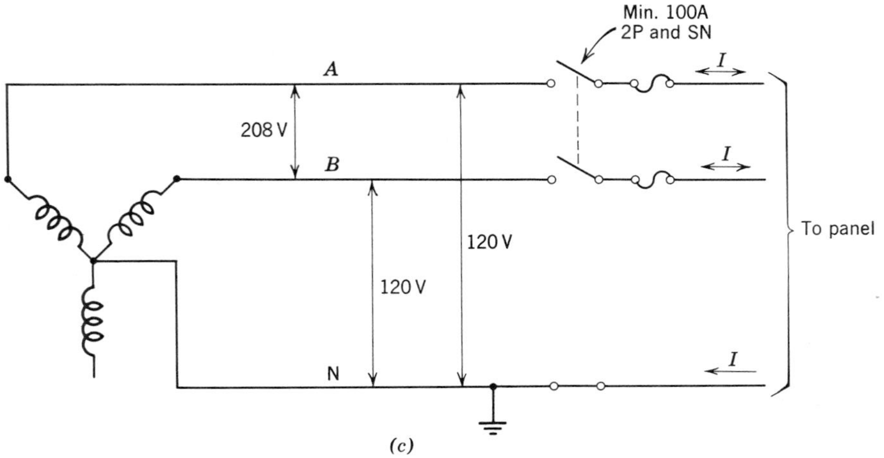

Fig. 17.1 (c) *120/208 V, single-phase, 3-wire service. This arrangement comprises two-thirds of the full 120/208 V, 3 phase, 4-wire connection shown in Fig. 17.1*(d-*1*), *below.*

ELECTRICITY

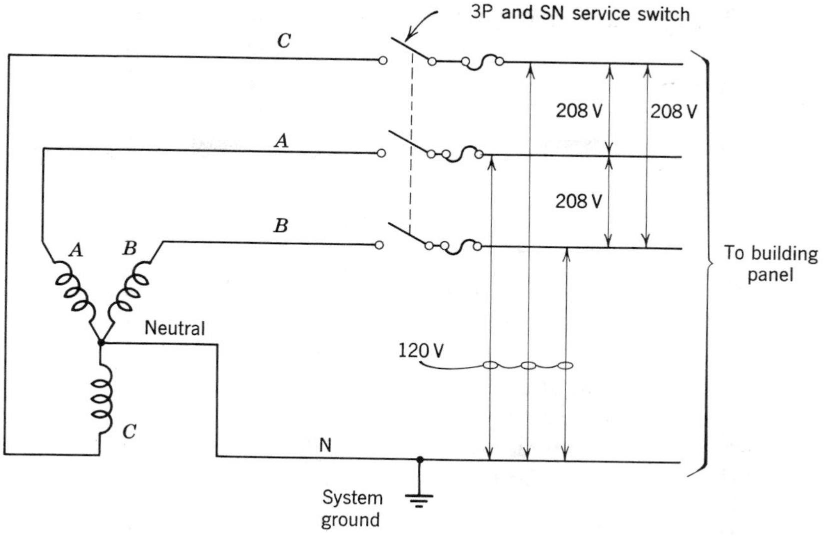

(d-1)

Fig. 17.1 (d-*1*) *120/208 V, 3-phase, 4-wire system. The neutral connection is connected to the system ground and is not broken by the service switch.*

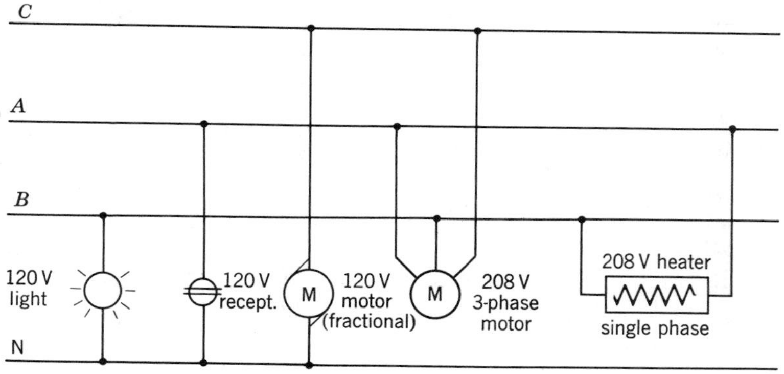

(d-2)

Fig. 17.1 (d-2) *The flexibility of the 120/208 V system is illustrated and accounts for its wide application. Loads are shown schematically. In practice, the loads are fed via protective devices in the panel. These are omitted here for clarity. A, B, C, and N represent the panel buses.*

As an example of the savings possible with this system, let us consider the wiring required for a 15 kW heater.

At 3 ϕ, 208 V:

$$I = \frac{15{,}000\ \text{W}}{\sqrt{3} \times 208\ \text{V}} = 42\ \text{A}$$

requiring No. 8 RHW wire (45-A capacity) at 208 V.

At 480 V:

$$I = \frac{15{,}000\ \text{W}}{\sqrt{3} \times 480\ \text{V}} = 18\ \text{A}$$

requiring only No. 12 RHW wire (20-A capacity). This system is also frequently referred to as 265/460 V and 255/440 V.

Voltage above 150 V to ground is generally avoided in residential branch circuits, but may be used in commercial and industrial facilities, within the guidelines established by the NEC.

6. 2400/4160 V, 3-phase, 4-wire systems are only used in very large commercial buildings or in industrial buildings with machinery requiring these voltages. The cost of running this voltage feeder within a building is high because of NEC

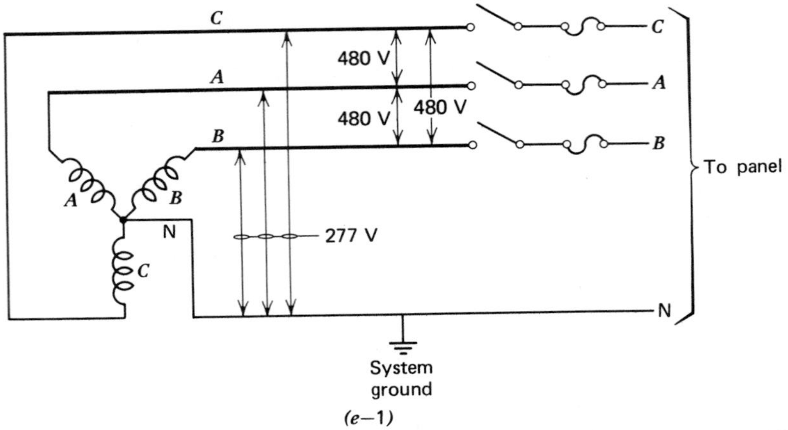

Fig. 17.1 (e-1) *277/480 V, 3-phase, 4-wire service system. The system is identical to the 120/208 V system shown in* (d-1) *except for voltages.*

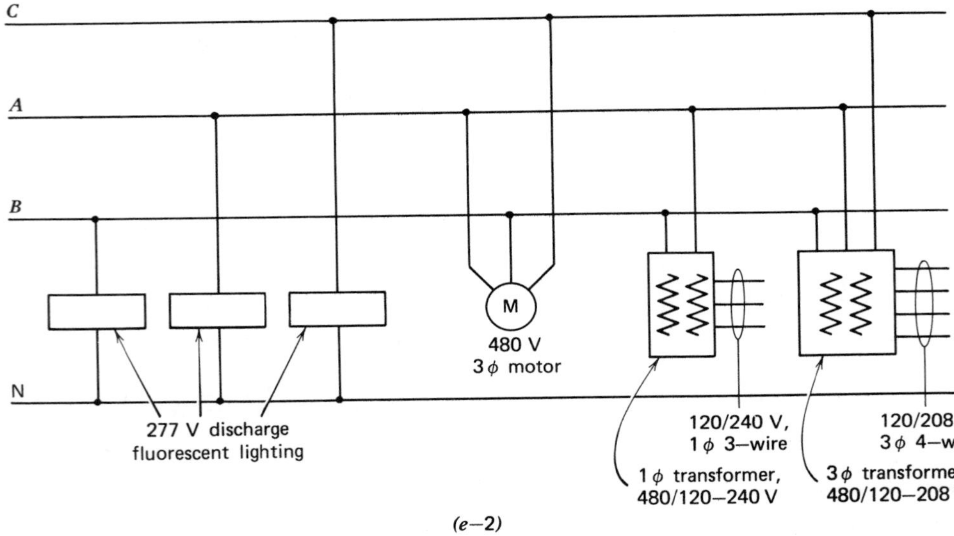

Fig. 17.1 (e-2) *Normal load arrangement for this system is illustrated. The lighting can be fluorescent or HID (mercury, metal-halide, sodium). Transformers, either single-phase or 3-phase, supply 120 V for receptacles and 208 V for loads requiring that voltage.*

requirements and the inherent high cost of 5-kV equipment. A detailed cost and engineering analysis by a competent engineer is required for each case. Voltages above this level are widely used in large industrial plants and beyond this text's consideration.

Reference was made above to the varied voltages assigned to the same voltage system. Thus, at the lowest level, we have 110, 115, and 120 V, at the next level 200, 208, 220, 230, and 240 V, at the next 255, 265, and 277 V, and finally 440, 465, and 480 V. These voltage differences arise because of the difference between transformer voltage standards, which establish the *system* voltage, and motor voltage standards, which govern *utilization* voltage level (see Table 17.3). The newer motor voltage standards are established at a level that is consonant with the system voltage. Note the close correspondence between motor standard voltage and system voltage with a

TABLE 17.3 **System and Utilization Voltages**[a]

Standard Voltages[b]			
System Voltage (Transformers)		*Utilization Voltage*[c] *(Motors)*	
Nominal	*With 4% Drop*[c]	*Current Standard*[d]	*Obsolete Standard*
120	115.2	115	110
208[d]	199.7	200	208
240[d]	230.4	230	220
480	460.8	460	440
600	576.0	575	550

[a]To eliminate any confusion between system and utilization voltages, the current NEMA standards are tabulated above.

[b]When specifying transformers, use system voltages; for motors, use utilization voltage.

[c]Note that utilization voltage corresponds to a 4% drop from system voltage, well within the normal motor tolerance.

[d]Motors for 208-V systems are rated 200 V. Motors for 240-V systems are rated 230 V. They cannot be used interchangeably without seriously affecting motor performance.

normal 4% feeder voltage drop. Thus we see that on 240- and 480-V systems, 230- and 460-V motors are completely suitable.

The difficulty that normally arises is in application of 230- and 240-V motors to 208-V systems. Despite the fact that motors will operate at plus or minus 10% voltage, 230- and 240-V motors *should not* be used on 208-V systems. Instead, motors specifically wound for 200 V should be specified. For a brief summary of the effects of undervoltage, refer to Table 17.4.

TABLE 17.4 **Effects of Undervoltage on Utilization Equipment**

Load	*10% Undervoltage*
Lighting	
Incandescent	Output reduced 30%
Fluorescent	Output reduced, poor start
Mercury	Low output, poor start
Motors	20% lower torque, hotter operation, reduced life, overloading
Heaters	20% reduction in output
Small tools	Stalling, low power

17.4 Grounding and Ground-Fault Protection

The vast majority of secondary wiring systems are grounded. The reasons for this arrangement are several and varied. Among them are

1. To prevent sustained contact between the low-voltage secondary system and the high-voltage primary system in the event of an insulation failure. Such contact could cause a breakdown of the secondary system insulation and severely endanger the system's users.
2. To prevent single grounds from going unnoticed until a second ground occurs, which would extensively disable the secondary system.
3. To permit locating ground faults with ease.
4. To protect against voltage surges.
5. To establish a neutral at zero potential for safety and for reference.

Points (2) and (4) are quite technical and a full explanation is beyond the purpose of this book. Point (5) requires that the neutral is

1. Never interrupted by switches or other devices.
2. Connected to ground only at one point—the service entrance.
3. Color coded white or natural gray for easy recognition.

A typical service-grounding diagram is given in Fig. 17.2. Universal acceptance and use of grounded secondary 120-V systems *introduces* another shock hazard while eliminating the dangers described above. This is shown in Fig. 17.3*a*. An accidental fault within an appliance can connect the metal case of the appliance to the line. This may readily occur with such common devices as an electric saw, clothes washer, dryer, or food mixer. A person contacting the appliance housing and simultaneously a ground, such as a water pipe, would receive a nasty 120-V shock. If the hands were wet, the shock could be fatal. Until such an incident occurred, however, the internal fault would remain an unnoticed but constant source of danger.

To eliminate this hazard, appliance manufacturers have always recommended that appliance housings be grounded to a cold water pipe and

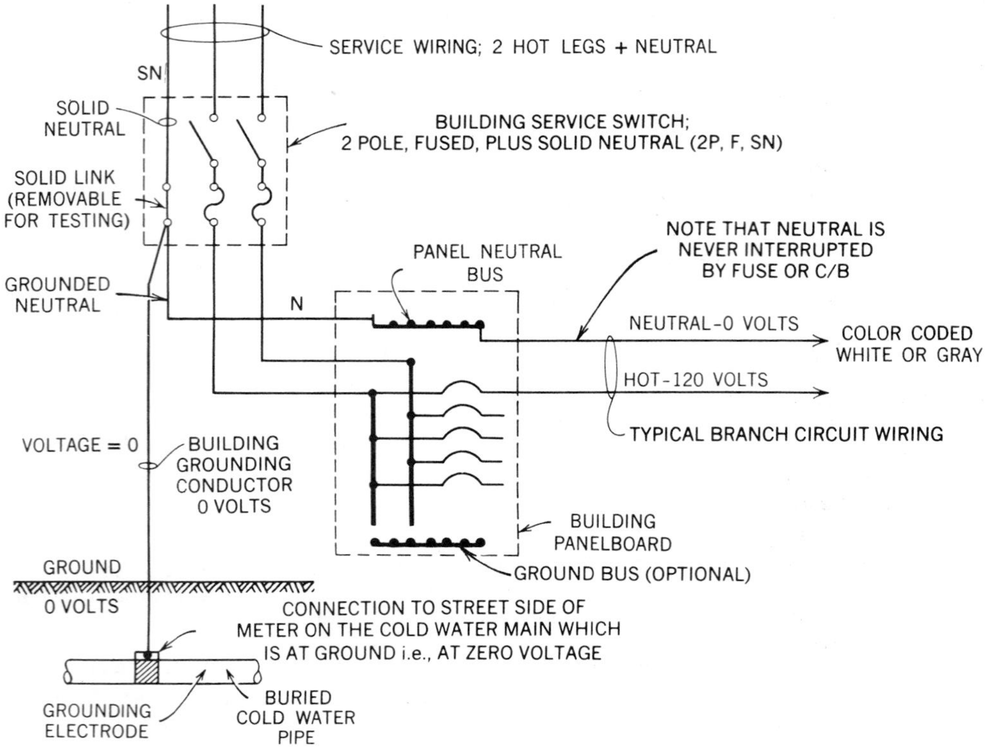

Fig. 17.2 *Typical service-grounding arrangement. Note that the grounded neutral is unbroken throughout. The ground bus, if present, is separate and distinct from the neutral bus, and both are grounded at the service entrance point.*

supply their appliances with 3-wire plugs: two wires connected to the appliance and the third wire to the housing. To accommodate such plugs and to provide a ground path, the National Electric Code requires all receptacles to be of the grounding type and all wiring systems to provide a ground path, separate and distinct from the neutral conductor (see Fig. 16.24). The result of such wiring is shown in Fig. 17.3*b*, where the ground current passes harmlessly through the internal fault, along the ground-wire path and back to the panel. A person contacting the appliance housing establishes a parallel ground path. However, since this path is usually of much higher resistance than the ground-conductor path, only a very small current will flow. Wet hands materially reduce contact resistance, and shock current can increase to a dangerous level. If the ground current is high, the branch circuit breaker or fuse will open, disconnecting the circuit.

When wiring systems are installed in metallic conduit, the conduit itself or the conduit plus a separate conductor within the conduit may be used as the grounding path. This latter method with the additional green ground wire is very much preferable, as explained in Fig. 17.3. When nonmetallic or flexible metallic wiring is used (''Romex'' or BX), a separate grounding conductor run with the regular circuit conductors *must* be used. All insulated ground conductors must have their covering colored green for identification as a grounding conductor. Many industrial installations install complete ''green-ground'' systems in an attempt to eliminate shock hazard and to reduce insulation failures. This has not been entirely successful for the reason alluded to above; that is, in order to clear the ground fault, current must be high enough to trip the branch circuit protective device. Otherwise, the ground fault continues to ''leak,'' unnoticed by the system's protective devices. Unfortunately, ground faults are by nature low-current, leak-type faults, because they result from weak

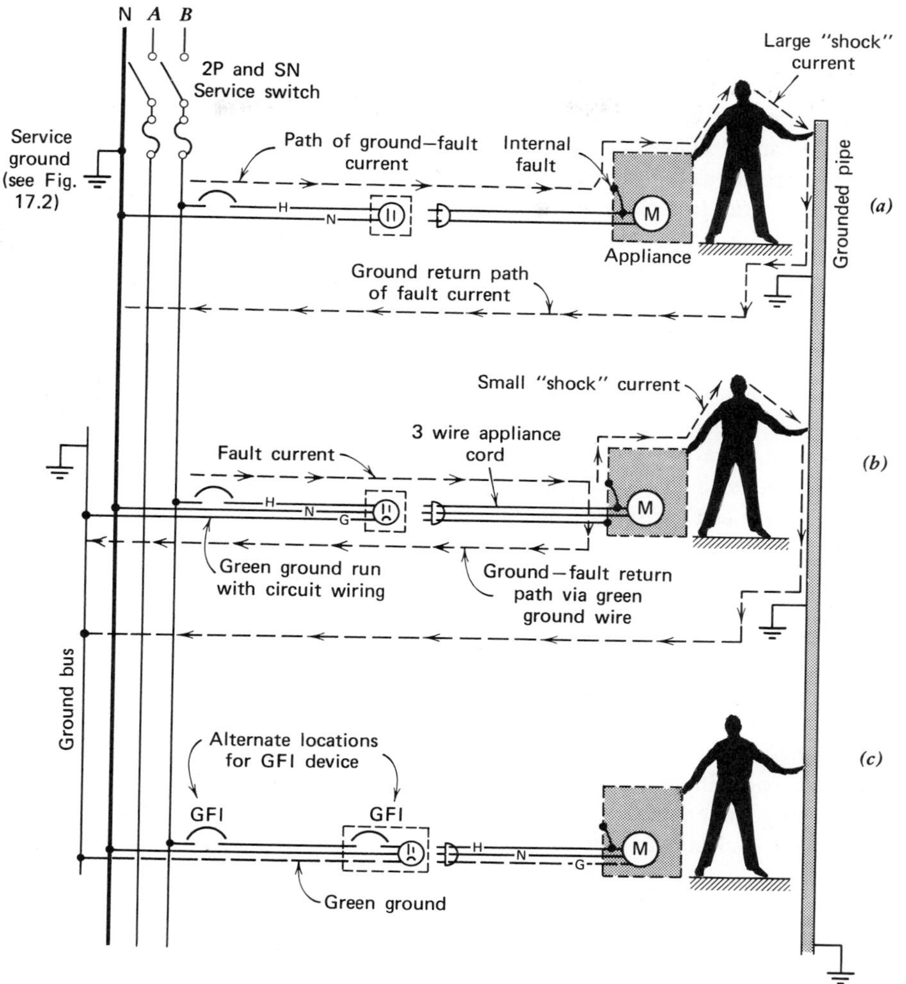

Fig. 17.3 Three types of circuit arrangements. (a) *Shows the conventional 2-wire, grounded neutral circuit, with no means of preventing shocks from ground faults.* (b) *Illustrates a similar arrangement but includes a green ground wire, which will considerably lessen the danger of ground-fault shocks. Note, however, that as long as the ground fault current is below the rating of the branch circuit protective device (that is, below 15 to 20 A), it will continue to flow within the appliance, causing overheating, arcing, and the eventual destruction of the appliance.* (c) *Illustrates the use of ground-fault circuit interrupter devices, now mandatory for bathrooms, outdoor locations, swimming-pool circuits, and other areas where ground faults are common and particularly hazardous.*

spots in insulation, dirt accumulation, and so on. Therefore, although the shock hazard is eliminated by the green-ground path, the fault continues to leak and arc until it becomes large enough to cause a major breakdown, frequently accompanied by fire.

To eliminate this dangerous situation, which occurs anytime there is a leak of current to ground in an electric circuit, the ground fault circuit interrupter (GFCI or GFI) was developed (see Fig. 17.3*c*). This device compares, with extreme precision, the current flowing in the hot and neutral legs of a circuit; if there is a difference, it indicates a ground fault and the device trips out. The rapidity of this action—approximately half a second—eliminates the possibility of dangerous shock hazard, which exists even in a properly grounded circuit, as in Fig. 17.3*b*, and, all the more so, for

ELECTRICITY

the circuit arrangement of Fig. 17.3*a*. The separate ground wire shown in Fig. 17.3*c*, and required by the NEC, serves to minimize the shock current taken by a person before the GFCI operates. The device can be applied at the panel to replace a normal circuit breaker (see Fig. 16.13*b*) or at an individual outlet to replace a normal receptacle device (see Fig. 16.26). The GFCI finds a ready application in the old 2-wire circuits illustrated in Fig. 17.3*a*. These aging circuit components are prone to ground faults and can best be protected with a GFI. It is also advisable to use GFI devices on all appliance circuits (see the NEC for locations where GFI use is mandatory, such as outdoors and in bathrooms). Application to lighting circuits is not essential, since fixtures are generally out of reach and are switch controlled. In mixed lighting and receptacle circuits, the GFCI is best applied at the outlet that is to be protected.

17.5 Energy Conservation Considerations

Before proceeding with a detailed description of design procedure, we will survey in this section many of the energy conservation ideas and techniques applicable to electrical distribution systems. This is done for ease of reference and cross-reference, since the individual items appear throughout the lengthy design procedure. ASHRAE standard 90, 1980 issue, which is mandated legally in most states, calls for a design of energy distribution systems "to conserve energy" (Section 8.0, Energy Distribution Systems). The following paragraphs are intended to fulfill that broad directive.

1. After establishing an energy budget based on projected loads and normal operation, set an energy reduction figure of 10 to 20% and accomplish it with the techniques given below. Annual energy consumption estimates are best made with the aid of the many computer programs available for that purpose.

2. Recognize the energy-use characteristics of all materials and systems specified (see Table 15.5 and Table 16.4. In general, select high-efficiency equipment (motors, transformers, etc.). If not available as such, use materials and equipment with the lowest temperature rise, since these have lowest losses. Economic justification can be established with life-cycle cost analysis. To avoid making a detailed cost analysis on every item, utilize one of the many available shortcut calculations for pay-back time. This will generally indicate the material of choice. When comparisons are close, a detailed analysis may be necessary.

3. Provide electric load control equipment (demand control), either as part of an overall building control system or separately (see Section 14.13).

4. In any multitenant residential building, provisions should be made to separately determine the energy consumed by each tenant. Where local codes and regulatory agencies permit, each tenant should be made financially responsible for the energy he uses. Exceptions to this rule would be made in the case of hotels, dormitories, and other transient facilities.

5. Where a choice of service voltages is available, select the highest available. Similarly, utilize the highest voltage in each class for interior distribution systems. This means 480 V in the 600-V class, 4 kV in the 5-kV class, and 13 kV in the 15-kV class. The result will be low line losses, small panelboards at the branch circuit level, and generally lower electrical contract cost (see Section 17.3).

6. In any building, the maximum total voltage drop *shall not* exceed 3% in branch circuits or feeders, for a total of 5% to the farthest outlet based on steady-state design load conditions.

7. Avoid the use of electric heating elements if alternatives are available. Electric heat is an inefficient use of national resources because of the low efficiency of fuel-to-electricity conversion.

8. Provide metering points (for fixed or plug-in meters) throughout the system to permit accurate analysis of power and energy use. Meters, both instantaneous reading and recording types, can provide data on equipment loading, load patterns, load coincidence, power factor, load voltage, power demand, and energy consumption. Analysis of these data will indicate how to program and shift loads for maximum operational efficiency. A flexible design that permits this load shifting is assumed [see item (10) below].

9. Include provisions for power factor correction in the system both at the device and at the feeder level. Then, if metering [see (8) above] indicates the necessity, add capacitors as required. High power factor reduces line losses, permits maximum utilization of equipment capacity, and avoids utility penalty charges. Utilization equipment rated greater than 1000 W and lighting equipment greater than 15 W, with an inductive reactance load component, should have a power factor of not less than 85% under rated load conditions. Utilization equipment with a power factor of less than 85% should be corrected to at least 90% under rated load conditions.

10. Size equipment as close as possible to the load. This normally results in maximum efficiency and high power factor. Where load varies considerably, for example, between day and night or weekend, consider splitting the loads so that part of the equipment can be switched off when load is low, and the remaining load can be fed from a "night and weekend" feeder. Such a design permits one to shut down whole systems rather than operate at very low load with concomitant high losses and low power factor. The design must be sufficiently flexible to permit shifting of loads between feeders if measurements on the operational facility indicate that this is desirable [see item (8) above]. The purpose would be to operate equipment near rated load and deenergize lightly loaded sections.

11. Use the most efficient type of control. This means solid-state control for motors, remote switch control for blocks of lighting, electronic control systems for elevators, and so on.

12. Arrange automatic time controls for 24-h loads such as vent fans, water coolers, vending machines, and calculators.

13. Seal all electric riser shafts to avoid heat loss by stack action.

14. Generally select the coolest possible locations for electrric equipment. Low ambient temperature (below 40°C) permits use of smaller equipment for the same load with concomitant lower cost and losses. Thus, if below-grade space is available, it is well suited for this purpose.

15. Provision for future expansion should be made by means of additional equipment in lieu of oversizing equipment initially. Here again the higher cost of two pieces of equipment can normally be justified by a detailed owning-and-operating cost analysis using realistic cost escalation and capital cost figures.

16. Energy-conservation techniques in lighting and lighting control are found in Sections 18.22, 18.23, 20.5, 20.7, 21.2, and 21.30.

17. Energy conservation techniques in vertical transportation are found in Sections 23.17, 24.1, 24.2, 24.3, and 25.12.

17.6 Terminology and Definitions

The succeeding discussions will utilize terms common in the electrical field. These terms are used in consonance with definitions given in Article 100 of the National Electrical Code. The reader is referred to that source for any necessary clarification.

17.7 Design Procedure

The steps involved in the electrical wiring design of any facility are outlined below. These may in some instances be performed in different order, or two or more steps may be combined, but the procedure normally used is that listed below.

1. Make an electrical load estimate based on areas involved, building data, and any other pertinent data (see Section 17.2).

2. In cooperation with the local electric utility, decide upon the point of service entrance, type of service run, service voltage, metering location, and building utilization voltage [see Sections 16.1 to 16.11 and Section 17.5, item (5)].

3. Determine with the client the usage of all areas, and type and rating of all client-furnished equipment including their specific electric ratings and service connection requirements.

4. Determine from other consultants such as HVAC, plumbing, elevators, kitchen, and the like, the exact electrical rating of all the equipment in their designs. This determination will often result

after conferences during which the electrical consultant makes valuable recommendations to these other specialists about the comparative characteristics and costs of equipment [see Section 17.5, items (2), (7), (9), and (10)].

5. Determine the location and estimate the size of all required electric equipment spaces including switchboard rooms, emergency equipment spaces, electric closets, and so forth. Panelboards are normally located in closets but may be located in corridor walls or elsewhere. This work is necessary at this point to enable the architect to reserve these spaces for electrical equipment. Once the design is accomplished in detail, the estimated space requirements can be checked and necessary adjustments made.

6. Design the lighting for the facility. This step, as will be discussed in detail in Chapters 18 to 21, is complex and involves a continued interaction between the architect and the lighting designer.

7. Depending on the type of facility, it may be necessary to separate the lighting layout from the receptacle and signal device layout for the sake of clarity of the plans. Once the decision has been made as to how this is to be handled, the lighting fixture layout can be made.

8. On the same plan, or on a separate plan, as decided, locate all electrical appratus including receptacles, switches, motors, and other power-consuming apparatus. Underfloor, under-carpet, and "over-ceiling" wiring and raceway systems would be shown at this stage. If extensive, a separate plan is made.

9. On the plans, locate signal apparatus such as phone outlets, speakers, microphones, TV outlets, fire and smoke detectors, and so on. At this stage provision is also made for load control wiring, building automatic control wiring, programmable control and computer wiring, and the like. Since many of these systems are covered by special drawings or by specification alone, this step may be limited to showing outlets only. The material that follows deals with wiring design only; all further discussion of signal equipment will be reserved for Chapter 22.

10. Circuit all lighting, devices, and power equipment to the appropriate panels, and prepare the panel schedules. Included in this step is the circuitry for emergency equipment.

11. Compute panel loads.

12. Prepare the riser diagram. This includes design of distribution panels, switchboards, and service equipment.

13. Compute feeder sizes and all protective equipment ratings.

14. Check the preceding work.

15. Coordinate the electrical work with the other trades and with the architectural plans. This is not really a separate step, but a continuing process starting at (9) above and covering all subsequent stages of the work.

The material for (1), (2), (3), (4), (6), (7), and (9) is covered elsewhere in this text. The remaining steps, that is, (5), (8), (10), (11), (12), (13), and (15) will be discussed in order below.

17.8 Electric Spaces

The spaces required for electric equipment in a facility vary radically, depending on the design and nature of the building. The spaces required around major pieces of electrical switchgear and transformers are discussed above (see Fig. 16.17). The NEC in Article 110 further specifies the minimum working spaces required in front of electrical equipment. In general, a minimum of 42 in. of clear space should be maintained in front of panels, switches, and other electrical apparatus.

(a) Residences. In private residences, the service equipment and the building panelboard are generally incorporated into a single unit. The main disconnect(s) is usually installed as the main switch/breaker of the panel. A number of typical residential service-panel arrangements are shown in Fig. 17.4. The panel is normally placed in the garage, utility room, or basement. The panel should be placed as close to the load center as practicable, without sacrificing valuable space or making the panel inaccessible. Frequently a smaller panel can be subfed from the main panel to feed the kitchen

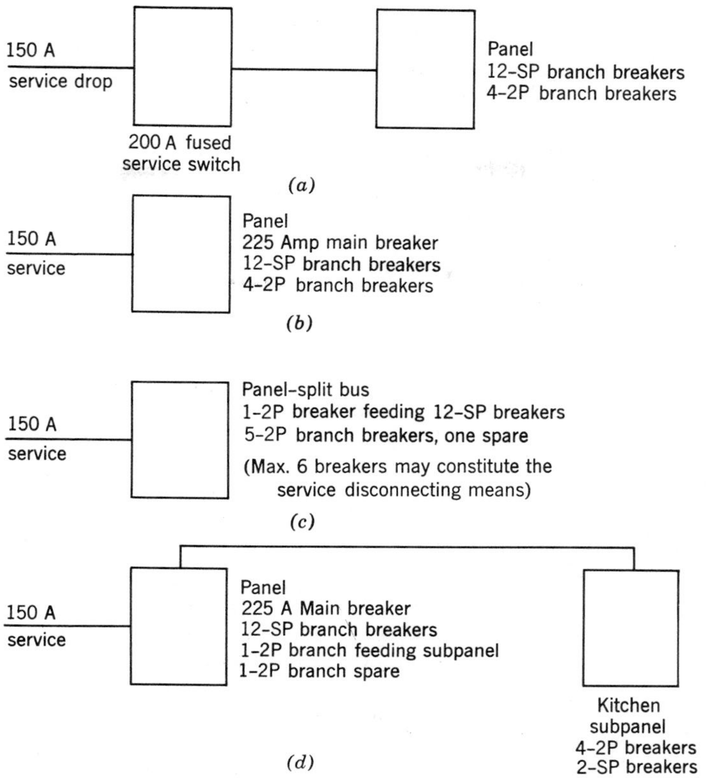

Fig. 17.4 *Typical 150-A service arrangements, applicable to residences and small commercial buildings. The service switch can be a separate unit* (a) *or combined with the branch circuit panelboard* (b–d). *The panel may be a single unit* (a, b), *two units in a single enclosure, fed by separate service switches* (c), *or a central panel and a subfed single use panel* (d)—*in this case a kitchen subpanel.*

and laundry loads. In apartments, panels are normally placed in the kitchen or the corridor immediately adjoining the kitchen. This location is chosen so that the panel circuit breaker may act as the required disconnecting means for fixed appliances larger than 300 V-A or ⅛ hp, as required by NEC.

(b) Commercial Spaces. The location of the required panelboards depends on their type and quantity, and on availability of space. In the research building of which Fig. 17.5 is a part plan, lighting panels are recessed into the corridor wall, since the building is only two stories high and the panels can be vertically stacked and fed by a single conduit. If this building were six or more stories high, an electric closet (see Section 17.9) would be advisable to accommodate the panel and riser conduits. Of course, when panels are installed in finished areas such as corridors, flush mounting is required.

To limit the voltage drop on a branch circuit in accordance with the Code requirements, panelboards should be located so that no circuit exceeds 100 ft in length. If circuits longer than this are unavoidable, No. 10 AWG wire should be used for runs of 100 to 150 ft and No. 8 AWG for longer circuits. These wire sizes apply to 15- or 20-A branch circuits, which are normally wired with No. 12 AWG wire.

The laboratory between the two offices of Fig. 17.5 is intended to function as a self-contained unit and is therefore equipped with its own panel. Multi-outlet assemblies, all wiring within the room, and the panel itself are surface-mounted to allow ready access to all components for the frequent rewiring encountered in laboratories. A main circuit breaker should be provided in such a panel to

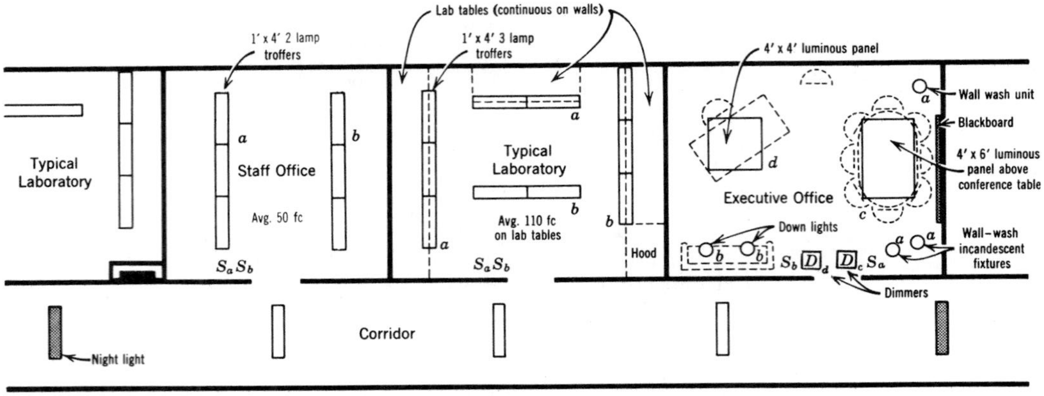

(a)

Lighting Plan

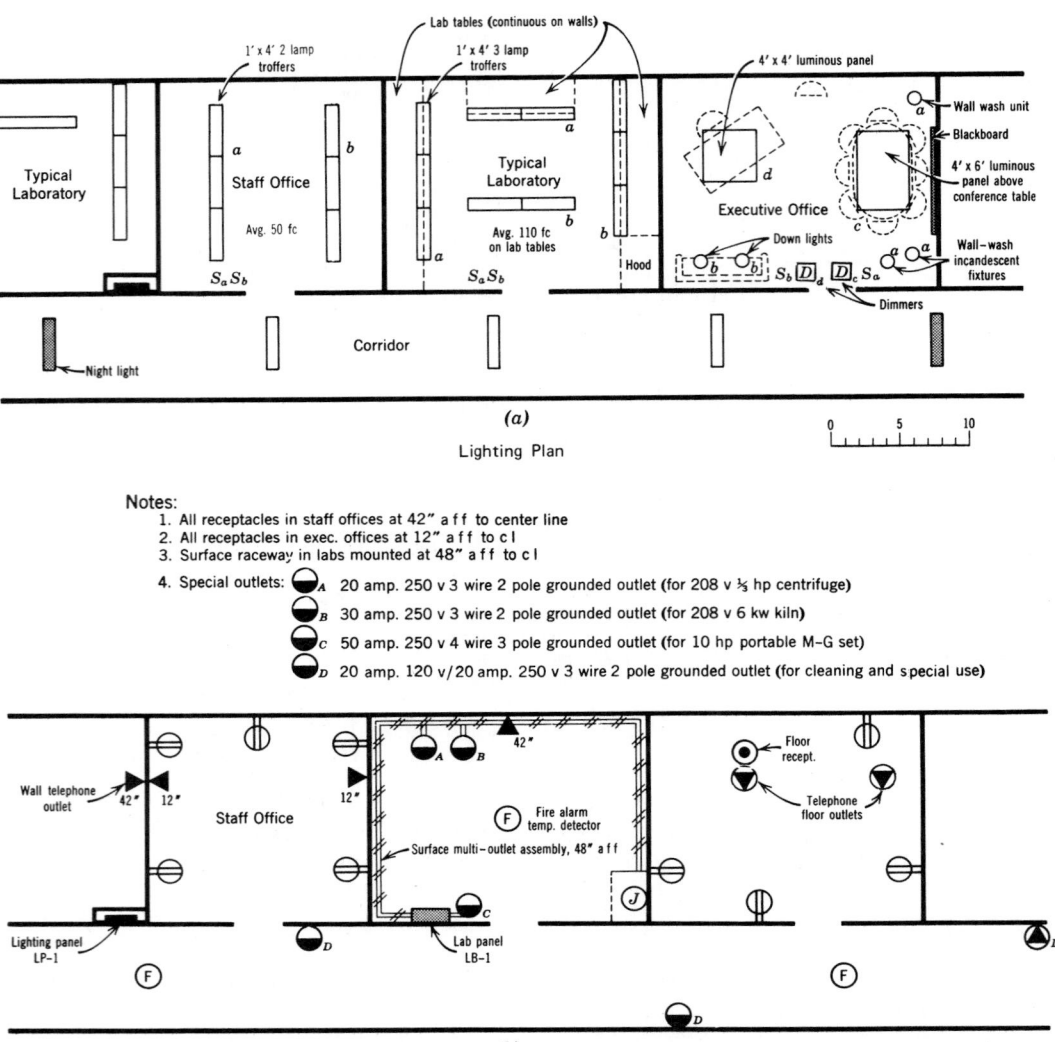

Notes:
1. All receptacles in staff offices at 42″ a f f to center line
2. All receptacles in exec. offices at 12″ a f f to c l
3. Surface raceway in labs mounted at 48″ a f f to c l
4. Special outlets: A 20 amp. 250 v 3 wire 2 pole grounded outlet (for 208 v ⅓ hp centrifuge)
   B 30 amp. 250 v 3 wire 2 pole grounded outlet (for 208 v 6 kw kiln)
   C 50 amp. 250 v 4 wire 3 pole grounded outlet (for 10 hp portable M-G set)
   D 20 amp. 120 v/20 amp. 250 v 3 wire 2 pole grounded outlet (for cleaning and special use)

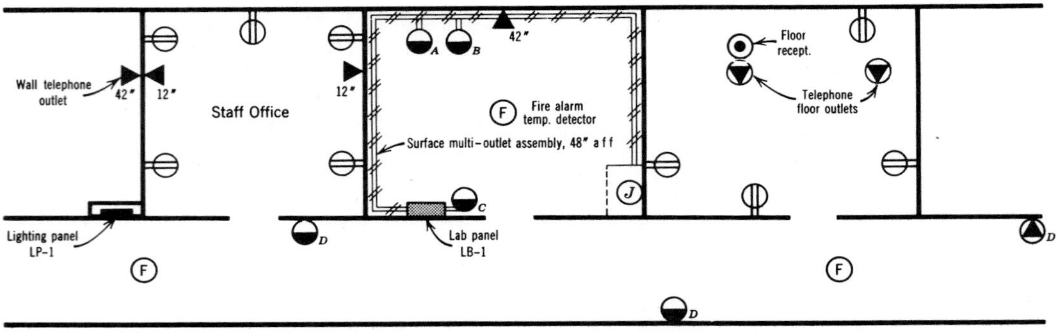

(b)

Power Plan

Fig. 17.5 *Typical floor plans for lighting and power for a section of an office-laboratory building. Separate lighting and power plans are drawn for the sake of clarity. For circuited plan see Fig. 17.17.*

act as a main disconnect, whether required by Code or not. Where panels are convenient to the load controlled, the panel circuit breakers may be used for switching.

Panels supplying large blocks of load simultaneously switched, such as auditorium house lights, lobby lights, large single-use office areas, store lighting, and the like, can be constructed with built-in contactors to switch the entire panel, with control at any desired remote location. These remote-control (RC) switches are discussed in Section 16.15. If only part of the panel's circuits is so arranged, a split bus panel is provided, partially contactor controlled (Fig. 17.6).

Small offices, stores, and other small buildings have lighting panels mounted in a convenient finished area and utilize the breakers for load switching. In large buildings, strategically located electric closets are provided to house all electrical supply equipment. Power panels and distribution panels are located as required by the loads fed through them.

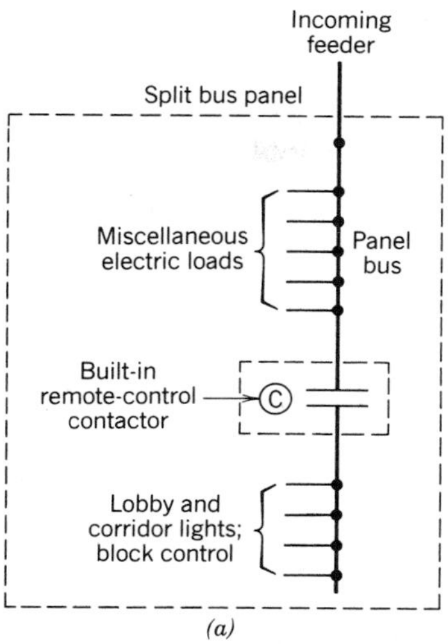

Fig. 17.6 (a) *Single-line diagram of a split bus panel. This design is used when a block of load is to be controlled as a unit. Control point is either remote, at the panel, or both.*

Fig. 17.6 (b) *Newly installed RC switches (mechanically held contactors) are shown to the left of a bank of existing panels. The contactors were installed to accomplish photocell and timer control of a department store's lighting. Courtesy of Automatic Switch Co.*

In general, branch circuit panels, distribution panels, and switchboards are best located near the electrical load center. This minimizes feeder length and reduces voltage drop, making it the most economical arrangement.

Every completely enclosed switchgear room, emergency generator room, or transformer vault should be equipped with an emergency light source. In generator rooms these should be battery operated to give illumination for generator repairs in the event of generator failure during a power outage.

17.9 Electric Closets

In the design of a building electric system, particularly in multistory construction, it is often advantageous and convenient to group the electrical equipment in a small room called an electric closet (see Fig. 17.7). The shape of this space can be varied to fit the architectural and electrical demands, but it should provide the following:

1. One or more locking doors.
2. Vertical stacking, above and below other electric closets and located so as not to block conduits entering or leaving horizontally. Thus locations on outside walls and adjoining shafts, columns, and stairs are poor.
3. Space free of other utilities such as piping or duct passing through the closet, either horizontally or vertically.
4. Sufficient wall space to mount all requisite and future panels, switches, transformers, telephone cabinets, and signal equipment.
5. Floor slots or sleeves of sufficient size for all present and future conduit or bus risers.
6. Sufficient floor space so that an electrician can work comfortably and safely on initial installation and repair.
7. Adequate illumination and ventilation.

17.10 Equipment Layout

Wiring devices, principally comprising receptacles and switches, are located as required by the equipment to be served and by the anticipated area use.

Switches for control of lighting or receptacles are normally placed on the strike side of the door. Other devices such as plug-in strips on walls and special-purpose receptacles are shown and identified. Signal outlet locations are often noted but generally remain uncircuited on floor plans, a riser being utilized to show interconnections. These include fire-alarm equipment, telephone and intercom equipment, radio and TV outlets, thermo-

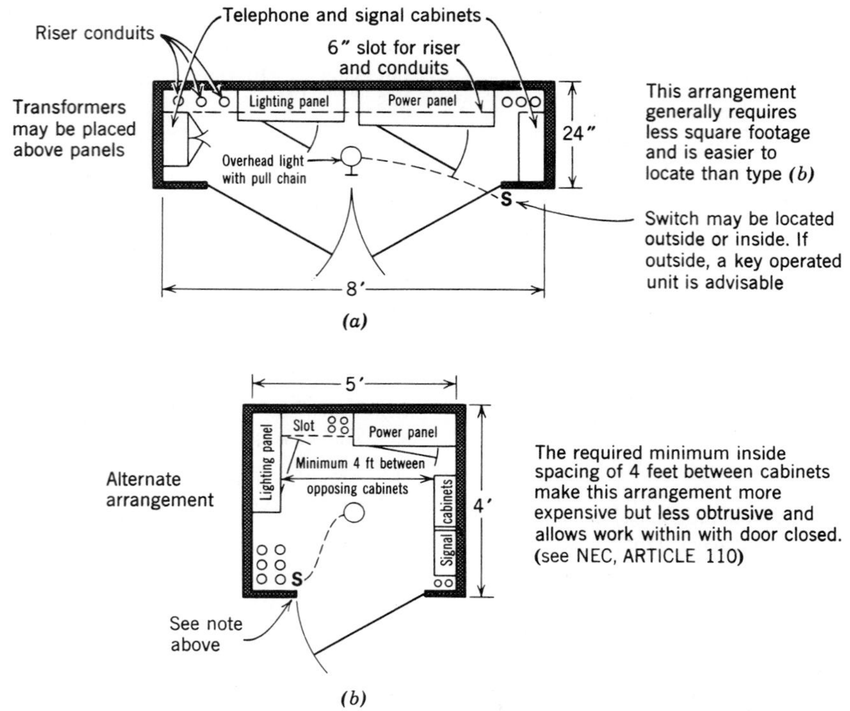

Fig. 17.7 *Typical electric closets with some usual equipment. If warranted by amount of equipment, separate closets may be used for signal and telephone conduits and cabinets.*

stats, and so on. These devices may be identified by a special symbol or note where a standard symbol is not available.

As mentioned above, lighting fixture outlets are normally placed on the same drawing as wiring devices unless the large number of the latter precludes showing the lighting, without undue cluttering of the drawings. In such an event, the lighting is shown on one drawing and receptacles on another, with signals shown on the one least occupied. A ceiling or underfloor wiring system would probably necessitate such separation. Motors, heaters, and other fixed and permanently wired equipment are shown and identified on the receptacle drawings (also called power drawings, in contradistinction to lighting drawings). Equipment furnished with a cord and plug is not normally shown. However, the receptacle intended for supplying such a device is shown and identified. A typical device layout is shown in Fig. 17.5. An abbreviated symbol list is given in Fig. 17.8. For a complete electrical symbol list, see McGuinness et al. (1980), pp. 665–672.

17.11 Application of Overcurrent Equipment

Before beginning an explanation of circuiting, it is necessary to explain the principles underlying overcurrent protection. As outlined in Chapter 16, the function of an overcurrent device is to open (interrupt) a circuit when the current rating of the equipment being protected is exceeded. These overcurrent devices are placed in circuits to protect wiring, transformers, lights, and all other equipment that can be damaged by excessive current. The following general rules govern the application of overcurrent protection:

1. Overcurrent devices must be placed on the line or supply side of the equipment being protected (see Fig. 17.9).
2. Overcurrent devices must be placed in all *ungrounded* conductors of the protected circuit (see Fig. 16.19*b*), page 794).
3. All equipment should be protected in accordance with its current-carrying capacity (see Fig. 17.10).

ELECTRICAL SYMBOL LIST

RACEWAYS

—///—	CONDUIT AND WIRING CONCEALED IN CEILING OR WALLS TICS INDICATE NO. OF CONDUCTORS EXCLUDING GROUNDS; 2 #12, ¾" CONDUIT UON.
— - —	CONDUIT AND WIRING CONCEALED IN OR UNDER FLOOR
- - - - -	CONDUIT AND WIRING EXPOSED
F-6	FEEDER F-6, SEE SCHEDULE, DWG NO.
— BX —	BX WIRING
— NM —	NONMETALLIC CABLE (ROMEX) WIRING
2, 4 / 2PLA / 3# 12, ¾" C	HOME RUN TO PANEL 2PLA—NUMERALS INDICATE CIRCUITS, 3 #12 AWG, ¾" RIGID STEEL CONDUIT
TC2A	HOME RUN TO TELEPHONE CABINET TC2A
— EC_T —	EMPTY CONDUIT, SUBSCRIPT INDICATES INTENDED USE T—TELEPHONE, IC—INTERCOM, FA—FIRE ALARM, ETC.
4" × 4" / A	SURFACE METAL RACEWAY, SEE NOTE ______, DWG ______ SIZE AND RECEPTACLES AS SHOWN

OUTLETS

CEILING	WALL	
(A) a	-(A) 3	OUTLET AND LIGHTING FIXTURE: LETTER IN CIRCLE INDICATES TYPE—SEE SCHEDULE. SUPERSCRIPT NO. INDICATES CIRCUIT. SUBSCRIPT LETTER INDICATES SWITCH CONTROL. RECTANGLE INDICATES FLUORESCENT
⊗ A	-⊗ B	OUTLET AND EXIT SIGN FIXTURE, UPPER CASE LETTER INDICATES TYPE, ARROWS INDICATE REQUIRED SIGN ARROWS
(E)	-(E)	OUTLET BOX, BLANK COVER—NOTE 2
(J)	-(J)	JUNCTION BOX, BLANK COVER

WIRING DEVICES

	DUPLEX CONVENIENCE RECEPTACLE OUTLET 15 AMP 2P 3W 125 V, GROUNDING, WALL MTD VERTICAL, ℄ 12" AFF.
GFCI	RECEPTACLE, RATED AS SHOWN. EQUIPPED WITH GROUND FAULT CIRCUIT INTERRUPTER
A	SPECIAL RECEPTACLE, LETTER DESIGNATES TYPE, SEE SCHED. DWG. NO. __ WALL MOUNTED
B	FLOOR OUTLET TYPE B, SEE DWG. NO. _____
or	MULTI-OUTLET ASSEMBLY, SEE DWG. NO. __ FOR SCHEDULE AND DETAILS (SEE SPEC.)
S_a	SINGLE POLE SWITCH, 15A 125 V, 50" AFF UON. SUBSCRIPT LETTER INDICATES OUTLETS CONTROLLED
$_aS_3$	SWITCH, 3 WAY, 15A 125 V, SEE SPEC.; CONTROLLING OUTLETS 'a'
S_K	SWITCH, KEY OPERATED, 15A 125 V
S_{SA}	SWITCH, SPECIAL PURPOSE, TYPE A, SEE SPEC.; SEE DWG. NO. _____
———	ABBREVIATIONS RELEVANT TO SWITCHES:

SP	—SINGLE POLE
DP	—DOUBLE POLE
SPDT	—SINGLE POLE DOUBLE THROW
DPDT	—DOUBLE POLE DOUBLE THROW
RC	—REMOTE CONTROL

[D]	OUTLET-BOX-MOUNTED DIMMER

Fig. 17.8 *Abbreviated electrical symbol list. For complete list see McGuinness et al. (1980), pp. 666–672.*

LIST OF ABBREVIATIONS

A,a	AMPERES
AFF	ABOVE FINISHED FLOOR
C/B	CIRCUIT BREAKER
CCT	CIRCUIT
EM	EMERGENCY
EC	EMPTY CONDUIT
F	FUSE
FA	FIRE ALARM
GND	GROUND
GFCI	GROUND FAULT CCT-INTERRUPTER
HP	HORSEPOWER
LTG	LIGHTING
MH	MOUNTING HEIGHT, MANHOLE
N	NEUTRAL
NC	NORMALLY CLOSED
NO	NORMALLY OPEN
NIC	NOT IN CONTRACT
OH	OVERHEAD
OC	ON CENTER
SW	SWITCH
T	THERMOSTAT
TEL	TELEPHONE
TV	TELEVISION
UON	UNLESS OTHERWISE NOTED
UG	UNDERGROUND
WP	WEATHERPROOF

EQUIPMENT

3 P SN 60/30 3R	SW., 3 POLE AND SN., 60A/30 AF. NEMA 3R ENCL., ND (NORMAL DUTY) 250 V
M OR RC 3 P, 60 A	CONTACTOR, ENCLOSED, ELECTRICALLY OPERATED, MECHANICALLY HELD, 3 POLE 60 A
	ITEM OF ELECTRICAL EQUIPMENT, AS INDICATED
LP-1	ELECTRIC PANELBOARD LP-1, RECESSED, SEE SCHEDULE ON DWG. _____
P-2	ELECTRIC PANEL (BOARD) P-2, SURFACE MTD
TEL. 18 × 24 × 6	CABINET, MOUNTING, SIZE, AND PURPOSE AS SHOWN

SIGNALING DEVICES

F 1, A	BELL OR GONG, INSCRIBED LETTER INDICATES SYSTEM (SEE BELOW) AND SUBSCRIPT LETTER OR NUMBER INDICATES TYPE e.g. A—8″ VIBRATING BELL, 12 V. DC; B—12″ WEATHERPROOF SINGLE STROKE GONG, 120 V. DC, etc.
F 1	FIRE DETECTOR, TYPE 1 (ETC.)
I 1	INTRUSION DETECTOR, TYPE 1 (ETC.)
SD 2	SMOKE DETECTOR, TYPE 2 (ETC.)
F	MANUAL STATION—WATCHMEN TOUR, FIRE ALARM, ETC. LETTER INDICATES SYSTEM (SEE BELOW)
SD A	CABINET OR CONTROL PANEL, SMOKE DETECTION, USE IDENTIFYING TYPE LETTER IF MORE THAN ONE TYPE IS USED ON THE PROJECT
BATT	AUXILIARY DEVICE, AS INDICATED
A, S 2	HORN OR LOUDSPEAKER, TYPE A, TYPE 2
B	TELEPHONE OUTLET, TYPE B
IC 2	INTERCOM OUTLET, TYPE 2
A	CLOCK SYSTEM OUTLET, TYPE A
TV	TV ANTENNA OUTLET
	OUTLET, LETTER INDICATES TYPE

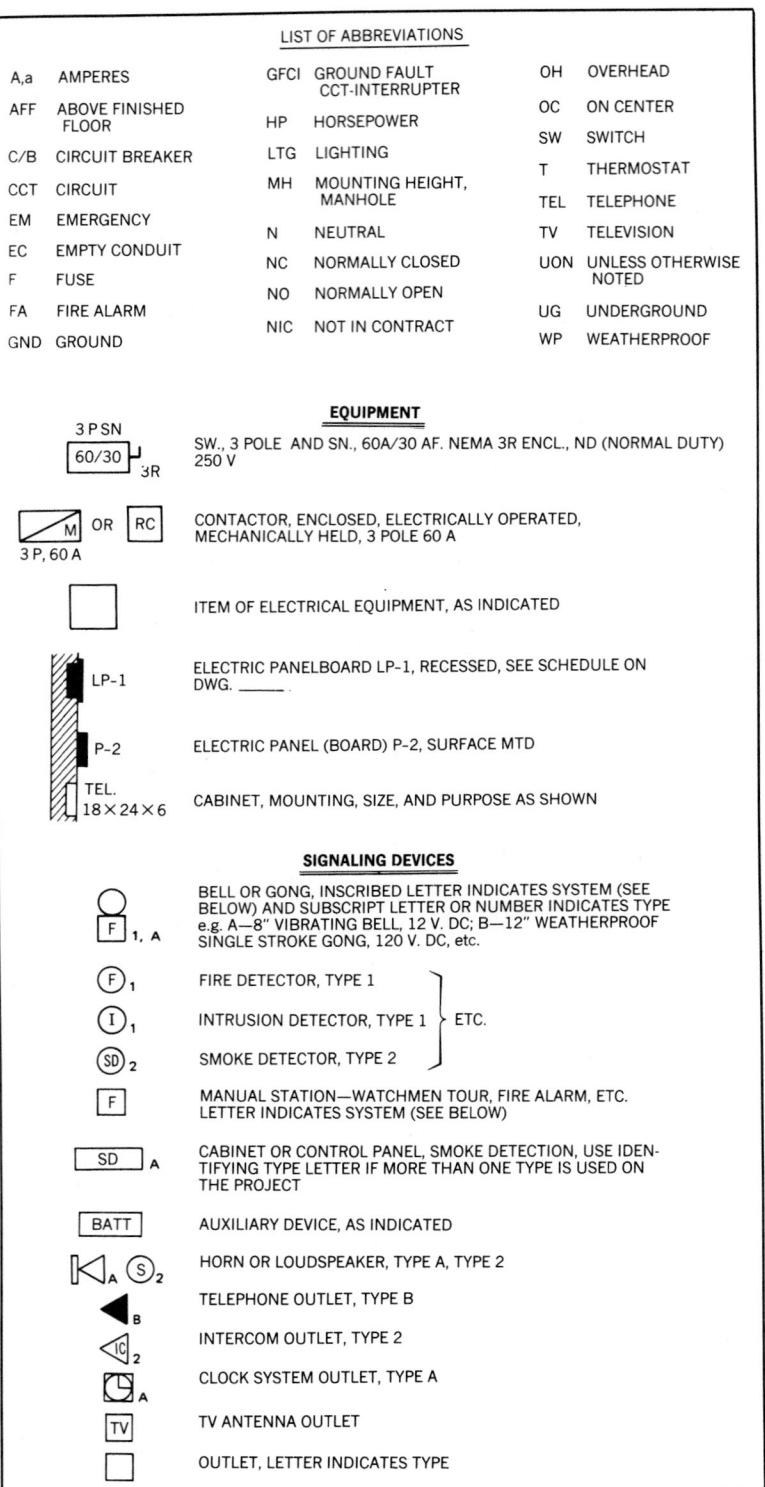

Fig. 17.8 (continued)

ELECTRICITY

SYSTEM TYPES

F, FA	FIRE ALARM
S, SD	SMOKE DETECTION
I, IA	INTRUSION ALARM
T, TEL	TELEPHONE
TV	TELEVISION
IC	INTERCOM

AUXILIARY DEVICES

BATT	BATTERY
S, SP	SPEAKER, LOUDSPEAKER
TC	TELEPHONE CABINET

MOTORS AND MOTOR CONTROL

	COMBINATION TYPE MOTOR CONTROLLER; ATL STARTER PLUS FUSED DISCONNECT SWITCH, NEMA SIZE II, SEE SCHEDULE DWG.___ ___ ___
	MOTOR #1, 5 HP, 3ϕ SQUIRREL CAGE UON
(T) OR [T]	DEVICE 'T' (SEE LIST OF ABBREVIATIONS)
S_T	MANUAL MOTOR CONTROLLER WITH THERMAL ELEMENT.
ATL	ACROSS THE LINE STARTER—MAGNETIC
FS	FUSED SWITCH
CB	CIRCUIT BREAKER
FV	FULL VOLTAGE
MCC	MOTOR CONTROL CENTER
S	START BUTTON—MOMENTARY CONTACT
ST	STOP BUTTON—MOMENTARY CONTACT

CONTROL DIAGRAMS AND WIRING DIAGRAMS

	MOMENTARY CONTACT PUSH BUTTON—NO—(START)
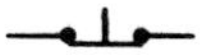	MOMENTARY CONTACT PUSH BUTTON—NC—(STOP)
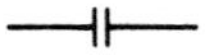	NORMALLY OPEN CONTACT—NO
	NORMALLY CLOSED CONTACT—NC
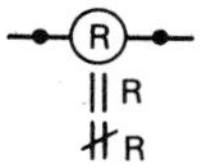	OPERATING COIL FOR RELAY OR OTHER MAGNETIC CONTROL DEVICE. WITH ONE NO AND ONE NC CONTACT. LETTERS NORMALLY USED ARE C, R FOR CONTROL COIL AND RELAY
	POWER WIRING
	CONTROL WIRING

ONE LINE DIAGRAMS

	CONDUCTOR, SIZE, AND TYPE INDICATED
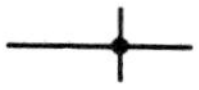	CONDUCTORS CROSSING, CONNECTED
	GROUND CONNECTION
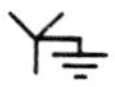	GROUNDED WYE
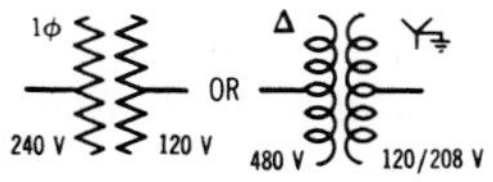	TRANSFORMER, 2 WINDING, 1ϕ INDICATES SINGLE PHASE; Δ-Y INDICATES 3 PHASE AND TYPE OF CONNECTION; NUMBERS INDICATE VOLTAGES.
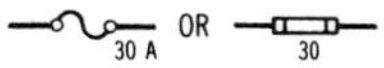	FUSE, 30 A TYPE - - -
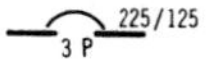	CIRCUIT BREAKER, MOLDED CASE, 3 POLE, 225 A FRAME/125 A TRIP
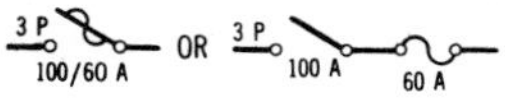	FUSED DISCONNECT SWITCH, 3 POLE, 100 A, 60 A FUSE
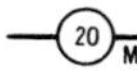	MOTOR, 3 PHASE SQUIRREL CAGE UON, 20 HP, MOTOR #3

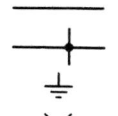

ELECTRICITY

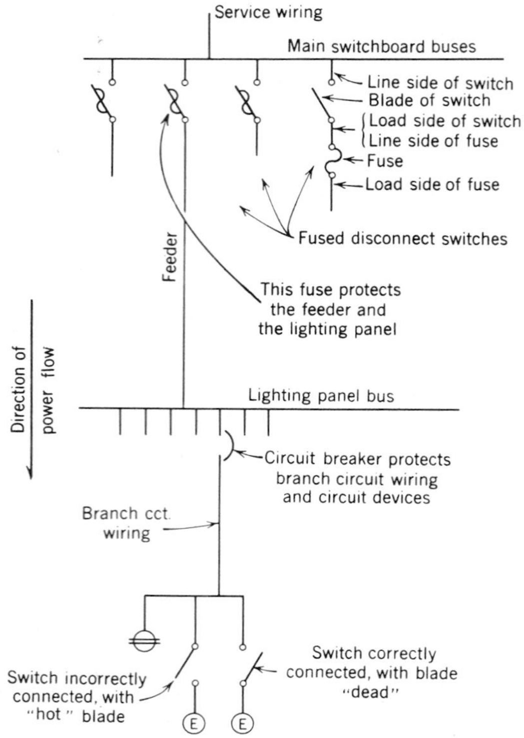

Fig. 17.9 Location of overcurrent protective equipment. Protective equipment should always be located at the point where the conductor receives its source of supply so that when it operates the current supply is cut off.

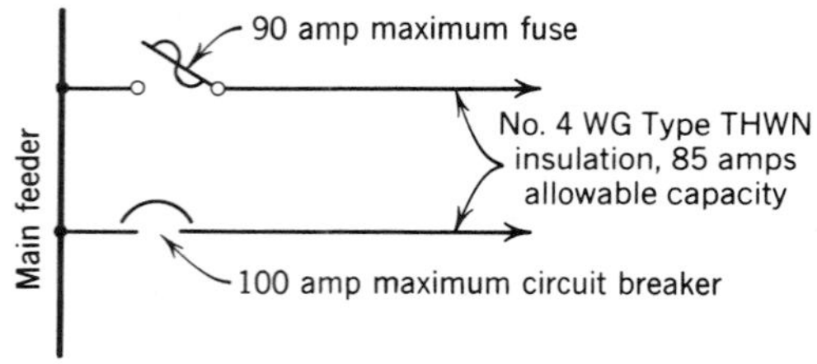

Fig. 17.10 The overcurrent protection must correspond to the rating of the protected equipment. Where ratings do not correspond exactly, the next larger standard size may be used.

4. Conductor sizes shall not be reduced in a circuit or tap, unless the smallest size wire is protected by the circuit overcurrent devices (see Fig. 17.11).
5. Overcurrent devices shall be located so as to be readily accessible.

17.12 Branch Circuit Design

A branch circuit by NEC definition refers only to the circuit conductors, although for our purposes and in trade parlance it includes the protective device and the outlets served. Such circuits may be multi-outlet general-purpose type (Fig. 17.12*a*), multi-outlet appliance type (Fig. 17.12*b*), or single-outlet type intended for a specific piece of equipment (Fig. 17.12*c*). The multi-outlet types are limited to 50 A in capacity, while the single-outlet type is governed in size only by the requirements of the item being served, and may be 200 or 300 A in size.

In its simplest form, a branch circuit comprises only two wires. However, multiwire branch circuits carrying 2- or 3-phase wires plus a neutral are also widely used. Generally, each branch circuit should be sized for the load connected to it, plus the load expansion that is expected. These general rules of good practice should be followed:

1. In all but the smallest installations, connect lighting, convenience receptacles, and appliances on separate groups of circuits, although this is not an NEC requirement.
2. General-purpose branch circuits should be 20 A and wired with No. 12 AWG wire. Switch legs may be No. 14 AWG if the lighting load permits.
3. Limit the circuit load on 15-A and 20-A circuits to the values shown in Table 17.5. This will provide the required building load expansion capability in the branch circuitry, that is, by bringing the loads on the branch circuits up to maximum, the additional building loads can be absorbed. However, since it is not always economical or feasible to expand existing circuits, panels are always equipped with spare circuit breakers and spaces for future circuit breakers. These then can alternatively be utilized to pick up building load expansion, or the user may use a combination of the two techniques, expanding some existing circuits and adding some new ones. (See Section 17.17 for a discussion of this subject.)

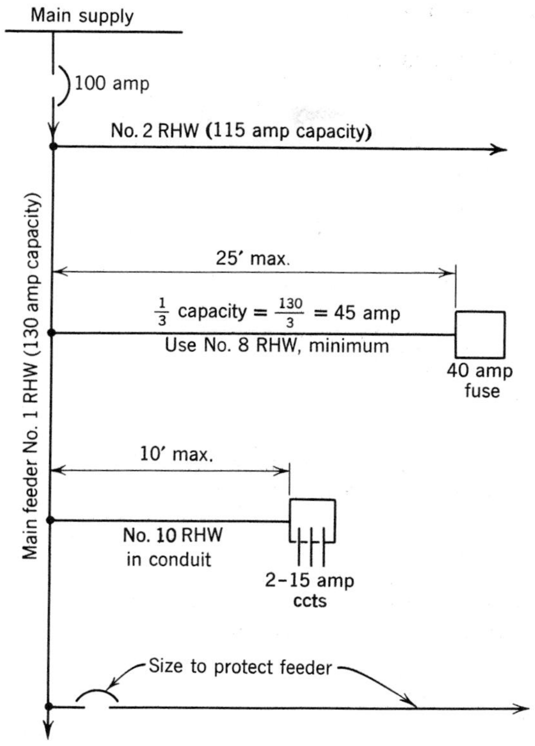

Taps may be made if smaller conductor is protected by main feeder protection (100 A C/B protects the No. 2).

Taps up to 25′ long are permitted it tap conductor has min. 1/3 ampacity of main, and terminates in a single C/B or set of fuses, which limits the current on the tap to the wire capacity. (40 A fuse limits current on 45 A capacity wire).

Taps up to 10′ long are permitted provided they have sufficient capacity for the circuits they feed; eg, if the wiring feeds 2–15 amp ccts, it must have 30 amp capacity and should be No. 10 AWG, RHW or equal.

Taps of any length may be made if the conductor is protected at the tap point by an appropriately sized overcurrent device.

Fig. 17.11 *Permissible tap arrangements. See NEC Article 240-21.*

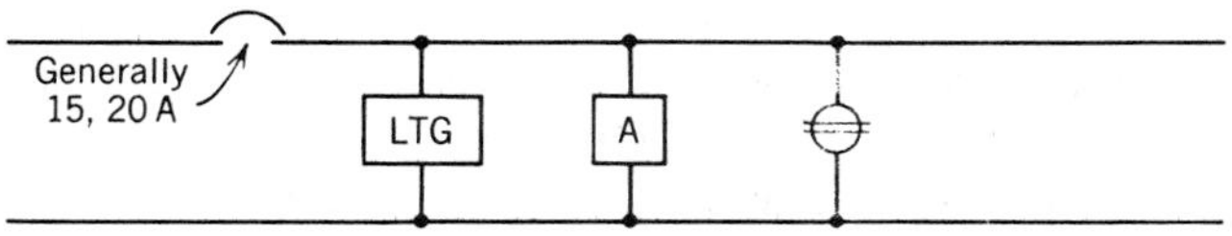

(a) General purpose branch circuit. Supplies outlets for lighting and appliances, including convenience receptacles.

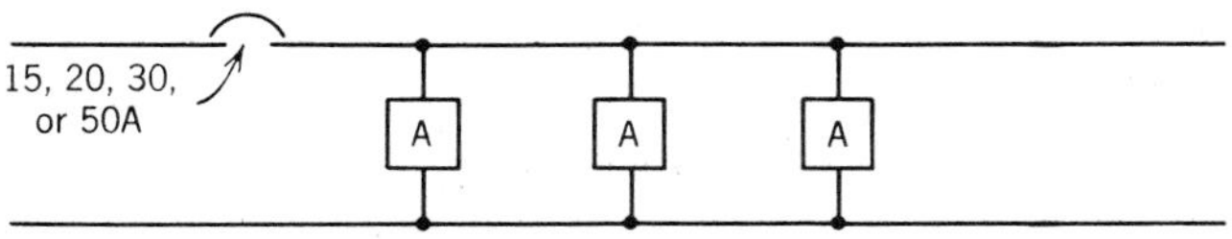

(b) Appliance branch circuit. Supplies outlets intended for feeding appliances. Fixed lighting not supplied.

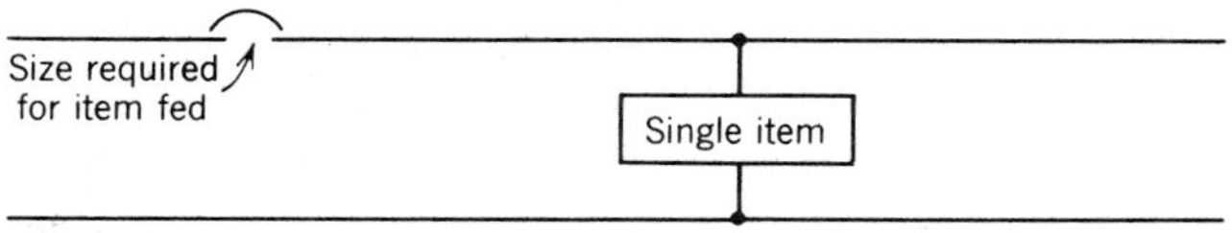

(c) Individual branch circuit, designed to supply a single, specific item.

Fig. 17.12 *Branch circuit types.*

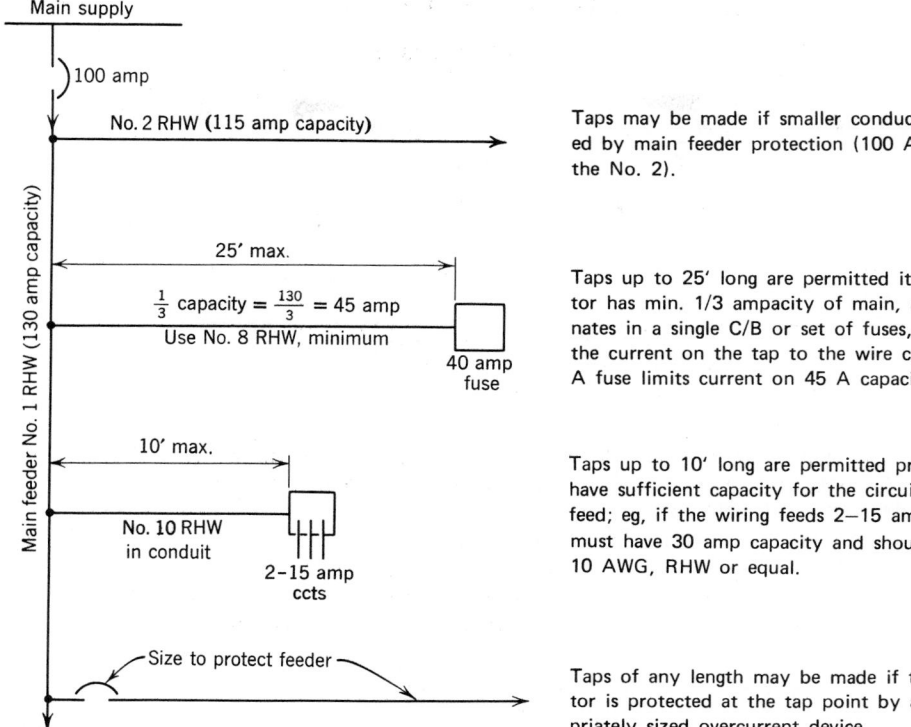

TABLE 17.5 **Recommended Branch Circuit Loads**[a]

Size Circuit	Circuit Amperes	Volt-Amperes at 120 V	Volt-Amperes at 277 V
		0% Expansion[b,c]	
15 A	12	1440	—
20 A	16	1920	4440
		25%[d] *Expansion*	
15 A	9.6	1150	—
20 A	12.8	1520	3600
		50%[e] *Expansion*	
15 A	8	960	—
20 A	11	1300	3000

[a]The loading shown will provide the needed expansion in the branch circuits. Where branch circuit expansion is not practical, the required expansion can be obtained from panel spares.

[b]See Table 17.1 for anticipated load growth.

[c]For 0% expansion, initial load = 80% of circuit rating, which is the maximum permissible for continuous loads. See Table 17.6, note 6.

[d]To accomplish 25% expansion, the circuit is derated to 80%, i.e., 80% of 12 A = 9.6 A, etc.

[e]To accomplish 50% expansion, the circuit is derated by ⅓, i.e., ⅔ of 12 A = 8 A, etc. (Then 50% expansion on 8 A yields 12 A.)

Since lighting and specific devices are circuited according to their nameplate rating, the only circuitry item left to the judgment of the designer is the number of convenience receptacles per circuit (cct). The NEC specifies that plug outlets (convenience receptacles) be counted, in totaling loads, at 1.5 A each unless included in the load for general lighting. Thus, following the guidelines stated above of 9- and 12-A loading on 15- and 20-A circuits, respectively, we would have by this method:

$$\text{15-A circuit } \frac{9}{1.5} = 6 \text{ outlets per cct}$$

$$\text{20-A circuit } \frac{12}{1.5} = 8 \text{ outlets per cct}$$

These figures must be used judiciously. If the devices to be energized are small, these quantities may be used. Such would be the case in a drafting room where only erasing machines (50 V-A) or a desk lamp (100 V-A) would be plugged in. However, for laboratory tables, office machines, or assembly benches, no more than two or three receptacles should be used or a 20-A circuit. Of course, diversity of use is an all-important factor and the more closely it can be estimated, the better the design result.

A further note of caution is in order. Receptacles should be arranged, if at all possible, so that the loss of a single circuit does not deprive an entire area of power. That is, for the sake of reliability, circuitry should be alternated to give each space parts of different circuits.

The Code further specifies certain requirements for conductors, devices, and loads permissible on general-purpose branch circuits. These are summarized in Table 17.6.

17.13 Alternative Wiring Techniques

The usual wiring system is a radial or ''tree'' type system in which conductors of progressively smaller size emanate radially from distribution points. These distribution points are the switchboards and panelboards throughout the system, which provide over-

TABLE 17.6 **Branch Circuit Requirements**

	Branch Circuit Size				
	15 A	*20 A*	*30 A*	*40 A*	*50 A*
Minimum size copper conductors	No. 14	12	10	8	6
Minimum size taps	No. 14	14	14	12	12
Overcurrent device rating	15 A	20	30	40	50
Lampholders permitted	Any type	Any type	Heavy duty	Heavy duty	Heavy duty
Receptacle rating permitted (see note 7)	15 A	15 or 20	30	40 or 50	50 A
Maximum load (see note 6)	15	20	30	40	50

NOTES:
1. Wiring shall be types RHW, RHH, T, THW, TW, THWN, THHN, XHHW in raceway or cable.
2. On 15-A circuit, maximum single appliance shall draw 12 A. On 20-A circuit, maximum single appliance shall draw 16 A. If combined with lighting or portable appliances, any fixed appliance shall not draw more than 7.5 A on a 15-A circuit, and 10 A on a 20-A circuit.
3. On a 30-A circuit, maximum single appliance draw shall be 24 A.
4. Heavy-duty lampholders are units rated not less than 750 W.
5. 30, 40, and 50-A circuits shall not be used for fixed lighting in residences.
6. When loads are connected for long periods, actual load shall not exceed 80% of the branch circuit rating. Conversely, continuous type loads shall be figured at 125% of actual load in all load calculations.
7. A single receptacle on an individual branch circuit shall have a rating not less than the circuit, for example, 15 A on a 15-A circuit, etc. 15-A receptacles on 20-A circuit shall not supply a load greater than 12-A for appliances. 20-A receptacles on 20-A circuit shall be limited to a 16-A load.

current protection for each of these radials (refer to Fig. 17.9). Note that at each step, power is tapped off in a smaller "package," and in accordance with the principles of overcurrent protection an appropriately sized protective device is placed at the point of tap. An alternate technique utilizes this tapping principle somewhat differently (refer to Fig. 17.11). The arrangement shown at the bottom of the figure—a tap protected at the source—is precisely what occurs at a panelboard, with the panel bus substituting for the feeder. This arrangement is shown in Fig. 17.13*a*. If we *eliminate* the panel and tap the feeder directly as in Fig. 17.13*b*, we have accomplished exactly the same end. The purpose of this second arrangement is to eliminate the branch circuit wiring, by placing the protective device at the load. Such an arrangement is shown in Fig. 17.14. The advantage of this system, in addition to the advantages inherent in a surface raceway wiring system, is that by eliminating branch circuit wiring, installation costs are reduced, voltage drop and energy losses in branch circuit conductors are negligible, and loads are individually protected. Obviously, this system finds its best application in areas where surface raceways are desirable for flexibility and loads would benefit from this type of individual control and protection. As an exercise, the reader might examine the possibility of utilizing this system for the layout of Fig. 17.5, justifying any conclusions from technical and economic points of view.

17.14 Branch Circuit Design Guidelines; Residential

1. The NEC (1981 edition) requires for residences sufficient circuitry to supply a load of 3 W/ft^2 in the building, excluding unfinished spaces such as porches, garages, and basements. Using Table 17.5, this works out to 480 ft^2 on a 15-A circuit (1440 V-A) and 640 ft^2 on a 20-A cct (1920 V-A). Allowing for some expansion (better practice) reduces these figures to 400 ft^2 on a 15-A cct and 500 on a 20-A cct.

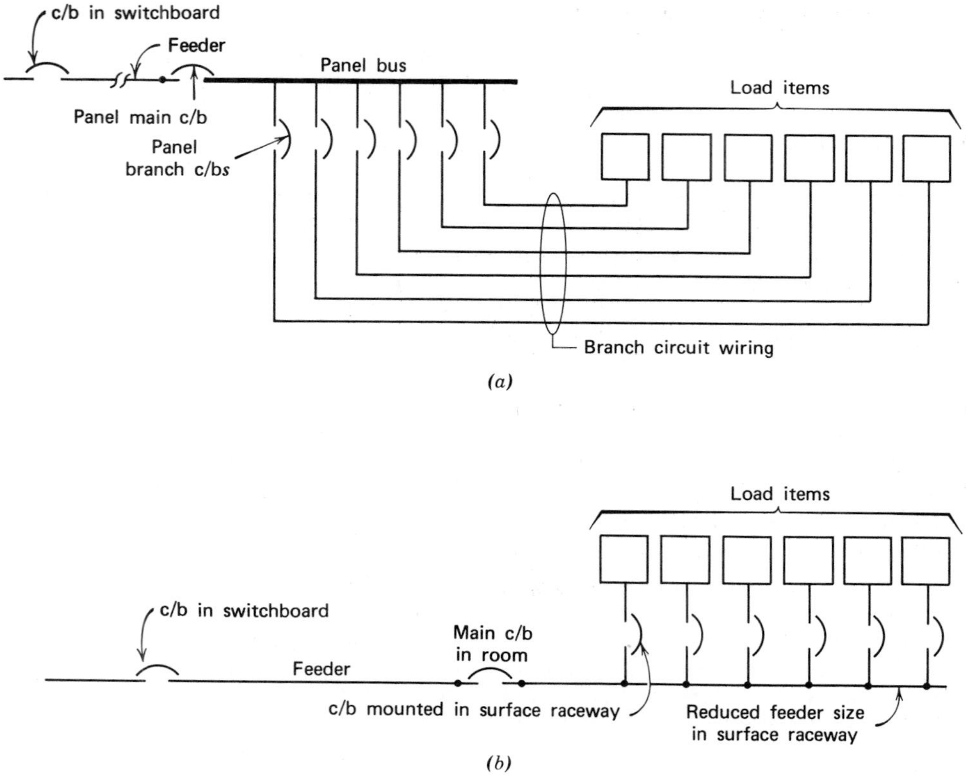

Fig. 17.13 (a) *Conventional radial wiring system. Branch circuits radiate from the panel to the loads and are protected at the panel by the panel circuit breakers. The panel itself is protected by a main c/b.* (b) *In this arrangement the heavy switchboard feeder is tapped by the room c/b that protects the room feeder. This feeder is in turn tapped at each load, thus eliminating a concentrated panelboard and almost all branch circuit wiring.*

2. The NEC requires (Article 220–3) a minimum of two 20-A appliance branch circuits (see Fig. 17.12*b*) to feed all of the small appliance outlets in the kitchen, pantry, dining room, and family room, and *only* these outlets. Furthermore, all kitchen outlets must be fed from at least two of these circuits (which may also feed other appliance outlets). This NEC requirement needs clarification. The NEC requires that at least two circuits be reserved for *appliance outlets,* but it does not specify what these appliance outlets are except to say, by inference, that all kitchen outlets are appliance outlets. In effect, then, according to the NEC all the receptacles are potential appliance outlets, and at least two circuits should be supplied to serve them. Good practice dictates that certain receptacles in each room be designated as appliance outlets even though they do not differ from the other outlets in appearance. These outlets are:

1. All kitchen receptacles.
2. One dining room receptacle.
3. One receptacle in the family (or living) room.

These receptacles should be circuited with preferably two, but no more than four, such outlets on a 20-A circuit, and the circuits should be arranged so that the kitchen has part of at least two circuits feeding its outlets.

3. Additional circuits similar to appliance circuits should be furnished to supply one outlet in each bedroom of a house that is not centrally air conditioned. Such outlets are intended for window air conditioners. (Good architectural and HVAC design will provide window arrangement, attic ventilation, insulation, sun-screening, and the like to obviate the necessity for these noisy energy users.) Also, on circuits of this kind place basement workbench outlets. These additional circuits

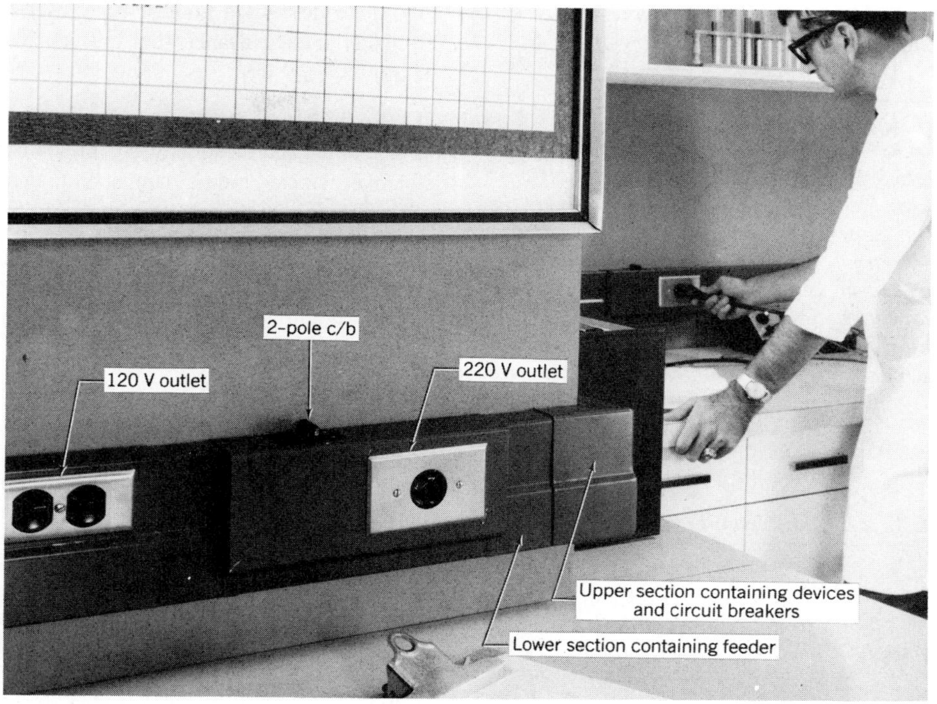

Fig. 17.14 *Typical room layout using the feeder tapping wiring technique. The room feeder is tapped from the main feeder and protected by a main disconnect that is either a circuit breaker or a fused switch (not shown). Beyond this point, the room feeder is run in surface raceway and is tapped at each device with a circuit breaker, which feeds an individual outlet. Overhead loads—such as lighting (not shown)—are fed from raceway circuit breakers. The equipment as actually installed in a laboratory facility (shown) is neat, flexible, and highly practical. Courtesy of Versa-Tek, Inc.*

must not be mixed with the appliance branch circuits discussed above, as they are not strictly appliance circuits by NEC definition.

4. The NEC requires that at least one 20-A circuit supply the laundry outlets. This requirement satisfies good practice. If an electric clothes dryer is anticipated (and it should be unless it is definitely known that a gas dryer will be used), an individual branch circuit should be supplied to serve this load, via a heavy-duty receptacle. (Obviously, facilities for hanging clothes must be provided for those who prefer not to waste energy.)

5. Lay out convenience receptacles so that no point on a wall is more than 6 ft from an outlet. Use 20-A, grounding-type receptacles only. Do not combine receptacles and switches into a single outlet except where convenience of use dictates high mounting of receptacles, as above counter spaces.

6. Circuit the lighting and receptacles so that each room has parts of at least two circuits. This includes basements and garages.

7. Avoid placing all the lighting in a building on a single circuit.

8. Supply at least one receptacle in the bathroom and one outside the house. Both must be GFCI types. This is an NEC requirement. An additional convenience is switch control of the outside receptacle from *inside* the house. Also, a timer-controlled outlet for a plug-in bathroom heater is a welcome convenience, but obviously means additional expense.

9. In rooms without overhead lights, provide switch control for one-half of a strategically located receptacle that is intended to supply a lamp. See Fig. 17.15 for the wiring arrangement in such a case.

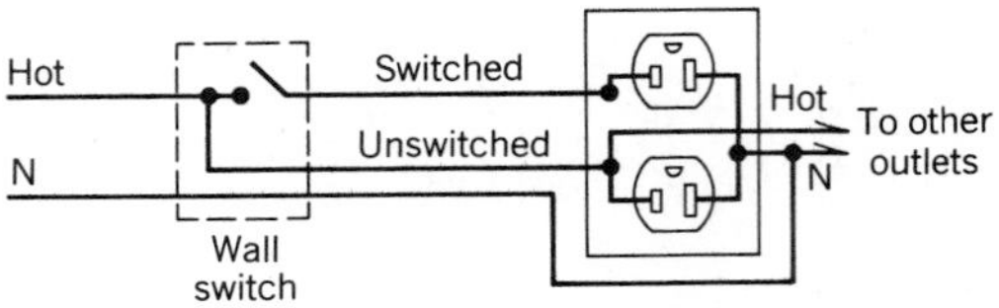

Fig. 17.15 *Split wiring of a duplex receptacle. Upper half is switch controlled; lower half is "hot" all the time. This allows wall switch control of a lamp or other device while maintaining part of receptacle live, for independent use. Notice that the receptacle is mounted with the grounding pole at the top. This is the safest way to install receptacles, since a metallic item such as a paper clip falling on to an inserted cap will contact the grounding pole only.*

10. Provide switch control for closet lights. Pull chains are a nuisance (but are considerably cheaper).

11. In bedrooms supply two duplex outlets at each side of the bed location to accommodate electric blanket, clocks, radios, lamps, and other such appliances.

12. Since receptacles are counted as part of general lighting and no additional load is included for them, no limit is placed on the number of receptacle outlets that may be wired to a circuit.

TABLE 17.7 **Load, Circuit, and Receptacle Chart for Residential Electrical Equipment**

Appliance	*NEC Type*[a]	*Typical Connected Volt-Amperes*	*Volts*	*Wires*[b]	*Circuit Breaker or Fuse*[c]	*Outlets on Circuit*	*NEMA Device*[d] *and Configuration See Fig. 16.24*
					Kitchen		
Range[e]	(F)	12,000	115/230	3 #6	60 A	1	14–60R
Oven (built in)[c]	(F)	4,500	115/230	3 #10	30 A	1	14–30R
Range top[c]	(F)	6,000	115/230	3 #10	30 A	1	14–30R
Dishwasher[c]	(F)	1,200	115	2 #12	20 A	1	5–20R
Waste disposer[c]	(F)	300	115	2 #12	20 A	1	5–20R
Broiler[e]	(P)	1,500	115	2 #12	20 A	1 or more	5–20R
Refrigerator[f]	(S)	300	115	2 #12	20 A	1 or more	5–20R
Freezer[f]	(S)	350	115	2 #12	20 A	1 or more	5–20R
					Laundry		
Washing machine	(S)	1,200	115	2 #12	20 A	1 or more	5–20R
Dryer[c]	(S)	5,000	115/230	3 #10	30 A	1	14–30R
Hand iron; Ironer	(P)	1,650	115	2 #12	20 A	1 or more	5–20R
					Living Areas		
Workshop	(P)	1,500	115	2 #12	20 A	1 or more	5–20R
Portable heater[g]	(P)	1,300	115	2 #12	20 A	1	5–20R

For good practice they should be limited to six on a 15-A circuit and eight on a 20-A circuit.

13. Kitchens should have a duplex appliance outlet every 36 in. of counter space, no less than two total. Each section of counter 12 in. (or more) wide must be supplied by at least one receptacle.

14. A disconnecting means, readily accessible, must be provided for electric ranges, cook tops, and ovens. It is good practice to utilize a small kitchen panel recessed into a corner wall to control the large kitchen appliances and to provide completely safe, accessible disconnecting means. Such an arrangement can also be cheaper if the length of run between the main panel and the kitchen is appreciable.

15. Perimeter lighting, inside controlled, can do much to lessen vandalism and discourage prowlers.

16. A tabulation of residential electrical equipment, including recommended circuits and receptacles, is shown in Table 17.7. A complete residential wiring plan for a small house is shown in Fig. 17.16. Although residential plans are normally left uncircuited, a completely circuited design is shown here for didactic purposes.

TABLE 17.7 **Load, Circuit, and Receptacle Chart for Residential Electrical Equipment (*Continued*)**

Appliance	*NEC Type*[a]	*Typical Connected Volt-Amperes*	*Volts*	*Wires*[b]	*Circuit Breaker or Fuse*[c]	*Outlets on Circuit*	*NEMA Device*[d] *and Configuration See Fig. 16.24*
Television[g]	(S)	300	115	2 #12	20 A	1 or more	5–20R
					Fixed Utilities		
Fixed lighting	(F)	1,200	115	2 #12	20 A	1 or more	—
Air conditioner ¾ hp[h]	(F)	1,200	115	2 #12	20 A or 30 A	1	5–20R
Central air conditioner[i]	(F)	5,000	115/230	3 #10	40 A	1	—
Sump pump[i]	(F)	300	115	2 #12	20 A	1 or more	5–20R
Heating plant, i.e., forced-air furnace[d,h,j]	(F)	600	115	2 #12	20 A	1	—
Attic fan[i]	(F)	300	115	2 #12	20 A	1 or more	5–20R

[a]Appliance types: (F) Fixed; (S) Stationary; (P) Portable.

[b]Number of wires does not include equipment grounding wires. Ground wire is No. 12 AWG for 20-A circuit and No. 10 AWG for 30-A and 50-A circuits.

[c]May be direct connected. For a discussion of disconnect requirements, see NEC Article 422.

[d]Equipment ground is provided in each receptacle.

[e]Heavy-duty appliances regularly used at one location should have a separate circuit. Only one such unit should be attached to a single circuit at the same time.

[f]Separate circuit serving only one other outlet is recommended.

[g]Should not be connected to circuit with appliances or other heavy loads.

[h]Separate circuit recommended.

[i]Recommended that all motor-driven devices be protected by a local motor-protection element unless motor protection is built into the device.

[j]Connect through disconnect switch equipped with motor-protection element.

ELECTRICITY

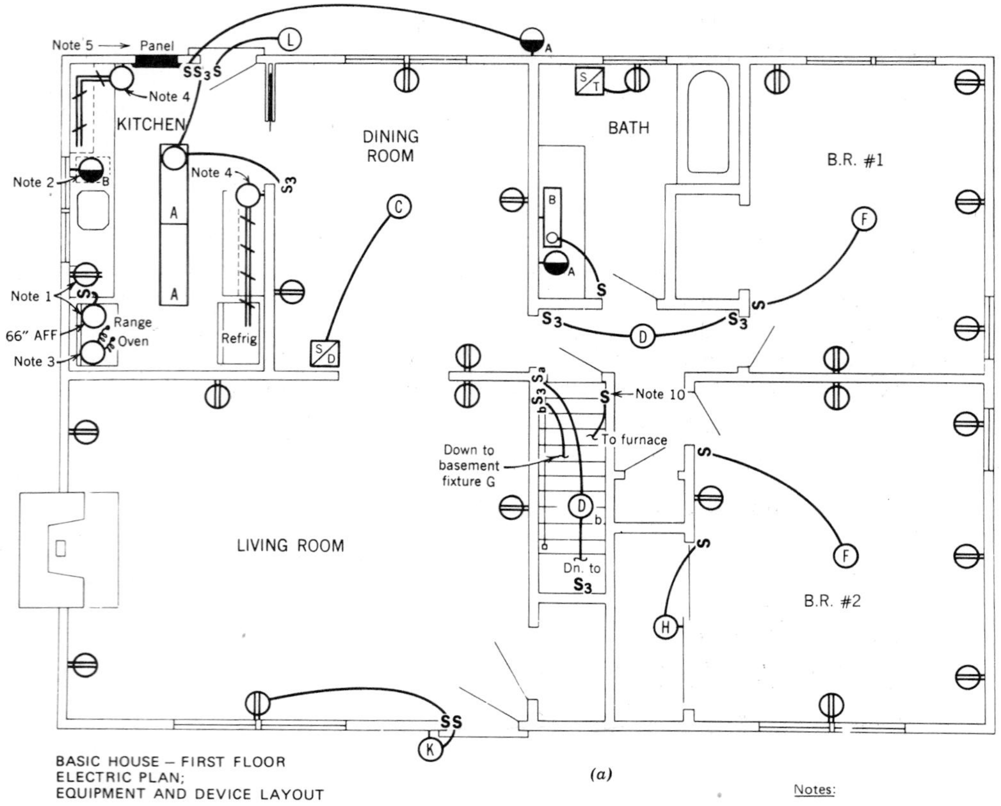

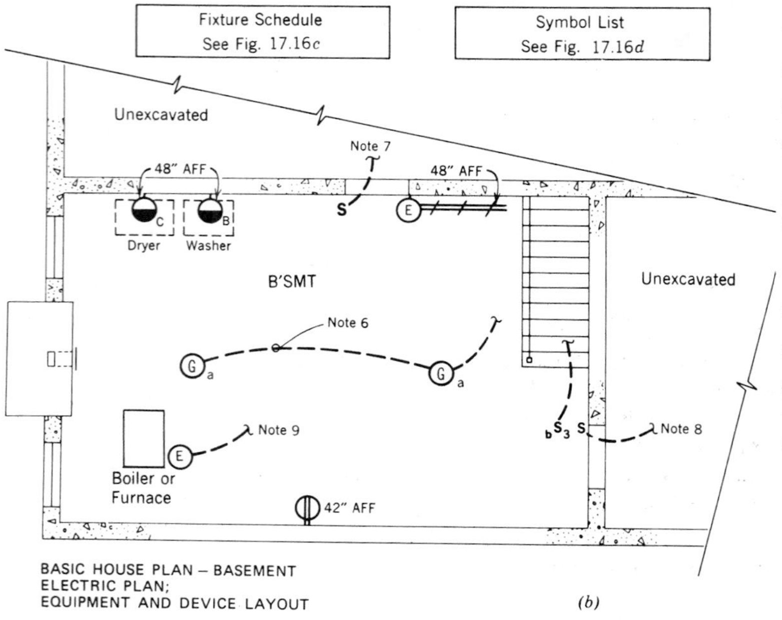

Notes:

1. Switch and outlet for exhaust fan. Switch wall mtd. above sink blacksplash. Outlet with blank cover mounted adjacent to fan wall opening. Separate switch may be omitted if fan is supplied with integral switch.
2. Dishwasher receptacle wall mtd. behind unit, 6" AFF.
3. Range and oven outlet boxes wall mtd., 36" AFF. Flexible connection to units.
4. Surface raceways outlet boxes to be flush mounted with flush covers, suitable for painting or other wall finish.
5. Max. ht. of top c/b to be 78" AFF.
6. Wiring shown as run exposed indicates absence of finished ceiling in basement level. All BX to be run through framing members. Attachment below ceiling joists not permitted.
7. Connect to 2–Type G fixtures ceiling mounted at 1/3 points.
8. Connect to 1–Type G fixture at at center.
9. Connect to shut–down switch at top of stairs. Boiler control wiring by others. See Note 10.
10. Boiler wiring safety disconnect. Provide RED wall plate, clearly marked "BOILER ON–OFF".

Fig. 17.16 *Electrical plan of a small house. Devices and lighting are shown on main floor and basement plans* (a) *and* (b). *A detailed fixture schedule* (c) *and symbol list* (d) *is also prepared. If circuitry is added, the completed plans for both levels will appear as in* (e) *and* (f). *A detailed panel schedule is given in* (g).

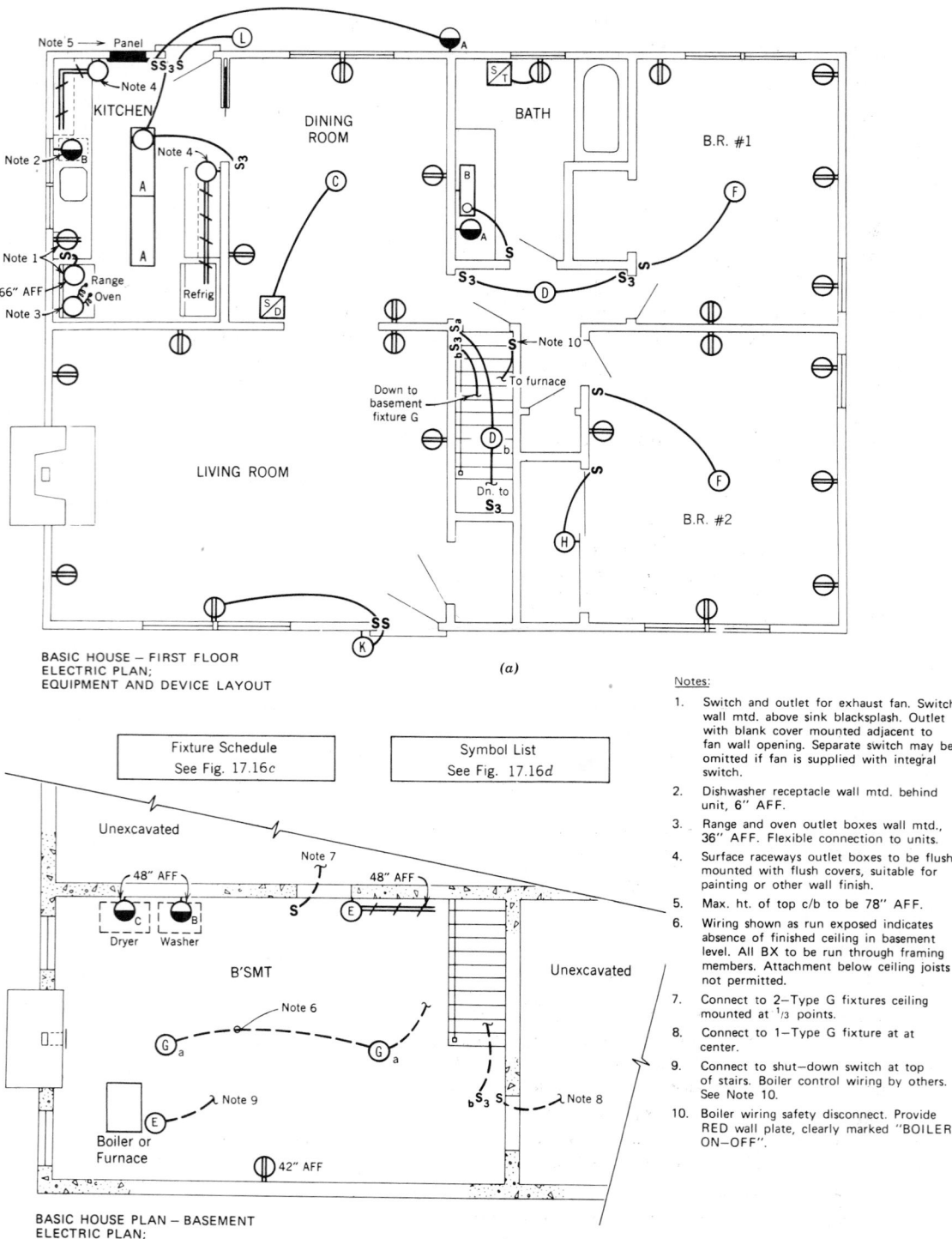

LIGHTING FIXTURE SCHEDULE			
TYPE	DESCRIPTION	MANUFACTURER	REMARKS
A	48" L X 12" W X 4" DEEP NOMINAL, 2 LAMP/FLUORES–CENT, WRAP–AROUND ACRYLIC LENS, F 40 WW/LAMPS. SURFACE MTD.	BRITE–LITE CO. CAT. #2/40/KFF OR EQUAL	4" DEPTH MAXIMUM
B	24" L, 1 LAMP 20 W FLUOR. FIXTURE, WRAP–AROUND WHITE DIFFUSER, WITH SINGLE–SWITCHED RECEPTACLE. MOUNT ABOVE MEDICINE CABINET.	BRITE–LITE CO. CAT. #1/20/BFF OR EQUAL	MAX. MTG. HT. 78" TO ℄.
C	ADJUSTABLE HEIGHT PENDANT INCANDESCENT, 3–75 W MAX., BUILT–IN 3–POSITION SWITCH.	HOMELAMP CO. CAT. #3/75/DRP OR EQUAL	———
D	10" D. DRUM–TYPE FIXTURE, WHITE GLASS DIFFUSER, CENTER LOCK–UP, 2–60 W INCAND. MAX., SURF. MTD.	BRITELITE CO. CAT. #2/60/HF OR EQUAL	6" MAX. DEPTH.
F	12" D. DRUM FIXTURE, CONCEALED HINGE ON OPAL GLASS DIFFUSER FOR RELAMPING WITHOUT GLASS REMOVAL, 2–75W INCAND. MAX. SURFACE MTD.	DENMARK LIGHTING SPECIAL UNIT #374821	NO SUBSTITUTION WILL BE ACCEPTED.
G	PORCELAIN LAMPHOLDER, PULL CHAIN WITH WIRE GUARD, 100 W. INCAND. SURF. MTD.	———	———
H	SAME AS TYPE G, EXCEPT W/O GUARD.	———	———
K	DECORATIVE OUTDOOR LANTERN, MAX. 150 W INCAND., WALL MTD. 84" AFF TO ℄.	TO BE CHOSEN BY OWNER	———
L	UTILITY OUTDOOR LIGHT, ANODIZED ALUMINUM BODY AND CYLINDRICAL OPAL GLASS DIFFUSER. 1–100 W INCAND. MAX. 84" AFF TO ℄.	UTIL–LITE CO. CAT. #1/100/BP OR EQUAL	IF VANDALISM IS OF CONCERN, SUBST. PLASTIC DIFFUSER.

(c)

SYMBOLS AND ABBREVIATIONS

BX CABLES RUN CONCEALED; TICS INDICATE NUMBER OF CONDUCTORS EXCLUDING GROUND WIRES. 2 #12 + BARE GROUND, UON

SAME AS ABOVE EXCEPT RUN EXPOSED.

WIRING RUN TURNING DOWN; WIRING TURNING UP

1 3 HOME RUN TO PANEL; ARROWS AND NUMERALS IDENTIFY CIRCUITS; TICS INDICATE WIRING – AS NOTED ABOVE.

OUTLET BOX AND FINAL CONNECTION TO EQUIPMENT WITH FLEXIBLE CONDUIT (OR BX).

OUTLET WITH SECTION OF SURFACE RACEWAY, 2 WIRE, SINGLE CIRCUIT, AND SEPARATE GREEN GND. 15A, 2P, 3 WIRE, RECEPTACLES ON 12" CENTERS.

(D)a CLG. OUTLET WITH INCANDESCENT LTG. FIXTURE D, SWITCH CONTROL – 'a'.

(H) WALL OUTLET W/INCAND. FIXT. 'H', M.HT. SHOWN.

A CLG. OUTLET W/FLUOR. FIXT. 'A'.

B WALL OUTLET W/FLUOR. FIXT. 'B', M.HT. SHOWN.

(J) JUNCTION BOX

DUPLEX CONVENIENCE RECEPTACLE, 15 A, 2 P, 3 W, 125 V. GROUNDING, WALL MTD., VERTICAL, ℄ 12" AFF NEMA 5–15 R.

A DUPLEX CONVENIENCE RECEPTACLES, 15A, 2P, 3W, 125V, GROUNDING. W/INTEGRAL GFCI AND GASKETED W.P. SELF–CLOSING COVER.

B SINGLE RECEPTACLE, 20 A, 2 P, 3 W, GND'G., NEMA 5–20 R.

C SINGLE RECEPTACLE, 30 A. 125/250 V. 3 POLE–4 WIRE GROUNDING NEMA 14–30 R; (NOTE 1)

S_a SINGLE POLE SWITCH, 15 A, 125 V, ℄ 50" AFF, UON, CONTROLLING OUTLET(S) 'a'.

$_aS_3$ SWITCH, 3 WAY, 15 A, 125 V, ℄ 50" AFF, UON, CONTROLLING OUTLETS 'a'.

S/T MANUAL TIMER SWITCH, 1 SET 15 AMP N.O. CONTACTS.

S/D OUTLET BOX MTD. SWITCH AND DIMMER, INCAND. LOAD ONLY, 600 WATTS MAXIMUM. ℄ 50" AFF.

FLUSH MTD. PANELBOARD;

AFF	ABOVE FINISHED FLOOR
MHT	MOUNTING HEIGHT
T	THERMOSTAT
UON	UNLESS OTHERWISE NOTED
GFCI	GROUND FAULT CIRCUIT INTERRUPTER
WP	WEATHER–PROOF
NO	NORMALLY OPEN

NOTE 1. CONTRACTOR TO SUPPLY MATCHING CAP.

(d)

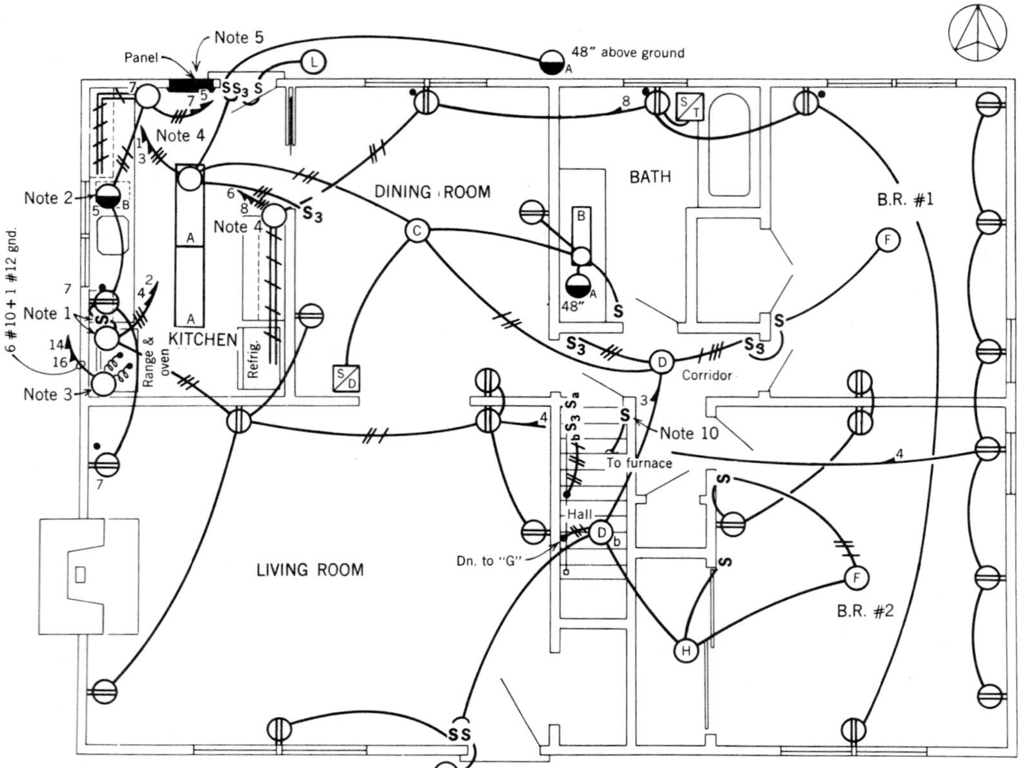

BASIC HOUSE – FIRST FLOOR
CIRCUITED ELECTRIC PLAN;

(e)

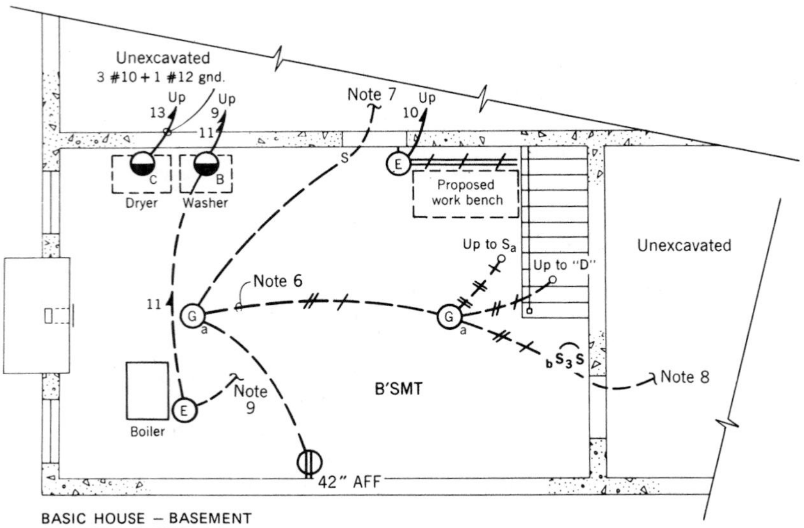

BASIC HOUSE – BASEMENT
CIRCUITED ELECTRIC PLAN,

(f)

***Fig. 17.16** (continued)*

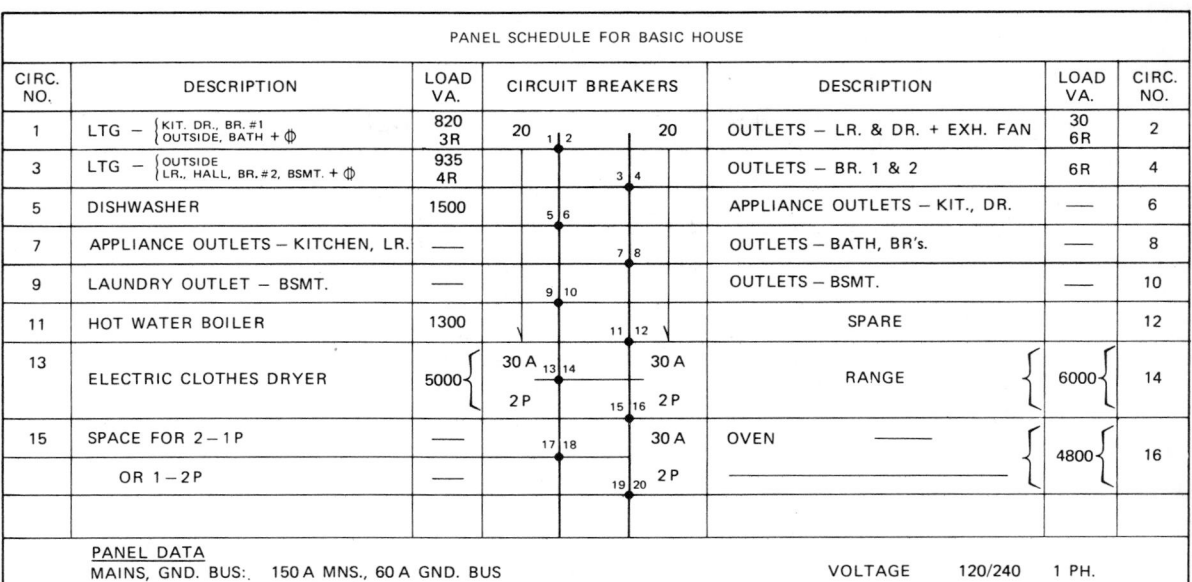

PANEL SCHEDULE FOR BASIC HOUSE

CIRC. NO.	DESCRIPTION	LOAD VA.	CIRCUIT BREAKERS			DESCRIPTION	LOAD VA.	CIRC. NO.
1	LTG — {KIT. DR., BR. #1 / OUTSIDE, BATH + ϕ}	820 3R	20	1 2	20	OUTLETS — LR. & DR. + EXH. FAN	30 6R	2
3	LTG — {OUTSIDE / LR., HALL, BR. #2, BSMT. + ϕ}	935 4R		3 4		OUTLETS — BR. 1 & 2	6R	4
5	DISHWASHER	1500		5 6		APPLIANCE OUTLETS — KIT., DR.	—	6
7	APPLIANCE OUTLETS — KITCHEN, LR.	—		7 8		OUTLETS — BATH, BR's.	—	8
9	LAUNDRY OUTLET — BSMT.	—		9 10		OUTLETS — BSMT.	—	10
11	HOT WATER BOILER	1300		11 12		SPARE		12
13	ELECTRIC CLOTHES DRYER	5000	30 A 2 P	13 14 / 15 16	30 A 2 P	RANGE	6000	14
15	SPACE FOR 2 — 1 P	—		17 18	30 A 2 P	OVEN	4800	16
	OR 1 — 2 P	—		19 20				

PANEL DATA
MAINS, GND. BUS: 150 A MNS., 60 A GND. BUS
MAIN C/B ~~OR SW/F~~ 150/100
BRANCH C/B INT. CAP. 5000 AMP.
MOUNTING — ~~SURF~~/RECESS
REMARKS: FRONT SUITABLE FOR PAINTING

VOLTAGE 120/240 1 PH.

(*g*)

17.15 Branch Circuit Design Guidelines; Nonresidential

(a) Schools. Since schools comprise an assembly of varied use spaces, including instruction, lab, shop, assembly, office, gymnasium, plus special areas such as swimming pools, photographic labs, and so on, it is not possible to generalize on branch circuit design considerations except for the following:

1. To accommodate the opaque and film projectors frequently used in the classroom, 20-A outlets wired two to a circuit are placed at the front and back of each such room. A similar receptacle, wired six or eight to a circuit is placed on each remaining wall.
2. Light switching should provide:
   (*a*) High–low levels for energy conservation and to permit low-level lighting for film viewing. With fluorescent lighting this can be accomplished by alternate ballast wiring and switching, thus avoiding the high cost of dimming equipment.
   (*b*) Separate switching of the lights on the window side of the room, which is often lighted sufficiently by daylight.
3. Provide appropriate outlets for all special equipment in labs, shops, cooking rooms, and the like.
4. Use heavy-duty devices and key-operated switches for public area lighting (corridors, etc.), plastic instead of glass in fixtures, and vandal-proof equipment wherever possible. All panels *must* be locked and should be in locked closets.
5. The NEC requires sufficient branch circuitry to provide a minimum of 3 W/ft^2 for general lighting in schools. Refer to the NEC Article No. 220. Unlike for residential occupancy, this figure does *not* include receptacles. Receptacles are calculated separately at 180 W each for ordinary convenience outlets.
6. Keep lighting and receptacles completely separate when circuiting.

(b) Office Space

1. In small office spaces (less than 400 ft^2) provide either one outlet for every 40 ft^2, or one outlet for every 10 linear ft of wall space, whichever is greater. In larger office

ELECTRICITY

spaces, provide one outlet every 100 to 125 sq ft beyond the initial 400 ft² (10 outlets). These should comprise wall outlets spaced as above plus floor outlets sufficient to make up the required total. In view of the increasingly heavy loads of office computers, these receptacles should be circuited at no more than six to a 20-A branch circuit, and less if the equipment to be fed so dictates. Figure 17.17 shows one possible circuiting arrangement for the room layouts shown in Fig. 17.5. Although other arrangements are possible, the net result is the same.

2. Corridors should have a 20-A, 120-V outlet every 50 ft, to supply cleaning and waxing machines.
3. As with all nonresidential buildings, convenience receptacles are figured at 180 W each.
4. Only specification grade equipment should be used.

(c) Industrial Spaces. These areas are so specialized that no meaningful guidelines can be given.

(d) Stores. In stores, good practice requires at least one convenience outlet receptacle for every 300 ft² in addition to outlets required for loads

Fig. 17.17 (a) *Alternate methods of circuiting are shown. Room 205 shows the actual junction box location, with flexible connections to the box at each fixture. Room 207 shows circuit numbers and switch designations only; the placement of junction boxes is understood, and conduit runs are omitted for clarity and because they most often are not representative of actual installation. Room 209 shows an outlet box at each fixture, with schematic conduit connections. All of these systems are in common use.*

(a)

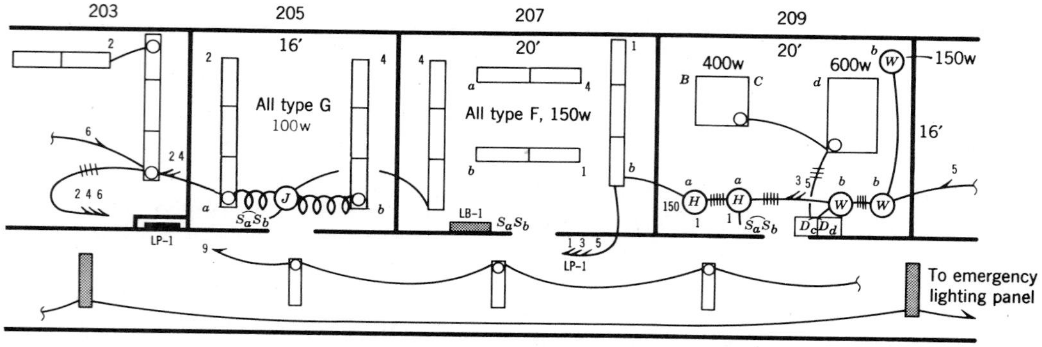

(b)

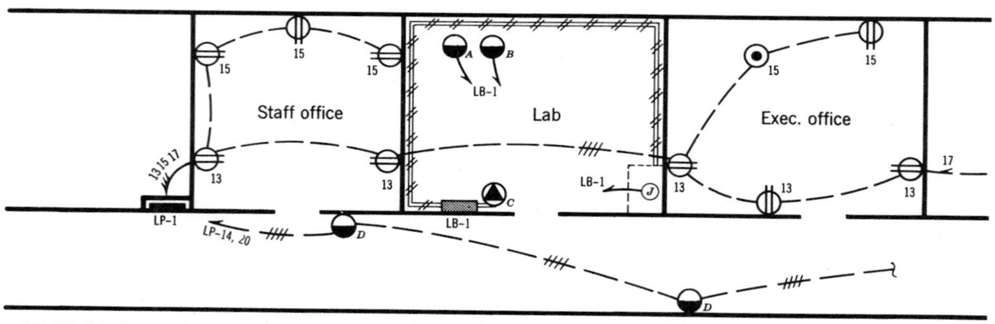

Fig. 17.17 (b) *Typical circuiting of several rooms in an office–lab building. Lighting and power (receptacles) are shown on separate plans (a) and (b) to avoid crowding (see Fig. 17.8 for symbols and Fig. 17.5 for notes). Lighting in offices is recessed; lighting in labs is surface mounted for flexibility. Note the double circuiting of the Type D receptacles.*

such as lamps, show windows, and demonstration appliances.

17.16 Load Tabulation

While circuiting the loads, a panel schedule is drawn up that lists the circuit numbers, load description, and wattage (actually volt-amperes), and the current rating and number of poles of the circuit-protective device feeding each circuit. Spare circuits are included to the extent that the designer considers them necessary and consonant with economy, but normally no less than 20% of the number of active circuits. Finally, spaces are left for future circuit breakers, in approximately the same quantity as the number of spare circuits, but always to round off the total number of circuits.

Panels are normally manufactured with an even number of poles. Thus if a panel had, with spares, 21 poles, the designer would probably require 3 spaces, to give a 24-circuit box. A typical panel schedule is shown in Fig. 17.18, which serves the laboratory of Fig. 17.17.

In calculating panel loads, the following rules apply:

1. Each specific appliance, device, lighting fixture, or other load is taken at its nameplate rating, except certain kitchen and laundry appliances for which the NEC allows a demand factor (see NEC Article 220).

2. Each convenience outlet, in other than residential spaces, is counted as 1.5 A (180 W).

3. Loads for special areas and devices such as show window lighting, heavy-duty lampholders, and multi-outlet assemblies are taken at the figures given in NEC Article 220.

4. Spare circuits are figured at approximately the same load as the average active circuits (1200 to 1500 W).

5. Spaces are not added into the load.

6. Continuous loads such as lighting are calculated at 125% of their actual value. See NEC Article 220-10(b), which applies to feeder calculations but applies equally to panel loading.

In calculating total panel loads, as shown in Fig. 17.18, no demand factor may be applied except as specifically stated in the NEC. This is true despite the knowledge that most often the usage will be such that average load will be lower than the maximum demand (see Section 17.20 for multipanel feeders). If it is known that certain loads will not or cannot be used simultaneously, the load total should reflect only the larger of the two. Thus, heating and cooling loads are *generally* not concurrent. Nor is building night floodlighting concurrent with the business office load, but it may be with general interior lighting. Note in Fig. 17.18

	ELEC. PANEL-LP-1 [1]			
	120/208 V 3ϕ 4W	LOAD IN WATTS		
No.	LOAD	ϕA	ϕB	ϕC
1	Lighting	1050		
2	Lighting	1050		
3	Lighting		1450	
4	Lighting		1050	
5	Lighting			1100
6	Lighting			1200
7	Lighting	800		
8	Lighting	1100		
9	Lighting		700	
10	Lighting		1050	
11	Lighting			1000
12	Lighting			1200
13	Receptacle [2]	900		
14	Receptacle [3]	900		
15	Receptacle		900	
16	Receptacle		900	
17	Spare			1200
18	Spare			1200
19	Spare	1200		
20	Receptacle [4]		1000	
				1000
21	Spare	1200		
			1200	
22-26	Spaces only			
	25% addition for			
	continuos loads [5]	1000	1063	1125
	Phase totals	9300	9313	9025
	Panel total		27638	V-A
	Max. ϕ current		78 amp.	
	25% spare capacity	~	20 amp.	(Future)
	Max. Phase I		98 amps	

Main breaker 225A 3pole (see text example A-1, section 17.19)
Trip 100 A
Feeder size 4 #2 THW in 2"C.

NOTES:
1. All C/B 1P; 50 A frame, 20 A trip except ccts 20 and 21, which are 2P, 50 AF/20 AT.
2. 5 receptacles @ 1.5 A each.
3. Corridor receptacle; 120 V section.
4. Corridor receptacle; 208 V section.
5. See Section 17.16.

Fig. 17.18 Schedule for lighting panel LP-1.

that 2-pole loads (208 single-phase) appear in two columns. Similarly, 3-phase loads would appear in three columns. Also note that the phase loads are *not equal*. It is the responsibility of the designer (or contractor) to circuit the loads so that the phases are as closely balanced in load as possible. If this is not done, one phase will carry considerably more current than the others. Since the panel feeder must be sized for the maximum phase current, this may lead to an oversized feeder and therefore a waste of money.

Having tablulated and balanced the loads and totaled them by phase, the maximum current is calculated. A portion of the spare capacity available in the branch circuits is added to the above total, as the basis for the calculation of the feeder load. This spare capacity, shown in Table 17.5, is something between 25 and 50%. The exact amount to be added initially in feeder sizing is developed below.

17.17 Spare Capacity

Load calculations for dwelling occupancies are detailed in NEC Article 220 and examples given in Chapter 9 of the NEC. Since these calculations are specialized but routine, and are covered there in detail, they will not be repeated herein. Again it is emphasized that possession of a current copy of the National Electrical Code (NFPA Bulletin 70) is a sine qua non of proper electrical design.

Having arrived at the panel load totals as detailed above, the next step is to size the conductors feeding the panel. To do this, an examination of the spare capacity of the panel and of the feeders is necessary, in order that the system design be consistent, giving equal capacity for future growth in all its components. Considering the panel circuitry first, let us examine the effect of load expansion, including spares and spaces (see Table 17.8).

As noted in Section 17.12 above, spare capacity is built into the branch circuitry *and* into the panels. Most often expansion is accomplished by additional loading on some circuits and by adding new circuits via spare circuit breakers in the panel. Table 17.8 gives the *ultimate* capacity of the panel, that is, fully loaded circuits and fully utilized spares and spaces. Since this ultimate capacity is rarely achieved, panel feeders need only be sized for initial loads as detailed in Section 17.16 above, and provision made for rewiring to meet anticipated expansion, by one of the techniques listed in Section 17.18 below.

These results can be summarized as follows: For panels in buildings expecting limited expansion, for which branch circuits are loaded to 80% of capacity (25% *branch circuit* expansion, see Table 17.5, note d), the ultimate panel load without new conduit work, i.e., merely by filling out circuits, is 1.5 L, or 50% beyond the initial load. By adding breakers in the spaces, this load can be expanded to 75% beyond the initial load. The corresponding figures for panels that are lightly loaded (66% capacity, i.e., 50% branch circuit expansion as in Table 17.5 note e), in anticipation of considerable load growth, are 80 to 110%, respectively. These results are summarized in Table 17.8.

TABLE 17.8 **Panel Initial and Expanded Loads**

	Panels in Facilities Expecting Limited Expansion; Circuits Initially Loaded to Give 25% Expandability (See Table 17.5)	*Panels in Facilities Expecting Extensive Expansion; Circuits Initially Loaded to Give 50% Expandability (See Table 17.5)*
Initial load	100%	100%
Initial plus spares	120%	120%
Load after all circuits including spares are loaded to maximum allowable	150%	180%
Load after utilizing 20% spaces also	175%	210%

NOTE: For development of these figures, see McGuinness et al. (1980), pp. 690–691.

17.18 Feeder Capacity

To achieve economy, the panel feeder must accommodate the initial load plus some portion of the future load. Spare capacity in feeders (to accommodate a considerable portion of the panel spare capacity as shown in Table 17.8), is provided by one or more of the following procedures:

1. Provide feeder (and conduit) capacity initially, to handle the entire eventual load. This method is most expensive—requiring initial outlay for no return—and is rarely used.

2. Provide feeder for initial plus spare with properly sized conduit. Conduit is sized for type THW or RHW without covering. This method, as we shall see, yields very limited spare capacity.

3. Provide feeder for initial load plus spare, with conduit oversized by one size. Size conduit for type THW wire, which is very widely used because of attractive price and excellent electrical properties. Some additional costs are entailed here.

If the initial wiring is done with type T or TW wire, the effect is approximately the same as having oversized the conduit by one size and wiring initially with THW wire. This is caused by the lower ampacity rating of T and TW as compared to THW, resulting in a larger conduit size for the same initial ampacity. The reader can work out exact figures in each case with the help of Tables 15.2*a* and 15.2*b*, and conduit capacity Table 17.9, or by using the NEC tables in Article 310 and Chapter 9.

TABLE 17.9 **Maximum Number of Conductors in Trade Sizes of Conduit or Tubing**

Type Letters	*Conductor Size AWG, MCM*	*Conduit Trade Size (Inches)*									
		1/2	*3/4*	*1*	*1 1/4*	*1 1/2*	*2*	*2 1/2*	*3*	*3 1/2*	*4*
TW, T, THW, RHW, and RHH (without outer covering)	6	1	2	4	7	10	16				
	4	1	1	3	5	7	12	17			
	2	1	1	2	4	5	9	13			
	1		1	1	3	4	6	9			
	0		1	1	2	3	5	8	12		
	00		1	1	1	3	5	7	10		
	000		1	1	1	2	4	6	9	12	
	0000			1	1	1	3	5	7	10	
	250			1	1	1	2	4	6	8	10
	300			1	1	1	2	3	5	7	9
	350				1	1	1	3	4	6	8
	400				1	1	1	2	4	5	7
THHN	6	1	4	6	11	15	26				
	4	1	2	4	7	9	16				
	2	1	1	3	5	7	11	16	25		
	1		1	1	3	5	8	12	18		
XHHW (4 to 500 MCM)	0		1	1	3	4	7	10	15		
	00		1	1	2	3	6	8	13		
	000		1	1	1	3	5	7	11	14	
	0000		1	1	1	2	4	6	9	12	
	250			1	1	1	3	4	7	10	12
	300			1	1	1	3	4	6	8	11
	350			1	1	1	2	3	5	7	9
	400				1	1	1	3	5	6	8

Source: National Electrical Code.

ELECTRICITY

4. Provide feeder for initial load plus spare and oversize conduit by *two* sizes. This will yield most of the capacity necessary in facilities anticipating large expansion.

5. Provide for initial load plus spare, with an empty conduit for future. This method is expensive because of high conduit cost, and it is infrequently advisable.

In (2), (3), and (4) the future capacity beyond that initially supplied is handled by the use of larger gauge wire in the existing conduit. To examine exactly what these alternatives provide in spare capacity, we have tabulated in Table 17.10 the maximum ampacity of various size conduits and the future capacity obtainable. Table 17.9 is taken directly from the NEC.

Note from Table 17.10 that simply by rewiring we can obtain only 10 to 15% additional capacity, whereas if the conduit had been oversized the additional capacity would be from 28 to 119%.

Returning now to the question of how large to make the feeder for a given panel load, we must balance future panel load, initial cost of feeder, and future capacity of existing conduit. It is best to avoid the installation of empty conduits since this is expensive. Rewiring, however, is relatively inexpensive and oversizing conduit is the method of choice if rapid expansion is anticipated.

Referring to Table 17.10, note that normal design uses THW cable. Design with T or TW is, in effect, a first step in oversizing conduit and is generally not economical. The second step is a deliberate oversizing of conduit that results in much increased conduit ampacity, as reflected by the figures. Using these figures in actual practice requires that the designer juggle cable and conduit cost against anticipated load growth to arrive at the most economical long-term solution. Applying these numbers to concepts previously developed we have:

1. Buildings designed for 25% branch circuit expansion (see Table 17.5) have a panel capacity of 1.75 times the load. If it is desired to design the panel feeder to carry the full expansion possible, then: Calculate the feeder on the basis of panel load plus 20% spares, and oversize the feeder by 20%. This gives a feeder capacity of

$$1.20 \times 1.2 = 1.44 \times \text{initial load}$$

Table 17.10 indicates that rewiring will add another 15% on the average. This gives

$$1.15 \times 1.44 = 1.66 \times \text{initial load}$$

which corresponds sufficiently closely to the 1.75 desired.

2. For a building with 50% branch circuit expansion, utilizing full panel space capacity gives us an ultimate panel capacity of 2.1 times the initial load (see Table 17.8). This is accomplished in the following way: Oversize the feeder by 15% and oversize the conduit by one size. This latter step gives an expansion of approximately 50%. Therefore,

$$\text{feeder capacity} = 1.15 \times 1.2 \times 1.5$$
$$= 2.07$$

which is the desired figure.

TABLE 17.10 **Maximum Wire and Ampacity of a Conduit, and Ampacity Gain on Rewiring**

Initial Conduit Size (Inches)	Initial Installation THW Cable: Max. Wire	Initial Installation THW Cable: Max Amperes	Rewiring with THHN or THWN — Using Original Conduit: Max. Wire	Max. Amperes	Capacity Increase	Capacity Increase Having Oversized Conduit: One Size	Two Sizes
1½	1	130	1/0	150	15%	76%	119%
2	3/0	200	4/0	230	15%	43%	90%
2½	250	255	300	285	12%	49%	80%
3	400	335	500	380	13%	37%	60%
			2–4/0	368			
3½	600	420	700	460	10%	28%	—

If, as in the case of laboratories, more than 100% expansion is anticipated (see Table 17.1), conduit should be oversized by two sizes and initial wiring oversized by approximately 25%. Feeders thus arranged will handle the new panels required to meet the anticipated expansion.

Two factors should be carefully noted here. First, note that the smaller conduits offer the largest expandability although, in dollars per amperes, they are more expensive. Second, in order to take advantage of spaces in a panel, conduit stubs should be taken from the panel and extended into hung ceilings or another procedure used to make the panel circuitry easily accessible in the future.

17.19 Panel Feeder Load Calculation

EXAMPLE 17.1. Refer to Fig. 17.18. The panel is for a laboratory/office area. Since large expansion is anticipated, circuitry follows the bottom section of Table 17.5. Ultimate panel load would be 26 cct at 1900 W = 50 kVA = 138 A. Thus the initial feeder is sized for 115 A but rewiring will allow as much as 235 A in a 2-in. conduit (oversized by 2 sizes). (See Table 17.10.) A 225-A frame c/b is chosen initially, since eventually the trip will be raised to 150 A.

The NEC in Article 220 specifies minimum watts-per-square foot (W/ft^2) figures for various occupancies, for lighting, and for miscellaneous power loads. Proper design procedure therefore requires that after detailed design of an area, the actual loading be compared to these minima, and the larger of the figures, as regards number of circuits and feeder load, be used. An example will help to make this clear.

EXAMPLE 17.2. Assume a single floor of an office building 100 × 200 ft. Assume also that 15% of the area is corridor and storage, equally divided between the two. Calculate the load and feeder size. Assume a good grade speculative construction venture.

SOLUTION

Office space	= 85% of 20,000 ft^2
	= 17,000 ft^2
Corridor and storage	= 15% of 20,000 ft^2
	= 3000 ft^2

The NEC specifies a minimum of 3½ W/ft^2 for lighting and 1 W/ft^2 for miscellaneous receptacles in office space. It further specifies ½ W/ft^2 minimum for corridors and ¼ W/ft^2 for storage. Therefore:

Lighting

Office load:	17,000 × 3½ =	59.5 kW
Corridor:	1500 × ½ =	0.75 kW
Storage:	1500 × ¼ =	0.38 kW
Minimum lighting load	=	60.6 kW

Receptacles

17,000 × 1 W/ft^2 = 17 kW

These figures would then be compared to the actual design loads. Receptacles are counted at 180 W each, as noted in Section 17.16, item 2. If the Code minima exceed the design load (as it well may if lighting is properly designed), panels must be equipped with sufficient additional circuits to make up the difference. The number of such circuits is up to the designer, since circuit loading is not specified.

For instance, suppose the lighting design was accomplished at 2 W/ft^2. Panel circuits would have to be provided for the additional 1½ W/ft^2, as follows:

$$17{,}000\ \text{ft}^2 \times 1.5\ \text{W/ft}^2 = 25.5\ \text{kW}$$

Assuming a 30 to 50% future load expansion, we would circuit the lighting loads at approximately 1300 W per circuit (see Table 17.5). Therefore, we would provide

$$\frac{25{,}500\ \text{W}}{1300\ \text{W}} = 20\ \text{additional circuits}$$

With respect to minimum feeder load, the Code specifies that it be increased by 25% if loads are continuous (three or more hours). This requirement allows for breakers to heat up in panels while carrying continuous load, and is waived for circuit breakers that are ambient compensated, that is, are rated to carry 100% load. Since we have established 80% of the breaker rating as maximum load (see Table 17.5), *we have already accounted for this factor in circuitry but must utilize it in feeder calculation*. Assuming that the Code minima are the design loads, then, for feeder calculation:

Lighting load = 60.6 kW × 125%	= 75.75 kW
Receptacle load as above	= 17.0 kW
Minimum feeder load	= 92.75 kW
25% future load	= 23.2 kW
Design feeder load	= 116 kW

ELECTRICITY

Since this load would be divided between several panels, the building electrical design might be such that the panels were not all fed by one feeder (see Fig. 17.19). However, assuming they *were*, the feeder would be calculated in terms of 3-phase current thus:

$$I = \frac{\text{kW}}{0.360} = \frac{116}{.360} = 322 \text{ A}$$

Usng THW cable, a minimum of 400 MCM would be required. Conduit would be a minimum of 3 in., but might be increased to 3½ or 4 in., according to the considerations of spare capacity discussed in the previous section.

As an aid in computing currents, Table 17.11 lists the relevant relations.

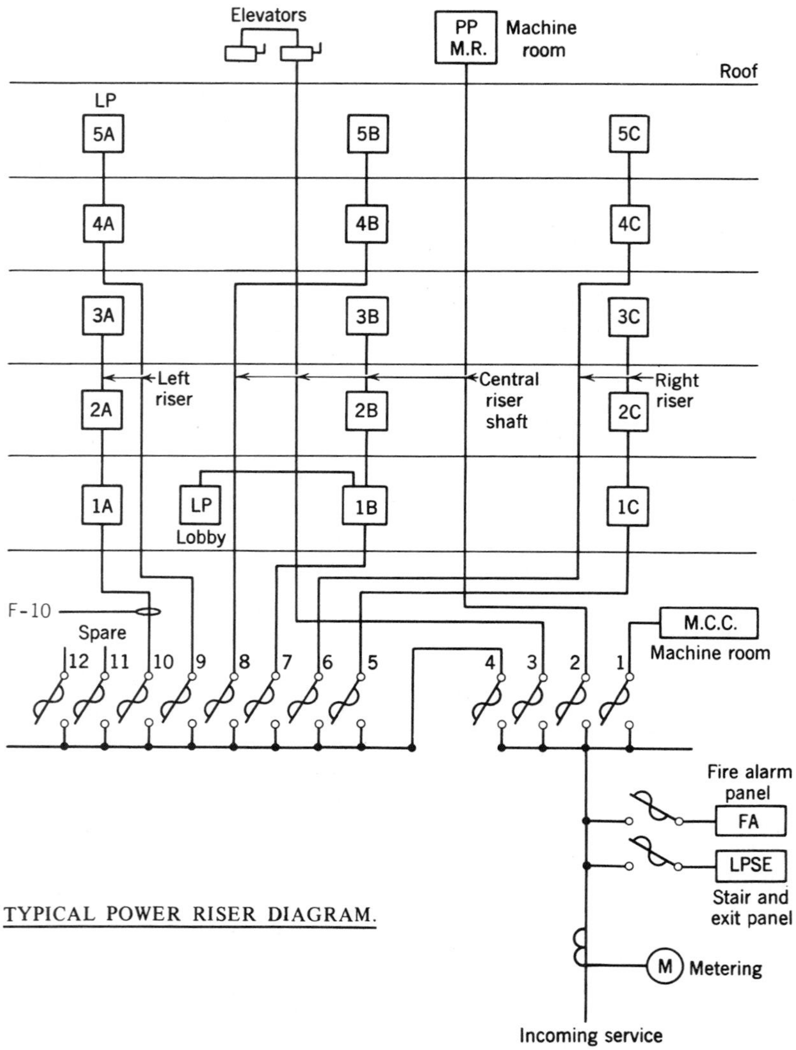

Fig. 17.19 Typical power riser diagram. Ordinarily the main switchboard would be shown as a large rectangle with the feeder emanating from it, and a switchboard schedule would detail the contents. Here, because of the unusual bus arrangement, the main switchboard appears as it would on a single-line diagram.

TABLE 17.11 **Current and Wattage Relationships**[a]

120-V Single-Phase	*120/240-V 3-Wire*	*120/208-V Single-Phase, 3-Wire*	*120/208-V 3-Phase*	*277/480-V 3-Phase*	*277-V Single-Phase*
$I = \frac{W}{120}$	$I = \frac{W}{240}$	$I = \frac{W}{208}$	$I = \frac{W}{360}$	$I = \frac{W}{830}$	$I = \frac{W}{277}$

[a]Assuming 100% power factor.

17.20 Riser Diagrams

When all devices are circuited and panels are located and scheduled, we are ready to prepare a riser diagram. A typical diagram, shown in Fig. 17.19, represents a block version of a single-line diagram except that, as the name implies, vertical relationships are shown. All panels, feeders, switches, switchboards, and major components are shown up to, but not including, branch circuiting. This diagram is an electrical version of a vertical section taken through the building.

EXAMPLE 17.3. Feeder F10 of Fig. 17.19 serves lighting panels 1A, 2A, and 3A. Calculate the required feeder size, considering loads, future expansion, and voltage drop.

SOLUTION. The loads on these panels have been computed in accordance with the above considerations and are:

Connected Load	LP-1A—100 A
	LP-2A—125 A
	LP-3A—110 A
	335 A

These figures include connected load, spares, and a 25% future factor. *The NEC requires that demand be taken per panel at 100% only for load calculated by square feet, which is lower than actual circuited loads.* If actual panel circuit loads are larger than the NEC minima, then they are utilized and any reasonable demand may be used, provided that at no time do the loads drop below the minima specified in the NEC. Therefore, we may apply diversity factors between panel loads in a judicious manner.

In office building work, typical diversity factors are

Lighting Panels Fed from a Single Feeder	*Diversity Factors*
1, 2	1.00
3, 4	1.09
5, 6, or 7	1.18
8, 9, or 10	1.33

Thus the load on feeder F-10, using 100% demand per panel and 1.09 diversity between panels, would be 335 × 1.0/1.09 = 307 A.

Methods for handling future expansion were discussed above. In this case the feeder, before voltage drop considerations, would be (from Tables 15.2 and 17.9) 4-350 MCM THW in 3 in. C. Note from Table 17.10, however, that 3 in. C gives a low percentage increase in rewiring. In this case, the wise choice would be to increase the feeder conduit to 3½ in., giving an eventual rewired capacity of 460 A, a 50% increase.

The final consideration in sizing a feeder is voltage drop. The recommended voltage drop from source to final panel should be no more than 1% for lighting and combination lighting and power feeders and 3% for power and heating load feeders. These figures should be adhered to as closely as possible. In instances of long runs where these restrictions will cause excessive cost, lighting feeders up to the branch panel may be run with 2% drop and power feeders up to 4% voltage drop. Total voltage drop to the last outlet in the circuit should not exceed 5%. Many tables and curves are published by manufacturers from which voltage drop can be obtained. Such a set of curves is shown in Fig. 17.20, which shows maximum length

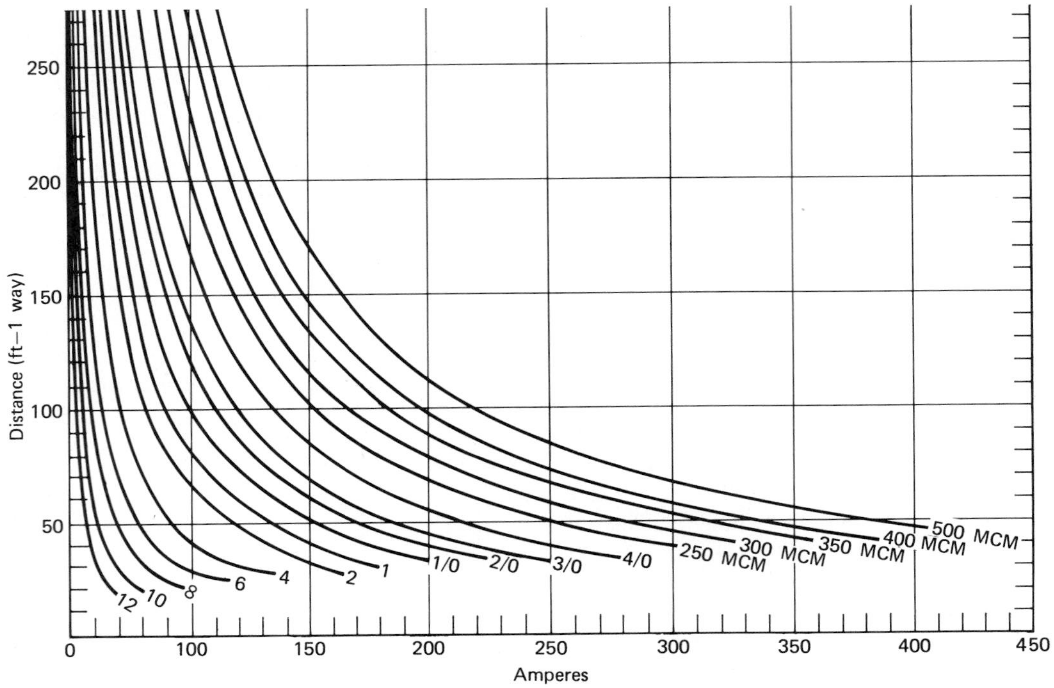

Fig. 17.20 Curves for determining voltage drop in copper cables. Curves show maximum one-way circuit length for 1% voltage drop.

of run for a 1% drop. Applying these curves to our last example:

Allowable voltage drop

= 1% of 208 V = 2.08 V

Distance—assume 80-ft run

From the curves, 307 A on 350 MCM cable will give a 1% drop in 50 ft. Therefore, the drop in 80 ft will be 1.6%, which is tolerable.

In summary, then, feeders are sized in accordance with load (actual or square feet, whichever is larger) and voltage drop. Conduit may be oversized for large future load expansion.

17.21 Service Equipment and Switchboard Design

The main switchboard shown in Fig. 17.19 constitutes a combination of service equipment and feeder switchboard. The service equipment portion of the board comprises the metering and the four main switches feeding risers, motor control center (MCC), roof machine room, and elevators. The feeder board comprises switches 5 through 12. Such an arrangement is permissible inasmuch as the NEC allows up to six fused switches or circuit breakers to serve as the service disconnect means. This arrangement was chosen in order to separate to the largest extent possible the motor loads (elevators, air-conditioning equipment, basement power, etc.) from the lighting. Such a procedure minimizes lighting fluctuations resulting from motor starting and yields simpler maintenance. Also, the size of the main switch is reduced. This switchboard would be of the metal-clad dead-front type with switches or circuit breakers, as desired.

Other considerations and general rules affecting service equipment are listed below.

1. A building may be supplied at one point by either a single set or parallel sets of service conductors.
2. Service drops may generally be not less than No. 8 AWG and service entrance conductors or underground service conductors not less than No. 6 AWG.

3. All equipment used for service including cable, switches, meters, and so on, shall be approved for that purpose.
4. It is recommended that a minimum of 100-A, 3-wire, 120/240-V service be provided for all individual residences.
5. No service switch smaller than 60 A or circuit breaker frame smaller than 50 A shall be used.
6. In multiple-occupancy buildings tenants must have access to their own disconnect means.
7. All building equipment shall be connected on the load side of the service equipment except that service fuses, metering, fire alarm and signal equipment, and equipment serving emergency systems may be connected ahead of the main disconnect (see Fig. 17.19).

17.22 Emergency Systems

Some of the considerations relevant to power reliability were discussed in Section 17.1(*b*), and a brief review of the equipment available to supply emergency power was presented in Section 16.29. Emergency lighting equipment is covered in Section 21.29. In this section, we will discuss possible arrangements of emergency power supply. The choice of arrangement and the size and type of equipment depends in large measure on the requirements of local codes, which determine the loads to be fed from the emergency system. The reader should note that although we are using the term *emergency,* the concepts involved are equally applicable to standby systems.

In general, when emergency power is discussed, it is assumed to be replacing "normal" power; that is, the assumption underlying most codes and ordinances is that power must be supplied to selected loads within the building because of a utility power outage. Cognizance must also be taken of situations in which normal power has not failed and the outage is localized because of an equipment failure. That aspect of design—reliability—is left to the designer. Some of the arrangements that will be discussed below differentiate between the nature of outages, that is, a utility or general outage versus an equipment or local outage. An exception to the above generalization occurs with health-care facilities, where the NEC in Article 517 specifies an internal electrical design that will in large measure cover both types of outages, down to the distribution level. The interested reader should refer to the referenced NEC article for further reading.

The emergency system includes all devices, wiring, raceways, and other electrical equipment, including the emergency source that is intended to supply electric power to the selected loads. These loads normally include egress lighting (stair, corridor, and exit and lobby lights), signal equipment such as public address and fire alarm that must remain functional during an emergency, and one or more elevators as required by Code. The recognized arrangements are discussed below.

An important change in Code requirements permits the use of a single power source for (1) emergency, (2) legally required standby, and (3) optional standby systems provided it is equipped with automatic selective load pickup and load-shedding equipment that will assure adequate power to the three types of systems in the order of priority stated (1, 2, 3).

1. Where emergency loads are light, a storage battery arranged to be connected automatically on power outage may be used. Where all emergency loads can be supplied with direct current, the arrangement of Fig. 17.21*a* is used. (Note that a-c lighting can accept d-c emergency power if equipped with a local inverter, as discussed in Section 21.29.) If a-c is required, the arrangement of Fig. 17.21*b* is utilized. When the emergency equipment is entirely separate from the normal equipment and is *normally deenergized,* the system of Fig. 17.21*c* is used. This arrangement is used in small facilities requiring egress lighting only, where it is found that supplying a completely separate emergency system is the preferred economic or engineering choice. Large battery installations are used where uninterrupted power is required, as is generally the case in computer installations where no power interruption, however short, can be tolerated. These systems are highly technical and beyond our scope here.

2. Where emergency loads are larger than can be supplied economically by batteries, and where 10-s start-up time is tolerable, a generator set is employed. The prime mover should preferably use

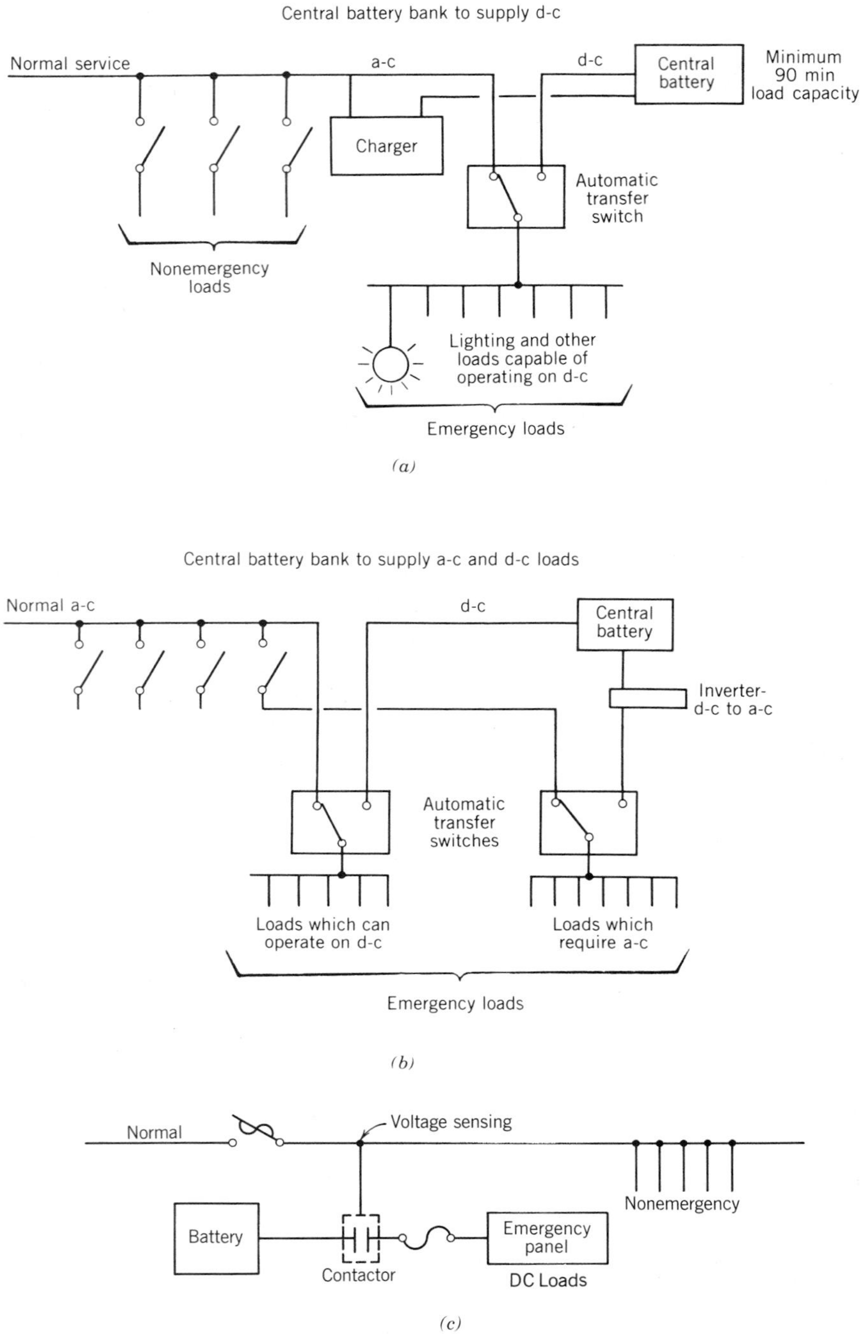

Fig. 17.21 *Use of a central battery for emergency power supply. Where all loads can be energized with d-c the arrangement of* (a) *is satisfactory. When a-c as well as d-c must be distributed, a central inverter is added as in* (b), *and the a-c and d-c emergency loads fed separately. In* (c) *the emergency loads are normally deenergized and activated through the contactor when it senses power loss.*

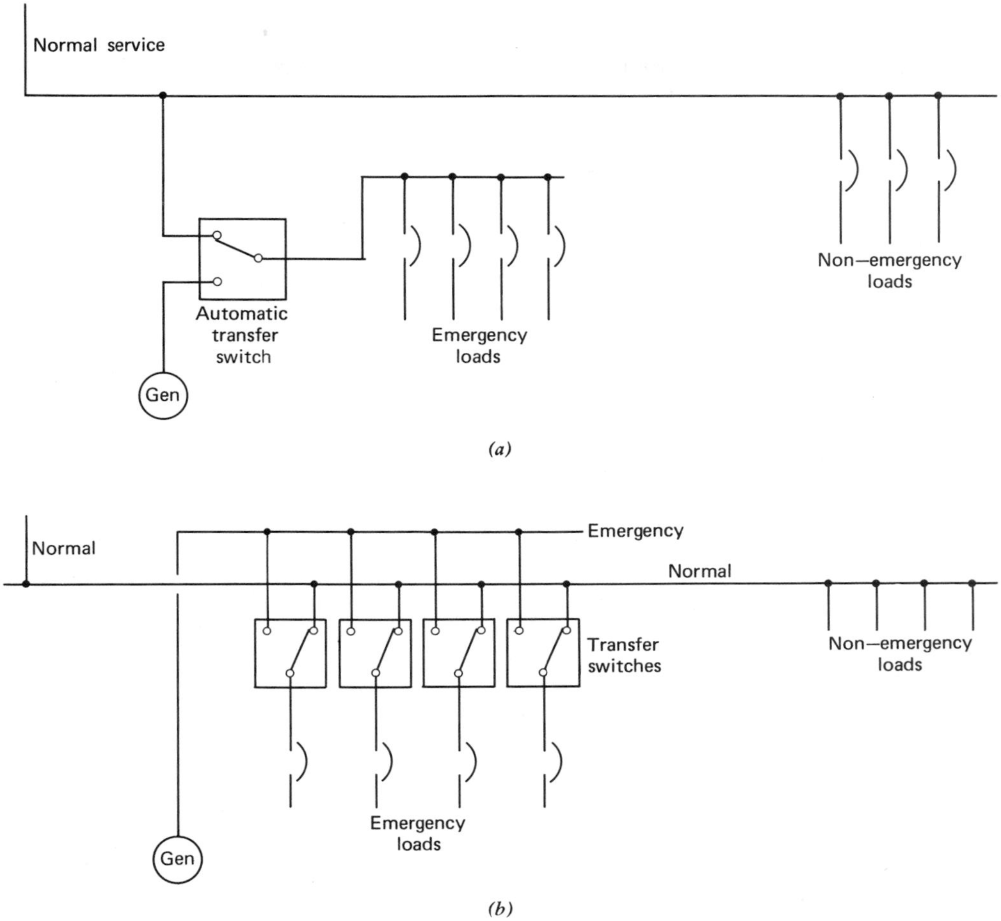

Fig. 17.22 *Alternate arrangements of emergency/normal power feed. In* (a) *a single transfer switch serves the normal power and transfers to the generator on power failure. In* (b) *the transfer switches are smaller, thus reducing the chance of a single equipment failure faulting out the entire emergency power system.*

on-site fuel (gasoline, diesel, bottled gas) for reliability. However, off-site fuel (utility gas, utility steam) may be used when the utility records indicate that simultaneous failure of electric and gas/steam service is rare. On-site fuel storage must be sufficient for 2 h of full-load operation. (It should be pointed out that a combination of sources can be used in a single building. For instance, a generator can supply bulk power loads and a battery installation selected lighting loads.)

The system can be arranged with a single transfer switch that senses normal power loss, as in Fig. 17.22*a*, or it can use multiple switches, each one of which will sense power loss *at its downstream location,* as in Fig. 17.22*b*. The latter system provides greater power reliability, provided that the design is such that the emergency power uses an independent power path to the transfer switches. Otherwise, a faulted piece of equipment that will interrupt normal power downstream will also prevent emergency power from reaching that point.

3. Many codes permit the use of two separate electric services in lieu of a normal service plus an emergency source, provided that the two sources are independent, that is, come from different utility transformers or feeders, enter the building at

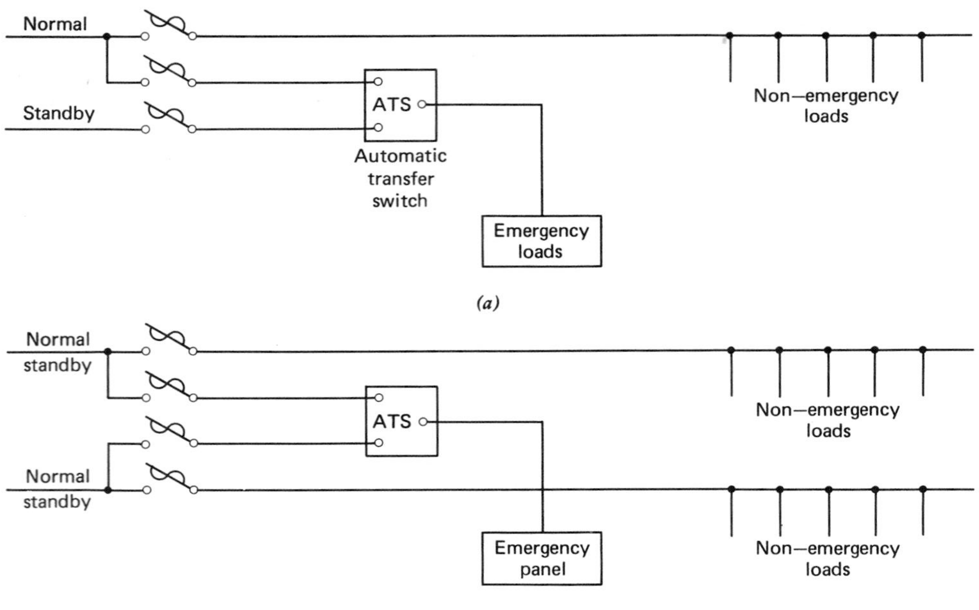

Fig. 17.23 *Emergency power is supplied by dual service. The arrangement shown in* (a), *where one service acts only as a standby, is more common than that of* (b), *where both supply normal loads and each acts as a standby for the other.*

different points and preferably from different directions, and use separate service drops or laterals. The point is, of course, that the type of reliability desired can be obtained only by minimizing the possibility of a single event interrupting both services. The usual arrangement is for one service to be ''normal'' and the other ''standby,'' as in Fig. 17.23*a*. A much less frequent case utilizes both feeders as ''normal,'' each carrying part of the normal load and each acting as a standby for the other. This is shown in Fig. 17.23*b*.

4. The least reliable arrangement is one in which the emergency loads are connected ahead of the main disconnects and are so arranged that a downstream fault within the building will not affect these items. This situation is illustrated in the riser diagram of Fig. 17.19 where the stair and exit panel, which supplies egress lighting, and the fire alarm panel are connected ahead of the building main disconnect and protected with their own fuses. Such an arrangement obviously can do nothing in the event of a power outage and, although once very popular, it is now falling into disuse as a result of more stringent codes.

5. The NEC recognizes a category of equipment for emergency illumination called ''unit equipment.'' These devices, discussed and illustrated in Section 21.29, consist of individual self-contained packages with battery, charger, and light source *permanently* mounted and wired at required locations. The panel device feeding these units should be capable of being locked or so arranged as to be accessible to authorized personnel only.

A portion of a typical stair and exit riser is shown in Fig. 17.24. Circuits connected to this panel are not switched, being of the constant burning type, unless switches are accessible to authorized persons only. Emergency system wiring (as the term is defined by Code) must be kept entirely independent of all other wiring and equipment and should not occupy the same enclosure or conduit as normal system wiring, except in dual-fed units

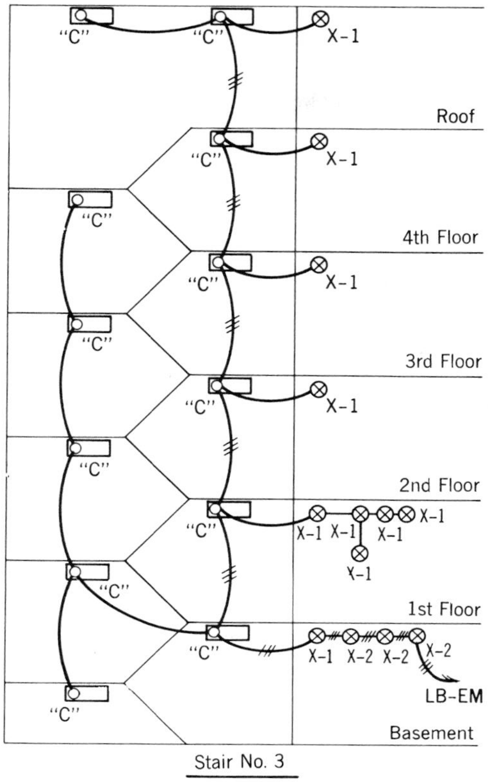

such as transfer switches. Standby system wiring is not subject to this limitation, and may share raceways and other enclosures with general wiring.

References

McGuinness, W. J., Stein, B., and Reynolds, J. S. (1980). *Mechanical Equipment for Buildings*, Wiley, New York.

Fig. 17.24 *Portion of a typical stair-and-exit riser. Panel LB-EM represents* Lighting panel, Basement, EM*ergency.*

PART VI
ILLUMINATION

It has been estimated that 90% of the information we obtain from all sources is received via our sense of sight. Vision, in turn, is made possible by the presence of light, the proper provision of which is discussed in this part of the book. Since architecture is a uniquely visually oriented profession, its practitioners must thoroughly understand the art and science of illumination in order to be able to integrate structure and light into a working whole that will appear, and function, according to the design intent.

Chapter 18, Lighting Fundamentals, introduces the subject with terminology, definitions, and basic characteristics and measures. A clear distinction is made between the physical and visual–photometric characteristics of light. The former treat light as a form of energy whereas the latter consider it as a visual phenomenon (i.e., the physiological response to a form of energy). Pursuing the subject of human response to light, the chapter continues with a discussion of factors in visual acuity. This subject is concluded with an explanation and tabulation of American and European lighting level recommendations.

The discussion continues with consideration of factors in lighting quality (as opposed to quantity), with particular emphasis on analysis and control of reflected glare. The chapter concludes with material on color vision, color in light, and color temperatures of light sources.

Chapter 19, Light Sources: Their Characteristics and Application, covers two basic subjects: daylighting and electric light sources. The daylighting section begins with a discussion of the characteristics of natural light and then proceeds to explain and demonstrate, with illustrative examples, a number of daylighting design methods. These are both graphic and analytic, American and international. The second subject, electric light sources, covers in detail incandescent, fluorescent, mercury, metal-halide, and sodium lamps. The discussions include a detailed description of construction and operating characteristics, including lamp accessories such as ballasts and lampholders. Emphasis is placed on luminous efficiency and overall life-cycle costs. The chapter concludes with a discussion of chromaticity of light sources, their spectral distribution, and color rendering index.

Chapter 20, Lighting Design, leads the reader step by step through the accepted design procedure methodology. The first section, on preliminary design, covers cost factors, power budgets, energy considerations, and illumination methods. This is followed by a section on detailed design, which discusses lighting fixtures characteristics and illumination calculation techniques. The chapter concludes with a discussion of lighting design evaluation. Illustrative examples are used to demonstrate and compare calculation techniques.

Chapter 21, Lighting Application, applies the knowledge and techniques explicated in the three preceding chapters to specific occupancies, each of which has its special needs, namely: residential, educational, commercial, and industrial facilities. In addition, important special topics are discussed in detail, including emergency lighting, exterior lighting, and lighting for computer terminals. Finally, lighting control strategies and techniques are covered in detail from the points of view of energy conservation, economics, and automatic controls.

18

LIGHTING FUNDAMENTALS

> Architecture is the masterly, correct and magnificent play of masses brought together in light. Our eyes are made to see forms in light; light and shade reveal these forms; cubes, cones, spheres, cylinders, or pyramids are the great primary forms which light reveals to advantage. *Le Corbusier*

18.1 Introductory Remarks

For many years an artificial division existed in the field of lighting design, dividing it into two disciplines: architectural lighting and utilitarian design. The former trend found expression in design that took little cognizance of visual task needs and displayed an inordinate penchant for incandescent wall washers and architectural lighting elements. The latter trend saw all spaces in terms of cavity ratios and performed its design function with footcandles and dollars as the ruling considerations. That both these trends have in large measure been eliminated is due largely to the efforts of thoughtful architects, engineers, and lighting designers, assisted in part by the energy consciousness that followed the 1973 Arab oil embargo. The latter spurred research into satisfying real vision needs within a framework of minimal energy use.

Another positive factor in the rationalization of lighting design has been the work of the Illuminating Engineering Society of North America (hereafter referred to as IES). Its activities in research, standardization, and publication have done much to place lighting design on a stable scientific basis while taking full cognizance of its essential artistic aspects. It is precisely this combination of science and art that makes interior lighting design an architectural-type discipline. The responsible lighting designer must consider quantitatively:

1. Daylight—its introduction and integration with artificial light.
2. The interrelationship between the energy aspects of artificial and natural lighting, heating, and cooling.
3. The effect of lighting on interior space arrangement and vice versa.
4. The characteristics, means of generation, and utilization techniques of artificial lighting.
5. Visual needs of specific tasks.
6. The effects of brightness patterns on visual acuity.

And qualitatively, the designer must consider:

7. The location, interrelationship, and psychological effects of light and shadow, that is, brightness patterns.
8. The use of color, both of light and of surfaces, and the effect of illuminant source on object color.
9. The artistic effects possible with color and patterns of light and shadow including the changes inherent in daylighting, and so on.

The list is almost endless because so much of the information we receive from our senses comes via our eyes, and what we see is a direct consequence of scene lighting.

As a result of the need to consider these and other interrelated factors, many of which are mutually incompatible, the lighting designer is faced with many difficult decisions. The purpose of the lighting chapters in this book is then twofold: to provide the background that will help the lighting designer make these decisions correctly, and to make him or her proficient in the use of lighting as a design material.

PHYSICS OF LIGHT

18.2 Light as Radiant Energy

The IES defines light as visually evaluated radiant energy or, more simply, a form of energy that permits us to see. If light is considered as a wave, similar to a radio wave or an alternating current wave, it has a frequency and a wave length. Figure 18.1 shows the position of light in the wave spectrum with relation to other wave phenomena of various frequencies.

From the chart we see that even the longest wavelength light (red) is a much higher frequency than radio and radar, and that visible light constitutes only a very small part of the wave energy spectrum.

Color is determined by wavelength. Starting at the longest wavelengths with red, we proceed through the spectrum of orange, yellow, green, blue, indigo, and violet to arrive at the shortest visible wavelengths (highest frequency).

When a light source produces energy over the entire visible spectrum in *approximately* equal quantities, the combination appears white (as is the case with daylight), whereas a source producing energy over only a small section of the spectrum produces its characteristic colored light. Examples are the blue-green clear mercury lamp and the yellow sodium lamp. Chromaticity of light sources is discussed in Chapter 19.

18.3 Transmittance and Reflectance

Lighting design is possible because light is predictable, that is, it obeys certain laws and exhibits certain fixed characteristics. Although some of these are so well known as to appear self-evident, a review is in order.

The *luminous transmittance* of a material such as a fixture lens or diffuser is a measure of its capability to transmit incident light. By definition, this quantity, known variously as *transmittance, transmission factor,* or *coefficient of transmission,* is the ratio of the total transmitted light to the total incident light. In the case of incident light containing several spectral components passing through a material that displays selective absorption, this factor becomes an average of the individual transmittances for the various components and must be used cautiously. A piece of frosted glass and a piece of red glass may both have a 70% transmission factor but obviously affect the incident light differently. In general then, transmission factors should be used only when referring to materials displaying nonselective absorption, that is, those that transmit the various component colors equally. Clear glass, for instance, displays a transmittance between 80 and 90%, frosted glass between 70 and 85%, and solid opal glass between 15 and 40%. The remainder is absorbed and reflected. See Table 19.5 for typical transmission factors.

Similarly, the ratio of reflected to incident light

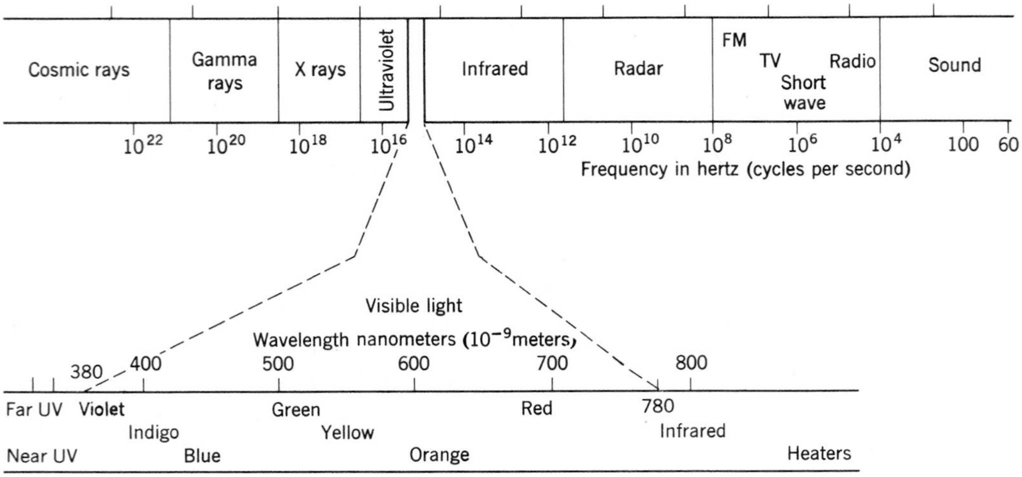

Fig. 18.1 Electromagnetic spectrum. See Chapter 19 for chromaticity diagrams of various light sources.

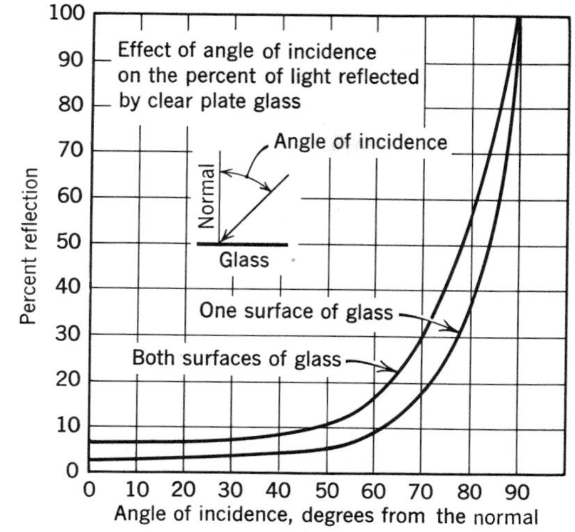

Fig. 18.2 Relation between angle of incidence and percent reflectance. This effect is important when considering the penetration of sunlight into interior spaces and, conversely, the exterior glare produced by reflection of the sun from building windows. See Fig. 3.11.

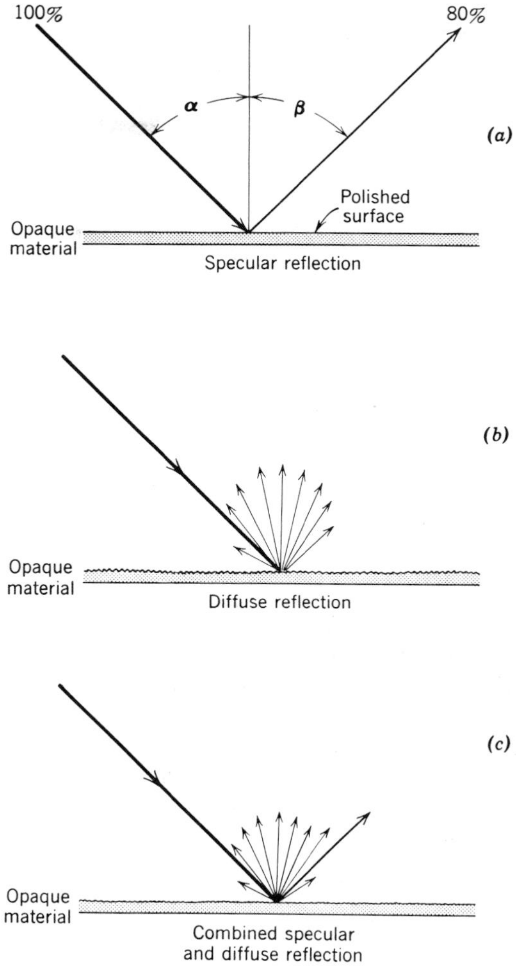

Fig. 18.3 *Reflection characteristics.* (a) *In specular reflection angle of incidence equals angle of reflection (α = β). Since 80% of light is reflected, reflectance is 80%; 20% of light is absorbed.* (b) *In diffuse reflection, incident light is spread in all directions by multiple reflections on the unpolished surface. Such surfaces appear equally bright from all viewing angles.* (c) *Most materials exhibit a combination of specular and diffuse reflection. Such a surface will mirror the source while producing a bright background.*

is variously called *reflectance, reflectance factor,* and *reflectance coefficient.* Thus if half the amount of light incident on a surface is bounced back, the reflectance is 50%, or 0.50. The remainder is absorbed, transmitted, or both. The amount of absorption and reflection depends on the type of material and the angle of light incidence (since light impinging upon a surface at grazing angles tends to be reflected rather than absorbed or transmitted; see Fig. 18.2). An example of almost perfect reflection from an opaque surface would be that from a well-silvered mirror, while almost complete absorption takes place on an object covered with lamp black or matte finish black paint. The effect of the material finish on reflection is shown in Fig. 18.3. See Table 19.7, and Table 20.3 for typical reflectance values. Reflectance measurement is discussed in Section 18.11.

If the reflection takes place on a smooth surface such as polished glass or stone, it is called specular reflection, as in Fig. 18.3*a*. If the surface is rough, multiple reflections take place on the many small projections on the surface, and the light is diffused as in Fig. 18.3*b*. Since the reflectance is a measure of total light reflected, it does not depend on whether the reflection is specular or diffuse, or a combination of both, as shown in Fig. 18.3*c*. Diffuse transmission takes place through any translucent source such as frosted glass, white glass, milky plexiglas, tissue paper, and so on. This diffusing principle is widely employed in

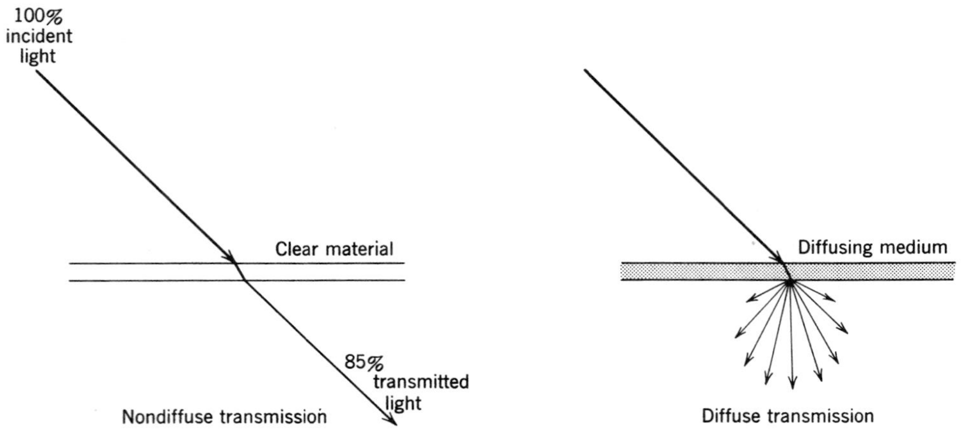

(*a*) (*b*)

Fig. 18.4 *Transmission characteristics.* (a) *In nondiffuse transmission, the light is refracted (bent) but emerges in the same beam as it enters. Clear materials such as glass, water, and certain plastics exhibit this type of transmission. In the instance illustrated the transmittance is 85% (the remaining 15% is reflected and absorbed.) The source of light is clearly visible through the transmitting medium.* (b) *With diffuse transmission, the source of light is not visible and, in the case of multiple sources, the diffusing surface will exhibit generally uniform brightness if the spacing between the light sources does not exceed 1½ times their distance from the material.*

lighting fixtures to spread the light generated by the bulb or tube within the fixture. Diffuse and nondiffuse transmission are illustrated in Fig. 18.4*a* and 18.4*b*.

18.4 Terminology and Definitions

Before beginning any discussion of lighting studies, techniques, and effects, it is important to have a basic understanding of the physical concepts and terminology involved and their interrelations. We will use and explain both the conventional American Standard (AS) and the SI systems of units. The latter, also called the metric system, is used as the basic system by the IES and in this book, whereas the lighting industry largely uses the AS system as its norm. To familiarize the reader with the SI system, we will frequently use duplicate units, with the second system's unit enclosed in square brackets [ ].

18.5 Luminous Intensity

The AS unit of *luminous intensity* is the candlepower [SI-candela], abbreviated cp [cd], and normally represented by the letter "I." It is analogous to pressure in a hydraulic system and voltage in an electric system and represents the force that generates the light that we see. An ordinary wax candle has a luminous intensity horizontally of approximately one candlepower [candela], hence the name. The candela and candlepower have the same magnitude. Luminous intensity is a characteristic of the source only; it is independent of the visual sense.

18.6 Luminous Flux

The unit of luminous flux, in both SI and conventional units, is the lumen, abbreviated lm. If we take a one candlepower [candela] source that radiates light equally in all directions and surround it with a transparent sphere of one foot [meter] radius (see Fig. 18.5*a*), then *by definition* the amount of luminous energy (flux) emanating from one square foot [meter] of surface on the sphere is one *lumen* [lumen]. Since there are 4π ft^2 [m^2] surface area in such a sphere, it follows that a source of one candlepower [candela] intensity produces 4π or 12.57 lm. The lumen, as luminous flux, or quantity of light, is analogous to flow in

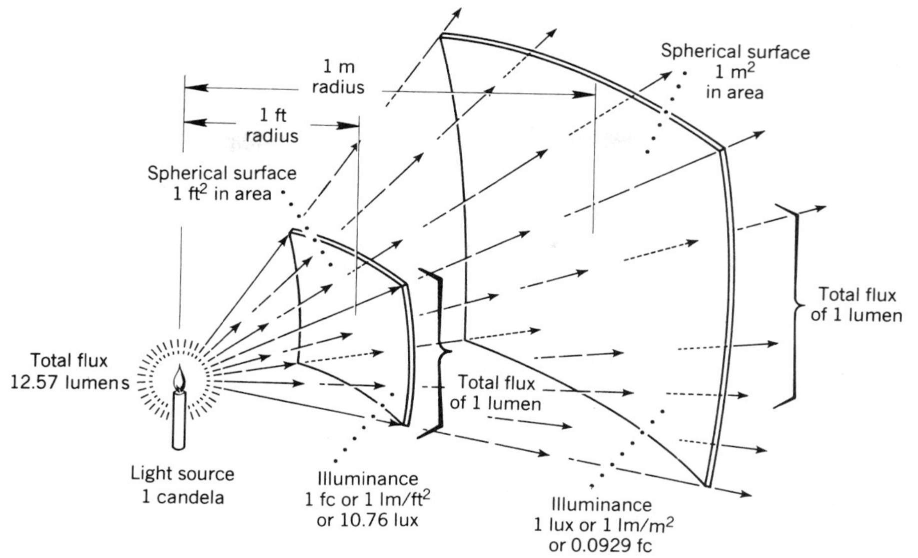

Fig. 18.5 A source of one candela intensity will produce 4Π (12.57) lumens of light flux. Thus each square foot [square meter] of (spherical) surface surrounding such a source will receive one lumen of light flux. This quantity of light will produce an illuminance of one footcandle [lux] on the spherical surface.

hydraulic systems and current in electric systems and is normally represented by the letter Φ.

In physical terms, the lumen is a unit of power, like the watt. However, unlike the watt, which is a radiometric unit directly convertible to other power units such as Btu/h, the lumen is a measure of *photometric* power. This means light power as perceived by the human eye and is therefore a function of human physiology. Since the visual response of the eye is frequency dependent, the apprehended light power is also frequency dependent and therefore varies with the spectral content of the impinging light. Figure 18.6 should clarify this concept. Figure 18.6*a* shows the spectral content of the visible energy produced by a 500-W incandescent lamp. Measured radiometri-

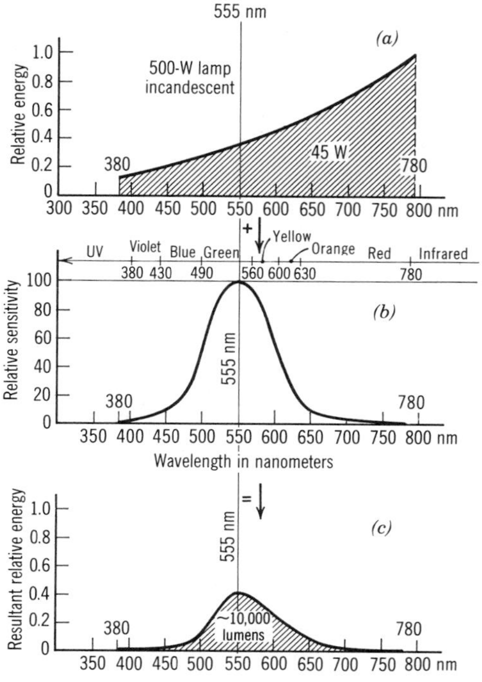

Fig. 18.6 Graphical demonstration of the method by which the unit of light flux is defined. (a) *Shows the spectrum of the light produced by a 500-W incandescent lamp. It amounts to approximately 45 W, measured radiometrically. When filtered by the human eye whose spectral sensitivity curve is given in* (b), *this light power is perceived as shown in* (c). *This new light power curve is expressed in lumens and indicates the quantity of light, as perceived by the eye.*

cally, it amounts to 45 W. However, when passed through a selective filter (Fig. 18.6*b*), which is effectively what happens when the light enters the eye, the resultant "understood" light power appears as in Fig. 18.6*c*, and therefore can no longer be measured in watts. Instead, we use a unit of eye-perceived, or photometric, power called the lumen. It should be obvious that if the spectral content curve in Fig. 18.6*a* were differently shaped, *even if the total radiometrically measured power were the same,* the resultant perceived power in Fig. 18.6*c* would be different.

A correlation can be made between photometric and radiometric power at the point of maximum response of the eye, which occurs at 555 nanometers wavelength. (A nanometer is 10^{-9} m. The older unit, now obsolete, is the Angstrom (Å), which is 10^{-8} cm, or 10^{-10} m, i.e., 1⁄10 of a nanometer.) (See Fig. 18.6*b*). One watt of monochromatic light at that wavelength will produce 683 lumens. However, since common light sources such as incandescent, fluorescent, mercury, etc., are not monochromatic, but produce light in many parts of the spectrum, (see Fig. 19.60), no single conversion factor between watts and lumens exists. Each source has its own luminous efficiency (lumens/watt), determined by its spectrum. For the 500-W lamp used as an illustration in Fig. 18.6, its luminous efficiency (efficacy) is 10,000 lm/500 watt or 20 lumens per watt (lpw).

ILLUMINATION

18.7 Illuminance

One lumen of luminous flux, uniformly incident on one square foot of area, produces an *illuminance* of one *footcandle* (fc). Illuminance is normally represented by the letter "*E*." Restated, illuminance is the density of luminous power, expressed in terms of lumens per unit area. If we were to consider a light bulb to be analogous to a sprinkler head, then the rate of water flow would be the lumens and the amount of water per unit time per square foot of floor area would be the footcandles. The acoustic analog is intensity in watts per area. In SI units, the area is expressed in square meters and the illuminance (illumination) in lux (lx). Thus the SI unit, lux, is smaller than the corresponding unit, footcandles, by the ratio of square feet to square meters. That is,

$$10.764 \text{ lux} = \text{one footcandle}$$

or multiply footcandle by 10.764 to obtain lux. These relations are shown in Fig. 18.5. Restating the above mathematically,

$$\text{footcandles} = \frac{\text{lumens}}{\text{square foot of area}}$$

$$\text{fc} = \frac{\text{lum}}{\text{ft}^2} \tag{18.1}$$

and

$$\text{lux} = \frac{\text{lumens}}{\text{square meters of area}}$$

$$\text{lx} = \frac{\text{lm}}{\text{m}^2} \tag{18.2}$$

As an *approximation* (8% error)

$$10 \text{ lx} \simeq 1 \text{ fc} \tag{18.3}$$

EXAMPLE 18.1. A 40-W, 430-mA (milliampere), 48-in. [122 cm] fluorescent tube produces 3200 lm. What is the illuminance on the floor of a 10-by-10. ft room assuming 40% overall efficiency and uniform illumination?

SOLUTION.

$$\text{Useful lumens} = 0.4 \times 3200 = 1280$$

$$\text{fc} = \frac{1280}{10 \times 10} = 12.8$$

$$\text{lx} = 12.8 \times 10.76 = 137.7$$

Calculating lux directly:

$$\text{lx} = \frac{1280}{10 \times 10} \times (3.28 \text{ ft/m})^2 = 137.7$$

By approximation:

$$\text{lx} \simeq 10 \cdot \text{fc} \simeq 128$$

Illuminance at a *point* can be computed from intensity, as shown in Section 18.12.

18.8 Luminance and Brightness

An object is perceived because light deriving from it enters the eye. The impression received is one of object *brightness*. This brightness sensation, however, is subjective, and depends not only on the object *luminance* (*L*) but also on the state of adaptation of the eye (see Sections 18.18 and 18.19). For this reason the physiological sensation is generally referred to in the literature as subjective or apparent brightness or simply *brightness*, whereas the measurable, reproducible state of object luminosity is its *luminance* (formerly "photometric brightness"). Luminance is normally defined in terms of intensity; it is the luminous intensity per unit *apparent* (projected) area of a primary (emitting) or secondary (reflecting) light source. Thus its units are candela per area. Specifically, the SI unit of luminance is candela per square meter (cd/m^2), sometimes referred to as the *nit*. The conventional (AS) unit is the footlambert, defined as

$$\text{fL} = \frac{1}{\pi}\frac{\text{cd}}{\text{ft}^2} \tag{18.4}$$

Other luminance terms such as stilb, apostilb, blondel, millilambert, and candela per square in. are best avoided. In this book, the term *luminance* will be used except where it is specifically intended to refer to the physiological sensation involved, in which case the terms *brightness*, *subjective brightness*, or *apparent brightness* will be used.

Luminance can also be defined in terms of luminous flux (see Fig. 18.7). A surface reflecting, transmitting, or emitting one lumen per square foot of area, *in the direction being viewed*, has a luminance of one footlambert (1 fL). This latter qualification is important since many surfaces (some fabrics, for instance) exhibit different luminances at different angles. Similarly, a surface emitting (directly or by reflection) one lumen per square meter has a luminance of $(1/\pi)(\text{cd/m}^2)$. Luminance has no readily conceivable mechanical or electrical analogy. Conversion factors for conventional (AS) and metric (SI) lighting units are given in Table 18.1.

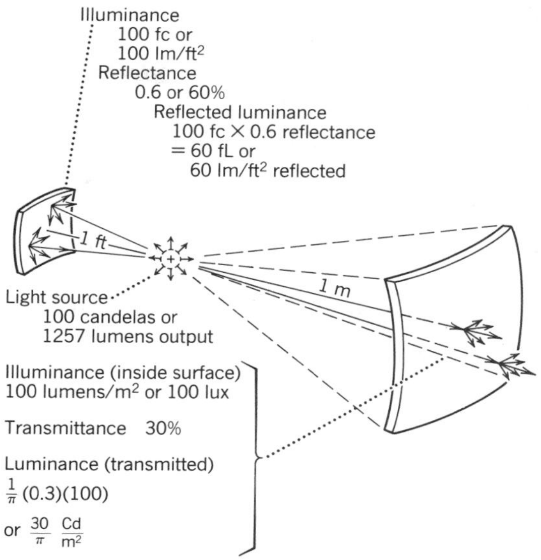

Fig. 18.7 Luminance may be either reflected or transmitted. In the former case it is calculated as the product of the incident lumens and the reflectance; in the latter, as the product of the incident lumens and the transmittance.

Since object luminance is that which is visually perceived and is a prime factor in visibility (and glare), it is important that the reader be able to perform basic luminance calculations. Although the eye does not differentiate between primary sources that generate and emit light, and secondary sources that derive their luminance from reflection or transmission, the differentiation is important in calculation procedures.

TABLE 18.1 **Lighting Units—Conversion Factors**

Unit	*Multiply*	*By*	*To Obtain*
Illuminance (*E*)	Lux	0.0929	Footcandle
	Footcandle	10.764	Lux
Luminance (*L*)	cd/m^2	0.2919	Footlambert
	Footlambert	3.4263	cd/m^2
Intensity (*I*)	Candela	1.0	Candlepower

Luminance of an emitting surface:

EXAMPLE 18.2.

(a) Calculate the luminance of a standard inside-frosted, 100-W incandescent light bulb (an A-19 bulb—see Fig. 19.32 for characteristic) with a maintained output of 1700 lm.

(b) Assume that an opal glass globe of 8 in. diameter (20 cm) and a transmittance of 35% surrounds the above bulb. Calculate the luminance of the globe. Use SI units throughout.

SOLUTION.

(a) Assume that the filament is a point source and that the inside frosting of the glass does not reduce the output (it does by about 1%). The definition of a point source tells us that one candela produces 4π lumens, distributed spherically. Therefore:

$$\frac{1 \text{ candela}}{4\pi \text{ lm}} = \frac{I \text{ candela}}{1700 \text{ lm}}$$

and the intensity, I, of the A-19 bulb is

$$I = 1700 \text{ lumens} \times \frac{1 \text{ candela}}{4\pi \text{ lm}} = 135 \text{ candela}$$

From Fig. 19.32, the A-19 bulb has a diameter of 2⅜″ [6 cm]. From the definition of luminance:

$$L = \frac{I}{A} \frac{\text{(intensity)}}{\text{(projected area)}}$$

The projected area becomes approximately a circle with a diameter of 6 cm (assuming a spherical bulb), whose area equals πr^2 or $\pi(.03 \text{ m})^2$. So:

$$L = \frac{135 \text{ candela}}{\pi(.03 \text{ m})^2} = 47{,}750 \frac{\text{candela}}{\text{m}^2}$$

This is a potential source of severe direct glare. See Section 18.25.

(b) We have already calculated the intensity of the source, but it is reduced to 35% by the globe. The projected area of the globe is larger than that of the bulb, that is,

$$\pi(\frac{.20}{2} \text{ m})^2$$

So the expression for luminance becomes:

$$L = \frac{135 \text{ candela } (0.35)}{\pi(.10 \text{ m})^2} = 1500 \frac{\text{candela}}{\text{m}^2}$$

This is no longer a potential source of direct glare.

EXAMPLE 18.3. Calculate the luminance of a standard 4-ft cool white fluorescent lamp, F40T12CW; see Table 19.11. Use SI units.

SOLUTION. Use the 2770 lm at 40% life as an average condition of bulb output. The luminous length is 4 ft (120 cm) and the diameter is 1½ in. or (3.8 cm). The luminous surface area of the tube is then:

$$A = L \times d \times \pi = (120 \text{ cm}) \times (3.8 \text{ cm}) \times \pi = (1.2 \text{ m})(.038 \text{ m})\pi$$

From the definition of lumen, we know that a source that emits one lumen per square meter has a luminance of $1/\pi$ candela/m^2. Therefore:

$$L = \frac{1}{\pi} \times \frac{2770 \text{ lm}}{(1.2 \text{ m})(.038 \text{ m})\pi} = 6155 \text{ candela/m}^2$$

Luminance for a reflecting surface:

EXAMPLE 18.4. Calculate the luminance of the page that you are now reading, assuming an illuminance of 500 lux and reflectance of 0.77 for the paper. Use SI units.

SOLUTION. From the definition of illuminance, one lux is produced by one lumen falling on one square meter. Therefore, the reflected lumens from the page are:

$$\text{reflected lumens} = 500 \frac{\text{lm}}{\text{m}^2} \times .77 = 385 \frac{\text{lm}}{\text{m}^2}$$

and as in Example 18.3 above, the luminance, L, is:

$$L = \frac{1}{\pi} \times \frac{385 \text{ lm}}{\text{m}^2} = 122.5 \frac{\text{candela}}{\text{m}^2}$$

(Typical luminances are given in Table 18.2.)

18.9 Illuminance Measurement

Field measurements of illuminance levels are most commonly made with a portable illuminance meter, two of which are illustrated in Fig. 18.8. These devices comprise a photoelectric material connected to a microammeter and are calibrated in

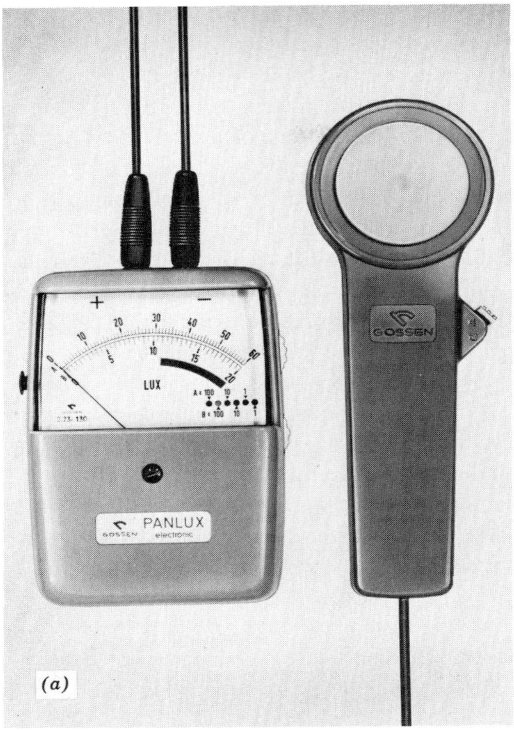

Fig. 18.8 (a) *Analog illuminance meter calibrated in lux and equipped with remote cable-connected cell unit, multiple ranges, and color and cosine correction. Ranges cover 0 to 120,000 lux with maximum ± 7.5% error. Unit is small and lightweight (380 grams). (Courtesy of Gossen.)*

Fig. 18.8 (b) *Portable, digital illuminance meter with a range of 0.01 to 200,000 lux (20,000 fc). The illustrated unit contains a microcomputer that permits integrating measurements, luminous intensity calculation (see Section 18.11), plus various other convenient measurement/calculation functions. (Courtesy of TOPCON.)*

footcandles or lux. The smaller units are convenient to use, but accuracy is generally not better than ± 5% for analog units and ± 2% for digital instruments.

As explained in Section 18.6, and as shown in Fig. 18.6, the human eye is not equally sensitive to the various wavelengths (colors). Maximum sensitivity is in the yellow-green area (wavelength of 555 nm) while sensitivity at the red and blue ends of the spectrum is quite low. This effect is so pronounced that 10 units of blue energy are required to produce the same visual effect as 1 unit of yellow-green. Therefore, if a meter is to be useful, its inherent response, which is quite different from that of the human eye, must be corrected to correspond to the eye. For this reason meters are ''color corrected.''

The cells (meters) must also be corrected for light incident at oblique angles that does not reach the cell due to reflection from the surface glass and shielding of the light-sensitive cell by the meter housing. This correction is known as cosine correction. A good meter must therefore be (and it will plainly so indicate) color and cosine corrected.

When taking actual readings, meters should be placed on a stable surface and readings taken after the needle stabilizes. For determining average room illuminance, a number of readings should be taken and an average computed. Where no definite height is specified, readings are taken at 30 in. above the floor. The meter must always be held with the cell parallel to the plane of the test. Thus to measure wall illuminance, the meter must be held with the cell parallel to the wall. If nighttime readings are desired and the test is being conducted during daylight hours, readings should be taken with and without the artificial illumination and the results subtracted. Detailed instructions for conducting field surveys are contained in the IES publication ''How to Make a Lighting Survey.'' Briefly, a survey of an existing indoor lighting installation should establish:

1. Type, rating, and age of sources.
2. Type, design, and model of luminaire.

3. Maintenance schedule.

It should also measure:

1. Mounting height.
2. Spacing and pattern of luminaires.
3. Reflectances of walls, floor, ceiling, and major items of furniture and equipment.
4. Illuminance levels throughout the area plus levels at walls and columns, all at the working plane elevation.

A special type of illuminance meter that measures Equivalent Spherical Illumination (ESI) is covered with the discussion of ESI in Section 18.26.

18.10 Luminance Measurement

In many respects the measurement of luminance is more important than that of illuminance (lux, footcandles), because it is brightness, not lux, that we see. That lux measurements are still much more common is due to the lower cost, better portability, and simplicity of use of lux meters as compared to luminance meters.

Luminance meters are of several types. Among them are an older comparator type that requires a brightness judgment to be made by the user (Fig. 18.9*a*), a direct-reading narrow-angle (spot) type (Fig. 18.9*b*), and a laboratory-grade unit that reads both luminance *and* contrast (Fig. 18.9*c*). This last is particularly useful in determining the effect of veiling reflection or contrast reduction, which is discussed in Section 18.26.

An approximation of the luminance of a reflecting or luminous source can be obtained using an illuminance (footcandle, lux) meter of the type shown in Fig. 18.8. For diffuse reflecting surfaces, the cell of the meter is placed against the surface and then slowly retracted 2 to 4 in. (5–10 cm) until a constant reading is obtained. The luminance, in footlamberts, is then 1.25 times the reading in footcandles, the 1.25 factor compensating for wide-angle losses. A device that uses

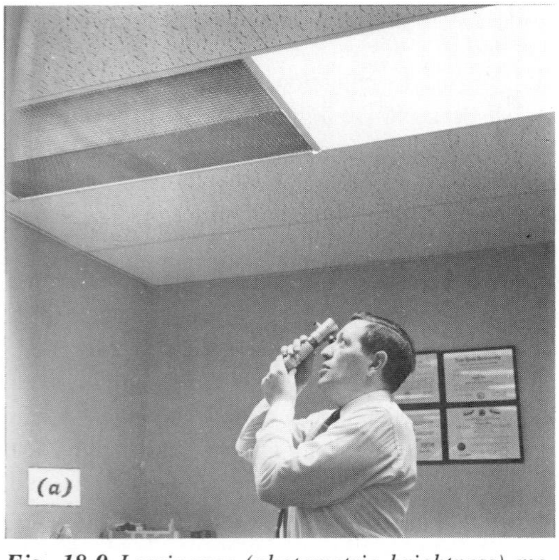

Fig. 18.9 *Luminance (photometric brightness) meters.* (a) *The comparator meter relies on the user to compare the brightness of the "scene" to a calibrated bright spot in the field of view. Accuracy depends on visual judgment and correct calibration, and rarely exceeds ± 10%. In the illustration the author is checking the luminance of a fluorescent fixture equipped with a miniature parabolic wedge louver, which has very low luminance above 45° (see Fig. 20.31, page 1033). This characteristic is readily apparent in comparison to the adjacent fixture, which utilizes a prismatic lens diffuser.*

Fig. 18.9 (b) *The illustrated "spot" meter displays digitally the luminance of the central 1° angle of view in an overall 9° scene. It is calibrated in footlamberts, with a range of 0 to ≈100,000 fL [~340,000 cd/m²]. The unit is readily portable. (Courtesy of Minolta.)*

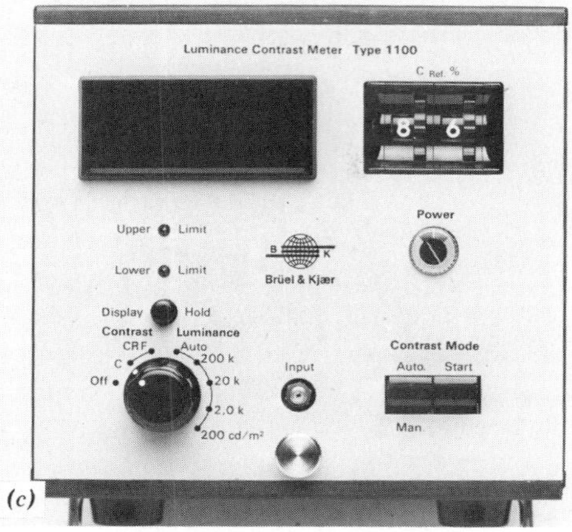

Fig. 18.9 (c) *This unit is a high-precision electronic digital instrument that measures luminance, luminance ratio, contrast, and contrast reduction. Use of the unit is explained in Fig. 18.30, page 896. Luminance measurement range is 0 to 200,000 cd/m² [56,400 fL] at high accuracy. (Courtesy of Brüel & Kjaer.)*

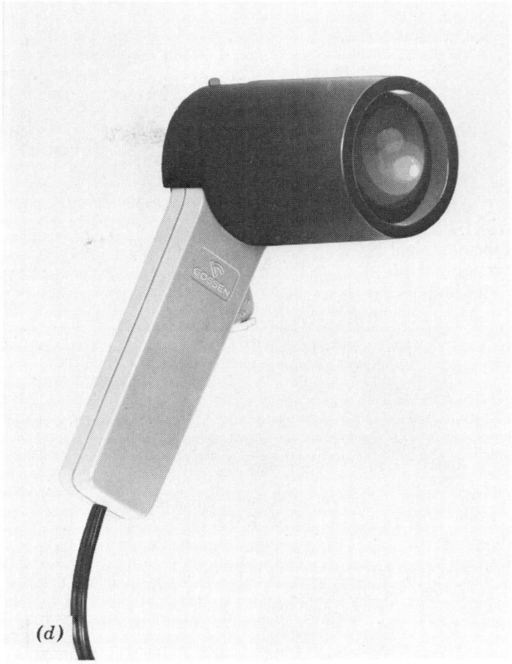

Fig. 18.9 (d) *This is an attachment to the lux meter of Fig. 18.8* (a) *and operates basically on the principal illustrated in Fig. 18.10, but with a somewhat higher accuracy. (Courtesy of Gossen.)*

this technique is shown in Fig. 18.9*d*. It is a direct-reading type. Basically, it is an illuminance meter equipped with a hooded cell arranged to block oblique light and calibrated in units of luminance.

For a luminous source, the cell of an illuminance meter is placed directly against the surface (see Fig. 18.10); the reading on the meter in footcandles is then the luminance in footlamberts (because footlamberts = lumens per area = footcandles). When using a meter calibrated in lux rather than footcandles, the readings must be divided by π to obtain the luminance in cd/m².

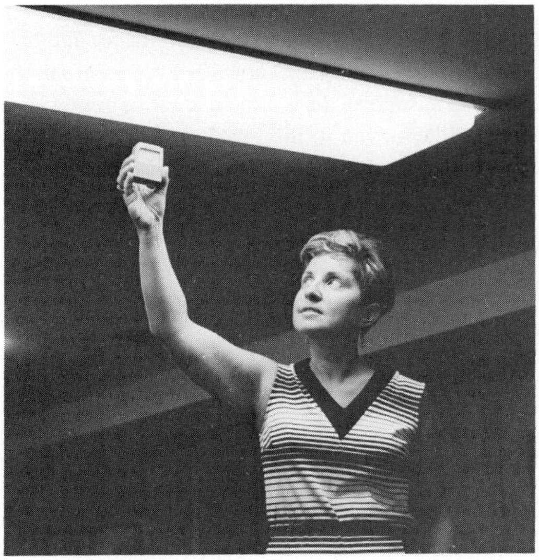

Fig. 18.10 *When the cell of a direct reading illuminance meter is held in contact with a luminous source, the surface luminance can be read directly or simply calculated. See text.*

18.11 Reflectance Measurements

It is often desirable to know the reflectance of a given surface since luminance can then be readily computed (see Fig. 18.7). Two methods of measuring reflectance are shown in Fig. 18.11: the known-sample method and the light-ratio method. If a sample of known reflectance factor (RF) is available, this method should be used since it yields more accurate results than the ratio method. The sample should be no smaller than 8 in. by 8 in.

It is a good idea for a budding lighting designer

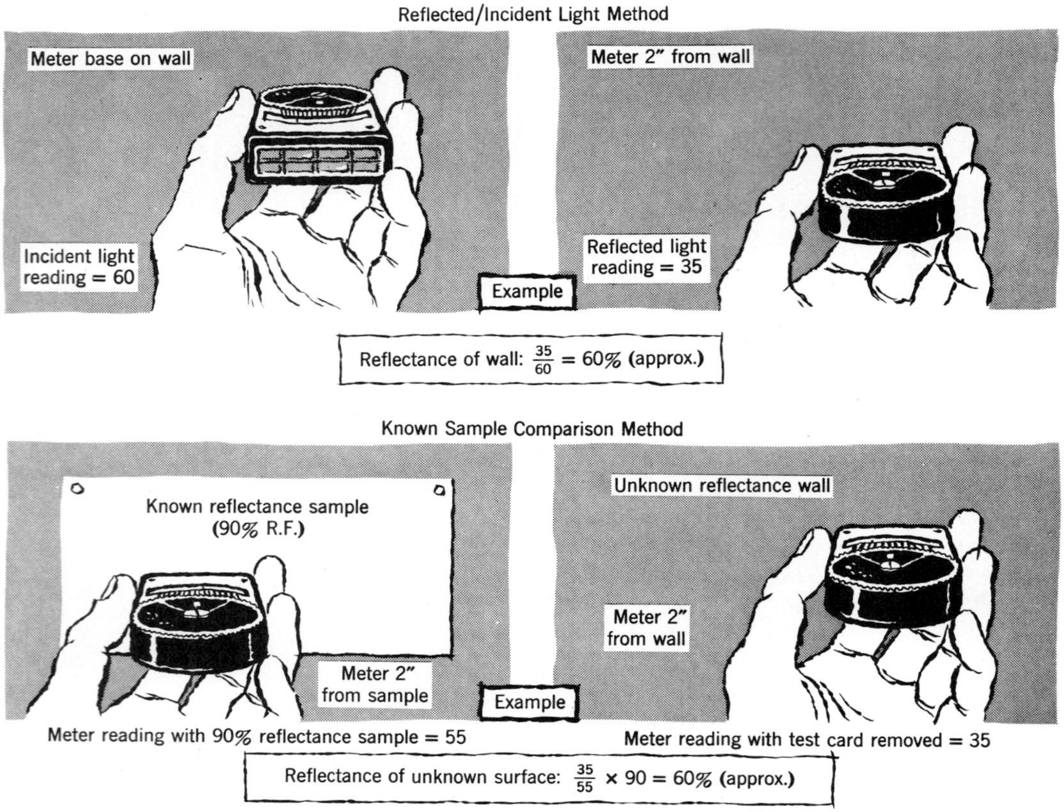

Fig. 18.11 *Two simple methods of measuring reflectance.*

to determine the reflectances, illuminance, and luminance levels of spaces and surfaces familiar to him or her, such as his office desk, adjoining wall, and the like—even to the extent of marking these figures on the respective surfaces in order to develop an appreciation of and a memory for these parameters. This will enable the designer to visualize the result of a lighting design and should be of considerable assistance. See Table 19.7 for typical reflectance values.

18.12 Inverse Square Law

We have already seen that, by definition, a point source of 1 cd produces an illumination of 1 lux on the inside surface of a surrounding sphere of 1 m radius (r). Since the surface area of this sphere is 4π m^2, a 1-cd source produces 4π lm of luminous flux. Now assume a sphere of 2-m radius surrounding this same source (see Fig. 18.12*a*). Since the same amount of flux is spread over a larger area, the illumination on the larger sphere is inversely proportional to the ratio of the two-sphere areas, that is,

$$\text{lux}_2 = \text{lux}_1 \times \frac{\text{area}_1}{\text{area}_2}$$

or

$$\text{lux}_2 = \text{lux}_1 \times \frac{4\pi r_1^2}{4\pi r_2^2}$$

$$= \text{lux}_1 \times \frac{r_1^2}{r_2^2} \tag{18.5}$$

In other words, the illumination is inversely proportional to the square of the distance from the source. In general terms

$$\text{lux [fc]} = \frac{\text{cp intensity}}{\text{distance}^2} \tag{18.6}$$

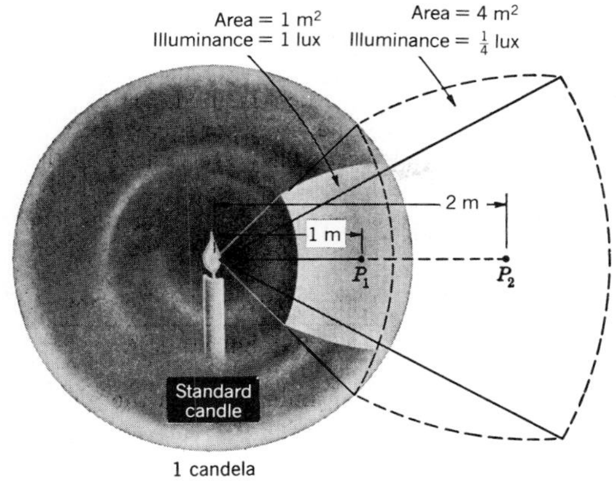

(a)

Fig. 18.12 (a) *Relations between candlepower, lumens, and lux defined with reference to a standard light source of 1 mean spherical candlepower (1 candela) located at the center of a sphere of one meter radius.*

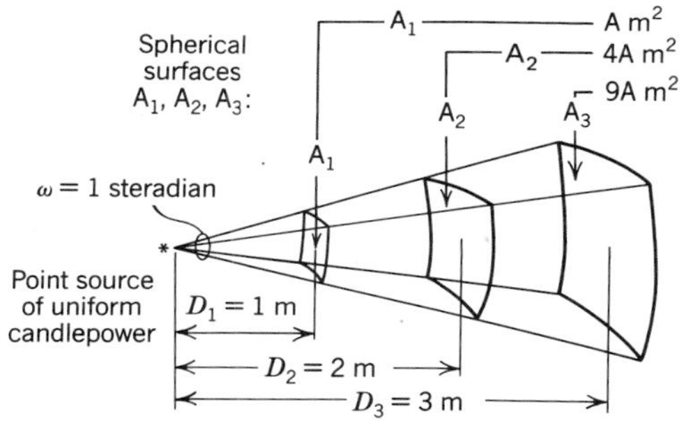

(b)

Fig. 18.12 (b) *Demonstration of inverse square law properties using a solid angle of unit size. Note that the surfaces are necessarily spherical, since points on a planar surface are not equidistant from the source.*

where distance is expressed in meters [feet]. (This holds true for surfaces normal to a source. For other situations see Section 20.38.)

This relationship can also readily be derived by using any solid angle and the area it intercepts, as in Fig. 18.12*b*. A glance at this figure shows clearly that the area intercepted is proportional to the square of the distance from the source and therefore the illumination is inversely proportional, as stated above.

18.13 Luminous Intensity (Candlepower) Measurements

Luminous intensity (candlepower) cannot be measured directly but must be computed from its illumination effects. The simplest way of doing this is to use the inverse square relationship developed in the preceding section. We measure the footcandle illumination produced on a plane at right angles to the source, at a known distance; then apply equation 18.6. For accurate measurement, the distance should be at least five and preferably ten times the maximum dimension of the source, since for anything other than a point source the equation is an approximation. The candlepower thus calculated is the luminous intensity in the direction being viewed. Since candlepower is not uniform in all directions for anything except an ideal point source, and since a single candlepower figure for a source is desirable for calculation purposes, the average of a number of candlepower figures taken from several directions is used. This average figure is called the mean spherical candlepower (mscp), and represents an equivalent point source that will produce 4π lm for every candela. Thus a 10-cp lamp will exhibit an average intensity of $\pm$ 10 cmp in all directions and will produce 40π lumens.

18.14 Candlepower Distribution Curves

If the candlepower figures calculated in the preceding section are plotted on polar coordinate axes, the resultant figure is called, logically, a candlepower distribution curve (CDC) for the particular source involved. The procedure for making this curve is straightforward. A photo cell is rotated around the source in a single plane, illuminance measured, and intensity (cp) calculated. Alternatively, the photo cell can be fixed and the source rotated. If the sources' distribution is symmetrical, as shown in Fig. 18.13, then only a single set of values is required, and the resultant plot is valid in all vertical planes through the source. Thus for incandescent lamps, downlights, open circular reflectors, and the like, only a single CDC is required. For a nonsymmetric source such as a fluorescent luminaire, CDC curves in several planes are required to define the fixture's distribution characteristic. Normally, manufacturers will pro-

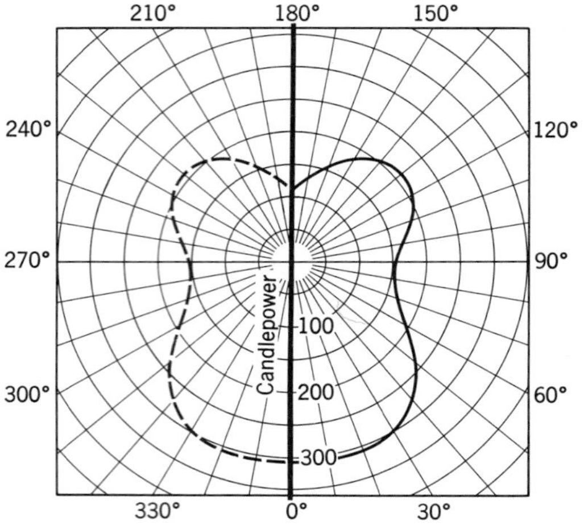

Fig. 18.13 Typical candlepower distribution curve for a general diffuse-type luminaire. Since the unit is symmetrical about its vertical axis, only one curve need be shown. Normally, only the right side of even this single curve is shown, due to symmetry, as in Fig. 18.14.

vide longitudinal and crosswise curves, plus a diagonal (45°) plane curve on request. This is illustrated in Fig. 18.14, where the three planes and typical resultant curves are shown. Most CDC plots are made on polar coordinates because such a plot clearly shows directions and magnitudes. Nevertheless, polar plots tend to crowd near the nadir and accurate magnitude readings at the cutoff angle are difficult to make. For this reason it is occasionally desirable to obtain a plot on rectangular coordinates. One such plot is shown in Fig. 18.15. The usefulness of candlepower distribution curves will become clear in our subsequent discussions on fixture diffusers (Section 20.19), point-by-point calculations (Section 20.39), and direct and reflected glare (Section 18.24 to 18.26).

It should be noted that the area of the CDC curve is not a measure of the lumen output, since a source that gives high candlepower near the vertical has its lumens in this direction spread over only a small area.

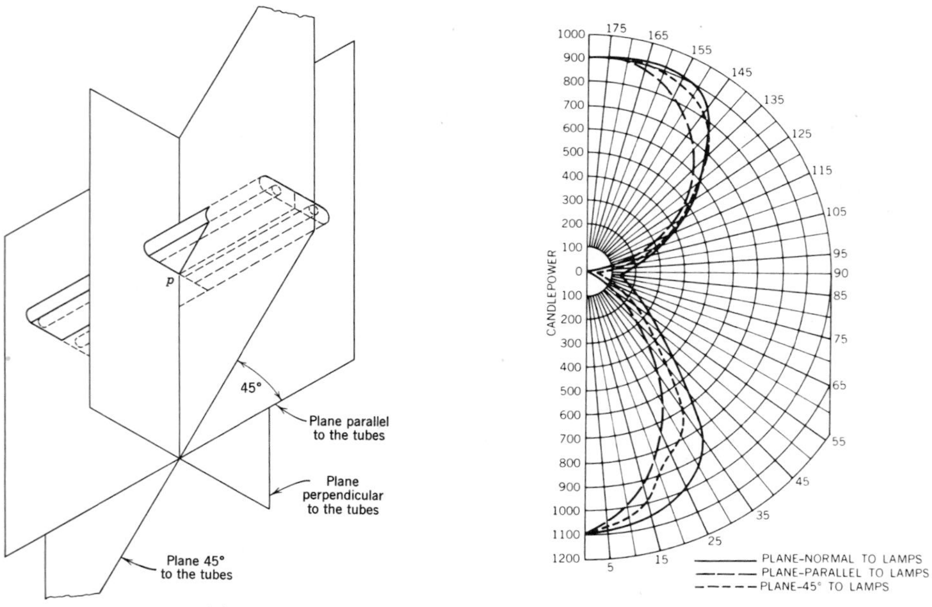

(a) (b)

Fig. 18.14 Because of asymmetry of fluorescent fixtures, candlepower distribution curves in three planes are plotted, as shown.

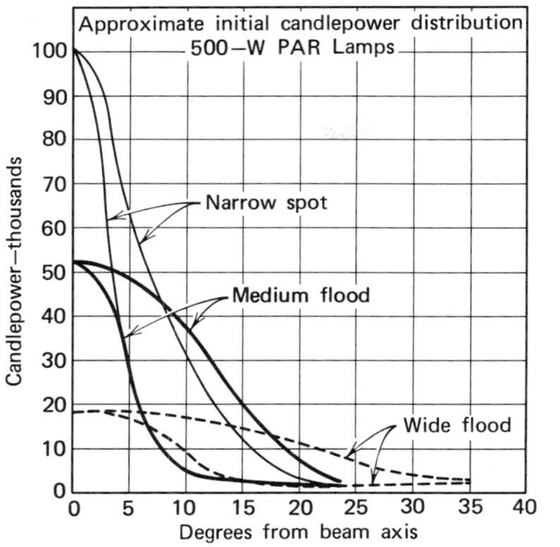

Fig. 18.15 Candlepower distribution curves plotted in rectangular coordinates. Note that candlepower values near the cutoff angles are easily read, which is not the case in polar plots.

LIGHT AND SIGHT

18.15 The Eye

Since all discussion of light and lighting techniques is irrelevant to our purposes unless ultimately related to vision, we turn to a cursory examination of the human eye before proceeding further with discussions of lighting.

Light impinging upon the eye enters through the pupil, the size of which is controlled by the iris, thereby controlling the amount of light entering the eye. The lens focuses the image on the retina, from which the optic nerve conveys the visual message by electric impulse to the brain. Figure 18.16 shows the structure of the eye and the parallel structure of a camera.

The central portion of the eye, near the fovea, contains light-sensitive cells called "cones" because of their shape. The cones are responsible for the ability to discriminate detail and also give us our sensation of color. As we proceed outward from the fovea, a second type of cell is encountered called a "rod" cell, also named after its shape. These cells are extremely light sensitive, giving response to light 1/10,000 as bright as that required by cone cells. However, rod cells lack color sensitivity, thus accounting for the fact that in dim light (rod vision), we have no color perception and all colors appear as varying shades of gray. Rod cells also lack detail discrimination, making "night vision" quite coarse. Finally, rod cells are slower acting than cone cells and therefore have a low degree of flicker fusion, or, stated conversely, they are highly motion sensitive. Since these cells occur at the outer portions of the retina, their motion sensitivity results in our being best

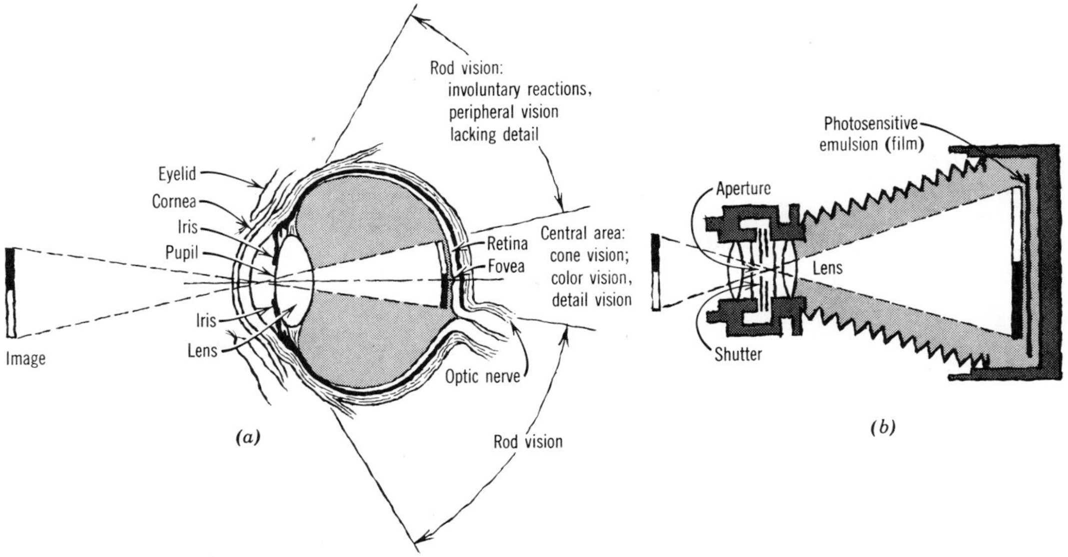

Fig. 18.16 The human eye and the camera operate on the same optic principles.

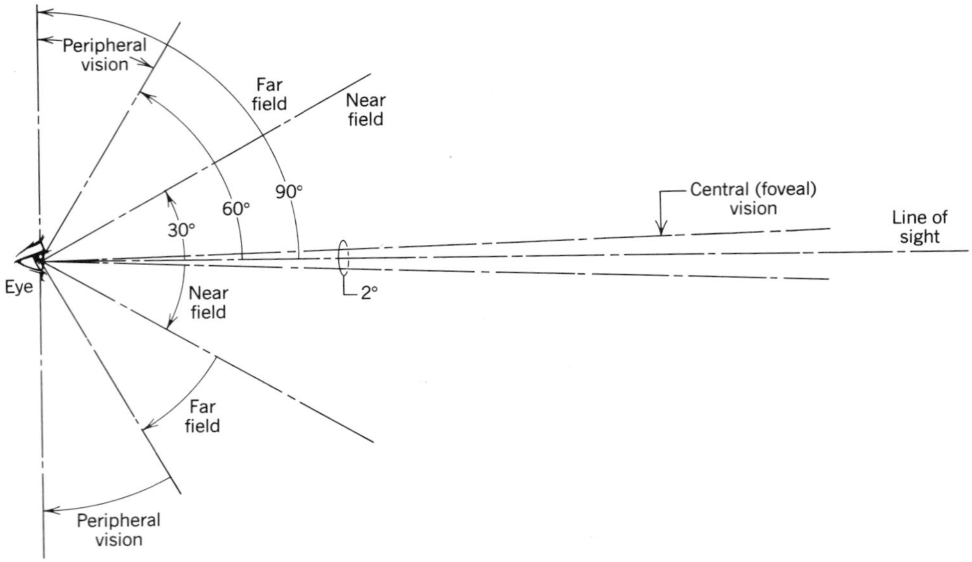

Fig. 18.17 The fields of vision of a normal pair of human eyes, and the angles subtended.

able to detect movement when looking out of the "corner of the eye." Looking at a fluorescent tube directly and then obliquely will demonstrate this effect.

Figure 18.17 is a sketch illustrating the angles involved in the field of vision. Of particular interest is the extreme narrowness of the cone of central (foveal) vision, in which acute perception of detail takes place. This area is so small that the eye must refocus on successive letters in the words you are now reading if you wish to examine each individually. Surrounding this central area is a cone of binocular vision of 30° half-angle, called the near field or surround, in which area most of coarser sight information is gathered. Beyond this cone we have far field and peripheral, primarily horizontal, monocular vision. It is this peripheral area that largely gives us our subjective, ambience-type reactions, which will be discussed below.

18.16 Factors in Visual Acuity

The three components of any seeing task are obviously the object or task, the lighting conditions, and the observer. Listed below are the variables affecting each of these three components. Based on the results of many investigations they can be categorized as of primary or secondary importance.

I. The Task
   Primary Factors
   (*a*) Size
   (*b*) Luminance (brightness)
   (*c*) Contrast
   (*d*) Exposure time—needed or given
   Secondary Factors
   (*e*) Type of object—required mental activity
   (*f*) Degree of accuracy required
   (*g*) Task; moving or stationary
   (*h*) Peripheral patterns

II. The Lighting Condition
   Primary Factors
   (*a*) Illumination level
   (*b*) Disability glare
   (*c*) Discomfort glare
   Secondary Factors
   (*d*) Luminance ratios
   (*e*) Brightness patterns
   (*f*) Chromaticity

III. The Observer
   Primary Factors
   (*a*) Condition of the eyes
   (*b*) Adaptation level
   (*c*) Fatigue level
   Secondary Factors
   (*d*) Subjective impressions; psychological reactions

Although in the following discussions these factors will be considered individually, many are interrelated. Thus luminance I(*b*) and adaptation III(*b*) result from the presence of illumination II(*a*); subjective impressions III(*d*) are dependent on brightness patterns II(*e*) and chromaticity II(*f*); fatigue III(*c*) results from a combination of many of the factors, and so on.

In the literature it is common to find reference to the quantity and quality of the lighting environment. In terms of the above factors, the quantity of light has reference to item II(*a*) and the quality to items II(*b*) through II(*f*).

The basic visual tasks are the perception of low contrast, fine detail, and brightness gradient. Assuming a good lighting environment, that is, low glare, acceptable brightness ratios, and white light plus a normal pair of unfatigued eyes, visual acuity is primarily dependent on items I(*a*) to I(*d*), the interrelated effects of which have been determined by a large number of field tests. Remember that the seeing task under discussion involves foveal vision focusing and concentrating on small-area detail. This is a vastly different task than reading, where the eye rapidly scans familiar images without focusing on details and the brain immediately understands even when much of the information is missing, as in poor reading copy. The task here discussed could be compared to reading an unfamiliar language written with unfamiliar signs and symbols, necessitating detailed examination of each symbol or part of symbol, individually.

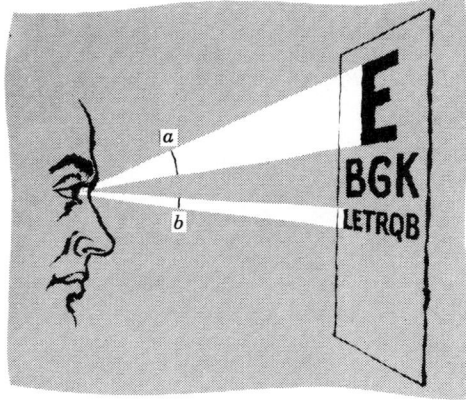

Fig. 18.18 *Relationship between object size and visibility is demonstrated by comparison of subtended angles* (a) *and* (b).

18.17 Size of Visual Object

Visual acuity is generally proportional to the physical size of the object being viewed given fixed brightness, contrast, and exposure time. Since the actual parameter is not physical size but subtended visual angle, visual ability can be increased by bringing the object nearer the eye (see Fig. 18.18). It is assumed that we are dealing with a pair of *young* eyes, since at ages above 40 the accommodation ability of the eye becomes limited and bringing the object closer blurs the focus.

18.18 Subjective Brightness

The sensation of vision, as explained above, is caused by light entering the eye. This light may be thought of as a group of convergent rays, each ray coming from a different point in space and therefore carrying different visual information. The composite of these rays comprises the entire visual picture that the eye sees and the brain comprehends. The individual rays differ from each other in intensity and chromaticity depending on the part of the viewed object from which they were reflected. The intensity of these cones of light determines and describes the perceived brightness of the object being viewed (see Fig. 18.19).

If the surface reflectance of the object being viewed is uniform and the illumination is also uniform, then the reflected rays of light will be equal in intensity and we will see an object of uniform luminance. If, however, as is generally the case, either the object or the illumination is nonuniform, we will see an object of varied luminance.

The human eye detects brightness over an astonishing range of more than 100 million to 1, the lower levels being accomplished after an adjustment period, called adaptation time. This period varies from 2 min for cone vision to up to 40 min for rod vision, for dark adaptation, but is much faster for both types for light adaptation (going from dark to light). The effects of adaptation on apparent (photometric) brightness are discussed in the following section. Table 18.2 lists some measured luminances of everyday visual tasks.

An interesting characteristic of light-level adaptation is a shift in the sensitivity curve of the eye (see Fig. 18.6*b*). Whereas for the light-adapted

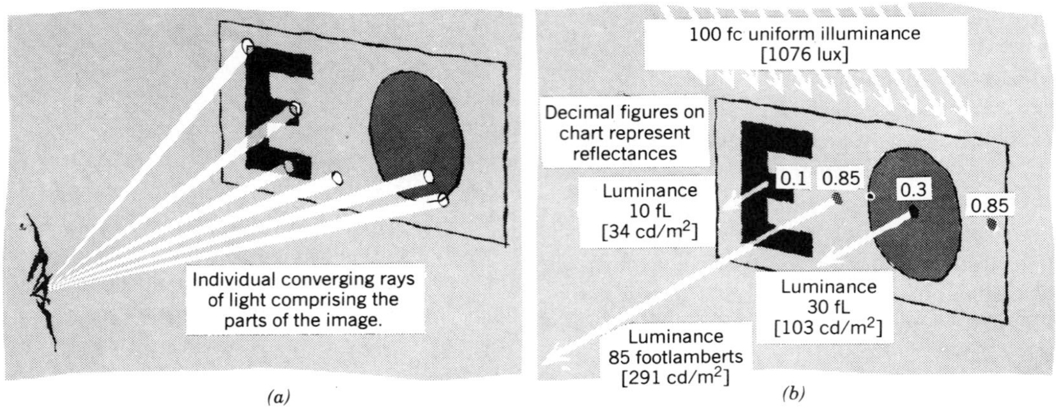

Fig. 18.19 (a) *Composition of the visual image.* (b) *Luminance of a nonuniform surface.*

TABLE 18.2*a* **Typical Luminance Values**[a]

Object	*Luminance* cd/m^2[b]	*Luminance* *Footlamberts*
Black glove on cloudy night	0.0003	0.0001
Wall brightness in a well-lighted office	100	30
This sheet of paper in an office	120	35
Green electroluminescent lamp	150	45
Asphalt paving—overcast day	1,300	380
North sky	3,500	1,000
Moon, candle flame	4,000–5,000	1,300
Fluorescent tube	7,000–8,000	2,200
Kerosene flame	8,500	2,500
Hazy sky or fog	15,000	4,400
Snow in sunlight	25,000	7,300
100-W inside frost incandescent lamp	50,000	14.600
Sun	2.3 E9	0.67 E9

[a]Values are rounded off.

[b]To obtain footlamberts; divide by 3.42.

TABLE 18.2*b* **Preferred and Permissible Luminances**

Item	*Luminance in* cd/m^2
Recommended road luminance	1–2
Minimum discernible	2–3
Clearly discernible human features	15–20
Preferred wall luminance	25–150
Preferred ceiling luminance	50–250
Preferred task luminance	100–500
Permissible luminaire luminance	1000–7000
Maximum contrast sensitivity	10,000

eye (photopic vision), maximum sensitivity is at 555 nm in the yellow-green region, the dark-adapted eye (scotopic vision) peaks at 520 nm in the blue-green region. This means that as the light dims, the warm colors—yellow, orange, red—become grayed, and the blues and violets stand out. This phenomenon can be important in lighting design of restaurants, where light levels generally vary inversely with restaurant quality. Very few foods are blue or violet.

Returning then to the primary consideration of visual acuity as affected by luminance, we can state that, in general, visual performance increases with object brightness. However, a great deal depends on the background against which an object is viewed and the consequent contrast in brightness between the object being viewed and its surroundings.

18.19 Contrast

In order to properly evaluate the effect of contrast (luminance ratio) on visibility, we must first determine the nature of the visual task or, more simply, exactly what it is that we are trying to see. As stated before, the basic visual tasks are detail discrimination and detection of low contrast. An example of the former task is reading of legible, good copy, fine print, while of the latter would be examination of surface textures or low contrast, washed out, poor copy.

Contrast is a dimensionless ratio, defined as

$$C = \frac{L_T - L_B}{L_B} \quad \text{or} \quad \frac{L_B - L_T}{L_B} \quad \text{or} \quad \left|\frac{L_B - L_T}{L_B}\right| \tag{18.7}$$

Where L_T and L_B are the luminance of the task and background respectively, in any units. Thus, C varies from 0 for no contrast to 1.0 for maximum contrast. In most situations, the illumination on the task and background is the same. Therefore, since luminance is the product of illuminance (lux) and reflectance, contrast can also be expressed as

$$C = \left|\frac{R_B - R_T}{R_B}\right| \tag{18.8}$$

that is, *contrast is generally independent of illumination* (ignoring specularity). Thus, for the black on white lettering that you are now reading,

$$C = \frac{0.77 - 0.045}{0.77} = 0.94$$

which accounts for the excellent legibility. (We are assuming no specularity.) See Table 18.9, for reflectance figures.

It is obvious that high contrast is the critical factor in visual appreciation of outline, silhouette, and size, which is involved in the task of reading. Thus black-on-white print can be read with ease even in moonlight, which is at best 0.1 lux illuminance. This is because the contrast is so high (94%, as above). An important conclusion can then be drawn about a *reading* task: With high contrast (clear, legible print), visibility is essentially *independent* of illumination above a certain minimum. Indeed, high illuminance values can be detrimental, since they generally go hand in hand with high luminance sources, and these in turn can cause veiling reflections (see Section 18.26). This indeed is the conclusion of some recent studies (Yonemura, 1981).

Now refer to Fig. 18.20. Note that as the contrast between the letters in the word ''performance'' and the background diminishes, the individual letters become harder to read. The end letters of the word require an illumination of up to 1000 lux, and that suffices only because we expect the letter ''*e*'' at the end. Were it an unknown sign, illumination of the magnitude of 10,000 lux + would be needed. These latter letters are an example of the second type of visual task mentioned above—low-contrast poor copy, or surface detail study.

The ''*e*'' at the end of ''performance'' in Fig. 18.20 is printed with the same density as the ''*c*'' in ''contrast.'' It exists, and with enough lighting, the negative effect of lack of contrast can be overcome. This is not so with a ninth-carbon copy or a washed-out photocopy. There, the data simply

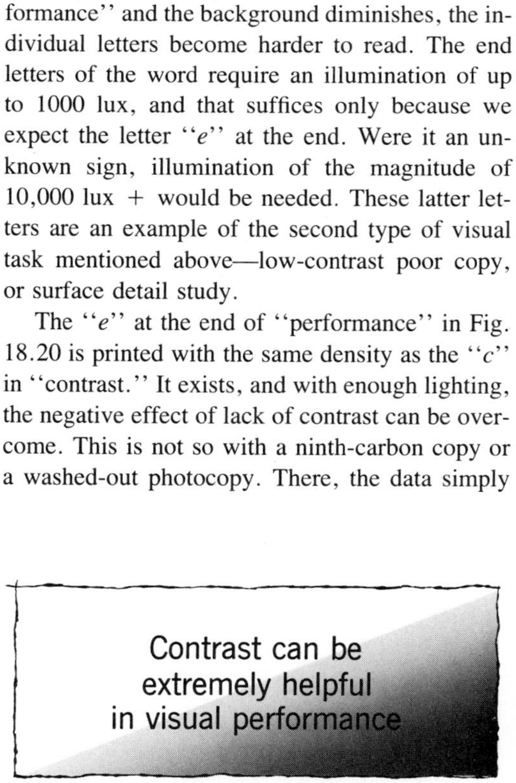

Fig. 18.20 *High contrast is helpful when the seeing task involves detection of silhouette detail.*

do not exist, and increased lighting will simply make this fact plainer.

High background luminance makes an object look darker and therefore assists in outline detail discrimination, which is precisely the visual task involved in reading. For this reason black on white is desirable for reading. Conversely, high background luminance makes surface examination more difficult. A simple experiment will demonstrate this effect. Stand near a window and hold your hand in front of you with the floor as background. The skin surface detail will be perfectly clear—in rough proportion to its luminance. Now hold your hand up against the window with the daytime sky as background. The hand outline will be clear, but the skin surface will appear dark—the brighter the sky, the darker the skin surface. The reason for this is that the eye automatically adjusts to the average brightness of the entire scene.

It is well known that when using an automatic exposure control camera to photograph a dark object on a light background such as a person in a snow scene, it is necessary to manually increase the camera aperture in order to obtain additional light to photograph the detail of the darker object. (In doing this we overexpose the rest of the scene.) Since we cannot easily control the aperture of our eyes, we must compensate for the detrimental effect of high background luminance in another way, by increasing the surface luminance of the visual task, for example. Indeed, this method is frequently employed. (See Section 18.27*b*.) Limited visual compensation can be made by squinting; this reduces the field of vision and the overall scene brightness. *Ideally, the luminance of a surface type task should be the same as that of the background,* but ratios of 3 : 1 are acceptable in most circumstances.

Another way of understanding the above is to consider the adaptation characteristic of the human eye (refer to Fig. 18.21). As stated, the eye adapts to the brightness level of the *overall* scene and sees each object in the scene with respect to that adaptation level. Thus at an adaptation level of 1 fL, a measured luminance ratio of 1 : 10 (horizontal scale on Fig. 18.21) appears to be only approximately 1 : 4 (vertical scale), that is, the apparent ratio is *smaller* than the actual. This is because the low level of eye adaptation causes the eye to "overload" easily and diminishes the difference between high brightnesses. At an adaptation level of 1000 fL (daylight conditions), the apparent and actual *ratios* correspond, that is, smaller ratios are recognizable. The second important conclusion that can be drawn is that at high adaptation levels apparent brightness is lower than actual and vice versa. Thus a shadowed object near a window will look *darker* than it actually is;

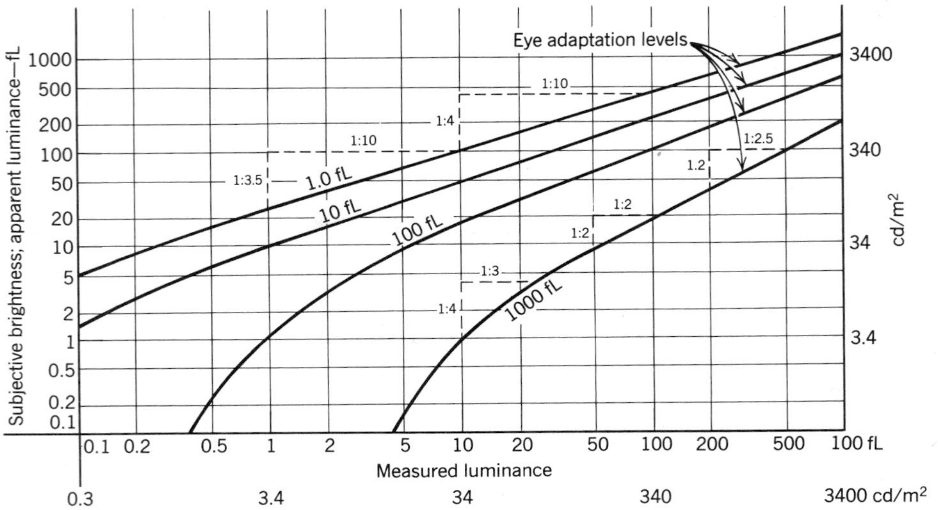

Fig. 18.21 Relation between subjective brightness and measured luminance levels. See text. (Adapted from H. Cotton, Principles of Illumination, *Wiley, New York, 1960.)*

contrary to first expectation, it must be better lighted than a similar object further inside the room for equal visibility.

That this effect (high level adaptation) is primarily important in daylight situations is also apparent from the curves, since at a 100-fL adaptation level, which is approximately the indoor condition, apparent and actual luminance levels coincide (within visual ability to recognize differences). However, the reverse effect, resulting from low adaptation levels, can be very important in design situations where low lighting levels are found, such as theaters, lecture halls, restaurants, and storage spaces. Sources of light that would be entirely acceptable at a higher adaptation level can easily become an annoying glare at a low level. Good examples are a theater usher's flashlight or the blinding glare of an oncoming car's headlights.

Reduction of contrast due to veiling reflections and the measurement of contrast and contrast reduction are discussed in Section 18.26.

18.20 Exposure Time

Registering a meaningful visual image is not an instantaneous process, but one that requires finite amounts of time. Just as a photograph can be taken in dim light by using a longer exposure, so can the human eye distinguish and discriminate fine detail in poor light given enough time (and neglecting eyestrain). Of course, the time needed depends on the type of task, but the principle of shorter time at higher illumination, within limits, remains the same. This is particularly true when the object being viewed is not static but in motion.

The phenomenon, however, is not linear. For one specific task tested, increasing the luminance by a factor of 6 halved the seeing time, whereas a further increase of sixfold in luminance reduced the time only approximately 20%. Thus, as in the case of improved contrast with increasing background brightness, we have a case of diminishing returns.

Exposure time has always been incorporated in British IES illuminance standards (see Table 18.3) and has recently been added to American IES illuminance tables. (See the discussion in Section 18.22.)

With the parameter of time, as with other parameters of visual acuity, the same qualification applies: When dealing with material that does not require detail discrimination, improved performance does not necessarily result from improved illumination. It has been amply demonstrated (*IES Lighting Handbook,* 1981) that speed of reading and comprehension are substantially independent of illumination levels above a minimum, but are very much dependent on the contrast quality of the material.

18.21 Other Factors in Visual Performance

The four factors considered in the preceding sectins were listed in Section 18.16 as the four primary task-oriented factors. Before proceeding with the lighting factors that occupy the remainder of this chapter, a few remarks bearing on the secondary task factors and the observer-oriented considerations are in order. The preceding discussion on visual performance factors implied that optimum performance is synonymous with optimal seeing conditions, that is, the lighting conditions that yield optimal performance of certain tasks under laboratory-controlled conditions will, when extrapolated to actual field conditions, yield an ideal visual environment. Unfortunately, this may *not* be the case since a direct correlation between optimal performance and minimum fatigue has not been established, nor is it possible to extrapolate the laboratory tests to the field condition without considerable inaccuracy. An observer performing a lab test has a different level of concentration and performance than a person at an 8-hour-a-day task. The latter will compensate for unsatisfactory seeing conditions by:

1. Moving the work to a better viewing angle.
2. Moving his or her head and eyes to a more comfortable position.
3. Reducing the distance between eyes and task, to the extent that the eyes accommodate.
4. Complaining about a poor contrast task so that something is done about it (such as fixing the photocopying machine).
5. Taking more time to perform the seeing task involved. (This item, if it affects produc-

tion, will frequently spur management to improve other factors.)

Furthermore, maximum performance may not be synonymous with maximum comfort or minimum fatigue. Indeed the reverse may be true. This field, which is under active research at this writing, has tentatively established that what is normally referred to as "eyestrain" is a condition of the eye muscles resulting from extensive and intensive eye use. [See Section 18.16, items I(*e, f*).] Thus, excellent performance under excellent lighting conditions can still produce fatigue because of the demanding nature of the task. Conditions will generally not produce fatigue if the observer is not "straining" to perform properly. In addition, as we discuss below, discomfort glare or even excessive lighting can cause fatigue without affecting performance. Thus the lighting designer who previously was concerned only with providing adequate uniform lighting levels should properly be concerned with all of the factors involved. Many are indeed beyond his or her control. Some, such as task contrast, previously thought to be outside the lighting designer's province, should be examined by the designer in the framework of an overall lighting plus task plus observer problem, and recommendations made. Acceptance of these recommendations is a management decision, based on cost effectiveness. This means that management must weigh the costs involved against the expenses, including the cost of errors caused by poor seeing conditions, and reach its conclusions accordingly.

QUANTITY OF LIGHT

18.22 Illuminance Levels

Returning to the list of factors in Section 18.16, and having discussed the task-oriented and observer-oriented items [except for item III(*d*), psychological reactions, which is covered in Section 18.32], we turn now to item II, *the lighting condition*. This is frequently, if somewhat inaccurately, divided into two groups—quantity and quality of lighting—with item II(*a*) representing quantity and items II(*b*) to II(*f*) quality. That such a division is not accurate will become clear in our discussion of glare in Sections 18.23 to 18.25.

The factors involved in visual acuity, as discussed briefly above, have been widely known and understood for many years. That knowledge, however, did not answer the most basic lighting design question, which is "How much light must I provide for the specific visual task at hand?" To provide an answer to that seemingly simple question, two approaches are possible, one pragmatic, one analytic. The former, which was the basic British approach, was to study specific tasks in actual and simulated field conditions. To these results, modifying factors of contrast, accuracy requirement, speed, and daylight provision were added. To a large extent, despite the influence of the American analytic approach, this is still the basic British IES recommendation, as shown in Table 18.3. The disadvantage of this approach was the necessity to study a large number of visual tasks individually.

The American approach to the problem of determining required lighting levels was analytic and based itself mainly on the work of Blackwell at Ohio University. The idea was to determine the conditions under which small differences in contrast could be detected for specific degrees of accuracy, with variable parameters of task size, luminance, and exposure time. These results were then related to field tasks by a process of extrapolation, using a device known as a Field Task Simulator. The results obtained were in large measure adapted by the American IES and became the basis of most American lighting design. The levels were higher than their British counterparts for a number of reasons, among which were that the American recommendations were based on maximum performance whereas the British IES uses a graded system, starting with a median or average requirement and increasing it only for tasks with stringent requirements or poor visual conditions.

Unfortunately, under pressure of cheap energy, a more-is-better philosophy, the interests of the not entirely objective lighting industry, and other similar pressures, the basically sound results of the analytic approach were pushed upward to such a point that, before the 1973 fuel crisis, some offices were being designed with 1000 fc [10,000 lux] illuminance levels. Fortunately, the situation has changed. The combination of the requirement for energy conservation plus the reaction of the reasoned portion of the lighting profession to previous excesses has had a profound and salutary effect on American lighting industry/IES illuminance recommendations. The most important result was the establishment in 1979 of a new procedure for determining illuminance requirements and its publication in 1981 in the sixth edition of the *IES Lighting Handbook*. (Details of the method are explained below.) Among other benefits resulting from the requirement for energy conservation are the following:

1. Extensive studies of veiling reflections were undertaken. These have led to the increased use of Equivalent Spherical Illumination with concomitant development of ESI meters and contrast meters. Another results has been the vastly increased use of computers in accurate lighting design.
2. Differentiation was made between the visual requirements of familiar and unfamiliar tasks. This led to a lowering of illumination requirements for ordinary reading and other common, repetitive tasks.
3. Governmental agencies commissioned their own studies (Yonemura, 1981) and trial construction buildings, on the basis of which they have issued their own illumination recommendations (see Table 18.8). A distinctive characteristic of these recommendations is the recognition that task duration as well as task nature is important; that is, a fatigue factor has been introduced into the recommendations.
4. Daylight is to be utilized to the maximum extent possible, (recognizing the inherent connection with heat loss and heat gain through windows) (see Section 19.2). This means that illumination levels *include* daylight contributions.
5. ASHRAE Standard 90, Section 9, has been legally adopted by numerous states. Among its provisions are a procedure for developing a power budgêt for a project (see Section 20.5). In calculating the lighting loads, this standard effectively requires that nonuniform layouts be used as follows:
   a *Task* lighting to accord with IES recommendations.
   b General area lighting to be one-third of task lighting, but not less than 215 lux [20 fc].
   c Noncritical (circulation) lighting to be one-third of general area lighting, but not less than 107 lux [10 fc].

   Thus, a room with task lighting of 750 lux [70 fc] would have area lighting of 250 lux [23 fc] and circulation lighting of 107 lux [10 fc]. Albeit this standard is binding only for budget calculation within which budget the designer is free to do as he or she pleases, it is difficult to see how a designer could avoid nonuniform design even if he so desired.

As a result of the multiplicity of sources, agencies, and authorities, the lighting designer should consult with the authorities having jurisdiction before establishing the criteria by which the job will be designed. To assist the designer, the following section is devoted to a study and comparison of the illuminance recommendations of the more important organizations.

18.23 Illuminance Recommendations

Visual task studies that have been performed by the researchers in this area have resulted in tabulations of required task luminances for hundreds of usual tasks. If these are categorized and acceptable luminances tabulated, the results are approximately these (assuming good quality, that is, high contrast):

Category of Visual Task	*Required Luminance (fL)*
Casual	5(3–6)
Ordinary	20(6–30)
Moderate	45(30–60)
Difficult	90(60–120)
Severe	Above 120

Since the relation between required luminance and incident illumination is known (fL = fc × reflectance), a determination of the required illumination for the listed tasks can readily be accomplished.

The crucial dependence of illumination on task reflectance (RF) can be seen by a glance at the tabulation below, which shows quantitatively the illumination requirements in the above categories

ILLUMINATION

TABLE 18.3 **Flowchart**

Task Group and Typical Task or Interior	Standard Service Illuminance Lux	Are Reflectances or Contrasts Unusually Low?	Will Errors Have Serious Consequences?	Is Task of Short Duration?	Is Area Windowless?	Final Service Illuminance, Lux
Storage areas and plant rooms with no continuous work	150	—	—	—	—	**150** (~15 fc)
Casual Work	200	—	—	—	no — 200 yes	**200** (~20 fc)
Rough work Rough machining and assembly	300	no — 300 yes	no — 300 yes	300	no — 300 yes	**300** (~30 fc)
Routine work Offices; control rooms, medium machining, and assembly	500	no — 500 yes	no — 500 yes	yes no — 500	500	**500** (~50 fc)
Demanding work Deep-plan, drawing or business machine offices. Inspection of medium machining	750	no — 750 yes	no — 750 yes	yes no — 750	750	**750** (~75 fc)
Fine work Color discrimination, textile processing, fine machining, and assembly	1000	no — 1000 yes	no — 1000 yes	yes no — 1000	1000	**1000** (~100 fc)
Very fine work Hand engraving, inspection of fine machining or assembly	1500	no — 1500 yes	no — 1500 yes	yes no — 1500	1500	**1500** (~150 fc)
Minute work Inspection of very fine assembly	3000	— 3000	— 3000	yes no — 3000	3000	**3000** (~300 fc)

Using local lighting, if necessary supplemented by use of optical aids, for example, binocular loupes, magnifiers, profile projectors, etc.

Source: British IES Code for Interior Lighting (1977).

NOTES

The standard service illuminance recommended for an application should be increased if:

1. unusually serious consequences, in terms of cost or danger, could result from mistakes in perception, or
2. unusually low reflectances or contrasts are present in the particular task, or
3. tasks for which the recommended standard service illuminance is less than 500 lx are carried out in windowless interiors.

The recommended standard service illuminance may be decreased if, in the judgment of the designer, the duration of the task is unusually short. This flowchart gives the steps by which the recommended standard service illuminance should be modified when one or more of these conditions apply. The resulting final service illuminance derived from the flowchart should then be used as the design value.

for tasks of radically different reflectance and points out why a single illumination scheme is often inadequate for an area containing widely differing visual tasks. Note that a 10% RF makes all tasks difficult and that casual seeing would be only outline recognition.

Category of Visual Task	Required fc [lux]* Reflectance 50%	10%
Casual	10 [100]	50 [550]
Ordinary	40 [400]	200 [2200]
Moderate	90 [100]	450 [4800]
Difficult	160 [1700]	900 [10000]
Severe	240 [2600]	1200+ [13000]

*Lux figures rounded.

(a) British IES. As stated above, the British IES recommendations consist of an extensive list of recommended ''service luminances'' for specific tasks, modified by considerations of contrast, daylight availability, accuracy, and speed requirements. The method of use is to consult the detailed table of service luminances (not reproduced here; see *British IES Code,* 1977) if the *specific* task is defined, or the flow chart (see Table 18.3) if only the type of task is known, in order to arrive at a recommended service luminance. This is then amended by the self-explanatory technique shown on the flowchart, to arrive at a final illuminance recommendation. For example, if an illuminance for planing and benchwork in a woodworking shop is desired, the designer would consult the detailed list (*British IES Code,* 1977) and would find a recommendation of 500 lux. He then enters the flowchart (Table 18.3) at 500 lux (routine work) and, based on a knowledge of the task and work conditions, would either remain at 500 lux, decrease to 300 lux, or increase to 750 lux. These illuminance levels (300, 500, and 750 lux) are chosen specifically to correspond to minimum significant level change; according to this standard (see Table 18.4). If, alternatively, the designer knows only that the space under consideration is to be used as a woodworking shop, he could select the *standard service luminance* directly from the general categories in the flowchart (Table 18.3).

The author has found this system of illuminance selection to be rapid, efficient, and reasonable and has used it effectively in many design projects.

(b) American IES. This new (1981) system is patterned after the British system. In lieu of the single luminance recommendation that characterized the early IES tables, this system provides a luminance range determined by task difficulty (contrast, size), within which range a specific illuminance is selected based on three weighting factors—age of the observer, the importance of speed and/or accuracy, and the reflectance of the background on which the task is seen. The procedure operates as follows:

1. An illuminance category (A through I) is initially selected, based on either a general description of the activity involved (see Table 18.5) or, if known, on a specific activity in a specific setting. (The extensive tables for this latter selection are found in the 1981 *IES Lighting Handbook* and are not reproduced here.) The category selection gives a three-number range of illuminances that correspond to the minimum significant changes shown in Table 18.4. For instance,

TABLE 18.4 **Schedule of Illumination Levels GSA**

Each level represents a significant subjective change. (Changes in levels of substantially lesser magnitude are of little consequence to the eye.)

Footcandles	*Lux*
0.2	2
0.5	5
1.0	10
2.0	20
3.0	30
5.0	50
7.5	75
10.0	100
15.0	150
20.0	200
30.0	300
50.0	500
75.0	750
100.0	1000
150.0	1500
200.0	2000
300.0	3000

Source: Adapted from Note B of the "General Schedule" of the *British IES Code for Interior Lighting* (1977).

TABLE 18.5 **Illuminance Categories and Illuminance Values for Generic Types of Activities in Interiors**

Type of Activity	Illuminance Category	Ranges of Illuminances	
		Lux	Footcandles
General lighting throughout spaces			
Public spaces with dark surroundings	A	20–30–50	2–3–5
Simple orientation for short temporary visits	B	50–75–100	5–7.5–10
Working spaces where visual tasks are only occasionally performed	C	100–150–200	10–15–20
Illuminance on task			
Performance of visual tasks of high contrast or large size	D	200–300–500	20–30–50
Performance of visual tasks of medium contrast or small size	E	500–750–1000	50–75–100
Performance of visual tasks of low contrast or very small size	F	1000–1500–2000	100–150–200
Illuminance on task, obtained by a combination of general and local (supplementary lighting)			
Performance of visual tasks of low contrast and very small size over a prolonged period	G	2000–3000–5000	200–300–500
Performance of very prolonged and exacting visual tasks	H	5000–7500–10000	500–750–1000
Performance of very special visual tasks of extremely low contrast and small size	I	10000–15000–20000	1000–1500–2000

Courtesy of Illuminating Engineering Society of North America.

TABLE 18.6 **Illuminance Values, Maintained, in Lux, for a Combination of Illuminance Categories and User, Room, and Task Characteristics (for Illuminance in Footcandles, Divide by 10)**

Part A ***General Lighting Throughout Room***

Weighting Factors		Illuminance Categories		
Average of Occupants Ages	Average Room Surface Reflectance (%)	A	B	C
Under 40	Over 70	20	50	100
	30–70	20	50	100
	Under 30	20	50	100
40–55	Over 70	20	50	100
	30–70	30	75	150
	Under 30	50	100	200
Over 55	Over 70	30	75	150
	30–70	50	100	200
	Under 30	50	100	200

TABLE 18.6 (***Continued***)

Part B **Illuminance on Task**

Weighting Factors			*Illuminance Categories*					
Average of Workers' Ages	*Demand for Speed and/or Accuracy*[a]	*Task Background Reflectance (%)*	*D*	*E*	*F*	*G*[b]	*H*[b]	*I*[b]
Under 40	NI	Over 70	200	500	1000	2000	5000	10000
		30–70	200	500	1000	2000	5000	10000
		Under 30	300	750	1500	3000	7500	15000
	I	Over 70	200	500	1000	2000	5000	10000
		30–70	300	750	1500	3000	7500	15000
		Under 30	300	750	1500	3000	7500	15000
	C	Over 70	300	750	1500	3000	7500	15000
		30–70	300	750	1500	3000	7500	15000
		Under 30	300	750	1500	3000	7500	15000
40–55	NI	Over 70	200	500	1000	2000	5000	10000
		30–70	300	750	1500	3000	7500	15000
		Under 30	300	750	1500	3000	7500	15000
	I	Over 70	300	750	1500	3000	7500	15000
		30–70	300	750	1500	3000	7500	15000
		Under 30	300	750	1500	3000	7500	15000
	C	Over 70	300	750	1500	3000	7500	15000
		30–70	300	750	1500	3000	7500	15000
		Under 30	500	1000	2000	5000	10000	20000
Over 55	NI	Over 70	300	750	1500	3000	7500	15000
		30–70	300	750	1500	3000	7500	15000
		Under 30	300	750	1500	3000	7500	15000
	I	Over 70	300	750	1500	3000	7500	15000
		30–70	300	750	1500	3000	7500	15000
		Under 30	500	1000	2000	5000	10000	20000
	C	Over 70	300	750	1500	3000	7500	15000
		30–70	500	1000	2000	5000	10000	20000
		Under 30	500	1000	2000	5000	10000	20000

NOTES:

1. Average weighted surface reflectances, including wall, floor, and ceiling reflectances, if they encompass a large portion of the task area or visual surround. For instance, in an elevator lobby, where the ceiling height is 25 ft, neither the task nor the visual surround encompasses the ceiling, so only the floor and wall reflectances would be considered.
2. In determining whether speed and/or accuracy is not important, important, or critical, the following questions need to be answered: What are the time limitations? How important is it to perform the task rapidly? Will errors produce an unsafe condition or product? Will errors reduce productivity and be costly? For example, in reading for leisure there are no time limitations and it is not important to read rapidly. Errors will not be costly and will not be related to safety; thus, speed and/or accuracy is not important. If, however, prescription notes are to be read by a pharmacist, accuracy is critical because errors could produce an unsafe condition and time is important for customer relations.
3. The task background is that portion of the task upon which the meaningful visual display is exhibited. For example, on this page the meaningful visual display includes each letter, which combines with other letters to form words and phrases. The display medium, or task background, is the paper, which has a reflectance of approximately 85%.

[a]NI = not important, I = important, and C = critical.

[b]Obtained by a combination of general and supplementary lighting.

Source: Courtesy of Illuminating Engineering Society of North America.

Category B is 50–75–100 lux and F is 1000–1500–2000 lux.

2. In the next step the three weighting factors are introduced. Age under 40 reduces the light requirement; unusual demand for speed or accuracy increases it, and particularly low or high background reflectance (below 30% or above 70%) increases or reduces it, respectively. This step is not a mechanical one in that at the higher luminances (categories D through I), the designer is expected to become thoroughly familiar with the importance, duration, and visual difficulty of the specific task involved, since the design aims at *task* lighting and not simply general room illumination (see Table 18.5). Having decided upon the weighting factor, the designer can now select the recommended illuminance from Table 18.6.

The resultant recommended illuminance is "raw" or conventional illumination, that is, average maintained lux [fc], *on the task* for categories D–I, and *in the room* for categories A–C. It is not Equivalent Spherical Illumination (see Section 18.26). Although the value of ESI as a concept has been amply demonstrated, and its use as a design tool is recommended, the IES felt at the time of publication of its method that it could not use ESI illuminance values and still achieve the necessary consensus. Therefore, the calculation procedure for the target illuminance is a lumen method for categories A–C and a point-by-point method for the *task* illumination of categories D–I.

For a space with several tasks of varying visual difficulty, the designer is expected to so design his lighting and controls that task requirements are met without overlighting. A uniform layout keyed to the most severe task is energy wasteful and is to be discouraged.

The IES recommended illuminance values are *not* applicable to installations where a visual task is not the deciding factor. Such installations include merchandising spaces, displays of all sorts, theatrical and artistic lighting, lighting for mood, lighting for safety, light used as part of an industrial process, and so on.

Recommended illuminances for selected exterior spaces are given in Table 21.5. For an extensive list including sport and recreation lighting see *IES Lighting Handbook* (1981).

TABLE 18.7*a* **Federal Energy Administration and General Services Administration Recommended Maximum Lighting Levels (1975)**

Task or Area	*Footcandle Levels*
Hallways or corridors	10 ± 5
Work and circulation areas surrounding work stations	30 ± 5
Normal office work such as reading and writing (on task only), store shelves, and general display areas	50 ± 10
Prolonged office work that is somewhat difficult visually (on task only)	75 ± 15
Prolonged office work that is visually difficult and critical in nature (on task only)	100 ± 20

(c) Other Recommendations. As mentioned in Section 18.22, the U.S. government commissioned a number of studies, the partial results of one of which is shown in Table 18.7. This system is the only one that uses fatigue, in terms of extended task duration, as a major weighting factor in determining required task illuminance. The American IES presupposes eight-hour exposure; the time factor in the British flow chart (Table 18.3) refers to short exposure tasks, not long-term fatigue. Table 18.7*a* shows the recommended illuminance levels resulting from these studies. The level for "normal" office work is 50 fc [500 lux] ±. This level is increased to either 75 fc or 100 fc on the basis of cumulative fatigue, operator age, or eyesight problems according to the criteria of Table 18.7*b*.

(d) Comparison of Methods. To give the reader a basis for comparison we selected three common tasks and used each of the above systems to arrive at an illuminance recommendation. The tasks are

1. School classroom; 3–4 h uninterrupted, age under 40, accuracy not critical.

TABLE 18.7*b* **Relative Visual Task Difficulty for Common Office Tasks**

Task Description	*Visual Difficulty Rating*
Book or magazine, printed matter, 8-point type and larger	2
Typed original	2
Ink writing (script)	3
Newspaper text	4
Shorthand notes, ink	4
Handwriting (script) in No. 2 pencil	5
Shorthand notes, No. 3 pencil	6
Poor copy from copying machine	7
Bookkeeping	8
Drafting	8
Telephone directory	12

NOTE.
Table 18.7*b* may be used as a guide in evaluating the degrees of visual difficulty for office work. It is based on the concept that visual difficulty for this kind of work is not only a function of the intrinsic characteristics of the task and the lighting system, but also of the length of time in which the task must be performed. To use this table, multiply the difficulty rating, as shown in the table, for each task performed at a given work place by a single worker by the number of decimal hours per day it is performed; for example: 3 hours, 15 minutes = 3.25 decimal hours. Add the products for each task. If the sum is greater than 40, provide 75 fc on the work station. If the sum is greater than 60, provide 100 fc on the work station. Multiply the difficulty factors by 1.5 if the operator is over 50 years of age, or if he has uncorrectable eyesight problems.

Source: Ross and Baruzzini, *Energy Conservation Applied to Office Lighting.* Federal Energy Administration, 1975.

2. General business office—varied tasks; 6–8 h of work, age 20–60, accuracy important.
3. Architects' office—drafting table work; 6–8 h of work, age 20–55, accuracy important.

The results, tabulated in Table 18.8, show that the three systems give similar results, with the American IES giving the greatest range of discretion to the designer.

QUALITY OF LIGHTING

18.24 Considerations of Lighting Quality

Quality of lighting is a term used to describe all of the factors in a lighting installation not directly connected with quantity of illumination. Certainly it is obvious that if two identical rooms are lighted to the same *average* illuminance, one with a single bare bulb and the second with a luminous ceiling, there is a vast difference in the two lighting systems. This difference is in the "quality" of the lighting, a term that describes the overall scene—that is, the luminances, diffusion, uniformity, and chromaticity of the lighting.

Excessive luminances and/or excessive luminance ratios in the field of vision are commonly referred to as glare. The quality of the lighting system must also include the visual comfort of the system, that is, the absence of glare. When the discomfort glare is caused by light sources in the field of vision, it is known as *direct* or *discomfort glare*. When the glare is caused by reflection of a light source in a viewed surface, it is known as *reflected glare* or *veiling reflection* (see Fig. 18.22).

ILLUMINATION

TABLE 18.8 **Comparison of Lighting Level Recommendations (Lux)**

	Source of Recommendation		
Activity	*British IES*	*American IES Handbook (1981)*	*U.S. Gov't Agency*
Classroom	500	200–500	400–600
Business office	500	300–750	500–750
Drafting room	750–1000	750–1500	800–1200

NOTES:
1. All levels are in lux. Divide by 10 for fc.
2. All levels refer to lighting on the task.

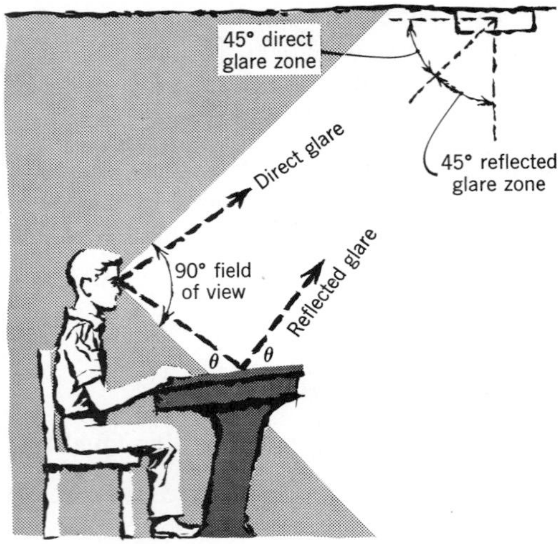

Fig. 18.22 Glare zones. The direct and reflected glare light paths are delineated on the diagram. Direct glare presupposes a head-up position, whereas reflected glare assumes eyes down, at a reading angle. Placement of lighting fixtures, room size, ceiling height, paint finishes, windows, and so on also affect luminance ratios and, therefore, glare.

18.25 Direct (Discomfort) Glare

The factors involved in producing discomfort glare are the luminance, size, and position of each light source in the vision field, plus the adaptation level of the eye. This last factor is also known as background brightness, because the eye adapts to the general or background brightness level. Obviously glare is proportional to source luminance and its size. It is necessary to consider these two parameters together to understand why a small, very bright source is not a serious problem, whereas a large, low brightness source (such as a luminous ceiling) may be. Indeed, a small bright source adds sparkle to the field of vision, and many observers find it pleasant in an otherwise monotonous lighting environment. Although discomfort glare from a scene is cumulative, source luminance is more important than the number of sources. For instance, if the luminance of a number of sources is halved, the reduction in glare is greater than is achieved by reducing the number of such sources by half. Indeed the latter procedure will have little effect on discomfort glare.

The remaining two factors are less self-evident. Glare decreases rapidly as the brightness source is moved away from the direct line of vision, and thus the glare produced depends on its position in the field of view. The amount of discomfort glare produced by a source is inversely proportional to the background brightness (eye adaptation level). Thus a ceiling fixture with a luminance of 1200 fL at 65° might easily constitute a source of discomfort glare in a space with an eye adaptation level of 50 fL. The same fixture would not be objectionable in a daylight condition, where the eye adaptation level might be 500 fL. A more striking example is that of an automobile's headlights, which at night are so severe a source of glare as to constitute "disabling glare," whereas in daylight, with its concommitant high eye adaptation level, these lights are noticeable, but not usually disturbing.

Keeping in mind the dependence of direct glare on eye adaptation level, a useful rule of thumb is that luminances of large sources should not exceed 2500 cd/m^2 [~ 700 fL] and small sources, 7500 cd/m^2 [~ 2200 fL]. The former is about the luminance of blue sky, the latter of a fluorescent lamp. The terms large and small depend not only on actual physical dimensions of the source but also on distance from the observer. That is, the actual criterion is apprehended size, or subtended visual angle, as shown in Fig. 18.18.

The sum of the individual glare source contributions is converted to a criterion called "visual comfort probability," or VCP, which is defined as the percentage of normal-vision observers who will be *comfortable* in that specific visual environment. The IES has established a set of standard conditions for which VCP of sources can be calculated. These include a 1000-lux illuminance, representative room dimensions, fixture height and observer position, and a head-up field of view limited to 53° above and directly forward from the observer (see Fig. 18.23).

With these conditions, direct glare will not be a problem if all three of the following conditions are satisfied:

1. The VCP is 70 or more.
2. The ratio of maximum-to-average luminaire luminance does not exceed 5 : 1 (preferably

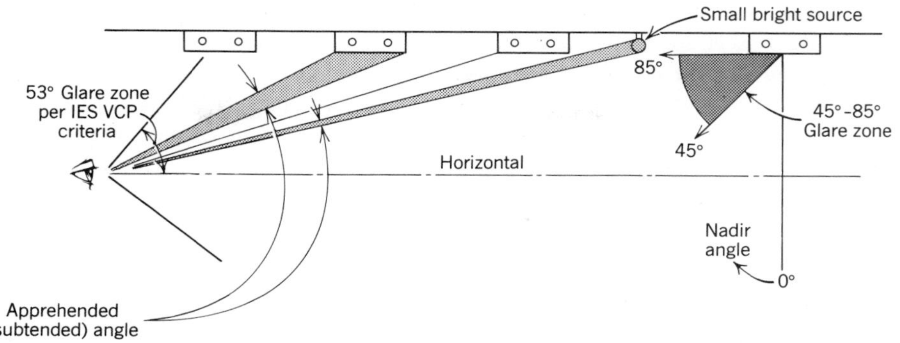

Fig. 18.23 *Glare determination. The glare contribution of each source depends on its size (subtended or apprehended* solid *angle), luminance, and location in the field of view. Note that the apprehended solid angle of a small source is such that even with high brightness, it is not objectionable. Such sources are normally called "sparkle." Glare will be much more objectionable with a dark background than with a light one; therefore, light-colored paints on ceilings and upper walls are recommended.*

3 : 1) at 45, 55, 65, 75, and 85° from the nadir, crosswise and lengthwise.

3. Maximum luminaire luminances crosswise and lengthwise do not exceed:

Angle Above Nadir, Degrees	*Maximum Luminance*	
	cd/m²	*fL*
45	7700	2250
55	5500	1605
65	3850	1125
75	2500	750
85	1700	495

A typical set of manufacturer's luminance and VCP data is shown in Fig. 18.24 for an actual ceiling-mounted fluorescent fixture with 4-40WT12 lamps. Note that all VCP values are considerably above the 70 minimum criteria. If full VCP data of this type are not available, an earlier criterion commonly called the "scissors curve criterion" can be used. This criterion, which is intended for application to a *single fixture,* by extrapolation can be applied to an entire space; that is, a room with fixtures that meet the scissors curve criterion will generally not fall below a VCP of 70. A luminance data check sheet using scissors curve criterion and typical fixture data is shown in Fig. 18.25. If we replace the scissors shown in the figure with the eye, this criterion tells us the limiting comfort brightnesses of the eye. That is:

Angle Above Nadir, Degrees	*Well-Tolerated Average Brightness, Footlamberts*
45	750
55	400
65	375
75	250
85	165

Despite the accuracy and excellence of the above VCP criterion, it is inherently limited by its own standard conditions, which are not easily applied to other situations, as follows:

1. In small spaces, VCP has little significance.
2. Tabulated VCP figures are given for the worst case in the room. Since VCP varies dramatically with observer position, the VCP values given are generally lower than actual.
3. VCP calculations are based on uniform layouts. Recent criteria effectively require nonuniform arrangements. To minimize veiling reflections, the closest fixtures in such arrangements are placed *outside* the frontal line of sight, thus decreasing direct glare and increasing VCP.

In view of the above (and other) reservations, most of which tend to make the actual direct glare situation better than the VCP calculation would indicate, it is recommended that layouts giving a VCP somewhat below 70 not be discarded out of hand. Instead, they should be carefully examined and, if possible, several observer positions calculated. By so doing the designer will develop a "feel" for glare sources that will enable him or her to make the intelligent judgments necessary. If the job warrants the expenditure of the additional computer time, there are programs available that will compute and plot VCP values in the space being studied and compute minimum or average VCP. With these as a guide, the designer can re-

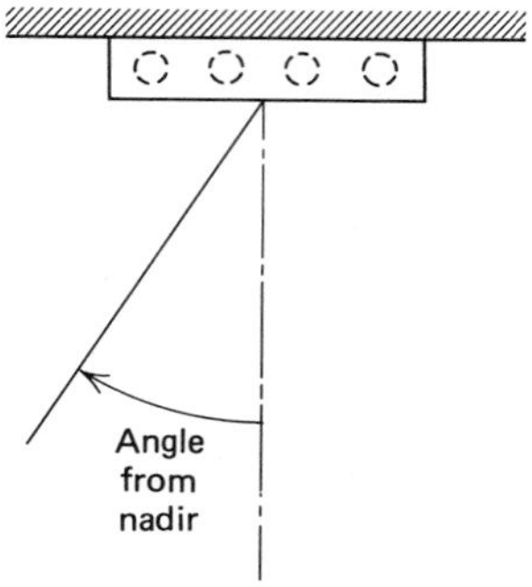

4–40-W Lamps
Prismatic Lens Diffuser

Average Luminance Data (fL)

Vertical Angles	Across Axes	45° Plane	Along Axes
60°	587	571	512
65°	310	302	317
70°	149	168	165
75°	121	120	112
80°	140	120	100
85°	162	136	122

IES Visual Comfort Probability Data

Room Size		Luminaires Lengthwise				Luminaires Crosswise			
W	L (in ft)	Ceiling Height (in ft)							
		8.5	10.0	13.0	16.0	8.5	10.0	13.0	16.0
20 ×	20	81	77	76	73	79	75	72	74
	30	81	78	76	74	79	76	72	70
	40	82	79	77	74	79	77	74	71
	60	81	80	78	76	79	77	75	72
30 ×	20	84	80	77	76	82	78	74	74
	30	83	80	77	73	82	79	74	70
	40	83	81	78	73	81	79	75	70
	60	82	80	78	74	80	78	76	72
	80	82	80	79	76	80	78	76	73

Reflectance:
Wall 50%
Ceiling cavity 80%
Floor cavity 20%
Work plane illumination: 100 fc

Fig. 18.24 *A typical set of manufacturers' published VCP and luminance data. (Data at 45°, 50°, and 55° are not presented.) (Courtesy of Holophane/Manville.)*

Crosswise and Lengthwise

Angle	Limit Value[1]	Maximum	Average	Ratio[2] Max/Avg
45°	2250			
55°	1605			
65°	1125			
75°	750			
85°	495			

[1]Three times the value of the sloped limiting line, in footlamberts
[2]Should not exceed 3.

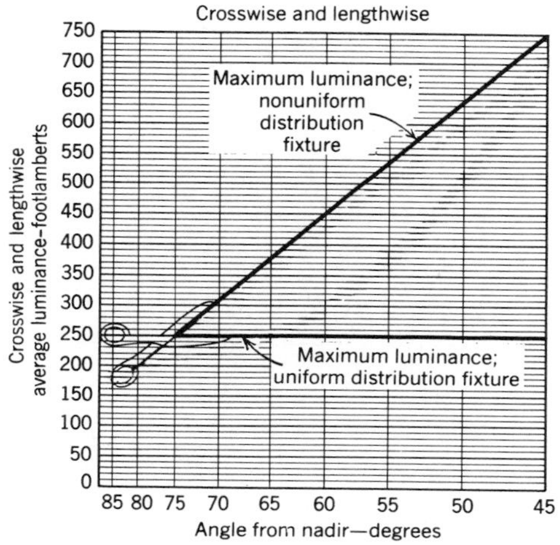

Fig. 18.25 *Fixture luminance (brightness) check sheet. Values of* average *luminance, in footlamberts, in both lengthwise and crosswise directions, are plotted on graphs of the type shown. All points must fall below the upper limit line for fixtures of nonuniform luminance and below the lower limit line (250 fL) for fixtures of uniform luminance (e.g., globes). Further, maximum fixture luminance should not exceed three times average at any angle, nor may the maximum exceed the limit values listed, which are three times the values of the sloping line.*

arrange and substitute equipment to obtain the desired condition.

See Section 20.26 for a comparison of the direct glare characteristics of lighting fixture diffusers.

18.26 Veiling Reflections and Reflected Glare

Although there is no generally accepted convention with respect to nomenclature, many people refer to *reflected glare* when dealing with specular (polished or mirror) surfaces and to *veiling reflections* when considering source reflections in dull or semimatte finish surfaces, which always exhibit some degree of specularity. We use the terms in that fashion in this book.

(a) Nature of the Problem. The problem of veiling reflections is much more complex than that of direct glare because it involves both the source *and* the task and is inherent in the act of seeing (refer to Fig. 18.26). Vision is produced by light being reflected from the object seen. The object mirrors the source(s) of the light into the eye of the observer. Thus if the object being viewed were replaced by a mirror, we would see the source(s) of light (see Fig. 18.26*a*). In an interior space there are multiple sources of light. The sources are usually one or more lighting fixtures near the observer. Other, more remote fixtures in the room are lesser sources of veiling reflections (see Fig. 18.26*b*). To the extent that the sources can be mirrored by the vision task, glare exists. (We are for the moment ignoring the sky as a light source and refer only to relatively small concentrated sources.) It is imperative to an understanding of this problem to appreciate the importance of the nature of the object being viewed, that is, the task. If the object were perfectly absorbent, that is, if it had a reflection coefficient of 0%, it would appear completely black as no light would be reflected into the eye (see Fig. 18.26*c*). Conversely, if the object were perfectly specular, like a clean mirror, and no light source were within the geometry of reflection, it too would appear black (see Fig. 18.26*d*). Thus if we took a mirror out on a cloudy night and shined a light on it from over our shoulder, it would be practically invisible since no light would be reflected into the eye.

The reader might try this experiment: in an inside space with a single overhead luminaire, try to examine the surface of a very clean, dust-free mirror. You will find that the best angle to hold it is *almost* at the angle at which the light source is seen. This is because the mirror is *almost* completely specular, and it is the slight diffuse reflection near the viewing angle that permits us to see the surface. Thus we understand that reflected glare is due to task surface specularity, whereas object definition, that is, the ability to see the task itself,

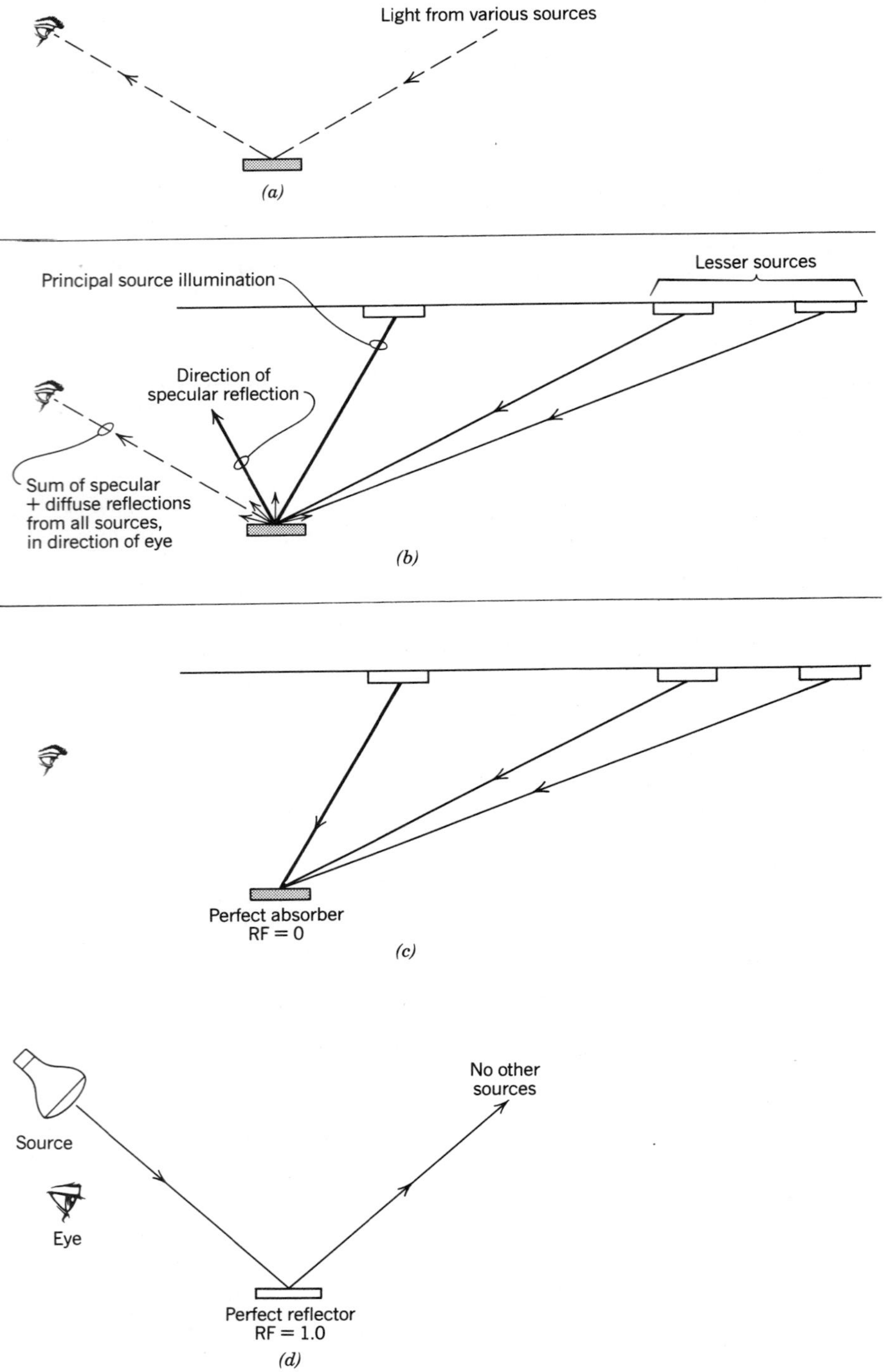

Fig. 18.26 (a) *The nature of the seeing process requires that light from the source(s) be reflected by the task into the eye.* (b) *The light entering the eye is the sum of all of the reflected light, specular and diffuse, from all sources, in the direction of the eye. If the task is specular, all of the sources will be seen reflected in the task.* (c) *A perfectly absorptive object is jet black since it reflects nothing.* (d) *A perfectly reflective one is also black since geometrically it cannot reflect light into the eye.*

is due to task surface diffuseness. A corollary of this conclusion is that veiling reflections, which are caused by mirroring of a source in the task, are proportional to source luminance and substantially independent of illumination level. The brighter the source, the more troublesome its reflection. These glare sources within the geometry of reflected vision are shown in Fig. 18.27 and the effects are shown in Fig. 18.28. Table 18.9 lists a few sample reflectance figures to demonstrate that most materials exhibit both a specular and a diffuse reflectance. In studying Fig. 18.27 it is important to note that a majority of visual work is done in the zone of 20 to 40° from the vertical, below the eye, and a maximum at the 25° angle shown in Fig. 18.29.

(b) Contrast Reduction. The principal effect of the reflection of a light source in a visual object is to reduce contrast between the object and its background, thus reducing visibility. It is as if a bright veil were spread over the object being viewed, which accounts for the nomenclature "veiling reflection." As the angle of the incident light approaches the viewing angle, the specularly reflected component of this light becomes more and more pronounced, and task contrast drops. This is clearly visible in Figs. 18.28 and 18.29. The worst situation occurs when the incident angle equals the viewing angle. When the specular reflectance of the task and background is high, as

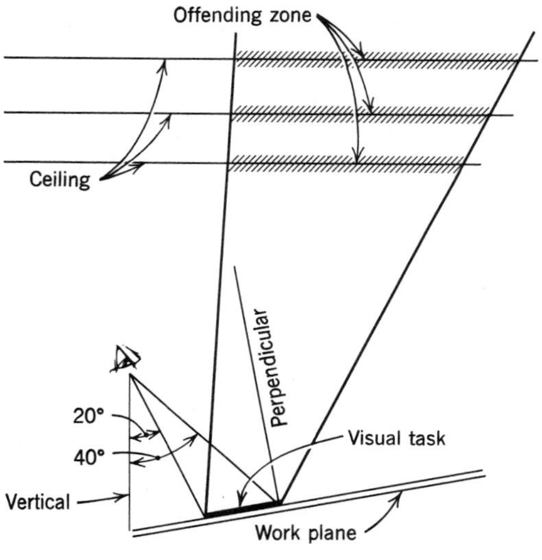

(a)

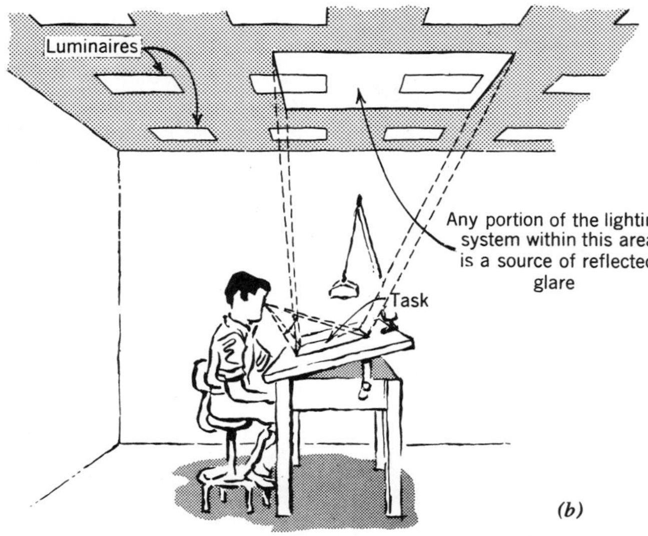

(b)

Fig. 18.27 *The geometry of reflected glare.* (a) *Since normal desk-top, head-down viewing angles vary from 20° to 40° from the vertical, the offending zone is the area on the ceiling corresponding to specular reflection between these two angles. (In specular reflection angle of incidence = angle of reflection.) Note that the higher the ceiling the larger this area becomes. In an office situation the draftsman in* (b) *would see ceiling fixtures in the offending zone reflected in his instruments, parallel straight edge, and work. Note the important fact that the offending zone moves back and becomes smaller as the table tilts up (see Fig. 18.34).*

ILLUMINATION

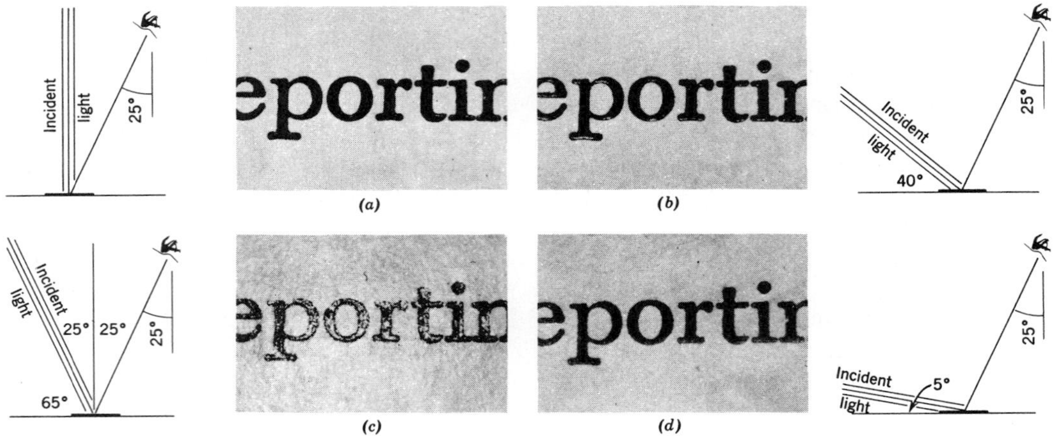

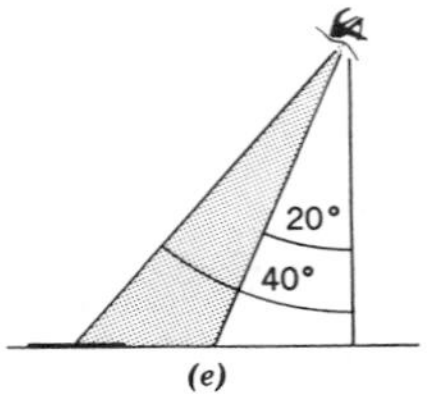

Fig. 18.28 Veiling reflection increases the difficulty of the seeing task. In (a), (b), *and* (d), *it is diffuse reflection that permits us to see, since the specular reflection angle does not correspond to the viewing angle except in* (c). *Veiling is pronounced at the 5° glancing angle even on a matte-finish surface. The normal viewing zone is from 20 to 40°, as shown in* (e). *(Courtesy of the IES of North America.)*

TABLE 18.9 **Typical Reflectances**

	Reflectance	
Material	*Specular*	*Diffuse*
Matte black paper	0.0005	0.04
Matte white paper	0.0030	0.77
Newspaper	0.0065	0.68
Very glossy white photo paper	0.048	0.83
Metallic paper—copper	0.11	0.28
Dull black ink	0.006	0.045
Super gloss black ink	0.039	0.016

Source: Courtesy of the IES.

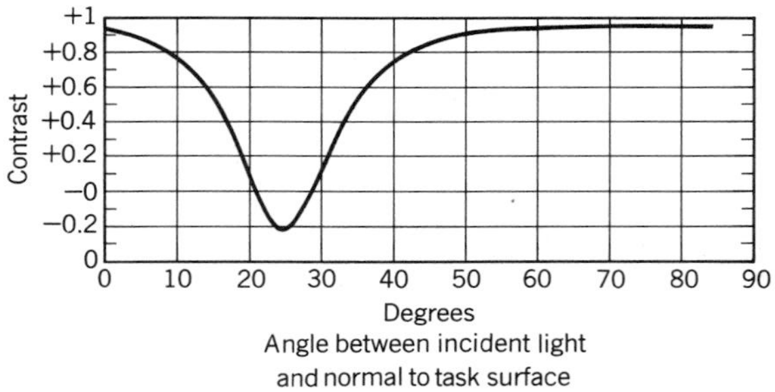

Fig. 18.29 Graph showing contrast reduction of a task with a specular background (such as a video monitor), as a function of the angle between incident light and the normal to the task surface. Viewing angle is assumed to be 25° from normal (see Fig. 18.28). Note that between 22° and 27° the contrast is negative. This indicates that background luminance exceeds that of the task, making the task essentially invisible. What is visible, is a reflected image of the source.

with the glass screen of a video display terminal, for instance, an image of the source is superimposed on the object, making viewing impossible (see Fig. 18.29). However, even with the high gloss finish of ''slick'' magazine paper, vision is still possible, although with much reduced efficiency.

When considering specular *and* diffuse reflectance, the equation for contrast given in Section 18.19 must be rewritten as

$$C = \frac{(L_{BD} + L_{BS}) - (L_{TD} + L_{TS})}{L_{BD} + L_{BS}} \qquad (18.8)$$

Where L_B and L_T are background and task luminances caused by diffuse (D) and specular (S) reflectance; that is, L_{BS} is the background luminance due to its specular reflectance, and so forth. If we were to rework the calculation of contrast of Section 18.19, including specularity, the result would be much different.

EXAMPLE 18.5. Assume an interior space lighted to an average illuminance of 75 fc [750 lux], using bare bulb fluorescent fixtures (luminance ≃ 7000 cd/m^2 or 2000 fL). The task is drafting with India ink on vellum. Reflectances are:

	Specular	*Diffuse*
Ink	0.021	0.038
Paper	0.018	0.71

Calculate the task contrast, without and with reflection of the 2000 fL source on the work.

SOLUTION.

(a) Without specularity, using equation 18.6 for diffuse reflection only:

$$C = \frac{R_B - R_T}{R_B}$$

$$= \frac{0.71 - 0.038}{0.71} = 0.946$$

(b) With specularity, using equation 18.8

$$C = \frac{(75 \text{ fc} \times 0.71 + 2000 \text{ fL} \times 0.018)}{L_{BD} + L_{BS}}$$

$$- \frac{(75 + 0.038) + (2000 \times 0.021)}{L_{BD} + L_{BS}}$$

$$C = \frac{(53.25 + 36) - (2.85 + 42)}{53.25 + 36} = 0.497$$

which is just over half of the previous contrast! If the contrast is normalized to the maximum contrast (as is usually done), the contrast reduction R can be expressed as

$$R = 1 - \frac{C}{C_{max}} \tag{18.9}$$

Thus, in this case, contrast reduction would be

$$R = 1 - \frac{0.497}{0.946} = 0.47$$

That is, contrast reduction would be 47%.

A similar calculation for clear, black typewritten material on good white bond paper yields a contrast reduction from 94% to 77% or R = 18%. In general, any contrast reduction above 15% is undesirable.

Since both specular and diffuse reflectances frequently vary with the angle of view, and exact figures are rarely available, accurate calculation is difficult. If a lighting system exists, or a mock-up can be made, measurements of contrast reduction can be accurately made with a contrast/luminance meter of the type shown in Fig. 18.9*c*. The method of use is shown in Fig. 18.30. A standard contrast device that is designed to correspond to a normal office (black typeface and white paper background) is positioned on the work surface and exposed to the ambient illumination. The task contrast is then measured at the same angle at which it would normally be viewed. Contrast reduction is automatically calculated and displayed.

A contrast reduction map of the work surface can thus easily be made (Fig. 18.30*b*). Thereafter, changes can be made in position of the work, viewer, or illumination sources to minimize contrast reduction. Note the pronounced effect of simply shifting the source out of the offending zone (see Fig. 18.27). Contrast reduction in the primary work area should not exceed 15% for good task visibility with specular work items such as glossy papers. If such a meter is not available, the system described in the next subsection can be used to predict lighting system performance as regards contrast reduction.

(c) Equivalent Spherical Illumination. Another way of approaching the problem of contrast reduction is to define a reference lighting system that is free of veiling reflections, and then relate actual lighting systems to it with a figure of merit. Conversely, one can measure the effectiveness of a given lighting system in terms of the equivalent glare-free system. Both of these ideas, which are essentially the same, are the basis of the concept of equivalent spherical illumination, normally known simply as ESI.

In order to achieve a lighting system almost free of reflected glare, it is necessary to construct an enclosed volume whose surfaces are uniformly reflective and whose primary source is obscured to the maximum extent possible. As illustrated (see Fig. 18.31), the integrating sphere is such a device. Light is introduced from the outside, split by a deflector, and evenly distributed throughout

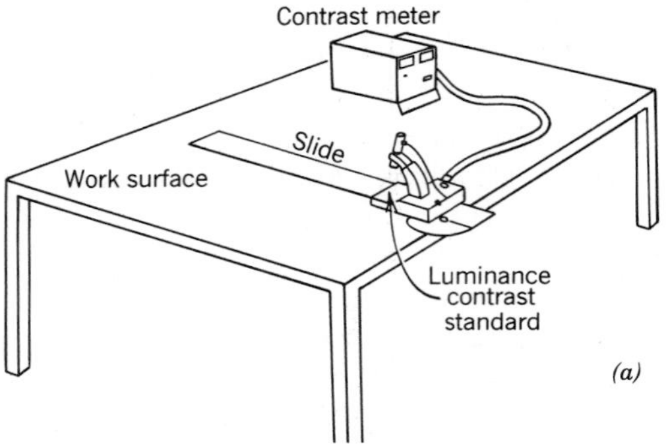

(a)

Fig. 18.30 The luminance contrast meter measures contrast reduction at the viewing angle. (a) *Since the sample task can be positioned at any point on the table, a chart of contrast reduction can be made for any viewing position and any task location.* (b) *Note that with a lighting fixture directly in the "offending zone," that is, above and in front of the viewer, a severe loss of contrast occurs over much of the work surface (see Fig. 18.27).* (c) *By shifting the relative position of the viewer and the source so that no source exists in the offending zone, contrast reduction is held to 3 to 4% over most of the work surface. Contrast reduction in the normal work area should not exceed 15%. (Courtesy of Brüel & Kjaer.)*

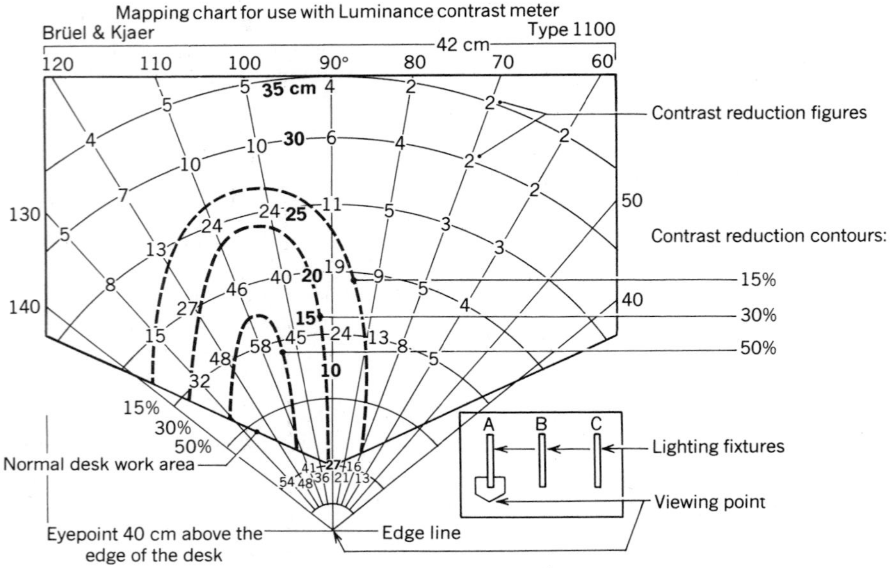

(b)

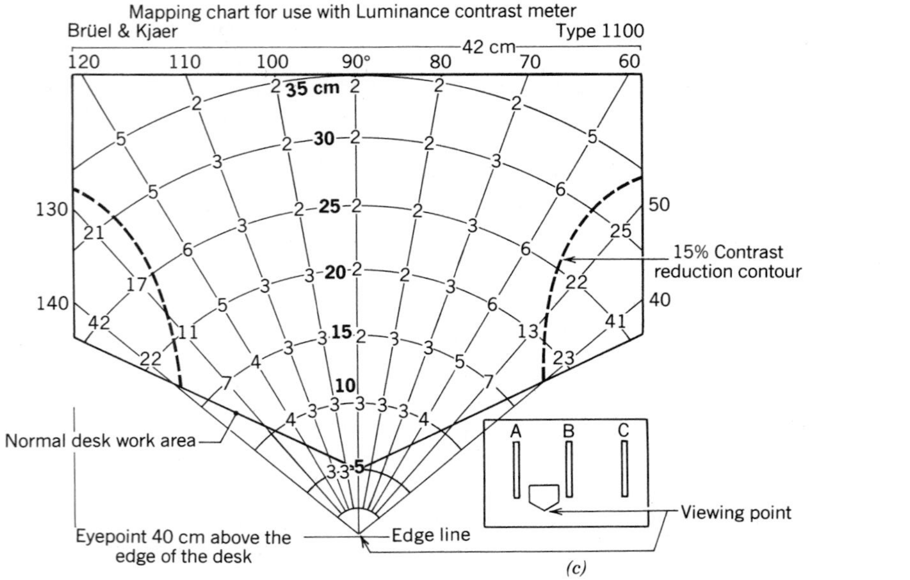

(c)

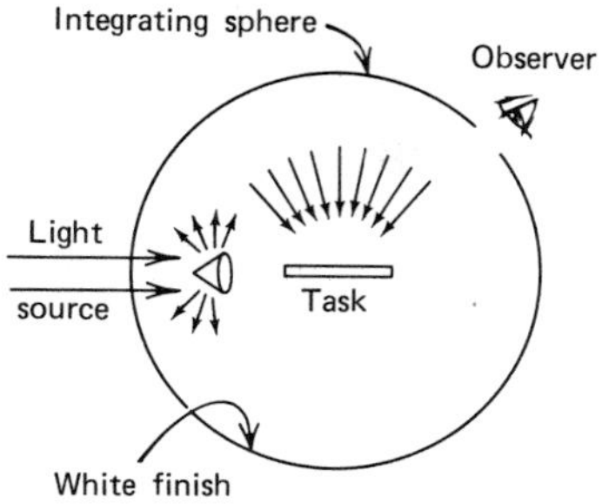

Fig. 18.31 Sphere illumination is produced by illuminating an object by reflection from the inside walls of an integrating sphere. The light source and observer are normally external.

the sphere by the multiple reflections from the white painted walls. The result is an evenly illuminated volume. When a task is introduced, the illumination falling on it is *entirely* uniform, that is, there are no high luminance sources reflected in it. It is therefore termed *spherically* illuminated. (Note the parallel to sky illumination.) The extent to which any other illumination system can duplicate this glare-free environment is that system's *equivalent* spherical illumination (ESI), and is simply the portion of its total illumination that is spherical, that is, diffuse, glare free.

Determination of ESI is accomplished by comparing contrast rendition in the spherical and test systems. A contrast rendition factor (CRF) of 1.00 would indicate that the system under test gives the same contrast rendition as the integration sphere and that all its illumination is spherical illumination. With a lower CRF, the ESI drops sharply. A study of school lighting (Sampson, 1970) gave the illustrated results for four viewing positions in a classroom lighted with ceiling-mounted continuous rows of 2 by 4 ft, four-lamp, 40-W fluorescent fixtures with lens type wraparound diffusers, on 10 ft centers (see Fig. 18.32). Carefully note that:

1. The CRF, and, therefore, the ESI, depends entirely on position and viewing angle, other factors in the space being equal.
2. In an ostensibly very well lighted (215 fc) position (M1), the useful illumination is only 28 fc!
3. The CRF can exceed 1.00, that is, the integrating sphere does not produce perfectly glare-free illumination but only nearly so. In such a location (M4), an increase in raw footcandles (118 to 236, i.e., double) results in a larger increase in ESI (i.e., 127.8 to 308.3, or 241%). Where CRF exceeds 1.0, ESI will exceed raw illuminance.

The results could have been anticipated, at least qualitatively, by examination of the observer positions vis-à-vis the layout. Positions M1 and M3 have bright sources in the offending zone, M1 more so than M3, as is borne out by the CRF figures. M2 is an excellent position in that it receives light contributions from the two sides; its footcandle value being lower than the others is due to wide row spacing. M4 is ideally placed with no glare sources in the offending zone but with a row of fixtures positioned behind it, so that it is geometrically impossible to act as a glare source. The CRF/ESI analysis gives quantitative expression to our qualitative prejudgment and as such is a valuable design tool. Note particularly that the CRF/ESI results shown in Fig. 18.32*a* clearly correspond to the results of a similar test made with the contrast meter, as shown in the charts of Fig. 18.30.

Figure 18.32*b* generalizes the conclusions drawn from Fig. 18.32*a*. Although it is drawn to reflect the ESI distribution of a single fixture (since there are an unlimited number of multiple-fixture arrangements), the principles shown are readily applicable to almost any layout. Recently, direct-reading ESI meters have come onto the lighting market and should prove valuable in the same fashion as the contrast meter described in the previous subsection (see Fig. 18.32*c*). However, it must be kept in mind that ESI is a device by which the essential characteristic that interests us, namely contrast reduction, can be determined. It seems to this writer that if measurements of actual systems are to be made, contrast reduction contour mapping would be more useful than ESI measurements, particularly if one remembers that the CRF/ESI relationship is nonlinear. See the numbers in Fig. 18.32*a*.

As with the VCP criteria for direct glare, so with the CRF/ESI criteria: there are ameliorating factors that generally make a given lighting system better than these criteria figures would indicate. Some of these factors have been mentioned already but bear repetition.

ILLUMINATION

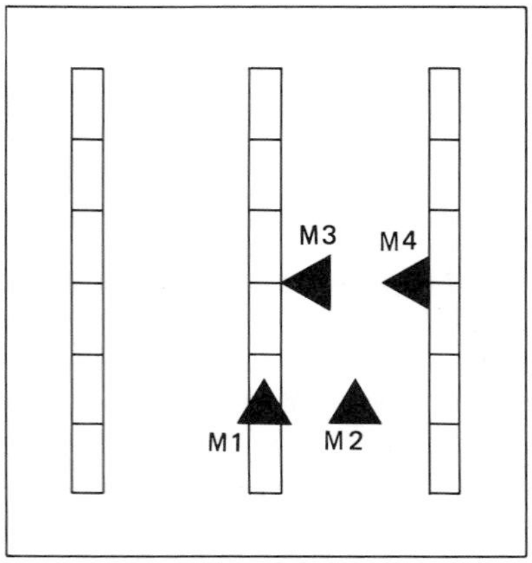

(a)

		Observer Position			
		M1	*M2*	*M3*	*M4*
TI	2L	108	92	125	118
	4L	215	185	250	235
CRF	2L	.75	1.00	.82	1.01
	4L	.76	1.00	.83	1.03
ESI	2L	17.8	91.9	31.5	127.8
	4L	28.4	185.3	58.1	308.3

TI–Task illumination
2L–2 lamps (inside pair)
4L–4 lamps

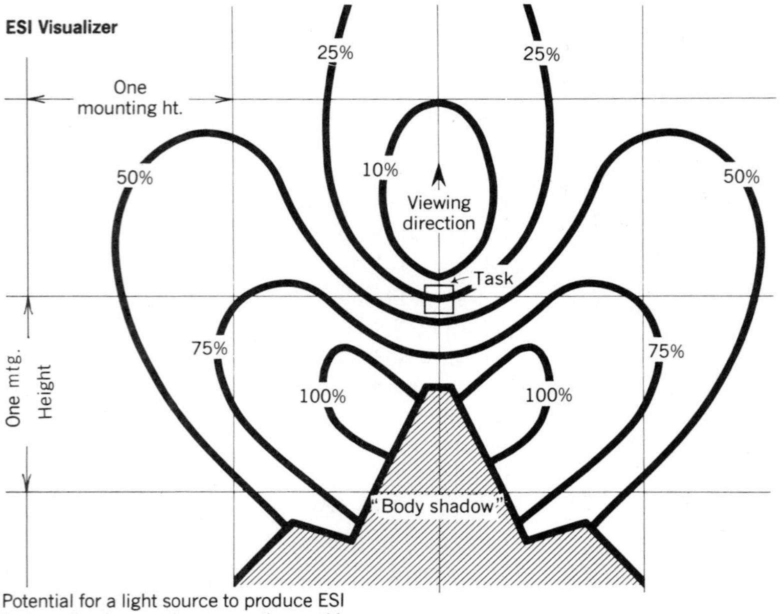

Potential for a light source to produce ESI footcandles from various mounting positions above a work plane.

Limitations of ESI Visualizer

- Relative ESI values only, not absolute. Do not use to calculate actual ESI.
- Applies only to single luminaire. Does not allow for effects of other luminaires in room (those effects combine nonlinearly, so cannot be simply summed).
- Applies to "standard" ESI conditions—pencil handwriting lying flat being viewed at 25°.

(b)

Fig. 18.32 (a) *A test classroom illuminated by three widely spaced rows of four-lamp fixtures with lens-type wraparound diffusers. Observer positions are shown by arrows. (Reproduced from Sampson, 1970.) The row of fixtures in front of position M4 are too far forward to be in the offending zone.* (b) *This diagram should assist the designer in predicting (and avoiding) reflected glare situations. Note that the 100% contour corresponds to position M4 in* (a), *the 10 to 25% contours cover positions M1 and M3 and the 75% contour roughly corresponds to the situation of position M2. In each case the designer must extrapolate by shifting the diagram to account for additional fixtures. (Courtesy of Widelight Corp.)*

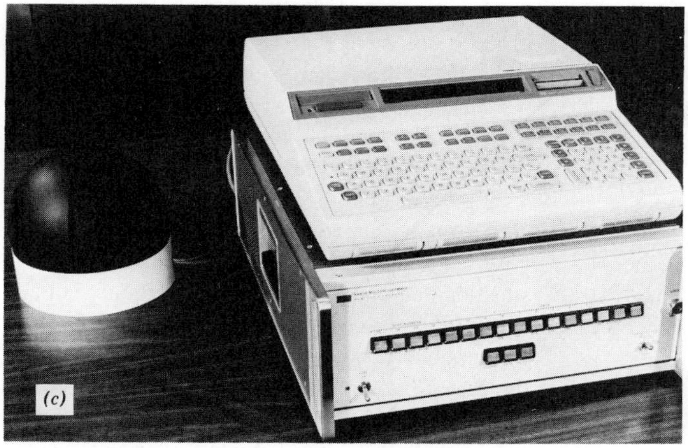

Fig. 18.32 (c) *The illustrated instrument is a practical, direct-reading ESI meter, suitable for field use. It measures light distribution at a given point, combines those data with task type and viewing direction, and calculates ESI illuminance, which is then digitally displayed. Stored within the device memory are the required data and calculation procedures. The black, slotted, hemispherical illuminance head is shown at the left. To its right are the calculation and display elements. (Courtesy of Holophane R & D/Manville Service Corp.)*

1. CRF and ESI are critically dependent on observer position and viewing angle. Although position is generally fixed by chair location, observers can and do change viewing angle and head aspect to correct for glare situations.
2. The nature of task (i.e., its specularity) is assumed to be fixed and unique. In many situations the task nature varies from hour to hour and thus also the CRF. When tasks *are* constant, a poor CRF will frequently lead to an improvement in the task, thus changing the CRF and ESI.

The procedure for calculating CRF and ESI requires a computer and presupposes that the detailed photometric characteristics of the lighting system and the task are available. A list of available computer programs can be found in Appendix **H.** To assist designers not equipped to run these programs, many luminaire manufacturers will make available computer printouts of both "raw" and ESI illuminance, point-by-point, for varying input data. See Section 20.38, Lighting Design Aids.

18.27 Control of Reflected Glare

Since the causes of veiling reflections are well understood, it would seem that a solution to the problem should by now have been adduced. Unfortunately, this is not the case. Although there is no known lighting method or material that will completely eliminate veiling reflections, there are a number of techniques that will minimize contrast loss due to veiling reflections while maintaining adequate illumination. These are:

Physical arrangement of sources, task, and observer so that reflected glare is minimal. See Section (*a*) below.

Adjusting brightnesses (eye adaptation level) so that objectionable brightness is minimized. See Section (*b*) below.

Design of the light source so that it causes minimal reflected glare. See Section (*c*) below.

Changing the task quality. See Section (*d*) below.

These techniques are discussed individually as follows:

(a) Physical Arrangement of System Elements. Since reflected glare is caused, as the name states, by reflection from a specular surface, the simplest and *most effective* technique is to arrange the geometry of the system so as to avoid the possibility of reflection. That is, we must remove the source from the offending zone, as in Fig. 18.33. Unfortunately this is totally effective only when a single luminaire is involved and when its placement with respect to the observer is completely adjustable—a rare combination. As should be clear from Figs. 18.30 and 18.32 and the related discussion, in a larger space utilizing multiple sources, particularly in continuous rows, placing the work between rows with the line of sight parallel to the long axis of the units is a very

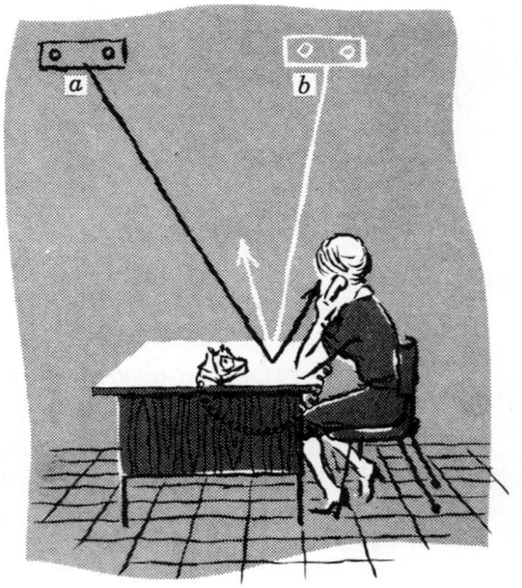

Fig. 18.33 Lighting fixture at (a) *will produce more glare than one at* (b) *because of the geometry of the light rays. Desk finish and luminaire brightness can be chosen to minimize loss of contrast.*

effective technique (see also Fig. 18.32, position M2). Position M4 is ''dangerous'' in that the center row can be a source of reflections. In this case it is not due to the wide 10-ft spacing (see Fig. 18.34*a*).

Remember also that the offending zone is dependent on the tilt of the desk, assuming the work is to be kept flat on it. Thus for a horizontal 3 × 5 ft standard desk, the offending zone is forward of the desk, as in Fig. 18.34*a*; with an elevated table, the ceiling glare source zone may well be behind the source, as in Fig. 18.34*b*. This being so, it is often possible to reduce glare simply by tilting the work and/or the work surface to such an angle that glare is eliminated.

All of the above geometric solutions presuppose a known detailed fixed-furniture layout, a situation that obtains in many but certainly not all cases. In the absence of such data, two alternatives are possible: to do a uniform layout and adjust the furniture to it, or vice versa. In practice, a combination of both is the most practical approach. Indeed, with the energy-oriented criteria discussed above requiring a nonuniform layout (to give task lighting rather than maximum levels throughout), after-the-fact rearrangement of fixtures is mandatory. Since low watts per square foot budgets have made ducted lighting fixture heat removal systems (air troffers) much less necessary, fixtures are easily shifted. This mobility is further enhanced by the extreme flexibility of lighting fixtures fed from ceiling plug-in raceways as discussed in Section 15.30. Figure 18.35 shows such a rearrangement, which results in saving five fixtures, a load reduction of 1 kW, and an *improvement* in visibility. The load density of 2.4 W/ft^2 is close to the recommended values (see Section 20.5).

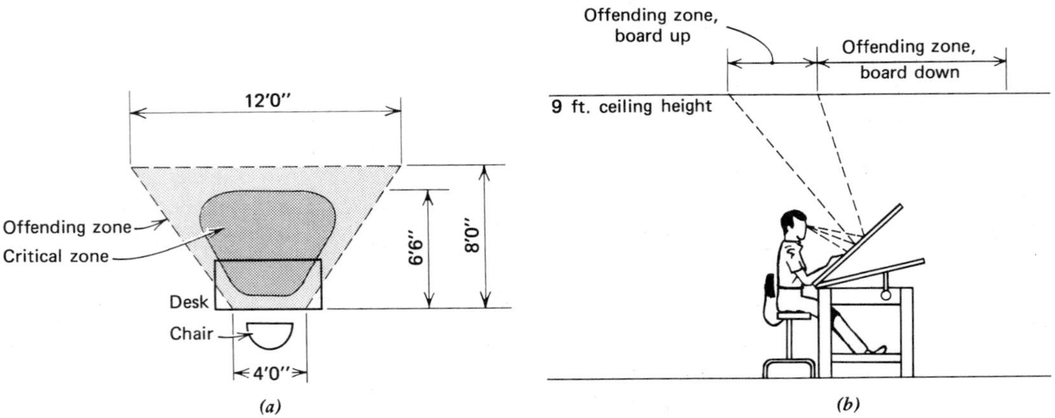

Fig. 18.34 (a) *If luminaires are kept out of the trapezoidal offending zone, a CRF of about 1.0 should be obtained. If the bulk of one or more luminaires projects into this zone and, in particular into the critical zone, CRF will drop sharply. These figures are for specular tasks. The dimensions shown are for a flat desk 3 ft × 5 ft and a 9-ft ceiling height. (From Ross and Barruzini, Lighting Systems Study, 1974.)* (b) *The dependence of glare zone on table tilt is illustrated. The offending zone becomes small as the table is raised so that with a table near vertical position glare is all but eliminated. See Fig. 18.27.*

ILLUMINATION

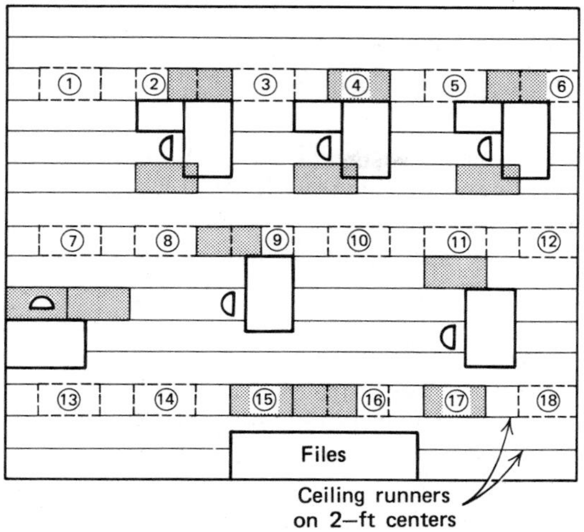

Fig. 18.35 The original uniform fixture layout utilized three rows of six 2 × 4, four-lamp fixtures, giving a total load of 3600 W, a load density of 3.3 W/sq ft, and a uniform illumination level of approximately 90 (raw) fc [900 lux]. The original layout is shown dotted and numbered. The rearranged layout uses 13 fixtures (shown shaded) for a total 2600 W, a load density of 2.4 W/sq ft, and better than 100 ESI fc [1000 lux] on each work surface. In addition, five fixtures are saved.

(b) Control of Area Brightness and Eye Adaptation Level. As discussed in Section 18.19, loss of contrast can be compensated for (and glare eliminated) by increased overall nonglare illumination. We are simply making the task brighter to override the detrimental veiling reflection. The problem with this technique, however, is that a large increase in illuminance is required to overcome the glare. This required increase can, in many instances, be most practically accomplished not by increasing overall room illumination with the associated extremely high energy consumption but by adding a supplementary source so arranged as to be free of reflected glare. By making this supplementary source's position adjustable (as in Figure 18.27*b*), we accomplish three things.

1. Veiling reflection is overcome.
2. The high level of illumination needed for exacting tasks is provided, with minimum energy expenditure.
3. The observer is granted complete control with resultant optimum lamp placement plus psychological satisfaction that will generally prevent worker complaints.

(Optimum position is generally to the left and slightly forward of the task).

We can demonstrate the effectiveness of a supplemental desk lamp by returning to Example 18.5.

EXAMPLE 18.6. Recalculate the contrast reduction of the ink on vellum visual task of Example 18.5, assuming that a desk lamp raises the illuminance to 200 fc [2000 lux] and is positioned so as to be glare free. (*Note:* An adjustable lamp with 2–15-W fluorescent tubes will produce about that illuminance on the task.)

SOLUTION. Contrast from Equation 18.8:

$$C = \frac{(L_{BD} + L_{BS}) - (L_{TD} + L_{TS})}{L_{BD} + L_{BS}}$$

$$C = 1 - \frac{200 \times 0.038 + 2000 \times 0.021}{200 \times 0.71 + 2000 \times 0.018}$$

$$C = 0.72$$

With this contrast (0.72), the contrast reduction from the original no-glare situation has been improved from the original 47% (0.946 to 0.497) to 24% (0.946 to 0.72). Since even a contrast reduction of 24% is unacceptable, a change in task-source geometry or a change in source luminance would be required, assuming that the task itself must remain unchanged.

(c) Control of Source Characteristics. The reflected brightness that causes loss of contrast is proportional to the luminaire luminance. It is apparent then that glare may be reduced by reducing luminaire luminance at the reflection angle. This can be accomplished in four ways:

Dimming or switching lamps. See number 1 below.

Using luminaires with lower overall luminance. See number 2 below.

Using the luminaire as a primary source to illuminate a large, low-brightness secondary source. See number 3 below.

Reduce the luminaire luminance *only at the offending angles*. See number 4 below.

1. Reducing the total output of a fixture will also reduce its output in the critical portion of the ceiling glare zone and can actually *increase* the ESI footcandles (i.e., improve task contrast).

2. In lieu of using a few small high-output sources, utilize larger-area, low-output sources (see Fig. 18.36). This has the effect of reducing the source luminance in the ceiling glare zone while increasing the illumination contribution from outside the glare zone, resulting in better contrast for the same or lower illumination level (lux). The disadvantage of this technique is increased fixture cost.

3. To overcome the economic disadvantage of multiple low-output, low-luminance sources, the ceiling can be used as a secondary source illuminated from high-output indirect or semi-indirect fixtures (see Section 20.11). These sources, which can be fluorescent or HID (mercury, metal-halide, sodium), have the advantage of high efficiency. The space's ceiling height must be sufficient to permit suspending the unit at least 18 in. (≈ 50 cm) down, to avoid ''hot spots'' on the ceiling. The minimum suspension length depends on the luminaire characteristic and is normally provided by the manufacturer. To assure high efficiency the ceiling should be painted with a high-reflectivity matte white paint and kept clean. The results obtained from a semi-indirect installation using 1500-mA, very high-output lamps are shown in Fig. 18.37*a*. Another, utilizing a 400-W indirect metal halide lighting unit, is illustrated in Fig. 18.37*b*. Of extreme importance is the CRF in excess of 1.00 in both installations with correspondingly high ESI. These results are typical of well-designed indirect lighting installations.

4. Since most vision takes place at about 25° (see Figs. 18.27 and 18.28), the light emitted by a fixture at 25° angle is the offending light. Therefore, any fixture that emits little or no light below 40° (which is the maximum vision angle) *cannot* produce veiling reflection regardless of its position in the field of view. This conclusion has been borne out by numerous studies that compared many different types of diffusers as to the direct and reflected glare produced. The results are shown in Fig. 18.38. As a result, diffuser manufacturers designed and produce a prismatic diffuser whose output below 30° and above 60° are diminished, in order to minimize reflected and direct glare, respectively. Due to the characteristic shape of the distribution curve, elements that are so designed are known industrywide as ''batwing'' diffusers or lenses. If observers can be positioned so that their sight lines are parallel to the longitudinal axis of the ceiling fixtures, lenses with linear (side-to-side) batwing characteristics will perform well. If the observing position varies in aspect with respect to the fixture, a radial batwing curve (in all directions) is required. Figure 18.39 shows batwing distribution curve and recommended luminaire arrangement. Diffusers and their characteristics are discussed in Section 20.26. There the reader will find the ''hardware'' that produces the batwing distribution, as well as a description of the glare-producing potential of common luminaire diffusers.

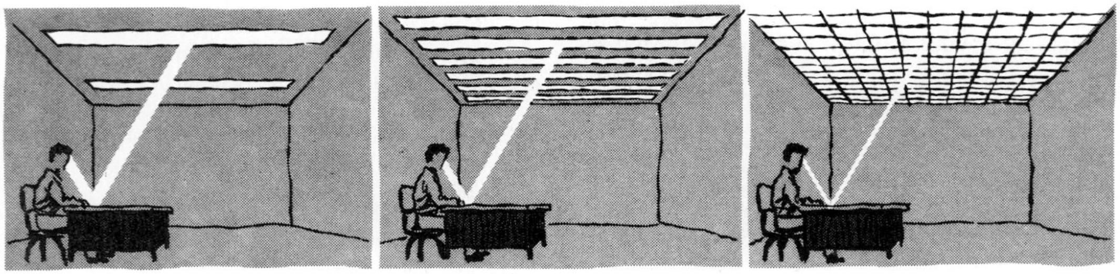

(*a*) (*b*) (*c*)

Fig. 18.36 *A concentration of light in the glaze zone* (a) *produces largest amount of reflected glare. As number of light sources is increased* (b) *in the glare zone and luminance is decreased, reflected glare is decreased. Least glare is from all-luminous ceiling, which also has lowest luminance* (c). *A similar result to* (c) *can be obtained with indirect lighting as in Fig. 18.37. (Courtesy of IES of North America.)*

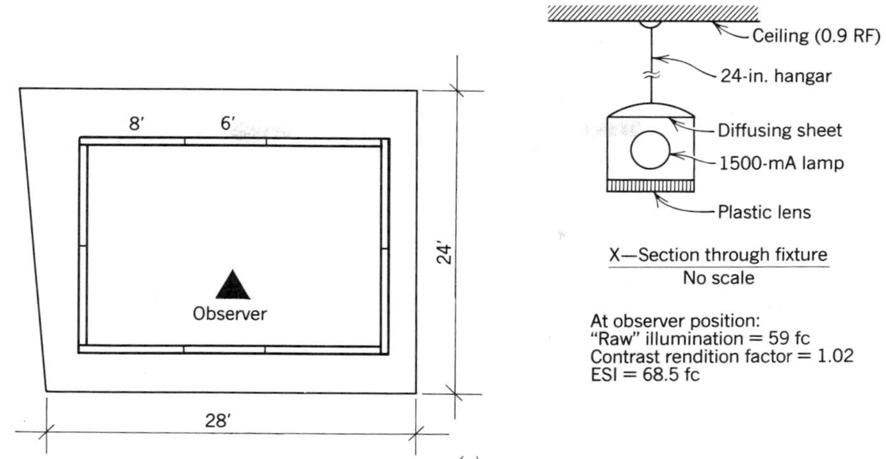

(a)

Fig. 18.37 (a) *With a high-reflectance, matte-finish ceiling this installation yields more ESI than raw footcandles (CRF > 1.0). Load is approximately 3 W/sq ft. (Reproduced from Sampson, 1970.)*

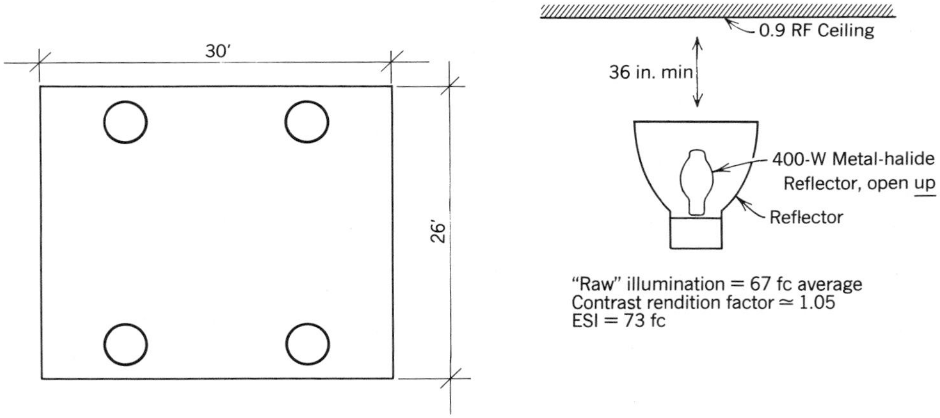

(b)

Fig. 18.37 (b) *Utilizing industrial reflectors to produce indirect lighting yields approximately the same results as in* (a) *above with only 2.2 W/sq ft. Due to compactness and high intensity of the source, a minimum of 3 ft between fixture and ceiling is desirable. Because dust accumulation in an open, upended reflector can cause considerable light reduction, this technique should be restricted to air-conditioned and other clean spaces.*

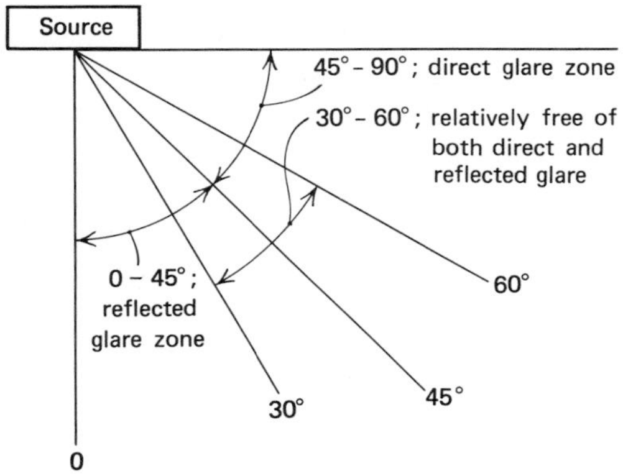

Fig. 18.38 *Glare zones are 0 to 45° and 45 to 90° for reflected and direct glare, respectively. Therefore, a diffuser that emphasizes the 30 to 60° zone will be least objectionable on both counts.*

ILLUMINATION

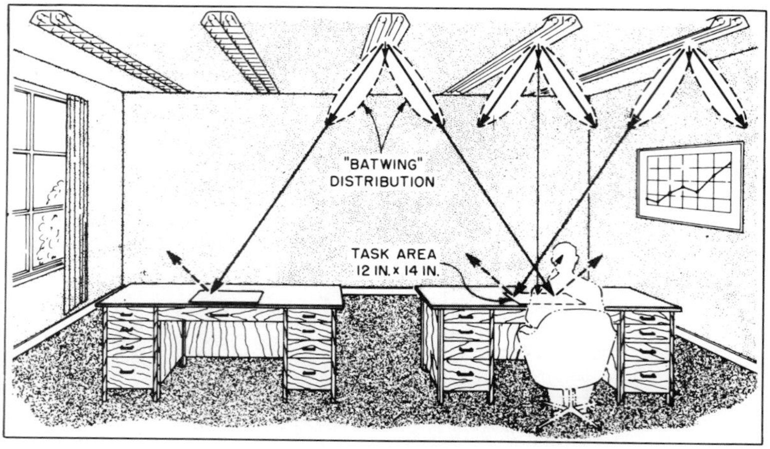

Fig. 18.39 *The action of the batwing is illustrated here. Note that the principal light output cannot produce veiling reflections. Compare this idealized characteristic with photometric test curves in Figs. 20.32, 20.33, and 20.34. (Courtesy of Illuminating Engineering Research Institute of the IES of North America.)*

(d) Changing the Task Quality. At this point it is abundantly clear that reducing the task specularity is at least as effective a means of reducing veiling reflections as changing the lighting system characteristics, if not more so. Traditionally, however, this option has been outside the purview of the lighting designer. With energy limitations having legal force, no area of consideration should be considered untouchable, least of all the nature of the work. It is therefore recommended that task contrast and specularity be actively considered and recommendations made in a framework of energy effectiveness and cost effectiveness. Thus, to produce adequate visibility it will often be cheaper to upgrade the task (in the visibility sense) than to upgrade the lighting system. Similarly, it will almost always be energy-economical to improve the task quality. A few suggestions follow:

1. Use felt pens and nonspecular paper for office tasks.
2. Use paper in lieu of pencil cloth, nylon, or photo reproductions in drafting work.
3. Provide means for tilting the work and changing the observers' viewing angle or position.
4. Institute a testing program to determine task characteristics and determine possible substitutions.

18.28 Luminance Ratios

As explained above, visual performance increases with contrast—that is, with difference in luminance between the object being viewed and its immediate surroundings. Conversely, however, the difference between the average luminance of the visual field (task) and the remainder of the field of vision should be low to avoid the discomfort of large rapid changes in eye adaptation level. Restated, *contrast is desirable in the object of view but undesirable in the field of view*. The room for which the basic scissors curve data were developed used reflectances of 50, 30, and 80% for walls, floor, and ceiling, respectively, and 35% for furniture, in order to establish a fairly high eye adaptation level so that direct glare that results from excessive luminances in the field of view is minimized. Recommendations for *maximum* luminance ratios to achieve a comfortable environment vary for different environments. Average figures for commercial interiors are presented in Table 18.10. To achieve these luminance ratios, it is obviously necessary to control carefully the reflectances of the major surfaces in a room. The 50, 30, 80, and 35% figures given above are recommended averages for work-type commercial and educational spaces. The marked difference between a background with proper reflectance and

TABLE 18.10 **Recommended Maximum Luminance Ratios**

To achieve a comfortable brightness balance, it is desirable and practical to limit luminance ratios between areas of appreciable size from normal viewpoints as follows:

1 to $\frac{1}{3}$	Between task and adjacent surroundings
1 to $\frac{1}{10}$	Between task and more remote darker surfaces
1 to 10	Between task and more remote lighter surfaces
20 to 1	Between luminaires (or fenestration) and surfaces adjacent to them
40 to 1	Anywhere within the normal field of view

These ratios are recommended as maximums; reductions are generally beneficial.

one with excessive brightness ratios caused by the low surrounding reflectances is shown in Fig. 18.40.

Fig. 18.40 *The reflected glare from luminaires disappears when a piece of light, diffuse linoleum is placed over the dark, polished desk top. Light-colored desk tops with a 35 to 50% reflectance result in task-to-background ratios within the 3:1 recommended range. Before the linoleum was placed, a reflection similar to the one seen above also existed on the left of the desk due to another luminaire. (Courtesy of the IES of North America.)*

18.29 Patterns of Luminance

Returning to the list of characteristics in Section 18.16, we note among the secondary factors in illumination the existence of "patterns of luminance." This is a way of describing the patterns of light and shadow in a space as they result from the method of illumination in that space. Thus a single source produces sharp shadows while a luminous ceiling or a completely indirect illumination system produces almost completely diffuse light (see Fig. 18.41). Diffusion is the degree to which light is shadowless and is therefore a function of the number of directions from which light impinges on a particular point and the relative intensities.

Perfect diffusion, rarely obtainable, would have equal intensities of light impinging from all directions, therefore yielding no shadows. Diffusion is generally judged by the depth and sharpness of

ILLUMINATION

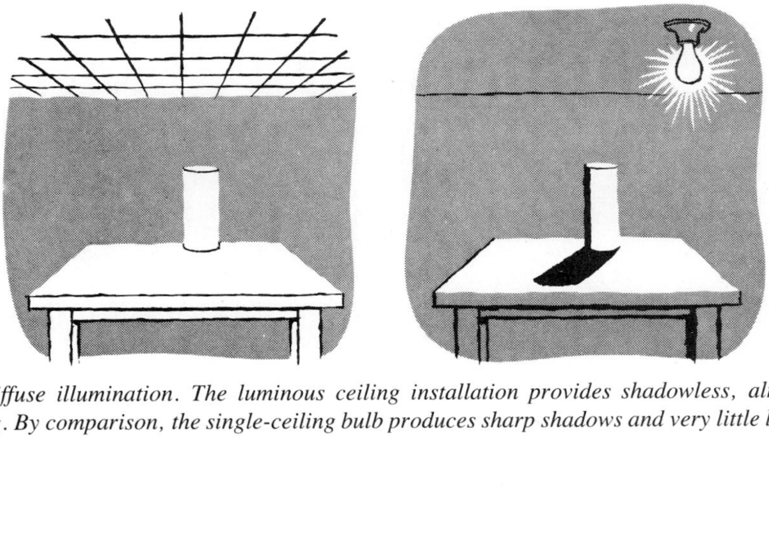

Fig. 18.41 *Diffuse illumination. The luminous ceiling installation provides shadowless, almost perfectly diffuse lighting. By comparison, the single-ceiling bulb produces sharp shadows and very little light diffusion.*

shadows. A room with well-diffused illumination resulting from multiple sources and high room surface reflectances yields soft multiple shadows that do not obscure the visual task. The only naturally occurring example of perfectly diffuse lighting is a daytime fog, which we know to be extremely disturbing to the eye, demonstrating that some directivity is desirable.

Some designers maintain that diffuse lighting is better than directional lighting for all installations. Although this is frequently true for offices, schoolrooms, machine shops, and drafting rooms where shadows would be highly disturbing and could be dangerous (as in the case of a machine shop), it is decidedly not the case where texture must be examined or surface imperfections detected by grazing angle reflections, or in any installation where the flat monotony of diffuse lighting is undesirable. For this reason, some directional lighting is often introduced as an adjunct to diffuse general lighting to lend interest by producing shadows and high brightness variations.

Indeed, as seen in Figure 18.42, directional light is what creates shape and is precisely the characteristic best used to influence architectural space and form.

Sections 20.10 through 20.15, which deal with systems of lighting, illustrate a few of the light/dark patterns produced by different lighting arrangements. The combinations of uplighting and downlighting, perimeter and ceiling lighting, are legion; each produces its own shadows and modeling, and each has a quality of its own. It is very much in the interest of the lighting designer to be familiar with these effects so that he or she can mentally visualize them as the design progresses. Indeed it would be well for a designer to prepare a reference sketchbook of such shadow diagrams. It is these patterns of light and darkness that give the ambience and the subjective reactions of sociability/isolation, clarity/fuzziness, spaciousness/crampedness, simplicity/clutter, formality/informality, boredom/excitement, definition/shapelessness, and so on. (Color has a great deal to do with subjective reactions and is discussed separately below.) The subject of psychological reactions to lighting environment is extensive and complex and can be only touched upon here to the extent of mentioning a few of the salient lighting techniques and their usual subjective responses.

In addition to modeling and texture accent, spots of high brightness, which can be called "glare" or "sparkle" depending on one's point of view, create interest and visual excitement. Lighting installations generally yield a sense of vividness or activity proportional to the level of illumination. However, this is not the case with very diffuse lighted areas, which even at high footcandle levels are tedious. This is particularly noticeable in large, low, luminous ceiling installations, which are

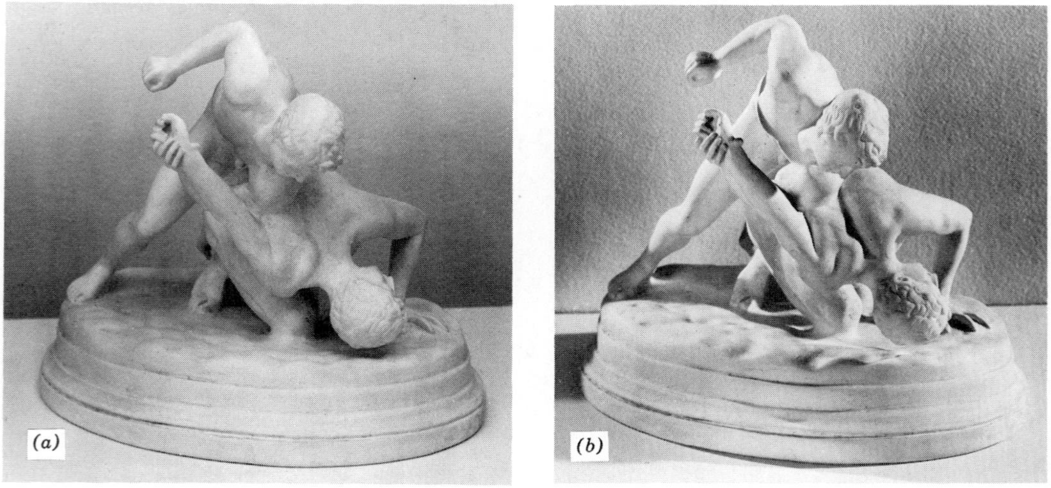

***Fig. 18.42** Totally diffuse lighting* (a) *destroys texture, whereas a combination of diffuse and directional lighting* (b) *produces the required modeling shadows. (Courtesy of Holophane/Manville.)*

completely lacking in visual interest. Small exposed incandescent lamps, a brightly lighted, rough-textured wall, and pendant fixtures with pierced reflectors are some of the techniques used to create this visual interest.

Visual attention can be drawn by high brightness. This well-known fact is used constantly in displaying merchandise. Note the following usual reactions.

A 3:1 luminance ratio will be noticed but will usually not affect behavior or draw attention.

A 10:1 luminance ratio will attract attention and, if interesting, will hold it.

A 50:1 luminance ratio or larger will highlight the object thus illuminated, practically to the exclusion of all else in the field of view.

Since brightnesses draw the eye's attention, all of the individual brightness sources in the field of view produce an overall impression. If there is some form or order or pattern to them (as a pattern of lighting fixtures) then the overall impression is not disturbing—it can be thought of as visually harmonious. If, on the other hand, they are in disarray, they produce a discordancy precisely as sound produces discordancy in the ear. This visual "noise" is frequently referred to as visual clutter and can be very disturbing. The designer is well advised to keep this important fact in mind when arranging light sources that are the primary sources of luminance in an enclosed space. A few aspects of fixture patterning are shown in Section 20.16.

18.30 Color Temperature

A light source is often designated with a "color temperature" such as 3400° K for quartz iodine lamps, 4200° K for cool white fluorescent tubes, and so on. This nomenclature derives from the fact that when a light-absorbing body (called a black body) is heated, it will first glow deep red, then cherry red, then orange until it finally becomes blue-white hot. The color of the light radiated—red from a red-hot body, white from a white-hot body—is thus related to its temperature. Therefore, by developing a black-body color temperature scale, we can compare the color of a light source to this scale and assign to it an approximate "color temperature," that is, the temperature to which a black body must be heated to radiate a light approximating the color of the source in question. Temperature is measured in degrees Kelvin, which is a scale that has its zero point at minus 460° F. Figure 18.43 shows the assigned color temperature of some common light sources.

Strictly speaking, a color temperature can be assigned only to a light source that produces light by heating, such as the incandescent lamp. Other sources, such as fluorescent lamps, produce light by processes that will be detailed in the next chapter. Such sources are assigned a *correlated color temperature,* which is the temperature of a black body whose chromaticity most nearly matches that of the light source. For such sources there is no relation whatever between their operating temperature and the color of the light produced.

As is well known, any nonspectral color illuminant is composed of two or more component color illuminants. When such a composite light,

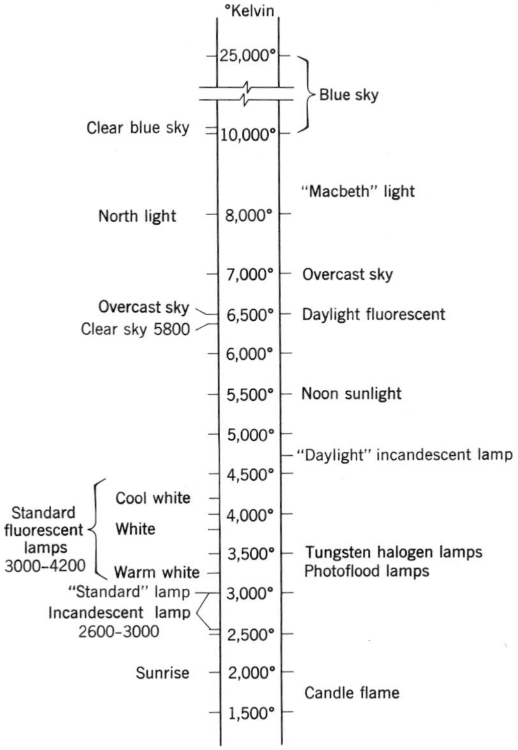

Fig. 18.43 Approximate color temperatures of common illuminants.

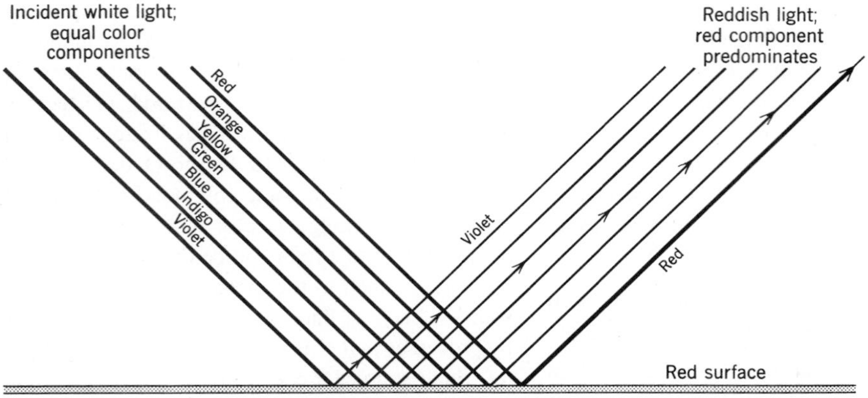

(a)

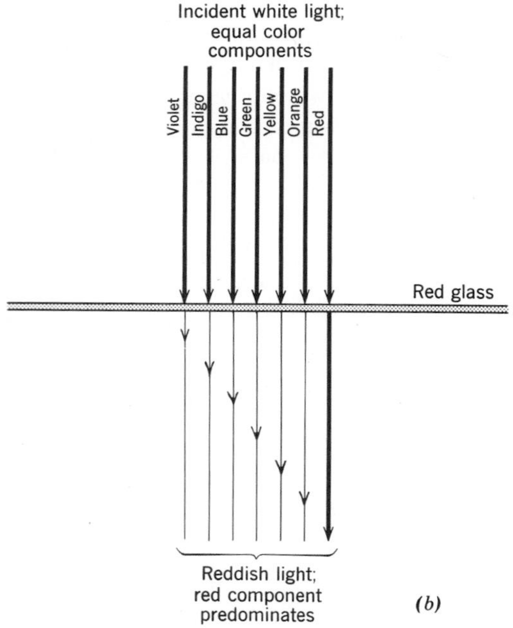

Fig. 18.44 (a) *Selective absorption of reflected light.*

Fig. 18.44 (b) *Selective absorption of transmitted light.*

for example, white, falls on a surface other than black or white, selective absorption occurs. The component colors are absorbed in different proportions so that the light reflected or transmitted is composed of a new combination of the same colors as had impinged on the surface. Thus a white light reflected from a red wall acquires a red tint since the component colors of the white light other than red were absorbed in greater proportion than the red. When reflected, the red light took prominence thus giving the reflected light a red tint. This is illustrated in Fig. 18.44*a*. Similarly, a white light when passed through a piece of red glass emerges as a reddish light since the other components were absorbed in much greater proportion than the red. This well-known phenomenon is illustrated in Fig. 18.44*b*.

It is this phenomenon that allows us to see color at all; the individual object pigmentation absorbs all other colors of light and reflects or transmits to the eye only its own hue.

18.31 Color

The color of the illuminant (light) and correspondingly the coloration of the objects within a space constitute an important facet of the lighting quality. The two factors, however, must not be considered separately since by definition the color of an object is its ability to modify the color of light incident upon it. It does this, as stated above, by a process of selective absorption, absorbing most of the light and reflecting or transmitting a spectrally modified light, rich in a single hue, as shown in Fig. 18.44*a*. The color reflected or transmitted is apprehended by the eye as the color of the object. An object is technically said to be colorless (not transparent) when it does not exhibit selective absorption, reflecting and absorbing the various components of the incident light nonselectively. Thus, white, black, and all shades of gray are colorless, neutral, achromatic or, more precisely, lack hue.

Hue is defined as that attribute by which we recognize and therefore describe colors as red, yellow, green, blue, and so on. Just as it is possible to form a series from white to black with the intermediate grays, so it is also possible to do the same with a hue.

The difference between the resultant colors of the same hue so arranged is called *brilliance* or *value*. White is the most brilliant of the neutral colors and black the least, pink is a more brilliant red hue than ruby, and golden yellow a more brilliant (lighter) yellow hue than raw umber.

Colors of the same hue and brilliance may still differ from each other in saturation, which is an indication of the vividness of hue or the difference of the color from gray. Thus pure gray has no hue; as we add color we change the saturation without changing the brilliance. The three characteristics then that define a particular coloration are *hue, brilliance,* and *saturation*. Using these terms we may define ''bay'' as a color red-yellow in hue of low brilliance and low saturation; while carmine is a color red in hue, of low brilliance and very high saturation.

Various systems of color classification have been devised, including the ISCC-NBC color system, the Munsell Color System, the Ostwald Color System, C.I.E., and the Chromaticity Diagram. In the Munsell color system (see Fig. 18.45) brilliance is referred to as *''value''* and saturation as *''chroma''*; thus a color is defined by hue, value, and chroma. The brilliance (value) of a pigment or coloration is related to its reflectance to white light. The higher the brilliance or value, the higher the reflectance factor, as might be predicted when one considers that white and black are the poles of brilliance. Chroma or saturation may be thought of as either the difference from gray or the purity of the color. Spectral colors have 100% purity and therefore maximum chroma.

When white is added to a pigment, it produces a tint; adding black produces a shade. When pigments are mixed to produce a particular color, we create this color by a subtractive process. That is, each pigment absorbs certain proportions of white light; when mixed, the absorptions combine to subtract (absorb) various colors of the white spectra, and leave only those colors that finally constitute the hue, value, and chroma of the pigment. This subtractive effect is also utilized when producing colors by filtering white light. The filter selectively absorbs component colors, transmit-

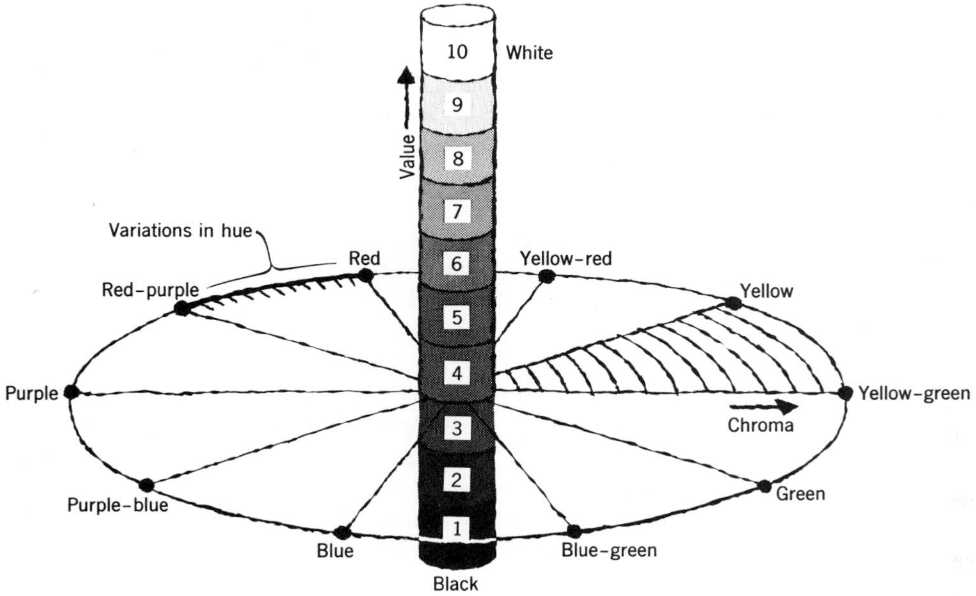

Fig. 18.45 *The Munsell color system defines a color by three characteristics: hue (color), chroma (saturation), and value (grayness).*

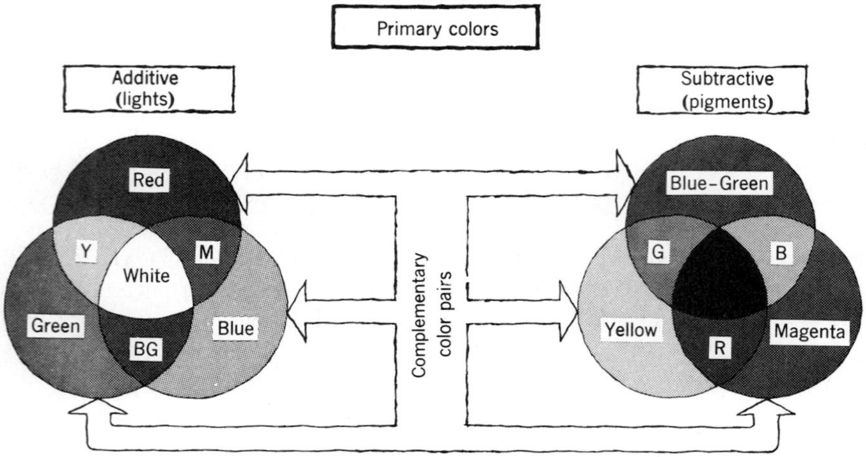

Fig. 18.46 *Primary and complementary colors. Complementary color pairs are shown by arrows. Pigments form color by an absorptive (subtractive) process; colored lights form colors by an additive process.*

ting only the component desired. Thus a blue filter transmits only blue and so on (see Fig. 18.44*b*).

Conversely, when light of the three primary colors of red, green, and blue are combined, they form white by an additive process (see Fig. 18.46).

The additive and subtractive primary colors are complementary; they combine to give a white or neutral gray, respectively. Thus, red and blue-green, blue and yellow, and green and magenta are complementary.

Therefore if a red object is illuminated with blue-green light, the object color appears gray, since the red pigment absorbs the blue-green and reflects nothing; hence the gray. This accounts for the common "lost red car" in parking lots illuminated with clear mercury lamps with their characteristic blue-green color. Similarly, a blue filter on a yellow light would transmit nothing.

18.32 Reactions to Color

Light of a particular hue (other than white) is rarely used for general illumination except to create a special atmosphere. When a space is lighted with colored light, the eye adapts by a phenomenon known as "color constancy" so that it can to a degree recognize colors of objects despite the spectral quality of the illuminant. However, the eyes become more sensitive to the missing colors that would make up white light. This phenomenon could be used to make meat look redder on a butcher counter by using blue-rich, red-poor, cool white lighting in the remainder of the store.

A similar phenomenon occurs when the eye is exposed to a monochromatic scene where the chromaticity is due to coloration of the objects, rather than the illumination. The eye in such a situation becomes sensitized to the complementary color; thus if after looking at a green surface one shifts the gaze to a white surface, one sees the complementary red color. Returning to our meat market, the use of green paint on the walls also enhances the redness of the meat. This effect in reverse also partly accounts for the extensive use of green for paints, linens, and gowns, and so on, in operating rooms. The eyes of the surgeons and nurses when diverted from the redness of the surgical area will be more comfortable seeing green on a green background than on a white one.

A similar effect is apparent object color differences when background color is changed. Thus a green object looks somewhat blue-green on a yellow background because the eye is supplying the complementary color to yellow—that is, blue. Similarly, the same green object looks slightly yellow-green when on a blue background, the eye supplying the yellow.

Apparent brightness of a color is a function of its hue, in that light colors appear lighter than dark

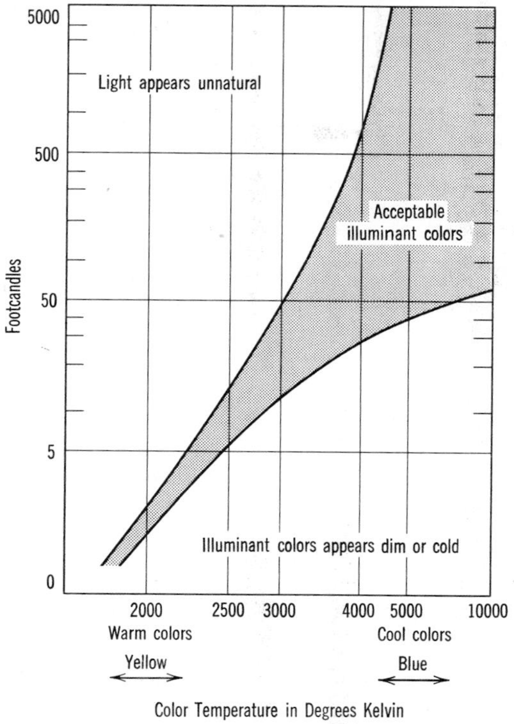

Fig. 18.47 Human preference for illuminant color as a function of light level. We like cool colors (blue) at high levels and warm colors (yellow-red) at low levels, which seems to connect with daylight and fire-light, respectively.

colors even when measured brightness is the same. Thus spaces may be defined by color within an area of equal illumination. Also, all colors tend to appear less saturated, that is, they appear "washed out" when illumination is high. Thus pigments of high saturation (chroma) must be used in well-lit spaces if they are to be effective, although extensive use of saturated colors is generally best avoided.

Furthermore, there is an observed relationship between the color of the *light* in a space and the range of acceptable levels of illumination. The curves of Fig. 18.47 indicate that cool illuminant color is desirable at high levels and warm at low levels, corresponding to daytime sky and night-time firelight, respectively.

Other well-known psychological effects of colors are the coolness of blues and greens and the warmth of reds and yellows. Similarly, red and yellow are "advancing" colors because objects lit with them tend to "advance" toward the observer, giving the appearance of becoming larger. The opposite effect is noted with blue and green, accounting for their being known as receding colors. Thus cool colors might well be used in a fur salon and warm colors in a display of summer wear. A practical, energy-saving application of these color phenomena would be to use warm colors to compensate somewhat for lowered thermostats in the winter, and cool colors for the opposite effect in summer. How to accomplish this without the expense of repainting twice a year is left to the ingenuity of the architect and interior designer. In an atmosphere designed to be calm and restful, greens should generally predominate either in illuminant color, object color, or both, except in eating areas, which should be lighted with reds and yellows since cool colors are generally unappetizing. Yellows and browns emphasize motion sickness, whereas blues and greens tend to the reverse. Warm and saturated colors produce activity; conversely cool, unsaturated colors are conducive to meditation. Cool colors also seem to shorten time passage and are well applied in areas of dull repetitive work.

A further discussion of color control, source colors, and color matching will be found in Section 19.25, dealing with spectral energy distribution of sources.

References

(British) *IES Code of Interior Lighting,* IES, London, 1977.

IES Lighting Handbook, 6th ed., IES of North America, 1981.

Sampson, F.K. (1970). *Contrast Rendition in School Lighting,* Educational Facilities Laboratories, New York.

Yonemura, G.T. and Tibbot, R.L. (1981). "Equal Apparent Conspicuity Contours with Five Bar Grating Stimuli," *Journal IES,* April, 1981, p. 155.

19

LIGHT SOURCES: THEIR CHARACTERISTICS AND APPLICATION

19.1 General Remarks

Long before the dawn of recorded history, the double blessing of fire, heat and light, was discovered. Even today in our sophisticated space age, fire is still used almost universally as the source of heat and, in a large proportion of the world's dwellings, as the source of light. Electrical lighting had its real beginning in about 1870 with the development of commercially usable arc lamps and was given greater impetus nine years later by Edison's first practical incandescent lamp. Today's practical electric light sources fall into two generic classifications: the incandescent lamp, including tungsten-halogen types and the gaseous discharge lamp, which includes fluorescent, mercury, metal-halide, and sodium lamps.

The efficiency of a light source is termed its *efficacy* and is measured in lumens per watt (lm/W). Table 19.1 lists efficacies of modern light sources, including ballast losses where applicable. It is misleading to use the efficacy of the light sources alone, as is often done in the literature, because the ballasts are inseparable from the lamp. Note that Table 19.1 gives a range of efficacies for each lamp type. In general, efficacy increases with wattage, as is clear from Fig. 19.1. It is therefore energy-economical to use a small number of higher wattage lamps than the reverse. (It is also usually more economical with respect to fixtures.)

Since electric lighting in American nonresidential buildings consumes 25 to 60% of the electric energy utilized, any attempt to reduce this must necessarily include integration of the cheapest (insofar as energy is concerned), most abundant, and, in many ways, most desirable form of lighting available—daylight.

DAYLIGHTING

19.2 Daylighting as a Lighting Design Factor

The provision of daylight in structures in the United States had in the recent past largely been considered an amenity rather than a necessity. As such its provision had been the province of architecture rather than lighting design. The reasons for this are clear. Daylight is indeed an amenity. Windows provide visual contact with the outside and the resultant daylight provides a bright, pleasant, airy ambience. When daylight enters through windows (side lighting, as opposed to toplighting), its horizontal directivity provides good modeling shadows, minimal veiling reflections, and excellent vertical surface illumination. Furthermore, the

TABLE 19.1 **Efficacy of Various Light Sources**[a]

Source	Efficacy (lm/W)
Candle	0.1
Oil lamp	0.3
Original Edison lamp	1.4
1910 Edison lamp	4.5
Modern incandescent lamp	8–20
Tungsten halogen lamp	16–20
Fluorescent lamp[b]	32–102
Mercury lamp[b]	30–60
Metal-halide lamp[b]	70–100
High-pressure sodium[b]	50–130
Low-pressure sodium	120–140

[a]See Fig. 19.1.

[b]Including ballast losses.

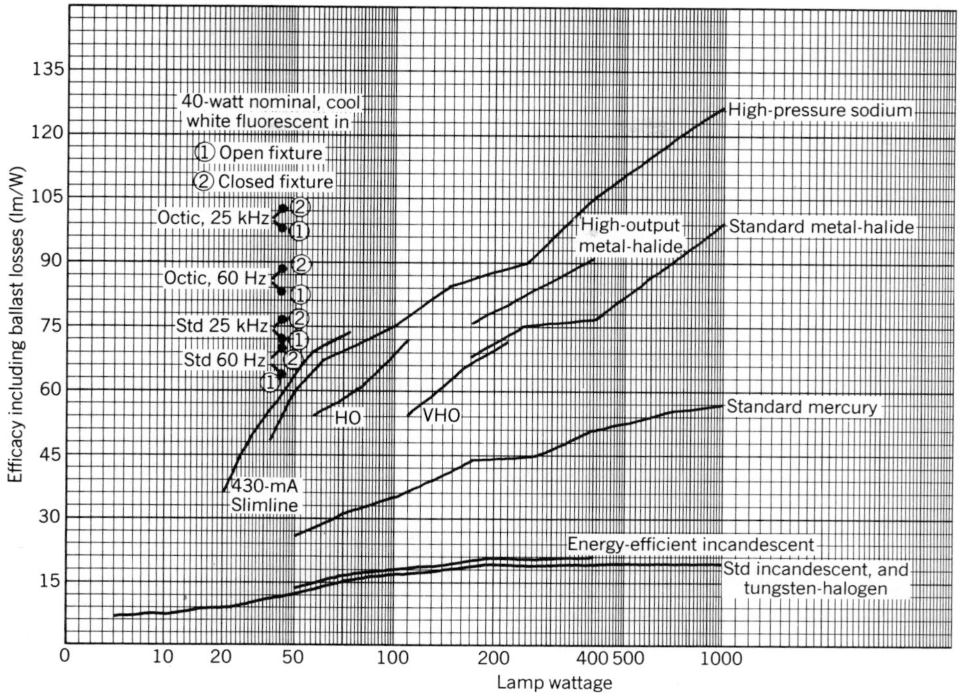

Fig. 19.1 Average light source efficacies as a function of lamp size. Ballast losses are included. Data from 1983 published manufacturers' data.

continual variation of daylight, which is one of its prominent characteristics, provides a constantly changing pattern of space illumination—one that is unattainable with artificial light. Since these changes are gradual, the eyes adapt easily and the effect is one of visual interest. Undoubtedly, as a result of these effects, numerous studies have conclusively demonstrated a marked preference for daylight over any other form of lighting.

On the other hand, although concern has been expressed about possible physical ill effects, none have *conclusively* been demonstrated to have been caused by lack of daylight, that is, by working in an artificially lighted space. Since an artificial lighting system had to be installed in any event, to furnish interior illumination during periods when daylight is insufficient, the practice arose in the United States of ignoring daylight and even of shutting it out deliberately. Careful design of an electric-lighting system can provide a satisfactory visual atmosphere, as these chapters explain. Furthermore, unlike daylight, control of such systems is relatively simple. Perhaps most important, an interior electric-lighting system has minimal impact on the building architecture, least of all on the all-important building facade. Finally, the energy to power electric-lighting systems *was* cheap.

The option of ignoring daylight in our high-energy-cost and energy-resource-poor society is no longer available. That being the case, the American designer must learn to cope with the special problems that daylight use presents in order to reap its benefits. Since daylight is variable, it creates special problems of glare control, direct sunlight control, and heat-gain limitation. In large measure the science (and art) of daylighting is not so much how to provide daylight as how to do so without the attendant undesirable effects. Put otherwise, the American designer must adapt and adopt the British technique of PSALI (Permanent Supplementary Artificial Lighting in Interiors), which is almost universally applied in Europe.

The material below, which discusses PSALI, sky luminance, outdoor illumination, and calculation of interior daylight levels, is necessarily brief, because of space considerations. Probably no subject bearing on architectural engineering has received more attention in the past decade than daylight, its source, and its ramifications. A bibliography of books and papers bearing on the subject would alone cover more space than is allotted to the subject itself in these pages. The interested reader is referred to the bibliography at the end of this chapter, which lists a few of the important sources in this field.

19.3 PSALI (Permanent Supplementary Artificial Lighting, Interiors)

This technique, which is really a design approach, views artificial lighting as supplementary to daylighting and not vice versa. The PSALI technique recognizes that nonresidential buildings are principally used during daylight hours and that sufficient daylight is generally available during these hours to provide much of the structure's lighting needs. The success of the PSALI design approach is predicated on three assumptions:

1. The large variation of interior daylight illuminance levels during the course of the day will not adversely affect visual performance, even when the actual levels fall below accepted recommendations. Stated otherwise, the same visual performance can be achieved with less daylight than artificial light, when compared on a footcandle (lux) basis. The rationale for this conclusion is based on two well-known phenomena of human vision: lightness constancy and Weber's law of visual contrast. Lightness constancy is the ability of the eye to distinguish between light and dark objects over an extremely wide range of illumination. Thus, even though a lump of coal in sunlight is brighter than this piece of paper in moonlight, we will correctly identify the one as black and the other as white. This is because the eye appreciates lightness or brightness not only by actual luminance of the object being viewed but also by the surrounding area. Thus, in a space with *slowly* changing illumination levels, as is the case with daylighted spaces, the eye will adapt to the altered level and see basically the same scene, in terms of lightness (and darkness) of objects. That is, the eye's appreciation of the pattern of brightnesses in the room will remain constant.

   Weber's law states that within the very wide illuminance range of lightness constancy (above 10 lux), contrast ratios remain constant. Since visual performance is closely related to contrast discrimination, it is also largely unaffected over the wide absolute range of daylight illuminance.
2. Illuminance recommendations to be used are the current ones as given in Section 18.22.
3. Daylight and artificial light can be readily and successfully combined, that is, artificial light can supplement daylight when the latter is insufficient (see Fig. 19.2).

We made particular reference above to nonresidential buildings because residences are occupied and extensively utilized at night when no daylight is available and PSALI is inapplicable. Obviously residences are also used during daylight hours, but the conditions of fixed work location and long-duration visual tasks generally do not apply. Nonresidential buildings that are regularly used at night as well as during daylight hours should be designed for both conditions, with appropriate controls.

Since interior daylighting is obviously directly dependent on exterior lighting levels, an understanding of the latter is a clear prerequisite to designing for the former. However, before beginning a discussion of exterior illuminance and the various methods by which the resultant interior daylight illuminance can be calculated, it is appropriate to ask several basic questions regarding daylight calculations in general:

1. Inasmuch as the majority of buildings must be designed for nighttime use as well, that is, with a complete system of electric lighting, what purpose does a knowledge of daylight distribution in such buildings serve?
2. Given the great variability of daylight with time of day, season, and weather, how can any meaningful calculations be made?
3. Furthermore, if daylight distribution is calculated for a variety of conditions, what use can be made of the results? What are the results we want, the definition of which will

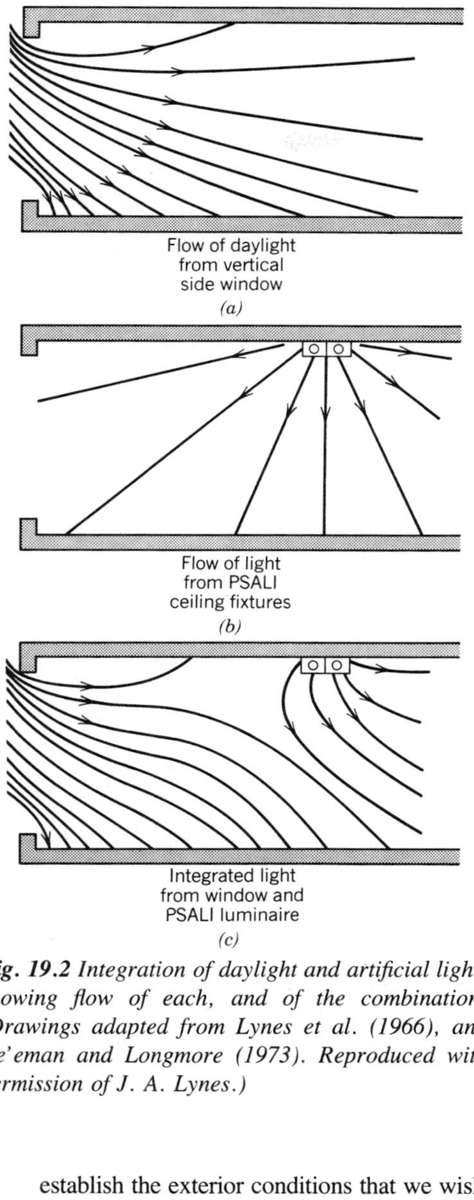

Fig. 19.2 *Integration of daylight and artificial light, showing flow of each, and of the combination. (Drawings adapted from Lynes et al. (1966), and Ne'eman and Longmore (1973). Reproduced with permission of J. A. Lynes.)*

establish the exterior conditions that we wish to study?

4. What degree of accuracy in calculation is desirable? Obtainable?

The answers to these questions are fairly clear and will define the direction that our study of daylight will take.

The purpose of daylight analysis is manifold.

1. Granted the variability of daylight, it is possible to calculate maximum, minimum, and, if enough data are available, statistical averages. Furthermore, there are extensive areas of the United States where for long periods of time the weather is constant—clear blue skies in the Southwest, solid overcast in the Northwest—and for these areas fairly accurate calculations of interior daylight (±25%) can be made.

2. Maximum daylight data, including direct sunlight, are needed for design of glare controls, shading devices, light shelves, and other types of reflectors, and for zoning of automatic daylight compensation controls (see Section 21.4). These data are also important in establishing a maximum cooling load.

3. Minimum daylight data are necessary to establish maximum PSALI (see Fig. 19.3) and to design the controls that will differentiate PSALI from nighttime lighting. This also defines maximum electric lighting loads, which are needed for heating and cooling calculations.

4. Average daylight data are necessary to establish percentiles on the basis of which annual energy budgets can be determined. Furthermore, average data are important in designing fenestration, since use of minimums will probably result in excessive glass area with the concomitant thermal problems discussed in Chapter 5. Thus it is reasonable to design fenestration on the basis of a 50 to 60% figure, which states that a given daylight level will be maintained for 50 to 60% of the daylight hours. In the analyses that follow, it is assumed that fenestration has been established. Obviously, the procedures that will be presented can also be used to establish fenestration, by starting with rules of thumb and utilizing a modified trial-and-error procedure to improve the initial design.

5. The problem of (in)accuracy in interior lighting calculations can be divided into two: that caused by inaccurate input data, and that which is inherent in all calculations based on statistical averages. Referring to the first, we will see below that hand-type, pencil-and-paper methods, and use of sky luminance data based on season, latitude, and solar altitude may introduce an error of 25% or more, depending on the type of sky. Furthermore, the calculation procedure itself involves approximations of considerable magnitude, so that results may be 30 to 50% from actual measured

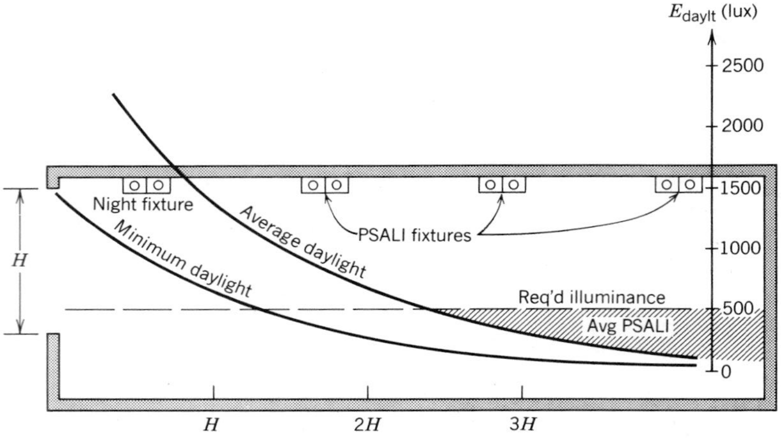

Fig. 19.3 Section through a room with daylighting from one side window. The typical average daylight curve shows that for this building, to a depth of about $2\frac{1}{2}$ times window height H*, daylight provides the necessary illuminance. Beyond that PSALI is required to maintain the required level. Even with minimum daylight conditions a depth of* H *into the room is daylighted, and PSALI lighting is required only beyond that depth. Thus the first (nearest the window) lighting fixture is used only in nighttime lighting, the second and third would be dimmed or multilevel switched, and the fourth (back wall) would be lighted during all hours of building use.*

Average daylight would represent the 85th percentile, that is, the level that (statistically) is maintained for 85% of the daylight hours. Setting of the minimum is flexible and depends on the type of control system. It usually represents the 95th percentile, that is, the level that is maintained for 95% of daylight hours.

levels. (That being so, the pointlessness of working to four decimal places should be obvious.) However, these *hand* calculation procedures are intended only to give the designer a ''feel'' for the daylight levels in the space and, more important, for the gradation of contours of daylight. Greater accuracy is possible only with computer-aided design procedures, which input more specific sky luminance data and reduce approximations.

For annual energy budget calculations, hour-by-hour daylight calculation is required, and, of course, this is done only by computer.

The second ''inaccuracy'' is a statistical deviation. Because of the infinite variability of weather in temperate climates, it is not possible to calculate the daylight condition for a specific time on a specific day, *nor is it necessary*. For this reason, the procedures shown below use single figures for whole seasons and rely on proper design of controls of both daylight and electric light, with PSALI, to furnish a well-lighted, low-energy-use interior.

We suggest that at this point the reader turn ahead to Section 21.4*e* and read through the section on daylight compensation, to understand how PSALI with proper controls will result in low energy use while maintaining the necessary lighting levels.

19.4 Characteristics of Outdoor Illumination

(a) Factors. The most prominent characteristic of daylight is its variability. Obviously the source of all daylight is the sun. The level of exterior illumination, at a particular place and time, depends on

1. Solar altitude and azimuth, which can be determined if latitude, date, and time of day are given. See Appendix B.
2. Weather conditions (cloud cover, smog).
3. Effects of local terrain (natural and man-made obstructions and reflections; see Fig. 3.7).

The position of the sun in the sky is expressed in terms of its altitude above the horizon and its azimuth angle. The latter is defined as its horizontal

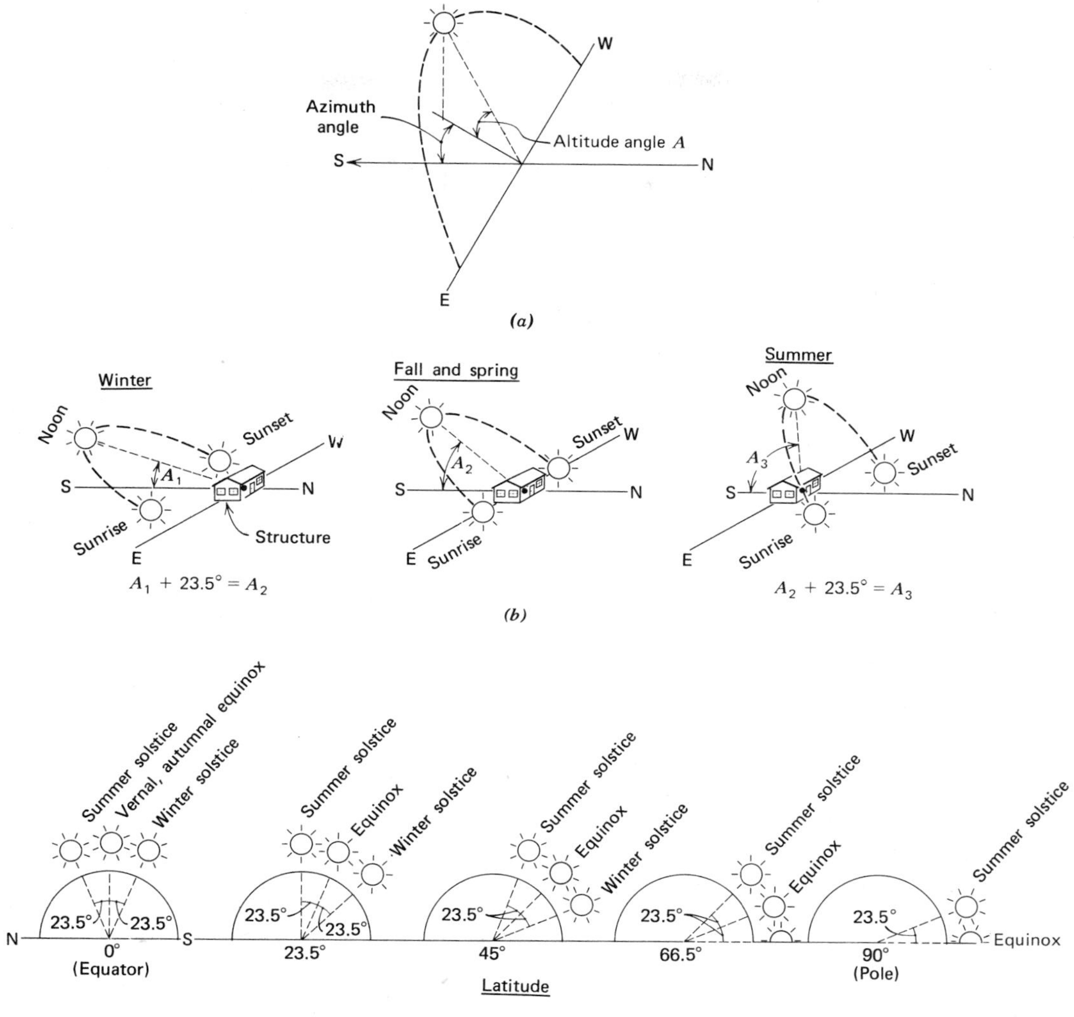

Fig. 19.4 (a) *The position of the sun is expressed in terms of vertical angle above the horizon (altitude) and horizontal angle, measured from the south (azimuth).* (b) *Approximate position of the sun in each of the seasons, at a midnorthern latitude (approximately 45°). Note that altitude angle is maximum in summer, minimum in winter, and in-between in spring and fall. Note too the length of daylight hours: maximum in summer, minimum in winter, and in-between in spring and fall.* (c) *Maximum sun altitude at various latitudes, for both soltices and equinoxes. Maximum summer sun altitude is 90° minus latitude plus 23.5°. Minimum winter sun altitude is 90° minus latitude minus 23.5°. Thus for all latitudes the yearly difference between maximum and minimum altitudes is twice 23.5°, or 47°, as shown.*

position angle, measured from the *south*. Both are normally expressed in degrees (see Fig. 19.4*a*). It is assumed that the reader is familiar with the basic astronomical phenomena governing the motion of the earth, which produce seasonal and latitudinal variations in the position of the sun. For our purpose, we will simply state the important facts, which are:

1. For all latitudes, the sun's altitude is highest in summer, lowest in winter, and in between in spring and fall (see Fig. 19.4*b*).

This fact, coupled with the shorter winter day, leads to the apparent rapid motion of the sun across the horizon in the winter and the apparent slow motion in the summer.

2. As a location approaches the equator (low latitude, either north or south), the sun's daily maximum altitude increases. The seasonal altitude *variation,* however, is the same for all latitudes (except at those extreme north and south latitudes where the sun is above or below the horizon for extended time periods; see Fig. 19.4*c*). This factor not only affects exterior illumination levels but also has a pronounced effect on the design and efficacy of sun-shading devices.
3. The sun's azimuth angle is entirely dictated by the time of day, since the sun by definition traverses the sky between sunrise and sunset. The principal significance of the azimuth angle is encountered when discussing building orientation, exposures, and shading angles. See Table 19.2.

The factor of cloud cover, unlike that of solar position, is predictable only statistically on the basis of extensive U.S. Weather Bureau observations at numerous weather stations throughout the United States. At locations other than those for which these data are available, an educated guess is necessary. Outside the United States, the designer must rely on locally available data, which are frequently meager or entirely unavailable. The third factor—that of local terrain and construction conditions that either reduce illumination by shadowing or increase it by reflection—is so particular and individual that it can be considered only on a case-by-case basis (see Fig. 3.7). Most of the calculation methods consider the effects of shadowing and obstruction and, to a lesser extent, the effects of reflections (see Sections 19.5 to 19.7).

(b) Sky Conditions. For hand calculation procedures, it is sufficient to establish four basic sky conditions. These are:

1. Completely overcast sky.
2. Clear sky, without sun (in the shade).
3. Clear sky, with sun.
4. Partly cloudy sky.

1. *Completely overcast sky*. This condition, which obtains for much of the year in northerly climates such as England, Scandinavia, and the Pacific Northwest, is also called the CIE sky, since it was adopted by the Commission Internationale de l'Éclairage (CIE) as the standard design sky for daylighting calculations (''Daylight,'' 1970). This sky as defined by the CIE has a nonuniform brightness distribution, increasing from horizon to zenith in approximately the ratio of 1:3. Sky luminance at any altitude angle above the horizon is defined as

$$L_A = L_z \frac{1 + 2 \sin A}{3} \tag{19.1}$$

where L_A = luminance at $A°$ above horizon
L_z = luminance at zenith

Thus at the horizon, where $A = 0°$,

$$L_A = \frac{L_z}{3}$$

as stated above. The illuminance (lux, fc) pro-

TABLE 19.2 **Typical[a] Solar Altitude and Azimuth Data as a Function of Date and Time of Day**

Latitude (°N)		Date	Solar Time A.M.: 6 / P.M.: 6	7 / 5	8 / 4	9 / 3	10 / 2	11 / 1	Noon
38[a]	Altitude	June 21	14	26	37	49	61	71	75
		Mar.–Sept. 21	—	12	23	34	43	50	52
		Dec. 21	—	—	7	16	23	27	28
	Azimuth	June 21	109	101	90	83	70	46	0
		Mar.–Sept. 21	90	81	71	58	43	24	0
		Dec. 21	—	—	54	43	30	16	0

[a]For data for other latitudes, see Appendix B.

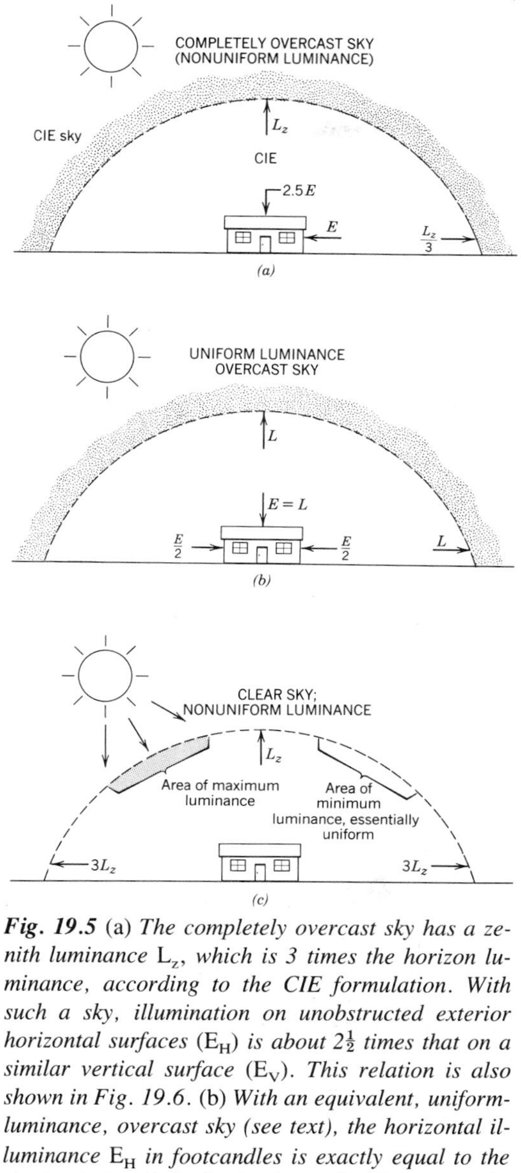

Fig. 19.5 (a) *The completely overcast sky has a zenith luminance* L_z, *which is 3 times the horizon luminance, according to the CIE formulation. With such a sky, illumination on unobstructed exterior horizontal surfaces* (E_H) *is about* $2\frac{1}{2}$ *times that on a similar vertical surface* (E_V). *This relation is also shown in Fig. 19.6.* (b) *With an equivalent, uniform-luminance, overcast sky (see text), the horizontal illuminance* E_H *in footcandles is exactly equal to the sky luminance* L *in footlamberts. In SI units, a sky luminance of* L *cd/m²* *will produce* πL *lux horizontal illuminance. Vertical surface illuminance from this distribution is obviously half the horizontal.* (c) *One widely accepted formulation for clear sky luminance is simply an inversion of the overcast distribution* (a); *horizon luminance is three times that of the zenith. Obviously the area around the sun is brightest. The area opposite the sun is darkest, and can be considered as essentially uniform at approximately 3500 cd/m² (1000 fL). Actual luminance varies from about 300 fL at the center to 2000 fL at the sides, averaging 1000 fL over the whole area.*

duced by this distribution on unobstructed exterior horizontal and vertical surfaces stands in the approximate ratio of 2.5:1 (see Fig. 19.5*a*). This results from an integration of Equation 19.1 over the whole sky. Pierpont (1983) derives a different vertical to horizontal illuminance ratio by utilizing an alternative sky distribution equation, and others have suggested different sky luminance distributions for the overcast sky.

There is, however, agreement among all sources that with an overcast sky, exterior horizontal illuminance varies directly with the sun's altitude, despite the overcast. Various formulations for the relation have been put forward. One formulation that gives good agreement with observations is given by Krochman (1963).

$$E_H = 300 + 21{,}000 \sin A \text{ (lux)} \qquad (19.2)$$

where E_H is exterior horizontal illuminance and A is the solar altitude, in degrees. Solar altitude (and azimuth) for a single latitude and times of day can be obtained from Table 19.2. Data for other latitudes are found in Appendix B. Figure 19.6 is a plot of year-round averages for both vertical and horizontal illuminance from an overcast sky as a function of solar altitude, based on U.S. Weather Bureau observations.

To perform accurate calculations of indoor daylight levels using a nonuniform sky luminance distribution, one requires some sort of sky protractor with which to measure sky angles from the interior observation point (see Fig. 19.7). With these angles, the sky luminance of each small patch of sky can be determined, and in turn the resultant interior daylight. Protractors are available from the Building Research Station, London, which simplify the work considerably, but it is still a time-consuming, laborious procedure. A common simplification is to establish an equivalent *uniform* sky brightness (luminance) for an overcast sky. The rationale for this procedure is that since both sky luminance and horizontal illumination vary with solar altitude, they vary with each other, though not directly. However, for a given solar altitude, E_H can be determined, and working back, an equivalent uniform sky luminance can be simply established, which is exactly equal, in footlamberts, to the previously established horizontal illuminance, E_H, in footcandles (see Fig. 19.6). The

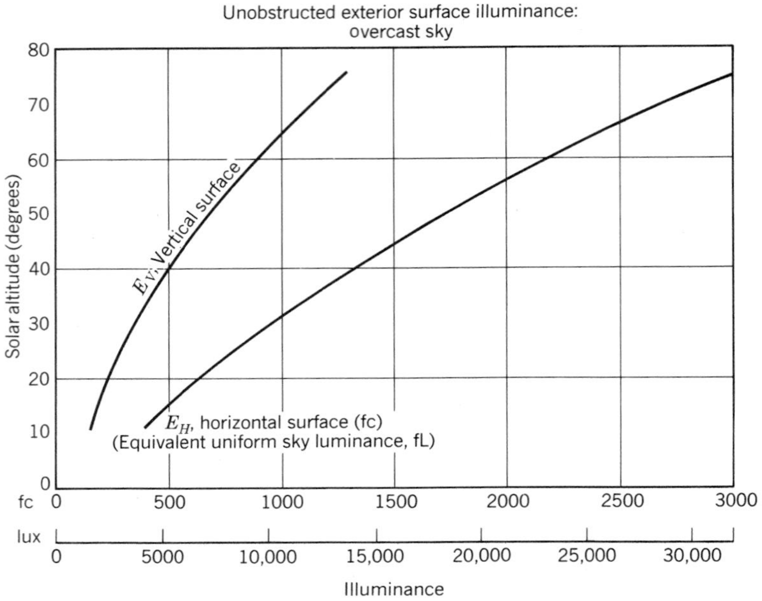

Fig. 19.6 *Curves giving unobstructed exterior surface illumination from an overcast sky, directly. The equivalent uniform sky luminance, in footlamberts, is equal to the horizontal illuminance, in footcandles. The ratio between horizontal and vertical surface illumination is 2.5 : 1 as shown in Fig. 19.5*a. *(Data based on U.S. Weather Bureau observations. Data from ''How to Predict. . .'' (1976), courtesy of Libbey-Owens-Ford.)*

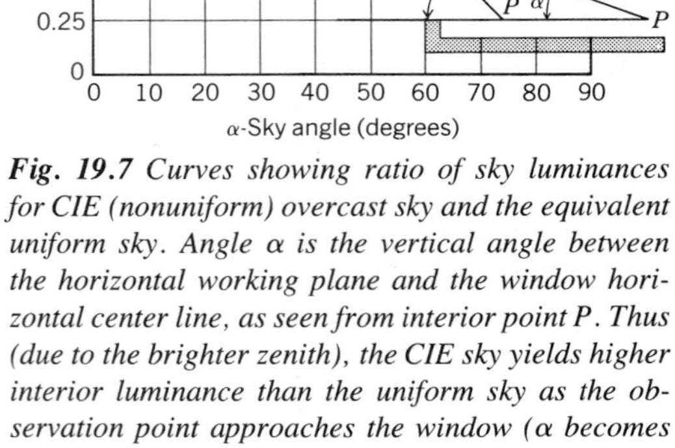

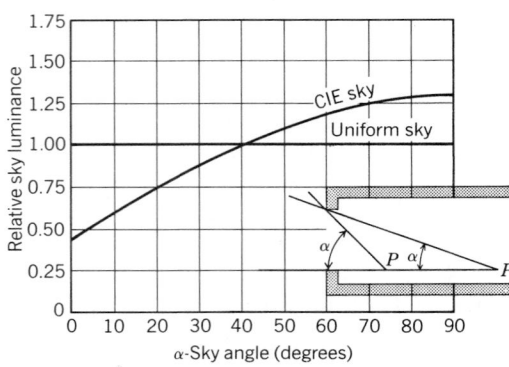

Fig. 19.7 *Curves showing ratio of sky luminances for CIE (nonuniform) overcast sky and the equivalent uniform sky. Angle α is the vertical angle between the horizontal working plane and the window horizontal center line, as seen from interior point P. Thus (due to the brighter zenith), the CIE sky yields higher interior luminance than the uniform sky as the observation point approaches the window (α becomes larger).*

advantage of this simplification is the elimination of altitude–luminance calculations.

For sidelighting calculations, the relationship between sky luminance for the CIE sky and the equivalent uniform sky must be taken into account. This is shown in Fig. 19.7. This plot shows that for points near a window, where sky angles are above 42°, the CIE sky yields greater daylight than the uniform sky. Therefore, uniform sky luminance values must be *increased* to yield a proper result for the actual nonuniform sky. The reverse is true for angles below 42°. Hopkinson et al. (1966) refer to this correction factor for converting from uniform sky to CIE sky as the ''Sky Luminance Ratio'' or simply the ''Z factor,'' and tabulates it.

It is interesting to compare exterior horizontal illumination obtained from the two sources given: Krochman's formula (Equation 19.2), and the observation based data of Fig. 19.6, for a few typical conditions. Solar altitude is obtained from Table 19.2.

Latitude: 38° N
Solar Time: 10 A.M.
Dates: Dec 21, March/Sept 21, June 21

	Eq. 19.2	Fig. 19.6
Dec. 21	790 fc	750 fc
Mar/Sept 21	1359 fc	1420 fc
June 21	1735 fc	2180 fc

The degree of agreement is generally satisfactory and either source will yield suitable results.

2. *Clear sky, horizontal illumination*. Exterior horizontal illumination on a cloudless day consists of two components: diffuse illumination from the entire sky plus the much larger component of direct sunlight. As with overcast sky, various empirical formulas for both components have been proposed, and here too, all sources agree that the total illumination, diffuse plus direct, varies directly with solar altitude. [The most widely accepted empirical formulation for clear sky luminance distribution is that of Kittler (see *IES Handbook,* 1981, Ref. Volume, pp. 9–84)]. Figure 19.8 gives values for both components of exterior horizontal illumination, based on weather bureau observations. The *sky only* values are used to determine shaded skylight illumination, or day-long ground illumination outside a shaded window, that is, a north-facing window, or an east/west window when the sun is on the opposite side of the building. In determining ground illumination, the values given in Fig. 19.8 must be reduced somewhat, as they represent unobstructed horizontal illumination, whereas the area outside a building window is partially obstructed from sky light by the building. If the building is so large that the ground outside the shaded window effectively receives diffuse radiation only from the half of the sky away from the sun, an average figure for E_H of 1000 fc can be used. This is because the luminance of the half of sky away from the sun varies from a minimum of approximately 300 fL for the deep blue patch directly opposite the sun to about 2000 fL at the sides, giving an average half-sky luminance of about 1000 fL. This, in turn, will give a horizontal illuminance E_H, diffuse, of about 1000 fc (see Fig. 19.5*c*).

Figure 19.8 also gives horizontal illuminance

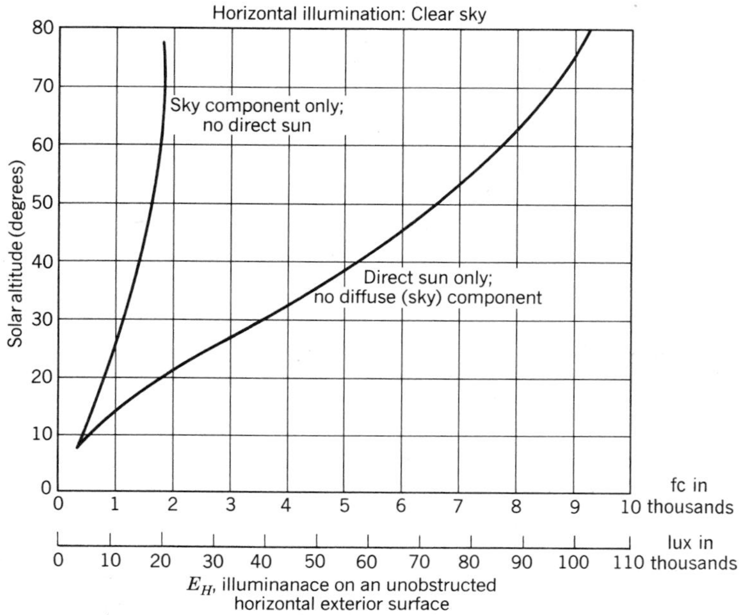

Fig. 19.8 Components of the exterior horizontal illumination on an unobstructed surface, from a clear sky, as a function of solar altitude. Total illumination E_H is the sum of the two components. (From data in Rennhackkamp, 1967.)

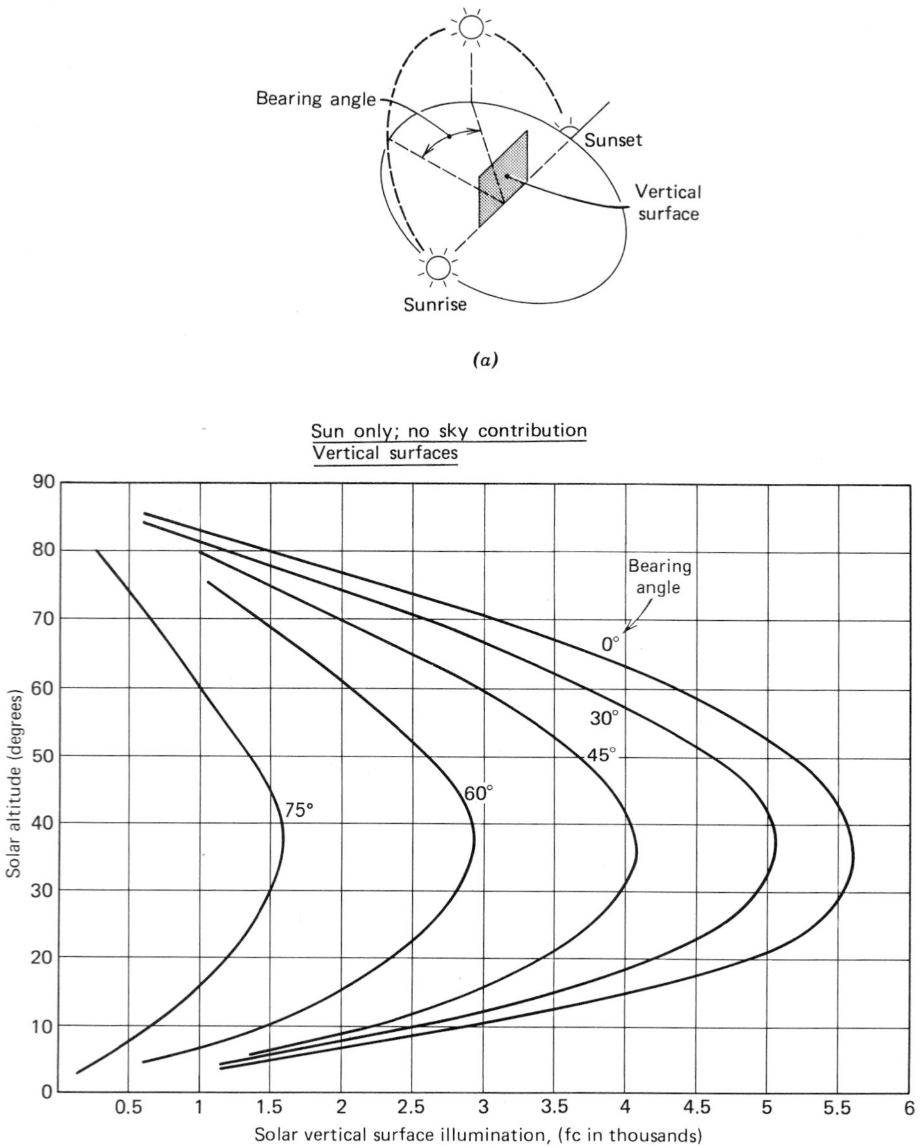

(b)

Fig. 19.9 (a, b) *Vertical surface illumination, year-long average, sun only, no sky contribution; bearing angle shown.*

from the sun only, as a function of solar altitude. This value, when combined with the proper portion of diffuse illumination as discussed above, is useful in determining ground illumination outside a sunny building exposure, or illumination on an unshaded skylight. The light incident on an external reflector or light shelf at a window can also be determined from these figures.

As with the overcast sky condition, an equivalent uniform sky luminance for clear sky can be derived from the horizontal illumination data, and such tables are available ("Recommended Practice," 1979). However, because such data are not readily usable with daylight factor criteria based on the overcast sky, as explained in Section 19.5, the tables are not presented here.

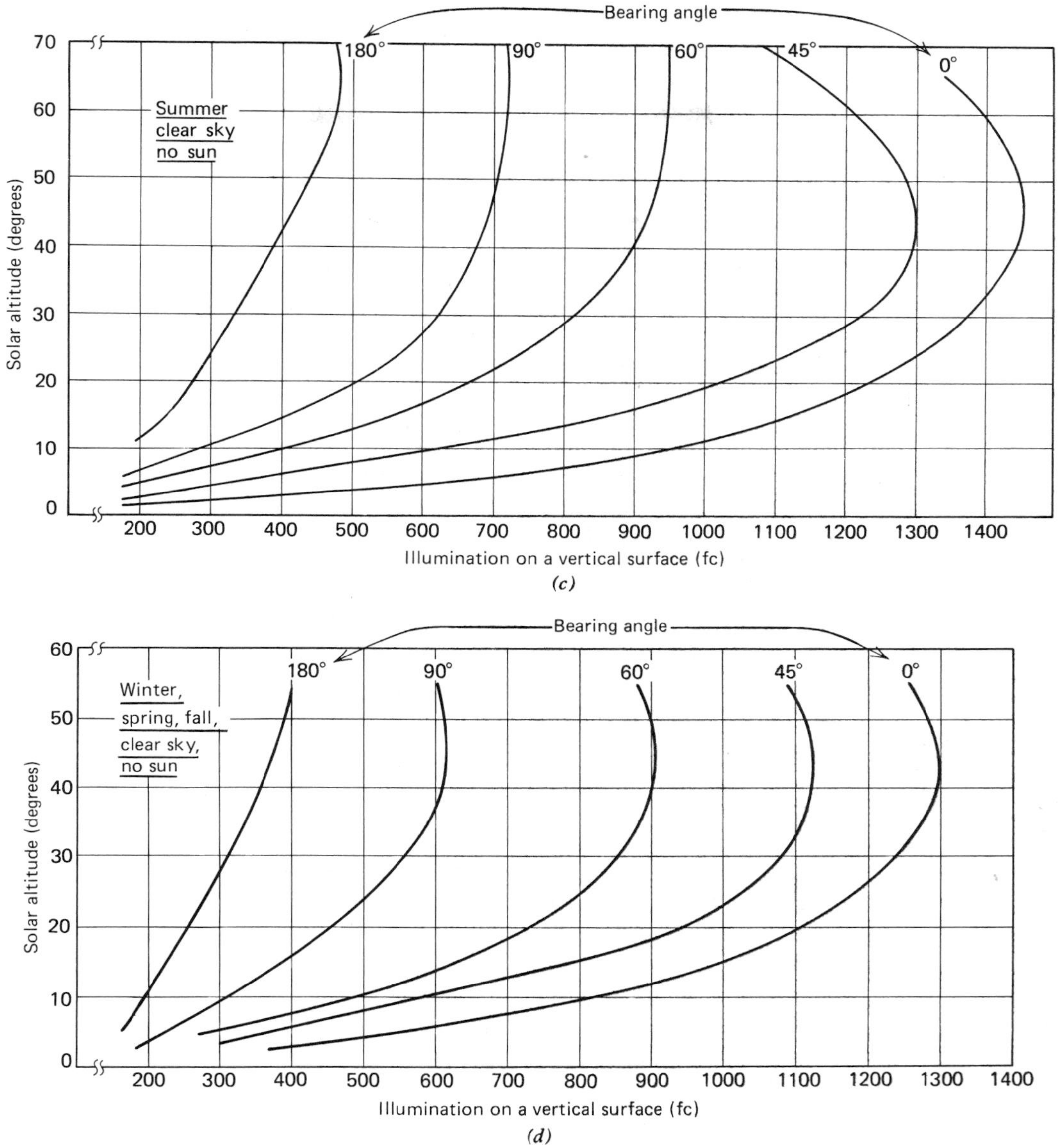

(c, d) *Vertical surface illumination from a clear sky, with no sun contribution, for seasons of the year. (Data from "How to Predict. . . ," 1976, courtesy of Libbey-Owens-Ford.)*

3. *Clear sky, vertical surface illumination.* Since most daylighting is accomplished via vertical fenestration, vertical surface illumination is the major component of interior daylight. It is also important for determining the daylight contribution of vertical elements in skylights. There is no simple relationship between horizontal and vertical illumination from a clear sky as there was for an overcast sky because the illumination on a vertical surface depends on solar azimuth as well as altitude. More specifically, it depends on the *bearing angle,* which is defined as the horizontal angle between vertical planes containing the sun and a perpendicular to the vertical surface in question. A bearing angle of 0° indicates that the sun plane is perpendicular to the vertical surface (see Fig. 19.9*a*). As with E_H, E_v (vertical illuminance) is divided into two components: sky only, and direct sun only, which are plotted in Fig. 19.9*b*–19.9*d* as a function of solar altitude and bearing angle.

ILLUMINATION

The *sky only* component is effectively for the half-sky, since a vertical surface can be exposed to a maximum of only half of the full sky.

The data are used as follows: Surfaces exposed to the sun are illuminated by the sun component at the relevant bearing angle (Fig. 19.9*b*) plus the sky component at the bearing angle (Fig. 19.9*c*–19.9*d*). Bearing angle for the sky component is important because of the nonuniform luminance of the clear sky—brightest near the sun and dullest opposite it. A vertical surface shaded from the sun is illuminated by the sky component, at the bearing angle, plus any reflected light from objects facing the window, at the bearing angle +180°. Thus a north-facing window at 180° bearing angle (noon solar time) is illuminated by the sky component at 180° plus reflected light from facing buildings *at 0° bearing angle*. These surfaces (at 0°) are illuminated by both sun and sky (see Fig. 19.22).

The data presented in the curves of Figs. 19.8 and 19.9 are based on year-long averages of observations taken by the U.S. Weather Bureau, modified by later observations of Rennhackkamp (1967). When more accurate observation data are available, they obviously should be used in preference. In many areas of the world, data on solar radiation are available, but illumination data are not (see Treado and Gillette, 1983). To translate these radiation data to illumination, an "efficiency" figure is required for solar energy, in units of lumens per watt of received radiation. Reliable average figures are:

Total radiation, sun and sky = 119 lm/W

Diffuse (sky) radiation only ≃ 14 lm/W

Direct solar radiation only ≃ 105 lm/W

Therefore, a surface receiving a total radiation of 750 W/m^2 is illuminated to

$$E = 750 \text{ W } (119 \text{ lm/W}) = 89{,}250 \text{ lux or } 8295 \text{ fc}$$

Obviously, the angle between the sun and the surface affects the amount of received radiation. Therefore, if only horizontal or perpendicular-to-the-sun radiation data are available, somewhat complex trigonometric calculations will give the radiation on a surface at any bearing angle. The relevant equations are found in many sources; see, for instance, Wieder (1982, pp. 35–37).

4. *Partly cloudy sky*. There is no way of expressing mathematically the sky luminance of a partly cloudy sky because of its infinite variability. However, statistical data on cloud cover are available from observations at many weather stations, and these data should be used in computer-calculated, hour-by-hour energy programs. For the purpose of illumination, it is important to note that the illumination from a partly cloudy sky is *higher* than that from a clear sky by 10 to 15%, because of the additional reflected sunlight from cloud edges. Several attempts have been made recently to account for this type of sky—including a formulation (Nakamura and Oki, 1983) for an average intermediate sky between clear and solid overcast—in terms of the effect on daylight factor within the room. As we will see in the next section, the concept of daylight factor was originally intended for overcast sky only, but, because of its great usefulness, it has recently been extended to clear sky (Bryan, 1980) and partly cloudy sky (Nakamura and Oki, 1983).

In the following sections several interior daylight analysis procedures will be presented. We specify "analysis" rather than design, because unlike the standard design procedure, we start with a daylight inlet (window, skylight, clerestory) and determine the daylight produced.

As discussed in detail in Section 19.3, the hand-calculation methods (including small programmable calculators) address maximum, minimum, and (possibly) average conditions. Large, mainframe computer-aided daylight analyses are beyond our scope, and the interested reader is referred to Appendix H. It is our feeling that interior daylight contours (isolux lines) are frequently more meaningful than isolated individual absolute numbers, and for this reason we will also review "hand" graphic methods that produce such contours.

19.5 Concept and Characteristics of Daylight Factor

Daylight factor (DF) is defined as

$$\text{DF} = \frac{\text{indoor illumination at a given point}}{\text{unobstructed exterior horizontal illumination } (E_H)} \tag{19.3}$$

expressed as a percentage. The concept is obviously applicable only where E_H can be calculated or measured, that is, where the sky luminance distribution is known or assumed. Thus it can be used with the CIE-defined overcast sky (Equation 19.1), uniform overcast sky, and clear sky whose distribution is fixed for the purpose of calculation. *Direct sun is excluded.* DF cannot readily be used for skies with constantly changing luminance (partly cloudy), since under such conditions the DF at a point also varies continuously, making the concept useless as a calculation tool for absolute daylight values. However, even with such a sky, plots of DF contours will show the *relative* daylight levels in an interior space. This is of great value to the lighting design in appreciating the patterns of brightness (luminance) within a room. As will become clear from an explanation of its components, a daylight factor associated with one type of sky luminance distribution cannot be used directly with another sky, without introducing considerable inaccuracy. As explained above (Section 19.4*b*) and shown in Fig. 19.7, *the sky component* of DF for uniform and nonuniform sky distributions can be converted to one another. However, this conversion is usually done only with overcast sky, where reflected components are small. Conversion of DF criteria between overcast and clear sky is unusual because of the great difference in importance of reflected light factors. To make such a conversion, DF must be divided into its components, the sky component (see Fig. 19.10, below) converted to uniform sky luminance, and the DF reconstituted with proper (and highly variable) reflected components. Since this is not readily accomplished, DF is almost exclusively used with overcast sky conditions.

The reason for the primary reliance on overcast sky conditions is historical. The method was developed in England (see Hopkinson, et al., 1966), where the solid overcast sky condition is predominant for much of the year. Considerable additional pioneer work on daylighting was done in Scandinavia, where the same sky condition prevails. As a result, the overcast sky became the basis for DF recommendations and remains so today. The validity of the DF concept as a method for expressing interior daylight conditions, both absolute and relative, should be obvious. With a given sky luminance distribution (which varies with solar altitude), variation in daylight *inside* will correspond exactly to variations outside for a given sky condition and the ratio (DF) remains the same. This assumes minimal effect from obstructions and ground reflections, as is indeed the case with a dull overcast sky. Then, as explained in Section 19.3, the phenomenon of lightness constancy and Weber's law of contrast constancy will smooth out the variations, making the *apparent* brightness due to daylight appear fairly uniform.

Thus the daylight factor method permits study of interior daylight distribution for varying fenestration, architectural arrangement, and building orientation. The overcast sky in the DF method, as stated, represents minimum exterior illumination, since the sky is frequently brighter. Hence actual interior daylight levels will frequently exceed the design minima. Since the overcast sky is *not* a glare source, the large window areas utilized in overcast sky design locales are not a glare source even when exterior levels are considerably above the design minima. However, such windows *are* a source of severe glare in clear sky conditions.

It is apparent that the DF varies within an interior space (see, for instance, Fig. 19.12, below). This variation, however, is constant for a given architectural configuration. Therefore, knowing the DF variation for a given space and the exterior illumination as derived from sky luminance data, the actual interior daylight levels throughout the space can easily be calculated. Going one step further, minimum DF levels (as well as minimum artificial illumination levels) can be *specified* for different occupancies. These assume a minimum average daylight figure in a given geographic location, and the architecture is then designed to meet these requirements. Thus a design based on a combination of minimum exterior brightness and minimum DF requirements will result in sufficient daylight for a large part of the working day under almost all exterior conditions.

The daylight at any point within an enclosed space is comprised of three components (see Fig. 19.10):

1. Sky component (SC).
2. Externally reflected component (ERC).
3. Internally reflected components (IRC_1 + IRC_2).

The *sky component* (SC) is that portion of the total

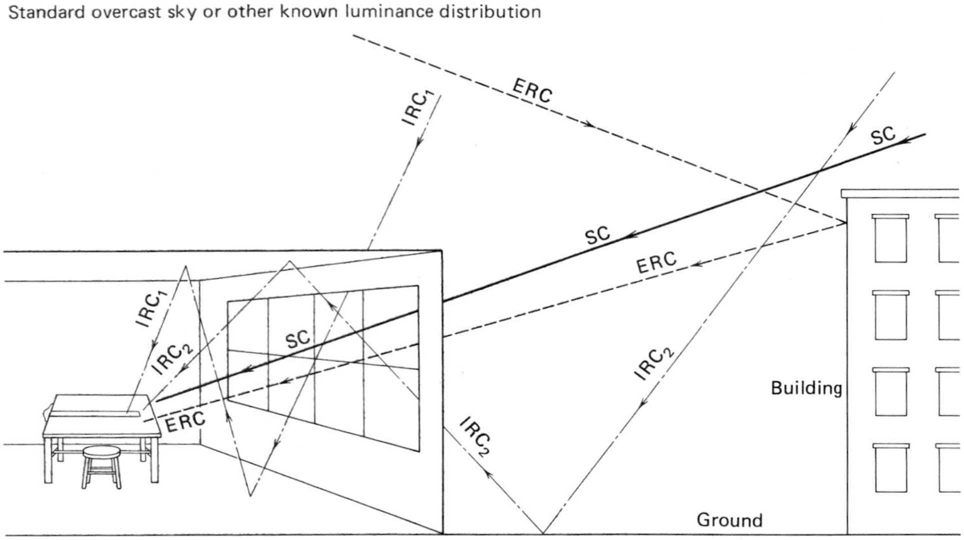

Fig. 19.10 Total daylight factor (DF) is composed of SC (sky component), ERC (externally reflected component), and IRC (internally reflected component). The latter in turn is subdivided into reflected skylight and reflected groundlight. Note that surfaces deep in the room are lighted in large part with rereflected light.

daylight at a point, which is received directly from the area of sky visible *through* the window. Since the sky component is *received* light, it takes into account light reduction due to window obstructions (mullions, etc.) and losses in transmission; that is,

$$\text{SC} = \text{incident skylight} - \text{window losses}$$

The externally reflected component (ERC) is light reflected from exterior obstructions onto the point under consideration. This *does not include ground-reflected light*. ERC is of significance only in built-up areas and can be estimated as the portion of sky component for that area of obstructed sky, reduced by the reflectance factor (RF) of the obstruction; that is,

$$\text{ERC} = \text{sky component} \times \text{RF (of obstruction)}$$

Thus if 25% of the sky is obstructed by a building with a 20% RF, we have

$$\text{ERC} = \text{SC} \times 0.25 \times 0.20$$

or

$$\text{ERC} = 5\% \text{ of SC, to be added to the remaining 75\% SC}$$

The *internally reflected component* (IRC) is the light received at the point under consideration that has been reflected from interior and exterior surfaces. Nevertheless, since IRC_2 is generally small, $\text{IRC} \cong \text{IRC}_1$ (see Fig. 19.10). IRC is therefore entirely dependent on surface reflectances and on the amount of window glazing and becomes a large portion of DF deep within the interior space (see Table 19.3 and Fig. 19.11). IRC is normally calculated using published interreflectance tables, as direct calculation is extremely complex. In summary, the DF then is the sum of the three components:

$$\text{DF} = \text{SC} + \text{ERC} + \text{IRC}$$

all calculated individually for each location being considered.

Typical curves for both horizontal and vertical daylight factor for a room with single sidelighting (windows on one side) are shown in Fig. 19.12. Note that DF_v is larger than DF_H, reversing the condition outside (see Fig. 19.5*a*). This is obviously due to the light coming from only the side. The reverse would be true for a toplighted space.

These curves are produced by a longhand daylight-protractor-aided technique (Building Re-

TABLE 19.3 **Effect of Wall Reflectance Factor on Proportion of Internally Reflected Component (IRC) in Daylight Factor**

Distance from Window, Feet	*30% Wall Reflectance*		*60% Wall Reflectance*	
	Total DF	$\frac{IRC}{DF}$ %	*Total DF*	$\frac{IRC}{DF}$ %
0	30	1	31	3.5
5	16	1.9	17	6.5
10	5.5	5.5	6.3	16.9
15	2.1	14.3	2.9	37.9
20	1.3	23	2.1	52.4

ROOM DATA:
Room 24 × 28 ft. — 70% ceiling RF
Window on 28-ft wall—one side only — 20% floor RF
Window area = 20% of floor area

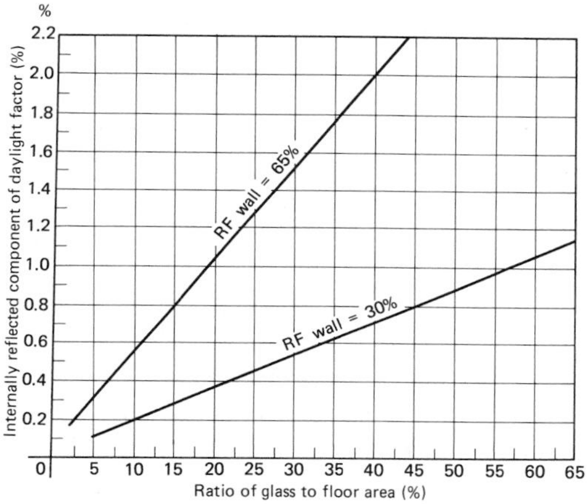

***Fig. 19.11** Plot of internally reflected component (IRC) of daylight factor (DF) as a function of amount of glazing, expressed as DF, as a percentage of exterior illuminance. As is expected, the effect of lighter wall finish becomes more pronounced as the fenestration area increases.*

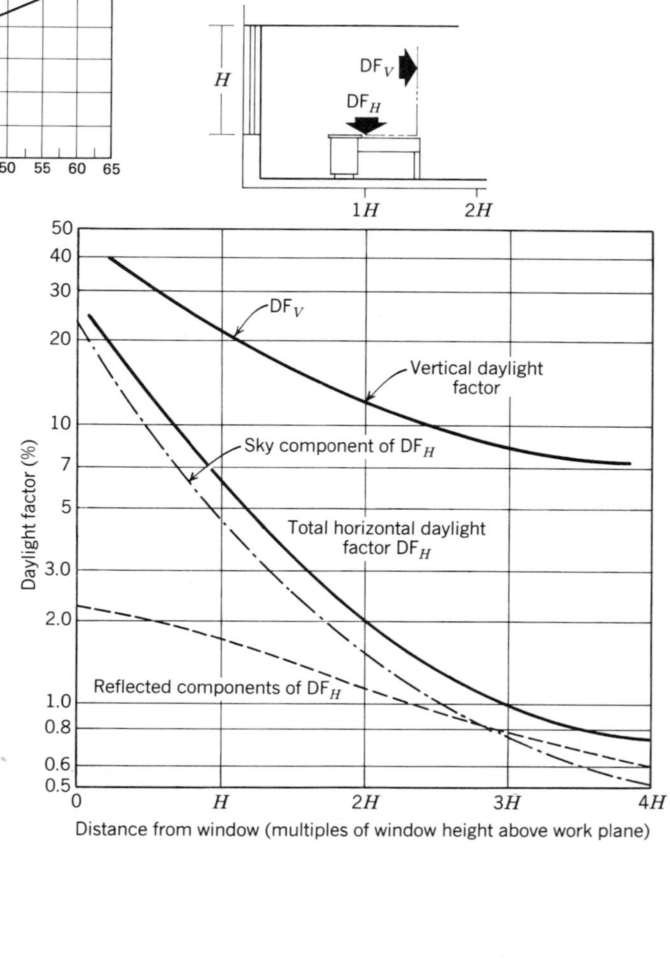

***Fig. 19.12** Typical daylight factors curves for horizontal (DF_H) and vertical (DF_V) illumination for a room with large windows on one side only. Note that the sky component represents almost the entire daylight factor near the window, but reduces its proportion at greater depths. There, interreflected light constitutes 50% of the available daylight. See also below, Fig. 19.24.*

search Station, London). The three components—SC, IRC, and ERC—are calculated separately and added to produce the total daylight factor. Any change in parameters, such as window dimensions or height above the working plane, ceiling height, surface reflectance, ground reflection, and obstructions, alters these curves and requires recalculation and replotting. Obviously, exact calculation of even a few variants in a space is a tedious, time-consuming procedure when done point by point, longhand. For this reason several alternative approaches are available:

1. Use of a computer program (see Appendix H).
2. Use of simplifications, such as standard curves, tabular data, or the CIE method described in Section 19.6 (see *Simplified Daylight Tables,* 1958).
3. Use of a library of graphic light distribution plots with varying parameters (see Section 19.7).

Space limitation prevents us from presenting complete data on each method. This is especially true with tabular data (item 2) and the library of graphic plots. We will therefore only survey those, while presenting the CIE sidelighting method in its entirety.

All the methods are primarily applicable to analysis rather than design, that is, calculating daylight, given the openings, and not vice versa. The architectural designer, however, must establish an initial design, and this is done either intuitively or, lacking experience, by following some rules of thumb. A few such rules are given in Table 5.3 *for overcast sky.*

Despite the fact the DF is a ratio and not an absolute value such as lux, it is applied as an absolute. That is, recommendations are made of minimum daylight factor for specific seeing tasks. Recommended minimum levels of daylight factor vary, but those listed in Table 5.2 are average British recommendations. These DF figures were established at a time when illuminance level recommendations were considerably lower than even the modest ones of today, as are given in Table 18.3. This is because on the basis of extensive observation, it was assumed that an overcast sky would give a minimum (85% of time) horizontal illuminance of 5000 lux (≈ 500 fc). Applying the DF recommendations of Table 5.3 to such a sky gives:

Ordinary tasks	75–125 lux
Moderate tasks	125–200 lux
Difficult tasks	200–320 lux

These figures are less than half of current recommendations. Obviously then, where exterior illuminance is indeed as low as 5000 lux, PSALI would be required for all areas beyond *H* feet, that is, one window height, from the window (see Fig. 19.12).

19.6 Daylight Analysis: CIE Method

(a) General. This method was the result of a search for a simple, rapid, straightforward, and fairly accurate daylight calculation method that would yield reliable results without the time-consuming constructions and calculations necessitated by other methods. After a study of considerable length and intensity, the CIE adapted and adopted a system developed in Australia by Dr. A. Dresler (*Daylight Design Diagrams,* 1963). The current CIE recommendations were published in 1970 (*Daylight,* 1970).

This system is based on the daylight factor described above as applied to the standard overcast CIE sky. Dr. Dresler developed a set of more than 100 curves covering rooms of varying proportions and fenestration. One such typical curve is shown in Fig. 19.13. The curves relate minimum DF (at a point 2 ft from the wall opposite the window) to the maximum permissible room depth, for given reflectances and a standard window design, thus establishing the room's properties. Depth, or width, is the dimension at right angles to the window wall.

The curves were calculated using the techniques developed in England, as described in the preceding section. The published CIE recommendations make a clear, strong case for the minimalist approach in selecting sky illumination and in design. They state simply that the number of design variables is so large and daylight itself so variable, that a simple routine method can be based on only minimal conditions for a given (selected) percentile. Therefore, the diagrams will give the

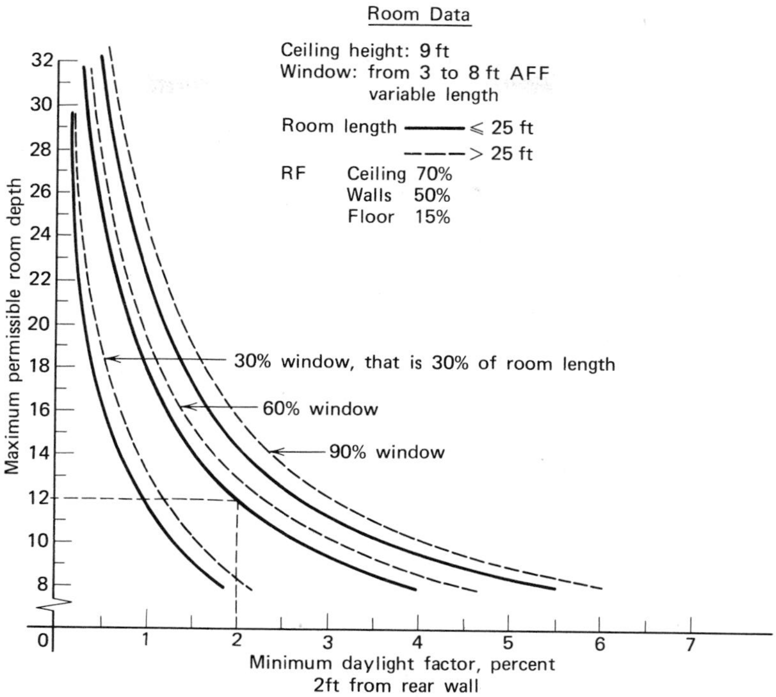

Fig. 19.13 Maximum room depth required to maintain minimum DF is proportional to window size. Thus for a room less than 25 ft long with a 5 ft high window for 60% of the room's length, depth cannot exceed 12 ft if 2% DF is to be maintained at a point 2 ft from the rear wall. (From Daylight Design Diagrams, *1963.)*

lowest level of daylight that can reliably be expected for a given percentage of normal working hours in sidelighted rooms, and the *average* level in toplighted spaces. Thus the designer can select the 95th percentile for minimum conditions, the 80th to 85th percentile for average conditions and the 60th percentile for maximum conditions. On the basis of these he can design maximum PSALI, estimate energy consumption, and establish the zones of lighting control.

(b) Characteristics of the Method. Advantages of the system are:

1. Consideration of obstructions, reflections, and interior reflections.
2. Use of overcast sky brightness that varies with latitude rather than a standard 5000 lux.
3. Applicability to a very wide range of side and top fenestration designs.
4. Establishment of required room proportions is architecturally more useful than solving for specific dimensions.

Limitations of the system are:

1. Inapplicable to clear-sky and sun conditions.
2. Inapplicable to other than rectangular-shaped rooms.
3. Unusable with sun-shading or high-reflectance ground.
4. Results give points of minimum, twice minimum, and four times minimum daylight only. Other points must be interpolated or extrapolated.
5. Window proportions and position in a wall are fixed.

Overall, the system accomplishes what it intended. The limitations listed are inherent in any rapid, simplified daylight calculation technique. They are listed here not as a criticism, but to advise the reader of these limitations. The standard itself, which should be in the library of every daylight designer, also clearly states these limiting characteristics.

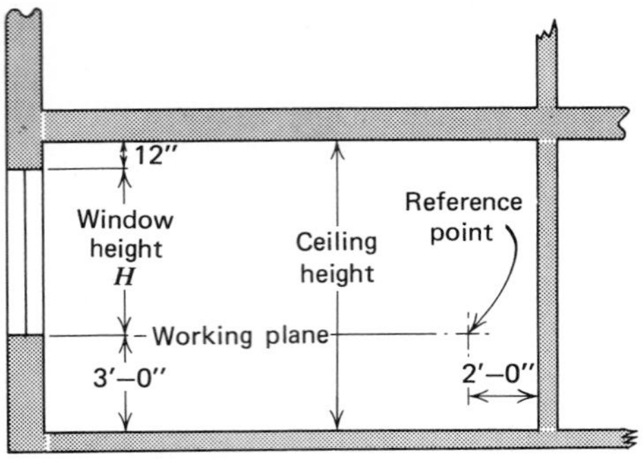

Section through a unilaterally lit room showing the assumed dimensions. These dimensions are the same for bilateral lighting except that the reference point is midway between the window walls.

(a)

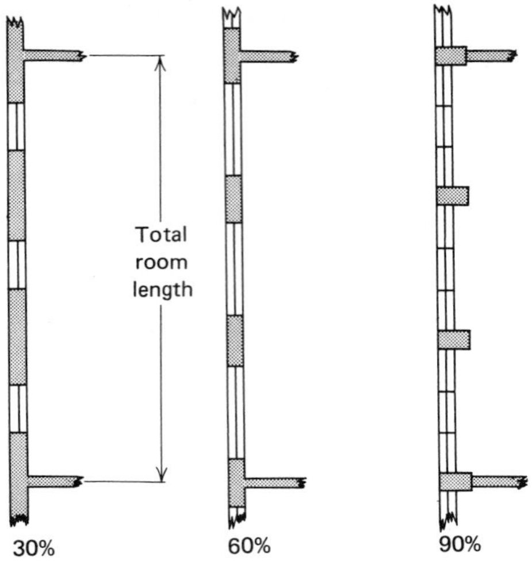

Plane of window walls showing window width expressed as a percentage of total room length.

(b)

***Fig. 19.14** Sketches indicating the parameters of the CIE calculation systems.* (a) *A vertical section through a room with dimensional data relevant to this system. Note that the sill height has been selected to coincide with a working plane at 90 cm (3 ft). Height of working plane varies between 30 and 36 in., the former being more common in American use, the latter in European use. A lower sill contributes only ground-reflected light at the working plane. Where window sill is much above the working plane (i.e., short windows, high on the wall), this design system is inapplicable.* (b) *Calculation of size (length) of windows with respect to overall room length. (From* Daylight, International Recommendations for the Calculation of Natural Daylight, *Publication CIE No. 16 (E-3.2), 1970; Commission Internationale de l'Éclairage. Reproduced with permission.)*

(c) Calculation Procedure. The CIE system is usable in two modes:

1. Given complete architectural data, find resultant daylight.
2. Given incomplete architectural data and required daylighting, find maximum room depth and/or other room proportions that will satisfy the daylighting requirement.

Obviously the former mode is simpler, since it leads directly to an answer. It will be demon-

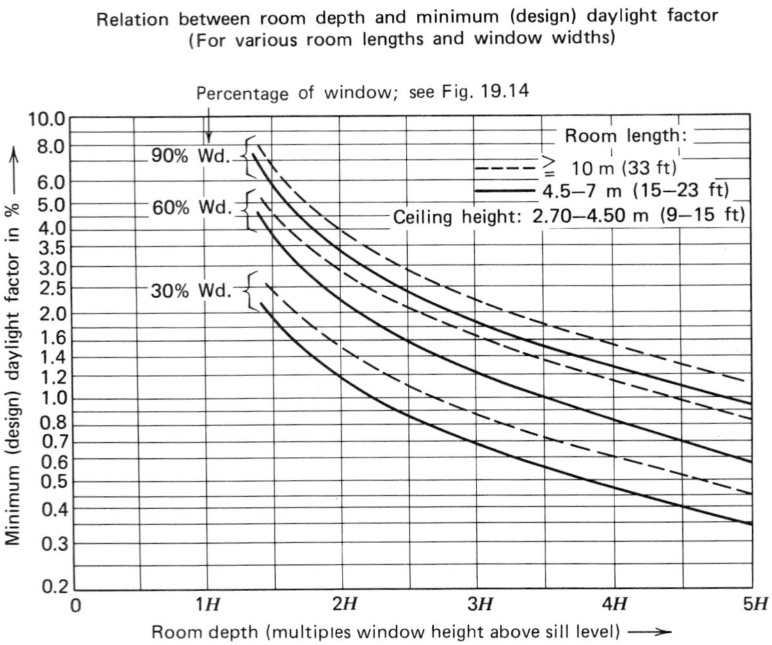

***Fig. 19.15** Basic design diagram that relates minimum daylight factor to room depth. Inasmuch as room depth is expressed in terms of window height, the curves effectively relate minimum daylight factor (2 ft from back wall) to room proportion. (From* Daylight, International Recommendations for the Calculation of International Daylight, *Publication CIE No. 16 (E-3.2), Commission Internationale de l'Éclairage. Reproduced with permission.)*

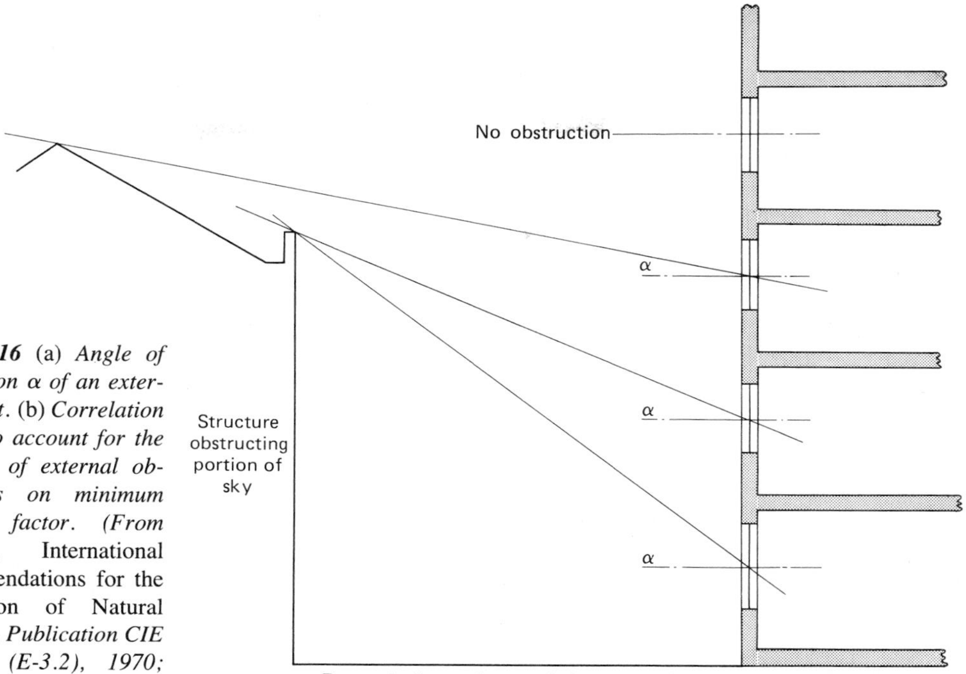

External obstructions and the angle of obstruction. The angle of obstruction α is measured from the center of the window to the outline of the opposite building.

(a)

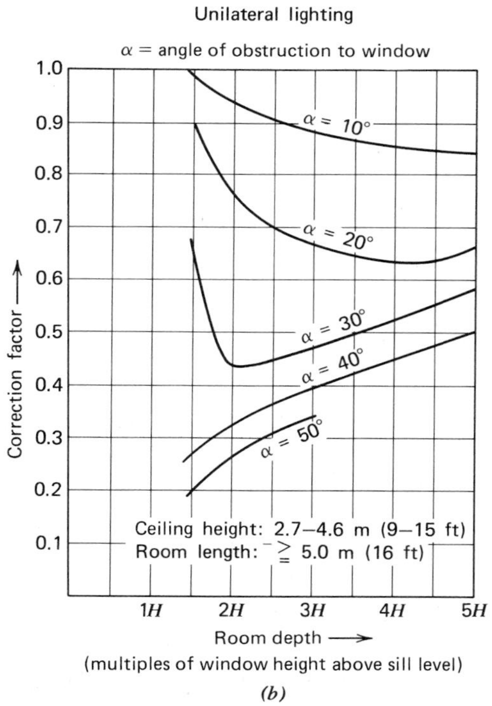

(b)

Fig. 19.16 (a) *Angle of obstruction α of an external object.* (b) *Correlation factors to account for the influence of external obstructions on minimum daylight factor. (From* Daylight, International Recommendations for the Calculation of Natural Daylight, *Publication CIE No. 16 (E-3.2), 1970; Commission Internationale de l'Éclairage. Reproduced with permission.)*

strated in an illustrative example below. The latter mode, since it has many answers, is more complex. That is, various combinations of fenestration and room dimensions will yield adequate daylight. For this reason, the designer should set room length (front to back) and percent fenestration of the window wall, leaving depth (width, side to side) as a variable, or set room length and depth, with percentage of fenestration as the variable. Ceiling height is usually fixed by other considerations. See Fig. 19.14 for sketches showing room parameters. The procedure in mode 1 is as follows (the reader can work out the mode 2 procedure):

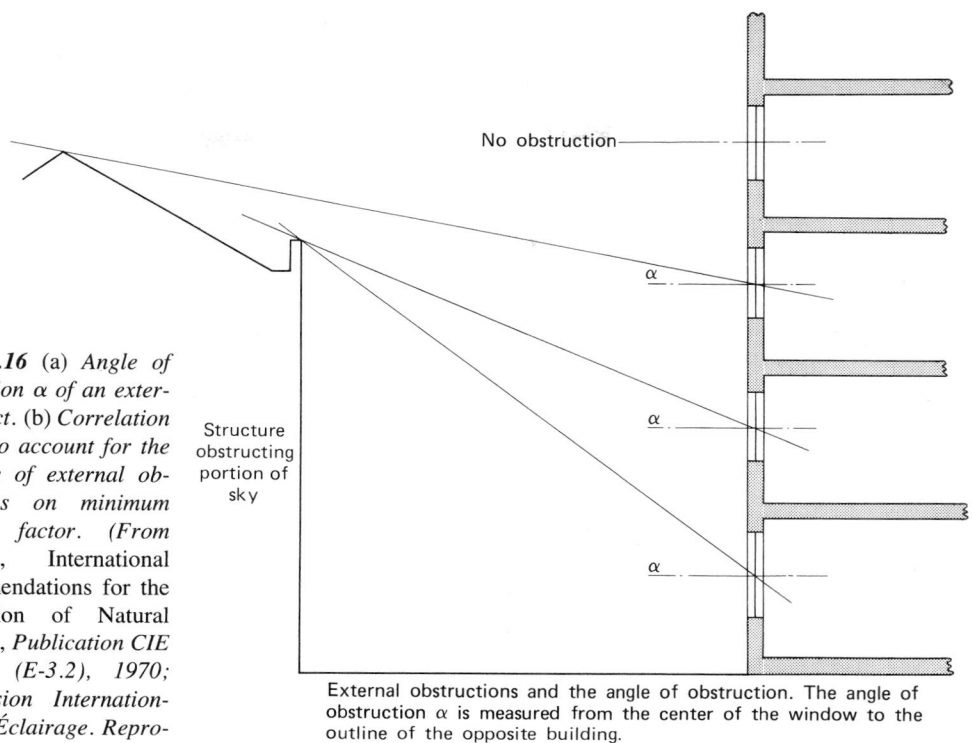

1. Express room depth in terms of window height. From Fig. 19.15 determine *design* daylight factor, that is, the daylight factor 2 ft from the wall opposite the window, on the window centerline.
2. The design daylight factor is larger than the actual service daylight factor because it is based on clear, clean glass and unobstructed horizon. Since this is rarely the case, the actual (service) daylight factor must be calculated using the correction factors of Fig. 19.16 and Table 19.4; that is,

TABLE 19.4 **Correction Factors to Be Used in CIE Daylight Calculations**

Diffuse Transmittance of Glass	*Correction Factor*
80%	0.95
70%	0.8
60%	0.7
50%	0.6
40%	0.45
30%	0.35

(*b*) Correction Factors to Allow for Dirt Accumulation on Glass

Locality	*Class of Industry*	*Angle of Slope (Measured to the Horizontal)* 90–75°	60–45°	30–0°
Country or outer-suburban area	Clean	0.9	0.85	0.8
	Dirty	0.7	0.6	0.55
Built-up residential area	Clean	0.8	0.75	0.7
	Dirty	0.6	0.5	0.4
Built-up industrial area	Clean	0.7	0.6	0.55
	Dirty	0.5	0.35	0.25

(*c*) Percentages to Use When Figure 19.17 Curves Are Applied to Periods Other Than 09.00–1700

Curve in Figure 19.17	*95%*	*90%*	*85%*	*80%*	*70%*	*60%*
Alternative period	Percentage of alternative period					
07.00–15.00	95	90	85	80	70	60
08.00–16.00	100	100	95	85	70	60
07.00–17.00	95	85	75	65	55	45
06.00–18.00	75	70	65	60	50	40

Source: From *Daylight, International Recommendations for the Calculation of Natural Daylight,* Publication CIE No. 16 (E-3.2), 1970; Commission Internationale de l'Éclairage.

$$\text{DF}_{\text{service}} = \text{DF}_{\text{design}} \times \text{correction factors}$$

3. Using the service daylight factor in Equation 19.4,

$$\text{Required exterior illumination} = \frac{\text{required interior illumination}}{\text{service DF}}$$

obtain required exterior illumination.

4. From the curves in Fig. 19.17, obtain the percentage of hours between 0900 and 1700 that the required illumination is maintained. For time periods other than this, see Table 19.4*c*.
5. From Fig. 19.18 determine the room positions at which illumination is twice or four times minimum.

The curves given here for unilateral sidelighting are typical of those found in the standard (*Daylight,* 1970), which covers, in addition, bilateral sidelighting and toplighting of the three principal varieties, that is, skylights, sawtooth roofs, and monitor roof.

An example of this simple, rapid technique should make its use clear. We have selected a geographic location appropriate to this method.

EXAMPLE 19.1. A classroom in a single-story Seattle elementary school is 25 ft long and 18 ft deep and has a 9 ft 6 in. ceiling. It receives daylight unilaterally from windows totaling 18 ft in length. See room sketch, Fig. 19.19. Window glazing is wired glass having a transmittance of 80%. The school is situated in a built-up residential area with a satisfactory cleaning schedule. Determine the portion of the year during which tasks requiring a minimum illumination of 150 lux can be carried out by daylight throughout the room. Also, determine what levels can be maintained for 85% of daylight hours, and at what distances from the window. (Note that 150 lux corresponds to a DF of 2% applied to an E_H of 7500 lux.)

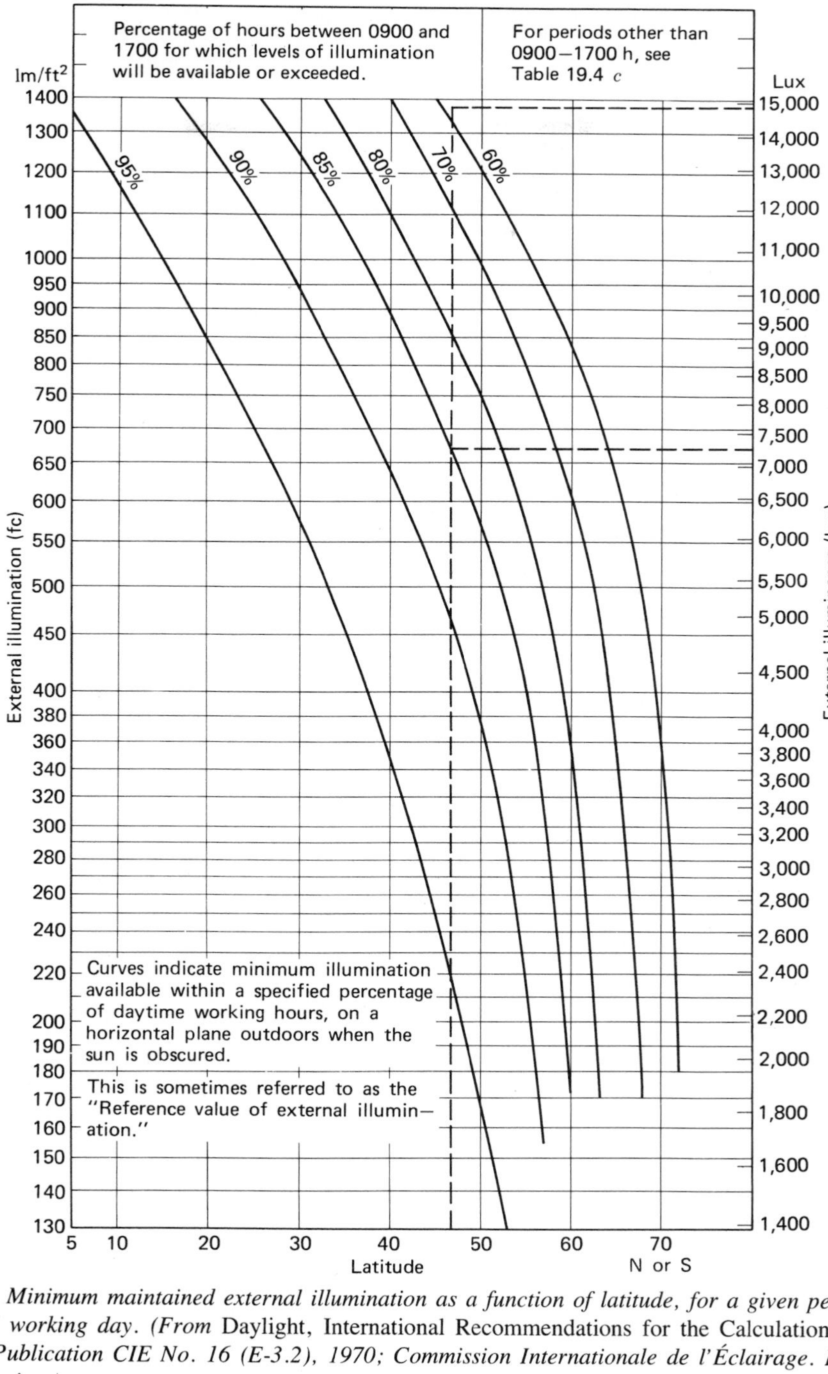

Fig. 19.17 Minimum maintained external illumination as a function of latitude, for a given percentage of the normal working day. (From Daylight, International Recommendations for the Calculation of Natural Daylight, *Publication CIE No. 16 (E-3.2), 1970; Commission Internationale de l'Éclairage. Reproduced with permission.)*

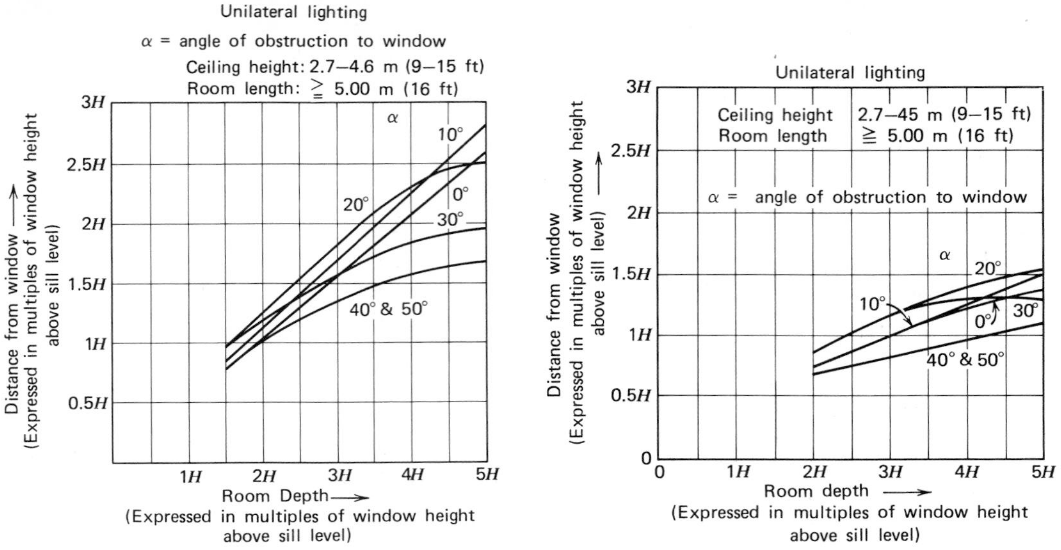

(a) (b)

Fig. 19.18 (a) *Distance from window at which daylight factor is twice minimum daylight factor*. (b) *Distance from window at which daylight factor is four times minimum daylight factor. (From* Daylight, International Recommendations for the Calculation of Natural Daylight, *Publication CIE No. 16 (E-3.2), 1970; Commission Internationale de l'Éclairage. Reproduced with permission.)*

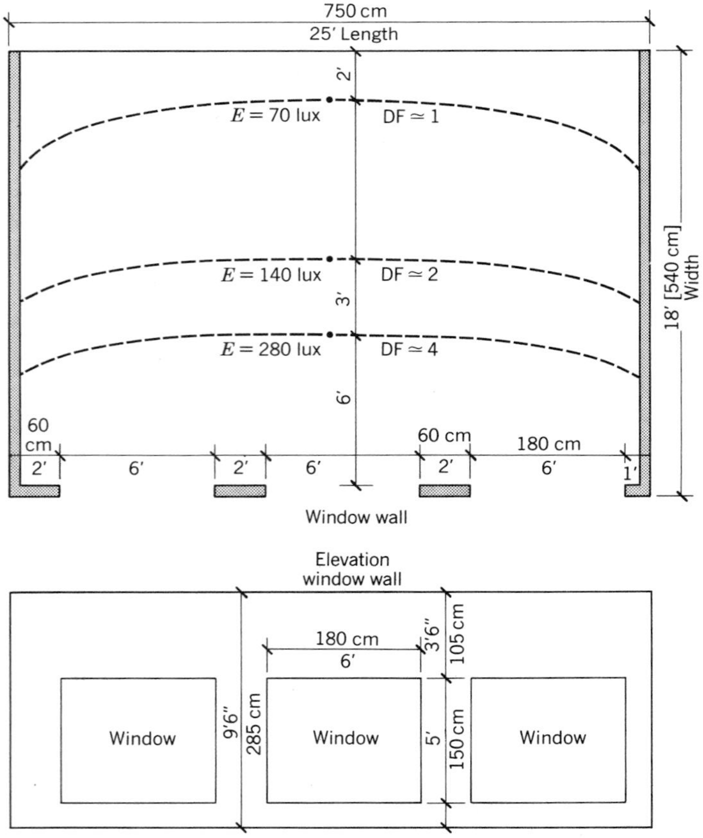

Fig. 19.19 Plan and window wall elevation of room of Example 19.1. The three daylight contours are estimated, based on the calculated centerpoint. They represent the levels maintained for 85% of daylight hours. Levels twice as high are maintained for 60% of the daylight hours.

SOLUTION

General. The latitude of Seattle is 47.6° N. The design condition for Seattle is solid overcast sky for 85% of the hours between 0900 and 1700, producing an unobstructed exterior horizontal illuminance E_H of 7000 lux.

1. Window height H is 5 ft (see Fig. 19.19) Room depth in terms of window height:

$$\frac{\text{18-ft depth}}{\text{5-ft window height}} = 3.6\ H$$

$$\text{Window coverage: } \frac{3 \times 6'}{25'} = 72\%$$

From Fig. 19.15, for room length of 25 ft, ceiling height 9 ft 6 in., window coverage 72%:

$$\text{Design DF at } 3.6H = 1.3$$

2. From Table 19.4 and Fig. 19.16*b* we obtain correction factors:

Glass transmission 0.95

Glass cleanliness 0.8

Therefore

$$\text{Service DF} = 1.3 \times 0.95 \times 0.8 \cong 1.00$$

3. Required exterior illumination:

$$E_H = \frac{\text{150 lux min.}}{\text{1.00 DF min}} \times 100 = 15{,}000 \text{ lux}$$

4. From Fig. 19.17 we see that with the given conditions, an illuminance of 150 lux will be maintained for less than 60% of the hours between 0900 and 1700. It was already apparent at step 2 that the desired 85th percentile of time would not be achieved, since the curves of Fig. 19.17 are based on a minimum DF of 2%, whereas our service DF is only 1.0, or half the required 2%. The level that *is* maintained for 85% of the hours is

$$E_{\min} = 7000 \text{ lux} \times 1.0 \text{ DF} \simeq 75 \text{ lux}$$

5. To complete the picture, we have from Fig. 19.18
   (a) DF doubles at $1.8H$ corresponding to 140 lux at a depth of 9 ft (1.8 × 5 ft).
   (b) DF quadruples at $1.2H$, corresponding to 280 lux at a depth of 6 ft (1.2 × 5′). These values (*a,b*) are maintained for 85% of daylight hours.
6. The calculated illuminance levels are accurate only at the centerline of the window wall. By visual estimate and extrapolation, rough contours can be drawn. They are shown on Fig. 19.19, and also in the next section on Fig. 19.21*c* for the purpose of comparing results with the graphic method discussed there.

In summary then, the CIE method is simple, rapid, and flexible, but provides only limited data. It is usable in both the design and analysis mode. Its exterior illuminance data (Fig. 19.17) seem to agree well with actual measurement (7250 lux on the figure as compared to 7000 lux measured average for Seattle). If contours are roughed in as shown on Fig. 19.19, guidelines for PSALI design and lighting control strategy can be drawn.

19.7 Graphic Daylighting Design Method (GDDM)

A method that applies the principle of daylight factor to overcast sky and shows results as DF (isolux) contours within a room (rather than individual DF factors at specific points) was developed by M. S. Millet and J. R. Bedrick (1980). Its primary advantage over the CIE method is that its results are a family of DF contours that are more meaningful to a lighting designer than either the isolated numbers of hand-calculation procedures or the rafts of numbers generated by computer programs. Its shortcomings are that it, too, is not readily applied to clear-sky conditions and that it requires the designer to acquire a "library" of 200 or so patterns that cover most design situations. For this latter reason, only the outline of the method is presented here. We strongly suggest that the reader acquire access to the catalog of patterns, so that he can apply this useful and relatively accurate method.

The authors of the method utilized a computer simulation program (UWLIGHT) to develop daylight distribution patterns in a room, resulting from either sidelights or skylights. To generalize the system, windows are identified by height to width

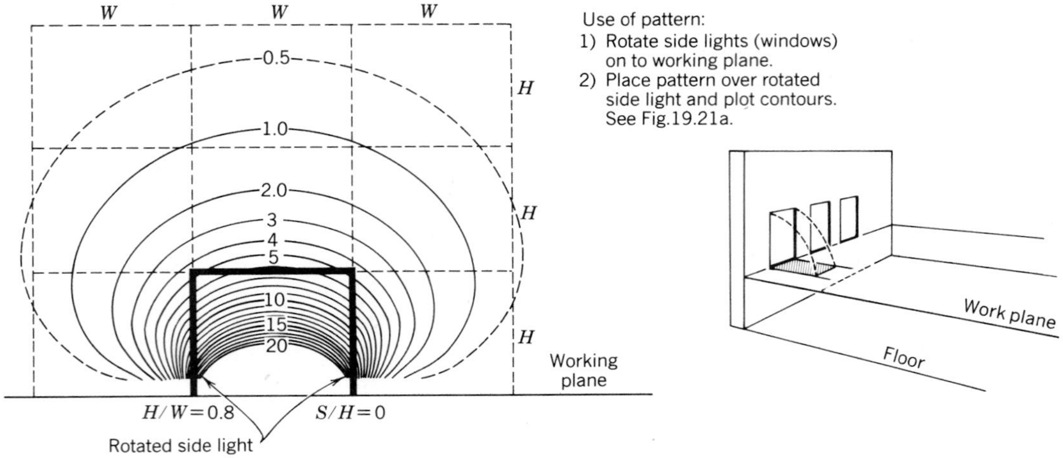

Fig. 19.20 Typical isolux contour map for a window with height-to-width ratio of 0.8, and sill at the working plane elevation. Numbers represent sky component of daylight factors, for overcast sky condition. The rectangles are the window outlines "rotated," i.e., projected, onto the work plane. See insert. (From Millet and Bedrick, 1980, p. 141.)

proportion (*H*/*W*), and the position of the isolux contours on the plan is determined by the ratio of the elevation of the sill over the work plane to the window height. This latter factor is a large advance over both the CIE system and the LOF/IES/lumen system (see Section 19.8), both of which are restricted to a sill height at the work plane. Thus the GDDM system can account for high windows, clerestories, and other designs intended to introduce daylight deep into the space—something that CIE and IES methods cannot do.

Figure 19.20 shows a typical isolux pattern for a window whose *H*/*W* ratio is 0.8 and whose sill is at the work plane. This particular window pattern was selected because it corresponds to the window of Example 19.1, Fig. 19.19, and will enable us to run through the same problem and compare results.

EXAMPLE 19.2. Repeat the design problem of Example 19.1 using the GDDM method.

STEP 1. Select the appropriate window pattern. Referring to Fig. 19.19, window proportion is

$$\frac{H}{W} = \frac{5}{6} = 0.83$$

We therefore select the pattern shown in Fig. 19.20 as the closest to our requirements.

STEP 2. On a plan of the room, trace the isolux pattern for each window. This is shown on Fig. 19.21*a*. *S*/*H* = 0 indicates that the pattern begins at the window wall, as shown. The patterns overlap, since the windows are close together. Where contours meet, the DF of the contours are added, producing points for new combined contours. The combined contours and their values, in DF, are shown in Fig. 19.21*b*.

STEP 3. In the final step the value of the combined contours is corrected to account for internally reflected components of daylight, plus light reduction due to glazing. The final contours are shown on Fig. 19.21*c*. Note that this diagram gives the designer a much more complete picture of the daylight contours than the results of the CIE method. For the purpose of comparison the three calculated DF factors of the CIE method are shown on Fig. 19.21*c*.

19.8 Daylight Design: LOF/IES/Lumen Method

A hand calculation procedure that addresses all types of sky conditions was developed in the early 1950s at Southern Methodist University (Dallas) by J. W. Griffith and Associates (Griffith, Arner,

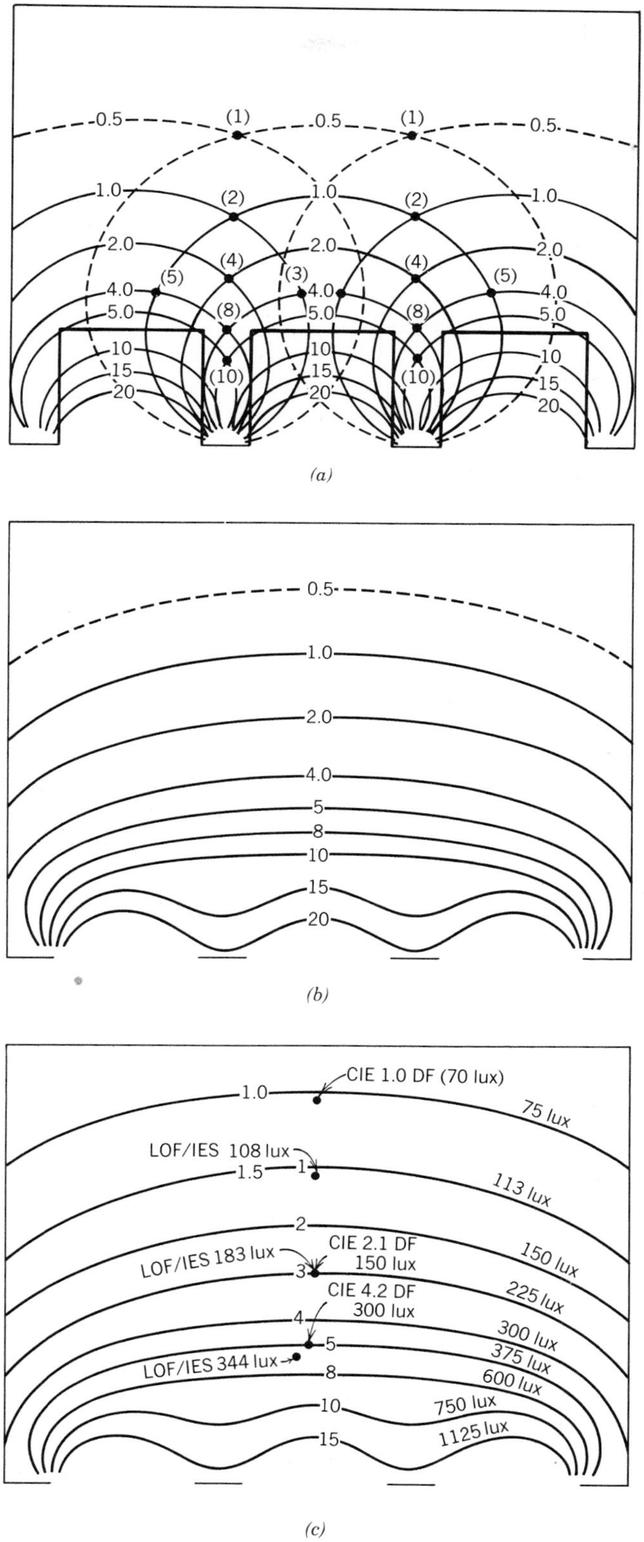

Fig. 19.21 (a) *Daylight contours for each window of Fig. 19.19 are plotted on the floor plan of the room being studied. Numbers in parentheses are combined sky components.* (b) *The isolux contours of* (a) *are combined to form new isolux contours, which represent the total sky component of daylight within the room.* (c) *The final isolux contours are calculated, including correction factors. The numbers represent daylight factors. Note the variance between these contours and the points calculated by the CIE method. The results of design by the LOF/IES method, discussed in the next section are also shown, to permit comparison with both the CIE and the GDDM methods.*

and Wenzler, 1957). The work was sponsored and published by the Libby-Owens-Ford Glass Co. (''How to Predict,'' 1976). The procedure was then adopted by the IES of North America for its Recommended Practice of Daylighting. This LOF/IES method is also frequently called the lumen method.

(a) General Remarks. This method is in large measure parallel to the IES method for calculating artificial lighting, in that it treats the window or toplight as a large area lighting source—essentially a transilluminated lighting fixture. It then applies coefficients to the light output of this ''daylight fixture,'' similar in usage to the familiar coefficient of utilization and light loss factor of a lighting fixture, and arrives at an interior daylight level. The lighting levels calculated for sidelighting are average levels, at predetermined points in the room. For toplighting it is an average overall level, assuming proper spacing of toplights. The method is completely described in the relevant current IES publication (*IES Recommended Practice*, 1979).

(b) Characteristics of the Method. Before entering into a detailed description of the technique, and demonstrating it with an illustrative example, a few general remarks are in order. The IES/LOF method is probably the most flexible manual technique available. It recognizes the importance of reflected light (see Fig. 19.22), can accommodate shading devices that will almost always be used on direct sun exposure, can readily account for various types of glazing—blinds, drapes, screens, and so on—is applicable to unilateral and bilateral sidelighting plus the usual varieties of toplighting, and is documented with fairly reliable tabular and graphic data. Nevertheless, it does have, to our mind, several distinct limitations or shortcomings of which the reader should be aware. These are:

1. The method is usable in only one mode, that is, given location and full architectural data, daylighting can be calculated. It cannot readily be used to determine room proportions, given the other data, as can the CIE method. Thus it is an *analysis* tool, not a *design* tool.
2. Since it considers the daylight source essentially as a lighting fixture with uniform luminance over its entire surface, *a single figure* of equivalent sky luminance must be used. (This introduces a degree of approximation, making the four-place precision of the coefficients given later in Table 19.6 questionable.)
3. A corollary to the above limitation is that obstructions such as buildings and trees are difficult to take into account. This leads to their being ignored, and a further reduction in accuracy.
4. With clear sky conditions, the northern exposure of a building receives up to 50% of its light from sunlight reflected from facing structures. This factor is ignored by the IES/LOF method, although it can be included as additional skylight or rereflected groundlight, depending on the solar altitude.
5. The standard unilaterally sidelighted room on which the calculations are based (Fig. 19.22*a*) calculates minimum lighting for a position 5 ft from the back wall. To our mind the CIE reference point of 2 ft from the back wall is a more useful figure.
6. The standard room assumes windows extending to the ceiling. This is infrequently the case, and it is important because the upper portion of a window contributes to deep daylight penetration much more than an equivalent lower area (see Fig. 19.22*b*). Thus minimum levels calculated with this window configuration are overoptimistic. Furthermore, the sill is assumed to be at the level of the working plane, that is, 30 to 36 in. [75–90 cm]. Note from Fig. 19.22*b* the effect of changes in sill height. Also, the method cannot accommodate high windows with sills more than 1 ft above the working plane (as clerestories). This is a serious deficiency.
7. The system calculates only three points in a room, on the window centerline. As stated in reference to the three points calculated by the CIE method, this is not normally sufficient to give a picture of the interior daylight distribution.

(c) Calculation Procedure. The calculation procedure for sidelighting will be discussed, since it is more frequently encountered than toplighting. The reader is referred to the current *IES Recom-*

Fig. 19.22 (a) *Standard room section for which LOF/IES daylight calculation method applies. Window is assumed to extend from a 36-in.-high sill to the ceiling. Daylight is calculated for three points in the room along a line from the center of the window to the center of the back wall. These points are indicated as Max, Mid, Min and are 5 ft from the window, at depth midpoint, and 5 ft from the back wall.* D *represents room depth from window to back wall. Room height from 8 to 14 ft can be accommodated.* (b) *The effect of window placement on interior daylight. Where skylight predominates* (b-*1*) *because of high window/high sill/low ground reflectance/strong reflected component from opposite structure, illuminance drops off rapidly as depth increases. When reflected groundlight is a major component, due to high ground reflectance* (b-2), *and/or strong rereflected sun from opposite structure, the window position on wall has minimum effect. Very low sill height will bring curve* B *above curve* A, *as in* (b-*3*). *Note that where skylight is blocked* (b-*3*), *lower sill height obviously results in better penetration. Rereflected sun can be added to both skylight and groundlight, as shown.*

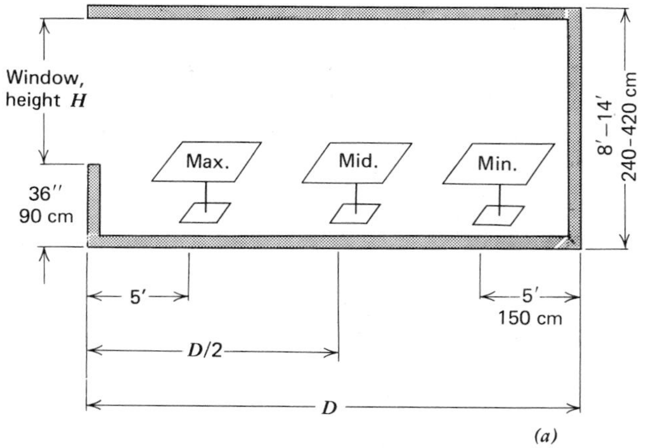

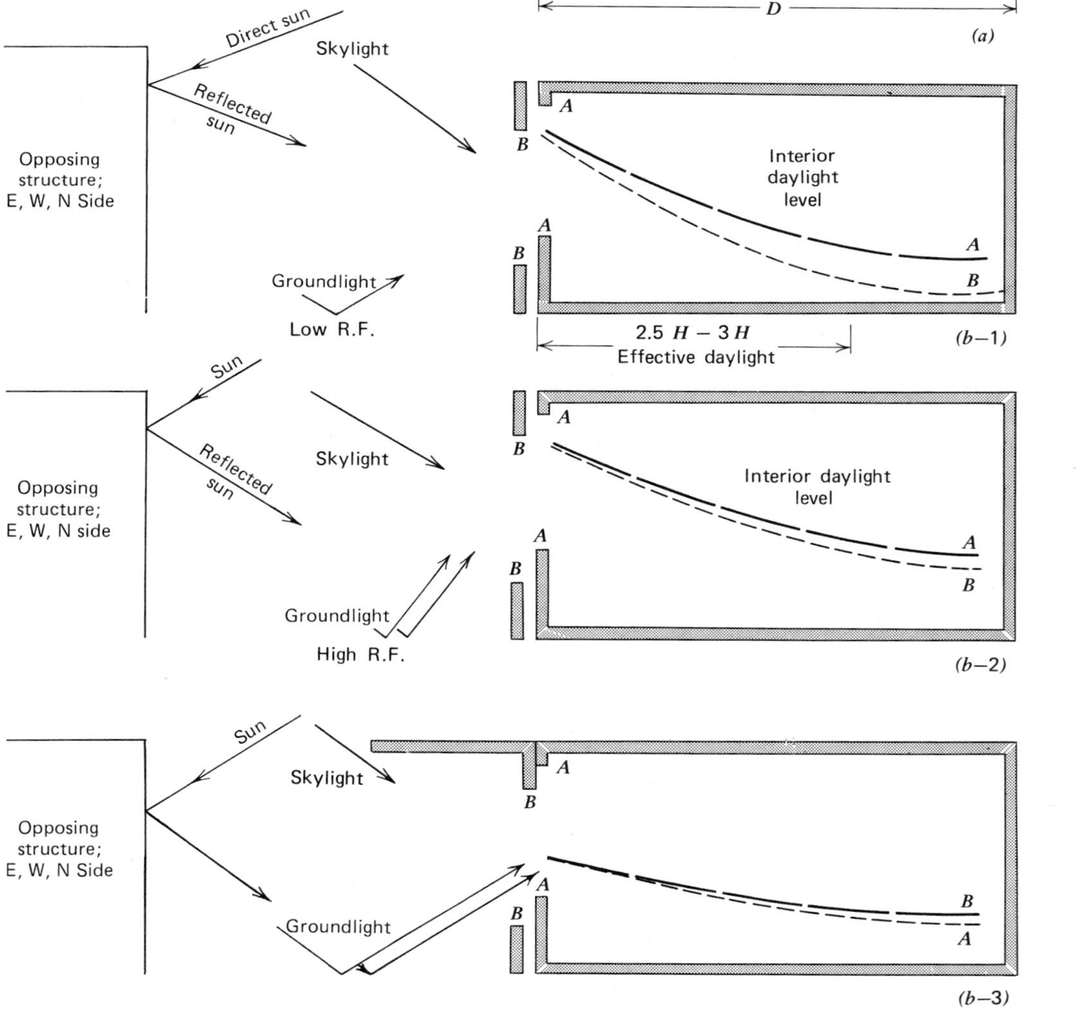

mended Practice of Daylighting for a discussion and presentation of the latter. It should present no difficulties since it is similar to, and simpler than, the sidelighting calculation procedure.

The daylight illumination on the work plane for the reference point(s) shown in Fig. 19.22*a* can be expressed as the sum of two components: that resulting from sky illumination, which we can call E_s, and that resulting from light reflected into the room from the ground, which we call E_G. Thus

$$E_D = E_S + E_G \qquad (19.4)$$

where

E_D = total daylight on the work plane, fc [lux]
E_S = work plane daylight resulting from sky light, fc [lux]
E_G = work plane daylight resulting from groundlight, fc [lux]

TABLE 19.5 **Transmission Data and Light Loss Factors for Windows—IES Method**

(a) Transmittance Data of Glass and Plastic Materials

Material	*Transmittance (percent)*
Polished plate window glass	80–90
Sheet drawn window glass	85–91
Heat-absorbing plate glass	70–80
Neutral low-transmission glass	10–60
Corrugated glass	80–85
Glass block	60–80
Clear plastic sheet	80–92
Colorless patterned plastic	80–90

Source: From *Recommended Practice of Daylighting,* IES.

(b) Average Window Maintenance Factors Expressed as Percentage of Clean Glass Transmission

	Office[a]	*Factory*[b]			
	Window Position				
	Vertical	*Vertical*	*30° from Vertical*	*60° from Vertical*	*Horizontal*
3-month cleaning cycle	82%	69%	62%	54%	50%
6-month cleaning cycle	73%	55%	45%	39%	34%

Source: From *Recommended Practice of Daylighting,* IES.
[a]Typical clean location.
[b]Typical dirty location.

Each of the two factors, E_S and E_G, is calculated as the window illumination, reduced by transmission losses (*AF, TF,* and *LLF*), and modified by room coefficients (*CU* and *K*), that is,

$$E_S = [E_W \times AF \times TF \times LLF] \times CU \times K \qquad (19.5)$$

where

E_S = daylight for the point selected (max, mid, min), from sky light, in fc [lux]
E_W = daylight incident on the window, from sky, in fc [lux] (see Figs. 19.6, 19.8, 19.9)
AF = net area of the window after subtracting mullions, glazing bars, etc., square feet
TF = transmission factor of the window, accounting for glazing plus shades, curtains, screens, and the like. (The individual factors of each are multiplied to obtain the overall factor.) Expressed as a dimensionless decimal; see Table 19.5*a*
LLF = light loss factor, expressing the degree of cleanliness of the glazing. Dimensionless decimal, see Table 19.5*b*
CU = coefficient of utilization selected for the point being calculated (max, mid, min), the sky condition (clear, overcast), the room characteristics, and the type of sun control device in use, if any; see Table 19.6
K = second utilization coefficient, selected with the same criteria as *CU* above; see Table 19.6

TABLE 19.6A **Coefficients of Utilization**[a]

Data: 75% Ceiling Reflectance[b], 30% Floor Reflectance, No Window Controls

		CU					
Length[c]:		*20 ft*		*30 ft*		*40 ft*	
Wall Reflectance:		*70%*	*30%*	*70%*	*30%*	*70%*	*30%*
(1) Overcast Sky							
Depth (ft)[c]							
	20	.0248	.0226	.0172	.0156	.0129	.0121
Max	30	.0245	.0223	.0169	.0155	.0123	.0118
	40	.0242	.0221	.0164	.0154	.0120	.0117
	20	.0143	.0105	.0091	.0078	.0073	.0064
Mid	30	.0052	.0045	.0049	.0036	.0031	.0030
	40	.0035	.0024	.0027	.0021	.0020	.0017
	20	.0078	.0048	.0057	.0039	.0045	.0033
Min	30	.0029	.0017	.0026	.0015	.0018	.0013
	40	.0017	.0008	.0014	.0008	.0011	.0007
(2) Clear Sky							
Depth (ft)							
	20	.0185	.0156	.0129	.0111	.0099	.0088
Max	30	.0183	.0156	.0123	.0108	.0088	.0083
	40	.0180	.0151	.0118	.0107	.0086	.0082
	20	.0138	.0094	.0090	.0071	.0075	.0060
Mid	30	.0074	.0049	.0056	.0039	.0041	.0033
	40	.0047	.0029	.0036	.0025	.0026	.0021
	20	.0095	.0054	.0071	.0044	.0060	.0039
Min	30	.0049	.0025	.0042	.0021	.0029	.0019
	40	.0028	.0013	.0024	.0012	.0019	.0011
(3) Uniform Brightness Ground							
Depth (ft)							
	20	.0132	.0101	.0092	.0079	.0073	.0064
Max	30	.0127	.0101	.0088	.0079	.0069	.0063
	40	.0123	.0101	.0084	.0077	.0065	.0062
	20	.0115	.0081	.0085	.0064	.0066	.0054
Mid	30	.0075	.0051	.0056	.0043	.0045	.0037
	40	.0050	.0033	.0040	.0030	.0038	.0023
	20	.0095	.0064	.0074	.0049	.0060	.0040
Min	30	.0046	.0023	.0037	.0021	.0030	.0019
	40	.0026	.0016	.0023	.0011	.0020	.0010

[a]Information in Table 19.6*A* was supplied by Prof. B. Evans of Virginia Polytechnic Institute and State University and is reproduced with his permission.

[b]To convert CU figures for a ceiling reflectance of 80%, divide by 0.9.

[c]Length is dimension along the window wall; depth is dimension at right angles to window.

TABLE 19.6B **K Factors**

80/75% CLG Reflectance, 30% Floor Reflectance

		K							
Ceiling Ht:		*8 ft*		*10 ft*		*12 ft*		*14 ft*	
Wall Reflectance:		*70%*	*30%*	*70%*	*30%*	*70%*	*30%*	*70%*	*30%*
(1) Overcast Sky									
Width (ft)									
	20	.125	.129	.121	.123	.111	.111	.0991	.0973
Max	30	.122	.131	.122	.121	.111	.111	.0945	.0973
	40	.145	.133	.131	.126	.111	.111	.0973	.0982
	20	.0908	.0982	.107	.115	.111	.111	.105	.122
Mid	30	.156	.102	.0939	.113	.111	.111	.121	.134
	40	.106	.0948	.123	.107	.111	.111	.135	.127
	20	.0908	.102	.0951	.114	.111	.111	.118	.134
Min	30	.0924	.119	.101	.114	.111	.111	.125	.126
	40	.111	.0926	.125	.109	.111	.111	.133	.130
(2) Clear Sky									
Width (ft)									
	20	.145	.155	.129	.132	.111	.111	.101	.0982
Max	30	.141	.149	.125	.130	.111	.111	.0954	.101
	40	.157	.157	.135	.134	.111	.111	.0964	.0991
	20	.110	.128	.116	.126	.111	.111	.103	.108
Mid	30	.106	.125	.110	.129	.111	.111	.112	.120
	40	.117	.118	.122	.118	.111	.111	.123	.122
	20	.105	.129	.112	.130	.111	.111	.111	.116
Min	30	.0994	.144	.107	.126	.111	.111	.107	.124
	40	.119	.116	.130	.118	.111	.111	.120	.118
(3) Uniform Brightness Ground									
Width (ft)									
	20	.124	.206	.140	.135	.111	.111	.0909	.0859
Max	30	.182	.188	.140	.143	.111	.111	.0918	.0878
	40	.124	.182	.140	.142	.111	.111	.0936	.0879
	20	.123	.145	.122	.129	.111	.111	.100	.0945
Mid	30	.0966	.104	.107	.112	.111	.111	.110	.105
	40	.0790	.0786	.0999	.106	.111	.111	.118	.118
	20	.0994	.108	.110	.114	.111	.111	.107	.104
Min	30	.0816	.0822	.0984	.105	.111	.111	.121	.116
	40	.0700	.0656	.0946	.0986	.111	.111	.125	.132

The quantity enclosed in square brackets represents the daylight entering the room through the window. Factor *CU* modifies this entering light in accordance with room *length* and depth. Factor *K* further accommodates the entering light in accordance with room *height* and depth. It requires two factors to introduce the effect of all room dimensions and proportions. The second factor, E_G, is similarly constituted and calculated, that is, window illumination resulting from reflected groundlight is reduced by transmission losses and room coefficients. Window illumination itself is calculated as $E_H/2$, that is, horizontal exterior illumination resulting from the *half-sky*, reduced by ground reflectance. Thus:

$$E_G = \frac{E_H \times RF}{2} \times [AF \times TF \times LLF] \times CU \times K \qquad (19.6)$$

where

E_G = daylight on the work plane, for the point selected (max, mid, min), from reflected groundlight, in fc [lux]
E_H = unobstructed horizontal illumination on the ground outside the window, in fc [lux] (see Figs. 19.6, 19.8)
RF = reflectance factor of the ground outside the window (see Table 19.7)
CU = coefficient of utilization for uniformly bright ground (see Table 19.6)
K = second utilization coefficient, for uniformly bright ground (see Table 19.6)
AF, TF, and LLF are same as above.

Due to space constraints, the coefficient tables for windows with sun controls (venetian blinds, shades) are not presented here. For these, as well as all toplighting data, refer to the LOF ("How to Predict," 1976) and IES (*Recommended Practice,* 1979) sources. Both these sources show a method for calculating the effect of overhangs above windows.

EXAMPLE 19.3. For the sake of comparison, we will use the building of Example 19.1. Find E_{max}, E_{mid}, and E_{min} for a spring day and a winter day, at 10 A.M. (2 P.M.). Assume the sky to be overcast so that a direct comparison with Examples 19.1 and 19.2 can be made. Reflectances: 75% ceiling, 30% floor, 50% ± wall. Assume large panes; 92% net glass area.

SOLUTION

A. *Sky Contribution E_S*

1. E_W: From Table 19.2 we find solar altitude to be

March 21—36°
Dec. 21—14°

From Fig. 19.6, vertical surface illumination E_V from overcast sky is

March 21—425 fc [4600 lux]
Dec. 21—175 fc [1900 lux]

TABLE 19.7 **Reflectances of Building Materials and Outside Surfaces**

Material	Reflectance (In percent)	Material	Reflectance (In percent)
Bluestone, sandstone	18	Asphalt (free from dirt)	7
Brick		Earth (moist cultivated)	7
light buff	48	Granolite pavement	17
dark buff	40	Grass (dark green)	6
dark red glazed	30	Gravel	13
Cement	27	Macadam	18
Concrete	55	Slate (dark clay)	8
Granite	40	Snow	
Marble (white)	45	new	74
Paint (white)		old	64
new	75	Vegetation (mean)	25
old	55		

Source: From *Recommended Practice of Daylighting,* IES.

ILLUMINATION

2. Net window area *AF*

window area = 3′ × 5′ × 6′ × 0.92 (net glass area) = 82.8 ft^2

Note: Since *CU* and *K* are based on dimensions in feet, window area must be expressed in square feet.)

3. $TF = 0.8$ (given)
4. $LLF = 0.9$ (given)
5. *CU*: From Table 19.6A(1)

$$\left.\begin{aligned} CU_{max} &= 0.021 \\ CU_{mid} &= 0.0117 \\ CU_{min} &= 0.0067 \end{aligned}\right\} \text{by interpolation and extrapolation}$$

6. *K*: From Table 19.6B(1) similarly,

$$K_{max} = 0.122$$
$$K_{mid} = 0.107$$
$$K_{min} = 0.094$$

We can now calculate the three levels of illumination for March 21, from sky light.

$$E_S\text{ (max)} = 425\text{ fc} \times [82.8 \times 0.8 \times 0.9] \times 0.021 \times 0.122 = 65\text{ fc}$$

Similarly, E_S (mid) = 32 fc

and

E_S (min) = 16 fc

The corresponding figures for Dec. 21 are 27 fc, 13 fc and 7 fc.

B. Groundlight Contribution (Equation 19.6)

1. From Fig. 19.6, horizontal surface illumination E_H from overcast sky is

   March 21—(36° solar alt)—1160 fc [12,500 lux]
   Dec. 21—(14° solar alt)—465 fc [5000 lux]

2. *RF:* Assume vegetation immediately outside the window. From Table 19.7, $RF = 0.25$
3. *TF, LLF,* are the same as in *A* above.
4. *CU:* From Table 19.6*A*(3)

$$CU_{max} = 0.0112$$
$$CU_{mid} = 0.0100$$
$$CU_{min} = 0.0085$$

5. *K;* from Table 19.6*B*(3)

$$K_{max} = 0.136$$
$$K_{mid} = 0.123$$
$$K_{min} = 0.108$$

6. From Equation 19.6

   for March 21:

$$E_{Gmax} = 13\text{ fc}$$
$$E_{Gmid} = 11\text{ fc}$$
$$E_{Gmin} = 8\text{ fc}$$

   for Dec. 21:

$$E_{Gmax} = 5\text{ fc}$$
$$E_{Gmid} = 4\text{ fc}$$
$$E_{Gmin} = 3\text{ fc}$$

C. Total Daylight (Equation 19.4)

March 21

$$E_{Dmax} = 65 + 13 = 78\text{ fc} = 840\text{ lux}$$
$$E_{Dmid} = 32 + 11 = 43\text{ fc} = 462\text{ lux}$$
$$E_{Dmin} = 16 + 8 = 24\text{ fc} = 258\text{ lux}$$

Dec. 21

$$E_{Dmax} = 27 + 5 = 32\text{ fc} = 344\text{ lux}$$
$$E_{Dmid} = 13 + 4 = 17\text{ fc} = 183\text{ lux}$$
$$E_{Dmin} = 7 + 3 = 10\text{ fc} = 108\text{ lux}$$

In order to compare the results with those of the CIE and GDDM method, the values calculated for December 21 are plotted on the room plan of Fig. 19.21(*c*). The December values are selected since they are minimum and would correspond most closely to the 85th percentile figures of both previous methods, and indeed the agreement is excellent. The figures for March 21 would correspond most closely to the 50th to 60th percentile of Fig. 19.17, that is, somewhat more than double the minimum figures—and here again the agreement is good.

In summary, then, it is our recommendation that each daylighting problem be worked through assuming three different sky conditions:

1. Completely overcast sky. Use sky luminances that are maintained for 85% of the time, that is, minimum luminance. Use the GDDM method if window patterns are available. If not, use the CIE or LOF/IES methods, and sketch in contours intuitively. For this sky condition, the need for computer assistance is minimal. As stated above,

these results will establish maximum PSALI needed and extent/type of control strategy.

2. Average sky, that is, overcast sky on March 21 or clear sky without sun, whichever is more prevalent in the geographic area. This condition, in the design mode, will assist in determining glass area and room proportions. In the analysis mode the results are usable for HVAC design and energy calculations. Hand calculation is possible; some form of computer assistance is desirable. For hour-by-hour energy calculations, computer calculation is required.
3. Clear sky with sun. These results will indicate need for shading and shielding devices and will give maximum HVAC loads.

In all studies, use the most accurate sky luminance data available, that is, if local data based on field observations are available, they are preferable to the generalized data presented here. Spaces lighted with multiple sources such as bilateral sidelighting, windows, and toplighting, or daylight plus artificial light are calculated by superposition (calculating the effect of each source separately and adding the results).

19.9 Other Daylight Analysis Procedures

Many hand-calculation techniques have been developed over the years, most of which are listed in *Daylight* (1970). A system similar to Millet's Graphic Design procedure is published by TNO (Holland). It is based on overcast sky and consists of a catalog of 200+ daylight diagrams for 40+ window designs and various obstruction angles. Output is shown as DF contours on a variable-density background, thus simulating the light effect of daylight penetration (see Fig. 19.23 and "TNO Daylight Diagrams"). (TNO also produces insolation and daylight meters, protractors, and design charts).

The restriction of the CIE/GDDM methods to overcast skies is a serious one, limiting their general application to relatively small portions of the United States where such weather prevails. On the other hand, this sky condition enormously simplifies the procedures, making the accuracy of hand-calculation methods reasonable. Clear-sky design requires addressing the problem of direct sun and, with it, the effect of sun control devices (shades, overhangs, etc.) on interior daylight. Attempts have been made, with some success, to apply the DF principle to clear sky conditions. Among these are Millet and Bedrick (1980) using an extension of the GDDM, Bryan (1980) using a modified Waldram diagram technique, and Farrell (1976). Other techniques not using DF are Jones (1983) using a simple graphic device called a "lune protractor," and Thrun and Jennings (1981) using the output of a large computer program to prepare families of illumination curves for various sky conditions, window direction, and simple sunshading. A recent useful addition to the library of daylighting design tools S. Selkowitz (1982) is a set of transparent overlays designed to be used with equidistant sun path projection charts (LOF type; see Appendix D). The overlays were generated using the CIE empirical formulations for clear sky with and without sun, and overcast sky. They give illumination on horizontal, vertical, and sloped surfaces. These data can then be used with DF or other calculation technique to give interior daylight levels.

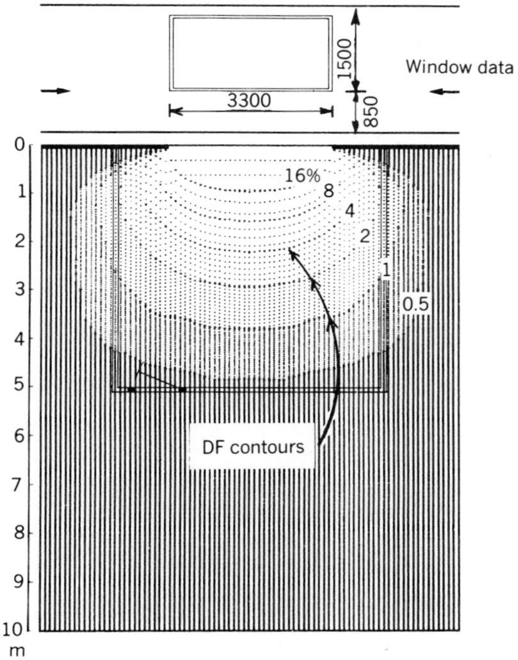

Fig. 19.23 A typical computer-produced daylight penetration diagram with DF contours. Dimensions are in millimeters. (Reproduced by permission of TNO Research Institute for Environmental Hygiene, Delft, The Netherlands.)

To handle sufficient data to arrive at relatively accurate results with clear skies, a computer-aided technique is required. The two avenues open are either to produce a "catalog" of curves or charts, which are then applied by hand (Millet, Thrun), or to use programmable calculators/computers in an interactive mode as an extension of hand methods. This latter seems clearly to be the trend, with particular emphasis on graphic output showing isolux contours, sun penetration, and so on. Because these methods and the associated hardware are being so rapidly developed, any listing here would undoubtedly be obsolete by the time the reader sees it. We recommend that the interested reader refer to the current daylighting literature for the most recent techniques, which can be applied with the resources available.

19.10 Additional Factors in Interior Daylighting

(a) Horizontal and Vertical Surfaces. Since the sky component of daylight enters side fenestration at an angle, it can be resolved into horizontal and vertical components, as shown in Fig. 19.24. The vertical component that illuminates horizontal surfaces is proportional to the sine of the angle of incidence, and the horizontal component that illuminates vertical surface is proportional to the cosine of this angle. Therefore, for horizontal tasks, windows should be *as high as possible* and, for vertical tasks, as low as possible. Since most tasks are horizontal, high windows will give better, deeper penetration than low windows of the same area.

(b) Window Details. The effect of window construction on total fenestration area reduction is often neglected. Even windows with narrow mullions and light metal frames have 8 to 10% obscuration; heavy window supports and small glass lights can result in 12 to 15% obscuration with proportional daylight reduction. Further obscuration readily results from dirt accumulation, wired glass, and mechanical system items such as pipes and ducts inside the room, adjacent to windows.

(c) Surface Reflectances. Interior reflectances are very important in daylight design. In addition to determining the magnitude of the internally reflected light component (IRC) within the room, they determine in large measure the eye adaptation level. A high adaptation level is desirable to avoid a sensation of glare when the window and its immediate surround are in the field of vision. Furthermore, the internally reflected light component contributes largely to the diffuseness of the room light. With low IRC, the sky component of daylight is the essential illuminant, and diffuseness and room penetration are reduced. Floors receive the sky component directly and should have at least 20% reflectance. Ceilings receive the ground component and should have at

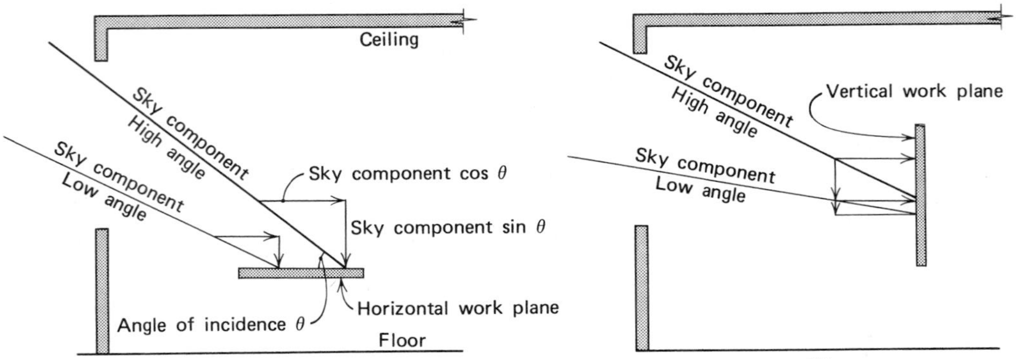

(a) *(b)*

Fig. 19.24 *Effect of daylight incidence angle on illumination components. High-angle skylight is more effective for illumination on the horizontal plane* (a), *whereas low-angle skylight is more effective on vertical surfaces* (b). *See also Fig. 19.12 for variation of horizontal and vertical daylight factor with depth.*

least 70% reflectance. Walls receive rereflected light. Since they are the surfaces seen at normal vision angles, they are responsible for eye adaptation levels and should have at least 50% reflectance.

Exterior surface reflection can provide the deep daylight penetration that is required for effective daylighting. Thus a concrete or light-painted ground surface (RF of 50 to 70%) will furnish one-quarter or two-thirds of the light incident on a window, depending on shading and orientation. When combined with a high-reflectance ceiling, optimal interior distribution is achieved.

(d) Glare and Heat Control (Sunshading). These are among the most difficult problems to overcome in daylighting. All exposures except north are exposed to direct sun during clear-sky weather, with the attendant glare and heat problem. Although a northern exposure does receive direct sun during the summer, the early (and late) hour, oblique angle, and low sun altitude combine to essentially eliminate all problems except possible glare. This subject was discussed at length in Chapters 5 and 6, primarily from the viewpoint of comfort. The reader is referred to Section 6.4*a* and to Table 6.3 for a description of adjustable sunshading devices. In addition to those devices, an effective technique that not only reduces glare, but also increases daylight penetration is the use of light-directing glass blocks (see Fig. 19.25).

We assume that it is understood that furniture be oriented so that windows are to the side or, if not possible, to the rear. Windows in the line of sight must be sufficiently distant as not to be a glare source and must be provided with some type of glare control, such as blinds, curtains, shades, or the like.

(e) Fixed Sunshades and Lightshelves (See Appendix D). Fixed exterior sunshades, particularly horizontal overhangs, can be arranged to function simultaneously as a light shelf, that is, as a reflector, to increase the flow of light to the *floor above* (see Fig. 19.26). If the overhang is moved down, as in Fig. 19.27, so that it shades the lower portion of the window, while reflecting light into the upper part of the same floor, it is usually referred to as a light shelf. Sunshades and light shelves have become common in recent ar-

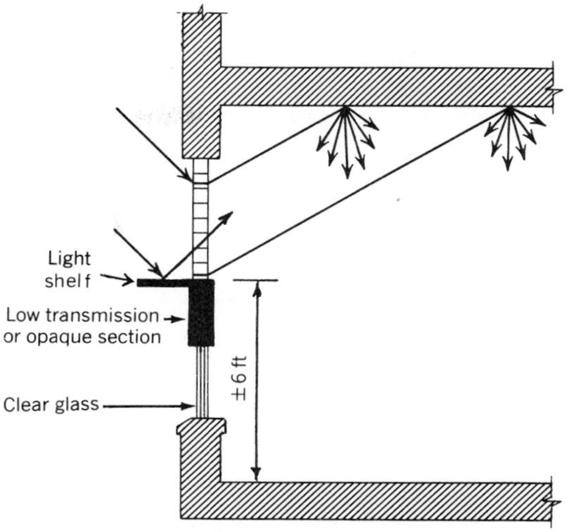

Fig. 19.25 *Arrows illustrate principal light paths into room with light-directing glass blocks above and clear glass below. Clear section permits unobstructed view outdoors. These blocks are particularly effective in low-ceilinged rooms, since the lower band of blocks near the 6-ft sill delivers most of the light to the deeper portions of the room. The use of a lightshelf below the blocks acts both as a reflector to increase the light transmission above and as a fixed horizontal sunshade for the window below.*

chitectural design because of the desire to increase the amount and penetration of daylight, while limiting the undesirable effects of direct insolation. That being so, it is necessary to understand the effect of fixed sunshades and lightshelves on the amount of daylight introduced, and its distribution for all types of sky conditions including overcast. Beam sunlighting is not included in this section.

In one interesting study, Millet, Lakin, and Moore (1981) compared various fixed shading devices that have the same shading mask (see Fig. 19.28) as to their effect on interior daylight distribution, for both direct sun and overcast sky conditions. (Readers unfamiliar with shading masks are referred to Olgyay and Olgyay, 1957; Ramsey/Sleeper, 1981; Lim et al., 1979; and Harkness and Mehta, 1978). The Millet et al. study conclusively indicates that devices with identical shading masks can yield vastly different interior daylight results. Therefore the lighting designer/architect would be well advised to test a model (either physically or mathematically) of any intended shading/reflect-

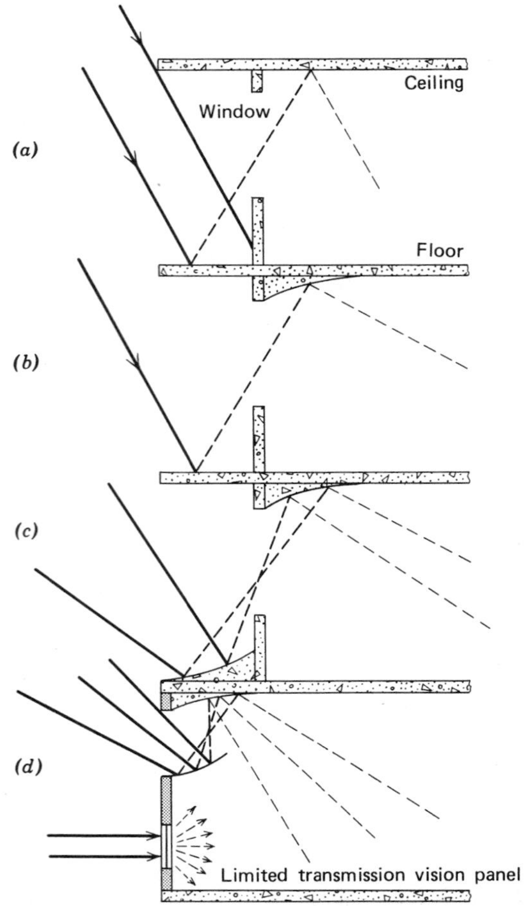

Fig. 19.26 *Window treatments that permit reflected daylight to enter the room but block direct sunlight.* (a) *Simple overhang acts as shield and reflector. Penetration is poor.* (b) *Addition of high-reflectance, curved section at ceiling increases room penetration.* (c) *Using a curved collector and reflector increases daylight factor deep into the room.* (d) *Reflector can be placed inside the building wall, using building wall for sun shield. Daylight penetration can be very deep, since sky factor angle is large.*

ing device under the anticipated sky conditions, before fixing his design in concrete.

Another study, Ander and Navvab (1983), used physical models to test a computer program output on the daylighting impact of fenestration controls. Angled fins, lightshelves, and an overhang were tested under simulated clear and overcast sky conditions. The study developed a ''fenestration coefficient'' that indicates the effect of the device in question on the daylight factor of specific interior areas. Other studies by Modest (1983) and Selkowitz, Navvab, and Mathews (1983) confirmed (among other conclusions) the intuitive notion that a lightshelf's principal action is to project daylight deep into a space, and so even out the level in the room. We trust that further studies will provide the designer with a useful ''catalog'' of exterior and interior shading devices and shelves, including reliable data on their daylighting (and thermal) impact.

(f) Suggested Daylighting Techniques. Some of the principles involved in the use of daylight in construction are demonstrated in Fig. 19.29. The reader is referred to the references at the end of the chapter, particularly *Daylight* (1970), Hopkinson et al. (1966), *Recommended Practice* (1979), *Daylight Design Diagrams* (1963), and ''How to Predict Interior Daylight Illumination'' (1976) for full coverage of the many aspects of daylighting that space limitations prevented our covering here. Among these are toplighting, treatment of lightshelves and overhangs, shading devices, daylight color, and similar topics.

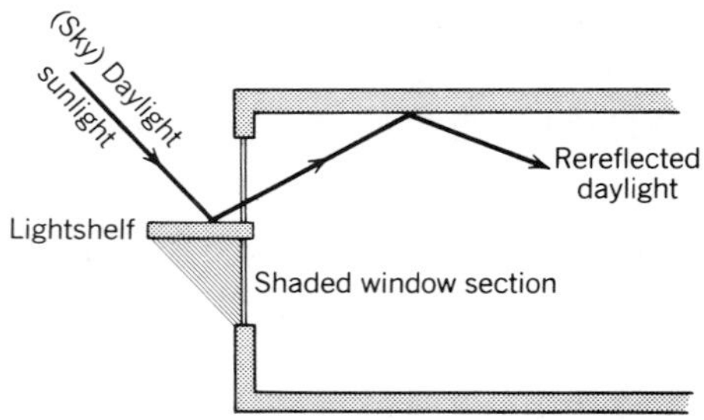

Fig. 19.27 *Lightshelf acts to reduce daylight near the window and increases it at greater depth. Shelf material (opaque, translucent) and angle of installation (horizontal, sloped up) markedly affect performance.*

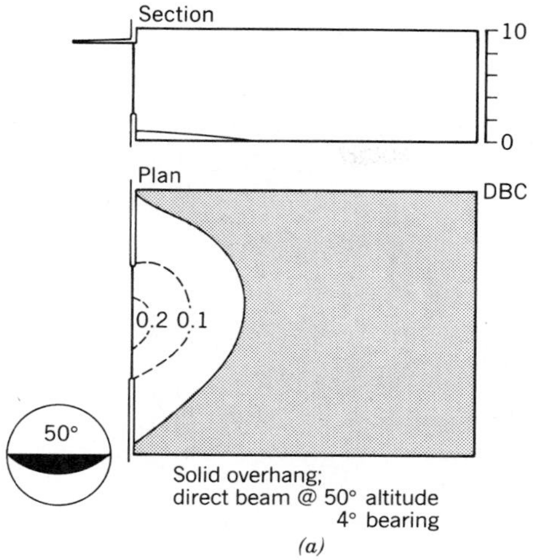

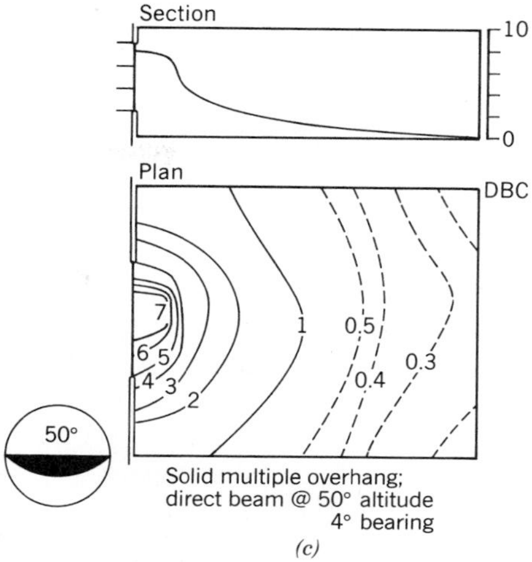

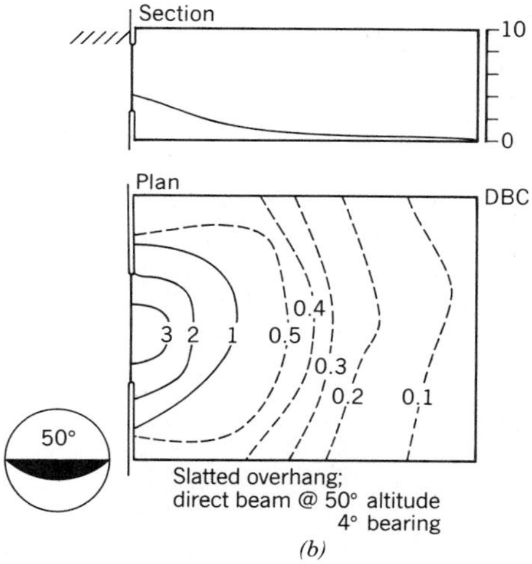

Fig. 19.28 *Three horizontal shading device designs with identical shading masks but resulting in entirely different interior daylight distributions under (simulated) direct sunlight conditions (labeled DBC). The solid overhang* (a) *provides complete (50°) window shading and results in a rapid dropoff of light in depth. The slanted slatted shade* (b) *acts both as shade and reflector, while the small stacked overhangs act as lightshelves, shading below and reflecting above. (Reproduced with permission, from Millet, Lakin, and Moore, 1981.)*

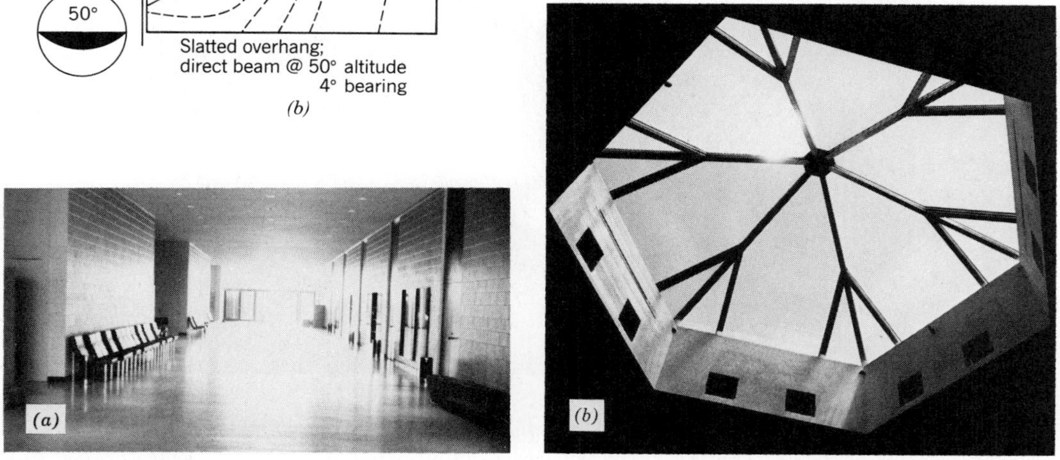

Fig. 19.29 (a) *When oriented east or west, expanses of glass must be provided with sun control devices. This unshielded, west-facing, 15-ft, glass-door facade creates intolerable disabling glare, the more so for the dimly lit interior corridor and associated low eye-adaptation level. (Photo by Stein.)* (b) *It is desirable from an architectural viewpoint to have the ceiling skylight area also incorporate artificial light sources, to be used when daylighting is insufficient. These alternate sources can be automatically modulated by the available daylight. This skylight is "glazed" with EXOLITE® double-skinned acrylic and polycarbonate sheet. (Photo courtesy of CYRO Industries.)* (c) *This shopping mall uses suspended, acrylic-enclosed mercury*

(Fig. 19.29 continued on next page)

ILLUMINATION

(Fig. 19.29 continued)

luminaires that blend with the large skylights. They illuminate the skylight well with light of similar color to daylight, giving the ceiling area a daylight appearance at all times. Note too the stepped ceiling construction, which serves to reduce brightness gradually. (Photo Courtesy of LD&A Magazine, *from article by J. Wilson, May 1974.)* (d) *Daylight factor here has a large ground component due to cutoff of the sky component by deep overhang, which also acts as a reflector for the ground light. Low-reflectance surfaces are suitable for a corridor that has no demanding visual task. (Photo courtesy of Libbey-Owens-Ford.)* (e) *Full-height clear glazing with negligible direct sun shading is applicable only on north elevations or*

(Fig. 19.29 continued)
climates with continual overcast sky, as in this Pacific Northwest installation. Note that work positions are not placed facing the glass, to avoid excessive brightness ratios in the field of view. Exterior planting reduces any incipient glare. (Photo courtesy of Libbey-Owens-Ford.) (f) *Unusual application of skylighting provides daylighting to an inner court of the International Monetary Fund Headquarters, Washington, D.C. Translucent acrylic panels give sun control while limiting heating and glare. (Photo courtesy of Roper IBG.)*

INCANDESCENT LAMPS

19.11 The Incandescent Filament Lamp

(a) Construction. The standard lamp consists simply of a tungsten filament inside a gas-filled, sealed glass envelope (see Fig. 19.30). Current passing through the high-resistance filament heats it to incandescence, producing light. Gradual evaporation of the filament causes the familiar blackening of the bulbs and eventual filament rupture and lamp failure. Incandescent lamps are available in many bulb and base types, with special designs for particular application (see Figs. 19.31 and 19.32). To diffuse the light, most bulbs are either etched on the inside (inside-frosted) or coated inside with white silica. The silica coating provides almost complete light diffusion at a cost of approximately 2 to 3% of the light output, whereas inside-frosted bulbs provide only partial diffusion but do not reduce light output. Inside-frosted bulbs are normally supplied for general service use unless other types are specified. Colored light is also readily available from either coated bulbs or bulbs of colored glass.

The lamp base is the means by which connection is made to the socket and thereby to the source of electric current. Most lamps are made with screw bases of various sizes, the most common being the medium screw base. General service lamps, of 300 W and larger, use the mogul screw base. Where exact positioning of the filament is important, as it is when lamps are placed in precise reflectors or in lens systems, a screw base cannot be used. Lamps designed for such use are furnished with one of the special bases illustrated in Fig. 19.31.

(b) Operating Characteristics. These are critically dependent on the volage at the lamps; therefore the life, output, and efficiency of a lamp can

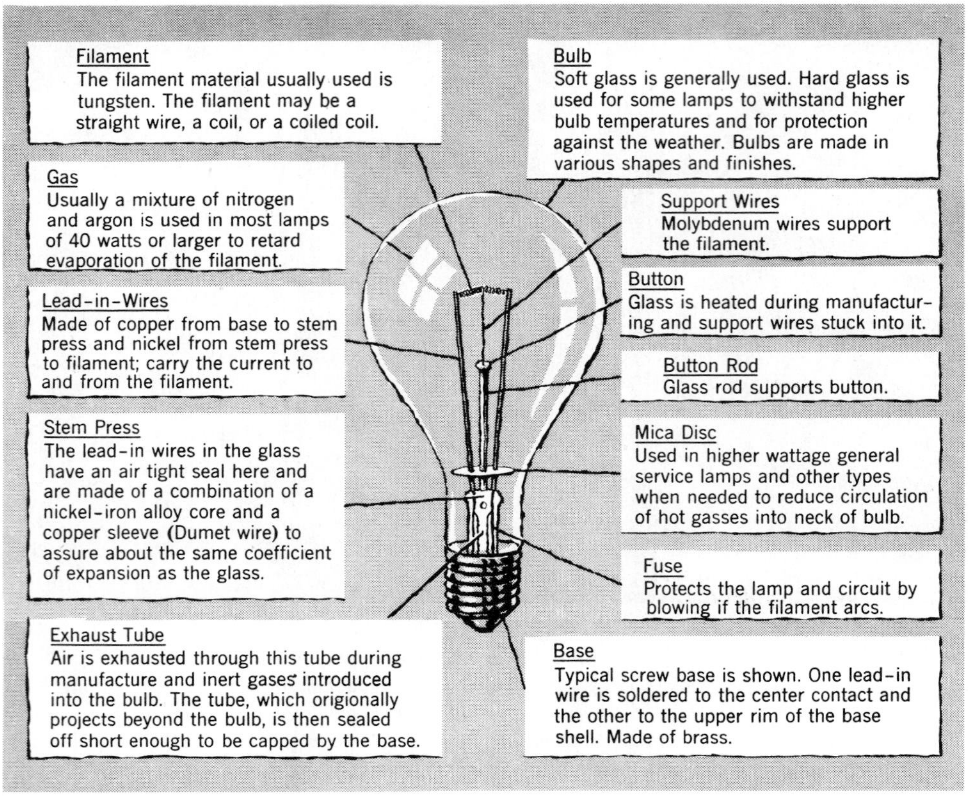
Filament
The filament material usually used is tungsten. The filament may be a straight wire, a coil, or a coiled coil.
Bulb
Soft glass is generally used. Hard glass is used for some lamps to withstand higher bulb temperatures and for protection against the weather. Bulbs are made in various shapes and finishes.
Gas
Usually a mixture of nitrogen and argon is used in most lamps of 40 watts or larger to retard evaporation of the filament.
Support Wires
Molybdenum wires support the filament.
Button
Glass is heated during manufacturing and support wires stuck into it.
Lead-in-Wires
Made of copper from base to stem press and nickel from stem press to filament; carry the current to and from the filament.
Button Rod
Glass rod supports button.
Stem Press
The lead-in wires in the glass have an air tight seal here and are made of a combination of a nickel-iron alloy core and a copper sleeve (Dumet wire) to assure about the same coefficient of expansion as the glass.
Mica Disc
Used in higher wattage general service lamps and other types when needed to reduce circulation of hot gasses into neck of bulb.
Fuse
Protects the lamp and circuit by blowing if the filament arcs.
Exhaust Tube
Air is exhausted through this tube during manufacture and inert gases introduced into the bulb. The tube, which origionally projects beyond the bulb, is then sealed off short enough to be capped by the base.
Base
Typical screw base is shown. One lead-in wire is soldered to the center contact and the other to the upper rim of the base shell. Made of brass.

***Fig. 19.30** Construction of a typical general-service incandescent lamp.*

be markedly altered by even a small change in operating voltage, as illustrated by Fig. 19.33.

For example, burning a 120-V lamp at 125 V (104.2%) means approximately:

16% more light (lumens)
7% more power consumption (watts)
8% higher efficacy (lumens per watt)
42% less life (hours)

Burning a 120-V lamp at 115 V (95.8%) means approximately:

15% less light (lumens)
7% less power consumption (watts)
8% lower efficacy (lumens per watt)
72% more life (hours)

Particular note should be taken of the effect of voltage on lamp life. In installations where lamp replacement is difficult and/or expensive, lamps may be burned slightly under voltage and life prolonged, thereby decreasing the frequency of replacement. However, since efficiency is decreased by this procedure and since energy cost is normally a major cost in any lighting installation over the life of the installation, a detailed cost analysis should be made by the consulting engineer involved. Conversely, where lamps are replaced before burnout on a group replacement system and initial installation cost per footcandle and/or energy costs are high, lamps may be burned overvoltage, thereby increasing output and efficiency but shortening life. This procedure is normal in sports-lighting installations because of the high cost of tower-mounted floodlights, making in mandatory to extract the maximum light from each unit. In stadium installations that have yearly burning schedules averaging less than 200 hours, 10% overvoltage operation doubles the light output but still allows a once-a-year, off-season relamping and is therefore a highly economical procedure.

Bulb shapes

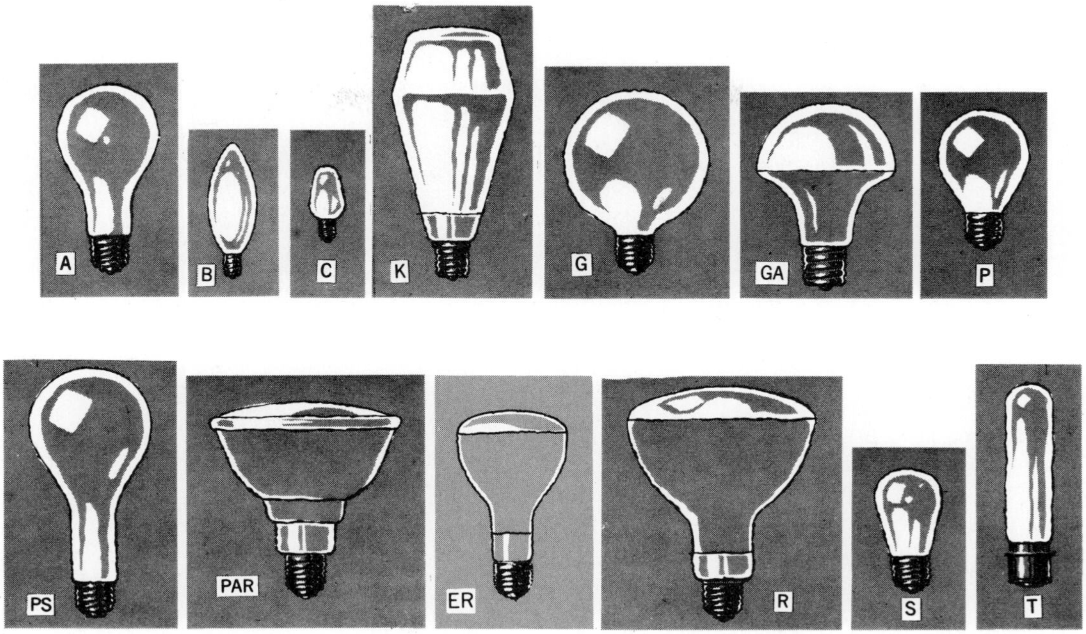

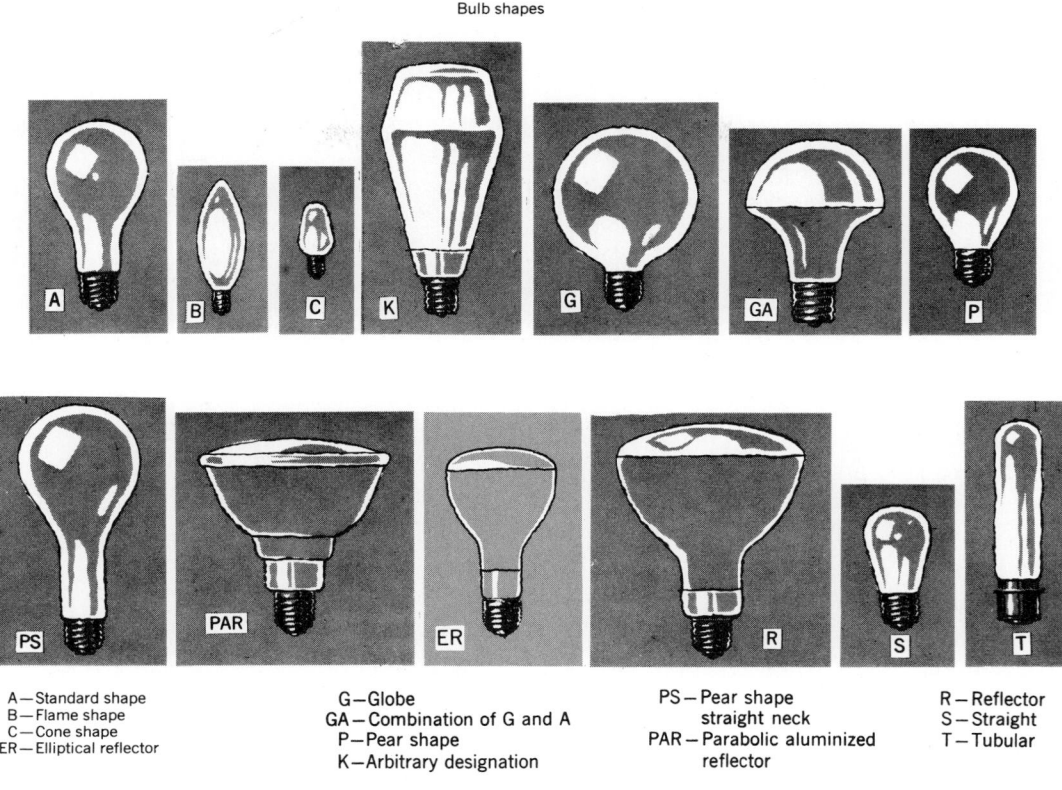

A—Standard shape
B—Flame shape
C—Cone shape
ER—Elliptical reflector

G—Globe
GA—Combination of G and A
P—Pear shape
K—Arbitrary designation

PS—Pear shape straight neck
PAR—Parabolic aluminized reflector

R—Reflector
S—Straight
T—Tubular

Base types

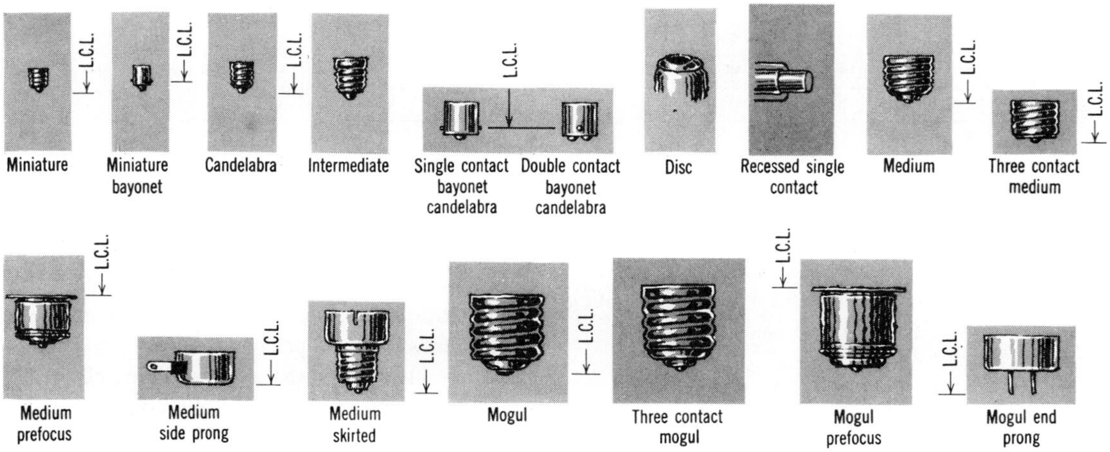

L.C.L.—Light center length

Fig. 19.31 *Common incandescent lamp bulb and base types with nomenclature. The bulb nomenclature indicates type and size; the letter being an abbreviation of the shape and the number equal to the maximum diameter in eighths of an inch. Thus a PS-52 is a pear-shaped bulb, 52/8 ($6\frac{1}{2}$) in. in diameter and an R-40 is a reflector lamp 40/8 (5) in. in diameter.*

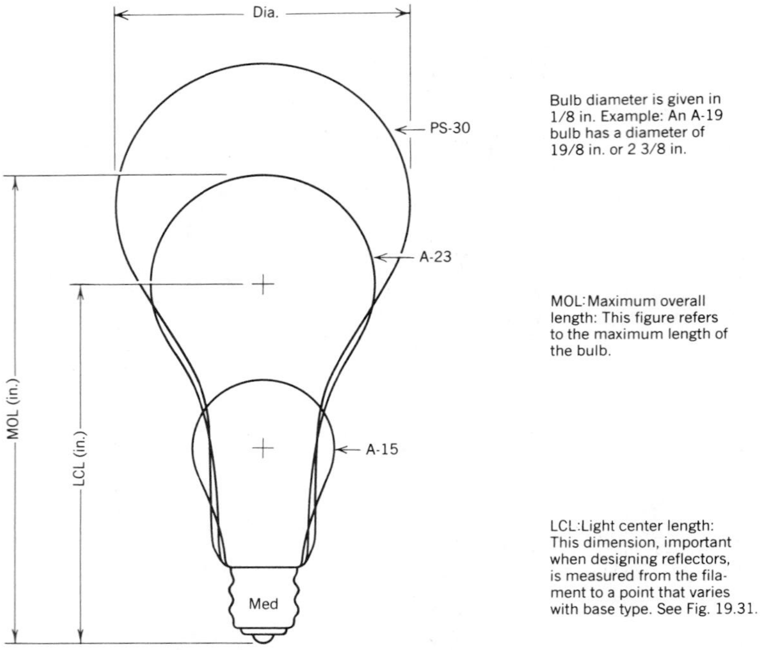

Dia.
PS-30
A-23
A-15
MOL (in.)
LCL (in.)
Med
Bulb diameter is given in 1/8 in. Example: An A-19 bulb has a diameter of 19/8 in. or 2 3/8 in.
MOL: Maximum overall length: This figure refers to the maximum length of the bulb.
LCL: Light center length: This dimension, important when designing reflectors, is measured from the filament to a point that varies with base type. See Fig. 19.31.

Lamps shown at slightly less than 1/2 actual size

	A – STANDARD SHAPE									PS – PEAR SHAPE							
WATTS	15	25	40	60	75	100	100	150	150	150	200	300	300	500	750	1000	1500
BULB	A–15	$A{-}19_1$	$A{-}19_2$	$A{-}19_3$	$A{-}19_3$	$A{-}19_3$	$A{-}21_1$	$A{-}21_2$	A–23	PS–25	PS–30	PS–30	PS–35	PS–40	PS–52	PS–52	PS–52
DIAMETER"	1 7/8	2 3/8	2 3/8	2 3/8	2 3/8	2 3/8	2 5/8	2 5/8	2 7/8	3 1/8	3 3/4	3 3/4	4 3/8	5	6 1/2	6 1/2	6 1/2
M.O.L."	3 1/2	3 7/8	4 1/4	4 7/16	4 7/16	4 7/16	5 1/4	5 1/2	6 3/16	6 15/16	8 1/16	8 1/16	9 3/8	9 3/4	13	13	13
L.C.L."	2 3/8	2 1/2	2 15/16	3 1/8	3 1/8	3 1/8	3 7/8	4	4 5/8	5 3/16	6	6	7	7	9 1/2	9 1/2	9 1/2
BASE	MED	MED	MED	MED	MED	MED	MED	MED	MED	MED	MED	MED	MOG	MOG	MOG	MOG	MOG
STANDARD FINISH	IF	IF	IF W	IF	IF	IF	IF	IF	CL IF	CL IF	IF	IF	IF	IF	CL IF	CL IF	CL IF

CL – CLEAR IF – INSIDE FROSTED

Fig. 19.32 *Typical dimensional data for common general-service incandescent lamps.*

In general, however, it is advisable to operate incandescent lamps at rated voltage, accepting balanced efficiency, output, and life.

(c) Other Characteristics.

1. *Lumen maintenance*. Light output decreases slowly with lamp life as the bulb blackens. Position during burning and bulb temperature affect this characteristic.
2. *Color*. White with a large yellow-red component and therefore highly flattering to the skin. As is explained in detail in Section 19.25, the spectral content of the light produced by a heated (incandescent) source depends on its temperature; high-wattage lamps are bluer, low-wattage lamps are yellower. Dimmed lamps give yellow-red light.
3. *Surroundings*. Generally, incandescent lamps are impervious to external heat, cold, or humidity. Starting of these lamps is completely unaffected by ambient conditions.
4. *Lamp efficiency*. Since incandescent lamps produce light as a by-product of heat, they are inherently inefficient. Luminous effi-

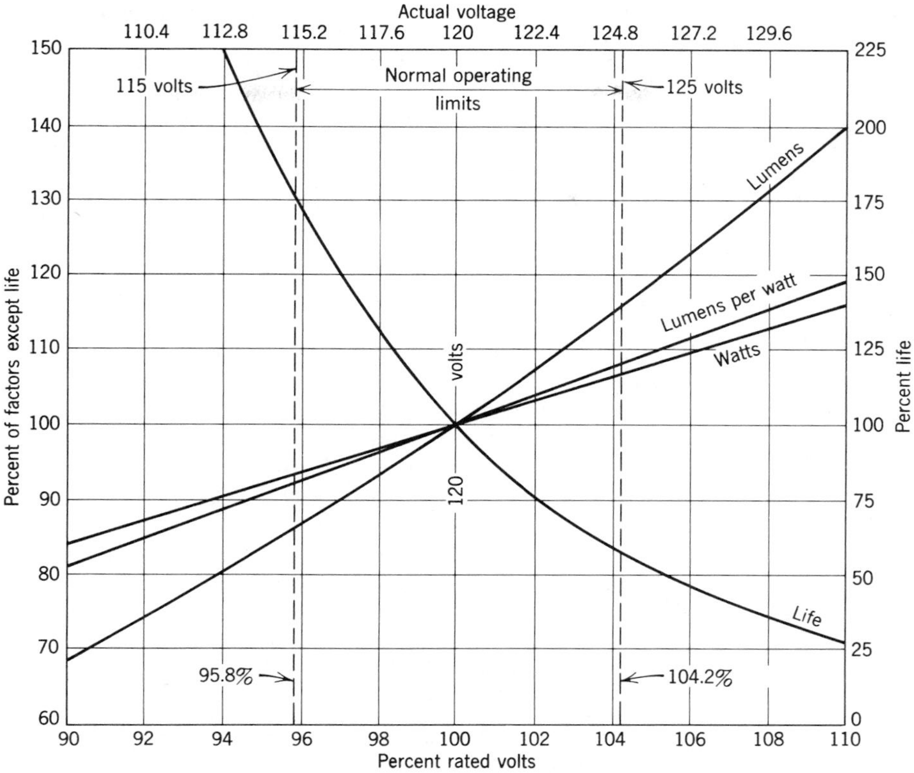

Fig. 19.33 Characteristics of a standard 120-V general-service incandescent lamp as a function of voltage.

ciency (efficacy) increases with wattage. Thus a 60-W general-service lamp produces 870 initial lumens or 14.5 lm/W, whereas a 100-W lamp produces the same light output as two 60-W lamps, or put otherwise, it results in a 20% energy savings. Refer back to Section 18.6 for a discussion of photometric efficiency, or efficacy.

(d) Summary. The principal advantages of incandescent lamps are low cost, instant start and restart, simple inexpensive dimming, simple compact installation requiring no accessories, cheap fixtures, focusable as a point source, high power factor, life independent of number of starts, and good color.

The principal disadvantages are low efficacy (see Fig. 19.1), short lamp life, and critical voltage sensitivity. Low efficacy results in a large number of fixtures, high maintenance costs, and large heat gain. Short lamp life results in high replacement labor cost. Voltage sensitivity requires careful and expensive circuit design. Also, light concentration of the filament (point source) requires careful fixture design in order to avoid glare and, if undesirable, sharp shadows. Because of the poor energy characteristic, incandescent lamp use should be limited to the following applications.

1. Infrequent or short-duration use.
2. Where low-cost dimming is required.
3. Where the point source characteristic of the lamp is important, as in focusing fixtures.
4. Where minimum initial cost is essential.
5. Where its characteristic color rendering is desired.

A brief list of conventional incandescent lamps and their physical and operating characteristics is given in Table 19.8. Lamp data for use in design

TABLE 19.8 **Typical Incandescent Lamp Data (Listing a Few of Many Sizes and Types of 115-, 120-, and 125-V Lamps)**

Watts and Life			*Lumens*		*Physical Data*		
Lamp Watts[a]	*Average Rated Life (h)*	*Approx. Color Temp.*[b] *(K)*	*Initial Lumens*	*Lumens per Watt*[c]	*Shape of Bulb*[d]	*Base*	*Description*
15	2500	—	126	8.4	A-15	Med	—
25	1000	—	228	9.1	A-19	Med	Rough service
25	2500	2550	357	14.3	A-19	Med	—
40	1500	—	460	11.5	A-19	Med	—
50	2000	—	525	10.5	ER30	Med	See Section 19.10*e*
60	1000	2800	870	14.5	A-19	Med	—
60	2500	—	775	12.9	A-19	Med	Long life
75	750	—	1190	15.9	A-19	Med	—
100	750	2870	1750	17.5	A-19	Med	—
100	750	—	1690	16.9	A-21	Med	—
100	2500	—	1460	14.6	A-19	Med	Long life
100	1000	—	1220	12.2	A-21	Med	Rough surface
150	750	2900	2810	18.7	A-21	Med	—
150	750	—	2680	17.9	PS-25	Med	—
200	750	2930	4000	20.0	A-23	Med	—
300	750	2940	6100	20.3	PS-25	Med	—
500	1000	3000	10850	21.7	PS-35	Mogul	—

[a]Figures in this column designate the input watts thus: 60 means 60 W. The letters identify the treatment of the glass bulb. All inside-frosted unless otherwise noted. Other letters have these meanings: W, white; SBIF, silver bowl, inside-frosted; Cl, clear; SW, soft white; Cer, ceramic.

[b]See Section 19.25.

[c]Efficacy (luminous efficiency), in lumens per watt, increases with filament temperature; therefore, with wattage.

[d]Bulb designations consist of a letter to indicate its shape and a figure to indicate the approximate maximum diameter in eighths of an inch (see Fig. 19.31).

should be taken from current manufacturers' literature. Data presented here are typical.

19.12 Special Incandescent Lamps, Including Energy-Saving Lamps

In the field of incandescent lamps other than the tungsten-halogen lamp, which is discussed separately, numerous special types are available. Some of the more important types are covered briefly in the following pages.

Rough service and *vibration* lamps are built to withstand rough handling and continuous vibration, respectively, both of which conditions are extremely hard on general-service lamp filaments. Neither type is intended for general use, and both types have lower luminous efficacy than a general-service lamp (see Table 19.8).

Extended-service lamps are designed for 2500-h life and are useful, as mentioned, in locations where maintenance is irregular and/or relamping is difficult. The lamp is really designed for slightly higher voltage than that at which it is applied, and therefore efficacy is reduced (see Table 19.8 and Fig. 19.33).

So-called long-life lamps, which are guaranteed to burn for 2, 3, or 5 years, are lamps designed for much higher voltages than that at which they operate. Since they normally sell at a high cost and are very inefficient, their use is seldom advisable. Before using such lamps, a cost comparison, including cost of lamps, energy, and relamping, should be made. See Appendix E.

(a) Reflector Lamps. These are made in "R" and "PAR" shapes (see Fig. 19.31) and contain a reflective coating on the inside of the glass en-

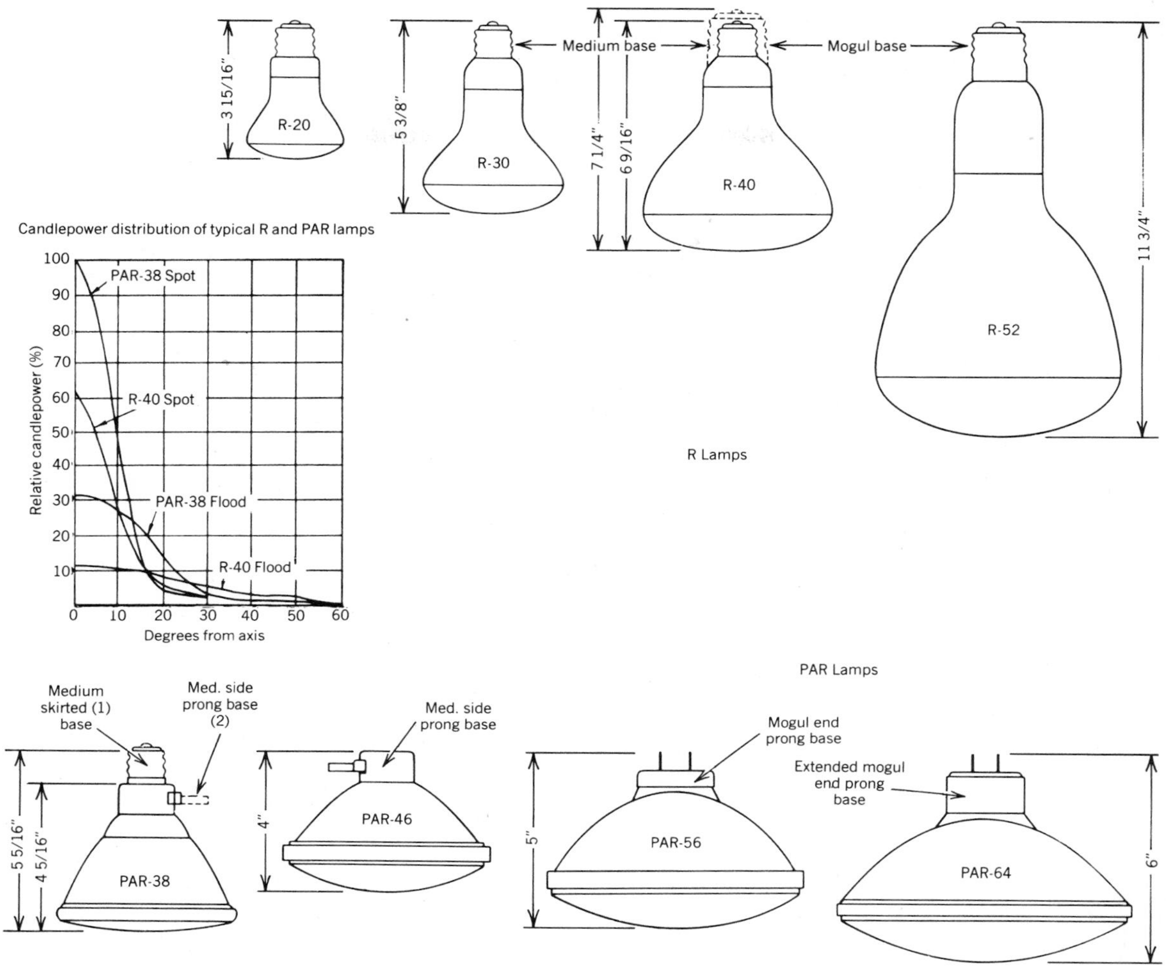

Fig. 19.34 Typical dimensional data for reflector spot and flood lamps. For photometric data, see manufacturers' catalogs.

velope; this gives the entire lamp accurate light beam control. Both types are available in narrow or wide beam design, commonly called *spot* and *flood,* respectively. R lamps are generally made in soft glass envelopes for indoor use, whereas PAR lamps are hard glass, suitable for exterior application.

Typical reflector lamp dimensional data and photometric data are given in Fig. 19.34. Illumination patterns resulting from typical PAR spot and flood lamps are shown in Fig. 19.35. When using R and PAR lamps, the fixture acts principally as a lampholder, since beam control is built into the lamp.

(b) Interference (Dichroic) Filters. Such filters, which had been previously used only in specialized applications such as projection lamps to remove heat from the light beam, are now available in PAR lamps. The basic filter is a thin film that operates on the interference principle rather than absorption. Thus the surface remains relatively cool.

In one design that is utilized to limit the heat in the light beam, the film is applied to the inside back of the lamp. It acts by transmitting infrared heat out the lamp back while reflecting light out the lamp front (see Fig. 19.36). Typical applications are in window displays, over food counters,

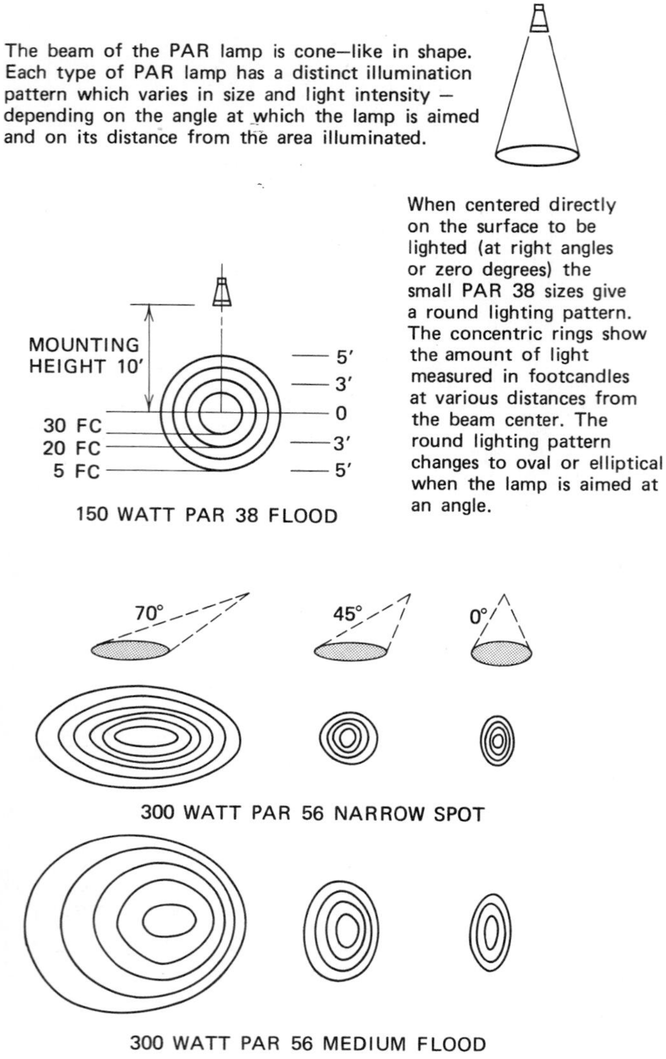
The beam of the PAR lamp is cone—like in shape. Each type of PAR lamp has a distinct illumination pattern which varies in size and light intensity — depending on the angle at which the lamp is aimed and on its distance from the area illuminated.
When centered directly on the surface to be lighted (at right angles or zero degrees) the small PAR 38 sizes give a round lighting pattern. The concentric rings show the amount of light measured in footcandles at various distances from the beam center. The round lighting pattern changes to oval or elliptical when the lamp is aimed at an angle.
MOUNTING HEIGHT 10′
30 FC
20 FC
5 FC
5′
3′
0
3′
5′
150 WATT PAR 38 FLOOD
70°
45°
0°
300 WATT PAR 56 NARROW SPOT
300 WATT PAR 56 MEDIUM FLOOD

The lighting pattern of the larger PAR 46, 56, and 64 lamps is oval or elliptical, whether centered directly on the surface or aimed at an angle. As shown in the two diagrams, aiming PAR lamps at progressively greater angles — proportionately increases both the length and width of the area illuminated. In general, for spotlights the length of the lighting pattern becomes proportionately greater — and in the case of floodlights, the width.

Fig. 19.35 Typical illumination patterns for PAR spot and flood lamps. (Courtesy of GTE/Sylvania, Inc.)

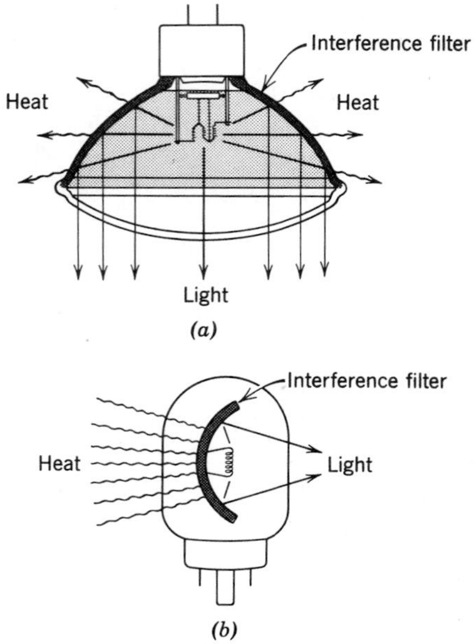

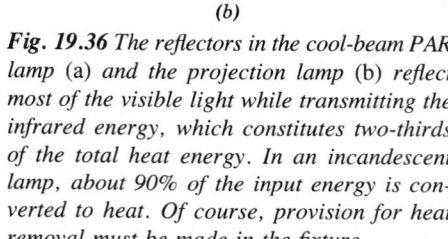

Fig. 19.36 The reflectors in the cool-beam PAR lamp (a) *and the projection lamp* (b) *reflect most of the visible light while transmitting the infrared energy, which constitutes two-thirds of the total heat energy. In an incandescent lamp, about 90% of the input energy is converted to heat. Of course, provision for heat removal must be made in the fixture.*

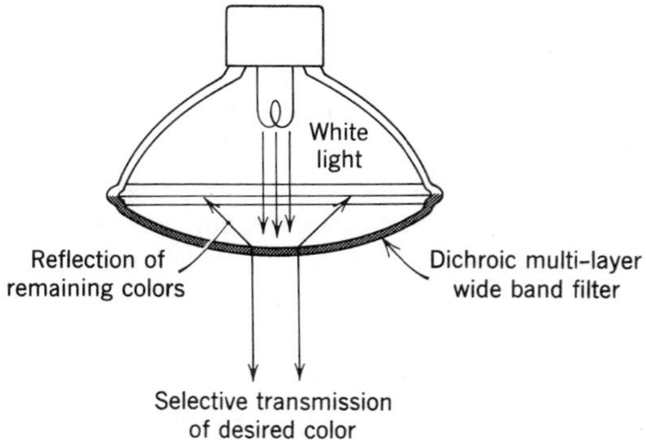

Fig. 19.37 Action of dichroic filter is one of selective interference rather than absorption. Each film layer transmits one color while reflecting its complement. Desired color is obtainable by action of several films.

and in any location where a "cool beam" is desirable. Of course, provision must be made for removal of the heat from the fixture if the lamp is housed.

In a second design, multiple-layer filters are applied to the front of the lamp. Each film acts to transmit one color and reflect is complement (two-color, hence dichroic). These dichroic filter lamps produce a purer, more saturated color at high efficacy than is possible with selective absorption filters (see Fig. 19.37).

(c) Low-Voltage Lamps. These lamps, in PAR shape and for 12-V operation, are available in extremely narrow beam spread (5–10°) for special precision control floodlighting. The low voltage makes their application to exterior work simpler.

(d) Heat-Mirror Lamps. Incandescent lamps are inherently inefficient because light is produced as a by-product of filament heating, and the heat, which comprises about 90% of the lamp's input energy, is wasted. A new design, developed by the Duro-Test Corp., utilizes a heat mirror coating on the inside of a spherical glass envelope, which transmits visible light but reflects heat. Thus the filament is heated to a large extent by reflected infrared energy rather than input energy, and efficiency is raised by 30 to 50%. These lamps are now becoming commercially available although no detailed characteristics are available at this writing. As with all nonstandard designs, the potential user should make an economic payback analysis before investing the additional first cost.

(e) Elliptic Reflector Lamps. These lamps, which are usually classed as energy-saving lamps, are simply an improved reflector design that increases the efficiency of the lamp–fixture combination over a similar installation using a conventional R reflector. The action of the reflector is shown in Fig. 20.18. This shape causes the beam to focus a few inches in front of the lamp, permitting high-efficiency application in pinhole downlights or deep baffle units, where use of ordinary R lamps causes trapping and loss of most of the lamps' output. This action *in a fixture* is illustrated in Fig. 20.19. Thus, although a 75-W ER (elliptical reflector) lamp will yield as much output from the above fixtures as a 150-W R lamp, it itself has no greater efficacy than the R lamp (see Table 19.8). In an open-type downlight or one designed for R-type reflectors, use of the (more expensive) ER lamp may actually *reduce* output.

(f) Energy-Saving Lamps. The requirement to conserve lighting energy has resulted in a line of energy-saving lamps being produced by every major manufacturer. These lamps, of all types, are known by trademarked names. Among them are General Electric's Watt-Miser™ and GTE/Sylvania's Super-Saver™. These lamps are rated at a wattage lower than the standard lamps they are intended to replace, and their catalog listing indicates this. Thus a 40A/34WM is a 34-W A-shape bulb Watt-Miser™ lamp, intended to replace a 40-W lamp. Its 410-lm output gives it an efficacy of 12.06 W as compared to the standard lamp's 455 lm and 11.38 lm/W. Thus it has 10% lower output (410/455) but 15% lower energy consumption (34/40). Hence its energy-saving characteristic. However, not all reduced-wattage lamps have higher efficacy than the corresponding standard lamps. Some simply reduce light output proportionately to the wattage, resulting in no saving at all, unless the system had been overdesigned to begin with. See, for instance the 150-W lamp in Table 19.9. Indeed, a *loss* may be incurred, since more fixture bodies are required to hold the lower output lamps, for the same total output.

At this writing, there are two types of energy-saving incandescent lamps. They are the krypton-filled lamps and the modified PAR lens-cover unit.

1. When the usual nitrogen-argon mixture is replaced with krypton, the filament remains hotter because of lower heat conduction by this gas, resulting in higher efficacy and longer life. Thus by juggling characteristics, the designer can have a low-wattage lamp with either long life and lower efficacy or standard life and high efficacy. See Table 19.9 for comparative ratings. Note that the efficacy of long-life lamps in both designs is about equal, making the choice a matter of wattage rating and not energy cost.
2. The 150-W PAR flood lens cover was redesigned to produce the same number of beam lumens with only 120 W. This was accomplished by redirecting stray light, otherwise

TABLE 19.9 **Comparison of Characteristics of Standard, Long-Life, and Two Types of Energy-Saving Lamps**

Standard Lamps	*40 W*	*60 W*	*75 W*	*100 W*	*150 W*
Initial lumens	455	870	1190	1750	2850
Life (hours)	1500	1000	750	1000	750
Efficacy (lm/W)	11.4	14.5	15.9	17.5	19.0
Long-Life Lamps	*40 W*	*60 W*	*75 W*	*100 W*	*150 W*
Initial lumens	420	775	1000	1490	2310
Life (hours)	2500	→	→	→	→
Efficacy (lm/W)	10.5	12.9	13.3	14.9	15.4
Energy-Saving Lamps, Normal Life	*34 W*	*52 W*	*67 W*	*90 W*	*135 W*
Initial lumens	410	800	1130	1620	2580
Life (hours)	1500	1000	750	750	750
Efficacy (lm/W)	12.1	15.4	16.9	18.0	19.1
Energy-Saving Lamps, Extended Life	*34 W*	*52 W*	*67 W*	*90 W*	*135 W*
Initial lumens	365	700	930	1360	2145
Life (hours)	2500	→	→	→	→
Efficacy (lm/W)	10.7	13.5	13.9	15.1	15.9

lost, into the beam and resulted in a 25% (150/120) increase in efficacy. The new design was then applied to another size lamp, which is now available to replace the 75-W, PAR 38 lamp as well.

As has been repeatedly stated, any additional first cost should be analyzed by a life-cycle cost analysis and the payback period calculated. The designer will find that a proper control system is often more economically attractive than low-wattage lamps, and that energy-saving lamps are primarily useful in retrofit work.

19.13 Tungsten-Halogen (Quartz-Iodine) Lamps

This lamp type, illustrated in Fig. 19.38, is similar to the standard incandescent lamp in that it produces light by heating a filament. It differs in that a small amount of halogen gas (usually iodine) is added to the inert gas mixture that fills the bulb. (Bulb material is quartz to withstand the high temperature of the halogen cycle.) This addition results in a retardation of filament evaporation. Filament evaporation is what normally occurs in the incandescent lamp, causing the familiar bulb blackening with attendant light reduction and eventual burnout. The mechanism of the regenerative halogen cycle (which retards evaporation) is shown in Fig. 19.38, along with a typical light-loss comparison chart. Although the lamp has approximately the same efficacy as an equivalent normal incandescent, it has the advantages of longer life, low lumen depreciation (98% output at 90% life), and a smaller envelope for a given wattage. Some typical lamp data are given in Table 19.10. Other lamp characteristics are similar in all respects to the incandescent lamp. Color temperature ranges between 2000 and 3400 K; spectral energy distribution is typical of a blackbody radiation, see Section 19.25), and dimming characteristics are similar as well.

Quartz-iodine lamps are available in tubular shape for use in reflector-type lamp holders or in PAR reflectors as an integral unit. They are not normally applied for general lighting because of their higher cost. They are applied where precise

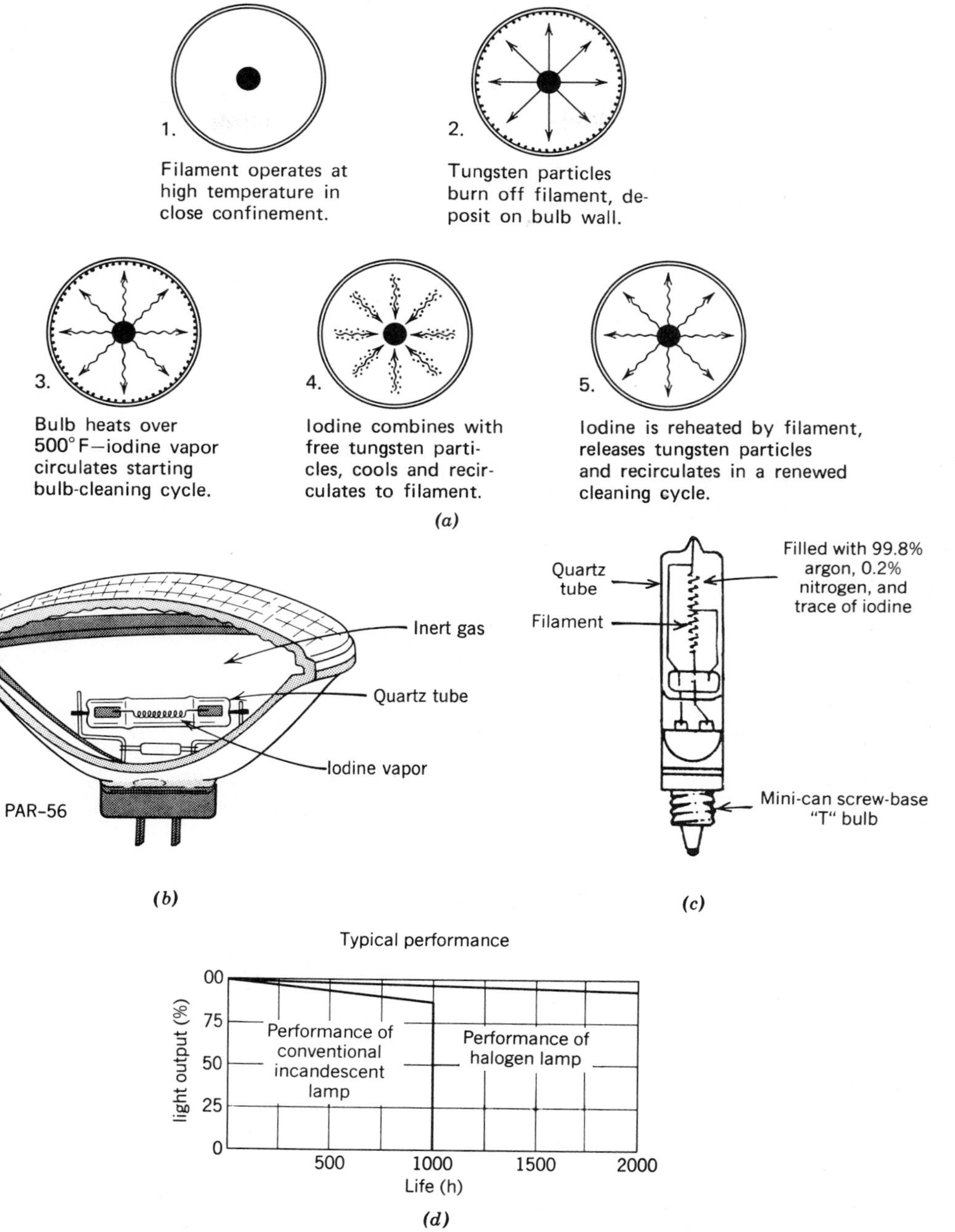

***Fig. 19.38** The self-regenerative halogen cycle* (a) *results in low light depreciation* (d) *whether the lamp is bare* (c) *or in an enclosure* (b).

beam characteristics are desired as in floodlighting, sports lighting, display lighting, and photographic lighting. As with conventional incandescent lamps, manufacturers are producing energy-saving designs that increase the attractiveness of these lamps for wider commercial application.

TABLE 19.10 **Typical Data for Quartz-Iodine (Tungsten-Halogen) Lamps**

Watts	Bulb	Maximum Overall Length (in)	Base	Rated Life (h)	Beam Type	Approximate Initial Total Lumens	Mean Lumens Through Life (percent)
100	T-4	2.81	Mini-can	750	—	1,800	93
150	T-4	2.81	Mini-can	2000	—	2,800	93
250	T-4	3.16	Mini-can	2000	—	5,000	97
400	T-4	3⅝	Mini-can	2000	—	8,250	97
500	T-4	6	Med-PF	2000	—	10,450	97
250	PAR-38	5.31	Medium skirted	6000	Spot	3,500	95
					flood	3,220	94
500	PAR-56	5	Mogul end prong	4000	Narrow spot	8,000	94
					Medium flood	8,000	94
					Wide flood	8,000	94
90[a]	PAR-38	5.31	Med. skirted	2000	15° spot 32°	—	—

[a]Intended to replace standard 150-W PAR Incandescent.

FLUORESCENT LAMPS

19.14 The Fluorescent Lamp: Construction

The second major category of light sources is that of electric discharge lamps, of which the fluorescent lamp is the best known and most widely used type. It has become so popular since its major introduction in 1937 that it has almost completely supplanted the incandescent lamp in all fields except specialty lighting and residential use. The typical fluorescent lamp comprises a cylindrical glass tube sealed at both ends and containing a mixture of an inert gas, generally argon, and *low-pressure* mercury vapor. Built into each end is a cathode that supplies the electrons to start and maintain the mercury arc, or gaseous discharge. The short-wave ultraviolet light, which is produced by the mercury arc, is absorbed by the phosphors with which the inside of the tube is coated and is reradiated in the visible light range. The fluorescent lamp is so called because its phosphors fluoresce, or radiate light, when exposed to ultraviolet light. The particular mixture of phosphors used governs the spectral quality of the light output.

The descriptions that follow cover *standard* lamps and circuits, that is, lamps operating at line frequency (60 H), with standard loading (200, 430, 800, 1500 mA), using core-and-coil magnetic ballasts, in common starting and operating arrangements. Special lamps, accessories, and circuits including low-wattage lamps and ballasts, high-frequency operation with electronic ballasts, 26-mm (T-8) tubes operating at 265 mA, modified waveform ballasts, triphospher lamps, and special-shape lamps are all discussed separately.

(a) Preheat Lamps. The original fluorescent lamp was a preheat design. Construction of a typical hot cathode lamp (preheat and rapid start) is shown in Fig. 19.39*a*; the basic preheat circuit is shown in Fig. 19.40*a*. The circuit utilizes a separate starter, which is a small cylindrical device that plugs into a preheat fixture. When the lamp circuit is closed, the starter energizes the cathodes; after a 2- to 5-s delay, it initiates a high-voltage arc across the lamp, causing it to start. Most starters are automatic, although in desk lamps the preheating is accomplished by depressing the start button for a few seconds and then releasing it. This closes the circuit and allows the heating current to flow; releasing the button causes the arc to strike.

Standard preheat lamps operate at 430 mA. All preheat lamps have bipin bases (see Fig. 19.39*b*). They range in wattage from 4 to 90 W and in

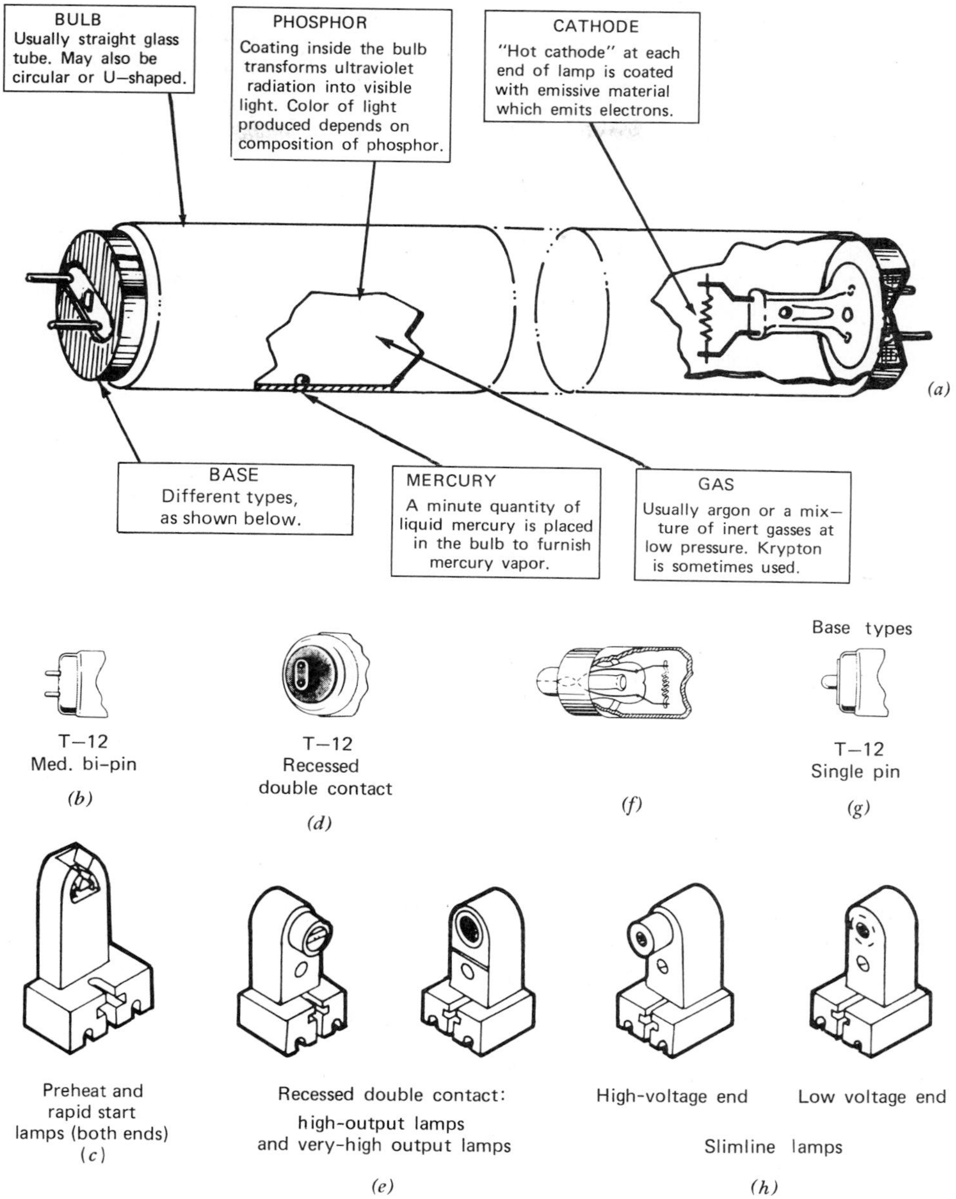
BULB
Usually straight glass tube. May also be circular or U—shaped.
PHOSPHOR
Coating inside the bulb transforms ultraviolet radiation into visible light. Color of light produced depends on composition of phosphor.
CATHODE
"Hot cathode" at each end of lamp is coated with emissive material which emits electrons.
(a)
BASE
Different types, as shown below.
MERCURY
A minute quantity of liquid mercury is placed in the bulb to furnish mercury vapor.
GAS
Usually argon or a mix—ture of inert gasses at low pressure. Krypton is sometimes used.
Base types
T—12
Med. bi-pin
(b)
T—12
Recessed
double contact
(d)
(f)
T—12
Single pin
(g)
Preheat and rapid start lamps (both ends)
(c)
Recessed double contact:
high-output lamps and very-high output lamps
(e)
High-voltage end
Low voltage end
Slimline lamps
(h)

Fig. 19.39 *Details of typical fluorescent lamps and associated lampholders.* (a) *Construction of preheat/rapid-start bipin base lamp. (Courtesy of GTE/Sylvania, Inc.) This type of lamp has type* (b) *base and is held in type* (c) *lampholder. High output HO and VHO rapid-start lamps use recessed d-c base* (d) *and lampholders* (e). *Instant-start lamps are similar in construction to* (a) *except with cathode construction* (f), *have a single pin base* (g), *and use single pin lampholders* (h), *which are different for each end.*

length from 6 to 90 in. A typical ordering abbreviation for a preheat lamp would be F15T12WW. This translates: fluorescent lamp, 15 W, tubular-shaped bulb, 12/8-in. diameter (number represents diameter in eighths of an inch), warm white color (see Table 19.11). In large measure preheat lamps have been supplanted by rapid-start and instant-start types.

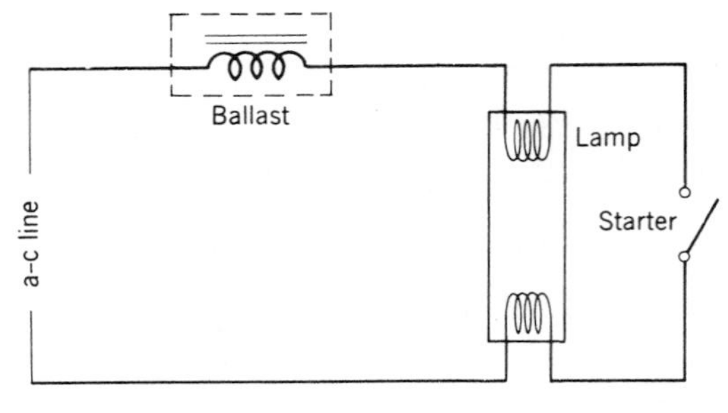

(a)

Fig. 19.40 *Basic circuits for preheat, rapid-start and instant-start (Slimline). For the sake of clarity, only single-lamp circuits are shown, and power-factor correcting capacitors, autotransformers, and compensators are omitted.*

(a) *Basic preheat circuit. Starter may be any of several types, manual or automatic. The circuit does not show compensators or other detailed elements for the sake of clarity. Most preheat lamps are T-12 and operate at 430 mA.*

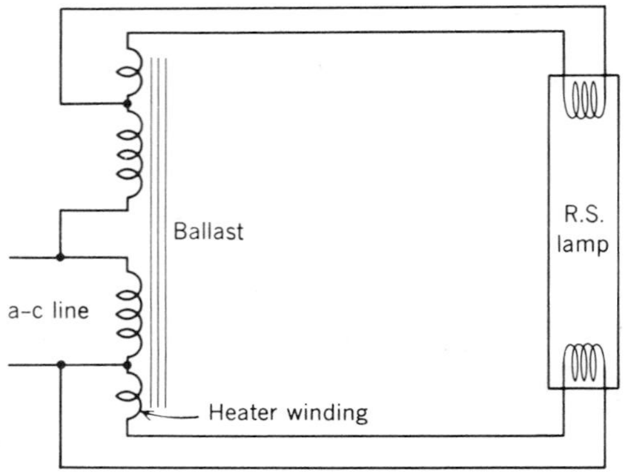

(b)

(b) *Basic rapid-start circuit. Note the special end windings used to supply voltage to heat the cathode continuously. To assure proper starting, all standard RS lamps must be mounted within ½ in. of a grounded metal strip extending the full length of the lamp (1 in. for HO and VHO lamps). Normal output lamps operate at 430 mA, HO at 800 mA, and VHO at 1500 mA.*

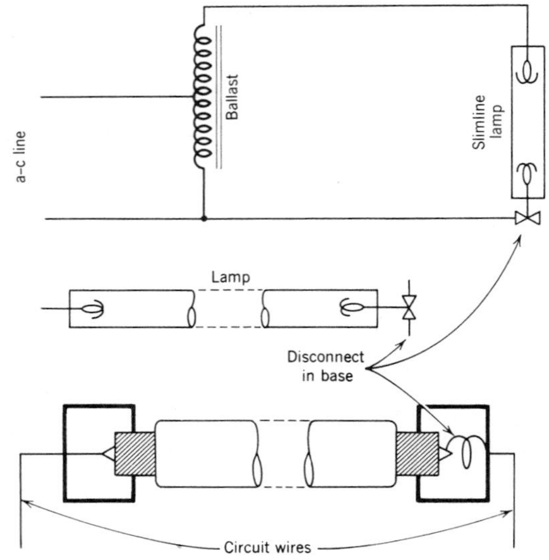

(c)

(c) *Basic instant-start circuit. Voltage from ballast transformer is high enough to strike an arc directly. Note that unlike preheat and rapid-start lamps, these are single pin, since cathodes are not preheated. T-6 and T-8 lamps normally operate at 200 mA, T-12 lamps at 430 mA.* (Lower portion of figure) *Because of the high voltage involved, the lampholder at one end is a disconnecting device that opens the circuit when the lamp is removed.*

TABLE 19.11 **Typical Fluorescent Lamp Data: Standard Lamps 60 Hz Standard Ballasts**

Lamp Abbreviation	Lamp Data: Lamp (W)	Lamp Data: Diameter (in)	Lamp Data: Length (in)	Lamp Current (mA)	Ballast (W)[b,c]	Total Watts[c]	Lamp Life (h)[d]	Initial Output (lm)[e]	Lumens at 40% Life	Initial Actual Efficacy (lm/W)[f]	Remarks
Preheat lamps[a]											
F-15 T-8 CW	15	8/8	18	430	8	23	7,500	870	765	38	Cool white
F-20 T-12 WW	20	12/8	24	430	10	30	9,000	1,300	1,155	43	Warm white
Rapid-start—preheat lamps[g]											
F40 T-12 CW	40	12/8	48	430	7.5	46	20,000+	3,150	2,770	68	
F40 T-12 WW	40	12/8	48	430	7.5	46	20,000+	3,200	2,815	70	Warm white
F40 T-12 CWX	40	12/8	48	430	7.5	46	20,000+	2,250	1,855	49	Cool white deluxe
F40 T-12 D	40	12/8	48	430	7.5	46	20,000+	2,600	2,290	57	Daylight
F40 T-12/C50	40	12/8	48	430	7.5	46	20,000+	2,200	1,890	48	5000 K color
F40 T-12/C75	40	12/8	48	430	7.5	46	20,000+	2,000	1,720	44	7500 K color
F40 T-12/U	40	12/8	—	430	7.5	46	12,000	2,900	2,525	55	"U" shape[h]
Rapid start—high output											
F48 T-12 CW/HO	60	12/8	48	800	12.5	72.5	12,000	4,300	3,740	55	
F60 T-12 CW/HO	75	12/8	60	800	15	90	12,000	5,400	4,700	60	
F72 T-12 CW/HO	85	12/8	72	800	22.5	107.5	12,000	6,650	5,785	62	
F96 T-12 CW/HO	110	12/8	96	800	18.5	128.5	12,000	9,200	8,005	72	
Rapid-start—very high output											
F48 PG-17 CW	110	12/8	48	1500	5	125	12,000	6,900	5,100	55	G.E. Power Groove®
F72 PG-17 CW	165	12/8	72	1500	10	175	12,000	11,500	8,510	66	G.E. Power Groove®
F96 PG-17 CW	215	12/8	96	1500	10	225	12,000	16,000	12,160	71	G.E. Power Groove®
Instant-start (Slimline) lamps											
F42 T-6 CW	25	6/8	42	200	10.5	35.5	7,500	1,750	1,490	49	
F64 T-6 CW	40	6/8	64	200	9	49	7,500	2,800	2,350	57	
F24 T-12 CW	20	12/8	24	430	14	34	7,500	1,150	1,035	34	
F36 T-12 CW	30	12/8	36	430	13	43	7,500	2,000	1,800	47	Warm white
F48 T-12 CW	40	12/8	48	430	12	52	9,000	3,000	2,760	58	
F72 T-12 CW	55	12/8	72	430	11	66	12,000	4,550	4,275	69	
F96 T-12 CW	75	12/8	96	430	13	85	12,000	6,300	5,800	74	

[a]Data given for a preheat circuit.

[b]Figures are for a two-lamp circuit.

[c]ANSI figures.

[d]Life figures are for 3-h burning per start.

[e]After 100-h burning.

[f]Includes ballast loss.

[g]Data given for lamps in a rapid-start circuit.

[h]U-shaped lamps available with 3⅝- or 6-in. leg spacing; all other characteristics equal.

(b) Rapid-Start Lamps. These are similar in construction to the preheat lamps; the basic difference is in the circuitry (see Fig. 19.40*b*). This circuit eliminates the delay inherent in preheat circuits by keeping the lamp cathodes constantly energized (preheated). When the lamp circuit is energized, the arc is struck immediately. No external starter is required. Because of this similarity of operation, rapid-start lamps will operate satisfactorily in a preheat circuit. The reverse is not true, because the preheat requires more current to heat the cathode than the rapid-start ballast provides (see Table 19.12 for interchangeability of lamps in the various circuits). By far the most popular lamp is the 40-W T-12 lamp. A standard ordering abbreviation for a lamp should be F40T12WW/RS, which indicates fluorescent, 40 W, T-12 bulb, warm white color, rapid start. However, this size is so common that the tube size and rapid-start designations are omitted and the lamp is simply F40WW.

Standard rapid-start lamps operate at 430 mA. If this current is increased, the output of the lamp also increases. Two special types of higher output rapid-start lamps are available. One operates at 800 mA and is called simply high output (HO). The second, which operates at 1500 mA (1.5 A), is called by different manufacturers very high output (VHO), superhigh output, or simply 1500-mA, rapid-start lamp. There is also a 1500-mA special lamp that uses what looks like a dented or grooved glass tube. This lamp, called Power Groove by General Electric, has somewhat higher output than the standard VHO tube. All high-output lamps use double contact bases and special ballasts (see Figs. 19.39*d* and 19.39*e*). This lamp is used in applications where high output is required from a limited size source such as outdoor sign lighting, street lighting, and merchandise displays. Because of the serious heat problems involved, VHO lamps are frequently operated without enclosing fixtures. Conversely, HO and VHO lamps are frequently used in cold environments that would prevent proper operation of standard output 430-mA lamps.

Most HO and VHO lamps are slightly less efficient than the standard 430-mA, rapid-start lamp and have considerably shorter life. Typical ordering abbreviations for high-output lamps are similar to the standard rapid-start lamps except that the number indicates length, not wattage. For instance, F72T12/CW/HO is fluorescent, 72 in. long, T-12 bulb, cool white, high output (800 mA). Similarly, F72T12/CW/VHO is fluorescent, 72 in. long, T-12 bulb, cool white, very high output (1500 mA). Typical characteristics for rapid-start lamps are given in Table 19.11.

A circuit known as trigger start uses a preheat lamp in a rapid-start type of circuit, but with higher strike voltage. It is used for lamps up to 20 W and does not use a starter. After the arc is struck, lamp voltage is reduced.

(c) Instant-Start Fluorescent Lamps. Slimline lamps are the best-known variety of instant-start fluorescent lamps. They use a high-voltage transformer to strike the arc without any cathode

TABLE 19.12 **Fluorescent Lamp Interchangeability: Standard Lamps and Ballasts Only**[a]

Lamp Type	*Ballast/Circuit Type*		
	Preheat	*Rapid-start*	*Instant-start*
Preheat	OK	Not good, poor starting	Not good, poor starting short life[b]
Instant-start (Slimline)	Won't start, not good[b]	Won't start, not good[b]	OK
Rapid-start	OK	OK	Not good, poor starting short life[b]
Preheat/rapid-start	OK	OK	Not good, poor starting short life[b]

[a]Special lamps such as low-wattage, high-wattage, 265-mA T-8 and so on must be used with matching ballast.

[b]Normally no possiblity of interchange. Instant-start lamp is single pin base; preheat/rapid-start lampholders are for bipin bases.

ILLUMINATION

preheating. These lamps have only a single pin at each end that also acts as a switch to break the ballast circuit when the lamp is removed, thus lessening the shock hazard (see Figs. 19.39*f*, 19.39*g*, 19.39*h*, and 19.40*c*). The lamps are generally operated in two-lamp circuits at various currents; normal currents are 200 and 430 mA, and normal lengths are 24, 36, 42, 48, 60, 64, 72, 84, and 96 in. These lamps are actually hot cathode instant-start lamps, which differentiates from the high-voltage cold cathode type. The high-voltage starting characteristic of instant-start circuits lowers lamp life to about half that of the corresponding rapid-start lamp.

Slimline lamps and ballasts are more expensive than rapid-start and are somewhat less efficient. However, they are manufactured in certain sizes and currents not made in rapid-start (e.g., 96 in., 430 mA), and they have the additional advantage of being able to start in much lower ambient temperatures (below 50 F) than rapid-start circuits. This starting characteristic makes the instant-start circuit particularly applicable to outdoor use. A typical ordering description for such a lamp would be F42T6CW Slimline, which means fluorescent, 42-in. length, tubular, 6/8-in. diameter, cool white, instant start. The T-6 narrow tube indicates a low-current, 200 mA lamp, in lieu of T-12 for the 430-mA lamp. Note also that in instant-start lamps the number following F indicates length, not wattage. This is true of all lamps that operate at other than 430 mA, which is the standard current. Typical characteristics appear in Table 19.11.

(d) Cold Cathode Tubes. The true cold cathode tube uses a large, thimble-shaped cathode and a high-voltage transformer that literally tears the electrons out of the large cathode to strike the arc. These lamps have a very long life that, in contradistinction to hot cathode lamps, is virtually unaffected by the number of starts. Cold cathode lamps have a lower overall efficiency than the hot cathode types and are normally used where long continuous runs are required, as in architectural-type lighting rather than in lighting fixtures. Cold cathode lamps are readily dimmed and also operate well at varying ambient temperatures.

19.15 The Fluorescent Lamp: Characteristics of Operation

(a) Lamp Life. The lamp life of a standard fluorescent tube is greatly dependent on the burning hours per start. One manufacturer now produces lamps whose life is independent of burning hours per start, but at the cost of overall life. The figures listed in Table 19.11 and in the lamp catalogs for lamp life are based on a burning cycle of 3 h per start and represents the average life of a group of lamps; that is, half the lamps of any group will have burned out at this time. Typical lamp mortality curves are shown in Fig. 19.41, and the effect of burning hours per start is shown in Fig. 19.42.

The significance of this item is connected with energy costs and utilization. From an energy source

ILLUMINATION

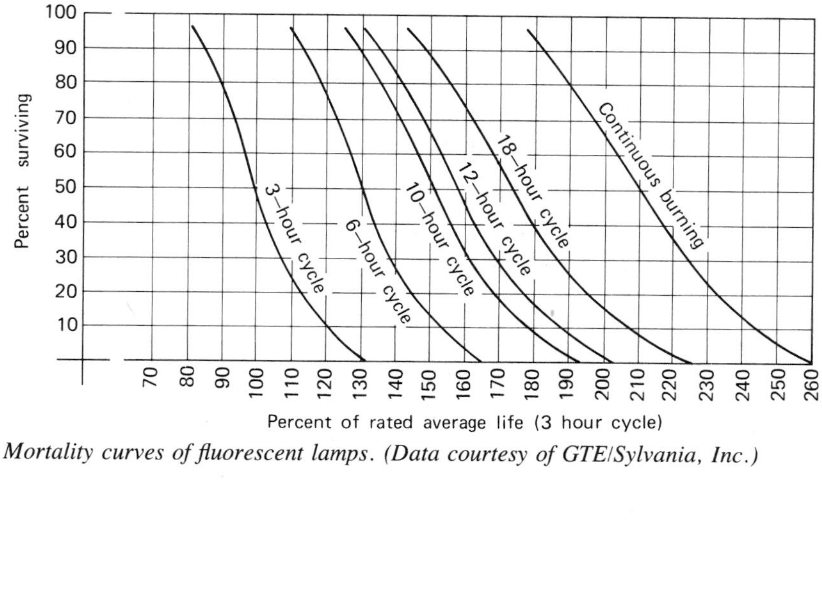

Fig. 19.41 *Mortality curves of fluorescent lamps. (Data courtesy of GTE/Sylvania, Inc.)*

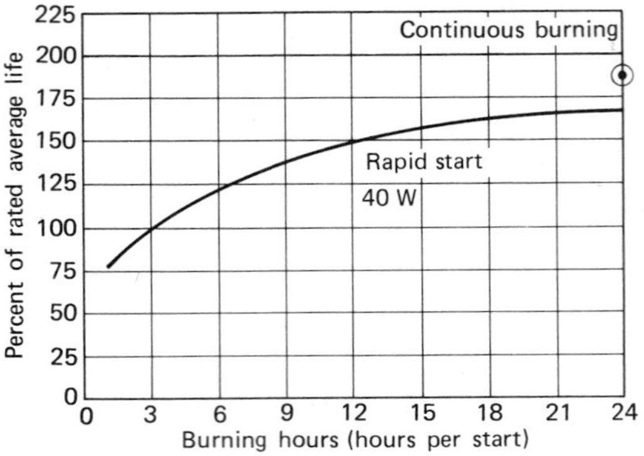

Fig. 19.42 Effect of burning hours on fluorescent lamp life. Note that at three burning hours per start, the average lamp life is 100% of the nominal catalog figure. (Data courtesy of GTE/Sylvania, Inc.)

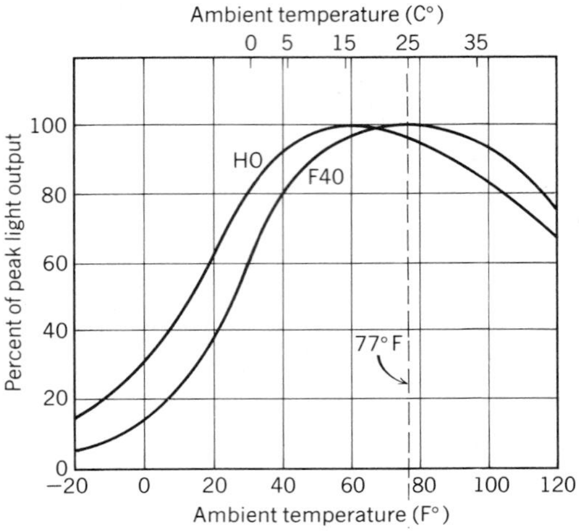

Fig. 19.43 Relation of light output to ambient temperature immediately around a fluorescent tube. Since the HO lamps run hotter, they require a lower ambient to maintain proper bulb wall temperature. (Courtesy of GTE/Sylvania, Inc.)

viewpoint, if an area is not utilized for periods of 10 min or more, fluorescent lamps should be shut off. This takes into account the resource energy required to replace a tube as a result of shortening its life. From a cost viewpoint, the break-even point depends on these factors:

1. Lamp life reduction as a function of burning hours per cycle.
2. Cost of energy.
3. Cost of lamp and lamp replacement.
4. Amount of time lamp remains off when shut off.
5. Cost of switching equipment (if any).
6. Life of the building.

With this number of variables it is not possible to give general solutions, and an individual analysis is required. However, several analyses by the author have shown that assuming ordinary office conditions, using lamp life data as given in Fig. 19.42, 20-yr fixture life, \$0.085/kWh energy cost escalating 3% annually, \$1.25 lamp cost, 15-min relamping time, and \$8 switching cost per lamp (one switch per two 2-lamp fixtures), lamps should be switched off any time they are not in use for 5 to 8 min or more. (The spread is caused primarily by variation in local labor rates.) It is thus clearly an economic fallacy to leave lamps burning to achieve longer lamp life.

(b) Effect of Temperature. The temperature of the coolest point on the lamp bulb wall determines the lamp's mercury vapor pressure, which in turn determines the lamp lumen output, wattage, and color. The bulb wall temperature itself is affected by room ambient tempertaure, airflow over the lamp (as in an air-return fixture), and the temperature of adjacent surfaces such as the ballast enclosure. These in turn are affected by such variables as time of day and weather and by construction details such as luminaire design and mounting method. Furthermore, lamp current depends largely on the particular type of ballast used. Thus it should be apparent that catalog data on lamp output and wattage, which are based on laboratory tests of bare bulbs at 25°C ambient temperature in still air (design conditions for maximum efficiency; see Fig. 19.43), may be very far from field operating conditions. For this reason, lamp (and ballast) manufacturers often list three values for lamp wattage in the catalog: laboratory conditions (ANSI), open luminaire (such as a parabolic reflector), and enclosed luminaire (such as a wraparound lens unit). Typical wattage/fixture values for two standard 40-W T-12 CW lamps with standard ballasts are 91, 88, and 78 W, respectively, that is, highest for laboratory conditions, with correspondingly high output, less for an open lamp fixture, and least for an enclosed fixture, because of its elevated lamp temperature.

The well-documented effect of increased light output in air-return luminaires is due to the cooling action of the air on overheated fluorescent tubes. However, overcooling, which may occur with outdoor fixtures or open suspended fixtures in air-conditioned rooms, also reduces output, as can be seen from Fig. 19.43. If knowledge of lamp wattage (and lumen output) is particularly important, as when selecting special ballasts for light reduction in retrofit work, field measurements on installed units must be made. For new construction, a full-size mockup with specific lamp and ballast combinations can be made and measurements of wattage and lumen output taken.

Special all-weather and jacketed lamps are available that will maintain fairly constant lumen output over a wide ambient temperature range. For outdoor use where starting below 50 F is necessary, rapid-start lamps require special low-temperature ballasts. Slimline lamps with normal ballasts will start readily down to 20 F and, by using the next higher voltage ballast, starting down to −20 F is possible.

(c) Voltage Effects. Voltage either above or below rating adversely affects life, unlike the effect of low voltage on the incandescent lamp. The results of operation at other than rated voltage are shown graphically in Fig. 19.44. Normal operating voltage range for ballasts is 110 to 125 V on 120-V circuits, 200 to 215 V on 208-V circuits, and 250 to 290 V on 277-V circuits.

(d) Lumen Maintenance. Lumen output of a fluorescent tube decreases rapidly during the first 100 h of burning and thereafter much more slowly. For this reason the tabulated initial lumen figures represent output after 100 h of burning. Data are also generally published on the lumen output at 40% of average rated life. This figure is approximately 85 to 90% of the 100-h initial value (see Fig. 19.45). Lumen maintenance is *not* affected

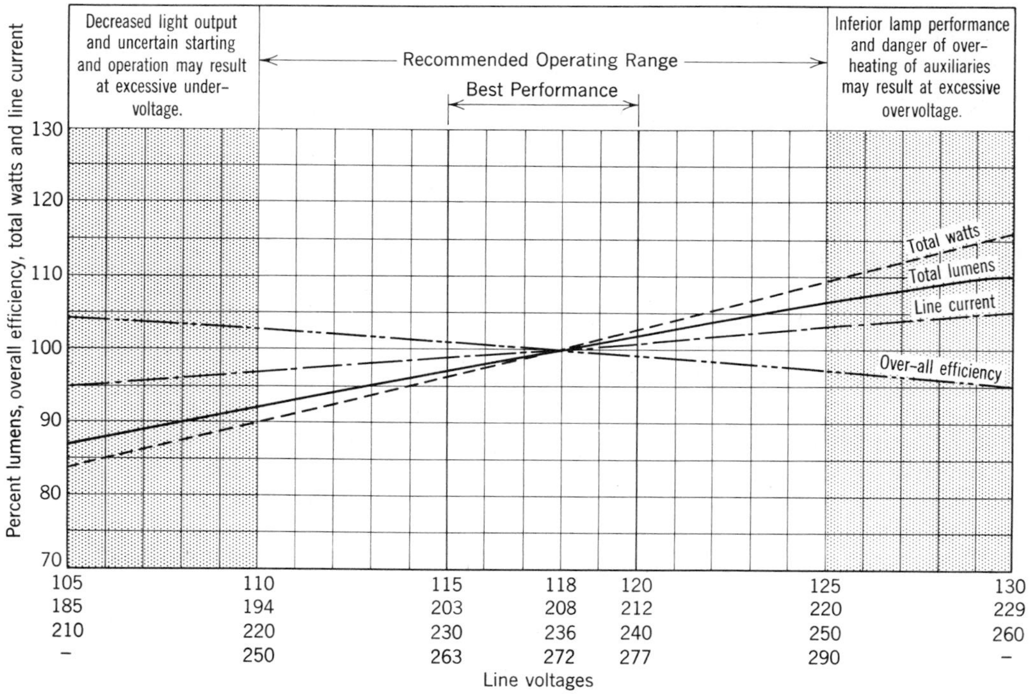

Fig. 19.44 Recommended operating ranges of circuit voltages for most satisfactory operation. The curves indicate the percentage changes in output lumens, efficiency, total watts, and current for line-voltage changes from the rated value. The nominal circuit voltages are 120, 208, and 277 V.

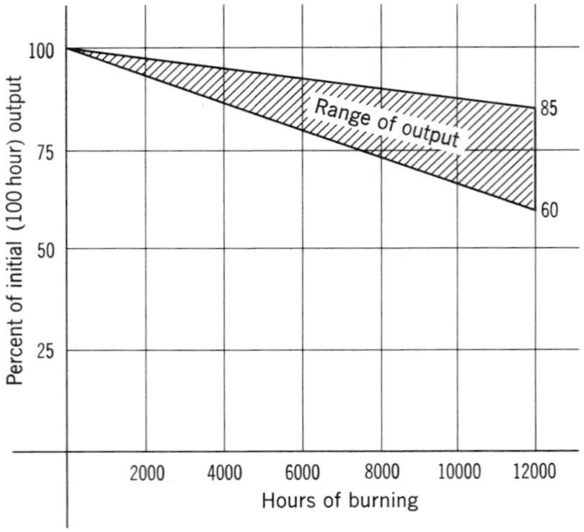

Fig. 19.45 Fluorescent lamp output depreciates with life. The majority of lamps fall in the upper part of the band. Lumen depreciation is unaffected by burning hours per start.

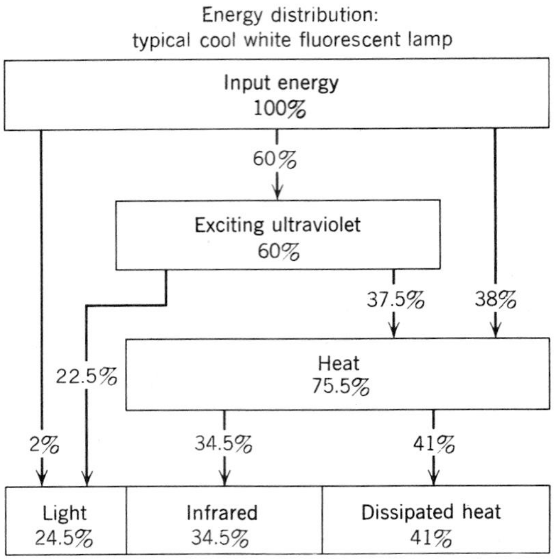

Fig. 19.46 Fluorescent lamps with efficacies of up to 85 lm/W are among the most efficient light sources available, yet still convert less than one-quarter of their energy to useful light. Ballast losses are not included in the percentages shown.

appreciably by the number of burning hours per starts. It is better for 200- and 430-mA lamps than for high-output (800- and 1500-mA) lamps.

(e) Efficacy. The design efficacy of a fluorescent lamp depends on operating current and the phosphors utilized. Figure 19.46 shows the energy distribution of a typical fluorescent lamp alone, not including ballast losses. Lamp current, in turn, is largely dictated by the ballast, although for normal design it is assumed that a ballast will be selected that will supply the design lamp current. This, however, is not always the case, since manufacturers now market a large array of ballasts intended to raise or lower normal lamp current (wattage) and alter other operating characteristics. However, all other conditions being equal, and assuming standard lamps and ballasts, warm white lamps are most efficient, followed closely by cool white, white, daylight, and colored lamps. Specialty colors or lamps designed to produce specific kelvin temperatures are low in output, with *lamp* efficacies in the 40–50 lm/W region. Triphosphor lamps, discussed below, are an exception. The range of efficacy for standard lamps is 40–85 lm/W, *including* ballast losses in the wattage figure (see Table 19.11). This is important, since discharge lamps are inoperative without ballasts, and neglecting ballast losses yields an artifically high and therefore misleading efficacy. It is important that the designer appreciate the significance of efficacy as compared to lamp wattage. In a normal design situation, an interior space requires a calculated quantity of light, in lumens. The designer usually attempts to deliver that light at minimum overall cost, that is, a combination of first cost and operating cost. Minimum first cost usually means *maximum light output* per fixture, consistent with good lighting design. Minimum operating costs means maximum *lumens per watt*. These two demands are not normally consonant. Over the life of the building, operating cost dominates, as life-cycle cost analyses show, and the lamp of choice is normally one with highest efficacy. Lamp *wattage,* in and of itself, is a meaningless quantity unless associated with a lumen output figure. Thus low-wattage or high-wattage lamps with their special ballasts are rarely the indicated choice in new design, because their efficacy does not usually justify the cost premium. These special lamps are useful in retrofit work. In such cases, the designer must make field measurements of actual lamp wattage and output, on the basis of which data a lamp/ballast selection can be made. Catalog data for these special lamp/ballast combinations are based on laboratory condition tests, which, as pointed out above, can vary widely from field conditions.

Generally, standard 430-mA lamps are most efficient, followed by HO 800-mA lamps, then VHO 1500-mA lamps. Specialty lamps such as reflector and low-wattage units are discussed in following paragraphs. Ballast losses, which constitute 5 to 16% of lamp wattage, depend on ballast type, circuit, manufacturer, type of fixture, ambient temperature, and number of lamps connected. Figures given in Table 19.11 are average.

To make a proper comparison between lamp types, it is not sufficient to compare cost to produce a given quantity of lumens, since high-output lamp installations use a smaller number of fixtures. A meaningful cost comparison requires a full life cycle or annual owning cost analysis.

(f) Dimming and Reduced Output. It is often desirable to reduce the output of a fluorescent lamp in order to reduce energy consumption, correct overlighting, compensate for daylight, change an area's function, and so forth. If it is anticipated that full output will again be required, dimming is indicated. If the change is long-term or permanent, a stepped or fixed reduced-output arrangement is probably indicated. Output reduction is accomplished by reducing effective lamp current. This can be accomplished by lowering primary voltage, adding impedance to the circuit, or shortening the time of current flow in each half cycle. The last method describes the gating action of a modern solid-state thyristor dimmer (SCR, triac).

1. *Full-range dimming*. When dimming is required to levels as low as 5% of full output, each individual lamp must be provided with a special dimming ballast. This ballast provides high voltage for starting and restrike even at low output levels and provides continuous cathode heating. This latter is important to assure that even at low lamp arc current, the lamp does not flicker and that lamp life is unaffected, because only arc current is reduced while cathodes remain fully heated. Finally, the dimming ballast permits gating of the lamp current in a "3-wire" control arrangement, which results in full-range dimming. In this circuit arrangement central control units can control the dimming function of the local ballasts, permitting dimmers to be preset, zoned, and timed as desired.

The lamp of choice in dimming circuits is the standard 430-mA rapid start, although 800- and 1500-mA can also be dimmed. An important characteristic of this dimming system is the linear relation between light output and power input, such that lamp efficacy is maintained down to about a third of output, after which efficacy drops and color shifts toward the blue-green portion of the spectrum, because of bulb cooling.

2. *Partial dimming*. The thyristor action shown in Fig. 19.47 can be applied to control the effective input voltage to a fluorescent luminaire with conventional ballasts, and by so doing dim the lamp. This is commonly known as "2-wire" dimming, since the controlled voltage is applied to the 2 input wires of the luminaire. However, since the high strike voltage and cathode heating circuit necessary for full-range dimming is absent, this arrangement is only useful down to 40 to 50% of full output. Below this level, light output drops sharply, lamp color changes, starting becomes unstable, and lamp life is affected. However, within the 100 to 50% range, the energy-output curve is linear, that is, efficacy is maintained, making this arrangement very desirable as an economic lighting-energy management system. This is explained fully in Sections 21.2 through 21.4, which covers lighting control. A typical controller of this design for dimming an entire 20-A 277-V circuit is shown in Fig. 21.8.

Low-energy lamps are generally *not* suitable for dimming service. This is because they operate with special ballasts that reduce lamp current in various ways. Some ballast designs change current

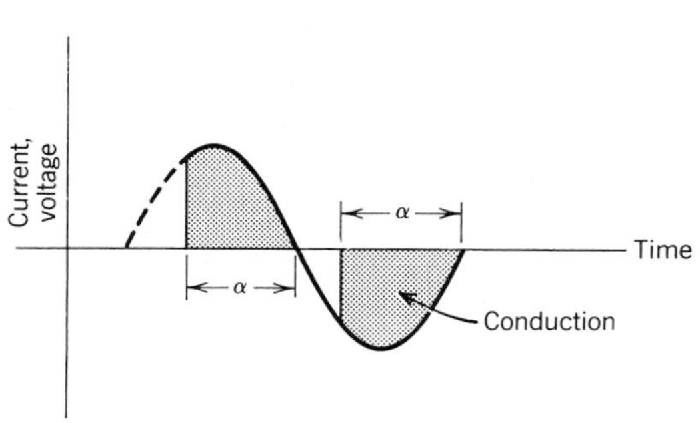

Fig. 19.47 Gating action of a solid-state dimmer. By controlling the gating, the conduction angle α can be adjusted, thus varying the effective voltage over the entire cycle.

waveform, others disconnect the cathode circuit, and still others change the circuit impedance and current-voltage characteristics. Since most of these circuit alterations would interfere with normal dimmer functioning, their use is inadvisable without a prior detailed engineering analysis. Also, low-energy lamps themselves are not operable with dimming ballasts, since these ballasts are designed for the parameters of standard lamps only.

3. *Output reduction*. The cost of dimming equipment is usually justified only if smooth continuous-level changes are required. See Sections 21.3 and 21.4. Where discrete-level switching is acceptable, one solution is the use of multilevel ballasts. Such ballasts, once quite popular, are used less today because local codes generally prohibit use of wall switches or other simple switching devices to change levels, instead requiring that an electrician perform the function. Obviously, this is inconvenient, time-consuming, and expensive. The reason for the prohibition is that the level change involves switching of ballast capacitors, with the possibility of high voltages (about 300 V) and current surges (up to 50 A). However, circuiting can be arranged so that switching is performed when the unit is de-energized, and if that is acceptable these ballasts can be very useful.

Where it is desired to reduce lamp output on a semipermanent basis while maintaining efficacy, impedance can be added to the circuit in various forms, one of which is shown in Chapter 21 (Fig. 21.7). When it is desired to restore full output, the device is removed. Alternatively, special lamps are available that contain integral impedances (Sylvania's Thrift-Mate™ is one) and that will operate at lowered output with standard ballasts. Both these solutions are best applied in enclosed fixtures, since the reduced lamp current may cause excessive lamp cooling in open fixtures with drop in efficacy and change in lamp color.

(g) Flashing. Fluorescent lamp flashing ballasts are similar to dimming ballasts in that the cathode circuit remains energized continuously and only the arc circuit is turned on and off. This permits flashing the lamps without affecting lamp life. Cold cathode, preheat, and rapid-start lamps are suitable for flashing service.

(h) High-Frequency Operation. It has long been known that operation of fluorescent lamps at frequencies above 60 Hz has many beneficial effects. Efficacy increases dramatically (see Fig. 19.48), the heavy and expensive magnetic core-and-coil ballast is replaced by small, light, quiet, and inexpensive solid-state components, heat loss is reduced by as much as 90%, maintenance is reduced, and dimming is greatly simplified. However, until recently, the drawback to all these desirable characteristics has been the difficulty in providing the requisite high-frequency power. Central static inverters are expensive, as is a high-frequency distribution power system. This situation has recently changed completely with the development of compact, reliable, and relatively inexpensive electronic ballasts. These devices, which are discussed further below in the ballast section, are actually small solid-state rectifier-inverter ballasts that change the incoming 60-Hz a-c to approximately 25 kHz. This local at-the-fixture conversion obviates the necessity for a high-frequency distribution system and makes high-frequency lamp operation with all its advantages economically attractive. Indeed, this is very clearly the present trend, and core-and-coil ballasts may soon be largely replaced for normal rapid-start installations. (Because of a high failure rate among solid-state ballasts of a specific type, doubt has been cast on their reliability generally. It is safe to say that solid-state devices are inherently stable and reliable, and that once these "startup" problems have been solved, electronic ballasts with all their inherent advantages will dominate the commercial market.)

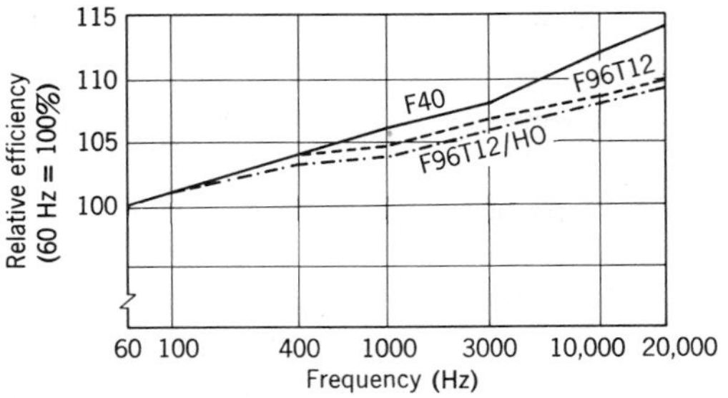

Fig. 19.48 *The efficiency of most fluorescent lamps increases when frequency increases. On this curve, the efficiency at 60 Hz is taken as the 100% value.*

ILLUMINATION

(i) Other Characteristics. Fluorescent light color is discussed in Sections 19.25 and 19.26. Fluorescent lamps are large and therefore necessitate a relatively expensive fixture both to hold the lamps and to control the light. Since the tubes emit light throughout their considerable length, accurate beam control is difficult, making fluorescent units best applicable to area lighting. The advantages of fluorescent lamps are long life, low cost, high output and efficacy, availability in an extremely wide range of sizes, colors, and brightnesses, and relative insensitivity to voltage fluctuation (important in brownout areas). Disadvantages are large size, which creates storage, handling, and relamping problems, and the fixture situation previously referred to.

19.16 Special Fluorescent Lamps

(a) U-Shaped Lamps. U-shaped lamps were developed to answer the need for a high-efficacy fluorescent source that could be utilized in a square fixture, since the normal fluorescent lamp shape is frequently not architecturally suitable (see Fig. 19.49). The U lamp is basically a standard 40-W, 48-in. fluorescent tube bent into a U shape and available with $3\frac{5}{8}$- or 6-in. leg spacing; the former can be accommodated three to a 2-ft square fixture

Fig. 19.49 Each 5-ft-square module uses a 2-ft-square fixture with two 40-W U-shaped fluorescent lamps, of the type shown in the insert. (Photo courtesy of GTE/Sylvania, Inc.)

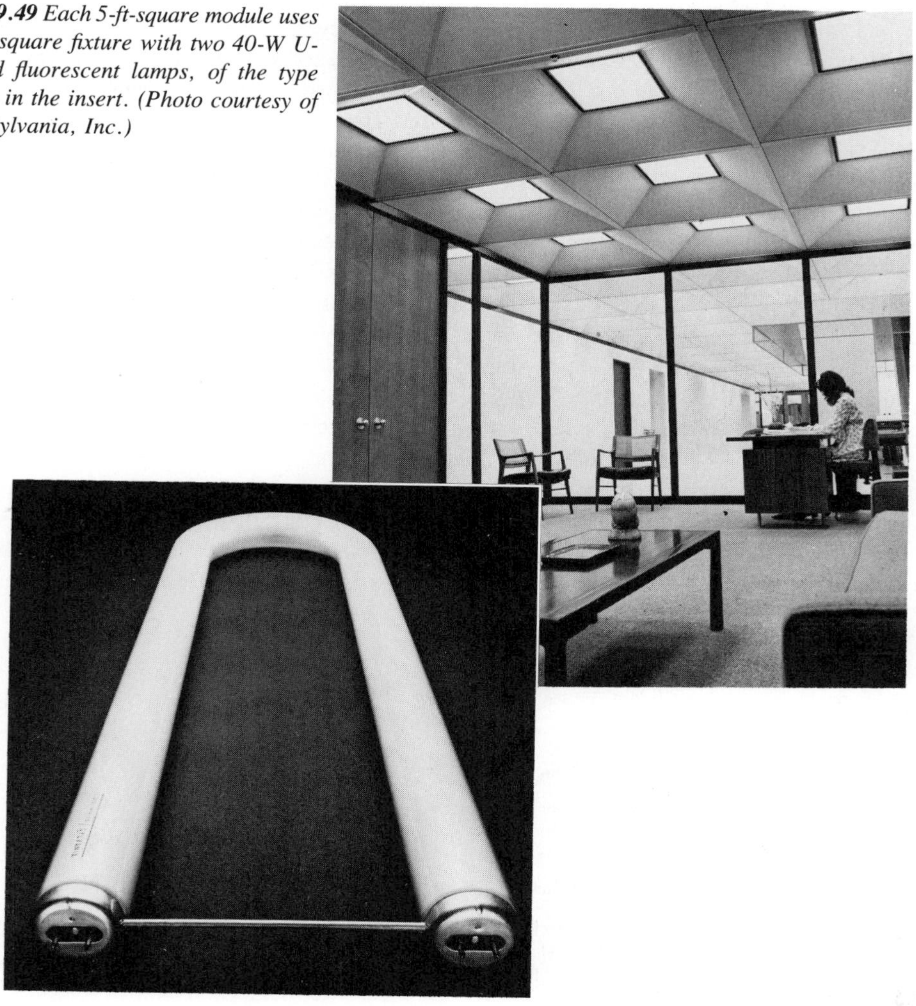

and the latter two to a 2-ft square fixture. The lamps operate on standard ballasts and have slightly lower output than the corresponding straight tube. Insofar as energy is concerned, their use is much more desirable than using 2-ft lamps, as can readily be seen from the following data.

Two foot square fluorescent fixture with

Four 2-ft CW lamps	110 W	5000 lm 9000-h life
Two U-shaped, CW lamps	100 W	5800 lm 12,000-h life

The panel fluorescent lamp that was produced in the late 1960s, for the same purpose, i.e., to provide a nonrectilinear source, is no longer available.

(b) Reflector and Aperture Lamps. These lamps, which are available in 800- and 1500-mA sizes, contain an internal reflector that performs in the same fashion as the more common reflector in the incandescent R and PAR lamps. The reflector lamp is completely phosphor coated, while the aperture lamp has a 30° clear "window" resulting in very high luminance of this slot (see Fig. 19.50). Both types have lower efficacy than a normal tube and are generally applied where an enclosing fixture is uneconomical or impractical, as in handrails or for sign illumination.

Tests using 135° and 235° reflector lamps in normal fluorescent fixtures intended for standard tubes indicate that the fixture coefficient of utilization increases up to 50%, depending on the fixture design. This is because the light normally trapped between the tubes and the fixture is saved, since almost no light is radiated above the cutoff of the internal reflector. Thus, using reflector lamps for general illumination can result in considerable savings in energy costs. If the use of these lamps is contemplated for normal lighting, the fixture and diffuser must be carefully selected to provide a sufficient degree of brightness control, since for a 800-mA, 235° reflector lamp, the aperature luminance is approximately 7400 fL, which is about three times the brightness of a normal 430-mA lamp (2400 fL).

(c) Low-Energy Lamps. The drive for energy conservation plus the desire to reduce lighting levels in existing overlighted spaces has resulted in a complete line of low-energy lamps. Their wattage ratings are lower than those of standard lamps because they are intended primarily as lower energy direct replacements for existing lamps. All such lamps are clearly marked by the manufacturer. They operate on lower current than standard lamps, require special matching ballasts for maximum effectiveness, and have an efficacy equal to, or somewhat higher than, standard lamps and ballasts. They have the disadvantage of higher cost, need for special ballasts where maximum energy reduction is desired, generally shorter life, inapplicability to dimming circuits, and problems of inventory and proper lamp replacement. To our mind, their use is indicated only where other light and wattage-reduction schemes such as circuit dimming and reduced wattage ballasts (see Section 19.14) are inapplicable. Here again, a proper economic analysis is indicated rather than relying on oversimplified charts and graphs, which show savings as a function of energy cost for the life of

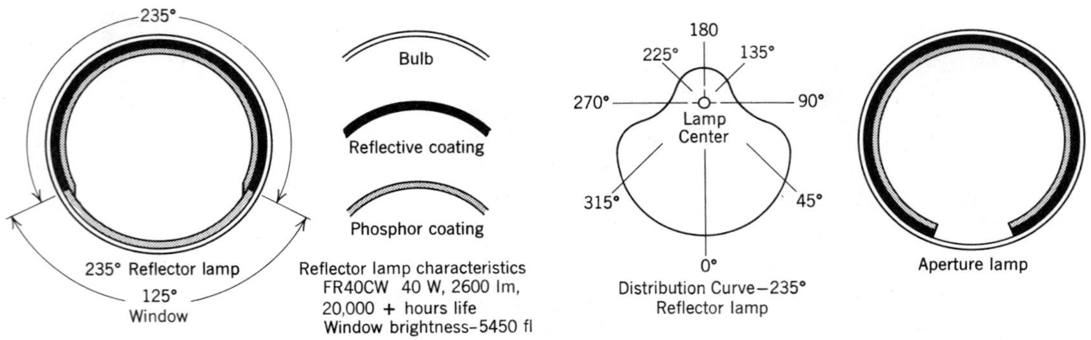

Fig. 19.50 Characteristics of reflector and aperture fluorescent lamps.

TABLE 19.13 **Comparative Life-Cycle Cost Analysis for Relamping a Single 2-Lamp, 8-Ft Slimline Fluorescent Fixture with Low-Energy Slimline Lamps (Retrofit Installation)**

Initial Costs	*75-W Standard*	*60-W Low Energy*
Two lamps	$9.00	$9.30
Annual costs (3000 h)		
Kilowatt-hours per year (incl. ballast loss)	528	438
Energy cost @ $0.085/kWh	44.88	37.23
Lamp replacement costs		
Lamp	2.25	2.33
Labor	3.50	3.50
	50.63	43.06
20-Year costs @ 8% interest rate and 3% annual escalation of energy costs	647.95	552.72
20-Year energy savings		1800 kWh

the lamps or per year. Most of these charts ignore relamping costs and interest rates, which results in a grossly distorted cost picture. A typical, proper life-cycle cost analysis is given in Table 19.13, with cost expressed in terms of present value. Since this analysis is predicated on the assumption that the application is a retrofit installation, it is not surprising that lower wattage lamps are cheaper to operate. In a new installation, where additional luminaire cost is incurred, the results may be entirely different, depending on the cost of luminaires, installation, and circuitry.

(d) "Octic" Fluorescent Lamps. Among the most interesting of recent developments in the fluorescent lamp field is a high-efficiency, 1-in. (26-mm) diameter lamp whose high-output phosphors have the additional desirable characteristic of excellent color rendering index (CRI). (See Section 19.26 for an explanation of this term.) These lamps are also called triphosphor lamps because they use phosphors that produce light primarily at three points; about 450 nm (blue), 540 nm (green), and 610 nm (red). Among them are Sylvania's Octron™, and Norelco's System 8™. They operate best with electronic ballasts at high frequency, although their performance with standard magnetic ballasts at 60 Hz is also an improvement on standard lamps. They are particularly temperature sensitive, making their use in dimming circuits inadvisable because of depreciation of output and color quality. Also, because of this characteristic, their performance in open-type luminaires such as parabolic reflectors should be carefully investigated before being specified. As with all special lamps, they are restricted to use with special ballasts and have shorter life than standard lamps. However, unlike the reduced-energy lamp, their high efficacy, particularly with electronic ballasts, makes their use in new construction attractive, if dimming is not intended. Typical average performance data are given in Table 19.14.

(e) Compact Fluorescent Lamps. The unwieldy linear shape of the standard fluorescent lamp, its double-ended connections, plus the need for a ballast have long restricted its use to specific commercial applications. Recently, however, new designs utilizing compact folded fluorescent tubes with integral ballast, terminating in a medium-screw Edison base have appeared on the market, as direct replacements for incandescent lamps. Although the compact shape reduces the lamp's output, its efficacy is still considerably higher than that of an equivalent wattage incandescent. That, coupled with the lamp's long life, makes these designs attractive for almost all uses to which incandescents are normally put. Light color may be problematic in instances where the red-orange characteristic of the incandescent is vital. However, in most applications this is not the case. These designs bear consideration for all applications involving at least three burning hours per start, and particularly where long life is advantageous, as in hard-to-reach locations. Typical designs are shown in Fig. 19.51.

TABLE 19.14 **Comparative Average Characteristics; 4-Ft Standard and "Octic" Lamps**

Lamp and Diameter	Nominal Lamp Wattage	Nominal Lamp Lumens	Nominal Lamp Current	Color Rendering Index	60-Hz Operation			
					Watts per Lamp (incl. ballast loss)		Efficacy (lm/W, incl. ballast loss)	
					Open Luminaire	Closed Luminaire	Open Luminaire	Closed Luminaire
F-40 T-12, 40 mm	40	3150	430 mA	50–65	47	43	64	70
"Octic" T-8, 26 mm	32	2900	265 mA	75	35	33	83	88

Lamp and Diameter	Nominal Lamp Wattage	Nominal Lamp Lumens	Nominal Lamp Current	Color Rendering Index	25-kHz Operation			
					Watts per Lamp (incl. ballast loss)		Efficacy (lm/W incl. ballast loss)	
					Open Luminaire	Closed Luminaire	Open Luminaire	Closed Luminaire
F-40 T-12, 40 mm	40	3150	430 mA	50–65	39	37	72	76
"Octic" T-8, 26 mm	32	2900	265 mA	75	29	28	98	102

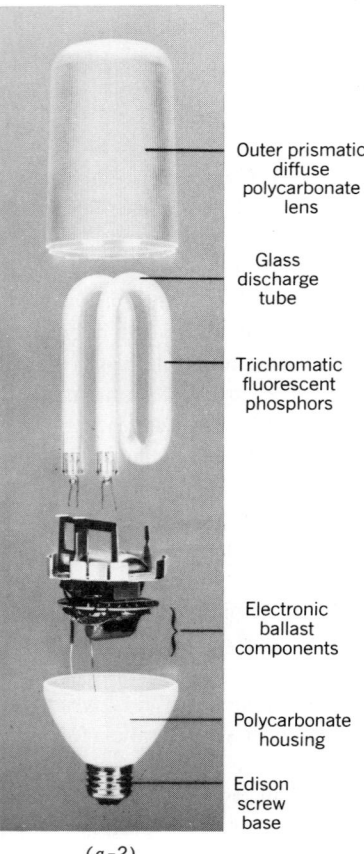

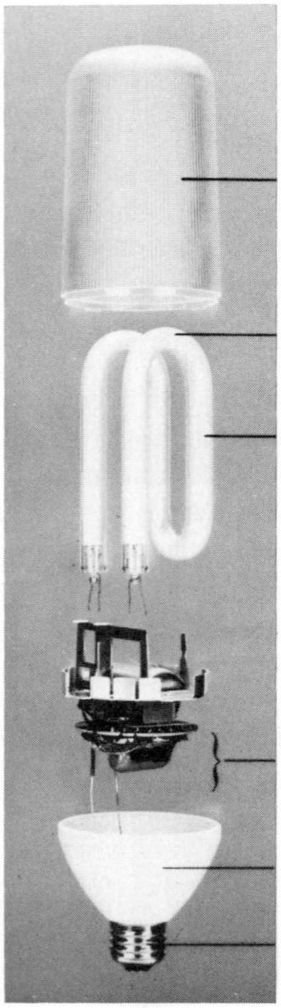

(a-2)

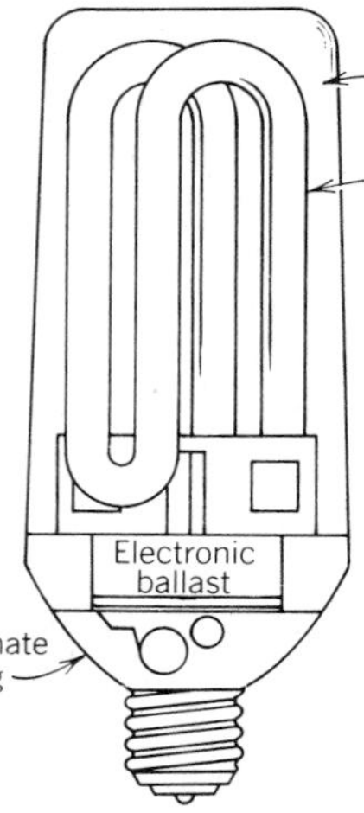

(a-1)

Fig. 19.51 (a) *The folded SL lamp is operated at high frequency to reduce heat build-up, improve output and efficacy, and produce light of acceptable color for an incandescent replacement. Lamp characteristics: MOL, 7+ in.; diameter, 3 in.; wt, <10 oz; color temperature, 2700°K; CRI, 81; power input, 18 W; life, 7500 h; efficacy, 60 lm/W; initial output, 1100 lm which is the equivalent of a 75-W incandescent. The assembled lamp is shown in* (a-1) *and the components in* (a-2).

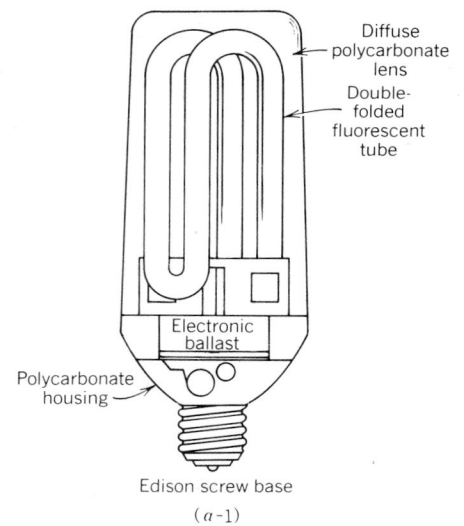

(b) *The ballast adapter shown in* (b-1) *accommodates either the PL7 or the PL9 lamp. The lighting characteristics of the PL7 (PL9) lamp are: output, 400 (600) lm, and efficacy, 57 (67) lm/W. Both lamps have correlated color temperature of 2700K (similar to an incandescent). See Section 19.25. Dimensions of the PL7 (PL9) lamp with ballast, as shown are: length 133 (165) mm; diameter of* each *tube is 12.5 mm. Lamp construction is shown in* (b-2). [*Photos* (a) *and* (b) *Courtesy of North American Phillips Lighting Corp.*] (c) *Lamp similar in design to that of* (a) *except with clear housing. Tubes are available in correlated color temperatures ranging from 2700 to 6500K, corresponding roughly to a range of warm white-incandescent to daylight.* (d) *Globe-shape lamp contains a double-folded fluorescent tube and accessories. Unit is designed to replace incandescent lamps in shape as well as base and, approximately, color.* [*Photos* (c) *and* (d) *courtesy of Toshiba Electric Equipment Co.*]

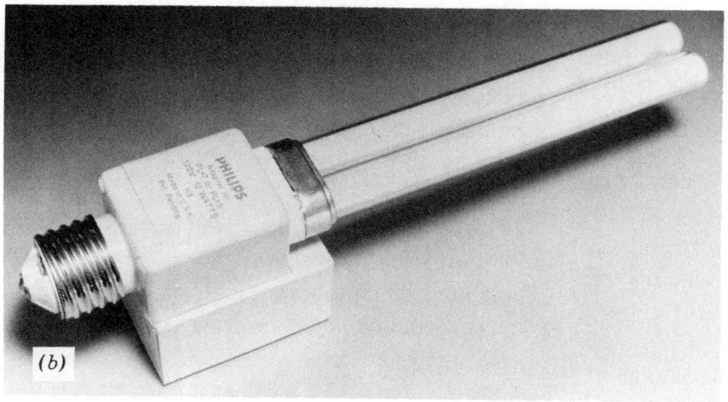

(b)

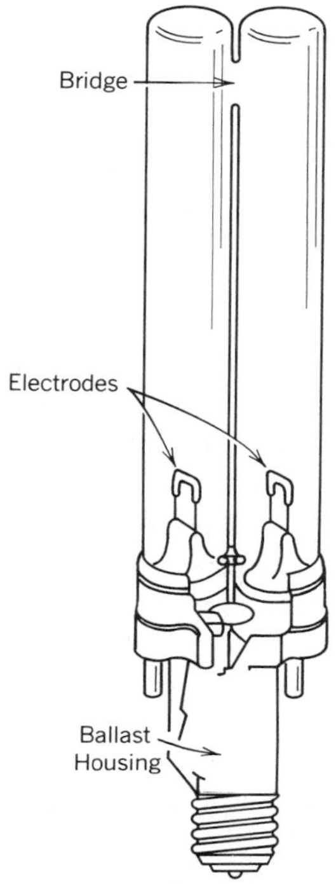

(b-2)

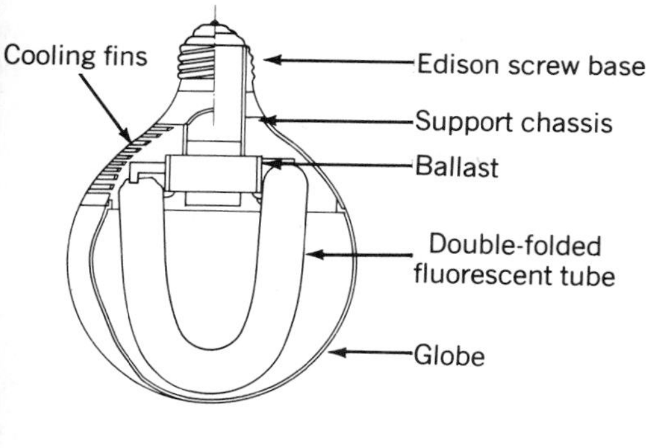

(d −2)

(c) ***(d −1)***

19.17 Core-and-Coil Ballasts for Fluorescent Lamps

The conventional, magnetic core-and-coil fluorescent lamp ballast is an assembly of electrical components responsible for the proper starting and continued operation of the fluorescent lamps in a fixture. By nature of its construction, it also produces considerable heat and noise. Some of the important considerations in selection and application of ballasts are discussed below.

(a) Ballast Temperature. Normal ballasts are designed to operate in the fixture at a maximum temperature of 90°C in a 25°C ambient. Ballast temperature is important, since ballast life is directly affected by it. At normal temperature a ballast life of 12 to 15 yr can be expected. A rule of thumb is that ballast life is halved for every 10°C above 90°C operation and conversely is doubled for every 10°C reduction in operating temperature. Ballast temperature rises 0.9°C per 1°C of ambient temperature above 25°C. Where ambient conditions above 25°C (77°F) are anticipated, special cool operation ballasts, suitable for use in ambients up to 50°C (122°F) should be specified, despite their higher cost.

(b) Ballast Labels. There are several organizations involved with ballast standards and testing. They are:

CBM—Certified Ballast Manufacturers Association.

ANSI—American National Standards Institute. Originates standards on a national level.

ETL—Electrical Testing Laboratories, Inc. A private, independent organization and a recognized authority in measurements and testing of lamps and lighting equipment.

UL—Underwriters Laboratories, Inc. An independent, nonprofit organization testing for public safety.

CSA—Canadaian Standards Association.

Ballasts should be UL labeled and CBM/ETL certified. The UL (Underwriters Laboratories) label assures intrinsic safety. CBM (Certified Ballast Manufacturers) establishes high-quality design criteria, and ETL (Electrical Testing Laboratories) tests the ballasts to determine that the design standards have been met.

(c) Ballast Protection. The National Electrical Code (NEC) requires that all ballasts for *indoor* fixtures be protected by an integral thermal-sensing device that will disconnect the ballast in the event of overheating. Overheating is caused by excessive voltage, excessive ambient temperature, or failure of a ballast component. These devices are either thermostatic (self-resetting) or fuse-type (self-destructive). Since two of the three conditions that cause overheating are usually correctable, our recommendation is to specify the self-resetting type of protector. Thermally protected ballasts are designated "type P" by the UL.

(d) Ballast Heat. Ballast heat is transferred to the fixture body by direct metal-to-metal contact (which must be unimpeded) and is then dissipated by radiation and convection from the fixture. Obviously, therefore, the location and method of fixture installation affect the heat transfer from the fixture and consequently the ballast temperature. Pendant fixtures (more than 6 in. below the ceiling) and fixtures recessed into ventilated suspended ceilings do not generally present a temperature problem. Fixtures mounted on insulating surfaces such as low-density acoustic tile, or into unventilated or heated ceiling spaces, or when boxed by a fire-rated enclosure and recessed into a fire-rated ceiling, *do* present serious heat-dissipation problems. For such installations, as for high-ambient-temperature installations, cool operation ballasts must be specified. For surface-mounted fixtures, an air space between the fixture and the ceiling material will markedly reduce ballast temperature. The space should be a minimum of 1.5 in. Since each installation situation represents almost a unique case because of the variables of fixture, ballast, ceiling material, and ambient temperature, the designer should require a temperature test of the specific fixture in the installation situation involved before acceptance of the unit. This can be specified along with shop drawings, photometric data, and a sample unit.

Normal ballasts are designed to start fluorescent lamps in an ambient temperature range of 50 to 105 F. If conditions outside these limits are expected, special ballasts must be specified. For outdoor installations, ballasts suitable for starting at +20, 0, and −20 F are available.

(e) Ballast Power Factor (pf). This factor is determined by unit design and is either high (above 0.9) or low (0.5–0.6). High-pf ballasts are more expensive than low-pf units, but the additional cost is readily repaid by lower line losses, smaller circuit conductors in long runs, and larger number of fixtures per circuit.

(f) Radio Noise. Frequently referred to as radio frequency interference (RFI), this noise is *not* produced by the ballast, but by the arc discharge in the fluorescent tube. (Occasionally a defective ballast does cause RFI.) To minimize RFI, ballasts are available with integral RF noise suppressors. In extreme cases additional suppression can be obtained by installation of RF noise attenuators in the fixture.

(g) Stroboscopic Effect. This is no longer a problem with single-lamp ballasts due to the use of long-persistence phosphors in all lamps.

(h) Noise. Ballasts produce a hum or buzz when operating that is transmitted to the fixture because of the integral contact required for heat transfer. This contact generally amplifies the noise. Total noise depends on sound rating of the ballast, fixture design, and acoustical characteristics of the room.

Ballasts are sound rated by a letter that indicates, not actual sound developed, but performance in a space. The ballast selected should be suitable for the lowest sound level likely to be encountered in the subject space. Generally, "A"-rated ballasts are selected for offices and other normally quiet areas where ballast noise would be objectionable. Since no manufacturer's standardization of ratings exists, recommendations should be obtained from ballast and fixture manufacturers or from an acoustic consultant. Where ballast noise or heat buildup may present a problem, ballasts can be remotely mounted if provision is made for heat dissipation and noise control (see Table 19.15). See Table 27.10 for acoustic criteria for selection of ballasts.

(i) Special Ballasts. All nonstandard-current lamps, including low-energy and high-energy units, require matching ballasts to supply the required current, waveform, and circuitry. These ballasts, unlike standard-current units, are not interchangeable. Mixing of lamps and ballasts, although possible, is not advised except on manufacturer's recommendation. Similarly, use of one maker's low-energy lamp with another's low-energy ballast is not suggested without prior testing or specific manufacturer's recommendation. It should be apparent that the stocking, inventory, and maintenance problems caused by nonstandard components are of sufficient magnitude to give pause before purchasing. Of course, if any entire facility is to be relamped and/or reballasted, the maintenance problem is reduced. There are four principal varieties of nonstandard ballast.

1. *Low-energy units* are intended to match specific low-energy lamps as explained above. Also, reduced current ballasts intended for use with *standard* lamps are readily available. The choice of lamp/ballast combination should be made only after careful analysis of the alternatives.

2. *High-energy units* are intended to be used with standard or high-output lamps. The purpose of this combination is either to increase output in an existing installation or to reduce the number of fixtures in a new installation. Among these units is General Electric's Maxi-Miser™.

3. *Multilevel ballasts* are useful when it is desired to change lighting levels evenly. The usual unit is two-level, that is, full output and 50%, but three-level units are available for full, two-thirds, and one-third output. As explained above, switching may be a problem because of unusually high voltage and current, but no-load switching can be arranged provided the intermediate OFF position is acceptable.

4. *Dimming ballasts* are made for standard lamps only and are normally applied one per lamp.

TABLE 19.15 **Magnetic 60-Hz, Fluorescent Lamp Ballasts**

Lamp Data			*Minimum Starting*		*Dimensions*		
Description	*Nominal Watts*	*Circuit Voltage*	*Temperature (°F)*	*Sound Rating*	*Length*	*Cross-Sect. Reference*	*Weight (lb)*
Rapid Start: 430 mA and Reduced Wattage							
(2) F30T-12/RS **Two-lamp**	30	120	0	A	9 1/2	A	3.3
		120	50	A	9 1/2	A	3.3
(2) F40T-12/RS **Two-lamp**	40	120	0	A	9 1/2	A	3.9
		120	50	A	9 1/2	A	3.3
		277	50	A	9 1/2	A	4.2
High Output: 800 mA							
(2) F36T-12/HO **Two-lamp**	45	120	50	B	11 3/4	B	8.3
(2) F48T-12/HO **Two-lamp**	60	120	50	B	11 3/4	B	8.3
(2) F72T-12/HO **Two-lamp**	87	120	−20	B	11 3/4	C	10.2
		277	−20	B	11 3/4	C	12.4
(2) F96T-12/HO **Two-lamp**	112	120	−20	B	11 3/4	C	10.2
		277	−20	B	11 3/4	C	10.8
Powergroove, SHO/VHO: 1500 mA							
(2) F72T-12 **Two-lamp**	168	120	−20	D	14 5/16	C	14.0
		277	−20	D	14 5/16	C	14.5
(2) F96PG-17 **Two-lamp** or (2) F96T-12 **Two-lamp**	218	120	−20	D	14 5/16	C	14.0
		277	−20	D	14 5/16	C	14.5

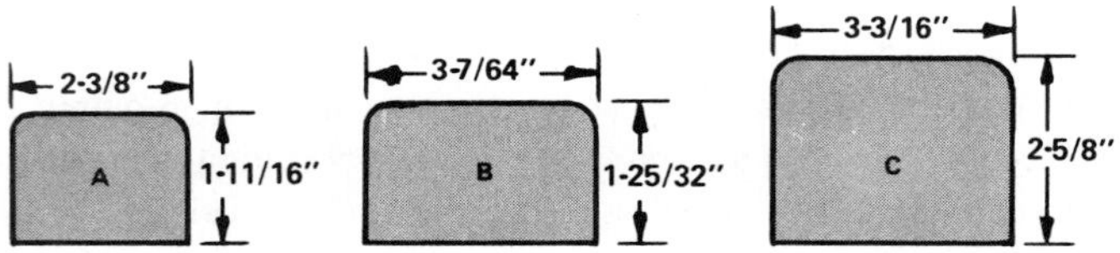

In summary, the course of least resistance, if not always least cost, is to use either standard lamps and ballasts or matched ballast/lamp combinations. However, careful analysis of each new or retrofit installation may yield a more desirable arrangement despite the confusing array of equipment and tendentious data sheets. Some typical ballast data are given in Table 19.15.

(j) Lamp Removal. In general, two-lamp ballasts operate at their lowest temperature with two active lamps, other conditions being equal. Ballasts are affected by lamp failure or removal as follows:

1. In *rapid-start* circuits, operation of a two-lamp ballast with one lamp removed or burned out will not damage the ballast. Ballast power loss will remain approximately the same.
2. In two-lamp *preheat* circuits, one lamp operation will cause ballast overheating and shortened life.

3. In two-lamp *instant-start* circuits, deactivation of one lamp will normally deactivate the entire ballast because the lamps are usually connected in series.

19.18 Electronic Ballasts for Fluorescent Lamps

We have already noted in Section 19.15*b* the advantages to be gained in high-frequency operation of fluorescent lamps. The electronic ballast is primarily a generator of high-frequency a-c and secondarily a ballast, since at these frequencies (25–30 kHz) all that is required to ''ballast'' the circuit, once the arc is formed, is a capacitor. The following is a brief comparison of the characteristics of fluorescent lamp operation with an electronic ballast at 25 to 30 kHz versus operation with conventional core-and-coil magnetic ballast, at 60 Hz. For the purpose of comparison, and where applicable, two 40-W RS lamps are assumed *and light output equalized* for the two situations.

(a) Circuit Wattage, Losses, and Component Temperatures. All these factors are intimately interrelated. In a typical open two-lamp fixture, average power consumption with standard ballast is 88 W; with electronic ballast 70 W. This 18-W (20%) reduction is primarily due to reduction in ballast heating. As is characteristic of solid-state equipment, the electronic ballast is almost lossless, and as a result runs at 50 to 60°C, as compared to 85 to 90°C for the magnetic unit. This has several additional desirable effects; the bulb runs cooler because fixture ambient temperature is lowered, and the fixture mounting constraints that are directed at controlling fixture ambient temperature are almost completely eliminated.

(b) Flicker. Conventional ballast circuits have 29% flicker, which is visible in peripheral vision and can be annoying. Flicker is substantially reduced with all electronic ballasts (10% maximum) and with most is entirely absent.

(c) Third-Harmonic Current. Conventional circuits have 5 to 15% third harmonic distortion, which appears as an undesirable current in the neutral of a 3-phase system. Electronic ballast circuits generally have higher third-harmonic currents, although stringent specifications can be met by premium units.

(d) Dimming. This capability can be much more easily and cheaply incorporated into the electronic design without appreciable increase in flicker; this is characteristic of conventional circuits. Local control at the electronic ballast permits individual fixture level adjustment—a ''tuning'' aspect absent from conventional circuits.

(e) Noise. Solid-state ballasts are almost completely silent and readily meet A rating requirements.

(f) Other Characteristics. Both standard and electronic ballasts are available with power factor and ballast factor above 90%. (Ballast factor is a figure of merit for light output, indicating the ratio between lamp output with this ballast as compared to lamp output with a reference ballast. It should exceed 90%.)

Standard ballasts will normally start a lamp at 50 F minimum. A special ballast is required for starting down to 0 F. Electronic ballasts normally start down to 0 F.

Standard circuits are voltage sensitive, as shown in Fig. 19.44. Electronic ballasts can easily be provided with a voltage-regulating circuit that will handle ±10% line voltage variation with no change in lamp performance.

Failure of one lamp in a two-lamp conventional circuit has the effects detailed in Section 19.17*j*. In electronic circuits, including Slimline, the lamps are operated in parallel. Failure of one lamp will cause the second either to be unaffected or to increase its output by 15 to 25%, depending on design.

Neither EMI (electromagnetic interference) nor RFI (radio frequency interference) is a problem with properly designed, shielded, and grounded equipment.

(g) Physical Characteristics. Electronic ballasts are smaller and much lighter than their standard counterparts because of the absence of the heavy, magnetic, iron core (see Fig. 19.52). As a result, handling, maintenance, and storage are easier.

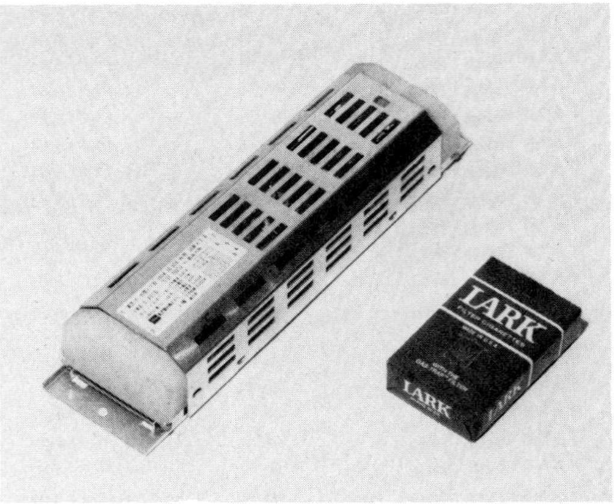

Fig. 19.52 *Typical two-lamp, 40-W, rapid-start electronic ballast with dimming capability. The physical data for this unit are: 260 × 66 × 44 mm (10.2 × 2.6 × 1.7 in.) and 0.47 kg (1 lb). This compares with 9.5 × 2.4 × 1.6 in. and 3.3 lb for a conventional unit (without dimming). (Photo courtesy of Toshiba Electric Equipment Co.)*

It should be apparent from the above that the electronic ballast has so many advantages over the standard unit that its general acceptance is only a matter of time. The duration of this transition period depends on removing "bugs" in design and reducing the price, which at this writing is two to three times that of the standard. Obviously, projects where first cost is the overriding consideration will use the cheapest unit, whatever that might be. The responsible lighting designer, however, will apply lighting, energy, and life-cycle cost criteria in reaching an equipment-selection decision.

HIGH-INTENSITY DISCHARGE (HID) LAMPS

Subsumed under this heading are mercury, metal-halide, and sodium-vapor lamps. These lamps have inherently high efficacy and, with appropriate color correction, can be utilized in any application, indoor or outdoor, that does not have critical color criteria.

19.19 Mercury Lamps

These lamps operate by passing an arc through a *high-pressure* mercury vapor contained in an arc tube made of quartz or glass (see Fig. 19.53). This action produces light in both the ultraviolet region (as in the low-pressure fluorescent lamp tube) and in the visible region, principally in the blue-green band. This color is characteristic of the clear mercury lamp. Details of spectral distribution are given in Section 19.25.

(a) Lamp Designations. Designations for mercury lamps are more complex than for incandescent or fluorescent lamps and have been a source of confusion for years, because manufacturers invented their own systems in an attempt to simplify specifications. ANSI adopted a simplified code some time ago that is now used by all manufacturers and is shown in Table 19.16. This code has five parts and is best illustrated by example. Lamp designation H 38 MP 100 DX indicates:

H—Mercury lamp
38—Ballast number
MP—Lamp physical characteristics
100—Lamp wattage
DX—Phosphor, glass coating, or coloring, optional with each manufacturer; lack of a letter indicates a clear lamp).

(b) Lamp Life. Lamp life is extremely long, averaging 24,000+ h based on 10 burning hours per start. Mercury lamps are not suitable for applications that are subject to constant switching; therefore, a relatively long period of burning per start was selected. A typical lamp mortality curve is shown in Fig. 19.54. Life is affected by ambient temperature, line voltage, and ballast design. Mercury lamps are not as sensitive to short burning cycles as fluorescents but, because of accelerated lumen depreciation near the end of life, they are normally replaced before burnout.

(c) Lumen Maintenance. This depends on the type of lamp and its burning position. Manufacturers publish data on *each* of their lamp types. In general, clear lamps have the best lumen main-

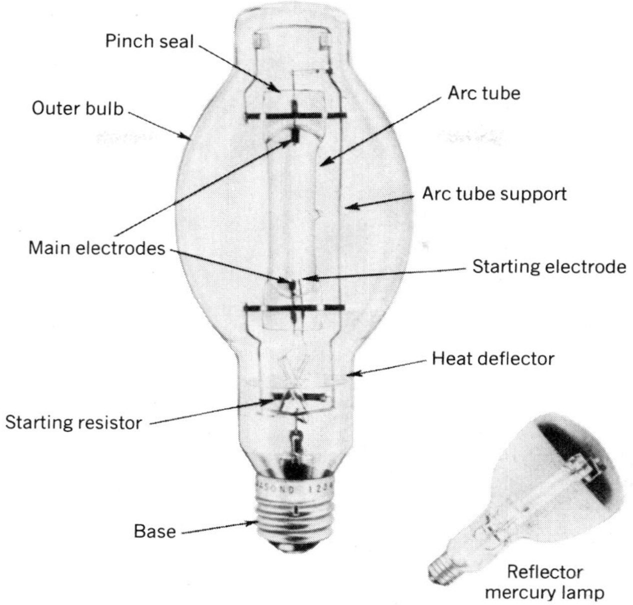

Fig. 19.53 *Typical construction of a clear mercury lamp.*

tenance, followed by color-improved and phosphor-coated units. A curve showing the average data is given in Fig. 19.55.

(d) Color Correction and Efficacy. These are added because the blue-green light distorts almost all colors. The outer bulb is coated with phosphors that are excited by the UV light and reradiate generally in the red band, which is entirely absent in the basic lamp color. Depending on the arc tube design and the phosphors used, the color of the emitted light can be corrected to make it acceptable for general indoor use. Lamps are available in clear, white, color-corrected, and white-deluxe, in ascending order of color improvement. The deluxe lamp also uses a stain on the envelope to filter out some of the blue-green, which obviously reduces lamp output. Efficacy of selected typical units, with and without ballast loss, is given in Table 19.16. Note that in general efficacy is lower than for fluorescent lamps.

(e) Ballasts. Ballasts are required, as with all arc discharge lamps, to start the lamp and thereafter to control the arc. The basic conventional ballast is simply a reactor that controls the arc after the discharge has been initiated. Three to six minutes is required for the lamp to reach full output, since heat must be generated by electron flow to vaporize the mercury in the arc tube before the arc will strike. Once extinguished, the lamp must cool and the pressure must be reduced before restrike is possible. This restart delay amounts to 3 to 8 min, depending on the ballast type, and is an important consideration in design, since a momentary outage will extinguish all lamps, leaving an interior area in the dark. Special fixtures now available utilize small quartz lamps to supply light during such outages. Alternatively, some incandescent lighting can be utilized that will maintain minimum illumination. Principal types of ballasts are:

1. Reactor ballast has low power factor, provides no voltage regulation, and should be used only where line voltage fluctuation does not exceed $\pm 5\%$.
2. Autotransformer ballast is a reactor unit with

TABLE 19.16 **Typical Data for Mercury Vapor Lamps**[a]

Lamp Watts	Bulb	Base	ANSI Ordering[b] Abbreviation	Light Center Length (in.)	Max Overall Length (in.)	Rated Average Life (h)	Approximate Lumens Initial	Approximate Lumens Mean[d]	Average[c] Ballast Loss (W)	Initial Efficacy (lm/W) Lamp and Ballast[e]
50	B-17	Med	H45AY-R40-50	3⅛	5⅛	16000+	1575	1250	10	26
75	E-17	Med	H43AV-R75/DX	3½	5 7/16	16000+	2800	2250	12	32
100	A-23	Med	H38MP-R100/DX	3½	5 7/16	18000+	4000	3050	15	35
	E-23½	Mogul	H38JA-R100/WDX	5	7½	24000+	3400	2600	15	30
	R-40	Med	H38BP-R100/DX		7½	24000+	2850	2000	15	25
175	E-28	Mogul	H39KC-R175/DX	5	8 5/16	24000+	8100	7200	20	44
	R-40	Med	H39B-R175		7½	24000+	5700	4800	20	29
250	E-28	Mogul	H37KB-250 H37KC-250/DX	5 5	8 5/16 8 5/16	24000+ 24000+	11200 12100	9850 9800	25	41 44
400	E-37	Mogul	H33GL-R400/DX	7	11½	24000+	22500	17200	40	51
	R-60	Mogul	H33FS-R400/DX		10⅛	24000+	15500	10800	40	35
700	BT-46	Mogul	H35N-R700/DX	9½	14½	24000+	41000	30250	70	55
1000	BT-56	Mogul	H36GW-R1000/DX	9½	15⅜	16000+	63000	40000	100	57

[a]For accurate, current data consult the manufacturers' catalogs. Data extracted from General Electric catalog.

[b]Explanation of color suffix in ordering abbreviations:

No letter clear lamp, CRI = 30
DX deluxe white, CRI = 45
WDX warm deluxe white, CRI = 54

[c]Losses vary widely with type of ballast; figures given are average.

[d]At two-thirds life.

[e]Using average ballast loss figures.

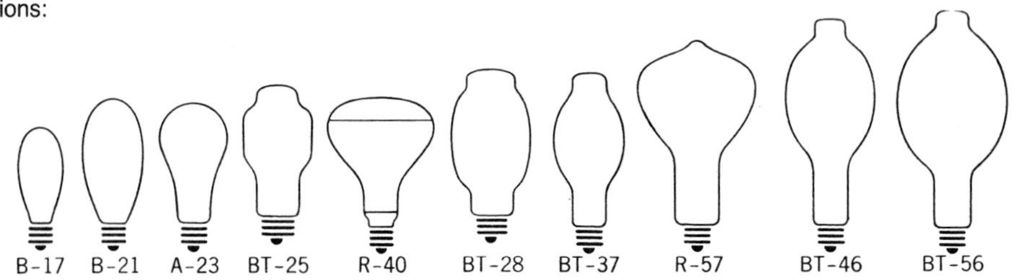

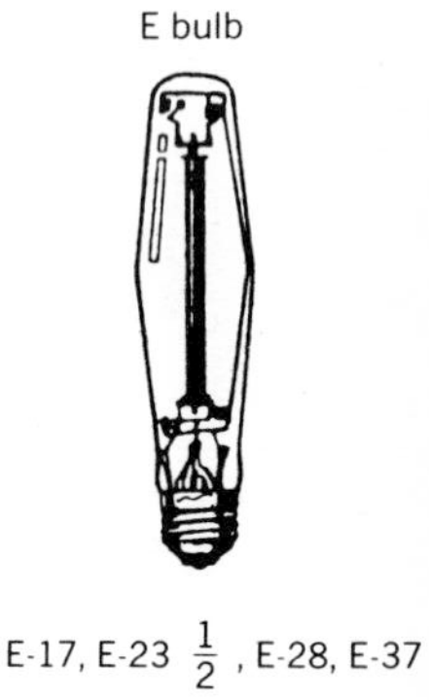

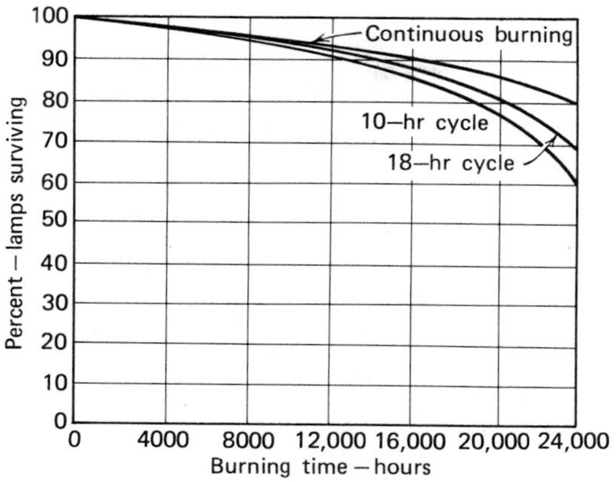

Fig. 19.54 *Typical life expectancy or survival curves for 175-, 400-, and 1000-W mercury lamps at various burning cycles.*

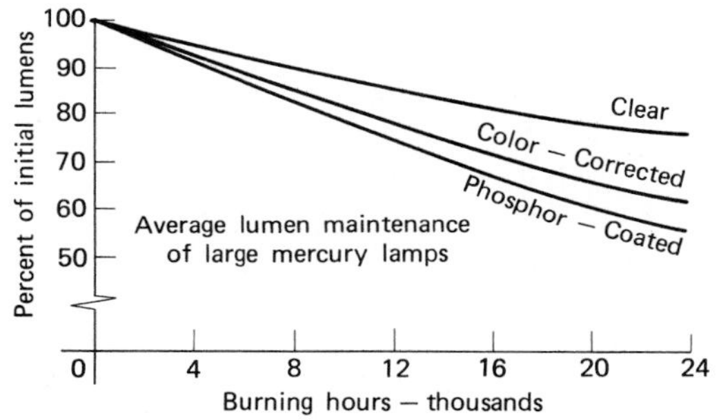

Fig. 19.55 *Average lumen maintenance for mercury lamps 175 W and larger.*

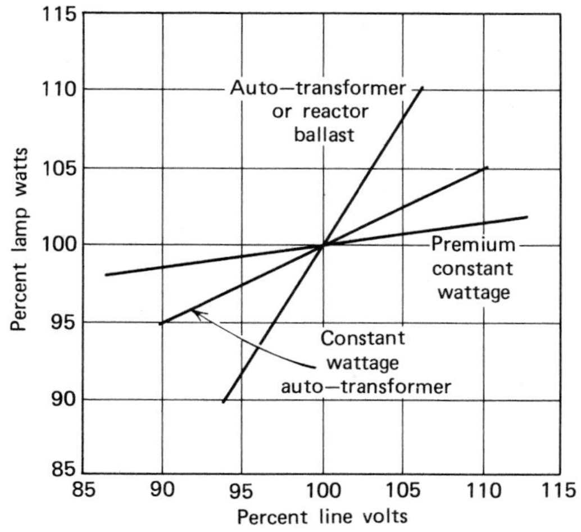

Fig. 19.56 *Effect of line-voltage variation on lamp watts with various ballast types.*

a transformer to match line voltage to lamp voltage. It also is low pf and nonregulating.

3. High-pf autotransformer ballast is the same as type 2, except with additional capacitor to improve pf.
4. Constant wattage autotransformer ballast, also called a lead circuit ballast, is a regulating high-pf unit that maintains lamp voltage, hence wattage and lumen output, fairly constant. Lamp wattage will vary 5% with a 10% voltage change.
5. Premium constant wattage or stabilized ballast provides lamp isolation, voltage regulation, high pf, low extinction voltage, and low inrush current.
6. Electronic ballasts are at this writing not readily available commercially, but probably will be in the near future. Their principal advantages are high efficacy, much lower lamp flicker, and drastic weight reduction—from 10 kg for a standard magnetic unit to less than 2 kg for an electronic unit. The principal drawback is, for the present, high cost.

Ballast prices increase from type 1 to type 5 in the ratio of 1.0 to approximately 2.5. When line voltage varies, lamp voltage is affected also, depending on the type of ballast (see Fig. 19.56). Lamp wattage and lumen output are directly proportional to lamp voltage. Lamp operation at overwattage is inadvisable, since lamp temperature increases while life and lumen maintenance decrease. Mercury ballasts are normally quite noisy. Where this may be a problem, remote mounting should be considered.

(f) Dimming. Dimming of mercury lamps is possible and entirely practicable with the use of a dimming ballast and solid-state dimming control. These are available for 400-, 700-, and 1000-W units. Mercury lamps have so large an output that shutting off a unit creates an imbalance in the lighting coverage—a problem readily solved by dimming. Cost analyses indicate the economic feasibility of such control, including sensing equipment that will automatically maintain illumination levels at preset levels. The payback period depends on power rates and local labor rates. One analysis performed by the author for a large installation, with 50–1000-W lamps, a $0.085/kWh energy rate, and 8% capital cost, indicated a 2½-yr payback period, assuming 4000-h annual operation.

A little-used but very effective and economical output-reduction technique is simply to change the circuit capacitance by an amount that depends on lamp size and ballast type. By doing this, the lamp wattage and output can be reduced by approximately 50% with no deleterious effect on lamp or ballast. This technique is by far the cheapest method of accomplishing an overall, even reduction in output. Relay switching of capacitors is the recommended technique.

(g) Application. Mercury-vapor lamps are applicable to indoor and outdoor use with proper attention to color and fixture brightness. Indoor application is generally limited to mounting 10-ft AFF or higher to avoid glare problems and to permit adequate area coverage. Use in industrial spaces and stores is common, as is discussed in detail in Chapter 21. However, because of low efficacy, mercury installations are being replaced by the more efficient metal-halide and sodium lamps.

Fig. 19.57 *Application of self-ballasted mercury lamp in a difficult access location.*

19.20 Special Mercury Lamps

In an attempt to satisfy the desire for a small lamp to take the place of incandescents in interior fixtures, manufacturers have made available mercury lamps in 40-, 50-, 75-, 100-, and 175-W sizes, in deluxe white and other color-corrected designs. For the smaller sizes, screw-in ballasts are available, so that replacing an incandescent is simply a matter of screwing in a ballast and a small mercury lamp. In the 175-W size the ballast must be separately mounted. Available also are self-ballasted lamps (require no separate ballast), which can be used where ballast mounting is inconvenient, expensive, or undesirable for other reasons. In both cases, there is no doubt that the extremely long life, good color, and reliability of these incandescent substitutes make their use attractive in locations where relamping is difficult and expensive (see Fig. 19.57). But it also is true that their efficacy is only slightly better than that of the incandescent, making their use, from an energy standpoint, questionable. Furthermore, the high cost of self-ballasted units makes this lamp questionable economically, except in very special cases. Similarly the long delay (5–10 min) in starting and restarting severely restricts applicability.

Whenever self-ballasted or small mercury units are contemplated, consideration should be given to fluorescents, metal-halide lamps, or low-pressure sodium. All these sources are more efficient than incandescent, self-ballasted mercury, and standard mercury (in that order; see Fig. 19.1), and the higher initial cost will be repaid rapidly. Table 19.17 gives an analysis of four alternatives for lighting an *existing,* difficult-access location, as in Fig. 19.57.

TABLE 19.17 **Cost Analysis of Four Alternative Methods of Relighting a Stairwell**[a,b]

Lamp Data and Fixture	*150-W Incandescent*	*75-W Mercury Separate Ballast*	*160-W Self-Ballasted Mercury*	*New Fluorescent Fixture with 40-W T-12 CW Lamp*
Initial lamp cost	$1.10	$18.00	$35.00	$1.67
Initial ballast/fixture cost incl. labor	—	24.00	—	60.00
Total initial cost	$1.10	$42.00	$35.00	$61.67
Lamp wattage incl. ballast	150W	90W	160W	48W
Lamp initial lumens	2850	2800	3000	3150
Lamp life-hours	750	24000+	20000	20000+
Annual data—4000 h				
Cost of lamp replacement				
Lamp	$5.90	$6.00	$4.50	$0.35
Labor	110.00	3.50	5.00	5.00
Total—Lamps	$116.90	$9.50	$9.50	$5.35
Cost of energy				
Kilowatt-hours	600	360	640	200
@ 0.085/kWh	$51.00	$30.60	$54.40	$17.00
Total annual cost	$167.90	$40.10	$63.90	$22.35
10-Yr Life Cycle Cost	$1190	$284	$453	$158
Relative LCC	7.5	1.8	2.9	1.0
Relative energy use	3.1	1.9	3.3	1.0

[a]Costs will vary with location and date, but the ratios should be approximately constant.

[b]Discount rate (interest rate, reflecting inflation) is assumed to be 10%. See Table E1.b, Appendix E. A 3% annual escalation of energy and labor costs is included.

1. Continue using 150-W incandescent lamps.
2. Use a screw-in ballast and a 75-W deluxe white mercury unit.
3. Install directly a self-ballasted 160-W mercury lamp.
4. *Replace* the existing fixture with a simple, 48-in. open-reflector strip fluorescent for one F-40 T-12 CW lamp.

From the table it is clear that alternative 4 is the economic choice and also, by far, the energy-resource choice. If installation of a fluorescent is impossible, the next choice is clearly the externally ballasted 75-W mercury unit.

19.21 Metal-Halide Lamps

The metal-halide lamp is basically a mercury lamp that has been altered by the addition to the arc tube of halides of metals such as thallium, indium, or sodium. The addition of these salts causes light to be radiated at frequencies other than the basic mercury colors and increases efficacy, but reduces the life and reduces lumen maintenance to 60% at two-thirds life. The color produced is much warmer than that of the mercury light. Clear lamps have a correlated color temperature of 3700 to 4700 K, and a CRI of about 65. Phosphor-coated lamps have a corresponding 3400 to 3800 K and CRI of about 70. A high-output version of the lamp is also available with better output and life characteristics. It is constructed with a special arc tube and is extremely sensitive to burning position. Clear lamps are recommended for exterior use and phosphor-coated units for all indoor application, including food displays.

A brief comparison of these two lamps can be seen on page 988.

ILLUMINATION

	Mercury	Metal-Halide	Special Higher Output Metal-Halide
Life	16,000 to 24,000 h	7500 to 15,000 h	10,000 to 20,000
Color	Poor to fair	Good to excellent	Good to excellent
Lamp only,[a] initial efficacy	50 to 60 lm/W	80 to 110 lm/W	85 to 125 lm/W

[a]Ballast loss approximately the same for all 3 types.

Since the color of the metal-halide lamp depends on the amount of ionized halide salt in the arc, lamp performance is extremely sensitive to voltage, temperature, and burning position. Mortality and lumen maintenance curves are similar to those for mercury lamps except for lower values as noted above. Strike time is shorter than for the mercury lamp, being 2 to 3 min, but restrike time is up to 10 min, making it necessary to supply an instant-start source in indoor areas lighted with these lamps. Certain metal-halides are usable with mercury ballasts, although generally they require special ballasts with higher voltage and better regulation than mercury lamps. Common trade names for metal halide lamps are Metalarc (Sylvania) and Multi-Vapor (G.E.). Typical data for standard metal halide are given in Table 19.18.

The continuing effort to find high-efficacy substitutes for incandescent lamps has produced a number of new metal-halide designs. One is an Edison screw-base lamp with built-in electronic ballast, intended to replace incandescents in residential and similar use. Its limitation, as with all sophisticated sources, is its initial cost. Another line of lamps, Sylvania's 3K(TM) lamps, have a correlated color temperature of 3200 K, which is very close to that of incandescents, and a CRI of 70. See Section 19.26 for a discussion of this important characteristic.

19.22 High-Pressure Sodium (HPS) Lamps

The highest efficacy general-purpose HID source is the high-pressure sodium lamp, marketed by its developer, General Electric, under the trade name *Lucalox,* by Norelco as *Son-Lux,* and by Sylvania under the name *Lumalux.*) Details of the lamp are shown in Fig. 19.58, and life and lumen maintenance data are given in Table 19.18. Construction is quite different from that of mercury and metal-halide lamps and, although it operates as an arc discharge unit, its excellent characteristics stem from the spectral absorption phenomenon of the contained sodium under high pressure. The resultant light is yellow tinted, similar to that of

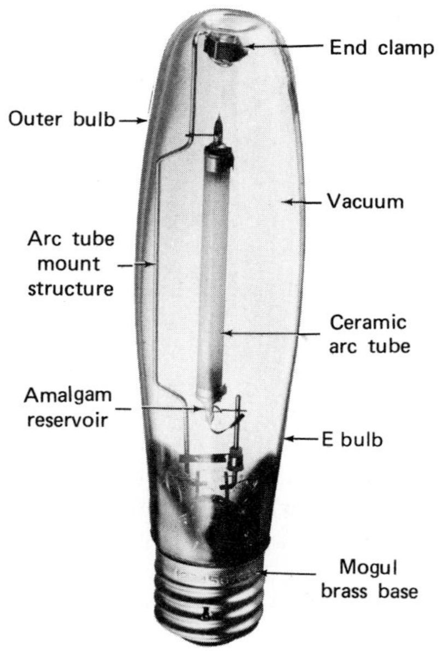

Fig. 19.58 *The main features of the HPS lamp are the alumina ceramic arc tube, amalgam reservoir, and rigid arc tube mount structure. (Photo courtesy of General Electric, Lighting Business Group.)*

TABLE 19.18 **Sylvania Metalarc™ Lamps**

Metalarc™ lamps: rated life and mean lumens based on minimum of 10 h per start operation. Rated life and mean lumens are reduced for shorter burning cycles.

Watts	Bulb[a]	Base	Description	Average Rated Hours Life	Approx. Vert. Lumens Initial	Approx. Vert. Lumens Mean[d]	Initial Efficacy Lamp	Initial Efficacy Lamp & Ballast[c]
175[b]	BT-28	Mogul	Clear Metalarc Phosphor coated	7,500 7,500	14,000 14,000	10,800 10,200	80	69
250[e]	BT-28	Mogul	Clear Metalarc Phosphor coated Metalarc	10,000 10,000	20,500 20,500	17,000 16,000	82	75
400[e]	BT-37	Mogul	Clear Metalarc Phosphor coated	15,000[f] 15,000[f]	34,000 34,000	25,600 24,600	85	77
1000[e]	BT-56	Mogul	Clear Metalarc Phosphor coated	12,000 12,000	110,000 110,000	88,000 84,000	110	100

Super-Metalarc™ Lamps: Lamps must be operated within ±15° of horizontal. Other lamps available for vertical operation.

Watts	Bulb	Base[g]	Description	Average Rated Hours Life	Approx. Horiz. Lumens Initial	Approx. Horiz. Lumens Mean[d]	Initial Efficacy Lamp	Initial Efficacy Lamp & Ballast
175	BT-28	Mogul	Clear Metalarc Phosphor coated	10,000	15,000 15,000	12,000 11,300	86	77
400	BT-37	Mogul	Clear Metalarc Phosphor coated	20,000	40,000	32,000 31,000	100	91

Super-Metalarc™ 3 K Lamps:[h] Lamps have correlated color temperature of 3200 K.

Watts	Bulb	Base	Description	Average Rated Hours Life	Initial	Mean	Lamp	Lamp & Ballast
175	BT-28	Mogul	Phosphor coated	10,000	14,000	10,500	80	69
250	BT-28	Mogul	Phosphor coated	10,000	21,500	16,500	86	77
400	BT-37	Mogul	Phosphor coated	20,000	37,000	28,000	92.5	84

Source: Courtesy of GTE/Sylvania, Inc.

[a]For diagram of bulb shape, see Table 19.16.

[b]Must be operated within ±15° of vertical.

[c]Calculated using average ballast loss figures.

[d]Taken at 40% of rated life.

[e]Any burning position.

[f]Life 20,000 h if operated within 15° of vertical.

[g]Position-oriented base.

[h]Horizontal operation only.

warm white fluorescent lamps. Typical characteristics are:

Lamp efficacy	80 to 140 lm/W
Efficacy, including ballast losses	57 to 128 lm/W
Life	16,000 to 24,000 h
Lumen maintenance	80 to 90% at 50% life
Warmup time	3 to 4 min
Restrike time	½ to 1½ min
Correlated color temp.	2100 K

Unlike the metal-halide lamp, the HPS unit is not voltage sensitive and is color constant. Its lumen maintenance is outstanding, as is its efficacy, both of which are the highest available for any general-use light source. Details of light color are discussed in Section 19.24, and comparative characteristics with other HID sources and fluorescent lamps are given in Section 19.26. As with all discharge lamps, a ballast is required to supply the high voltage to strike the arc and to control the arc once struck. HPS ballasts are quite different from those for mercury or metal-halide lamps because of the high voltages necessary. In addition the ballast lamp power is a function of lamp voltage. As with other HID sources, electronic ballasts are now becoming available. However, to make the changeover from existing mercury installations to HPS attractive financially, a line of special HPS units is available that can be substi-

TABLE 19.19 **Sylvania High-Pressure Sodium Lamps—Lumalux™ (HPS Lamps)**

Watts	Bulb	Base[a]	Average Rated Hours Life[b]	Approx. Lumens Initial	Approx. Lumens Mean	Description	Lamp Efficacy (lm/W)	Lamp and Ballast[c] Efficacy
35	B-17	Med	16,000	2,250	2,025	Clear	64	48
				2,150	1,935	Coated[d]	61	46
50	B-17	Med	24,000+	4,000	3,600	Clear	80	60
				3,800	3,420	Coated	76	57
70	B-17	Med	24,000+	6,300	5,670	Clear	90	67
				5,985	5,387	Coated	86	64
100	BT-17	Med	24,000+	9,500	8,850	Clear	95	74
				8,800	7,920	Coated	88	69
150	BT-17	Med	24,000+	16,000	14,400	Clear	106	85
				15,000	13,500	Coated	100	80
250	E18	Mogul	24,000+	27,500	24,750	Clear	110	90
	BT-28			26,000	23,400	Coated	104	85
400	E-18	Mogul	24,000+	50,000	45,000	Clear	125	105
	BT-37			47,500	42,750	Coated	119	100
1000	E-25	Mogul	24,000+	140,000	126,000	Clear	140	128

Source: Data courtesy of GTE/Sylvania, Inc.

[a]All bases are screw type. Lamps of 50-, 70-, 100-, and 150-W ratings are available in either medium or mogul screw base.

[b]Based on operation on proper auxiliary equipment for 10 h or more per start.

[c]Conventional core-and-coil ballast.

[d]Use in open-bottomed fixtures or where glare is a problem.

tuted directly for existing mercury lamps. This gives considerable improvement in illumination and reduction in energy costs, since on the average HPS lamps will supply double the efficacy of mercury lamps. These direct mercury replacements are marketed under the trade names of E-Z Lux and Unalux by General Electric and Sylvania, respectively.

HPS lamps are available in clear and coated designs. The former is effectively a point source and, because of its extreme brightness, must be enclosed in a fixture. The latter is intended to substitute photometrically for coated mercury lamps and to constitute a lesser glare source, since lamp surface brightness is correspondingly reduced. Table 19.19 gives characteristics of typical HPS lamps that are presently available. Because of high efficacy, their energy-saving implications are obvious. Therefore, this lamp should be considered primarily, or as a substitute, for all existing or new HID applications. Its yellow color, which becomes less noticeable as the eye color adapts, can be made even more acceptable by using white sources in conjunction.

19.23 Low-Pressure Sodium (SOX) Lamps

This lamp, also referred to as SOX, produces light of sodium's characteristic monochromatic deep yellow color, making it inapplicable for general lighting. Because of its very high efficacy of over 150 lm/W *including* ballast loss (cf. HPS, Table 19.19), it can be applied wherever color is not an important criterion. Thus SOX is widely used for street, road, and area lighting, as well as for emergency or after-hours indoor lighting. Another desirable aspect of SOX lamps is their 100% lumen maintenance. This, coupled with the discharge lamp's typically long life (18,000+ h), makes SOX lamps the most economical source available today in terms of cost per million lumens produced.

19.24 Chromaticity

The CIE (Commission Internationale de l'Éclairage) color system is the internationally accepted standard for designating illuminant color. In this system the relative proportions of each of the three primary colors (red, green, blue) required to produce a given illuminant color are calculated, and the result is plotted on a standard chromaticity diagram (see Figure 19.60 in the next section). To calculate these proportions, measurements are made across the entire spectrum of the illuminant under test. These figures are then weighted in terms of the three primary colors. The resulting figures represent the proportions of red, green, and blue required to produce the spectrum color at that wavelength. These values are called the *tristimulus values* for that color and are designated by capital letters: X (red), Y (green), and Z (blue). The Y (green) value is also proportional to that color's luminosity.

19.25 Spectral Distribution of Light Sources

In addition to providing sufficient light of adequate quality, the lighting designer must also be concerned with the spectral content of the selected illuminant, since object color depends heavily on the illuminant. As shown earlier in Fig. 18.44*a*, perceived object color is the result of selective absorption and reflection of components of the illuminating light by the pigments of the object being viewed. It is therefore obvious that the *illuminant* must contain the color of the *object,* in order for us to see the object's color. It is not so obvious that the relative energy of an illuminant at a particular wavelength determines the saturation and brilliance with which we see a color. To understand this, refer to Fig. 19.59. In graphic form, the relative spectral energy of a few common light sources have been plotted, as a function of wavelength, that is, color. If we compare graphs 19.59*g* and 19.59*h*, which show the spectral content of two of the most common light sources, cool white and warm white fluorescent lamps, respectively, we note that the principal difference lies in the amount of blue and orange in their spectrum. Cool white has 1.3 blue units and 1.8 orange; warm white has 0.3 blue units and 2.7 orange. As a result, a blue object will be bright under cool white light and dull (grayed) under warm white, whereas

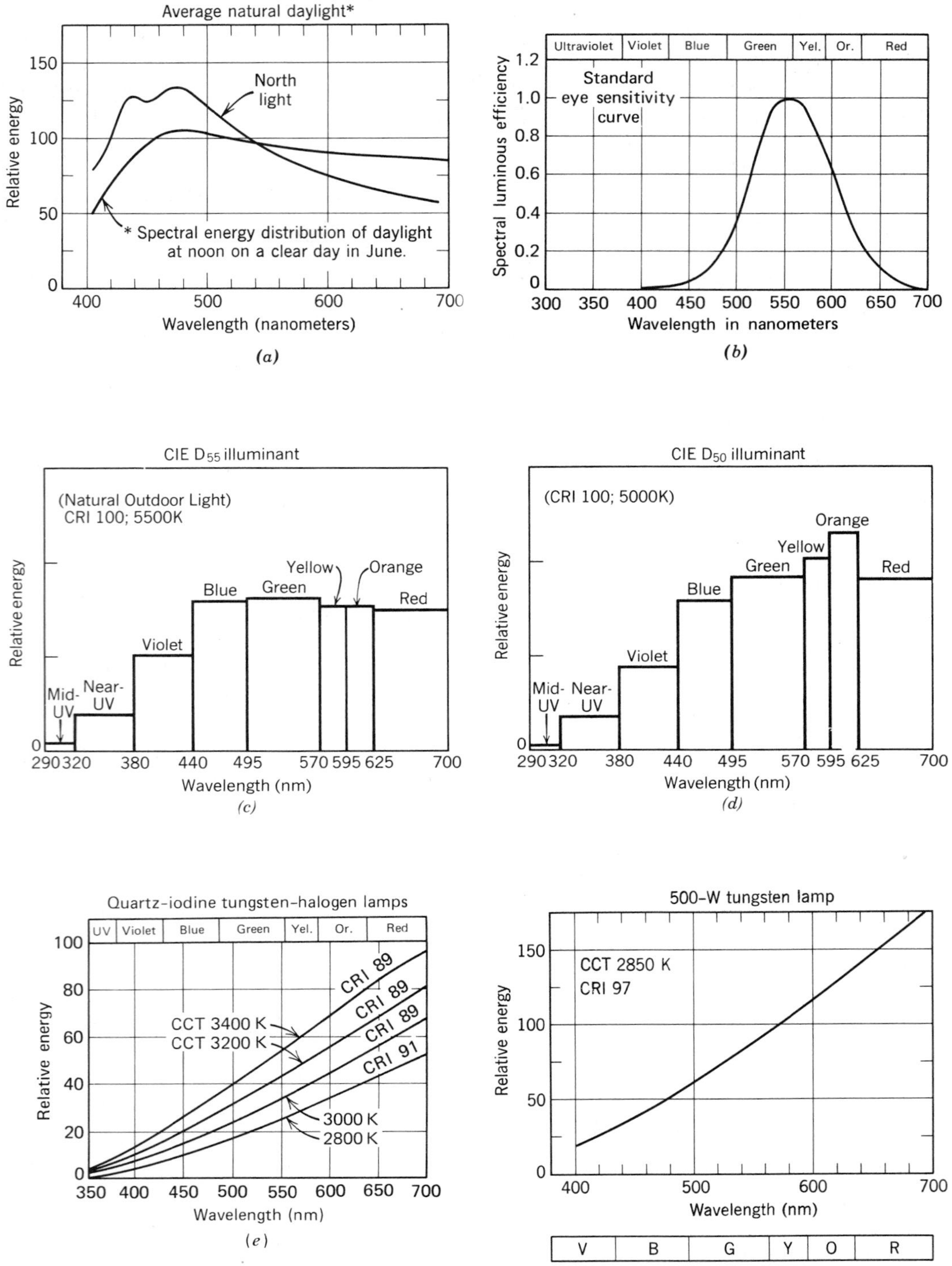

Fig. 19.59 Spectral energy distribution curves for typical illuminants, plus their correlated color temperature (CCT) and color rendering indices (CRI). Since only a radiating blackbody has a true color temperature, a source with mixed color illuminants is assigned a correlated color temperature, which is the temperature of a blackbody radiator whose chromaticity most nearly matches that of the light source. See Section 18.30.

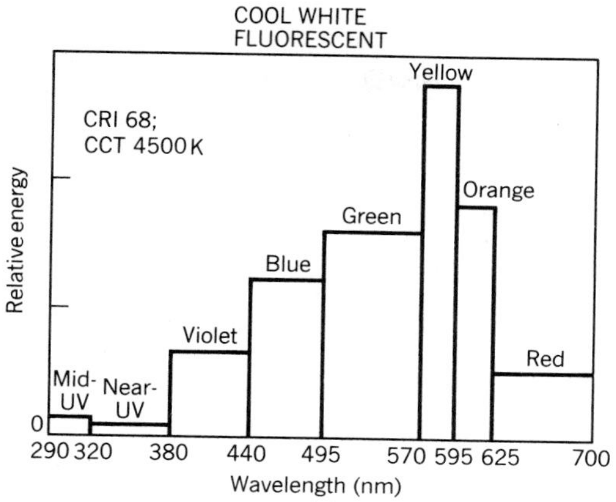
COOL WHITE FLUORESCENT
CRI 68;
CCT 4500 K
Yellow
Orange
Green
Blue
Violet
Red
Mid-UV
Near-UV
Relative energy
0
290 320 380 440 495 570 595 625 700
Wavelength (nm)

(g)

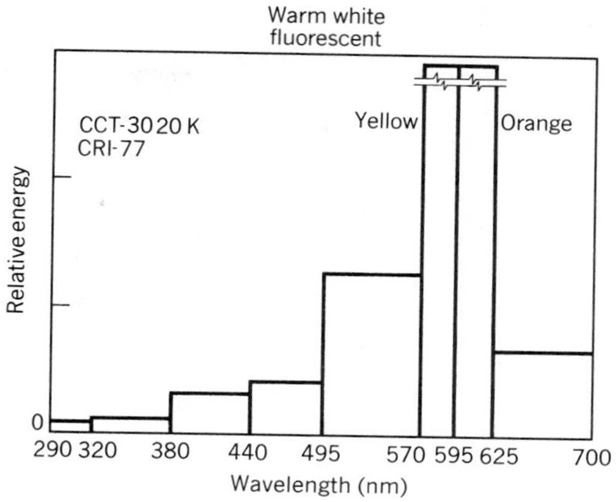
Warm white fluorescent
CCT-3020 K
CRI-77
Yellow
Orange
Relative energy
0
290 320 380 440 495 570 595 625 700
Wavelength (nm)

(h)

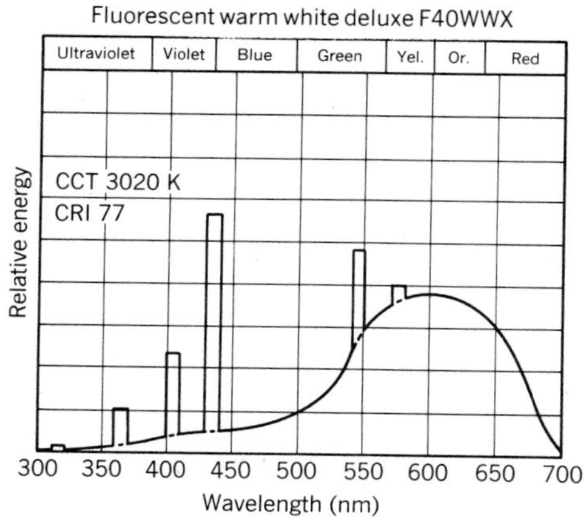
Fluorescent warm white deluxe F40WWX
Ultraviolet
Violet
Blue
Green
Yel.
Or.
Red
CCT 3020 K
CRI 77
Relative energy
300 350 400 450 500 550 600 650 700
Wavelength (nm)

(i)

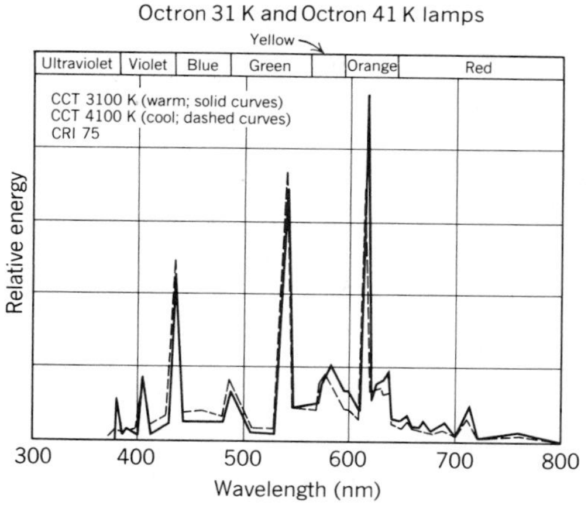
Octron 31 K and Octron 41 K lamps
Yellow
Ultraviolet
Violet
Blue
Green
Orange
Red
CCT 3100 K (warm; solid curves)
CCT 4100 K (cool; dashed curves)
CRI 75
Relative energy
300 400 500 600 700 800
Wavelength (nm)

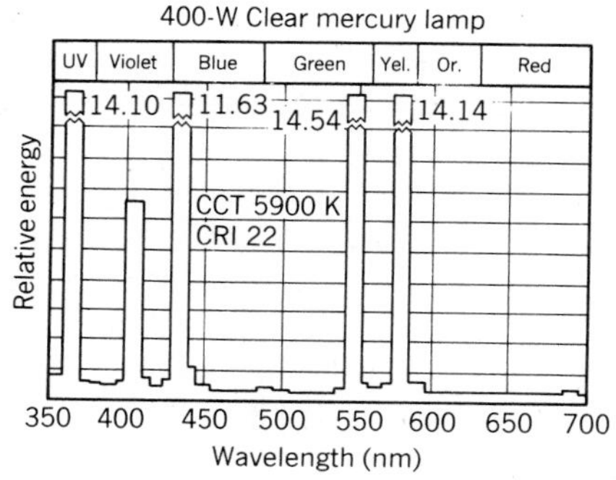
400-W Clear mercury lamp
UV
Violet
Blue
Green
Yel.
Or.
Red
14.10
11.63
14.54
14.14
CCT 5900 K
CRI 22
Relative energy
350 400 450 500 550 600 650 700
Wavelength (nm)

(k)

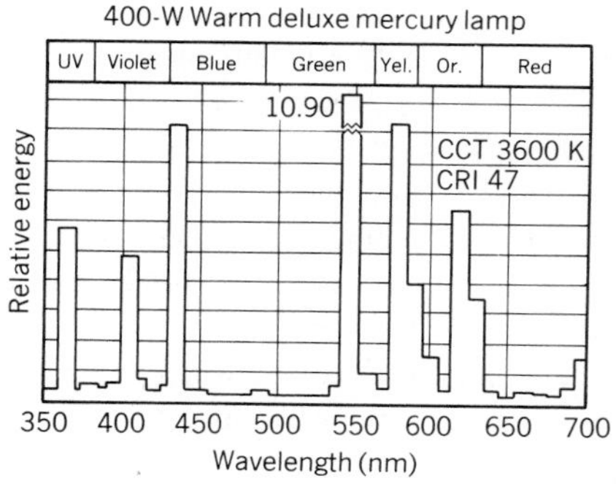
400-W Warm deluxe mercury lamp
UV
Violet
Blue
Green
Yel.
Or.
Red
10.90
CCT 3600 K
CRI 47
Relative energy
350 400 450 500 550 600 650 700
Wavelength (nm)

(l)

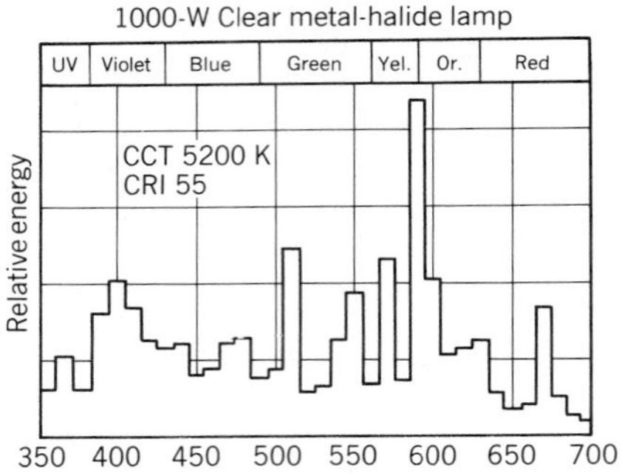

(m)

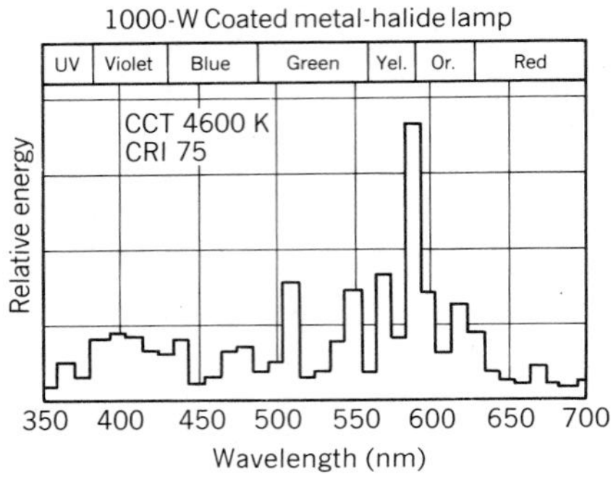

(n)

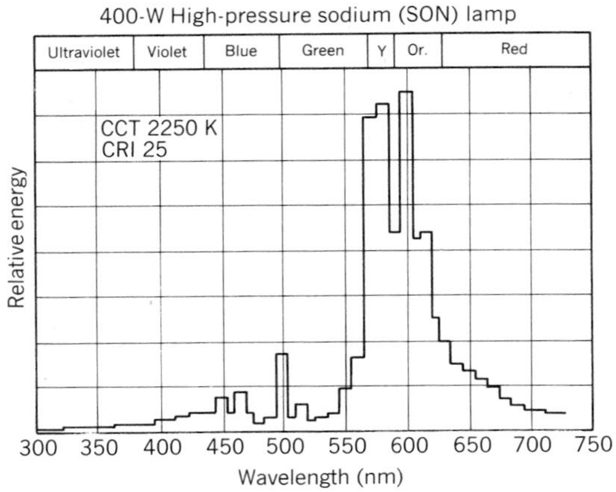

(o)

the opposite will occur with an orange object. The situation is more pronounced with the clear mercury lamp (Fig. 19.59*k*) as compared to the clear metal-halide lamp (Fig. 19.59*m*). A red object will be gray under the mercury lamp and unrecognizable as red, whereas under the metal-halide its redness will show clearly, and so on.

This concern for perceived object color, which relates not only to furnishings but also to paints and prefinished construction materials such as carpets or floor tile, is quite properly the province of the architect and lighting designer, who in turn must possess the necessary knowledge and information to make the appropriate choices of both illuminant *and* object color. Spectral composition graphs of the type shown in Fig. 19.59 are available from manufacturerers for all light sources, and they should be examined when considering the characteristics of a particular light source.

One of the best ways to compare illuminants is first to expose a dull white surface to the illuminants, side by side but separated by an opaque divider, to get an impression of the illuminate color, and then to expose a series of colored chips, again side by side, to see which colors are brightened and which are grayed. The intensity of illumination also influences the appearance of colors, and it must be considered in choosing object colors. As intensity is increased, reflection increases, particularly with pale tints (high value) that contain much white pigment and thus tend to wash out color. Therefore, with high-intensity lighting, saturation of colors should be high for true, brilliant color rendition.

Refer again to Fig. 19.59. Note from diagrams *a, e,* and *f* that the spectrum of a light source that produces light as a result of heating is continuous. Sunlight is equal in spectrum to a blackbody radiator at 5500 K; north light equal to one at about 8000 to 10,000 K; a 500-W incandescent lamp approximately equal to one at 2850 K; and so on. If the spectrum of a blackbody radiator is plotted on a chromaticity diagram, its locus is a continuous curved line as seen in Fig. 19.60. The chromaticity of all true blackbody radiators will fall exactly on this line, with the location depending on temperature. Daylight, for most purposes, falls on this locus, although because of selective atmospheric absorption and other phenomena, it is actually slightly off. Incandescent lamp chromaticity is very close to this locus because it is also a heat–light radiator.

A source that produces light by means of individual phosphors can also have chromaticity on this locus if the phosphors will, in effect, produce a continuous spectrum similar to that of a blackbody radiator. Thus we see that CIE standard illuminants (see Figs. 19.59*c* and 19.59*d*) have spectral components over the entire spectrum, yielding an exact equivalent of blackbody radiator at a particular color temperature. Therefore their color temperature and their correlated color temperature are the same. For other sources, their correlated color temperature is established by their chromaticity locus in relation to the diagonal lines crossing the blackbody locus, as in Fig. 19.60. Each of these lines is isothermal, that is, all chromaticities on it have the *same correlated color temperature*. Thus the reader can see readily that two sources with widely differing spectral content

ILLUMINATION

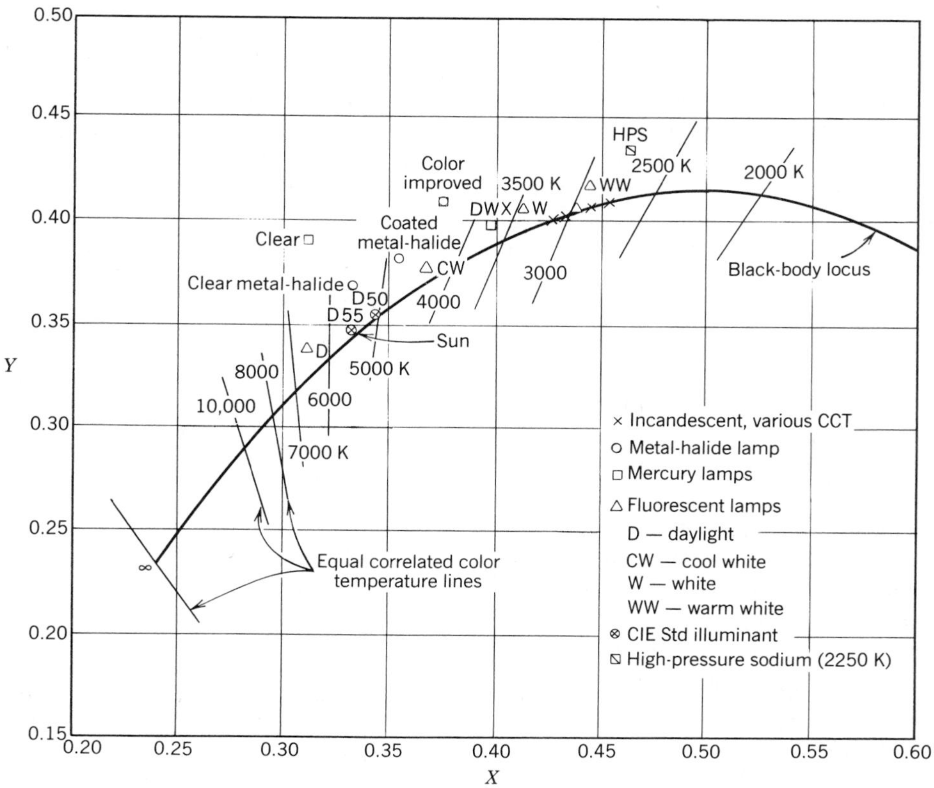

Fig. 19.60 *Basic chromaticity diagram showing relation of common illuminant chromaticities to that of the blackbody locus. Illuminants whose coordinates fall on the same line crossing the blackbody locus have the same correlated color temperature.*

and therefore object color rendering can have the same correlated color temperature.

19.26 Color Rendering (Index), CRI

Color rendering is defined as the degree to which perceived colors of objects illuminated by a test source conform to the colors of the same objects as illuminated by a reference source. The *color rendering index* (CRI) of a source is a two-part concept, comprising a color temperature that establishes the reference standard and a number that indicates how closely the illuminant approaches the standard. *The standard is always daylight at that color temperature*. Therefore, the color rendering index of a light is really a measure of how closely it approximates daylight of the same color temperature. Two lights cannot be compared unless their color temperatures are equal or quite close. A color rendering index of 100 indicates an illuminant whose spectral content is equal to daylight of that temperature. Color rendering indices for typical common lamps are given in Fig. 19.59.

Table 19.20 lists the color characteristics of a few of the major sources. An illuminant's own color appearance on a neutral surface will depend on its own spectral content, but if the observer is placed in a space illuminated with this source, the eye after a short exposure time will become adapted to the source color and will detect only a degree of whiteness rather than an actual tint.

Where it is necessary to detect small color differences between two objects, a light poor in object color, or complementary to the object color, should be used, at a relatively high illumination level, followed by a light high in object color, at the same illumination level. If this is not possible, two widely different but broad-spectrum illuminants should be used, preferably at the same illumination level. Another technique is the use of a special, fixed color source. For a full discussion, see the *IES Handbook*. It should be remembered in all considerations of color comparison, matching, and rendering that object color depends on the spectral energy distribution of the light source (illuminant), and therefore any change in the spectral content will change the object appearance. Two sources of the same color temperature and therefore apparent whiteness can have quite different spectral content and will therefore render object colors differently. A case in point would be a 3000 K warm white fluorescent tube and an incandescent lamp (500-W photoflood) of approximately the same color temperature. Color temperature is an expression of dominant color, not spectral distribution.

References

Ander, G. E., and Navvab, M. (1983). *Daylight Impacts of Fenestration Controls,* American Solar Energy Society, Washington, D.C.

Bryan, H. J. (1980). "A Simplified Procedure for Calculating the Effects of Daylight from Clear Skies," *Journal of the IES of North America,* April, pp. 142–151.

Daylight Design Diagrams (1963). Published by Service Division, Commonwealth Department of Labor and National Service, Melbourne, Australia.

Daylight, International Recommendations for the Calculation of Natural Daylight. Publication CIE No. 16 (E-3.2), 1970; Commission Internationale de l'Éclairage.

Farrell, R. (1976). "Calculating Direct Illumination from the Sky Under Clear Sky Conditions." *Journal of the IES of North America,* July, pp. 218–222.

Griffith, J. W., Arner, W. J., and Wenzler, (1957). "Practical Daylighting Prediction," IES National Technical Conference, Atlanta.

Harkness, E., and Mehta, M. (1978). "Solar Radiation Control in Buildings," Applied Science Publ., London.

Hopkinson, R. G., Longmore, J., and Graham, A. M. (1958). *Simplified Daylight Tables,* National Building Studies, Report No. 26, HMSO, London.

Hopkinson, R. G., Pethebridge, R., and Longmore, J. (1966). *Daylighting,* Heinemann, London.

"How to Predict Interior Daylight Illumination," Pamphlet, Libbey-Owens-Ford Co., 1976.

Jones, B. F. (1983). "Very Very Simple Hand Calculations for Daylighting," *Proceeding of 1983 International Daylighting Conference,* Phoenix, pp. 87–92.

Kimball, H. H., and Associates (1923). "Daylight Illumination on Horizontal, Vertical and Sloping Surfaces," *U.S. Weather Review,* Vol. 50, p. 615.

Krochman, J. (1963). "Über die horizontal Beleuchtungsstarke der Tagesbeleuchtung," *Lichtechnik,* Vol. 15, No. 11.

TABLE 19.20 **A Guide for Lamp Selection Based on General Color Rendering Properties**

Type of Lamp	*Lamp Efficacy (lm/W)*	*Lamp Appearance Effect on Neutral Surfaces*	*Effect on "Atmosphere"*	*Colors Strengthened*	*Colors Grayed*	*Effect on Complexions*	*Remarks*
			Fluorescent Lamps				
Cool[a] white CW	High	White	Neutral to moderately cool	Orange, yellow, blue	Red	Pale pink	Blends with natural daylight—good color acceptance
Deluxe[a] cool white CWX	Medium	White	Neutral to moderately cool	All nearly equal	None appreciably	Most natural	Best overall color rendition; simulates natural daylight
Warm[b] white WW	High	Yellowish white	Warm	Orange, yellow	Red, green, blue	Sallow	Blends with incandescent light—poor color acceptance
Deluxe[b] warm white WWX	Medium	Yellowish white	Warm	Red, orange, yellow, green	Blue	Ruddy	Good color rendition; simulates incandescent light
Daylight	Medium-high	Bluish white	Very cool	Green, blue	Red, orange	Grayed	Usually replaceable with CW
			Incandescent Lamps, Tungsten-Halogen				
Incandescent filament	Low	Yellowish white	Warm	Red, orange, yellow	Blue	Ruddiest	Good color rendering
			High-Intensity Discharge Lamps				
Clear mercury	Medium	Greenish blue-white	Very cool, greenish	Yellow, blue, green	Red, orange	Greenish	Very poor color rendering
Deluxe white[a] mercury	Medium	Purplish white	Warm, purplish	Red, blue, yellow	Green	Ruddy	Color acceptance similar to CW fluorescent
Metal-halide[a]	High	Yellowish white	Moderately cool	Yellow, green, blue	Red	Grayed	Color acceptance similar to CW fluorescent
High-pressure sodium[b]	High	Yellowish	Warm, yellowish	Yellow, green, orange	Red, blue	Yellowish	Color acceptance approaches that of WW fluorescent

Source: Courtesy of General Electric Co., Lighting Business Group.

[a]Greater preference at higher levels.

[b]Greater preference at lower levels.

Lim, Rao, et al. (1979). ''Environmental Factors in the Design of Building Fenestration,'' Applied Science Publ., London.

Lynes, J. A., Burt, W., Jackson, G. K., and Cuttle, C. (1966). ''The Flow of Light into Buildings,'' *Transactions of Illuminating Engineering Society (London),* Vol. 31, No. 3, pp. 65–91.

Millet, M. S., and Bedrick, J. R. (1980). *Graphic Daylighting Design Method,* LBL/DOE.

Millet, S. M., Bedrick, J. R., and Adams, C. (1980). ''Graphic Daylight Design Method,'' paper presented at International Symposium on Daylighting, Berlin, July.

Millet, S. M., Lakin, J. E., and Moore, J. (1981). ''Rainy Day Shading; The Effect of Shading Devices on Daylighting and Thermal Performance,'' University of Washington, Department of Architecture.

Modest, F. F. (1983). ''Daylighting Calculations for Non-Rectangular Interior Spaces with Shading Devices,'' *Journal of the IES of North America,* July.

Nakamura, H., and Oki, M. (1983). ''Calculation of Daylight Factor Dominated by Intermediate Sky,'' *Proceedings of International Daylighting Conference,* Phoenix, pp. 57–59.

Ne'eman, E., and Longmore, J. (1973). ''The Integration of Daylight with Artificial Light,'' paper presented at CIE Symposium, Istanbul.

Olgyay, A., and Olgyay, V. (1957). *Solar Control and Shading Devices,* Princeton University Press, Princeton, N.J.

Pierpont, W. (1983). ''A Simple Sky Model for Daylighting Calculations,'' *Proceedings of International Daylighting Conference,* Phoenix, pp. 47–51.

Ramsey, C. G., Sleeper, H. R., and AIA (1981). *Architectural Graphic Standards,* 7 ed., Wiley, New York, pp. 76–86.

Recommended Practice of Daylighting. (1962, current edition, 1979). IES of North America, New York.

Rennhackkamp, W. M. H. (1967). ''Sky Luminance Distribution on Warm Arid Climates,'' *Proceedings of 16th International Conference on Illumination,* Washington, D.C.

Selkowitz, S. (1982). ''Daylight Design Overlays for Equidistant Sun-Path Projections,'' LBL.

Selkowitz, S., Navvab, M., and Mathews, S. (1983). ''Design and Performance of Light Shelves,'' *Proceedings of the International Daylighting Conference,* Phoenix.

Szokolay, S. V. (1980). *Environmental Science Handbook,* Halsted/Wiley, New York, p. 100.

Thrun, E., and Jennings, R. (1981). ''Levels of Performance Related Illumination Available from Daylight in Typical Office/School Interiors,'' *Journal of the IES of North America,* October, pp. 7–14.

TNO Daylight Diagrams. Available from TNO Research Institute for Environmental Hygiene, POB 214, 2600 AE Delft, The Netherlands.

Treado, S., and Gillette, G. (1983). ''Measurement of Sky Luminance, Sky Illuminance and Horizontal Solar Radiation,'' *Journal of the IES of North America,* April, pp. 130–135.

Wieder, S. (1982). *An Introduction to Solar Energy for Scientists and Engineers.* Wiley, New York.

20

LIGHTING DESIGN

20.1 General

Lighting design is a combination of applied art and applied science. There can be many solutions to the same lighting problem, all of which will satisfy the minimum requirements, yet some will be dull and pedestrian while others will display ingenuity and resourcefulness. The competent lighting designer approaches each problem afresh, bringing to it a knowledge of current technology and years of background and experience, yet rarely being satisfied with a carbon copy of a previous design. And it is these years of background with their successful and not-so-successful designs coupled with a constant striving for improvements that are the characteristics differentiating the competent lighting consultant, designer, or engineer from the person who attempts to force each new job into the unwilling mold of a previous design.

Because of the large number of interrelated factors in lighting, no single design is the correct one, and for this very reason it is not entirely desirable to solve a lighting problem with a step-by-step technique. However, since this technique is a good avenue of approach for the uninitiate who lacks the experience necessary to view an entire solution, we have adopted it.

20.2 Goals of a Lighting Design

Simply stated, the goal of lighting is to create an efficient and pleasing interior. These two requirements, that is, the utilitarian and esthetic, are not antithetical as is demonstrated by every good lighting design. Light can and should be used as a primary architectural material. We elaborate on these goals below.

1. Lighting levels should be adequate for efficient seeing of the particular task involved. Variations within acceptable luminance ratios in a given field of view are desirable to avoid monotony and to create perspective effects.

2. Lighting equipment should be unobtrusive, but not necessarily invisible. Fixtures can be chosen and arranged in various ways to complement the architecture or to create dominant or minor architectural features or patterns. Fixtures may also be decorative and thus enhance the interior design.

3. Lighting must have the proper quality as discussed previously. Accent lighting, directional lighting, and other highlighting techniques increase the utilitarian as well as architectural quality of a space.

4. The entire lighting design must be accomplished efficiently in terms of capital and energy resources, the former determined principally by life-cycle costs and the latter by operating energy costs and resource-energy usage. Both the capital and energy limitations are, to a large extent, outside the control of the designer, who works within constraints in these areas. Obviously, these constraints are maxima.

With these goals before us we can write a lighting design procedure, keeping in mind that the order of steps shown is not necessarily the same in each lighting problem and that, since all of the factors are closely interrelated, it is often necessary to apprehend several of the stages simultaneously before arriving at a decision.

20.3 Lighting Design Procedure

For the benefit of the reader whose background and study framework is purely architectural, it may be helpful to draw parallels between the purely architectural approach to lighting design and the more structured approach described below. The terminology used in the text *InsideOut* (see Brown et al., 1982) is typical of an architect's approach. There the authors speak of a four-stage plan: conceptualizing, scheming, developing, and finalizing. The *conceptualizing* stage is general and

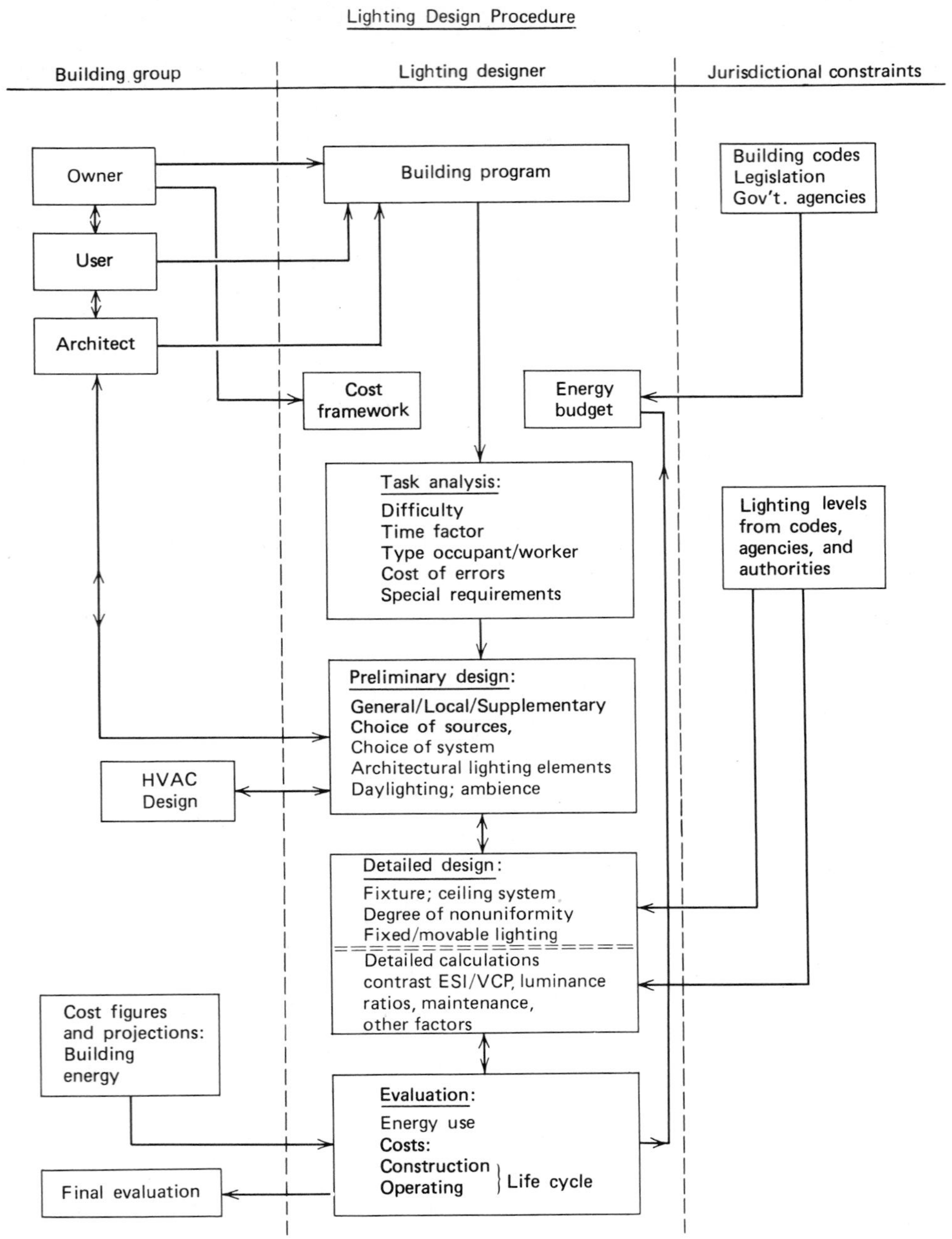

Fig. 20.1 Lighting design procedure chart.

overall; it considers the occupant in the visual environment, and his visual relation to task, immediate environment, and surroundings. It corresponds, in the flowchart of Fig. 20.1, to the background thought and ideas for the two boxes labeled "task analysis" and "preliminary design." *Scheming* is the on-paper preliminary design stage; 1/16- or 1/32-in. scale, corresponding to parts of the "preliminary design" and "detailed design" stages. *Developing* is essentially the detailed design stage, with input from (the boxes on Fig. 20.1 labeled) costs, energy, codes, regulations, and other disciplines. *Finalizing* is comparing, checking, evaluating, and comparing design results to goals and requirements.

(a) Project Constraints. *Refer to Fig. 20.1.* This flowchart, which represents the design procedure and its interactions, should be referred to throughout the necessarily lengthy discussion that follows, in order to maintain perspective. It is important that the reader be aware of job constraints and of the interactions between the lighting designer and the remainder of the design group. We deliberately emphasize this to demonstrate the interdisciplinary nature of lighting design in general and its particular connection with HVAC and daylighting (fenestration). This approach, which is most often referred to as the systems design approach, will be followed throughout the discussion.

Item 4 of the list in Section 20.2 referred to constraints. These can alternatively be classified as the owner-architect-user team and the jurisdictional authorities. In some detail these are:

1. *Owner-designer-user group*. The owner establishes the cost framework, both initial and operating. A part of both of these may be a rent structure, which in turn determines and is determined by the space usage. If the owner is also the occupant, the cost factors change somewhat but remain in force. The architect determines the amount and quality of daylighting and the architectural nature of the space to be lighted. Many of these data are detailed in the building program. Obviously the architect and lighting designer (who may be one person) should interact in this aspect of building design.

2. The jurisdictional authorities *may* include:

FEA—Federal Energy Administration

GSA—General Services Administration

NCSBCS—National Conferences of States for Building Codes and Standards

ASHRAE—American Society of Heating, Refrigerating and Air Conditioning Engineers

IES—Illuminating Engineering Society

NBS—National Bureau of Standards

Most of these are jurisdictional by reference, that is, the actual authorities will specify that the lighting system meet the requirements of ASHRAE, IES, and so on. If public money is involved, FEA/GSA standards will probably be involved. The principal area of involvement is that of energy budgets and lighting levels, both of which affect every aspect of lighting design including source type, fixture selection, lighting system, fixture placement, and even maintenance schedules. For this reason the first step in the lighting design procedure is to establish the *project lighting cost framework and the project energy budget*.

(b) Task Analysis. As shown in Fig. 20.1, this step essentially determines the needs of the task. Factors to be considered in addition to the nature of the task are its repetitiveness, variability, who is performing it (i.e., condition of the occupant's eyes), task duration, cost of errors, and special requirements. Several of these factors have been discussed in the preceding sections dealing with quality of light. The reader will be referred to the appropriate sections in the following analysis.

(c) Design Stage. This is the active consideration stage during which detailed suggestions will be raised, considered, modified, accepted, or rejected. This is also the most interactive stage as is clearly seen from Fig. 20.1. At its completion, a detailed, workable design is in hand. The critical interactions here are with the architect in daylighting and with the HVAC group in power loads. The former may result in relocating a space within the building, the latter in making a change in a lighting system or HVAC system. In brief, this stage consists of the following steps:

1. Select the lighting system. Select type of light source and distribution characteristic of fixture(s) or area source and consider effects of daylighting, economics, and electric loads.
2. Calculate the lighting requirements. Use the applicable calculation method and establish the fixture pattern, considering the architectural effects.
3. Design the supplemental decorative and architectural lighting.
4. Review the resultant design. Check the design for quality, quantity, esthetic effect, and originality.

(d) Evaluation Stage. With the design on paper, it can now be analyzed for conformance to the principal constraints of cost and energy. If the design stage has been carefully accomplished, with due attention to these factors, the result of the final evaluation should be gratifying. The results of this stage are fed to the architectural group for use in the final overall project evaluation. In the following sections we consider in detail each of the steps in the design procedure.

20.4 Cost Factors

This is a particularly difficult item for a novice lighting designer because it requires experience in the field and an acquaintance with commercially available equipment. Also, the inevitable tradeoffs between first cost and operating cost cannot intelligently be made unless the cost structure is clearly understood. The following guidelines should be of considerable assistance both in avoiding unpleasant surprises when a job is estimated and in preparing cost analyses:

1. Decide at the outset what cost criteria will be applied, that is, the relative importance of first cost, operating costs, annual owning costs, and life-cycle costs.

2. Tradeoff decisions are required between first cost and operating costs. For example, incandescent lamps and fixtures are low in first cost and high in operating cost, and so on (see Table 19.17). Dimming and control equipment falls into this area of decisions.

3. Manufacturer's catalog items are *always* cheaper than specials, and can be priced more readily.

4. Compare annual owning costs of two systems or methods. Conversion of these data to life-cycle cost comparisons is straightforward. With the continued instability in energy costs, it is suggested that even in an *annual* owning cost comparison, two different energy costs be used and the impact of a sharp increase studied.

5. The impact of lighting energy on the operating cost of the entire building must be studied, and the apportionment of costs determined. The only practical means of accomplishing this is by computer program. Programs can be readily adjusted to reflect the impact of the lighting system on building costs, and in particular on HVAC first cost and operating costs. It is incorrect to artificially separate the lighting system from the HVAC system with which it intimately interacts.

The lower the lighting system's energy level, the lower the building's overall operating cost. The argument that heat from a lighting system is fully utilized to heat the building and is therefore not wasted is a specious one, which has been refuted on many counts:

HVAC system first cost is higher.

HVAC year-round cost is higher.

Lighting energy cost is higher.

Life-cycle costs are higher.

Energy resource use is higher.

This subject is discussed in detail in Section 21.22.

20.5 Power Budgets

The requirement to establish a project's lighting power budget in accordance with a specified procedure has now been incorporated into the building codes of many states.

The purpose of this budget determination procedure is not to dictate design procedure. Indeed all standards explicitly so state. Instead, the purpose is to develop an overall maximum power budget *within* which the designer is free to do as he or she wishes. Obviously, prodigality in one area is necessarily at the expense of another area, since maximum power is inflexible, and the entire budget is built on reasonable design techniques. Still, there is enough leeway in the budget and enough exceptions so that the designer is not overly restricted. It is suggested that the reader obtain and scan through the ANSI/ASHRAE/IES and DOE material at this point [see below], but reserve detailed study of their application guidelines, since they contain references to terminology and techniques with which he or she will not be familiar until completion of the chapter. However, this material is presented here to maintain the proper order of events in the design procedure.

The standards that define the establishment of a lighting power budget are:

ANSI/ASHRAE/IES 90A, *Energy Conservation in New Building Design* (current edition, 1980), Section 9.

IES Standard LEM-I, *Recommended Proce-*

dure for Lighting Power Limit Determination (current edition, 1982).

Proposed Energy Conservation Standards for New Commercial Buildings, Department of Energy (DOE), 1983.

The IES standard, also called the unit power density (UPD) procedure, utilizes precalculated unit power densities in watts per square meter (foot) for specific types of visual task areas and contains an extensive table of values. The DOE standard uses a similar approach, except that its tabulated unit power densities are generally lower than those of the IES. See Table 5.4 for the DOE list.

The procedure by which the unit power densities are calculated is given in the ASHRAE standard. Simply stated, it assumes minimum coefficients of utilization, lamp efficacy, and light loss factor according to stated criteria and then calculates the maximum power density as:

$$PD = \frac{P}{A} = \frac{E}{CU \times LE \times LLF}$$

where

P = lighting power budget of a space, in watts
PD = lighting power density in watts per square meter [ft^2]
E = illuminance in lux [fc]; see Table 18.5
A = area in square meters [ft^2]
CU = coefficient of utilization (see Section 20.27)
LE = light source efficacy
LLF = light loss factor (see Section 20.30)

Since the room's area is unknown, and since the coefficient of utilization depends on area, an "area factor" is included in the UPD list (see Table 5.4, Part B).

Thus the designer can establish the proposed facility's lighting power budget by using the tabulated UPD values, and where these are insufficient or unavailable, he can calculate the required UPD by using the following criteria, which are excerpted from the ASHRAE standard. For recommended illuminance values, see Table 18.5.

Coefficient of Utilization (CU)

	Room Cavity Ratio						
Task Criteria	*0*	*1*	*2*	*3*	*4*	*5*	*6*
For spaces with tasks subjected to veiling reflections and where visual comfort is important	.55	.55	.50	.45	.40	.36	.32
For spaces without tasks, or with tasks not subjected to veiling reflections, but where visual comfort is important	.63	.63	.57	.51	.46	.41	.37
For spaces without tasks and where visual comfort is not a criterion	.70	.70	.63	.57	.51	.46	.41

Lamp Efficacies (LE) to Be Used for Applications[a]

Efficacy Criteria	*Efficacy (lm/W)*
Where moderate color rendition is appropriate and for all outdoor spaces	55
Where good color rendition is appropriate	40
Where high color rendition is appropriate, and total space is less than 4.6 m² (50 ft²)	25

[a]Based on initial lumen output and including ballasts losses.

Light Loss Factor (LLF) for Task/Areas

Clean	*Light Dirt*	*Medium Dirt*	*Dirty*	*Very Dirty*
.70	.65	.60	.55	.50

ILLUMINATION

Because of the multiplicity of standards and guidelines, the lighting designer may find a contradiction between authorities. In such an event, written clarification from the appropriate authorities is advisable. Lest the designer feel that these power maxima place excessive limitations on design choice, it should be pointed out that

1. The IES/ASHRAE technique does not consider daylight contributions at all. This is an area that can be used to great advantage by the lighting designer to achieve flexibility and ''breathing space'' in other areas.
2. Exceptions to the budget criteria include residential, theater, and exhibition spaces. Since these are specifically areas that utilize low-efficiency sources, no difficulties should be encountered in their lighting design.

By applying knowledge gained in these pages, and in actual design, low-wattage, high-lighting-quality designs can be accomplished that will readily meet the IES/ASHRAE power limitations. Then, proper application of lighting energy management techniques (see Sections 21.2–21.4) will result in a low-energy-use structure.

20.6 Task Analysis

Refer to Fig. 20.1. This is the stage at which the quantity and quality of lighting required for the tasks are decided. The factors affecting this choice as shown in the figure are difficulty, time factor, occupant, cost of errors, and special requirements.

(a) Difficulty. The components of visual difficulty were discussed at length in Sections 18.16 to 18.21, and the results in terms of lighting levels appear in Tables 18.3 to 18.8. Essentially, the designer will examine the type of task involved, and after determining the applicable authority he will select the required illuminance.

In the absence of specific instructions to the contrary, the designer will use IES recommendations. If there ar several tasks to be performed at the same point, the most difficult, subject to time considerations, will be selected; that is, if the more difficult task occurs infrequently, it may be reasonable to provide supplementary portable lighting or even to suggest moving to another brighter location. If it is the major task, lighting should be based on it and provision made for intensity reduction for less demanding work.

Variation in task difficulty is particularly common in spaces in public buildings. Thus a school gym can be used for athletics, band concerts (despite the acoustics), and town meetings—three totally disparate lighting requirements. In these and similar instances it is common practice to treat the space as essentially three different spaces and to design lighting for each with a careful eye to maximum common equipment usage. Similar problems are encountered in basements, multipurpose rooms, and conference/meeting/lecture/exhibition rooms. Fortunately, most such spaces do not have severe seeing tasks.

The task variation referred to here is the variation that occurs in one very specific location and is not to be confused with task variation in an area, however restricted. Thus a small private office of, say, 8 × 8 ft has a desk, file cabinet, and circulation space, three tasks of differing but constant difficulty in one small space. The corresponding lighting for these is also fixed and varies with the task severity. The values listed in Tables 18.3 to 18.8 represent the required illumination on the surface in question whether horizontal, vertical, or in between. Since the flux method of calculating illumination normally yields the 30-in., horizontal-plane illumination level, it is necessary to be cognizant of the ratio of horizontal to vertical illumination for various lighting systems. This ratio is approximately 3:1 for narrow-distribution direct and semi-direct lighting; 2.5:1 for wide-distribution luminaires of the same type; and 1.5:1 for indirect, general diffuse lighting.

Since the illumination values listed assume adherence to both recommended luminance ratios and reflectances (see Section 18.28), it is necessary to select, in conjunction with the interior designer, finishes and reflectances for surfaces within the area. If, for instance, in a private office a dark wall finish of 10% reflectance is chosen, it will be necessary for the lighting designer to compensate for this by additional wall lighting to maintain the recommended maximum 10:1 brightness ratio (see Table 18.10 and the discussion of point f in Section 20.7). The atmosphere created by vertical surface luminances will be discussed below. Table 20.1 lists reflectances of some common interior paint finishes.

TABLE 20.1 **Approximate Reflection Factors**

Medium Value Colors	
White	80–85
Light gray	45–70
Dark gray	20–25
Ivory white	70–80
Ivory	60–70
Pearl gray	70–75
Buff	40–70
Tan	30–50
Brown	20–40
Green	25–50
Olive	20–30
Azure blue	50–60
Sky blue	35–40
Pink	50–70
Cardinal red	20–25
Red	20–40

(b) Time Factor. As discussed in Sections 18.20 and 18.22, the length of time for which the task must be accomplished is important in difficult work. Beginning with moderately difficult tasks, that is, luminance of about 50 fL [170cd/m^2], prolonged intensive application, or rapidly changing tasks, would require illumination to be raised one (or more) levels. Alternatively, quality could be improved by increasing daylight or task contrast.

If, on the other hand, this worker readily learns the work and it becomes routine, illumination can be reduced one level because, effectively, the degree of difficulty has been reduced.

(c) Occupant. Since ordinarily the age and other specific characteristics of the worker are not known, a standard distribution is assumed and the recommendations as tabulated take account of this. On the other hand, if there is a high percentage of older workers, as is the case in certain industries, lighting should be raised one level. This compensates for inability of the eyes to accommodate and for the tendency to tire easily.

(d) Cost of Errors. This involves an economic tradeoff between cost of improving seeing accuracy as against the cost of the improved lighting. Performance can be brought close to perfection, but the cost of so doing increases much more rapidly than the proportional increase in performance. Standard lighting should provide 90% ± work accuracy. Thus this step is basically an economic calculation, the criteria for which must come from the owner or user. Tasks in which this problem is encountered include all types of inspection, proofreading, textile matching, very fine machining, and jewelry manufacturing.

(e) Special Requirements. These include any nonstandard task lighting requirements. Some of these are specific illuminant color, directionality for shadowing, and reflections as required for inspection, polarization, and controlled variations, as required in a space with varied tasks or varying daylight factor. In addition to these the physical dimensions of the task often create special requirements of their own. We tend to assume a small object in the horizontal plane, since that is the normal office task. However, there are exceptions such as a drafting board, a large machine, an inspection bench, or a cutting table. Consequently, these special requirements arise:

1. *Large tasks.* In large tasks, the angle of seeing changes from 20 to 70° from the vertical, resulting in radically changed glare angles and reflection from the task.
2. *Three-dimensional tasks.* These tasks shadow themselves, particularly when containing undercuts and reveals. An architect's model shop presents such tasks. When it is necessary to see into an opening, an intense narrow beam is required.
3. *Tools.* Tools cast shadows below and in front when lighted from above and behind. A fabric cutter must see ahead of and below the cutting machine.
4. *Nonhorizontal tasks.* These must be calculated for the plane in which they stand. As stated earlier, the ratio between horizontal and vertical illumination varies between 1.5:1 to 3:1, depending on the system. Task lighting requirements are stated in the plane of the task. This can have a pronounced effect on the lighting system selected and its arrangement.
5. *Task observed from various positions.* There are instances where a fixed task is observed from several angles, such as a drawing in a conference room or a wall display. Illumination must be adequate for all viewing angles.

PRELIMINARY DESIGN

20.7 Energy Considerations

Energy considerations must pervade every aspect of the design process. Some background material is in order here, to place the lighting energy subject in proper perspective. Best estimates indicate that as of 1985 lighting consumed approximately 25% of the electric power generated in the United States. In terms of *resources* this amounts to approximately 4 million barrels (bbl) of oil per day. From these estimates, the usage by occupancy is approximately:

Residential—20%

Industrial—20%

Stores—20%

Schools and offices—15%

Outdoor and other—25%

In commercial buildings lighting consumes about 20 to 30% of the building's electric energy, more in residences and less in industrial facilities. By judicious design a reduction of 40 to 50% in lighting energy is attainable. Translated into resources, this reduction can readily amount to more than 1 million bbl of oil per day. Few will disagree that such a goal is well worth the effort. Every watt per square foot reduction in lighting energy results in at least 1.25 W/ft^2 [m^2] savings in air-conditioned buildings. It has been demonstrated by actual designs that offices and schools can be *well* lighted with less than 2.5 W/ft^2 [27 W/m^2]. The question to be answered then is: What design guidelines can be followed to effect this energy-conscious design? In 1983 the U.S. Department of Energy drafted a set of guidelines for energy-efficient lighting for new construction. They are presented below in a group, followed by more detailed recommendations. The same recommendations will also be found at appropriate points throughout the description of the lighting design process.

1. *DOE guidelines for energy-efficient lighting:* Use DOE recommended lighting power densities; see Table 5.4.

Use dimmers in any area over 400 ft^2 with a power density greater than 1.0. The dimmers should be able to reduce general lighting by at least 50% (for periods of reduced lighting demand). Where possible, use task lighting, daylight and daylighting controls, more efficient lamps and fixtures, and lighting units with heat-removal and heat-recovery capabilities.

In all rooms where use can be made of daylight, design lighting systems so that lighting units in portions of the room where daylight is available can be controlled separately from the rest of the room. At a minimum, the row of lighting units nearest the source of daylight should comply. In general, an area twice the window height from all windowed walls should be considered for daylighting control.

Four-lamp and three-lamp fluorescent lighting units with multiple ballasts should have at least two separate control devices.

Install separate control devices where spaces are served by more than one lighting circuit. Each should be accessible from within the space.

Install individual lighting controls for each partitioned room.

2. IES Standard LEM-3, *Lighting Design Criteria for Effective Energy Utilization;* provides the reader with appropriate criteria for energy-conscious lighting design.

(a) *Design lighting for expected activity.* This is the task lighting approach. It is wasteful of energy to light any surface to a higher level than it requires. Since many spaces contain varied seeing tasks, nonuniform lighting is recommended. One way to accomplish this for areas where exact furniture layout is not available is to use readily movable fixtures. Providing overall high-level illumination with provision for switching to reduce levels is not advisable because of increased first cost and the psychological impetus to operate at maximum levels. Another solution is fixed fixtures for general low-level lighting and supplementary task lighting. Other factors and techniques to be borne in mind are:

(1) Grouping of tasks with similar lighting requirements.

(2) Place most severe seeing tasks at best daylight locations.

(3) Heat-removal fixtures (air troffers) increase efficiency of the units 10 to 20% but make the fixtures immobile when ducted. Consider venting into ceiling plenum.

(4) Advantages of nonuniform lighting increase as the space between work stations increases.

(b) *Design with effective, high-quality, efficient, low-maintenance, thermally controlled luminaires. Effective* means providing useful light, with a high ESI component, and minimum direct glare. In cases where much of the viewer's time is spent in a head-up position, as in schools, or where the viewer can compensate for veiling reflections, the decision should lean toward high VCP. (See Section 18.25.) Where work and viewing position are fixed and most of the viewer's time is spent head down, the decision should lean toward low reflected glare. (See Sections 18.26 and 18.27).

(1) *A high-quality* luminaire is made with permanent finishes such as Alzac or multicoat baked enamel, so that its performance after 8 to 10 years of service will be comparable to the original.

(2) An *efficient* luminaire is one with a high CU (coefficient of utilization; see Section 20.27), that is, minimum light loss within the fixture.

(3) A *low-maintenance* luminaire remains clean for extended periods and is designed so that all reflecting surfaces can be easily and rapidly cleaned, without demounting. Enclosed fixtures should be gasketed. Nongasketed units collect and retain dust and cause rapid output depreciation. Relamping should be simple and rapid, to encourage group relamping programs that are energy efficient and cost effective. A 20% increase in maintained light is possible if lamps are replaced at the end of their *useful* life, that is, when output is down to 70% of initial maintained lumens, and fixtures are cleaned and maintained on a fixed schedule. No cost tradeoff is generally involved, since periodic maintenance and relamping is normally cheaper than one-at-a-time maintenance and burnout replacement, *and* yields 20% higher average lumens. A tradeoff is involved between the additional cost of more frequent maintenance and the energy savings produced. Fixtures in relatively inaccessible locations such as high ceilings must be designed for low maintenance, and maintenance should be on a fixed schedule. Tradeoff here is between higher cost of low-maintenance units, and high maintenance costs. Use of high-lumen-maintenance sources is assumed.

(4) A *thermally controlled luminaire* is one that controls the heat generated by the light source. This item depends to a large extent on the type of HVAC system, the lighting heat load, and the type of fixtures employed. Detailed analysis of this point involves HVAC considerations and the overall impact of lighting energy on the building. However, remember that the method of fixture installation controls, in large measure, the utilization of lighting heat. See Fig. 20.2. As noted above, return-air troffers raise light output by 10 to 15% (by lowering bulb temperature), in addition to reducing the a-c load.

(c) *Use efficient light sources and accessories.* This point is self-explanatory. Tradeoffs involved here are:

(1) Between first cost and life-cycle costs.

(2) Between desired illuminant color and efficiency. Some of the highest efficacy sources have less than ideal color but, since the eye adapts quickly, off-white color, as from HPS lamps, should be considered for indoor uses.

(3) Between light control and efficiency. Fluorescent sources, which are highly efficient, do not lend themselves to good beam control and are principally useful for area coverage.

(4) Between architectural considerations and efficiency. Fluorescent sources are ef-

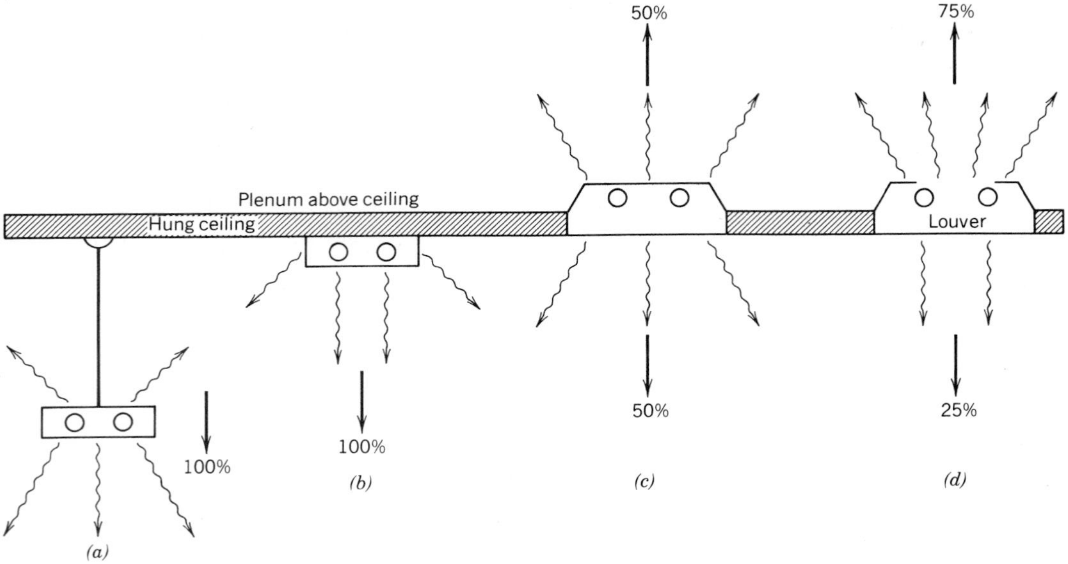

Fig. 20.2 Method of fixture installation controls the transfer of lighting heat. (a) *Suspended units contribute all their heat to the space, and themselves remain cool.* (b) *Surface-mounted fixtures also place all their heat in the space, but because of blocked transfer upwards, run hot.* (c) *Static recessed units transfer about 50% of their heat to the plenum whereas louvred units* (d) *transfer about 75%. When ducted, heat transfer up can be as high as 85%.*

ficient and have good color but require a large fixture, which tends to dominate the space. A possible compromise is the U-shaped lamp in a modular ceiling (see Fig. 19.49).

Figure 20.3 will assist the designer in determining roughly what the various sources represent in terms of watts per square foot [meter] load. See also Fig. 19.1 for a source efficacy comparison. Thus, with a target of 2 to 2.5 W/ft^2 for office-building lighting, the use of incandescent downlights is obviously severely restricted.

Spill light and borrowed light are often neglected sources. Glass in upper wall sections can provide sufficient corridor lighting from borrowed office lighting. Sources with high lumen *maintenance* such as tungsten-halogen and high-pressure sodium should be given preference.

In the category of accessories, ballasts are the largest energy users. For fluorescent lamps use electronic or low-loss conventional ballasts. Avoid single-lamp ballasts as they are less efficient than multilamp types. See Sections 19.17 and 19.18.

For HID sources, match ballast to the lamp, use solid-state types where available, and use the less efficient regulating ballast only where required.

For all sources, use high-power-factor ballasts only.

(d) *Use daylight, properly.* Daylight must be considered as a regular light source subject to weather variations and time of building use. Obviously, a three-shift industrial plant cannot use daylight on all shifts, but it can for at least one shift, and design should reflect this fact. Part of proper daylight design is control of window brightness. Excessive brightness causes severe and even disabling glare. A corollary of excessive brightness is excessive heat gain. Both are manageable with common control devices, manual and automatic.

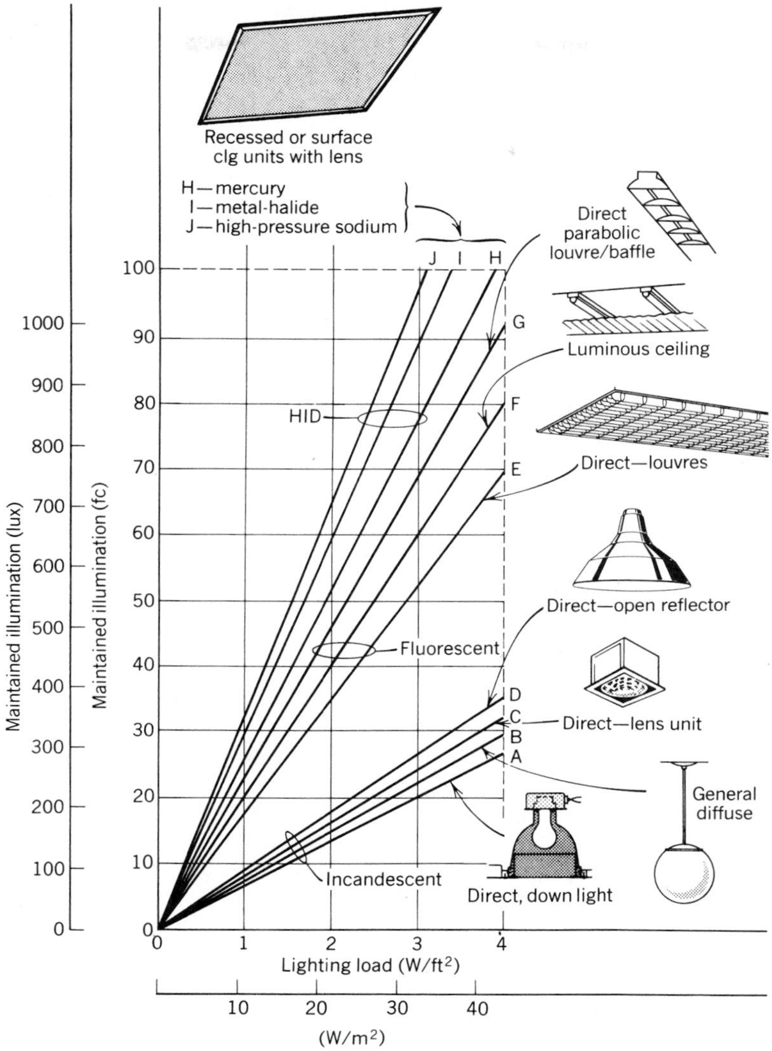

Fig. 20.3 Estimating chart for lighting load and related illumination levels for different light sources. The chart was calculated for a fairly large room—approximately classroom size. Although all the figures are necessarily approximate because of the variables involved, the chart gives figures close enough for a first approximation. Notice the increase in output as the sources change from incandescent to fluorescent to HID.

(e) *Use energy-efficient lighting control strategies*. The subject of lighting control, including manual and automatic switching, dimming, sensing, and intensity control is covered in Sections 21.2 through 21.4, for both new installations and retrofit work. Proper design of controls can reduce energy consumption over a noncontrolled installation by as much as 60% without reducing lighting effectiveness. The economic trade-off in new construction is between the initial cost of control equipment on the one hand and the initial lighting equipment savings plus annual energy savings on the other hand. Proper decisions require life-cycle cost analysis. See Appendix E, Economic Analysis.

(f) *Use light finishes on ceilings, walls, floors, and furnishings*. This point is self-explanatory and is examined in a number of sections. A brief summary of recommendation ranges would be:

Ceilings—80 to 92%

Walls—40 to 60%

Furniture, office machines, and equipment—25 to 45%

Floors—20 to 40%

In addition to higher illumination levels in the room, high reflectances minimize uncomfortable luminance ratios, as between fixture and upper wall or task and background. With respect to these ratios (see Table 18.10):

(1) Between task and near surround—aim for 3:1, accept 5:1.
(2) Between task and immediate area—aim for 10:1, accept 20:1.
(3) Between luminaires and their background—aim for 20:1, accept 40:1.
(4) Anywhere in the normal field of view—aim for 40:1, accept 80:1.

Note that the targets themselves are maximums and acceptance of values above maximum should only be for very good reasons.

20.8 Preliminary Design

Again referring to Fig. 20.1, the preliminary design phase is the time during which ideas crystallize, but in terms of areas and patterns, as well as light and shadow, not yet in terms of hardware. At this stage the quality of the system is decided on, that is, the luminances, diffuseness, chromaticity, and proportion of vertical to horizontal lighting are determined. The latter establishes in large measure the room ''mood'' or lighting ambience. In preceding sections these items were discussed in some detail. In the sections that follow on lighting systems (direct, indirect, etc.), the quality of each will be considered and applications suggested. In the overall view, however, the ultimate quality of the lighting system, its visual pleasantness, centers of visual attention, highlights and shadows, as well as texture and forms, will be a deft and perhaps artistic combination of the above considerations, and will establish, as the term implies, the quality of the lighting design. A few observations, not covered elsewhere, are mentioned below.

Planes other than the ''working plane'' must always be considered. The ratio of vertical to horizontal illumination of the chosen lighting system will determine wall luminance, while the floor finish will have a pronounced effect on the ceiling illumination for direct lighting systems.

The chromaticity of the room lighting depends primarily on the source but secondarily on the luminaire and surface finishes. A ''white'' source can be tinted slightly by the use of a colored reflector in the luminaire. Of course, the effect on luminaire output of such a change must be considered. In the case of semi-indirect and indirect lighting this same effect can be accomplished by the use of colored ceiling and upper wall surfaces, which serve as secondary reflectors and become the actual luminous source for the room. Recommendations in Table 19.20 cover choice of source.

20.9 Illumination Methods

There are three methods of illumination: general, local and supplementary, and combined general and local.

(a) General Lighting. This is a system designed to give uniform and generally, though not necessarily, diffuse lighting throughout the area under consideration. The method of accomplishing this result varies from the use of luminous ceiling to properly spaced and chosen downlights, but the resultant lighting on the *horizontal working plane* must be the same, that is, reasonably uniform. It may be, but is not necessarily, task lighting.

(b) Local and Supplementary Lighting. These are two terms that are used interchangeably but have slightly different meanings. By definition, *local lighting* provides a small, high-level area of lighting without contributing to the general lighting. *Supplementary lighting* also provides a restricted area of high intensity, but supplements the general lighting. In actual practice it is difficult to differentiate between the two. A desk lamp, a high-

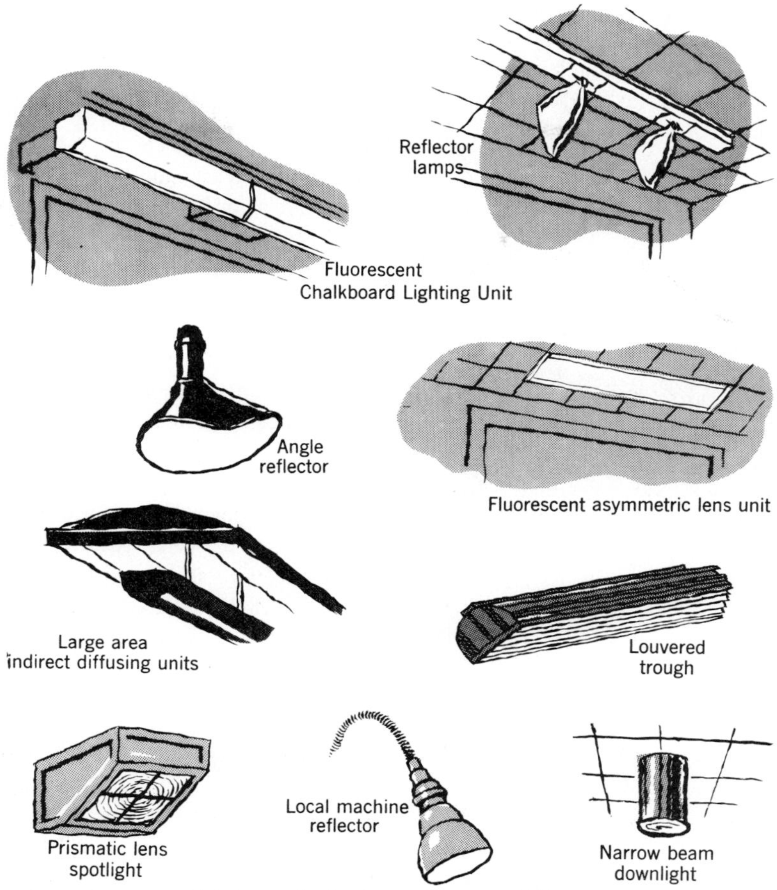

Fig. 20.4 *Typical supplementary lighting units for incandescent and fluorescent sources.*

intensity downlight on a merchandising display, and a track light illuminating wall displays all seem to answer both definitions and in practice are referred to as local, supplementary, or local-supplementary lights. Typical of this genre are the units illustrated in Fig. 20.4.

(c) Combined General and Local Lighting. This illumination method is used in spaces where the general visual task is low, but supplementary lighting is required in a limited area for a particular task.

These three *methods* of illumination can be accomplished in many ways by the use of luminaires and luminous sources of different types, since the illumination method is a function of *both* luminaire arrangement and the luminaire's inherent lighting distribution. The term used to describe the effect of the combination of a particular fixture type applied in a particular way is the *lighting system*. Thus a reflector-type fixture when aimed down gives *direct* light. The same fixture beamed up at the ceiling gives *indirect* light. The following section describes the systems that constitute the vast majority of lighting installations.

20.10 Types of Lighting Systems

No one lighting system can be said to be the single choice in a given instance; on the contrary, the designer will normally have a choice of at least two systems that will, if utilized properly, yield illumination of adequate quantity and good quality. However, other factors, such as harmonization with the architecture and economics, will usually tip the balance in favor of one or the other. The five generic types of lighting systems are indirect, semi-indirect, diffuse or direct-indirect, semidirect, and direct.

20.11 Indirect Lighting

See Fig. 20.5*a*. Between 90 and 100 percent of the light output of the luminaires is directed to the ceiling and upper walls of the room. The system is called indirect because practically all the light reaches the horizontal working plane indirectly, that is, via reflection from the ceiling and upper walls. *Therefore, the ceiling and upper walls in effect become the light source* and, if these surfaces have a high-reflectance finish, the room illumination is quite diffuse (shadowless). Since the source must be suspended at least 18 in. [45 cm] from the ceiling (depending on the unit's output) to avoid ceiling "hot spots," this system requires a minimum ceiling height of 9 ft, 6 in. [~3 m].

In addition to diffuseness, the resultant illumination is generally uniform, and direct and reflected glare are low. (A CRF in excess of 1.0 is common, with associated high ESI footcandles.) See Fig. 18.37*b* and associated text.

To avoid an excessive luminance ratio between the luminaire and its surrounding field, the luminaire can be made translucent, at least on the bottom surfaces and sometimes on the sides. Approximately 750 lux [75 fc] is the maximum horizontal-plane illumination attainable without excessive *overall* ceiling luminance of about 2500 cd/m² [730 fL]. See Section 18.25. With practically no veiling reflections, that is, a CRF in excess of 1.0, this is sufficient for all but the most difficult tasks. The lack of shadow, low source brightness, and highly diffuse quality created by

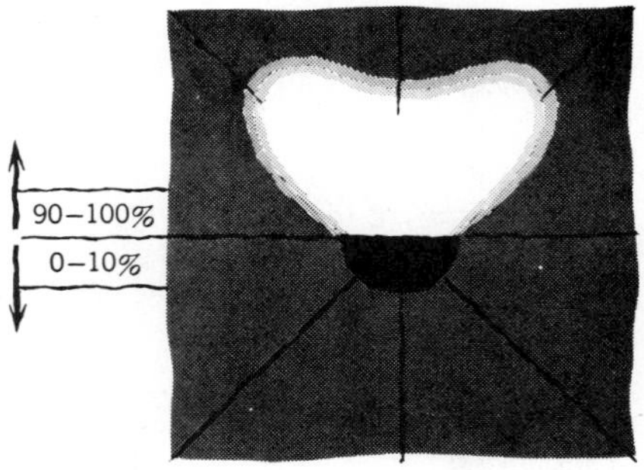

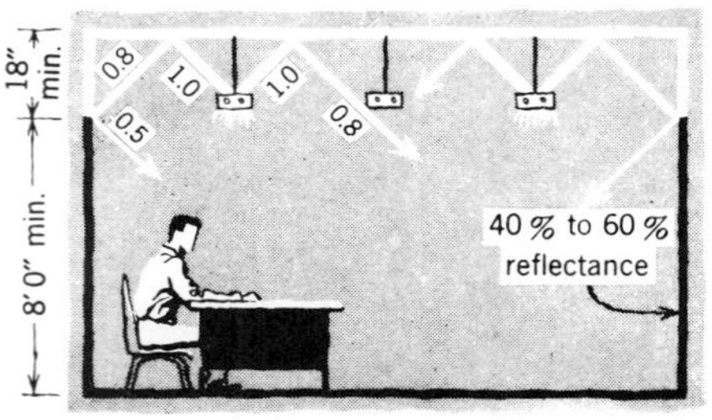

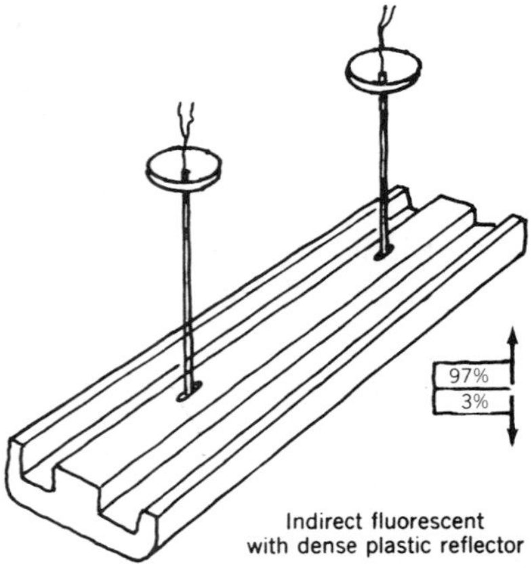

(*a*)

(*b*)

Fig. 20.5 (a) *Indirect lighting. The white surfaces of the drawings represent the areas that are illuminated by the indirect fixtures, and in turn illuminate the room. This type of installation gives a uniformly bright ceiling.* (b) *Use of architectural coves gives brightness gradient on ceiling and, if properly designed, nearly uniform illumination in the room.*

indirect lighting give a very quiet, cool ambience to this type of lighted space, suitable for private offices, lounges, and plush waiting areas. Areas having specular visual tasks use this system to advantage.

When properly designed, particularly when the source of light is architectural coves (see Fig. 20.5*b*), the ceiling has a floating, almost infinitely deep or skyline quality, which is pleasant and can be used to give an impression of height in a large room of low ceiling. This system is not to be confused with the self-luminous transilluminated ceiling, which is a direct lighting system of entirely different quality and effect. A further characteristic of the indirect lighting system is loss of texture on vertical surfaces, as is common to all fully diffuse lighting.

Indirect lighting is inherently inefficient, since much of the useful light reaches the working plane only after double reflection—within the fixture and off the ceiling. Although to a considerable extent this is offset by the absence of veiling reflections, applications to difficult seeing tasks normally require supplementary lighting. Thus an indirectly lighted drafting room having tables equipped with supplementary lamps would take advantage of both systems—the local high-intensity light at a maximum of 200 fc [~ 2000 lux] for the restricted area being worked on, and overall table lighting of 40 to 50 fc ESI. The latter would also solve any reflected glare problems arising from the many viewing angles required by large tasks such as drawings.

20.12 Semi-indirect Lighting

Between 60 and 90 percent of the light is directed upward to the ceiling and upper walls. This distribution is similar to that of indirect, except that it is somewhat more efficient and allows higher levels of illumination without undesirable brightness contrast between fixture and surroundings along with lower ceiling brightness. A typical fixture employs a translucent diffusing element through which the downward component shines and is illustrated in Fig. 20.6. The ceiling remains the principal radiating source and the diffuse character of room lighting remains. Direct and reflected glare are both very low, as with indirect lighting. See Fig. 18.37*a*. In both indirect and semi-indirect systems, it is often desirable to add accent lighting or downlighting, to break the monotony inherent in these systems and to establish a visual point of interest, or to create required modeling shadows.

The quality of the semi-indirect system is somewhat different than indirect because attention is not drawn to the fixtures, since they exhibit less contrast with the background ceiling brightness.

In both indirect and semi-indirect lighting systems, the light undergoes a number of ceiling and wall reflections before reaching the horizontal working plane. The use of colored paints serves to tint the room illumination slightly because of selective absorption.

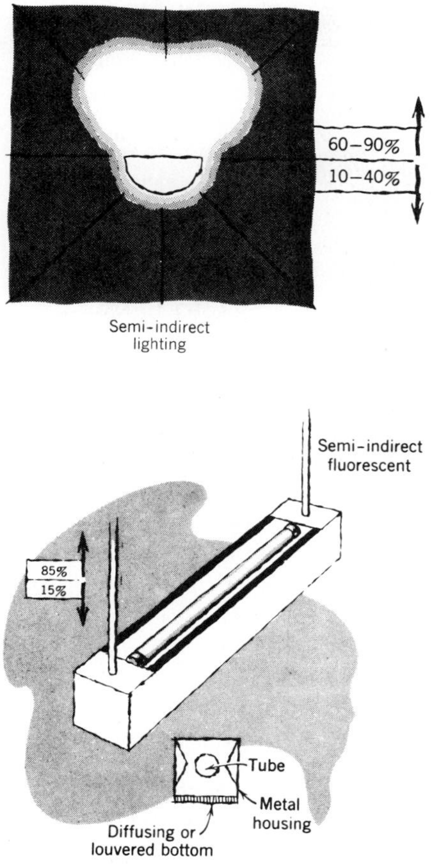

Fig. 20.6 *Semi-indirect lighting.*

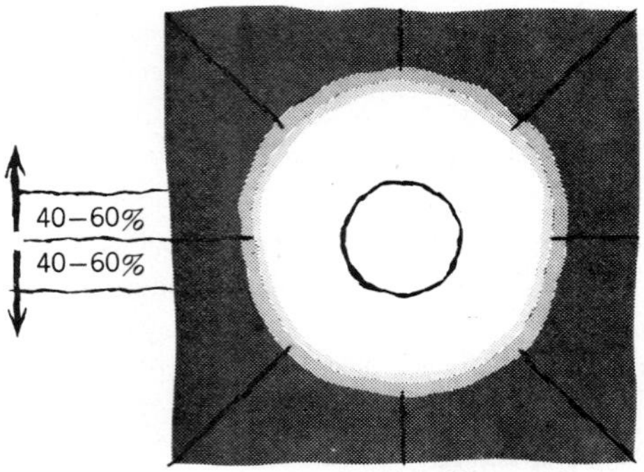

General diffuse

(a)

(b)

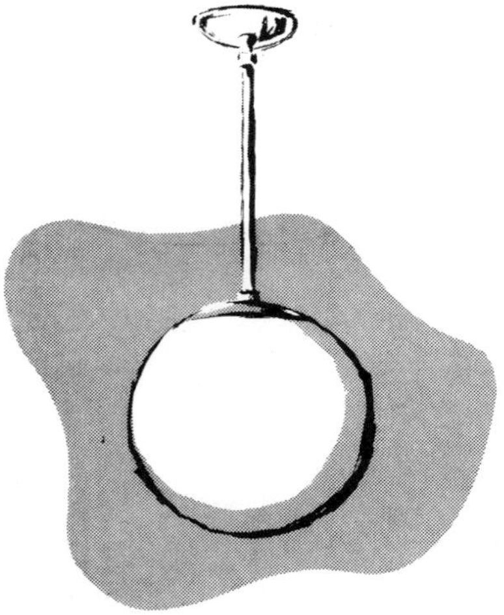

General diffuse incandescent

(c)

Fig. 20.7 (a) *General diffuse lighting. Note that all room surfaces are illuminated and become secondary sources* (b); *the primary source if illuminant is the direct radiation from the fixture* (c). *The floor contribution is low due to its normally low reflectance.*

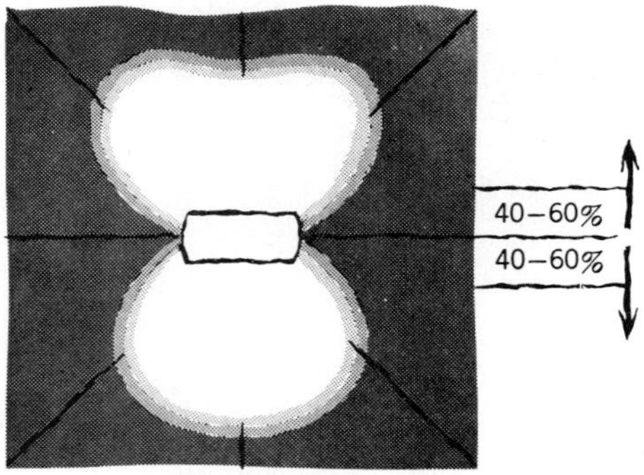

Direct-indirect

(a)

(b)

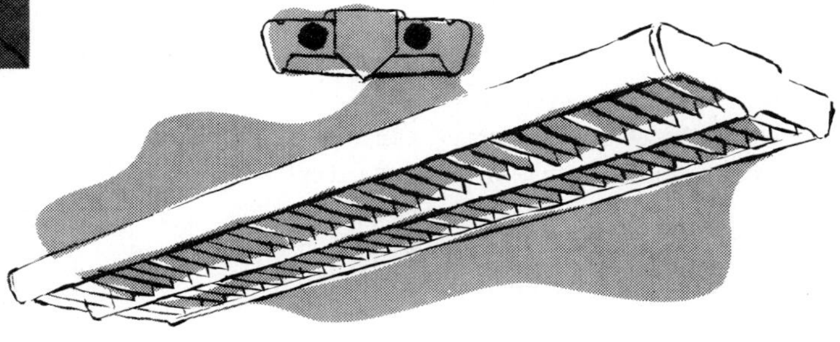

Direct-indirect fluorescent

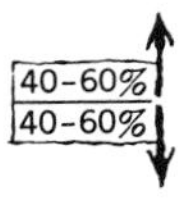

(c)

Fig. 20.8 (a) *Direct-indirect lighting. Upper and lower room surfaces are luminous* (b), *but center of walls is not because of the lack of horizontal light from fixtures* (c). *Principal light on working plane comes directly from the luminaire.*

20.13 General Diffuse and Direct-Indirect Lighting

This type provides approximately equal distribution of light upward and downward, resulting in a bright ceiling and upper wall. For this reason brightness ratios in the upper-vision zone are usually not a problem. Since the ceiling is a major though secondary source of room illumination, diffuseness will be good, with resultant satisfactory vertical-plane illumination.

Diffuse fixtures give light in all directions, whereas direct-indirect have little horizontal component (see Figs. 20.7 and 20.8). Stems should be of sufficient length to avoid excessive ceiling brightness, generally not less than 12 in. [30 cm].

Since the impression of illumination depends to a large extent on *wall brightness* because this is the surface we see most often, a space with general diffuse illumination will *appear* lighter than one with direct-indirect because of the darker walls in the latter (see Figs. 20.8*b* and 20.12).

By avoiding excessively bright luminaires and giving attention to positioning of sources and viewing angles, direct and reflected glare can both be kept low. Fixture brightnesses are interest points and the space will not appear dull and monotonous. Efficiency of these two systems is good. Both are well applied in spaces requiring overall uniform lighting at moderate levels such as classrooms, standard office work spaces, and merchandising areas.

20.14 Semidirect Lighting

With this type of lighting system 60 to 90% of the luminaire output is directed downward and the remaining upward component serves to illuminate the ceiling (see Fig. 20.9). If the ceiling has a high reflectance, this upward component will normally be sufficient to minimize direct glare, depending upon eye adaptation level. The degree of diffuseness will depend in large measure on the reflectances of room furnishings and of the floor. Shadowing should not be a problem when upward components are at least 25% and ceiling reflectance not less than 70%. With smaller upward components the system is essentially direct lighting (see Section 20.15). The system is inherently efficient. Reflected glare can be controlled by the methods discussed in Section 18.26. With adequate wall illumination, the quality of the lighting gives a pleasant working atmosphere. It is applicable to offices, classrooms, shops, and other working areas.

20.15 Direct Lighting

Since essentially all the light is directed downward, ceiling illumination is entirely due to light reflected from the floor and room furnishings. This system then, more than any other, requires a light, high-reflectance, diffuse floor unless a dark ceiling is desired from an architectural or decorative

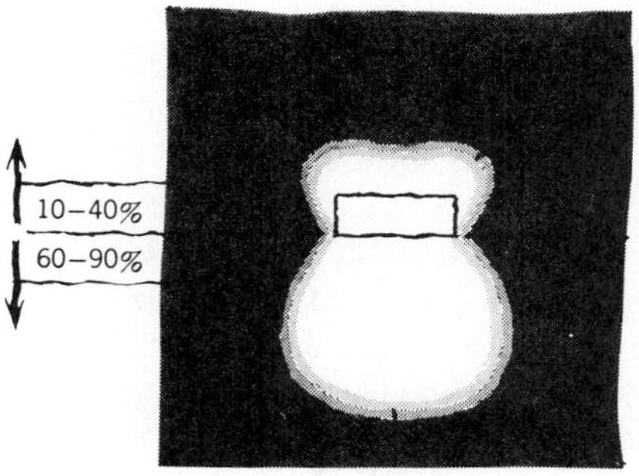

(*a*)

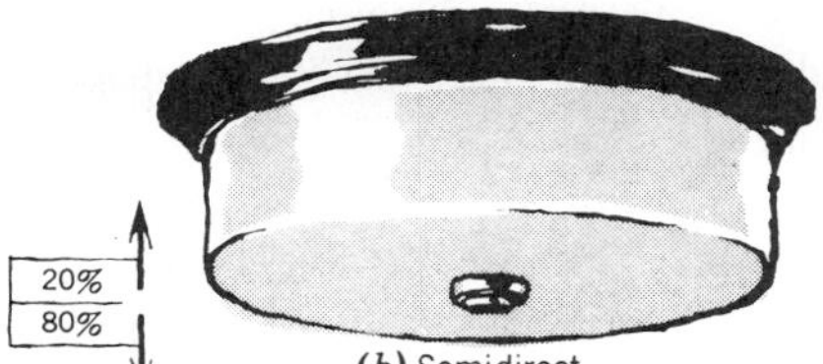

(*b*) Semidirect

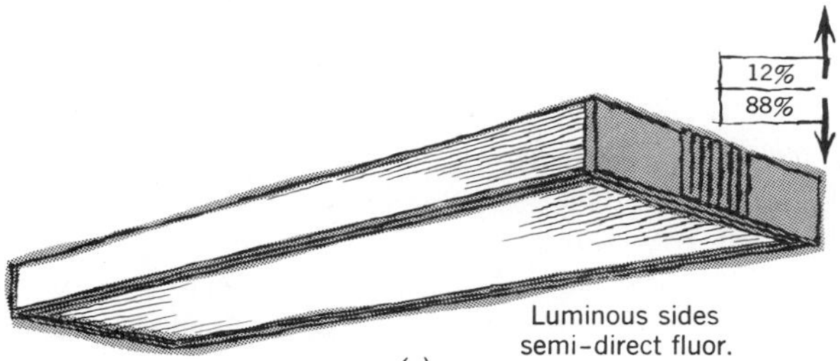

(*c*)

Fig. 20.9 Semidirect lighting provides its own ceiling brightness (a), *with surface-mounted fixtures* (b), *or pendant/surface units* (c). *Other characteristics are similar to direct lighting.*

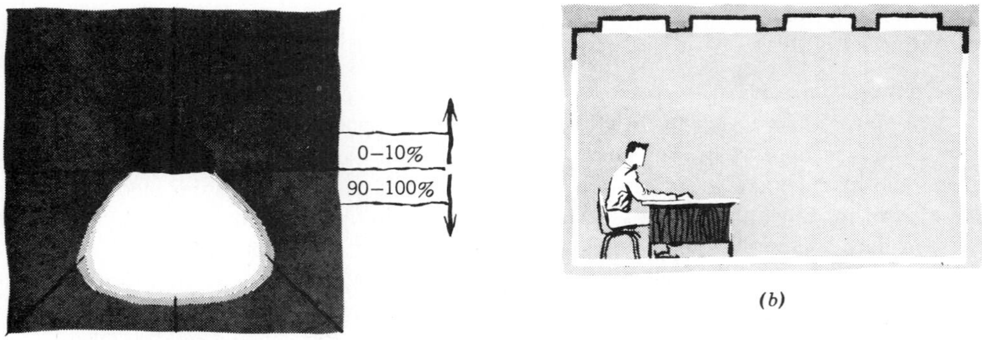

(a) Direct lighting—spread

(b)

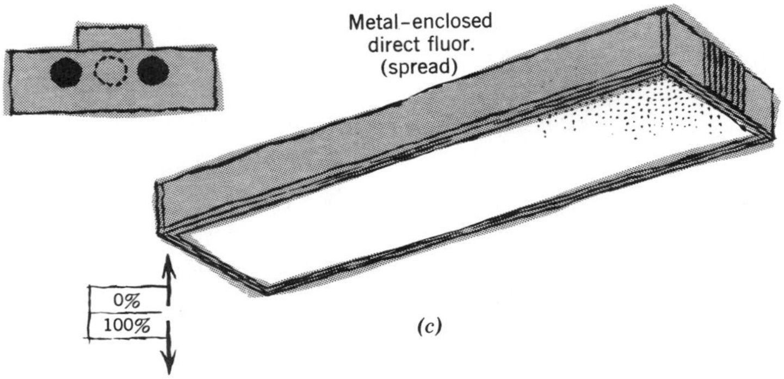

(c)

Fig. 20.10 *Spread-type direct lighting* (a), *illuminates all room surfaces except the ceiling* (b), *which is only illuminated by reflection from the floor. Some diffuseness is evident. The most common type of unit in this category is the direct fluorescent, either surface-mounted* (c) *or troffer type recessed in a hung ceiling.*

viewpoint. Occasionally the ceilings are deliberately painted a dark color and pendant direct fixtures used to lower the apparent ceiling of a poorly proportioned room or to hide unsightly piping, ductwork, and so on.

The effect of direct lighting depends greatly on whether the luminaires are spread or concentrating (Figs. 20.10 and 20.11). In the former case considerable diffusion of light emitted at high angles from the nadir results from reflections on floor, furniture, and walls. The result is a working atmosphere with slightly darkened walls and ceiling. This type of lighting, which is most widely represented by the recessed fluorescent troffer in a suspended ceiling, is common for general office lighting. The fixtures themselves form a ceiling surface of light and dark areas, and the quality of the entire system is not unpleasant. Difficulties associated with direct glare and veiling reflections can be controlled by proper use of reflectances, use of low-brightness units, and judicious arrangement of viewing positions (see Fig. 18.32*a*). When direct lighting units are used in a uniform pattern, this latter option disappears, and the need for particularly low-brightness units and high ceiling reflectivity, or specialty diffusers like the batwing design, is increased.

Direct lighting gives little vertical surface illumination, requiring the addition of perimeter lighting in business atmospheres (see Fig. 20.12).

Concentrating downlights create sharp shadows and a theatrical atmosphere, not appropriate to a working commercial space. See Fig. 21.27, below. When used alone, they can be appropriate in restaurants and other areas where the privacy type of atmosphere generated by limited-area horizontal illumination and minimal vertical-surface illumination is desired. When a fixture is designed with a black cone or baffle or other device that is nonreflecting at the viewing angle, the fixture ap-

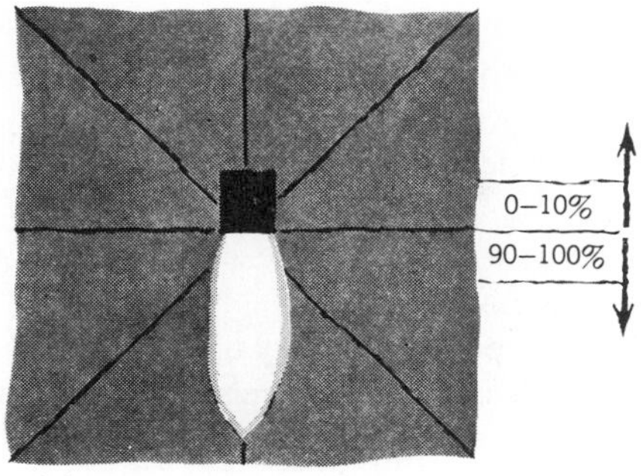

(a) Direct lighting—concentrating

(b)

Fig. 20.11 *With concentrating direct distribution* (a), *the floor is the only luminous surface* (b), *other than the ceiling fixture. Diffuseness is absent. Walls are dark. Incandescent downlights* (c) *are of this type unless equipped with spread-type lenses.*

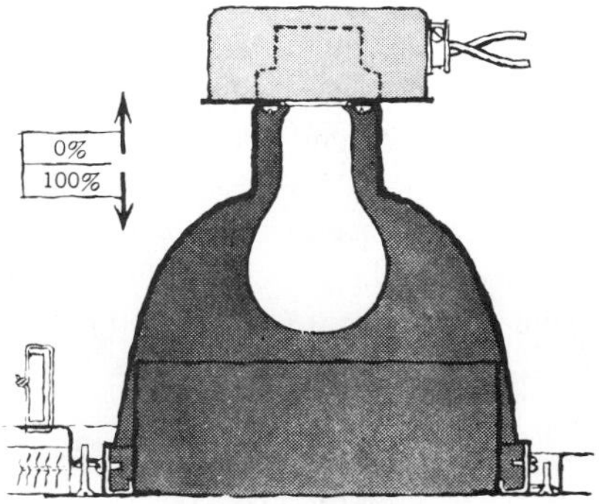

(c) Direct concentrating

pears dark. It is our opinion that installations providing high-horizontal-surface illumination, with no apparent source of brightness, such as those using black-cone downlights, are disturbing to our normal bright-sun-and-sky orientation and should therefore be used cautiously and only in limited areas. This same comment, but to a lesser extent, is applicable to very low-brightness diffusers such as the parabolic wedge type (see Fig. 18.9*a*). There, however, the unit has the redeeming characteristic of low reflected glare, which is not the case with downlights.

In summary then, spread direct lighting is suitable for general lighting while concentrated direct lighting, which reduces vertical illumination, is appropriate for highlights, local and supplementary lighting, and specialized viewing.

Fig. 20.12 *The use of large-dimension lighting fixtures in a low-ceiling room is possible if the apparent size of the unit is reduced. Here at a mounting height of 7 ft, 6 in., 4 × 4 ft units are acceptable because the lattice on the face of each unit gives the impression of reduced fixture size. Note also that the apparent illumination in* (b) *is greater than in* (a) *(though both are exactly equal on the table surface) due to the wall wash in the background. The eye perceives vertical surface illumination more readily than horizontal and retains the impression for the entire space.*

20.16 Size and Pattern of Luminaires

Because of its luminance, each luminaire or other luminous source is a point of visual attention. To the extent that luminaires are numerous, large, very bright, or arranged in striking patterns, attention will be drawn to them and away from other surfaces. Furthermore, color elements or accent lighting can be added deliberately to draw attention. Rigid rules cannot be set down covering these criteria, but examples can demonstrate the principles.

Luminaire size could correlate with room size and ceiling height. Fluorescent fixtures larger than 2 × 4 ft [60 × 120 cm] should not be used in ceilings below 10 ft unless their size is minimized by some sort of surface pattern (see Fig. 20.12). Transilluminated ceilings are *all* fixture and therefore require a minimum of 12 ft [3.66 m] mounting height. When installed below this level, particularly in large rooms, the effect is oppressive, as if the sky were lowered on us. To offset this effect, the use of colored, shaped, or dark panels is of some help. In place of a luminous ceiling (Fig. 20.13), a large-area, coffer-type fixture can be utilized, which gives the impression of great depth (see Fig. 20.14).

To achieve the uniformity of illumination necessary for general lighting, regular spacing is desirable. However, various effects may be obtained within the regularity, to accomplish an architectural purpose, as shown in Fig. 20.15. The pattern of lights must never be at cross-purposes with any dominant architectural pattern; rather it should either reinforce an architectural form or be neutral. If a strong architectural element is absent, a dominant lighting pattern may be desirable. Conversely, a strong architectural element can either be reinforced (Fig. 20.15*p*-1) or utilized to carry a neutral lighting pattern (Fig. 20.15*p*-2, 3).

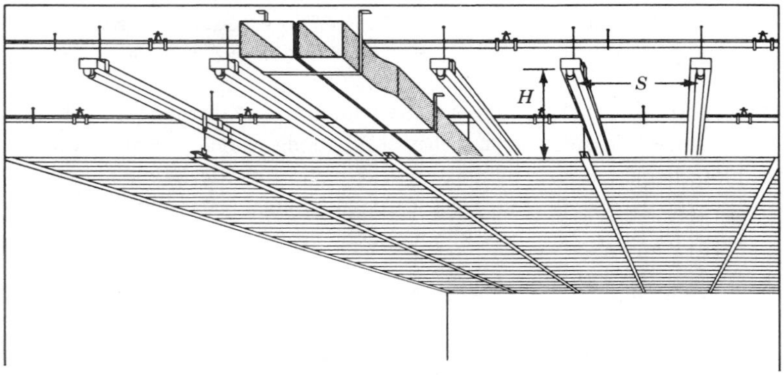

***Fig. 20.13** Luminous ceiling installation. Note that in properly designed installation, piping and ductwork do not affect the light distribution. As a rule of thumb, fixture spacing* S *should not exceed 1.5 times fixture height* H *above the diffusing element, for uniform illumination. This type of installation is useful for specular tasks, if supplemental lighting is impractical.*

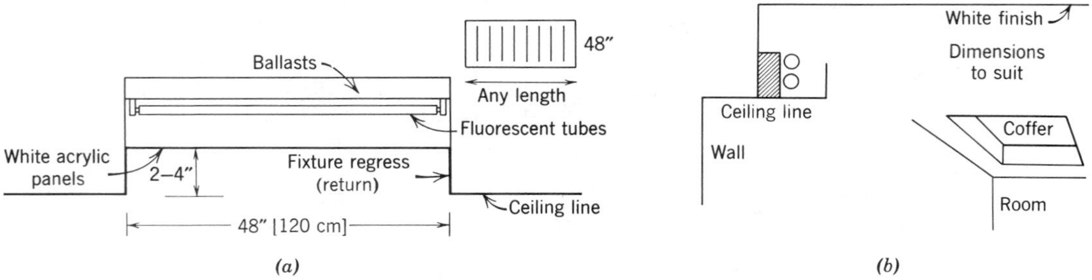

***Fig. 20.14** Types of coffer.* (a) *Coffer fixture, for direct diffuse lighting, generally available in 4 ft [120 cm] width and any length.* (b) *Architectural coffer, for indirect lighting, designed as required. (See Chapter 21, Fig. 21.17), for dimensional data. Both types give an illusion of great depth and soft, glare-free illumination. Coffers are also ideal daylight sources when designed in conjunction with skylights as in Fig. 19.29.*

(*a*) *Longitudinal lines increase apparent length, direct traffic flow, decrease direct glare.*

(*b*) *Horizontal lines shorten and widen a space, but also increase direct glare.*

(*c*) *Diagonal lines minimize shadows and break rectangular patterns. They are architecturally dominant.*

(*d*) *Rectangular pattern is architecturally dominant. It is a poor choice in stores where attention downward is desired.*

(*e*) *Cornices, valances, and coves are luminous ceiling borders. In large rooms suspended coves achieve uniform ceiling brightness and when designed with a downward component or combined with local lighting, as illustrated, give a pleasant, intimate atmosphere.*

(*f*) *Coffers create a decorative architectural effect and can be designed to resemble skylights or can be built into actual skylights.*

(*g*) *Luminous ceiling system (see Fig. 20.13) provides high illumination, low brightness, and high diffusion. Some accent of either color or lighting is desirable to relieve monotony.*

Fig. 20.15 *(Continued on next page)*

(h) *Downlights are architecturally neutral and may therefore be spaced evenly . . . or unevenly.*

(k) *The large can-shaped surface-mounted downlights dominate the area's appearance despite the high ceiling. The pimpled look could be eliminated by recessing the units. Or, taking advantage of the prominence of these units, interest could be added by altering the regular pattern.* [*See* ***(l)***.]

(l) *The lines of lights in the center converge optically to produce a directional flow of traffic toward the escalator. The remaining floor lighting is provided by visually pleasing circular patterns around columns (Penn Station, New York). (Photo by B. Stein.)*

(m) *That patterns of lighting are plainly visible even during the daytime is apparent from this photo. The attractiveness of uniformity of fixture pattern can readily be seen. (Photo by B. Stein.)*

Fig. 20.15 *(continued)*

*(**n**) Since fixtures are readily visible even when unlit during daylight hours, their outline can be accentuated and the resultant pattern utilized as an architectural motif. (Courtesy of Welton Becket & Assoc.)*

*(**o**) Lighting can be utilized as a medium to connect the inside and outside of a building. The simple expedient of continuing the lighting pattern beyond the window or wall glass provides visibility from inside out as well as outside in. Care must be exercised to avoid fixture placement that will reflect in the glass. (Photo by Marc Neuhof.)*

Continuous row installations eliminate the dominating checkerboard effect of closely spaced individual luminaires (and are cheaper). Coves and cornices give the ceiling a floating or light effect. Geometric patterns can be used to add interest or break monotony of large areas, such as department stores. Generally, incandescent downlights are not dominant, and regularity of placement is not essential. Nonuniform layouts with large sources create a distinct pattern problem inasmuch as they are too large to be neutral, and the nonuniformity can create visual confusion (see Fig. 20.16). The only cure for this problem is to minimize the source brightness by using low-brightness luminaires. (See Fig. 21.33 and Section 20.27.)

The incorporation of daylight into the luminous sources of a space should take cognizance of size and pattern as well. Large windows are not consonant with small ceiling sources, whereas skylights are readily integrated with other ceiling units. A frequently neglected consideration is the appearance of a source when de-energized. With proper daylight and energy-conserving design, many sources will be unlighted during the normal-use hours of the space. Obviously, low-brightness sources will change least in appearance, which is another factor in their favor.

FACTORS IN DESIGN

20.17 Design Decisions

Refer again to Fig. 20.1. At this point in the design process the lighting hardware is chosen on the basis of the considerations adduced in the preliminary design stage and the appropriate calculations performed. Some spaces will require overall, uniform illumination. These spaces are calculated by the lumen method, which yields average illumination. Other spaces will utilize local lighting alone, or local lighting in addition to general, requiring point-by-point illumination calculations or some other method for restricted-area calculation. Additional considerations at this design stage are the control strategy (see Sections 21.2–21.4), type of ceiling system, for example, modular, movable fixture, and integrated service, and ancillary considerations of ballast noise, fixture heat distribution, and maintenance. Also de-

(**p-1**) *Treatment of dominant architectural patterns. In each case the lighting designer was faced with essentially the same problem: a low-level seeing task in a large space with a dominant architectural ceiling. Different solutions were arrived at, each of which is consonant with the architectural ceiling, (*p-1) and (p-2) *accomplish this by following the dominant line;* (p.3) *follows it with a neutral pattern. (Photo by M. B. Warren.)*

(**p-2**) *Floor reflection and daylight provide the required ceiling and wall illumination. (Photo by L. Reens.)*

(**p-3**) *The lighting in this extremely strong architectural pattern was harmonized deftly by recessing fixtures deeply into the lattice motif. Metal-halide HID units and tungsten-halogen units were used. (Courtesy of GTE/Sylvania, Inc.)*

Fig. 20.15 *(continued)*

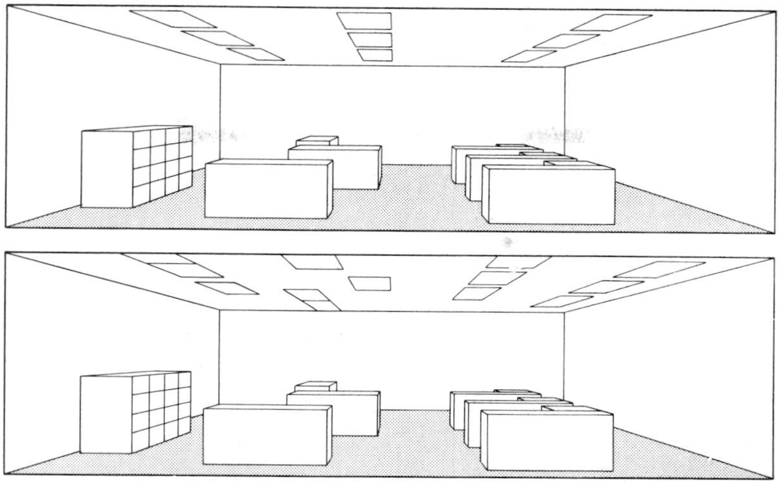

Fig. 20.16 (a) *The uniform layout of Fig. 18.35 page 901, shown here in perspective, is neutral in that it does not dominate the space or draw the eye.* (b) *The nonuniform layout can be dominant in the pejorative sense if the eye is drawn by the lack of pattern or symmetry.*

cided here is whether to utilize work-station-mounted or built-in lighting, both of which are principally applicable to open-plan spaces. The sections immediately following are a discussion of the photometric and other characteristics of lighting fixtures, without which proper equipment selection is obviously impossible.

20.18 Direct-Lighting Luminaire Characteristics

It is important to understand the action of luminaire reflectors. The basic shapes and beam patterns are illustrated in Figs. 20.17 to 20.21, while shielding methods are shown in Fig. 20.22. See also Fig. 21.15. Although most of the illustrations use point sources (incandescent or HID lamps), the principle illustrated is applicable to reflectors for linear (fluorescent) sources, when considered in section.

Note from Fig. 20.18*a* and 20.18*b* that the so-called pinhole downlight requires an elliptic reflector to focus the light through this hole at point "*x*" in order to maintain even minimal fixture efficiency. Elliptic reflectors are large and frequently space above the ceiling is too restricted for their use. A lamp with an integral elliptic reflector, which can therefore be utilized in a standard baffled reflector without severe losses, is illustrated in Fig. 20.23.

20.19 Lighting Fixture Distribution Characteristics

When selecting a ceiling fixture for general coverage, the designer is faced with a choice of quite literally hundreds of units, most of which purport to be ideal for the purpose intended. Efficiency and economics will be discussed below; glare considerations were covered previously; distribution will be our concern in this section. A review at this point of Section 18.14, by the reader, would be useful.

The two distribution curves shown in Fig. 20.24 are actual test results of two 2-lamp, 1 ft wide by 4 ft long, semidirect fluorescent fixtures, with prismatic enclosure. The flat bottom of the curve in Fig. 20.24*a* indicates even illumination over a wide area, therefore permitting a high spacing-to-mounting height ratio (1.4) for uniform illumination. The rounded bottom of the curve in Fig. 20.24*b* indicates uneven illumination and closer required spacing for horizontal uniformity (spacing-to-mounting height ratio of 1.2). Uniformity is defined in the next section. The straight sides of the curve in Fig. 20.24*a* show a fairly sharp cutoff, and the small amount of light above 45°

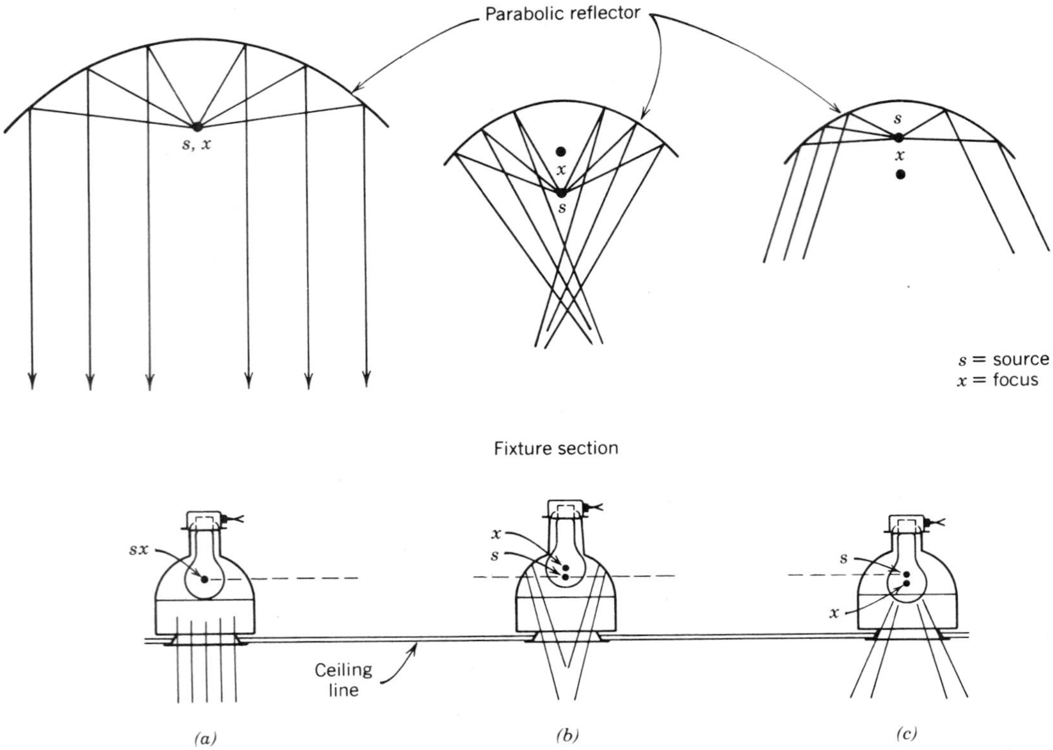

***Fig. 20.17** Parabolic reflector action:* (a) *with source at focal point rays are parallel;* (b) *with source below focal point they converge;* (c) *with source above focal point they diverge. This focusing action is illustrated by fixtures correspondingly designated. Note that type* (c) *requires a large ceiling opening to achieve even minimal efficiency.*

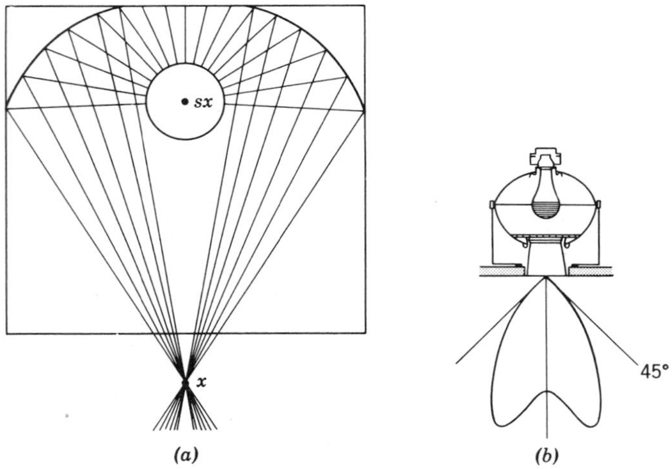

Fig. 20.18 (a) *Action of elliptical reflector section. With the light source at one focal point, the light converges at the other focal point. This effect is useful in fixture design as in* (b). *By projecting light up only, through use of silvered bowl lamp, the output light can be redirected through a constricted aperture at the other focal point, with little loss. Beyond the aperture, the light diverges. The cone in the bottom of the fixture provides cutoff at high angles of viewing (see Fig. 20.23).*

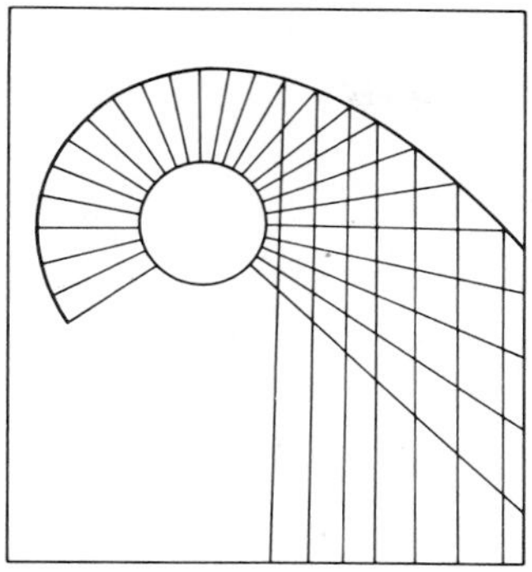

Fig. 20.19 The extended section reflector allows the source to be concealed (shielded) while projecting the light downward.

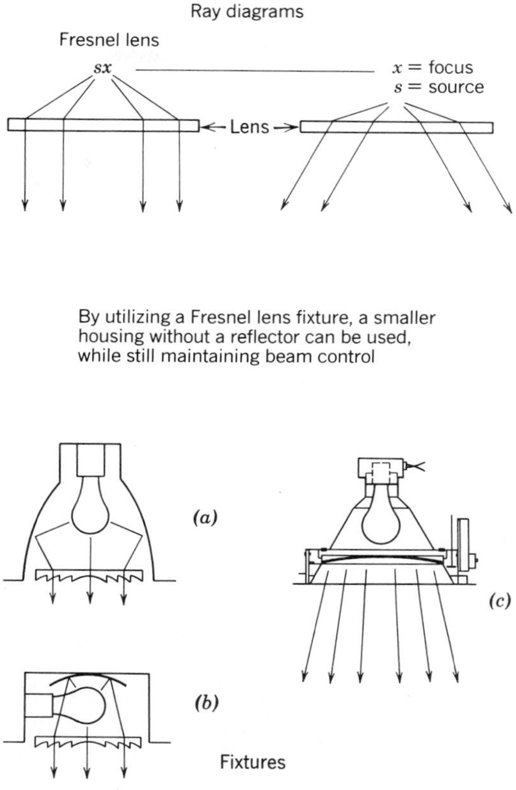

Fig. 20.20 *Action of a Fresnel lens. The lens performs the same function as a reflector, controlling the beam as a function of source placement. By utilizing a lens fixture, the curved reflector can be largely eliminated yielding a smaller fixture, while maintaining accurate beam control. Common design* (c) *uses regressed lens to provide shielding, although lens brightness is not normally objectionable.*

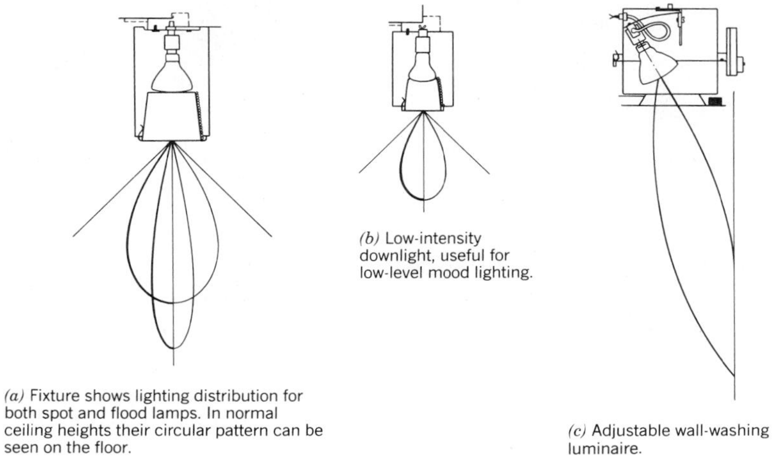

(a) Fixture shows lighting distribution for both spot and flood lamps. In normal ceiling heights their circular pattern can be seen on the floor.
(b) Low-intensity downlight, useful for low-level mood lighting.
(c) Adjustable wall-washing luminaire.

Fig. 20.21 (a)–(c) *Fixtures utilizing lamps with integral reflectors serve generally as lamp holders and to provide beam shielding.*

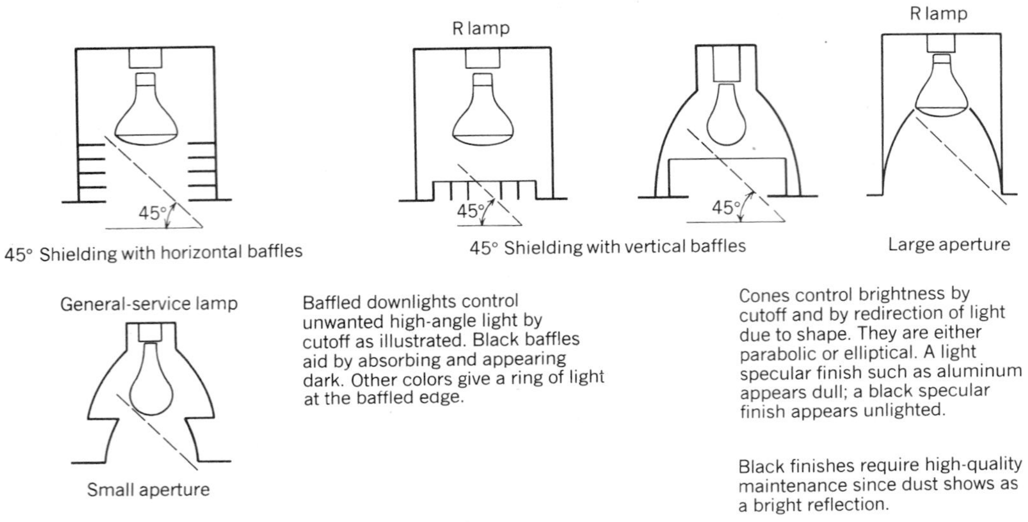
R lamp
R lamp
45°
45°
45°
45° Shielding with horizontal baffles
45° Shielding with vertical baffles
Large aperture
General-service lamp
Baffled downlights control unwanted high-angle light by cutoff as illustrated. Black baffles aid by absorbing and appearing dark. Other colors give a ring of light at the baffled edge.
Cones control brightness by cutoff and by redirection of light due to shape. They are either parabolic or elliptical. A light specular finish such as aluminum appears dull; a black specular finish appears unlighted.
Black finishes require high-quality maintenance since dust shows as a bright reflection.
Small aperture

Fig. 20.22 *Methods for shielding the light source.*

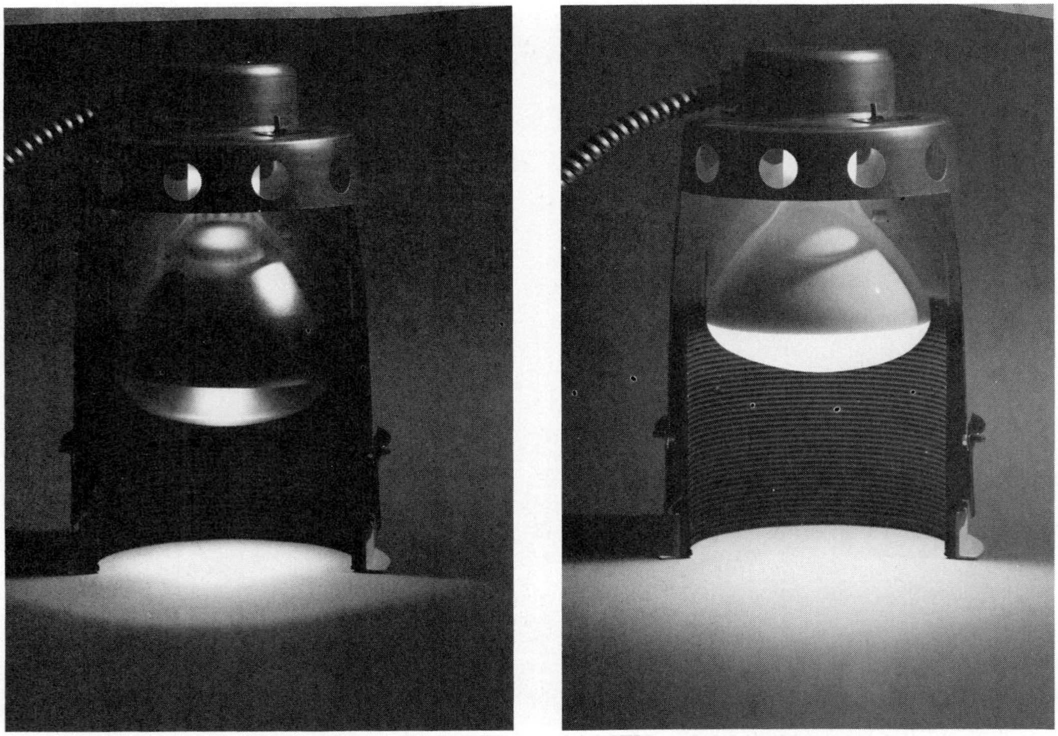

Fig. 20.23 *Baffled downlights are low in brightness at normal viewing angles because the black baffles trap and absorb the lamp's output. The result is low efficiency, as in the fixture on the right. The lamp on the left contains an elliptical reflector that focuses the light below the fixture, resulting in even lower fixture brightness with high efficiency. This principle is useful in pinhole downlights as illustrated in Fig. 20.18. (Photos courtesy of General Electric, Lighting Business Group.)*

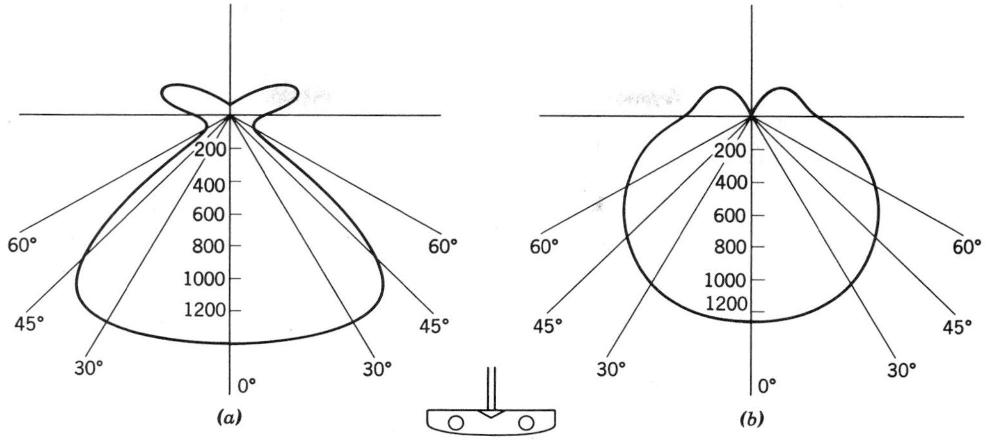

Fig. 20.24 Semidirect fluorescent crosswise fixture distribution (two lamps, 40-W, prismatic enclosure). Note the sharp cutoff and wide horizontally even distribution of (a) *in contrast to the diffuse, broad, and horizontally uneven distribution of* (b).

means high efficiency, sufficient wall lighting, adequate diffuseness, and very little direct glare problem, but a distinct possibility of veiling reflections. Conversely, the curve in Fig. 20.24*b* shows a large amount of horizontal illumination (above 45°), with resultant direct glare, diffuseness, and relative inefficiency, since horizontal light is attenuated by multiple reflections before reaching the horizontal working plane. Here, however, low output below 45° minimizes reflected glare. The uplight component of fixture (*a*) is directed outward to cover the ceiling and will not cause hot spots; the corresponding light from fixture (*b*) is concentrated above the fixture and will give uneven illumination of the ceiling. Thus we see that a rapid inspection of a fixture curve performed by an informed person can yield a large amount of information on the fixture's performance.

20.20 Uniformity of Illumination

In any space intended to be lighted uniformly with multiple, discrete, ceiling-mounted direct-lighting system light sources, it is necessary to establish a fixture spacing that will give acceptable uniformity of illumination. A ratio of maximum to minimum illumination on the working plane of 1.2 to 1.3 is readily acceptable since lesser ratios are not easily noticed. See Table 18.4. For general background or circulation lighting, up to 1.5 is acceptable. The recommended spacing-to-mounting-height ratio given by manufacturers (see figures immediately above the distribution curves for each fixture in Table 20.2) are generally based on a 1.2 figure. See Fig. 20.25. We mentioned above that the fixture of Figure 20.24*a* had a high spacing-to-mounting-height ratio (*S/MH*) because of its flat-bottomed curve. This ratio, when not given by the manufacturer, may be approximated from Fig. 20.26.

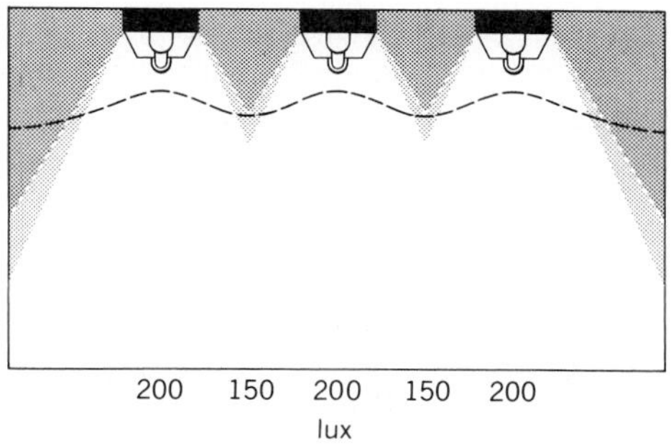

Fig. 20.25 The ratio of maximum to minimum illumination should not exceed 1.30 in areas requiring uniform illumination.

An accurate method of calculating maximum to minimum illumination ratios is available (see *IES Handbook,* 1981).

It is well known that illumination levels near walls drop off at least 30% even in a well-designed installation. To counteract this effect, particularly

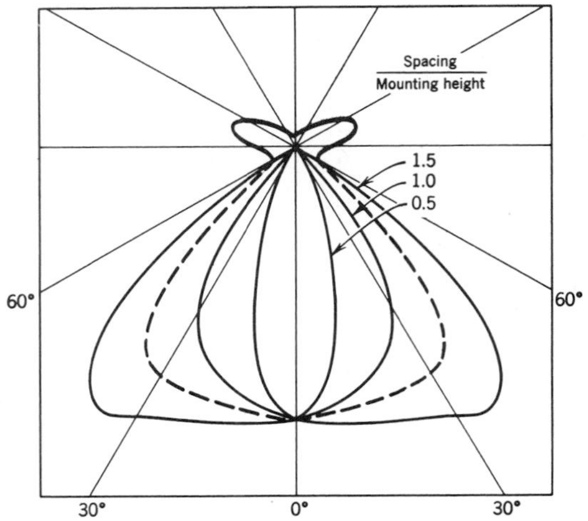

Fig. 20.26 *Typical distribution curves for approximating ratio of fixture spacing to the mounting height* (S/MH) *above the working plane, for* direct lighting *luminaires. To determine* S/MH *for a specific unit, compare its curve to those shown and interpolate. Although these curves were developed for* direct lighting *point sources such as incandescent (and HID), they can also be used with asymmetric distribution luminaires such as fluorescent. In such cases the curves will give the* S/MH *ratio in the plane whose characteristic is used, that is, lengthwise or crosswise. The dotted curve was transferred from Fig. 20.24*a. *Its permissible* S/MH *is somewhat higher than the curves indicate because it is a* semidirect *distribution, and its ceiling light component permits wider spacing between units. (After Odle and Smith, from* IES Journal, *January 1963.)*

when placement of furniture is such that visual tasks will occur near walls, the designer should arrange to provide additional illumination in these areas. This may readily be accomplished by additional fixtures, higher output units, perimeter lighting, or some type of wall-washing arrangement. Particular stress should be placed on this type of local lighting where wall reflectances are low, such as at walls covered with book shelves, equipment racks, low-reflectance paint, or dark

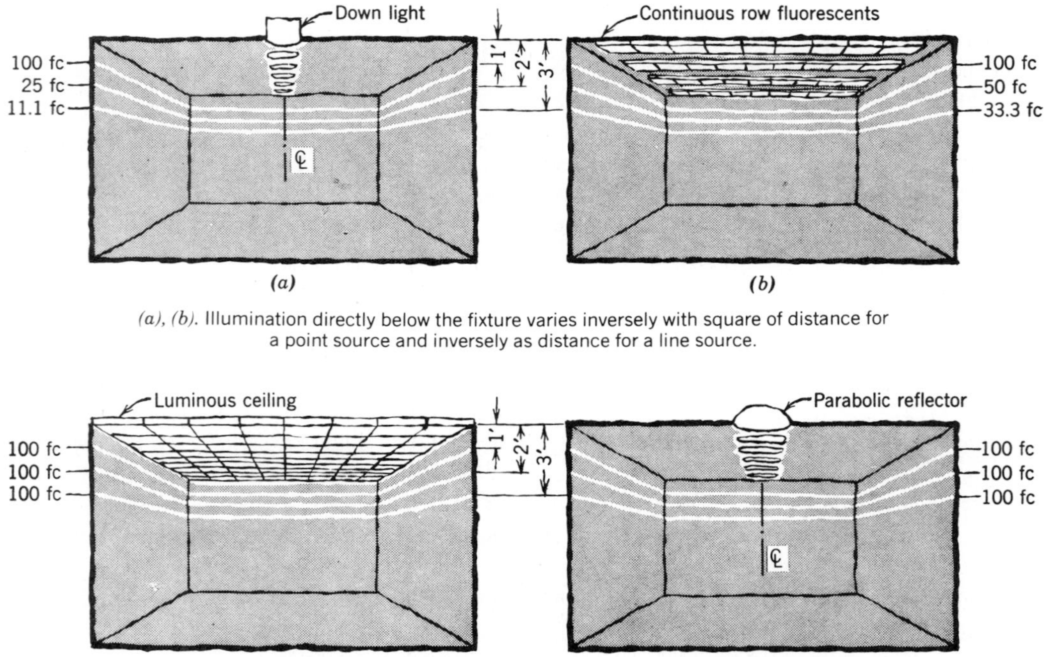

(a), (b). Illumination directly below the fixture varies inversely with square of distance for a point source and inversely as distance for a line source.

(c), (d) Illumination remains constant at all distances from either an infinite (or nearly) source or a parabolic reflector.

Fig. 20.27 Variation of illuminance vertically, directly below the fixtures, for different source types.

wood paneling. Fixture end should be no more than 1 ft, and fixture sides no more than 2 ft from walls.

The foregoing discussion of illumination uniformity concerned itself with uniformity on a horizontal plane. Occasionally, it is necessary to know the degree of uniformity vertically, that is, on horizontal planes at different elevations *directly below the fixtures*. Four different lighting situations are normally encountered. They are point sources such as incandescent downlights, line sources such as continuous-row fluorescent fixtures, infinite sources such as luminous ceilings—whether trans-illuminated or indirect—and parabolic reflector beams such as from PAR lamps. The vertical uniformity of each type is shown graphically in Fig. 20.27.

20.21 Luminaire Mounting Height

The mounting height of luminaires is normally established before spacing. In arriving at a mounting height for fixtures with an upward component, a balance must be struck between low mounting, which will control ceiling brightness and give good utilization of light, and the reticence to dominate an area, particularly a large room, by using such a low mounting height that the apparent ceiling height is affected (see Fig. 20.28). Rules of thumb for mounting height are:

1. Indirect and semi-indirect luminaires should be suspended no less than 18 in. from ceiling, and preferably 24 to 36 in.
2. Direct-indirect and semidirect fluorescent fixtures should be suspended not less than 12 in. for two-lamp units and 18 in. for four-lamp units.

The effect of pendant length on coefficient of utilization (efficiency; see Section 20.27) is given in Fig. 20.29.

20.22 Lighting Fixtures

Before proceeding further with design, we wish to discuss the principal item of lighting hardware, the fixture itself. This and the following sections cover fixture construction, installation, and appraisal. The architect need simply stop to consider that lighting fixtures constitute 25 to 30% of the electrical budget or 4 to 5% of the overall building budget to appreciate their importance. Since the difference between a quality unit and an inferior one is often not readily visible to the casual observer, particular care must be taken in the specification of lighting fixtures and in examination of shop drawings *and* samples. All fixtures if applied properly will give a sufficient quantity of light, but only a good unit will combine quantity with good quality, ease of installation, facility of maintenance, and indefinite life. In addition, installation must be proper to ensure mechanical rigidity and safety, electrical safety, freedom from excessive temperatures, and requisite accessibility of component parts and of the fixture outlet box. The material below is a combination of NEC minimum requirements plus factors that the author has found important beyond these minima.

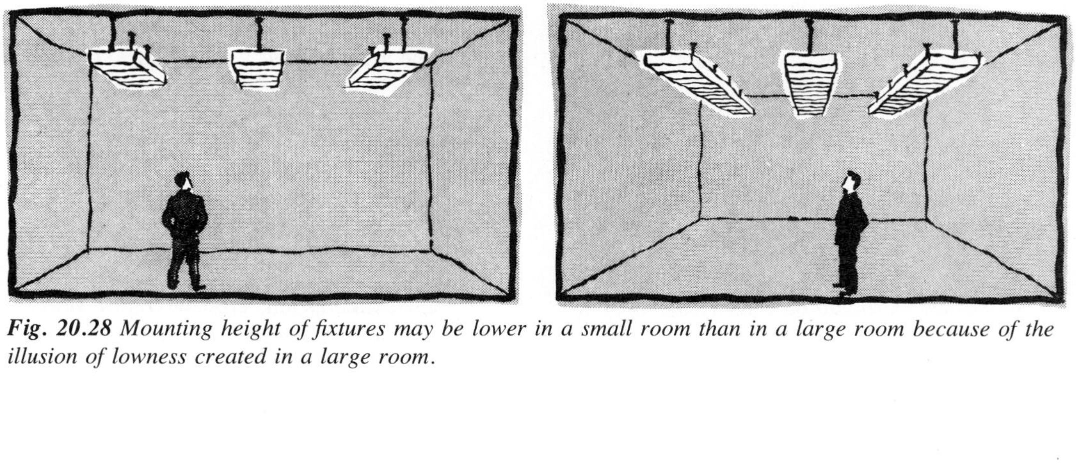

Fig. 20.28 *Mounting height of fixtures may be lower in a small room than in a large room because of the illusion of lowness created in a large room.*

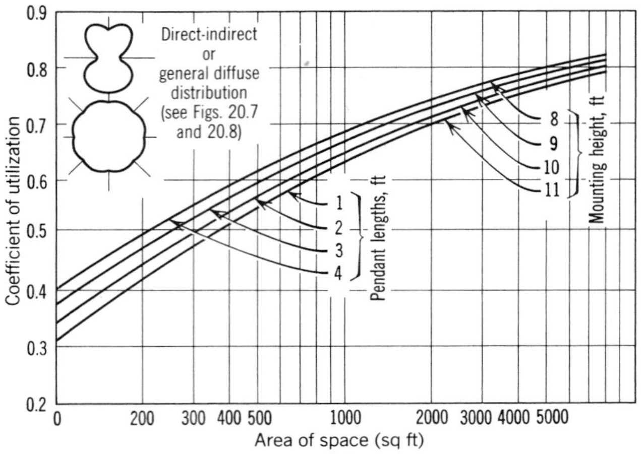

(a)

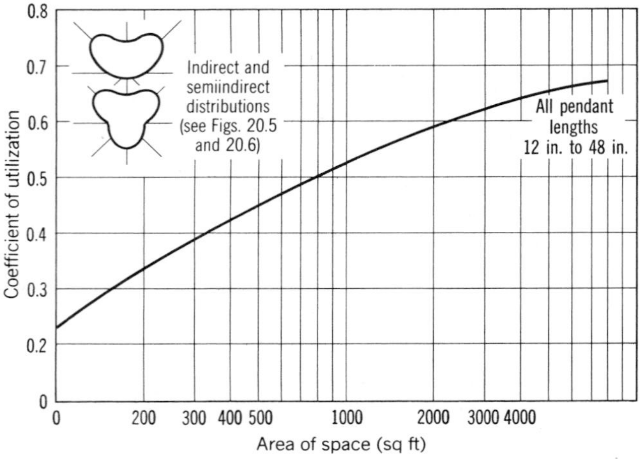

(b)

***Fig. 20.29** Coefficient of utilization (lighting system efficiency) as a function of pendant length for various distributions. With a substantial downward component as in direct-indirect or general diffuse* (a)*, system efficiency rises slowly as the fixture descends (pendant length increases). Maximum differentials occur in small rooms and can reach 20%. Where the ceiling is the light source as in indirect and semi-indirect systems* (b)*, the pendant length does not change the room illumination (see also Fig. 20.27c). This curve can be used to estimate CU for indirect and semi-indirect, where not found in Table 20.2 or manufacturers' catalogs.*

20.23 Lighting Fixture Construction

1. All fixtures should be wired and constructed to comply with local codes, NEC and Underwriters Laboratories Standard for Lighting Fixtures, and shall bear the Underwriters Label, where label service is available. Reflector Luminaire Manufacturers (RLM) standards should be adhered to for all porcelain enameled fixtures.

2. Fixtures should generally be constructed of 20 gauge (0.0359 in.) thick steel minimum. Cast portions of fixtures should be no less than 1⁄16 in. thick.

3. All metals should be coated. The final coat should be a baked-enamel white paint of minimum 85% reflectance, except for anodized or ''Alzac'' surfaces.

4. No point on the outside surface of any fixture should exceed 90° C.

5. Each fixture should be identified by a label carrying the manufacturer's name and address and the fixture catalog number.

6. Glass diffuser panels in fluorescent fixtures should be mounted in a metal frame. Plastic diffusers should be suitably hinged. ''Lay-in'' plastic diffusers should not be used.

7. Plastic diffusers should be of the slow-burning or self-extinguishing type with low smoke-density rating and low heat-distortion tempera-

tures. This latter should be low enough so that the plastic diffuser will distort sufficiently to drop out of the fixture before reaching ignition temperature.

8. It is *imperative* that plastics used in air-handling fixtures be of the noncombustible, low-smoke-density type. These requirements also apply to other nonmetallic components of such fixtures.

9. All plastic diffusers should be clearly marked with their composition material, trade name, and manufacturer's name and identification number. Results of ASTM combustion tests should be submitted with fixture shop drawings. The characteristics of many plastic diffusers change radically with age and exposure to ultraviolet light. Glass and acrylic plastic are stable in color and strength. Other plastics may yellow and even turn brown, thus diminishing light transmission radically as well as changing the fixture appearance. Some plastics that are initially very tough and "vandalproof" embrittle with age and exposure to weather or the ultraviolet light of a mercury or fluorescent source. Thus the long-range as well as initial characteristics of all diffuser elements must be investigated before specification and approval.

10. Ballasts should be mounted in fixtures with captive screws on the fixture body, to allow ballast replacement without fixture removal.

11. All fixtures mounted outdoors, whether under canopies or directly exposed to the weather, should be constructed of appropriate weather-resistant materials and finishes, including gasketing to prevent entrance of water into wiring, and should be marked by the manufacturer, "Suitable for Outdoor Use."

20.24 Lighting Fixture Installation

1. Fixtures mounted on combustible surfaces should not subject these surfaces to a temperature exceeding 90°C, in a 40°C ambient space. To this end insulating-material spacers or simply an air space may be used.

2. Although most codes allow fluorescent fixtures weighing less than 40 lb to be mounted directly on the horizontal metal members of hung-ceiling systems, experience has shown that vibration, member deflection, routine maintenance operation on equipment in hung ceilings, and poor workmanship can cause such fixtures to fall, endangering life. It is therefore strongly recommended that all fixtures, surface, pendant, or recessed, whether mounted individually or in rows, be supported from the black channel iron supporting the ceiling system (purlins) or directly from the building structure, but in no case by the ceiling system itself. This is particularly important in the case of an exposed "Z" spline ceiling system.

3. Fixtures installed in bathrooms should *not* have an integral receptacle and when installed on walls, shall have nonmetallic bodies. These are safety precautions.

20.25 Lighting Fixture Appraisal

As mentioned above, the intense competition in the lighting field necessitates close scrutiny of the characteristics of fixtures. To compare the relative merits of similar lighting fixtures as manufactured by different companies, complete test data plus a sample in a regular shipping carton from a normal manufacturing run are necessary.

The following list should be used as a basic guide, with additional items added according to job requirements:

1. *Photometric and design data*. Manufacturers should furnish complete test data, including candlepower distribution curve(s), coefficients of utilization (CU), wall and ceiling luminance coefficients, luminance data from 45 to 85°, a table of visual comfort probability (VCP), and recommended spacing-to-mounting-height ratio (*S/MH*). In addition, many manufacturers either regularly publish, or make available on request, various design aids such as isolux (isocandle) curves, ESI charts, and point-by-point computer printouts for different layouts. These very useful data and their application are covered in Section 20.38.

2. *Construction and installation*. The designer should check the sample for workmanship, rigidity, quality of materials and finish, and ease of installation, wiring, and leveling. Installation instruction sheets should be sufficiently detailed. Results

of actual operating temperature tests in various installation modes should be included. Air-handling fixtures should be furnished with heat-removal data, pressure-drop curves, air-diffusion data, and noise-criteria (NC) data for different airflow rates.

3. *Maintenance*. Luminaires should be simply and quickly relampable, resistant to dirt collection, and simple to clean. Replacement parts must be readily available.

20.26 Luminaire Diffusers

The purpose of the diffusing element is to screen the source and to diffuse and redirect the light. The four types in common use are translucent glass and plastics, various sizes and shapes of louvres and baffles including miniature parabolic section louvres, prismatic plastic and glass sheets, and "batwing" characteristic diffusers. Each of these types must be considered on its merits, and a decision arrived at based on photometric characteristics, cost, ease of maintenance, appearance, and fire safety. A rapid review of the photometric characteristics of this important element of a lighting fixture follows.

(a) Translucent Diffusers. Because these do not redirect the light but merely diffuse it, the distribution characteristic is as in Fig. 20.30*a*. Typical examples of this type are white opal glass, frosted glass, and white plastics such as Plexiglas, polystyrene, vinyl, and polycarbonates. The distribution is basically the same as it would be for bare lamps. Diffusion is good. Depending on the material, direct glare can be a problem (VCP is poor). Veiling reflections are high, that is, ESI is low. Spacing-to-mounting-height ratio does not exceed 1.5. The fixture is generally inefficient. Wall illumination is good because of a large component of high-angle light (which reduces VCP).

A special type of flat plastic panel that polarizes the transmitted light was introduced some years ago. These panels held great promise because they produce a marked decrease of veiling reflection at an angle of 60°, but much less at other angles. Since most viewing is in the 20–40° range, using these panels does not result in any appreciable reduction in reflected glare, in normal office work situations. From experience in a drafting room equipped with luminaires utilizing high-efficiency multilayer polarizers, it can be stated that visual discomfort from reflected glare, as personally experienced, and as reported by a large staff, is not noticeably reduced.

(b) Louvres and Baffles. These are generally rectangular section metal or plastic and serve basically to shield the source (see Fig. 20.22). Typical

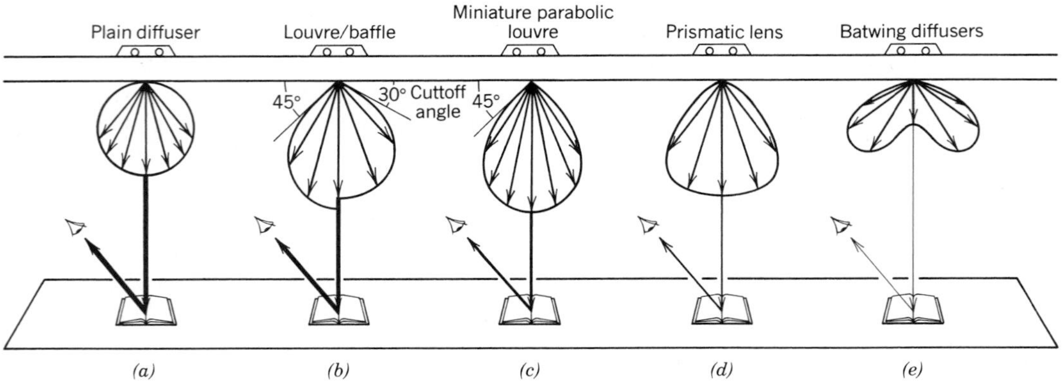

Fig. 20.30 *Comparison of typical candlepower distribution curves for common (fluorescent) luminaire diffuser elements. For a full description see text. Note that for a given geometry of viewer and luminaire, the severity of veiling reflections depends entirely on the fixture's photometric characteristic. In the individual figures, the potential to produce the reflected glare is indicated by line weight. Note that the batwing distribution* (e) *concentrates its output in the 30–60° range, which minimizes both direct and reflected glare.*

candlepower distribution curves are shown in Fig. 20.30*b*. The exact shape depends on the cutoff angle, design of the louvre, and its finish. Louvres finished in specular Alzac or dark colors yield low direct glare. The large downward light component can cause serious veiling reflections. Overall fixture efficiency is fair. The value of *S/MH* does not exceed 1.5 and varies inversely with cutoff angle. This is because the basic circular distribution is changed to egg shaped by cutoff and redirection, reducing the high-angle light. Thus, a 45° cutoff has lower direct glare but requires closer spacing.

A special design of this category is the miniature egg-crate parabolic wedge louvre shown in Fig. 20.31. These units redirect a large portion of the light directly downward and because of their specular finish appear completely dark—darker indeed than the rest of the ceiling, when viewed obliquely. See Figs. 18.9*a*, 21.33. Fixtures using these louvres have low efficiency due to the trapped light, with a maximum CU of about 0.5. VCP is very high, but veiling reflections can be troublesome; *S/MH* is low, varying between 1.0 and 1.5. The candlepower distribution curve is shown in Fig. 20.30*c*. When using these units, additional wall lighting is almost always required.

(c) Prismatic Lens. There are many designs available with varying distribution characteristics. Figures 20.30*d* and 20.24*a* can be taken as typical of this genre. They produce an efficient fixture (high coefficient of utilization), good diffusion, wide permissible spacing—*S/MH* as high as 2.0—and low direct glare (high VCP). Veiling reflections can be troublesome, depending on viewing angles and position (see Fig. 18.32*a*).

(d) Batwing Diffusers. The theory of this type of diffuser is covered in paragraph 4 of Section 18.27*c*. A typical characteristic is shown in Fig. 20.30*e*. There are two fluorescent luminaire designs that produce the batwing shape distribution characteristic—a prismatic lens and parabolic reflectors and baffles.

1. *Prismatic batwing diffusers*. These are either linear or radial, that is, they produce the batwing distribution either in one direction or in all directions. Typical characteristics of both types are shown in Fig. 20.32. Note that the characteristic shape is more pronounced in the linear diffuser,

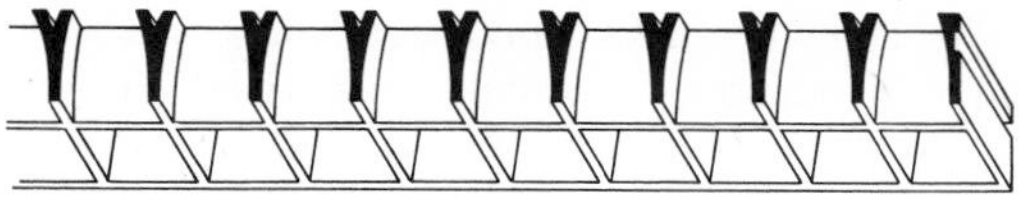

Fig. 20.31 *Section through an egg-crate type of parabolic wedge louvre. This louvre gives exceptionally low brightness and high VCP. Most such units are made of aluminized plastic.*

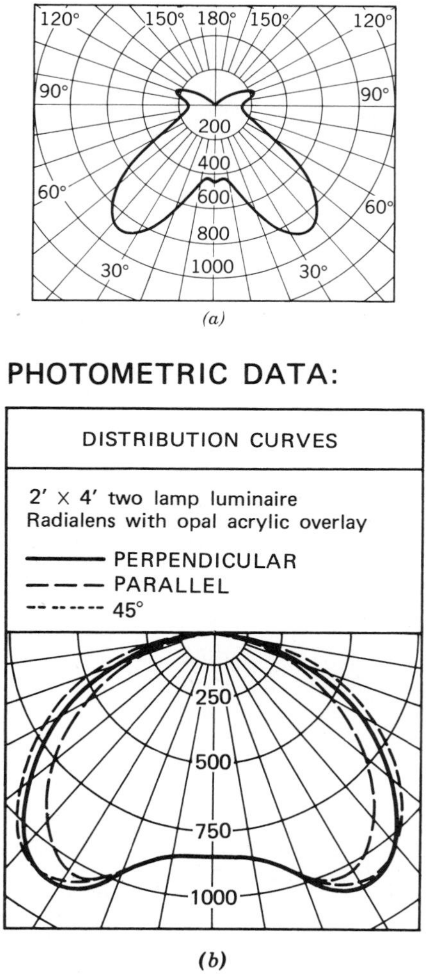

Fig. 20.32 (a) *Linear batwing distribution with extremely sharp cutoff in the upper and lower ranges. Curve taken across lamp axis for Holophane Percepta (registered trademark) single lamp unit. (Courtesy of Holophane/Mansville.)* (b) *Distribution curves for a radial batwing distribution lens. Note that the perpendicular, parallel, and diagonal curves are almost identical. Zonal flux is maximum in the 30–60° range and drops off at both extremes, as desired. (Courtesy of J. W. Carroll and Sons.)*

which indicates better control of veiling reflections in that direction (normally crosswise). Fixtures equipped with these diffusers have good efficiency, low direct and reflected glare, and good diffusion. As with all enclosed ungasketed fixtures, the lens acts as a dust trap, necessitating frequent cleaning to maintain high output.

2. *Parabolic reflectors*. These luminaires produce the batwing characteristic in the crosswise direction only. The lengthwise characteristic is circular. Typical units are shown in Figs. 20.33 and 20.34. The principal advantage of the batwing characteristic, that of veiling reflection control, is shown in Fig. 20.35*a*. Note that the fixtures are

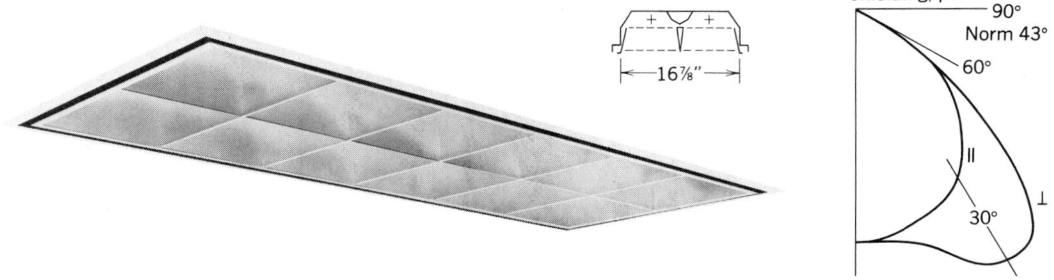

Fig. 20.33 Distribution characteristic of a typical two-lamp baffled, parabolic reflector fixture. All surfaces are semispecular aluminum. The crosswise direction has a typical batwing characteristic, which gives good ESI and VCP. (Courtesy of Columbia Lighting.)

Luminaire Luminance Data

	Average Footlamberts	
Deg	*Parallel*	*Crosswise*
0	720	720
45	583	1608
55	508	1949
65	146	209
75	40	63
85	32	56

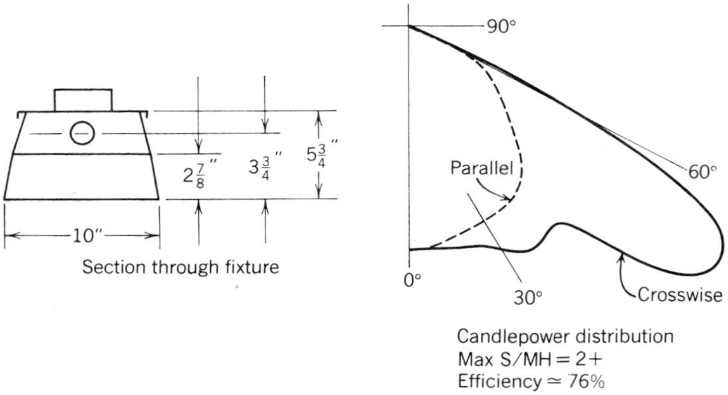

Coefficients of Utilization; Zonal Cavity Method

Reflectance		*Room Cavity Ratios*										
Floor	*Wall*	*0*	*1*	*2*	*3*	*4*	*5*	*6*	*7*	*8*	*9*	*10*
.30	.60	100	90	81	72	65	57	51	46	41	37	33
.30	.50	100	88	77	68	60	53	46	41	36	32	29
.30	.40	100	86	75	65	56	49	42	37	32	28	25
.30	.30	100	84	72	62	53	46	39	34	29	25	22
.20	.60	90	83	75	68	61	55	49	44	40	36	32
.20	.50	90	81	73	65	58	51	45	40	35	31	28
.20	.40	90	80	70	62	54	47	41	36	31	27	24
.20	.30	90	79	68	60	52	44	38	33	29	25	22

NOTE: Ceiling cavity reflectance ρ_{CC} = 80%.

Fig. 20.34 Photometric characteristic of the single-lamp, baffled, parabolic reflector luminaire used in the installation shown in Figure 20.35. The pronounced batwing characteristic (crosswise) accounts for the reduced veiling reflections from specular items in the field of view, as seen in Fig. 20.35. (Courtesy of Peter L. Goodman/Edison Price.)

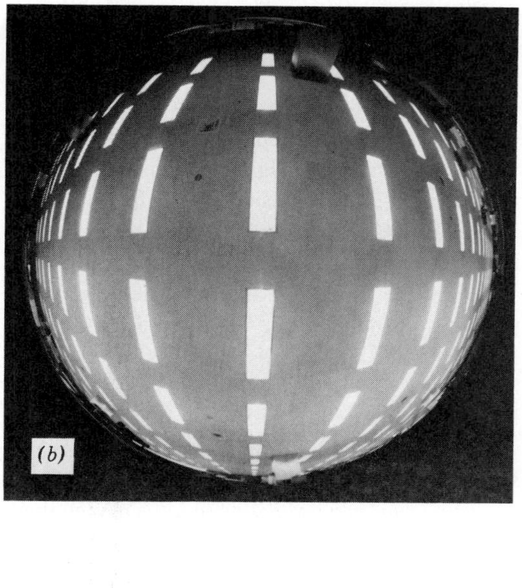

Figure 20.35 (a) *Drafting room lighted with* single-lamp *parabolic reflector baffled fixtures whose characteristics are given in Fig. 20.34. Note the absence of reflection in specular items on the drafting table. Light from other fixtures* (b) *causes high overall luminance that eliminates the problem of light coming from the "offending zone" on the ceiling.* (b) *Fish-eye view of the ceiling, taken from the center of the front table of* (a). *Note that considerable light is contributed from adjacent rows, due to the batwing distribution. (Photos courtesy of Peter L. Goodman/Edison Price.)*

installed as shown graphically in Fig. 18.39, above, and that reflections, even on the specular surfaces shown, are absent. This is so for two reasons: the light from the offending zone is limited by the fixture characteristic, and the light from adjoining luminaires increases background luminance to the point that the ''offending'' light is unnoticed. (Reread the first paragraph of Section 18.27b.) Figure 20.35*b* shows this effect clearly. Luminaires of this design are very efficient, exhibit low direct glare in addition to low reflected glare, and because of the open design, remain clean for extended periods. The *S/MH* is high—between 1.5 and 2.0.

20.27 Luminaire Efficiency: Coefficient of Utilization

A luminaire, variously called a fixture, lighting unit, or reflector, comprises a device for physically supporting the light source and usually for directing or controlling the light output of this source. Because of internal reflections, some of the generated lumen output of the lamp is lost within the fixture. The ratio of output lumens to lamp (input) lumens, expressed as a percentage, represents the luminous efficiency of the fixture. This information is normally available from the manufacturer. This characteristic has little meaning by itself, however, since the actual overall efficiency of a luminaire depends on the space in which it is used.

To illustrate, let us consider the case of a large room with a high, dark ceiling. If we were to use a high-efficiency (say 80%) indirect lighting unit in such a room, most of the light directed upward would be lost (absorbed), and the actual lighting on the working plane would be very low. If, however, the same room were illuminated with 50% efficiency direct-lighting units utilizing the same wattage, the illumination on the working plane would be considerably higher than in the first case.

Similarly, if we consider a small room with dark walls and ceiling, lighted alternatively by diffuse lighting and by direct lighting units of the same wattage and unit efficiency, the horizontal-plane illumination will be higher in the case of the direct units because of the large loss of the horizontal and upward components of the diffuse lighting on the walls and ceiling. It should be obvious then that the fixture efficiency *alone* is not sufficient but *the overall luminous efficiency of a particular unit in a particular space* is the required figure of merit. This number, since it describes the utilization of the fixture output in a specific space, is known as the *coefficient of utilization* (CU). It is defined as the ratio between the lumens reaching the horizontal working plane and the generated lumens. Since each luminaire will have a different coefficient for every different space in which it is used, a system of standardization has been evolved utilizing room cavities (explained below) of certain proportions and various surface reflectances. The fixture coefficients are then computed and tabulated as shown in Table 20.2. It should be emphasized that the figures given in this table are for the generic fixture type only; in an actual job, actual fixture data should be used. When fixture data are not available, as in preliminary estimates and calculations, use the coefficients listed in Section 20.5. To summarize, CU is a factor that combines fixture efficiency and distribution with room proportions, mounting height, and surface reflectances.

DETAILED DESIGN PROCEDURES

20.28 Calculation of Average Illuminance

Having selected a luminaire on the basis of all the foregoing criteria, it remains only to calculate the number of such units required in each space, for *general* illumination, and to arrange them properly. Although a number of calculation methods are available, the lumen (flux) method is simplest and most applicable to our needs for area lighting calculations. Illuminance calculation from point, line, or area sources is covered in Sections 20.38 to 20.41. Luminance (photometric brightness) calculations are covered in Section 20.42.

Before beginning a detailed description of the zonal cavity calculation method, a general comment on precision and accuracy is in order. The degree of precision of any calculation need not exceed either the accuracy required or the precision of the data available. Thus there is no point in working to three decimal places if a ±10% rounding of the result is common and acceptable, or if the data available are only accurate to one decimal place. As the reader will see, the lumen (lighting flux) method of average illuminance calculation is replete with assumptions and estimates.

TABLE 20.2 **Coefficients of Utilization for Typical Luminaires with Suggested Maximum Spacing Ratios**

To obtain a coefficient of utilization:

1. Determine cavity ratios for the room, ceiling, and floor.
2. Determine the effective ceiling and floor cavity reflectances from Table 20.3. Use initial ceiling, floor, and wall reflectances.
3. Obtain coefficient of utilization (CU) for 20% effective floor cavity reflectance from appropriate table below for luminaire type to be used. Interpolate, when necessary, to obtain CU for exact room cavity ratio for nearest effective ceiling cavity reflectances above and below reflectance obtained in step 2; interpolate between these CUs to obtain CU for step 2 ceiling cavity reflectance.
4. If effective floor cavity reflectance differs significantly from 20%, obtain multiplier from Table 20.4 and apply this to the CU obtained in step 3.
5. To obtain CU for a ceiling cavity reflectance (ρ_{CC}) of 30 or 10%, multiply the figure for ρ_{CC} = 50% by 0.85 and 0.70, respectively. This is an approximation. For exact figures see *IES Handbook* (1981).
6. Use the figure in the last column ($\rho_{CC} = 0$; $\rho_W = 0$) for outdoor lighting, i.e., no walls or ceiling.
7. Legend:

ρ_{CC} = percent effective ceiling cavity reflectance
ρ_W = percent wall reflectance
RCR = room cavity ratio
Maximum *S*/*MH* guide = ratio of maximum luminaire spacing to mounting above work plane.

NOTE: In some cases, luminaire data in this table are based on an actual typical luminaire; in other cases, the data represent a composite of generic luminaire types. Therefore, whenever possible, specific luminaire data should be used in preference to this table of typical luminaires.

The polar intensity sketch (candlepower distribution curve) and the corresponding spacing-to-mounting height guide are representative of many luminaires of each type shown.

Typical Luminaire	Typical Distribution and Percent Lamp Lumens: Maintenance Category	Maximum S/MH Guide	ρ_{CC} → / ρ_W → / RCR	80 / 50	80 / 30	80 / 10	70 / 50	70 / 30	70 / 10	50 / 50	50 / 30	50 / 10	0 / 0
				Coefficients of Utilization for 20% Effective Floor Cavity Reflectance (ρ_{FC} = 20)									
1	V	1.5	0	.87	.87	.87	.81	.81	.81	.69	.69	.69	.44
			1	.71	.67	.63	.66	.62	.59	.56	.53	.50	.31
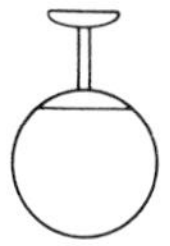	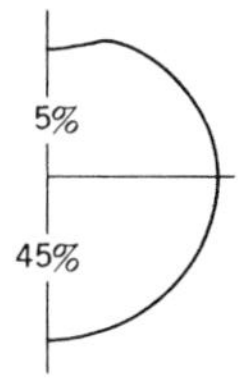		2	.61	.54	.49	.56	.50	.46	.47	.43	.39	.23
			3	.52	.45	.39	.48	.42	.37	.41	.36	.31	.18
			4	.46	.38	.33	.42	.36	.30	.36	.30	.26	.15
			5	.40	.33	.27	.37	.30	.25	.32	.26	.22	.12
			6	.36	.28	.23	.33	.26	.21	.28	.23	.19	.10
			7	.32	.25	.20	.29	.23	.18	.25	.20	.16	.09
Pendant diffusing sphere with incandescent lamp			8	.29	.22	.17	.27	.20	.16	.23	.17	.14	.07
			9	.26	.19	.15	.24	.18	.14	.20	.15	.12	.06
			10	.23	.17	.13	.22	.16	.12	.19	.14	.10	.05

(*continued*)

TABLE 20.2 **Coefficients of Utilization for Typical Luminaires with Suggested Maximum Spacing Ratios** (*continued*)

Typical Luminaire	Typical Distribution and Percent Lamp Lumens: Maintenance Category	Maximum S/MH Guide	ρ_{CC}:	80			70			50			0
			ρ_W:	50	30	10	50	30	10	50	30	10	0
			RCR	Coefficients of Utilization for 20% Effective Floor Cavity Reflectance (ρ_{FC} = 20)									
3	IV	1.3	0	.99	.99	.99	.97	.97	.97	.92	.92	.92	.83
	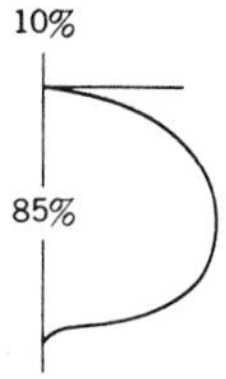		1	.88	.85	.82	.86	.83	.81	.83	.80	.78	.72
			2	.78	.73	.68	.76	.72	.67	.73	.69	.66	.61
			3	.69	.62	.57	.67	.61	.57	.65	.60	.56	.52
			4	.61	.54	.49	.60	.53	.48	.58	.52	.48	.45
			5	.54	.47	.41	.53	.46	.41	.51	.45	.41	.38
			6	.48	.41	.35	.47	.40	.35	.46	.39	.35	.32
Porcelain-enameled ventilated standard dome with incandescent lamp			7	.43	.35	.30	.42	.35	.30	.41	.34	.30	.28
			8	.38	.31	.26	.38	.31	.26	.37	.30	.26	.24
			9	.35	.28	.23	.34	.27	.23	.33	.27	.23	.21
			10	.31	.25	.20	.31	.24	.20	.30	.24	.20	.18
7	IV	0.7	0	.52	.52	.52	.51	.51	.51	.48	.48	.48	.44
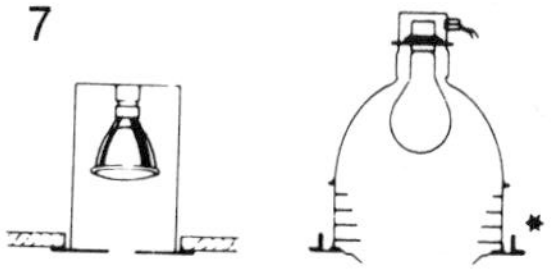	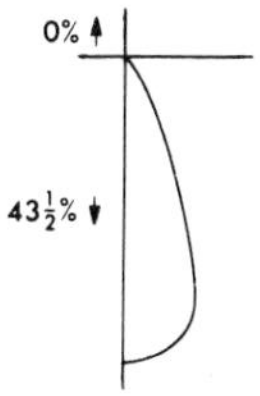		1	.49	.48	.48	.48	.48	.47	.47	.46	.46	.42
			2	.47	.46	.45	.46	.45	.44	.45	.44	.43	.41
			3	.45	.44	.43	.45	.43	.42	.44	.42	.42	.40
			4	.43	.42	.41	.43	.41	.40	.42	.41	.40	.38
EAR-38 lamp above 51 mm (2″) diameter aperture (increase efficiency to 54½% for 76 mm (3″) diameter aperture)			5	.42	.40	.39	.41	.40	.38	.41	.39	.38	.37
			6	.40	.39	.37	.40	.38	.37	.39	.38	.37	.36
			7	.39	.37	.36	.39	.37	.36	.38	.37	.35	.35
			8	.37	.36	.34	.37	.35	.34	.37	.35	.34	.33
			9	.36	.34	.33	.36	.34	.33	.35	.34	.33	.32
			10	.35	.33	.32	.35	.33	.32	.34	.33	.32	.31

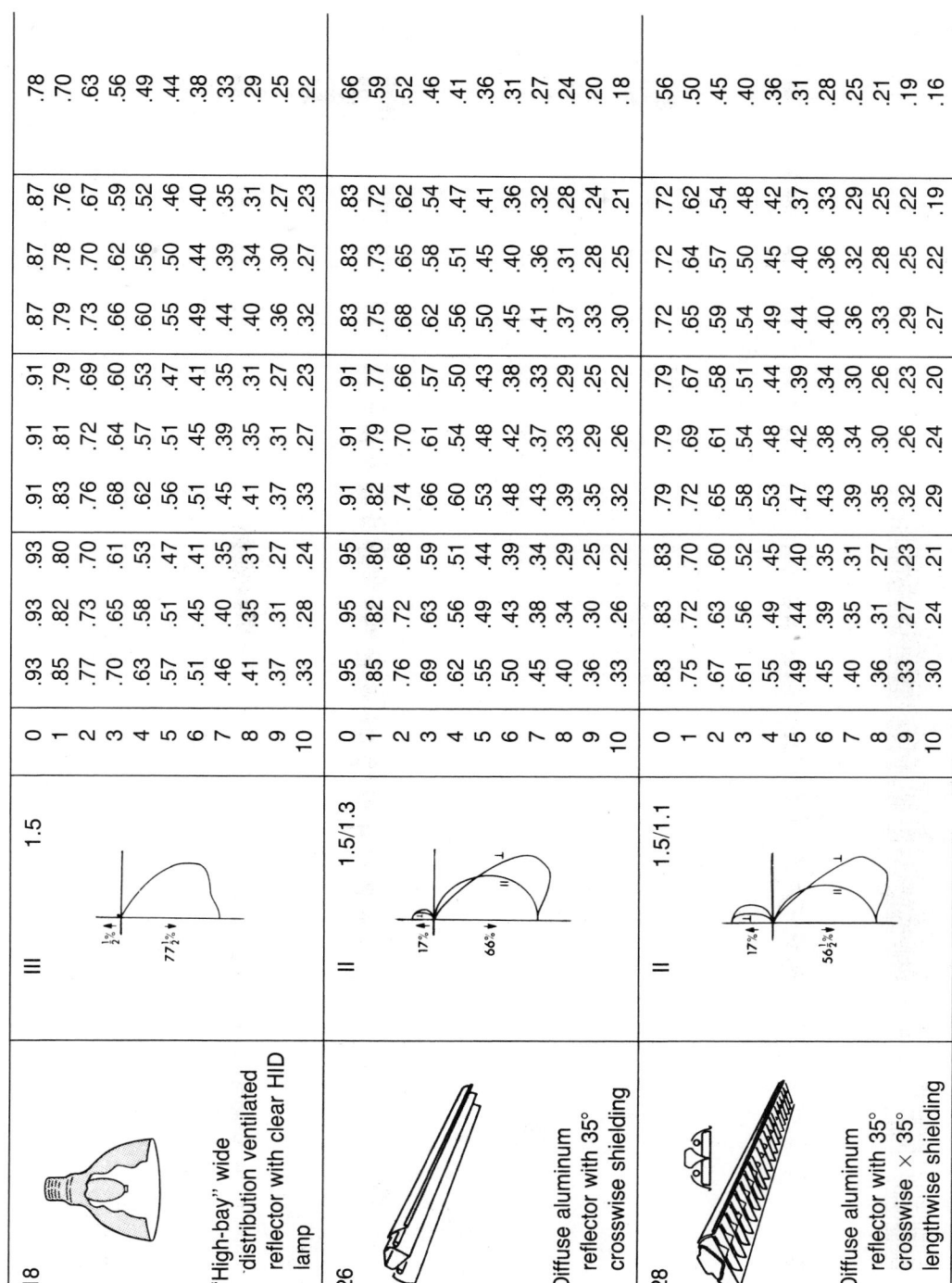

18 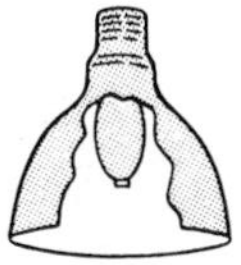 "High-bay" wide distribution ventilated reflector with clear HID lamp	III 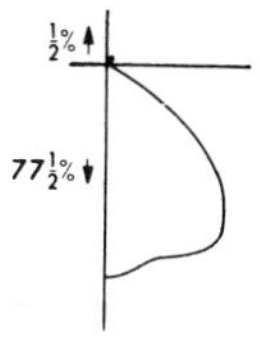	1.5	0	.93	.93	.93	.91	.91	.91	.87	.87	.87	.78
			1	.85	.82	.80	.83	.81	.79	.79	.78	.76	.70
			2	.77	.73	.70	.76	.72	.69	.73	.70	.67	.63
			3	.70	.65	.61	.68	.64	.60	.66	.62	.59	.56
			4	.63	.58	.53	.62	.57	.53	.60	.56	.52	.49
			5	.57	.51	.47	.56	.51	.47	.55	.50	.46	.44
			6	.51	.45	.41	.51	.45	.41	.49	.44	.40	.38
			7	.46	.40	.35	.45	.39	.35	.44	.39	.35	.33
			8	.41	.35	.31	.41	.35	.31	.40	.34	.31	.29
			9	.37	.31	.27	.37	.31	.27	.36	.30	.27	.25
			10	.33	.28	.24	.33	.27	.23	.32	.27	.23	.22
26 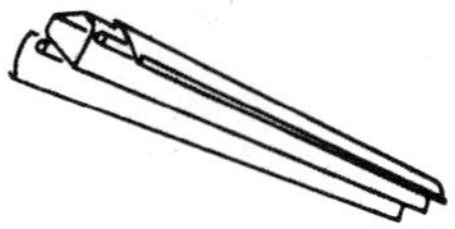 Diffuse aluminum reflector with 35° crosswise shielding	II 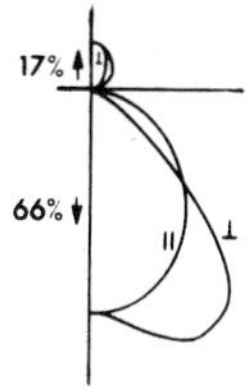	1.5/1.3	0	.95	.95	.95	.91	.91	.91	.83	.83	.83	.66
			1	.85	.82	.80	.82	.79	.77	.75	.73	.72	.59
			2	.76	.72	.68	.74	.70	.66	.68	.65	.62	.52
			3	.69	.63	.59	.66	.61	.57	.62	.58	.54	.46
			4	.62	.56	.51	.60	.54	.50	.56	.51	.47	.41
			5	.55	.49	.44	.53	.48	.43	.50	.45	.41	.36
			6	.50	.43	.39	.48	.42	.38	.45	.40	.36	.31
			7	.45	.38	.34	.43	.37	.33	.41	.36	.32	.27
			8	.40	.34	.29	.39	.33	.29	.37	.31	.28	.24
			9	.36	.30	.25	.35	.29	.25	.33	.28	.24	.20
			10	.33	.26	.22	.32	.26	.22	.30	.25	.21	.18
28 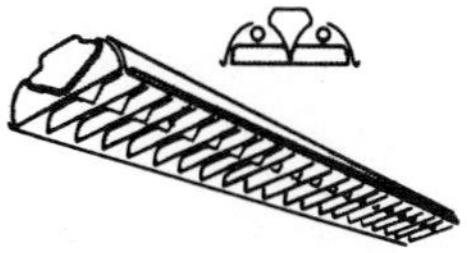 Diffuse aluminum reflector with 35° crosswise × 35° lengthwise shielding	II 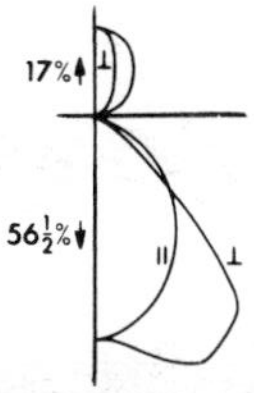	1.5/1.1	0	.83	.83	.83	.79	.79	.79	.72	.72	.72	.56
			1	.75	.72	.70	.72	.69	.67	.65	.64	.62	.50
			2	.67	.63	.60	.65	.61	.58	.59	.57	.54	.45
			3	.61	.56	.52	.58	.54	.51	.54	.50	.48	.40
			4	.55	.49	.45	.53	.48	.44	.49	.45	.42	.36
			5	.49	.44	.40	.47	.42	.39	.44	.40	.37	.31
			6	.45	.39	.35	.43	.38	.34	.40	.36	.33	.28
			7	.40	.35	.31	.39	.34	.30	.36	.32	.29	.25
			8	.36	.31	.27	.35	.30	.26	.33	.28	.25	.21
			9	.33	.27	.23	.32	.26	.23	.29	.25	.22	.19
			10	.30	.24	.21	.29	.24	.20	.27	.22	.19	.16

(continued)

TABLE 20.2 **Coefficients of Utilization for Typical Luminaires with Suggested Maximum Spacing Ratios (*continued*)**

Typical Luminaire	Typical Distribution and Percent Lamp Lumens: Maintenance Category	Maximum S/MH Guide	ρ_{CC}: / ρ_W: / RCR	80 / 50	80 / 30	80 / 10	70 / 50	70 / 30	70 / 10	50 / 50	50 / 30	50 / 10	0 / 0
				Coefficients of Utilization for 20% Effective Floor Cavity Reflectance (ρ_{FC} = 20)									
33	VI	N.A.	0	.77	.77	.77	.68	.68	.68	.50	.50	.50	.12
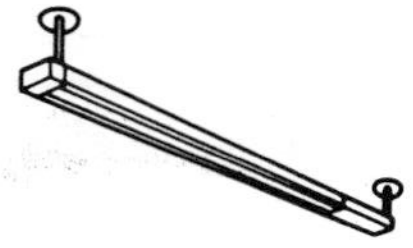			1	.67	.64	.62	.59	.57	.54	.44	.42	.41	.10
	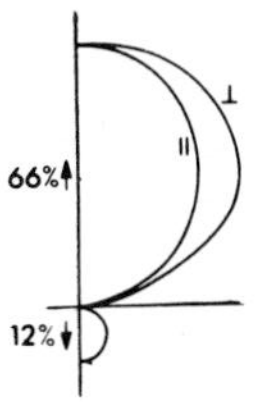		2	.59	.54	.50	.52	.48	.45	.38	.36	.34	.09
			3	.51	.46	.42	.45	.41	.37	.34	.31	.28	.07
			4	.45	.40	.35	.40	.35	.31	.30	.27	.24	.06
			5	.40	.34	.30	.35	.30	.27	.26	.23	.20	.05
			6	.36	.30	.26	.32	.27	.23	.24	.20	.18	.05
			7	.32	.26	.22	.28	.23	.20	.21	.18	.15	.04
Luminous bottom suspended unit with extra-high-output lamp			8	.29	.23	.19	.25	.21	.17	.19	.16	.13	.03
			9	.26	.20	.17	.23	.18	.15	.17	.14	.12	.03
			10	.24	.18	.15	.21	.16	.13	.16	.12	.10	.03
35	V	1.5/1.2	0	.81	.81	.81	.78	.78	.78	.72	.72	.72	.59
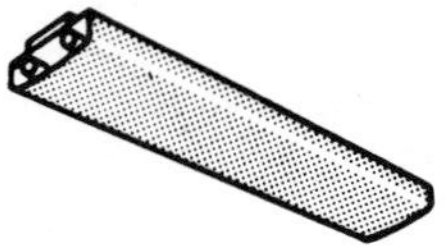			1	.71	.69	.66	.69	.66	.64	.64	.62	.60	.50
	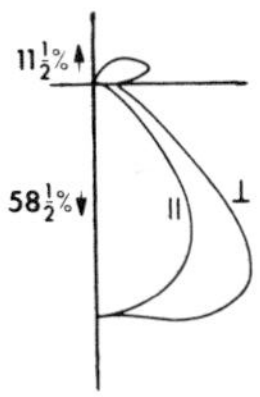		2	.64	.59	.56	.61	.58	.54	.57	.54	.51	.44
			3	.57	.52	.48	.55	.50	.47	.51	.48	.45	.38
			4	.51	.46	.41	.49	.44	.41	.46	.42	.39	.34
			5	.46	.40	.36	.44	.39	.35	.41	.37	.34	.29
			6	.41	.35	.31	.40	.35	.31	.38	.33	.30	.26
			7	.37	.31	.27	.36	.31	.27	.34	.29	.26	.23
Two-lamp prismatic wraparound; multiply by 0.95 for four lamps			8	.33	.28	.24	.32	.27	.23	.30	.26	.22	.19
			9	.30	.24	.20	.29	.24	.20	.27	.23	.19	.17
			10	.27	.22	.18	.26	.21	.18	.25	.20	.17	.15

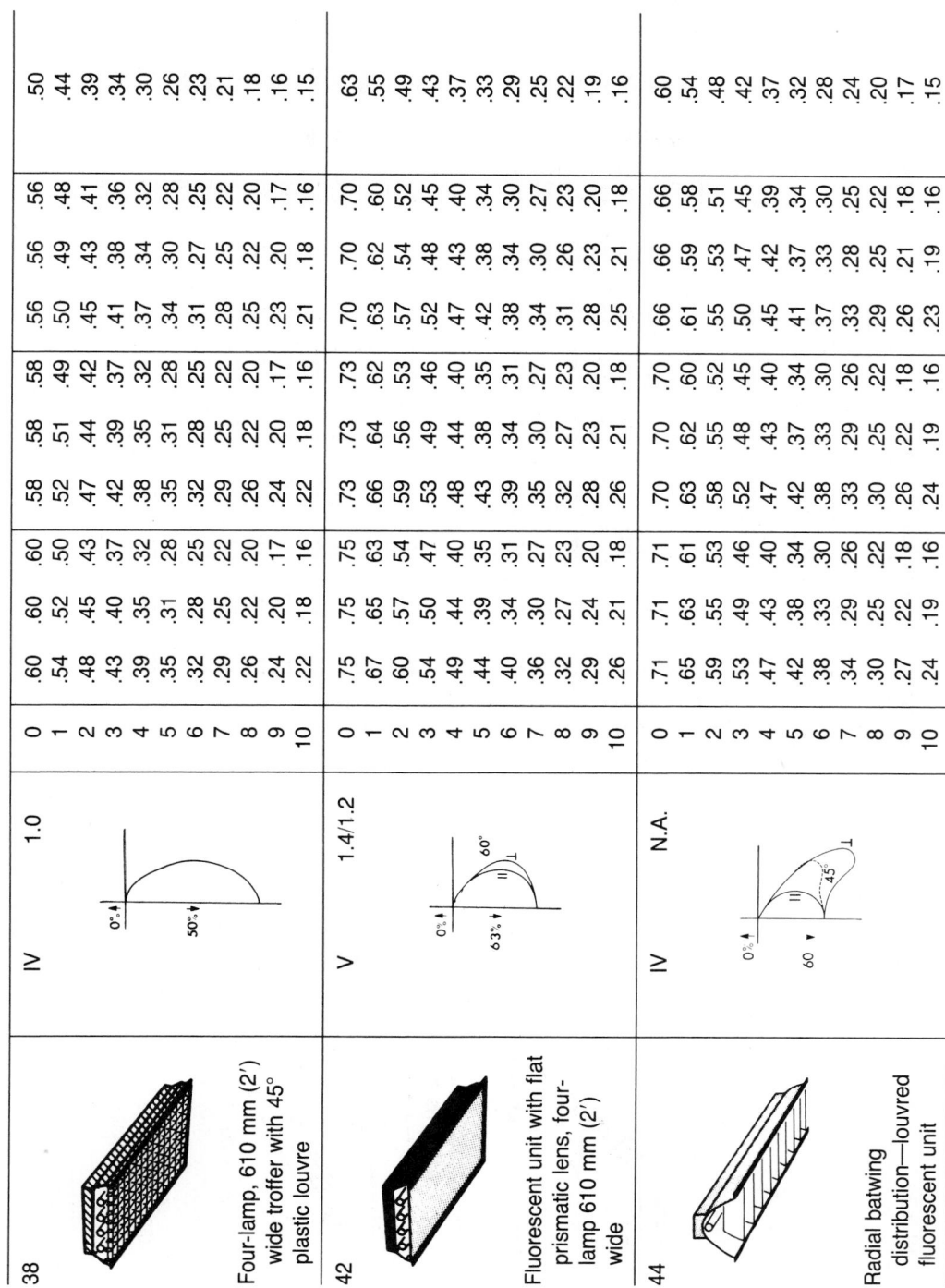

38	IV	1.0	0	.60	.60	.60	.58	.58	.58	.56	.56	.56	.50
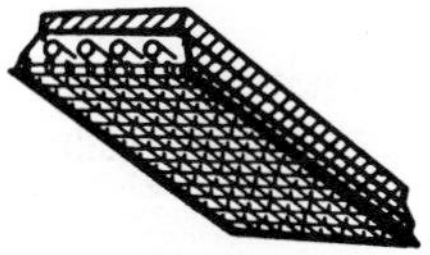	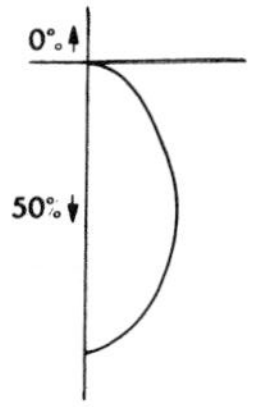		1	.54	.52	.50	.52	.51	.49	.50	.49	.48	.44
			2	.48	.45	.43	.47	.44	.42	.45	.43	.41	.39
			3	.43	.40	.37	.42	.39	.37	.41	.38	.36	.34
			4	.39	.35	.32	.38	.35	.32	.37	.34	.32	.30
			5	.35	.31	.28	.35	.31	.28	.34	.30	.28	.26
			6	.32	.28	.25	.32	.28	.25	.31	.27	.25	.23
			7	.29	.25	.22	.29	.25	.22	.28	.25	.22	.21
Four-lamp, 610 mm (2′) wide troffer with 45° plastic louvre			8	.26	.22	.20	.26	.22	.20	.25	.22	.20	.18
			9	.24	.20	.17	.24	.20	.17	.23	.20	.17	.16
			10	.22	.18	.16	.22	.18	.16	.21	.18	.16	.15
42	V	1.4/1.2	0	.75	.75	.75	.73	.73	.73	.70	.70	.70	.63
	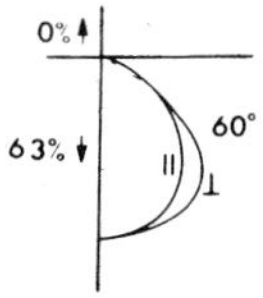		1	.67	.65	.63	.66	.64	.62	.63	.62	.60	.55
			2	.60	.57	.54	.59	.56	.53	.57	.54	.52	.49
			3	.54	.50	.47	.53	.49	.46	.52	.48	.45	.43
			4	.49	.44	.40	.48	.44	.40	.47	.43	.40	.37
			5	.44	.39	.35	.43	.38	.35	.42	.38	.34	.33
			6	.40	.34	.31	.39	.34	.31	.38	.34	.30	.29
Fluorescent unit with flat prismatic lens, four-lamp 610 mm (2′) wide			7	.36	.30	.27	.35	.30	.27	.34	.30	.27	.25
			8	.32	.27	.23	.32	.27	.23	.31	.26	.23	.22
			9	.29	.24	.20	.28	.23	.20	.28	.23	.20	.19
			10	.26	.21	.18	.26	.21	.18	.25	.21	.18	.16
44	IV	N.A.	0	.71	.71	.71	.70	.70	.70	.66	.66	.66	.60
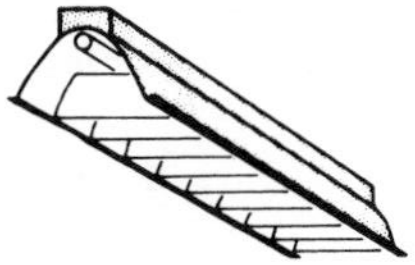	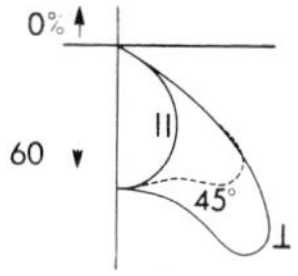		1	.65	.63	.61	.63	.62	.60	.61	.59	.58	.54
			2	.59	.55	.53	.58	.55	.52	.55	.53	.51	.48
			3	.53	.49	.46	.52	.48	.45	.50	.47	.45	.42
			4	.47	.43	.40	.47	.43	.40	.45	.42	.39	.37
			5	.42	.38	.34	.42	.37	.34	.41	.37	.34	.32
			6	.38	.33	.30	.38	.33	.30	.37	.33	.30	.28
			7	.34	.29	.26	.33	.29	.26	.33	.28	.25	.24
Radial batwing distribution—louvred fluorescent unit			8	.30	.25	.22	.30	.25	.22	.29	.25	.22	.20
			9	.27	.22	.18	.26	.22	.18	.26	.21	.18	.17
			10	.24	.19	.16	.24	.19	.16	.23	.19	.16	.15

(*continued*)

TABLE 20.2 **Coefficients of Utilization for Typical Luminaires with Suggested Maximum Spacing Ratios (*continued*)**

Typical Luminaire	Typical Distribution and Percent Lamp Lumens: Maintenance Category	Maximum S/MH Guide	ρ_{CC}: / ρ_W: / RCR	80 / 50	80 / 30	80 / 10	70 / 50	70 / 30	70 / 10	50 / 50	50 / 30	50 / 10	0 / 0
				Coefficients of Utilization for 20% Effective Floor Cavity Reflectance (ρ_{FC} = 20)									
45 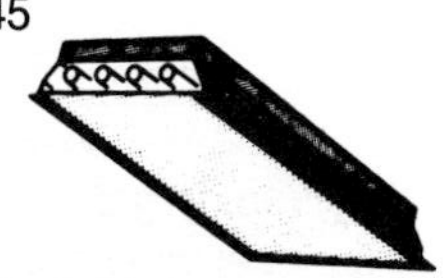	V	N.A.	0	.57	.57	.57	.56	.56	.56	.53	.53	.53	.48
			1	.50	.48	.47	.49	.47	.46	.47	.46	.44	.41
	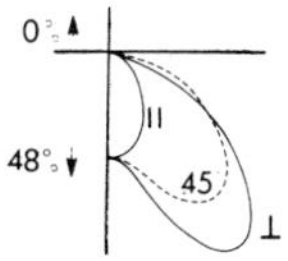		2	.44	.41	.38	.43	.40	.38	.41	.39	.37	.34
			3	.39	.35	.32	.38	.34	.31	.37	.33	.31	.29
			4	.34	.30	.27	.33	.29	.26	.32	.29	.26	.24
Radial batwing			5	.30	.25	.22	.29	.25	.22	.28	.24	.22	.20
distribution—four lamp,			6	.26	.22	.19	.26	.22	.18	.25	.21	.18	.17
610 mm (2′) wide			7	.23	.19	.16	.23	.19	.16	.22	.18	.16	.14
fluorescent unit with			8	.21	.16	.13	.20	.16	.13	.19	.16	.13	.12
flat prismatic lens and			9	.18	.14	.11	.18	.14	.11	.17	.14	.11	.10
overlay			10	.16	.12	.09	.16	.12	.09	.16	.12	.09	.08
46 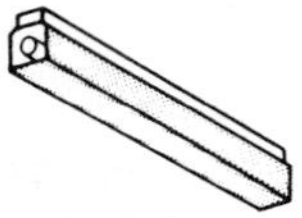	V	N.A.	0	.87	.87	.87	.84	.84	.84	.77	.77	.77	.64
			1	.76	.73	.70	.73	.70	.67	.67	.65	.63	.53
	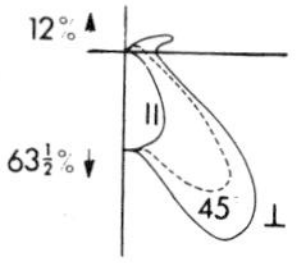		2	.66	.61	.57	.64	.59	.56	.59	.56	.52	.44
			3	.59	.53	.48	.56	.51	.47	.53	.48	.44	.38
			4	.52	.45	.40	.50	.44	.40	.47	.42	.38	.32
Bilateral batwing			5	.46	.39	.34	.44	.38	.33	.41	.36	.32	.27
distribution—one lamp,			6	.41	.34	.29	.39	.33	.29	.37	.31	.27	.23
surface-mounted			7	.36	.30	.25	.35	.29	.24	.33	.27	.23	.20
fluorescent with			8	.32	.26	.21	.31	.25	.21	.29	.24	.20	.17
prismatic wraparound			9	.29	.22	.18	.28	.22	.18	.26	.21	.17	.14
lens			10	.26	.20	.16	.25	.19	.15	.23	.18	.15	.12

47  Radial batwing distribution—four lamp, 610 mm (2′) wide fluorescent unit with flat prismatic lens	V	1.7	0	.71	.71	.71	.69	.69	.69	.66	.66	.66	.60
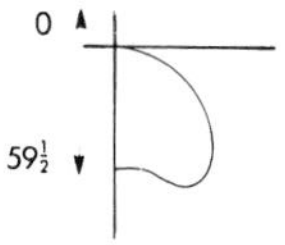			1	.62	.60	.58	.61	.59	.57	.59	.57	.55	.51
			2	.55	.51	.47	.53	.50	.47	.51	.48	.46	.42
			3	.48	.43	.39	.47	.43	.39	.45	.41	.38	.36
			4	.42	.37	.33	.41	.37	.33	.40	.36	.32	.30
			5	.37	.32	.27	.36	.31	.27	.35	.30	.27	.25
			6	.33	.27	.23	.32	.27	.23	.31	.26	.23	.21
			7	.29	.24	.20	.29	.24	.20	.28	.23	.20	.18
			8	.26	.21	.17	.25	.20	.17	.25	.20	.17	.15
			9	.23	.18	.14	.23	.18	.14	.22	.17	.14	.13
			10	.21	.16	.12	.20	.16	.12	.20	.15	.12	.11
48 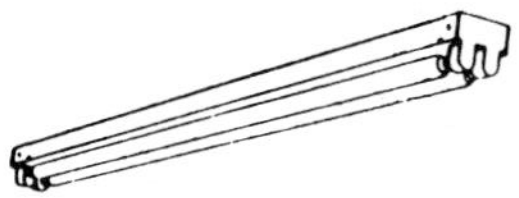 Two-lamp fluorescent strip unit	I	1.6/1.2	0	1.01	1.01	1.01	.96	.96	.96	.87	.87	.87	.68
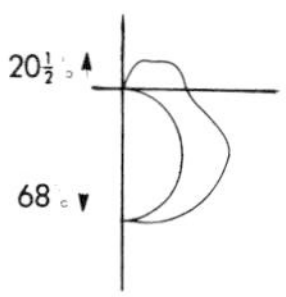			1	.85	.81	.77	.81	.77	.73	.73	.70	.67	.53
			2	.73	.66	.61	.69	.63	.58	.63	.58	.54	.42
			3	.63	.56	.50	.60	.53	.48	.55	.49	.44	.35
			4	.56	.47	.41	.53	.46	.40	.48	.42	.37	.29
			5	.49	.40	.34	.46	.39	.33	.42	.36	.31	.24
			6	.43	.35	.29	.41	.34	.28	.38	.31	.26	.20
			7	.39	.31	.25	.37	.29	.24	.34	.27	.23	.17
			8	.34	.27	.21	.33	.26	.21	.30	.24	.19	.15
			9	.31	.23	.18	.30	.23	.18	.27	.21	.17	.12
			10	.28	.21	.16	.27	.20	.16	.25	.19	.15	.11
49 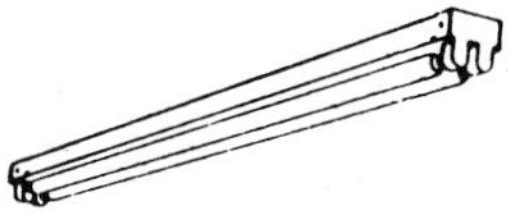 Two-lamp fluorescent strip unit with 235° reflector fluorescent lamps	I	1.4/1.2	0	1.13	1.13	1.13	1.09	1.09	1.09	1.01	1.01	1.01	.85
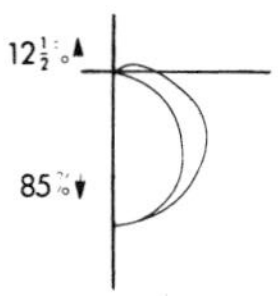			1	.96	.92	.88	.93	.89	.85	.87	.83	.80	.68
			2	.83	.76	.70	.80	.74	.68	.75	.69	.65	.55
			3	.73	.65	.58	.70	.63	.57	.66	.59	.54	.46
			4	.64	.55	.49	.62	.54	.48	.58	.51	.46	.39
			5	.56	.47	.41	.55	.46	.40	.51	.44	.38	.33
			6	.50	.41	.35	.49	.40	.34	.46	.38	.33	.28
			7	.45	.36	.30	.44	.35	.30	.41	.34	.28	.24
			8	.40	.32	.26	.39	.31	.25	.37	.30	.25	.21
			9	.36	.28	.22	.35	.27	.22	.33	.26	.21	.18
			10	.33	.25	.20	.32	.24	.19	.30	.23	.19	.15

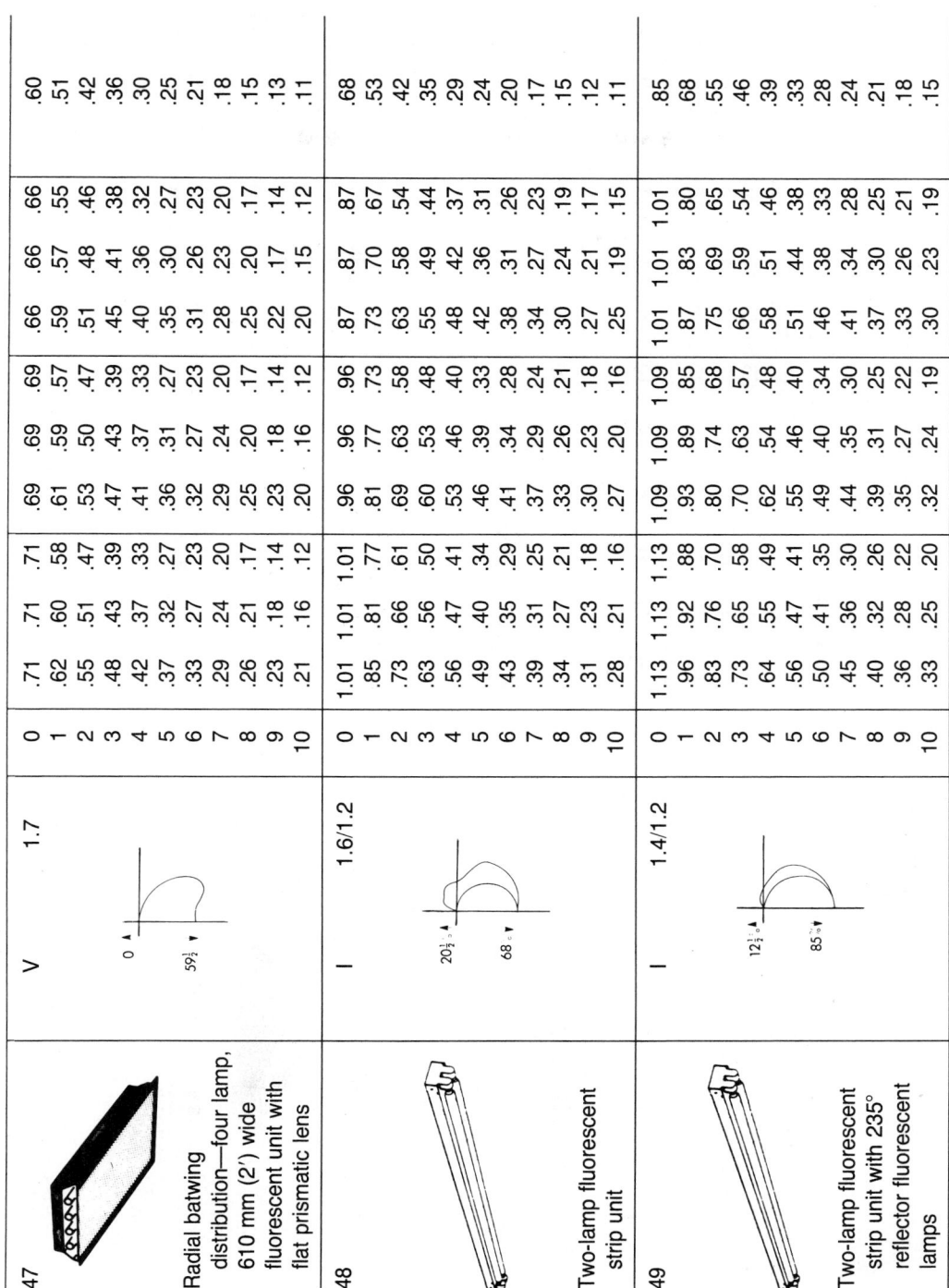

(continued)

TABLE 20.2 **Coefficients of Utilization for Typical Luminaires with Suggested Maximum Spacing Ratios** (*continued*)

Typical Luminaire	Typical Distribution and Percent Lamp Lumens — Maintenance Category	Maximum S/MH Guide	ρ_{CC}:	80			70			50			0		
			ρ_W:	50	30	10	50	30	10	50	30	10	0		
			RCR	Coefficients of Utilization for 20% Effective Floor Cavity Reflectance (ρ_{FC} = 20)											
50			1	.42	.40	.39	.36	.35	.33	.25	.24	.23			
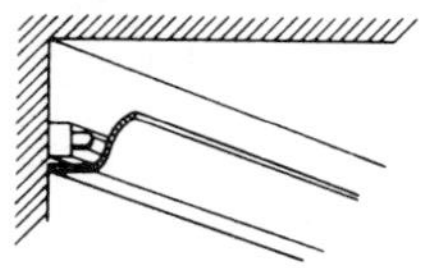			2	.37	.34	.32	.32	.29	.27	.22	.20	.19	Coves are not recommended for lighting areas having low reflectances		
			3	.32	.29	.26	.28	.25	.23	.19	.17	.16			
			4	.29	.25	.22	.25	.22	.19	.17	.15	.13			
			5	.25	.21	.18	.22	.19	.16	.15	.13	.11			
			6	.23	.19	.16	.20	.16	.14	.14	.12	.10			
			7	.20	.17	.14	.17	.14	.12	.12	.10	.09			
Single-row fluorescent lamp cove without reflector, multiplied by 0.93 for two rows and by 0.85 for three rows.			8	.18	.15	.12	.16	.13	.10	.11	.09	.08			
			9	.17	.13	.10	.15	.11	.09	.10	.08	.07			
			10	.15	.12	.09	.13	.10	.08	.09	.07	.06			
													ρ_{CC} = 10%		
													ρ_W = 50	30	10
53 ρ_{CC} from below ~45%			1							.51	.49	.48	.47	.46	.45
			2							.46	.44	.42	.43	.42	.40
			3							.42	.39	.37	.39	.38	.36
			4							.38	.35	.33	.36	.34	.32
			5							.35	.32	.29	.33	.31	.29
			6							.32	.29	.26	.30	.28	.26
Louvred ceiling. Ceiling efficiency ~50%; 45° shielding opaque louvres of 80% reflectance. Cavity with minimum obstructions and painted with 80% reflectance paint—use ρ_{CC} = 50.			7							.29	.26	.23	.28	.25	.23
			8							.27	.23	.21	.26	.23	.21
			9							.24	.21	.19	.24	.21	.19
			10							2	.19	.17	.22	.19	.17

NOTES:
1. Data extracted from *IES Handbook* (1981), Reference Volume, with permission.
2. Multiply by 1.05 for three lamps and 1.1 for two lamps.

Among these are:

1. It is assumed that the space is empty. This is not normally the case.
2. It is assumed that all surfaces are perfect diffusers. This is not the case.
3. All surface reflectances are estimates, ±10%.
4. Maintenance conditions are estimates, at best ±10%.
5. No allowance is made for deviation of the performance of an individual product from its specification.

Any attempt to account accurately for these approximations would enormously complicate the calculation and would serve no useful purpose for this type of *average* illuminance calculation. For this reason, the procedure presented below has introduced some approximations (which we feel are well justified), in the interest of simplicity. These approximations are noted wherever used.

20.29 Calculation of Horizontal Illuminance by the Lumen (Flux) Method

The lumen method of calculation is a procedure for determining the *average* maintained illuminance (footcandles, lux) on the working plane in a room. The method presupposes that luminaires will be spaced so that uniformity of illumination is provided in order that an *average* calculation have validity. The method is based on the definition of one foot-candle [lux] as one lumen incident on one square foot [meter] of area, that is,

$$\text{footcandles [lux]} = \frac{\text{lumens}}{\text{area in ft}^2\ [\text{m}^2]}$$

As explained above, the ratio between the lumens reaching the working plane and the lumens generated is the coefficient of utilization, CU. Or

$$\text{lumens on the working plane} = \text{lamp lumens} \times \text{CU}$$

Therefore

$$\text{illuminance } E = \frac{\text{lamp lumens} \times \text{CU}}{\text{area}}$$

The coefficient CU is selected from Table 20.2 of generic fixture types or similar tables provided by the luminaire manufacturer, by a method known as the zonal cavity method. It is explained in Section 20.31.

The illuminance figure so calculated is *initial average* illumination. This initial level is reduced by the effect of temperature and voltage variations, dirt accumulation on luminaires and room surfaces, lamp output depreciation, and maintenance conditions. All of these effects are cumulatively referred to as the light loss factor (LLF):

$$\text{maintained } E = \text{initial } E \times \text{LLF}$$

(This factor was previously known as the maintenance factor, MF.) The procedure required to arrive at this factor is explained in the following section.

Our final expression for maintained illuminance E as calculated by the lumen method is, therefore,

$$E = \frac{\text{lamp lumens} \times \text{CU} \times \text{LLF}}{\text{area}} \tag{20.1}$$

where E is footcandles if area is expressed in square feet, or lux, where area is in square meters. Lamp lumens is the total within the space, and equal to

$$\text{no. of fixtures} \times \text{lamps per fixture} \times \text{lumens per lamp}$$

Our formula then becomes

$$E = \frac{\text{luminaires} \times \text{lamps per luminaire} \times \text{lumens per lamp} \times \text{CU} \times \text{LLF}}{\text{area}} \tag{20.2}$$

or, conversely, solving for the number of luminaires required to achieve a target maintained luminance E,

$$\text{No. of luminaires} = \frac{E \times \text{area}}{\text{lamps per luminaire} \times \text{lumens per lamp} \times \text{CU} \times \text{LLF}} \tag{20.3}$$

If we wish to determine the area per luminaire that is required to give a specified illumination, we have

$$\text{area per luminaire} = \frac{\text{lamps per luminaire} \times \text{lumen per lamp} \times \text{CU} \times \text{LLF}}{E} \tag{20.4}$$

This is usually a much more useful procedure in large spaces than calculating for the entire room. For instance, it is much more convenient to know that to maintain, say, 60 fc with a given luminaire, 70 ft^2 per unit is required than it is to know that for an 18,000 ft^2 floor, 257 fixtures are necessary. The former figure allows us to establish a pattern, say 7 × 10; the latter figure is too large to be immediately useful. Therefore, for rooms requiring more than a nominal number of luminaires, the latter calculation should always be used.

20.30 Calculation of Light Loss Factor (LLF)

The light loss factor is composed of elements that can be categorized as recoverable and nonrecoverable. The former can be improved by maintenance; the latter cannot. The total LLF is the product of all the individual factors. The survey that follows includes approximations. For more precise data, see *IES Handbook* (1981).

Among the nonrecoverable loss factors are the following.

(a) Luminaire Ambient Temperature. Light output changes when a fixture operates at other than its design temperature. With normal indoor installation use 1.0, that is, no depreciation. For air troffers *increase* output by 10%, that is, use a factor of 1.10. For other conditions, refer to technical data on the source involved.

(b) Voltage. When the lamp operates at rated voltage, use 1.0. Details of source sensitivity to voltage are given in Chapter 19.

(c) Luminaire Surface Depreciation. This factor is proportional to age and depends on type of surface involved.

(d) Components. Losses can be due to use of nonstandard components such as ballasts, reflectors, louvres, and so on. When no exact data are available and no special conditions apply, use 0.9 for the product $a \times b \times c \times d$.

The factors below are *recoverable,* that is, can be improved to initial conditions by maintenance.

(e) Room Surface Dirt. This factor is self-explanatory. Obviously, lighting systems that depend heavily on surface reflections, such as indirect, are more heavily affected than systems that deliver most of their useful light directly. Assuming a 24-month cleaning cycle, and normal conditions of cleanliness, use the appropriate base factor below. Alter it for other conditions such as infrequent maintenance and unusual cleanliness.

Direct lighting: 0.92 ±5%
Semidirect lighting: 0.87 ±8%
Direct-indirect lighting: 0.82 ±10%
Semi-indirect lighting: 0.77 ±12%
Indirect lighting: 0.72 ±17%

(f) Lamp Lumen Depreciation. This factor depends on the type of lamp and the replacement schedule. Use the following when exact data are unavailable:

	Group Replacement	*Replacement on Burnout*
Incandescent	0.94	0.88
Tungsten-halogen	0.98	0.94
Fluorescent	0.90	0.85
Mercury	0.82	0.74
Metal-halide	0.87	0.80
High-pressure sodium	0.94	0.88

(g) Burnouts. This self-explanatory factor depends on maintenance schedules and method of replacement. Use the following:

Group replacement procedures: 1.0
Individual replacement on burnout: 0.95

(h) Luminaire Dirt Depreciation (LDD). This factor depends on luminaire design, atmosphere conditions in the space, and maintenance schedule. The maintenance category is obtained from manufacturer's data or from Table 20.2. The type of atmosphere is determined by considering the space involved. Assuming a 12-month cleaning schedule and normal room cleanliness, use the base number in Fig. 20.36 and change it to match con-

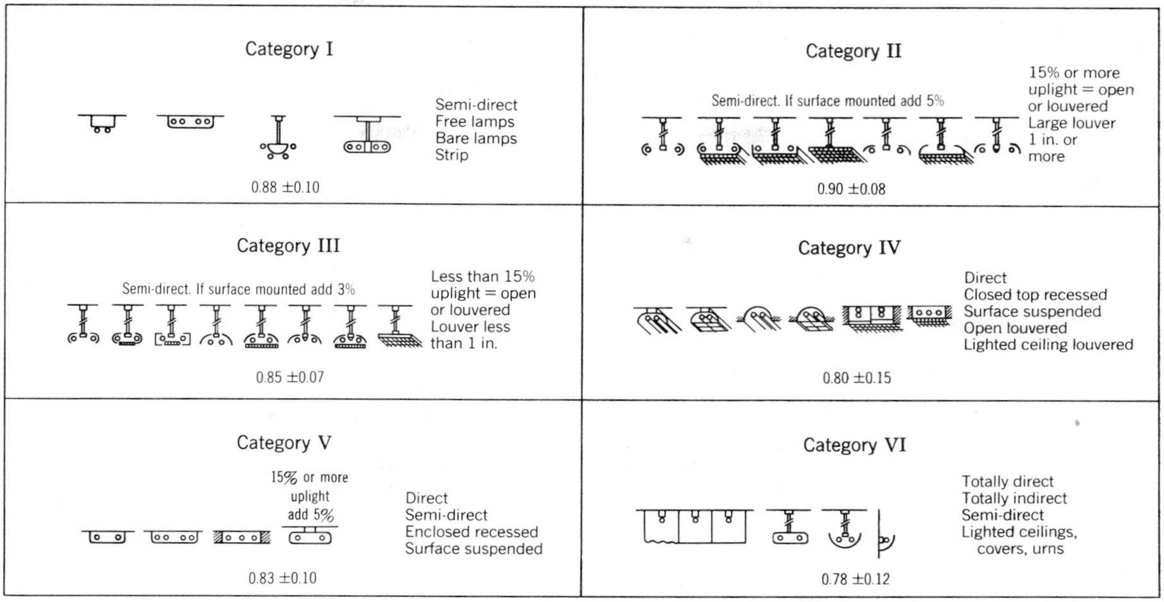

***Fig. 20.36** The LDD factor is determined from the category of luminaire, which is an indication of its proneness to dirt accumulation, plus a knowledge of room ambient conditions.*

ditions of dirt and maintenance. The categories correspond to those of the IES.

Total LLF is the product of all the depreciation factors above, that is,

$$\text{LLF} = a \times b \times c \times d \times e \times f \times g \times h$$

For example, a fluorescent air troffer in a regularly maintained group-lamp-replacement, air-conditioned office might typically have an LLF of

$$\text{LLF} = 1.1 \times 1 \times 0.92 \times 1 \times 0.95 \times 0.9 \times 1.0 \times 0.93 = 0.80$$

The same fixture in the same office, but with walls and fixture cleaned only when replacing burned-out lamps would typically have an LLF of

$$\text{LLF} = 1.1 \times 1 \times 0.92 \times 1 \times 0.87 \times 0.85 \times 0.95 \times 0.78 = 0.55$$

Thus, if in the first case the maintained illumination is E fc, in the second case it is 0.55/0.80 or 0.69Efc, that is, a reduction of 31% as a result of poor maintenance. When a detailed determination of light loss factor is not possible, use the factors given in Section 20.33. They are somewhat more conservative than those given in Section 20.5c.

20.31 Determination of Coefficient of Utilization (CU) by the Zonal Cavity Method

The coefficient of utilization connects a particular fixture to a particular space, by relating the luminaire's light distribution characteristic to the room size and its surface reflectances. To account for the luminaire's mounting height and its relationship to the working plane, the space is divided into three cavities: the ceiling cavity above the fixture, the floor cavity below the working plane, and the room cavity between the two (see Fig. 20.37). Given the surface reflectances, the effective reflectances of the floor and ceiling cavities can be obtained. With these, the CU can be selected from the tables (either Table 20.2 or manufacturer's data) and the lumen formula (equation 20.3 above) applied to arrive at average illuminance. A step-by-step explanation of the method plus illustrative examples will demonstrate the procedure. The reader should follow the steps with the flow chart in Fig. 20.38 and the calculation form in Fig. 20.39.

STEP 1. First, dimensional data are recorded. In offices, schools, and many other occupancies

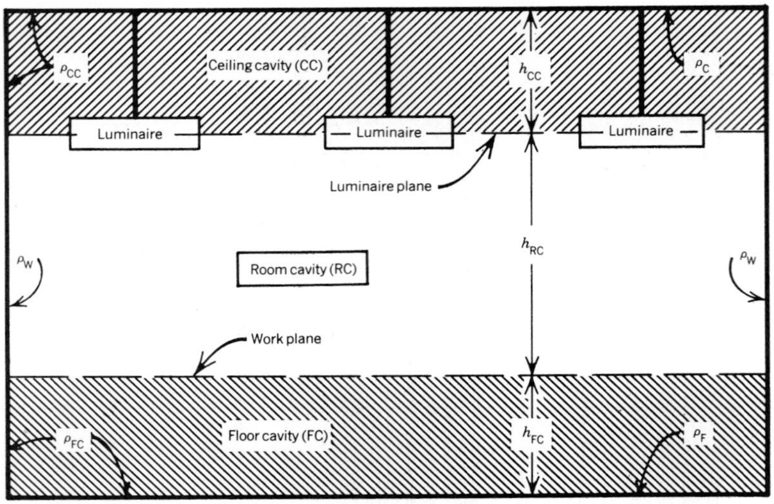

Notes:

1. Obtain effective ceiling cavity reflectance (ρ_{CC}) for combination of ceiling and wall reflectance from Table 20.3. Note that for surface-mounted or recessed luminaires, CCR = 0 and the ceiling reflectance may be used as the effective cavity reflectance. Since LLF accounts for room surface depreciation, use initial reflectances.
2. Obtain effective floor cavity reflectance (ρ_{FC}) for combination of floor and wall reflectances from Table 20.3.
3. Obtain coefficient of utilization for 20% effective floor cavity reflectance condition from table for luminaire, interpolating between tabulated values as required to match room size and ceiling and wall reflectance combinations. If effective floor cavity reflectance (ρ_{FC}) obtained in step 3 differs significantly from 20%, obtain the multiplier from Table 20.3. Multiply the coefficient of utilization by this multiplier.
4. If *initial illumination* is desired, CU selection should be obtained by above procedure and the light loss factor LLF omitted from the formula.
5. Legend:

ρ_C = ceiling reflectance
ρ_{CC} = ceiling cavity reflectance
ρ_W = wall reflectance
ρ_F = floor reflectance
ρ_{FC} = floor cavity reflectance
h = height in feet or meters
h_{RC} = height of room cavity

Fig. 20.37 *Room cavities as used in the zonal cavity method.*

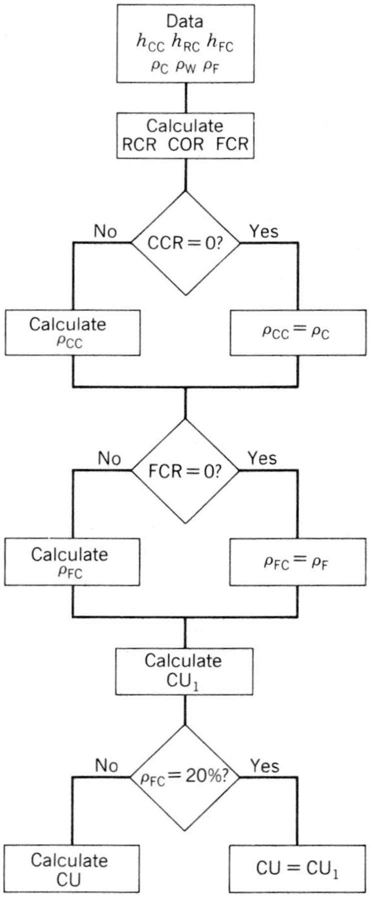

Fig. 20.38 *Zonal cavity method flow chart.*

GENERAL INFORMATION

1. Project identification: ______________________________

   (Give name of area and/or building and room number)

2. Average maintained illumination for design: ______ lux [footcandles]

Luminaire data:

3. Manufacturer: ______________________
4. Catalog number: ______________________

Lamp data:

5. Type and color: ______________________
6. Number per luminaire: ______________________
7. Total lumens per luminaire: ______________________

SELECTION OF COEFFICIENT OF UTILIZATION

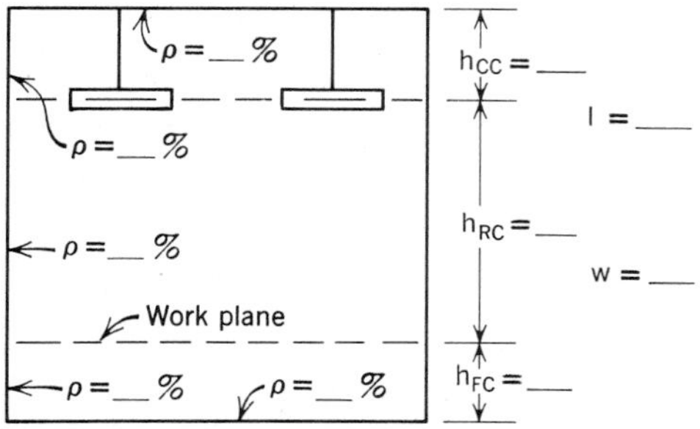

8. Step 1: Fill in sketch at right.
9. Step 2: Determine Cavity Ratios by formulas.

   9a. Room cavity ratio, RCR = ______________________

   9b. Ceiling cavity ratio, CCR = ______________________

   9c. Floor cavity ratio, FCR = ______________________

10. Step 3: Obtain effective ceiling cavity reflectance (ρ_{CC}) from Table 20.3. ρ_{CC} = ______
11. Step 4: Obtain effective floor cavity reflectance (ρ_{FC}) from Table 20.3. ρ_{FC} = ______
12. Step 5: Obtain coefficient of utilization (CU) from manufacturer's data. CU = ______

SELECTION OF LIGHT LOSS FACTORS

Unrecoverable

13. Luminaire ambient temperature ______
14. Voltage to luminaire ______
15. Luminaire surface depreciation ______
16. Other factors ______

Recoverable

17. Room surface dirt depreciation ______
18. Lamp lumen depreciation ______
19. Lamp burnouts factor ______
20. Luminaire dirt depreciation LDD ______

21. Total light loss factor, LLF (product of individual factors above): ______

CALCULATIONS

(Average maintained illumination level)

$$\text{Number of luminaires} = \frac{(\text{Illuminance}) \times (\text{Area})}{(\text{Lumens per luminaire}) \times (\text{CU}) \times (\text{LLF})}$$

22. = ______________________ =

$$\text{Lux [footcandles]} = \frac{(\text{number of luminaires}) \times (\text{lumens per luminaire}) \times (\text{CU}) \times (\text{LLF})}{(\text{area})}$$

23. = ______________________ =

24. Calculated by: ______________________ Date: ______________

***Fig. 20.39** Zonal cavity method calculation form. (Courtesy of IES of North America.)*

TABLE 20.3 **Percent Effective Ceiling or Floor Cavity Reflectance (ρ_{CC}, ρ_{FC}) for Various Reflectance Combinations**

Percent Ceiling ρ_C or Floor Reflectance ρ_F:	90				80				70			50			30				10		
Percent Wall Reflectance ρ_W:	90	70	50	30	80	70	50	30	70	50	30	70	50	30	65	50	30	10	50	30	10
Ceiling or Floor Cavity Rates CCR or FCR																					
0	90	90	90	90	80	80	80	80	70	70	70	50	50	50	30	30	30	30	10	10	10
0.2	89	88	86	85	79	78	77	76	68	67	66	49	48	47	30	29	29	28	10	10	9
0.4	88	86	83	81	78	76	74	72	67	65	63	48	46	45	30	29	27	26	11	10	9
0.6	88	84	80	76	77	75	71	68	65	62	59	47	45	43	29	28	26	25	11	10	9
0.8	87	82	77	73	75	73	69	65	64	60	56	47	43	41	29	27	25	23	11	10	8
1.0	86	80	74	69	74	71	66	61	63	58	53	46	42	39	29	27	24	22	11	9	8
1.2	86	78	72	65	73	70	64	58	61	56	50	45	41	37	29	26	23	20	12	9	7
1.4	85	77	69	62	72	68	62	55	60	54	48	45	40	35	28	26	22	19	12	9	7
1.6	85	75	66	59	71	67	60	53	59	52	45	44	39	33	28	25	21	18	12	9	7
1.8	84	73	64	56	70	65	58	50	57	50	43	43	37	32	28	25	21	17	12	9	6
2.0	83	72	62	53	69	64	56	48	56	48	41	43	37	30	28	24	20	16	12	9	6
2.2	83	70	60	51	68	63	54	45	55	46	39	42	36	29	28	24	19	15	13	9	6
2.4	82	68	58	48	67	61	52	43	54	45	37	42	35	27	28	24	19	14	13	9	6
2.6	82	67	56	46	66	60	50	41	53	43	35	41	34	26	27	23	18	13	13	9	5
2.8	81	66	54	44	66	59	48	39	52	42	33	41	33	25	27	23	18	13	13	9	5
3.0	81	64	52	42	65	58	47	38	51	40	32	40	32	24	27	22	17	12	13	8	5
3.5	79	61	48	37	63	55	43	33	48	38	29	39	30	22	26	22	16	11	13	8	5
4.0	78	58	44	33	61	52	40	30	46	35	26	38	29	20	26	21	15	9	13	8	4
4.5	77	55	41	30	59	50	37	27	45	33	24	37	27	19	25	20	14	8	14	8	4
5.0	76	53	38	27	57	48	35	25	43	32	22	36	26	17	25	19	13	7	14	8	4

Extracted from *IES Handbook;* reprinted with permission; for more complete data see *IES Handbook* (1981).

the work plane is 30 in. In drafting rooms it is 36 to 38 in., in shops 42 to 48 in., in carpet stores and sail-cutting rooms the work plane is at the floor level. The three "h" terms are the heights of the various cavities. In this step also, identify the initial reflectance of the room surfaces and fill in the sketch in Fig. 20.39. Similarly, establish wall and floor reflectances. Utilize the nearest reflectance to those given in Table 20.3. Interpolation between values is *not* required.

STEP 2. See Fig. 20.39. This step involves determining the cavity ratios of the room, by calculation. The basic expression for a cavity ratio (CR) is

$$\text{CR} = 2.5 \times \frac{\text{area of cavity wall}}{\text{area of work plane}} \quad (20.5)$$

In a rectangular space the area of the cavity wall is $h \times (2l \times 2w)$ or $2h(l + w)$; therefore

$$\text{CR} = \frac{2.5 \times 2h \times (l + w)}{\text{area of work plane}}$$

or

$$\text{CR} = 5h \times \frac{l + w}{l \times w} \quad (20.6)$$

For other than rectangular rooms, the area can be calculated as required. For instance, in a circular room, the cavity wall area $= h \times 2\pi r$ and the work plane area is πr^2. Thus

$$\text{CR} = \frac{2.5 \times h \times 2\pi r}{\pi r^2} = \frac{5h}{r} \quad (20.7)$$

For each of the cavities in a rectangular room we have:

Room Cavity Ratio

$$\text{RCR} = 5h_{\text{RC}} \frac{l + w}{l \times w} \quad (20.8)$$

Ceiling Cavity Ratio

$$\text{CCR} = 5h_{\text{CC}} \frac{l + w}{l \times w} \quad (20.9)$$

Floor Cavity Ratio

$$\text{FCR} = 5h_{\text{FC}} \frac{l + w}{l \times w} \quad (20.10)$$

For reference, since all cavity ratio figures are related, having determined one, the others are

$$\text{CCR} = \text{RCR} \frac{h_{\text{CC}}}{h_{\text{RC}}} \quad (20.11)$$

$$\text{FCR} = \text{RCR} \frac{h_{\text{FC}}}{h_{\text{RC}}} \quad (20.12)$$

and

$$\text{CCR} = \text{FCR} \frac{h_{\text{CC}}}{h_{\text{FC}}} \quad (20.13)$$

STEP 3. See Table 20.3 and Figs. 20.38 and 20.39. This step involves obtaining the effective ceiling reflectance (ρ_{CC}) from Table 20.3. Note that the wall reflectance remains as selected in step 1. If the fixtures are surface mounted or recessed, then CCR $= 0$ and ρ_{CC} = selected ceiling reflectance.

STEP 4. See Table 20.3 and Figs. 20.38 and 20.39. This step involves obtaining the effective floor reflectance ρ_{FC} as above in step 3, for ρ_{CC}. If the floor is the working plane, FCR $= 0$ and ρ_{FC} = selected floor reflectance.

STEP 5. Select CU from manufacturer's data. Note that interpolation may be necessary for ceiling cavity reflectance (ρ_{CC}), which is between the figures in the CU table. See Example 20.1 in the next section. Also, factors for CU correction for ρ_{FC} other than 20% (standard in CU tables) are given in Table 20.4.

TABLE 20.4[a] **Factors for Effective Floor Cavity Reflectances Other Than 20% (Any Wall Reflectance)[b]**

For 30% effective floor cavity reflectance, *multiply* by appropriate factor below.
For 10% effective floor cavity reflectance, *divide* by appropriate factor below.

Room cavity ratio	*Percent Effective Ceiling Cavity Reflectance, ρ_{CC}* 80	70	50	10
1	1.08	1.06	1.04	1.01
2	1.06	1.05	1.03	1.01
3	1.04	1.04	1.03	1.01
4	1.03	1.03	1.02	1.01
5	1.03	1.02	1.02	1.01
6	1.02	1.02	1.02	1.01
7	1.02	1.02	1.01	1.01
8	1.02	1.02	1.01	1.01
9	1.01	1.01	1.01	1.01
10	1.01	1.01	1.01	1.01

[a]Extracted from *IES Handbook* (1981), and reprinted with permission.

[b]For more precise data, for varying ρ_W, see *IES Handbook* (1981).

STEP 6. Calculate illuminance and number of fixtures or area per luminaire in the usual fashion.

Illustrative examples and shortcut methods are demonstrated below. CU coefficients are listed in Table 20.2.

20.32 Zonal Cavity Calculations: Illustrative Examples

EXAMPLE **20.1.** It is suggested that the reader photocopy Fig. 20.39 and fill it in as the solution progresses.

Given. Classroom: 6 m × 8 m × 3.70 m height, elementary school. Initial reflectances: ceiling 80%, entire wall 50%, floor 20%. (Note that the sketch on Fig. 20.39 can accommodate different reflectances for the upper, center, and lower wall sections.) Provide adequate lighting using fluorescent lamp fixtures. Assume yearly maintenance, lamp replacement at burnout, proper voltage and ballasts, medium clean atmosphere.

SOLUTION:

(a) Illuminance Level. Refer to Table 18.5. Depending on the class grade, the illuminance category could be either D or E. We will use E, which gives an illuminance range of 500–750–1000 lux. In Table 18.6 under category E we find: age—under 40, speed/accuracy are important, and background is over 70% (white paper). This yields a recommendation of 500 lux. Note that the British IES recommendation in Table 18.3 also gives 500 lux. The daylight contribution in much of the space is normally considerable in schools because of the hours of use. It will frequently exceed this 500-lux level, but to demonstrate the calculation procedure, we must assume an absence of daylight. Lines 1, 2, and 8 of Fig. 20.39 and the room sketch can be filled in.

(b) Luminaire Selection. The criteria we have previously developed for a classroom situation requires an installation that will yield

1. Low direct glare (high VCP), because students spend a large proportion of their time in a heads-up position.
2. Low veiling reflections (high CRF), because much of the seeing task involves high reflectance materials, occasionally specular.
3. High efficiency and low energy use to meet IES/ASHRAE and most governmental requirements.
4. Minimum required maintenance, in view of the poor cleaning and maintenance situation that obtains in many schools.

Although ceiling height is sufficient to permit use of indirect lighting with all its distinct advantages, it is not chosen, because

5. The low CU would not meet IES/ASHRAE requirements, which are often mandatory in public buildings (see Table 20.2, luminaire 33, and Section 20.5.
6. The luminaire maintenance category is bad, and given the type of maintenance expected, light reduction would be serious.
7. Indirect lighting depends on a highly reflective ceiling, requiring yearly cleaning and repainting at intervals not exceeding 5 years. This is not generally the case in public schools. As a result we select luminaire no. 44 from Table 20.2. This unit is a parabolic aluminum reflector with louvres, and exhibits a 45° cutoff and crosswise batwing distribution. (Figures 20.33, 20.34 and 20.35*a* are units of this design.) This unit meets requirements 1 through 4, above.

Although its distribution curve (see Table 20.2) shows no upward component, most commercial units of this design do have slots in the top of the reflector and show 5 to 8% uplight. This avoids an excessively dark ceiling and undesirable luminance ratios between fixture and background. We would select such a unit. Mounting height should be about 270 cm [~9 ft] to permit easy maintenance, and good row spacing. For this design *S/MH* is between 1.5 and 2.0. Work plane height is 75 cm [30 in.].

(c) Calculations

STEP 1. The required data that should appear on the sketch in Fig. 20.39 are

$h_{CC} - 1.0$ m $h_{RC} = 1.95$ m $h_{FC} = 0.75$ m $l = 8$ m

$\rho_C = 80\%$ $\rho_W = 50\%$ $\rho_F = 20\%$ $w = 6$ m

STEP 2. From equations 20.8, 20.9, and 20.10

$$\text{RCR} = 5h_{RC}\frac{l + w}{l \times w} = 2.84,$$

$$\text{CCR} = 5\,(1)(0.29) = 1.46,$$

$$\text{FCR} = 5\,(0.75)(0.29) = 1.09$$

STEP 3. From Table 20.3 obtain effective reflectances:

For ρ_{CC} use $\rho_C = 0.8$, $\rho_W = 0.5$, and CCR $= 1.46$.

$$\therefore \rho_{CC} = 0.62$$

STEP 4. For ρ_{FC} use $\rho_F = 0.2$, $\rho_W = 0.5$, and FCR $= 1.09$.

$$\therefore \rho_{FC} = 0.18 \text{ by interpolation}$$

STEP 5. Interpolation between RCR of 2.8 and 3.0 is necessary. The coefficient of utilization for the selected luminaire—no. 44 of Table 20.2—can now be obtained by double interpolation.

$\rho_W = 0.50$		*CU*	
$\rho_{CC} \rightarrow$	*0.70*	*0.62*	*0.50*
RCR			
2.	0.58		0.55
2.84		?	
3.	0.52		0.50

Required CU = 0.52. No correction from Table 20.4 is required, as ρ_{FC} is close to 20%. Note that this CU is above the minimum recommended by IES/ASHRAE 90 of 0.45 for a room with RCR of 3.0. (See Section 20.5.) At this stage lines 9 through 12 of Fig. 20.39 can be filled in.

STEP 6. The light loss factor (see Section 20.30 above) results from establishing items 13 to 20 of Fig. 20.39. These are:

Items 13–16	0.9
Item 17	0.95
Item 18	0.85
Item 19	0.95
Item 20	0.80

Item 21: LLF =

$$(0.9)(0.95)(0.85)(0.95)(0.80) = 0.55$$

STEP 7. *Lumen calculation.* A typical classroom is naturally divided into a large student seating area and a teacher's area. The illuminance requirement is approximately the same for both so that the room can be treated as a single unit for visual task purposes. A good lighting design will include chalkboard lighting, the spill light from which is usually sufficient to illuminate the very front of the room. Treating the entire space as one visual task area, and assuming two nominal 40-W cool white fluorescent lamps per fixture, the number N of luminaires required is, from Equation 20.3,

$$N = \frac{(500\ \text{lux})\ (6 \times 8\ \text{m})}{2(3200\ \text{lm})(0.52)(0.55)} = 13.11$$

The remaining lines of Fig. 20.39 can now be completed.

Refer now to Fig. 20.40 for the layout. Two lengthwise rows of fixtures, spaced 300 cm apart, will give excellent coverage, as shown. The spacing-to-mounting height ratio is only 1.54, well below the maximum of 1.7 to 2.0. The sixth fixture in the outside (window) row provides illumination for the teacher's desk. Two *single-lamp* batwing distribution or asymmetric distribution luminaires provide chalkboard and front-of-room lighting. The window-wall row is separately switched and is further from the wall than the inside row, due to the usual presence of daylight. Point-by-point il-

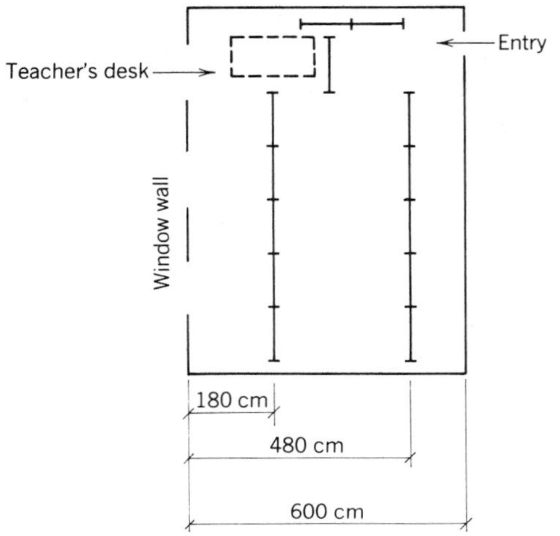

***Fig. 20.40** Layout of batwing distribution pendant luminaires in a typical classroom. Note that the outside row is further from the wall than the inside row because of daylight contribution. It is also extended to illuminate the teacher's desk.*

luminance including raw lux and ESI lux can be computed by one of the available computer programs. Daylight contribution can also be included. Actual *average* illuminance in the room would be

$$E = \frac{11 + \frac{1}{2} + \frac{1}{2}}{13.11}(500 \text{ lux}) \simeq 450 \text{ lux}$$

assuming similar characteristics for the one- and two-lamp fixtures.

EXAMPLE 20.2. Large clerical business office 60 × 100 × 8 ft; initial reflectances 80, 30, 30. Provide 50 fc general lighting with fluorescent, single-lamp F-40 T-12 CW troffers. Space is air conditioned. Lamps are replaced on a burnout basis and fixture then cleaned.

SOLUTION. We select the fixture of Fig. 20.34 because of its high *S/MH* ratio and good CU and VCP. (This fixture also appears in Fig. 20.35.)

Calculations. (The reader can fill in a copy of Fig. 20.39.) Assume working plane at 36 in. AFF; although usually at 30 in., the large number of business machines in use raises the elevation here.

STEP 1. $h_{CC} = 0$; $h_{RC} = 5$; and $h_{FC} = 3$.

STEP 2. CCR = 0; RCR = 0.666; and FCR = 0.4.

STEP 3. $\rho_{CC} = 80\%$ (recessed fixture; use ceiling reflectance).

STEP 4. From Table 20.3, $\rho_{FC} = 27\%$ and $\rho_W = 30\%$.

STEP 5. From Fig. 20.34, by double interpolation:

	CU		
RCR	$\rho_{FC} = 20\%$	$\rho_{FC} = 27\%$	$\rho_{FC} = 30\%$
0	0.90		1.00
0.666		?	
1.0	0.79		0.84

$\rho_{CC} = 80\%$ $\rho_W = 30\%$

which results in CU = 0.87.

STEP 6. Light loss factors, per Fig. 20.39:

1.1 0.92

1.0 0.85

0.9 0.95

1.0 0.88

$$\text{LLF} = 0.9(0.92)(0.85)(0.95)(0.88) = 0.59$$

STEP 7.

$$\text{area/luminaire} = \frac{1 \times 3200 \times 0.87 \times 0.59}{50 \text{ fc}} = 32.85 \text{ ft}^2$$

Fixtures can be mounted on a 6 × 6 grid (approximately as seen in Fig. 20.35), or in continuous rows spaced 8 ft on centers. Since the maximum spacing-to-mounting-height ratio is more than 2, an 8-ft side-to-side spacing is well within limits. Note the very important fact that a level of 50 fc (500 lux) of very high quality can be achieved with a power density of less than 1.5 W/ft² [16 W/m²].

20.33 Zonal Cavity Calculation by Approximation (based on a method developed by B. F. Jones)

For a first approximation, make the following assumptions:

1. $\rho_{CC} = 80$ for a large room $\simeq$ 100 ft sq (~ 10,000 ft², ~ 1000 m²)

   $\rho_{CC} = 70$ for a medium room $\simeq$ 30 ft sq (~ 1000 ft², ~ 100 m²)

   $\rho_{CC} = 60$ for a small room $\simeq$ 12 ft sq (~ 150 ft², ~ 15 m²)
2. $\rho_{FC} = 20$
3. Assume all rooms are square. To do this for a rectangle, take one-third the difference in dimensions and add to the smaller dimension to obtain equivalent width *w*. Then, for square rooms

$$\text{RCR} = \frac{10 h_{RC}}{w}$$

4. Assume LLF = 0.65 for very good conditions, 0.55 for average, and 0.45 for poor.

EXAMPLE 20.3 (by approximation). Classroom as in Example 20.1.

$\rho_{CC} = 70$ $\rho_W = 50$ $\rho_{FC} = 20$ (assumptions)

$$\text{RCR} = \frac{10 \times 1.95}{6 + \frac{2}{3}} = \frac{19.5}{6.66} = 2.93$$

Obtain CU from Table 20.2, fixture no. 44.

$$CU = 0.52 \text{ by visual inspection}$$

$$N = \frac{500(6 \times 8 \text{ m})}{2(3200)(0.52)(0.55)}$$

$$= 13.11 \text{ fixtures}$$

Thus, the result is the same as the accurate calculation. Let us check Example 20.2.

EXAMPLE 20.4 (by approximation).

$$\rho_{CC} = 80 \qquad \rho_W = 30 \qquad \rho_{FC} = 20$$

$$RCR = \frac{10 \times 5}{60 + 40/3} = \frac{50}{73} = \frac{2}{3} = 0.66$$

CU = 0.90 by inspection (mental interpolation)
LLF = 0.55

$$\text{area/luminaire} = \frac{1 \times 3200 \times 0.90 \times 0.55}{50}$$

$$= 31.7 \text{ sq ft}$$

This result is within 5% of the accurate calculation. We thus see that these simple approximations give answers sufficiently accurate for most uses and are therefore recommended.

In conclusion, then, with respect to zonal cavity calculations, we can make the following statements:

1. For everyday calculations of rectangular rooms, with assumed reflectances, use the assumptions listed as a first approximation in 1 to 4 above.
2. For rooms where a high degree of accuracy is desired and actual reflectances are known, use the long method, with visual interpolation.
3. For rooms of unusual shape or rooms with special conditions such as coffered ceilings, mixed-material walls, and partial height partitions, use the long method. This should not be necessary for more than 5% of calculations.

20.34 Effect of Cavity Reflectances on Illuminance

It should be obvious to every reader at this point in our study that the reflectances of the various room cavities have a marked effect on the CU, because of light reflections within the room. To demonstrate this graphically, we have plotted in Figs. 20.41, 20.42, and 20.43 the effect of varying cavity reflectances on the three principal types of fixture distribution: semi-indirect, direct-indirect, and direct-spread. Note that, as expected, the ceiling cavity reflectance has the most pronounced effect with indirect fixtures, and the floor reflectance with direct units. Since lighting costs amount to 3 to 5% of *total construction* cost for many types of buildings such as offices, a 20% differential in lighting units can amount to as much as 1% of the total cost of the facility. This amount

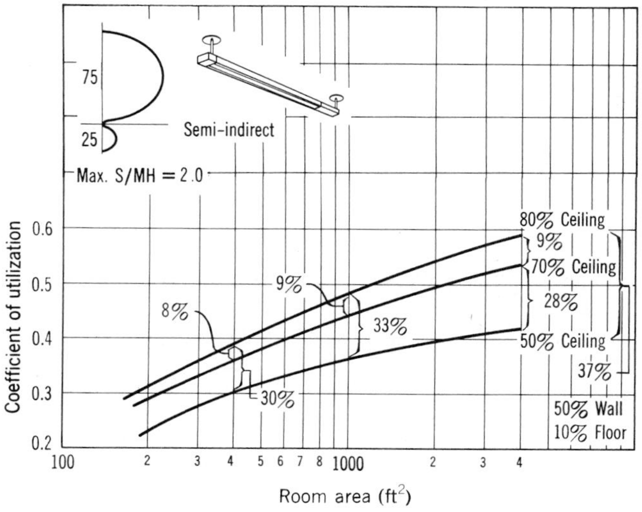

Fig. 20.41 Effect of surface reflectances on the CU of a luminaire with semi-indirect distribution. As expected, since the ceiling becomes the light source, its reflectance has the most pronounced effect. With this particular unit having a 25% downward component, floor finish also has an appreciable effect, increasing CU by an average 10% for a 30% reflectance floor. The effect of wall reflectance naturally increases as rooms become smaller and the proportion of wall surface is larger. The change in CU between a 30% and a 50% reflectance wall varies from 15% for a 400-square-foot room to 5% for a 4000-square-foot room.

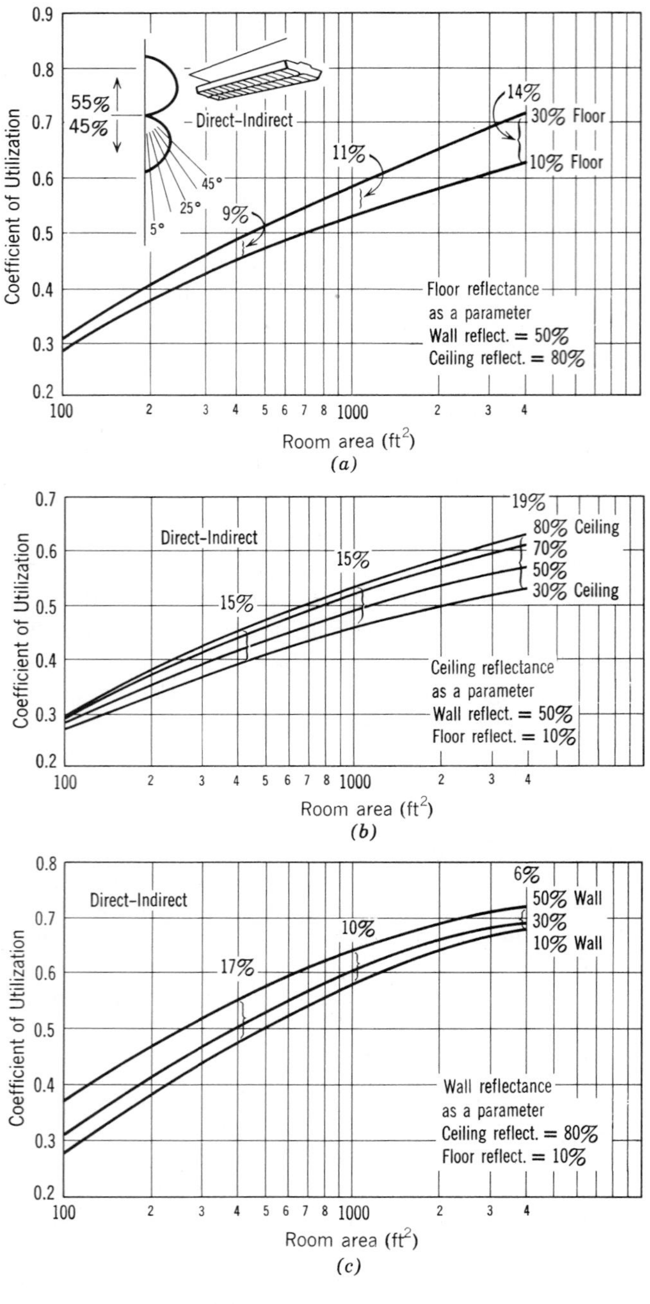

Fig. 20.42 Effect of surface reflectances on the CU of a luminaire with direct-indirect distribution. With this distribution, the effects of ceiling and floor are most pronounced, with a wall effect only in small rooms.

would not only pay for the increased cost of higher reflectance finishes and materials but also would reduce both initial and operating cost. These data clearly indicate the necessity for the lighting designer to have considerable influence on the choice of room materials and finishes, a situation which unfortunately does not usually occur.

20.35 Modular Lighting Design

An increasingly large number of buildings are being designed on a modular system resulting in a need for flexible lighting to fit the module utilized. In such buildings, once the general lighting scheme and the fixture involved are established, it is con-

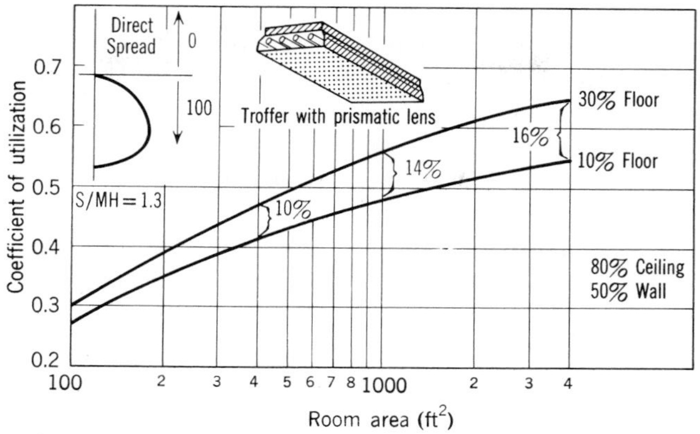

Fig. 20.43 Effect of surface reflectances on the CU of a luminaire with direct (spread) distribution. Floor finish is most important, with wall reflectance important only in small rooms. Since these fixtures have no upward component, all ceiling illumination is derived from reflection. Thus in room with floor reflectance less than 20%, ceiling finish has no effect on room illumination.

venient to draw a family of curves for the fixture chosen, thereby facilitating the utilization of the modular unit in various spaces. ''Area'' may readily be replaced with multiples of modular areas, as shown in Fig. 20.44.

20.36 Computer-Aided Design

As pointed out in preceding sections, the use of a computer in lighting calculation is a necessity rather than a convenience if accurate results are to be produced. Once the calculation is done, only a computer analysis will give us the VCP and ESI values for selected viewing locations and directions. In addition, the computer gives the designer a degree of flexibility not otherwise possible in that:

1. It performs the calculations accurately and rapidly.
2. It frees the designer for other, less routine work.
3. It gives the designer the ability to change parameters repeatedly, without making the amount of calculation excessive, as would be the case in hand calculation.

It is this last characteristic that gives the de-

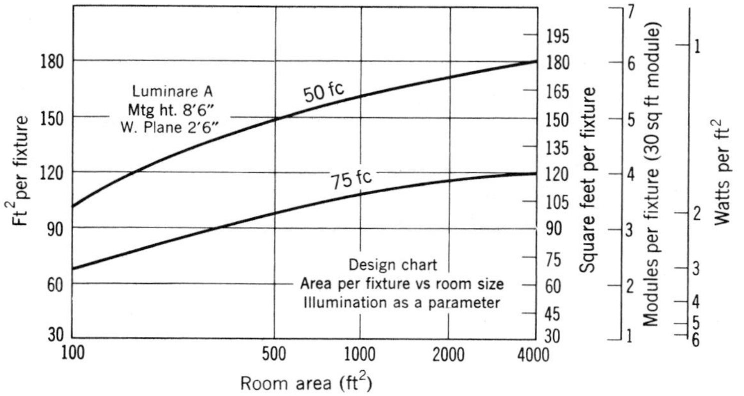

Fig. 20.44 Luminaire design chart. For frequently used fixtures, this type of chart gives a rapid design figure for various size rooms. As seen from the ordinates, the figures can be translated into number of modules and watts per square foot.

signer greatest flexibility. The ability to run a series of calculations for a pendant fixture with pendant lengths varying from 12 to 36 in. in 3-in. intervals, or to change paint colors and reflectance on various surfaces and to note the effect, gives the designer a freedom urgently needed. In addition, the computer can consider related items such as first costs, energy use, operating costs, and impact on HVAC, items whose complexity because of interrelations put them well beyond the pencil-and-hand calculator's ability. A few of the currently available programs are listed in Appendix H.

20.37 Calculating Illuminance at a Point

The lumen (flux) method of horizontal illuminance calculation explained above is appropriate for spaces in which illumination is essentially uniform throughout. However, even in such a space, illuminance varies at least ± 10%, and near columns, walls, windows, bookcases, and the like, considerably more. Therefore, to answer the often asked question "How much light will I have on my desk?", the designer must turn to other methods. Three are available:

1. Calculation of illuminance at selected points by computer, as explained in the preceding section.
2. Utilization of one of the design aids explained in the following section.
3. Longhand calculation by one of the methods presented in Sections 20.39 to 20.41.

These methods will also yield results where the lumen method is simply inapplicable. Among these applications are layouts that are intentionally nonuniform, calculation of illumination on planes other than horizontal (e.g., wallwashers), calculation of illuminance resulting from architectural lighting elements such as coves, valances, and the like, and illuminance calculations for nonstandard light sources for which coefficient data of the type given in Table 20.2 are not available.

20.38 Design Aids

By this term we mean any of the various curves, charts, plots, or tables either prepared by the designer or made available by luminaire manufacturers, the purpose of which is to simplify and speed lighting design when using that particular lighting fixture. The reliability of data so obtained depends entirely on that of the manufacturer involved, and their use should be governed accordingly. More recently, it has become customary for major manufacturers to provide computer output charts and tables based on the designers' proposed layout(s). When using these, it behooves the designer to study the data carefully in the computer program being used, since the fixture supplier is certainly not a disinterested party, and the program may be "weighted" accordingly. A brief description of the common design aids follows.

(a) Isolux Charts. These charts, also called isofootcandle charts, are based on the type traditionally supplied by manufacturers of outdoor lighting equipment such as street lights and floodlights, but equally applicable to interior lighting. Their use is illustrated in Fig. 20.45. The basic tool is an isolux diagram for a single luminaire. This is either calculated (see Sections 20.39–20.41), measured from a full-scale mockup (the most accurate and simplest method), or obtained from the manufacturer. Since the relative positions of source and illuminated point are reversible, that is, if a source at A causes illuminance E at point B, then the source at B will cause the same illuminance E at point A, placing the center of the isolux chart at the point in question will permit direct reading of the illumination contribution of each luminaire. It then remains simply to sum the individual contributions to obtain the (scalar) illumination at the desired point. The method is shown on Fig. 20.45.

(b) Illuminance "Cone" Charts. See Fig. 20.46. When the light distribution of a direct downlight is symmetrical, as is generally the case with recessed incandescent and HID downlights, a cone can be drawn showing illuminance directly under the fixture at various distances. The projected circles are defined by maximum illuminance at the center, and half of this illuminance at the edge. This projected circle can be used in the same fashion as the isolux chart in Section 20.38a, except that only two values are given—that at the center and that on the circumference.

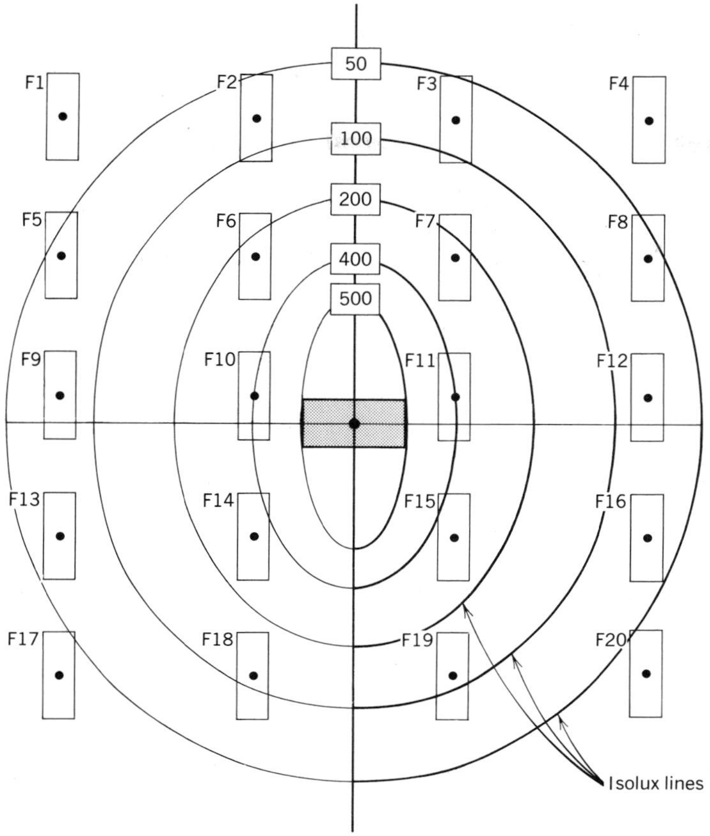

Luminaire	*Contribution (Lux)*
F2, F3	75 × 2
F5, F8	50 × 2
F6, F7	225 × 2
F9, F12	75 × 2
F10, F11	400 × 2
F13, F16	60 × 2
F14, F15	300 × 2
F18, F19	100 × 2
	2570 lux total

Fig. 20.45 *The ellipses represent isolux lines for a single luminaire at a given height above the work plane. They are centered on the point (the work area of a desk) for which it is desired to determine the illuminance. The total illuminance is the sum of the individual luminaire contributions. The center of the luminaire is the point of reference. Therefore when two or more isolux lines pass through a fixture, its contribution is determined by the interpolated isolux line passing through its center. Note the symmetry around the vertical axis, necessitating a plot of only half the ellipses.*

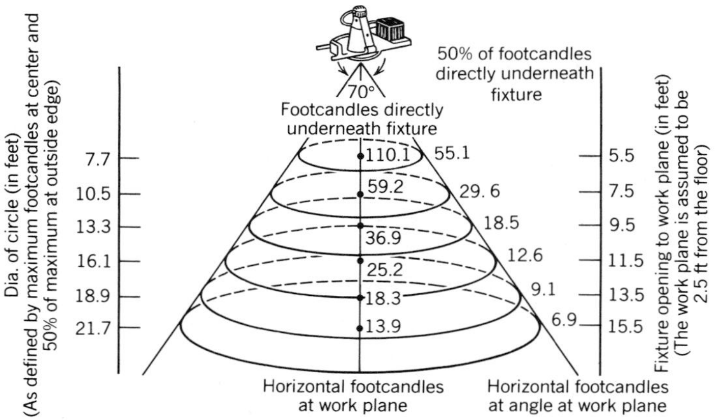

Fig. 20.46 *For downlights with symmetrical distribution, a "cone of light" as shown can be drawn, giving the illuminance directly below the luminaire, and a circle at the circumference of which illuminance is half this maximum. For more accurate work, additional horizontal circles can be added. (Courtesy of Halo Lighting.)*

ILLUMINATION

Luminaire Height Above Working Plane	Horizontal Distance from Luminaire Centerline (ft) 0	2	4	6	8	10
	Initial footcandles					
6 ft	6.9	8.4	6.2	3.9	2.4	1.5
	9.4	9.8	7.2	4.7	2.7	1.6
	6.7	7.4	5.3	3.2	1.9	1.1
	7.5	7.7	5.5	3.2	1.8	1.1
8 ft	3.6	4.4	4.0	3.2	2.1	1.5
	4.9	5.2	4.7	3.5	2.5	1.7
	3.5	3.8	3.5	2.5	1.8	1.2
	4.0	4.1	3.6	2.6	1.8	1.2
10 ft	2.2	2.7	2.6	2.3	1.8	1.4
	3.0	3.1	3.1	2.6	2.0	1.5
	2.2	2.3	2.4	2.0	1.5	1.1
	2.4	2.5	2.5	2.0	1.5	1.1

(a)

Average illumination on wall (footcandles)
Horizontal Distance from Fixtures (ft)

Distance from ceiling (feet)	4	3	2	1	0	1	2	3	4
1	0.2	1.3	3.3	9.0	13.2	9.0	3.3	1.3	0.2
2	0.9	2.6	6.6	15.0	20.5	15.0	6.6	2.6	0.9
3	1.2	3.4	7.6	15.4	20.5	15.4	7.6	3.4	1.2
4	1.5	3.4	7.3	12.8	16.0	12.8	7.3	3.4	1.5
5	1.6	3.2	6.1	9.3	10.7	9.3	6.1	3.2	1.6
6	1.6	2.8	5.0	7.0	7.8	7.0	5.0	2.8	1.6
7	1.6	2.8	4.3	5.4	6.1	5.4	4.3	2.8	1.6

Single unit installation 150W R40 FL 1950 lumens
Values based on one fixture mounted 3 ft from wall

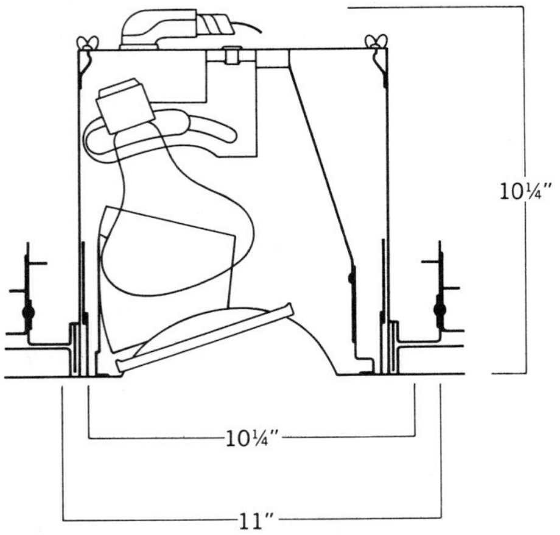

(b)

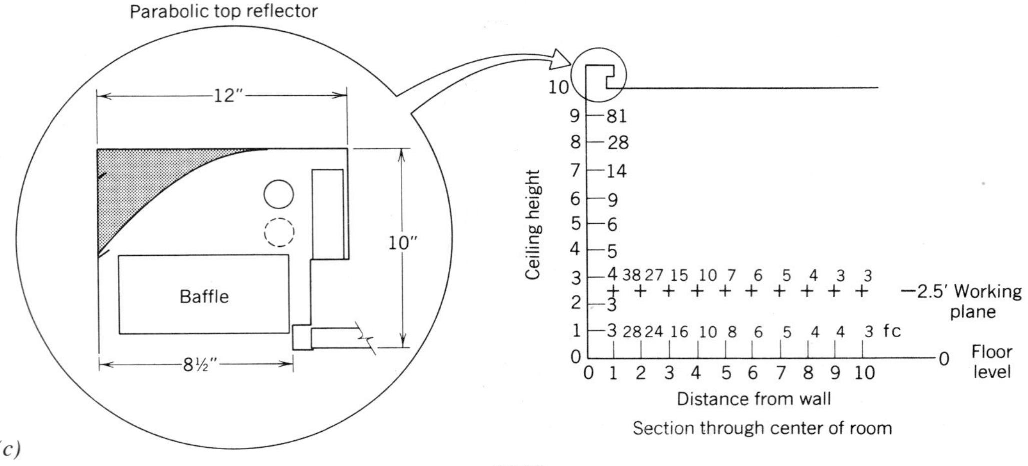

(c)

Fig. 20.47 *Illuminance charts and tables take various forms.* (a) *Tabulation of illuminance from an overhead symmetrical distribution luminaire, in terms of vertical and horizontal distance from the unit. This is a meld of the isolux lines of Fig. 20.45 and the cone chart of Fig. 20.46.* (b) *Chart of vertical surface illumination for a single wall-wash unit, a given distance (3 ft) from the wall. For a multiple-fixture installation, copies of the chart can be overlaid at the proper spacing. Alternatively, the data can be converted to isofootcandle lines and these overlaid to obtain a graphic chart of the entire wall illuminance.* (c) *For lighting design with architectural lighting elements, charts in both plan and section are required, to give wall (vertical) and working plane (horizontal) illuminance data. Numbers shown are footcandles at room center. [Charts courtesy of Marvin Elec., subsidiary of Kidde* (b), *and Columbia Lighting* (c)*].*

(c) Illuminance Tables and Charts. These take various forms as shown in Fig. 20.47. All, however, give specific illuminance data, in numerical form, for specific points. The values are usually obtained from computer printout and may be raw or ESI illuminance, or both.

20.39 Calculating Illuminance from a Point Source

It is well to note at the outset that all the methods discussed below calculate illumination at *a point*. The answer to the constant query "How much light will I have on my desk from a luminaire at this location?" is arrived at by taking several points on the desk and calculating illumination at each one. Shortcuts can be made for symmetry and so on. The reader should also understand that most of these techniques are laborious and are generally performed by the consulting engineer or lighting specialist rather than the architect. They are presented here as source material for the technically oriented reader.

The basis of point source calculations is the inverse square law developed in Section 18.12:

$$fc = \frac{cp}{D^2} \qquad (18.6)$$

where fc, cp, and D are footcandle illumination, candlepower intensity, and distance, respectively. Refer to Fig. 20.48. The horizontal illumination at a point P as shown in Fig. 20.48*a* is

$$\text{horizontal } fc_P = \frac{cp}{D^2}\cos\theta \qquad (20.14)$$

and the vertical illumination at that same point is

$$\text{vertical } fc_P = \frac{cp}{D^2}\sin\theta \qquad (20.15)$$

however, since

$$\cos\theta = \frac{H}{D} \quad \text{and} \quad \sin\theta = \frac{R}{D}$$

we have then at point P

$$\text{horizontal illumination} = \frac{cp}{H^2}\cos^3\theta \qquad (20.16)$$

and

$$\text{vertical illumination} = \frac{cp}{R^2}\sin^3\theta \qquad (20.17)$$

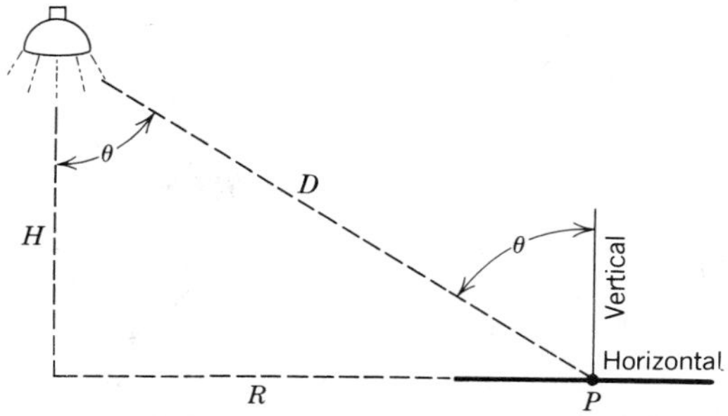

Horizontal footcandles at point P

$$= \frac{\text{Candlepower} \times \cos\theta}{D^2} = \frac{CP}{H^2}\cos^3\theta$$

Vertical footcandles at point P

$$= \frac{\text{Candlepower} \times \sin\theta}{D^2} = \frac{CP}{R^2}\sin^3\theta$$

Fig. 20.48 *Relationship between intensity and illuminance when source can be considered a point source, that is, when inverse square law applies. Source major dimension must not exceed 0.2D to be considered a point source.*

Since the candlepower intensity in the direction of point P is normally taken from a candlepower distribution curve, θ is normally known, and these expressions are readily usable. Very few commercial sources are actually point sources. However, *when the maximum dimension of the source is less than five times the distance to point P*, the equations will give satisfactory results. Note that these equations can be used to calculate and plot isolux diagrams for point sources, of the type shown in Figs. 20.45 and 20.46.

EXAMPLE 20.5. Referring to Fig. 20.48 and the candlepower distribution curve of Fig. 20.49, find the horizontal and vertical illumination at a point P, which is 10 ft below and 12 ft away from the source.

SOLUTION

$$H = 10 \text{ ft} \qquad R = 12 \text{ ft}$$

$$\theta = \tan^{-1}\frac{12}{10} = 50°$$

$$\sin\theta = 0.766 \qquad \cos\theta = 0.643$$

$$cp \text{ at } 50° = 6600$$

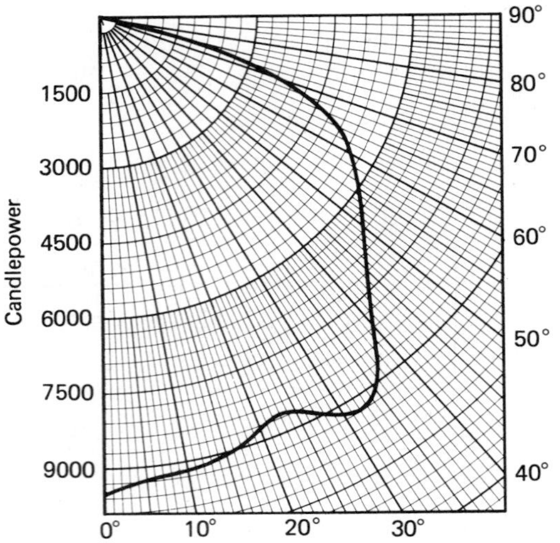

Fig. 20.49 Typical candlepower distribution plot for use in inverse square law calculation.

$$\text{horizontal illumination} = \frac{6600}{10^2}(0.643)^3 = 17.5 \text{ fc}$$

$$\text{vertical illumination} = \frac{6600}{12^2}(0.766)^3 = 20.8 \text{ fc}$$

20.40 Calculating Illuminance from a Line Source

When the maximum source dimension exceeds 20% of the distance to the point illuminated, it can no longer be considered a point source. A source that is long and narrow such as a fluorescent tube can be considered as a line source. For a line source where length is much greater than width, the illumination varies *inversely with distance* as opposed to inversely with the square of the distance for a point source, that is,

$$fc = \text{constant} \times \frac{cp}{D} \qquad (20.18)$$

Thus if illumination at point P_1 is known, that at P_2 can be calculated as

$$fc_2 = fc_1 \times \frac{D_1}{D_2} \qquad (20.19)$$

When the point being considered is directly below the source and the source is very long, Equation 20.18 becomes

$$fc = \frac{fL \times W}{2D} \qquad (20.20)$$

where

fc = illumination at distance D in footcandles
fL = source brightness in footlamberts
D = distance in feet
W = width of source

When the point is not directly below the source, the equation becomes complex (see *IES Handbook*, 1981). When the source is very long with respect to width, the effect of longitudinal offset (in the same direction as the length of the source) diminishes and the only offset that need be considered is a lateral one—away from the axis of the source. Table 20.5 provides illumination values for this situation for several specific luminaire types. Obviously the longer the row, the more accurate the results. Lumens per foot data for fluorescent lamps are available from Table 19.11 by simply dividing lumen output by length.

EXAMPLE 20.6. Find the horizontal illumination on a line 8 ft below and 2 ft offset from a continuous row of two 800-mA HO lamps in a white-enameled reflector.

SOLUTION. Refer to Fig. 20.50 and Table 20.5, Part C. Note that the maximum vertical distance

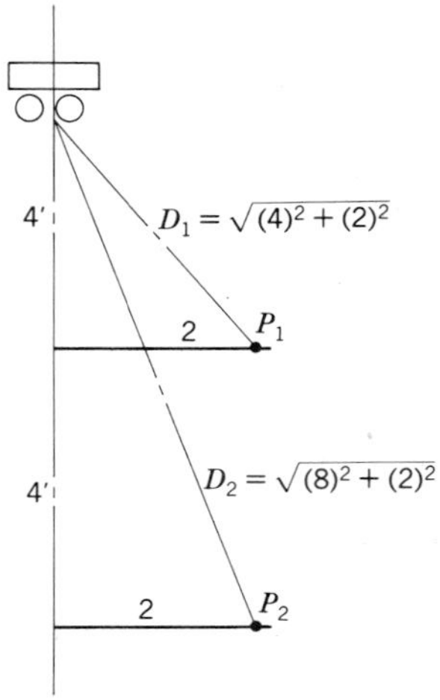

Fig. 20.50 For long linear sources, the ratio of illuminanes is $\frac{E_{P_2}}{E_{P_1}} = \frac{D_1}{D_2}$.

TABLE 20.5 K_H and K_V Values

Part A. ***Horizontal*** $F_C = K_H \times$ ***Lamp Lumens per Foot***
Broad Distribution—White-Enameled Channel

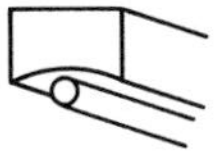

Vertical Distance from Centerline of Unit	K_H Values			
	Horizontal Distance from Centerline of Unit			
	0 ft	*1 ft*	*2 ft*	*3 ft*
1 ft	0.087	0.045	0.017	0.007
2 ft	0.040	0.033	0.020	0.011
3 ft	0.025	0.022	0.017	0.011
4 ft	0.018	0.015	0.014	0.009

Part B. ***Vertical*** $F_C = K_V \times$ ***Lamp Lumens per Foot***
Broad Distribution—White-Enameled Channel

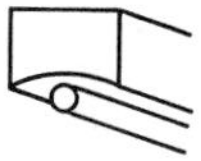

Vertical Distance from Centerline of Unit	K_V Values				
	Horizontal Distance from Centerline of Unit				
	3 in.	*6 in.*	*9 in.*	*12 in.*	*18 in.*
1 ft	0.014	0.030	0.037	0.043	0.041
2 ft	—	0.007	0.017	0.014	0.019
3 ft	—	—	—	0.005	0.008
4 ft	—	—	—	—	—

Part C. ***Horizontal*** $F_C = K_H \times$ ***Lamp Lumens per Foot***
Broad Distribution—White-Enameled Reflector

Vertical Distance from Centerline of Unit	K_H Values			
	Horizontal Distance from Centerline of Unit			
	0 ft	*1 ft*	*2 ft*	*3 ft*
1 ft	0.438	0.127	0.008	0.001
2 ft	0.233	0.150	0.061	0.017
3 ft	0.145	0.120	0.077	0.041
4 ft	0.106	0.095	0.072	0.048

Part D. ***Vertical*** $F_C = K_V \times$ ***Lamp Lumens per Foot***
Broad Distribution—White-Painted Cornice, No Reflector

Vertical Distance from Centerline of Unit	K_V Values				
	Horizontal Distance from Centerline of Unit				
	3 in.	*6 in.*	*9 in.*	*12 in.*	*18 in.*
9 in.	0.185	0.159	0.175	0.165	0.129
1 ft, 9 in.	0.011	0.028	0.044	0.057	0.068
2 ft, 9 in.	0.004	0.010	0.017	0.023	0.032
3 ft, 9 in.	0.002	0.005	0.008	0.012	0.018

Source: Reproduced by permission of General Electric, Lighting Business Group.

given in the table is 4 ft, whereas we require 8 ft. However, if we calculate the illuminance at 4 ft below and 2 ft offset (point P_1 in Fig. 20.50), then by Equation 20.19 we can calculate the illuminance at point P_2. From Table 19.11, the output of an HO lamp is approximately 1100 lumens per foot.

Illuminance at point P_1 is

$$E_{P1} = K \times \frac{\text{lumens}}{\text{foot}} = 0.072 \text{ (2 lamps) (1100 lm/ft)} = 158.4 \text{ fc}$$

From Equation 20.19, the illuminance at point P_2 is related to that at P_1 inversely as their distance from the source, that is,

$$\frac{E_{P_2}}{E_{P_1}} = \frac{D_1}{D_2} = \frac{\sqrt{4^2 + 2^2}}{\sqrt{8^2 + 2^2}} = 0.542$$

and

$$E_{P_2} = 0.542\ (E_{P_1}) = 0.542(158.4 \text{ fc}) = 85.9 \text{ fc}$$

20.41 Calculating Illuminance from Linear and Area Sources

When the source is too wide to be considered a line source (the definition is relative and depends on distance to the illuminated surface), it is referred to as a linear source. The *direct component* of the illuminance at a point, resulting from such a source, or from a large area source, can be calculated by manual graphical or analytical means, both of which are based on an assumed distribution, generally lambertian (diffuse). Since most fixtures being applied today do *not* have lambertian characteristics (e.g., parabolic reflectors, prismatic diffusers), the results of such calculations are necessarily approximate. (Skylights and luminous ceilings *do* have a lambertian distribution, and there these calculation methods do give reliable results.) In addition to the direct component of illuminance, a *reflected component* must be added that depends on the point's location in the room and the room's characteristics. This calculation involves charts, diagrams, and tables. The interested reader will find a full description of these manual methods in Section 9, Reference Volume, 1981 *IES Lighting Handbook*.

However, because these manual methods are laborious, and frequently inapplicable, they are not presented here. Further, the increasing use of desktop computers that can handle detailed input for the specific light source involved, without broad approximations, is making these manual procedures inefficient and obsolete. We recommend that when point-by-point illuminance calculation is desired, such a detailed-input program be used. Alternatively, one of the design aids described above, based on a specific light source, can be used.

20.42 Average Luminance Calculations

The basic equations relating luminance to candela intensity and to illuminance are covered in Section 18.8. That section also deals with the calculation of source luminance, and with reflected luminance when the illuminance (lux) level is known. Thus, once horizontal illuminance has been calculated by any of the methods described above (Sections 20.28 through 20.41), and knowing the reflectance of an object, its horizontal luminance can readily be calculated (see Section 18.8). However, as explained in detail in Sections 18.28 and 18.29, which discuss lighting quality, and in Sections 20.10 through 20.16, which deal with lighting systems and patterns, the luminance impression of a visual environment is affected more by vertical surface luminance than by horizontal. (See also Table 18.2*b* and Fig. 20.12.) For this reason it is important to be able to calculate *average* vertical surface (wall) luminance in the same simple, straightforward fashion that *average* horizontal illumination is calculated. In addition, it is useful to know the average luminance of the ceiling cavity in a space, in order to judge the contrast between all luminous objects, including luminaires, that have the ceiling cavity as background.

Straightforward calculation of both wall and ceiling cavity luminance (L_W and L_{CC}) is possible through the use of luminance coefficients that are similar in application and calculation to utilization coefficients. These coefficients are listed in Table 20.6 for the same generic fixture types as given in Table 20.2. For actual installation calculations, it is preferable to obtain coefficients from luminaire manufacturers. The average luminance calculations are parallel to those for illuminance.

TABLE 20.6 **Wall Luminance Coefficients and Ceiling Cavity Luminance Coefficients for Typical Luminaires**

To obtain a luminance coefficient, follow the procedure detailed at the head of Table 20.2 to find a coefficient of utilization. Data should be obtained from manufacturers of actual luminaires used.

	ρ_{CC}:	80			70			50			ρ_{CC}:	80			70			50		
	ρ_W:	50	30	10	50	30	10	50	30	10	ρ_W:	50	30	10	50	30	10	50	30	10
Typical Luminaire	RCR	*Wall Luminance Coefficients for* ρ_{FC} = 20									RCR	*Ceiling Cavity Luminance Coefficient* ρ_{FC} = 20								
1											0	.42	.42	.42	.36	.36	.36	.25	.25	.25
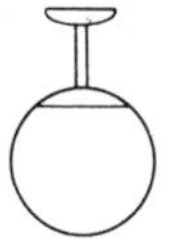	1	.32	.18	.06	.30	.17	.05	.27	.15	.05	1	.42	.40	.37	.36	.34	.32	.25	.23	.22
	2	.27	.15	.05	.25	.14	.04	.23	.13	.04	2	.42	.38	.35	.36	.33	.30	.24	.23	.21
	3	.24	.13	.04	.22	.12	.04	.20	.11	.03	3	.41	.37	.33	.35	.32	.29	.24	.22	.20
	4	.21	.11	.03	.20	.10	.03	.17	.09	.03	4	.41	.36	.32	.35	.31	.28	.24	.22	.20
	5	.19	.10	.03	.18	.09	.03	.16	.08	.02	5	.40	.35	.31	.34	.30	.27	.24	.21	.19
	6	.18	.09	.03	.16	.08	.02	.14	.07	.02	6	.39	.34	.31	.34	.30	.27	.23	.21	.19
	7	.16	.08	.02	.15	.08	.02	.13	.07	.02	7	.39	.34	.31	.33	.29	.27	.23	.21	.19
	8	.15	.07	.02	.14	.07	.02	.12	.06	.02	8	.38	.34	.30	.33	.29	.26	.23	.20	.19
Pendant diffusing sphere	9	.14	.07	.02	.13	.06	.02	.12	.06	.02	9	.38	.33	.30	.33	.29	.26	.23	.20	.19
with incandescent lamp	10	.13	.06	.02	.12	.06	.02	.11	.05	.01	10	.37	.33	.30	.32	.29	.26	.22	.20	.19
3											0	.15	.15	.15	.13	.13	.13	.09	.09	.09
	1	.23	.13	.04	.23	.13	.04	.21	.12	.04	1	.15	.13	.11	.13	.11	.10	.09	.08	.07
	2	.22	.12	.04	.22	.12	.04	.21	.11	.04	2	.14	.11	.08	.12	.09	.07	.08	.07	.05
	3	.21	.11	.03	.21	.11	.03	.20	.11	.03	3	.13	.10	.06	.12	.08	.06	.08	.06	.04
	4	.20	.10	.03	.19	.10	.03	.19	.10	.03	4	.13	.08	.05	.11	.07	.04	.08	.05	.03
	5	.19	.09	.03	.18	.09	.03	.18	.09	.03	5	.12	.08	.04	.11	.07	.04	.07	.05	.02
	6	.18	.09	.03	.17	.09	.03	.17	.09	.02	6	.12	.07	.03	.10	.06	.03	.07	.04	.02
	7	.17	.08	.02	.16	.08	.02	.16	.08	.02	7	.11	.06	.03	.10	.05	.02	.07	.04	.02
Porcelain-enameled	8	.16	.08	.02	.15	.08	.02	.15	.07	.02	8	.11	.06	.02	.09	.05	.02	.06	.04	.01
ventilated standard dome	9	.15	.07	.02	.15	.07	.02	.14	.07	.02	9	.10	.05	.02	.09	.05	.02	.06	.03	.01
with incandescent lamp	10	.14	.07	.02	.14	.07	.02	.13	.07	.02	10	.10	.05	.02	.09	.04	.02	.06	.03	.01
7											0	.08	.08	.08	.07	.07	.07	.04	.04	.04
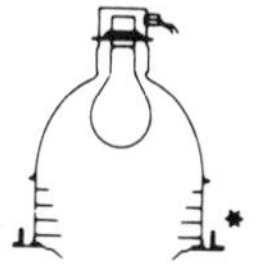	1	.06	.03	.01	.06	.03	.01	.05	.03	.01	1	.08	.07	.07	.06	.06	.06	.04	.04	.04
	2	.05	.03	.01	.05	.03	.01	.05	.03	.01	2	.07	.06	.05	.06	.05	.05	.04	.04	.03
	3	.05	.03	.01	.05	.03	.01	.04	.02	.01	3	.06	.05	.04	.05	.04	.04	.04	.03	.03
	4	.05	.02	.01	.05	.02	.01	.04	.02	.01	4	.05	.04	.04	.05	.04	.03	.03	.03	.02
	5	.05	.02	.01	.04	.02	.01	.04	.02	.01	5	.05	.04	.03	.04	.03	.03	.03	.02	.02
	6	.04	.02	.01	.04	.02	.01	.04	.02	.01	6	.05	.03	.03	.04	.03	.02	.03	.02	.02
Reflector downlight with	7	.04	.02	.01	.04	.02	.01	.04	.02	.01	7	.04	.03	.02	.04	.03	.02	.03	.02	.01
baffles and inside frosted	8	.04	.02	.01	.04	.02	.01	.04	.02	.01	8	.04	.03	.02	.03	.02	.02	.02	.02	.01
lamp. See note on this unit	9	.04	.02	.01	.04	.02	.01	.04	.02	.01	9	.04	.02	.02	.03	.02	.01	.02	.02	.01
in Table 20.2.	10	.04	.02	.00	.04	.02	.01	.04	.02	.01	10	.04	.02	.01	.03	.02	.01	.02	.01	.01

(*continued*)

TABLE 20.6 **Wall Luminance Coefficients and Ceiling Cavity Luminance Coefficients for Typical Luminaires** (***continued***)

	ρ_{CC}:	80			70			50			ρ_{CC}:	80			70			50		
	ρ_W:	50	30	10	50	30	10	50	30	10	ρ_W:	50	30	10	50	30	10	50	30	10
Typical Luminaire	*RCR*	*Wall Luminance Coefficients for ρ_{FC} = 20*									*RCR*	*Ceiling Cavity Luminance Coefficient ρ_{FC} = 20*								
18											0	.15	.15	.15	.13	.13	.13	.08	.08	.08
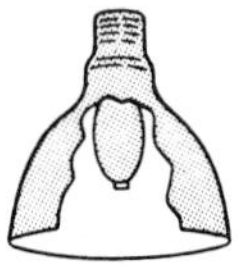	1	.18	.10	.03	.17	.10	.03	.16	.09	.03	1	.14	.13	.11	.12	.11	.10	.08	.08	.07
	2	.17	.09	.03	.16	.09	.03	.15	.09	.03	2	.13	.11	.09	.11	.09	.08	.08	.06	.05
	3	.16	.09	.03	.16	.08	.03	.15	.08	.02	3	.12	.09	.07	.11	.08	.06	.07	.06	.04
	4	.16	.08	.02	.15	.08	.02	.14	.08	.02	4	.12	.08	.06	.10	.07	.05	.07	.05	.03
	5	.15	.08	.02	.15	.07	.02	.14	.07	.02	5	.11	.07	.05	.10	.06	.04	.07	.04	.03
	6	.14	.07	.02	.14	.07	.02	.14	.07	.02	6	.11	.07	.04	.09	.06	.03	.06	.04	.02
	7	.14	.07	.02	.14	.07	.02	.13	.07	.02	7	.10	.06	.03	.09	.05	.03	.06	.04	.02
High-bay wide-distribution	8	.13	.07	.02	.13	.06	.02	.13	.06	.02	8	.10	.06	.03	.09	.05	.03	.06	.04	.02
ventilated reflector with	9	.13	.06	.02	.13	.06	.02	.12	.06	.02	9	.10	.05	.03	.08	.05	.02	.06	.03	.02
clear HID lamp	10	.12	.06	.02	.12	.06	.02	.12	.06	.02	10	.09	.05	.02	.08	.04	.02	.06	.03	.01
26											0	.28	.28	.28	.24	.24	.24	.16	.16	.16
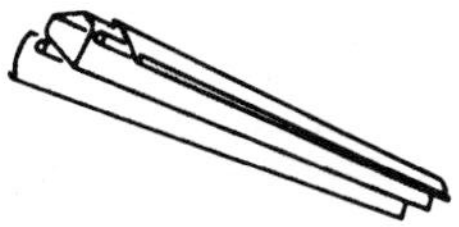	1	.20	.11	.04	.19	.11	.03	.17	.10	.03	1	.27	.26	.24	.23	.22	.21	.16	.15	.15
	2	.19	.10	.03	.18	.10	.03	.16	.09	.03	2	.27	.24	.22	.23	.21	.19	.16	.14	.13
	3	.18	.10	.03	.17	.09	.03	.15	.08	.03	3	.26	.22	.20	.22	.19	.17	.15	.13	.12
	4	.17	.09	.03	.16	.09	.03	.15	.08	.02	4	.25	.21	.18	.22	.18	.16	.15	.13	.11
	5	.16	.08	.02	.16	.08	.02	.14	.07	.02	5	.25	.21	.17	.21	.18	.15	.15	.12	.11
	6	.15	.08	.02	.15	.07	.02	.14	.07	.02	6	.24	.20	.17	.21	.17	.15	.14	.12	.10
	7	.15	.07	.02	.14	.07	.02	.13	.07	.02	7	.24	.19	.16	.20	.17	.14	.14	.12	.10
Diffuse aluminum reflector	8	.14	.07	.02	.13	.07	.02	.12	.06	.02	8	.23	.19	.16	.20	.16	.14	.14	.11	.10
with 35° crosswise	9	.13	.06	.02	.13	.06	.02	.12	.06	.02	9	.23	.19	.15	.20	.16	.13	.14	.11	.10
shielding	10	.13	.06	.02	.12	.06	.02	.11	.06	.02	10	.23	.18	.15	.20	.16	.13	.14	.11	.09
28											0	.27	.27	.27	.23	.23	.23	.16	.16	.16
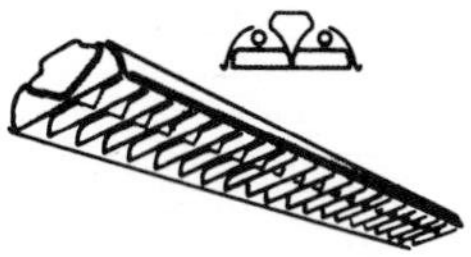	1	.17	.10	.03	.16	.09	.03	.14	.08	.03	1	.26	.24	.23	.22	.21	.20	.15	.14	.14
	2	.16	.09	.03	.15	.09	.03	.14	.08	.02	2	.25	.23	.21	.21	.20	.18	.15	.14	.13
	3	.15	.08	.02	.15	.08	.02	.13	.07	.02	3	.24	.21	.19	.21	.19	.17	.14	.13	.12
	4	.15	.08	.02	.14	.07	.02	.13	.07	.02	4	.24	.21	.18	.20	.18	.16	.14	.12	.11
	5	.14	.07	.02	.13	.07	.02	.12	.06	.02	5	.23	.20	.17	.20	.17	.15	.14	.12	.11
	6	.13	.07	.02	.13	.06	.02	.11	.06	.02	6	.23	.19	.17	.20	.17	.14	.14	.12	.10
	7	.13	.06	.02	.12	.06	.02	.11	.06	.02	7	.23	.19	.16	.19	.16	.14	.13	.11	.10
Diffuse aluminum reflector	8	.12	.06	.02	.12	.06	.02	.11	.05	.02	8	.22	.18	.16	.19	.16	.14	.13	.11	.10
with 35° crosswise and 35°	9	.12	.06	.02	.11	.05	.02	.10	.05	.01	9	.22	.18	.15	.19	.16	.13	.13	.11	.10
lengthwise shielding	10	.11	.05	.01	.11	.05	.01	.10	.05	.01	10	.22	.18	.15	.19	.15	.13	.13	.11	.09

33 											0	.65	.65	.65	.56	.56	.56	.38	.38	.38
	1	.20	.12	.04	.18	.10	.03	.13	.08	.02	1	.65	.63	.62	.55	.54	.53	.38	.37	.37
	2	.19	.10	.03	.17	.09	.03	.12	.07	.02	2	.64	.61	.59	.55	.53	.51	.38	.37	.36
	3	.17	.09	.03	.15	.08	.02	.11	.06	.02	3	.64	.60	.58	.55	.52	.50	.37	.36	.35
	4	.16	.08	.02	.14	.07	.02	.11	.06	.02	4	.63	.59	.57	.54	.51	.49	.37	.36	.35
	5	.15	.08	.02	.13	.07	.02	.10	.05	.02	5	.63	.59	.56	.54	.51	.49	.37	.36	.34
	6	.14	.07	.02	.12	.06	.02	.09	.05	.01	6	.60	.58	.55	.54	.50	.48	.37	.35	.34
	7	.13	.07	.02	.12	.06	.02	.09	.04	.01	7	.62	.58	.55	.53	.50	.48	.37	.35	.34
Luminous bottom suspended unit with extra-high-output lamp	8	.12	.06	.02	.11	.05	.02	.08	.04	.01	8	.61	.57	.55	.53	.50	.48	.37	.35	.34
	9	.12	.06	.02	.10	.05	.01	.08	.04	.01	9	.61	.57	.55	.53	.50	.48	.36	.35	.34
	10	.11	.05	.01	.10	.05	.01	.07	.04	.01	10	.61	.57	.54	.52	.49	.47	.36	.35	.34
35 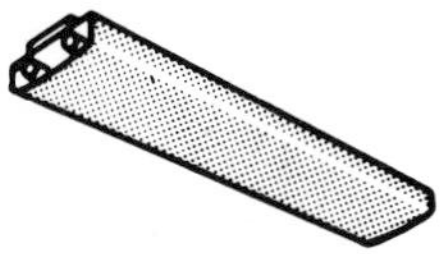											0	.22	.22	.22	.19	.19	.19	.13	.13	.13
	1	.19	.11	.03	.18	.10	.03	.17	.10	.03	1	.21	.20	.18	.18	.17	.16	.12	.12	.11
	2	.18	.10	.03	.17	.09	.03	.15	.09	.03	2	.21	.18	.16	.18	.16	.14	.12	.11	.10
	3	.16	.09	.03	.16	.08	.03	.14	.08	.02	3	.20	.17	.14	.17	.15	.13	.12	.10	.09
	4	.15	.08	.02	.15	.08	.02	.14	.07	.02	4	.19	.16	.13	.17	.14	.12	.11	.10	.08
	5	.14	.07	.02	.14	.07	.02	.13	.07	.02	5	.19	.15	.13	.16	.13	.11	.11	.09	.08
	6	.14	.07	.02	.13	.07	.02	.12	.06	.02	6	.18	.15	.12	.16	.13	.10	.11	.09	.07
	7	.13	.06	.02	.12	.06	.02	.12	.06	.02	7	.18	.14	.11	.16	.12	.10	.11	.09	.07
Two-lamp prismatic wraparound—multiply by 0.95 for four lamps	8	.12	.06	.02	.12	.06	.02	.11	.05	.02	8	.18	.14	.11	.15	.12	.10	.11	.08	.07
	9	.12	.06	.02	.11	.05	.02	.11	.05	.01	9	.17	.13	.11	.15	.12	.09	.10	.08	.07
	10	.11	.05	.01	.11	.05	.01	.10	.05	.01	10	.17	.13	.11	.15	.11	.09	.10	.08	.07
38 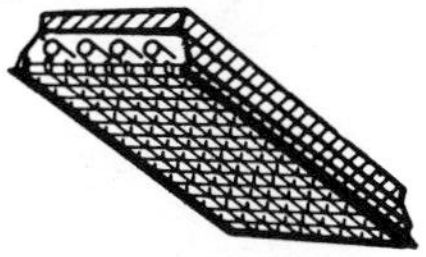											0	.09	.09	.09	.08	.08	.08	.05	.05	.05
	1	.13	.07	.02	.12	.07	.02	.12	.07	.02	1	.09	.08	.07	.08	.07	.06	.05	.05	.04
	2	.12	.07	.02	.12	.06	.02	.11	.06	.02	2	.08	.07	.05	.07	.06	.05	.05	.04	.03
	3	.11	.06	.02	.11	.06	.02	.11	.06	.02	3	.08	.06	.04	.07	.05	.03	.05	.03	.02
	4	.11	.06	.02	.10	.05	.02	.10	.05	.02	4	.07	.05	.03	.06	.04	.03	.04	.03	.02
	5	.10	.05	.01	.10	.05	.01	.10	.05	.01	5	.07	.04	.03	.06	.04	.02	.04	.03	.02
Four-lamp, 2-ft-wide troffer with 45° plastic louvre—multiply by 1.05 for three lamps and 1.1 for two lamps.	6	.10	.05	.01	.09	.05	.01	.09	.05	.01	6	.07	.04	.02	.06	.04	.02	.04	.02	.01
	7	.09	.04	.01	.09	.04	.01	.09	.04	.01	7	.06	.04	.02	.06	.03	.02	.04	.02	.01
	8	.09	.04	.01	.08	.04	.01	.08	.04	.01	8	.06	.03	.02	.05	.03	.01	.04	.02	.01
	9	.08	.04	.01	.08	.04	.01	.08	.04	.01	9	.06	.03	.01	.05	.03	.01	.04	.02	.01
	10	.08	.04	.01	.08	.04	.01	.07	.04	.01	10	.06	.03	.01	.05	.03	.01	.03	.02	.01
42 											0	.12	.12	.12	.10	.10	.10	.07	.07	.07
	1	.16	.09	.03	.16	.09	.03	.15	.09	.03	1	.11	.10	.09	.09	.08	.07	.06	.06	.05
	2	.15	.08	.03	.15	.08	.03	.14	.08	.02	2	.10	.08	.06	.09	.07	.06	.06	.05	.04
	3	.15	.08	.02	.14	.08	.02	.14	.07	.02	3	.10	.07	.05	.08	.06	.04	.06	.04	.03
	4	.14	.07	.02	.13	.07	.02	.13	.07	.02	4	.09	.06	.04	.08	.05	.03	.05	.04	.02
	5	.13	.07	.02	.13	.07	.02	.12	.06	.02	5	.09	.06	.03	.08	.05	.03	.05	.03	.02
Fluorescent unit with flat prismatic lens; four-lamp, 2 ft wide—multiply by 1.05 for three lamps and 1.10 for two lamps.	6	.12	.06	.02	.12	.06	.02	.12	.06	.02	6	.09	.05	.03	.07	.04	.02	.05	.03	.02
	7	.12	.06	.02	.12	.06	.02	.11	.06	.02	7	.08	.05	.02	.07	.04	.02	.05	.03	.01
	8	.11	.05	.02	.11	.05	.02	.11	.05	.02	8	.08	.04	.02	.07	.04	.02	.05	.03	.01
	9	.11	.05	.01	.10	.05	.01	.10	.05	.01	9	.08	.04	.02	.07	.03	.01	.05	.02	.01
	10	.10	.05	.01	.10	.05	.01	.10	.05	.01	10	.07	.04	.01	.06	.03	.01	.04	.02	.01

(continued)

TABLE 20.6 **Wall Luminance Coefficients and Ceiling Cavity Luminance Coefficients for Typical Luminaires (*continued*)**

	ρ_{CC}:	80			70			50			ρ_{CC}:	80			70			50		
	ρ_W:	50	30	10	50	30	10	50	30	10	ρ_W:	50	30	10	50	30	10	50	30	10
Typical Luminaire	*RCR*	*Wall Luminance Coefficients for ρ_{FC} = 20*									*RCR*	*Ceiling Cavity Luminance Coefficient ρ_{FC} = 20*								
44 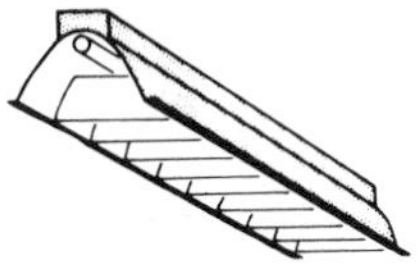	0										0	.114	.114	.114	.097	.097	.097	.066	.066	.066
	1	.137	.078	.025	.133	.076	.024	.125	.072	.023	1	.105	.094	.084	.089	.080	.072	.061	.055	.050
	2	.131	.072	.022	.128	.070	.022	.121	.067	.021	2	.097	.079	.064	.083	.068	.055	.057	.047	.038
	3	.127	.068	.020	.124	.066	.020	.118	.064	.019	3	.092	.068	.049	.079	.059	.043	.054	.041	.030
	4	.123	.064	.019	.120	.063	.019	.115	.061	.018	4	.087	.060	.039	.075	.052	.034	.052	.036	.024
	5	.119	.060	.018	.116	.060	.017	.112	.058	.017	5	.084	.053	.031	.072	.046	.027	.050	.032	.019
	6	.114	.057	.016	.112	.056	.016	.108	.055	.016	6	.080	.048	.025	.069	.042	.022	.048	.029	.016
	7	.110	.054	.015	.108	.054	.015	.104	.053	.015	7	.078	.044	.021	.067	.038	.018	.046	.027	.013
Radial batwing	8	.106	.052	.015	.104	.051	.014	.101	.050	.014	8	.075	.041	.018	.065	.036	.016	.045	.025	.011
distribution—louvred	9	.102	.049	.014	.100	.049	.014	.097	.048	.014	9	.073	.038	.015	.063	.033	.013	.043	.023	.009
fluorescent unit	10	.097	.047	.013	.096	.046	.013	.093	.046	.013	10	.070	.036	.013	.060	.031	.012	.042	.022	.008
45 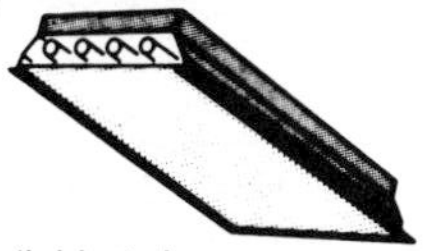	0										0	.092	.092	.092	.078	.078	.078	.053	.053	.053
	1	.143	.081	.026	.139	.079	.025	.133	.076	.024	1	.086	.075	.065	.074	.065	.056	.051	.044	.039
	2	.135	.074	.023	.132	.073	.022	.127	.071	.022	2	.082	.064	.048	.070	.055	.041	.048	.038	.029
	3	.127	.067	.020	.124	.066	.020	.119	.065	.020	3	.079	.055	.036	.067	.047	.031	.046	.033	.022
	4	.119	.062	.018	.117	.061	.018	.112	.059	.018	4	.075	.049	.028	.065	.042	.025	.044	.029	.017
Radial batwing	5	.113	.057	.017	.111	.057	.017	.107	.056	.016	5	.073	.044	.022	.062	.038	.019	.043	.026	.014
distribution—four lamp,	6	.106	.053	.015	.104	.053	.015	.101	.052	.015	6	.070	.040	.018	.060	.034	.016	.041	.024	.011
610 mm (2′) wide	7	.100	.049	.014	.098	.049	.014	.095	.048	.014	7	.067	.037	.015	.057	.032	.013	.040	.022	.010
fluorescent unit with flat	8	.094	.046	.013	.093	.046	.013	.090	.045	.013	8	.064	.034	.013	.055	.029	.012	.038	.021	.008
prismatic lens and	9	.089	.043	.012	.088	.043	.012	.085	.042	.012	9	.062	.032	.011	.053	.027	.010	.037	.019	.007
overlay—see note 2	10	.084	.040	.011	.083	.040	.011	.081	.039	.011	10	.059	.030	.010	.051	.026	.009	.035	.018	.006
46 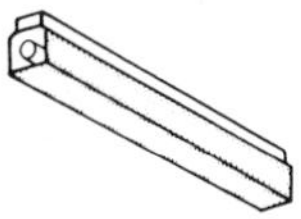	0										0	.236	.236	.236	.201	.201	.201	.138	.138	.138
	1	.234	.133	.042	.225	.128	.041	.208	.119	.038	1	.229	.210	.194	.196	.181	.167	.134	.124	.115
	2	.213	.117	.036	.205	.113	.035	.190	.106	.033	2	.222	.193	.168	.190	.166	.145	.130	.115	.101
	3	.195	.104	.031	.188	.101	.030	.175	.095	.029	3	.216	.180	.151	.185	.155	.131	.127	.108	.092
	4	.181	.094	.028	.175	.091	.027	.162	.086	.026	4	.211	.170	.139	.181	.147	.121	.124	.102	.085
	5	.170	.087	.025	.164	.084	.025	.153	.080	.023	5	.206	.163	.131	.177	.141	.114	.122	.098	.080
Bilateral batwing	6	.159	.080	.023	.153	.078	.022	.143	.073	.021	6	.201	.157	.124	.173	.136	.108	.119	.095	.077
distribution—one lamp,	7	.149	.074	.021	.144	.072	.020	.135	.068	.020	7	.197	.152	.120	.169	.131	.105	.117	.092	.074
surface-mounted	8	.141	.069	.019	.137	.067	.019	.128	.064	.018	8	.193	.148	.117	.166	.128	.102	.115	.090	.072
fluorescent, with prismatic	9	.134	.065	.018	.129	.063	.018	.121	.060	.017	9	.189	.144	.114	.163	.125	.099	.113	.088	.071
wraparound lens	10	.126	.061	.017	.122	.059	.016	.115	.056	.016	10	.185	.141	.112	.159	.122	.098	.111	.086	.069

47	0										0	.114	.114	.114	.097	.097	.097	.066	.066	.066
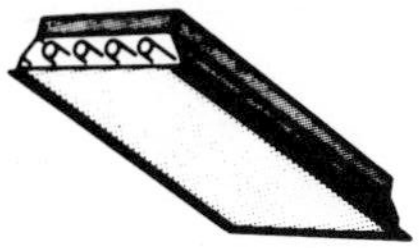	1	.175	.100	.032	.171	.098	.031	.163	.094	.030	1	.107	.093	.081	.091	.080	.069	.063	.055	.048
	2	.168	.092	.028	.164	.090	.028	.157	.087	.027	2	.102	.079	.059	.087	.068	.051	.060	.047	.036
	3	.157	.083	.025	.153	.082	.025	.147	.080	.024	3	.097	.068	.045	.083	.059	.039	.057	.041	.027
	4	.147	.076	.022	.144	.075	.022	.138	.073	.022	4	.093	.060	.035	.080	.052	.030	.055	.036	.021
	5	.139	.071	.021	.136	.070	.020	.131	.068	.020	5	.090	.054	.028	.077	.047	.024	.053	.033	.017
Radial batwing	6	.130	.065	.019	.128	.065	.019	.123	.063	.018	6	.086	.049	.023	.074	.043	.020	.051	.030	.014
distribution—four lamp,	7	.122	.060	.017	.120	.060	.017	.116	.059	.017	7	.082	.045	.019	.071	.039	.017	.049	.027	.012
610 mm (2′) wide	8	.115	.056	.016	.114	.056	.016	.110	.055	.016	8	.079	.042	.016	.068	.036	.014	.047	.025	.010
fluorescent unit with flat	9	.109	.053	.015	.108	.052	.015	.104	.052	.015	9	.076	.039	.014	.065	.034	.013	.045	.024	.009
prismatic lens—see note 2	10	.103	.049	.014	.102	.049	.014	.099	.048	.014	10	.072	.037	.013	.062	.032	.011	.043	.022	.008
48	0										0	.325	.325	.325	.278	.278	.278	.189	.189	.189
	1	.316	.180	.057	.304	.173	.055	.280	.161	.051	1	.320	.295	.273	.274	.253	.235	.187	.174	.162
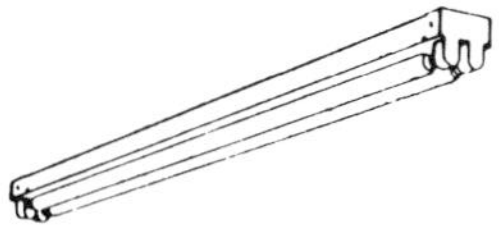	2	.282	.155	.047	.271	.149	.046	.250	.139	.043	2	.314	.275	.242	.269	.237	.209	.184	.164	.146
	3	.252	.134	.040	.242	.130	.039	.223	.121	.037	3	.307	.261	.223	.264	.225	.193	.181	.156	.136
	4	.229	.119	.035	.220	.115	.034	.202	.107	.032	4	.301	.250	.210	.258	.216	.183	.178	.150	.129
	5	.212	.108	.031	.203	.104	.030	.187	.098	.029	5	.295	.241	.201	.253	.209	.175	.175	.146	.124
	6	.195	.098	.028	.188	.095	.027	.173	.089	.026	6	.289	.234	.195	.248	.203	.170	.171	.142	.120
	7	.181	.090	.025	.174	.087	.025	.161	.081	.023	7	.283	.228	.190	.244	.198	.166	.168	.139	.117
	8	.169	.083	.023	.163	.080	.023	.150	.075	.021	8	.278	.224	.186	.239	.194	.163	.165	.136	.115
Two-lamp fluorescent strip	9	.159	.077	.021	.153	.074	.021	.141	.070	.020	9	.273	.220	.184	.235	.191	.160	.163	.134	.114
unit	10	.149	.071	.020	.143	.069	.019	.132	.065	.018	10	.268	.216	.182	.231	.188	.158	.160	.132	.113
49	0										0	.280	.280	.280	.239	.239	.239	.163	.163	.163
	1	.334	.190	.060	.324	.185	.059	.303	.174	.056	1	.273	.247	.224	.234	.212	.192	.160	.146	.133
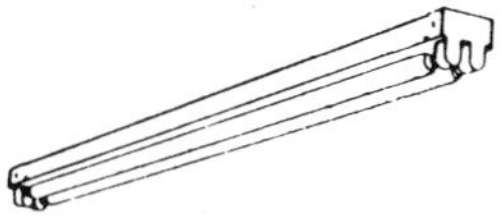	2	.302	.166	.051	.292	.161	.050	.274	.153	.047	2	.266	.225	.189	.228	.193	.164	.156	.134	.114
	3	.272	.145	.043	.263	.141	.042	.247	.134	.041	3	.259	.208	.168	.222	.179	.145	.152	.125	.102
	4	.248	.129	.038	.241	.126	.037	.226	.120	.036	4	.251	.196	.153	.216	.169	.133	.148	.118	.094
	5	.231	.118	.034	.224	.115	.033	.210	.110	.032	5	.245	.186	.142	.210	.161	.124	.145	.112	.087
	6	.213	.107	.031	.207	.105	.030	.195	.100	.029	6	.238	.178	.135	.205	.154	.118	.141	.108	.083
	7	.199	.098	.028	.193	.096	.027	.182	.092	.026	7	.232	.172	.130	.199	.149	.113	.138	.104	.080
Two-lamp fluorescent strip	8	.186	.091	.026	.181	.089	.025	.170	.085	.024	8	.226	.166	.125	.194	.144	.109	.134	.101	.078
unit with 235° reflector	9	.175	.085	.024	.170	.083	.023	.161	.079	.023	9	.220	.162	.122	.190	.140	.107	.131	.099	.076
fluorescent lamps	10	.164	.079	.022	.160	.077	.022	.151	.074	.021	10	.215	.158	.120	.185	.137	.104	.128	.096	.074

NOTES:

1. Data extracted, with permission, from *IES Handbook* (1981), Reference Volume. Coefficients for fixtures 1, 3, 7, 18, 26, 28, 33, 35, 38, and 42 have been rounded to two decimal places. For three decimal figures, see *IES Handbook* (1981).
2. Multiply coefficients by 1.05 for three lamps and 1.1 for two lamps.

Average initial wall luminance (cd/m^2):

$$L_W = \frac{\text{lamp lumens} \times \text{wall luminance coefficient}}{\pi \times \text{floor area in m}^2} \quad (20.21)$$

and average initial ceiling cavity luminance in cd/m^2:

$$L_{CC} = \frac{\text{lamp lumens} \times \text{ceiling cavity luminance coefficient}}{\pi \times \text{floor area in m}^2} \quad (20.22)$$

If area is expressed in square feet and π omitted from the equation, L is expressed in footlamberts.

To obtain *maintained* luminance values, a light loss factor similar to that explained in Section 20.30 is introduced. It is similarly calculated except that item 17 (see Fig. 20.39), room surface dirt, is calculated using the following figures:

Lighting System	*Wall Luminance*	*Ceiling Luminance*
Direct	0.82 ±10%	0.75 ±10%
Semidirect	0.87 ± 7%	0.82 ±10%
Direct-indirect	0.92 ± 5%	0.85 ± 8%
Semi-indirect	0.87 ± 7%	0.88 ± 7%
Indirect	0.82 ±10%	0.90 ± 5%

For ceiling mounted or recessed luminaires, L_{CC} is the average luminance of the ceiling between luminaires. For pendant luminaires, the calculated L_{CC} is that of an imaginary plane at the height of the luminaires. It is useful in determining brightness ratios when compared to luminaire luminance at the seeing angle involved. The ceiling cavity, as the wall, is assumed to have a lambertian characteristic, that is, perfect diffuseness, making luminance independent of viewing angle.

It would be instructive to calculate the wall luminance of the office of Example 20.2. The photometric data in Fig. 20.34 do not include the wall luminance coefficient, since luminance coefficients are not normally published by luminaire manufacturers. However, based on other data available, a figure of 0.22 is appropriate given RCR, ρ_{CC}, and ρ_W of 0.66, 80%, and 30%, respectively. Initial wall luminance is then

$$L_W = \frac{3200 \times 0.22}{32 \text{ ft}^2} = 22 \text{ fL} = 75.5 \text{ cd/m}^2$$

This is within the preferred range of 50 to 200 cd/m^2. See Table 18.2*b*. In actuality the average wall luminance would probably be higher because of the practice of placing the last row of luminaires quite close to the wall.

EVALUATION STAGE

20.43 Design Evaluation

The design stage having been completed, the final step is evaluation of the design from the three aspects—lighting, costs, and energy. The lighting aspects include quantity, quality, ESI and/or contrast reduction calculations, luminance ratios, mood, ambience, texture, color, variation, psychological impressions, orientations, and daylight use—in short, a review of all the lighting factors discussed in detail above. Since a good deal of experience is required to visualize the actual results from the drawings, the novice designer would do well to have the review done, at least in part, by one having such experience. The other two aspects of evaluation, cost and energy, can be performed readily with the aid of contractor's estimating figures for cost and a straightforward calculation for energy. The results are compared to the budget figures developed at the preliminary design stage. As we have repeatedly stressed, the useful cost figures are life cycle, annual operating and first cost, for economic comparisons, operating budgets, and construction budgets, respectively. In the following chapter we will present recommendations for specific occupancies, accompanied by actual cost studies and energy analyses. Detailed cost studies including the impact of lighting on air conditioning, the proportional cost of the wiring system, and the proper apportionment of costs involves the entire building and can be accurately performed only by computer. Studies of this type are generally made by consulting engineers rather than architects, and then only after initial, operating, and total costs have been set in proper perspective for the particular job, by the architect and client. This is necessary because often, as in the case of speculative construction, the client's overriding consideration is first cost, thereby rendering a complete cost analysis unnecessary. Any attempt to completely separate costs for lighting, HVAC, structure, and so on is arbitrary because of the intimate interaction between these factors. Lighting designers are well advised to keep themselves and the construction team aware of this, if they are to fully fulfill their responsibility.

References

Brown, G. Z., Reynolds, J. S., and Ubbelohde, M. S. (1982). *InsideOut*. Wiley, New York.

Elmer, W. B. (1980). *The Optical Design of Reflectors*. Wiley, New York.

Illuminating Engineering Society of North America (1981). *IES Handbook,* 6th ed. IES of North America, New York.

21

LIGHTING APPLICATION

21.1 Introduction

In the preceding three chapters, we examined lighting fundamentals, sources, and design procedures. In this final lighting chapter we shall consider application of lighting principles to specific situations. The facilities covered in some detail include residential, educational, commercial, institutional, and industrial occupancies. Each is examined from the viewpoint of its special requirements and suggested approach. The latter includes lighting materials, sources, as well as comparative economics and energy considerations. The chapter concludes with consideration of special types of indoor lighting plus a short section on exterior lighting.

LIGHTING CONTROL

21.2 Requirement for Lighting Control

The term *lighting control* means all the techniques by which the lighting system will be operated and covers both manual and automatic controls. The control strategy must be decided on simultaneously with the lighting design because the control scheme must be appropriate to the light source. In turn, the system accessories and arrangement depend on the control scheme. For instance, if dimming is decided on, using a fluorescent light source, then the range of dimming determines the type of ballast, the ballast switching points, and the degree of dimming flexibility.

The primary purposes of lighting control are flexibility and economy: flexibility to provide the modifications of brightness and pattern desired by the designer, and economy of both energy resources and monetary resources. (See Appendix E for treatment of economic analyses.) A properly designed lighting control system will reduce energy usage over a noncontrolled installation by 10 to 60% (see Peterson and Rubenstein, 1982; Rubenstein, 1981; Turner, 1982). In addition, financial operating economies will result from:

1. Reduced energy use.
2. Reduced air conditioning costs as a result of lower lighting waste heat.
3. Longer lamp and ballast life due to lower operating temperatures and lower output.
4. Lower maintenance costs due to longer equipment life.

21.3 Elements of Lighting Control

(a) Switching. There are two basic types of control functions—switching and dimming. Switching is an on–off function. By selecting the number of lighting elements to be switched in each switching action, the designer can establish the number of control levels. The more levels, the finer the control. Thus, in a space requiring several levels of uniform illumination for different functions, the designer has many control alternatives. He can switch entire fixtures, but this adversely affects uniformity. Assuming three-lamp fluorescent fixtures as an example, the designer can obtain better uniformity and four levels of illumination by switching the ballasts (assuming one 2-lamp and one single lamp or half of a 2-lamp ballast):

All ballasts on	100% illumination
2-Lamp ballast on	66% illumination
1-Lamp ballast on	33% illumination
All ballasts off	0 illumination

This type of switching has the advantage of light reduction in relatively small steps, at low cost. A typical arrangement is shown in Fig. 21.1. Use of split-wired two-lamp ballasts rather than one-lamp units is advantageous from the cost and energy

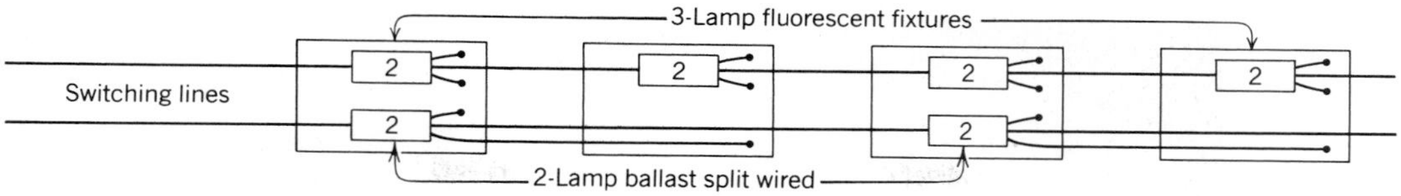

Fig. 21.1 *Multilevel switching arrangement of three-lamp fluorescent fixtures. Note that only two lamp ballasts are used in the interest of energy and financial economy. Alternatively, two level ballasts can be used.*

viewpoints. Even more uniform light reduction and finer control are possible with two-level ballasts (at increased cost). There, each lamp remains lighted but at either full or half output. Thus the designer could have 0, 50, and 100% levels, or by combining alternate ballast switching with two-level ballasts, he could have 0, 17, 33, 50, 67, 83, and 100%. However, if that degree of control is desirable, dimming is probably preferable. The choice will depend on the type of space, and on the situation economics, as discussed below.

(b) Dimming. The techniques and equipment required for dimming each of the different light sources, as well as the effect on the color of the light produced and on the lamp, are discussed in Chapter 19. Figure 21.2 shows typical lumen output versus power input curves for common light sources. Note that for fluorescent lamps, *even with conventional ballasts,* dimming down to approximately 40% of output is possible without reducing efficacy. This desirable characteristic can be exploited in control schemes where it is desired to gradually change light output without sacrificing efficiency. Since below 40% output efficacy drops off, an economical and efficient control scheme combines dimming and switching of multilamp fluorescent fixtures to yield an almost stepless output range of 13 to 100% output. Continuous dimming over a 10 to 100% range is practicable with special dimming ballasts or with electronic ballasts. As discussed in Chapter 19, electronic ballasts are much more energy efficient than conventional ones, and must be considered for all *new* installations, dimmed or not. (Because of high failure rates of some electronic ballasts, a few government agencies and large private firms will not consider their use at present. We assume that when the deficiencies are removed, the prohibition will be lifted also.) For retrofit work, SCR (silicon-controlled rectifier) or triac dimmers will give excellent results with existing conventional core-and-coil ballasts, down to 40% output, as noted above.

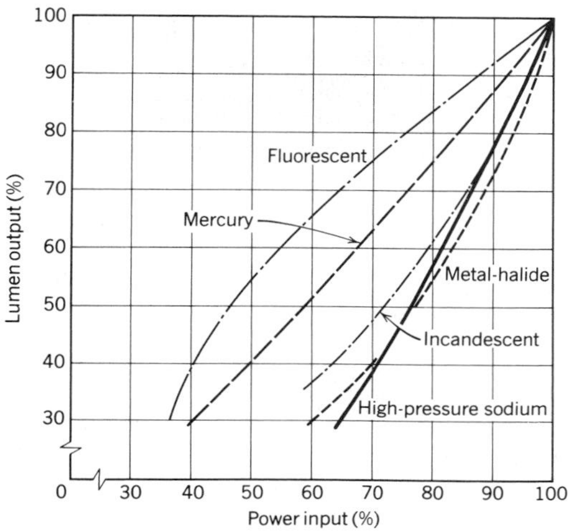

Fig. 21.2 *Typical dimming curves for generic light source types. Note that fluorescent is efficient and approximately linear, down to 40% output. All other sources have reduced efficacy when dimmed.*

(c) Control Initiation. The control operation is either manual or automatic. Manual operation is usually applicable only to a small number of simple functions such as on–off or level switching. Even then, however, the tendency is to leave lights on at the maximum level and not to shut them off when leaving a space. Numerous studies have demonstrated that no lasting appreciable energy economy is possible when the control initiation is entirely manual and relies on a facility's personnel, however well intentioned. Even a system of rewards for energy conservation rarely has long-term effects. Limited energy economy is pos-

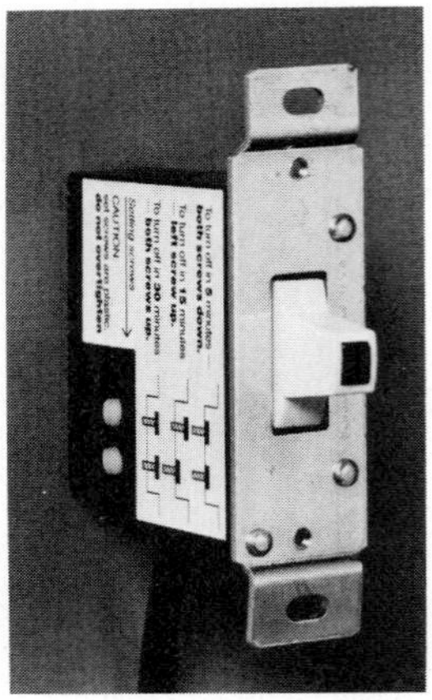

Fig. 21.3 *Time-out switch replaces ordinary wall switch and is adjustable to turn off in 5, 15, or 30 minutes. Typical applications are closets, stock rooms, and other spaces with short time occupancy. (Photo courtesy of Enertron.)*

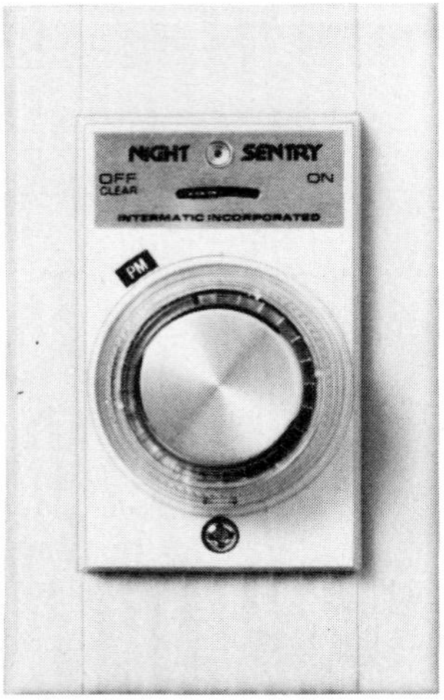

Fig. 21.4 *Programmable electronic 24-h repeating timer, fits into a standard wall box and provides up to 24 on–off operations per day, plus dimming and switching of noninductive (incandescent lighting) loads. (Photo courtesy of Intermatic.)*

sible when the turn-off function is automated by the use of ''time-out'' switches, which open after a preset interval (see Fig. 21.3). Significant energy reduction and use limitation is best achieved with automatic control initiation.

Automatic controls are of two types: an open-circuit type and a closed-loop feedback type. The former initiates a control function that is independent of the actual lighting situation, whereas the latter reacts to the condition of the lighting situation it controls via a feedback loop. The most common type of open-circuit lighting control is the programmable time controller. These vary from the small simple unit of Fig. 21.4 to the much more sophisticated unit shown earlier in Fig. 16.10*b*. These devices are available in a myriad of designs and capacities, but all perform the same basic function—(remote) control of devices and circuits on a preprogrammed time base. The programming in turn is determined after analysis of operating schedules, task requirements, and field conditions. With ''tight'' programming, energy savings of up to 50% over an uncontrolled installation are possible (see Peterson and Rubenstein, 1982). The control signals may be line voltage, low voltage, on–off signals, or dimmer control signals. All are time based, and in this control configuration, all are insensitive to the actual field condition. (Some systems are ''wireless'' and use high-frequency signals impressed on the power wiring system to transmit control signals. This arrangement is particularly useful in retrofit work. See Section 14.14.) Because these devices act only on a time base, an override feature must be incorporated to permit accommodation of special conditions. Thus if the timer is arranged to shut off the row of lights adjacent to the windows between 10 A.M. and 3 P.M., local override must be provided to accommodate dark rainy days and the like. Similarly, if lights are shut off during nonworking hours, provision must be made for persons working overtime. The override arrangement can be entirely local, in which case it may lead to energy waste because it depends on local cancellation; it can be local with time-out, which can be a nuisance to a person working for an extended period; or it can incorporate an override feedback link to the controller, usually operated by telephone lines. In general, programmable time controls are best applied to facilities with regular, repetitive schedules and few exceptional situations.

An alternative form of this device is one for which the control function is primary and the time-based repetitive functions are secondary. These devices are generally applied in energy management systems (see Section 14.14). They differ from the above in that they control many more points,

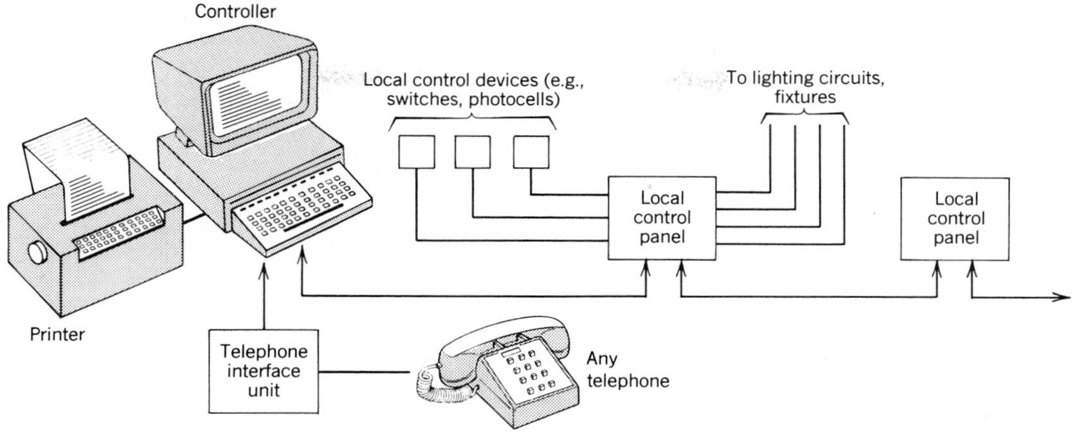

Fig. 21.5 *Schematic arrangement of a large lighting control system. The controller can schedule and supervise more than 5000 control points via the local control panels. The local panel accepts coded commands from the controller and operates individual devices and circuits via local relays. It also accepts local control signals from manual (switches) and automatic devices. Override is provided via telephone command to the controller, and all functions, including overrides, appear as "hard copy" on the printer.*

without necessarily having reference to a time base, that is, the condition of a function may be long-term and nonrepetitive. These devices are applicable to large systems for which (remote) tuning, level switching, and centralized, supervised, automatic control are desirable. Although referred to in the manufacturers' literature as lighting controllers, identical units are equally useful in building automation control, HVAC and solar energy control, energy management, and the like. A schematic arrangement of such a system is shown in Fig. 21.5.

The most flexible and useful (and expensive) of the automatic control devices is called a programmable controller (PC). It differs from a programmable time controller in that it interacts with the controlled environment, although the dividing line between a programmable controller and a large flexible time-based controller of the type shown in Fig. 21.5 is not a sharp one. The principal difference is that a programmable controller, by definition, contains a central processor, input/output (I/O) interface, memory, and a programming device. An operational block diagram is shown in Fig. 21.6. In addition to the usual time-based signal function, the controller accepts information (feedback) from field devices via its I/O device, "processes" the signal in the central processor, which consists of logic and storage memory, and

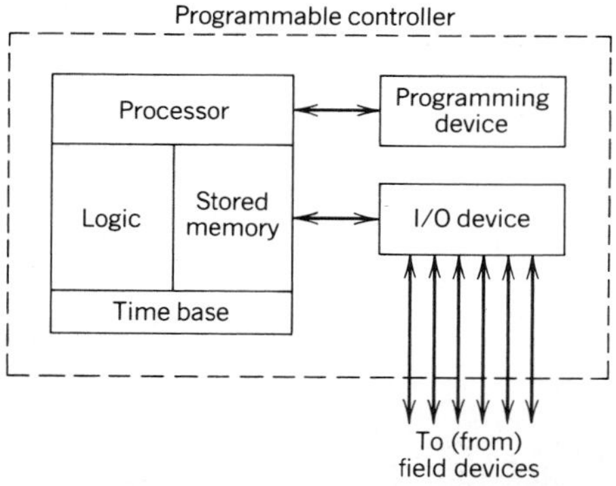

Fig. 21.6 *Block diagram of a programmable controller. These devices are designed to continuously supervise field conditions and are therefore used in industrial applications as well as lighting and environmental system control.*

sends out a resultant processed control signal. These devices are used in lighting control where active continuous, nonrepetitive control functions are required. (See Fig. 21.11 for a typical application.)

21.4 Lighting Control Strategy

It is apparent that a good lighting control system will match the lighting supplied to the lighting required, *as the requirement varies,* and thus overlighting and underlighting are avoided. In addition, the control system must be capable of permitting initial adjustments (see Section 21.4a, below) and external, nonlighting-connected constraints such as commands from a peak-demand controller. The common lighting system situations addressed by the control system are listed below.

(a) System "Tuning." In every lighting installation there is a difference between the design intent and the field result. This comes about from assumptions and imprecision in calculation, differences between specified and installed equipment, equipment location changes, and so forth. The responsible lighting designer will "tune" the lighting system in the field to attain the intended design. This usually means *reducing* levels in nontask areas, since spill light is frequently sufficient for circulation, rough material handling, and the like. This tuning can result in energy reduction of 20 to 30% (see Rubenstein, 1981) depending on the control technique. The smaller the group of light sources controlled, the more accurate the tuning and the larger the energy saving. Lighting system retuning is also required when the function of an entire space is changed, or when a single area is altered by a furniture move or by a task change. Tuning is a one-step function in the sense that once accomplished, it does not require change unless the space function changes. Therefore, it should also be reversible.

As stated, the action most often required is a reduction in illuminance. This can be accomplished by:

1. Replacement of lamps with others of lower wattages, and replacement of fluorescent tubes with low-wattage tubes or "phantom" tubes. See Section 19.16.
2. Replacement of fluorescent ballast with low-output (low-current) ballasts. See Section 19.17.
3. Addition of output reducing device to fluorescent fixtures. See Fig. 21.7.

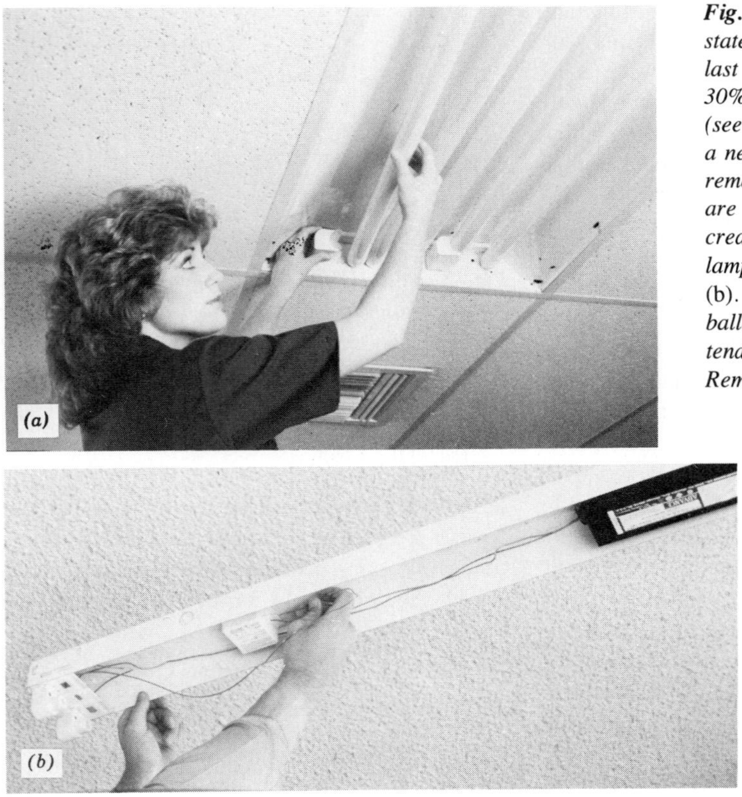

Fig. 21.7 *This type of compact solid-state electronic circuit reduces ballast power input by approximately 30% and lamp output by about 28% (see curve in Fig. 21.2), resulting in a net* gain *in efficacy. Power factor remains above 90%. Similar units are available for larger power decrease. They can be mounted on the lamp end* (a) *or in the future channel* (b). *An ancillary benefit is cooler ballast operation, resulting in extended life. (Photos courtesy of Remtec Systems.)*

Fig. 21.8 *Typical single-circuit fluorescent dimmer. This unit is rated at 277 V and will adjust the energy input to all the fluorescent ballasts on a single 20-A circuit, from 40 to 100% of full capacity. Light output is reduced somewhat less than energy, improving efficacy. Level adjustment is manual, at the dimmer, and uniform for the entire circuit. Groups of such dimmers can be remotely adjusted from a central controller. The illustrated unit is 15′ × 9′ × 8¾ in. deep, and weighs 20 lb. Similar units are available for other currents and voltages, and for HID light sources. Their action is shown in Fig. 21.2. (Photo courtesy of Lutron.)*

4. Ballast switching or use of multilevel ballasts. See Section 21.3a.
5. Dimming by adjustment of potentiometer at individual fixtures.
6. Dimming of groups of fixtures from centralized dimmers. See Fig. 21.8.

Items 4 and 6 can be accomplished locally or from a remote central controller.

(b) Variable Time Schedule. No normal task area has a constant 24-hour, 365-day, lighting requirement. In commercial and industrial spaces, work areas have regularly scheduled periods during which task lighting is not required. These include coffee and lunch breaks, cleaning periods, shift changes, and unoccupied periods. Programmed time controls can readily save 10 to 25% of the energy use as compared to relying on occupants to manually operate the controls. The action of such a controller in an actual installation is shown in Fig. 21.9. ''Tight'' scheduling took account of lunch hour and provided lighting only in restricted areas being cleaned rather than whole floors. Payback period for the investment in control equipment varies between 1½ and 5 years.

(c) Variable Space Occupancy. Within a normal 9 A.M. to 5 P.M. working schedule, offices in commercial spaces are unoccupied for 30 to 60% of the time. The reasons are manifold—coffee break, conference, work assignment, illness, vacation, and reassignment to a different work location are a few. Occupancy sensors can turn off lights after a preset minimum period of about 10 minutes or can dim the light level to a minimum in areas such as corridors, which always require some light. (They can also turn off other energy consumers such as fan coil units, air conditioners, fans.) Reestablishment of the original lighting level can be instantaneous, delayed, or manual on the action of the occupant. Another useful function of

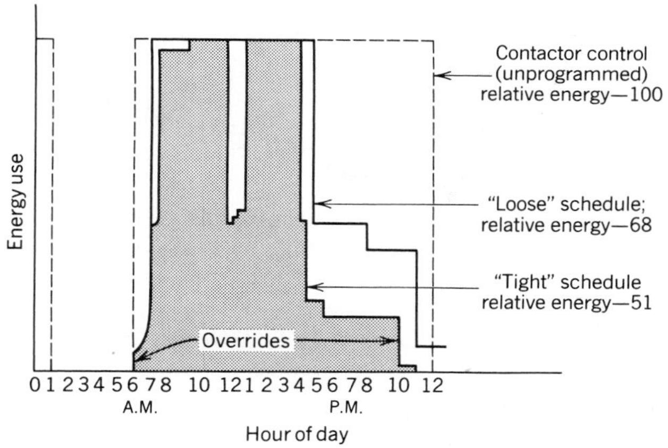

Fig. 21.9 *Actual plot of energy usage on the 58th floor of the World Trade Center in New York, with 3 degrees of lighting control. Note the importance of override by occupants, when using ''tight'' scheduling. Override was provided on a 1000-sq. ft zone basis. (Data extracted from Peterson and Rubenstein, 1982.)*

ILLUMINATION

an occupancy sensor is to provide automatic override in scheduled systems, thus both relieving the occupant of the necessity of using a manual override and limiting the energy use to actual occupancy time. It can also light the occupant's way into a space, and shut off the system after he has left.

Occupancy sensors that react to motion are of two types—ultrasonic and passive infrared. Both cover a maximum area of about 500 ft^2 per unit. Payback for this equipment runs between 6 months and 3 years depending on the type of space and the degree of control already existing. They are best applied in areas that are divided into individual rooms and work spaces. Building codes that require dual switching in rooms above a certain size will usually accept occupancy sensors as an alternative system, since they have been demonstrated to be much more effective in energy conservation. (See also Section 22.2, which discusses the security application of motion detectors and illustrates their operation.)

Another type of occupancy detector reacts to audible sound. This design is lower in first cost than motion detectors but may present problems of sensitivity adjustment in spaces with highly variable noise levels. A typical unit is shown in Fig. 21.10.

(d) Deliberate Overdesign. Referring to Section 20.30, we see that in order to maintain a minimum lighting level to the end of a maintenance period, we deliberately overdesign initially. The extent of the overdesign is the reciprocal of the light loss factor (LLF). With an average LLF of 0.60, this initial overdesign amounts to (1/0.60) or 66%. Assuming a linear light falloff over a 2-year maintenance period, this overdesign results in an average of 33% energy waste. (In the next 2-year maintenance cycle the savings will be slightly less due to a small amount of unrecoverable loss.) Because the light depreciation is a continuous and very gradual process over the maintenance period, the most appropriate control strategy is one that will reduce the initial overlighting by the required amount (as measured in the field) and gradually restore it, as the system ages. This is best accomplished by a centrally controlled dimming system, operating in conjunction with local light sensors (photocells). The photocells measure ambient light, and in response to their signals the controller operates the dimming units to raise (or lower) the light output. Depending on the size of the installation, the dimmers (of the type shown in Fig. 21.8) can either be dispersed and controlled from a central point, or the entire system can be installed at one central location. The modulating action of such a system over the maintenance period is shown in Fig. 21.11.

In a new installation, the choice of whether to use electronic ballasts and full-range dimming, conventional ballasts and partial dimming, or a system of multilevel switching is an economic one and depends on many factors, the most important of which is the cost of energy. See the economic discussion below. Often a combined system is advisable. In interior zones initial lighting reduction does not exceed 40%, and full-range dimming is not required. In perimeter zones, daylight often provides all of the required light, and either full-range dimming, SCR dimming plus switching, or a multilevel switching system is required. (SCR

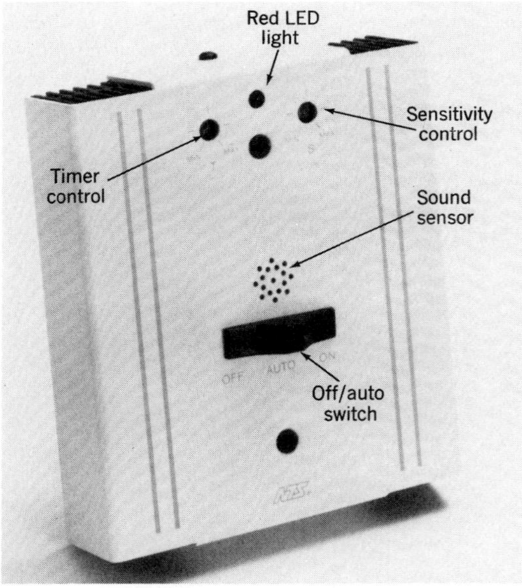

Fig. 21.10 *Typical audible-sound acoustic occupancy sensor. Such units are arranged to react to selected frequencies that exist in an occupied space. Absence of these sounds causes the unit to time-out and lights to be turned off. Conversely, the unit will re-energize the lighting when occupancy sounds are detected. (Photo courtesy of National Technical Systems.)*

dimmers are also referred to as lighting energy adjusters, reducers, or modulators in the manufacturers' literature.) See Section 21.3a above and the discussion of daylight compensation below. Payback period for a light reduction/compensation installation of this type varies from 1 to 5 years. Shorter payback periods can be obtained by using multilevel switching rather than dimming, as discussed in Section 21.3a. Although dimming gives stepless control and therefore higher energy savings, the lower cost of switching reduces the payback period.

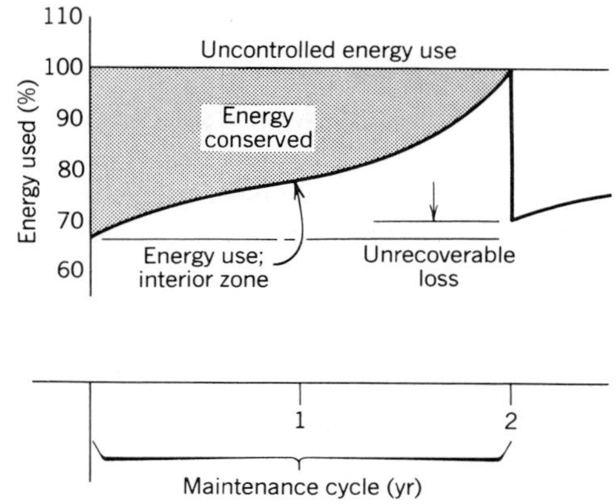

(a)

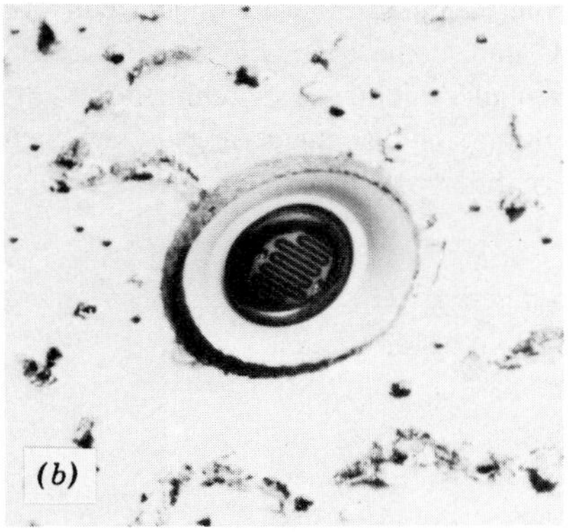

(b)

Fig. 21.11 (a) *Graph of energy use by a system that reduces initial lighting level to compensate for initial overdesign and gradually increases level as system output depreciates. In subsequent cycles the energy savings are reduced slightly because of unrecoverable output loss.* (b) *Typical ceiling-mounted photocell. The location of this small and unobtrusive device must be carefully selected so that the unit's signal corresponds accurately to the lighting level at the point (Photo courtesy of WideLite.)*

An additional favorable effect of initial light reduction is the lengthening of effective lamp life and a reduction in the rate of its lumen depreciation. When a lamp is operated at rated voltage, its lumen output drops during its life, according to the type of lamp. (See, for instance, Sections 19.11, 19.15, and 20.30.) However, if lamps are operated at reduced output, as would be the case if lamps are dimmed to compensate for initial overlighting, the lamp life cycle is greatly extended, lumen depreciation is reduced, and lamp energy consumption is linearly reduced. Typical life extension (to economic replacement) figures are:

Fluorescent	80%
Metal-halide	40%
High-pressure sodium	20%

These figures are for interior zones; for perimeter zones with ambient daylight compensation they are higher.

(e) Daylight Compensation. It should be apparent that a control system arranged for continuous ambient light compensation, as described in the preceding subsection, is automatically arranged to compensate for ambient daylight. The difference is that ambient compensation for deliberate overdesign is a very gradual process of *increasing* output, whereas daylight compensation can be a minute-by-minute variation, and generally in the direction of *decreased* artificial light. Because of these possible rapid variations, most switching systems are undesirable, as the constant on–off or level switching of lamps can be very annoying to occupants. A system that switches in small increments by use of multilevel ballasts in multiballast fixtures can be acceptable if combined with a fairly wide ''dead band'' between ON and OFF (as with a thermostat) to prevent ''hunting'' and excessive switching. (Indeed some variations in light levels as occur with daylight can be considered desirable. It is not necessary to *exactly* compensate for ambient daylight.)

Automatic dimming is the system of choice. Since in perimeter areas daylight often supplies all the required light, the system must either be full-range dimming with electronic ballasts (for fluorescent installations) or partial dimming with conventional core-and-coil ballast, plus switching. Here again the switching action can be disturbing

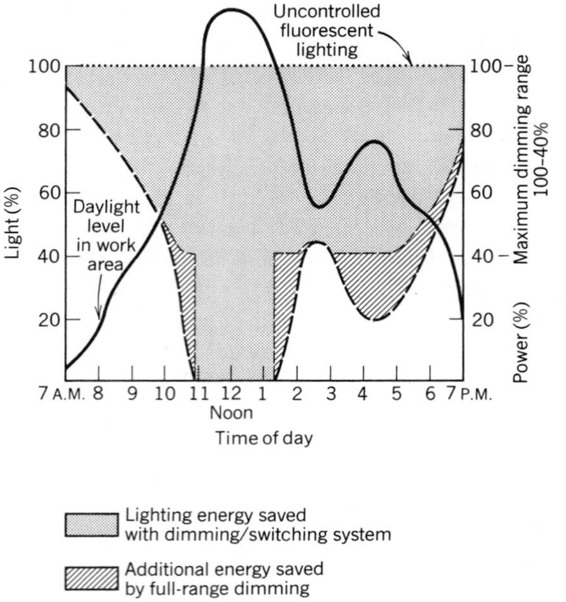

Fig. 21.12 Typical graph of energy savings with daylight-compensating lighting control. A full-range dimming system is more effective than one that dims down only to 40%, since daylight often supplies most of the required lighting. An economic analysis is required to determine whether the additional cost of such a system is justified.

and must be carefully designed. Figure 21.12 shows the action of both types of dimming system in a fluorescent installation with daylight compensation. The crucial design element in a daylight-compensating system is the establishment of zone areas. Depending on latitude and climate, the southern and possibly the east and west exposures can have an interior (second) perimeter zone that receives sufficient daylight for a large enough portion of the year to economically warrant dimming. The northern exposure has only a narrow perimeter zone (see Fig. 21.13). As a rule of thumb, the size of the zones is established by determining the maximum room depth that receives at least half its illuminance from daylight for several hours a day. A computerized economic study with perimeter zone depth as a variable is the proper method of determining zone size. See Matsuura (1979).

Placement of the control photocells depends on the control system. Where daylight compensation is desired in conjunction with overdesign compensation (preceding subsection), area photocells are desirable, since they give a feedback control signal for the specific area involved. Alternatively, a daylight-factor (DF) map of a space can be made, preferably after the installation is complete and the furniture in place, and photocells located at one point, based on this map.

Refer to Fig. 21.14. We know from our study of daylight in Chapter 19 that the DF at any inside point is constant. Therefore, by measuring actual daylight level at a point either immediately inside or outside the window, shielded from direct sun and from inside artificial lighting, we can relate it to any area in the room by the ratio of daylight factors and establish a switching level at which the lighting for that inside area is switched, partially or fully. An example will make this clear.

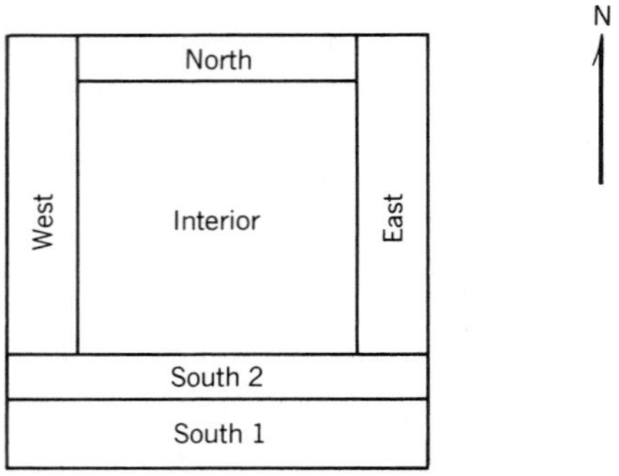

Fig. 21.13 Typical building section showing approximate daylight perimeter zones. Exact delineation of zones depends on latitude, climate, window design, and cost of electric energy.

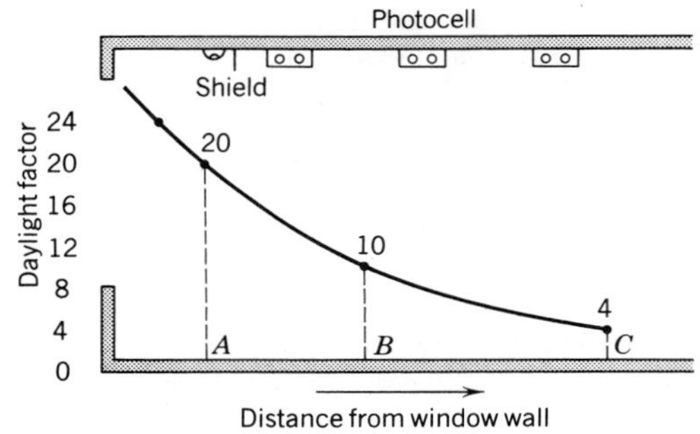

Fig. 21.14 Typical daylight factor curve plotted on a room section with one side window. The photo cell control technique for daylight compensation described in the text assumes a constant D.F. distribution indoors, that is, fixed luminous sky and no direct sun.

In Fig. 21.14, the photocells are mounted at point *A*. The ratio of daylight between points *A* and *B* is the ratio of their daylight factors, that is, 20/10 or 2.0. Therefore, if 500 lux of daylight is required at point *B* before switching lights in that area, 2.0 (500) or 1000 lux is required at point *A*. Similarly, a 500-lux requirement at point *C* corresponds to (20/4) (500) or 2500 lux at point *A*. Therefore switching can be initiated by two single-level photocells or a dual-level unit at point *A*, with settings at 1000 and 2500 lux. Other switching arrangements can be made on the same principle. Dimming initiation can also be arranged in this fashion.

Daylight compensation can reduce energy use in perimeter areas by up to 60% depending on latitude, climate, hours of building use, initial power density, and so forth. The amount saved for this entire building obviously depends on the building configuration, that is, the ratio of perimeter to total area. Payback time is usually in the range of 3 months to 3 years.

Summarizing, a well-designed lighting control system can provide energy conservation of up to 60%, extremely long lamp life, reduced cooling costs, extended ballast life, and reduced maintenance costs. Investment payback period (see Appendix E), when considering all these factors, is always short and therefore financially attractive. An aspect of a centralized lighting control system not discussed above, since it is not directly concerned with lighting, is its use in connection with peak demand reduction (see Section 14.13). When interconnected with a demand controller, this approach acts to reduce electric lighting load in accordance with a predetermined preferential-load schedule. Significant savings can thereby be achieved, as explained in Chapter 14.

RESIDENTIAL OCCUPANCIES

21.5 Residential Lighting

Residential lighting offers to the lighting designer great opportunity for originality and ingenuity, since a residence combines more functions and needs than any other building. Furthermore, it generally requires that all work be done at minimal cost, and that the end result please a range of tastes. The designer approaches the problem with a list of requirements, a perception of the space, and two basic tools: the fixture and the architectural lighting element. The former was covered in Sections 20.18 through 20.26 and is augmented by the fixture data in Fig. 21.15. The latter is discussed below. The reader is referred to *Design Criteria* (1980) for additional details on residential lighting.

(a) Energy Considerations. In determining a power budget (see Section 20.5), use the following unit power densities:

Bath	4.3 W/sq ft
Bedroom	1.4 W/sq ft
Finished spaces	2.2 W/sq ft
Kitchen	4.0 W/sq ft
Laundry	1.0 W/sq ft
Garage, unfinished spaces	0.5 W/sq ft

1. Provide means for reducing light levels in all areas. A kitchen during food preparation does not have the same lighting requirements as a kitchen being entered for a "refrigerator raid." Low-level lighting provision should be made in *all* rooms, including bathrooms. To accomplish this use high–low switches, simple dimmers, multilevel ballasts, and multilevel switching. An ancillary benefit is that ambience can be changed thereby in multiuse rooms such as dining rooms, family rooms, and finished basements.
2. Provide local task lighting for difficult tasks such as the location at which family accounts are handled.
3. Provide dimming and switching for accent lighting.
4. Provide appropriate controls for exterior lights.
5. In large residences consider programmable time switches and low-voltage control for ease of remote control and energy savings.
6. Use daylight in areas normally occupied during daylight hours such as kitchens and living rooms. Consider skylights with built-in artificial lighting for these areas.

Wide Profile

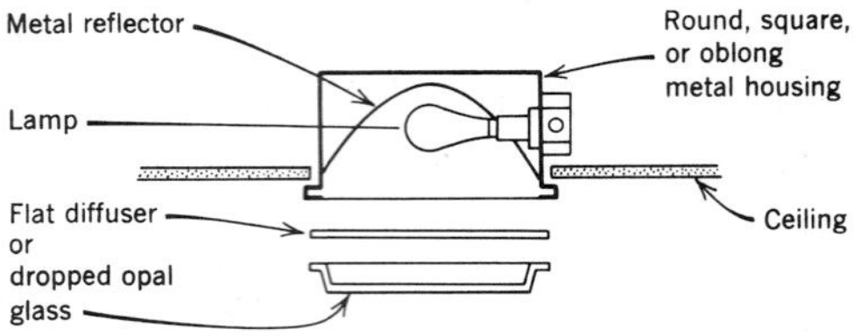

For general illumination (almost always used in multiple). Basement, reacreation rooms, kitchens, laundries, halls (service). Used singly or in small groups for small areas such as walk-in closets, garage, entry doors, overhangs in porches. Because of high luminance of diffuser, seldom used in living or social areas.

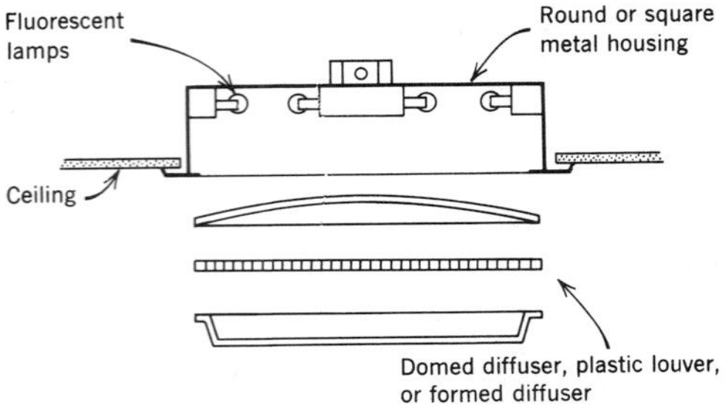

Very wide distribution, excellent as general lighting for kitchen, laundry, recreation room and bath. Because of large size and high lumen output, fewer units required. Often used signly or in pairs for entry halls and foyers and for skylight effect in interior halls and stairways.

Narrow Profile

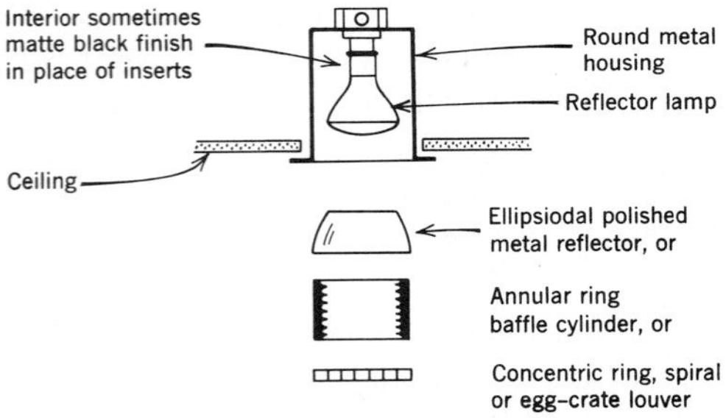

Accent lighting over plants, cocktail tables, etc. Wall lighting–mounted close to textured surfaces such as brick, stone, rough wood and fabrics. Task lighting–food preparation areas (may cause specular reflections), in multiple on quite close spacing for general lighting. Most effective when used near perimeter of room so that some light spills onto wall. Dramatic effects for family rooms, recreation rooms, and formal living areas. Supplementary stair lighting–shadow patterns define treads and risers. Dining tables–provide functional light on dining table to supplement decorative effect from hanging luminaire.

Medium Profile

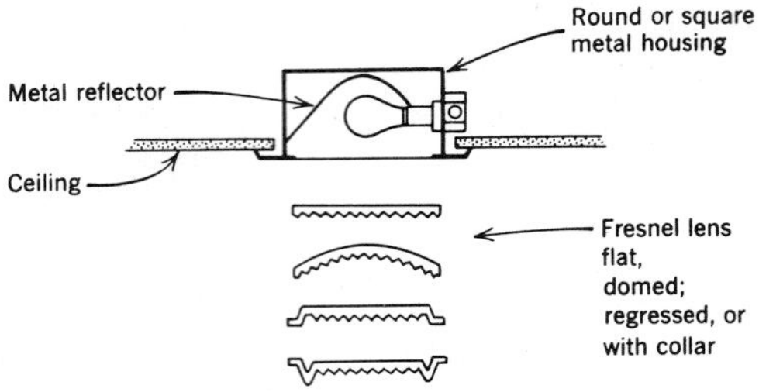

Used for specific task lighting where task area is large, such as kitchen sink, kitchen island counter or range, laundry tubs and ironing, game table, workbench, hobby area. Used for general lighting in restricted areas such as halls, entries and baths. Multiple groupings are satisfactory for the general lighting of kitchens and recreation rooms. If weatherproof, appropriate for outdoor uses, including overhangs, porches, entries.

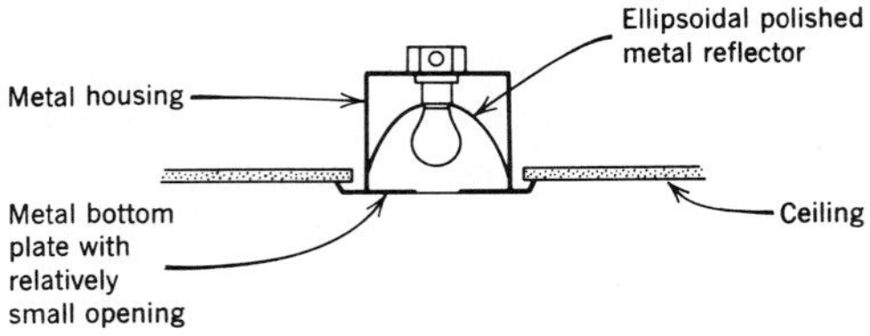

Uses basically the same as the Fresnel unit listed above, except that the lower luminance makes this type of equipment more usable in living and dining areas.

Special Asymmetric Profile

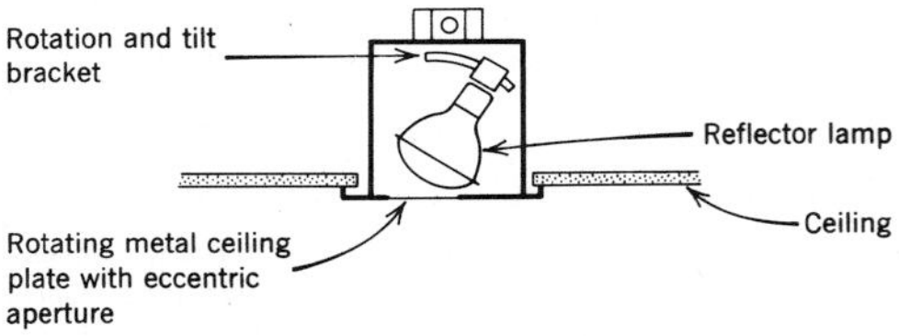

Useful for gallery or picture lighting and to light sculpture. If scalloping effect is acceptable can be used for wall lighting and to accent fireplace surfaces. Large size of bottom aperture sometimes make this unacceptable for highly styled interiors. May also be used for lighting piano music and sewing machines.

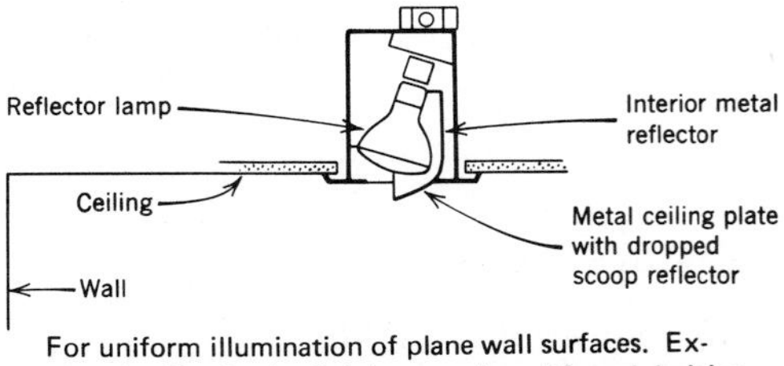

For uniform illumination of plane wall surfaces. Extremely effective for lighting murals and for minimizing wall imperfections. *Not* generally to be used for lighting textured wall surfaces because it directs no grazing light at wall. Spacing of these units is critical–follow manufacturer's recommendations closely.

Fig. 21.15 *Design features of recessed luminaires having wide, medium, narrow, and asymmetric profiles. See also Figs. 20.17 through 20.23, pages 1024 to 1026. (Courtesy of the IES of North America.)*

(b) Sources. When using fluorescent, choose proper color for space (see Table 19.20). Despite their lower efficacy, use of 5000 K and daylight fluorescents as the artificial source in lighted skylights is very effective. Listed below are the appropriate sources to be used in different areas of a residence:

1. Work and utility areas including kitchens, laundry, and workshop—fluorescent of appropriate color or metal-halide.
2. Built-in architectural elements—fluorescent.
3. Bedrooms, portable lamps, accent lights—incandescent, tungsten-halogen, compact fluorescent, circular fluorescent.
4. Circulation areas, stairwells, closets—incandescent.
5. Exterior: for short periods—incandescents; for long periods—HID.
6. Bathrooms: general; incandescent or warm white fluorescent; mirror lighting—incandescent.
7. All rooms—daylight where possible.
8. All spaces—use incandescent when source is turned on and off frequently or lighted for short periods only.

(c) Recommendations. These design recommendations are applicable to residential occupancies of all types.

1. Use general/task-lighting concept with recommended levels as in Tables 21.1 and 21.2.
2. Provide luminance ratios as in Fig. 21.16.
3. Provide general lighting in all spaces, sufficient for movement and casual seeing. Hallways require little lighting; stairs require more. Light stairs from directly above or ahead to create a shadow directly below the tread front. Lighting from the front eliminates shadows and can create a safety hazard.
4. Do *not* avoid ceiling lights as is so frequently done. Wide-profile ceiling fixtures provide general lighting; switch-controlled table lamps do not.

(d) Fixtures and Luminous Elements. See Fig. 21.15.

1. Utilize diffuse distribution for general lighting, narrow-distribution downlight for area and furniture accents, and narrow-distribution, ceiling-recessed incandescent wall-washers for accenting surfaces such as brick walls.
2. Use built-in lighting to the extent possible, including architectural lighting elements (see next section). We believe that this demonstrates integrity of concept. For this reason, we recommend that the flexibility of track lighting be utilized for accent and task lighting but not for general lighting.
3. Private residences are the exception to the rule of selecting off-the-shelf items in preference to specials. The lighting should complement the architecture and furnishings, and frequently this can best be accomplished by original designs.

21.6 Architectural Lighting Elements

Architectural lighting elements are coves, cornices, valances, coffers, skylights, and other luminous constructions not normally comprising a

TABLE 21.1 **Current Illuminance Recommendations for General Lighting**

Activity or Area	*Typical American Recommendation: Average Lux*	*Other Authorities: Average Lux*
Conversation and relaxation	50–100[a]	50–100
Passage areas	50–100[a]	50–100
Areas other than kitchen	200–500	100–200
Kitchen	500–1000	300–500

[a]General lighting in these areas need not be uniform.

ILLUMINATION

TABLE 21.2 **Current Illuminance Recommendations for Specific Visual Tasks**[a]

Seeing Task	*Typical American Recommendation: Average Lux*[b]	*Other Authorities: Average Lux*[b]
Dining	100–200	100–150
Grooming, makeup	200–500	500
Handcraft		
Ordinary seeing tasks	200–500	200–500
Difficult seeing tasks	500–1000	500–750
Critical seeing tasks	1000–2000	>1250
Kitchen duties		
Food preparation and cleaning involving difficult seeing tasks	500–1000	750–1000
Serving and other noncritical tasks	200–500	200–300
Laundry tasks	200–500	100–300
Reading and writing		
Handwriting, reproductions, and poor copies	500–1000	750
Books, magazines, and newspapers	200–500	300
Sewing, hand or machine		
Dark fabrics	1000–2000	>1250
Medium fabrics	500–1000	700–1000
Light fabrics	200–500	300–500
Table games	200–300	300

[a]Selection of illuminance within the given range is made on the basis of criteria given in Section 18.23.
[b]Divide by 10 to get footcandles. Due to the range of values, use of the exact 10.76 figure is unnecessary.

Zone 2 The immediate surroundings (area adjacent to the visual task)	
Desirable ratio	1/3 to equal to task*
Minimum acceptable ratio	1/5 to equal to task*

Zone 3 The general surroundings (not immediataly adjacent to task)	
Desirable ratio	1/5 to 5 times task*
Minimum acceptable ratio	1/10 to 10 times task*

*Typical task luminance range is 40 to 120 cd/m^2[12–35 fL] and seldom exceeds 200 cd/m^2[60 fL].

Fig. 21.16 *Seeing zones and recommended luminance ratios for residential visual tasks.*

lighting fixture. Although such units are normally less efficient than lighting fixtures, their use is often indicated by architectural considerations. Empirical design data are given in Fig. 21.17.

Using fluorescent tubes, it is possible to avoid dark spots between lamps by placing lamps at a slight angle rather than end-to-end, thus enabling ends to overlap. Reflectors increase installation efficiency. When used, they should be aimed 15 to 25° above the horizontal and field-adjusted for best ceiling coverage. When using double strips, they should be stacked vertically as shown in Fig. 21.17*m*. Light output for double-lamp installation rarely exceeds 1.75 times the single-lamp output.

(a) Lighted Cornices

Cornices direct all their light downward to give dramatic interest to wall coverings, draperies, murals, etc. May also be used over windows where space above window does not permit valance lighting. Good for low-ceilinged rooms.

(b) Lighted Valances

Valances are always used at windows, usually with draperies They provide up-light which reflects off ceiling for general room lighting and down-light for drapery accent. When closer to ceiling than 10 inches use closed top to eliminate annoying ceiling brightness.

(c) Lighted Coves

Coves direct all light to the ceiling. Should be used only with white or near-white ceilings. Cove lighting is soft and uniform but lacks punch or emphasis. Best used to supplement other lighting. Suitable for high-ceilinged rooms and for places where ceiling heights abruptly change.

(d) Lighted High Wall Brackets

High wall brackets provide both up and down light for general room lighting. Used on interior walls to balance window valance both architecturally and in lighting distribution. Mounting height determined by window or door height.

(e) Lighted Low Wall Brackets

Low brackets are used for special wall emphasis or for lighting specific tasks such as sink, range, reading in bed, etc. Mounting height is determined by eye height of users, from both seated and standing positions. Length should relate to nearby furniture groupings and room scale.

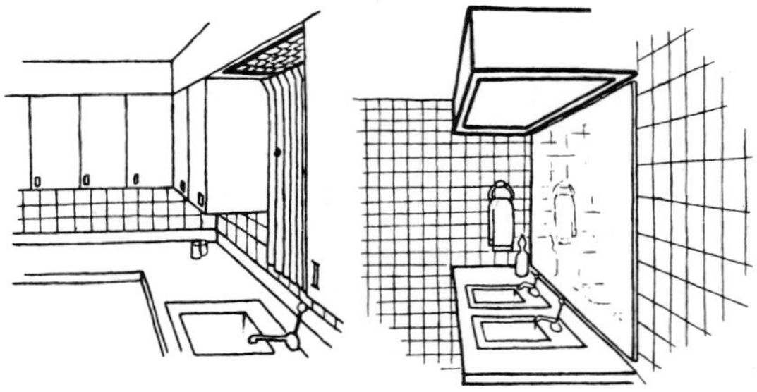

(f) Lighted Soffits

Soffits over work rease are designed to provide higher level of light directly below. Usually they are easily installed in furred-down area over sink in kitchen. Also are excellent for niches over sofas, pianos, built-in desks, etc.

Bath or dressing room soffits are designed to light user's face. They are almost always used with large mirrors and counter-top lavatories. Length usually tied to size of mirror. Add luxury touch with attractively decorated bottom diffuser.

Fig. 21.17 *Residential architectural lighting elements. (Courtesy of the IES of North America except as otherwise noted.)*

(g) Lighted Canopies

The canopy overhang is most applicable to bath or dressing room. It provides excellent general room illumination as well as light to the user's face.

(h) Luminous Ceilings

Totally luminous ceilings provide skylight effect very suitable for interior rooms or utility spaces, such as kitchens, baths, laundries. With attractive diffuser patterns, more decorative supports, and color accents they become acceptable for many other living spaces such as family rooms, dens, etc. Dimming controls desirable.

(j) Luminous Wall Panels

Luminous wall panels create pleasant vistas; are comfortable background for seeing tasks; add luxury touch in dining areas, family rooms and as room dividers. Wide variety of decorative materials available for diffusing covers.

(k) Typical Valance

This "typical" dimensional drawing applies only to commonly encountered window valance situations. Obviously, other window treatments could necessitate modifications in these critical dimensions; i.e., vertical blinds, double track situations, curved bay windows, etc.

The same "job-tailored" variations can occur in the design of any type of structural lighting device. Therefore no other dimensional drawings have been included here.

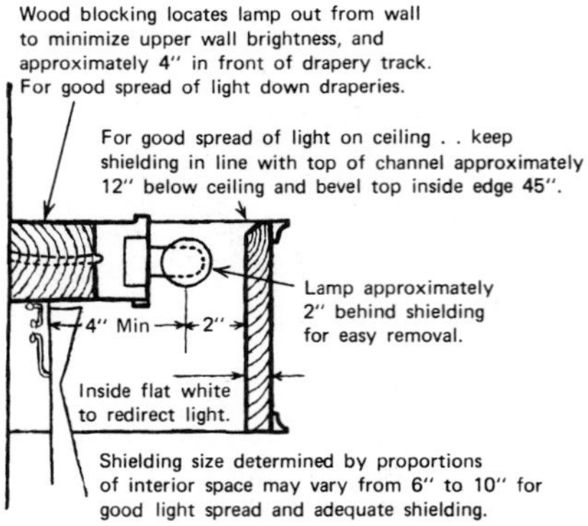
Wood blocking locates lamp out from wall to minimize upper wall brightness, and approximately 4" in front of drapery track. For good spread of light down draperies.
For good spread of light on ceiling . . keep shielding in line with top of channel approximately 12" below ceiling and bevel top inside edge 45".
Lamp approximately 2" behind shielding for easy removal.
4" Min
2"
Inside flat white to redirect light.
Shielding size determined by proportions of interior space may vary from 6" to 10" for good light spread and adequate shielding.

Fig. 21.17 (continued)

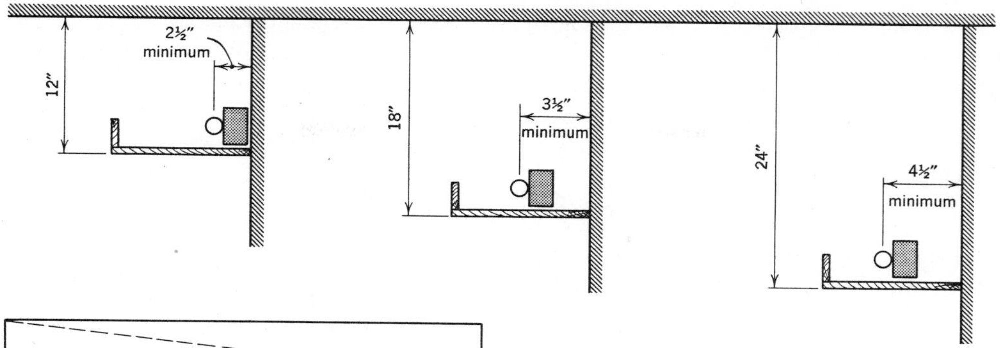

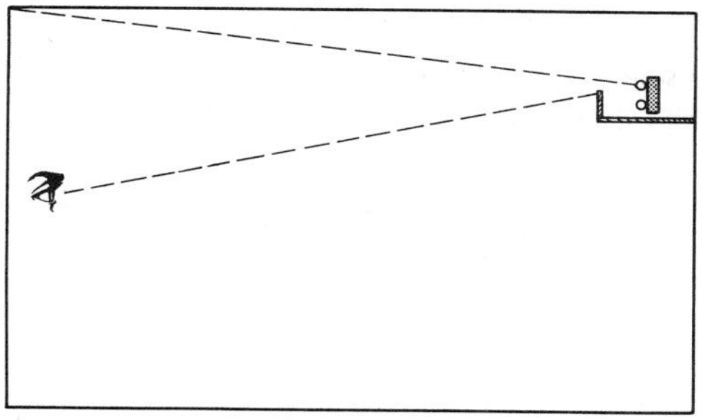

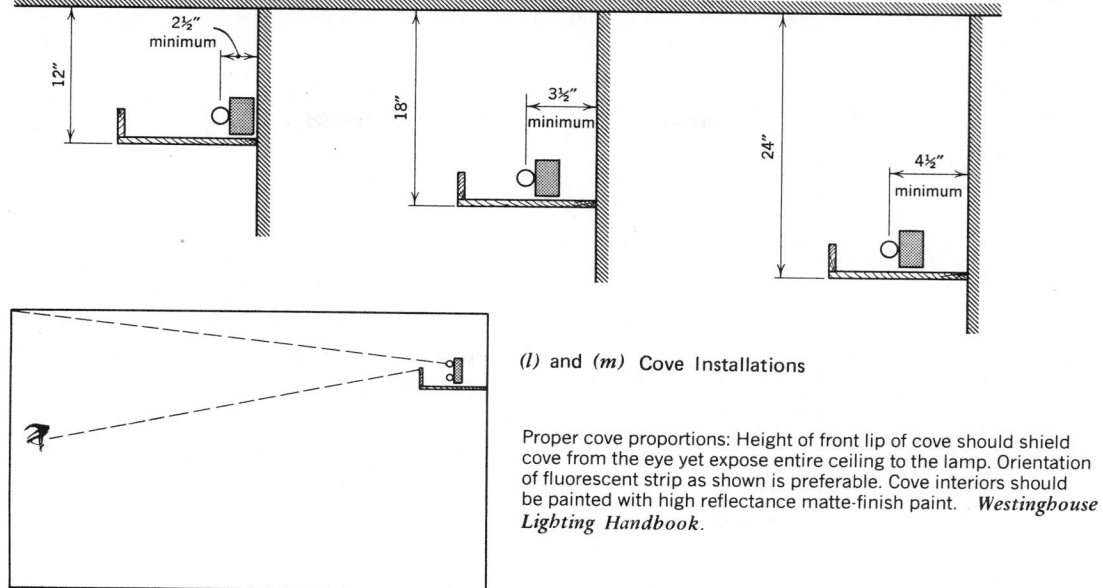

(l) and *(m)* Cove Installations

Proper cove proportions: Height of front lip of cove should shield cove from the eye yet expose entire ceiling to the lamp. Orientation of fluorescent strip as shown is preferable. Cove interiors should be painted with high reflectance matte-finish paint. *Westinghouse Lighting Handbook*.

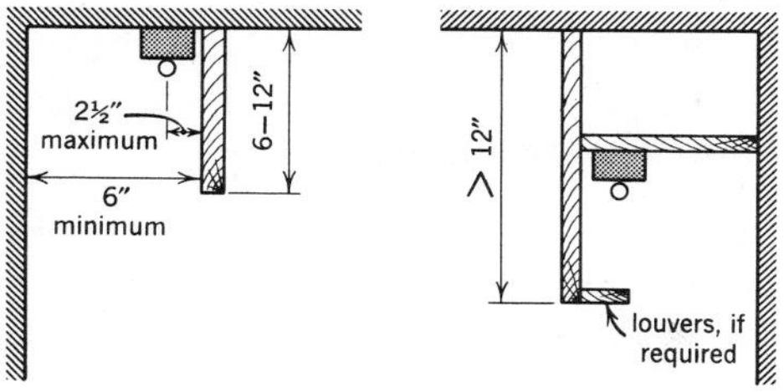

(n) Typical Cornices

Wall washing equipment mounted in valances and cornices provide improved brightness ratios and may be used for lighting desks against walls, or vertical illumination of walls and objects mounted thereon. *Westinghouse Lighting Handbook*.

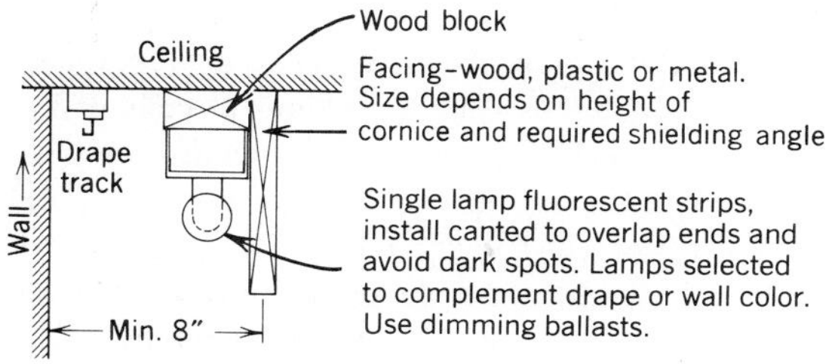

Paint all surfaces matte white.

***Fig. 21.17** (continued)*

EDUCATIONAL FACILITIES

21.7 Institutional and Educational Buildings

The lighting requirements for some spaces in educational facilities coincide with requirements for commercial (office) and institutional buildings. To that extent the remarks herein are applicable there also. Generally, school buildings are maintained from operating funds obtained from taxes, and the budget is *always* tight. Therefore, all equipment in public buildings must be extremely hardy, vandal-proof, as maintenance-free as possible, and low in energy consumption. Maintenance in such buildings is generally poor and on a repair rather than preventive basis. With these overall criteria in mind, the following remarks apply to lighting equipment.

1. Use source with highest possible efficacy. Remember that daylight has the highest efficacy, followed by HPS, fluorescent, and other HID sources.

2. Where specific color lamps are called for, such as deluxe white, the requirement should be permanently stenciled in large letters on the lighting fixture.

3. Long-life sources should always be given preference because of lower maintenance. Thus corridor and stair lighting should be fluorescent or HID. This is also important in locations where relamping is difficult, as in high-ceiling rooms such as gyms and assembly rooms. In such spaces, extended-life lamps are recommended, with preference to HID sources.

4. In calculating levels, low figures for LLF (light loss factors) should be used to allow for aging of paints and dirt accumulation. Cleaning of lighting fixtures in schools is virtually unknown. A figure of 0.5 to 0.6 is reasonable. This being so, provision should be made to reduce initial overlighting. See Section 21.4.

5. Most schools are *not* air conditioned. The masking air noise being absent, careful control must be exercised on noise and vibration from ballasts, diffusers, and so on. Ballast noise increases with current rating. Therefore, 800-mA high-output and 1500-mA very-high-output lamps must be used with caution, particularly in locations that amplify sounds, or where low NC (Noise Criteria; see chapter 26) obtains.

6. Lighting equipment must be designed for an absolute minimum of maintenance. This means captive screws, rust-preventive plated parts, captive-hinged diffusers requiring only one man to maintain, ballast replacement without demounting fixtures (plug-in ballasts are available), nonyellowing plastics, and high-quality finish and assembly.

For recommended illuminance levels in all spaces, see *IES Lighting Handbook* (1981) and Section 18.23.

21.8 Art Rooms

The primary requirement here is for constant color daylight. Thus north windows and skylights are highly desirable. For artificial lighting, since color is so important, high-CRI fluorescent tubes are recommended. General illumination should be augmented by user-adjustable supplementary lighting. If modeling is anticipated, spotlights for this purpose are required. For display of artwork, adjustable wall illumination is required. Ceiling track-mounted units are an excellent choice (see Fig. 21.18). These ceiling tracks can also be utilized to provide outlets for desk lamps (see Section 15.30).

21.9 Assembly Rooms, Auditoriums, and Multipurpose Spaces

The varied activities in these rooms make flexible lighting imperative. For performances, low-level dimmed incandescent lighting is required. Here incandescent is the recommended source because of the lower cost of dimming and short burning periods. For assembly, this can be augmented by architectural elements along walls and drapes, and in the ceiling. For study, additional ceiling fluorescents or HID units can be switched on. The combinations are legion; the different usages are the critical consideration (see Figs. 21.19 and 21.20). Acoustic considerations are acute because of the low NC criteria. Thus the generally noisy ballasts of HID sources should be located with care. An additional consideration is step lighting. These units should be mounted to the side of, or in risers, to illuminate the tread, and particularly its leading edge.

21.10 Gymnasium Lighting

Gyms present a situation similar to auditoriums in that they have widely varying usages. All fixtures

Fig. 21.18 Art exhibition room, illustrating good and bad lighting techniques. Upper wall fenestration is excellent for deep daylight penetration. Track lighting is ideal for display of art. The mixture of incandescent downlights for general lighting is excessive and an eyesore. Also, the positioning of the track lights can create both direct and reflected glare problems and annoying shadows unless the sources are selected properly and ceiling height is at least 10 ft.

Fig. 21.19 Schools frequently utilize spaces for multiple functions. This space, normally used as a dining area, doubles as an assembly room. The architecture did not lend itself to conventional fixtures. High-intensity, indirect tungsten-halogen units, in concrete beam junctures, provide sufficient light for both uses.

Fig. 21.20 Institutional cafeteria illuminated by cove lighting in deep pyramidal coffers. Lighting is even, glare free, soft in quality, and pleasant, yet of sufficient intensity to permit using the cafeteria as a working-meeting space. (Photo by Stein.)

Fig. 21.21 This indoor tennis court uses 1000-W metal-halide luminaires directed at the ceiling and reflecting shadow-free light of over 1000 lux maintained on to the playing surface. Fixtures are mounted 20 ft above the floor. Sealed luminaires were chosen for both photometric performance and the low maintenance provided by their dustproof construction. (Photo courtesy of WideLite Corp.)

should be sturdy and guarded. Phosphor-coated mercury, HPS and high CRI metal-halide are excellent choices for color, life, control, and efficiency. Multiple levels should be available by switching or dimming. For dance and assembly use, other fixtures can be lamped with long-life incandescent or tungsten halogen, which provide good color for low-intensity lighting and also provide illumination during HID startup or restart after an outage. All fixtures should be designed for relamping from the floor. Locker rooms should use guarded-strip fluorescents. An interesting application of HID lighting in an indirect system is shown in Fig. 21.21. Recognition of a possible problem with dirt accumulation and relamping is necessary with such an arrangement.

21.11 Classrooms

The essential room in the school is the classroom. Refer to Section 18.28 for recommended surface reflectances. The modern classroom utilizes extensive audiovisual teaching aids and therefore requires multiple lighting levels. See Sections 21.3 and 21.4 for control techniques and Section 21.20 for lighting in areas using video display units.

(a) Energy Considerations and Sources. Use standard or energy-efficient fluorescent lamps, cool white or high-CRI white for direct and direct-indirect fixtures; 800- and 1500-mA lamps, same colors for semi-indirect and indirect lighting. Incandescent sources should not be used. Daylight, to the maximum extent, is desirable. Some form of daylight compensation is mandatory.

(b) Choice of Lighting System. With proper design, adequate lighting can be provided with 2 W/ft^2 or less. Since ESI is so sensitive to viewing position, the designer must be inherently familiar with the type of system, fixtures, and arrangements that will minimize reflected glare without exceeding energy limitations. To this end, careful rereading of Sections 18.26 and 18.27 will be of considerable assistance.

(c) Costs. Refer to Appendix E. Convert costs to life-cycle costs using data given there. *Do not* compare costs on the basis of dollars per footcandle (i.e., by dividing maintained illumination by cost), since this leads to preference for higher illumination levels and consequently higher wattage levels. *Do* compare life-cycle costs of alternate adequate illumination systems.

21.12 Lecture Hall Lighting

Lecture hall lighting is similar, with respect to sources and other considerations, to illumination for classrooms. Adjustable-level fluorescent lighting is necessary for demonstrations, films, and the like. Auxiliary lighting for demonstration table and chalkboard completes the design. High-ceiling installations can utilize color-corrected mercury or metal-halide for general lighting. Controls for lighting should be at the demonstration table. See Fig. 21.22.

21.13 Laboratory Lighting

Laboratories differ from classrooms in that tables are fixed, bench surfaces are frequently very dark, many of the items used exhibit specular reflection, vertical surface illumination is important, and visual tasks are not normally prolonged or severe. With low ceilings, use direct fixtures with small uplight component, run crosswise to the tables. Luminaires with batwing distribution will minimize reflected glare from specular equipment. See Fig. 21.23. If ceiling height is sufficient, indirect lighting is highly desirable for the same reason. Indirect lighting will also provide a high degree of diffuseness necessary for vertical surface illumination. See Fig. 21.24 for suggested layouts.

21.14 Library Lighting

Libraries comprise several different seeing tasks, each of which requires its own lighting solution.

(a) General Reading Room. Here two solutions are possible and both are in common use. In the first, general lighting is supplied over the entire area, which is sufficient for reading tasks. For this purpose fluorescent or color-improved HID sources are normally applicable, the latter with ceiling heights of at least 10 ft. The long life and high efficacy of these sources are suited to the long burning hours found in libraries. The second, and

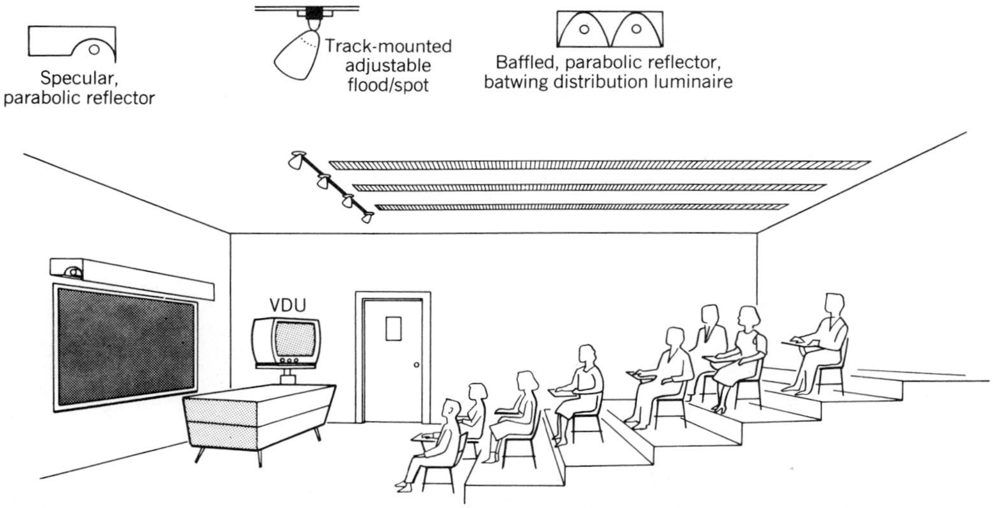

Fig. 21.22 *Typical lecture room lighting utilizes 45° cutoff baffled parabolic reflector troffers for minimum direct and reflected glare, adjustable track lights for demonstration table illumination, and an asymmetric reflector for chalkboard lighting. The large visual display unit in the room will not have veiling reflections with the illustrated lighting arrangement. See Section 21.20 for a discussion of lighting in areas using visual display devices.*

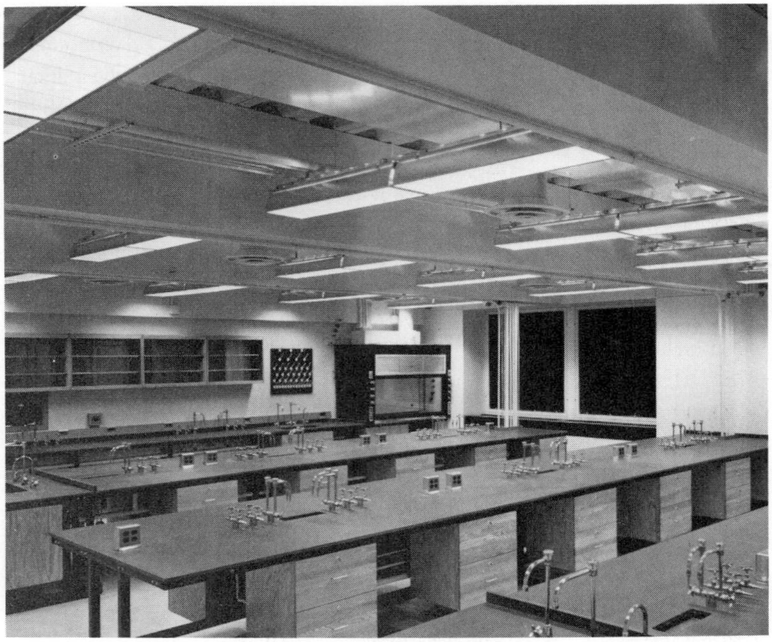

Fig. 21.23 *Baffled luminaires with specular parabolic reflectors are run crosswise to minimize veiling reflections. Note the uplight component, which serves to illuminate the ceiling and minimize excessive luminance ratios. (Photo courtesy of Peter Goodman/Edison Price.)*

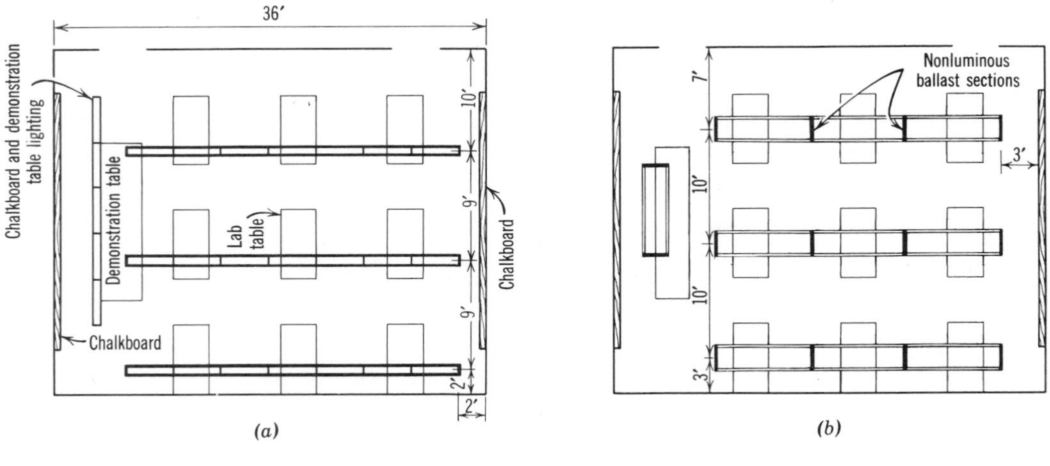

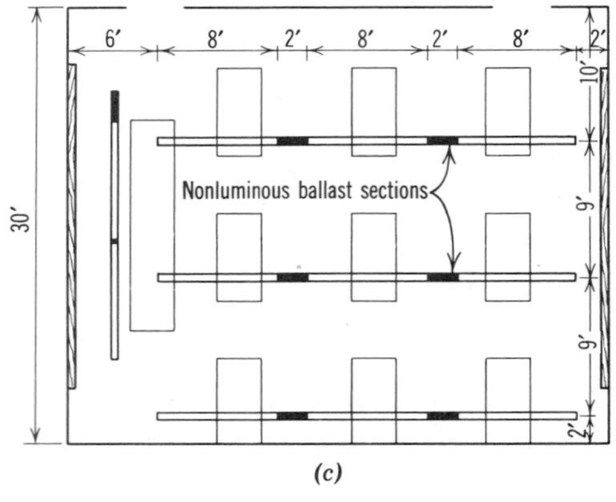

Fig. 21.24 *Laboratory lighting schemes. Running lights across tables or in aisles is preferable from the aspect of reflected glare.* (a) *Pendant direct-indirect units.* (b) *and* (c) *Variations of the single, semi-indirect HO and VHO unit design illustrated in Table 20.2, fixture no. 33. Relatively high noise level of ballasts is generally not objectionable in labs.*

more energy-efficient solution, involves low-level general lighting supplemented by local reading lighting on the tables or in carrels. This solution is consonant with task-lighting orientation and is to be preferred. Reading lights should be fluorescent, user adjustable if possible, and arranged to avoid veiling reflections when not user adjustable (see Section 21.19).

Wherever HID sources are used, an instant restart source must be available to supply minimal lighting after an outage. Many commercial HID luminaires contain a small tungsten-halogen source for this purpose. Ballast noise can be a problem in low-NC-criteria spaces such as libraries. Special low-noise ballasts and enclosures are available and should be employed.

(b) Stack Areas. In stack areas the required vertical surface illumination is best supplied by one of the special fluorescent units designed for this purpose. These are mounted between stacks and no higher than 24 in. above them for best results (see Fig. 21.25).

21.15 Specialty Room Lighting

Rooms with specific tasks such as sewing and typing rooms and shops must be carefully considered. Where moving parts are present, safety is a prime consideration and adjustable local lighting is often necessary. No general rules can be suggested other than those previously explained in detail. Particular attention must be given to often-forgotten vertical illumination requirements, which are vital in areas such as storage, stacks, and shops. Kitchens and cafeterias are color sensitive, and sources must be carefully chosen.

21.16 Corridors and Stairways: All Buildings

Corridors intended only for circulation need only be lighted to 10 fc ± unless a specific seeing task requires higher levels, for example, bulletin boards and lockers (see Fig. 21.26). Lighting can also be used to give direction by longitudinal arrangement. Wall-mounted or recessed wall lighting is particularly effective in corridors, giving walk illumination plus lighting for posters, bulletins, and so on. Fluorescent luminaires mounted across corridors, particularly when corridors are long, are effective in reducing the tunnel impression. Incandescent sources are not recommended, because of low efficacy, high maintenance, and frequency of relamping. Fluorescent and HID sources are also suggested for stairwells. Care must be exercised here, however, to avoid direct glare, which causes attention to shift from the stairs to the light and may thereby cause a hazard.

ILLUMINATION

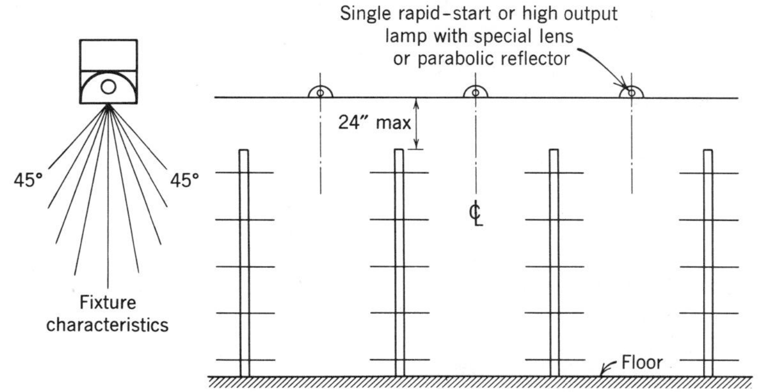

Fig. 21.25 Stack lighting is best accomplished by fixtures with lenses specifically designed for the purpose. Fixtures with baffles and plastic diffusers generally do not give adequate vertical surface illumination.

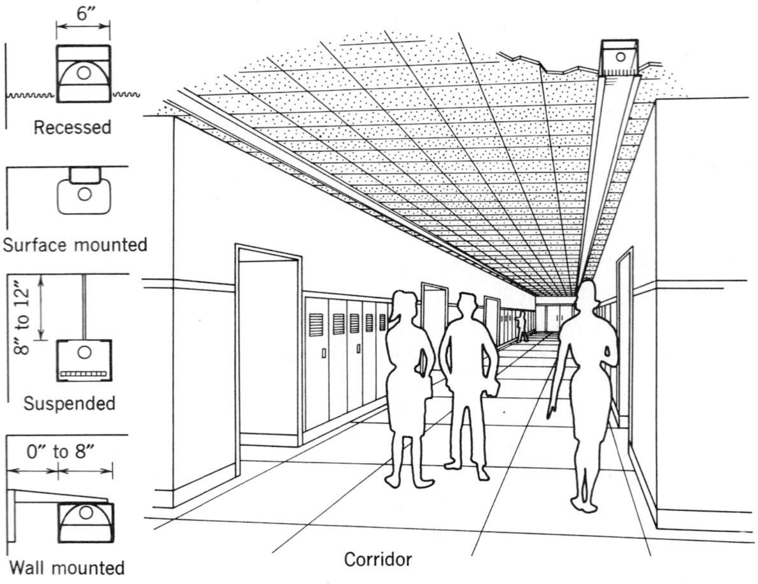

Wall lighting with single lamp units

Fig. 21.26 *Lighting of school corridors. High-reflectance walls, floor, and ceiling improve utilization of light and increase the feeling of cheerfulness. The lighting technique illustrated is appropriate for school corridors. The rows of luminaires at each sidewall illuminate bulletin boards, special displays, and the faces and interiors of lockers more effectively than units centered in the ceiling.*

COMMERCIAL INTERIORS

21.17 Office Lighting

The following information applies primarily to offices in commercial buildings and secondarily to similar spaces in other occupancies, such as in educational and industrial buildings. In the latter cases, the general remarks applicable to facilities of those types take precedence. The special problems associated with visual display units (CRTs, computer screens, word processors) are discussed in Section 21.20. Task-ambient (nonuniform) lighting is covered in Sections 21.18 and 21.19. The reader is referred to IES Standard Practice for Office Lighting (1983) for a full discussion of office lighting.

(a) Light Sources. The most commonly used source is fluorescent. HID can be used to advantage in spaces with at least 10 ft (~3 m) ceiling height. When using HPS it is advisable to mix it with color-improved mercury or metal-halide to avoid possible objections to its yellow color. All HID lamps are possible glare sources. This problem can be minimized by use of indirect lighting where ceiling height permits.

In most office applications, color is not critical but it is important. Source color must be coordinated with the color scheme of room surfaces and furnishings. In areas with large daylight contribution, source correlated color temperature should be at least 4000 K. Use low-wattage, energy-saving lamps and accessories throughout. Incandescents may be used for storage areas, closets, and other short-burning-period uses. Incandescent and tungsten-halogen track lighting is used to advantage to illuminate displays of all sorts. Low-voltage units are useful in this application. See Section 19.13.

(b) Illumination Levels. These are to be found in Section 18.23 for the particular type of activity involved. See also IES Standard Practice for Office Lighting (1983). Recommended reflectances for room surfaces are:

Ceiling	80% minimum
Walls	50–70%

Partitions	40–70%
Floor	20–40%
Desktops, furniture	25–45%
Window blinds	40–60%

Frequently, the upper wall section is painted to match the ceiling, that is, with a lighter finish than the remainder of the wall. This serves the double function of increasing ceiling cavity brightness, particularly with suspended fixtures, and increasing vertical illumination due to reflection from this surface.

(c) Vertical Surface Illumination. This is required for visual tasks in offices, such as files, desk drawers, card files, and secretarial copy stands. Large area luminaires and a high degree of diffuseness are desirable. This is especially true in large offices where wall reflections are absent. Light-finish furniture surfaces, surface-mounted fixtures that yield an illuminated ceiling, and high-reflectance floors will also assist in this.

Fig. 21.27 Pendant two-lamp fluorescent fixture supplies high-level task lighting for the work area and enough spill light for circulation in this small room. The 20% up-component assists in this and provides requisite ceiling illumination. When using the VDU, a lower level of illuminance is preferable, obtainable by dimming, or switching a two-level ballast. Switching off one lamp entirely is not advisable. (Courtesy of Armstrong World Industries.)

(d) Private Offices. Here a task-ambient approach is indicated. Frequently, spill light from the task area will suffice for general illumination in small rooms. See Figs. 21.27 and 21.28. In larger rooms, provide downlights in sitting areas, wallwash for accent and brightening dark walls, special lighting for display boards, paintings, and so on. If the ceiling is high, pendant fixtures can create a horizontal plane to correct poor room proportions.

(e) General Offices. The two basic approaches to general office lighting are a uniform layout that provides task-level lighting in the entire area, and a task-ambient design. The former, discussed here, is most appropriate to speculative type construction, where the furniture layout is unknown but a complete job is required. (Frequently, however, the construction contractor only provides sufficient electric power for lighting and

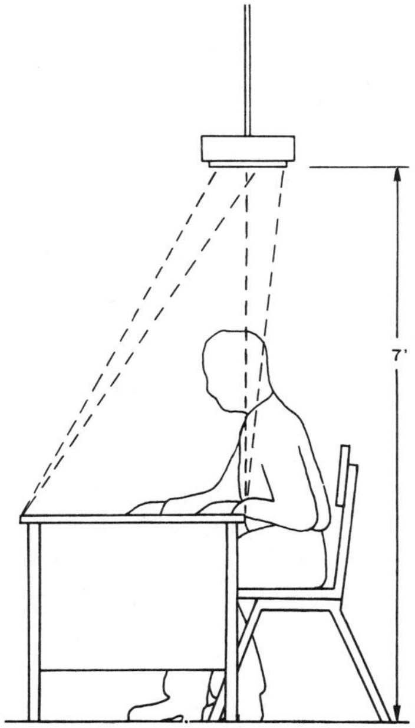

Fig. 21.28 Geometry of the task-ambient fixture of Fig. 21.27. When arranged as shown, veiling reflections are minimized and glare on the VDU screen is essentially eliminated. (Courtesy of Armstrong World Industries.)

ILLUMINATION

miscellaneous power, and the lighting of the space is designed *after* the space has been rented, that is, as ''tenant'' work. In such cases, a task-ambient design is possible since tenant work is tailored to the user's needs).

As has been repeatedly pointed out, a task-level overall layout is wasteful of energy and will create problems with building energy budgets. Therefore, a uniform layout when used should be designed so that levels can be lowered easily in areas not requiring task lighting. See Sections 21.2 to 21.4. Furthermore, since no control of the geometry of direct and reflected glare is possible in such layouts, the luminaire selected must give minimum average glare in *all* viewing directions. The system that will best meet these requirements is one that uses a maximum number of low-brightness luminaires in a dense layout. Modular ceilings, with a single-lamp luminaire in each module, are a feasible, economic approach. See Fig. 21.34. A structurally coffered ceiling with a luminaire in each coffer (see Fig. 21.30*a*, below) is another viable alternative, since the higher cost of luminaires is offset by the absence of a hung ceiling. The luminaire of choice would be some type of low-brightness lens or louvre unit. Batwing distribution is not indicated because of high direct glare (low VCP) and strong veiling reflections at work locations whose line of vision is crosswise to a lighting fixture. Where the viewing direction is established, but exact furniture layout is unknown, a uniform layout with low-brightness parabolic reflectors with batwing characteristic can be used to advantage, as in Fig. 20.35. There, an initial level of 750 lux is achieved with a power density of less than 16 W/sq m [1.5 W/sq ft], and an excellent VCP of 88.

(f) Office Lighting Equipment. Office lighting equipment is generally not handled roughly. Fixtures may have touch latches, light hinges, and adjustable devices without fear of breakage or vandalism.

(g) Maintenance. In most offices maintenance is provided on a trouble call basis. Lamps are replaced on burnout, and the fixture is then cleaned. Because of the long life of fluorescents and HID sources, this generally means a 3- to 5-year cleaning cycle. Since most offices use lay-in troffers, an LLF of 0.65 to 0.70 is reasonable in air-conditioned spaces, lower in open-window offices.

(h) Noise. Noise from air conditioning plus adjacent street, traffic, and process noise makes the use of higher noise level ballasts, as are found on HO and VHO lamps, frequently feasible. For private offices, A-rated ballasts for 430-mA lamps are the best choice. (For more detail, see Chapter 27.)

(i) Fenestration. When fenestration is absent, a lighted valance is recommended around the room. This will remove the wall–ceiling line and will partially compensate for the lack of windows. It will also brighten the walls and increase illumination on desks placed adjacent to the walls.

(j) Supplementary Lighting. Supplementary lighting can be mounted on the ceiling or on ceiling-track-fed poles of the type shown in Fig. 15.30. These are, in effect, track lights, mounted on vertical tracks.

(k) Control. Control strategy (see Sections 21.2–21.4) should *minimally* provide (after tuning):

1. Daylight compensation.
2. Small groups of lights to remain lighted while the remainder are off to permit off-hours work.
3. Path lighting through large spaces to permit traverse without turning on all lights.
4. Careful scheduling with supervised local override.

With these general guidelines in mind, the following sections discuss specific topics in office lighting.

21.18 Nonuniform (Task-Ambient) Office Lighting Design Using Ceiling-Mounted Units

Efficient modern office lighting, like other work area lighting, is predicated on a task-ambient design. This approach has the advantages of minimum contrast reduction (high ESI) at the task, and minimum power use, the latter resulting in low operating cost. There are four applicable design methods:

1. Uniform luminaire layout with appropriate controls that will result in the desired task-ambient layout. Since luminaire output will vary radically from one to another, only very low brightness units of the types shown elsewhere (Fig. 18.9*a*, or Fig. 21.33) should be used. With these units, the large differences in output will not be noticeable, and the desirable symmetry of ceiling appearance will be maintained. The disadvantage of this approach is the high first cost for luminaires and controls.

2. Uniform ceiling layout that provides low-level lighting for circulation and miscellaneous easy visual tasks, plus work-station-mounted task lighting. This latter can either be integral with the furniture (see the following section) or a separate, generally adjustable unit mounted on, or standing on, the desk. This latter is particularly useful for very severe seeing tasks when adjustability of angle and distance between light and work is vital. Integral furniture units are not adjustable, and are therefore limited to a maximum of about 1000 lux on the task. Higher levels would generate excessive heat and noise and, almost certainly, glare.

3. Nonuniform (asymmetrical) luminaire layout, with luminaires located as required to suit the tasks. Obviously, a detailed furniture layout is required. See Fig. 18.35.

A more detailed study is presented in Figure 21.29. This is an office layout for a small technical office. In accordance with current recommendations, 50 fc is to be furnished on all work surfaces except the drafting table, DT, and two small offices where close work on poor copy will be performed. Figure 21.29*b* shows a standard uniform layout generally using three-lamp, 2 × 4 fluorescent troffers, with high-quality lens diffusers. VCP is good due to low lens brightness. Illumination on the drafting table is also good because of the fortuitous position of the fixture above it. Other work surfaces receive more than adequate raw footcandles but in many instances low ESI due to bad veiling reflections. In Fig. 21.29*c*, fixtures have been arranged to supply task lighting. Ambient lighting comes from spill light and three fixtures placed in a relatively large, open circulation space. The results are good VCP, much higher ESI footcandles on the tasks, and lower energy use. As pointed out in Fig. 20.16, a disadvantage of this system is the jumbled appearance of the ceiling. For this reason, nonuniform layouts are well suited to coffer-type ceilings, as in Fig. 21.30, where the presence or absence of a fixture is not as prominent. This is all the more important where glass outer wall areas make the ceilings readily visible from the outside (see Fig. 21.36 below). This appearance problem can be mitigated considerably by using a uniform arrangement with adequate controls, as discussed in paragraph 1, above. Another method of minimizing the visual disarray of a nonuniform luminaire layout is to use smaller luminaires. Thus 2 × 2 ft fluorescents with U tubes would be a better choice than 1 × 4 ft, with essentially the same output. Even better from the appearance viewpoint would be the use of HID units that are smaller yet. Finally, as mentioned in paragraph 1, use of very low brightness parabolic diffusers causes the fixtures to blend with the ceiling and minimizes the negative effect of nonuniformity of layout.

4. Pendant units of the type shown in Fig. 21.28 for ceiling mounted task oriented lighting. This system is usable for small areas only, since a proliferation of stem-mounted units in a large area would create an unacceptable visual jumble. A viable application would be a room with a center aisle and one row of desks on each side of the aisle.

21.19 Nonuniform Office Lighting Using Furniture-Integrated Luminaires

In lieu of ceiling-mounted fixtures, ambient and task lighting can be supplied by furniture-mounted units and free-standing, indirect HID units. Advantages include the following:

1. The problem of furniture layout and layout changes is eliminated.
2. Initial construction cost is reduced.
3. Energy requirements are lowered because of short distances between light source and task.
4. Each occupant has complete control of his or her task lighting including, in some designs, positioning control.
5. Maintenance is very much simplified, since fixtures are readily accessible from the floor.

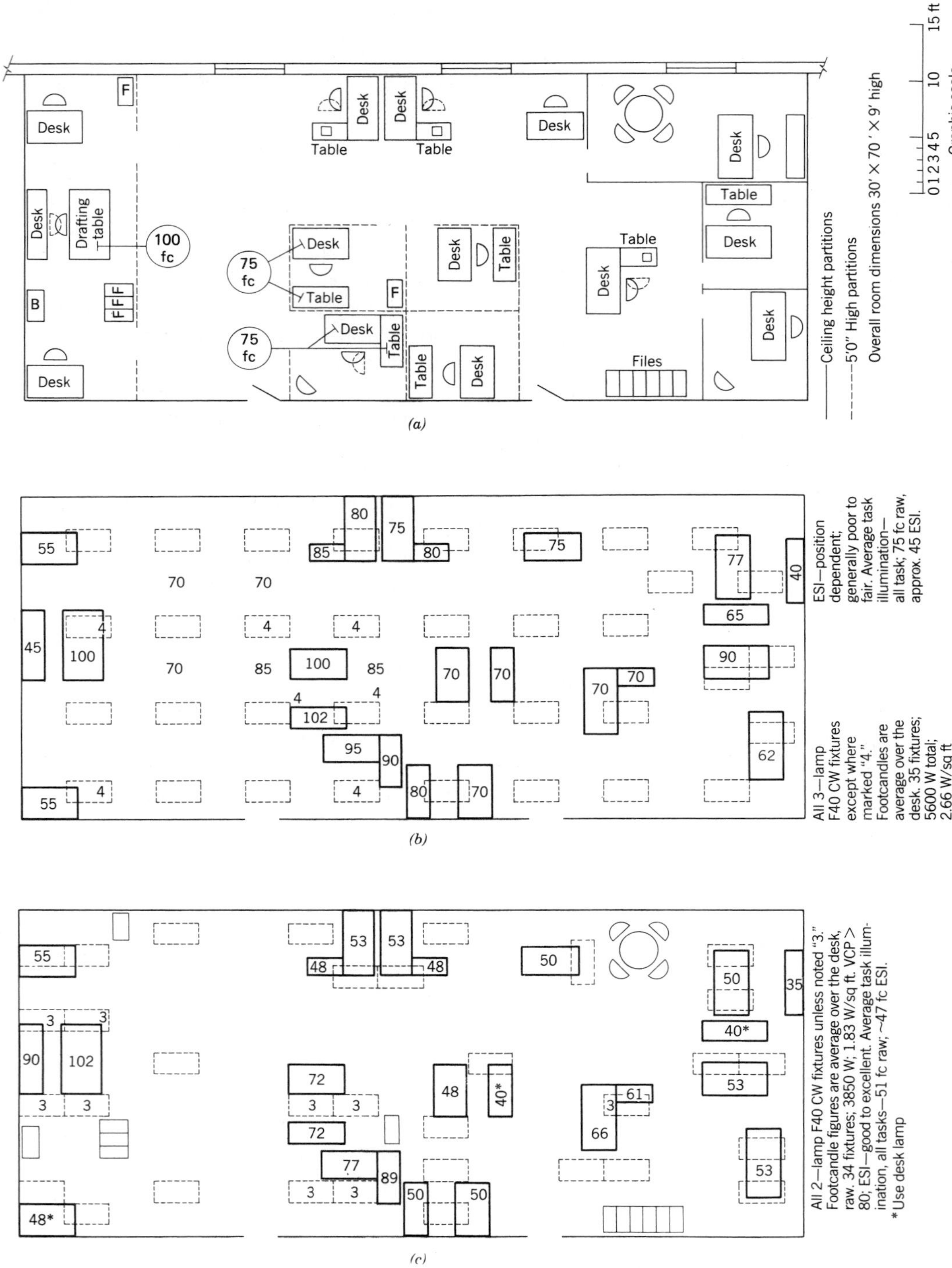

Fig. 21.29 (a) *Layout for a small technical office. Illuminance required is 50 fc (raw) except where otherwise noted.* (b) *Uniform lighting layout generally using three-lamp 2 ft × 4 ft troffers (four-lamp where indicated) gives generally satisfactory raw illuminance levels but considerable reflected glare at many locations.* (c) *Nonuniform lighting layout generally using two-lamp 2 ft × 4 ft troffers (three-lamp where indicated) gives lower raw footcandles but* higher *ESI illuminance. All figures shown are raw footcandles calculated by the point-by-point method. Layout* (a) *taken from* Lighting System Study, *1974, Ross and Baruzzini, GSA, PBS.*

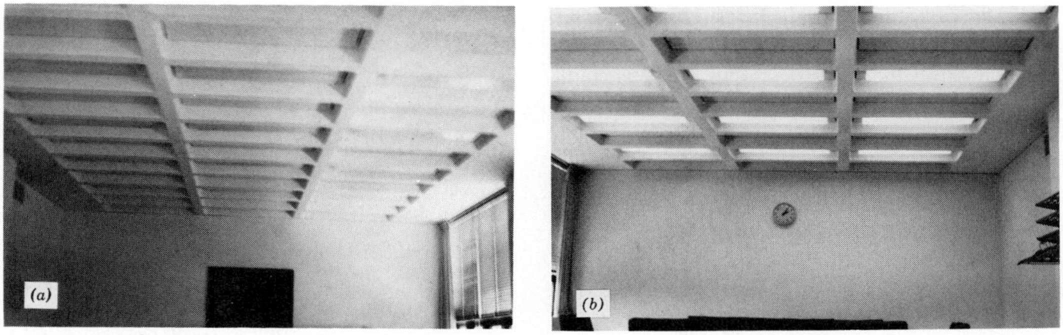

Fig. 21.30 A coffer ceiled room (a) *is illuminated with 1 × 4, two-lamp units in alternate coffers. The appearance both unlighted* (a) *and lighted* (b) *is symmetrical and pleasing. A nonuniform layout would be less objectionable here than in a flat-hung ceiling. (Photos by Stein.)*

6. Floor-to-floor height can frequently be reduced.
7. Tax advantages normally accrue due to higher depreciation rates on furniture than on the building.

Disadvantages include the following:

1. Difficulty in dissipating heat and minimizing ballast noise due to proximity of sources to user.
2. Veiling reflections can be severe.
3. Luminance ratios in the near and far surround may exceed recommended levels.
4. Difficulty in lighting a free-standing open desk, since most of the fixture types are undercounter or sidewall mounted.
5. Difficulty in evenly lighting large table or L-shaped desk areas because of the concentrating nature of the lighting units.
6. Not readily applicable to automatic switching and dimming schemes.

Figure 21.31 graphically shows the problem of local desk lighting. The author's own experience has been that only systems that permit user positioning adjustment have a high degree of user acceptability. However, many such systems are available, and the lighting designer is advised to personally test a system before specifying it.

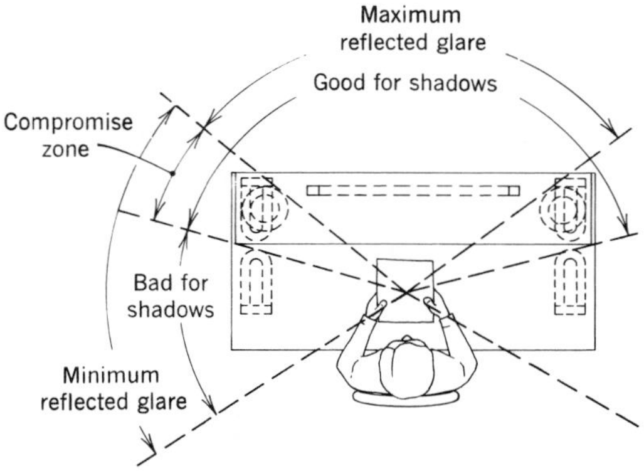

Fig. 21.31 Task lighting can create severe reflected glare or hand shadows if the luminaire location is not carefully selected. Further constraints are: location in elevation for shielding (needs to be low) and for good distribution (needs to be high); appearance considerations, and compatibility with the worker's physical movements. Light source position adjustability removes most user objections to this type of task lighting.

21.20 Lighting for Visual Display Unit (VDU) Areas

One result of the emergence of the computer era has been the almost explosive proliferation of visual display units in business offices. These devices, variously called VDUs, CRT screens, monitors or monitor screens, video displays, or simply TV screens, will very soon become standard desktop items. Because of its specular surface, the VDU creates a special problem in office lighting that becomes the deciding consideration in selecting the lighting system. Simply stated, the primary problem is to avoid reflection in the screen of any luminous source in the area including luminaires, windows, illuminated walls, and even light-colored clothing. Any such reflection makes reading of the data on the screen difficult, and even impossible. A secondary problem is avoidance of reflections on the usually specular keyboard (and other specular objects in the vicinity). (Matte-finish keyboards are now available; nonspecular screens are being developed.)

It is undoubtedly obvious (see Fig. 21.32) that the problem is one of geometry, complicated by the fact that the screen is often up to 20° from the vertical and the keyboard is up to 30° from the horizontal. A further difficulty lies in the contradictory nature of adjacent visual tasks: Viewing of the screen calls for a low (vertical) ambient light level (75–125 lux), whereas the reading and writing work done in conjunction with the screen data requires a much higher (horizontal) level (300–700 lux). A full solution to the problem therefore involves careful attention to the selection and location of the VDU equipment, and control of room surface reflectances, in addition to proper lighting design. The following recommendations refer to all these considerations, with the knowledge that some may not be in the purview of the architect/lighting designer.

(a) Equipment.

1. The screen should be recessed as deeply as possible into the VDU. This will reduce the ambient light level at the screen and make reading easier. It will also reduce glare by restricting the room surface area that is reflected in the screen.
2. All parts of the VDU, including keys, should have a matte finish.
3. The VDU should be mounted on a swivel that will permit changing the geometry of reflection. If possible, obtain a VDU with adjustable tilt.

(b) Location.

1. Avoid locations where the VDU *viewer* faces a window or other large bright area, to avoid excessive luminance ratios.
2. Equip windows within the geometry of reflection with low-reflectance vertical blinds. Blackout curtains are not required. Remember that windows can become a secondary source of reflection, that is, luminaire to window to VDU screen. See Fig. 21.32(*b*).
3. Test each VDU location by running a small mirror over the face of the screen to find glare sources, and correct each.

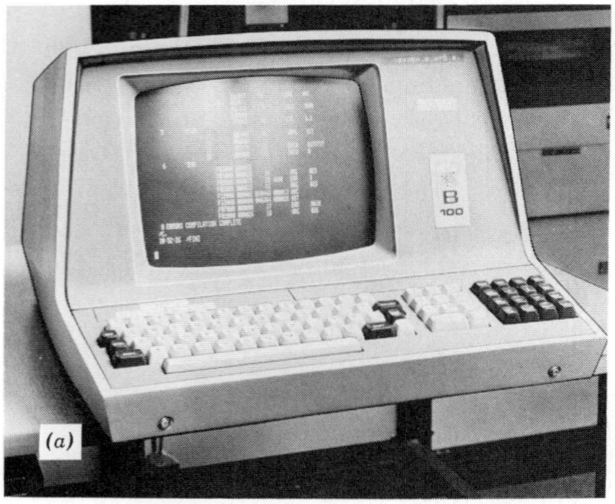

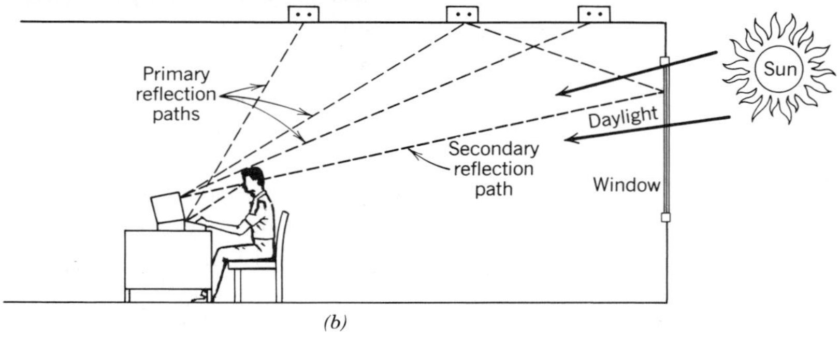

Fig. 21.32 (a) *Primary sources of reflection in the VDU screen and keyboard are luminous areas such as lighting fixtures and windows. (Photo courtesy of Brüel and Kjaer.)* (b) *Specular surfaces such as window glass can become secondary reflection sources via rereflection of light from luminaires (or other light sources), as shown.*

Fig. 21.33 The lighting system of choice uses 45° cutoff parabolic wedge louvres, shown here in a pyramidal 5-ft-square modular ceiling. The window would require treatment to reduce brightness except on a solidly overcast sky. Also, a darker desktop surface would be preferable. [*Photo courtesy of Armstrong World Industries (ceiling system supplier).*]

(c) Lighting. The system needed is one that will prevent reflection of bright luminaires on the screen, provide sufficient task light for standard office tasks, and completely shield the light source. Two lighting solutions are available:

1. Indirect lighting giving a horizontal working plane illuminance of about 300 lux. Higher levels will result in an unacceptably bright ceiling, which will cause a haze of glare over the upper portion of the VDU screen. Such an installation should be dimmer controlled. Rooms with very high ceilings (above 10 ft) can tolerate brighter ceilings and therefore higher horizontal plane illuminance.
2. Fluorescent fixtures with 45° cutoff parabolic wedge louvres (see Figs. 20.31 and 21.33). The 45° cutoff will cause any lamp image to be reflected below the viewing angle *in a vertical screen,* so that there is no necessity to place a diffusing element between the lamp and the louvre. Such a requirement *does* exist if louvre cutoff is less than 45° or if a canted screen is used. Also, such a diffuser is suggested if keys and other VDU terminal accessories are specular. Otherwise the fluorescent tube will be reflected in the key, which can readily obscure the key identification. The dark appearance of the louvre must be maintained

by frequent cleaning, since dust shows up as bright spots on the louvre.

(d) Finishes. Walls, floors, and furniture should be finished in low-chroma, low-value colors, with a maximum reflectance of 50%. Very dark and very light desktops are to be avoided, because of excessive luminance ratios and the latter also because of reflections. Similarly, operators should be advised to avoid white clothing and specular clothing accessories.

21.21 Integrated and Modular Ceilings

The cost, appearance, and design-flexibility advantages of an integrated ceiling design over field-assembled and coordinated systems have long been known. As a result ceiling systems with integrated lighting, acoustic control, and air-handling capabilities are commercially available in modular sizes, among which are 60 in. square, 48 in. square, and 30 × 60 in. Modules are made in flat and pyramidal shapes, the latter having several distinct advantages over the flat:

1. More interesting and esthetically pleasing.
2. More acoustic absorbency due to ceiling angles and large surface area.
3. Recessed center provides visual baffling, permitting use of higher brightness sources while maintaining high VCP.

A typically equipped pyramidal module is shown in Fig. 21.34. Possible luminaire arrangements for both flat and pyramidal shapes are given in Fig. 21.35, and examples of installation are shown in

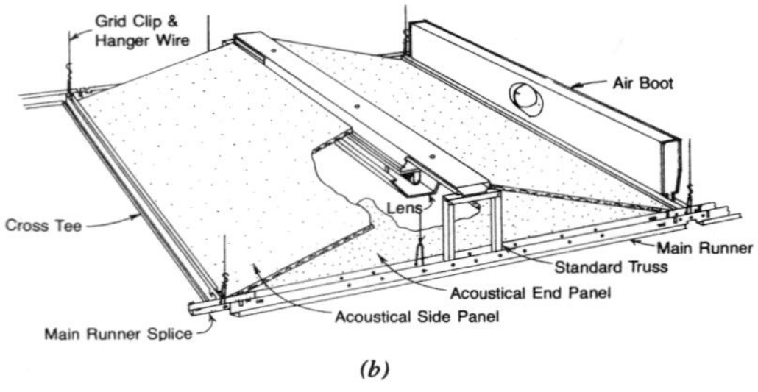

Fig. 21.34 (a) *Integrated ceiling based on a 5-ft-square module. (Photo courtesy of Armstrong World Industries.) Each module* (b) *contains a one-lamp, low-brightness, wraparound lens luminaire in the center, surrounded by acoustical panels in a pyramidal shape. Air supply and return is handled at the runners. Other designs pass return air through the luminaire. The installation shown is also fire-rated, and sprinklers can be installed at runner intersections.*

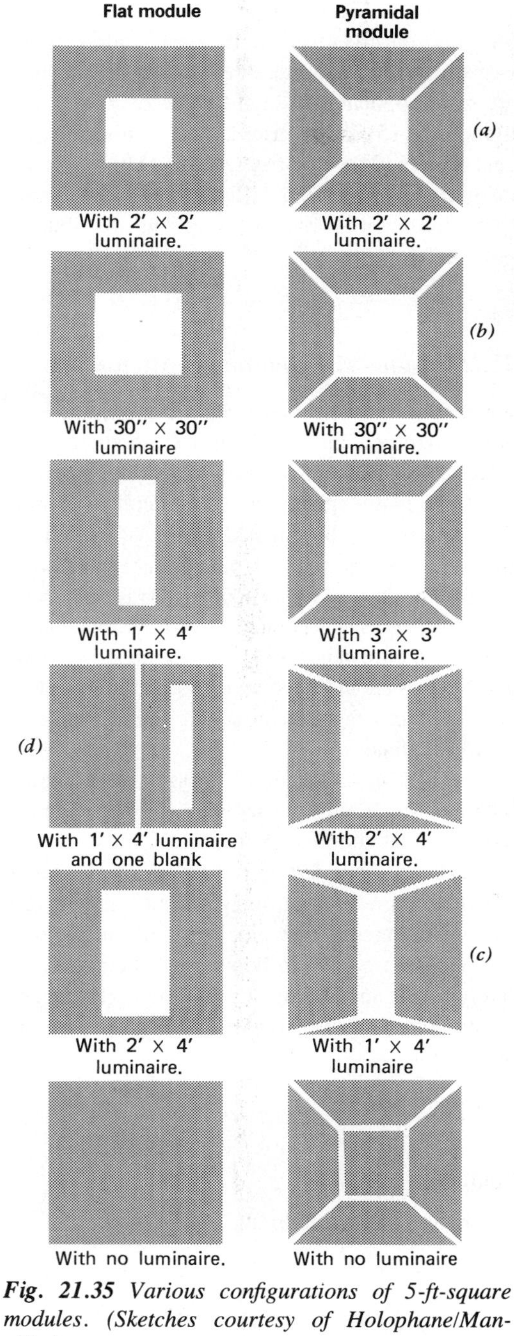

Fig. 21.35 *Various configurations of 5-ft-square modules. (Sketches courtesy of Holophane/Manville.)*

Figs. 21.33 and 21.36. In addition to the design flexibility available, electrified track can be integrated into the system runners, to supply both the lighting fixtures and power poles. This increases

Fig. 21.36 *A square modular ceiling utilizing square lighting units demonstrates architectural symmetry, blending and, incidentally, that lighting is easily visible from the outside, even during the day. (Photo by Stein.)*

the mobility of the fixtures, so necessary for adequate task lighting.

21.22 Lighting and Air Conditioning

The reduction of lighting power density levels to below 2 W/sq ft in all but special areas has considerably reduced the impact of lighting-generated heat on a building's HVAC system. In non-air-conditioned buildings, the lighting heat contribution is directly applicable to building heating. Fixture efficiency is directly affected by its temperature. Fluorescent units operate at an optimum temperature of 77 F. Temperatures above and below this decrease output and fixture efficiency. Thus heat removal from units is desirable even at low lighting-energy levels. The most effective method of fixture heat removal is by duct connection to the unit itself. This method, however, is relatively expensive and immobilizes the fixture. Alternatively the plenum can be exhausted with air passing through the fixtures, picking up excess heat. These details are essentially part of the HVAC design and are covered in Part Two.

INDUSTRIAL LIGHTING

21.23 General

In industrial lighting the prime and overriding consideration of all work, lighting included, is its profitability; that is, its economic impact on the company. Given acceptable standards of comfort and safety for the working staff, additional costs for lighting must be self-justifying economically. In one case a good lighting installation was improved at considerable cost. Production jumped 15%, of which 3% was sufficient to amortize the cost of the lighting alteration. In another case an outlay for new inspection lighting reduced product failures and proved economically sound. In a third, improved lighting reduced accidents, improved employee morale, and consequently improved production. The cases studied are far too numerous to mention; general principles will be adduced instead. For control strategies, see Sections 21.2 to 21.4.

21.24 Levels and Sources

Levels of illumination are detailed in the IES, *American National Standard Practice for Industrial Lighting* (A 11.1), latest issue. Other applicable recommendations are found in Section 18.23. Where levels above 50 to 75 fc are required, ambient illumination must be supplemented by local illumination. For details on supplementary lighting the IES *Recommended Practice for Supplementary Lighting* may be consulted, which also contains data on inspection lighting. Both these subjects are too specialized to be treated here. Industrial facilities lend themselves readily to daylighting, since many are one-story structures. Thus roof monitors, skylights, and clerestories are readily applicable and extremely desirable. However, since industrial facilities are frequently sited in industrial areas with attendant heavy atmospheric soot and dirt, a frequent cleaning and maintenance program is necessary if the LLF is to be kept at reasonable levels. This observation is obviously also applicable to indoor light facilities (to the extent that the indoor activity warrants).

Sources for industrial application should be high-efficacy, low-maintenance types, that is, HID and fluorescent. Where color is not critical, HPS is the recommended source. Adaptation to its warm yellow color is rapid and, as stated above, if it is mixed with metal-halides or mercury sources, no problem should be encountered. HID sources are easier to maintain, store, clean, and relamp than fluorescent and have equal or better efficacy, but have the disadvantage of delayed restrike time. Because of their concentrating nature, HID sources are more applicable to high-bay (>25 ft) and medium-bay (15–25 ft) installations, while fluorescent are suited for low-bay (<15 ft), although specially designed low-bay HID fixtures are available that minimize the inherent glare and distribution problems involved.

21.25 Industrial Luminance Ratios

For reasons explicated at length in preceding sections, brightness ratios in industrial situations must be controlled. Recommendations are given in Table 18.10. In many situations it is difficult to control the surrounding brightness. Ceilings, which so frequently are covered with piping, ducts, and other equipment, must be *light*. In other words, the above equipment must be painted with high-reflectance finishes, maintenance and cleaning must be good, and fixtures should have an upward component of light to avoid more than a 20:1 ratio of task to ceiling luminance.

Use of bright saturated colors for general surface painting should be avoided, however, since they draw attention and frequently have special significance. In addition to color-coded piping (banding is preferable), red frequently means fire equipment; green, first aid; orange, danger; and so on. White is also to be avoided, being excessively bright and susceptible to dirt. Light, unsaturated colors are preferable. Recommended minimum reflectances are:

Ceiling	80–90%
Walls	40–60%
Equipment	25–45%
Floors	20% minimum

21.26 Industrial Lighting Glare

The problem of direct glare can be acute in low-bay installations, and that of reflected glare in high-bay designs, when both use point sources. One method of reducing direct glare is the use of low-brightness prismatic lens units that utilize a black aluminum reflector behind the prismatic lens. The

Fig. 21.37 By utilizing a black reflector surface behind the prismatic glass lining in lieu of a polished aluminum one, fixture brightness can be dramatically reduced with less than 10% overall light loss. (Photo courtesy of Holophane/Manville.)

pronounced reduction in high-angle brightness of such luminaires (shown with 400-W mercury lamp) is shown in Fig. 21.37. This reduction in brightness is accomplished with only a 10% reduction in useful light. Methods of minimizing veiling reflection from all sources have been discussed previously.

21.27 Industrial Lighting Equipment

The cost of maintenance increases with labor rates. For this reason, high-quality lighting equipment will yield lowest owning and life-cycle costs. For instance, the cost of replacing a ballast for an HID lighting unit frequently *exceeds* the cost of the ballast. It is thus obvious that it is more economical to utilize long-life, high-quality ballasts, particularly where luminaires are not readily accessible.

Other suggestions for lowering costs, both initial and operating, include using ventilated luminares that tend to be self-cleaning by convection (see Fig. 21.38) in addition to giving the needed upward light component, using bus-mounted fixtures for rapid installation and repair (see Fig. 21.39), using lowering mechanisms on high-bay units to avoid catwalk or platform relamping with concomitant *extremely* high cost, using fixtures arranged for "stick" relamping from the floor in medium- and low-bay work, and generally incorporating the most modern equipment into the plant.

Proper maintenance is of paramount importance in industrial facilities because of the prevalence of dirt, vibration, and rough service. Under "maintenance" is subsumed cleaning, relamping, inspection, and preventive maintenance. Relamping on a burnout basis is extremely uneconomical because of disruption of production and lowered production due to lumen depreciation before burnout. Relamping is normally done on a planned group basis. Similarly, if the specific facility has a high dirt accumulation rate, cleaning must also be done on a planned group basis, rather than only at relamping time.

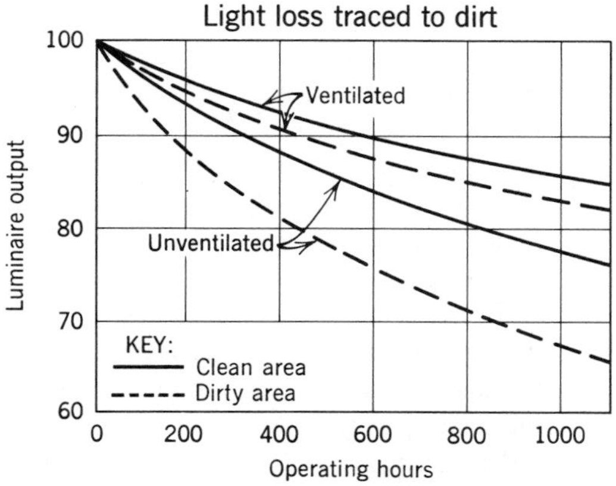

Fig. 21.38 Graph demonstrating the advantage of ventilated fixtures.

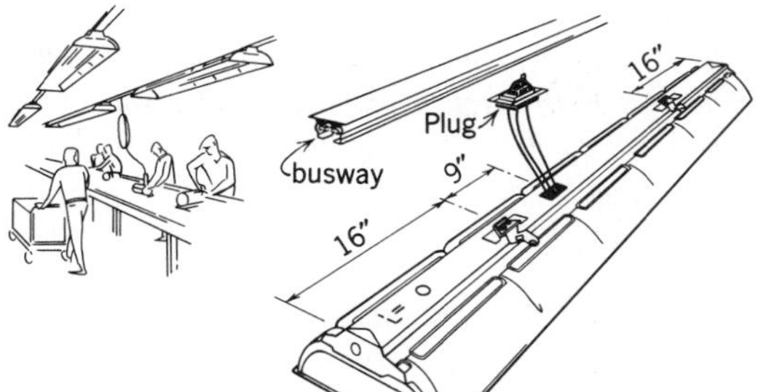

Fig. 21.39 Industrial lighting design using plug-in lighting bus to feed and support fixtures. This allows fixtures to be installed, moved, and maintained with ease, thus reducing initial and operating costs. (Courtesy of Day-Brite Lighting Division, Emerson Electric Co.)

ILLUMINATION

Ballast noise, including the high levels of HID ballasts, is not usually a factor in industrial facilities because of high ambient noise. In relatively quiet installations and/or where fluorescent fixtures are mounted a short distance above the work bench as in inspection and fine assembly, this is not true and noise ratings and conditions must be examined carefully.

21.28 Vertical-Surface Illumination

In industrial facilities more than any other, the illumination of vertical surfaces is crucial. This is a result of the nature of the work; machines, storage, gauges, and so on, all require high-level vertical-surface illumination. Examining Fig. 21.40 and the derivation given there, we note that maximum vertical illumination (illumination resulting from the horizontal-lighting component) is obtained when the angle between the fixture's vertical axis and the work is approximately 35°. Hence, we should select a fixture whose candlepower distribution curve demonstrates such a characteristic. Of course, the derivation is for a single location and fixture. For good vertical and angular illumination over a large area, arrange fixtures with considerable overlap.

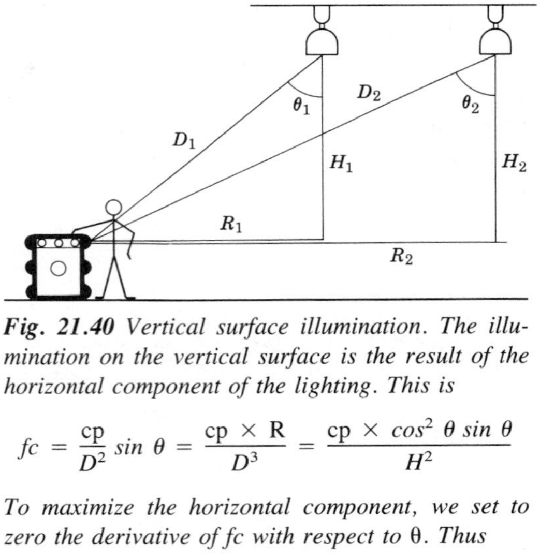

Fig. 21.40 Vertical surface illumination. The illumination on the vertical surface is the result of the horizontal component of the lighting. This is

$$fc = \frac{\text{cp}}{D^2} \sin\theta = \frac{\text{cp} \times R}{D^3} = \frac{\text{cp} \times \cos^2\theta \sin\theta}{H^2}$$

To maximize the horizontal component, we set to zero the derivative of fc with respect to θ. Thus

$$\frac{dfc}{d\theta} = \frac{\text{cp}}{H^2}(-2\cos\theta\sin^2\theta + \cos^3\theta) = O$$

or

$$2\sin^2\theta = \cos^2\theta \qquad \tan^2\theta = \tfrac{1}{2}$$

$$\tan\theta = 0.707 = \frac{R}{H} \qquad \theta \cong 35°$$

SPECIAL LIGHTING APPLICATION TOPICS

21.29 Emergency Lighting

Emergency lighting is required when the normal lighting is extinguished for any of three reasons:

1. General power failure.
2. Failure of the building's electrical system.
3. Interruption of current flow to a lighting unit, even as a result of inadvertent or accidental operation of a switch or circuit disconnect.

As a result of the third requirement, sensors must be installed at the most localized level, i.e., at the lighting fixture (voltage sensor) or in the lighted space (photocell sensor).

(a) Codes and Standards. Since emergency lighting is a safety-related item, it is covered by various codes, all of which have jurisdiction. In addition, there are widely accepted technical society and industry standards whose recommendations normally exceed the minima legally required by the codes.

1. Life Safety Code, ANSI/NFPA 101, current issue—1981. This code defines the locations within specific types of structure requiring emergency lighting and specifies the level and duration of the lighting.
2. National Electric Code, NFPA 70, current issue—1984. This code deals with system arrangements for emergency light (and power) circuits, including egress and exit lighting. It discusses power sources and system design.
3. OSHA regulations. These are primarily safety oriented, and in the area of emergency lighting discuss primarily exit and egress lighting requirements.
4. Industry standards. These include the publications of the IES and the IEEE (see Tables 21.3 and 21.4 below). Because codes and

standards are constantly being revised and updated, the designer in any actual construction project must determine which codes have jurisdiction, obtain current editions, and fulfill their requirements. The material presented below is in the nature of general information and good practice, but is not intended to take the place of construction and safety codes.

(b) Amount and Duration of Emergency Lighting. The codes, and many authorities, require a minimum of 1.0 fc (10.76 lux) at floor level throughout the means of egress, as sufficient to permit orderly egress. The *IES Handbook* (1981) recommends that escape routes be lighted to not less than 1% of their normal illuminance, but in no case less than 5 lux (0.5 fc). Where this IES recommendation falls below 1 fc (normal illumination less than 100 fc) we would abide by the 1 fc minimum required by the Life Safety Code.

Duration of emergency lighting is normally specified to be a minimum of 90 minutes for egress, and indefinite periods (also at higher levels) for facilities that cannot be evacuated. Tables 21.3 and 21.4 give general criteria and typical recommendations for emergency lighting for egress as well as for other purposes.

(c) Types of Emergency Lighting Systems. Sections 16.29 and 17.22 cover emergency power systems, including generators and central battery systems. The former supplies selected portions of the normal lighting system through special emergency-lighting panels. The latter can supply a d-c distribution system or, if equipped with an inverter, can supply a-c as well. The availability of efficient inverters has practically eliminated central d-c systems. D-c systems have the added disadvantage that the incandescent fixtures they supply are not part of the normal system and may obtrude on the architecture, even when recessed and attractively finished. This is also true, to a lesser extent, of package units with spotlight-type heads (see Fig. 21.41). Such units are best applied in individual rooms and isolated locations.

The central battery with inverter, like the central generator, supplies lighting units that are usually (although not necessarily) part of the normal system. This arrangement has the advantages of economy, neatness, ability to use a-c sources (fluorescent), and reliability (see Fig. 21.42). Reliability can be further enhanced by using completely integral battery-charger-inverter packages mounted entirely within a fluorescent fixture (see Fig. 21.43). These units are usually designed to provide the 1½ h of illumination required by the code and to be completely maintenance free for 5 to 10 years, after which the battery is simply replaced. Since high temperatures seriously affect battery life, mounting these integral packages in a location other than that occupied by the fixture ballast will appreciably extend their life (see Fig. 21.42).

(d) Emergency Lighting Design Considerations. The level of required emergency illumination in specific areas should be related to the area's level of normal illumination and the degree of hazard in the area. Therefore, we would provide:

Exit area	5 fc [50 lux]
Stairs	3.5–5 fc [35–50 lux]
Hazard areas, such as machinery room	2–5 fc [20–50 lux]
Other spaces	1.0 fc [10 lux]

When the illumination level in an interior space drops sharply from a level of 30 to 100 fc to 1 to 5 fc, the eyes require up to 5 min to fully accommodate. During this long period the space's occupants are essentially sightless—a condition that lends itself readily to panic. For this reason, bright, spotlight-type heads must be *very carefully arranged*. Otherwise they can create disabling glare and distorting shadows, and impede eye accommodation.

Although it is customary to furnish the required emergency illumination from ceiling- or wall-mounted fixtures, consideration must be given to the code requirement that specified emergency lighting levels be maintained at floor level. Since heavy smoke can readily obscure light from overhead fixtures, it may be advisable to install some emergency lighting fixtures near the floor level.

(e) Exit Lighting. Most codes require either self-illuminated signs or 5 fc [50 lux] on non-illuminated exit signs. Some exit signs are equipped with a battery and controls that will provide 1½ h

TABLE 21.3 **Condensed General Criteria for Preliminary Consideration for Emergency Lighting Applications**

Specific Need	*Maximum Tolerance Duration of Power Failure*	*Recommended Minimum Auxiliary Supply Time*	*Type of Auxiliary Power System*[a]		*System Justification*
			Emergency	*Standby*	
Evacuation of personnel	Up to 10 seconds, preferably not more than 3 seconds	2 hours	×		Prevention of panic, injury, loss of life Compliance with building codes and local, state, and federal laws Lower insurance rates Prevention of property damage Lessening of losses due to legal suits
Perimeter and security	10 seconds	10 to 12 hours during all dark hours	×	×	Lower losses from theft and property damage Lower insurance rates Prevention of injury
Warning	From 10 seconds up to 2 or 3 minutes	To return to prime power source	×		Prevention or reduction of property loss Compliance with building codes and local, state, and federal laws Prevention of injury and loss of life

Restoration of normal power system	1 second to indefinite depending on available light	Until repairs completed and power restored	×	×	Risk of extended power and light outage due to a longer repair time
General lighting	Indefinite; depends on analysis and evaluation	Indefinite: depends on analysis and evaluation		×	Prevention of loss of sales Reduction of production losses Lower risk of theft Lower insurance rates
Hospitals and medical areas	Uninterruptible to 10 s. NFPA 76A-1977 and 101-1976 allow 10 seconds for alternate power source to start and transfer	To return of prime mover	×	×	Facilitate continuous care to patients by surgeons, medical doctors, nurses, and aides Compliance with all codes, standards, and laws Prevention of injury or loss of life Lessening of losses due to legal suits
Orderly shutdown time	0.1 seconds to 1 hour	10 minutes to several hours	×		Prevention of injury or loss of life Prevention of property loss by a more orderly and rapid shutdown of critical systems Lower risk of theft Lower insurance rates

Source: Extracted with permission from "Recommended Practice for Emergency and Standby Power Systems," ANSI/IEEE Standard 446-1980, © 1980 by the Institute of Electrical and Electronics Engineers, Inc.

[a]See Section 16.29 for definition of these terms.

TABLE 21.4 **Typical Emergency and Standby Lighting Recommendations**

Standby[a]	*Immediate, Short-Term*[b]	*Immediate, Long-Term*[c]
Security Lighting	Evacuation Lighting	Hazardous Areas
Outdoor perimeters	Exit signs	Laboratories
Closed circuit TV	Exit lights	Warning lights
Night lights	Stairwells	Storage areas
Guard stations	Open areas	Process areas
Entrance gates	Tunnels	
	Halls	Warning Lights
Production Lighting		Beacons
Machine areas	Miscellaneous	Hazardous areas
Raw materials storage	Standby generator areas	Traffic signals
Packaging	Hazardous machines	
Inspection		Health-Care Facilities
Warehousing		Operating rooms
Offices		Delivery rooms
		Intensive care areas
Commercial Lighting		Emergency treatment areas
Displays		
Product shelves		Miscellaneous
Sales counters		Switchgear rooms
Offices		Elevators
		Boiler rooms
Miscellaneous		
Switchgear rooms		
Landscape lighting		
Boiler rooms		
Computer rooms		

Source: Extracted with permission from "Recommended Practice for Emergency and Standby Power Systems," IEEE Standard 446-1980 © 1980, Institute of Electrical and Electronics Engineers, Inc.

[a]An example of a standby lighting system is an engine-driven generator.

[b]An example of an immediate short-term lighting system is the common unit battery equipment.

[c]An example of an immediate long-term lighting system is a central battery bank rated to handle the required lighting load only until a standby engine-driven generator is placed on line.

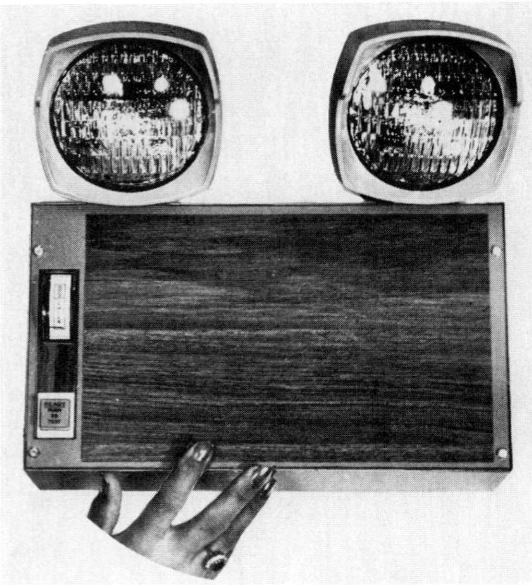

Fig. 21.41 *Self-contained, maintenance-free package, good for at least 90 min of emergency light. The package contains a battery charger, inverter, and controls. Heads, which are normally 6- or 12-V tungsten-halogen lamps, are either integral or remotely mounted. Similar units are available for recessed mounting. Heads with various photometric characteristics are available. (Photo courtesy of TORK.)*

of illumination on loss of utility power. Others are arranged to illuminate the exit area beneath the sign (see Fig. 21.44) and still others are equipped with a flasher and/or an audible beeper that will assist people to find the exit in a light-obscuring, smoke-filled room. Finally, some exit signs are nonelectrical self-illuminating, requiring no elec-

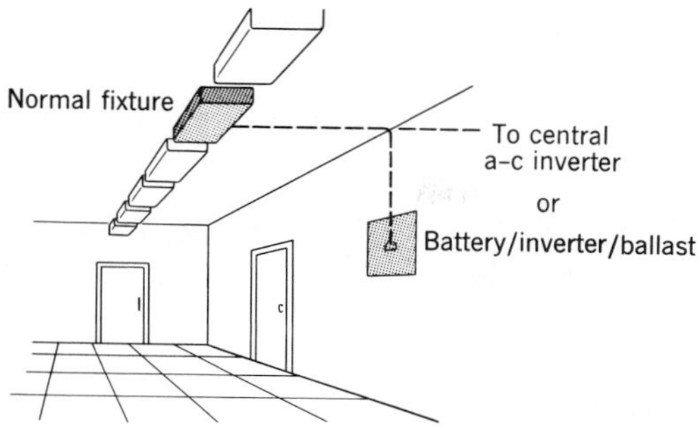

Fig. 21.42 The use of an integral or remote concealed battery/inverter (shown) allows instantaneous emergency lighting of areas blacked out by local or general power failure. The use of fluorescent sources provides seven times the illumination possible with incandescent, for the same battery size.

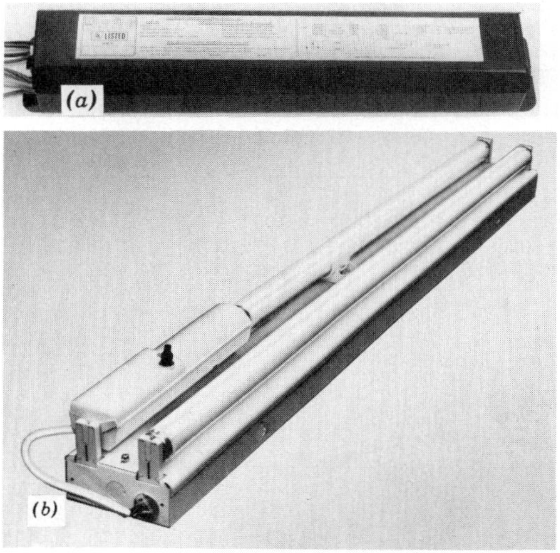

Fig. 21.43 Compact (13 in. long) ballast-shaped enclosure (a) *containing battery, charger, inverter, and controls can be mounted in the ballast chamber, between lamps, or on the fixture housing. This unit produces approximately 1000 lm from one lamp only, regardless of type of lamp or number of lamps in luminaire. (Photo courtesy of Bodine.) Another design* (b) *is mounted in tandem with a 36-in. lamp, replacing a 48-in. lamp. (Photo courtesy of Dual-Lite.)*

trical wiring connection. All exit lights require continuous illumination and are part of the emergency lighting system.

Fig. 21.44 Typical application of exit light with built-in battery, charger, and controls. Note that the bottom of the unit illuminates the area immediately in front of the exit. (Photo by Stein.)

21.30 Building Retrofit

Existing building lighting systems can be modified to increase efficiency, decrease glare, and decrease energy consumption by applying the procedures, techniques, and knowledge gained in the preceding four chapters.

(a) Sources

1. Replace standard fluorescent tubes with low-energy units.
2. Replace continuous-burning incandescent sources with high-efficacy fluorescent or HID.
3. Replace general-service incandescent lamps in downlights with lower wattage R, PAR, or elliptical reflector or low-voltage lamps.

4. Replace existing HID sources with higher-efficacy HID units such as HPS and metal-halide. Use those that can be operated on mercury-lamp ballasts, to avoid the cost of ballast replacement.
5. Increase daylight use—add reflectors to increase room penetration.

(b) Fixtures

1. Replace diffusers with more efficient ones and reduce glare. This will permit reducing lighting levels.
2. Install an appropriate control and daylight compensating system.
3. Institute a program of maintenance that will permit decreasing energy use by at least 20% while maintaining output.
4. Modify fixture layout in accordance with a task-ambient lighting philosophy and add task lighting.

(c) Other

1. Rearrange tasks so that the most difficult ones benefit most from daylight.

TABLE 21.5 **Lighting Application Guide**

		Watts per Square Foot Generally Required			
Application	*Minimum Footcandles Maintained*[a]	*Tungsten-Halogen*	*Mercury Units*	*Metal-Halide*	*High-Pressure Sodium*
Automobile Parking					
Attendant parking	2	0.38	0.17	0.11	0.075
Industrial lots	1	0.13–0.15	0.06–0.07	0.037–0.044	0.026–0.03
Self-parking lots	1	0.13–0.15	0.06–0.07	0.037–0.044	0.026–0.03
Shopping Centers					
Neighborhood	1	0.13–0.19	0.06–0.09	0.037–0.055	0.026–0.038
Average commercial	2	0.26–0.3	0.12–0.135	0.075–0.087	0.052–0.06
Heavy traffic	5	0.65	0.29	0.19	0.13
Automobile Sales Lots					
Front row (Front 20 ft)	50	10.	4.5	2.9	2.0
Remainder	10	1.5–1.8	0.68–0.81	0.44–0.52	0.3–0.36
Building					
Construction	10	1.5–1.8	0.68–0.81	0.44–0.52	0.3–0.36
Excavation	2	0.26–0.3	0.12–0.14	0.075–0.09	0.052–0.06
Buildings up to 50 ft High	***Adj. Area***				
	Light Dark				
Light surfaces	15 5	3.3 1.2	1.5 0.54	0.96 0.35	0.66 0.24
Medium light	20 10	4.3 2.2	1.94 1.0	1.25 0.64	0.86 0.44
Dark surfaces	50 20	10.0 4.3	4.5 1.94	2.9 1.25	2.0 0.86
Billboards and Signs	***Adj. Area***				
	Light Dark				
Good contrast	50 20	10.0 4.3	4.5 1.94	2.9 1.25	2.0 0.86
Poor contrast	100 50	20.0 10.0	9.0 4.5	5.8 2.9	4.0 2.0
Protective Lighting					
Gates and vital area	5	1.2	0.54	0.35	0.24
Building surrounds	1	0.15–0.19	0.07–0.09	0.044–0.055	0.03–0.04
Roadways					
Along buildings	1	0.24	0.11	0.07	0.05
Open areas	0.5	0.08–0.1	0.036–0.045	0.023–0.029	0.02
Storage yards (active)	20	3.6–4.3	1.6–1.94	1.04–1.25	0.72–0.86
Storage yards (inactive)	1	0.15–0.19	0.07–0.09	0.044–0.055	0.03–0.04
Shopping Centers					
Parking areas (attraction)	5	0.65	0.29	0.19	0.13
Buildings (attraction)			(See Buildings)		
Used Car Lots			(See Automobile Parking)		

[a]All footcandle levels for ground area applications are *horizontal* values.

2. Repaint to give requisite reflectances.

Install an appropriate lighting control system as described in Sections 21.2 to 21.4 above.

21.31 Floodlighting

Floodlighting, both interior and exterior, is extensively used for such diverse locations as are listed in Table 21.5, in addition to the more common sports lighting, which is not listed. At the designer's disposal are a variety of sources with respect to output, color, life, efficiency, and wattage (see Chapter 19).

Although a detailed floodlighting design involves complex calculations beyond the scope of this work, it is often sufficient for the designer to utilize a watts per square foot table such as Table 21.5 to determine the approximate floodlighting requirements.

Thus, if one is concerned with lighting a self-service parking lot at a neighborhood shopping center, and metal-halide is selected, Table 21.5 tells us that approximately 0.055 W/sq ft will suffice. If the lot is 200 × 500 ft or 100,000 sq ft, then 0.055 × 100,000 = 5500 W is required.

Arrangement and choice of equipment remains then, before the problem can be considered solved. Considerable assistance on this score can be obtained from either the lighting engineer involved or from representatives of the equipment manufacturers.

Although most floodlight installations use a single type, the installation of Fig. 21.45*a* used a combination of metal-halide and HPS to obtain the desired effect.

Fig. 21.45 (a) *The Statue of Liberty was relighted for the American Bicentennial. It was found that fifty-eight 1000-W metal-halide units give 40 to 50 fc of white light on the statue. Eleven 400-W high-pressure sodium units were selected to complement the color of the granite base, which they light to 10 fc. (Photo courtesy of Crouse-Hinds Co.)* (b) *Floodlighted section of wall surrounding the Old City of Jerusalem, Israel, adjacent to the Jaffa gate. Light sources are 400-W, HPS units, giving an average illumination level of 50 lux. Sodium source was chosen to enhance the yellow-red color of the stone. (Photo courtesy of City of Jerusalem and J. Stroumsa, Chief Engineer.)* (c) *Church of All Nations, Mount of Olives, Jerusalem, Israel. Floodlight sources are 250- and 400-W mercury and metal-halide units, giving an average illumination of 70 lux. Sources were selected to complement the colors in the mosaic at the top of the facade. (Photo courtesy of City of Jerusalem and J. Stroumsa, Chief Engineer.)*

Fig. 21.46 (a) *Concrete pole with aluminum arm and mercury luminaire.* (b) *Other concrete pole sections. (Photos courtesy of Union Metal Corp.)*

21.32 Street Lighting

Although detailed street-lighting calculations and design considerations are beyond our scope (see appropriate IES standards), a few remarks are in order. New installations now use HID sources almost exclusively. The low efficacy and short life of incandescent sources and the bulkiness of fluorescents make them obsolete. Furthermore, high street-lighting levels reduce vandalism and crime, improve night merchandising, and add to an area's attractiveness. Some typical designs of street lighting and other outside luminaires are shown in Figs. 21.46 and 21.47.

Fig. 21.47 *The ''lollypop'' fixture, even if esthetically pleasing, gives poor illumination downward (note large collar). The narrow pole is weakened by the large handhole. (Photo by Stein.)*

References

American National Standard Practice for Office Lighting, RP-1 (1983). IES of North America, New York.

ANSI/IEEE Standard 446, *Recommended Practice for Emergency and Standby Power,* (1980). IEEE/Wiley, New York.

Design Criteria for Lighting Interior Living Spaces, Publication RP-11 (1980). IES of North America, New York.

IES Lighting Handbook, 6th ed. IES of North America, New York.

Matsuura, K. (1979). Turning-off Line in Perimeter Areas for Saving Lighting Energy in Side-Lit Offices, *Energy and Buildings,* 2.

Peterson, D., and Rubenstein, F. (1982). *Energy Savings Through Effective Lighting Control,* Laurence Berkeley Laboratories, February.

Pierpoint, W. (1981). ''Equi-visibility Lighting Control Principles,'' *Journal of the IES of North America,* October.

Rubenstein, F. (1981). *Energy Saving Benefits of Automatic Lighting Controls,* Laurence Berkeley Laboratories, September.

Turner, W. C. (ed.). (1982). *Energy Management Handbook,* Wiley, New York.

PART VII

SIGNAL EQUIPMENT

Modern buildings, from the simplest residence to the most complex industrial facility, depend completely on electrical, signal, alarm, and communication systems for normal functioning. The systems discussed in this part of the book are security, music/sound, intercom, clock and program, and paging. Fire and smoke detection and alarms are covered completely in Part IV (Chapter 13), Fire Protection. This discussion is from the point of view of a user—that is, we emphasize system application rather than design. Operating principles are adduced, and application data are related to available equipment. Thus, although equipment is constantly being improved, operating and application principles remain substantially constant, thereby permitting the designer to readily adapt new hardware to existing system arrangements.

22

SIGNAL SYSTEMS

22.1 Introduction

No area of equipment design and application to buildings has seen such great and rapid changes as the field of signal equipment. Under this title is subsumed all signal, communication, and control equipment, the function of which is to assist in effecting proper building operation. Included are surveillance equipment such as fire and interior alarm; audio and visual communication equipment such as telephone, intercom, and television, both public and closed circuit; and timing equipment such as clock and program. These systems are no longer limited in application. Clock and program equipment, which once were the exclusive interest of schools and some industrial facilities, are now incorporated into building mechanical equipment control systems. Closed-circuit TV, which was once limited to classroom and college use, is commonplace in mercantile areas as part of surveillance systems. The hundreds of signals generated throughout a large facility are logged, channeled, and controlled by means of specially programmed computers and microprocessors. All the signal systems that once were separate and distinct are now frequently combined and serve multiple purposes.

Obviously a detailed study of the design and application of such equipment is beyond the scope of this book or, for that matter, of any single book. We shall attempt, however, to discuss the basic operation of the various systems, some of the equipment available, application to different types of facilities, and the impact of these systems on the spaces within a structure. The types of facility considered are single and multiple residences, schools, stores, office buildings, and industrial facilities. Hospitals and laboratories are combinations of the area types above plus facilities too highly specialized to be discussed herein.

The systems covered include surveillance, communication, and time-based signal arrangements. Antenna systems are discussed as a special case of a communication (reception) system.

22.2 Principles of Intrusion Detection

To understand the design of security systems, it is necessary first to understand the characteristics of the commonly available intrusion detectors (sensors) on which these systems are based. Once the intrusion alarm has been given by a device (as with fire detection), the signal must be processed and appropriate measures taken. This may include sounding loud alarms, turning on lights, sending signals to private surveillance services or police, and so forth.

(a) Normally Open (NO) Contact Devices. See Fig. 22.1*a*. Primary among these are switch mats. Contact is made as a result of pressure on the mat due to an intruder's weight. These devices, normally used in residences, are applied where only a single path of entry exists to the protected area, so that the mat cannot easily be circumvented, if it is not noticed. However, the device can be readily defeated by simply moving the mat or stepping over or around it. Other disadvantages are that the switch mat restricts free movement by authorized personnel (residents), and it can be set off by small animals. Finally, being an open-circuit device, its circuit is unsupervised.

(b) Normally Closed (NC) Contact Devices. See Fig. 22.1*b*. These come in a variety of designs, the most common of which are magnetic contacts for doors and windows, spring-loaded plunger contacts for doors and windows, window foil, and trip wires. These devices, though not easily defeated, also restrict use of the protected opening. This is not a problem with doors, but it can be inconvenient with windows in residences, where open windows are desired for ventilation.

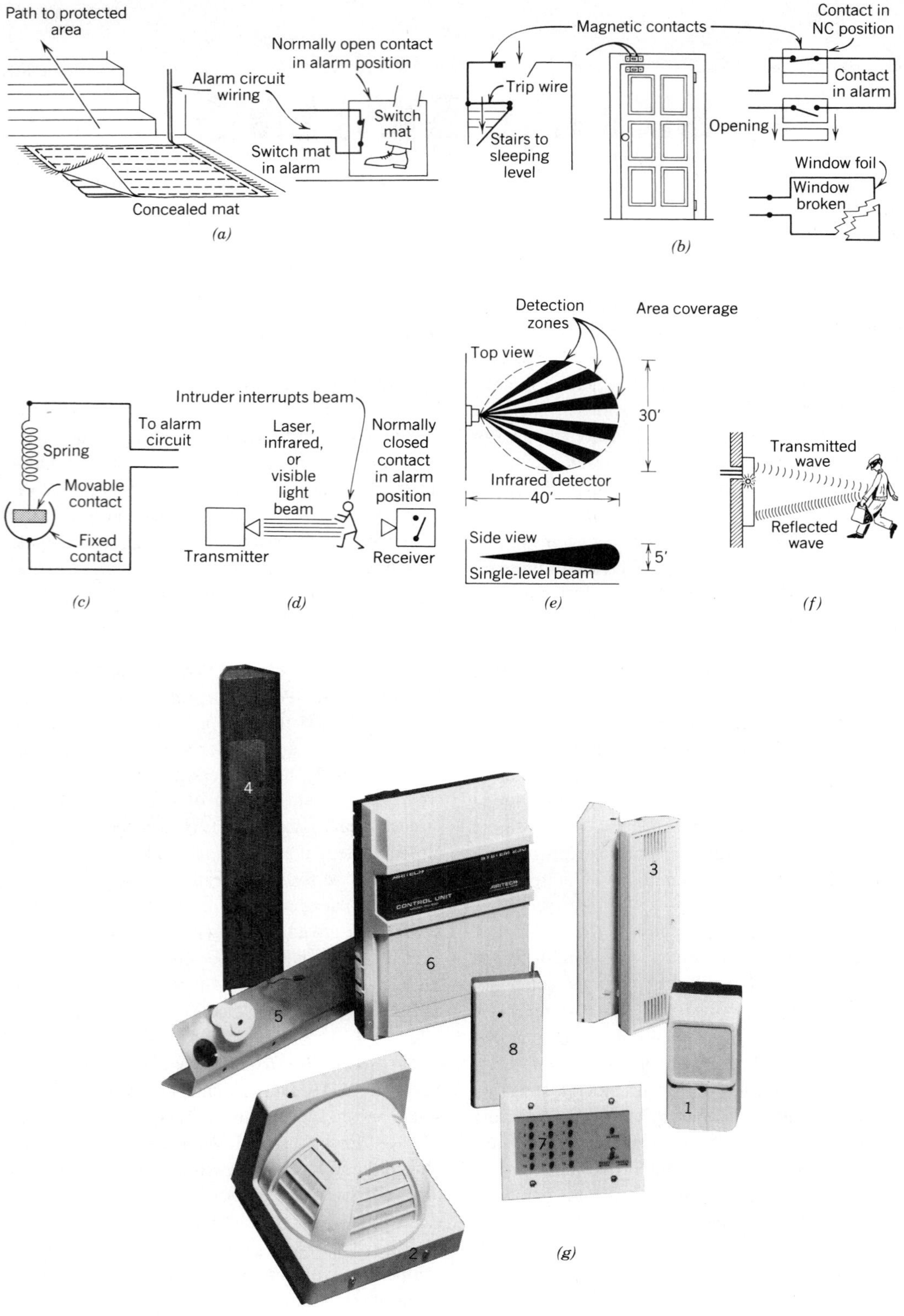
Path to protected area
Normally open contact in alarm position
Alarm circuit wiring
Switch mat
Switch mat in alarm
Concealed mat
(a)
Magnetic contacts
Contact in NC position
Trip wire
Stairs to sleeping level
Contact in alarm
Opening
Window foil
Window broken
(b)
Detection zones
Area coverage
Top view
30′
Intruder interrupts beam
To alarm circuit
Spring
Movable contact
Fixed contact
Laser, infrared, or visible light beam
Normally closed contact in alarm position
Transmitter
Receiver
Infrared detector
40′
Side view
Single-level beam
5′
Transmitted wave
Reflected wave
(c)
(d)
(e)
(f)
CONTROL UNIT
(g)

Fig. 22.1 Typical intrusion detectors. (a) *Normally open contact device such as a switch mat is operated by the force (weight) of the intruder.* (b) *Magnetic door contacts are the first line of intrusion alarm at the house. The second may be a low-level trip wire at the base of the stairs.* (c) *Vibration detectors are very sensitive to motion and can be used on multipaned windows that cannot be conductive-strip foiled without spoiling their appearance.* (d) *Photoelectric beams form an effective intrusion barrier if placed so that their detection or avoidance is difficult.* (e) *Passive infrared detectors give basically 30 × 40 ft oval protective zone, starting as a narrow beam and widening with distance. Focusability permits exact coverage of any area in a space. Units are also available with multilevel beams that give vertical as well as horizontal coverage.* (f) *Motion detectors depend on the Doppler effect. They detect changes in the frequency of a signal reflected from a moving object. Sensitivity is highest when relative motion is greatest, that is, when the intruder is moving directly toward (or away from) the detector.* (g) *Typical intrusion detection equipment: (1) Passive infrared detector. Maximum sensitivity is 2°C differential between target and background, and target motion of 2½ in./s. Unit is approximately 6 in. high, 3½" wide, and 4 to 6 in. deep. They are available with different coverage patterns, with a maximum coverage of 2000 sq ft. (2) High-sensitivity multizone wide-angle passive IR detector gives coverage of 17 zones horizontally and vertically. (3) Ultrasonic detectors arranged to give broad coverage of approximately 20 × 50 ft. Each unit is 11 × 5½ × 3½ in. (4) High-sensitivity balanced signal type of ultrasonic detector differentiates between intrusion signal and random environmental signals, thus reducing false alarms. (5) Unobtrusive, button-type ultrasonic sensor intended for ceiling mounting. (6–8) System control panel, annunciator, and mechanical alarm interface unit. (Photo courtesy of Aritech Corp.)*

The problem with windows can be overcome by using multiple sets of contacts. Trip wires can be used only where traffic in the area is cut off during protected hours. (It is advisable to "exercise" magnetic contacts periodically to prevent their freezing in position after a long period of disuse.)

(c) Mechanical Motion Detectors. See Fig. 22.1*c*. Where window foil or fixed contacts are impractical, a mechanical motion detector can be used. This device is basically a spring-mounted contact suspended inside a second contact surface. Any appreciable motion of the surface on which the device is placed will cause the contacts to make *momentarily,* turning in an alarm. These devices are very sensitive and can be activated by sonic booms, wind, and even a heavy truck passing by. Because of this, most such units are provided with sensitivity adjustment.

(d) Photoelectric Devices. See Fig. 22.1*d*. These devices operate on the simple principle of beam interruption. When the beam is received, an alarm contact in the receiver is normally closed. Interruption of the beam causes the contact to open, setting off the alarm. Older devices of this design use a visible light beam, and rely, for concealment, on the fact that light is invisible except when reflected from an intervening object. This is quite effective indoors, but outside dust, insects, birds, and so forth will show the location of the beam, permitting it to be circumvented. Birds and small animals will set it off, too. Dispersion of visible light also limits the throw of the devices when used outside. Modern units use lasers or infrared beams, which are less easily detected and can be arranged to distinguish between intruder and other disturbance. These latter devices have a longer effective range for exterior use. When using a laser beam, the signal can be picked up, amplified, and retransmitted in a different direction, thus establishing a perimeter security "fence" from a single source.

(e) Passive Infrared "Presence" Detector. See Fig. 22.1*e*. This device acts on the principle that all objects emit infrared radiation, or heat. The amount radiated depends primarily on the object's temperature and secondarily on its material, color, and texture. The IR passive sensor uses a lens or mirror that focuses on a small area and concentrates the IR radiation collected from that area into a sensor. Since IR radiation in an area that is undisturbed (not necessarily unoccupied) changes very slowly, because object temperature changes very slowly, any rapid change in the IR reading of that area indicates an object entering (or leaving) the space, causing an alarm. (We have already seen how these detectors can be used as occupancy sensors, to turn off lights when a space's occupant leaves. See Section 21.4.) The ability to focus on a particular area is utilized to cover areas both horizontally and vertically. Also, since the IR detector is not sensitive to motion, but only to heat, it is usable where motion in the monitored area is unavoidable. The principal disadvantage to passive IR detectors is that rapid temperature changes caused by direct insolation, a cold breeze, a heater turning on, and the like can cause false alarms.

(f) Motion Detectors. See Fig. 22.1*f*. These devices, which operate either at microwave frequencies or at ultrasonic frequencies, detect motion in the protected area by the Doppler effect. (This is the same effect that changes the perceived sound of a car horn or train whistle as the vehicle passes.) Any moving body will change the received frequency of the signal it reflects, and an alarm will sound. However, because the Doppler effect depends on relative motion between the source and the moving body, an intruder moving laterally may go undetected, if sensitivity has been reduced to avoid false alarms. Therefore, units should be located so that the path of an intruder is as nearly as possible directly toward or away from the detector. Ultrasonic units are cheaper than microwave but can be disturbed by strong air turbulence and very loud noises. Microwave units are undisturbed by air or noise, but because they penetrate solids (like ordinary TV signals), they can be affected by motion outside the protected area.

(g) Acoustic Detectors. These units alarm when the noise level exceeds a preset minimum. Alternatively, they can be arranged to respond to a particular range of frequencies, corresponding to the noise of breaking glass, forced entry, or whatever is desired. Although applied principally in security systems, they can also be used as occupancy sensors for switching of lighting. Such an application is described in Section 21.4c and illustrated in Fig. 21.10.

A variety of the above-described devices are illustrated in Fig. 22.1*g*.

In summary, as with fire detectors, no single unit will serve all situations, and the advice of an expert is desirable in determining which unit or combination of units to specify.

PRIVATE RESIDENTIAL SYSTEMS

22.3 General

Modern private residences utilize a variety of signal apparatus that greatly enhance their functional value. Figure 22.2 shows a residence that has been provided with what would be considered adequate but by no means excessive sound and signal equipment for a house of its size. In general, all signal systems require a source of signal, equipment to process the signal, including transmitting it, and finally a means of indicating the signal, either audibly, visually, or on permanent record "hard copy." A complex system still falls into this threefold category except that the individual items of equipment and their functions become more sophisticated.

In connection with complexity of equipment, the reader is undoubtedly familiar with the microprocessor controls now on the residential market that will handle security, fire alarm, time functions, lighting, and so forth. These are basically time-based programmable switches, of the type described in Sections 14.14 and 16.15. They function on dedicated wiring or on powerline carrier signals (PLC), that is, high-frequency signals impressed on electric power wiring. Since it is not our purpose to discuss automation or central microprocessor control, but rather the basic systems, the following descriptions consider the systems separately, despite the clear trend to consolidated control.

Residential fire alarm systems are covered in Section 13.16. The components and design of such a system, including detectors, circuit arrangement, and the like are also discussed in Chapter 13.

We have listed in Table 22.1 the systems and equipment found in the residence of Fig. 22.2, by the threefold classification. Note that the fire alarm, smoke detection, and intrusion alarm systems have been combined into a single system. This simplifies operation and avoids unnecessary equipment duplication. As we discuss the more complex systems, it will be seen that the basic functions remain unchanged.

As shown in Fig. 22.2, a single control panel can serve a multiplicity of residential systems. A typical unit of this type is illustrated in Fig. 13.38. The panel is designed to display the nature and location of the alarm device that has "tripped." A riser diagram for this residence is shown in Fig. 22.2*d*. The alarm devices themselves are not shown, because they appear on the plans and duplication serves no useful purpose.

22.4 Residential Intrusion Alarm Systems

Although any or all of the devices described in Section 22.2 may be used, residences normally utilize door and window contact switches, and possibly one ultrasonic motion detector, as shown in Fig. 22.2. A manual switch at the end of a long cord is also often provided so that the resident may at will set off the alarm if an intruder is heard. The system may employ the same audible signals as the fire system or its own components. Although done infrequently, intrusion alarm systems can be continuously supervised by connection with central stations of companies whose business is such supervision and who will either respond directly to an alarm call or notify local police authorities of any illegal entry.

22.5 Residential Television Antenna Systems

The large number of multiset American homes has made the central television antenna system a desirable feature of the modern residence. Systems with more than two outlets generally require a booster amplifier (except in strong signal areas) and are known as amplified systems.

The function of the system is to supply a television signal at each wall outlet, so that a receiver may be operated at any location and so that two or more receivers may operate simultaneously.

The functioning of the system is simply to amplify the signal received by the antenna and by means of special cable to distribute these amplified signals in a concealed cable to the various wall outlets. The type and location of antenna, gain (amplification) of the amplifier, and type of cable are variables that, being dependent on the specific installation, are best left to a competent and reliable local television company or design engineer. If a dish-type or other antenna is considered, provision must be made for cable entry via wall sleeves, sealed inside and out.

22.6 Residential Intercom Systems

The public demand for step-saving conveniences has resulted in the wide acceptance of the home "intercom" (see Fig. 22.3). Although available with various features, the basic system comprises one or more masters and several remote stations, one of which monitors the front door, allowing it to be answered from various points within the home. In general, master stations allow selective calling, whereas remote stations operating through the masters are nonselective. The systems are particularly useful when left in the open (monitor) position for remote "baby-sitting." The applicability of such systems to residences with outbuildings should be immediately apparent. Since wiring is low voltage and low power, multiconductor color-coded intercom cable is generally used, run concealed within walls, attics, and basements.

Systems are also available that impose the signals onto the house power wiring. This has the advantage of eliminating separate wiring and making remote stations portable—they are connected simply by plugging into a power outlet.

Fig. 22.2 *(see overleaf) Signal plans for typical residence.* (a) *Lower level.* (b) *Upper level.* (c) *Symbol key.* (d) *Riser diagram.*

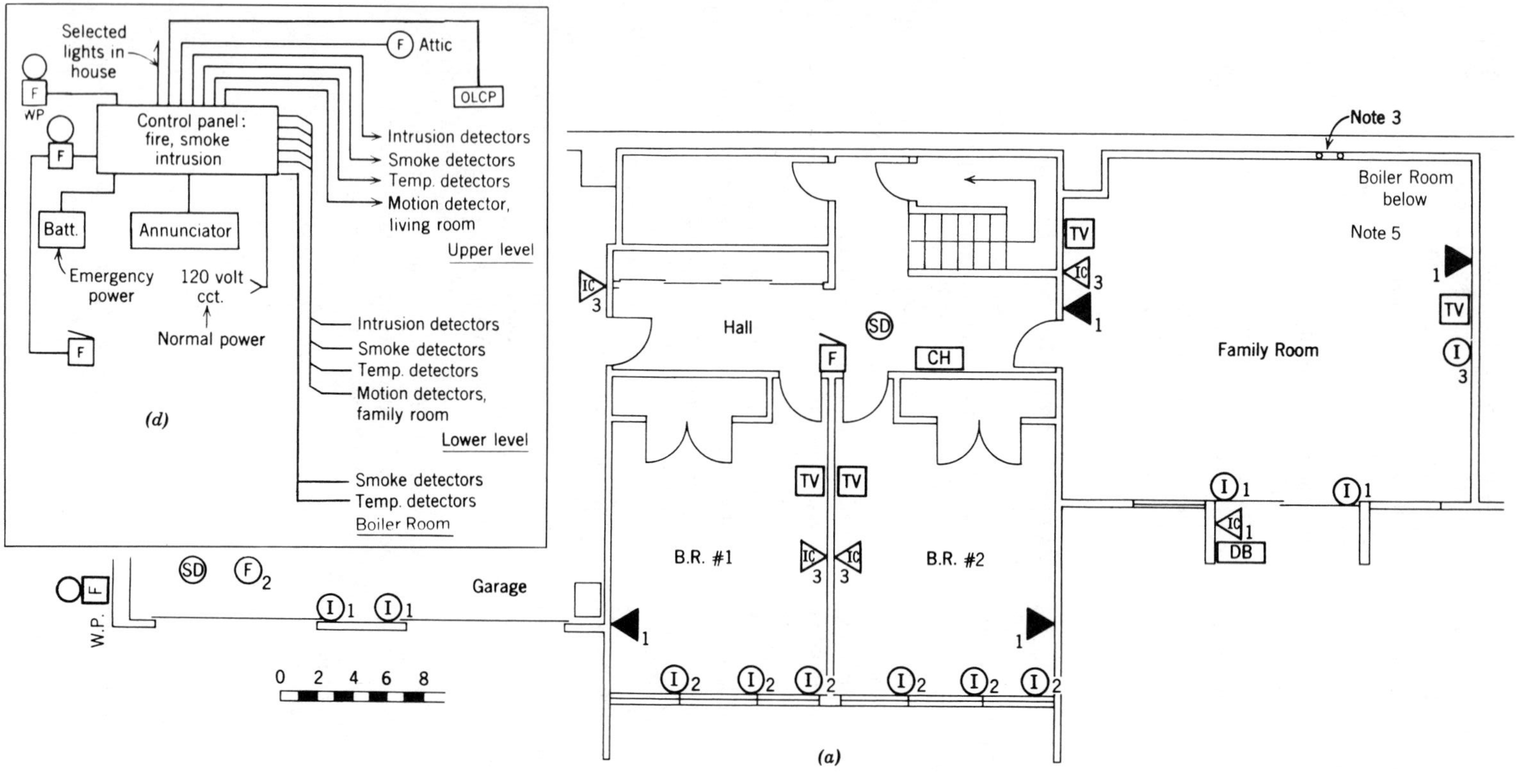

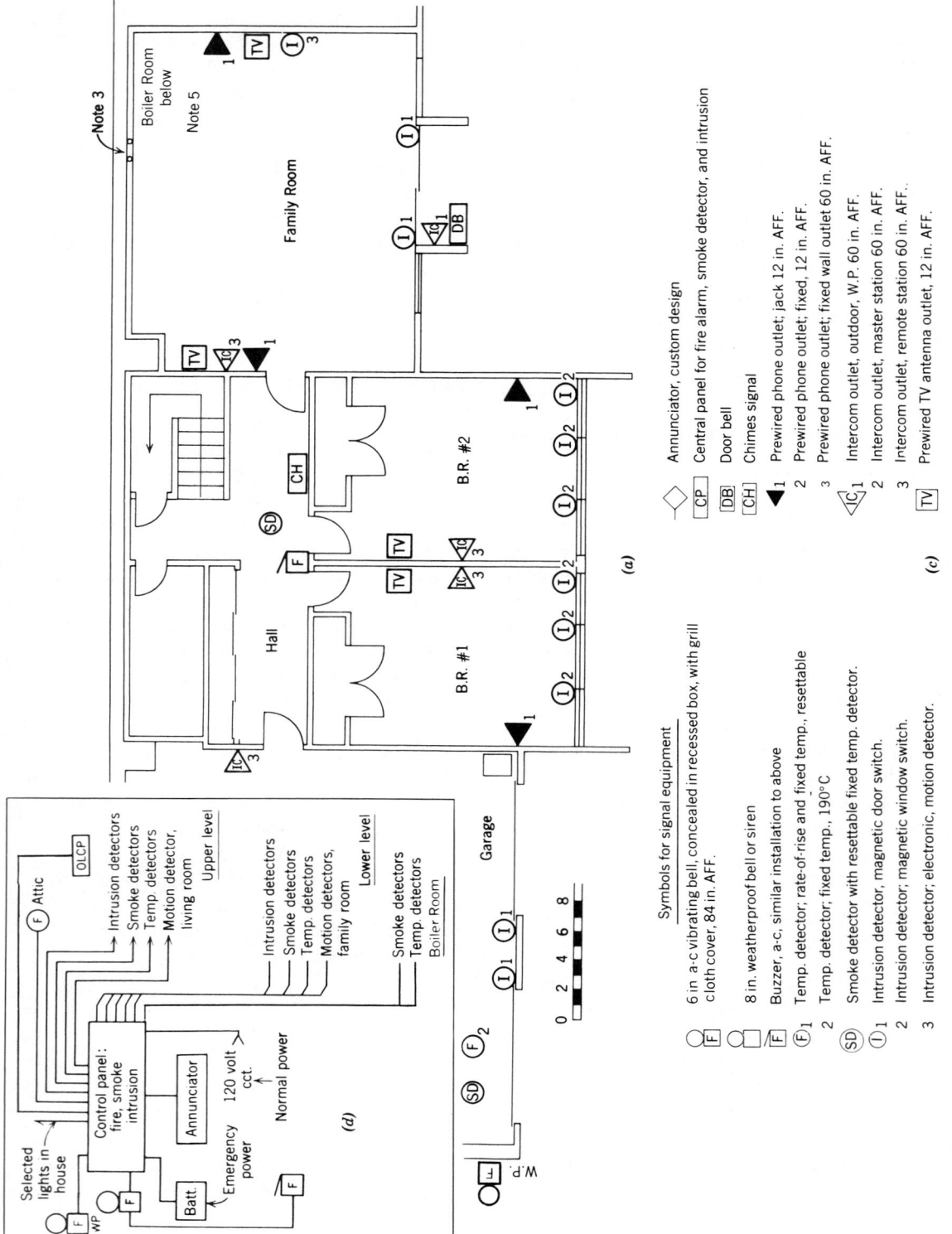

Symbols for signal equipment

○ F	6 in a-c vibrating bell, concealed in recessed box, with grill cloth cover, 84 in. AFF.
○ □	8 in. weatherproof bell or siren
F	Buzzer, a-c, similar installation to above
(F) 1	Temp. detector; rate-of-rise and fixed temp., resettable
2	Temp. detector; fixed temp., 190°C
(SD)	Smoke detector with resettable fixed temp. detector.
(I) 1	Intrusion detector, magnetic door switch.
2	Intrusion detector; magnetic window switch.
3	Intrusion detector; electronic, motion detector.

◇	Annunciator, custom design
CP	Central panel for fire alarm, smoke detector, and intrusion
DB	Door bell
CH	Chimes signal
◀ 1	Prewired phone outlet; jack 12 in. AFF.
2	Prewired phone outlet; fixed, 12 in. AFF.
3	Prewired phone outlet; fixed wall outlet 60 in. AFF.
◁IC 1	Intercom outlet, outdoor, W.P. 60 in. AFF.
2	Intercom outlet, master station 60 in. AFF.
3	Intercom outlet, remote station 60 in. AFF.
TV	Prewired TV antenna outlet, 12 in. AFF.

(c)

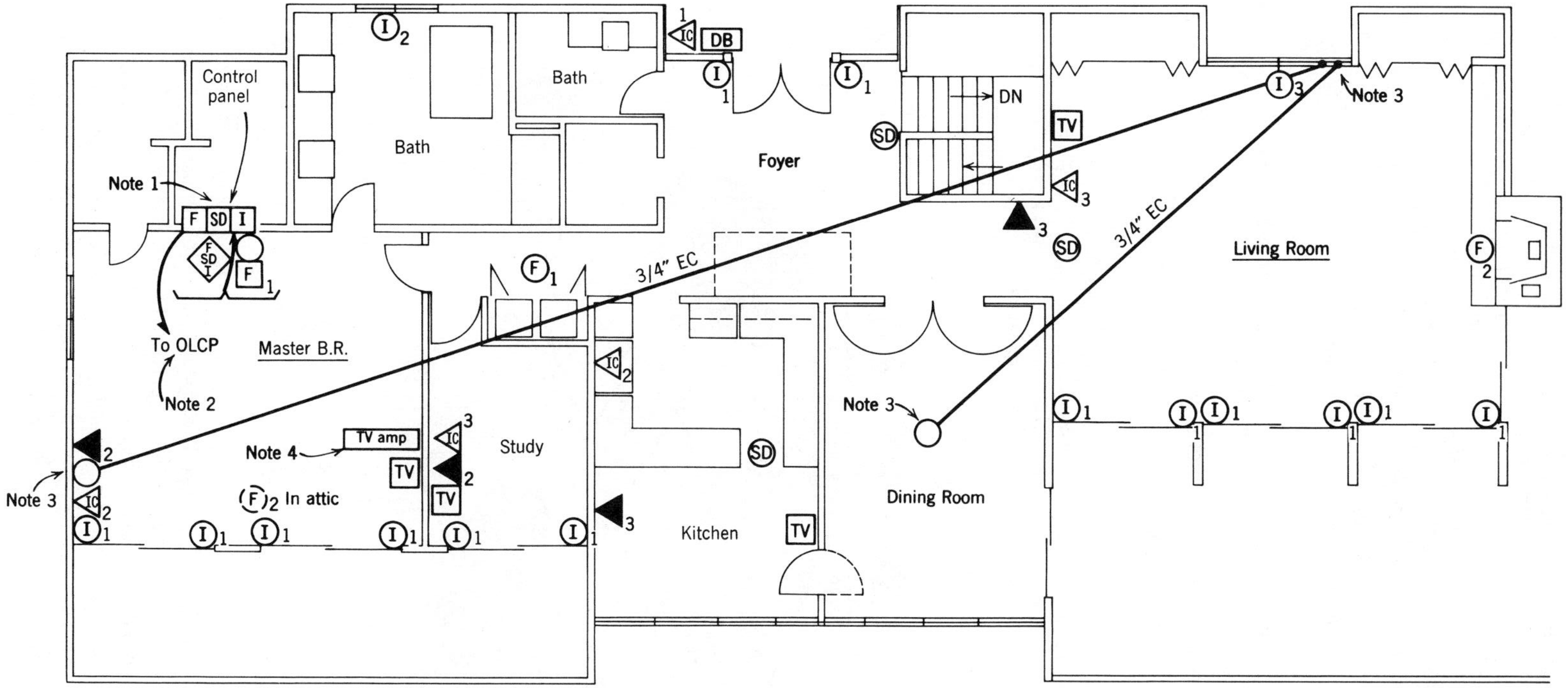

(b)

Notes:

1. The fire detection, smoke detection, and intrusion alarm devices all operate from a single control panel, see (d). The alarm bell is common. The annunciator indicates the device operated and its location.
2. The connection between the signal control panel and OLCP (outside lighting control panel) activates all outside lights when a signal device trips. Selected lights inside the house can also be connected to go on. See riser diagram (d).
3. Two 3/4-in. empty plastic conduits, extending from 4-in. boxes in living room wall down to family room and terminating in 4-in. flush boxes. Boxes to be 18 in. AFF and fitted with blank covers. Also, extend a 3/4-in. plastic EC from one 4-in. box in living room to 12-in. speaker backbox recessed in dining room ceiling. Locate in the field. From the second 4-in. box in living room extend a 3/4-in. empty plastic conduit to an empty 4-in. box in the master bedroom, 18-in. AFF. Finish with blank cover.
4. Provide television antenna amplifier, recessed in wall box, with hinged ventilated cover, 18-in. AFF. Connections to antenna and to all television outlets by television antenna subcontractor. Provide 120-V outlet at the amplifier, with switch to disconnect.
5. Boiler room to contain smoke detector, fixed 190 F heat detector, and remote station intercom outlet.

TABLE 22.1 **Elements of Residential Signal Systems**

System Type	*Signal Generator*	*Signal Processor*[a]	*Signal Transducer*
Fire alarm	Temperature and smoke detectors	Control cabinet(s)	Bells, annunciator, buzzer, siren
Intrusion alarm	Door and window switches, motion detector	Control cabinet	Bells, buzzer, annunciator, siren
Emergency call system	Pull, push button	Control cabinet	Bells, annunciator, corridor lights
Door bell	Push button	Transformer	Buzzer, chime
TV antenna	TV station and house antenna	Amplifier	TV set
Intercom	Microphone, speaker—mike	Amplifier	Speakers in various stations

[a]The proper wiring and switching is included under this title in all cases.

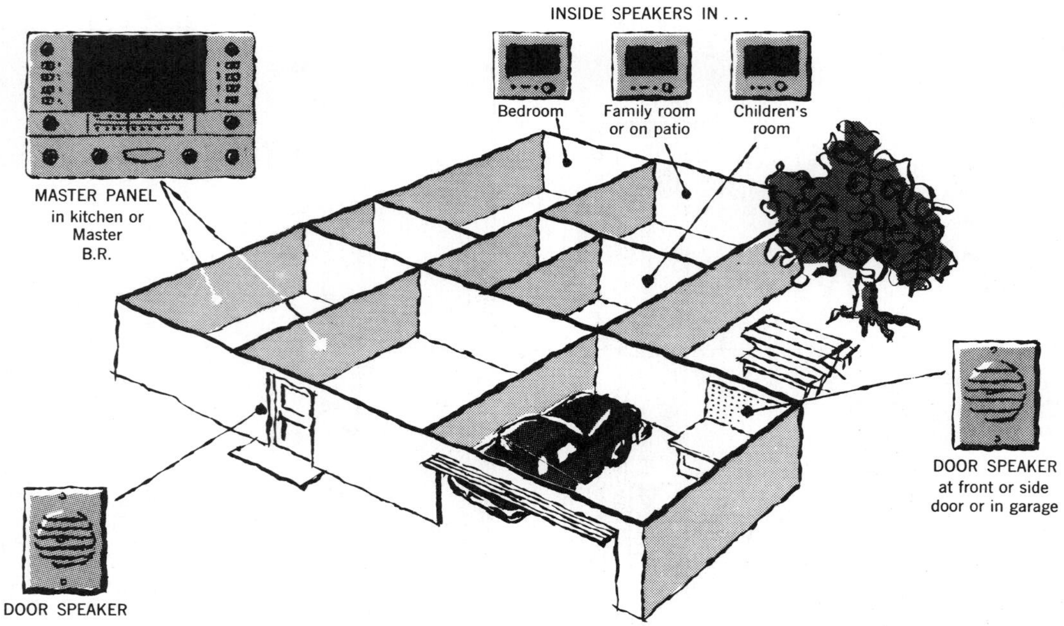

Fig. 22.3 Typical residential intercom equipment.

22.7 Residential Telephone Systems

Prior to relatively recent court decisions permitting users to install their own telephone equipment, the actual wiring within a structure was done only by the utility, in the user's raceway system. Today, work beyond the service entrance may be done by the owner, in fashion similar to other signal work. In residential work the telephone company normally follows the route of the electric service, entering the building overhead or underground as desired. In both cases a separate service entrance means must be provided: if aerial, a sleeve through the wall; if underground, a separate entrance conduit. Unless a residence has many entering lines, no source of power is required for the telephone equipment.

Wiring of telephone instruments when installed *after* completion of the residence consists of a single surface-mounted cable that, even if skillfully installed, is unsightly at best and completely objectionable at worst. Prewiring consists of running the cables on the wall framing and into empty device boxes to which instruments are later connected. Instruments can be wall or desk type, the latter also being available for jacking into outlets around the house.

MULTIPLE-DWELLING SYSTEMS

See Section 13.17 for a description of multiple-occupancy residential fire detection and alarm systems.

22.8 Multiple-Dwelling Lobby Intercom and Security Systems

Apartment houses, residences, and hotels combine the functions of the intrusion alarm and doorbell systems in the familiar lobby-to-apartment communication system. The most basic system is a series of push buttons in the lobby and an intercom speaker or telephone with which to communicate with residents. At the other end, the tenant has a speaker microphone plus a lobby door-opener button (see Fig. 22.4). This system can also be arranged to utilize the tenants' regular telephones. When the number of tenants is large, an alphabetical roster is added to the apartment-button panel to avoid the nuisance of scanning all the apartment names when the sought party's apartment number is not known. When the number is larger yet, a simple push button per apartment arrangement becomes cumbersome, and is usually replaced by an alphabetical tenant register plus a dial or button phone. Closed-circuit television can be added to the lobby–tenant system, enabling the occupant not only to converse with, but actually to see, the caller. Such a system will increase (by 7 to 10%) the electrical contract cost for an average apartment house.

In addition to the security provisions provided by the apartment-to-lobby audio and video connections, additional security and alarm devices have been used such as emergency call buttons within the apartment at desired locations. These will light alarm lights, ring bells, and perform any other alarm functions required, to cover the situation of an intruder who manages to bypass the lobby security check. In geriatric designs, these buttons serve to *unlock* the apartment door to allow aid to enter, if summoned by lights and alarms. In luxury apartments, apartment doors can be monitored from a central security desk and any unscheduled door movement subjected to immediate investigation. These systems are custom designed to the needs and requirements of the owner.

A security problem applicable to all facilities, including residential ones, deals with the problem of limiting entry in unsupervised areas to authorized persons. The problem of keys and locks is well known, despite advances in that field. More sophisticated means include magnetic cards and electronic combination locks, which, because of

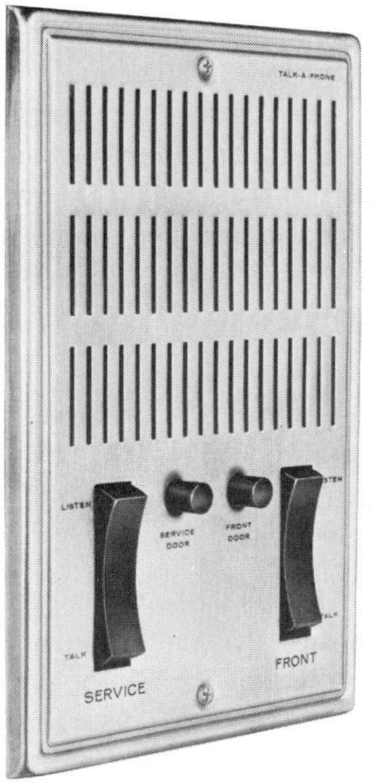

Fig. 22.4 *Apartment house communication–security equipment. Typical apartment unit contains intercom controls and door-opening buttons for main and auxiliary entrances. (Photo courtesy of Talk-a-Phone Co.)*

ease of code change, are particularly useful in residential facilities that cater to transients.

Another aspect of security that is particularly appropriate to housing for the elderly and handicapped, although it is applicable to all housing installations, is the emergency call system. The purpose of this system is to alert *outsiders* to an emergency situation *inside* a closed apartment. This alarm system is essentially a way to call for help in time of illness or other distress. Many construction and housing codes include descriptions of the required equipment. Most often they prescribe a call initiation button in each bedroom *and* bathroom, which will register an audible (alarm) and visible (annunciated) signal at a location that is monitored, locally or remotely, 24 hours a day. Additional signals are required in the floor corridor and at the apartment, the purpose of which is to alert immediate neighbors to the distress call.

22.9 Multiple-Dwelling Television Antenna Systems

All modern multiple residences supply each room with one or more TV/FM jack outlets. The master antenna systems feeding these are similar to the residential type except for size and electronic design of the components. The antenna should be placed only after a survey has been made to determine signal strength patterns on the roof. Space below the roof and a single 15-A circuit should be allowed for the amplifier equipment. Coaxial cables are run through floor sleeves and tapped at various points to provide good signals in each apartment.

22.10 Multiple-Dwelling Telephone Systems

As in the small residence, the telephone service normally follows the same entrance path and method of entrance as the electric power service. For the sake of economy in underground construction, the two services often share the same trench, albeit in different raceways, and utilize twin manholes where such are required. Typical entrance arrangements for any large building, residential or other, are shown in Fig. 22.5.

The service entrance space requirements vary with the size of the building and telephone capacity. For a small apartment house of the garden or three-story type, a clear wall space of 2 to 4 ft is sufficient. A terminal room is required only in very large buildings.

Apartment buildings and dormitories differ from commercial structures in that the floor plans of all floors are similar, so that the arrangement of risers is relatively simple. It is common practice to utilize cable only, in risers that extend through vertically aligned closets in apartments. To accommodate these cables, a sleeve through the floor between closets is necessary. If a riser is located

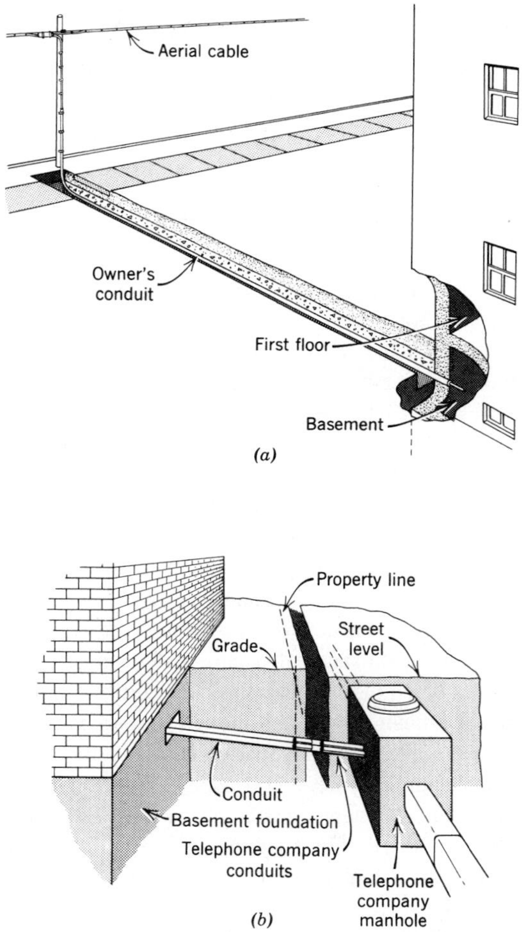

Fig. 22.5 *Telephone cable may enter a building underground, originating on overhead lines* (a) *or in manholes* (b).

in a shaft other than a closet, conduit is normally utilized to allow for easy installation, protection, and repair. If the location is accessible, as in an alcove, only a sleeve is provided. When the riser is located outside the apartment, each dwelling unit is connected to the riser by a conduit with a junction box at either end. These conditions are illustrated in Fig. 22.6.

Beyond the apartment service point, the individual rooms can normally be prewired entirely without conduit, or with only a few short sleeves.

SCHOOL SYSTEMS

22.11 General

The proper operation of a modern school requires that flexible and efficient signal and communications equipment be available to the administrative and teaching staff. Such equipment, engineered to meet the needs of the individual institutions, will do much toward optimum utilization of staff and student time. School fire detection and alarm systems are discussed in Section 13.18.

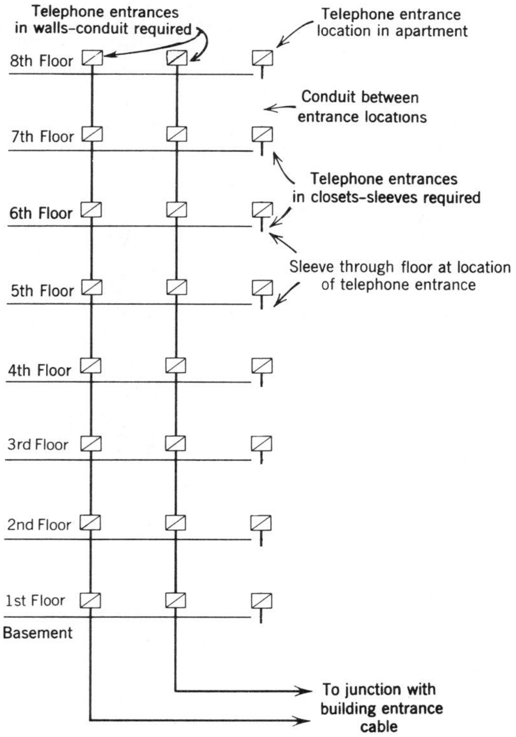

Fig. 22.6 *Typical telephone riser diagram. Note the need for conduit between apartments when installation is made inaccessible, as in a wall.*

22.12 School Security Systems

Although intrusion alarms and security systems were not historically normal school requirements, this situation has unfortunately changed. Sensing devices on doors and windows can be arranged both to trip local alarm devices and, via auxiliary circuits, to notify police headquarters. Often, vandals can be frightened off by having the alarm system actuate a protective lighting system that will illuminate the building exterior and any interior areas desired, such as record rooms. A perimeter alarm detection system of the types described in Section 22.2 can be installed in particularly vandalism-prone areas to assist in preventing entry to the school premises after hours. Though expensive, they are very frequently cost effective.

22.13 School Clock and Program Systems

The clock system and the program system were at one time separate and distinct, sharing only the timekeeping facilities provided by the master clock. Now that electromechanical programming devices are becoming obsolete, the two systems are actually one, but the traditional two-system name remains. Referring to Fig. 22.7, we see that the heart of the system is the time base (electronic clock). This is precisely the same device that provides timing for all programmable switches and controllers, as for instance in Figs. 14.14 and 16.10. In those instances, the timer controls various switching events; in a clock and program device it controls clock signals and various audible devices.

The clock system may use conventional clocks (with hands) or digital units. The former are usually locally powered, and a correction signal is periodically transmitted via dedicated wiring from the master clock at the controller. This same controller continuously transmits a binary-coded sig-

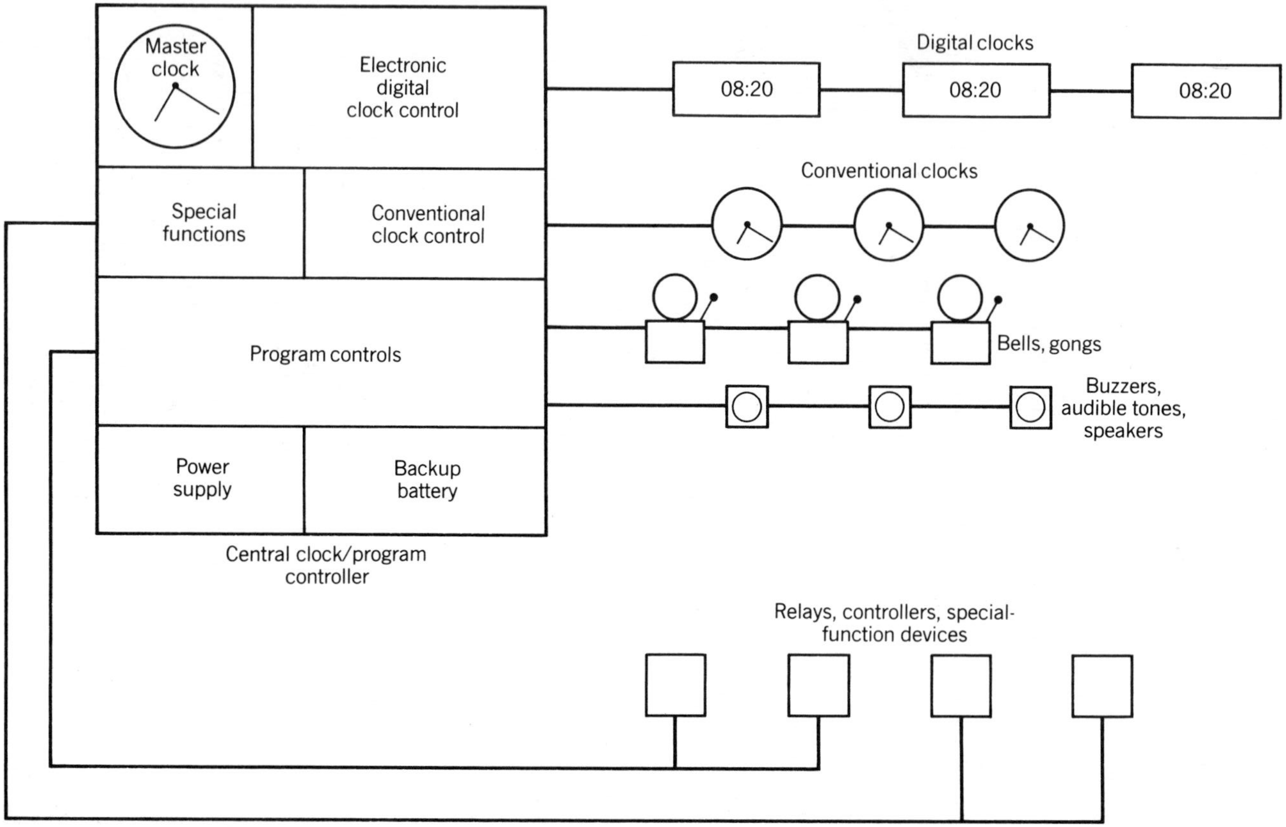

Fig. 22.7 The central clock and program device uses its electronic time base to control clocks of all types, time-based audible signals, and event controllers such as relays and switches.

nal to digital ''slave'' clocks, which can be either of the self-illuminated LED (light-emitted diode) type or of the LCD (liquid-crystal display) type. LCD units are easily visible only in high ambient illumination and when viewed directly (not at an acute angle). LED units are best applied in areas of low ambient illuminance. Conventional large-face clocks are easily visible in all ambient light situations.

The programming function of the controller serves to delineate audibly the various time periods into which the school day and week are divided. A single-circuit unit is utilized in the instance of an institution that operates entirely on one schedule, such as an elementary school on a morning period–lunch–afternoon period regimen. For a school employing different schedules for its various parts, the controller can provide multiple program schedules on different circuits, depending on its design. Controllers are user programmable and are normally provided with a local, visible, master clock, crystal control conversion to assure accurate timekeeping regardless of line frequency variation, backup power source to maintain user programming and master clock local display, and various conveniences such as daylight saving time correction, security arrangements, event timers, and so on. If desired, the clock and program controllers can also be used for mechanical system control, by the simple expedient of adding relays and switching devices, as shown in Fig. 22.7.

The audible devices in a program system may be bells, gongs, buzzers, horns, or a tone reproduced on a classroom loudspeaker. The latter system has the following advantages:

1. Clear audibility in each classroom, with adjustable volume to cover quiet and noisy areas.
2. No possibility of confusion between program tone and other signals such as fire alarm gongs.
3. Multiple use of the speaker unit, for classroom sound as well as program tone.
4. Complete flexibility of programming that is not possible with hall gongs. This is particularly desirable in schools with special programs for groups of students.

22.14 School Intercom Systems

Various types of intercom systems are available depending on the needs of the school building involved. In small schools, a simple wired intercom system connecting the various offices is usually sufficient. This is supplemented by outside telephones in the administrative offices and a functional paging system that is normally part of the school's sound system arrangement. With larger buildings and correspondingly larger numbers of extensions and multiple-function demands, more sophisticated equipment is required. The unit illustrated in Fig. 22.8 is typical of modern school intercom equipment, which is in actuality a private telephone system of considerable flexibility. Such a system is generally interfaced with the school sound system, and provides these functions:

1. Intercom between staff members and offices.
2. Direct communication with classrooms, including selective and all-call capability.
3. Zone, group call, and conference call functions.
4. Interconnection with the outside phone system.

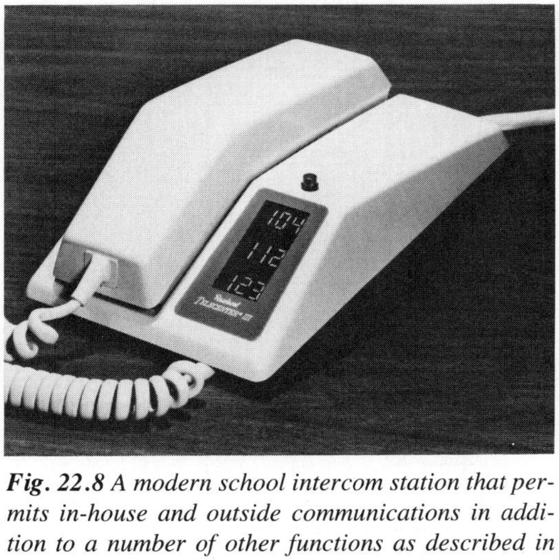

Fig. 22.8 *A modern school intercom station that permits in-house and outside communications in addition to a number of other functions as described in the text. Switchboards are eliminated, and all control and switching equipment is solid state. (Photo courtesy of Rauland-Borg.)*

These systems use direct push-button "dialing," eliminating the necessity for switchboards and operators. All stations are coded with three-digit codes, and all switching is solid state, minimizing maintenance problems. Such systems are adequate for all but the largest institutions.

22.15 School Sound Systems

The integrated sound-paging-radio system designed for school use offers several modes of operation and considerable flexibility. Its function is to provide a means for distributing recorded (records, tapes), broadcast (AM/FM), or live sound to preselected areas of the school. Thus, a simple system might provide a record player and single microphone input, and a single channel to all the speakers in the school, whereas a complex system can be arranged to operate with three simultaneous input signals distributed to six different areas of the school. As might be expected, increasing flexibility appreciably increases the cost of the system (see Fig. 22.9).

The system consists of a desk-size control console containing most of the input units, amplifiers, and switching devices, connecting to the remote loudspeakers, with related volume controls where required. The input units may comprise one or more AM/FM tuners, multispeed turntables or record changers, tape deck, and microphones. One microphone is normally located at the console with others in the principal's office, auditorium, school office, or other selected location. If desired, mike outlets can be spotted around the school and a spare mike and stand supplied to be plugged in at any of these points.

Loudspeakers, located in classrooms, gymnasium, auditorium, cafeteria, and outdoors receive the amplified signal through the switching mechanisms located in the console. It is the function of these switches to deliver the program material to the various loudspeaker circuits, which are also called program lines. Thus, using a system with multiple amplifiers, music can be piped to the cafeteria, an important radio address to senior classrooms and teachers' lounge, and instructions to an outdoor gym class or team practice. An all-call feature also allows announcements to reach all speakers in the system simultaneously. The intercom system discussed above can be incorporated into the sound system to allow conversation between classrooms and the console or other points, although it is often kept entirely separate.

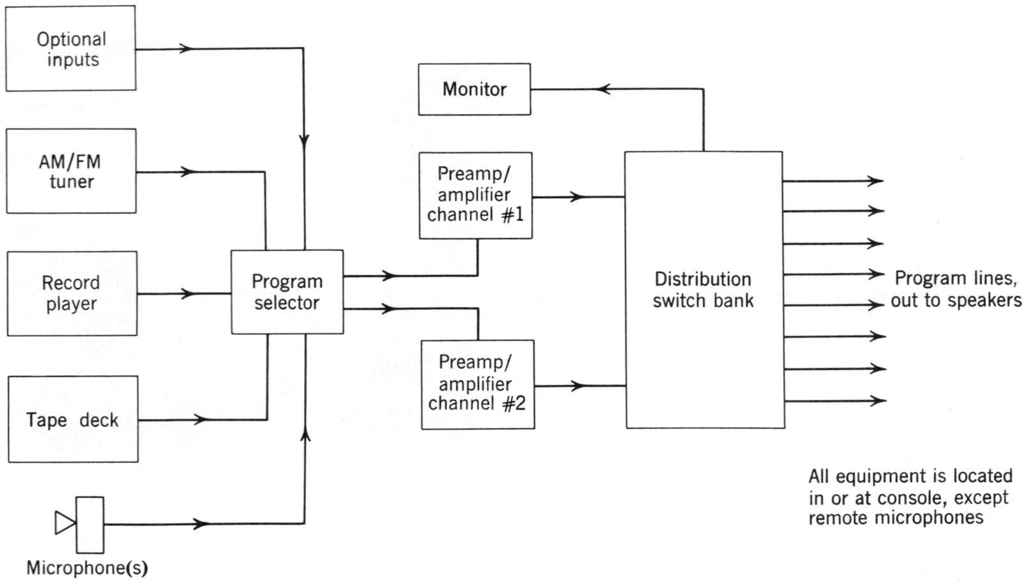

Fig. 22.9 Block diagram of a two-channel sound system. Optional facilities that can be added to the console are tape deck, private telephone communications, and equipment for rebroadcast of signals between areas. Intercom and master clock and program can also be incorporated.

The console is usually built in a desk arrangement, and it is advisable to provide sufficient space for it and for the person who operates it. Often an alcove of 30 to 50 sq ft is reserved for it and a library of recordings.

Loudspeakers may be in flush or surface baffles, at the discretion of the designer. Gymnasium, cafeteria, and auditorium units are normally flush-mounted in the ceiling. For large areas such as these it is well to provide a volume control, enclosed in a recessed wall box with a locking cover. A common variation of the above-described system uses separate subsystems for the cafeteria, auditorium, ball field, or other areas utilizing sound equipment frequently. These smaller systems have their own input, amplification, and control devices but utilize speakers in common with the central console. Normally the console has an override feature that allows it to override local systems.

For a discussion of high-quality sound systems required for recital halls in music schools and the like, refer to Sections 26.27 through 26.29.

22.16 School Electronic Teaching Equipment

The hardware consists of a library of recordings, a closed-circuit TV terminal, and an interactive computer terminal programmed with learning material. The first two devices are useful with a student in a passive mode. The computer terminal permits the student to actively participate in the lesson on a one-to-one basis, with the terminal acting as tutor. An auxiliary function of the computer terminal is that of information retrieval for more advanced students working on projects.

The entire subject of electronic teaching aids is complex and rapidly developing. The well-designed school will have adequate space and electrical and HVAC provisions for the extensive expansion into this field, which will undoubtedly continue.

OFFICE BUILDING SYSTEMS

22.17 General

Under this category we include systems found in all office, professional, and sales-type buildings. Such buildings house tenants with varying schedules and requirements and, unless large, do not have a full-time custodian. These factors must then be considered in the design of the signal systems for such buildings.

Although in many of the medium-sized and large buildings control, alarm, and security functions are combined in multiuse apparatus and consoles, which we will discuss, the basis systems are essentially separate despite shared-use equipment. We discuss them individually to demonstrate function and equipment, and combined to show economies and modern practice.

22.18 Office Building Security Systems

Although automatic surveillance systems *are* applicable to office and mercantile occupancies, they are more frequently found in industrial facilities and are discussed under that heading. Office buildings normally utilize some type of manual watchman's tour system so that surveillance of unoccupied areas is conducted on some regular basis.

The simplest type is nonelectric and comprises a number of small cabinets, each containing a key, placed at intervals around the interior and exterior of the building. The watchman uses these keys to operate a special clock that he carries about, thus recording the exact time at which he "clocked in" at any specific location. Alternatively, the clock is wall mounted and the guard carries only a key (see Fig. 22.10).

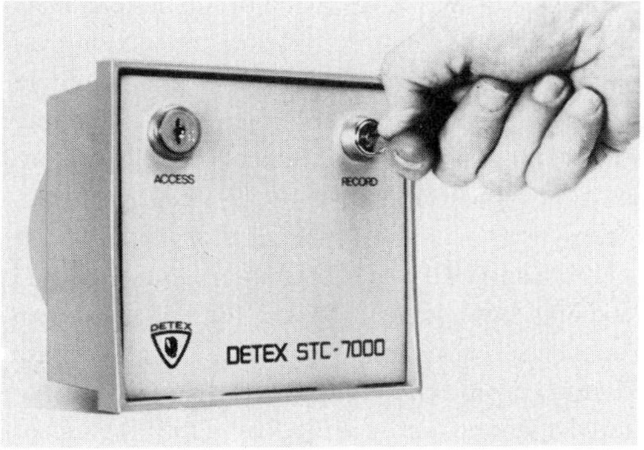

Fig. 22.10 *Watchman's tour station. Here the clock and recording tape are inside. The station is operated by simply inserting a key, as shown. (Photo courtesy of Detex Corp.)*

Electrical systems are available that permit constant supervision and are particularly effective where more than one person is on duty. Such systems show on a panel the location and progress of the watchman, by means of lights that glow when the device at each location is operated. Since part of the effectiveness of these systems lies in the timing of the tour, a system can be arranged to sound an alarm if a particular station is not operated within a specific time period. Telephone jacks spaced at points along the guard's route allow him to communicate with the supervising office or other point, without interrupting the scheduled tour. For protection of areas housing extremely valuable items or documents, an intrusion alarm system may be employed.

22.19 Office Building Communications Systems

This planning item is composed of three parts, which are frequently melded into a single network:

1. Intraoffice voice communication, or intercom.
2. Interoffice and intraoffice data communication using telephone lines.
3. Outside-the-building communication via phone company lines.

Prior to court decisions permitting user equipment connection to the telephone company network, items 1 and 3 were entirely separate unless the utility supplied the intercom service as well. Today, in most locations, the user has the choice of purchasing or leasing as much of the system as he desires. In other words, the communication system including cables, instruments, switching equipment, and so on can either be all privately owned, all supplied by the local telephone company, or almost any division desired. However, except for a small interoffice intercom, duplication is eliminated. That is, the same instruments and switching equipment are used for both intercom and outside connection. What to buy, rent, or lease constitutes an economic decision, since the required functions are satisfied by either private or telephone company equipment.

The item of data handling is highly specialized and depends on the type of business being conducted. In modern offices separate raceways or expanded telephone raceways are provided for the relatively heavy cables required for this use. Although fiberoptic cables and multiplexing increase the efficiency of cable use, requirements are also continually increasing, and raceway space requirements remain high. The field of office communications is highly specialized, and the building designer should consult with prospective tenants to determine their requirements. Where this is not possible, as in speculative office building construction, the guidelines given below will be helpful.

22.20 Office Building Communications Planning

Planning for the telephone and other communications equipment in an office building is of prime importance because of the large amounts and critical locations of required space. Therefore it must be done simultaneously with other space planning. Exact requirements for office space are generally unknown at design time and, even if they were known, planning would have to account for changes in space usage as well as increased communications and data transmission services. For this reason, all planning is based on square foot areas. Planning is essentially for spaces only, from incoming service to final instrument, since cabling and equipment are furnished and installed either by a private telephone equipment supplier or by the telephone company. Space is required for:

1. Service entrance including terminal space, cabinet, or room.
2. Riser spaces, shafts, conduits, and cabinets.
3. Apparatus closets for equipment.
4. Satellite locations for interconnections.
5. Equipment rooms for specific-use equipment.
6. Distribution system including conduits, boxes and underfloor duct, and raceway systems.

When multiplexing, solid-state switching, or computer data control is used, space requirements for items 3, 4, and 5 can often be reduced. It is in the interest of the client to be aware of this, in order that precious rental space not be wasted.

(a) Service Entrance. Adjacent to the service conduit a spare sleeve placed in the foundation wall and sealed will provide for future service expansion. Inside the building the telephone cable is terminated in a wall box, cabinet, or terminal room depending on the cable size. For buildings up to about 70,000 sq ft of rentable area (depending on the type of anticipated tenant use), wall-mounted terminal cabinets are normally sufficient. In any case, the area should be dry, and it requires light, ventilation, one or more 120-V outlets on a separate circuit, and a good ground connection. In special cases more power is required.

Surface-type wall installations comprise simply a sheet of ¾-in. marine plywood on to which the phone company mounts its cabinet. Flush installations in finished areas require a cabinet with ¾-in. plywood back.

(b) Riser Shafts. These can accept the cables extending beyond the terminal room and carry them vertically through the building. Connection between terminal space and risers is preferable in conduit.

The riser shafts provide means for the cables to extend vertically and to terminate at each floor. Ideally, the risers comprise a series of vertically aligned closets connected by 3½- or 4-in. sleeves set in the floors and extending 1 in. above the floor. It is preferable to separate communications closets from electrical power closets. Where multiple risers are used, shafts should be interconnected by several 2-in. conduits to allow for interconnection of systems.

(c) Riser Closets. Here, cables from the riser system are interconnected to switching and power equipment, as well as to the cables that radiate from the closet to station locations throughout the floor. These closets may also be called *zone closets* or *apparatus closets,* particularly if they function with an underfloor raceway system. Walls of the closet should be lined with plywood at least ¾-in. thick to support the weight of switching and connection panels, power equipment, terminals, connecting blocks, and other hardware. Each riser (apparatus) closet must be provided with a switched ceiling light, and a separate 20-A, 120-V circuit with two duplex receptacles. A source of emergency power is desirable to avoid curtailment of telephone service during outages.

(d) Satellite Closets. Unlike riser (apparatus) closets, satellite closets do not contain switching and power equipment. Their primary use is to provide cable-connecting and -terminating facilities in large complex facilities, where riser closet space is insufficient.

(e) Equipment Rooms. Where extensive cross-connection is required or where tenants utilize PBX (private switchboard) equipment, the required equipment is placed in equipment rooms. These spaces, which are actually small closets or alcoves, should contain a 20- to 30-A, 208-V circuit, a 20-A, 120-V outlet, a grounding point, good lighting, and, obviously, sufficient equipment space. Space requirements vary with each installation and are obtainable from the equipment supplier. Ventilation is essential, as is absorptive acoustic material on the ceiling and at least one wall. Connections between these spaces and other communication equipment closets should be via floor or ceiling ducts or multiple 3½-in. conduits.

(f) Horizontal Distribution. Cabling from riser, satellite, apparatus, and equipment closets to individual outlets and instruments can be underfloor (Sections 15.24–15.28), in ceilings and plenum spaces (Section 15.30), under carpet (Section 15.29), or in surface raceways (Section 15.23). Because of the large raceway volumes required, conduit is infrequently used. When using underfloor raceways a header capacity of 2 sq in. per work station is reasonable, based on one multiline telephone and one video display unit per station.

(g) Fiber Optic Cables. In installations with very heavy data transmission loads, in those using video systems, and/or in applications for which the high-security, low-noise, and broad-bandwidth characteristics of optic fiber cables are desirable, they can be used in lieu of copper cabling. Space requirements and connection accessories are obtainable from manufacturers.

22.21 Office Building Supervisory Control Center

As the modern office building's mechanical and electrical systems increased in complexity, the need arose for a central point of supervision, control, and data collection from which to survey and control an entire building's functioning. From such a point the water, air conditioning, heating, ventilating, electrical, and other systems could be controlled manually and automatically with much greater accuracy than if no such central control point were available. Data on temperatures, pressures, flow, current, voltage, and all of the many parameters of mechanical systems could be made instantly available, so that operational decisions could be made more accurately. Also, all systems could be monitored here, and all alarms instantly acted upon, automatically or manually.

Such control centers, generally called *supervisory control and data centers,* are now installed as a matter of course in office buildings. They are equipped with microprocessors or computers that process the huge amount of data received to arrive at operational decisions intended to optimize system performance. As such, they result in a considerable savings in operating costs, in addition to the work force savings generated by their remote monitoring and control functions. These units normally perform these functions:

1. Remote monitoring, recording, logging.
2. Remote start–stop.
3. Remote controls, resets, changes.
4. Alarm functions.

The unit is programmed to respond to the data input of item 1 under normal circumstances and item 4 under unusual ones, with the responses of items 2 and 3 being automatic. Operator intervention is needed to change programming, to override automatic responses, to check alarms, and to periodically run through the system.

Energy management functions of such a unit would be:

1. Programmed lighting—reduce levels, turn off unused area lights.
2. Electric load control—as described in Section 14.14.
3. Optimized HVAC operation—based on preprogrammed procedures and continuous response control.

These supervisory control centers are very similar to the energy-management controllers described and illustrated in Section 14.14, with additional control, recording, display, measuring, and dis-

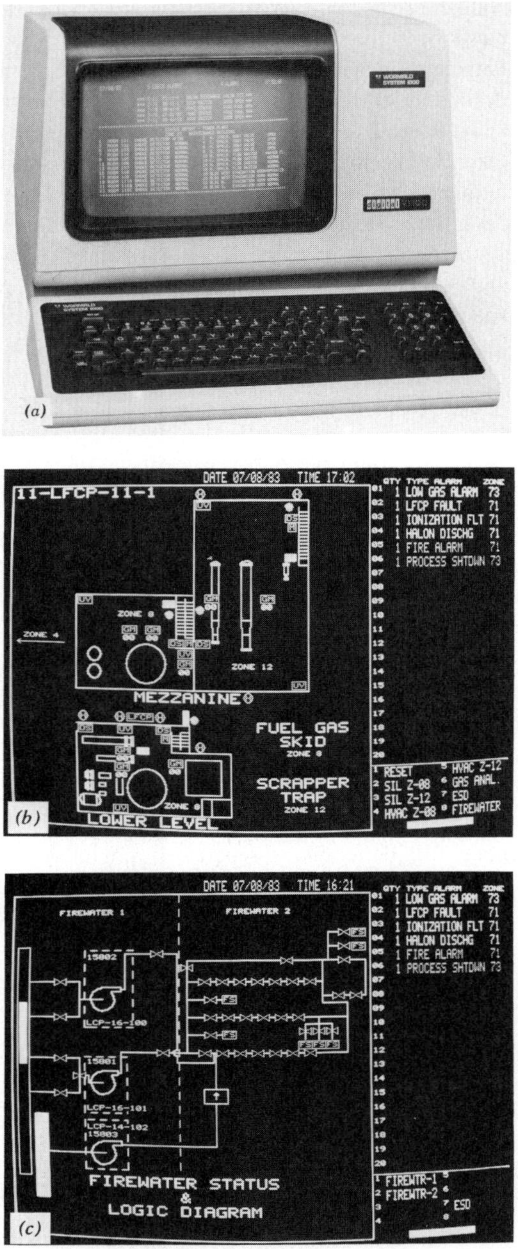

Fig. 22.11 (a) *The keyboard and display unit of a large, computerized central control system. Displayed on the screen is a status report for a variety of detectors, valves, switches, and other devices for a specific zone in a building. The unit can also display the location of devices, and their status, on a floor plan* (b) *or a schematic of a particular mechanical system* (c). *All remote devices can be monitored, logged, and controlled. (Photo and computer graphics courtesy of Wormald.)*

play functions as desired, as in Fig. 14.13. Figure 22.11 illustrates a large, computerized central alarm and control system applicable to any type of facility. Economic studies readily demonstrate the financial justification for the considerable first cost involved in acquiring a control unit both for new construction and for retrofitting an existing building.

Architecturally the unit requires good lighting and ventilation, extensive raceway space, but little area. Because systems are tailored to the building, no guidelines can be stated for space requirements.

INDUSTRIAL BUILDING SYSTEMS

22.22 General

All industrial facilities ranging from the taxpayer loft, housing a small hand assembly plant, to the immense steel manufacturing plant, require a variety of signal and alarm equipment. Although, as in the case of commercial structures, a detailed analysis of the equipment is out of place here, it behooves the building designer to know generally the function, operation, and availability of this type of equipment.

Fire alarm systems for industrial buildings are discussed in Section 13.20. Audible alarms in industrial facilities for any of the building security systems must be selected with the usually high ambient noise level in mind. See Figs. 22.12 and 22.13 for recommendations.

22.23 Industrial Building Security Systems

Among the most important signal functions in this type of facility is protection. Although the control point may be a common one, these systems are varied and perform separate functions.

(a) Door and Exit Controls. Outside doors and doors to restricted areas are supervised by electrified security hardware that triggers an alarm when a door is opened without authorization. The alarm mechanism may be concealed, or openly displayed, to act as a deterrent. Annunciation can be provided by a separate panel or tied into a central alarm panel.

(b) Personnel Entry Control. Several levels of security are available and can be applied in accordance with security needs. Beyond a simple lock, the first level is a card reader that grants entry to the card holder. The next level requires the encoding of a number simultaneously with presentation of a card, thus barring entry to unauthorized card holders (found or stolen cards). The third level involves more sophisticated identifcation procedures such as voice prints, or verification by an attendant. The attendant can compare card data and the caller's appearance with stored data displayed on an adjacent screen as in Fig. 22.14, providing a triple check and negating the effect of forged cards.

(c) Watchman's Tour Equipment. When tour equipment is used, it is frequently of the combination-alarm type. This type of station allows guards to call in, alerts them to waiting calls, permits a general alarm to be turned in by key operation, and is available as a manual fire alarm station. The station may be coded or noncoded as desired. The tour is normally timed so that any delay automatically sends an alarm and summons help.

(d) Intrusion Security Systems. A combination of the systems described in Section 22.2 is normally employed. Perimeter systems are normally arranged to sound an audible alarm and to energize all security lighting.

(e) Proximity Alarm. This system utilizes the electrical capacitive effect of a human body to trip an alarm. It is particularly useful in protecting a single item, for example, a file or safe within a room. Since the protective field extends only a few inches, normal use of the space can be maintained while protecting a single unit within the area.

All the above alarms, plus others not discussed, can be connected to a central monitoring console. If the console is unattended, it is normally auxiliarized, so as to alert security personnel. Usually, however, the alarm signal is transmitted to a central security station—normally a constantly attended guard post. In all cases, a per-

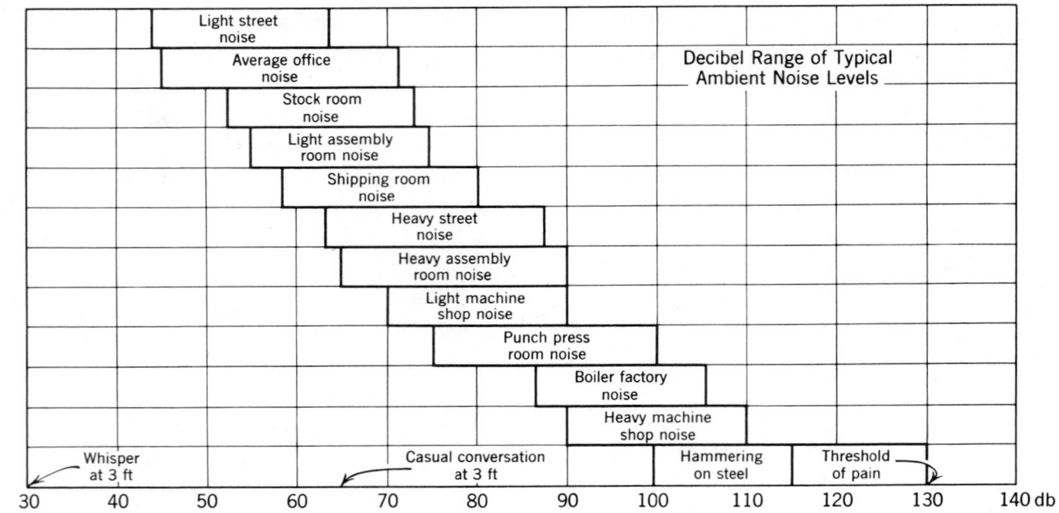

(a)

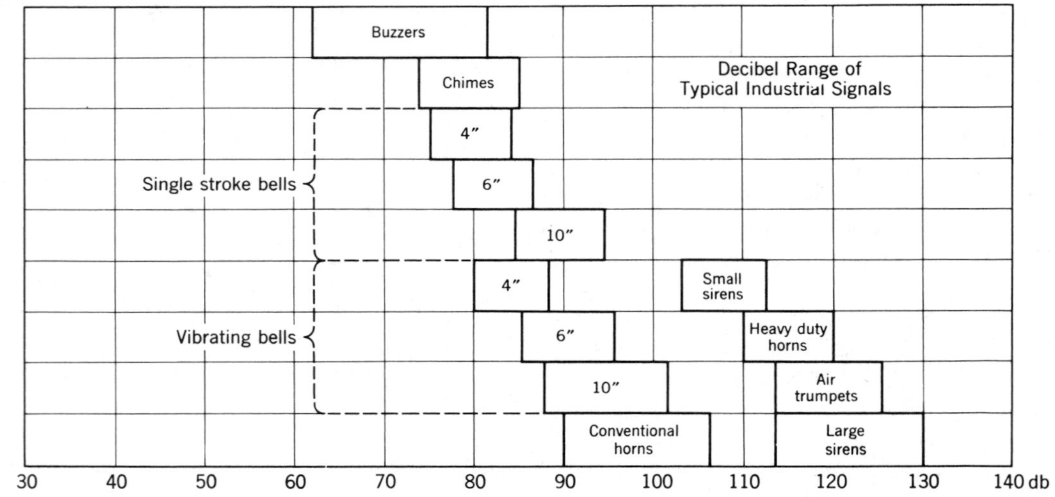

(b)

Fig. 22.12 (a) *Decibel range of typical industrial noise levels.* (b) *Decibel rating of typical industrial signals may be compared with ambient noise levels* (a) *to facilitate selection.*

manent printed record of every alarm should be made.

22.24 Industrial Building Paging Systems

All the time saving and efficiency potential that results from the proper use of the various signal and communication equipment available can be lost if there is no recipient for the information transmitted. Furthermore, sometimes a decision must be reached quickly to avoid costly delay, reruns, and so on. These factors combine to make rapid and accurate paging an extremely important function in a manufacturing operation, particularly of the larger variety.

Paging systems fall into two general categories and several subcategories: they are either visual, audible, or both, and are either common or selective. The simplest visual and audiovisual types

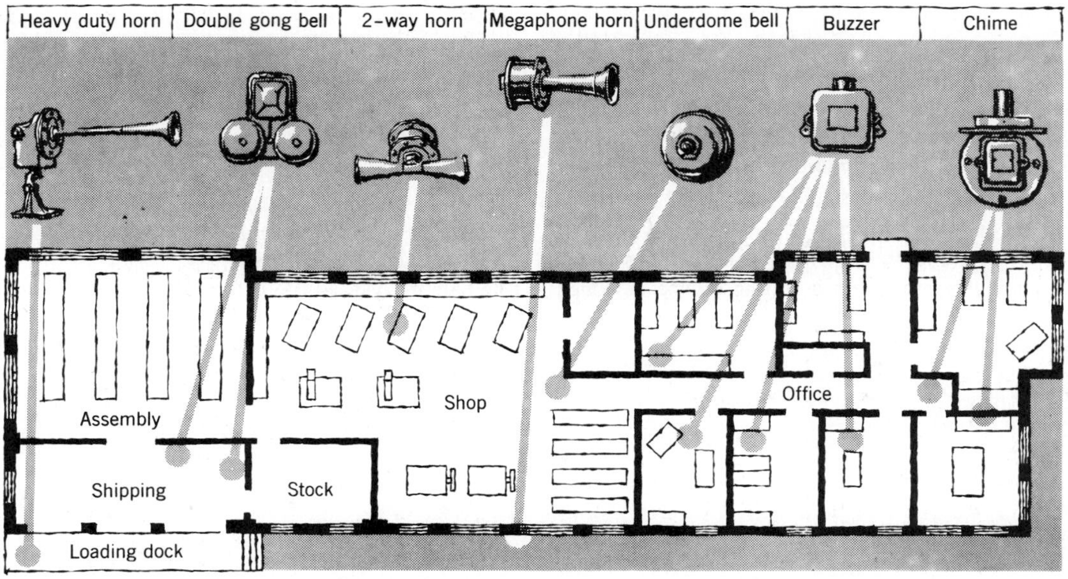

Fig. 22.13 *Plan of a small industrial facility showing suggested locations of typical signals.*

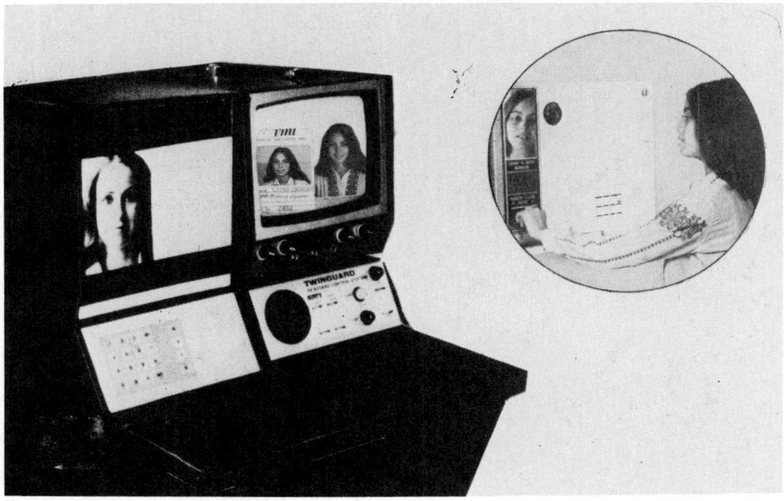

Fig. 22.14 *A person desiring entry inserts her identification card and is viewed by a camera at the entrance (inset). At a remote location, images of the subject and her ID card are displayed on a screen (right), while ID data from a memory bank are displayed for comparison (left). (Photos courtesy of Visual Methods, Inc.)*

comprise flashing lights, which may be combined with buzzers or bells, either or both of which are generally coded. Such systems are nonselective in that they impinge on the senses of all the building occupants—an obvious disadvantage.

More sophisticated systems utilize a small pocket device that is carried by each person likely to be paged—maintenance personnel, plant engineers, executives, and so forth. By means of either direct radio transmission or of electric fields induced by induction loops installed throughout the building, an individual pocket device can be alerted by a buzz. In some systems, the alerted person then listens to the message directly. On others, it is necessary for a person, once having been paged, to go to a phone and call in to a central paging desk to receive the message. Others utilize small hand-held, two-way radio transmitters with paging, to enable conversation between the page originator and the recipient.

In any of these systems, it is necessary to have a paging operator and a coding device at which the paging calls originate. Often, in a small factory, paging is handled by the regular phone operator.

References

Publications of the National Fire Protection Assocation, Boston: *Fire Protection Handbook,* 15th ed., 1981.

Life Safety Code, NFPA 101.

PART VIII

TRANSPORTATION

No facilities in a multistory structure are taken for granted more consistently than elevators and escalators, which are relied on to move people quickly and safely under all conditions, including emergencies. This movement should be quiet, trouble free, and economical. Furthermore, since vertical transportation accounts for 10 to 15% of a construction budget, somewhat less of the building area, and somewhat more of the operating cost, and is a determining factor in building shape, core layout, and lobby design, we see that elevator and escalator selection is a major task for the architectural designer.

Chapter 23, Passenger Elevators, introduces the subject with a description of the components of such facilities, including traction equipment, cars, safety devices, and systems of control and supervision. Special topics such as codes and standards, requirements for the handicapped, and solid-state controls are also covered. The discussion then turns to actual selection of units. Criteria are established for elevator performance, including interval, waiting time, handling capacity, average trip time, and round-trip time. Extensive car performance data (curves) are presented, which can be used with the other criteria to select the size, number, and speed of elevators to fulfill the requirements. The discussion continues with principles of elevator zoning and a comparison of single-zone and multizone systems. The chapter concludes with an analysis of spatial requirements for elevators, including shafts, lobbies, and core arrangement. Throughout, application of principles and data is demonstrated through the use of examples.

Chapter 24, Special Topics, addresses important passenger elevator considerations other than elevator selection. These include power and energy requirements, with emphasis on energy conservation, fire-emergency operation, emergency power requirements, and passenger security. The next part of the chapter is devoted to special cases: freight elevators, hydraulic elevators, nonvertical designs, residential elevators, and chair lifts, with discussion of equipment, capacities, selection, and economy, and comparison (where applicable) to conventional design. The chapter concludes with a section on material-handling devices: dumbwaiters, horizontal and vertical conveyors, pneumatic systems, container delivery systems, and automated self-propelled vehicles. This section answers the need to plan for material handling in commercial and institutional buildings rather than treating the subject as an afterthought, with resultant inefficiency, overloading of freight facilities, and misappropriation of passenger elevators.

Chapter 25, Moving Stairways and Walks, treats these subjects with a discussion similar to the coverage of passenger elevators. Descriptions of equipment components, capacities, sizes, speeds, and methods of selection are followed by special topics including lighting, fire protection, power and energy requirements, and budget estimating. Information on inclined walks (ramps) is also presented.

23

VERTICAL TRANSPORTATION: PASSENGER ELEVATORS

GENERAL INFORMATION

23.1 Introduction

Among the many decisions that must be reached by the designer of a multistory building, probably none is more important than the selection of the vertical transportation equipment, that is, the passenger, service, and freight elevators and the escalators. Not only do these items represent a major building expense, being in the case of a 25-story office building as much as 10% of the construction cost, but also the quality of elevator service is an important factor in a tenant's choice of space in competing buildings.

Although the final decision as to the type of equipment rests with the architect, the factors affecting it are so numerous that it behooves the building designer to consult with an elevator expert. Such consultation service is readily available from consultants in the field and to an extent from the major elevator and escalator manufacturers. It is the function of this chapter to familiarize the architect and engineer with the nature and application of vertical transportation equipment, to enable them to make preliminary design decisions before turning to the consultants.

23.2 Passenger Elevators

Our discussion is principally concerned with the general-purpose traction elevator. Hydraulic elevators are covered in Section 24.10.

Ideal performance of an elevator installation provides minimum waiting time for a car at any floor level, comfortable acceleration, rapid transportation, smooth and rapid retardation, automatic leveling at landings, and rapid loading and unloading at all stops. Furthermore, the system must provide quick and quiet power operation of doors, good visual floor indication both in the cars and at landings, easily operated car and landing call buttons (or other devices), smooth, quiet, and safe operation of all mechanical equipment for all conditions of loading, comfortable lighting, and generally pleasant car atmosphere.

In addition to the passenger-oriented service considerations above, the elevators have architectural aspects as well. The cars and shaftway doors must be treated in a manner consonant with the architectural unity of the building. More important, though, the shaftways are major space elements whose integration into the building is a prime factor in composition.

23.3 Codes and Standards

Perhaps more than any other item of construction, elevators are governed by strict installation codes. The ''bible'' of the industry is the American National Standards Institute (ANSI) code A 17.1, *Safety Code for Elevators, Dumbwaiters, Escalators and Moving Walks,* an up-to-date copy of which should be an integral part of every architect's and engineer's working library. This code has legal force in most parts of the United States. As with other items, some states and municipalities have their own elevator codes (Massachusetts, Wisconsin, Pennsylvania, New York City, Seattle, Boston, among others) that are generally based on, and more stringent than, the ANSI code.

In addition to the elevator code, other construction and installation codes have an influence on elevator work. Thus NFPA No. 101, *Life Safety Code,* states certain fire safety requirements, NFPA No. 70 (the National Electric Code) governs some of the electrical aspects of elevator construction, and state and local laws add a multitude of requirements and restrictions bearing on fire safety, emergency power, security regulations, and special accommodations for handicapped persons. Provisions for the handicapped are covered by ANSI A 117.1 (Barrier-free), a special industry code, and in most locations by local law. As with most large industries, the elevator industry is self-regulating and standardized. The National Elevator Industry, Inc. publishes standard elevator layouts for traction and hydraulic installations, as well as its own elevator standard for the handicapped, "Suggested Minimum Passenger Requirements for the Handicapped" (NEII, 600 Third Avenue, New York, NY 10016). Elevator consultants and elevator companies are normally knowledgeable as to all the codes and standards in force, but this does not relieve the architect-engineer of legal responsibility for the installation. Therefore, we strongly recommend that in the preliminary planning stage all pertinent regulations concerning vertical transportation be acquired and studied.

ELEVATOR EQUIPMENT

23.4 Arrangement of Principal Parts

The car, cables, elevator machine, control equipment, counterweights, hoistway, rails, penthouse, and pit make up the principal parts in any traction elevator installation. An idea of the functioning and orientation of these units of equipment can be obtained from an inspection of Fig. 23.1. Specific installations will vary somewhat; in newer systems microprocessor logic modules may be used in lieu of the older electromechanical, relay-operated control, and dispatch panels and solid-state equipment may be used in lieu of a motor-generator (m-g) set. On the whole, however, the basic functional components remain the same for all installations.

The cars, with their equipment for safety, convenience, comfort, and finish, are the only items with which the average passenger is familiar. Indeed, some of the building's prestige depends on proper design of the car. Essentially, the cab is a cage of some fire-resistant material supported on a structural frame, to the top member of which the cables are fastened. By means of guide shoes on the side members, the car is guided in its vertical travel in the shaft. The car is provided with safety doors, operating-control equipment, floor-level indicators, illumination, emergency exits, and ventilation. It is designed for long life, quiet operation, and low maintenance.

The cables (ropes) that are connected to the cross-head (top beam of the elevator) and carry the weight of the car and its live load are made of groups of traction steel wires especially designed for this application. Four to eight cables, depending on car speed and capacity, are placed in parallel. Although multiple ropes are used primarily to increase traction area, they also increase the elevator safety factor, since each rope is normally capable of supporting the entire load. The minimum factor of safety varies from 7.6 to 12 for passenger elevators and 6.6 to 11 for freight elevators. The cables from the top of the car pass over a motor-driven cylindrical sheave at the traction machine (grooved for the cables) and then downward to the counterweight.

The counterweight is made up of cut steel plates stacked in a frame that is attached to the opposite ends of the cables to which the car is fastened. It is guided in its travel up and down the shaft by two guide rails typically installed on the back wall of the shaft. (Obviously the counterweight travels in the reverse direction to the car.) See Fig. 23.1.

The weight of the counterweight is equal to the weight of the empty car plus 40% of the live load. It has several purposes: to provide adequate traction at the sheave for car lifting, to reduce the size of the traction machine, and to reduce power demand and energy cost. Approximately 75% of the energy expended in lifting a load is returned to the system by regeneration when the load is lowered. The "lost" energy appears as heat, primarily in the machine room. See Section 24.2 for a discussion of system energy requirements.

To compensate for hoist rope weight, which in high-rise elevators becomes an important factor, cables are attached to the bottom of the car and the counterweight, thus equalizing loads regard-

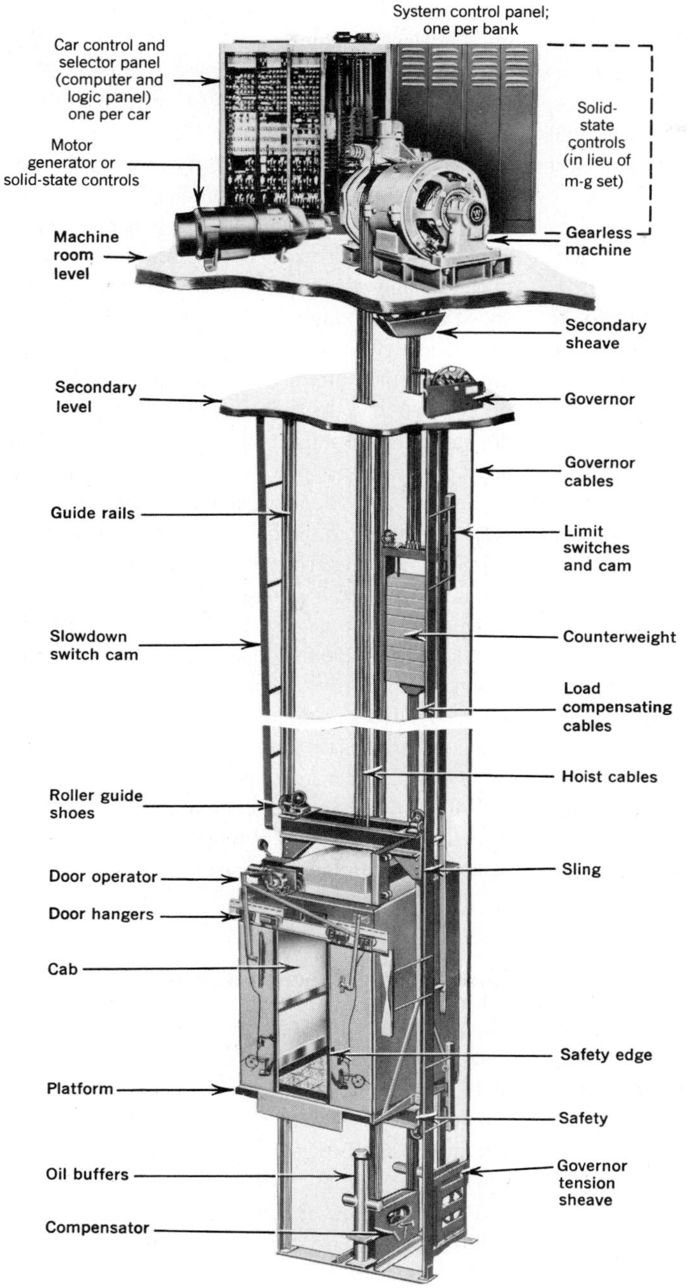

Fig. 23.1 Principal components of a typical gearless elevator installation. All new installations use solid-state controls in lieu of an m-g set. See sections 23.14 to 23.16.

less of the cab position. These cables can be seen in Fig. 23.1.

The elevator machine turns the sheave and lifts or lowers the car. It consists of a heavy structural frame on which are mounted the sheave and driving motor, the gears (if any), the brakes, the magnetic safety brake, and certain other auxiliaries. In most existing installations the elevator-driving motor receives its energy from a separate m-g set, which is in operation during the period that the particular elevator is available for handling traffic. This m-g set is properly considered a part of the elevator machine, although it may be located some distance from it. In new installations, solid-state power and control equipment replaces the traditional m-g set because of its inherent advantages, as discussed in Section 23.17. The governor that limits the car to safe speeds is mounted on or near the elevator machine.

The control equipment, in a general sense, is the combination of push buttons, contacts, electronic equipment, relays, solid-state switching, cams, and devices that are operated manually or automatically to initiate the door operation, starting, acceleration, retardation, leveling, and stopping of the car. These auxiliaries are interrelated in such a way that the major apparatus functions to produce the maximum of safety, comfort, and convenience. Electrical limit switches automatically stop the car from overrunning at the top and bottom of the hoistway. The well-known floor indicators, floor pilot lights, car panels, lobby control panel, call buttons at floor levels, floor-leveling devices, and up and down indicating lamps are all parts of the coordinated control equipment.

The shaft, or hoistway, is the vertical passageway for the car and counterweights. On its sidewalls are the car guide rails and certain mechanical and electrical auxiliaries of the control apparatus. At the bottom of the shaft are the car and counterweight buffers. At the top is the structural platform on which the elevator machine rests. The elevator machine room (which may be on one or two levels) is usually directly above the hoistway. It contains the traction motor, the m-g set or solid-state control that supplies energy to the elevator machine, the control board, and other control equipment. All machinery and control equipment are designed for quiet, vibration-free operation.

23.5 Gearless Traction Machines

A gearless traction machine consists of a d-c motor, the shaft of which is directly connected to the brake wheel and driving sheave. The elevator hoist ropes are placed around this sheave. The absence of gears means that the motor must run at the same relatively low speed as the driving sheave. Since it is not practical to build d-c motors for operation at very low speeds, this type of machine is utilized for medium- and high-speed elevators, that is, 400 fpm and above. The motors range from 20 to 375 hp. Gearless machines are generally utilized for passenger service, with usual car capacities of 2000 to 4000 lb, although specials of up to 10,000 lb, such as at the World Trade Center, have been built. Below 400 fpm, geared machines are used. In the range of 400 to 700 fpm, a 2:1 roping arrangement (see Section 23.7) is generally used. This reduces motor size and increases sheave speed, thus reducing the cost. Above 800 fpm, motor speed is high enough for 1:1 roping to be applied economically. At this writing maximum car speeds are 1600 fpm, although 2000-fpm systems and faster have already been developed and will undoubtedly be an important factor in the development of ever taller, practical, workable buildings.

The gearless traction machine is generally considered superior to the geared machine. It is more efficient, is quieter in operation, requires less maintenance, and has longer life. The decision as to whether these advantages are worth the additional cost involved is made only after a careful analysis. Generally, a gearless machine is chosen where rise is more than 250 ft and smooth, high-speed operation is desired. In the intermediate range of rise and speeds, that is, 150 to 250 ft height and 400 to 500 fpm, excellent equipment, both geared and gearless, is available. The choice depends on performance characteristics and cost.

23.6 Geared Traction Machines

The geared traction machine (Fig. 23.2) employs a worm and gear interposed between the driving motor and the hoisting sheave. The driving motor can therefore be a smaller, cheaper high-speed unit. The motor itself may be a-c in new installations, whereas the gearless unit is always d-c.

Prior to the development of a solid-state, thyristor-controlled a-c drive, a-c motors were applicable only to very-low-speed elevators (25–150 fpm), utilizing rheostatic control, which yields a relatively rough ride. Smooth operation and higher speeds required the use of an m-g set to drive a d-c motor. Today, however, many of the major manufacturers can supply an a-c, geared traction machine for installations up to 350 fpm with thyristor or variable-frequency, solid-state drive, and at least one manufacturer makes units to 500 fpm.

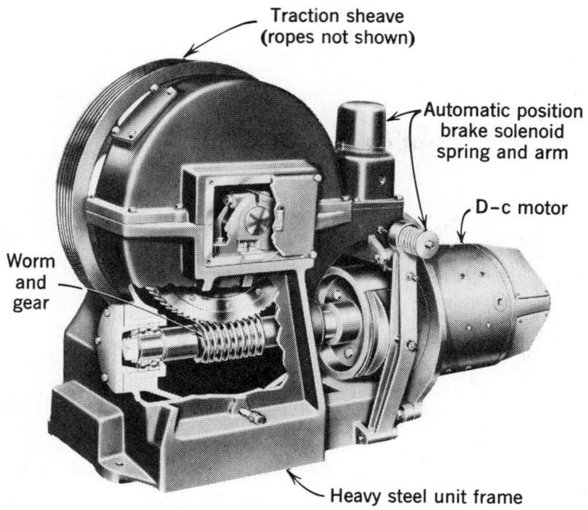

Fig. 23.2 *Cutaway of a typical d-c geared traction machine. Note the grooves for multiple ropes in the traction sheave. (Photo courtesy of Westinghouse Elevator Co.)*

These installations provide acceleration and speed control characteristics of the same high quality as d-c units, at lower cost and higher efficiency. Geared traction machines are used for some passenger service and most freight service. Table 23.1 summarizes the characteristics of presently available geared and gearless elevators.

23.7 Arrangement of Elevator Machines, Sheaves, and Ropes

The simplest method of arranging vertical travel of a car is to pass a rope over a sheave and counterbalance the weight of the car by a counterweight. Then, by rotating the sheave, the car will move up or down and require very little energy to do it. This is essentially the scheme that is used on a majority of high-speed passenger elevators, as illustrated in Fig. 23.3*a*.

When the four or more supporting ropes merely pass over the sheave T and connect directly to the counterweights, the lifting power is exerted by the sheave through the traction of the ropes in the parallel grooves on the sheave. This system is referred to as the single-wrap traction elevator machine. The function of sheave S is merely that of a guide pulley; it is called the deflector sheave.

In Fig. 23.3*b* the ropes from the car are first wrapped over the traction sheave T, then around the secondary or idler sheave S, once more around sheave T, and back over S to the counterweights. This arrangement is characteristic of the one-to-one, double-wrap traction machine. It provides greater traction than the single-wrap machine and is used in many automatic high-speed installations.

A 1:1 roping arrangement (Fig. 23.3*a, b,* and *d*) gives no mechanical advantage; that is, the drive must supply sufficient power to move the unbalanced load. The 2:1 roping (Fig. 23.3*c*) has a mechanical advantage of 2, which results in a high-speed, low-power, and, therefore, low-cost traction machine. This arrangement is used for a wide variety of installations varying from medium-speed (500–700 fpm) gearless passenger elevators to low-speed, heavy-duty freight units.

In types *a, b,* and *c* in Fig. 23.3, the elevator machines are located at the top of the hoistway.

TABLE 23.1 **Comparative Table of Geared and Gearless Elevators**

Type	*Max Rise (ft)*	*Speed (fpm)*	*Control*	*Life*	*Cost of Mainten-ance*	*Initial Cost*	*Smooth-ness*
Geared a-c	150	50–200	Rheostatic	30–40 years for gear and worm ↓	Medium ↓	Low	Poor
	300	150–500	Thyristor			Medium	Excellent
Geared d-c	175	50–400	Variable voltage			Medium	Fair
	250	350	Variable frequency			Medium	Excellent
Gearless a-c	Unlimited	400–2000	Solid state, variable voltage	Indefinite	High	High	Excellent

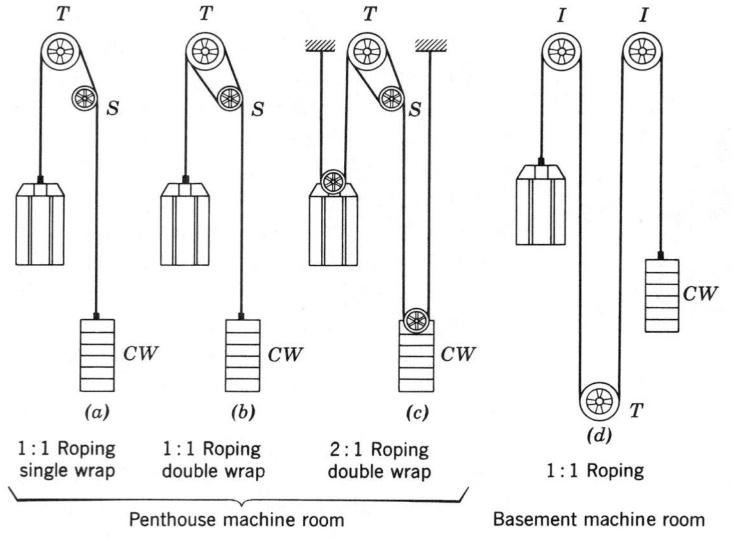

***Fig. 23.3** Elevator roping and sheave arrangement:* T = *traction sheave,* S = *secondary sheave,* I = *idler sheave,* CW = *counterweight.*

When the elevator machines are placed in the basement, a very different arrangement of cables and sheaves must be utilized to secure the same results (see Fig. 23.3*d*). Much more rope is required in such an arrangement, and consequently the problems of rope maintenance are increased. These systems, however, obviate the necessity for a tall penthouse; and where this is desirable for architectural or other reasons, a basement machine is used (see Fig. 23.3). This arrangement uses geared traction equipment, with speeds up to 400 fpm. All the illustrated ropings are applicable to the full range of car capacities up to 4000 lb.

23.8 Safety Devices

The main brake of an elevator is mounted directly on the shaft of the elevator machine (see Fig. 23.2). The elevator is first slowed by dynamic braking of the motor, and the brake then operates to clamp the brake drum, thus holding the car still at the floor.

A dual safety system, designed to stop an elevator car automatically before the car's speed becomes excessive, is normally used. The device that acts first is a centrifugal governor, which is independent of the other elevator machinery, and at normal speeds has no effect on the operation of the elevator. In the event of a limited overspeed, the governor will cut off the power to the d-c motor and set the brake. This usually stops the car, but should the speed still increase, the governor actuates the two safety rail clamps, which are mounted at the bottom of the car, one on each side. These devices clamp the guide rails by wedging action, bringing the car to a smooth stop (see Fig. 23.4).

Oil or spring buffers are usually placed in the elevator pit. Their purpose is not to stop a falling car but to bring it to a partially cushioned stop if it should overtravel the lower terminal. Electrical final-limit switches are located a few feet below and above the safe travel limits of the elevator car. If the car overtravels (down or up), these switches de-energize the traction motor and set the main brake. Safety arrangements under emergency condition of fire or power failure are discussed in Section 24.4.

23.9 Elevator Doors

The choice of car and hoistway door affects the speed and quality of elevator service considerably. Doors for passenger elevators are power operated and are synchronized with the leveling controls so that the doors are fully opened by the time a cab comes to a complete stop at the landing. The closing time, however, varies with the type of door and size of opening. For safety reasons, the kinetic

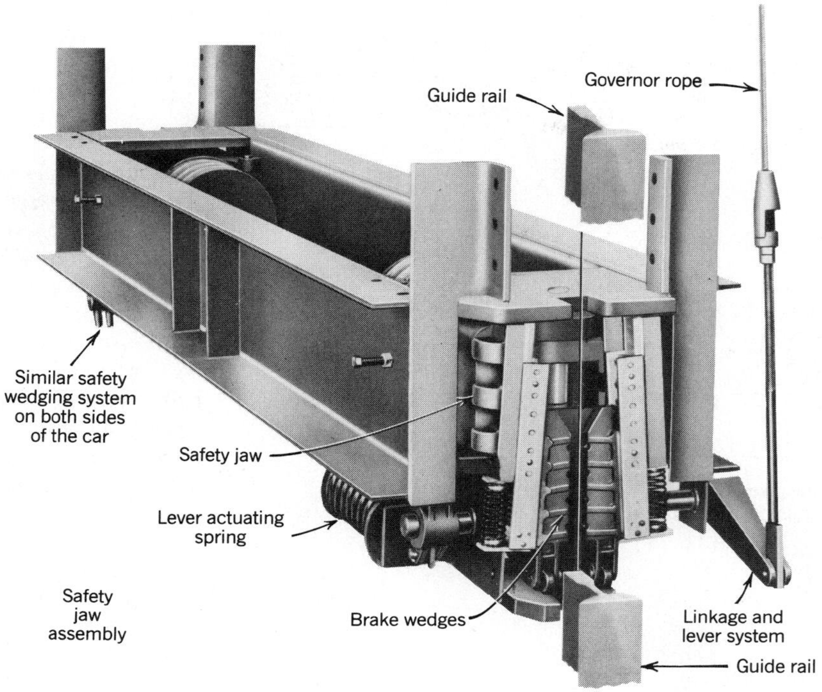

Fig. 23.4 Elevator safety devices. The governor trips, clamping the governor rope and releasing the safety jaws that exert a constant retarding force on the car rails, thus bringing the car to a gradual and safe stop. (Photo courtesy of Westinghouse Elevator Co.)

energy of an automatic door is limited to 7 ft-lb, and its closing pressure to 30 lb. To provide fastest closing within this energy limitation, a center-opening door is used. Also, to reduce passenger transfer time and avoid discomfort, a clear opening of 3 ft, 6 in. (106.7 cm) is used in most commercial installations, which permits simultaneous loading and unloading without undue passenger contact. (Some consultants feel that simultaneous passenger transfer is practical only with a 48-in. clear opening. See Strakosch, 1983.) See Fig. 23.5. When an opening narrower than 36 in. is used, loading will be delayed until unloading is complete; therefore, speed and quality of service will be markedly reduced. Such small doors are applied only in residential or small, light-traffic buildings. The available types are shown in Fig. 23.6.

A two-speed door design is used where space conditions dictate or where a wide opening is required. The two-speed nomenclature reflects the fact that the two halves of the door must travel at different speeds to complete their travel simultaneously (see Fig. 23.6*c*).

Installations can be equipped with an electronic sensing device that detects passengers in a wide area on the landing in front of the car door, rather than only directly in the door path. Such detection, often accompanied by an audible signal, causes the car door to remain open for a predetermined length of time, or a closing door to reverse. These devices are particularly useful in installations where passengers cannot approach the entrance or cannot enter the car quickly, for example, riders with baggage or holding children, people in wheelchairs, and employees moving bulky objects such as beds or carts in hospitals or wheeled objects in office and industrial facilities.

All automatic elevators, whether equipped with detection beams or not, are required by ANSI to have a safety edge device on the car doors that will cause the car and hoistway doors, which operate in synchronism, to reopen when the safety edge meets any obstruction.

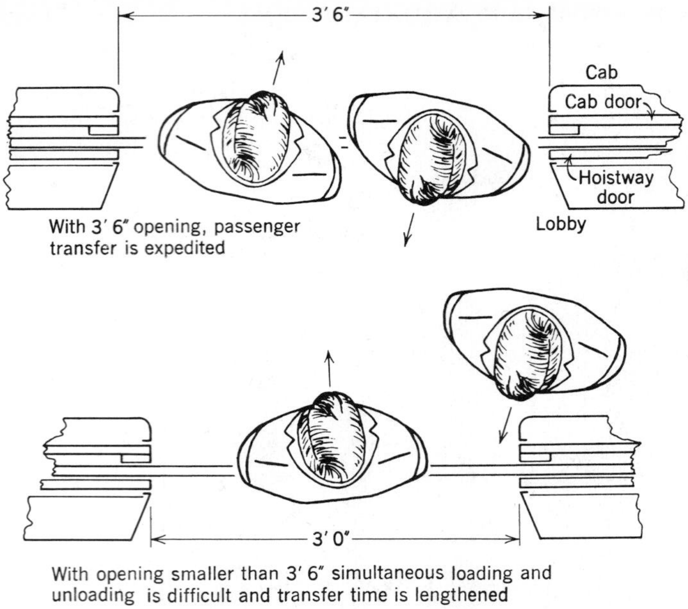

Fig. 23.5 *Transfer of passengers with door openings of different widths. With openings smaller than 3 ft, 6 in. (42 in.), simultaneous loading and unloading is difficult and transfer time is lengthened. With a 42-in. opening, large men or people with bulky outerwear may brush against each other in passing. For complete isolation of passengers, a 48-in. opening may be necessary.*

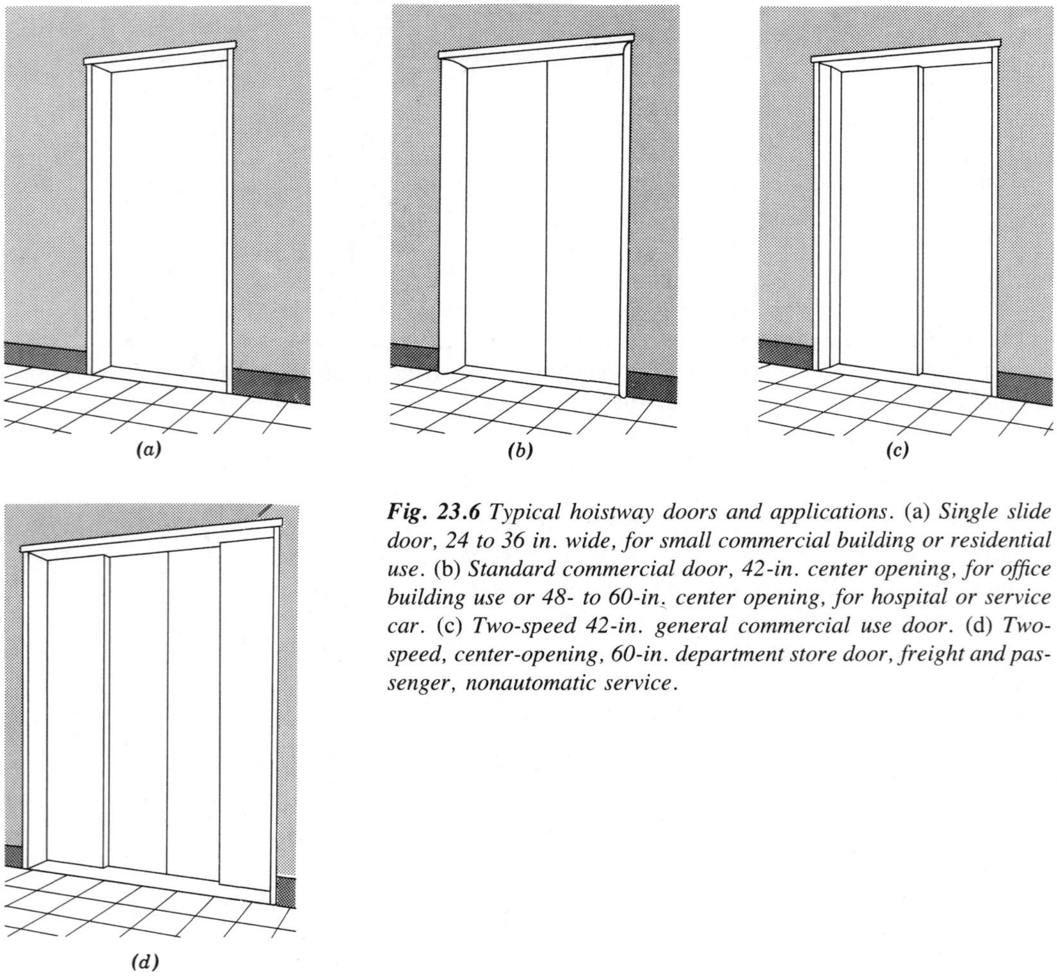

Fig. 23.6 *Typical hoistway doors and applications.* (a) *Single slide door, 24 to 36 in. wide, for small commercial building or residential use.* (b) *Standard commercial door, 42-in. center opening, for office building use or 48- to 60-in. center opening, for hospital or service car.* (c) *Two-speed 42-in. general commercial use door.* (d) *Two-speed, center-opening, 60-in. department store door, freight and passenger, nonautomatic service.*

23.10 Cabs and Signals

Possibly the only area in which the architect has a free hand in selection of equipment is in the decor of the cabs and the styling of hallway and cab signals. A normal elevator specification is a functional one that describes the intended operation of the equipment and normally includes an amount to cover optional decor of the cabs. The type and functioning of signal equipment is also specified, but finish and styling are optional. Cab interiors may be finished in wood paneling, plastic (Micarta or Formica), stainless steel, or almost any material desired. Floors may be tile, wood, or carpeting as selected. Illumination may be from ceiling fixtures, coves, or completely illuminated luminous ceiling, of standard or special design. For each bank of elevators, it is wise to furnish at least one set of wall mats, to protect wall finishes when cars are being used to move tenant furniture. This is especially important where no separate service car has been provided.

Car and hallway signals and lanterns should be designed to fulfill their basic functions, take proper consideration of the needs of the handicapped, and be consonant with the decor of the cabs and the corridors. (For a discussion of cab control panel, see Section 23.25.) The hall buttons (Fig. 23.7*a*) should indicate desired direction of travel and by visual means indicate that a call has been placed. The hall lantern (Fig. 23.7*b*) located at each car entrance must visually indicate the direction of travel of an arriving elevator and preferably its present location. This latter feature allows waiting passengers to move to the next arriving car in a bank and thereby speeds service. Hall stations can be equipped with special switches for fire, priority, and limited access service, as required (Fig. 23.7*c*).

Within the cab, indication of travel direction and present location can be accomplished either with separate fixtures or by indicators built into the car panel. Several manufacturers now market voice synthesizers built into the car panel that announce the floor, direction of travel, and any other desired message such as safety or emergency messages.

(a) Basic hall station

(b) Combination hall lantern and car position display

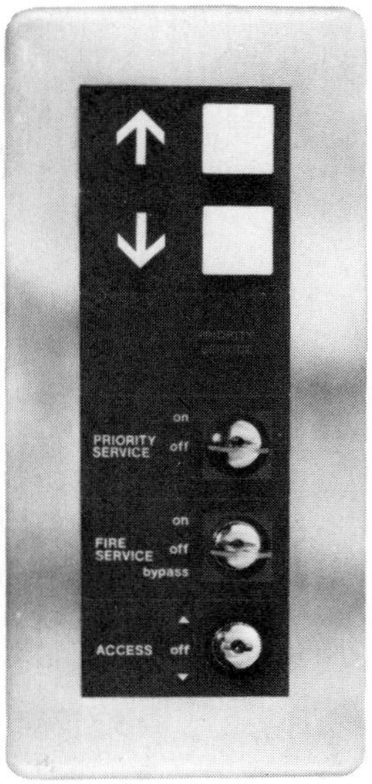

(c) Expanded hall station

Fig. 23.7 *Typical hall fixtures. Buttons* (a) *must show direction and have visual indication. Lantern* (b) *must show travel direction and preferably also present car location. Special hall stations* (c) *may incorporate priority, emergency, and security features. (Photos courtesy of Otis Elevator.)*

23.11 Requirements for the Handicapped

Although the recommendations of the NEII special standard "Suggested Minimum Passenger Elevator Requirements for the Handicapped" are not mandatory except where local building codes make them so, it is certainly advisable to adopt them in consonance with the modern concept of barrier-free buildings.

The basic physical limitations addressed are those of ambulation and sight. Thus to ease access for passengers in wheelchairs (or with walking aids), the standard requires excellent car leveling, 42-in. minimum clear door opening, delayed door closing, detection beams that will reopen the door *without contact* on sensing a passenger, inside car dimensions that will permit turning a wheelchair, buttons and emergency controls within easy reach, and appropriate car furnishings. For those with sight impairment, the standard calls for audible signals in addition to easily seen and recognized visual ones, both in the car and at landings, to indicate call registration, car approach, car landing, direction of travel, floor, car position, and so on. In this connection, a voice synthesizer is of invaluable assistance. Additionally, Braille plates adjacent to car floor buttons and large, easily recognizable symbols adjacent to passenger-controlled emergency controls are to be used.

Many of these ideas will make life easier for

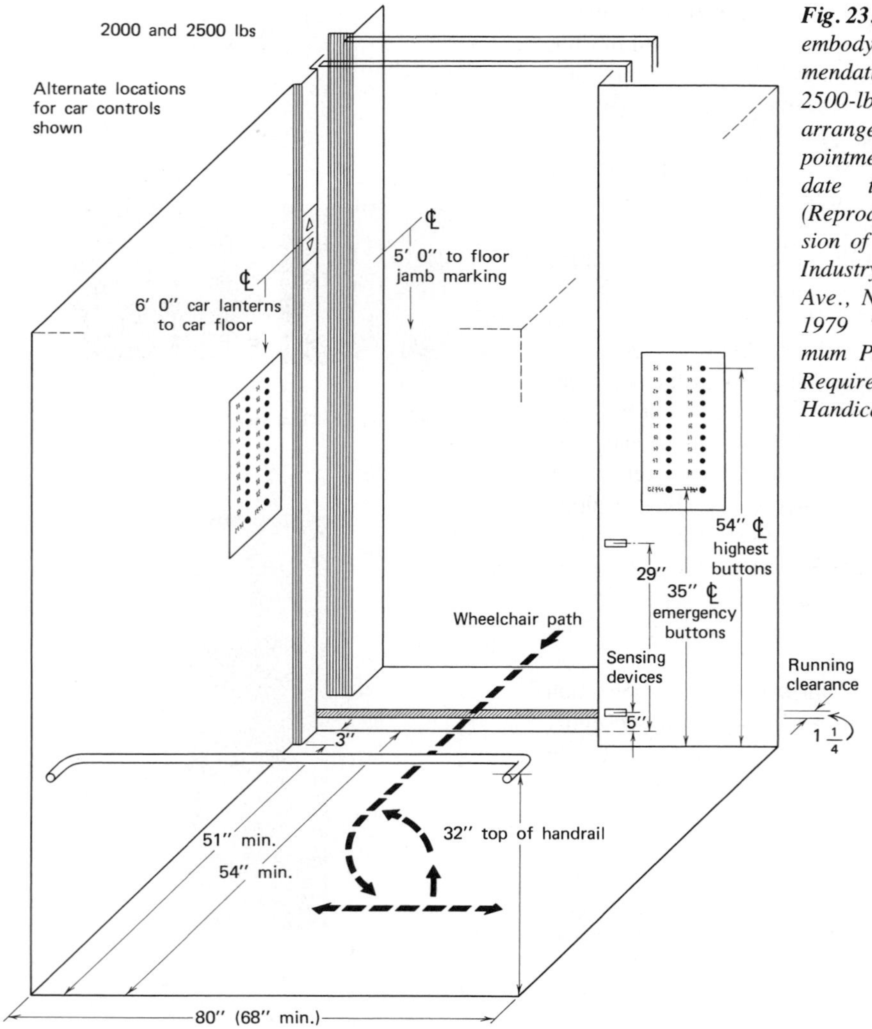

Fig. 23.8 Composite sketch embodying NEII recommendations for 2000- and 2500-lb elevator car sizes, arrangements, and appointments to accommodate the handicapped. (Reproduced with permission of National Elevator Industry, Inc., 600 Third Ave., New York, N.Y. © 1979 "Suggested Minimum Passenger Elevator Requirements for the Handicapped.")

all passengers with full faculties. A note of caution, however, is in order. Delayed door closing may increase travel time appreciably. In buildings with traffic peaks, it may be necessary to designate one or more specific elevators for use by the handicapped during these periods. Figure 23.8 is a composite sketch showing the application of some of these recommendations.

ELEVATOR CONTROL

23.12 Elevator Control Systems

The equipment, arrangement, and interconnections that determine the movement and performance of a single car are designated by us as elevator control. Subsumed under this heading is the equipment that controls travel, door operation, leveling, call buttons, and floor signals (lanterns). This control is separate and distinct from the control that governs the interaction of a group of cars—the latter is designated as operating or supervisory systems. To an extent, and particularly with a single car, the two designations overlap.

Elevator car acceleration and deceleration are accomplished by controlling the *speed* of the motor that drives the elevator traction machine. This speed control is performed in a number of ways, which are described below. The extent to which this motor speed control can be performed smoothly, rapidly, and accurately is the principal figure of merit of the control system. The other functions of the control system mentioned above are, in general, independent of the motor control system and are normally furnished to match the quality of the drive system. Thus a highly sophisticated leveling system would not usually be found with a rheostatic motor control system, and so forth.

23.13 Rheostatic Elevator Motor Control

In installations where economic considerations, low rise and low speeds (25–150 fpm), or the type of traffic do not justify the use of the more costly variable-voltage control or variable-frequency control, and where a hydraulic unit is not desirable, the traction machine may be driven by an a-c motor whose speed is controlled by rheostat (variable resistance; see Fig. 23.9*a*). Such elevators are usually operated by an operator's hand wheel or lever in the car rather than being automatic. The ride is not smooth. Use of a two-speed motor can improve this deficiency somewhat, although it remains much inferior to variable-voltage installations. Automatic leveling is available on some a-c, two-speed, rheostatically controlled elevator motors, operating at car speeds up to 100 fpm. New installations of this type are unusual for passenger service. Many existing ones are being converted to automatic because of the efficiency and better service available with solid-state electronic control.

23.14 Variable-Voltage Elevator Motor Control

Before the development of electronic motor control, the only practical, economical method of obtaining the precise motor speed control necessary for smooth elevator operation was to impress a variable d-c voltge on a d-c (traction) motor. Such fine control was not possible with an a-c motor without extremely expensive and complex drive equipment. All existing gearless installations, and most good-quality geared ones, use this system. The variable d-c voltage required to drive the d-c traction machine was traditionally obtained from an auxiliary m-g set comprising an a-c motor and a d-c generator (see Fig. 23.9*b*). This arrangement is known as a Ward–Leonard system or as unit multivoltage (umv) drive.

Today, the variable d-c voltage required to drive a gearless d-c traction motor can be obtained from a completely solid-state electronic control system, which replaces the traditional m-g set (see Fig. 23.9*c*). Since the traction motor itself operates equally well with either type of d-c supply, many retrofit installations utilize the original long-life gearless traction motors with new solid-state car and group supervisory control systems. The advantages of solid-state control are discussed in Section 23.17.

As noted in Section 23.6, and shown in Table 23.1, the development of solid-state controls has made possible the use of a-c motors in mid-rise (300-ft maximum) and medium-speed (500-fpm

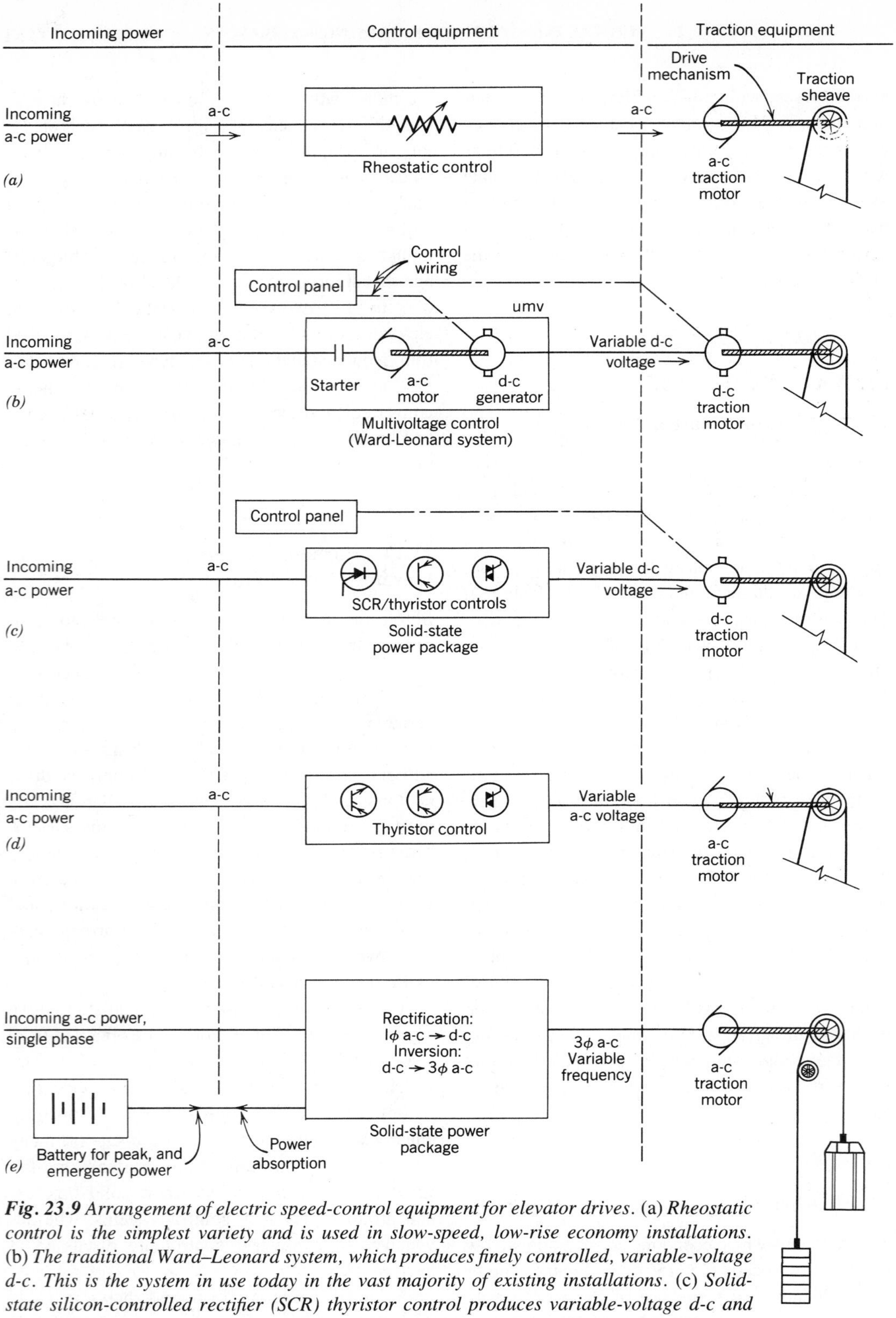

Fig. 23.9 *Arrangement of electric speed-control equipment for elevator drives.* (a) *Rheostatic control is the simplest variety and is used in slow-speed, low-rise economy installations.* (b) *The traditional Ward–Leonard system, which produces finely controlled, variable-voltage d-c. This is the system in use today in the vast majority of existing installations.* (c) *Solid-state silicon-controlled rectifier (SCR) thyristor control produces variable-voltage d-c and finds application in new and retrofit medium- and high-speed installations.* (d) *Thyristor-produced variable-voltage a-c is used in high-quality, mid-rise, medium-speed geared installations.* (e) *Variable-frequency a-c is energy efficient, operates from a light-duty electric service, and is applicable to mid-rise installations.*

maximum) applications. These controls supply variable-voltage (chopped-wave) a-c to geared a-c traction motors with an accuracy and flexibility that rivals umv d-c motor control. The control system is variously known as servo, or feedback control, via a thyristor bank. A single-line representation of this type of control is given in Fig. 23.9*d*.

23.15 Variable-Frequency, Alternating-Current Control

The most recent advance in the application of electronic controls to elevator motors is a system that supplies variable *frequency* to an a-c motor (see Fig. 23.9*e* and Table 23.1). This technique is ideal for a-c motors because it is highly energy efficient, is capable of extreme accuracy, and lends itself well to central supervisory control. The Otis Elevator Co. design for this system, known as MRVF (medium-rise variable frequency), utilizes a battery for peak power, emergency power, and regenerative power absorption, resulting in an excellent energy efficiency characteristic (see Section 24.2). It also has the distinct advantage of utilizing a light-duty, *single-phase* a-c line and by means of rectification and inversion, supplies three-phase (3ϕ) a-c power to the traction motor.

23.16 Elevator Operation

When considering operation, whether the variable voltage is supplied by an m-g set or a solid-state installation is immaterial. Differences in other areas are discussed below. The following is a description of conventional umv elevator operation.

Assuming the system to be energized and at rest, registration of a call from a station in the lobby or an upper floor corridor will activate the system. The particular elevator that will answer the call is selected by the supervisory system. The m-g set related to that elevator car is started and comes up to a speed slowly. In the case of solid-state control, voltage is available immediately, and the m-g set starting time is eliminated. This difference between m-g set and solid-state control disappears in the more usual case of an m-g set that is already running. When the d-c generator has developed full voltage, the elevator control is cut in and the car is ready for operation. Automatic signals initiate the automatic sequence for accelerating. Various relay devices in the car control panel release the brake and energize the elevator motor, and the elevator accelerates to its rated speed. Reverse operations are initiated when decelerating and finally stopping (landing) the car. When the car stops, the brake holds the sheave and elevator stationary.

The motion of a single car is determined by the action of three principal items of equipment: the car controller, the motor controls, and the system supervisory equipment. The action of this latter equipment is discussed in Sections 23.18 to 23.25.

The function of the car controller panel is generally to duplicate the action of the elevators and to anticipate hall calls. This is accomplished in older controllers by the action of a sort of miniature elevator system that contains sets of contacts over which small sliding brushes move in synchronism with the corresponding car movements in the hoistway. New systems utilize an electronic or optical position transducer, which is at least as accurate as its mechanical forerunner, but does not require periodic alignment to maintain accuracy.

This information, that is, the car position and the waiting calls, is then fed to the (supervisory or relay) panel, which initiates the proper procedures. In the case of group (system) control, the system panel operates to control the individual car control panel. In the final analysis, however, the control panel controls the direction of car travel, starting, running, leveling, and stopping, and operation of doors. In the newer systems, solid-state switching devices may be substituted for conventional electrical relays, but the action remains essentially the same.

23.17 A Comparison: Motor-Generator Sets and Solid-State Control

Solid-state equipment will in time undoubtedly replace the traditional m-g set and the associated electrical and electromechanical controls, because of inherent advantages. The principal drawback at present is the higher initial cost of solid-state equipment. However, when considering life-cycle cost, the relative amounts may be reversed.

(a) Energy Considerations. One of the principal advantages of the solid-state elevator control lies in the energy savings possible. A considerable portion of the energy consumed by an elevator system is wasted when it is not used to perform the fundamental traction work of the elevator system. Sources of energy waste are:

1. The m-g set idling while the elevator is at standstill—during loading, unloading, and waiting periods. Studies indicate this to be approximately 50% of the total time the m-g set is operating.
2. The m-g set startup after being automatically shut down when a car is not in use.
3. The essential inefficiency of an m-g set, whose overall losses run approximately 20% of input power.
4. Forced-ventilation equipment in machine rooms, necessitated by the high heat losses of the m-g sets. (See Sections 24.1 and 24.2.)

Since the m-g set is eliminated with solid-state (thyristor) control, all m-g set losses are eliminated and ventilation requirements are sharply reduced. Depending on the particular elevator design, electrical energy costs are reduced by 10 to 25%. The actual cost impact can be seen in Section 23.17*e* below, "Operating Costs."

(b) Physical Considerations. A solid-state control panel occupies approximately the same space as a standard m-g set control board, including the motor starter panel. The car controller (one per car) is considerably smaller in its solid-state configuration (see Table 23.2) than its electromechanical counterpart, as is the group supervisory panel (one per bank).

Elimination of the m-g set plus smaller control equipment thus reduces the required machine room size, with concomitant construction costs savings and increase in usable space. Also, the absence of m-g sets reduces the machine room floor loading, yielding possible savings in construction, and in some instances permits elimination of the secondary machine room level (see Fig. 23.1). Comparative layouts for the same elevator installation are shown in Fig. 23.10, graphically illustrating the space economy involved.

(c) Car Operation. Appreciably better passenger service is obtainable when using electronic equipment for a number of reasons:

1. Better group control. See Section 23.22.
2. Better car control, due to faster response of components and finer control, particularly of jerk (rate of change of acceleration). This is the principal factor in passenger discomfort.

The exact degree of improvement possible when changing to solid state equipment in a retrofit job depends on the quality of both the old and the new systems. It is greatest when a microcomputer group control system replaces relay-logic terminal dispatch system.

TABLE 23.2 **Comparative[a] Size and Weights of Elevator Control Equipment: Conventional[b] versus Solid State**

	Gearless Traction Motor: 500 to 1200 fpm					
	System Processor (Group Controller)		*Car Controller and Selector*		*Motor Drive*	
	Conventional	*Solid State*	*Conventional*	*Solid State*	*Conventional*	*Solid State*
Width	70 in.	24 in.	70 in.	43 in.	30–49 in.	48 in.
Height	100 in.	100 in.	84–100 in.	84 in.	63–84 in.	84 in.
Depth	24 in.	24 in.	24 in.	24 in.	24 in.	24 in.
Weight	1500 lb	600 lb	1200–1400 lb	1000 lb	500–1200 lb	1600 lb

[a]Dimensions for other manufacturers will vary from these.

[b]M-g set, required in the conventional system for each car, measures 65–70 in. long by 15–24 in. wide and weighs 1200–4000 lb.

Source: Courtesy of Westinghouse Elevator Co.

TRANSPORTATION

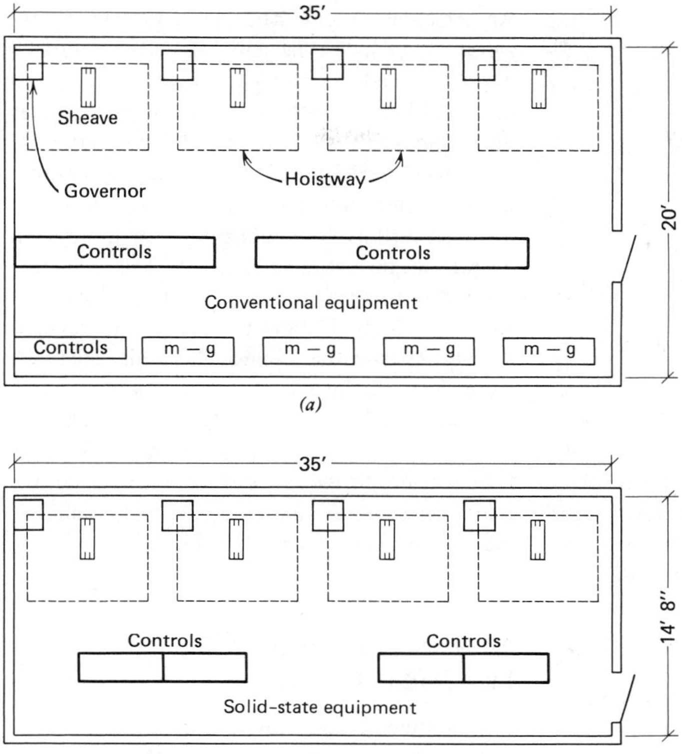

Fig. 23.10 *Space savings possible with solid-state control* (b) *over conventional controls* (a) *for a bank of four 3500-lb, 800-fpm, gearless traction elevators. Actual dimensions vary from one manufacturer to another.*

(d) Other Factors. Elimination of the m-g set and use of electronic control have other desirable effects:

1. Improvement of the facility power factor, since the m-g set power factor is low due to starting and no-load operation. This may result in a reduction in the facility's electric energy bill.
2. Elimination of current and kW demand peaks, which are characteristics of motor starting operation. This will result in a reduction in electric bills. The overall reduction, including energy and demand changes, can reach 35%.
3. Reduction in noise and vibration in the machine room, which frequently is transmitted to other parts of the building.
4. Reduction in maintenance costs, because the periodic maintenance required for continuously rotating and sliding equipment is eliminated. Solid-state equipment maintenance is minimal.
5. Possible reduction in the size of the emergency generator required to operate an elevator, with concomitant cost reduction. This reduction is possible since the very high current inrush taken on m-g starting is eliminated. The unit must, however, be large enough to handle the regenerative braking current produced by a full car traveling down.

(e) Operating Costs. A brief life-cycle cost analysis of a five-car elevator bank can serve as an example of the cost figures involved. Input data to the study are:

Five 3500-lb, 600-fpm, gearless cars, group supervisory control operation: 10 h a day, 22 days per month.

Overall system losses at 30%.

25-year equipment life.

Power cost $0.08 per kWh, increasing 8% per year.

Maintenance cost $400 per car month for m-g set equipment, escalating 6% per year.

Note: In major cities this cost will run nearer to $550 per month.

Cost of capital—11%.

Analysis. The two major operating costs that can be substantially reduced by the use of solid-state control are energy costs and maintenance costs. (For the sake of simplicity, we neglect kW demand costs.) Overall savings possible are the sum of these two factors.

1. *Energy Cost*. From Section 24.2, the average energy monthly cost of this *bank* is $630, whereas with solid-state equipment it is $360. This is a yearly difference of $3240. The present value of the electric energy savings over 25 years, assuming 8% annual escalation, can be calculated (see Appendix E).

$$\textit{present value} = 18.15 \times \$3240 = \$58{,}800$$

2. *Maintenance Cost*. Assume that only a 15% economy in maintenance costs is possible. We then have

$$\text{annual savings per car} = \frac{\$400}{\text{month}} \times 15\% \times 12 \text{ months} = \$720$$

$$\text{annual savings per 5-car bank} = 5 \times \$720 = \$3600$$

Present value (*PV*) of life-cycle maintenance cost *savings* is

$$PV = 14.5 \times \$3600 = \$52{,}209$$

The total (rounded) present value of combined savings for energy and maintenance is therefore $111,000. To reach a decision, compare this amount to the increased first cost of the solid-state equipment.

SYSTEMS OF ELEVATOR OPERATION AND SUPERVISION

23.18 System Control Requirements

An operating system is that control, as stated above, that governs the automatic (operatorless) response of a single car or a group of cars to calls for service. An effective system therefore must take cognizance of all hall calls and car calls, car travel directions, and car positions—in relation to each other and in relation to the call requirements, plus the relationships between up and down traffic and the trends of traffic. This last is required in order that the system anticipate demand rather than react to it, since only by anticipation can the system operate at maximum efficiency. The operating system considers all the above data and dispatches cars accordingly. Obviously, since traffic and calls are never static, the control that will satisfy all these demands in a large elevator system is an extremely sophisticated one. Without such control, however, large elevator systems will render poor service, to the dissatisfaction of owner and rider. On small systems the operating control is much simpler, as described below.

23.19 Single Automatic Push-Button Control

This system is the simplest of the passenger-operated automatic control schemes. It handles only one call at a time, providing an uninterrupted trip for each call. A single corridor button at each level can register a call only when the car is not in motion. To indicate availability of the car, an "in use" light is placed over the hall call button. Calls can be placed when the light is off. This control scheme is applicable only to a short-rise, inactive elevator, that is, one making five or fewer trips per hour. Such elevators are found in small apartment houses, residences, small professional buildings, and industrial buildings using freight elevators.

23.20 Collective Control

This system is no longer used in new installations in the United States, although it is common in other countries. With only a slight increase in traffic beyond very light, single automatic push-button control becomes unsatisfactory, because no call storage provision is made and waiting periods can become extremely long, particularly with slow cars and moderately high rise. A control system therefore evolved that provides a single call button at each landing. The elevator stops at each floor that has registered a call to ''collect'' the waiting passenger. Hence the ''collective'' nomenclature. The car cannot differentiate between up and down calls since only a single button is provided. A car direction light is frequently placed adjacent to the call button. This will tell the prospective passengers whether the car that stops in response to their call is going in the right direction. As can readily be seen, this type of operating system is acceptable only for light residential or industrial service.

23.21 Selective Collective Operation

This type of collective operation is ''selective'' in that it is arranged to collect all waiting ''up'' calls on the trip up and all hall ''down'' calls on the trip down. The control system ''stores'' all calls until they are answered, and automatically reverses the direction of travel at the highest and lowest calls. When all calls have been cleared, the car will remain at the floor of its last stop awaiting the next call, and its m-g set will stop after several minutes. Pressure on any ''up'' or ''down'' button at any landing, or on any floor button in the car, will start the m-g set and set the car into operation.

Selective collective control is standard in locations where service requirements are moderate such as in apartment houses, small offices, and hospitals. Since these locations often require more than one car, a control scheme is available for groups of one to three cars. It automatically assigns each landing call to the car best situated to answer it, prevents more than one car from answering a call, allows one car to be detached for freight duty while others serve passengers, and automatically parks all but one car at the ground floor, that car acting as a free car until service calls require the use of the parked cars.

Although selective collective control is standard for most residential and other light-to-moderate service requirement buildings, its inherent characteristics frequently result in long waiting periods for elevators. These characteristics are:

1. Highest call reversal.
2. Shutdown of the m-g set (where used) after calls have been answered.
3. Strong tendency toward bunching of cars.

The last characteristic is particularly annoying in groups of three cars. Frequently a passenger will arrive at a landing to find that all three cars have just passed, going in the same direction. The result is that service is only slightly better than that which would be rendered by a single car, except that load (handling) capacity is greater. For this reason, operation of more than two cars with this system is not recommended and operation of more than three cars is not feasible.

Although collective control furnishes adequate elevator service where requirements are light to moderate, it is still basically a signal-controlled system that weighs all calls equally and takes no cognizance of traffic patterns.

To overcome this shortcoming, elevator engineers over the years 1950 to 1980 developed numerous supervisory control systems, among which were multiple zoning arrangements and programmed traffic patterns. The former divided the building into zones and dispatched elevators within these zones, thus reducing waiting time. The latter recognized that most buildings exhibit repetitive traffic patterns—such as morning up-peak, evening down-peak, lunch hour up and down peaks, and so on—and arranged car travel accordingly to give optimum service. Choice of the mode of operation was generally left to the lobby attendant, whose judgment was relied on to select the mode appropriate to the need. This was frequently less than satisfactory and resulted in the operational mode being selected on a clock-time basis. The basic problem with all these older preprogrammed operating systems is that they can react in only a few modes because of physical limitations. The control system consists of relays, contacts, and hardwiring. As the system becomes more complex, the number of these devices increases enormously and places a practical limit on the sophistication of the system. This limitation disappeared with the advent of the computerized, microprocessor-controlled operating system.

23.22 Computerized System Control

The most advanced type of control system continuously monitors demand and controls car motions in response to demand only; that is, it analyzes all the possibilities and answers each call in optimum fashion. The term "optimum" depends, of course, on the system design strategy, and this varies between manufacturers. Such a system is possible only with the aid of a central computer combined with programmable microprocessor-controlled peripherals, since the amount of data that must be collected and processed is enormous. The factors that must be considered in determining which car will answer which call are:

1. Condition of each car, including its load, registered calls, priority status, if any, and status of its m-g set, if any.
2. Distance of each car from a registered hall call.
3. Car unloading calls ahead of and behind each hall call.
4. Special conditions, including delays, priority calls, and timing of registered lobby calls.

On the basis of all these data, the computer calculates not only which car is in the best position to answer a call but also travel time for the hall call *and* the calls already registered in that car. The fundamental criterion is optimum service, that is, minimum total trip time including waiting. The computer also analyzes the traffic trends many times each second and reevaluates the importance of all factors in order to anticipate service requirements.

The actual program logic for even a small system is beyond the scope of this book, but its guidelines are not.

1. The system should be initially programmed for the anticipated service needs.
2. It should be readily reprogrammable to meet changes in building needs *at nominal cost,* and with minimum shutdown time. It must be possible to detach cars from the system during testing, reprogramming, and routine maintenance to provide minimal service even during off-hours.
3. In the case of retrofit work, the system should be adaptable not only to existing traction equipment but also to anticipated modernization, again without extensive outlay. When an existing group operating system is modernized by installation of a microcomputer group controller, it is almost always also necessary to change the car controls to programmable solid-state equipment.
4. The system should be self-diagnostic so that maintenance is simplified and "bugs" are easily located. In this area, a printout of performance in response to actual and test signals and calls can be of considerable assistance, and is recommended.

Proper operation of the system should result in:

1. The closest car responding to a call.
2. All floors getting equal service, including the basement if desired.
3. Reduced waiting time in the halls by preferential treatment to bypassed calls.
4. Proper handling of multifloor tenants in office buildings, to permit efficient interfloor traffic without going to the lobby.
5. Reduced-length floor stop, by proper sensing and door control.
6. Protection against nuisance calls and abnormal door operation.
7. Appropriate action in emergencies such as power failure to one or more cars or abnormal car operation.
8. Minimum electric demand and energy costs. In the case of rehabilitation work, this should be readily demonstrable.

23.23 Rehabilitation Work: Performance Prediction

The primary reason for rehabilitating an existing elevator system is to improve its operating performance. As mentioned above, the traction machinery has extremely long life, particularly in the gearless configuration. Thus the usual rehabilitation project consists of replacing the m-g set, car control panel, and group controller with solid-state, microprocessor-controlled programmable equipment, while retaining the original traction equipment. The car door operator, which controls door action, is usually also replaced, since door opening and closing characteristics are an important factor in overall trip time, and therefore system performance.

Using large computer installations, major manufacturers have developed elevator system simulators. These devices enable architects and owners to input proposed (or existing) building data, and to receive visual and printed readout on the performance of proposed systems (see Fig. 23.11). Since such programs are interactive, the operator can change the input data and the characteristics of the proposed equipment until the de-

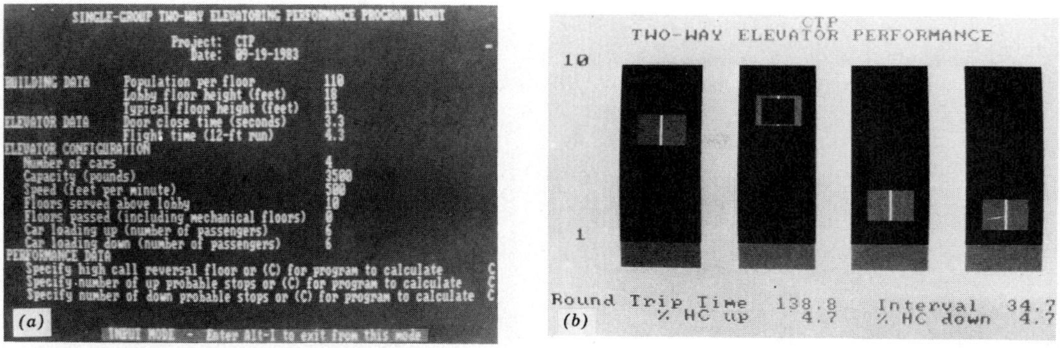

Fig. 23.11 Elevator system simulation. A computerized mathematic model of an automated elevator system permits the operator to enter building data (a) *and vary elevator control strategy. The results, displayed graphically* (b) *show the system response. (Courtesy of Otis Elevator.)*

sired performance level is reached. These devices are particularly useful in modernization work, since the owner will see in advance the operation of a proposed system *in his building* and can make the required decisions on this basis.

23.24 Lobby Control Panel

A lobby control panel (also called a traffic director's panel or an information control center) is furnished with each bank of elevators and is usually mounted in a readily accessible and easily visible position in the building lobby.

The contents of the panel are generally these:

1. An information module that gives information on corridor calls, special status of cars, car movement direction, and, optionally, location of cars. Information on location is optional because some feel that it leads to the undesirable "tracking" of cars by persons waiting in the lobby, with consequent group movement, crowding, and impatience. The information can be displayed in tabular form (Fig. 23.12*a*) or graphically (Fig. 23.12*b*).

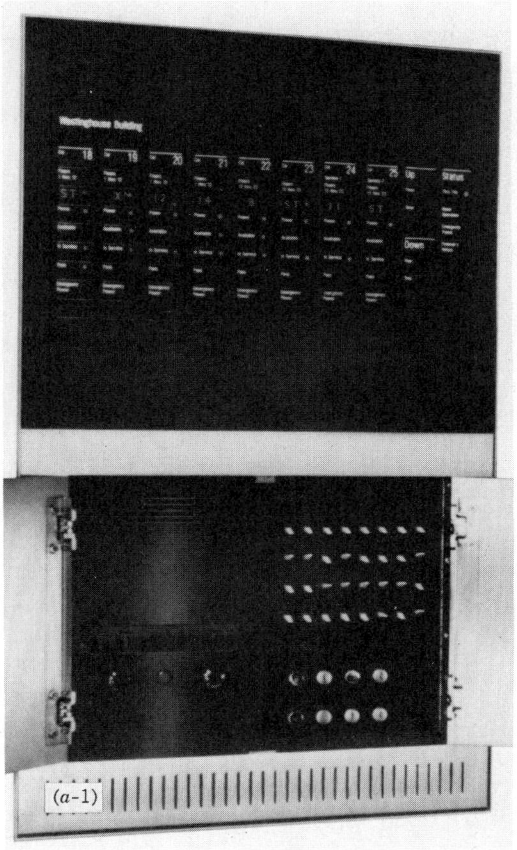

Fig. 23.12 (a-*1*) *Typical traffic director's station for a modern group supervisory system (Westinghouse Mark V.) The upper section shows, in tabular form, status information on all cars in the bank: position, direction of travel, power, car availability, whether in service, passing status, and availability of emergency power. In addition, data on upper floor hall calls are shown. This section also contains the fireman's lobby return switch, which can recall all cars to the lobby and can restore normal service on termination of the emergency (or false alarm). If desired, the same tabular data can be presented to another location. The lower section of the traffic director's panel contains power and control switches, plus communication equipment. These controls are normally behind locked cabinet doors.*

TRANSPORTATION

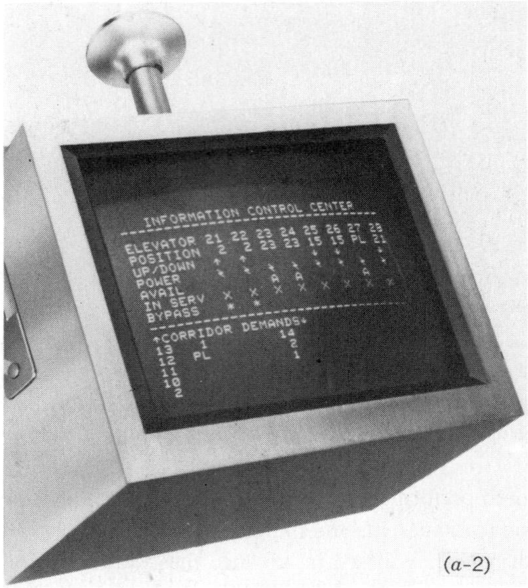

Fig. 23.12 (continued) (a-2) *A ceiling-mounted location display monitor.* (b) *The lobby monitor screen can have a graphic display showing car status and hall call data. This unit, the Westinghouse Micro-Scan, is used in system modernization work. (Photos and diagram courtesy of Westinghouse Elevator Co.)*

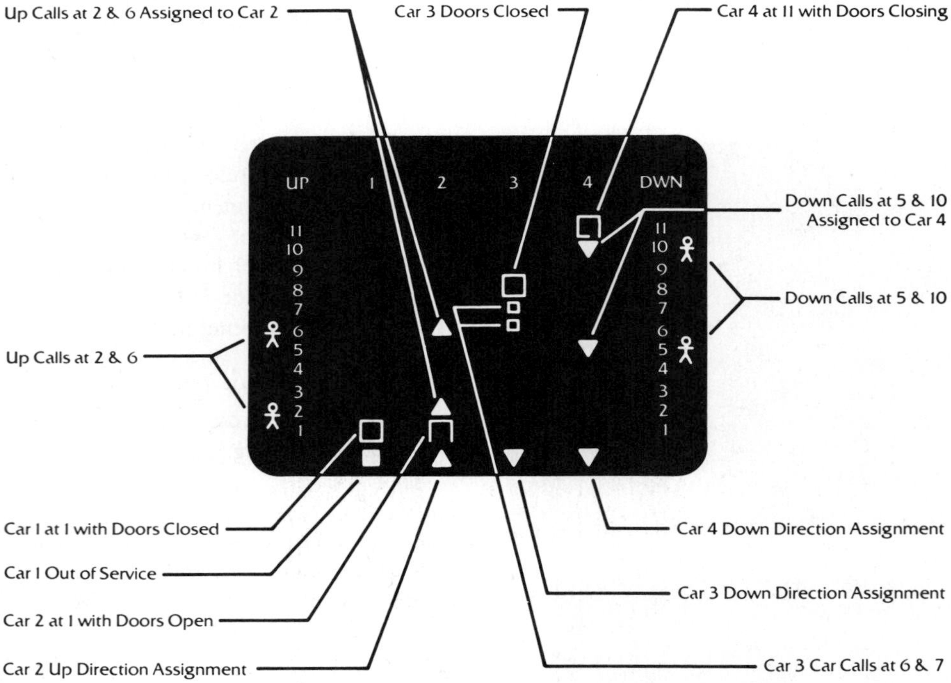

(b)

2. A communications compartment that permits two-way communication with cars and other selected locations.
3. A switch compartment, behind locked doors, containing a "power-on" and "in-service" switch for each car in the bank, plus special switches.

Once the system has been activated by operating the switches in the locked compartment, no manual intervention is necessary to keep the system in operation.

The panel does provide for manual intervention to permit special types of operation. Thus a car can be

1. Arranged to travel without operating the usual audible and visual signals (*inconspicuous riser*).
2. Taken out of supervisory control and operated manually (*attendant or independent service*).
3. Selected for night or weekend service while the other cars are shut down.
4. Assigned to a particular floor on a fixed- or priority-basis call (*convention feature or priority*).

Other operational features that can be provided are concerned with emergency service, including the "fireman's return" feature required by ANSI and many local fire codes (Section 24.5) and the controls related to switching of power between cars in the event that operation on an emergency generator is necessary.

23.25 Car Panel

A typical modern car panel is illustrated in Fig. 23.13. Every car panel (station) is equipped with full-access buttons for call registry, door-open and alarm, and switches for emergency stop and fireman control. Also always provided behind an unlocked door is a telephone or intercom device that permits communication with the lobby panel. A door-close button is sometimes provided if extensive hand operation of the car is anticipated. It is activated only when the car is under manual control. Controls that do not concern the normal passenger are grouped in a locked compartment in the car station. These include hand-operation switch, light, fan and power switches, and any special controls such as security and emergency devices. Finally, a compartment accessible only to technicians contains the devices controlling door motion, car signals, door and car position transducers, load-weighing control, door and platform detection beam equipment, speech synthesizer (if used), and visual displays.

Fig. 23.13 Car panel ergonometrically designed for convenient use by normal as well as handicapped passengers. Each floor button is lighted and adjoins a Braille plate (raised numbers). Four special buttons—door-open, door-close, alarm, and emergency stop—are placed at the bottom of the panel and marked with internationally recognized symbols. Fireman's controls and special access switches are at the top of the panel, along with speaker/microphone. This panel (Otis Elevonic 401) is also equipped with a voice synthesizer and an upper panel that can display any desired information, in addition to showing the car location. Note also the arrow on the door jamb, which clearly indicates the direction of car travel. (Photo courtesy of Otis Elevator.)

ELEVATOR SELECTION

23.26 General Considerations

The selection of elevators for any but the simplest buildings requires the simultaneous consideration of several factors: adequate elevator service for the intended building usage, the economics of elevator selection, and the architectural integration of spaces assigned to elevators, including lobbies, shafts, and machine rooms. As must be obvious, these factors are interdependent; therefore, in large complex buildings many combinations are possible. The selection of a single optimum system for such cases is most practical with the aid of a computer, or simulator. For most buildings, however, certain guidelines can yield entirely satisfactory results with hand computation. These guidelines will be developed and explained below. The criteria of elevator service quality are:

1. Interval and average waiting time.
2. Handling capacity.
3. Travel time.

23.27 Definitions

A clear definition of terms, including variant usages, is imperative for the proper study of a subject. To that end, an abbreviated list of important definitions for our study follows.

Interval (I) or lobby dispatch time is the average time between departure of cars from the lobby.

Average lobby time or average waiting time is the average time spent by a passenger between arriving in the lobby and leaving the lobby in a car.

Registration time is the waiting time at an upper floor after registering a call.

Round-trip time (RT) is the average time required for a car to make a round trip, starting from the lower terminal and returning to it. The time includes a statistically determined number of upper floor stops in one direction and an express return trip.

Travel time or average trip time (AVTRP) is the average time spent by passengers from the moment they arrive in the lobby to leaving the car at an upper floor.

Handling capacity (HC) is a figure of merit for an elevator system, indicating the maximum number of passengers that can be handled in a given period, usually 5 min; thus "5-min handling capacity." It is expressed as a percentage of the building population.

Zone is a group of floors in a building that is considered as a group, with respect to elevator service. It may consist of a physical entity—a group of upper floors above and below which are blind shafts—or it may be a product of the elevator group control system, changing with system needs.

23.28 Interval or Lobby Dispatch Time, and Waiting Time

In an ideal installation, at least from the riding public's point of view, a car would be waiting at the lower terminal on the rider's arrival or would be available after a short wait. Since cars leave the lobby separated in time by the *interval (I)* and passengers arrive at the lobby in random fashion, the average waiting time in the lobby should be half the interval. Field tests show, however, that it is actually higher than this, and the figure most used in the industry is 60%; that is,

$$\text{average waiting time} = 0.6I$$

Excellent office building design provides an average wait of 15 to 18 s during up peaks, with 22 s considered good and 26 borderline.

Since modern control systems automatically zone the building with the result that some cars do not return to the lobby, the interval as a figure of merit can be somewhat misleading. As a basis of comparison, however, a 30-s interval is excellent, 35 borderline, and 40 just acceptable in other than center-city office spaces. Table 23.3 lists acceptable intervals for various building types.

With intervals in this range, riders will not be conscious of any irksome delay in elevator service. Consciousness of delay is considered a major drawback in rental desirability and should be sedulously avoided for all conditions of traffic except during morning and evening peaks, when a certain delay is expected and therefore tolerated, however grudgingly. Even in both these cases, a modern group supervisory system will recognize any *timed-out* call, that is, a call with registration time

TABLE 23.3 **Suggested Elevator Intervals**

Facility	*Interval (s)*
Office Buildings	
Center city	25–30
Investment	30–35
Suburban	35–40
Residential	
Prestige apartments	50–70
Middle-income apartments	60–80
Low-income apartments	80–120
Dormitories	60–80
Hotels—1st quality	40–60
Hotels—2nd quality	50–70

TABLE 23.4 **Car Passenger Capacity (p)**

Elevator Capacity (lb)	*Maximum Passenger Capacity*	*Normal Passenger[a] Load per Trip*
2000	12	10
2500	17	13
3000	20	16
3500	23	19
4000	28	22

[a]The number of passengers carried on a trip during peak conditions is approximately 80% of the car capacity.

exceeding 50 s, as a priority call. Priority calls are answered by the first available car, usually within 15 s. If a considerable amount of interfloor traffic is expected during peak periods, as may be the case when a large company occupies several upper floors, elevator service must be increased by 20 to 40% over that calculated to maintain proper intervals.

23.29 Handling Capacity

The frequency, or interval, with which a car appears at the lobby is one of the two factors that determine the passenger capacity of an elevator system. The other is obviously the size or capacity of the elevator car. The system's *handling capacity* is completely determined by these two factors—car size and interval, and is independent of the number of cars. This can be best understood by visualizing the system as a single set of doors that open periodically (interval) to remove a given number of passengers (car capacity) from the waiting group. Whether that set of doors represents a single car or many cars that take turns is immaterial. The only factors that fix handling capacity are passenger load (car capacity) and frequency of loading (interval). See Table 23.4.

Note that cognizance is taken of the fact that during peak traffic periods cars are not loaded to maximum capacity, but only to about 80%—a figure determined by actual count in many existing installations.

As a convenient measure of capacity, the handling capacity of a system for 5 min is taken as a standard. This is because a 5-min rush period is used as a measure of a system's ability to handle traffic. This may be expressed thus:

$$\text{handling capacity } (HC) = \frac{300 \text{ s} \times \text{passengers/car}}{\text{interval}}$$

or

$$HC = \frac{300p}{I} \tag{23.1}$$

where p is individual car loading. When the interval is 30 s, the system handling capacity is $10p$, a convenient figure to remember.

To establish a figure of merit for building service, system capacity HC must be related to building size. This is normally done by establishing a minimum percentage of building population that the system will handle in 5 min. A good system for a diversified office building will handle no less than 12% of the building population. Similar figures are shown in Table 23.5 for various types of facilities.

In planning a building the population must, of course, be estimated. This is particularly difficult in speculative-type, diversified-use buildings. However, based on rental cost, area, and building type, a fair estimate can be made. Population estimates for office buildings are based on net area, that is, actual available area for tenancy. Table 23.6 gives suggested density figures, while Table 23.7 gives average office building efficiency figures for use in calculating net area.

TABLE 23.5 **Minimum Handling Capacities (HC)**

Facility	*Percent of Population to be Carried in 5 Min*
Office Buildings	
Center city	12–14
Investment	11.5–13
Single purpose	14–16
Residential	
Prestige	5–7
Other	6–8[a]
Dormitories	10–11
Hotels—1st quality	12–15
Hotels–2nd quality	10–12

[a]Due to more urgent traffic demands, particularly at the school and work exodus.

TABLE 23.6 **Population of Typical Buildings for Estimating Elevator and Escalator Requirements**

Building Type	*Net Area*
Office Buildings	*Square feet per person*
Diversified	
Large lower floors	125–140[a]
Upper floors	140–160
Average use	150
Single purpose	130
Hotels	*Persons per sleeping room*
Normal use	1.3
Conventions	1.9
Hospitals	*Visitors and staff per bed*[b]
General private	3
General public (large wards)	3–4
Apartment Houses	*Persons per bedroom*
High-rental housing	1.5
Moderate-rental housing	2.0
Low-cost housing	2.5–3.0

[a]Density may vary for different floors. Clerical and stenographic area may have a population density as high as 70 square feet per person.

[b]If visiting hours are restricted, visitor population will determine elevator requirements. If visiting is not restricted to a certain few hours, staff requirements may determine elevator design. Where traffic is heavy, a combination of passenger cars and larger "hospital" cars should be used to provide optimum service.

23.30 Travel Time or Average Trip Time

The average trip time or time to destination is the sum of the lobby waiting time plus travel time to the median floor stop. Car round-trip time is also used as a criterion but is not as relevant or meaningful as trip time. In a commercial atmosphere a trip of less than 1 min is highly desirable, a 75-s trip acceptable, a 90-s trip annoying, and a 120-s trip the limit of toleration. Obviously in the more relaxed atmosphere of a residence, where interval alone can account for a minute or more of trip time, these maxima are revised upward. From Fig. 23.14 we see that the 2000-lb and 2500-lb cars used in residential buildings can have a 17-story rise, even with 60-s interval, without excessive trip time. On the other hand, the 3500-lb car, which is almost universal in office buildings (Fig. 23.15), is limited to a maximum of 16 floors local run before exceeding the 90-s limit and about 6 to 8 floors to stay within the 75-s criterion.

An important reservation on the foregoing statements must be noted. The curves of Figs. 23.14 and 23.15 are based on statistical calculations, empirical data, and field observations, as discussed in the next section. This being so, the average values that these figures give should be considered to be $\pm 10\%$, and borderline cases can be shifted either way. Designs that show high travel

TABLE 23.7 **Office Building Efficiency**

Building Height	*Net Rentable Area as Percentage of Gross Area*
0–10 floors	Approximately 80%
0–20 floors	Floors 1–10 approximately 75%
	11–20 approximately 80%
0–30 floors	Floors 1–10 approximately 70%
	11–20 approximately 75%
	21–30 approximately 80%
0–40 floors	Floors 1–10 approximately 70%
	11–20 approximately 75%
	21–30 approximately 80%
	31–40 approximately 85%

NOTE: Applicable to buildings with 15,000 to 20,000 gross square feet per floor.

Source: Reprinted from G. R. Strakosch, *Vertical Transportation, Elevators and Escalators,* 2nd ed. New York: Wiley, 1983.

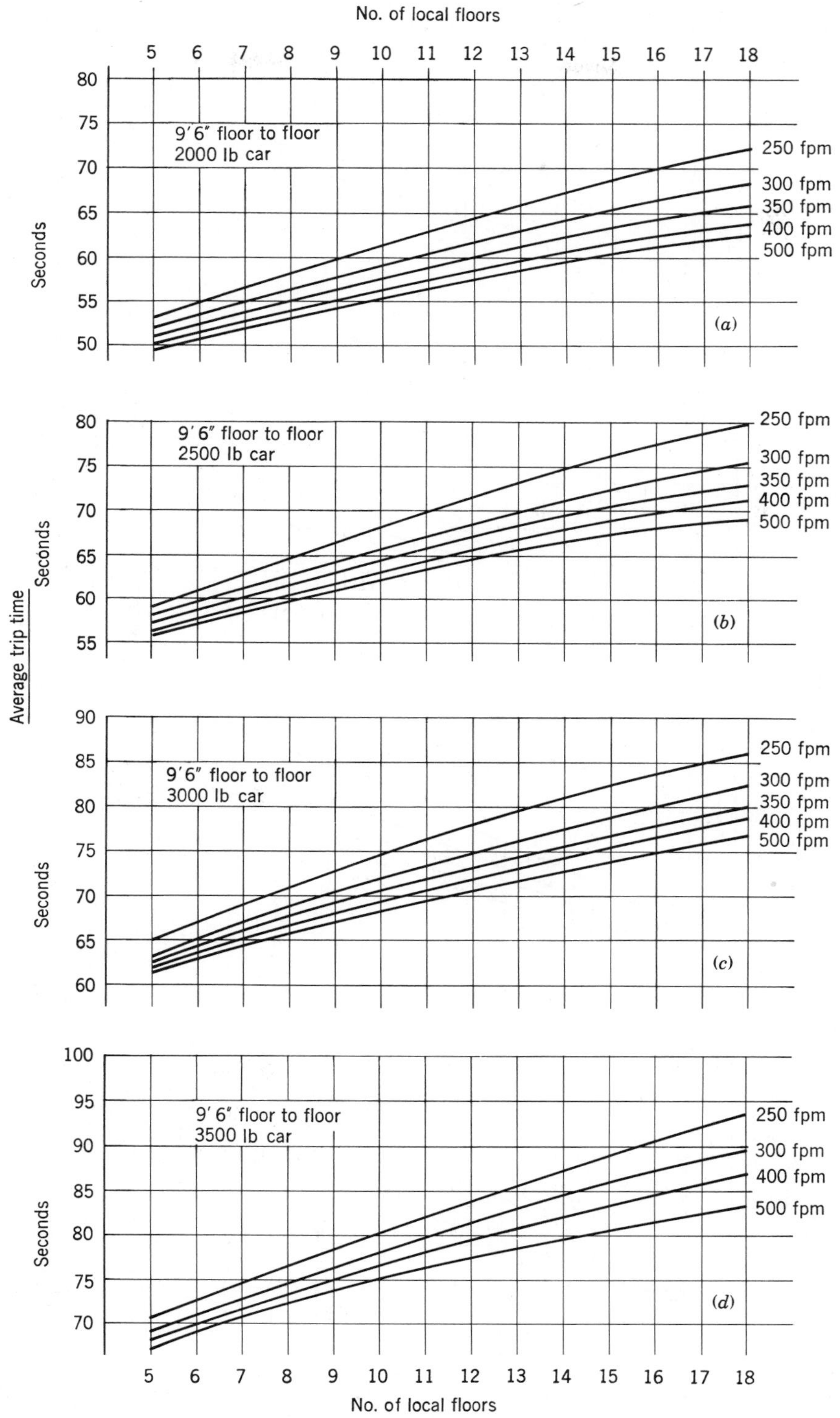

Fig. 23.14 *Plots of average trip time for various car speeds and capacities with 9-ft, 6-in. floor height, and 30-s interval.*

TRANSPORTATION

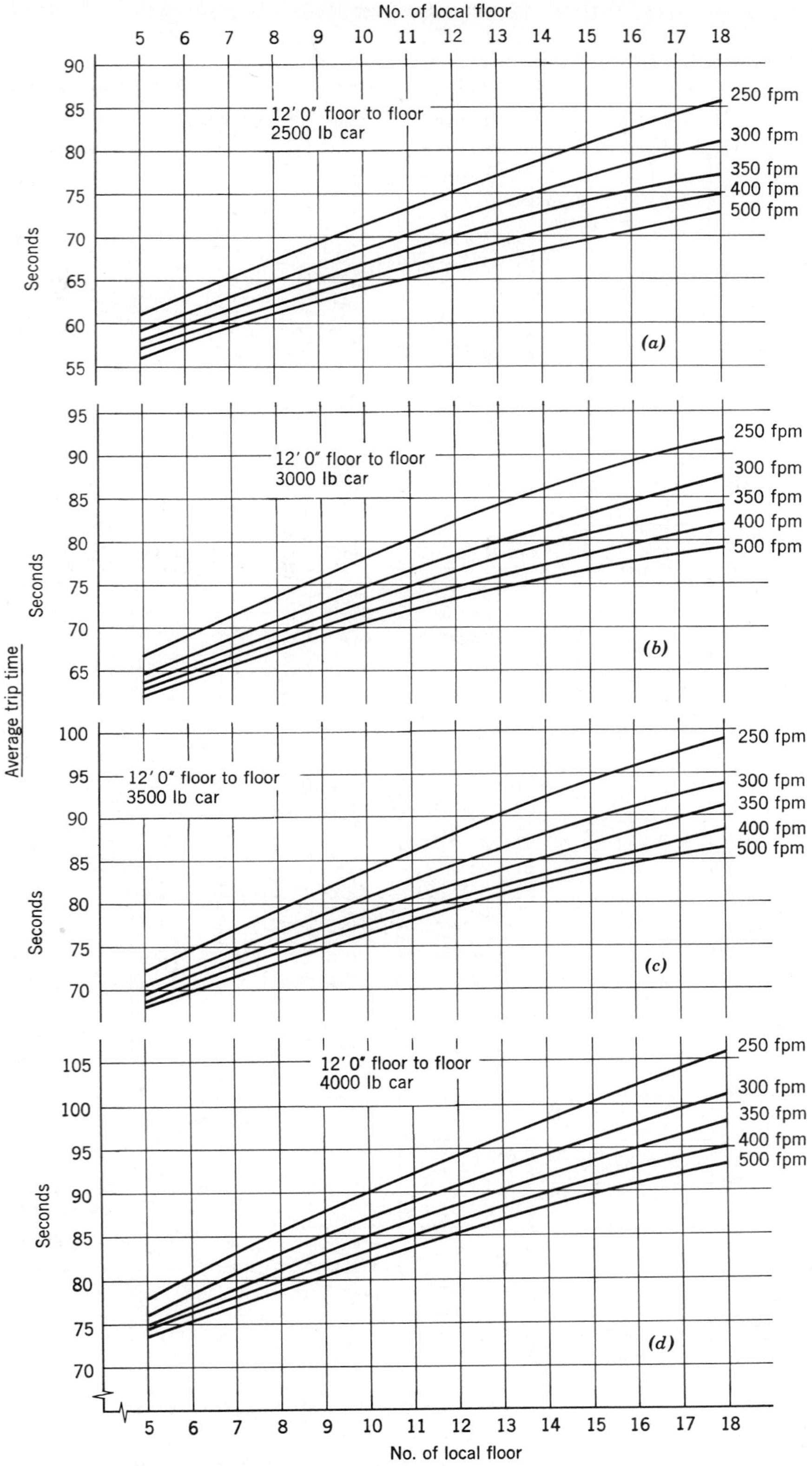

Fig. 23.15 Plots of average trip time for various car speeds and capacities for 12-ft floor height and 30-s interval.

time on paper frequently work out well in the field, because lobby loading is often less than 80%, upper floor stops are less than the statistical figure due to groups of people going to the same floor, and staggered working hours relieve traffic peaks. Also, a feature called high-call reversal takes account of the fact that cars do not travel to the top on each trip, but reverse at the top call. This can reduce average trip time by 5 to 10%. Finally, solid-state controls allow more rapid acceleration and deceleration without discomfort, further reducing trip time by several percent.

23.31 Round-Trip Time

The figure for round-trip time is composed of the sum of four factors. They are: time to accelerate and decelerate, time to open and close doors at all stops, time to load and unload, and running time (see Figs. 23.16 to 23.18). It is physically the time consumed by a car from door opening at the lower terminal to door opening at the same terminal at the end of a round trip. Since the actual number of stops made by a car is unknown, a statistical probability figure is used, based on the passenger capacity of the car and number of local floors above the lower terminal. In calculating this round-trip time (*RT*), it is assumed that a car will depart the lower terminal when loaded. No intentional delay is included at either lower or upper terminal. The *RT* thus calculated is a median figure, with any single actual round trip taking more or less time. In detail, round-trip time consists of the time expended in:

1. Loading at the lobby.
2. Door closing at the lobby.
3. Accelerating from the terminal and from each stop.
4. Decelerating at each stop.
5. Passenger transfer at each stop.
6. Door operation at each stop.
7. Running time at rated speed between stops.
8. Return express run from the last stop.

These figures are obtained as follows:

1. *Field observations*—items 1 and 5 are based on a 3 ft, 6 in. door opening. A smaller door opening increases passenger transfer time.
2. *Calculations*—items 2, 3, 4, 6, 7, and 8.

Door-closing time is based on a 3 ft, 6 in. center-opening door with adjustable speed.

Acceleration and deceleration times are calculated with a maximum of 4 ft/s/s because anything beyond that results in physical discomfort to the passengers.

Running time at rated speed takes place after the car has accelerated and before it begins to decelerate. If we consider that it takes between 20 and 30 ft to accelerate to 700 fpm, depending on rate of acceleration, we see that in local runs the car never gets to rated speed. It simply accelerates and decelerates. The higher speed equipment with its larger motor accelerates more quickly and gives some time advantage on the return express run, but has no great time advantage overall. This accounts for the small reduction above 500 fpm in Figs. 23.16 and 23.17.

In calculating round-trip time for cars in upper zones, it is necessary to know the time required to traverse the express floors. This may be obtained from Fig. 23.18. The times given therein are for *one-way* express runs. Thus to calculate *RT* for an upper zone car, take the *RT* corresponding to the upper local floors and add *twice* the figure obtained for express run time from Fig. 23.18.

23.34 Single-Zone Systems

Having established the relationships that govern the design and performance of an elevator system comprising a single zone, it would be helpful to follow through an example.

EXAMPLE 23.1. Office building, downtown, diversified use, 14 rentable floors above lobby, each 12,000 sq ft net. Floor-to-floor height—12 ft. Determine a workable elevator system.

SOLUTION. From Table 23.5, average *HC* is 13%. From Table 23.3, maximum control interval is 30 s. From Table 23.6, average population density is 150 sq ft per person.

Building population:

$$\frac{\text{14 floors at 12,000 sq ft}}{\text{150 sq ft per person}} = \text{1120 persons}$$

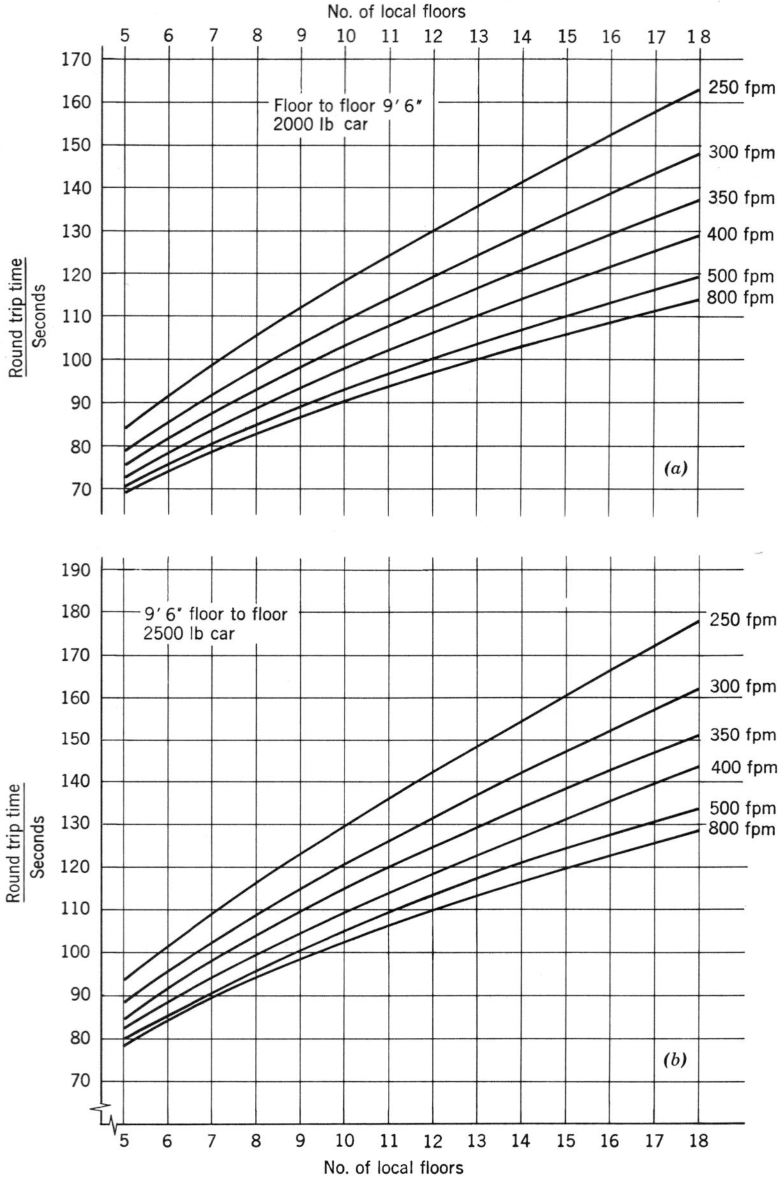

Fig. 23.16 Plots of round-trip time for various car speeds and capacities with 9-ft, 6-in. floor height and 30-s interval.

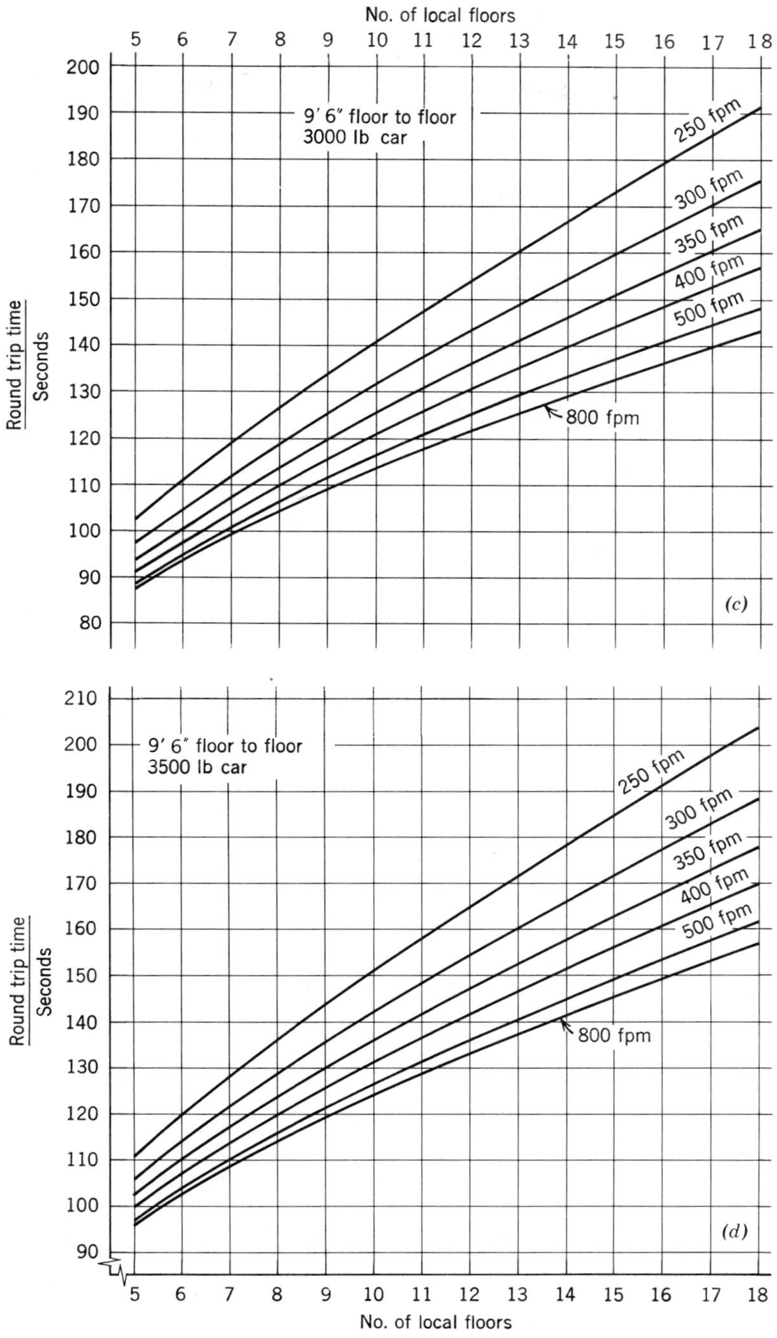

Fig. 23.16 (continued)

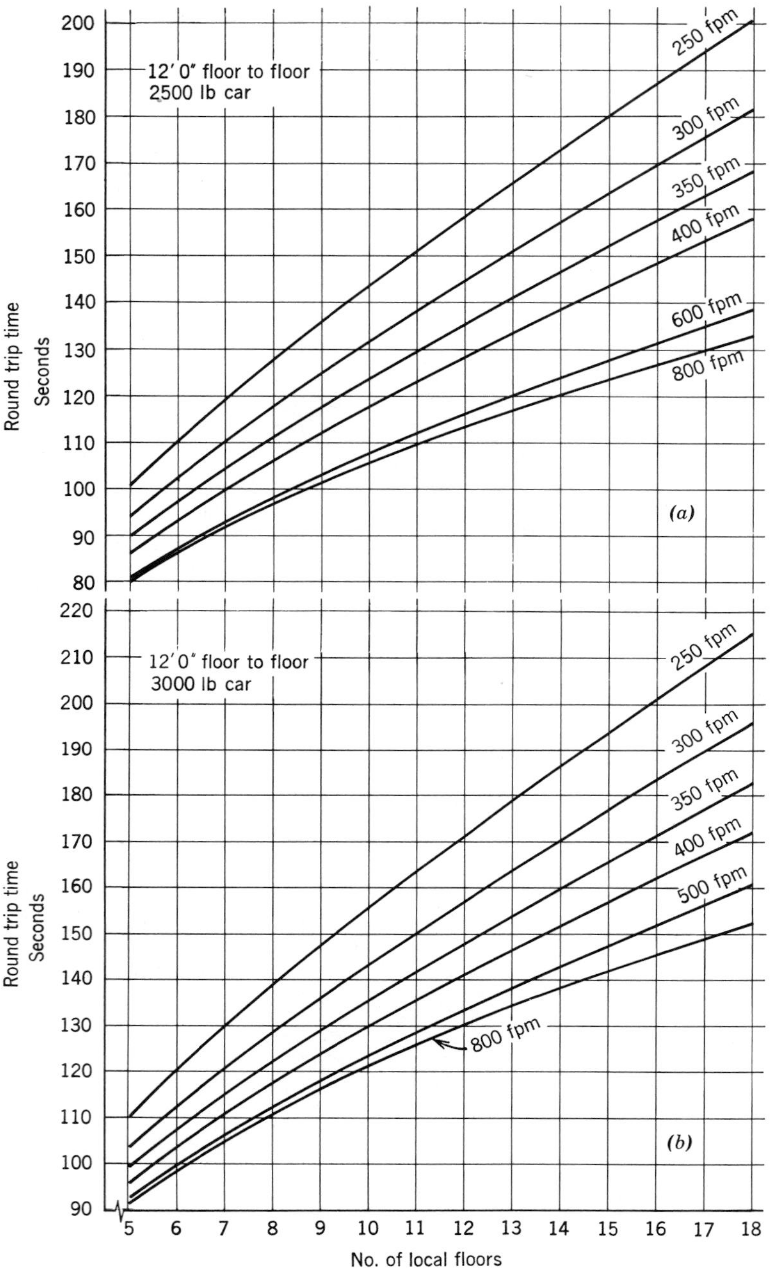

Fig. 23.17 Plots of round-trip time for various car speeds and capacities with 12-ft floor height and 30-s interval.

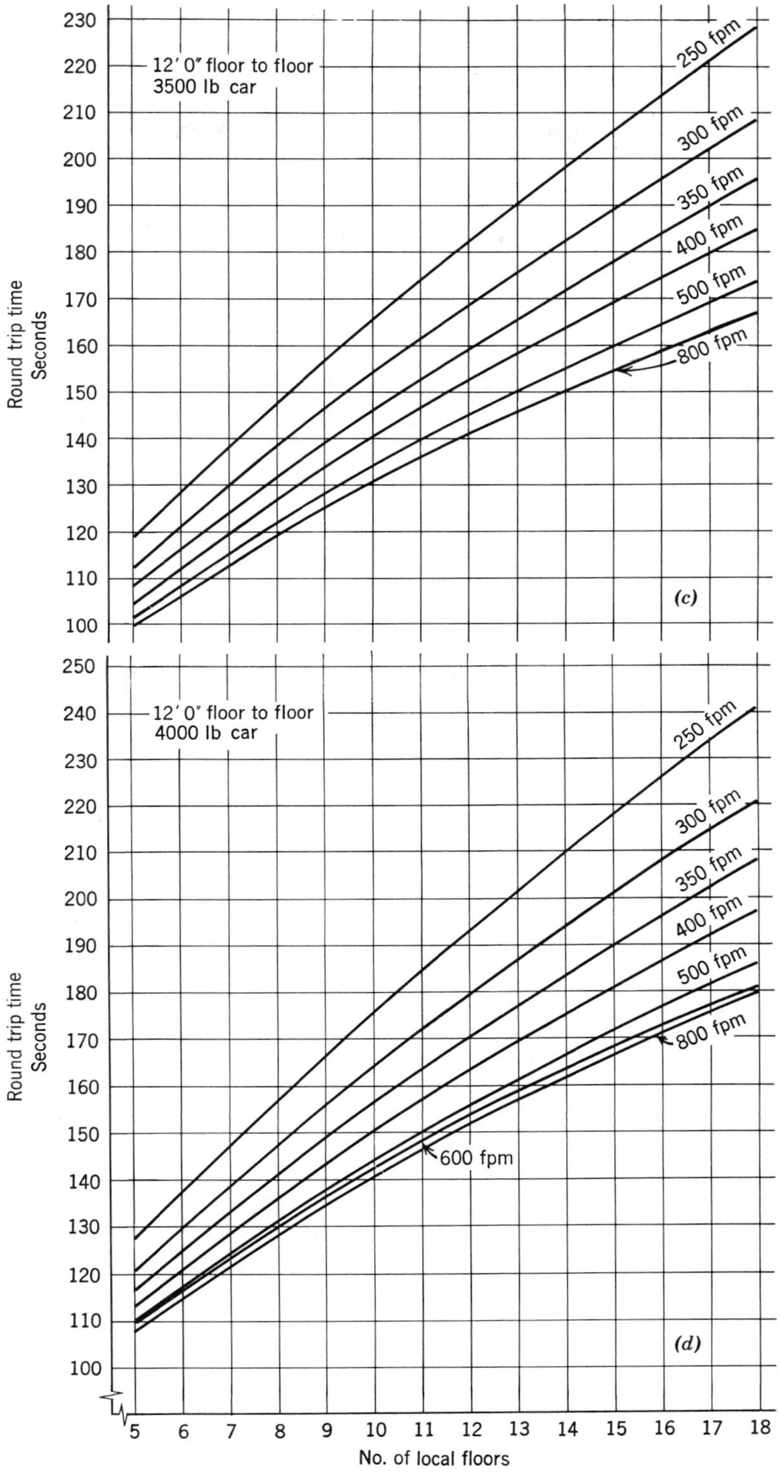

Fig. 23.17 (continued)

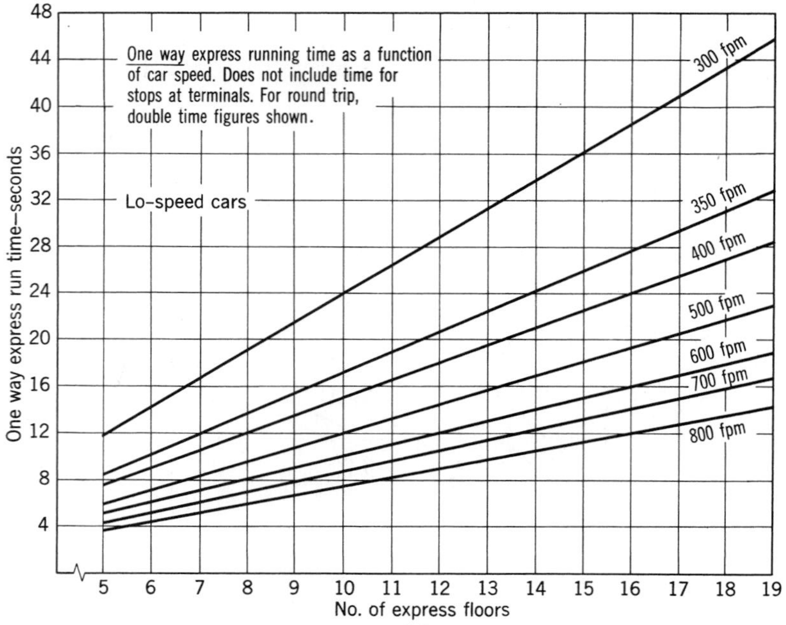

(a)

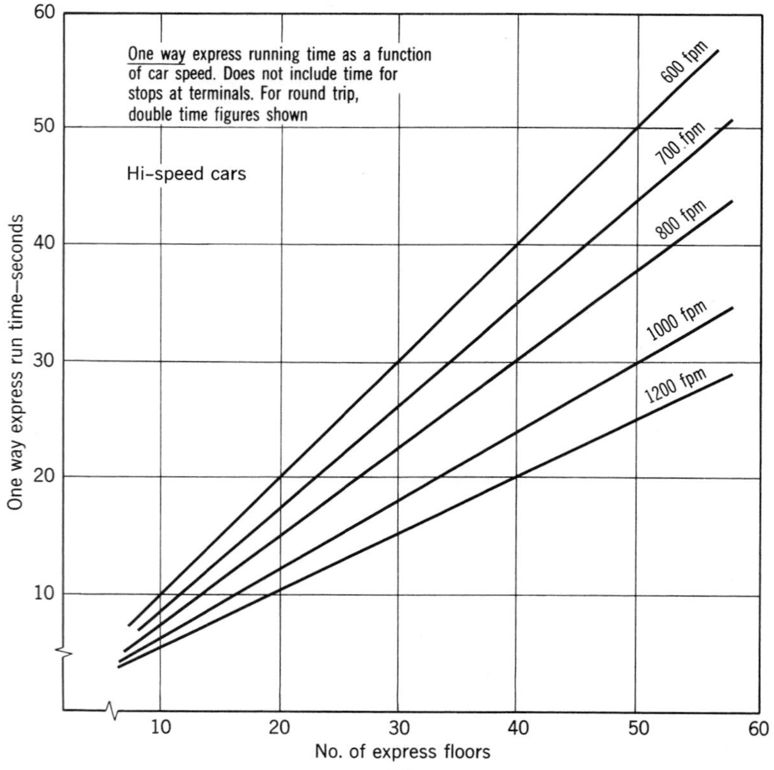

(b)

Fig. 23.18 One-way express running time, not including terminal time.

Suggested minimum handling capacity:

$$PHC = 13\%$$
$$HC = 0.13 \times 1120 = 146 \text{ persons}$$
$$\text{rise} = 14 \text{ floors at } 12 \text{ ft} = 168 \text{ ft}$$

From Table 23.8 we select a recommended car size of 3000 lb at 500 fpm.

3000 lb
500 fpm

Then from Figs. 23.17*b* and 23.15*b*:

$$RT = 143 \text{ s}, \qquad AVTRP = 76 \text{ s}$$

Single-car capacity $h = 300p/RT$ (see Table 23.4 for p):

$$h = \frac{300(16)}{143} = 33.5 \text{ persons}$$

$$N = \frac{HC}{h} = \frac{146}{33.5} = 4.4 \text{ cars, say } 5$$

(four cars would given an excessive interval)

$$I = \frac{RT}{N} = \frac{143}{5} = 28.3 \text{ s}$$

$$\text{actual } PHC = \frac{5(13)}{4.4} = 14.77\%$$

These figures are excellent.

We should also try faster cars, to reduce the number of cars, or slower cars, to reduce car cost. We select 700 fpm and 400 fpm.

3000 lb
700 fpm

$$RT = 140, \qquad AVTRP = 73 \text{ s}$$

TABLE 23.8 **Elevator Equipment Recommendations**

Usage		*Car Capacity (lb)*	*Minimum Car Speed (fpm)*	*Car Travel (feet)*
Office buildings			350–400	0–125
	Small building	2500	500–600	126–225
	Medium building	3000	700	226–275
	Large building	3500	800	276–375
			1000	Above 375
Hotels		2500		
		3000	As above	As above
Hospitals			150	0–60
			200–250	61–100
		3500	250–300	101–125
		4000	350–400	126–175
			500–600	176–250
			700	Above 250
Apartment houses[a]			100	0–75
		2000	200	76–125
		2500	250–300	126–200
			350–400	Above 200
Retail stores			200	0–100
		3500	250–300	101–150
		4000	350–400	151–200
		5000	500	Above 200

[a]FHA minimum requirements call for full-collective variable voltage control, minimum of two cars, and approximately 120 bedrooms per car, for all buildings exceeding seven stories in height.

$$h = \frac{300(16)}{140} = 34.3 \text{ persons}$$

$$N = \frac{146}{34.3} = 4.3, \text{ say 4 cars}$$

$$I = \frac{137}{4} = 34.2$$

$$\text{actual } PHC = \frac{4(13\%)}{4.3} = 12.1\%$$

This solution, that is, 4 cars at 700 fpm, 3000 lb, is marginally acceptable. Interval is somewhat high, but handling capacity is acceptable. Now, trying the slower car:

3000 lb
400 fpm

$$RT = 152, \qquad AVTRP = 77 \text{ s}$$

$$h = \frac{300(16)}{152} = 31.6 \text{ persons}$$

$$N = \frac{146}{31.6} = 4.62, \text{ say 5 cars}$$

$$I = \frac{152}{5} = 30.4 \text{ s}$$

$$\text{actual } PHC = \frac{5(13\%)}{4.62} = 14.1\%$$

This is also an acceptable arrangement. Thus, of the three arrangements, the two most likely choices are

4 cars, 3000 lb, 700 fpm
5 cars, 3000 lb, 400 fpm

To reduce the interval of the four-car solution, we can try a high-speed smaller car, since a larger car will *increase* the interval.

2500 lb
700 fpm

$$RT = 122 \text{ s}, \qquad AVTRP = 66 \text{ s}$$

$$h = \frac{300(13)}{122} = 32 \text{ persons}$$

$$N = \frac{146}{32} = 4.56 \text{ cars}$$

Using four cars:

$$I = \frac{122}{4} = 30.5 \text{ s}$$

$$PHC = \frac{4}{4.56}(13) = 11.4\%$$

Handling capacity is insufficient. Using five cars:

$$I = \frac{122}{5} = 24.4 \text{ s}$$

$$PHC = \frac{5}{4.56}(13\%) = 14.25\%$$

Of these two solutions only the five-car arrangement is satisfactory, since four cars yield insufficient *PHC*. However, because interval is low and *PHC* high, we can drop the speed and still obtain a workable arrangement:

2500 lb
400 fpm

$$RT = 139 \text{ s}, \qquad AVTRP = 71 \text{ s}$$

$$h = \frac{300(13)}{139} = 28.0 \text{ persons}$$

$$N = \frac{146}{28.0} = 5.2, \text{ say 5 cars}$$

$$I = \frac{139}{5} = 27.8 \text{ s}$$

$$PHC = \frac{5}{5.2}(13\%) = 12.5\%$$

which is acceptable, and obviously cheaper than five cars of 3000 lb capacity and 400 fpm speed.

We thus have two solutions to compare:

1. Four 3000-lb cars at 700 fpm, gearless, giving an interval of 34.2 s and a handling capacity of 12.4%.
2. Five 2500-lb cars at 400 fpm, geared, giving an interval of 27.8 s and a handling capacity of 12.5%.

The final selection should at this point be made on the basis of cost and core layout. When considering cost, note that first cost is the governing factor only in a speculative venture. With an owner–operator building, the cost comparison should be on a life-cycle basis. Cost figures must

TABLE 23.9 **Relative First Cost[a] Figures for Passenger Elevators at Various Speeds (fpm)**

Car Size (lb)	Hydraulic	Geared Traction			Gearless Traction			
	100	200	350	500	500	700	1000	1200
2000	40	80	100	130	165	170	220	235
2500	43	85	115	145	175	180	235	250
3000	50	90	120	150	180	185	250	265
3500	58	95	125	155	190	195	265	275
4000	60	100	135	165	200	205	280	300
4500[b]	70	120	150	185	225	230	300	325
5000[b]	75	130	160	200	240	250	330	350

[a]Costs are ±10%; based on standard fixtures, cabs, and entrances, and average rise for the speed indicated.

[b]Service elevator or hospital elevator.

NOTE: 1. See Table 23.1 for geared/gearless comparison.
See Table 23.8 for speed/rise recommendation.

reflect the impact of elevator space requirements on net rentable area in the building. Comparative relative cost figures are given in Table 23.9. Utilizing these data we have

1. Four 3000-lb cars at 700 fpm = 4 × 185 = 740 cost units.
2. Five 2500-lb cars at 400 fpm = 5 × 125 = 625 cost units.

The choice, *on a first-cost basis,* is clearly in favor of the geared 400-fpm units. However, the costs are sufficiently close that in a life-cycle cost analysis, even without considering the comparative space cost, which favors the four-car bank, the lower maintenance and energy costs of the gearless machine might reverse the results.

The recommendations given in Tables 23.8 and 23.9 are based on extensive field surveys and experience.

23.35 Multizone Systems

To properly understand the method by which zoning can be established and utilized, it must be remembered that handling capacity HC is fully established when size of car and interval are fixed, that is,

$$HC = \frac{300p}{I} \tag{23.1}$$

By definition, PHC is a percentage of the population; therefore

$$HC = PHC \times \text{population}$$

If we assume a population density D, we have

$$HC = PHC \times \frac{\text{area}}{D} \tag{23.5}$$

Substituting Equation 23.1

$$\frac{300p}{I} = PHC \times \frac{\text{area}}{\text{density}}$$

$$\text{area} = \frac{300 \times p \times D}{PHC \times I} \tag{23.6}$$

This equation has been plotted in curves of Fig. 23.19, with area as a function of interval and PHC, for a population density of 150 sq ft per person, average.

A glance at these curves shows the area that can be handled by a specific combination of car size and interval. In the problem of the preceding paragraph, an area of 168,000 sq ft with a minimum PHC of 13% and a maximum 30-s I is readily located on the set of curves for a 2500-lb car (Fig. 23.19*b*). A 3000-lb car (Fig. 23.19*c*) will necessitate excessive interval, and a 2000-lb car (Fig. 23.19*a*) cannot readily handle the load. Thus an immediate selection of a 2500-car can be made. The remaining calculations are routine. Use of these curves allows immediate car size selection, thus

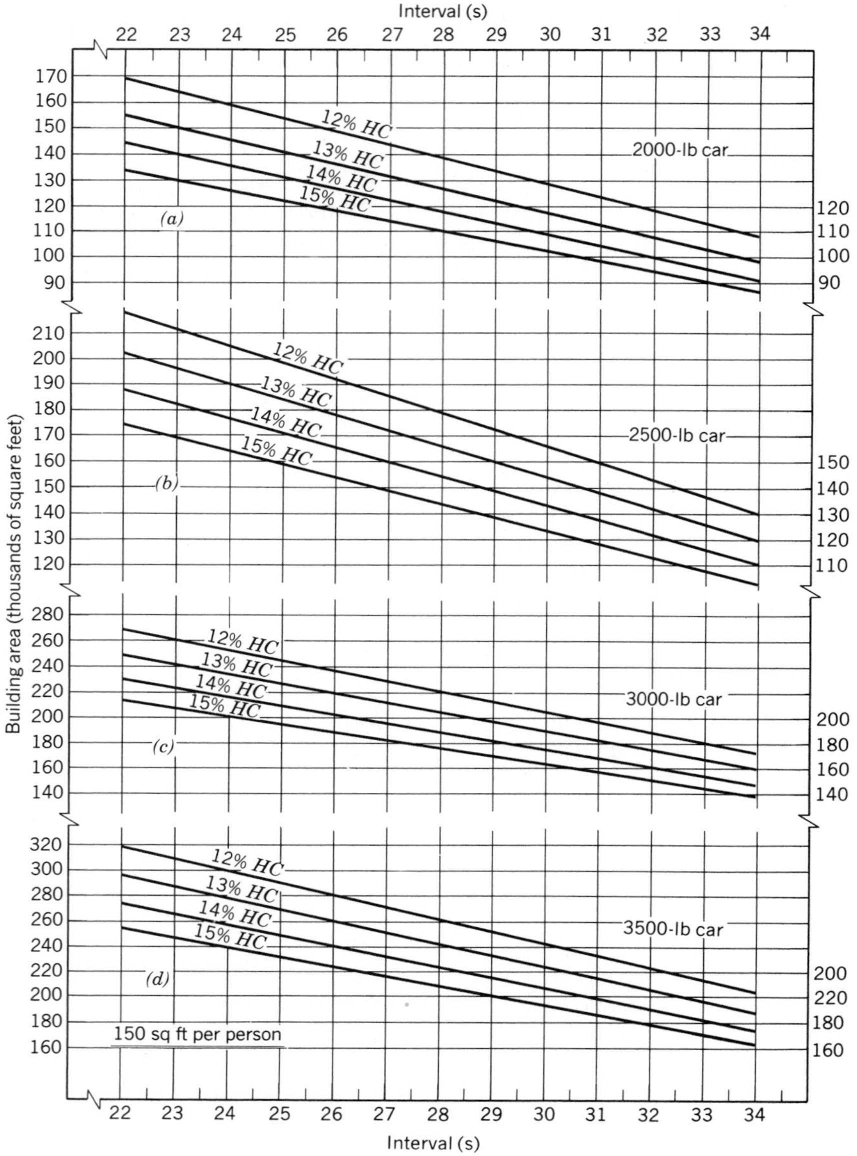

Fig. 23.19 Curves show the (zones) areas of a building that can be handled by cars of various specific capacities, as a function of interval and percent handling capacity. Curves are based on an average population density of 150 sq ft per person.

eliminating much of the trial-and-error method of the preceding section, the purpose of which was to show the calculation technique and demonstrate the effect of changing car capacity and speed.

Applying these curves to the results of the preceding problem, we can see immediately that zoning increases handling capacity, but at the cost of longer interval. In practice a 14-story building would not normally be split into upper and lower zones; it would be elevatored as one zone, as calculated above. However, to demonstrate the effect of zoning, we will attempt to solve the same problem using a two-zone solution.

The upper zone in a two-zone building is frequently smaller than the lower, to allow for the additional time of express run to the upper zone. As a trial here, we can split the building as follows:

Lower zone 8 floors, 96,000 sq ft

Upper zone 6 floors, 72,000 sq ft

Figure 23.19*a* shows immediately that even with 2000-lb cars, the upper zone would have too long an interval and too high an *HC*. The curve also tells us that this building is not a candidate for zoning, although a solution is theoretically possible using two equal seven-story zones of 84,000 sq ft each and 2000-lb cars. However, 2000-lb cars are not suggested for office building use because of door size limitation; they take a maximum 36-in. door whereas, as explained above, a 42-in. door is the minimum required for commercial use.

If we increase the building height to 18 stories, which is the usual limit for a single-zone solution, we can demonstrate the effect of zoning by providing both a single-zone and a two-zone solution and comparing them.

Single-Zone Solution. Eighteen stories above the lobby at 12,000 sq ft = 216,000 sq ft.

From Fig. 23.19*c*, a 3000-lb car will give acceptable handling capacity and interval.

$$\text{rise} = 18 \times 12 = 216 \text{ ft}$$

$$\text{population} = \frac{216{,}000 \text{ sq ft}}{150 \text{ sq ft/person}} = 1440 \text{ persons}$$

Suggested 5-min handling capacity is 13%, that is,

$$HC = 0.13(1440) = 187 \text{ persons}$$

3000 lb
500 fpm

$$RT = 161 \text{ s}, \qquad AVTRP = 78 \text{ s}$$

$$h = \frac{300(16)}{161} = 29.8 \text{ persons}$$

$$N = \frac{HC}{h} = \frac{187}{29.8} = 6.28, \text{ say 6 cars}$$

$$\text{actual } PHC = \frac{6.00}{6.28}(13\%) = 12.42\%$$

$$I = \frac{161}{6} = 26.8 \text{ s}$$

Increasing the car speed will not reduce the number of cars but will increase the price sharply. Thus the one-zone solution is:

6 cars, 3000 lb, 500 fpm (geared)

From Table 23.9 the relative first cost is

$$6(150) = 900 \text{ cost units}$$

Two-Zone Solution. From Fig. 23.19*b* we see that for a car of minimum size (2500 lb), a minimum zone area of approximately 110,000 sq ft is indicated. Therefore, we divide the building into two equal nine-story zones of 108,000 sq ft each, and use 2500-lb cars.

Upper zone:
2500 lb
800 fpm

$$AVTRP = \text{express time} + \text{local time} = 6.5 + 62 = 68.5 \text{ s}$$

$$RT = 2(\text{express time}) + \text{local time} = 2(6.5 \text{ s}) + 101 = 114 \text{ s}$$

$$h = \frac{300(13)}{114} = 34.2 \text{ persons}$$

$$N = \frac{HC}{h} = \frac{187/2}{34.2} = 2.73, \text{ say 3 cars}$$

$$I = \frac{114}{3} = 38.0 \text{ s}$$

$$\text{Actual } PHC = \frac{3}{2.73}(13) = 14.28\%$$

This is an acceptable solution despite the somewhat high interval. Increasing car speed to 1000 fpm would only marginally reduce interval while sharply increasing cost.

Lower zone:
2500 lb
500 fpm

$$RT = 105 \text{ s}, \qquad AVTRP = 63 \text{ s}$$

$$h = \frac{300(13)}{105} = 37.14 \text{ persons}$$

$$N = \frac{187/2}{37.14} = 2.52, \text{ say 3 cars}$$

$$I = \frac{105}{3} = 35 \text{ s}$$

$$PHC = \frac{3}{2.52}(13\%) = 15.5\%$$

Therefore, we have a two-zone solution of

3 cars, 2500 lb, 800 fpm, gearless, 555 cost units and

3 cars, 2500 lb, 500 fpm, geared, 435 cost units.

Total cost—990 cost units, as compared to the 900 cost units, and better interval, of the single-zone solution.

These results indicate that zoning is advisable where high handling capacity is required, that is, where population density is high, for example, 120 sq ft per person. In such a case, for a similar 18-story building, a two-zone solution would clearly be preferred. It is left to the reader as an exercise to work through such a problem.

Summarizing the comparison between one- and two-zone systems, we find that the two-zone system has higher *HC* and interval, but lower trip time; it also has fewer shaftway door openings and lower maintenance cost. We are assuming no machine room in the middle of the building. If one were contemplated there, its cost impact on the building would be compared to the additional rental space saved on the floors above. Also, the architectural effect of an oversize floor in the building center would have to be considered.

Where faced with the problem of elevatoring a large high-rise building, all the foregoing curves and tables can be utilized to arrive at a solution by hand calculation. However, the factors involved are so numerous and so closely interrelated that a proper solution requires the use of computer calculation, simulation, or both. Among these factors are zoning, rental rates for different floors, machine room space, and structural reactions. It is always advisable to confer with an elevator consultant on high-rise buildings.

23.36 Other Elevator Selection Considerations

As in every design discipline, certain generalizations, or rules of thumb, have grown up in elevator design. For commercial buildings rough costs can be estimated from Tables 23.9 and 23.10.

A good rule of thumb for cost is 11% of total construction cost. Special considerations for various building types are given below.

(a) Office Buildings, Hotels, and Industrial Buildings. The expected population may be estimated from Table 23.6. An interval of 30 s is desirable, although slightly longer is acceptable in all except large structures in congested areas. A 5-min capacity of 13% of building population is safe. The basic type of control is automatic electronic supervisory. Table 23.8 lists the characteristics of equipment usually used in these buildings.

Approximately one service car per 10 passenger cars should be provided, or alternatively one car for every 300,000 sq ft of net area. Service cars should be 4000 lb or larger without dropped ceiling and, if also used for passenger service, equipped with wall pads. An oversized door (e.g., 4 ft, 0 in. or 4 ft, 6 in.) is particularly useful in handling furniture. Service elevators should have a shaftway door at every level plus easy access to the truck dock or other freight entrance as well as the lobby. These cars operate as service cars normally but can serve as passenger cars in peak periods to alleviate congestion and delay. This fact is particularly useful in marginal service designs.

(b) Apartment Houses. Studies indicate that apartment house traffic depends not only on population but also on location and type of tenant. Houses with many children experience a school-

TABLE 23.10 **Office Buildings: Cost of Elevator and Electric Work**

	Number of Stories		
Item	*20*	*35*	*60*
Elevator work	10.9%	11.9%	12.2%
Electric work	13.3%	12.6%	12.2%

hour peak; houses in midtown with predominantly adult tenancy exhibit evening peaks due to the homecoming working group and outgoing amusement traffic. Normally a single elevator will suffice, although a second car functioning as a service and/or passenger car is sometimes indicated, particularly in buildings taller than six stories. The cars may be banked or separated, as desired. If a single car is used, it should be service elevator size.

Self-service collective control is the general choice, with provision for attendant control in prestige buildings. With smaller cars and short rise, a swing-type manual corridor door is acceptable; in larger installations both the cab and corridor door should be the power-operated sliding type.

Service elevators must be large enough to handle bulky furniture and should therefore be at least 2500 lb and preferably 3500 lb, with a 48-in. door and high ceiling. Hoistways must be isolated from sleeping rooms by lobbies or other space. Similarly, machine rooms must be isolated, since the starting and stopping of motors and other machine room noises are an effective barrier to sound sleep. Security arrangements are discussed in Section 24.6.

(c) Hospitals. As mentioned in Table 23.6, the governing factor in the determination of elevator requirements may be either normal hospital traffic or visitor traffic, depending on the visiting-hour schedule. Due to the large amount of vehicular traffic such as stretcher carts, wheelchairs, beds, linen carts, and laundry trucks, the elevator cabs are much deeper than the normal passenger type. This type of car when used for passenger service holds more than 20 persons and therefore gives slow service. For this reason, it is occasionally advisable to utilize some normal passenger cars in addition to hospital-size cars, particularly in large hospitals.

The use of tray and bulk carts in food service imposes a considerable load on the elevators before, during, and after meals, and passenger service is seriously disrupted. To alleviate this congestion and delay, many architects and hospital administrators prefer the use of dumbwaiter cars or another of the many types of materials-handling system that will handle a 15½ × 20 in. food tray. These systems can also be used for transporting pharmaceuticals and other items and are discussed in Sections 24.19 and 24.20.

Elevators should be grouped centrally, although separated by type of use. Car control is normally self-service collective.

Population of the hospital may be estimated from Table 23.6. Experience has shown that a carrying capacity of 45 passengers in a 5-min period is adequate (estimating each vehicle as equivalent to 9 passengers).

Intervals should not exceed 1 min. All recommendations for service to the handicapped should be adopted. See Section 23.11.

(d) Retail Stores. Retail stores present a unique problem in vertical transportation inasmuch as the objective is partially to transport persons to specific levels and partially to expose the passengers to displayed merchandise. For this reason modern stores rely almost exclusively on escalators, with one or two small elevators intended for use by staff and handicapped persons. When for some reason it is desired to equip a store with elevators, use the recommendations shown in Table 23.8, calculated for a load of between 10 and 20% of store population. Control should be automatic, selective collective. Cars are arranged in a straight line to facilitate loading and waiting.

THE PHYSICAL PROPERTIES AND SPATIAL REQUIREMENTS OF ELEVATORS

23.37 Shafts and Lobbies

The elevator lobbies and shafts form one of the major space factors with which the architect is concerned. The elevator lobby on each floor is the focal point from which the corridors radiate for access to all rooms, stairways, service rooms, and so forth. Such lobbies obviously must be located one above the other. The ground-floor elevator lobby (also called the lower terminal) must be conveniently located with respect to main entrances. The modern equipment within or placed adjacent to this area should include public telephones, building directory, elevator indicators, and control panels.

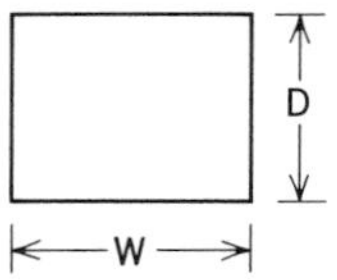

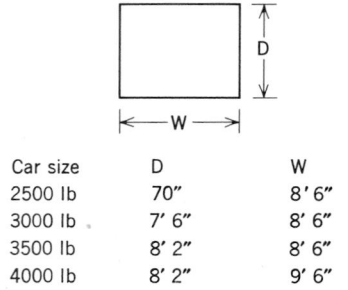

Car size	D	W
2500 lb	70″	8′6″
3000 lb	7′ 6″	8′ 6″
3500 lb	8′ 2″	8′ 6″
4000 lb	8′ 2″	9′ 6″

Fig. 23.20 *Rough hoistway dimensional data to be used in architectural single-line planning stage.*

All lobbies should be adequate in area for the peak-load gathering of passengers to ensure rapid and comfortable service to all. The number of people contributing to the period of peak load (15- to 20-min peak) determines the required lobby area on the floor.

Approximately 5 sq ft of floor space per person should be provided at peak periods for waiting passengers at a given elevator or bank of elevators. The hallways leading to such lobbies should also provide about 5 sq ft per person, approaching the lobby. Under self-adjusting relaxed conditions, density is about 7 sq ft per person. However, in peak periods, crowding occurs, reducing

Fig. 23.21 *Lobby groupingsforsinglezones:three-, four-, six-, and eight-car groups.*

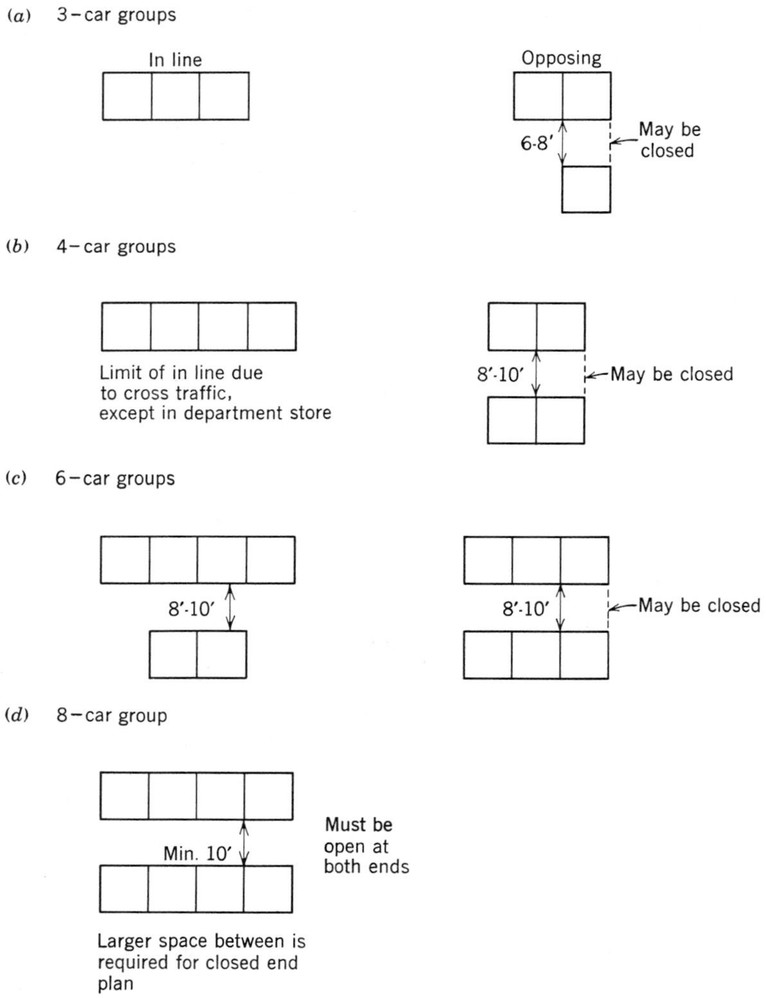

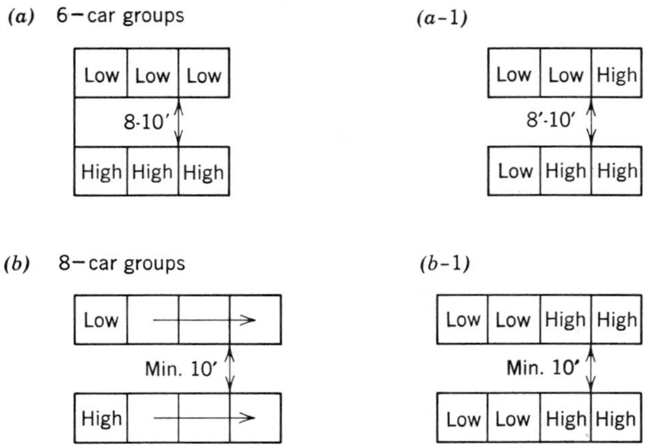

Fig. 23.22 *Lobby groupings for multiple zones. Arrangement* (a) *is preferable to* (a-*1*), *and* (b) *to* (b-*1*).

this to 3 to 4 sq ft. An acceptable compromise is 5 sq ft per person.

The main lower terminal of elevator banks is generally on the street-floor level, although it may be on the mezzanine level, when the elevations of the street entrances vary so that one side of the building is at mezzanine level, while another entrance is lower. Such a situation is ideal for the use of escalators, which will economically and rapidly carry large numbers of persons between levels, thus making practical and efficient a single main lower elevator terminal. The upper terminal is usually the top floor of the building. Typical dimensional data and lobby arrangements are shown in Figs. 23.20 to 23.22.

23.38 Dimensions and Weights

Manufacturers will supply standard layouts for elevators including dimensions, weights, and structural loads. Furthermore, to assist in preliminary design, major manufacturers have agreed upon, and publish, a set of Standard Elevator Layouts via their trade organization, the National Elevator Industry, Inc. See Section 23.11 for NEII recommendation for the handicapped. One such standard, with applications, is reproduced in Fig. 23.23 for 500- to 700-fpm gearless units in the full range of car capacities. These standards are available from the NEII.

As may be seen from Fig. 23.23, it is necessary, in providing for an elevator installation, to consider such factors as the depth of the pit, the dimensions of the hoistway, the clearance from the top of the hoistway to the floor of the penthouse, the size of the penthouse, and the loads that must be carried by the supporting beams.

The penthouse floor and secondary level floor are located above the shaft of each elevator and need approximately 1½ stories of additional height above the top of the support beam of a given elevator when it is standing at its top-floor location. The actual floor area required by the elevator traction machine and its controls is roughly two times the area of the elevator shaft itself. The required area of the floor of the secondary level is no larger than the elevator shaft it serves. As seen in Fig. 23.23, the machine room contains the bulk of the elevator machinery. Since some of this equipment will have to be moved for maintenance, it is advisable to furnish an overhead trolley beam that can be used during installation as well. The maximum beam load will be supplied by the elevator manufacturer.

When penthouse space is not available and a hydraulic unit is not desired, a basement traction unit, also referred to as an underslung arrangement, can be used. These units are low speed (100–350 fpm) and are therefore applicable where rise is limited and traffic light to medium. Figure 23.24 shows a shaft section with car and dimensional data.

Vertical Transportation Standards
Electric Passenger Elevators

Single slide entrance

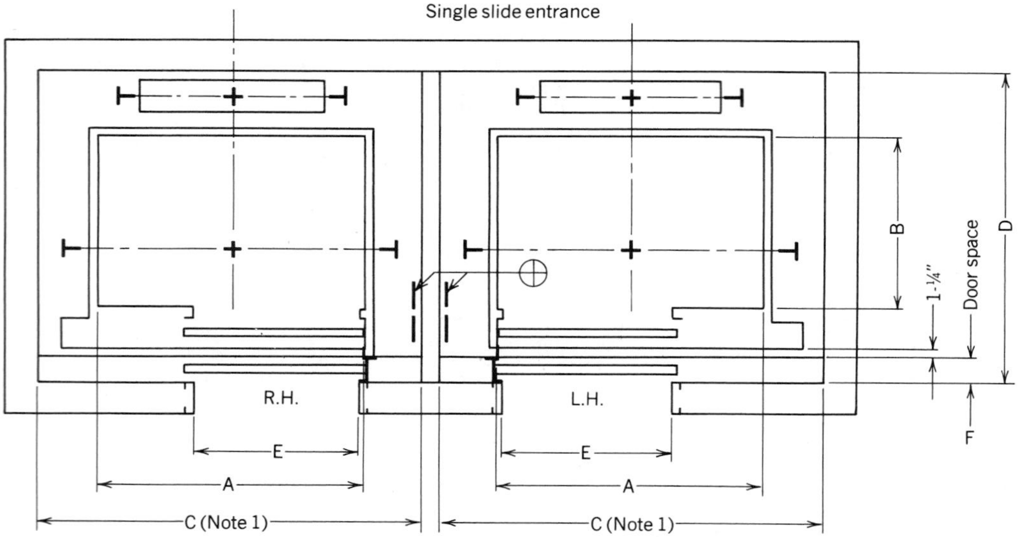

Plan view

Note 1. "C" dimension does not include any allowance for rail backing.

Rated Speeds 500–700 fpm

Rated Load (lb)	Area[a]	Dimensions (ft–in.) +A	+B	C	D	E	Entrances Type	F min (in.)
Application: Office Bldg. Apartment Houses, Hotels, Banks, Etc.								
2000	24.2	5–8	4–3	7–5	6–11	3–0	Single slide[b] Center opening	5
2500	29.1	6–8	4–3	8–6	6–11	3–6	Single slide[b] Center opening	5
3000	33.7	6–8	4–7	8–6	7–4	3–6	Single slide[b] Center opening	5
3500	38.0	6–8	5–3	8–6	8–0	3–6	Single slide[b] Center opening	5
4000	42.2	7–8	5–3	9–6	8–0	4–0	Center opening[b]	5

[a]Maximum allowable inside car area in square feet.

[b]These car sizes and entrance types comply with NEII suggested minimum passenger elevator requirements for the handicapped; 2000 lb rated load will not accommodate ambulance-type stretcher.

Fig. 23.23 *Typical elevator installation dimensional data. (Reproduced with permission of National Elevator Industry Inc., 600 Third Ave., New York, N.Y.* © Vertical Transportation Standards, *1983 edition.)*

Vertical Transportation Standards
Electric Passenger Elevators

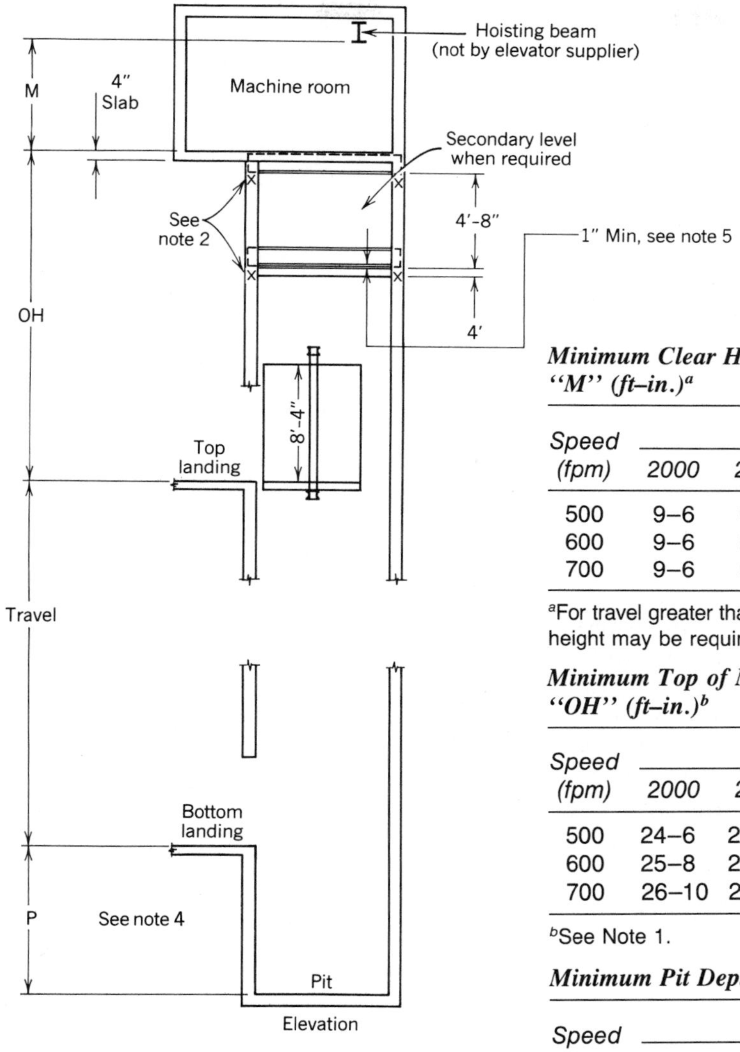

Fig. 23.23 (cont.)

Minimum Clear Height of Machine Room "M" (ft–in.)[a]

Speed (fpm)	Rated Load (lb) 2000	2500	3000	3500	4000
500	9–6	9–6	9–6	9–6	9–6
600	9–6	9–6	10–0	10–6	10–6
700	9–6	9–6	10–6	10–6	10–6

[a]For travel greater than 500 ft, increased machine room height may be required.

Minimum Top of Machine Room Floor "OH" (ft–in.)[b]

Speed (fpm)	Rated Load (lb) 2000	2500	3000	3500	4000
500	24–6	24–6	24–6	24–6	24–6
600	25–8	25–8	25–8	25–8	25–8
700	26–10	26–10	26–10	26–10	26–10

[b]See Note 1.

Minimum Pit Depth "P" (ft–in.)[c]

Speed (fpm)	Rated Load (lb) 2000	2500	3000	3500	4000
500	11–0	11–0	11–0	11–0	11–0
600	12–9	12–9	12–9	12–9	12–9
700	12–9	12–9	12–9	12–9	12–9

[c]See Note 4.

Notes

1. "OH" dimensions are based on an 8 ft, 4 in. overall car height.
2. Supports for elevator machine beams at X-X in elevation, not by elevator supplier.
3. Machine room floor slab is flush with or above top of machine beams.
4. For travel greater than 400 ft, increased pit depth may be required. Consult elevator supplier.
5. Top of support to be 1 in. above top of slab.

TRANSPORTATION

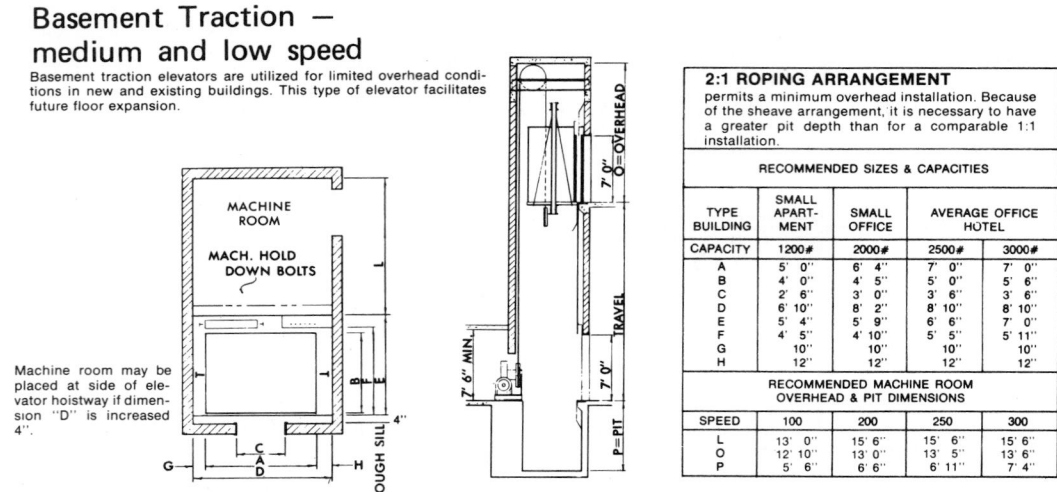

Basement Traction — medium and low speed

Basement traction elevators are utilized for limited overhead conditions in new and existing buildings. This type of elevator facilitates future floor expansion.

Machine room may be placed at side of elevator hoistway if dimension "D" is increased 4".

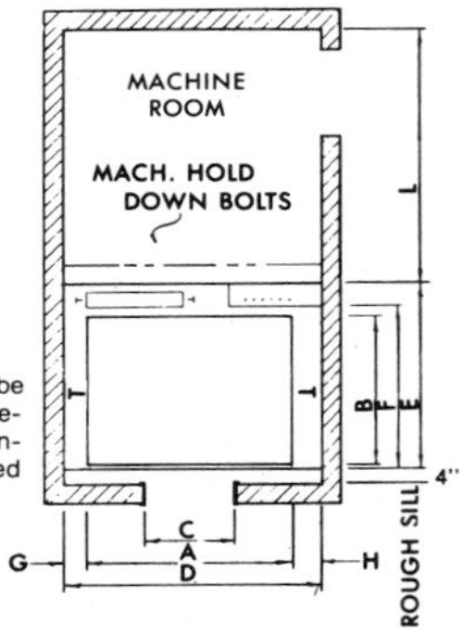

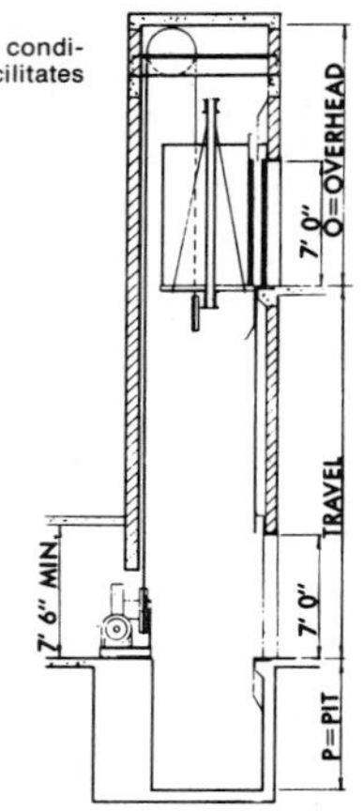

2:1 ROPING ARRANGEMENT permits a minimum overhead installation. Because of the sheave arrangement, it is necessary to have a greater pit depth than for a comparable 1:1 installation.

RECOMMENDED SIZES & CAPACITIES				
TYPE BUILDING	SMALL APARTMENT	SMALL OFFICE	AVERAGE OFFICE HOTEL	
CAPACITY	1200#	2000#	2500#	3000#
A	5' 0"	6' 4"	7' 0"	7' 0"
B	4' 0"	4' 5"	5' 0"	5' 6"
C	2' 6"	3' 0"	3' 6"	3' 6"
D	6' 10"	8' 2"	8' 10"	8' 10"
E	5' 4"	5' 9"	6' 6"	7' 0"
F	4' 5"	4' 10"	5' 5"	5' 11"
G	10"	10"	10"	10"
H	12"	12"	12"	12"
RECOMMENDED MACHINE ROOM OVERHEAD & PIT DIMENSIONS				
SPEED	100	200	250	300
L	13' 0"	15' 6"	15' 6"	15' 6"
O	12' 10"	13' 0"	13' 5"	13' 6"
P	5' 6"	6' 6"	6' 11"	7' 4"

***Fig. 23.24** Typical data for basement traction machine (underslung) arrangement, used where penthouse is unavailable or undesirable. (Courtesy of Montgomery Elevator Co.)*

23.39 Structural Stresses

For the purpose of structural design it is necessary to know the overhead load that must be supported by the foundations, by structural columns extending upward to the penthouse, and by the main beams that support the penthouse floor and subfloor. These loads (reactions) are supplied by manufacturers and usually include the actual dead weights of equipment when the elevator is not in motion, plus the added weight caused by the momentum of all moving parts and passengers when the elevator is at top speed and is suddenly caused to stop rapidly by the safety devices.

References

Publications of the National Elevator Industry, Inc., New York: "Suggested Minimum Passenger Elevator Requirements for the Handicapped," 1979. *Vertical Transportation Standards,* 1983 ed.

Strakosch, G. R. (1983). *Vertical Transportation, Elevators and Escalators,* 2nd ed., Wiley, New York.

24

VERTICAL TRANSPORTATION: SPECIAL TOPICS

POWER AND ENERGY

24.1 Power Requirements

The power required by an elevator is that amount necessary to perform the necessary traction work and to overcome friction. Since power is equal to the *rate* at which work is done the elevator motor size is directly proportional to the speed of the system. In other words, it requires proportionately more power to lift a 3000-lb car at 700 fpm than at 200 fpm. This relationship is shown in Fig. 24.1, which shows the minimum size of a d-c elevator traction motor as a function of speed, for different capacity cars. This motor supplies the system friction in addition to the required traction power. (For power data on hydraulic elevators, see Section 24.10.) Since friction is higher in geared machines than in gearless units, the traction motor must be larger. The m-g set motor size (where used) is approximately 20% larger than the value shown, to compensate for the set losses.

When solid-state control is used (and the m-g

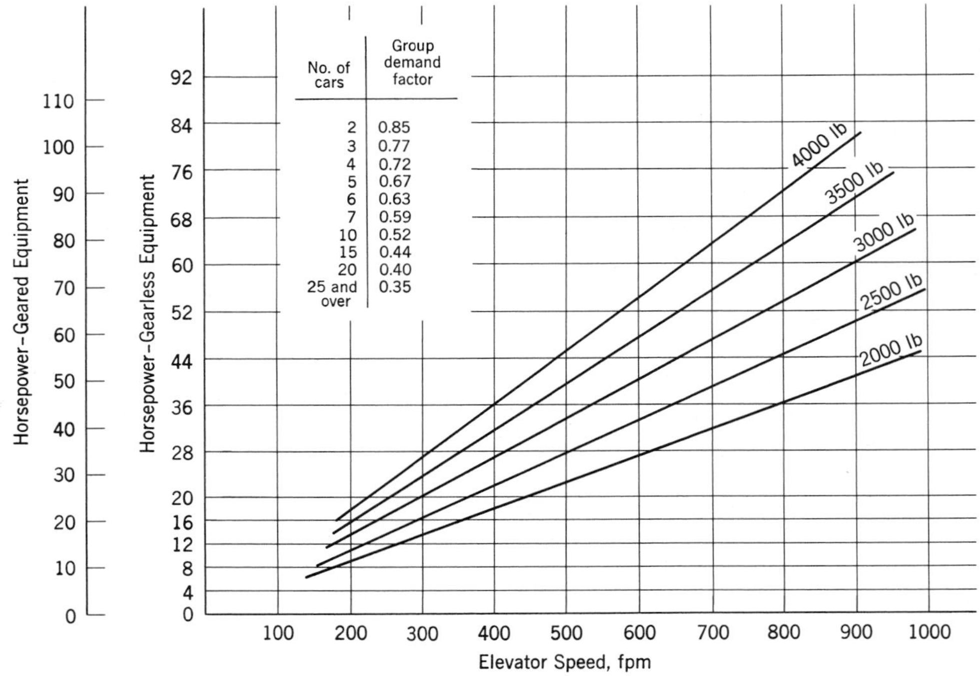

No. of cars	Group demand factor
2	0.85
3	0.77
4	0.72
5	0.67
6	0.63
7	0.59
10	0.52
15	0.44
20	0.40
25 and over	0.35

Fig. 24.1 *Elevator traction motor power requirements per car. The m-g set motor (if used) is approximately 20% larger than the traction machine.*

set eliminated), the values shown in Fig. 24.1 are still valid inasmuch as the size of the traction motor is unaffected by the type of variable-voltage or -frequency supply.

An elevator moves only about 50% of the time, the remainder being spent standing at various landings. As the number of cars in a bank increases, the probability of *all* the cars being in operation simultaneously decreases, resulting in a system demand factor less than 1.0. The factor is shown directly in Fig. 24.1.

As an example of the use of the curves, consider a bank of five 3500-lb, 600-fpm units. From Fig. 24.1, each car requires 48 hp

$$\text{group demand factor} = 0.67$$

$$\text{total instantaneous power required} = 5 \times 48 \times 0.67 = 160 \text{ hp}$$

Note that this is the *traction motor* power requirements. Assuming an overall efficiency (eff) of 80% for an m-g set, the elevator system power requirement is

$$\text{system power} = \frac{160 \text{ hp}}{80\% \text{ eff}} = 200 \text{ hp}$$

which would have to be provided by the building power system. When using solid-state control, with an efficiency in excess of 90%, system power requirements would be reduced proportionately.

24.2 Energy Requirements

The energy used by an elevator is essentially the system friction, including the heat generated by the brakes plus the electrical losses in the traction motor and power supply equipment—rotary or solid state. The energy expended in raising the car and its passengers is simply stored as potential energy. It is *returned to the power system* when the car and passengers descend, via the system of regenerative braking in use in almost all elevator systems. Refer to Fig. 24.2, which shows the approximate efficiencies of the components of a typical system. With these data, we are able to calculate a system's energy consumption.

EXAMPLE 24.1 Given a system of five 3500-lb, 600-fpm (gearless) cars calculate:

(a) Heat generated in the machine room during peak periods. Assume solid-state control.
(b) Approximate monthly energy cost using a combined demand/energy rate of $0.08/kWh.

SOLUTION

(a) During peak periods the traction motor operates approximately 50% of the time and is at standstill the other half. Assume that while operating it draws 90% of full load. Therefore, for one car, from Fig. 24.1

$$\text{traction motor} = 48 \text{ hp}$$

Total loss per machine:
In controls:

$$\frac{48 \text{ hp}}{0.9 \text{ eff}} \times 90\% \text{ load} \times 50\% \text{ operation} \times 10\% \text{ loss} = 2.4 \text{ hp}$$

In traction motor:

$$48 \text{ hp} \times 90\% \text{ load} \times 50\% \text{ operation} \times 20\% \text{ loss} = 4.32 \text{ hp}$$

$$\text{Total} = 6.72 \text{ hp} = 17{,}100 \text{ Btu/h}$$

Since there are five elevators operating, the total heat generated is

$$5 \times 17{,}000 \text{ Btu/h} = 85{,}500 \text{ Btu/h}$$

This is the rating of a home furnace. No diversity is taken since all the machines are operating and the heating is additive; diversity is applicable only in calculating instantaneous load. In view of the large heat gain in an elevator machine room, it is well to equip the room with thermostatically controlled forced ventilation arranged to keep the room temperature from exceeding 90F. Occasionally spill air from an air conditioning system can be utilized for cooling purposes. In older designs using m-g sets, the losses would be 30,000 Btu/h per machine or 150,000 Btu/h for the bank.

(b) To calculate the monthly energy cost, an estimate must be made of the total usage of the system. Assuming the system to be in an office building, a reasonable breakdown of operation during a 24-h day would be

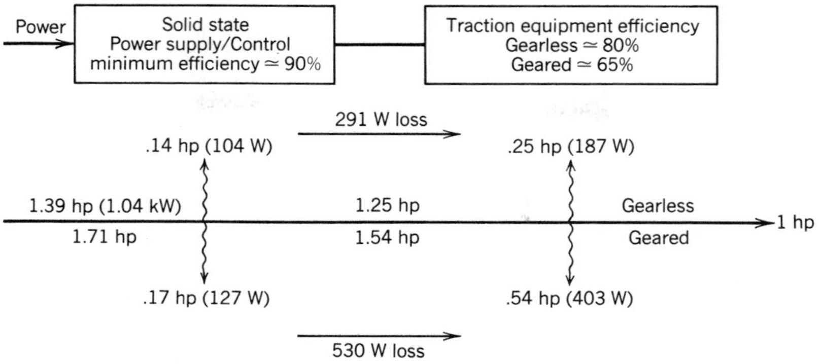

Fig. 24.2 Block diagram showing losses in the system per horsepower delivered to the elevator car, and the equivalent wattages. Note that the losses in a geared system are almost double those of a gearless one. Figures shown are for solid-state controls.

2 h peak use
2 h 70% of peak
6 h 50% of peak
14 h 10% of peak

This gives an average of 30% of peak load for the bank. Therefore, per car

$$\begin{aligned} \text{energy} &= 30\% \times \text{total losses} \times 24 \text{ h} \\ &= 0.3 \times 6.72 \text{ hp} \times 24 \text{ h} \\ &= 48 \text{ hp-h} = 36 \text{ kWh/day/car} \end{aligned}$$

Monthly cost would be

$$36 \frac{\text{kWh}}{\text{day}} \times 25 \text{ days} \times \$0.08$$

$$\simeq \$72\text{/month/car}$$

$$\simeq \$360\text{/month for the bank}$$

In a design using m-g sets in lieu of the solid-state controls, the monthly cost would be $630 for the bank.

24.3 Energy Conservation

A reduction in energy consumption can be accomplished by implementing the following recommendations:

For Existing Elevators

1. Increase interval during nonpeak hours.
2. Replace m-g sets with thyristor controls.
3. Recycle machine room waste heat.
4. Shut down some units completely during off hours.

For Building in the Planning Stage

1. Design for maximum trip time.
2. Use the lowest speeds possible, within a type—that is, geared and gearless.
3. Use gearless equipment whenever possible (see Fig. 24.2).
4. After construction, implement the recommendations for existing elevators above.

Since elevator shafts have a powerful stack effect, measures should be taken to counteract this during the heating season.

24.4 Emergency Power

Major power failures and local "brownouts" have demonstrated forcefully the need for a standby or emergency power source of adequate size to operate a building's elevators. Few experiences are so harrowing as being trapped in the crowded confines of a small box suspended in a long vertical shaft, with little or no light, and complete strangers for companions.

A common misconception relative to elevators is that on failure of power, the cars will automatically descend to the nearest landing where exit will be possible. In actuality, the brake is set immediately on power outage and the car remains stationary. Hydraulic cars can be lowered by operation of a manual valve; *small* traction cars can be cranked to a landing by hand; but large cars are fixed in position. This is particularly bad for cars in blind shafts, that is, express shafts with no shaftway doors. In such cases escape from the cars via hatchway is not practicable and, when emergency power is not available, breaking through the shaftway walls is the only recourse.

In addition to simple inconvenience, loss of elevator service in facilities such as hospitals, mental institutions, and jails constitutes a danger to life. For this reason most codes require that emergency power be available in public buildings to operate at least one elevator at a time, and for lighting and communications. Most installations separate the emergency power functions, providing a diesel generator for traction power and separate individual elevator battery packs for communications, lighting, and preferably, also the car fan. The latter two items can be furnished by the elevator manufacturers with the cars, as an option. Typical self-contained emergency lighting units are shown in Fig. 24.3.

The generator is normally sized to supply one elevator motor at a time, with manual or automatic switching arranged between unit controllers. Thus each car in turn can be brought to a landing and thereafter a single car retained in service. Obviously, if it is desired to operate more than one car, a larger generator can be installed. This might well be the case in a multiwing building with critical service requirements, such as a hospital.

The amount of power required, the size of the emergency generator, and the equipment size necessary to absorb regenerative power are all data that can be furnished by the consulting engineer and the elevator manufacturer.

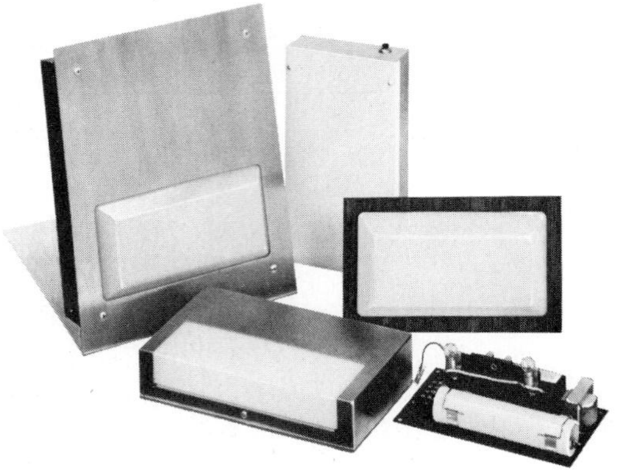

Fig. 24.3 *Typical self-contained emergency lighting units. Each contains, in addition to the light source, a battery and the requisite controls, which in the case of a flourescent source includes a d-c/a-c inverter. (Photo courtesy of Nylube Products.)*

OTHER CONSIDERATIONS

24.5 Fire Safety

Most codes specify the procedure that the elevator control equipment must implement once a fire emergency has been initiated. Details vary somewhat, but in general the actions are these:

1. A fire signal may be introduced manually at a fire station or at the lobby control panel, or automatically by the action of a detector or water-flow switch.
2. The fire signal will cause all cars to close doors and immediately return to the lower terminal, without stopping. All car floor buttons and emergency stop switches are disabled, as are all hall call buttons.
3. All controls that can be affected by smoke or heat are disabled.
4. Once recalled, the cars park with doors open and are thereafter operable manually only, by activating a fireman's key in the car panel. See Section 23.25. The cars can then be used by trained personnel to transport fire personnel and equipment, and for evacuation. In the event of a false alarm, the emergency procedure can be overridden at the lobby panel and the system returned to normal while the source of the alarm is located. (This is a particularly important feature in large buildings with automatic fire alarm systems containing hundreds of fire, smoke, and water-flow detectors. See Section 13.19.)

24.6 Elevator Security

The problem of physical security in elevators is a serious one, inasmuch as the traveling elevator is

Man Trying To Hide In Corner Under Camera

Fig. 24.4 *Wide-angle TV camera intended for elevator cab surveillance. A prominent, printed warning in the cab is an integral part of the system's effectiveness. (Photo courtesy of Visual Methods, Inc.)*

an enclosed space that can be rendered inaccessible simply by pressing the emergency stop button. Thereafter, an attacker can escape at the floor of his choice. To ameliorate this danger to an extent, elevators are equipped with alarm buttons, which alert residents and security personnel, if any. Every elevator, by code, must be equipped with communication equipment, but a hand-held phone is not practical in a security-emergency situation. More effective is an open two-way communication system with "no-hands" operation, in the car. When a closed-circuit TV monitor is added, utilizing a wide-angle camera in each car (see Fig. 24.4), the security problem will have been addressed to a considerable extent. Obviously, using a communication and TV system presupposes continuous manning of the building security desk.

At least one major manufacturer now markets a device that will alarm automatically on detecting sudden violent motions or a sharp pointed instrument. The matter of handling the alarm is problematic, since an automatically locked door can be forced open manually, and, furthermore, the advisability of locking a violent person *in* the elevator with potential victims is questionable.

Security in buildings is often a matter of restricting access to (and from) a floor or car. This can be accomplished by push-button combination locks or coded cards, the proper use of which will permit access (see Chapter 22). However, if a second person happens to accompany the authorized person, the effectiveness of this type of access barrier is seriously compromised. In sum, the most effective security system is a combination of automatic monitoring and access devices, coupled with continuous supervision by persons who know the appropriate action to take in an emergency.

24.7 Elevator Noise

As already noted, elevator operation, with its rotating, sliding, and vibrating masses, can be a cause of serious noise disturbance to quiet areas such as sleeping rooms, libraries, and certain types of office space. Noise can be reduced by the appropriate application of vibration isolators (e.g., between guide rails and the structure) and by proper control, such as door operation only when required, and not when parking, but primarily by placing noise-sensitive areas away from shafts and machine rooms. Furthermore, the clatter and whirring sound of the older machine room, caused by relays, step switches, m-g sets, and sliding contacts, can be entirely eliminated by the use of solid-state equipment.

24.8 Elevator Specifications

Two basic types of specification for elevator equipment, as for other types of equipment, are utilized. The performance specifications describe job conditions and invite contractors to submit detailed proposals including full engineering. The burden of comparing proposals then falls on the owner, who—if competent to properly perform such an evaluation—would most probably do better to utilize the equipment-type specification.

In recent years the use of performance specifications has increased because of the advent of preengineered, premanufactured systems. These are supplied by the major manufacturers and have these advantages:

1. Approximately 10% lower cost than a custom-designed system.
2. A complete engineered and tested system

whose performance and cost are known exactly.
3. Rapid delivery.
4. Minimum supervision required by the owner and architect.

If architects decide to use a custom-designed system, they must prepare detailed drawings and specifications. The specifications must include:

Elevator type, rated load and speed.

Maximum travel.

Number of landings and openings.

Type of control and supervisory system.

Details of car and shaft doors.

Signal equipment.

Characteristics of power supply.

Finishes.

This last item can be left as a dollar allowance for architectural treatment of the car interior. Since the selection of, and technical specifications for, elevators are specialized and complex, the services of an elevator consultant are usually required.

In addition to the technical portions of the specifications, it is imperative that the following items be covered in detail.

(a) Owner's Responsibility. The *construction* contractor normally provides:

1. The hoistway, including properly designed, lighted, drained, waterproofed, and ventilated machine room and pit.
2. Access doors, ladders, and required guards.
3. Guide rail bracket supports, and support for machine and sheave beams.
4. Electric feeder terminating in switch in machine room.
5. Hoistway outlets for light, power, and telephone.
6. Temporary light and power during construction.
7. Concrete machine foundations.
8. Vents, holes, and other work to satisfy fire codes.
9. All cutting, patching, and chasing of walls, beams, masonry, and so on.
10. Coordination of all work.
11. Any special work, as negotiated.

(b) Elevator Contractor's Responsibility. Provide complete, working, tested, and approved system in accordance with specifications, plus any special work such as painting, special tests, scheduling of work, and temporary elevator service.

(c) Special Job Conditions. These include work restrictions, scheduling, penalties or bonuses, test reports, and the like.

In alteration and modernization work, the problems of coordination are more complex, and an elevator contractor experienced in this type of work must be selected. To this end, in all elevator contract work, bids should be solicited from parties named on qualified bidder lists. Part of an elevator contract comprises maintenance of the installation for a specific period after completion. Contractors with poor maintenance facilities should be avoided.

SPECIAL ELEVATORS

24.9 Unique Traction Designs

Elevator engineers have always provided interesting and original solutions to unusual problems. A detailed analysis of these solutions is beyond our scope here; however, a rapid review is of interest. For more details see Strakosch (1983).

(a) Sky-Plaza System. For skyscraper buildings such as the World Trade Center in New York and multiple-use buildings such as the John Hancock Center in Chicago—both of which are, in effect, stacked multiple buildings—the elevator solution involves transporting large groups of people from the street lobby to an upper lobby called a sky plaza. At this point the passengers transfer to another elevator to continue their upward journey. The advantages of this system are

1. Reduction in the space occupied by elevators, since the shafts do not extend the entire height of the buildings.
2. Interrupting the otherwise lengthy vertical trip by the horizontal break at the sky lobby.

(b) Double-Deck Elevators. This is an old technique, recently revised to answer the needs of tall buildings such as the Sears and Citicorp Tow-

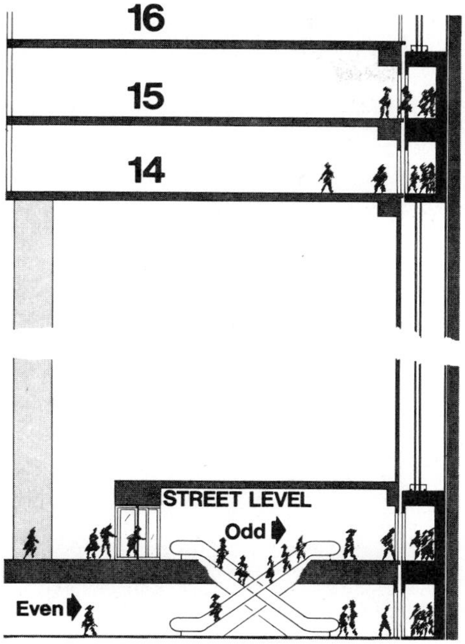

(a)

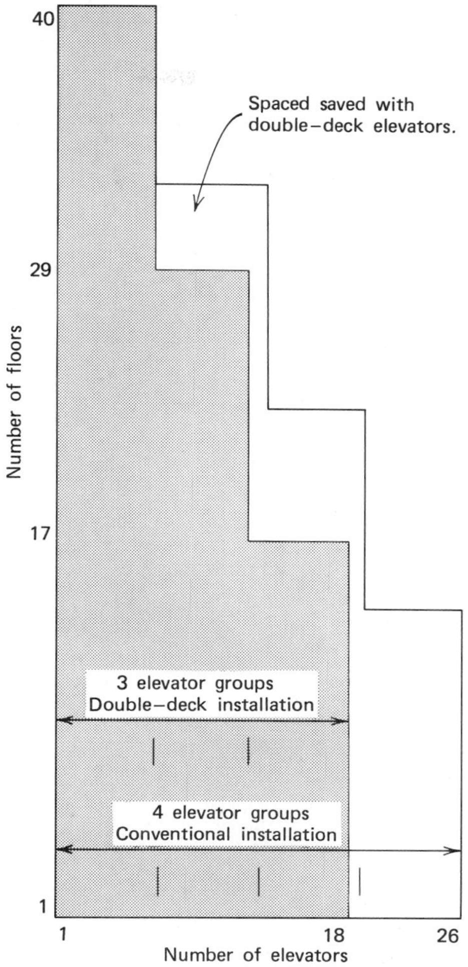

(b)

Fig. 24.5 (a) *The double-deck car serves to increase cab capacity and decrease shaft space. Coincidence of calls in the upper and lower cabs reduces the number of local stops made by the double car.* (b) *Graphical representation of the space saved in a 40-story building by the use of double-deck elevators. (Courtesy of Otis Elevator.)*

ers. The principal purpose is to limit the otherwise prohibitively large amount of space occupied by elevator shafts (see Fig. 24.5). The double-deck car increases shaft capacity, decreases the number of local stops, and increases the rental area available. This technique can also be combined with sky lobbies for further space economy, as was done in the Sears Tower.

(c) Observation Car Elevators. By placing the traction lifting mechanism *behind* the car, attaching the car at the back, and using a glass-enclosed, observation-style car, a spectacular unit can be constructed that becomes an attraction in itself. The basic construction is shown in Fig. 24.6, and a well-designed example is seen in Fig. 24.7. If the back screen is treated properly, the car gives the impression of movement without any apparent motive force or machinery.

(d) Slant Elevators. Although elevators are normally conceived as traveling vertically, this is not necessarily so. Slant or inclined elevators have been constructed in numerous locations. The design varies depending on the angle of incline. Figure 24.8 shows a well-known application of an inclined elevator.

24.10 Hydraulic Elevators

All the foregoing elevators are traction types; that is, they are raised and lowered as a result of the tractive force of cables attached to, or passing under, the car. In contrast to these, the hydraulic or plunger elevator is raised and lowered quite simply, by means of a movable rod (plunger) rigidly

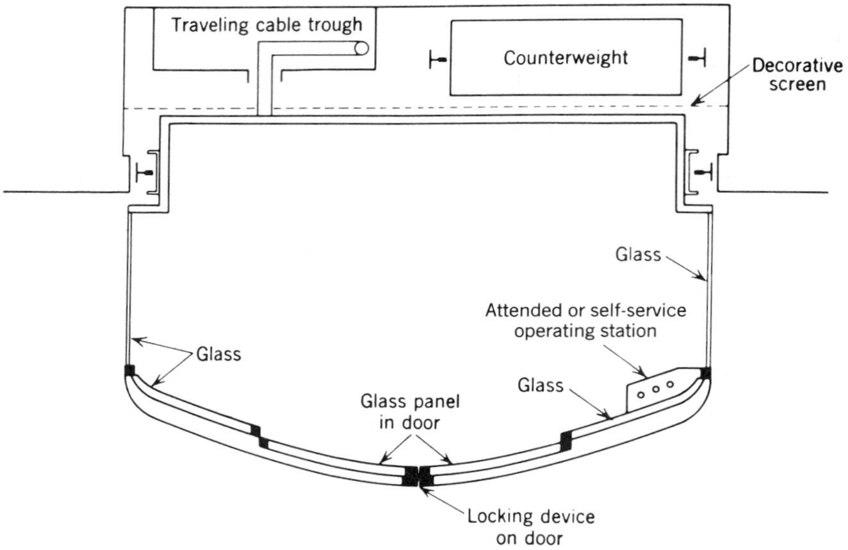

Fig. 24.6 *Basic design of a traction-type observation elevator. Cars are small to permit all the occupants to enjoy the view through the cab's glass walls. An inside railing (not shown) is frequently used to prevent passengers from contacting the glass cab walls. (From Strakosch, 1983. Reprinted with permission of John Wiley & Sons.)*

Fig. 24.7 *A twin installation with unusual car design, at the Hilton Hotel in Atlanta. Note the effectiveness of the screening, making the traction equipment barely visible. (Photo courtesy of Hauenstein and Burmeister, Inc., which supplied the cars for this installation.)*

fixed to the bottom of the elevator car. The absence of cables, drums, m-g sets, elaborate controllers and safety devices, and penthouse equipment makes this system inherently inexpensive and often the indicated choice for low-speed (up to 200 fpm), low-rise (up to 65 ft) applications, where construction of the plunger pit does not present difficulties and absence of a penthouse is desirable.

The components of a typical hydraulic unit are shown in Fig. 24.9. This system operates very much the same way as a hydraulic automobile jack. Oil from a reservoir is pumped under the plunger thereby raising it and the car. The pump is stopped during downward motion, the car being lowered by gravity and controlled by the action of bypass valves, which also control the positioning of the car during upward motion. Control systems normally used are similar to those for traction types, for example, single push-button, collective, and selective collective. Similarly, door arrangements are the same as in traction types, that is, single slide, center opening, and two speed. Automatic leveling is readily available and is standard on all automatic units. Typical layout and dimensional data for standard plunger units are given in Fig.

Fig. 24.8 The St. Louis Gateway Arch has a 10-passenger, inclined elevator in each leg. Placement of doors, arrangement of counterweight, and size of shaft all depend on the angle of incline. The car moves 82 ft horizontally during its 386 ft of total travel, at an incline of approximately 12°. (Photo courtesy of Bethlehem Steel Corporation, which supplied the elevator rope for this installation.)

24.10, along with capacities and application recommendations.

Where drilling a plunger hole presents difficulties, a hydraulic installation using a telescoping plunger can be installed. These cars are very limited in rise and speed and are applicable only to small two- to three-story buildings. A cutaway for this type of unit is given in Fig. 24.11.

From the point of view of the construction, the major inherent advantage of hydraulic units is the absence of overhead traction equipment. In Fig. 24.9 we see that only the guide rails project above the car and, if these are camouflaged, the impression of a free-standing elevator car is given. This effect can be used to good advantage *inside* large, open spaces such as exist in shopping malls and stores; when combined with glass-enclosed, observation-type cabs, the effect is striking (see Figs. 24.12 and 25.4).

The major inherent *disadvantage* of the hydraulic elevator is its operating expense. Since it is not counterweighted, it requires a relatively large motor to drive the oil pump, and *all* the energy is lost in heat. As an example of the operating cost consider a 3500-lb, 125-fpm hydraulic unit in a department store. Such a unit requires a 40-hp motor. Assuming the unit to be in operation 10 h/day, 6 days a week, and assuming a normal 60%

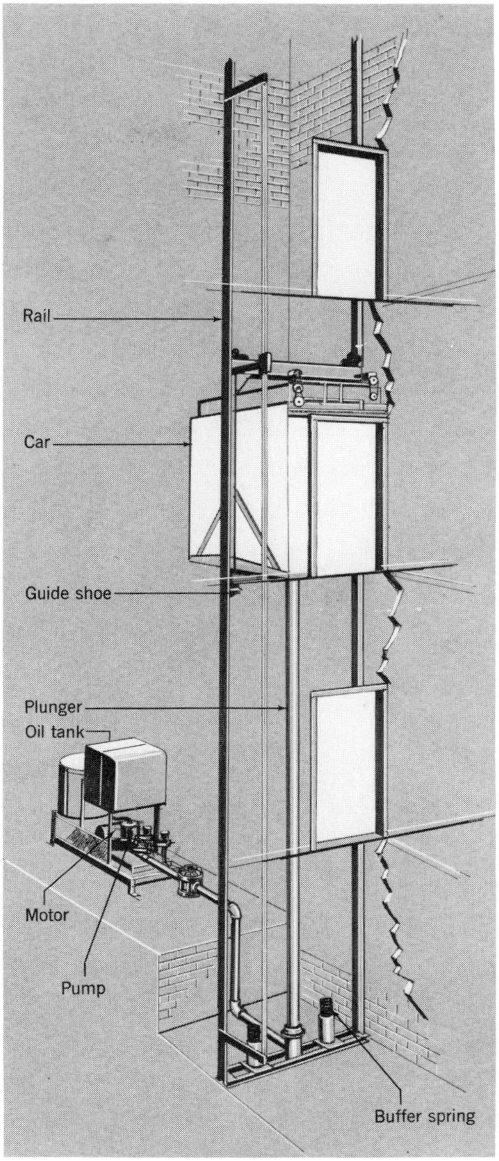

Fig. 24.9 Phantom view of a typical plunger-type hydraulic elevator installation. (The oil pump is actually inside the tank, but it is shown outside for clarity.) (Courtesy of Westinghouse Elevator Co.)

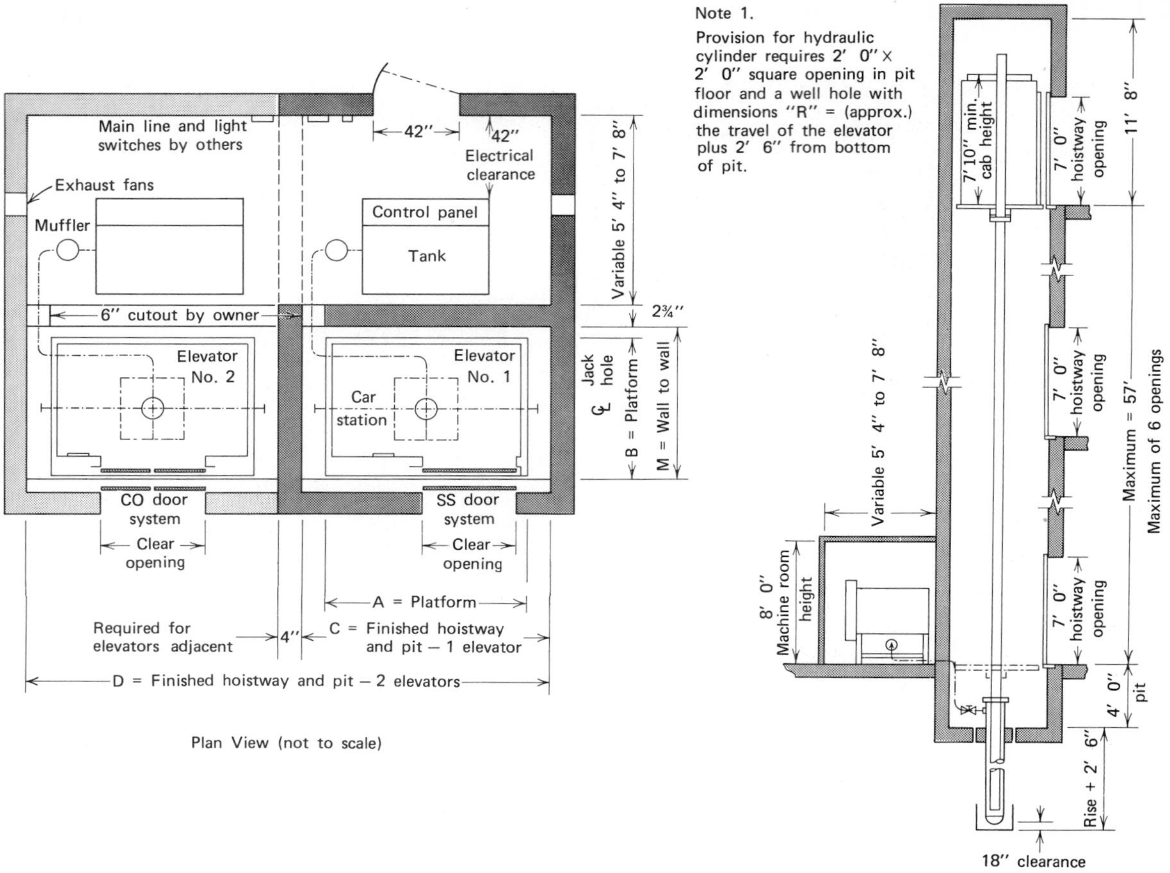

Capacity (lb)	Standard Speed (fpm)	Door System	Maximum Standard Openings	Maximum Standard Rise	Clear Opening	Platform A Width	Platform B Depth	Hoistway C Clear Hatch	Hoistway D Clear Hatch	Hoistway M Wall to Wall	Motor hp	Application
1500	75	SS	5	57′–0″	2′–8′	5′–0″	5′–0″	6′–6″	13′–4″	5′–9″	15–30	Residential
2500	125	SS	6	57′–0″	3′–0″	6′–2″	5′–0″	7′–4″	15′–0″	5′–9″	20–40	Residential,
2500	125	SS	6	57′–0″	3′–6′	7′–2″	5′–0″	8′–4″	17′–0″	5′–9″	20–40	small office
2500	125	CO	6	57′–0″	3′–6″	7′–2″	5′–0″	8′–4″	17′–0″	5′–9″	20–40	Residential, office, hotel
3000	100	CO	6	57′–0″	3′–6″	7′–2″	5′–6″	8′–4″	17′–0″	6′–3″	25–50	Store, office
3500	100	CO	6	57′–0″	3′–6″	7′–2″	6′–2″	8′–6″	17′–4″	6′–11″	30–60	Store, office
4000	100	CO	6	57′–0″	4′–0″	8′–2″	6′–2″	9′–6″	19′–4″	6′–11″	30–60	Store, office

***Fig. 24.10** Typical dimensional, capacity, and layout data for conventional plunger-type hydraulic elevator. Door systems are single slide (SS) or center opening (CO). (Extracted with permission from published data of Westinghouse Elevator Co.)*

time-in-operation figure, we have (remembering that the motor operates only in the up direction),

$$\text{energy used/day} = \frac{40\text{ hp}}{0.82\text{ eff}} \times \frac{0.746\text{ kW}}{\text{hp}} \times 60\% \times 10\text{ h} \times 1/2 = 110\text{ kWh/day}$$

At \$0.08/kWh we have

$$\text{monthly energy cost} = 110 \times \frac{6\text{ days}}{\text{week}} \times \frac{4.33\text{ weeks}}{\text{month}} \times \$0.08 = \$229/\text{month}$$

Compare this to the previously calculated (Section 24.2) monthly energy cost of \$72 when using solid-state equipment and \$126 with m-g set power

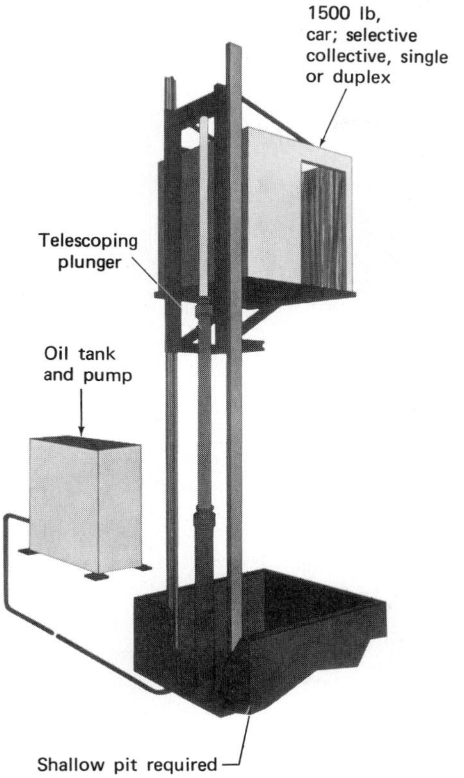

Fig. 24.11 *"Holeless" hydraulic elevator is similar in construction to the standard plunger-type, except limited to 22-ft rise, 1500-lb car, and 75 fpm. (Courtesy of Otis Elevator.)*

Fig. 24.12 *This four-story-high, exterior-plunger elevator in the Los Angeles rapid transit system is practical and fulfills viewing function. (Photo courtesy of Montgomery Elevator Co.)*

supply, for a 3500-lb, 600-fpm traction car, for an appreciation of the value of a counterweight.

24.11 Residential Elevators and Chair Lifts

Although the special needs of the handicapped have only lately been recognized in legislated requirements, the elevator industry has been providing for the handicapped for years, on a private basis. Chair lifts shown in Fig. 24.13, wheelchair lifts as in Fig. 24.14, and private elevators as in Fig.

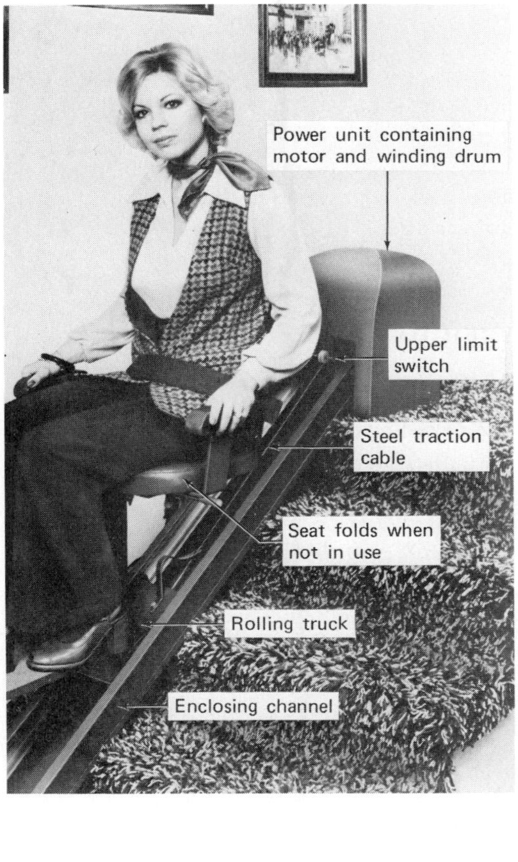

Fig. 24.13 *Typical layout for a single-seat, folding chair lift. The seat is rigidly attached to a rolling truck mounted inside an enclosed steel track. The truck is pulled by a steel cable operated from a winding drum in the power unit at the top of the stairs. (Photo courtesy of the Inclinator Company of America.)*

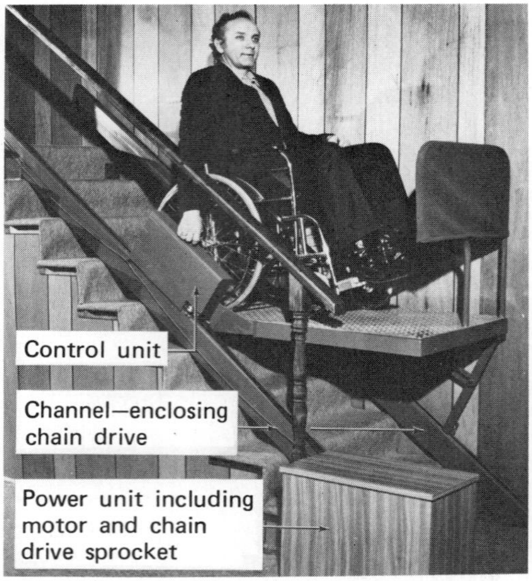

Fig. 24.14 *A wheelchair lift installed relatively unobtrusively on a stair. The platform forms a bottom step, leaving the stairway open for normal use. This unit is arranged to descend to the lower terminal in the event of a power failure. It is roller-chain driven and operates at 22 fpm. (Photo courtesy of Flinchbaugh/Murray Corp.)*

Fig. 24.15 *A 450-lb, 30-fpm residential elevator in an open installation. The operating mechanism is similar to that of the chair lift in Fig. 24.13. The cab is rigidy attached to a rolling truck that is lifted by a winding drum. The track, within which the truck rolls, is easily seen here, although it can readily be concealed. The power unit and drum can be located at the top, bottom, or center of the installation. Limit switches prevent overrun. Control is manual or automatic, as selected, with call and send buttons at each landing. A 6-in. pit is required below the bottom landing. (Photo courtesy of Inclinator Company of America.)*

24.15 are widely used to overcome the stair barrier in private homes. These items are also covered by the elevator code and must be equipped with the safety devices there specified. All units operate on household electric current and require minimal maintenance.

FREIGHT ELEVATORS

24.12 Freight Elevators: General

The preceding material, which dealt with passenger traffic, had as its prime consideration the most effective solution to the problem of vertically transporting a given number of persons. The problem with respect to freight elevators is similar: to transport a given tonnage of freight efficiently, economically, and quickly. The service car in a facility can be considered to be a freight car but, if utilized for passenger duty at all, it must meet passenger service requirements. If passenger duty is not required, or if much freight is to be handled, a car designed specifically for freight is used.

Factors to be considered in freight elevator selection, in addition to tonnage movement per hour, are size of load, method of loading, travel, type of load, type of doors, and speed and capacity of cars. Since these factors are interrelated, the actual process of selection involves making assumptions on the basis of recommendation and then arriving at a solution, very much as was done for passenger elevators.

It is beyond the scope of this book to discuss in detail the selection of material-handling elevators because of the large number of considerations involved. Therefore we shall restrict ourselves in the following paragraphs to descriptive material and recommendations. Also, since freight elevators form such an important link in industrial processes, a careful and detailed material-flow study should be made before freight elevators are selected. Elevator manufacturer's representatives and materials-handling consultants can be very helpful in this regard.

24.13 Freight Car Capacity

Figure 24.16 is a section through a typical freight car shaft. Capacities corresponding to a specific platform size (Tables 24.1 and 24.2) are due to the varying square foot loads that are permissible. Cognizance of this is taken by the ANSI Code for

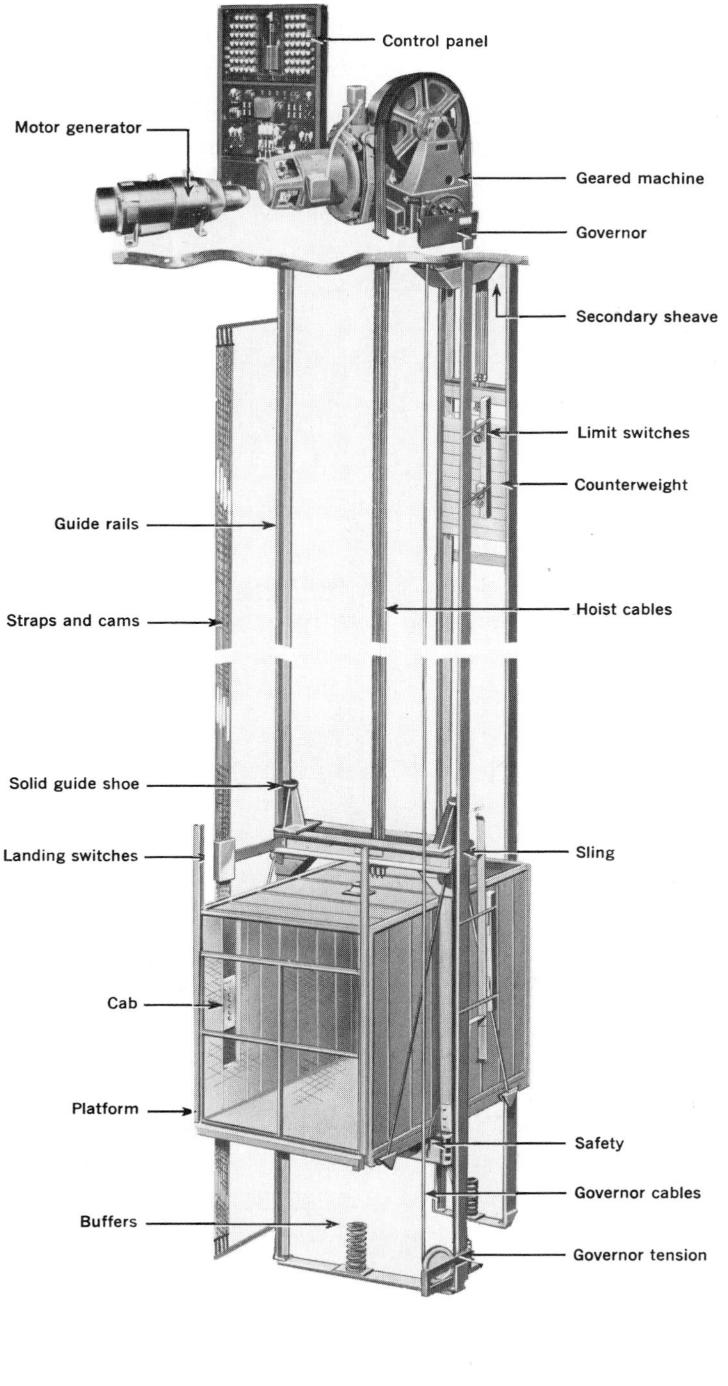

Fig. 24.16 *Components of a typical freight elevator installation utilizing a variable-voltage controlled geared traction machine. The sling that lifts the car is frequently arranged with double sheaves over which the hoisting ropes pass. This roping arrangement increases the mechanical advantage of the lifting ropes. The control panel shown is applicable to a one- or two-car installation. (Photo courtesy of Westinghouse Elevator Co.)*

TABLE 24.1 **Loading by Hand or by Hand Truck**

Capacity (lb)	*Platform Width*	*Size (Depth)*
2500	5′4″	7′0″
3000	6′4″	8′0″
3500	6′4″	8′0″
4000	6′4″	8′0″
5000	8′4″	10′0″
6000	8′4″	10′0″
8000	8′4″	12′0″
10,000[a]	8′4″	12′0″
12,000[a]	10′4″	14′0″

[a]Elevators of this size should always be considered for industrial truck loading.

TABLE 24.2 **Loading by Industrial Trucks**

Capacity (lb)	*Platform Width*	*Size (Depth)*
10,000	8′4″	12′0″
12,000	10′4″	14′0″
16,000	10′4″	14′0″
18,000	10′4″	16′0″
20,000	12′4″	20′0″

Elevators, which has established three load classifications for freight elevators.

Class A. General Freight Loading, by hand truck. Single items may not exceed 25% of the car-rated load. Rated load is based on 50 pounds per square foot (psf) of net inside platform area.

Freight Elevators — Traction

LIGHT AND MEDIUM DUTY FREIGHT ELEVATORS						
CAPACITY	2500#	3000#	4000#	6000#	8000#	10,000#
A	5′-4″	6′-4″	6′-4″	8′-4″	8′-4″	10′-4″
B	7′-0″	8′-0″	8′-0″	10′-0″	10′-0″	14′-0″
C	5′-0″	6′-0″	6′-0″	8′-0″	8′-0″	10′-0″
D	7′-4″	8′-4″	8′-4″	10′-4″	10′-10″	12′-10″
L	13′-0″	14′-0″	14′-0″	14′-0″	14′-0″	15′-0″

HEAVY DUTY POWER TRUCK FREIGHT ELEVATORS					
CAPACITY	10,000#	12,000#	16,000#	18,000#	20,000#
A	8′-4″	10′-4″	10′-4″	10′-4″	12′-4″
B	12′-0″	14′-0″	14′-0″	16′-0″	20′-4″
C	8′-0″	10′-0″	10′-0″	10′-0″	12′-0″
D	11′-4″	13′-6″	14′-0″	14′-2″	16′-6″
L	14′-0″	15′-0″	15′-0″	17′-0″	21′-0″

MINIMUM PIT & OVERHEAD DIMENSIONS FOR LIGHT & MEDIUM DUTY FREIGHT ELEVATORS				
CAR SPEED	50	75	100	200
O	16′-0″	16′-0″	16′-0″	16′-0″
P	5′-6″	5′-6″	5′-6″	6′-0″

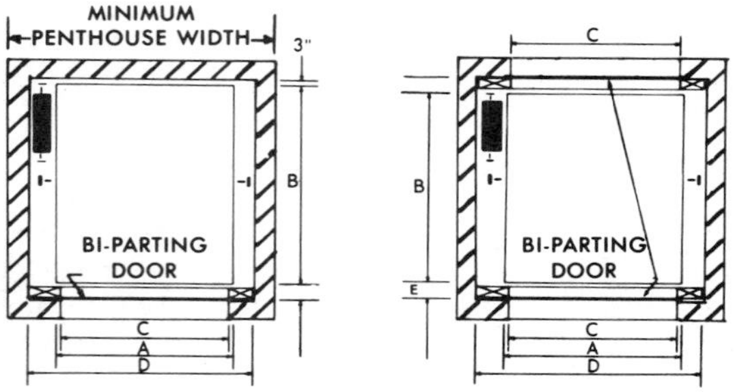

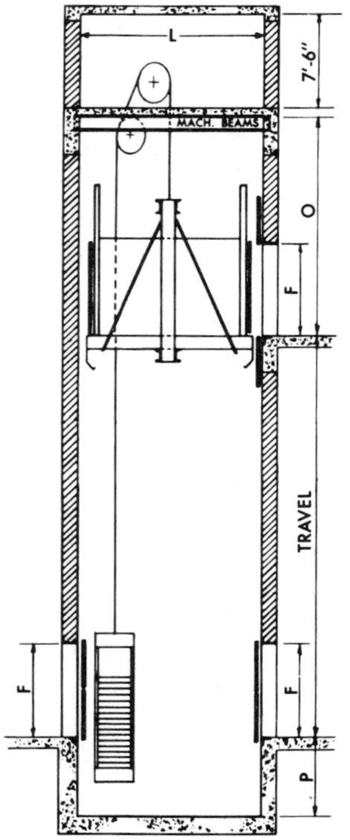

Fig. 24.17 Typical dimensional data for traction-type freight elevators. (Courtesy of Montgomery Elevator Co.)

Class B. Motor Vehicle Loading. Car will carry automobiles or automobile trucks. Rating is based on a load of 30 psf of net inside platform area.

Class C. Industrial Truck Loading. Rated load is based on 50 psf of net inside platform area. Car must have automatic leveling.

24.14 Freight Elevator Description

Since speeds are generally between 50 and 200 fpm, a geared-type machine is used almost universally.

The preferred system of control is collective, with variable-voltage, d-c supply. However, if the car is used infrequently (fewer than five trips a day), economy is very important, accurate leveling is not essential, and a rougher ride is tolerable, then a-c rheostatic control may be used. For low-rise jobs, a hydraulic unit is most often employed. These, as with the umv traction units, provide accurate control, smooth operation, and very accurate automatic leveling. Hydraulic units rarely exceed 60 ft in height, and operate at speeds up to 125 fpm. Accessories such as governors, safeties, and brakes are similar to those for passenger elevators previously described.

General-purpose freight elevators, whether traction (Fig. 24.17) or hydraulic (Fig. 24.18), in load ranges up to 20,000 lb, are standard design items, applicable to all types of commercial and industrial buildings. Heavier units are individually engineered. As with passenger elevators, structural reactions are supplied by the manufacturer to the architect, who is responsible for adequate

Freight Elevators — Oil Hydraulic

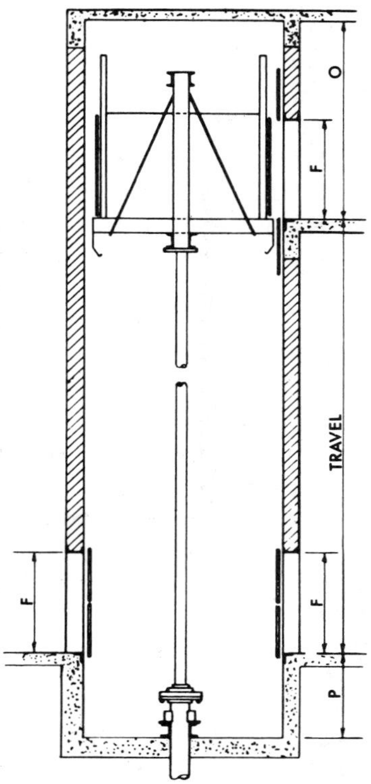

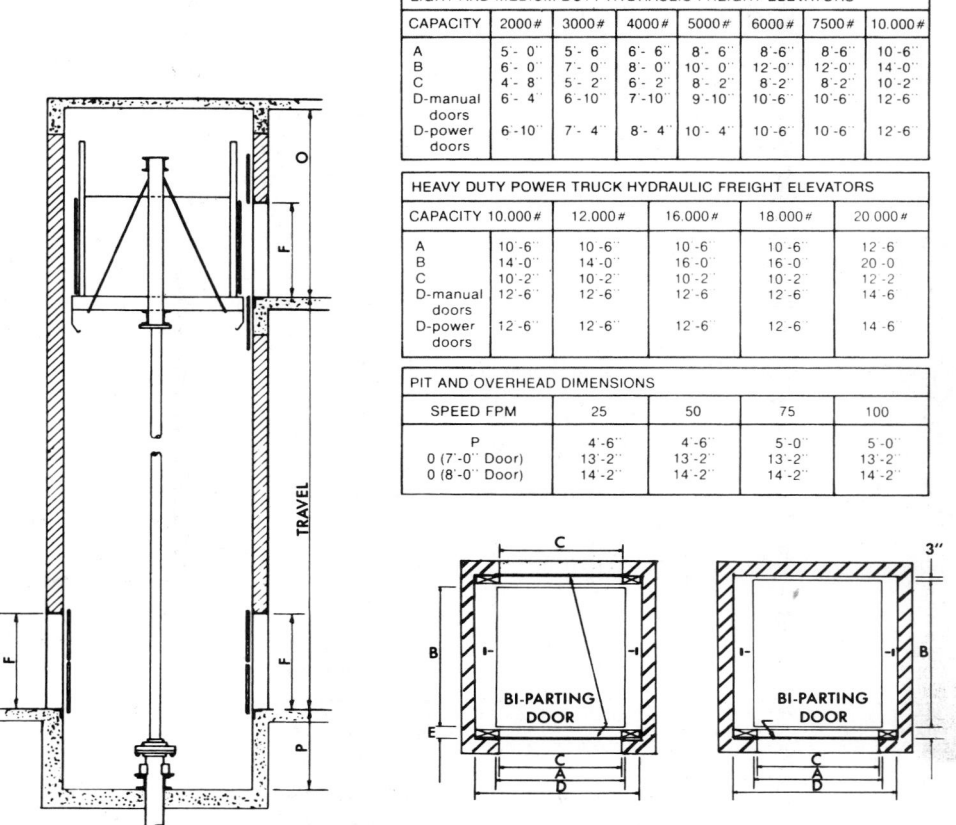

LIGHT AND MEDIUM DUTY HYDRAULIC FREIGHT ELEVATORS							
CAPACITY	2000#	3000#	4000#	5000#	6000#	7500#	10.000#
A	5'-0"	5'-6"	6'-6"	8'-6"	8'-6"	8'-6"	10'-6"
B	6'-0"	7'-0"	8'-0"	10'-0"	12'-0"	12'-0"	14'-0"
C	4'-8"	5'-2"	6'-2"	8'-2"	8'-2"	8'-2"	10'-2"
D-manual doors	6'-4"	6'-10"	7'-10"	9'-10"	10'-6"	10'-6"	12'-6"
D-power doors	6'-10"	7'-4"	8'-4"	10'-4"	10'-6"	10'-6"	12'-6"

HEAVY DUTY POWER TRUCK HYDRAULIC FREIGHT ELEVATORS					
CAPACITY	10.000#	12.000#	16.000#	18.000#	20.000#
A	10'-6"	10'-6"	10'-6"	10'-6"	12'-6
B	14'-0"	14'-0"	16'-0"	16'-0"	20-0
C	10'-2"	10'-2"	10'-2"	10'-2"	12'-2
D-manual doors	12'-6"	12'-6"	12'-6	12'-6"	14'-6
D-power doors	12'-6"	12'-6"	12'-6"	12'-6	14'-6

PIT AND OVERHEAD DIMENSIONS				
SPEED FPM	25	50	75	100
P	4'-6"	4'-6"	5'-0"	5'-0"
0 (7'-0" Door)	13'-2"	13'-2"	13'-2"	13'-2"
0 (8'-0" Door)	14'-2"	14'-2"	14'-2"	14'-2"

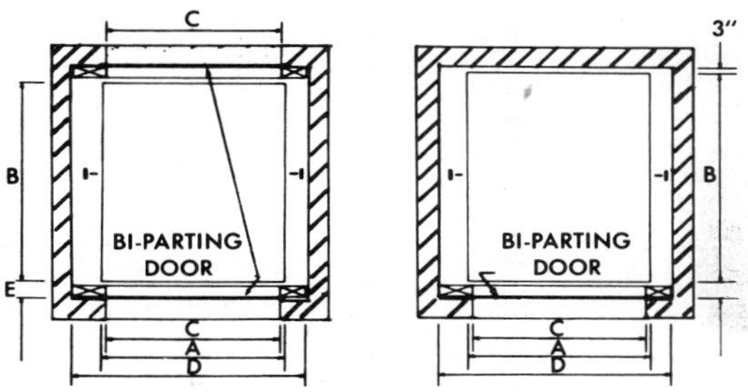

Fig. 24.18 Typical dimensional data for hydraulic-type freight elevators. (Courtesy of Montgomery Elevator Co.)

structional supports. This item is of great importance in larger car installations, since traction unit rails must be supported every few feet and additional steel provided to accomplish this.

24.15 Freight Elevator Cabs, Gates, and Doors

Cabs for freight service are normally built of heavy-gauge steel with a multilayer wooden floor, the entire unit being designed for hard service. Guarded ceiling light fixtures are required. Cab gates slide up vertically and are a minimum of 6 ft high. Hoistway doors are normally vertical lift, center-opening, manual or power operated. Both cab gate and hoistway doors are counterweighted and open fully to give complete floor and head clearance (see Fig. 24.19).

24.16 Freight Elevator Cost Data

The cost of a freight elevator installation, as with passenger elevator installation, is dependent on many factors, principally: capacity, type of control, use, and type of door operation.

Fig. 24.19 A pair of large hydraulic freight elevators for automotive use. These units are automatic, self-leveling, and equipped with power-operated biparting shaftway doors and cab gate. As with many freight installations, both ends of the cab are open. Note that two control stations are provided in the cab—one at each gate. (Photo courtesy of Harris-Preble Co., which supplied the doors in this installation. Hydraulic lifts are by Becker.)

Since exact pricing, like actual selection, is outside our scope, it is our recommendation that a reputable manufacturer or elevator consultant be consulted for such information. We can, however, make some general remarks on pricing, as follows:

1. Variable-voltage controlled equipment, depending on type, costs 20 to 50% more than rheostatically controlled equipment.
2. Above a basic two-floor rise, the cost increases linearly with rise.
3. Electric door operation can increase the cost of a car installation 10 to 25%.

As an example of *comparative* pricing, using a nominal 100 for an 8000-lb, 75-fpm, four-floor, manual door car with a-c rheostatic control and automatic leveling, the same car with variable-voltage control, 150 fpm, and electrically operated doors would cost approximately 180.

24.17 Garage Car-Parking Facility

Multistory automobile parking facilities that increase parking space by replacing interfloor ramps with freight elevators have the characteristic of very slow service. Since this is obviously unacceptable in a downtown commercial application, an alternative arrangement using a materials-handling approach is available. One such system consists of a central structural tower that supports a continuous loop of car-carrying "baskets," that is, a sort of Ferris wheel. An automobile enters a cage at a terminal, which can be anywhere in the loop—top, bottom, center—and is "parked" by the simple expedient of operating the controls at the terminal. The loop rotates, the car remains in its cage, and an empty cage is presented at the terminal. A car is retrieved by calling back the appropriate cage. In principle, the system is similar to the vertical conveyor described in Section 24.22.

The advantages of this parking system are as follows.

1. Very high space utilization efficiency; approximately 40 cu m (1400 cu ft) per mid-size car.
2. Minimal horizontal space requirement.

3. Simplicity of operation with minimum labor.
4. High degree of security to parked vehicles and to drivers.
5. Rapid operation.

Where height of the structure is limited but horizontal size is not, a similar arrangement can be constructed horizontally. In effect the vertical structure described above is lying on its side and is two ''stories'' high, and as wide as necessary to accommodate the required number of vehicles.

This parking system is used principally overseas because of the prevalence of small cars, which readily fit into the carrying cage. With the increasing number of compact cars in the United States, it may find increasing application in congested areas of American cities, as well.

MATERIAL-HANDLING EQUIPMENT

24.18 Material Handling: General

The material-handling equipment discussed briefly here is that which finds application in commercial and institutional buildings. Industrial materials handling is an entirely separate subject not germane to our purpose. The need to transport material within a building has always existed and until approximately a decade ago was done largely manually, with mechanical assistance. Thus offices used messengers; hospitals used dumbwaiters, service elevators, conveyors, and chutes. The single exception to this situation was (and still is) the extensive use of pneumatic tube systems in large stores. Today's systems accomplish the same end—that is, the transfer of materials—but automatically and, in general, much more rapidly. First cost of these systems is frequently high, but the reduction in labor and increase in speed generally yield a short payback period combined with a marked rise in efficiency.

Modern, commercial material-handling systems can be grouped roughly into four categories:

1. *Elevator-type systems*. These are vertical lift, car-type systems including the common dumbwaiter and ejection lifts, which are basically automated dumbwaiters.
2. *Conveyor-type systems*. These include horizontal and vertical conveyors.
3. *Pneumatic systems*. These include sophisticated pneumatic tube systems and pneumatic trash and linen systems.
4. *Other systems*. Systems that do not fit easily into any of the above categories include automated messenger carts and automatic track-type container delivery systems.

24.19 Manual Load/Unload Dumbwaiters

The use of dumbwaiters in structures of various types often provides the most convenient and economical means of transporting relatively small articles between levels. In department stores, such units transport merchandise from stock areas to selling or pickup counters; in hospitals dumbwaiters often transport food, drugs, linens, and other necessary small items. In multilevel restaurants, office dining rooms, and the like, dumbwaiters are almost always used for delivery of food from the kitchen and for return of soiled dishes.

Dumbwaiter cars are limited to a platform area of 9 sq ft and a maximum height of 4 ft. The car may be, and frequently is, compartmented by shelves. Normal speed ratings are 45 to 150 fpm, with a capacity of up to 500 lb. Cars may be of the traction (counterweighted) or drum (direct pickup) type. Control is normally ''call and send'' between two floors, although multibutton selector switch or central dispatching arrangements are available for applications with more than two floors. Loading may be floor, counter, or any other specified height (see Fig. 24.20 for typical layouts).

24.20 Automated Dumbwaiters

These units are also known as ejection lifts because of the method of delivery (see Fig. 24.21). They find their best application in institutions and other facilities that require rapid scheduled vertical movement of relatively large items. Thus, the device is ideally suited for delivery of food carts, linens, dishes, bulk liquids containers, and so on. The load can be a cart (Fig. 24.21*a* and 24.21*b*) or a basket (Fig. 24.21*c*) containing the items being transported. At the delivery terminal the item must be picked up and transferred horizontally to its final destination if remote from the delivery point.

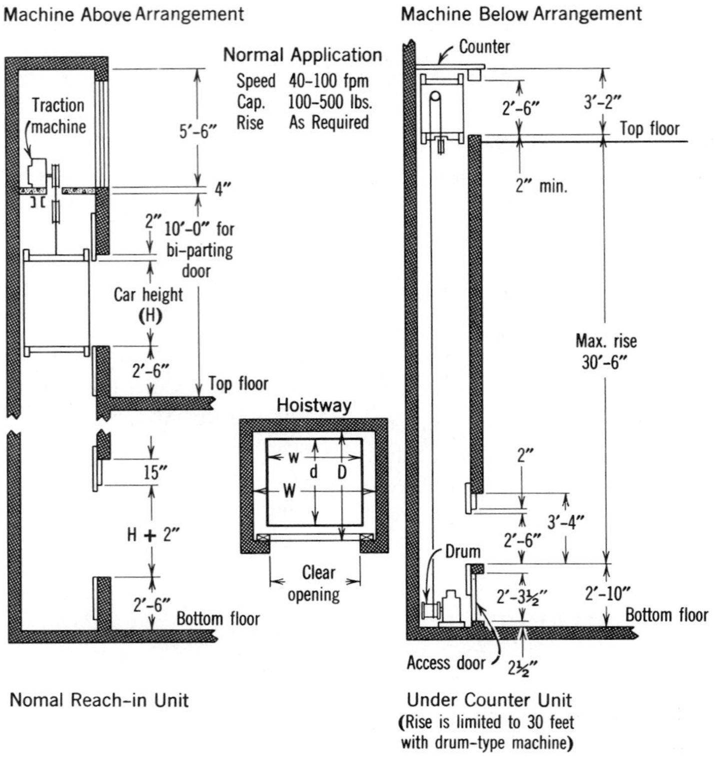

RECOMMENDED SIZES OF DUMBWAITERS

MAX. DUTY			CAR		HOISTWAY		
traction type machine 1:1 roping	traction type machine 2:1 roping	*drum type machine 2:1 roping	(w) width	(d) depth	(W) width	(D) depth without car gate	(D) depth with car gate
400 lb @ 100 fpm	500 lb @ 50 fpm	400 lb @ 45 fpm	2′-0″	2′-6″	3′-2″	2′-11″	3′-0¼″
↑	↑	↑	2′-0″	3′-0″	3′-2″	3′-5″	3′-6¼″
			2′-6″	2′-6″	3′-8″	2′-11″	3′-0¼″
			2′-6″	3′-0″	3′-8″	3′-5″	3′-6¼″
			3′-0″	2′-6″	4′-2″	2′-11″	3′-0¼″
↓	↓	↓	3′-0″	3′-0″	4′-2″	3′-5″	3′-6¼″
400 lb @ 100 fpm	500 lb @ 50 fpm	400 lb @ 45 fpm	3′-6″	2′-6″	4′-8″	2′-11″	3′-0¼″
Under-Counter Dumbwaiter		300 lb @ 50 fpm	2′-6″	1′-8½″	3′-5″	2′-1½″	——

Standard car heights—3′-0″, 3′-6″, 4′-0″

Fig. 24.20 Typical layout data for manual-load dumbwaiters.

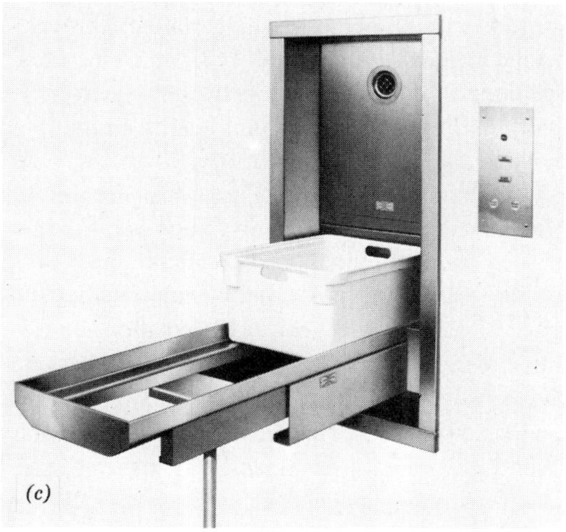

Fig. 24.21 (a) *Open ejection-lift unit showing cart ejection mechanism. Shaftway doors are vertical bi-parting.* (b) *The same unit being loaded with food carts and dispatched to the various floors of the hospital. At the upper floors the carts are rolled away by the attendants. Later the lifts are used to return soiled dishes and trays. (Photos courtesy of Courion Industries, Inc.)* (c) *Baskets can be used in a similar system for handling smaller loads. (Photo courtesy of AMSCO/American Sterilizer Co., Erie, Pa.)*

Payload capacity is available up to 1000 lb and car speeds up to 300 fpm. Round-trip time for a 200-fpm unit with 5 loading stations is approximately 2 min, with 10 stations about 2.5 min. Major considerations for these units are their relatively high cost and the large shaft area required.

24.21 Horizontal Conveyors

Although horizontal conveyors find their best application in industrial facilities, they are also usable in commercial buildings such as mail order houses, which require a continuous flow of material. Restrictions in application stem from inflexible right-of-way requirements, noise generation, and a degree of danger if left unprotected or exposed to unauthorized persons. Cost is relatively low, and capacity is virtually unlimited.

24.22 Vertical Conveyors

The action of this system is similar to the automated dumbwaiter in that the system transfers vertically and automatically loads and unloads, but the similarity ends there. Vertical conveyors are constructed with a moving continuous-loop chain

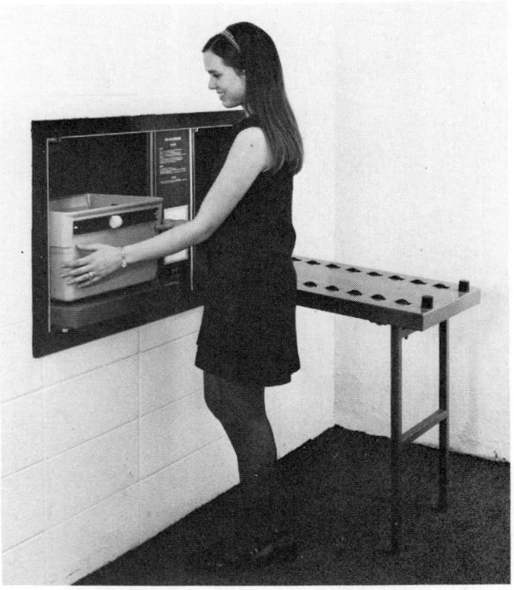

Fig. 24.22 Selective vertical conveyor. The basket placed into the sending terminal will be picked up and delivered to its address. At the right is the receiving terminal. (Photo courtesy of Translogic Corp.)

to which are attached carriages that pick up and deliver tote boxes. At sending and receiving stations the operator places the items to be moved (up to 40 lb) in the tote box, ''addresses'' the box in one of several ways depending on the system, and places it at a pickup point (see Fig. 24.22). The first empty carriage on the chain will pick up the box and deliver it to its address. Drawbacks of this system are the large shaft required, noise, and cumbersome arrangements when interfacing with horizontal conveyors. Cost is moderate.

24.23 Pneumatic Tubes

This well-tried system will undoubtedly continue in wide use until electronic data reproduction completely replaces the transfer of pieces of paper between two points. Pneumatic tube systems are available with 2¼- to 6-in. ranges of tube diameters (special shapes are also used) and with single or multiple loops. Older systems were positive pressure; newer systems are generally vacuum. Compressors run 25 to 35 hp, require substantial machine room and air manifold space, and are noisy. Piping of the standard 2¼-in. lines is not difficult, although larger size systems present some space problems due to large minimum bending radii. Overall, pneumatic tube systems (also called pneumatic dispatch systems) are cheap, reliable, and very fast.

24.24 Pneumatic Trash and Linen Systems

Rapid movement of bagged or packaged trash and linen from numerous outlying stations to a central collecting point is usually handled by this system. Linen systems are found generally in hospitals; trash systems in various facilities, frequently in conjunction with compactors. The system is basically a network of large pipes, negatively pressurized, with numerous loading stations throughout the building. Pipes are 16-, 18-, or 20-in. sizes, operating at high static pressure. A system normally can handle only one unit load at a time, but moves it so quickly (20 to 30 fps) that system capacity is large and delays are not encountered. Material placed into a loading station is picked up as soon as the previous load clears. Compressors are large and very noisy, requiring considerable space allocation and acoustical isolation. In addition to the main vacuum system, a high-pressure air line is required to operate the doors, and sprinkler heads must be installed every few floors. Overall costs are low to moderate. For the specific task performed, a cheaper and more efficient transfer technique is difficult to find.

24.25 Automated Container Delivery Systems

This ingenious arrangement employs captive and secure containers locked onto a motorized carriage that, in turn, is locked onto the track system. Power for the motor in the carriage is picked off a third rail at 24 V d-c. The entire assembly moves horizontally or vertically with equal ease, at a constant 100 fpm. The container, which can be opened only with a proper key and only at a station, carries a maximum load of 20 lb, which is adequate

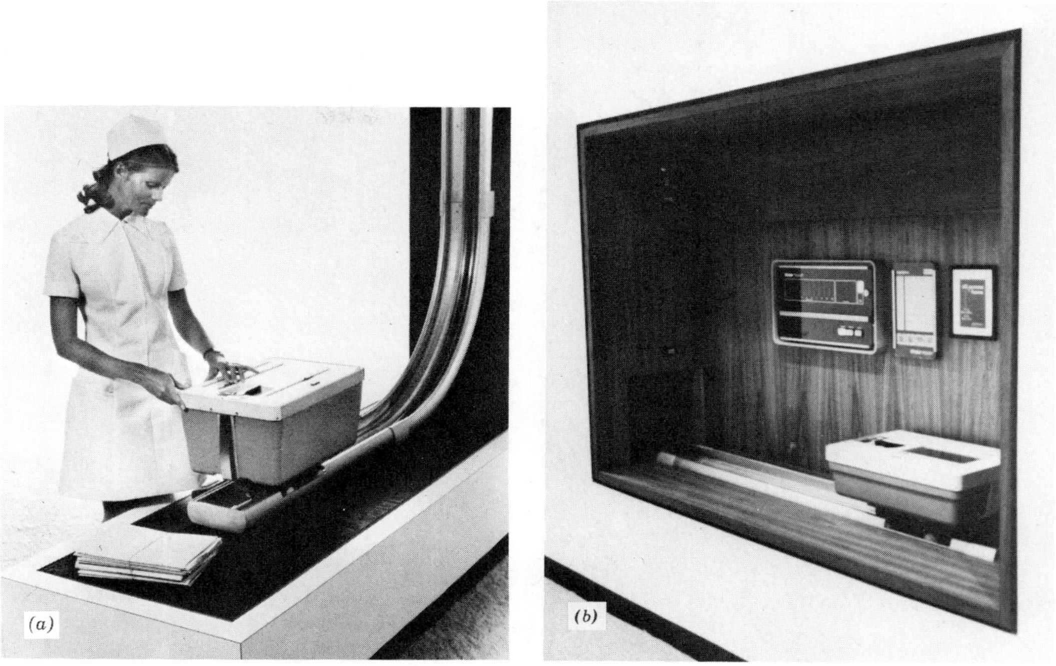

Fig. 24.23 (a) *Adjustment of the encoding magnets that address the car in a sending/receiving station for Mosler's "Material Distribution System." This car has a capacity of 1700 cu in. and 20 lb.* (b) *A smaller car, with a capacity of 850 cu in. and 11 lb. Mounted on the wall above the car in this library installation are a directory of stations and a network monitor panel that controls and monitors the entire system. (Photos courtesy of Mosler Systems Division.)*

for most uses. Addressing the unit is done simply by moving magnetic devices on the container. Right-of-way conflicts occasionally arise but, because of the ease with which the unit moves and the small cross section of the container (about 13 in. × 14 in.), the problems are not serious. The equipment is quiet, unobtrusive, easy to use, and attractive. It is also relatively expensive. See Fig. 24.23.

24.26 Automated Self-Propelled Vehicles

Robot self-powered vehicles follow a predetermined route on a single level. One type stops at scheduled points for pickup and delivery of interoffice papers. In effect, it is an automated office boy. It can be held at a stop as long as desired and, when released, will continue its scheduled trip. For offices with a large volume of incoming and outgoing mail and interdepartmental paper, this device works admirably. Another type (see Fig. 24.24) is an automated distribution system. Since it is restricted to a single level, it interfaces well with a selective vertical conveyor (Section 24.22), to give both horizontal and vertical capability.

24.27 Conclusion

The foregoing very brief summary simply describes the types of equipment available. For each facility being planned, the architect must study the material transfer problems, remembering that buildings not only handle and process but also generate material; an office building generates about a pound of waste per 100 square feet per day—a prodigious amount in today's large office structures. This type of dry waste can be compacted,

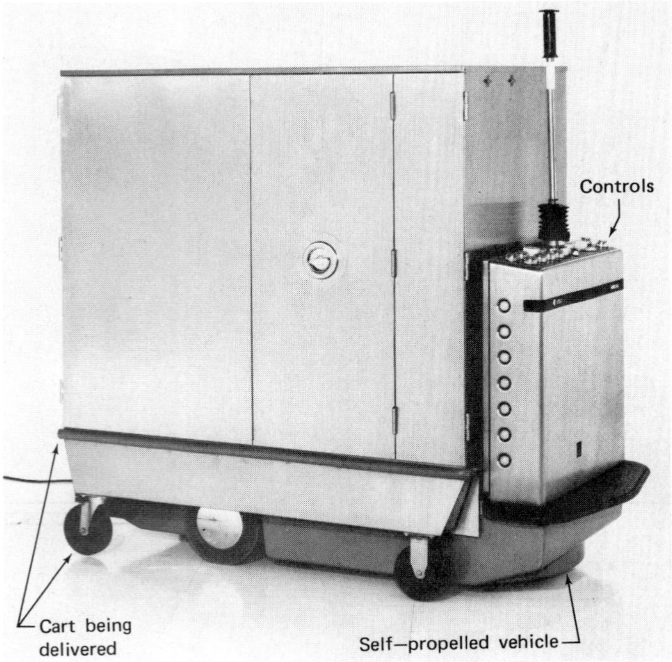

Fig. 24.24 *This automated, self-propelled vehicle follows a predetermined route, guided by wires buried in the floor. The bulk cart being delivered was picked up by the unit from an automated dumbwaiter system (see Section 24.20 and Fig. 24.21). Such vehicles can also be operated manually. (Photo courtesy of AMSCO/American Sterilizer Co., Erie, Pa.)*

bailed, and sold, unlike garbage and wet waste. In addition to considerations of the type of material being handled, there are factors of speed, scheduling, location of stations, labor and material costs, space requirements, noise generation, and energy requirements. To consider and evaluate all these factors in a large and complex facility is generally beyond the ability of the architect alone. Thus expert advice from consultants who specialize in materials handling and from manufacturers' representatives must be sought.

Reference

Strakosch, G. R. (1983). *Vertical Transportation, Elevators and Escalators,* 2nd ed., Wiley, New York.

25

MOVING STAIRWAYS AND WALKS

MOVING ELECTRIC STAIRWAYS

25.1 General

The moving stairway is also referred to as an escalator or as an electric stairway. This section uses all three names. The escalator was first operated at the Paris Exposition in 1900. Its successors not only deliver passengers comfortably, rapidly, and safely, but also continously receive and discharge their live loads at a constant speed with practically no delay at any landing. The annoyance of waiting for elevators is not present. Also, time is not lost by acceleration, retardation, leveling, or door operation nor by pressing hall buttons, by passenger interferences in getting in or out of the cars, and so on. One seldom sees a waiting passenger or congestion of passengers at the lighted combplate of an escalator.

Instead of formal lobbies and hallways leading to a bank of elevators on each floor, the electric stairway is always in motion, inviting passengers to ride. The corridors, aisles, and other passageways in existing buildings usually provide space for floor openings adequate for the installation of escalators. In contrast, it would in most cases be almost impossible to install an adequate bank of elevators in an existing building to meet the need for vertical transportation. Elevator hoistways must be vertical from bottom to top floors; an escalator installation can be ''staggered'' at various appropriate locations (see Fig. 25.1). Figure 25.2 shows a schematic view of a modular escalator. These are designed differently from standard units and are discussed separately in Section 25.6.

A standard stairway is assembled from three separate sections of structural truss—an upper section, a middle section, and a lower section similar to the upper one (see Fig. 25.6*a*). The middle straight section may be any desired length to provide rises for floor heights from 10 to 23 ft, for example. When the rise exceeds 20 ft, an intermediate support is located between the two end supports of the stairway. Generally, the upper corners of the bottom and top ends of the truss, after assembly, carry the complete weight of the stairway mechanism and its live load.

25.2 Location

Because escalators are constantly moving and are generally part of a horizontal and vertical trip, they must be placed directly in the main line of circulation. This is in contrast to the elevator bank, which, being a vertical transportation unit, can be set off as an element on its own, for people to approach and utilize. Escalators must therefore be placed in the area served, and with a total and even dominating view of it. This allows potential riders to immediately:

1. Locate the escalators.
2. Recognize the individual escalator's destination.
3. Easily and comfortably move toward the escalator.

One of the most effective ways to disorient traffic movement is to inadequately or poorly mark escalator destinations. The resultant milling about, false starts, and general unhappiness can be observed in numerous otherwise well-designed buildings.

Sufficient lobby space must be provided at the base for queuing where anticipated, and most particularly at discharge points. A restricted or poorly marked area here will cause passenger hesitation and traffic backup. Since the escalator discharges continuously, backup of traffic is dangerous and

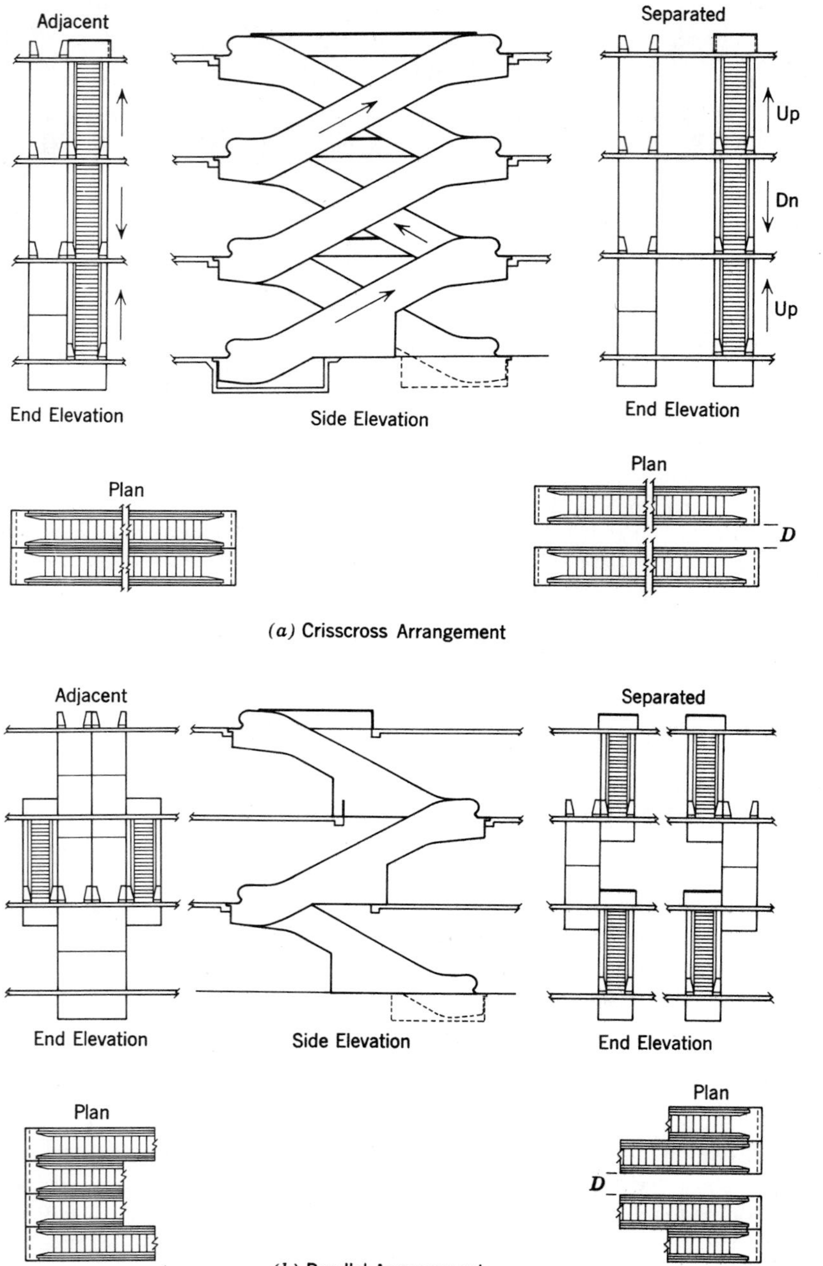

(*a*) Crisscross Arrangement

(*b*) Parallel Arrangement

Multiple Escalator Arrangement

Fig. 25.1 *Plan views, side and end elevations of escalator banks in* (a) *crisscross and* (b) *parallel arrangements. Distance* D *between stairs may be selected as desired.*

Fig. 25.2 Schematic of a modular design escalator. Note that length is varied by inserting additional center sections between the top and bottom sections, with additional drive units as required. (Courtesy of Westinghouse Elevator Co.)

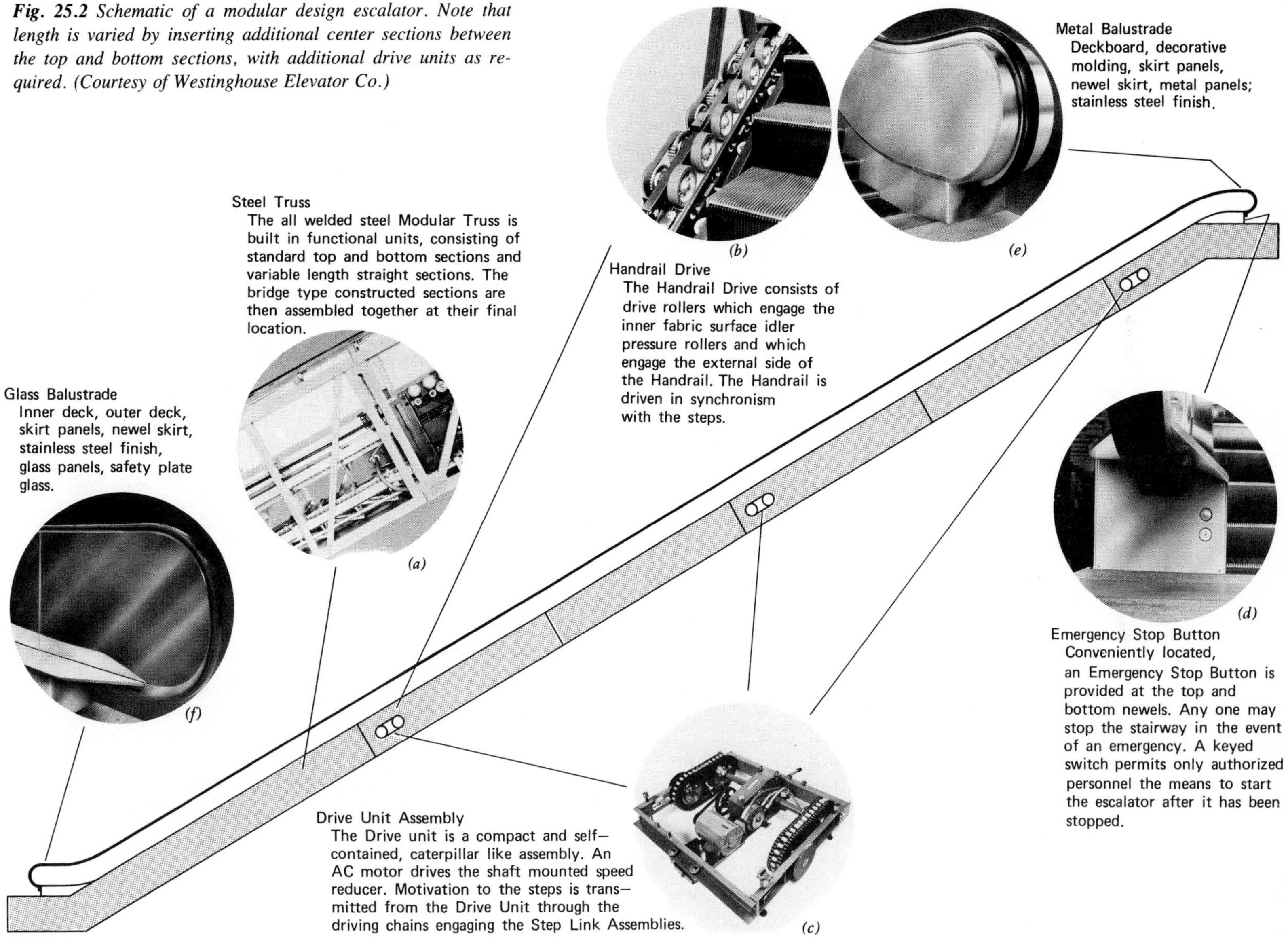
Steel Truss
The all welded steel Modular Truss is built in functional units, consisting of standard top and bottom sections and variable length straight sections. The bridge type constructed sections are then assembled together at their final location.
Handrail Drive
The Handrail Drive consists of drive rollers which engage the inner fabric surface idler pressure rollers and which engage the external side of the Handrail. The Handrail is driven in synchronism with the steps.
Metal Balustrade
Deckboard, decorative molding, skirt panels, newel skirt, metal panels; stainless steel finish.
Glass Balustrade
Inner deck, outer deck, skirt panels, newel skirt, stainless steel finish, glass panels, safety plate glass.
Emergency Stop Button
Conveniently located, an Emergency Stop Button is provided at the top and bottom newels. Any one may stop the stairway in the event of an emergency. A keyed switch permits only authorized personnel the means to start the escalator after it has been stopped.
Drive Unit Assembly
The Drive unit is a compact and self—contained, caterpillar like assembly. An AC motor drives the shaft mounted speed reducer. Motivation to the steps is trans—mitted from the Drive Unit through the driving chains engaging the Step Link Assemblies.
(a)
(b)
(c)
(d)
(e)
(f)

therefore intolerable. This is particularly crucial in theaters and stadiums, where even momentary hesitation during peak traffic periods could be disastrous. To avoid this, three design steps, in descending order of importance, are taken:

1. Provide well-marked escalators with sufficient traffic-carrying capacity.
2. Provide collecting space at intermediate landings so that pressure can be relieved.
3. Provide a slight setback for the next escalator so that the necessary 180° turn can be readily negotiated (see Fig. 25.3).

At the landing, an escalator should discharge into an open area with no turns or choice of direction necessary. Where such is unavoidable, *large* clear signs should make hesitation unnecessary. Landing space in front of an escalator terminal should be 6 to 8 ft *minimum* for a 90-fpm unit, and 10 to 12 ft for a 120-fpm unit. For escalators that will be reversed to accommodate change in traffic direction, this landing space must be provided at top and bottom.

25.3 Parallel and Crisscross Arrangements

Escalators may be installed so that the up and down stairs crisscross each other as in Fig. 25.1*a*, or they may be arranged in parallel as in Fig. 25.1*b*. Both arrangements may have the up and down stairs separated by any desired distance. Separating the stairways gives the advantage of easier mixing of riders entering at the various levels with riders making a continuous trip. Also, by separating the escalators in the crisscross arrangement or by stacking them in the parallel arrangement (see Fig. 25.4), passengers making a multifloor trip can be forced to traverse a specific area on each floor. This area can obviously be used to advantage for display of impulse-buying merchandise, and this is indeed the major consideration in favor (from the store's point of view) of these arrangements. A negative reaction can be produced with these arrangements when:

1. Insufficient floor space is provided for the transit between escalators, causing crowding, pushing, and general annoyance.
2. Insufficient elevator service is provided for passengers wishing to travel at least three floors. This forces people to make a multistory escalator trip, which in itself can be wearying, particularly when carrying parcels. If such a trip is further lengthened by an enforced walk-around at each floor, it becomes a source of severe irritation, often sufficient to keep customers away from the store.

Fig. 25.3 *Single, 32-in. crisscross escalator in Lord & Taylor's Oakbrook Center store (Chicago). Note the setback of the descending escalator, which is helpful in making a smooth turn to the next escalator. By separating the two escalators horizontally, an enforced walk-through can be created. (Photo courtesy of Montgomery Elevator Co.)*

Fig. 25.4 The stacked parallel arrangement of escalators at Stix, Baer & Fuller's Independence Mall store (Kansas City, Mo.), requires riders to walk around when traveling more than one floor. To avoid the annoyance of excessively long trips, riders are provided with the alternative of glass-cab hydraulic elevators. The glass escalator balustrades are architecturally consonant with the glass cab and the glass-protective barriers around the center well of the store. (Photo courtesy of Montgomery Elevator Co.)

The crisscross arrangement is generally favored because of lower cost, minimum floor space occupied, and lowest structural requirements. The parallel arrangement, being less efficient and more expensive, has as its only virtue a very impressive appearance that strongly draws people to it. For this reason it is frequently employed, particularly in multiple banks of three or four, in transportation terminals (see Fig. 25.5). In such installations, flexibility is maintained by operating all but one

Fig. 25.5 The largest moving stairways in Western Europe are installed in the Stockholm subway system. This parallel bank of three escalators is 70-m (230-ft) long with a 33-m (108-ft) rise. Unlike American standards of 30° incline, 90- or 120-fpm speed, and 32- or 48-in. width, these units are at 27.3° incline, 45-m/min (147.6-fpm) speed, and 1000-m (40-in) width. (Photo courtesy of Orenstein & Koppel, West Germany.)

escalator in the direction of heaviest traffic. Reversibility of escalators provides this most desirable feature.

25.4 Size, Capacity, and Speed

All escalators in the United States are installed at an angle of 30 degrees from the horizontal. Thus the rise is equal to 57% of the projected floor area. The safety code limits escalator speed to 125 fpm along the axis of rise. In actual practice two speeds are available: 90 fpm and 120 fpm. Installations are occasionally two-speed, with the higher speed utilized during rush hours and the lower speed during "off" hours. Where no rush is encountered, the lower speed is utilized, since the 120-fpm speed presents some difficulty to the less agile passenger.

Moving stairways are generally available in widths of 32 and 48 in., both measured at the hip level between the balustrades. These two principal sizes correspond to tread widths of 24 and 40 in., respectively. All treads have a 16 in. depth and 8 in. rise. Table 25.1 lists the maximum and actual capacities of escalators and includes the equivalent metrically sized units. The 32-in. unit is rated for 1¼ persons per step. A nominal figure assumes one person per step, or about 75% of the maximum figure. The 48-in. width assumes 2 persons per step. Here also a derating to about 75% of maximum is more realistic, taking into account empty steps, briefcase- and package-carrying persons, and so on. The table also gives observed capacity, which is considerably lower. Only peak load in transportation terminals and stadiums approaches even nominal capacity.

25.5 Components

The major components of a standard (as opposed to a modular) escalator installation are shown in Fig. 25.6. Safety devices are discussed in Section 25.7.

The truss (Fig. 25.6*a*) is a welded steel frame that supports the entire apparatus. The tracks are steel angles attached to the truss on which the step rollers are guided, thus controlling the motion of the steps. The sprocket assemblies, chains, and machine provide the motive power for the unit, much like the simple chain drive of a bicycle. The controller, which consists of contactors, relays, and a circuit breaker, is normally located near the drive machine. An emergency stop button wired to the controller and placed near or on the escalator housing will stop the drive machine and apply the brake (see lower left of escalator, Fig. 25.3).

Key-operated control switches at the top and bottom newels will start, stop, and reverse the stairway. The handrail is driven by two sheaves and is powered from the top sprocket assembly. It is synchronized with the tread motion to provide stability to riding passengers and a support for entering and leaving passengers. Handrails disappear at inaccessible points at newels. The balustrade assembly is designed for maximum safety of persons stepping on or off the escalators.

A particularly attractive design utilizing a transparent balustrade made of tempered glass is illustrated in Fig. 25.4. They are frequently referred to as crystal balustrades. In these units, the handrail is pinch-driven within the truss. In addition to metal and glass as balustrade materials, back-illuminated fiberglass and wood are also used.

TABLE 25.1 **Escalator Passenger Capacity**

Step Width	Speed	Passengers per Hour		
		Maximum	Nominal	Observed
English Units; American Manufacture				
32 in.	90 fpm	5,000	3750	2100
	120 fpm	6,666	5025	2800
48 in.	90 fpm	8,000	6000	4000
	120 fpm	10,665	8025	5500
Metric Units; Foreign Manufacture				
800 mm	30 m/min	6,000	—	—
1200 mm	30 m/min	9,000	—	—

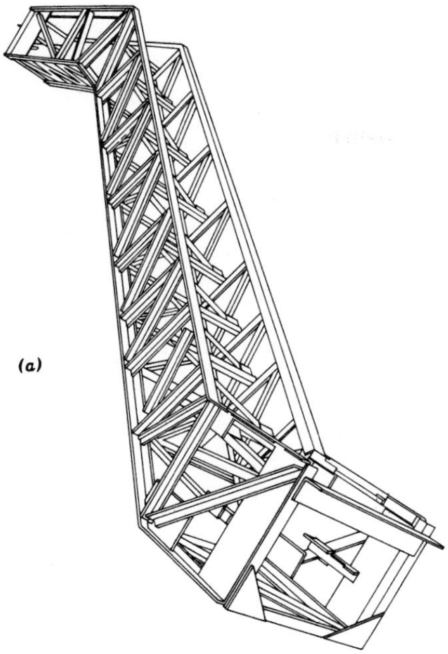

(a)

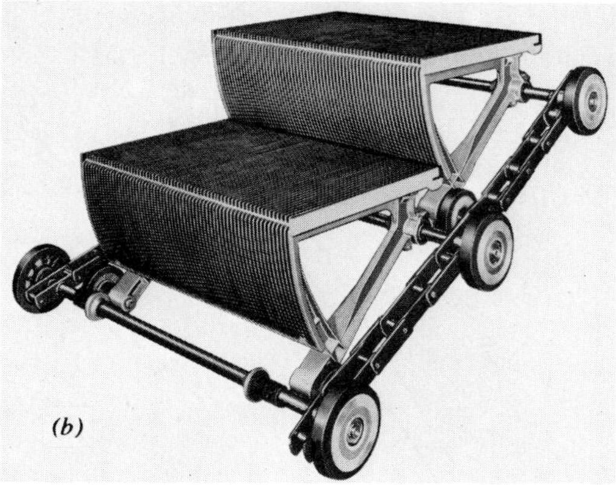

(b)

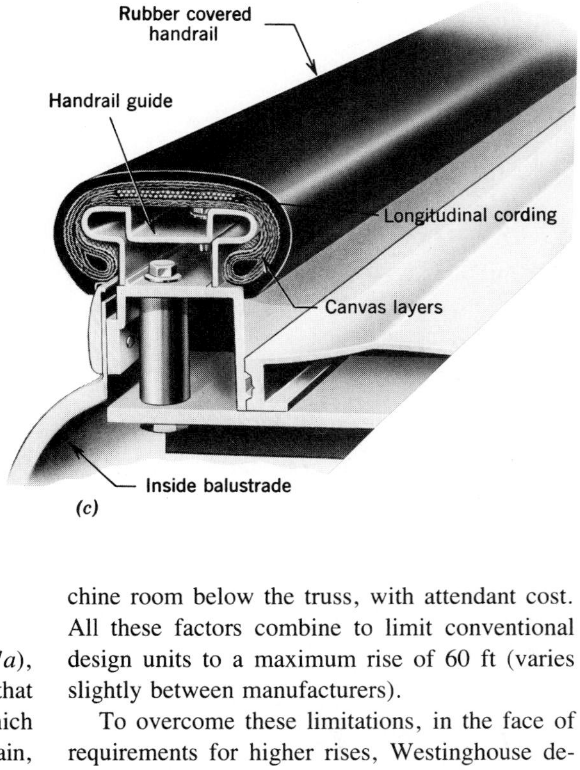

(c)

Fig. 25.6 (a) *The truss is a three-section, built-up, welded steel unit that supports the moving stairway equipment.* (b) *Typical step assembly for electric stairway. Riser is 8 in. high. Comb pattern is designed to prevent materials from being caught between riser and tread.* (c) *Constant flexure of the handrail necessitates layered construction. (Illustrations courtesy of Westinghouse Elevator Co.)*

25.6 Standard and Modular Designs

In conventional escalator design (see Fig. 25.7*a*), all the motive power is delivered at one point; that is, the drive motor drives the main chain, which drives the top sprocket, which drives the step chain, which pulls up the steps, causing the entire assembly to move. This arrangement is suitable for moderate rises of up to approximately 25 ft. Beyond that the design becomes increasingly inefficient. As the rise increases, the loads on all the drive components including chains and sprockets increase sharply. Furthermore, to accommodate the heavier equipment necessitated, truss width increases, as does wellway size and balustrade decks. For rises about 25 ft the drive motor is too large to fit inside the truss and requires a separate machine room below the truss, with attendant cost. All these factors combine to limit conventional design units to a maximum rise of 60 ft (varies slightly between manufacturers).

To overcome these limitations, in the face of requirements for higher rises, Westinghouse developed a modular design that was introduced in 1973 under the trade name Modular Escalators. This design has unlimited rise capability because it is constructed with additional drive motors along the length of the unit, in a modular design pattern (see Figs. 25.7*b* and 25.2). By spreading the drive load throughout the length of the unit, the inherent limitations listed above, which are caused by a single drive location, are eliminated (refer to Fig. 25.7*c*). Note that this distributed-drive principle simplifies the mechanism considerably and in-

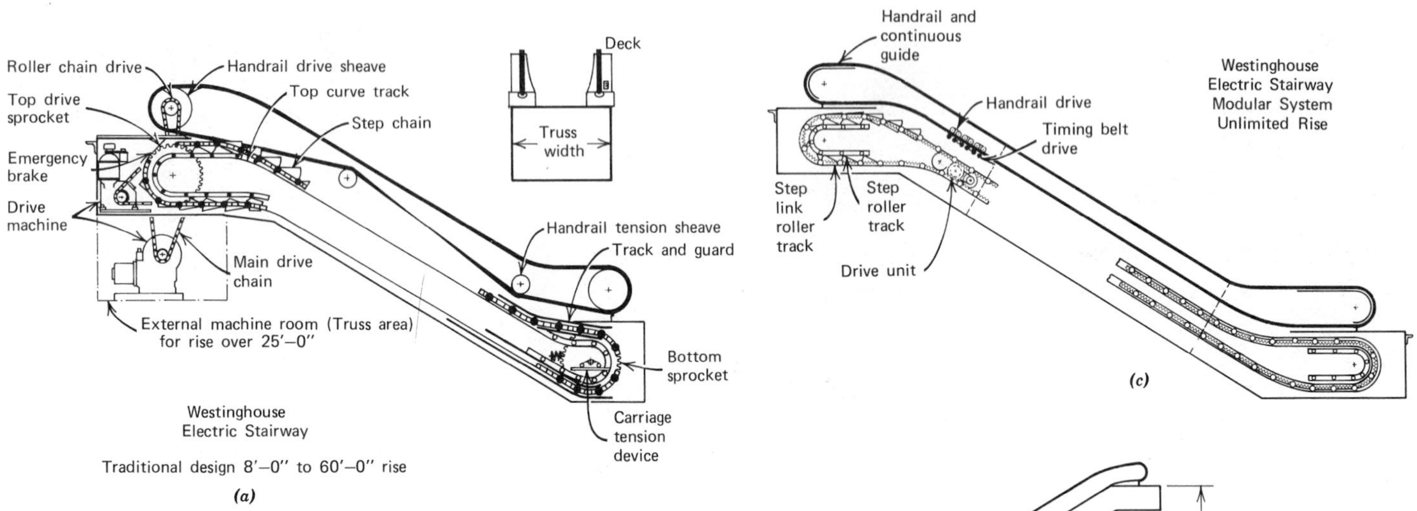

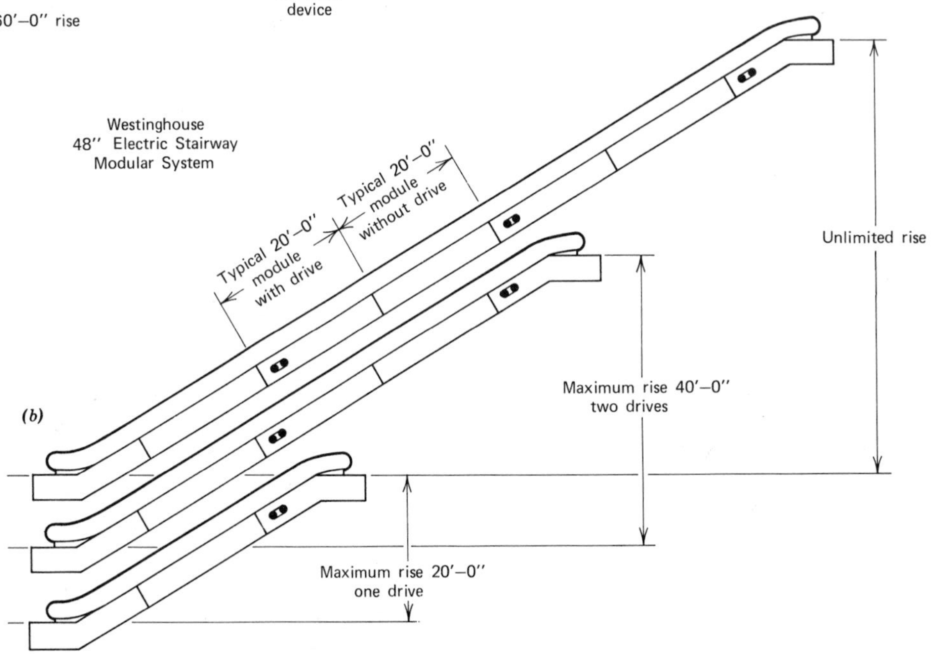

Fig. 25.7 (a) *Operating mechanism of a traditionally designed electric stairway. Note that all motive power comes from the drive at the top of the stair via main chain, top sprocket, and stair chain.* (b) *Schematic sectional drawing showing modular design. The drive system is distributed along the length of the stair, which comprises a top section, a bottom section, and as many modular intermediate sections as are necessary to accomplish the requisite length and rise (see also Fig. 25.2).* (c) *Operating mechanism of the modular escalator. The distributed drive adds motive power throughout the length of the drive chain. (Diagrams courtesy of Westinghouse Elevator Co.)*

TABLE 25.2 **Modular Escalator Motor Drives**

	Escalator Size	
	32 in.	*48 in.*
Speed (fpm):	*90–120*	*90–120*
Motor (hp):	*10*	*10*
One drive, nominal, max rise	30 ft, 0 in.	20 ft, 0 in.
Two drives, nominal, max rise	60 ft, 0 in.	40 ft, 0 in.
Three drives, nominal, max rise	90 ft, 0 in.	60 ft, 0 in.

Source: Data courtesy of Westinghouse Elevator Co.

creases efficiency greatly. In a high-rise application this results in much lower total motor horsepower, with attendant energy and cost savings. Chain loads are constant regardless of length, as is truss size. The modular drive machine units are all identical and a machine room is never necessary. Since tensions are held to low levels, maintenance is low. Furthermore, the helical gear drive in the modular units is 10 to 15% more efficient than the worm gear of the traditional design. In summary, then, the Westinghouse modular design offers distinct advantages for high-rise escalator installations. Table 25.2 lists the necessary drives for various rises and stair widths.

25.7 Safety Features

Protection to passengers during normal operation is assured by a number of safety features associated with moving stairways:

1. Handrails and steps travel at exactly the same speed (90 or 120 fpm) to assure steadiness and balance and to aid stepping on or off the combplates.
2. The steps are large and steady, and are designed to prevent slipping.
3. Step design and step leveling with the combplates at each landing prevent tripping upon entering or leaving the escalator. This is accomplished with two or three (depending on manufacturer) horizontal steps at each end of the escalator.
4. The balustrade includes all enclosures as furnished by the escalator manufacturer, including the deckboards, inside panels, skirt guards, handrail guards, handrails, and combplates. Details of these parts are designed to prevent the catching of passengers' clothing or packages. Close clearances provide safety near the combplates and step treads.
5. An automatic service brake will bring the stairway to a smooth stop if:
   (*a*) The drive chain or the step chain is broken or abnormally stretched.
   (*b*) A foreign object is jammed into the handrail inlet, between the skirt guard and step or between steps, causing them to separate.
   (*c*) A power failure occurs.
   (*d*) The emergency stop button is operated (one is located at each end of the escalator).
   (*e*) Any of the fire safety system devices operates. See Section 25.8.

   Passengers would then walk the steps as they would any stationary stairway.
6. In case of overspeed or underspeed, an automatic governor shuts down the escalator, prevents reversal of direction (up or down), and operates the service brake.
7. Adequate illumination must be provided by the building at all landings, at the combplates, and completely down all stairways. Some escalator designs provide built-in lighting, as discussed in Section 25.9.

25.8 Fire Protection

Four methods of affording protection in case of fire near escalators are available: the rolling shutter, the smoke guard, the spray-nozzle curtain, and the sprinkler vent. One of these methods is required by code when more than two floors are pierced. Figure 25.8 illustrates clearly how the wellway at a given floor level may be entirely closed off by the fire shutter, thus preventing draft and the spread of fire upward through escalator

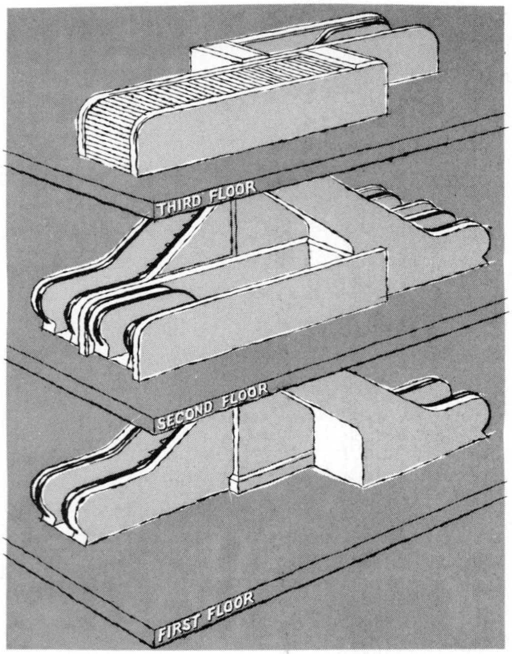

Fig. 25.8 Rolling-shutter method of wellway fire protection.

wells. The movement is actuated by temperature and smoke relays that automatically actuate the motor-driven shutters. The shutter in Fig. 25.8 is shown at the third floor level, but other shutters may be installed at the tops of horizontal wellway openings at any floor.

Figure 25.9 illustrates the smoke-guard method of protection. It consists of fireproof baffles surrounding the wellway and extending downward about 20 in. below the ceiling level. Smoke and flames rising upward to the escalator floor opening meet a curtain of water automatically released from sprinkler heads of the usual type, shown at the ceiling level. The baffle is a smoke and flame deflector. The vertical shields between adjacent sprinklers ensure that the spray from one will not cool the nearby thermal fuses, preventing the opening of adjacent sprinklers.

The spray-nozzle curtain of water (not shown) is quite similar to the smoke-guard protection. Here closely spaced, high-velocity water nozzles form a compact water curtain to prevent smoke and flames from rising through the wellways. Automatic thermal or smoke relays open all nozzles simultaneously.

The sprinkler-vent fire control is shown in Fig. 25.10. The fresh air intake housed on the roof contains a blower to drive air downward through escalator floor openings, while the exhaust fan on the roof creates a strong draft upward through an exhaust duct; this duct in turn draws air from the separate ducts just under the ceiling of each moving stairway floor opening. Three such separate wellway ducts are shown. Each duct has a number of smoke-pickup relays that automatically start the fresh air fans. The usual spray nozzles on the ceiling near the stairways aid in quenching the fire.

25.9 Lighting

Adequate illumination of a moving stairway, particularly at the landings, is important from decorative as well as safety standpoints, since it is usually desirable to highlight the moving stair installation. In a stairwell-type installation, where general-area lighting does not provide sufficient illumination for the escalator, lighting consonant with the adjacent illumination is installed on the ceiling above the stairway, with special emphasis on lighting the combplate. Thus, in Fig. 25.3 the general illumination is supplemented by down-lights above the stairs. In Fig. 25.5 banks of fluorescent lights are placed across the escalator bank at frequent intervals along the rise. Note the additional concentrated light at the combplate. Two different lighting treatments of similar type installations are shown in Fig. 25.11.

25.10 Application

1. Main floor locations should be chosen in the direct flow of traffic to assure maximum use.

2. Vertical arrangements should be made to accomplish specific purposes, such as exposure of merchandise, maximum passenger capacity, and maximum accessibility to various areas.

3. The aspect of reversibility of an electric stairway should be considered in applications where major traffic flow is unidirectional. Light traffic in

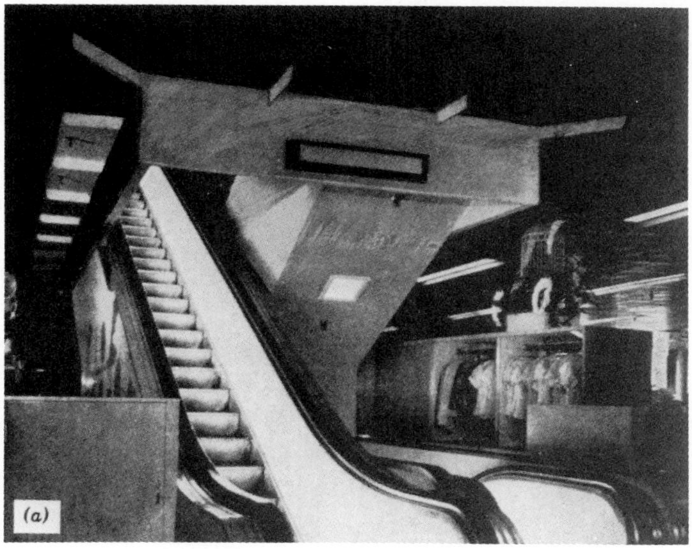

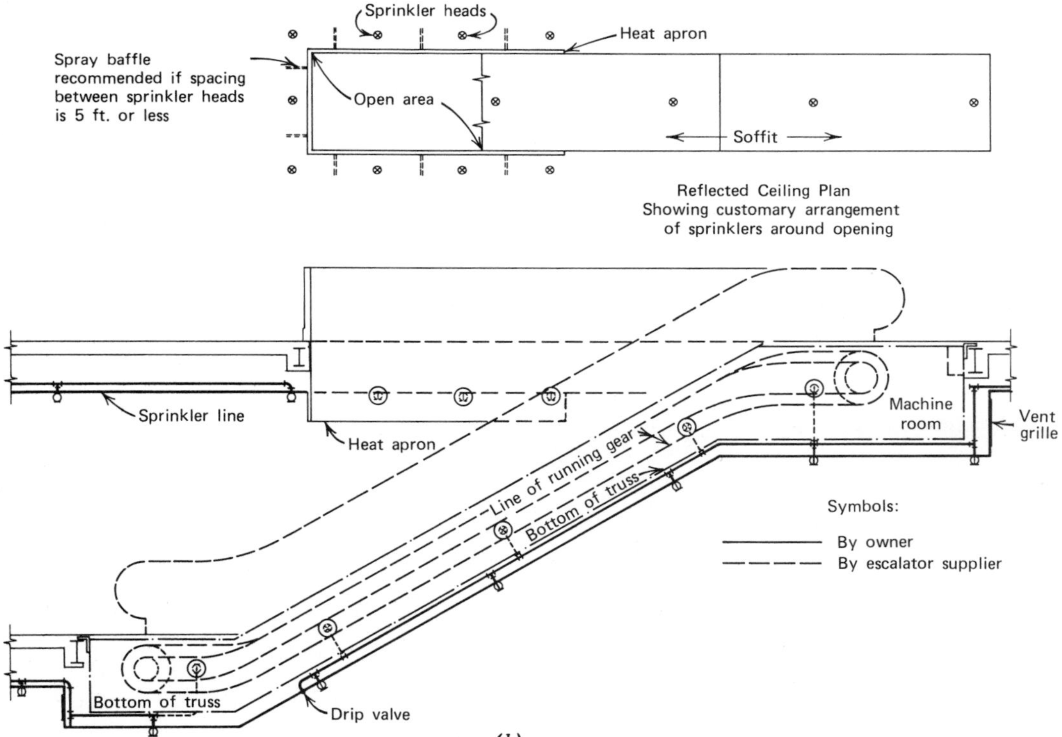

(b)

Fig. 25.9 (**a**) *Smoke-guard method of fire protection for a 32-in. moving stairway, crisscross type. The escalator floor opening (per floor) is approximately 4 ft, 4 in. by 14 ft, 8 in.* (**b**) *Reflected ceiling plan and section showing baffle and sprinkler layout. (Courtesy of Otis Elevator.)*

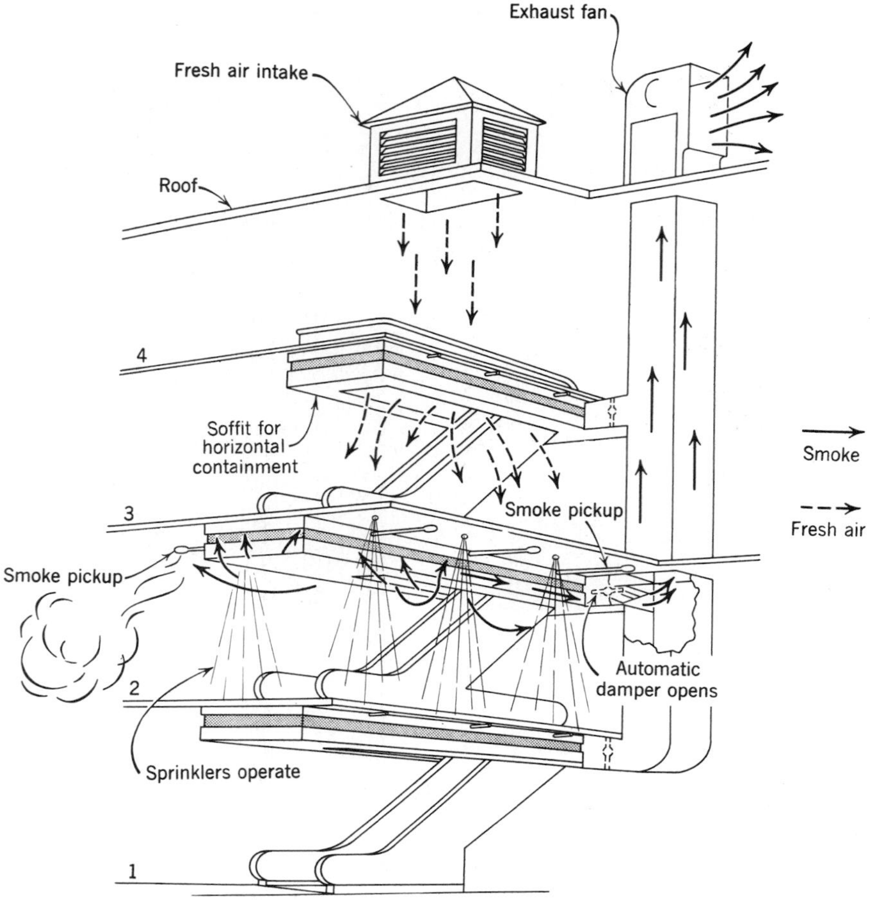

Fig. 25.10 Sprinkler vent fire protection for escalator openings. An exception (with control) to the rule against perforations in floors.

Fig. 25.11 *The balustrade section of an escalator can be used to supply the required task (stair) lighting.* (a) *A continuous flourescent strip is installed at the balustrade base at Fox Hills Mall (Culver City, Calif.) (Photo courtesy of Montgomery Elevator Co.)* (b) *A similar source is placed under the handrail of the crystal balustrade in this covered mall in West Germany. Note the additional light at the base for illuminating the combplate. (Photo courtesy of Orenstein & Koppel, West Germany.)*

Fig. 25.12 *A conventional arrangement provides a parallel bank of two electric stairways separated by a fixed stair in this Canadian application (Place des Jardins, Montreal). Reversibility of the escalators provides the desired flexibility, although normal operation is one up and one down. The fixed stair accommodates those who cannot or will not use a moving stair, as well as reverse traffic when both escalators are operating in the same direction. (Photo courtesy of Montgomery Elevator Co.)*

the reverse direction can be handled by a normal fixed stair, adjacent to the escalator (see Fig. 25.12). Similarly, a bank of two escalators can operate either both up, both down, or one up and one down to handle variable traffic conditions in such areas as office buildings and transportation terminals.

4. Exterior escalators can provide an attractive, interesting, and economical solution to transporting people to selected entry points in a building without the necessity of extending the building to cover the entrance (see Fig. 25.13).

25.11 Elevators and Escalators

The use of elevators and escalators should be considered together, as a single problem in vertical transportation as applied to the particular facility being designed. In this connection, particularly in case of modernization, Fig. 25.14 provides an interesting comparison.

In certain facilities there are times during which demand for vertical transportation is so large that elevators are not a feasible solution. A prime example is the school building. During class change, virtually the entire building population moves, with as much as 80% moving between floors. Since class change time is at most 10 min, the only reasonable solution is the combined use of fixed and moving stairs. In other buildings such as multifloor stores, the escalator provides for short trips of one or two floors, and the elevator generally transports passengers traveling three or more stories.

A comparison of travel time between escalator and elevator is of interest. Using a normal speed of 90 fpm, a 12-ft floor requires 16 s for travel plus about 5 to 6 s to turn, for a total of approximately 22 s. Thus a four-story trip would take approximately 88 s. A similar elevator trip would take at most 60 s. The additional escalator time is not noticed in the activity of boarding, turning, and riding. However, a trip of more than four stories becomes tiresome, and all the more so with the enforced ''walkaround'' at each floor in the separated arrangement. See Section 25.3.

25.12 Electric Power Requirements

Standard American electric stairways are driven by three-phase, 60-Hz, a-c induction motors at standard voltages (208, 230, and 460 V). Horsepowers of driving motors are shown in Table 25.3. A comparison with the figures in Table 25.2 shows that the power requirements for standard elevators in rises beyond 20 ft exceed those of the modular

Fig. 25.13 An attractive exterior escalator installation avoids the necessity of interior stairs and escalators, while providing an item of architectural interest at San Francisco's Candlestick Park. (Photo courtesy of Montgomery Elevator Co.)

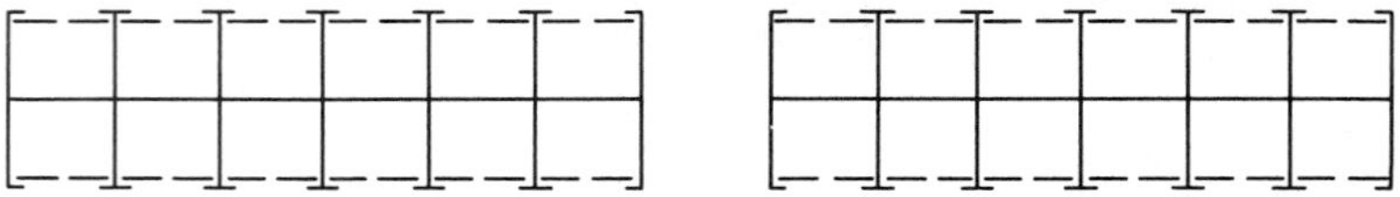

Elevators

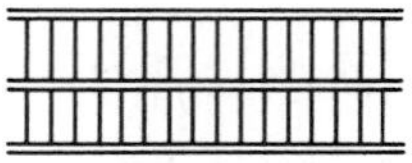

Electric stairways

Fig. 25.14 Comparative space requirements for equivalent passenger handling capacity. Note the marked space savings offered by escalators.

TABLE 25.3 **Typical Escalator Motor Sizes**

Size (in.)	*Speed (fpm)*	*Rise (ft)*	*Motor (hp)*
32	90/120	14	5
	90/120	17	7½
48	90	17	7½
	90	21	10
	90/120	25	15

design. This is due to the higher friction losses in the gears and chain mechanisms of the standard drive.

It is recommended that no more than four escalators be served by a single electric feeder, and further that not all the escalators of an installation, whatever the number, be served from the same feeder.

Since obviously one cannot be trapped on an escalator, emergency power is rarely required. Ventilation for the machinery should be supplied for approximately 40% of the power to be dissipated as heat. Thus a 10-hp motor would require the dissipation of $0.40 \times 10 \times 2500$ Btu/hp or approximately 10,000 Btu/h.

25.13 Structural Design and Installation Data

The architect and the engineers must design the floor openings, stairway supports, and other structural work and finishes. A typical moving stairway drawing, arranged for the information of the architect and trained stairway erectors, is shown in Fig. 25.15. A careful review of all the details shown on these plans and indicated specifications exhibits

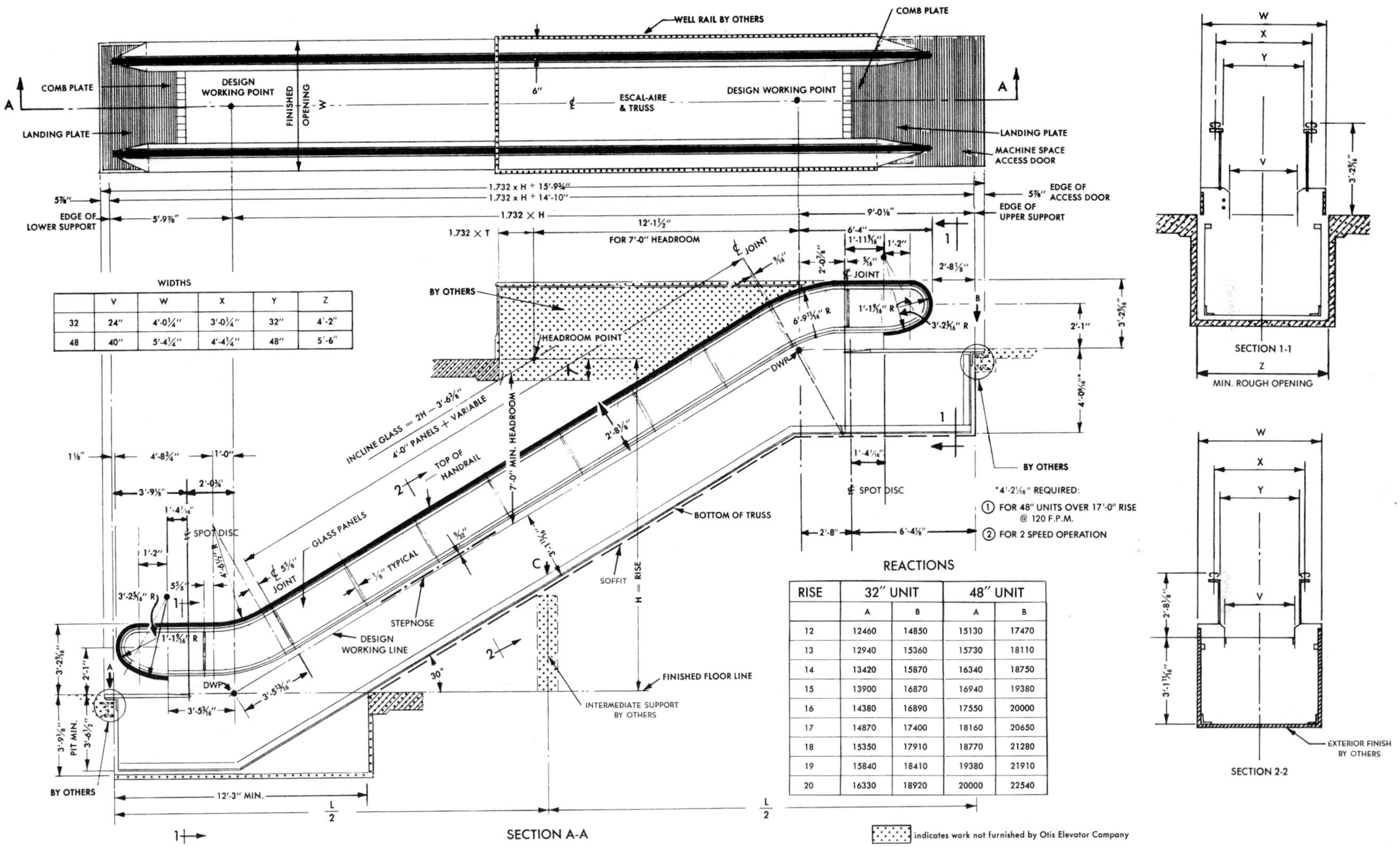

WIDTHS

	V	W	X	Y	Z
32	24"	4'-0¼"	3'-0¼"	32"	4'-2"
48	40"	5'-4¼"	4'-4¼"	48"	5'-6"

REACTIONS

RISE	32" UNIT		48" UNIT	
	A	B	A	B
12	12460	14850	15130	17470
13	12940	15360	15730	18110
14	13420	15870	16340	18750
15	13900	16870	16940	19380
16	14380	16890	17550	20000
17	14870	17400	18160	20650
18	15350	17910	18770	21280
19	15840	18410	19380	21910
20	16330	18920	20000	22540

Fig. 25.15 *Typical escalator layout with dimensions and reactions, as furnished by the manufacturer. This Otis glass balustrade unit is an Escal-aire type, J series. (Courtesy of Otis Elevator.)*

the coordination that is necessary among the architect, engineer, and erection superintendent.

Outlined on these plans are the details for which the architects and engineers are responsible, including structural, mechanical, and electrical features. It will be seen that two ''working points'' are identified, between which a very strong steel wire is tightly stretched. From these two points all other measurements are made, that is, locating the centerline of the truss sections, placing the lower and upper landing truss support beams, and so on. Such plans as this one are available from all escalator manufacturers for any standard type of stairway.

TABLE 25.4 **Relative[a] Escalator Prices**

Rise (ft)	Width: 32 in., Speed 90 fpm	32 in., 120 fpm	48 in., 90 fpm	48 in., 120 fpm
14	100[a,b]	106	118	129
16	103	109	121	133
18	105	111	124	136
20	108	114	127	140
22	111	117	130	143
24	113	120	133	146
26	115	122	135	150
28	118	125	138	155
30	120	127	141	160

[a]Base price figure is 100 for the shortest, narrowest, slowest unit (i.e., 14-ft rise, 32 in. wide, 90 fpm).
[b]Add 7–8% for glass balustrade for any unit.

25.14 Budget Estimating for Escalators

The cost of an escalator includes the cost of the associated mechanical and electrical equipment, plus the shipping installation charges. The manufacturer provides expert engineering and a union field erector who supervises the installation, which is done by unionized elevator and escalator mechanics.

The 32- and 48-in. electric stairways are considered standard production models. These may be furnished from a 10- to 25-ft rise, operating at 90 or 120 fpm. On special orders, other rises and speeds may be obtained.

A tabulation of *relative* escalator prices on a base of 100, is given in Table 25.4. Prices for units with rise above 35 ft rise very rapidly and depend on the type of unit used. The designer is referred to the suppliers for quotes on all units. To these figures must be added the cost of builders' work, wellway protection, lighting, outside balustrades, and plaster.

MOVING WALKS AND RAMPS

25.15 General

Moving walks and ramps are different from moving stairways in application, function, construction, and capacity. Escalators have as their primary function the movement of large numbers of people vertically, when such vertical distance does not exceed approximately five stories, as noted above. This specific transportation function the moving stair performs extremely well with minimum cost, space, and maintenance.

When vertical transportation of wheeled vehicles and large parcels is required, the use of an electric stairway is awkward, if not entirely impossible. For such functions and others discussed below, the moving ramp may best be utilized.

Unlike the elevator and escalator, the moving walk or ramp serves a dual function, that is, horizontal transportation only, or a combined function of horizontal and vertical transportation. For the purpose of our discussions, we will define a moving *walk* as one with an incline not exceeding 5° where the principal function is horizontal motion and inclined motion is incidental to the horizontal. A moving *ramp* is a device with an incline limited by code to 15°, where vertical motion is as important as or more important than the horizontal component. It should be understood that the walk and ramp are physically the same device, differently applied.

25.16 Application of Moving Walks

The principal uses of moving walks or moving sidewalks, as they are sometimes called, are to:

1. Eliminate and/or accelerate burdensome walking.
2. Eliminate congestion.

3. Force movement.
4. Easily transport large, bulky objects.

Anyone who has walked the seemingly endless distances in a major airport, carrying a heavy suitcase, can appreciate the near-absolute necessity for a moving walkway. It is for this reason that transportation terminals hve become major users of this item (see Fig. 25.16). Other transportation terminals, such as rail and ship terminals, also can often find excellent applications for the moving walk, since much heavy and bulky luggage is moved in these areas.

Application of the apparent distance compression that moving walks provide permits placement of parking areas remote from the pedestrians' destination. Thus, a store can extend its parking area with no annoyance to patrons who must make the long trip to their cars with bulky packages or shopping carts. These advantages are all the more appreciated by persons with a walking impediment.

A second application of walks, as noted above, is the routing of traffic to avoid congestion, milling about, and lost time and motion. This is particularly applicable in transportation terminals, where persons are always traveling in opposite directions through the same and often restricted area,

Fig. 25.16 *Twin autowalks at Manchester Airport, England. These are the pallet type with glass balustrades and continuous built-in fluorescent lighting. These walks, which are each 1 m (40 in.) wide and 55 m (180 ft) long, travel in opposite directions. Because a single-loop drive runs both walks, they cannot both be operated in the same direction. (Photo courtesy of Orenstein & Koppel, West Germany.)*

such as in "fingers" leading from airplanes to the main air terminal.

Movement of persons past a display window or some other point is used where the moving walk prevents congestion by preventing stopping. The "movement of objects" application demonstrates clearly that the moving walk is in function simply a large conveyor belt, regardless of its construction.

25.17 Application of Moving Ramps

The moving ramp that combines horizontal and vertical movement is principally applicable as follows:

1. To move persons and wheeled vehicles vertically.
2. To move persons who lack the agility required to use an escalator.
3. To vertically move large, bulky objects.

Ramps have found a fertile field of application in multilevel stores where escalators are not feasible for shopping-cart users. Such stores may also utilize rooftop parking that is made accessible to cart users via a moving ramp. Since luggage carriers do not easily adapt to usage on escalators, transportation terminals, which are almost always multilevel, also find extensive application for the moving ramps (see Fig. 25.17).

25.18 Size, Capacity, and Speed

The speed, physical dimensions, and therefore passenger capacity of walks and ramps are not standardized as is the case with escalators. Manufacturers utilize many different tread widths, combined with various speeds and ramp angles of incline. The combinations are designed to suit the situation. Most installations, however, are 26 in. (660 mm) one passenger or 40 in. (1000 mm) two passenger, these figures being derived from stair pallet dimensions. The code allows wider units on horizontal runs.

Furthermore, since the maximum ramp speed varies with angle of slope, and with design of entering point, passenger capacity ratings vary with each design. Higher speeds are allowed by code

Fig. 25.17 Moving ramps in weatherproof design provide direct access to upper floors. This 12° ramp leads to the second floor of a store in Copenhagen. Note the glass balustrade and continuous flourescent lighting fixtures below the handrail. (Photo courtesy of Orenstein & Koppel, West Germany.)

for level entrance than with sloping entrance, for the obvious reason that the level entrance is easier to board. Table 25.5 and Fig. 25.18 give maximum speeds and typical passenger capacity for specific units.

Since capacity varies with width, speed, and type of entrance, exact capacity figures must be obtained for each specific design. Maximum practical lengths at present are approximately 1000 ft, with longer units in design.

TABLE 25.5 **Maximum Operating Speeds of Moving Ramps**

	Maximum Speed (fpm)	
Angle of Incline	*Level Entrance*	*Sloping Entrance*
0–3°	180	180
3–5°	180	160
5–8°	180	140
8–12°	140	130
12–15°	140	125

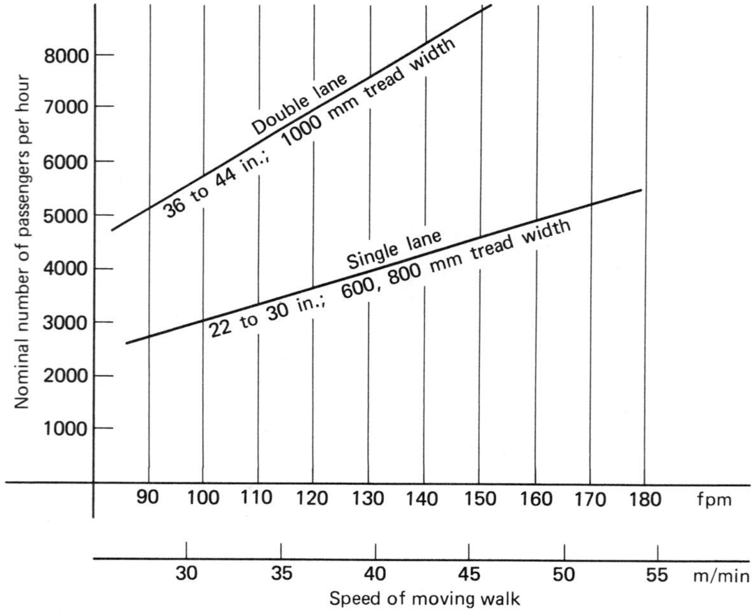

Fig. 25.18 The capacity of moving walks varies with speed, angle of incline, and tread width. Capacity shown is for maximum incline permitted at that walk speed. Because of the requirement for handrail support, tread widths greater than double lane are not utilized.

25.19 Components

Moving walks are manufactured in two separate and entirely distinct designs. The first, which is derivative of the escalator, uses a flattened pallet in place of a step. In all other respects—the drive mechanism, safeties, brake, handrails, and so on—the unit is similar to an escalator. The second design is based on a conveyor belt and utilizes a continuous belt, constructed of rubber-covered steel or fabric, supported on idler sheaves or slider bed, and driven by a typical roller-drive mechanism at the ends of the belt. As with escalators, a wide choice of materials and colors is available for side panels, drum, and balustrades.

25.20 Safety Devices

Normal safety considerations on all walk and ramp installations include:

1. Thin groove belting, to provide sure footing, plus comb action at the ends to avoid jamming.
2. Handrails synchronized with tread motion, extending beyond the treadway to ease entry and exit on to a normally level area.
3. Emergency stop devices, both hand operated and automatic as with escalators. The latter are activated by disturbances on the tread or in the drive machinery, causing the walk to shut down. Also as with escalators, handrails disappear at inaccessible points at newels, and special lighting is generally provided at combplates.

25.21 Conclusion

Since moving walks and ramps, unlike moving stairs, are nonstandard and are specifically tailored to each application, no general data can be given on power or space requirements and structural design. Each job must be referred to the manufacturers and the details developed.

PART IX
ACOUSTICS

The best (and simplest) distinction between the terms ''sound'' and ''noise'' is a subjective one—the former is desirable, the latter is not. This definition does not consider the content of the sound. For example, speech, which is a desirable sound in most instances, can become a ''noise'' when emanating from a neighbor's apartment or an adjoining office. In such a case an air conditioner, which is most often considered to be a ''noise,'' can provide a desired ''sound'' by masking the intruding speech. The function of architectural acoustics is simply to reinforce this definition; that is, to enhance desired sound and to maximally attenuate noise.

Chapter 26, Fundamentals of Architecture/Acoustics, introduces the subject with a discussion of physical sound theory and physiological hearing phenomena. The latter include the negative effects of noise, which are primarily psychological (annoyance) at low levels but become physical and result in hearing damage at high levels. Criteria of the U.S. Occupational Health and Safety Administration (OSHA) are given. The discussion moves to the core of the subject—room acoustics—with an explanation of absorption and reverberation, and develops acoustic design criteria for various indoor activities. The chapter concludes with a description of high-grade sound reinforcement systems.

Chapter 27 is devoted to noise control. Noise within a space is controlled by absorption, whereas control of noise transference between spaces is a matter of isolation. The chapter begins with a discussion of the application of absorptive materials for room noise reduction. It treats the problem of interspace noise conduction, dividing the problem into two parts (airborne noise and structure-borne noise), since the solutions are different. Relevant criteria, including sound transmission class (STC) and impact isolation class (IIC), are adduced and explained, and solutions to noise transference problems of different types are suggested. The chapter offers sections on mechanical system noise control and on acoustic recommendations and criteria, ending with a section on exterior acoustics.

26

FUNDAMENTALS OF ARCHITECTURAL ACOUSTICS

SOUND THEORY AND HEARING PHENOMENA

26.1 General

Architectural acoustics may be defined as the technology of designing spaces, structures, and mechanical systems to meet hearing needs. With proper design, ''wanted'' sounds can be heard properly and ''unwanted'' sounds, or ''noise,'' can be attenuated or masked to the point where it does not cause annoyance. However, achieving good acoustics has become increasingly more difficult for a variety of reasons. To cut costs, the weight of construction materials used in many of today's buildings is reduced. Since light structures generally transmit more sound than heavy ones, this practice poses major acoustical problems. Forty percent or more of a building budget may be allocated for mechanical systems—most of which make noise. Outside noise sources such as cars, trucks, trains, and airplanes present problems in isolating interior spaces from exterior sound.

Building owners and tenants are aware that good acoustic environments in buildings are possible, and the architect is expected to provide them. A clear understanding of the principles explained in this and the following chapter will assist the architect in accomplishing straightforward designs alone and, in more complex instances, cooperating knowledgeably with the project's acoustic consultant. The importance of proper acoustic design is all the more critical, since after-the-fact acoustic ''repair'' is often difficult and frequently impossible without substantial structural alterations.

All acoustics situations have three common elements—source, transmission path, and receiver. The source can be made louder or quieter and the path can be made to transmit more or less sound. The listener's reception of sound also may be influenced. This chapter presents essential aspects of architectural acoustics to assist a designer in defining acoustic goals. Moreover, it describes the achievable goals as well as various methods for reaching them through design.

26.2 Sound: Definition and Generation

Sound can be defined in a number of different ways depending on the aspect we desire to study. Thus sound is a physical wave, or a mechanical vibration, or simply a series of pressure variations, in an elastic medium. For airborne sound, the medium is air. For structure-borne sound, the medium is concrete, steel, wood, glass, and combinations of all these materials. A much more limited definition is probably more appropriate to our study, that is, to define sound simply as an audible signal. This does not mean that subsonic or supersonic signals are not sound, nor does it mean that we are taking a stand on the existential question of whether unheard sound exists. It simply means that the science of architectural acoustics is concerned with the building occupant, and sounds that he or she cannot detect are generally not our concern. To further clear the air, it is always assumed that the hearer has a pair of healthy young ears with a detection range of 20 to 20,000 Hz. With these givens, it is probably best to view sound as a series of pressure variations. In air, these pressure variations take the form of periodic compres-

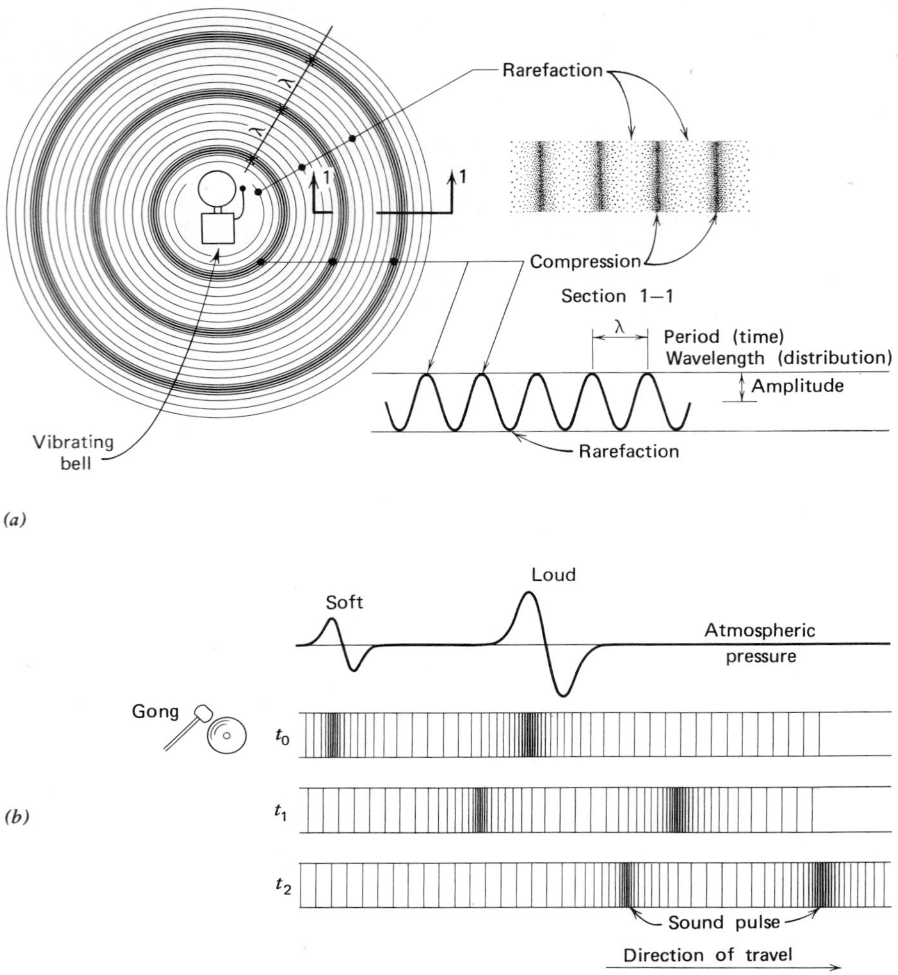

Fig. 26.1 *Sound pressure waves.* (a) *This continuous vibration from a vibrating bell causes a series of compressions and rarefactions to travel outward longitudinally from the source. Amplitude information is carried by pressure; that is, greater amplitude means greater compression and greater rarefaction. (Compression and rarefaction are expressed diagramatically as line density, although they are actually molecular phenomena as shown in the upper drawing).* (b) *Two single impulses of different magnitude (amplitude) traveling away from the source. Note how amplitude information is carried by difference in pressure.*

sions and rarefactions (see Fig. 26.1). The bell radiates a pure tone in all directions equally, that is, it creates a circular wave front. As the bell vibrates, it sets up vibrations in the air, of the same frequency, which can best be visualized in the form of a sectional view. Notice that the pressure changes containing the sound information travel in the same direction as the wave front—longitudinally. This is unlike a radio signal, for instance, in which the wave travels longitudinally but the information—that is, the modulation, is expressed in wave height and shape—is transverse. Sound is therefore *longitudinal* mechanical wave motion.

26.3 Frequency

The number of times the cycle of compression and rarefaction of air occurs in a given unit of time is described as the frequency of a sound. For example, if there are 1000 cycles in one second, the frequency of the sound is 1000 cps [1000 hertz (Hz) in the standard nomenclature]. Thus in Fig. 26.1, higher frequencies would be shown by compressions and rarefactions that are closer together and lower frequencies by those that are further apart. In sound, the concept of frequency is often referred to by a term borrowed from music—pitch. The higher the frequency, the higher the pitch, and vice versa. As stated, the approximate frequency range of a healthy young person's hearing is 20 to 20,000 Hz. The upper limit decreases with age as a result of a process called presbycusis. Recognition of this phenomenon can be of importance in schools, since very high-pitched sounds that are inaudible to most adults (more pronounced in men than women) can be a source of extreme annoyance to students. For example, dentists report that high-speed turbine drills and teeth-cleaning devices cause extreme auditory discomfort to many young patients. These devices produce sounds in the 15- to 20-kilohertz (kHz) range.

The human speaking voice has a range of approximately 100 to 600 Hz in *fundamentals,* but harmonics (overtones) reach to approximately 7500 Hz. Most speech information, as discussed below, is carried in the upper frequencies, while most *energy* exists in the lower frequencies. The critical range of speech communication is 300 to 4000 Hz. Overtones outside these frequencies give the voice its characteristic sound and specific identity. (See Figs. 26.14 and 26.19.)

A sound composed of only one frequency is a pure tone. Except for the sound generated by a tuning fork, few sounds are truly pure. Musical sounds are composed of a fundamental frequency and integral multiples of the fundamental frequency (harmonics). Most common sounds are complex combinations of frequencies. Figure 26.2 shows examples of pure tones, musical notes, and common sounds; Fig. 26.3 shows the frequency ranges of some common devices and phenomena. The frequencies in the scale of Fig. 26.3 all stand in the ratio of 2:1 to each other, that is, 16:32:63:125:250, and so on. Borrowing again from musical terminology, they are one *octave* apart. These particular frequencies are also accepted internationally as the center frequencies of octave bands used for the purpose of sound specification.

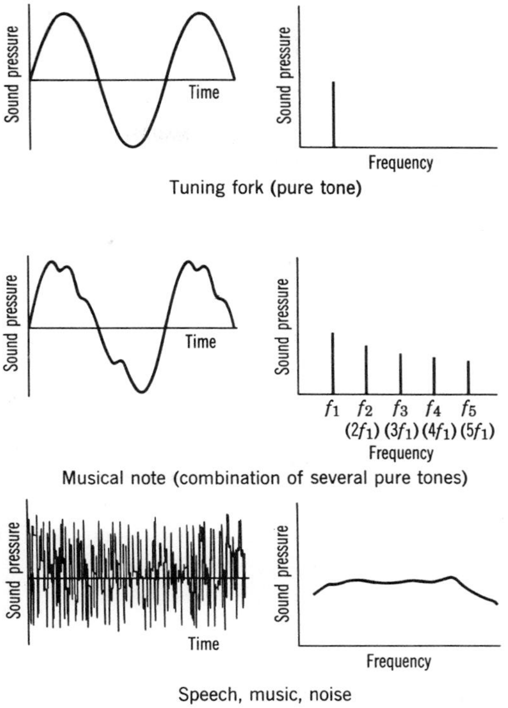

Fig. 26.2 *Schematic representations of a pure tone, a musical note, and more complex sounds (speech, music, and noise), showing the variation of sound pressure with time and frequency.*

For technical reasons, a geometric mean is used. Thus 250 is the center frequency of an octave band ranging from $250/\sqrt{2}$ to $250\sqrt{2}$ and that particular octave is known as the 250-Hz octave. If finer division is required, ½-octave and ⅓-octave bands are used.

26.4 Velocity of Propagation

Sound travels at different velocities depending on the medium. In air, at sea level, sound velocity is 344 m/s or 1130 fps. This corresponds to 770 miles per hour or 1239 kilometers per hour—slow indeed when compared to light, which has a velocity of 186,000 miles per *second.* Since sound travels not only in air but also through parts of a structure, it is of interest to know the velocities in other media. See Table 26.1. For our purposes, velocity changes due to temperature and altitude (atmos-

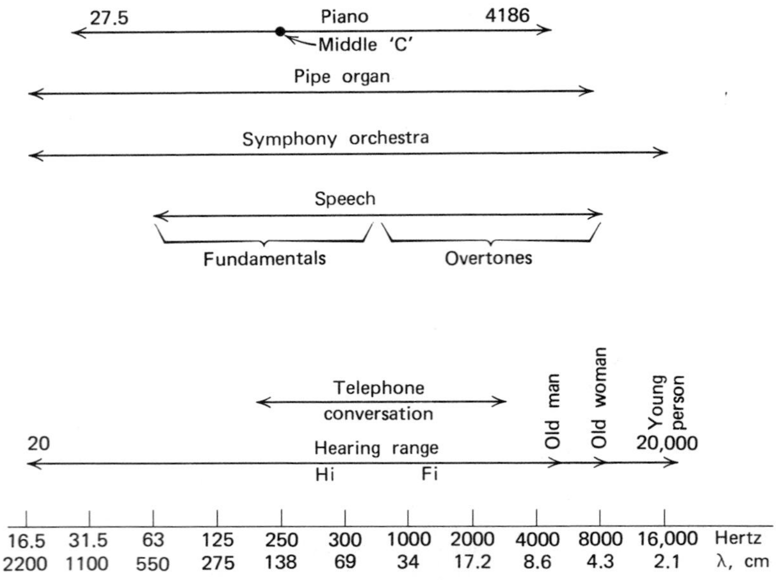

Fig. 26.3 *Frequency ranges of common instruments. Wavelength is calculated on an assumed propagation velocity of 344 m/s.*

pheric pressure) may be ignored and, for rough calculation, 1100 fps and 350 m/s may be used as velocity in air, since both are within 3% error.

26.5 Wavelength and Types of Propagation

The wavelength of a sound may be defined as the distance between similar points on successive waves or the distance the sound travels in one cycle. The relationship between wavelength, frequency, and velocity of sound is expressed as

$$\lambda = \frac{c}{f} \tag{26.1}$$

where

λ = wavelength, ft (m)
c = velocity of sound, fps (m/s)
f = frequency of sound, Hz

Low-frequency sounds are characterized by long wavelengths and high-frequency sounds by short wavelengths. Sounds with wavelengths ranging from ½ in. to 50 ft can be heard by humans. A simple nomograph presented later in this chapter (see Fig. 26.31) permits rapid determination of wavelength, given the frequency, and vice versa.

TABLE 26.1 **Sound Propagation Velocity in Various Media**

	Velocity	
Medium	*Meters per Second*	*Feet per Second*
Air	344	1130
Water	1410	4625
Wood	3300	10,825
Brick	3600	11,800
Concrete	3700	12,100
Steel	4900	16,000
Glass	5000	16,400
Aluminum	5800	19,000

NOTE: These figures are approximate, since the listed materials vary in density. Average frequency is used.

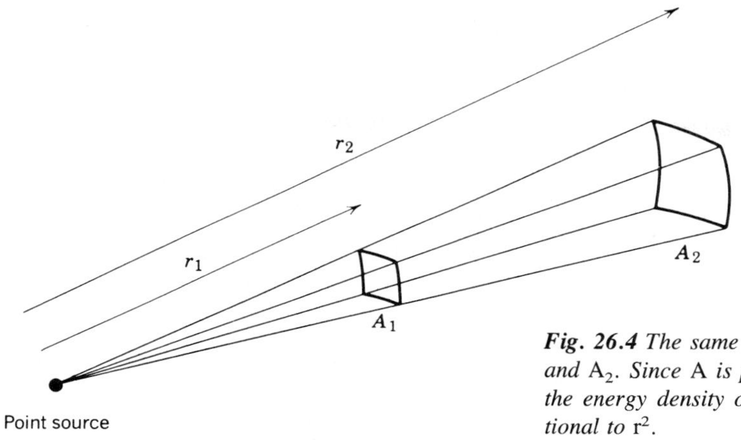

Fig. 26.4 The same total energy passes through A_1 *and* A_2*. Since* A *is proportional to the square of* r*, the energy density or intensity is inversely proportional to* r^2.

26.6 Sound Magnitude

The magnitude of a sound signal is a much more complex subject because of the different terms in use and because of the great range of numbers involved. When we speak of sound magnitude, we think of loudness, which is a subjective, ear-oriented reaction not linearly related to the physical quantity of sound. The level (quantity) of sound is described variously as sound power, sound pressure, sound pressure level (SPL), sound intensity, and sound intensity level (IL), all of which differ from each other, and from subjective loudness. To clearly understand these concepts, and it is imperative that they be so understood, a comprehension of how we hear and how sound is propagated in free space is necessary.

26.7 Sound Intensity; Free Field Propagation

A point sound source of constant power (watts) radiating in free space, that is, at a distance far from the effects of any reflecting surface, is represented in the drawing of Fig. 26.4. The *sound intensity* at any distance from the source is expressed as

$$I = \frac{P}{A} \qquad (26.2)$$

where

I = sound (power) intensity, W/cm² (W/m²)
P = acoustic power, W
A = area, cm² (m²)

It is traditional in architectural acoustics to express area in square centimeters, although the MKS (SI) system would require area to be stated in square meters. We will continue to use square centimeters. Conversion data are supplied in Table 27.15. See also Table 27.16 for a listing of symbols and abbreviations. Since the sound radiates freely in all directions,

$$I = \frac{P}{4\pi r^2} \quad \text{W/cm}^2 \qquad (26.3)$$

where r is the radius of an imaginary enclosing sphere. (In English units this is

$$I = \frac{P}{930 \times 4\pi r^2} \quad \text{W/ft}^2 \qquad (26.4)$$

since 1 ft² = 930 cm².) The intensities at distances r_1 and r_2 from the source stand in the ratio of

$$\frac{I_1}{I_2} = \frac{r_2^2}{r_1^2} \qquad (26.5)$$

which is the formula for the classic *inverse square law,* stating that *intensity is inversely proportional to (distance)² from the source*. Note the exact correspondence of these relations to the derivations for illumination from a point source, found in Section 18.12 and Fig. 18.12. Figure 26.5 shows graphically how a sound pulse is attenuated in *strength* (but not in waveform) as it travels outward from the source by action of distance.

The preceding derivation is based on a *point* source, that is, a source that is small relative to the wavelength produced. This type of source pro-

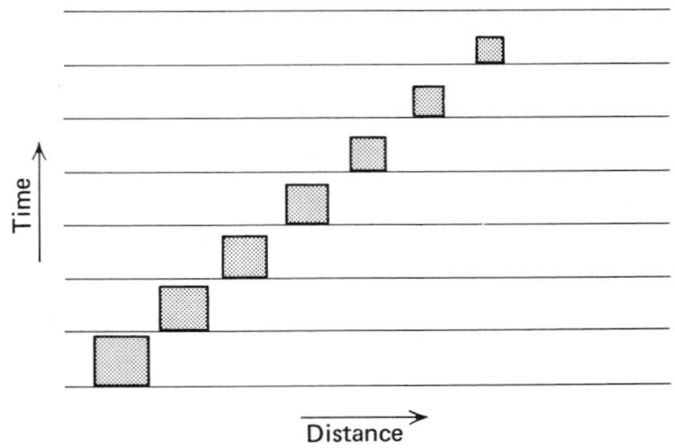

Fig. 26.5 *Attenuation of a sound signal as it travels away from the source. The shape (information) remains constant when traveling in a nondispersive medium such as air. This is not the case with travel in solids, where different frequencies travel at different velocities, and therefore cause a wave-shape change with time and distance.*

duces spherical waves. Line sources such as strings produce cylindrical waves. Large vibrating surfaces such as walls produce plane waves. The importance of these facts will become clear in the discussions on barriers and diffraction in Chapter 27.

The threshold of hearing, that is, the minimum sound power intensity (I) that a normal ear can detect, is 10^{-16} W/cm^2. (Actually, the ear responds to pressure, as explained below.) The maximum sound intensity that the ear can accept without damage is approximately 10^{-3} W/cm^2. This gives a range of 10^{13}, or 10 million million to 1 (10,000,000,000,000:1). Table 26.2 gives an idea of the physical significance of these numbers. Two problems arise immediately when dealing with quantities of this type; the numbers themselves are very small and the ratios are very large. Furthermore, the human ear responds logarithmically, not arithmetically to sound pressure (and intensity); that is, doubling the intensity of a sound does not double its loudness—the change is barely perceptible. To solve these problems it would be much more convenient if we were to construct a scale that:

1. Started at zero for the minimum sound (intensity or pressure) we can hear.
2. Used whole numbers rather than negative powers of 10.
3. Had some fixed relationship between an arithmetic difference and a loudness change; say, 10 units equals a doubling (or halving) of loudness. Thus, on such a scale, the difference between 20 and 30, and 60 and 70, would always be a doubling of loudness.

Such a scale exists. It is the decibel scale.

26.8 Intensity Level (IL) and the Decibel (db)

The word "level" indicates a quantity relative to a base quantity. Intensity *level* is the ratio between a given intensity and a base intensity. If we express intensity level as

TABLE 26.2 **Comparison of Decimal, Exponential, and Logarithmic Statements of Various Acoustic Intensities**

Intensity (W/cm^2)		*Intensity Level—Logarithmic Notation*	*Examples*
Decimal Notation	*Exponential Notation*		
0.001	10^{-3}	130 db	Painful
0.0001	10^{-4}	120 db	
0.00001	10^{-5}	110 db	75-piece orchestra
0.000001	10^{-6}	100 db	
0.0000001	10^{-7}	90 db	Shouting at 5 ft
0.000000001	10^{-9}	70 db	Speech at 3 ft
0.00000000001	10^{-11}	50 db	Average office
0.0000000000001	10^{-13}	30 db	Quiet unoccupied office
0.00000000000001	10^{-14}	20 db	Rural ambient
0.000000000000001	10^{-15}	10 db	
0.0000000000000001	10^{-16}	0 db	Threshold of hearing

$$IL = 10 \log \frac{I}{I_0} \tag{26.6}$$

where

IL = intensity level, db
I = intensity, W/cm^2
I_0 = base (i.e., 10^{-16} W/cm^2, the threshold of hearing)
log = logarithm to base 10

then we have established a scale that satisfies the three conditions set forth above. The quantity IL, intensity level, is dimensionless, since it indicates simply a ratio between two numbers. It is measured in decibels (db) for convenience in expressing the large numbers involved. Table 26.2 shows the great convenience of using the logarithmic decibel scale as compared to either decimal notation or exponential notation. Table 26.3 gives a short listing of subjective loudness changes expressed in decibels. Note that 10 db indicates a doubling of loudness, as specified; 20 db is loudness doubled twice, that is, four times as loud. The *difference* (Δ) between any two intensity levels

$$\Delta IL = IL_2 - IL_1 = 10 \log \frac{I_2}{I_0} - 10 \log \frac{I_1}{I_0}$$

$$\therefore \Delta IL = 10 \log \frac{I_2}{I_1} \text{ db} \tag{26.7}$$

A few examples using db notation and logarithmic calculations should help the reader become familiar with this useful system.

EXAMPLE 26.1. Two sound sources produce intensity levels of 50 and 60 db, respectively, at a point. When these are functioning simultaneously, what is the total sound intensity level? (We assume identical frequency content and random phase; that is, the phase relationship between the two sources changes in a random manner.)

TABLE 26.3 **Subjective Loudness Changes and Corresponding Intensity Level Changes**

Change in Level (db)	*Subjective Change in Loudness*
3	Barely perceptible
6[a]	Perceptible
7	Clearly perceptible
10	Twice or half as loud
20	Four times or one-quarter as loud

[a]Six decibels corresponds to the change encountered when distance to the source in a free field is doubled (halved).

SOLUTION. Note that we are dealing with intensity level, not intensity, since intensity itself has little significance for us. The technique involved in adding two sound intensity levels has three steps:

1. Convert both to actual intensity.

$$IL = 10 \log \frac{I}{I_0}$$

so

$$60 = 10 \log \frac{I_1}{10^{-16}}$$

or

$$6.0 = \log \frac{I_1}{10^{-16}}$$

Then, using the definition of a base 10 logarithm:

$$10^6 = \frac{I_1}{10^{-16}}$$

$$I_1 = (10^{-16})\, 10^6 = 10^{-10} \text{ W/cm}^2$$

By similar calculation,

$$I_2 = 10^{-11} \text{ W/cm}^2$$

2. Add arithmetically.

$$I_1 + I_2 = 10^{-10} + 10^{-11}$$
$$= (10 \times 10^{-11}) + 10^{-11}$$
$$I_{tot} = 11 \times 10^{-11} \text{ W/cm}^2$$

3. Reconvert to decibels. To find the intensity level IL corresponding to the new total intensity $I_1 + I_2$, we simply apply Equation 26.6:

$$IL_{tot} = 10 \log \frac{I_{tot}}{I_0}$$
$$= 10 \log \frac{11 \times 10^{-11}}{10^{-16}}$$
$$= 10 (\log 11 + \log 10^5)$$
$$= 10 (1.04 + 5)$$
$$= 60.4 \text{ db}$$

which is only a fraction larger than the original 60 db of the stronger sound.

EXAMPLE 26.2. Assume two noise signals of 60 db each. What is the combined strength in decibels?

SOLUTION. One method would be to calculate levels as in Example 26.1. A shorter method is to find the difference between the two signals and to add it to either one. Using Equation 26.7,

$$\Delta IL = IL_{\text{comb}} - IL_1 = 10 \log \frac{I_{\text{comb}}}{I_1}$$

$$= 10 \log \frac{2I_1}{I_1}$$

$$= 10 \log 2$$

$$= 10\ (0.30)$$

$$= 3 \text{ db}$$

This answer (which is independent of any particular level) gives us the extremely important fact that doubling a signal intensity raises the intensity level by 3 *db*. (In our case, the combined intensity level would obviously be 60 db + 3 db or 63 db). Similarly, quadrupling a signal's intensity raises the received level by 6 db. This is because

$$\Delta IL = 10 \log \frac{4I}{I}$$

$$= 10 \log 4$$

$$= 10\ (0.60)$$

$$= 6 \text{ db}$$

This technique is very useful when combining a large number of identical levels, as in the following example.

EXAMPLE 26.3. A factory will contain 20 identical machines, each of which generates a noise level of 80 db. What will be the combined noise level? (Ignore problems of frequency content, phase, and fields.)

SOLUTION.

$$\Delta IL = IL_{\text{tot}} - IL_{\text{single}}$$

$$= 10 \log \frac{I_{\text{tot}}}{I_{\text{single}}}$$

$$= 10 \log \frac{20\ I_{\text{single}}}{I_{\text{single}}}$$

$$= 10 \log 20$$

$$\Delta IL = 10(1.3) = 13 \text{ db}$$

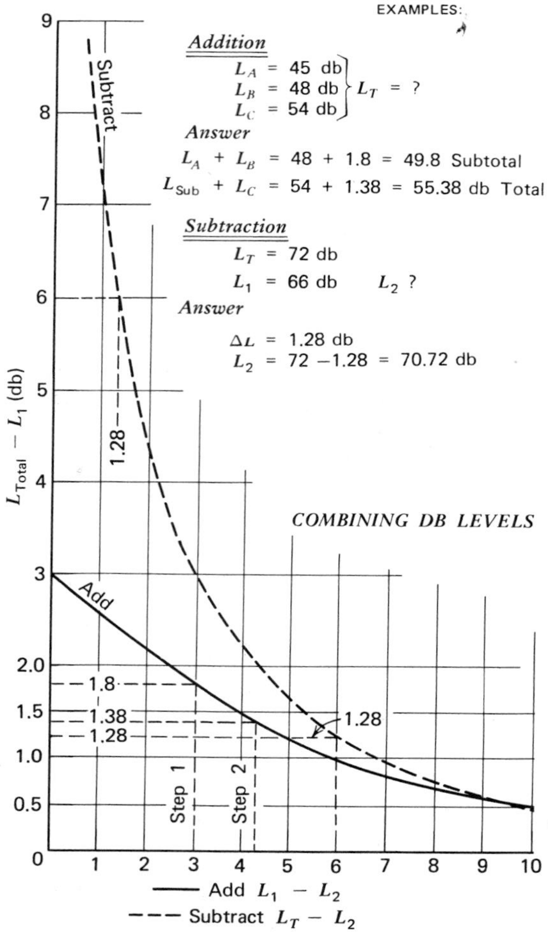

Fig. 26.6 Curves for combining decibel levels of uncorrelated sound sources.

Therefore the total noise level will be

$$IL_{\text{tot}} = 80 \text{ db} + 13 \text{ db} = 93 \text{ db}$$

A chart for combining decibel levels of two sources is given below in Fig. 26.6, which eliminates the somewhat lengthy procedure detailed in Example 26.1. Referring to Table 26.3, we note that the human ear is not responsive to fractional db changes; indeed even a 3-db change is barely perceptible. This being so, we recommend that the detailed calculation and the chart in Fig. 26.6 be reserved for situations where a high degree of accuracy is required. For everyday calculations, the following rules of thumb may be used to combine db levels of two sources:

1. When the difference between two sources is 1 db or less, add 3 db to the higher level to obtain the total.
2. When the difference is 2 to 3 db, add 2 db.
3. When the difference is 4 to 8 db, add 1 db.
4. When the difference is 9 db or more, ignore the lower source.

A comparison of addition by these rules of thumb and by an accurate method shows that at usual levels, the error involved is always less than 1%.

db Level		*Sum*	
Lower	*Higher*	*Approximate*	*Accurate*
60	60	63	63
60	62	64	64.4
60	64	65	65.5
60	66	67	67
60	68	69	68.7
60	70	70	70.5

Returning to the inverse square law expressed in Equation 26.5, we are now in a position to determine the effect on sound intensity level of moving away from the sound source.

EXAMPLE 26.4. Given a sound source that produces sound intensity IL at a distance d_1 from the source (the reader can substitute any numbers desired, or follow the problem with symbols), what are the intensities at twice the distance? Four times?

SOLUTION. We know from Equation 26.7 that

$$\Delta IL = 10 \log \frac{I_2}{I_1}$$

and we also know from Equation 26.5 that

$$\frac{I_1}{I_2} = \frac{d_2^2}{d_1^2}$$

or

$$\frac{I_1}{I_2} = \frac{(2d_1)^2}{(d_1)^2} = 4$$

Substituting in Equation 26.7, we have

$$\Delta IL = 10 \log \frac{I_2}{I_1}$$

$$= 10 \log 1/4$$

$$= 10\,(-0.6)$$

$$= -6 \text{ db}$$

which tells us that sound intensity level (not pressure) is reduced by 6 db. Similarly, when distance is quadrupled, it is reduced by 12 db.

To summarize, then, intensity level increases (decreases) 3 db with every doubling (halving) of power and decreases (increases) 6 db with every doubling (halving) of distance. Figures 26.6 through 26.9 illustrate these relationships.

The ear responds to sound pressure, not inten-

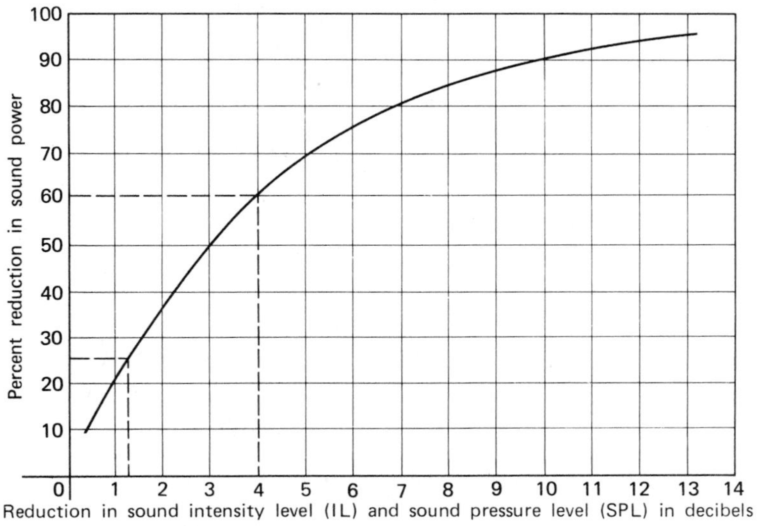

Fig. 26.7 Decibel level decrease as a function of percentage power (intensity) decrease.

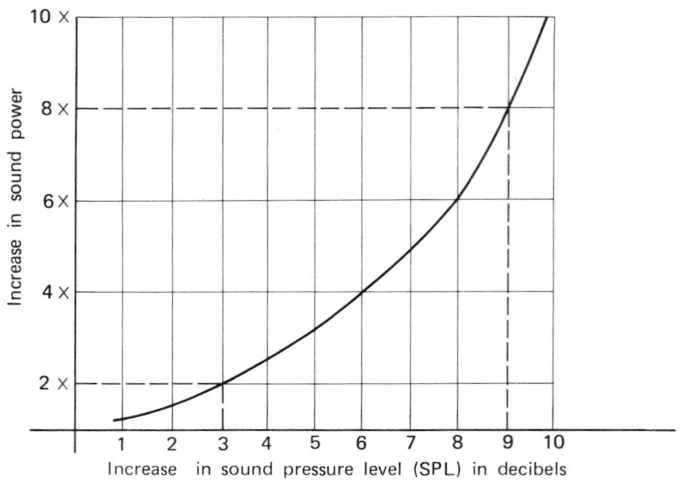

Fig. 26.8 Decibel level increase as a function of power (intensity) increase.

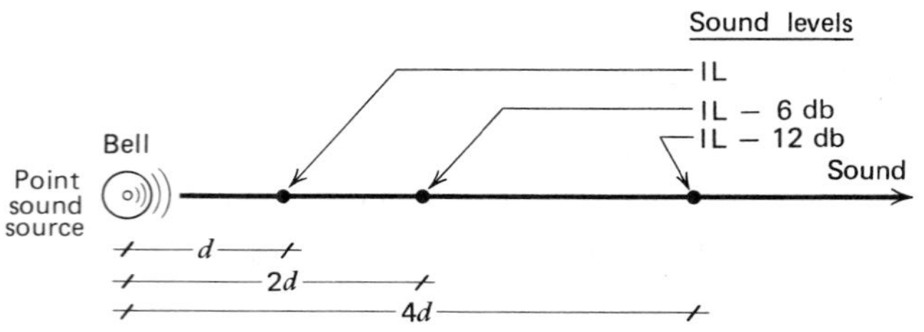

Fig. 26.9 Sound energy levels at varying distances from the source. Each doubling of distance reduces the level by 6db. These relationships hold true only in a free field.

sity. As explained below, sound *pressure level* (SPL) is equal numerically to *intensity level* (IL) (at least for normal temperature and pressure, i.e., for our use in architectural acoustics). Therefore, the foregoing examples and manipulations of intensity level are equally applicable to sound pressure level.

As explained in some detail in the following section, the combined effect of two sounds depends on their frequency content. In the examples above we assumed signals either of identical frequency and random phase or of very-wide-frequency spectrum—so wide that phase phenomena are not significant. In architectural acoustics work, such an assumption is generally valid.

26.9 Human Hearing

Refer to Fig. 26.10, which is a *schematic diagram* of the human ear, since explanation of the functioning would be more difficult on an anatomical drawing. The outer ear is funnel shaped and serves as a sound-gathering input terminal to the auditory system. Sound energy travels through the auditory canal (outer ear) and sets in motion the components of the middle ear, comprising the eardrum, hammer, anvil, and stirrup. The stirrup acts as a piston to transmit vibrations into the fluid of the inner ear. This fluid motion causes movement of hair cells in the cochlea, which in turn stimulates nerves at their bases, which in turn transmit electrical impulses along the eighth cranial nerve, to the brain. These impulses we understand as sound.

Frequency recognition is accomplished in the cochlea, by the basilar membrane. This membrane resonates at one end at about 20 Hz and at the other at 20 kHz, giving the ear its frequency range. The ear hears and recognizes distinct frequencies, yet the hearing mechanism has the ability, apparently as directed by the brain, either to hear individual frequencies *or* to combine them into a single sound. Thus when we hear a string quartet we can, *at will,* hear either the entire quartet or each instrument individually. With concentration

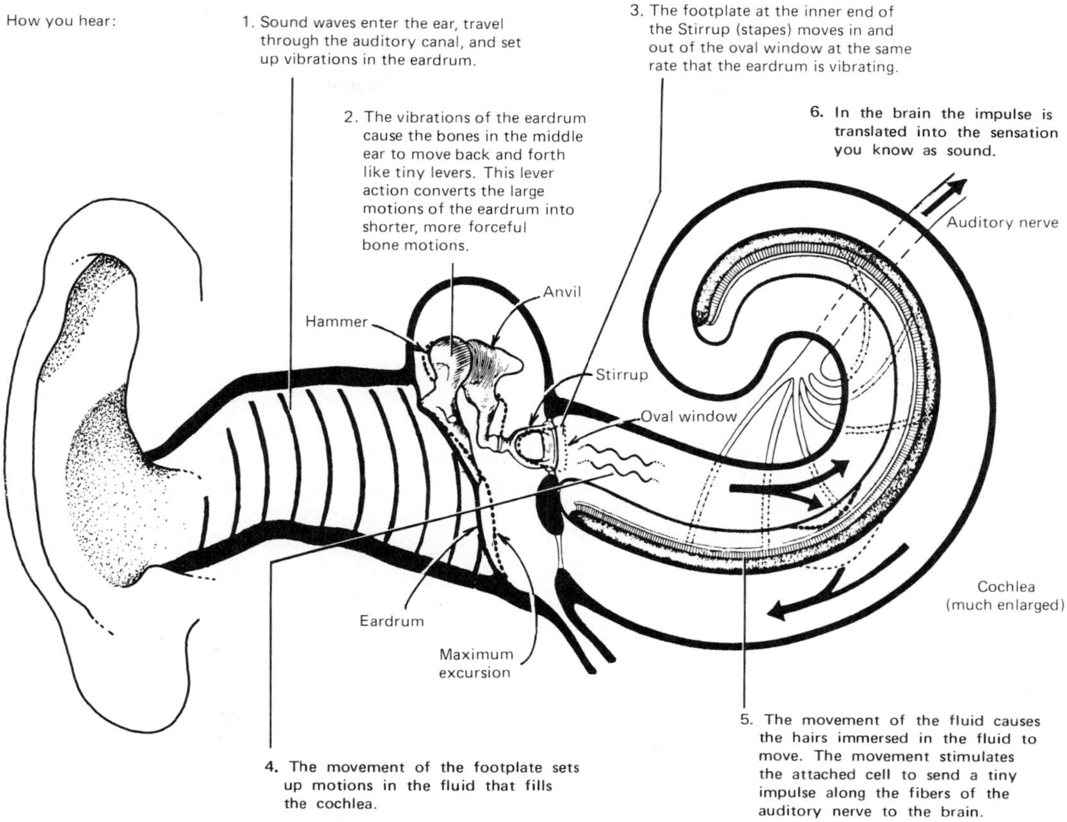
How you hear:
1. Sound waves enter the ear, travel through the auditory canal, and set up vibrations in the eardrum.
2. The vibrations of the eardrum cause the bones in the middle ear to move back and forth like tiny levers. This lever action converts the large motions of the eardrum into shorter, more forceful bone motions.
3. The footplate at the inner end of the Stirrup (stapes) moves in and out of the oval window at the same rate that the eardrum is vibrating.
4. The movement of the footplate sets up motions in the fluid that fills the cochlea.
5. The movement of the fluid causes the hairs immersed in the fluid to move. The movement stimulates the attached cell to send a tiny impulse along the fibers of the auditory nerve to the brain.
6. In the brain the impulse is translated into the sensation you know as sound.
Auditory nerve
Anvil
Hammer
Stirrup
Oval window
Eardrum
Maximum excursion
Cochlea (much enlarged)

Fig. 26.10 *Schematic diagram showing functioning of the human ear. The dotted lines adjacent to each movable element indicate the extent of its movement. (From* Quieting: A Practical Guide to Noise Control. *NBS Handbook 119, July 1976.)*

(vision helps in this), a ''trained'' ear can pick out a single instrument in an orchestra of 120 pieces, even if there is more than one such instrument in the group. Conductors do this regularly. Similarly, the ear can perform the selection known as the ''cocktail party effect,'' that is, pick out one voice in background noise that may be 20 db louder than the wanted signal. In effect, the ear is attenuating the unwanted signals. Normally, however, the ear does precisely the opposite. It combines sounds that are clearly distinct from each other in frequency and phase. The three tones in a chord struck on a piano are different in frequency and, if played as a very rapid triplet, out of phase. Yet the ear combines them and hears a single sound, despite the fact that the maxima of the three tones do not occur simultaneously. In the preceding decibel calculations (Section 26.8), as well as those that will follow, it is assumed that the ear ignores phase differences and combines frequencies. This may not always be the case, particularly when the frequencies are very far apart. For this reason the type of information found in Table 26.4 (i.e., single-number representations of complex sounds) can be misleading and must be used with caution. This problem is discussed in the next section and in Section 26.12, which explains measurements and weighting networks.

At the threshold of hearing (approximately 0 db), the displacement of air molecules impinging on the eardrum, and the eardrum excursion, are approximately one angstrom unit (1 Å = 10^{-8} cm), which is approximately the diameter of an *atom*. Were the ear an order of magnitude more sensitive, it would hear thermal noise. The human ear is thus close to the practical limit of sensitivity.

At the other end of the noise spectrum, the threshold of pain corresponds to 130 db, and a molecular and eardrum motion of approximately 0.25 mm. An astonishing range indeed!

As is obvious from the brief description of the hearing process above, movement of the eardrum is caused by air pressure. Therefore the quantities that interest us more than intensity (*power density*) are pressure, which is *force density*, and pressure level, which is the ratio of sound pressure to a base level, expressed in db.

26.10 Sound Pressure and Sound Pressure Level (SPL)

The usual base, reference, pressure corresponds to the threshold of hearing and is taken to be 20 μPa or 2 × 10^{-4} microbar (μbar). See Table 27.15. As with intensity, this pressure is established as 0 db for the purpose of sound pressure *level* calculation. Since the ear responds logarithmically to intensity and since pressure varies as the square root of intensity, we can write the expression

$$\text{SPL} = 10 \log \frac{p^2}{p_0^{\,2}}$$

or

Fig. 26.11 Common sound sources plotted at their dominant frequencies and levels as typically heard by the observer. The equal loudness curves (see Section 26.11) show why certain sounds seem louder than others despite the pressure levels that would indicate the contrary. (From Quieting: A Practical Guide to Noise Control. *NBS Handbook 119, July 1976.)*

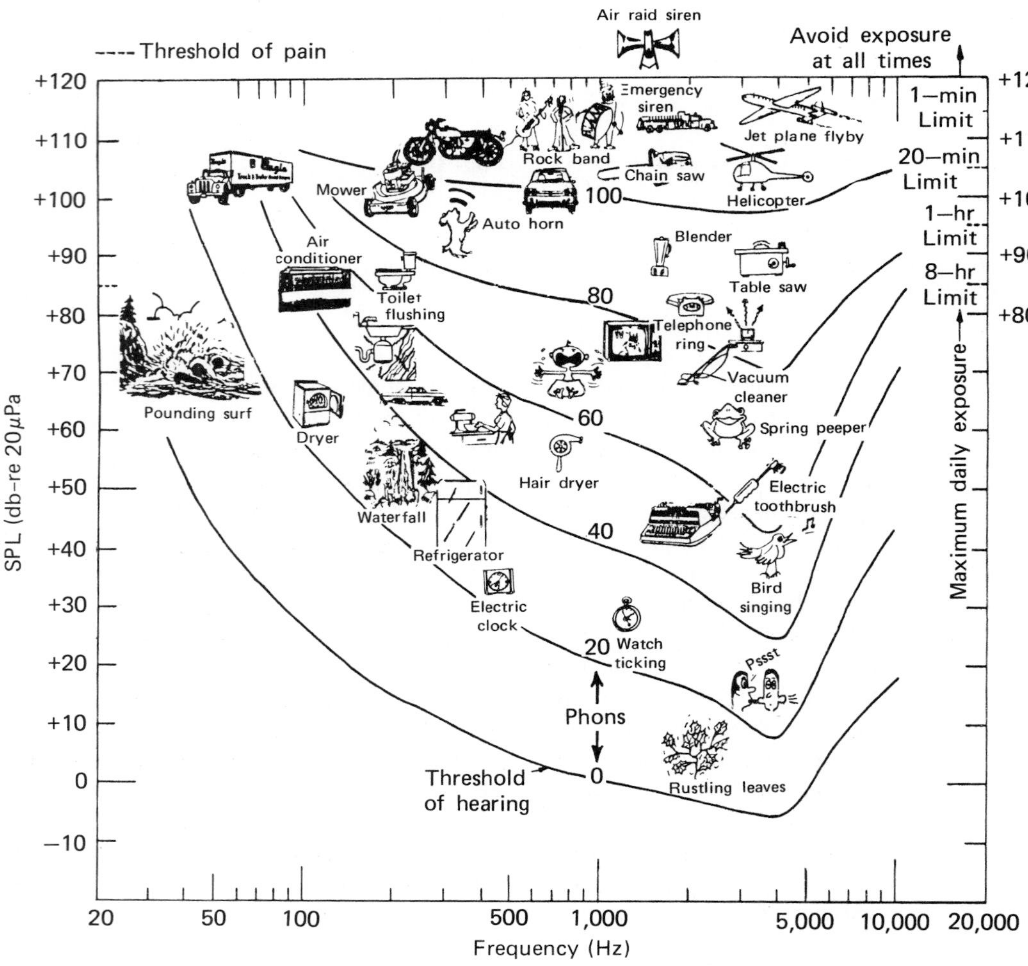

$$SPL = 20 \log \frac{p}{p_0} \qquad (26.8)$$

where

SPL = sound pressure level, db
p = pressure, Pa (bars)
p_0 = reference base pressure, Pa (20 μPa or $2E^{-4}$ μbar).

Since for both intensity level and sound pressure level, the 0-db base corresponds to the hearing threshold, we have equalized the db scales for sound pressure level (SPL) and intensity level (IL) *and the db values of the two can be used interchangeably*. Obviously, though, the actual intensity and the actual pressure corresponding to a particular decibel level are different—completely different, in magnitude and units. For instance, 70 db equals 10^{-9} W/cm intensity, and 0.063 Pa pressure. However, the important fact is that 70 db corresponds approximately to a particular noise level. We say "approximately" because assigning a single-number decibel level to a sound presents two difficulties:

1. The sound pressure level varies with time, except for a pure steady tone.
2. The different components of most common sounds vary in pressure level.

To overcome this problem two techniques are used. If a sound has a dominant frequency, that frequency's level can be used (see Fig. 26.11). This would be the case for a relatively constant sound such as a motor, or a fan or blower. Other sounds that vary widely in level and frequency can be plotted on an octave-band chart with maximum level for a minimum percent of time (see Fig. 26.12). Thus vehicular traffic is represented by levels maintained for 90% of the time. Speech sounds are those that will be exceeded only 1% of the time. The single-number levels in Table

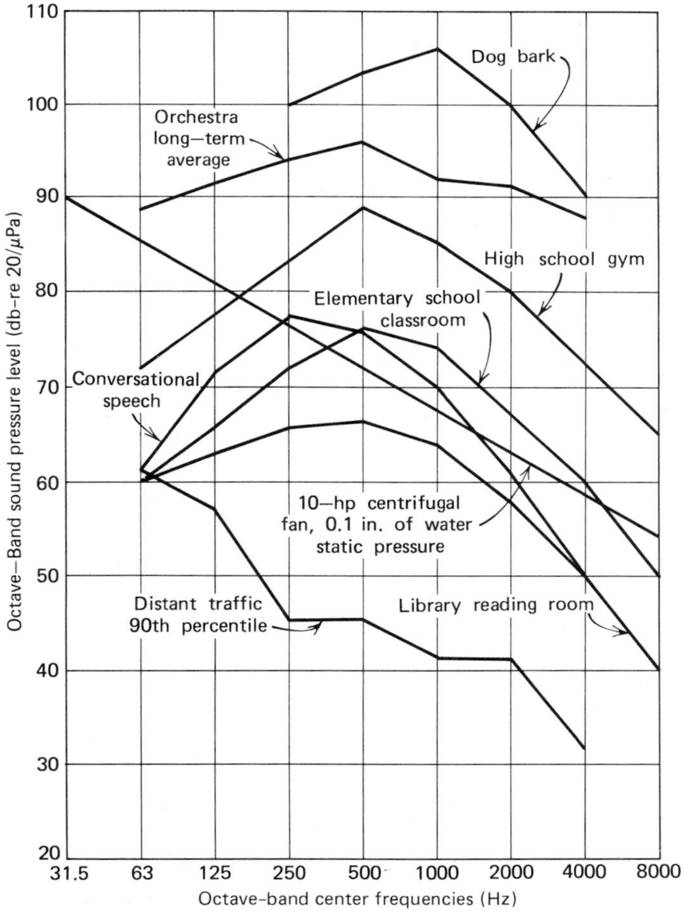

Fig. 26.12 Sound pressure levels of eight typical sounds with different time characteristics.

TABLE 26.4 **Common Noise Levels**

Sound Pressure Level, SPL (db)	*Typical Sound*	*Subjective Impression*
150		(Short exposure can cause hearing loss)
140	Jet plane takeoff	
130	Artillery fire, riveting, machine gun	(Threshold of pain)
120	Siren at 100 ft, jet plane (passenger ramp), thunder, sonic boom	Deafening
110	Woodworking shop, hard-rock band, accelerating motorcycle	Sound can be felt (Threshold of discomfort)
100	Subway (steel wheels), loud street noise, power lawnmower, outboard motor	Very loud, conversation difficult; ear protection required for sustained occupancy
90	Noisy factory, truck unmuffled, train whistle, machine shop, kitchen blender, pneumatic jackhammer	
80	Printing press, subway (rubber wheels), noisy office, supermarket, average factory	(Intolerable for phone use)
70	Average street noise, quiet typewriter, freight train at 100 ft, average radio, department store	Loud, noisy; voice must be raised to be understood
60	Noisy home, hotel lobby, average office, restaurant, normal conversation	Usual background; normal conversation easily understood
50	General office, hospital, quiet radio, average home, bank, quiet street	
40	Private office, quiet home	Noticeably quiet
30	Quiet conversation, broadcast studio	
20	Empty auditorium, whisper	
10	Rustling leaves, soundproof room, human breathing	Very quiet
0 db		Intolerably quiet Threshold of audibility

26.4 have been assigned on this basis and are primarily useful to establish a mental-aural comparison base and to use in maximum exposure calculations, as are discussed in Section 26.21. Where the position of the listener is not specified in the table, it is assumed to be at normal *close* distances, that is, 10 to 20 ft from a train, 3 to 5 ft from a radio, and the like.

26.11 Loudness Level: The Phon Scale

The human ear is not uniformly sensitive over its entire frequency range of 20 Hz to 20 kHz. The 120- to 130-db upper limit pain threshold occurs at all frequencies. However, at the lower limit, the 0-db threshold occurs only at 1000 Hz. The ear is in fact most sensitive at 3000 to 4000 Hz, at which frequencies the threshold is about −5 db. This nonlinear response exists throughout the ear's range. To determine the nature of this nonlinearity, a great number of tests were conducted with pure tones at different frequencies, in which listeners were asked to equate subjective loudness of signals. These tests resulted in a family of curves called *equal loudness level contours* or, alternatively, *Fletcher-Munson equal loudness contours*

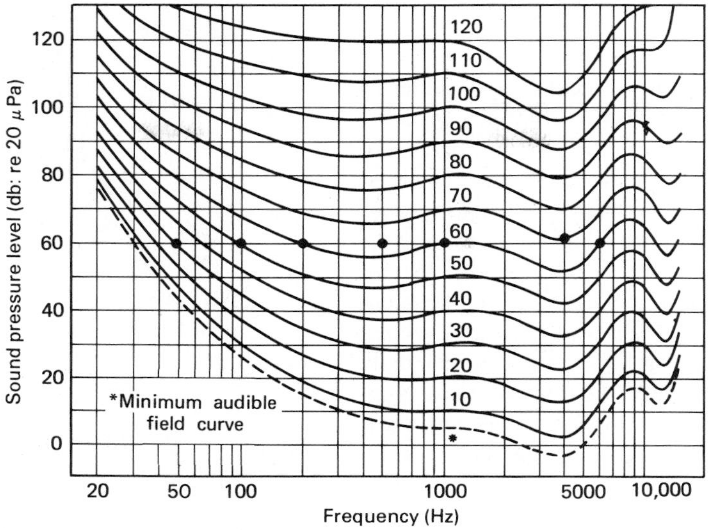

Fig. 26.13 *Standard equal loudness contours. These curves are accurate for a listener with normal binaural hearing, situated in the near field of a source producing pure tones directly ahead of the listener. Note that an SPL of 60 db corresponds to 30 phons at 50 Hz, 50 phons at 100 Hz, 63 phons at 500 Hz, 60 phons at 1 kHz, 70 phons at 4 kHz, and 60 phons at 6 kHz. This indicates the relative flatness of the ear's response in the central frequency–loudness range, and the sharp drop at low frequencies.*

(after two of the principal researchers). These curves (see Fig. 26.13) are internationally recognized and standardized, and are used as the reference for normal hearing response. They are also used to ''weight'' measuring devices, as is explained in Section 26.12. Note that *by definition:*

1. All points on a single contour have the same subjective sensation of loudness.
2. The loudness level in *phons* of the *entire* contour is defined as the db level of that contour at 1000 Hz.

The curves demonstrate some interesting phenomena:

1. Sensitivity drops off sharply at low frequencies, particularly at low-db levels. (For this reason most high-fidelity amplifiers provide automatic bass boost at low-volume levels.)
2. Maximum sensitivity occurs between 3 and 4 kHz—precisely the frequencies that contain most information in human speech. (Note the dip between 3 and 4 kHz). See Fig. 26.14.
3. In the normal listening range of 45 to 85 db, and the normal frequency range of 150 Hz to 6 kHz, the contour is substantially flat; that is, the ear's response is effectively linear. It is only at extremes of sound level and frequency that nonlinearity occurs.

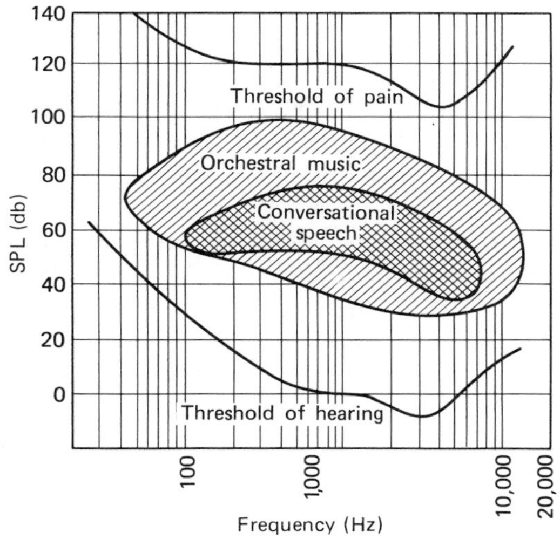

Fig. 26.14 *The positions of speech and wide-range music in the human ear's aural field are illustrated. Speech is in the nominally linear response area of the ear as is most music (see Fig. 26.13). Beyond these frequencies, the ear's action is effectively to attenuate the signal.*

26.12 Measurement of Sound Pressure Level

The need for a means of measuring sound levels is obvious. One such instrument is the integrating sound-level meter illustrated in Fig. 26.15*a*. To correlate meter readings with subjective loudness impressions, most such single-reading instruments are furnished with three weighting networks, the characteristics of which are given in Fig. 26.15*b*. The A network corresponds to the 40-phon contour and discriminates against low frequencies (see Fig. 26.13). The B network corresponds similarly to the 70-phon contour. The C network gives substantially flat uniform response. In practice it was found that the B and C networks did not correspond well to subjective loudness reports. This was because the loudness contours on which they were based were taken for pure tones, whereas most field measurements are of very complex sounds. The A network, however, did correspond well, and has therefore become the standard single-number measuring scale for loudness of sounds at any frequency and intensity. Measurements taken in this way should be labeled ''dbA'' to identify the scale. The sound meter is equipped with two different damping characteristics—fast and slow. The former will follow reasonably rapid fluctuations. The latter is more heavily damped and will give an average reading of sounds with rapid changes of more than 4 db. Extremely rapid noise pulses are impulse noises, requiring special instruments.

It is known that the ear ''averages'' sounds over a minimum period, and that sound impulses shorter than this period sound quieter than they would sound as steady-state noise. This minimum sound length is taken to be between 50 and 200 milliseconds (1/20–1/5 s) depending on which researcher's work is used. A compromise figure of 70 ms (1/14 s) is frequently found in the literature. This figure becomes important when considering the effect of echoes. See Sections 26.23 through 26.25.

More accurate measurements of complex sounds than are possible with a standard meter are accomplished with sophisticated instruments that measure intensity in octave bands and plot the results as in Fig. 26.12. Such measurements are necessary for accurate application of sound absorption and attenuation materials whose characteristics are also

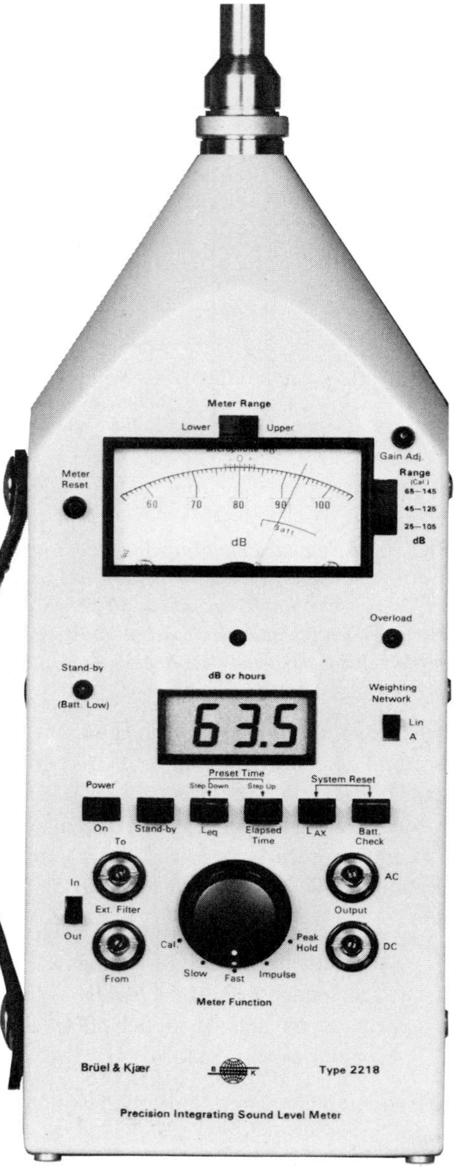

(a)

Fig. 26.15a *Precision-integrating sound level meter. This unit will measure either instantaneous sound level or L_{eq} (equivalent continuous sound level). The latter is important when sound at a location varies with time, and an equivalent level for design purposes is required. The unit takes level and duration into account and displays L_{eq} directly. It is arranged to measure dbA or linear (dbC), fast or slow time constant. It adapts with filter networks to read octave or 1/3-octave bands, and reads out digitally. It also has peak-hold settings, which measure maximum sound during a time exposure. (Courtesy of Brüel and Kjaer.)*

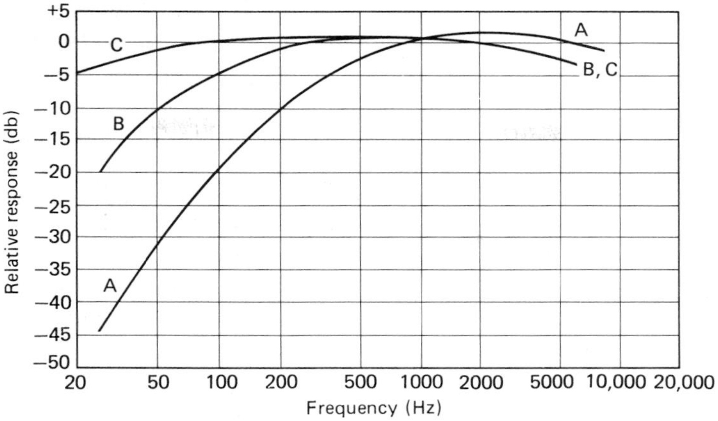

(b)

Fig. 26.15b *Frequency–response characteristics in the ANSI Standard for Sound-Level Meters. (Courtesy of the American National Standards Institute.)*

nonlinear over the frequency spectrum. Single-number dbA readings are known as *overall* levels and are useful as preliminary data and for broad-spectrum design.

26.13 Sound Absorption

When sound energy impinges on a material, part is reflected and the remainder is absorbed, in the sense that it is not reflected. (Some of the energy is transmitted, as discussed in the next chapter; see Fig. 27.2.) Most materials are neither perfect reflectors nor perfect absorbers. The coefficient of absorption (α) is defined as

$$\alpha = \frac{I_a}{I_i} \qquad (26.9)$$

where

I_a = sound power density (intensity) absorbed by the material, W/cm^2
I_i = intensity impinging on the material, W/cm^2
α = absorption coefficient (no units)

Thus α is a measure of absorption efficiency. When $\alpha = 1.0$ all the impinging energy is absorbed. Since open space has this characteristic, α has also been defined as the ratio between the absorption of a given material and that of an *open window* of the same area. Obviously then for an opening (window, door, etc.)

$$\alpha = 1.0$$

The total absorption (A) provided by a surface is proportional to its area and its absorption coefficient, that is,

$$A = S\alpha \qquad (26.10)$$

where

A = total absorption in *sabins*
S = surface area, square feet or meters
α = coefficient of absorption

Since α is a ratio and thus unitless, and S is a unit of area, $S\alpha$ should be area as well. Instead, sound absorption units are called sabins in honor of W. C. Sabine, a pioneer in architectural acoustics. One sabin (m^2) is the sound absorption equivalent to an open window one square meter in area. Similarly, one sabin (ft^2) is equivalent to 1 square foot of open window. Obviously, 1 sabin (m^2) equals 10.76 sabin (ft^2). When not specified, a sabin (ft^2) is intended.

Most rooms are constructed of several materials, each having a different absorption coefficient α. Thus to determine the total absorption of the room, it is necessary to sum the component absorptions, that is,

$$\Sigma S\alpha = S_1\alpha_1 + S_2\alpha_2 + \cdots + S_n\alpha_n$$

or

$$\Sigma\overline{A} = \overline{A}_1 + \overline{A}_2 + \cdots + \overline{A}_n \qquad (25.11)$$

ACOUSTICS

where

$\Sigma S\alpha$ = total absorption in the room, sabins
S_1, S_2, etc. = area of each material
α_1, α_2, etc. = coefficient of each material
$\overline{A}_1$, $\overline{A}_2$, etc. = total absorption of each different material

If S is expressed in square feet, then $\overline{A}$ will be in sabins (ft^2); if S is in square meters then $\overline{A}$ is sabins (m^2).

The mechanics of absorption and the use of absorptive materials are discussed in Sections 27.3 and 27.4 in connection with noise reduction techniques. Absorption coefficients for common materials, acoustic materials, and auditorium furnishings are tabulated in Table 27.1. It is important to note that for most common materials, absorption (and therefore α) varies with frequency. Thus absorption calculations must be made individually for the frequencies being studied.

26.14 Reverberation

Reverberation is the persistence of sound after the cause of sound has stopped—a result of repeated reflections. Reverberation time (T_R) describes the period required for the sound level to decrease 60 db after the sound source has stopped producing sound. For rooms of usual size and shape, the reverberation time at a specific frequency may be found by the formula:

$$T_R = K \times \frac{V}{\Sigma\overline{A}}, \quad \text{seconds} \qquad (26.12)$$

where

K = a constant, equal to 0.05 when measurements are in feet and 0.16 when in meters
V = room volume, ft^3 (m^3)
$\Sigma\overline{A}$ = total absorption, sabins (ft^2 or m^2) at that frequency

(For spaces of unusual shapes, see Beranek, 1971; Embelton, 1971; Gomperts, 1965; Young, 1959.) Subjectively, reverberation is one of the most pronounced hearing reactions in an enclosed space. Simply, it is the ear's reaction to echoes, giving an impression of "liveness" or "deadness."

In most room acoustics studies, reverberation times are calculated at 125, 500, 1000, and 2000 Hz. The midfrequency (500–1000 Hz) range is generally the reference used in specifying the reverberation time of a room when studying the speech characteristic of the space.

Reverberation can be considered as a mixture of previous and more recent sounds. The converse of reverberation or reverberance is articulation. An articulate environment keeps each sound event separate rather than running them together. Spaces for speech activities should be less reverberant—more articulate—than those designed for performance of music. See Sections 26.23 and 26.24 for reverberation criteria for speech and music rooms, respectively.

26.15 Sound Fields in an Enclosed Space

The inverse square law described in Section 26.7 holds true for the acoustic *far field,* which is the sound field sufficiently far from the source that intensity is proportional to pressure squared. This field is developed in open, obstruction-free circumstances. Propagation in an enclosed space is quite different. There, when a sound reaches a wall or other large (with respect to wavelength) obstruction, part of the sound energy is reflected and part absorbed. Thus the sound at any point in the room is a combination of direct sound from the source plus reflected sound from walls and other obstructions. If the reflections are so large that the sound level becomes uniform throughout the room, the field within the room is termed a *diffuse* one (no shadows), and intensity measurements with respect to a specific source are meaningless. Of course, if it is our intention to measure sound pressure level at a specific point, such as a seat in an auditorium, the type of field in the room is irrelevant.

Most rooms do not have such a high level of reflection that a diffuse field is created. Instead, there is a *near field* near the source, a *free field* beyond the near field, and a *reverberant field* near the walls (see Fig. 26.16). They can be recognized as follows:

1. The near field is generally within one wavelength of the lowest frequency of sound produced by the source. Within this distance

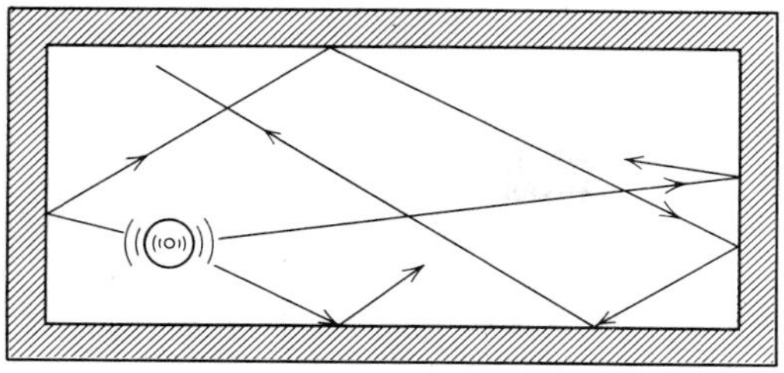

(a)

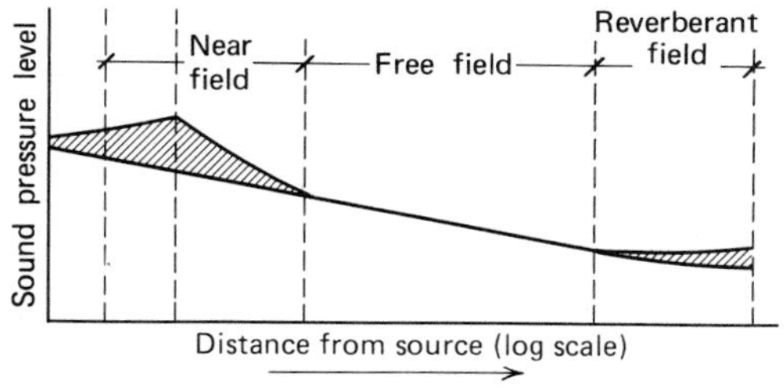

(b)

Fig. 26.16 *Type of sound field in an enclosed space depends in large measure on reflections (reverberation) and absorption. In a typical room* (a) *there is a near field adjacent to the sound, a free field beyond that, and a reverberant field adjacent to the walls.* (b) *In a large hall or auditorium the reverberant field dominates and level remains approximately constant.*

SPL measurements vary widely and are not meaningful. (The maximum wavelength for the human male voice is about 11 ft.)

2. Near large obstructions such as walls, the *reverberant field* is dominant and approaches a diffuse condition. In well-designed auditoriums the reverberant (diffuse) field predominates and sound pressure level remains relatively constant beyond the free field area.
3. The *free* (far) *field* exists between the near and reverberant fields, and there intensity varies as pressure squared and inversely with distance. In this field, sound pressure level drops 6 db with each doubling of distance from the source, and it is in this field that meaningful SPL measurements can be made, with respect to a specific small source.

Since it is desirable to be able to predict the noise level (SPL) in a space during the design stage, and also because of the difficulty of SPL measurement in rooms with various types of sound field, as explained above, it is useful to have equations that relate sound *power* level (PWL) to sound *pressure* level. Sound power level is supplied by manufacturers with each item of equipment, at either octave or ⅓-octave points.

To avoid confusion, it will be helpful to the reader to think of sound pressure level (SPL) as the resultant ''noise'' or sound in an enclosed space, resulting from a source in that space, and affected by the characteristics of the space and the position of the listener. It is thus an end effect. The sound power level (PWL) is a measure of the amount of sound generated by a source, independent of its environment. In free space the two quantities are simply related by the inverse square law, whereas in enclosed spaces the room characteristics come into play. Roughly speaking, an analogy to lighting can be drawn: SPL corresponds to room illumination, that is, footcandles, and PWL corresponds to the lumen output of the source. The basic relationship between sound pressure level and sound power level for a single, small, nondirective source in a room large enough to have both free and reverberant fields is:

$$SPL = PWL + 10 \log \left(\frac{1}{4\pi r^2} + \frac{4}{R} \right) \qquad (26.13)$$

where

SPL = sound pressure level, db
PWL = sound power level, db
r = distance from source, m
R = room constant, m^2

The factor R can be calculated from

$$R = \frac{S\overline{\alpha}}{1 - \overline{\alpha}}$$

where

S = total room surface area, m^2
$\overline{\alpha}$ = average absorption coefficient of all materials in the room

When calculating in square feet, add 10.5 to the right-hand side of Equations 26.13 and 26.15. The average absorption $\overline{\alpha}$ is calculated by first calculating the room's entire absorption in sabins, and then dividing by total area:

$$\overline{\alpha} = \frac{S_1\alpha_1 + S_2\alpha_2 + \cdots + S_n\alpha_n}{S_1 + S_2 + \cdots + S_n} \qquad (26.14)$$

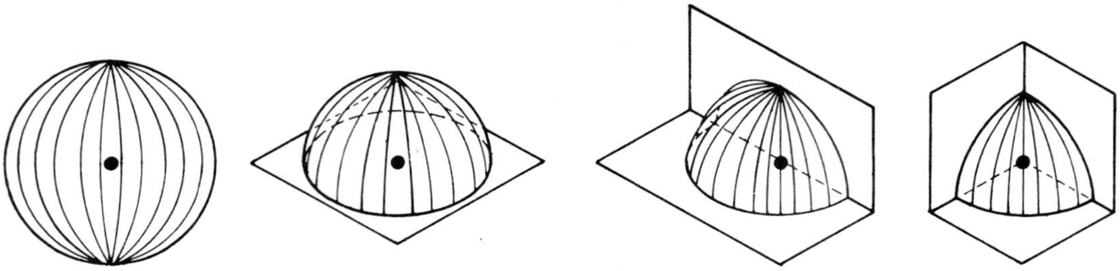

Fig. 26.17 Diagrams illustrating directivity factors for either inherently directive sources or nondirective sources placed adjacent to large reflecting surfaces. (Courtesy of Barry Blower Co.)

For a sound source with directional characteristics, the applicable equation is

$$SPL = PWL + 10 \log \left(\frac{Q}{4\pi r^2} + \frac{4}{R} \right) \quad (26.15)$$

where Q is a directivity factor and all other factors are as in Equation 26.13. The directional constant is either inherent in the sound source, and as such will be part of the given data, or can be obtained from Fig. 26.17 for a nondirectional source made directional by adjacent reflecting surfaces. In a room containing more than one sound source, the sound pressure levels can be combined as explained in Section 26.8 above. To understand the application of these equations, and the effect of absorptive material in a space, see Example 27.1 in Section 27.7, Noise Reduction by Absorption.

To understand the effect of room absorption on the development of the reverberant field in a large space, refer to Fig. 26.18. The field is plotted from Equation 26.15, with correction for units in feet, for a room 80 ft × 120 ft × 30 ft high, using average absorptions $\bar{\alpha}$ ranging from 0.05 to 0.7, that is, from very live to very dead. Note particularly that the addition of sufficient absorbent material can entirely prevent the development

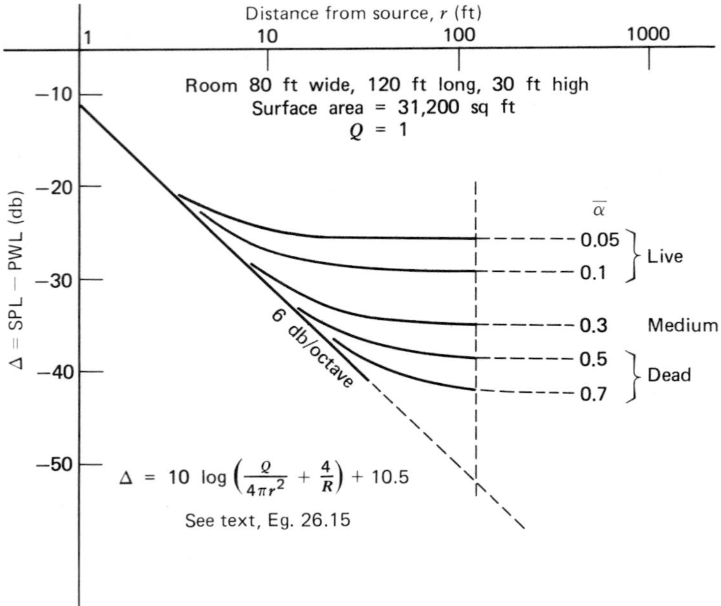

Fig. 26.18 Curves show that development of reverberant field (constant SPL, or flat curve) is dependent on the room's absorption characteristics. In a live room, ($\bar{\alpha}$ = 0.1) the reverberant field begins 20 ft from the source. In a dead room ($\bar{\alpha}$ = 0.5) it begins at 120 ft from the source, that is, at the back wall. Effectively then, there is no reverberant field in such a room. Slope of asymptote is 6 db per octave, that is, inverse square attentuation.

of a reverberant (diffuse) field. This is particularly important in noisy industrial interiors where the building materials commonly employed are generally nonabsorbent, and the noise sources are both numerous and loud. For further discussion see Sections 27.2 through 27.8. An empirical expression relating power level to *intensity* level is given in Section 27.7.

26.16 Other Factors in Hearing

(a) Masking. When two separate sources of sound are perceived simultaneously, the perception of each is made more difficult by the presence of the other. This is known as masking, which is defined as the number of decibels a sound has to be raised above its threshold when perceived alone, to be perceived in the presence of another sound. Effectively then, masking is an upward shift of audibility threshold. Masking is greatest when two sounds are close in frequency or frequency content, since the ear has greater difficulty separating them. Also, a low frequency will more effectively mask a high frequency than the reverse, for the same decibel levels. With broad-frequency sounds the masking effect is difficult to predict, since it depends in part on how ''hard'' the listener is listening. Masking is an extremely important technique in noise control where background noise levels are deliberately manipulated to mask unwanted sounds. The background noises used for this purpose are of the broadband continuous variety, which are themselves noninformation bearing. They serve to obliterate lower level, information-bearing sounds that would cause annoyance.

(b) Time. As stated above, impulse sounds are apprehended at lower levels than the same sound intensity over a longer period. Similarly, because of the time constant of the ear's mechanical linkages, sounds closer than 10 ms apart cannot be distinguished from each other, and those up to 50 ms apart are poorly distinguished (see Section 26.23). Beyond this point differentiation becomes increasingly clear. This effect is of particular importance in the design of halls and auditoriums, with respect to reception of echoes.

(c) Directivity. The exact mechanism by which the binaural aspect of hearing detects direction is not clearly understood. The single ear is not phase sensitive, but it may be that binaurally, phase sensitivity exists, at least at low frequencies, and that this assists in detection of direction. At high frequencies, phase detection is clearly nonexistent and sense of direction may be due to diffraction effects around the head. These effects would also explain the accuracy of detection in a horizontal plane, which research indicates to be in the order of 5° change. It would not explain how the ears detect changes in the vertical plane immediately in front of the listener. This latter situation, although much less accurate than horizontal plane detection, definitely exists. In enclosed space, reverberation will blur most phase differences and ''stereo'' information will be almost completely dependent on high frequencies in the near field.

SOUND SOURCES

26.17 Speech

As can be seen from Fig. 26.14, the ear's sensitivity is maximum in the speech frequency and normal energy range. Speech sounds vary in time length between 30 and 300 ms so that the ear perceives them individually and clearly. Speech is comprised of *phonemes,* which are individual and distinctive sounds that to an extent vary from language to language. That is, certain ones exist in one language and not in another. Since certain phonemes carry more information than others, it is these that good architectural acoustics must be particularly careful to preserve intact, to preserve intelligibility. In English, consonants carry much more information than vowels, as can readily be demonstrated by writing a sentence first without consonants and then without vowels:

Most speech energy is concentrated in the 100- to 600-Hz range.

o ee ee i oeae i e 100–600 e ae

Mst spch nrgy s cncntrtd n th 100–600 Hrtz rng. (See Fig. 26.19.)

The male voice centers its energy around 500 Hz; the female about 900 Hz. It is, however, in the high frequencies that consonants have most of their energy. Phonemes such as *s* and *sh* have most of

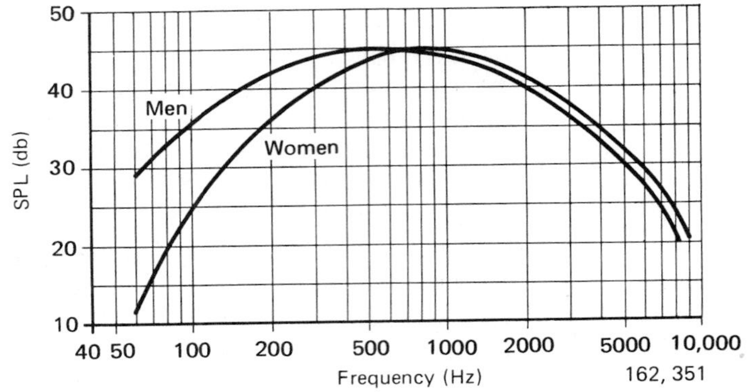

Fig. 26.19 Frequency distribution of typical male and female voices. (From Brüel and Kjaer, 1963.)

their energy above 2 kHz and both are particularly important in conveying intelligence.

Normal speech averages between 40 and 50 db sound pressure level at 3 to 4 ft, with a dynamic range of from about 30 db for a soft speech to about 65 db for loud speech at the same distance. Extremes of speech are 10 db for a whisper and 80 db for a shout, but in both these instances intelligibility is sharply reduced because of lack of consonant power. Indeed, in shouting, emphasis is necessarily on vowels, so that it is generally accepted that 70 db SPL is about the upper limit of fully intelligible human speech. (Singers who frequently exceed 90 db do so at great loss of intelligibility.) Another result of the high-frequency content of consonants, hence intelligibility, is its directiveness. The higher the frequency, the greater its directivity and the less its diffraction (ability to turn corners). Therefore, intelligibility of speech is greatest directly in front of the speaker and least behind him. The high-frequency tones are most easily absorbed and least reflected and diffracted.

26.18 Other Sounds

Music is much broader and complex than speech in frequency and dynamic range. It has no direct parallel to intelligibility. "Reception" of music is a combination of physiological and psychological phenomena. As such, it is beyond most of the purposes of this study but is briefly examined in our discussion of room acoustics, auditoriums, and halls. *Noise* is variously defined as unwanted sound, sound with no intelligence content, and broadband sound, depending on the listener and the situation. All three definitions are correct at various times and situations. For our purpose we must assume that any sound can be referred to as noise.

NOISE CRITERIA

26.19 Negative Effects of Noise

Although noise effects and their control are the subject of Chapter 27, noise criteria and their background are discussed here as part of our overall study of hearing and sound sources. There are two basic approaches to the negative effects of noise, a pschological-practical one and a purely physiological one. The latter is concerned with the physical impact of noise on the body including hearing loss and other deleterious conditions (see Section 26.22). The former is concerned with noise levels that cause annoyance and disturbance to daily activities including work, relaxation, and rest, and is discussed in the following sections.

26.20 Noise and Annoyance

Research has developed accurate data on loudness. The concept of annoyance, however, being primarily subjective and psychological, is much more elusive. Tests have shown that *in general,* annoyance as a result of noise is:

1. Proportional to the loudness of the noise.
2. Greater for high-frequency than low-frequency noise.

3. Greater for intermittent than continuous noise.
4. Greater for pure-tone than for broad-band noise.
5. Greater for moving or unlocatable (reverberant) noise than for a fixed-location sound.
6. Much greater for an intelligence-bearing noise (neighbor's radio) than for a no-sense noise.

To establish criteria for acceptable background noise, certain of these effects must be neglected for the sake of simplicity. [They can be and are considered in construction design and in establishing levels of masking noise (see Chapter 27)]. We thus ignore:

Factor 3, since design is based on continuous sounds.

Factor 4, since broadband noise is assumed.

Factor 5, since the noise source is assumed to be fixed in location.

Factor 6, since we consider noise level rather than content.

Thus the particularity and special characteristics of noises such as a barking dog (3), a whistle (4), a single passing vehicle (5), and intelligible sounds (6) are not considered. Consideration of the remaining factors (1 and 2) on interference with speech communication resulted in concepts called the Articulation Index (AI) and the Speech Interference Level (SIL), determined by reading a carefully selected set of phonetically balanced nonsense syllables to a test audience in the presence of different levels and compositions of background noise. The ratio of correct answers to total syllables was the Articulation Index. An Articulation Index (Beranek, 1949) in excess of 0.5 indicated a condition in which perfect intelligibility could be expected. A simplified version of the AI called the Speech Interference Level (SIL) was devised by Beranek. The SIL consists simply of the arithmetic average in decibels of the background-noise sound levels in three octave bands, 600 to 1200, 1200 to 2400, and 2400 to 4800 Hz, since it was found that correlation between intelligibility and the sound power in these three bands could be established.

Beranek developed the well-known and widely accepted noise criteria (NC) curves shown in Fig. 26.20 on the basis of SIL data and loudness level (LL) information (*Fundamentals of Noise,* 1971; Pearsons and Bennet, 1974; *Procedure for Computation of Loudness of Noise,* 1968). (Loudness level gives information similar to that provided by equal loudness contour, as shown in Fig. 26.13, but it uses broadband noise instead of pure tones.) The NC curves take cognizance of the field-determined fact that most people prefer to speak at a level no greater than 22 phons above the background noise. By combining the SIL levels in decibels with this fact, the NC curves are derived, that is, they represent a loudness level 22 phons higher than the SIL in db. These contours represent then the maximum *continuous* background noise that will be considered acceptable in the environment specified and correspond fairly accurately to background noise level in commercial environments. A similar set of curves called noise rating (NR) curves, proposed by the International Standards Organization (ISO), finds considerable application outside the United States. These curves are less stringent than NC curves in the low frequencies and more stringent in the high frequencies.

In application, the octave-band spectrum of a noise over the range of 63 to 8000 Hz is measured and plotted on an NC curve sheet. The lowest NC curve that is not exceeded by any portion of the plot is the NC rating of the noise. Thus a specification of maximum noise levels of NC-30 means that no portion of sound power level of any continuous background noise in the area may cross the NC-30 contour. Conversely, a piece of equipment rated NC-35 has an octave-band spectrum completely below NC-35. A fan rating of NC-53 would indicate that at some point on its frequency-spectrum plot the fan exceeded NC-50 by 3 db (see Fig. 26.20).

The NC number of a noise varies between 5 and 10 db below the measured dbA. NR numbers correspond closely to NC numbers. The virtue of the NC curve system is that it is a single-number specification for an entire frequency spectrum. Its disadvantage is that it was derived for, and is primarily accurate for, speech conversation against a continuous, non-intelligence-bearing noise. This is not the situation that is most troublesome in an office; indeed, continuous machine-type noise may be *helpful* in masking unwanted speech interfer-

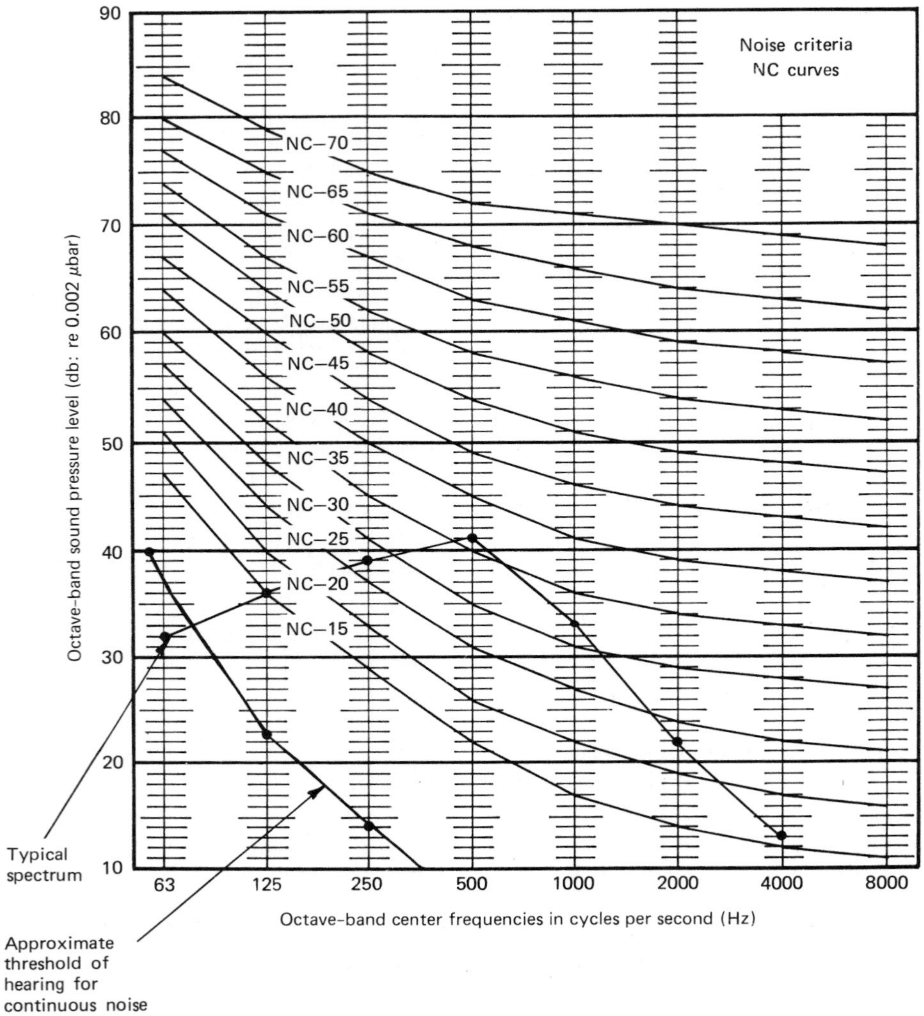

Fig. 26.20 Noise criteria curves. Typical spectrum plotted would be rated NC-36 as it exceeds NC-35 at 500 Hz by 1 db (see Table 27.11 for specific recommendations for interior spaces).

ence. See Sections 27.21 and 27.22, and Fig. 27.33.

It was found, when introducing background noise to provide masking, that a noise with a frequency spectrum conforming in shape to an NC curve was neither pleasant nor unnoticed (neutral), but instead sounded emphasized at both ends; that is, it rumbled and hissed. To avoid this problem (since background noise should be unnoticed), a new set of curves was introduced by Beranek in 1971, called preferred noise criteria (PNC) curves, which overcame the foregoing objections. They are intended primarily to be applied as a criterion for deliberate white noise, that is, for noise whose frequency characteristic roughly corresponds to a PNC curve. NC curves remain the commonly used yardstick for continuous, nonintentional background noise. Table 27.11 gives both NC and PNC recommendations for different environments.

26.21 High Noise Levels; Hearing Protection; OSHA

It has long been recognized that continuous exposure to high noise levels causes a degree of temporary deafness in a majority of people and that long periods of such exposure, even on an inter-

§ 1910.95 Occupational noise exposure.

(a) Protection against the effects of noise exposure shall be provided when the sound levels exceed those shown in Table G-16 when measured on the A scale of a standard sound level meter at slow response. When noise levels are determined by octave band analysis, the equivalent A-weighted sound level may be determined as follows:

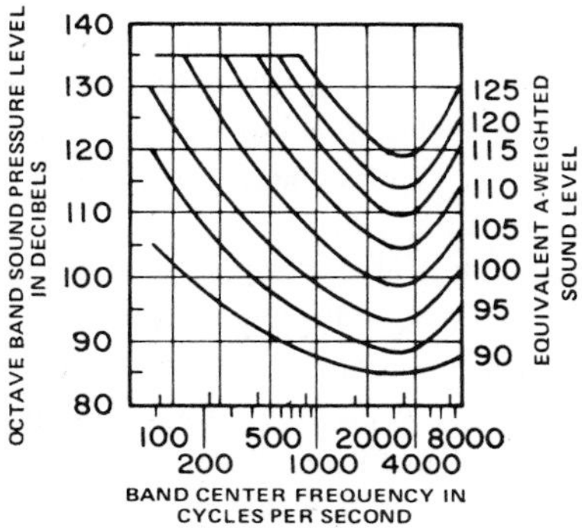

Equivalent sound level contours. Octave band sound pressure levels may be converted to the equivalent A-weighted sound level by plotting them on this graph and noting the A-weighted sound level corresponding to the point of highest penetration into the sound level contours. This equivalent A-weighted sound level, which may differ from the actual A-weighted sound level of the noise, is used to determine exposure limits from Table G-16.

[1910.95 amended at 39 FR 19468, June 3, 1974]

(b)(1) When employees are subjected to sound exceeding those listed in Table G-16, feasible administrative or engineering controls shall be utilized. If such controls fail to reduce sound levels within the levels of Table G-16, personal protective equipment shall be provided and used to reduce sound levels within the levels of the table.

(2) If the variations in noise level involve maxima at intervals of 1 second or less, it is to be considered continuous.

(3) In all cases where the sound levels exceed the values shown herein, a continuing, effective hearing conservation program shall be administered.

TABLE G-16—PERMISSIBLE NOISE EXPOSURES[1]

Duration per day, hours	Sound level dBA slow response
8	90
6	92
4	95
3	97
2	100
1½	102
1	105
½	110
¼ or less	115

[1] When the daily noise exposure is composed of two or more periods of noise exposure of different levels, their combined effect should be considered, rather than the individual effect of each. If the sum of the following fractions: $C_1/T_1+C_2/T_2+\cdots C_n/T_n$ exceeds unity, then, the mixed exposure should be considered to exceed the limit value. Cn indicates the total time of exposure at a specified noise level, and Tn indicates the total time of exposure permitted at that level.

[1910.95 Table G-16 amended at 39 FR 19468, June 3, 1974]

Exposure to impulsive or impact noise should not exceed 140 dB peak sound pressure level.

Fig. 26.21 The standard for exposure to noise in the work place. OSHA.

mittent 8-h workday basis, appears to produce permanent hearing impairment. Many experts place the safe upper limit at 85 db. In addition, studies have indicated that continual exposure to noise levels as low as 75 to 85 dbA level can produce or contribute to numerous physical and psychological ailments including headache, digestive problems, tachycardia, high blood pressure, anxiety, and nervousness—an almost complete catalog of human illnesses. Since continuous noises are most severe in industry, regulatory legislation in the United States has been most stringent and effective in this area. In 1969 the Walsh–Healy Public Contracts Act was passed, and thereafter its provision for maximum permissible exposure to noise levels was incorporated into the Occupational Safety and Health Act. Both the act and the associated regulatory agents, the Occupational Safety and Health Administration, are known as OSHA. The relevant provisions of this act are reproduced in Fig. 26.21. To avoid complex regulations, limitations of exposure are given in terms

TABLE 26.5 **Typical Industrial Noise Levels**[a]

Equipment	*dbA*
Printing press plant (medium size automatic)	86
Heavy diesel-propelled vehicle (about 25 ft away)	92
Heavy-duty grinder	93
Air compressor	94
Plastic chipper	96
Cutoff saw	97
Multiple spot welder	98
Turbine condenser	98
15 cu ft air compressor	100
Drive gear	103
Banging of steel plate	104
Magnetic drill press	106
Air chisel	106
Positive displacement blower	107
Air hammer	107
Vacuum pump	108
Jolt squeeze hammer	122

[a]These are approximate values for typical generic equipment types and should not be used as design values.

of single-number dbA values. The value is determined either by direct A-scale measurement, or by octave-band measurements. In the latter case, SPL readings are taken with a linear C-scale instrument and plotted on the curves given in Fig. 26.21. These curves represent A-scale response. That is, a meter with an A-weighted scale would read the values shown when exposed to a given sound pressure level at a given octave-band center frequency. The maximum penetration of the octave-band plot represents the dbA value of that noise, which is then applied as in Table G-16 of Fig. 26.21. Typical industrial noise levels are given in Table 26.5. When these levels are exceeded, management must take steps to reduce the exposure, either by reducing the noise or by providing hearing protectors. Typical characteristics of a few types are given in Fig. 26.22.

OSHA does not deal extensively with impulse noises, except to state that they shall not exceed 140-db peak sound pressure level. The area of impulse noise is quite different from continuous noise, since apprehended levels depend on duration, and specification is difficult. Much work has been done in this area by the military for obvious reasons, and the interested reader is referred to the literature (Magrab, 1975; Ward et al., 1968), since this subject is substantially outside the area of architectural acoustics.

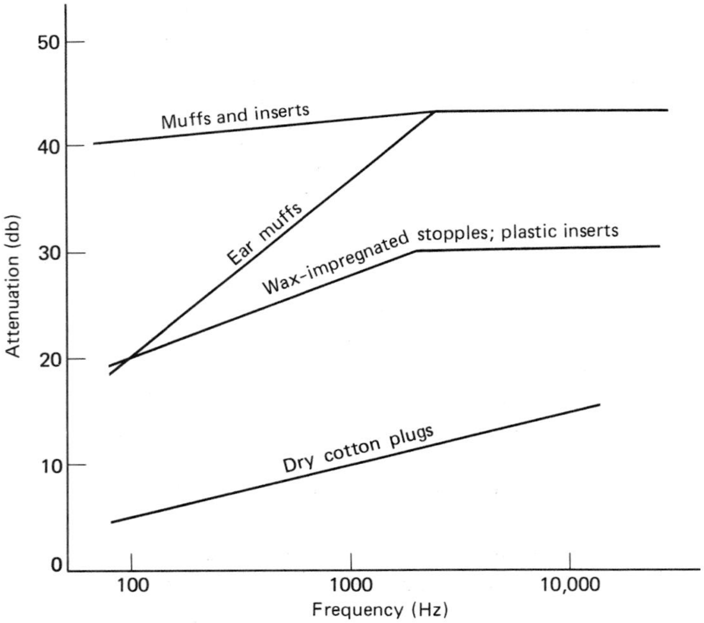

Fig. 26.22 *Sound attenuation characteristics of various types of ear protectors. (From* Quieting: A Practical Guide to Noise Control, *p. 13.)*

ROOM ACOUSTICS

26.22 Sound in Enclosures

When a continuous sound is generated in an enclosure, fields are set up as described in Section 26.15. When the sound is not a continuous tone or noise but a series of discrete sounds, following one upon the other and containing intelligence, as in speech or music, the room must be designed to maintain and enhance this intelligibility. That is what is meant by design of room acoustics.

The generated sound radiates out from the source until it strikes a room boundary or other large surface. Before reaching this surface the sound intensity is attenuated by distance (inverse square law) and by absorption in the air. This latter is appreciable only in large rooms, and at frequencies above 2000 Hz. (For detailed data on air absorption, see Magrab, 1975.) In the absence of exact data, assume a figure of 0.40 sabin (m^2)/100 m^3 and see Table 27.1.) When the sound reaches the wall, it is partially reflected and partially absorbed. A small portion is also transmitted into adjoining spaces. The energy transmitted is so small that it has little effect on the space within which the sound originates although, as discussed in Chapter 27, it may be very important in the surrounding spaces. The ratio between the energy absorbed and the energy reflected will significantly affect what one hears within the space. Specifically, if little energy is absorbed and much is reflected, two effects will be noticeable. Intermittent sounds will be mixed together (which may make speech *less* intelligible or music *more* pleasant), and steady sounds will accumulate into a reverberant field, making the space "noisy." Conversely, if much energy is absorbed and little reflected, the room will sound quiet to speech and "dead" to music. At this point the reader may wish to review the technicalities of absorption and reverberation as discussed in Sections 26.13 through 26.15 above.

26.23 Criteria for Speech Rooms

The overriding criterion for speech is intelligibility. Since speech consists of short disconnected sounds 30 to 300 ms in length (see Section 26.17), among which are high-frequency, low-energy phonemes, the ideal room must assure the ear's reception of these phonemes as they are given. Since a slow decay rate is the equivalent of a masking noise, reverberation must be kept to a minimum. We can obtain a good approximation of the subjective feeling of liveness of a room, for purposes of *speech*, from the relation

$$T_R = 0.3 \log \frac{V}{10} \qquad (26.14)$$

where

T_R = optimum reverberation time in seconds, for speech
V = room volume, m^3

For instance, a typical classroom might have a volume of 150 m^3 (5300 ft^3). Optimum reverberation time is

$$T = 0.3 \log 15 = 0.35 \text{ s}$$

Reverberation times longer than this would sound live, shorter ones dead and flat. Indeed an increase of 20% in reverberation time would make the room excessively live and boomy and would negatively affect speech intelligibility. Figure 26.23 gives optimum mid-frequency reverberation times as a function of room size and use. Figure 26.24 gives *maximum* reverberation time figures for speech, in large rooms. The reflected sounds associated with reverberation have either a salutary or a deleterious effect. As explained in Section 26.16b, the ear cannot distinguish between sounds that arrive within a maximum of 50 ms (1/20 s) of each other. (Some authorities use 40 ms, i.e., 1/25 s.) Sounds arriving within this time *reinforce* the direct path signal and appear to come from the source. Sounds arriving after this time are apprehended as a fuzzy echo or elongation of the sound, reducing intelligibility and directiveness. Since the range of 40 to 50 ms corresponds at 344 m/s to 13.76 to 17.2 m, a speech room must be so arranged that the difference between the first reflection path and the direct path is no greater than 17 m and preferably 14 m or less. (See Fig. 26.25.) For more detail concerning this and related factors on reflected paths, refer to Section 26.25.

Too *low* a reverberation time (very high absorption, minimum reflection) is also undesirable because:

1. It limits the size of the room to that which can be covered by direct sound only.

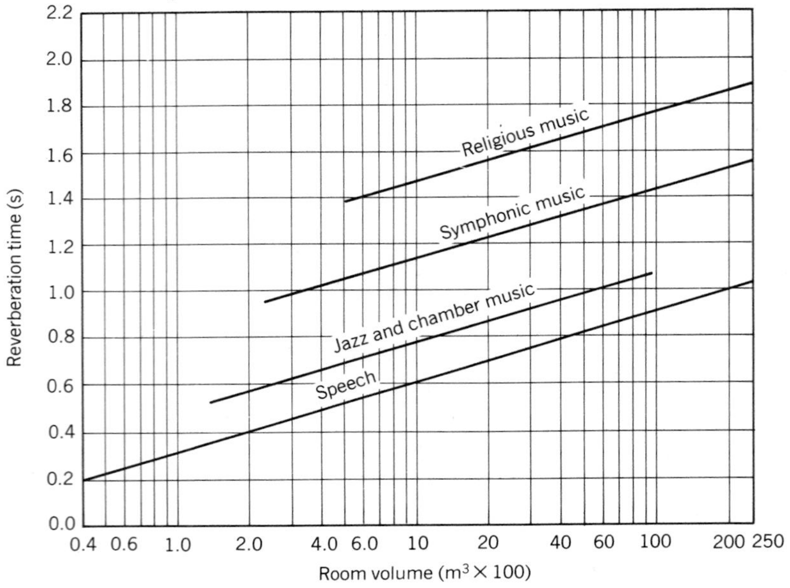

Fig. 26.23 Optimum reverberation times, for the frequency range of 500–1000 Hz. (Reprinted by permission, from E.B. Magrab, Environmental Noise Control, Wiley, New York, 1975, p. 206.)

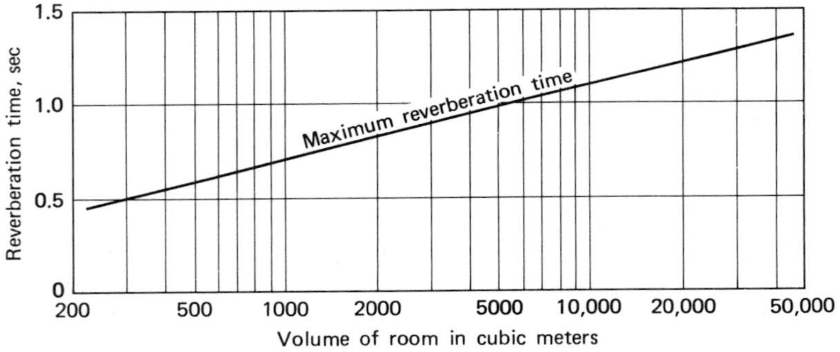

Fig. 26.24 *Maximum recommended reverberation time for speech in auditoriums and lecture halls. (From P. V. Brüel,* Sound Insulation and Room Acoustics, *1951.)*

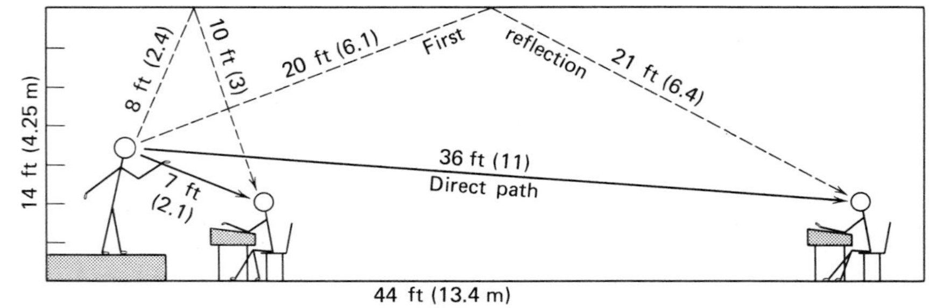

Fig. 26.25 *Sound paths in a typical medium-sized lecture room. Note that for both extremes of listener position, the maximum path-length difference between direct and first reflection is 11 ft. Thus signal is reinforced and intelligibility should be excellent if room absorption is provided to limit reverberation time to about ½ s maximum (see Fig. 26.24). Numbers in parentheses are dimensions in meters.*

2. It is disturbing to the speaker, since absence of reflection prevents him or her from gauging proper voice level and tends to cause excessive effort (shouting, as when outdoors).

Thus proper design of a room for speech is a compromise between the need for some reflection and the desire to minimize reflection to preserve intelligibility.

26.24 Criteria for Music Performance

Adequate design for a music space requires recognition of the following:

1. Large-volume spaces require direct-path sound reinforcement by reflection.
2. Relatively long reverberation time is needed to enhance the music—the exact amount depending on the type of music (see Fig. 26.23). Designers should keep in mind that recommendations vary as much as 100% between respected sources. In addition to an optimum reverberation time in the central 500-Hz to 1-kHz range, the reverberations at other frequencies should exhibit a slight drop in the higher frequencies and a rise at the lower frequencies to compensate for the sharp drop in ear sensitivity. Thus T_R at 100 Hz should be, according to most authors, between 25 and 50% longer than T_R at the center frequencies. As a matter of interest, concert halls judged to be excellent by musical experts have a center frequency reverberation time between 1.6 and 1.8 s.
3. Directivity declines if the reinforcing signal is excessively delayed. With large ensembles, directivity gives the sense of depth and instrument location necessary for proper appreciation. This is often referred to as clarity or definition in music. With a solo instrument this problem is diminished.
4. Brilliance of tone is primarily a function of high-frequency content. Since these frequencies are most readily absorbed, a good direct path must exist between sound source and listener. Since our eyes and ears are close together, a good sound path exists when a good vision path exists. At the other end of the spectrum, lack of sufficient bass expresses itself as a loss of ''fullness,'' which is often caused by resonant absorption (see Section 27.3).

The actual design of music performance space is a very complex procedure involving extensive calculations of absorption, reverberation time and ray diagraming, and juggling of materials, dimensions, and wall angles. Simulation techniques and acoustic models are also employed. Most modern design also uses movable reflector panels and other active variables. After construction is completed, extensive tests are conducted and field adjustments are made.

26.25 Ray Diagrams and Sound Paths

Ideally, every listener in a lecture hall, theater, or concert hall should hear the speaker or performer with the same degree of loudness and clarity. Since this is obviously impossible by direct-path sound, the essential design task is to plan methods for reinforcing desirable reflections and minimizing and controlling undesirable ones. Normally only the first reflection is considered in ray diagraming, since it is strongest. Second and subsequent reflections are usually attenuated to the point that they need not be considered except for the special situations of flutter, echoes, and standing waves discussed below.

(a) Specular Reflection. Specular reflection occurs when sound reflects off a hard polished surface. This characteristic can be used to good advantage to create an effective image source. In ancient Greek and Roman theaters, seats were arranged on a steep, conical surface around the performers. The virtue of the arrangement (see Fig. 26.26*a*) is that the sound power travels to each location with minimal attenuation. The same effect can be accomplished by placing the sound source above the seats. This is not practical physically, but it can be accomplished effectively by the use of a reflecting panel (see Fig. 26.26*b*). Panel dimension must be at least one wavelength at the lowest frequency under consideration. Figure 26.27 is a conversion chart from frequency to wavelength in feet and meters.

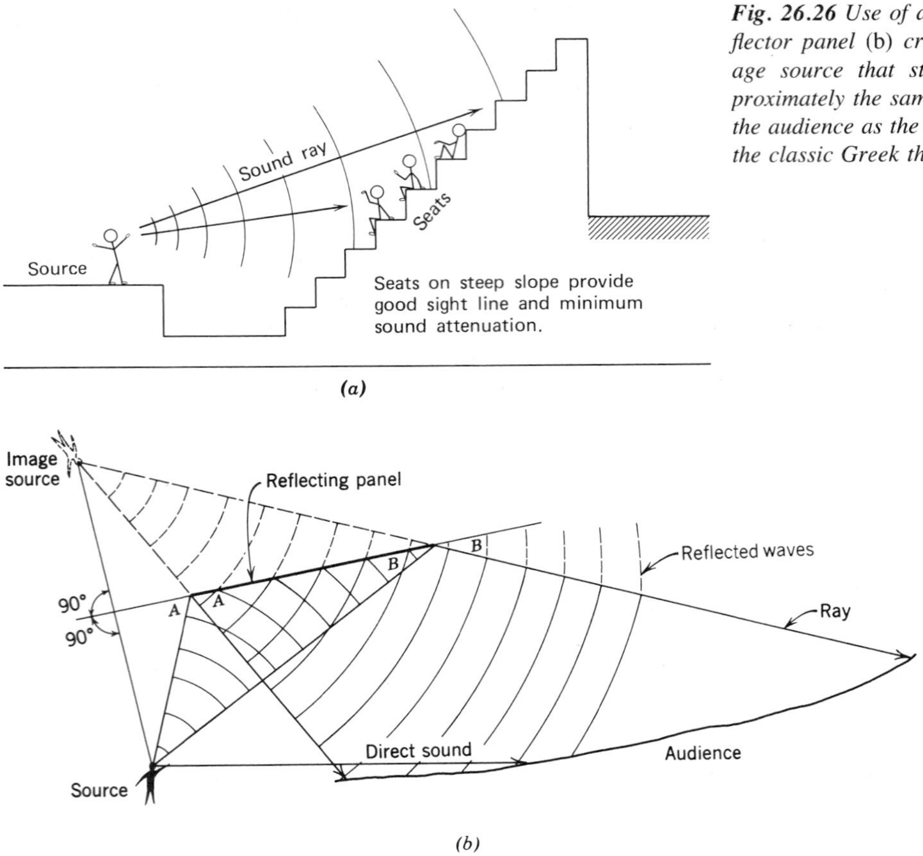

Fig. 26.26 *Use of an angled reflector panel* (b) *creates an image source that stands in approximately the same relation to the audience as the performer in the classic Greek theater* (a).

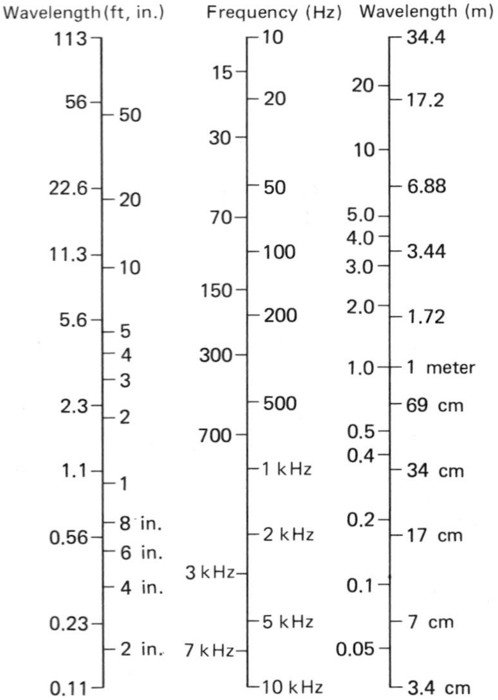

Fig. 26.27 *A nomograph for determining wavelength in feet or meters, given frequency in hertz, or vice versa. Speed of sound is taken as 344 m/s and 1128 ft/s. To use, hold a straightedge horizontally across the nomograph, and read the figures directly.*

(b) Ray Diagrams. Ray diagraming is a design procedure for analyzing reflected sound distribution throughout a hall, using the first reflection only. Figure 26.28 shows a ray diagram. The rays are drawn normal (perpendicular) to the spherically propagating sound waves. Specular reflection is assumed. That is, the angles between the reflecting panel and the incident and reflected rays are always equal. Thus, in addition to the direct sound, each listener is receiving reflected sound energy. It is as though there were additional sound sources, the *real* one and numerous image sources. Figure 26.28 shows the application of a ray diagram to a lecture hall. In Fig. 26.28*a* the

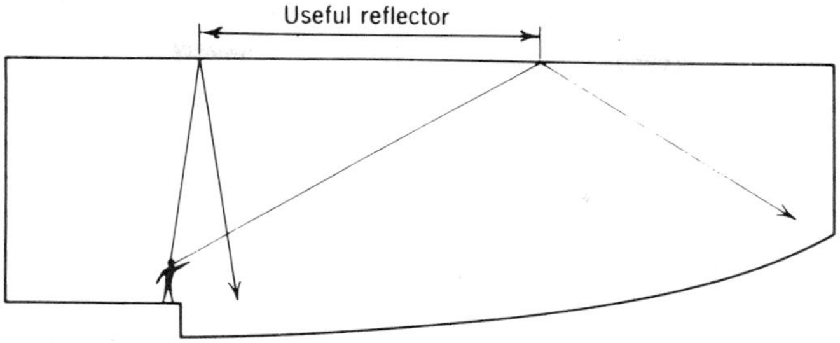

(a) Flat ceiling

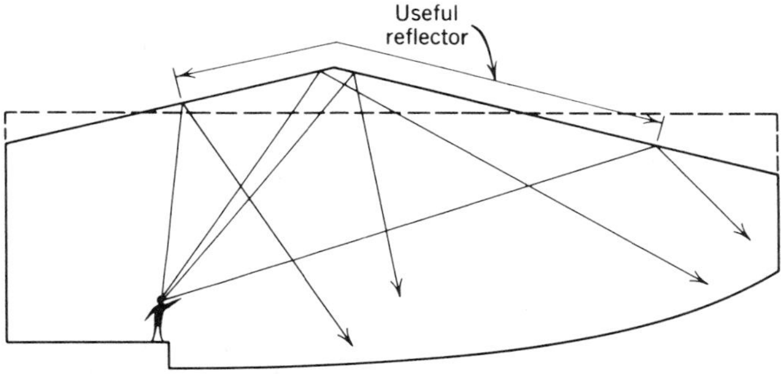

(b) Two panel ceiling increases useful reflecting area

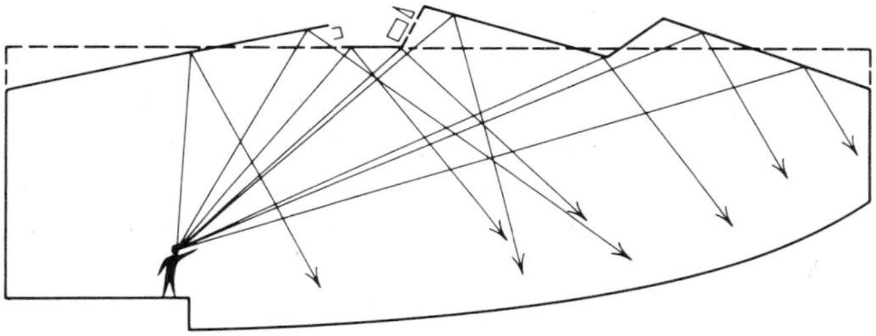

(c) Multifaceted ceiling incorporates lights and loudspeakers

Fig. 26.28 *Section through a typical lecture room showing use of ray diagrams.*

stage height and seating slope are arranged to provide good sight lines, and the ceiling height is established by reverberation requirements, esthetics, cost, and so on. It can be seen that less than half the ceiling is providing useful reflection. By dividing the ceiling into two panels (Fig. 26.28*b*), people in the rear of the room perceive the direct source plus two image sources, and the useful reflecting area is increased by 50%. In Fig. 26.28*c*, the shape has been further refined to include a lighting slot and a loudspeaker grille.

Although they are a useful design tool, ray diagrams have certain restrictions. For example, the hall is three dimensional and the diagrams two dimensional, that is, sectional. To properly ray diagram, depth (width) must also be considered, and this unduly complicates the diagrams. Also, design must always be a compromise between ray diagrams for various "speaking positions" on the stage. Thus a paraboloid may be a perfect shape for one source position but will be very poor for other positions.

(c) Echoes. As explained in Section 26.23, a clear echo is caused when reflected sound *at sufficient intensity* reaches a listener more than 50 ms after he has heard the direct sound. Echoes, even if not distinctly discernible, are undesirable in

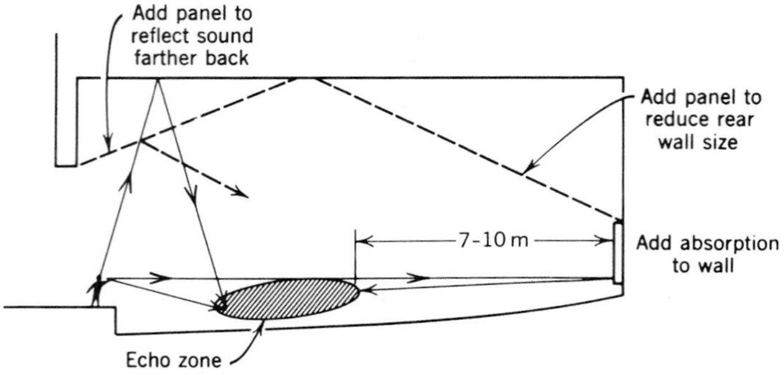

Fig. 26.29 Auditorium section showing the causes and remedies for two typical echoes.

rooms. They are annoying and make speech less intelligible. The relative annoyance is dependent on the time delay and loudness relative to the direct sound, which, in turn, are dependent on the size, position, shape, and absorption of the reflecting surface.

Typical echo-producing surfaces in an auditorium are the back wall and the ceiling above the proscenium. Figure 26.29 shows these problems and suggests remedies. Note that the energy that produced the echoes can be redirected to places where it becomes useful reinforcement. If echo control by absorption alone were used on the ceiling and back wall, that energy would be wasted. The rear wall, since its area cannot be reduced too far, may have to be made more sound absorptive to reduce the loudness of the reflected sound.

(d) Flutter. A flutter, perceived as a buzzing or clicking sound, is comprised of repeated echoes traversing back and forth between two nonabsorbing parallel (flat or concave) surfaces. Flutters often occur between shallow domes and hard, flat floors. The remedy for a flutter is either to change the shape of the reflectors or their parallel relationship, or to add absorption. The solution chosen will depend on reverberation requirements, cost, or esthetics.

(e) Focusing. Concave domes, vaults, or walls will focus reflected sound into certain areas of rooms. This has several disadvantages. For example, it will deprive some listeners of useful sound reflections and cause hot spots at other audience positions (see Fig. 26.31*a*).

(f) Diffusion. This is the converse of focusing and occurs primarily when sound is reflected from convex surfaces. A degree of diffusion is also provided by flat horizontal and inclined reflectors (see Fig. 26.30). In a diffuse sound field the sound level remains relatively constant throughout the space, an extremely desirable property for musical performances.

(g) Creep. This describes the reflection of sound along a curved surface from a source near the surface. Although the sound can be heard at points along the surface, it is inaudible away from the surface. Creep is illustrated in Fig. 26.31*b*.

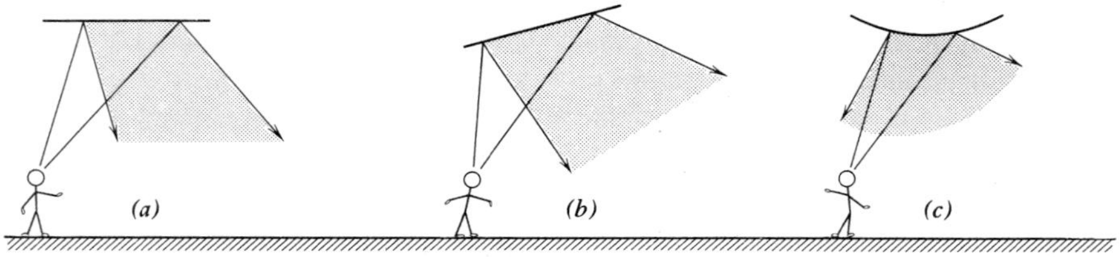

Fig. 26.30 *Sound diffusion can be created with reflectors of different shapes, ranging from the horizontal flat* (a), *inclined flat* (b), *or convex* (c). *Diffusion improves from* (a) *to* (c).

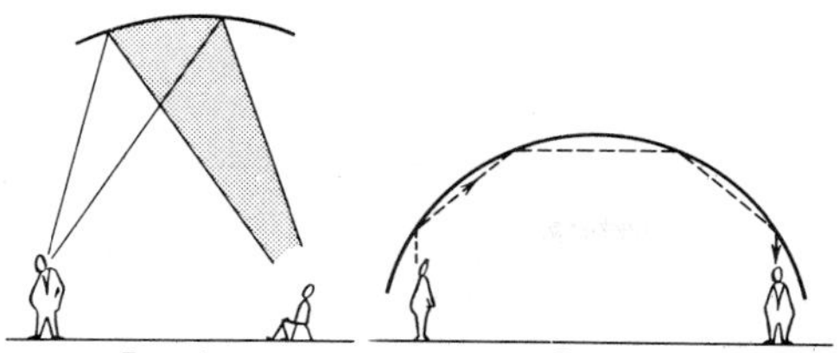

Fig. 26.31 Two undesirable phenomena in room acoustics.

(h) Standing Waves. Standing waves and flutters are very similar in principle and cause, but are heard quite differently. When an impulse (such as a hand clap) is the energy source, a flutter will occur between two parallel walls. When a steady, pure tone is the source, a standing wave will occur, but only when the parallel walls are spaced apart at some integral multiple of a half-wavelength.

When the parallel walls are exactly one-half wavelength apart, the tone will sound very loud near the walls and very quiet halfway between them. This is because at the center, the reflected waves traveling in one direction are exactly one-half wavelength away from those traveling in the other, and thus equal *but opposite in pressure,* which results in total cancellation. In other rooms standing waves are noted as points of quiet and maximum sound in the room. Standing waves are important only in rooms small with respect to the wavelengths generated (smallest room dimension <30 ft for music or <15 ft for speech).

Another effect of standing waves, *resonance,* is the accentuation of the particular frequency that will cause a standing wave in a room of that dimension. Thus, if one speaks (or plays a musical instrument) standing near a wall of a room, about 8 ft by 8 ft, one will notice an abnormal and sometimes unpleasant loudness in the sound at about 280 Hz.

Thus, when a musician plays a scale, one note may seem far louder than the adjacent ones, and listeners in one section of the room will hear a quality of sound different from that heard by those in other sections. This effect *must* be avoided for music performance but is merely an annoyance in rooms designed for speech use. This is one of the reasons that one finds music rehearsal rooms, broadcast studios, and so on with nonparallel walls and undulating ceilings. These irregularities direct sound energy toward the absorbing materials of the room and cause the standing waves to degenerate.

26.26 Auditorium Design

This is a general term used to describe a space where people sit and listen to speech or music. A large lecture hall, a multipurpose space, and a concert hall are auditoriums. Before beginning design of an auditorium, its potential use must be determined. If planned activities range from lectures to symphony orchestra concerts, the design approach for acoustics will differ, significantly, from a design approach for a space that would house only one of these activities. Therefore, the first step in the acoustical and architectural design must be determining the program. If the program for the auditorium includes activities that need different acoustical environments, it must be decided early whether the acoustics will be a compromise between the program extremes or adjustable for various activities. Acoustical environments can be altered by changing volume, moving reflecting surfaces, and adding or subtracting sound-absorbing treatment. Figure 26.32 illustrates several examples of acoustical adjustability.

Factors that influence the acoustical design include audience size, range of performance activities, and sophistication of the potential audience. Obviously, an acoustically good 1200-seat theater is more difficult to design than an acoustically good 400-seat theater. In addition, the caliber of performance production and audience expectations are important design considerations. For example, a small school auditorium and a professional theater will have widely divergent demands from both audience and performers.

Acoustical design of an auditorium includes room acoustics, noise control, and sound system design. Noise control is covered in Chapter 27.

The audience size determines the basic floor area of an auditorium. Once this area has been fixed, the volume of the room is developed according to reverberation requirements of the space.

Figure 26.33 shows a typical auditorium in plan

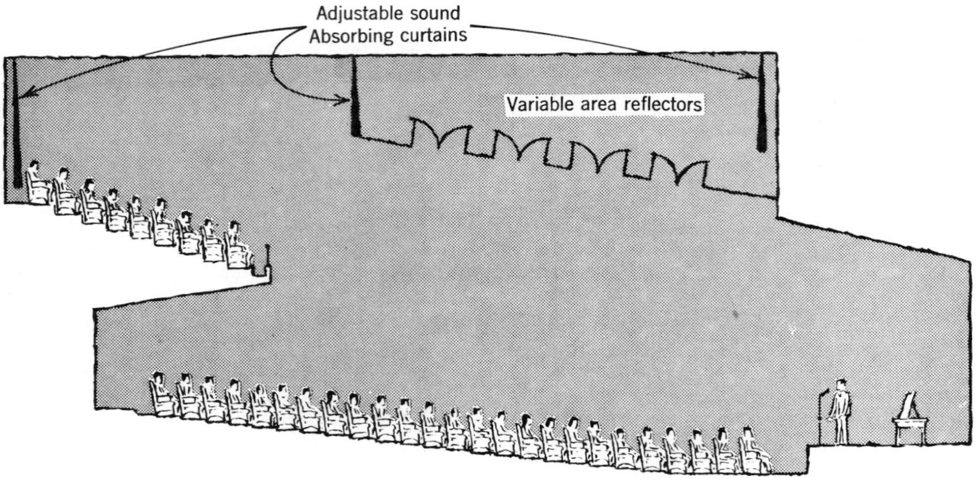

Fig. 26.32 *Adjustable acoustic elements in an auditorium.*

and section. The shape of wall and ceiling surfaces is developed to give proper distribution of sound and eliminate focusing or echoes. Essential characteristics of the design include:

1. Ceiling and side walls at the front of the auditorium distribute sound to the audience. These surfaces must be close enough to the performers to minimize time delays between natural sound and reflected sound.
2. Ceiling and side walls provide diffusion.

Acoustics must be considered in selection of materials used in an auditorium. All auditoriums use both sound-reflecting and sound-absorbing materials. Since the largest area of sound-absorbing material in any auditorium is the audience, the difference in acoustical characteristics that occur without an audience may be minimized by using fully upholstered seating.

Chairs with fully upholstered seats and backs, covered in an open-weave material, will have absorption characteristics closely approximating an audience. Using the auditorium in Fig. 26.33 as an example, the reverberation characteristics of an auditorium with various materials may be examined. In the first example, the room use is assumed to be for music performance. The only sound absorption is that provided by the audience and seating. In the second set of calculations absorptive curtains were installed along the rear wall and a portion of the side wall. This configuration might be used for lectures in a room that is adjustable between speech and music configurations. A third configuration might use permanent sound-absorbing treatment installed on the ceiling and rear and side walls. Because of its low reverberation time, this configuration would be appropriate only for movies and lectures, not for music activities.

These simple examples indicate the effect of changes in the amount of absorption on the characteristics of a room. Adjustable treatments permit the characteristics of the room to be modified to any point between the extremes to meet the program acoustic requirements of a multipurpose hall.

Existing spaces may require remedial treatment to eliminate unwanted phenomena such as focusing and echoes, as shown in Fig. 26.34. In the first example, the surface of the dome was covered with sound-absorbing material to eliminate focusing; in the second, sound-absorbing treatment was applied to a curved rear wall to eliminate an echo. Such treatment also will affect the reverberation characteristics.

SOUND REINFORCEMENT SYSTEMS

26.27 Objectives and Criteria

The purpose of a sound reinforcement system is just what the name indicates—to reinforce the sound, which would otherwise be inadequate. Thus

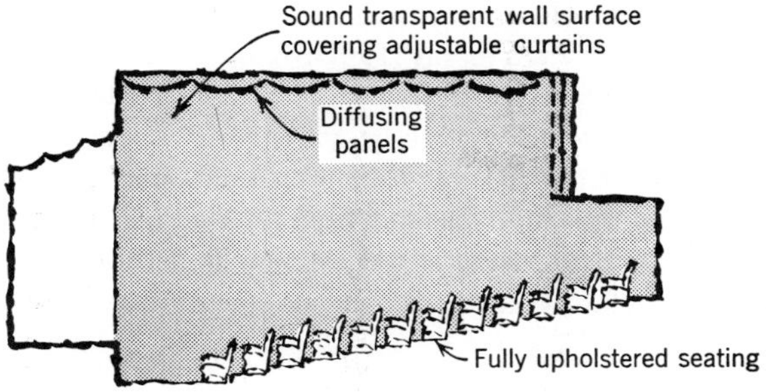

Section

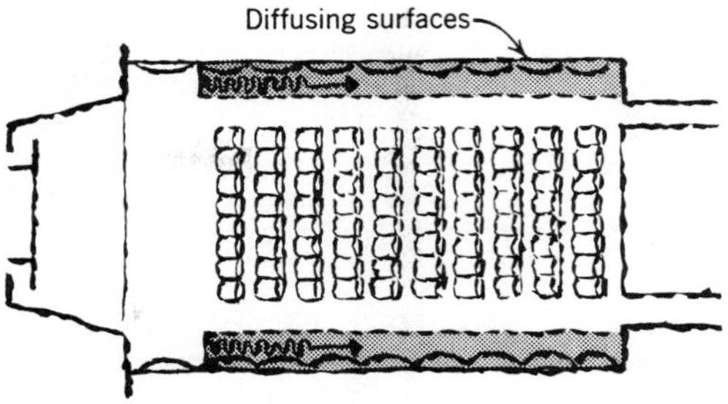

Plan

Simplified Calculations of Midfrequency (500 and 1000 Hz) Average Reverberation Times

$$\text{reverberation time } (RT) = \frac{0.05 \times \text{volume (cu ft)}}{\text{total absorption (sabins)}}$$

$$\text{volume} = 155{,}500 \text{ cu ft}$$

More Reverberant Condition (curtains retracted)

	Area	*a*	*Absorption*
Seating and stage (with audience and performers)	3323	.92	3060
Wall area Concrete block	8000	.2	1600
Lower rear wall Permanent sound-absorbing treatment	450	.88	396
Total absorption More reverberant condition			5056

$$RT = \frac{0.05 \times 155{,}500}{5056} = 1.5 \text{ s}$$

Less Reverberant Condition (curtains exposed)

	Area (sq ft =)	*a*	*Absorption*
Seating and stage (with audience and performers)	3323	.92	3060
Wall area Concrete block (balance covered by curtains)	3600	.2	720
Curtains	4400	.45	1970
Lower rear wall Permanent sound-absorbing treatment	450	.88	396
Total absorption Less reverberant condition			6146

$$RT = \frac{0.05 \times 155{,}500}{6146} = 1.2 \text{ s}$$

Fig. 26.33 *Auditorium with surface treatments for control of reflections and reverberation.*

an ideal sound system will give the listener the same loudness, quality, directivity, and intelligibility as if the source of sound were immediately adjacent to him—a distance of 2 to 3 ft for speech and farther for music, depending on type and number of instruments. This situation must obtain for every position in the space except that a variation of ±3 db for loudness is tolerable. The other factors should remain constant. Of these factors, loudness and intelligibility have been discussed at length and should be well understood. By quality we mean that frequency response should be linear

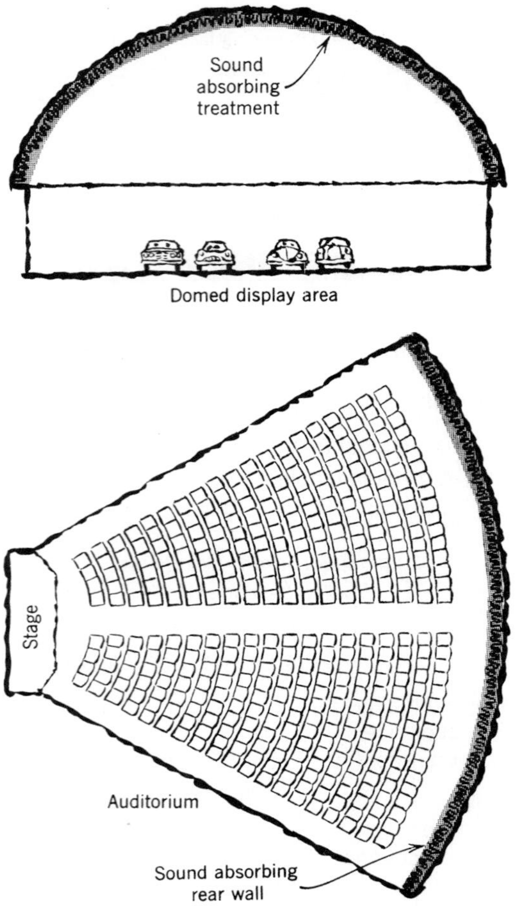

Fig. 26.34 Sound-absorbing treatment used to eliminate focusing from dome and curved auditorium wall.

so that reproduced sound bears the same relation between its frequency components as the original sound. (Quality is then field-adjusted by "voicing" or "equalization," as discussed below.)

Directivity is the characteristic whereby the sound appears to be coming from the originating source, that is, the loudspeakers should be directionally "invisible," and the listener must have the impression of actually hearing the source. It should be emphasized that sound systems cannot correct poor acoustic design completely, although they can improve a bad situation.

Generally, sound systems will be required in spaces larger than 50,000 ft^3 ($\sim$1400 m^3). In terms of population, this volume translates as 550 persons in lecture rooms (15 ft average ceiling height and 6 ft^2 per person) and 325 persons in theaters (20 ft average ceiling height and 7.5 ft^2 per person). In such a room (50,000 ft^3) a normal speaking voice can maintain a volume level of only 55 to 60 db, depending on room design and voice strength. With background noise at NC 30 (see Table 27.11), the speaker will be heard; at higher noise levels intelligibility will suffer.

26.28 Components and Specifications

All sound systems consist of three basic elements: input devices, amplifier(s), and loudspeaker systems.

(a) Input. Input is usually a microphone, a radio source, or some form of prerecorded material.

(b) Amplifier and Controls. Amplifiers must be rated to deliver sufficient power to produce intensity levels of 80 db for speech, 95 db for light music, and 105 db for symphonic music. This assumes a *maximum* background noise level of 60 dbA. Thus 80-db speech intensity will be 20 db higher, or four times as loud as the noise level. If noise level is *known* to be below 60 db maximum, amplifier and loudspeaker power ratings can be reduced accordingly. The amplifier should carry technical specifications for signal-to-noise ratio, linearity, and distortion. Exact figures here depend on application and are left to the acoustics specialist or sound engineer to supply.

In addition to the usual volume, tone mixing, and input–output selector controls, the amplifier *must* contain special equalization controls for signal shaping. These are highly critical filter networks that are used to *voice* or *equalize* a system after installation. Equalization is a sine qua non of a good sound system; without it the system will howl, sound rough, give insufficient and poorly distributed gain and sound level, and generally sound poor. Essentially, voicing tailors the system to the acoustic properties of the space. A system not equipped for equalization is not a professional system, and results will verify it. Furthermore, the specification must provide for the services of a

Fig. 26.35 *Loudspeaker system using delayed signal to underbalcony area.*

competent sound engineer to perform the equalization after installation and construction is complete.

Another control frequently required in theater systems is a delay mechanism or circuit that can introduce a time delay into a signal being fed to a loudspeaker. Figure 26.35 shows a sound system that covers a majority of an auditorium from a central-loudspeaker cluster. The underbalcony seating areas are hidden from the central cluster and receive the reinforced sound from distributed loudspeakers in the underbalcony soffit. To provide directional realism, the signal to the underbalcony loudspeakers must be delayed to allow the weaker signal from the central speakers to arrive first. Delay is necessary because electrical signals travel at the speed of light, whereas sound is much slower (one-millionth the speed, approximately). With this arrangement, sound will seem to come from the source, and the directivity so necessary to realism is maintained.

(c) Loudspeakers. These are the heart of any sound system and obviously must be of the same high quality as the remainder of the system. Indeed, system economies will show up much more quickly in loudspeaker performance than in any other component. Selection of speakers is a complex, technical task not within our scope. Nevertheless, a few general remarks are in order. The best systems use central-speaker arrays consisting of high-quality, sectional (multicell), directional, high-frequency horns, and large-cone woofers. These assemblies are very large, and the architect should be aware of the dimensions that must be accommodated. Units 6 ft wide, 8 ft high, and 3 ft deep, with a weight of 1000+ lbs are common in a large lecture hall or small theater. This size is necessitated by the large wavelength of low-frequency sound (see Fig. 26.27). Smaller units with folded horns can be used, at a sacrifice in low-frequency response. If only speech is to be reproduced, these will perform adequately. Distributed systems use small (4–12 in. diameter) low-level speakers, ceiling mounted, and firing directly down. To give adequate response, these units must be mounted in at least a 2-ft^3 enclosure. Smaller enclosures will usually seriously compromise performance.

26.29 Loudspeaker Considerations

Loudspeaker system design and placement must be coordinated with the architectural design. The two principal types of loudspeaker system are central and distributed. The loudspeakers in a central system are a carefully designed array of directional high-frquency units combined with less directional low-frequency units, placed above and slightly in front of the primary speaking position. In most theaters, this location is just above the proscenium on the centerline of the room. Located

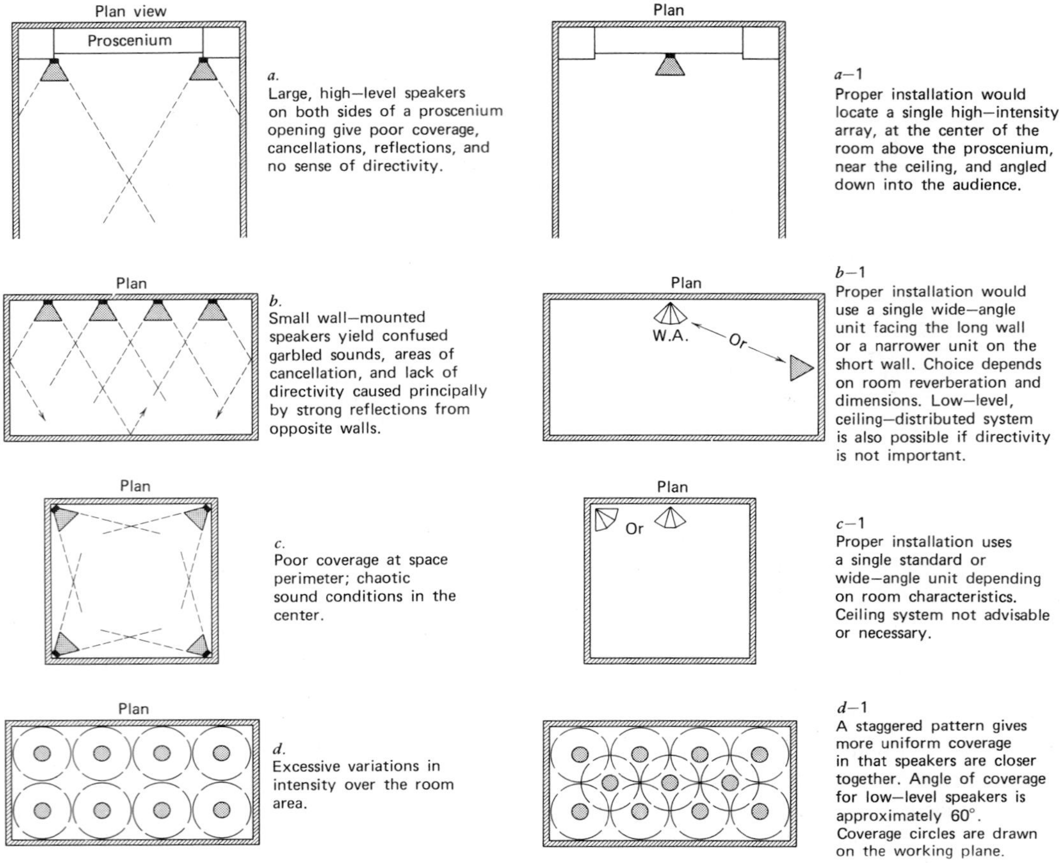

Fig. 26.36 *Poor layouts* (a) *to* (d) *and good layouts* (a-1) *to* (d-1). *Wide-angle speakers are available to fit most needs. Ceiling speakers give effective coverage on a 60° cone. Working plane is taken to be 4 ft AFF for seated audience and 6 ft AFF for standing listeners.*

in this position, the system provides directional realism and is simple in its design.

The distributed loudspeaker system consists of a series of low-level loudspeakers located overhead throughout the space. Each loudspeaker covers a small area, in a manner similar to downlights. This type of system is used in low-ceilinged areas where a central-loudspeaker cluster could not provide proper coverage. It also can be used for public address functions if directional realism is not essential, in spaces such as exhibition areas, airline terminals, and offices. In addition, distributed loudspeaker systems provide flexibility for use in spaces where source and listener locations vary according to the use of the space, since loudspeakers can easily be switched to provide proper coverage. Combination loudspeaker systems that include both central and distributed units are used to solve special problems, as discussed earlier and illustrated in Fig. 26.35. In general, a listening position should receive sound from only one loudspeaker. Systems that cover seating areas with signals from several scattered loudspeakers usually increase the loudness of the sound but tend to produce garbled speech. This rule is the principal reason that the arrangements shown in Fig. 26.36*a–d* will guarantee a bad job. The common practice of placing one loudspeaker on either side of a proscenium opening (Fig. 26.36*a*), or rows of speakers on one or both sides of a room (Fig. 26.36*b*), is particularly to be deplored.

Essential parameters for location and design of

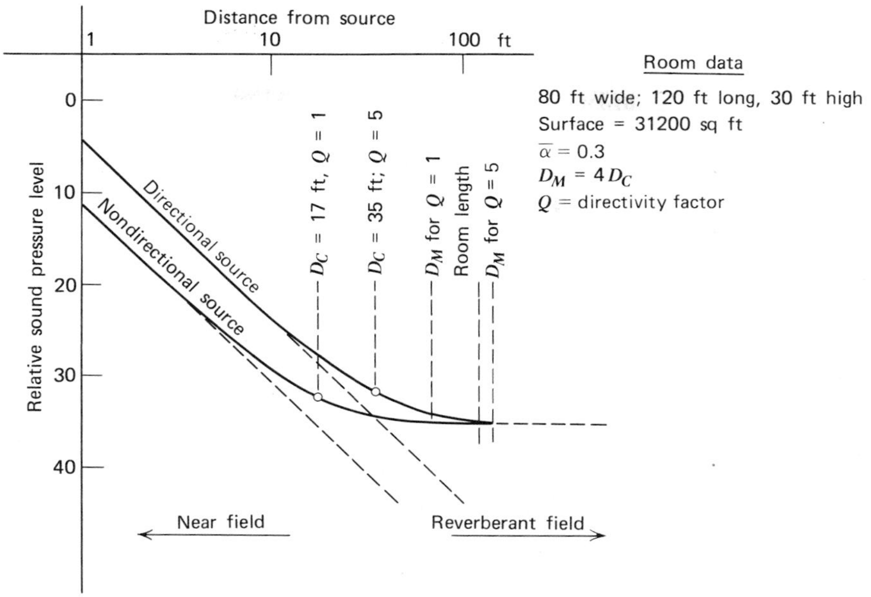

Fig. 26.37 *Plot of relative sound pressure level versus distance from sound source in a large, medium-absorption room. Lower curve is for a nondirectional source. Upper curve is for a directional loudspeaker.* D_C *is critical distance. Maximum distance* D_M *(throw) is four times* D_C.

the control position create problems for the architect. The sound system operator must be within the coverage pattern of the loudspeakers. For proper operation, he or she should be able to hear the sound as it is heard by the audience. Some current auditorium designs locate sound system controls within the audience seating pattern (see Fig. 27.47). Others place a control room with a completely open wall or a large window at the rear of the auditorium. Monitor loudspeakers and earphones are inadequate substitutes for actual listening within the auditorium. In churches the simple control equipment can be located at the rear of the congregation area.

To understand selection and positioning of a high-level directional loudspeaker, it is necessary first to understand its action in a space. Refer to Fig. 26.37. We have transferred the curve $\overline{\alpha}$ = 0.3, that is, a medium (neither live nor dead) room, from Fig. 26.18 to Fig. 26.37 and have added a loudspeaker curve calculated from Equation 26.15 for $Q = 5$, which is a common figure for multicell directional speakers. Note the result. First the level in the near field and free field areas has been raised 7 db. Second, the critical distance (D_C) has been doubled, that is, increased from 17 to 35 ft. *Critical distance* is defined as that point at which the distance between the asymptote and actual curve is 3 db. *At that point the direct and reverberant components are equal.* Beyond that point the reverberant field predominates. We have already stated that to maintain good directivity and clarity, the direct component should predominate or at least constitute a large portion of the sound. It is widely accepted that *maximum* distance from loudspeaker to listener should not exceed four times the critical distance for good quality reproduction and comprehension (preferably less). Note then on Fig. 26.37 that with D_C at 35 ft, the maximum distance becomes 140 ft. Since the room is only 120 ft long, a single high-level directional array is satisfactory. If the room were longer, the rear portions would require additional coverage. Alternatively, room characteristics could be changed by adding absorption to reduce reverberance, or a more highly directive loudspeaker could be used. If it is not possible to use a central system, distributed speakers will perform adequately but without directivity (see Table 26.6). These decisions are the province of the acoustics expert, without whose

TABLE 26.6 **Area of Coverage; 60° Cone Loudspeaker Firing Directly Down**

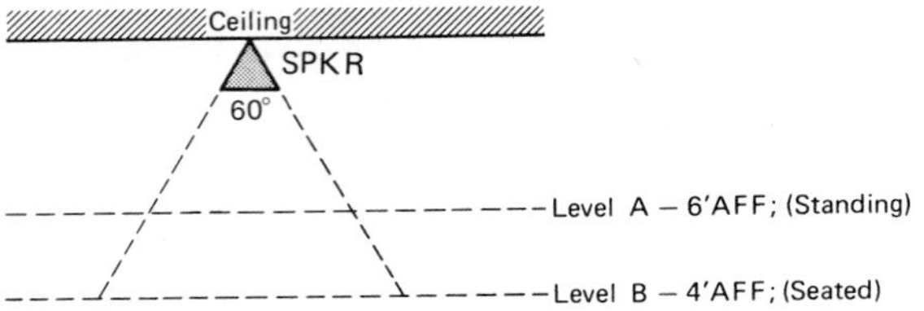

Ceiling Height, Feet	*At Level A*		*At Level B*	
	Diameter[a]	*Area*[b]	*Diameter*	*Area*
8	1.15	4.2	1.7	17
9	1.73	9.4	2.3	26
10	2.31	17	2.9	38
11	2.9	26	3.5	51
12	3.5	38	4.0	67
13	4.0	51	4.6	85
14	4.6	67	5.8	105
15	5.2	85	6.4	127
16	5.8	105	6.9	151

[a]Diameter of coverage circle (see Fig. 26.36).
[b]Area of coverage circle (see Fig. 26.36).

expert advice complex or expensive installations should not be designed.

References cited in this chapter are given next. For related bibliography, see the References for Chapter 27.

References

Beranek, L. L. (1949). *Acoustic Measurements,* Wiley, New York.

Beranek, L. L. (1971). *Noise and Vibration Control,* McGraw-Hill, New York.

Brüel, P. V. (1951). *Sound Insulation and Room Acoustics,* Chapman & Hall, London.

Brüel, P. V., and Kjaer, (1963). *Architectural Acoustics,* Denmark.

Embelton, T. (1971). ''Absorption Coefficients of Surfaces Calculated from Decaying Sound Fields,'' *Journal of the Acoustical Society of America,* Vol. 50, No. 3.

Fundamentals of Noise. (1971). Report NTID 300.15, U.S. Environmental Protection Agency, Washington, D.C., pp. 55–56.

Gomperts, M. (1965/1966). ''Do the Classical Reverberation Formulas Still Have a Right for Existence?'' *Acustica,* Vol. 16.

Magrab, E. B. (1975). *Environmental Noise Control,* Wiley, New York.

Pearsons, K. S., and Bennet, R. L. (1974). *Handbook of Noise Rating,* National Aeronautics and Space Administration Report NASA CR-2376, Washington, D.C., pp. 55–56.

Procedure for Computation of Loudness of Noise. (1968). ANSI, S3.4, New York.

Quieting: A Practical Guide to Noise Control. (1976). National Bureau of Standards Handbook No. 119, NBS, Washington, D.C.

Terminology, Laws and Application of Fan Sound Data. (1972). Publication SD100-1972, Barry Blower Co. Minneapolis.

Ward, W. D., et al. (1968). *Proposed Damage-Risk Criterion for Impulse Noise,* Report of Working Group 57, National Research Council–National Academy of Sciences Committee on Hearing, Bioacoustics and Biomechanics (CHABA).

Young, R. W. (1959). ''Revision of the Speech Privacy Calculation,'' *Journal of the Acoustical Society of America,* Vol. 31, No. 7.

27

BUILDING NOISE CONTROL

27.1 Introduction

Noise control in buildings is comprised of three components:

1. Reduction of noise generation at the source by proper selection and installation of equipment.
2. Reduction of noise transmission from point to point (along the transmission path) by proper selection of construction materials and appropriate construction techniques.
3. Reduction of noise at the receiver through acoustical treatment of the relevant spaces to meet NC criteria developed in Chapter 26.

Assurance of speech privacy is achieved by manipulation of all of the above plus the use of masking noise where necessary.

NOISE REDUCTION

27.2 Principles of Noise Reduction

Noise reduction is essentially the science of converting acoustical energy into another, less disturbing form of energy—heat. Since the amounts of energy involved are minute—130 db corresponds to 1/1000 of a watt or 0.003 Btu/h—the heat produced is completely negligible. This conversion is by absorption, by the room contents and wall coverings, and by the structure itself. The former controls noise levels *within* a space and the latter noise transmission between spaces. The reasons for this will become clear as our discussion proceeds, but it is important that the principle be stated at the outset. Noise control treatment of a room will affect the reverberant noise level within that room but will have minimal effect on the noise level in adjoining spaces. Refer to Fig. 27.1 for a graphic presentation of this fundamental fact. The best that can be accomplished with acoustic room treatment is elimination of the reverberant field, that is, to make the intensity at the room boundaries what it would have been in free space, as in Fig. 27.1*d*. (Even this is extremely difficult; the actual field at the wall would be above 72 db, except in a completely anechoic chamber.) Adding further wall or other acoustic absorbent as in Fig. 27.1*e* does nothing in the room itself and has minimal effect on the overall transmission loss, since the transmission loss of the acoustic material itself is very low, as can be seen in Fig. 27.2*b*.

ABSORPTION

27.3 Mechanics of Absorption

We have already learned in Sections 26.13 and 26.14 the definition of sound absorption and its relevance to room reverberation characteristics. At this point it is appropriate to reexamine absorption as an acoustic phenomenon so that we may understand the application of absorptive material. Refer to Fig. 27.2*a*. In an untreated room of normal construction, when the sound waves strike the walls or ceiling, a small portion is transmitted, a small portion is absorbed, and most of the sound is reflected. The exact proportions obviously depend on the nature of construction. When acoustical treatment is applied to the room surfaces as in Fig. 27.2*c*, some of the energy in the sound waves is dissipated before the sound reaches the wall. The transmitted portion is slightly reduced, but the reflection is greatly reduced. The difference between the two situations is shown graphically in Fig. 27.3. Refer back to Fig. 26.18, where the result of adding absorptive material to a room is shown in greater detail. The difference between a room with average absorption of 0.1 and the same room with $\overline{\alpha}$ of 0.7 is 15 db, which is a reduction in *loudness* of approximately 1.5 times (see Table 26.3). We will now examine the acoustic materials themselves and the effect of varying type, quantity, thickness, and installation methods.

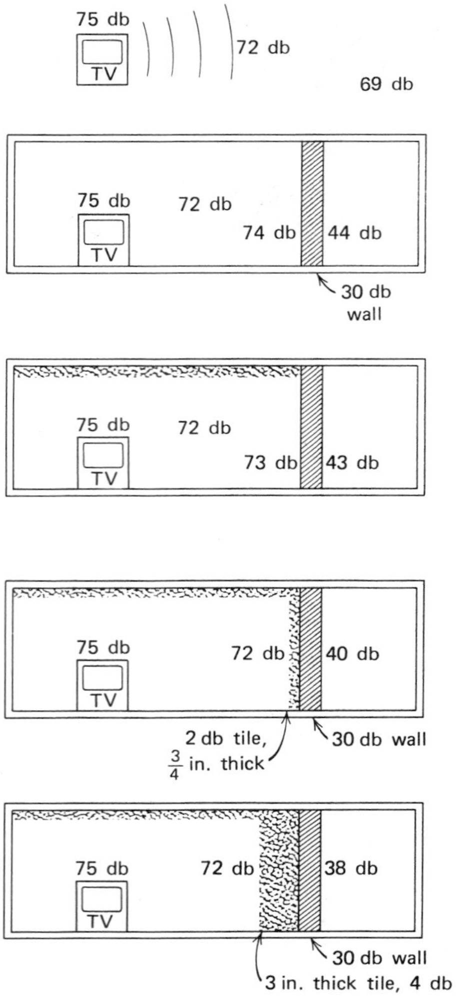

(a) TV set in free space produces 75-db sound level, which drops 6 db for each doubling of distance. Attenuation by inverse square law (see Section 26.7).

(b) TV still produces 75 db. In the free field, sound drops to 72 db but builds up to 74 db at the wall due to reverberant field reinforcement (see Figs. 26.16 and 26.18). Wall attenuation is 30 db. Sound on other side of the wall is 74 − 30 = 44 db.

(c) Acoustic tile ceiling acts to reduce room reverberant field. Free field is extended. Level at wall is 73 db. Level in second space is 73 − 30 = 43 db.

(d) Entire room is acoustically treated, effectively eliminating reverberant field. Room is "dead." Level on second side of wall is 72 db less acoustic tile loss, less wall loss (that is, 72 − 2 − 30 = 40 db).

(e) Add another 2¼ in. of acoustic wall treatment. Room is "dead." Level at wall 72 db. Level in second space = 72 − 4 − 30 = 38 db.

Fig. 27.1 *Action of acoustic absorbent material.*

27.4 Absorptive Materials

There are three families of devices for sound absorption—fibrous materials, panel resonators, and volume resonators. All types absorb sound by changing sound energy into heat energy. Only fibrous materials and panel resonators are used commonly in buildings. Volume resonators, also known as Helmholtz resonators, after their originator, are used principally as enclosures for absorbing a narrow band of frequencies. The following discussion refers to fibrous absorbers—the other two types are discussed separately in Section 27.8.

The fibrous or porous materials absorb by the frictional drag produced by moving the air in small spaces within the material. The absorption provided by a specific material depends on its thickness, density, porosity, and resistance to airflow. For example, materials must be thick to absorb low-frequency sounds effectively. Since the action depends on absorbing energy by "pumping" air through the material, the air paths *must extend from one side to the other*. A fibrous material with sealed pores is useless as an acoustic absorbent. (Therefore, painting will generally ruin a porous absorber.) A simple test is to blow smoke through the material. If the smoke passes through freely and the material is porous, fibrous, and thick, it should be a good sound absorbent. Porosity, provided it is above 70%, does not much affect absorption. Below this figure sound absorbency decreases as porosity decreases. Table 27.1 gives

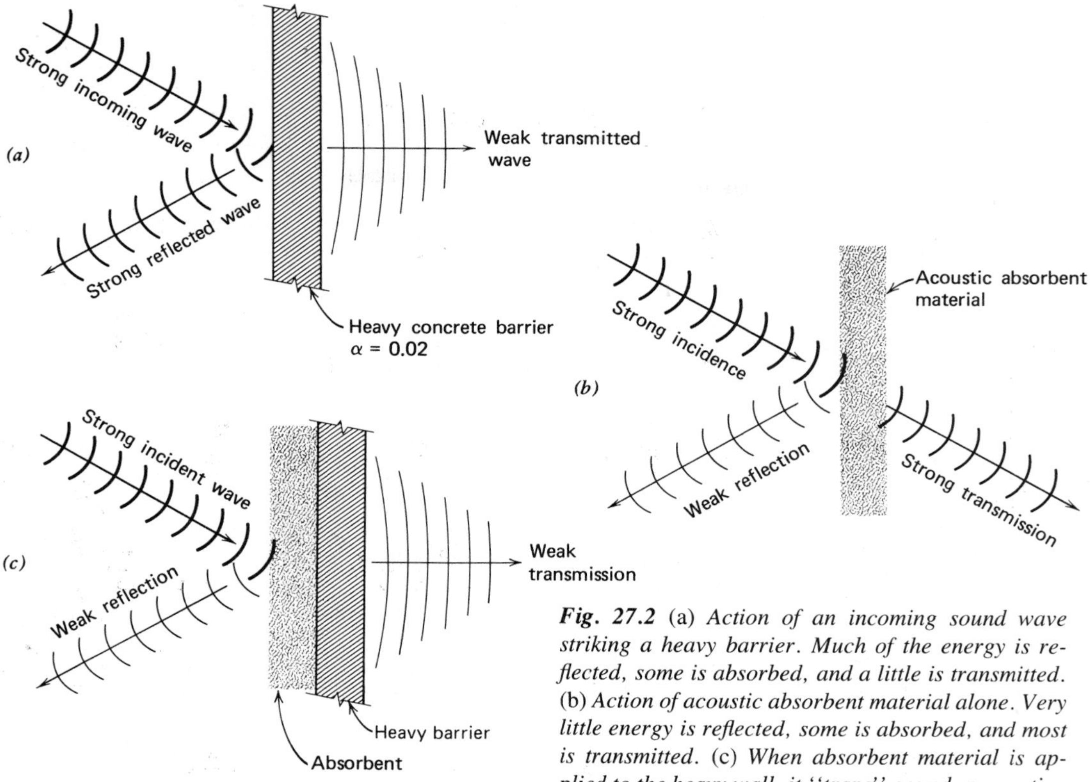

Fig. 27.2 (a) *Action of an incoming sound wave striking a heavy barrier. Much of the energy is reflected, some is absorbed, and a little is transmitted.* (b) *Action of acoustic absorbent material alone. Very little energy is reflected, some is absorbed, and most is transmitted.* (c) *When absorbent material is applied to the heavy wall, it "traps" sound, preventing reflection, while wall mass acts to reduce transmission.*

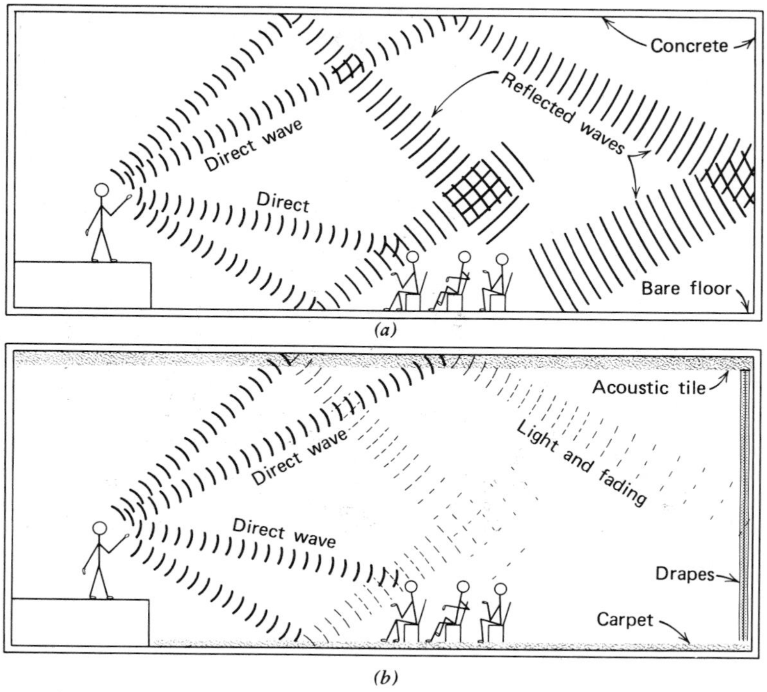

Fig. 27.3 *In the untreated space* (a) *reverberant (reflected) sound constitutes the greater portion of received sound in much of the room. These reflections are largely eliminated in* (b) *by wall and ceiling absorption. Note that direct wave is completely unaffected.*

TABLE 27.1 **Coefficients of Absorption[a]—α**

General Building Materials and Furnishings	125 Hz	250 Hz	500 Hz	1000 Hz	2000 Hz	4000 Hz	NRC[b]
	Absorption Coefficients						
Air, sabins per 1000 ft³ at 50% rh				0.9	2.3	7.2	—
Brick, unglazed	0.03	0.03	0.03	0.04	0.05	0.07	0.005
Brick, unglazed, painted	0.01	0.01	0.02	0.02	0.02	0.03	0.00
Carpet, heavy, on concrete	0.02	0.06	0.14	0.37	0.60	0.65	0.29
Carpet, heavy, on 40-oz hairfelt or foam rubber	0.08	0.24	0.57	0.69	0.71	0.73	0.55
Concrete block, coarse	0.36	0.44	0.31	0.29	0.39	0.25	0.35
Concrete block, painted	0.10	0.05	0.06	0.07	0.09	0.08	0.05
Fabrics							
Light velour, 10 oz/yd², hung straight, in contact with wall	0.03	0.04	0.11	0.17	0.24	0.35	0.15
Medium velour, 14 oz/yd², draped to half area	0.07	0.31	0.49	0.75	0.70	0.60	0.55
Heavy velour, 18 oz/yd², draped to half area	0.14	0.35	0.55	0.72	0.70	0.65	0.60
Floors							
Concrete or terrazzo	0.01	0.01	0.015	0.02	0.02	0.02	0.00
Linoleum, asphalt, rubber, or cork tile on concrete	0.02	0.03	0.03	0.03	0.03	0.02	0.05
Wood	0.15	0.11	0.10	0.07	0.06	0.07	0.10
Glass							
Large panes of heavy plate glass	0.18	0.06	0.04	0.03	0.02	0.02	0.05
Ordinary window glass	0.35	0.25	0.18	0.12	0.07	0.04	0.15
Gypsum board, ½ in. nailed to 2 × 4's 16 in. o.c.	0.10	0.08	0.05	0.03	0.03	0.03	0.05
Marble or glazed tile	0.01	0.01	0.01	0.01	0.02	0.02	0.00
Openings							
Stage, depending on furnishings			0.25–0.75				
Deep balcony, upholstered seats			0.50–1.00				
Grilles, ventilating			0.15–0.50				
Plaster, gypsum or lime, smooth finish on tile or brick	0.013	0.015	0.02	0.03	0.04	0.05	0.05
Plaster, gypsum or lime, on lath	0.14	0.10	0.06	0.05	0.04	0.03	0.05
Plywood paneling, ⅜ in. thick	0.28	0.22	0.17	0.09	0.10	0.11	0.15
Rough wood as tongue and groove cedar	0.24	0.19	0.14	0.08	0.13	0.10	0.14
Water surface, as in a swimming pool	0.008	0.008	0.013	0.015	0.020	0.025	0.00
Absorption of Seats and Audience[c]	*125 Hz*	*250 Hz*	*500 Hz*	*1000 Hz*	*2000 Hz*	*4000 Hz*	*NRC[b]*
Audience, in upholstered seats, per ft² of floor area	0.60	0.74	0.88	0.96	0.93	0.85	—
Unoccupied cloth upholstered seats, per ft² of floor area	0.49	0.66	0.80	0.88	0.82	0.70	—
Wooden pews, occupied, per ft² of floor area	0.57	0.61	0.75	0.86	0.91	0.86	—
Students in tablet-arm chairs, per ft² of floor area	0.30	0.42	0.50	0.85	0.85	0.84	—

Acoustic Absorptive Materials	Absorption Coefficients 125 Hz	250 Hz	500 Hz	1000 Hz	2000 Hz	4000 Hz	NRC[b]
Fiberglass painted ceiling boards[e]							
Textured: 5/8 in. thick	0.68	0.88	0.70	0.91	0.97	0.93	0.85
3/4 in. thick	0.66	0.85	0.72	0.94	0.99	0.98	0.90
1 in. thick	0.69	0.91	0.79	0.99	0.99	0.99	0.90
Random fissured, 5/8 in.	0.64	0.82	0.68	0.86	0.83	0.57	0.80
Perforated, 5/8 in.	0.71	0.89	0.68	0.90	0.96	0.98	0.85
Fiberglass glass cloth ceiling board[e]							
Nubby: 3/4 in. thick	0.75	0.91	0.70	0.93	0.99	0.99	0.90
1 in. thick	0.68	0.93	0.77	0.99	0.99	0.99	0.90
Fiberglass prefinished ceiling tile, 3/8 in. thick[e]	0.70	0.83	0.62	0.78	0.91	0.92	0.80
Celotex mineral fiber tile[f]							
Natural fissured, 3/4 in. thick (Fig. 27.10*a*)	0.47	0.49	0.51	0.75	0.86	0.80	0.65
Textured, 3/4 in. thick (Fig. 27.10*b*)	0.49	0.55	0.53	0.80	0.94	0.83	0.70
Plaid design, 3/4 in. thick (Fig. 27.10*c*)	—	—	—	—	—	—	0.70
LeBaron design, 3/4 in. thick (Fig. 27.10*d*)	—	—	—	—	—	—	0.70
Striated design, 3/4 thick (Fig. 27.10*e*)	—	—	—	—	—	—	0.70
Perforated lay-in panel, 5/8 in. thick (Fig. 27.10*f*)	0.27	0.26	0.52	0.75	0.68	0.53	0.55
Gold Bond, National Gypsum[g] mineral fiber tiles and panels[g]							
Solitude panels ("Fire Shield"), washable acrylic finish[g]							
Perforated, 5/8 in. thick	0.25	0.29	0.60	0.83	0.71	0.53	0.60
Fissured, 5/8 in. thick	0.28	0.32	0.65	0.73	0.73	0.75	0.60
Textured, 5/8 in. thick	0.28	0.36	0.65	0.62	0.44	0.33	0.50
Perforated asbestos panels, 1 in. thick[g]							
Uniform	0.60	0.65	0.49	0.71	0.73	0.51	0.65
Random	0.56	0.51	0.49	0.68	0.60	0.31	0.60
Perforated metal panel ("Acoustimetal"),[g] enameled, 1 9/16 in. thick							
Square pattern	0.59	0.85	0.88	0.99	0.97	0.79	0.90
Diagonal pattern	0.63	0.84	0.86	0.99	0.99	0.91	0.90
Sound blocks ("Tectum"),[g,h]							
3 in. thick × 15½ in. square	0.32	0.60	1.43	2.36	2.32	2.41	1.68

[a]Complete tables of coefficients of the various materials that normally constitute the interior finish of rooms may be found in books on architectural acoustics. This table will be useful in making simple calculations.

[b]Noise reduction coefficient: the arithmetic average of the α values at 250, 500, 1000, and 2000 Hz.

[c]When the audience is randomly spaced, use an average of 5.0 sabins (ft^2) per person.

[d]Installed in hung ceiling with at least 16 in. to slab.

[e]Courtesy of Owens-Corning Fiberglas.

[f]Courtesy of Celotex-Jim Walter Co.

[g]Courtesy of Gold Bond/National Gypsum.

[h]Clipped or glued to wall; minimum 24 in. o.c.

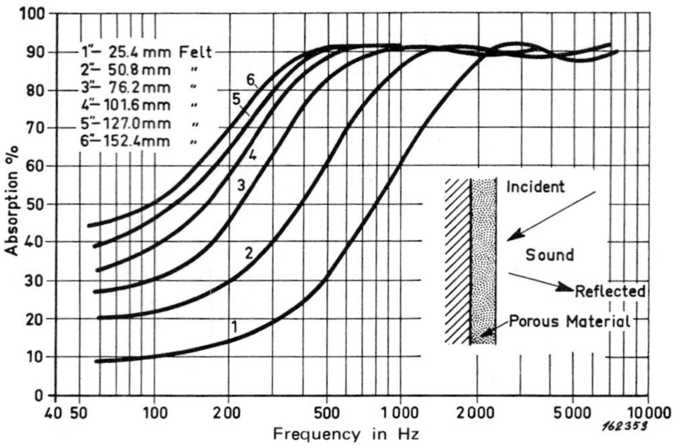

Fig. 27.4 Variation of absorption coefficient with thickness of felt absorbent. Note particularly that beyond 1 kHz, all thicknesses give the same α, whereas at low frequencies the absorption is proportional to thickness. Furthermore it requires a very heavy layer to give appreciable absorption at low frequency. (Courtesy of Brüel and Kjaer, Sound Insulation and Room Acoustics, *Chapman and Hall, London, 1951.)*

absorption coefficients for fibrous absorbent materials and for building materials and furnishings. Several important conclusions can be drawn from examination of this table.

1. For absorbent materials, absorption is normally higher at high frequencies than at low.

2. Absorption is not always proportional to thickness, but depends on the type material being used and the method of installation (see Fig. 27.4). It is clear from this figure that beyond a nominal thickness, little is to be gained by adding thickness except at very low frequencies, or when installed discontinuously, as in item 3, below.

3. It is possible to obtain an α greater than 1.0 by using very thick blocks. See "Sound blocks" in Table 27.1. These are installed at a distance from each other. Edge absorption is very large, particularly at high frequencies.

4. Installation methods have a pronounced effect as discussed in the following section.

27.5 Installation of Absorptive Materials

Coefficient ratings for absorptive materials are always given with mountings corresponding to ASTM requirements. The most common standardized mounting methods are given in Fig. 27.5. Installation of absorbent directly on a wall or ceiling is the least effective means, since exposure to sound energy is minimal (see Fig. 27.6). When an air gap is left between the porous layer and the rigid surface, the combination acts almost as well, in midfrequencies, as a layer of absorbent equivalent in thickness to the air plus porous material. One problem with this technique is that at the λ/2 node of a standing wave there is a severe drop in absorption, as can be seen from Fig. 27.6*c*. Thus, for instance, at 1000 Hz, one-half wavelength is approximately 7 in. At that distance, α drops severely, but is a maximum at λ/4 or 3½ in. For ceiling tile hung at 16 in. below the slab (Fig. 27.5, method 7), the drop in absorption occurs at

$$\lambda/2 = 16 \text{ in.}$$

$$\lambda = 32 \text{ in.} = 2.67 \text{ ft}$$

$$f = \frac{1128}{2.67} = 422 \text{ Hz}$$

which is midfrequency. This factor should be considered in applying absorptive material; that is, avoid a spacing corresponding to a drop in absorption at a sensitive frequency. To obtain good low-frequency absorption, it is essential that a deep air space be provided behind the porous absorbent

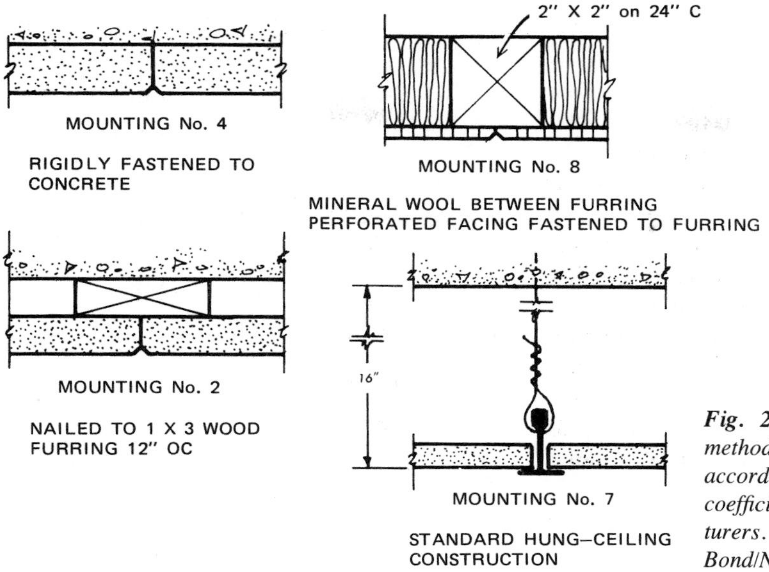

Fig. 27.5 Standard test-mounting methods for absorptive material, in accordance with which absorptive coefficients are given by manufacturers. (Diagrams courtesy of Gold Bond/National Gypsum.)

material, and that walls be treated in addition to the ceiling.

In increasing order of effectiveness, absorbent material can be applied:

1. Directly to room surface.
2. Hung below ceiling and supported away from the walls.
3. Hung from the ceiling as louvres or baffles.
4. Made up into shapes such as cubes, or tetrahedrons, and suspended from the ceiling.

The last two techniques are extremely effective because they expose a very large surface of porous material—much larger than could be accomplished with wall or ceiling covering. Of course, these suspended objects become architectural elements and must be handled accordingly. In contrast, surface coverings are relatively neutral. In general, treatment should not be limited to one room surface such as the ceiling. All three principal surfaces in the direction of sound propagation, that is, ceiling, floor, and back wall, should be treated *approximately equally* for best results. The common practice of treating the ceiling only is generally inadvisable, since high frequencies are highly directive and may not reach the ceiling until the third reflection. See specific recommendations in Section 27.34.

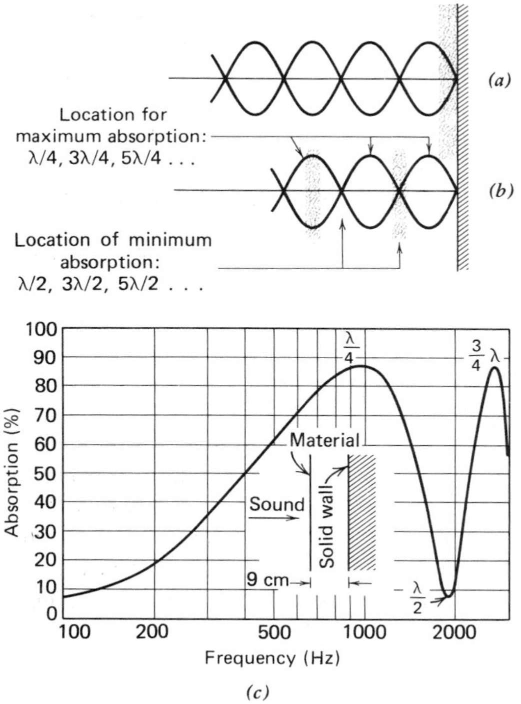

Fig. 27.6 Sound waves striking a surface of large mass will create standing waves at certain frequencies, depending on room dimensions (a). *If insulation is placed at the nodes of these waves* (b), *attenuation is drastically reduced* (c). *(Illustration* (c) *from Brüel and Kjaer,* Sound Insulation and Room Acoustics, *Chapman and Hall, London, 1951.)*

27.6 Noise Reduction Coefficient

The last column in Table 27.1 is labeled NRC—noise reduction coefficient. This figure is the arithmetic average of the absorption coefficients at 250, 500, 1000, and 2000 Hz. The name is ill chosen inasmuch as it cannot be used as the words would seem to imply. It is useful as a single-number criterion for the *midband* effectiveness of a porous absorber. Obviously if careful design is required or high- and low-end frequencies are of interest, NRC is nearly useless, and detailed calculations over the entire frequency range must be made. Similarly, two materials with the same NRC can perform quite differently, since the NRC is an average and few materials have a flat characteristic.

27.7 Noise Reduction by Absorption

The reader is referred back to Sections 26.13 through 26.15 and Fig. 26.37 for a review of the characteristics of absorption, reverberation, and fields in an enclosed space, as background material for a discussion of noise reduction by absorption. Equation 26.13 relates sound pressure level (SPL) and sound power level (PWL) as a function of distance from the source and room absorption. The following example shows the application of this equation and the result of adding absorption to the space.

EXAMPLE 27.1 An open blower is installed on the floor, away from the walls, in a large enclosed space that is 6 m long, 12 m wide, and 4 m high. The floor is concrete ($\alpha = 0.01$), walls and ceiling painted block ($\alpha = 0.07$). The sound power levels (PWL) supplied by the manufacturer are 90 db at 500 Hz and 87 db at 2000 Hz. Calculate the noise level SPL at distances of 5 m and 10 m from the blower outlet:

(*a*) In the original room.
(*b*) With double the absorption.
(*c*) With quadruple the absorption.

SOLUTION.

1. From Equation 26.15, we have the expression

$$SPL = PWL + 10 \log_{10}\left[\frac{Q}{4\pi r^2} + \frac{4}{R}\right]$$

In our example, $Q = 2$ (see Fig. 26.17).

2. The first step is to calculate room factor R for the three situations of absorption.

$$R = \frac{S\bar{\alpha}}{1 - \bar{\alpha}}$$

$$S = 2(48) + 2(24) + 2(72) = 288 \text{ m}^2$$

from Equation 26.14,

$$\bar{\alpha} \text{ original} = \frac{3(48)(0.07) + 1(72)(0.07) + (72)(0.01)}{288} = 0.055$$

$\bar{\alpha}_a = 0.055$ $\qquad R_a = 16.76$

$\bar{\alpha}_b = 2(0.055) = 0.11$ $\qquad R_b = 35.6$

$\bar{\alpha}_c = 4(0.055) = 0.22$ $\qquad R_c = 81.23$

3. Calculating SPL values and tabulating, we obtain

	500 Hz		2000 Hz	
SPL for:	*5 m*	*10 m*	*5 m*	*10 m*
Original room	83.9	83.8	80.9	80.8
Double $\bar{\alpha}$	80.8	80.6	77.8	77.6
Quadruple $\bar{\alpha}$	77.5	77.1	74.5	74.1

The results indicate the very important fact that *doubling the absorption decreases the noise level by only 3 db*. Therefore it requires a quadrupling of absorption to make the decrease noticeable (see Table 26.3). This is obviously an expensive procedure with diminishing returns.

If the entire space is considered to be a reverberant field, then the noise (sound) intensity level (*IL*) can be expressed as follows:

$$IL = \text{PWL} - 10 \log \Sigma S\alpha + 6 \text{ db}$$

or

$$IL = PWL - 10 \log \Sigma A + 6 \text{ db} \qquad (27.1)$$

where

$\Sigma S\alpha = \Sigma A$
$\quad$ = total absorption in room, sabins (ft^2, m^2)

ACOUSTICS

IL = intensity level, db
PWL = sound power level, db

Although increasing absorption decreases the noise level, the level never is reduced below the free field level for that distance from the source (see Fig. 27.1).

The amount of noise reduction provided by additional absorption may be determined by

$$\text{noise reduction} = IL_1 - IL_2$$
$$= 10 \log \Sigma A_2 - 10 \log \Sigma A_1$$
$$\therefore NR = 10 \log \frac{\Sigma A_2}{\Sigma A_1} \qquad (27.2)$$

where

ΣA_1 = total absorption, initial condition
ΣA_2 = total absorption, final condition

From Equation 27.2 it is obvious that doubling the absorption results in a noise reduction of 3 db, since $10 \log_{10} 2 = 3$ (db). It is of interest to work out a practical example using equations 27.1 and 27.2 and comparing the results with the more precise relation given in Equation 26.15.

EXAMPLE 27.2. Referring to Fig. 27.7, calculate the original noise level and the subsequent noise reduction by three steps of sound absorption treatment, assuming a completely reverberant field in the space, that is, SPL independent of location.

SOLUTION.

(a) Original Condition. Painted concrete block chamber, 10 × 10 × 10 ft.
Fan power level

At 500 Hz = 88 db
At 2000 Hz = 78 db

Absorption	*Area*	α	*Total Absorption* ($\Sigma S\alpha$)
500 Hz	600 ft^2	0.06	36 sabins (ft^2)
2000 Hz	600 ft^2	0.09	54 sabins (ft^2)

Sound intensity level before treatment

At 500 Hz:

$$\begin{aligned} IL &= \text{sound power} - 10 \log \Sigma S\alpha + 6 \text{ db} \\ &= 88 \text{ db} - 10 \log 36 + 6 \text{ db} \\ &= 88 \text{ db} - 15.6 \text{ db} + 6 \text{ db} \\ &= 78.4 \text{ db} \end{aligned}$$

At 2000 Hz:

$$\begin{aligned} IL &= 78 \text{ db} - 10 \log 54 + 6 \text{ db} \\ &= 78 \text{ db} - 17.3 \text{ db} + 6 \text{ db} \\ &= 66.7 \text{ db} \end{aligned}$$

(b) Ceiling Treatment Only

At 500 Hz:

$$\begin{aligned} \alpha &= 0.82 \\ \text{additional absorption} &= 100\,(0.82 - 0.06) \\ &= 76 \text{ sabins} \\ NR &= 10 \log \frac{76 + 36}{36} \\ NR &= 4.9 \text{ db} \end{aligned}$$

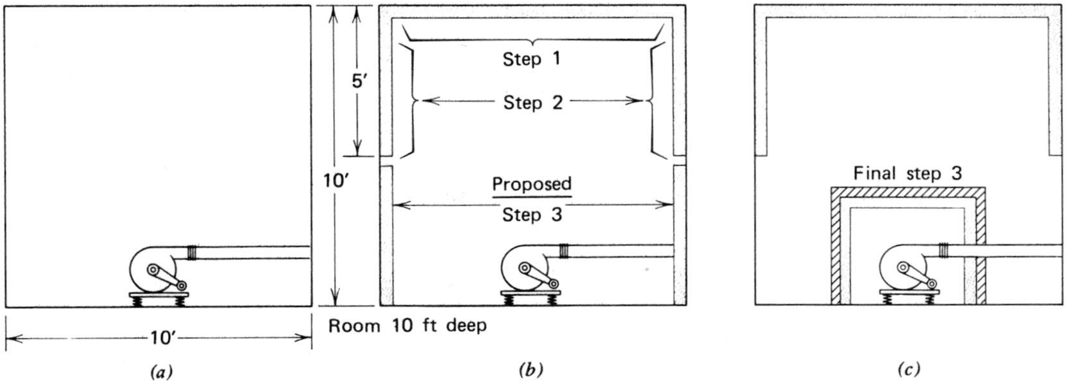

Fig. 27.7 Quieting the room (a) *by addition of absorptive material is cost effective only through step 2* (b). *Further quieting would be accomplished locally* (c), *which might obviate the necessity for wall treatment (step 2). Ceiling treatment should remain.*

At 2000 Hz:

$$\alpha = 0.94$$
$$\Delta A = 100\ (0.94 - 0.09) = 85$$
$$NR = 10 \log \frac{85 + 54}{54}$$
$$NR = 4.1 \text{ db}$$

(c) Ceiling and Half-Wall Treated

At 500 Hz:

$$\begin{aligned}\text{added absorption} &= 300\ (0.82 - 0.06)\\ &= 228 \text{ sabins}\\ NR &= 10 \log \frac{228 + 36}{36}\\ &= 8.7 \text{ db}\end{aligned}$$

At 2000 Hz:

$$\begin{aligned}\text{added absorption} &= 300\ (0.94 - 0.09)\\ &= 255 \text{ sabins}\\ NR &= 10 \log \frac{255 + 54}{54}\\ &= 7.5 \text{ db}\end{aligned}$$

(d) Ceiling and Full-Wall Treatment

At 500 Hz:

$$\begin{aligned}\Delta A &= 500\ (0.82 - 0.06)\\ &= 380 \text{ sabins}\\ NR &= 10 \log \frac{380 + 36}{36}\\ &= 10.6 \text{ db}\end{aligned}$$

At 2000 Hz:

$$\begin{aligned}\Delta A &= 500\ (0.94 - 0.09)\\ &= 425 \text{ sabins}\\ NR &= 10 \log \frac{425 + 54}{54} = 9.5 \text{ db}\end{aligned}$$

Summary

	IL (SPL)	
	500 Hz	*2000 Hz*
Bare room	78.4	66.7
Ceiling treated	−4.9 db (−4.9)	−4.1 db (−4.2)
Half-wall treatment	−8.7 db (−8.7)	−7.5 db (−7.7)
Full-wall treatment	−10.6 db (−10.7)	−9.5 db (−9.6)

The numbers in parentheses are the SPL differences as calculated from Equation 26.15, using $Q = 2$ and $r = 5$ ft.

We would conclude here that the third step is really not worthwhile, since only a negligible additional decibel of quieting is accomplished. This example is intended to indicate the law of diminishing returns in quieting by absorption. Starting with a live room, the initial application is effective. Beyond that, additional quieting by absorption is not economical, and the same outlay would be better applied in quieting the machine itself, probably with a machine enclosure as indicated in Fig. 27.7*c*.

27.8 Panel and Cavity Resonators

Panel resonators are built with a membrane such as thin plywood or linoleum in front of a sealed air space generally containing absorbent material. The panel is set in motion by the alternating pressure of the impinging sound wave. The sound energy is converted into heat through internal viscous damping. Panel resonators are used where efficient low-frequency absorption is required and middle- and high-frequency absorption is unwanted or provided by another treatment (see Fig. 27.8). Panel resonators are often used in recording studios.

A *volume or cavity resonator* (Helmholtz resonator) is an air cavity within a massive enclosure, connected to the surroundings by a narrow neck opening. The impinging sound causes the air in the neck to vibrate, and the air mass behind causes the entire construction to resonate at a particular frequency. At that frequency absorption is very great—approaching unity, and dropping sharply above and below this frequency (see Fig. 27.8). By adjusting neck opening and cavity dimensions, the unit can be tuned to resonate at any desired frequency. This makes it extremely useful when a major single frequency is present, as with 120-Hz transformer hum. Such an installation is shown in Fig. 27.9 along with a typical concrete block resonator. Fibrous filler can be used in the block to increase high-frequency absorption.

27.9 Acoustically Transparent Surfaces

The soft porous material of which most absorbents are constructed is covered with perforated metal or wood panels, which act as physical protection and as stiffeners. These coverings are acoustically

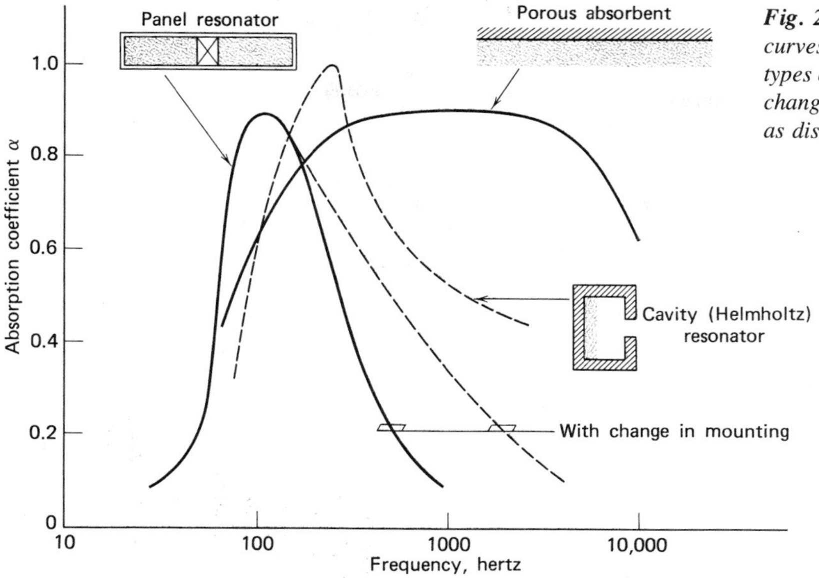

Fig. 27.8 *Typical absorption curves for the three major types of absorbers. All can be changed by varying design, as discussed in the text.*

transparent except at higher frequencies. The frequency at which a noticeable reduction in absorption occurs can be estimated for circular holes as

$$f = \frac{40p}{d} \qquad (27.3)$$

where

f = frequency, Hz
p = percentage of open area
d = diameter of holes, in.

Thus for ¼-in. holes and 60% open area, which is a typical commercial construction,

$$f = \frac{40(6)}{0.25} = 9600 \text{ Hz}$$

which is very high and generally not of major concern. It is always desirable for a given percentage of open area to have a larger quantity of small holes since, as is obvious from the formula, this raises the interference frequency. It is also desirable to stagger the holes as this improves absorption. An open-weave fabric is almost completely transparent to sound and can be used as a decorative cover on absorbent wall coverings (see Fig. 27.10).

Fig. 27.9 *Typical application of cavity resonators to absorb transformer hum. Wall is constructed of masonry units constructed to act as resonators (see insert). (This installation is made with "SOUND-BLOX" as manufactured by the Proudfoot Co. through whose courtesy photo is supplied.)*

27.10 Absorption Recommendations

To summarize the discussion above, absorption techniques are useful and effective:

1. To change room reverberation characteristics.
2. In spaces with distributed noise sources such as offices, schools, restaurants, and machine shops.

TABLE 27.2 **Typical Recommended Acoustical Treatment**

Material Characteristics	*Acoustical Tile*	*Metal-Faced Acoustical Units*	*Acoustical Plaster*	*Acoustical Form Board*
Size and form	Square tiles 12 in. × 12 in.; some up to 48 in. × 96 in. Roof deck 2 ft × 8 ft—most common thicknesses, 5/8 and 3/4 in.	12 in. × 24 in. up to 24 in. × 120 in. Thickness is controlled by acoustical pad backing. Either paper-wrapped mineral wool or cut or roll glass fiber. Units run 2 to 3 in. total thickness.	A plasterlike material of special fibrous or particulate aggregate, 1/2 to 1 1/2 in. thick.	Sizes and thickness vary to fit structural requirements (floor and roof slabs only) 1 to 2 1/2 in. for wall application.
Surface	Wide variety of perforated, textured, and sculptured surfaces with white, painted, vinyl, and glass cloth finishes.	Baked-enamel finish usually white, but available in color, perforated or slotted.	Fine-grain white texture, but may be spray painted.	"Shredded Wheat" or smooth pattern.
Method of installation	Adhesive, nailing, or stapling to wood furring; lay-in grids; concealed spline-grid.	Attached to metal supports nailed to wood furring or hung in a proprietary metal suspension system.	Trowel or spray.	According to structural design (floor and roof slabs only). Adhesive or mechanical application on walls.
Major area of application	All interior applications. Check specifications for application in high-humidity areas.	All interior applications. Check specifications for application in high-humidity areas.	Most interior applications. Especially useful for large curved surfaces. Not satisfactory in high humidity.	Floor and roof decks wherever applicable. Wall surfaces in most interior spaces.
Advantages	Provide widest range of finishes in high-absorption units. Mineral fiber tiles are rated incombustible. Lay-in units provide access to plenum above.	Incombustible. Easily maintained. Can be washed or painted; replacement units will match original job. Permits easy access to plenum.	Rapid, low-cost installation on large irregular or curved surface. Fine-grain texture. Incombustible—can be applied with (or as part of) fire protection.	Combines form board with thermal insulation and acoustical absorption.

Limitations	Allows transmission of sound over partition in hung ceiling unless selected for high-attenuation factor.	Allows transmission of sound over partition in hung ceiling unless backed by impervious layer.	Easily abused. Difficult to match finish in patching or repairing. Does not always perform as advertised. Performance limited to high-frequency absorption. Dusty in application.	Not advisable for high-humidity areas.

Material Characteristics	*Unit Sound Absorbers*	*Carpeting*	*Drapery*
Size and form	From 12 in. × 12 in. × 2 in. units to 24 in. × 48 in. plastic coated glass fiber baffles.	Sized or seamed to fit any floor or wall area.	Acoustically transparent fabrics ranging from opaque velours to transparent/translucent glass fibers.
Surface	12 in. × 12 in. units in smooth to rough surface. Baffles are vinyl wrapped.	Cut, looped, or combination to achieve any degree of "fuzzing."	Velveteen to bouclé (rough).
Method of installation	Units may be applied using special wall clips, pendents (ceiling only), or adhesive. Baffles are wire supported.	Tacking (floor only) or adhesive.	According to function and esthetics.
Major area of application	12 in. × 12 in. units are useful in all interior areas. Baffles in industrial areas.	Floors and walls.	Window or opaque wall areas and room dividers.
Advantages	Permit maximum flexibility. Hung units add sound absorption without requiring lowering lights or sprinkler heads.	Provides a relative degree of mid- and high-frequency absorption and luxurious appeal.	Provides a relative degree of overall sound absorption and luxurious appeal.
Limitations	Each application must be designed individually.	Acoustical absorption increases with pile height, pad, and fiber density.	Acoustical absorption increases with fabric density percent, fold when drawn, and air space behind.

Source: Reprinted with permission of the authors from Goodfriend and Sulewsky, "A Guide to Acoustic Materials," *Architectural and Engineering News*, February 1970.

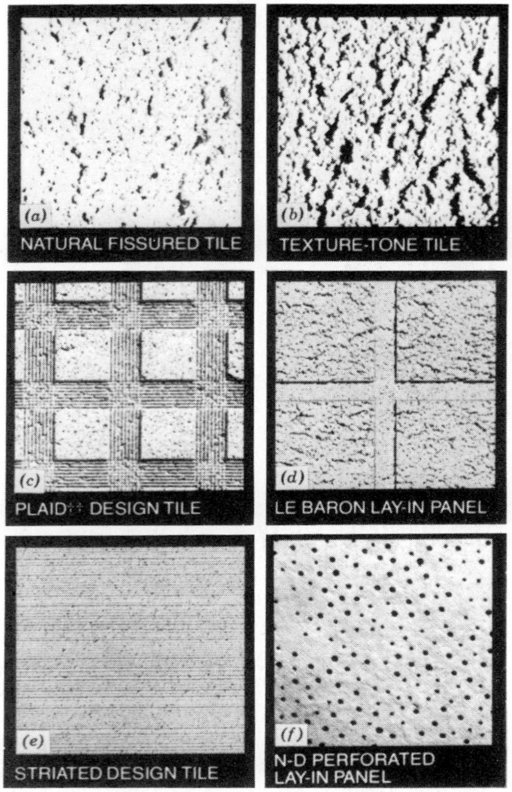

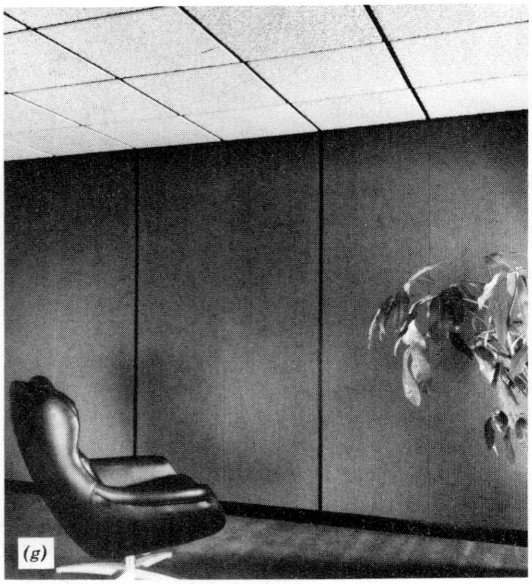

Fig. 27.10 (a–f) *Acoustic ceiling tiles and panels of varying designs, with varying types of perforations. (Courtesy of Celotex Building Products Division, Jim Walter Corp.)* (g) *A fabric-covered mineral-fiber, acoustic wall covering, coordinated with the acoustic ceiling tile pattern. (Photo, courtesy of Armstrong World Industries, illustrates their "Sound-Soak" Panels.)*

3. In spaces with a hard surface and little absorptive content.
4. Where listeners are in the reverberant field. (No amount of absorptive material can reduce intensity levels in the direct field.)

Concentrated noise sources are better handled by individual equipment enclosures than by room treatment, since enclosures reduce direct field noise, which, as stated above, room surface treatment cannot do. Typical application recommendations are given in Table 27.2.

SOUND ISOLATION

27.11 Airborne and Structure-Borne Sound

In contradistinction to the preceding sections, which were concerned with the phenomenon of in-room sound reduction by absorption, the following sections will discuss the characteristics of sound *transmission* between enclosed spaces. The distinction is often made between airborne and structure-borne sound, although in reality they differ only in sound origin. *Airborne sound* originates in a space with any sound-producing source and changes to *structure-borne* sound when the sound wave strikes the room boundaries. Structure-borne sound is generally understood as direct impact caused by a vibrating or impacting source directly contacting the structure. Hence a child crying in the adjoining apartment is contributing airborne sound; the same child bouncing a ball on the floor is creating structure-borne sound, in this case by impact. The building heating pumps that were installed without proper damping mounts create structure-borne sound by vibration.

In reality, all sound transmission is both airborne and structure-borne since, once having entered the structure, the sound travels along the structure and causes the structure to vibrate, generating airborne sound. Figure 27.11 should assist in understanding this action. In Fig. 27.11*a* the

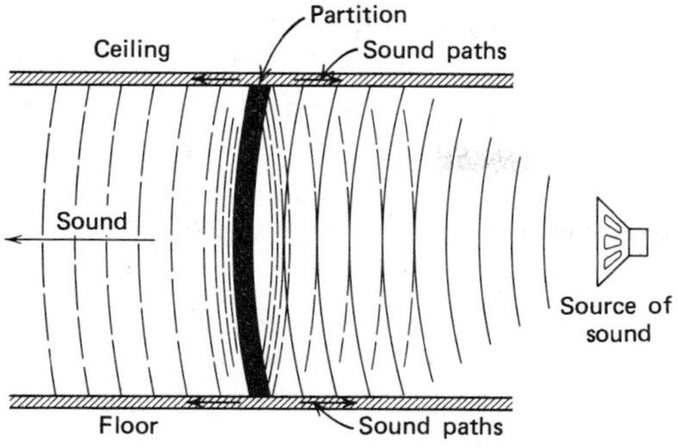

(a)

Fig. 27.11 (a) *Airborne sound is so called because it originates in the air.* (b) *Structure-borne sound originates from mechanical contact between structure and vibrating or impacting sources. As such its energy level is usually much higher than airborne sound.* (c) *Some techniques for controlling airborne sound. (Data from* A Guide to Airborne, Impact and Structure-Borne Noise Control in Multi-family Dwellings, *1968.)*

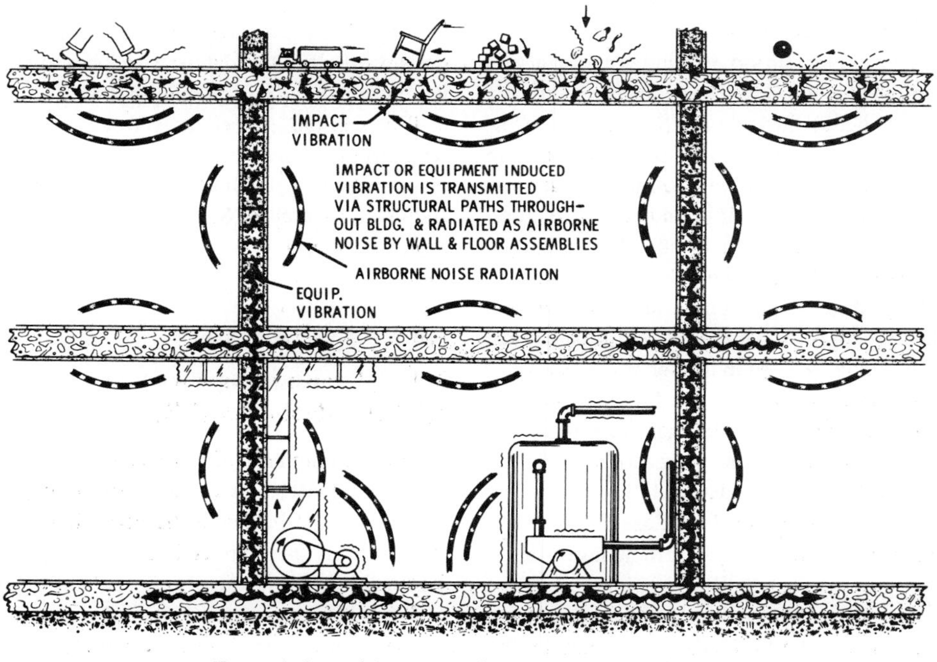

Transmission of Impact and Structure—Borne Noise.

(b)

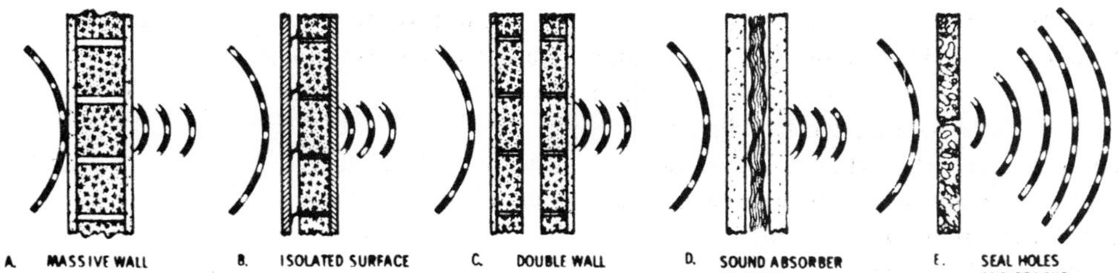

(c)

sound is airborne, originating in the air at one side of the partition. The incidence of sound energy *causes the partition to vibrate,* generating sound on the other side. Sound does not "pass through"—it causes the structure to become a secondary source. The partition vibrates primarily in the direction of the sound, that is, in the vertical plane. It also vibrates in other modes causing some sound energy to pass into the floor and ceiling, depending on the mode of attachment. This energy becomes structure-borne sound.

In Fig. 27.11*b* the process is similar, but reversed. Energy is introduced into the structure directly and efficiently by mechanical contact, that is, vibration and impact. Sound travels along the structure, as shown and by causing the structure to vibrate, creates airborne sound. In a structure with rigid wall-to-floor connections, these sounds are clearly heard throughout the structure. Airborne sound is generally much less disturbing than structure borne, since its initial power is very low and it attenuates rapidly at boundaries. Structure-borne sound is generally at a much higher energy level initially and attenuates slowly as it travels through the structure, thereby causing disturbance over large sections of a building. This disturbance is magnified by the "sounding board effect."

We are all familiar with the fact that a tuning fork must be held up to the ear to be heard directly, but if its handle is placed on a table the sound is amplified. This action is not really amplification but an increase in the efficiency of energy transfer. In general, the efficiency of a radiator is proportional to the ratio of its surface dimensions to the sound wavelength. A tuning fork vibrating at concert A (440 Hz) of wavelength 2½ *feet* cannot couple its energy into the air. It is simply too small. By placing the instrument on a table whose dimensions are approximately one wavelength, we permit it to transfer its energy efficiently; hence the "amplification." The same effect can be extremely troublesome in structure-borne sound. A vibrating pump itself makes little sound. However, it transfers a large amount of energy into the structure, which will appear as audible sound at each partition, floor, and wall that is rigidly coupled to the structure. Soft (damping) connections prevent energy transfer, thereby greatly attenuating the transmission of sound energy into connecting efficient radiating surfaces; hence the desirability of such flexible connections.

Airborne sound changes direction easily, with low frequencies most flexible in this regard. Thus sound can travel through long distances in corridors and ventilating ducts with attenuation of 6 db per doubling of distance. Structure-borne sound travels much more rapidly (see Table 26.1) and with attenuation as low as 1 db per kilometer. Mass of the structure is an effective attenuator only in the direction of radiation. That is, a sound traveling along a massive structure will radiate little outward from the structure (although enough to be very annoying) because the large mass minimizes vibration in that direction. Thus in Fig. 27.11*b*, noise from impact on the floor above will probably be louder than noise from machines below because the former generates sound directly downward while the latter introduces energy into the entire network of parallel paths.

The sections immediately following will deal with airborne sound and the means for controlling it (see Fig. 27.11*c*). Impact noise (structure-borne sound) is covered in Sections 27.23 to 27.25.

AIRBORNE SOUND

27.12 Transmission Loss (TL) and Noise Reduction (NR)

The transmission loss (TL) of a barrier is the ratio, expressed in decibels, of the acoustic energy reradiated by the barrier to the acoustic energy incident on it. This number is a figure of merit for the sound-isolating quality of the wall itself and is obtained from controlled laboratory tests. (In Europe transmission loss is referred to as sound reduction index, *R*.) The number that is of greater importance to the building designer is the actual noise reduction between two spaces separated by a barrier of transmission loss (TL), that is, the action of the barrier in situ. This noise reduction is defined as the difference between the intensity levels in the two rooms, that is,

$$NR = IL_{\text{room 1}} - IL_{\text{room 2}} \tag{27.4}$$

and is related to the TL of the barrier by the expression

$$NR = TL - 10 \log \frac{S}{A_R} \tag{27.5}$$

where

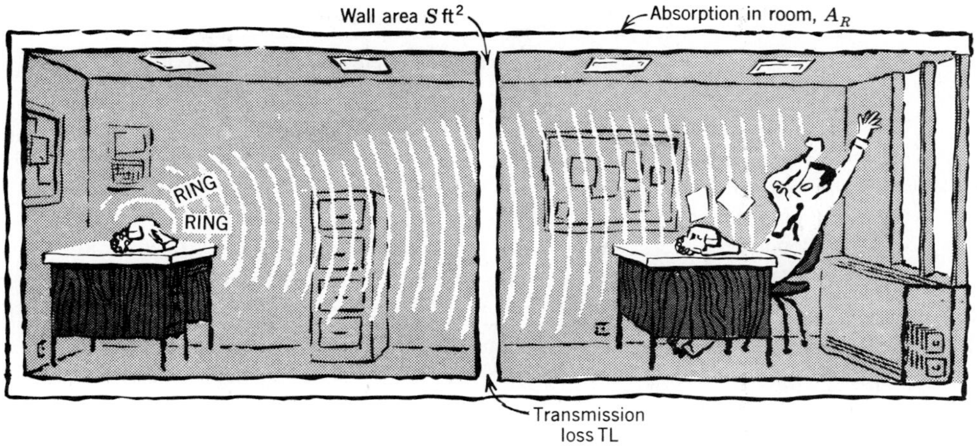

Fig. 27.12 Illustration of the simple case of airborne sound transmission between adjacent rooms through a common barrier. With a sound source in one room, the transmitted sound level is dependent not only on the transmission loss of the barrier but also on the area of the barrier and the receiving-room absorption. The background noise level determines whether the transmitted sound will be heard. See Section 27.21 and Figs. 27.32, 27.33, and 27.35.

NR = noise reduction, db
TL = barrier transmission loss, db
S = area of barrier wall (ft^2, m^2)
A_R = total absorption of *receiving* room, sabins (ft^2, m^2)

We see, therefore, that noise reduction and transmission loss are not equal but are related by the size of the dividing wall S and the absorption characteristic of the receiving room, A_R. A moment's thought will confirm the logic of this relation. When sound energy impinges on the wall, the wall in turn becomes the sound source, radiating into the receiving room. Therefore, the amount of sound energy transferred is proportional to the area S of the common wall between the two spaces.

The sound level in the receiving room is related to its own reverberance (absorption characteristic A) as we have seen repeatedly. (Refer back to Equation 26.13, for example.) Thus if the receiving room is a reverberant, live space, A is low and NR is less than TL. Conversely, if the receiving room is dead, A is large and noise reduction can be greater than transmission loss, depending on the ratio of barrier wall size to room area. In lieu of calculating, the following rule of thumb can be used:

1. For a live receiving room,
   $NR = TL - 1$ db
2. For a medium receiving room,
   $NR = TL + 4$ db
3. For a dead receiving room,
   $NR = TL + 7$ db

The extreme case of "deadness" of the receiving room is one with no walls, that is, sound transmission from inside to the exterior. In such cases, NR exceeds TL by 10 to 15 db, depending on the point outside where IL is measured. Figure 27.12 illustrates these relationships. To acquire facility with sound-isolation techniques, the reader must become familiar with the relationship of transmission loss to the barrier wall's physical characteristics, its mass, rigidity, material of construction, and method of construction and attachment. These considerations are the subject of the following sections.

Note: The reader is cautioned to be careful with the term *noise reduction* (NR), since a term noise reduction coefficient (NRC) also exists, which is completely unrelated. The latter, as pointed out in Section 27.6, is very poorly named.

27.13 Barrier Mass

Sound transmission requires that the barrier be set into vibration by the incident sound energy. Al-

TABLE 27.3 **Typical Average[a] Transmission Loss**

Barrier Construction	Surface Weight (lb/ft²)	Transmission Loss (db)
3⁄16-in. plate glass	2.5	20
1⁄4-in. asbestos—cement sheet	3.0	21
2-in. plaster on wire lath	15	34
3-in. cinder block	25	39
4-in. concrete	50	45
8-in. brick, plastered	80	50
12-in. brick, plastered	120	53

[a]Over the frequency range of 150 to 3000 Hz.

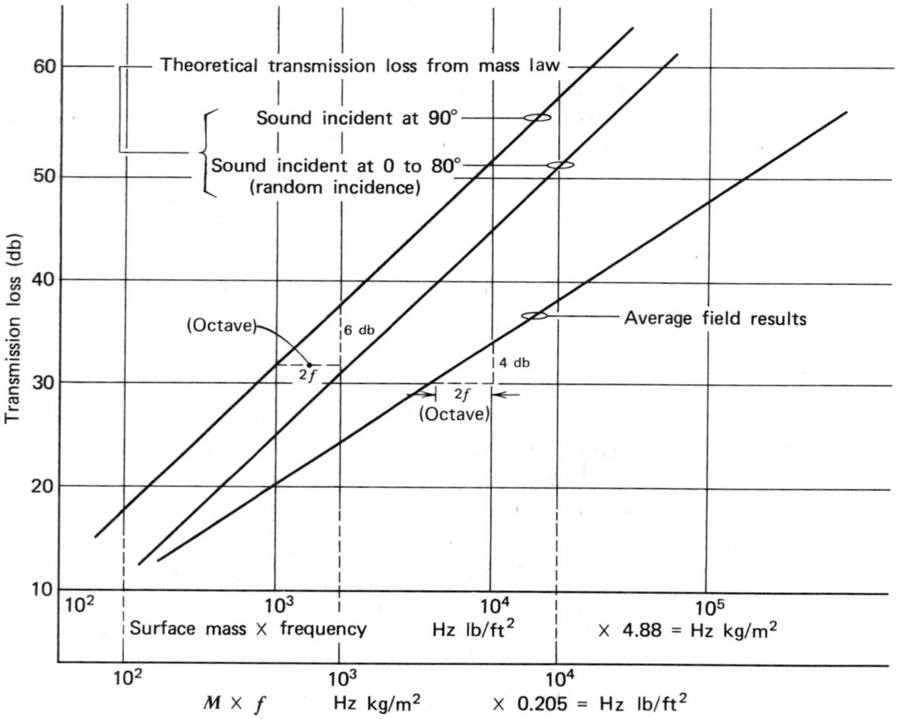

Fig. 27.13 *Graphic representation of mass law action in attenuation of transmitted sound. Normal (90°) incidence results in maximum transmission loss. Field or random incidence (0 to 80°) is approximately 6 db lower, but maintains the 6 db per octave slope. Field results are much lower due to flanking and stiffness effects, and average 4 db per octave (frequency doubling).*

though this was stated above, we repeat it here to emphasize the fundamental importance of this simple statement. (We are of course referring to a barrier that is impervious to air, i.e., a solid barrier. Otherwise the moving air molecules carrying the sound will simply pass through with minimal transmission loss.) The impinging energy acts as a force on the wall. Since $F = MA$, the larger the mass, the less it will vibrate. When other factors are taken into account, the resultant relationship is known in acoustics as the mass law. It states that for a nonporous, homogeneous structure of low stiffness, the sound transmission loss is proportional to the logarithm of surface mass (the weight of the wall per square foot of area) and to the frequency of vibration (acceleration factor). Thus doubling the mass or frequency will, theoretically, cause an increase of 6 db in the transmission loss or, stated otherwise, the slope of *TL* versus the frequency–mass (*fM*) curve is 6 db (see Fig. 27.13). Table 27.3 lists the average transmission loss of some common barriers, confirming the overall operation of the mass law. Figure 27.13 is a graphic representation of mass law operation. With sound incident at 90°, maximum energy is imparted to the barrier and the entire mass resists, resulting in maximum transmission loss. In practice, however, sound is incident from 0 to 80° (called field or random incidence) reducing the mass effect, but keeping the slope at 6 db per octave. Due to nonhomogeneity, porousness, and stiffness, actual field results indicate transmission losses nearer 4 db per octave as shown by the lower curves in the figure.

27.14 Stiffness and Resonance

The stiffness of a barrier is a function of its material composition and the rigidity of its mounting. The former is dependent on its internal cohesiveness, that is, its modulus of elasticity, and the latter depends on its boundary restraints—whether the barrier is tightly or loosely held. A homogeneous material of high Young's modulus has great cohesiveness between its molecules. As soon as one molecule is set in motion by the incident sound, the motion is passed to the next, and so on, making the material an excellent sound conductor. Homogeneous materials with low modulus of elasticity have high internal damping (motion of molecules is not transmitted well), and they are good sound insulators. Composite materials such as concrete and organic materials such as wood do not conform to these general rules.

The rigidity of mounting can be likened to a drumhead—the tighter it is stretched, the better it resounds. Ridigity (stiffness) in a panel barrier resists damping and assists vibrations, making it a poor insulator. To recapitulate, stiffness, both internal and external, is an undesirable characteristic in sound isolation. A material such as lead, which has high mass and low modulus of elasticity, and resists rigid mounting, is an excellent sound attenuator.

The effects of stiffness and mass both vary with frequency, unfortunately in opposite directions. Stiffness acts to reduce transmission loss as frequency increases while, as we have seen, mass acts to increase it; the combined effect is shown in Fig. 27.14. At very low frequencies the mass and stiffness effects negate each other, giving the resonance dips shown. Beyond approximately 200 Hz most common wall construction enters the mass law range and continues with it until the critical frequency. Deviations from a smooth 4 to 6 db per octave slope are due to the nonhomogeneous nature of most wall construction. At the critical frequency the phase of incident sound waves corresponds or "coincides" with the phase of vibration (shear wave) of the barrier in such a way as to pass a large portion of the incident energy. See insert on Fig. 27.14. This shows in Fig. 27.14 as the "coincidence dip." This effect is most pronounced in thin homogeneous partitions and light stiff ones.

Critical frequency f_c, as a function of panel thickness for common materials, is plotted in Fig. 27.15. To avoid a coincidence dip in the audible range, partitions can be either very heavy and stiff, which greatly decreases the critical frequency, or heavy and limp (resilient), which greatly increases the critical frequency. In practical terms, cost effectiveness is heavily in favor of the latter alternative. Thus, for instance, the TL of a wood partition can be improved by grooving it to increase flexibility. The dramatic improvement in TL, resonances, and coincidence dip achieved by use of resilient mounting of a simple masonry partition is shown in Fig. 27.16. Both walls have the same weight—21 + lb/ft^2. The solid wall A has better attenuation below 200 Hz in the stiffness-controlled range. Above that frequency the resilient-mounted partition is 10 db better. The sound transmission class (STC) (see Section 27.16) of wall A is 40; that of wall B is 51. Both show a coincidence dip at approximately 250 Hz (see Fig. 27.15) but that of wall B is shallow, that of A deep and wide. Furthermore, wall B is consistently better than mass law attenuation, wall A consistently worse (due to stiffness).

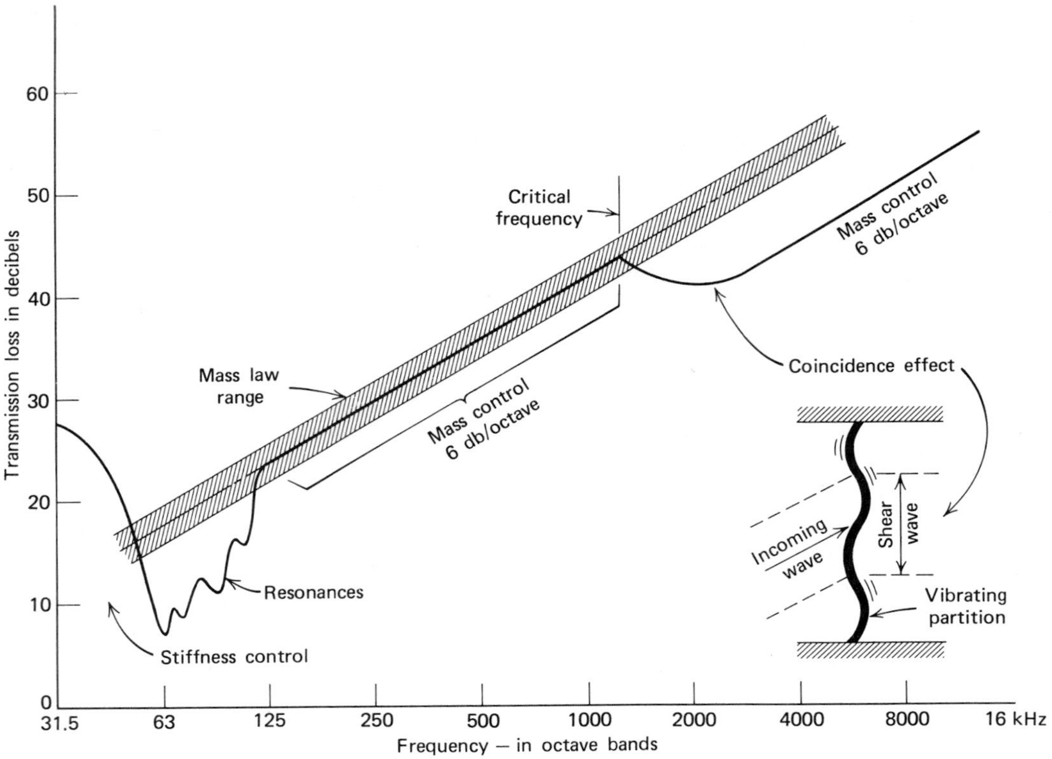

Fig. 27.14 Transmission loss as a function of frequency, assuming constant surface mass.

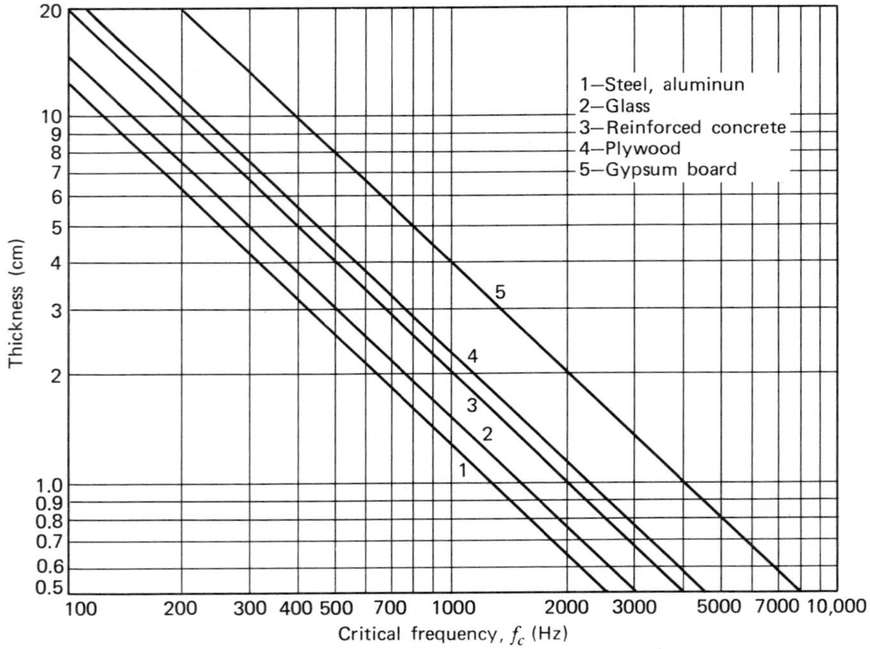

Fig. 27.15 Critical frequency as a function of thickness for several common materials. (Reprinted with permission from E. B. Magrab, Environmental Noise Control, *Wiley, New York, 1975, p. 262.)*

ACOUSTICS

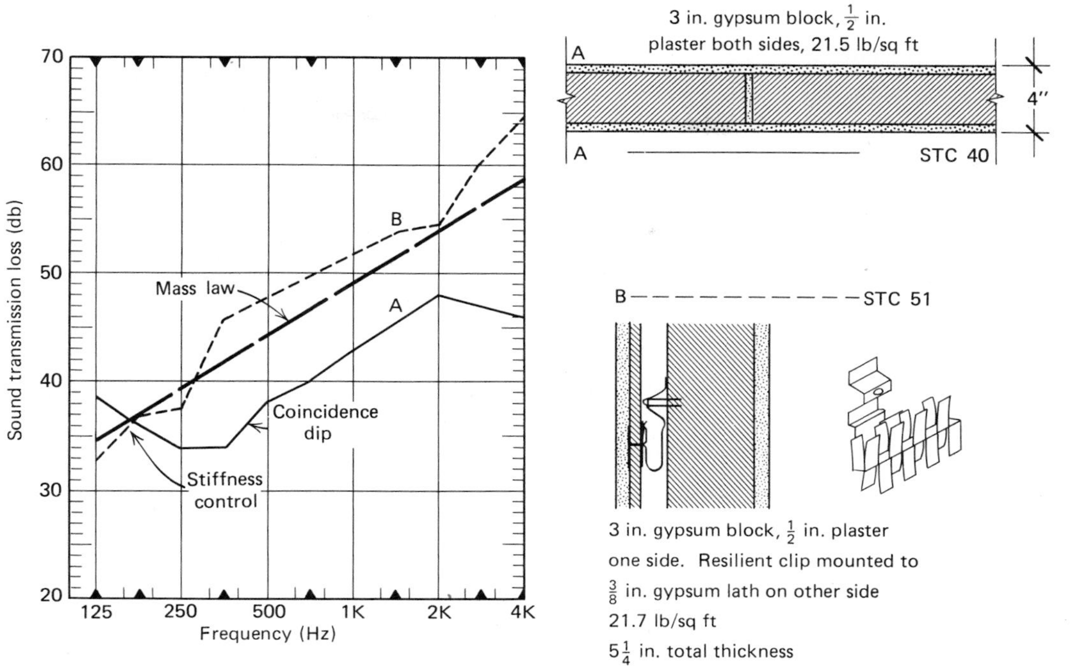

Fig. 27.16 Transmission loss characteristics of two equal weight partitions with similar boundary constraints. The solid partition A is worse than the mass law due to stiffness. The resilient-mounted wall B performs better than the mass law and much better than wall A, except at the lowest frequencies. (Data extracted from A Guide to Airborne; Impact and Structure-Borne Noise Control in Multi-family Dwellings, *1968, pp. 18, 19.)*

27.15 Compound Barriers (Cavity Walls)

Since the maximum theoretical increase in transmission loss with mass increase is 6 db per doubling of mass, it is apparent that this method of transmission loss improvement rapidly reaches the limits of practicality. Indeed as we have seen, actual single homogeneous walls fall below the mass law curve. This is because mass increase brings with it stiffness increase, which as we have seen acts to *reduce* transmission loss. If, however, a barrier is constructed of two separate layers without rigid interconnection, its performance is better than the calculated transmission loss based on mass alone. Note that even the nonrigid wire ties of wall B in Fig. 27.17 lower its STC by five points. At low frequencies, where stiffness controls the transmission loss (see Fig. 27.14), the cavity in *C* acts as a rigid connection between the layers, adding stiffness and increasing transmission loss. At higher frequencies, in the mass law range, the air in the cavity acts as a damping coupling to reduce stiffness. The net result is an improvement in performance throughout the frequency range.

Transmission loss for the entire cavity wall increases with the width of the air space at the rate of approximately 5 db per doubling. Performance can be improved still further by filling the void with porous, sound-absorbent material. This acts to further decrease the stiffness of the compound structure *and* to absorb sound energy that reflects back and forth between the two inside surfaces. The performance of cavity walls is reduced by any rigid interconnections between leaves. Thus a common stud wall with frequent rigid interconnections acts little better than a single homogeneous wall. However, a stud wall with staggered studs exhibits much improved performance over a

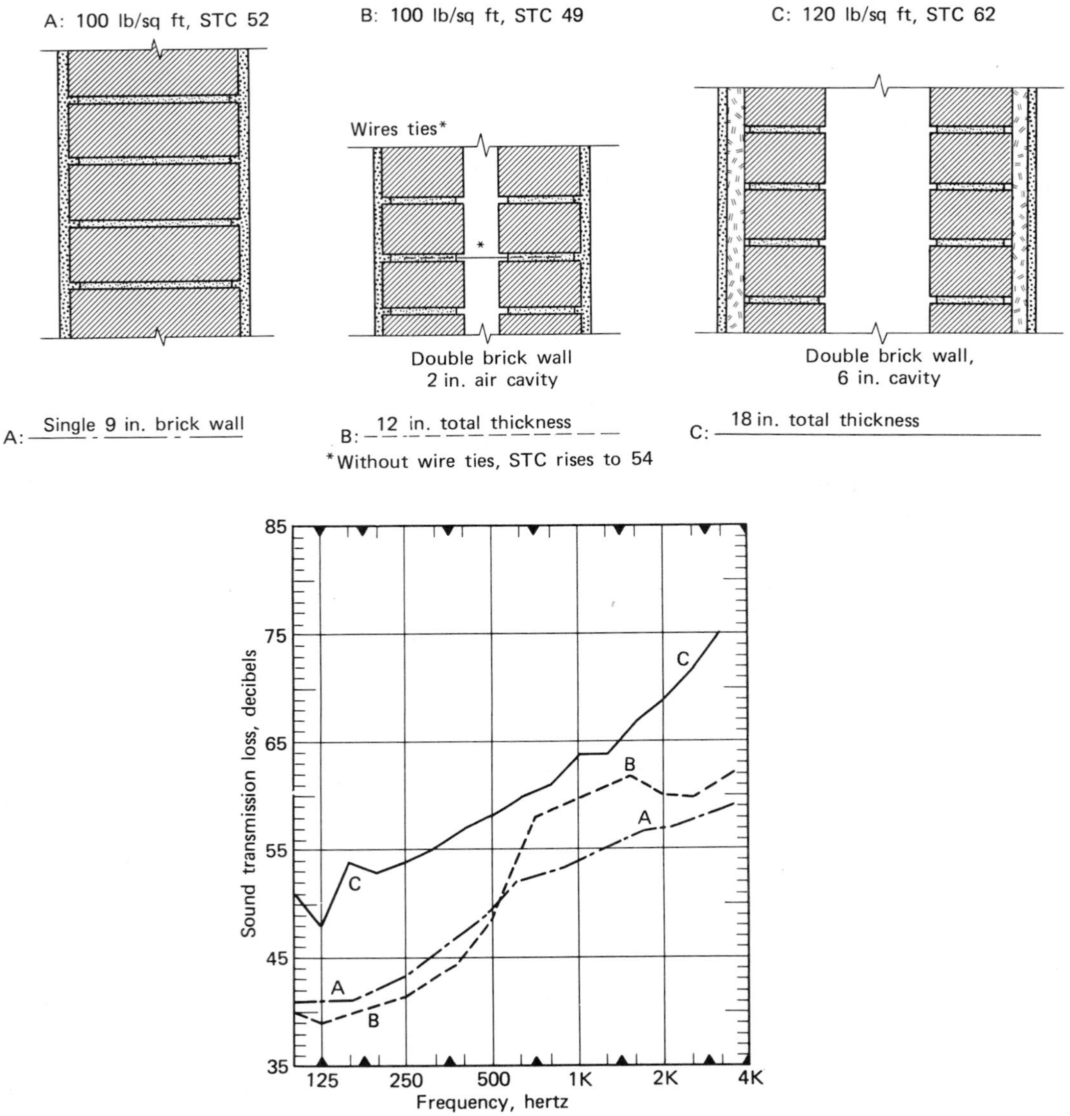

Fig. 27.17 Transmission loss curves showing the effect of air space on heavy wall construction. All three walls are approximately the same mass. The 2-in. air space in wall B is not significant until the higher frequencies, whereas the large, 6-in. air space of wall C is effective throughout the frequency spectrum. Data from A Guide to Airborne, Impact and Structure-Borne Noise Control in Multi-family Dwellings, *1968, pp. W6, 22, 23).*

single-material wall or a common stud wall. The above effects are illustrated in Figs. 27.17 and 27.18 (see also Appendix F).

27.16 Average Transmission Loss (TL); Sound Transmission Class (STC)

Various attempts at using a single-number average transmission loss to describe a barrier's characteristics have been tried but with little success. Indeed these averages can be misleading, since they ignore both deficiencies and proficiencies at particular frequencies. Their use, therefore, in all but rough work is to be discouraged.

To avoid the shortcomings of averages and yet

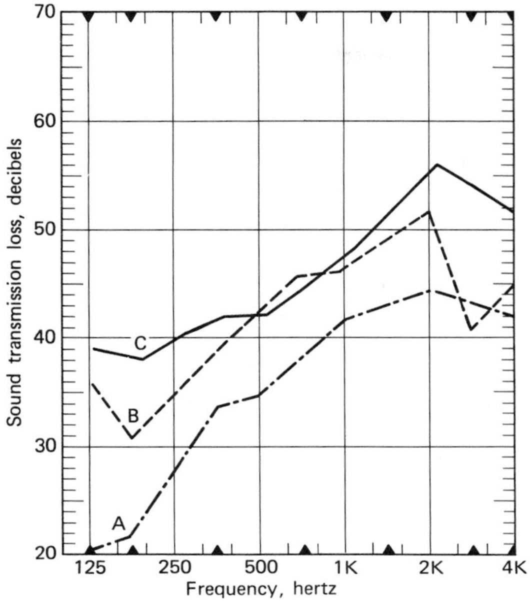

Fig. 27.18 Transmission loss curves illustrating the effect on lightweight walls of stiffness reduction and addition of absorptive material in the cavity. Curve A is a standard stud wall as found in frame construction. Curve B shows the advantage of staggered studs, over the entire frequency range. The dip at 3 kHz is a coincidence dip for a single leaf. Addition of absorption material C improves the attentuation characteristic at both ends and is particularly useful for its low end improvement. (Data from A Guide to Airborne, Impact and Structure-Borne Noise Control in Multi-family Dwellings, *1968, pp. W20, 34, 37.)*

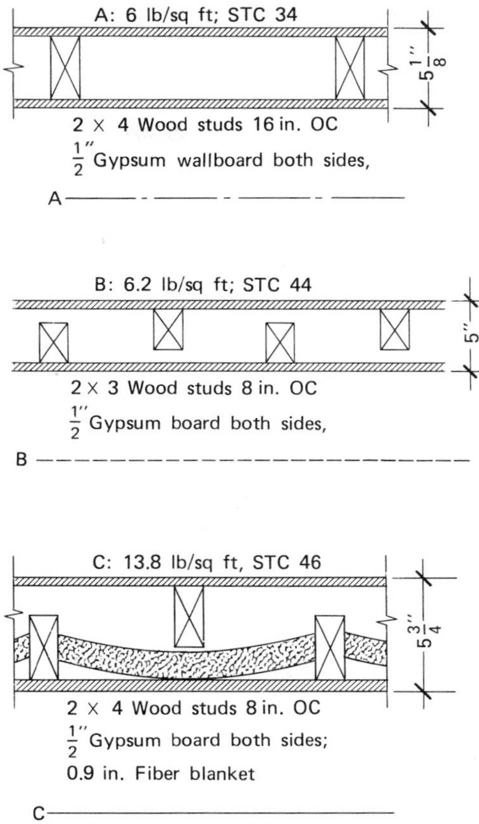

to benefit from the indisputable convenience of single-number ratings, a system of standard contours was developed, called sound transmission class (STC) contours. Actual test results for a given construction, measured in a series of sixteen ⅓-octave bands, are compared to the standard STC contours according to a fixed procedure, and the STC number for that barrier is derived. The technique is illustrated in Fig. 27.19. Figure 27.20 shows two *TL* curves and the STC ratings of each. The nine-frequency average *TL* for each is approximately 40. However, there is a major dip in transmission loss of curve *B*. The STC of *A* is 42 and of *B*, 33. Thus, the STC approach considers a flaw in performance that is ignored by average *TL*. Nevertheless, since STC fails to give credit for performance *above* the established requirements, octave-band transmission-loss data, rather than STC ratings, should be used in all critical work such as music rooms or mechanical rooms where certain particular frequencies may be dominant.

Figure 27.21 gives three standard STC contours that are of interest because they are used by the FHA to specify grades of construction. The criteria for their application are found in Section 27.33. An appreciation of the degree of speech sound isolation provided by walls with different STC ratings is given in Section 27.21 (see Table 27.7). Since the subjective reaction on the quiet side depends on the background noise level there, the table gives this reaction for two NC curve lev-

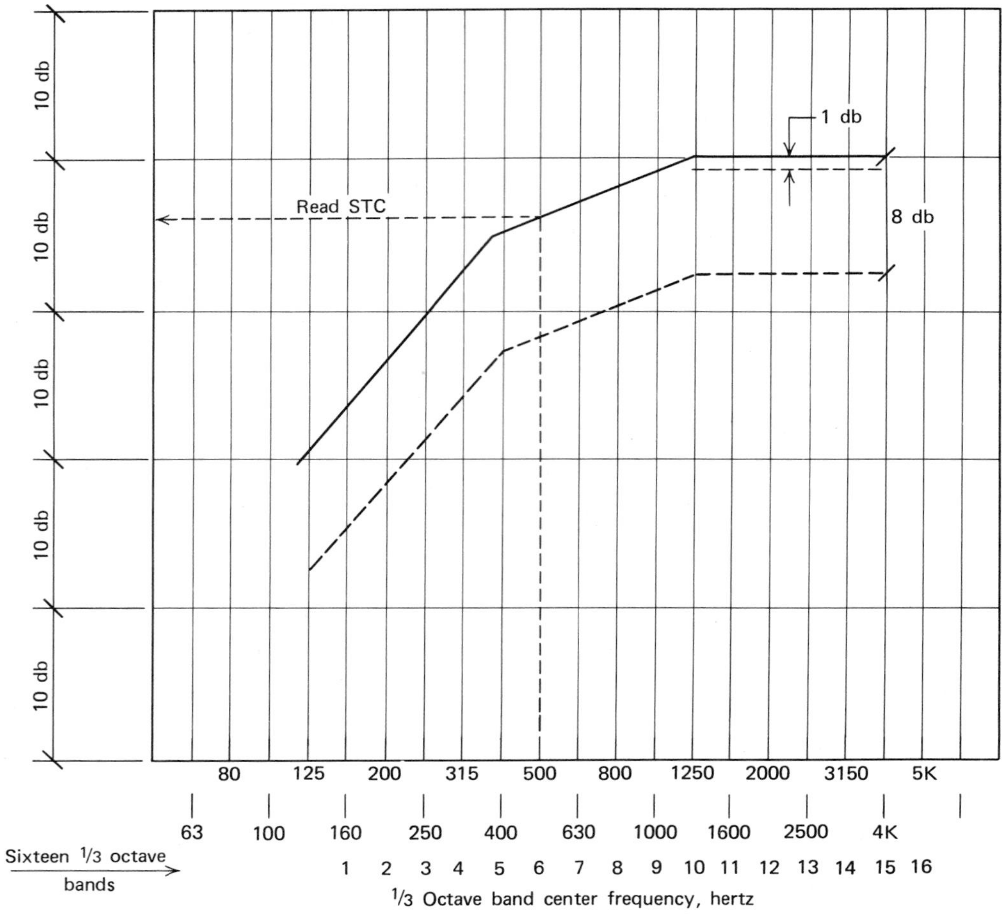

Fig. 27.19 Overlay from which sound transmission class is determined graphically.

els. To assist the designer, extensive sound transmission testing has been performed on most types of standard wall and partition construction and the results published. Table 27.4 and Appendix F give descriptions of construction with typical details, transmission loss data, STC ratings, and other pertinent data.

27.17 Composite Walls and Leaks

It is frequently necessary to determine the transmission loss of a composite wall, that is, a wall with a window, door, vent opening, and the like. It should be clearly appreciated that the two materials are "in parallel," to borrow an electrical concept, and the behavior is similar to that situation. That is, the overall performance will be strongly affected by the poorer of the two, with some tempering of the degradation for small areas. Figure 27.22 enables us to calculate situations of this type.

Since an opening in a wall is effectively a second material of $TL = 0$, the curves of Fig. 27.22 can be replotted as in Fig. 27.23. Note carefully that the curves very rapidly flatten out; thus any wall with 1% open area will have a *maximum* transmission loss of 20 db, which is all but useless as sound insulation. For this reason it is absolutely imperative that all openings be completely sealed, particularly those around doors and windows. A hairline crack degrades a wall 6 db; a keyhole will

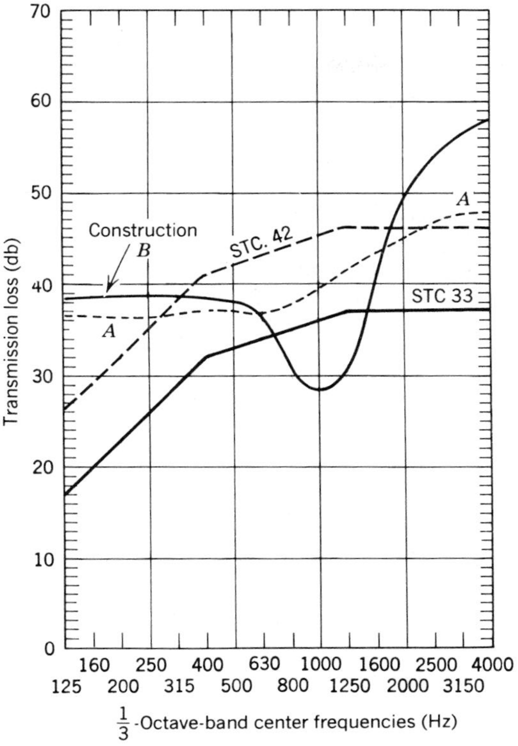

Fig. 27.20 Curves A *and* B *are two different construction types with the same average TL. However, because of its center dip, construction* B *has an STC of only 33, as compared to STC of construction* A.

degrade a door 3 db, and so forth. Special considerations for doors and windows are discussed below. Care must also be taken with such common acoustic leaks as back-to-back electric outlets, pipes passing through walls, and medicine cabinets—in fact any break in the integrity of a partition. All such openings must be caulked to make an airtight joint if any degree of sound isolation is to be maintained. Examples 27.3 and 27.4 show how barriers of high *TL* are degraded by standard openings. To maintain the integrity of a barrier, special care must be taken in the design of windows and doors, as explained below.

EXAMPLE 27.3. Given a wall 9 ft × 18 ft with a transmission loss of 52 db at 1000 Hz, containing a 3 ft × 7 ft 6-in. hollow core door of 22-db

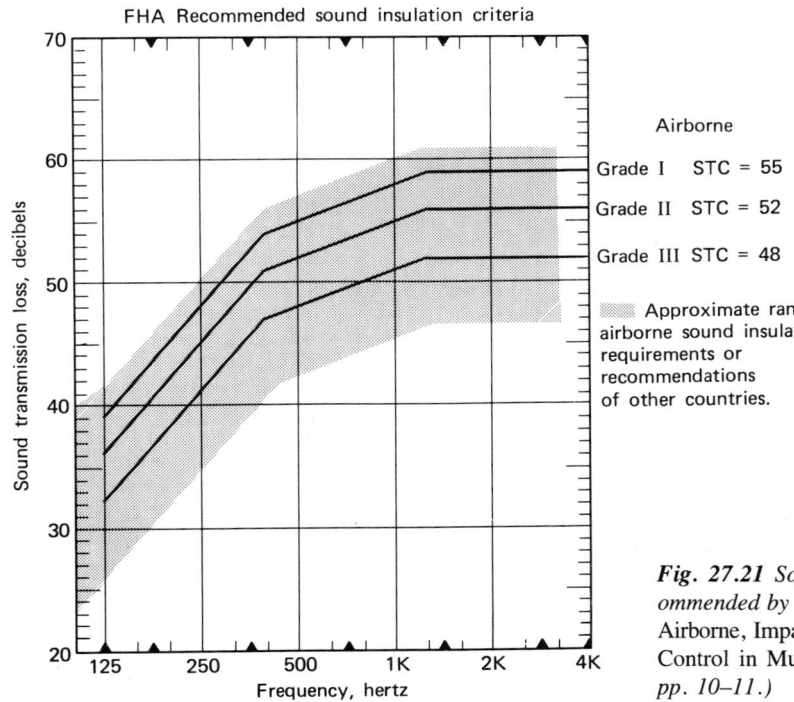

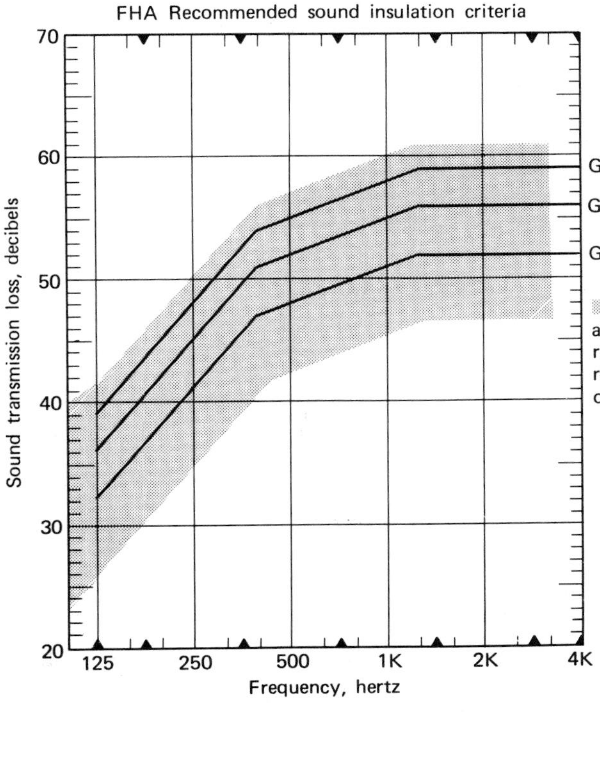

Fig. 27.21 Sound insulation criteria recommended by the FHA. (From A Guide to Airborne, Impact and Structure-Borne Noise Control in Multi-family Dwellings, *1968, pp. 10–11.)*

TABLE 27.4a **Improvements in STC Rating of Stud[a] Partitions[b]**

Description	*STC*[c]
Basic partition: single wood studs, 16 in. on centers, ½-in. gypsum board each side, air cavity	35
Add to basic partition	
Double gypsum board, one side	+2
Double gypsum board, both sides	+4
Single-thickness absorbent material in air cavity	+3
Double-thickness insulation	+6
Resilient channel supports for gypsum board	+5
Staggered studs	+9
Double studs	+13

[a]For application to metal stud partitions, use adders as in note *b*, but begin with STC = 40 for 3⅝-in. basic partition.

[b]When using two improvements add additional +2; for three improvements add +3.

EXAMPLE: Improvements to 35 STC basic partition:

Staggered wood studs	+9
Double gypsum board, one side	+2
Single insulation	+3
Adder	+3
Total	+17
Total STC	35 + 17 = 52

[c]The STC figures are conservative. Other sources list the same constructions with 1 to 5 points higher STC.

TABLE 27.4b **STC Ratings of Masonry Walls**

Description	*STC*[a]
4-in. lightweight[b] hollow block	36
4 in. dense hollow block	38
6-in. lightweight hollow block	41
6-in. dense hollow block	43
8-in. lightweight hollow block	46
8-in. dense hollow block	48
12-in. lightweight hollow block	51
12-in. dense hollow block	53
4-in. brick	41
6-in. brick	45
8-in. brick	49
12-in. brick	54
6-in. solid concrete	47
8-in. solid concrete	50
10-in. solid concrete	53
12-in. solid concrete	56

[a]See note *c*, Table 27.4*a*.

[b]All ratings of lightweight block assume sealing with paint. Note that this reduces absorption.

MODIFICATIONS

Add sand to cores of hollow blocks	+3
Add plaster to one side	+2
Add plaster to both sides	+4
Add furring strips, lath and plaster:	
1 side	+6
2 sides	+10
Add plaster via resilient mounting:	
1 side	+10
2 sides	+15

transmission loss. Find the overall *TL* of the composite wall.

SOLUTION. Refer to Fig. 27.22.

$$TL_1 - TL_2 = 30 \text{ db}$$

$$\frac{S_2}{S_1} = \frac{3 \times 7\frac{1}{2}}{9 \times 18} \times 100 = 13.9\%$$

From curves:

$$TL_1 - TL_c = 21.5$$

$$TL_c = 52 - 21.5 = 30.5 \text{ db}$$

That is, a door with an area of only 14% of the entire wall reduces the *TL* of the structure from 52 to 30.5 db—that is, from excellent to *very poor*.

EXAMPLE 27.4. An exterior brick/frame wall having a *TL* of 54 at 1000 Hz, measuring 8 ft × 16 ft, is pierced by two wood frame windows, each of area 3 ft × 4 ft, with single ⅛-in. glass, and has a *TL* of 34 at 1000 Hz. Find the combined transmission loss.

SOLUTION.

$$TL_1 - TL_2 = 54 - 34 = 20 \text{ db}$$

$$\frac{S_2}{S_1} = \frac{2 \times 3 \times 4}{8 \times 16} \times 100 = 18.75\%$$

From Fig. 27.22,

$$TL_1 - TL_c = 12.5 \text{ db}$$

$$TL_c = 41.5$$

Again the result is a reduction from an excellent wall to a poor one.

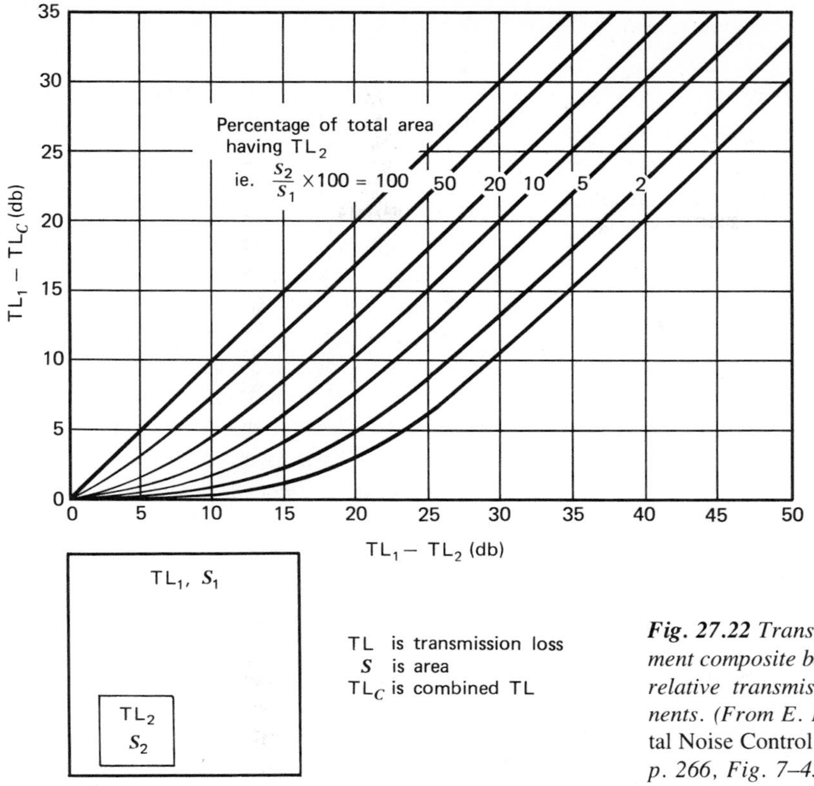

Fig. 27.22 Transmission loss of a two-element composite barrier as a function of the relative transmission loss of the components. (From E. B. Magrab, Environmental Noise Control, *Wiley, New York, 1975, p. 266, Fig. 7–45.)*

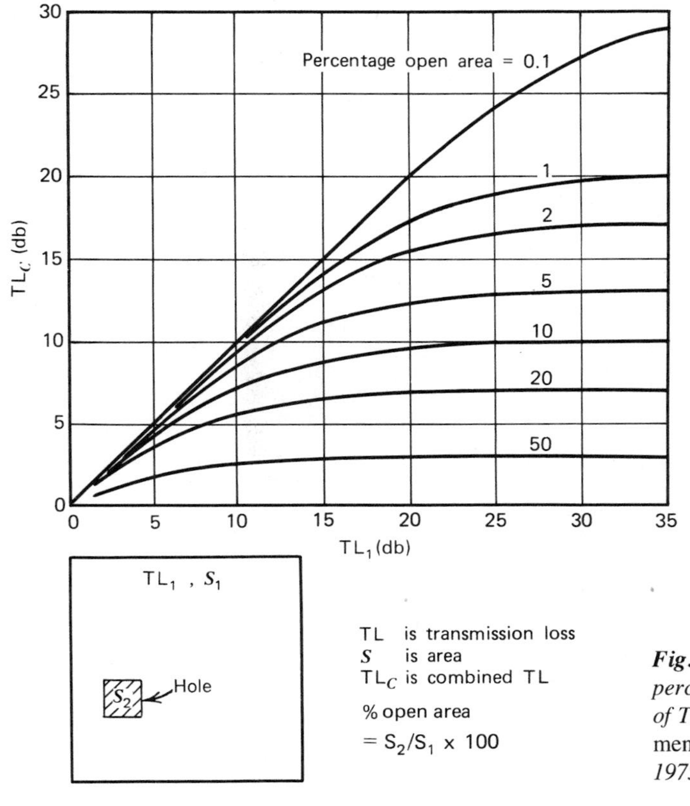

Fig. 27.23 The effect of a hole of a given percentage of the total area on a partition of TL, db. (From E. B. Magrab, Environmental Noise Control, *Wiley, New York, 1975, p. 268, Fig. 7–47.)*

27.18 Doors and Windows

As can be appreciated from the preceding section, doors and windows in large measure determine the overall transmission loss of a wall. Since in almost every instance doors and windows are of lower acoustic rating than that of the wall in which they are mounted, particular care must be taken not to further degrade performance with leaks.

(a) Doors. Table 27.5 gives average *TL* figures for common wooden doors and for doors arranged in a sound lock, that is, two doors with a sufficient space between to allow door swing. As with all transmission loss averages, the number is usable only as an approximate figure of merit. STC is a better criterion; a frequency analysis is best. The important conclusions to be drawn from this table are:

1. Louvered doors and doors undercut to permit air movement are useless as sound barriers.
2. The most important step in sound-proofing doors is complete sealing around the opening. The door in the closed position should exert pressure on these gaskets, making the joints airtight.
3. Doors constructed of two leaves separated by a sound-absorbing material (e.g., steel sheets with fiber batting inside) are more effective than a single solid material door of the same weight.

Commercial products are available with STC rating up to 65 for application in sensitive locations.

(b) Windows. Windows are critically important to block exterior noise, and all the more so, since exterior wall construction is generally of high STC, making the window the deciding factor in the composite transmission loss. Sound leaks through cracks in operable windows will normally establish the windows' rating regardless of type of glazing. Fortunately, the attention now given to the sealing of windows for thermal purposes has had a salutary effect on their acoustic properties. As with doors, the importance of proper gasketing and sealing cannot be overstressed. Double glazing is effective only when the two panes are separated by a wide air gap (see Table 27.6). A narrow air gap acts as a stiff spring between the panes. The resultant TL is approximately that of a single pane of double weight.

In addition to a window's sound transmission characteristic when closed, it is important to consider the TL when open, because of ventilation and natural cooling requirements. The sound attenuation between the center of a room with a clear-through open window and a point some distance outside, is 5 to 15 db. This drops to 3db ± as the receiver–observer approaches the open element. By making the path from inside to outside indirect, the "open" window attenuation can be increased to as much as 25 db, but with consid-

TABLE 27.5 **Average Sound Transmission Loss of Doors**

Construction	*Average TL (db)*	*STC*
Louvered door	10	15
Any door, 2-in. undercut	12	17
$1\frac{1}{2}$ in. hollow core door, no gasketing	15	22
$1\frac{1}{2}$ in. hollow core door, gaskets and drop closure	20	25
$1\frac{3}{4}$ in. solid wood door, no gasketing	25	30
$1\frac{3}{4}$ in. solid wood door, gaskets and drop closure	30	35
2 hollow core doors, gasketed all around with sound lock	38	45
2 solid core doors, gasketed all around, with sound lock	48	55
Special commercial construction, with lead lining and full sealing		45–65

TABLE 27.6 **Sound Transmission Class (STC) of Window Construction**

Window Construction	*STC*
Operable wood sash, ⅛-in. glass, unsealed	25
Operable wood sash, ¼-in. glass, unsealed	25
Operable wood sash, ¼-in. glass, gasketed	30
Operable wood sash, laminated glass, unsealed	28
Operable wood sash, double-glazed, ⅛-in. panes, ⅜-in. air space, gasketed	29
Fixed sash, double ⅛-in. panes, 3-in. air space, gasketed	44
Fixed sash, double ¼-in. panes, 4-in. air space, gasketed	48

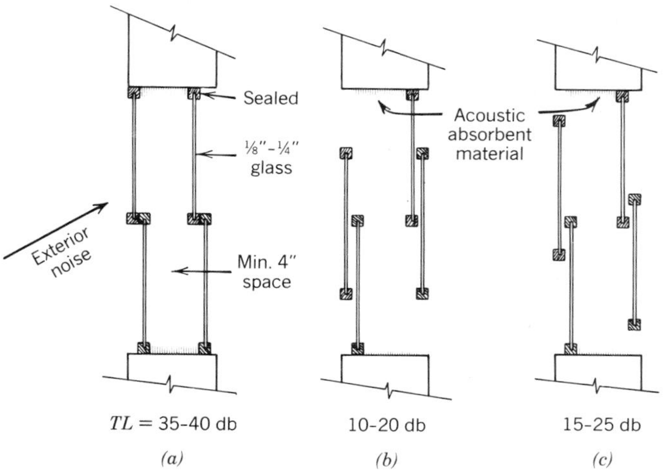

Fig. 27.24 The degree of attenuation of external noise can be regulated when using pairs of double-hung (or horizontally sliding) windows with acoustic sealant and absorbent materials. Ventilation airflow varies inversely with transmission loss.

erable reduction of airflow, hence ventilation. Several possible arrangements with approximate TL figures are given in Fig. 27.24. This principle can be applied advantageously when exterior noise reduction is important, but sealed windows are undesirable. Window opening style and placement can also have an effect on the amount of exterior noise admitted, as shown in Figs. 27.25 and 27.26.

27.19 Diffraction; Barriers

The physical process by which sound passes around obstructions and through very small openings is called diffraction. Simply stated, diffraction is a process whereby any point on a wave front establishes a new wave front when passing an obstacle. Thus, although much of a sound wave is blocked by a small opening, the portion that does get through establishes a new wave front. The *amplitude* of the diffracted wave is affected by the relationship between the size of the opening and the wavelength. Hence for a small hole, short wavelengths (high frequencies) are attenuated less than long wavelengths (low frequencies) (see Fig. 27.27).

When sound encounters a finite-length barrier, it diffracts around and over it, approximately as shown in Fig. 27.28. The attenuation of the diffracted sound depends on frequency, type of source, and the dimensions of the barrier. For a point

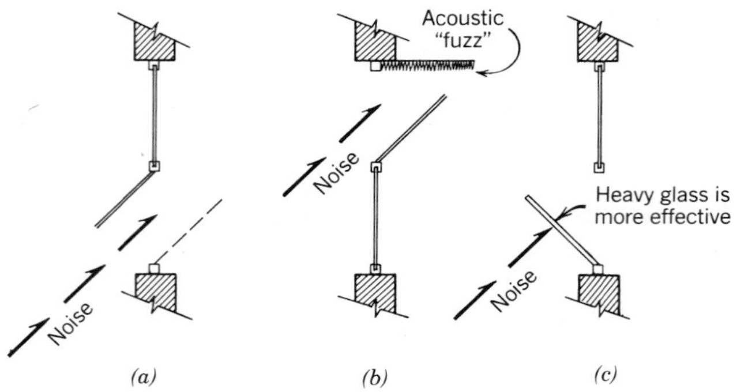

Fig. 27.25 *Alternative arrangements of the same basic "hopper" window design can yield results differing by up to 10 db. Design* (a) *is entirely open, and the noise path is unobstructed deep into the room. Design* (b) *is about 5 db better than* (a) *at frequencies above 1 kHz because of higher absorption and less diffraction. Lower frequencies diffract readily around the window leaf and are less affected by absorptive material. Design* (c) *can be 10 db better than* (a) *because it interposes a rigid barrier into the noise path, particularly at high frequencies. In this arrangement, the glass thickness is important.*

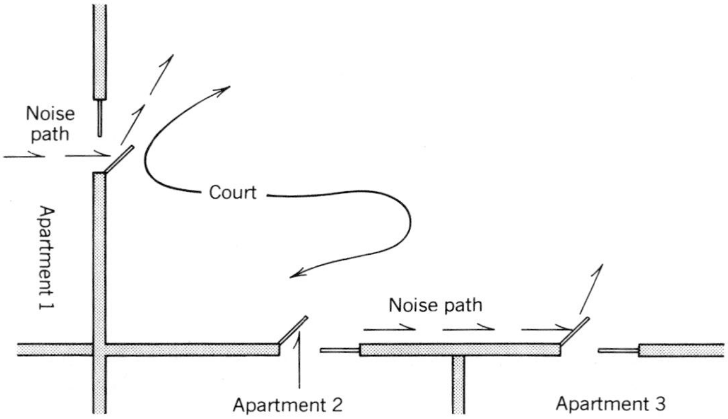

Fig. 27.26 *Noise transfer between continuous corner spaces, as in apartments 1 and 2, can be particularly severe if windows are improperly designed. Swinging windows, as shown, are preferable to double-hung or hopper, because they act to reflect sound away from the adjacent space. Similarly, adjacent spaces on the same wall, as for apartments 2 and 3, can benefit from this type of swinging arrangement, which is preferable to sliding or double-hung designs.*

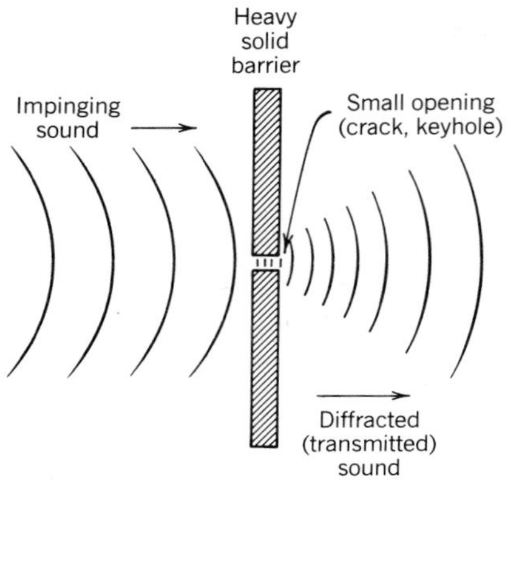

Fig. 27.27 *Sound passes through small openings by diffraction. The intensity of the transmitted sound is proportional both to frequency and to the size of the opening. It is always less than the intensity of the impinging sound.*

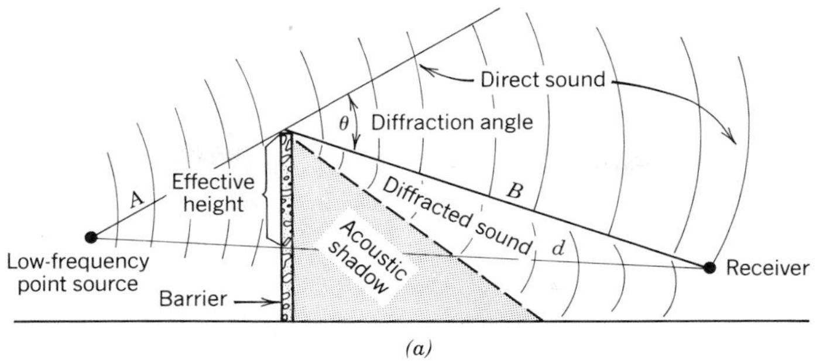

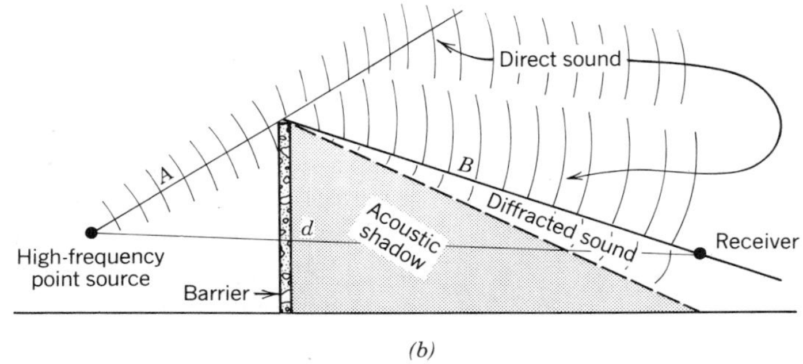

Fig. 27.28 Comparison of the effect of a barrier on sources of different frequencies. The low-frequency sound (a) *diffracts more readily over the barrier than the high-frequency sound* (b) *because of its longer wavelength. Thus the lower the frequency, the smaller the "acoustic shadow" and the lower the barrier attenuation. Of course the "shadow" is not sharply defined; it represents increasing attenuation as the observer approaches the barrier. See Fig. 27.29.*

source, with only a single *practical* path around the obstruction (barrier) the noise reduction in db can be calculated from Maekawa's equation:

$$NR = 20 \log \left[\frac{\sqrt{2\pi N}}{\tanh \sqrt{2\pi N}} \right] + 5 \text{ db} \qquad (27.6)$$

where

$N = (F/565)(A + B - d)$
NR = noise reduciton, db
F = frequency, Hz
$A + B$ = shortest path length around the barrier, ft (over or around)
d = straightline distance, source-to-receiver, ft

Note that this equation:

1. Is applicable only to exterior barriers where sound passing over the barrier is partially diffracted and partially attenuated by distance. In an interior situation, sound passing over the barrier strikes the ceiling and is reflected down, increasing the received sound and effectively reducing barrier attenuation. Maximum exterior barrier attenuation is 24 db as compared to about 15 db for a partial height interior partition.
2. Assumes that the barrier is very long, so that only one sound path exists. In practice, a barrier whose length is at least four times the distance between the source and the wall is sufficient, if *the barrier is close to the source*. If the barrier is close to the receiver, it must be longer still.
3. Assumes a point source. Line sources (such as traffic) show 20 to 25% less attenuation for the same barrier.

The equation will, however, give reliable, usable results when the dimensions of the source are small with respect to the barrier, as is the case for speech; individual motors, fans, engines, and other

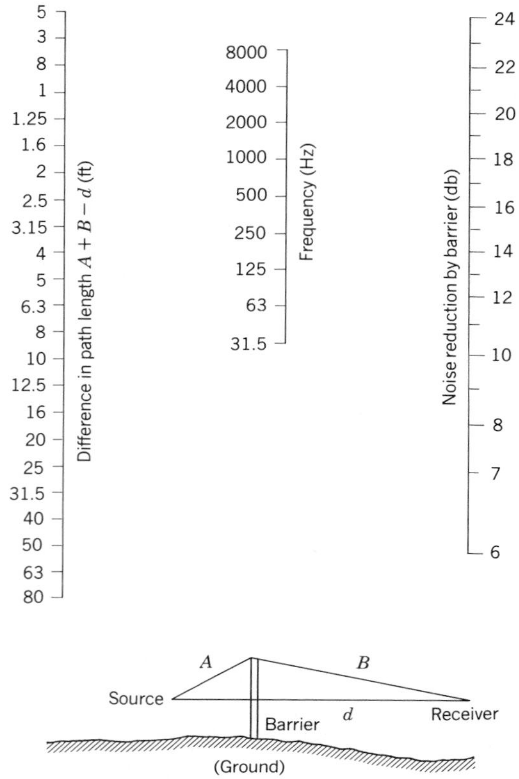

Fig. 27.29 This nomograph to estimate the noise reduction afforded by a barrier is based on Equation 27.6 and assumes a point (small) source and only a single path around the barrier. The dimensions A, B, *and* d *are taken from the insert sketch.* A *plus* B *represent the shortest path around the barrier—which may be over* or around it. *(Reprinted with permission from B. Fader,* Industrial Noise Control. *Wiley, New York, 1981, pp. 148–149.)*

mechanical devices; and individual motor vehicles. The chart in Fig. 3.15 relates barrier dimensions and position to traffic noise reduction. Note there that frequency is not a variable, since the chart has been plotted for an average attenuation at 220 Hz, which is a center traffic frequency.

It should be apparent that the best location for a barrier is either very close to the source or very close to the receiver. The worst position for attenuation is halfway between. All effective barriers are assumed to be opaque and to have a minimum surface density of 5 lb/ft^2 (~20 kg/m^2). The inherent TL of the barrier need not be very high; a massively thick barrier has only marginally higher attenuation than an opaque light one, within the above minimum. Although maximum noise reduction of an *exterior* barrier is about 24 db, in practice it rarely exceeds 20 db. Figure 27.29 is a nomograph based on Equation 27.6.

27.20 Flanking

Just as sound will pass through the acoustically weakest part of a composite wall, so it will also find parallel or flanking paths, that is, an acoustic short-circuit. Proper design of window locations to avoid flanking paths has already been shown in Fig. 27.26. The same situation obtains with respect to doors and any other openings between spaces. Thus in Fig. 27.30 a high STC wall between the two spaces is in large measure defeated by flanking paths F5, F6, and F7. In office spaces the most common flanking path is via the plenum, as in Fig. 27.30, path F1, and Fig. 27.31*b* and *d*. Ductwork with registers or grilles in various rooms acts as an excellent intercom system unless completely lined with sound-absorptive material (see Section 27.28). Even then, low-frequency sound is only minimally attenuated, and special measures must be employed if good transmission loss is required. This subject is discussed further below, under Mechanical System Noise Control.

27.21 Noise Reduction and Background Noise

Referring back to Section 27.12, we saw that

$$IL_2 = IL_1 - NR \qquad (27.4)$$

where NR is noise reduction and IL_2 and IL_1 are sound intensity levels in the receiving and source rooms, respectively. If the resulting IL_2 is below the background noise level, it will not be a source of annoyance; conversely, if it is greater than the background noise it will be heard and, depending on intensity and content, could be a source of disturbance. See Fig. 27.32 for a graphic representation of this idea.

With this in mind, we can also tabulate the performance of a sound barrier, in terms of receiving-room background levels (see Table 27.7).

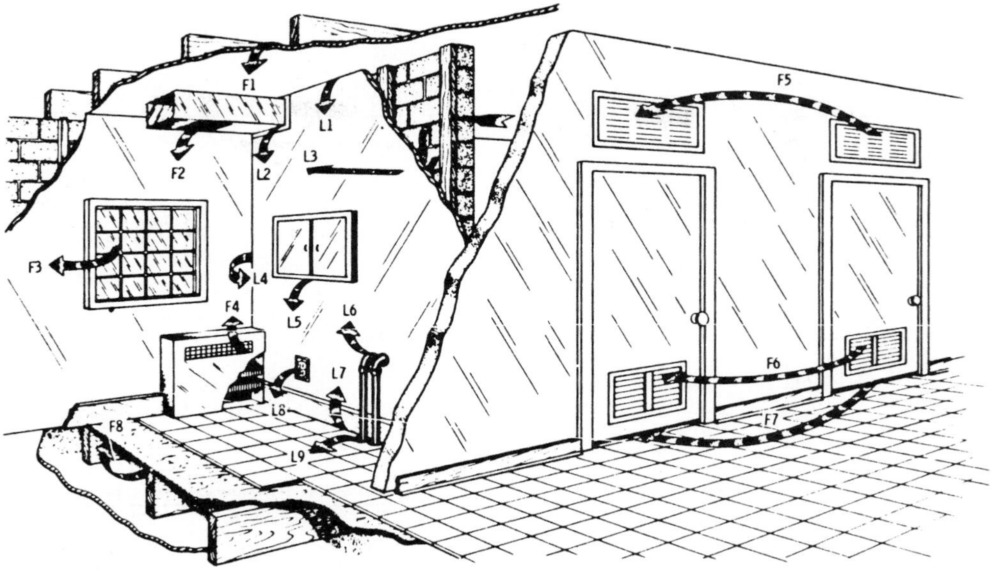

FLANKING NOISE PATHS

F1 OPEN PLENUMS OVER WALLS, FALSE CEILINGS
F2 UNBAFFLED DUCT RUNS
F3 OUTDOOR PATH, WINDOW TO WINDOW
F4 CONTINUOUS UNBAFFLED INDUCTOR UNITS
F5 HALL PATH, OPEN VENTS
F6 HALL PATH, LOUVERED DOORS
F7 HALL PATH, OPENINGS UNDER DOORS
F8 OPEN TROUGHS IN FLOOR-CEILING STRUCTURE

NOISE LEAKS

L1 POOR SEAL AT CEILING EDGES
L2 POOR SEAL AROUND DUCT PENETRATIONS
L3 POOR MORTAR JOINTS, POROUS MASONRY BLK
L4 POOR SEAL AT SIDEWALL, FILLER PANEL ETC.
L5 BACK TO BACK CABINETS, POOR WORKMANSHIP
L6 HOLES, GAPS AT WALL PENETRATIONS
L7 POOR SEAL AT FLOOR EDGES
L8 BACK TO BACK ELECTRICAL OUTLETS
L9 HOLES, GAPS AT FLOOR PENETRATIONS

OTHER POINTS TO CONSIDER, RE: LEAKS ARE (A) BATTEN STRIP A/O POST CONNECTIONS OF PREFABRICATED WALLS, (B) UNDER FLOOR PIPE OR SERVICE CHASES, (C) RECESSED, SPANNING LIGHT FIXTURES, (D) CEILING & FLOOR COVER PLATES OF MOVABLE WALLS, (E) UNSUPPORTED A/O UNBACKED WALL BOARD JOINTS (F) EDGES & BACKING OF BUILT-IN CABINETS & APPLIANCES, (G) PREFABRICATED, HOLLOW METAL, EXTERIOR CURTAIN WALLS.

Fig. 27.30 *Flanking transmission of airborne noise. (Reprinted from* A Guide to Airborne, Impact and Structure-Borne Noise Control in Multi-family Dwellings, *1968.)*

Stated otherwise, the apparent isolation provided by a barrier may be greater than the actual noise reduction. Figure 27.33 shows two conditions of adjacent spaces. Although the source room level is uniform and partitions on both sides of the source room are identical, the background noise in the two receiving rooms is different. In *A*, the background is NC-35; in *B*, it is NC-25. The occupant of room *A* hears nothing from the source room. The occupant of room *B* hears clearly. Occupant *A* probably will praise the partition while occupant *B* will complain. Although the levels of reradiated sound are identical in the two receiving rooms, the intruding signal is masked by the background noise in *A* but it is clearly audible in *B*. Thus, the apparent noise reduction is substantially higher in *A* than in *B*. Section 27.22 investigates this phenomenon of speech privacy.

Sound isolation may be achieved very economically by careful planning. Storage and circulation can serve as buffers for noise-sensitive areas. Physical separation of noisy areas from quiet ones often eliminates the need for complicated and expensive compound barriers.

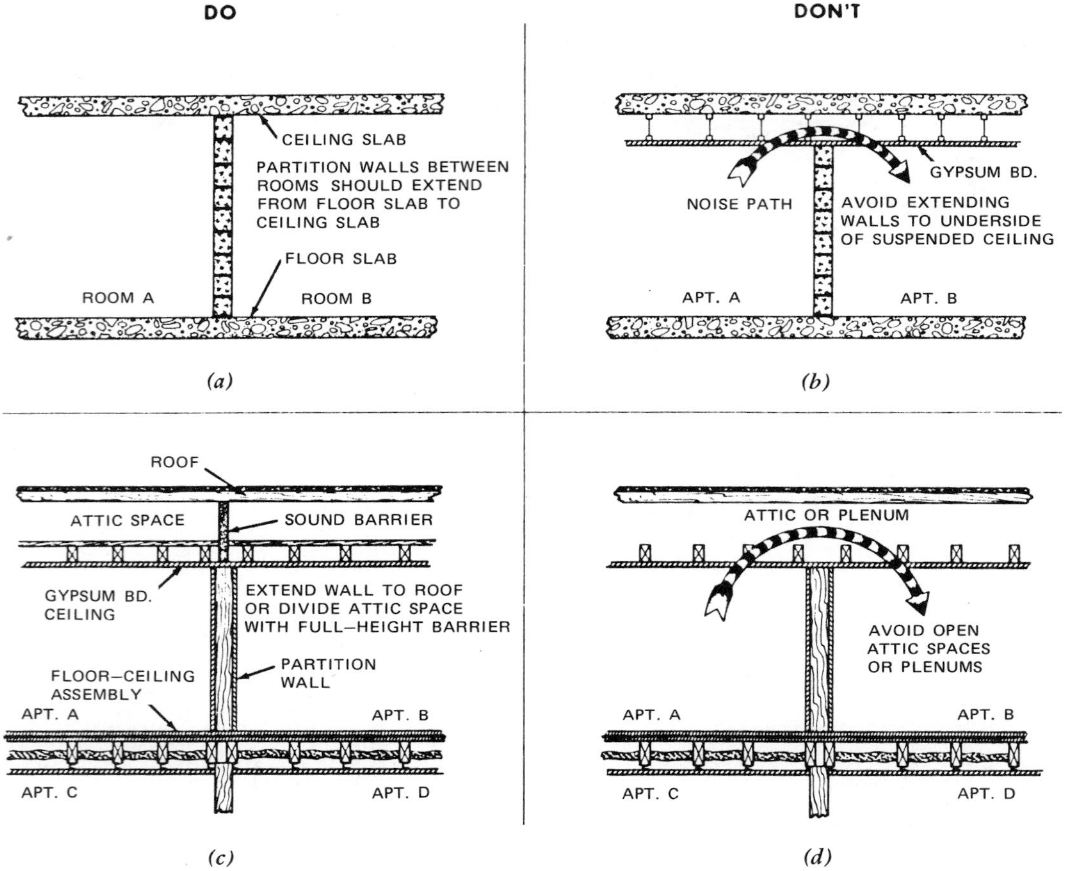

***Fig. 27.31** Construction, techniques to avoid flanking paths. (Reprinted from* A Guide to Airborne, Impact and Structure-Borne Noise Control in Multi-family Dwellings, *1968.)*

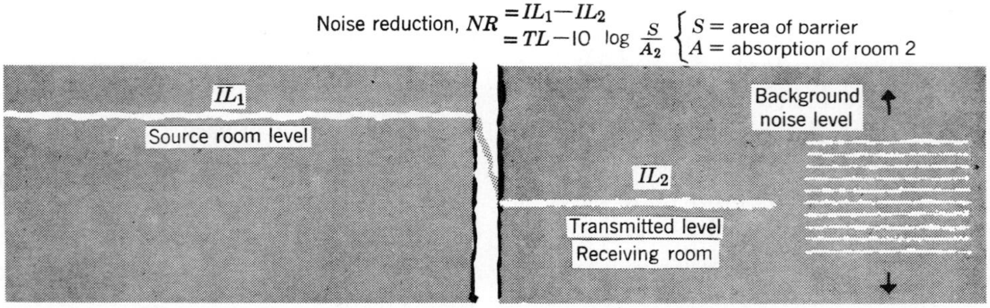

***Fig. 27.32** The background noise level determines whether the transmitted sound will actually be heard.*

TABLE 27.7 **Relations Between Barrier STC Rating and Hearing Condition on the Receiving Side**

Barrier STC[a]	*Hearing Condition*	*Description*	*Application*
25	Normal speech can be understood quite easily and distinctly through the wall.	Poor	Space divider
30	Loud speech can be understood fairly well. Normal speech can be heard but not easily understood.	Fair	Room divider where concentration not essential
35	Loud speech can be heard, but is not easily intelligible. Normal speech can be heard only faintly, if at all.	Good	Suitable for offices next to quiet spaces
42–45	Loud speech can be faintly heard but not understood. Normal speech is inaudible.	Very good	For dividing noisy and quiet areas; party wall between apartments
46–50	Very loud sounds, such as loud singing, brass musical instruments, or a radio at full volume can be heard only faintly or not at all.	Excellent	Music room, practice room, sound studio, bedrooms adjacent to noisy areas

[a]Assuming a background noise level of NC-25. With higher background, (e.g., NC-35), STC can be degraded one step.

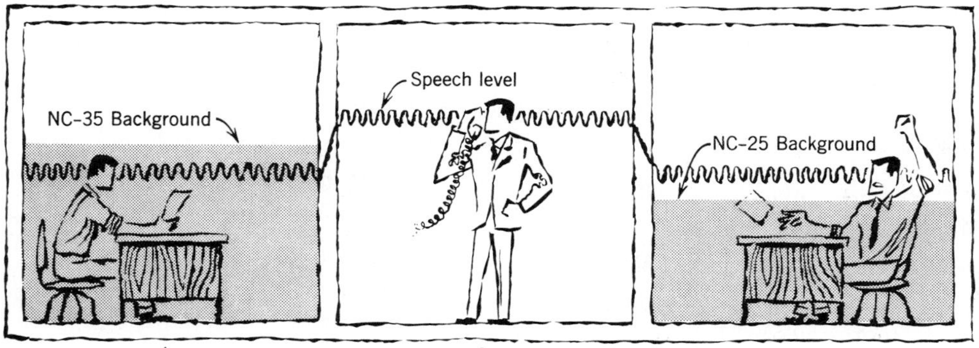

A Background noise masks intruding speech
C
B Low background noise allows speech to intrude

Fig. 27.33 *The occupant in room A with background noise NC-35 (≈ 45 dbA) is unaware of the noise that is so disturbing to occupant B, whose NC-25 (≈ 36 dbA) is insufficient to mask C's loud speech.*

27.22 Speech Privacy

The subject of speech privacy received considerable additional study and emphasis with the advent of open-plan offices (office landscaping), although the same problem prevails in all spaces. Essentially the purpose of the investigations was to determine the factors affecting speech privacy and to quantify them with a degree of accuracy sufficient for design purposes. These studies indicate that there are six factors involved, which can be subsumed under two headings:

1. Speech rating (source room, No. 1)
   a. Speech effort—a measure of the loudness of speech.
   b. Source room floor area—gives the approximate effect of room absorption.
   c. Privacy allowance—What is the measure of privacy required?
2. Isolation rating (receiving room, No. 2)
   d. STC—rating of the barrier.
   e. Noise reduction factor A_2/S is an indication of receiving-room absorption, that is, the difference between NR and TL; A_2 is area of receiving room; S the area of the barrier between rooms. Absorption is assumed to be average. For live rooms raise this factor two points; for a dead room lower two points.
   f. Background noise level—in receiving room.

An analysis sheet for enclosed spaces is provided in Fig. 27.34. See also Cavanaugh et al. (1962) and Young (1965). The two examples of this analysis that follow should clarify its use. The reader should follow the analysis with Fig. 27.34 in hand. The numbered steps in the examples correspond to the numbers in the figure.

EXAMPLE 27.5

(*a*) Source room:

General clerical office, 40 × 60 × 9 ft

16-ft partition, STC 40

(*b*) Receiving room:

Conference rooms, 16 × 24 ft

Background noise level, 40 dbA (NC-30)

Privacy analysis

(*a*)	1. Speech effort: raised	66
	2. Area $A_1 > 1000$ ft^2	0
	3. Privacy—normal	9
	Speech rating:	75
(*b*)	4. STC	40
	5. A_2/S	2
	[(16 × 24)/(16 × 9) = 2.6, corresponding to 2.0]	
	6. Noise level	40
	Isolation rating:	82
	$a - b =$	−7

Therefore the STC rating of the partition can be reduced to 33 without affecting speech privacy.

EXAMPLE 27.6

(*a*) Source Room:

Drafting room 20 × 30 ft
Common wall 12 × 8 ft high
STC: 26 (half glass, with door)

(*b*) Receiving room:

Supervisor's office, 12 × 14 × 8 ft
Background noise level, 35 dbA

Privacy analysis:

(*a*)	Speech effort—conversational	60
	Source room area A_1	2
	Privacy—confidential	13
		75
(*b*)	STC	26
	A_2/S	2
	[(12 × 14)/(12 × 8) = 1.8, corresponds to 1.6; use 2.0, since system uses whole numbers only.]	
	Noise level	35
		63
	$a - b =$	12
	or strong dissatisfaction	

The suggested corrections here are to increase STC to 36 by gasketing the door and increasing background noise level in the receiving room to 40 dbA (NC-30). This should result in a satisfactory con-

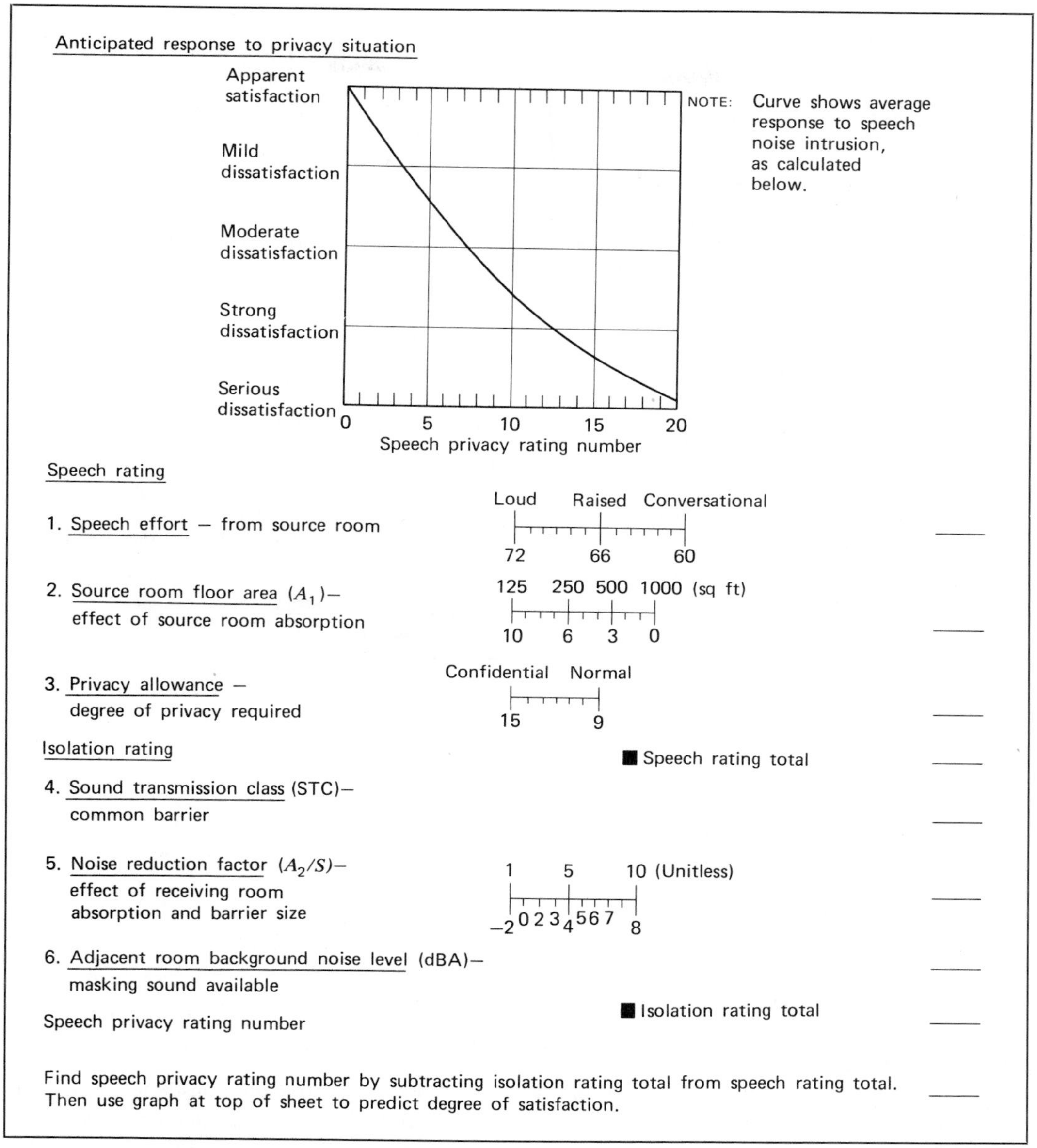

Fig. 27.34 *Speech privacy analysis sheet. (Based on W. J. Cavanaugh et al., 1962.)*

dition. If still unsatisfactory, the glazing could be doubled.

A form similar to Fig. 27.34 is available for open plan offices. We have not included it because these spaces require critically accurate design best left to an acoustics expert. In these open spaces where partial height partitions can provide a *maximum* of 15 db isolation, masking noise of fairly high level must be introduced. Too high a level will result in the noise itself being disturbing; too low a level results in speech interference, extreme dissatisfaction, and impediments to work. Since in such areas, control of background noises such as office machines is extremely difficult, success depends in large measure on proper physical

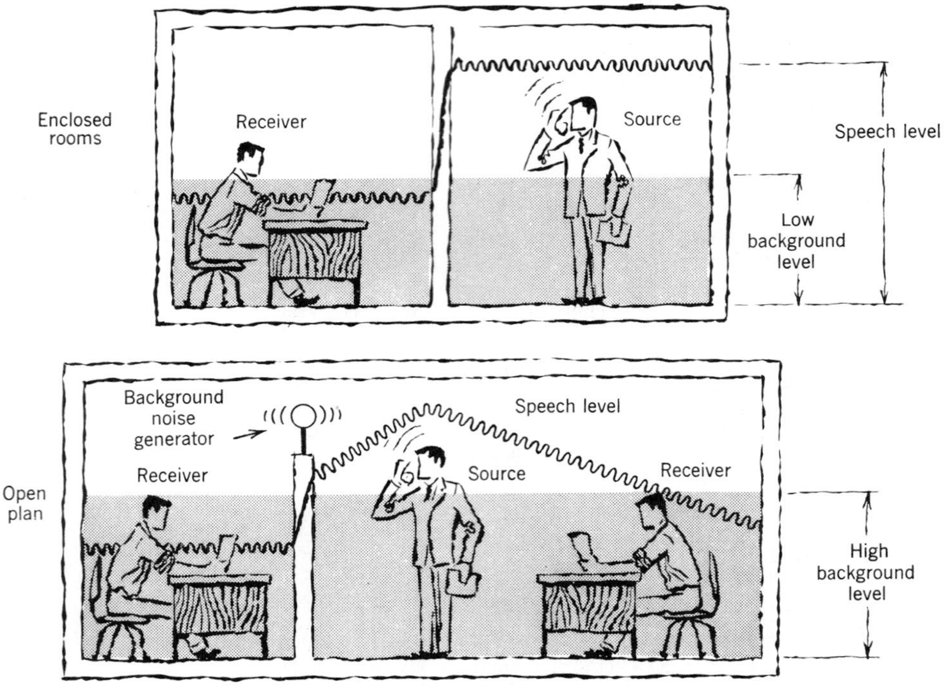

Fig. 27.35 The open-plan background noise level must be high enough to mask adjacent speech but not so loud that it constitutes a disturbance itself.

placement of noise sources and grouping of common acoustic need areas. The indiscriminate reliance on noise generators is to be discouraged (see Fig. 27.35). Generally speaking, open areas should use barriers at least 5 × 8 ft of very high absorbency and acoustic ceilings no higher than 9 ft with a minimum absorption coefficient of 0.7, backed by a plenum at least 30 in. deep. The partial height barriers must be set so that the listener is shielded from line-of-sight noise, either direct or on first reflection. Obviously the taller the barrier, the better.

STRUCTURE-BORNE SOUND

27.23 Structure-Borne Impact Noise

As stated in Section 27.11, structure-borne noise is at least as serious a problem as airborne noise because:

1. There is no air cushion between the source and the structure. Thus high-intensity energy is introduced into the structure, through which it travels with minimum attenuation and at great speed.

2. Sound, once introduced into the structure, is attenuated well only by discontinuities in the structure. Since the structure must have integrity to carry the loads, discontinuities of the type that will stop noise are complex and expensive.

3. The entire structure constitutes a network of parallel paths for sound. Therefore, partial solutions are useless, since sound will find flanking paths. The entire structure must be soundproofed to yield good results.

4. Unlike the case of airborne noise, additional mass does not usually alleviate floor-borne noise, particularly in long spans where the floor acts as a diaphragm.

5. The increasing use of exposed structural ceiling eliminates the attenuation that can be introduced into a plenum above a hung ceiling. This is particularly bad, since most structure-borne noise

is carried by floor structures (rather than walls), which radiate sound up and down.

The discussion that follows will be limited to impact noise. Refer to Section 27.27 for a brief treatment of vibration that is felt rather than heard, and is in effect a very low-frequency noise. Many of the recommendations and techniques that will minimize impact noise will also reduce vibration.

27.24 Control of Impact Noise

Impact noise problems can be controlled in two ways—by preventing or minimizing the impact, and by attenuating it once it has occurred. Since we are concerned with structure, the latter problem is discussed first. The former problem is covered in Section 27.26. Impact on floors is more serious than wall impact because the latter will be partially attenuated at the wall/floor joint, whereas the former is introduced directly into the building framework.

The discussion below addresses each of the solutions shown in Fig. 27.36.

(a) Cushion the Impact. See Fig. 27.36*a*. This obvious solution will frequently eliminate all but severe problems. The resilient materials in common use are floor tiles of vinyl, rubber, and cork, or carpeting on pads, in ascending order of impact insulation. See Section 27.25 and Appendix G for quantitative data on impact isolation.

(b) Float the Floor. See Fig. 27.36*b*. Since the key to elimination of structure-borne sound is *isolation,* separating the impacted floor from the structural floor by a resilient element is extremely effective. This element can be rubber or mineral wool pads, or blankets, or special spring metal sleepers. The effectiveness depends on the mass of the floating floor, compliance of the resilient support, and degree of isolation of the floating floor. This last element is extremely important, since flanking paths via end contacts with walls can short-circuit the floating element's sound impedance and defeat the system. With floating floors it is important that:

1. Mass of the floating floor be large enough to properly spread the loads. Otherwise, the pad will compress and deform sufficiently to transmit the impact.
2. Total construction must be airtight. Airtight is soundtight.
3. Particular care be exercised where partitions rest on the floating floor (see Fig. 27.37).
4. Short-circuits at walls or by penetrations be avoided; see Fig. 27.11*b*. Details of proper construction techniques are given in *A Guide to Airborne, Impact and Structure-borne Noise Control in Multi-family Dwellings* (see References).
5. Construction throughout be consistent. Mixed construction types invite flanking noise paths. (See Fig. 27.38).

(c) Suspend the Ceiling—and Use Absorber in Cavity. See Fig. 27.36(*c,d*). As stated, the most disturbing noise is radiated down from the ceiling. A flexibly suspended ceiling with an acoustic absorbent layer suspended in it can be very effective if not flanked by paths leading into the walls and from there reradiating into the space

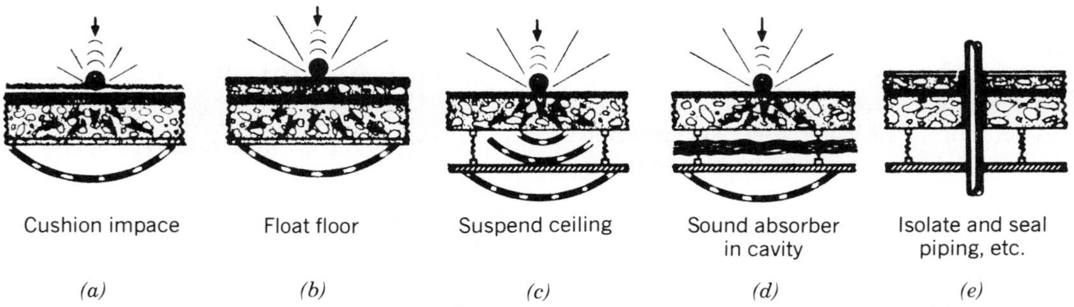

Fig. 27.36 Methods of controlling impact sound transmission through floors. (Reprinted Quieting: A Practical Guide to Noise Control, *1976, p. 50.)*

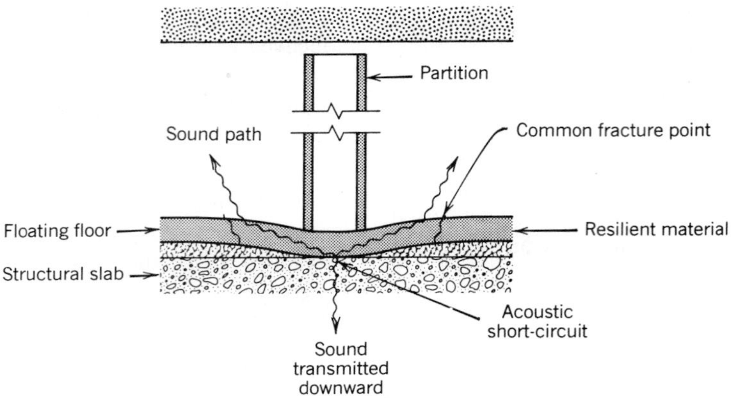

Fig. 27.37 Caution must be exercised when supporting partitions on floating floors to prevent structural failures or short-circuiting of the floating element, as illustrated. (Reprinted from A Guide to Airborne, Impact and Structure-Borne Noise Control in Multi-family Dwellings, *1968.)*

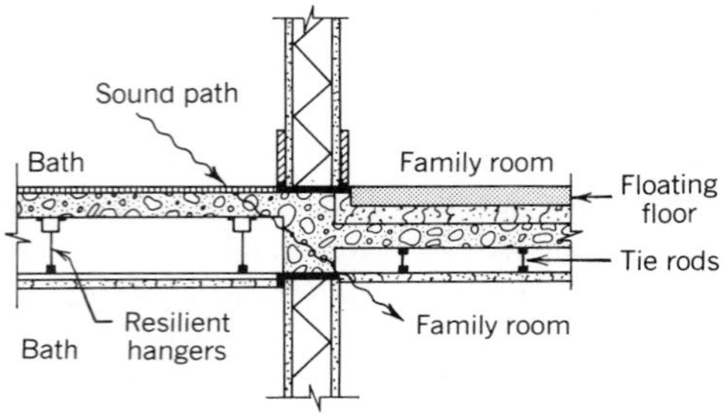

Fig. 27.38 Flanking paths in mixed-construction type floors. FHA does not recommend mixing construction types unless provisions (e.g., expansion joints or breaks in all structural paths between each space) have been made to prevent flanking. (From A Guide to Airborne, Impact and Structure-Borne Noise Control in Multi-family Dwellings, *1968.)*

below. It is imperative that the entire floor slab above be decoupled from the walls below by resilient separators.

(d) Isolate all Piping. See Fig. 27.36*e*. All rigid structures such as piping must be isolated so as not to form a flanking path, and caulked with resilient sealing so as not to constitute an air–sound leak.

27.25 Impact Isolation Class (IIC)

This is a single-number, impact isolation rating for floor construction, similar in intent and derivation to STC wall ratings. Tests are made with a standard tapping machine and the noise levels measured in ⅓-octave bands. These are plotted and compared to a standard contour, exactly as with sound transmission class. Figure 27.39 gives the standard contour. Details of typical floor constructions along with IIC ratings are given in Appendix G. Resilient floor finishes on any of the floor constructions not provided with them will add to the IIC ratings approximately as follows:

1/16-in. vinyl tile	0
⅛-in. linoleum or rubber tile	4 ± 1
¼-in. cork tile	10 ± 2
Low-pile carpet on fiber pad	12 ± 2
Low-pile carpet on foam rubber pad	18 ± 3
High-pile carpet on foam rubber pad	24 ± 3

MECHANICAL SYSTEM NOISE CONTROL

27.26 Mechanical Noise Sources

Mechanical devices obviously make noise. And, generally, the more power they consume, the more noise they make. In many of today's buildings, 40% of the total cost is spent on mechanical systems. These systems are located throughout a building.

In most buildings, the primary sources of me-

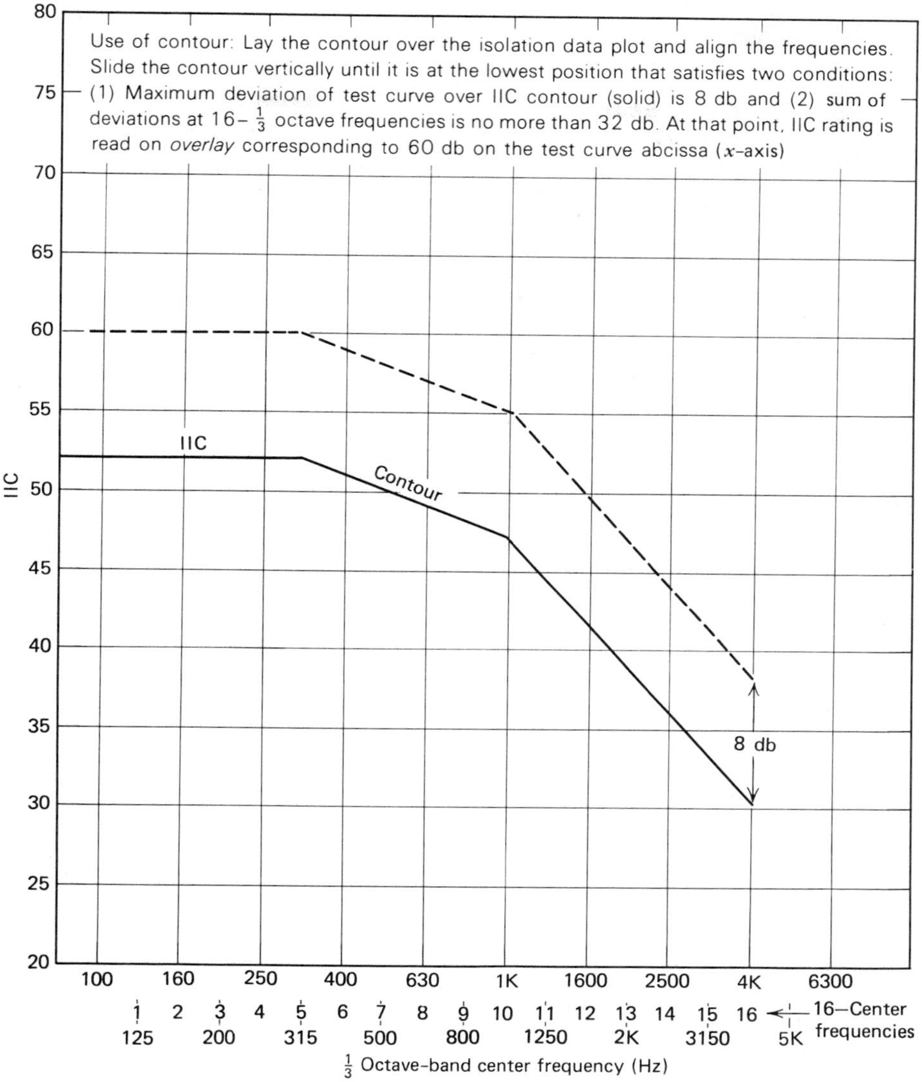

Fig. 27.39 Overlay from which impact isolation class can be determined graphically. (Reprinted from A Guide to Airborne, Impact and Structure-Borne Noise Control in Multi-family Dwellings, *1968.)*

chanical noise are the components of the air conditioning and air-handling systems such as fans, compressors, cooling towers, condensers, ductwork, dampers, mixing boxes, induction units, and diffusers. The curve of Fig. 27.40 depicts typical air-handling system noise and indicates the portions of the spectrum produced by each group of components.

Pumps are another source of mechanical noise.

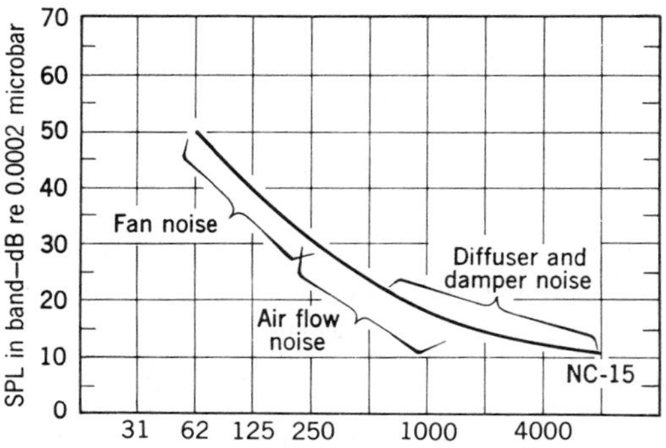

Fig. 27.40 Frequency spectrum of noise of HVAC system components.

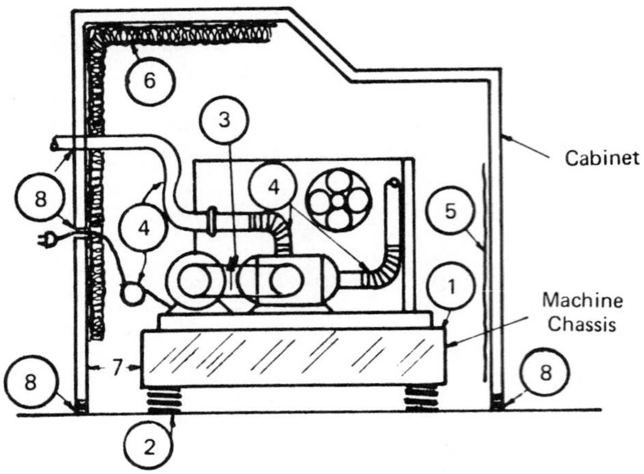

Fig. 27.41 *Techniques to reduce the generation of airborne and structure-borne noise in machines and appliances. (Reprinted from* A Guide to Airborne, Impact and Structure-Borne Noise Control in Multi-family Dwellings, *1968.)*

1. Install motors, pumps, fans, etc. on most massive part of the machine.
2. Install such components on resilient mounts or vibration isolators.
3. Use belt drive or roller drive systems in place of gear trains.
4. Use flexible hoses and wiring instead of rigid piping and stiff wiring.
5. Apply vibration damping materials to surfaces undergoing most vibration.
6. Install acoustical lining to reduce noise buildup inside machine.
7. Minimize mechanical contact between the cabinet and the machine chassis.
8. Seal openings at the base and other parts of the cabinet to prevent noise leakage.

Pump noise is frequently transmitted along pipes to remote points.

Elevators, escalators, and freight elevators also introduce mechanical noise into buildings. Escalators and freight elevators pose few problems, since they are localized in a specific area and have low operation speeds. However, elevator car operation is rapid, and it affects large areas. In addition, the motors and switchgear are located on or above the prime upper floors of a building. Motor, shaftway, and switching noise must be properly controlled to prevent annoyance to building tenants located near the shaftways or mechanical penthouses. Vibration isolation of these major components is a specialized problem beyond the scope of this text.

27.27 Quieting of Machines

Machines cause noise by vibration. This noise is imparted directly to the surrounding air and by vibrational contact to the surrounding structure. Therefore, there are three ways to reduce this noise:

1. Reduce the vibration itself.
2. Reduce the airborne noise by decoupling the vibration from efficient radiating sources.
3. Decouple the vibrating source from the structure.

Refer to Fig. 27.41: Items 1, 3, and 4 reduce vibration; items 4, 5, 6, and 7 reduce and decouple the vibration from the radiating cabinet; and items 2 and 8 decouple the vibrating source from the structure. Once the noise becomes airborne or structure borne, the isolation techniques studied above are employed.

Vibration reduction takes two forms, damping and isolation. Damping is accomplished by rigidly coupling the vibrating source to a large mass, frequently called an inertia block. Much of the energy is absorbed and dissipated as friction; the remainder results in lower-amplitude vibration (see Fig. 27.42). *Isolation* is accomplished by supporting the vibrating mass on resilient supports. These take many forms and are used in tandem. Thus machines are supported on fibrous, rubber, or spring steel vibration isolators, and the entire mass can be supported on a floating floor, which in turn rests on resilient vibration isolators as in Fig. 27.42. Large machines are supported on special commercial "sandwiches" of asbestos, lead, cork, and other strong resilient materials. Piping is supported on cork pads and hung on resilient hangers.

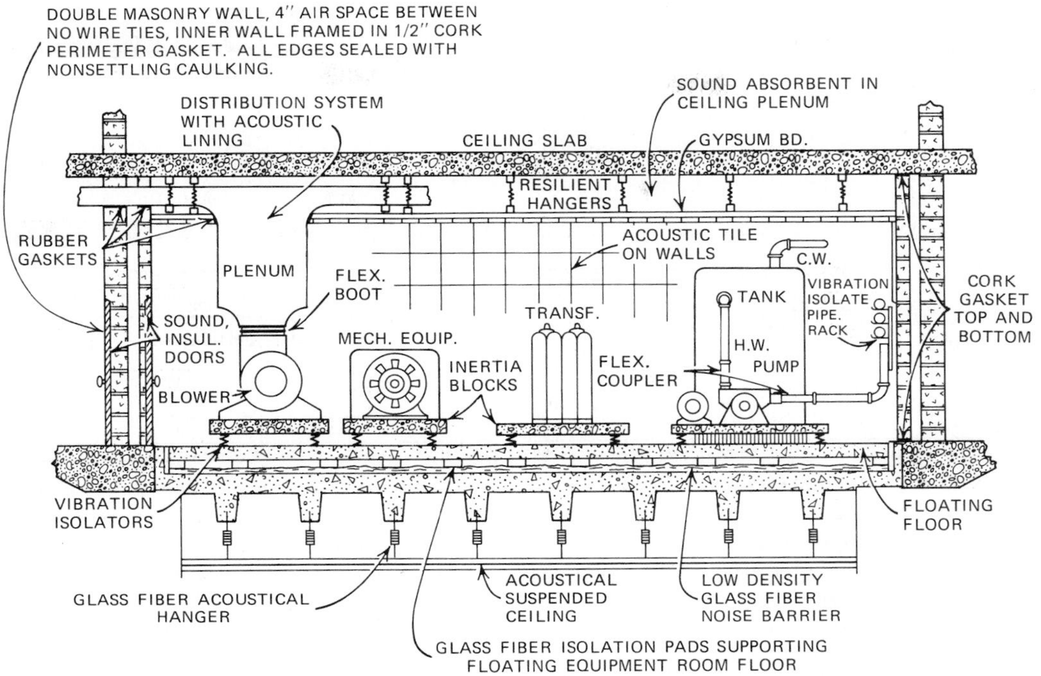

Fig. 27.42 *Soundproofing a mechanical equipment room. (Reprinted from* A Guide to Airborne, Impact and Structure-Borne Noise Control in Multi-family Dwellings, *1968.)*

The recent emphasis on emergency electric generators has caused a serious noise and vibration problem due to the large mass and extremely high noise levels. For such units, complete enclosures are frequently the best approach. Flexible joints in all pipes and ducts connected to vibrating machines are mandatory. This includes flexible conduit connectors to all motors, transformers, and lighting fixtures using ballasts.

27.28 Duct System Noise Reduction

Design of quiet duct system operation entails more than specification of duct lining. Air turbulence generates noise. Turbulence and noise levels increase as the velocity of airflow increases. Table 27.8 gives recommended design velocities for various distances from the terminal devices. Velocities should be as low as possible, since sound

TABLE 27.8 **Air Speeds in Ducts to Yield NC-15 or NC-25 Background Levels**[a]

	Supply		*Return*	
Location	*NC-15*	*NC-25*	*NC-15*	*NC-25*
Slot speed at min. ½ in. opening	250 fpm	350 fpm	300 fpm	420 fpm
10 ft of duct before opening	300	420	350	490
Next 20 ft	400	560	450	630
Next 20 ft	500	700	570	800
Next 20 ft	640	900	700	980
Next 20 ft	800	1120	900	1260
Next 20 ft	1000	1400	1100	1540
Next 20 ft	1300	1820	1450	2030
Next 20 ft	1600	2240	1800	2520

[a]Ducts with 1- to 2-in. inside duct lining, all duct sizes.

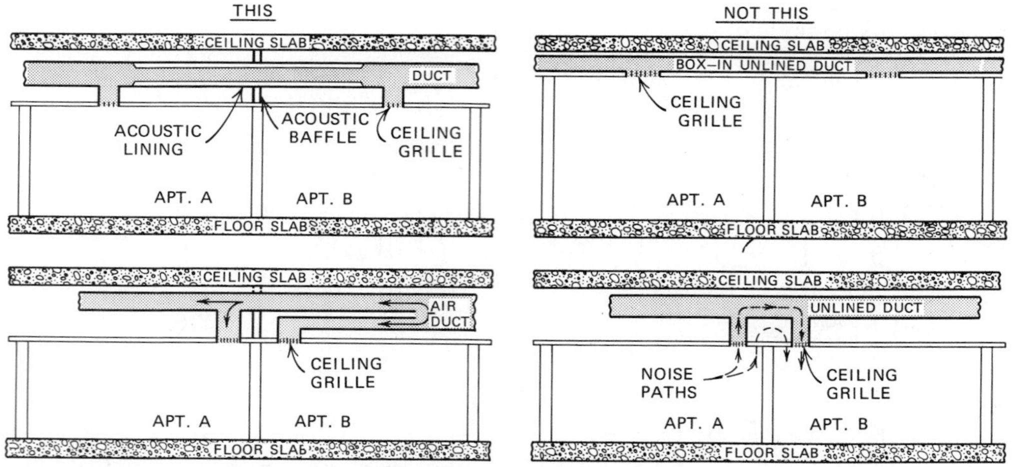

Fig. 27.43 *Since ducts are extremely efficient sound transmission paths, considerable precaution must be taken to avoid cross-talk and ventilation, combustion, and equipment noise. Avoid running ducts as a common supply or return between rooms, unless such systems are properly baffled and lined with sound-absorbing material. The common practice, in wood frame structures, of using troughs between joists as a common return duct between dwellings results in very serious noise transmission problems. Caulk or seal around ducts at all points of penetrations through partitions. Use double-wall ducts, acoustical lining, flexible boots, and resilient hangers where required. Dwelling units should be serviced by separate supply and return ducts, which branch off a main duct system. (Reprinted from* A Guide to Airborne, Impact and Structure-Borne Noise Control in Multi-family Dwellings, *1968, pp. 8-7,8-8.)*

increases exponentially with velocity. Allowable velocities increase as the distance from the terminal device becomes greater. Return velocities may be slightly higher than supply.

Sound travels as easily against as with the airflow in ductwork. Therefore, both supply and return systems must be lined to control transmission of fan noise. Maximum noise reduction occurs at bends in the ducts. For maximum reduction in short runs, a pair of 90° bends often is designed into a system, since lining is not effective in runs shorter than 50 to 60 ft.

Other design approaches that are part of a quiet system include smooth transitions at changes of duct size and large radius bends with turning vanes. Attenuation drops rapidly with increasing size of duct, and therefore ducts should not be deliberately oversized. Cross-talk between rooms and between ducts can be minimized by using lined ducts, separating adjacent ducts as much as possible, and gluing damping material on the outside and lining on the inside. Damping material is particularly effective in preventing the thin metal walls of the ducts from resonating. Mufflers and silencers are effective in reducing fan noise. The pressure drop they introduce must be compensated for in the fan. The *ASHRAE Guide and Data Book*'s chapter on noise control in fans, plenums, housings, and ducts should be consulted for their recommendations. Figure 27.43 shows some of the ways by which cross-talk and flanking noises can be reduced. Figure 27.44 shows some of the techniques employed for quieting duct noise.

27.29 Piping System Noise Reduction

As with airflow, noise increases exponentially with velocity. Piping is not a major noise source normally, since the radiating diameter is small, although flow velocities much in excess of 8 fps, where the pipe is in contact with the structure, can create noise problems, particularly when passing through NC-15 to NC-25 areas (see Table 27.11). Domestic water system mains should be limited to 50 psi in other than tall buildings and pressure

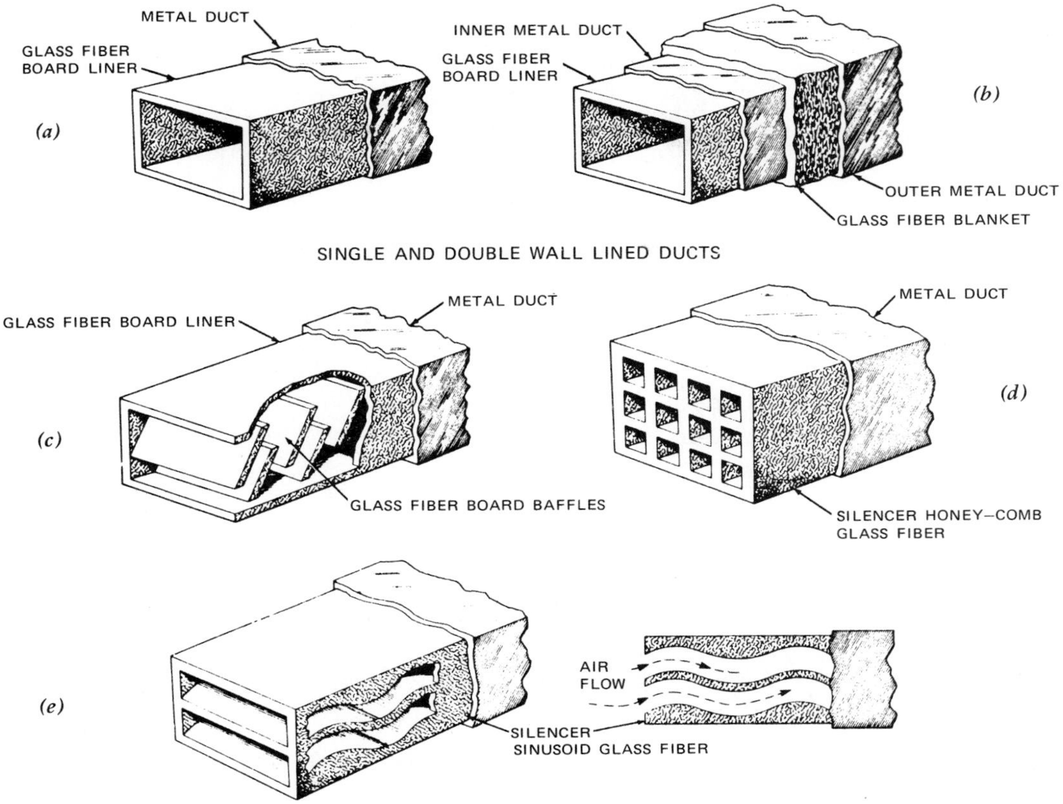

Fig. 27.44 *Unlined duct has negligible sound attenuation, and acts as an excellent speaking tube. Inside lining* (a) *gives 2 to 3 db attenuation per foot in the 1–2 kHz range, dropping rapidly above and below those frequencies, and giving negligible low-frequency attenuation.* (b) *Double lining gives higher attenuation and reduces cross-talk between ducts. Duct silencers and baffles* (c–e) *give high broadband attenuation—maximum of 10 to 12 db/ft in the 1–2 kHz range, and lower above and below. They are useful to reduce fan noise in short runs but cause considerable pressure drop. (Reprinted from* A Guide to Airborne, Impact and Structure-Borne Noise Control in Multi-family Dwellings, *1968, Figure 8-66.)*

in branches to 35 psi. In high-rise structures pressure-reducing valves will be required in high-pressure mains to meet these recommendations. Obviously piping must be designed to prevent water hammer, and noise sources must be located away from quiet areas.

Pumps, as with all rotating equipment, are sources of vibration and noise and should be treated as described in Section 27.27. Figure 27.45 shows a typical pump installation with appropriate noise reduction measures. For at least a distance of 100 pipe diameters beyond the pump, resilient pipe hangers should be used. With centrifugal pumps as with fan and blowers, machine sound concentrates in narrow bands and, if extremely disturbing, can be attenuated with resonant filters. Reciprocating pumps are more difficult to control as the pulsations are more vibration than noise. Flexible connections in the piping and U-joints in the piping will absorb much of this vibration.

27.30 Electrical Equipment Noise

Electrical equipment is generally overlooked as a noise source and this is unwise. Most electrical noise is 120-Hz hum. This can be very disturbing because it is so low a frequency and, as we have

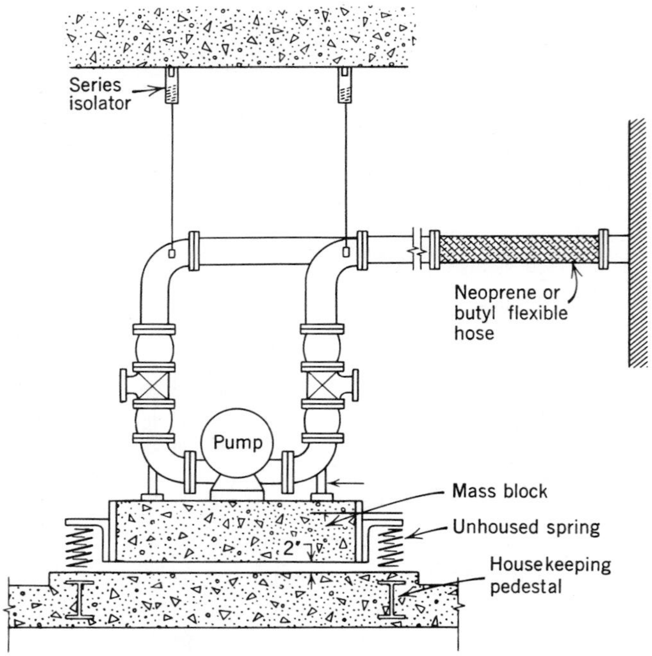

Fig. 27.45 *Typical pump installation with appropriate isolation and damping measures.*

TABLE 27.9 **Maximum Sound Levels: Dry-Type Transformers**

kVA	*Decibels (NEMA Standard)*
0–9	40
10–50	45
51–150	50
151–300	55
301–500	60

noted repeatedly, low-frequency noise is difficult to attenuate. Transformer noise levels are dictated by NEMA and ANSI standards. For a premium price lower noise units are obtainable. Table 27.9 lists maximum noise levels for dry-type units. Most manufacturers warranty noise levels below these. Oil- and silicone-filled units are normally quieter than dry-type transformers, as are units designed for lower temperature rise. Transformer noise can be minimized by these steps:

1. Mount unit on vibration isolators.
2. If transformer is wall hung, use resilient hanging. If it is floor mounted, place on as massive a slab as possible.
3. Locate the unit so that reflections do not amplify the sound. Sound-absorbent material on the walls behind the units is not useful at 120 Hz. Only cavity resonators will absorb appreciable amounts of sound at that frequency.
4. Use only flexible conduit connections.
5. Avoid locating transformers adjacent to, or immediately outside, quiet areas. A common error in this regard is placing a transformer pad immediately below the window of an NC 15–25 area.

The second major source of 120-Hz hum is conventional, core-and-coil discharge light fixture ballasts. (Electronic ballasts are practically noiseless.) This includes fluorescent plus all the HID sources. Table 27.10 lists the recommended application of conventional fluorescent ballasts. HID ballasts were not a problem until recently when mercury, metal–halide, and sodium lamps moved from their former noisy industrial surroundings into quiet office spaces. These ballasts can be very noisy, and care must be exercised in their placement. With all ballasts, the method of mounting has a marked effect on the radiated noise. As pointed out earlier, when a small vibrating source is coupled rigidly to a larger body, noise is amplified

TABLE 27.10 **Acoustic Criteria for Selection of Conventional Core-and-Coil Fluorescent Lamp Ballasts**

For an Installation in	*Use of Ballasts with This Rating Will Usually Be Satisfactory*
TV or radio station, church, synagogue	A
Offices, residence, night school	B
Library, reception or reading rooms, school study hall	C
Noisy office, doctor's or dentist's office, classroom	D

because of increased source-to-air coupling. Since fluorescent ballasts are necessarily closely coupled to large metal fixtures for heat dissipation purposes, the sound radiation is much amplified. A large number of fluorescent fixtures mounted in a plenum can create a serious problem. Solution of the problem lies in use of absorptive material in plenums, flexible conduit connection to fixtures, and resilient fixture hanging. In severe cases ballasts can be remote-mounted. HID ballasts are inherently noisier than fluorescent but are generally less troublesome, being coupled to small radiating bodies and generally mounted higher.

27.31 Noise Problems Due to Equipment Location

Roof-mounted HVAC units have proven themselves very economical and very noisy. The vibration, short duct runs, and sound reflections are serious problems that can be solved with vibration isolators, sound mufflers, and careful location of equipment. Roof-mounted cooling towers are a particular problem when located adjacent to a taller building. That problem has led to a spate of lawsuits and noise control legislation in many cities. For this reason particular attention should be paid in design to all exterior equipment.

In high-rise buildings, problems are engendered by the conflicts between the stringent noise requirements of the prime upper floor space and the near presence of elevator machine rooms, mechanical equipment rooms, and cooling towers. These problems are almost impossible to solve after construction and require the services of an acoustics expert during design.

27.32 Commercial Sound-Isolating Enclosures

In areas such as industrial plants with multiple high noise sources, it is desirable to reduce noise at its source. To accomplish this, many commercial products in the form of curtains, panels, and partial and full enclosures are available that will effectively isolate the noise source. Most of these operate on the mass principle and comprise sandwich construction of materials frequently including lead sheet (see Fig. 27.46). These enclosures are tailored to the specific noise and are not normally the responsibility of the building designer. It is, however, important to know that they will be used, as well as their characteristics, so that appropriate isolation can be designed into the structure for the sound that is radiated from the enclosures.

Fig. 27.46 *Commercial multilayer acoustic noise barrier material composed of high-mass vinyl and absorbent blanket. It acts as both an absorbent and an isolating barrier. When mounted on ¼-in. plywood it forms a barrier of STC 33 rating. (Photo courtesy of Ferro Corp.)*

TABLE 27.11 **Recommended Category Classification and Suggested Noise Criteria Range for Steady Background Noise**[a]

Type of Space (and Acoustical Requirements)	*PNC*[a,b] *Curve*	*NC Curve*	*Equivalent*[c] *dbA*
Concert halls, opera houses, and recital halls (for listening to faint musical sounds).	10–20	10–20	20–30
Broadcast and recording studios (distant microphone pickup used).	10–20	15–20	25–30
Large auditoriums, large drama theatres and houses of worship (for excellent listening conditions).	20 max.	20–25	30–35
Broadcast, television, and recording studios (close microphone pickup only).	25 max.	20–25	30–35
Small auditoriums, small theatres, small churches, music rehearsal rooms, large meeting and conference rooms (for good listening), or executive offices and conference rooms for 50 people (no amplification).	Not to exceed 35	25–30	35–40
Bedrooms, sleeping quarters, hospitals, residences, apartments, hotels, motels, and so forth (for sleeping, resting, relaxing).	25–40	25–35	35–45
Private or semiprivate offices, small conference rooms, classrooms, libraries, and so forth (for good listening conditions).	30–40	30–35	40–45
Living rooms and similar spaces in dwellings (for conversing or listening to radio and TV).	30–40	35–45	45–55
Large offices, reception areas, retail shops and stores, cafeterias, restaurants, and so forth (for moderately good listening conditions).	35–45	35–50	45–60
Lobbies, laboratory work spaces, drafting and engineering rooms, general secretarial areas (for fair listening conditions).	40–50	40–45	50–55
Light maintenance shops, office and computer equipment rooms, kitchens, and laundries (for moderately fair listening conditions).	45–55	45–60	55–70
Shops, garages, power-plant control rooms, and so forth (for just acceptable speech and telephone communication). Levels above PNC-60 are not recommended for any office or communication situation.	50–60	—	—
For work spaces where speech or telephone communication is not required, but where there must be no risk of hearing damage.	60–75	—	—

[a]After Beranek et al. (1971).

[b]Available in Magrab (1975) and Beranek et al. (1971).

[c]For information only.

Source: Reprinted with permission from E. B. Magrab, *Environmental Noise Control,* Wiley, New York, 1975.

RECOMMENDATIONS AND CRITERIA

Recommendations for background noise levels (NC criteria) are given in Table 27.11. Criteria for partition isolation (STC) and impact isolation (IIC) are given in the following sections and in Table 27.12.

27.33 Multioccupancy Residential STC/IIC Criteria

The most important and binding criteria for residential work in the United States are issued by the Department of Housing and Urban Development in conjunction with the Federal Housing Admin-

TABLE 27.12 **Recommended STC for Partitions; Specific Occupancies**

	Wall, Partition, or Panel Between		*Sound Isolation Requirement: Background Level in Room Being Considered*	
Type of Occupancy	*Room Being Considered AND*	*Adjacent area*	*Quiet*	*Normal*
Normal school buildings without extraordinary or unusual activities or requirements	Classrooms	Adjacent classrooms	STC 42	STC 40
		Corridor or public areas	STC 40	STC 38
		Kitchen and dining areas	STC 50	STC 47
		Shops	STC 50	STC 47
		Recreation areas	STC 45	STC 42
		Music rooms	STC 55	STC 50
		Mechanical equipment rooms	STC 50	STC 45
		Toilet areas	STC 45	STC 42
	Music practice rooms	Adjacent practice rooms	STC 55	STC 50
		Corridor and public areas	STC 45	STC 42
Executive areas, doctors' suites; confidential privacy requirements	Office	Adjacent offices	STC 50	STC 45
		General office areas	STC 48	STC 45
		Corridor or lobby	STC 45	STC 42
		Washrooms and toilet areas	STC 50	STC 47
Normal office; normal privacy requirements	Office	Adjacent offices	STC 40	STC 38
		Corridor, lobby, exterior	STC 40	STC 38
		Washrooms, kitchen, dining	STC 42	STC 40
Any occupancy, using rooms for group meetings	Conference rooms	Other conference rooms	STC 45	STC 42
		Adjacent offices	STC 45	STC 42
		Corridor or lobby	STC 42	STC 40
		Exterior of building	STC 40	STC 38
		Kitchen and dining areas	STC 45	STC 42
Large offices, drafting areas, banking floors, etc.	Large general office areas	Corridors, lobby, exterior	STC 38	STC 35
		Data-processing area	STC 40	STC 38
		Kitchen and dining areas	STC 40	STC 38
Motels and urban hotels, hospitals and dormitories	Bedrooms	Adjacent bedrooms[a]	STC 52	STC 50
		Bathroom[a]	STC 50	STC 45
		Living rooms[a]	STC 45	STC 42
		Dining areas	STC 45	STC 42
		Corridor, lobby, or public spaces	STC 45	STC 42

[a]Separate occupancy.

Source: Courtesy of U.S. Gypsum.

istration (HUD/FHA). The reader is referred to the latest issue of *A Guide to Airborne, Impact and Structure-Borne Noise Control* for the full text. Tables 27.13 and 27.14 give the essential data presented there.

The recommendations are divided into Grades I, II, and III. Grade II is the most important category and is applicable primarily in residential urban and suburban areas considered to have the ''average'' noise environment. The nighttime exterior noise levels might be about 40 to 45 dbA, and the permissible interior noise environment should not exceed NC 25–30 characteristics. Grade I is suburban, with a ''quiet'' noise environment characterized by a nighttime exterior noise level of about 35 to 40 dbA. *Grade I STC/IIC criteria are three points higher than those of Grade II.*

Grade III recommendations are minimal and can be characterized as ''noisy,'' with an average nighttime exterior noise level of about 55 dbA or higher. *Grade III STC/IIC recommendations are four points lower than those of Grade II.*

The fundamental criteria of airborne sound insulation between dwelling units are, for Grade II:

$$\text{Wall partitions} \quad \text{STC} \geq 52$$

$$\text{Floor–ceiling assemblies} \quad \text{IIC} \geq 52$$

These apply where similar function spaces are contiguous, as bedroom to bedroom and living room to living room. Where this is not the case, the isolation must be revised up or down to meet the maximum sensitivity requirement.

TABLE 27.13 **Criteria for Airborne Sound Insulation of Wall Partitions Between Dwelling Units**

Partition Function Between Dwellings			
Apt. A		*Apt. B*	*Grade II STC*
Bedroom	to	Bedroom	52
Living room	to	Bedroom[a]	54
Kitchen[b]	to	Bedroom[a]	55
Bathroom	to	Bedroom[a]	56
Corridor	to	Bedroom[a,c]	52
Living room	to	Living room	52
Kitchen[b]	to	Living room[a]	52
Bathroom	to	Living room	54
Corridor	to	Living room[a,c,d]	52
Kitchen	to	Kitchen[e]	50
Bathroom	to	Kitchen	52
Corridor	to	Kitchen[a,c,d]	52
Bathroom	to	Bathroom	50
Corridor	to	Bathroom[a,c]	48

[a]Whenever a partition wall might serve to separate several functional spaces, the highest criterion must prevail.

[b]Or dining or family or recreation room.

[c]It is assumed that there is no entrance door leading from corridor to living unit.

[d]Criterion applies to the partition. Doors in corridor partitions must have the rating of the partition, not vice versa.

[e]Double wall construction is recommended to minimize kitchen impact noises.

Source: Reprinted from *A Guide to Airborne, Impact, and Structure-borne Noise Control in Multifamily Dwellings* (1968). For Grade I, add 3 points; for Grade III, subtract 4 points.

27.34 Specific Occupancies

(a) Schools. School buildings house spaces of many kinds—classrooms, auditoriums, gymnasiums, cafeterias, shop areas, and music suites—that pose acoustics problems.

1. *Auditoriums.* All auditoriums require a sound system for some of the activities accommodated. Figure 27.47 indicates the microphones, loudspeakers, and control locations in a typical auditorium sound system. The most difficult aspect, architecturally, is integration of the loudspeaker system into the design. To provide proper sound reinforcement, loudspeakers must be located properly without obstructions. To accomplish this, the loudspeaker system should be incorporated in the earliest drawings.

In general, a school auditorium must be a multipurpose facility. It should be designed to meet speech requirements and also should be suitable for the school's music activities.

Often a modified gymnasium (gymnatorium) or cafeteria (cafetorium) functions as an auditorium. Obviously, acoustics compromises occur in such facilities. The large areas of sound-absorbing treatment in either kind of space make them un-

TABLE 27.14 **Criteria for Airborne and Impact Sound Insulation of Floor–Ceiling Assemblies Between Dwelling Units**

Partition Function Between Dwellings			*Grade II*	
Apt. A		*Apt. B*	*STC*	*IIC*
Bedroom	above	Bedroom	52	52
Living room	above	Bedroom[a]	54	57
Kitchen[b]	above	Bedroom[a,c]	55	62
Family room	above	Bedroom[a,d]	56	62
Corridor	above	Bedroom[a]	52	62
Bedroom	above	Living room[e]	54	52
Living room	above	Living room	52	52
Kitchen	above	Living room[a,c]	52	57
Family room	above	Living room[a,d]	54	60
Corridor	above	Living room[a]	52	57
Bedroom	above	Kitchen[c,e]	55	50
Living room	above	Kitchen[c,e]	52	52
Kitchen	above	Kitchen[c]	50	52
Bathroom	above	Kitchen[a,c]	52	52
Family room	above	Kitchen[a,c,d]	52	58
Corridor	above	Kitchen[a,c]	48	52
Bedroom	above	Family room[e]	56	48
Living room	above	Family room[e]	54	50
Kitchen	above	Family room[e]	52	52
Bathroom	above	Bathroom[c]	50	50
Corridor	above	Corridor	48	48

[a]This arrangement requires greater impact sound insulation than the converse, where a sensitive area is above a less sensitive area.

[b]Or dining or family or recreation room.

[c]It is assumed that plumbing fixtures, appliances, and piping are installed with proper vibration isolation.

[d]The airborne STC criteria in this table apply as well to vertical partitions between these two spaces.

[e]This arrangement requires equivalent airborne sound insulation and perhaps less impact sound insulation than the converse.

Source: Reprinted from *A Guide to Airborne, Impact, and Structure-borne Noise Control in Multifamily Dwellings* (1968). For Grade I, add 3 points; for Grade II, subtract 4 points.

suitable as auditoriums and most events that require speech amplification.

2. *Classrooms*. Typical classrooms are approximately 30 ft square with 10-ft ceilings. Adequate speech communication is easily achieved in a room of this size. Classroom acoustic design usually involves:

a. Locating sound-absorbing treatment to reduce classroom noise levels.
b. Ensuring adequate privacy between adjacent spaces.
c. Control of air-handling system noise.

Acoustic tile ceilings yield adequate sound absorption for most classrooms. An NRC of 0.7 is

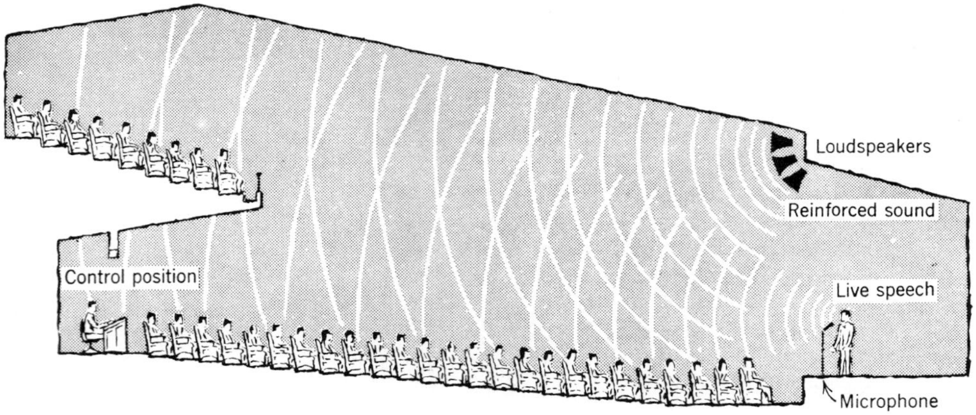

Fig. 27.47 Auditorium sound system elements.

recommended. See Section 27.6. An alternate method involves a combination of sound-absorbing wall treatment and floor carpeting, which also produces adequate noise reduction. Carpeting, alone, is generally insufficient.

Partition systems must produce sufficient isolation to prevent disturbance from activities in other classrooms and corridors. Such partitions should run full height from floor to ceiling slab or roof construction. If return air transfer ducts are needed, their noise reduction characteristics must be as good as the walls or doors that they penetrate (for NC data, see Table 27.11). Unit ventilators commonly used for classrooms produce approximately this level of background noise.

3. *Music suites.* School music programs usually range from individual instruction to band and choral concerts. The teaching spaces required for such a program include practice rooms, ensemble rooms, and large rehearsal spaces. Both room acoustics design and sound isolation are critical in music suites. Privacy between adjacent spaces is particularly critical, since simultaneous use is necessary.

4. *Dining areas.* The activity in cafeterias or lunchrooms generates noise. The kitchen and serving areas should be separated from the eating spaces. Ceilings and available wall areas in the cafeteria should be treated with sound-absorbing material such as acoustic tiles to reduce potential buildup of sounds produced by conversation, traffic, moving chairs, and dishes. Ideally, the area should be carpeted. Moreover, unless the ceiling is completely treated with a highly efficient sound absorbing material, the environment will be unsatisfactory. Minimum NRC should be 0.8.

5. *Gymnasiums.* Activities in gymnasiums create so much noise that even extensive treatment will not quiet these spaces. A quiet gymnasium probably would be unsatisfactory, since spectators are conditioned to consider the "noise" as an enjoyable aspect of athletic events. However, to provide a proper environment for normal sports activities, the ceiling area should be sound absorbing. In addition, if a sound amplification system is to be used, sound-absorbing wall treatment may be required to eliminate echoes that would reduce intelligibility of announcements. An NRC of 0.7 is suggested with sound-absorbent material to be ceiling mounted.

If a gymnasium will also serve as an auditorium, loudspeaker system placement needs special consideration. For example, the loudspeakers should be located above the source location for speeches and plays.

6. *Swimming pools.* The acoustic environment of swimming pools is often chaotic. Most sound-absorbing materials disintegrate in the high-humidity conditions prevalent in pool areas. Special combinations of materials, as well as some recently developed prefabricated sound-absorbing units, have moisture-resistant properties.

7. *Shops.* Metal, woodworking, and scenery shops in schools contain many noise sources—

saws, planers, drill presses, and manual tools. Each generates high airborne and structure-borne noise levels. Consolidating noisy areas and maximizing the distance between them and critically quiet spaces are essential. Ceiling and wall absorptive treatment with an NRC of at least 0.75 is recommended.

(b) Churches. The basic activities of most churches combine speech and music. Thus, the church environment must be hospitable, acoustically, to both. A nonreverberant chapel meets the requirement for speech communication but it is unsatisfactory for music. A cathedral usually will provide a magnificent acoustical environment for organ and choral works, while it is nearly unusable for intelligible speech communication. The architectural plan also must respond to religious requirements including the relative positioning of pulpits, lecterns, the altar, and choir.

Successful church acoustics can be achieved by designing the overall environment for music and providing special assistance for speakers. Figure 27.48 illustrates a church that is adequately reverberant for music and includes a sound-reflecting canopy over the pulpit to direct the minister's voice to the congregation. In some larger churches, a loudspeaker located above the canopy further reinforces speech from the pulpit. The choir and organ communicate with the entire volume of the church and, therefore, benefit from the reverberant environment.

(c) Offices. Although office buildings contain public spaces, auditoriums, and restaurants, prime occupancy is in small office areas. Most acoustics problems in office buildings relate to privacy—either between spaces within a single firm or between adjacent firms. Speech privacy has been discussed at length in Section 27.22, including consideration of open-plan offices.

Satisfying speech privacy requirements does not necessarily create a good acoustical environment in an office building. As already pointed out, the operation of air-handling equipment, pumps, elevators, plumbing, and escalators can create excessive noise and vibration, particularly since executive offices or restaurants often are located on the top floors of multistory office buildings.

(d) Apartment Buildings. Large apartment buildings sometimes house thousands of residents. Privacy and freedom from annoyance are high on the list of tenant requirements. See HUD/FHA criteria in Section 27.33.

The performance of the partitions is compromised in many designs by careless planning of convenience outlets, medicine cabinets, and mechanical services. Moreover, direct-exhaust duct connections between apartments and back-to-back placement of medicine cabinets result in loss of privacy. Back-to-back convenience outlets must be avoided.

Installation of rugs or carpeting provides the best protection against footfall noise. Some leases now require that a tenant provide such impact-reducing floor covering over most of the floor area in his apartment. Good design also dictates that similar spaces in adjacent apartments be grouped—bedrooms next to bedrooms, for example. Ab-

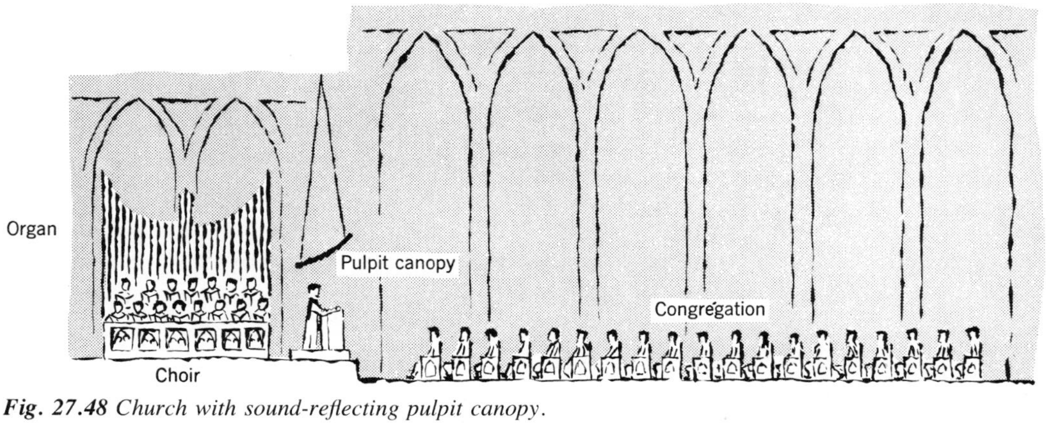

Fig. 27.48 *Church with sound-reflecting pulpit canopy.*

sorptive material in bedrooms should be ceiling mounted. An NRC of 0.6 is recommended.

Apartment house site selection seldom includes consideration of acoustics. Nevertheless, truck routes, superhighways, and airports can be annoying "neighbors." As mentioned above, cooling towers serving adjacent buildings must be considered during planning stages.

OUTDOOR ACOUSTIC CONSIDERATIONS

27.35 Sound Power and Pressure Levels in Free Space (Outdoors)

The equations of Section 26.15 are not applicable to outdoor propagation, in which the large reflective component of the indoor condition is absent. Although the propagation of sound outdoors may not appear to be of immediate importance in architectural acoustics, outdoor noise sources such as traffic, cooling tower, and aircraft are frequently loud enough to disturb activities within or immediately adjacent to a building. Conversely, the noise made by building equipment such as cooling towers, heat pumps, and even window air conditioners may be loud enough to disturb neighbors in a nearby building. For this reason it is desirable to have some basic understanding of outdoor sound propagation.

For preliminary evaluation of an outdoor noise problem, assuming a nondirectional source, one needs to know only the power level radiated by the source as a function of frequency and time; from this one can establish the level of sound at the appropriate distance as follows:

$$SPL = PWL - 20 \log r + (Q - 1) \text{ db} \quad (27.6)$$

where

SPL = sound pressure level

PWL = equipment power level

r = distance from the source, ft

Q = directivity factor (see Fig. 26.17)

This formula is fairly accurate for a *small* source. For large sources such as cooling towers and traffic, which do not exhibit inverse square properties, sound level estimates are best made on the basis of experience and empirical data beyond the scope of this book (see Magrab, 1975; Schaudinischky, 1976). For small outdoor sources, the equipment power level can be estimated by measuring the sound power level at 5 ft and adding 15 db. Other factors, such as moisture in the air, the presence of trees, wind, and temperature gradients, will affect outdoor sound propagation to some extent, but they can be ignored except when great distances (i.e., over 1000 ft) are involved. Barriers, which are most effective outdoors, were discussed in Section 27.19.

27.36 Building Siting

As important as interior structural design is building siting vis-à-vis exterior noise sources. See also Section 3.5 and Figs. 3.14 and 3.15. Since this subject is somewhat beyond our scope, the discussion is brief. Buildings should be sited, with respect to noise sources:

1. To use natural terrain noise barriers (Fig. 27.49*a*).
2. With respect to trees as noise barriers, rely only on thick wooded areas (Fig. 27.49*b*).
3. To avoid naturally poor sites (Fig. 27.49*c*).
4. To avoid sound reflection from other buildings (Fig. 27.49*d*).

Factor (4) is also important in a multiwing building; in avoiding U shapes or other configurations where a central court becomes an echo chamber.

Where avoidance of an exterior noise source is impossible, quiet zones can be buffered from the noise by placing higher noise areas on the noisy side of the building. Thus, in a school, classrooms and offices can be buffered by a cafeteria and gym; in a residence, bedrooms by living rooms and corridors; in an office building, private offices by noisier clerical offices; and so on.

REFERENCE MATERIAL

27.37 Definitions

A-Scale. Identifies the filtering system that has characteristics that roughly match the response

THIS
NOT THIS

(a)

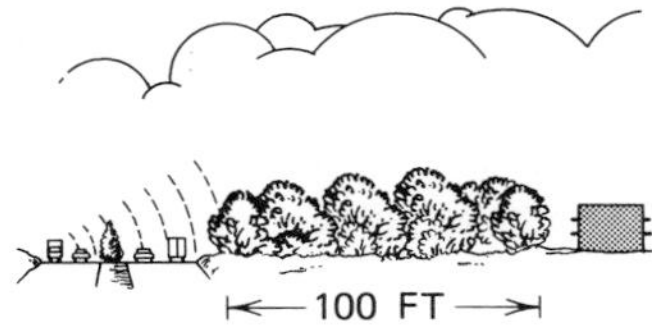

100 FT
A THICK GROWTH OF LEAFY TREES AND UNDERBRUSH REDUCES NOISE ABOUT 6 to 7 DB/100 FT (AVERAGE OVER AUDIBLE FREQ. RANGE)
LOW–FREQ. LOSS: 3–4 dB
HIGH–FREQ. LOSS: 10–12 dB

(b)

HIGH FREQ. REDUCTION 3–4 dB
SINGLE ROW OF TREES IS WORTHLESS AS NOISE BARRIER. DUE TO INTER–REFLECTION MULTI–ROWS OF TREES ARE MORE EFFECTIVE

(c)

AVOID HOLLOWS OR DEPRESSIONS. THEY ARE GENERALLY NOISIER THAN FLAT OPEN LAND.

(d)

BUILDING SITES IN OPEN AREAS ARE LESS NOISY THAN SITES IN CONGESTED BUILDING AREAS

TRAFFIC ARTERIES BETWEEN TALL BUILDINGS ARE QUITE NOISY.

Fig. 27.49 (a) *Use of natural noise barriers.* (b) *Effectiveness of wooded areas as noise barriers. Noise reduction of trees.* (c) *An example of a poor building site.* (d) *Building sites near traffic arteries and other buildings. (Reprinted from* A Guide to Airborne, Impact and Structure-Borne Noise Control in Multi-family Dwellings, *1968, p. 52.)*

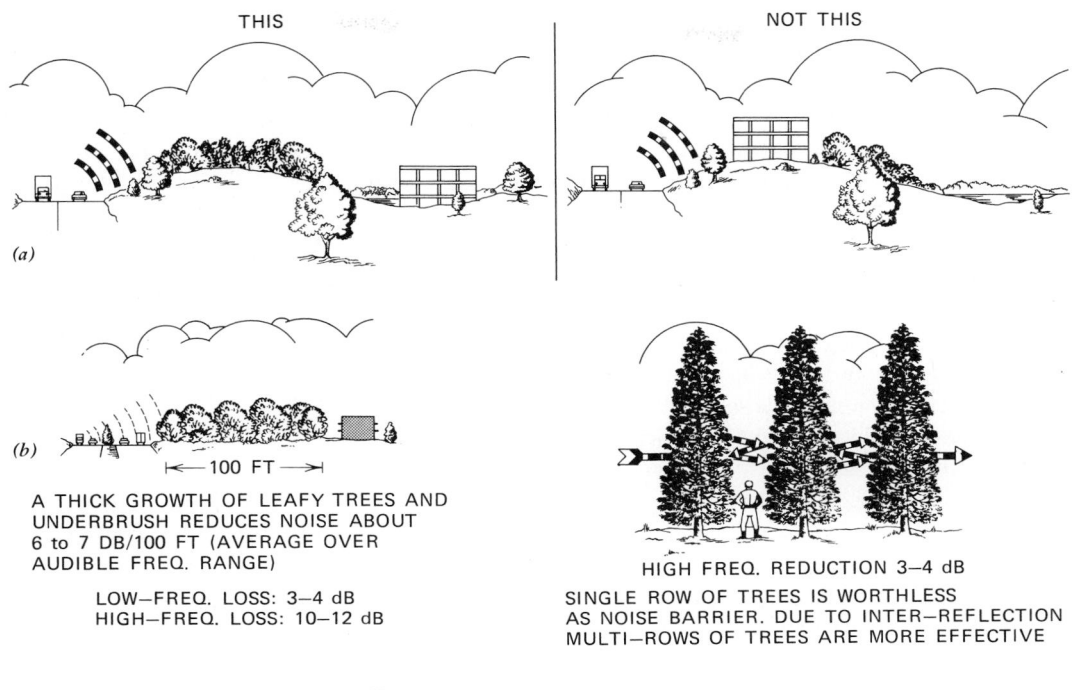

characteristics of the human ear at low sound levels (below 55 db sound pressure level, but frequently used to gauge levels to 85 db). A-scale measurements are often referred to as dbA.

Absorber, Sound. A device, panel, or material specifically designed to absorb sound energy. Such devices are usually constructed of porous materials composed of organic or mineral fibers.

Absorption Coefficient (α). The absorption coefficient of a material or sound-absorbing device is the ratio of the sound absorbed to the sound incident on the material or device. The sound absorbed by a material or device is usually taken as the sound energy incident on the surface minus the sound energy reflected.

Anechoic Room. An anechoic room provides a free field acoustic testing environment like the out-of-doors. All the sound emanating from a source is essentially absorbed at the walls of the anechoic room. Hence, there are no reflections and the spatial sound radiation pattern of a source may be determined. An anechoic room may be described as "echoless" or acoustically "dead."

Baffle or Barrier, Sound. A shielding structure or partition used to increase the effective length of a sound transmission path between two locations. Such structures often are constructed or surfaced with sound-absorbing materials and are frequently used to seal open plenums above ceilings or below floors.

Critical Frequency. The lowest frequency at which the wavelength of a bending wave, traveling in a structure, is the same as the wavelength in air at that frequency. Coupling between the air and the structure is very good at this point, and sound waves may move from the structure to the air and vice versa with ease.

Damping. Dissipation of structure-borne noise (usually traveling bending waves) by conversion to some other form of energy, usually heat. For the most part, this is accomplished by using a material with a high internal energy-absorbing capacity (i.e., high internal damping).

Decibel (db). See "Sound Pressure Level."

Diffuse Sound Field. A sound field that at any given point is made up of sound waves of all angles of incidence.

Direct Field. The sound in a region in which all or most of the sound arrives directly from the source without reflection.

Flanking Transmission. The transmission of sound or noise from one room to another by indirect paths, rather than directly through an intervening partition.

Free Sound Field (Free Field). A field in a homogeneous, isotropic medium free from boundaries. In practice, it is a field in which the effects of the boundaries are negligible over the region of interest. In the free field, the sound pressure level decreases 6 db for a doubling of distance from a point source.

Impact Isolation Class (IIC). A single-figure rating that provides an estimate of the impact sound-isolating performance of a floor–ceiling assembly.

Intensity Level (IL). A measure of the acoustic power passing through a unit area expressed on a decibel scale referenced to some standard (usually 10^{-12} W/cm^2).

Loudness. The subjective human definition of the intensity of a sound. Human reaction to sound is highly dependent on the sound pressure and frequency.

Masking. The presence of a background noise increases the level to which a sound signal must be raised in order to be heard or distinguished. If the level of background noise is significantly higher than that of the sound signal (e.g., a sound transmitted from another room), the transmitted sound signal cannot be heard. This effect is known as masking.

Mass Law. The law relating to the transmission loss of walls, which says that in a part of frequency range, the magnitude of the loss is controlled entirely by the mass per unit area of the panel. The law also says that the transmission loss increases 6 db for each doubling of frequency or each doubling of the panel mass per unit area.

Noise. Any undesired sounds, usually of different frequencies, resulting in an objectionable or irritating sensation.

Noise Reduction (NR).

1. The reduction in sound pressure level caused by making some alteration to a sound source.
2. The difference in sound pressure level

measured between two adjacent rooms caused by the transmission loss of the intervening wall.

Octave Band (OB). A range of frequency where the highest frequency of the band is double the lowest frequency of the band. The band is usually specified by the center frequency.

Phon. Loudness level, at a particular frequency, equal to the 1000-Hz decibel level of that equal-loudness contour.

Pink Noise. Wide frequency spectrum noise, whose amplitude drops 3 db per octave with increasing frequency. Particularly useful for masking.

Reverberation. The persistence or echoing of previously generated sound caused by reflection of acoustic waves from the surfaces of enclosed spaces.

Sabin. The unit of acoustic absorption. One sabin (ft^2) is the absorption of one square foot of perfect sound-absorbing material.

Sound-Level Meter. An instrument for the direct measurement of sound pressure level. Sound-level meters often are made with various filtering networks that measure the sound directly on A, B, C, etc., scales. Sound-level meters may also incorporate octave-band filters for measuring sound directly in octave bands.

Sound Power Level (PWL). A measure of the total airborne acoustic power generated by a noise source, expressed on a decibel scale referenced to some standard (usually 10^{-12} W).

Sound Pressure Level (SPL). A measure of the air pressure change caused by a sound wave. Expressed on a decibel scale referenced to some standard (usually 0.0002 μbar).

Transmission Loss (TL). The reduction of airborne sound power that is caused by placing a wall or barrier between the reverberant sound field of a source and its receiver. Transmission loss is a property of the wall or barrier.

White Noise. Noise of wide frequency range in which the amplitude of the noise is essentially the same in all frequency bands.

27.38 Units and Conversions

See Table 27.15.

TABLE 27.15 **Acoustic Units and Conversions**

Units	*MKS*	*CGS*
Force	kilogram-meter/s^2 = newton	gram-cm/s^2 = dyne
Intensity	watts/meter^2	watts/cm^2
Pressure	newtons/meter^2 = pascals	dynes/cm^2 = microbars

	In Conversion:		
Quantity	*Multiply*	*By*	*To Obtain*
Force	newtons	10^5	dynes
	dynes	10^{-5}	newtons
Intensity	watts/cm^2	10^4	watts/m^2
	watts/m^2	10^{-4}	watts/cm^2
Pressure	pascals	10	microbars
	microbar	10^{-1}	pascals

NOTE. One atmosphere = 1 bar = 10^6 μbar.

27.39 Symbols

See Table 27.16.

References

A Guide to Airborne, Impact and Structure-Borne Noise Control in Multi-family Dwellings. (1968). U.S. Department of Housing and Urban Development, Washington, D.C. (1968).

Application of Sound Power Level Ratings. (June 1973). AMCA publication 303, Arlington Heights, Ill.

Beranek, L. L. (1949). *Acoustic Measurements,* Wiley, New York.

Beranek, L. L. (1954). *Acoustics,* Mc-Graw-Hill, New York.

Beranek, L. L. (1960). *Noise Reduction,* McGraw-Hill, New York.

Beranek, L. L. (1962). *Music, Acoustics and Architecture,* Wiley, New York.

Beranek, L. L. (1971). *Noise and Vibration Control,* McGraw-Hill, New York.

Beranek, L. L., Blazer, W. E., and Figwer, J. J. (1971). ''PNC Curves and Their Application to Rooms,'' *Journal of the Acoustical Society of America,* pp. 1223–1228.

Broch, J. T. (1971). *Application of the B & K Measuring Systems to Acoustic Noise Measurements,* 2nd ed., Naerum, Denmark.

Brüel, P. V. (1951). *Sound Insulation and Room Acoustics,* Chapman & Hall, London.

ACOUSTICS

TABLE 27.16 **Symbols and Abbreviations Used Commonly in Acoustics**

A =	Total absorption, sabins; area in square inches or square feet
A_R =	Absorption in receiving room, sabins
$\overline{A}_{1,2\cdots}$ =	Total absorption of each different material in a space, sabins
c =	Velocity of sound, feet per second
d =	Distance from source, meters or feet
db =	Decibel
f =	Frequency of sound, hertz (cps)
I =	Intensity, watts per square centimeter
I_a =	Absorbed energy, watts per square centimeter
I_i =	Incident energy, watts per square centimeter
I_0 =	Reference intensity, 10^{-16} watt per square centimeter
IIC =	Impact insulation class, no units
IL =	Intensity level, decibels
NC =	Noise criterion, no units
NRC =	Noise reduction coefficient, no units
NR =	Noise reduction, decibels
p =	Pressure, pascals or microbars
p_0 =	Reference base pressure 2×10^{-5} Pa
P =	Acoustic power, watts
Pa =	Pascal, unit of pressure (SI)
PWL =	Sound pressure level, decibels
R =	Room constant, square feet
r =	Distance from source, meters or feet
S =	Surface area, square feet
SPL =	Sound pressure level, decibels
STC =	Sound transmission class, no units
T_R =	Reverberation time, seconds
TL =	Transmission loss, decibels
V =	Volume, cubic feet
W,P =	Sound power, watts
W_0 =	Reference base sound power 10^{-12} watt
a, α =	Absorption coefficient (no units)
$\bar{a}, \bar{\alpha}$ =	Average absorption coefficient (no units)
λ =	Wavelength, feet or meters
Σ =	Sum of, or total (no units)
$\Sigma S_\alpha = \Sigma\overline{A}$ =	Total absorption, sabins
Δ =	Change in a quantity or difference between two quantities

NOTE. Where definitions are expressed in feet, centimeters or meters is also understood, with proper conversion factors, and vice versa.

Brüel, P. V., and Kjaer (1963). *Application of Sound Power,* Denmark.

Brüel, P. V., and Kjaer. *Measuring Sound,* Denmark.

Cavanaugh, W. J., Farrell, W. R., Hirtle, P. W., and Watters, B. G. (1962). "Speech Privacy in Buildings," *Journal of the Acoustical Society of America,* Vol. 34, No. 4.

Embelton, T. (1971). "Absorption Coefficients of Surfaces Calculated from Decaying Sound Fields," *Journal of the Acoustical Society of America,* Vol. 50, No. 3.

Fundamentals of Noise. (1971). Report NTID 300.15, U.S. Environmental Protection Agency, Washington, D.C., pp. 55–56.

Gomperts, M. (1965/1966). "Do the Classical Reverberation Formulas Still Have a Right for Existence?" *Acustica,* Vol. 16.

Harris, C. (1957). *Handbook of Noise Control,* McGraw-Hill, New York.

Information on Levels of Environmental Noise. (1974). Report 550/9-74-004, U.S. Environmental Protection Agency, Washington, D.C.

Magrab, E. B. (1975). *Environmental Noise Control,* Wiley, New York.

Pearsons, K. S., and Bennet, R. L. (1974). *Handbook of Noise Rating,* National Aeronautics and Space Administration Report NASA CR-2376, Washington, D.C., pp. 55–56.

Procedure for Computation of Loudness of Noise. (1968). ANSI, USAS 53.4, New York.

Quieting: A Practical Guide to Noise Control. (1976). National Bureau of Standards Handbook No. 119, NBS, Washington, D.C.

Schaudinischky, L. H. (1976). *Sound, Man and Building,* Applied Science Publishers, London.

Schiff, M. L. (1975). "Noise Control for Draft Fans," *Specifying Engineer,* pp. 62–70.

Terminology, Laws and Application of Fan Sound Data. (1972). Publication SD100-1972, Barry Blower Co.

Ward, W. D., et al. (1968). *Proposed Damage-Risk Criterion for Impulse Noise,* Report of Working Group 57, NAS-NRC Committee on Hearing, Bioacoustics and Biomechanics (CHABA).

Young, R. W. (1959). "Sabine Reverberation Equation and Sound Power Calculations," *Journal of the Acoustical Society of America,* Vol. 31, No. 7.

Young, R. W. (1965). "Revision of the Speech Privacy Calculation," *Journal of the Acoustical Society of America,* Vol. 38, No. 4.

PART X
APPENDICES

The purpose of this part is to present extensive important reference and ancillary data in a separate, easily accessed section. The appendices are entitled:

A

CLIMATIC CONDITIONS FOR THE UNITED STATES, CANADA, AND MEXICO

Prepared by ASHRAE Technical Committee 4.2, Weather Data, from data compiled from official weather stations where hourly weather observations are made by trained observers.

Station Designations

AP = airport

CO = urban areas (influenced by surroundings)

No designation = semirural, similar to airport locations

S = solar data available

Winter Design Conditions

For United States: percentage of the three-month period, December through February.

For Canada: percentage for January.

Summer Design Conditions

For United States: percentage of the four-month period, June through September.

For Canada: percentage for July.

Other stations: Many more stations in each state or province are found in the *ASHRAE 1981 Handbook of Fundamentals,* Chapter 24.

Interpretations Between Stations. As a general rule, weather data from listed stations can be adjusted for nearby locations in these ways:

*Adjustment for elevation.** For lower elevations, which tend to be warmer, *increase* the design temperatures; for higher elevations, *decrease* the design temperatures:

Dry-bulb temperature	1F° per 200 ft elevation
Wet-bulb temperature:	1F° per 500 ft elevation

Adjustment for air mass (at coastlines). Along the West Coast, both dry-bulb and wet-bulb design temperatures increase with distance from the ocean. Along the Gulf coast, dry-bulb temperatures increase for the first 200 to 300 miles inland, while wet-bulb temperatures decrease slightly. (Beyond this 200- to 300-mile belt, both dry-bulb and wet-bulb values decrease.)

*Elevations for many locations are found in Appendix B, Table B.15.

TABLE A.1 **Outside Design Conditions: United States and Canada**

	Winter	Summer		
State and Station	*Design Dry-Bulb (97.5%)*	*Design Dry-Bulb and Mean Coincident Wet-Bulb (2.5%)*	*Mean Daily Range*	*Design Wet-Bulb (2.5%)*
ALABAMA				
Auburn	22	93/76	21	78
Birmingham AP	21	94/75	21	77
Huntsville AP	16	93/74	23	77
Mobile AP	29	93/77	18	79
Montgomery AP	25	95/76	21	79
ALASKA				
Anchorage AP	−18	68/58	15	59
Fairbanks AP (S)	−47	78/60	24	62
Juneau AP	1	70/58	15	59
Nome AP	−27	62/55	10	56
ARIZONA				
Flagstaff AP	4	82/55	31	60
Phoenix AP (S)	34	107/71	27	75
Prescott AP	9	94/60	30	65
Tuscon AP (S)	32	102/66	26	71
Winslow AP	10	95/60	32	65
Yuma AP	39	109/72	27	78
ARKANSAS				
Fayetteville AP	12	94/73	23	76
Fort Smith AP	17	98/76	24	79
Little Rock AP (S)	20	96/77	22	79
Texarkana AP	23	96/77	21	79
CALIFORNIA				
Bakersfield AP	32	101/69	32	71
Barstow AP	29	104/68	37	71
Burbank AP	39	91/68	25	70
Eureka/Arcata AP	33	65/59	11	60
Fresno AP (S)	30	100/69	34	71
Long Beach AP	43	80/68	22	69
Los Angeles AP (S)	43	80/68	15	69
Los Angeles CO (S)	40	89/70	20	71
Needles AP	33	110/71	27	75
Oakland AP	36	80/63	19	64
Pomona CO	30	99/69	36	72
Redding AP	31	102/67	32	69
Sacramento AP	32	98/70	36	71
San Diego AP	44	80/69	12	70
San Francisco AP	38	77/63	20	64
San Francisco CO	40	71/62	14	62
San Luis Obispo	35	88/70	26	71
Santa Barbara AP	36	77/66	24	67
Santa Maria AP (S)	33	76/63	23	64
Santa Monica CO	43	80/68	16	69
COLORADO				
Alamosa AP	−16	82/57	35	61

TABLE A.1 **Outside Design Conditions: United States and Canada (Continued)**

	Winter	*Summer*		
State and Station	*Design Dry-Bulb (97.5%)*	*Design Dry-Bulb and Mean Coincident Wet-Bulb (2.5%)*	*Mean Daily Range*	*Design Wet-Bulb (2.5%)*
Boulder	8	91/59	27	63
Colorado Springs AP	2	88/57	30	62
Denver AP	1	91/59	28	63
Fort Collins	−4	91/59	28	63
Grand Junction AP (S)	7	94/59	29	63
Leadville	−4	81/51	30	55
Pueblo AP	0	95/61	31	66
CONNECTICUT				
Bridgeport AP	9	84/71	18	74
Hartford	7	88/73	22	75
New Haven AP	7	84/73	17	75
DELAWARE				
Wilmington AP	14	89/74	20	76
DISTRICT OF COLUMBIA				
Washington National AP	17	91/74	18	77
FLORIDA				
Daytona Beach AP	35	90/77	15	79
Fort Myers AP	44	92/78	18	79
Gainesville AP (S)	31	93/77	18	79
Jacksonville AP	32	94/77	19	79
Key West AP	57	90/78	9	79
Miami AP (S)	47	90/77	15	79
Miami Beach CO	48	89/77	10	79
Orlando AP	38	93/76	17	78
Panama City, Tyndall AFB	33	90/77	14	80
Pensacola CO	29	93/77	14	79
Tallahassee AP (S)	30	92/76	19	78
Tampa AP (S)	40	91/77	17	79
West Palm Beach AP	45	91/78	16	79
GEORGIA				
Athens	22	92/74	21	77
Atlanta AP (S)	22	92/74	19	76
Augusta AP	23	95/76	19	79
Columbus, Lawson AFB	24	93/76	21	78
Macon AP	25	93/76	22	78
Rome AP	22	93/76	23	78
Savannah-Travis AP	27	93/77	20	79
Valdosta-Moody AFB	31	94/77	20	79
HAWAII				
Hilo AP (S)	62	83/72	15	74
Honolulu AP	63	86/73	12	75
Wahiawa	59	85/72	14	74
IDAHO				
Boise AP (S)	10	94/64	31	66

TABLE A.1 **Outside Design Conditions: United States and Canada (Continued)**

	Winter	Summer		
State and Station	Design Dry-Bulb (97.5%)	Design Dry-Bulb and Mean Coincident Wet-Bulb (2.5%)	Mean Daily Range	Design Wet-Bulb (2.5%)
Lewiston AP	6	93/64	32	66
Moscow	0	87/62	32	64
Pocatello AP	−1	91/60	35	63
ILLINOIS				
Carbondale	7	93/77	21	79
Champaign/Urbana	2	92/74	21	77
Chicago, O'Hare AP	−4	89/74	20	76
Chicago CO	2	91/74	15	77
Moline AP	−4	91/75	23	77
Peoria AP	−4	89/74	22	76
Rockford	−4	89/73	24	76
Springfield AP	2	92/74	21	77
INDIANA				
Evansville AP	9	93/75	22	78
Fort Wayne AP	1	89/72	24	75
Indianapolis AP (S)	2	90/74	22	76
Lafayette	3	91/73	22	76
Muncie	2	90/73	22	76
South Bend AP	1	89/73	22	75
Terre Haute AP	4	92/74	22	77
IOWA				
Ames (S)	−6	90/74	23	76
Burlington AP	−3	91/75	22	77
Des Moines AP	−5	91/74	23	77
Dubuque	−7	88/73	22	75
Iowa City	−6	89/76	22	78
Mason City AP	−11	88/74	24	75
Sioux City AP	−7	92/74	24	77
Waterloo	−10	89/75	23	77
KANSAS				
Dodge City AP (S)	5	97/69	25	73
Goodland AP	0	96/65	31	70
Manhattan, Fort Riley (S)	3	95/75	24	77
Topeka AP	4	96/75	24	78
Wichita AP	7	98/73	23	76
KENTUCKY				
Bowling Green AP	10	92/75	21	77
Lexington AP (S)	8	91/73	22	76
Louisville AP	10	93/74	23	77
LOUISIANA				
Alexandria AP	27	94/77	20	79
Baton Rouge AP	29	93/77	19	80
Lafayette AP	30	94/78	18	80
Lake Charles AP (S)	31	93/77	17	79

TABLE A.1 **Outside Design Conditions: United States and Canada (Continued)**

State and Station	Winter: Design Dry-Bulb (97.5%)	Summer: Design Dry-Bulb and Mean Coincident Wet-Bulb (2.5%)	Summer: Mean Daily Range	Summer: Design Wet-Bulb (2.5%)
Monroe AP	25	96/76	20	79
New Orleans AP	33	92/78	16	80
Shreveport AP (S)	25	96/76	20	79
MAINE				
Bangor, Dow AFB	−6	83/68	22	71
Caribou AP (S)	−13	81/67	21	69
Portland (S)	−1	84/71	22	72
MARYLAND				
Baltimore AP	13	91/75	21	77
Baltimore CO	17	89/76	17	78
Cumberland	10	89/74	22	76
Frederick AP	12	91/75	22	77
Salisbury (S)	16	91/75	18	77
MASSACHUSETTS				
Boston AP (S)	9	88/71	16	74
Pittsfield AP	−3	84/70	23	72
Worcester AP	4	84/70	18	72
MICHIGAN				
Alpena AP	−6	85/70	27	72
Detroit	6	88/72	20	74
Escanaba	−7	83/69	17	71
Flint AP	1	87/72	25	74
Grand Rapids AP	5	88/72	24	74
Muskegon AP	6	84/70	21	73
Sault Ste. Marie AP (S)	−8	81/69	23	70
Traverse City AP	1	86/71	22	73
MINNESOTA				
Bemidji AP	−26	85/69	24	71
Duluth AP	−16	82/68	22	70
International Falls AP	−25	83/68	26	70
Minneapolis/St. Paul AP	−12	89/73	22	75
Rochester AP	−12	87/72	24	75
MISSISSIPPI				
Columbus AFB	20	93/77	22	79
Jackson AP	25	95/76	21	78
Meridian AP	23	95/76	22	79
Vicksburg CO	26	95/78	21	80
MISSOURI				
Columbia AP (S)	4	94/74	22	77
Kansas City AP	6	96/74	20	77
St Louis AP	6	94/75	21	77
St Louis CO	8	94/75	18	77
Springfield AP	9	93/74	23	77
MONTANA				
Billings AP	−10	91/64	31	66
Bozeman	−14	87/60	32	62

TABLE A.1 **Outside Design Conditions: United States and Canada (Continued)**

	Winter	*Summer*		
State and Station	*Design Dry-Bulb (97.5%)*	*Design Dry-Bulb and Mean Coincident Wet-Bulb (2.5%)*	*Mean Daily Range*	*Design Wet-Bulb (2.5%)*
Cut Bank AP	−20	85/61	35	62
Glasgow AP (S)	−18	89/63	29	66
Great Falls AP (S)	−15	88/60	28	62
Helena AP	−16	88/60	32	62
Kalispell AP	−7	87/61	34	63
Lewiston AP	−16	87/61	30	63
Miles City AP	−15	95/66	30	68
Missoula AP	−6	88/61	36	63
NEBRASKA				
Grand Island AP	−3	94/71	28	74
Lincoln CO (S)	−2	95/74	24	77
North Platte AP (S)	−4	94/69	28	72
Omaha AP	−3	91/75	22	77
Scottsbluff AP	−3	92/65	31	68
NEVADA				
Elko AP	−2	92/59	42	62
Ely AP (S)	−4	87/56	39	59
Las Vegas AP (S)	28	106/65	30	70
Lovelock AP	12	96/63	42	65
Reno AP (S)	10	92/60	45	62
Tonopah AP	10	92/59	40	62
Winnemucca AP	3	94/60	42	62
NEW HAMPSHIRE				
Berlin	−9	84/69	22	71
Concord AP	−3	87/70	26	73
Portsmouth, Pease AFB	2	85/71	22	74
NEW JERSEY				
Atlantic City CO	13	89/74	18	77
Newark AP	14	91/73	20	76
Trenton CO	14	88/74	19	76
NEW MEXICO				
Albuquerque AP (S)	16	94/61	27	65
Farmington AP	6	93/62	30	65
Las Cruces	20	96/64	30	68
Los Alamos	9	87/60	32	61
Raton AP	1	89/60	34	64
Roswell, Walker AFB	18	98/66	33	70
Silver City AP	10	94/60	30	64
Tucumcari AP	13	97/66	28	69
NEW YORK				
Albany AP (S)	−1	88/72	23	74
Binghamton AP	1	83/69	20	72
Buffalo AP	6	85/70	21	73
Ithaca (S)	0	85/71	24	73
Massena AP	−8	83/69	20	72
NYC—Central Park (S)	15	89/73	17	75

TABLE A.1 **Outside Design Conditions: United States and Canada (Continued)**

	Winter	*Summer*		
State and Station	*Design Dry-Bulb (97.5%)*	*Design Dry-Bulb and Mean Coincident Wet-Bulb (2.5%)*	*Mean Daily Range*	*Design Wet-Bulb (2.5%)*
NYC—Kennedy AP	15	87/72	16	75
Rochester AP	5	88/71	22	73
Syracuse AP	2	87/71	20	73
NORTH CAROLINA				
Asheville AP	14	87/72	21	74
Charlotte AP	22	93/74	20	76
Greensboro AP (S)	18	91/73	21	76
Raleigh/Durham AP (S)	20	92/75	20	77
Wilmington AP	26	91/78	18	80
NORTH DAKOTA				
Bismarck AP (S)	−19	91/68	27	71
Fargo AP	−18	89/71	25	74
Minot AP	−20	89/67	25	70
Williston	−21	88/67	25	70
OHIO				
Akron-Canton AP	6	86/71	21	73
Cincinnati CO	6	90/72	21	75
Cleveland AP (S)	5	88/72	22	74
Columbus AP (S)	5	90/73	24	75
Dayton AP	4	89/72	20	75
Mansfield AP	5	87/72	22	74
Toledo AP	1	88/73	25	75
Youngstown AP	4	86/71	23	73
OKLAHOMA				
Norman	13	96/74	24	76
Oklahoma City AP (S)	13	97/74	23	77
Stillwater (S)	13	96/74	24	76
Tulsa AP	13	98/75	22	78
OREGON				
Astoria AP (S)	29	71/62	16	63
Bend	4	87/60	33	62
Eugene AP	22	89/66	31	67
Medford AP (S)	23	94/67	35	68
Pendleton AP	5	93/64	29	65
Portland AP	23	85/67	23	67
Portland CO	24	86/67	21	67
Salem AP	23	88/66	31	68
PENNSYLVANIA				
Allentown AP	9	88/72	22	75
Erie AP	9	85/72	18	74
Harrisburg AP	11	91/74	21	76
Philadelphia AP	14	90/74	21	76
Pittsburgh AP	5	86/71	22	73
Pittsburgh CO	7	88/71	19	73
Scranton/Wilkes-Barre	5	87/71	19	73
State College (S)	7	87/71	23	73

TABLE A.1 **Outside Design Conditions: United States and Canada (Continued)**

State and Station	Winter: Design Dry-Bulb (97.5%)	Summer: Design Dry-Bulb and Mean Coincident Wet-Bulb (2.5%)	Summer: Mean Daily Range	Summer: Design Wet-Bulb (2.5%)
RHODE ISLAND				
Newport (S)	9	85/72	16	75
Providence AP	9	86/72	19	74
SOUTH CAROLINA				
Anderson	23	92/74	21	76
Charleston AFB (S)	27	91/78	18	80
Charleston CO	28	92/78	13	80
Columbia AP	24	95/75	22	78
Spartanburg AP	22	91/74	20	76
SOUTH DAKOTA				
Huron AP	−14	93/72	28	75
Pierre AP	−10	95/71	29	74
Rapid City AP (S)	−7	92/65	28	69
Sioux Falls AP	−11	91/72	24	75
TENNESSEE				
Bristol–Tri City AP	14	89/72	22	75
Chattanooga AP	18	93/74	22	77
Knoxville AP	19	92/73	21	76
Memphis AP	18	95/76	21	79
Nashville AP (S)	14	94/74	21	77
TEXAS				
Abilene AP	20	99/71	22	74
Amarillo AP	11	95/67	26	70
Austin AP	28	98/74	22	77
Brownsville AP (S)	39	93/77	18	79
Bryan AP	29	96/76	20	78
Corpus Christi AP	35	94/78	19	80
Dallas AP	22	100/75	20	78
Del Rio, Laughlin AFB	31	98/73	24	77
El Paso AP (S)	24	98/64	27	68
Forth Worth AP (S)	22	99/74	22	77
Galveston AP	36	89/79	10	80
Houston AP	32	94/77	18	79
Houston CO	33	95/77	18	79
Laredo AFB	36	101/73	23	78
Lubbock AP	15	96/69	26	72
Lufkin AP	29	97/76	20	79
Midland AP (S)	21	98/69	26	72
Port Arthur AP	31	93/78	19	80
San Angelo, Goodfellow AFB	22	99/71	24	74
San Antonio AP (S)	30	97/73	19	76
Sherman Perrin AFB	20	98/75	22	77
Waco AP	26	99/75	22	78
Wichita Falls AP	18	101/73	24	76
UTAH				
Cedar City AP	5	91/60	32	63

TABLE A.1 **Outside Design Conditions: United States and Canada (Continued)**

	Winter	*Summer*		
State and Station	*Design Dry-Bulb (97.5%)*	*Design Dry-Bulb and Mean Coincident Wet-Bulb (2.5%)*	*Mean Daily Range*	*Design Wet-Bulb (2.5%)*
Salt Lake City AP (S)	8	95/62	32	65
Vernal AP	0	89/60	32	63
VERMONT				
Barre	−11	81/69	23	71
Burlington AP (S)	−7	85/70	23	72
VIRGINIA				
Charlottesville	18	91/74	23	76
Norfolk AP	22	91/76	18	78
Richmond AP	17	92/76	21	78
Roanoke AP	16	91/72	23	74
WASHINGTON				
Bellingham AP	15	77/65	19	65
Olympia AP	22	83/65	32	66
Seattle CO (S)	27	82/66	19	67
Seattle-Tacoma AP (S)	26	80/64	22	64
Spokane AP (S)	2	90/63	28	64
Walla Walla AP	7	94/66	27	67
Yakima AP	5	93/65	36	66
WEST VIRGINIA				
Beckley	4	81/69	22	71
Charleston AP	11	90/73	20	75
Huntington CO	10	91/74	22	77
Wheeling	5	86/71	21	73
WISCONSIN				
Ashland	−16	82/68	23	70
Eau Claire AP	−11	89/73	23	75
Green Bay AP	−9	85/72	23	74
La Cross AP	−9	88/73	22	75
Madison AP (S)	−7	88/73	22	75
Milwaukee AP	−4	87/73	21	74
WYOMING				
Casper AP	−5	90/57	31	61
Cheyenne AP	−1	86/58	30	62
Lander AP (S)	−11	88/61	32	63
Laramie AP (S)	−6	81/56	28	60
Rock Springs AP	−3	84/55	32	58
Sheridan AP	−8	91/62	32	65
Province and Station				
ALBERTA				
Calgary AP	−23	81/61	25	63
Edmonton AP	−25	82/65	23	66
Medicine Hat AP	−24	90/65	28	68
BRITISH COLUMBIA				
Dawson Creek	−33	79/63	26	64

TABLE A.1 **Outside Design Conditions: United States and Canada (Continued)**

	Winter	*Summer*		
State and Station	*Design Dry-Bulb (97.5%)*	*Design Dry-Bulb and Mean Coincident Wet-Bulb (2.5%)*	*Mean Daily Range*	*Design Wet-Bulb (2.5%)*
Nanaimo (S)	20	80/65	21	66
Prince George AP (S)	−28	80/62	26	64
Prince Rupert CO	2	63/57	12	58
Vancouver AP (S)	19	77/66	17	67
Victoria CO	23	73/62	16	62
MANITOBA				
Flin Flon	−37	81/66	19	68
Winnipeg AP (S)	−27	86/71	22	73
NEW BRUNSWICK				
Fredericton AP (S)	−11	85/69	23	71
Saint John AP	−8	77/65	19	68
NEWFOUNDLAND				
Corner Brook	0	73/63	17	66
St. John's AP (S)	7	75/65	18	67
NORTHWEST TERR.				
Fort Smith AP (S)	−45	81/64	24	66
Inuvik (S)	−53	77/60	21	62
Yellowknife AP	−46	77/61	16	63
NOVA SCOTIA				
Halifax AP (S)	5	76/65	16	67
Sydney AP	3	80/68	19	70
Yarmouth AP	9	72/64	15	66
ONTARIO				
Kitchener	−2	85/72	23	74
Ottawa AP (S)	−13	87/71	21	73
Sudbury AP	−19	83/67	22	70
Toronto AP (S)	−1	87/72	20	74
PRINCE EDWARD ISLAND				
Charlottetown AP (S)	−4	78/68	16	70
QUEBEC				
Chicoutimi	−22	83/68	20	70
Montréal AP (S)	−10	85/72	17	74
Québec AP	−14	84/70	20	72
Sept Iles AP (S)	−21	73/61	17	65
Val d'Or AP	−27	83/68	22	70
SASKATCHEWAN				
Prince Albert AP	−35	84/66	25	68
Regina AP	−29	88/68	26	70
Saskatoon AP (S)	−31	86/66	26	68
YUKON TERRITORY				
Whitehorse AP (S)	−43	77/58	22	59

[a]S = solar data available. U.S. solar data available from the National Climatic Data Center, Federal Building, Ashville, NC 28801. Canadian solar data available from The Canadian Climate Center, Atmospheric Environment Service, 4905 Dufferin St., Downsview, Ontario M3H 5T 4.

APPENDICES

TABLE A.2 **Outdoor Design Conditions: Mexico and Puerto Rico**

	Winter	*Summer*		
Country and Station	*Design Dry-Bulb 97.5%*	*Design Dry-Bulb 2.5%*	*Mean Daily Range*	*Design Wet-Bulb 2.5%*
MEXICO				
Guadalajara	42	91	29	67
Mérida	61	95	21	79
Mexico City	39	81	25	60
Monterrey	41	95	20	78
Vera Cruz	62	89	20	83
PUERTO RICO				
San Juan	68	88	11	80

B

SOLAR DATA

This appendix presents both *clear day* and *average day* data.

Solar Intensity and Solar Heat Gain Factors (Tables B.1–B.10).

These are listed for five latitudes; solar heat gain factors occur through one layer of double-strength sheet (DSA) glass. The data are based on an "average cloudless" day at each latitude. Tables B.1 through B.10 are copyrighted by the American Society of Heating, Refrigerating, and Air Conditioning Engineers, Inc., Atlanta, GA., and are reprinted by permission from *ASHRAE 1981 Handbook of Fundamentals*.

Solar Position and Clear Day Global Irradiance (Insolation) (Tables B.11–B.14).

These are listed for four latitudes. The data are presented in conventional units for the following tilt angles (degrees above horizontal): latitude $-10°$, latitude, latitude $+10°$, latitude $+20°$, and vertical. To convert to SI units: Btu/h $ft^2 \times 0.317 = W/m^2$.

Elevation, Latitude, Average Horizontal Insolation, Average Vertical Insolation, Average Temperature, and Heating Degree Days (Table B.15).

These are listed in conventional units for January and July, as well as yearly totals, for a number of locations (Fig. B.1). To convert horizontal surface (HS) and vertical surface (VS) to SI units: Btu/h $ft^2 \times 0.317 = W/m^2$. To convert average temperature (TA) to Celsius: $(°F - 32)/1.8 = °C$. To convert DD(F) to DD(C): DD(F) $\times$ 9/5 = DD(C) or DD(K).

APPENDICES

TABLE B.1 **Solar Intensity and Solar Heat Gain Factors[a] for 16° N Latitude (Conventional Units)[b]**

Date	Solar Time am	Direct Normal Btuh ft²	N	NNE	NE	ENE	E	ESE	SE	SSE	S	SSW	SW	WSW	W	WNW	NW	NNW	HOR	Solar Time pm
			Solar Heat Gain Factors, Btuh ft²																	
Jan 21	7	141	5	6	44	92	124	134	126	96	49	6	5	5	5	5	5	5	14	5
	8	262	14	15	55	147	210	240	233	189	114	25	14	14	14	14	14	14	79	4
	9	300	21	21	32	122	200	244	251	219	152	58	22	21	21	21	21	21	150	3
	10	317	26	26	27	66	150	209	233	223	178	102	31	26	26	26	26	26	203	2
	11	325	29	29	29	31	77	148	195	210	194	146	75	31	29	29	29	29	236	1
	12	327	30	30	30	30	32	73	139	184	199	184	138	72	32	30	30	30	248	12
HALF DAY TOTALS			110	112	196	461	760	1000	1096	1020	781	426	211	127	111	110	110	110	805	
Feb 21	7	182	8	17	84	138	169	172	150	103	36	8	8	8	8	8	8	8	25	5
	8	273	17	19	96	180	231	247	224	166	77	18	17	17	17	17	17	17	101	4
	9	305	23	24	64	153	214	242	233	188	110	30	23	23	23	23	23	23	174	3
	10	319	28	29	33	92	161	202	211	188	134	61	30	28	28	28	28	28	229	2
	11	326	32	32	32	37	83	136	167	172	149	102	49	33	32	32	32	32	263	1
	12	328	33	33	33	33	34	60	107	142	154	142	106	60	34	33	33	33	275	12
HALF DAY TOTALS			124	137	321	609	865	1023	1034	885	582	287	174	132	124	124	124	124	930	
Mar 21	7	201	11	53	124	172	192	183	145	82	15	10	10	10	10	10	10	10	40	5
	8	272	20	50	140	205	239	235	195	123	35	19	19	19	19	19	19	19	120	4
	9	299	26	35	109	179	218	225	197	138	57	27	26	26	26	26	26	26	192	3
	10	312	31	33	61	120	165	182	172	134	76	34	32	31	31	31	31	31	247	2
	11	318	34	35	36	53	87	114	125	116	89	55	36	35	34	34	34	34	280	1
	12	320	35	35	36	36	37	47	69	87	93	86	68	47	37	36	36	35	291	12
HALF DAY TOTALS			141	226	494	755	928	975	879	643	319	187	153	142	139	139	139	139	1025	
Apr 21	6	14	2	8	12	14	14	12	8	2	1	1	1	1	1	1	1	1	1	6
	7	197	24	94	153	187	191	167	117	45	14	13	13	13	13	13	13	13	53	5
	8	256	27	99	172	216	227	204	150	69	24	22	22	22	22	22	22	22	131	4
	9	280	31	79	149	193	208	193	147	77	31	29	29	29	29	29	29	29	197	3
	10	293	35	54	102	141	158	151	120	73	37	34	33	33	33	33	33	33	249	2
	11	299	38	40	54	72	86	88	78	60	43	38	38	36	36	36	36	37	279	1
	12	301	39	39	39	40	40	41	43	45	45	45	43	41	40	39	39	39	289	12
HALF DAY TOTALS			179	403	674	859	922	851	653	352	174	159	157	156	155	155	155	156	1057	
May 21	6	44	14	30	41	45	43	34	19	4	3	3	3	3	3	3	3	3	5	6
	7	193	50	120	168	191	185	150	92	24	16	16	16	16	16	16	16	17	62	5
	8	244	52	132	189	218	215	179	115	38	25	24	24	24	24	24	24	25	135	4
	9	268	49	116	171	198	197	167	109	45	32	30	30	30	30	30	30	32	197	3
	10	280	47	89	130	151	150	126	84	44	37	35	35	35	35	35	35	37	245	2
	11	286	47	63	79	87	83	70	52	41	40	39	38	38	38	39	39	41	273	1
	12	288	46	46	44	43	42	41	41	41	41	41	41	41	42	43	44	46	282	12
HALF DAY TOTALS			283	575	804	916	897	748	493	217	172	167	167	167	167	168	169	176	1058	
Jun 21	6	53	20	39	52	55	51	39	20	4	4	4	4	4	4	4	4	4	7	6
	7	188	62	128	172	190	179	141	80	20	16	16	16	16	16	16	16	18	64	5
	8	238	66	142	194	217	207	167	99	31	25	25	25	25	25	25	25	27	135	4
	9	261	63	130	178	198	190	154	93	37	31	31	31	31	31	31	31	33	194	3
	10	273	59	104	140	154	145	115	70	39	37	36	36	36	36	36	36	38	241	2
	11	279	57	76	90	92	82	63	46	41	40	39	39	39	39	40	41	43	268	1
	12	281	57	55	50	45	43	42	41	41	41	41	41	42	42	45	50	55	277	12
HALF DAY TOTALS			356	648	850	929	876	700	430	194	174	171	171	171	172	173	176	190	1049	
Jul 21	6	41	14	29	39	42	40	31	18	4	3	3	3	3	3	3	3	3	6	6
	7	184	51	118	164	185	179	145	88	23	16	16	16	16	16	16	16	17	62	5
	8	236	55	132	187	214	210	174	111	37	25	25	25	25	25	25	25	26	133	4
	9	259	52	117	170	196	193	163	106	44	32	31	31	31	31	31	31	33	194	3
	10	272	50	92	131	151	148	123	81	44	38	36	36	36	36	36	36	38	241	2
	11	278	49	66	81	88	83	69	52	42	41	40	39	39	39	40	40	42	269	1
	12	279	49	48	46	44	43	42	42	42	42	42	42	42	43	44	46	48	277	12
HALF DAY TOTALS			296	580	799	903	878	729	478	215	176	172	171	171	171	172	173	182	1043	
Aug 21	6	11	2	7	10	12	12	10	6	2	1	1	1	1	1	1	1	1	1	6
	7	180	26	92	145	176	180	156	109	42	15	14	14	14	14	14	14	14	53	5
	8	240	30	100	168	209	219	196	143	65	25	23	23	23	23	23	23	23	128	4
	9	266	33	82	148	190	203	187	142	74	33	30	30	30	30	30	30	30	193	3
	10	279	37	58	104	140	155	147	117	71	39	36	35	35	35	35	35	35	243	2
	11	285	40	43	57	75	86	87	76	59	44	40	39	38	38	38	38	39	273	1
	12	287	41	41	41	42	42	43	44	45	46	45	44	43	42	41	41	41	282	12
HALF DAY TOTALS			191	410	666	837	891	817	624	339	180	167	165	164	163	163	163	164	1033	
Sep 21	7	179	12	50	114	158	176	168	133	76	15	11	11	11	11	11	11	11	39	5
	8	253	21	49	134	196	227	224	186	119	36	20	20	20	20	20	20	20	116	4
	9	281	28	36	106	173	211	217	191	134	57	28	27	27	27	27	27	27	185	3
	10	295	32	34	61	118	161	178	168	132	76	35	33	32	32	32	32	32	238	2
	11	302	35	36	37	54	86	113	123	114	88	56	38	36	35	35	35	35	271	1
	12	304	36	36	37	38	39	49	69	86	93	86	69	48	39	38	37	36	282	12
HALF DAY TOTALS			146	226	475	722	885	931	842	622	319	192	159	148	145	144	144	144	991	
Oct 21	7	166	8	18	79	128	156	159	139	95	33	9	8	8	8	8	8	8	25	5
	8	259	17	20	95	174	223	237	215	159	74	19	17	17	17	17	17	17	99	4
	9	292	24	25	65	150	209	235	225	182	106	31	24	24	24	24	24	24	170	3
	10	307	29	30	34	92	158	197	205	183	130	60	31	29	29	29	29	29	224	2
	11	314	32	32	33	39	83	133	163	167	145	100	49	34	32	32	32	32	258	1
	12	316	33	33	33	34	35	60	105	139	150	138	104	60	35	34	33	33	270	12
HALF DAY TOTALS			127	141	318	592	836	986	996	852	563	283	175	136	128	127	127	127	911	
Nov 21	7	134	5	6	43	89	119	129	120	92	47	6	5	5	5	5	5	5	14	5
	8	255	15	15	55	145	206	235	228	185	111	25	15	15	15	15	15	15	78	4
	9	295	21	21	33	121	197	241	247	215	150	57	22	21	21	21	21	21	149	3
	10	312	26	26	28	67	147	206	230	220	176	100	31	26	26	26	26	26	201	2
	11	320	29	29	29	31	77	146	192	207	191	144	74	31	29	29	29	29	234	1
	12	322	30	30	30	30	32	72	137	181	196	181	137	72	32	30	30	30	246	12
HALF DAY TOTALS			112	113	197	456	749	983	1077	1001	767	420	210	128	112	112	112	112	799	
Dec 21	7	118	4	5	30	72	101	112	107	85	48	7	4	4	4	4	4	4	10	5
	8	255	13	14	41	132	198	233	231	193	124	33	13	13	13	13	13	13	69	4
	9	297	20	20	25	108	191	241	254	227	165	72	21	20	20	20	20	20	138	3
	10	315	25	25	26	56	144	208	239	233	192	117	35	25	25	25	25	25	191	2
	11	323	28	28	28	29	73	150	202	221	207	161	86	30	28	28	28	28	223	1
	12	325	29	29	29	29	30	77	149	197	212	196	149	76	30	29	29	29	234	12
HALF DAY TOTALS			104	105	159	402	710	975	1099	1050	836	484	228	125	105	104	104	104	748	
			N	NNW	NW	WNW	W	WSW	SW	SSW	S	SSE	SE	ESE	E	ENE	NE	NNE	HOR	PM

[a]Total solar heat gains for DSA glass (based on a ground reflectance of 0.20).

[b]Half-day totals computed by Simpson's rule, time interval = 10 min.

APPENDICES

TABLE B.2 **Solar Intensity and Solar Heat Gain Factors[a] for 32° N Latitude (Conventional Units)[b]**

Date	Solar Time am	Direct Normal Btuh/ft²	N	NNE	NE	ENE	E	ESE	SE	SSE	S	SSW	SW	WSW	W	WNW	NW	NNW	HOR	Solar Time pm
			Solar Heat Gain Factors, Btuh · ft²																	
Jan 21	7	1	0	0	0	1	1	1	1	1	0	0	0	0	0	0	0	0	0	5
	8	203	9	9	29	105	160	189	189	159	103	28	9	9	9	9	9	9	32	4
	9	269	15	15	17	91	175	229	246	225	169	82	17	15	15	15	15	15	88	3
	10	295	20	20	20	41	135	209	249	250	212	141	46	20	20	20	20	20	136	2
	11	306	23	23	23	24	68	159	221	249	238	191	110	29	23	23	23	23	166	1
	12	310	24	24	24	24	25	88	174	228	246	228	174	88	25	24	24	24	176	12
HALF DAY TOTALS			79	79	107	284	570	856	1015	1014	853	553	264	112	80	79	79	79	512	
Feb 21	7	112	4	7	47	82	102	106	95	67	26	4	4	4	4	4	4	4	9	5
	8	245	13	14	65	149	205	228	216	170	95	17	13	13	13	13	13	13	64	4
	9	287	19	19	32	122	199	242	248	216	149	55	20	19	19	19	19	19	127	3
	10	305	24	24	25	62	151	213	241	232	189	112	31	24	24	24	24	24	176	2
	11	314	26	26	26	28	76	156	208	227	212	165	87	28	26	26	26	26	207	1
	12	316	27	27	27	27	29	79	155	204	221	204	155	79	29	27	27	27	217	12
HALF DAY TOTALS			100	103	201	445	735	978	1080	1010	780	452	228	122	100	100	100	100	691	
Mar 21	7	185	10	37	105	153	176	173	142	88	20	9	9	9	9	9	9	9	32	5
	8	260	17	25	107	183	227	237	209	150	62	18	17	17	17	17	17	17	100	4
	9	290	23	25	64	151	210	237	227	183	107	30	23	23	23	23	23	23	164	3
	10	304	28	28	30	87	158	202	215	195	144	70	29	28	28	28	28	28	211	2
	11	311	31	31	31	34	82	142	179	188	168	120	59	32	31	31	31	31	242	1
	12	313	32	32	32	32	33	66	122	162	176	162	122	66	33	32	32	32	252	12
HALF DAY TOTALS			124	162	359	629	875	1033	1041	888	589	326	193	136	125	124	124	124	874	
Apr 21	6	66	9	35	54	65	66	56	38	12	4	3	3	3	3	3	3	3	7	6
	7	206	17	80	146	188	200	182	136	65	16	14	14	14	14	14	14	14	61	5
	8	255	23	61	144	200	227	219	177	107	30	22	22	22	22	22	22	22	129	4
	9	278	28	36	103	168	206	212	187	133	58	29	28	28	28	28	28	28	188	3
	10	290	32	34	52	108	155	177	172	141	87	39	33	32	32	32	32	32	233	2
	11	295	35	35	36	47	83	118	135	132	108	70	40	36	35	35	35	35	262	1
	12	297	36	36	36	37	38	53	82	106	115	106	82	53	38	37	36	36	271	12
HALF DAY TOTALS			161	296	550	792	952	992	889	645	360	228	177	157	153	152	152	152	1015	
May 21	6	119	33	77	108	121	116	94	56	13	8	8	8	8	8	8	8	9	21	6
	7	211	36	111	170	202	204	174	118	42	19	18	18	18	18	18	18	19	81	5
	8	250	29	94	165	208	220	199	149	73	27	25	25	25	25	25	25	25	146	4
	9	269	33	61	128	177	198	190	155	93	37	32	31	31	31	31	31	31	201	3
	10	280	36	40	76	121	150	156	138	99	54	37	35	35	35	35	35	35	243	2
	11	285	38	39	42	59	83	99	102	90	68	47	40	39	37	37	37	37	269	1
	12	286	38	39	40	40	41	47	59	70	74	70	59	47	41	40	40	39	277	12
HALF DAY TOTALS			222	438	702	900	985	933	747	447	250	199	183	177	175	174	174	175	1098	
Jun 21	6	131	44	92	123	135	127	99	55	12	10	10	10	10	10	10	10	11	28	6
	7	210	47	122	176	204	201	168	108	35	20	20	20	20	20	20	20	21	88	5
	8	245	36	106	171	208	214	189	135	60	28	27	27	27	27	27	27	27	151	4
	9	264	35	74	137	178	193	180	139	77	35	32	32	32	32	32	32	32	204	3
	10	274	38	47	86	125	146	145	123	83	45	38	36	36	36	36	36	36	244	2
	11	279	40	41	47	64	82	91	89	75	56	43	41	40	39	39	39	39	269	1
	12	280	41	41	41	42	42	46	52	58	60	58	52	46	42	42	41	41	276	12
HALF DAY TOTALS			261	504	762	935	985	897	678	372	225	197	189	185	184	184	183	186	1122	
Jul 21	6	113	34	76	105	117	113	90	53	12	9	9	9	9	9	9	9	9	22	6
	7	203	38	111	167	198	198	169	114	41	20	19	19	19	19	19	19	19	81	5
	8	241	31	95	163	204	215	194	145	70	28	26	26	26	26	26	26	26	145	4
	9	261	34	64	129	175	195	186	150	90	37	32	32	32	32	32	32	32	198	3
	10	271	37	42	78	121	148	153	134	96	53	38	36	36	36	36	36	36	240	2
	11	277	39	40	43	60	83	98	99	88	66	47	41	40	38	38	38	38	265	1
	12	279	40	40	41	41	42	48	58	68	72	68	58	48	42	41	41	40	273	12
HALF DAY TOTALS			231	444	701	890	967	912	726	433	248	202	187	182	180	179	179	180	1088	
Aug 21	6	59	10	33	50	60	60	51	34	11	4	4	4	4	4	4	4	4	8	6
	7	190	19	79	141	179	190	172	128	61	17	15	15	15	15	15	15	15	61	5
	8	240	25	63	141	195	219	210	170	102	31	23	23	23	23	23	23	23	128	4
	9	263	30	39	104	166	200	206	181	127	57	31	29	29	29	29	29	29	185	3
	10	276	34	36	55	109	153	173	167	136	84	40	35	34	34	34	34	34	229	2
	11	282	36	37	39	50	84	116	131	127	104	69	41	38	36	36	36	36	256	1
	12	284	37	37	37	39	40	54	81	103	111	103	81	54	40	39	37	37	265	12
HALF DAY TOTALS			171	303	546	774	922	955	854	618	352	231	184	166	162	161	160	160	999	
Sep 21	7	163	10	35	96	139	159	156	128	80	20	10	10	10	10	10	10	10	31	5
	8	240	18	26	103	173	215	224	198	143	60	19	18	18	18	18	18	18	96	4
	9	272	24	26	64	146	202	227	218	177	105	31	24	24	24	24	24	24	158	3
	10	287	29	29	32	86	154	196	208	189	141	70	31	29	29	29	29	29	204	2
	11	294	32	32	32	36	81	139	174	182	163	118	59	34	32	32	32	32	234	1
	12	296	33	33	33	33	35	66	120	158	171	158	120	66	35	33	33	33	244	12
HALF DAY TOTALS			130	164	345	598	831	982	993	852	574	325	197	142	130	129	129	129	845	
Oct 21	7	99	4	7	43	74	92	96	85	60	24	5	4	4	4	4	4	4	10	5
	8	229	13	15	63	143	195	217	206	162	90	17	13	13	13	13	13	13	63	4
	9	273	20	20	33	120	193	234	239	208	144	54	21	20	20	20	20	20	125	3
	10	293	24	24	26	62	147	207	234	225	183	109	32	24	24	24	24	24	173	2
	11	302	27	27	27	29	76	152	203	221	207	160	85	29	27	27	27	27	203	1
	12	304	28	28	28	28	30	78	151	199	215	199	151	78	30	28	28	28	213	12
HALF DAY TOTALS			103	106	200	433	708	941	1038	972	753	441	226	125	104	103	103	103	679	
Nov21	7	2	0	0	0	1	1	1	1	1	1	0	0	0	0	0	0	0	0	5
	8	196	9	9	29	103	156	184	184	155	100	27	9	9	9	9	9	9	32	4
	9	263	16	16	17	90	173	225	241	221	166	80	17	16	16	16	16	16	88	3
	10	289	20	20	21	41	134	206	245	246	209	138	45	21	20	20	20	20	136	2
	11	301	23	23	23	24	67	157	218	245	234	188	109	29	23	23	23	23	165	1
	12	304	24	24	24	24	25	87	171	224	243	224	171	87	25	24	24	24	175	12
HALF DAY TOTALS			80	81	108	282	561	841	996	995	838	544	261	113	81	80	80	80	509	
Dec 21	8	176	7	7	19	84	135	163	166	143	97	31	7	7	7	7	7	7	22	4
	9	257	14	14	15	77	162	218	238	222	171	89	15	14	14	14	14	14	72	3
	10	288	18	18	18	34	127	204	246	251	216	148	52	19	18	18	18	18	119	2
	11	301	21	21	21	22	63	157	222	252	243	197	116	29	21	21	21	21	148	1
	12	304	22	22	22	22	23	89	177	232	252	232	177	89	23	22	22	22	158	12
HALF DAY TOTALS			71	71	84	227	500	792	965	986	852	578	275	107	71	71	71	71	440	
			N	NNW	NW	WNW	W	WSW	SW	SSW	S	SSE	SE	ESE	E	ENE	NE	NNE	HOR	PM

[a]Total solar heat gains for DSA glass (based on a ground reflectance of 0.20).

[b]Half day totals computed by Simpson's rule, time interval = 10 min.

TABLE B.3 **Solar Intensity and Solar Heat Gain Factors[a] for 40° N Latitude (Conventional Units)[b]**

Date	Solar Time am	Direct Normal Btuh/ft²	N	NNE	NE	ENE	E	ESE	SE	SSE	S	SSW	SW	WSW	W	WNW	NW	NNW	HOR	Solar Time pm
			Solar Heat Gain Factors, Btuh/ft²																	
Jan 21	8	142	5	5	17	71	111	132	133	114	75	22	6	5	5	5	5	5	14	4
	9	239	12	12	13	74	154	205	224	209	160	82	13	12	12	12	12	12	55	3
	10	274	16	16	16	31	124	199	241	246	213	146	51	17	16	16	16	16	96	2
	11	289	19	19	19	20	61	156	222	252	244	198	118	28	19	19	19	19	124	1
	12	294	20	20	20	20	21	90	179	234	254	234	179	90	21	20	20	20	133	12
HALF DAY TOTALS			61	61	73	199	452	734	904	932	813	561	273	101	62	61	61	61	354	
Feb 21	7	55	2	3	23	40	51	53	47	34	14	2	2	2	2	2	2	2	4	5
	8	219	10	11	50	129	183	206	199	160	94	18	10	10	10	10	10	10	43	4
	9	271	16	16	22	107	186	234	245	218	157	66	17	16	16	16	16	16	98	3
	10	294	21	21	21	49	143	211	246	243	203	129	38	21	21	21	21	21	143	2
	11	304	23	23	23	24	71	160	219	244	231	184	103	27	23	23	23	23	171	1
	12	307	24	24	24	24	25	86	170	222	241	222	170	86	25	24	24	24	180	12
HALF DAY TOTALS			84	86	152	361	648	916	1049	1015	821	508	250	114	85	84	84	84	548	
Mar 21	7	171	9	29	93	140	163	161	135	86	22	8	8	8	8	8	8	8	26	5
	8	250	16	18	91	169	218	232	211	157	74	17	16	16	16	16	16	16	85	4
	9	282	21	22	47	136	203	238	236	198	128	40	22	21	21	21	21	21	143	3
	10	297	25	25	27	72	153	207	229	216	171	95	29	25	25	25	25	25	186	2
	11	305	28	28	28	30	78	151	198	213	197	150	77	30	28	28	28	28	213	1
	12	307	29	29	29	29	31	75	145	191	206	191	145	75	31	29	29	29	223	12
HALF DAY TOTALS			114	139	302	563	832	1035	1087	968	694	403	220	132	114	113	113	113	764	
Apr 21	6	89	11	46	72	87	88	76	52	18	5	5	5	5	5	5	5	5	11	6
	7	206	16	71	140	185	201	186	143	75	16	14	14	14	14	14	14	14	61	5
	8	252	22	44	128	190	224	223	188	124	41	22	21	21	21	21	21	21	123	4
	9	274	27	29	80	155	202	219	203	156	83	29	27	27	27	27	27	27	177	3
	10	286	31	31	37	92	152	187	193	170	121	56	32	31	31	31	31	41	217	2
	11	292	33	33	34	39	81	130	160	166	146	102	52	35	33	33	33	33	243	1
	12	293	34	34	34	34	36	62	108	142	154	142	108	62	36	34	34	34	252	12
HALF DAY TOTALS			154	265	501	758	957	1051	994	782	488	296	199	157	148	147	147	147	957	
May 21	5	1	0	1	1	1	1	1	0	0	0	0	0	0	0	0	0	0	0	7
	6	144	36	90	128	145	141	115	71	18	10	10	10	10	10	10	10	11	31	6
	7	216	28	102	165	202	209	184	131	54	20	19	19	19	19	19	19	19	87	5
	8	250	27	73	149	199	220	208	164	93	29	25	25	25	25	25	25	25	146	4
	9	267	31	42	105	164	197	200	175	121	53	32	30	30	30	30	30	30	195	3
	10	277	34	36	54	105	148	168	163	133	83	40	35	34	34	34	34	34	234	2
	11	283	36	36	38	48	81	113	130	127	105	70	42	38	36	36	36	36	257	1
	12	284	37	37	37	38	40	54	82	104	113	104	82	54	40	38	37	37	265	12
HALF DAY TOTALS			215	404	666	893	1024	1025	881	601	358	247	200	180	176	175	174	175	1083	
Jun 21	5	22	10	17	21	22	20	14	6	2	1	1	1	1	1	1	1	2	3	7
	6	155	48	104	143	159	151	121	70	17	13	13	13	13	13	13	13	14	40	6
	7	216	37	113	172	205	207	178	122	46	22	21	21	21	21	21	21	21	97	5
	8	246	30	85	156	201	216	199	152	80	29	27	27	27	27	27	27	27	153	4
	9	263	33	51	114	166	192	190	161	105	45	33	32	32	32	32	32	32	201	3
	10	272	35	38	63	109	145	158	148	116	69	39	36	35	35	35	35	35	238	2
	11	277	38	39	40	52	81	105	116	110	88	60	41	39	38	38	38	38	260	1
	12	279	38	38	38	40	41	52	72	89	95	89	72	52	41	40	38	38	267	12
HALF DAY TOTALS			253	470	734	941	1038	999	818	523	315	236	204	191	188	187	186	188	1126	
Jul 21	5	2	1	2	2	2	2	1	1	0	0	0	0	0	0	0	0	0	0	7
	6	138	37	89	125	142	137	112	68	18	11	11	11	11	11	11	11	12	32	6
	7	208	30	102	163	198	204	179	127	53	21	20	20	20	20	20	20	20	88	5
	8	241	28	75	148	196	216	203	160	90	30	26	26	26	26	26	26	26	145	4
	9	259	32	44	106	163	193	196	170	118	52	33	31	31	31	31	31	31	194	3
	10	269	35	37	56	106	146	165	159	129	81	41	36	35	35	35	35	35	231	2
	11	275	37	38	40	50	81	111	127	123	102	69	43	39	37	37	37	37	254	1
	12	276	38	38	38	40	41	55	80	101	109	101	80	55	41	40	38	38	262	12
HALF DAY TOTALS			223	411	666	885	1008	1003	858	584	352	248	204	186	181	180	180	181	1076	
Aug 21	6	81	12	44	68	81	82	71	48	17	6	5	5	5	5	5	5	5	12	6
	7	191	17	71	135	177	191	177	135	70	17	16	16	16	16	16	16	16	62	5
	8	237	24	47	126	185	216	214	180	118	41	23	23	23	23	23	23	23	122	4
	9	260	28	31	82	153	197	212	196	151	80	31	28	28	28	28	28	28	174	3
	10	272	32	33	40	93	150	182	187	165	116	56	34	32	32	32	32	32	214	2
	11	278	35	35	36	41	81	128	156	160	141	99	52	37	35	35	35	35	239	1
	12	280	35	35	35	36	38	63	106	138	149	138	106	63	38	36	35	35	247	12
HALF DAY TOTALS			164	273	498	741	928	1013	956	751	474	296	205	166	157	156	156	156	946	
Sep 21	7	149	9	27	84	125	146	144	121	77	21	9	9	9	9	9	9	9	25	5
	8	230	17	19	87	160	205	218	199	148	71	18	17	17	17	17	17	17	82	4
	9	263	22	23	47	131	194	227	226	190	124	41	23	22	22	22	22	22	138	3
	10	280	27	27	28	71	148	200	221	209	165	93	30	27	27	27	27	27	180	2
	11	287	29	29	29	31	78	147	192	207	191	146	77	31	29	29	29	29	206	1
	12	290	30	30	30	30	32	75	142	185	200	185	142	75	32	30	30	30	215	12
HALF DAY TOTALS			119	142	291	534	787	980	1033	925	672	396	222	137	119	118	118	118	738	
Oct 21	7	48	2	3	20	36	45	47	42	30	12	2	2	2	2	2	2	2	4	5
	8	204	11	12	49	123	173	195	188	151	89	18	11	11	11	11	11	11	43	4
	9	257	17	17	23	104	180	225	235	209	151	64	18	17	17	17	17	17	97	3
	10	280	21	21	22	50	139	205	238	235	196	125	38	22	21	21	21	21	140	2
	11	291	24	24	24	25	71	156	212	236	224	178	101	28	24	24	24	24	168	1
	12	294	25	25	25	25	27	85	165	216	234	216	165	85	27	25	25	25	177	12
HALF DAY TOTALS			88	89	152	351	623	878	1006	974	791	493	247	117	89	88	88	88	540	
Nov 21	8	136	5	5	18	69	108	128	129	110	72	21	6	5	5	5	5	5	14	4
	9	232	12	12	13	73	151	201	219	204	156	80	13	12	12	12	12	12	55	3
	10	268	16	16	16	31	122	196	237	242	209	143	50	17	16	16	16	16	96	2
	11	283	19	19	19	20	61	154	218	248	240	194	116	28	19	19	19	19	123	1
	12	288	20	20	20	20	21	89	176	231	250	231	176	89	21	20	20	20	132	12
HALF DAY TOTALS			63	63	75	198	445	721	887	914	798	551	269	101	63	63	63	63	354	
Dec 21	8	89	3	3	8	41	67	82	84	73	50	17	3	3	3	3	3	3	6	4
	9	217	10	10	11	60	135	185	205	194	151	83	13	10	10	10	10	10	39	3
	10	261	14	14	14	25	113	188	232	239	210	146	55	15	14	14	14	14	77	2
	11	280	17	17	17	17	56	151	217	249	242	198	120	28	17	17	17	17	104	1
	12	285	18	18	18	18	19	89	178	233	253	233	178	89	19	18	18	18	113	12
HALF DAY TOTALS			52	52	56	146	374	649	822	867	775	557	276	94	53	52	52	52	282	
			N	**NNW**	**NW**	**WNW**	**W**	**WSW**	**SW**	**SSW**	**S**	**SSE**	**SE**	**ESE**	**E**	**ENE**	**NE**	**NNE**	**HOR**	**PM**

[a]Total solar heat gains for DSA glass (based on a ground reflectance of 0.20).

[b]Half day totals computed by Simpson's rule, time interval = 10 min.

TABLE B.4 **Solar Intensity and Solar Heat Gain Factors[a] for 48° N Latitude (Conventional Units)[b]**

Date	Solar Time am	Direct Normal Btuh/ · ft^2	N	NNE	NE	ENE	E	ESE	SE	SSE	S	SSW	SW	WSW	W	WNW	NW	NNW	HOR	Solar Time pm
			Solar Heat Gain Factors, Btuh · ft^2																	
Jan 21	8	37	1	1	4	18	29	34	35	30	20	6	1	1	1	1	1	1	2	4
	9	185	8	8	8	53	118	160	176	166	129	69	10	8	8	8	8	8	25	3
	10	239	12	12	12	22	106	175	216	223	195	136	50	12	12	12	12	12	55	2
	11	261	14	14	14	15	53	144	208	239	233	190	116	26	14	14	14	14	77	1
	12	267	15	15	15	15	16	86	171	226	245	226	171	86	16	15	15	15	85	12
HALF DAY TOTALS			43	43	46	117	316	567	729	776	701	512	259	85	43	43	43	43	203	
Feb 21	7	4	0	0	1	3	3	3	3	2	1	0	0	0	0	0	0	0	0	5
	8	180	8	8	36	103	149	170	166	136	82	17	8	8	8	8	8	8	25	4
	9	247	13	13	16	90	168	216	230	209	155	71	14	13	13	13	13	13	66	3
	10	275	17	17	17	38	131	203	242	244	207	138	44	18	17	17	17	17	105	2
	11	288	19	19	19	20	65	158	221	249	239	192	113	27	19	19	19	19	130	1
	12	292	20	20	20	20	22	89	176	231	250	231	176	89	22	20	20	20	138	12
HALF DAY TOTALS			68	68	107	274	541	816	968	967	813	531	261	104	68	68	68	68	395	
Mar 21	7	153	7	22	80	123	145	145	123	80	23	7	7	7	7	7	7	7	20	5
	8	236	14	15	76	154	204	222	206	158	82	15	14	14	14	14	14	14	68	4
	9	270	19	19	3	121	193	234	239	207	142	52	20	19	19	19	19	19	118	3
	10	287	23	23	24	58	146	208	237	231	189	115	33	23	23	23	23	23	156	2
	11	295	25	25	25	26	74	156	210	232	218	172	94	28	25	25	25	25	180	1
	12	298	26	26	26	26	27	83	161	211	228	211	161	83	27	26	26	26	188	12
HALF DAY TOTALS			100	118	250	494	775	1012	1100	1014	767	465	244	126	101	100	100	100	636	
Apr 21	6	108	12	53	86	105	107	93	64	23	6	6	6	6	6	6	6	6	15	6
	7	205	15	61	132	180	199	189	148	84	18	14	14	14	14	14	14	14	60	5
	8	247	20	32	111	179	219	225	196	138	55	21	20	20	20	20	20	20	114	4
	9	268	25	26	60	141	197	223	215	176	106	33	25	25	25	25	25	25	161	3
	10	280	28	28	31	77	148	193	209	194	150	80	31	28	28	28	28	28	196	2
	11	286	31	31	31	33	78	140	181	193	177	133	69	33	31	31	31	31	218	1
	12	288	31	31	31	31	34	71	131	172	186	172	131	71	34	31	31	31	226	12
HALF DAY TOTALS			147	242	461	724	957	1098	1081	895	605	370	226	156	141	140	140	140	875	
May 21	5	41	17	31	40	42	39	29	14	3	3	3	3	3	3	3	3	3	5	7
	6	162	35	97	141	162	160	133	85	24	12	12	12	12	12	12	12	13	40	6
	7	219	23	90	158	200	212	191	142	68	21	19	19	19	19	19	19	19	91	5
	8	248	26	54	132	190	218	214	178	113	38	25	25	25	25	25	25	25	142	4
	9	264	29	32	82	151	194	208	192	147	77	32	29	29	29	29	29	29	185	3
	10	274	33	34	39	90	145	178	184	163	116	57	35	33	33	33	33	33	219	2
	11	279	35	35	36	40	79	126	155	160	142	101	54	37	35	35	35	35	240	1
	12	280	35	35	35	36	38	63	107	139	150	139	107	63	38	36	35	35	247	12
HALF DAY TOTALS			215	388	645	893	1065	1114	1007	749	483	316	225	184	174	173	173	174	1045	
Jun 21	5	77	35	61	76	80	72	53	24	6	5	5	5	5	5	5	5	8	12	7
	6	172	46	110	155	175	169	138	84	22	14	14	14	14	14	14	14	16	51	6
	7	220	29	101	165	204	211	187	135	60	23	21	21	21	21	21	21	21	103	5
	8	246	29	64	139	191	215	206	168	101	34	27	27	27	27	27	27	27	152	4
	9	261	31	36	91	153	190	199	180	133	66	33	31	31	31	31	31	31	193	3
	10	269	34	36	45	94	143	169	171	148	101	50	36	34	34	34	34	34	225	2
	11	274	36	36	38	44	79	118	142	145	126	88	49	38	36	36	36	36	246	1
	12	275	37	37	37	38	40	60	96	124	134	124	96	60	40	38	37	37	252	12
HALF DAY TOTALS			257	459	722	955	1095	1102	955	678	436	299	228	197	189	188	188	191	1108	
Jul 21	5	43	18	33	42	45	41	30	15	3	3	3	3	3	3	3	3	4	6	7
	6	156	37	96	138	159	156	129	82	24	13	13	13	13	13	13	13	14	41	6
	7	211	25	90	156	196	207	186	138	66	22	20	20	20	20	20	20	20	92	5
	8	240	27	56	132	187	214	209	174	110	38	26	26	26	26	26	26	26	142	4
	9	256	30	34	83	149	191	204	187	143	75	33	30	30	30	30	30	30	184	3
	10	266	34	35	41	90	143	174	180	158	113	56	36	34	34	34	34	34	217	2
	11	271	36	36	37	42	79	124	151	156	138	99	54	38	36	36	36	36	237	1
	12	272	36	36	36	37	39	63	104	136	146	136	104	63	39	37	36	36	244	12
HALF DAY TOTALS			223	395	646	886	1050	1092	983	730	474	315	229	190	181	179	179	180	1042	
Aug 21	6	99	13	51	81	98	100	87	60	22	7	7	7	7	7	7	7	7	16	6
	7	190	17	61	128	172	190	179	141	79	19	15	15	15	15	15	15	15	61	5
	8	232	22	34	110	174	211	216	188	132	53	23	22	22	22	22	22	22	114	4
	9	154	27	28	63	139	192	216	108	169	102	34	27	27	27	27	27	27	159	3
	10	266	30	30	33	78	145	188	203	188	144	78	33	30	30	30	30	30	193	2
	11	272	32	32	32	36	78	137	175	187	171	129	68	35	32	32	32	32	215	1
	12	274	33	33	33	33	36	71	128	167	189	167	128	71	36	33	33	33	223	12
HALF DAY TOTALS			157	251	459	709	929	1060	1040	862	587	366	231	165	151	149	149	149	869	
Sep 21	7	131	8	21	71	108	128	128	108	71	21	8	7	7	7	7	7	7	20	5
	8	215	15	16	72	144	191	207	193	148	77	16	15	15	15	15	15	15	65	4
	9	251	20	20	34	116	184	223	227	197	136	52	21	20	20	20	20	20	114	3
	10	269	24	24	25	58	141	200	228	221	182	112	34	24	24	24	24	24	151	2
	11	278	26	26	26	28	73	151	203	223	210	166	92	29	26	26	26	26	174	1
	12	280	27	27	27	27	29	82	156	204	220	204	156	82	29	27	27	27	182	12
HALF DAY TOTALS			105	121	240	465	729	953	1040	963	737	453	243	131	106	105	105	105	614	
Oct 21	7	4	0	0	2	3	4	4	3	2	1	0	0	0	0	0	0	0	0	5
	8	165	8	9	35	96	139	159	155	126	77	16	8	8	8	8	8	8	25	4
	9	233	14	14	16	88	161	207	220	199	148	68	15	14	14	14	14	14	66	3
	10	262	18	18	18	39	128	196	233	234	199	133	43	18	18	18	18	18	104	2
	11	274	20	20	20	21	64	153	213	241	231	186	109	27	20	20	20	20	128	1
	12	278	21	21	21	21	23	87	171	223	242	223	171	87	23	21	21	21	136	12
HALF DAY TOTALS			71	71	108	266	519	780	925	925	779	513	256	106	72	71	71	71	391	
Nov 21	8	36	1	1	4	18	29	34	35	30	20	6	1	1	1	1	1	1	2	4
	9	179	8	8	9	52	115	156	171	161	125	67	10	8	8	8	8	8	26	3
	10	233	12	12	12	22	104	172	212	218	191	133	49	13	12	12	12	12	55	2
	11	255	15	15	15	15	52	142	204	234	228	186	114	26	15	15	15	15	77	1
	12	261	15	15	15	15	17	85	168	222	240	222	168	85	17	15	15	15	85	12
HALF DAY TOTALS			44	44	47	117	310	555	713	760	686	502	255	85	44	44	44	44	204	
Dec 21	9	140	5	5	6	36	86	120	133	127	100	56	8	5	5	5	5	5	13	3
	10	214	10	10	10	16	91	156	194	201	179	126	49	10	10	10	10	10	38	2
	11	242	12	12	12	13	46	134	195	225	220	180	111	25	12	12	12	12	57	1
	12	250	13	13	13	13	14	81	163	215	233	215	168	81	14	13	13	13	65	12
HALF DAY TOTALS			33	33	34	73	233	458	610	665	616	468	247	76	34	33	33	33	141	
			N	NNW	NW	WNW	W	WSW	SW	SSW	S	SSE	SE	ESE	E	ENE	NE	NNE	HOR	PM

[a]Total solar heat gains for DSA glass (based on a ground reflectance of 0.20).

[b]Half day totals computed by Simpson's rule, time interval = 10 min.

TABLE B.5 **Solar Intensity and Solar Heat Gain Factors[a] for 64° N Latitude (Conventional Units)[b]**

Date	Solar Time am	Direct Normal Btuh ft²	Solar Heat Gain Factors, Btuh · ft² N	NNE	NE	ENE	E	ESE	SE	SSE	S	SSW	SW	WSW	W	WNW	NW	NNW	HOR	Solar Time pm
Jan 21	10	22	1	1	1	1	9	16	20	21	19	13	5	1	1	1	1	1	1	2
	11	81	3	3	3	3	15	45	67	77	75	62	38	8	3	3	3	3	6	1
	12	100	3	3	3	3	4	33	67	89	96	89	67	33	4	3	3	3	8	12
HALF DAY TOTALS			5	5	5	6	25	79	121	142	141	119	75	23	5	5	5	5	11	
Feb 21	8	18	1	1	3	10	15	17	17	14	9	2	1	1	1	1	1	1	1	4
	9	134	5	5	6	43	89	118	128	119	90	45	6	5	5	5	5	5	13	3
	10	190	8	8	8	18	87	144	176	180	157	108	38	9	8	8	8	8	28	2
	11	215	10	10	10	11	44	122	177	202	197	160	97	20	10	10	10	10	41	1
	12	222	11	11	11	11	12	73	147	194	210	194	147	73	12	11	11	11	45	12
HALF DAY TOTALS			29	30	33	89	244	446	578	617	560	411	212	66	30	29	29	29	106	
Mar 21	7	95	4	11	47	74	90	91	79	53	17	4	4	4	4	4	4	4	9	5
	8	185	9	10	46	113	158	177	170	135	78	14	9	9	9	9	9	9	32	4
	9	227	13	13	16	88	159	203	215	194	143	64	14	13	13	13	13	13	59	3
	10	249	16	16	16	35	122	190	226	228	194	130	42	16	16	16	16	16	84	2
	11	260	17	17	17	18	60	148	209	236	228	184	109	25	17	17	17	17	99	1
	12	263	18	18	18	18	19	85	168	221	239	221	168	85	19	18	18	18	105	12
HALF DAY TOTALS			68	74	150	334	596	854	984	958	779	504	257	104	68	68	68	68	335	
Apr 21	5	27	8	18	24	27	26	20	12	2	1	1	1	1	1	1	1	1	2	7
	6	133	12	59	102	127	132	118	84	35	8	8	8	8	8	8	8	8	21	6
	7	194	14	41	113	163	189	185	153	96	25	13	13	13	13	13	13	13	51	5
	8	228	17	19	79	153	201	217	201	153	79	19	17	17	17	17	17	17	85	4
	9	248	21	21	32	111	180	219	225	197	138	55	22	21	21	21	21	21	116	3
	10	260	23	23	24	51	134	194	225	221	185	118	38	24	23	23	23	23	140	2
	11	266	24	24	24	26	68	148	202	225	214	171	99	29	24	24	24	24	155	1
	12	268	25	25	25	25	27	83	159	208	224	208	159	83	27	25	25	25	160	12
HALF DAY TOTALS			131	218	410	671	943	1150	1186	1036	763	487	273	149	121	120	120	120	651	
May 21	4	51	30	44	51	51	43	28	8	3	3	3	3	3	3	3	3	10	6	8
	5	132	48	95	125	135	125	96	50	11	9	9	9	9	9	9	9	11	26	7
	6	185	28	97	150	181	183	158	109	40	15	15	15	15	15	15	15	15	55	6
	7	218	21	63	138	189	211	201	161	94	24	19	19	19	19	19	19	19	90	5
	8	239	23	28	97	167	209	220	198	146	68	25	23	23	23	23	23	23	124	4
	9	252	26	27	45	122	183	215	215	184	123	46	27	26	26	26	26	26	152	3
	10	261	28	28	30	61	135	188	212	205	167	102	36	28	28	28	28	28	174	2
	11	265	30	30	30	32	72	141	188	207	195	154	87	33	30	30	30	30	188	1
	12	267	30	30	30	30	33	78	146	189	204	189	146	78	33	30	30	30	192	12
HALF DAY TOTALS			247	425	680	950	1177	1291	1218	985	708	465	288	191	169	168	168	176	911	
Jun 21	4	93	53	83	96	94	78	50	14	7	7	7	7	7	7	7	7	21	16	8
	5	154	62	114	148	158	145	110	55	14	12	12	12	12	12	12	12	14	39	7
	6	194	36	107	162	191	192	163	110	39	18	17	17	17	17	17	17	18	71	6
	7	221	24	71	145	193	213	200	158	89	25	22	22	22	22	22	22	22	105	5
	8	239	25	33	104	170	208	216	192	139	62	27	25	25	25	25	25	25	137	4
	9	251	28	29	51	124	181	210	208	175	115	43	29	28	28	28	28	28	165	3
	10	258	30	30	32	65	134	183	204	195	157	94	36	30	30	30	30	30	186	2
	11	262	32	32	32	34	72	137	180	196	184	144	82	35	32	32	32	32	199	1
	12	263	32	32	32	32	35	76	138	179	193	179	138	76	35	32	32	32	203	12
HALF DAY TOTALS			322	533	801	1061	1253	1317	1195	946	679	455	296	211	192	190	191	213	1021	
Jul 21	4	53	32	47	55	54	46	29	9	4	4	4	4	4	4	4	4	11	8	8
	5	128	49	94	123	133	124	95	50	11	10	10	10	10	10	10	10	11	28	7
	6	179	30	96	148	177	180	155	106	39	16	15	15	15	15	15	15	15	57	6
	7	211	22	64	137	186	207	197	157	92	25	20	20	20	20	20	20	20	92	5
	8	231	24	30	97	165	205	215	193	142	67	26	24	24	24	24	24	24	124	4
	9	245	27	28	47	121	180	211	211	179	120	46	28	27	27	27	27	27	152	3
	10	253	29	29	31	62	134	185	208	200	164	100	37	29	29	29	29	29	174	2
	11	257	31	31	31	33	72	139	185	202	191	151	86	34	31	31	31	31	187	1
	12	259	31	31	31	31	34	78	143	185	200	185	143	78	34	31	31	31	192	12
HALF DAY TOTALS			258	434	684	946	1163	1269	1193	965	697	462	292	198	177	175	175	185	918	
Aug 21	5	29	9	20	27	30	28	22	13	2	2	2	2	2	2	2	2	2	3	7
	6	123	13	58	97	121	125	111	80	34	9	9	9	9	9	9	9	9	23	6
	7	181	15	42	109	157	180	176	145	92	26	14	14	14	14	14	14	14	53	5
	8	214	19	21	78	148	193	208	192	147	76	21	19	19	19	19	19	19	87	4
	9	234	22	22	34	109	174	211	217	189	133	55	23	22	22	22	22	22	117	3
	10	246	25	25	26	52	131	188	217	214	178	114	39	25	25	25	25	25	140	2
	11	252	26	26	26	28	69	144	196	217	207	166	97	31	26	26	26	26	154	1
	12	254	27	27	27	27	29	82	155	201	217	201	155	82	29	27	27	27	159	12
HALF DAY TOTALS			142	226	410	657	914	1109	1141	997	740	478	275	158	131	130	130	130	656	
Sep 21	7	77	4	10	39	62	74	75	65	44	15	4	4	4	4	4	4	4	8	5
	8	163	10	10	43	103	143	160	154	123	71	14	10	10	10	10	10	10	31	4
	9	206	14	14	17	83	148	189	200	181	133	61	15	14	14	14	14	14	57	3
	10	229	16	16	17	35	116	179	213	214	183	123	41	17	16	16	16	16	81	2
	11	240	18	18	18	19	59	141	198	224	216	174	104	26	18	18	18	18	96	1
	12	244	19	19	19	19	21	82	160	209	227	209	160	82	21	19	19	19	101	12
HALF DAY TOTALS			71	77	142	307	547	787	910	891	731	480	249	106	72	71	71	71	324	
Oct 21	8	17	1	1	3	10	14	16	16	13	8	2	1	1	1	1	1	1	1	4
	9	122	5	5	6	40	82	109	118	110	83	42	6	5	5	5	5	5	13	3
	10	176	9	9	9	18	83	135	165	169	147	102	36	9	9	9	9	9	29	2
	11	201	11	11	11	11	43	116	167	191	186	152	92	20	11	11	11	11	41	1
	12	208	11	11	11	11	13	70	140	184	199	184	140	70	13	11	11	11	46	12
HALF DAY TOTALS			31	31	34	86	231	420	542	580	527	388	202	66	32	31	31	31	108	
Nov 21	10	23	1	1	1	1	10	17	21	22	20	14	5	1	1	1	1	1	1	2
	11	79	3	3	3	3	15	44	65	75	74	61	37	8	3	3	3	3	6	1
	12	97	4	4	4	4	4	32	66	87	93	87	66	32	4	4	4	4	8	12
HALF DAY TOTALS			5	5	5	6	26	79	120	141	140	117	74	23	6	5	5	5	11	
Dec 21	11	4	0	0	0	0	1	2	3	4	4	3	2	0	0	0	0	0	0	1
	12	16	0	0	0	0	1	5	11	14	15	14	11	5	1	0	0	0	1	12
HALF DAY TOTALS			0	0	0	0	1	5	9	11	11	10	7	3	0	0	0	0	1	
			N	NNW	NW	WNW	W	WSW	SW	SSW	S	SSE	SE	ESE	E	ENE	NE	NNE	HOR	PM

[a]Total solar heat gains for DSA glass (based on a ground reflectance of 0.20).

[b]Half day totals computed by Simpson's rule, time interval = 10 min.

APPENDICES

TABLE B.6 **Solar Intensity and Solar Heat Gain Factors[a] for 16° N Latitude (SI Units)[b]**

Date	Solar Time am	Direct Normal W/m²	N	NNE	NE	ENE	E	ESE	SE	SSE	S	SSW	SW	WSW	W	WNW	NW	NNW	HOR	Solar Time pm
			Solar Heat Gain Factors, W/m²																	
Jan 21	7	445	17	19	138	291	390	424	397	303	155	19	17	17	17	17	17	17	43	5
	8	827	45	48	174	463	662	757	734	596	359	79	45	45	45	45	45	45	249	4
	9	948	67	67	102	384	630	770	791	690	481	183	69	67	67	67	67	67	472	3
	10	1001	82	82	86	209	474	658	737	704	563	321	97	82	82	82	82	82	640	2
	11	1025	92	92	92	96	242	467	614	663	612	462	236	96	92	92	92	92	745	1
	12	1032	95	95	95	95	100	228	438	580	628	580	438	228	100	95	95	95	782	12
HALF DAY TOTALS			348	352	618	1453	2398	3153	3458	3217	2465	1344	666	401	350	348	348	348	2539	
Feb 21	7	575	24	55	265	435	532	544	474	326	113	26	24	24	24	24	24	24	80	5
	8	862	53	60	304	567	729	778	706	525	244	56	53	53	53	53	53	53	319	4
	9	961	74	77	202	482	676	763	733	592	347	96	74	74	74	74	74	74	549	3
	10	1006	90	91	104	292	508	636	665	593	423	193	94	90	90	90	90	90	722	2
	11	1027	99	99	102	118	262	428	527	542	471	323	154	103	99	99	99	99	831	1
	12	1034	103	103	103	105	108	189	336	448	487	448	336	189	108	105	103	103	868	12
HALF DAY TOTALS			390	431	1013	1922	2730	3228	3263	2792	1836	906	547	417	393	390	390	390	2933	
Mar 21	7	634	36	167	393	544	606	578	458	260	47	33	33	33	33	33	33	33	126	5
	8	857	63	157	442	648	752	741	615	390	111	61	61	61	61	61	61	61	380	4
	9	943	84	110	343	565	689	709	622	435	180	86	82	82	82	82	82	82	606	3
	10	983	98	103	191	379	519	575	543	424	240	107	100	98	98	98	98	98	778	2
	11	1003	108	110	113	166	273	361	395	366	281	173	114	110	108	108	108	108	885	1
	12	1008	111	111	112	114	117	149	216	273	295	273	216	149	117	114	112	111	919	12
HALF DAY TOTALS			443	712	1558	2383	2926	3076	2773	2028	1006	588	483	448	440	483	437	437	3233	
Apr 21	6	44	7	24	37	43	43	37	24	7	2	2	2	2	2	2	12	2	4	6
	7	622	75	298	482	589	604	528	369	141	45	42	42	42	42	42	42	42	169	5
	8	807	85	312	543	682	718	644	473	217	74	70	70	70	70	70	70	70	413	4
	9	885	97	248	469	610	657	608	465	244	97	90	90	90	90	90	90	90	623	3
	10	924	112	171	321	444	499	476	380	231	118	109	166	106	106	106	106	106	784	2
	11	942	120	127	169	228	270	276	245	189	136	121	118	115	115	115	115	118	882	1
	12	948	123	123	124	125	126	129	135	141	143	141	135	129	126	125	124	123	911	12
HALF DAY TOTALS			565	1272	2127	2711	2909	2684	2059	1111	547	503	494	491	490	489	490	492	3333	
May 21	6	138	43	94	128	141	134	106	59	11	9	9	9	9	9	9	9	10	195	6
	7	608	157	378	531	603	583	474	290	76	49	49	49	49	49	49	49	55	17	5
	8	771	165	415	598	689	677	564	362	121	78	76	76	76	76	76	76	80	425	4
	9	845	156	366	539	626	622	526	344	141	100	96	96	96	96	96	96	100	621	3
	10	883	149	281	410	478	474	398	264	139	116	111	111	111	111	111	111	116	772	2
	11	902	147	198	248	273	262	220	166	130	126	123	120	120	120	122	124	128	862	1
	12	907	146	144	140	134	132	131	130	129	129	129	130	131	132	134	140	144	890	12
HALF DAY TOTALS			893	1813	2537	2891	2829	2360	1555	685	541	528	527	526	527	529	533	557	3338	
Jun 21	6	168	64	124	163	175	162	123	64	13	12	12	12	12	12	12	17	13	24	6
	7	593	195	404	543	598	565	445	252	63	52	52	52	52	52	52	52	57	203	5
	8	750	209	449	612	684	653	526	313	98	79	79	79	79	79	79	79	84	425	4
	9	823	199	409	560	626	601	487	294	118	98	98	98	98	98	98	98	105	613	3
	10	861	187	329	441	486	459	363	222	125	116	113	113	113	113	113	113	121	759	2
	11	879	181	241	283	290	258	200	146	130	126	122	122	122	122	125	128	136	847	1
	12	885	179	173	158	142	134	132	130	129	129	129	130	132	134	142	158	173	873	12
HALF DAY TOTALS			1122	2043	2683	2930	2763	2207	1357	612	548	540	540	540	542	545	555	599	3308	
Jul 21	6	128	43	90	122	134	127	99	55	11	9	9	9	9	9	9	9	10	17	6
	7	579	161	373	518	585	564	456	277	74	51	51	51	51	51	51	51	55	194	5
	8	743	173	415	590	676	661	549	349	110	78	78	78	78	78	78	78	83	419	4
	9	818	165	371	537	618	610	513	334	138	102	99	99	99	99	99	99	104	611	3
	10	857	157	289	413	475	467	389	257	138	118	113	113	113	113	113	113	120	759	2
	11	876	154	207	255	277	262	218	164	133	128	125	122	122	122	125	128	131	848	1
	12	882	154	151	145	138	135	134	133	132	132	132	133	134	135	138	145	151	875	12
HALF DAY TOTALS			933	1829	2521	2848	2770	2298	1507	680	554	541	539	539	540	542	547	575	3289	
Aug 21	6	136	7	21	31	37	36	31	20	6	2	2	2	2	2	2	2	2	4	6
	7	569	81	289	458	555	567	493	343	131	48	45	45	45	45	45	45	45	167	5
	8	757	94	315	531	660	691	617	451	206	79	74	74	74	74	74	74	74	404	4
	9	838	104	258	467	598	640	589	448	235	103	95	95	95	95	95	95	95	608	3
	10	879	118	183	328	443	490	464	368	224	122	114	110	110	110	110	110	110	766	2
	11	899	127	136	180	235	271	273	240	185	138	127	124	120	120	120	120	124	860	1
	12	905	129	130	130	131	132	134	139	143	145	143	139	134	132	131	130	130	889	12
HALF DAY TOTALS			603	1294	2099	2640	2810	2578	1969	1069	568	528	519	516	515	514	515	518	3258	
Sep 21	7	565	38	157	360	497	554	529	419	240	48	35	35	35	35	35	35	35	122	5
	8	797	67	156	424	618	716	705	587	374	113	64	64	64	64	64	64	64	367	4
	9	887	87	114	335	547	665	684	602	424	181	90	86	86	86	86	86	86	585	3
	10	931	101	107	193	372	507	560	529	415	240	111	103	101	101	101	101	101	752	2
	11	952	111	114	118	169	272	356	389	361	279	176	118	114	111	111	111	111	856	1
	12	958	114	114	116	118	121	153	217	272	293	272	217	153	121	118	116	114	889	12
HALF DAY TOTALS			461	712	1500	2276	2791	2937	2658	1963	1007	605	501	466	457	455	454	454	3126	
Oct 21	7	524	25	56	249	404	492	502	437	300	105	27	25	25	25	25	25	25	79	5
	8	816	55	64	299	548	702	747	677	502	234	59	55	55	55	55	55	55	313	4
	9	920	76	80	204	473	659	741	711	573	336	97	76	76	76	76	76	76	537	3
	10	968	92	94	108	291	499	621	647	577	412	189	97	92	92	92	92	92	707	2
	11	991	102	105	122	261	420	515	528	459	315	154	106	102	102	102	102	102	814	1
	12	997	105	105	105	107	111	189	330	437	474	437	330	189	111	107	105	105	850	12
HALF DAY TOTALS			402	444	1002	1869	2637	3110	3141	2689	1776	891	553	428	404	402	402	402	2872	
Nov 21	7	423	17	19	134	280	375	406	379	289	147	19	17	17	17	17	17	17	43	5
	8	806	46	49	174	456	651	742	719	583	350	78	46	46	46	46	46	46	247	4
	9	929	67	67	105	382	623	779	779	679	472	180	70	67	67	67	67	67	468	3
	10	984	83	83	87	210	470	651	727	694	554	316	98	83	83	83	83	83	635	2
	11	1009	92	92	92	98	242	462	607	654	603	455	234	98	92	92	92	92	740	1
	12	1016	96	96	96	96	101	227	432	572	619	572	432	227	101	96	96	96	775	12
HALF DAY TOTALS			352	357	620	1438	2363	3101	3396	3159	2421	1324	664	404	355	352	352	352	2520	
Dec 21	7	372	13	14	93	228	318	354	339	268	150	21	13	13	13	13	13	13	31	5
	8	803	42	43	129	416	625	734	728	609	390	105	42	42	42	42	42	42	219	4
	9	936	63	63	78	341	604	761	800	716	520	226	66	63	63	63	63	63	436	3
	10	993	78	78	81	178	445	657	753	735	605	369	112	79	78	78	78	78	602	2
	11	1019	87	87	87	92	231	474	638	698	654	506	270	94	87	87	87	87	704	1
	12	1026	91	91	91	91	95	241	469	620	670	620	469	241	95	91	91	91	739	12
HALF DAY TOTALS			328	330	501	1269	2241	3076	3468	3312	2638	1527	721	393	331	328	328	328	2361	
			N	NNW	NW	WNW	W	WSW	SW	SSW	S	SSE	SE	ESE	E	ENE	NE	NNE	HOR	PM

[a]Total solar heat gains for DSA glass (based on a ground reflectance of 0.20).

[b]Half day totals computed by Simpson's rule, time interval = 10 min.

APPENDICES

TABLE B.7 **Solar Intensity and Solar Heat Gain Factors[a] for 32° N Latitude (SI Units)[b]**

Date	Solar Time am	Direct Normal W/m²	Solar Heat Gain Factors, W/m²																	Solar Time pm
			N	NNE	NE	ENE	E	ESE	SE	SSE	S	SSW	SW	WSW	W	WNW	NW	NNW	HOR	
Jan 21	7	4	0	0	1	3	4	4	4	3	2	0	0	0	0	0	0	0	0	5
	8	640	28	29	93	330	505	597	596	502	326	88	29	28	28	28	28	28	102	4
	9	849	48	48	53	286	553	721	775	711	534	258	52	48	48	48	48	48	278	3
	10	931	63	63	64	129	427	659	784	788	670	444	144	64	63	63	63	63	430	2
	11	967	71	71	71	75	213	502	698	784	750	602	347	90	71	71	71	71	524	1
	12	977	74	74	74	74	79	277	548	718	777	718	548	277	79	74	74	74	556	12
HALF DAY TOTALS			249	250	338	897	1797	2700	3201	3198	2692	1746	831	353	252	249	249	249	1614	
Feb 21	7	352	13	23	148	258	323	336	299	212	83	14	13	13	13	13	13	13	30	5
	8	771	41	43	204	469	646	719	683	538	300	53	41	41	41	41	41	41	200	4
	9	905	60	60	101	386	627	764	783	680	471	174	63	60	60	60	60	60	402	3
	10	964	75	75	78	195	475	670	760	733	595	352	99	75	75	75	75	75	556	2
	11	990	83	83	83	88	240	491	656	717	670	519	275	89	83	83	83	83	652	1
	12	998	86	86	86	86	91	250	489	644	696	643	489	250	91	86	86	86	684	12
HALF DAY TOTALS			314	324	635	1405	2318	3086	3408	3188	2461	1426	718	384	316	314	314	314	2181	
Mar 21	7	583	30	116	332	483	556	545	447	276	63	29	29	29	29	29	29	29	100	5
	8	821	54	78	339	576	717	746	661	473	195	57	54	54	54	54	54	54	315	4
	9	914	74	77	203	475	662	746	716	578	339	95	74	74	74	74	74	74	516	3
	10	959	88	88	96	273	499	637	677	615	456	221	93	88	88	88	88	88	667	2
	11	980	96	96	98	108	258	447	564	592	529	380	185	101	96	96	96	96	762	1
	12	987	99	99	99	99	105	209	386	511	554	511	386	209	105	99	99	99	795	12
HALF DAY TOTALS			393	512	1131	1985	2761	3258	3284	2801	1858	1030	609	429	394	391	391	391	2757	
Apr 21	6	210	29	110	172	205	207	177	119	38	11	11	11	11	11	11	11	11	23	6
	7	649	53	253	462	593	631	575	428	204	49	45	45	45	45	45	45	45	191	5
	8	804	73	192	453	632	715	689	559	337	95	69	69	69	69	69	69	69	408	4
	9	876	89	113	324	532	649	669	591	418	183	92	87	87	87	87	87	87	593	3
	10	913	101	106	165	342	490	599	543	445	275	123	104	101	101	101	101	101	736	2
	11	932	109	109	115	149	263	371	426	415	340	222	126	113	109	109	109	109	825	1
	12	937	112	112	112	116	120	167	260	335	363	335	260	167	120	116	112	112	854	12
HALF DAY TOTALS			508	932	1734	2500	3003	3129	2806	2033	1135	721	559	496	482	479	479	479	3203	
May 21	6	374	104	244	340	381	367	295	175	39	26	26	26	26	26	26	26	28	67	6
	7	666	112	350	535	638	643	550	374	134	60	57	57	57	57	57	57	59	256	5
	8	787	93	295	519	655	694	629	470	229	86	80	80	80	80	80	80	80	461	4
	9	849	104	194	404	558	625	601	488	294	117	100	97	97	97	97	97	97	633	3
	10	882	114	127	240	383	473	491	435	313	170	117	110	110	110	110	110	110	766	2
	11	898	121	124	132	186	260	312	322	285	215	147	126	122	118	118	118	118	847	1
	12	904	121	123	125	127	130	148	186	220	223	220	186	148	130	127	125	123	873	12
HALF DAY TOTALS			700	1382	2214	2840	3106	2943	2356	1409	788	627	577	558	551	550	548	551	3464	
Jun 21	6	412	140	289	388	425	401	314	174	39	32	32	32	32	32	32	32	35	89	6
	7	662	148	384	556	644	634	529	342	109	62	62	62	62	62	62	62	65	279	5
	8	773	115	335	540	656	677	597	427	188	89	84	84	84	84	84	84	84	476	4
	9	831	110	234	431	563	609	567	440	244	111	101	101	101	101	101	101	101	642	3
	10	863	120	150	272	395	461	458	387	261	143	120	114	114	114	114	114	114	770	2
	11	880	126	130	148	202	258	288	280	237	177	135	128	125	122	122	122	122	847	1
	12	885	128	129	130	132	134	144	164	183	191	183	164	144	134	132	130	129	871	12
HALF DAY TOTALS			824	1589	2403	2950	3108	2829	2137	1174	710	621	595	585	582	579	579	586	3538	
Jul 21	6	358	106	240	332	370	355	285	168	39	27	27	27	27	27	27	27	30	70	6
	7	640	118	349	526	623	626	534	361	129	62	59	59	59	59	59	59	61	257	5
	8	761	98	300	515	645	680	613	456	221	89	82	82	82	82	82	82	82	457	4
	9	823	107	202	405	553	615	588	475	284	117	100	100	100	100	100	100	100	626	3
	10	856	118	133	247	383	467	481	424	303	167	120	113	113	113	113	113	113	757	2
	11	873	124	128	137	191	261	308	314	277	210	147	129	125	121	121	121	121	836	1
	12	879	126	127	128	130	133	150	184	214	226	214	184	150	133	130	128	127	861	12
HALF DAY TOTALS			728	1402	2210	2809	3052	2877	2290	1365	783	637	591	574	568	566	565	568	3431	
Aug 21	6	187	30	103	159	188	189	162	108	35	12	12	12	12	12	12	12	12	25	6
	7	599	59	250	444	565	598	542	402	192	53	49	49	49	49	49	49	49	192	5
	8	756	79	200	446	614	690	662	536	321	96	74	74	74	74	74	74	74	403	4
	9	830	95	124	328	523	632	648	570	402	179	98	93	93	93	93	93	93	583	3
	10	869	106	113	175	344	482	544	526	429	266	126	110	106	106	106	106	106	722	2
	11	889	115	117	122	157	264	365	414	401	328	217	131	119	115	115	115	115	808	1
	12	894	117	117	117	122	126	170	255	324	350	324	255	170	126	122	117	117	837	12
HALF DAY TOTALS			539	957	1721	2443	2910	3014	2694	1950	1109	729	581	524	510	507	506	506	3152	
Sep 21	7	514	32	109	302	437	502	492	405	252	62	30	30	30	30	30	30	30	97	5
	8	758	57	81	324	546	679	706	626	450	190	61	57	57	57	57	57	57	304	4
	9	857	77	82	201	459	637	717	688	557	330	99	77	77	77	77	77	77	498	3
	10	905	91	91	101	270	485	617	655	596	444	220	97	91	91	91	91	91	645	2
	11	928	100	100	102	113	257	437	549	576	516	373	187	106	100	100	100	100	737	1
	12	935	103	103	103	103	110	210	379	499	540	499	379	210	110	103	103	103	769	12
HALF DAY TOTALS			409	518	1089	1888	2621	3097	3132	2689	1812	1025	620	446	411	408	408	408	2664	
Oct 21	7	312	13	24	136	233	291	301	268	190	75	14	13	13	13	13	13	13	30	5
	8	724	42	46	200	450	616	684	649	511	285	54	42	42	42	42	42	42	198	4
	9	862	63	63	104	378	608	738	755	655	454	170	65	63	63	63	63	63	395	3
	10	923	77	77	81	197	465	652	737	711	577	342	101	77	77	77	77	77	546	2
	11	952	86	86	86	91	239	481	639	697	651	505	269	93	86	86	86	86	639	1
	12	960	89	89	89	89	95	247	478	626	677	626	278	247	95	89	89	89	671	12
HALF DAY TOTALS			325	335	632	1365	2234	2968	3275	3067	2376	1390	714	394	328	325	325	325	2143	
Nov 21	7	5	0	0	1	3	4	5	4	3	2	0	0	0	0	0	0	0	0	5
	8	619	29	29	93	323	493	581	579	488	316	86	29	29	29	29	29	29	102	4
	9	829	49	49	55	284	545	709	761	697	523	253	49	49	49	49	49	49	277	3
	10	912	64	64	65	131	423	650	772	775	659	437	142	65	64	64	64	64	428	2
	11	949	72	72	72	76	213	496	688	773	739	593	342	91	72	72	72	72	521	1
	12	959	75	75	75	75	80	274	541	708	766	708	541	274	80	75	75	75	553	12
HALF DAY TOTALS			253	254	342	890	1769	2653	3143	3138	2643	1717	824	355	256	253	253	253	1607	
Dec 21	8	556	22	22	59	265	426	515	523	452	305	98	23	22	22	22	22	22	69	4
	9	812	43	43	46	244	512	686	751	701	539	282	48	43	43	43	43	43	228	3
	10	908	57	57	57	106	402	642	777	792	683	466	163	59	57	57	57	57	375	2
	11	949	66	66	66	69	200	497	700	795	766	620	367	92	66	66	66	66	468	1
	12	960	69	69	69	69	73	281	559	733	794	733	559	281	73	69	69	69	500	12
HALF DAY TOTALS			223	223	265	717	1577	2499	3043	3111	2687	1823	868	339	225	223	223	223	1389	
			N	NNW	NW	WNW	W	WSW	SW	SSW	S	SSE	SE	ESE	E	ENE	NE	NNE	HOR	PM

[a]Total solar heat gains for DSA glass (based on a ground reflectance of 0.20).

[b]Half day totals computed by Simpson's rule, time interval = 10 min.

APPENDICES

TABLE B.8 **Solar Intensity and Solar Heat Gain Factors[a] for 40° N Latitude (SI Units)[b]**

Date	Solar Time am	Direct Normal W/m²	Solar Heat Gain Factors, W/m²																	Solar Time pm
			N	NNE	NE	ENE	E	ESE	SE	SSE	S	SSW	SW	WSW	W	WNW	NW	NNW	HOR	
Jan 21	8	446	17	17	55	223	350	417	420	358	236	60	17	17	17	17	17	17	44	4
	9	753	37	37	41	233	485	648	706	658	504	260	42	37	37	37	37	37	173	3
	10	865	51	51	51	97	390	627	761	776	671	460	161	53	51	51	51	51	303	2
	11	913	59	59	59	62	193	493	699	796	769	623	372	89	59	59	59	59	390	1
	12	926	62	62	62	62	66	293	563	740	802	740	563	283	62	62	62	62	419	12
HALF DAY TOTALS			194	194	231	628	1426	2316	2852	2941	2566	1770	860	318	196	194	194	194	1117	
Feb 21	7	175	6	10	71	127	160	167	150	107	43	6	6	6	6	6	6	6	11	5
	8	692	33	35	158	407	576	651	628	505	296	56	33	33	33	33	33	33	136	4
	9	854	52	52	70	337	587	738	773	689	496	209	55	52	52	52	52	52	309	3
	10	926	65	65	67	155	450	666	777	768	641	408	120	66	65	65	65	65	451	2
	11	958	73	73	73	77	224	504	690	769	730	579	325	86	73	73	73	73	538	1
	12	967	76	76	76	76	80	271	536	702	760	702	536	271	80	76	76	76	568	12
HALF DAY TOTALS			267	271	478	1140	2043	2888	3308	3202	2591	1602	790	361	269	267	267	267	1730	
Mar 21	7	540	27	92	295	441	514	509	425	271	69	26	26	26	26	26	26	26	83	5
	8	789	50	57	288	534	686	732	665	494	232	54	50	50	50	50	50	50	268	4
	9	889	67	70	147	429	640	749	744	625	404	127	69	67	67	67	67	67	450	3
	10	938	80	80	85	226	482	653	722	682	539	299	91	80	80	80	80	80	587	2
	11	961	88	88	88	94	247	476	623	673	622	473	244	94	88	88	88	88	673	1
	12	968	91	91	91	91	97	238	458	602	650	602	458	238	97	91	91	91	702	12
HALF DAY TOTALS			358	440	954	1777	2626	3265	3429	3055	2191	1270	694	417	360	357	357	357	2411	
Apr 21	6	282	36	144	228	275	279	241	164	56	16	15	15	15	15	15	15	15	34	6
	7	651	50	223	442	584	633	588	451	255	51	45	45	45	45	45	45	45	193	5
	8	795	69	140	402	601	706	703	594	391	130	69	67	67	67	67	67	67	389	4
	9	865	84	91	254	488	638	691	640	494	260	91	84	84	84	84	84	84	557	3
	10	901	96	99	117	291	480	589	608	538	381	177	101	96	96	96	96	96	685	2
	11	920	104	107	107	122	255	411	506	522	459	323	164	109	104	104	104	104	766	1
	12	926	106	106	106	109	114	196	341	448	486	448	341	196	114	109	106	106	794	12
HALF DAY TOTALS			487	835	1580	2390	3020	3314	3135	2466	1539	935	628	495	467	464	464	464	3020	
May 21	5	3	1	2	3	3	3	2	1	0	0	0	0	0	0	0	0	0	0	7
	6	453	113	284	403	458	446	364	223	56	33	33	33	33	33	33	33	35	96	6
	7	681	90	320	520	638	659	580	412	172	63	59	59	59	59	59	59	59	276	5
	8	787	86	230	471	629	694	655	519	295	92	80	80	80	80	80	80	80	461	4
	9	844	99	131	330	518	620	632	551	382	168	101	96	96	96	96	96	96	616	3
	10	875	107	114	171	332	467	529	513	419	262	127	111	107	107	107	107	107	737	2
	11	891	115	115	121	152	256	357	409	400	331	222	133	120	115	115	115	115	812	1
	12	896	117	117	117	121	126	171	258	329	355	329	258	171	126	121	117	117	836	12
HALF DAY TOTALS			679	1275	2102	2818	3231	3232	2778	1897	1129	780	630	568	554	550	550	552	3418	
Jun 21	5	68	32	54	68	70	63	46	20	5	4	4	4	4	4	4	4	7	8	7
	6	488	150	329	450	501	478	380	222	53	39	39	39	39	39	39	39	43	126	6
	7	681	118	355	543	648	654	562	385	145	68	65	65	65	65	65	65	67	306	5
	8	776	94	268	492	633	680	626	480	252	93	85	85	85	85	85	85	85	484	4
	9	829	105	160	359	524	607	601	507	332	142	104	100	100	100	100	100	100	633	3
	10	859	112	121	197	345	457	500	468	366	218	122	115	112	112	112	112	112	750	2
	11	874	119	123	128	166	254	332	367	347	279	188	130	124	119	119	119	119	821	1
	12	879	121	121	121	127	131	163	227	281	301	281	227	163	131	127	121	121	844	12
HALF DAY TOTALS			799	1483	2314	2968	3275	3151	2580	1649	995	743	642	602	592	588	587	594	3551	
Jul 21	5	7	3	5	7	7	6	5	2	0	0	0	0	0	0	0	0	1	1	7
	6	435	116	281	395	447	433	352	216	55	34	34	34	34	34	34	34	37	100	6
	7	656	95	321	513	625	643	564	400	166	66	62	62	62	62	62	62	62	278	5
	8	762	90	236	468	620	680	639	505	285	94	83	83	83	83	83	83	83	459	4
	9	818	102	138	333	513	610	618	537	371	165	104	99	99	99	99	99	99	611	3
	10	850	110	118	177	333	462	519	501	407	255	129	114	110	110	110	110	110	729	2
	11	866	117	120	125	157	256	352	400	389	321	217	135	123	117	117	117	117	802	1
	12	871	120	120	120	125	130	172	253	320	345	320	253	172	130	125	120	120	826	12
HALF DAY TOTALS			705	1296	2102	2792	3180	3164	2707	1842	1110	783	643	586	572	568	567	570	3395	
Aug 21	6	255	38	137	214	256	259	223	151	52	18	17	17	17	17	17	17	17	38	6
	7	603	55	223	426	557	602	557	426	222	55	49	49	49	49	49	49	49	196	5
	8	747	75	149	397	584	681	676	569	374	128	74	72	72	72	72	72	72	386	4
	9	819	89	97	259	481	621	669	618	475	251	97	89	89	89	89	89	89	549	3
	10	857	102	105	126	294	472	574	590	519	567	173	107	102	102	102	102	102	674	2
	11	876	109	109	113	130	257	403	492	505	443	313	165	116	109	109	109	109	753	1
	12	882	112	112	112	115	120	197	333	434	470	434	333	197	120	115	112	112	780	12
HALF DAY TOTALS			518	861	1571	2338	2929	3196	3015	2370	1496	932	647	524	496	492	492	492	2983	
Sep 21	7	472	28	87	265	395	460	456	381	244	66	27	27	27	27	27	27	27	80	5
	8	725	52	61	275	504	646	689	626	467	224	57	52	52	52	52	52	52	258	4
	9	830	71	73	148	413	613	717	712	599	391	129	73	71	71	71	71	71	434	3
	10	882	84	84	89	224	468	631	697	659	522	293	96	84	84	84	84	84	567	2
	11	906	92	92	92	99	245	463	604	652	603	461	242	99	92	92	92	92	651	1
	12	914	95	95	95	95	101	237	446	584	631	584	446	237	101	95	95	95	679	12
HALF DAY TOTALS			374	447	917	1683	2484	3092	3257	2918	2119	1250	699	432	376	373	373	373	2329	
Oct 21	7	153	6	10	64	113	142	148	132	94	38	7	6	6	6	6	6	38	12	5
	8	644	35	37	155	387	545	615	592	476	280	56	35	35	35	35	35	35	136	4
	9	811	54	54	73	329	567	710	743	661	476	202	57	54	54	54	54	54	305	3
	10	884	67	67	70	157	439	646	752	742	619	395	120	69	67	67	67	67	443	2
	11	917	76	76	76	80	223	491	670	745	707	562	317	89	76	76	76	76	529	1
	12	927	78	78	78	78	84	267	521	681	737	681	521	267	84	78	78	78	558	12
HALF DAY TOTALS			277	282	479	1106	1965	2771	3173	3074	2494	1555	780	369	280	277	277	277	1704	
Nov 21	8	430	17	17	55	217	339	404	406	346	228	66	18	17	17	17	17	17	44	4
	9	733	38	38	42	231	476	634	691	643	492	254	43	38	38	38	38	38	173	3
	10	846	52	52	52	99	385	617	748	763	659	145	159	54	52	52	52	52	303	2
	11	894	60	60	60	63	192	486	688	783	757	613	367	89	60	60	60	60	388	1
	12	908	63	63	63	63	67	280	555	728	789	728	555	280	67	63	63	63	418	12
HALF DAY TOTALS			197	198	235	623	1403	2273	2797	2884	2516	1738	850	319	199	197	197	197	1115	
Dec 21	8	279	9	9	25	129	212	259	264	230	157	52	10	9	9	9	9	9	20	4
	9	685	31	31	33	188	427	584	646	611	477	260	40	31	31	31	31	31	124	3
	10	825	45	45	45	78	358	594	732	755	661	462	173	47	45	45	45	45	244	2
	11	882	53	53	53	55	177	477	685	786	765	624	379	90	53	53	53	53	327	1
	12	898	56	56	56	56	60	281	560	736	798	736	560	281	60	56	56	56	356	12
HALF DAY TOTALS			165	165	178	461	1180	2046	2594	2736	2446	1757	870	298	167	165	165	165	891	
			N	NNW	NW	WNW	W	WSW	SW	SSW	S	SSE	SE	ESE	E	ENE	NE	NNE	HOR	PM

[a]Total solar heat gains for DSA glass (based on a ground reflectance of 0.20).

[b]Half day totals computed by Simpson's rule, time interval = 10 min.

APPENDICES

TABLE B.9 Solar Intensity and Solar Heat Gain Factors[a] for 48° N Latitude (SI Units)[b]

Date	Solar Time am	Direct Normal W/m²	Solar Heat Gain Factors, W/m²: N	NNE	NE	ENE	E	ESE	SE	SSE	S	SSW	SW	WSW	W	WNW	NW	NNW	HOR	Solar Time pm
Jan 21	8	116	4	4	13	57	90	109	110	94	63	18	4	4	4	4	4	4	7	4
	9	584	24	24	26	168	371	505	555	523	406	217	30	24	24	24	24	24	79	3
	10	754	37	37	37	69	333	554	682	702	615	429	159	39	37	37	37	37	174	2
	11	823	45	45	45	47	166	455	656	753	734	598	365	83	45	45	45	45	244	1
	12	842	48	48	48	48	51	271	541	713	772	713	541	271	51	48	48	48	269	12
HALF DAY TOTALS			134	134	144	368	996	1788	2298	2448	2212	1616	817	267	136	134	134	134	639	
Feb 21	7	11	0	1	5	8	10	11	10	7	0	0	0	0	0	0	0	0	1	5
	8	568	24	25	114	324	470	537	524	428	259	52	24	24	24	24	24	24	78	4
	9	780	42	42	49	284	530	683	727	660	488	225	45	42	42	42	42	42	210	3
	10	869	54	54	55	120	415	641	764	768	653	434	137	55	54	54	54	54	331	2
	11	908	61	61	61	64	205	498	696	787	755	607	355	84	61	61	61	61	409	1
	12	920	63	63	63	63	68	280	555	728	790	728	555	280	68	63	63	63	435	12
HALF DAY TOTALS			214	216	337	864	1708	2575	3054	3051	2565	1677	825	328	216	214	214	214	1246	
Mar 21	7	482	23	71	253	387	458	458	388	253	71	23	22	22	22	22	22	22	64	5
	8	744	44	48	239	486	644	701	651	498	257	49	44	44	44	44	44	44	214	4
	9	853	60	60	104	381	609	738	753	652	449	164	62	60	60	60	60	60	371	3
	10	906	72	72	75	183	460	656	749	728	596	363	104	72	72	72	72	72	493	2
	11	932	79	79	79	83	232	492	663	731	689	541	297	88	79	79	79	79	568	1
	12	939	81	81	81	81	86	261	509	666	720	666	509	261	86	81	81	81	594	12
HALF DAY TOTALS			317	372	790	1558	2446	3193	3471	3199	2420	1466	769	399	319	316	316	316	2006	
Apr 21	6	340	39	167	271	330	337	294	203	74	20	19	19	19	19	19	19	19	46	6
	7	646	49	191	417	567	628	595	468	264	56	45	45	45	45	45	45	45	189	5
	8	779	64	99	350	566	690	709	619	435	173	67	64	64	64	64	64	64	359	4
	9	847	79	83	191	444	621	703	679	554	335	105	79	79	79	79	79	79	507	3
	10	884	90	90	97	244	466	610	660	613	472	251	97	90	90	90	90	90	618	2
	11	902	97	97	97	105	245	443	570	609	558	419	217	103	97	97	97	97	689	1
	12	908	99	99	99	99	106	224	414	543	587	543	414	224	106	99	99	99	713	12
HALF DAY TOTALS			463	765	1454	2285	3019	3465	3411	2825	1908	1169	713	492	445	442	442	442	2761	
May 21	5	129	52	97	125	133	122	91	44	9	8	8	8	8	8	8	8	10	16	7
	6	511	112	305	443	512	504	418	267	75	38	38	38	38	38	38	38	40	125	6
	7	690	73	283	498	631	668	604	448	214	66	61	61	61	61	61	61	61	287	5
	8	782	83	170	418	599	689	675	562	358	120	79	79	79	79	79	79	79	449	4
	9	834	93	101	259	475	611	656	605	463	243	100	93	93	93	93	93	93	585	3
	10	864	103	106	124	283	458	561	579	513	366	179	109	103	103	103	103	103	690	2
	11	879	109	109	113	127	249	396	488	505	447	320	170	116	109	109	109	109	756	1
	12	884	111	111	111	114	120	198	336	439	474	439	336	198	120	114	111	111	778	12
HALF DAY TOTALS			679	1224	2035	2817	3360	3515	3176	2363	1525	997	709	579	550	546	546	549	3297	
Jun 21	5	243	111	192	241	252	228	166	75	19	17	17	17	17	17	17	17	24	38	7
	6	544	146	348	488	552	534	434	266	70	46	46	46	46	46	46	46	49	162	6
	7	693	93	317	521	642	467	591	427	188	72	67	67	67	67	67	67	67	324	5
	8	775	90	203	440	604	678	651	529	320	107	84	84	84	84	84	84	84	479	4
	9	822	98	114	286	482	600	629	567	418	208	105	98	98	98	98	98	98	610	3
	10	849	108	113	142	296	450	534	540	465	319	157	114	108	108	108	108	108	711	2
	11	864	114	114	120	138	248	373	449	456	396	279	156	121	114	114	114	114	775	1
	12	869	116	116	116	120	126	189	303	391	423	391	303	189	126	120	116	116	796	12
HALF DAY TOTALS			811	1446	2279	3013	3454	3477	3013	2140	1376	942	718	620	597	593	592	602	3495	
Jul 21	5	135	57	104	133	141	129	96	46	11	9	9	9	9	9	9	9	12	18	7
	6	492	116	302	436	501	492	407	259	75	40	40	40	40	40	40	40	43	130	6
	7	666	78	285	492	619	653	588	436	207	69	63	63	63	63	63	63	63	290	5
	8	757	86	176	416	590	675	660	547	348	119	81	81	81	81	81	81	81	448	4
	9	809	96	106	263	471	601	643	591	450	237	104	96	96	96	96	96	96	582	3
	10	839	106	109	130	285	453	550	566	500	356	177	112	106	106	106	106	106	684	2
	11	855	112	112	117	132	249	390	477	492	435	312	169	119	112	112	112	112	749	1
	12	859	114	114	114	117	124	198	329	428	462	428	329	198	124	117	114	114	771	12
HALF DAY TOTALS			705	1247	2039	2795	3311	3446	3101	2303	1496	993	721	598	570	565	565	569	3287	
Aug 21	6	311	42	161	256	310	316	273	190	70	22	21	21	21	21	21	21	21	51	6
	7	599	53	193	403	543	598	565	444	250	59	49	49	49	49	49	49	49	193	5
	8	732	69	108	347	549	665	681	593	417	168	73	69	69	69	69	69	69	358	4
	9	801	84	90	198	437	605	681	655	534	323	108	84	84	84	84	84	84	502	3
	10	839	95	95	104	247	453	593	639	592	456	245	104	95	95	95	95	95	610	2
	11	858	102	102	102	112	247	433	553	589	539	406	215	110	102	102	102	102	679	1
	12	864	104	104	104	104	113	224	404	526	568	526	404	224	113	104	104	104	702	12
HALF DAY TOTALS			496	790	1449	2237	2929	3343	3282	2720	1852	1156	728	521	475	471	471	471	2741	
Sep 21	7	414	24	66	224	342	403	403	342	224	66	24	23	23	23	23	23	23	62	5
	8	678	46	51	228	455	602	654	608	467	244	52	46	46	46	46	46	46	206	4
	9	792	63	63	107	366	581	702	716	621	430	163	66	63	63	63	63	63	358	3
	10	848	75	75	79	182	444	630	719	699	573	353	107	75	75	75	75	75	476	2
	11	876	82	82	82	88	230	476	639	704	664	524	292	93	82	82	82	82	549	1
	12	884	84	84	84	84	91	257	494	643	695	643	494	257	91	84	84	84	574	12
HALF DAY TOTALS			332	381	758	1467	2300	3007	3280	3039	2324	1429	766	412	334	331	331	331	1937	
Oct 21	7	12	0	1	5	9	11	12	11	8	3	0	0	0	0	0	0	0	1	5
	8	522	25	27	111	304	439	501	489	398	242	51	25	25	25	25	25	25	79	4
	9	734	44	44	52	276	508	652	693	629	466	216	47	44	44	44	44	44	208	3
	10	825	56	56	58	122	403	618	736	739	628	418	136	58	56	56	56	56	327	2
	11	866	64	64	64	67	203	483	673	760	729	587	345	86	64	64	64	64	403	1
	12	878	66	66	66	66	71	274	538	704	763	704	538	274	71	66	66	66	429	12
HALF DAY TOTALS			223	225	340	838	1637	2461	2918	2918	2459	1619	808	334	226	223	223	223	1233	
Nov 21	8	115	4	4	13	57	90	108	109	93	62	18	4	4	4	4	4	4	7	4
	9	565	25	25	27	165	363	492	540	509	394	211	31	25	25	25	25	25	81	3
	10	735	38	38	38	70	328	543	668	688	602	420	156	40	38	38	38	38	175	2
	11	804	46	46	46	48	165	448	645	739	720	587	358	83	46	46	46	46	244	1
	12	823	48	48	48	48	53	267	531	700	758	700	531	267	53	48	48	48	269	12
HALF DAY TOTALS			138	138	147	368	979	1752	2250	2396	2165	1583	804	268	140	138	138	138	642	
Dec 21	9	442	16	16	18	113	272	378	420	401	316	176	25	16	16	16	16	16	42	3
	10	676	30	30	30	51	286	491	613	636	563	398	156	32	30	30	30	30	119	2
	11	765	38	38	38	40	146	421	615	708	694	569	351	80	38	38	38	38	181	1
	12	789	41	41	41	41	44	256	514	679	734	679	514	256	44	41	41	41	204	12
HALF DAY TOTALS			105	105	107	231	735	1444	1924	2098	1944	1477	778	239	107	105	105	105	444	
			N	NNW	NW	WNW	W	WSW	SW	SSW	S	SSE	SE	ESE	E	ENE	NE	NNE	HOR	PM

[a]Total solar heat gains for DSA glass (based on a ground reflectance of 0.20).

[b]Half day totals computed by Simpson's rule, time interval = 10 min.

TABLE B.10 **Solar Intensity and Solar Heat Gain Factors[a] for 64° N Latitude (SI Units)[b]**

Date	Solar Time am	Direct Normal W/m^2	Solar Heat Gain Factors, W/m^2																	Solar Time pm
			N	NNE	NE	ENE	E	ESE	SE	SSE	S	SSW	SW	WSW	W	WNW	NW	NNW	HOR	
Jan 21	10	69	2	2	2	4	29	51	63	66	58	41	16	2	2	2	2	2	4	2
	11	256	9	9	9	9	46	143	210	242	238	196	120	24	9	9	9	9	18	1
	12	316	11	11	11	11	12	104	211	280	302	280	211	104	12	11	11	11	24	12
HALF DAY TOTALS			16	16	16	18	80	250	382	449	446	374	237	72	17	16	16	16	33	
Feb 21	8	56	2	2	9	31	46	53	53	44	27	6	2	2	2	2	2	2	3	4
	9	422	16	16	18	136	280	373	403	375	285	143	18	16	16	16	16	16	41	3
	10	601	26	26	26	57	276	453	554	567	495	341	119	28	26	26	26	26	90	2
	11	679	32	32	32	34	139	386	558	638	622	506	307	64	32	32	32	32	128	1
	12	701	34	34	34	34	37	231	464	613	662	613	464	231	37	34	34	34	143	12
HALF DAY TOTALS			93	93	103	279	770	1408	1822	1947	1767	1298	668	210	[illegible]	93	93	93	334	
Mar 21	7	300	12	34	147	235	284	288	249	168	54	13	12	12	12	12	12	12	27	5
	8	582	29	31	147	357	499	559	536	427	246	44	29	29	29	29	29	29	100	4
	9	717	41	41	51	277	502	642	679	613	450	203	44	41	41	41	41	41	187	3
	10	786	49	49	50	111	386	598	714	719	613	410	132	51	49	49	49	49	264	2
	11	821	54	54	54	57	190	468	658	746	718	579	344	80	54	54	54	54	314	1
	12	831	56	56	56	56	61	269	530	696	754	696	530	269	61	56	56	56	331	12
HALF DAY TOTALS			213	234	473	1052	1879	2695	3103	3021	2457	1591	811	330	215	213	213	213	1057	
Apr 21	5	85	24	56	77	85	81	64	36	6	4	4	4	4	4	4	4	5	8	7
	6	419	38	187	320	400	416	371	266	111	26	24	24	24	24	24	24	24	67	6
	7	613	43	129	355	516	595	584	482	303	80	41	41	41	41	41	41	41	162	5
	8	720	54	60	249	483	633	685	634	484	250	60	54	54	54	54	54	54	269	4
	9	783	65	65	100	351	567	691	710	620	436	174	68	65	65	65	65	65	367	3
	10	820	72	72	76	161	421	613	709	698	582	372	119	74	72	72	72	72	441	2
	11	839	77	77	77	82	215	468	637	710	676	540	312	93	77	77	77	77	448	1
	12	846	79	79	79	79	85	262	503	655	708	655	503	262	85	79	79	79	504	12
HALF DAY TOTALS			413	687	1294	2116	2974	3628	3741	3268	2408	1537	861	471	381	378	378	378	2053	
May 21	4	160	94	139	162	161	135	87	25	10	10	10	10	10	10	10	11	31	20	8
	5	416	152	298	393	425	395	302	158	34	29	29	29	29	29	29	29	33	81	7
	6	584	90	304	474	570	579	499	343	126	49	46	46	46	46	46	46	46	175	6
	7	688	65	199	436	596	665	634	507	297	78	60	60	60	60	60	60	60	284	5
	8	754	72	89	307	527	660	694	623	459	215	78	72	72	72	72	72	72	390	4
	9	796	82	85	143	384	577	678	679	579	389	147	85	82	82	82	82	82	481	3
	10	822	89	89	94	193	427	593	668	645	528	322	114	89	89	89	89	89	549	2
	11	836	93	93	93	100	226	446	594	652	614	485	275	105	93	93	93	93	592	1
	12	841	95	95	95	95	103	248	460	597	644	597	460	248	103	95	95	95	606	12
HALF DAY TOTALS			780	1341	2145	2998	3712	4073	3841	3108	2234	1467	909	603	533	529	530	554	2875	
Jun 21	4	294	181	263	302	297	247	156	43	21	21	21	21	21	21	21	23	67	50	8
	5	485	195	360	466	498	458	346	175	44	39	39	39	39	39	39	39	45	124	7
	6	614	113	338	510	603	605	516	347	122	57	55	55	55	55	55	55	57	223	6
	7	698	74	225	457	610	672	632	498	282	80	68	68	68	68	68	68	68	331	5
	8	754	79	105	327	535	657	682	605	438	197	85	79	79	79	79	79	79	433	4
	9	791	89	93	161	393	572	662	655	552	362	137	92	89	89	89	89	89	520	3
	10	814	96	96	102	205	423	577	642	615	497	297	114	96	96	96	96	96	585	2
	11	827	100	100	100	108	228	431	568	619	581	456	258	110	100	100	100	100	627	1
	12	831	101	101	101	101	110	240	436	566	610	566	436	240	110	101	101	101	641	12
HALF DAY TOTALS			1016	1680	2526	3347	3953	4154	3769	2985	2142	1436	935	667	605	601	603	673	3219	
Jul 21	4	168	101	149	173	172	144	93	27	12	12	12	12	12	12	12	12	35	24	8
	5	405	154	297	389	420	390	298	156	36	32	32	32	32	32	32	32	36	88	7
	6	564	94	303	467	560	567	488	335	124	51	49	49	49	49	49	49	49	181	6
	7	665	69	201	431	585	652	620	495	290	80	63	63	63	63	63	63	63	289	5
	8	730	75	94	307	520	648	680	609	449	211	81	75	75	75	75	75	75	393	4
	9	771	85	88	148	382	567	665	664	566	380	146	88	85	85	85	85	85	481	3
	10	797	92	92	98	196	422	583	655	631	516	315	116	92	92	92	92	92	548	2
	11	812	96	96	96	104	226	439	582	638	601	475	271	108	96	96	96	96	590	1
	12	816	98	98	98	98	107	246	452	585	630	585	452	246	107	98	98	98	604	12
HALF DAY TOTALS			814	1370	2157	2985	3669	4004	3763	3046	2198	1457	920	626	557	553	554	582	2985	
Aug 21	5	92	28	62	85	94	89	71	40	8	6	6	6	6	6	6	6	6	11	7
	6	388	42	182	306	380	395	352	252	107	30	27	27	27	27	27	27	27	73	6
	7	570	48	132	344	494	567	555	458	289	81	45	45	45	45	45	45	45	168	5
	8	675	59	66	247	468	609	657	607	464	241	66	59	59	59	59	59	59	273	4
	9	737	70	70	107	345	549	666	683	597	420	172	74	70	70	70	70	70	368	3
	10	775	78	78	82	165	412	594	685	674	562	361	122	80	78	78	78	78	440	2
	11	794	82	82	82	89	216	456	617	686	653	523	305	99	82	82	82	82	485	1
	12	801	84	84	84	84	92	260	489	635	684	633	489	260	92	84	84	84	501	12
HALF DAY TOTALS			447	714	1293	2073	2884	3498	3600	3147	2333	1509	869	500	413	409	409	410	2069	
Sep 21	7	242	12	30	122	194	234	238	206	139	47	13	12	12	12	12	12	12	26	5
	8	513	30	33	136	324	451	505	484	387	225	45	30	30	30	30	30	30	97	4
	9	651	43	43	54	261	468	596	631	570	420	193	47	43	43	43	43	43	181	3
	10	722	52	52	53	111	366	563	672	677	577	388	131	54	52	52	52	52	255	2
	11	758	57	57	57	61	185	445	623	706	680	549	329	83	57	57	57	57	303	1
	12	769	59	59	59	59	65	260	504	660	715	660	504	260	65	59	59	59	320	12
HALF DAY TOTALS			224	241	448	968	1727	2484	2871	2810	2305	1513	787	336	227	224	224	224	1021	
Oct 21	8	54	2	2	10	30	45	52	51	42	26	6	2	2	2	2	2	2	4	4
	9	383	17	17	19	127	259	345	372	346	263	133	19	17	17	17	17	17	42	3
	10	556	28	28	28	58	261	426	520	532	465	321	115	29	28	28	28	28	91	2
	11	633	34	34	34	36	135	367	528	603	588	479	292	64	34	34	34	34	130	1
	12	655	36	36	36	36	40	222	441	581	628	581	441	222	40	36	36	36	144	12
HALF DAY TOTALS			98	98	109	273	728	1324	1711	1829	1661	1225	638	209	100	98	98	98	339	
Nov 21	10	72	2	2	2	5	31	53	67	69	62	43	17	3	2	2	2	2	4	2
	11	250	9	9	9	10	46	140	207	238	234	192	118	24	9	9	9	9	19	1
	12	307	11	11	11	11	13	102	207	274	295	274	207	102	13	11	11	11	25	12
HALF DAY TOTALS			17	17	17	19	81	248	378	444	440	369	234	71	18	17	17	17	35	
Dec 21	11	12	0	0	0	0	2	7	10	12	11	9	6	1	0	0	0	0	1	1
	12	51	1	1	1	1	2	16	34	45	48	45	34	16	2	1	1	1	3	12
HALF DAY TOTALS			1	1	1	1	3	16	28	35	36	32	22	8	1	1	1	1	2	
			N	NNW	NW	WNW	W	WSW	SW	SSW	S	SSE	SE	ESE	E	ENE	NE	NNE	HOR	PM

[a]Total solar heat gains for DSA glass (based on a ground reflectance of 0.20).

[b]Half day totals computed by Simpson's rule, time interval = 10 min.

TABLE B.11 **Solar Position and Clear Day Insolation, 32° N Latitude**

Date	Solar Time A.M.	Solar Time P.M.	Solar Position Alt	Solar Position Azm	Direct Normal	Global Irradiance (Btu/h ft²) Horiz	South-Facing Elevation Angle 22	32	42	52	90
Jan 21	7	5	1.4	65.2	1	0	0	0	0	1	1
	8	4	12.5	56.5	203	56	93	106	116	123	115
	9	3	22.5	46.0	269	118	175	193	206	212	181
	10	2	30.6	33.1	295	167	235	256	269	274	221
	11	1	36.1	17.5	306	198	273	295	308	312	245
	12		38.0	0.0	310	209	285	308	321	324	253
	Surface Daily Totals				2458	1288	1839	2008	2118	2166	1779
Feb 21	7	5	7.1	73.5	121	22	34	37	40	42	38
	8	4	19.0	64.4	247	95	127	136	140	141	108
	9	3	29.9	53.4	288	161	206	217	222	220	158
	10	2	39.1	39.4	306	212	266	278	283	279	193
	11	1	45.6	21.4	315	244	304	317	321	315	214
	12		48.0	0.0	317	255	316	330	334	328	222
	Surface Daily Totals				2872	1724	2188	2300	2345	2322	1644
Mar 21	7	5	12.7	81.9	185	54	60	60	59	56	32
	8	4	25.1	73.0	260	129	146	147	144	137	78
	9	3	36.8	62.1	290	194	222	224	220	209	119
	10	2	47.3	47.5	304	245	280	283	278	265	150
	11	1	55.0	26.8	311	277	317	321	315	300	170
	12		58.0	0.0	313	287	329	333	327	312	177
	Surface Daily Totals				3012	2084	2378	2403	2358	2246	1276
Apr 21	6	6	6.1	99.9	66	14	9	6	6	5	3
	7	5	18.8	92.2	206	86	78	71	62	51	10
	8	4	31.5	84.0	255	158	156	148	136	120	35
	9	3	43.9	74.2	278	220	225	217	203	183	68
	10	2	55.7	60.3	290	267	279	272	256	234	95
	11	1	65.4	37.5	295	297	313	306	290	265	112
	12		69.6	0.0	297	307	325	318	301	276	118
	Surface Daily Totals				3076	2390	2444	2356	2206	1994	764
May 21	6	6	10.4	107.2	119	36	21	13	13	12	7
	7	5	22.8	100.1	211	107	88	75	60	44	13
	8	4	35.4	92.9	250	175	159	145	127	105	15
	9	3	48.1	84.7	269	233	223	209	188	163	33
	10	2	60.6	73.3	280	277	273	259	237	208	56
	11	1	72.0	51.9	285	305	305	290	268	237	72
	12		78.0	0.0	286	315	315	301	278	247	77
	Surface Daily Totals				3112	2582	2454	2284	2064	1788	469
Jun 21	6	6	12.2	110.2	131	45	26	16	15	14	9
	7	5	24.3	103.4	210	115	91	76	59	41	14
	8	4	36.9	96.8	245	180	159	143	122	99	16
	9	3	49.6	89.4	264	236	221	204	181	153	19
	10	2	62.2	79.7	274	279	268	251	227	197	41
	11	1	74.2	60.9	279	306	299	282	257	224	56
	12		81.5	0.0	280	315	309	292	267	234	60
	Surface Daily Totals				3084	2634	2436	2234	1990	1690	370

TABLE B.11 **Solar Position and Clear Day Insolation, 32° N Latitude (Continued)**

						Global Irradiance (Btu/h ft²)					
	Solar Time		Solar Position				South-Facing Elevation Angle				
Date	A.M.	P.M.	*Alt*	*Azm*	*Direct Normal*	*Horiz*	*22*	*32*	*42*	*52*	*90*
Jul 21	6	6	10.7	107.7	113	37	22	14	13	12	8
	7	5	23.1	100.6	203	107	87	75	60	44	14
	8	4	35.7	93.6	241	174	158	143	125	104	16
	9	3	48.4	85.5	261	230	220	205	185	159	31
	10	2	60.9	74.3	271	274	269	254	232	204	54
	11	1	72.4	53.3	277	302	300	285	262	232	69
	12		78.6	0.0	279	311	310	296	273	242	74
	Surface Daily Totals				3012	2558	2422	2250	2030	1754	458
Aug 21	6	6	6.5	100.5	59	14	9	7	6	6	4
	7	5	19.1	92.8	190	85	77	69	60	50	12
	8	4	31.8	84.7	240	156	152	144	132	116	33
	9	3	44.3	75.0	263	216	220	212	197	178	65
	10	2	56.1	61.3	276	262	272	264	249	226	91
	11	1	66.0	38.4	282	292	305	298	281	257	107
	12		70.3	0.0	284	302	317	309	292	268	113
	Surface Daily Totals				2902	2352	2388	2296	2144	1934	736
Sep 21	7	5	12.7	81.9	163	51	56	56	55	52	30
	8	4	25.1	73.0	240	124	140	141	138	131	75
	9	3	36.8	62.1	272	188	213	215	211	201	114
	10	2	47.3	47.5	287	237	270	273	268	255	145
	11	1	55.0	26.8	294	268	306	309	303	289	164
	12		58.0	0.0	296	278	318	321	315	300	171
	Surface Daily Totals				2808	2014	2288	2308	2264	2154	1226
Oct 21	7	5	6.8	73.1	99	19	29	32	34	36	32
	8	4	18.7	64.0	229	90	120	128	133	134	104
	9	3	29.5	53.0	273	155	198	208	213	212	153
	10	2	38.7	39.1	293	204	257	269	273	270	188
	11	1	45.1	21.1	302	236	294	307	311	306	209
	12		47.5	0.0	304	247	306	320	324	318	217
	Surface Daily Totals				2696	1654	2100	2208	2252	2232	1588
Nov 21	7	5	1.5	65.4	2	0	0	0	1	1	1
	8	4	12.7	56.6	196	55	91	104	113	119	111
	9	3	22.6	46.1	263	118	173	190	202	208	176
	10	2	30.8	33.2	289	166	233	252	265	270	217
	11	1	36.2	17.6	301	197	270	291	303	307	241
	12		38.2	0.0	304	207	282	304	316	320	249
	Surface Daily Totals				2406	1280	1816	1980	2084	2130	1742
Dec 21	8	4	10.3	53.8	176	41	77	90	101	108	107
	9	3	19.8	43.6	257	102	161	180	195	204	183
	10	2	27.6	31.2	288	150	221	244	259	267	226
	11	1	32.7	16.4	301	180	258	282	298	305	251
	12		34.6	0.0	304	190	271	295	311	318	259
	Surface Daily Totals				2348	1136	1704	1888	2016	2086	1794

NOTE: 1 Btu/h ft² = 3.152 W/m².

Source: Reprinted by permission from Peter J. Lunde, *Solar Thermal Engineering*, © 1980, John Wiley & Sons Inc., New York.

TABLE B.12 **Solar Position and Clear Day Insolation, 40° N Latitude**

	Solar Time		Solar Position			Global Irradiance (Btu/h ft²)					
					Direct		South-Facing Elevation Angle				
Date	A.M.	P.M.	Alt	Azm	Normal	Horiz	30	40	50	60	90
Jan 21	8	4	8.1	55.3	142	28	65	74	81	85	84
	9	3	16.8	44.0	239	83	155	171	182	187	171
	10	2	23.8	30.9	274	127	218	237	249	254	223
	11	1	28.4	16.0	289	154	257	277	290	293	253
	12		30.0	0.0	294	164	270	291	303	306	263
	Surface Daily Totals				2182	948	1660	1810	1906	1944	1726
Feb 21	7	5	4.8	72.7	69	10	19	21	23	24	22
	8	4	15.4	62.2	224	73	114	122	126	127	107
	9	3	25.0	50.2	274	132	195	205	209	208	167
	10	2	32.8	35.9	295	178	256	267	271	267	210
	11	1	38.1	18.9	305	206	293	306	310	304	236
	12		40.0	0.0	308	216	306	319	323	317	245
	Surface Daily Totals				2640	1414	2060	2162	2202	2176	1730
Mar 21	7	5	11.4	80.2	171	46	55	55	54	51	35
	8	4	22.5	69.6	250	114	140	141	138	131	89
	9	3	32.8	57.3	282	173	215	217	213	202	138
	10	2	41.6	41.9	297	218	273	276	271	258	176
	11	1	47.7	22.6	305	247	310	313	307	293	200
	12		50.0	0.0	307	257	322	326	320	305	208
	Surface Daily Totals				2916	1852	2308	2330	2284	2174	1484
Apr 21	6	6	7.4	98.9	89	20	11	8	7	7	4
	7	5	18.9	89.5	206	87	77	70	61	50	12
	8	4	30.3	79.3	252	152	153	145	133	117	53
	9	3	41.3	67.2	274	207	221	213	199	179	93
	10	2	51.2	51.4	286	250	275	267	252	229	126
	11	1	58.7	29.2	292	277	308	301	285	260	147
	12		61.6	0.0	293	287	320	313	296	271	154
	Surface Daily Totals				3092	2274	2412	2320	2168	1956	1022
May 21	5	7	1.9	114.7	1	0	0	0	0	0	0
	6	6	12.7	105.6	144	49	25	15	14	13	9
	7	5	24.0	96.6	216	114	89	76	60	44	13
	8	4	35.4	87.2	250	175	158	144	125	104	25
	9	3	46.8	76.0	267	227	221	206	186	160	60
	10	2	57.5	60.9	277	267	270	255	233	205	89
	11	1	66.2	37.1	283	293	301	287	264	234	108
	12		70.0	0.0	284	301	312	297	274	243	114
	Surface Daily Totals				3160	2552	2442	2264	2040	1760	724
Jun 21	5	7	4.2	117.3	22	4	3	3	2	2	1
	6	6	14.8	108.4	155	60	30	18	17	16	10
	7	5	26.0	99.7	216	123	92	77	59	40	14
	8	4	37.4	90.7	246	182	159	142	121	97	16
	9	3	48.8	80.2	263	233	219	202	179	151	47
	10	2	59.8	65.8	272	272	266	248	224	193	74
	11	1	69.2	41.9	277	296	296	278	253	221	92
	12		73.5	0.0	279	304	306	289	263	230	98
	Surface Daily Totals				3180	2648	2434	2224	1974	1670	610

TABLE B.12 **Solar Position and Clear Day Insolation, 40° N Latitude (Continued)**

Date	Solar Time A.M.	Solar Time P.M.	Solar Position Alt	Solar Position Azm	Direct Normal	Global Irradiance (Btu/h ft²) Horiz	South-Facing Elevation Angle 30	40	50	60	90
Jul 21	5	7	2.3	115.2	2	0	0	0	0	0	0
	6	6	13.1	106.1	138	50	26	17	15	14	9
	7	5	24.3	97.2	208	114	89	75	60	44	14
	8	4	35.8	87.8	241	174	157	142	124	102	24
	9	3	47.2	76.7	259	225	218	203	182	157	58
	10	2	57.9	61.7	269	265	266	251	229	200	86
	11	1	66.7	37.9	275	290	296	281	258	228	104
	12		70.6	0.0	276	298	307	292	269	238	111
	Surface Daily Totals				3062	2534	2409	2230	2006	1728	702
Aug 21	6	6	7.9	99.5	81	21	12	9	8	7	5
	7	5	19.3	90.0	191	87	76	69	60	49	12
	8	4	30.7	79.9	237	150	150	141	129	113	50
	9	3	41.8	67.9	260	205	216	207	193	173	89
	10	2	51.7	52.1	272	246	267	259	244	221	120
	11	1	59.3	29.7	278	273	300	292	276	252	140
	12		62.3	0.0	280	282	311	303	287	262	147
	Surface Daily Totals				2916	2244	2354	2258	2104	1894	978
Sep 21	7	5	11.4	80.2	149	43	51	51	49	47	32
	8	4	22.5	69.6	230	109	133	134	131	124	84
	9	3	32.8	57.3	263	167	206	208	203	193	132
	10	2	41.6	41.9	280	211	262	265	260	247	168
	11	1	47.7	22.6	287	239	298	301	295	281	192
	12		50.0	0.0	290	249	310	313	307	292	200
	Surface Daily Totals				2708	1788	2210	2228	2182	2074	1416
Oct 21	7	5	4.5	72.3	48	7	14	15	17	17	16
	8	4	15.0	61.9	204	68	106	113	117	118	100
	9	3	24.5	49.8	257	126	185	195	200	198	160
	10	2	32.4	35.6	280	170	245	257	261	257	203
	11	1	37.6	18.7	291	199	283	295	299	294	229
	12		39.5	0.0	294	208	295	308	312	306	238
	Surface Daily Totals				2454	1348	1962	2060	2098	2074	1654
Nov 21	8	4	8.2	55.4	136	28	63	72	78	82	81
	9	3	17.0	44.1	232	82	152	167	178	183	167
	10	2	24.0	31.0	268	126	215	233	245	249	219
	11	1	28.6	16.1	283	153	254	273	285	288	248
	12		30.2	0.0	288	163	267	287	298	301	258
	Surface Daily Totals				2128	942	1636	1778	1870	1908	1686
Dec 21	8	4	5.5	53.0	89	14	39	45	50	54	56
	9	3	14.0	41.9	217	65	135	152	164	171	163
	10	2	20.7	29.4	261	107	200	221	235	242	221
	11	1	25.0	15.2	280	134	239	262	276	283	252
	12		26.6	0.0	285	143	253	275	290	296	263
	Surface Daily Totals				1978	782	1480	1634	1740	1796	1646

NOTE: 1 Btu/h ft^2 = 3.152 W/m^2.

Source: Reprinted by permission from Peter J. Lunde, *Solar Thermal Engineering*, © 1980, John Wiley & Sons Inc., New York.

TABLE B.13 **Solar Position and Clear Day Insolation, 48° N Latitude**

Date	Solar Time A.M.	Solar Time P.M.	Solar Position Alt	Solar Position Azm	Direct Normal	Global Irradiance (Btu/h ft²) Horiz	South-Facing Elevation Angle 38	48	58	68	90
Jan 21	8	4	3.5	54.6	37	4	17	19	21	22	22
	9	3	11.0	42.6	185	46	120	132	140	145	139
	10	2	16.9	29.4	239	83	190	206	216	220	206
	11	1	20.7	15.1	261	107	231	249	260	263	243
	12		22.0	0.0	267	115	245	264	275	278	255
	Surface Daily Totals				1710	596	1360	1478	1550	1578	1478
Feb 21	7	5	2.4	72.2	12	1	3	4	4	4	4
	8	4	11.6	60.5	188	49	95	102	105	106	96
	9	3	19.7	47.7	251	100	178	187	191	190	167
	10	2	26.2	33.3	278	139	240	251	255	251	217
	11	1	30.5	17.2	290	165	278	290	294	288	247
	12		32.0	0.0	293	173	291	304	307	301	258
	Surface Daily Totals				2330	1080	1880	1972	2024	1978	1720
Mar 21	7	5	10.0	78.7	153	37	49	49	47	45	35
	8	4	19.5	66.8	236	96	131	132	129	122	96
	9	3	28.2	53.4	270	147	205	207	203	193	152
	10	2	35.4	37.8	287	187	263	266	261	248	195
	11	1	40.3	19.8	295	212	300	303	297	283	223
	12		42.0	0.0	298	220	312	315	309	294	232
	Surface Daily Totals				2780	1578	2208	2228	2182	2074	1632
Apr 21	6	6	8.6	97.8	108	27	13	9	8	7	5
	7	5	18.6	86.7	205	85	76	68	59	48	21
	8	4	28.5	74.9	247	142	149	141	129	113	69
	9	3	37.8	61.2	268	191	216	208	194	174	115
	10	2	45.8	44.6	280	228	268	260	245	223	152
	11	1	51.5	24.0	286	252	301	294	278	254	177
	12		53.6	0.0	288	260	313	305	289	264	185
	Surface Daily Totals				3076	2106	2358	2266	2114	1902	1262
May 21	5	7	5.2	114.3	41	9	4	4	4	3	2
	6	6	14.7	103.7	162	61	27	16	15	13	10
	7	5	24.6	93.0	219	118	89	75	60	43	13
	8	4	34.7	81.6	248	171	156	142	123	101	45
	9	3	44.3	68.3	264	217	217	202	182	156	86
	10	2	53.0	51.3	274	252	265	251	229	200	120
	11	1	59.5	28.6	279	274	296	281	258	228	141
	12		62.0	0.0	280	281	306	292	269	238	149
	Surface Daily Totals				3254	2482	2418	2234	2010	1728	982
Jun 21	5	7	7.9	116.5	77	21	9	9	8	7	5
	6	6	17.2	106.2	172	74	33	19	18	16	12
	7	5	27.0	95.8	220	129	93	77	59	39	15
	8	4	37.1	84.6	246	181	157	140	119	95	35
	9	3	46.9	71.6	261	225	216	198	175	147	74
	10	2	55.8	54.8	269	259	262	244	220	189	105
	11	1	62.7	31.2	274	280	291	273	248	216	126
	12		65.5	0.0	275	287	301	283	258	225	133
	Surface Daily Totals				3312	2626	2420	2204	1950	1644	874

TABLE B.13 **Solar Position and Clear Day Insolation, 48° N Latitude (Continued)**

Date	Solar Time A.M.	Solar Time P.M.	Solar Position Alt	Solar Position Azm	Direct Normal	Global Irradiance (Btu/h ft²) Horiz	South-Facing Elevation Angle 38	48	58	68	90
Jul 21	5	7	5.7	114.7	43	10	5	5	4	4	3
	6	6	15.2	104.1	156	62	28	18	16	15	11
	7	5	25.1	93.5	211	118	89	75	59	42	14
	8	4	35.1	82.1	240	171	154	140	121	99	43
	9	3	44.8	68.8	256	215	214	199	178	153	83
	10	2	53.5	51.9	266	250	261	246	224	195	116
	11	1	60.1	29.0	271	272	291	276	253	223	137
	12		62.6	0.0	272	279	301	286	263	232	144
	Surface Daily Totals				3158	2474	2386	2200	1974	1694	956
Aug 21	6	6	9.1	98.3	99	28	14	10	9	8	6
	7	5	19.1	87.2	190	85	75	67	58	47	20
	8	4	29.0	75.4	232	141	145	137	125	109	65
	9	3	38.4	61.8	254	189	210	201	187	168	110
	10	2	46.4	45.1	266	225	260	252	237	214	146
	11	1	52.2	24.3	272	248	293	285	268	244	169
	12		54.3	0.0	274	256	304	296	279	255	177
	Surface Daily Totals				2898	2086	2300	2200	2046	1836	1208
Sep 21	7	5	10.0	78.7	131	35	44	44	43	40	31
	8	4	19.5	66.8	215	92	124	124	121	115	90
	9	3	28.2	53.4	251	142	196	197	193	183	143
	10	2	35.4	37.8	269	181	251	254	248	236	185
	11	1	40.3	19.8	278	205	287	289	284	269	212
	12		42.0	0.0	280	213	299	302	296	281	221
	Surface Daily Totals				2568	1522	2102	2118	2070	1966	1546
Oct 21	7	5	2.0	71.9	4	0	1	1	1	1	1
	8	4	11.2	60.2	165	44	86	91	95	95	87
	9	3	19.3	47.4	233	94	167	176	180	178	157
	10	2	25.7	33.1	262	133	228	239	242	239	207
	11	1	30.0	17.1	274	157	266	277	281	276	237
	12		31.5	0.0	278	166	279	291	294	288	247
	Surface Daily Totals				2154	1022	1774	1860	1890	1866	1626
Nov 21	8	4	3.6	54.7	36	5	17	19	21	22	22
	9	3	11.2	42.7	179	46	117	129	137	141	135
	10	2	17.1	29.5	233	83	186	202	212	215	201
	11	1	20.9	15.1	255	107	227	245	255	258	238
	12		22.2	0.0	261	115	241	259	270	272	250
	Surface Daily Totals				1668	596	1336	1448	1518	1544	1442
Dec 21	9	3	8.0	40.9	140	27	87	98	105	110	109
	10	2	13.6	28.2	214	63	164	180	192	197	190
	11	1	17.3	14.4	242	86	207	226	239	244	231
	12		18.6	0.0	250	94	222	241	254	260	244
	Surface Daily Totals				1444	446	1136	1250	1326	1364	1304

NOTE: 1 Btu/h ft² = 3.152 W/m².

Source: Reprinted by permission from Peter J. Lunde, *Solar Thermal Engineering,* © 1980, John Wiley & Sons Inc., New York.

APPENDICES

TABLE B.14 **Solar Position and Clear Day Insolation, 64° N Latitude**

	Solar Time		Solar Position			Global Irradiance (Btu/h ft²)					
							South-Facing Elevation Angle				
Date	A.M.	P.M.	Alt	Azm	Direct Normal	Horiz	54	64	74	84	90
Jan 21	10	2	2.8	28.1	22	2	17	19	20	20	20
	11	1	5.2	14.1	81	12	72	77	80	81	81
	12		6.0	0.0	100	16	91	98	102	103	103
	Surface Daily Totals				306	45	268	290	302	306	304
Feb 21	8	4	3.4	58.7	35	4	17	19	19	19	19
	9	3	8.6	44.8	147	31	103	108	111	110	107
	10	2	12.6	30.3	199	55	170	178	181	178	173
	11	1	15.1	15.3	222	71	212	220	223	219	213
	12		16.0	0.0	228	77	225	235	237	232	226
	Surface Daily Totals				1432	400	1230	1286	1302	1282	1252
Mar 21	7	5	6.5	76.5	95	18	30	29	29	27	25
	8	4	12.7	62.6	185	54	101	102	99	94	89
	9	3	18.1	48.1	227	87	171	172	169	160	153
	10	2	22.3	32.7	249	112	227	229	224	213	203
	11	1	25.1	16.6	260	129	262	265	259	246	235
	12		26.0	0.0	263	134	274	277	271	258	246
	Surface Daily Totals				2296	932	1856	1870	1830	1736	1656
Apr 21	5	7	4.0	108.5	27	5	2	2	2	1	1
	6	6	10.4	95.1	133	37	15	9	8	7	6
	7	5	17.0	81.6	194	76	70	63	54	43	37
	8	4	23.3	67.5	228	112	136	128	116	102	91
	9	3	29.0	52.3	248	144	197	189	176	158	145
	10	2	33.5	36.0	260	169	246	239	224	203	188
	11	1	36.5	18.4	266	184	278	270	255	233	216
	12		37.6	0.0	268	190	289	281	266	243	225
	Surface Daily Totals				2982	1644	2176	2082	1936	1736	1594
May 21	4	8	5.8	125.1	51	11	5	4	4	3	3
	5	7	11.6	112.1	132	42	13	11	10	9	8
	6	6	17.9	99.1	185	79	29	16	14	12	11
	7	5	24.5	85.7	218	117	86	72	56	39	28
	8	4	30.9	71.5	239	152	148	133	115	94	80
	9	3	36.8	56.1	252	182	204	190	170	145	128
	10	2	41.6	38.9	261	205	249	235	213	186	167
	11	1	44.9	20.1	265	219	278	264	242	213	193
	12		46.0	0.0	267	224	288	274	251	222	201
	Surface Daily Totals				3470	2236	2312	2124	1898	1624	1436
Jun 21	3	9	4.2	139.4	21	4	2	2	2	2	1
	4	8	9.0	126.4	93	27	10	9	8	7	6
	5	7	14.7	113.6	154	60	16	15	13	11	10
	6	6	21.0	100.8	194	96	34	19	17	14	13
	7	5	27.5	87.5	221	132	91	74	55	36	23
	8	4	34.0	73.3	239	166	150	133	112	88	73
	9	3	39.9	57.8	251	195	204	187	164	137	119
	10	2	44.9	40.4	258	217	247	230	206	177	157
	11	1	48.3	20.9	262	231	275	258	233	202	181
	12		49.5	0.0	263	235	284	267	242	211	189
	Surface Daily Totals				3650	2488	2342	2118	1862	1558	1356

TABLE B.14 **Solar Position and Clear Day Insolation, 64° N Latitude (Continued)**

	Solar Time		Solar Position			Global Irradiance (Btu/h ft²)					
							South-Facing Elevation Angle				
Date	A.M.	P.M.	Alt	Azm	Direct Normal	Horiz	54	64	74	84	90
Jul 21	4	8	6.4	125.3	53	13	6	5	5	4	4
	5	7	12.1	112.4	128	44	14	13	11	10	9
	6	6	18.4	99.4	179	81	30	17	16	13	12
	7	5	25.0	86.0	211	118	86	72	56	38	28
	8	4	31.4	71.8	231	152	146	131	113	91	77
	9	3	37.3	56.3	245	182	201	186	166	141	124
	10	2	42.2	39.2	253	204	245	230	208	181	162
	11	1	45.4	20.2	257	218	273	258	236	207	187
	12		46.6	0.0	259	223	282	267	245	216	195
	Surface Daily Totals				3372	2248	2280	2090	1864	1588	1400
Aug 21	5	7	4.6	108.8	29	6	3	3	2	2	2
	6	6	11.0	95.5	123	39	16	11	10	8	7
	7	5	17.6	81.9	181	77	69	61	52	42	35
	8	4	23.9	67.8	214	113	131	123	112	97	87
	9	3	29.6	52.6	234	144	190	182	169	150	138
	10	2	34.2	36.2	246	168	237	229	215	194	179
	11	1	37.2	18.5	252	183	268	260	244	222	205
	12		38.3	0.0	254	188	278	270	255	232	215
	Surface Daily Totals				2808	1646	2108	2008	1860	1662	1522
Sep 21	7	5	6.5	76.5	77	16	25	24	24	23	21
	8	4	12.7	62.6	163	51	92	92	90	85	81
	9	3	18.1	48.1	206	83	159	159	156	147	141
	10	2	22.3	32.7	229	108	212	213	209	198	189
	11	1	25.1	16.6	240	124	246	248	243	230	220
	12		26.0	0.0	244	129	258	260	254	241	230
	Surface Daily Totals				2074	892	1726	1736	1696	1608	1532
Oct 21	8	4	3.0	58.5	17	2	9	9	10	10	10
	9	3	8.1	44.6	122	26	86	91	93	92	90
	10	2	12.1	30.2	176	50	152	159	161	159	155
	11	1	14.6	15.2	201	65	193	201	203	200	195
	12		15.5	0.0	208	71	207	215	217	213	208
	Surface Daily Totals				1238	358	1088	1136	1152	1134	1106
Nov 21	10	2	3.0	28.1	23	3	18	20	21	21	21
	11	1	5.4	14.2	79	12	70	76	79	80	79
	12		6.2	0.0	97	17	89	96	100	101	100
	Surface Daily Totals				302	46	266	286	298	302	300
Dec 21	11	1	1.8	13.7	4	0	3	4	4	4	4
	12		2.6	0.0	16	2	14	15	16	17	17
	Surface Daily Totals				24	2	20	22	24	24	24

NOTE: 1 Btu/h ft² = 3.152 W/m².

Source: Reprinted by permission from Peter J. Lunde, *Solar Thermal Engineering,* © 1980, John Wiley & Sons Inc., New York.

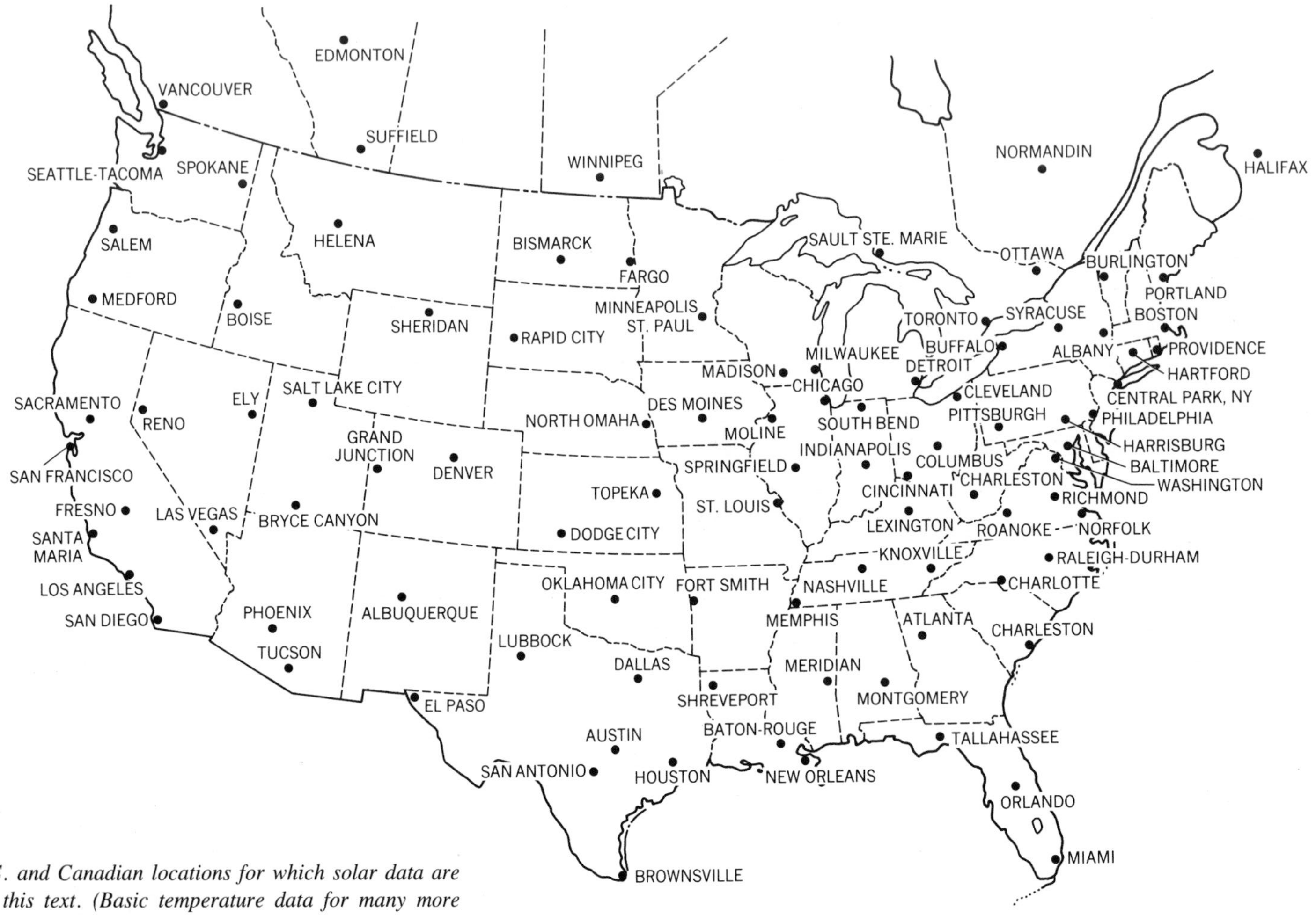

Fig. B.1 *U.S. and Canadian locations for which solar data are published in this text. (Basic temperature data for many more locations are found in Appendix A.) (Reprinted by permission from J.D. Balcomb et al;* Passive Solar Design Handbook *Volume 3, ©1983, American Solar Energy Society, Inc., Boulder.)*

TABLE B.15 **Average Insolation, Temperature, and DD Data**

Elevation in feet, latitude in degrees north latitude, HS (horizontal surface) and VS (vertical surface) in Btu/day ft^2, TA in degrees F, D50 through D65 in degree days F.

United States

MONTGOMERY, ALABAMA Elev 203 Lat 32.3

	HS	VS	TA	D50	D55	D60	D65
Jan	752	896	48	148	256	394	556
Jul	1841	820	81	0	0	0	0
Yr	1390	946	65	445	866	1474	2269

PHOENIX, ARIZONA Elev 1112 Lat 33.4

	HS	VS	TA	D50	D55	D60	D65
Jan	1021	1462	51	78	162	285	428
Jul	2486	964	91	0	0	0	0
Yr	1371	1326	70	187	459	919	1552

TUCSON, ARIZONA Elev 2556 Lat 32.1

	HS	VS	TA	D50	D55	D60	D65
Jan	1099	1539	51	80	166	292	442
Jul	2341	922	86	0	0	0	0
Yr	1874	1307	68	214	525	1036	1752

FORT SMITH, ARKANSAS Elev 463 Lat 35.3

	HS	VS	TA	D50	D55	D60	D65
Jan	744	996	39	346	497	651	806
Jul	2065	908	82	0	0	0	0
Yr	1406	1013	61	996	1622	2405	3336

FRESNO, CALIFORNIA Elev 328 Lat 36.8

	HS	VS	TA	D50	D55	D60	D65
Jan	657	886	45	176	308	457	611
Jul	2685	1076	81	0	0	0	0
Yr	1714	1210	62	507	1021	1741	2650

LOS ANGELES, CALIFORNIA Elev 105 Lat 33.9

	HS	VS	TA	D50	D55	D60	D65
Jan	926	1293	55	21	83	186	331
Jul	2307	942	69	0	1	5	19
Yr	1596	1157	62	64	299	849	1819

SACRAMENTO, CALIFORNIA Elev 26 Lat 38.5

	HS	VS	TA	D50	D55	D60	D65
Jan	597	829	45	183	315	464	617
Jul	2688	1131	75	0	0	0	0
Yr	1646	1198	60	554	1097	1871	2843

SAN DIEGO, CALIFORNIA Elev 30 Lat 32.7

	HS	VS	TA	D50	D55	D60	D65
Jan	976	1325	55	9	58	160	314
Jul	2186	902	70	0	0	1	6
Yr	1600	1151	63	23	170	623	1507

SAN FRANCISCO, CALIFORNIA Elev 16 Lat 37.6

	HS	VS	TA	D50	D55	D60	D65
Jan	708	1023	48	82	210	363	518
Jul	2392	1034	63	0	2	21	93
Yr	1556	1156	57	202	705	1643	3042

SANTA MARIA, CALIFORNIA Elev 236 Lat 34.9

	HS	VS	TA	D50	D55	D60	D65
Jan	854	1198	51	51	150	296	450
Jul	2341	965	62	0	3	30	112
Yr	1610	1172	57	155	624	1604	3053

DENVER, COLORADO Elev 5331 Lat 39.7

	HS	VS	TA	D50	D55	D60	D65
Jan	840	1465	30	623	778	933	1088
Jul	2273	1053	73	0	0	0	0
Yr	1570	1334	50	2592	3588	4733	6016

GRAND JUNCTION, COLORADO Elev 4839 Lat 39.1

	HS	VS	TA	D50	D55	D60	D65
Jan	791	1296	27	726	880	1035	1190
Jul	2465	1094	79	0	0	0	0
Yr	1661	1346	53	2514	3412	4434	5605

HARTFORD, CONNECTICUT Elev 180 Lat 41.9

	HS	VS	TA	D50	D55	D60	D65
Jan	477	694	25	781	936	1091	1246
Jul	1649	861	73	0	0	0	0
Yr	1060	835	49	2971	3948	5075	6350

WASHINGTON, D.C. Elev 289 Lat 38.9

	HS	VS	TA	D50	D55	D60	D65
Jan	572	793	32	555	710	865	1020
Jul	1817	883	75	0	0	0	0
Yr	1210	912	54	2004	2869	3864	5010

MIAMI, FLORIDA Elev 7 Lat 25.8

	HS	VS	TA	D50	D55	D60	D65
Jan	1057	1121	67	1	4	18	53
Jul	1763	787	82	0	0	0	0
Yr	1474	941	76	3	14	55	206

ORLANDO, FLORIDA Elev 118 Lat 28.5

	HS	VS	TA	D50	D55	D60	D65
Jan	999	1151	60	13	42	105	197
Jul	1801	795	81	0	0	0	0
Yr	1488	984	72	39	126	348	733

TALLAHASSEE, FLORIDA Elev 69 Lat 30.4

	HS	VS	TA	D50	D55	D60	D65
Jan	877	1033	53	73	150	256	408
Jul	1748	786	81	0	0	0	0
Yr	1434	969	68	215	501	951	1563

ATLANTA, GEORGIA Elev 1033 Lat 33.6

	HS	VS	TA	D50	D55	D60	D65
Jan	718	884	42	246	393	546	701
Jul	1812	821	78	0	0	0	0
Yr	1347	941	61	758	1362	2150	3095

BOISE, IDAHO Elev 2867 Lat 43.6

	HS	VS	TA	D50	D55	D60	D65
Jan	485	770	29	651	806	961	1116
Jul	2613	1309	75	0	0	0	0
Yr	1499	1255	51	2420	3395	4536	5833

CHICAGO, ILLINOIS Elev 623 Lat 41.8

	HS	VS	TA	D50	D55	D60	D65
Jan	507	756	24	797	952	1107	1262
Jul	1944	984	75	0	0	0	0
Yr	1217	960	51	2954	3881	4940	6127

MOLINE, ILLINOIS Elev 594 Lat 41.4

	HS	VS	TA	D50	D55	D60	D65
Jan	535	803	22	884	1039	1194	1349
Jul	1939	974	75	0	0	0	0
Yr	1226	973	50	3191	4117	5178	6395

SPRINGFIELD, ILLINOIS Elev 614 Lat 39.8

	HS	VS	TA	D50	D55	D60	D65
Jan	585	852	27	723	877	1032	1187
Jul	2058	984	76	0	0	0	0
Yr	1304	1003	53	2558	3434	4425	5558

TABLE B.15 **Average Insolation, Temperature, and DD Data (Continued)**

INDIANAPOLIS, INDIANA Elev 807 Lat 39.7

	HS	VS	TA	D50	D55	D60	D65
Jan	496	668	28	685	840	995	1150
Jul	1806	891	75	0	0	0	0
Yr	1167	873	52	2511	3403	4421	5577

SOUTH BEND, INDIANA Elev 774 Lat 41.7

	HS	VS	TA	D50	D55	D60	D65
Jan	416	566	24	806	961	1116	1271
Jul	1852	944	72	0	1	4	6
Yr	1140	864	49	3112	4084	5199	6462

DES MOINES, IOWA Elev 965 Lat 41.5

	HS	VS	TA	D50	D55	D60	D65
Jan	581	912	19	949	1104	1259	1414
Jul	2097	1037	75	0	0	0	0
Yr	1314	1065	49	3491	4435	5510	6710

DODGE CITY, KANSAS Elev 2582 Lat 37.8

	HS	VS	TA	D50	D55	D60	D65
Jan	827	1303	31	596	750	905	1060
Jul	2295	1013	79	0	0	0	0
Yr	1562	1232	55	2132	2980	3945	5046

TOPEKA, KANSAS Elev 886 Lat 39.1

	HS	VS	TA	D50	D55	D60	D65
Jan	681	1033	28	683	837	992	1147
Jul	2128	993	78	0	0	0	0
Yr	1387	1036	54	2325	3175	4137	5243

LEXINGTON, KENTUCKY Elev 988 Lat 38.0

	HS	VS	TA	D50	D55	D60	D65
Jan	546	714	33	531	685	840	995
Jul	1850	881	76	0	0	0	0
Yr	1221	892	55	1865	2686	3632	4729

BATON ROUGE, LOUISIANA Elev 75 Lat 30.5

	HS	VS	TA	D50	D55	D60	D65
Jan	785	889	51	90	174	294	451
Jul	1746	786	82	0	0	0	0
Yr	1380	913	67	232	530	1006	1670

NEW ORLEANS, LOUISIANA Elev 10 Lat 30.0

	HS	VS	TA	D50	D55	D60	D65
Jan	835	950	53	73	150	252	403
Jul	1813	801	82	0	0	0	0
Yr	1438	943	68	197	465	887	1465

SHREVEPORT, LOUISIANA Elev 259 Lat 32.5

	HS	VS	TA	D50	D55	D60	D65
Jan	762	920	47	154	264	403	552
Jul	2014	864	83	0	0	0	0
Yr	1428	973	66	428	832	1415	2167

PORTLAND, MAINE Elev 62 Lat 43.6

	HS	VS	TA	D50	D55	D60	D65
Jan	450	689	22	884	1039	1194	1349
Jul	1659	894	68	0	1	5	27
Yr	1052	857	45	3648	4758	6039	7498

DETROIT, MICHIGAN Elev 627 Lat 42.4

	HS	VS	TA	D50	D55	D60	D65
Jan	417	585	26	760	915	1070	1225
Jul	1835	951	73	0	0	0	0
Yr	1122	869	50	2931	3890	4986	6228

SAULT STE. MARIE, MICHIGAN Elev 725 Lat 46.5

	HS	VS	TA	D50	D55	D60	D65
Jan	325	492	14	1110	1265	1420	1575
Jul	1835	1045	64	1	7	33	96
Yr	1044	861	40	4969	6198	7607	9193

MINNEAPOLIS, MINNESOTA Elev 837 Lat 44.9

	HS	VS	TA	D50	D55	D60	D65
Jan	464	768	12	1172	1327	1482	1637
Jul	1970	1071	72	0	1	5	11
Yr	1172	996	44	4584	5631	6824	8159

MERIDIAN, MISSISSIPPI Elev 308 Lat 32.3

	HS	VS	TA	D50	D55	D60	D65
Jan	744	883	47	163	274	413	575
Jul	1823	815	81	0	0	0	0
Yr	1371	933	65	510	955	1582	2388

SAINT LOUIS, MISSOURI Elev 564 Lat 38.7

	HS	VS	TA	D50	D55	D60	D65
Jan	627	898	31	581	735	890	1045
Jul	2049	959	79	0	0	0	0
Yr	1329	1006	56	1961	2762	3686	4750

HELENA, MONTANA Elev 3898 Lat 46.6

	HS	VS	TA	D50	D55	D60	D65
Jan	419	719	18	989	1144	1299	1454
Jul	2334	1312	68	1	3	12	33
Yr	1266	1134	43	4151	5342	6689	8190

NORTH OMAHA, NEBRASKA Elev 1325 Lat 41.4

	HS	VS	TA	D50	D55	D60	D65
Jan	634	1034	20	924	1079	1234	1389
Jul	2106	1038	75	0	1	3	7
Yr	1323	1078	49	3369	4309	5381	6601

ELY, NEVADA Elev 6253 Lat 39.3

	HS	VS	TA	D50	D55	D60	D65
Jan	819	1380	24	818	973	1128	1283
Jul	2447	1094	67	0	2	11	23
Yr	1675	1391	44	3716	4922	6291	7814

LAS VEGAS, NEVADA Elev 2178 Lat 36.1

	HS	VS	TA	D50	D55	D60	D65
Jan	978	1553	44	216	346	493	645
Jul	2588	1039	90	0	0	0	0
Yr	1866	1431	66	631	1129	1788	2601

RENO, NEVADA Elev 4400 Lat 39.5

	HS	VS	TA	D50	D55	D60	D65
Jan	800	1345	32	561	716	871	1026
Jul	2692	1167	69	0	1	5	17
Yr	1764	1439	49	2292	3345	4590	6022

ALBUQUERQUE, NEW MEXICO Elev 5312 Lat 35.0

	HS	VS	TA	D50	D55	D60	D65
Jan	1016	1562	35	459	614	769	924
Jul	2489	995	79	0	0	0	0
Yr	1830	1379	57	1497	2292	3216	4292

ALBANY, NEW YORK Elev 292 Lat 42.7

	HS	VS	TA	D50	D55	D60	D65
Jan	456	674	22	884	1039	1194	1349
Jul	1725	908	72	0	1	3	9
Yr	1068	843	48	3424	4428	5586	6888

TABLE B.15 **Average Insolation, Temperature, and DD Data (Continued)**

BUFFALO, NEW YORK Elev 705 Lat 42.9

	HS	VS	TA	D50	D55	D60	D65
Jan	349	465	24	815	970	1125	1280
Jul	1776	935	70	0	0	3	12
Yr	1037	780	47	3322	4363	5551	6927

NEW YORK, NEW YORK Elev 187 Lat 40.8

	HS	VS	TA	D50	D55	D60	D65
Jan	500	708	32	552	707	862	1017
Jul	1688	861	77	0	0	0	0
Yr	1101	849	55	1931	2759	3737	4848

SYRACUSE, NEW YORK Elev 407 Lat 43.1

	HS	VS	TA	D50	D55	D60	D65
Jan	385	538	24	818	973	1128	1283
Jul	1758	931	72	0	1	3	11
Yr	1037	791	48	3215	4218	5366	6678

CHARLOTTE, NORTH CAROLINA Elev 768 Lat 35.2

	HS	VS	TA	D50	D55	D60	D65
Jan	719	944	42	255	402	555	710
Jul	1831	841	79	0	0	0	0
Yr	1346	981	61	828	1451	2257	3218

RALEIGH-DURHAM, NORTH CAROLINA Elev 440 Lat 35.9

	HS	VS	TA	D50	D55	D60	D65
Jan	694	924	41	300	451	605	760
Jul	1776	832	78	0	0	0	0
Yr	1297	955	59	990	1659	2509	3514

BISMARCK, NORTH DAKOTA Elev 1647 Lat 46.8

	HS	VS	TA	D50	D55	D60	D65
Jan	467	847	8	1296	1451	1606	1761
Jul	2184	1241	71	0	2	8	18
Yr	1251	1145	41	5235	6364	7627	9044

FARGO, NORTH DAKOTA Elev 899 Lat 46.9

	HS	VS	TA	D50	D55	D60	D65
Jan	415	720	6	1367	1522	1677	1832
Jul	2120	1210	71	0	1	5	13
Yr	1206	1075	41	5485	6607	7858	9271

CINCINNATI, OHIO Elev 889 Lat 39.1

	HS	VS	TA	D50	D55	D60	D65
Jan	500	659	31	587	741	896	1051
Jul	1771	869	76	0	0	0	0
Yr	1160	858	54	2117	2973	3951	5070

CLEVELAND, OHIO Elev 804 Lat 41.4

	HS	VS	TA	D50	D55	D60	D65
Jan	388	507	27	716	871	1026	1181
Jul	1828	929	71	0	1	4	9
Yr	1093	808	50	2804	3768	4879	6154

COLUMBUS, OHIO Elev 833 Lat 40.0

	HS	VS	TA	D50	D55	D60	D65
Jan	459	606	28	670	825	980	1135
Jul	1755	876	74	0	0	0	0
Yr	1128	834	52	2524	3438	4491	5702

OKLAHOMA CITY, OKLAHOMA Elev 1302 Lat 35.4

	HS	VS	TA	D50	D55	D60	D65
Jan	801	1114	37	413	565	719	874
Jul	2128	925	82	0	0	0	0
Yr	1463	1073	60	1232	1903	2734	3695

MEDFORD, OREGON Elev 1299 Lat 42.4

	HS	VS	TA	D50	D55	D60	D65
Jan	407	565	37	417	571	725	880
Jul	2475	1207	72	0	1	3	11
Yr	1356	1033	53	1576	2505	3621	4930

SALEM, OREGON Elev 200 Lat 44.9

	HS	VS	TA	D50	D55	D60	D65
Jan	332	471	39	348	502	657	812
Jul	2142	1154	67	0	1	7	43
Yr	1130	897	52	1265	2220	3411	4852

HARRISBURG, PENNSYLVANIA Elev 348 Lat 40.2

	HS	VS	TA	D50	D55	D60	D65
Jan	536	763	30	617	772	927	1082
Jul	1764	883	76	0	0	0	0
Yr	1152	887	53	2221	3093	4086	5224

PHILADELPHIA, PENNSYLVANIA Elev 30 Lat 39.9

	HS	VS	TA	D50	D55	D60	D65
Jan	555	792	32	549	704	859	1014
Jul	1758	876	77	0	0	0	0
Yr	1170	905	55	1935	2775	3749	4865

PITTSBURGH, PENNSYLVANIA Elev 1224 Lat 40.5

	HS	VS	TA	D50	D55	D60	D65
Jan	424	553	28	679	834	989	1144
Jul	1689	857	72	0	1	3	7
Yr	1071	793	50	2635	3574	4669	5930

PROVIDENCE, RHODE ISLAND Elev 62 Lat 41.7

	HS	VS	TA	D50	D55	D60	D65
Jan	506	750	28	670	825	980	1135
Jul	1695	878	72	0	0	0	0
Yr	1114	884	50	2566	3543	4669	5972

CHARLESTON, SOUTH CAROLINA Elev 39 Lat 32.9

	HS	VS	TA	D50	D55	D60	D65
Jan	744	904	49	120	222	360	521
Jul	1799	813	80	0	0	0	0
Yr	1346	940	65	360	756	1355	2146

RAPID CITY, SOUTH DAKOTA Elev 3169 Lat 44.0

	HS	VS	TA	D50	D55	D60	D65
Jan	542	928	22	871	1026	1181	1336
Jul	2223	1161	73	1	2	8	13
Yr	1344	1177	47	3681	4749	5965	7324

KNOXVILLE, TENNESSEE Elev 981 Lat 35.8

	HS	VS	TA	D50	D55	D60	D65
Jan	621	785	41	302	449	602	756
Jul	1804	839	78	0	0	0	0
Yr	1275	909	60	1018	1671	2504	3478

MEMPHIS, TENNESSEE Elev 285 Lat 35.0

	HS	VS	TA	D50	D55	D60	D65
Jan	683	870	41	312	455	606	760
Jul	1972	879	82	0	0	0	0
Yr	1368	965	62	988	1588	2357	3227

NASHVILLE, TENNESSEE Elev 591 Lat 36.1

	HS	VS	TA	D50	D55	D60	D65
Jan	580	721	38	369	519	673	828
Jul	1891	869	80	0	0	0	0
Yr	1272	891	59	1195	1874	2720	3696

TABLE B.15 **Average Insolation, Temperature, and DD Data (Continued)**

AUSTIN, TEXAS — Elev 620 Lat 30.3

	HS	VS	TA	D50	D55	D60	D65
Jan	864	1008	50	116	207	333	483
Jul	2105	865	85	0	0	0	0
Yr	1478	974	68	289	602	1088	1737

BROWNSVILLE, TEXAS — Elev 20 Lat 25.9

	HS	VS	TA	D50	D55	D60	D65
Jan	913	923	60	18	51	116	225
Jul	2212	867	84	0	0	0	0
Yr	1550	917	74	44	127	325	650

DALLAS, TEXAS — Elev 489 Lat 32.8

	HS	VS	TA	D50	D55	D60	D65
Jan	821	1035	45	189	312	457	608
Jul	2122	890	86	0	0	0	0
Yr	1470	1014	66	505	943	1543	2290

EL PASO, TEXAS — Elev 3917 Lat 31.8

	HS	VS	TA	D50	D55	D60	D65
Jan	1125	1572	44	210	355	509	663
Jul	2450	934	82	0	0	0	0
Yr	1901	1327	63	561	1102	1810	2678

HOUSTON, TEXAS — Elev 108 Lat 30.0

	HS	VS	TA	D50	D55	D60	D65
Jan	772	852	52	71	150	263	416
Jul	1828	805	83	0	0	0	0
Yr	1353	884	69	161	409	825	1434

LUBBOCK, TEXAS — Elev 3241 Lat 33.6

	HS	VS	TA	D50	D55	D60	D65
Jan	1031	1497	39	343	494	648	803
Jul	2412	956	80	0	0	0	0
Yr	1768	1279	60	1069	1739	2582	3545

SAN ANTONIO, TEXAS — Elev 794 Lat 29.5

	HS	VS	TA	D50	D55	D60	D65
Jan	895	1026	51	93	179	302	451
Jul	2121	863	85	0	0	0	0
Yr	1501	973	69	213	490	941	1570

BRYCE CANYON, UTAH — Elev 7588 Lat 37.7

	HS	VS	TA	D50	D55	D60	D65
Jan	914	1511	20	936	1091	1246	1401
Jul	2424	1044	62	1	8	47	128
Yr	1742	1404	40	4693	6106	7675	9133

SALT LAKE CITY, UTAH — Elev 4226 Lat 40.8

	HS	VS	TA	D50	D55	D60	D65
Jan	639	1017	28	683	837	992	1147
Jul	2590	1186	77	0	0	0	0
Yr	1606	1301	51	2648	3612	4725	5983

BURLINGTON, VERMONT — Elev 341 Lat 44.5

	HS	VS	TA	D50	D55	D60	D65
Jan	385	572	17	1029	1184	1339	1494
Jul	1721	941	70	0	1	4	20
Yr	1023	815	44	4142	5230	6464	7876

NORFOLK, VIRGINIA — Elev 30 Lat 36.9

	HS	VS	TA	D50	D55	D60	D65
Jan	678	932	41	300	450	605	760
Jul	1853	868	78	0	0	0	0
Yr	1327	990	59	974	1646	2489	3488

RICHMOND, VIRGINIA — Elev 164 Lat 37.5

	HS	VS	TA	D50	D55	D60	D65
Jan	632	863	38	390	543	698	853
Jul	1774	849	78	0	0	0	0
Yr	1250	936	58	1296	2021	2909	3939

ROANOKE, VIRGINIA — Elev 1175 Lat 37.3

	HS	VS	TA	D50	D55	D60	D65
Jan	660	911	36	423	577	732	887
Jul	1796	854	75	0	0	0	0
Yr	1271	958	56	1486	2277	3211	4307

SEATTLE-TACOMA, WASHINGTON — Elev 400 Lat 47.4

	HS	VS	TA	D50	D55	D60	D65
Jan	262	378	38	367	521	676	831
Jul	2248	1299	65	0	2	16	80
Yr	1056	857	51	1386	2393	3662	5185

SPOKANE, WASHINGTON — Elev 2365 Lat 47.6

	HS	VS	TA	D50	D55	D60	D65
Jan	315	496	25	763	918	1073	1228
Jul	2357	1368	70	0	2	7	21
Yr	1227	1068	47	3061	4150	5411	6835

CHARLESTON, WEST VIRGINIA — Elev 951 Lat 38.4

	HS	VS	TA	D50	D55	D60	D65
Jan	498	638	35	483	636	791	946
Jul	1682	827	75	0	0	0	0
Yr	1125	822	55	1726	2540	3488	4590

MADISON, WISCONSIN — Elev 860 Lat 43.1

	HS	VS	TA	D50	D55	D60	D65
Jan	515	822	17	1029	1184	1339	1494
Jul	1934	1009	70	0	1	5	14
Yr	1193	978	45	4086	5143	6352	7730

MILWAUKEE, WISCONSIN — Elev 692 Lat 42.9

	HS	VS	TA	D50	D55	D60	D65
Jan	479	731	19	949	1104	1259	1414
Jul	1962	1017	70	0	1	4	15
Yr	1194	957	46	3774	4833	6045	7444

SHERIDAN, WYOMING — Elev 3966 Lat 44.8

	HS	VS	TA	D50	D55	D60	D65
Jan	517	900	21	899	1054	1209	1364
Jul	2329	1237	70	1	2	9	28
Yr	1333	1170	45	3860	4991	6279	7708

Canada

EDMONTON, ALBERTA — Elev 2220 Lat 53.6

	HS	VS	TA	D50	D55	D60	D65
Jan	324	746	4	1421	1574	1728	1883
Jul	1977	1378	62	1	7	38	117
Yr	1114	1205	36	6317	7563	9016	10650

SUFFIELD, ALBERTA — Elev 2549 Lat 50.3

	HS	VS	TA	D50	D55	D60	D65
Jan	433	937	7	1333	1486	1640	1794
Jul	2173	1377	67	0	2	11	49
Yr	1239	1269	40	5500	6637	7923	9393

VANCOUVER, BRITISH COLUMBIA — Elev 310 Lat 49.3

	HS	VS	TA	D50	D55	D60	D65
Jan	254	395	37	425	572	724	878
Jul	2021	1239	63	0	1	13	82
Yr	1060	916	50	1791	2781	4041	5588

WINNIPEG, MANITOBA — Elev 820 Lat 49.9

	HS	VS	TA	D50	D55	D60	D65
Jan	461	1011	0	1588	1740	1893	2047
Jul	2025	1264	67	0	2	9	45
Yr	1190	1199	36	6925	8062	9338	10790

TABLE B.15 **Average Insolation, Temperature, and DD Data (Continued)**

HALIFAX, NOVA SCOTIA				Elev 136	Lat 44.6		
	HS	VS	TA	D50	D55	D60	D65
Jan	456	737	26	752	900	1051	1204
Jul	1694	929	65	0	1	8	57
Yr	1076	907	46	3457	4500	5746	7211

TORONTO, ONTARIO					Elev 443	Lat 43.7	
	HS	VS	TA	D50	D55	D60	D65
Jan	487	777	22	891	1041	1194	1348
Jul	1958	1035	70	0	0	3	18
Yr	1171	948	46	3842	4853	6013	7343

OTTAWA, ONTARIO				Elev 377	Lat 45.5		
	HS	VS	TA	D50	D55	D60	D65
Jan	510	914	13	1169	1320	1473	1627
Jul	1875	1040	69	0	1	4	23
Yr	1158	1015	43	4912	5965	7158	8529

NORMANDIN, QUEBEC					Elev 450	Lat 48.8	
	HS	VS	TA	D50	D55	D60	D65
Jan	454	921	0	1564	1719	1873	2028
Jul	1707	1031	62	1	7	40	118
Yr	1092	1053	34	7037	8308	9762	11376

Source: Reprinted by permission from J. D. Balcomb et al., *Passive Solar Design Handbook,* Vol. 3, © 1983, American Solar Energy Society, Inc., Boulder.

APPENDICES

C

ANNUAL SOLAR PERFORMANCE

This material is abridged from J. D. Balcomb et al., *Passive Solar Design Handbook,* Volume 3, © 1983, American Solar Energy Society, Inc., Boulder.

This appendix lists 30 passive solar heating systems (5 water wall, 10 Trombe wall, 5 direct gain, and 10 sunspace) for the U.S. and Canadian locations shown in Fig. B.1. The *Passive Solar Design Handbook* lists 94 systems, for about twice as many locations. The specifications for all 94 systems are listed here, so that the original source may be consulted if your passive system is not represented by one of those in this appendix.

TABLE C.1 **Characteristics of Selected Passive Solar Heating Systems**

Part A. ***Water Wall Systems***

Designation	*Thermal Storage Capacity*[a] *(Btu/ft² F)*	*Wall Thickness (in.)*	*No. of Glazings*	*Wall Surface*	*Night Insulation*
WW-A1	15.6	3	2	Normal	No
WW-A2	31.2	6	2	Normal	No
WW-A3[b]	46.8	9	2	Normal	No
WW-A4	62.4	12	2	Normal	No
WW-A5	93.6	18	2	Normal	No
WW-A6[b]	124.8	24	2	Normal	No
WW-B1	46.8	9	1	Normal	No
WW-B2[b]	46.8	9	3	Normal	No
WW-B3	46.8	9	1	Normal	Yes
WW-B4[b]	46.8	9	2	Normal	Yes
WW-B5	46.8	9	3	Normal	Yes
WW-C1	46.8	9	1	Selective	No
WW-C2	46.8	9	2	Selective	No
WW-C3[b]	46.8	9	1	Selective	Yes
WW-C4	46.8	9	2	Selective	Yes

Part B. ***Trombe Wall Systems: Vented***

Designation	*Thermal Storage Capacity*[c] *(Btu/ft² F)*	*Wall Thickness*[c] *(in.)*	ρck *(Btu²/h ft⁴ F²)*	*No. of Glazings*	*Wall Surface*	*Night Insulation*
TW-A1	15	6	30	2	Normal	No
TW-A2	22.5	9	30	2	Normal	No
TW-A3[b]	30	12	30	2	Normal	No
TW-A4	45	18	30	2	Normal	No
TW-B1	15	6	15	2	Normal	No
TW-B2	22.5	9	15	2	Normal	No
TW-B3	30	12	15	2	Normal	No
TW-B4	45	18	15	2	Normal	No
TW-C1	15	6	7.5	2	Normal	No
TW-C2	22.5	9	7.5	2	Normal	No
TW-C3	30	12	7.5	2	Normal	No
TW-C4	45	18	7.5	2	Normal	No
TW-D1	30	12	30	1	Normal	No
TW-D2	30	12	30	3	Normal	No
TW-D3	30	12	30	1	Normal	Yes
TW-D4	30	12	30	2	Normal	Yes
TW-D5	30	12	30	3	Normal	Yes
TW-E1	30	12	30	1	Selective	No
TW-E2	30	12	30	2	Selective	No
TW-E3[b]	30	12	30	1	Selective	Yes
TW-E4	30	12	30	2	Selective	Yes

TABLE C.1 **Characteristics of Selected Passive Solar Heating Systems (Continued)**

Part C. ***Trombe Wall Systems: Unvented***

Designation	*Thermal Storage Capacity*[c] *(Btu/ft² F)*	*Wall Thickness*[c] *(in.)*	*ρck (Btu²/h ft⁴ F²)*	*No. of Glazings*	*Wall Surface*	*Night Insulation*
TW-F1	15	6	30	2	Normal	No
TW-F2	22.5	9	30	2	Normal	No
TW-F3[b]	30	12	30	2	Normal	No
TW-F4[b]	45	18	30	2	Normal	No
TW-G1	15	6	15	2	Normal	No
TW-G2	22.5	9	15	2	Normal	No
TW-G3[b]	30	12	15	2	Normal	No
TW-G4[b]	45	18	15	2	Normal	No
TW-H1	15	6	7.5	2	Normal	No
TW-H2	22.5	9	7.5	2	Normal	No
TW-H3	30	12	7.5	2	Normal	No
TW-H4	45	18	7.5	2	Normal	No
TW-I1	30	12	30	1	Normal	No
TW-I2[b]	30	12	30	3	Normal	No
TW-I3[b]	30	12	30	1	Normal	Yes
TW-I4[b]	30	12	30	2	Normal	Yes
TW-I5	30	12	30	3	Normal	Yes
TW-J1	30	12	30	1	Selective	No
TW-J2[b]	30	12	30	2	Selective	No
TW-J3	30	12	30	1	Selective	Yes
TW-J4	30	12	30	2	Selective	Yes

Part D. ***Direct-Gain Systems***

Designation	*Thermal Storage Capacity*[c] *(Btu/ft² F)*	*Mass Thickness*[c] *(in.)*	*Ratio of Mass to Glazing Area*	*No. of Glazings*	*Night Insulation*
DG-A1[b]	30	2	6	2	No
DG-A2[b]	30	2	6	3	No
DG-A3[b]	30	2	6	2	Yes
DG-B1	45	6	3	2	No
DG-B2[b]	45	6	3	3	No
DG-B3	45	6	3	2	Yes
DG-C1	60	4	6	2	No
DG-C2	60	4	6	3	No
DG-C3[b]	60	4	6	2	Yes

TABLE C.1 **Characteristics of Selected Passive Solar Heating Systems (Continued)**

Part E. ***Sunspace Systems***

Designation	*Type*[d]	*Tilt (Degrees)*	*Common Wall*	*End Walls*	*Night Insulation*
SS-A1[b]	Attached	50	Masonry	Opaque	No
SS-A2[b]	Attached	50	Masonry	Opaque	Yes
SS-A3	Attached	50	Masonry	Glazed	No
SS-A4	Attached	50	Masonry	Glazed	Yes
SS-A5[b]	Attached	50	Insulated	Opaque	No
SS-A6	Attached	50	Insulated	Opaque	Yes
SS-A7	Attached	50	Insulated	Glazed	No
SS-A8	Attached	50	Insulated	Glazed	Yes
SS-B1[b]	Attached	90/30	Masonry	Opaque	No
SS-B2	Attached	90/30	Masonry	Opaque	Yes
SS-B3[b]	Attached	90/30	Masonry	Glazed	No
SS-B4	Attached	90/30	Masonry	Glazed	Yes
SS-B5[b]	Attached	90/30	Insulated	Opaque	No
SS-B6	Attached	90/30	Insulated	Opaque	Yes
SS-B7	Attached	90/30	Insulated	Glazed	No
SS-B8	Attached	90/30	Insulated	Glazed	Yes
SS-C1	Semi-enclosed	90	Masonry	Common	No
SS-C2	Semi-enclosed	90	Masonry	Common	Yes
SS-C3	Semi-enclosed	90	Insulated	Common	No
SS-C4[b]	Semi-enclosed	90	Insulated	Common	Yes
SS-D1[b]	Semi-enclosed	50	Masonry	Common	No
SS-D2[b]	Semi-enclosed	50	Masonry	Common	Yes
SS-D3	Semi-enclosed	50	Insulated	Common	No
SS-D4	Semi-enclosed	50	Insulated	Common	Yes
SS-E1[b]	Semi-enclosed	90/30	Masonry	Common	No
SS-E2	Semi-enclosed	90/30	Masonry	Common	Yes
SS-E3	Semi-enclosed	90/30	Insulated	Common	No
SS-E4	Semi-enclosed	90/30	Insulated	Common	Yes

[a]Per unit of projected area.

[b]Listed in this text.

[c]The thermal storage capacity is per unit of projected area, or, equivalently, the quantity ρct. The wall thickness is listed only as an appropriate guide by assuming ρc = 30 Btu/ft^3 F (ρ of 150 lb/ft^3 and c of 0.2 Btu/lb F).

[d]See Fig. C.1 for additional description.

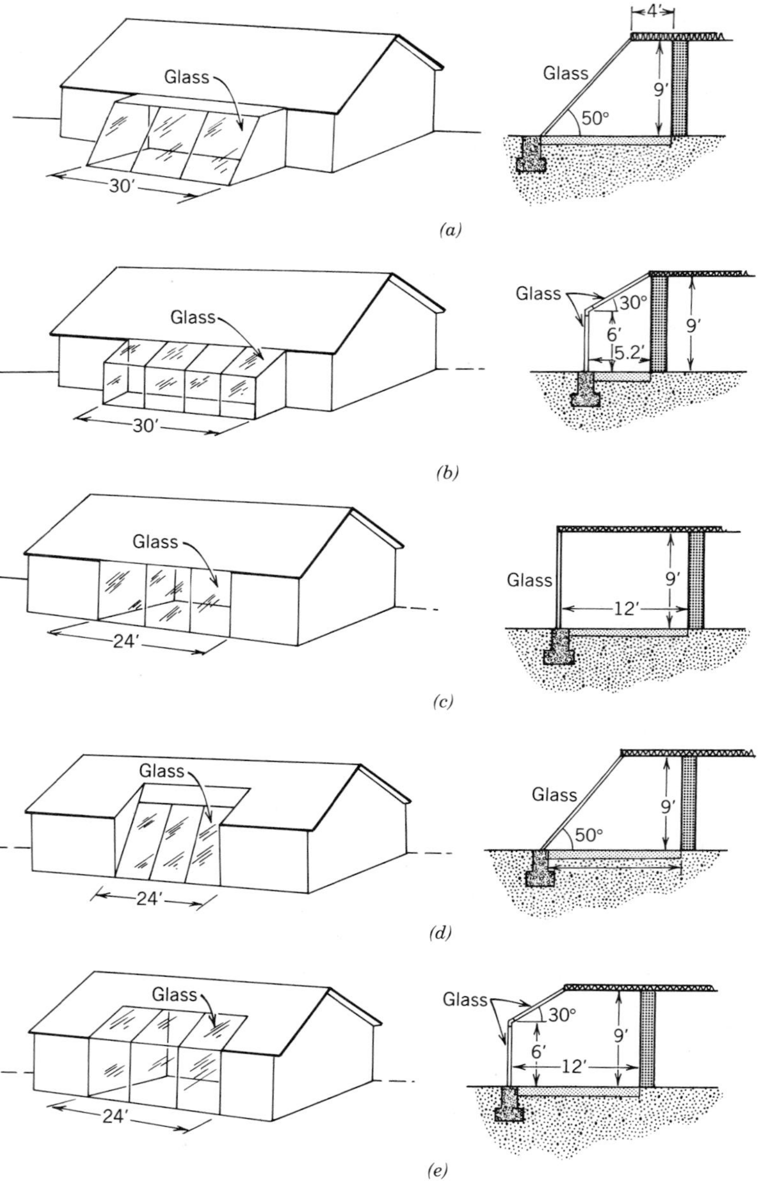

Fig. C.1 *Types of sunspaces, described in Table C.1, Part E.* (a) *and* (b) *Considered* attached *to the building;* (c), (d), *and* (e) *considered* semi-enclosed *by the building. The architectural detail at the sides of type* (d) *is insignificant; no shading of the sunspace by the building was accounted for in the performance estimates of Table C.2. (Reprinted by permission from J.D. Balcomb et al.,* Passive Solar Design Handbook, *Volume 3, ©1983, American Solar Energy Society, Inc., Boulder.)*

TABLE C.2 **Annual Passive Heating Performance: LCR**

For each location, DD listed are DD65. For a description of SSF (Solar Savings Fraction) see Fig. 5.15, page 181; LCR (load–collector ratio) is described in Section 5.6. LCR units are Btu/DD ft^2.

United States

MONTGOMERY, ALABAMA — *2269 DD*

SSF =	.10	.20	.30	.40	.50	.60	.70	.80	.90
WW-A3	340	162	98	66	47	34	25	18	12
WW-A6	287	162	106	74	54	40	30	22	15
WW-B2	303	161	103	71	51	38	28	20	14
WW-B4	303	183	124	89	66	50	38	28	19
WW-C3	320	206	144	106	80	61	46	35	24
TW-A3	334	144	84	55	39	28	20	15	10
TW-E3	356	211	142	101	75	56	42	31	21
TW-F3	245	117	71	48	34	25	18	13	9
TW-F4	201	104	66	45	32	24	18	13	8
TW-G3	180	91	56	38	27	20	15	11	7
TW-G4	144	75	47	32	23	17	12	9	6
TW-I2	248	128	80	55	40	29	22	16	10
TW-I3	262	140	90	62	45	33	25	18	12
TW-I4	259	149	99	70	51	38	29	21	14
TW-J2	282	159	104	73	53	40	30	22	15
DG-A1	247	109	62	39	24	14	—	—	—
DG-A2	261	119	72	49	34	24	16	10	5
DG-A3	312	145	90	63	46	34	25	18	10
DG-B2	268	124	76	52	38	29	21	14	8
DG-C3	361	173	106	75	56	44	34	26	18
SS-A1	687	214	113	70	47	33	24	16	11
SS-A2	606	260	152	99	69	50	36	26	17
SS-A5	1155	215	103	62	40	28	20	14	9
SS-B1	473	163	88	55	38	27	19	13	9
SS-B3	442	152	81	51	35	24	17	12	8
SS-B5	618	147	73	44	29	20	14	10	6
SS-C4	313	149	90	61	43	31	23	17	11
SS-D1	633	255	144	93	64	46	33	23	15
SS-D2	535	282	178	122	87	64	47	34	23
SS-E1	479	202	116	75	52	37	27	19	12

FORT SMITH, ARKANSAS — *3336 DD*

SSF =	.10	.20	.30	.40	.50	.60	.70	.80	.90
WW-A3	230	109	66	44	31	23	17	12	8
WW-A6	196	110	72	50	37	27	20	15	10
WW-B2	211	112	71	49	36	26	20	14	10
WW-B4	218	132	90	65	48	36	27	20	14
WW-C3	235	151	106	78	59	45	35	26	18
TW-A3	227	97	56	37	26	19	14	10	6
TW-E3	258	153	103	74	55	41	31	23	16
TW-F3	166	79	48	32	23	16	12	9	6
TW-F4	137	71	44	30	22	16	12	9	6
TW-G3	122	61	38	26	18	13	10	7	5
TW-G4	99	51	32	22	16	11	8	6	4
TW-I2	172	89	56	38	28	20	15	11	7
TW-I3	182	98	62	43	31	23	17	12	8
TW-I4	186	107	71	50	37	28	21	15	10
TW-J2	201	113	74	52	38	29	21	16	11
DG-A1	156	66	35	19	—	—	—	—	—
DG-A2	179	81	49	32	22	14	9	5	—
DG-A3	224	104	64	45	33	24	18	12	6
DG-B2	183	84	51	35	26	19	13	8	3
DG-C3	261	124	77	54	41	32	25	19	13
SS-A1	453	139	73	45	30	21	15	11	7
SS-A2	416	179	104	68	48	34	25	18	17
SS-A5	753	138	66	39	25	17	12	8	5
SS-B1	311	106	57	36	24	17	12	8	5
SS-B3	284	96	51	32	21	15	11	7	5
SS-B5	400	94	46	28	18	13	9	6	4
SS-C4	218	103	63	42	30	22	16	12	8
SS-D1	418	167	94	61	42	30	21	15	10
SS-D2	369	195	123	85	61	45	33	24	16
SS-E1	311	130	74	48	33	23	17	12	8

PHOENIX, ARIZONA — *1552 DD*

SSF =	.10	.20	.30	.40	.50	.60	.70	.80	.90
WW-A3	766	369	229	157	113	84	62	45	30
WW-A6	639	367	245	175	130	98	73	54	36
WW-B2	659	356	231	163	119	89	67	49	33
WW-B4	633	387	266	194	145	111	84	62	42
WW-C3	647	420	297	221	168	129	99	73	50
TW-A3	751	327	196	132	94	69	51	37	24
TW-E3	729	437	297	215	160	122	92	68	46
TW-F3	553	269	167	115	83	61	45	33	22
TW-F4	449	237	152	107	78	58	43	31	21
TW-G3	404	207	131	91	66	49	37	27	18
TW-G4	321	169	108	76	55	41	31	22	15
TW-I2	541	283	181	127	93	69	51	37	25
TW-I3	564	307	199	140	103	77	57	42	28
TW-I4	542	316	212	152	113	85	64	47	32
TW-J2	595	338	224	160	119	89	67	49	33
DG-A1	590	273	168	114	80	57	40	26	14
DG-A2	571	268	168	117	85	63	46	32	20
DG-A3	640	302	193	137	103	78	58	42	26
DG-B2	582	275	176	126	95	73	56	41	27
DG-C3	738	349	225	163	124	97	77	59	41
SS-A1	1458	459	247	157	107	76	55	38	25
SS-A2	1234	533	314	209	147	107	78	55	36
SS-A5	2440	470	230	141	94	66	47	33	21
SS-B1	1011	353	195	126	87	62	45	32	21
SS-B3	952	332	183	118	81	58	42	29	19
SS-B5	1326	325	166	103	70	49	35	25	16
SS-C4	679	325	201	138	100	74	54	39	26
SS-D1	1344	547	315	207	145	104	76	54	35
SS-D2	1084	575	367	254	184	136	101	73	48
SS-E1	1035	443	260	173	122	88	64	46	30

LOS ANGELES, CALIFORNIA — *1819 DD*

SSF =	.10	.20	.30	.40	.50	.60	.70	.80	.90
WW-A3	869	418	256	174	125	92	69	50	34
WW-A6	725	414	273	194	143	108	81	60	41
WW-B2	748	401	258	180	131	98	73	54	37
WW-B4	713	432	294	213	159	121	92	68	47
WW-C3	737	473	333	246	186	143	110	82	57
TW-A3	851	371	219	147	104	76	56	41	27
TW-E3	833	495	333	239	178	135	102	76	52
TW-F3	628	304	187	127	92	68	50	37	25
TW-F4	510	268	170	118	86	64	48	35	24
TW-G3	459	234	146	101	73	54	40	30	20
TW-G4	364	190	121	84	61	45	34	25	17
TW-I2	614	320	202	140	102	76	57	41	28
TW-I3	649	350	225	157	115	86	64	47	32
TW-I4	611	353	235	167	124	94	71	52	36
TW-J2	671	379	249	176	130	98	74	54	37
DG-A1	679	310	189	127	90	64	45	31	16
DG-A2	655	303	187	130	94	69	51	36	23
DG-A3	734	341	213	150	113	86	64	47	30
DG-B2	678	314	195	138	104	80	62	46	31
DG-C3	853	397	248	177	134	106	85	66	46
SS-A1	1909	601	319	202	138	98	70	49	32
SS-A2	1604	688	401	265	187	135	98	70	46
SS-A5	3188	619	300	182	122	85	60	42	27
SS-B1	1312	457	250	160	110	79	57	40	26
SS-B3	1261	439	239	153	105	75	54	38	25
SS-B5	1719	423	213	132	89	63	45	31	20
SS-C4	773	370	226	154	110	81	60	44	30
SS-D1	1762	713	407	266	186	134	97	69	45
SS-D2	1404	738	467	323	233	172	127	92	61
SS-E1	1347	573	333	220	154	111	81	58	38

TABLE C.2 **Annual Passive Heating Performance: LCR (Continued)**

SACRAMENTO, CALIFORNIA									2843 DD
SSF =	.10	.20	.30	.40	.50	.60	.70	.80	.90
WW-A3	373	173	103	68	47	33	24	16	10
WW-A6	311	173	111	77	55	40	29	20	12
WW-B2	329	172	108	74	52	38	27	19	12
WW-B4	325	194	130	92	67	50	37	26	17
WW-C3	342	217	150	109	81	61	45	33	21
TW-A3	369	155	89	58	39	28	19	13	8
TW-E3	383	224	148	105	76	56	41	29	19
TW-F3	269	126	75	50	34	24	17	12	7
TW-F4	219	112	69	46	33	23	17	11	7
TW-G3	197	97	59	40	28	20	14	10	6
TW-G4	157	80	49	33	23	17	12	8	5
TW-I2	270	137	85	57	40	29	21	14	9
TW-I3	285	150	94	64	46	33	24	16	10
TW-I4	279	158	103	72	52	38	28	20	13
TW-J2	304	168	109	75	54	40	29	20	13
DG-A1	276	119	66	40	24	13	—	—	—
DG-A2	289	129	77	50	34	23	15	9	2
DG-A3	343	156	96	66	47	34	24	16	8
DG-B2	298	134	82	56	39	28	20	13	6
DG-C3	400	184	114	80	59	44	33	24	15
SS-A1	798	237	121	72	46	31	21	13	8
SS-A2	684	284	161	102	69	48	33	22	14
SS-A5	1375	241	112	64	40	26	17	11	6
SS-B1	542	179	94	57	37	25	17	11	6
SS-B3	508	167	87	52	34	22	15	9	6
SS-B5	721	163	79	46	29	19	13	8	5
SS-C4	343	159	95	63	44	31	22	15	10
SS-D1	722	280	153	95	63	42	29	19	11
SS-D2	594	305	187	124	86	61	43	30	18
SS-E1	542	220	123	77	51	34	23	15	9

SAN DIEGO, CALIFORNIA									1507 DD
SSF =	.10	.20	.30	.40	.50	.60	.70	.80	.90
WW-A3	968	466	287	196	141	105	78	57	38
WW-A6	805	461	305	218	161	122	92	68	46
WW-B2	830	446	287	202	148	111	83	61	42
WW-B4	788	479	327	238	178	136	103	77	53
WW-C3	811	523	369	273	208	160	123	92	64
TW-A3	948	413	245	165	117	86	64	46	31
TW-E3	918	547	370	267	199	151	115	85	58
TW-F3	699	339	209	143	103	77	57	42	28
TW-F4	567	298	190	133	97	72	54	40	27
TW-G3	510	260	164	113	82	61	46	34	23
TW-G4	405	212	135	94	69	51	38	28	19
TW-I2	681	355	226	157	115	85	64	47	32
TW-I3	718	388	251	176	129	97	73	53	36
TW-I4	676	392	261	187	139	105	80	59	40
TW-J2	743	420	277	197	146	110	83	61	42
DG-A1	757	348	213	145	103	74	53	37	22
DG-A2	727	337	210	146	107	79	58	42	27
DG-A3	809	376	237	168	126	96	73	53	34
DG-B2	747	347	218	155	117	91	70	52	35
DG-C3	934	436	275	197	150	119	95	74	52
SS-A1	2065	655	353	224	153	109	78	55	36
SS-A2	1730	750	442	293	207	149	109	78	51
SS-A5	3469	673	331	202	136	95	67	47	30
SS-B1	1425	500	276	178	123	88	63	45	29
SS-B3	1367	480	265	170	117	83	60	42	28
SS-B5	1875	462	236	147	99	70	50	35	23
SS-C4	858	410	252	172	124	92	68	50	34
SS-D1	1904	779	450	295	206	149	108	76	50
SS-D2	1520	808	515	357	258	190	140	101	67
SS-E1	1463	629	369	245	172	124	90	65	43

SAN FRANCISCO, CALIFORNIA									3042 DD
SSF =	.10	.20	.30	.40	.50	.60	.70	.80	.90
WW-A3	506	247	152	103	73	53	38	27	18
WW-A6	428	247	163	115	84	62	46	33	22
WW-B2	444	241	155	108	78	58	42	30	20
WW-B4	436	267	182	131	97	73	55	40	27
WW-C3	461	298	210	154	116	88	66	49	33
TW-A3	493	219	130	86	61	44	32	22	15
TW-E3	514	309	208	149	110	82	61	45	30
TW-F3	365	179	111	75	53	39	28	20	13
TW-F4	299	159	101	70	50	37	27	19	13
TW-G3	268	138	87	60	43	31	23	16	11
TW-G4	214	113	72	50	36	26	19	14	9
TW-I2	363	192	122	84	61	44	33	23	15
TW-I3	388	212	137	95	69	51	37	27	18
TW-I4	372	217	145	103	76	56	42	30	20
TW-J2	407	232	153	108	79	59	44	31	21
DG-A1	382	177	106	69	46	31	19	10	—
DG-A2	382	181	112	76	54	39	27	18	10
DG-A3	438	212	133	93	68	51	37	26	15
DG-B2	390	189	118	82	61	45	34	24	14
DG-C3	505	248	158	111	83	64	50	38	25
SS-A1	1129	369	198	122	81	55	38	25	15
SS-A2	984	434	254	165	113	79	56	38	24
SS-A5	1843	375	184	110	71	47	32	21	13
SS-B1	772	278	153	96	64	44	31	21	13
SS-B3	739	266	145	91	60	41	28	19	12
SS-B5	990	254	129	79	51	35	24	16	10
SS-C4	457	222	137	93	66	48	35	25	16
SS-D1	1058	441	252	161	109	75	52	35	22
SS-D2	878	469	296	200	141	101	72	50	32
SS-E1	799	349	203	131	89	62	43	29	18

SANTA MARIA, CALIFORNIA									3053 DD
SSF =	.10	.20	.30	.40	.50	.60	.70	.80	.90
WW-A3	523	261	164	114	82	61	45	33	22
WW-A6	447	263	177	128	95	72	54	40	27
WW-B2	462	256	168	119	88	66	49	36	25
WW-B4	455	283	197	144	109	83	63	47	32
WW-C3	483	318	228	170	130	100	77	57	40
TW-A3	508	231	140	95	68	50	37	27	18
TW-E3	537	328	225	164	123	94	71	53	36
TW-F3	378	190	120	83	60	45	33	24	16
TW-F4	311	169	110	77	57	42	32	23	16
TW-G3	278	147	94	66	48	36	27	20	13
TW-G4	222	120	78	55	40	30	22	16	11
TW-I2	377	203	132	93	68	51	38	28	19
TW-I3	404	226	149	106	78	58	44	32	22
TW-I4	387	230	156	113	84	64	48	36	24
TW-J2	422	246	165	119	88	67	50	37	25
DG-A1	392	188	117	79	55	38	26	16	7
DG-A2	389	190	121	85	61	45	32	22	14
DG-A3	443	220	142	102	77	58	44	31	19
DG-B2	394	196	127	91	69	53	40	29	19
DG-C3	509	254	167	121	92	73	58	45	31
SS-A1	1171	396	220	142	98	69	49	34	22
SS-A2	1028	468	283	190	135	98	71	50	33
SS-A5	1896	400	204	127	86	60	42	29	18
SS-B1	802	298	170	111	77	55	40	28	18
SS-B3	770	287	163	106	74	52	37	26	17
SS-B5	1022	271	144	91	62	44	31	22	14
SS-C4	473	235	147	102	74	55	41	30	20
SS-D1	1107	477	282	188	132	95	68	48	31
SS-D2	928	511	332	232	169	125	92	66	43
SS-E1	838	378	227	153	108	78	56	40	26

TABLE C.2 **Annual Passive Heating Performance: LCR (Continued)**

DENVER, COLORADO — *6016 DD*

SSF =	*.10*	*.20*	*.30*	*.40*	*.50*	*.60*	*.70*	*.80*	*.90*
WW-A3	224	108	66	45	32	24	18	13	9
WW-A6	193	111	73	52	38	29	22	16	11
WW-B2	207	112	72	51	37	28	21	15	10
WW-B4	216	132	91	66	49	38	29	21	15
WW-C3	233	152	108	80	61	47	36	27	19
TW-A3	219	96	57	38	27	20	14	10	7
TW-E3	255	154	104	75	56	43	32	24	17
TW-F3	161	79	48	33	24	17	13	9	6
TW-F4	134	71	45	31	23	17	13	9	6
TW-G3	120	61	38	26	19	14	11	8	5
TW-G4	97	51	32	22	16	12	9	7	4
TW-I2	169	89	56	39	29	21	15	12	8
TW-I3	180	98	63	44	32	24	18	13	9
TW-I4	183	107	72	51	38	29	22	16	11
TW-J2	198	113	75	53	39	30	22	16	11
DG-A1	150	66	37	21	15	—	—	—	—
DG-A2	172	80	49	33	23	15	10	6	—
DG-A3	216	103	65	46	34	25	19	13	7
DG-B2	177	83	52	36	27	20	14	9	4
DG-C3	253	121	77	55	42	33	26	20	14
SS-A1	444	140	74	46	31	22	16	11	7
SS-A2	413	180	105	70	49	35	26	18	12
SS-A5	714	139	67	40	26	18	13	9	5
SS-B1	303	106	57	36	25	18	13	9	6
SS-B3	276	96	51	32	22	15	11	8	5
SS-B5	380	93	47	28	19	13	9	6	4
SS-C4	213	103	63	43	31	23	17	12	8
SS-D1	415	168	96	62	43	30	22	15	10
SS-D2	370	197	125	86	62	46	34	24	16
SS-E1	307	130	75	49	34	24	17	12	8

MIAMI, FLORIDA — *206 DD*

SSF =	*.10*	*.20*	*.30*	*.40*	*.50*	*.60*	*.70*	*.80*	*.90*
WW-A3	3514	1741	1086	746	538	396	293	213	143
WW-A6	2951	1722	1152	822	608	457	343	252	172
WW-B2	2983	1640	1068	751	549	410	306	223	152
WW-B4	2809	1733	1191	865	647	491	371	274	188
WW-C3	2835	1849	1310	969	735	562	429	319	220
TW-A3	3416	1539	929	627	447	327	241	174	117
TW-E3	3212	1944	1323	955	711	537	405	299	205
TW-F3	2542	1268	794	546	394	290	215	156	105
TW-F4	2066	1114	719	503	367	273	203	148	100
TW-G3	1857	973	620	431	313	232	172	126	85
TW-G4	1472	789	508	355	259	192	143	105	71
TW-I2	2444	1305	839	586	427	317	236	172	117
TW-I3	2557	1416	924	651	476	355	265	194	132
TW-I4	2406	1418	953	683	506	381	286	211	144
TW-J2	2647	1526	1015	723	533	400	300	220	150
DG-A1	2824	1351	848	588	425	314	231	167	111
DG-A2	2597	1254	793	557	409	305	227	165	112
DG-A3	2799	1372	869	615	461	351	265	194	126
DG-B2	2641	1306	833	589	446	346	269	203	141
DG-C3	3185	1586	1019	720	543	429	340	264	187
SS-A1	6675	2252	1249	806	558	399	289	206	138
SS-A2	5630	2548	1534	1031	731	532	390	282	190
SS-A5	10965	2307	1177	733	497	351	251	178	118
SS-B1	4704	1744	990	647	451	324	236	169	113
SS-B3	4512	1680	954	623	434	312	226	162	109
SS-B5	6088	1610	851	539	368	262	188	134	89
SS-C4	3045	1498	932	639	460	339	251	182	123
SS-D1	6282	2699	1597	1063	749	543	397	286	193
SS-D2	5041	2762	1788	1250	909	674	502	367	250
SS-E1	4936	2219	1333	895	634	461	338	244	164

WASHINGTON, D.C. — *5010 DD*

SSF =	*.10*	*.20*	*.30*	*.40*	*.50*	*.60*	*.70*	*.80*	*.90*
WW-A3	123	55	32	20	14	10	7	4	3
WW-A6	106	58	37	25	17	13	9	6	4
WW-B2	121	63	39	26	19	13	10	7	4
WW-B4	135	81	54	39	28	21	16	11	8
WW-C3	153	98	68	50	38	28	22	16	11
TW-A3	122	49	28	17	12	8	5	4	2
TW-E3	164	97	65	46	34	25	19	14	9
TW-F3	88	40	23	15	10	7	5	3	2
TW-F4	74	37	22	15	10	7	5	3	2
TW-G3	65	31	19	12	8	6	4	3	2
TW-G4	54	27	16	11	7	5	4	3	2
TW-I2	98	49	30	20	14	10	7	5	3
TW-I3	106	56	35	24	17	12	9	6	4
TW-I4	114	65	43	30	22	16	12	9	6
TW-J2	122	68	44	30	22	16	12	8	6
DG-A1	65	—	—	—	—	—	—	—	—
DG-A2	100	43	24	14	7	—	—	—	—
DG-A3	140	64	39	27	19	13	9	6	—
DG-B2	102	45	26	17	11	6	—	—	—
DG-C3	166	78	48	33	25	19	14	10	6
SS-A1	269	77	38	22	14	9	6	4	2
SS-A2	264	110	63	40	27	19	14	9	6
SS-A5	442	74	32	18	11	7	4	2	1
SS-B1	179	57	29	17	11	7	5	3	2
SS-B3	158	48	24	13	8	5	3	2	—
SS-B5	226	48	21	12	7	4	2	1	—
SS-C4	127	58	34	23	16	11	8	6	4
SS-D1	245	92	50	30	20	13	9	6	3
SS-D2	235	121	75	51	36	26	18	13	8
SS-E1	173	67	36	21	14	9	6	4	2

ORLANDO, FLORIDA — *733 DD*

SSF =	*.10*	*.20*	*.30*	*.40*	*.50*	*.60*	*.70*	*.80*	*.90*
WW-A3	1105	545	341	235	171	127	95	69	47
WW-A6	931	542	364	261	195	148	112	83	57
WW-B2	950	521	340	241	178	134	101	74	51
WW-B4	909	561	387	283	214	163	125	93	64
WW-C3	928	606	432	322	246	190	146	110	76
TW-A3	1076	482	291	198	142	105	78	57	38
TW-E3	1045	632	432	314	236	180	137	102	70
TW-F3	799	397	249	172	125	93	69	51	34
TW-F4	651	350	226	159	117	88	66	48	33
TW-G3	585	305	195	136	100	75	56	41	28
TW-G4	465	248	160	113	83	62	47	34	23
TW-I2	778	414	267	188	138	103	78	57	39
TW-I3	816	450	295	209	154	116	87	64	44
TW-I4	777	458	309	223	167	126	96	71	49
TW-J2	853	491	328	235	175	132	100	74	51
DG-A1	863	410	257	177	127	93	67	47	29
DG-A2	817	392	249	176	129	96	71	51	34
DG-A3	905	436	280	200	152	116	88	64	41
DG-B2	838	403	259	187	142	111	86	65	44
DG-C3	1043	501	325	236	181	143	115	89	63
SS-A1	2147	701	385	250	174	126	92	66	44
SS-A2	1819	806	484	327	234	172	127	92	62
SS-A5	3554	717	361	225	154	110	79	56	37
SS-B1	1501	541	305	200	141	102	75	54	36
SS-B3	1431	517	291	191	134	97	71	51	34
SS-B5	1951	498	260	165	114	82	59	42	28
SS-C4	972	476	297	205	148	110	82	60	41
SS-D1	2001	839	494	330	235	172	127	91	61
SS-D2	1615	874	566	399	293	220	164	120	82
SS-E1	1557	685	410	276	198	145	107	78	52

TABLE C.2 **Annual Passive Heating Performance: LCR (Continued)**

TALLAHASSEE, FLORIDA — *1563 DD*

SSF =	.10	.20	.30	.40	.50	.60	.70	.80	.90
WW-A3	522	248	151	102	74	54	40	29	20
WW-A6	436	246	162	115	85	64	48	35	24
WW-B2	454	242	154	108	79	59	44	32	22
WW-B4	442	267	181	131	98	75	57	42	29
WW-C3	458	293	206	152	116	89	68	51	35
TW-A3	513	221	129	86	61	45	33	24	16
TW-E3	515	304	205	147	110	83	63	47	32
TW-F3	377	180	110	75	54	40	29	21	14
TW-F4	306	159	101	70	51	38	28	21	14
TW-G3	276	139	87	60	43	32	24	17	12
TW-G4	220	114	72	50	36	27	20	15	10
TW-I2	373	192	121	84	61	45	34	25	17
TW-I3	391	209	134	94	69	51	38	28	19
TW-I4	379	218	144	103	76	58	44	32	22
TW-J2	415	233	152	108	80	60	45	33	23
DG-A1	395	177	105	69	47	32	21	12	4
DG-A2	396	180	111	76	55	40	28	19	12
DG-A3	459	209	131	92	69	52	39	28	17
DG-B2	411	185	115	82	61	47	35	26	16
DG-C3	537	244	152	110	83	66	52	40	28
SS-A1	1043	325	171	108	74	53	38	27	18
SS-A2	897	383	223	147	104	76	56	40	27
SS-A5	1733	332	159	96	65	45	32	23	15
SS-B1	723	249	135	86	60	43	31	22	15
SS-B3	682	234	127	81	56	40	29	21	13
SS-B5	942	228	114	70	48	34	24	17	11
SS-C4	470	222	135	92	66	49	36	26	18
SS-D1	963	386	219	143	100	73	53	38	25
SS-D2	787	412	260	180	131	97	73	53	35
SS-E1	738	310	179	118	83	61	44	32	21

BOISE, IDAHO — *5833 DD*

SSF =	.10	.20	.30	.40	.50	.60	.70	.80	.90
WW-A3	190	85	49	30	20	13	8	5	3
WW-A6	160	86	54	36	24	17	11	7	4
WW-B2	176	90	55	37	25	17	12	8	5
WW-B4	185	109	72	50	36	26	19	13	8
WW-C3	202	127	88	63	46	35	25	18	12
TW-A3	188	76	42	26	17	11	7	4	2
TW-E3	222	129	84	59	42	31	22	16	10
TW-F3	136	62	35	22	14	9	6	4	2
TW-F4	112	55	33	21	14	10	6	4	2
TW-G3	101	48	28	18	12	8	5	3	2
TW-G4	81	40	24	15	10	7	5	3	2
TW-I2	144	72	43	28	19	13	9	6	3
TW-13	154	79	49	32	22	15	11	7	4
TW-I4	158	89	57	39	28	20	14	10	6
TW-J2	171	93	59	40	28	20	14	10	6
DG-A1	124	46	16	—	—	—	—	—	—
DG-A2	152	66	37	22	13	6	—	—	—
DG-A3	197	89	53	36	25	17	12	7	—
DG-B2	157	69	40	26	16	10	5	—	—
DG-C3	233	107	66	45	32	24	17	12	7
SS-A1	403	115	55	31	19	11	7	4	2
SS-A2	371	151	83	52	34	23	15	10	6
SS-A5	675	115	49	26	14	8	4	2	—
SS-B1	270	85	42	24	14	9	5	3	1
SS-B3	244	75	36	19	11	6	2	—	—
SS-B5	349	75	33	18	10	5	2	—	—
SS-C4	184	84	49	31	21	15	10	6	4
SS-D1	366	136	70	41	25	16	10	6	3
SS-D2	324	163	98	64	43	30	21	14	8
SS-E1	265	102	53	31	18	11	6	3	1

ATLANTA, GEORGIA — *3095 DD*

SSF =	.10	.20	.30	.40	.50	.60	.70	.80	.90
WW-A3	245	114	68	46	32	23	17	12	8
WW-A6	207	116	75	52	38	28	21	15	10
WW-B2	222	117	74	51	37	27	20	14	10
WW-B4	228	137	93	67	49	37	28	21	14
WW-C3	246	157	110	81	61	47	35	26	18
TW-A3	242	102	59	38	27	19	14	10	6
TW-E3	271	160	107	76	57	42	32	23	16
TW-F3	176	83	50	33	23	17	12	9	6
TW-F4	145	74	46	31	22	16	12	9	6
TW-G3	130	64	40	27	19	14	10	7	5
TW-G4	104	53	33	23	16	12	9	6	4
TW-12	182	93	58	40	28	21	15	11	7
TW-I3	193	103	65	45	32	24	18	13	8
TW-I4	195	111	73	52	38	28	21	16	10
TW-J2	211	118	77	54	39	29	22	16	11
DG-A1	168	70	38	21	8	—	—	—	—
DG-A2	191	85	51	33	23	15	9	5	—
DG-A3	236	109	67	46	34	25	18	12	7
DG-B2	195	89	54	37	26	19	13	8	4
DG-C3	277	130	80	56	42	33	25	19	13
SS-A1	516	154	80	49	33	23	16	11	7
SS-A2	462	195	113	74	51	37	26	19	12
SS-A5	877	154	73	43	28	19	13	9	6
SS-B1	351	116	62	39	26	18	13	9	6
SS-B3	324	107	56	35	23	16	11	8	5
SS-B5	460	104	51	30	20	14	9	6	4
SS-C4	232	108	65	44	31	22	16	12	8
SS-D1	470	184	103	66	45	32	23	16	10
SS-D2	405	212	133	91	65	48	35	25	17
SS-E1	349	143	81	52	36	25	18	13	8

CHICAGO, ILLINOIS — *6127 DD*

SSF =	.10	.20	.30	.40	.50	.60	.70	.80	.90
WW-A3	92	38	20	12	7	4	2	—	—
WW-A6	80	41	25	16	10	7	4	3	1
WW-B2	95	47	28	19	13	9	6	4	2
WW-B4	112	65	43	30	22	16	12	8	5
WW-C3	129	82	56	41	30	23	17	12	8
TW-A3	93	35	18	10	6	3	2	—	—
TW-E3	138	80	53	37	27	20	14	10	7
TW-F3	66	27	14	8	5	3	1	—	—
TW-F4	55	26	15	9	6	4	2	1	—
TW-G3	49	22	12	7	5	3	2	—	—
TW-G4	41	19	11	7	4	3	2	1	—
TW-I2	78	37	22	14	10	7	4	3	2
TW-I3	85	43	26	17	12	8	6	4	2
TW-I4	94	52	34	23	17	12	9	6	4
TW-J2	101	54	34	23	17	12	9	6	4
DG-A1	31	—	—	—	—	—	—	—	—
DG-A2	79	31	15	—	—	—	—	—	—
DG-A3	119	53	31	21	14	10	6	3	—
DG-B2	81	33	18	10	—	—	—	—	—
DG-C3	143	65	39	27	19	14	11	7	4
SS-A1	217	56	25	13	7	4	2	—	—
SS-A2	220	88	48	30	20	14	9	6	4
SS-A5	374	52	19	8	—	—	—	—	—
SS-B1	140	39	18	9	4	—	—	—	—
SS-B3	119	31	11	—	—	—	—	—	—
SS-B5	181	31	11	—	—	—	—	—	—
SS-C4	102	45	25	16	11	7	5	3	2
SS-D1	193	67	33	18	11	6	3	1	—
SS-D2	193	96	58	38	26	18	13	9	6
SS-E1	128	44	20	10	—	—	—	—	—

TABLE C.2 **Annual Passive Heating Performance: LCR (Continued)**

MOLINE, ILLINOIS — *6395 DD*

SSF =	.10	.20	.30	.40	.50	.60	.70	.80	.90
WW-A3	92	38	20	12	7	4	2	1	—
WW-A6	80	41	25	16	11	7	5	3	2
WW-B2	96	47	28	19	13	9	6	4	2
WW-B4	112	65	43	30	22	16	12	9	6
WW-C3	129	82	56	41	30	23	17	13	8
TW-A3	93	35	18	10	6	4	2	1	—
TW-E3	138	80	52	37	27	20	15	10	7
TW-F3	66	27	15	9	5	3	2	—	—
TW-F4	55	26	15	9	6	4	2	1	—
TW-G3	49	22	12	7	5	3	2	1	—
TW-G4	41	19	11	7	4	3	2	1	—
TW-I2	78	37	22	14	10	7	5	3	2
TW-I3	85	43	26	17	12	8	6	4	2
TW-I4	94	52	34	23	17	12	9	6	4
TW-J2	101	54	34	23	17	12	9	6	4
DG-A1	31	—	—	—	—	—	—	—	—
DG-A2	79	31	15	5	—	—	—	—	—
DG-A3	119	53	31	21	14	10	6	3	—
DG-B2	81	33	17	10	4	—	—	—	—
DG-C3	144	65	39	27	19	14	11	7	4
SS-A1	213	55	25	13	7	4	2	—	—
SS-A2	218	87	48	30	20	14	9	6	4
SS-A5	363	51	19	8	—	—	—	—	—
SS-B1	138	39	18	9	5	2	—	—	—
SS-B3	116	30	11	—	—	—	—	—	—
SS-B5	177	31	11	3	—	—	—	—	—
SS-C4	102	45	25	16	11	7	5	3	2
SS-D1	190	66	33	18	11	6	4	2	—
SS-D2	191	95	58	38	26	19	13	9	6
SS-E1	127	43	20	10	4	—	—	—	—

INDIANAPOLIS, INDIANA — *5577 DD*

SSF =	.10	.20	.30	.40	.50	.60	.70	.80	.90
WW-A3	88	36	19	11	6	4	2	—	—
WW-A6	76	39	23	15	10	6	4	3	1
WW-B2	92	45	27	18	12	8	6	4	2
WW-B4	109	64	42	29	21	16	11	8	5
WW-C3	127	80	55	40	30	22	17	12	8
TW-A3	89	33	16	9	5	3	2	—	—
TW-E3	135	78	51	36	26	19	14	10	7
TW-F3	63	26	13	8	4	2	1	—	—
TW-F4	53	24	14	8	5	3	2	1	—
TW-G3	47	21	11	7	4	3	1	—	—
TW-G4	39	18	10	6	4	3	2	1	—
TW-I2	75	36	21	13	9	6	4	3	2
TW-I3	82	41	25	16	11	8	5	4	2
TW-I4	92	51	33	22	16	12	8	6	4
TW-J2	98	53	33	23	16	11	8	6	4
DG-A1	—	—	—	—	—	—	—	—	—
DG-A2	76	29	14	—	—	—	—	—	—
DG-A3	116	51	30	20	14	9	6	3	—
DG-B2	78	31	16	9	—	—	—	—	—
DG-C3	140	64	38	26	19	14	10	7	4
SS-A1	214	56	25	13	7	4	2	—	—
SS-A2	218	87	48	30	20	14	9	6	4
SS-A5	358	51	19	8	—	—	—	—	—
SS-B1	138	39	17	9	5	2	—	—	—
SS-B3	118	30	11	—	—	—	—	—	—
SS-B5	175	31	11	—	—	—	—	—	—
SS-C4	98	43	24	15	10	7	5	3	2
SS-D1	191	66	33	18	11	6	4	2	—
SS-D2	192	96	58	38	26	18	13	9	6
SS-E1	128	43	20	10	4	—	—	—	—

SPRINGFIELD, ILLINOIS — *5558 DD*

SSF =	.10	.20	.30	.40	.50	.60	.70	.80	.90
WW-A3	121	52	29	19	12	8	6	4	2
WW-A6	103	55	34	23	16	11	8	5	3
WW-B2	119	60	37	25	17	12	9	6	4
WW-B4	133	78	52	37	27	20	15	11	7
WW-C3	150	95	66	48	36	27	20	15	10
TW-A3	121	47	26	16	10	7	5	3	2
TW-E3	162	94	62	44	32	24	18	13	9
TW-F3	86	38	21	13	9	6	4	3	2
TW-F4	72	35	21	13	9	6	4	3	2
TW-G3	64	30	17	11	8	5	4	2	1
TW-G4	53	25	15	10	7	5	3	2	1
TW-I2	97	47	29	19	13	9	7	5	3
TW-I3	104	53	33	22	15	11	8	6	4
TW-I4	113	63	41	28	21	15	11	8	5
TW-J2	121	66	42	29	21	15	11	8	5
DG-A1	62	—	—	—	—	—	—	—	—
DG-A2	100	41	22	12	5	—	—	—	—
DG-A3	141	63	37	25	18	13	9	5	—
DG-B2	102	43	24	15	10	5	—	—	—
DG-C3	168	77	46	32	23	18	13	10	6
SS-A1	269	72	35	20	12	8	5	3	2
SS-A2	261	105	59	37	25	18	12	9	5
SS-A5	465	69	29	15	9	5	3	2	—
SS-B1	177	53	26	15	9	6	4	2	1
SS-B3	155	44	20	11	6	4	2	—	—
SS-B5	231	44	19	10	6	3	2	—	—
SS-C4	126	56	33	21	15	10	7	5	3
SS-D1	240	86	45	27	17	11	8	5	3
SS-D2	228	115	70	47	33	24	17	12	8
SS-E1	167	61	31	18	11	7	4	3	1

DES MOINES, IOWA — *6710 DD*

SSF =	.10	.20	.30	.40	.50	.60	.70	.80	.90
WW-A3	107	45	25	15	10	7	4	3	2
WW-A6	92	48	29	19	13	9	7	4	3
WW-B2	107	54	33	22	15	11	8	5	3
WW-B4	122	72	48	34	25	18	13	10	6
WW-C3	140	88	61	44	33	25	19	14	9
TW-A3	108	41	22	13	8	6	4	2	1
TW-E3	150	87	57	40	29	22	16	12	8
TW-F3	76	32	18	11	7	5	3	2	1
TW-F4	64	30	18	11	8	5	4	2	1
TW-G3	57	26	15	9	6	4	3	2	1
TW-G4	47	22	13	8	6	4	3	2	1
TW-I2	88	42	25	17	11	8	6	4	2
TW-I3	95	48	29	19	14	10	7	5	3
TW-I4	104	58	37	26	19	14	10	7	5
TW-J2	111	60	38	26	19	14	10	7	5
DG-A1	48	—	—	—	—	—	—	—	—
DG-A2	90	36	19	10	—	—	—	—	—
DG-A3	131	58	34	23	16	11	7	4	—
DG-B2	92	38	21	13	7	—	—	—	—
DG-C3	157	71	43	29	21	16	12	9	5
SS-A1	238	63	29	16	10	6	4	2	1
SS-A2	237	95	52	33	22	15	11	7	5
SS-A5	405	59	23	11	6	3	1	—	—
SS-B1	155	45	21	12	7	4	2	1	—
SS-B3	133	36	15	7	3	—	—	—	—
SS-B5	200	37	15	7	3	—	—	—	—
SS-C4	114	50	29	19	13	9	6	4	3
SS-D1	212	75	38	22	14	9	6	3	2
SS-D2	207	104	63	42	29	21	15	10	7
SS-E1	144	51	25	14	8	5	2	1	—

TABLE C.2 **Annual Passive Heating Performance: LCR (Continued)**

DODGE CITY, KANSAS — 5046 DD

SSF =	.10	.20	.30	.40	.50	.60	.70	.80	.90
WW-A3	223	104	63	42	30	22	16	11	8
WW-A6	189	106	69	48	35	26	20	14	10
WW-B2	204	108	69	48	34	26	19	14	9
WW-B4	211	127	87	62	47	35	27	20	14
WW-C3	228	147	103	76	58	44	34	25	17
TW-A3	220	93	54	35	25	18	13	9	6
TW-E3	251	148	100	72	53	40	30	22	15
TW-F3	160	75	46	31	22	16	12	8	6
TW-F4	132	68	42	29	21	15	11	8	6
TW-G3	118	59	36	25	18	13	10	7	5
TW-G4	95	49	30	21	15	11	8	6	4
TW-I2	167	85	54	37	27	20	15	11	7
TW-I3	177	94	60	42	30	22	17	12	8
TW-I4	180	103	68	49	36	27	20	15	10
TW-J2	195	109	71	50	37	28	21	15	10
DG-A1	149	62	33	17	—	—	—	—	—
DG-A2	174	77	47	31	21	14	9	4	—
DG-A3	218	100	62	43	32	24	17	12	6
DG-B2	177	80	49	34	25	18	13	8	3
DG-C3	256	119	74	52	40	31	24	18	12
SS-A1	457	135	70	43	29	20	14	10	6
SS-A2	414	173	100	65	46	33	24	17	11
SS-A5	770	134	63	37	24	16	11	8	5
SS-B1	309	101	54	34	23	16	11	8	5
SS-B3	281	91	48	30	20	14	10	7	4
SS-B5	402	90	44	26	17	12	8	6	4
SS-C4	212	99	60	41	29	21	16	11	8
SS-D1	415	161	90	57	39	28	20	14	9
SS-D2	363	188	118	81	58	43	31	23	15
SS-E1	305	124	70	45	31	22	16	11	7

LEXINGTON, KENTUCKY — 4729 DD

SSF =	.10	.20	.30	.40	.50	.60	.70	.80	.90
WW-A3	123	53	30	19	13	9	6	4	2
WW-A6	105	56	35	23	16	11	8	6	4
WW-B2	121	61	37	25	17	12	9	6	4
WW-B4	135	79	53	37	27	20	15	11	7
WW-C3	152	96	67	49	36	27	21	15	10
TW-A3	123	48	26	16	10	7	5	3	2
TW-E3	165	96	63	44	32	24	18	13	9
TW-F3	88	38	22	14	9	6	4	3	2
TW-F4	73	35	21	14	9	6	4	3	2
TW-G3	66	30	18	11	6	5	4	2	2
TW-G4	54	26	15	10	7	5	3	2	1
TW-I2	99	48	29	19	13	9	7	5	3
TW-I3	107	54	33	22	16	11	8	6	4
TW-I4	114	64	41	29	21	15	11	8	5
TW-J2	123	67	43	29	21	15	11	8	5
DG-A1	65	—	—	—	—	—	—	—	—
DG-A2	102	42	23	13	6	—	—	—	—
DG-A3	143	64	38	26	18	13	9	5	—
DG-B2	105	44	25	15	10	5	—	—	—
DG-C3	171	78	47	32	24	18	13	10	6
SS-A1	282	77	37	21	13	9	6	4	2
SS-A2	272	110	62	39	26	18	13	9	6
SS-A5	474	74	31	16	10	6	4	2	1
SS-B1	186	56	28	16	10	6	4	3	1
SS-B3	165	48	22	12	7	4	3	1	1
SS-B5	240	48	20	11	6	4	2	1	1
SS-C4	128	57	33	21	15	10	7	5	3
SS-D1	254	92	48	29	18	12	8	5	3
SS-D2	239	120	74	49	34	25	18	12	8
SS-E1	177	66	34	20	12	8	5	3	2

TOPEKA, KANSAS — 5243 DD

SSF =	.10	.20	.30	.40	.50	.60	.70	.80	.90
WW-A3	163	74	43	28	19	14	10	7	4
WW-A6	139	76	48	33	24	17	13	9	6
WW-B2	154	80	50	34	24	18	13	9	6
WW-B4	165	99	66	47	35	26	19	14	10
WW-C3	182	116	81	59	44	34	26	19	13
TW-A3	162	66	37	24	16	11	8	6	4
TW-E3	199	116	77	55	40	30	23	16	11
TW-F3	117	53	31	20	14	10	7	5	3
TW-F4	97	48	30	20	14	10	7	5	3
TW-G3	87	42	25	17	12	8	6	4	3
TW-G4	70	35	21	14	10	7	5	4	2
TW-I2	126	63	39	26	19	13	10	7	5
TW-I3	134	70	44	30	21	16	11	8	5
TW-I4	141	80	52	37	27	20	15	11	7
TW-J2	152	84	54	38	27	20	15	11	7
DG-A1	100	36	—	—	—	—	—	—	—
DG-A2	131	56	32	20	12	7	—	—	—
DG-A3	173	78	48	33	23	17	12	8	3
DG-B2	134	59	34	23	16	11	6	—	—
DG-C3	205	95	58	40	30	23	18	13	8
SS-A1	341	98	49	29	19	13	9	6	3
SS-A2	321	132	75	48	33	23	17	12	7
SS-A5	572	96	42	24	15	10	6	4	2
SS-B1	229	73	37	22	14	10	7	4	3
SS-B3	205	63	32	18	12	8	5	3	2
SS-B5	296	63	29	16	10	7	4	3	2
SS-C4	162	74	44	29	20	15	11	7	5
SS-D1	309	116	63	39	26	18	12	8	5
SS-D2	282	144	89	60	42	31	22	16	10
SS-E1	223	87	47	29	19	13	9	6	4

NEW ORLEANS, LOUISIANA — 1465 DD

SSF =	.10	.20	.30	.40	.50	.60	.70	.80	.90
WW-A3	534	253	153	104	74	54	40	29	19
WW-A6	445	251	164	116	85	64	48	35	24
WW-B2	464	247	157	109	79	59	44	32	22
WW-B4	451	271	184	132	99	75	57	42	29
WW-C3	467	298	209	154	117	89	68	51	35
TW-A3	524	225	132	87	62	45	33	24	16
TW-E3	525	310	207	149	110	84	63	47	32
TW-F3	385	184	112	76	54	40	29	21	14
TW-F4	313	163	102	70	51	38	28	20	14
TW-G3	282	142	88	60	43	32	24	17	12
TW-G4	224	116	73	50	36	27	20	15	10
TW-I2	381	196	123	85	61	46	34	25	17
TW-I3	400	214	136	95	69	51	38	28	19
TW-I4	387	222	146	104	77	58	44	32	22
TW-J2	423	237	155	109	80	60	45	33	22
DG-A1	405	181	107	70	48	32	21	12	4
DG-A2	406	185	113	77	55	40	28	19	11
DG-A3	471	214	133	93	69	52	39	28	17
DG-B2	423	190	117	82	62	47	35	25	16
DG-C3	552	251	155	111	84	66	52	40	27
SS-A1	1072	335	176	111	76	54	39	27	18
SS-A2	922	394	229	151	106	77	57	41	27
SS-A5	1777	342	164	99	66	46	33	23	15
SS-B1	744	257	139	88	61	44	32	22	15
SS-B3	703	242	131	83	57	41	29	21	14
SS-B5	967	236	118	72	49	34	25	17	11
SS-C4	480	227	138	93	66	49	36	26	18
SS-D1	990	398	225	147	103	74	54	38	25
SS-D2	809	423	267	184	134	99	74	53	36
SS-E1	760	320	184	121	85	62	45	32	21

TABLE C.2 **Annual Passive Heating Performance: LCR (Continued)**

SHREVEPORT, LOUISIANA — *2167 DD*

SSF =	.10	.20	.30	.40	.50	.60	.70	.80	.90
WW-A3	360	170	103	69	49	36	26	19	13
WW-A6	302	170	111	78	57	42	32	23	16
WW-B2	319	169	107	74	54	40	30	22	14
WW-B4	317	191	129	93	69	52	39	29	20
WW-C3	334	214	150	110	83	64	48	36	25
TW-A3	354	151	88	58	41	30	22	15	10
TW-E3	372	220	147	105	78	59	44	33	22
TW-F3	260	124	75	50	36	26	19	14	9
TW-F4	212	110	69	47	34	25	19	13	9
TW-G3	190	95	59	40	29	21	16	11	7
TW-G4	152	78	49	34	24	18	13	10	6
TW-I2	261	134	84	58	42	31	23	17	11
TW-I3	275	147	94	65	47	35	26	19	13
TW-I4	271	156	103	73	54	40	30	22	15
TW-J2	296	166	108	76	56	42	31	23	15
DG-A1	263	115	66	41	26	16	7	—	—
DG-A2	277	125	76	51	36	25	17	11	5
DG-A3	330	151	94	65	48	36	27	19	11
DG-B2	286	129	79	55	41	30	22	15	9
DG-C3	384	179	110	78	59	46	36	27	19
SS-A1	715	222	117	73	49	35	25	17	11
SS-A2	629	270	156	103	72	52	38	27	18
SS-A5	1202	225	107	64	42	29	21	14	9
SS-B1	494	170	91	58	39	28	20	14	9
SS-B3	461	158	85	53	36	26	18	13	8
SS-B5	645	154	76	46	31	22	15	11	7
SS-C4	330	156	94	63	45	33	24	18	12
SS-D1	659	265	149	96	67	48	35	24	16
SS-D2	555	292	183	126	91	67	49	36	24
SS-E1	501	210	120	78	54	39	28	20	13

BOSTON, MASSACHUSETTS — *5621 DD*

SSF =	.10	.20	.30	.40	.50	.60	.70	.80	.90
WW-A3	101	43	24	15	10	6	4	3	1
WW-A6	87	46	28	19	13	9	6	4	3
WW-B2	103	52	31	21	15	10	7	5	3
WW-B4	118	70	46	33	24	18	13	9	6
WW-C3	136	86	60	44	33	25	19	14	9
TW-A3	102	39	21	13	8	5	3	2	1
TW-E3	146	84	56	39	29	21	16	11	8
TW-F3	72	31	17	11	7	4	3	2	1
TW-F4	61	29	17	11	7	5	3	2	1
TW-G3	54	25	14	9	6	4	3	2	1
TW-G4	45	21	12	8	5	4	3	2	1
TW-I2	84	41	24	16	11	8	6	4	2
TW-I3	91	46	28	19	13	9	7	5	3
TW-I4	100	56	36	25	18	13	10	7	5
TW-J2	107	58	37	25	18	13	10	7	4
DG-A1	43	—	—	—	—	—	—	—	—
DG-A2	85	34	18	9	—	—	—	—	—
DG-A3	125	56	33	22	16	11	7	4	—
DG-B2	87	36	20	12	7	—	—	—	—
DG-C3	151	68	41	28	21	16	12	8	5
SS-A1	230	61	29	16	10	6	4	2	1
SS-A2	231	93	52	33	22	15	11	7	5
SS-A5	389	58	23	11	6	3	—	—	—
SS-B1	151	44	21	12	7	4	2	1	—
SS-B3	129	36	16	8	3	—	—	—	—
SS-B5	193	36	15	7	3	—	—	—	—
SS-C4	109	48	28	18	12	9	6	4	3
SS-D1	206	73	38	22	14	9	5	3	2
SS-D2	203	103	63	42	29	21	15	10	6
SS-E1	140	51	25	14	8	4	2	—	—

PORTLAND, MAINE — *7498 DD*

SSF =	.10	.20	.30	.40	.50	.60	.70	.80	.90
WW-A3	69	26	13	6	—	—	—	—	—
WW-A6	61	30	17	10	6	4	2	—	—
WW-B2	77	37	22	14	9	6	4	3	1
WW-B4	95	55	36	25	18	13	10	7	4
WW-C3	113	71	49	36	26	20	15	11	7
TW-A3	71	24	11	5	—	—	—	—	—
TW-E3	120	69	45	32	23	17	12	9	6
TW-F3	49	19	9	4	—	—	—	—	—
TW-F4	42	18	10	5	3	—	—	—	—
TW-G3	37	16	8	4	2	—	—	—	—
TW-G4	32	14	8	4	3	1	—	—	—
TW-I2	63	29	17	11	7	5	3	2	1
TW-I3	69	34	20	13	9	6	4	3	2
TW-I4	80	44	28	19	14	10	7	5	3
TW-J2	85	45	28	19	14	10	7	5	3
DG-A1	—	—	—	—	—	—	—	—	—
DG-A2	62	22	8	—	—	—	—	—	—
DG-A3	102	44	26	17	12	8	5	—	—
DG-B2	64	24	12	—	—	—	—	—	—
DG-C3	125	55	33	23	16	12	9	6	3
SS-A1	180	44	18	8	—	—	—	—	—
SS-A2	192	75	41	25	17	11	7	5	3
SS-A5	302	39	12	—	—	—	—	—	—
SS-B1	114	29	12	—	—	—	—	—	—
SS-B3	93	20	—	—	—	—	—	—	—
SS-B5	144	21	—	—	—	—	—	—	—
SS-C4	83	35	20	12	8	5	4	2	1
SS-D1	160	52	24	12	5	—	—	—	—
SS-D2	168	83	50	32	22	15	11	7	4
SS-E1	101	31	10	—	—	—	—	—	—

DETROIT, MICHIGAN — *6228 DD*

SSF =	.10	.20	.30	.40	.50	.60	.70	.80	.90
WW-A3	69	25	11	—	—	—	—	—	—
WW-A6	60	29	16	9	4	—	—	—	—
WW-B2	77	37	21	13	8	5	3	2	1
WW-B4	95	55	36	25	18	13	9	6	4
WW-C3	113	71	49	35	26	19	14	10	7
TW-A3	71	24	10	—	—	—	—	—	—
TW-E3	120	68	45	31	22	16	12	8	5
TW-F3	49	18	7	—	—	—	—	—	—
TW-F4	42	18	9	4	—	—	—	—	—
TW-G3	37	15	7	3	—	—	—	—	—
TW-G4	31	14	7	4	2	—	—	—	—
TW-I2	62	29	16	10	6	4	2	1	—
TW-I3	69	34	20	12	8	5	4	2	1
TW-I4	80	44	28	19	13	10	7	5	3
TW-J2	85	45	28	19	13	9	6	4	3
DG-A1	—	—	—	—	—	—	—	—	—
DG-A2	62	22	—	—	—	—	—	—	—
DG-A3	103	45	26	17	11	7	4	—	—
DG-B2	64	24	11	—	—	—	—	—	—
DG-C3	125	56	33	22	16	11	8	5	2
SS-A1	180	43	17	7	—	—	—	—	—
SS-A2	191	75	40	25	16	11	7	5	3
SS-A5	307	38	9	—	—	—	—	—	—
SS-B1	113	29	10	—	—	—	—	—	—
SS-B3	92	19	—	—	—	—	—	—	—
SS-B5	144	21	—	—	—	—	—	—	—
SS-C4	83	35	19	12	7	5	3	2	—
SS-D1	158	51	23	10	—	—	—	—	—
SS-D2	167	82	49	32	21	15	10	7	4
SS-E1	100	29	—	—	—	—	—	—	—

TABLE C.2 **Annual Passive Heating Performance: LCR (Continued)**

SAULT ST. MARIE, MICHIGAN								9193 DD	
SSF =	.10	.20	.30	.40	.50	.60	.70	.80	.90
WW-A3	38	—	—	—	—	—	—	—	—
WW-A6	34	—	—	—	—	—	—	—	—
WW-B2	54	22	9	—	—	—	—	—	—
WW-B4	76	42	26	17	12	8	5	3	2
WW-C3	95	58	39	28	20	14	10	7	4
TW-A3	43	—	—	—	—	—	—	—	—
TW-E3	99	55	35	23	16	11	8	5	3
TW-F3	26	—	—	—	—	—	—	—	—
TW-F4	24	—	—	—	—	—	—	—	—
TW-G3	21	—	—	—	—	—	—	—	—
TW-G4	19	—	—	—	—	—	—	—	—
TW-I2	44	17	—	—	—	—	—	—	—
TW-I3	51	22	10	—	—	—	—	—	—
TW-I4	64	34	20	13	8	5	3	2	1
TW-J2	67	34	20	12	8	5	3	1	—
DG-A1	—	—	—	—	—	—	—	—	—
DG-A2	43	—	—	—	—	—	—	—	—
DG-A3	85	36	20	12	7	3	—	—	—
DG-B2	45	10	—	—	—	—	—	—	—
DG-C3	105	46	27	17	11	7	4	2	—
SS-A1	138	23	—	—	—	—	—	—	—
SS-A2	159	59	29	16	9	4	2	—	—
SS-A5	224	—	—	—	—	—	—	—	—
SS-B1	82	—	—	—	—	—	—	—	—
SS-B3	60	—	—	—	—	—	—	—	—
SS-B5	100	—	—	—	—	—	—	—	—
SS-C4	61	23	9	—	—	—	—	—	—
SS-D1	120	26	—	—	—	—	—	—	—
SS-D2	139	65	36	21	13	8	4	2	1
SS-E1	63	—	—	—	—	—	—	—	—

MINNEAPOLIS–ST. PAUL, MINNESOTA								8159 DD	
SSF =	.10	.20	.30	.40	.50	.60	.70	.80	.90
WW-A3	64	22	8	—	—	—	—	—	—
WW-A6	56	26	14	7	—	—	—	—	—
WW-B2	73	34	20	12	8	5	3	1	—
WW-B4	91	53	34	24	17	12	9	6	4
WW-C3	109	68	47	34	25	19	14	10	6
TW-A3	66	21	8	—	—	—	—	—	—
TW-E3	115	66	43	30	21	15	11	8	5
TW-F3	45	15	—	—	—	—	—	—	—
TW-F4	39	16	7	—	—	—	—	—	—
TW-G3	34	13	6	—	—	—	—	—	—
TW-G4	29	12	6	3	—	—	—	—	—
TW-I2	59	27	15	9	6	3	2	—	—
TW-I3	65	31	18	11	7	5	3	2	—
TW-I4	77	42	26	18	13	9	6	4	3
TW-J2	81	43	27	18	12	9	6	4	2
DG-A1	—	—	—	—	—	—	—	—	—
DG-A2	59	20	—	—	—	—	—	—	—
DG-A3	99	43	25	16	10	7	4	—	—
DG-B2	61	22	9	—	—	—	—	—	—
DG-C3	122	54	32	21	15	11	8	5	2
SS-A1	158	35	12	—	—	—	—	—	—
SS-A2	174	67	36	22	14	9	6	4	2
SS-A5	269	29	—	—	—	—	—	—	—
SS-B1	98	23	—	—	—	—	—	—	—
SS-B3	76	—	—	—	—	—	—	—	—
SS-B5	124	12	—	—	—	—	—	—	—
SS-C4	79	33	18	11	7	4	2	1	—
SS-D1	139	42	16	—	—	—	—	—	—
SS-D2	153	74	44	28	19	13	9	6	3
SS-E1	84	19	—	—	—	—	—	—	—

MERIDIAN, MISSISSIPPI								2388 DD	
SSF =	.10	.20	.30	.40	.50	.60	.70	.80	.90
WW-A3	312	149	90	61	43	31	23	16	11
WW-A6	264	150	98	69	50	37	28	20	14
WW-B2	279	150	95	66	48	35	26	19	13
WW-B4	282	171	116	84	62	47	35	26	18
WW-C3	299	193	135	100	75	57	44	33	22
TW-A3	306	132	78	51	36	26	19	13	9
TW-E3	331	197	133	95	70	53	40	29	20
TW-F3	225	108	66	44	31	23	17	12	8
TW-F4	184	96	61	42	30	22	16	12	8
TW-G3	165	84	52	35	25	19	14	10	7
TW-G4	133	69	43	30	21	16	12	8	6
TW-I2	228	119	75	51	37	27	20	15	10
TW-I3	242	130	83	58	42	31	23	17	11
TW-I4	240	139	92	65	48	36	27	20	13
TW-J2	261	148	97	68	50	37	28	20	14
DG-A1	223	99	56	34	21	11	—	—	—
DG-A2	240	110	67	45	31	21	14	9	4
DG-A3	288	135	84	59	43	32	24	16	9
DG-B2	245	114	70	49	35	26	19	13	7
DG-C3	333	161	100	70	53	41	32	24	16
SS-A1	628	197	104	65	44	31	22	15	10
SS-A2	560	242	142	93	65	47	34	24	16
SS-A5	1050	198	96	57	37	26	18	12	8
SS-B1	432	150	81	51	35	25	18	12	8
SS-B3	403	139	75	47	32	22	16	11	7
SS-B5	561	135	68	41	27	19	13	9	6
SS-C4	289	138	84	56	40	29	21	15	10
SS-D1	580	236	134	86	59	42	30	21	14
SS-D2	497	263	167	114	82	60	45	32	21
SS-E1	439	186	107	70	48	34	25	17	11

SAINT LOUIS, MISSOURI								4750 DD	
SSF =	.10	.20	.30	.40	.50	.60	.70	.80	.90
WW-A3	156	70	40	26	18	12	9	6	4
WW-A6	132	72	45	31	22	16	11	8	5
WW-B2	148	76	47	32	22	16	12	8	5
WW-B4	159	95	63	45	33	25	18	13	9
WW-C3	177	112	78	57	42	32	24	18	12
TW-A3	155	62	35	22	15	10	7	5	3
TW-E3	192	112	74	52	38	29	21	16	10
TW-F3	112	50	29	19	13	9	6	4	3
TW-F4	92	46	28	18	13	9	6	4	3
TW-G3	83	39	23	15	11	8	5	4	2
TW-G4	67	33	20	13	9	7	5	3	2
TW-I2	121	60	37	24	17	12	9	6	4
TW-I3	129	67	42	28	20	14	10	7	5
TW-I4	136	76	50	35	25	19	14	10	7
TW-J2	146	80	51	36	26	19	14	10	7
DG-A1	93	32	—	—	—	—	—	—	—
DG-A2	126	53	30	18	11	5	—	—	—
DG-A3	168	76	46	31	22	16	11	7	—
DG-B2	129	56	32	21	14	9	5	—	—
DG-C3	199	92	56	38	28	22	17	12	8
SS-A1	335	94	46	27	17	12	8	5	3
SS-A2	314	128	72	46	31	22	16	11	7
SS-A5	569	92	40	22	13	9	6	4	2
SS-B1	224	70	35	21	13	9	6	4	2
SS-B3	200	61	30	17	11	7	5	3	2
SS-B5	292	60	27	15	9	6	4	2	1
SS-C4	156	71	42	27	19	13	10	7	4
SS-D1	302	112	59	36	24	16	11	8	5
SS-D2	275	140	86	57	40	29	21	15	10
SS-E1	215	83	44	27	18	12	8	5	3

TABLE C.2 **Annual Passive Heating Performance: LCR (Continued)**

HELENA, MONTANA 8190 DD

SSF =	.10	.20	.30	.40	.50	.60	.70	.80	.90
WW-A3	108	46	25	14	8	3	—	—	—
WW-A6	93	48	29	18	11	7	4	2	—
WW-B2	108	54	33	21	14	9	6	4	2
WW-B4	123	73	48	33	24	17	12	8	5
WW-C3	141	89	61	44	33	24	18	13	8
TW-A3	108	42	22	12	7	3	—	—	—
TW-E3	151	88	58	40	29	21	15	11	7
TW-F3	77	33	18	10	5	—	—	—	—
TW-F4	64	31	17	10	6	3	1	—	—
TW-G3	57	26	15	9	5	3	—	—	—
TW-G4	47	23	13	8	5	3	1	—	—
TW-I2	88	43	25	16	11	7	4	3	1
TW-I3	95	48	29	19	13	9	6	3	2
TW-14	104	58	37	26	18	13	9	6	4
TW-J2	112	61	38	26	18	13	9	6	4
DG-A1	51	—	—	—	—	—	—	—	—
DG-A2	90	37	19	9	—	—	—	—	—
DG-A3	131	59	35	23	16	11	7	3	—
DG-B2	92	40	22	12	6	—	—	—	—
DG-C3	155	72	44	30	21	15	11	7	4
SS-A1	240	64	29	14	7	—	—	—	—
SS-A2	238	96	53	32	21	14	9	6	3
SS-A5	405	60	23	8	—	—	—	—	—
SS-B1	156	46	21	10	—	—	—	—	—
SS-B3	134	36	14	—	—	—	—	—	—
SS-B5	200	37	14	—	—	—	—	—	—
SS-C4	115	51	29	18	12	8	5	3	2
SS-D1	214	76	37	20	10	—	—	—	—
SS-D2	209	105	63	41	27	19	13	8	5
SS-E1	146	52	23	9	—	—	—	—	—

NORTH OMAHA, NEBRASKA 6601 DD

SSF =	.10	.20	.30	.40	.50	.60	.70	.80	.90
WW-A3	119	52	29	19	13	9	6	4	2
WW-A6	102	55	34	23	16	12	8	6	4
WW-B2	118	60	37	25	17	13	9	6	4
WW-B4	132	78	52	37	27	20	15	11	7
WW-C3	149	95	66	48	36	27	21	15	10
TW-A3	120	47	26	16	11	7	5	3	2
TW-E3	161	94	62	44	32	24	18	13	9
TW-F3	85	37	21	13	9	6	4	3	2
TW-F4	71	34	21	13	9	7	5	3	2
TW-G3	64	30	17	11	8	5	4	3	2
TW-G4	52	25	15	10	7	5	3	2	1
TW-I2	96	47	29	19	13	10	7	5	3
TW-I3	104	53	33	22	16	11	8	6	4
TW-I4	112	63	41	28	21	15	11	8	5
TW-J2	120	65	42	29	21	15	11	8	5
DG-A1	61	—	—	—	—	—	—	—	—
DG-A2	99	41	22	12	6	—	—	—	—
DG-A3	140	62	37	25	18	13	9	5	—
DG-B2	101	42	24	15	10	5	—	—	—
DG-C3	168	76	46	32	23	18	14	10	6
SS-A1	261	70	33	19	12	8	5	3	2
SS-A2	255	103	57	36	25	17	12	8	5
SS-A5	447	67	28	14	8	5	3	2	—
SS-B1	171	51	25	14	9	6	4	2	1
SS-B3	148	42	19	10	6	3	2	—	—
SS-B5	222	43	18	9	5	3	2	—	—
SS-C4	125	56	33	21	15	10	7	5	3
SS-D1	233	84	43	26	17	11	7	5	3
SS-D2	223	112	69	46	32	23	17	12	8
SS-E1	161	59	30	18	11	7	5	3	1

ELY, NEVADA 7814 DD

SSF =	.10	.20	.30	.40	.50	.60	.70	.80	.90
WW-A3	185	88	54	36	26	19	14	10	7
WW-A6	159	91	60	42	31	23	17	13	8
WW-B2	173	93	60	42	30	23	17	12	8
WW-B4	184	112	77	56	42	32	24	18	12
WW-C3	203	132	93	69	53	40	31	23	16
TW-A3	181	78	46	31	22	16	11	8	5
TW-E3	221	132	90	65	48	37	28	20	14
TW-F3	133	64	39	26	19	14	10	7	5
TW-F4	110	58	37	25	18	13	10	7	5
TW-G3	99	50	31	21	15	11	8	6	4
TW-G4	80	42	26	18	13	10	7	5	3
TW-I2	141	73	47	32	23	17	13	9	6
TW-I3	152	82	53	37	27	20	15	11	7
TW-I4	156	91	61	43	32	24	18	13	9
TW-J2	168	96	63	45	33	25	19	14	9
DG-A1	117	49	25	10	—	—	—	—	—
DG-A2	143	66	40	26	18	11	7	—	—
DG-A3	184	87	55	39	29	21	15	10	5
DG-B2	145	68	42	29	21	15	10	6	—
DG-C3	217	103	66	47	36	28	22	17	11
SS-A1	406	125	66	41	27	19	13	9	5
SS-A2	378	163	95	63	44	31	22	16	10
SS-A5	668	123	59	35	23	15	10	7	4
SS-B1	270	92	50	31	21	15	10	7	4
SS-B3	246	83	45	28	18	13	9	6	3
SS-B5	344	81	40	24	16	11	7	5	3
SS-C4	179	86	52	36	26	19	14	10	7
SS-D1	375	150	85	55	37	26	18	12	8
SS-D2	337	178	113	78	56	41	30	21	14
SS-E1	270	113	65	42	29	20	14	10	6

LAS VEGAS, NEVADA 2601 DD

SSF =	.10	.20	.30	.40	.50	.60	.70	.80	.90
WW-A3	522	250	154	105	75	55	41	30	20
WW-A6	437	250	165	117	87	65	49	36	24
WW-B2	455	245	158	110	81	60	45	33	22
WW-B4	444	270	184	134	100	76	58	43	29
WW-C3	458	296	209	154	117	90	69	51	35
TW-A3	511	222	132	88	63	46	34	24	16
TW-E3	513	306	207	149	111	84	64	47	32
TW-F3	376	182	112	76	55	41	30	22	14
TW-F4	307	161	103	71	52	32	29	21	14
TW-G3	276	140	88	61	44	33	24	18	12
TW-G4	220	115	73	51	37	27	20	15	10
TW-I2	373	194	124	86	62	46	35	25	17
TW-I3	389	211	136	95	70	52	39	28	19
TW-I4	380	220	147	105	78	59	44	33	22
TW-J2	415	235	155	110	81	61	46	34	23
DG-A1	393	179	108	71	49	33	22	13	4
DG-A2	394	182	113	78	56	41	29	20	12
DG-A3	454	211	133	94	70	53	40	28	17
DG-B2	405	188	118	84	63	48	36	26	16
DG-C3	527	246	156	112	86	67	53	41	28
SS-A1	967	304	162	102	69	49	35	24	16
SS-A2	833	361	212	140	98	71	51	37	24
SS-A5	1628	309	151	91	60	42	30	20	13
SS-B1	670	233	128	82	56	40	29	20	13
SS-B3	623	217	119	75	52	37	26	18	12
SS-B5	877	213	108	67	45	31	22	15	10
SS-C4	468	224	137	94	67	50	37	27	18
SS-D1	890	362	208	135	94	67	48	34	22
SS-D2	736	390	248	171	123	91	67	48	32
SS-E1	682	292	170	112	78	56	41	29	19

TABLE C.2 **Annual Passive Heating Performance: LCR (Continued)**

ALBUQUERQUE, NEW MEXICO — *4292 DD*

SSF =	.10	.20	.30	.40	.50	.60	.70	.80	.90
WW-A3	300	146	90	62	45	33	24	18	12
WW-A6	256	148	99	70	52	39	30	22	15
WW-B2	270	147	96	67	49	37	28	20	14
WW-B4	275	169	116	85	64	49	37	28	19
WW-C3	293	191	136	101	77	59	46	34	24
TW-A3	293	129	77	52	37	27	20	15	10
TW-E3	323	195	133	96	72	55	42	31	21
TW-F3	216	106	66	45	33	24	18	13	9
TW-F4	178	95	61	42	31	23	17	13	9
TW-G3	159	82	52	36	26	20	15	11	7
TW-G4	128	68	43	30	22	16	12	9	6
TW-I2	221	117	75	52	38	28	21	16	11
TW-I3	234	128	83	59	43	32	24	18	12
TW-I4	234	137	92	66	49	37	28	21	14
TW-J2	254	146	97	69	51	39	29	22	15
DG-A1	211	97	57	36	22	13	5	—	—
DG-A2	227	107	67	46	32	23	16	10	5
DG-A3	274	131	83	59	44	34	25	18	10
DG-B2	232	110	69	49	37	28	21	14	8
DG-C3	318	155	98	71	54	43	34	26	18
SS-A1	591	187	101	64	44	31	22	16	10
SS-A2	531	232	137	92	65	47	34	25	16
SS-A5	980	187	92	56	37	26	18	13	8
SS-B1	403	141	78	50	35	25	18	13	8
SS-B3	372	130	71	46	31	22	16	11	7
SS-B5	518	127	65	40	27	19	13	9	6
SS-C4	277	135	83	57	41	31	23	17	11
SS-D1	548	225	130	85	59	43	31	22	14
SS-D2	474	253	162	113	82	61	45	33	22
SS-E1	410	176	103	68	48	35	25	18	12

NEW YORK (CENTRAL PK), NEW YORK — *4848 DD*

SSF =	.10	.20	.30	.40	.50	.60	.70	.80	.90
WW-A3	104	45	25	16	10	7	5	3	2
WW-A6	90	48	30	20	14	10	7	5	3
WW-B2	105	54	33	22	15	11	8	5	3
WW-B4	121	72	48	34	25	18	14	10	7
WW-C3	139	88	61	45	33	25	19	14	10
TW-A3	104	41	22	13	9	6	4	2	1
TW-E3	148	87	57	41	30	22	16	12	8
TW-F3	74	33	18	11	7	5	3	2	1
TW-F4	63	30	18	12	8	5	4	2	1
TW-G3	56	26	15	10	6	4	3	2	1
TW-G4	46	22	13	9	6	4	3	2	1
TW-I2	86	42	25	17	12	8	6	4	3
TW-I3	93	48	29	20	14	10	7	5	3
TW-I4	102	58	37	26	19	14	10	7	5
TW-J2	109	60	38	26	19	14	10	7	5
DG-A1	47	—	—	—	—	—	—	—	—
DG-A2	86	36	19	10	—	—	—	—	—
DG-A3	127	57	34	23	16	11	8	4	—
DG-B2	89	37	21	13	8	—	—	—	—
DG-C3	151	70	43	29	22	16	12	9	5
SS-A1	229	64	31	17	11	7	4	3	1
SS-A2	233	96	54	34	23	16	11	8	5
SS-A5	374	60	25	13	7	4	2	—	—
SS-B1	152	47	23	13	8	5	3	2	—
SS-B3	131	38	17	9	5	2	—	—	—
SS-B5	190	38	16	8	4	2	—	—	—
SS-C4	111	50	29	19	13	9	6	4	3
SS-D1	209	77	40	24	15	10	6	4	2
SS-D2	207	106	65	43	30	22	16	11	7
SS-E1	145	54	27	16	9	6	3	2	—

BUFFALO, NEW YORK — *6927 DD*

SSF =	.10	.20	.30	.40	.50	.60	.70	.80	.90
WW-A3	34	—	—	—	—	—	—	—	—
WW-A6	32	—	—	—	—	—	—	—	—
WW-B2	52	22	10	—	—	—	—	—	—
WW-B4	74	41	26	18	12	8	6	4	2
WW-C3	93	57	39	28	20	15	11	8	5
TW-A3	40	—	—	—	—	—	—	—	—
TW-E3	97	54	34	23	16	12	8	6	4
TW-F3	23	—	—	—	—	—	—	—	—
TW-F4	22	—	—	—	—	—	—	—	—
TW-G3	19	—	—	—	—	—	—	—	—
TW-G4	18	—	—	—	—	—	—	—	—
TW-I2	42	17	7	—	—	—	—	—	—
TW-I3	49	21	11	5	—	—	—	—	—
TW-I4	62	33	20	13	9	6	4	3	2
TW-J2	65	33	19	12	8	5	4	2	1
DG-A1	—	—	—	—	—	—	—	—	—
DG-A2	41	—	—	—	—	—	—	—	—
DG-A3	84	34	19	11	7	4	—	—	—
DG-B2	43	—	—	—	—	—	—	—	—
DG-C3	104	44	26	16	11	8	5	3	—
SS-A1	139	24	—	—	—	—	—	—	—
SS-A2	158	58	30	17	10	6	4	2	1
SS-A5	246	—	—	—	—	—	—	—	—
SS-B1	81	—	—	—	—	—	—	—	—
SS-B3	58	—	—	—	—	—	—	—	—
SS-B5	104	—	—	—	—	—	—	—	—
SS-C4	59	22	10	4	—	—	—	—	—
SS-D1	117	28	—	—	—	—	—	—	—
SS-D2	137	64	37	23	15	9	6	4	2
SS-E1	60	—	—	—	—	—	—	—	—

CHARLOTTE, NORTH CAROLINA — *3218 DD*

SSF =	.10	.20	.30	.40	.50	.60	.70	.80	.90
WW-A3	256	120	72	48	34	25	18	13	8
WW-A6	217	121	79	55	40	30	22	16	11
WW-B2	232	123	78	54	39	28	21	15	10
WW-B4	238	143	97	70	52	39	29	21	15
WW-C3	254	163	114	84	63	48	37	27	19
TW-A3	253	107	62	41	28	20	15	10	7
TW-E3	281	166	111	79	59	44	33	24	17
TW-F3	185	87	52	35	25	18	13	9	6
TW-F4	151	78	49	33	24	17	13	9	6
TW-G3	136	68	42	28	20	15	11	8	5
TW-G4	109	56	35	24	17	12	9	7	4
TW-I2	190	97	61	42	30	22	16	12	8
TW-I3	201	107	68	47	34	25	18	13	9
TW-14	202	116	77	54	40	30	22	16	11
TW-J2	220	123	80	56	41	31	23	17	11
DG-A1	178	75	41	23	11	—	—	—	—
DG-A2	199	89	54	35	24	16	10	6	—
DG-A3	245	113	70	48	35	26	19	13	7
DG-B2	203	93	56	39	28	20	14	9	4
DG-C3	287	135	84	59	44	34	27	20	13
SS-A1	524	157	82	50	34	23	17	11	7
SS-A2	469	198	114	75	52	37	27	19	13
SS-A5	886	157	74	44	28	19	13	9	6
SS-B1	358	119	64	40	27	19	13	9	6
SS-B3	330	109	58	36	24	17	12	8	5
SS-B5	468	106	52	31	20	14	10	7	4
SS-C4	241	113	68	46	32	24	17	12	8
SS-D1	478	187	105	67	46	32	23	16	10
SS-D2	412	215	135	92	66	48	35	26	17
SS-E1	356	146	83	53	37	26	18	13	8

TABLE C.2 **Annual Passive Heating Performance: LCR (Continued)**

FARGO, NORTH DAKOTA — *9271 DD*

SSF =	*.10*	*.20*	*.30*	*.40*	*.50*	*.60*	*.70*	*.80*	*.90*
WW-A3	55	14	—	—	—	—	—	—	—
WW-A6	49	20	—	—	—	—	—	—	—
WW-B2	66	30	16	9	5	—	—	—	—
WW-B4	86	49	31	21	15	10	7	5	3
WW-C3	104	64	44	31	23	17	12	9	6
TW-A3	58	15	—	—	—	—	—	—	—
TW-E3	109	62	39	27	19	14	10	7	4
TW-F3	39	8	—	—	—	—	—	—	—
TW-F4	34	12	—	—	—	—	—	—	—
TW-G3	30	9	—	—	—	—	—	—	—
TW-G4	26	10	—	—	—	—	—	—	—
TW-I2	54	23	12	6	3	—	—	—	—
TW-I3	60	28	15	9	5	3	1	—	—
TW-I4	72	39	24	16	11	8	5	4	2
TW-J2	76	40	24	16	10	7	5	3	2
DG-A1	—	—	—	—	—	—	—	—	—
DG-A2	54	16	—	—	—	—	—	—	—
DG-A3	95	40	23	14	9	5	—	—	—
DG-B2	56	18	—	—	—	—	—	—	—
DG-C3	116	51	30	19	13	9	6	4	—
SS-A1	142	28	—	—	—	—	—	—	—
SS-A2	162	61	31	18	11	7	4	3	1
SS-A5	241	18	—	—	—	—	—	—	—
SS-B1	87	14	—	—	—	—	—	—	—
SS-B3	63	—	—	—	—	—	—	—	—
SS-B5	108	—	—	—	—	—	—	—	—
SS-C4	73	29	15	8	4	2	—	—	—
SS-D1	123	33	—	—	—	—	—	—	—
SS-D2	141	67	39	24	16	10	7	4	2
SS-E1	70	—	—	—	—	—	—	—	—

CLEVELAND, OHIO — *6154 DD*

SSF =	*.10*	*.20*	*.30*	*.40*	*.50*	*.60*	*.70*	*.80*	*.90*
WW-A3	57	16	—	—	—	—	—	—	—
WW-A6	50	21	8	—	—	—	—	—	—
WW-B2	68	30	16	9	5	3	—	—	—
WW-B4	87	49	32	21	15	11	8	5	3
WW-C3	106	65	44	32	23	17	13	9	6
TW-A3	61	16	—	—	—	—	—	—	—
TW-E3	112	63	40	28	20	14	10	7	5
TW-F3	40	10	—	—	—	—	—	—	—
TW-F4	35	12	—	—	—	—	—	—	—
TW-G3	31	10	—	—	—	—	—	—	—
TW-G4	27	10	4	—	—	—	—	—	—
TW-I2	55	24	12	7	4	2	—	—	—
TW-I3	62	29	16	9	6	3	2	1	—
TW-I4	74	39	24	16	11	8	5	4	2
TW-J2	78	40	24	16	11	7	5	3	2
DG-A1	—	—	—	—	—	—	—	—	—
DG-A2	56	16	—	—	—	—	—	—	—
DG-A3	98	41	23	14	9	6	2	—	—
DG-B2	58	18	—	—	—	—	—	—	—
DG-C3	120	52	30	20	14	10	7	4	—
SS-A1	177	38	11	—	—	—	—	—	—
SS-A2	186	70	37	22	14	9	6	4	2
SS-A5	314	32	—	—	—	—	—	—	—
SS-B1	108	23	—	—	—	—	—	—	—
SS-B3	87	—	—	—	—	—	—	—	—
SS-B5	142	11	—	—	—	—	—	—	—
SS-C4	75	30	15	8	5	3	1	—	—
SS-D1	151	44	16	—	—	—	—	—	—
SS-D2	161	77	44	28	19	13	8	6	3
SS-E1	91	17	—	—	—	—	—	—	—

CINCINNATI, OHIO — *5070 DD*

SSF =	*.10*	*.20*	*.30*	*.40*	*.50*	*.60*	*.70*	*.80*	*.90*
WW-A3	99	42	23	14	9	6	4	2	1
WW-A6	85	44	27	17	12	8	6	4	2
WW-B2	101	50	30	20	14	10	7	5	3
WW-B4	117	69	45	32	23	17	13	9	6
WW-C3	135	85	59	43	32	24	18	13	9
TW-A3	100	38	20	12	7	5	3	2	—
TW-E3	144	84	55	39	28	21	15	11	7
TW-F3	70	30	16	10	6	4	2	1	—
TW-F4	59	28	16	10	7	4	3	2	1
TW-G3	53	24	13	8	5	4	2	1	—
TW-G4	44	21	12	8	5	3	2	1	—
TW-I2	82	40	24	15	10	7	5	3	2
TW-I3	90	45	27	18	13	9	6	4	3
TW-I4	99	55	35	24	18	13	9	7	4
TW-J2	105	57	36	25	18	13	9	7	4
DG-A1	40	—	—	—	—	—	—	—	—
DG-A2	84	33	17	8	—	—	—	—	—
DG-A3	124	55	33	22	15	10	7	4	—
DG-B2	86	35	19	11	6	—	—	—	—
DG-C3	148	68	41	28	20	15	11	8	5
SS-A1	233	63	29	16	9	6	3	2	1
SS-A2	234	95	52	33	22	15	11	7	4
SS-A5	390	59	23	11	6	3	—	—	—
SS-B1	153	45	21	11	7	4	2	1	—
SS-B3	132	36	15	7	3	—	—	—	—
SS-B5	194	37	14	6	3	—	—	—	—
SS-C4	107	47	27	17	12	8	6	4	2
SS-D1	210	75	38	22	13	9	5	3	2
SS-D2	206	104	63	42	29	20	15	10	6
SS-E1	143	51	25	13	8	4	2	—	—

COLUMBUS, OHIO — *5702 DD*

SSF =	*.10*	*.20*	*.30*	*.40*	*.50*	*.60*	*.70*	*.80*	*.90*
WW-A3	75	29	14	7	2	—	—	—	—
WW-A6	66	32	18	11	7	4	2	1	—
WW-B2	82	40	23	15	10	7	4	3	2
WW-B4	99	58	38	26	19	14	10	7	5
WW-C3	118	74	51	37	27	20	15	11	7
TW-A3	77	27	12	6	2	—	—	—	—
TW-E3	125	72	47	33	23	17	13	9	6
TW-F3	53	21	10	5	—	—	—	—	—
TW-F4	45	20	11	6	3	2	—	—	—
TW-G3	40	17	9	5	2	—	—	—	—
TW-G4	34	15	8	5	3	2	—	—	—
TW-I2	66	31	18	11	7	5	3	2	1
TW-I3	73	36	21	14	9	6	4	3	2
TW-I4	84	46	29	20	14	10	7	5	3
TW-J2	89	47	30	20	14	10	7	5	3
DG-A1	—	—	—	—	—	—	—	—	—
DG-A2	67	24	10	—	—	—	—	—	—
DG-A3	107	47	27	18	12	8	5	—	—
DG-B2	69	26	13	5	—	—	—	—	—
DG-C3	130	59	35	23	17	12	9	6	3
SS-A1	192	48	21	10	5	—	—	—	—
SS-A2	202	80	44	27	18	12	8	5	3
SS-A5	324	44	14	—	—	—	—	—	—
SS-B1	123	33	14	6	—	—	—	—	—
SS-B3	103	24	—	—	—	—	—	—	—
SS-B5	155	25	5	—	—	—	—	—	—
SS-C4	88	38	21	13	8	6	4	2	1
SS-D1	172	58	27	14	8	4	1	—	—
SS-D2	177	88	53	34	23	16	11	8	5
SS-E1	111	35	14	—	—	—	—	—	—

TABLE C.2 **Annual Passive Heating Performance: LCR (Continued)**

OKLAHOMA CITY, OKLAHOMA									*3695 DD*
SSF =	*.10*	*.20*	*.30*	*.40*	*.50*	*.60*	*.70*	*.80*	*.90*
WW-A3	241	113	68	46	32	24	17	12	8
WW-A6	204	114	75	52	38	28	21	15	10
WW-B2	219	116	74	51	37	27	20	15	10
WW-B4	226	136	92	67	50	37	28	21	14
WW-C3	242	156	109	81	61	47	36	27	18
TW-A3	238	101	58	38	27	19	14	10	7
TW-E3	267	158	106	76	56	43	32	24	16
TW-F3	173	82	49	33	24	17	13	9	6
TW-F4	142	73	46	31	23	17	12	9	6
TW-G3	128	64	39	27	19	14	10	7	5
TW-G4	103	53	33	23	16	12	9	6	4
TW-I2	180	92	58	40	29	21	16	11	8
TW-I3	190	101	64	45	32	24	18	13	9
TW-I4	192	110	73	52	38	29	22	16	11
TW-J2	209	117	76	54	39	29	22	16	11
DG-A1	164	69	38	21	9	—	—	—	—
DG-A2	187	84	50	33	23	15	10	5	—
DG-A3	233	107	66	46	34	25	18	13	7
DG-B2	192	87	53	37	27	20	14	9	4
DG-C3	273	127	79	56	42	33	26	20	13
SS-A1	488	146	76	47	32	22	16	11	7
SS-A2	440	186	108	71	49	36	26	18	12
SS-A5	824	146	69	41	27	18	13	9	6
SS-B1	332	111	59	37	25	18	13	9	6
SS-B3	304	100	53	33	22	16	11	8	5
SS-B5	434	98	48	29	19	13	9	6	4
SS-C4	228	107	65	44	31	23	17	12	8
SS-D1	445	175	98	63	43	31	22	16	10
SS-D2	387	202	128	88	63	46	34	25	16
SS-E1	330	136	77	50	34	24	18	12	8

SALEM, OREGON									*4852 DD*
SSF =	*.10*	*.20*	*.30*	*.40*	*.50*	*.60*	*.70*	*.80*	*.90*
WW-A3	170	71	38	23	14	8	4	1	
WW-A5	145	71	42	26	17	11	6	3	1
WW-A6	140	71	43	27	18	11	7	4	2
WW-B2	158	77	46	29	20	13	8	5	3
WW-B4	167	96	62	43	30	22	15	10	6
WW-C3	184	114	77	55	40	30	22	15	10
TW-A3	171	64	33	20	12	7	3	1	—
TW-E3	204	115	74	51	36	26	19	13	8
TW-F3	122	51	28	16	10	6	3	—	—
TW-F4	99	46	26	16	10	6	3	2	—
TW-G3	89	40	22	14	8	5	3	1	—
TW-G4	72	34	19	12	8	5	3	1	—
TW-I2	130	61	36	23	15	10	6	4	2
TW-I3	139	68	41	26	18	12	8	5	3
TW-I4	143	78	49	33	23	16	12	8	5
TW-J2	155	82	51	34	24	17	11	8	4
DG-A1	107	30	—	—	—	—	—	—	—
DG-A2	140	56	29	16	8	—	—	—	—
DG-A3	186	79	46	30	20	14	9	5	—
DG-B2	145	59	33	20	11	5	—	—	—
DG-C3	224	96	57	39	27	20	14	9	5
SS-A1	402	106	48	26	14	7	3	—	—
SS-A5	683	106	42	20	10	—	—	—	—
SS-B1	268	78	36	19	10	5	—	—	—
SS-B3	244	68	30	15	6	—	—	—	—
SS-B5	352	69	28	13	6	—	—	—	—
SS-C4	169	73	41	26	17	11	7	4	2
SS-D1	357	124	61	34	20	11	5	2	—
SS-D2	311	150	88	57	38	26	17	11	6
SS-E1	256	91	44	24	12	—	—	—	—

MEDFORD, OREGON									*4930 DD*
SSF =	*.10*	*.20*	*.30*	*.40*	*.50*	*.60*	*.70*	*.80*	*.90*
WW-A3	174	75	41	25	16	9	5	2	—
WW-A6	144	75	46	30	20	13	8	4	2
WW-B2	162	80	48	31	21	14	9	6	3
WW-B4	171	99	65	45	32	23	16	11	7
WW-C3	189	117	80	57	42	31	23	16	10
TW-A3	175	67	36	21	13	8	4	2	—
TW-E3	208	118	77	53	38	27	20	13	8
TW-F3	125	54	30	18	11	7	3	1	—
TW-F4	102	48	28	18	11	7	4	2	1
TW-G3	92	42	24	15	9	6	3	2	—
TW-G4	74	35	20	13	8	5	3	2	—
TW-I2	133	64	38	24	16	11	7	4	2
TW-I3	142	71	43	28	19	13	9	5	3
TW-I4	146	80	51	35	25	17	12	8	5
TW-J2	158	85	53	36	25	18	12	8	5
DG-A1	110	35	—	—	—	—	—	—	—
DG-A2	142	58	32	18	9	—	—	—	—
DG-A3	187	82	48	32	22	15	9	5	—
DG-B2	146	61	35	21	13	7	—	—	—
DG-C3	224	98	60	41	29	21	14	10	5
SS-A1	415	110	51	28	16	9	4	2	—
SS-A2	372	146	79	48	31	21	14	8	5
SS-A5	732	111	45	23	12	5	—	—	—
SS-B1	274	81	38	21	12	6	3	—	—
SS-B3	249	71	33	17	8	—	—	—	—
SS-B5	368	72	30	15	7	—	—	—	—
SS-C4	173	76	43	27	18	12	8	5	3
SS-D1	367	129	65	37	22	13	7	3	1
SS-D2	318	156	92	60	40	27	18	12	7
SS-E1	262	95	48	27	15	7	—	—	—

HARRISBURG, PENNSYLVANIA									*5224 DD*
SSF =	*.10*	*.20*	*.30*	*.40*	*.50*	*.60*	*.70*	*.80*	*.90*
WW-A3	106	47	26	17	11	7	5	3	2
WW-A6	92	50	31	21	14	10	7	5	3
WW-B2	107	55	34	23	16	11	8	6	4
WW-B4	122	73	49	35	25	19	14	10	7
WW-C3	140	89	62	45	34	26	20	14	10
TW-A3	106	42	23	14	9	6	4	3	2
TW-E3	150	88	58	41	30	23	17	12	8
TW-F3	76	34	19	12	8	5	4	2	1
TW-F4	64	31	19	12	8	6	4	3	2
TW-G3	57	27	16	10	7	5	3	2	1
TW-G4	47	23	14	9	6	4	3	2	1
TW-I2	87	43	26	17	12	9	6	4	3
TW-I3	94	49	30	20	14	10	7	5	3
TW-I4	103	58	38	27	19	14	10	8	5
TW-J2	110	61	39	27	19	14	10	7	5
DG-A1	49	—	—	—	—	—	—	—	—
DG-A2	88	37	20	11	—	—	—	—	—
DG-A3	127	58	35	24	17	12	8	5	—
DG-B2	90	39	22	14	8	3	—	—	—
DG-C3	152	71	44	30	22	17	13	9	5
SS-A1	231	65	31	18	11	7	5	3	1
SS-A2	235	97	55	35	24	17	12	8	5
SS-A5	373	61	26	13	8	4	2	1	—
SS-B1	153	47	23	13	8	5	3	2	1
SS-B3	133	39	18	10	5	3	1	—	—
SS-B5	191	39	17	8	5	2	1	—	—
SS-C4	113	51	30	20	13	9	7	5	3
SS-D1	212	78	41	25	16	10	7	4	2
SS-D2	209	107	66	44	31	22	16	11	7
SS-E1	146	55	28	16	10	6	4	2	1

TABLE C.2 **Annual Passive Heating Performance: LCR (Continued)**

PHILADELPHIA, PENNSYLVANIA									*4865 DD*
SSF =	*.10*	*.20*	*.30*	*.40*	*.50*	*.60*	*.70*	*.80*	*.90*
WW-A3	125	57	33	21	14	10	7	5	3
WW-A6	109	59	38	26	18	13	9	7	4
WW-B2	123	64	40	27	19	14	10	7	5
WW-B4	137	82	55	39	29	22	16	12	8
WW-C3	155	99	69	51	38	29	22	16	11
TW-A3	125	51	28	18	12	8	6	4	2
TW-E3	167	98	66	47	34	26	19	14	9
TW-F3	90	41	24	15	10	7	5	3	2
TW-F4	75	38	23	15	10	7	5	4	2
TW-G3	67	32	19	13	9	6	4	3	2
TW-G4	55	27	17	11	8	5	4	3	2
TW-I2	100	50	31	21	15	11	8	5	3
TW-I3	108	57	35	24	17	12	9	6	4
TW-I4	116	66	43	30	22	16	12	9	6
TW-J2	125	69	45	31	22	17	12	9	6
DG-A1	67	16	—	—	—	—	—	—	—
DG-A2	102	44	25	14	8	—	—	—	—
DG-A3	142	65	40	27	19	14	10	6	—
DG-B2	104	46	27	17	11	7	—	—	—
DG-C3	168	79	49	34	25	19	15	11	7
SS-A1	266	77	38	23	14	9	6	4	2
SS-A2	263	110	63	40	28	19	14	10	6
SS-A5	432	74	33	18	11	7	4	3	1
SS-B1	178	57	29	17	11	7	5	3	2
SS-B3	157	48	24	14	8	5	3	2	
SS-B5	224	48	22	12	7	4	3	2	—
SS-C4	129	59	35	23	16	11	8	6	4
SS-D1	244	93	50	31	20	13	9	6	4
SS-D2	234	121	75	51	36	26	19	13	8
SS-E1	173	67	36	22	14	9	6	4	2

CHARLESTON, SOUTH CAROLINA									*2146 DD*
SSF =	*.10*	*.20*	*.30*	*.40*	*.50*	*.60*	*.70*	*.80*	*.90*
WW-A3	363	173	105	70	50	36	27	19	13
WW-A6	306	173	113	79	58	43	32	23	16
WW-B2	322	172	110	76	55	41	30	22	15
WW-B4	321	194	132	95	70	53	40	30	20
WW-C3	338	217	152	112	84	64	49	37	25
TW-A3	357	154	90	59	42	30	22	16	10
TW-E3	376	223	150	107	79	60	45	33	23
TW-F3	262	126	76	51	36	27	19	14	9
TW-F4	214	112	70	48	35	25	19	14	9
TW-G3	192	97	60	41	29	22	16	11	8
TW-G4	154	80	50	34	25	18	13	10	6
TW-I2	264	137	86	59	42	31	23	17	11
TW-I3	278	149	95	66	48	35	26	19	13
TW-I4	274	158	105	74	54	41	31	22	15
TW-J2	299	168	110	77	57	42	32	23	16
DG-A1	266	118	68	43	27	16	8	—	—
DG-A2	279	127	77	52	36	26	18	11	6
DG-A3	331	154	95	66	49	37	27	19	11
DG-B2	287	132	81	56	41	31	23	16	9
DG-C3	384	182	113	79	60	46	37	28	19
SS-A1	718	225	119	74	50	35	25	18	11
SS-A2	633	273	159	104	73	53	38	27	18
SS-A5	1205	227	109	65	43	30	21	15	9
SS-B1	497	172	93	59	40	28	20	14	9
SS-B3	464	160	86	54	37	26	19	13	8
SS-B5	647	156	78	47	31	22	16	11	7
SS-C4	333	158	96	65	46	34	25	18	12
SS-D1	663	268	152	98	68	49	35	25	16
SS-D2	560	295	186	128	92	68	50	36	24
SS-E1	505	214	123	80	55	40	29	20	13

PITTSBURGH, PENNSYLVANIA									*5930 DD*
SSF =	*.10*	*.20*	*.30*	*.40*	*.50*	*.60*	*.70*	*.80*	*.90*
WW-A3	65	22	7	—	—	—	—	—	—
WW-A6	57	26	13	7	—	—	—	—	—
WW-B2	74	35	19	12	7	5	3	2	—
WW-B4	93	53	34	24	17	12	9	6	4
WW-C3	111	69	47	34	25	19	14	10	7
TW-A3	68	21	7	—	—	—	—	—	—
TW-E3	117	67	43	30	21	15	11	8	5
TW-F3	46	16	—	—	—	—	—	—	—
TW-F4	39	16	7	—	—	—	—	—	—
TW-G3	35	13	6	—	—	—	—	—	—
TW-G4	30	12	6	3	—	—	—	—	—
TW-I2	60	27	15	9	5	3	2	1	—
TW-I3	67	32	18	11	7	5	3	2	1
TW-I4	78	42	27	18	13	9	6	4	3
TW-J2	83	43	27	18	12	9	6	4	3
DG-A1	—	—	—	—	—	—	—	—	—
DG-A2	60	20	—	—	—	—	—	—	—
DG-A3	101	43	25	16	10	7	4	—	—
DG-B2	62	22	9	—	—	—	—	—	—
DG-C3	124	55	32	21	15	11	8	5	2
SS-A1	183	43	16	5	—	—	—	—	—
SS-A2	193	74	40	24	15	10	7	4	3
SS-A5	321	38	—	—	—	—	—	—	—
SS-B1	115	28	9	—	—	—	—	—	—
SS-B3	94	17	—	—	—	—	—	—	—
SS-B5	149	20	—	—	—	—	—	—	—
SS-C4	81	33	18	11	6	4	2	1	—
SS-D1	160	51	22	9	—	—	—	—	—
SS-D2	168	82	48	31	21	14	10	7	4
SS-E1	100	28	—	—	—	—	—	—	—

RAPID CITY, SOUTH DAKOTA									*7324 DD*
SSF =	*.10*	*.20*	*.30*	*.40*	*.50*	*.60*	*.70*	*.80*	*.90*
WW-A3	134	60	35	22	15	10	7	5	3
WW-A6	116	63	39	27	19	13	9	6	4
WW-B2	130	68	42	28	20	14	10	7	4
WW-B4	144	86	57	41	30	22	16	12	8
WW-C3	161	103	71	52	39	29	22	16	11
TW-A3	133	54	30	19	12	8	6	4	2
TW-E3	174	102	68	48	35	26	19	14	9
TW-F3	96	44	25	16	11	7	5	3	2
TW-F4	80	40	24	16	11	7	5	4	2
TW-G3	72	34	20	13	9	6	4	3	2
TW-G4	59	29	17	11	8	6	4	3	2
TW-I2	106	53	32	22	15	11	8	5	3
TW-I3	114	60	37	25	17	13	9	6	4
TW-I4	122	69	45	31	23	17	12	9	6
TW-J2	131	72	46	32	23	17	12	9	6
DG-A1	75	22	—	—	—	—	—	—	—
DG-A2	109	47	26	15	8	—	—	—	—
DG-A3	150	69	41	28	20	14	10	6	—
DG-B2	112	49	28	18	12	7	—	—	—
DG-C3	177	83	51	35	26	19	15	11	6
SS-A1	282	81	39	22	14	9	5	3	2
SS-A2	275	114	64	40	27	19	13	9	5
SS-A5	473	78	33	17	10	6	3	2	—
SS-B1	187	59	29	17	10	6	4	2	1
SS-B3	165	50	24	13	7	4	2	—	—
SS-B5	239	50	22	12	6	4	2	—	—
SS-C4	136	63	37	24	17	12	8	6	4
SS-D1	257	96	50	30	19	12	8	5	3
SS-D2	243	124	76	51	35	25	18	12	8
SS-E1	181	70	36	21	13	8	5	3	1

TABLE C.2 **Annual Passive Heating Performance: LCR (Continued)**

NASHVILLE, TENNESSEE									*3696 DD*
SSF =	*.10*	*.20*	*.30*	*.40*	*.50*	*.60*	*.70*	*.80*	*.90*
WW-A3	161	72	42	27	18	13	9	6	4
WW-A6	137	74	47	32	22	16	12	8	5
WW-B2	152	78	48	33	23	17	12	9	6
WW-B4	164	97	65	46	34	25	19	14	9
WW-C3	181	115	79	58	43	33	25	18	12
TW-A3	160	65	36	23	15	11	8	5	3
TW-E3	197	115	76	54	39	29	22	16	11
TW-F3	116	52	30	19	13	9	7	5	3
TW-F4	96	47	28	19	13	9	7	5	3
TW-G3	86	41	24	16	11	8	6	4	2
TW-G4	69	34	21	14	10	7	5	3	2
TW-I2	125	62	38	25	18	13	9	7	4
TW-I3	133	69	43	29	20	15	11	8	5
TW-I4	139	78	51	35	26	19	14	10	7
TW-J2	150	82	53	36	26	19	14	10	7
DG-A1	98	34	—	—	—	—	—	—	—
DG-A2	130	55	31	19	11	6	—	—	—
DG-A3	173	78	47	32	23	16	11	7	2
DG-B2	134	58	33	22	15	10	6	—	—
DG-C3	205	94	57	39	29	22	17	12	8
SS-A1	351	100	50	29	19	13	9	6	4
SS-A2	328	135	76	49	33	24	17	12	8
SS-A5	595	98	43	24	15	10	7	4	2
SS-B1	236	74	38	23	15	10	7	5	3
SS-B3	212	65	32	19	12	8	5	3	2
SS-B5	307	65	30	17	10	7	4	3	2
SS-C4	160	73	43	28	19	14	10	7	5
SS-D1	317	119	64	39	26	18	13	8	5
SS-D2	287	147	90	61	43	31	23	16	10
SS-E1	229	89	48	29	19	13	9	6	4

MEMPHIS, TENNESSEE									*3227 DD*
SSF =	*.10*	*.20*	*.30*	*.40*	*.50*	*.60*	*.70*	*.80*	*.90*
WW-A3	227	105	62	41	29	21	15	11	7
WW-A6	191	106	68	47	34	25	19	13	9
WW-B2	207	108	68	47	33	25	18	13	9
WW-B4	214	128	86	62	46	34	26	19	13
WW-C3	231	147	103	75	57	43	33	24	17
TW-A3	224	93	54	35	24	17	12	9	6
TW-E3	254	149	100	71	52	39	29	22	15
TW-F3	163	76	45	30	21	15	11	8	5
TW-F4	134	68	42	28	20	15	11	8	5
TW-G3	120	59	36	24	17	12	9	6	4
TW-G4	97	49	30	20	15	11	8	6	4
TW-I2	169	86	53	36	26	19	14	10	7
TW-I3	179	95	60	41	29	22	16	11	8
TW-I4	182	104	68	48	35	26	20	14	10
TW-J2	198	110	71	50	36	27	20	15	10
DG-A1	153	62	32	16	—	—	—	—	—
DG-A2	178	78	46	30	20	13	8	3	—
DG-A3	223	102	62	43	31	23	17	11	6
DG-B2	182	81	49	33	24	17	12	7	2
DG-C3	261	122	75	52	39	30	23	18	12
SS-A1	469	138	71	44	29	20	14	10	6
SS-A2	422	177	102	66	46	33	24	17	11
SS-A5	802	138	64	37	24	16	11	8	5
SS-B1	318	104	55	34	23	16	11	8	5
SS-B3	291	95	49	30	20	14	10	7	4
SS-B5	419	93	45	26	17	12	8	6	3
SS-C4	216	100	60	40	28	20	15	11	7
SS-D1	425	165	92	58	39	28	20	14	9
SS-D2	371	192	120	82	58	43	31	22	15
SS-E1	315	128	72	46	31	22	16	11	7

KNOXVILLE, TENNESSEE									*3478 DD*
SSF =	*.10*	*.20*	*.30*	*.40*	*.50*	*.60*	*.70*	*.80*	*.90*
WW-A3	195	90	53	35	24	17	12	9	6
WW-A6	166	92	59	40	29	21	16	11	7
WW-B2	181	95	59	40	29	21	15	11	7
WW-B4	190	114	77	55	40	30	23	16	11
WW-C3	208	132	92	68	51	38	29	22	15
TW-A3	193	80	46	29	20	14	10	7	5
TW-E3	228	134	89	63	46	35	26	19	13
TW-F3	140	65	38	25	18	12	9	6	4
TW-F4	116	59	36	24	17	12	9	6	4
TW-G3	104	51	31	20	14	10	7	5	3
TW-G4	84	42	26	17	12	9	6	5	3
TW-I2	148	75	46	31	22	16	12	8	6
TW-I3	158	83	52	36	25	19	14	10	6
TW-I4	162	92	60	42	31	23	17	12	8
TW-J2	175	97	63	44	32	23	17	13	8
DG-A1	127	50	23	—	—	—	—	—	—
DG-A2	155	68	40	25	16	10	5	—	—
DG-A3	198	91	55	38	27	20	14	9	4
DG-B2	159	71	42	28	20	14	9	5	—
DG-C3	234	109	67	46	34	26	20	15	10
SS-A1	414	122	62	37	24	17	12	8	5
SS-A4	388	157	69	57	39	27	20	14	9
SS-A5	697	121	55	32	20	13	9	6	4
SS-B1	281	91	48	29	10	12	9	6	4
SS-B3	256	82	42	25	17	11	8	5	3
SS-B5	364	81	38	22	14	9	6	4	3
SS-C4	189	88	52	35	24	17	13	9	6
SS-D1	377	145	80	50	34	23	16	11	7
SS-D2	334	173	107	73	51	37	27	19	13
SS-E1	277	111	61	38	26	18	13	9	5

AUSTIN, TEXAS									*1737 DD*
SSF =	*.10*	*.20*	*.30*	*.40*	*.50*	*.60*	*.70*	*.80*	*.90*
WW-A3	455	217	132	89	64	47	35	25	17
WW-A6	381	216	142	100	74	55	42	30	21
WW-B2	399	213	136	95	69	51	38	28	19
WW-B4	391	237	161	116	87	66	50	37	25
WW-C3	408	262	184	136	103	79	60	45	31
TW-A3	446	193	113	75	53	39	28	20	14
TW-E3	457	271	182	131	97	74	56	41	28
TW-F3	328	157	96	65	47	34	25	18	12
TW-F4	268	139	88	61	44	33	24	18	12
TW-G3	240	121	76	52	37	28	21	15	10
TW-G4	192	99	63	43	31	23	17	13	8
TW-I2	327	169	107	74	53	40	29	21	14
TW-I3	344	185	118	82	60	45	33	24	16
TW-I4	335	193	128	91	67	51	38	28	19
TW-J2	366	206	135	95	70	53	40	29	20
DG-A1	340	152	90	58	39	26	16	8	—
DG-A2	346	158	97	66	47	34	24	16	9
DG-A3	404	186	116	82	61	46	34	24	14
DG-B2	358	163	101	71	53	40	30	22	13
DG-C3	472	218	136	97	74	58	46	35	24
SS-A1	906	286	152	95	65	47	33	24	15
SS-A2	790	341	199	131	93	67	49	35	23
SS-A5	1493	291	140	85	57	40	28	20	13
SS-B1	629	219	119	76	52	37	27	19	12
SS-B3	592	206	111	71	49	35	25	18	11
SS-B5	812	200	100	62	42	29	21	15	9
SS-C4	412	196	119	81	58	43	31	23	15
SS-D1	841	341	194	127	88	64	46	33	22
SS-D2	698	368	233	161	117	87	64	47	31
SS-E1	643	273	158	104	73	53	38	27	18

TABLE C.2 **Annual Passive Heating Performance: LCR (Continued)**

BROWNSVILLE, TEXAS — *650 DD*

SSF =	.10	.20	.30	.40	.50	.60	.70	.80	.90
WW-A3	955	455	279	190	137	101	75	54	37
WW-A6	790	450	297	211	156	117	88	65	44
WW-B2	814	435	280	195	143	107	80	59	40
WW-B4	773	468	319	231	173	131	100	74	51
WW-C3	790	509	358	265	201	154	118	88	61
TW-A3	941	404	239	160	114	83	61	44	30
TW-E3	895	532	359	258	192	146	110	82	56
TW-F3	690	331	204	139	100	74	55	40	27
TW-F4	557	291	185	129	94	70	52	38	26
TW-G3	502	254	160	110	80	59	44	32	22
TW-G4	398	207	131	91	66	49	37	27	18
TW-I2	669	346	220	152	111	82	62	45	30
TW-I3	700	377	243	170	125	93	70	51	35
TW-I4	664	383	255	182	135	102	77	57	39
TW-J2	730	411	270	192	141	106	80	59	40
DG-A1	749	340	208	140	99	71	51	34	21
DG-A2	718	328	204	142	103	76	56	40	25
DG-A3	799	369	231	164	122	93	70	51	32
DG-B2	733	340	212	150	114	88	68	50	33
DG-C3	916	431	269	192	146	116	92	71	50
SS-A1	1954	615	334	214	148	106	76	54	35
SS-A2	1626	707	421	281	199	145	106	76	51
SS-A5	3342	630	313	193	130	92	66	46	30
SS-B1	1357	473	263	171	119	85	62	44	29
SS-B3	1295	453	253	164	114	81	59	42	28
SS-B5	1810	436	225	141	96	68	49	34	23
SS-C4	845	400	245	167	120	88	66	48	32
SS-D1	1788	733	427	283	199	144	105	75	50
SS-D2	1427	765	491	343	249	185	137	100	67
SS-E1	1385	596	353	235	166	121	88	63	42

EL PASO, TEXAS — *2678 DD*

SSF =	.10	.20	.30	.40	.50	.60	.70	.80	.90
WW-A3	457	224	139	95	68	51	38	27	18
WW-A6	388	225	150	107	79	60	45	33	23
WW-B2	403	220	143	100	74	55	41	30	21
WW-B4	399	245	168	123	92	70	53	40	27
WW-C3	416	271	192	143	109	84	64	48	33
TW-A3	445	198	119	80	57	42	31	22	15
TW-E3	463	279	190	137	103	78	59	44	30
TW-F3	330	163	101	69	50	37	27	20	14
TW-F4	270	144	93	65	47	35	26	19	13
TW-G3	242	126	80	55	40	30	22	16	11
TW-G4	194	103	66	46	34	25	19	14	9
TW-I2	329	174	112	78	57	42	32	23	16
TW-I3	347	191	124	87	64	48	36	26	18
TW-I4	340	199	134	96	71	54	41	30	21
TW-J2	371	213	141	101	25	56	42	31	21
DG-A1	339	158	96	63	43	29	19	11	—
DG-A2	343	162	102	71	51	37	26	18	11
DG-A3	399	191	121	86	65	49	37	26	16
DG-B2	351	168	106	76	57	44	33	24	15
DG-C3	460	224	142	102	78	62	49	38	26
SS-A1	876	286	155	98	67	48	34	24	16
SS-A2	774	343	203	135	95	69	51	36	24
SS-A5	1425	289	143	87	58	41	29	20	13
SS-B1	604	218	121	78	54	38	28	20	13
SS-B3	565	203	112	72	50	35	26	18	12
SS-B5	771	197	101	63	42	30	22	15	10
SS-C4	412	201	124	85	61	45	34	25	17
SS-D1	822	343	198	130	91	65	48	34	22
SS-D2	692	372	238	165	119	89	66	48	32
SS-E1	624	273	160	106	75	54	39	28	19

DALLAS, TEXAS — *2290 DD*

SSF =	.10	.20	.30	.40	.50	.60	.70	.80	.90
WW-A3	349	166	101	69	49	36	27	19	13
WW-A6	294	167	110	78	57	43	32	24	16
WW-B2	310	166	106	74	54	40	30	22	15
WW-B4	310	188	128	93	69	53	40	30	20
WW-C3	327	210	148	109	83	64	49	36	25
TW-A3	342	148	87	58	41	30	22	16	10
TW-E3	363	216	146	105	78	59	45	33	23
TW-F3	251	121	74	50	36	26	19	14	9
TW-F4	206	107	68	47	34	25	19	14	9
TW-G3	185	93	58	40	29	21	16	11	8
TW-G4	148	77	48	33	24	18	13	10	7
TW-I2	254	132	83	58	42	31	23	17	11
TW-I3	267	144	92	64	47	35	26	19	13
TW-I4	265	153	102	72	54	40	30	22	15
TW-J2	288	163	107	76	56	42	31	23	16
DG-A1	252	112	65	41	26	16	8	—	—
DG-A2	266	122	75	51	36	25	17	11	6
DG-A3	318	147	92	65	48	36	27	19	11
DG-B2	274	126	78	55	41	31	23	16	9
DG-C3	373	174	108	78	59	46	37	28	19
SS-A1	675	214	114	71	49	35	25	18	11
SS-A2	603	261	153	101	71	52	38	27	18
SS-A5	1103	216	104	63	42	29	21	14	9
SS-B1	469	164	89	57	39	28	20	14	9
SS-B3	436	152	82	52	36	26	18	13	8
SS-B5	600	148	74	45	31	21	15	11	7
SS-C4	320	152	93	63	45	33	25	18	12
SS-D1	629	256	146	95	66	48	35	25	16
SS-D2	536	284	180	124	90	67	50	36	24
SS-E1	479	203	118	77	54	39	28	20	13

HOUSTON, TEXAS — *1434 DD*

SSF =	.10	.20	.30	.40	.50	.60	.70	.80	.90
WW-A3	468	217	131	89	64	47	34	25	16
WW-A6	387	216	141	100	73	55	41	30	20
WW-B2	407	213	135	94	69	51	38	28	19
WW-B4	395	236	160	116	86	66	50	37	25
WW-C3	410	261	183	135	103	79	60	45	31
TW-A3	462	194	113	75	53	39	28	20	13
TW-E3	462	271	181	130	97	73	55	41	28
TW-F3	337	158	96	65	46	34	25	18	12
TW-F4	273	140	88	61	44	33	24	17	12
TW-G3	246	122	75	52	37	28	20	15	10
TW-G4	196	100	62	43	31	23	17	12	8
TW-I2	334	169	106	73	53	39	29	21	14
TW-I3	350	185	118	82	60	45	33	24	16
TW-I4	340	193	127	91	67	50	38	28	19
TW-J2	372	206	134	95	70	52	39	29	19
DG-A1	351	152	90	58	39	26	16	8	—
DG-A2	357	159	97	66	47	34	24	16	9
DG-A3	417	186	116	81	60	46	34	24	14
DG-B2	370	162	101	71	53	40	30	21	13
DG-C3	491	217	136	97	74	58	45	35	24
SS-A1	954	288	151	96	66	47	34	24	15
SS-A2	815	341	198	132	93	68	49	35	23
SS-A5	1602	295	140	85	57	40	28	20	13
SS-B1	661	221	119	76	53	38	27	19	12
SS-B3	622	208	112	71	49	35	25	18	11
SS-B5	869	203	100	62	42	29	21	15	9
SS-C4	424	197	119	80	58	42	31	23	15
SS-D1	871	341	194	127	89	64	46	33	21
SS-D2	708	367	233	161	117	87	64	46	31
SS-E1	666	274	158	104	73	53	39	27	18

TABLE C.2 **Annual Passive Heating Performance: LCR (Continued)**

LUBBOCK, TEXAS									*3545 DD*
SSF =	*.10*	*.20*	*.30*	*.40*	*.50*	*.60*	*.70*	*.80*	*.90*
WW-A3	340	164	102	69	50	37	27	20	14
WW-A6	289	166	110	79	58	44	33	24	17
WW-B2	304	164	106	75	55	41	31	23	15
WW-B4	305	187	128	94	70	54	41	30	21
WW-C3	323	210	149	111	84	65	50	37	26
TW-A3	333	146	87	58	42	31	23	16	11
TW-E3	357	215	146	106	79	60	46	34	23
TW-F3	245	119	74	51	37	27	20	15	10
TW-F4	201	106	68	48	35	26	19	14	10
TW-G3	180	93	58	41	29	22	16	12	8
TW-G4	145	76	49	34	25	18	14	10	7
TW-I2	248	130	83	58	42	32	24	17	12
TW-I3	262	143	93	65	48	36	27	20	13
TW-I4	260	152	102	73	54	41	31	23	16
TW-J2	283	161	107	76	57	43	32	24	16
DG-A1	244	111	66	42	27	17	9	—	—
DG-A2	258	120	75	51	36	26	18	12	6
DG-A3	306	145	92	65	49	37	28	20	12
DG-B2	262	124	78	55	42	32	24	17	10
DG-C3	356	170	109	78	60	47	38	29	20
SS-A1	681	212	114	73	50	35	25	18	12
SS-A2	600	260	154	103	72	53	39	28	18
SS-A5	1149	214	105	64	43	30	21	15	10
SS-B1	464	161	89	57	39	28	20	14	10
SS-B3	431	149	82	52	36	26	19	13	9
SS-B5	606	145	74	46	31	22	16	11	7
SS-C4	313	150	93	64	46	34	25	18	12
SS-D1	626	254	147	97	67	49	35	25	17
SS-D2	531	283	181	126	92	68	50	37	25
SS-E1	470	200	118	78	55	40	29	21	14

BURLINGTON, VERMONT									*7876 DD*
SSF =	*.10*	*.20*	*.30*	*.40*	*.50*	*.60*	*.70*	*.80*	*.90*
WW-A3	31	—	—	—	—	—	—	—	—
WW-A6	30	—	—	—	—	—	—	—	—
WW-B2	50	21	10	—	—	—	—	—	—
WW-B4	72	41	26	17	12	9	6	4	2
WW-C3	91	56	38	27	20	15	11	8	5
TW-A3	36	—	—	—	—	—	—	—	—
TW-E3	94	53	34	23	16	12	8	6	4
TW-F3	21	—	—	—	—	—	—	—	—
TW-F4	20	—	—	—	—	—	—	—	—
TW-G3	18	—	—	—	—	—	—	—	—
TW-G4	17	—	—	—	—	—	—	—	—
TW-I2	40	16	7	—	—	—	—	—	—
TW-I3	47	21	11	5	—	—	—	—	—
TW-I4	60	32	20	13	9	6	4	3	2
TW-J2	63	32	19	12	8	5	4	2	1
DG-A1	—	—	—	—	—	—	—	—	—
DG-A2	38	—	—	—	—	—	—	—	—
DG-A3	80	33	18	11	7	4	—	—	—
DG-B2	39	—	—	—	—	—	—	—	—
DG-C3	100	43	25	16	11	8	5	3	—
SS-A1	123	19	—	—	—	—	—	—	—
SS-A2	146	54	28	16	10	6	3	2	
SS-A5	214	—	—	—	—	—	—	—	—
SS-B1	70	—	—	—	—	—	—	—	—
SS-B3	45	—	—	—	—	—	—	—	—
SS-B5	88	—	—	—	—	—	—	—	—
SS-C4	56	21	10	4	—	—	—	—	—
SS-D1	103	23	—	—	—	—	—	—	—
SS-D2	127	60	34	21	14	9	6	4	2
SS-E1	48	—	—	—	—	—	—	—	—

SALT LAKE CITY, UTAH									*5983 DD*
SSF =	*.10*	*.20*	*.30*	*.40*	*.50*	*.60*	*.70*	*.80*	*.90*
WW-A3	198	92	54	35	24	17	12	8	5
WW-A6	169	93	60	41	29	21	15	10	6
WW-B2	184	96	60	41	29	21	15	11	7
WW-B4	193	116	78	55	41	30	22	16	11
WW-C3	211	134	93	68	51	39	29	21	14
TW-A3	196	82	47	30	20	14	10	7	4
TW-F3	231	136	90	64	47	35	26	18	12
TW-F3	143	66	39	26	18	12	9	6	4
TW-F4	118	60	37	24	17	12	9	6	4
TW-G3	105	52	31	21	14	10	7	5	3
TW-G4	85	43	26	18	12	9	6	4	3
TW-I2	150	76	47	32	22	16	12	8	5
TW-I3	160	84	53	36	26	18	13	9	6
TW-I4	164	94	61	43	31	23	17	12	8
TW-J2	178	99	64	44	32	23	17	12	8
DG-A1	131	52	24	—	—	—	—	—	—
DG-A2	157	69	40	26	16	10	5	—	—
DG-A3	200	92	56	39	28	20	14	9	4
DG-B2	161	73	43	29	20	14	9	4	—
DG-C3	237	110	69	48	35	27	20	15	9
SS-A1	421	124	62	37	23	15	10	6	4
SS-A2	388	161	91	58	39	27	19	13	8
SS-A5	695	124	55	31	19	12	8	5	2
SS-B1	283	92	47	28	18	12	8	5	3
SS-B3	257	82	41	24	15	10	6	4	2
SS-B5	362	82	38	21	13	8	5	3	2
SS-C4	191	89	53	35	24	17	12	9	5
SS-D1	386	147	79	49	32	21	14	9	5
SS-D2	342	174	107	71	50	36	25	17	11
SS-E1	281	111	60	37	24	16	11	7	4

NORFOLK, VIRGINIA									*3488 DD*
SSF =	*.10*	*.20*	*.30*	*.40*	*.50*	*.60*	*.70*	*.80*	*.90*
WW-A3	233	109	66	44	31	22	16	12	8
WW-A6	198	111	72	50	37	27	20	14	10
WW-B2	213	113	72	49	36	26	19	14	9
WW-B4	220	133	90	65	48	36	27	20	14
WW-C3	237	152	107	79	59	45	34	26	17
TW-A3	229	97	57	37	26	18	13	9	6
TW-E3	261	154	104	74	55	41	31	23	15
TW-F3	167	79	48	32	22	16	12	8	6
TW-F4	138	71	44	30	22	16	12	8	6
TW-G3	123	62	38	26	18	13	10	7	5
TW-G4	100	51	32	22	16	11	8	6	4
TW-I2	174	89	56	38	27	20	15	11	7
TW-I3	184	98	63	43	31	23	17	12	8
TW-I4	187	108	71	50	37	28	21	15	10
TW-J2	203	114	74	52	38	28	21	15	10
DG-A1	158	67	36	19	—	—	—	—	—
DG-A2	181	82	49	32	22	14	9	4	—
DG-A3	226	105	65	45	33	24	18	12	6
DG-B2	186	85	52	35	25	19	13	8	3
DG-C3	265	125	78	55	41	32	25	19	12
SS-A1	470	143	74	46	31	21	15	10	6
SS-A2	428	182	105	69	48	34	25	17	11
SS-A5	786	142	67	39	25	17	12	8	5
SS-B1	321	108	58	36	24	17	12	8	5
SS-B3	295	98	52	32	21	15	10	7	4
SS-B5	416	96	47	28	18	13	9	6	4
SS-C4	221	104	63	42	30	22	16	11	8
SS-D1	431	170	95	61	42	29	21	15	9
SS-D2	378	198	125	85	61	45	33	23	15
SS-E1	321	132	75	48	33	23	16	11	7

TABLE C.2 **Annual Passive Heating Performance: LCR (Continued)**

ROANOKE, VIRGINIA *4307 DD*

SSF =	.10	.20	.30	.40	.50	.60	.70	.80	.90
WW-A3	177	83	50	33	23	16	12	8	5
WW-A6	152	85	55	38	28	20	15	11	7
WW-B2	167	88	56	38	28	20	15	11	7
WW-B4	177	107	73	52	39	29	22	16	11
WW-C3	195	126	88	65	49	37	28	21	14
TW-A3	175	74	43	28	19	14	10	7	4
TW-E3	212	126	85	61	45	34	25	18	12
TW-F3	127	60	36	24	17	12	9	6	4
TW-F4	106	54	34	23	16	12	9	6	4
TW-G3	94	47	29	19	14	10	7	5	3
TW-G4	77	39	24	17	12	9	6	4	3
TW-I2	136	70	44	30	21	15	11	8	5
TW-I3	145	77	49	34	24	18	13	9	6
TW-I4	150	87	57	41	30	22	17	12	8
TW-J2	162	91	60	42	31	23	17	12	8
DG-A1	112	45	20	—	—	—	—	—	—
DG-A2	140	62	37	24	15	9	5	—	—
DG-A3	181	85	52	36	26	19	14	9	4
DG-B2	142	65	39	27	19	13	9	4	—
DG-C3	213	101	63	45	33	26	20	15	10
SS-A1	368	111	57	35	23	16	11	7	5
SS-A2	347	147	85	56	38	27	20	14	9
SS-A5	602	109	51	29	19	12	8	6	3
SS-B1	250	83	44	27	18	12	9	6	4
SS-B3	227	74	39	24	15	10	7	5	3
SS-B5	317	73	35	21	13	9	6	4	2
SS-C4	173	81	49	33	23	17	12	9	6
SS-D1	340	133	74	47	32	22	15	11	7
SS-D2	308	161	101	69	49	36	26	19	12
SS-E1	248	101	57	36	24	17	12	8	5

SPOKANE, WASHINGTON *6835 DD*

SSF =	.10	.20	.30	.40	.50	.60	.70	.80	.90
WW-A3	112	43	19	—	—	—	—	—	—
WW-A6	93	44	23	9	—	—	—	—	—
WW-B2	112	52	29	17	9	4	—	—	—
WW-B4	126	71	45	30	20	14	9	6	3
WW-C3	143	88	59	41	29	21	15	10	6
TW-A3	115	40	17	—	—	—	—	—	—
TW-E3	155	87	55	37	26	18	12	8	5
TW-F3	80	30	13	—	—	—	—	—	—
TW-F4	66	28	14	—	—	—	—	—	—
TW-G3	59	24	11	—	—	—	—	—	—
TW-G4	49	21	11	4	—	—	—	—	—
TW-I2	91	41	22	13	7	—	—	—	—
TW-I3	99	47	26	16	9	5	2	—	—
TW-I4	107	57	35	23	15	10	7	4	2
TW-J2	115	60	36	23	15	10	6	4	2
DG-A1	54	—	—	—	—	—	—	—	—
DG-A2	98	37	16	—	—	—	—	—	—
DG-A3	141	60	34	21	13	8	4	—	—
DG-B2	102	40	19	—	—	—	—	—	—
DG-C3	170	74	44	29	19	13	8	5	—
SS-A1	267	64	24	—	—	—	—	—	—
SS-A2	256	97	50	29	17	10	6	3	2
SS-A5	465	62	17	—	—	—	—	—	—
SS-B1	173	45	16	—	—	—	—	—	—
SS-B3	150	35	—	—	—	—	—	—	—
SS-B5	229	37	—	—	—	—	—	—	—
SS-C4	120	50	26	15	8	4	—	—	—
SS-D1	233	75	31	—	—	—	—	—	—
SS-D2	220	104	59	36	23	14	9	5	3
SS-E1	158	48	—	—	—	—	—	—	—

SEATTLE-TACOMA, WASHINGTON *5185 DD*

SSF =	.10	.20	.30	.40	.50	.60	.70	.80	.90
WW-A3	151	61	31	16	—	—	—	—	—
WW-A6	125	62	35	20	10	—	—	—	—
WW-B2	143	69	39	24	15	9	4	1	—
WW-B4	154	87	55	37	26	18	12	8	4
WW-C3	171	105	70	49	35	25	18	12	8
TW-A3	152	56	27	14	—	—	—	—	—
TW-E3	188	105	67	45	31	22	15	10	6
TW-F3	108	44	22	11	—	—	—	—	—
TW-F4	89	40	21	12	5	—	—	—	—
TW-G3	80	35	18	10	4	—	—	—	—
TW-G4	65	29	16	9	5	—	—	—	—
TW-I2	117	54	30	18	11	6	3	—	—
TW-I3	126	61	35	22	14	8	5	2	—
TW-I4	132	71	44	29	20	13	9	6	3
TW-J2	142	74	45	29	20	13	9	5	3
DG-A1	92	—	—	—	—	—	—	—	—
DG-A2	127	50	24	11	—	—	—	—	—
DG-A3	172	74	42	26	17	11	6	—	—
DG-B2	133	53	28	15	—	—	—	—	—
DG-C3	207	91	53	35	24	16	11	7	3
SS-A1	358	95	40	19	6	—	—	—	—
SS-A2	332	129	67	39	24	15	9	5	2
SS-A5	595	94	34	13	—	—	—	—	—
SS-B1	238	69	29	13	—	—	—	—	—
SS-B3	216	59	23	—	—	—	—	—	—
SS-B5	307	60	22	—	—	—	—	—	—
SS-C4	153	65	35	21	13	7	4	1	—
SS-D1	322	110	51	25	10	—	—	—	—
SS-D2	286	137	78	48	31	20	13	7	4
SS-E1	228	78	34	12	—	—	—	—	—

CHARLESTON, WEST VIRGINIA *4590 DD*

SSF =	.10	.20	.30	.40	.50	.60	.70	.80	.90
WW-A3	113	48	27	16	11	7	5	3	2
WW-A6	97	51	31	20	14	10	7	5	3
WW-B2	113	57	34	23	16	11	8	5	3
WW-B4	128	75	50	35	25	19	14	10	7
WW-C3	145	92	63	46	34	26	19	14	9
TW-A3	114	44	23	14	9	6	4	2	1
TW-E3	157	91	60	42	30	22	17	12	8
TW-F3	81	35	19	12	8	5	3	2	1
TW-F4	68	32	19	12	8	5	4	2	1
TW-G3	60	28	16	10	6	4	3	2	1
TW-G4	50	24	14	9	6	4	3	2	1
TW-I2	92	45	27	17	12	8	6	4	3
TW-I3	100	51	31	20	14	10	7	5	3
TW-I4	108	60	39	27	19	14	10	7	5
TW-J2	116	63	40	27	19	14	10	7	5
DG-A1	55	—	—	—	—	—	—	—	—
DG-A2	95	38	20	10	—	—	—	—	—
DG-A3	136	60	36	24	17	12	8	4	—
DG-B2	97	41	22	13	8	—	—	—	—
DG-C3	162	74	45	30	22	16	12	9	5
SS-A1	264	71	34	19	11	7	5	3	2
SS-A2	258	104	58	36	24	17	12	8	5
SS-A5	443	68	28	14	8	4	2	1	—
SS-B1	174	52	25	14	8	5	3	2	1
SS-B3	153	43	20	10	5	3	1	—	—
SS-B5	224	43	18	9	5	2	1	—	—
SS-C4	120	53	31	20	13	9	6	4	3
SS-D1	237	85	44	26	16	11	7	4	2
SS-D2	226	114	69	46	32	23	16	11	7
SS-E1	165	60	30	17	10	6	4	2	1

TABLE C.2 **Annual Passive Heating Performance: LCR (Continued)**

MADISON, WISCONSIN 7730 DD SSF =	.10	.20	.30	.40	.50	.60	.70	.80	.90
WW-A3	77	30	15	8	3	—	—	—	—
WW-A6	67	34	19	12	7	4	2	—	—
WW-B2	84	41	24	15	10	7	5	3	2
WW-B4	101	59	39	27	19	14	10	7	5
WW-C3	119	75	51	37	28	21	15	11	7
TW-A3	79	28	13	7	3	—	—	—	—
TW-E3	126	72	47	33	24	18	13	9	6
TW-F3	55	22	11	5	—	—	—	—	—
TW-F4	47	21	11	7	4	2	—	—	—
TW-G3	42	18	9	5	3	—	—	—	—
TW-G4	35	16	9	5	3	2	—	—	—
TW-I2	68	32	18	12	8	5	3	2	1
TW-I3	75	37	22	14	10	7	4	3	2
TW-I4	85	47	30	21	15	11	8	5	3
TW-J2	91	48	30	21	14	10	7	5	3
DG-A1	—	—	—	—	—	—	—	—	—
DG-A2	68	25	11	—	—	—	—	—	—
DG-A3	109	47	28	18	12	8	5	—	—
DG-B2	70	27	14	6	—	—	—	—	—
DG-C3	133	59	35	24	17	13	9	6	3
SS-A1	192	47	20	9	3	—	—	—	—
SS-A2	200	78	42	26	17	12	8	5	3
SS-A5	329	42	13	—	—	—	—	—	—
SS-B1	122	32	13	5	—	—	—	—	—
SS-B3	100	22	—	—	—	—	—	—	—
SS-B5	156	24	—	—	—	—	—	—	—
SS-C4	90	38	22	13	9	6	4	2	1
SS-D1	169	56	26	13	6	—	—	—	—
SS-D2	175	86	51	34	23	16	11	7	5
SS-E1	108	34	12	—	—	—	—	—	—

EDMONTON, ALBERTA 10645 DD SSF =	.10	.20	.30	.40	.50	.60	.70	.80	.90
WW-A3	88	32	12	—	—	—	—	—	—
WW-A6	76	36	18	—	—	—	—	—	—
WW-B2	93	44	25	14	7	—	—	—	—
WW-B4	109	63	40	27	18	13	8	5	3
WW-C3	126	78	53	37	27	20	14	10	6
TW-A3	89	31	12	—	—	—	—	—	—
TW-E3	134	76	49	33	23	16	11	7	4
TW-F3	63	24	—	—	—	—	—	—	—
TW-F4	53	23	10	—	—	—	—	—	—
TW-G3	47	19	8	—	—	—	—	—	—
TW-G4	39	17	8	—	—	—	—	—	—
TW-I2	76	35	19	11	5	—	—	—	—
TW-I3	81	39	22	13	7	3	—	—	—
TW-I4	92	50	31	21	14	9	6	4	2
TW-J2	98	52	32	20	13	9	6	3	2
DG-A1	—	—	—	—	—	—	—	—	—
DG-A2	78	29	11	—	—	—	—	—	—
DG-A3	118	52	30	19	12	7	3	—	—
DG-B2	81	33	15	—	—	—	—	—	—
DG-C3	143	65	39	26	17	12	7	4	—
SS-A1	180	43	—	—	—	—	—	—	—
SS-A2	193	75	39	22	13	7	4	2	—
SS-A5	288	38	—	—	—	—	—	—	—
SS-B1	117	29	—	—	—	—	—	—	—
SS-B3	95	—	—	—	—	—	—	—	—
SS-B5	144	20	—	—	—	—	—	—	—
SS-C4	98	42	22	13	7	—	—	—	—
SS-D1	162	50	13	—	—	—	—	—	—
SS-D2	170	82	47	28	18	11	7	4	2
SS-E1	106	25	—	—	—	—	—	—	—

SHERIDAN, WYOMING 7708 DD SSF =	.10	.20	.30	.40	.50	.60	.70	.80	.90
WW-A3	126	56	32	20	13	9	6	4	2
WW-A6	108	58	37	24	17	12	8	6	3
WW-B2	123	63	39	26	18	13	9	6	4
WW-B4	137	82	55	39	28	21	15	11	7
WW-C3	155	99	68	50	37	28	21	15	10
TW-A3	125	50	28	17	11	7	5	3	2
TW-E3	167	98	65	46	33	25	18	13	9
TW-F3	90	40	23	14	10	6	4	3	1
TW-F4	75	37	22	14	10	7	5	3	2
TW-G3	67	32	19	12	8	5	4	2	1
TW-G4	55	27	16	11	7	5	3	2	1
TW-I2	101	50	30	20	14	10	7	5	3
TW-I3	108	56	35	23	16	12	8	6	4
TW-I4	116	66	43	30	21	16	11	8	5
TW-J2	125	69	44	30	22	16	11	8	5
DG-A1	68	—	—	—	—	—	—	—	—
DG-A2	103	44	24	14	7	—	—	—	—
DG-A3	143	65	39	27	19	13	9	5	—
DG-B2	105	46	26	17	10	6	—	—	—
DG-C3	170	79	49	34	25	18	14	10	6
SS-A1	273	76	36	20	12	7	4	2	1
SS-A2	265	109	61	38	26	18	12	8	5
SS-A5	463	73	31	16	8	4	2	—	—
SS-B1	180	55	27	15	9	5	3	2	—
SS-B3	157	46	21	11	6	2	—	—	—
SS-B5	232	47	20	10	5	2	—	—	—
SS-C4	129	59	35	23	15	11	8	5	3
SS-D1	246	91	47	28	17	11	7	4	2
SS-D2	234	119	73	48	33	23	16	11	7
SS-E1	172	65	33	19	11	6	3	2	—

SUFFIELD, ALBERTA 9391 DD SSF =	.10	.20	.30	.40	.50	.60	.70	.80	.90
WW-A3	112	47	25	14	8	4	—	—	—
WW-A6	95	49	29	18	12	7	4	2	—
WW-B2	112	56	33	21	14	10	6	4	2
WW-B4	126	74	48	34	24	17	12	9	5
WW-C3	143	90	62	44	33	24	18	13	8
TW-A3	112	43	22	12	7	3	—	—	—
TW-E3	154	88	58	40	29	21	15	11	7
TW-F3	80	34	18	10	6	2	—	—	—
TW-F4	67	31	18	11	6	4	2	—	—
TW-G3	59	27	15	9	5	3	1	—	—
TW-G4	49	23	13	8	5	3	2	—	—
TW-I2	91	44	26	16	11	7	5	3	1
TW-I3	97	49	29	19	13	9	6	4	2
TW-I4	107	59	38	26	18	13	9	6	4
TW-J2	114	62	39	26	18	13	9	6	4
DG-A1	54	—	—	—	—	—	—	—	—
DG-A2	94	38	19	9	—	—	—	—	—
DG-A3	135	60	36	23	16	11	7	3	—
DG-B2	97	41	22	13	6	—	—	—	—
DG-C3	162	73	45	30	22	16	11	7	4
SS-A1	227	58	25	12	—	—	—	—	—
SS-A2	228	90	49	30	19	12	8	5	3
SS-A5	379	54	19	—	—	—	—	—	—
SS-B1	149	42	18	8	—	—	—	—	—
SS-B3	125	31	9	—	—	—	—	—	—
SS-B5	190	34	11	—	—	—	—	—	—
SS-C4	118	52	30	19	12	8	5	3	2
SS-D1	202	68	32	16	7	—	—	—	—
SS-D2	199	98	58	38	25	17	12	7	4
SS-E1	137	46	20	—	—	—	—	—	—

TABLE C.2 **Annual Passive Heating Performance: LCR (Continued)**

VANCOUVER, BRITISH COLUMBIA — *5592 DD*

SSF =	.10	.20	.30	.40	.50	.60	.70	.80	.90
WW-A3	154	63	32	16	—	—	—	—	—
WW-A6	128	64	36	20	11	—	—	—	—
WW-B2	146	70	40	24	15	9	4	1	—
WW-B4	157	89	57	38	26	18	12	8	4
WW-C3	173	107	71	50	36	26	18	12	7
TW-A3	155	58	28	14	—	—	—	—	—
TW-E3	191	107	68	46	32	22	15	10	6
TW-F3	110	46	23	11	—	—	—	—	—
TW-F4	91	41	22	12	5	—	—	—	—
TW-G3	81	36	19	10	—	—	—	—	—
TW-G4	66	30	16	9	5	—	—	—	—
TW-I2	119	56	31	19	11	6	3	—	—
TW-I3	128	63	36	22	14	8	4	2	—
TW-I4	134	72	45	29	20	13	9	5	3
TW-J2	144	76	46	30	20	13	9	5	3
DG-A1	95	—	—	—	—	—	—	—	—
DG-A2	129	52	25	12	—	—	—	—	—
DG-A3	174	76	43	27	17	11	6	—	—
DG-B2	134	55	29	15	—	—	—	—	—
DG-C3	208	93	54	36	24	16	11	6	3
SS-A1	348	93	40	18	—	—	—	—	—
SS-A2	326	127	66	39	23	14	8	5	2
SS-A5	574	93	34	12	—	—	—	—	—
SS-B1	232	68	29	12	—	—	—	—	—
SS-B3	210	58	22	—	—	—	—	—	—
SS-B5	298	60	21	—	—	—	—	—	—
SS-C4	155	67	36	21	13	7	4	1	—
SS-D1	314	108	50	24	8	—	—	—	—
SS-D2	282	136	77	48	30	19	12	7	4
SS-E1	224	78	33	—	—	—	—	—	—

HALIFAX, NOVA SCOTIA — *7210 DD*

SSF =	.10	.20	.30	.40	.50	.60	.70	.80	.90
WW-A3	95	40	22	13	8	5	2	—	—
WW-A6	82	43	26	17	12	8	5	3	—
WW-B2	98	49	30	20	14	10	7	4	2
WW-B4	114	67	45	31	23	17	12	9	6
WW-C3	132	84	58	42	32	24	18	13	9
TW-A3	96	36	19	11	7	4	2	—	—
TW-E3	141	82	54	38	28	21	15	11	7
TW-F3	68	29	16	10	6	3	—	—	—
TW-F4	57	27	16	10	6	4	3	1	—
TW-G3	51	23	13	8	5	3	2	—	—
TW-G4	42	20	12	7	5	3	2	1	—
TW-I2	80	39	23	15	10	7	5	3	2
TW-I3	87	44	27	18	12	9	6	4	2
TW-I4	96	54	35	24	17	13	9	7	4
TW-J2	103	56	35	24	17	13	9	6	4
DG-A1	36	—	—	—	—	—	—	—	—
DG-A2	80	32	16	7	—	—	—	—	—
DG-A3	121	53	32	22	15	10	7	3	—
DG-B2	82	34	19	11	6	—	—	—	—
DG-C3	146	66	40	28	20	15	11	8	4
SS-A1	227	61	29	16	9	5	—	—	—
SS-A2	230	93	52	33	22	15	10	7	4
SS-A5	374	57	23	11	4	—	—	—	—
SS-B1	148	43	20	11	6	—	—	—	—
SS-B3	128	35	15	6	—	—	—	—	—
SS-B5	187	35	14	6	—	—	—	—	—
SS-C4	104	46	27	17	12	8	5	4	2
SS-D1	205	73	37	21	13	7	3	—	—
SS-D2	202	102	62	41	29	20	14	9	6
SS-E1	138	50	24	13	6	—	—	—	—

WINNIPEG, MANITOBA — *10789 DD*

SSF =	.10	.20	.30	.40	.50	.60	.70	.80	.90
WW-A3	71	28	13	—	—	—	—	—	—
WW-A6	63	31	17	10	4	—	—	—	—
WW-B2	79	39	23	14	9	6	3	2	—
WW-B4	97	56	37	26	18	13	10	7	4
WW-C3	114	72	49	36	26	20	15	10	7
TW-A3	73	25	12	—	—	—	—	—	—
TW-E3	120	69	45	32	23	17	12	8	5
TW-F3	50	20	9	—	—	—	—	—	—
TW-F4	43	19	10	5	—	—	—	—	—
TW-G3	38	16	8	3	—	—	—	—	—
TW-G4	32	15	8	4	2	—	—	—	—
TW-I2	64	30	17	11	7	4	2	1	—
TW-I3	70	34	20	13	8	5	3	2	1
TW-I4	81	45	29	20	14	10	7	5	3
TW-J2	86	46	29	20	14	10	7	4	3
DG-A1	—	—	—	—	—	—	—	—	—
DG-A2	64	24	9	—	—	—	—	—	—
DG-A3	103	45	27	17	12	8	4	—	—
DG-B2	65	26	13	—	—	—	—	—	—
DG-C3	125	56	34	23	17	12	8	6	2
SS-A1	154	35	12	—	—	—	—	—	—
SS-A2	172	67	36	22	14	9	6	3	2
SS-A5	257	29	—	—	—	—	—	—	—
SS-B1	97	23	—	—	—	—	—	—	—
SS-B3	74	—	—	—	—	—	—	—	—
SS-B5	121	14	—	—	—	—	—	—	—
SS-C4	84	36	20	12	8	5	3	2	—
SS-D1	136	42	17	—	—	—	—	—	—
SS-D2	151	74	44	28	19	13	9	5	3
SS-E1	83	21	—	—	—	—	—	—	—

OTTAWA, ONTARIO — *8735 DD*

SSF =	.10	.20	.30	.40	.50	.60	.70	.80	.90
WW-A3	88	36	19	11	6	3	—	—	—
WW-A6	76	39	23	15	10	6	4	2	—
WW-B2	92	46	27	18	12	8	6	4	2
WW-B4	109	64	42	29	21	16	11	8	5
WW-C3	126	79	55	40	30	22	17	12	8
TW-A3	90	33	17	9	5	3	—	—	—
TW-E3	135	78	51	36	26	19	14	10	
TW-F3	63	26	14	8	4	2	—	—	—
TW-F4	53	24	14	8	5	3	2	—	—
TW-G3	47	21	11	7	4	2	1	—	—
TW-G4	39	18	10	6	4	3	2	—	—
TW-I2	75	36	21	14	9	6	4	3	1
TW-I3	82	41	25	16	11	8	5	3	2
TW-I4	92	51	33	23	16	12	8	6	4
TW-J2	98	53	33	23	16	12	8	6	4
DG-A1	—	—	—	—	—	—	—	—	—
DG-A2	76	29	14	—	—	—	—	—	—
DG-A3	117	51	30	20	14	9	6	3	—
DG-B2	78	31	17	9	—	—	—	—	—
DG-C3	142	63	38	26	19	14	10	7	4
SS-A1	204	50	22	11	5	—	—	—	—
SS-A2	209	82	45	28	18	12	8	5	3
SS-A5	343	46	16	—	—	—	—	—	—
SS-B1	130	35	15	7	—	—	—	—	—
SS-B3	108	25	6	—	—	—	—	—	—
SS-B5	166	27	8	—	—	—	—	—	—
SS-C4	99	43	24	15	10	7	5	3	2
SS-D1	180	60	29	15	8	4	—	—	—
SS-D2	182	90	54	35	24	17	12	8	5
SS-E1	117	38	16	—	—	—	—	—	—

TABLE C.2 **Annual Passive Heating Performance: LCR (Continued)**

TORONTO, ONTARIO									*6827 DD*
SSF =	*.10*	*.20*	*.30*	*.40*	*.50*	*.60*	*.70*	*.80*	*.90*
WW-A3	91	38	21	12	8	5	3	—	—
WW-A6	79	41	25	16	11	7	5	3	1
WW-B2	95	47	29	19	13	9	6	4	3
WW-B4	111	65	43	31	22	16	12	9	6
WW-C3	129	82	56	41	31	23	17	13	9
TW-A3	92	35	18	11	7	4	2	—	—
TW-E3	138	80	53	37	27	20	15	11	7
TW-F3	65	27	15	9	5	3	2	—	—
TW-F4	55	26	15	9	6	4	3	1	—
TW-G3	49	22	12	8	5	3	2	1	—
TW-G4	41	19	11	7	5	3	2	1	—
TW-I2	77	37	22	14	10	7	5	3	2
TW-I3	84	42	26	17	12	8	6	4	2
TW-I4	94	52	34	23	17	12	9	6	4
TW-J2	100	54	34	24	17	12	9	6	4
DG-A1	30	—	—	—	—	—	—	—	—
DG-A2	78	31	15	6	—	—	—	—	—
DG-A3	118	52	31	21	14	10	6	3	—
DG-B2	79	32	18	10	5	—	—	—	—
DG-C3	143	64	39	27	19	15	11	8	4
SS-A1	215	56	25	14	8	4	2	—	—
SS-A2	219	88	48	30	20	14	10	6	4
SS-A5	364	52	20	9	—	—	—	—	—
SS-B2	173	72	40	26	17	12	8	6	—
SS-B4	165	68	38	24	16	11	7	5	3
SS-B5	177	31	12	4	—	—	—	—	—
SS-C4	101	44	25	16	11	8	5	4	2
SS-D1	192	67	33	19	11	7	4	1	—
SS-D2	192	96	58	38	27	19	13	9	6
SS-E1	128	44	21	10	4	—	—	—	—

NORMANDIN, QUEBEC									*10528 DD*
SSF =	*.10*	*.20*	*.30*	*.40*	*.50*	*.60*	*.70*	*.80*	*.90*
WW-A3	65	25	11	—	—	—	—	—	—
WW-A6	58	29	16	8	—	—	—	—	—
WW-B2	73	36	21	13	8	5	2	1	—
WW-B4	92	54	36	25	18	13	9	6	4
WW-C3	110	70	48	35	26	19	14	10	7
TW-A3	65	23	10	—	—	—	—	—	—
TW-E3	115	67	44	31	22	16	12	8	5
TW-F3	46	18	7	—	—	—	—	—	—
TW-F4	40	18	9	—	—	—	—	—	—
TW-G3	35	15	7	—	—	—	—	—	—
TW-G4	30	13	7	3	—	—	—	—	—
TW-I2	59	28	16	10	6	3	2	—	—
TW-I3	65	33	19	12	8	5	3	1	—
TW-I4	77	43	28	19	13	9	7	4	3
TW-J2	81	44	28	19	13	9	6	4	2
DG-A1	—	—	—	—	—	—	—	—	—
DG-A2	58	22	—	—	—	—	—	—	—
DG-A3	96	43	26	17	11	7	4	—	—
DG-B2	60	24	11	—	—	—	—	—	—
DG-C3	117	54	33	23	16	11	8	5	2
SS-A1	148	36	13	—	—	—	—	—	—
SS-A2	169	68	37	22	14	9	6	3	2
SS-A5	231	30	—	—	—	—	—	—	—
SS-B1	94	24	—	—	—	—	—	—	—
SS-B3	74	—	—	—	—	—	—	—	—
SS-B5	111	15	—	—	—	—	—	—	—
SS-C4	78	34	19	12	7	4	2	—	—
SS-D1	135	44	18	—	—	—	—	—	—
SS-D2	151	75	45	29	19	13	8	5	3
SS-E1	83	22	—	—	—	—	—	—	—

D

SUNSHADING

D.1 Introduction

This appendix presents a graphic-analytic method for the design of fixed sunshading devices. Although it is assumed that the reader is to an extent familiar with the principles of sunshading and shading mask construction, a brief review of these topics is presented, to develop the necessary background for the design methods that will be demonstrated. It is further assumed that the reader recognizes the importance of sunshading and the importance of the effect of insolation on thermal gain (and loss), daylighting, glare, and so on. A full discussion of these topics is found in Chapters 1 and 6 of this volume.

The shadow cast by a projection from a wall depends on:

1. The shape and dimensions of the projection.
2. The direction in which the wall faces.
3. The position of the sun.

The first two factors are fixed; the third varies with the hour of the day and the date, that is, with the apparent motion of the sun.

D.2 Solar Position: Altitude and Azimuth

Refer to Fig. 19.4*a* and Tables B.11 to B.14. The position of the sun at any instant with respect to an observer on the ground is defined by its *altitude angle A* and its *azimuth angle Az*. Note that in our discussion *azimuth is measured from south*. (In other texts it is measured from north; there is no consensus on this point.) The altitude of the sun varies during the day, starting and ending at 0° at sunrise and sunset, and reaching a daily maximum at *solar* noon. (See Section D.3 for explanation of solar and clock time.) The altitude angle at noon varies from day to day, reaching a yearly maximum at the summer solstice on June 21, a minimum at the winter solstice on December 21 and a point halfway between on the vernal and autumnal equinoxes of March 21 and September 21. These seasonal changes are shown graphically on Fig. 19.4*b* and Fig. D.1. Note that the altitude change is approximately 23½° every three months, making the daily change approximately a quarter-degree and the maximum difference in altitude an-

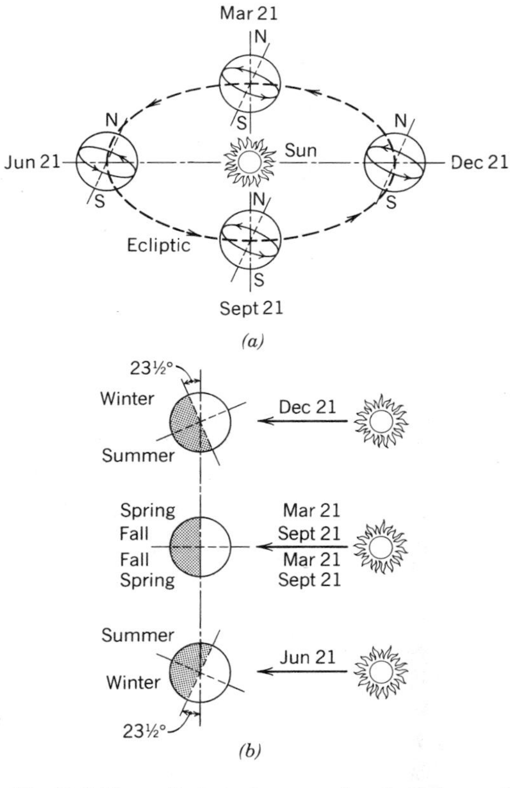

***Fig. D.1** The ecliptic is the annual path of the earth around the sun. The tilt of the earth's axis to the plane of the ecliptic results in the seasonal variations. Note in* (a) *and* (b) *that the northern hemisphere tilts away from the sun in December resulting in winter's low sun altitude and cold weather. In June the effect is reversed.*

APPENDICES

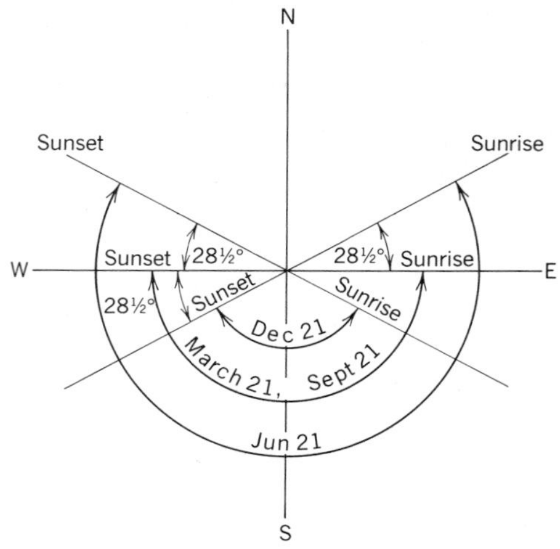

Fig. D.2 Projected sunpaths for solstices and equinoxes at latitude 32° N (Jerusalem, Savannah, Tucson). The sun's azimuth angle, like its altitude, varies with time of day and date. The sun's azimuth path from sunrise to sunset, for this latitude is shown in horizontal projection. Sunrise on June 21 is 28½° North of East, and sunset is correspondingly North of West. On December 21 sunrise and sunset are 28½° South of East and West.

gle about 47°, between the summer and winter solstices. See Tables B.11 to B.14 for accurate altitude/azimuth figures.

Referring to the ecliptic (Fig. D.1*a*), we can see that the height of the sun (its altitude angle), depends on the observer's position on the earth, specifically, his latitude. This can be expressed as follows:

$$\text{maximum solar altitude} = 113.5° - \text{latitude}$$
$$\text{equinox solar altitude} = 90° - \text{latitude}$$
$$\text{minimum solar altitude} = 66.5° - \text{latitude}$$

These relations are shown graphically in Fig. 19.4*c*.

On the equinoxes (March 21 and September 21), the sun rises due east and sets due west. In the northern hemisphere, the sun rises north of east between March 22 and September 20 and sets north of west. This is shown graphically on Fig. 19.4*b* and is projected, in Fig. D.2, for one specific latitude. Note the positions of sunrise and sunset with respect to the E–W line, to understand the explanation above.

D.3 Sun Time and Clock Time

Although for the purpose of sunshading it is not normally important to differentiate between sun time and clock time, it is important that the reader know that such a difference exists, and can be as much as an hour. The following is a brief discussion of timekeeping. The interested reader can refer to the References for a more detailed explanation. (In our discussion we will refer to the sun's motion rather than apparent motion, since relative to the observer, the sun does indeed move.)

The basis of all timekeeping is the length of the solar day, that is, one full rotation of the earth on its axis. More rigidly defined, it is the interval between two sun passages across the observer's meridian. However, because the sun moves about 1° per day on the ecliptic (360° in 365 days), it must move approximately 361° in a solar day. Furthermore, the sun's motion on the ecliptic is not uniform—it moves more rapidly when further from the earth. As a result, the actual length of the solar day varies. Since it is impractical for us to use timepieces that need daily adjustment, we use an average, or *mean solar day,* as the basis for timekeeping. The difference between mean (clock) and apparent (actual) solar time is called the equation of time and can be determined from a curve called the *analemma* or from tabulations in various sources. Furthermore, the 360° earth circumference is divided into 24 one-hour time zones, each therefore separated from the adjacent zone by 15° of longitude, which corresponds to approximately 1000 miles. Therefore an observer at any point other than directly on a time meridian (which is normally in the center of the time zone) must make a time correction of up to 30 min to correspond to actual solar time. (In some countries a reference meridian is more than 30 min from the time zone extremes. In such places, the longitude correction plus the equation of time correction can total well over an hour, and one is no longer able to neglect it. This situation does *not* exist in the continental United States.) Although not vital, because the length of shaded time is the same, it is good practice to make a table of mean time versus apparent solar time for any project where shading and insolation are important. In any case, the designer must remember that unless otherwise specifically noted (and this is rare), all sun charts show sun

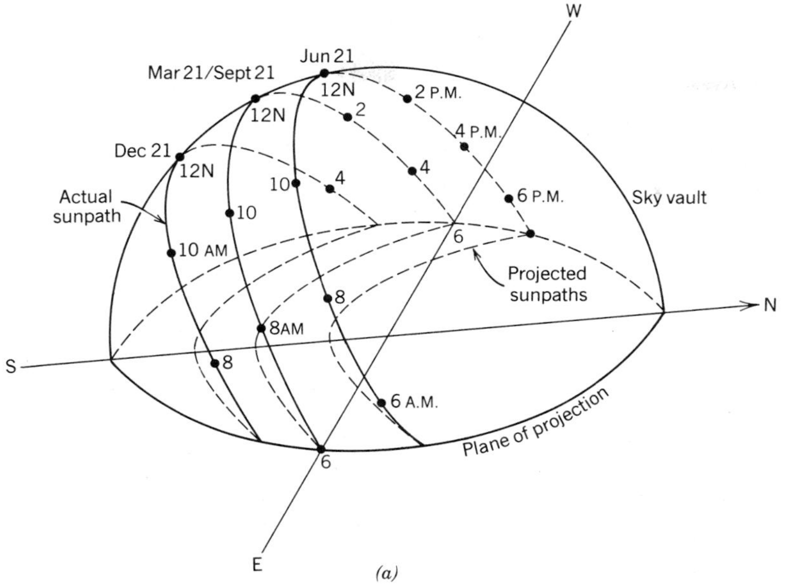

(a)

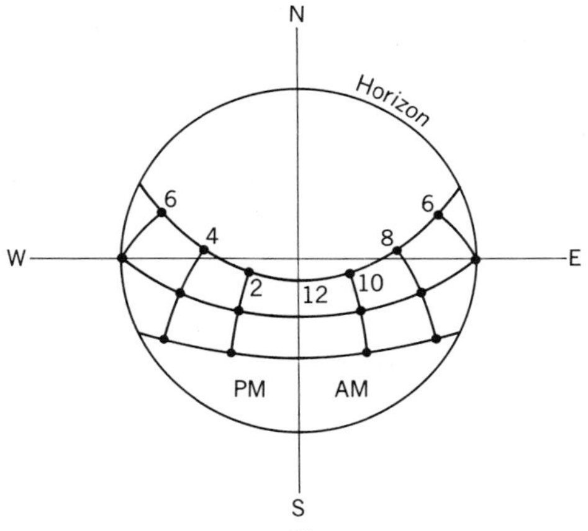

(b)

***Fig. D.3** Basis of horizontal projection of the sunpaths. The sky vault* (a) *is represented as a hemisphere, which projects as a circle* (b) *on the projection plane. The sunpaths in* (a) *project as curved lines on the projection plane* (b)*, the exact shape of which depends on the projection method (Fig D.4). The numbers on the sunpaths in* (a) *are hours—6 A.M., 8 A.M., 10 A.M., 12 Noon, 2 P.M., 4 P.M., 6 P.M. When hour points are connected between sunpaths* (b)*, they form hour lines, which are useful for interpolation in completed diagrams.*

time, that is, actual solar time. This includes the well-known Libbey-Owens-Ford (LOF) *Sun Angle Calculator* (1975).

D.4 Sunpath Projection

The design of sunshading devices necessitates projecting the three-dimensional solar path onto a two-dimensional surface. There are basically two types of projection: vertical and horizontal. A version of the former is familiar to users of Waldram diagrams in daylight design. Horizontal projections are more common. The basic idea behind the horizontal projection is shown in Fig. D.3. The sky vault is represented as a hemisphere that projects as a circle, the center of which is the zenith. The observer looks down onto the hemispherical sky vault and projects the sunpaths onto the two-dimensional, projected circle. The four methods of projection are:

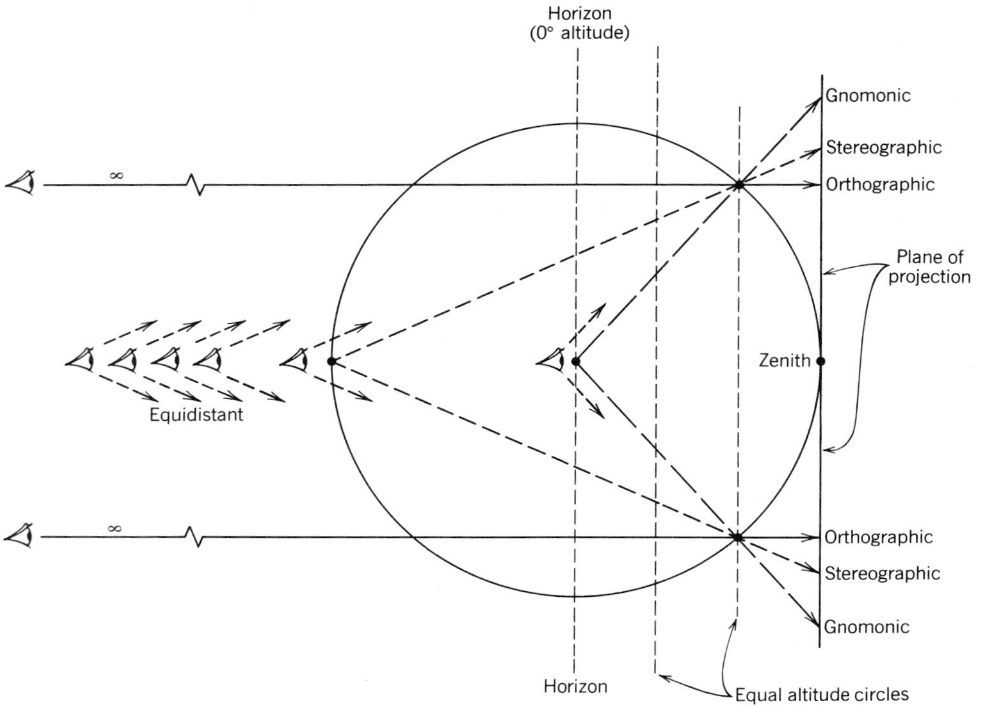

Fig. D.4 *The four standard horizontal projection methods, with observer position and projected point. Note that in the equidistant projection the observer position varies and therefore the projected point also varies. Thus the projected point not shown. Note also that as the altitude drops and approaches the horizon, the gnomonic projection extends out to infinity. Typical projections are given in Fig. D.5.*

Gnomonic.
Orthographic.
Stereographic.
Equidistant.

The position of the observer's eye for each of these projections is shown in Fig. D.4, and projected circles of equal solar altitude are shown in Fig. D.5 for all four methods. Spacing between circles of equal altitude is the basic difference between the four methods. In all of them, lines of equal azimuth project as straight lines emanating from the center of the circle.

1. The *gnomonic projection* is derived from the sundial. The observer is at the center of the projection; therefore low sun angles (sunrise, sunset) extend to infinity. For this reason, this method is rarely used for solar charts but is used frequently for buildings shadow studies. The method is illustrated in the sun peg shadow charts of Fig. 3.8.

2. The *orthographic projection* views the sky vault from infinity and projects as in a standard architectural plan, that is, with all projection lines parallel. Circles of equal altitude simply "drop" onto the projection plane. This method has the advantage of *not* requiring a special protractor (see below). Since it is geometrically correct, altitude angles can be plotted directly on the diagram. Its disadvantage is that the area of 0 to 20° is highly compressed (see Fig. D.5*b*). It is therefore applicable to low latitudes, where the sun rises rapidly, as for instance in Jerusalem, Savannah, and Tucson latitude 32°N, where the sun reaches 20° altitude at 06:20 on June 21, 07:30 on March 21/September 21, and 08:40 on December 21.

3. The *stereographic projection* is used widely throughout the world, except in the United States. The observer is fixed at the top of the hemisphere, resulting in wide-spaced, low-angle altitudes. This makes it particularly useful for northern latitudes,

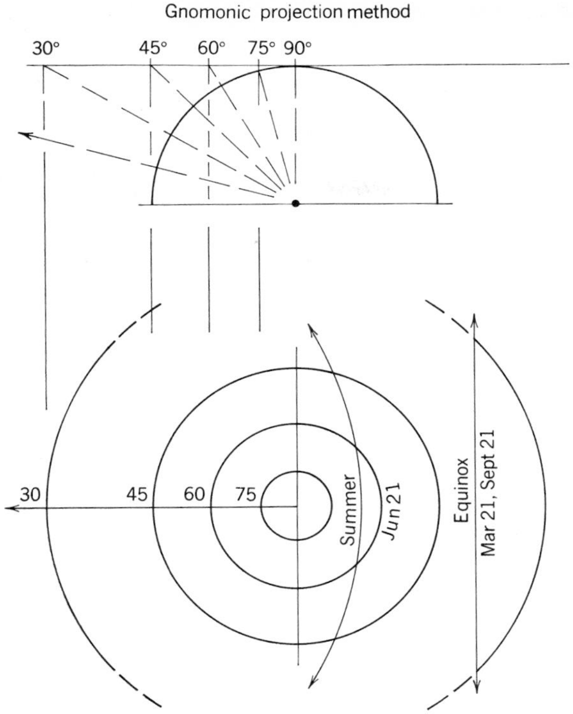

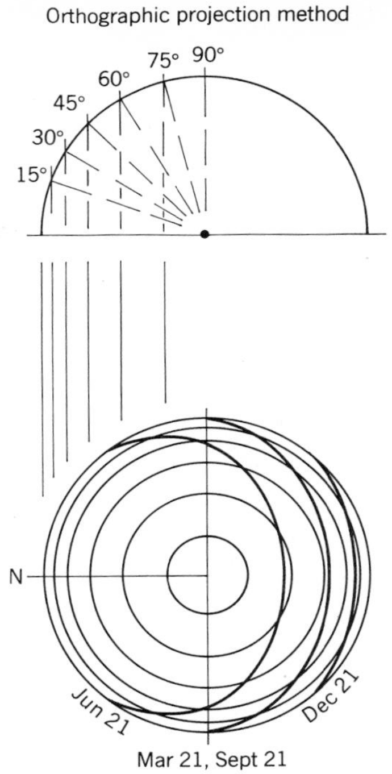

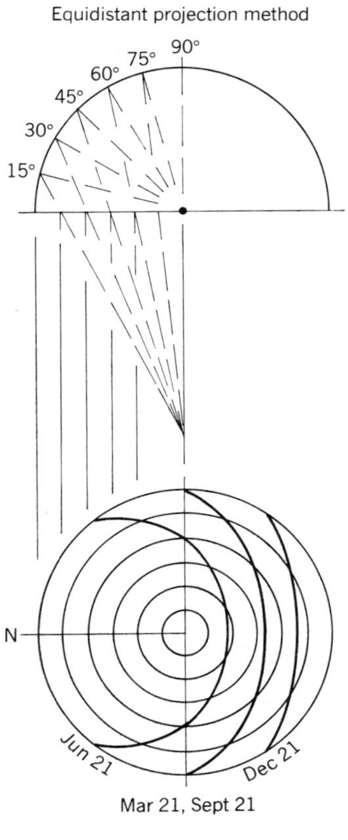

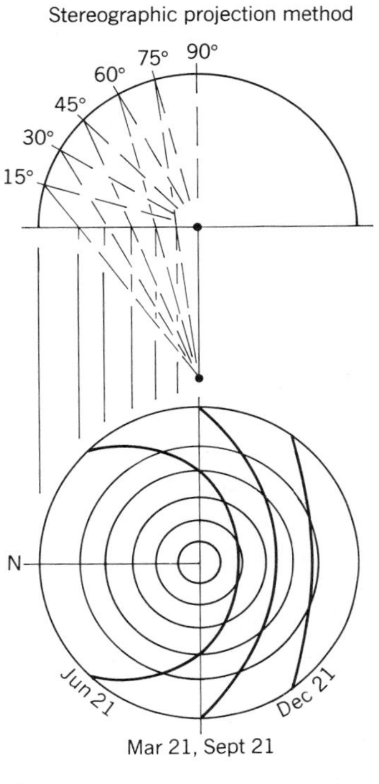

Fig. D.5 *In all projections the solstice and equinox sunpaths are shown, since (see Fig. D.3a) the solstices are the outer limits and the equinox(es) line is the center. For the gnomonic projection, the winter solstice line cannot be shown because it is too far out, being at low altitude. (Reproduced, with permission from S.V. Szokolay,* Environmental Science Handbook, *1980, Halsted/Wiley, New York.*

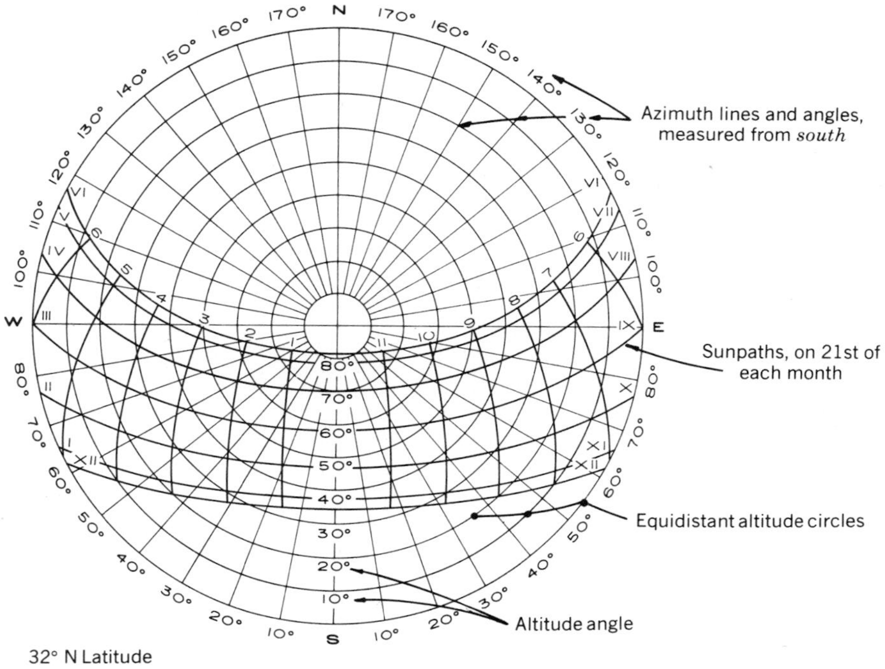

Fig. D.6 *Typical equidistant sunpath projection for 32° N latitude. This latitude is approximately that of Savannah, Georgia, Tucson, Arizona, San Diego, California, and Jerusalem, Israel. In all of these locations sun shading is desirable at least from the vernal to the autumnal equinox. (Diagram after Libbey-Owens-Ford, Sun Angle Calculator, 1975.)*

that is, above 40°N, since the sun at these latitudes is usually low in the sky. A special protractor is required; (see Section D.5). The principal advantage of this projection is that it is very simple to draw sunpaths at any scale and for any latitude without having to rely on published sunpaths. Thus for any serious sunshading design, the designer can prepare a large-scale sunpath chart at the exact latitude of the project and do accurate graphical analysis. For a detailed description of the method of preparing the charts and protractor, see Szokolay (1980, pp 314–317).

4. The *equidistant projection* is used almost exclusively in the United States because of the ready availability and wide acceptance of the LOF sunpath charts (*Sun Angle Calculator,* 1975). These charts, one of which is illustrated in Fig. D.6, show the altitude circles as spaced equidistantly, a projection that requires moving the observer during projection, as shown in Fig. D.4. This method is equally applicable to all latitudes. Its disadvantage is the difficulty of preparing sunpaths and the required shadow angle protractor. Thus the user is essentially limited to published material, with its limitations of size and latitude. The LOF diagrams are available only in steps of 4° of latitude.

D.5 Shadow Angles and Shading Masks

The sun's altitude and azimuth angles are very useful in developing the sunpath diagrams but are much less so in defining the shadow angles cast by a projection, on a wall exposed to the sun. Since there is no universal agreement on angle nomenclature, we will very carefully define the angles involved. Refer to Fig. D.7, which shows the shadow cast by a horizontal overhang on a wall exposed to sunlight. Note that the shadow is defined by two angles: *the vertical shadow angle (VSA),* which indicates the position of the leading edge of the shadow as defined from the leading edge of the overhang, and the *horizontal shadow angle (HSA),* which defines the leading edge of a shadow cast by a vertical element (absent here, and indicated by a dotted line), as defined with respect to that element's leading edge. The terms

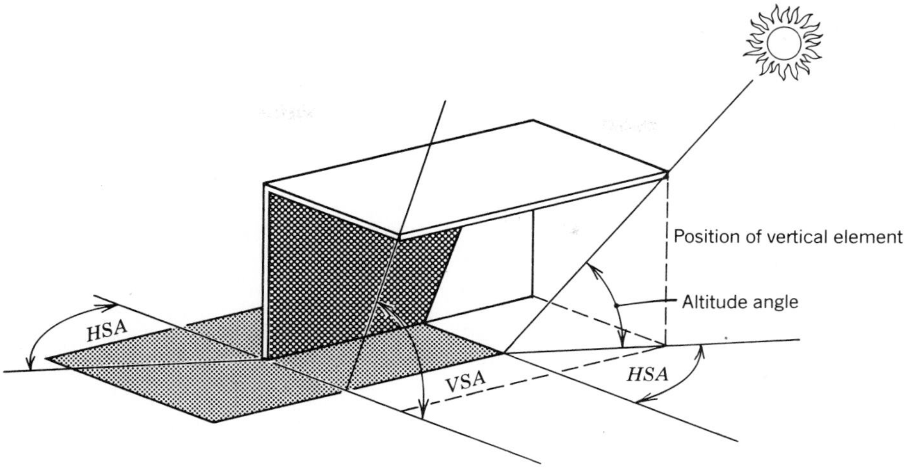

Fig. D.7 The shadow cast by a horizontal overhang is best defined by angles VSA and HSA, the vertical shading angle and horizontal shading angles, respectively. VSA is also known in the United States as the profile angle.

VSA and HSA are in use throughout the world, although in the United States, the vertical shadow angle is also known widely as the *profile angle* because it is so designated on the LOF sun angle calculator. Figure D.8 shows the same information in slightly different form; line *DA* is the shadow

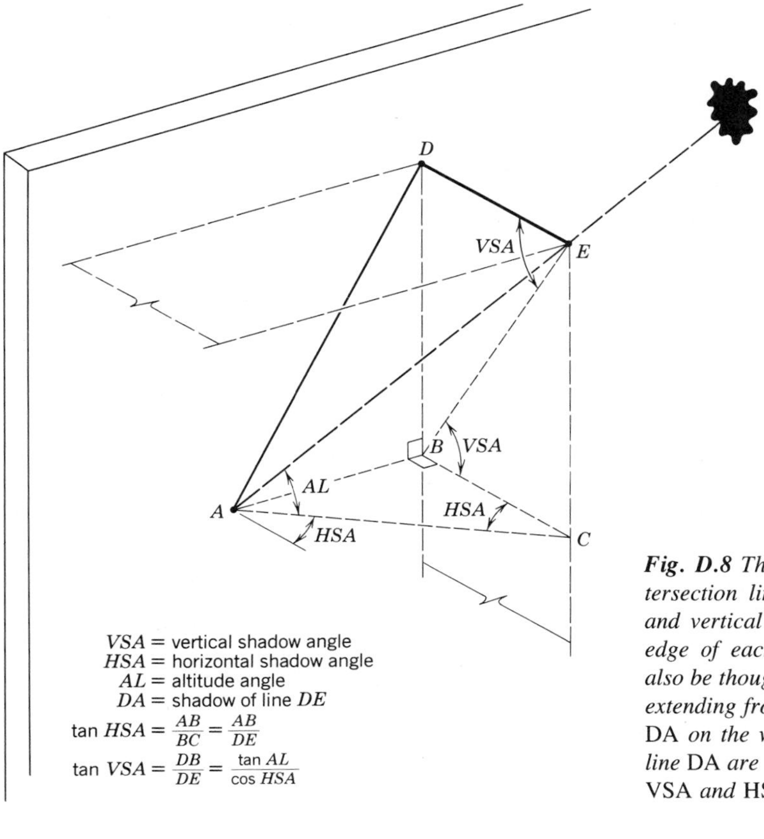

$$\tan HSA = \frac{AB}{BC} = \frac{AB}{DE}$$

$$\tan VSA = \frac{DB}{DE} = \frac{\tan AL}{\cos HSA}$$

Fig. D.8 The shadow DA *cast by the intersection line* DE, *between a horizontal and vertical shading element, defines the edge of each of these shadows.* DE *can also be thought of as a pin, normal to, and extending from the wall, casting a shadow* DA *on the wall. The location and size of line* DA *are best defined in terms of angles* VSA *and* HSA.

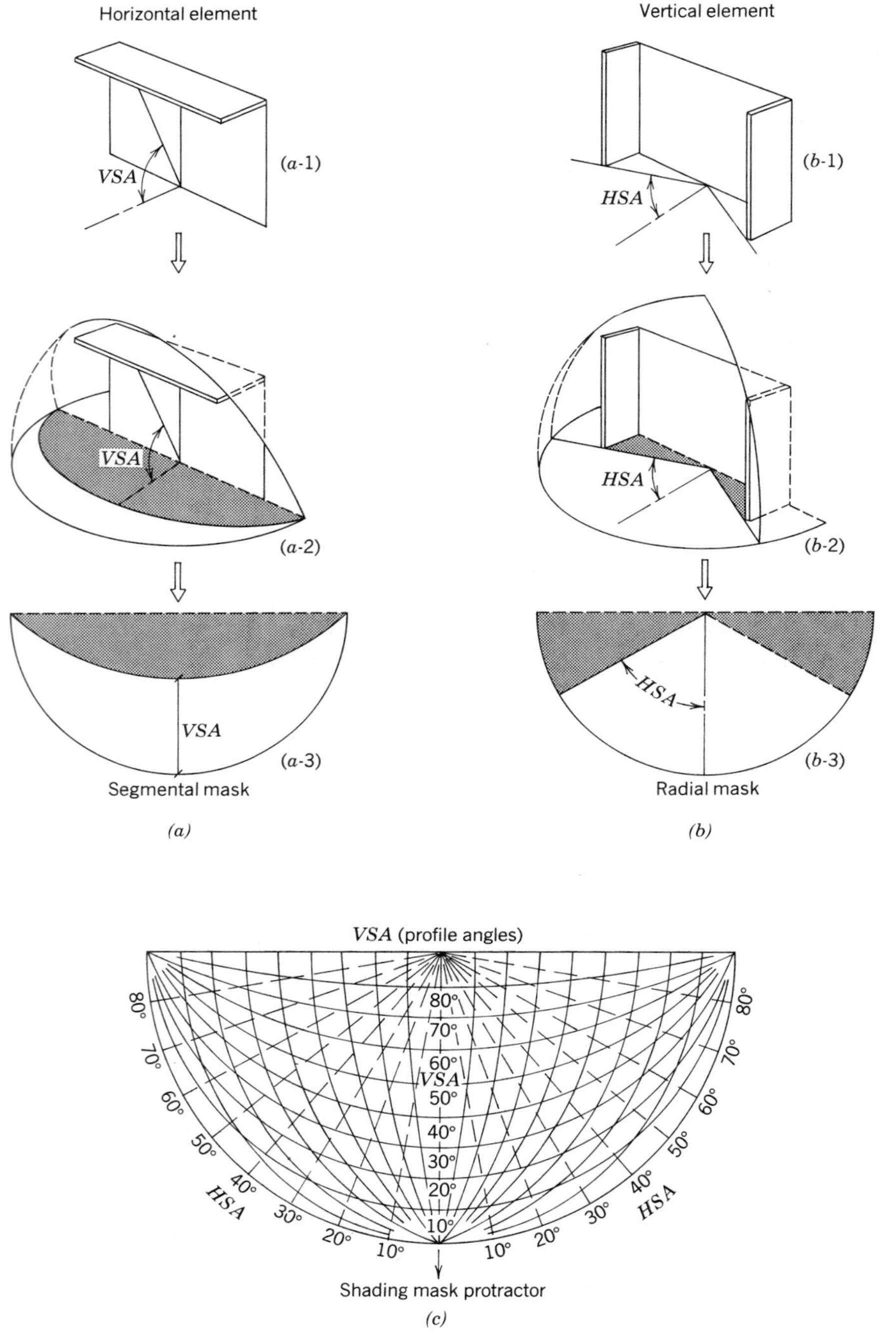

Fig. D.9 A shading mask is the horizontal projection of the shadow cast by elements with the same VSA or HSA. Thus any infinitely long *horizontal element with the VSA shown in* (a-*1*) *will have a shading mask as shown in* (a-*3*). *Similarly, any* infinitely high *vertical elements with the HSA shown in* (b-*1*) *will have the shadow mask shown in* (b-*3*). *A protractor of the type shown in* (c) *is required to properly draw the segmental horizontal element mask. This device must be drawn to the same scale and with the same projection method (equidistant, stereographic) as the sunpath diagram on which it will be used. The protractor of* (c) *matches the sunpath diagram of Fig. D.6 and is drawn with equidistant projection.*

cast by line *DE*, which is the intersection between a horizontal projection and a vertical projection. This line is particularly important in determining the required extent of a shading element, as will be shown below.

Refer now to Fig. D.9. The locus of the leading edge of the shadow cast by all horizontal elements with the same VSA (profile angle) (Fig. D.9*a*-1) projects as a segmental line (Fig. D.9*a*-2), and when plotted on the projection plane (similar to the sunpath projection in Fig. D.3), shows as a segmental mask (Fig. D.9*a*-3). To draw this segment, a protractor is required, as shown in Fig. D.9*c*. Note that the angular notation on the segmental section of the protractor corresponds to VSA as shown on Fig. D.9*a*-3. Users of the LOF sun angle calculator can use the profile angle overlay to draw the required segment. It is important not to confuse the VSA with its complement. A simple way to remember this is to consider the limits of VSA. With no shading, VSA is 90° and the segment on Fig. D.9*a*-3 disappears. With a very deep overhang, VSA approaches 0° and the *unshaded* area in Fig. D.9*a*-3 shrinks. This action shows clearly on the protractor of Fig. D.9*c*.

Refer now to Fig. D.9*b*. Vertical shading elements Fig. D.9*b*-1 with the same HSA show as radii (*b*-2) and project as radial lines from the center, on the radial mask, at angle HSA from normal to the wall (*b*-3). These radial lines can be drawn with the assistance of an ordinary protractor, or by use of the protractor in Fig. D.9*c*. The reader must bear in mind that full segments and full radial lines are the shading masks of *infinitely long elements,* which, of course, do not actually exist. Shading masks that appear in the literature (Olgyay, 1975; Ramsey and Sleeper, 1981) are frequently drawn as if for infinite elements, and the reader must know how to truncate them if accurate shading design is desired.

D.6 Use of Shading Masks

Once a shading mask has been drawn, it is applicable to any latitude and exposure direction. The use of the mask is straightforward. It is drawn on some transparent medium (e.g., paper or plastic sheet), laid on top of a sunpath diagram of the proper latitude, *and drawn to the same scale*. Its centerpoint is placed on the centerpoint of the sunpath diagram, and it is rotated until its facing direction (the direction of a normal to the wall) is aligned with the appropriate azimuth line on the sunpath diagram. Assuming that the shading mask has been correctly drawn to provide 100% shading for the *entire window,* the shaded hours for the window in question can then be read directly from the sunpath diagram underneath. (This is why the shading mask must be drawn on a transparent medium.) See Fig. D.10.

It is extremely important to keep in mind that a horizontal overhang shading element the same width as the window can provide full shade for the window only when the sun is exactly opposite the window, that is, when the bearing angle is 0°. (See Fig. 19.9*a*.) This situation obtains for only an instant. See Fig. D.11. At all other times, some part of the window will be exposed to the sun. Therefore, to provide full shading for more than an instant with only a horizontal element and no vertical (side) elements, *the overhang must extend beyond the sides of the window*. The amount of such extension can be determined both graphically and analytically. Since graphic solutions are amply treated in the literature (Harkness and Mehta, 1978; Lim et al., 1979; Cowan, 1980) we will restrict ourselves to the analytic solution, leaving to the reader the parallel graphic solution. In doing this work, the reader should keep in mind the requirement that shading masks for 100% coverage must be prepared in order that the extremities of the opening be shaded.

D.7 Finite Horizontal Elements

The first design step is to establish the required *depth* of the unit, that is, the amount it projects from the wall. Although any percent coverage desired is possible, most elements are designed to give either 50 or 100% coverage. The reason for this is explained in the illustrative design example below. Figure D.12 shows how the required depth and the corresponding segmental mask are determined. The next step is to determine the required side extensions beyond the window's sides. This can best be illustrated by a practical example, using the analytic technique. First however, a few remarks about the example presented are in order.

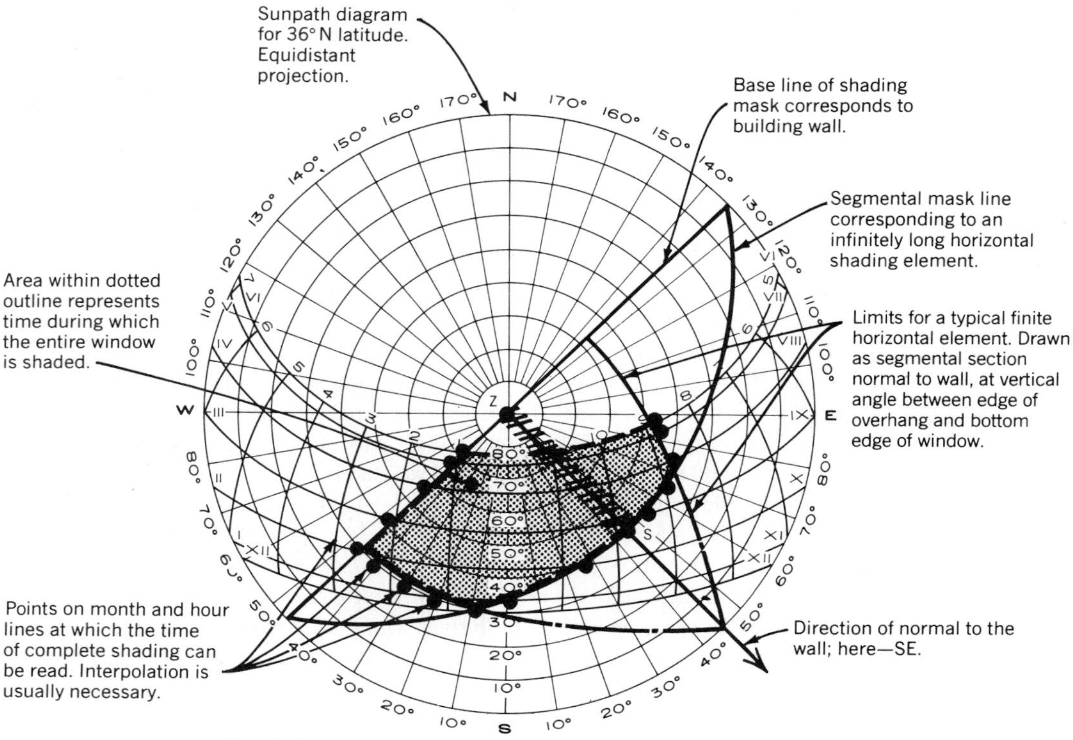
Sunpath diagram for 36° N latitude. Equidistant projection.
Base line of shading mask corresponds to building wall.
Segmental mask line corresponding to an infinitely long horizontal shading element.
Limits for a typical finite horizontal element. Drawn as segmental section normal to wall, at vertical angle between edge of overhang and bottom edge of window.
Area within dotted outline represents time during which the entire window is shaded.
Points on month and hour lines at which the time of complete shading can be read. Interpolation is usually necessary.
Direction of normal to the wall; here—SE.
N
S
E
W
Z
36° N Latitude

Fig. D.10 A typical shading mask for a horizontal overhang on a window is laid over a sunpath diagram, to permit determination of the shaded hours. The mask is drawn to provide full shading for the entire window. The hours during which the entire window is shaded are determined by the intersection of the mask perimeter with the date/hour lines of the sunpath diagram, and are tabulated on the figure. However, only at the hours falling along the normal-to-the-wall line ZS (where the bearing angle is 0°) and within the dotted area, is a window fully shaded by an overhang element only as wide as the window. *Indeed, line ZS is the shading mask of such an element with respect to 100% shading of the entire window (see Fig. D. 11). Note that the element shown here is* not *symmetrical about the center, indicating an extension to the left, to provide a larger shade period after noon. This is typical with east-facing windows, and is reversed for western exposures.*

Time of Day When Window is Fully Shaded

With Required Overhang Extensions			With Element Same Width as Window
Date	*From*	*To*	*Time*
21 June	0900	1240	1115
21 July/May	0850	1250	1100
21 Aug./Apr.	0845	1330	1035
21 Sept./Mar.	0940	1400	0955
21 Oct./Feb.	1040	1410	—
21 Nov./Jan.	12N	1300	—
21 Dec.	1230		—

Due to the symmetry of solar motion about the ecliptic, the position of the sun is symmetrical on both sides of its maximum/minimum positions, that is, the solstices. Thus the sun's position on May 21 is the same as on July 21, since both dates are one month from the summer solstice. As a result, any *fixed* shading device will give the same shade in the spring as in the summer. Since in many locations spring is cool, the desired summer shading will produce spring shading that *may not* be

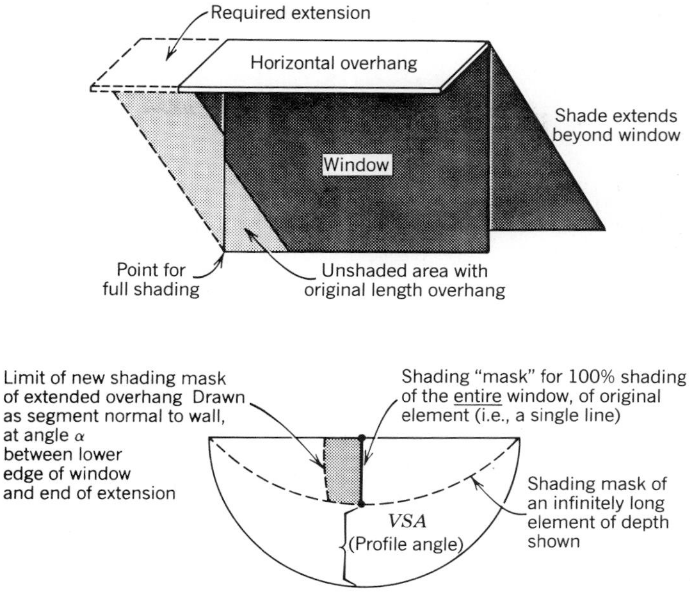

Fig. D.11 *A horizontal overhang with sufficient depth to provide 100% shading (i.e., to cast a shadow reaching to the window sill line, but only as wide as the window) can provide full shading only when the sun is exactly opposite the window. At any other time (bearing angle ≠ 0), part of the window will be exposed to the sun. To compensate for this, if full shading is desired, an entension is required, as shown. For south-facing windows the extension is symmetrical on both sides. For east-facing windows en extension to the left is usually used to provide full shading for hot afternoon hours; for west-facing windows, the reverse is customary. Shading from the low early morning or late afternoon sun is not practical with horizontal elements.*

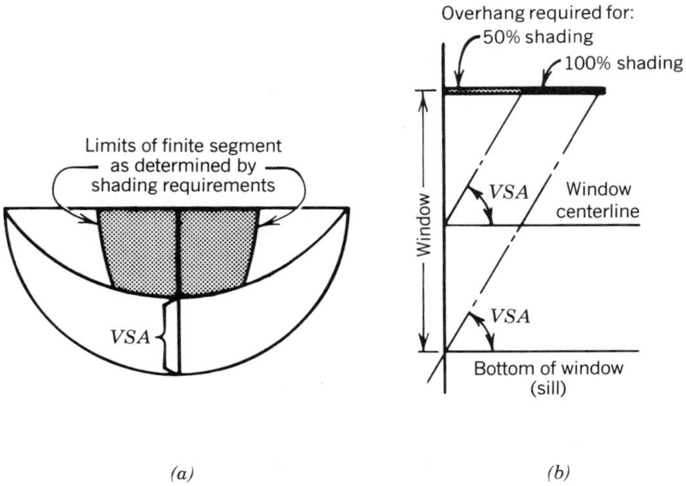

Fig. D.12 *To find the required depth of a horizontal overhang, having established the required segmental shading mask* (a) *it is only necessary to read* VSA *(profile angle) off the protractor in Fig. D.9* (c) *that corresponds to the segment in* (a) *and plot this angle as shown in* (b) *(wall section). Angle* VSA *can be plotted for 50% coverage, 100% coverage, or any other figure. The depth of the overhang can then be measured directly from the section, which is drawn to scale. Alternatively, if the overhang depth is known, the corresponding segment can be drawn.*

desirable. The solutions to this apparent dilemma are either to use a movable or variable size fixed shading device, or to compromise with the amount of shading, that is, to design for 50% shading for late summer, giving 50% insolation in early spring and 100% shading in early summer (and late spring).

The example that follows requires 100% shading for most of the summer (spring) for two reasons:

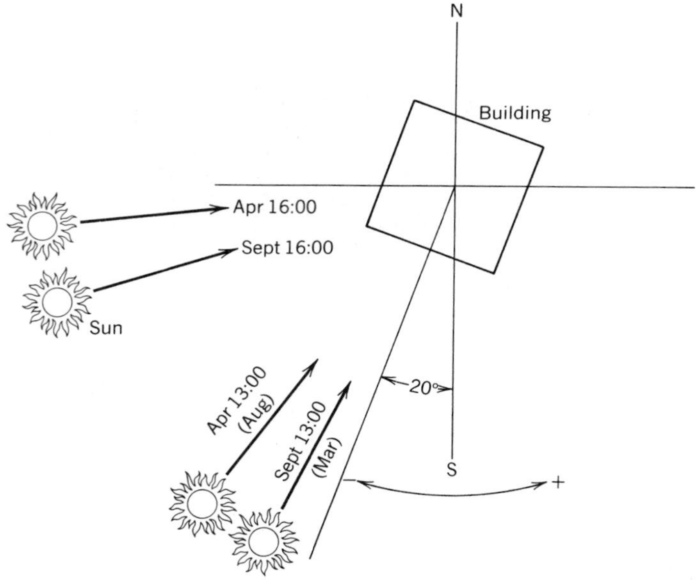

(a)

Date	Time	Sun Azm	Shade direction	*VSA*	*HSA*[a]	*DE*[b] (depth, cm)	*AB*[c] (extension, cm)
Apr 21	13:00	−37°	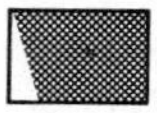	67°	−17°	42	−40
	16:00	−84°	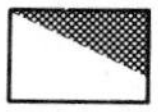	54.5°	−64°	71	−266
Sept 21	13:00	−27°	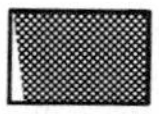	55°	−7°	70	−16
	16:00	−73°	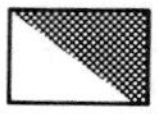	37.5°	−53°	130*	−172

[a] HSA = (Sun azm) − (Bldg. bearing), i.e., (Azm) − (20°)
[b] DE = Window height/tan VSA = 100/VSA
[c] AB = DE tan HSA = overhang depth × tab HSA = 130 tan HSA; the negative sign indicates an extension to the left.

(b)

***Fig. D.13** Sketches for solution of Example D.1 (See text.)*

1. At the selected latitude of 32°N, sunshading is frequently desirable in the spring.
2. We wish to show that 100% shading using only a horizontal overhang can result in a very large element and that the shading characteristic of such an element can be duplicated with a much smaller combined horizontal–vertical element.

EXAMPLE D-1. Design a simple horizontal over-the-window shading element that will provide 100% window shading from 1 P.M. to 4 P.M., from April 21 to September 21 for a window 100 cm [39 in.] high and 150 cm [59 in.] wide, in a wall facing 20° west of south. The building is at latitude 32°N.

Analytic Solution

STEP 1. The first step is to draw a sketch (Fig. D.13*a*) showing the orientation of the building. All the sketches connected with this example are shown in Fig. D.13.

STEP 2. Using either the LOF sun angle cal-

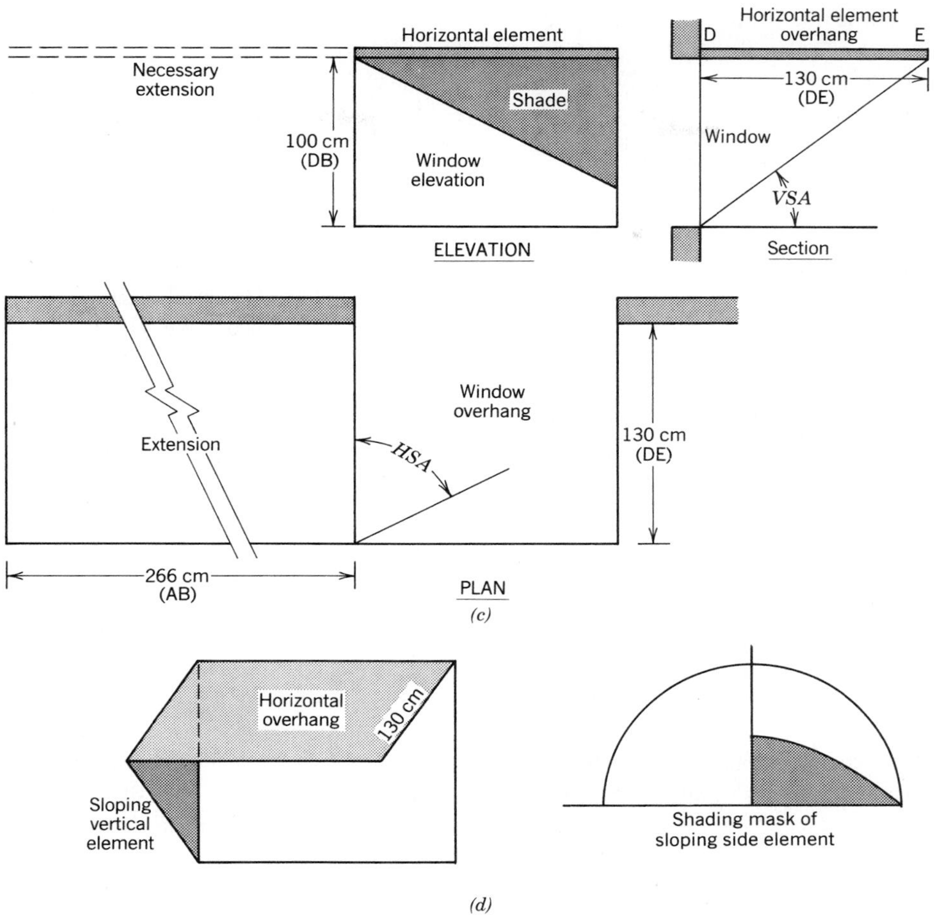

Fig. D.13 (continued)

culator, or Fig. D.6, tabulate the sun positions for the four times that bracket the shading period. Tabulate them as shown in Fig. D.13*b* and show sun directions as in Fig. D.13*a*. The purpose of establishing sun direction is to be able to immediately determine the required shading element end treatment, by sketching the direction in which the shade falls. This is shown in the fourth column of the tabulation of Fig. D.13*b*.

STEP 3. By inspection of the solar azimuth and the bearing angle (the angle between perpendicular to the wall and the vertical plane through the sun), it is apparent that all shadows fall to the right, and therefore the horizontal element must be lengthened to the left. Maximum extension is required for shading at 1600 on April 21. The next step is to establish the shadow angles *VSA* and *HSA* for each condition. These can be read directly from the LOF sun chart (profile angle = *VSA*; bearing from normal line = *HSA*) or by overlaying the protractor of Fig. D.9*c* on Fig. D.6. *HSA* need not be read off: It is simply the difference between sun azimuth and building bearing (20°) and so can be calculated directly. Fill in the required angles in the *VSA* and *HSA* columns of the table in Fig. D.13*b*.

STEP 4. At this point it is advisable to draw sketches of the window in elevation, section, and plan to understand the calculations. See Fig. D.13*c*. It is necessary for us first to establish the depth of the overhang, and with that known, to then calculate the required extension of the left. Refer now to Fig. D.8. Note carefully that the first item that we wish to calculate is the depth of the overhang, *DE*. Since

$$\tan VSA = \frac{DB}{DE}$$

and we know the window height, *DB*, to be 100 cm [39 in.], we have

$$DE = \frac{100 \text{ cm}}{\tan VSA}$$

We perform this simple calculation for every VSA, and we will select the deepest overhang as the one required. The figures are listed under column *DE* (depth) of Fig. D.13*b*.

STEP 5. Having found the overhang depth required, we can calculate the end treatment needed, that is, the amount that the overhang must be extended to cover the window. Study Fig. D.8. The extension needed is obviously *AB*, since the area *ABD* is unshaded. Therefore, if we move *D* a distance of *AB* (on the figure—to the right; in our example—to the left), point *A* will move to *B*, and shading will be complete. Since

$$\tan HSA = \frac{AB}{DE}$$

and we have already established that *DE* is 130 cm [51 in.], we have

$$AB = 130 \tan HSA$$

These four values are calculated and tabulated in Fig. D.13*b*. We find that the required extension to the left, beyond the window, is 266 cm [105 in.]. At this point we can complete the sketch in Fig. D.13*c* with dimensions.

STEP 6. This step is the evaluation stage that follows every design. Our evaluation would be that if, indeed, the stated shading hours must be maintained, an overhang 130 cm [51 in.] deep and 416 cm [164 in.] wide (266 + 150) is excessively large, and we would recommend a vertical element on the left side. The type of vertical element recommended is shown in Fig. D.13*d*; it is a sloping element that reaches from the front edge of the overhang to the window. To understand how such a unit shades, refer again to Fig. D.8. Note very carefully that because *DA* is the shadow of *DE*, triangle *DAB* is the shadow area corresponding to a vertical side element *DEB*. That is, if *DEB* were solid, *DAB* would be its shade. We thus see the extremely important and useful fact that *a sloping side element like DEB is equivalent to an infinitely long extension on that side*. Thus the full segmental areas shown in shading masks can be established not by an impossible infinitely long horizontal element, but by a horizontal element the width of the window and two sloping elements. In our case, such an element would eliminate the necessity for a massive 266-cm [164 in.] side extension.

References

Cowan, H. J., Editor. (1980). *Solar Energy Applications in the Design of Buildings,* Applied Science Publishers, London.

Harkness, E., and Mehta, M. (1978). *Solar Radiation Control in Buildings,* Applied Science Publishers, London.

Lim, B. P., Rao, K. R., Tharmaratnam, K., and Mattar, A. M. (1979). *Environmental Factors in the Design of Building Fenestration,* Applied Science Publishers, London.

Olgyay, A., and Olgyay, V. (1951). *Solar Control and Shading Devices,* Princeton University Press, Princeton, N.J.

Ramsey, C. G., and Sleeper, H. R. (1981). *Architectural Graphic Standards,* 7th ed., AIA/Wiley, New York.

Sun Angle Calculator. (1975). Libbey-Owens-Ford. Available from LOF, 811 Madison Ave., Toledo, OH 43695.

Szokolay, S. V. (1980). *Environmental Science Handbook,* Halsted/Wiley, New York.

E

ECONOMIC ANALYSIS

E.1 Economic Decisions

As has been pointed out repeatedly in this text, economic factors are a major consideration in almost every decision in the building design process. The discussion below is necessarily brief but covers the essential principles. For in-depth study of economic analyses in the design and construction process, the reader is referred to the references at the end of the appendix. Publications of the National Bureau of Standards, (Marshall and Ruegg, 1980a, 1980b, and Ruegg et al., 1978), are the sources of some of the principles explicated below, and are so referenced.

The most common economic questions are;

1. How can the cost of different systems *that produce the same end result* be compared? This may apply to the purchase of a single item such as a motor, to the lighting of a room, or to the choice of the type of HVAC installation for an entire system. The principle involved is the same.

2. Assuming that improving a system, existing or proposed, will reduce operating costs and that many different types of improvement are possible, what preliminary economic guidelines can be established to determine whether any proposed investment appears cost effective? Stated otherwise, how can the initial, simple rate of return of a proposed investment be determined so that it can be compared to the *minimum expected rate of return* on investment?

3. Having determined in answer to question 2 that a number of different proposals, each with its own investment and payback (savings) figures, are apparently cost effective, how can these be compared to determine which is *most* cost effective? Put another way, which proposal has the maximum *net benefit or savings*? This is the type of decision that must be made when comparing energy conservation projects, such as the lighting control systems described in Section 21.4.

4. Having established which of the proposals is most cost effective, how can the *actual* rate of return of the investment be determined? This figure, often referred to as *internal rate of return (IRR),* can then be compared to the required rate of return, to determine whether the investment is economically desirable.

5. What is the *payback period* of the investment?

These questions are considered individually in the discussion below.

E.2 Life-Cycle Cost (LCC)

Cost comparisons made on the basis of first cost are legitimate only when operating costs do not bear consideration, either because there are none or because the builder is not the owner. Proposals for pouring a concrete slab might represent the former situation, and speculative construction for immediate sale the latter. In such cases the technique of cost comparison is simple. However, when operating costs are involved that vary from system to system, the only reasonable way to determine and compare system costs is by *life-cycle costing*. The basic theory and calculation techniques are well known (see AIA, 1977; Marshall and Ruegg, 1980a, 1980b; Ruegg, 1978), so they will be reviewed only briefly here.

Stated simply, a life-cycle cost represents the total cost of an item or system over its entire life cycle, that is, the sum of first cost and all future costs, less salvage. When the life cycle of the system being studied does not correspond to that of the building in which it is installed (as is usually the case), the designer must decide which life cycle to use. For instance, a lighting system has an estimated life of 15 to 20 years, whereas the building life is usually at least double that. It is usually wiser to make the study over the shorter period, since almost certainly, changes in technology will present an entirely different picture when the time comes to replace the system. This is true of most mechanical/electrical systems being designed today.

Life-cycle cost, in equation form is

$$LCC = IC + MC + AC + OC \quad \text{(E.1)}$$

where

LCC = life-cycle cost
IC = investment cost, (i.e., first cost minus salvage)
MC = maintenance and repair cost
AC = amortization (replacement) cost (not always included in life-cycle cost calculation)
OC = operating costs, including labor and energy

All costs are expressed in terms of present value, which is explained below. Costs can also be annualized. We use the present value format because we feel it gives a clearer answer to the economic decision questions posed above. When calculating costs, the designer must include related ancillary costs. For instance, in comparing lighting systems, the cost figures must reflect the impact on the HVAC system and the wiring system. (For lighting system cost comparisons, see IES of North America, 1981.)

When the life cycle is short—say two years maximum, the last three components of Equation E.1 can be assumed to be constant without introducing an intolerable error. However, for longer periods this is obviously not a valid assumption because of escalating labor and energy costs. Since these two factors weigh heavily in all economic comparisons, and since they generally exist in different proportions in the systems being considered, escalation in their costs will obviously change the economic balance when costs are projected over the equipment life. It is therefore necessary to change the usual present value formula to reflect this escalation.

The present value of future payments of 1.00, made at the *end* of a series of *n* periods, is

$$PV = \frac{(1 + i)^n - 1}{i\,(1 + i)^n} \quad \text{(E.2)}$$

where

PV = present value
i = interest (discount) rate expressed as a decimal (e.g., 8% = 0.08)
n = number of periods

The factor i is more accurately the discount rate rather than the simple interest rate, the difference being that the latter reflects inflation, either actual or anticipated. If, however, the future payments escalate, the expression becomes:

$$PV = \frac{1 - \left(\frac{K}{1 + i}\right)^n}{\frac{1 + i}{K} - 1} \quad \text{(E.3)}$$

where K is the escalation rate per period, and the remaining items are as in Equation E.2. Thus for a 5% annual escalation, $K = 1.05$, and so on. The values of PV are tabulated in Table E.1 for interest (discount) rates of 8, 10, and 12%, and escalation rates of 3, 5, 8, and 10%. Other values can be calculated from the equations. Table 15.5, Table 16.4, Table 19.13, and Table 19.17 were calculated assuming a 3% annual energy cost escalation.

It is apparent that the result of a life-cycle-cost analysis depends heavily on accurate forecasting when using escalation and that comparative results can readily shift not only with escalation estimates of the various components but also with the length of the life cycle.

E.3 Initial (Simple) Rate of Return

Addressing the second question in Section E.1, we wish to determine whether to examine a proposed cost-saving investment in detail. That is, is the initial (simple) rate of return sufficient? The difference between initial and *actual* rates of return is that the former is simply

$$\text{initial rate of return} = \frac{\text{annual savings}}{\text{investment}}$$

whereas the actual (internal) rate of return depends on the life of the project and the *total* savings accrued during this life cycle. The calculation method for the latter is given in Section E.5.

Initial rate of return is a useful criterion in that it can be very easily calculated and then compared

TABLE E.1*a* **Present Value of n Future Payments Beginning at the End of the First Period**

	Interest (Discount) Rate per Period						
n	*6%*	*8%*	*10%*	*12%*	*15%*	*20%*	*25%*
1	0.94	0.93	0.91	0.89	0.87	0.83	0.80
2	1.83	1.78	1.74	1.69	1.63	1.53	1.44
3	2.67	2.58	2.49	2.40	2.28	2.11	1.95
4	3.47	3.31	3.17	3.04	2.86	2.59	2.36
5	4.21	3.99	3.79	3.61	3.35	2.99	2.69
6	4.92	4.62	4.36	4.11	3.78	3.33	2.95
7	5.58	5.21	4.87	4.56	4.16	3.61	3.16
8	6.21	5.75	5.34	4.97	4.49	3.84	3.33
9	6.80	6.25	5.76	5.33	4.77	4.03	3.46
10	7.36	6.71	6.15	5.65	5.02	4.19	3.57
11	7.89	7.14	6.50	5.94	5.23	4.33	3.66
12	8.83	7.54	6.81	6.19	5.42	4.44	3.73
13	8.85	7.90	7.10	6.42	5.58	4.53	3.78
14	9.30	8.24	7.37	6.63	5.72	4.61	3.82
15	9.71	8.56	7.61	6.81	5.85	4.68	3.86
16	10.11	8.85	7.82	6.97	5.95	4.73	3.89
17	10.48	9.12	8.02	7.12	6.05	4.78	3.91
18	10.83	9.37	8.20	7.25	6.13	4.81	3.93
19	11.16	9.60	8.37	7.37	6.20	4.84	3.94
20	11.47	9.82	8.51	7.47	6.26	4.87	3.95
21	11.76	10.02	8.65	7.56	6.31	4.89	3.96
22	12.04	10.20	8.77	7.65	6.36	4.91	3.97
23	12.30	10.37	8.88	7.72	6.40	4.93	3.98
24	12.55	10.53	8.99	7.78	6.43	4.94	3.98
25	12.78	10.68	9.08	7.84	6.46	4.95	3.99

to the *minimum* desired rate of return on investment. We emphasize *minimum,* since actual rate of return is always lower than the simple, initial rate of return, except where savings escalate during the project life cycle. If this preliminary comparison indicates that the simple rate of return exceeds the minimum by a margin of a few percent, it is likely that the project is economically viable, and a detailed analysis can be undertaken.

E.4 Cost-Effectiveness Comparisons

This analysis is undertaken in answer to question 3, that is, several projects are being considered, all of which meet the stated criterion. Which is most desirable from a cost-effectiveness viewpoint? (If only one project is under consideration, the analysis proceeds directly to the IRR calculation described in the next section.)

The difference between life-cycle project cost and life-cycle project savings, that is, net lifetime savings, is a measure of cost effectiveness. Therefore, to compare the cost effectiveness of two projects, it is necessary simply to calculate this differential. To express this as an equation, we write:

$$\text{net savings} = \text{life-cycle savings} - \text{life-cycle cost}$$

Since this is true for all projects, a comparative differential can be calculated by using the differential in each term of the life-cycle cost equation, that is,

TABLE E.1*b* **Present Value of *n* Future Payments Beginning at the End of the First Period and Escalating at *K* per Period**

	Interest (Discount) Rate per Period 8%			
	Annual (Periodic) Escalation Rate K			
n	*1.03*	*1.05*	*1.08*	*1.10*
1	0.95	0.97	1.00	1.02
2	1.86	1.92	2.00	2.06
3	2.73	2.84	3.00	3.11
4	3.56	3.73	4.00	4.19
5	4.35	4.60	5.00	5.28
6	5.10	5.44	6.00	6.40
7	5.82	6.26	7.00	7.54
8	6.50	7.06	8.00	8.70
9	7.15	7.84	9.00	9.88
10	7.78	8.59	10.00	11.08
11	8.37	9.33	11.00	12.30
12	8.94	10.04	12.00	13.55
13	9.48	10.73	13.00	14.82
14	9.99	11.41	14.00	16.11
15	10.48	12.06	15.00	17.43
16	10.95	12.70	16.00	18.77
17	11.40	13.32	17.00	20.13
18	11.82	13.92	18.00	21.53
19	12.23	14.51	19.00	22.94
20	12.62	15.08	20.00	24.39
21	12.99	15.63	21.00	25.86
22	13.34	16.17	22.00	27.35
23	13.68	16.69	23.00	28.88
24	14.00	17.20	24.00	30.43
25	14.30	17.69	25.00	32.01

TABLE E.1*c* **Present Value of *n* Future Payments Beginning at the End of the First Period and Escalating at *K* per Period**

	Interest (Discount) Rate per Period 10%			
	Annual (Periodic) Escalation Rate K			
n	*1.03*	*1.05*	*1.08*	*1.10*
1	0.94	0.96	0.98	1.00
2	1.81	1.87	1.95	2.00
3	2.63	2.74	2.89	3.00
4	3.40	3.57	3.82	4.00
5	4.12	4.36	4.73	5.00
6	4.80	5.12	5.63	6.00
7	5.43	5.84	6.51	7.00
8	6.02	6.53	7.37	8.00
9	6.57	7.18	8.22	9.00
10	7.09	7.81	9.05	10.00
11	7.58	8.41	9.87	11.00
12	8.03	8.98	10.67	12.00
13	8.46	9.53	11.46	13.00
14	8.85	10.05	12.23	14.00
15	9.23	10.55	12.99	15.00
16	9.58	11.27	13.74	16.00
17	9.90	11.48	14.47	17.00
18	10.21	11.91	15.19	18.00
19	10.50	12.32	15.90	19.00
20	10.76	12.72	16.59	20.00
21	11.02	13.09	17.27	21.00
22	11.25	13.45	17.94	22.00
23	11.47	13.80	18.59	23.00
24	11.68	14.12	19.24	24.00
25	11.87	14.44	19.87	25.00

TABLE E.1*d* **Present Value of *n* Future Payments Beginning at the End of First Period and Escalating at *K* per Period**

	Interest (Discount) Rate per Period 12%			
	Annual (Periodic) Escalation Rate K			
n	*1.03*	*1.05*	*1.08*	*1.10*
1	.92	.94	.96	.98
2	1.77	1.82	1.89	1.95
3	2.54	2.64	2.79	2.89
4	3.26	3.41	3.66	3.82
5	3.92	4.14	4.49	4.74
6	4.52	4.82	5.29	5.64
7	5.08	5.45	6.07	6.52
8	5.59	6.05	6.82	7.38
9	6.06	6.61	7.54	8.23
10	6.49	7.13	8.23	9.07
11	6.89	7.63	8.90	9.89
12	7.26	8.09	9.55	10.69
13	7.59	8.52	10.17	11.49
14	7.90	8.92	10.77	12.26
15	8.19	9.30	11.35	13.03
16	8.45	9.66	11.91	13.78
17	8.69	9.99	12.45	14.51
18	8.91	10.31	12.97	15.23
19	9.11	10.60	13.47	15.95
20	9.30	10.87	13.95	16.64
21	9.47	11.13	14.42	17.34
22	9.63	11.37	14.87	18.00
23	9.78	11.60	15.30	18.66
24	9.91	11.81	15.72	19.31
25	10.04	12.01	16.12	19.95

$$\Delta_{net} = \Delta_{savings} - \Delta_{costs}$$

Taking a very simple example, assume the following conditions.

	Project A	*Project B*
Investment	$10,000	$15,000
Life cycle	10 years	8 years
Annual savings	$2,000	$3,200
Initial rate of return	20%	21.3%
Discount rate	12%	12%

Since project B has an apparently higher rate of return, we would formulate the differential as B minus A. Using Table E.1*a,* the present worth of lifetime savings for project B (without escalation) is $3,200 × (4.97) or $15,904, and that of project A is $2,000 × (5.65) or $11,120. Therefore the differential cost benefit (CB) of B minus A is $\Delta_{life\ savings}$ minus $\Delta_{project\ costs}$, or

$$\Delta_{CB} = (\$15{,}904 - \$11{,}120) - (\$15{,}000 - \$10{,}000) = \$4{,}784 - 5{,}000 = -\$216$$

indicating that project A is more cost effective despite the apparently higher rate of return of project B. If we include escalation in costs, obviously the life savings of the longer life project (A) will shrink more than the savings of shorter life project B. Using Table E.1*d,* we find that the projects are about equal, with 3% annual escalation. Above that, project B is more cost effective.

E.5 Internal Rate of Return (IRR)

This figure represents the rate of return at which lifetime savings are exactly equal to lifetime costs. It cannot be calculated directly, but only by trial and error. We must assume a rate of return, calculate present value of savings, and compare PV to lifetime costs. Because of this difficulty, IRR is frequently neglected, and initial rate of return used instead, although as we saw above, this can readily give misleading results. To illustrate calculation of IRR, we will return to our very simple example and calculate IRR for projects A and B.

For project A, we need a present worth factor of $10,000/$2,000 = 5, in 10 years. Referring to Table E.1*a,* we see that at a discount (interest) rate of 15%, the present worth factor is 5.019, indicating that IRR is slightly above this. (It is actually 15.1%.) For project B we need an 8-year present worth factor of $15,000/$3,200 or 4.69. Referring again to Table E.1*a,* we see that this factor falls between 12 and 15%. By a series of trials with Equation E.2, we arrive at 13.7%, which is 1.4% *less* than the same factor for project A, confirming the conclusion reached in the preceding step.

E.6 Payback Period

The payback period referred to in most proposals is the reciprocal of the initial rate of return and is usually referred to as the *simple* payback period. However, just as we have seen that the actual rate of return differs from the simple rate (and is usually less than it), so the actual or *discounted payback period* is different, and usually longer than the simple one. This period is, logically, the period of time required for the accumulated net savings to equal the initial investment, with all figures expressed in present value dollars. To illustrate, let us return to the example given above.

1. *Proposal A.* As in the IRR calculation, we need a present worth factor of $10,000/$2,000, or 5, at a 12% discount rate. From Table E.1*a* we find this to be about 8.1 years. This compares to a simple payback period of 5 years. Note that this is *not* the reciprocal of internal rate of return.
2. *Proposal B.* Here we need a present worth factor of $15,000/$3,200 or 4.7. At 12% discount rate, from Table E.1*a,* this is about 7.3 years. This compares to a simple payback period of 4.7 years.

Note from these results the very important fact that a shorter payback period does not necessarily indicate a more cost-effective investment. Here, proposal A has a better return precisely because it yields good savings for a longer period. Summing up the results of our simple study, we have:

	Project A	*Project B*
Initial rate of return	20%	21.3%
Internal rate of return	15.1%	13.7%
Simple payback period	5 years	4.7 years
Discounted payback period	8.1 years	7.3 years

This shows very clearly that although initial rate of return and simple payback period are indications of a proposal's value, they are useless in comparative studies.

References

American Institute of Architects. (1977). *Life Cycle Cost Analysis—A Guide for Architects,* AIA, Washington, D.C.

Edison Electric Institute. *Interest Tables, 0–25%,* EEI, New York.

Illuminating Engineering Society of North America. (1981). *Lighting Handbook, Application* Volume, 6th ed., Section 3, IES of North America, New York.

Marshall, H. E., and Ruegg, R. T. (1980a). *Energy Conservation in Buildings: An Economics Guidebook for Investment Decisions,* National Bureau of Standards Handbook 132, NBS, Washington, D.C.

Marshall, H. E., and Ruegg, R. T. (1980b). *Simplified Energy Design Economics,* National Bureau of Standards Special Publication 544, NBS, Washington, D.C.

Ruegg, R. T., et al. (1978). *Life Cycle Costing,* National Bureau of Standards 113, NBS, Washington, D.C.

F

SOUND TRANSMISSION DATA FOR WALLS

To use Appendix F:

1. Find desired construction Type Code in Index F.1.
2. Find desired STC corresponding to the selected construction Type Code.
3. Refer to Table F.1 for details of wall construction.
4. Refer to Index F.2 for wall thickness weight, STC, fire rating, and the transmission losses at standard octave midpoints.

EXAMPLE. An interior masonry partition with an STC between 50 and 55 is desired. Of particular interest is the TL at 1000 Hz.

1. From Index F.1, we find Type Code "c."
2. From Index F.1, we note that constructions W6, W12b, W15b, and W22b have appropriate STC ratings.
3. From Table F.1 we decide that construction W15b is most suitable for the proposed use.
4. From Index F.2 we find that construction W15b has a transmission loss of 55 db at 1000 Hz.

APPENDICES

INDEX F.1 **Sound Transmission Class: Walls**

Type Code:

a. *Wooden stud* **b.** *Metal stud* **c.** *Masonry* **d.** *Concrete* **e.** *Staggered stud* **f.** *Plaster* **g.** *Gypsum wallboard* **h.** *With resilient element* **i.** *Absorbent blankets or fill* **j.** *Fiber board* **k.** *Lead* **l.** *Gypsum core board* **m.** *Double wall*

STC	Type	Item No.	STC	Type	Item No.
63	d,f	W4	46	a,f	W30a
62[a]	c,f,j,m	W23	46	a,f	W31a
56	c	W7	46	a,f,k	W31c
56[a]	c,f	W8	46	a,e,g,i	W37
55[a]	b,g,i	W63	45	c,d	W10b
54[a]	c,f,m	W22b	45	a,f,i	W38
54	b,f,h	W52	45	b,f,h	W50a
53[a]	d,f	W2	45	g,l,m	W85a
53	c,f,h	W15b	44	c	W12a
52[a]	c,f	W6	44	a,e,g	W34a
51	b,f	W44a	44	a,e,g	W35a
51	b,g,h,i	W67	44	a,f,h	W40a
50	c,k	W12b	43	c,d	W10a
50	b,g,i	W60	43[a]	d,j,m	W21
50	g,i,l,m	W85b	43	a,e,f	W35a
49[a]	c,f,m	W22a	43	a,e,f	W36a
48	c,d	W9	43	b,f,k	W43b
48	a,e,f,i	W36b	42[a]	c,f	W5
48	b,f,k	W43c	42	a,f	W14a
47	d	W1	41	b,f	W43a
47	a,g,k	W28b	41	b,g	W55a
47	a,f,k	W31b	40	c,f	W13
47	a,g,h	W39a	40	a,g	W32a
46	c,f	W11a	39	a,g	W28a
46	c,f,h	W15a	36	g,l	W80

[a]Field measurement.

INDEX F.2 **Sound Transmission Loss: Walls**

Designation	Thickness (in.)	Weight (lb/ft²)	Transmission Loss (db) 125	250	500	1K	2K	4K	STC	Fire Rating hr
W1	3	39	35	40	44	52	58	64	47	½
W2	7	80	39	42	50	58	64	—	53	3
W4	Approx. 16	184	50	54	59	65	71	68	63	4+
W5	5½	55	34	34	41	50	66	—	42	2.5
W6	10	100	41	43	49	55	57	—	52	4+
W7	12	121	45	45	53	58	60	61	56	4+
W8	25	280	50	53	52	58	61	—	56	4+
W9	12	79	46.5	44	46	52	54	56	48	4
W10a	6	34	32	33	40	47	51	48	43	1
W10b	6	34	37	36	42	49	55	58	45	1
W11a	5¼	35.8	36	37	44	51	55	62	46	2
W12a	3¾	26.1	40	40	40	48	55	56	44	1.5
W12b	5	31	41	46	46	56	63	67	50	1.5
W13	4	21.5	39	34	38	43	48	46	40	3
W14a	5	23.4	37	42	39	44	49	49	42	4
W15a	5	27	38	37	44	51	56	59	46	3
W15b	6	31	45	44	50	55	56	59	53	4
W20	7	22.9	32	46	49	53	58	66	52	3
W21	10¼	37	41	42	46	51	52	—	43	Not available
W22a	12	100	37	41	48	60	60.5	—	49	4+
W22b	12	100	40	44	55	67.5	70	—	54	4+
W23	18	120	48	54	58	64	69	—	62	4+
W28a	5	6	21	28	35	42	45	41	39	0.5
W28b	Approx. 5⅛	12	27	37	43	52	56	—	47	0.5
W30a	5¾	13.4 to 15.7	32	37	42	47	47	63	46	0.75
W31a	5¾	13.4 to 15.7	32	37	42	48	48	63	46	0.75
W31b	Approx. 5⅞		33	41	45	52	55	65	47	0.75
W31c	Approx. 5⅞	17–19	32.5	40	43	47	50	62	46	0.75
W32a	5½	8.2	27	31	39	45	52.5	48	40	1
W34a	5	6.2	36	36	40	47	52	45	44	0.5
W35a	6½	13.4	41	41	46	49	41	54	44	1.5
W35b	5¾	15.6	48	46	48	48	48	59	43	1
W36a	6¼	11.1	36	33	42	42	41	51	43	0.75
W36b	6¼	12.8	37	37	49	50	52	66	48	1
W37	5¾	13.8	39	40	42	47.5	55	51.5	46	0.5
W38	5⅜	14.2	39	45	48	50	44	54	45	1
W39a	6¼	6.7	30	40	46	50	49	49	47	1
W40a	Approx. 6½	14.4	43	41	48	50	42	56	44	1
W43a	3⅜	12.3	27	37	43	46	39	47	41	0.75
W43b	3½	15.2	35	43	45	47	48	58	43	0.75
W43c	Approx. 3½	18.2	36	45	47	50	53	61	48	0.75
W44a	5	15.7	34	38	47	50	52	58	51	1
W50a	5¼	13	35	46	48	51	48	43	45	0.75
W52	5	19	50	52	55	56	52	60	54	1
W55a	4⅞	6	29	36	40	46	40	46	41	1
W60	3½	5.4	34	40	47	50	53	54	50	1
W63	6⅛	11.5	36	47	51	57	57	62	55	2
W67	6½	11.3	41	46	49	51	50	60	51	2
W80	2¼	10.2	34	34	37	38	39	45	36	1
W85a	5⅛	14.6	36	35	45	51	53	57	45	3
W85b	6	12.8	37	37	54	56	56	62	50	3

TABLE F.1 **Acoustic Characteristics of Walls: For Other Data See Index F.2 (Continued)**

Designation	Description	Section Sketch
Solid Concrete		
W1	3-in.-thick solid concrete wall poured in situ in test opening. All surface cavities were sealed with thin mortar mix.	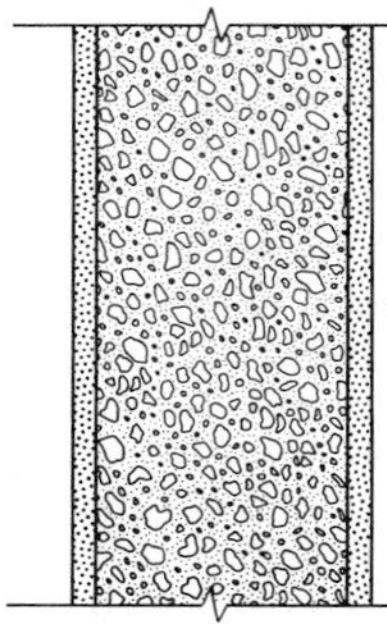 W 1, W 2
W2	6-in.-thick concrete wall with ½-in.-thick layer of plaster on both sides.	
W4	Wall of 4, 6, and 8 × 8 × 16 in. sand and gravel aggregate solid concrete blocks; on each side, ¼- to ½-in.-thick layer of cement gypsum plaster and sand.	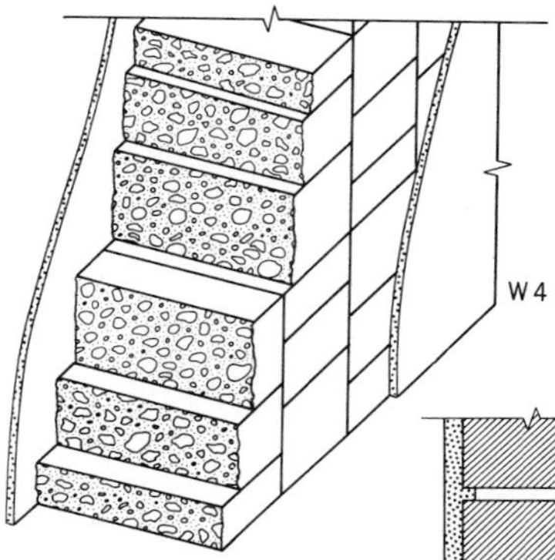 W 4
W5	4½-in.-thick brick wall with ½-in.-thick layer of plaster on each side.	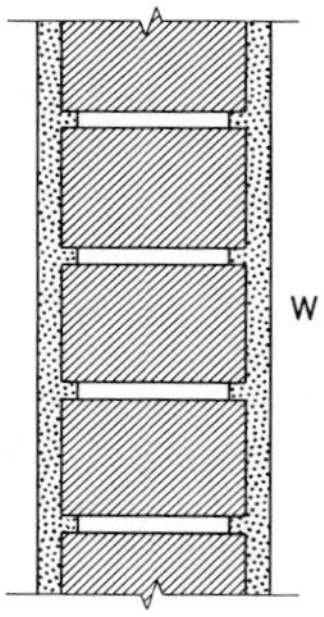 W 5, W 6
W6	9-in.-thick brick wall with ½-in.-thick layer of plaster on each side	
W7	12-in.-thick brick wall.	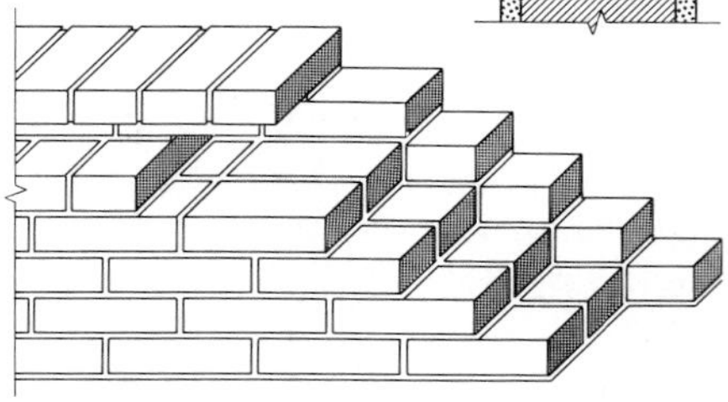 W 7
W8	24-in.-thick stone wall with ½-in.-thick layer of plaster on both sides.	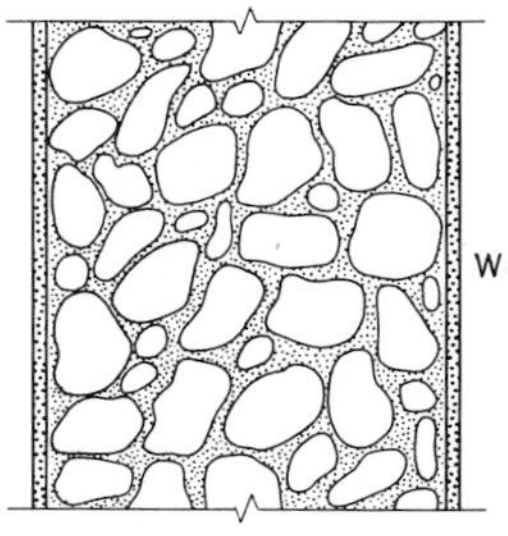 W 8

TABLE F.1 **Acoustic Characteristics of Walls: For Other Data See Index F.2 (Continued)**

Designation *Description* *Section Sketch*

Hollow Concrete Block

W9 12-in. wall made of hollow 8 × 8 × 12 in. and 8 × 4 × 16 in. concrete blocks.

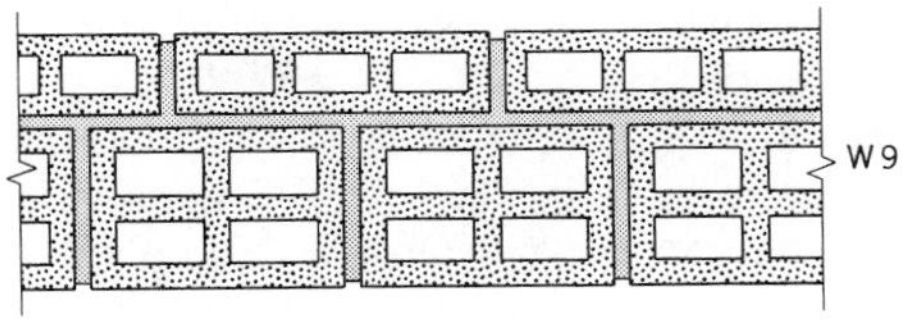

W10a 6-in. hollow concrete blocks constructed with vertical mortar joints staggered.

W10b Similar to W10a except wall painted.

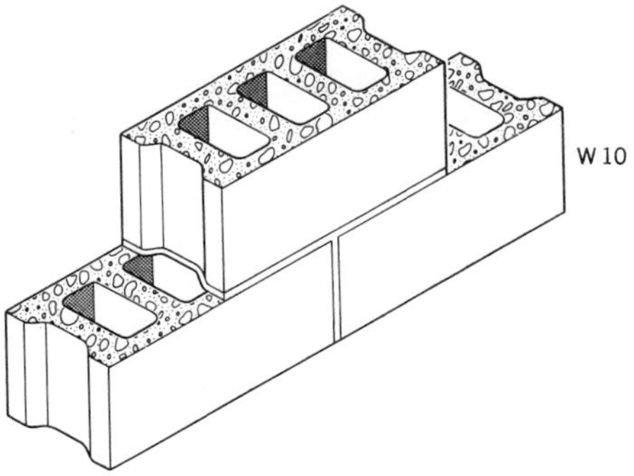

Cinder Block

W11a 4 × 8 × 16 in. hollow cinder blocks; on each side, 5/8 in. of sanded gypsum plaster.

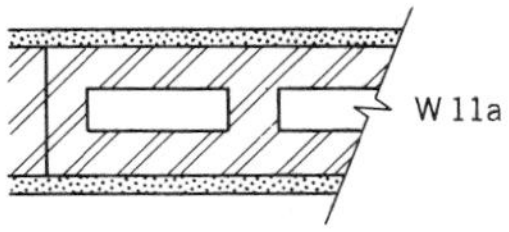

Cement Block

W12a 3 5/8 × 7 3/4 × 13 1/2 in. lightweight-aggregate cement blocks with 1/2-in. mortar joints; three coats of masonry paint applied to each side of partition.

W12b Same as W12a except 1 × 2-in. furring strips were nailed vertically to partition on one side; 1/16-in. layer of lead, 3.94 lb/ft^2, nailed to furring strips, 1/4-in. plywood-covered lead with joints caulked.

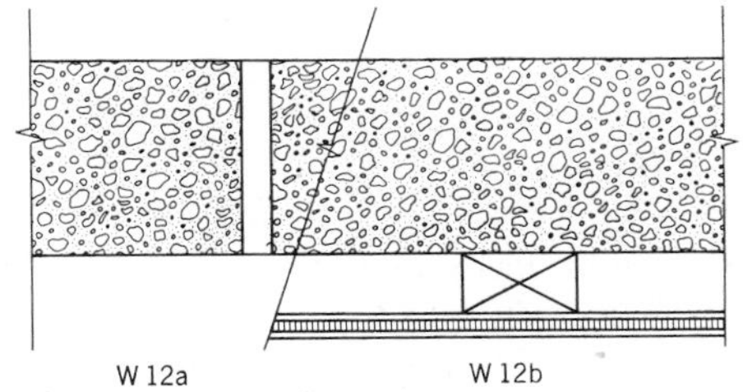

Hollow Gypsum Block

W13 3-in. hollow gypsum blocks cemented together with 3/8-in. mortar joints; on each side, 1/2-in. sanded gypsum plaster.

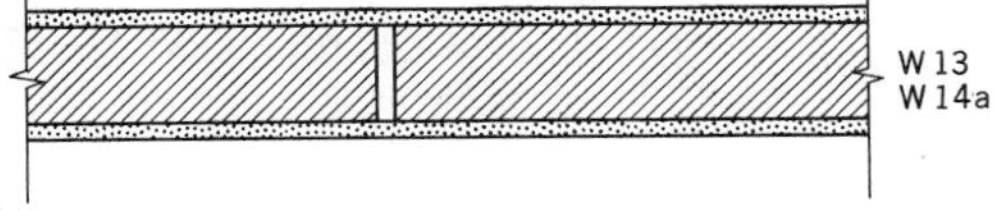

W14a 4-in. hollow gypsum blocks cemented together with 3/8-in. mortar joints; on each side, 1/2-in. sanded gypsum plaster.

TABLE F.1 **Acoustic Characteristics of Walls: For Other Data See Index F.2 (Continued)**

Designation	Description	Section Sketch
Hollow Gypsum Block, Resilient One Side, Plaster Both Sides		
W15a	3 × 12 × 30 in. hollow gypsum blocks with $\frac{1}{2}$-in. mortar joints. On one side $\frac{7}{16}$-in. sanded gypsum plaster; on the other side resilient clips, spaced 18 in. on centers vertically and 16 in. on centers horizontally, held to $\frac{3}{4}$-in. metal channels 16 in. on centers, to which expanded metal lath was wire-tied; $\frac{11}{16}$-in. sanded gypsum plaster. $\frac{1}{16}$-in. white-coat finish applied to both sides.	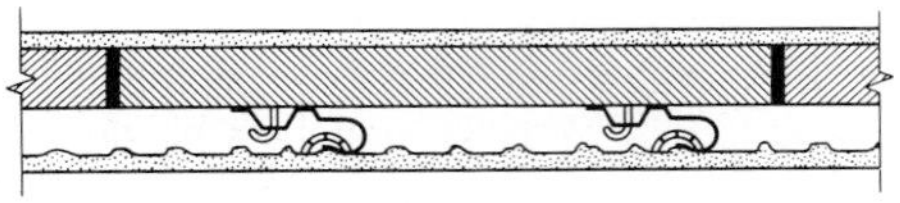 W 15,
W15b	Similar to W15a except 4 × 12 × 30 in. gypsum blocks were used.	
Hollow Concrete		
W21	Precast concrete hollow wall panels with in situ concrete posts and beams. Panels have $1\frac{1}{2}$-in.-thick concrete shells with $6\frac{1}{4}$-in. air space between them. Layer of fiberboard $\frac{1}{2}$-in. thick is adhered to the exposed surfaces of the panel.	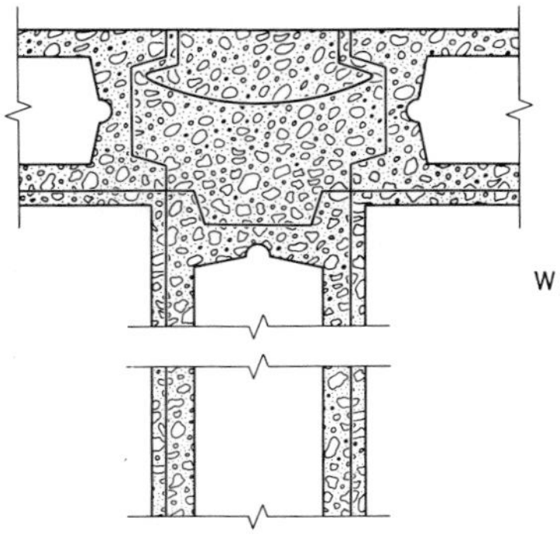 W 21
Double Walls		
W22a	Double wall with $4\frac{1}{2}$-in.-thick brick leaves separated by a 2-in. cavity (wire ties between leaves); $\frac{1}{2}$-in. plaster on exposed sides.	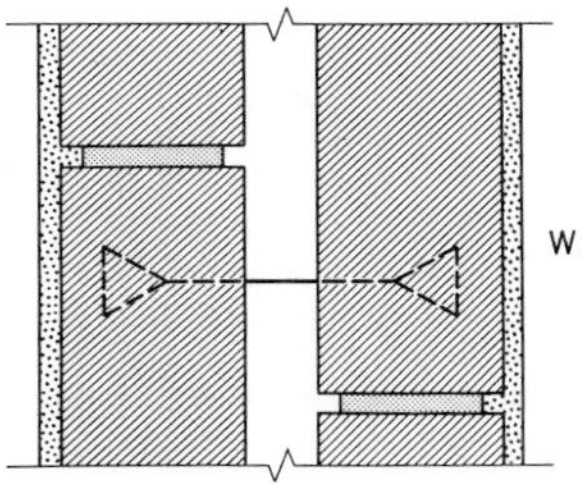 W 22
W22b	Similar to W22a but no wire ties between the leaves.	
W23	Double wall with $4\frac{1}{2}$-in.-thick brick leaves, 6-in. cavity (no ties); on exposed sides, $\frac{1}{2}$-in. plaster on 1-in.-thick wood-wool slabs mortared to the brick walls.	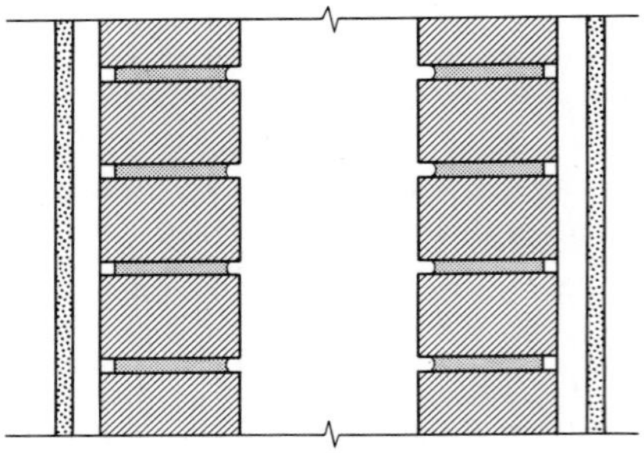 W 23
Wood Stud Walls		
W28a	2 × 4 in. wooden studs, 16 in. on centers, $\frac{1}{2}$-in. gypsum wallboard nailed to each side. All joints taped and finished.	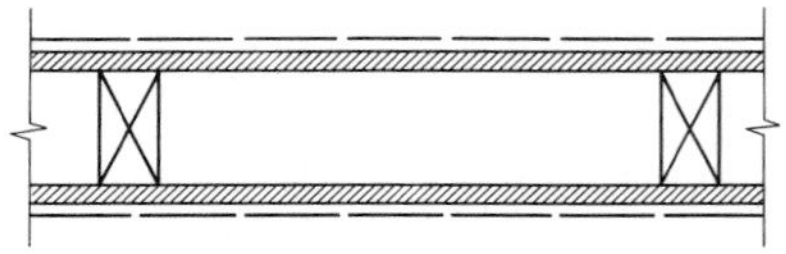 W 28
W28b	Similar to W28a except a layer of lead, 2.95 lb/ft^2, was laminated to each side of panel.	

TABLE F.1 **Acoustic Characteristics of Walls: For Other Data See Index F.2 (Continued)**

Designation	*Description*	*Section Sketch*
W29	2 × 4 in. wooden studs 16 in. on centers attached to 2 × 4 in. wooden floor and ceiling plates, ⅝-in. tapered-edge gypsum wallboard nailed 7 in. on centers to both sides of studs. All joints taped and finished.	
W30a	2 × 4 in. wooden studs, 16 in. on centers, attached to 2 × 4 in. wooden floor and ceiling plates, ⅜-in. gypsum lath nailed to studs on both sides, ½-in. sanded plaster with white-coat finish.	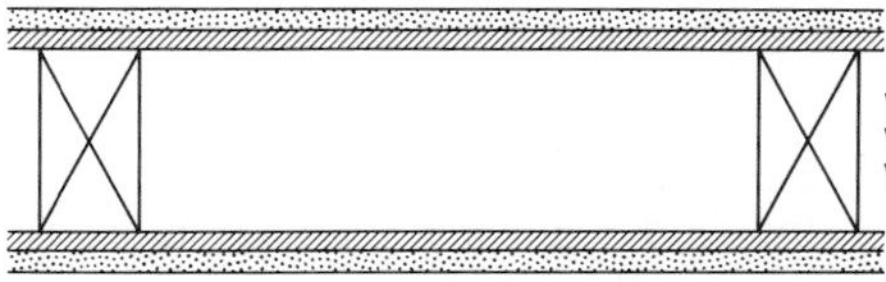 W 30 W 31 W 32
W31a	2 × 4 in. wooden studs, 16 in. on centers, ⅜-in. gypsum lath nailed to studs on both sides, ½-in. sanded plaster with white-coat finish.	
W31b	Similar to W31a except a 0.065-in.-thick layer of lead weighing 3.85 lb/ft^2 was laminated to each side of panel.	
W31c	Similar to W31a except a 0.13-in.-thick layer of lead weighing 7.9 lb/ft^2 was laminated to one side of panel.	
W32a	2 × 4 in. wooden studs, 16 in. on centers; on each side two layers of ⅜-in. gypsum wallboard cemented together; joints in exposed surfaces taped and finished.	
Staggered Wood Stud Walls		
W34a	2 × 3 in. wooden studs, 16 in. on centers, staggered 8 in. on centers, attached to 2 × 4 in. wooden plates at ceiling and floor; ½-in. gypsum wallboard nailed 7 in. on centers on both sides to studs. All joints taped and finished.	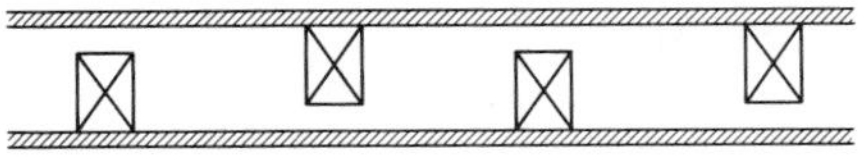 W 34a
W35a	2 × 3 in. wooden studs, 16 in. on centers, staggered 8 in. on centers (attached to 2 × 4 in. wooden plates at floor and ceiling) two layers of ⅝-in. tapered-edge gypsum wallboard, first layer nailed 7 in. on centers, second layer nailed 16 in. on centers. All exposed joints taped and finished.	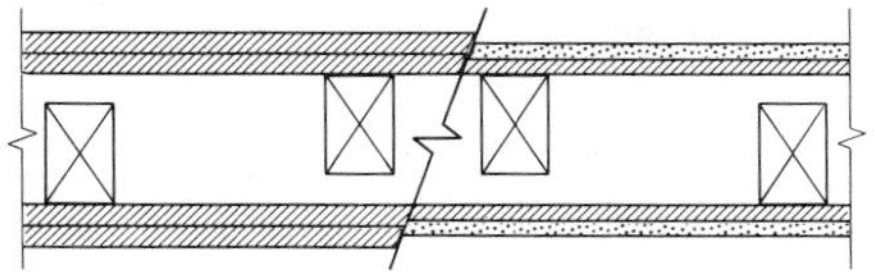 W 35a W 35b
W35b	Similar to W35a except the wall was constructed with ⅜-in. perforated gypsum lath and ½-in. sanded gypsum plaster with white-coat finish.	

TABLE F.1 **Acoustic Characteristics of Walls: For Other Data See Index F.2 (Continued)**

Designation	*Description*	*Section Sketch*
W36a	2 × 4 in. wooden studs, 16 in. on centers, staggered 8 in. on centers, and offset ½ in. On each side ⅜-in. gypsum lath nailed to studs, ½-in. gypsum vermiculite plaster, machine applied, and a hand-applied white-coat finish.	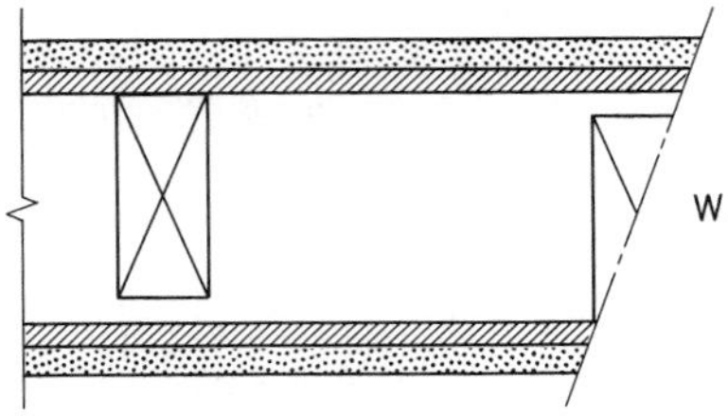 W 36a
W36b	Same as W36a except the space between the studs contained vermiculite fill with a density of 6.3 lb/ft³.	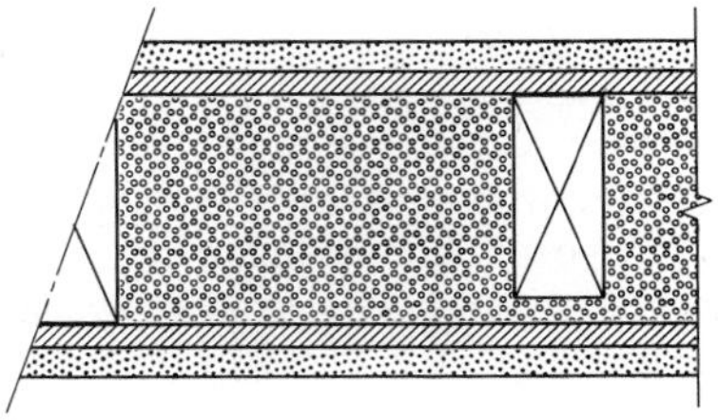 W 36b
W37	2 × 4 in. wooden studs, 16 in. on centers, staggered 8 in. on centers, attached to 2 × 4¾ in. wooden floor and ceiling plates; ½-in. gypsum wallboard nailed on both sides to studs, 0.9-in. wood-fiber wool blanket stapled on the inside of one side of the wall. All joints taped and finished.	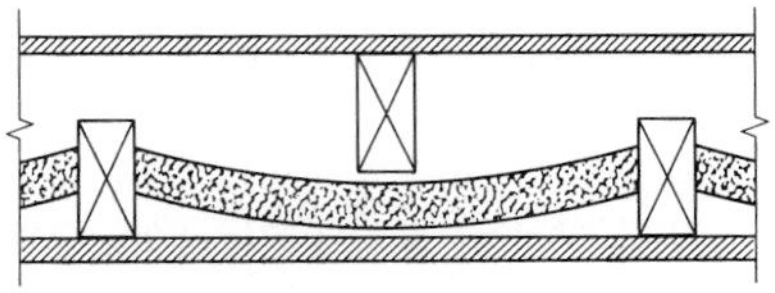 W 37
Slotted Wood Studs		
W38	2 × 4 in. slotted wooden studs, 16 in. on centers, attached to 2 × 4 in. wooden floor and ceiling plates; ⅜-in. gypsum lath nailed 7 in. on centers to studs, ½-in. gypsum plaster with white-coat finish applied to both sides. 3-in. mineral fiber batts stapled between studs.	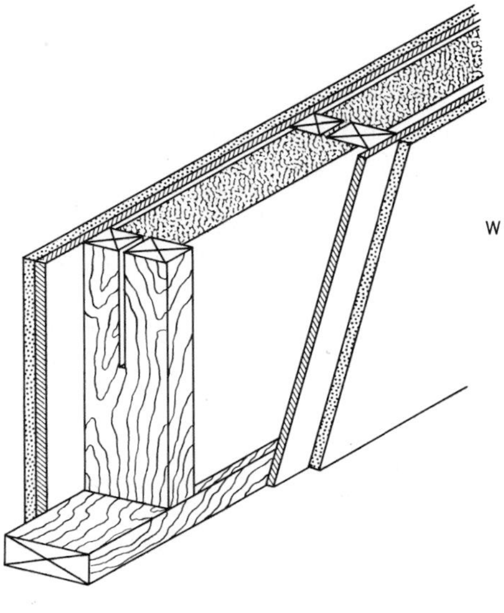 W 38

TABLE F.1 **Acoustic Characteristics of Walls: For Other Data See Index F.2 (Continued)**

Designation	*Description*	*Section Sketch*
Wood Studs; Resilient Mounting		
W39a	2 × 4 in. wooden studs, 16 in. on centers, attached to 2 × 4 in. wooden floor and ceiling plates; resilient channels nailed horizontally to both sides of studs 24 in. on centers, ⅝-in. gypsum wallboard screwed 12 in. on centers to channels. All joints taped and finished.	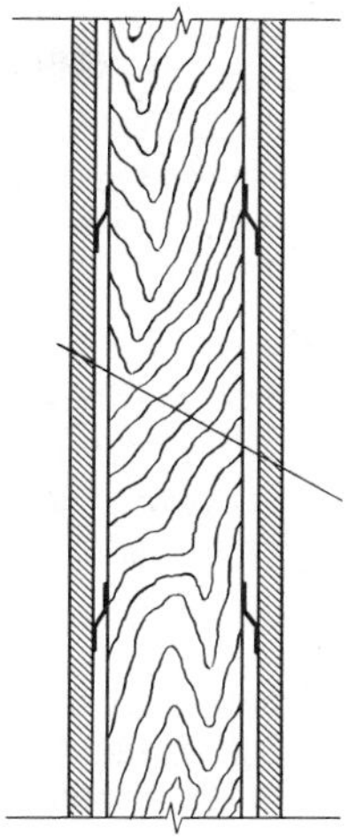 W 39a
W40a	2 × 4 in. wooden studs, 16 in. on centers; resilient clips, nailed to studs on both sides, held ⅜-in. gypsum lath, ½-in. sanded gypsum plaster with white-coat finish.	 W 40a
Steel Truss Stud Wall		
W43a	1⅝-in. steel truss studs; ⅜-in. gypsum lath, ½-in. plaster on both sides.	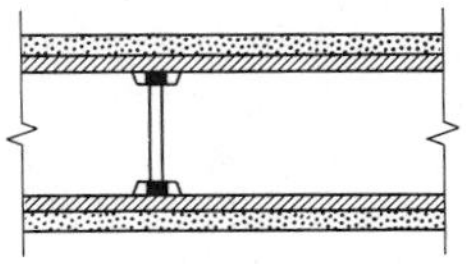 W 43 W 44
W43b	Similar to W43a except a layer of lead, 2.95 lb/ft^2, was laminated to one side of partition.	
W43c	Similar to W43a except a layer of lead, 2.95 lb/ft^2, was laminated to each side of partition.	
W44a	3¼-in. steel truss studs, 24 in. on centers, attached to metal floor and ceiling tracks; on both sides ⅜-in. perforated gypsum lath attached with wire clips wire-tied to studs, ½-in. sanded gypsum plaster.	
Steel Truss Studs; Resilient Mountings		
W50a	2½-in. steel truss studs 16 in. on centers, ⅜-in. gypsum lath attached with resilient clips to studs, ½-in. plaster applied to both sides.	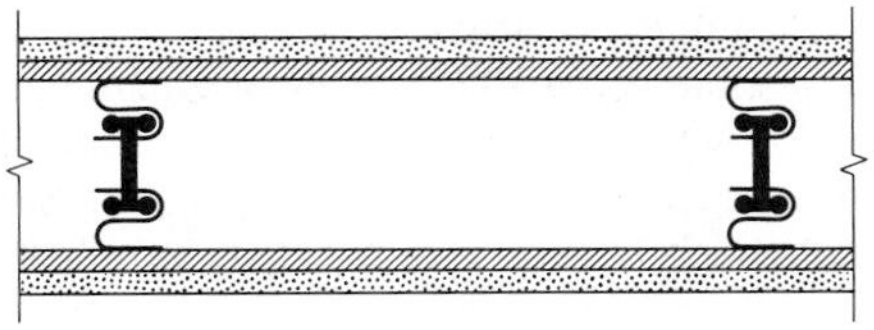 W 50a

TABLE F.1 **Acoustic Characteristics of Walls: For Other Data See Index F.2 (Continued)**

Designation	*Description*	*Section Sketch*
W52	3¼-in. steel truss studs, 16 in. on center; on each side resilient clips fastened 16 in. on centers to studs, ¼-in. metal rod wire-tied to clips, diamond mesh metal lath wire-tied to metal rods, ¾-in. sanded gypsum plaster.	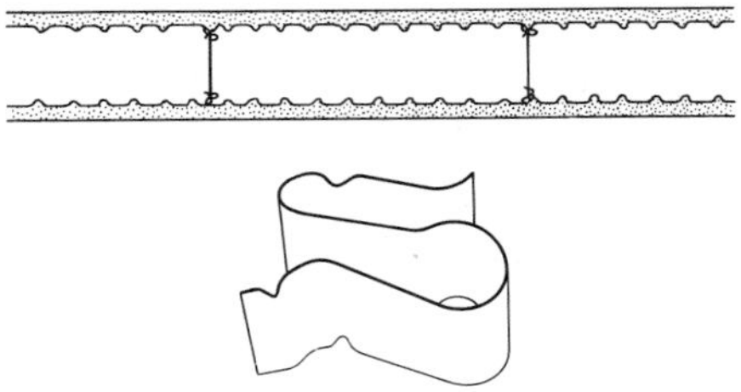 W 52
Metal Channel Stud Wall		
W55a	3⅝-in. metal channel studs, 24 in. on centers, set into 3⅝-in. metal floor and ceiling runners; ⅝-in. gypsum wallboard screwed to studs on both sides. All joints taped and finished.	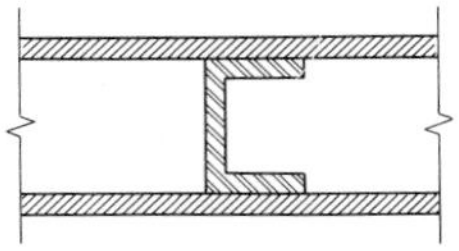 W 55a
W60	2½-in. metal channel studs, 24 in. on centers, set in 2½-in. metal floor and ceiling runners; ½-in. vinyl-coated gypsum wallboard adhesively attached and screwed to studs on both sides. All joints sealed with caulking compound. Aluminum batten strips screwed 12 in. on centers to gypsum board at joints; top and bottom finished with aluminum ceiling and base trim. 2-in. mineral fiber blankets hung between studs.	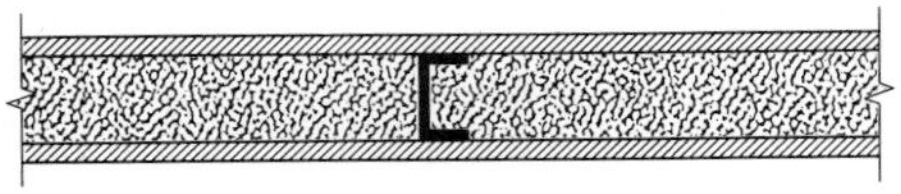 W 60
W63	3⅝-in. metal channel studs, 24 in. on centers, set into 3⅝-in. metal runners, which were attached through continuous beads of nonsetting resilient caulking compound to floor and ceiling, respectively. Two layers of ⅝-in. gypsum wallboard attached to both sides of studs. First layer screwed 8 in. on centers at joints and 12 in. on centers in field; second layer laminated and screwed 24 in. on centers to first layer, with joints staggered 24 in. 1½-in.-thick mineral fiber felt, 3 lb/ft^3, stapled between studs. All exposed joints taped and finished. The ¼-in. clearance around the perimeter closed with a nonsetting resilient caulking compound.	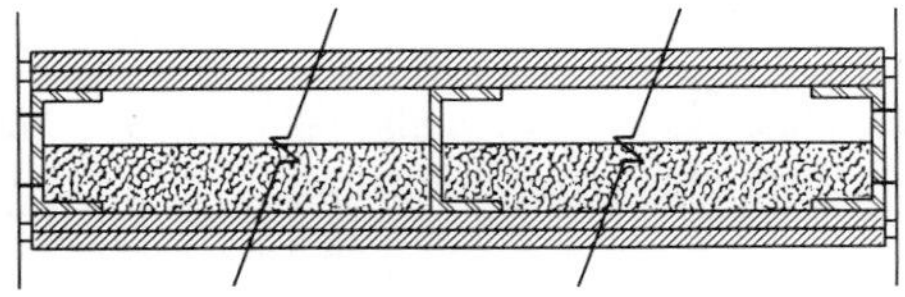 W63
W67	3⅝-in. metal channel studs, 24 in. on centers, set in 3⅝-in. metal floor and ceiling runners; ⅝-in. gypsum wallboard screwed to studs on both sides. On one side, resilient channels screwed horizontally, 24 in. on centers to inner layer; ⅝-in. gypsum wallboard screwed to channels. On the other side, ⅝-in. gypsum wallboard laminated directly to inner layer. 3-in. mineral fiber blankets hung between studs. All exposed joints taped and finished.	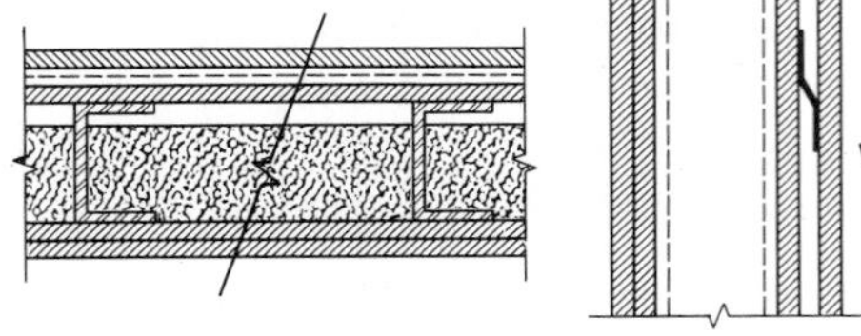 W 67

TABLE F.1 **Acoustic Characteristics of Walls: For Other Data See Index F.2 (Continued)**

Designation	*Description*	*Section Sketch*
Gypsum Partitions		
W80	24-in.-wide panels constructed of 1 × 24 in. gypsum core board offset 1½ in. at edges to form tongue-and-groove edge; ⅝-in. vinyl-faced gypsum wallboard laminated to both sides of core board. Panels inserted into two-piece metal floor and ceiling tracks. Gypsum to gypsum screws at ¼ and ½ points along vertical edges of face boards.	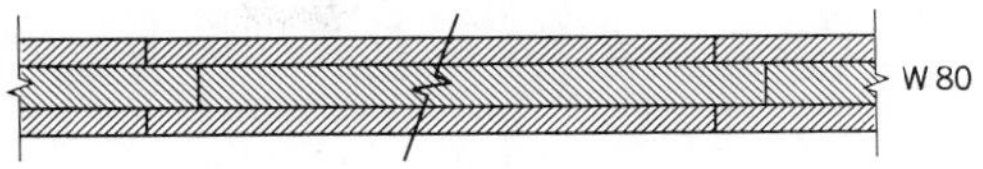
W85a	Double wall with 1⅜-in. air space. Each leaf consisted of 24-in.-wide panels of ⅝-in. gypsum core board strips, 7½ and 4⅜ in. wide, offset 1½ in. at edges to form tongue and groove; ⅝-in., vinyl-faced, gypsum wallboard laminated to both sides of core board strips. Panels screwed 12 in. on centers to 1¼ × 1 in. angle floor and ceiling runners.	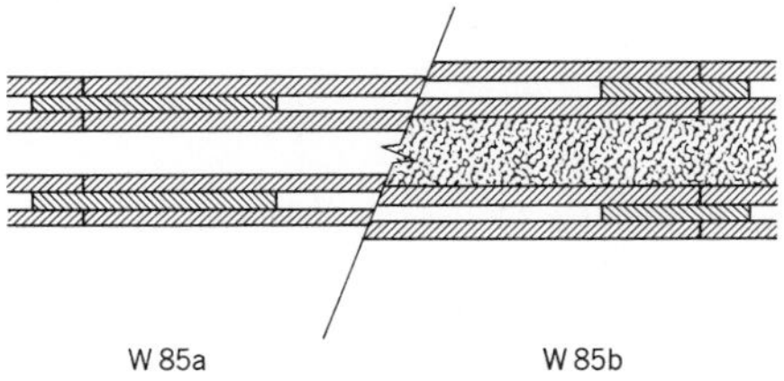
W85b	Similar to W85a except space between leaves was 2⅛ in. and contained 2-in. mineral fiber blankets stapled to one leaf. ¼-in. perimeter clearance closed with a nonsetting resilient caulking compound. Vertical face layer joints sealed with joint compound.	

G

SOUND TRANSMISSION AND IMPACT INSULATION DATA FOR FLOOR/CEILING CONSTRUCTION

To use Appendix G:

1. Find desired construction Type Code in Index G.1 or G.2. Index G.1 lists the constructions by STC. Index G.2 lists the constructions by IIC.
2. Find desired STC/IIC ratings corresponding to selected Type Code.
3. Refer to Table G.1 for details of construction.
4. Refer to Index G.3 for thickness weight, STC, IIC, fire rating, and transmission loss at standard octave midpoints.

INDEX G.1 **Sound Transmission Class: Floor/Ceiling Constructions**

Type Code:

a. *Wooden joist*
b. *Metal joist*
c. *Concrete or masonry*
d. *Plaster ceiling*
e. *Gypsum board ceiling*
f. *With resilient elements*
h. *With carpeting*
i. *With absorbent blankets*
j. *With separate ceiling joists*

STC	*IIC*	*Type*	*Item No.*	*STC*	*IIC*	*Type*	*Item No.*
55[a]	57[a]	c,d,f	F14	46[a]	42[a]	c,d	F23
54[a]	64[a]	c,d,f	F17b	45	44	a,e,f	F44
52	80	a,e,i,j	F48	44	42	c,f	F3-2d
51[a]	53[a]	c,d,f	F10	44	41	c	F3-1a
51[a]	48[a]	c,d	F7a	44	80	c,h	F2-1a
50[a]	53[a]	c,e,f	F25	44	29	c	F1-c
50[a]	51[a]	c,e	F27	44	25	c	F1a
50[a]	48[a]	c,d	F16	43[a]	43[a]	a,d	F32b
49[a]	48[a]	c,d	F9	42[a]	32[a]	c	F22
48[a]	47[a]	c,d	F12	40	32	a,e,i	F40a
48	33	b,c,d	F60a	39[a]	37[a]	a,e	F34
47	62	b,c,e	F58	37	33	a,e	F38a
47[a]	42[a]	c,d	F24	37	32	a,e	F39a
47	59	b,c,d,h	F57b	34[a]	32[a]	a,e	F30
47	37	b,c,d	F57a	29[a]	32[a]	a,e	F35a
46	74	b,c,d	F60c	29	56	a,e,h	F35b

[a]Field measurement.

Use of Appendix G

EXAMPLE. A standard (simple) wooden joist floor–ceiling construction is required with a minimum IIC of 55. Any STC above 35 is acceptable.

1. Since IIC is the determining factor, refer to Index G.2. Basic Type Code is "a."
2. From Index G.2 we note that none of the listed "a" constructions gives an IIC of 55. However, from Section 27.25 we note that the addition of carpeting will add 10 to 27 points to the IIC, making items F34, F38a, F39a, and F30 all suitable. We have discounted special and resilient constructions, because of the requirement for standard (simple) construction.
3. From Table G.1 we select construction F30 as being most appropriate for application of carpeting.
4. From Index G.3 we have (without carpet): thickness 9½ in., STC 34, IIC 32 (+ carpet), fire rating ¼ h. With high pile carpeting on foam rubber pad, this construction will have IIC > 55, STC > 35.

INDEX G.2 **Impact Insulation Class: Floor/Ceiling Constructions**

Type Code:

a. *Wooden joist*
b. *Metal joist*
c. *Concrete or masonry*
d. *Plaster ceiling*
e. *Gypsum board ceiling*
f. *With resilient ceiling element*
g. *With resilient floor element*
h. *With carpeting*
i. *With absorbent blankets*
j. *With separate ceiling joists.*

IIC	*STC*	*Type*	*Item No.*	*IIC*	*STC*	*Type*	*Item No.*
80	52	a,e,h,i,j	F48	44	45	a,e,f	F44
80	44[b]	c,h	F2-1a	43[a]	43[a]	a,d,g	F32b
74	46	b,c,d,h	F60c	42[a]	47[a]	c,d	F24
64[a]	54[a]	c,d,g	F17b	42[a]	46[a]	c,d	F23
62	47	b,c,e,h	F58	42	44[b]	c,g	F3-2(d)
59	47[b]	b,c,d,h	F57b	41	44[b]	c	F3-1(a)
57[a]	55[a]	c,d,g	F14	37	47	b,c,d	F57a
56[a]	29[b]	a,e,h	F35b	37[a]	39[a]	a,e	F34
53[a]	51[a]	c,d,g	F10	33	48	b,c,d	F60a
53[a]	50[a]	c,e,g	F25	33	37	a,e	F38a
51[a]	50[a]	c,e	F27	32[a]	42[a]	c	F22
48[a]	51[a]	c,d	F7a	32	40	a,e,i	F40a
48[a]	50[a]	c,d	F16	32	37	a,e	F39a
48[a]	49[a]	c,d,g	F9	32[a]	34[a]	a,e	F30
47[a]	48[a]	c,d	F12	32[a]	29[a]	a,e	F35a
				29	44[b]	c	F1c
				25	44	c	F1a

[a]Field measurement.
[b]Estimated on the basis of similar structures.

INDEX G.3 **Floor/Ceiling Sound Transmission and Construction Data**

Designation[a]	*Thickness (in.)*	*Weight (lb/ft²)*	*Transmission Loss (db)*						*STC*	*IIC*	*Fire Rating (h)*
			125	*250*	*500*	*1K*	*2K*	*4K*			
F1a	4	53	59	67	75	78	77	75	44	25	1
F1c	4⅛	54	59	66	73	76	74	—	44	29	1
F2-1(a)	4⅜	54	38	36	32	23	—	—	44	80	1
F3-1(a)	4½	55	59	64	58	42	26	—	44	41	1
F3-11(d)	4⅛	53	65	67	72	67	48	40	44	42	1
F7a	5½	61	66	70	68	66	59	37	51	48	2
F9	6⅝	65	63	69	71	68	55	38	49	48	2
F10	8¼	90	60	61	59	54	45	—	51	53	2½
F12	10	62	68	65	67	66	61	50	48	47	3
F14	9½	83	59	58	55	48	42	—	55	57	3
F16	8⅛	65	56	60	60	62	54	—	50	48	2
F17b	9¼	57	56	50	44	40	30	—	54	64	2
F22	6¼	28	76	69	80	79	75	—	42	32	¾
F23	9½	45	71	73.5	72	71	66	54	46	42	¾
F24	10¼	65	67	66	67	66	62	—	47	42	¾
F25	10	45	65	61	57	51	39	—	50	53	¾
F27	7⅝	50	62	66	61	58	46	—	50	51	¾
F30	9½	7	79	79	80	76	69	—	34	32	¼
F32b	11	12	69	72	68	64	55	—	43	43	¾
F34	10¼	9.9	83	80	71	57	52	45	39	37	1
F35a	10	9.2	83	83	72	54	52	43	29	32	—
F35b	10	9.2	58	44	38	24	23	—	29	56	—
F38a	11¾	9	86.5	85.5	85	80	68.5	—	37	33	½
F39a	11⅞	9.5	80	80	81	71	66	58	37	32	1
F40a	11⅞	10	78	84	80	70	65	58	40	32	1
F44	10½	10.1	71	74	70	63.5	61	—	45	44	¾
F48	12⅜	10.7	34	28	20	17	18	—	52	80	¾
F57a	18 9/16	23.2	79	72	72	70	63.5	—	47	37	3
F57b	18 9/16	23.2	59	43	32	18	—	—	47	59	3
F58	21½	20.4	56	41	33	23	17	—	47	62	1
F60a	11	38.2	66	68.5	74	76	75	—	48	33	1½
F60c	11⅝	39	46	38	33.5	25	20	9	46	74	1½

[a]Material extracted from *A Guide to Airborne, Impact and Structure-Borne Noise Control in Multi-Family Dwellings,* HUD/FHA/NBS, 1971.

TABLE G.1 **Acoustic Characteristics of Floors: For Acoustic Data see Index G.3**

Code[a]	Description	Section Sketch

Reinforced Concrete Slab

F1a 4-in.-thick reinforced concrete slab, isolated from support structure. Concrete was reinforced with 6 × 6 in. No. 6 AWG reinforcing mesh placed at the centerline horizontal plane of the slab. All surface cavities were sealed with a thin mortar mix.

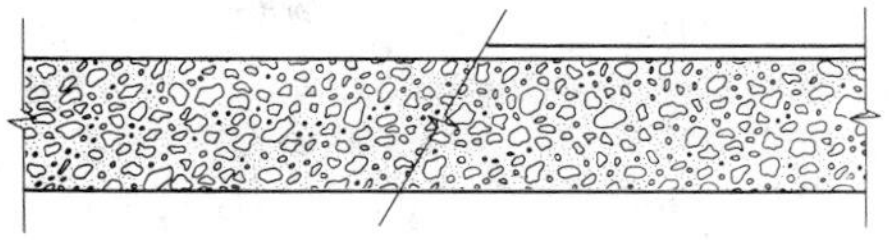

F 1a F 1c

F1c Same as F1a except ⅛-in.-thick vinyl tile was adhered to concrete.

Reinforced Concrete with Floor Coverings

See also F1c above.

F2-1(a) 4-in.-thick reinforced concrete slab with carpeting and pad. The carpeting was of ¼-in. wool loop pile with ⅛-in. woven jute backing, 0.49 lb/ft^2; the foam rubber pad was ¼ in. thick and weighed 0.53 lb/ft^2.

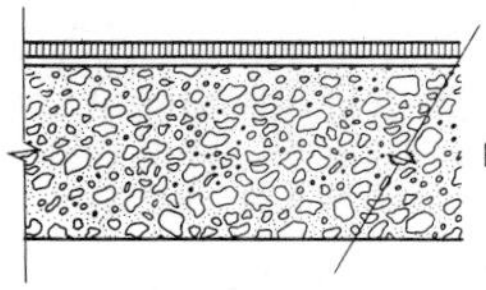

F 2-I(a)

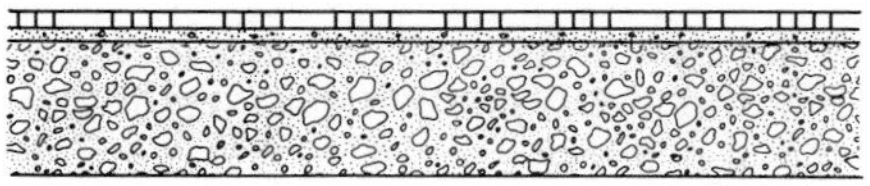

F 3

F3-1(a) 4-in. reinforced concrete slab with ½ × 9 × 9 in. oak blocks, 1.8 lb/ft^2, set in mastic.

F3-2(d) 4 in. concrete slab with ⅛-in. cork.

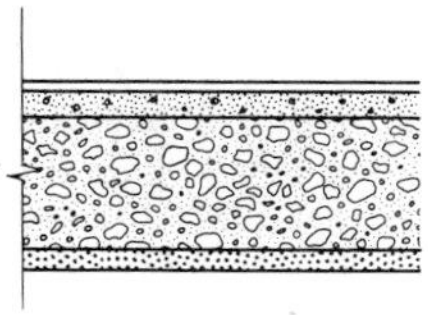

F 7a

F7a 4⅜-in.-thick reinforced concrete slab. On the floor side, ¾-in.-thick, sand-cement screed with ⅛-in. linoleum floor covering. On the ceiling side, ⅜-in. layer of plaster.

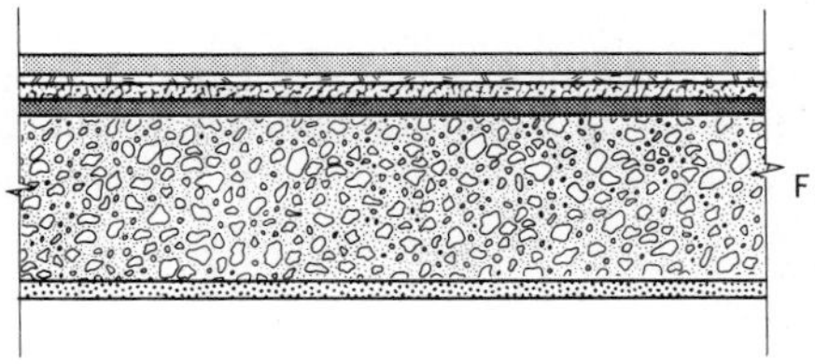

F 9

F9 4⅜-in.-thick reinforced concrete slab. On the floor side, ½-in.-thick layer of bitumen with ½-in.-thick soft wood fiberboard, which was covered with a thin layer of bitumen with sand and a ¾-in.-thick sand-cement screed. On the ceiling side ⅜-in. layer of plaster.

Reinforced Concrete Slab, Floating Floor

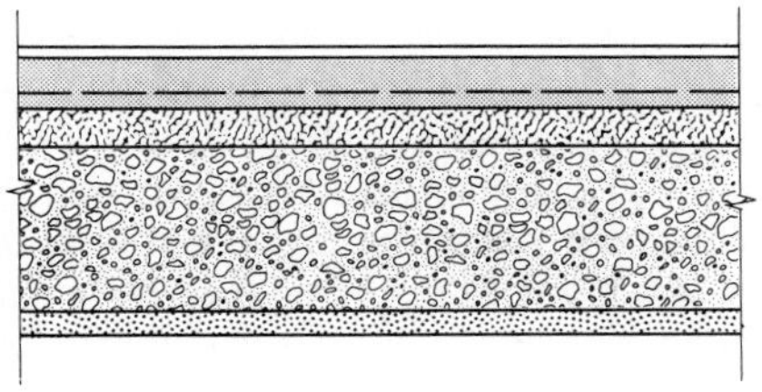

F 10

F10 5-in.-thick reinforced concrete. On the floor side, 1½-in.-thick wire mesh reinforced sand-cement screed floating on ½-in.-thick bitumen-bonded, glass-wool quilt covered with building paper. On the screed, ½-in.-thick pitch-mastic with a linoleum floor covering. On the ceiling side, ½-in. layer of plaster.

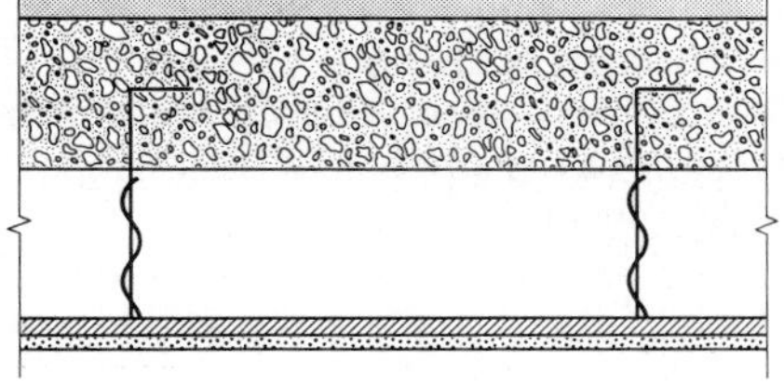

F 12

F12 4⅜-in.-thick reinforced concrete slab. On the floor side, ¾-in.-thick sand-cement screed. On the ceiling side, brick wire mesh, suspended 4 in. with wire hangers, held ⅞-in. gypsum plaster.

TABLE G.1 **Acoustic Characteristics of Floors: For Acoustic Data see Index G.3 (Continued)**

Code[a]	*Description*	*Section Sketch*
F14	6-in.-thick reinforced concrete slab. On the floor side, $\frac{3}{4}$-in.-thick tongue-and-groove wood flooring nailed to $1\frac{1}{2} \times 2$ in. wooden battens, 16 in. on centers, floating on 1-in.-thick, glass-wool quilt. On the ceiling side, $\frac{1}{2}$-in. layer of plaster.	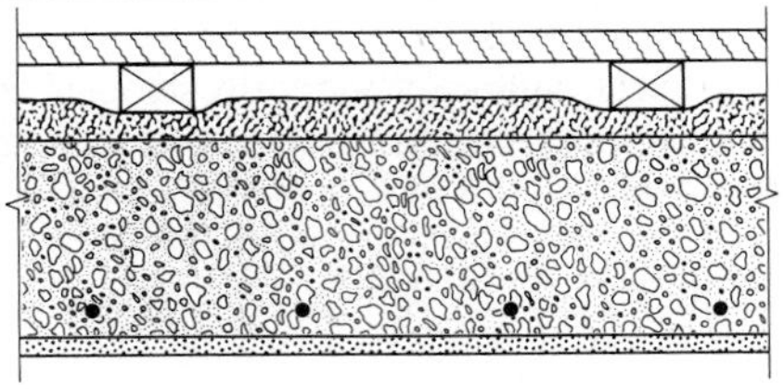 F 14
Concrete with Hollow Blocks		
F16	5×10 in. hollow masonry blocks, 14 in. on centers, with spaces between blocks filled with 5-in.-thick reinforced concrete. On the floor side, $\frac{7}{8}$-in.-thick wood blocks adhered to $1\frac{1}{2}$-in.-thick, sand-cement screed. On the ceiling side, $\frac{3}{4}$-in. layer of plaster.	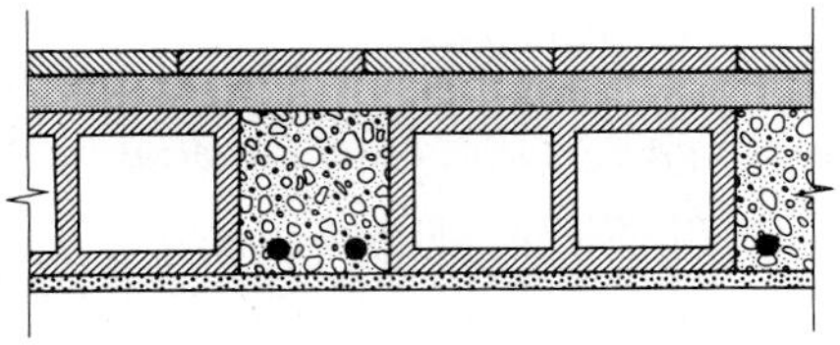 F 16
F17b	$4 \times 12\frac{1}{2}$ in. hollow masonry blocks, $15\frac{1}{2}$ in. on centers, with spaces between blocks filled with 4-in.-thick reinforced concrete. On the floor side, 2-in.-thick, sand-cement screed; linoleum on 1-in.-thick wood flooring nailed to 1×2 in. wooden battens, spaced $15\frac{1}{2}$ in. on centers, floating on glass-wool quilt, approximately 1 in. thick. On the ceiling side, $\frac{3}{4}$-in. layer of plaster.	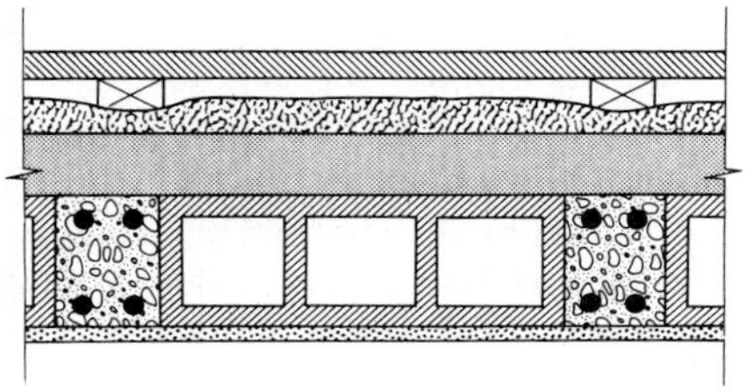 F 17b
Concrete Channel Slab		
F22	Prefabricated concrete channel slabs mortared together 20 in. on centers. Each slab had a 3-in.-deep trapezoidal channel with bases of 11 and $14\frac{3}{4}$ in. On the floor side, $\frac{3}{4}$-in.-thick, sand-cement finish.	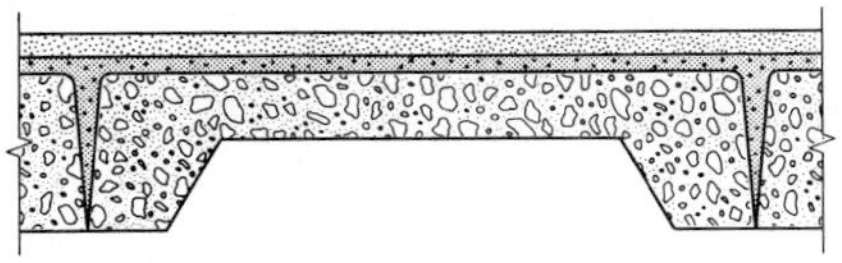 F 22
Ribbed Concrete		
F23	$7\frac{1}{4}$-in. ribbed concrete floor. Ribs were $5\frac{1}{4} \times 3\frac{3}{4}$ in., spaced 21 in. on centers, with 1×2 in. wooden nailing strips cast into ends. On the floor side, the slab was 2 in. thick with a $\frac{3}{4}$-in.-thick, sand-cement screed. On the ceiling side, $\frac{5}{8}$-in.-thick wooden laths nailed to nailing strips, held $\frac{5}{8}$-in.-thick plaster.	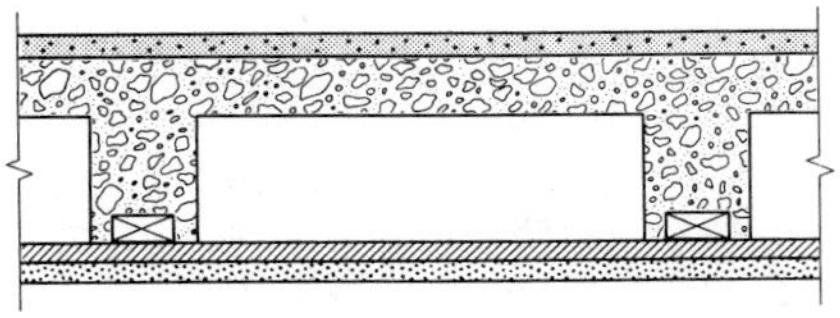 F 23
Concrete Channel Beam		
F24	7-in. precast trapezoidal concrete channel beams, 14 in. on centers, with spaces between beams filled with sand-cement mix. On the floor side, $1\frac{1}{2}$-in.-thick, sand-cement screed with 1-in.-thick, wood-block floor	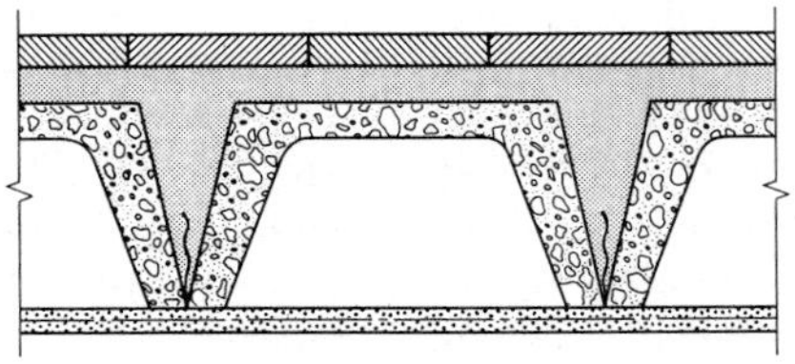 F 24

TABLE G.1 **Acoustic Characteristics of Floors: For Acoustic Data see Index G.3 (Continued)**

Code[a] *Description* *Section Sketch*

covering. On the ceiling side, approximately ¾-in.-thick layer of plaster on expanded metal lath.

Precast Concrete Beam, Floating Floor

F25 5-in. precast concrete channel beams, 14½ in. on centers, with spaces between beams filled with a sand-cement mix. On the floor side, ⅞-in.-thick, tongue-and-groove wood flooring nailed to 1 × 2 in. wooden battens, 20 in. on centers, on approximately 1-in.-thick, glass-wool quilt on ¾-in.-thick, sand-cement screed. On the ceiling side, ⅛-in. layer of plaster on ⅜-in. gypsum wallboard nailed to 1 × 2 in. wooden battens spaced 14½ in. on centers.

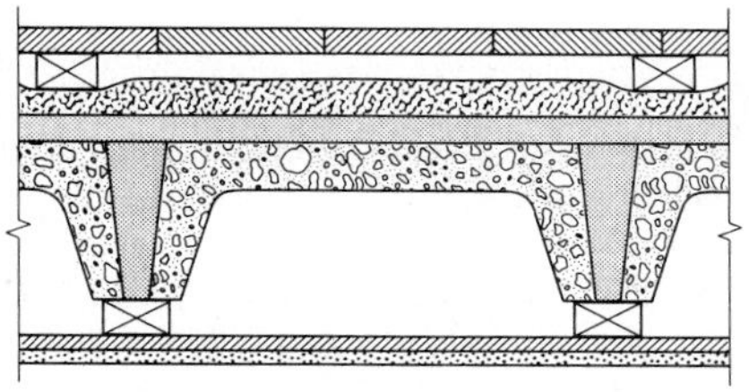

F 25

Hollow Concrete Beam

F27 5-in. precast trapezoidal hollow concrete beams, 14½ in. on centers, with bases of 14 and 12½ in. Spaces between beams filled with sand-cement mix. On the floor side, 1-in.-thick, sand-cement screed with 3/16-in. cork tile floor covering. On the ceiling side, ⅜-in.-thick gypsum wallboard attached to 1 × 2 in. wooden battens held by metal clips.

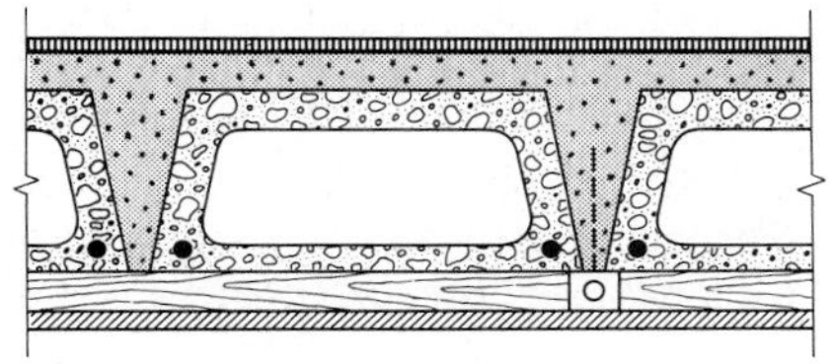

F 27

Wooden Joist

F30 2 × 8 in. wooden joists 16 in. on centers. On the floor side, ⅞-in., tongue-and-groove flooring nailed to joists; on the ceiling side, ⅜-in. gypsum wallboard nailed to joists with joints sealed.

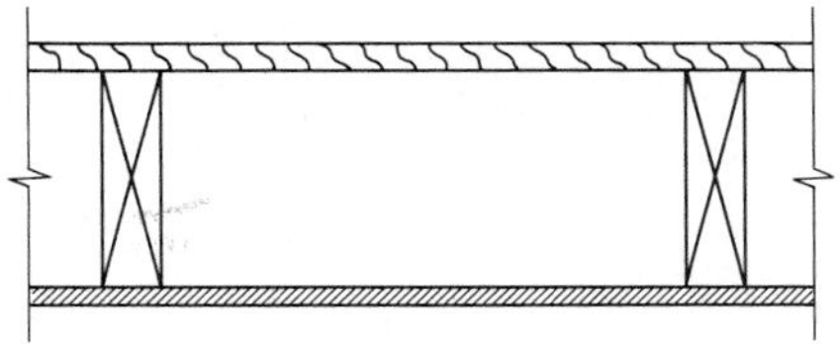

F 30

F32b 2 × 8 in. wooden joists 18 in. on centers. On the floor side, ⅞-in., tongue-and-groove wood flooring nailed to joists. On the ceiling side, 1-in. battens nailed through glass-wool quilt, approximately 1 in. thick; ½-in. layer plaster on ¼-in.-thick wood lath.

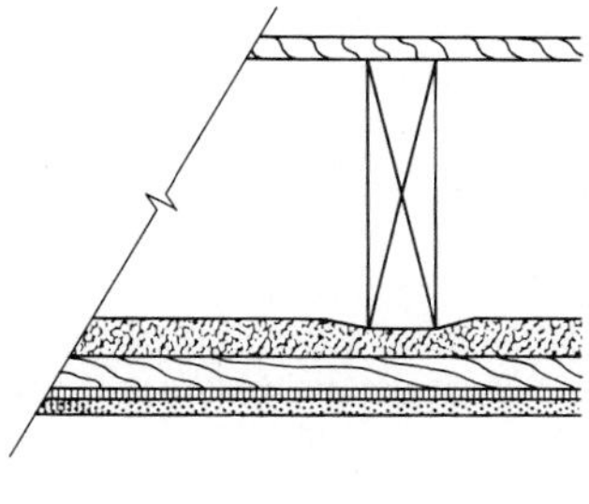

F 32b

F34 2 × 8 in. wooden joists, 16 in. on centers. On the floor side, ½-in.-thick C-D plywood nailed 8 in. on centers to joists, 25/32-in.-thick hard wood flooring on plywood. On the ceiling side, ½-in.-thick gypsum wallboard nailed 6 in. on centers to joists. All joints taped and finished; ceiling tile adhered to gypsum board.

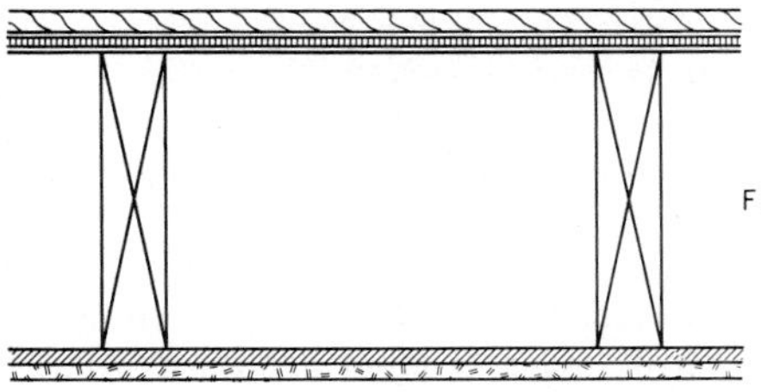

F 34

TABLE G.1 **Acoustic Characteristics of Floors: For Acoustic Data see Index G.3 (Continued)**

Code[a]	*Description*	*Section Sketch*
F35a	2×8 in. wooden joists, 16 in. on centers. On the floor side, $1\frac{1}{2}$-in.-thick, tongue-and-groove wood fiberboard nailed to joists, vinyl tile floor covering. On the ceiling side, $\frac{1}{2}$-in.-thick gypsum wallboard nailed 6 in. on centers to joists. All joints taped and finished.	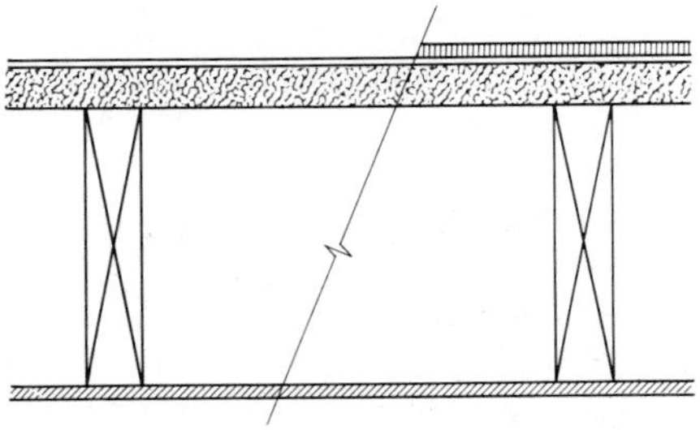 F 35a F 35b
F35b	Similar to F35a except fiberboard covered with carpet and pad.	
F38a	2×10 in. wooden floor joists spaced 16 in. on centers. $\frac{5}{8}$-in. fir plywood subfloor nailed to joists 8 in. on centers; $\frac{1}{2}$-in. plywood underlayment nailed to subfloor with joints staggered to miss joints of the subfloor; $\frac{1}{8} \times 9 \times 9$ in. vinyl asbestos tile glued to underlayment. On the ceiling side, $\frac{1}{2}$-in. gypsum wallboard nailed 12 in. on centers with all joints and nailheads taped and finished.	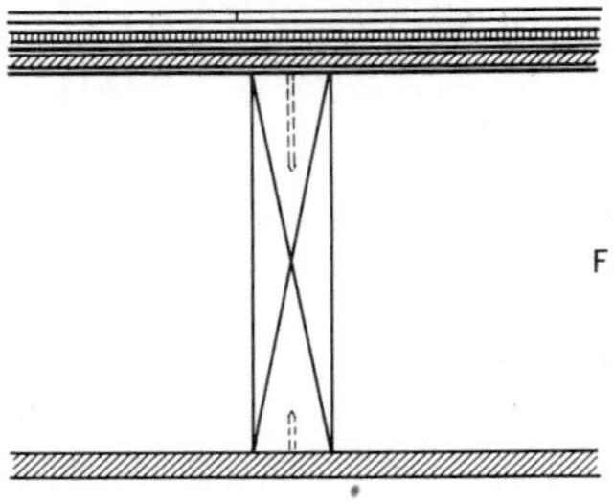 F 38a
F39a	2×10 in. wooden joists 16 in. on centers. On the floor side, $\frac{1}{2}$-in.-thick plywood subfloor nailed 6 in. on centers along edges and 10 in. on centers in field, building paper underlayment, $\frac{25}{32} \times 2\frac{1}{4}$ in. oak wood flooring nailed at each joist intersection and midway between joists. On the ceiling side, $\frac{5}{8}$-in.-thick gypsum wallboard, nailed 6 in. on centers to joists. All joints taped and finished.	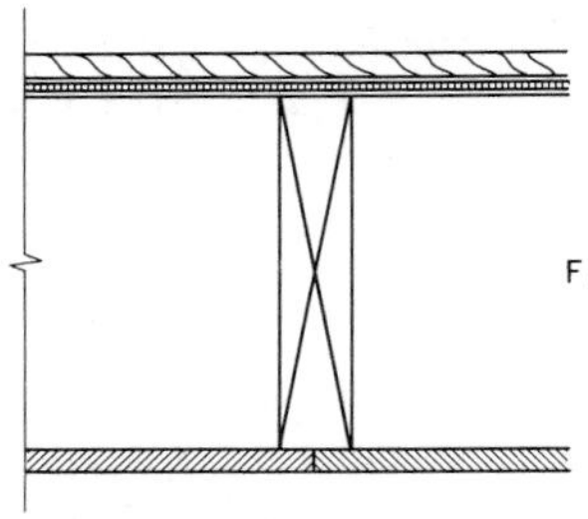 F 39a
Wooden Joist, Resilient Ceiling		
F44	2×8 in. wooden joists 16 in. on centers. On the floor side, $\frac{3}{4}$-in.-thick wood subfloor, layer of building paper, and $\frac{3}{4}$-in.-thick tongue-and-groove fir finish flooring. On the ceiling side, resilient runners bridged across joists and nailed 12 in. on centers to joists; $\frac{5}{8}$-in.-thick gypsum wallboard screwed to resilient runners. All joints taped and finished.	F 44 
Wooden Joist with Insulation		
F40a	2×10 in. wooden joists 16 in. on centers with 3-in.-thick mineral fiber batts stapled between joists. On the floor side, $\frac{1}{2}$-in.-thick plywood subfloor nailed 6 in. on centers along edges and 10 in. on centers in field, building paper underlayment, $\frac{25}{32}$ ×	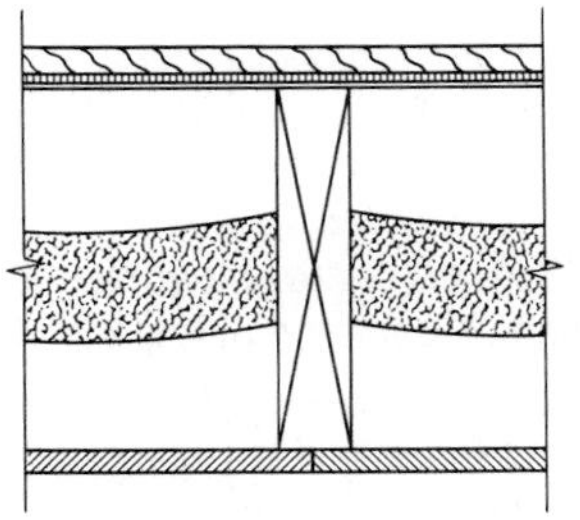 F 40a

APPENDICES

TABLE G.1 **Acoustic Characteristics of Floors: For Acoustic Data see Index G.3 (Continued)**

Code[a]	*Description*	*Section Sketch*
	$2\frac{1}{4}$ in. oak wood flooring nailed at each joist intersection and midway between joists. On the ceiling side, $\frac{5}{8}$-in.-thick gypsum wallboard nailed 6 in. on centers to joists. All joints taped and finished.	
F48	2×8 in. wooden joists, 16 in. on centers. On the floor side, $1\frac{1}{8}$-in.-thick regular C-D rough plywood nailed 6 in. on centers along periphery and 16 in. on centers at other bearings, plywood covered with an all-hair pad (40 oz/yd^2) and all-wool pile (44 oz/yd^2) carpet. The total weight of the carpet was 4.14 lb/yd^2 and the total thickness was $\frac{3}{8}$ in. On the ceiling side, 2×4 in. wooden joists, 16 in. on centers, staggered 8 in. on centers relative to the floor joists, 3-in.-thick fibered glass blankets stapled between ceiling joists, $\frac{5}{8}$-in.-thick gypsum wallboard nailed to ceiling joists. All joints taped and finished; entire periphery of panel caulked and sealed. The ceiling was supported independently from the floor structure.	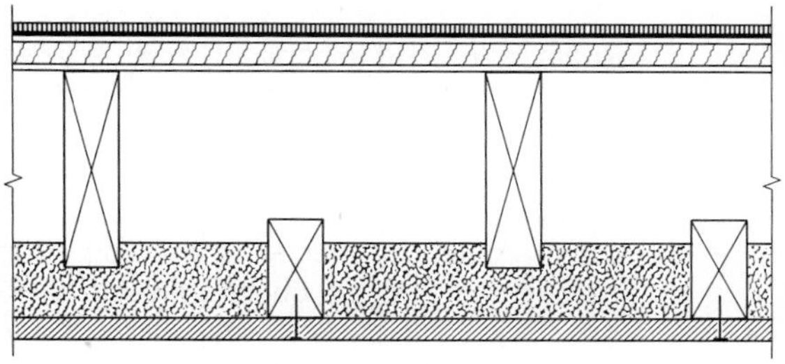 F 48
	Steel Joist with Concrete Floor	
F57a	$2\frac{1}{2}$-in.-thick perlite concrete, 72 lb/ft^3 on 28-gauge corrugated steel units supported by 14-in. steel bar joists; $\frac{1}{8}$-in.-thick asphalt tile cemented to concrete. On the ceiling side, $\frac{3}{4}$-in. furring channels, $13\frac{1}{2}$ in. on centers, wire-tied to joists, 3.4 lb/yd^2 diamond mesh metal lath wire-tied to furring channels, $\frac{9}{16}$-in. coat of plaster with $\frac{1}{16}$-in., white-coat finish.	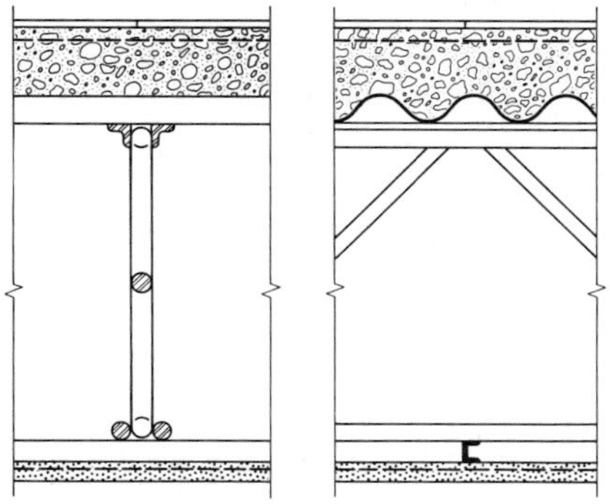 F 57
F57b	Same as F57a except carpet and pad in lieu of asphalt tile.	
F58	18-in. steel joists 16 in. on centers. On the floor side, $\frac{5}{8}$-in.-thick C-D rough plywood nailed to joists, $1\frac{5}{8}$-in.-thick foamed concrete, 100 lb/ft^3, slab constructed on the plywood; concrete covered with an all-hair pad (40 oz/yd^2) and an all-wool pile (44 oz/yd^2) carpet. Total weight of the carpet, 4.14 lb/yd^2, total thickness, $\frac{3}{8}$ in. On the ceiling side, $\frac{5}{8}$-in.-thick gypsum wallboard nailed to joists. All joints taped and finished; entire periphery of panel caulked and sealed.	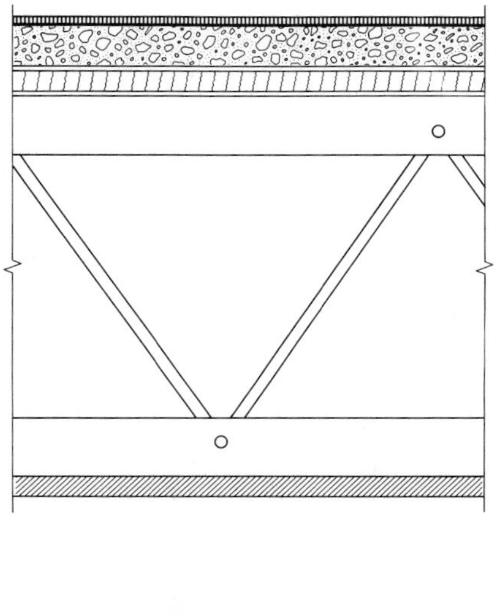 F 58

F60a 7-in. steel bar joists spaced 27 in. on centers. On the floor side, 3⁄8-in. metal rib lath attached to top of joists, and 2-in.-thick poured concrete floor. On the ceiling side, 3⁄4-in. metal furring channels wire-tied to joists 16 in. on centers; 3⁄8 × 16 × 48 in. plain gypsum lath held with wire clips and sheet metal end joint clips; 7⁄16-in. sanded gypsum plaster and 1⁄16-in., white-coat finish.

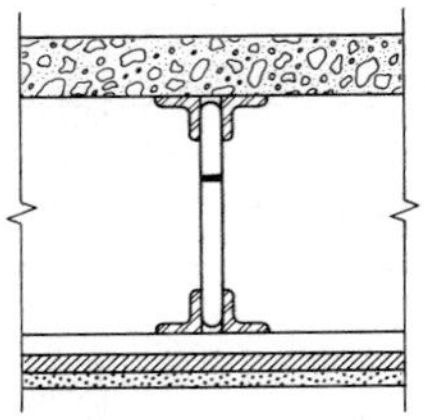

F60c Structure F60a with nylon carpeting and foam rubber pad placed on the floor. The carpet pad had an uncompressed thickness of 1⁄4 in., backed with a woven jute fiber cloth. The carpet had 1⁄8-in. woven backing and 1⁄4-in. looped pile spaced 7 loops per inch with a total thickness of 3⁄8 in.

[a]Material extracted from *A Guide to Airborne, Impact and Structure-Borne Noise Control in Multi-Family Dwellings,* HUD/FHA/NBS, 1971.

H

OPTIONS FOR COMPUTATION ASSISTANCE

This book presents ''manual'' calculation procedures—simple enough to require only a hand-held calculator and an informed and patient user. It is by now obvious, however, that some very time-consuming procedures are required for the detailed environmental analysis of buildings. If a machine can be made to automatically perform repetitious calculations, its human user can be freed to do higher grade tasks. Three degrees of such automatic computation assistance are available. In order of increasing complexity, they are the *programmable calculator,* the *micro-* (or ''personal'') *computer,* and the *mainframe computer*. It is important to realize that with each step toward greater complexity, the user gets faster and more thorough analysis. But each step also requires a greater purchase cost for the hardware (machine) and greater training time for the user—whether to learn how to run the software (prepared programs, which are themselves more expensive with each step) or to learn how to program the machine. There is also decreasing mobility with each step. Hand-held calculators may be small enough to carry in a wallet, and programmable calculators (which should be accompanied by a printer unit) can be stuffed into a briefcase. However, most personal computers are not yet really portable, although some can be installed in an automobile. Mainframe computers may be accessible by telephone line, but printed analysis requires access to fixed-in-place equipment. Finally, with each step, a typical user of prepared software becomes more remote from the calculation process—for better or for worse.

The point of this introduction, then, is to help you select the appropriate tool for the task. Quick approximations require no more than a hand-held calculator; thorough analysis of many alternative designs for a given problem will be more efficiently done with automatic computation.

There is a rapidly growing array of software for the automatic calculations associated with building energy performance. Table H.1 presents a guide to the selection of software for building thermal analysis. If your architectural idea cannot be reduced to numbers, then it cannot be assisted by computer analysis.

(a) Hand-Held Calculators. This book presents manual procedures that require a common device, the hand-held calculator. It should have trigonometric functions and one or two memories. For acoustics, logarithms are essential as well. Powered by batteries or by photovoltaic cells, hand-held calculators are truly portable, and inexpensive enough for every professional to own at least one.

(b) Programmable Calculators. The key to the successful use of most of these devices is the *magnetic card* that carries the instructions for automatic computation. This tiny card may carry a program developed by its user or a program that has been developed elsewhere and purchased for use. A program is the encoded equivalent of the long series of keystrokes that would be necessary if a given procedure were to be done on a hand-held calculator. Prepared programs are frequently ''protected'' so that their set of instructions cannot be printed out for inspection by the user; you get step-by-step results, but not the calculation details of the steps themselves.

Any programmable calculator can be used to develop your own programs; be prepared to spend at least one work day learning the intricacies of your chosen machine and for trial-and-error time.

Table H.1 **Considerations in Choosing Computer Software for Heating or Cooling Analysis**

Part A. ***Facts and Figures***

- Package name and version
- Price
- Systems available for (applications)
- Required supporting software
- Memory requirements
- Diskette capacity required
- Utility programs included
- Internal format, records, types
- Portability
- User skill level required
- System upgrade policy and cost

Part B. ***Qualitative Factors***

- Documentation
  - Organization for learning—tutorials, examples
  - Organization for reference
  - Readability
  - Includes all needed information
  - Table of contents and index
- Menus (optional)
  - Ease of use
  - Includes all needed information
- Ease of use
  - Initial startup and installation
  - Conversion of external data
  - Operator use
  - Application implementation (speed, accuracy, etc.)
- Error recovery
  - From input error
  - Restart from interruption
  - From data media damage (bad disk)
- Support
  - For initial startup
  - For system improvement

Part C. ***Technical Factors***

- General technical user concerns
  - Speed and convenience
  - Validity and accuracy
  - Ability to modify analysis
  - Source code availability
- Building types
  - Residential buildings
  - Commercial and nonresidential buildings
  - Underground or below-grade considerations
  - Special conditions
- Weather and climate
  - Weather data source
  - Availability of weather data files
  - Ease of generating weather data
- Building or system loads analysis
  - Method of analysis
  - Orientation and shading
  - Number of building zones
  - Building schedules and internal gains
  - Temperature setpoints and deadbands
- Solar features or systems analysis
  - Method of analysis
  - Orientation and shading
  - System types modeled
  - Component properties and ranges
  - Multiple solar features
  - System limitations
- Report generation
  - Summary and detail report formats
  - Flexibility of report formats
  - Ability to generate custom reports
  - Ability to print reports to disk files
- Special features
  - Supports graphics output
  - Terminals supported
  - Printers or plotters supported
  - Other special features

Source: M. Steven Baker, *Computer-Aided Solar Design* presentation at Oregon Solar Conference, September 1984.

In general, these machines will *not* tell you when mistakes are made; as with hand-held calculators, you must be able to recognize mistakes by answers that you see to be unreasonably large or small. For this reason, it is nearly essential to purchase a printer unit to accompany your programmable calculator; errors are much more easily traced when you have a printed record of the process. Including the printer generally doubles total cost. Since prepackaged software can be purchased for only a few brands [Texas Instruments (TI) and Hewlett Packard (HP) are most common], strongly consider purchasing a machine for which such a ''library'' of software exists.

For those who would rather prepare their own magnetic cards, an excellent reference book is entitled *Building Systems Design with Programmable Calculators* (Daryanani, 1980). It offers a catalog of complete programs for the TI-59 and the HP-97 for water and waste pipe sizing, air duct design, electric lighting design, heat transmission coefficients, solar shading and average insolation on tilted surfaces, life-cycle costing, and general planning for space and distribution system layout for preliminary design.

Detailed programs for solar procedures such as passive solar heating performance, active collector system sizing, passive cooling by night ventilation of thermal mass, and so on can be found in publications of the American Solar Energy Society (ASES), which contain the proceedings of national solar conferences. However, these are usually presented without the patient development of user instructions that accompany the programs in Daryanani's book. Some books on solar design include programs for these calculators; *Solar Architecture: The Direct Gain Approach,* (Johnson, 1981) gives program codes for the TI-59, for determining interior air temperature swing in passively solar heated buildings, for solar angles and clear sky radiation, and for average daily radiation. A more expensive book is *Solar Energy Programs* (F-Chart Software, 1980). It contains 13 programs for the HP-41 and the TI-59, basically for active solar system design. For daylight analysis with a TI-59, see both the August and September 1981 issues of *Solar Age* magazine, for the program for "Quicklite 1" developed by Prof. Harvey Bryan, MIT, Cambridge, Massachusetts.

Prepared software is available from a variety of sources. The manufacturers of the calculators provide a wide range of software, only some of which is applicable to building design. Preprinted magnetic cards for the TI-59 and the HP-97 are available for passive and active solar performance analysis from such sources as Total Environmental Action (TEA: Church Hill, Harrisville, NH 03450), and Princeton Energy Group (PEG, 575 Ewing St., Princeton, NJ 08540). Since available programs—and their prices—change rapidly, the latest catalogs from these suppliers should be consulted.

Although programmable calculators promise a saving of the user's time, there are several cautions. First, the data required as input (e.g., areas and U values for thermal calculations) must still be generated. Next, the user must spend time to become familiar with the equipment and then with the details of the software. For example, to gain confidence that the program for the monthly solar savings fraction is being correctly run, a first-time user should expect to invest at least a half a day reading the manual and running example programs. When months will elapse between uses of a specific program, some retraining time should be scheduled. Some of the more complicated programs need periodic feeding of data by the user; this may cause frequent interruptions of the work you were "freed" to do, to keep the program going. And finally, there is the syndrome of a user, entranced by the effortless generation of data, spending as much time running many alternative calculations as would have been spent manually calculating the few most logical choices.

The programmable calculator is threatened by competition from microcomputers. When these newer, more numerically powerful devices achieve the size and price of the programmable calculator, such calculators will join the slide rule as historic aids to computation.

(c) Microcomputers. These are becoming commonplace in architectural offices, since their services include word processing, graphics, billing, and other project management functions in addition to their role in a building's technical analysis. A dizzying array of software has been developed, and very lively competition among hardware manufacturers has already eliminated some early promising contenders. The recommended method is to choose the software you like and then buy the hardware for it. Despite the dangers of immediate obsolescence of such a list, the following are a few early-established sources of thermal and daylighting analysis software and some of the equipment on which these programs will run.

CALPAS series of programs for daylighting, passive and other thermal analysis, developed by the Berkeley Solar Group (3140 Grove St., Berkeley, CA 94703). Available for the IBM PC. The buyer receives the program, user's manual, one year of user support services, a full-day seminar, five weather files, and a newsletter. (A mainframe computer subscription service is also available.)

F-CHART series for active solar heating design, available for most U.S. types of microcomputer; from F-Chart Software (4406 Fox Bluff Rd., Middleton, WI 53562). The programs allow analysis of four collector types and eight system types (including passive direct gain and storage wall systems).

SOLARSOFT (Box 124, Snowmass, CO 81654), has a series of solar analysis programs for Apple computers, for both passive and active solar heating systems.

MICROPAS, written by Enercomp (757 Russell Blvd, Davis, CA 95616), is an hourly simulation of energy performance for six seasons (seven consecutive days per season). It runs on most microcomputers.

The Designers Software Exchange, Laboratory of Architecture and Planning at MIT (Cambridge, MA 02139) offers a rapidly growing array of programs for a variety of calculation devices.

A bibliography of computer programs in the general area of HVAC is available from ASHRAE (1791 Tullie Circle NE, Atlanta, GA 30329). Similarly, a listing of available computer programs in all areas of lighting design, arranged by program capability, type of hardware, and vendor, is published periodically in *Lighting Design and Application* (see June 1984, for example). An up-to-date list is available from the Computer Committee, IES of North America (345 East 47 St., New York, NY 10017).

Advertisements and articles about late developments in computer programs are especially numerous in magazines such as *Solar Age,* and in the Solar Conference Proceedings published yearly by the American Solar Energy Society.

(d) Mainframe Computers. Most universities, many municipal governments, and some larger architectural firms own these machines. Many other professionals buy access to big computers through time-sharing. The more complicated thermal analysis programs require extensive training for operators, and as much as several days to generate the input data. When a detailed performance comparison is required for several alternative designs on larger buildings, these programs are appropriate.

Most mainframe computer programs for thermal analysis require hourly data for input; this usually requires the purchase of a tape of data for a typical meteorological year (TMY), available from the U.S. Department of Commerce, National Climatic Center (Asheville, NC 28801) for a variety of locations. The resulting energy analysis is thus based on a very detailed, sequential analysis of the building's use of energy for all purposes. As can be imagined, considerable detail about the building's configuration and equipment must be included as input for this analysis.

The most widely used of the early mainframe computer programs include:

DOE series, often considered the standard evaluation technique; the user's manual is available from the Los Alamos Scientific Laboratory.

BLAST series (Building Loads Analysis and System Thermodynamics); the user's manual is available from the U.S. Army Construction Engineering Research Laboratory. (These programs have been compared in an analysis done at the Solar Energy Research Institute, Golden, Colorado.)

CALPAS series, developed by the Berkeley Solar Group, at the address listed above.

TRNSYS, a detailed simulation program for active solar systems, developed at the University of Wisconsin, Madison.

SUPERLITE, which includes daylighting in overall energy analysis; developed at the Lawrence Berkeley Laboratory, Berkeley, California.

Again, these large, detailed, complex programs are not for the casual user. It is more likely that a consulting mechanical engineer rather than an architect will utilize them. They represent a sophisticated opportunity for analyzing energy performance and choosing between alternative designs.

References

Daryanani, Sital (1980). *Building Systems Designs with Programmable Calculators,* McGraw-Hill, New York.

F-Chart Software. (1980). *Solar Energy Programs,* F-Chart Software, 4406 Fox Bluff Rd., Middleton, WI 53562.

Johnson, Timothy. (1981). *Solar Architecture: The Direct Gain Approach,* McGraw-Hill, New York.

I

METRICATION; SI UNITS; CONVERSIONS

I.1 General Comments on Metrication

The building profession has been slower than most in adopting the metric (more accurately the Système International, i.e., SI) system of units for many reasons, a discussion of which is not relevant here. The change will come. Many of the major professional societies use a mixture of both systems because certain units are so entrenched that changing them is an almost insurmountable task. However, the advantages of the SI system are well known to all and need no justification here. In this book we have followed current industry practice—a mixture of the two systems. For this reason tables of conversions and approximations are presented below to enable the reader to live with both systems. Also given below are some useful facts that should make use of the SI (metric) system a bit easier.

I.2 SI Nomenclature, Symbols

For a full discussion of the SI system the reader is referred to *AIA Metric Building and Construction Guide*, edited by S. Braybrooke, (Wiley, New York, 1980). The units in common use include the basic SI units; meter, kilogram, and second (MKS), plus a host of derived, supplementary, and non-SI units such as pascal (pressure), radian (solid angle), and kilowatt-hour (energy), respectively. Also, multiple and submultiple units such as liter, metric ton, and millibar are so common that they stand as separate units instead of being expressed as $10^{-3}m^3$, 10^3kg and $10^{-2}Pa$. For this reason we omit a detailed analysis of the subject and simply supply data that by experience we have found useful.

Table I.1 lists prefixes with their accepted symbols. Symbols do not change in plural. That is, 6

TABLE I.1 **These Prefixes May Be Applied to All SI Units**

Multiples and Submultiples	*Prefixes*	*Symbols*
1 000 000 000 = 10^9	giga	G
1 000 000 = 10^6	mega	M[a]
1 000 = 10^3	kilo	k[a]
100 = 10^2	hecto[b]	h
10 = 10	deka	da
0.1 = 10^{-1}	deci	d
0.01 = 10^{-2}	centi	c[a]
0.001 = 10^{-3}	milli	m[a]
0.000 001 = 10^{-6}	micro	μ[a]
0.000 000 001 = 10^{-9}	nano	n
0.000 000 000 001 = 10^{-12}	pico	p

[a]Most commonly used.
[b]A hectare is a square hectometer (i.e., 10^4 m^2).

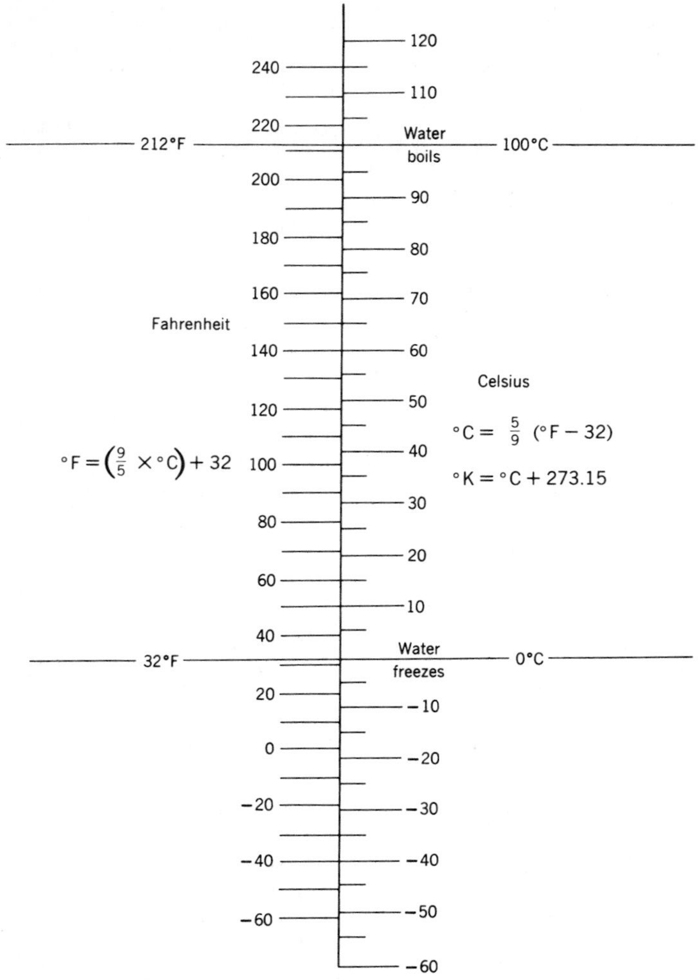

Fig. I.1 Conversion, Fahrenheit degrees–Celsius degrees.

millimeters is written 6 mm and 20 kilograms is written 20 kg. All units and prefixes except Fahrenheit and Celsius are uncapitalized when written out, as in megaton or meter.

I.3 Common Usage

1. Length: meter (m), kilometer (km), millimeter (mm), micrometer (μm), nanometer (nm).
2. Area: square meter (m^2), hectare (ha).
3. Volume: cubic meter (m^3), liter (L).
4. Flow: cubic meters per second (m^3/s).
5. Velocity, airflow: meters per second (m/s).
6. Weight: kilogram (kg), gram (g).

The SI system clearly differentiates between mass (kg) and force ($kg \cdot m/s^2$), the latter being given a separate name and symbol, newton (N). Weight is not used, because it is a force that depends on acceleration and is therefore variable. However, the construction industry in large measure continues to use the terms mass, weight, and force interchangeably.

7. Force: newton (N), kilonewton (kN). A newton is the force required to accelerate 1 kg at 1.0 m/s^2.

8. Pressure: pascal (Pa), kilopascal (kPa). A pascal is a newton per square meter (N/ m^2).
9. Energy, work, quantity of heat: joule (J), kilojoule (kJ), megajoule (MJ). A joule is a watt-second (W·s).
10. Temperature: degree Celsius (°C), degree kelvin (°K).

The SI unit (Kelvins) is not used as commonly as °C. (Celsius is the accepted term; ''centigrade'' is obsolete). The Celsius and Kelvin scales are subdivided equally but start at different points, that is, 0° K is −273.15°C. Therefore, to determine Kelvin from Celsius, simply add 273.15. Increments are equal because of equal subdivision; that is, a change of 10K° is the same as a 10C° change. Because of its special importance, a separate Fahrenheit/Celsius conversion chart is given in Fig. I.1.

11. Illumination: see Table 18.1.
12. Acoustics: see Table 27.15.
13. CGS/MKS conversions: see Table 27.15.
14. Abbreviations: see Table I.2.
15. Approximations: see Table I.3.

I.4 Conversion Factors

Table I.4 is an alphabetized list of useful conversion factors. Because normal use involves a hand calculator, we have avoided scientific notation and used decimal notation instead; that is, we write 0.00378, not 3.78×10^{-3} or 3.78E-3.

TABLE I.2 **Typical Abbreviations: All Systems of Units**

atmospheres	atm	kilopascals	kPa
British thermal units	Btu	kilowatts	kW
British thermal units per hour	Btu/h	kilowatt-hours	kWh
calorie	cal	liters	L
cubic feet	cf, ft^3	liters per second	L/s
cubic feet per minute	cfm, ft^3/min	megajoules	MJ
cubic feet per second	cfs, ft^3/s	meganewtons	MN
cubic meters	m^3	megapascals	MPa
feet	ft	meters	m
feet per second	fps, ft/s	meters per second	m/s
gallons	gal	miles per hour	mph
gallons per hour	gph, gal/h	millimeters	mm
gallons per minute	gpm, gal/min	millimeters of mercury	mm Hg
grams	g	newtons	N
hectares	ha	ounces	oz
horsepower	hp	pounds	lb
inches	in.	pounds of force	lbf
inches of mercury	in. Hg	pounds per cubic foot	lb/ft^3
joules	J	second	sec, s
kilocalories	kcal	square feet	ft^2
kilograms	kg	square inches	in.^{2}
kilograms per second	kg/s	square meters	m^2
kilojoules	kJ	watts	W
kilometers	km	watts per square meter	W/m^2
kilometers per hour	kph, km/h	yards	yd
kilonewtons	kN		

TABLE I.3 **Common Approximations**

Approximate Common Equivalents			
1 inch	= 25 millimeters	1 millimeter	= 0.04 inch
1 foot	= 0.3 meter	1 liter	= 61 cubic inches
1 yard	= 0.9 meter	1 meter	= 3.3 feet
1 mile	= 1.6 kilometers	1 meter	= 1.1 yards
1 square inch	= 6.5 square centimeters	1 kilometer	= 0.6 mile
1 square foot	= 0.09 square meter	1 square centimeter	= 0.16 square inch
1 square yard	= 0.8 square meter	1 square meter	= 11 square feet
1 acre	= 0.4 hectare	1 square meter	= 1.2 square yards
1 cubic inch	= 16 cubic centimeters	1 hectare	= 2.5 acres
1 cubic foot	= 0.03 cubic meter	1 cubic centimeter	= 0.06 cubic inch
1 cubic yard	= 0.8 cubic meter	1 cubic meter	= 35 cubic feet
1 quart	= 1 liter	1 cubic meter	= 1.3 cubic yards
1 gallon	= 0.004 cubic meter	1 liter	= 1 quart
1 ounce	= 28 grams	1 cubic meter	= 250 gallons
1 pound	= 0.45 kilogram	1 kilogram	= 2.2 pounds
1 horsepower	= 0.75 kilowatt	1 kilowatt	= 1.3 horsepower

TABLE I.4 **Useful Conversion Factors: Alphabetized**

Multiply	*By*	*To Get*
acres	4047	square meters
atmospheres	33.93	feet of water
atmospheres	29.92	inches of mercury
atmospheres	760.0	millimeters of mercury
Btu (energy)	0.252	kilocalories
	1.055	kilojoules
Btu/h (power)	0.2928	watts
Btu/h/ft^2 (energy transfer)	3.152	watts per square meter
BtuF (heat capacity)	1.897	kilojoules per kelvin[a]
Btu/lb°F (specific heat)	4.182	kilojoules per kilogram per kelvin[a]
Btu/h/Fft (thermal conductivity[b])	1.729	watts per kelvin[a] per meter
Btu/hFft^2 (conductance[c])	5.673	watts per kelvin[a] per square meter
Btu/Fday (building load coefficient, BLC)	0.022	watts per kelvin[a]
Btu/Fdayft^2 (load–collector ratio, LCR)	0.236	watts per kelvin[a] per square meter
cubic feet	0.028	cubic meters
cubic feet	7.481	gallons
cubic feet	28.32	liters
cubic feet per minute	0.472	liters per second
cubic feet per second	2.832	liters per second
cubic inches	16.39	cubic centimeters
cubic meters	35.32	cubic feet
cubic meters	1.308	cubic yards
cubic meters	264.2	gallons
cubic yards	0.765	cubic meters
feet	0.305	meters
feet	304.8	millimeters
feet per second	0.3048	meters per second
foot-pounds of force per second	1.356	watts

TABLE I.4 **Useful Conversion Factors: Alphabetized (Continued)**

Multiply	*By*	*To Get*
gallons	3.785	liters
gallons her hour	0.00152	liters per second
gallons per minute	0.0022	cubic feet per second
gallons per minute	0.06308	liters per second
grams	0.035	ounces (avoirdupois)
hectares	2.471	acres
horsepower	0.746	kilowatts
horsepower	746	watts
inches	25.4	millimeters
inches of mercury	0.033	atmospheres
inches of mercury	1.133	feet of water
inches of mercury (60°F)	3377	newtons per square meter
inches of mercury	0.491	pounds per square inch
inches of water	0.002458	atmospheres
inches of water	0.036	pounds per square inch
inches of water (60°F)	248.8	newtons per square meter
kilocalories	3.968	British thermal units
kilocalories	4190	joules
kilograms	2.205	pounds
kilograms per cubic meter	1.686	pounds per cubic yard
kilograms per square meter	0.0033	feet of water
kilograms per square meter	0.0029	inches of mercury
kilograms per square meter	0.205	pounds per square foot
kilograms per square meter	0.001422	pounds per square inch
kilojoules	0.948	British thermal units
kilojoules per kilogram	0.430	British thermal units per pound
kilometers	0.621	miles
kilometers per hour	0.621	miles per hour
kilonewtons	0.1004	tons of force
kilonewtons	224.8	pounds of force
kilopascals	20.89	pounds of force per square foot
kilowatts	1.341	horsepower
kilowatt-hours	3.6	megajoules
liters	0.03532	cubic feet
liters	61.02	cubic inches
liters	0.2642	gallons
liters	1.057	quarts
liters per second	2.119	cubic feet per minute
liters per second	951.0	gallons per hour
liters per second	15.85	gallons per minute
megajoules	0.278	kilowatt-hours
meganewtons	100.36	tons of force
megapascals	145.04	pounds of force per square inch
megapascals	9.324	tons of force per square foot
meters	3.281	feet
meters	1094	yards
meters per second	196.86	feet per minute
meters per second	2.237	miles per hour
miles	1.609	kilometers
miles per hour	1.609	kilometers per hour
miles per hour	0.447	meters per second

TABLE I.4 **Useful Conversion Factors: Alphabetized (Continued)**

Multiply	*By*	*To Get*
milliliters	0.061	cubic inches
milliliters	0.035	fluid ounces
millimeters	0.039	inches
millimeters of mercury	133.3	newtons per square meter
million gallons per day	18.94	cubic meters per hour
newtons	0.225	pounds of force
ounces (avoirdupois)	28.35	grams
ounces (fluid)	28.41	milliliters
pounds	0.454	kilograms
pounds of force	4.448	newtons
pounds of force per square foot	47.88	pascals
pounds of force per square inch	6.895	kilopascals
pounds per cubic foot	16.02	kilograms per cubic meter
pounds per cubic yard	0.593	kilograms per cubic meter
pounds per square foot	4.882	kilograms per square meter
square feet	0.0929	square meters
square inches	645.2	square millimeters
square kilometers	0.386	square miles
square meters	10.76	square feet
square meters	1.196	square yards
square miles	2.590	square kilometers
square yards	0.836	square meters
tons of force	9.964	kilonewtons
tons of force per square foot	107.25	kilopascals
tons of force per square inch	15.44	megapascals
torr (millimeters of mercury at 0°C)	133.3	newtons per square meter
watts	3.412	British thermal units per hour
watts	0.738	foot-pounds of force per second
watts per square meter	0.317	British thermal units per square foot
yards	0.914	meters

Add your own conversion factors here.

Multiply	*By*	*To Get*
Multiply	*By*	*To Get*
Multiply	*By*	*To Get*

[a] °K or °C.
[b] Thermal conductivity (K).
[c] Thermal conductance (C) or transmittance (U).

INDEX